Lecture Notes in Computer Science 3316

Commenced Publication in 1973
Founding and Former Series Editors:
Gerhard Goos, Juris Hartmanis, and Jan van Leeuwen

Editorial Board

David Hutchison
  Lancaster University, UK
Takeo Kanade
  Carnegie Mellon University, Pittsburgh, PA, USA
Josef Kittler
  University of Surrey, Guildford, UK
Jon M. Kleinberg
  Cornell University, Ithaca, NY, USA
Friedemann Mattern
  ETH Zurich, Switzerland
John C. Mitchell
  Stanford University, CA, USA
Moni Naor
  Weizmann Institute of Science, Rehovot, Israel
Oscar Nierstrasz
  University of Bern, Switzerland
C. Pandu Rangan
  Indian Institute of Technology, Madras, India
Bernhard Steffen
  University of Dortmund, Germany
Madhu Sudan
  Massachusetts Institute of Technology, MA, USA
Demetri Terzopoulos
  New York University, NY, USA
Doug Tygar
  University of California, Berkeley, CA, USA
Moshe Y. Vardi
  Rice University, Houston, TX, USA
Gerhard Weikum
  Max-Planck Institute of Computer Science, Saarbruecken, Germany

Nikhil R. Pal   Nikola Kasabov
Rajani K. Mudi   Srimanta Pal
Swapan K. Parui (Eds.)

Neural
Information Processing

11th International Conference, ICONIP 2004
Calcutta, India, November 22-25, 2004
Proceedings

 Springer

Volume Editors

Nikhil R. Pal
Srimanta Pal
Indian Statistical Institute
Electronics and Communications Sciences Unit
203 B. T. Road, Calcutta 700 108, India
E-mail: {nikhil,srimanta}@isical.ac.in

Nikola Kasabov
Auckland University of Technology
Knowledge Engineering and Discovery Research Institute (KEDRI)
Private Bag 92006, Auckland, New Zealand
E-mail: nik.kasabov@aut.ac.nz

Rajani K. Mudi
Jadavpur University
Department of Instrumentation and Electronics Engineering
Salt-lake Campus, Calcutta 700098, India
E-mail: rkmudi@iee.jusl.ac.in

Swapan K. Parui
Indian Statistical Institute
Computer Vision and Pattern Recognition Unit
203 B. T. Road, Calcutta 700 108, India
E-mail: swapan@isical.ac.in

Library of Congress Control Number: 2004115128

CR Subject Classification (1998): F.1, I.2, I.5, I.4, G.3, J.3, C.2.1, C.1.3, C.3

ISBN 978-3-540-23931-4   ISBN 978-3-540-30499-9 (eBook)
DOI 10.1007/978-3-540-30499-9

This work is subject to copyright. All rights are reserved, whether the whole or part of the material is concerned, specifically the rights of translation, reprinting, re-use of illustrations, recitation, broadcasting, reproduction on microfilms or in any other way, and storage in data banks. Duplication of this publication or parts thereof is permitted only under the provisions of the German Copyright Law of September 9, 1965, in its current version, and permission for use must always be obtained from Springer. Violations are liable to prosecution under the German Copyright Law.

springeronline.com

© Springer-Verlag Berlin Heidelberg 2004
Originally published by Springer-Verlag Berlin Heidelberg New York in 2004

Typesetting: Camera-ready by author, data conversion by Olgun Computergrafik
Printed on acid-free paper   SPIN: 11359166   06/3142   5 4 3 2 1 0

Preface

It is our great pleasure to welcome you to the *11th International Conference on Neural Information Processing* (ICONIP 2004) to be held in Calcutta. ICONIP 2004 is organized jointly by the Indian Statistical Institute (ISI) and Jadavpur University (JU). We are confident that ICONIP 2004, like the previous conferences in this series, will provide a forum for fruitful interaction and the exchange of ideas between the participants coming from all parts of the globe. ICONIP 2004 covers all major facets of computational intelligence, but, of course, with a primary emphasis on neural networks. We are sure that this meeting will be enjoyable academically and otherwise.

We are thankful to the track chairs and the reviewers for extending their support in various forms to make a sound technical program. Except for a few cases, where we could get only two review reports, each submitted paper was reviewed by at least three referees, and in some cases the revised versions were again checked by the referees. We had 470 submissions and it was not an easy task for us to select papers for a four-day conference. Because of the limited duration of the conference, based on the review reports we selected only about 40% of the contributed papers. Consequently, it is possible that some good papers are left out. We again express our sincere thanks to all referees for accomplishing a great job. In addition to 186 contributed papers, the proceedings includes two plenary presentations, four invited talks and 18 papers in four special sessions. The proceedings is organized into 26 coherent topical groups. We are proud to have a list of distinguished speakers including Profs. S. Amari, W.J. Freeman, N. Saitou, L. Chua, R. Eckmiller, E. Oja, and T. Yamakawa. We are happy to note that 27 different countries from all over the globe are represented by the authors, thereby making it a truly international event.

We are grateful to Prof. A.N. Basu, Vice-Chancellor, JU and Prof. K.B. Sinha, Director, ISI, who have taken special interest on many occasions to help the organizers in many ways and have supported us in making this conference a reality. Thanks are due to the Finance Chair, Prof. R. Bandyopadhyay, and the Tutorial Chair, Prof. B.B. Chaudhuri.

We want to express our sincere thanks to the members of the Advisory Committee for their timely suggestions and guidance. We sincerely acknowledge the wholehearted support provided by the members of the Organizing Committee. Special mention must be made of the organizing Co-chairs, Prof. D. Patranabis and Prof. J. Das for their initiative, cooperation, and leading roles in organizing the conference. The staff members of the Electronics and Communication Sciences Unit of ISI have done a great job and we express our thanks to them. We are also grateful to Mr. Subhasis Pal of the Computer and Statistical Services Center, ISI, for his continuous support. Things will remain incomplete unless we mention Mr. P.P. Mohanta, Mr. D. Chakraborty, Mr. D.K. Gayen and Mr. S.K. Shaw without whose help it would have been impossible for us to make this conference a success. We must have missed many other colleagues and friends who have helped us in many ways; we express our sincere thanks to them also.

We gratefully acknowledge the financial support provided by different organizations, as listed below. Their support helped us greatly to hold this conference on this scale.

Last, but surely not the least, we express our sincere thanks to Mr. Alfred Hofmann and Ms. Ursula Barth of Springer for their excellent support in bringing out the proceedings on time.

November 2004

Nikhil R. Pal
Nikola Kasabov
Rajani K. Mudi
Srimanta Pal
Swapan K. Parui

Funding Agencies

- Infosys Technologies Limited, India
- IBM India Research Lab, India
- Department of Science and Technology, Govt. of India
- Council of Scientific and Industrial Research, Govt. of India
- Reserve Bank of India
- Department of Biotechnology, Govt. of India
- Defence Research and Development Organization, Govt. of India
- Department of Higher Education, Govt. of West Bengal, India
- Jadavpur University, Calcutta, India
- Indian Statistical Institute, Calcutta, India

Organization

Organizers
Indian Statistical Institute, Calcutta, India
Jadavpur University, Calcutta, India
Computational Intelligence Society of India (CISI), India

Chief Patrons
K.B. Sinha, Indian Statistical Institute, India
A.N. Basu, Jadavpur University, India

Honorary Co-chairs
S. Amari, Riken Brain Science Institute, Japan
T. Kohonen, Neural Networks Research Centre, Finland

General Chair
N.R. Pal, Indian Statistical Institute, India

Vice Co-chairs
E. Oja, Helsinki University of Technology, Finland
R. Krishnapuram, IBM India Research Lab, India

Program Chair
N. Kasabov, University of Otago, New Zealand

Organizing Co-chairs
D. Patranabis, Jadavpur University, India
J. Das, Indian Statistical Institute, India

Joint Secretaries
S.K. Parui, Indian Statistical Institute, India
R.K. Mudi, Jadavpur University, India
S. Pal, Indian Statistical Institute, India

Tutorials Chair
B.B. Chaudhuri, Indian Statistical Institute, India

Finance Chair
R. Bandyopadhyay, Jadavpur University, India

Technical Sponsors
Asia Pasific Neural Network Assembly (APNNA)
World Federation on Soft Computing (WFSC)

Advisory Committee

A. Atiya, Cairo University, Egypt
Md.S. Bouhlel, National Engineering School of Sfax, Tunisia
S. Chakraborty, Institute of Engineering and Management, India
G. Coghill, University of Auckland, New Zealand
A. Engelbrecht, University of Pretoria, South Africa
D. Fogel, Natural Selection Inc., USA
K. Fukushima, Tokyo University of Technology, Japan
T. Gedeon, Australian National University, Australia
L. Giles, NEC Research Institute, USA
M. Gori, Università di Siena, Italy
R. Hecht-Nielsen, University of California, USA
W. Kanarkard, Ubonratchathani University, Thailand
O. Kaynak, Bogazici University, Turkey
S.V. Korobkova, Scientific Centre of Neurocomputers, Russia
S.Y. Lee, Korea Advanced Institute of Science and Technology, Korea
C.T. Lin, National Chiao-Tung University, Taiwan
D.D. Majumder, Indian Statistical Institute, India
M. Mitra, Jadavpur University, India
D. Moitra, Infosys Technologies Limited, India
L.M. Patnaik, Indian Institute of Science, India
W. Pedrycz, University of Alberta, Canada
S.B. Rao, Indian Statistical Institute, India
V. Ravindranath, National Brain Research Centre, India
J. Suykens, Katholieke Universiteit Leuven, Belgium
A.R. Thakur, Jadavpur University, India
S. Usui, Neuroinformatics Lab., RIKEN BSI, Japan
L. Wang, Nanyang Technological University, Singapore
L. Xu, Chinese University of Hong Kong, Hong Kong
T. Yamakawa, Kyushu Institute of Technology, Japan
Y.X. Zhong, University of Posts and Telecommunications, China
J. Zurada, University of Louisville, USA

Track Chairs

Quantum Computing	E. Behrman (USA)
Bayesian Computing	Z. Chan (New Zealand)
Bio-informatics	J.Y. Chang (Taiwan)
Support Vector Machines and Kernel Methods	V.S. Cherkassky (USA)
Biometrics	S.B. Cho (Korea)
Fuzzy, Neuro-fuzzy and Other Hybrid Systems	F.K. Chung (Hong Kong)
Time Series Prediction and Data Analysis	W. Duch (Poland)
Evolutionary Computation	T. Furuhashi (Japan)
Neuroinformatics	I. Hayashi (Japan)
Pattern Recognition	R. Kothari (India)
Control Systems	T.T. Lee (Taiwan)
Image Processing and Vision	M.T. Manry (USA)
Robotics	J.K. Mukherjee (India)
Novel Neural Network Architectures	M. Palaniswami (Australia)
Brain Study Models	V. Ravindranath (India)
Brain-like Computing	A. Roy (USA)
Learning Algorithms	P.N. Suganthan (Singapore)
Cognitive Science	R. Sun (USA)
Speech and Signal Processing	H. Szu (USA)
Computational Neuro-science	S. Usui (Japan)
Neural Network Hardware	T. Yamakawa (Japan)

Organizing Committee

K. Banerjee	U. Garai	P. Pal
J. Basak	Karmeshu	S. Raha
U. Bhattacharya	R. Kothari	Baldev Raj
B.B. Bhattacharya	S. Kumar	K. Ray
B. Chanda	K. Madhanmohan	K.S. Ray
N. Chatterjee	K. Majumdar	B.K. Roy
B.N. Chatterji	A.K. Mandal	K.K. Shukla
B. Dam	M. Mitra	B.P. Sinha
A.K. De	D.P. Mukherjee	B. Yegnanarayana
K. Deb	K. Mukhopadhyay	
U.B. Desai	P.K. Nandi	
B.K. Dutta	U. Pal	

Reviewers

S. Abe
A. Abraham
M. Alcy
A.S. Al-Hegami
N.M. Allinson
E. Alpaydin
L. Andrej
A. Arsenio
A. Atiya
M. Atsumi
M. Azim
T. Balachander
J. Basak
Y. Becerikli
R. Begg
L. Behera
Y. Bengio
U. Bhattacharya
C. Bhattacharyya
A.K. Bhaumik
A. Biem
Z. Bingul
S. Biswas
S.N. Biswas
M. Blumenstein
S. Buchala
A. Canuto
A.C. Carcamo
F.A.T. Carvaho
K.-M. Cha
D. Chakraborty
Debrup Chakraborty
S. Chakraborty
U. Chakraborty
C.-H. Chan
B. Chanda
H. Chandrasekaran
K. Chang
B.N. Chatterjee
B.B. Chaudhuri
H.-H. Chen
J. Chen
K. Chen
T.K. Chen

V. Cherkassky
W. Cheung
E. Cho
S. Cho
S.B. Cho
K. Chokshi
N. Chowdhury
A.A. Cohen
R.F. Correa
S. Daming
J. Das
N. Das
C.A. David
A.K. Dey
R. De
K. Deb
W.H. Delashmit
G. Deng
B. Dhara
G. Dimitris
M. Dong
Y. Dong
T. Downs
K. Doya
W. Duch
A. Dutta
D.P.W. Ellis
M. Er
P. E'rdi
M. Ertunc
A. Esposito
E.C. Eugene
X. Fan
Z.-G. Fan
O. Farooq
S. Franklin
M. Fukui
T. Furuhashi
M. Gabrea
M. Gallagher
U. Garain
A. Garcez
S.S. Ge
T. Gedeon

T.V. Geetha
A. Ghosh
S. Ghosh
B.G. Gibson
K. Gopalsamy
K.D. Gopichand
R. Gore
M. Grana
L. Guan
C. Guang
H. Guangbin
A. Hafez
M. Hagiwara
M. Hattori
I. Hayashi
G. Heidemann
G.Z. Chi
S. Himavathi
P.S. Hiremath
A. Hirose
L. Hongtao
C.-H. Hsieh
W.-Hsu
B. Huang
H.-D. Huang
Y.K. Hui
M.F. Hussin
S. Ikeda
T. Inoue
H. Ishibuchi
P. Jaganathan
M. Jalilian
M. Jalili-Kharaajoo
G. Ji
X. Jiang
L. Jinyan
T. Kalyani
T. Kambara
C.-Y. Kao
K. Karim
N. Kasabov
U. Kaymak
O. Kaynak
S.S. Khan

J.Y. Ki
J. Kim
J.-H. Kim
K. Kim
K.B. Kim
H. Kita
A. Koenig
M. Koppen
K. Kotani
R. Kothari
R. Kozma
K. Krishna
R. Krishnapuram
S.N. Kudoh
C. Kuiyu
A. Kumar
A.P. Kumar
S. Kumar
M.K. Kundu
Y. Kuroe
S. Kurogi
J. Kwok
H.Y. Kwon
J. Laaksonen
S. LaConte
A. Laha
D. Lai
S. Laine
R. Langari
J.-H. Lee
K.J. Lee
V.C.S. Lee
J. Li
P. Li
S. Lian
C. Lihui
C.-J. Lin
C.T. Lin
C. Liu
J. Liu
P. Lokuge
R. Lotlikar
M. Louwerse
C.L. Lu
T. Ludermir

C.K. Luk
P.-C. Lyu
Y. Maeda
S. Maitra
S.P. Maity
K. Majumdar
M. Mak
F.J. Maldonado
A.K. Mandal
J. Mandziuk
N. Mani
D.H. Manjaiah
M. Matsugu
Y. Matsuyama
B. McKay
O. Min
M. Mitra
S. Mitra
B.M. Mohan
P.P. Mohanta
R.K. Mudi
S. Mukherjea
A. Mukherjee
D.P. Mukherjee
J.K. Mukherjee
K. Mukhopadhyaya
A.K. Musla
W. Naeem
P. Nagabhushan
H. Najafi
T. Nakashima
P.K. Nanda
P. Narasimha
M. Nasipura
V.S. Nathan
G.S. Ng
A. Nijholt
G.S. Nim
D. Noelle
A. Ogawa
E. Oja
H. Okamoto
P.R. Oliveira
T. Omori
Y. Oysal

G.A.V. Pai
N.R. Pal
S. Pal
U. Pal
S. Palit
R. Panchal
J.-A. Park
K.R. Park
S.K. Parui
M.S. Patel
S.K. Patra
M. Perus
T.D. Pham
A.T.L. Phuan
H. Pilevar
M. Premaratne
S. Puthusserypady
S. Qing
M. Rajeswari
K.S. Rao
F. Rashidi
M. Rashidi
V. Ravindranath
B.K. Rout
A. Roy
P.K. Roy
R. Roy
J. Ruiz-del-Solar
S. Saha
S. Saharia
A.D. Sahasrabudhe
S. Sahin
J.S. Sahmbi
M. Sakalli
A.R. Saravanan
S.N. Sarbadhikari
P. Sarkar
P.S. Sastry
A. Savran
H. Sawai
A. Saxena
C.C. Sekhar
A. Sharma
C. Shaw
B.H. Shekar

P.K. Shetty
Z. Shi
C.N. Shivaji
P. Shivakumara
K.K. Shukla
A.P. Silva Lins
M.J. Silva Valenca
J.K. Sing
R. Singh
S. Singh
S. Sinha
M. Sirola
G. Sita
K.R. Sivaramakrishnan
J. Sjoberg
K. Smith
X. Song
M.C.P. de Souto
A. Sowmya
R. Srikanth
B. Srinivasan
P.N. Suganthan
C. Sun
R. Sun
E. Sung
V. Suresh
J. Suykens

R. Tadeusiewicz
P.K.S. Tam
H. Tamaki
C.Y. Tang
E.K. Tang
P. Thompson
K.-A. Toh
A. Torda
V. Torra
D. Tran
J. Uchida
S. Usui
P. Vadakkepat
B. Valsa
D. Ventura
A. Verma
B. Verma
J. Vesanto
E. Vijayakumar
D. Wang
J. Wang
L. Wang
S. Wang
J. Watada
O. Watanabe
J. Wei
A. Wichert

R.H.S. Wong
K.-W. Wong (Kevin)
B. Xia
C. Yamaguchi
Y. Yamaguchi
T. Yamakawa
L. Yao
S. Yasui
Z. Yeh
D.C.S. Yeung
H. Yigit
C.-G. Yoo
N.M. Young
C. Yu
M. Yunqian
C. Zanchettin
Z. Zenn Bien
B.-T. Zhang
D. Zhang
L. Zhang
Q. Zhang
Y. Zhang
L. Zhiying
S. Zhong
D. Zhou
H. Zujun

Table of Contents

Computational Neuroscience

Complex-Valued Neural Networks

Self-organizing Maps

Evolutionary Computation

Control Systems

Cognitive Science

Biometrics

Adaptive Intelligent Systems

Brain-Like Computing

Learning Algorithms

Novel Neural Networks

Image Processing

Pattern Recognition

Neuroinformatics

Fuzzy Systems

Neuro-fuzzy Systems

Hybrid Systems

Ant Colony

Neural Network Hardware

Robotics

Signal Processing

Support Vector Machine

Time Series Prediction

Bioinformatics

Neurobiological Foundation
for the Meaning of Information

Walter J. Freeman

Department of Molecular and Cell Biology
University of California
Berkeley CA 94720-3206 USA
http://sulcus.berkeley.edu

Abstract. Brains create meaning and express it in information. They select and pre-process the information carried by sensory stimuli as sense data, from which they construct meaning. They post-process cognitive meaning into informative commands that control the goal-directed actions that express meaning. Meaning exists in the interaction of subjects with their environments. The process of perception by which brains construct meaning from information can be explained by analyzing the neural activity in human and animal brains as subjects engage in meaningful behaviors. Measurement is followed by decomposition and modeling of the neural activity in order to deduce brain operations. Brains function hierarchically with neuronal interactions within and between three levels: microscopic of single neurons, mesoscopic of local networks forming modules, and macroscopic of the global self-organization of the cerebral hemispheres by the organic unity of neocortex. Information is carried in continuous streams of microscopic axonal pulses. Meaning is carried in mesoscopic local mean fields of dendritic currents in discontinuous frames resembling cinemas, each frame having a spatial pattern of amplitude modulation of an aperiodic carrier wave.

1 Introduction

James Barham [2] laid a foundation in physics for a theory of meaning in terms of nonequilibrium thermodynamics and the nonlinear dynamics of coupled oscillators. He described these oscillators as self-governed by attractors in phase space. He proposed that a biological system should be characterized as a generalized nonlinear oscillator that is stabilized far from thermodynamic equilibrium by means of successive phase transitions. The stability is achieved by effective interaction of the high-energy system with other high-energy oscillators sharing the environment that serve as constraints. Effective interactions are by thermodynamic engagement of the inner and outer high-energy oscillators (for example, attack and consumption by a predator or evasion and escape by its prey). Both predator in search of food and its prey in search of shelter are high-energy oscillators. The predator is stabilized when it captures and consumes its prey. The prey is stabilized when it escapes and finds shelter (Fig. 1).

Information exists in low-energy environmental energy fluxes that are correlated with the high-energy fluxes so as to serve as signals of distant events. Examples are the sights, sounds and odors of both predator and prey. He called the brain counterpart of an environmental low-energy flux an "epistemon" and identified it with a

N.R. Pal et al. (Eds.): ICONIP 2004, LNCS 3316, pp. 1–9, 2004.
© Springer-Verlag Berlin Heidelberg 2004

chaotic attractor in the attractor landscape of a sensory system. He interpreted the meaning of the information in the low-energy flux as the "prediction of successful functional action" [2, p. 235], so that information could be either correct or wrong. In biological terms, the meaning of a stimulus for an organism is demonstrated by the use to which it is put, which was described by J. J. Gibson [21] as its 'affordance'.

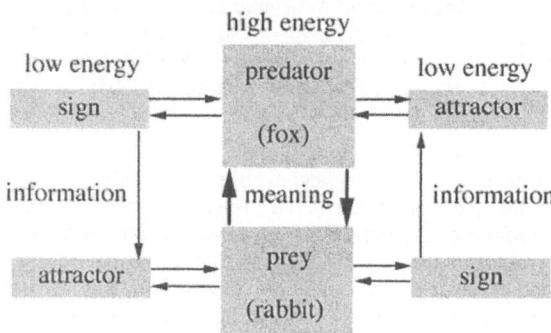

Fig. 1. The dynamic relation between meaning and information is illustrate

The meaning is valid when it leads to successful adaptation to environmental constraints through the action-perception cycle [27, 26]. The problem to be solved in order to apply this theory is to translate Barham's "epistemon" (sign) and the attractor and its basin of attraction into terms of neurodynamics. In order to test and validate the theory, it must also be recast in engineering terms. For example, the meaning for robots of low-energy environmental constraints would be observed in the successful adaptation of autonomous intentional robots to cope successfully with environmental challenges without the intervention of observers or controllers [25].

2   Experimental Foundation

A major constraint in devising a theory of meaning is that no physical or chemical measurement of brain activity is a direct measure of meaning. Meaning can be experienced subjectively in oneself, and one can infer it in other subjects from the behavioral context in which measurements are made, but one cannot measure it to express it in numbers. An equally important constraint is that no measurement of brain activity makes sense unless the investigator has sufficient control over the behavior of a subject to be able to infer the teleology of the subject, whether animal or human, at the time of measurement. The teleology includes the history, intention, expectation, motivation, and attentiveness of the subject. For this reason, all of the data on brain function must be accumulated from studies in which the subjects are carefully trained or coached to enter and maintain overt states of normal behavior that can be reproduced and measured along with the measurements of brain activity.

Yet another requirement is stability. Bak, Tang and Wiesenfeld [1] proposed that a complex system such as a brain evolves by self-organization to a critical state at the edge of chaos, by which it maintains a readiness to adapt rapidly to unpredictable changes in its environment and thereby maintain its integrity in accord with Barham's [2] theory. Adaptation is by repetitive phase transitions; the space-time patterns of its state variables re-organize themselves abruptly and repeatedly. His prime example was the performance of a sand pile, in which a steady drip of grains of sand onto the central peak gave the pile the shape of a cone. The slope of the cone increased to a maximum that was maintained by repeated avalanches as sand continued to pour onto the apex. The avalanches had fractal distributions in size and time intervals. Bak called this a state of "self-organized criticality" (SOC), and he characterized it by the fractal distributions and the $1/f^{\alpha}$ form of the temporal spectra of the avalanches with α as the critical exponent. He concluded that the $1/f^{\alpha}$ spectra were explained by the self-similarity of the recurrent events over broad scales of time and space.

Recent advances in technology have made it possible to for neurobiologists to observe the electrochemical oscillations of energy that enable brains to maintain their states far from equilibrium at the edge of stability [9]. Interactive populations of neurons are nonlinear oscillators that create and maintain landscapes of chaotic attractors. Their oscillatory activity in primary sensory cortices can be observed, measured and analyzed by simultaneously recording the EEG [3] of multiple populations with high-density arrays of electrodes placed epidurally over the cortex. This technique is feasible, because the main source of EEG potentials is the sum of the dendritic currents of the neurons in local neighborhoods that control the firing rates of the action potentials. That sum is accompanied by extracellular potential differences giving access to spatiotemporal patterns of the local mean fields [8, 9, 10].

The $1/f^{\alpha}$ form of the EEG PSD (Fig. 2) has been repeatedly demonstrated in both temporal spectra [4, 23, 31] and spatial spectra of EEG recorded intracranially in animals [4, 14] and neurosurgical patients [20].

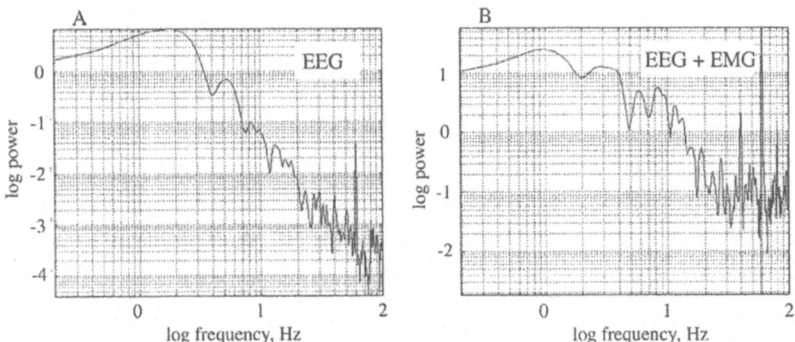

Fig. 2. Upper frame: Examples of scalp EEG PSD from frontal cortex of a subject at rest, eyes open. Lower frame: EEG + EMG from tensing scalp muscles. 1/f holds for EEG, not for EMG (electromyogram means muscle potentials, which approximate white noise)

Although EEG and EMG (the electromyogram from the action potentials of scalp muscles) appear to be similar in their time series, they differ significantly, because the spectrum of EMG tends to be flat like that of white noise, not $1/f^{\alpha}$ of EEG [17]. However, the temporal EEG spectra usually include prominent peaks in the clinical bands of theta (3-7 Hz), alpha (7-12 Hz), beta (12-30 Hz), and gamma (30-80 Hz), so that SOC cannot explain all of cortical dynamics. In particular, inclusion of the limbic and thalamic controls of cortical function is essential for modeling brain function [5,

30, 32], but the focus here is on the intrinsic macroscopic properties of cortical activity, the self-organized properties of which are modulated and controlled by the brain stem nuclei, basal ganglia and thalamus.

These spatial patterns reveal "itinerant trajectories" through successions of chaotic attractors, which begin to dissolve into "attractor ruins" as soon as they are accessed [34]. The patterns are recorded with high-density electrode arrays, intracranially on or in the brains of cats, rabbits, and neurosurgical patients, and from the scalps of normal volunteers. Each attractor forms during training of the subjects by reinforcement to discriminate sensory stimuli. An attractor that is selected by the information in sensory input is realized in the spatial pattern of amplitude modulation (AM) of a chaotic carrier wave. Spatial AM patterns form repeatedly with exposure to the conditioned stimulus that is paired with an unconditional stimulus (Fig. 3). Each pattern is measured with an array of 64 electrodes on the cortical surface. Signal identification and pattern classification have been done with high temporal resolution using Fourier decomposition [4, 15], wavelets [18] and the Hilbert transform [3, 19]. The differences among the 64 EEGs across the array for each AM pattern are expressed by the 64 amplitudes that specify a 64x1 column vector and a point in 64-space [9].

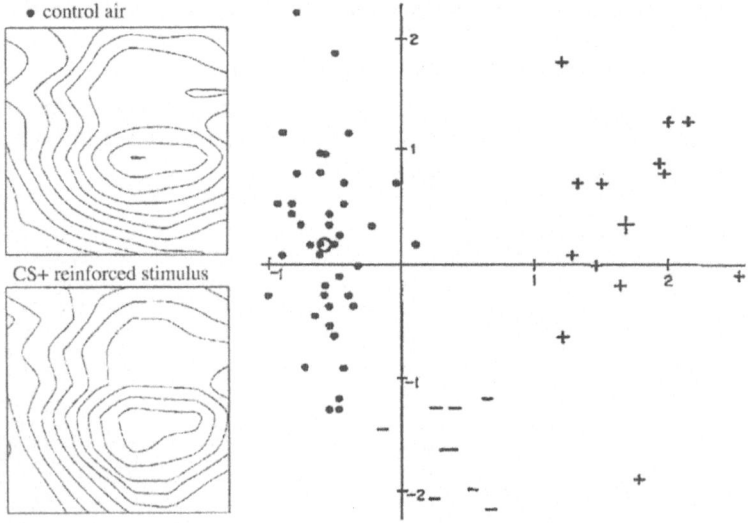

Fig. 3. Left frames: Examples of RMS patterns of amplitude modulation of gamma bursts. Right frame: Classification by stepwise discriminant analysis of bursts from trials with conditioned stimuli, reinforced (+) or not (-) compared with bursts breathing control air

The AM patterns are never twice identical; however, because of their similarity within the class, they form a cluster in 64-space, in which the center of gravity specifies an average AM pattern. A cluster forms for each stimulus that a subject learns to discriminate. Statistical classification of EEG patterns is done by assigning membership to clusters on the basis of minimizing Euclidian distances in 64-space. The basin of attraction provides for abstraction and generalization that is needed to define the class of each stimulus. The site of the cluster in 64-space changes whenever the meaning of the stimulus changes, revealing the adaptability of the mechanism for classification, and the unique dependence on the cumulative experience of each subject.

3 Phase Transitions
Constituting 'Cinematographic' Cortical Dynamics

A unique2-D phase gradient in the form of a cone has been found to accompany each AM pattern in the olfactory bulb and also in the visual, auditory and somatic primary sensory cortices. The phase velocities were commensurate with the distribution of conduction velocities of intrabulbar and intracortical axons running parallel to the surfaces [15, 13]. As a result, the modal half-power diameter (15 mm) and the 95% upper inclusion range (28 mm) of neocortical AM patterns were substantially larger than the bulbar surface (10 mm). Unlike the bulbar EEG in which the phase velocity was invariant with gamma frequency [14], in the neocortical EEG the phase velocity co-varied with gamma frequency, but the half-power diameter did not.

The conclusion was drawn that visual, auditory, somatosensory, and olfactory receiving areas had the capacity for input-dependent gain increase [7] leading to destabilization. Emergence of self-organized mesoscopic patterns was by a 1st order phase transition that was completed within 3-7 ms depending on the center carrier frequency, independently of the radius of the conic section that accompanied the AM pattern. The location, time of onset, size and duration of each wave packet were demarcated by the phase, whereas its perceptual content was expressed in an AM pattern, which appeared within 25-35 ms of the wave packet onset [11, 12]. The content, as defined by classification with respect to CSs, was context-dependent, unique to each subject, and it was distributed with long-range correlations over delimited domains of both the cortical surface and the gamma spectrum. Clearly the content did not come directly from the stimulus, nor was it imported from other parts of the brain. It was realized in the landscape by the selection of an appropriate attractor by the stimulus. It was the phase transition that released the cortex from its existing state, giving it the degree of freedom necessary to advance to a new state along the chaotic itinerant trajectory.

4 Transition by Anomalous Dispersion

The high phase velocities were of exceptional interest, because they greatly exceeded the velocities of serial synaptic transmission across the bulb and neocortical sensory areas. For example, the modal radius of the axon collaterals parallel to the surface from bulbar excitatory neurons was about .5 mm, and the time for transmission of an impulse input by synaptic relay over the entire bulb (about 10 mm) by convolution would require about 100 ms, about 20-fold greater than the observed time required [7]. A comparable distinction is made between group velocity and phase velocity in media that conduct light [22, p. 42 and p. 205]. The transmission of energy and information in such media can never exceed the speed of light, but when the frequency of the carrier light is close to an absorption or resonance band of a medium, the phase velocity can appear to exceed the group velocity. The excess in apparent velocity above the speed of light manifests "anomalous dispersion". It does not carry information.

By analogy, the maintenance in cortex by self-organized criticality of a resonance band might correspond to an absorption band in light media. Whereas the group velocity would correspond to the average rate of serial synaptic transmission of information by action potentials from one cortical area to another, the phase velocity would correspond to the spatial rate of spread of a phase transition across the cortex. Anomalous dispersion in the bulb or neocortex could not carry information at the phase velocity exceeding the limiting velocity of group (serial synaptic) transmission, but it might trigger the expression of information previously stored in synaptic weights into the spatial AM patterns of gamma oscillations, with negligible time lags between widely separated areas. The phase transitions clearly involve thalamic controls, but the global coordination of the timing and content of beta and gamma oscillations, even over the entire extent of both cerebral hemispheres, may be an intrinsic property of the neocortex viewed as an integrated tissue. The high speed of the phase transition can be explained by the small number of long axons in bulb and cortex with high conduction velocities. Because cortex is maintained at the edge of stability, only a small push is needed to cross over a separatrix to a new basin. The long axons can provide the push by virtue of small world effects [36] and the scale-free dynamics of neocortex [35, 12].

Evidence for global phase transitions has now been accrued from scalp EEG recording in normal human volunteers at recurrence rates in the alpha range. They are most clearly seen in the time derivative of the instantaneous phase of the EEG calculated from the Hilbert transform [16] (Fig. 4). The distances of distribution of the

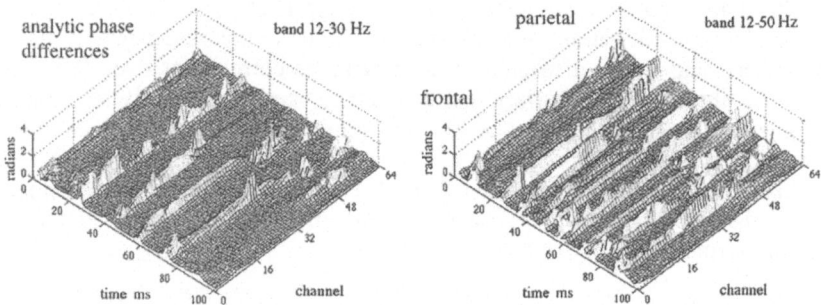

Fig. 4. Coordinated analytic phase differences (CAPD) were calculated for two pass bands of the EEG. The scalp EEG was recorded with a 1x64 electrode array spaced 3 mm from the forehead to the occiput. The derivative of the analytic phase from Hilbert transform of the beta EEG was approximated by time differences [16]. This function revealed an abrupt change in phase that took place over the frontal lobe and the parietal lobe but not at the same times. The time differences of the jump were under the 5 ms resolution of the digitizing. At other times in other subjects the phase locking was over the parietal or occipital area, or over all areas

The statistical relations between the unfiltered EEG averaged across the 64 channels and the standard deviation (SD) of the coordinated analytic phase differences were investigated by cross-correlating the two time series from the array over a time period of 5 sec (1000 time points at 5 ms sample interval) and calculating its power spectral density using the FFT. In subjects with eyes closed and at rest a prominent peak appeared in the autospectrum of the EEG and also in the cospectrum of the EEG and the CAPD cross-correlation (Fig. 5). When the subjects opened their eyes or engaged in the intentional action of tensing their scalp muscles to produce controlled amounts of muscle noise (EMG), the alpha peak (8-12 Hz) disappeared from the autospectrum and cospectrum, and a theta peak (3-7 Hz) often appeared. The results indicate that the CAPD are manifestations of the cinematographic dynamics of neocortex, which is constituted by successive frames formed by phase transitions, and that the theta and alpha rhythms are the macroscopic manifestations of this on-going process. The disappearance of alpha waves ('alpha blocking') appears not to be 'desynchronization' but 'deperiodicization' owing to an increase in the mean wave length of the oscillation and an increase in its variance; hence the appearance of alpha blocking is the result of applying linear analysis to the output of a nonlinear system.

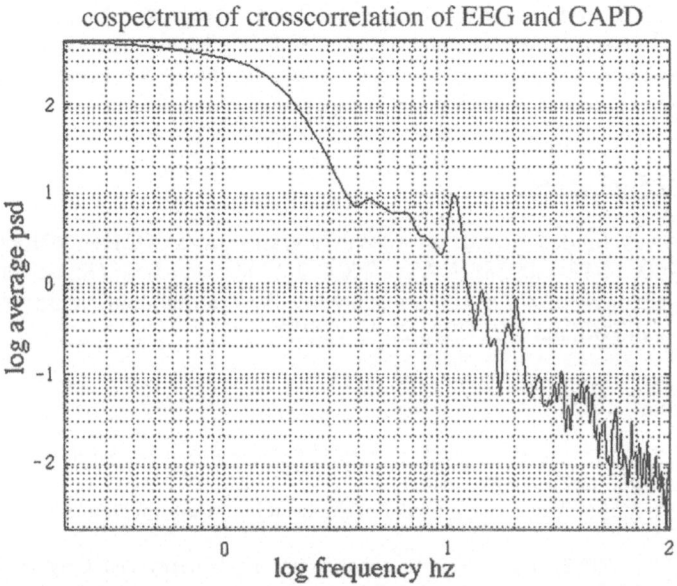

Fig. 5. The cospectrum of the crosscorrelation between the EEG and the coordinated analytic phase differences revealed a peak in the alpha range (8-12 Hz) or in the theta range (3-7 Hz). This finding indicated that the recurrence rate of the global state transitions was most prominent in the alpha band (as shown in Fig. 5 in a subject with eyes closed), and in the theta band in a subject with eyes open (F9g. 2). The close relationship documented in [17] implies that the theta and alpha rhythms may manifest the frame rates at which AM patterns succeed one another at rest and during the course of thought

A starting point is to visualize the receptor input from the retinas that enables a person to recognize a familiar face breaking into a smile [10, 12]. Sensory receptors are selective for types of environmental energy, not for information. The cortex must constantly receive an enormous barrage of action potentials that induces in the visual cortex explosions of action potentials from all of the motion, edge, and color detectors in the visual field. Experimental evidence summarized here indicates that any heightening of activity as during saccades can destabilize the cortex and induce the formation of sequences of brief spatial patterns of neural activity. The phenomenon of coalescence or condensation may resemble the formation by fish, birds and insects of schools, flocks and swarms [6] or by water molecules into rain drops and snowflakes. Individuals synchronize their activities to conform to the whole, yet they retain their autonomy. In the sensory cortices the patterns bind only a small fraction of the total variance of "neural swarms", so the patterns may not be observable in recording from one or a few neurons. They can be observed by multichannel EEG recording with high-density arrays of electrodes placed epidurally over the cortex. This is because the main source of EEG potentials is the sum of the dendritic currents of the neurons in local neighborhoods that control the firing rates of the action potentials. That sum is accompanied by extracellular potential differences that give access to the local mean fields of activity governing the collective behavior [8, 9].

By these techniques meaning is seen to exist in the relations between each animal and human individual and its environment that is shared with others. Thus meaning is ontological; it is understood epistemologically in three ways: by phenomenological experience, by seeing goal-directed actions, and now for the first time by witnessing the patterns of neural activity that form within brains in the normal course of the creation and exercise of meaning.

Acknowledgements

This work was funded in part by research grants from NIMH (MH06686), ONR (N63373 N00014-93-1-0938), NASA (NCC 2-1244), and NSF (EIA-0130352). This report was adapted from a lecture presented to the International Neural Network Society IJCNN in Portland OR, USA, 22 July 2003.

References

1. Bak, P., Tang, C. and Wiesenfeld, K. (1987) Self-organized criticality: an explanation of 1/f noise. *Phys. Rev. Lett.* **59**: 364-374.
2. Barham, J. (1996) A dynamical model of the meaning of information. *Biosystems* **38**: 235-241.
3. Barlow, J.S. (1993) The Electroencephalogram: Its Patterns and Origins. (MIT Press, Cambridge MA).
4. Barrie, J.M., Freeman, W.J. and Lenhart, M.D. (1996) Spatiotemporal analysis of prepyriform, visual, auditory and somesthetic surface EEG in trained rabbits. *J. Neurophysiol.* **76**: 520 539.
5. Destexhe, A. (2000) Modeling corticothalamic feedback and the gating of the thalamus by the cerebral cortex. *J. Physiol.-Paris* **94**: 91-410.
6. Edelstein-Keshet, L., Watmough, J. and Grunbaum, D. (1998) Do traveling band solutions describe cohesive swarms? An investigation for migratory locusts. *J. Math. Biol.* **171**: 515-549.
7. Freeman, W.J. (1975) *Mass Action in the Nervous System.* (Academic Press, New York).
8. Freeman, WJ (1992) Tutorial in Neurobiology: From Single Neurons to Brain Chaos. *Int. J. Bifurc. Chaos* **2**: 451-482.
9. Freeman, W.J. (2000) Neurodynamics. An Exploration of Mesoscopic Brain Dynamics. (Springer-Verlag, London.
10. Freeman, WJ. (2003a) A neurobiological theory of meaning in perception. Part 1. Information and meaning in nonconvergent and nonlocal brain dynamics. Int. J. Bifurc. Chaos**13**: 2493-2511.
11. Freeman WJ. (2003b) A neurobiological theory of meaning in perception. Part 2. Spatial patterns of phase in gamma EEG from primary sensory cortices measured by nonlinear wavelets. Intern J Bifurc. Chaos**13**: 2513-2535.

12. Freeman WJ. (2004a) Origin, structure and role of background EEG activity. Part 1. Analytic amplitude. Clin. Neurophysiol. 115: 2077-2088.
13. Freeman WJ. (2004b) Origin, structure and role of background EEG activity. Part 2. Analytic phase. Clin. Neurophysiol. 115: 2089-2107.
14. Freeman, W.J. and Baird, B. (1987) Relation of olfactory EEG to behavior: Spatial analysis. *Behav. Neurosci.* **101**: 393-408.
15. Freeman, W.J. and Barrie, J.M. (2000) Analysis of spatial patterns of phase in neocortical gamma EEG in rabbit. *J. Neurophysiol.* **84**: 1266-1278.
16. Freeman, W.J., Burke, BC and Holmes, M.D. (2003) Application of Hilbert transform to scalp EEG. Human Brain Mapping 19(4):248-272.
17. Freeman, W.J., Burke, B.C., Holmes, M.D. and Vanhatalo, S. (2003) Spatial spectra of scalp EEG and EMG from awake humans. Clin. Neurophysiol. **114**: 1055-1060..
18. Freeman, W.J. and Grajski, K.A.(1987) Relation of olfactory EEG to behavior: Factor analysis. *Behav. Neurosci.* **101**: 766-777.
19. Freeman, W.J. and Rogers, L.J. (2002) Fine temporal resolution of analytic phase reveals episodic synchronization by state transitions in gamma EEG. *J. Neurophysiol.* **87**, 937-945.
20. Freeman, W.J., Rogers, L.J., Holmes, M.D. and Silbergeld, D.L. (2000) Spatial spectral analysis of human electrocorticograms including the alpha and gamma bands. *J. Neurosci. Meth.* **95**: 111-121.
21. Gibson JJ (1979) The Ecological Approach to Visual Perception. Boston: Haughton Mifflin.
22. Hecht, E. and Zajac, A. (1974) *Optics.* (Addison-Wesley, Reading MA), pp. 38-42, 205-205.
23. Hwa, R.C. and Ferree, T. (2002) Scaling properties of fluctuations in the human electroencephalogram. *Physical Rev. E* **66**: 021901.
24. Kozma, R. and Freeman WJ (2001) Chaotic Resonance: Methods and applications for robust classification of noisy and variable patterns. *Int. J. Bifurc. Chaos* **10**: 2307-2322.
25. Kozma, R., Freeman, W.J. and Erdí, P. (2003) The KIV model - Nonlinear spatiotemporal dynamics of the primordial vertebrate forebrain. Neurocomputing **52**: 819-826.
26. Merleau-Ponty; M. (1945/1962) *Phenomenology of Perception.* (C Smith, Trans.). (Humanities Press, New York).
27. Piaget, J. (1930) *The Child's Conception of Physical Causality.* (Harcourt, Brace, New York).
28. Principe, J.C., Tavares, V.G., Harris, J.G. and Freeman, W.J. (2001) Design and implementation of a biologically realistic olfactory cortex in analog VLSI. *Proc. IEEE* **89**: 1030-1051.
29. Robinson, P.A., Wright, J.J. and Rennie, C.J. (1998) Synchronous oscillations in the cerebral cortex. *Phys. Rev. E* **57**: 4578-4588.
30. Robinson, P.A., Loxley, P.N., O'Connor, S.C. and Rennie, C.J. (2001) Modal analysis of corticothalamic dynamics, electroencephalographic spectra, and evoked potentials. *Physical Rev. E* **6304**: #041909.
31. Srinivasan, R., Nunez, P. L. and Silberstein, R. B. (1998) Spatial filtering and neocortical dynamics: estimates of EEG coherence. *IEEE Trans. Biomed Engin.* **45**: 814-826.
32. Steriade, M. (2000) Corticothalamic resonance, states of vigilance and mentation. *Neurosci.* **101**: 243-276.
33. Taylor, J.G. (1997) Neural networks for consciousness. *Neural Networks* **10**: 1207-1225.
34. Tsuda, I. (2001) Toward an interpretation of dynamics neural activity in terms of chaotic dynamical systems. Behav. Brain Sci. 24: 793-847.
35. Wang XF, Chen GR. (2003) Complex networks: small-world, scale-free and beyond. IEEE Trans. Circuits Syst. 2003, 31: 6-20.
36. Watts, D.J. and Strogatz, S.H. (1998) Collective dynamics of 'small world' networks. *Nature* **394**: 440-442.

Neural Information Processing Efforts to Restore Vision in the Blind

Rolf Eckmiller, Oliver Baruth, and Dirk Neumann

Department of Computer Science University of Bonn, 53117 Bonn, Germany
eckmiller@nero.uni-bonn.de

Abstract. Retina Implants belong to the most advanced and truly 'visionary' man-machine interfaces. Such neural prostheses for retinally blind humans with previous visual experience require technical information processing modules (in addition to implanted microcontact arrays for communication with the remaining intact central visual system) to simulate the complex mapping operation of the 5-layered retina and to generate a parallel, asynchronous data stream of neural impulses corresponding to a given optical input pattern. In this paper we propose a model of the human visual system from the information science perspective. We describe the unique information processing approaches implemented in a learning Retina Encoder (RE), which functionally mimics parts the central human retina and which allows an individual optimization of the RE mapping operation by means of iterative tuning using learning algorithms in a dialog between implant wearing subject and RE.

1 Introduction

Blindness in humans and some animals can be caused by retinal degenerative defects (especially: retinitis pigmentosa, RP and macular degeneration, MD), which gradually lead to total blindness. Degeneration of the light-sensitive photoreceptor layer (Fig. 1) typically triggers various pathological processes, which irreversibly destroy the physiological intra-retinal structure and the complex retinal mapping function [2], [19]. However, a significant number of retinal ganglion cells (Fig. 1) forming the optic nerve at the 'retinal output' as well as subsequent parts of the central visual system often remain intact. Several studies have shown that electrical stimulation with microcontact foils at the ganglion cell layer elicit evoked responses in the monkey visual cortex [6] and even visual sensations in humans [8]. More recently, it could even be demonstrated [11] that local electrical stimulation with chronically implanted electrodes at the retinal ganglion cell layer in blind RP-patients yielded certain localized visual sensations.

Let us assume here that the microcontact -, biocompatibility-, and implantation issues, which absorb most of the current research efforts towards visual implants are being solved. We are then confronted with the following information processing challenges:

A- What are the optimal stimulation signal time courses for an implanted microcontact foil with 20 or 100 stimulation electrodes, which may stimulate more than one ganglion cell per electrode?

B- How will the central visual system respond if only a small fraction of presumably several thousand ganglion cells, which participate with its spatio-temporal filter properties (see below) in the mapping operations of the central retina under physiological conditions, will receive stimulation patterns?

N.R. Pal et al. (Eds.): ICONIP 2004, LNCS 3316, pp. 10–18, 2004.
© Springer-Verlag Berlin Heidelberg 2004

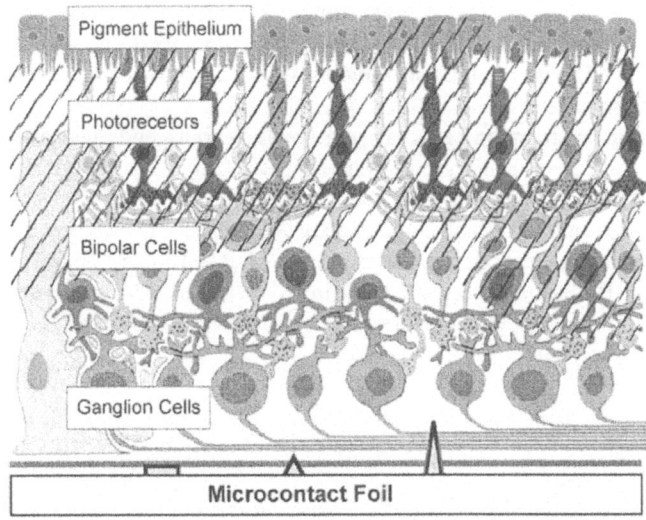

Fig. 1. Schematic cross section through part of a vertebrate retina with partial degeneration. The retina covers large portions of the inside of the eye ball. Counter-intuitively, light reaches the photoreceptors from below, which requires that the physiological retina consists of five layers o translucent neurons.

Top layer: retinal pigment epithelium. The neural retina below consists of the photoreceptor layer with rods and cones, the horizontal cell layer (not marked); the layer of bipolar cells, which form connections between photoreceptors above and ganglion cells below; the amacrine cell layer (not marked), and the layer of ganglion cells with thin nerve fibers (axons) extending horizontally to the right, which eventually form the optic nerve as the exclusive communication link to the central visual system. The diagonally hatched region indicates the assumed degenerated / dysfunctional part of the retina including part of the pigment epithelium. The microcontact foil at the bottom is depicted schematically with three possible shapes of microcontacts for highly localized electrical stimulation of ganglion cells and / or fibers.

C- How can a human subject after for example 10 years of blindness 're-activate' the central visual system and help to optimize the required RE mapping operation from an optical pattern to a corresponding stimulation data stream for the individually implanted set of microcontacts with an unpredictable positioning of electrodes relative to ganglion cells, from which the central visual system may still 'expect and remember' unique (pattern-specific) activity time courses as part of the retina mapping and pre-processing function?

These fundamental questions are being addressed in the following paragraphs.

2 Concept of the Visual System with Two Mapping Functions

From the information science perspective, the primate visual system (Fig. 2) consists of a retina module as a large ensemble of spatio-temporal (ST) filters [5] represented by the receptive field (RF) properties of mostly P- and M-type ganglion cells [3], [13], [21]. Based on the neurobiological data on structure and function of these P- and M-cells, the ST filters can be simulated with concentric spatial Difference-of-Gaussian (DoG)- or wavelet-type characteristics, which generate non-invertible, ambiguous output signals in that a given output signal can be caused by a number of different input signals. These M- and P-type ST filters feed into a central visual system module (VM) with the task of inducing visual percepts P2 corresponding to optical input patterns P1.

Human visual perception [12], [16], [17] which transcends neuroscience and biophysics, is considered in this visual system model (Fig. 2) as the result of a sequence of two unidirectional mapping operations: mapping 1 of an optical pattern P1 in the physical domain onto a retinal output vector $\underline{R1}$(t) in the neural domain by means of the retina module as encoder and mapping 2 of $\underline{R1}$(t) in the neural domain onto a visual percept P2 in the perceptual domain by means of VM. At present, neither the neurophysiological correlates nor the neuroanatomical location of the visual perceptual domain are certain.

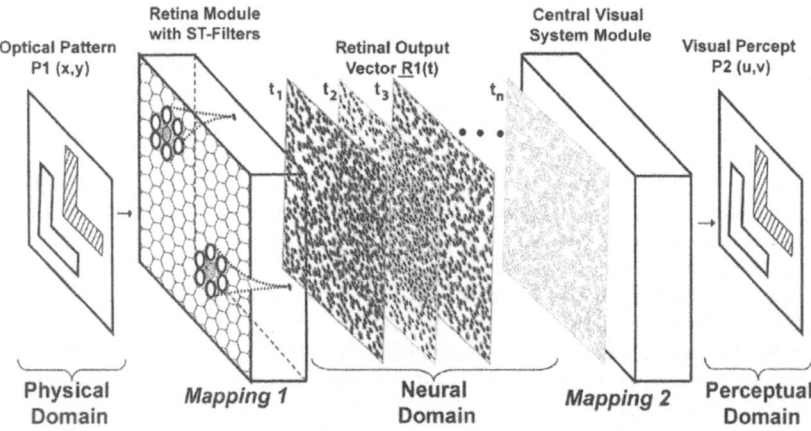

Fig. 2. Schema of the primate visual system as sequence of two mapping operations. Left: Optical pattern P1 in the physical domain as input pattern; retina module with spatio-temporal ST filters to perform mapping 1 from the physical domain at the input onto the neural domain at the output. Middle: retinal output vector $\underline{R1}$(t) as output pattern represented by a parallel data stream of neural impulse activity depicted by data frames from t_1 to t_n traveling along the optic nerve. Right: central visual system module representing all participating visual and oculomotor brain -structures and -functions to perform mapping 2 from a $\underline{R1}$(t) at the retinal output onto a visual percept P2 in the perceptual domain

The model of the visual system (Fig. 2) is based on the following assumptions:

(A) The neural activity data stream as retinal output vector $\underline{R1}$(t), which is generated for a given optical pattern P1 by the ensemble of about 1 million retinal ganglion cells as a result of a complex intra-retinal mapping 1 operation must in principle be invertible back into the originating input pattern P1 in all geometrical detail. This postulate implies that the human visual system, in contrast to a pattern recognition module, is capable of capturing all features of optical patterns in order to subsequently perform different mapping operations with selected emphasis on movement, contrast, shape, recognition, similarity, etc.

(B) The typical receptive field properties of primate retinal ganglion cells (P- or M-cells) are highly ambiguous for most input configurations and represent non-invertible ST filter operations.

(C) The central visual system represents a kind of 'Gate' between the neural domain within the objectively accessible realm of information science, neuroscience, and physics and the perceptual domain within the subjectively accessible realm of psychophysics, psychology, and philosophy [15]. Although mapping 2 of VM (Fig. 2) is by no means understood, it is treated in our model as the mathematically exact inverse of mapping 1. This is because any mapping 2 generating an exact, unique representation of pattern P1 has to resolve the ambiguity in $\underline{R1}$(t) and has to 'reconstruct' all geometrical (and even dynamic) details of P1. In other words, mapping 2, which leads into the perceptual domain has to be in a way at least with regard to certain pattern properties equivalent to an inverse of mapping 1. This postulate implies that a given visual percept P2 – for the purpose of this model – has a unique, unambiguous correspondence to one optical pattern P1.

(D) The ambiguity of many components of the retinal output vector $\underline{R1}$(t) can be partly removed by logical, decision tree-type information processing [9] and partly by micro eye movements under control of the central visual system during fixation [14], [18] to shift the projection of P1 on the retina by a very small defined amount. A detailed description of a novel retina encoder (RE*) based on this model is presented elsewhere [9], [10].

3   Retina Implant: Functional Replacement of the Human Retina

Fig. 3 depicts the retina implant schematically [4]. The retina encoder (RE) [7], [5] (Fig. 4), which can be located in a frame of glasses, has a photosensor array with well over 100,000 pixels at the input and about 100 to 1,000 technical ganglion cell out-

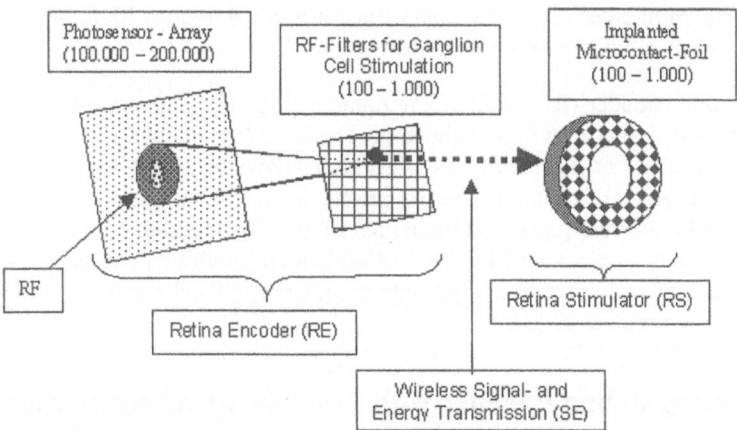

Fig. 3. Principle of a learning retina implant. Inside the eye, implanted adjacent to the retinal ganglion cell layer is the retina stimulator (RS) to elicit neural activity sequences. Outside the eye is the retina encoder (RE), composed of a number of tunable spatio-temporal (ST) filters with receptive field (RF) properties. A wireless signal- and energy transmission module (SE) communicates the stimulation sequences to the implanted, individually addressable electrodes of RS

puts generating impulse sequences for elicitation of spike trains. Information processing within RE simulates the receptive field type filter operations for each ganglion cell individually. The ganglion cell output is subsequently encoded and transmitted via an electromagnetic and/or optoelectronic wireless transmission channel (SE) to the implanted retina stimulator (RS). RS will be implanted adjacent to the retinal ganglion cell layer and consists of an array of 100 to 1.000 microcontacts for localized, bi-phasic electrical stimulation of ganglion cells/fibers as well as a receiver and signal distributor.

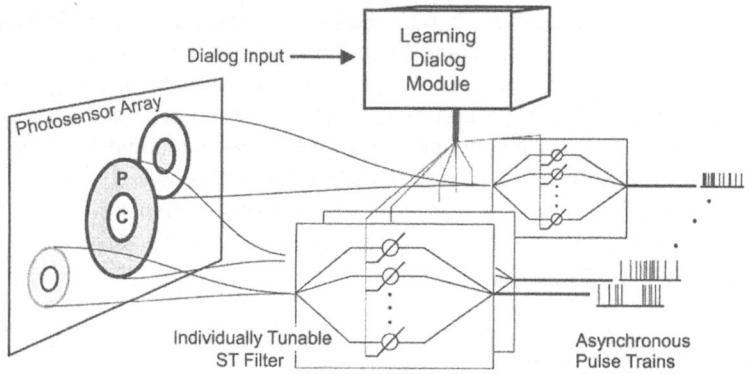

Fig. 4. Schema of information processing in the retina encoder (RE) to mimic part of the human retina. RE consists of spatio-temporal (ST) filters to provide the implanted electrodes with different stimulation pulse time courses. ST filter inputs, which simulate the receptive fields (RF) of M- or P-type ganglion cells are arranged as concentric areas each with a distinct central region C and peripheral region P within the photosensor array. The photosensor array represents the photoreceptor layer whereby the ST filter outputs represent retinal ganglion cells. Spatial and temporal ST filter parameters are individually tunable (via the learning dialog module, see Fig. 5) in order to find the optimal match of the respective RF-location, -size, and functional properties for the electrode-defined input of the central visual system

The retina encoder (RE) under development (Fig. 4) simulates about 200 tunable ST filters in real time on a digital signal processor, each with an array of several thousand photosensors at the input. The algorithm allows to adjust various spatial and temporal properties of each ST filter by means of several parameters. Various spatio-temporal patterns were processed by the ST filters. The functional range of operation of the ST filters as measured by the spatiotemporal amplitude spectra, included the spatial and temporal properties of primate retinal M and P cells with good approximation in the photopic range [5].

4 Learning Retina Encoder with Tunable Spatio-temporal Filters

Our concept of a 'Learning Retina Implant' [5], [7] with tunable ST filters is based on the assumption that a given blind human subject will be able based on the still intact capabilities of the central visual system and corresponding visual perception to "tell" the retina encoder (RE) (Fig. 4) by means of dialog-based tuning (see Fig. 5) what the best parameter settings are for the different ST filters. This tuning task in a dialog has to be based on the actual, re-gained visual perception of the initially blind subject. We developed a procedure to test alternative tuning algorithms in subjects with normal vision. For this purpose the mapping function of the unknown central visual system of a future blind subject was simulated by a neural network, which had been trained to perform an approximate inverse mapping of a RE mapping with a selected parameter vector. Fig. 5 gives a schema of the proposed dialog concept. The monitor picture on the left depicts pattern P1 as letter L moving upwards to the right. The right monitor shows an ellipse moving downwards to the right as simulation of the assumed initially perceived pattern P2. RE is depicted as a module of tunable ST filters (RE has been typically implemented with more than 200 ST filters) with receptive field (RF) input surfaces as open circles on the left. The central visual system with contacted ganglion cells represents two structures for two different cases: case 1) simulation of the central visual system for a dialog with a subject with normal vision, and case 2) real central visual system of an implant-wearing subject perceiving pattern P2 during presentation of pattern P1.

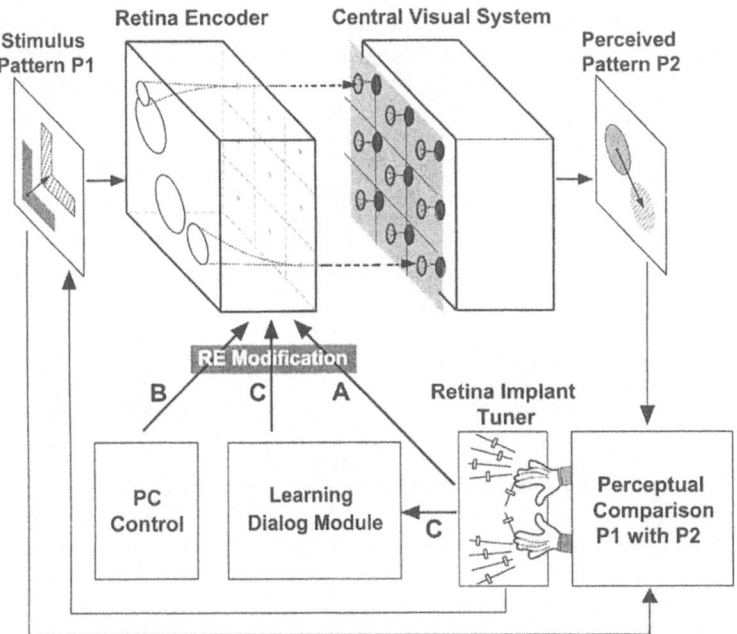

Fig. 5. Dialog-based RE tuning schema for normally sighted (central visual system as mapping inverter module of RE) and blind subjects. Retina encoder output signals reach the electrode array of an implanted microcontact foil (gray rhombic structure with open ellipses as microcontacts). Subject compares visual percept P2 with input pattern P1. RE modification methods: A-direct manual modification of parameter values, B- sweep through a large parameter space under PC control, C- dialog-based tuning by means of a learning dialog module

Three possible methods for retina encoder parameter changes (RE Modification) have been studied (Fig. 5): A- direct manual modification of parameter values, B- sweep through large parameter space under PC control, C- dialog-based tuning by means of a learning dialog module. Typically, a subject with normal vision was asked to compare a given current visual percept P2 with the corresponding input pattern P1.

In the future, in case 2), an implant-wearing, initially blind subject will hopefully receive an electrically induced visual percept P2. During dialog-based RE tuning, the subject provides the input to the dialog module (Fig. 5) based on a comparison between the actually perceived pattern P2 and the desired pattern P1, which in case 2) is not visible on a monitor but made known to the subject via another sensory channel.

Accordingly, the envisioned dialog may be thought of as a combination of 'tuning' a multi-dial radio to a station and producing desired visual patterns with a functionally cryptic visual pattern generator.

The learning RE dialog module in Fig. 5 has been implemented by a combination of a neural network and a knowledge based component. In both cases, a human subject suggests changes of the RE function. For this purpose, the dialog module generates a set of for example six RE parameter vectors, which lead to six different percepts P2'. By selecting for example the three "best matching" out of six percepts, the subject informs the dialog module, which then generates another six percepts based on the perceptual evaluation of the subject. In an iterative process, subject and dialog module jointly 'find' the RE parameter vector, which is optimal for this individual. We assume that this dialog-based tuning process will also help implanted-wearing subjects to re-gain a modest amount of vision including Gestalt-perception of larger objects (table, window, door, etc.).

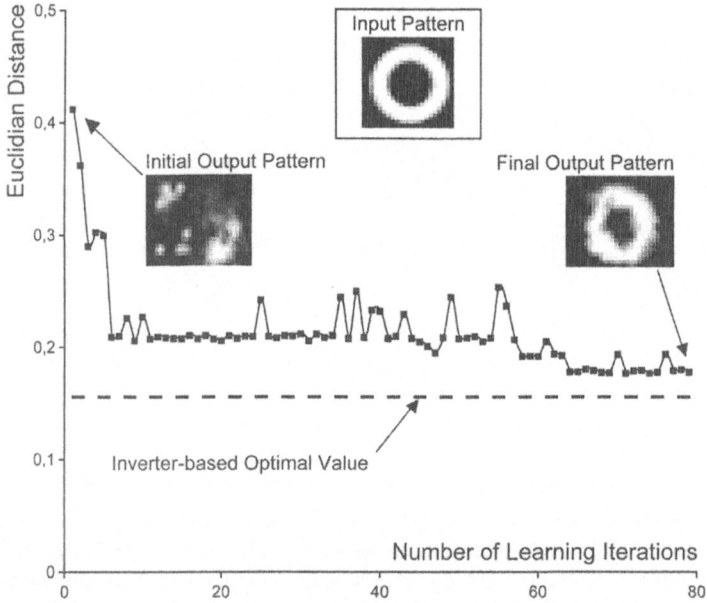

Fig. 6. Typical improvement of the learning retina encoder (RE) tuning in an iterative dialog between a RE with a learning dialog module and a subject with normal vision. The inverter-based optimal value was limited by the type of pre-trained neural network to simulate the central visual system module as inverter and by various pre-set RE parameters.

Fig. 6 gives a typical example for dialog-based tuning of the learning retina encoder (RE). RE was implemented by 256 ST filters consisting of four separate filter-types with 11 tunable parameters (partly spatial, partly temporal) each. Both input-

and output pattern arrays consisted of 32 x 32 pixels. Previously, the central visual system as inverter module had been trained to approximately invert the 'reference mapping' of RE defined by an arbitrarily selected parameter vector.

The inverter simulation did not achieve a perfect mapping inversion but only an 'optimal' value as marked as broken line in Fig. 6. For the purpose of this dialog-based tuning experiment, the RE parameter vector was arbitrarily modified so as to drastically change the RE mapping function relative to the reference mapping, which was used for the inverter training before. A subject with normal vision was looking at the input pattern P1 (slowly horizontally moving white ring against a black background) on one monitor and at the current simulated output pattern P2 on another monitor. P2 was generated by the sequence of two mapping operations: mapping 1 by RE with sub-optimally tuned filter parameters and mapping 2 by the inverter module. Initially, P2 had no clear resemblance with P1 as shown in Fig. 6 (left inset). The dialog-module generated six alternative versions of P2. The subject selected three out of six. Subsequently, the dialog module generated another six slightly 'improved' versions of P2 by using a combination of learning algorithms [20] including evolutionary algorithms (EA). Each iteration took less than one minute. As shown in Fig. 6, after 80 iterations, the Euclidian distance between P2 and P1 was significantly reduced and the P2 could be clearly recognized as a white ring (right inset).

5 Discussion and Conclusions

Besides the largely unresolved microcontact-, biocompatibility-, and implantation issues and certain ethical questions, there are still major information processing challenges left. The visual system occupies a major portion of the human brain and appears to be a highly parallel structure of mostly unknown function. Advances in visual neuroscience seem to add continuously to our appreciation of its complexity rather than to an understanding of its function [1]. On top of this, we do not understand visual perception [16] in terms of an approachable part of the brain. In that regard it seems far easier to develop autonomous vehicles with sensory and motor facilities rather than an artifact that reliably mimics your or my visual perception. We are left then with the hope that the central visual system of a blind subject with extensive previous visual experience will accept our simplistic offers presented as an electrically induced data stream of stimulation pulses along the optic nerve. Will the central visual system refuse or accept these offers? The key element of our learning retina encoder approach is its property to be functionally modified by the individual's visual perception. Within the near future there may be opportunities to successfully test this intelligent neuroprosthetic technology in blind individuals to their satisfaction.

References

1. Chalupa, L.M., Werner, J.S.: The Visual Neurosciences, Volume 1+2. MIT Press Cambridge (2004)
2. Cuenca, N., Pinilla, I., Sauve, Y., Lu, B., Wang, S., Lund, R.D.: Regressive and Reactive Changes in the Connectivity Patterns of Rod and Cone Pathways of P23H Transgenic Rat Retina. Neuroscience 127 (2004) 301-317
3. Dacey, D.M., Peterson, M.R.: Dendritic Field Size and Morphology of Midget and Parasol Ganglion Cells of the Human Retina. Proc. Natl. Acad. Sci. 89 (1992) 9666-9670

4. Eckmiller, R.: Learning Retina Implants with Epiretinal Contacts. Ophthalmic Res 29 (1997) 281-289
5. Eckmiller, R., Hünermann, R., Becker, R.: Exploration of a Dialog-Based Tunable Retina Encoder for Retina Implants. Neurocomputing. 26-27 (1999) 1005-1011
6. Eckmiller, R., Hornig, R., Gerding, H., Dapper, M., Böhm, H.: Test Technology for Retina Implants in Primats. ARVO, Invest. Ophthal. Vis. Sci. 41 (2001) 942
7. Eckmiller, R.: Adaptive Sensory-Motor Encoder for Visual or Acoustic Prosthesis. US Patent 6,400,989 (2002)
8. Eckmiller, R., Hornig, R., Ortmann, V., Gerding, H.: Test Technology for Acute Clinical Trials of Retina Implants. ARVO, Invest. Ophthal. Vis. Sci. 43 (2002) 2848
9. Eckmiller, R., Neumann, D., Baruth, O.: Specification of Single Cell Stimulation Codes for Retina Implants. ARVO Conf. Assoc. Res. Vis. Ophthal. (2004) 3401
10. Eckmiller, R. Neumann, D., Baruth, O.: Tunable Retina Encoders for Retina Implants: Why and How. J. Neural Eng. submitted (2004)
11. Humayun, M.S., Weiland, J.D., Fujii, G.Y. et al.: Visual Perception in a Blind Subject with a Chronic Microelectronic Retinal Prosthesis. Vision Res. 43 (2003) 2573-2581
12. Humphreys, G.W., Bruce, V.: Visual Cognition: Computational, Experimental, and Neuropsychological Perspectives. Lawrence Erlbaum Publ., London (1989)
13. Lee, B.B., Pokorny, J., .Smith, V.C., Kremers, J.: Responses to Pulses and Sinusoids in Macaque Ganglion Cells. Vision Res. 34 (1994) 3081-3096
14. Martinez-Conde, S., Macknik, S.L., Hubel, D.H.: The Role of Fixational Eye Movements in Visual Perception. Nature Rev. Neurosci. 5 (2004) 229-240
15. Noe, A.: Action in Perception. MIT Press. Cambridge (2004)
16. Noe, A., O'Regan, J.K.: Perception, Attention and the Grand Illusion. Psyche 6 (2000) 6-15
17. O'Regan, J.K.: Solving the "Real" Mysteries of Visual Perception: The World as an Outside Memory. Can. J. Psychol. 46 (1992) 461-488
18. Rucci, M., Desbordes, G.: Contributions of Fixational Eye Movements to the Discrimination of Briefly Presented Stimuli. J. Vision 3 (2003) 852-864
19. Santos, A., Humayun, M.S., de Juan, E.J., Greenberg, R.J., Marsh, M.J., Klock, I.B.,et al.: Preservation of the Inner Retin in Retinitis Pigmentosa. A Morphometric Analysis. Arch. Ophthalmol. 115 (1997) 511-515
20. Schwefel, H.-P., Wegener, I., Weinert, K.: Advances in Computational Intelligence, Theory and Practice. Springer Publisher, Berlin - Heidelberg (2003)
21. Watanabe, M. Rodieck, R.W.: Parasol and Midget Ganglion Cells of the Primate Retina. J. Comp. Neurol. 289 (1989) 434-454

Synchronous Phenomena for Two-Layered Neural Network with Chaotic Neurons

Katsuki Katayama[1], Masafumi Yano[1], and Tsuyoshi Horiguchi[2]

[1] Tohoku University
Research Institute of Electrical Communication
Sendai 980-8577, Japan
[2] Tohoku University
Department of Computer and Mathematical Sciences, GSIS
Sendai 980-8579, Japan

Abstract. We propose a mathematical model of visual selective attention using a two-layered neural network, based on an assumption proposed by Desimone and Duncan. We use a spiking neuron model proposed by Hayashi and Ishizuka, which generates periodic spikes, quasi-periodic spikes and chaotic spikes. The neural network consists of a layer of hippocampal formation and that of visual cortex. In order to clarify an attention shift, we solve numerically a set of the first-order ordinary differential equations, which describe a time-evolution of neurons. The visual selective attention is considered as the synchronous phenomena between the firing times of the neurons in the hippocampal formation and those in a part of the visual cortex in the present model.

1 Introduction

Recently, studies of visual selective attention have been performed actively as a starting point to understand various higher functions of brain including attention and consciousness. The visual selective attention is considered as the function by which an amount of the visual information can be reduced in order to perform effective processing of information in the brain. A spot-light hypothesis was proposed for the visual selective attention by Crick [1]. In his hypothesis, the visual information is searched item by item sequentially such as a spot-light does, and only a part of the visual information is selected by the spot-light of the attention. An alternative hypothesis was proposed by Desimone and Duncan [2]. They proposed that the visual selective attention is an emergent property of competitive interactions which work across the visual cortex. The visual information competes with each other, and only necessary visual information is obtained in the brain.

Wu and Guo [3] proposed a simple mathematical model of the visual selective attention, based on the assumption proposed by Desimone and Duncan [2] and a neurophysiological experiment performed by Iijima et al [4]. Their model is constructed by using a two-layered neural network which consists of a layer

N.R. Pal et al. (Eds.): ICONIP 2004, LNCS 3316, pp. 19–30, 2004.
© Springer-Verlag Berlin Heidelberg 2004

of the hippocampal formation and that of the visual cortex using neurons described by a phase oscillator. In their model, the state of a neuron is described by a single phase variable. They investigated an attention shift by numerical calculations. They pointed out that the function of the visual selective attention can be considered from the obtained results as synchronous phenomena of the frequency between the phase oscillators in the hippocampal formation and those in a part of the visual cortex.

Hayashi and Ishizuka proposed a spiking neuron model which can generate not only periodic spikes but also quasiperiodic spikes and chaotic spikes [5]. We consider that the spiking neuron model proposed by Hayashi and Ishizuka are biologically more plausible than that described by the phase oscillator. It is important to investigate whether the idea given by Wu and Guo works even for biologically plausible neurons. We hence propose a mathematical model of the visual selective attention by using a two-layered neural network which consists of a layer of the hippocampal formation and that of the visual cortex using spiking neurons proposed by Hayashi and Ishizuka, and investigate it by means of numerical calculations.

In §2, we formulate a mathematical model of the visual selective attention. In §3, we investigate the model by numerical calculations, and discuss the visual selective attention from the obtained results. Concluding remarks are given in § 4.

2 Formulation

In this section, we formulate a mathematical model of the visual selective attention using a two-layered neural network with spiking neurons proposed by Hayashi and Ishizuka [5], based on the neurophysiological experiment performed by Iijima et al [4]. We show schematically an architecture of the neural network in our model in Fig.1; the two-layered neural network consists of a layer of the hippocampal formation and that of the visual cortex. We assume that there exist N_H neurons in the hippocampal layer, and N_C neurons in the visual cortical layer; N_C neurons in the visual cortical layer are divided into L_G groups according to the values of the external input and each group has N_G neurons. Let $\mathcal{N}_H$ denote a set of the neurons in the hippocampal layer, $\mathcal{N}_C$ a set of the neurons in the visual cortical layer and $\mathcal{N}_{G_k}$ a set of the neurons in the visual cortical layer in a group to which neuron k belongs; hence we have $|\mathcal{N}_H| = N_H$, $|\mathcal{N}_C| = N_C$ and $|\mathcal{N}_{G_k}| = N_G$. We consider that (1) each neuron in the hippocampal layer is connected with all other neurons in the same layer and with all neurons in the visual cortical layer by excitatory synapses, and (2) each neuron in the visual cortical layer is connected by excitatory synapses with all other neurons in the same layer, even though the neurons are divided into L_G groups, and with all neurons in the hippocampal layer. We use 8 variables in order to express a state of a neuron proposed by Hayashi and Ishizuka [5]; a state of neuron i is then denoted by $\boldsymbol{X}_i^H = (V_i^H, m_i^H, h_i^H, n_i^H, m_{si}^H, h_{si}^H, n_{si}^H, n_{ii}^H)$, where V_i^H, m_i^H, h_i^H, n_i^H, m_{si}^H, h_{si}^H, n_{si}^H and n_{ii}^H correspond to a membrane potential and gating variables, respectively, where s means slow and i inward rectification. The first-order ordinary

Hippocampal Formation (H)

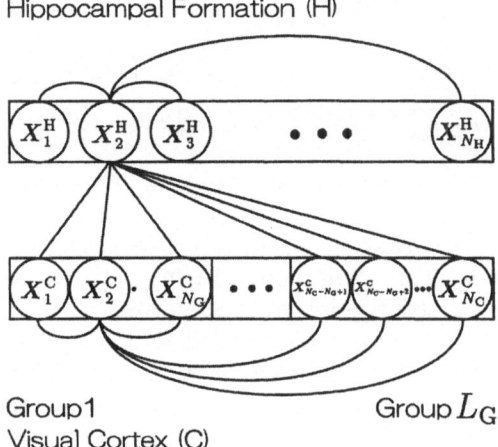

Group1   Group L_G
Visual Cortex (C)

Fig. 1. An architecture of the mathematical model of the visual selective attention used in the present paper.

equation of the time-evolution for the membrane potential, V_i^H ($i = 1, \cdots, N_H$), in the hippocampal layer is given as follows:

$$C\frac{dV_i^H}{dt} = I_i^{\text{ion,H}} + I_i^{\text{syn,H}} + I_i^{\text{ext,H}}, \tag{1}$$

$$I_i^{\text{ion,H}} = g_{\text{Na}}\left(m_i^H\right)^3 h_i^H \left(V_{\text{Na}} - V_i^H\right) + g_K \left(n_i^H\right)^4 \left(V_K - V_i^H\right)$$

$$+ g_L(V_L - V_i^H) + g_{\text{Na}_s} m_{si}^H h_{si}^H \left(V_{\text{Na}} - V_i^H\right)$$

$$+ g_{K_s} n_{si}^H \left(V_K - V_i^H\right) + g_{K_i} n_{ii}^H \left(V_K - V_i^H\right) - I_p, \tag{2}$$

where C represents a capacitance, and g_{Na}, g_K, g_L, g_{Na_s}, g_{K_s} and g_{K_i} represent conductances of each current. The details of first-order ordinary differential equations for m_i^H, h_i^H, n_i^H, m_{si}^H, h_{si}^H, n_{si}^H and n_{ii}^H are described in Ref.5, and we omit them here for lack of space. The relaxation constants of the gating variables m_i^H, h_i^H and n_i^H are set to 1, those of the gating variables m_{si}^H, h_{si}^H and n_{si}^H are set to 10, and that of the gating variable n_{ii}^H are set to 80, respectively. $I_i^{\text{ext,H}}$ is an external input, and is considered as an input from the amygdala to the hippocampal formation, for example. The state of neuron k in the visual cortical layer is denoted by $\boldsymbol{X}_k^C = (V_k^C, m_k^C, h_k^C, n_k^C, m_{sk}^C, h_{sk}^C, n_{sk}^C, n_{ik}^C)$, where V_k^C, m_k^C, h_k^C, n_k^C, m_{sk}^C, h_{sk}^C, n_{sk}^C and n_{ik}^C correspond to a membrane potential and gating variables, respectively. The first-order ordinary differential equation of the time-evolution for the membrane potential, V_k^C ($k = 1, \cdots, N_C$), in the visual cortical layer is given as follows:

$$C\frac{dV_k^C}{dt} = I_k^{\text{ion,C}} + I_k^{\text{syn,C}} + I_k^{\text{ext,C}}, \tag{3}$$

$$I_k^{\text{ion,C}} = g_{\text{Na}} \left(m_k^C\right)^3 h_k^C \left(V_{\text{Na}} - V_k^C\right) + g_{\text{K}} \left(n_k^C\right)^4 \left(V_{\text{K}} - V_k^C\right)$$

$$+ g_{\text{L}} (V_{\text{L}} - V_k^C) + g_{\text{Na}_s} m_{sk}^C h_{sk}^C \left(V_{\text{Na}} - V_k^C\right)$$

$$+ g_{\text{K}_s} n_{sk}^C \left(V_{\text{K}} - V_k^C\right) + g_{\text{K}_i} n_{ik}^C \left(V_{\text{K}} - V_k^C\right) - I_{\text{p}}, \tag{4}$$

where $I_k^{\text{ext,C}}$ is an external input. The details of first-order ordinary differential equations for m_k^C, h_k^C, n_k^C, m_{sk}^C, h_{sk}^C, n_{sk}^C and n_{ik}^C are described in Ref.5, and we omit them here. The relaxation constants of the gating variables m_k^C, h_k^C and n_k^C are set to 1, those of the gating variables m_{sk}^C, h_{sk}^C and n_{sk}^C are set to 10, and that of the gating variable n_{ik}^C are set to 80, respectively.

We notice that N_C neurons are divided into L_G groups according to the values of the external input. We consider that each group receives the same visual information from the retina. We assume that the synaptic connections are described as follows:

$$I_i^{\text{syn,H}} = -\frac{g_{\text{H}}}{N_{\text{H}} - 1} \sum_{i \neq j} \left(V_i^H - V_j^H\right) - \frac{g_{\text{R}}}{N_{\text{C}}} \sum_{k=1}^{N_{\text{C}}} \left(V_i^H - V_k^C\right), \tag{5}$$

$$I_k^{\text{syn,C}} = -\frac{g_{\text{C}_1}}{N_{\text{C}} - 1} \sum_{l \in \mathcal{N}_{G_k} \backslash \{k\}} \left(V_k^C - V_l^C\right)$$

$$-\frac{g_{\text{C}_2}}{N_{\text{C}} - 1} \sum_{l \in \mathcal{N}_{\text{C}} \backslash \mathcal{N}_{G_k}} \left(V_k^C - V_l^C\right) - \frac{g_{\text{R}}}{N_{\text{H}}} \sum_{i=1}^{N_{\text{H}}} \left(V_k^C - V_i^H\right), \tag{6}$$

where g_{H}, g_{C_1}, g_{C_2} and g_{R} represent the strengths of connections between the neurons in the hippocampal layer, between the neurons within each group in the visual cortical layer, between the neurons among each group in the visual cortical layer, and between a neuron in the hippocampal layer and a neuron in the visual cortical layer, respectively.

The firing time for neuron i in the hippocampal layer and that for neuron k in the visual cortical layer are defined by times given by T_i^H and T_k^C, respectively:

$$T_i^H = \left\{ t \mid \left(V_i^H = 0\right) \wedge \left(\dot{V}_i^H > 0\right) \right\}, \tag{7}$$

$$T_k^C = \left\{ t \mid \left(V_k^C = 0\right) \wedge \left(\dot{V}_k^C > 0\right) \right\}, \tag{8}$$

where the time derivatives of V_i^H and V_k^C are denoted by $\dot{V}_i^H$ and $\dot{V}_k^C$, respectively. The number of firing times for neuron i in the hippocampal layer and that for neuron k in the visual cortical layer during a time-window $\tau(m-1) \leq t < \tau m$ ($m = m_{\text{i}}, m_{\text{i}} + 1, \cdots, m_{\text{f}} - 1, m_{\text{f}}$) are defined, respectively, as follows:

$$F_i^H(m) = \left| T_i^H \cap \{t \mid \tau(m - 1) \leq t < \tau m\} \right|, \tag{9}$$

$$F_k^C(m) = \left| T_k^C \cap \{t \mid \tau(m - 1) \leq t < \tau m\} \right|. \tag{10}$$

We define an inter-spike interval (ISI) by a difference between a certain firing time and a firing time just before it. The ISI for neuron i in the hippocampal layer and that for neuron k in the visual cortical layer are defined, respectively, as follows:

$$\Lambda_i^{\mathrm{H}}(t) = \min\left(T_i^{\mathrm{H}} \cap \{t_2 \mid t_2 > t\}\right)$$
$$-\max\left(T_i^{\mathrm{H}} \cap \{t_1 \mid t_1 \le t\}\right), \tag{11}$$

$$\Lambda_k^{\mathrm{C}}(t) = \min\left(T_k^{\mathrm{C}} \cap \{t_2 \mid t_2 > t\}\right)$$
$$-\max\left(T_k^{\mathrm{C}} \cap \{t_1 \mid t_1 \le t\}\right). \tag{12}$$

We assume that a frequency of the firing for neuron i in the hippocampal layer and that for neuron k in the visual cortical layer are obtained, respectively, as follows:

$$f_i^{\mathrm{H}}(t) = 1/\Lambda_i^{\mathrm{H}}(t), \tag{13}$$
$$f_k^{\mathrm{C}}(t) = 1/\Lambda_k^{\mathrm{C}}(t). \tag{14}$$

We consider a correlation of the firing time among the neurons by using the number of times for the firing of neurons $F_i^{\mathrm{H}}(m)$ and $F_k^{\mathrm{C}}(m)$. We define time-dependent correlations $\kappa^{\mathrm{H}}(t)$, $\kappa^{\mathrm{C}}(t)$ and $\kappa^{\mathrm{HC}(n)}(t)$ of the firing time among the neurons; $\kappa^{\mathrm{H}}(t)$ for the neurons in the hippocampal layer, $\kappa^{\mathrm{C}}(t)$ for the neurons in the visual cortical layer and $\kappa^{\mathrm{HC}(n)}(t)$ between the neurons in the hippocampal layer and those for each group in the visual cortical layer, respectively, as follows:

$$\kappa^{\mathrm{H}}(t) = \frac{1}{N_{\mathrm{H}}(N_{\mathrm{H}}-1)} \sum_{i \ne j} \sum$$

$$\times \frac{\displaystyle\sum_{m=\left[\frac{t}{\tau}\right]}^{\left[\frac{t+T}{\tau}\right]} F_i^{\mathrm{H}}(m) F_j^{\mathrm{H}}(m)}{\sqrt{\displaystyle\sum_{m=\left[\frac{t}{\tau}\right]}^{\left[\frac{t+T}{\tau}\right]} \left\{F_i^{\mathrm{H}}(m)\right\}^2 \sum_{m=\left[\frac{t}{\tau}\right]}^{\left[\frac{t+T}{\tau}\right]} \left\{F_j^{\mathrm{H}}(m)\right\}^2}}, \tag{15}$$

$$\kappa^{\mathrm{C}}(t) = \frac{1}{N_{\mathrm{C}}(N_{\mathrm{C}}-1)} \sum_{k \ne l} \sum$$

$$\times \frac{\displaystyle\sum_{m=\left[\frac{t}{\tau}\right]}^{\left[\frac{t+T}{\tau}\right]} F_k^{\mathrm{C}}(m) F_l^{\mathrm{C}}(m)}{\sqrt{\displaystyle\sum_{m=\left[\frac{t}{\tau}\right]}^{\left[\frac{t+T}{\tau}\right]} \left\{F_k^{\mathrm{C}}(m)\right\}^2 \sum_{m=\left[\frac{t}{\tau}\right]}^{\left[\frac{t+T}{\tau}\right]} \left\{F_l^{\mathrm{C}}(m)\right\}^2}}, \tag{16}$$

$$\kappa^{\mathrm{HC}(n)}(t) = \frac{1}{N_{\mathrm{H}} N_{\mathrm{G}}}$$

$$\times \sum_{i=1}^{N_{\mathrm{H}}} \sum_{k=N_{\mathrm{G}}(n-1)+1}^{N_{\mathrm{G}} n} \frac{\displaystyle\sum_{m=\left[\frac{t}{\tau}\right]}^{\left[\frac{t+T}{\tau}\right]} F_i^{\mathrm{H}}(m) F_k^{\mathrm{C}}(m)}{\sqrt{\displaystyle\sum_{m=\left[\frac{t}{\tau}\right]}^{\left[\frac{t+T}{\tau}\right]} \left\{F_i^{\mathrm{H}}(m)\right\}^2 \sum_{m=\left[\frac{t}{\tau}\right]}^{\left[\frac{t+T}{\tau}\right]} \left\{F_k^{\mathrm{C}}(m)\right\}^2}}$$

$$(n = 1, \cdots, L_{\mathrm{G}}). \tag{17}$$

3   Results by Numerical Calculations

We investigate the system formulated in §2 by solving Eqs.(1)-(6) by numerical integrations, and discuss the visual selective attention from the obtained results. The parameters in Eqs.(1)-(6) are fixed to $C = 2$, $V_{\mathrm{Na}} = 50$, $V_{\mathrm{K}} = -70$, $V_{\mathrm{L}} = -70$, $g_{\mathrm{Na}} = 60$, $g_{\mathrm{K}} = 10$, $g_{\mathrm{Na_s}} = 1.4$, $g_{\mathrm{K_s}} = 0.18$, $g_{\mathrm{K_i}} = 0.2$, $g_{\mathrm{L}} = 0.063$ and $I_{\mathrm{p}} = -3$ in the present paper. We assume that there exist 100 neurons in the hippocampal layer and 100 neurons in the visual cortical layer, and the visual cortical neurons are divided into 5 groups with equal size in the present paper: $N_{\mathrm{H}} = 100$, $N_{\mathrm{C}} = 100$, $L_{\mathrm{G}} = 5$ and $N_{\mathrm{G}} = 20$. The external input to the neurons in the hippocampal layer is assumed to be $I_i^{\mathrm{ext,H}} = A + B u_i$ for $i = 1, \cdots, N_{\mathrm{H}}$, where A and B are constant and u_i is taken from $U[-1, 1]$, where we denote a uniform distribution in an interval $[\alpha, \beta]$ as $U[\alpha, \beta]$ hereafter. Thus, the external input is randomly distributed with a deviation $B/\sqrt{3}$ around a mean A for the neurons in the hippocampal layer.

Here we notice that a frequency of a neuron proposed by Hayashi and Ishizuka [5] depends on the strength of the external input. In order to show this, we give the frequency f and Lyapunov exponents as a function of the external input I^{ext} for a neuron proposed by Hayashi and Ishizuka [5] after enough time has been passed in Fig.2. Thus, we find that the neuron proposed by Hayashi and Ishizuka [5] can generate periodic spikes, quasiperiodic spikes and chaotic spikes by changing the strength of the external input in the range $I^{\mathrm{ext}} \geq -3.73$; in particular, it can generate chaotic spikes in the range $-2.40 \leq I^{\mathrm{ext}} \leq -2.27$, because the maximal Lyapunov exponent becomes positive.

We investigate the attention shift for two cases; one of them is the attention shift from a state with a periodic spike to one with another periodic spike (case 1), and another is the attention shift from a state with a chaotic spike to one with a periodic spike (case 2). We fix to $g_{\mathrm{H}} = 0.015$, $g_{\mathrm{C_1}} = 0.015$, $g_{\mathrm{C_2}} = 0.01$ and $g_{\mathrm{R}} = 0.02$ hereafter in the present paper.

We now explain an attention shift for the case 1. We assume that the external inputs to the neurons in the visual cortical layer are given by $I_k^{\mathrm{ext,C}} = 1.0 + 0.5\,(n-1) + 0.05 u_k$ for $k = 20\,(n-1) + 1, \cdots, 20n$ for group n $(n = 1, \cdots, 5)$, where u_k is taken from $U[-1, 1]$; we note that the essential behavior for the system does not depend on how to select the order of the external input as for the neuron groups. The parameters of the correlation of the firing time among the neurons are fixed to $\tau = 5$ and $T = 50$. We change $A = 1.5$ for $0 \leq t < 5000$

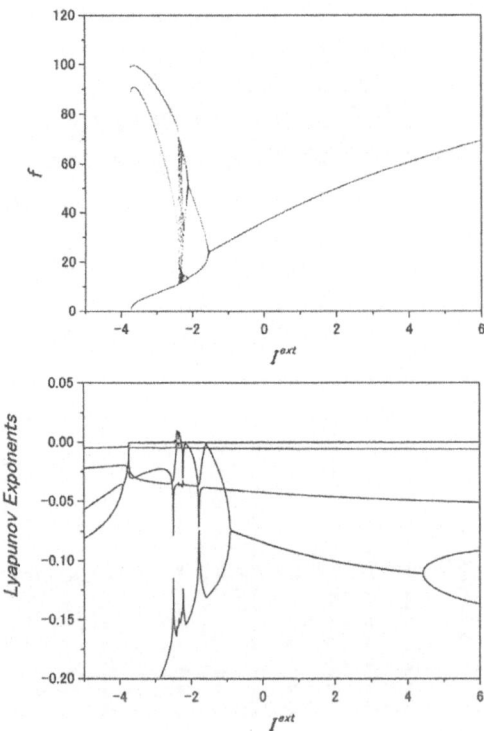

Fig. 2. The frequency f and Lyapunov exponents as a function of the external input I^{ext} after enough time has been passed for a neuron proposed by Hayashi and Ishizuka [5].

to $A = 2.5$ at $t = 5000$ and keep $A = 2.5$ for $t \geq 5000$; we set $B = 0.05$ for $t \geq 0$. In Fig.3, we show the firing times of the neurons in both the hippocampal layer and the visual cortical layer for $4000 \leq t \leq 4200$ and $6000 \leq t \leq 6200$. The firing times of all the neurons in the hippocampal layer synchronize with each other, although the strength A of the external inputs has been changed at $t = 5000$. For $0 \lesssim t < 5000$, only the firing times of the neurons of $k = 21, \cdots, 40$ in the visual cortical layer synchronize with the firing times of the neurons in the hippocampal layer. On the other hand, by switching the value of A from 1.5 to 2.5 at $t = 5000$, only the firing times of the neurons of $k = 61, \cdots, 80$ in the visual cortical layer synchronize with the firing times of the neurons in the hippocampal layer for $t \gtrsim 5000$. We show the time-dependent correlations of the firing time among the neurons, $\kappa^H(t)$, $\kappa^C(t)$ and $\kappa^{HC(n)}(t)$, in Fig.4. We see from $\kappa^H(t)$ that the firing times of all the neurons in the hippocampal layer almost synchronize with each other. We also see that the firing times of all the neurons in the hippocampal layer almost synchronize with those of the neurons in group 2 in the visual cortical layer from $\kappa^{HC(2)}(t)$ for $0 \lesssim t < 5000$, and that the firing times of all the neurons in the hippocampal layer almost synchronize with those

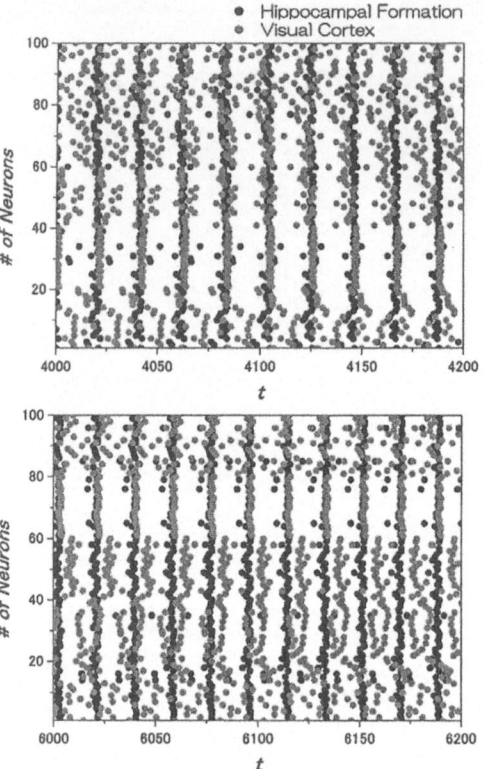

Fig. 3. Firing times of the neurons in both the hippocampal layer and the visual cortical layer in the case of $A = 1.5$ and $B = 0.05$ for $0 \leq t < 5000$ and $A = 2.5$ and $B = 0.05$ for $t \geq 5000$.

of the neurons in group 4 in the visual cortical layer from $\kappa^{\text{HC}(4)}(t)$ for $t \gtrsim 5000$. We show the frequencies $f_i^{\text{H}}(t)$ and $f_k^{\text{C}}(t)$ as a function of time t in Fig.5. All the neurons in the hippocampal layer are entrained to the state with the frequency of about 48 for $0 \lesssim t < 5000$. Only the neurons of $k = 21, \cdots, 40$ in the visual cortical layer are entrained to the state with the frequency of about 48 occurring in the hippocampal layer for $0 \lesssim t < 5000$. On the other hand, the values of the frequency of other neurons in the visual cortical layer oscillate strongly without being entrained to the state with the frequency of about 48 in the hippocampal layer. By switching the value of A from 1.5 to 2.5 at $t = 5000$, all the neurons in the hippocampal layer are entrained to a state with the frequency of about 54 for $t \gtrsim 5000$. Only the neurons of $k = 61, \cdots, 80$ in the visual cortical layer are entrained to the state with the frequency of about 54 in the hippocampal layer for $t \gtrsim 5000$. On the other hand, the values of the frequency of other neurons in the visual cortical layer oscillate strongly without being entrained to the state with the frequency of about 54. This result for the frequency is in qualitative

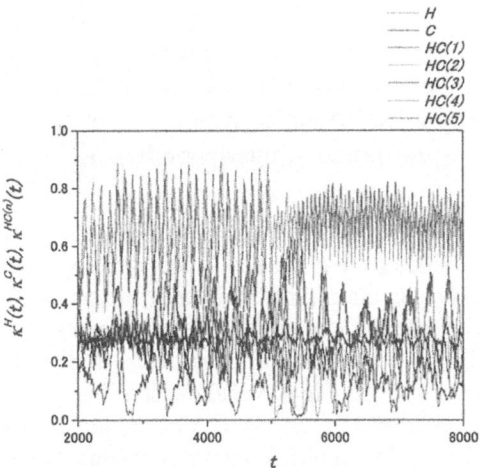

Fig. 4. Time-dependent correlations $\kappa^{\mathrm{H}}(t)$, $\kappa^{\mathrm{C}}(t)$ and $\kappa^{\mathrm{HC}(n)}(t)$ of the firing time among the neurons in the case of $A = 1.5$ and $B = 0.05$ for $0 \le t < 5000$ and $A = 2.5$ and $B = 0.05$ for $t \ge 5000$.

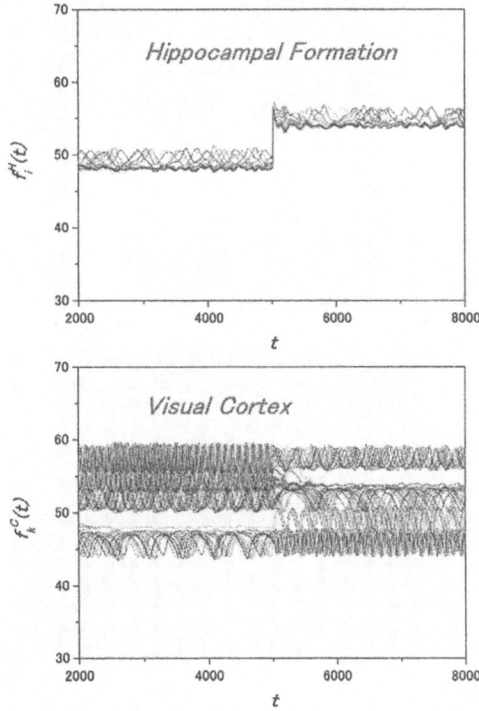

Fig. 5. Frequencies $f_i^{\mathrm{H}}(t)$ and $f_k^{\mathrm{C}}(t)$ as a function of the time t in the case of $A = 1.5$ and $B = 0.05$ for $0 \le t < 5000$ and $A = 2.5$ and $B = 0.05$ for $t \ge 5000$.

28   Katsuki Katayama, Masafumi Yano, and Tsuyoshi Horiguchi

agreement with that obtained for the phase oscillatory neurons by Wu and Guo
[3]. It turns out that the attention shift has been performed quickly in our model.

Next, we explain an attention shift for case 2. We assume that the external
inputs to the neurons in the visual cortical layer are given by $I_k^{\text{ext,C}} = -3.34 +$
$0.05u_k$ for $k = 1, \cdots, 20$ (group 1, quasiperiodic), $I_k^{\text{ext,C}} = -2.34 + 0.05u_k$ for
$k = 21, \cdots, 40$ (group 2, chaotic), $I_k^{\text{ext,C}} = -1.34 + 0.05u_k$ for $k = 41, \cdots, 60$
(group 3, periodic), $I_k^{\text{ext,C}} = -0.34 + 0.05u_k$ for $k = 61, \cdots, 80$ (group 4, periodic)
and $I_k^{\text{ext,C}} = 0.66 + 0.05u_k$ for $k = 81, \cdots, 100$ (group 5, periodic), respectively.
The parameters of the correlation of the firing time among the neurons are fixed
to $\tau = 20$ and $T = 200$. We change $A = -2.34$ (chaotic) for $0 \le t < 5000$ to
$A = -0.34$ (periodic) at $t = 5000$ and keep $A = -0.34$ (periodic) for $t \ge 5000$;
we set $B = 0.05$ for $t \ge 0$. In Fig.6, we show the firing times of the neurons in
both the hippocampal layer and the visual cortical layer for $4000 \le t \le 4500$ and
$6000 \le t \le 6500$. The firing times of all the neurons in the hippocampal layer
synchronize with each other, although the strength A of the external inputs has

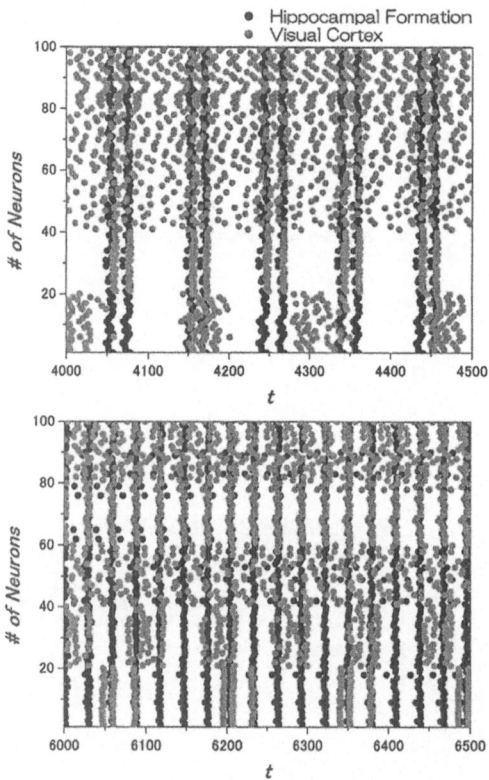

Fig. 6. Firing times of the neurons in both the hippocampal layer and the visual cortical
layer in the case of $A = -2.34$ and $B = 0.05$ for $0 \le t < 5000$ and $A = -0.34$ and
$B = 0.05$ for $t \ge 5000$.

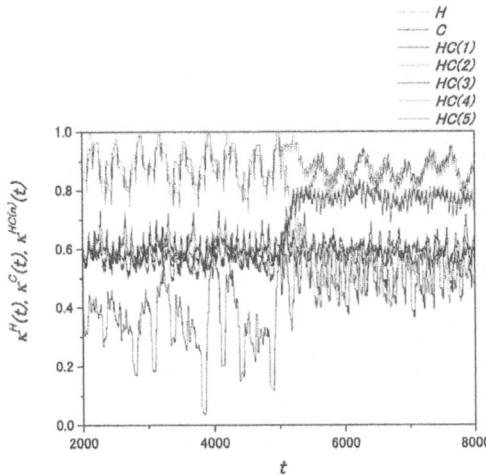

Fig. 7. Time-dependent correlations $\kappa^H(t)$, $\kappa^C(t)$ and $\kappa^{HC(n)}(t)$ of the firing time among the neurons in the case of $A = -2.34$ and $B = 0.05$ for $0 \le t < 5000$ and $A = -0.34$ and $B = 0.05$ for $t \ge 5000$.

been changed at $t = 5000$. For $0 \lesssim t < 5000$, only the firing times of the neurons of $k = 21, \cdots, 40$ in the visual cortical layer synchronize with the firing times of the neurons in the hippocampal layer, although the behavior of the neuronal spikes is chaotic. On the other hand, by switching the value of A from -2.34 to -0.34 at $t = 5000$, only the firing times of the neurons of $k = 61, \cdots, 80$ in the visual cortical layer synchronize with the firing times of the neurons in the hippocampal layer for $t \gtrsim 5000$. We show the time-dependent correlations of the firing time among the neurons, $\kappa^H(t)$, $\kappa^C(t)$ and $\kappa^{HC(n)}(t)$, in Fig.7. We see from $\kappa^H(t)$ that the firing times of all the neurons in the hippocampal layer almost synchronize with each other. We also see that the firing times of all the neurons in the hippocampal layer almost synchronize with those of the neurons in group 2 in the visual cortical layer from $\kappa^{HC(2)}(t)$ for $0 \lesssim t < 5000$, and that the firing times of all the neurons in the hippocampal layer almost synchronize with those of the neurons in group 4 in the visual cortical layer from $\kappa^{HC(4)}(t)$ for $t \gtrsim 5000$. Thus, the attention shift has been performed quickly in our model.

4   Concluding Remarks

We have proposed a mathematical model for the visual selective attention using the two-layered neural network, based on the assumption proposed by Desimone and Duncan [2] and the neurophysiological experiment performed by Iijima et al [4]. We have used the spiking neuron model proposed by Hayashi and Ishizuka, which generates the periodic spikes, the quasiperiodic spikes and the chaotic spikes [5]. The neural network has been constructed by the layer of the hippocampal formation and that of the visual cortex. We have solved by the numerical calculations the set of the first-order ordinary differential equations, which describe the time-evolution of each neuron, in order to clarify the attention shift. The visual selective attention is considered as the synchronous phenomena between the firing times of the neurons in the hippocampal formation and those in a part of the visual cortex in our model.

K. K. was partially supported by Research Fellowships of the Japan Society for the Promotion of Science for Young Scientists.

References

1. Crick, F.: Function of the Thalamic Reticular Complex: The searchlight hypothesis. Proceedings of the National Academy of Sciences USA **81** (1984) 4586-4590
2. Desimone, R., Duncan, J.: Neural Mechanisms of Selective Visual Attention. Annu. Rev. Neurosci. **18** (1995) 193-222
3. Wu, Z., Guo, A.: Selective Visual Attention in a Neurocomputational Model of Phase Oscillators. Biol. Cybern. bf 80 (1999) 205-214
4. Iijima, T., Witter, M.P., Ichikawa, M., Tominaga, T., Kajiwara, R., Matsumoto, G.: Entorhinal-Hippocampal Interactions Revealed by Real-Time Imaging. Science **272** (1996) 1176-1179
5. Hayashi, H., Ishizuka, S.: Chaotic Nature of Bursting Discharges in the *Onchidium* Pacemaker Neuron. J. theor. Biol. **156** (1992) 269-291

Influence of Dendritic Spine Morphology on Spatiotemporal Change of Calcium/Calmoduline-Dependent Protein Kinase Density

Shuichi Kato[1], Seiichi Sakatani[1,2], and Akira Hirose[1]

[1] Department of Frontier Informatics, Graduate School of Frontier Sciences,
The University of Tokyo, 7-3-1 Hongo, Bunkyo-ku, Tokyo, 113-8656 Japan
{kato,sakatani}@eis.t.u-tokyo.ac.jp, ahirose@ee.t.u-tokyo.ac.jp
[2] Japan Society for the Promotion of Science

Abstract. Glutamic acid sensitivity of AMPA (α-amino-3-hydroxy-5-methyl-4-isoxazole propionic acid) receptor at spine is an important factor that determines excitatory synaptic weight. It has been suggested that the weight is determined by spine morphology. It was also reported that the thick spines (mushroom type) whose head is fully developed remain mostly stable. Therefore, the memory in the cerebral neuronal circuit is possibly stored in the spine morphology. However, it is still unclear how glutamic acid sensitivity varies with the change in spine morphology. In this paper, we show the relation between the spine morphology and the glutamic acid sensitivity. We classify the spines into 3 types (mushroom, thin and stubby) according to physiological observation. We analyze the spatiotemporal dynamics of CaMKII-CaMCa$_4$ concentration, that increases the glutamic acid sensitivity, when calcium ions are injected at the spine head. We find that CaMKII-CaMCa$_4$ concentration in the mushroom-type spine is much higher than those in the others and that its decay is much slower. This result is the first finding that shows the relation between the change in the spine morphology and the glutamic acid sensitivity, which connects synaptic characteristics that determines the brain functions and the spine morphology.

1 Introduction

In recent years, the development of measurement such as electron microscope [1, 2] and two-photon laser scanning microscope [3, 4] enables us to observe spine morphology and spatiotemporal dynamics of the calcium concentration in the spine. As a result, it has been found that both the absolute number [2, 5–8] and the shape [5, 9–13] of spines can change sometimes quite drastically in young or mature animals.

As for the association of the spine structure with the human brain function, it was reported that, in mental retardation cases, we often observe the thin spines caused by a dysgenesis [14]. It was also reported that the spine neck, if any,

N.R. Pal et al. (Eds.): ICONIP 2004, LNCS 3316, pp. 31–36, 2004.
© Springer-Verlag Berlin Heidelberg 2004

defilades metabolic factors such as calcium ions and that it adjusts the spine function individually [1, 4, 15, 16]. Recently, it was reported that glutamic acid sensitivity at hippocampus CA1 pyramidal cell was strongly enhanced if the size of spine head is larger, while it was not for headless spines. Since the spine glutamic acid sensitivity of α-amino-3-hydroxy-5-methyl-4-isoxazole propionic acid (AMPA) receptor is an important factor that determines the excitatory synaptic weight, it is suggested that the synaptic weight is determined by the spine morphology and that furthermore the memory is stored in the spine structure. That is to say, it is considered that the state of synaptic function that determines the brain function is related to the spine morphology.

Therefore the fundamental mechanism to be clarified is how glutamic acid sensitivity varies with the change in spine morphology. To reveal it, we need to find out the relation between the spine morphology and the CaMKII-CaMCa$_4$ concentration that changes the glutamic acid sensitivity. It is very difficult to observe physiologically the concentrations of multiple chemicals at once by using fluorescence indicators. In addition, in such a reaction and diffusion of multi-chemical case, the changes of concentrations cannot be solved analytically.

In this paper, we present the relation between spine morphology and glutamic acid sensitivity. We analyze the concentration change of CaMKII-CaMCa$_4$, that increases the glutamic acid sensitivity, by using numerical calculation for chemical reaction and chemical diffusion. As a result, we show that CaMKII-CaMCa$_4$ concentration in the mushroom-type spine becomes much higher than those in others. Its decay is also found much slower.

2   Calculation Method

In chemical diffusion and reaction calculation, the concentration change of a chemical C$_i$ is expressed with a reaction-related time-dependent function $R_i(t)$, that is determined by material concentrations, as

$$\frac{\partial [C_i]}{\partial t} + D_i\left(\frac{\partial^2 [C_i]}{\partial x^2} + \frac{\partial^2 [C_i]}{\partial y^2} + \frac{\partial^2 [C_i]}{\partial z^2}\right) = R_i(t) \tag{1}$$

where i denotes material index, D_i is diffusion coefficient of each material and $[C_i]$ denotes each chemical concentration.

In this paper, we evaluate the degree of rise in the glutamic acid sensitivity by the increase of CaMKII-CaMCa$_4$ concentration. CaMKII-CaMCa$_4$ is generated through the reaction paths in Fig.1(b). Figure 1(a) shows the kinetic constant of the Ca$^+$ pump and the other chemicals. Figure 1(c) shows the diffusion coefficient of each chemicals. Calmoduline (CaM) is capable to combine with up to 4 calcium ions. CaMKII-CaMCa$_4$ is generated by the combination of CaMCa$_3$ or CaMCa$_4$ with calcium/calmoduline-dependent protein kinase II (CaMKII).

A spine contains many chemicals. Thereby, in Fig.1, B stands for buffer proteins except for CaM. The chemical reaction is expressed as

$$B + Ca^{2+} \underset{K_{Bb}}{\overset{K_{Bf}}{\rightleftarrows}} B \cdot Ca^{2+} \tag{2}$$

where $B \cdot Ca^{2+}$ denotes the compound made up of buffer proteins and Ca^{2+}.

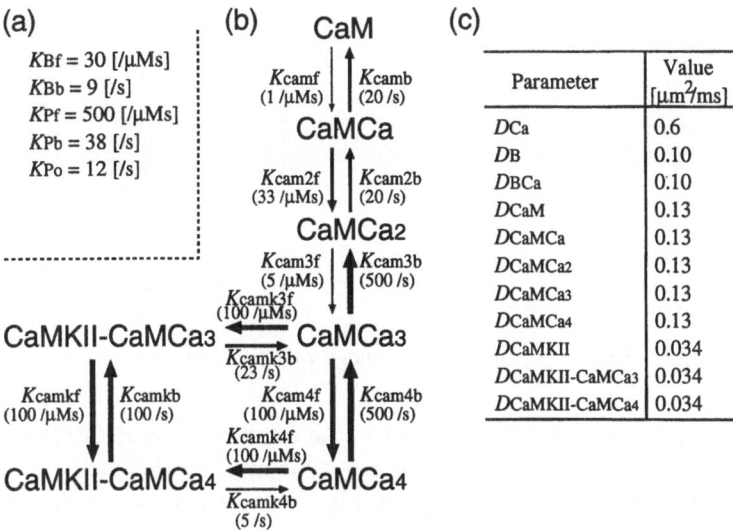

Fig. 1. (a)Kinetic constants (b)reaction paths and (c)diffusion coefficients.

The pumps discharge Ca^{2+} from the spine surface. The equivalent chemical reaction by the pump is expressed as

$$P + Ca^{2+}_i \underset{K_{Pb}}{\overset{K_{Pf}}{\rightleftarrows}} P \cdot Ca^{2+} \overset{K_{Po}}{\rightarrow} Ca^{2+}_o \qquad (3)$$

where P, Ca^{2+}_i and Ca^{2+}_o denote Pump/Exchanger, calcium ion in spine and extra-spine calcium ion, respectively.

3   Spine Model

Figure 2(a) shows 3 kinds of typical spine morphology. We name the thick spine whose head is fully developed the mushroom type, the thin spine the thin type and head-less spine the stubby type. The spine morphology is characterized by the radius of spine neck r_n and the total spine length L which are shown at the bottom of Fig.2(a). The spine morphology is 3 dimensional. However, we can drop it to 2 dimensional model, as shown in Fig.2(b), by considering the vertical axial symmetry. Each spine receives the Ca^{2+} input from the spine head and Ca^{2+} is always discharged by the pump on the side surface.

The rise in intracellular Ca^{2+} concentration is caused mainly through 3 paths. One is the action-potential induced Ca^{2+} transient, another is the Ca^{2+} influx from N-methyl-D-aspartate (NMDA) receptor and the last one is the Ca^{2+} release from intracellular Ca^{2+} store. In the spine, Ca^{2+} influx from NMDA receptor is dominant and, therefore, we consider only the Ca^{2+} influx from NMDA receptor. The influx of calcium ion J_{Ca} is expressed as

$$J_{Ca} = g_{const} \cdot t \cdot e^{-\frac{t}{\tau}} \qquad (4)$$

34   Shuichi Kato, Seiichi Sakatani, and Akira Hirose

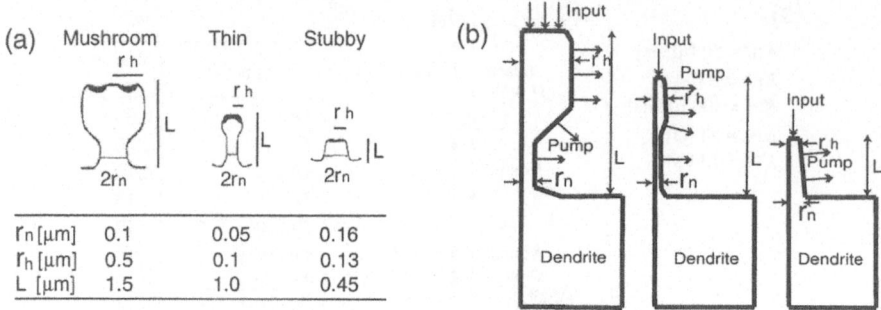

Fig. 2. (a) Spine morphology and (b) spine model.

where g_{const} ($= 5e \times 10^{-2}[\text{mol/m}^2\text{s}]$) is the constant that determines the amplitude of Ca^{2+} influx, t is time and τ is time constant(5.0[ms]). Ca^{2+} inputs, which are illustrated as input arrows in Fig.2, are input to whole the spine head.

4   Calculation Results

Figure 3 shows the spatiotemporal change of Ca^{2+} concentration: (a) mushroom-type, (b) thin type and (c) stubby type. The concentration decay in the mushroom-type and the thin-type is slower than that in the stubby-type since the spine neck chokes the diffusion to the dendrite. This fact is consistent with experimental knowledge [1, 4, 15, 16]. By comparing mushroom-type spine result with that of the thin-type, the concentration decay by Ca^{2+} diffusion is slower since the proportion of spine body to the spine neck of the mushroom-type is larger than that of the thin-type. However, the Ca^{2+} emission from the spine

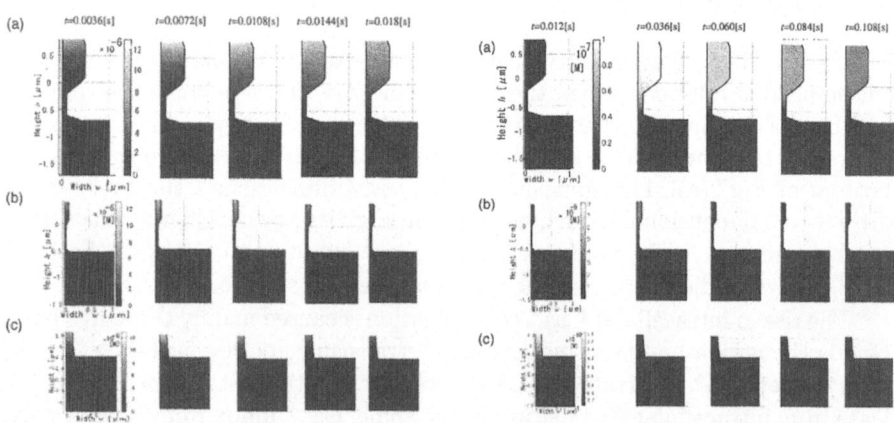

Fig. 3. Spatiotemporal change of Ca^{2+} concentration (a)Mushroom type (b)Thin type (c)Stubby type.

Fig. 4. Spatiotemporal change of CaMKII-CaMCa4 concentration (a) Mushroom type (b) Thin type (c) Stubby Type.

surface by the pump is rather small so that only a small concentration gradient is caused. The larger the spine-head area and the thicker the spine is, the Ca^{2+} pumping has less influence on the Ca^{2+} concentration since the Ca^{2+} emission by the pump occurs only on the side surface.

Figure 4 shows the spatiotemporal changes of CaMKII-CaMCa$_4$ concentration: (a) mushroom-type spine, (b) thin one and (c) stubby one. Figure 4 shows that the CaMKII-CaMCa$_4$ concentration gets very higher than those in others. The decay time of CaMKII-CaMCa$_4$ concentration in mushroom-type spine is also 10–100 times larger than those in others. CaMKII-CaMCa$_4$ diffusion to the dendrite as well as Ca^{2+} outflow is restricted by the neck. CaMKII-CaMCa$_4$ concentrations are about one tenth of Ca^{2+} ones and the decay is 10 times slower. The slower decay is caused by 20 times smaller diffusion coefficient than that of Ca^{2+}.

5   Discussion on the Results and the Relation to LTPs

Surprisingly, CaMKII-CaMCa$_4$ concentration at the spine head is the same as that at the internal part, although Ca^{2+} concentration at the spine head is higher than that at the internal part. This phenomenon is caused as follows. 95% Ca^{2+} combines with the buffer proteins, and it is highly resistant to generation of CaMCa since K_{camf} is 20 times smaller than K_{camb}. Contrarily, CaMKII-CaMCa$_4$ concentration decays at a slow speed since K_{camk4f} is 20 times larger than K_{camk4b}. That is, the higher forward reaction than the backward one makes the decay time of forward-generated material longer. At the spine neck, the gradient of CaMKII-CaMCa$_4$ concentration occurs, but it diffuses only slowly since the diffusion coefficient is small.

Accordingly, it is needed for incidence of long-term potentiation (LTP) that CaMKII-CaMCa$_4$ concentration get almost homogeneous because, otherwise, CaMKII-CaMCa$_4$ concentrations rapidly decrease and they cannot activate AMPA receptors at the head.

Additionally we can partly explain the reason why we observe many thin spines in mentally retardate persons as follows. When Ca^{2+} influxes through the NMDA receptor, Ca^{2+}, CaM and their compound divide CaMKII from the bound actin filaments (F-actin) and make CaMKII combined with the postsynaptic density (PSD) [29]. At the same time, the rest of F-actin plays polymerization reaction and make the spine thicker [30]. At the thin spines the LTP less occurs and, therefore, the spines cannot become thicker but remain unstable, which will be related to the possible mental retardation through spine dysgenesis [14].

6   Conclusion

We have shown the relation between the spine morphology and the glutamic acid sensitivity. We have classified the spines into 3 types (mushroom, thin and stubby) according to physiological observation. We have analyzed the spatiotemporal dynamics of CaMKII-CaMCa$_4$ concentration, that increases the glutamic acid sensitivity, when calcium ions are injected at the spine head. We have found that CaMKII-CaMCa$_4$ concentraion in the mushroom-type spine is much higher than those in others and that its decay is much slower. This result is the first finding that shows the relation between the change in the spine morphology and the glutamic acid sensitivity, which connects synaptic characteristics that determines the brain functions and the spine morphology.

Acknowledgement

This work was partly supported by the Ministry of Education, Culture, Sports, Science and Technology, Grant-in-Aid for Scientific Research on Grant-in-Aid for JSPS Fellows 15-11148 (Sakatani).

References

1. K. M. Harris et al., editors, *Intrinsic Determinants of Neuronal From and Function*, 179–199, Alan Liss, New York, 1988.
2. K. M. Harris and J. K. Stevens. *J. Neurosci.*, 9, 4455–4469, 1989.
3. W. Denk, J. H. Strickler and W. W. Webb. *Science*, 248, 73–6, 1990.
4. R. Yuste and W. Denk. *Nature*, 375, 682–4, 1995.
5. K. M. Harris, F. E. Jensen, and B. Tsao. *J. Neurosci.*, 12, 2685–2705 1992.
6. C. S. Woolley et al. *J. Neurosci.*, 10, 4035–4039, 1990.
7. J. T. Trachtenberg et al. *Nature*, 420, 788–94, 2002.
8. J. Grutzendler, N. Kasthuri and W.-B. Gan. *Nature*, 420, 812–816, 2002.
9. A. Van Harrefeld and E. Fidková. *Exp. Neurol.*, 49, 736–749, 1975.
10. K. S. Lee et al. *J. Neurophysiol.*, 44, 247–258, 1980.
11. W. T. Greenough et al., editor, *Synaptic Plasticity*, 335–372, Guilford Press, New York, 1985.
12. N. L. Desmond and W. B. Levy. *Synapse*, 5, 39–43, 1990.
13. R. K. S. Carverly and D. G. Jones. *Brain Res. Rev.*, 15, 215–249, 1990.
14. D. P. Purpura. *Science*, 186, 1126–8, 1974.
15. K. Svoboda, D. W. Tank and W. Denk. *Science*, 272, 716–719, 1996.
16. G. M. Shepherd. *J. Neurohysiol.*, 75, 2197–2210, 1996.
17. M. Matsuzaki et al. *Nature Neurosci.*, 4, 1086–1092, 2001.
18. P. Ascher and L. Nowak. *J. Physiol.*, 399, 247–266, 1988.
19. C. Jahr and C. Stevens. *J. Neurosci.*, 10, 1830–1837, 1990.
20. T. Nishizaki et al. *Biochem. Biophy. Res. Commun.*, 254, 446–449, 1999.
21. C. F. Stevens et al. *Curr. Biol.*, 4, 687–693, 1994.
22. P. M. Lledo et al. *Proc. Natl. Acad. Sci. USA*, 92, 11175–11179, 1995.
23. A. Barria et al. *Science*, 276, 2042–2045, 1997.
24. I. Yamaguchi and K. Ichikawa. *Neurosci. Res.*, 30, 91–98, 1998.
25. A. Zhou and E. Neher. *J. Physiol.*, 480, 123–136, 1993.
26. C. Koch. *Biophysics of Computaion.* Oxford University Press, New York, 1999.
27. W. R. Holmes. *J. Comput. Neurosci.*, 8, 65–85, 2000.
28. C. J. Coomber. *Neural Comput.*, 10, 1653–1678, 1998.
29. K. Shen and T. Meyer. *Science*, 284, 162–167, 1999.
30. Y. Fukazawa et al. *Neuron*, 38, 447–460, 2003.

Memory Modification Induced by Pattern Completion and STDP in Hippocampal CA3 Model

Toshikazu Samura and Motonobu Hattori

Interdisciplinary Graduate School of Medicine and Engineering,
Computer Science and Media Engineering, University of Yamanashi
4-3-11 Takeda, Kofu-shi, Yamanashi, 400-8511 Japan
{toshikazu,hattori}@as.media.yamanashi.ac.jp

Abstract. One of the role of the hippocampus is viewed as modifying episodic memory so that it can contribute to form semantic memory. In this paper, we show that the pattern completion ability of the hippocampal CA3 and symmetric spike timing-dependent synaptic plasticity induce memory modification so that the hippocampal CA3 can memorize invariable parts of repetitive episode as essential elements and forget variable parts of those as unnecessary ones.

1 Introduction

Since human brain's high-level functions entirely depend on memories stored there, understanding human memory is essential for elucidation of the brain. Memories are classified broadly into two types: declarative memory and non-declarative memory [1]. Moreover, declarative memory is classified into episodic memory and semantic memory. Recent study has suggested that semantic memory which corresponds to knowledge derives from schematization of substantial parts in repetitive episodic memory. Therefore, episodic memory is essential to declarative memory.

It is well known that the hippocampus plays an important role in the acquisition of episodic memory [2]. In addition to this function, Niki has indicated that hippocampus modifies episodic memory [2]. Eichenbaum has also suggested that each episodic coding by the hippocampus is added to others to compose a final stage of semantic knowledge destined for consolidation in the cerebral cortex [3]. However, it is not clear how the hippocampus modifies episodic memory. Our fundamental interest is how semantic memory is formed from daily experiences that occur in a temporal context: especially, how substantial parts are extracted from repetitive episodes in the hippocampus so that they can contribute to be stored as semantic memory. Moreover, from the engineering point of view, modeling the hippocampal function of knowledge acquisition is indispensable to develop brain-like intelligent systems.

Introspecting our memories, we tend to forget frequent episodes than infrequent episodes. Especially, we easily forget variable parts of frequent episodes.

N.R. Pal et al. (Eds.): ICONIP 2004, LNCS 3316, pp. 37–43, 2004.
© Springer-Verlag Berlin Heidelberg 2004

For example, we can remember stores and houses on the daily commuter way. Buildings, which are not often changed, correspond to invariable parts in the frequent episodes and we can recall them anytime. While we can also memorize pedestrians who are there by chance then and recall them within a few days – this means hippocampus memorizes such pedestrians –, we cannot recall them after a week. Thus, the hippocampus may memorize invariable parts in frequent episodes as essential core elements, and conversely it may forget variable pars in those as unnecessary ones little by little. Notice that here we focus on the memory modification in only frequent episodes because an infrequent episode often accompanies emotion and is mediated by not only the hippocampus but also other subcortical areas such as the amygdala.

In this paper, we propose a hippocampal CA3 model based on recent physiological findings: spike timing-dependent plasticity (STDP) and pattern completion. We show that the above mentioned memory modification is induced by the pattern completion ability of the CA3 and the symmetric STDP.

2   Physiological Background

2.1   Spike Timing-Dependent Synaptic Plasticity

Spike timing-dependent synaptic plasticity (STDP) is observed from a hippocampal slice and so on [4]. STDP is a rule that modifies a synaptic weight according to the interval between pre- and postsynaptic spikes. The modification rate of a weight and its polarity (potentiation or depression) depend on the interval. STDP has two types: asymmetric one [4] and symmetric one [5]. Symmetric STDP works as a coincidence detector. If both pre- and postsynaptic neurons fire simultaneously, the synapse between them is potentiated. Conversely, if they fire separately, the synapse between them is depressed.

Many experimental results of STDP are obtained from CA3–CA1 synapses, which are called schaffer collaterals. However, Debanne et al. have suggested that CA3–CA3 synapses appear to be identical to CA3–CA1 ones as regards both their basic properties and mechanisms of synaptic plasticity [6]. In addition, several inhibitory interneurons exist in the CA3 region [7], and they seem to be similar to the CA1 region. Kobayashi has indicated that asymmetric STDP alters to symmetric one by taking effect of inhibitory interneurons into account [5].

Thus, we extend the results of STDP obtained from CA3–CA1 synapses to CA3–CA3 ones. That is, we employ symmetric STDP in the proposed hippocampal CA3 model.

2.2   Pattern Completion

Pattern Completion is a function involved in memory retrieval. In the brain, a stimulus is expressed by the pattern of neuronal activity. So memory retrieval means reconstruction of the same pattern. In our memory retrieval, the complete pattern can be recalled from an incomplete pattern. This function of complementing patterns is called pattern completion and it is commonly accepted that hippocampal CA3 supports this function [8]. The major feature of CA3 is that there exist recursive axons that return part of CA3 output to CA3 itself. This recursive axon is called recurrent collateral and contributes to recall the complete pattern from incomplete patterns. Thus, CA3 functions as an autoassociative memory that leads to pattern completion.

3   Hippocampal CA3 Model

3.1   Spiking Neuron Model

The proposed hippocampal CA3 model consists of spiking neurons. This neuron model sums excitatory postsynaptic potential (EPSP) which is caused by presynaptic spikes. The following equation shows the EPSP at time r since a spike arrived at the neuron,

$$\varepsilon\left(r\right) = R_{\varepsilon} \frac{r}{r_{\varepsilon}} \exp\left(-\frac{r}{r_{\varepsilon}}\right) \tag{1}$$

where r_{ε} denotes the time until EPSP reaches maximum and R_{ε} denotes the magnitude of the EPSP. R_{ε} is chosen so that the max value of $\varepsilon\left(r\right)$ approaches 1.

The membrane potential of the ith neuron at time t, $h_i^{\mathrm{mem}}\left(t\right)$ is affected by two sources: excitations $h_i\left(t\right)$ and refractories $h_i^{\mathrm{ref}}\left(t - t_i^{\mathrm{last}}\right)$,

$$h_i^{\mathrm{mem}}\left(t\right) = h_i\left(t\right) + h_i^{\mathrm{ref}}\left(t - t_i^{\mathrm{last}}\right) \tag{2}$$

$h_i\left(t\right)$ is defined as a weighted sum of EPSP caused by all presynaptic spikes to the ith neuron, and t_i^{last} shows the latest firing time of the ith neuron. The refractoriness of the ith neuron is defined as follows:

$$h_i^{\mathrm{ref}}\left(s\right) = \begin{cases} -\infty & \text{for } s \leqq \delta^{\mathrm{abs}} \\ -R_{\mathrm{ref}} \exp\left(-\frac{s - \delta^{\mathrm{abs}}}{r_{\mathrm{ref}}}\right) & \text{for } s > \delta^{\mathrm{abs}} \end{cases} \tag{3}$$

where s shows the elapsed time from the latest firing time. For a short period δ^{abs} after the ith neuron firing, the neuron has absolute refractory. That is, the neuron can't fire during this period. Following the absolute refractory, the refractoriness decays and converges to 0.

The behavior of each neuron is stochastic and the firing of the ith neuron depends on the following probability,

$$P_i\left(t\right) = \frac{1}{2}\left(1 + \tanh\left(\beta\left(h_i^{\mathrm{mem}}\left(t\right) - T\right)\right)\right) \tag{4}$$

where β determines the randomness of the firing and T denotes the threshold for the ith neuron.

3.2   Synaptic Plasticity

In the proposed CA3 model, each synaptic weight is modified according to symmetric STDP. Spike interval Δt between the ith postsynaptic neuron and the jth presynaptic neuron is given by

$$\Delta t_{ij} = (T_i - T_j) \tag{5}$$

where T_i and T_j denote the spike time of the ith postsynaptic neuron and that of the jth presynaptic one, respectively.

In this study, we employ the *semi-nearest-neighbor* manner for pairing spikes [9]. That is, for each presynaptic spike, we consider only one preceding postsynaptic spike and ignore all earlier spikes. All postsynaptic spikes subsequent to the presynaptic spike are also considered. For each pre-/postsynaptic spike pair, the synaptic weight is updated as follows:

$$\Delta w_{ij} = 0.81 \left(1 - 0.65 \left(0.1 \Delta t_{ij} \right)^2 \right) e^{-(0.1 \Delta t_{ij})^2 / 2} + 1 \tag{6}$$

$$w_{ij} \left(t + \Delta t \right) = m_i \Delta w_{ij} w_{ij} \left(t \right) \tag{7}$$

where

$$m_i = \frac{C}{\sum_j^n \Delta w_{ij} w_{ij} \left(t \right)} \tag{8}$$

and C is a normalizing constant. Owing to the coefficient m_i, the sum of synaptic weights is conserved. We defined (6) by tracing experimental results [5].

3.3   Structure of Hippocampal CA3

The proposed hippocampal CA3 model is composed of n spiking neurons which have recurrent connections. Therefore, the neurons in CA3 receive external inputs and feedback inputs from CA3. The membrane potential of the ith neuron at time t is given by

$$h_i^{\text{ca3}} \left(t \right) = W^{\text{ext}} \sum_{t_i^{\text{ext}}} \varepsilon \left(t - t_i^{\text{ext}} - \Delta_{\text{axon}} \right)$$

$$+ \sum_j^N \sum_{t_j^{\text{ca3}}} w_{ij}^{\text{rc}} \left(t_j^{\text{ca3}} + \Delta_{\text{axon}} \right) \varepsilon \left(t - t_j^{\text{ca3}} - \Delta_{\text{axon}} \right) \tag{9}$$

where W^{ext} shows the synaptic weight for external inputs, Δ_{axon} shows axonal delay of a presynaptic neuron, and t_i^{ext} and t_j^{ca3} denote the spike time of an external input and that of the jth neuron, respectively. $w_{ij}^{\text{rc}} \left(t_j^{\text{ca3}} + \Delta_{\text{axon}} \right)$ shows the synaptic weight when the spike of the jth neuron at time t_j^{ca3} arrived at the ith neuron with axonal delay Δ_{axon}. w_{ij}^{rc} is initialized as follows:

$$w_{ij}^{\text{rc}} \left(0 \right) = \begin{cases} 0 & i = j \\ \frac{C}{(n-1)} & i \neq j \end{cases} \tag{10}$$

and is modified by STDP.

In the proposed CA3 model, frequent episodes are memorized by potentiating the synaptic weights w_{ij}^{rc} between neurons firing concurrently. However, since variable parts of an episode are seldom repeated, they are initially complemented by invariable parts of that. This causes the spike time difference between neurons which represent variable and invariable parts, and as a result, the synaptic weights between them are depressed by STDP. Eventually, their synaptic weights becomes small enough not to activate neurons corresponding to variable parts after repeated exposure to similar episodes. Thus, we can expect that the proposed CA3 model can retain essential elements of an episode and forget unnecessary ones in this way.

4   Computer Simulation Results

In this section, we demonstrate the memory modification by the proposed hippocampal CA3 model. In the simulation, we set the parameters as follows: $r_\varepsilon = 4.0$, $R_\varepsilon = 2.7$, $\Delta t = 1$, $\delta^{abs} = 3$, $R_{ref} = 16.0$, $r_{ref} = 7.0$, $T = 0.85$, $\beta = 60.0$, $C = 1.19$, $n = 30$, $W^{ext} = 1.0$ and $\Delta_{axon} = 12$. We applied two patterns to the proposed CA3 model: a complete pattern (CP) and an incomplete pattern (IP). The CP is represented by the activation of the 1st, 11th, 19th, 25th and 28th neurons from the left and the IP lacks the 11th neuron of the CP. First, the CP was applied to the CA3 model once every 200 unit time for 17 times (CP phase). Then, the IP was exposed in the same way for 85 times (IP phase). Namely, we applied input patterns 102 times in total. We regarded representation overlapped between the CP and IP as the invariable part of an episode, whereas we regarded the difference between them as variable part.

4.1   Pattern Completion

Here we examined the ability of pattern completion of the proposed CA3 model. Fig.1(a) shows the result when the IP was applied right after the CP phase. As shown in this figure, since the synaptic weights had been potentiated in the CP phase so that the CP could be recalled, the lack (the activation of the 11th neuron) was correctly complemented by the proposed model. That is, the proposed CA3 model shows the ability of pattern completion properly.

4.2   Memory Modification

After the CP phase, while pattern completion persists for some time, some synaptic weights connected to the 11th neuron gradually decrease during the IP phase because of the spike timing and STDP. Fig.1(b) shows the result when the 93th input was applied to the proposed model. Although the membrane potential of the 11th neuron was still not small, we could not observe pattern completion from this point in time. Fig.2 shows the transition of the synaptic weights to the 11th neuron. We can see that the synaptic weights to the 11th neuron were gradually depressed during the IP phase, especially that from the 1st neuron. As a result, the proposed CA3 model forgot the variable part of the episode and modified memory.

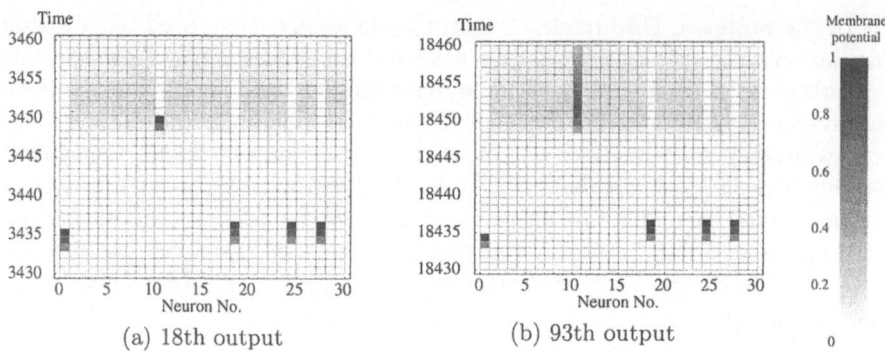

(a) 18th output   (b) 93th output

Fig. 1. CA3 output (*black: cell firing; gray level: membrane potential*).

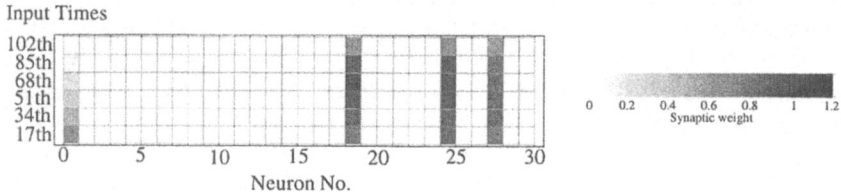

Fig. 2. Transition of synaptic weights to the 11th neuron (*gray level: strength*).

5   Conclusions

In this paper, we have proposed a hippocampal CA3 model which consists of spiking neurons and employs symmetric STDP as the rule of synaptic modification. Computer simulation results have shown the ability of pattern completion of the proposed model. Moreover, they have also shown that synaptic weights corresponding to the complemented part gradually decrease and eventually it comes not to be activated because of the pre-/postsynaptic spike timing and STDP. Namely, we have shown that pattern completion and STDP induce memory modification so that the CA3 model can memorize invariable parts of frequent episodes as essential elements and forget variable parts of those as unnecessary ones.

In the future research, we will develop the entire hippocampal model and study how this fundamental memory modification can contribute to form semantic memory by using more complicated temporal patterns.

Acknowledgment

The authors would like to thank Dr. Morisawa and Dr. Kobayashi for valuable discussions.

References

1. Nakazawa, K.: An interdisciplinary approach to cognitive neuroscience. Cell Technology **21** (2002) 982–985
2. Niki, K., Jing, L.: Hi-level cognition and hippocampus. Technical Report of IEICE **NC2002-103** (2003) 1–6
3. Eichenbaum, H.: How does the hippocampus contribute to memory. Trends in Cognitive Sciences **7** (2003) 427–429
4. Bi, G.: Spike timing-dependent synaptic plasticity. Cell Technology **21** (2002) 986–990
5. Kobayashi, Y., Shimazaki, H., Aihara, T., Tsukada, M.: Spatial distributions of hippocampal LTP/LTD induced electrically from schaffer collaterals and stratum oriens with relative timing. The brain and neural networks **8** (2001) 57–64
6. Debanne, D., Gähwiler, B.H., Thompson S.M.: Long-term synaptic plasticity between pairs of individual CA3 pyramidal cells in rat hippocampal slice cultures. Journal of Physiology **507.1** (1998) 237–247
7. Gulyás, A.I., Miles, R., Hájos, N., Freund, T.F.: Precision and variability in postsynaptic target selection of inhibitory cells in the hippocampal CA3 region. European Journal of Neuroscience **5** (1993) 1729–1751
8. Nakazawa, K., Quirk, M.C., Chitwood, R.A., Watanabe, M., Yeckel, M.F., Sun, L.D., Kato, A., Carr, C.A., Johnston, D., Wilson, M.A., Tonegawa, S.: Requirement for hippocampal CA3 NMDA receptors in associative memory recall. Science **297** (2002) 211–218
9. Izhikevich, E.M., Desai, N.S.: Relating STDP to BCM. Neural Computation **15** (2003) 1511–1523

Neural Mechanism of Binding ITD Information with IID One for Generating Brain Map of Sound Localization

Kazuhisa Fujita[1], ShungQuang Huang[1],
Yoshiki Kashimori[1,2], and Takeshi Kambara[1,2]

[1] Department of Information Network Science, School of Information Systems,
University of Electro-Communications, Chofu, Tokyo, 182-8585, Japan
k-z@nerve.pc.uec.ac.jp, {kashi,kambara}@pc.uec.ac.jp
[2] Department of Applied Physics and Chemistry,
University of Electro-Communications, Chofu, Tokyo, 182-8585, Japan

Abstract. Barn owls perform sound localization based on analyses of interaural differences in arrival time and intensity of sound. Two kinds of neural signals representing the interaural time difference (ITD) and the interaural intensity difference (IID) are processed in parallel in anatomically separate pathway. We explain briefly the neural models of pathways of ITD and IID detection which we have already presented [1, 2]. We present a neural network model of ICc ls in which the signals representing ITD and IID are first combined with each other. It is shown using our neural models how the neural map, in which ITD and IID are represented along the axes being perpendicular mutually, can be generated in ICc ls by the excitatory inputs from ICc core representing ITD and the inhibitory inputs from bilateral VLVps representing IID. We show that the firing rates of ICc ls neuron are well represented by a suppressive multiplication of ITD and IID inputs. This seems to be the origin of the multiplicative binding found in ICx by Pena and Konishi [3].

1 Introduction

Barn owls perform sound localization based on analyses of interaural differences in arrival time and intensity of sound. Two kinds of neural signals representing the interaural time difference(ITD) and the interaural intensity difference(IID) are processed in anatomically separate pathways that start from the cochlea nuclei in both ears. Both the signals are processed in parallel along the neural pathways to the higher sensory brain modalities. ITD is mainly used for detecting the horizontal direction of sound source and IID for the vertical direction. The neural map for detecting the spatial direction of sound source is formed in the brain of barn owls, based on the interaural arrival time and intensity information[4].

The sound localization of barn owl with respect to horizontal direction has been observed to be made with a remarkable high accuracy based on analysis of ITD in arrival time [4]. We have shown [1] that this hyper acuity for detection of

N.R. Pal et al. (Eds.): ICONIP 2004, LNCS 3316, pp. 44–49, 2004.
© Springer-Verlag Berlin Heidelberg 2004

ITD is accomplished through a combination of four kinds of functional factors:(1) the functional structure of neural unit(CD unit) for coincidence detection; (2) projection of CD units randomly arranged in every array of the units in nucleus laminaris (NL) to ITD sensitive neurons arranged randomly and densely in a few single arrays in the core of central nucleus of the inferior colliculus(ICc core); (3)convergence of outputs of all the arrays tuned in to a single sound frequency in ICc core into ITD sensitive neurons arranged regularly in a single array in the lateral shell of central nucleus of the inferior colliculus (ICc ls); and(4) integration of outputs of frequency-tuned arrays in ICc ls over all frequencies in external nucleus of inferior colliculus(ICx).

The sound localization of owl with respect to vertical direction is made based on analysis of IID. The neural information processing for detection of IID is made on neural pathway parallel with the pathway of ITD detection before both the signals arrive at ICc ls. The pathway for the IID detection is the angular nucleus in the cochlear nucleus → the nucleus ventralis lemnisci lateralis pars posterior (VLVp) → ICc ls(the first site of convergence of ITD and IID information) → ICx [4, 5].

In order to clarify the neural mechanism of detection of IID, we presented a neural model of VLVp [2] in which the signals of sound intensities coming from both the ears are combined to compare with each other. In the model, the information of IID is represented by position of the firing zone edge in the chains of IID sensitive neurons in both right and left VLVp units. We have shown [2] that the mutual inhibitory coupling between both VLVp units [5, 6] can induce the cooperative formation of clear firing zone edge in both VLVp units so that the firing zone in both units do not over lap with each other but the firing zone gap becomes as narrow as possible.

In the present paper, we investigated using the neural models of ICc core [1] and VLVp [2] for detection of ITD and IID, respectively, how the two kinds of signals coming from ICc core and VLVp are combined in ICc ls to make the neural map representing the spatial direction of sound source in ICx.

2   Neural Model of ICc ls Binding ITD Information with IID One

Two kinds of neural signals representing ITD and IID are converged firstly in the lateral shell of ICc ls. The maps representing ITD and IID, which are tuned in to single sound frequencies, are formed in ICc ls. The value of ITD is represented by the position of firing neuron in the chain of ITD sensitive neurons within ICc core [1, 4]. The axones of ITD sensitive neurons in ICc core make excitatory synaptic connections with the neurons in ICc ls, whereas the axons of main neurons in R- and L- VLVp units make inhibitory synaptic connections with the main neurons in ICc ls [6, 7].

We proposed in the present paper the model (Fig.1) showing how the map can be generated in ICc ls by the excitatory inputs from ICc core and the inhibitory inputs from right and left VLVp units. Then, values of ITD and IID are

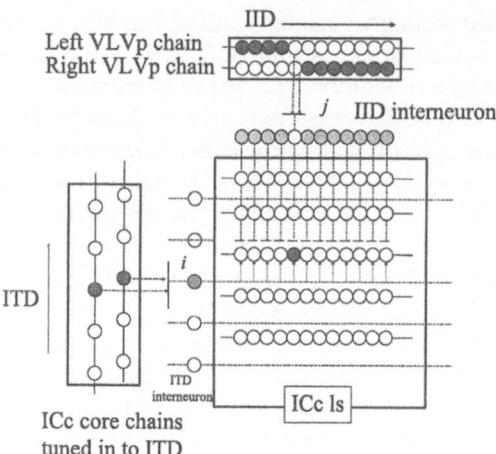

Fig. 1. Schematic description of functional connections of the lattice of main neurons in ICc ls with the chains of ITD sensitive neurons within ICc core and with the chains of IID sensitive neurons within right and left VLVp units. Those connections are made through a single ITD interneuron for each value of ITD and through a single IID interneuron for each value of IID. Arrows in the ICc ls network denote the excitatory synapses and short bars denote the inhibitory synapses.

represented along the axes perpendicular mutually as shown in Fig.1. The main neurons in ICc ls are arranged in the form of a lattice. The neurons in each row are received excitatory inputs from an interneuron gathering outputs of relevant neurons in ICc core chains, which is tuned in to a single value of ITD, as shown in Fig.1. The neurons in each column are received inhibitory inputs from an interneuron gathering bilaterally the outputs of right and left VLVp units at the relevant position through excitatory synapses as shown in Fig.1.

The membrane potential $V_s(i,j;t)$ and the output $U_s(i,j;t)$ of ICc ls neuron at (i,j) site of the lattice are determined by

$$\frac{dV_s(i,j;t)}{dt} = \frac{1}{\tau_m}(-(V_s(i,j;t) - V_{\text{rest}}) + w_{\text{ITD}} \times U_{\text{ITD}}(i;t) - w_{\text{IID}}U_{\text{IID}}(j;t)), \quad (1)$$

$$U_s(i,j;t) = \frac{1}{1 + \exp[-(V_s(i,j;t) - V_{\text{thr}})/T]}, \quad (2)$$

where V_{rest} is the potential in the resting state, $U_{\text{ITD}}(i;j)$ and $U_{\text{IID}}(j;t)$ are the outputs of ith ITD and jth IID interneurons, respectively, w_{ITD} and w_{IID} are the strength of relevant synaptic connections, and V_{thr} and T are the threshold value of membrane potential and the rate constant, respectively. The membrane potentials $V_{\text{ITD}}(i;t)$ and $V_{\text{IID}}(j;t)$ of ith ITD and jth IID interneurons, respectively, are also obtained using the leaky integrator neuron model used in Eq.(1). The output of each interneuron is represented by 1 or 0, the probability that $U_{\text{ITD}}(i;t)$ is 1 is given by

$$Prob(U_{\mathrm{ITD}}(i;t) = 1) = \frac{1}{1 + \exp[-\frac{(V_{\mathrm{ITD}}(i;t) - V_{\mathrm{thr}})}{T}]}. \tag{3}$$

$U_{\mathrm{IID}}(j;t)$ is also obtained from $V_{IID}(j;t)$ using the equivalent equation.

Thus, the lattice array of main neurons in ICc ls functions as the map in which the value of ITD is represented along the column direction and the value of IID is represented along the row direction as shown in Fig.1. Under application of binaural sound stimulus, only one neuron group in the chains within ICc core fire, where the ITD selectivity of the neuron group corresponds to the value of ITD of the stimulus. Therefore, the neurons in only the row of the ICc ls lattice corresponding to the ITD value receive the excitatory inputs through the relevant interneuron. On the other hand, each neuron in the row receives inhibitory inputs from almost all of IID interneurons as seen in Fig.1. The interneuron receiving signals from VLVp neurons in the narrow gap, whose position corresponds to the value of IID of the stimulus, dose not fire as shown in Fig.1. Therefore, the neurons in the column corresponding to the value of IID is not inhibited by the outputs of R- and L- VLVp units. Thus, the neuron in the lattice, which is firing under application of the sound with a pair of definite values of ITD and IID, can represent the value of ITD by its position along the column direction and the value of IID by its position along the row direction.

3   Ability of the ICc ls Model to Represent Values of ITD and IID as a Map

3.1   Response Properties of ICc ls Model

To investigate how ICc ls neurons bind values of IID represented by VLVp pair with values of ITD represented by ICc core, we calculated the firing rates of ICc ls neurons using our neural models of VLVp pair [2] and ICc core [1]. The result is shown in Fig.2. Figure 2a shows the firing probability $U_{\mathrm{ITD}}(i)$ ($i = 1 \sim 61$) of ITD interneuron i which is receiving the outputs of ICc core neurons tuned to ITD during a finite stimulation period. Figure 2b shows the firing probability $U_{\mathrm{ITD}}(j)$ ($j = 1 \sim 31$) of IID interneuron j which is receiving the outputs of IID sensitive neurons in both VLVp chains during the some period. Figure 2c shows the firing frequency of the ICc ls network. The peak of firing frequency appears at the position (center) corresponding to ITD=0 and IID=0. However, the representation of ITD and IID in ICc ls is not unique, because there exist the other two peaks. This problem is solved by integration of the outputs over ω in ICx [4].

3.2   Multiplicative Binding of ITD and IID Information

Pena and Konishi [3] have recently shown observing the firing frequency of neurons sensitive both to ITD and IID in ICx that the neurons bind ITD and IID information in a multiplicative manner. However, the mechanism generating the

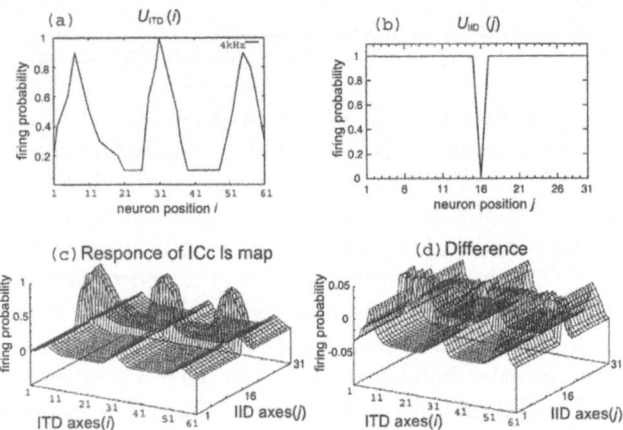

Fig. 2. Response probability of ICc ls to a pair of ITD and IID inputs (ITD=0,IID=0). Firing probabilities of (a) ITD interneurons ($i = 1 \sim 61$) (b) IID interneurons ($j = 1 \sim 31$) (c) Firing probability of ICc ls neurons (d) difference between the firing pattern (c) and the ideal multiplicative output $f(i,j)$ (Note the scale of perpendicular axis).

multiplicative binding is not clear yet. In order to clarify the mechanism, we compared the firing patterns of ICc ls map calculated with the ideal multiplicative output. The ideal output $f(i,j)$ is represented according to Pena and Konishi [3] by using the firing frequencies $U_{ITD}(i)$ and $U_{IID}(j)$ as

$$f(i,j) = f_0 + \lambda U_{ITD}(i) \times (1 - U_{IID}(j)), \tag{4}$$

where f_o and λ are constant adjustable parameters. Note that the multiplicative term of $f(i,j)$ contributes to the suppression of ICc ls activity due to inhibitory connections between VLVp and ICc ls. In Fig.2d, we showed the difference between the firing frequency of ICc ls shown in Fig.2c and $f(i,j)$. The difference is quite small compared with the firing frequency shown in Fig.2c.This led us to conclude that ICc ls neurons bind multiplicatively ITD and IID information. The present model provides a novel multiplicative binding, that is, ICc ls encodes ITD and IID information as a suppressive multiplication between them.

3.3   Origin of the Multicative Binding Between ITD and IID Information

The origin of the multiplicative binding in ICc ls is quite clear if our model of ICc ls network is adopted. The firing probability of each ICc ls neuron is given by the product of the two kinds of probabilities, where one is the probability of relevant ITD interneuron firing and the other is that of relevant IID interneurons resting. Then, the ITD and IID interneurons fire according to the probabilities determined by the outputs of relevant ICc core neurons and VLVp neurons, respectively, as shown in Eq.(3).

Pena and Konishi [3] found that the multiplicative binding is made by neurons sensitive to a pair of ITD and IID stimuli in ICx. The neurons sensitive to a pair of specific values of ITD and IID are distributed regularly in the similar manner to the map of ICc ls shown in Fig.1 [1]. Therefore, it is a quite reasonable assumption that the outputs of (i,j) neuron in ICx are induced by the inputs given by integrating the outputs of (i,j) neurons in ICc ls network tuned in to each frequency over all frequencies. Then, the relation of outputs of ICx neurons to the outputs of ICc core and VLVp pair is equivalent to the relevant relation for ICc ls neurons.

4   Conclusion

In order to clarify the neural mechanism by which ITD information is bound with IID information so that the neural map for sound localization is generated in ICc ls, we presented a neural model of ICc ls network. The ICc ls neuron sensitive specifically to a pair of specific values of ITD and IID is made by excitatory inputs from ICc core encoding ITD and the inhibitory inputs from bilateral VLVp encoding IID. We showed how effectively the fire gap made in bilateral VLVps transmits IID information to ICc ls by the inhibitory signals. We showed also that the firing rates of ICc ls neuron are well represented by a suppressive multiplication of ITD and IID inputs as expected in the experimental result.

References

1. Inoue, S., Yoshizawa, T., Kashimori, Y., Kambara, T.: An essential role of random arrangement of coicidence detection neurons in hyper accurate sound location of owl. Neurocomputing, 38-40 (2001) 675-682
2. Fujita, K., Huang, S.-Q., Kashimori, Y., Kambara, T.: Neural mechanism of detecting interaural intensity differences in the owl's auditory brainstem for sound localozation. Neurocomputing, in press (2004).
3. Pena, J.L., Konishi, M.: Auditory spatial receptive field created by multiplication, Science, 292 (201) 249-252
4. Konishi, M.: Listening with two ears. Scientific American, 268 (1993) 66-73
5. Takahashi, T.T., Keller, C.H.: Commissural connections mediate inhibition for the computation of interaural level difference in the barn owl. J.comp.Physiol.A, 170 (1992) 161-169
6. Adolphs, R.: Bilateral inhibition generates neuronal responses tuned to interaural level differences in the auditory brainstem of the barn owl. J.Neurosci, 13 (1993) 3647-4668.
7. Mordans, J., Knudsen, E.I.: Representation of interaural level difference in the VLVp, the first site of binaural comparison in the barn owl's auditory system. Hearing Res, 74 (1994) 148-164

The Spatiotemporal Dynamics
of Intracellular Ion Concentration and Potential

Seiichi Sakatani[1,2] and Akira Hirose[1]

[1] Department of Frontier Informatics, Graduate School of Frontier Sciences,
The University of Tokyo, 7-3-1 Hongo, Bunkyo-ku, Tokyo, 113-8656 Japan
sakatani@eis.t.u-tokyo.ac.jp, ahirose@ee.t.u-tokyo.ac.jp
[2] Japan Society for the Promotion of Science

Abstract. It is well known that membrane potential in neuron is determined by the ion concentrations and ion permeability. However, the ion
concentrations are determined by both the ion flows through ion channels
and intracellular and extracellular ion flows. Therefore, it is needed to
solve the ion concentrations and potential distribution simultaneously to
analyze the spatiotemporal change in membrane potential. In this paper,
we construct the theory for spatiotemporal dynamics of ion concentration and potential. We adopt Hodgkin–Huxley-type nonlinear conductance to express the ion permeability and use Nernst–Planck equation
to denote the ion concentrations. We also assume that the electric charge
is conserved at intra- and extra-cellular space and at the boundary. By
using the theory, we numerically analyze the distribution of intracellular
ion concentrations and potential. As a result, the following phenomena
are revealed. When the cell depolarized, firstly ions flow into (or out
to) only to the thin space very adjacent to the membrane by rapid ion
flows through ion channels. Secondly, ions slowly diffuse far away by ion
concentration gradients. The movement speeds are determined only by
their diffusion coefficients and almost independent of membrane potential. This theory has a high degree of availability since it is extendable
to cells with various types of ion channels.

1 Introduction

The dynamics of membrane potential is explained by the changes of ion conductance on membrane. On the other hand, the propagation dynamics of membrane
potential at the dendrites has been analyzed in one dimension by using the cable
theory[6] and compartmental models[6–8]. Recently, the theory that expresses
two dimensional propagation was proposed since the membrane expands in two
dimension[9]. However, in these conventional research, the intracellular potential
in the depth direction and the influence of the distribution of intracellular ion
concentration on membrane potential propagation have not been studied.

It is well known that the origin of membrane potential lies in ion concentration and ion permeability (Goldman–Hodgkin–Katz voltage equation[10, 11]).
Each ion concentration is determined by the ion flow through channels and the

N.R. Pal et al. (Eds.): ICONIP 2004, LNCS 3316, pp. 50–56, 2004.
© Springer-Verlag Berlin Heidelberg 2004

flow in both the intracellular space and extracellular one. On the other hand, ion flow at ion channel depends on membrane potential or ion concentration. The ion movement in the intracellular and extracellular space is followed by the concentration gradient as well as potential gradient. That is to say, there is interdependence among membrane potential and ion concentration. Consequently, it is needed to solve ion concentration and potential distribution inside and outside of the cell at the same time for the analysis of spatiotemporal change in membrane potential.

In this paper, we construct the spatiotemporal dynamic theory of ion concentration and potential (Ion Concentration and Potential (ICP) theory). For the construction, we express ion permeability on the membrane by nonlinear conductance where we treat each ion conductance density as a variable. We use Nernst–Planck equation[2–5] to express the intra- and extra-cellular ion concentration. We set that the electric charge is conserved at intra- and extra-cellular space and at the boundary. By using this theory, we numerically analyze the spatiotemporal dynamics of intracellular ion concentration and potential. We show that the phenomenon and the ion diffusion speed are consistent with the observed results in the physiological experiment.

2   Theory

Here we construct spatiotemporal dynamic theory of intracellular ion concentration and potential. To analyze the spatiotemporal distribution at the action potential initiation, we consider only rapid temporal change. The intracellular ion movement is followed by the concentration and potential gradients. By using Nernst–Planck equation, the concentration of an ion A at any intracellular points is expressed as

$$\frac{\partial [A]_i}{\partial t} - \nabla \cdot (\mu_A z_A [A]_i \nabla V_i + D_A \nabla [A]_i) = \frac{d[A]_{inf}}{dt} \tag{1}$$

where $[A]_i$ is intracellular ion A concentration, t is time, μ_A is mobility, z_A is valence, V_i is intracellular potential, D_A is diffusion coefficient and $[A]_{inf}$ is ion influx. We consider only potassium, sodium and chloride ions among the intracellular ions. Each ion concentration is expressed as

$$\frac{\partial [K^+]_i}{\partial t} - \nabla \cdot (\mu_K z_K [K^+]_i \nabla V_i + D_K \nabla [K^+]_i) = \frac{d[K^+]_{inf}}{dt} \tag{2}$$

$$\frac{\partial [Na^+]_i}{\partial t} - \nabla \cdot (\mu_{Na} z_{Na} [Na^+]_i \nabla V_i + D_{Na} \nabla [Na^+]_i) = \frac{d[Na^+]_{inf}}{dt} \tag{3}$$

$$\frac{\partial [Cl^-]_i}{\partial t} - \nabla \cdot (-\mu_{Cl} z_{Cl} [Cl^-]_i \nabla V_i + D_{Cl} \nabla [Cl^-]_i) = \frac{d[Cl^-]_{inf}}{dt} \tag{4}$$

where $[K^+]_i$, $[Na^+]_i$ and $[Cl^-]_i$ are the intracellular potassium, sodium and chloride concentrations, respectively. μ_K, μ_{Na} and μ_{Cl} are their mobilities. z_K, z_{Na} and z_{Cl} are their valences, while D_K, D_{Na} and D_{Cl} are their diffusion coefficients.

$[K^+]_{inf}$ denotes the potassium ion influx. The rapid influx at the boundary between the membrane and intracellular space occurs that through the potassium channel. That is, the potassium ion influx $[K^+]_{inf}$ is expressed as

$$\frac{d[K^+]_{inf}}{dt} = \begin{cases} 0 & \text{(Intracellular)} \\ J_K^{ch} = -\frac{\bar{g}_K n^4 (V_m - E_K)}{F} & \text{(Boundary)} \end{cases} \quad (5)$$

where J_K^{ch} is the potassium ion influx per unit time through the potassium channel, $\bar{g}_K$ is the maximum conductance of the channel, n is the activation variable which Hodgkin and Huxley[1] introduced, E_K is the equilibrium potential and F is Faraday's constant. V_m is membrane potential and is defined as $V_m \equiv V_{ib} - V_{eb}$ in terms of the potential at the boundary between intracellular space and membrane V_{ib} and that at the boundary between extracellular space and membrane V_{eb}.

$[Na^+]_{inf}$ also denotes sodium ion influx which occurs through the channel. The sodium ion influx $[Na^+]_{inf}$ is also followed by

$$\frac{d[Na^+]_{inf}}{dt} = \begin{cases} 0 & \text{(Intracellular)} \\ J_{Na}^{ch} = -\frac{\bar{g}_{Na} m^3 h (V_m - E_{Na})}{F} & \text{(Boundary)} \end{cases} \quad (6)$$

where J_{Na}^{ch} is the sodium ion influx per unit time through sodium channel, $\bar{g}_{Na}$ is the maximum conductance, m and h are the activation and inactivation variable which Hodgkin and Huxley[1] expressed, respectively and E_{Na} is the equilibrium potential.

In like wise, $[Cl^-]_{inf}$ also denotes chloride ion influx through the chloride channel. The chloride ion influx $[Cl^-]_{inf}$ is also described as

$$\frac{d[Cl^-]_{inf}}{dt} = \begin{cases} 0 & \text{(Intracellular)} \\ J_{Cl}^{ch} = \frac{g_{Cl}(V_m - V_{rest})}{F} & \text{(Boundary)} \end{cases} \quad (7)$$

where J_{Cl}^{ch} is the chloride ion influx per unit time by the chloride channel, $\bar{g}_{Cl}$ is the channel conductance and V_{rest} is the resting potential.

The law of conservation of electric charge in the cell is expressed as,

$$\nabla \cdot (\sigma_i \nabla V_i) = 0 \quad (8)$$

where V_i is intracellular potential and σ_i is intracellular conductivity. We assume that intracellular charged particles are restricted to potassium, sodium and chloride ions. Therefore, intracellular conductivity σ_i is expressed as follow.

$$\sigma_i = (\mu_K z_K [K^+]_i + \mu_{Na} z_{Na} [Na^+]_i + \mu_{Cl} z_{Cl} [Cl^-]_i) F. \quad (9)$$

The influx current is the sum of capacitive current and ion current, and that is equal to the current flowing at the intracellular space at the boundary, i.e.,

$$\nabla \cdot (\sigma_i \nabla V_{ib}) = -C_m \frac{dV_m}{dt} - I_{ion} \quad (10)$$

where V_{ib} is the potential at the boundary between intracellular space and membrane and C_m denotes membrane capacitance per unit area. I_{ion} is the total of ion current and expressed as

$$I_{ion} = \bar{g}_K n^4 (V_m - E_K) + \bar{g}_{Na} m^3 h (V_m - E_{Na}) + g_{Cl}(V_m - V_{rest}) \qquad (11)$$

3   Numerical Analysis and Results

We use the finite-element-analysis software (COMSOL, FEMLAB) for the numerical analysis. We focus on only potassium and sodium ions which play a dominant role on the spatiotemporal change in membrane potential, and neglect other ion effects. We also assume that voltage-dependent potassium channel and sodium channel are distributed uniformly on the membrane of the soma and the dendrite. We consider only the effects of fast changes, shorter than 10 ms and neglect slow ones. For the calculation simplicity, we treat the extracellular potential as the ground and analyze the intracellular ion concentration and potential.

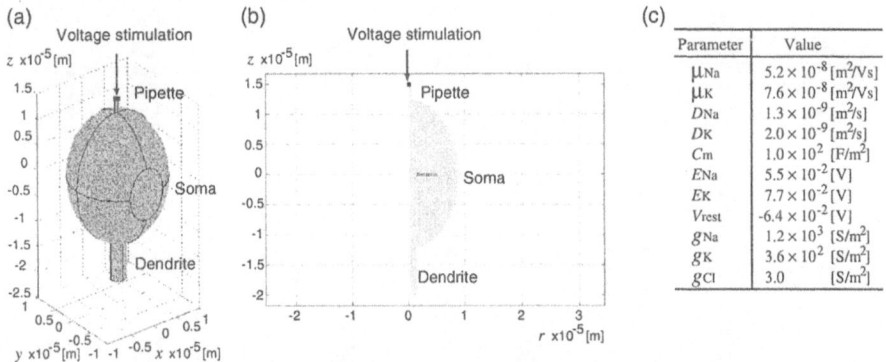

Fig. 1. (a) 3 dimensional cell model, (b) z axial symmetrical 2 dimensional model and (c) physical parameters.

Figure 1 shows the cell model which consist of a soma, a dendrite and a pipette for stimulation. We can drop 3 dimensional cell in Fig.1(a) to 2 dimensional model in Fig.1(b) by considering the vertical axial symmetry. Figure 1(c) shows physical parameters.

The potential at the platinum electrode in the patch-clamping pipette is set at 0[mV]. The potential in the soma near the pipette is then driven from the resting voltage to the depolarized one for 5[ms]. At the initial state, the potential is set uniformly at −64[mV] (resting potential), potassium ion concentration at 130[mM] and sodium ion concentration at 10[mM].

Figure 2(a) shows the spatiotemporal change of intracellular potential for 5 ms. The intracellular potential rises to 25[mV] after 1 ms and drop to −20[mV]

54   Seiichi Sakatani and Akira Hirose

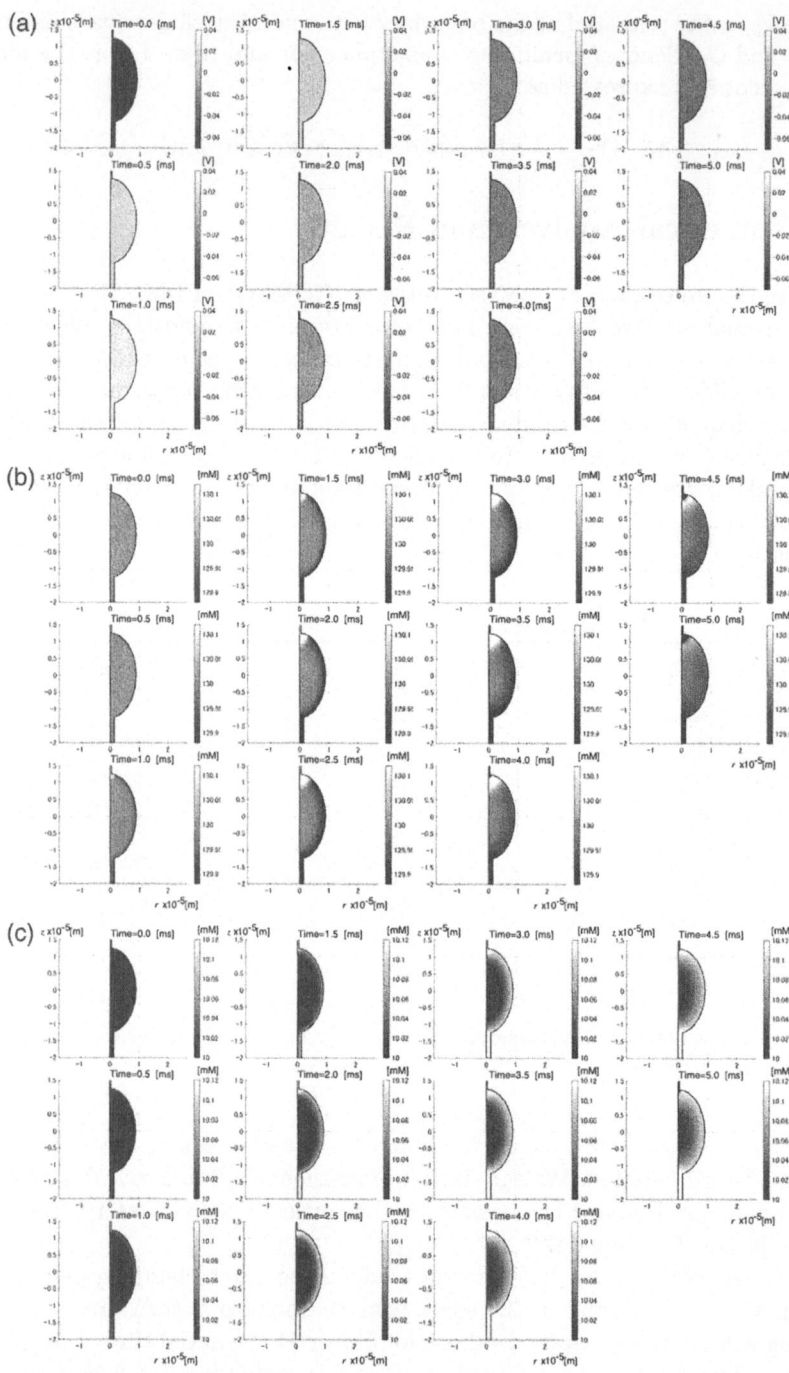

Fig. 2. (a) The spatiotemporal changes of (a) intracellular potential, (b) intracellular potassium ion and (c) intracellular sodium ion, respectively.

after 3 ms. The temporal changes in potential occur uniformly and potential gradients don't exist in the intracellular space. This result is consistent with conventional knowledge that the surface potentials on membrane rapidly become homogeneous by the transient current.

Figure 2(b) also shows the spatiotemporal change in intracellular potassium ion concentration for 5 ms. The cell is depolarized, and the potassium ion concentrations just below the membrane decrease. This is because the potassium ions flow out of the cell through voltage-dependent potassium channel. The potassium ion concentrations in the soma are higher than those in the pipette at 0.5–2.5 ms, and vice versa at 3.0–5.0 ms. When the soma depolarizes over 0 mV at 0.5–2.5 ms, the potential in the soma is higher than those in the pipette (Fig.2(a)). That is, potassium ions move from the soma to the pipette according to the potential gradient. Consequently, the concentrations in the pipette are higher. On the contrary, when the soma depolarizes under 0 mV at 3.0–5.0 ms, the potential in the soma is lower than those in the pipette (Fig.2(a)). Again, potassium ions move from the pipette to the soma according to the potential gradient. Therefore, the concentrations in the pipette are lower.

We estimate the migration speed of potassium ion from Fig.2(b). The time when the potassium ion concentration becomes minimum value at $(9 \times 10^{-6}, 0)$, $(8 \times 10^{-6}, 0)$ and $(7 \times 10^{-6}, 0)$ are 2.1[ms], 2.7[ms] and 3.3[ms], respectively and their intervals are constant. The diffusion distance of potassium ion is expressed as $\sqrt{D_K t}$, and the distance that the ions move for 0.6 ms is 1.1 μm. Since this is almost the same as each distance $(1.0[\mu m])$ among points, the migration speed of intracellular potassium ion is determined only by the diffusion coefficient and the drift have little effect on it. This is consistent with the result of potential in Fig.2(a).

Figure 2(c) also shows the spatiotemporal change in intracellular sodium ion concentration for 5 ms. The cell is depolarized, the sodium ion concentrations rise. That's because sodium ions flow into the cell through the voltage-dependent sodium channel. The change in sodium ion concentration is also governed by the concentration gradients since the potential gradients don't exist in the cell. The high concentration layer of sodium ion on the r axis moves through a distance of 1 μm in the negative direction in 1 ms from 3 ms to 4 ms. This distance is almost the same as the diffusion distance of sodium ion in 1 ms that is 1.2 μm $(\sqrt{D_{Na} t} = 1.2[\mu m])$. That is, the migration speed of intracellular sodium ion is also determined only by the diffusion coefficient. It is confirmed that this speed is the same as that observed in the cultured cell[13].

4   Conclusion

The results are summarized as follows. (1) When the cell is depolarized, firstly ions flow into (or out to) the region very adjacent to the membrane by rapid flows through ion channels. (2) Secondly, ions slowly move far away by the ion concentration gradients without drift. The speeds are determined by their diffusion coefficients. These two phenomena are consistent with the results in the physiological experiments. The Ion Concentration and Potential (ICP) theory is the quantitative theory that expresses the spatiotemporal dynamics of ion concentration and potential.

Acknowledgement

This work was partly supported by the Ministry of Education, Culture, Sports, Science and Technology, Grant-in-Aid for Scientific Research on Grant-in-Aid for JSPS Fellows 15-11148 (Sakatani).

References

1. Hodgkin A. L. and Huxley A. F. *J. Physiol.*, 117, 500–544, 1952.
2. Nernst W. *Z. Phys. Chem.*, 3, 613–637, 1888.
3. Nernst W. *Z. Phys. Chem*, 4, 129–181, 1889.
4. Planck M. *Ann. Phys. Chem.*, *Neue Folge* 39, 161–186, 1890.
5. Planck M. *Ann. Phys. Chem.*, *Neue Folge* 40, 561–576, 1890.
6. Rall W. In Reiss R. F., editor, *Neural theory and modeling*, 73–94, Standord University Press, Stanford, CA, 1964.
7. Perkel D. H. et al. *Neurosci.*, 6, 823–837, 1981.
8. Holmes W. R. and Rall W. *J. Neurophysiol.*, 68, 1421–1437, 1992.
9. Hirose A. and Murakami S. *Neurocomp.*, 43, 185–196, 2002.
10. Goldman D. E. *J. Gen. Physiol.*, 27, 37–60, 1943.
11. Hodgkin A. L. and Katz B. *J. Physiol.*, 108, 37–77, 1949.
12. Rall W. *Science*, 126, 454, 1957.
13. Sakatani S. PhD thesis, Univ. of Tokyo, 2003.

A Model That Captures Receptive Field Properties of Orientation Selective Neurons in the Visual Cortex

Basabi Bhaumik*, Alok Agarwal, Mona Mathur, and Manish Manohar

Department of Electrical Engineering, Indian Institute of Technology, Delhi,
Hauz Khas, New Delhi-110016, India
{bhaumik,alok}@ee.iitd.ac.in, msmathur@hotmail.com,
maniishmanohar2002@yahoo.co.in

Abstract. A purely feedforward model has been shown to produce realistic simple cell receptive fields (RFs). The modeled cells capture a wide range of receptive field properties of orientation selective cortical cells in the primary visual cortex. We have analyzed the responses of 72 nearby cell pairs to study which RF properties are clustered. Orientation preference shows strongest clustering and RF phase the least clustering. Our results agree well with experimental data (DeAngelis et al, 1999, Swindale et al, 2003).

Keywords: visual cortex; orientation selectivity; receptive field; neuron

1 Introduction

Ever since the discovery of orientation selective cells by Hubel and Wiesel (Hubel and Wiesel, 1962) a number of models have been proposed. These models can be broadly classified into two categories. (i) In these models receptive fields of simple cells is generated by construction. These models deal with the origin of orientation selectivity either through feedforward connections (Hubel and Wiesel, 1962; Ferster, 1987) or inhibitory intracortical connections (Koch and Poggio, 1985; Carandini and Heeger, 1994, Wörgötter and Koch, 1991) or recurrent connections (Somers et al., 1995, Douglas et al., 1995). The primary assumption in these models is that initial orientation specificity in the cortex is generated by converging thalamic inputs. However, these models do not attempt to answer how this bias is generated. (ii) Models (von der Malsburg, 1973; Linsker, 1986; Miller, 1994, Bhaumik and Mathur, 2003) that have tried to address underlying developmental mechanisms that lead to formation of simple cell receptive fields with segregated ON and OFF regions. In all the models in this category the receptive fields size i.e the arbor size of all cortical cells are taken to be equal. In this paper we extend our earlier work (Bhaumik and Mathur, 2003) to include variations in arbor sizes in the cortical cells, thereby removing the constraints of fixed arbor size. Initially a cortical cell receives very small weights ($\sim 10^{-6}$) from the entire arbor window and arbor sizes were chosen from a uniform random distribution. As receptive fields develop, significant weights are found only in a smaller region and an effective arbor size emerges. With this model we study the micro organizations of near neighbour cortical cell properties such as orientation selectivity, orientation tuning and magnitude, receptive field phase and size. Our

---

* Corresponding author. This work is sponsored by Department of Science and Technology, Ministry of Science and Technology, India.

N.R. Pal et al. (Eds.): ICONIP 2004, LNCS 3316, pp. 57–63, 2004.
© Springer-Verlag Berlin Heidelberg 2004

simulated results match closely with experimental results (DeAngelis et al, 1999; Freeman, 2003: Swindale et al., 2003).

2 The Feedforward Model for the Formation of Simple Cell RFs

2.1 Model Assumptions

For the formation of simple cell RFs, we have proposed a model based on competition for neurotrophic factors and cooperation among near neighbors through diffusion (Bhaumik and Mathur, 2003). The model is based on biologically plausible assumptions of: (a) Competition for a pre-synaptic resource where a pre-synaptic cell has a fixed amount of resource to distribute among its branches. This would constrain the number of axonal branches a neuron can maintain; (b) Competition between axons for target space. The axons are competing for neurotrophic factors, growth or survival promoting factors, released by the postsynaptic cells upon which the axons innervate. Competition for target space or post-synaptic competition is used in all models for development of ocular dominance and; (c) Diffusive cooperation between near neighbor (i) cortical cells and (ii) same type of i.e. ON-ON and OFF-OFF LGN cells. Studies on Long-term potentiation LTP have shown the generation of diffusible signals at the active synapses leads to strengthening of the nearby synapses. Diffusive spread of signals also provides local anti-correlation between ON and OFF LGN cells. No synaptic weight normalization is used. Both cooperation and competition determine the strength of synapses in the model. Fixed resources are used for both pre and postsynaptic competition. Such competition among synapses for finite resources, such as receptor or a trophic factor controlling the number of synapses have been observed (Xiong et al, 1994). The cooperation among neighboring cells can occur through release and uptake of diffusible factors (Bonhoeffer, 1996).

2.2 Model Architecture and Equations

The model consists of three hierarchical layers: retina, LGN and cortex. All the layers are modeled as two-dimensional arrays. Both retina and LGN comprise of two distinct (ON and OFF) layers of size 30x30. Cortex consists of one layer of 50x50 spiking cells. Retinal and LGN cells are modeled as center surround gaussian filters with fixed one to one connectivity from retina to LGN. For the details of retinal cell's spatial receptive field, temporal response functions and mechanism for generation of spikes we have used the model in (Wörgötter and Koch, 1991). The response of cortical cell at a given time is calculated using SRM model (Spike response model) (Gerstner, 1999). A cortical cell receives thalamic projections (both ON and OFF) from a PxP region centered symmetrically about its corresponding retinotopic position in the LGN. Initial synaptic strengths are very weak and randomly organized. Time evolution of synaptic strengths represents cortical development and is achieved through the following differential equation for weight updation.

$$\frac{\partial W_{IJ}^+}{\partial t} = (\gamma_1 - K_1)(\gamma_2 - K_2) \mathbf{A}_R \ W_{IJ}^+ + D_L \frac{\partial^2 W_{IJ}}{\partial J^2} + D_C \frac{\partial^2 W_{IJ}}{\partial I^2} \tag{1}$$

where, W^+_{IJ} (W^-_{IJ}) represents the strength of the connection from the ON-(OFF) center LGN cell at position J=(J_1,J_2) in LGN layer to the cortical cell at position I=(I_1,I_2) in the cortical layer. $W_{IJ} \in \{W^+_{IJ}, W^-_{IJ}\}$. Since we are considering only simple cells, from any location, either ON connection $W^+_{I,J}$ or OFF connection $W^-_{I,J}$ exists. $K^2_1 = \Sigma^{NXN}_{P=1}(W^+_{PJ})^2$, is the sum square of synaptic strength of all branches emanating from the LGN cell at the location J. γ_1 represent fixed presynaptic resources available in the LGN cell at location J. The term $(\gamma_1 - K_1)$ enforces competition for resources among axonal branches in a LGN cell. Similarly the term $(\gamma_2 - K_2)$ enforces competition among LGN cells for target space in the cortex. $K^2_2 = \Sigma^{MXM}_{P=1}(W_{IP})^2$ is the sum of the square of synaptic strength of all branches of LGN cells converging on the cortical cell at location I. γ_2 represent fixed postsynaptic resources available in the cortical cell at location I. A_R is the arbor function. The arbor function defines the region from where a cortical cell receives its initial unorganized thalamic afferents. The amount of afferents a cell receives is determined by the arbor window size PxP. The PxP regions were chosen from a uniform random distribution in which the sizes varied uniformly between 9x9 and 21x21 (mean 14.5, std. 3.44). The type of function (square, circular, Gaussian etc) used within the arbor window defines the arbor window type. A trapezoidal window has been used, where in the window height reduces as one moves towards the periphery of the window. D_L and D_C are the diffusion in the LGN and cortex respectively. MxM and NxN are the sizes of the LGN and the cortex respectively. A similar equation is used for updating W^-_{IJ}.

3 Results

Weight update is simulated for a 50x50 cortical layer and two overlapping 30x30 LGN layers using circular boundary conditions. Initial weights of the order of 10^{-6} are picked up from a uniform random distribution. The differential equation for weight updating is simulated in difference mode using synchronous weight update. Out of phase update is carried out for ON and OFF weights.

3.1 Receptive Fields

The contour plots of every third cell from a 15x15 cross-section of a 50x50-simulated cortex are shown in Fig 1(a). Initially a cortical cell receives very small weights (~ 10^{-6}) from the entire arbor window. As receptive fields develop, significant weights are found only in a smaller region and an effective arbor size (mean12.6 x12.6, std. 1.76) emerges (by ignoring all the weights less than 1/20th of the maximum weight). The weight development equation given in (1) tends to minimize the variations in arbor sizes of the nearby cortical cells.

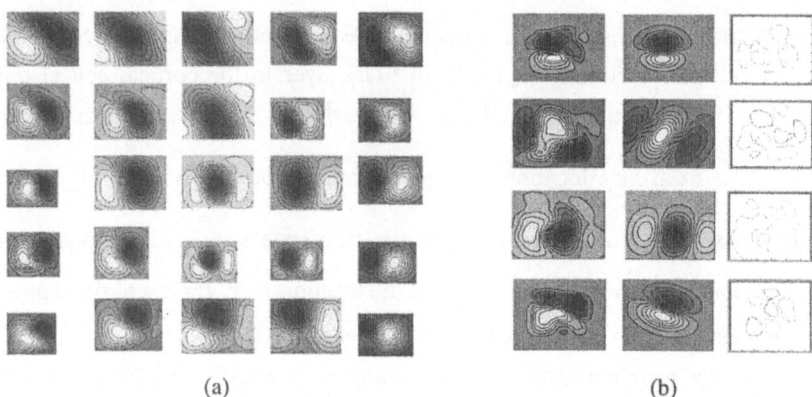

(a)   (b)

Fig. 1. (a) A 50x50 cortex was simulated using eqn.(1) (D_C=0.3, D_L=0.075, $\gamma_1 = 5, \gamma_2 = 1$). Every third cell from a 15x15 section of this simulated cortex is shown. The contours within the sub-fields indicate the strength of the synaptic connections. Shades of white (black) indicate synapses with ON (OFF) center LGN cells. (b) Simulated RF, 2D Gabor fit and residual error for four simulated cells

3.2   Simulated Cell Characterization

To obtain response of simulated cortical cells the retina in the three-layer visual pathway model was stimulated with sinusoidal grating of 50% contrast at 0.5 cycles/degree spatial frequency and moving at a velocity of 2 deg/sec, unless specified otherwise. The spike response of the cortical cells to various input stimuli was calculated. The orientation of the input sinusoidal grating was varied from 0° to 180° in steps of 18°. The direction of motion of the grating was always orthogonal to the orientation of the grating orientation was presented to the retina for thirty times and peristimulus histogram (PSTH) is made for each of these thirty presentation with a bin width of 100 ms. Spike rates per second were computed for individual bins. The cell spike response for each orientation of input stimulus is the average response of the cell over these thirty PSTH. Ten responses were obtained for ten orientations of input stimulus. These ten responses are then converted into vectors having magnitude equal to response amplitude and angle equal to twice the angle of the grating. Half the angle of the resultant of these vectors gives the orientation preference of the cell. To calculate tuning hwhh (half width at half height) a third order spline curve was fitted through the ten responses for each cell. To obtain RF phase of nearby neurons the simulated RFs were fitted with Gabor functions in the least square sense using an unconstrained minimization algorithm in Matlab (fminsearch, Mathworks). In Figure1(b) we show the spatial RFs and the 2D Gabor filter fits for four simulated cells. The plots are shown in three columns. First column plots the RF of the cell as obtained through simulations, the second column gives the corresponding 2D Gabor fit for the cells and the residual error obtained after subtracting the fitted Gabor function from the simulated RF is plotted in the third column.

3.3 Nearby Neurons

De Angelis et al, (De Angelis et al., 1999; Freeman, 2003) reported in detail RFs parameters of nearby neurons in adult cat and kitten visual cortex. In their work 'nearby pairs of simple cells' had extensive overlap between their RFs but the cells were not adjacent cells. Cells with positional offsets (measured in the direction orthogonal to the orientation of the cell) as large as 1.9 degrees between the RF centers have also been included in their study. However, of the 66 cells pair for which phase differences have been plotted 62 (i.e. 97%) cell pairs had a separation of less than 1deg between their RF centers. In our model the average positional offset between the RF centers of adjacent cells is 6 minutes. If a cell is chosen and a circle of radius 1 deg is drawn around it, then on an average along the x and y direction there will lay 10 cells on each side of the cell and not less than 7 cells in any other direction. To compare simulated results with the experimental data (De Angelis et al., 1999; Freeman, 2003) we have presented results for nearby cells pairs whose RF centers are separated by no more than one degree in the direction orthogonal to the orientation of the cell.

3.4 Clustering of Receptive Field Parameters

To address which aspects of receptive field structure are clustered within primary visual cortex we have selected 72 pairs of nearby cells from our simulated cortex. The positional offsets of the nearby cells chosen for this study are less than one degree as shown in Figure 2(a). A scatter plot showing the value of a particular RF parameter for one neuron (cell 2, vertical axis) plotted against nearby neuron (cell 1, horizontal axis) is used to study clustering of RF parameters. From the orderly map of orientation preference (Blasdel, 1992) and from the simultaneous electrode measurements (DeAngelis et al, 1999; Freeman 2003) orientation preference is reported to be strongly clustered. In Figure 2(b) we find that in our simulated cells orientation preference in nearby cells are tightly grouped around a diagonal line of unity slope. The tuning in terms of hwhh and tuning height of nearby cells is shown in Figure 2(c) and Figure 2(d) respectively. In the cell pairs indicated by circle with a dot inside either one of the cell in the pair or both are located at or near pinwheels in the orientation map (not shown here) of the simulated cortex. The cells at or near pinwheels show very poor tuning and low tuning height. However, narrow tuning width could be associated with small as well as large tuning heights. Our result is consistent with single-unit studies (Maldonado et al., 1997) and optical imaging studies (Swindale et al, 2003). Spatial phase of nearby cell pairs are plotted Figure2(e). Because phase is circular variable, the largest possible phase difference between a pair of RFs is 180°. The dashed lines in the figure denotes phase difference of 180°. A large number of cell pairs show phase differences in the range from 75° to 145°. Our results agree with results (DeAngelis et al., 1999) that phase is not clustered like orientation preference. Fig 2(f) shows difference in RF size. The cell pairs with larger size differences are generally located about 1 degree apart.

4 Conclusions

We have presented a model that captures in sufficient detail the receptive field properties of orientation selective cells in primary visual cortex. We have studied (i) how similar are the RFs of nearby neurons in the simulated cortex, (ii) which RF parameters are clustered within a cortical column and (iii) if RF parameters differ, which are those parameters. Our results confirm experimental data that orientation preference is most clustered and spatial phase accounts for most of the difference between receptive fields of nearby neurons. We also show that orientation tuning and height of nearby cell pairs depends on the location of cells in the orientation map.

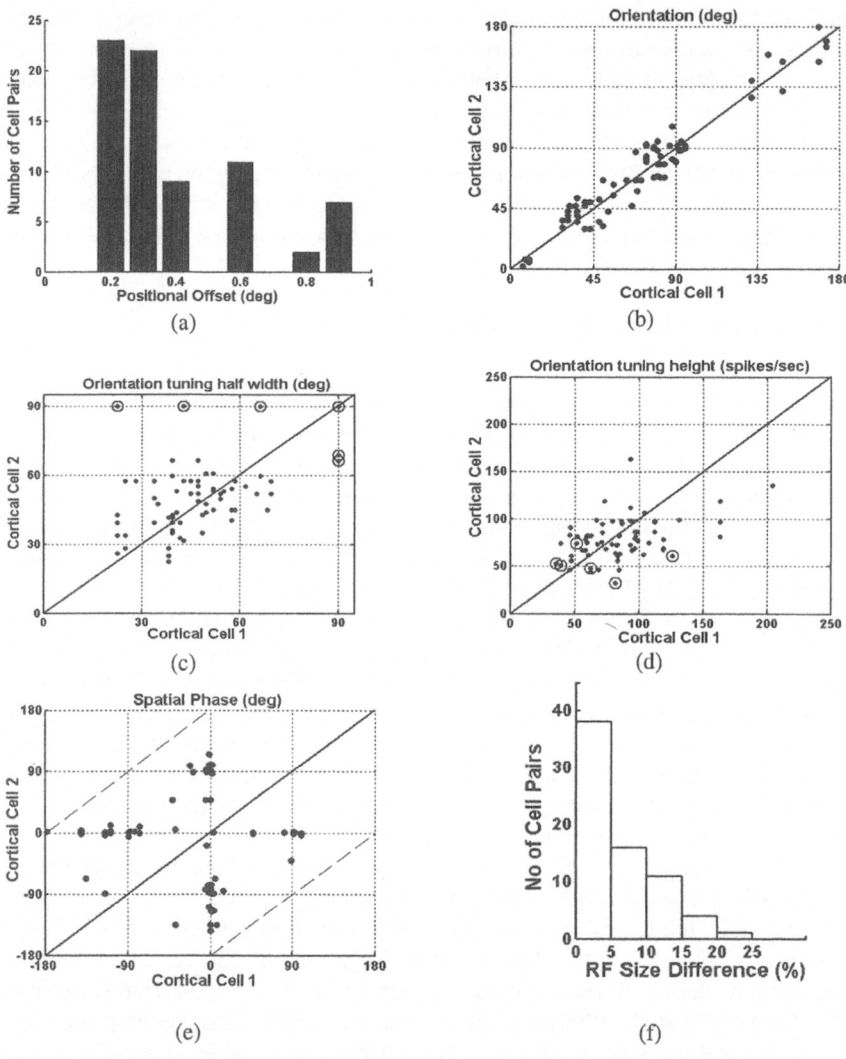

Fig. 2. Correlations in various RF parameters in the selected 72 pairs of nearby cells from our simulated cortex. (a) Positional offsets of the nearby cell pairs chosen for this study; (b) Orientation preference in nearby cells shows strong clustering; (c) Tuning hwhh of nearby cell pair; (d) Orientation tuning height in spikes/sec; (e) Phase differences between cells and (f) Difference in RF size in the cell pairs

References

1. Bhaumik B, Mathur M.: A Cooperation and Competition Based Simple Cell Receptive Field Model and Study of Feed-Forward Linear and Nonlinear Contributions to Orientation Selectivity. Journal of Computational Neuroscience, Vol. 14 (2003) 211-227.
2. Blasdel GG.: Orientation selectivity, preference, and continuity in monkey striate cortex. J Neurosci, Vol. 12 (1992) 3139 –3161.
3. Bonhoeffer T.: Neurotrophins and activity dependent development of the neocortex. Curr.: Opin. Neurobiol, Vol. 6 (1996) 119-126.
4. Carandini M, Heeger D.: Summation and division by neurons in the visual cortex. Science, Vol. 264 (1994) 1333-1336.
5. DeAngelis GC, Ghose GM, Ohzawa I, Freeman RD.: Functional micro-organization of primary visual cortex: receptive field analysis of nearby neurons. Journal of Neuroscience, Vol. 19 (1999) 4046-4064.
6. Douglas RJ, Koch C, Mahowald M, Martin KAC, Suarez HH.: Recurrent excitation in neocortical circuits. Science,Vol. 269 (1995) 981-985.
7. Ferster D.: Origin of orientation selective EPSPs in simple cells of the cat visual cortex. Journal of Neuroscience, Vol. 7 (1987) 1780-1791.
8. Freeman RD.: Cortical columns: A multi-parameter examination. Cereb Cortex, Vol. 13 (2003) 70-72.
9. Gerstner W.: Spiking Neurons. In Pulsed Neural Networks, (Eds. Mass W, Bishop CM), pp. 3-54, MIT Press, Cambridge.Grinvald A, Lieke E, Frostig RD, Gilbert CD, Wiesel TN (1986) Functional architecture of cortex revealed by optical imaging of intrinsic signals. Nature,Vol. 324 (1999) 361-364.
10. Hubel DH, Wiesel TN.: Receptive fields, binocular interaction and functional architecture in the cat's visual cortex. Journal of Physiology, Vol. 160 (1962) 106-154.
11. Koch C, Poggio T.: The synaptic veto mechanism: does it underlie direction and orientation selectivity in the visual cortex? In Models of the visual cortex, (Eds. Rose DR, Dobson VG), John Wiley, New York (1985) 408-419
12. Linsker R.: From basic network principles to neural architecture: Emergence of spatial-opponent cells. Proceedings of National Academy of Sciences, USA, Vol. 83 (1986) 7508-7512.
13. Maldonado PE, Gödecke I, Gray CM, Bonhoeffer T.: Orientation selectivity in pinwheel centers in cat striate cortex. Science, Vol. 276 (1997) 1551–1555.
14. Miller KD.: A model for the development of simple cell receptive fields and the ordered arrangement of orientation columns through activity-dependent competition between ON and OFF center inputs. Journal of Neuroscience, Vol. 14 (1994) 409-441.
15. Somers DC, Nelson SB, Sur M.: An emergent model of orientation selectivity in cat visual cortical simple cells. Journal of Neuroscience, Vol. 15 (1995) 5448-5465.
16. Swindale NV, Amiram ,Amir Shmuel A.: Spatial pattern of response magnitude and selectivity for orientation and direction in cat visual cortex. Cerebral Cortex, Vol. 13 (2003) 225–238.
17. Von der Malsburg C.: Self Organization of orientation selective cells in the striate cortex. Kybernetik, Vol. 14 (1973) 85-100.
18. Wörgötter F, Koch C.: A detailed model of the primary visual pathway in the cat: Comparison of afferent excitatory and intracortical inhibitory connection schemes for orientation selectivity. Journal of Neuroscience, Vol. 11(7) (1991) 1959-1979.
19. Xiong M, Pallas SL, Lim S, Finlay BL.: Regulation of retinal ganglion cell axon arbor size by target availability: Mechanism of compression and expansion of the retinotectal projection. J. Comp. Neurol, Vol. 344 (1994) 581-597.

Development of a Simple Cell Receptive Field Structure: A Model Based on Hetero-synaptic Interactions

Akhil R. Garg[1], Basabi Bhaumik[2], and Klaus Obermayer[3]

[1] Department of Electrical Engineering, J.N.V. University
Jodhpur 342001, India
garg_akhil@yahoo.com
[2] Department of Electrical Engineering, I.I.T Delhi
New Delhi 110016, India
bhaumik@ee.iitd.ac.in
[3] Department of Computer Science and Electrical Engineering
Technical University Berlin, Germany
oby@cs.tu-berlin.de

Abstract. Recent experimental studies of hetero-synaptic interactions in various systems have shown the role of spatial signaling in plasticity, challenging the conventional understanding of Hebb's rule. It has also been found that activity plays a major role in plasticity, with neurotrophins acting as molecular signals translating activity into structural changes. Furthermore, role of synaptic efficacy in biasing the outcome of competition has also been revealed recently. Motivated by these experimental findings we present a model for the development of a simple cell receptive field structure based on competitive and cooperative hetero-synaptic interaction in the spatial domain. We find that with proper balance of competition and cooperation, the inputs from the two populations (ON/OFF) of LGN cells segregate starting from the homogeneous state. We obtain segregated ON and OFF regions in simple cell receptive field.

1 Introduction

Simple cells in layer IV of mammalian primary visual cortex show strong preferences for oriented bars and edges. These cells are found to have spatial receptive fields(RFs) composed of segregated elongated ON/OFF subfields [1][2] [3]. The process of development of orientation selectivity or the formation of RF structure involves both the formation of new connections, strengthening and elimination of some of the already existing connections [4]. It has been shown that correlation based rules of synaptic modifications combined with the constraints to ensure competition provide a reasonable account of development of RF structure of simple cells in visual cortex [4][5]. The constraints used in many of these models of Hebbian learning are based on the idea of imposing competition among synapses dependent on some form of global intracellular signal reflecting the state of many

N.R. Pal et al. (Eds.): ICONIP 2004, LNCS 3316, pp. 64–69, 2004.
© Springer-Verlag Berlin Heidelberg 2004

synapses [6][7]. Typically these constraints keep the sum of synaptic strength received by a cell or mean activity of the cell constant using process of normalization. The process of normalization is not biologically realistic and leaves an open issue as to how the competition is actually implemented biologically. Here we explore an entirely different approach for the development of simple cell receptive field structure based on the following experimental findings : (a) Activity and synaptic efficacy dependent hetero-synaptic competition among axons for limited amount of common resource [8][9] . (b) Cooperative hetero-synaptic interactions in spatial domain [10]. In the following using a computational model for the formation of single cell receptive field structure we find that the model based on the above mentioned findings, is sufficient for inputs to segregate and to maintain this segregation: starting from homogeneous state and as found experimentally, segregated ON and OFF regions in simple cell receptive field can be obtained. Furthermore there is no requirement to include additional constraints, such as any form of explicit normalization, fixed intra-cortical synaptic strengths and hard bounds on synaptic strengths for formation of RF structure.

2   Material and Methods

In this section we describe the basic architecture of the model, its underlying biological assumptions and basic computational strategy. We assume a two layer structure, the output layer composed of a two-dimensional $m \times m$ sheet consisting of m^2 cortical cells which shall represent layer IV C of cat primary visual cortex. Input layer which represents the corresponding LGN layers is subdivided into two two-dimensional $n \times n$ sheets with each consisting of n^2 cells. One sheet labeled "ON" consists of ON-type LGN cells and the other sheet labeled "OFF" consists of OFF-type LGN cells. All sheets are assumed to be regular, and periodic boundary conditions are enforced for computational convenience. Cells in LGN layer are labeled by letters such as i,j,... denoting two dimensional position vectors in the input layer sheets. Cortical cells are labeled by letters x,y .. denoting two dimensional position vectors in the output layer sheet. Each LGN cell is constrained to always arborize over a fixed, topographically appropriate circular patch of cortical cells. The value of the synaptic strength at time t between LGN cell i in sheet "ON" and cortical cell x is described by its peak synaptic conductance $g_{xi}^{ON}(t)$. Similarly the value of the synaptic strength at time t between a LGN cell i in sheet OFF and cortical cell x is given by $g_{xi}^{OFF}(t)$. Initially all the synaptic strengths are chosen randomly and independently from a gaussian distribution described by a mean value g and variance $\pm \alpha$. Furthermore, initially, each cortical cell has synaptic connections from both type of LGN populations (ON/OFF). These connections are of excitatory and modifiable nature. We assume that each model cortical cell also receives fixed inhibitory input from N inhibitory interneurons, termed as untuned global inhibition and synaptic strength of connection between each inhibitory interneurons and cortical cell is given by g_{in} (inhibitory peak synaptic conductance). The membrane potential of the model neuron is thus determined by

$$\tau_m \frac{dV}{dt} = V_{rest} - V + G_{ex}(t)(E_{ex} - V) + G_{in}(t)(E_{in} - V) \qquad (1)$$

with τ_m=20ms, V_{rest}=-70mV, E_{ex}=0mV and $E_{in} = -70mV$ [7]. E_{ex} and E_{in} are the reversal potentials for the excitatory and inhibitory synapses. When the membrane potential reaches the threshold value of -54mV, the neuron fires an action potential and the membrane potential is reset to -60mV. The synaptic conductances G_{ex} and G_{in} and their related peak conductances are measured in the units of the leakage conductance of the neuron and are thus dimensionless. Whenever a particular ON/OFF type LGN cell fires, the corresponding peak synaptic conductance contributes towards the value of total excitatory conductance G_{ex}. Similarly, whenever any of the inhibitory cell receives input in the form of spike it contributes towards the value of total inhibitory conductance G_{in}. Otherwise, both excitatory and inhibitory synaptic conductances decay exponentially i.e. $\tau_{ex} \frac{dG_{ex}}{dt} = -G_{ex}$ and $\tau_{in} \frac{dG_{in}}{dt} = -G_{in}$, where $\tau_{ex} = \tau_{in} = 5mS$. Every time the postsynaptic cell fires an action potential there is a possibility that the peak synaptic conductances of synapses connected to that particular cortical cell may change. We in our model assume the changes are because of competition which is dependent upon the existing value of synaptic strengths of two population of LGN cells from same location in LGN sheets, the activity of presynaptic and postsynaptic cell we call this change as C_{xi}^{ON} and C_{xi}^{OFF}, given by equations (2) and (3) respectively. In addition to the local competition, the modifications in synaptic strengths are also dependent upon cooperative hetero-synaptic interactions, i.e. modifications at one set of synapse are often accompanied by changes at nearby synapses. These interactions may be because of extracellular diffusion of diffusible substances or because of some long range intracellular signal [10]. The pre- and post-hetero synaptic interactions are implemented in a way that every local alteration of synaptic strength is propagated to nearby synapses of the same pre- or postsynaptic cell. A similar analysis has previously been used by [11] [12] in the context of orientation selectivity and ocular dominance. We describe these changes by following equations:

$$C_{xi}^{ON} = (g_{xi}^{ON} A_i^{ON} - g_{xi}^{OFF} A_i^{OFF}) \cdot y_x \qquad (2)$$
$$C_{xi}^{OFF} = (g_{xi}^{OFF} A_i^{OFF} - g_{xi}^{ON} A_i^{ON}) \cdot y_x \qquad (3)$$

$$\frac{dg_{xi}^{ON}}{dt} = eps[C_{xi}^{ON} + K_1 \sum_{i'}(\delta ii' + \rho h_{ii'})C_{xi'}^{ON} + K_2 \sum_{x'}(\delta xx' + \rho h_{xx'})C_{x'i}^{ON}] \qquad (4)$$

$$\frac{dg_{xi}^{OFF}}{dt} = eps[C_{xi}^{OFF} + K_1 \sum_{i'}(\delta ii' + \rho h_{ii'})C_{xi'}^{OFF} + K_2 \sum_{x'}(\delta xx' + \rho h_{xx'})C_{x'i}^{OFF}]$$
$$(5)$$

Here, A_i^{ON}, A_i^{OFF} are the activities of LGN cells we generate these activities by an independent Poisson's process at each time step. y_x is the activity of a model cortical cell, it has value either 1 or 0 depending upon the value of its membrane potential i.e. whether it is above or below threshold at each time step. K1,K2,eps and ρ are constants, δ is Kronecker delta and $h_{\alpha\alpha'}$ is the distance dependent interaction function. In the above equations (4) and (5), besides C_{xi}^{ON} and C_{xi}^{OFF} there are two terms on the right hand side of the equality, describing the redistribution of change in peak synaptic conductances of connections between different presynaptic cells and the same postsynaptic cell (left term) and the redistribution of change in peak synaptic conductances of connections between different postsynaptic cells and the same presynaptic cell (right term). After the above mentioned changes, the new values of each peak synaptic conductance is recalculated as

$$g_{xi}^{ON} = g_{xi}^{ON} + \frac{dg_{xi}^{ON}}{dt} \qquad (6)$$

$$g_{xi}^{OFF} = g_{xi}^{OFF} + \frac{dg_{xi}^{OFF}}{dt} \qquad (7)$$

if now the value of either g_{xi}^{ON} or g_{xi}^{OFF} becomes less than zero then it is made equal to zero. In biological terms this means that now there is no connection between those pre- and post synaptic cell where the value of peak synaptic conductance is zero. The whole process of calculation of synaptic conductance, membrane potential of each cortical cell, changes in individual peak synaptic conductance depending upon post synaptic cell firing is repeated till the time the sum total of peak synaptic conductances reaching a particular cortical cell do not exceed a predefined constant value S.

3 Results

Simulations show that if the development is allowed to take place without any competition then no segregation in the regions of receptive field takes place. K1 is controlling the rate at which the effect of change taking place at one location is transferred to nearby locations due to post hetero-synaptic interactions. To study the role of K1 in development of RF structure, we keep the value of all the parameters except K1 fixed and vary the value of K1. The value of K1 is varied from 0.001 to 0.08 and as can be seen from figure (1a) and (1b) for many values of K1 we get the development of RF structure. In this and all the other figures showing RF structure we have plotted contour plot of $gD_{xi} = g_{xi}^{ON} - g_{xi}^{OFF}$, the difference in the final peak synaptic conductances of connections between ON/OFF LGN cells and each cortical cell at every spatial location in RF region.

To study the role of the pre hetero-synaptic interactions we repeat the simulations with more number of cells in cortex and by keeping the values of all the variables except K2 fixed for which we had obtained segregated regions in RF structure.

As shown in figure(2a) with incorporation of pre hetero-synaptic interactions, nearby cells in the cortical sheet tend to have similar RF structure, more increase in K2 makes more cells to have similar RF structure. The orientation tuning curves are as shown in figure(2b) as can be seen nearby cells have similar orientation tuning curves.

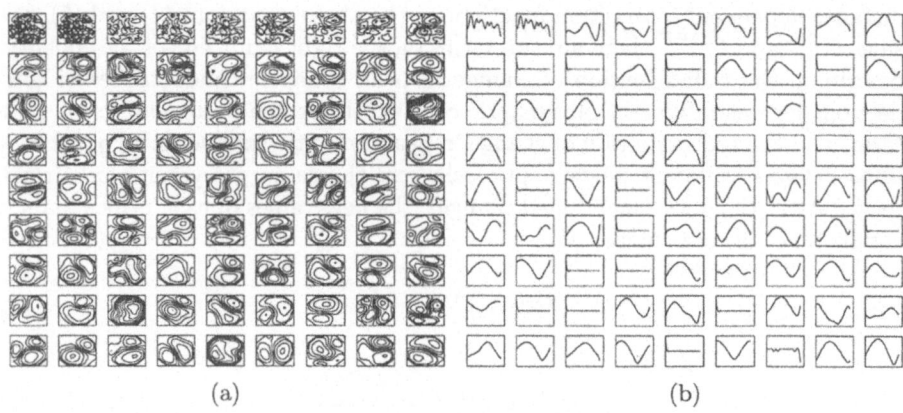

(a)   (b)

Fig. 1. (a) Contour plot of RF structure of a single cells in cortex developed with competitive and post hetero-synaptic cooperative interactions, each subplot is for different value of K1. The value of K1 increases row wise with leftmost subplot in first row obtained having lowest value of K1. (b) Orientation tuning curve of each subplot shown in (a).

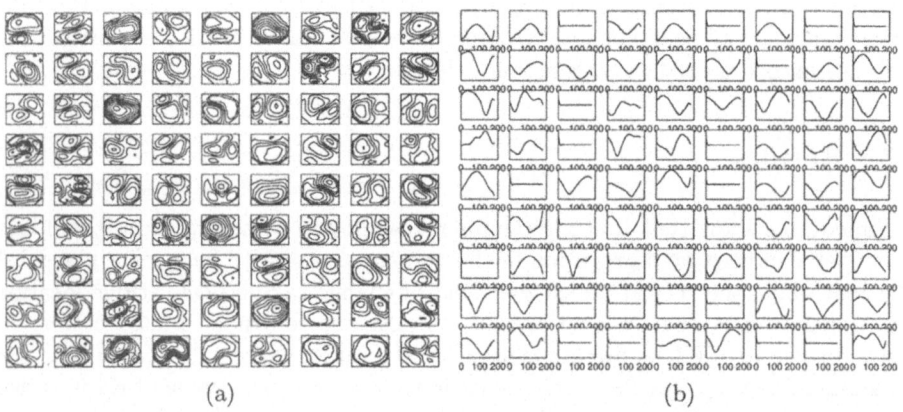

(a)   (b)

Fig. 2. (a) Contour plot of RF structure of 9x9(81) cells in cortex developed with competitive pre and post hetero-synaptic cooperative interactions. (b) Orientation tuning curve of RF structure of 9x9(81) cells in cortex shown in (a).

During our simulation study we observe that development in synaptic strength takes place only for those firing rates for which either the excitation is in balance condition or more than the inhibition. All these simulations suggest that competition combined with post hetero-synaptic interactions are needed for segregation of input population to take place and formation of subregions in RF structure. While competition combined with both post and pre hetero-synaptic interactions are needed for the formation of RF structure having subregions combined with similar RF structure for neighboring cells.

4   Conclusion

We have presented a model for the development of simple cell RF structure based on the mechanism of hetero-synaptic competitive and cooperative mechanism. The results suggest that both mechanisms are not only necessary but should be properly coupled for the formation of RF structure similar to what is found experimentally. Synaptic normalization for which there are no biological evidences is essential in most of the previous models for incorporating competition, the competition incorporated by us eliminates need of any such normalization mechanism. Instead of putting a hard bound on individual synaptic strength, we have imposed a constrain in which the sum total of all synaptic strength reaching a cortical cell is not allowed to exceed beyond certain value, this may be true as the cell may have some physical limits. This is also important as because of this all the synapses do not have equal synaptic strengths as have been observed experimentally [13].

References

1. Hubel D H and Wiesel T N 1962 *J.Physiol.* **160** p 106-154
2. Reid R C and Alonso J M 1995 *Nature* **378** p 281-284
3. Chung S and Ferster D 1998 *Neuron* **20** p 1177-1189
4. Miller K D 1994 *J.Neurosci.* **14** p 409-441
5. Miller K.D. 1996*Models of Neural Networks III eds Domany E., van Hemmen J.L and Shulten K.* (New York: Springer) p 55-78.
6. Miller K D and Mackay D J C 1994 *Neural Comput.* **6** p100-126.
7. Song S, Miller K D and Abbott L F 2000 *Nature Neurosci.* **3** p 919-926.
8. Katz L C and Shatz C J 1996 *Science* **274** p 1133-1138
9. Poo M M 2001 *Nature Rev. Neurosci.* **2** p 24-32
10. Bi G 2002 *Biol. Cybern.* **87** p 319-332.
11. Bhaumik B. and Mathur M 2003 *J. of Comput. Neurosci.* **14** p 211-227.
12. Stetter M, Lang E W and Obermayer K 1998 *Neuroreport* **9** p 2697-2702.
13. Kara P,Pezaris J S,Yurgenson S and Reid R C 2002 *PNAS* **99** p 16261-16266

The Role of the Basal Ganglia in Exploratory Behavior in a Model Based on Reinforcement Learning

Sridharan Devarajan, P.S. Prashanth, and V.S. Chakravarthy

Department of Aerospace Engineering and Department of Electrical Engineering,
Indian Institute of Technology, Madras, India
schakra@ee.iitm.ernet.in

Abstract. We present a model of basal ganglia as a key player in exploratory behavior. The model describes exploration of a virtual rat in a simulated "water pool" experiment. The virtual rat is trained using a reward-based or reinforcement learning paradigm which requires units with stochastic behavior for exploration of the system's state space. We model the STN-GPe system as a pair of neuronal layers with oscillatory dynamics, exhibiting a variety of dynamic regimes like chaos, traveling waves and clustering. Invoking the property of chaotic systems to explore a state space, we suggest that the complex "exploratory" dynamics of STN-GPe system in conjunction with dopamine-based reward signaling present the two key ingredients of a reinforcement learning system.

1 Introduction

The basal ganglia (BG), a group of sub-cortical nuclei, including the corpus striatum, subthalamic nucleus (STN), the substantia nigra (SN), have long been afforded the role of a gate or a selector among action representations in the cortex competing for limited resources. In fact some have called these nuclei the Vertebrate Solution to the Selection Problem [1]. They have also been implicated in sequence generation [2] and working memory [3]. Their dysfunction in motor disorders such as Parkinsons disease has been well documented [4]. In the present work we assign yet another role to the basal ganglia (specifically to the STN-GPe segment within the basal ganglia) – as a system that provides the exploratory drive needed in activities like navigation, foraging etc.

Reinforcement learning is that form of unsupervised learning where the training signal is in the form of a global scalar known as the *reward*. Neural network models of reinforcement learning use stochastic output units for exploration of output state space, i.e the only way the network can know the correct response to an input is by "guessing." The probabilistic output neurons ensure that the system thoroughly explores the space of responses to a certain input so that, the correct response when it occurs can be reinforced. Chaotic systems have been known to exhibit exhaustive exploration of their state-space. It is well-known that a network of non-linear oscillators is intrinsically chaotic [5]. Recently a network of oscillators has been proposed as a model of motor unit recruitment in skeletal muscle. The complex dynamics of the network is used to model desynchronized activity of motor neurons in healthy muscle [6]. Oscillatory neural activity is known to exist in several structures in the brain including the basal ganglia, hippocampus, sensory cortices etc. Oscillatory dynamics in the basal ganglia have been observed at the level of the Sub-Thalamic Nucleus –

N.R. Pal et al. (Eds.): ICONIP 2004, LNCS 3316, pp. 70–77, 2004.
© Springer-Verlag Berlin Heidelberg 2004

Globus Pallidus network ([7],[4],[8]). The STN-GPe network, depending on the patterns of the interconnections and values of the interconnecting weights, has been shown to support three general classes of sustained firing patterns: clustering, propagating waves, and repetitive spiking that may show little regularity or correlation [8]. Furthermore, it has also been demonstrated that each activity pattern can occur continuously or in discrete episodes. The mesencephalic dopaminergic input to the basal ganglia might help modulate the activity of the STN-GPe loop by serving as a reward signal to these units in the so-called *indirect pathway* in the basal ganglia [2]. We hypothesize that the complex oscillations of the STN-GPe segment within the basal ganglia provide the exploratory dynamics necessary for reward-based or reinforcement learning.

The paper is organized as follows: In Section 2 we elaborate on the exact role of the basal ganglia in exploratory behavior. We then present a computational model of the STN-GPe segment with a network of oscillatory neurons. Next in Section 3 we evaluate this network in the context of a simulated "waterpool experiment". In real versions of these experiments a rat has to learn the location of a submerged (invisible) platform in a pool of water based on spatial cues placed around the pool. Finally we conclude with a discussion on the unique dynamics exhibited by the oscillatory network and its significance in a biological context, the possible reasons for the validity and means of validation of our hypothesis, scope for further work and the possible alternatives to our model.

2   The Model

2.1   Description of the Basal Ganglia Model

For a general consensus regarding signal flow in basal ganglia see ref. [9] and [10]. Further, it has been suggested that, the STN-GPe system is ideally placed to produce oscillations [7]. We propose a simplified three layer architecture (Fig. 1) for the basal ganglia consisting of an input layer representing visual input, a hidden layer representing the oscillatory STN_GPe system, and an output layer representing the selected movement of the animal for the given input.

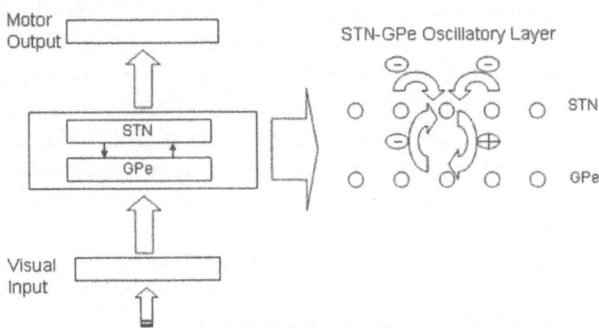

Fig. 1. Overall architecture of the network is shown on the left. The visual input is presented to the STN-GPe layer through a set of weights. Interactions between STN and GPe layers produce complex oscillatory activity. Activity of STN layer is fed to the output layer. A detailed diagram of connectivity in the STN-GPe layer is shown on the right

In neurophysiological terms, the input to the first layer represents the unresolved/competing visual representations arising from the cortex reaching the striatum (caudate/putamen). The activity of the hidden layer consisting of oscillatory units corresponds to the *lumped* activity of the STN-GPe oscillatory network. The output at the final layer corresponds to the motor output at the level of GPi (or EP in the case of the rat) to which the STN projects.

We suppose that the reward signal received from the limbic system is translated into fluctuations from baseline levels of dopamine secreted by the mesencephalic dopaminergic system comprising of the Ventral Tegmental Area (VTA) and the SNc (substantia nigra pars compacta) nucleus of the basal ganglia [11]. There is an increase in the overall level of dopmine when there is a positive reward and a corresponding decrease in case of a negative reward, with the magnitudes of these fluctuations being correlated with the magnitudes of the rewards. This reward signal (i.e, level of dopamine) is propagated as a global reinforcement signal that serves to modify the synaptic weights both among striato-pallidal (STR-GPe) and pallidopeduncular (STN-GPe to GPi) projections thereby leading to learning, i.e the generation of a potentially rewarding motor output based on the current sensory input.

We now present a concise summary of the operation of our model, due to constraints on space. For a detailed mathematical description the interested reader may consult [12]. The "visual input", presented to the input layer of Fig. 1, is forwarded to the the STN-GPe oscillatory layer via a weight stage. The STN-GPe layer is a actually a double layer: interaction between its sublayers produce oscillations. Equations similar to ours have been used by Gillies et al (2001) to describe their model of STN-GPe interaction. The activity of STN layer is controlled by a parameter D ($0 < D< 100$) which represents dopamine levels. D determines the percentage of ON neurons in STN layer. Another factor that crucially controls STN layer activity is the pattern of lateral connectivity, determined by the parameter ε. Thus each STN has a negative center and a positive surround; the relative sizes of center and surround are determined by ε. Smaller ε implies, more negative lateral STN connections, which tends to de-correlate oscillations of STN neurons. In the absence of input from the input layer, as ε is varied from 0 to 2, the activity of STN-GPe system exhibits three different regimes: 1) chaos, 2) traveling waves, and 3) clustering [6]. Operation of the network in the first regime – chaos – is most crucial since it is the chaotic dynamics in the STN-GPe layer that makes the network extensively explore the output space. The output of STN layer is forwarded further to the output layer where it is translated into "motor output". Reward resulting from this motor output is communicated back as fluctuations in D of STN layer (high reward implies high D). The weight stage connecting input layer to STN-GPe layer is trained by reinforcement learning [13]. The weight stage between STN-GPe layer and the output layer is trained by a form of competitive learning [12].

3 The Simulated Waterpool Experiment

The above network is used to drive exploratory behavior in a simulated version of the so-called "water pool experiments" [14]. In real water pool experiments, a rat is made to explore a pool of water searching for a submerged platform, which is invisible since the water is muddy. The rat attempts to navigate with the help of landmarks placed around the pool. On reaching the platform the rat receives an intrinsic reward (relief) or an external reward administered by the experimenter. The experiment is repeated by placing the rat at various locations in the pool.

The setup used in our simulation is depicted in Fig. 2. The large circle represents the water pool. The small segment on the right of the pool is the submerged platform. Eight landmarks are placed around the periphery of the circle with uniform spacing. The landmarks are vertical poles with specific height and are assumed to be uniquely identifiable by some property other than the height, such as, for instance, color. The model rat has an angle of vision of 180 degrees. The rat is also assumed to have a position (point size) and an orientation in the water pool at any instant. From a given viewing point the rat can "see" a segment of the fringe of the pool containing a subset of the landmarks present around the pool. More details of the setup can be obtained from [12].

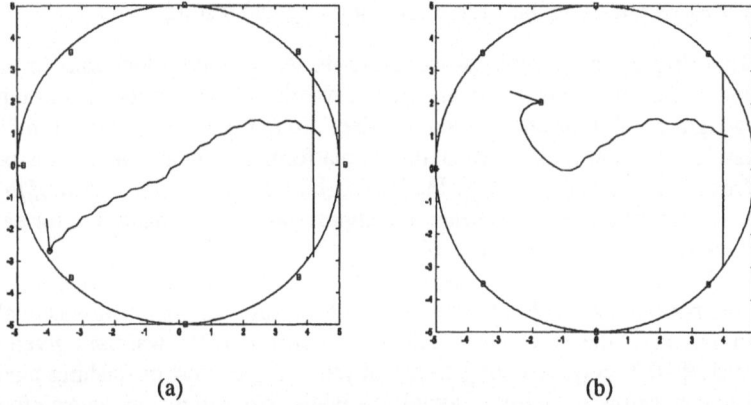

(a)   (b)

Fig. 2. The water pool experimental setup – the waterpool is in the form of a circle centered at the origin. The poles around the rim of the pool are represented by squares. The dark line at x = 4 represents the edge of the platform. Various trajectories taken by the rat for different initial locations (depicted as dark circles) and orientations (depicted by short line emanating from initial location). Cases a & b show navigation to the platform when the platform is within and outside initial range of vision respectively

(i) View-matrix: This is a vector that encodes the animal's "view" of the waterpool. (ref. [12] for details). The view vector is presented as input to the network of Fig. 2.

(ii) Output representation: The view-vector is presented as an input to the 3-layer network of Fig. 1. Activity of the output layer represents the rat's motion in response to the current view. Each node in the output layer represents a displacement direction. The rat is displaced in the direction corresponding to the encoded direction of the "winning" node in the output layer. The rat moves a fixed distance of 'd' per time step. The rat's motion results in a change in the view and the cycle continues.

3.1 Simulation Results

In this section we will briefly describe the experimental parameters and various outcomes of the computer simulation experiments based on the reinforcement learning framework discussed above.

- STN-GPe layer: A 10x10 grid of oscillatory neurons is chosen for the STN-GPe layer.
- Output layer: The output layer consists of K (=5) neurons which produce output by a "winner-take-all" mechanism and map onto K distinct output states. The maximum movement deviation angle, $T_{max}/2$, is 30 degrees, i.e 30 degrees on either side of the current orientation.
- Fluctuations in D: The dopamine level D providing the reward signal is assumed to be linearly proportional to the reward obtained with a maximum/minimum value for the fluctuations that is approximately 50% of the baseline value. Thus for a baseline value of, say 50, the dopamine level varies between 25 for -1 reward and 75 for +1 reward.

There are two phases in the simulation – training and testing.

3.1.1 Training. In the training phase the rat is set at random locations in the water pool at random orientations and is allowed to wander. Based on its current input vector and weight configurations, the rat wanders around the water pool, initially in an almost random fashion until it hits upon the platform by chance. When this occurs, a positive reward of +1 is provided to the rat *based on the input and output of the previous step,* so that the rat learns to select the appropriate output maneuver for each kind of visual input *before* entering the pool.

During its wandering in the water-pool the rat often comes into contact with the walls of the pool i.e its trajectory often attempts to cross the pool's dimensional limits. In the simulation, at these instances, the rat is bounced off the wall and given a negative reward of -0.3, corresponding to the physical discomfort of dashing against the wall, so that it learns to actively avoid the walls. No weight update/reinforcement occurs during the wandering motion of the rat (except negative reinforcement at the walls) until the rat reaches the platform.

The rat, after several iterations learns to head directly to the platform with minimal wandering as shown, for example in Figures 2a & 2b, for extreme orientations of the rat towards the platform. A plot of the mean number of steps to platform vs. training time corresponding to one set of training trials can be found in Figure 3.

It is clear from this figure that as training progresses, the rat learns to effectively navigate towards the platform in fewer steps on an average.

3.1.2 Testing. This phase involves testing the rat's movements without any update to the neural network weights, i.e without providing any form of reinforcement, neither positive reward at the platform nor negative reward at the walls. This phase, though not biologically very realistic, is a standard practice in neural network literature to evaluate network performance.

While testing, the rat learns to efficiently avoid the walls, and once its view is within a reasonable range of the platform, heads in an almost straight line for the platform. In this phase we find interesting dynamics of the oscillatory STN-GPe layer as shown in Figures 4a,b & c. As the rat approaches the platform, the STN-GPe layer

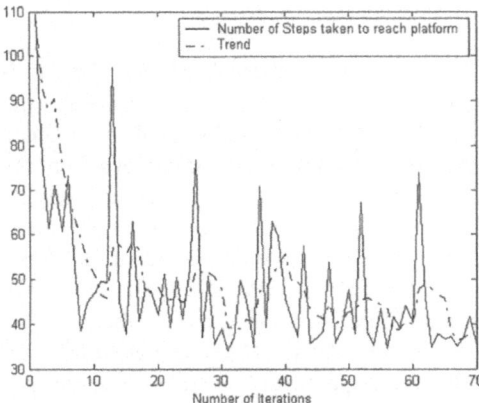

Fig. 3. A plot of mean number of steps to reach platform and wall bounces versus training cycle number. It can be seen that the rat learns to navigate effectively with minimal wandering and wall bounces

settles into a bistable state and each of the neurons enter into a periodic alteration or a sustained maintenance of their respective outputs (Fig. 4a &b). On the contrary, when the rat is looking away from the platform, and exploring the other parts of the pool, the dynamics of the STN-GPe layer exhibit wandering activity (Fig. 4c).

In order to characterize the observed dynamics of the STN-GPe layer for each of the turning "toward" and turning "away" cases, two measures – effective dimension[1] and average correlation coefficient[2] are computed.

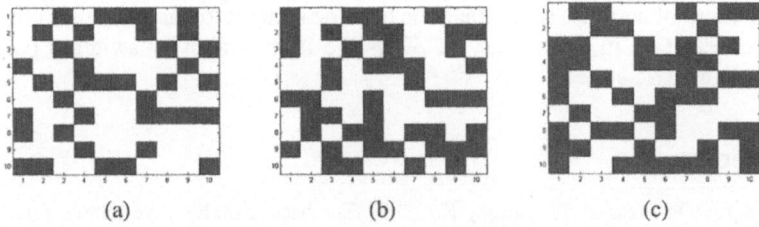

(a)   (b)   (c)

Fig. 4. (a and b) The two-states characterizing the oscillatory dynamics of the hidden layer when the rat is heading toward the platform. The effective dimension and correlation value for these states were evaluated and found to be 2 and -0.4432 respectively. (c) A snapshot of STN layer state when the rat is turning away from the platform. The effective dimension and correlation value for these states were evaluated and found to be 6 and 0.3743 respectively. Therefore, the effective dimension of the STN layer activity is high when the animal is exploring the pool for the platform and low when it is heading towards it

---

[1] Effective dimension is a measure of effective number the degrees of freedom of the activity, $v(t)$, of a system. Let λ_k and λ_{max} the kth and the highest eigenvalues of the autocorrelation matrix of the activity, $v(t)$, over a duration of interest, such that $\lambda_k / \lambda_{max} = \frac{1}{2}$, then 'k' is the effective dimension.

[2] Average correlation coefficient is the average value of correlation between pairs of components of, $v(t)$, say, $vi(t)$ and $vj(t)$. The averaging is performed over a large number of randomly chosen pairs of components.

4 Discussion

Mechanism for the learning driven by the level of dopamine in the STN-GPe loop:
The D-level learning by the STN-GPe loop was achieved in the network model by a
oscillatory neural network model. We hypothesize the neural correlate of this network
to be pallido-nigro-striatal circuit involving projections from the GPe to the SNc and
back to the striatum. The detailed mechanics of this feedback learning are as follows:
The STN-GPe oscillator attempts to learn the level of Dopamine which signal is fed
in via the SNc-Striosomal projections through the striatal Matrix into the GPe. The
error is back propagated through the GPe-SNc (pallidal-nigral) projection (of which
there is little documented literature), and it is hypothesized that neurons in the SNc
compute the difference between the actual percentage of GPe units that are active (D_a)
and the percentage that are required to be active (as given by their own activity reflecting the amount of dopamine, D, thereby feeding the error signal back into the
striatum and further to the GPe-STN loop where it may modulate the percentage of
active neurons according to eqns. (8) & (9). This idea of error-back propagation is
similar to that proposed by Berns & Sejnowski (1998) wherein the error signal, computed by hypothetical projections from the striatum and the GPi to the SNc/VTA,
modulates the STN-GPe synaptic weights.

Discriminating novel from familiar odors – the case of olfactory bulb: Studies with
the olfactory systems in rabbits show that, when a familiar odor is presented to the
animal, the olfactory bulb responds with a rhythmic waveform; however, when the
stimulus is novel or unfamiliar, activity in the bulb exhibited chaotic wandering [15].
This is analogous to the STN-GPe layer in our model. When the rat is turned away
from the platform and is searching for it, the STN-GPe layer exhibited desynchronized pattern of activity; when the rat is heading straight towards the platform, there is
a sharp reduction in the activity of STN-GPe layer – activity switches periodically
between only two states.

References

1. Redgrave P, Prescott TJ, Gurney K (1999). The basal ganglia: a vertebrate solution to the
   selection problem? Neuroscience, 89, pp.1009-1023.
2. Berns GS, Sejnowski TJ (1998). A computational model of how the Basal Ganglia produce
   sequences. Journal of Cognitive Neuroscience, 10:1, pp.108-121
3. Houk, J. C., J. L. Davis, and D. G. Beiser (1995). Models of Information Processing in the
   Basal Ganglia. Cambridge, MA, MIT Press
4. Bevan MD, Magill PJ, Terman D, Bolam JP, Wilson CJ (2003). Move To The Rhythm:
   Oscillations In The Subthalamic Nucleus-External Globus Pallidus Network. Trends in
   Neuroscience (in press)
5. Chirikov, B., (1979), A universal instability of many-dimensional oscillator systems, Phys.
   Rev., 52:263-379.
6. Chakravarthy VS, Thomas ST, Nair N (2003). A model for scheduling motor unit recruitment in skeletal muscle. International Conference on Theoretical Neurobiology, National
   Brain Research Center, Gurgoan, February, 24-26.
7. Gillies, A., Willshaw, D., and Li, Z., (2002) Subthalamic-pallidal interactions are critical in
   determining normal and abnormal functioning of the basal ganglia *Proc R Soc Lond B Biol
   Sci.* 2002 Mar 22;269(1491):545-51.

8. Terman, D., Rubin, J.E., Yew, A.C., and Wilson, C.J. (2002) Activity Patterns in a Model for the Subthalamopallidal Network of the Basal Ganglia. J Neurosci. 2002 Apr 1:22(7):2963-2976.
9. Harner AM (1997). An Introduction to Basal Ganglia Function. Boston University, Boston, Massachusetts.
10. Obeso JA, Rodriguez-Oroz MC, Rodriguez M, Arbizu J, Gimenez-Amaya JM (2002). The Basal Ganglia and Disorders of Movement: Pathophysiological Mechanisms. News Physiol Sci, 17, pp.51-55
11. Montague, Dayan & Sejnowski (1996). A Framework for Mesencephalic Dopamine Systems Based on Predictive Hebbian Learning, *The Journal of Neuroscience*, 16(5):1936-1947.
12. Sridharan, D., (2004) "Human Factors in Aviation: Willed action and its disorders," MTech Thesis, Department of Aerospace Engineering, IIT, Madras, India.
13. Barto AG, (1999). Reinforcement Learning. M.A. Arbib (ed.) The Handbook of Brain Theory and Neural Networks (1st Edition). Cambridge, MA: MIT Press.
14. Morris, R.G.M., Garrud, P., Rawlins, J.N.P., O'Keefe, J. (1982). Place navigation impaired in rats with hippocampal lesions. *Nature*, 297, 681-683.
15. Skarda, C. A. and Freeman, W. J. (1987). How brain makes chaos in order to make sense of the world. Behavioral and Brain Sciences, 10:161-195.

A Functional Role of FM Sweep Rate of Biosonar in Echolocation of Bat

Kazuhisa Fujita[1], Eigo Kamata[1], Satoru Inoue[2],
Yoshiki Kashimori[1,3], and Takeshi Kambara[1,3]

[1] Department of Information Network Science, School of Information Systems,
University of Electro-Communications, Chofu, Tokyo, 182-8585, Japan
{k-z,kamata}@nerve.pc.uec.ac.jp, {kashi,kambara}@pc.uec.ac.jp
[2] Department of Computer Science, Faculty of Engineering,
Saitama Institute of Technology, Okabe, Saitama, 369-0293, Japan
[3] Department of Applied Physics and Chemistry,
University of Electro-Communications, Chofu, Tokyo, 182-8585, Japan

Abstract. Most species of bats making echolocation use frequency modulated(FM) ultrasonic pulses to measure the distance to targets.These bats detect with a high accuracy the arrival time differences between emitted pulses and their echoes generated by targets. In order to clarify the neural mechanism for echolocation, we present neural model of inferior colliculus(IC). The bats increase the downward frequency sweep rate of emitted FM pulse as they approach the target. The functional role of this modulation of sweep rate is not yet clear. In order to investigate the role, we calculated the response properties of our models of IC changing the target distance and the sweep rate. We found based on the simulations that the distance of a target in various ranges may be encoded the most clearly into the activity pattern of delay time map network in IC, when bats adopt the observed FM sweep rate for each range of target distance.

1 Introduction

Mustached bats emit ultrasonic pulses and listen to returning echoes for orientation and hunting flying insects. The bats analyze the correlation between the emitted pulse and their echoes and extract the detailed information about flying insects based on the analysis. This behavior is called echolocation. The neural circuits underlying echolocation detect the velocity of target with accuracy of 1 cm/sec and the distance of target with accuracy of 1 mm. To extract the various information about flying insects, mustached bats emit complex biosonar that consists of a long constant-frequency (CF) component followed by a short frequency-modulated(FM) component. Each pulse contains four harmonics and so eight components represented by (CF_1,CF_2,CF_3,CF_4 and FM_1, FM_2, FM_3,FM_4) as shown in Fig.1 [4]. The information of target distance and velocity are processed separately along the different pathways in the brain by using four FM components and four CF components, respectively [5].

N.R. Pal et al. (Eds.): ICONIP 2004, LNCS 3316, pp. 78–83, 2004.
© Springer-Verlag Berlin Heidelberg 2004

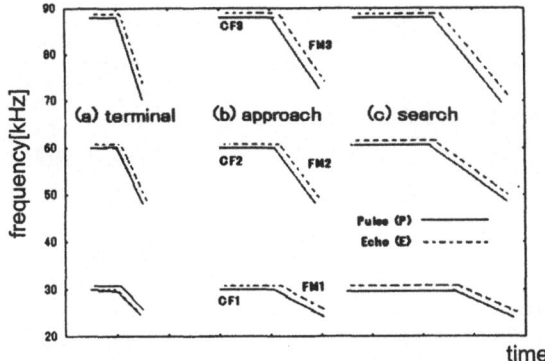

Fig. 1. Schematized sonagrams of mustached bat biosonar pulses emitted during three phases of target-oriented flight (terminal, approach, and search phases). The three harmonics of the pulses each contain a long CF component (CF_{1-3}) followed by a short FM component (FM_{1-3}). (a), Terminal phase; CF and FM durations are 5 and 2ms. (b), Approach phase, CF and FM durations are 15 and 3ms. (c), Search phase, CF and FM durations are 30 and 4ms [3]. Thus,the frequency downward sweep rate of FM_n pulse (n = 1 -4) is 6n/2, 6n/3,and 6n/4 kHz/ms for terminal, approach, and search phases, respectively.

In the present paper, we consider the neural mechanism detecting the target-distance information. A primary cue for measuring the distance(target range) is the time intervals between the emitted pulses and the echoes returned by targets. A 1.0ms delay of the echo corresponds to a 17.3 cm target distance at 25 °C. When the bat emits a pulse, its ears are stimulated by the pulse FM_1-FM_4 at the same time and by the echo FM_2-FM_4 after a delay time, because FM_1 component in the pulse is very weak compared with FM_2-FM_4 components [6]. There exist FM-FM neurons in the auditory cortex of mustached bat, which respond strongly when the pulse and echo are combined with a particular echo delay [3]. That is, they can decode target range.

We consider here the problem what is real reason for which the bats change the repetition rate of the pulse depending on the target range during their foraging flight. In order to solve the problem, we present neural network model of inferior colliculus (IC) and show how the subcortical signal processing is made so that the specific functions of those neurons can be generated.

2   Neural Network Model of Inferior Colliculus(IC)

The network model of IC consists of multiple layers, each of which is a linear array of delay-tuned(DT) neurons as shown in Fig.2a. The DT neurons are tuned in to specific echo delay times ranging from 0.4 to 18 msec. These DT neurons in a single layer are also tuned in to a specific pair of pulse PFM_1 from 30kHz to 24kHz. The bat uses the echo sound EFM_2 of second harmonics pulse PFM_2 for the detection of target distance, whose frequency is swept downward from 60kHz to 48kHz.

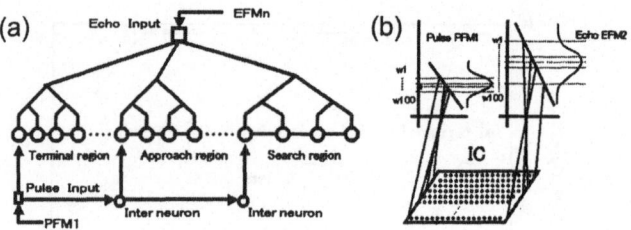

Fig. 2. (a) Neural model of time disparity detection(TD) layer consisting of signal array of delay-tuned neurons in IC. The array consists of three regions, terminal, approach, and search regions whose neurons are tuned in to delay times within the ranges(0-3ms), (3-8ms), and (8-18ms), respectively. The interneuron distance is random, but the average distance is increased as the region is changed from terminal to search. (b)Neural model of multiple time disparity detection(TD) layers each of which is tuned in to a combination of specific frequencies of pulse PFM_1 and echo EFM_n. The frequencies of PFM_1 and EFM_n are downward swept from 30 to 24kHz and from 30n to 24n kHz, respectively. Signal propagation pathes from IC to MGB and from MGB to AC are also shown.

In the present model, we divide both the frequency range for pulse and echo into 10 subranges, and consider 100 linear arrays of DT neurons(time difference detection layer(TD layer)) as shown schematically in Fig.2b. That is, DT neurons in each TD layer are specifically tuned in to a specific pair of one frequency subrange for pulse and one frequency subrange for echo. More detailed description of neural network model of IC has been given in the previous paper [1, 2].

First, we described the structure and function of the single TD layer, and secondly those of multiple layers.

2.1   Model of Single TD Layer

The structure of a single TD layer is illustrated in Fig.2a, where each circle denotes a delay-tuned(DT) neuron and the delays are created by different lengths of delay line and interneurons. The distance between adjacent DT neurons in a TD layer is changed randomly along the linear array. In order to see more clearly the dependence of response of IC networks on the frequency sweep rate of PFM_1 and EFM_2, we divided the linear array of DT neurons into three regions as shown in Fig.1. Although the interneuron distances are random, the average of the distances in each region is constant and different between three regions. The average value is increased as the region changes form terminal to search region.

The membrane potential of DT neuron is determined by

$$\frac{du(k,r,i;t)}{dt} = -\frac{1}{\tau}\left(u(k,r,i;t) - u_{rp}\right) + I_P(t) + I_E(t), \tag{1}$$

where $u(k, r, i; t)$ is the membrane potential of i th DT neuron in region r of k th TD layer, τ is the relaxation time, u_{rp} is the resting potential, and $I_P(t)$ and $I_E(t)$ are the input impulse trains of biosonar pulse and its echo, respectively.

2.2 Conditions for Simulating the Neural Processes of Detection of Echo Delay Time

The situation of simulation is described by the five quantities; (1) the initial distance d from the bat to a target, (2) the velocity Vr of the bat relative to the target, (3) the initial frequency ω_{P0} of the pulse, (4) the initial frequency ω_{E0} of the echo, and (5) frequency sweep rate Rs. Echo delay time T is determined by these quantities as,

$$T = \frac{2d}{C + V_r} - \frac{(C - V_r)\omega_{P0}}{(C + V_r)R_S} - \frac{(\omega_P - \omega_{P0})}{Rs} + \left(\frac{C - V_r}{C + V_r}\right)^2 \left(\frac{\omega_{P0}}{\omega_{E0}}\right)\frac{\omega_E}{R_S}, \quad (2)$$

where C is the sound velocity, and ω_P and ω_E are the relevant frequencies of the pulse and echo, respectively, which take one of 10 subranges of the pulse frequency range(30kHz to 24kHz) and one of 10 subranges of echo frequency range(60-48kHz).

When k th TD layer is tuned in to a pulse frequency ω_P and an echo frequency ω_E, $I_E(t)$ in Eq.(1) is given by $I_P(t - T)$.

2.3 Model of Multiple TD Layers

To know target distance, bats use a FM component of emitted ultrasonic pulse and its echo. The pulse and echo are decomposed into their Fourier components in the ears. Then, the components are processed separately for each frequency to detect the time difference between the pulse and the echo, until finally the results for each frequency are integrated. The process is illustrated schematically in Fig.2b.

The membrane potential of each neuron in each TD layer is calculated by Eq.(1), where the timings of input $I_P(t)$ and $I_E(t)$ are changed depending on the values of ω_P and ω_E to which the TD layer is tuned in. That is, $t_p = (\omega_{P0} - \omega_P)/R_S$ and $t_E = t_P + T$, where t_P and t_E are the input timings of pulse and echo impulse, respectively.

The output $U(k, r, i; t)$ of each neuron is given by

$$U(k, r, i; t) = \frac{1}{1 + exp\left(-(u(k, r, i; t) - \theta_{IC})/h_{IC}\right)}, \quad (3)$$

where θ_{IC} and h_{IC} are the threshold and the variation rate, respectively.

3   Result

3.1   Situation Used in Our Simulation of Detecting Echo Delay Time

Initial Distance d from the Bat to a Target
We consider three cases where a target is in the three ranges, terminal, approach, and search ranges. The value of d in each range is $0 < d \leq 50$ cm for the terminal range, 50 cm$< d \leq 140$ cm for the approach range, and 140 cm $< d \leq 300$ cm for the search range [3]. The echo delay time T corresponding to those values of d is $0 < T \leq 3$ ms for the terminal range, 3 ms $< T \leq 8$ ms for the approach range, and 8 ms $< T \leq 18$ ms for the search range. Thus, we designed the models of terminal, approach, and search regions in a TD layer shown in Fig.2a so that delay-tuned neurons in each region can respond to a pair of pulse and echo whose delay time is within the relevant time range.

V_r, ω_{P0}, and ω_{E0}
We chose $C/40$ for the velocity V_r of the bat relative to the target and 30kHz and 60kHz for the initial frequency ω_{P0} and ω_{E0} of pulse and echo, respectively.

3.2   Response Properties of Multiple TD Layers in IC Induced by Three Kinds of FM Pulses

We studied the response properties of IC stimulated by three kinds of FM pulses whose frequency sweep rate is $6/2, 6/3$ and $6/4$ kHz/ms. We consider their echoes generated by a target which is within the terminal, approach, and search range, respectively. It has been observed [3] that when the target is in the terminal range(d=0-50cm), the bat emits the FM pulse whose downward sweep rate is around $6/2$ kHz/ms. We calculated the temporal variation of firing pattern integrated over multiple TD layers in IC in the four cases where a target is at d = 10cm, 30cm, 70cm, and 170 cm. The result is shown in Fig.3. When the target is within the terminal range, the peak of firing pattern appears correctly at the neuron whose delay time T tuned is 0.6ms for d = 10 cm and 2ms for d = 30cm. The maximum firing frequency is high and the width of firing pattern is narrow. We show in Figs.3c and d the response of the multiple TD layers induced by a target within the approach range (d = 70cm) and within the search range (d = 170cm), respectively. Although the peak appears around the neuron with T= 4 and 10ms, respectively, the peak height is quite low compared with Figs.3a and b.

Therefore, FM pulse with $R_S = 6/2$ kHz/ms is the most suitable for detecting a target within the terminal range. That sweep rate is really used by the bat. The similar results were obtained for R_S=6/3 and 6/4 kHz/ms. FM pulse with R_S=6/3 and 6/4 kHz/ms are the most suitable for detecting a target within the approach and search region, respectively.

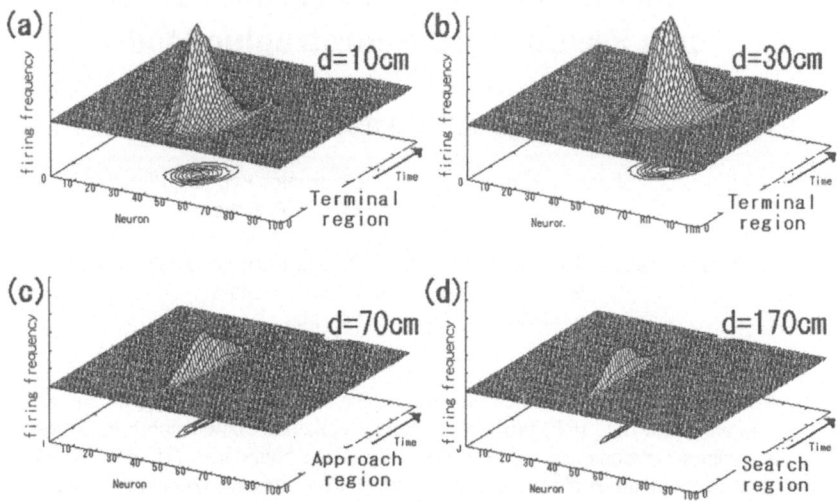

Fig. 3. Temporal variations of firing patterns in IC integrated over multiple TD layers stimulated by PFM with $R_S = 6/2$ kHz/ms and its echo generated by a target whose distance is (a) 10cm, (b) 30cm, (c) 70cm, and (d) 170cm.

4   Concluding Remarks

We have presented here the functional model of delay-time map constructed with delay-tuned neurons in auditory cortex. We showed that the model of delay-time map has the essential functions required for detection of target distance and may reproduce well the observed results. Especially, we have clarified the functional role of changing frequency downward sweep of FM pulse, when the bat pursues insects without colliding with obstacles and being attacked by predators.

References

1. Inoue,S., Kimyou,M., Kashimori, Y., Hoshino, O., Kambara, T.: A neural model of medial geniculate body and auditory cortex detecting target distance independently of target velocity in echolocation, Neurocomputing **32-33** (2000) 833-841
2. Inoue,S., Kimyou,M., Kashimori, Y., Hoshino, O., Kambara, T.: A basic neural mechanism for acoustic imaging, In: K.Yasue, M.Jibu, and T.D. senta (Eds.) No matter, Never Mind, Advances in consciousness research, John Benjamins, Amsterdam, vol.**33** (2002) pp.281 - 288
3. O'Neill, W. E., Suga, N.: Target range - sensitive neurons in the auditory cortex of the mustached bat, Science, **203** (1979) 67 -73
4. O'Neill, W. E., Suga,N.: Encoding of target range and its representation in the auditory cortex of the mustached bat, J.Neurosci. **2** (1982) 17-31
5. Suga, N: Cortical computational maps for auditory imaging, Neural Networks **3** (1990) 3-21
6. Suga, N., O'Neill, W.E.: Neural axis representingtarget range in the auditory cortex of the mustached bat. Science **206** (1979) 351-353

Orientation Map Emerges in Parallel with the Formation of Receptive Fields in a Feedforward Neurotrophic Model

Mona Mathur[1,*] and Basabi Bhaumik[2]

[1] Advanced Systems Laboratory, ST Microelectronics Pvt. Ltd, Plot Nos-2&3,
Sector-16A, Noida - 201301, UP, India
mona.mathur@st.com
[2] Department of Electrical Engineering, Indian Institute of Technology,
Hauz Khas, New Delhi – 110016, India
bhaumik@ee.iitd.ac.in

Abstract. A feed-forward neurotrophic model has been shown to generate realistic receptive field (RF) profiles for simple cells that show smooth transitions between subregions and fade off gradually at the boundaries [1]. RF development in the neurotrophic model is determined by diffusive cooperation and resource limited competition guided axonal growth and retraction in the geniculocortical pathway. Simple cells developed through the model are selective for orientation (OR) [1] and capture a wide range of spatial frequency properties of cortical cells [2]. Here, we show that the development of spatial receptive structure of the cells through the phenomena of competition and cooperation is also accompanied with formation of an orientation map (ORmap). Once these maps appear they remain stable.

1 Introduction

The *ORmap*, depicting gradually changing OR preferences across the cortical surface is a characteristic feature of the cortical organization of OR selective cells. Emergence of these maps parallels the development of OR selectivity in the visual cortex. Both the OR selectivity and the OR maps can be observed as early as the first cortical responses can be measured [3] i.e. at birth or before eye opening. At birth thalamocortical connections are well developed, while the horizontal connections are still clustered [4], indicating thereby that the OR selectivity observed at birth is a manifestation of the RF structure of the cells. These findings indicate towards the existence of common biological mechanisms responsible for the emergence of RF structure and thus OR selectivity and ORmaps in the visual cortex.

Many groups have tried to explore the principles underlying the development of these ORmaps (for recent reviews see: [5],[6]). Some of these have modeled the formation of ORmaps along with the development of OR selective cells from a competition of ON-center and OFF-center cell responses in the LGN [7],[8],[9]. These models are based on one or more of the following assumptions: (a) use of synaptic normalization to bring in competition, (b) existence of fixed Mexican hat intracortical connections, and (c) use of an input stimulus to stimulate development. These

---

* Corresponding Author.

N.R. Pal et al. (Eds.): ICONIP 2004, LNCS 3316, pp. 84–89, 2004.
© Springer-Verlag Berlin Heidelberg 2004

assumptions are not supported by experimental findings [1] and a need for introducing models that are closer to biological processes has been stressed [10],[11].

Neurotrophic factors (NTFs) or neurotrophins are found to play important role in neuronal survival and/or differentiation [12],[13]. Competition among growing axons for NTFs has also been reported [14]. Models based on such competition have been proposed for the development of neuromuscular junctions [15] and ocular dominance columns [11]. We have proposed a model based on such competition for NTFs for the development of thalamocortical connections. The growth of thalamic afferents leads to the formation of subfields in the RFs of the modeled cells [1]. These RFs resemble experimentally measured RFs for simple cells and exhibit OR selectivity [1] and spatial frequency selectivity [2]. Here, we show that formation of subregions within the RFs occurs in such a manner that the OR preferences of nearby cells change smoothly across the simulated cortical surface forming an ORmap. This paper examines the development and characteristics of ORmaps formed through the model.

2 Development of Orientation Map in the Neurotrophic Model

A feedforward model consisting of three hierarchical layers: retina, LGN and cortex, has been used to model the formation of RFs and OR map in the visual cortex. The development of synaptic strengths is modeled through diffusive cooperation and resource limited competition for pre and postsynaptic resources. The axons compete for pre-synaptic resources (e.g. receptor molecules) that are present in limited amount in the LGN cells. A role for pre-synaptic resource was first suggested for elimination of polyneuronal innervations in neuromuscular system [16]. Competition also exists among axons for the post-synaptic resources (NTFs) that are present in limited amount at the post-synaptic sites i.e. the cortical cells. The model and its underlying assumptions have been discussed at length in [1]. All the layers are modeled as regular two-dimensional arrays. Both retina and LGN comprise of two distinct (ON and OFF) layers of size 30x30. Cortex consists of one layer of 50x50 spiking cells. Retinal and LGN cells are modeled as center surround gaussian filters with fixed one to one connectivity from retina to LGN. A cortical cell receives thalamic projections (both ON and OFF) from a 13x13 region centered symmetrically about its corresponding retinotopic position in the LGN. Initial synaptic strengths are very weak and randomly organized. Time evolution of synaptic strengths represents cortical development and is achieved through the following differential equation for weight updation

$$\frac{\partial W_{IJ}^{+}}{\partial t} = (\gamma_1 - K_1)(\gamma_2 - K_2) A_R(I, J) W_{IJ}^{+} + D_L \frac{\partial^2 W_{IJ}}{\partial J^2} + D_C \frac{\partial^2 W_{IJ}}{\partial I^2} \tag{1}$$

W_{IJ}^{+} (W_{IJ}^{-}) represents the strength of the connection from the ON-(OFF) center LGN cell at position J in LGN layer to the cortical cell at position I in the cortical layer. $W_{IJ} \in \{W_{IJ}^{+}, W_{IJ}^{-}\}$. $K_1^2 = \Sigma_{P=1}^{NXN} (W_{PJ}^{+})^2$, is the sum square of synaptic strength of all branches emanating from the LGN cell at the location J. γ_1 represent fixed presynaptic resources available in the LGN cell at location J. The term $(\gamma_1 - K_1)$ enforces competition for resources among axonal branches in a LGN cell.

Similarly the term $(\gamma_2 - K_2)$ enforces competition among LGN cells for target space in the cortex. $K_2^2 = \Sigma_{P=1}^{MXM} (W_{IP})^2$ is the sum of the square of synaptic strength of all branches of LGN cells converging on the cortical cell at location I. γ_2 represent fixed postsynaptic resources available in the cortical cell at location I. A_R is the arbor function. D_L and D_C are the diffusion constants in the LGN and cortex respectively. MxM and NxN are the sizes of the LGN and the cortex respectively. A similar equation is used for updating W_{IJ}^-.

We show that in this neurotrophic model the ORmap emerges in parallel with the formation of simple cell RFs. Initial synaptic strengths are randomly distributed; the cells have no RF structure and little or no OR bias. This results in randomly distributed OR preferences on the cortical surface. As the synapses grow and mature, cells develop RFs, become OR selective and OR preferences are also found to change smoothly across the simulated cortical surface forming an OR map.

3 Simulated Maps

Fig.1(a) shows an ORmap for one of the simulated cortices. The preferred orientations of the cells are represented linearly in 16 colors (red–yellow–green–blue–magenta); areas responding best to a horizontal stimulus are coded in red; areas responding best to a vertical stimulus are coded in green and so on. All the salient regions of an experimentally observed ORmap can be seen in the simulated map. The pinwheels are well distributed and connected by regions of fast OR change called fractures. Saddle regions and linear regions can also be seen. For the simulated cortices we also plotted the angle-magnitude (polar) maps. One such polar map is shown in Fig. 1(b).

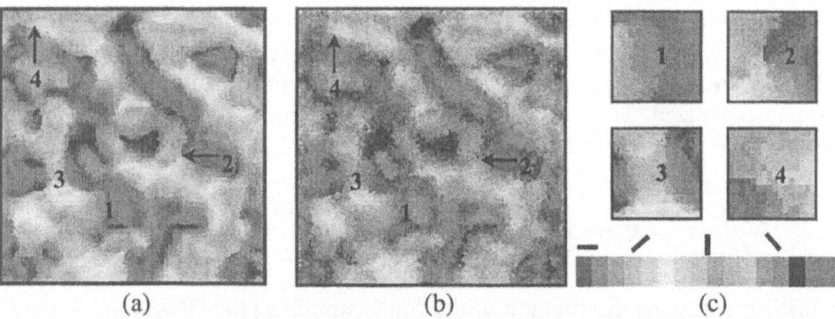

(a)   (b)   (c)

Fig. 1. (a) Orientation map (angle map) of one of the simulated cortex. Preferred orientation of the cells as computed by vector addition of the responses obtained for 10 different stimuli are represented linearly in 16 colors (red–yellow–green–blue–magenta) (Colored figures are available at http://www.geocities.com/monasurimathur/mona.html). Marked regions indicate: (1) Linear zones, (2) Singularities, (3) saddle regions, (4) fractures. (b) Polar map where the brightness of the colors codes for the magnitude of the resultant vector. The brighter is the color the larger is the magnitude of the resultant vector. (c) A close up of the marked regions

3.1 Stability of Orientation Maps with Time

Chapman et al [3] observed the development of OR preference maps in the ferret primary visual. They observed that OR preferences emerge very early in development and once established; the maps remain stable. The initially recorded OR activity maps are of low contrast. Over a period of several days, the maps mature into high-contrast, adult-like maps. During this period of maturation, the features of the developing maps were found to be very stable.

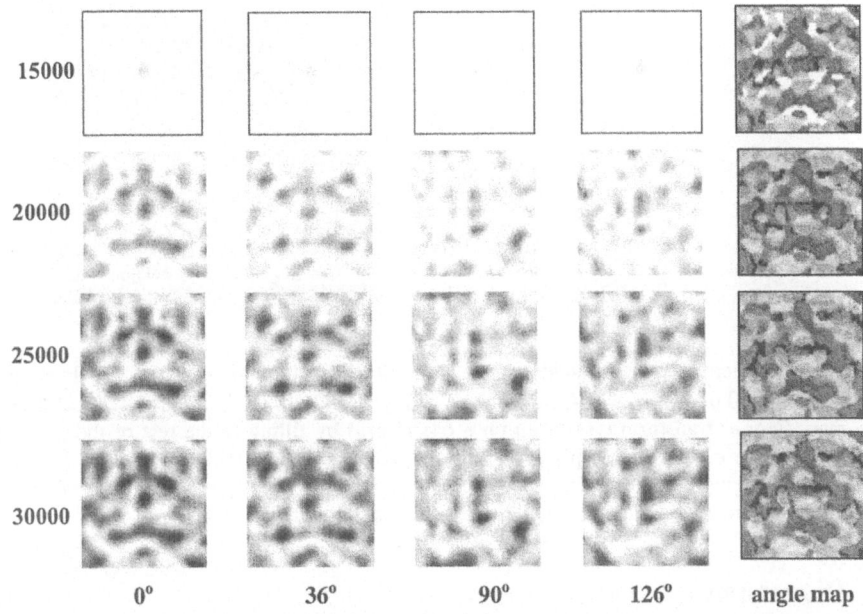

Fig. 2. A 50x50 cortex was simulated at D_C=0.1, D_L=0.075, γ_1=γ_2=1, with a time step of 0.001 for 30,000 iterations. Single condition maps and OR (angle) maps are shown for four iteration steps (15000, 20000, 25000, 30000) during the development process. Each column of the single condition maps shows iso-orientation maps recorded in response to a particular OR of a moving sinusoidal grating and the row gives the iteration number. Initially not all cells are responsive to input stimuli. The white regions in the OR map (angle map) obtained at iteration 15000 correspond to cells that did not respond to any input stimulus

In our model, the growth of synaptic weights leads to formation of RFs that leads to emergence of OR selective responses. We studied the change in the OR preference of the cells with the growth of their synaptic strengths. Responses of the cortical cells were obtained at different time steps during the development process. Fig. 2 shows the ORmaps (angle maps) and single condition maps recorded at four different time steps. Initially the cells have very weak synaptic strengths and do not respond to the input stimuli shown. Even at around 15000 iterations, some of cells had very weak synapses and therefore did not respond. The white regions in the ORmap obtained at iteration 15000 correspond to such cells. A comparison of the maps obtained at different time steps shows that the maps once formed are very stable.

3.2 Orientation Map and Its FFT

The power spectrum of OR preference maps gives a clear indication of how the OR preferences repeat. The OR maps obtained in monkeys are isotropic and this gives an annulus shaped spectrum [18], while for other species like cats the Fourier spectra are hemicircular [6] depicting the anisotropy found in these maps. The maps obtained through the model do not have the same periodicity in all directions and this anisotropy is reflected in the half moon shaped Fourier power spectrum of the modeled maps (Fig. 3).

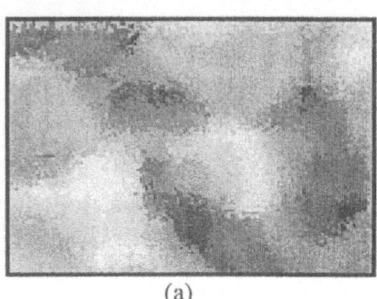 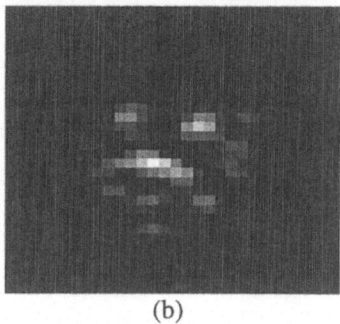

(a)   (b)

Fig. 3. (a) Orientation map (angle map) of a 20x30 section of a cortex simulated with DC=0.09, DL=0.075, $\gamma1=\gamma2=1$. Color scheme of Fig. 1 is used. (b) Fourier power spectrum of the map. Fourier transform was taken on a 64x64 grid by filling in the rest of the grid with zeros. Only the central 26x26 pixels of full spectra are depicted in the figure, as the power outside this region was zero. The spectrum is shown on a linear gray scale with black hue representing zero power and white hue representing maximum power

4 Discussion

We have shown that the neurotrophic model based on diffusive cooperation and re-source- limited competition can achieve the desired spatial layout of OR preferences on the cortical surface. The cortical diffusion constant is the key model parameter that determines the average density of the pinwheels on the simulated map. All the salient regions of the ORmap namely, linear regions, singularities, saddle regions and fractures could be seen in the simulated maps. In our simulations, ORmap emerges in parallel with the formation of RFs of the cells. Once the maps appear they remain stable. The Fourier spectrum of our maps is half moon shaped resembling the spectrum of ORmaps obtained in cats.

References

1. Bhaumik, B., Mathur, M.: A Cooperation and Competition Based Simple cell Receptive Field Model and Study of Feed-forward Linear and Nonlinear Contributions to orientation selectivity. Journal of Computational Neuroscience, Vol. 14 (2003) 211-227
2. Mathur, M., Bhaumik, B.: Study of Spatial frequency selectivity and its spatial organization in the visual cortex through a feedforward model. In Computational Neuroscience Meeting (CNS), Baltimore, MD, USA (2004)

3. Chapman, B., Stryker, M.P., Bonhoeffer, T.: Development of orientation preference maps in ferret primary visual cortex. Journal of Neuroscience, Vol. 16 (1996) 6443-6453
4. Sur, M., Leamey, C.A.: Development and plasticity of cortical areas and networks Nature Reviews Neuroscience, Vol. 2 (2001) 251:262
5. Erwin, E., Obermayer, K., Schulten, K.: Models of orientation and ocular dominance columns in visual cortex: a critical comparison. Neural Computation, Vol. 7 (1995) 425-468
6. Wörgötter, F.: Comparing different modeling approaches of visual cortical cell characteristics. Cerebral Cortex, (Eds. Ulinski et al.), Vol. 13, Kluwer Academic, Plenum Publishers, New York (1999)
7. Miller, K.D.: A model for the development of simple cell receptive fields and the ordered arrangement of orientation columns through activity-dependent competition between ON and OFF center inputs. Journal of Neuroscience, Vol. 14 (1994) 409-441
8. Stetter, A., Müller, A., Lang, E.W.: Neural network model for the coordinated formation of orientation preference and orientation selectivity maps. Physical Review E, Vol. 50, No. 5 (1994) 4167-4181
9. Miyashita, M., Tanaka, S.: A mathematical model for the self-organization of orientation columns in the visual cortex. NeuroReport, Vol. 3 (1992) 69-72
10. Miller, K.D.: Equivalence of a sprouting-and-retraction model and correlation-based plasticity models of neural development. Neural Computation, Vol. 10 (1998) 529-547.
11. Elliott, T., Shadbolt, N.R.: Competition for Neurotrophic factors: Ocular dominance Columns. Journal of Neuroscience, Vol. 18, No. 15, (1998) 5850-5858.
12. Cellerino, A., Maffei, L.: The action of neurotrophins in the development and plasticity of the visual cortex. Progress in Neurobiology, Vol. 49, (1996) 53-71.
13. McAllister, A.K., Katz, L.C., Donald, C.Lo.: Neurotrophins and synaptic plasticity. Annual Review Neuroscience, Vol.22 (1999) 295-318.
14. Purves, D.: *Neural activity and the growth of the brain*. Cambridge University Press, Cambridge (1994)
15. Rasmussen, C.E., Willshaw, D.J.: Presynaptic and postsynaptic competition in models for the development of neuromuscular connections. Biol. Cybernetics, Vol.68 (1993) 409-419
16. Willshaw, D.J.: The establishment and the subsequent elimination of polyneural innervation of developing muscle: theoretical considerations. Proc. R. Soc. B, Vol. 212, (1981) 233-252
17. Bonhoeffer, T., Grinvald, A.: Iso-orientation domains in cat visual cortex are arranged in pinwheel-like patterns. Nature, Vol. 353 (1991) 429-431.
18. Niebur, E., Wörgötter, F.: Design Principle of Columnar Organization in Visual Cortex. Neural Computation, Vol. 6, (1994) 602-614.

The Balance Between Excitation and Inhibition Not Only Leads to Variable Discharge of Cortical Neurons but Also to Contrast Invariant Orientation Tuning

Akhil R. Garg[1], Basabi Bhaumik[2], and Klaus Obermayer[3]

[1] Department of Electrical Engineering, J.N.V. University
Jodhpur 342001, India
garg_akhil@yahoo.com
[2] Department of Electrical Engineering, I.I.T Delhi
New Delhi 110016, India
bhaumik@ee.iitd.ac.in
[3] Department of Computer Science and Electrical Engineering
Technical University Berlin, Germany
oby@cs.tu-berlin.de

Abstract. The orientation tuning width of the spike response of neuron in layer V1 of primary visual cortex does not change with contrast of input signal. It is also known that cortical neurons exhibit tremendous irregularity in their discharge pattern which is conserved over large regions of cerebral cortex. To produce this irregularity in responses the neurons must receive balanced excitation and inhibition. By a modeling study we show that if this balance is maintained for all levels of contrast it results in variable discharge patterns of cortical neurons at all contrast and also in contrast invariant orientation tuning. Further, this study supports the role of inhibition in shaping the responses of cortical neurons and we also obtain changes in circular variance with changing contrast, similar to what is observed experimentally.

1 Introduction

Simple cells in layer IV of mammalian primary visual cortex show contrast invariant orientation tuning [1][2]. Many approaches for achieving contrast invariance have been proposed, some authors [3] suggest that contrast changes simply multiplies the response of a cell obtained using linear model by a constant changing this gain value gives rise to contrast invariant orientation tuning. Others [4][5] suggest that width of orientation tuning is the emergent property of intra-cortical circuitry and is independent of the parameters of input stimulus, including stimulus contrast. Using narrowly tuned excitatory and broadly tuned inhibitory intra-cortical interactions, they showed that a cell emerges to be sharply tuned even when it receives weakly tuned LGN(Lateral geniculate nucleus) inputs. On the other hand Troyer et. al have used phase specific feedforward inhibition,

N.R. Pal et al. (Eds.): ICONIP 2004, LNCS 3316, pp. 90–95, 2004.
© Springer-Verlag Berlin Heidelberg 2004

which has the effect at only non-preferred orientations [6], while McLaughlin et. al have used isotropic cortico-cortical inhibition for sharpening of orientation tuning [7].

Another important fact is that cortical neurons exhibit tremendous variability in the number and temporal distribution of spikes in their discharge patterns [8][9] and it has been reported that noise contributes to contrast invariance [2][10]. Recent review article [11] suggest that feedforward excitation provides the orientation preference to simple cells, and there is global(untuned) inhibition which sharpens the orientation selectivity. Experimental studies[15] show the presence of of both tuned and untuned inhibitory cortical cells. In the present study we have tried to investigate how the balance of feedforward excitation by global untuned inhibition contribute to contrast invariant orientation tuning. It is interesting to report that if this balance is maintained for all contrast levels then it leads to variable discharge patterns of cortical cell at all contrast level and to contrast invariant orientation tuning. As is known that with increasing contrast the LGN firing rate increases resulting in the increase of feedforward excitation, if simultaneous increase in global untuned inhibition is there then the balance between excitation and inhibition is maintained. In our model we achieve this by making inhibition dependent upon the average firing rates of the population of LGN cells for every stimulus condition, therefore with increase in contrast inhibition increases automatically. Our model resembles other models using inhibition for increasing orientation selectivity but differs from such models in that we use feedforward untuned inhibition rather than broadly tuned cross-orientation or feedforward phase specific tuned inhibition. As found in experimental studies [12], we also get contrast dependent changes in circular variance, with increase in contrast there is decrease in circular variance.

2   Material and Methods

A model of visual pathway used was developed elsewhere and has been shown to produce realistic responses to visual stimuli [5]. Simple cells RFs(Receptive Fields) were modeled as Gabor functions [13], which is a two dimensional Gaussian multiplied by a sinusoid. The positive values of Gabor function were taken to be ON subregion yielding connection from ON type LGN cells, and negative values of Gabor function were taken to be OFF subregion yielding connection from OFF type LGN cells. The model simple cell in cortical layer was a single compartment, integrate and fire neuron that received synaptic input in the form of transient conductance changes at both excitatory and inhibitory synapses. The membrane potential of the model neuron is determined by

$$\tau_m \frac{dV}{dt} = V_{rest} - V + G_{ex}(t)(E_{ex} - V) + G_{in}(t)(E_{in} - V) \qquad (1)$$

with τ_m=20ms, V_{rest}=-70mV, E_{ex}=0mV and $E_{in} = -70mV$ [14]. E_{ex} and E_{in} are the reversal potentials for the excitatory and inhibitory synapses. When the membrane potential reaches the threshold value of -54mV, the neuron fires an action potential and the membrane potential is reset to -60mV. The synaptic conductances G_{ex} and G_{in} and their related peak conductances g_{ex} and g_{in} are measured in the units of the leakage conductance of the neuron and are thus dimensionless. The values of individual peak excitatory conductance g_{ex} is taken from Gabor function, representing the receptive field of a cortical cell. In other words, individual peak synaptic conductance represents the synaptic strength between a LGN cell of particular type located at a particular location and cortical cell. Whenever a particular ON/OFF type LGN cell fires, the corresponding peak synaptic conductance contributes towards the value of total excitatory conductance. Each cortical cell also receives feedforward inhibition from N inhibitory interneurons, termed as untuned global inhibition, synaptic strength of connection between each inhibitory interneurons and cortical cell is given by g_{in}(inhibitory peak synaptic conductance). In our model we incorporate this by making each of these inhibitory cells receive input in the form of spikes generated by independent Poisson's process. The firing frequency used for generating spikes is proportional to the LGN firing rates. Additionally, whenever any of the inhibitory cell receives input in the form of spike it contributes towards the value of total inhibitory conductance. Otherwise, both excitatory and inhibitory synaptic conductances decay exponentially i.e. $\tau_{ex}\frac{dG_{ex}}{dt} = -G_{ex}$ and $\tau_{in}\frac{dG_{in}}{dt} = -G_{in}$, where $\tau_{ex} = \tau_{in} = 5mS$.

2.1   Balance of Excitation and Inhibition

The condition of balance between excitation and inhibition is when the mean value of the excitatory current and mean value of inhibitory current input to a cortical cell are same. Therefore, from equation(1) for a particular value of $V = V_m$ we get,

$$< G_{ex} > (E_{ex} - V_m) = - < G_{in} > (E_{in} - V_m) \qquad (2)$$

$$< G_{in} >= - < G_{ex} > \frac{E_{ex} - V_m}{(E_{in} - V_m)} \qquad (3)$$

where, $< G_{ex} >$ and $< G_{in} >$ are the excitatory and inhibitory synaptic conductances which are temporally averaged for every stimulus condition and which depend upon mean firing rate, the number and the value of peak synaptic conductances of excitatory and inhibitory inputs respectively, since E_{ex} and E_{in} are constants, on replacing $-\frac{(E_{ex}-V_m)}{(E_{in}-V_m)}$ by K in equation(3) we get $< G_{in} >= K* < G_{ex} >$ Also,

$$< G_{ex} >= n_{ex} < f_{ex} > g_{ex} \qquad (4)$$

and

$$< G_{in} >= n_{in} < f_{in} > g_{in} \qquad (5)$$

where n_{ex}, n_{in} are the total number of excitatory and inhibitory inputs received by the cell, $< f_{ex} >$, $< f_{in} >$ are the average firing rates of these inputs and g_{ex}, g_{in} are the values of the excitatory and inhibitory peak synaptic conductances representing the connection strength between the excitatory and inhibitory input cells to a cortical cell in question. If n_{ex}, n_{in}, g_{ex} and g_{in} are constants then we get $< f_{in} >= K_1 < f_{ex}$. This gives the relationship between average firing rates of excitatory and inhibitory inputs, to maintain the balance between excitation and inhibition.

2.2   Circular Variance

Circular variance(CV) has been used as a measure to quantify the effects of contrast on orientation tuning [12]. CV is calculated using the mean firing rate of the neuron according to following $CV = 1 - |R|$, where

$$R = \frac{\sum_k r_k e^{i2\theta_k}}{\sum_k r_k} \tag{6}$$

In the above equation, r_k is the mean firing rate at orientation k and θ_k is the orientation in radians.

3   Results

Moving sinusoidal grating of different orientation and particular spatial frequency was used as an input stimuli. We accumulated the number of spikes for a trial period of 1000msec for each orientation and repeated the process 100 times, and then obtained the average number of spikes for each orientation. Figure(1a) shows the orientation tuning curve of a cortical cell obtained at a contrast of 40 percent. This curve is a plot between orientation of the input stimuli and response of a cell in form of average number of spikes per second for each orientation. In actual we showed 10 different orientation of input stimuli

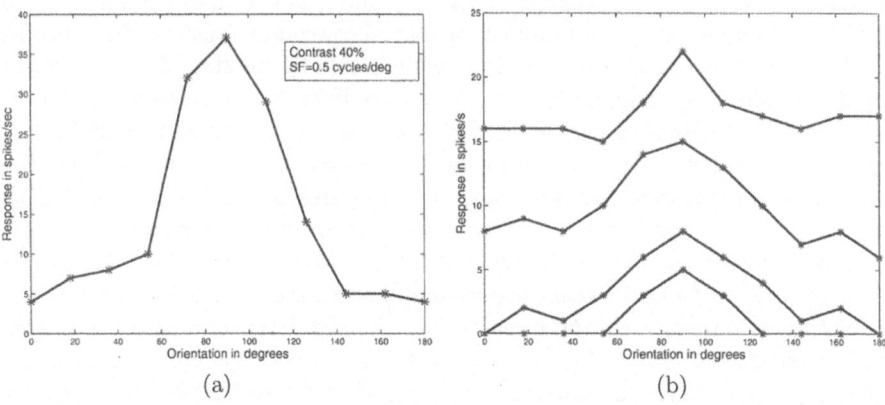

(a)   (b)

Fig. 1. (a) Orientation tuning curves with balanced excitation and inhibition for 40 percent Contrast(b) Orientation tuning curves for different level of contrast obtained with very less inhibition.

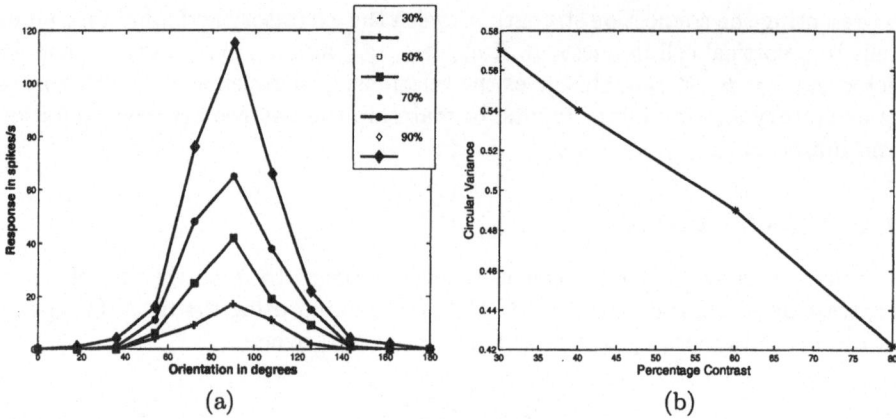

Fig. 2. (a) Orientation tuning curve for different contrast, with balanced excitation and inhibition in each case(b) Plot between CV and Contrast.

starting from 0 degrees to 162 degree with an increment of 18 degrees. As can be seen from the plot that the cell has a preference to a stimulus of an orientation of 90 degrees. The HWHH (half width at half height) obtained from the orientation tuning curve was 26 degrees. At a particular time and preferred orientation the bright and dark portions of the grating stimuli align with the cortical cells ON and OFF subregions simultaneously. When the matching between the two is perfect, the total number of LGN cells which are in position to contribute towards raising the membrane potential of cortical cell are much more than when the grating is in null position this makes the cell orientation selective. If we keep the inhibition unchanged with changes in contrast then as shown in figure(1b), with increase in contrast the selectivity of the cell changes from sharply tuned to untuned at higher contrast levels. To obtain contrast invariant tuning at all contrast, there should be change in the net inhibition with change in contrast of input stimuli so as to maintain the balance between excitation and inhibition at all contrast levels. In our model we achieve this by making the firing rate of inputs to inhibitory cells dependent on the feedforward excitatory input firing rate, which changes with change in contrast. If we do this so as to maintain the balance in excitation and inhibition at all contrast levels, we indeed get contrast invariant orientation tuning as shown in figure(2a). It can be seen that the hwhh for all contrast levels is almost same and on calculations we found it to be approximately 26 degrees for all contrasts, quite similar to what is observed in experimental studies. Recent experimental studies [12] found an inverse relationship between contrast and circular variance. To determine whether or not a similar relationship holds for our modeling results, we calculated circular variance (see methods) for the same results as shown in figure(2a). We also obtain an inverse relationship between contrast and circular variance shown in figure(2b). This is due to null orientation suppression in general at all contrast levels.

4   Conclusion

We have constructed a simple model that accounts for contrast invariant orientation tuning. In this model untuned inhibition, dependent on feedforward input plays a vital role in shaping the response of a cortical cell. Role of feedforward dependent inhibition in shaping the response of a cortical cell has been shown previously in modeling study [6]. In their model tuned inhibition was necessary. Experimental studies [15] supports the presence of both types of inhibitory cells to exist in layer IV of primary visual cortex. It would be interesting to study the combined effect of both in shaping the response of a cortical cell. Also, using this model we show that balance in excitation and inhibition leads to not only irregular discharge pattern of cortical neurons but also to contrast invariant orientation tuning. We have analytically derived the relationship needed between the firing rates of excitatory and inhibitory inputs to maintain this balance.

References

1. Sclar G. and Freeman R. D. 1982 *Exp. Brain Res.* **46** p 457-461
2. Anderson J.S.,Lampl I.,Gillespie D.C., Ferster D. 2000 *Science* **290** p 1968-1972
3. Carandini M,Heeger D and Movshon J 1997 *J Neurosci.* **17** p 8621-8644
4. Ben-Yishai R,Bar Or R and Sompolinsky H 1995 *Proc. Nat. Acad Sci USA* **92** p 3844-3848
5. Somers D, Nelson S and Sur M 1995 *J. Neurosci.* **15** p 5448-5465
6. Troyer T. W.,Krukowski A. E.,Priebe N. J. and Miller K. D. 1998 *J. Neurosci.* **18** p 5908-5927
7. McLaughlin D., Shapley R.,Shelley M and Wielaard J. 2000 *Proc. Natl. Acad. Sci. USA* **99** p 1645-1650
8. Shadlen M.N and Newsome W. T 1994 *Current Biology* **4** p 569-579
9. Shadlen M. N. and Newsome W. T 1998 *The Journal of Neurosci.* **18** p 3870-3896
10. Hansel D and Vreeswijk C. 2002 it The Journal of Neurosci. bf 22 p 5118-5128
11. Shapley R., Hawken M. and Ringach D. L. 2003 *Neuron* **38** p 689-699
12. Alitto H. J and Usrey W. M. 2004 *Articles in press J. Neurophysiol.* 10.1152/jn.00943,2003
13. Jones J P and Palmer L A 1987 *J. Neurophysiol.* **58** p 1187-1211
14. S. Song, K. D. Miller and L. F. Abbott 2000 *Nature Neurosci.* **3** p 919-926
15. Hirsch J.A., Martinez L.M, Pillai C., Alonso J.M., Wang Q. and Sommer F. T. 2003 *Nature Neurosci.* **12** p 1300-1308.

Stochastic Resonance Imaging – Stochastic Resonance Therapy: Preliminary Studies Considering Brain as Stochastic Processor

Prasun Kumar Roy

National Brain Research Centre, National Highway-8, Manesar, Gurgaon,
Haryana 122 050, India
pkroy@nbrc.ac.in

Abstract. The novel field of Stochastic Resonance effect (SR) uses optimized pertur-bation or statistical fluctuation (so-called 'noise') to critically enhance the behaviour or sensitivity of a system to an input parameter or signal. We use SR to explore a new paradigm for increasing the efficiency of diagnostic and therapeutic radiology for the brain. Firstly, we demonstrate experimentally how SR enhances the neuroimaging process in MRI/fMRI, utilizing stochastic enhancement of paramagnetic relaxation of organometallic compounds as gadolinium-pentatate and deoxyhaemoglobin for imaging brain lesions or cognitive activation respectively. Case study on using SR for differentiating MRI scans of brain tumour recurrence versus necrosis is presented. Secondly, we show how SR can enhance gadolinium radiotherapy, or electrotherapy (deep brain electrostimulation). We present a model for therapeutic SR for neoplastic or degenerative lesions. We outline the prospect of developing the emerging field of stochastic neuroscience, inspired by constructive noise-tissue interaction.

1 Introduction

A promising prospect to enhance the efficiency of neuroradiological processes, whether diagnostic or therapetic, is offered by the recently discovered unitary phenomenon of stochastic resonance (SR) effect [1], an emerging research field in computational neuro-science and bioengineering. SR is a general principle of nonlinear dynamics applicable to various systems, whether physical, chemical, biological, computational or quantum. Simply stated, the SR process arises since positive peaks of a weak signal, under opti-mized conditions, adds to positive peaks of the perturbation or noise, resulting in some amplitudes which become considerably higher (crossing the threshold), thus helping detection or target action. Studies carried out by various investigators, including the author [2-5], show that, for such biological systems, a small optimized stochastic pertur-bation added to the input or signal appreciably increase the responsivity or signal:noise ratio in the system; in other words, noise is added to decrease the noisiness of a system, that is, noise is used to counter noise. Though this may seem counterintuitive, SR has been used to enhance various processes relevant to neurobiologists, such as:

- x-ray/γ-ray/raman spectra, electron paramagnetic resonance of organic chemicals [6-8].
- peptide-induced neuromodulation, or electrostimulation, for neuronal signal transduction, and treatment in cerebral stroke, neural injury or sleep apnoea [9-11].

N.R. Pal et al. (Eds.): ICONIP 2004, LNCS 3316, pp. 96–103, 2004.
© Springer-Verlag Berlin Heidelberg 2004

However the *practical* application of SR effect as a novel technique in neuroimaging or therapy has not been systematically pursued, and the applicability is the aim of our study.

Rationale. Factors necessary for SR to occur are: (i) a source of input signal, (ii) a *noise* source (iii) a threshold, viz. the minimum signal strength needed for a process to be distinctly effective. A radiological process is basically a threshold process for a signal (a faint input signal, whether diagnostic or therapeutic, does not have the desired effect). Routine radiological operations thus satisfy the factors (i) and (iii); we now need to devise the factor (ii) for SR to occur. SR's amplification is shown by the SR output equation, implying a peak at an optimal noise level (fig. 1a):

$$SnR = (a/v)^2 \, exp \, (-2 \, H/v),$$

where a, H, v and SnR denote input signal amplitude, threshold, noise power and signal-noise ratio respectively [1]. The peak response is evident if fig 1a. To induce SR, noise power is varied (optimized) for maximal enhancement. We now elucidate how one can devise SR enhancement is diagnostic and therapetic neuroradiology.

2 Stochastic Resonance Imaging (SRI): Paramagnetic Noise

In magnetic resonance imaging (MRI) or functional MRI (fMRI) of brain, involving signal of blood vasculature or tumours, the differential MR signal is related to a paramagnetic substance in blood. In fMRI, the signal relates to change of [oxyhaemoglobin (OxHb) level + deoxyhaemoglobin (DeHb) level] in blood produced by a cognitive task: the paramagnetic substance is DeHb, an organo-metallic compound of a transition series metal (Fe). DeHb is internally produced from the anoxic reduction of OxHb which is diamagnetic. On other hand, in structural MRI of brain pathology involving tumour or blood vessel, the paramagnetic substance is externally-administered intravenously and is gadolinium meglumine, an organometallic compound of another transition series metal (Gd). MRI output depends on relationship between the response of gadolinium and solvent water, a diamagnetic substance. We now probe the scope of giving noise through paramagnetic contrast in body fluid (having water as basic medium, as blood or cerebro-spinal fluid). Using Bloch NMR formalism, quantum/nuclear stochastic resonance pheno-mena has been theoretically predicted, and verified by computational simulation [12].

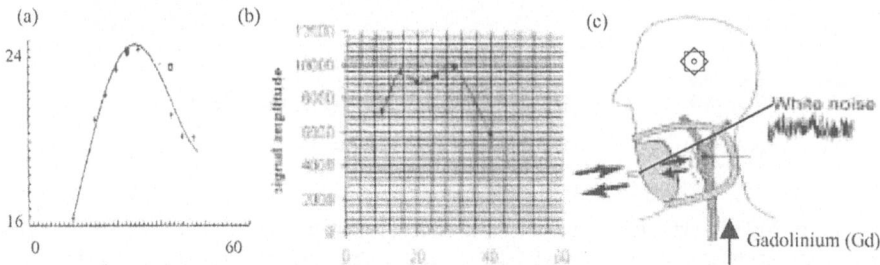

Fig. 1. (a) Pattern of stochastic resonance (SR). Noise power (x-axis) vs. SnR (y-axis, in dB). (b) SR enhancement of MRI signal using kinetic noise of contrast agent. Perturbation intensity (x-axis) vs. signal amplitude (y-axis). (c) Cervical chamber for giving perturbation to Gd input to lesion

Dynamics. Relaxation processes occur, due to microscopic effects as stochastic fluc-tua-tion of local dipolar field actuated by the stochastic kinetic motions of the nuclei [13]. We consider magnetization of water medium as a dependent variable of the stochastic nature of nuclear relaxation process, namely the noise intensity inducing the relaxation process. The aqueous medium can be sent to a regime where transverse and longitudinal relaxation times T_1 and T_2 equalize, if one adds paramagnetic agent. At a definite range of concentration of the agent, the stochastic collision events of the water molecules with the molecules of the agent, dominate the nuclear relaxation dynamics. Thereby, the system transits to a regime of faster motion; here the correla-tion time of local magnetic fields sensed by the nuclei is very short (in comparison with the time period of Larmor frequency), this ensures that $T_1 = T_2$ (= T^* say) [13]. Below the said range of concentra-tion, the collision events or noise will not domi-nate. On other hand, larger concentration of the agent higher than the definitive range, induces too swift relaxation rate. This indicates effective tuning of noise strength induced by the agent (the kinetic 'noise' source) if it has optimum concentration level, where relaxation noise dynamics is valid.

Experimental Corroboration. An experiment was carried out using paramagnetic agent cupric sulphate pentahydrate in aqueous solution using a radiological phantom (artificial model of polyurethane or glass-ware that would model fluid of blood ves-sel) under a 1.5 Tesla Siemens Magnetom® MRI scanner. Phantom's image was obtained by the scout sequence. Spin echo sequence was used to measure T_1 and T_2 periods at TR = 300 ms and 2000 ms respectively, using dynamic analysis option. Localizer sequence was given, and then long external radiofrequency pulse of width 200 ms was administered, with TR = 500 ms; 128 acquisitions were taken and fre-quency was adjusted (receiver gain = 64). The spectra and post-processed image was taken, and the amplitude observed. Amplitude was measured for the agent at concen-trations of 10, 15, 20, 15, 30, 40 mili-moles [mM] (fig 1b). Thus we discern that SR peak occurs at kinetic noise intensity induced by 24 mM of the paramagnetic agent; there is amplitude decrease on higher or lower noise intensity (i.e. paramagnetic agent concentration). This is characteristic of SR (fig 1a).

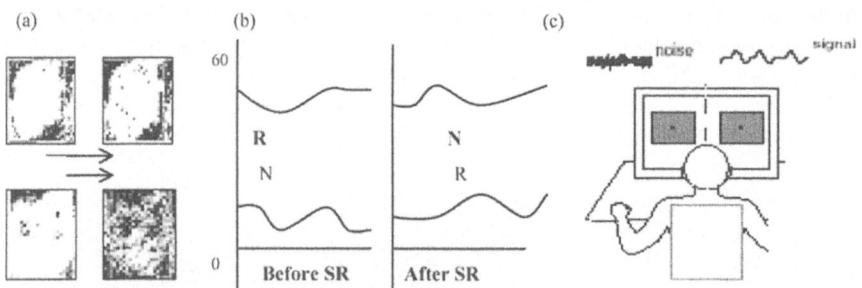

Fig. 2. (a) MRI image before and after SR enhancement in brain tumour recurrence (*upper panel*) and necrosis (*lower panel*). By SR, necrotic areas clearly show up as patches, while recurrence as an uniformity without patches. (b) Variance Z of the proton spin density images (y-axis) plotted against position (x-axis). Necrosis image N is below Recurrence image R, as necrosis has more proton leak and homogenization due to cell death. After SR, the strong varie-gated structuration of necrosis increases Z. (c) SR scheme adaptable for fMRI. Signal and noise given by separate eyes

Neuroimaging Implementation. Magnetic moment of cupric ion and gadolinium ion are 1 and 9 Bohr magnetons, and the molarity of cupric and gadolinium agents are 249 and 928. So we deduce, as first order approximation, that SR for Gd would occur around 2.67 mM concentration. Clinical Gd injectable solution is 0.5 M, and dilution at cerebral artery after systemic circulation, is about 0.5-1%, i.e. 2.5-5 mM. This matches the theore-tically predicted value of 2.67 mM; hence the SR for Gd can be feasible, the experiments for which we are pursuing. Note the importance of variation in contrast agent concentra-tion required to obtain SR maxima. Our aim is to induce such a variation of cerebral arterial concentration of gadolinium, administered intra-venously in the forearm. This variation can be done by using a standard clinical cervi-cal pneumatic chamber around the neck connected to a pneumatic pump that can administer a sinusoidal variation in chamber pressure, with zero mean and increasing variance (fig 1c). Such chambers are used to administer a desired modulation of cere-bral arterial (intra-carotid) blood pressure and volume for cerebrovascular research [14]. Using a sinusoidal variation, one induces an alteration of Gd concentration that would induce a maximal response at a specific Gd concentration that acts as kinetic noise function ξ apropos fig. 1b.

There are other routes to administer SR noise to the MRI imaging system, e.g. by giving a gaussian variation (white noise function) ζ through at any of several levels: the signal generation level (via pneumatic chamber), or signal detection level, or ac-cessed signal level, etc. In fig. 2a we present SR enhancement of internal structure of brain tumour recurrence vis-à-vis post-radiotherapy necrosis, their differential diagno-sis is a major challenge in contemporary neuroimaging. Fig 2b shows a quantitative index to gauge SR enhanced images, namely an entropic index that characterizes the image texture. This is done by scaling the image from 256×256 pixel size to 30×30 pixel size with 5 pixel as step size. Thus we have 51 vertical strips of image. The y-axis plots the variance Z (of each vertical strip of the new image), calculated using the deviation of each pixel from the average intensity of the strip; while the image scale (strip position) forms the x-axis. Entropic graph of the raw image and of the SR image are compared. Results are striking; the entropic values increase where structuration has been revealed by SR (e.g. necrosis), but decreases where there is no revealed structuration (e.g. recurrence) (fig 2b).

2.1 Functional Neuroimaging: Stochastic Resonance Enhancement

The fMRI signal is actuated by the cerebral blood flow and hence a stochastic noise α to the cerebrovascular flow using a pneumatic carotid collar (fig 1c) can enhance the fMRI signal, this noise would impress a stochastic function on the endogenous para-magnetic agent deoxyhaemoglobin. Further, for fMRI cognitive experiment with visual/auditory stimulus, an alternative SR route is to give the experimental stimulus to one eye/ear. To the other eye/ear we give noise function β via a screen with sto-chastically varying illumination, or a earphone with stochastically varying hiss (fig 2c). Respective noise functions are generated by a light-emitting diode or a sonic piezo cell, whose current is controlled by gaussian noise generator. The noise power (gaussian variance) β is increased until maximum enhancement is obtained in the fMRI signal. For linguistic application, we are testing SR on cortical activation areas in trilingual subjects, speaking pre-Aryan (Malayali), Aryan (Hindusthani) and post-

Aryan (English), so as to study differential cognitive activation, semantic processing or anatomical localization. It may be mentioned that if two noises ψ and ϕ are simultaneously used, there is, under specific optimality, a double stochastic resonance with a notable enhancement of signal. Double resonance has been already noted in physical, biological and information systems [15], and we have demonstrated its applicability to signal processing in brain [2], the system dynamics depending on the parameter 'noise-to-noise ratio', namely NNR = ψ / ϕ.

3 Stochastic Resonance Therapy (SRT): Noise Through Input Agent

Using stochastic perturbation of the input flow of a therapetic agent, SR enhancement can occur in the target activity of the agent (e.g. radiotherapy, drug therapy or physical therapy). Stochasticity has been used to enhance biochemical effects, e.g. signal trans-duction across neuronal ion-channels, photochemical reactions (involving photon-biochemical interaction) and free-radical induced bimolecular reactions (fig 3a), applicable to dimerization, polymerization and catalysis [9,18-19]. To propose the SRT concept, we consider gadolinium again, but as an agent in therapeutic radiology, namely gadolinium therapy for brain tumours, especially in florid glioblastoma where mean survival is 54 weeks. Here gadolinium meglumine is chelated with a chromophore moeity (Magnevist-pBR322®, Hoechst), this chelate has affinity for DNA minor groove in tumour cell [20]. Under neutron radiotherapy of gadolinium-infused tumour [21], high energy photons (γ-ray) generate locally by gadolinium-neutron interaction. Such photons produce free-radical induced DNA dimerization, lethal to brain tumour cells [17]. SR can be induced by giving noise through any of the reactants, gadolinium or neutrons, viz. (1) modulating the flow of gadolinium compound using a stochastic cervical pneumatic noise, or (2) modifying the neutron flux by a stochastic noise that alters the radio-frequency (RF) current of the proton generator used to produce neutrons (fig. 3b).

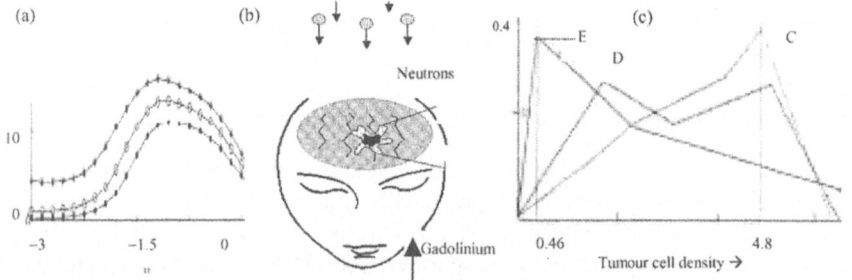

Fig. 3. (a) SR enhancement of bimolecular reactions; perturbation (log noise) of flow of a reactant substance forms x-axis, while y-axis is output concentration. The curves show situation at 10, 20, 50 sec time (b) SR enhancement for brain tumour radiotherapy: perturbation at two routes: Gd inflow (by cervical chamber) or neutron input (radiofrequency generator). (c) Mathematical model for SR enhanced therapetic effect of cytomodulative agent. X-axis: Tumour cell density; y-axis: probability function. As perturbation rises, cell population shifts from macro-lesion C (low noise, $\sigma = 0.5$) to meso-lesion D (medium noise, $\sigma = 0.86$) to micro-lesion E (high noise, $\sigma = 2.8$)

Experimental Corroboration. By stochastic differential equation, we have analyzed the effect of a stochastic fluctuation or noise perturbation impressed on tumour cell death rate r, induced by a therapetic agent [16]. We know that variation in rate of input flow or flux of a therapetic agent (radiation, cytomodulative drugs) is reflected as random varia-tion of the cytolytic rate r that gives it a stochastic character [3, 18]. As noise or fluctua-tion σ of cytolytic rate r increases (fig 3c), the tumour cell density moves from macro-lesion state C (large lesion, many neoplastic cells, progressing tumour) to meso-lesion D (inter-mediate lesion, metastable tumour) to micro-lesion E (small lesion, few tumour cells, regressing tumour). Asymmetry of the 3 graphs of probability distributions can be characterized by the normalized third moment (skew-ness) index Γ. Statistically, Γ of a distribution varies between -1 (right skewed) to 0 (symmetrical) to $+1$ (left skewed).

From fig 3c, the skewness coefficient Γ of the probability distribution curve for state C is around -1, while Γ for equipoised state D is around zero, and Γ for state E around $+1$. Stage D is the critical homeostatic state where tumour cell birth (by mitotic cell-division) balances tumour cell death (by therapetic agent). The states can be char-acterized quantitatively by homeostaticity coefficient $H = \sqrt{(1 - \Gamma^2)}$. Fig 4a plots the H value against the noise variation σ; note that at intermediate values of σ, there is maximization of H value, indicating stochastic resonance; the curve matches the theo-retical pattern of SR (fig 1a) as also the experimental pattern of SR enhancement of a bi-molecular/ dimerzation chemical reaction induced by reactant flux or photon radia-tion (fig 3a).

Towards Therapetic Implementation. We elucidate that SR enhancement of thera-petic application can be done by perturbation of a parameter that modulates the cyto-lytic rate r, namely by perturbing the radiation level, free radical or oxidative level, or cytotoxic drug level. As a case study we consider applicability to a common clinical constellation, the neurofibromatosis-neurofibroma-neurofibrosarcoma gamut. Fi-brosarcoma experi-mental findings are shown in Fig 4b, there is arrest of the im-planted tumour cells in mice, using oxyaemic perturbations induced by endostatin agent, that induces oxyaemic variation between 11 mmHg to 2.5 mmHg of pO_2 level. Note that σ of this variation is 4.86, that exceeds the critical value σ value 2.83 of state E (fig 3c), thus ensuring that SR regression will occur in the present case. It is also observed that without endostatin perturbation the lesion grows progressively and that the cells are resistant to standard chemotherapetic agent as cyclophosphamide. Indeed the proposed SR formalism pro-vides a robust elucidation of therapeutic effect of cytomodulative perturbation on drug-resistant tumour cells. It is well known that conventional theory cannot explain satisfac-torily the increased efficiency of cyto-modulative perturbations on neoplastic cells [22].

Such experimental SR approach would be applicable to therapetic modalities that per-turb the r variable, i.e. radiotherapy, chemotherapy etc. There is also indirect empirical evidence justifying a stochastic perturbative approach to radiotherapy as per fig 4c. This shows the clinically-observed pattern of increase of therapetic efficiency as radiothera-petic perturbation (beam fractionation) is increased. This rising graph corresponds to ascending left-side graph in fig 1a. For fig 4c, if perturbation consid-erably exceeds the hyper-fractionation level, there is decrease in therapetic efficiency with radiotoxicity (corresponding to post-optimal descending arm of fig 1a; however the post-optimal descending stage is not shown in fig 4c. Indeed these experimental observations corro-borate our proposed principle of stochastic resonance radiotherapy using perturbation of either gadolinium or neutron flow, via a pneumatic collar or radiofrequency modulation respectively. SR approach may also be used for enhancing the radiofrequency controlled electron path in standard intensity-modulated radiotherapy, used for brain tumours, cerebrovascular malformation and some refractory neurodegenerative disorders.

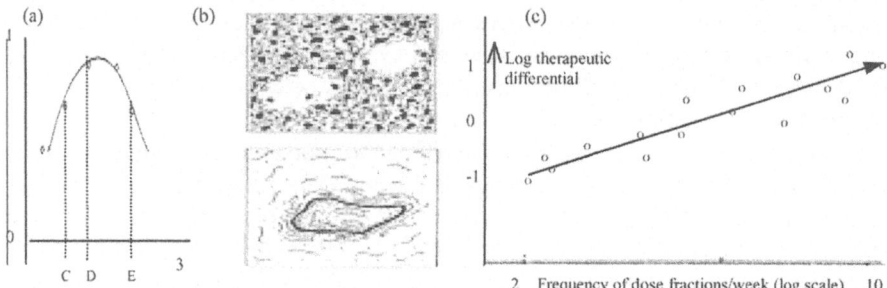

Fig. 4. (a) Homeostaticity index of lesion (y-axis) as one increases the stochastic perturbation noise σ of flow rate of therapeutic agent (x-axis), confirming SR behaviour. C, D, E corresponds to the states of fig 3c given above. (b) The cells of the lesion before and after the perturbation induced by cytomodulative agent (*top and bottom panels* respectively). Top panel shows characteristic neoplastic cells, while the bottom panel a circumscribed micro-focus in arrest state. (c) Increased effectivity of therapeutic radiology outcome as radiation perturbation increases from hypofractionation to normofractionation to hyperfractionation. X-axis is the perturbation intensity, and y-axis the radiobiological effect (therapeutic differential, or normalized log iso-effective dose)

3.1 Physical Therapy and Drug Therapy: Stochastic Resonance Enhancement

One can also consider SR to boost (i) physical therapy (electrostimulation) in neural diseases as parkinsonism, neuralgia etc, (ii) drug therapy, especially where perturbation can be given to the drug flow rate near its target, the brain, so as to prevent damping of perturbations. Drugs given intra-arterially or intrathecally via cerebrospinal fluid, are obvious candidates for study (e.g. arabinose to open blood-brain barrier, opioids etc).

4   Conclusion: Towards Stochastic Processing and Neuroengineering

Neuroscience and medicine has been historically dominated by the deterministic New-tonian model, where noise or fluctuations have been looked upon as inevitable evils, to be neglected or lessened. Nevertheless, the remarkable discovery of noise-enhanced operations by electronic engineers and then by biophysicists, has snow-balled into a technological paradigm shift from a deterministic to a stochastic and probabilistic world-view. We showed that stochastic resonance can be utilized as a new diagnostic and therapetic approach to the brain, when conventional procedures are deficient. Though used for radiology, the overall SR approach of this paper can be generalized for enhancing effectivity of biological signals in general, whether for diagnosis or treatment, e.g. electrodiagnosis (using EEG, EVP) or pharmacotherapy (using drugs). A collective effort among neuroscientists, radiologists, pharmacologists and computer engineers is imperative for ushering in the promising modality of stochastic neuroengineering.

References

1. Gammaitoni, L et al: Stochastic Resonance. Rev Mod Phys. 70:1 (1998) 223-272
2. Roy P et al: Cognitive stochastic resonance, J. Inst Electronics & Telecom Engr 49 (2003) 183-95
3. Roy P et al: Neurocomputation: fluctuation instability. IEEE Trans Evol Comp 6 (2002) 292-305
4. Roy P: Stochastic resonance in language or EEG, J. Quantitative Linguistics 11 (2004) in press
5. Upadyay A, Roy P: MRI-Stochastic Resonance, Tech Rept, Nat Brain Res Ctr, Gurgaon (2002)
6. Alibegov, B: Stochastic Resonance in x-ray, Astron. Lett, 22:4 (1996) 564-66
7. Vaudelle, F: Stochastic Resonance: Raman scattering. J. Opt Soc Amer. B-15 (1998) 2674-80
8. Pardi, L: Stochastic resonance: electron paramagnetic system. Phys Rev Lett 67 (1991) 1799-02
9. Bezrukov, S: Noise-induced enhancement in signal transduction, Nature 378 (1995) 362-64
10. Glanz, J: Shaping the senses with neural noise, Science, 277 (1997) 1759-60
11. Suki, B et al: Life-support systems benefit from stochastic noise, Nature 393 (1998) 127-28
12. Makarov, D: Stochastic resonance in quantum well structure, Phys Rev E 52 (1995) R2257-60
13. Abragam, A: Principles of nuclear magnetism, Oxford University Press, Oxford (1961)
14. Eckberg, D et al: Neck device for carotid activation, J Lab Clin Med 85 (1975) 167-73
15. Kurths, J et al: Additive noise in non-equilibrium transitions, Chaos, 11 (2002) 570-80
16. Roy P et al: Tumor stability analysis, Kybernetes: Int J System Sc & Engg, 29 (2002) 896-927
17. Perez, C, Brady, L: Principles of radiotherapy of brain tumors, Lippincott, New York (1997)
18. Horsthemke, W et al: Noise-induced transitions: Physics and biology, Springer, Berlin (1994)
19. Liu, Z et al: Noise-induced enhancement of chemical reactions, Chaos, 12:2 (2002) 417-25
20. DeStasio, G et al: Gadolinium in human glioblastoma, Cancer Research, 61 (2001) 4272-77
21. Culbert C: Computational gadolinium therapy, Phys in Bio & Med. 48 (2003) 3943-59
22. Kerbel, R: Resistance to drug resistance, Nature, 390 (1997) 335-37

Ultra-wideband Beamforming by Using a Complex-Valued Spatio-temporal Neural Network

Andriyan B. Suksmono[1] and Akira Hirose[2]

[1] Department of Electrical Engineering
Institut Teknologi Bandung, Jl. Ganesha No. 10 Bandung, Indonesia
suksmono@ltrgm.ee.itb.ac.id, suksmono@yahoo.com
[2] Department of Electrical and Electronics Engineering
The University of Tokyo
ahirose@ee.t.u-tokyo.ac.jp

Abstract. We propose an ultra-wideband (UWB) beamforming technique by using a spatio-temporal complex-valued multilayer neural network (ST-CVMLNN). The complex-valued backpropagation through time (CV-BPTT) is employed as a learning algorithm. The system is evaluated with an ultra-wideband monocycle signal. Preliminary simulation results in suppression of UWB interferer and in steering for desired UWB signal, demonstrating the applicability of the proposed system.

1 Introduction

Conventionally, a multilayer neural-network utilizing backpropagation algorithm learns input-output pattern in a static manner. The network maps an input vector X to an output vector Y. It is well suited for pattern recognition applications, and also applicable to a time series with stationary statistics, where the input vector is defined in term of past samples $\{x(n), x(n-1), \ldots, x(n-k-1), x(n-k)\}$. Whenever the signal is non-stationary, a multilayer neural network possessing memory is necessary.

In [1] we presented a multilayer CVNN that is capable to deal with multiple desired signals-multiple interference case. The incoming signals are assumed to be a monochromatic (or narrowband). To extend its capability in handling wideband signals, the structure of the processor should be modified. Based on wideband array [2], a tapped-delay-line (TDL) network should be applied. It also can be regarded as a processing-with-memory system. Therefore, a natural extension of the multilayer neural network for an array is a spatio-temporal network.

In the view of neurobiology, the use of time delay is motivated by the omnipresence of signal delays in the brain. Action potential's pulse transmission along an axon can also be digitally represented as TDL. The delays (and memory) also play an important role in neurobiological information processing [3].

2 Spatio-temporal Processing in Biological Neuron and the Spatio-temporal Neural Network

Neuron is regarded as the most elemental computing unit in live beings. The artificial neural network is inspired by the biological neuron. Fig. 1 shows analogy of spatio-

N.R. Pal et al. (Eds.): ICONIP 2004, LNCS 3316, pp. 104–109, 2004.
© Springer-Verlag Berlin Heidelberg 2004

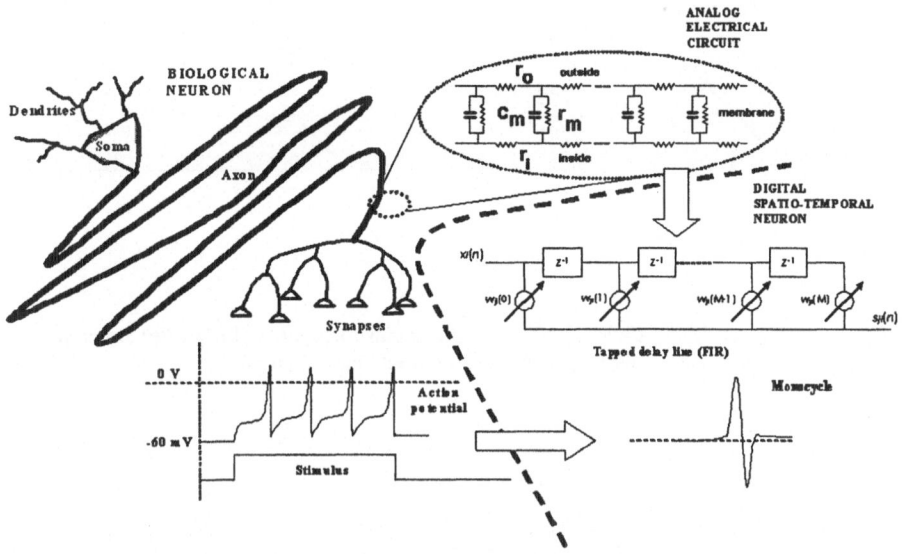

Fig. 1. Spatio-temporal processing in biological and artificial neural network

temporal processing between the biological neuron and our proposed digital spatio-temporal neural network system. The upper left part of the figure shows a generic neuron modeled after spinal motor's; consisting of dendrites, cell body (soma), axon and synapses. Neuron is a cell; it has a nucleus and related metabolic apparatus. The neuron receives input from other cells through dendrites, processes the information and integrating the inputs in dendrites and soma, and yields output results in a form of electrical spikes called action potentials (lower left part of the figure) whose frequency is depending on the amplitude of the stimulus. The information is delivered through the axon to synapses, which is in turn connected to other neurons/cells.

Axon is the transmission line for neurons to communicate with others. It is long and thin and has a very high resistance and large capacitance. Electrical voltage will drop very rapidly along the axon if the line was simply linear. However, an actual axon is equipped with nonlinear ion channels. They compensate the decay and keep the pulse shape unchanged so that they communicate with among neurons.

Learning in the neuron is achieved by modification of synaptic strength. Analogous to that of muscles, the synaptic becomes stronger when it is used frequently. In the computational model of a neuron (see [3] for detail examples), the synaptic strength is analog to a weighting factor. Along with RC that forms filter, in the digital domain we can construct a TDL/FIR (Finite Impulse Response) filters. The set of delayed pulses is then summed and fed into a (nonlinear) activation function yielding a spatio-temporal processor.

Elemental unit of the spatio-temporal digital neuron is a TDL shown in Fig.2. It consists of an array of memories and adjustable weighting devices, whose value is adjusted during adaptation process. For simplification, TDL in Fig. 2.a is represented by a double-edged triangle shown in Fig. 2.b. The multilayer structure shown in Fig.3, consisting of one input layer, one hidden layer and one output layer, is constructed from the elemental unit. When quadrature demodulation is not employed, a

106   Andriyan B. Suksmono and Akira Hirose

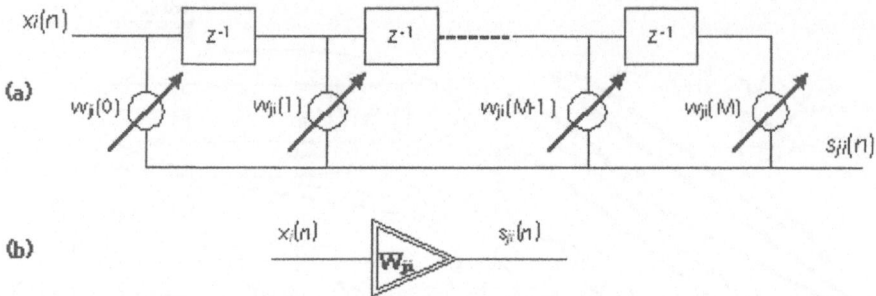

(a)

(b)

Fig. 2. (a) Tapped Delay Line (TDL) or Finite Impulse Response (FIR) filter structure representing processing with memory and (b) its simplified block diagram

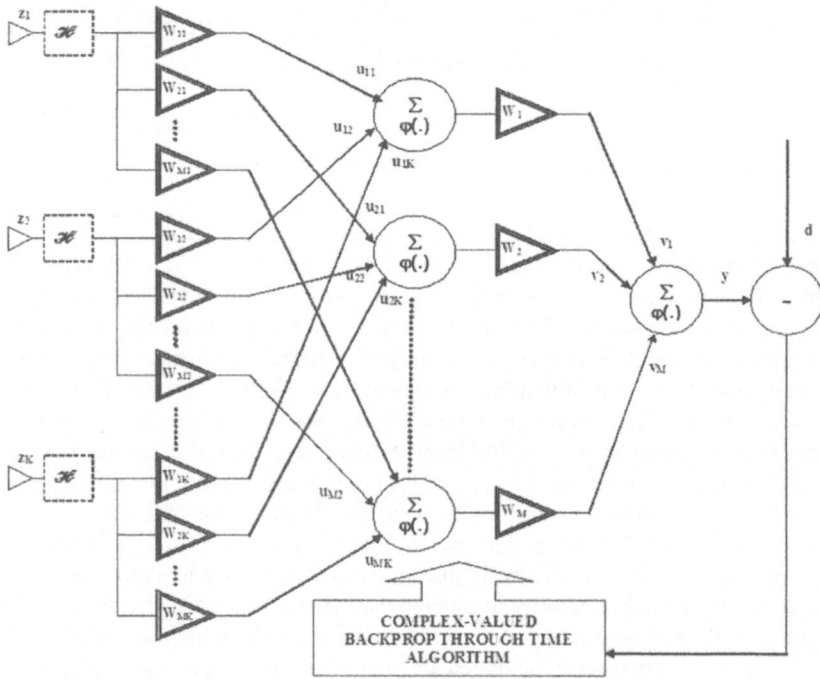

Fig. 3. Block diagram of the complex-valued spatio-temporal multilayer neural network for ultra-wideband beamforming application. In the figure, $\mathcal{H}$ is a Hilbert transform operator and $\varphi(.)$ is an activation function

Hilbert transformer should be applied to produce complex-valued signals. In the proposed system, we adopt Widrow's two-mode adaptation process where a reference signal is provided.

The BPTT is an algorithm to update weights of a multilayer neural network with memory [3, 4]. The algorithm computes local gradient of error to be used in updating synaptic strength. Following the complex-valued backpropagation derivation explained in [5], we generalize real-valued BPTT to CV-BPTT. The algorithm is as follows:

1. Propagate the input signal through the network in the forward direction, layer by layer, then compute the error signal $e_j(n)$ for neuron j at the output layer.
2. For neuron j at the output layer, compute

$$\delta_j(n) = e_j(n)\left(\varphi'_j(n)\right)^H \tag{1}$$

$$\mathbf{w}_{ji}(n) = \mathbf{w}_{ji}(n) + \eta\delta_j(n)\mathbf{x}_i^H(n) \tag{2}$$

where $\mathbf{x}_i(n)$ is the state of synapse i of a hidden neuron connected to output neuron j, $\varphi'_j(n)$ is derivative of the activation function and $(\cdot)^H$ is Hermitian (complex-conjugate transpose).
3. For neuron j in a hidden layer, compute

$$\delta_j(n-lp) = \left(\varphi'\left(v_j(n-lp)\right)\right)^H \sum_{r \in A}\Delta_r^T(n-lp)\mathbf{w}_{rj} \tag{3}$$

$$\mathbf{w}_{ji}(n+1) = \mathbf{w}_{ji}(n) + \eta\delta_j(n-lp)\mathbf{x}_i^H(n-lp) \tag{4}$$

where $\Delta_r(n-p) = \left[\delta_r(n-p), \delta_r(n+1-p), ..., \delta_r(n)\right]^T$, p is the order of each synaptic FIR filter, the index l identifies the corresponding hidden layer, and A is a set of indices of neurons that contributes to forward-propagated error.

In the simulation experiment, the learning factor η is decayed as $\eta(n+1) = \eta(n)/(n+1)$ to achieve stability and convergence of the neural network.

3 Experiments and Preliminary Results

In the simulation, we demonstrate the applicability of ST-CVMLNN in UWB beamforming of a 5 element uniform linear array (ULA). The network consists of three layers with 25 tapped delay line in each neuron. The input layer consists of 5 neurons, the hidden layer has 3 neurons, 1 neuron at the output layer and we employ linier activation function. The UWB signal being used is a monocycle.

The system is trained to reject signals coming from $0°$ and steering its beams toward desired UWB signals coming from $40°$. Firstly, the weighting factors are set to unity, meaning that the array received strongly from $0°$. The resulting beampatterns are shown in Fig.4 (a) on a linear plot and (b) on a polar plot.

Next, the weights are randomized and the array is trained to reject interference and to steer beam direction. Initially, the learning speed is 1.25×10^{-3} and it is decayed during iterations. The resulting beampattern is displayed in Fig. 4 (c) on a linier plot and (d) on a polar plot. From these figures, we observe that the beampattern have been reformed, it placed nulls in $0°$ and maximum at around $30°$ and at $-20°$. Error curve in Fig.5 shows convergence of the system at around 10 iterations.

From the simulation results we observe that the null has been placed correctly, but the maximum beam is only approaching the intended direction at $40°$. Additionally, there is an extra maximum at $-20°$. Some further tuning is necessary to adjust the array so that it works properly.

4 Conclusions and Further Direction

Preliminary construction, simulation and results for a spatio-temporal complex-valued multilayer neural network for beamforming the UWB signal have been

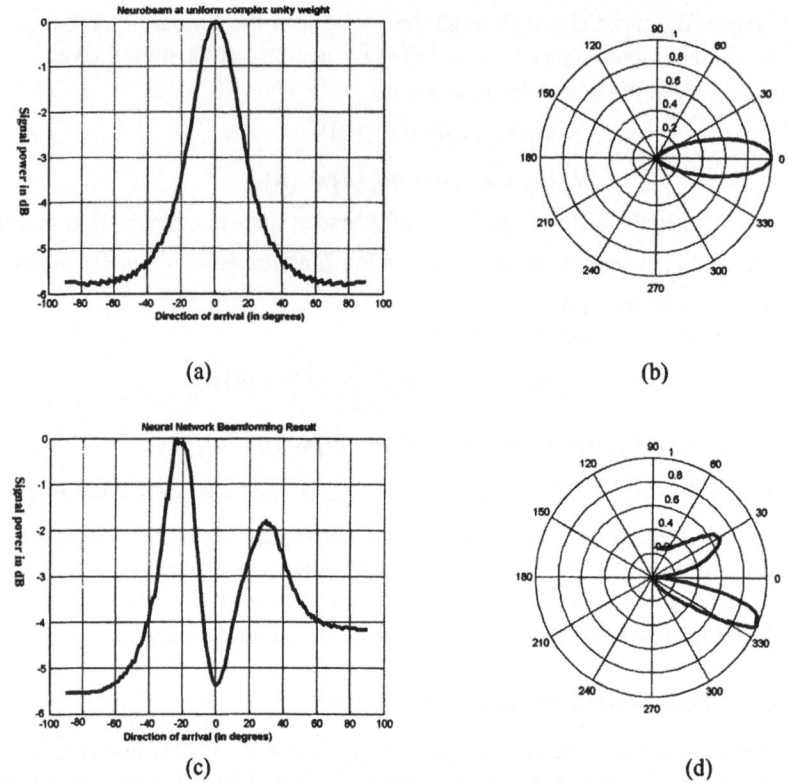

(a)   (b)

(c)   (d)

Fig. 4. Beampatterns of a 5 elements uniform linear array (ULA) utilizing ST-CVMLNN before (a,b) and after (c,d) beamforming. The UWB interfering signal comes from $0°$, while the desired one comes from $40°$. In (b) and (d) beampatterns are shown in polar plots format

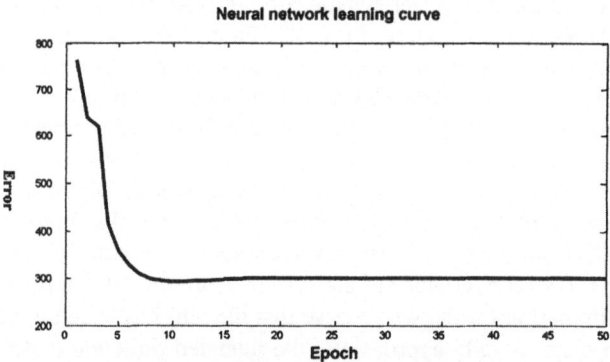

Fig. 5. Error curve during CV-BPTT training showing convergence at around 10 iterations. The learning speed η is decayed during iterations for stability

presented. It is found that the UWB interferer has been suppressed successfully, although the maximum beam is only approaching the desired direction. In the forthcoming works, more improvement will be done.

Acknowledgement

The monocycle signal was generated and captured by the IRCTR, TU-Delft. The authors are indebted for a permission of using the signal.

References

1. A.B. Suksmono & A. Hirose, "Adaptive beamforming by using complex-valued multilayer perceptron", *LNCS*, Vol. 2714/2003, Springer-Verlag, (Proc. ICANN/ICONIP-2003, pp. 959-966.
2. B. Widrow, P.E. Mantey, L.J. Griffiths, and B.B. Goode, "Adaptive Antenna Systems," *Proc. of the IEEE*, No. 55, 1967, pp. 2143-2159.
3. S. Haykin, *Neural Network: A Comprehensive Foundation*, Prentice Hall International.
4. P.J. Werbos, "Backpropagation Through Time: What It does and How to Do It," *Proc. Of the IEEE*, Vo. 78, No. 10, October 1990, pp. 1550-1560.
5. A. Hirose, "Coherent neural networks and their applications to control and signal processing," in "Soft Comp. in Syst. and Control Tech.", S.G. Tzafestas, ed., *Series in Robotics and Intel. Syst.*,WSPC, 1999.
6. F. Anderson, W. Christiensen, L. Fullerton and B. Kortegaard, "Ultra-wideband beamforming in sparse arrays," *IEE Procedings-H*, Vol. 138, No.4, Aug. 1991, pp.342-346.

A Model of Hopfield-Type Quaternion Neural Networks and Its Energy Function

Mitsuo Yoshida[1], Yasuaki Kuroe[2], and Takehiro Mori[1]

[1] Department of Electronics and Information Science,
Kyoto Institute of Technology
Matsugasaki, Sakyo-ku, Kyoto 606-8585, Japan
yoshid7m@ics.dj.kit.ac.jp, mori@dj.kit.ac.jp
[2] Center for Information Science,
Kyoto Institute of Technology
Matsugasaki, Sakyo-ku, Kyoto 606-8585, Japan
kuroe@dj.kit.ac.jp

Abstract. Recently models of neural networks that can directly deal with complex numbers, complex-valued neural networks, have been proposed and several studies on their abilities of information processing have been done. Furthermore models of neural networks that can deal with quaternion numbers, which is an extension of complex numbers, have also been proposed. However they are all multilayer quaternion neural networks. This paper proposes a model of recurrent quaternion neural networks, Hopfield-type quaternion neural networks. We investigate dynamics of the proposed model from the point of view of the existence of an energy function and derive its condition.

1 Introduction

In recent years, there have been increasing research interests of artificial neural networks and many efforts have been made on applications of neural networks to various fields. Furthermore models of neural networks that can deal with multidimensional data, complex neural networks and quaternion neural networks, have been proposed and their abilities of information processing have been investigated. In complex domain, the models of recurrent type complex-valued neural networks have been proposed and investigated their dynamics based on its energy function [1, 2]. In quaternion domain, the models of quaternion neural networks investigated are all multilayer quaternion neural networks [3–5].

In this paper we propose a class of recurrent type quaternion neural networks which are a quaternion-valued extension of the Hopfield-type neural networks [6], and investigate dynamics of the proposed model from the point of view of the existence of an energy function and derive its condition.

A quaternion number is defined by

$$x = x^{(0)} + i x^{(1)} + j x^{(2)} + k x^{(3)} \tag{1}$$

where $x^{(0)}, x^{(1)}, x^{(2)}, x^{(3)}$ are real numbers, $\{i, j, k\}$ are imaginary units for which the following relations hold.

N.R. Pal et al. (Eds.): ICONIP 2004, LNCS 3316, pp. 110–115, 2004.
© Springer-Verlag Berlin Heidelberg 2004

$$i^2 = -1, \quad j^2 = -1, \quad k^2 = -1,$$
$$ij = -ji = k, \quad jk = -kj = i, \quad ki = -ik = j. \tag{2}$$

The multiplication of quaternion is performed according to the above relations. Note that quaternion numbers are non-commutative on multiplication, that is $xy \neq yx$. The quaternion conjugate is defined by $x^* = x^{(0)} - ix^{(1)} - jx^{(2)} - kx^{(3)}$. Using this, the real part of x, that is $x^{(0)}$, is represented by $\mathcal{R}e(x) = \frac{1}{2}(x + x^*)$ and the vector part of x, that is $(ix^{(1)} + jx^{(2)} + kx^{(3)})$, is represented by $\mathcal{V}e(x) = \frac{1}{2}(x - x^*)$. The norm of quaternion number is defined by $|x|^2 = x^*x = xx^* = x^{(0)^2} + x^{(1)^2} + x^{(2)^2} + x^{(3)^2}$. The multiplicative commutative property of conjugate products holds: $(xy)^* = y^*x^*$. In the following, the set of quaternion (real) numbers is denoted by $\mathbb{H}$ $(\mathbb{R})$. The n-dimensional quaternion (real) space is denoted by $\mathbb{H}^n$ $(\mathbb{R}^n)$ and the set of $n \times m$ quaternion (real) matrices is denoted by $\mathbb{H}^{n \times m}$ $(\mathbb{R}^{n \times m})$. For $A \in \mathbb{H}^{n \times m}$ $(a \in \mathbb{H}^n)$, A^t (a^t) denotes the transpose of A (a). $\{a_{pq}\} \in \mathbb{H}^{n \times m}$ $(\mathbb{R}^{n \times m})$ denotes the $n \times m$ quaternion (real) matrix whose (p, q) element is denoted by $a_{pq} \in \mathbb{H}$ $(\mathbb{R})$.

2   Model of Hopfield-Type Quaternion Neural Networks

We consider a class of quaternion neural networks, that is a quaternion-valued extension of the Hopfield neural networks, described by differential equations of the form:

$$\begin{cases} \tau_p \dfrac{du_p}{dt} = -u_p + \displaystyle\sum_{q=1}^{n} w_{pq} v_q + b_p \\[2mm] v_p = f(u_p) \end{cases} \quad (p = 1, 2, \cdots, n) \tag{3}$$

where n is the number of neurons, $\tau_p \in \mathbb{R}$ is the time constant of the pth neuron (positive real number), $u_p \in \mathbb{H}$, $v_p \in \mathbb{H}$ are the state and the output of the pth neuron at time t, respectively, $b_p \in \mathbb{H}$ is the constant input, $w_{pq} \in \mathbb{H}$ is a connection weight from the qth neuron to the pth neuron, $f(\cdot)$ is an activation function which is a nonlinear quaternion-valued function from $\mathbb{H}$ into $\mathbb{H}$, and $du_p/dt := du_p^{(0)}/dt + i\,(du_p^{(1)}/dt) + j\,(du_p^{(2)}/dt) + k\,(du_p^{(3)}/dt)$.

In the real-valued neural networks, an activation function $f(\cdot)$ is usually chosen to be a smooth and bounded function such as sigmoidal functions. In the quaternion region, let us express an activation function $f : \mathbb{H} \to \mathbb{H}$ as:

$$f(u) = f^{(0)}(u^{(0)}, u^{(1)}, u^{(2)}, u^{(3)}) + i\, f^{(1)}(u^{(0)}, u^{(1)}, u^{(2)}, u^{(3)})$$
$$+ j\, f^{(2)}(u^{(0)}, u^{(1)}, u^{(2)}, u^{(3)}) + k\, f^{(3)}(u^{(0)}, u^{(1)}, u^{(2)}, u^{(3)}) \tag{4}$$

where $f^{(l)} : \mathbb{R}^4 \to \mathbb{R}$ $(l = 0, 1, 2, 3)$. We assume the following conditions corresponding to the above situations as an activation function of quaternion neuron.

Assumption 1

(i) $f^{(l)}(\cdot)$ are continuously differentiable with respect to $u^{(m)}$ $(l, m = 0, 1, 2, 3)$.
(ii) There exists some $M > 0$ such that $|f(\cdot)| \leq M$.

With this assumption, we can define the Jacobi matrix of an activation function f at $u = u_p$, denoted by $\boldsymbol{J}_f(u_p) = \{\alpha_{lm}(u_p)\} \in \mathbb{R}^{4 \times 4}$ as

$$\alpha_{lm}(u_p) = \left. \frac{\partial f^{(l)}}{\partial u^{(m)}} \right|_{u=u_p} \tag{5}$$

In order to write (3) in an abbreviated form, we define vectors $\boldsymbol{u} \in \mathbb{H}^n$, $\boldsymbol{v} \in \mathbb{H}^n$, $\boldsymbol{b} \in \mathbb{H}^n$ and $\boldsymbol{f} : \mathbb{H}^n \to \mathbb{H}^n$ by $\boldsymbol{u} := (u_1, u_2, \cdots, u_n)^t$, $\boldsymbol{v} := (v_1, v_2, \cdots, v_n)^t$, $\boldsymbol{b} := (b_1, b_2, \cdots, b_n)^t$ and $\boldsymbol{f}(\boldsymbol{u}) := (f(u_1), f(u_2), \cdots, f(u_n))^t$, respectively, and matrices $\boldsymbol{T} \in \boldsymbol{R}^{n \times n}$ and $\boldsymbol{W} \in \mathbb{H}^{n \times n}$ by $\boldsymbol{T} = \mathrm{diag}(\tau_1, \tau_2, \cdots, \tau_n)$ and $\boldsymbol{W} = \{w_{pq}\}$. With this notation, the model of quaternion neural networks (3) is rewritten as

$$\begin{cases} \boldsymbol{T}\dfrac{d\boldsymbol{u}}{dt} = -\boldsymbol{u} + \boldsymbol{W}\boldsymbol{v} + \boldsymbol{b} \\ \boldsymbol{v} \ = \boldsymbol{f}(\boldsymbol{u}) \end{cases} \tag{6}$$

3   Existence Conditions of Energy Function

We define an energy function for (6) on the analogy of that for the real-valued Hopfield neural networks as follows.

Definition 1. $E(\boldsymbol{v})$ *is an energy function of the quaternion-valued neural network* (6) *if* $E(\boldsymbol{v})$ *is a mapping* $E : \mathbb{H}^n \to \mathbb{R}$ *and the derivative of* E *along the trajectories of* (6), *denoted by* $\left.\frac{dE(\boldsymbol{v})}{dt}\right|_{(6)}$, *satisfies* $\left.\frac{dE(\boldsymbol{v})}{dt}\right|_{(6)} \le 0$. *Furthermore* $\left.\frac{dE(\boldsymbol{v})}{dt}\right|_{(6)} = 0$ *if and only if* $\frac{d\boldsymbol{v}}{dt} = 0$.

We assume the following conditions on the weight matrix and the activation functions of (6).

Assumption 2

   (i) *The connection weights matrix* $\boldsymbol{W}$ *satisfies* $w_{qp} = w_{pq}^*$.
   (ii) *The activation function* f *is injective function.*
   (iii) $\boldsymbol{J}_f(u_p)$ *is symmetric matrix for all* $u_p \in \mathbb{H}$.
   (iv) $\boldsymbol{J}_f(u_p)$ *is positive definite matrix for all* $u_p \in \mathbb{H}$.

Under the condition (ii) of Assumption 1 and the condition (ii) of Assumption 2, there exists the inverse function of $f : f(\mathbb{H}) \to \mathbb{H}$, and we express this as $u = g(v)$ where

$$\begin{aligned} g(v) &= g^{(0)}(v^{(0)}, v^{(1)}, v^{(2)}, v^{(3)}) + \boldsymbol{i}\, g^{(1)}(v^{(0)}, v^{(1)}, v^{(2)}, v^{(3)}) \\ &\quad + \boldsymbol{j}\, g^{(2)}(v^{(0)}, v^{(1)}, v^{(2)}, v^{(3)}) + \boldsymbol{k}\, g^{(3)}(v^{(0)}, v^{(1)}, v^{(2)}, v^{(3)}). \end{aligned} \tag{7}$$

Then, the following lemma on $g(v)$ holds.

Lemma 1. *If* f *satisfies the conditions* $(ii) - (iv)$ *of Assumption 2, there exists a function* $G(v^{(0)}, v^{(1)}, v^{(3)}, v^{(3)}) : \mathbb{R} \times \mathbb{R} \times \mathbb{R} \times \mathbb{R} \to \mathbb{R}$, *such that*

$$\frac{\partial G}{\partial v^{(l)}} = g^{(l)}(v^{(0)}, v^{(1)}, v^{(2)}, v^{(3)}) \qquad (l = 0, 1, 2, 3). \tag{8}$$

Proof. We define the Jacobi matrix of g at $v = v_p$ as $\boldsymbol{J}_g(v_p) = \{\beta_{lm}\} \in \mathbb{R}^{4\times 4}$ where $\beta_{lm} = \partial g^{(l)}/\partial v^{(m)}|_{v=v_p}$. By partially differentiating both sides of the equations $u^{(l)} = g^{(l)}(v^{(0)}, v^{(1)}, v^{(2)}, v^{(3)})$ with respect to $u^{(m)}$ $(l, m = 0, 1, 2, 3)$, respectively, the relation $\boldsymbol{I} = \boldsymbol{J}_g(v_p)\boldsymbol{J}_f(u_p)$ is obtained for all $u = u_p$, where $\boldsymbol{I} \in \mathbb{R}^4$ is identity matrix. From the condition $(iii)(iv)$ of Assumption 2 and this relation, it holds that $\boldsymbol{J}_g(v_p) = \{\boldsymbol{J}_f(u_p)^{-1}\}^t = \boldsymbol{J}_g(v_p)^t$, hence,

$$\frac{\partial g^{(l)}}{\partial v^{(m)}}(v_p) = \frac{\partial g^{(m)}}{\partial v^{(l)}}(v_p) \qquad (l, m = 0, 1, 2, 3) \tag{9}$$

hold for all $v_p \in \mathbb{H}$. Let define a function G:

$$G(v^{(0)}, v^{(1)}, v^{(3)}, v^{(3)})$$

$$:= \int_0^{v^{(0)}} g^{(0)}(\rho, 0, 0, 0)d\rho + \int_0^{v^{(1)}} g^{(1)}(v^{(0)}, \rho, 0, 0)d\rho$$

$$+ \int_0^{v^{(2)}} g^{(2)}(v^{(0)}, v^{(1)}, \rho, 0)d\rho + \int_0^{v^{(3)}} g^{(3)}(v^{(0)}, v^{(1)}, v^{(2)}, \rho)d\rho. \tag{10}$$

Using equations (9), the function G satisfies equations (8).   □

The following theorem shows the existence of an energy function for the quaternion neural network (6) under the previous assumptions. In the proof of this theorem, we will propose an energy function for (6) and prove that this function satisfies Definition 1 using the above lemma.

Theorem 1. *Suppose that the quaternion neural network (6) satisfies Assumption 1. If the weight matrix $\boldsymbol{W}$ and the activation functions satisfy Assumption 2, then there exists an energy function for (6).*

Proof. The proposed function E of (6) is as follows:

$$E(v) = -\sum_{p=1}^n \sum_{q=1}^n \left\{ \frac{1}{2} v_p^* w_{pq} v_q + \mathcal{R}e\left(b_p^* v_p\right) - G(v_p^{(0)}, v_p^{(1)}, v_p^{(3)}, v_p^{(3)}) \right\} \tag{11}$$

where $G(\cdot)$ is defined by equation (10). ¿From the condition (i) of Assumption 2, the equations $(v_p^* w_{pq} v_q + v_q^* w_{qp} v_p) = 2\mathcal{R}e(v_p^* w_{pq} v_q)$ hold, then the value of function E is always real number, that is $E : \mathbb{H}^n \to \mathbb{R}$.

Let us define the gradient operator in the quaternion domain as:

$$\nabla_{v_p}^{\mathbb{H}} E(v) = \frac{\partial E}{\partial v_p^{(0)}} + i \frac{\partial E}{\partial v_p^{(1)}} + j \frac{\partial E}{\partial v_p^{(2)}} + k \frac{\partial E}{\partial v_p^{(3)}}, \tag{12}$$

which are calculated as

$$\nabla_{v_p}^{\mathbb{H}} E(v) = -\left(\sum_{q=1}^n w_{pq} v_q + b_p - u_p \right) \tag{13}$$

Hence the derivative of energy function (11) along the trajectories of (6) is calculated as follows.

$$
\begin{aligned}
\frac{dE(\boldsymbol{v})}{dt}\bigg|_{(6)} &= \sum_{p=1}^{n}\sum_{l=0}^{3}\frac{\partial E}{\partial v_p^{(l)}}\frac{dv_p^{(l)}}{dt} = \sum_{p=1}^{n}\mathcal{R}e\left\{\nabla_{v_p}^{\mathrm{H}}E(\boldsymbol{v})^*\frac{dv_p}{dt}\right\}\\
&= \sum_{p=1}^{n}\mathcal{R}e\left\{-\left(\sum_{q=1}^{n}w_{pq}v_q + b_p - u_p\right)^*\frac{dv_p}{dt}\right\}\\
&= -\sum_{p=1}^{n}\mathcal{R}e\left\{\left(\frac{du_p}{dt}\right)^*\tau_p\left(\frac{dv_p}{dt}\right)\right\}\\
&= -\sum_{p=1}^{n}\left(\frac{d\hat{v}_p}{dt}\right)^t\tau_p\boldsymbol{J}_g(v_p)^t\left(\frac{d\hat{v}_p}{dt}\right) \leq 0
\end{aligned}
\tag{14}
$$

where $\hat{v}_p = (v_p^{(0)}, v_p^{(1)}, v_p^{(2)}, v_p^{(3)})^t \in \mathbb{R}^4$, Since $\tau_p > 0$ and $\boldsymbol{J}_g(v_p)$ are positive definite for any v_p (for $p = 1, 2, \cdots, n$), function (11) satisfies the required conditions for energy function (Definition 1). Furthermore, $\frac{dE(\boldsymbol{v})}{dt}\big|_{(6)} = 0$ if and only if $\frac{d\hat{v}}{dt} = 0 \Leftrightarrow \frac{d\boldsymbol{v}}{dt} = 0$. $\square$

Note that as the functions which satisfy Assumption 1 and 2, for instance, $f(u) = \tanh(u^{(0)}) + i\tanh(u^{(1)}) + j\tanh(u^{(2)}) + k\tanh(u^{(3)})$ and $f(u) = u/(1+|u|)$ can be considered. For all the quaternion parameters of (6), when the coefficients of all or two imaginary units, (i, j, k) or (j, k), are 0, (11) is reduced to the energy function for the real-valued [6] or complex-valued [1] neural networks. Therefore Theorem 1 is an extension for the results of those networks.

4   Qualitative Analysis of Quaternion Neural Networks Based on Energy Function

The qualitative behavior of the quaternion neural network (6) can be studied via the derived energy function. A point $\tilde{v} \in \mathbb{H}^n$ is an equilibrium point if and only if $\nabla_{v_p}^{\mathrm{H}}E(\tilde{v}) = 0$ $(p = 1, 2, \cdots, n)$, that is, the stationary points of the energy function coincide with the equilibrium points of (6). We define the Hessian matrix of the energy function by $\boldsymbol{H}(\boldsymbol{v}) = \{\boldsymbol{\Pi}_{pq}(\boldsymbol{v})\} \in \mathbb{R}^{4n \times 4n}$ as

$$
\boldsymbol{\Pi}_{pq}(\boldsymbol{v}) = \{\Gamma_{lm}^{pq}(\boldsymbol{v})\} \in \mathbb{R}^{4\times 4}, \quad \Gamma_{lm}^{pq}(\boldsymbol{v}) = \left\{\frac{\partial^2 E}{\partial v_p^{(l)}v_q^{(m)}}(\boldsymbol{v})\right\} \in \mathbb{R}
\tag{15}
$$

which is calculated as $\boldsymbol{H}(\boldsymbol{v}) = -\{\hat{\boldsymbol{W}}_{pq}\} + \mathrm{blockdiag}\{\boldsymbol{J}_g(v_p)\}$ where $\hat{\boldsymbol{W}}_{pq}$ given by

$$
\hat{\boldsymbol{W}}_{pq} := \begin{pmatrix}
w_{pq}^{(0)} & -w_{pq}^{(1)} & -w_{pq}^{(2)} & -w_{pq}^{(3)}\\
w_{pq}^{(1)} & w_{pq}^{(0)} & -w_{pq}^{(3)} & w_{pq}^{(2)}\\
w_{pq}^{(2)} & w_{pq}^{(3)} & w_{pq}^{(0)} & -w_{pq}^{(1)}\\
w_{pq}^{(3)} & -w_{pq}^{(2)} & w_{pq}^{(1)} & w_{pq}^{(0)}
\end{pmatrix} \in \mathbb{R}^{4\times 4} \quad (p, q = 1, 2, \cdots, n).
\tag{16}
$$

The following theorems about the equilibrium points of (6) will help to consider their applications.

Theorem 2. *Suppose that the quaternion neural network* (6) *satisfies Assumption* 1 *and* 2. *If there is no* $v \in \mathbb{H}^n$ *satisfying* $\nabla_{v_p}^{\mathrm{H}} E(v) = 0$ $(p = 1, 2, \cdots, n)$ *and* $\det(H(v)) = 0$, *simultaneously, then there no nontrivial periodic solutions exist and each non-equilibrium solution converges to an equilibrium point of* (6) *as* $t \rightarrow \infty$.

Theorem 3. *Suppose that the quaternion neural network* (6) *satisfies Assumption* 1 *and* 2. *A point* $\tilde{v} \in \mathbb{H}^n$ *is an asymptotically stable equilibrium point if and only if* $\nabla_{v_p}^{\mathrm{H}} E(\tilde{v}) = 0$ $(p = 1, 2, \cdots, n)$ *and* $H(\tilde{v}) > 0$.

Those theorems are derived based on the results of reference [7], identifying the quaternion neural networks with $4n$ dimensional real-valued neural networks.

5 Conclusion

In this paper we proposed a model of recurrent type quaternion neural networks and investigated existence conditions of energy function for the network. The existence conditions include those of real-valued or complex-valued neural networks. Furthermore, we considered for qualitative behavior of the network that was derived from the analysis based on the energy function. Those results are obtained along the line of authors' previous study for complex-valued recurrent neural networks. Further work is underway on possibilities of their application.

References

1. Y. Kuroe, N. Hashimoto, "On energy function for complex-valued Hopfield-type neural networks," KES 2002, E. Damiani et al. ed., pp. 623–627, IOS Press, 2002.
2. Y. Kuroe, M. Yoshida and T. Mori, "On activation functions for complex-valued neural networks – existence of energy functions –," in LNCS 2714, ICANN/ICONIP 2003, O. Kaynak et al. ed., pp. 985–992, Springer-Verlag, 2003.
3. T.Nitta, "A quaternary version of the back-propagation algorithm," Proc. IEEE Int. Conf. on Neural Networks, vol. 5, pp. 2753–2756, Perth, Australia, 1995.
4. P. Arena, L. Fortuna, G. Muscato and M. G. Xibilia, "Multilayer perceptrons to approximate quaternion valued functions," Neural Networks, vol. 10, no. 2, pp. 335–342, 1997.
5. T. Isokawa, T. Kusakabe, N. Matsui and F. Peper, "Quaternion neural network and its application," in LNAI 2774, KES2003, V. Palade et al. ed., pp. 318–324, Springer-Verlag, 2003.
6. J. J. Hopfield, "Neurons with graded response have collective computational properties like those of two-state neurons," Proc. Natl. Acad. Sci. USA, Vol. 81, pp. 3088–3092, 1984.
7. J. H. Li, A. N. Michel and W. Porod, "Qualitative Analysis and Synthesis of a Class of Neural Networks," IEEE Trans. on Circuits and Systems, Vol. 35, No. 8, pp. 976–986, 1988.
8. I. L. Kantor, A. S. Solodovnikov, "Hypercomplex numbers," Springer-Verlag, 1989.

Mode-Utilizing Developmental Learning Based on Coherent Neural Networks

Akira Hirose[1,2], Yasufumi Asano[2], and Toshihiko Hamano[1]

[1] Department of Frontier Informatics,
The University of Tokyo, 7-3-1 Hongo, Bunkyo-ku, Tokyo 113-8656, Japan
ahirose@ee.t.u-tokyo.ac.jp
http://www.eis.t.u-tokyo.ac.jp
[2] Department of Electronic Engineering,
The University of Tokyo, 7-3-1 Hongo, Bunkyo-ku, Tokyo 113-8656, Japan

Abstract. We propose a mode-utilizing developmental learning method. Thereby a system possesses a mode parameter and learns similar or advanced tasks incrementally by using its cumulative skill. We construct the system based on the coherent neural network where we choose its carrier frequency as the mode parameter. In this demonstration, we assume two tasks: basic and advanced. The first is to ride a bicycle as long as the system can before it falls. The second is to ride as far as possible. It is demonstrated that the system finds self-organizingly a suitable value of the mode parameter in the second task learning. The learning is performed efficiently to succeed in riding for a long distance.

1   Introduction

There have been several researches reported to realize modulated or context-dependent behavior by preparing internal state in neural networks. They include, for example, the PATON concerning association tasks [1],[2] and the switched multiple network module systems where change their mode by using a mode-dependent switch or a weighting factor for modules [3],[4]. One of the most meaningful evaluations of the behavioral modulation performance is the flexibility to environmental demands. Another is the generalization characteristic of the behavioral modulation versus the variation of the modulation key parameters.

Previously we proposed a behavioral modulation in an lightwave coherent neural network [5]. We utilize its optical carrier frequency f as the mode parameter. Its advantages in general lie in the high degree of freedom in modulation and the flexibility in the generalization characteristics [6], [7]. These merits arise from the orthogonality and the completeness of the basis function against one of the neural connection variables τ, the 'delay time variable' [5]. In other words, the trigonometric function $\cos 2\pi f\tau + i \sin 2\pi f\tau = e^{i2\pi f\tau}$ yields the advantages.

This paper proposes a mode-utilizing developmental learning method based on the coherent neural networks by using the carrier frequency as the key parameter to change the behavioral mode. The system learns a similar but new task or a advanced one self-organizingly with a behavioral modulation. The network

N.R. Pal et al. (Eds.): ICONIP 2004, LNCS 3316, pp. 116–121, 2004.
© Springer-Verlag Berlin Heidelberg 2004

utilizes its cumulative experience to accelerate the learning process. We assume the following two tasks in the present demonstration. The first is to ride as long as the system can before it falls (basic task). The second is to ride as far as possible (advanced task). The mode key is found adjusted self-organizingly in the developmental learning. It is demonstrated that the learning is performed efficiently to show a successful long distance ride.

2   System Construction and Human-Bicycle Model

Network Construction: Figure 1 shows our complex-valued neuron where x_m, w_{nm}, y_n are complex-valued input signals, connection weights and output signals, respectively. The activation function is the amplitude-phase type sigmoid function that treats the amplitude and the phase separately as [8]

$$y_n = A \, \tanh \left(g \left| \sum_m w_{nm} x_m \right| \right) \exp \left(i \, \arg \left\{ \sum_m w_{nm} x_m \right\} \right) \quad (1)$$

where A and g stand for saturation amplitude and small signal gain, respectively.

The activation function transforms the amplitude in a saturation manner, just like the ordinary real-valued sigmoid function, while it leaves the phase unchanged. A complex-valued network works as a coherent neural network when it is fed with input signals that has a carrier frequency f (non-baseband signals). Thereby we express the neural weight w_{nm} by an attenuation / amplification factor $|w_{nm}|$ and a delay time τ_{nm} with the carrier frequency f as [8]

$$w_{nm}(f) = |w_{nm}| \exp[i2\pi f \tau_{nm}] \quad (2)$$

That is to say, the coherent network behavior, such as self-organization, learning and task processing, depends on the carrier frequency f. If we prepare various τ_{nm}, then w_{nm} has completeness and orthogonality. Therefore, by choosing f as the modulation key, we realize a more flexible behavioral modulation than when we with a set of conventional sigmoid network modules.

Figure 2 shows the single-layered and forward coherent network construction that interacts with human and bicycle. The variables are explained below.

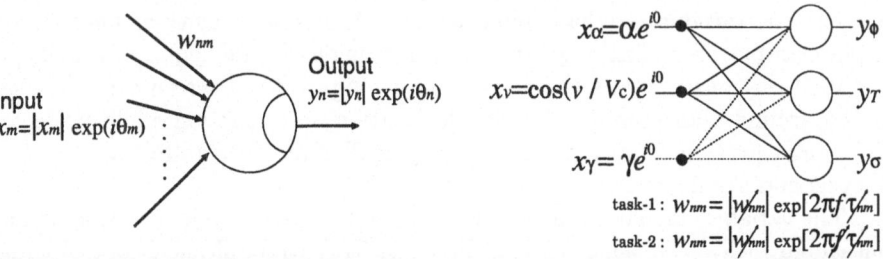

Fig. 1. Amplitude-phase-type complex-valued neuron.

Fig. 2. Network construction.

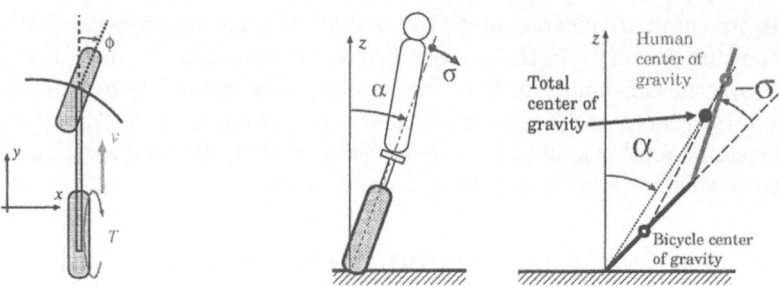

Fig. 3. Physical model and parameters of a person on a bicycle.

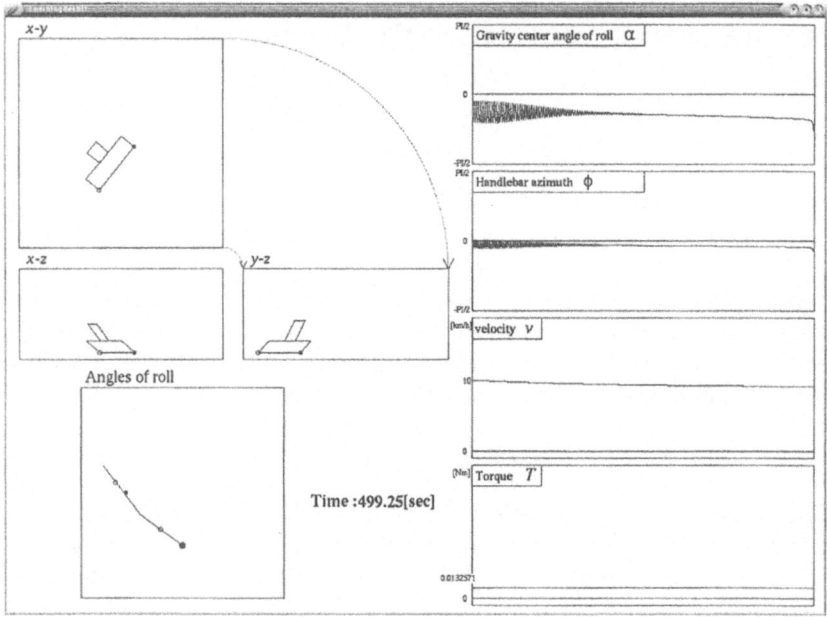

Fig. 4. Captured simulator screen.

Human and Bicycle Model: Figure 3 presents the human and bicycle physical model. The variables are handlebar azimuth ϕ, bicycle velocity v, wheel torque T, human-bicycle rolling angle σ and rolling angle of total center of gravity α.

We have constructed a physics simulator. Figure 4 is a screen capture where $x - y$ ground plan presents the bicycle location and direction while $y - z$ and $x - z$ are elevation. The angles-of-roll graphic illustrates the rolling angles of the bicycle and the human.

The variables are shown also in Fig.2 as well as the bicycle direction angle γ measured with a zero angle of y-axis direction. We extract physical control signals from the complex-valued network outputs as $\phi = \phi_c \cdot \mathbf{Im}[y_\phi]$, $T = T_c \cdot \mathbf{Re}[y_T]$, $\sigma = \sigma_c \cdot \mathbf{Im}[y_\sigma]$ where ϕ_c, T_c and σ_c are gain constants.

3   Experiment

There are two stages in the learning to ride a bicycle. First the network attempts random trials for a certain number of times. Secondly it selects the best trial set of weights as the initial state and starts a hill climbing with a random fractions $\Delta|w|$ and $\Delta\tau$, i.e., $|w_{nm}| \longleftarrow |w_{nm}| + \Delta|w|$ and $\tau_{nm} \longleftarrow \tau_{nm} + \Delta\tau$. These stages correspond to trials and errors that we, human being, really experience when we learn bicycle riding.

Task 1 (Basic Task): Riding as Long as Possible. In this mode, we do not use the bicycle direction information γ (blind condition). We assign this condition and state "Mode 1", and fix the carrier frequency f in (2) as $f_0 = 100[\text{Hz}]$.

Figure 5 gives a typical result: (a)Riding time t_R versus random trial steps, (b)that versus hill climbing learning steps after the random trial in (a) to choose the best weight set as the initial, and (c)riding locus after the hill climbing learning is completed. The aim to ride for a long time is achieved, but we find the locus is cyclic and the bicycle does not go forward.

Task 2 (Advanced Task): Ride as Far as Possible. We make the carrier frequency variable, which enables a self-organizing mode search. The system is fed with the bicycle direction information γ (seeing condition). The frequency self-organization is performed in the same manner as $f \longleftarrow f + \Delta f$ with a small fraction Δf. We call this condition and state "Mode 2" (meta-learning).

In Task 2, we choose a score function S to evaluate the achievement as $S = \sum_{t_i=0}^{t_R}(S_x(t_i) + S_y(t_i)) + C_{S \leftarrow t} \times t_R$ for discrete time steps (unit t_i=50[ms]) where S_x is higher when the bicycle is near to $x = x_0$ (initial x location) line, while S_y becomes higher as it go further. The constant $C_{S \leftarrow t}$ is a coefficient to convert the riding time to the present score scale. The detail will be given in the presentation.

Figure 6 presents the mode-utilizing developmental learning results. The system takes the best result in Task 1 as the initial state and, then, conducts a hill climbing learning for Task 2. We find in Fig.6(a) that the score S increases quickly and monotonically. Figure 6(b) shows the carrier frequency change which is equivalent to the behavioral mode change. Starting with f_0, the system finds

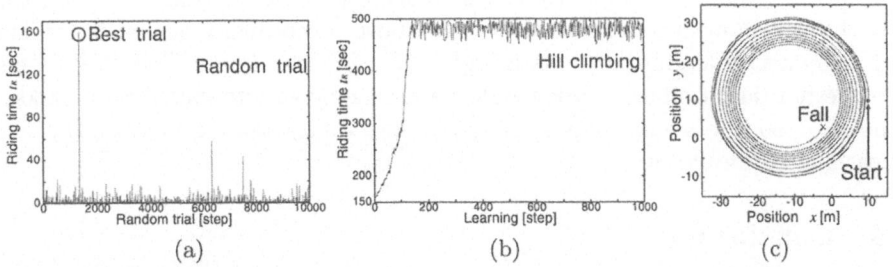

Fig. 5. Typical results: (a)Riding time t_R versus random trial steps, (b)that versus hill climbing learning steps for the best weight-set point in (a), and (c)riding locus after the learning is completed.

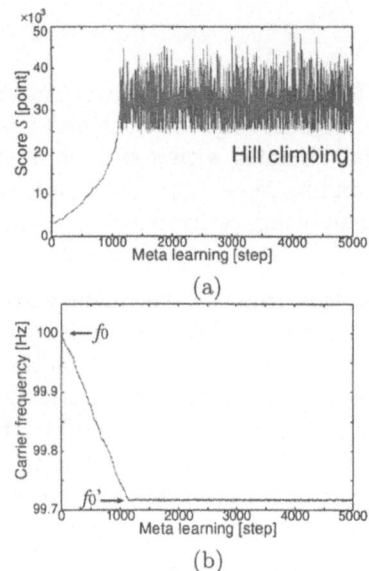

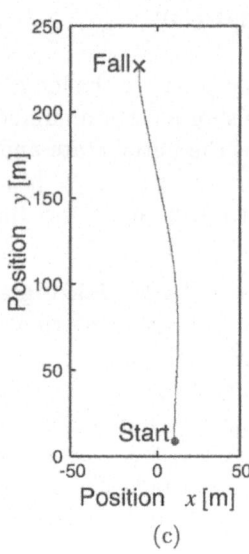

Fig. 6. Results of the mode-utilizing learning: Evolutions of (a) score S for each riding and (b)carrier frequency f, and (c) riding locus after the learning. (Score example: $\sum S_x = 7{,}900$, $\sum S_y = 2{,}400$, $C_{S \leftarrow t} \cdot t_R = 27{,}400$ which reflect the contribution ratio.)

the frequency f'_0 optimal for Task 2 self-organizingly. In 6(c), we find the locus is almost straight, resulting in a further riding. That is, the system learns quickly an optimal set of far riding weights by utilizing the mode and modulating the behavior by beginning with the initial weight set obtained for Task 1.

Comparison Experiments for Task 2: For comparison, Fig.7 shows typical results when we tried 5,000-times hill climbing after (a)1,000- or (b)10,000-times random trials, respectively. The direction information γ is fed to the system (seeing state, again).

In Fig.7, we find the following three facts. (1)On the random trial stage, the probability to obtain a high score is very small as it ever is. (2)On the hill climbing learning stage, we cannot find a score increase, which is very different from the result in Fig.6(a). (3)The final score is mostly much lower than that in the developmental learning mentioned above. Accordingly, the mode-utilizing developmental learning realizes a highly efficient learning in Task 2. It is also suggested that a difficult task should be broken down into incremental tasks, if possible, so that we can adopt the developmental learning to improve the total learning performance.

4   Conclusion

We have proposed the mode-utilizing developmental learning. The system is based on the coherent neural network with the carrier frequency as the mode-selecting key. We assumed two tasks to ride a bicycle. The first is to ride as long

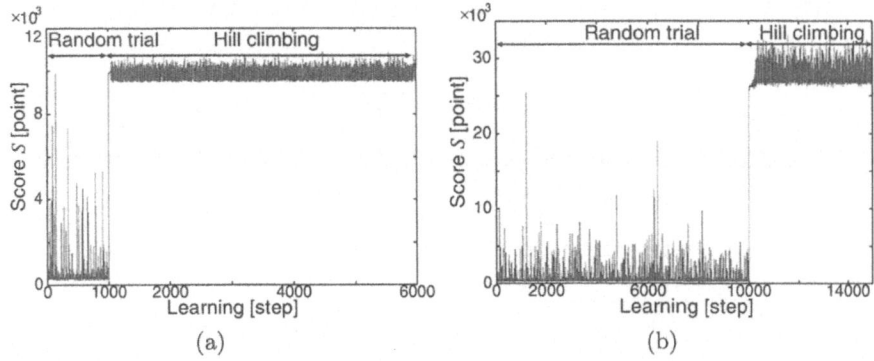

Fig. 7. Typical example of scores versus learning steps for best learning results after (a)1,000 or (b)10,000 random trials.

as the system can before it falls. The second is an advanced one, i.e., to ride as far as possible. The mode key has been found self-organizing in the developmental learning. It has been demonstrated that the total learning is performed efficiently to show a successful ride for a long distance.

Acknowledgment

The authors acknowledge the help of Sotaro Kawata in software preparation.

References

1. Omori, T., Mochizuki, A., Mizutani, K.: "Emergence of symbolic behavior from brain like memory with dynamic attention," Neural Networks, 12 (1999) 1157–1172
2. Omori, T., Mochizuki, A., "PATON: A model of context dependent memory access with an attention mechanism," Brain Processes, Theories and Models, MIT Press (1995) 134-143
3. Wolpert, D.M., Kawato, M., "Multiple paired fotward and inverse models for motor control," Neural Networks, 11 (1998) 1317–1329
4. Hartono, P., Hashimoto, S., "Temperature switching in neural network ensemble," J. Signal Processing, 4 (2000) 395-402
5. Hirose, A., Eckmiller, R., "Coherent optical neural networks that have optical-frequency-controlled behavior and generalization ability in the frequency domain" Appl. Opt., 35, 5 (1996) 836-843
6. Hirose, A., Tabata, C., Ishimaru, D., "Coherent neural network architecture realizing a self-organizing activeness mechanism," Proc. of Int'l Conf. on Knowledge-based Eng. Sys. (KES) 2001, (Sept. 6-8, 2001, Osaka) 576-580
7. Hirose, A., Ishimaru, D., "Context-dependent behavior of coherent neural systems based on self-organizing mapping of carrier frequency values," Proc. of Int'l Conf. on Knowledge-based Engineering Systems (KES) 2002 (Sept. 16-18, 2002, Crema) 638-642
8. Kawata, S., Hirose, A., "Coherent lightwave neural network systems," *in* "Complex-Valued Neural Networks: Theories and Appliactions,"* A.Hirose, ed., The Series on Innovative Intelligence, World Scientific Publishing Co., (2003)

Dynamics of Complex-Valued Neural Networks and Its Relation to a Phase Oscillator System

Ikuko Nishikawa[1] and Yasuaki Kuroe[2]

[1] College of Information Science and Engineering, Ritsumeikan University
Shiga, 525-8577, Japan
nishi@ci.ritsumei.ac.jp
[2] Center for Information Science, Kyoto Institute of Technology
Kyoto, 606-8585, Japan
kuroe@dj.kit.ac.jp

Abstract. A network of complex-valued rotating neurons is proposed and its dynamics is compared with that of a coupled system of phase oscillators. The dynamics of the phases of neurons is shown to be described as a system of the phase oscillators with a pair-wise sinusoidal coupling. The amplitudes affect the phase dynamics in a way to modulate the coupling strength. As the system of phase oscillators is known to be effective for an area-wide signal control of an urban traffic network, the complex-valued neural network is also useful to the signal control. The similarity and the difference caused by the existence of the amplitude are discussed from the point of view of the flow control.

1 Introduction

Various models of complex-valued neural network are recently studied[1], and one typical example is the fully connected neural network as the mathematical extension of an ordinary real-valued neural network such as Hopfield network[2]. In this paper, the dynamics of the complex-valued neural network is decomposed into the dynamics of the amplitude and the phase of each neuron. Then the dynamics of the phases is compared with a coupled system of phase oscillators, and the effect of the existence of the amplitudes is discussed.

In the followings, some fundamental characteristics of a system of phase oscillators are briefly reviewed in Section 2, and then a network of complex-valued rotating neurons is introduced in Section 3 and the relation to the phase oscillator is clarified. The latter half of the paper discusses the effectiveness of both systems from the viewpoint of the application to the traffic flow control. Section 4 explains the framework of the flow control by a decentralized system of signals, and points out the correspondence between the dynamics of a signal pair and a phase oscillator pair. The formula for the offset control is shown in Section 5, and the amplitude effect in the complex-valued neural network is discussed.

2 Entrainment in a Coupled System of Phase Oscillators

2.1 A Phase Oscillator

A phase oscillator is originally introduced to describe a periodic motion on a limit cycle in a dissipative dynamical system. Its dynamics is simply expressed

N.R. Pal et al. (Eds.): ICONIP 2004, LNCS 3316, pp. 122–129, 2004.
© Springer-Verlag Berlin Heidelberg 2004

by a non-dimensional phase valuable $\phi \in [0, 2\pi)$, which indicates a state on a limit cycle. A phase $\phi(t)$ at time t is given by a constant angular velocity motion with period T, by an appropriate definition of the phase according to the motion. Thus, the dynamics of $\phi(t)$ is written by the equation $d\phi/dt = \omega$, where $\omega = 2\pi/T$ is the frequency [3].

2.2   1 Pair of Interacting Phase Oscillators

Let us consider the following coupled system of 2 phase oscillators ϕ_1 and ϕ_2, where both dynamics are described by the above equation with each frequency ω_1 and ω_2, and their interaction is given by the phase difference at each time t;

$$\frac{d\phi_1}{dt} = \omega_1 + \Gamma(\phi_2 - \phi_1), \quad \frac{d\phi_2}{dt} = \omega_2 + \Gamma(\phi_1 - \phi_2) \tag{1}$$

$$\Gamma(\phi) = K \cdot \sin(\phi - \delta), \quad \text{where } K \text{ is a coupling coefficient} \tag{2}$$

As an arbitrary form of 2π-periodic function is possible for a coupling function $\Gamma(\phi)$, the simplest sinusoidal type of Eq.(2) has been often used for the mathematical analysis. Then the following results are known[3]. If the original frequencies for 2 systems are different ($\omega_1 \neq \omega_2$), then the frequency entrainment emerges according to a coupling strength $K > 0$. On the other hand, if the original frequencies are same ($\omega_1 = \omega_2$), then there is always a unique stable solution of the phase entrainment with a constant phase difference $\phi_1 - \phi_2$.

The examples below show some fundamental characteristics of the system (1) with the same frequency $\omega_1 = \omega_2 \equiv \omega$. A simple analysis shows that the following coupled system always possesses a unique stable solution $\Delta\phi_0$ for a phase difference $\Delta\phi = \phi_1 - \phi_2$.

$$\frac{d\phi_1}{dt} = \omega + K_1 \cdot \sin(\phi_2 - \phi_1 - \delta), \quad \frac{d\phi_2}{dt} = \omega + K_2 \cdot \sin(\phi_1 - \phi_2 - \delta) \tag{3}$$

$$\text{(where } K_1, K_2 > 0)$$

Especially,

– If an interaction is one-way, then $\Delta\phi_0 = \delta$ for $K_1 = 0$, while $\Delta\phi_0 = -\delta$ for $K_2 = 0$. This shows that the interaction term of ϕ_1 sets $\Delta\phi = -\delta$ as a target, while the interaction term of ϕ_2 sets $\Delta\phi = \delta$ as a target. In general, each interaction term works to attain the phase difference which makes an argument of $\sin(\cdot)$ be zero.
– If both interaction strengths are equal ($K_1 = K_2$), then

$$\Delta\phi_0 = 0 \text{ for } 0 \leq |\delta| < \pi/2, \quad \text{while } \Delta\phi_0 = \pi \text{ for } \pi/2 < |\delta| \leq \pi, \tag{4}$$

while $\forall \Delta\phi$ is neutral stable for $\delta = \pi/2$.

Thus, in general, the stable phase difference $\Delta\phi_0$ takes a value between $\Delta\phi = -\delta$ and δ, which are target phase differences for ϕ_1 and ϕ_2, respectively. $\Delta\phi_0$ changes continuously according to the coupling strengths K_1 and K_2. That is, if 2 coupling strengths are equal, then $\Delta\phi_0$ takes 0 or π as the middle of $-\delta$ and δ, and the closer one is chosen. If 2 coupling strengths differ, $\Delta\phi_0$ becomes closer to the target of the stronger, and in the limit of the strength difference where one is negligible, the stronger attains the target.

2.3   A System of Phase Oscillators with a Pair-Wise Interaction

A system of N phase oscillators with a pair-wise sinusoidal interaction are described by the following equation;

$$\frac{d\phi_i}{dt} = \omega_i + \sum_{j=i, j\neq i}^{N} K_{ij} \cdot \sin(\phi_j - \phi_i - \delta_{ij}), \quad i = 1, \ldots, N \tag{5}$$

(K_{ij} : a coupling coefficient from j to i)

As the special case, the previous discussion also directly leads to the following results on a phase oscillator system. Let a phase oscillator be located at each node of a 2-dimenstional square lattice, and interacts each other with adjacent oscillators by Eq.(3) with a common coupling strength K. Then, a unique stable steady state is whether a phase entrainment to the same phase for all oscillators, if $0 \le |\delta| < \pi/2$ (in-phase type), or a phase entrainment to an anti-phase, where all adjacent pairs have half a cycle difference, if $\pi/2 < |\delta| \le \pi$ (anti-phase type).

3   Network of Complex-Valued Rotating Neurons

3.1   Complex-Valued Neural Network

Several models of complex-valued neural networks have been proposed and studied in recent years. Among those researches, a class of fully connected complex-valued neural networks is investigated as a natural extension of the Hopfield-type neural networks to complex-valued neurons[2]. The dynamics is described by

$$\tau_i \frac{du_i}{dt} = -u_i + \sum_{j=1}^{N} w_{ij}x_j + \theta_i, \quad x_i = f(u_i), \quad i = 1, \ldots, N \tag{6}$$

where N is the number of neurons, $x_i \in C$, $u_i \in C$ and $\theta_i \in C$ are the output, the internal state and the external input of the i-th nrueon, respectively. $\tau_i(> 0) \in R$ is the time constant, and $w_{ij} \in C$ is the connection weight from the j-th neuron to the i-th neuron. The activation function $f(\cdot)$ is common to all neurons and is a nonlinear complex function; $f : C \to C$. In the followings, let $\forall i, \theta_i = 0$ for the simplisity.

3.2   Complex-Valued Rotating Neuron

Let us review the motion of a single neuron, which is described by a first order linear differential equation $du_i/dt = -\alpha \cdot u_i$, according to the 3 cases of α value.

1. **A real α:** Each neuron i is described by $\tau_i du_i/dt = -u_i, \tau_i \in R$. The solution is $u_i(t) = u_i(0)\exp(-t/\tau_i)$, with a unique stable point $u(t) = 0$ if $\tau_i > 0$.
2. **A purely imaginary α:** The equation is $du_i/dt = \imath\omega_i \cdot u_i$, where $\imath$ is an imaginary unit and $\omega_i \in R$. The solition is $u_i(t) = u_i(0)\exp(\imath\omega_i \cdot t)$, which has a neutrally stable periodic solution with a frequency ω_i.
3. **A general complex-value α:** The equation is $du_i/dt = (-1/\tau_i + \imath\omega_i) \cdot u_i$, and the solution is $u_i(t) = u_i(0)\exp\{(-1/\tau_i + \imath\omega_i) \cdot t\}$, with a unique stable point $u(t) = 0$ if $\tau_i > 0$. This case is led to the case 1 by a transformation $u_i(t) \to u_i{}'(t) = u_i(t)\exp(-\imath\omega_i t)$, which depends on ω_i.

3.3   Decomposition into the Dynamics of Phase and Amplitude

Let us consider the fully connected network of the case 3 neurons, to explicitly consider the rotating motion of a single neuron. Namely, we propose the dynamics;

$$\frac{du_i}{dt} = (-\frac{1}{\tau_i} + \imath \omega_i)u_i + \sum_{j=1}^{N} w_{ij}x_j, \quad x_i = f(u_i), \quad i = 1,\ldots,N \qquad (7)$$

as a network of the compled-valued rotating neurons. However, in the rest of the paper we only consider a homogeneous network with $\forall i, \tau_i = 1, \omega_i = \omega$, which is led to Eq.(6) under the following restriction (8) on the activation function $f(\cdot)$.

First, let us rewrite the complex variables $u_i(t)$ and w_{ij} by the amplitude and the phase, namely, $u_i(t) = r_i(t)\exp(\imath \phi_i(t))$ and $w_{ij} = \kappa_{ij}\exp(-\imath \delta_{ij})$.

Next, we restrict the form of the activation function $f(\cdot)$ to the following;

$$f(u_i(t)) = f_R(r_i(t)) \cdot \exp(\imath \phi_i(t)) \qquad (8)$$

$f_R(\cdot)$ is a nonlinear real function and here we assume $f_R : R_+ \to R_+$, where $R_+ = \{r \geq 0, r \in R\}$. Moreover, it is proved in [2] that when $f_R(r)$ is bounded and continuously differentiable with respect to $r \in R_+$, then a sufficient condition for the existence of an energy function of the network is

$$\forall r \in R_+, \frac{df_R(r)}{dr} > 0 \quad \text{and} \quad \lim_{r \to 0} \frac{f_R(r)}{r} > 0 \qquad (9)$$

with the condition on the weight matrix $W = (w_{ij})$ that W be Hermitian:

$$W = W^* \qquad (10)$$

where $A^* = \overline{A^t}$ is a conjugate transposed matrix. Eqs.(9) and (10) directly correspond to the existence conditions of the energy function in the real-valued Hopfiled network that $f(\cdot)$ should be bounded, continuously differentiable and monotonically increasing, and that W is symmetric, $W = W^t$. We assume $\forall i, w_{ii} = 0$ for the simplicity in the following discussion.

Then, the real and imaginary parts of Eq.(7) are decomposed into the following 2 equations (11) for the amplitude $r_i(t)$, and (12) for the phase $\phi_i(t)$.

$$\frac{dr_i(t)}{dt} = -r_i(t) + \sum_{j=1,j\neq i}^{N} \kappa_{ij}f_R(r_j(t)) \cdot \cos\big(\phi_j(t) - \phi_i(t) - \delta_{ij}\big) \qquad (11)$$

$$\frac{d\phi_i(t)}{dt} = \omega + \frac{1}{r_i(t)} \sum_{j=1,j\neq i}^{N} \kappa_{ij}f_R(r_j(t)) \cdot \sin\big(\phi_j(t) - \phi_i(t) - \delta_{ij}\big) \qquad (12)$$

$$i = 1,\ldots,N$$

Eq.(12) takes the same form as Eq.(5) with the homogeneous frequency $\forall i, \omega_i = \omega$. Only difference is that the coupling coefficient κ_{ij} is multiplied by $f_R(r_j)/r_i$ as the effect of the amplitudes.

As is explained in 2.2, each interaction term in Eq.(12) works to attain the phase difference $\phi_j - \phi_i$ which makes the argument of $\sin(\cdot)$ be zero. Therefore, if the coupled system (11),(12) converges to any stable solution $r_i^*, \phi_i^*(t)$, $\sin(\phi_j^*(t) - \phi_i^*(t) - \delta_{ij}) \simeq 0$ for most pairs of i and j. Then, Eq.(11) is led to

$$r_i^* = \sum_{j=1, j \neq i}^{N} \kappa_{ij} f_R(r_j^*) \cdot \cos(\phi_j^*(t) - \phi_i^*(t) - \delta_{ij}) \simeq \sum_{j=1, j \neq i}^{N} \kappa_{ij} f_R(r_j^*) \quad (> 0)$$

which is simply the case for the conventional real-valued Hopfiled network with $f_R(\cdot) : R_+ \to R_+$. Hence, the amplitude r_i is the internal state, or activity, of the i-th neuron activated by the input from the other neurons. Consequently, using the modified constant $\kappa_{ij}\prime = \kappa_{ij} f_R(r_j^*)$, the coupling coefficient in Eq.(5) is approximately given by $K_{ij} \simeq \kappa_{ij}\prime / r_i^* \simeq \kappa_{ij}\prime / \sum_{k=1}^{N} \kappa_{ik}\prime$, which shows that the bigger is r_i, the less ϕ_i is affected by the interaction from the other neurons, and the more ϕ_i affects the phase dynamics of the other neurons.

According to (9), the existence of the energy function is assured with the activation functions $f_R(r) = \tanh(r)$ and $f_R(r) = r/(c + 1/\gamma \cdot r)$ $(c, \gamma > 0)$, but is not assured with $f_R(r) = 1$, as is discussed in [2].

Finally, let us consider the case that the interaction strength is small compared with the original dynamics of a single neuron. Namely, if the second term on the right hand side of Eq.(7) is multiplied by $\varepsilon \ll 1$, then $r_i^* = O(\varepsilon)$ from Eq.(11). However the interaction term in Eq.(12) is $O(1)$. This is because $f_R(r_j^*) = O(\varepsilon)$ from the latter of Eq.(9), and thus $\kappa_{ij}\prime = O(\varepsilon)$, but $K_{ij} = O(1)$.

4   Control of a Flow Network

4.1   Traffic Network as Flow Network

We consider a finite network of flows, which is modeled by a finite lattice space. A node represents an intersection of multiple flows, and a gate is located at each node to control the traffic of the flows. The function of the gate is to give the right of way to only one flow at a time. Here we assume the gate dynamics is periodic for constant flows. Each flow enters the network from one boundary and goes out from another boundary after a travel within the network. There is neither sink nor source and the flow is conserved, in a sense that amounts of inflow and outflow are equal at any point on the network.

One typical example is an urban traffic network. Let us consider a finite area of a traffic network. A traffic signal is equipped at each intersection of the streets. The area is modeled by a 2-dimensional lattice space, where intersection C_i is a node and signal i corresponds to a gate at the node. A lattice is a topological representation of a traffic network.

The dynamics of a traffic signal is periodic, with a phase of green, yellow and red light in one cycle. Periodic dynamics of a signal is given by 3 control parameters, namely, a cycle length, an offset and a split. An offset is a phase difference between adjacent signals, and a split is a ratio of a green light time. In many conventional approaches, a cycle length T is usually set common for an area-wide traffic control, while a split is sometimes given point-wise for each intersection. Therefore, an offset is taken as the most important parameter for an area-wide control and many researches have been reported. Here we also focus on the offset control of traffic signals.

4.2   Objective of a Traffic Flow Control

The following simplified model has been conventionally used for traffic flow. If a traffic is less stopped by signals at intersections, then its traveling time is kept shorter. Therefore, the smoothest travel can be attained by setting all adjacent signals along the street have a time delay equal to the traveling time. The traveling time from C_j to an adjacent C_i with distance L_{ij} is L_{ij}/v, where a constant velocity v is assumed for a steady flow. Thus, let us define L_{ij}/v as a desired offset value between j and i. The non-dimensional normalized quantity is obtained as $2\pi \cdot L_{ij}/Tv$ for 2π-periodic motion.

Here we assume the symmetric network $\forall i, j, L_{ij} = L_{ji}$ for the simplicity, though all the following discussion is applicable to any asymmetric case. Then, the desired offsets Δ_{ij} and Δ_{ji} in both directions are given by $|\Delta_{ij}| = |\Delta_{ji}| = 2\pi \cdot L_{ij}/Tv$ (mod 2π). Now we recognize from $|\Delta_{ij}| = |\Delta_{ji}|$, namely $\Delta_{ij} = -\Delta_{ji}$ (mod 2π), that this equation is realized only for $\Delta_{ij} = 0$ or π. This means that the above desired offset can not be attained even for a pair of signals in general. Therefore as the additional criteria, for the effective transportation over the area, we set such control objective that a directed pair with higher flow density $\mathcal{K}_{ij}$ for $C_j \to C_i$ takes precedence in attaining Δ_{ij}.

4.3   Optimal Offset for a Symmetric Traffic Flow

The following is known for the i, j-pair of traffic signals for the symmetric flow $\mathcal{K}_{ij} = \mathcal{K}_{ji}$. Let denote a desired offset $\Delta \equiv |\Delta_{ij}| = |\Delta_{ji}|$. A simple average of both desired offsets are 0 or π, noticing that an offset is considered by mod 2π. Between 0 and π, the closer to the original desired offsets $\pm\Delta$ is

$$0 \text{ for } 0 \leq |\Delta| < \pi/2, \quad \pi \text{ for } \pi/2 < |\Delta| \leq \pi \tag{13}$$

The following result is well known in the transportation system research for the symmetric flow. Based on the model of single saturated flow, as the most simplified model for a traffic flow, the total waiting time, as a function of the difference from a desired offset, becomes minimum at the offset value of Eq.(13). Moreover, a recent traffic simulation with individual vehicle motion shows the same result for an optimal offset[6]. Thus, both from a simple analysis and a computer simulation, it seems appropriate to give the optimal signal offset by Eq.(13) for a symmetric traffic flow.

In a square street network with the same distances L between all adjacent intersections, the desired offsets between any pair of adjacent signals are all given by $\Delta = 2\pi \cdot L/Tv$. The optimal offset (13) in the above sense leads to 0 offset for all adjacent signal pairs when $|\Delta| < \pi/2$, and half a period offset for all adjacent signal pairs when $\pi/2 < |\Delta| \leq \pi$. Let us call the former offset pattern as a simultaneous type, while the latter pattern as an alternate type.

4.4   Correspondence Between the Dynamics of a Traffic Signal Pair and a Phase Oscillator Pair

Eq.(4) in Section 2 is obtained as a steady phase difference $\Delta\phi_0$ for an oscillator pair given by Eq.(3) with equal coupling strengths. This formula is the same as

Eq.(13) in 4.3 which is the optimal offset of a signal pair for a symmetric traffic flow $\mathcal{K}_{ij} = \mathcal{K}_{ji}$. This also leads to the coincidence of in-phase (or anti-phase) type entrainment in a oscillator system, with simultaneous (or alternate) type offset of a signal system, under a common coupling strength and a common flow density, respectively. Moreover, when the coupling strengths differ in an oscillator pair, it is shown that a unique steady solution of a phase difference is closer to the target of the stronger, and this characteristic coincides with the control criteria described in 4.2. This leads to the offset control method by a phase oscillator system, which is described in the next section.

5   Signal Offset Control by a Coupled System of Phase Oscillators and by a Complex-Valued Neural Network

5.1   Basic Formula for Offset Control

The signals are modeled by phase oscillators with a common frequency $\omega = 2\pi/T$. As is shown in 2.2, the interaction by $\sin(\phi_j - \phi_i - \delta)$ causes a phase entrainment with a phase difference δ. We utilize this character for an offset control and set the constant δ equal to a desired offset Δ_{ij} for a signal pair. In addition, a coupling coefficient between an oscillator pair is given by a monotonically increasing function $F(\mathcal{K}_{ij}) > 0$ of a flow density $\mathcal{K}_{ij}$. Thus, the basic formula for an offset control is given by the following equation for signal i [4];

$$\frac{d\phi_i}{dt} = \omega + \epsilon \cdot \frac{1}{|\mathcal{N}_i|} \sum_{j \in \mathcal{N}_i} F(\mathcal{K}_{ij}) \cdot \sin(\phi_j - \phi_i - \Delta_{ij}) \tag{14}$$

where $\mathcal{N}_i$ is a set of signals adjacent to i, $|\mathcal{N}_i|$ is the number of its elements, and $\epsilon > 0$ is a common coupling coefficient. In the rest of this section, we take $\epsilon = 1.0$ and $F(\mathcal{K}) = \mathcal{K}$. Then, the flow density $\mathcal{K}_{ij}$ directly corresponds to the coupling coefficient $\mathcal{K}_{ij}$.

Complex-valued neural network described by Eqs.(11) and (12) is equally applicable to the offset control. In this case, the weight is given by $w_{ij} = \mathcal{K}_{ij} \exp(-\imath \Delta_{ij})$. The only difference caused by the existence of the amplitude $r_i(t)$ is that the coupling coefficient $\mathcal{K}_{ij}$ is multiplied by $f_R(r_j)/r_i$ as is shown in 3.3. $r_i(t)$ is the internal state which is activated by the input from the other neurons, and in the present case of a traffic network, it corresponds to the total incoming traffic flow (or equivalently, the total outgoing traffic flow, because of the flow conservation) at the intersection C_i. As the result, the signal at the intersection with heavier traffic affects the adjacent signals stronger and is less affected. Conversely, the signal that copes with less traffic tends to follow. The nonlinear function $f_R(r)$ has the saturation effect on this tendency.

5.2   Extension to Bi-directional Coupling

Eq.(14) has an interaction term only for a downstream signal i to attain a desired offset Δ_{ij} for a flow from C_j to C_i. Let us denote this form of interaction as a uni-directional coupling. On the other hand, the following formula gives the interaction equally to both downstream signal i and upstream signal j, to control the offset Δ_{ij} for a flow from C_j to C_i;

$$\frac{d\phi_i}{dt} = \omega + \frac{1}{|\mathcal{N}_i|} \sum_{j \in \mathcal{N}_i} \left\{ \frac{\mathcal{K}_{ij}}{2} \sin(\phi_j - \phi_i - \Delta_{ij}) + \frac{\mathcal{K}_{ji}}{2} \sin(\phi_j - \phi_i + \Delta_{ji}) \right\}$$

which is denoted as a bi-directional coupling.

For the complex-valued neural network, this corresponds to $w_{ij} = \mathcal{K}_{ij} \exp(-\imath \Delta_{ij}) + \mathcal{K}_{ji} \exp(\imath \Delta_{ji}) = \overline{w_{ji}}$, which meets the Hermitian condition Eq.(10) on W.

5.3   Further Extensions and Evaluation by a Traffic Simulator

A system is further generalized to a non-straight flow, to a network with multi-leg intersections with arbitrary number of legs, to a signal which has a different phase to each flow with a different outgoing direction, and to a signal with a right-turn-only phase [5]. The interaction remains pair-wise between the adjacent signals throughout these extensions.

The effectiveness of the proposed method is evaluated by a traffic simulator. In the computer simulation, first, a system of differential equations is numerically solved and a set of offsets is obtained as a steady solution. Then the obtained offset values are given to a traffic simulator, which explicitly describes a motion of individual vehicle and dynamics of each signal [6]. Statistics such as an average travel time and an average number of waiting vehicles are used for a quantitative evaluation. The several simulations show the effective control by the phase model under various conditions. The detailed simulation results on the complex-valued neural network with various activation functions will be reported in the future.

6   Summary

A network of complex-valued rotating neurons is proposed, and it is shown that the phase dynamics of the neurons is simply described by a system of rotating phase oscillators with a pair-wise sinusoidal interaction. The effect of the amplitude and the application to the traffic flow control are also discussed.

References

1. Ed. A. Hirose, "Complex-Valued Neural Networks", Series on Innovative Intelligence, vol. 5, World Scientific, 2003.
2. Y. Kuroe et al., "On activation functions for complex-valued neural networks – existence of energy functions –", *ICANN/ICONIP 2003*, ed., Okyay Kaynak et.al., Lecture Notes in Computer Science 2714, pp. 985–992, Springer, 2003.
3. Y. Kuramoto, "Chemical Oscillations, Waves and Turbulence", Springer-Verlag, 1984.
4. I. Nishikawa et al., "Area-wide Control of Traffic Signals by a Phase Model," *Trans. Society of Instrument and Control Engineers*, vol. 39, no. 2, pp. 199–208, 2003.
5. I. Nishikawa and H. Kita, "Information Flow Control by Autonomous Adaptive Nodes and Its Application to an On-line Traffic Control", *Proceedings of the 6th JCIS* , Sept. 2003. I. Nishikawa, "Dynamics of Oscillator Network and Its Application to Offset Control of Traffic Signals", *Proceeding of IJCNN2004*, July 2004.
6. Goto, Thesis, Kyoto University, 2000. (In Japanese)

Two Models for Theta Precession Generation Using the Complex Version of the Nagumo-Sato Neuron Model and the Hodgkin-Huxley Equations

Iku Nemoto

Tokyo Denki University School of Information Environemt
Inzai, Chiba, 270-1382, Japan
nemoto@sie.dendai.ac.jp

Abstract. The firing of the so-called place cells in the rat hippocampus shows theta precession, meaning that the probability of firing follows the local theta potential with certain phase lags which depend on the instantaneous position of the animal during running in a given field. In this paper, the precession is modeled by two models, one is the complex version of the Nagumo-Sato single neuron and the other the well-known Hodgkin-Huxley equations.

1 Introduction

The hippocampus plays important roles in many aspects of memory. So-called place cells in the rat hippocampus show firing synchronized with the local theta potential. Place cells are activated when the animal runs over a certain small area of its running field, and the timing of their spikes with respect to the local theta potential changes as the animal passes over it. This shift of phase is called theta precession (O'Keefe and Recce, 1993) and is considered to give the animal some clue about its position.

A theory of hippocampal memory was developed based on theta phase precession (Yamaguchi, 2004). In it, theta precession generation is modeled by a set of differential equations. We here show that theta precession can be modeled easily by the complex neuron model (Nemoto and Saito, 2002). We next show that it can be modeled by simply driving the Hodgkin-Huxley equations by a sinusoidal current superimposed on a bias. We connect the well-known synchronization of the H-H membrane (Nemoto et al., 1975) with theta precession and show that synchronization can be obtained at such low frequencies as 8 Hz by a simple change of a parameter.

2 Theta Precession

It was found that the so-called place cells in the rat hippocampus fire impulses in synchrony with the local theta rhythm measured in the hippocampus and their relative phase with respect to the theta rhythm advances as the animal runs through the designated course(O'Keefe and Recce, 1993). Fig. 1 is a schematic diagram showing firing patterns of several units having different places in the running field for optimal activation. As the animal passes through each of these places, the corresponding unit fires at successively earlier phases of the theta wave, and if we look at the firing pattern of all the units in one theta cycle, it may be considered to reproduce the whole running course in a compressed time period so that the animal may know its present position relative to the whole running course.

N.R. Pal et al. (Eds.): ICONIP 2004, LNCS 3316, pp. 130–135, 2004.
© Springer-Verlag Berlin Heidelberg 2004

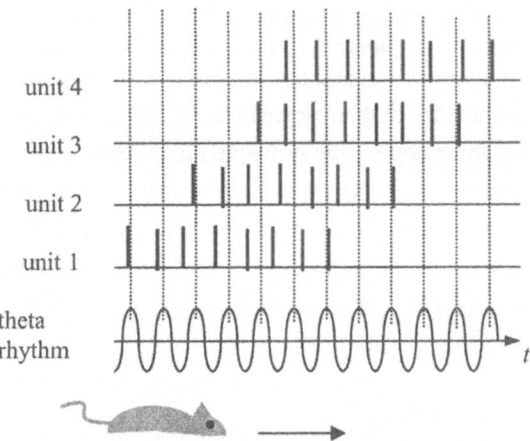

unit 4

unit 3

unit 2

unit 1

theta
rhythm

Fig. 1. Schematic drawing of theta rhythm precession, adapted from Yamaguchi, 2002. See text for detail.

3 Complex Version of the Nagumo-Sato Model

We proposed a complex version of the Nagumo-Sato single neuron model (Nemoto and Saito, 2002) and extended it to a network (Nemoto, 2003). It is given by the equations:

$$\eta_i(n) = A_i - \frac{1}{\alpha_i}\sum_{l=0}^{n}\beta_i^l\xi_i(n-l)$$

$$\xi_i(n+1) = \Theta(\eta_i(n)) = \begin{cases} 0 & (|\eta_i(n)| < 1) \\ \dfrac{\eta_i(n)}{|\eta_i(n)|} & (|\eta_i(n)| \geq 1) \end{cases} \quad i = 1,2,\ldots,N \tag{1}$$

where $\eta_i(n), \xi_i(n), A_i$ are respectively, the complex-valued membrane potential, output and the input to the i-th neuron. β_i is a complex parameter less than 1 in modulus and represents the degree of influence of the past (up to n time steps) output of the unit on its own present membrane potential. We here con-sider $N = 2$ completely separate neurons 1 and 2. After the variable conversion:

$$z_i(n) = 1 + \alpha_i\beta_i A_i - \sum_{l=0}^{n-1}\beta_i\xi_i(n-l), \quad c_i = 1 - \alpha_i A_i(1-\beta_i) \tag{2}$$

we get the dynamical system:

$$z_i(n) - 1 = \begin{cases} \beta_i(z_i(n-1) - c_i), & |z_i(n-1) - c_i| < \alpha_i \\ \beta_i(z_i(n-1) - c_i) - \dfrac{z_i(n-1) - c_i}{|z_i(n-1) - c_i|}, & |z_i(n-1) - c_i| \geq \alpha_i \end{cases} \tag{3}$$

for $i = 1,2$. We consider $z_i(n)$ to represent the state of the neurons. When $|z_i(n) - c_i| \geq \alpha_i$, the i-th neuron is considered to fire with the phase of the impulse train being $\arg(z_i(n)-1)$, and if $|z_i(n) - c_i| < \alpha_i$, it is in the resting state. We want to see how the amplitude of the input to the neuron affects the phase of the output. For the purpose, we use the first neuron as a reference. We change the amplitude of the input to the second neuron $|A_2|$ and see the phase difference $\arg z_2 - \arg z_1$ which is considered to be the amount of theta precession. Therefore we neglect the value $\arg z_1$ itself. Fig. 2 shows an example. The horizontal axis represents $|A_2|$ and the vertical axis shows the advance of phase $\arg z_2 - \arg z_1$ at the 50-th impulse. The value of $\arg \beta$ is taken as a parameter. When $\arg \beta > 0$, the phase of z_2 advances relative to that of z_1 when $|A_2| > |A_1|$. As impulses keep on firing, the phase continues to advance for constant $|A_2| > |A_1|$. The other parameter values are: $\alpha_i = 0.1, |\beta_i| = 0.8$. Fig. 3 shows the time course of the phase difference for

$$|A_2| = \begin{cases} |A_1| = 4, & n < 50, n > 75 \\ 6.5, & 50 \le n \le 75 \end{cases} . \tag{4}$$

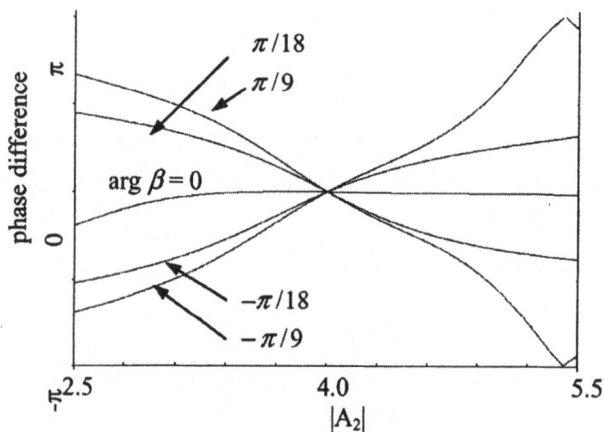

Fig. 2. $\arg z_2 - \arg z_1$ as a function of $|A_2|$. $|A_1| = 4$.

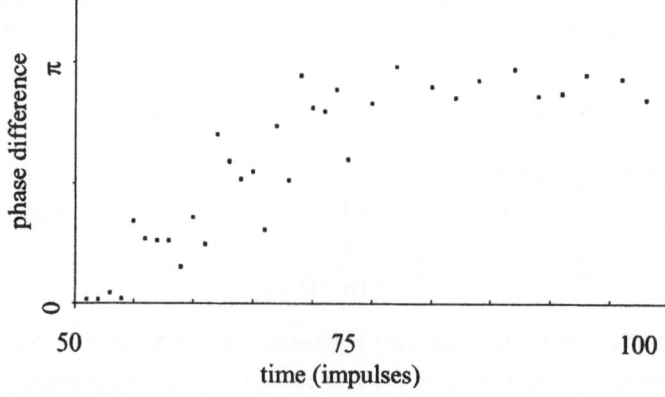

Fig. 3. Time course of the phase difference for the inputs described by eq. (4).

There is a trend of increase of phase advance from 50-th to 75-th impulses and then the phase difference becomes constant on the average as expected because the input $|A_2|$ drops to the original value of 4. The behavior shown in this figure can probably be utilized in modeling the theta precession in a very simple way.

4 Modeling by the Hodgkin-Huxley Equations

We do not give the equations here because of lack of space but we use the Hodgkin-Huxley equations in their original form (Hodgkin and Huxley, 1952). The current density I through the membrane (outward positive) is considered to be the input and is composed of a DC bias I_b and a sinusoidal current:

$$I = I_b + A \sin(2\pi f t) \tag{4}$$

As is well known, when $A = 0$, the H-H membrane fires with the frequency determined by the value of I_b. Therefore, we consider the frequency determined by I_b the intrinsic frequency of a particular place cell. It was assumed that during the activation period of the place cell, there is a constant input current which makes a constant increase of the intrinsic impulse frequency (Yamaguchi, 2003). Our assumption may be more realistic, because an increase in the bias current may well happen as the animal nears the place most favored by the cell.

Fig. 4 shows an example. The parameter values are exactly the same as the original H-H model. The bias current was set according to:

$$I_b(t) = \begin{cases} I_0 + 2t(I_{max} - I_0)/T_d, & (t < T_d/2) \\ I_{max} - 2(t - T_d/2)(I_{max} - I_0)/T_d, & (t > T_d/2) \end{cases} \tag{5}$$

T_d is the period of running of the animal. I_b increases until $t = T_d/2$ when the animal is assumed to reach the most favored place by the particular place cell, and it decreases as the animal gets away. The parameter values are: $I_0 = 2\mu A/cm^2$, $I_{max} = 12, T_d = 5$ sec, $f = 70 Hz$. The horizontal line represents the time and the vertical line the phase delay (contrary to Figs. 2 and 3) with respect to the reference sinusoid. The four curves correspond to four values of A: 0.5, 0.8, 1, and 2 $\mu A/cm^2$. It is seen that when the bias current I_b becomes maximum at $T_d/2$, the phase delay becomes minimum and as I_b decreases afterwards, it again increases. The larger amplitudes of the sinusoidal wave tend to synchronize the H-H membrane for a larger range of I_b with the phase shift less dependent on I_b. When the firing becomes out of synchrony, the firing patterns become quite irregular due to pseudoperiodic and partly chaotic behavior of the impulses (Aihara, 1985), but this is not the topic in our present discussion. The phase increases faster after $T_d/2$ than it decreases before $T_d/2$. Actual place cells do not seem to see their phase increase after the animal reaches the favored point. Instead, firiring simply stops when it leaves the place. Therefore, we may adopt a bias current which does not change after the animal reaches the point. The question is whether it is natural or not.

Although the above qualitative behavior may be suitable for a model of theta precession, the frequency involved is much too high for the 8 Hz theta rhythm. We next

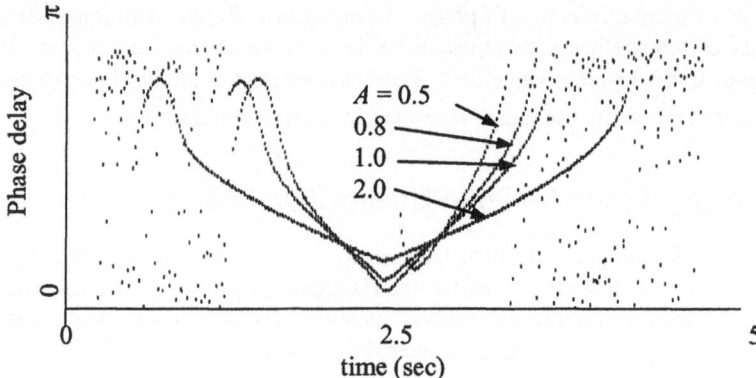

Fig. 4. Phase delay in the H-H membrane with respect to the reference for bias current described by eq. (5) with $T_d = 5$ s.

try to lower the whole frequency by changing the parameters in the dynamics of Na⁺ concentration. Fig. 5 shows an example. The voltage dependency of one of the constants for the differential equation for h which is a variable describing Na⁺ conductance is described by

$$\beta_h(V) = \left(\exp\left\{ -\frac{V + V_h}{10} + 1 \right\} \right)^{-1} \tag{6}$$

where usually, $V_h = 30$. Here we use $V_h = 50$, which yields quite low frequencies of impulse firing. All other equation parameters are exactly the same as the original.

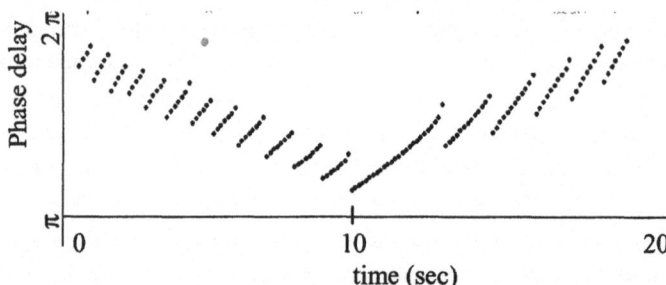

Fig. 5. Phase delay in the H-H membrane with respect to the basic rhythm. See text.

The current parameters are: $I_0 = 0.77$, $I_{max} = 0.805$, $T_d = 20$, $A = 0.05$ and the driving frequency is set at 8 Hz. The response takes a zigzag course but the behavior seems to be suitable for a possible model of theta precession.

5 Conclusion

Theta rhythm precession in a running rat's hippocampus was modeled by a complex neuron model and by the Hodgkin-Huxley equations. At present we have only shown that these models can behave similarly to the place cells in their synchronous behaviors and more detailed analysis of the models is necessary before applying to physiological data.

References

1. Aihara, K. Matsumoto, G. (1984). Periodic and non-periodic responses of a periodically forced Hodgkin-Huxley oscillator. *J. Theor.Biol.* 109, 249-269.
2. Hodgkin, A.L. & Huxley, A.F. (1952). A quantitative description of membrane current and its application to conduction and excitation in nerve. *J. Physiol.* 117, 500-544.
3. Nemoto, I..Miyazaki, S., Saito, M. & Utsunomiya, T. (1975). Behavior of solutions of the Hodgkin-Huxley equations and its relation to properties of mechanoreceptors, *Biophys. J.* 15, 469-479.
4. Nemoto, I. & Saito, K. (2002). A complex-valued version of Nagumo-Sato model of a single neuron and its behavior. *Neural Networks*, 15, 833-853.
5. Nemoto, I. (2003). Some properties of the network consisting of two complex-valued Nagumo-Sato Neurons, *Proc. 7th Int'l Conf KES2003*, 351-357.
6. O'Keefe, J. & Recce, M. (1993). Phase relationship between hippocampal place units and the EEG theta rhythm. *Hippocampus*, 3, 317-330.
7. Yamaguchi, Y.(2002). Neural mechanism of hippocampal memory organized by theta rhythm. *Jpn J. Neuropsychopharmacol.*, 22, 169-173.
8. Yamaguchi, Y. (2003). A theory of hippocampal memory based on theta phase precession. Biological Cybernetics 89, 1-9.

Using Self-organizing Map
in a Computerized Decision Support System

Miki Sirola, Golan Lampi, and Jukka Parviainen

Helsinki University of Technology, Laboratory of Computer and Information Science,
P.O.Box 5400 FIN-02015 HUT, Finland
{miki.sirola,golan.lampi,jukka.k.parviainen}@hut.fi
http://www.cis.hut.fi

Abstract. Modern computerized decision support systems have developed to their current status during many decades. The variety of methodologies and application areas has increased during this development. In this paper neural method Self-Organizing Map (SOM) is combined with knowledge-based methodologies in a rule-based decision support system prototype. This system, which may be applied for instance in fault diagnosis, is based on an earlier study including compatibility analysis. A Matlab-based tool can be used for example in fault detection and identification. We show with an example how SOM analysis can help decision making in a computerized decision support system. An error state model made in Simulink programming environment is used to produce data for the analysis. Quantisation error between normal data and error data is one significant tool in the analysis. This kind of decision making is necessary for instance in state monitoring in control room of a safety critical process in industry.

1 Introduction

The beginning of the decision support system development dates back to the 1960s. The development began from two rather different directions: expert systems and simple information systems. Knowledge-based techniques were introduced about 20 years later. During these decades the variety of methodologies as well as different application areas has increased a lot. Today a wide range of methodologies from classical information science into artificial intelligence and modern data mining including all possible decision making and problem solving techniques are in use. Decision support systems are introduced comprehensively in [1] and [2].

Self-Organizing Map (SOM) [3] is an artificial neural networks algorithm based on competitive learning. It is effective in analysis and visualisation of multi-dimensional data. The SOM helps in mapping nonlinear statistical dependencies from multi-dimensional measurement data into simple geometrical relations, usually into two-dimensional space. The map maintains roughly the most important topological and metric relations from the original measurement element, and clusters the data. Clustering is needed for instance in data analysis

N.R. Pal et al. (Eds.): ICONIP 2004, LNCS 3316, pp. 136–141, 2004.
© Springer-Verlag Berlin Heidelberg 2004

of complex processes and systems. The self-organizing map has been used in many engineering applications [4].

Knowledge-based methodologies can be used in decision support systems in many ways. Some possibilities in use are presented in [5]. In these systems advice is given to control room personnel or people working with maintenance of a power plant. Hierarchically organized rule base is used in ennobling process data, and decision analytic approach in choosing correct actions when there exist several alternatives. Simulators are used to give predictions of calculated process quantities. Information and instructions are revised in suitable form for the user by an advanced Man-Machine Interface (MMI) system.

The role of a Decision Support System (DSS) in a process monitoring task is the following. Measurement data from the process is used either directly in the DSS or after some kind of preprocessing phase. The output of the decision support system is presented to the operator via an MMI system. The operator can use this information or possible advice when control actions are decided. Process is never controlled directly by a decision support system.

In this paper we combine self-organizing map and knowledge-based techniques, and build a prototype of a decision support system based on an earlier compatibility analysis [6]. The possibilities to use neural methods in computerized decision support systems are also studied on more general basis, although the focus is in the new DSS application. We demonstrate how SOM analysis can help decision making in a computerized decision support system. This kind of decision making is needed for example in state monitoring of a safety critical process in industry. Fault diagnosis including both detection and identification of faults is one suitable application area. The problem here is to combine in an efficient way these two different approaches to produce a more competent computerized decision support system than those that have been built before. A small process example built with Simulink program package is included in this study. SOM was chosen as a DSS tool because of its ability to do a two-dimensional visualisation of multidimensional data. It is also faster than some other clustering algorithms [7].

A compatibility analysis about the usage of SOM analysis with knowledge-based decision support system was done already earlier [6]. It was studied how SOM analysis can help in detecting faults in a process example. A leak in the primary circuit of a BWR nuclear power plant was simulated with a simplified model. These findings have also been utilized when the prototype was built.

The SOM analysis offer results of statistical analysis for the rule-based reasoning hierarchy in a computerized decision support system. The utilization of correlations, clusters, shape of SOM map, U-matrix and trajectories all add new features into the information content that DSS can ennoble further. Quantisation error, which is calculated between normal data and error data including simulated fault, turned out to be the most useful tool found in this analysis. If there are a limited number of possible error cases, even an atlas of error maps is possible to produce. The prototype called DERSI is presented in [8], including more detailed technical description.

2   Prototype Description

The prototype built according to the explained principles is called DERSI. DERSI is a DSS that utilizes SOM. It can be used to monitor an arbitrary process that can supply data in matrix form. DERSI is a Matlab software program and it is built on top of a Matlab extension named SOMToolbox [9]. SOMToolbox is a result of an earlier software development project in the Laboratory of Computer and Information Science in Helsinki University of Technology.

DERSI can identify process states. It is based mainly on supervised learning. Data of a process state is in a matrix where columns represent a specific process variable and rows represent a specific sample. Data of different process states has to be available separately (so that measurements of state 1 are in matrix 1, measurements of state 2 are in matrix 2, etc.). For correct operation, it is important to have enough samples in the matrices.

It is possible to do preprosessing and calculate additional variables like differences from the existing variables and concatenate the corresponding variable column vectors horizontally with the process state data matrix. This is done, if the information value of the calculated variables is significant in this context. The result is a complete state data matrix. Every complete state data matrix is taught to a separate state SOM. All these matrices are also concatenated vertically and a SOM U-matrix is formed from this data.

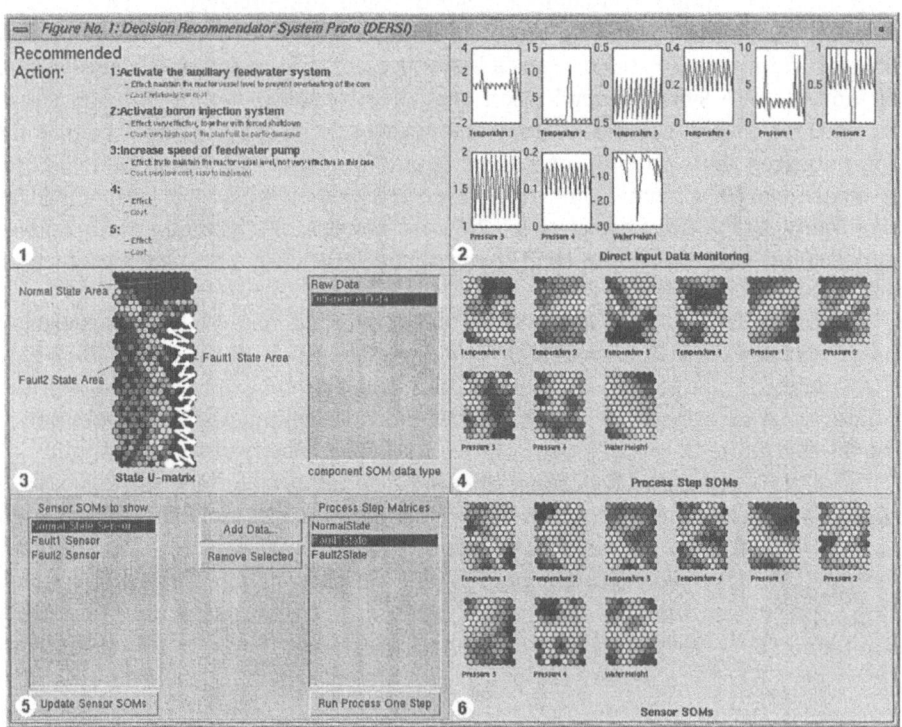

Fig. 1. DERSI GUI.

A recommendation engine implements the DSS functionality. It is programmed in object oriented way and has four important classes: Decision, QErrorSensor, DecisionDatabase and DecisionRecommendator.

DERSI MMI is a GUI that is shown in Figure 1. It has six different fields. Field 1 has recommended decisions. Field 3 has the U-Matrix on the left side. Former studies have shown that it is possible for different process states to be mapped to different clusters in the U-Matrix [6]. A trajectory is formed in the U-matrix from input data. Field 4 has SOM input data component planes. Differences between input data and sensor data can be found by comparing them to the sensor data component planes in Field 6. In the figure an imaginary leak of a BWR nuclear power plant is simulated.

3   Testing Prototype

The plan is to make a simulated process example and corresponding specific rule base into DERSI. Before this we have tested DERSI with artificial data made by simpler means. In this paper we present an error state model made in Simulink, which is a simulation tool in Matlab programming environment. A similar but simpler model is seen in Figure 2. Although the model is not a representation of some physical process, it has some features that are similar to those found in real processes.

Five variables of data were generated with the model mentioned before. The used model has 5 differential equations and correspondingly 5 measured vari-

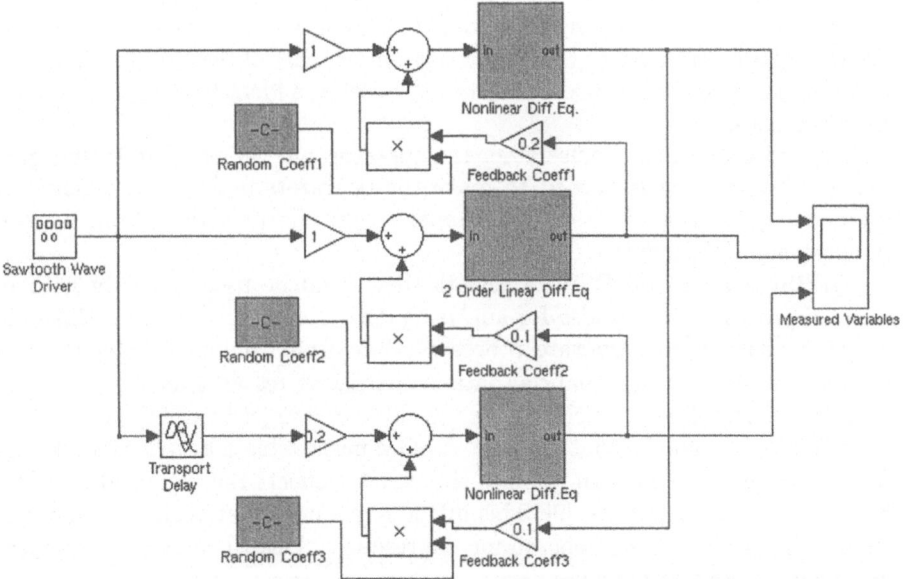

Fig. 2. A simplified version of the Simulink model used in testing DERSI.

ables. Complexity has been introduced by using nonlinear and second order differential equations. If specific values are set to different coefficients in the model, some variables begin to diverge rapidly. An analogy can be drawn with this unstability property and a real industrial process in a serious fault state.

Interdependencies between variables have been created by feeding part of a differential equation output to the drive of another differential equation. The model process has also stochastic properties. Every differential equation feedback has a random coefficient. Every time the simulation is run, the random coefficient gets a random value in the range $[1 - a, 1 + a]$ (a of 0.05 was used).

Four states were generated with the model. Four DERSI sensors were each taught with data of one of these states. These sensors were able to recognize 3 of the four states with 16 different simulation runs.

Although the quantisation error method was succesful, the U-matrix visualisation was not very utilizable in this simulation. Because fault state data still had outliers after initial preprocessing, the U-matrix visualisation did not show clusters in the simulation data. It seems that data preprocessing is significant especially in U-matrix visualisations.

4   Discussion

A decision support system using self-organizing map has been developed based on an earlier compatibility analysis. SOM methodology is used together with knowledge-based methods in DERSI prototype. An error state model was used to simulate faults in process data. Our future plan is to construct more realistic process example.

We have been able to demonstrate how self-organizing map can be utilized in a computerized decision support system. The results of statistical data analysis are given for the rule base of the knowledge-based part of the system. Especially quantisation error has turned out to be very useful in identifying fault situations from the data.

SOM visualisations are useful for the operator, but the problem is that quite much extra training is needed to make the operators understand enough the methodology, so that they are able to make right interpretations from the two-dimensional SOM maps.

DERSI is a general DSS framework that could be used not only with industrial process data but also for analysing data from other domains. Examples might be data from corporate processes, data from computer networks (e.g. Congestion monitoring), financial data or marketing research data.

It is important that which process variables are taught to a SOM and what is the SOM mask value for that variable. A SOM mask value is between 0 and 1 and it tells how much the distribution of the variable effects the organization of the SOM in the training phase [9]. This introduces a new interesting problem: how should the mask values be chosen for the results of DERSI to be as informative and unambiguous as possible?

DERSI is especially useful when there are so many input variables that it is hard to comprehend with variable plots or even SOM component planes, what is happening in the process. It is here that the U-matrix and DSS unit of DERSI show their strength.

Our error state model can somewhat simulate process characteristics. Still to use real data or data from a more realistic process model is more challenging, and it would make the whole study more plausible. The simulated stochastic features need also better comparisons to stochastic properties of a real process. However, many interesting features could be demonstrated already now.

The SOM part of this DSS is not able to handle semi-structured or unstructured data, but the knowledge-based part can. The idea to add self-organizing map technique into a DSS is not to replace something, but to complement. Our approach and suggested combination of methologies has the advantage of using all possibilities in data analysis in addition, compared to traditional knowledge-based decision support systems.

References

1. Turban E., Aronson J.: Decision support systems and intelligent systems. Prentice Hall (1998)
2. Marakas G.: Decision support systems in the 21st century. Prentice Hall (1999)
3. Kohonen T.: The self-organizing map. Springler, Berlin, Heidelberg (1995)
4. Simula O., et.al.: The self-organizing map in industry analysis. Industrial applications of neural networks. CRC Press (1999)
5. Sirola M.: Computerized decision support systems in failure and maintenance management of safety critical processes. VTT Publications 397. Espoo, Finland (1999)
6. Sirola M., Vesanto J.: Utilization of neural methods in knowledge-based decision support systems - state monitoring as a case example. IASTED International Conference for Modelling, Identification and Control. Innsbruck, Austria (2000)
7. Vesanto J., Alhoniemi E.: Clustering of the Self-Organizing Map. IEEE Transactions on Neural Networks, Volume 11, Number 3, pp. 586-600 (2000)
8. Sirola M. et.al.: Neuro computing in knowledge-based decision support systems. EHPG-Meeting of OECD Halden Reactor Project. Sandefjord, Norway (2004)
9. Vesanto J., et.al.: Technical report on SOM Toolbox 2.0. Espoo, Finland (2000)

An Empirical Study on the Robustness of SOM in Preserving Topology with Respect to Link Density

Arijit Laha

Institute for Development and Research in Banking Technology
Castle Hills, Hyderabad 500 057 India
alaha@idrbt.ac.in

Abstract. Practical implementations of SOM model require parallel and synchronous operation of the network during each iteration in the training stage. However this implicitly implies existence of some communication link between the winner neuron and all other neurons so that update can be induced to the neighboring neurons. In the current paper we report the results of an empirical study on the retention of topology preservation property of the SOM when such links become partially absent, so that during a training iteration not all the neighbors of the winner may be updated. We quantify our results using three different indexes for topology preservation.

Keywords: SOM, topology preservation, link density

1 Introduction

The Self-organizing Map (SOM) [1] introduced by Kohonen is a self organizing network based on competitive learning. However SOM stands apart from other competitive learning networks due to its unique property of "topology preservation". This is achieved through arrangement of the neurons in SOM in a regular lattice structure in (usually) two dimensional plane known as the "output plane" or "viewing plane" and incorporation of "neighborhood update" strategy during the training process. The SOM is originally inspired by the discovery of different spatially ordered maps in brain [2], [3], [4]. Many of them are found in cerebral cortex area for different perceptual tasks. Originally Kohonen modelled the SOM algorithm in form of a system of coupled differential equations [5], [6]. The computer simulation using them treated each neuron independently and did not demand any synchronization among the neurons. The topological ordering is achieved due to various lateral feedback connections among the neighbors. However such a simulation (mimicking the biological neurons) is computationally intensive and ill-suited for practical applications. For practical purposes a simpler algorithm [1] is used which leads to functional appearance of the topologically ordered maps.

The simplified SOM algorithm (hereafter referred as SOM algorithm unless stated otherwise) is almost invariably implemented by simulating it on serial

N.R. Pal et al. (Eds.): ICONIP 2004, LNCS 3316, pp. 142–149, 2004.
© Springer-Verlag Berlin Heidelberg 2004

computers. During the training, in each iteration the update of the winner and its neighbors are done in a way equivalent to a parallel operation. For quick learning, initially the neighborhood is defined large enough to cover almost whole network [1]. Thus there is an implicit assumption at work that every neuron (a possible winner) is connected to all the other neurons by some kind of direct communication link, so that the winner can induce the update of another neuron if it happens to fall within the neighborhood of the winner. Such a scenario of complete connection among biological neurons is highly unlikely. In the present paper we investigate the effect of partial absence of such connections on the topology preservation property of the practical SOM algorithm. To measure the topology preservation property we use three indexes, 1) the well-known "topographic product" [7], 2) a measure of topology preservation proposed by Su et. al [8] and 3) a rank correlation based measure introduced by Laha and Pal in [9].

2   SOM and Topology Preservation

2.1   The Self-organizing Map

SOM is formed of neurons located on a regular (usually)1D or 2D grid. Thus each neuron is identified with a index corresponding to its position in the grid (the viewing plane). Each neuron i is represented by a weight vector $\mathbf{w}_i \in \Re^p$ where p is the dimensionality of the input space. In t-th training step, a data point $\mathbf{x}$ is presented to the network. The winner node with index r is selected as

$$r = \underbrace{arg\ min}_{i}\{\|\mathbf{x} - \mathbf{w}_{i,t-1}\|\}$$

and $\mathbf{w}_{r,t-1}$ and the other weight vectors associated with cells in the spatial neighborhood $N_t(r)$ are updated using the rule:

$$\mathbf{w}_{i,t} = \mathbf{w}_{i,t-1} + \alpha(t)h_{ri}(t)(\mathbf{x} - \mathbf{w}_{i,t-1}),$$

where $\alpha(t)$ is the learning rate and $h_{ri}(t)$ is the neighborhood kernel (usually Gaussian). The learning rate and the radius of the neighborhood kernel decreases with time. During the iterative training the SOM behaves like a flexible net that folds onto the "cloud" formed by the input data.

2.2   Topology Preservation in SOM and Its Measures

In any map the topology preservation refers to the preservation of neighborhood relation from the input space to output space. Thus if topology is preserved, nearby features in the input space are mapped onto neighboring locations in the output space. In context of SOM this translates to its property that nearby points in the input space, when presented to the SOM, activates same node or nodes those are close in the viewing plane. Performance of many SOM-based applications depend crucially on the extent to which the map is topologically ordered. However, since the SOM implements a nonlinear and usually dimension reducing mapping, measuring the topology preservation is often very important. We describe three quantitative measures bellow.

Topographic Product. The topographic product [7] is by far most well known and widely used measure of topology preservation in SOM. It considers all order relationships between neuron pairs. For each neuron j the sequences $n_k^U(j)$ and $n_k^V(j)$ are determined such that $n_k^U(j)$ is the k-th nearest neighbor of j in the output space while $n_k^V(j)$ is the k-th nearest neighbor of j in the input space. Then an intermediate quantity $P_3(j,k) = \left(\prod_{l=1}^{k} \frac{d^V(\mathbf{w}_j, \mathbf{w}_{n_l^U})}{d^V(\mathbf{w}_j, \mathbf{w}_{n_l^V})} \cdot \frac{d^U(j, n_l^U)}{d^U(j, n_l^V)}\right)^{1/2k}$ is computed, where $d^U(.)$ and $d^V(.)$ denote the Euclidean distances computed in the output (viewing) space and the input space respectively. The topographic product is computed by averaging over all neurons j and all orders k of neighborhood as:

$$P = \frac{1}{N(N-1)} \sum_{j=1}^{N} \sum_{k=1}^{N-1} \log(P_3(j,k)) \qquad (1)$$

The value $P = 0$ signifies perfect topology preservation. Non-zero values signify mismatch between input and output space and hence lack of topology preservation.

A Measure of Topology Violation. In [8] Su et. al proposed a measure of topology violation based on the observation that if a map is topologically ordered then the weight vector of each node should be more similar to the weight vectors of its immediate neighbors (8 neighbors for a 2-D SOM) on the lattice than to the weight vectors of its non-neighbors. The measure for 2-D SOM can be formulated as follows:

Let Λ_r be the set containing the immediate 8 neighbors of node r and Ω_r denote the set containing the nodes which are not immediate neighbors of node r. Let the size of the map is $m \times n$. Consider a node $i \in \Omega_r$ and another node $i_r \in \Lambda_r$ such that $i_r = \underset{k \in \Lambda_r}{argmin} \|\mathbf{p}_i - \mathbf{p}_k\|$, where, $\mathbf{p}_i = (p_{i1}, p_{i2})$ is the position vector of the node i in the lattice plane and $\|\mathbf{p}_i - \mathbf{p}_k\|$ is the Euclidean distance between the nodes i and k. Since node r is closer to the neighboring node i_r than to i in the lattice plane, the weight vector of node r should be more similar to the weight vector of the node i_r than to the weight vector of the node i. Therefore, if the map is preserving the topology then for each node r the following relation should hold:

$$\|\mathbf{w}_i - \mathbf{w}_r\| \geq \|\mathbf{w}_{i_r} - \mathbf{w}_r\| \text{ for } 1 \leq r \leq m \times n, i_r \in \Lambda_r \text{ and } i \in \Omega_r. \qquad (2)$$

Now the quantitative measure of topology violation V is defined as:

$$V = \sum_{r=1}^{m \times n} \sum_{i \in \Theta_r} \left[1 - \exp^{-\|\mathbf{p}_i - \mathbf{p}_r\|^2}\right] \frac{\|\mathbf{w}_{i_r} - \mathbf{w}_r\| - \|\mathbf{w}_i - \mathbf{w}_r\|}{\|\mathbf{w}_{i_r} - \mathbf{w}_r\|}, \qquad (3)$$

where $\Theta_r = \{i : \|\mathbf{w}_i - \mathbf{w}_r\| < \|\mathbf{w}_{i_r} - \mathbf{w}_r\| \text{ for } i \in \Omega_r \text{ and } i_r \in \Lambda_r\}$ is the set of nodes in Ω_r those violate condition 2 with respect to node r. The measure of violation V has the properties: 1) $V = 0$ if $\Theta_r = \emptyset$, i.e., the topology is perfectly preserved, 2) the larger the value of V the greater is the violation and 3) if $i \in \Theta_r$ and the nodes r and i is far apart in the lattice plane, their contribution to V will be high due to the factor $(1 - \exp^{-\|\mathbf{p}_i - \mathbf{p}_r\|^2})$.

A Measure of Topology Preservation Using Rank Correlation Coefficient. When objects are arranged in order according to some quality which they all possess to a varying degree, they are said to be ranked with respect to that quality. In SOM for a node j the sequences of neighborhood indexes $n_k^U(j)$ and $n_k^V(j)$ (as defined in section 2.2.1) produce two rankings of the neighboring neurons with respect to their proximity to j in the output plane and the input space respectively. If topology is preserved, these two rankings should show similarity, i.e., they should be correlated. Kendall's τ [10] coefficient is a measure of the intensity of rank correlation between two rankings. Kendall's τ coefficient is computed as follows:

Let R_1 and R_2 be two rankings of a set of n objects. Define the natural order 1,2,... as direct order (i.e., the pair, say, 2,3 is said to be in direct order and 3,2 is said to be in inverse order). Now for every distinct pair of objects from the set of n objects, set the value $v_1 = +1$ if they are in direct order in R_1, set $v_1 = -1$ if they are in inverse order. Similarly set v_2 according to the order in R_2. Multiply v_1 and v_2 to obtain the score for the pair of the objects. Let S be the sum of the scores for all pairs of objects (total $\frac{n(n-1)}{2}$ pairs). Then τ is defined as, $\tau = \frac{2S}{n(n-1)}$.

Major properties of τ are (a) if the rankings are in perfect agreement, i.e., every object has the same rank in both, τ is $+1$, indicating perfect positive correlation, (b) if the rankings are in perfect disagreement, i.e., one ranking is the inverse of other, τ is -1, indicating perfect negative correlation and (c) for other arrangements τ should lie between these limiting values. Increase of values from -1 to $+1$ corresponds to increasing agreements between the ranks.

However, it may happen that several objects possess a quality to same degree. This is the case of tied rank. The common practice is to mark such objects in the rankings and make their contribution to the score 0 (thus, the score due to a tied pair in any of the ranking becomes 0). If there are u objects tied among themselves in R_1, then $\frac{u(u-1)}{2}$ pairs will contribute to zero to the score S. Similarly v tied objects in R_2 will cause $\frac{v(v-1)}{2}$ pairs to contribute 0 to S. So total number of tied pairs in R_1 is $U = \frac{1}{2}\sum u(u-1)$ and in R_2 is $V = \frac{1}{2}\sum v(v-1)$ where the $\sum$ the summation is over all tied scores in respective ranking. Thus τ for tied rankings is defined as

$$\tau = \frac{S}{\sqrt{[\frac{1}{2}n(n-1) - U][\frac{1}{2}n(n-1) - V]}}. \tag{4}$$

Thus for each node j in a $m \times n$ SOM the rank correlation coefficient τ_j can be computed using eq. 4. The aggregate of the τ_js defined

$$T = \frac{1}{mn} \sum_{j=1}^{m \times n} \tau_j \tag{5}$$

can be utilized as a good measure of topology preservation in the SOM. However due to large number of ties in the sequences $n_k^U(j)$ for SOMs with output dimensions more than 1, T does not attain the value 1 even for perfect topology preservation.

3   Link Density

When the SOM algorithm is implemented on a serial computer, during the training, in each iterative step, the computations are arranged to reflect parallelism in winner search and updates. In biological systems or hardware implementations, where true parallelism can be achieved, this requires for each neuron to be directly connected to the all other neurons those are part of the map. However such complete connections are unlikely in biological systems and hard to achieve and maintain in electronic hardware. So the question is worth asking that whether the SOM algorithm can withstand less than complete connections? If it can then to what extent? The absence of a connection manifests itself in *not allowing a neuron to update* though it falls within the neighborhood of the winner. In other words, the absence of the connection between a neuron (say a) and another (say b) will result in no update of b when a is a winner, even if b falls within the neighborhood of a. To express the concept of partial connection between one neuron and its neighbors we use the term "link density", measured in percentage values. The value of 100% denotes complete connection as in usual SOM, and 0% denotes no connection, i.e., no neighborhood update. Though the link density, as defined here is the property of each individual node, we refer to an SOM having a value of link density to indicate that all the nodes in the network have the same value of link density.

Again we can think about two different types of deviations from full connectivity. The first case can be thought as natural one. A neuron is more likely to be connected to the nearby neurons than to those far away. This can be modelled using a probability distribution such that the probability of a link between two nodes being present is inversely proportional to the Euclidean distance in the output space between the nodes. This situation is analogous to what can be expected in biological networks or may be desirable for designing hardware implementations economically.

The other situation involves random absence of links. This can be modelled by a uniform distribution over the interneuron distances. The situation is analogous to damage caused in the biological maps due to accident and diseases or random failures of components in hardware implementations. In the current paper we investigate empirically both the situations separately with different levels of link densities. We use three quantitative indexes described in the previous section to evaluate the topology preservation properties.

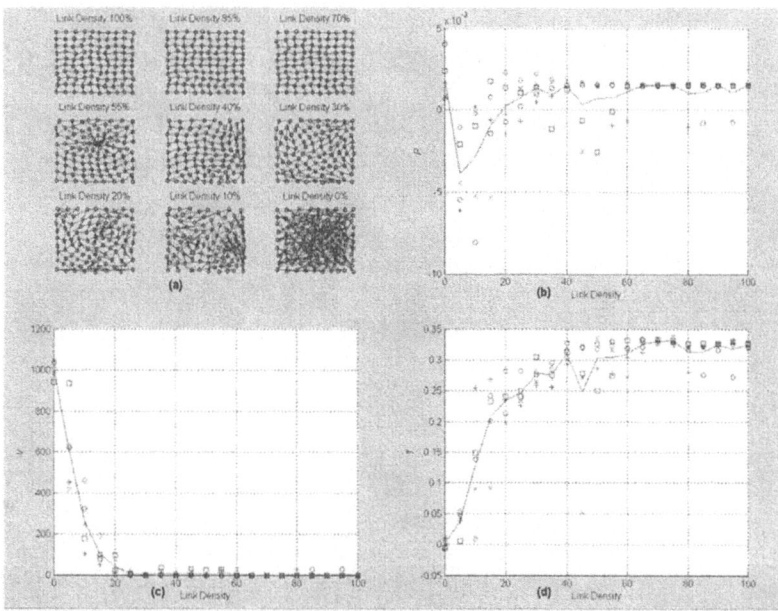

Fig. 1. The graphical results when the probability of absence of a link is proportional to the interneuron distance

4   Experimental Results and Discussion

The data set used in all experimental studies reported here contain ten thousand 2D points uniformly distributed over a square. We have studied both the cases described above, 1) when the link density of the winner neuron decreases in proportion to its distance from other neurons and 2) when the absence of links are random. For both the cases we have experimented with link densities varying from 0% to 100% in steps of 5%, for each value of link density 5 SOMs are trained. All the SOMs are of size 10 × 10 and all the parameters other than the link densities and randomization seeds for different runs for a particular link density, are identical. The results for case 1 and 2 are summarized graphically in Figures 1 and 2 respectively. Each of the figures is divided into four panels. The panel (a) contains the view of the maps for some selected SOMs with different link densities. The panels (b)-(d) depict the variation of topology preservation property w.r.t the link density measured in topographic product P (Eq. 1), index of topology violation V (Eq. 3) and rank correlation based index of topology preservation T (Eq. 5) respectively. In each of these panels the measurements for 5 SOMs for each link density is marked with the symbols ○, +, ×, ◇ and □ respectively and the solid line represents the averages of 5 measurements.

As expected, in both of the cases, with decrease of link density the topology preservation suffers. All the three indexes agree over that. However the topological product values are difficult to interpret in this context since its deviation

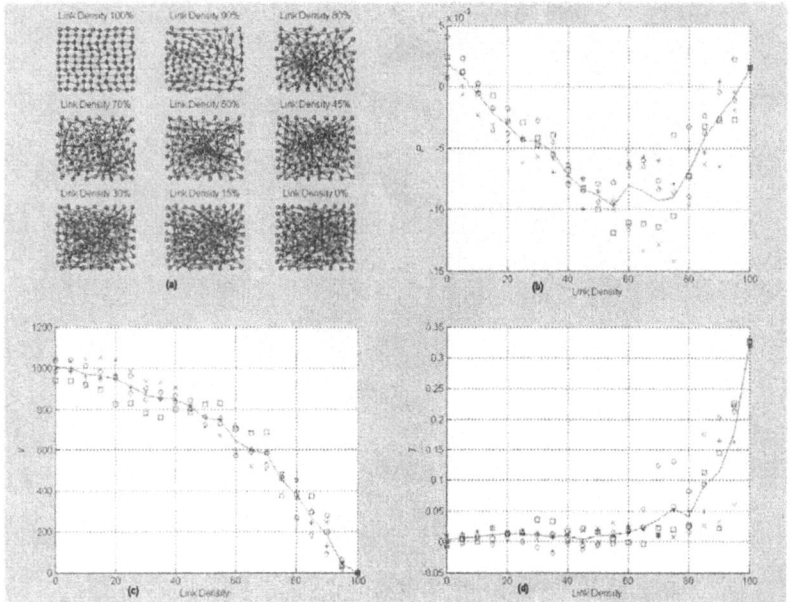

Fig. 2. The graphical results when the probability of absence of a link is random

from 0 (the perfect preservation) is more of a indicative of dimensional mismatch between input and output spaces. The other two indexes reflect the situations better, though V shows relative insensitivity to lack of topology preservation due cases like twisting of the map and below a threshold it increases rapidly. The third coefficient T shows the change of value as can be expected comparing with the visual inspection of the maps. It is also sensitive to the changes due to twisting of map.

All three measurements indicate good robustness of the SOM algorithm in case 1. It can be seen from figure 1 that the significant decrease of topology preservation can not be detected until the link density falls to nearly 30%. In contrast, in case 2 the drastic decrease of topology preservation can be observed even at link density as high as 90%.

5   Conclusion

The empirical study reported in this paper demonstrate the robustness of SOM even when the link density is quite low provided the absence of links are more likely to occur between distant neurons. This emphasizes the localized learning in SOM. However it is also found that in case absence of link is equally likely at all interneuron distance, the system degrades very quickly. These results are in agreement with the biological analogues we mentioned earlier. The study also provide a comparison regarding the usefulness of three quantitative measures for topology preservation in SOM.

References

1. T. Kohonen, "The self-organizing map," *Proc. IEEE*, vol. 78, no. 9, pp. 1464-1480, 1990.
2. T. Kohonen, *Self-Organization and Associative Memory*, Springer-Verlag, 1989.
3. J. A. Anderson, A. Pellionisz and E Rosenfeld (eds.)", *Neurocomputing 2: Directions for Research*, MIT Press, 1990.
4. E. I. Knudsen, S. Du Lac and S. D Esterly, "Computational maps in brain", *Ann. Rev. Neurosci.*, vol. 10, pp. 41-65, 1987.
5. T. Kohonen, "Automatic formation of topological maps of patterns in a self-organizing sysyem", *Proc. 2nd Scandinavian Conf. on Image Analysis*, pp. 214-220, 1981.
6. T. Kohonen, "Self-organized formation of topologically correct feature maps", *Biol. Cybern.*, vol. 43, pp. 59-69, 1982.
7. H. Bauer and K. R. Pawelzik, "Quantifying the Neighborhood Preservation of Self-Organizing Feature Maps," *IEEE Trans. on Neural· Networks*, vol. 3, no. 4, pp. 570-579, 1992.
8. M. C. Su, H. T. Chang and C. H. Chou, "A novel measure for quantifying the topology preservation of self-organizing feature maps", *Neural Processing Letters*, vol. 15, no. 2, pp. 137-145, 2002.
9. A. Laha and N. R. Pal, "On different variants of self-organizing feature map and their property", *Proc. of the IEEE Hong Kong Symposium on Robotics and Controls*, Volume 1 pp I-344 - I-349, 1999.
10. M. Kendall and J. D. Gibbons, *Rank Correlation Coefficient*, Edward Arnold, 1990.

Extending the SOM Algorithm
to Non-Euclidean Distances via the Kernel Trick

Manuel Martín-Merino[1] and Alberto Muñoz[2]

[1] University Pontificia of Salamanca, C/Compañía 5, 37002 Salamanca, Spain
mmerino@ieee.org
[2] University Carlos III, C/Madrid 126, 28903 Getafe, Spain
alberto.munoz@uc3m.es

Abstract. The Self Organizing Map is a nonlinear projection technique that allows to visualize the underlying structure of high dimensional data. However, the original algorithm relies on the use of Euclidean distances which often becomes a serious drawback for a number of real problems. In this paper, we present a new kernel version of the SOM algorithm that incorporates non-Euclidean dissimilarities keeping the simplicity of the classical version. To achieve this goal, the data are nonlinearly transformed to a feature space taking advantage of Mercer kernels, while the overall data structure is preserved.
The new SOM algorithm has been applied to the challenging problem of word relation visualization. We report that the kernel SOM improves the map generated by other alternatives for certain classes of kernels.

1   Introduction

The Self Organizing Map (SOM) [7] is a non-linear projection technique that helps to discover the underlying structure of high dimensional data. It has been applied to a wide variety of practical problems [7, 8] with remarkable results.

However, the algorithm originally proposed in [7] relies on the use of the Euclidean distance. Therefore, the performance is not satisfactory when the object relations can not be accurately modeled by Euclidean dissimilarities.

There are a large variety of applications for which the Euclidean distance fails to reflect the object proximities. This is the case for sparse high dimensional data [1]. In particular, the Euclidean distance is not appropriate to model semantic relations between terms, in the context of text mining [11, 9].

In this paper, we propose a new kernel version of the SOM Batch [7] algorithm that transforms nonlinearly the data to a feature space [15] where the SOM prototypes are organized. This is analogous to organize the network in input space but measuring the object proximities with a non-Euclidean dissimilarity induced by appropriate kernels. The dissimilarities induced are expected to reflect more accurately the object proximities. The new algorithm is derived from the minimization of an error function and avoids the need to solve complex nonlinear optimization problems.

N.R. Pal et al. (Eds.): ICONIP 2004, LNCS 3316, pp. 150–157, 2004.
© Springer-Verlag Berlin Heidelberg 2004

The main contribution of this paper is to extend the SOM algorithm to deal with non-Euclidean dissimilarities but keeping the simplicity of the SOM originally proposed by [7]. The new model takes advantage of kernel techniques and is derived from the optimization of an error function. Therefore it has a solid theoretical foundation. Finally the algorithm is compared rigorously with some alternatives using textual data.

This paper is organized as follows: Section 2 introduces the SOM Batch algorithm. Section 3 presents a kernelized version of the SOM Batch algorithm. In section 4 the new model is applied to the problem of word relation visualization. Finally, section 5 draws some conclusions and outlines future research trends.

2   The Self-organizing Map

The SOM algorithm [7] is a nonlinear projection technique that allow us to visualize the underlying structure of high dimensional data. Input vectors are represented by neurons arranged along a regular grid (usually 1D-2D) in such a way that similar vectors in input space become spatially close in the grid.

From a practical point of view, the SOM algorithm originally proposed by [7] is analogous to the algorithm that results from the minimization of the following quantization error [5]:

$$E(\mathcal{W}) = \sum_r \sum_{x_\mu \in V_r} \sum_s h_{rs} D(x_\mu, w_s) \qquad (1)$$

where w_s denotes the prototype associated to neuron s in input space and V_r the corresponding Voronoi region. D is the square Euclidean distance and h_{rs} is a neighborhood function (for instance a Gaussian of width σ) that performs as a smoothing kernel over the grid [10]. The smoothing parameter σ is adapted iteratively from an initial value σ_i (usually large) to a final value σ_f that determines the variance of the principal curve [10]. σ is adapted in each iteration using for instance the following rule: $\sigma(t) = \sigma_i(\sigma_f/\sigma_i)^{t/N_{iter}}$ [10].

The function error (1) is minimized when nearby prototypes according to the Euclidean distance are represented by neighboring neurons in the grid.

The optimization can be carried out by a simple iterative algorithm made up of two steps: First a quantization algorithm is run that represents the dataset by a certain number of prototypes. Next, the prototypes are organized by minimizing equation (1). This results in a simple updating rule for the network prototypes [7, 10].

3   Kernel SOM Algorithm

In this section we propose a new SOM algorithm that incorporates non-Euclidean dissimilarities keeping the simplicity of its original version. To achieve this goal, the dataset is non-linearly transformed to a feature space, where the object proximities are measured by the Euclidean distance. However considering this measure in feature space is analogous to work with a non-Euclidean dissimilarity in the original input space induced by appropriate kernels. Besides, as we explain next, the interpretation of the algorithm as a kernel method significantly simplifies the resulting optimization problem.

Let $k(\boldsymbol{x}_i, \boldsymbol{x}_j)$ be a Mercer kernel [15]. That is, there exists a non-linear map ϕ such that $k(\boldsymbol{x}_i, \boldsymbol{x}_j) = \phi(\boldsymbol{x}_i)^T \phi(\boldsymbol{x}_j)$. The SOM error function in feature space can be written in terms of ϕ as:

$$E(\mathcal{W}) = \sum_r \sum_{\phi(\boldsymbol{x}_\mu) \in V'_r} \sum_t h_{rt} (\phi(\boldsymbol{x}_\mu) - \boldsymbol{w}_t)^T (\phi(\boldsymbol{x}_\mu) - \boldsymbol{w}_t) \qquad (2)$$

where $\boldsymbol{w}_t$ are the SOM prototypes in feature space that can be written as $\boldsymbol{w}_t = \sum_i \alpha_{ti} \phi(\boldsymbol{x}_i)$ [15] and V'_r denotes the corresponding Voronoi region computed in feature space. Finally, h_{rt} is a neighborhood function that determines the degree of smoothing as in the classic version.

The minimization of the function error (2) is done through an iterative algorithm (inspired in the classic version [7,5]) made up of two steps:

1. *Voronoi Tessellation*: Each pattern is assigned to the nearest neighbor prototype according to the Euclidean distance. Fortunately, the Euclidean distance factorizes in terms of scalar products only. Therefore it can be written exclusively in terms of kernel evaluations:

$$\begin{aligned} d^2(\phi(\boldsymbol{x}_\mu), \boldsymbol{w}_t) &= (\phi(\boldsymbol{x}_\mu) - \boldsymbol{w}_t)^T (\phi(\boldsymbol{x}_\mu) - \boldsymbol{w}_t) \\ &= [\phi(\boldsymbol{x}_\mu) - \sum_i \alpha_{ti} \phi(\boldsymbol{x}_i)]^T [\phi(\boldsymbol{x}_\mu) - \sum_j \alpha_{tj} \phi(\boldsymbol{x}_j)] \\ &= k(\boldsymbol{x}_\mu, \boldsymbol{x}_\mu) - 2 \sum_i \alpha_{ti} k(\boldsymbol{x}_\mu, \boldsymbol{x}_i) + \sum_{ij} \alpha_{ti} \alpha_{tj} k(\boldsymbol{x}_i, \boldsymbol{x}_j) \qquad (3) \end{aligned}$$

Notice that the above distance in feature space induces a non-Euclidean dissimilarity in input space. This feature help to remove the hypothesis of normal distribution for the data under the principal curve which is assumed by several SOM like algorithms [13].

2. *Quantization Error Optimization*: This step adapts the network prototypes $\boldsymbol{w}_t$ to minimize the error function (2). This can be done easily considering that the quantization error can be written exclusively in terms of kernel evaluations. In fact, substituting the expression for the Euclidean distance in feature space, the error function (2) can be written as:

$$E(\mathcal{W}) = \sum_r \sum_{\phi(\boldsymbol{x}_\mu) \in V'_r} \sum_t h_{rt} \left[k(\boldsymbol{x}_\mu, \boldsymbol{x}_\mu) - 2 \sum_i \alpha_{ti} k(\boldsymbol{x}_\mu, \boldsymbol{x}_i) + \sum_{ij} \alpha_{ti} \alpha_{tj} k(\boldsymbol{x}_i, \boldsymbol{x}_j) \right]$$
$$(4)$$

Notice that if k is a Mercer kernel, the minimization of the above error function in dual space is equivalent to a quadratic optimization problem. Computing the first derivative and equating to zero, we get the following system of linear equations for the α_{ti} coefficients:

$$\mathbf{a} = \frac{1}{2} \mathbf{k\alpha N}, \qquad (5)$$

where **k** is the kernel matrix and

$$\mathbf{a} = (a_{it}) = \sum_r h_{rt} \sum_{\phi(x_\mu) \in V'_r} k(x_i, x_\mu), \tag{6}$$

$$\mathbf{N} = (N_{it}) = diag(\sum_r N_r h_{rt}). \tag{7}$$

Now, the $\boldsymbol{\alpha}$ matrix that minimizes (4) can be obtained solving explicitly the equation (5)

$$\boldsymbol{\alpha} = 2\mathbf{k}^\dagger \mathbf{a} \mathbf{N}^{-1}, \tag{8}$$

where $\mathbf{k}^\dagger$ denotes the pseudoinverse of the kernel matrix.

We next discuss briefly an important issue regarding the kind of kernels to be considered by the new model.

Notice that the nonlinear mapping ϕ induced by the associated kernel should not change drastically the underlying structure of the original dataset. Otherwise the relations suggested by the grid of neurons would become meaningless. Therefore, we are interested in kernels that give rise to maps that preserve roughly the ordering of neighbors in input space. Fortunately, it is justified next that both, the RBF and certain polynomial kernels verify this property.

Let δ_{ij} be the Euclidean distance in feature space. It can be written just in terms of kernel evaluations as [15]:

$$\delta_{ij}^2 = d^2(\phi(x_i), \phi(x_j)) = k(x_i, x_i) + k(x_j, x_j) - 2k(x_i, x_j) \tag{9}$$

A simple calculation shows that for the RBF kernel $\delta_{ij}^2 = f(d_{ij}^2) = 2(1 - e^{-d_{ij}^2/\sigma_k^2})$, where d_{ij}, δ_{ij} refer to the dissimilarities in input and feature spaces respectively. The nonlinear function f is monotonically increasing and hence it preserves the neighbor's ordering induced by d_{ij} in input space.

Similarly, it can be justified that the nonlinear maps induced by polynomial kernels of the form $k(x_i, x_j) = (x^T y)^k$ with k an odd integer, preserve the data structure. Indeed, any dissimilarity in feature space verifies that [4]

$$\tilde{\delta}_{ij}^2 = \tilde{c}_{ii} + \tilde{c}_{jj} - 2\tilde{c}_{ij}, \tag{10}$$

where $\tilde{c}_{ij}$ denotes the similarity between the patterns i, j. Equating the right hand side of (9) and (10) it can be deduced that the similarities in feature space verify $\tilde{c}_{ij} = k(x_j, x_j) = (x_i^T x_j)^k = (c_{ij})^k$, where c_{ij} denotes the similarities in input space. This suggests again that for k an odd integer the nonlinear mapping ϕ preserves the underlying data structure suggested by the similarity considered in input space.

We next comment shortly how to reconstruct the prototypes in input space for the kernel SOM which may be interesting for some practical applications. Our experience suggests that the centroid of the Voronoi regions defined by equation (11) can be a good approximation if the smoothing parameter σ_f for the neighborhood function h_{rt} is small.

$$w_r = \frac{1}{N'_r} \sum_{s \in V'_r} x_s, \tag{11}$$

where V_r' refers to the Voronoi region in feature space and x_s is the vectorial representation of pattern s in input space.

Alternatively the method proposed by [14] may be used to get approximate preimages of the network prototype in input space regardless of the value for the smoothing parameter.

Finally we finish this section with a concise remark about the computational complexity of the kernel SOM algorithm. The computational burden of the first step *(Voronoi Tesselation)* is determined by the computation of the kernel matrix and the Euclidean distances in feature space (3). This is roughly equal to $\mathcal{O}(dN^2)$, where N denotes the number of patterns and d the dimension of the vector space representation. The second step *(Quantization error optimization)* involves the computation of a kernel pseudoinverse matrix through an SVD. This linear algebra operation does not usually involve more computation than the previous step particularly when the kernel matrix is sparse [3,15]. Therefore, the kernel SOM does not increase the computational burden of its classic version [8].

4   Experimental Results

In this section, the kernel SOM is applied to generate a visual representation of term relationships in a textual collection. This is a challenging problem in which the Euclidean distance often fails to model the term proximities [11,9].

To guarantee an objective evaluation of the maps, we have built a collection made up of 2000 documents recovered from three commercial databases "LISA", "INSPEC" and "Sociological abstract". The terms can be classified in seven main topics, according to the database thesaurus.

To evaluate the mapping algorithms we check if words nearby in the map are associated in the thesaurus. To this aim, the word map is first partitioned into 7 groups using a clustering algorithm such as PAM [6]. Then, the maps are evaluated from different view points by means of the following objective measures: The F measure [2] determines if words grouped together in the map are related in the thesaurus. It is a measure of the overall word map quality. The entropy measure [16] E gives an idea of the overlapping between different topics in the map. Small values are preferred. Finally, the mutual information I [16] is a nonlinear correlation measure between the classifications induced by the thesaurus and the word map respectively. It gives valuable information about the position of less frequent terms [17].

Words are clustered for SOM maps using the following procedure. First the SOM prototypes are projected using Sammon mapping [12,7] (see figure 1). Next, they are clustered together and each term is assigned to the group of the nearest neighbor prototype.

The network topology for SOM algorithms has been chosen linear because the organization is usually easier.

The parameters of the kernel SOM have been set up experimentally as for the SOM algorithm. Hence, $(\sigma_i \approx \text{Nneur}/2)$ and $\sigma_f \approx 2$ give good experimental results for the problem at hand. However, as the kernels considered become strongly non-linear, larger values for the smoothing parameter (σ_f) help to improve the network organization.

Regarding the kernel parameters, k denotes the degree of the polynomial kernel considered and σ_k the width of the Gaussian kernel. σ_k is fixed experimentally ranging from the 50 percentile of the dissimilarity histogram (corresponding to linear kernels) to the 4 lower percentile which corresponds to strongly non-linear kernels.

Table 1. Kernel SOM versus Kernel PCA and SOM for our textual collection.

	F	E	I
[1] SOM	0.70	0.38	0.23
[2a] *Polynomial kernel SOM* ($k = 1$)	0.72	0.32	0.26
[2b] *Polynomial kernel SOM* ($k = 3$)	0.57	0.41	0.19
[2c] *RBF kernel SOM* ($\sigma_k = 0.8$)	0.69	0.29	0.26
[2d] *RBF Kernel SOM* ($\sigma_k = 0.5$)	0.47	0.35	0.15
[3a] Polynomial kernel PCA ($k = 1$)	0.59	0.46	0.20
[3b] Polynomial kernel PCA ($k = 3$)	0.55	0.45	0.17
[3c] RBF kernel PCA ($\sigma_k = 0.8$)	0.56	0.42	0.19
[3d] RBF kernel PCA ($\sigma_k = 0.5$)	0.38	0.61	0.08

Parameters: [1] Nneur = 88, Niter = 30, $\sigma_i = 30$, $\sigma_f = 2$; Nneur=40; [2a] Niter = 20, $\sigma_i = 25$, $\sigma_f = 2$; [2b-c] Niter = 15, $\sigma_i = 25$, $\sigma_f = 2$; [2d] Niter = 20, $\sigma_i = 25$, $\sigma_f = 4$

Table 1 compares empirically the kernel SOM (rows 2) with two popular alternatives such as kernel PCA [15] (rows 3) and the SOM considered in the WEBSOM project [8] (row 1). The most important results are the following:

The kernel SOM algorithm with slightly nonlinear RBF or polynomial kernels ($2c, 2a$) outperforms the SOM algorithm (1). In particular the position of less frequent terms is improved up to 12% (I) and the overlapping between different topics in the map is significantly reduced ($\Delta E = 24\%$ for the RBF kernel). However, table 1 shows that strongly nonlinear kernels such as $2d$, $2b$ have a negative impact on the word maps generated by kernel SOM. Notice that this problem is observed for kernel PCA ($3b, 3d$) as well which suggests that strongly nonlinear RBF or polynomial kernels are not suitable to handle textual data. It is worth noting that both, our kernel SOM (2) and SOM (1) improve significantly the maps generated by Kernel PCA (3) regardless of the parameters considered.

Finally, figure 1 illustrates the performance of the kernel SOM algorithm from a qualitative point of view. The kernel SOM prototypes have been reconstructed (see section 3) and projected to $\mathbb{R}^2$ using Sammon mapping [7, 12]. For the sake of clarity only a small subset of words have been shown. Notice that the organization of the kernel SOM prototypes is satisfactory as well as the term relationships induced by the network.

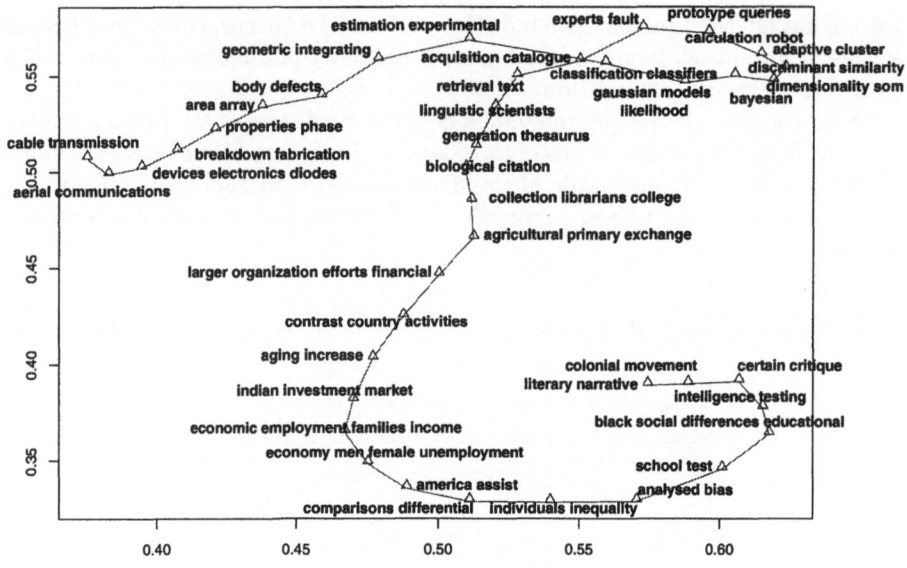

Fig. 1. Word map generated by kernel SOM for a subset of terms.

5   Conclusions and Future Research Trends

In this paper we have extended the SOM algorithm to deal with non-Euclidean dissimilarities. The kernel SOM presented transforms non-linearly the data to feature space taking advantage of certain class of Mercer kernels that preserve the data structure. The proposed algorithm has been tested in a challenging problem such as word relation visualization. Our algorithm has been compared objectively with widely known alternatives such as SOM and kernel PCA.

The experimental results suggest that the kernel SOM algorithm with slightly nonlinear kernels improves significantly the maps generated by SOM and particularly by kernel PCA. Our model improves considerably the position of less frequent terms and achieves a remarkable reduction of the overlapping between different topics in the map.

Future research will focus on the development of new specific kernels for the problem under consideration.

References

1. C. C. Aggarwal, A. Hinneburg and D. A. Keim. On the surprising behavior of distance metrics in high dimensional spaces. In Proc. of the International Conference on Database Theory (ICDT), 420-434, London, UK., January 2001.
2. R. Baeza-Yates and B. Ribeiro-Neto. Modern Information Retrieval. Addison-Wesley, New York, 1999.
3. M. W. Berry, Z. Drmac, and E. R. Jessup. Matrices, vector spaces and information retrieval. SIAM review, 41(2):335-362, 1999.

4. T. F. Cox and M. A. A. Cox. Multidimensional Scaling. Chapman & Hall/CRC Press, New York, second edition, 2001.
5. T. Heskes. Energy functions for self-organizing maps. In E. Oja and S. Kaski editors, Kohonen Maps, chapter 6, 303-315. Elsevier, Amsterdam, 1999.
6. L. Kaufman and P. J. Rousseeuw. Finding groups in Data; An Introduction to Cluster Analysis. John Wiley and Sons, USA, 1990.
7. T. Kohonen. Self-Organizing Maps. Springer Verlag, Berlin, second edition, 1995.
8. T. Kohonen, S. Kaski, K. Lagus, J. Salojarvi, J. Honkela, V. Paatero, and A. Saarela. Organization of a massive document collection. IEEE Transactions on Neural Networks, 11(3):574-585, May 2000.
9. M. Martín-Merino and A. Muñoz. Self organizing map and Sammon mapping for asymmetric proximities. In Lecture Notes on Computer Science (2130), 429-435, Springer Verlag, Berlin, 2001.
10. F. Mulier and V. Cherkassky. Self-organization as an iterative kernel smoothing process. Neural Computation, 7:1165-1177, 1995.
11. A. Muñoz. Compound key word generation from document databases using a hierarchical clustering art model. Journal of Intelligent Data Analysis, 1(1), 1997.
12. A. Muñoz. Self-organizing Maps for outlier detection. Neurocomputing, 18:33-60, 1998.
13. A. Ruiz and P. E. López de Teruel. Nonlinear kernel-based statistical pattern analysis. IEEE Transactions on Neural Networks, 12(1):16-32, January 2001.
14. B. Schölkopf, S. Mika, C. J. C. Burges and P. Knirsch. Input space versus feature space in kernel-based methods. IEEE Transactions on Neural Networks, 10(5):1000-1017, 1999.
15. Scholkopf, B. and A. J. Smola. Learning with Kernels, MIT Press, Cambridge, 2002.
16. A. Strehl, J. Ghosh, and R. Mooney. Impact of similarity measures on web-page clustering. Workshop of Artificial Intelligence for Web Search , Austin, Texas, USA, 58-64, July 2000.
17. Y. Yang and J. O. Pedersen. A comparative study on feature selection in text categorization. In Proc. of the International Conference on Machine Learning, 412-420, Nashville, Tennessee, USA, July 1997.

An Efficient Two-Level SOMART Document Clustering Through Dimensionality Reduction

Mahmoud F. Hussin[1], Mohamed S. Kamel[2], and Magdy H. Nagi[1]

[1] Dept. of Computer Science & Automatic Control, University of Alexandria,
Alexandria, Egypt
mfarouk@pami.uwaterloo.ca
[2] Dept. of Electrical and Computer Engineering, University of Waterloo,
Waterloo, Ontario, Canada
mkamel@uwaterloo.ca

Abstract. Document Clustering is one of the popular techniques that can unveil inherent structure in the underlying data. Two successful models of unsupervised neural networks, Self-Organizing Map (SOM) and Adaptive Resonance Theory (ART) have shown promising results in this task. The high dimensionality of the data has always been a challenging problem in document clustering. It is common to overcome this problem using dimension reduction methods. In this paper, we propose a new two-level neural network based document clustering architecture that can be used for high dimensional data. Our solution is to use SOM in the first level as a dimension reduction method to produce multiple output clusters, then use ART in the second level to produce the final clusters using the reduced vector space. The experimental results of clustering documents from the RETURES corpus using our proposed architecture show an improvement in the clustering performance evaluated using the entropy and the f_measure.

1 Introduction

Document clustering attempts to organize documents into groups where each group represents some topic that is different than those topics represented by the other groups. It has been used in presenting organized and understandable results to the user of a search engine query, and creating document taxonomies. Document clustering has also been used in efficient information retrieval by focusing on relevant subsets (clusters) rather than the whole data collection.

Unsupervised artificial neural networks are widely used for document clustering. Neural networks are highly suited to textual input, being capable of identifying structure of high dimensions within a body of natural language text. They work better than other methods even when the data contains noise, has a poorly understood structure and changing characteristics.

The self-organizing map (SOM) [1] is a neuro-computational algorithm to map high-dimensional data to a two-dimensional space through a competitive and unsupervised learning process. It takes a set of objects (e.g. documents), each object represented by a vector of terms (keywords) and then maps them onto the nodes of a two-dimensional grid. The Adaptive Resonance Theory (ART) [2], is another type of

N.R. Pal et al. (Eds.): ICONIP 2004, LNCS 3316, pp. 158–165, 2004.
© Springer-Verlag Berlin Heidelberg 2004

unsupervised neural networks that possesses several interesting properties that make it appealing in the area of text clustering. It allows for plastic yet stable learning, and known for its ability to perform on-line and incremental clustering. Typically, ART has the ability to create new category dynamically, and can develop input categories at a given level of specificity, which depends on a vigilance parameter.

Representative work on document clustering based on unsupervised neural networks includes the hierarchical SOM (HSOM) [3], tree-based SOM (TS-SOM)[4], WEBSOM project [5], growing hierarchical SOM (GHSOM) [6], and hierarchical ART (HART) [7][8]. In previous work [9][10], we also considered phrase based document clustering representation with flat SOM and HSOM, and proposed a combination between SOM and ART in HSOMART [11][12].

In the text domain, documents are often represented by a vector of word counts in a vector space model of documents. The dimension of a word vector is the size of the vocabulary of a document collection (tens of thousands of words are not unusual). This high dimensionality of data poses two challenges. First the presence of irrelevant and noisy features can mislead the clustering algorithm. Second, in high dimensions data may be sparse (the curse of dimensionality), making it difficult for an algorithm to find any structure in the data [13]. To ameliorate these problems, a large number of dimension reduction approaches have been developed and tested in different application domains and research communities. The main idea behind these techniques is to map each text document into a lower dimensional space. The associations present in the lower dimensional representation can be used to perform clustering and classification more efficiently.

These dimension reduction techniques can be classified into three categories. One refers to the set of techniques based on feature selection schemes that reduce the dimensionality by selecting a subset of the original features, and techniques that derive new features by clustering the terms. These dimension reduction techniques aim to minimize the information loss compared to the original data or to maintain the similarity distance found in the data set. The second type of dimension reduction techniques is based on feature transformations which project high dimensional data onto interesting subspaces, Latent Semantic Indexing (LSI), Principle Component Analysis (PCA), and Multidimensional Scaling (MDS) are examples of this type. The third class of dimension reduction techniques is the self-organizing map (SOM) that uses a neuro-compuational approach.

In this paper, we propose a two-level document clustering architecture based on neural network, by using SOMs in the first level as a dimension reduction technique to map the original high dimension vector space to a reduced one based on its output clusters, and using the ART in the second level as a clustering algorithm to improve the quality of the document clustering process.

Our motivation behind this work is based on: first, the capability of SOM with fast learning process to project the high dimension input space on prototypes of a two dimensional grid that can be effectively utilized to reduce the dimension. Second, a considerable good quality clustering results using ART with the reduced dimension.

For a given high dimension data collection, we use multiple SOMs with different map sizes to output clusters, then a new vector space based on the resulted SOMs clusters was constructed and ART is used to produce the final clusters.

The remainder of this paper is organized as follows. Section 2 presents the vector space model representation. Section 3 describes the two-level SOMART document clustering method. In section 4 we show experimental results and their evaluation. Conclusion and future work are given in section 5.

2   The Vector Space Model for Document Representation

Most document clustering methods use the vector space model to represent document objects, where, each document is represented by a word vector. Typically, the input collection of documents are represented by a word-by-document matrix A:

$$A = (a_{ik}),$$

Where, a_{ik} is the weight of word i in document k. There are three different main weighting schemes used for the weight a_{ik}. Let f_{ik} be the frequency of word i in document k, N the number of documents in the collection, and n_i the total number of times word i occurs in the whole collection. Table 1 summarizes the three weighting schemes that will be used in our experiments.

Table 1. Three different weighting schemes used in the vector space model

Method	Weighting Scheme	Comments
Binary	$a_{ik} = \begin{cases} 1 & \text{if } f_{ik} > 0 \\ 0 & \text{otherwise} \end{cases}$	The simplest approach is to let the weight be 1 if the word occurs in the document and 0 otherwise.
Term Frequency	$a_{ik} = f_{ik}$	Another simple approach is to use the frequency of the word in the document.
Term Frequency Inverse Document Frequency	$a_{ik} = f_{ik} * \log\left(\dfrac{N}{n_i}\right)$	A well-known approach for computing word weights, which assigns the weight to word i in document k in proportion to the number of occurrences of the word in the document, and in inverse proportion to the number of documents in the collection for which the word occurs at least once.

3   Two-Level SOMART Document Clustering

The key idea of the two-level SOMART document clustering approach is to combine the information of multiple runs of different sizes SOMs in the first level to construct a new reduced vector space based on output clusters, then to apply an ART clustering algorithm to produce a final set of clusters in the second level as shown in Fig. 1. This idea is based on, combining the fast learning capability of SOM to map a high dimension space onto low dimension (typically two) space with the accuracy of the clusters produced by ART neural network[11].

The clustering process of this approach starting by applying the vector space model of the document collection to output different clusters. We apply different SOMs sizes to produce different features representing the document in the collection. Here, SOM

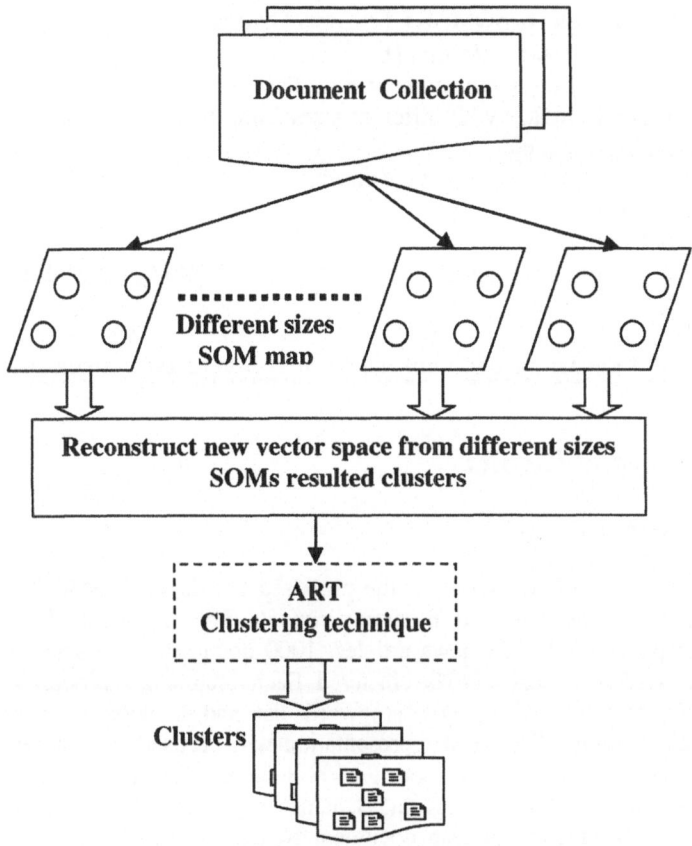

Fig. 1. Two-level SOMART Document Clustering

is used for dimensionality reduction belonging to topology preserving algorithms. This kind of algorithms aim to represent high dimensional data spaces in a low dimensional space while preserving as much as possible the structure of the data in the high dimensional data space.

Let $C = \{c1,c2,\ldots\ldots,cj\}$, where j is the total number of output clusters. Now the new reduced vector space representing the set of document collection is constructed. Each document is represented by a cluster vector instead of word vector. Typically the input collection of documents are represented by a binary cluster-by-documents matrix B, instead of a word-by-document matrix:

$$B = (b_{jk}), \text{ where } b_{jk} = \begin{cases} 1 & \text{if } document\ k \in \text{cluster j} \\ 0 & \text{if } document\ k \notin \text{cluster j} \end{cases}$$

Finally, to produce the final clusters using the reduced vector space B, we apply an ART clustering algorithm. The algorithm for two-level SOMART document clustering is summarized in algorithm1.

Algorithm 1: Two-level SOMART Document Clustering

Given a set of document collection D,
1. Prepare the word-by-document matrix *A* of set *D*.
2. Apply multiple SOMs with different sizes using matrix *A* to project *D* onto *j* clusters, $(c_1, c_2, ..., c_j)$.
3. Construct the cluster-by-document matrix *B* of set *D*.
  For each document *k*
    For each cluster *j*
      If document *k* belongs to cluster *j* **then** $b_{jk} = 1$ **else** $b_{jk} = 0$ **end if**
    End for
  End for
4. Apply ART clustering algorithm using matrix *B* to produce final clusters.

---

4 Experimental Results

4.1 Experimental Setup

To demonstrate the effectiveness of the proposed two-level SOMART method, the REUTERS test corpus was used in our experiments. This is a standard text clustering corpus, composed of 21,578 news articles. 1000 documents are selected from this corpus and used as the test set to be clustered. Each document is processed by removing a set of common words using a "stopword" list, and the suffixes are removed using a Porter stemmer. The word representation will be used as features to make a word-by-document matrix of the set of 1000 document, so the vector space size is 1000X7293. Three different weighting schemes were used: binary, term frequency (TF), and term frequency inverse document frequency (TF/IDF). In the two-level SOMART, the reduced cluster-by-document matrix size is 1000X379, 1000X383, and 1000X373, respectively.

The documents were clustered using SOM, Two-level SOM, and Two-level SOMART techniques implemented using the SOM-PAK package developed by Kohonen et. al. [14], and the ART gallery simulation package developed by Liden [15]. The two-level SOM clustering method uses SOM to produce the final clusters instead of using ART as in the two-level SOMART. The configurations of these document clustering techniques were as follows:

- The two-level SOMART, uses SOMs with different sizes ranging from 4 units (2X2) to 100 units (10X10), with 0.02 learning rate, and the second level using ART1 with vigilance value ranging from 0.02 to 0.16 and 0.9 learning rate.
- The two-level SOM, the same first level as in two-level SOMART, and the second level using SOM with map sizes ranging from 16 units (4X4) to 100 units (10X10) with learning rate 0.02.
- The SOM used with dimensions ranging from 16 units (4X4) to 100 units (10X10) with learning rate 0.02.

4.2 Quality Measures

Two measures are widely used in text mining literature to evaluate the quality of the clustering algorithms: cluster entropy and F-measure [16]. Both of these techniques rely on labeled test corpus, where each document is assigned to a class label.

The F-measure combines the precision and recall concepts from information retrieval, where each class is treated as the desired result for the query, and each cluster is treated as the actual result for the query, for more details refer to [9]. Cluster entropy uses the entropy concept from information theory and measures the "homogeneity" of the clusters. Lower entropy indicates more homogeneous clusters and vice versa.

In the test corpus, some documents have multiple classes assigned to them. This does not affect the F-measure, however the cluster entropy can no longer be calculated. Instead of cluster entropy, we define the class-entropy[9], which measures the homogeneity of a class rather than the homogeneity of the clusters.

4.3 Results

Basically we would like to maximize the F_measure, and minimize the Class entropy of clusters to achieve high quality clustering. The class entropy and F-measure results are shown in Figures 2-4 for the vector space with three different weighting schemes applied to SOM, two-level SOM (TSOM), and two-level SOMART (TSOMART) techniques. In most cases the Two-level SOM performed better in the clustering task than the SOM, also the Two-level SOMART performed better than the Two-level SOM and SOM for different number of output clusters. The average reduction of the class entropy using the two-level SOM technique relative to the SOM technique is 4.7%, 8.5%, and 2.6% for binary, term frequency, and tf/idf respectively. The corresponding improvement in the F-measure is 15.9%, 8.9%, and 5.5% for binary, term frequency, and tf/idf respectively. Similarly, the average reduction of the class entropy using two-level SOMART instead of SOM is 41.8%, 42.8%, and 48.6% for binary, term frequency, and tf/idf, respectively, while the corresponding improvement in F-measure is 40.7%, 48.2%, and 24.4%.

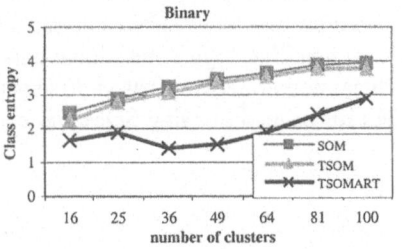 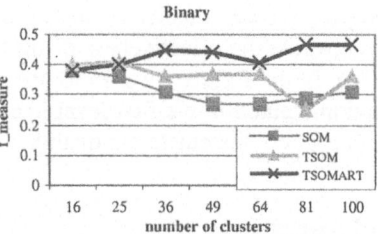

Fig. 2. Class entropy and F-measure for 1000 document set, using binary representation and clustered by SOM, TSOM, and TSOMART techniques

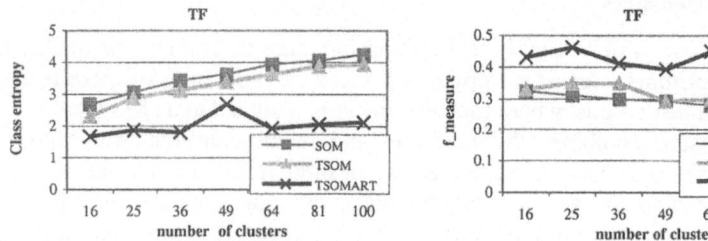

Fig. 3. Class entropy and F-measure for 1000 document set, using TF representation and clustered by SOM, TSOM, and TSOMART techniques

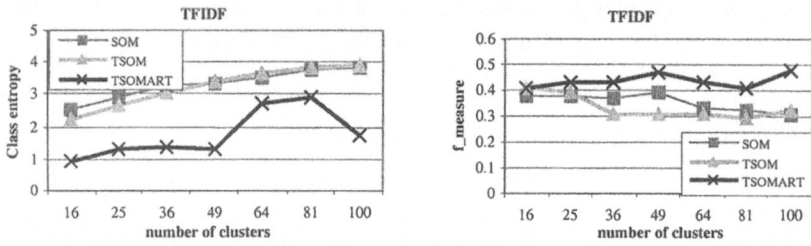

Fig. 4. Class entropy and F-measure for 1000 document set, using TFIDF representation and clustered by SOM, TSOM, and TSOMART techniques

5   Conclusion and Future Work

In this paper, we proposed a new two-level SOMART document clustering method. It uses SOM as a dimension reduction technique in the first level to produce multiple clusters in order to construct the reduced cluster-by-document matrix, then ART is applied to achieve an improved final clustering in the second level. The performance was evaluated by testing the two-level SOMART on the REUTRES test corpus, and comparing it to classical SOM and two-level SOM using both class entropy and the F_measure. The experimental results demonstrate that the proposed clustering method achieves a better quality clustering than SOM, and two-level SOM. The results also demonstrate the effectiveness of the method in handling high dimensional data clustering. In the future, we plan to use hierarchical neural network based dimension reduction method in the first level like HSOM [10] instead of using SOM and evaluate its impact on the clustering quality.

References

1. T. Kohonen, "Self-organizing maps," Springer Verlag, Berlin, 1995.
2. G.A. Carpenter and S. Grossberg, "A massively parallel architecture for a self-organizing neural pattern recognition machine.," Computer Vision, Graphics, and Image processing, vol. 34, pp. 54-115, 1987.

3. J. Lampinen and E. Oja, "Clustering properties of hierarchial self-organizing maps," Journal of Mathematical Imaging and Vision, pp. 261-272, 1992.
4. P. Koikkalainen, "Fast deterministic self-organizing maps," In Proc Int'l Conf Neural Networks, Paris, France, pp. 63-68, 1995.
5. T. Kohonen, S. Kaski, K. Lagus, J. Saloja, V. Pattero, and A. Saarela, "Organization of a massive document collection. IEEE Transactions on Neural Networks, Special Issue on Neural Networks for Data Mining and Knowledge Discovery, 11(3), 574–585, 2000.
6. M. Dittenbach, and D. Merkl, and A. Rauber, "Hierarchical Clustering of Document Archives with the Growing Hierarchical Self-Organizing Map," In Proc of the Int'l Conf on Artificial Neural Networks (ICANN01), Vienna, Austria, pp. 21-25, August 2001.
7. G. Bartfai, "An ART-based Modular Architecture for Learning Hierarchical Clusterings," Neurocomputing, Vol.13, pp.31-45, September 1996.
8. G. Bartfai, and R. White, "Adaptive Resonance Theory-based Modular Networks for Incremental Learning of Hierarchical Clusterings," Connection Science, Vol.9, No.1, pp.87-112, 1997.
9. J. Bakus, M. F. Hussin, and M. Kamel, "A SOM-Based Document Clustering using Phrases," In Proc of the 9th Int'l Conf on neural information processing, Singapore, pp. 2212-2216, November 2002.
10. M.F. Hussin, J. Bakus, and M. Kamel, "Enhanced phrase-based document clustering using Self-Organizing Map (SOM) architectures", Book Chapter in: Neural Information Processing: Research and Development, Springer-Verlag, pp. 405-424, May, 2004.
11. M.F. Hussin, M. Kamel "Document clustering using hierarchical SOMART neural network". In Proceedings of the 2003 Int'l Joint Conf on Neural Network, Portland, Oregon, USA, pp. 2238-2242, July, 2003.
12. M.F. Hussin, M. Kamel, "Integrating Phrases to Enhance HSOMART based Document Clustering," In Proc of the 2004 Int'l Joint Conf on Neural Network, Budapest, Hungry, Vol. 3, pp. 2347-2352, July 2004.
13. Xiaoli Fern, and Carla Brodley, "Random Projection for High Dimensional Data Clustering: A Cluster Ensemble Approach", In proc. Of The Twentieth International Conference on Machine Learning (ICML-2003), Washington, DC USA, August 2003.
14. T. Kohonen, J. Kangas and J. Laaksonen, "SOM-PAK: the self-organizing map program package ver.3.1," SOM programming team of Helsinki University of Technology, Apr. 1995.
15. L. Liden, "The ART Gallery Simulation Package ver.1.0," Dept. of cognitive and neural systems, Boston University, 1995.
16. M. Steinbach, G. Karypis, and V. Kumar, "A comparison of document clustering techniques," KDD'2000, Workshop on Text Mining, 2000.

Color Image Vector Quantization
Using Wavelet Transform
and Enhanced Self-organizing Neural Network

Kwang Baek Kim[1] and Dae Su Kim[2]

[1] Dept. of Computer Engineering, Silla University, Korea
[2] Dept. of Computer Science, Hanshin University, Korea
gbkim@silla.ac.kr

Abstract. This paper proposes a vector quantization using wavelet transform and enhanced SOM algorithm for color image compression. To improve the defects of SOM algorithm, we propose the enhanced self-organizing algorithm, which, at first, reflects the error between the winner node and the input vector in the weight adaptation by using the frequency of the winner node, and secondly, adjusts the weight in proportion to the present weight change and the previous weight one as well. To reduce the blocking effect and improve the resolution, we construct vectors by using wavelet transform and apply the enhanced SOM algorithm to them. The simulation results show that the proposed method energizes the compression ratio and decompression ratio.

1 Introduction

Computer Graphics and Imaging applications have started to make inroads into our everyday lives due to the global spread of Information technology. This has made the image compression an essential tool in computing with workstations, personal computers and computer networks. Videoconferencing, desktop publishing and archiving of medical and remote sensing images all entail the use of image compression for storage and transmission of data [1]. Compression can also be viewed as a form of classification, since it assigns a template or codeword to a set of input vectors of pixels drawn from a large set in such a way as to provide a good approximation of representation. A color image is composed of three primary components. The most popular choices of color primaries are (R, G, B), (Y, I, Q), and (Y, U, V), etc. In this paper, we considered color images in the (R, G, B) domain with a color of one pixel determined by three primary components, the red(R), green (G) and blue (B). Each component is quantized to 8 bits, hence 24 bits are needed to represent one pixel. The number of palette elements to be represented by 24 bits is 2^{24}, but all of the colors are not used to represent one image. So it is possible to compress pixel colors of the real image. Also, the compression of pixel colors is necessarily needed because of the limitation of disk space and the transmission channel bandwidth [2][3]. In the compression methods currently used, the image compression by Vector Quantization(VQ) is most popular and shows a good data compression ratio. Most methods by VQ use the LBG algorithm developed by Linde, Buzo, and Gray [4][5]. However this algorithm reads the entire image several times and moves code vectors into optimal position in each step. Due to the complexity of algorithm, it takes considerable time to execute. The vector quantization for color image requires the analysis of image pixels

N.R. Pal et al. (Eds.): ICONIP 2004, LNCS 3316, pp. 166–171, 2004.
© Springer-Verlag Berlin Heidelberg 2004

to determine the codebook previously not known, and the self-organizing map(SOM) algorithm, which is the self-learning model of neural network, is widely used for the vector quantization. However, the vector quantization using SOM shows the underutilization that only some code vectors generated are heavily used [6,7]. This defect is incurred because it is difficult to estimate correctly the center of data with no prior information of the distribution of data.

In this paper, we propose the enhanced SOM algorithm, which, at first, reflects the error between the winner node and the input vector in the weight adaptation using the frequency of the winner node, and secondly, adjusts the weight in proportion to the current and the previous changes of weight. By using the wavelet transform and the proposed SOM algorithm, we implement and evaluate the vector quantization. The evaluation result shows that the proposed VQ algorithm reduces the requirement of computation time and memory space, and improves the quality of the decompressed image decreasing the blocking effect.

2 Enhanced Self-organizing Neural Network

In this paper, we improved the SOM algorithm by employing three methods for the efficient generation of the codebook. First, the error between winner node and input vector and the frequency of the winner node are reflected in the weight adaptation. Second, the weight is adapted in proportion to the present and the previous changes of weight at the same time. Third, in the weight adaptation for the generation of initial codebook, the weight of the adjacent pixel of the winner node is adapted together.

In the proposed method, the codebook is generated by scanning the entire image only two times. In the first step, the initial codebook is generated to reflect the distribution of the given training vectors. The second step uses the initial codebook and regenerates the codebook by moving to the center within the decision region. To generate the precise codebook, it needs to select the winner node correctly and we have to consider the real distortion of the code vector and the input vector. For this management, the measure of frequency to be selected as winner node and the distortion for the selection of the winner node in the competitive learning algorithm are needed. We use the following equation in the weight adaptation.

$$w_{ij}(t+1) = w_{ij}(t) + \alpha(x_i - w_{ij}(t))$$

$$\alpha = f(e_j) + \frac{1}{f_j} \qquad (1)$$

Where α is the learning factor between 0 and 1 and is set between 0.25 and 0.75 in general. $(x_i - w_{ij}(t))$ is an error value and represents the difference between the input vector and the representative code vector. This means weights are adapted as much as the difference and it prefers to adapt the weight in proportion to the size of the difference. Therefore, we use the normalized value for the output error of the winner node that is converted to the value between 0 and 1 as a learning factor. The larger the output error is, the more the amount for the weight adaptation is. So, the weight is adapted in proportion to the size of the output error. $f(e_j)$ is the normalization function that converts the value of e_j to the value between 0 and 1, e_j is the output error of the j-th neuron, and f_j is the frequency for the j-th neuron as the winner.

The above method considers only the present change of weight and does not consider the previous change. So in the weight adaptation, we consider the previous weight change as well as the present ones. This concept corresponds to the momentum parameter of BP. We will also call this concept as a momentum factor. Based on the momentum factor, the equation for the weight adaptation is as follows:

$$w_{ij}(t+1) = w_{ij}(t) + \delta_{ij}(t+1) \tag{2}$$

$$\delta_{ij}(t+1) = \alpha(x_i - w_{ij}(t)) + \alpha\delta_{ij}(t) \tag{3}$$

In equation (3), the first term represents the effect of the present change of weight and the second term is the momentum factor representing the previous ones.

3   Application of Wavelet Transform

In this paper, for the proposed SOM algorithm, we apply a wavelet transform to reduce the block effect and to improve the decompression quality. After the wavelet transforms the color image, the color image is compressed by applying the vector quantization using the enhanced SOM algorithm to each separated RGB values. That is, by applying the wavelet transforms to the image, input vectors are generated, and the enhanced SOM algorithm are applied to the input vectors. If the index of winner node corresponding to the input vector is found, the original image vector corresponding to the transformed input vector is stored in the codebook. Wavelet transform is applied to the original image in the vertical and horizontal direction of a low frequency prior to the codebook generation. Specially, the image information of the original resolution is maintained without the down sampling used in the existing wavelet transform. Using the low frequency pass filter of wavelet emphasizes the strong areas of image and attenuates weak areas, have an equalization effect and remove the noise. Fig. 1 shows the structure of wavelet transform [8].

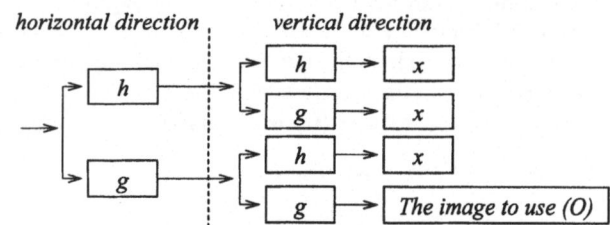

h : high frequency band pass   g : low frequency band pass

Fig. 1. The structure of wavelet transforms

4   Simulation

An experiment environment was implemented on an IBM 586 Pentium III with C++ Builder. Color bitmap images of 128×128 pixel size were used in the experiment. One image is divided into blocks of 4×4 size and each block is represented by the vector of 16 bytes, which constitutes the codebook. In this paper, the proposed VQ algorithm and LBG algorithm are compared in performance. In the case of the codebook generation and the image compression, the vector quantization using the enhanced SOM algorithm improves 5 times in the computation time than LBG algorithm and generates the codebook by scanning all image vectors only two times. This reduces the requirement of memory space. The application of the wavelet transform lightens the block effect and improves the recovery quality. Fig. 2 shows color images used in the experiment. Although the proposed algorithm can be applied to grayscale images, we selected various color images for this experiment because the proposed vector quantization algorithm is for the color image.

(a) Image 1: Album cover   (b) Image 2: Endoscopes

Fig. 2. Image samples used for experiment

Table 1 shows the size of codebooks generated by SOM Algorithm, enhanced SOM and the integration of wavelet transform and enhanced SOM for images in Fig. 2. In Table 1, the proposed integration of wavelet transform and enhanced SOM algorithm shows a more improved compression ratio than other methods. In the case of image 2, which the distribution of color is various, the compression ratio is low compared with different images. For the comparison of decompression quality, we measure the mean square error (MSE) between the original image and the recovered image, and presented in Table 2 the MSE of each image in the three algorithms.

Table 1. Size of codebook by VQ (unit: byte)

Algorithms Images	SOM	Enhanced SOM	Wavelet and Enhanced SOM
Image1	48816	33672	26528
Image2	54081	53649	28377

Table 2. Comparison of MSE (Mean Square Error) for compressed images

Algorithms Images	SOM	Enhanced SOM	Wavelet and Enhanced SOM
Image1	14.2	13.1	11.3
Image2	13.8	12.7	10.6

As shown in Table 2, the integration of wavelet transform and enhanced SOM algorithm shows the lowest MSE. Also, for images shown in Fig. 3, the decompression quality of LBG algorithm is worse than the above three algorithms.

LBG algorithm generates 10's temporary codebooks until the creation of the optimal codebook and requires a high computation time for codebook generation. Oppositely, the proposed algorithm generates only one codebook in the overall processing and reduces greatly the computation time and the memory space required for the

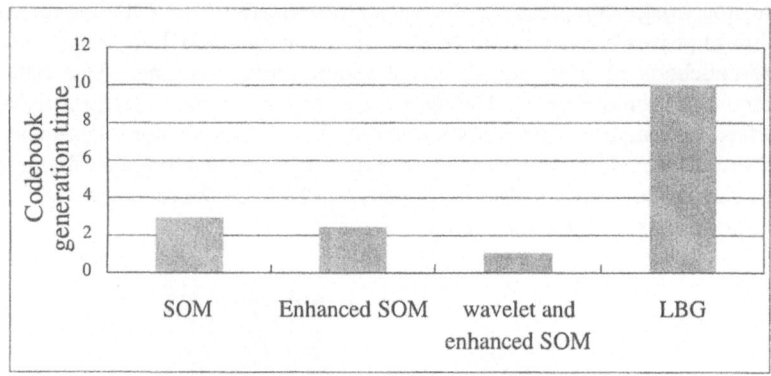

Fig. 3. Comparison of processing time for codebook generation

codebook generation. Fig.4 and Fig.5 show recovered images for original images of Fig.2 respectively. The improved SOM algorithm improves the compression ratio and the recovery quality of images by the codebook dynamic allocation more than the conventional SOM algorithm.

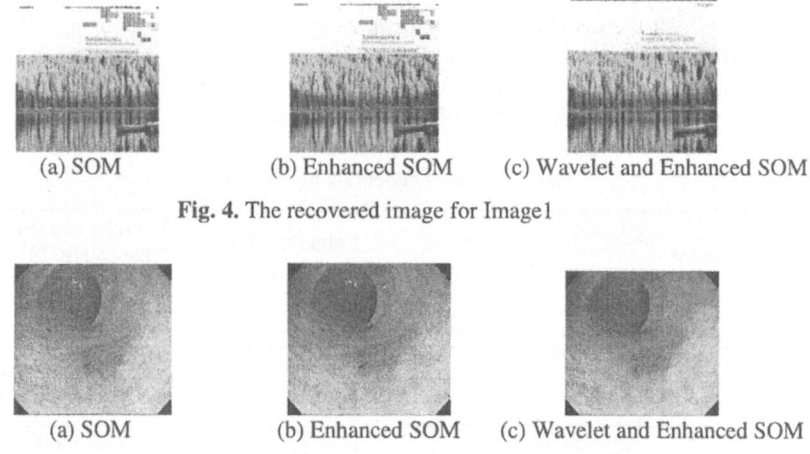

(a) SOM   (b) Enhanced SOM   (c) Wavelet and Enhanced SOM

Fig. 4. The recovered image for Image1

(a) SOM   (b) Enhanced SOM   (c) Wavelet and Enhanced SOM

Fig. 5. The recovered image for Image2

5 Conclusion

The proposed method in this paper can be summarized as follows: Using the enhanced SOM algorithm, the output error concept is introduced into the weight adaptation and the momentum factor is added. The simulation results show that the enhanced SOM algorithm for the color image compression produces a major improvement in both subjective and objective quality of the decompressed images. LBG algorithm is traditionally used for the codebook generation and requires considerable time especially for large size images, since the codebook is generated by repetitive scanning of the whole image. The proposed method is apt to real time application because the codebook is created by scanning the whole image only twice. The enhanced SOM algorithm performs the learning in two steps and total learning vectors are used only once in each step. In the first step, it produces the initial codebook by reflecting the distribution of learning vectors well. In the second step, it produces the optimal codebook by shifting to the current center of each code group based on the initial codebook. For reducing the memory space and the computation time for the codebook generation, we construct vectors by using wavelet transform and we apply the enhanced SOM algorithm to them. The simulation results showed that the integration of the wavelet transform and the enhanced SOM algorithm improves the defects of vector quantization such as the time and memory space caused by the complex computation and the block effect.

References

1. Rabbani, M. and Jones, P. W.: Digital Image Compression Technique. *Spie Optical Engineering Press* (1991) 144-169.
2. Orchard, M. T. and Bouman, C. A.: Color Quantization of Images. IEEE Trans. On Sp, Vol.39, No.12 (1991) 2677-2690.
3. Godfrey, K. R. L. and Attikiouzel, Y.: Self-Organized Color Image Quantization for Color Image Data Compression. Proc. of ICNN, Vol.3 (1993) 1622-1626.
4. Gersho, A. and Gray, R. M.: Vector Quantization and Signal Compression. *Kluwer Academic Publishers* (1992).
5. Oehler, K. L. and Gray, R. M.: Combining Image Compression and Classification using Vector Quantization. *IEEE Multimedia* (1997) 36-45.
6. Kim, K. B. and Cha, E. Y.: A Fuzzy Self-Organizing Vector Quantization For Image. *Proc. of IIZUKA*, Vol.2 (1996) 757-760.
7. Madeiro, F., Vilar, R. M., Fechine, J. M. and Aguiar Neto, B. G.: A Slef-Organizing Algorithm for Vector Quantizer Design Applied to Signal Processing. *Int. Journal of Neural Systems*, Vol.9, No.3 (1999) 219-226.
8. Strang, G. and Nguyen, T.: Wavelets and Filter Banks, Wellesley-Cambridge Press (1996).

Using SOM-Based Data Binning
to Support Supervised Variable Selection

Sampsa Laine and Timo Similä

Helsinki University of Technology, Laboratory of Computer and Information Science,
P.O. Box 5400, FI-02015 HUT, Finland
{sampsa.laine, timo.simila}@hut.fi

Abstract. We propose a robust and understandable algorithm for supervised variable selection. The user defines a problem by manually selecting the variables Y that are used to train a Self-Organizing Map (SOM), which best describes the problem of interest. This is an illustrative problem definition even in multivariate case. The user also defines another set X, which contains variables that may be related to the problem. Our algorithm browses subsets of X and returns the one, which contains most information of the user's problem. We measure information by mapping small areas of the studied subset to the SOM lattice. We return the variable set providing, on average, the most compact mapping. By analysis of public domain data sets and by comparison against other variable selection methods, we illustrate the main benefit of our method: understandability to the common user.

1 Introduction

Selection of data is a central task of any data analysis effort. Its importance is explicated in many data analysis models, such as, the Knowledge Discovery from Databases (KDD) [1] and the CRoss-Industry Standard Process for Data Mining (CRISP) [2]. This paper shows how the Self-Organizing Map (SOM) [3] can support this task. The user creates a SOM that well describes the problem. This is an exploratory data analysis task, and it has been extensively discussed e.g. in [4]. The main result of this paper is an algorithm that looks for other variables, which relate to this SOM-based definition of the problem. Already established methods are based on, e.g. finding the variable set that has the highest mutual information with the variables that define the problem [5]. Our method resembles such an approach. The key difference is that we do not measure distances in the original space, but on the lattice of the SOM.

We show with theoretical discussion and practical experiments that the proposed method compares favorably with common statistical methods. In conceptual comparison, we use the criteria of supervised operation, robustness and understandability. According to [6] these criteria define tools that appeal to the common user.

Supervised variable selection algorithms allow, and require, the user to define what he/she considers interesting [7]. A simple example is regression analysis.

N.R. Pal et al. (Eds.): ICONIP 2004, LNCS 3316, pp. 172–180, 2004.
© Springer-Verlag Berlin Heidelberg 2004

Stepwise methods can find the variables that create models, which best estimate the variable selected by the user. An unsupervised algorithm has no such information and must resort to the study of general properties of the variables. They may choose a variable subspace that minimizes entropy [8], select variables from variable clusters [9], or, as PCA, create a mapping which reduces dimensionality but retains variance [10]. However, all unsupervised methods share the same problem. As they do not know the goals of the user, they may select statistically significant variables that are irrelevant given the problem. Supervised methods, for example, the method proposed in this paper, can focus onto the variables that are relevant to the user's current task.

Robust algorithms allow straightforward analysis of versatile data. Consider a counter example: an algorithm that makes several statistical assumptions on the inputted data. For example, methods that assume Gaussian distributions are biased by outliers: a data point gains weight according to its squared distance from the main body of data. Another example is the assumption of linearity: nonlinear dependencies in the data may mislead the model to state that there is no information. Generally, violating the assumptions of an algorithm leads to biased results. These problems can be avoided by a careful pretreatment of the data. Outliers can be removed and variables can be transformed to obey the assumptions. However, a user without an extensive statistical background is unlikely to start this task. We propose a nonparametric method that makes few assumptions of the data. This promotes the usability of this tool to the common engineer.

According to [7], understandability, to the user, is often more important than accuracy. We consider understandability the most important of the three criteria discussed in the paper. People accept only the results that they find understandable and justifiable. The method presented in this paper is based on visualization and it is conceptually simple.

2   The Proposed Method

In the sense of supervised operation, our method resembles regression analysis. In both cases, the user selects the variables Y that are the targets of the study and, moreover, selects the variables X among which the best covariates are sought from. Then, manual or algorithmic methods are used to find the subset of X that allows the identification of the most accurate model. Consider, for example, an industrial plant. The analyst might look for the process control variables X, which best estimate a set of product quality parameters Y. The difference of our method to regression is that we do not look for a model. While we lose the benefit of gaining a model of this phenomenon, we have the benefit of relaxing the restrictions imposed by the selected regression model structure. Reconsidering the industrial plant, the user is provided with a more general characterization of the mapping from the selected subset of X to Y. Instead of creating a mathematical model, the user is offered a set of relevant control variables and visualization of their impact on product quality.

Regression model construction is guided by prediction error. While operating without such a model, we need a criterion for measuring information between X and Y. Our criterion resembles mutual information, which has earlier been used for variable selection [5]. However, mutual information between X and Y is symmetric, whereas we consider Y to be dependent on X. Neighboring data points from X should be mapped onto an undivided local area in Y. On the other hand, the inverse mapping from Y to X may be one-to-many. The most particular property of the criterion is the measurement of the dispersion in the mapping. Instead of measuring it in the Y-space, we measure it on the SOM lattice trained into the Y-space. While this mapping distorts the true distances between the data points, it allows a nonparametric binning of the data and visualization in the Y-space to the user.

Suppose that we have n measurements of the target variables y_i and the corresponding measurements of potentially related variables x_i. A SOM is trained into the Y-space to illustrate the distribution of possibly multivariate target data. The user can reselect the target variables, or, if necessary, change the training parameters of the SOM, until the problem is well described. See [3, 4] for more information of this phase. The target data is binned by assigning each measurement onto the nearest SOM neuron according to the best matching unit (BMU) rule [3]. We denote the neuron, which according to the Euclidean distance is closest to the measurement y_i in the Y-space, by $\text{BMU}(y_i)$. Moreover, we denote the indices of k nearest measurements of x_i in the X-space by $I(x_i)$, again, by using the Euclidean distance. The compactness of the mapping from X to Y is measured by the cost function

$$c(X) = \frac{1}{n} \sum_{i=1}^{n} \frac{1}{k} \sum_{j \in I(x_i)} d(\text{BMU}(y_i), \text{BMU}(y_j)), \qquad (1)$$

where d is the same distance function that was used in the neighborhood kernel of the SOM during the training phase. Distances are calculated using the positions of the neurons on the regular lattice (output space), when, for instance, all distances between adjacent neurons are the same. The inner sum calculates the average distance between the mapping of x_i and the mappings of its k nearest neighbors. The outer sum averages over all data points.

The variable selection algorithm browses various subsets of X and calculates the cost for each of them. Changing the subset of X affects the neighborhoods $I(x_i)$. The subset with the smallest cost will be selected. Any combinatorial optimization algorithm may be incorporated to the cost function minimization. Due to the combinatorial problem, we may have to resort to a local minimum.

The above approach has two sets of free parameters. Firstly, the generally well understood training parameters of the SOM [3]. Secondly, the sole parameter related to the variable selection, which is the size of the neighborhood in the X-space, denoted by k. Selecting a large neighborhood leads to global analysis focusing on continuous mappings. A smaller neighborhood allows more diverse mappings. For example, neighboring parts of a cluster in X may be mapped onto distinct areas of Y. Due to the visual nature of our approach, the user can study the results created by various values of this parameter. There hardly exists a single rule, which fits in every case. However, we performed the experiments of this paper with the neighborhood size of one third of the size of the whole data set.

Robustness of the approach may be analyzed from three points of view. Firstly, the SOM is a robust method of modeling probability distribution of the data in the Y-space. As the method is nonparametric, it does not fit a prototypical distribution model, but hosts a variety of distributions. The SOM performs a sort of equal mass binning by aiming to position an equal amount of data into each bin. Thus, outliers are unlikely to trouble the analysis. Secondly, the X-space is also studied without assuming any properties of the distribution of the data. The k nearest neighbor procedure focuses on the areas of dense data. Thirdly, the dependence of Y on X is measured by the compactness criterion (1) without assuming any specific model structure between the spaces. Even complex mappings can be studied by varying a single free parameter k without extensive model construction.

Understandability ensues from efficient visualizations and conceptual simplicity of the method. Consider, for example, that Y is high-dimensional. Another binning method, such as, the k-means algorithm [11], may be theoretically more solid than the SOM. However, to accept the results, the user needs to understand, how the bins reside in the Y-space. We display the results on the same SOM lattice that was defined by the user to describe the problem. Many other methods can handle multivariate, complex data and even allow robust processing, but most of them require adequate skills as regards the selected method. Any user familiar with the SOM can use the variable selection algorithm presented in this paper.

The proposed method has two drawbacks: a lack of mathematical rigor and high computational complexity. The SOM has eluded full theoretical justification, and also the k nearest neighbor algorithm is more difficult to analyze formally than parametric methods. However, both of the methods have proven their capability by a large number of users. The computational complexity of the variable selection is $\mathcal{O}(n^2)$. This is due to the computation of all interpoint distances when looking for the k nearest neighbors. If the number of data points is high, the user must apply a sampling procedure to reduce the computation.

3 Experiments

This section analyses two data sets that are publicly available with documentations at the UCI Machine Learning Repository[1].

3.1 Boston Housing Data

The data is about the median value of homes and 13 accompanying sociodemographic variables in 506 census tracts in Boston area in 1970. We preprocessed

---

[1] http://www.ics.uci.edu/~mlearn/MLRepository.html

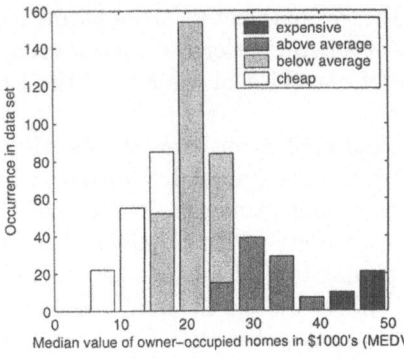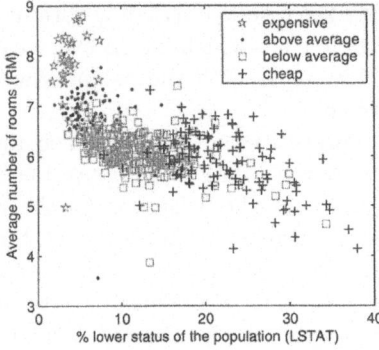

Fig. 1. A histogram and division of the Y-data into four clusters (*left*). The best X-variable pair found by the algorithm to explain the Y-data (*right*).

the data by scaling the variables to zero mean and unit variance. The first step in our method is to select the variables of interest. We used the median value of homes, MEDV, which then became the sole variable of Y. X was selected to contain the rest of the variables. Then, we used the variable selection algorithm to find the subset of X that best describes Y. The algorithm returned the variables RM, the average number of rooms per homes, and LSTAT, the perceived social status of the population.

A histogram of the Y-data is presented in Fig. 1. No SOM is shown, since there is a single target variable. The data has been clustered into four clusters to support the study of the found variables. We used hierarchical clustering with the error sum of squares-criterion [12]. Note that we performed the clustering only to support the study of the results. The locations of the four prices clusters in the found subspace of X are also presented in Fig. 1. Homes are cheap if the proportion of people with a lower social status in the neighborhood is high. Expensive homes are best indicated by the number of rooms: large homes are expensive.

To evaluate the results, we use the three criteria discussed above. The analysis is supervised: the user could define the median value of homes as the target of the analysis. An unsupervised tool, for example, correlations analysis would have emphasized the statistically most significant correlations between the sociodemographic indicators. The robustness of our method was not clearly manifested: this data set has reliably been analyzed, in other studies, with less robust methods. However, the results are understandable: they provide a clear description of the relation of the two found variables to the value of homes.

3.2   Automobile Data

The data is from 1985 Ward's Automotive Yearbook. It has 26 variables and 206 instances. As in the above example, we scaled each variable to zero mean and unit variance, but did not make any other preprocessing. In this case, we

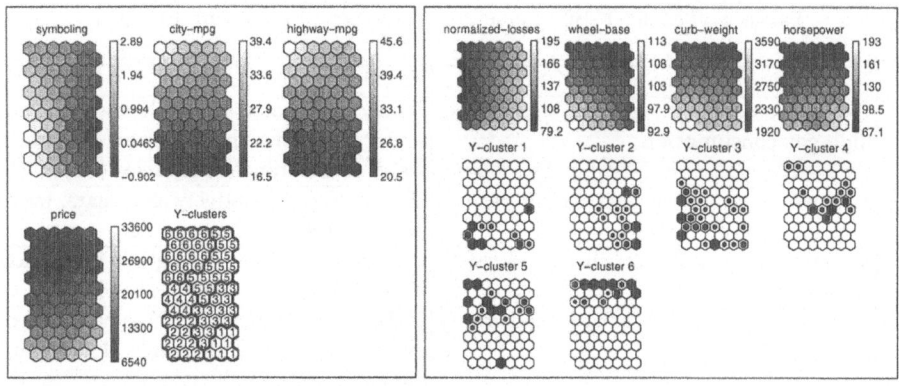

Fig. 2. The Y-SOM to describe the user's interests and a clustering of the lattice into six clusters (*left*). The X-SOM and the histograms of data points assigned to the Y-SOM clusters on the lattice (*right*).

decided to perform a study of the safety and economic aspects of the cars. Thus, we defined our problem with the SOM presented in Fig. 2. The first variable of Y, *symboling*, describes the insecurity of a car: a high value states that the car is less safe than other cars of the same price group. *city-mpg* and *highway-mpg* describe miles-per-gallon information: a high number indicates high consumption of gas. The fourth variable is the price of the car in USD. We clustered the surface of the Y-SOM according to the method of [13] using the error sum of squares-criterion [12]. The properties of these clusters are discussed in Table 1.

We used the variable selection algorithm to find the four variable set that contains the best information of the problem defined by the Y-SOM. The found subspace is illustrated in Fig. 2. The first variable is *normalized-losses*, which is the average loss payment (USD) per vehicle per insured vehicle year. The number has been normalized for all autos within a particular size classification (two-door small, station wagon, etc.). *wheel-base* is the distance between front and back wheels, in inches; *curb-weight* is the weight of a passenger-less car, in pounds; *horse power* is the power of the engine.

We illustrate the locations of the Y-SOM clusters with the hit-histograms shown in Fig. 2. The larger the dot in the histograms is, the higher is the number of data points classified into the respective X-SOM-neuron. The core of the analysis is to study where the clusters reside on the X-SOM, which is trained using the informative variables. Consider e.g. cluster 1, that is, the most expensive cars. According to the first hit-histogram, the cluster resides on two areas, bottom left and bottom right of the X-SOM. A common denominator of these areas is a high curb-weight. The cars at bottom left have high wheel-base, that is, they are long. The cars on bottom right have high horse power. The latter type causes high losses to insurance companies. Table 1 provides a similar description for all of the clusters.

Table 1. The six Y-SOM clusters characterized according to Fig. 2.

	Properties according to the Y-SOM	Properties according to the X-SOM
1	Expensive cars with good safety record, high gas consumption.	Heavy cars, some of which are long, and some that have high power and cause losses to insurance companies (IC).
2	Expensive cars with bad safety record, high gas consumption.	High powered, relatively short cars, which cause significant losses to IC.
3	Moderately priced cars with good safety record.	Cars with average or high weight. Most cars do not cause high losses to IC.
4	Moderately priced cars with bad safety record.	Compared to cluster 3, these cars are, on average, shorter and lighter. They cause higher losses to IC.
5	Cheap cars with moderately good safety record.	Comparably short, light and low-powered cars, which cause low losses to IC.
6	Cheap cars with moderately bad safety record, the lowest gas consumption.	The shortest cars. Compared to the cars of cluster 5, on average, these cars cause more losses to IC.

The above analysis provides good insight to the general factors related to the safety and economics of various cars. We illustrated that the Y- and X-spaces can be high-dimensional without major loss of understandability.

3.3   Comparison with Other Methods

We verified the above results by analyzing the data sets with other variable selection algorithms: the median k-NN method and Multi-Layered Perceptron (MLP) fit [14]. The median k-NN method is similar to our method. The difference is that the distances are not measured on the Y-SOM lattice, but directly in the Y-space. Instead of the mean of the distances from the center point to its neighbors, the median distance is used, which diminishes the adverse influence of outliers. The goal of this analysis is to study whether the mapping onto the SOM distorts the analysis. The MLP approach involves fitting of a MLP with a hidden layer of 3 neurons. The training set contains 60% and both the validation and test set 20% of data points. The training of the MLP stops when the validation error starts to rise. The cost is the mean absolute prediction error of the test set. In order to reduce the effect of random division of the data into different types of sets and the effect of random initialization, the MLP is fitted 20 times for each combination of covariates.

In the case of the Boston data set the median k-NN method returned the same variables as the proposed method. The MLP method returned a large group of variables, but when explicitly requesting the algorithm to return a two variable subset, the variables RM and LSTAT were again selected. This variable pair has been identified the most important subset also by [15].

In the Automobile data case we requested all methods to retrieve four covariates. The k-NN median method returned the same variables as our method.

However, the MLP approach suggested a varying set of four covariates in different runs of the variable selection. This is probably due to the random initialization of the MLP weights and entrapments to local minima. In this sense, the median- and SOM-based methods seemed more robust.

4   Summary and Conclusions

In this paper we have proposed a variable selection method that, according to our discussion, is supervised, robust and understandable. The user supervises the method by creating a SOM that describes the problem of interest, and then uses the proposed algorithm to find variables that contain relevant information. The algorithm is nonparametric and the dependence between the two variable spaces is studied in a model-free sense, which allows robust performance. The understandability ensues both from the visualization capabilities of the SOM and from the conceptual simplicity of the method. The experiments show that the proposed method works well with real analysis tasks. The comparisons confirm that the results are in line with other methods and studies made by other authors.

Future research focuses on selecting the neighborhoods in the X-space. Instead of using a hard-limited neighborhood, a smoother version can be obtained by calculating the rank ordering of the distances from the studied point to all other data points. Terms in the compactness criterion (1) can then be weighted based on the rank ordering. Promising results have been obtained by applying this technique.

References

1. Fayyad, U., Piatetsky-Shapiro, G., Smyth, P.: From Data Mining to Knowledge Discovery in Databases. AI Magazine **17** (1996) 37-54
2. Chapman, P., Clinton, J., Khabaza, T., Reinartz, T., Wirth, R.: CRISP-DM 1.0 Step-by-Step Data Mining Guide. Technical report, CRISM-DM Consortium, http://www.crisp-dm.org (2000)
3. Kohonen, T.: Self-Organizing Maps. 3rd edn. Springer-Verlag (2001)
4. Vesanto, J.: Data Exploration Process Based on the Self-Organizing Map. PhD thesis, Helsinki University of Technology, http://lib.hut.fi/Diss/2002/isbn9512258978/ (2002)
5. Bonnlander, B., Weigend, A.: Selecting Input Variables Using Mutual Information and Nonparametric Density Estimation. In: Procoodings of the International Symposium on Artificial Neural Networks (ISANN). (1994) 42-50
6. Laine, S.: Using Visualization, Variable Selection and Feature Extraction to Learn from Industrial Data. PhD thesis, Helsinki University of Technology, http://lib.hut.fi/Diss/2003/isbn9512266709/ (2003)
7. Glymour, C., Madigan, D., Pregibon, D., Smyth, P.: Statistical Themes and Lessons for Data Mining. Data Mining and Knowledge Discovery **1** (1997) 11-28
8. Dash, M., Liu, H., Yao, J.: Dimensionality Reduction for Unsupervised Data. In: Proceedings of the 9th International Conference on Tools with Artificial Intelligence (ICTAI). (1997) 532-539

9. Lagus, K., Alhoniemi, E., Valpola, H.: Independent Variable Group Analysis. In: Proceedings of the International Conference on Artificial Neural Networks (ICANN). (2001) 203-210
10. Jolliffe, I.T.: Principal Component Analysis. Springer-Verlag (1986)
11. Selim, S.Z., Ismail, M.A.: K-Means-Type Algorithms: A Generalized Convergence Theorem and Characterization of Local Optimality. IEEE Transactions on Pattern Analysis and Machine Intelligence **6** (1984) 81-87
12. Ward, J.H.: Hierarchical Grouping to Optimize an Objective Function. Journal of the American Statistical Association **58** (1963) 236-244
13. Vesanto, J., Alhoniemi, E.: Clustering of the Self-Organizing Map. IEEE Transactions on Neural Networks **11** (2000) 586-600
14. Haykin, S.: Neural Networks: A Comprehensive Foundation. 2nd edn. Prentice Hall (1998)
15. Doksum, K., Samarov, A.: Nonparametric Estimation of Global Functionals and a Measure of the Explanatory Power of Covariates in Regression. The Annals of Statistics **23** (1995) 1443-1473

Packing Bins Using Multi-chromosomal Genetic Representation and Better-Fit Heuristic

A.K. Bhatia* and S.K. Basu**

Department of Computer Science
Banaras Hindu University
Varanasi-221005, India
Fax: +91-542-2368285
swapankb@bhu.ac.in

Abstract. We propose a multi-chromosome genetic coding and set-based genetic operators for solving bin packing problem using genetic algorithm. A heuristic called *better-fit* is proposed, in which a left-out object replaces an existing object from a bin if it can fill the bin better. Performance of the genetic algorithm augmented with the better-fit heuristic has been compared with that of hybrid grouping genetic algorithm (HGGA). Our method has provided optimal solutions at highly reduced computational time for the benchmark uniform problem instances used. The better-fit heuristic is more effective compared to the best-fit heuristic when combined with the coding.

Keywords: Genetic Algorithm, Bin Packing Problem, Heuristics.

1 Introduction

We apply genetic algorithm (GA) [8] to solve bin packing problem (BPP) which is NP-hard. For a given list of objects and their sizes, BPP consists of finding a packing of the objects using the minimum number of bins of the given capacity. Many online and offline heuristics [2] such as Next-fit (NF), First-fit (FF), Best-fit (BF), First-fit-decreasing (FFD), and Best-fit-decreasing (BFD) have been devised for the BPP.

Existing GAs for the BPP use single-chromosome codings. *Binary coding* requires a chromosome of length $n * q$ bits, where n is the number of objects and q is the upper bound on the number of bins [10]. It forms lengthy chromosomes and the operators create infeasible strings requiring use of penalty terms during fitness evaluation. In *Object membership coding*, objects are assigned a bin number in the range $[1, q]$. The operators produce infeasible chromosomes and so penalty terms are used to take care of infeasibility. *Object permutation coding* defines the chromosomes as permutations of the object indices. It requires specialized crossover operators such as PMX, OX, CX, CI [8, 10]. It is difficult to incorporate heuristics in the coding.

Grouping representation [6] constructs a chromosome in two parts. The first part consists of the object membership coding. The second part consists of the groups present in the first part. The groups are termed as the genes and the genetic operators work on

---

* Present address: National Bureau of Animal Genetic Resources, Karnal - 132001 (India).
** Corresponding author.

N.R. Pal et al. (Eds.): ICONIP 2004, LNCS 3316, pp. 181–186, 2004.
© Springer-Verlag Berlin Heidelberg 2004

the grouping part. Falkenauer hybridizes dominance criterion [9] with this coding [4, 5]. The combined method has been called the hybrid grouping genetic algorithm (HGGA) and the reported results are the best obtained with a genetic algorithm, in the authors' knowledge.

Grouping representation requires fixation of an upper limit on the number of bins. One has to keep track of the bin labels for the working of the genetic operators, which totally disrupt the chromosomes and induce infeasibility. We propose an alternative multi-chromosome genetic representation, which wards off these problems. A heuristic for BPP called *better-fit* is also proposed. The proposed GA has been executed in combination with both the best-fit and the proposed better-fit heuristics. The results obtained on a benchmark data set are compared with those obtained with the HGGA.

We organise the paper as: section 2 explains the multi-chromosomal coding, section 3 introduces the better-fit heuristic, section 4 describes our experimental setup, section 5 analyses the results obtained, and section 6 contains our concluding remarks.

2   Multi-chromosomal Grouping Genetic Algorithm (MGGA)

2.1   Multi-chromosomal Genetic Representation

We term objects as genes and bins as chromosomes. A gene is the index number of an object. A chromosome consists of a subset of genes, which can be accommodated in it. An individual is a set of such chromosomes. Individual construction from different object indices provides different individuals. Figure 1 shows two individuals constructed with the next-fit heuristic for the following BPP instance with a bin capacity of 10. *IndivI* starts from index '1' and *IndivII* from index '4'. Each individual contains chromosomes shown by 'chr-j'.

Object index(o_i): 1 2 3 4 5 6 7 8 9 10
Object size (s_i): 8 3 5 2 1 6 3 5 4   7

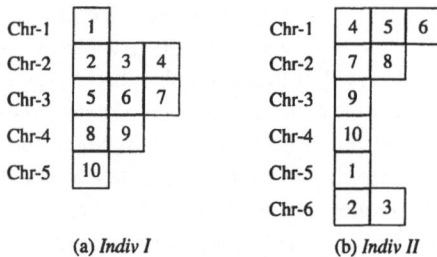

(a) *Indiv I*   (b) *Indiv II*

Fig. 1. Two individuals with multi-chromosome coding.

2.2   Genetic Operators

Crossover: Randomly extract a chromosome from each of the two selected individuals. Let chromosome $A \in IndivI$ and $B \in IndivII$. Define three sets: $(A - B), (B - A)$ and $(A \cup B)$. Remove the objects in $(B - A)$ from the *IndivI* and those in $(A - B)$ from the *IndivII*. Insert objects of $(A \cup B)$ into the chromosomes in both *IndivI* and *IndivII*. Build chromosome(s) from the remaining objects and append to the individual.

Mutation: Remove a chromosome from an individual with the given mutation probability. Insert the left-out genes into the chromosomes in the individual. Build a chromosome from the remaining genes and append it to the individual.

Translocation: We apply the operator at the gene level. Extract each gene with a given translocation probability. If it is the only gene, remove the chromosome. Try to insert the extracted gene into the chromosomes of the individual. If the gene does not fit in any of the chromosomes, define a new chromosome with the gene and append it to the individual.

3 Hybridization with Problem Heuristics

Packing of the *left-out* genes after the application of the genetic operators requires problem related heuristics. We introduce a new heuristic for one-dimensional BPP and name it **better-fit**. It packs a *left-out* gene in the first bin that it can fill better than any of the existing genes. The existing gene is then removed and the *left-out* gene is inserted into the bin. The replaced gene is then tried to be packed with the better-fit heuristic, starting with the first bin. The process continues till a replaced gene cannot be better-fitted in any of the chromosomes in the individual. The last gene is then packed with the best-fit heuristic.

Figure 2 demonstrates working of the better-fit heuristic. Gene '10' having object size equal to 7 is left out after the action of a genetic operator. It replaces gene '6' having an object size 6 from the chromosome-1 and so on. Gene '7' cannot be better-fitted in any of the chromosomes. So it gets packed with the gene '8' by using the best-fit heuristic.

We hybridize MGGA with the best-fit heuristic and also with the better-fit heuristic.

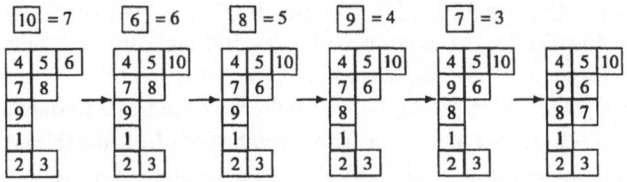

Fig. 2. Packing of a *left-out* gene into an individual with Better-Fit heuristic.

4 Experimental Setup

4.1 Test Data Sets

We use BPP 'uniform' data sets available from Beasley's OR-Library [1], which consists of uniform[20,100] integer object sizes with a bin capacity equal to 150. Twenty instances are available for each of the problem sizes 120, 250, 500 and 1000.

4.2   Genetic Parameters

Genetic parameters are determined using several methods like parameter-tuning, self-adaptation, etc. [3]. We decide the parameter values based on tuning and heuristics.

The initial population is generated with the first-fit heuristic. Population size has been fixed at a value that provided good quality solution with the least computational time using the combination MGGA+BetterFit. The parameter values are 4, 6, 6, and 8 for problem sizes of 120, 250, 500 and 1000 respectively.

Probability of mutation(p_m) varies with the remaining gap in a bin and the progress of the GA run. The p_m is calculated as:

$[p_{m_j}]_t = [MF]_t * [1 - \frac{F_j}{C}]$ where $[p_{m_j}]_t$ denotes the p_m of jth chromosome at generation t. $[MF]_t = (t/T) * (100 - leeway\%)$ is the mutation factor at generation t. F_j is the sum of object sizes in the jth bin. C is the bin capacity. t and T are the current generation and total generations. $leeway\% = (L_1 * C - \sum s_i)/(\sum s_i) * 100$, $i = 1, n$. $L_1 = \lceil (\sum s_i)/(C) \rceil$ is a lower bound for the BPP [9, 7]. s_i is size of the ith object.

Translocation probability (p_r) is higher for small size objects. The operator proves disruptive for well-filled bins at later generations. Therefore, we use a decay function to control the parameter.

$[p_{r_i}]_t = [TF]_t * (1 - \frac{s_i}{C})$ where $[p_{r_i}]_t$ denotes the p_r of the ith object and $[TF]_t = 0.1 * e^{-\frac{t}{100}}$ is the translocation factor at generation t.

We adopt the evaluation function $f_{indiv} = \frac{\sum_{j=1}^{z}(F_j/C)^2}{z}$ [5], where f_{indiv} is the fitness of an individual. F_j is the sum of the object sizes in the jth chromosome. z is the number of bins in a feasible individual.

We use tournament selection with a size two and parameter value 0.8, crossover probability equal to 0.8, and total number of generations equal to 5000.

5   Results

We hybridize MGGA with the best-fit as well as with the better-fit heuristics. Each problem instance of sizes 120, 250, 500 and 1000 is solved ten times with the two combinations. Results for the instances of sizes 500 and 1000 are only shown in the Tables 1and 2 respectively for the reason of space limitation.

The indices in the vector $(i_0, i_1, i_2, \ldots)$ in the tables indicate frequency of a solution equal to L_1, $L_1 + 1$, $L_1 + 2$ and so on. The lower bound, L_1 is the theoretical optimum. 'Result' column under the Hybrid GGA shows the solution in a single run [5]. 'Mean-time' depicts the average time in seconds on a PC (Pentium III 733 MHz). 'Time' under Hybrid GGA shows the time in seconds for a single run on R4000 Silicon Graphics Workstation under IRIX5.1[5].

MGGA+BetterFit can find optimal solutions in all the 200 runs of the problem instances of size 120 while the MGGA+BestFit can find optimal solutions in 70 runs only. MGGA+BetterFit could find optimal solutions in 175 runs while the MGGA+BestFit could find optimal solutions in 14 runs out of 200 runs for the problem instances of size 250. For the problem instances of size 500, MGGA+BetterFit found optimal solutions in 193 runs while the MGGA+BestFit failed to do so in all the 200 runs. MGGA+BetterFit found optimal solution in all the 200 runs while the MGGA+BestFit

Table 1. Performance of the MGGA on uniform instances of problem size=500.

Instance_id	Optimum(L_1)	MGGA+BestFit		MGGA+BetterFit		Hybrid GGA	
		Results	Mean-time	Results	Mean-time	Result	Time
u500_00	198	(,,1,4,3,1,1)	158.75	(10)	1.74	198	480.5
u500_01	201	(,,,1,4,3,1,1)	182.76	(10)	2.41	201	177.7
u500_02	202	(,,2,2,3,,1,2)	159.59	(10)	0.61	202	347.9
u500_03	204	(,,4,1,1,3,1)	168.48	(10)	6.55	204	11121.2
u500_04	206	(,1,1,3,2,2,1)	125.19	(10)	0.63	206	267.6
u500_05	206	(,,2,2,3,3)	121.55	(10)	0.57	206	129.7
u500_06	207	(,1,1,4,1,3)	170.48	(4,6)	131.15	207	1655.5
u500_07	204	(,,4,3,2,,1)	165.05	(9,1)	50.78	204	1834.7
u500_08	196	(,1,1,3,2,3)	156.36	(10)	1.19	196	501.5
u500_09	202	(,2,1,,4,2,1)	111.19	(10)	0.55	202	92.5
u500_10	200	(,,1,3,2,3,1)	115.47	(10)	0.53	200	106.2
u500_11	200	(,2,2,2,3,,1)	145.41	(10)	0.77	200	152.3
u500_12	199	(,1,,1,3,3,2)	173.95	(10)	1.72	199	1019.3
u500_13	196	(,,,,1,3,1,5)	177.77	(10)	0.59	196	135.5
u500_14	204	(,,,3,,5,1,1)	110.28	(10)	0.58	204	951.7
u500_15	201	(,2,1,2,3,2)	124.77	(10)	0.64	201	375.2
u500_16	202	(,,1,2,,3,1,3)	104.11	(10)	0.55	202	162.6
u500_17	198	(,1,3,1,4,,1)	148.27	(10)	0.71	198	336.8
u500_18	202	(,3,2,1,1,1,1,1)	140.61	(10)	0.64	202	143.9
u500_19	196	(,1,3,1,1,,1,2,1)	158.97	(10)	0.85	196	306.8
Overall values		(,15,30,36,46, 35,21,15,2)	143.95	(193,7)	10.19		1015

Table 2. Performance of the MGGA on uniform instances of problem size=1000.

Instance_id	Optimum(L_1)	MGGA+BestFit		MGGA+BetterFit		Hybrid GGA	
		Results	Mean-time	Results	Mean-time	Result	Time
u1000_00	399	(,,,3,,2,1,1,,,2,1)	554.43	(10)	3.19	399	2924.7
u1000_01	406	(,,1,,2,1,1,,,,2,2,1)	527.04	(10)	2.79	406	4040.2
u1000_02	411	(,,1,2,3,,2,1,,,1)	463.54	(10)	2.92	411	6262.1
u1000_03	411	(,,1,,1,1,3,2,,1,1)	630.84	(10)	66.0	411	32714.3
u1000_04	397	(,1,,1,2,1,1,1,1,1,,1)	590.48	(10)	14.92	397	11862.0
u1000_05	399	(,,,,2,1,1,2,1,1,1,1)	584.20	(10)	4.63	399	3774.3
u1000_06	395	(,,,1,1,,1,1,,1,4,,,,1)	506.34	(10)	2.74	395	3033.2
u1000_07	404	(,,,,,2,1,,,3,,3,,,,1)	492.10	(10)	2.94	404	9878.8
u1000_08	399	(,,,,2,2,2,1,,1,1,1)	571.05	(10)	4.09	399	5585.2
u1000_09	397	(,,,1,,3,1,2,1,,1,,1)	635.48	(10)	25.13	397	8126.2
u1000_10	400	(,,,1,,,2,2,,3,1,,,1)	572.68	(10)	2.70	400	3359.1
u1000_11	401	(,1,,1,3,,2,1,2)	547.66	(10)	7.02	401	6782.3
u1000_12	393	(,,,,,2,1,1,3,1,,,2)	519.67	(10)	2.73	393	2537.4
u1000_13	396	(,,,1,1,2,2,,2,,,1,,1)	521.30	(10)	3.12	396	11828.8
u1000_14	394	(,,,2,3,,,1,1,1,,1,,1)	624.22	(10)	31.60	394	5838.1
u1000_15	402	(,,,2,,3,,3,,1,,1)	615.15	(10)	13.89	402	12610.8
u1000_16	404	(,,,,,,2,3,1,3,1)	371.95	(10)	2.64	404	2740.8
u1000_17	404	(,,1,,2,2,,2,,1,1,,,1)	647.63	(10)	14.83	404	2379.4
u1000_18	399	(,,,,1,1,3,,1,2,,,1,,,1)	497.82	(10)	2.79	399	1329.7
u1000_19	400	(,1,,1,,4,4)	511.82	(10)	2.71	400	3564.2
Overall results		(,3,4,16,23,23,28, 27,15,17,18,13,7,3,,3)	549.27	(200)	10.67		7058.6

failed to do so in all the 200 runs for the problem instances of size 1000. HGGA failed to find the optimal solutions in the five problem instances $u120\_8$, $u120\_19$, $u250\_07$, $u250\_12$, $u250\_13$ while the MGGA+BetterFit failed to find optimal solutions in the two problem instances $u250\_12$, $u250\_13$.

MGGA+BetterFit takes overall average time equal to 0.05, 12.04, 10.19 and 10.67 seconds while the HGGA takes 381.2, 1336.6, 1015 and 7058.6 seconds for the instances of size 120, 250, 500 and 1000 objects respectively. The small computational time with our method is attributed to three factors: it uses an efficient multi-chromosome genetic coding, Better-fit heuristic is gene-based, while the dominance criterion used in the HGGA is a combinatorial heuristic, and lastly genetic operators in the HGGA produce infeasible individuals that require repair while this is not the case with the MGGA.

6   Conclusion

Our approach in combination with the 'better fit' is superior to its combination with the 'best fit', as observed on the benchmark problem instances and provides results of matching quality compared to those obtained with the hybrid grouping genetic algorithms (HGGA), but MGGA+BetterFit requires very small computational time. Although we have demonstrated effectiveness of our method for the bin packing problem, it can be used in solving other grouping problems also.

References

1. J. E. Beasley. OR-Library: Distributing test problems by electronic mail. *Journal of Operational Research Society*, 41(11):1069–72, 1990.
2. E. G. Coffman,Jr, G. Galambos, S. Martello, and D. Vigo. Bin packing approximation algorithms: combinatorial analysis. In D.-Z. Du and P. Pardalos, editors, *Handbook of Combinatorial Optimization*. Kluwer Academic Publishers, 1998.
3. A. E. Eiben, R. Hinterding, and Z. Michalewicz. Parameter control in evolutionary algorithms. *IEEE Transactions on Evolutionary Computation*, 3(2):124–141, 1999.
4. E. Falkenauer. A hybrid genetic algorithm for bin packing. *Journal of Heuristics*, 2(1):5–30, 1996.
5. E. Falkenauer. *Genetic Algorithms and Grouping Problems*. John Wiley & Sons, 1998.
6. E. Falkenauer and A. Delchambre. A genetic algorithm for bin packing and line balancing. In *Proc. IEEE Int. Conf. on Robotics and Automation, France*, pages 1186–1192, 1992.
7. S. P. Fekete and J. Schepers. New classes of fast lower bounds for bin packing problem. *Mathematical Programming Series A*, 91(1):11–31, 2001.
8. D. E. Goldberg. *Genetic Algorithms in Search, Optimization and Machine Learning*. Addison Wesley, Reading MA., 1989.
9. S. Martello and P. Toth. Lower bounds and reduction procedures for the bin packing problem. *Discrete Applied Mathematics*, 28:59–70, 1990.
10. C. Reeves. Hybrid genetic algorithms for bin-packing and related problems. *Annals of Operations Research*, 63:371–396, 1996.

Data Association for Multiple Target Tracking: An Optimization Approach

Mukesh A. Zaveri, S.N. Merchant, and Uday B. Desai

SPANN Lab, Electrical Engineering Dept., IIT Bombay - 400076
{mazaveri,merchant,ubdesai}@ee.iitb.ac.in

Abstract. In multiple target tracking the data association, observation to track fusion, is crucial and plays an important role for success of any tracking algorithm. The observation may be due to true target or may be clutter. In this paper, data association problem is viewed as an optimization problem and two methods, (i) using neural network and (ii) using the evolutionary algorithm, have been proposed and compared.

1 Introduction

Various approaches have been proposed for multiple target tracking and data association in literature [1]. The most common method for data association is nearest neighbor method. But there is always uncertainty about the origin of observation and hence, it may result in a false track. To avoid this uncertainty about the origin of an observation, joint probabilistic data association filter (JPDA) [2] and probabilistic multiple hypotheses tracking (PMHT) algorithm [3] have been proposed. Though the JPDA algorithm has excellent performance, the computational burden is too heavy with large number of targets and observations. On the other hand PMHT operates in batch mode and uses the centroid of observations to update state, and hence it is sensitive to the size of validation region and the amount of clutter falling in it. To overcome the above problems with data association and tracking arbitrary trajectory, earlier we have proposed multiple model based tracking algorithm using Expectation-Maximization (EM) algorithm [4]. It gives excellent performance, nevertheless due to the iterative nature of the algorithm, the computational burden is quite heavy.

In this paper, we propose approaches based on the neural network and an evolutionary (genetic) algorithm for data association, which reduces the number of computations and provides robust data association. With the evolutionary algorithm there is no guarantee of obtaining the optimal solution. But it does provide a set of potential solutions in the process of finding the best solution. The idea behind using the neural network is to extract the stored pattern in a form of observation to track association from the given set of observations (data). The novelty of the proposed algorithm is that no observation is assigned to any target implicitly but an assignment weight is calculated for each validated observation for a given target and it is used by tracking algorithm. In the proposed algorithm an observation likelihood given a target state is treated as mixture probabilistic

N.R. Pal et al. (Eds.): ICONIP 2004, LNCS 3316, pp. 187–192, 2004.
© Springer-Verlag Berlin Heidelberg 2004

density function (pdf). It allows one to incorporate multiple models for target dynamics and consequently, is able to track maneuvering and non-maneuvering targets simultaneously in absence of any apriori knowledge about the target dynamics in the presence of clutter.

2 Data Association: Problem Formulation

In this section, the problem is described in multimodel framework for data association and tracking. Let N_t be the number of targets at time k, and it may vary with time. $\boldsymbol{\Phi}_k$ represents concatenated combined state estimates for all targets $t = 1, \ldots, N_t$, i.e. $\boldsymbol{\Phi}_k = (\Phi_k(1), \Phi_k(2), \ldots, \Phi_k(N_t))^T$ where $\Phi_k(t)$ is combined state estimate at time instant k for target t. The state at time instant k by model m for target t is represented by $\phi_k^m(t)$. Let the observation process $\mathcal{Y}$ and its realization, the observations at time instant k be denoted by a vector $\mathbf{Y}_k = (y_k(1), y_k(2), \ldots, y_k(N_k))^T$, where N_k denotes the number of observations obtained at time k, N_k may also vary with time. To assign observations to targets, an association process defined as $\mathcal{Z}$ is formed. It is used to represent the true but unknown origin of observations. $\mathbf{Z}_k$ is a realization of an association process at time instant k, and it is referred to as an association matrix. For each model, a validation matrix (association matrix) is defined. With multiple models, we represent $\mathbf{Z}_k$ as combined (logically OR operation) realization of $\mathcal{Z}$ and is defined as, $\mathbf{Z}_k = \mathbf{z}_{k,1} + \mathbf{z}_{k,2} + \ldots + \mathbf{z}_{k,M}$ where $\mathbf{z}_{k,m}$ is the association matrix at time instant k for model m. Here M is the total number of models used for tracking. Each $\mathbf{z}_{k,m}$ is $(N_t + 1) \times (N_k + 1)$ matrix, (t, i)-th element of association matrix is $z_{km}(t, i)$ and it is 1 if observation $y_k(i)$ falls in validation gate of target t otherwise it is 0.

2.1 Neural Network Based Data Association

The data association problem is treated as incomplete data in absence of any information about an origin of an observation. EM algorithm is used to evaluate MAP estimation of target state. EM algorithm estimates the assignment probabilities and the assignment weights as a by product [4]. EM algorithm calculates assignment weights and updates state vector for the targets in an iterative mode, which is computationally more demanding. To speed-up the computations, in the proposed approach assignment weights are obtained using Hopfield neural network [5] and these are directly used in a single step to update state for the targets.

Finding the assignment weights $\hat{\mathbf{Z}}_k(t, i)$ from the likelihoods $p(y_k(i)|\Phi_k(t))$ is a similar to the traveling salesman problem. If the output voltages $V_i(t), i = 0, 1, \ldots, N_k, t = 0, 1, 2, \ldots, N_t$ of a $(N_t + 1) \times (N_k + 1)$ matrix of neurons are defined as assignment weights, then rows indicate targets and columns represent observations. The constraints for the data association are as follows. Each row sum of voltages must be unity. At most one large entry is to be favored in every row and column. We assume that no two observations come from the same target and no observation can come from two targets. All these requirements are fulfilled by minimizing the energy function with respect to the voltages as described in [5] and can be written as,

$$
\begin{aligned}
\mathcal{E} = {} & \tfrac{A}{2} \sum_{i=l}^{N_k} \sum_{t=1}^{N_t} \sum_{\substack{s=1 \\ s \neq t}}^{N_t} V_i(t) V_i(s) + \tfrac{B}{2} \sum_{i=1}^{N_k} \sum_{t=1}^{N_t} \sum_{\substack{j=1 \\ j \neq i}}^{N_k} V_i(t) V_j(t) \\
& + \tfrac{C}{2} \sum_{t=1}^{N_t} \left(\sum_{i=1}^{N_k} V_i(t) - 1 \right)^2 + \tfrac{D}{2} \sum_{i=1}^{N_k} \sum_{t=1}^{N_t} (V_i(t) - \rho_i(t))^2 \\
& + \tfrac{E}{2} \sum_{i=1}^{N_k} \sum_{t=1}^{N_t} \sum_{\substack{s=1 \\ s \neq t}}^{N_t} \left(V_i(t) - \sum_{\substack{j=1 \\ j \neq i}}^{N_k} \rho_j(s) \right)^2
\end{aligned}
\tag{1}
$$

For our proposed algorithm to incorporate *likelihood of an observation due to each model* the normalized likelihoods, $\rho_i(t)$ is given by,

$$
\rho_i(t) = \frac{\sum_{m=1}^{M} p_m[y_k(i)|\phi_k^m(t)]\mu_k^m(t)}{\sum_{j=1}^{N_k} \sum_{m=1}^{M} p_m[y_k(j)|\phi_k^m(t)]\mu_k^m(t)}
\tag{2}
$$

where $i = 1, 2, \ldots, N_k$ and $t = 1, 2, \ldots, N_t$. It can be shown that evaluation of assignment weights based on an energy function (1) can be obtained using the difference equation in an iterative mode [5]. The initial conditions $V_i(t) = \rho_i(t)$ provide a convenient alternative to the nearly uniform and randomized initial values. Once the assignment weights are calculated by neural net, these weights are used to update the state vector of each model for a given target.

2.2 Evolutionary Algorithm Based Data Association

In this approach, the evolutionary algorithm is used by exploiting JPDA approach to evaluate assignment weights for data association. JPDA evaluates all feasible events of observation-to-track association and hence, computationally it is expensive. In our approach, we use all the best solutions (tuples) from all generations given by the evolutionary algorithm to calculate the assignment weights. These best tuples act as most likely feasible events. Once the evolutionary algorithm is over, the assignment weight for an observation to the given target is calculated by summing over all *feasible* joint events in which the marginal event of interest occurs. The number of best tuples found using the evolutionary algorithm is much less compared to the number of feasible events used in the JPDA. It reduces the amount computations to a great extent.

Using this combined association matrix $\mathbf{Z}_k$, a combined likelihood measure matrix $\mathcal{E}$ is formed, where each entry $\mathcal{E}(t, i)$ is given by

$$
\mathcal{E}(t, i) = \begin{cases} \dfrac{p(y_k(i)|\Phi_k(t))}{\sum_{j=1}^{N_k} p(y_k(j)|\Phi_k(t))} & \text{if } \mathbf{Z}_k(t, i) = 1 \\ 0 & \text{if } \mathbf{Z}_k(t, i) = 0 \end{cases}
\tag{3}
$$

where $p(y_k(i)|\Phi_k(t)) = \sum_{m=1}^{M} p_m(y_k(i)|\phi_k^m(t))\mu_k^m(t)$ represents the likelihood of the observation given a combined state estimate $\Phi_{(k|k-1)}(t)$ for target t at time k, and it is treated as mixture probability. The combined likelihood measure matrix $\mathcal{E}$, given by (3), is used by the evolutionary algorithm.

Evolutionary algorithm is based on salient operators like crossover, mutation and selection. Initially, a random set of population of elements that represent the candidate solutions is created. Crossover and mutation operations are applied on the set of population elements to generate a new set of offsprings which serve as new candidate solutions. Each element of the population of elements is assigned a fitness value (quality value) which is an indication of the performance measure. In our formulation the likelihood measure $\mathcal{E}(t, i)$ is considered as a fitness value while designing the fitness function. We form a string consisting of target number as a symbol. It represents a solution for data association problem i.e. observation to track (target) pairing. This string is also called as tuple. If tuple is indicated by symbol n then the quality of solution is represented by fitness function $f(n)$ and it is defined as, $f(n) = \sum_i \mathcal{E}(t, i)$ where i is the observation index and t represents target number from the given tuple n. The details of various operators of the evolutionary algorithm for data association can be found in [6].

Let the assignment weight matrix be $\hat{\mathbf{Z}}_k$, each entry $\hat{\mathbf{Z}}_k(t, i)$ in this matrix indicates assignment weight for assigning an observation i to target (track) t.

$$\hat{\mathbf{Z}}_k(t, i) = \begin{cases} p\{\theta_{it}|\mathbf{Y}^k\} = \sum_{\theta \in \mathcal{G}} \frac{1}{c} \left\{ \prod_{j=1}^{m_k} p[y_k(j)|\theta_{it}(k), \mathbf{Y}^{k-1}] \right\} & \text{if } \mathbf{Z}_k(t, i) = 1 \\ 0 & \text{if } \mathbf{Z}_k(t, i) = 0 \end{cases} \tag{4}$$

where $\mathcal{G}$ represents a set of best solutions from all generation obtained using the evolutionary algorithm. $p\{\theta_{it}|\mathbf{Y}^K\}$ denotes probability of the event in which observation i is assigned to target t, and is obtained by summing over all events in which observation i is assigned to track t. where c is the normalization constant. Here, θ represents an event (tuple). $p[y_k(j)|\theta_{it}(k), \mathbf{Y}^{k-1}]$ is the likelihood of each observation j for a given event θ_{it}, in which observation i is assigned to target t.

$$p[y_k(j)|\theta_{it}(k), \mathbf{Y}^{k-1}] = \begin{cases} N_{s_j}[y_k(j)] & \text{if}[\theta_{it}(k, j)] = s \neq 0 \\ V^{-1} & \text{if}[\theta_{it}(k, j)] = 0 \end{cases} \tag{5}$$

where $\theta_{it}(k, j)$ represents the jth entry in tuple θ_{it}. At the jth index in a tuple if we have a non zero entry s, it means that observation j is assigned to target s in a given event θ_{it}. For our case, likelihood of an observation is treated as mixture pdf and with evolutionary algorithm, we have used normalized likelihood, which is defined in (3), i.e. $N_{s_j}[y_k(j)] = \mathcal{E}(s, j)$. Observation not associated to any target is assumed uniformly distributed with probability value equal to $1/V$. For simulation, the value for V^{-1} is set to 0.01. After calculating assignment weights for each target t and observation i for which $\mathbf{Z}_k(t, i) = 1$, the entries in assignment matrix $\hat{\mathbf{Z}}_k$ are normalized so that in each row sum of assignment weights equals to 1.0. This normalized assignment matrix is used to update target state.

After calculation of assignment weights it is followed by a tracking algorithm. In our proposed algorithm the observation and target state are assumed to be Gaussian distributed. With the independence assumption for each target and for each model, it can be shown that the state estimate for each model can be

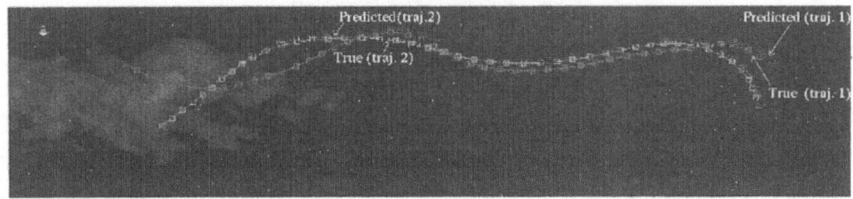

Fig. 1. Tracked trajectories at frame number 44 - ir50 clip (0.05% clutter).

obtained by standard Kalman equations using interacting multiple model based algorithm [4]. Detail mathematical steps are not described here due to space limitation.

3   Simulation Results

We are interested in tracking of airborne targets using InfraRed (IR) sensors. In order to validate our proposed approaches we have used synthetic IR images which are generated using real time temperature data. For simulation, the generated frame size is 1024 × 256 and very high target movement of ±20 pixels per frame. Maneuvering trajectories are generated using the B-Spline function. It is important to note that these generated trajectories do not follow any specific model. In our simulations, we have used constant acceleration (CA) model and Singers' maneuver model (SMM) for tracking. For all simulated trajectories, tracking filters are initialized using positions of the targets in the first two frames.

In our investigations the various parameters used for simulation are set as follows. For neural network based data association and tracking, the maximum number of iterations are kept fixed to 500 to evaluate energy function (1) using the difference equation in [5]. Values $A = 1000, B = 1000, C = 1000, D = 100$, and $E = 100$ are found to be appropriate with 0.05% clutter level. Similarly for evolutionary algorithm based data association and tracking, the number of generations is set to 20. By default, the number of solutions is set to 8. If the number of possible tuples (solutions) are less than the specified number, it is set to the minimum of these two numbers. The initial crossover and mutation probability are set to 0.650 and 0.010 respectively. The choice of these parameters for the evolutionary algorithm is application dependent and in our case we have obtained these after extensive simulations.

Due to space limitation it is not possible to present all the results and therefore the output pertaining to a sample clip titled "ir50" is shown in Figure 1. This figure indicates the presence of two targets. Both, the true trajectory and the predicted trajectory are indicated in Figure 1. Using the proposed tracking algorithms, mean prediction error in position (x-y plane) is depicted in Table 1 for different trajectories for different clips without clutter and with clutter for both the proposed algorithms. We also compared our proposed tracking algorithms with earlier proposed method [4]. The mean prediction error in position

Table 1. Mean Prediction Error.

	Data association using			
	Neural network		Evolutionary algorithm	
Traj. No.	no clutter	0.05% clutter	no clutter	0.05% clutter
ir44 clip				
1	1.8730	1.9147	1.7698	1.6468
2	3.6895	3.9563	2.6986	2.8329
ir49 clip				
1	3.1750	3.5274	2.1925	2.2708
2	3.3655	3.3619	2.4040	2.4397
ir50 clip				
1	3.6960	3.7714	2.7013	2.6985
2	3.5550	3.6575	2.4110	2.7067

for different trajectories were found to be less in most of the cases with the proposed methods compared to earlier method. Earlier method even fails to track targets in clips ir49 and ir50 with 0.05% clutter level.

4   Conclusion

From the extensive simulations it is concluded that both the proposed methods, (i) using neural network and (ii) using the evolutionary algorithm, for data association are robust and performs equally well in the presence of multiple target and clutter. It does assign an observation to a track implicitly and avoids uncertainty about the origin of an observation. The inclusion of multiple models allows us to track an arbitrary movement of target successfully.

References

1. Bar-shalom, Y., Fortmann, T.E.: Tracking and Data Association. Academic Press (1989)
2. Gad, A., Majdi, F., Farooq, M.: A Comparison of Data Association Techniques for Target Tracking in Clutter. In: Proceedings of 5th International Conference on Information Fusion. (2002) 1126–1133
3. P. Willett, Y. Ruan and R. Streit: PMHT: Problems and Some Solutions. IEEE Transactions on Aerospace and Electronic Systems **38** (2002) 738–754
4. Zaveri, M.A., Desai, U.B., Merchant, S.: Interacting Multiple Model Based Tracking of Multiple Point Targets using Expectation Maximization Algorithm in Infrared image sequence. In: Proceedings of SPIE: Visual Communications and Image Processing (VCIP) 2003. Volume 5150. (2003) 303–314
5. Sengupta, D., Iltis, R.A.: Neural Solution to the Multitarget Tracking Data Association Problem. IEEE Transactions on Aerospace and Electronic Systems **25** (1989) 96–108
6. Zaveri, M.A., et al.: Genetic Algorithm Based Data Association and Tracking of Multiple Point Targets. In: Proceedings of 10th National Conference on Communications (NCC - 2004), Banglore, India (2004) 414–418

Expected Running Time Analysis of a Multiobjective Evolutionary Algorithm on Pseudo-boolean Functions

Nilanjan Banerjee and Rajeev Kumar

Department of Computer Science and Engineering
Indian Institute of Technology Kharagpur
Kharagpur, WB 721 302, India
rkumar@cse.iitkgp.ernet.in

Abstract. In this paper we suggest a multiobjective evolutionary algorithm based on a restricted mating pool (REMO) with a separate archive for storing the remaining population. Such archive based algorithms have been used for solving real-world applications, however, no theoretical results are available. In this paper, we present a rigorous expected running time complexity analysis for the algorithm on two discrete pseudo boolean functions. We use the well known linear function LOTZ (Leading Zeros : Trailing Ones) and a continuous multiobjective quadratic function which is adapted to the discrete boolean space, for the analysis. The analysis shows that the algorithm runs with an expected time of $O(n^2)$ on LOTZ. Moreover, we prove that the bound holds with an overwhelming probability. For an unary encoding of the multiobjective quadratic function $((x-a)^2, (x-b)^2)$ in the boolean space, the expected running time of REMO is found to be $O(n\log n)$. A simple strategy based on partitioning of the decision space into fitness layers is used for the analysis.

1 Introduction

Evolutionary Algorithms (EAs) are randomized search heuristics that try to imitate the process of natural evolution. They are a broad class of heuristics that are applied to optimization problems. There are countless reports on the successful application of EAs with a huge number of empirical results but theoretical results are very few.

In case of single objective optimization theoretical analyses on running time complexity for few functions are available. Droste et al. [1] provided a rigorous theoretical analysis of the so called $(1 + 1)$ EA on a wide variety of functions, especially linear functions. ONE-MAX function [2] has also been studied on the (1+1) EA. Results on the time bounds of algorithms in the discrete search space is available [3]. Rigorous proof on analysis of runtime in the continuous search space has been obtained only recently [4]. An analysis of Evolutionary Algorithms on whether crossover is essential is shown in [5]. However, work on theoretical analysis of multiobjective optimization is rare. Deriving sharp asymptotic bounds for multiobjective optimizers was started by Laumanns et al. [6] and is further extended by them in [7]. They defined a group of multiobjective pseudo-boolean functions which they analyze on their MOEAs.

The algorithms used in most of the work above use a single member population or an unbounded population. There is another group of genetic algorithms which use

N.R. Pal et al. (Eds.): ICONIP 2004, LNCS 3316, pp. 193–198, 2004.
© Springer-Verlag Berlin Heidelberg 2004

an archive. Such algorithms have been empirically demonstrated to work efficiently but no theoretical analysis of such algorithms is available. In this work we define a simple archive based algorithm and perform an expected running time analysis of the algorithm on a linear and a quadratic function. We prove that for the linear function the expected running time is better than those obtained for the same function in [6,7]. We also show that the algorithm performs efficiently for the well known bi-objective quadratic function. Thus, the work aims at the study of the behavior of an archive based algorithm on a few simple multiobjective functions.

The rest of the paper is organized as follows. Section 2 discusses the new algorithm REMO. Section 3 and 4 analyses the algorithm on the LOTZ and the quadratic multiobjective function, respectively. Section 5 concludes the paper.

2   Algorithm

Restricted Evolutionary Multiobjective Optimizer (REMO)

---

1. Initialize two sets $P = \phi$ and $A = \phi$, where P is the mating pool and A is an archive.
2. Choose an individual x uniformly at random from $Y = \{0,1\}^n$.
3. $P = \{x\}$.
4. **loop**
5. Select an individual y from P at random.
6. Apply mutation operator on y by flipping a single randomly chosen bit and create y'.
7. $P = P \setminus \{l \in P \mid l \prec y'\}$.
8. $A = A \setminus \{l \in A \mid l \prec y'\}$.
9. **if** there does not exist $z \in P \cup A$ such that $z \succ y'$ or $f(z) = f(y')$ **then** $P = P \cup \{y'\}$
10. **end if.**
11. **if** cardinality of P is greater than 2 **then**
12. **Handler Function.**
13. **end if.**
14. **end loop.**

Handler Function

---

1. For all the members x of $P \cup A$ calculate a fitness function $F(x, P \cup A) = H(x)$ where $H(x)$ denotes the number of Hamming neighbors of x in $P \cup A$.
2. Select the two individuals with the minimum $F(x, P \cup A)$ values into P and put the rest of the individuals in the archive A. In case of equal $F(x, P \cup A)$ values the selection is made at random.

---

3   LOTZ : Leading Ones Trailing Zeros

The LOTZ is a bi-objective linear function that was formulated and analyzed by Laumanns et al. [6, 7]. The authors proved expected runtimes of $\theta(n^3)$ and $\theta(n^2 log n)$ for their algorithms SEMO (Simple Evolutionary Multiobjective Optimizer) and FEMO (Fair Evolutionary Multiobjective Optimizer) respectively on LOTZ. We show that REMO (algorithm above) has an expected running time of $O(n^2)$ for LOTZ and moreover prove that the above bound holds with an overwhelming probability.

3.1   Problem Definition

The Leading Ones (LO), Trailing Zeros (TZ) and the LOTZ problems can be defined as follows where the aim is to maximize both the objectives:

$$LO(x) = \sum_{i=1}^{n} \prod_{j=1}^{i} x_j, \qquad TZ(x) = \sum_{i=1}^{n} \prod_{j=i}^{n} (1 - x_j)$$
$$LOTZ(x) = (LO(x), TZ(x)).$$

Proposition 1. The Pareto-optimal front for LOTZ can be represented as a set $S = \{(i, n - i) \mid 0 \leq i \leq n\}$ and the Pareto set consists of all bit vectors belonging to the set $P = \{ 1^i 0^{n-i} \mid 0 \leq i \leq n \}$.

Proof. In the first part of the proof we aim to show that corresponding to any arbitrary non-Pareto optimal bit vector in the decision space we can always find an individual in P which dominates it. Let us consider any arbitrary individual $Y = 1^i 0\{0, 1\}^{n-(i+j+2)} 10^j$ and an optimal bit string $X = 1^k 0^{n-k}$ where $0 \leq k \leq n$. It is clear that if $k = i+1$ then $LO(Y) < LO(X)$. In the string Y, there is a 0 following the i Leading-ones and a 1 preceding the j Trailing-zeros. This implies that $j \leq n - (i + 2)$ therefore $TZ(Y) < TZ(X)$, thus proving that X dominates Y.

Now we need to show that any bit string of the form $X = 1^k 0^{n-k}$ where $0 \leq k \leq n$ cannot be dominated by any other individual in the decision space. It is clear that if $LO(Y)$ or $TZ(Y)$ (where Y is defined in the above paragraph) is greater than that of X for a certain choices of i and j, the other objective is bound to be less for Y, thus implying that in the worst case the two strings are incomparable. The same argument holds for any string $Z = 1^l 0^{n-l}$ where $l \neq k$; thus proving the proposition.

3.2   Analysis

The expected running time analysis of the function above is divided into two distinct phases. Phase 1 ends with the first Pareto-optimal point in the population P, and Phase 2 ends with the entire Pareto-optimal set in $P \cup A$.

Theorem 1. The expected running time of REMO on LOTZ is $O(n^2)$. The above bound holds with a probability of $1 - e^{-\Omega(n)}$.

Proof. We partition the decision space into fitness layers defined as (i, j), $(0 \leq i, j \leq n)$ where i refers to the number of Leading-ones and j is the number of Trailing-zeros in a chromosome.

For LOTZ, in phase 1 the population cannot contain more than one individual for REMO because a single bit flip will create a child that is either dominating or is dominated by its parent. Phase 1 begins with an initial random bit vector in P. An individual can climb up a fitness layer (i, j) by a single bit mutation if it produces the child $(i+1, j)$ or $(i, j+1)$. The probability of flipping any particular bit in the parent is $\frac{1}{n}$, thus the probability associated with such a transition is $\frac{2}{n}$. The factor of 2 is multiplied because we could either flip the leftmost 0 or the rightmost 1 for a success. Therefore, the expected waiting time for such a successful bit flip is at most $\frac{n}{2}$. If we pessimistically assume that Phase 1 begins with a random individual in the population then algorithm would require at most n successful mutation steps till the first Pareto-optimal point is found. Thus, it takes $\frac{n^2}{2}$ steps for the completion of Phase 1. To prove that the above bound holds with an overwhelming probability let us consider that the algorithm is run for n^2 steps. The expected number of successes for these n^2 steps is at least $2n$. If S denotes the number of successes, then by Chernoff's bounds :

$$P[S \le (1 - \tfrac{1}{2}) \cdot 2n] = P[S \le n] \le e^{-\frac{n}{4}} = e^{-\Omega(n)}.$$

Phase 2 begins with an individual of the form $I = (i, n - i)$ in P. A success in Phase 2 is defined as the production of another Pareto-optimal individual. The first successful mutation in Phase 2 leads to production of the individual $I_{+1} = (i + 1, n - i - 1)$ or $I_{-1} = (i - 1, (n - i + 1))$ in the population P. The probability of such a step is given by $\frac{2}{n}$. Thus, the waiting time till the first success occurs is $\frac{n}{2}$. If we assume that after the first success I and I_{-1} are in P (without loss of generality), then the Pareto-optimal front can be described as two paths from $1^{i-1}0^{n-i+1}$ to 0^n and 1^i0^{n-i} to 1^n. At any instance of time T, let the individuals in P be represented by $L = (l, n - l)$ and $K = (k, n - k)$ where $0 \le k < l \le n$. As the algorithm would have followed the path from $(i - 1, n - i + 1)$ to $(k, n - k)$ and $(i, n - i)$ to $(l, n - l)$ to reach the points L and K, it is clear that at time T all the individuals of the form $S = (j, n - j)$ with $l < j < k$ have already been found and form a part of the archive A. Moreover, the handler function, assures that L and K are farthest in Hamming distance . At time T the probability of choosing any one individual for mutation is $\frac{1}{2}$. Let us assume, without loss of generality, that the individual selected is $(k, n - k)$. The flipping of the left most 0 produces the individual $K_{+1} = (k + 1, n - k - 1)$ and the flipping of the rightmost 1 produces the individual $K_{-1} = (k - 1, n - k + 1)$. Since, the algorithm does not accept weakly dominated individuals and K_{+1} is already in A, the production of K_{-1} can only be considered as a success. Thus, the probability of producing another Pareto-optimal individual at time T is $\frac{1}{4n}$. The expected waiting time of producing another Pareto-optimal individual is at most $4n$. Since, no solutions on the Pareto-optimal front is revisited in Phase 2, it takes a maximum of $n + 1$ steps for its completion. Therefore, REMO takes $O(n^2)$ for Phase 2. By arguments similar to Phase 1, it can be shown that the bound in phase 2 holds with a probability $1 - e^{-\Omega(n)}$.

Altogether considering both the phases, REMO takes n^2 steps to find the entire Pareto-optimal set for LOTZ. For the bound on the expected time we have not assumed anything about the initial population. Thus, the above bound on the probability holds for the next n^2 steps. Since the lower bound on the probability that the algorithm will find the entire Pareto set is more than $\frac{1}{2}$ (in fact exponentially close to 1) the expected number of times the algorithm has to run is bounded by 2. Combining the results of both the phases 1 and 2 yields the bounds in the theorem.

4 Quadratic Function

We use a continuous bi-objective minimization function $((x - a)^2, (x - b)^2)$ and adapt it to the discrete boolean decision space in the following manner:

if $\|x\| = \sum_{i=1}^n x_i$, QF : $((\|x\| - a)^2, (\|x\| - b)^2)$

Proposition 2. *The Pareto-optimal front of QF is the set* $F = \{(i^2, (i - (b - a))^2) \mid a \le i \le b\}$ *and consists of individuals where* $a \le \|x\| \le b$. .

Proof. First, we aim to prove that corresponding to any arbitrary individual in the decision space (which is not in F) we can always find a bit vector in F which dominates it. We represent any arbitrary individual with i number of ones as X_i. If $i < a$ then the objective value for $X_i = ((i - a)^2, (i - b)^2)$. In the best case $i = a - 1$. The value of $QF(X_i) = (1, (b - a + 1)^2)$. We can find the individual (in F) with $\|x\| = a$ to have both the objectives less than X_i, thus proving that X_i is dominated. A similar proof can be proposed for any bit vector in which $\|x\| > b$. Now we need to prove that there does not exist any bit vector that can dominate an individual in F. Let us represent an individual in F as X_f. It is obvious to see that any other bit vector in F is incomparable to X_f. An individual with $\|x\| > b$ or $\|x\| < a$ in the best case can have any one objective value lower than that of X_f. Therefore, either the individual is dominated or is incomparable to X_f; thus proving the proposition.

4.1 Analysis

Theorem 2. *The expected running time of REMO on QF is $O(n\log n)$ for any value of a and b.*

Proof. We partition the analysis into two phases. Phase 1 ends with the first Pareto-optimal point in P and the second phase continues till all the Pareto-optimal bit vectors are in $P \cup A$. In phase 1 there can be a maximum of 2 individuals in $P \cup A$. Thus, the archive A is empty. This is because a single bit mutation of a parent with $\|x\| < a$ or $\|x\| > b$ will produce an individual which is dominated by or dominates its parent. We partition the decision space into sets with individuals having the same number of ones. Let us consider a bit vector represented as I_d where d represents the number of ones in the individual. A single bit mutation of I_d is considered to be a success if the number of ones increases (decreases) when $d < a(d > b)$. Therefore a success S requires the flipping of any one of the d 1-bits ($n - d$ 0-bits) when $d < a$ ($d > b$). The probability of a successful mutation $P(S) = \frac{d}{2n} (or \frac{n-d}{2n})$. The expected waiting time of S is given by $E(S) \le \frac{2n}{d} (or \frac{2n}{n-d})$. The total expected time till the first Pareto optimal individual arrives in the population is at most $\sum_{i=1}^n \frac{2n}{d} = 2nH_n = 2n\log n + \theta(2n) = O(n\log n)$, where H_n stands for the nth harmonic number, by the linearity of expectations.

  Phase 2 works with the assumption that $b - a > 1$ or else there would be no second phase. The number of individuals in the population is bounded by 2. The selection mechanism ensures that they are the bit vectors that are most capable of producing new individuals. The Pareto-front can be visualized as a path of individuals with number of ones varying from a to b or b to a. Let us represent any individual with $a < \|x\| < b$ as I_k where k represents the number of ones in the bit vector. Such a bit vector can be produced either by an individual with $k + 1$ ones or $k - 1$ ones. The associated probability for such a successful mutation is at least $\frac{k+1}{2n}$ and $\frac{n-k+1}{2n}$ respectively. Hence, the expected waiting time till the I_k^{th} Pareto optimal point is in the population (assuming that its parent is in the population) is $E(I_k) \leq \frac{2n}{k+1}$ and $\frac{2n}{n-k+1}$ for the two cases above. Thus, the total expected time till all the Pareto points are in $P \cup A$ is at most $\sum_{k=a}^{b} E(I_k) \leq \sum_{k=a}^{b} \frac{2n}{k+1} \leq \sum_{k=0}^{b-a} \frac{2n}{k+1} = 2nH_{b-a}$. Therefore, the expected time for Phase 2 is at most $2nelog(b - a) + \theta(2ne) = O(nlog(b - a))$. Since a and b can have a maximum value of n the expected running time for REMO on QF is $O(nlogn)$.

5 Discussion and Conclusions

In this paper, an archive based multiobjective evolutionary optimizer (REMO) is presented and a rigorous runtime complexity analysis of the algorithm on a linear multiobjective function and one multiobjective quadratic function is shown. The key feature of REMO is its special restricted population for mating and a separate archive. The idea is to restrict the mating pool to a constant c. The value of 2 for c is sufficient for most linear and quadratic functions. In case of certain linear functions a single individual population with a similar selection scheme as REMO may suffice. However, two bit vectors may be required for functions where the Pareto front can be reached via two paths as is the case of the quadratic function. The bounds for REMO presented in the paper are better to those found on the linear function LOTZ earlier in [6, 7].

References

1. Droste, S., Jansen T., Wegener, I.: On the Analysis of the $(1 + 1)$ Evolutionary Algorithm. Theoretical Computer Science, **276** (2002) 51 - 81.
2. Garnier, J. , Kallel, L., Schoenauer, M. : Rigourous Hitting Times for Binary Mutations. Evolutionary Computation, **7(2)** (2002) 167 - 203.
3. Droste, S. , Jansen, T., Wegener, I.: On the Optimization of Unimodal Functions with the $(1 + 1)$ Evolutionary Algorithm. Proceedings of the 5th Conference of Parallel Problem Solving from Nature (PPSN V), LNCS **1498** (1998) 13 - 22.
4. Jagerskupper, J. : Analysis of Simple Evolutionary Algorithm for Minimization in Euclidean Spaces. Proceedings of the 30th International Colloquium on Automata, Languages and Programming. LNCS **2719** (2003) 1068 - 1079.
5. Jansen,T. ,Wegener, I. : On the Analysis of Evolutionary Algorithms, a Proof that Crossover Really Can Help. Proceedings of the 7th Annual European symposium of Algorithms(ESA 99). LNCS **1643** (1999) 184 - 193.
6. Laumanns, M., Thiele, L., Zitzler, E., Welzl. E., Deb, K. : Running Time Analysis of Multiobjective Evolutionary Algorithms on a Discrete Optimization Problem. Parallel Problem Solving from Nature (PPSN VII), LNCS **2439** (2002) 44 - 53.
7. Laumanns, M., Thiele, L., Zitzler, E.: Running Time Analysis of Evolutionary Algorithms on Vector-Valued Pseudo-Boolean Functions. IEEE Transactions on Evolutionary Computation, 2004.

The Influence of Gaussian, Uniform, and Cauchy Perturbation Functions in the Neural Network Evolution

Paulito P. Palmes and Shiro Usui

RIKEN Brain Science Institute
2-1 Hirosawa, Wako, Saitama 351-0198, Japan
ppalmes@brain.riken.jp
usuishiro@riken.jp

Abstract. Majority of algorithms in the field of evolutionary artificial neural networks (EvoANN) rely on the proper choice and implementation of the perturbation function to maintain their population's diversity from generation to generation. Maintaining diversity is an important factor in the evolution process since it helps the population of ANN (Artificial Neural Networks) to escape local minima. To determine which among the perturbation functions are ideal for ANN evolution, this paper analyzed the influence of the three commonly used functions, namely: Gaussian, Cauchy, and Uniform. Statistical comparisons were conducted to examine their influence in the generalization and training performance of EvoANN. Our simulations using the glass classification problem indicated that for mutation-with-crossover-based EvoANN, generalization performance among the three perturbation functions were not significantly different. On the other hand, mutation-based EvoANN that used Gaussian mutation performed as good as that with crossover but it performed worst when it used either Uniform or Cauchy distribution function. These observations suggest that crossover operation becomes a significant operation in systems that employ strong perturbation functions but has less significance in systems that use weak or conservative perturbation functions.

1 Introduction

There are two major approaches in evolving a non-gradient based population of neural networks, namely: Mutation-based approach using EP (Evolutionary Programming) or ES (Evolutionary Strategies) concepts and Crossover-based approach which is based on GA (Genetic Algorithm) implementation. While the former relies heavily on the mutation operation, the latter considers the crossover operation to be the dominant operation of evolution. Common to these approaches is the choice of the perturbation function that is responsible for the introduction of new characteristics and information in the population. Since the selection process favors individuals with better fitness for the next generation, it is important that the latter generation will not be populated by individuals

N.R. Pal et al. (Eds.): ICONIP 2004, LNCS 3316, pp. 199–204, 2004.
© Springer-Verlag Berlin Heidelberg 2004

that are too similar to avoid the possibility of being stuck in a local minimum. The way to address this issue is through the proper choice and implementation of the perturbation function, encoding scheme, selection criteria, and the proper formulation of the fitness function. In this study, we are interested in the first issue.

The SEPA (Structure Evolution and Parameter Adaptation) [4] evolutionary neural network model is chosen in the implementation to ensure that the main driving force of evolution is through the perturbation function and the crossover operation. The SEPA model does not use any gradient information and relies only in its mutation's perturbation function and crossover operation for ANN evolution.

2   Related Study

Several studies have been conducted to examine the influence of the different perturbation functions in the area of optimization. While Gaussian mutation is the predominant function in numerical optimization, the work done by [7] indicated that local convergence was similar between Gaussian and spherical Cauchy but slower in non-spherical Cauchy. Studies done by [8] in evolutionary neural networks found that Cauchy mutation had better performance than Gaussian mutation in multimodal problems with many local minima. For problems with few local minima, both functions had similar performance.

A study conducted by [1] combined both the Gaussian and Cauchy distributions by taking the mean of the random variable from Gaussian together with the random variable from Cauchy. Preliminary results showed that the new function performed as good or better than the plain Gaussian implementation.

Common to these approaches is the reliance of the system to the perturbation function to effect gradual changes to its parameters in order for the system to find a better solution. In a typical implementation, the perturbation function undergoes adaptation together with the variables to be optimized. Equations (1) and (2) describe a typical implementation using Gaussian self-adaptation [1]:

$$\eta' = \eta + \eta N(0,1) \tag{1}$$
$$x' = x + \eta' N(0,1) \tag{2}$$

where x is the vector of variables to be optimized; η is the vector of search step parameters (SSP), each undergoing self-adaptation; N is the vector of Gaussian functions with mean 0 and the standard deviation controlled by their respective SSPs;

The typical implementations in evolutionary neural networks also follow similar formulation for the mutation of weights:

$$w = w + \mathbf{N}(0, \alpha\epsilon(\varphi)) \ \ \forall w \in \varphi$$

where $\mathbf{N}(0, \alpha\epsilon(\varphi))$ is the gaussian perturbation with mean 0 and standard deviation $\alpha\epsilon(\varphi)$; w is a weight; and $\epsilon(\varphi)$ is an error function of network φ (e.g. mean-squared error) which is scaled by the user-defined constant α.

Unlike in a typical function optimization problem where the main goal is to optimize the objective function, the goal of neural network evolution is to find the most suitable architecture with the best generalization performance. Good network training performance using a certain perturbation function does not necessarily translate into a good generalization performance due to overfitness. It is important, therefore, to study the influence of the different perturbation functions in the training and the generalization performances of ANN. Moreover, knowing which combination of mutation and adaptation strategies are suited for a particular perturbation function and problem domain will be a big help in the neural network implementation. These issues will be examined in the future. In this paper, our discussions will only be limited to the performance of EvoANN in the glass classification problem taken from the UCI repository [2].

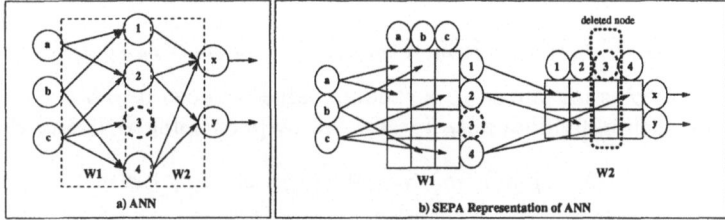

Fig. 1. Evolutionary ANN.

3   Evolutionary ANN Model

Neural Network implementation can be viewed as a problem in optimization where the goal is to search for the best network configuration having good performance in training, testing, and validation. This is achieved by training the network to allow it to adjust its architecture and weights based on the constraint imposed by the problem. The SEPA model (Fig. 1) used in this study addresses this issue by making weight and architecture searches become a single process that is controlled by mutation and crossover. Changes caused by mutation and crossover induce corresponding changes to the weights and architecture of the ANN at the same time [3]. In this manner, the major driving force of evolution in SEPA is through the implementation of the crossover and mutation operations. This makes the choice of the perturbation function and the implementation of adaptation, mutation, and crossover very important for the successful evolution of the network. Below is a summary of the SEPA approach:

1. At iteration t=0, initialize a population $P(t) = \{net_1^t, ..., net_\mu^t\}$ of μ individuals randomly:

$$net_i = \{W1_i, W2_i, \theta_{w1}^i, \theta_{w2}^i, \rho(pr_i, m_i, \sigma_i)\}$$

where: $W1, W2$ are the weight matrices; θ_{w1}, θ_{w2} are the threshold vectors; ρ is the perturbation function; pr is the mutation probability; m is the strategy parameter; and σ is the step size parameter (SSP).

2. Compute the fitness of each individual based on the objective function Q_{fit} [5]:

$$Q_{fit} = \alpha * Q_{acc} + \beta * Q_{nmse} + \gamma * Q_{comp}$$

where: Q_{acc} is the percentage error in classification; Q_{nmse} is the percentage of normalized mean-squared error (NMSE); Q_{comp} is the complexity measure in terms of the ratio between the active connections c and the total number of possible connections c_{tot}; α, β, and γ are constants used to control the strength of influence of their respective factors.

3. Using rank selection policy, repeat until there are μ individuals generated:
   - Rank-select two parents, net_k and net_l, and apply crossover operation by exchanging weights between $W1_k$ and $W1_l$ and weights between $W2_k$ and $W2_l$:

   $$\forall (r,c) \in W1_k \wedge W1_l, \text{if rand}() < \Theta, swap(W1_k[r][c], W1_l[r][c])$$

   $$\forall (r,c) \in W2_k \wedge W2_l, \text{if rand}() < \Theta, swap(W2_k[r][c], W2_l[r][c])$$

   where Θ is initialized to a random value between 0 to 0.5.

4. Mutate each individual net_i, $i = 1, ..., \mu$, by perturbing $W1_i$ and $W2_i$ using:

$$\delta_i = \rho(\sigma_i); \; m'_i = m_i + \rho(\delta_i); \; w'_i = w_i + \rho(m'_i)$$

where: σ is the SSP (step size parameter); δ is mutation strength intensity; ρ is the perturbation function; m is the adapted strategy parameter, and w is the weight chosen randomly from either W1 or W2.

5. Compute the fitness of each offspring using Q_{fit}.
6. Using elitist replacement policy, retain the best two parents and replace the remaining parents by their offsprings.
7. Stop if the stopping criterion is satisfied; otherwise, go to step 2.

4 Experiments and Results

Two major SEPA variants were used to aid in the analysis, namely: mutation-based (mSEPA) and the mutation-crossover-based (mcSEPA or standard SEPA). Furthermore, each major variant is divided into three categories, namely: mSEPA-c (Cauchy-based); mSEPA-g (Gaussian-based); and mSEPA-u (Uniform-based). Similarly, mcSEPA follows similar categorization, namely: mcSEPA-c, msSEPA-g, and mcSEPA-u which is based on the type of perturbation function used.

Table 1 summarizes the important parameters and variables used by the different variants. The glass problem was particularly chosen because its noisy data made generalization difficult which was a good way to discriminate robust variants. The sampling procedure divided the data into 50% training, 25% validation, and 25% testing [6]. The objective was to forecast the glass type (6 types) based on the results of the chemical analysis (6 inputs) using 214 observations.

Table 2 shows the generalization performance of the different SEPA variants. The posthoc test in Table 2 uses the Tukey's HSD wherein average error results

Table 1. Feature Implemented in SEPA for the Simulation.

SEPA Main Features		
Features	**Implemented**	**Comment**
selection type	rank	rank-sum selection
mutation type	gaussian-cauchy-uniform	depends on the variant
mutation prob	0.01	
SSP size	$\sigma = 100$	Uniform range is U(-100,100)
crossover type	uniform	randomly assigned between (0,0.5)
replacement	elitist	retains two best parents
population size	100	
no. of trials	30	
max. hidden units	10	
max. generations	5000	
stopping criterion	validation sampling	evaluated at every 10th generation
fitness constants	$\alpha = 1.0, \beta = 0.7, \gamma = 0.3$	
classification	winner-takes-all	

Table 2. ANOVA of Generalization Error in Glass Classification Problem.

Gaussian vs Uniform vs Cauchy		
Variants	**Average Error**	**Std Dev**
mSEPA-g	0.3912*	0.0470
mcSEPA-u	0.4006*	0.0380
mcSEPA-g	0.4031*	0.0516
mcSEPA-c	0.4113*†	0.0626
mSEPA-u	0.4194†	0.0448
mSEPA-c	0.4453†	0.0649
Linear-BP [6]	0.5528	0.0127
Pivot-BP [6]	0.5560	0.0283
NoShortCut-BP [6]	0.5557	0.0370

*, † (Tukey's HSD posthoc test classification using $\alpha = 0.05$ level of significance)

that are not significantly different are indicated by the same label (* or †). Table 2 indicates that for mutation-based SEPA (mSEPA), Gaussian perturbation is significantly superior than the Uniform and Cauchy functions. For the mutation-crossover-based SEPA (mcSEPA), there is no significant difference among the three perturbation functions.

Furthermore, the table also indicates that any SEPA variant has superior generalization than any of the Backpropagation variants tested by Prechelt [6]. Since these results are only limited to the glass classification problem and BP can be implemented in many ways, the comparison of SEPA with the BP variants are not conclusive and requires further study. Moreover, Figure 2 and Table 2 suggest that even though the Uniform perturbation has the best training performance in mSEPA, it has the worst generalization performance.

5   Conclusion

This preliminary study suggests that for evolutionary neural networks that rely solely in mutation operation, Gaussian perturbation provides a superior generalization performance than the Uniform and Cauchy functions. On the other hand, introduction of crossover operation helps to significantly improve the performance of the Cauchy and Uniform functions. It also suggests that in order to manage complexity provided by more chaotic perturbation functions such

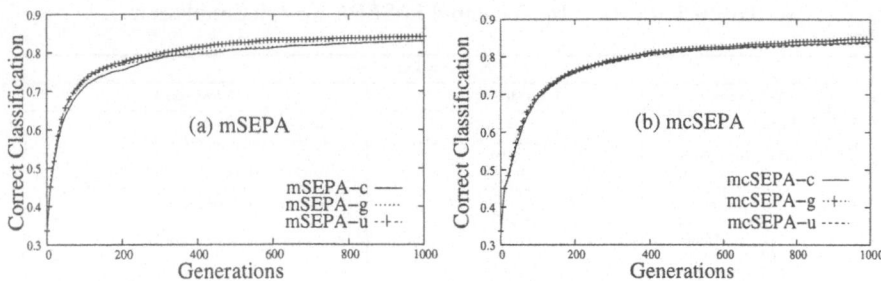

Fig. 2. Training Performance of the Different SEPA Variants.

as that of the Uniform and Cauchy perturbations, a proper crossover operation must be introduced to leverage and exploit the wider search coverage introduced by these functions. The simulation also indicates that that superior performance in training for mutation-based evolution does not necessarily imply a good generalization performance. It may even worsen the generalization performance due to too localized searching.

References

1. K. Chellapilla and D. Fogel. Two new mutation operators for enhanced search and optimization in evolutionary programming. In B.Bosacchi, J.C.Bezdek, and D.B.Fogel, editors, *Proc. of SPIE: Applications of Soft Computing*, volume 3165, pages 260–269, 1997.
2. P. M. Murphy and D. W. Aha. *UCI Repository of machine learning databases*. University of California, Department of Information and Computer Science, Irvine, CA, 1994.
3. P. Palmes, T. Hayasaka, and S. Usui. Evolution and adaptation of neural networks. In *Proceedings of the International Joint Conference on Neural Networks, IJCNN*, volume II, pages 397–404, Portland, Oregon, USA, 19-24 July 2003. IEEE Computer Society Press.
4. P. Palmes, T. Hayasaka, and S. Usui. SEPA: Structure evolution and parameter adaptation. In E. Cantu Paz, editor, *Proceedings of the Genetic and Evolutionary Computation Conference*, volume 2, page 223, Chicago, Illinois, USA, 11-17 July 2003. Morgan Kaufmann.
5. P. Palmes, T. Hayasaka, and S. Usui. Mutation-based genetic neural network. IEEE Transactions on Neural Network, 2004. article in press.
6. L. Prechelt. Proben1–a set of neural network benchmark problems and benchmarking rules. Technical Report 21/94, Fakultat fur Informatik, Univ. Karlsruhe, Karlsruhe, Germany, Sept 1994.
7. G. Rudolph. Local convergence rates of simple evolutionary algorithms with cauchy mutations. *IEEE Trans. on Evolutionary Computation*, 1(4):249–258, 1997.
8. X. Yao, Y. Liu, and G. Liu. Evolutionary programming made faster. *IEEE Trans. on Evolutionary Computation*, 3(2):82–102, 1999.

Closest Substring Problem –
Results from an Evolutionary Algorithm

Holger Mauch

University of Hawaii at Manoa, Dept. of Information and Computer Science,
1680 East-West Road, Honolulu, HI 96822
hmauch@hawaii.edu

Abstract. The closest substring problem is a formal description of how
to find a pattern such that from a given set of strings a subregion of
each string is highly similar to that pattern. This problem appears frequently in computational biology and in coding theory. Experimental results suggest that this NP-hard optimization problem can be approached
very well with a custom-built evolutionary algorithm using a fixed-length
string representation, as in the typical genetic algorithm (GA) concept.
Part of this success can be attributed to a novel mutation operator introduced in this paper. For practical purposes, the GA used here seems to
be an improvement compared to traditional approximation algorithms.
While the time complexity of traditional approximation algorithms can
be analyzed precisely, they suffer from poor run-time efficiency or poor
accuracy, or both.

Keywords: Genetic Algorithm, Closest String Problem, Closest Substring Problem, Radius of Code

1 Introduction

The goal of this paper is to provide a means to solve large instances of the
closest substring problem as they arise in practical applications in a time efficient
manner.

In the following let Σ be a fixed finite alphabet. For example in computational
biology terms, one can think of as $\Sigma = \{A, C, G, T\}$ representing the 4 nucleotide
bases when working on the DNA level, or $\Sigma = \{Ala, \ldots, Val\}$ representing the
20 amino acids when working on the protein level. The set of natural numbers
is denoted as $\mathbb{N}$. If t is a length l substring of s, that is there are (potentially
empty) strings u, v such that $s = utv$ and $|t| = l$, we will write $t \triangleleft_l s$ for short.

Definition 1. *The* closest substring problem *(CSSP) takes as input*

- *a set $S = \{s_1, \ldots, s_m\}$ of m strings where $s_i \in \Sigma^n$ for $1 \leq i \leq m$, and
  $n, m \in \mathbb{N}$ and*
- *the substring length $l \in \mathbb{N}$ where $l \leq n$.*

The desired optimal output is a string $t^* \in \Sigma^l$ (called a closest substring*), such
that*

$$z = \max_{s \in S} \min_{t \triangleleft_l s} \{d(t^*, t)\} \tag{1}$$

is minimal, where d denotes the Hamming distance between two strings.

N.R. Pal et al. (Eds.): ICONIP 2004, LNCS 3316, pp. 205–211, 2004.
© Springer-Verlag Berlin Heidelberg 2004

Note that while

$$z^* = \min_{u \in \Sigma^l} \max_{s \in S} \min_{t \triangleleft_l s}\{d(u,t)\}$$

is unique, t^* is not necessarily unique.

One can determine a lower bound for z^* by constructing a $m(n-l+1) \times m(n-l+1)$-matrix that records the distance of every length-l substring of s_i to every length-l substring of s_j for all $i,j \in \{1,\dots,m\}$. (Entries comparing length-l substrings of s_i with length-l substrings of the same s_i can be ignored.) This is accomplished by determining the minimum entries within those $(n-l+1) \times (n-l+1)$-submatrices that record all the substring distances between a pair of strings. Then calculate the maximum value d_{max} among these minimum entries. A triangular inequality argument establishes

$$z^* \geq \lceil d_{max}/2 \rceil. \qquad (2)$$

This can be done in $O(m^2n^2)$ time. For some problem instances, this expense might be worthwhile, because it allows the CSSP algorithm to take an early exit in case equality holds in (2). Indeed, the suggested GA takes advantage of this fact when tested on the problem instance used in section 4.

Definition 2. *The* closest string problem *(CSP) is a special case of the closest substring problem with $n = l$.*

The CSSP and the more specific CSP appear frequently in computational biology, e.g. in the context of deriving artificial transgenes to create virus-resistant plants [1], in the context of primer design, or in the context of finding similar regions in a given set of protein sequences. In terms of coding theory the CSP is referred to as the problem of finding the (minimum) radius of a code [2], also known as the Hamming radius 1-clustering problem [3].

2   Complexity of the Problem and Related Work

Frances and Litman showed in [2], that the minimum radius decision problem of arbitrary binary codes is NP-complete. Based on this fact it can easily be shown that the CSP and the CSSP are NP-complete.

An exhaustive search of the space Σ^l is easy to program but inefficient. Calculating the distance according to (1) from each of the $|\Sigma|^l$ points in the search space to all $(n-l+1)m$ input regions in order to find a closest substring takes $l(n-l+1)m|\Sigma|^l$ pairwise base comparisons, since it takes l base comparisons to calculate the distance between a pair of substrings. This method is not practical for large l.

Branch and bound techniques allow to prune the search space and to improve the brute force approach, but the improvements are not significant enough to implement an algorithm with sufficient efficiency for real world sized CSSPs.

The customized GA on the other hand has acceptable time requirements and outputs results within a reasonable amount of time.

One direction of previous work established polynomial time algorithms for restricted cases of the CSP. Gąsienic et al. [3] described an algorithm for the CSP that runs in $n^{O(m)}$ time – only practical for a small number m of strings. Gramm et al. [4] give efficient solutions to the CSP in case of a small optimal value z^* and also for the case of $m = 3$.

Another direction of research focused on the design and analysis of approximation algorithms for the CSSP and the CSP. Lanctot et al. [5] established a polynomial-time $(4/3 + \epsilon)$-approximation algorithm for the CSP and a heuristic for the CSSP based upon it. Li et al. [6] improved these results by providing a polynomial time approximation scheme (PTAS) for the CSP and also for the CSSP. The problem is that for accurate approximations the running time gets too large for real-world sized problems.

GAs cannot guarantee to find a closest substring either, but they are more robust, less sensitive to larger input parameter values and have a good chance to produce a result that is good enough to be useful in a practical setting, as is demonstrated by experimental results in section 4.

3   Genetic Algorithm Approach

3.1   Introduction to Genetic Algorithms

Genetic algorithms (GA) [7–9], inspired by biological systems, mimic the Darwinian evolution process. GAs tend to make more copies of individuals (fixed-length character strings), which exhibit higher fitness, as measured by a suitable fitness function. Over time individuals in the population evolve because of natural selection and because genetic operations (mutation, recombination) modify individuals.

After the random generation of an initial population, the GA enters an evaluation - selection - alteration - cycle until the termination criterion (e.g. maximum number of generations, perfect individual sighted, etc.) is satisfied. GAs are a robust search technique and they are widely used in optimization. The discrete search space and the lack of further constraints indicate that the CSSP should be a good application area for GAs.

3.2   Genetic Algorithm Design for the Closest Substring Problem

"A representation should always reflect fundamental facts about the problem at hand" [10, p.97]. For the CSSP the most natural representation for candidate solutions are strings over the alphabet Σ. Therefore the population in the GA is a collection of strings from Σ^l.

The fitness function f used to evaluate an individual string $\tilde{t}$ is simply based on the objective function (1), i.e.

$$f(\tilde{t}) = \max_{s \in S} \min_{t \lhd_l s} \{d(\tilde{t}, t)\}$$

Note that a lower fitness value is considered better and that the "fittest" individuals that can ever evolve are closest substrings – they have a fitness of z^*. The fitness evaluation of an individual is the most time consuming part of the GA – it requires $l(n - l + 1)m$ character comparisons per individual per generation. Future improvements of the GA could aim at designing a fitness function that can be evaluated faster.

The procedures employed for random initialization (draw a character for each locus from the uniform discrete density function $p(x) = 1/|\Sigma| \quad \forall x \in \Sigma$), selection (tournament style), recombination (uniform crossover), and mutation (uniform mutation probability θ_M for each locus) are widely used in the GA community. The only operator that has been designed specifically for the CSSP is a second type of mutation operator as described in the following subsection.

3.3   The Shift-Mutation Operator

In addition to the ordinary mutation operator that substitutes a character in a string with a given small probability it looks desirable to have a mutation operator that shifts all characters of an individual to the left or right. A slightly alternative design would be to rotate the characters, i.e. reinsert the character that gets pushed off one end at the other end. Similar mutation operators have been used previously, e.g. in [11–13]. Due to the nature of the CSSP problem an individual mutated in such a way would still have a fitness close to its original value, but would increase the genetic variety of the population dramatically. This is because the other genetic operators (ordinary mutation, crossover) are designed such as not to disturb the value of a certain locus (its absolute position in the string) by too much. However, the shift-mutation operator causes all characters to be relocated to a new absolute position, but preserves their ordering, which makes it particularly suited for the CSSP. Or, as Whitley describes this interdependent relationship in [7, p.246]:

> ... one cannot make *a priori* statements about the usefulness of a particular mutation operator without knowing something about the type of problem that is to be solved and the representation that is being used for that problem ...

The suggested shift-mutation operator is applied to an individual string of the population with probability θ_{M2}. There is an equal chance for a left or right shift. An original string $s = \sigma_1 \ldots \sigma_n$ is left-shifted to $s' = \sigma_2 \ldots \sigma_n \tau$ or right-shifted to $s' = \tau \sigma_1 \ldots \sigma_{n-1}$ where the new random symbol τ is drawn uniformly from Σ. A generalization of this mutation operator might shift by more than one position at a time and will be taken into consideration for future experiments.

The experimental results in section 4 indicate that the addition of the special mutation operator leads to slightly better results.

4   Experimental Results of the GA Approach

The real-world problem instance studied consists of $m = 116$ viral sequences over the alphabet $\{A, C, G, T\}$ each having a sequence length of $n = 626$. The desired substring length is $l = 50$. The GA parameters are set as follows.

- Maximum number of generations: 100.
- Population size: 500.
- Recombination parameter (crossover rate) $\theta_R = 0.6$.
- Ordinary mutation parameter $\theta_M = 0.02$. (This equals $1/l$, i.e. one can expect one base mutation per individual per generation).
- Special shift-mutation parameter $\theta_{M2} = 0.03$. (That is, one can expect a total of 15 individuals to get shifted, left or right, per generation).
- Selection parameter $\theta_S = 1$. Nonoverlapping generations. Tournament selection with tournament size 2.

The statistics of a typical GA run are graphed in figure 1. Most of the optimization progress happens within the first 80 generations. A GA run for 100 generations takes about 40 minutes on a Intel Celeron 2.0 GHz with 256MB RAM.

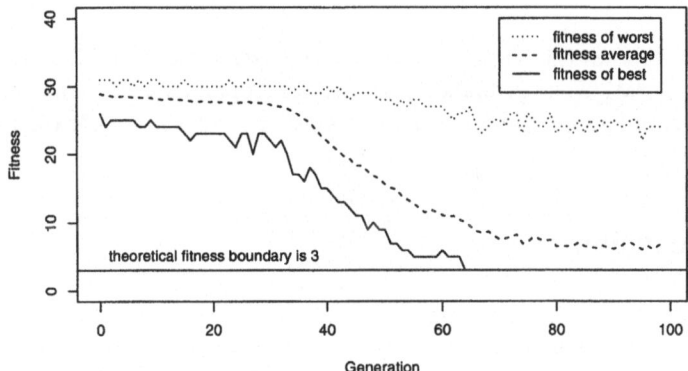

Fig. 1. Fitness Statistics for a Sample GA Run.

The interpretation of the distance matrix leads to $d_{max} = 5$ and thus it follows from inequality (2) that $z^* \geq 3$. Indeed, 2 out of 20 GA runs found closest substrings at a distance of $z^* = 3$. In the other 18 runs substrings at a distance of 4 were found. Running the same experiment without the special shift-mutation operator lead to 20 runs producing a substring at a distance of 4. While this does not prove that the improvements due to the shift-mutation operator are statistically significant, a beneficial effect has been observed.

For many practical purposes, a substring at distance 4 still provides sufficient homology in biological applications of this instance. Therefore, for the creation of the success statistics (figure 2), a sequence with a distance (and therefore fitness) of 4 was considered a success. If after 100 generations no individual with fitness 3 or 4 evolved, the GA run is considered a failure. No failure was observed in a sample of 20 GA runs. Note that the trivial approximation algorithm that works for the CSP (pick the most suitable string from S and designate it as the desired closest string) does not work for the CSSP and therefore even GA runs that yield an individual with fitness $d_{max} = 5$ or worse (i.e. failures) could be of some value in practice.

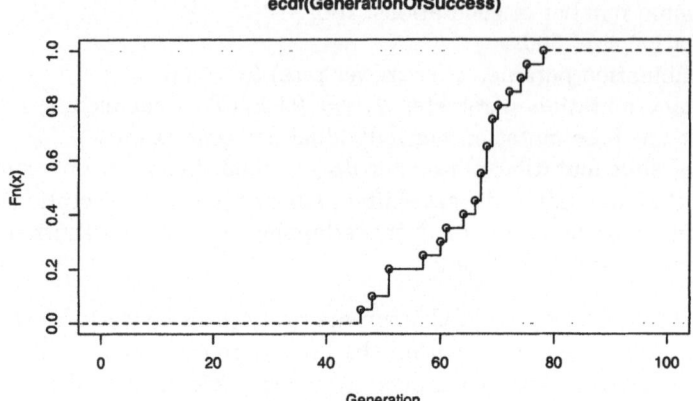

Fig. 2. Success Statistics.

Figure 2 shows the empirical cumulative distribution function which assigns to every generation the probability of success. Apparently the "fine-tuning" – i.e. the discovery of an individual with fitness 3 or 4 – takes place between generation 40 and 80.

5   Conclusion

Traditional approximation algorithms for the CSSP have the drawback to be either inefficient or inaccurate when applied to the large input sizes arising in reality. The genetic algorithm with a specifically designed mutation operator as described here is capable of combining efficiency with accuracy to solve large instances of the CSSP. In comparison with existing commercial GA software products it has been determined that the custom-built GA operates more efficiently and produces higher success rates than off-the-shelf GA software products, which cannot be adjusted easily to perform well on the CSSP. Future research will look into faster, heuristic ways of computing the fitness function and use more specialized genetic operators.

References

1. Mauch, H., Melzer, M.J., Hu, J.S.: Genetic algorithm approach for the closest string problem. In: Proceedings of the 2003 IEEE Bioinformatics Conference (CSB2003), Stanford, California, August 11–14, 2003, IEEE Computer Society Press (2003) 560–561
2. Frances, M., Litman, A.: On covering problems of codes. Theory of Computing Systems **30** (1997) 113–119
3. Gasienic, L., Jansson, J., Lingas, A.: Approximation algorithms for hamming clustering problems. In: CPM 2000. Volume 1848 of LNCS., Springer-Verlag (2000) 108–118

4. Gramm, J., Niedermeier, R., Rossmanith, P.: Exact solutions for closest string and related problems. In Eades, P., Takaoka, T., eds.: ISAAC 2001. Volume 2223 of LNCS., Springer-Verlag (2001) 441–453
5. Lanctot, J.K., Li, M., Ma, B., Wang, S., Zhang, L.: Distinguishing string selection problems. Information and Computation **185** (2003) 41–55
6. Li, M., Ma, B., Wang, L.: On the closest string and substring problems. Journal of the ACM **49** (2002) 157–171
7. Bäck, T., Fogel, D.B., Michalewicz, Z., eds.: Evolutionary Computation 1 - Basic Algorithms and Operators. Institute of Physics Publishing, Bristol, UK (2000)
8. Goldberg, D.: Genetic Algorithms in Search, Optimization, and Machine Learning. Addison-Wesley, Reading, MA (1989)
9. Holland, J.H.: Adaptation in Natural and Artificial Systems. University of Michigan Press, Ann Arbor, MI (1975)
10. Banzhaf, W., Nordin, P., Keller, R.E., Francone, F.D.: Genetic Programming - An Introduction: On the Automatic Evolution of Computer Programs and its Applications. Morgan Kaufmann Publishers, Inc., San Francisco, CA (1998)
11. Gen, M., Cheng, R.: Genetic Algorithms and Engineering Design. John Wiley and Sons, Inc., New York, NY (1996)
12. Ono, I., Yamamura, M., Kobayashi, S.: A genetic algorithm for job-shop scheduling problems using job-based order crossover. In: Proceedings of IEEE International Conference on Evolutionary Computation (ICEC'96). (1996) 547–552
13. Tavares, J., Pereira, F.B., Costa, E.: Evolving Golomb rulers. In: Proceedings of Genetic and Evolutionary Computation (GECCO 2004). (2004) 416–417

Quantum-Inspired Evolutionary Algorithms and Its Application to Numerical Optimization Problems

André V. Abs da Cruz[1], Carlos R. Hall Barbosa[1], Marco Aurélio C. Pacheco[1], and Marley Vellasco[1,2]

[1] ICA – Applied Computational Intelligence Lab,
Electrical Engineering Department
Pontifícia Universidade Católica do Rio de Janeiro
{andrev,hall,marco,marley}@ele.puc-rio.br
[2] Department of Computer Science
University College of London, UK

Abstract. This work proposes a new kind of evolutionary algorithm inspired in the principles of quantum computing. This algorithm is an extension of a proposed model for combinatorial optimization problems which uses a binary representation for the chromosome. This extension uses probability distributions for each free variable of the problem, in order to simulate the superposition of solutions, which is intrinsic in the quantum computing methodology. A set of mathematical operations is used as implicit genetic operators over those probability distributions. The efficiency and the applicability of the algorithm are demonstrated through experimental results using the F6 function.

1 Introduction

Many research efforts in the field of quantum computing have been made since 1990, after the demonstration that computers based on principles of quantum mechanics can offer more processing power for some classes of problems. The principle of superposition, which states that a particle can be in two different states simultaneously, suggests that a high degree of parallelism can be achieved using this kind of computers. Its superiority was shown with few algorithms such as the Shor's algorithm [1, 2], for factoring large numbers, and the Grover's algorithm [3], for searching databases. Shor's algorithm finds the prime factors of a n-digit number in polynomial time, while the best known classical algorithm has a complexity of $O(2^{n^{1/3}} log(n)^{2/3})$. On the other hand, Grover's algorithm searches for an item in a non-ordered database with n items with a complexity of $O(\sqrt{n})$ while the best classical algorithm has a complexity of $O(n)$.

Research on merging evolutionary algorithms with quantum computing has been developed since the end of the 90's. This research can be divided in two different groups: one that, motivated by the lack of quantum algorithms, focus on developing new ones by using techniques for automatically generating programs

N.R. Pal et al. (Eds.): ICONIP 2004, LNCS 3316, pp. 212–217, 2004.
© Springer-Verlag Berlin Heidelberg 2004

[4]; and another which focus on developing quantum-inspired evolutionary algorithms [5–7]. The latter approach, in which this work could be included, allows the algorithm to be executed on classical computers.

This paper is organized as follows: section 2 describes the proposed quantum-inspired evolutionary algorithm; section 3 describes the experiments; section 4 presents the results obtained; and finally section 5 draws some conclusions regarding the work.

2   Quantum-Inspired Evolutionary Algorithm

Quantum-inspired evolutionary algorithms rely on the concepts of "quantum bits", or qubits, and on superposition of states from quantum mechanics [5,6]. The state of a quantum bit can be represented as:

$$|\varphi> = |\alpha> + |\beta>$$

(1)

Where α and β are complex numbers that represent probability amplitudes of the corresponding states. $|\alpha|^2$ and $|\beta|^2$ give the probability of the qubit to be in state 0 and in state 1, respectively, when observed. The amplitude normalization guarantees that:

$$|\alpha|^2 + |\beta|^2 = 1$$

(2)

The quantum-inspired evolutionary algorithm with binary representation [5, 6] works properly in problems where this kind of representation is more suited. But, in some specific situations, representation by real numbers is more efficient (for instance, in function optimization, where one wants to find a maximum or minimum by adjusting some variables). The question then is: how to implement this representation using the quantum-inspired paradigm?

To answer that question it is important to consider the following questions:

- How to represent a superposition of states, since in this kind of problem the genes can assume values in a continuum interval between the variables' limits?
- How to update those values so that the algorithm converges towards an optimal or sub-optimal value?

For the first question the answer is very simple: instead of using probabilities of observing a particular state, a *probability distribution function* is defined for each variable, allowing a random selection of values in the variable's universe of discourse. In this work, in order to avoid an exponential growth in storage needs and to reduce computational cost a set of rectangular pulses has been employed to represent the distributions. This approach provides two major advantages: only the centre and the width of each pulse must be stored; and it simplifies the calculation of cumulative distribution functions, which are needed in the drawing of the random numbers used in the algorithm.

Therefore, the algorithm's initialization procedure begins with the definition of a value N that indicates how many pulses will be used to represent each variable's probability distribution function. Then, for each single pulse used in each variable, it must be defined:

- The pulse centre in the mean point of the variable domain;
- The pulse height as the inverse of the domain length divided by N.

At the end of this process, the sum of the N pulses related to a variable will have a total area of 1.

Suppose, for instance, that one wishes to initialize a variable with an universe of discourse equals to the interval $[-50, 50]$ and to use 4 rectangular pulses to represent the probability distribution function for this variable; in this case, each pulse would have a width equal to 100 and height equal to $1/100/4 = 0.0025$.

The set of probability distribution functions for each variable (genes) related to the problem creates a superposition $Q_i(t)$ for each variable i of the problem. From this $Q_i(t)$ distribution, a set of n points are randomly drawn, which will form the population $P(t)$.

After choosing the individuals that will form the population $P(t)$, it is necessary to update the probability distribution $Q_i(t)$, in order to converge to the optimal or sub-optimal solution, similarly to the conventional crossover from classical genetic algorithms. The method employed in this work consists of choosing randomly m individuals from the population $P(t)$ using a roulette method identical to the one used in classical genetic algorithms. Then, the central point of the first pulse is redefined as the mean value of those m chosen individuals. This process is repeated for each one of the N pulses that define the distribution $Q_i(t)$. The value m is given by:

$$m = \frac{n}{N} \tag{3}$$

Where N is the number of pulses used to represent the probability distribution function and n is size of the population $P(t)$.

In addition, after each generation, the pulses' width is contracted symmetrically regarding its center. This contraction is made following an exponential decay, according to the following formula:

$$\sigma = (u - l)^{(1 - \frac{t}{T})^\lambda} - 1 \tag{4}$$

Where σ is the pulse width, u is the domain's upper limit, l is the lower limit, t is the current algorithm generation, T is the total number of generations and λ is a parameter that defines the decay rate for the pulse width.

It is important to notice that as the pulses have their widths contracted and their mid-points changed, their sums will look less like a rectangular signal and will start to have several different shapes.

Although this algorithm is able to, intrinsically, recombine existing solutions (by summing up the pulses and using those sums as probability distribution functions for future drawing new individuals randomly), an operator similar to the mutation from classical genetic algorithms is also used in the quantum-inspired algorithm. This operator makes small random moves in the pulses mid-points by summing small random numbers to the center's value. This operator is used to avoid premature convergence to local minima or maxima.

3   Experiments

To evaluate the performance of the proposed algorithm a benchmark problem was used. The problem consists of optimizing the F6 function, a hard-to-optimize function due to the presence of several local maxima very close to each other. The equation that defines this function is shown below:

$$F6(x,y) = 0.5 - \frac{\left(\sin \sqrt{x^2 + y^2}\right)^2 - 0.5}{1.0 + 0.001(x^2 + y^2)^2} \tag{5}$$

The global maximum of this function is in the (0,0) point.

A classical genetic algorithm was used to compare the results. The parameters of this algorithm are shown in Table 1.

Table 1. Parameters for the classical genetic algorithm.

Mutation Rate	8%
Crossover Rate	80%
Gap	20%
Population Size	100
Generations	40
Number of Evaluations	4000
Genetic Operators	Arithmetical Crossover, Uniform and Creep Mutation
Domain	$x, y \in [-100, 100]$
Selection Method	Roulette with steady state

Table 2. Parameters for the quantum–inspired evolutionary algorithm.

Mutation Rate	2%
Pulses per Variable	4
Pulse Width Decay Rate (λ)	20%
Number of Observations $P(t)$	100
Generations	40
Number of Evaluations	4000

For the quantum-inspired evolutionary algorithm the set of parameters in Table 2 was used.

Those values have provided the best results and were obtained after systematic experiments, with several different configurations. For each experiment 20 rounds were made and the mean value for the evaluation was calculated.

4   Results

The results presented in Figure 1 were obtained through experiments using a domain for the x, y variables in the interval $[-100, 100]$.

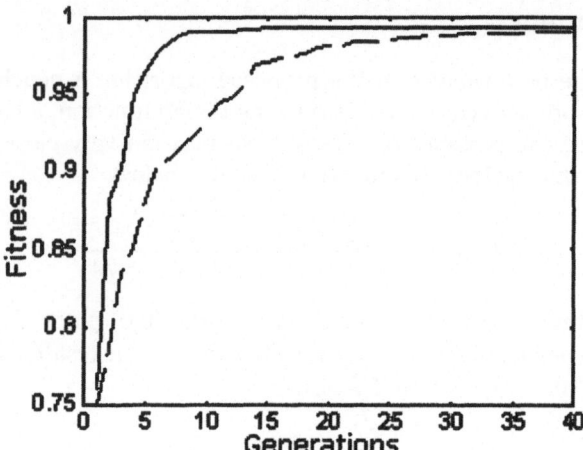

Fig. 1. Comparison between the classical (dashed line) genetic algorithm and the quantum–inspired (solid line) evolutionary algorithm.

This plot shows that the quantum-inspired evolutionary algorithm presented better performance regarding the necessary number of generations to reach the best solutions. Additionally, the final result obtained with the quantum-inspired algorithm is slightly better than the one obtained by the traditional genetic algorithm.

When the domain bounds for x and y variables are increased, the performance graphs are modified as in figure 2.

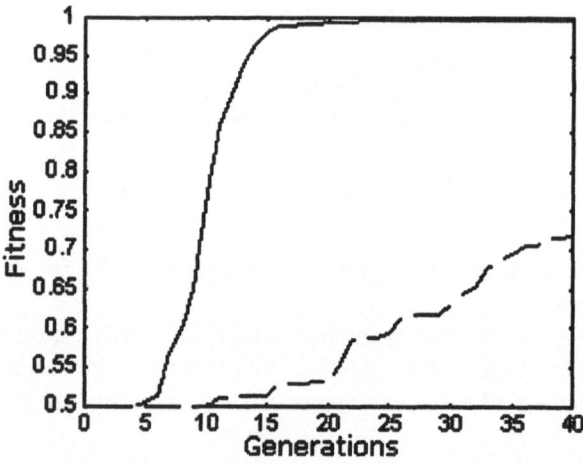

Fig. 2. Comparison between the classical (dashed line) genetic algorithm and the quantum–inspired (solid line) evolutionary algorithm after increasing the domain bounds.

As can be observed from Figure 2, the quantum-inspired algorithm did not suffer significant loss in its performance after augmenting the domain bounds. This shows that the algorithm might be used as a robust method for optimizing problems where the domain's size is critical.

5 Conclusions

This paper presented a new quantum inspired evolutionary algorithm suitable for problems that demand chromosomes with real number representation. The paper showed that the proposed algorithm is very efficient for solving the presented problem. Also, it has been demonstrated that the algorithm is robust for larger domains, which suggests the need for further investigations in order to determine the applicability of this algorithm for problems where domain's size is critical. It is necessary however, to use other benchmark functions in order to fully evaluate the algorithm's performance. It would be interesting to use several other functions with different characteristics (many local maxima, very flat surfaces near the optimal solution, etc).

References

1. Shor, P.W.: Algorithms for quantum computation: Discrete log and factoring. In: Foundations of Computer Science, Proc. 35th Ann. Symp., IEEE Computer Society Press (1994) 124–134
2. Shor, P.W.: Quantum computing. Documenta Mathematica (1998) 467–486
3. Grover, L.K.: A fast quantum mechanical algorithm for database search. In: Proceedings of the 28th Annual ACM Symposium on the Theory of Computing (STOC), ACM Press (1996) 212–219
4. Spector, L., Barnum, H., Bernstein, H.J., Swami, N.: Finding a better-than-classical quantum AND/OR algorithm using genetic programming. In: Proceedings of the Congress on Evolutionary Computation. Volume 3., IEEE Press (1999) 2239–2246
5. Han, K.H., Kirn, J.H.: Genetic quantum algorithm and its application to combinatorial optimization Problem. In: Proceedings of the 2000 Congress on Evolutionary Computation, IEEE Press (2000) 1354–1360
6. Han, K.H., Kirn, J.H.: Quan uminspired t evolutionary algorithm for a class of combinatorial optimization. IEEE Transactions on Evolutionary Computation 6 (2002) 580–593
7. Narayanan, A., Moore, M.: Genetic quantum algorithm and its application to combinatorial optimization problem. In: Proceedings of the 1996 IEEE International Conference on Evolutionary Computation (ICEC96), IEEE Press (1996) 61–66

Multiobjective Genetic Search
for Spanning Tree Problem

Rajeev Kumar, P.K. Singh, and P.P. Chakrabarti

Department of Computer Science and Engineering
Indian Institute of Technology Kharagpur
Kharagpur, WB 721 302, India
{rkumar,pksingh,ppchak}@cse.iitkgp.ernet.in

Abstract. A major challenge to solving multiobjective optimization problems is to capture possibly all the (representative) equivalent and diverse solutions at convergence. In this paper, we attempt to solve the generic multi-objective spanning tree (MOST) problem using an evolutionary algorithm (EA). We consider, without loss of generality, edge-cost and tree-diameter as the two objectives, and use a multiobjective evolutionary algorithm (MOEA) that produces diverse solutions without needing *a priori* knowledge of the solution space. We test this approach for generating (near-) optimal spanning trees, and compare the solutions obtained from other conventional approaches.

1 Introduction

Computing a minimum spanning tree (MST) from a connected graph is a well-studied problem and many fast algorithms and analytical analyses are available [1–5]. However, many real-life network optimization problems require the spanning tree to satisfy additional constraints along with minimum edge-cost. There are many engineering applications requiring MST problem instances having a bound on the degree, a bound on the diameter, capacitated trees or bounds for two parameters to be satisfied simultaneously [1]. Finding spanning trees of sufficient generality and of minimal cost subject to satisfaction of additional constraints is often NP-hard [1, 2].

Many such design problems have been attempted and approximate solutions obtained using heuristics. For example, the research groups of Deo et al. [3–5] and Ravi et al. [1, 2] have presented approximation algorithms by optimizing one criterion subject to a budget on the other. In recent years, evolutionary algorithms (EAs) have emerged as powerful tools to approximate solutions of such NP-hard problems. For example, Raidl & Julstorm [6, 7] and Knowles & Corne [8, 9] attempted to solve diameter and degree constrained minimum spanning tree problems, respectively using EAs. All such approximation and evolutionary algorithms yield a *single* optimized solution subject to satisfaction of the constraint(s).

We argue that such constrained MST problems are essentially multiobjective in nature. A multiobjective optimizer yields a set of all representative equivalent and diverse solutions rather a single solution; the set of all optimal solutions is called the Pareto-front. Most conventional approaches to solve network design problems start with a minimum spanning tree (MST), and thus effectively minimize the cost. With some variations induced by ϵ-constraint method, most other solutions obtained are located near the

N.R. Pal et al. (Eds.): ICONIP 2004, LNCS 3316, pp. 218–223, 2004.
© Springer-Verlag Berlin Heidelberg 2004

minimal-cost region of the Pareto-front, and thus do not form the complete (approximated) Pareto-front.

In this paper, we try to overcome the disadvantages of conventional techniques and single objective EAs. We use multiobjective EA to obtain a (near-optimal) Pareto-front. For a wide-ranging review, a critical analysis of evolutionary approaches to multiobjective optimization and many implementations of multiobjective EAs, see [10]. These implementations achieve diverse and equivalent solutions by some diversity preserving mechanism but they do not talk about convergence. Kumar & Rockett [11] proposed use of Rank-histograms for monitoring convergence of Pareto-front while maintaining diversity without any *explicit* diversity preserving operator. Their algorithm is demonstrated to work for problems of *unknown* nature. Secondly, assessing convergence does not need *a priori* knowledge for monitoring movement of Pareto-front using rank-histograms. Some other recent studies have been done on combining convergence with diversity. Laumanns et al. [12] proposed an ϵ-dominance for getting an ϵ-approximate Pareto-front for problems whose optimal Pareto-set is *known*.

In this work, we use the Pareto Converging Genetic Algorithm (PCGA) [11] which has been demonstrated to work effectively across complex problems and achieves diversity without needing *a priori* knowledge of the solution space. PCGA excludes any explicit mechanism to preserve diversity and allows a natural selection process to maintain diversity. Thus multiple, equally good solutions to the problem, are provided. We consider edge-cost and diameter as the two objectives to be minimized though the framework presented here is generic enough to include any number of objectives to be optimized. The rest of the paper is organized as follows. In Section 2, we include a brief review of the multiobjective evolutionary algorithm (MOEA). We describe, in Section 3, the representation scheme for the spanning tree and its implementation using PCGA. Then, we present results in Section 4 along with a comparison with other approaches. Finally, we draw conclusions in Section 5.

2   Multiobjective Evolutionary Algorithms: A Review

Evolutionary/Genetic Algorithms (EAs/GAs) are randomized search techniques that work by mimicking the principles of genetics and natural selection. (In this paper, we use the term EA and GA interchangeably.) EAs are different from traditional search and optimization methods used in engineering design problems. Most traditional optimization techniques used in science and engineering applications can be divided into two broad classes: direct search algorithms requiring only the objective function; and gradient search methods requiring gradient information either exactly or numerically. One common characteristic of most of these methods is that they all work on a point-by-point basis. An algorithm begins with an initial solution (usually supplied by the user) and a new solution is calculated according to the steps of the algorithm. These traditional techniques are apt for well-behaved, simple objective functions, and tend to get stuck at sub-optimal solutions. Moreover, such approaches yield a *single* solution. In order to solve complex, non-linear, multimodal, discrete or discontinuous problems, probabilistic search heuristics are needed which may work with a set of points/initial solutions, especially for multiobjective optimization which yields a set of (near-) optimal points, instead of a single solution.

Mathematically, in a maximization problem of m objectives, an individual objective vector F_i is partially less than another individual objective vector F_j (symbolically represented by $F_i \prec F_j$) iff:

$$(F_i \prec F_j) = (\forall_m)(f_{mi} \leq f_{mj}) \wedge (\exists_m)(f_{mi} < f_{mj})$$

Then F_j is said to dominate F_i where $F = (f_1, ..., f_m)$ is a vector-valued objective function. If an individual is not dominated by any other individual, it is said to be non-dominated. The notion of Pareto-optimality was introduced to assign equal probabilities of regeneration to all the individuals in the population, and the advantage of the Pareto rank-based research is that a multiobjective vector is reduced to a scalar fitness without combining the objectives in any way.

Almost all the multiobjective evolutionary algorithms/implementations have ignored the issue of convergence and are thus, unsuitable for solving *unknown* problems. Another drawback of most of these algorithms/implementations is the explicit use of parameterized sharing, mating restriction and/or some other diversity preserving operator. Any explicit diversity preserving mechanism method needs prior knowledge of many parameters and the efficacy of such a mechanism depends on successful fine-tuning of these parameters. It is the experience of almost all researchers that proper tuning of sharing parameters is necessary for effective performance, otherwise, the results can be ineffective if parameters are not properly tuned. In particular to MOST problem where we use a special encoding [7], incorporation of such knowledge is not an easy task.

3   Design and Implementation

Evolutionary algorithm operators namely mutation and crossover imitate the process of natural evolution, and are instrumental in exploring the search space. The efficiency of the genetic search depends how a spanning tree is represented in a chromosome. There are many encoding schemes available in literature see [7] for a detailed review and comparison. One classic representation scheme is Prüfer encoding which is used by Zhou & Gen [13]. The scheme is shown to be time and space efficient and able to represent all feasible solutions. However, Raidl & Julstorm [7] and Knowles & Corne [9] have pointed out that Prüfer numbers have poor locality and heritability and are thus unsuitable for evolutionary search. There are many other variants of Prüfer mappings too [5]. Recently, Raidl & Julstorm [7] proposed representing spanning trees directly as sets of their edges and have shown locality, heritability and computational efficiency of the edge sets for genetic search. In this work, we use their scheme for representing spanning trees to explore the search space.

We generate initial population based on random generation of spanning trees; we do not choose the cheapest edge from the currently eligible list of edges (as per Prim's algorithm) rather we select a random edge from the eligible list. The other variants of generating initial trees are based on One- Time-Tree Construction (OTTC) [4] and Randomized Greedy Heuristics (RGH) [6] algorithms. We select crossover operation to provide strong habitability such that the generated trees consist of the parental edges as far as possible. For generating valid trees, we include non-parental edges into the offspring tree. The mutation operator generates the valid spanning trees. We use the Pareto-rank based EA implementation, Roulette wheel selection for selecting the parents and rank-histogram for assessing the convergence [11].

4   Results

We tested generation of dual objective spanning tree using our MOEA framework and selected benchmark data taken from Beasley's OR library[1]. For comparison, we also include results obtained from two well-known diameter constrained algorithms, namely, One-Time Tree Construction (OTTC) [4] and Randomized Greedy Heuristics (RGH) [6] algorithms. Both algorithms are single objective algorithms and generate a single tree subject to the diameter constraint. Our MOST algorithm simultaneously optimizes both the objectives and generates a (near-optimal) Pareto-front which comprises a set of solutions. Therefore, we iteratively run both the OTTC and RGH algorithms by varying the value of the diameter constraint and generate sets of solutions to form the respective Pareto-fronts, for comparison with the Pareto-front obtained from the proposed multiobjective evolutionary algorithm. We have included results obtained from 50 and 100 node data in Figures 1 and 2, respectively.

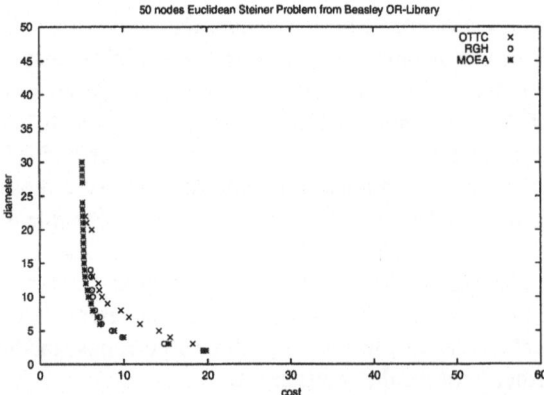

Fig. 1. Pareto front generated, for a 50 node data, from OTTC, RGH and Evolutionary algorithms.

It can be observed from Figures 1 and 2, that this is indeed difficult to find the solutions in the higher range of diameter. In fact, RGH algorithm could not find any solutions in this range of diameter; we generated multiple sets of solutions with multiple runs of RGH algorithm with different initial values but none of the run could generate any solution in this range of diameter. It can also be observed from Figures 1 and 2 that the solutions obtained form OTTC algorithm are good in lower and higher range of diameter, however, the results obtained from RGH are good only in the lower range of the diameter. Contrary to this, EA is able to locate solutions in the higher range of the diameter with almost comparable quality of the solutions obtained by OTTC. The solutions obtained by OTTC in the middle range are much sub-optimal and are inferior to the solutions obtained by EA. In the upper-middle range of diameters, RGH could not locate solutions at all, and the solutions located in this range by the OTTC are much

---

[1] http://mscmga.ms.ic.ac.uk/info.html

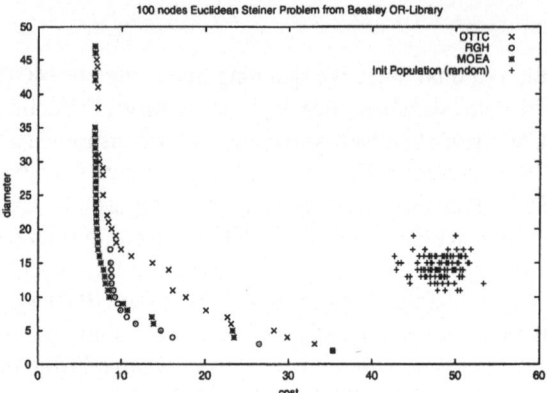

Fig. 2. Pareto front generated, for a 100 node data, from OTTC, RGH and Evolutionary algorithms. Initial population used for evolutionary algorithm is also shown.

inferior to the solutions obtained by EA. Thus, the quality of solutions obtained by EA is much superior in this range, and comparable in higher range to those of OTTC. Solutions obtained from EA are marginally inferior compared to RGH algorithm in very low-range of diameter; these solutions can be improved by fine-tuning of some parameters and procedures. This is an active area of research and is being investigated.

These are interesting observations, and are partly contrary to those reported by Raidl & Julstorm [6]. Raidl & Julstorm have shown that their technique works the best over all the other such techniques including OTTC. We reiterate that their conclusions were based on the experiments which they did for a particular value of the diameter; they could not observe the results over the entire range of diameter. We are currently investigating the empirical behavior shown by these three algorithms, and how this knowledge can be used to further improve the solution-set.

5   Conclusions

In this work, we demonstrated generating spanning trees subject to their satisfying the twin objectives of minimum cost and diameter. The obtained solution is a set of (near-optimal) spanning trees that are non-inferior with respect to each other.

The work presented in this paper presents a generic framework which can be used to optimize any number of objectives simultaneously for spanning tree problems. The simultaneous optimization of objectives approach has merits over the constrained-based approaches, e.g., OTTC and RGH algorithms. It is shown that the constrained-based approaches are unable to produce quality solutions over the entire range of the Pareto-front. For example, the best known algorithm of diameter-constrained spanning tree is RGH which is shown to be good for smaller values of diameters *only*, and is unable to produce solutions in the higher range. Similarly, the other well-known OTTC algorithm produces sub-optimal solutions in the middle range of the diameter. EA could obtain superior solutions in the entire range of the objective-values. The solutions obtained by EA may further be improved marginally by proper tuning of evolutionary operators for the specific values of the objectives by introducing problem specific knowledge while designing evolutionary operators; such type of improvement, is however, difficult with an approximation algorithm.

References

1. Marathe, M.V., Ravi, R., Sundaram, R., Ravi, S.S., Rosenkrantz, D.J., Hunt, H.B.: Bicriteria Network Design Problems. J. Algorithms **28** (1998) 142 – 171
2. Ravi, R., Marathe, M.V., Ravi, S.S., Rosenkrantz, D.J., Hunt, H.B.: Approximation Algorithms for Degree-Constrained Minimum-Cost Network Design Problems. Algorithmica **31** (2001) 58 – 78
3. Boldon, N., Deo, N., Kumar, N.: Minimum-Weight Degree-Constrained Spanning Tree Problem: Heuristics and Implementation on an SIMD Parallel Machine. Parallel Computing **22** (1996) 369 – 382
4. Deo, N., Abdalla, A.: Computing a Diameter-Constrained Minimum Spanning Tree in Parallel. In: Proc. 4th Italian Conference on Algorithms and Complexity (CIAC 2000), LNCS 1767. (2000) 17 – 31
5. Deo, N., Micikevicius, P.: Comparison of Prüfer-like Codes for Labeled Trees. In: Proc. 32nd South-Eastern Int. Conf. Combinatorics, Graph Theory and Computing. (2001)
6. Raidl, G.R., Julstrom, B.A.: Greedy Heuristics and an Evolutionary Algorithm for the Bounded-Diameter Minimum Spanning Tree Problem. In: Proc. 18th ACM Symposium on Applied Computing (SAC 2003). (2003) 747 – 752
7. Julstrom, B.A., Raidl, G.R.: Edge Sets: An Effective Evolutionary Coding of Spanning Trees. IEEE Trans. Evolutionary Computation **7** (2003) 225 – 239
8. Knowles, J.D., Corne, D.W.: A New Evolutionary Approach to the Degree-Constrained Minimum Spanning Tree Problem. IEEE Trans. Evolutionary Computation **4** (2000) 125 – 133
9. Knowles, J.D., Corne, D.W.: A Comparison of Encodings and Algorithms for Multiobjective Minimum Spanning Tree Problems. In: Proc. 2001 Congress on Evolutionary Computation (CEC-01). Volume 1. (2001) 544 – 551
10. Deb, K.: Multiobjective Optimization Using Evolutionary Algorithms. Chichester, UK: Wiley (2001)
11. Kumar, R., Rockett, P.I.: Improved Sampling of the Pareto-front in Multiobjective Genetic Optimization by Steady-State Evolution: A Pareto Converging Genetic Algorithm. Evolutionary Computation **10** (2002) 283 – 314
12. Laumanns, M., Thiele, L., Deb, K., Zitzler, E.: Combining Convergence and Diversity in Evolutionary Multiobjective Optimization. Evolutionary Computation **10** (2002) 263 – 282
13. Zohu, G., Gen, M.: Genetic Algorithm Approach on Multi-Criteria Minimum Spanning Tree Problem. European J. Operations Research **114** (1999) 141–152

A Partheno-genetic Algorithm
for Combinatorial Optimization

Maojun Li[1,2,*], Shaosheng Fan[1], and An Luo[3]

[1] College of Electrical & Information Engineering,
ChangSha University of Science & Technology, ChangSha 41 00 76, P.R. China
{famgge,fanss508}@sina.com
[2] College of Information Science & Technology, Central South University,
ChangSha 41 00 83, P.R. China
[3] College of Electrical & Information Engineering, Hunan University,
ChangSha 41 00 82, P.R. China
an_luo@hotmail.com

Abstract. Genetic Algorithms (GA) Using ordinal strings for combinatorial optimization must use special crossover operators such as PMX, OX and CX, instead of general crossover operators. Considering the above deficiency of GA
using ordinal strings, a Partheno-Genetic Algorithm (PGA) is proposed that
uses ordinal strings and repeals crossover operators, while introduces some particular genetic operators such as gene exchange operators, which have the same
function as crossover operators. The genetic operation of PGA is simpler and its
initial population need not be varied and there is no "immature convergence" in
PGA. The schema theorem of PGA was analyzed. Similarly with TGA, by genetic operators processing schemas, the individuals in the population continually move towards optimal individual in PGA, finally the optimal solution can
be gained. The global convergence of PGA was studied. It was also proved that
optimal maintaining operation is the key operation to make the algorithm global
convergent.

1 Introduction

The chromosomes of Genetic Algorithms (GA) fall into two groups: one is ordinal
string and the other non-ordinal string. While solving the combinatorial optimization
problems such as Traveling Salesman Problem [1], Train Line Holding Problem [2]
and Job-shop Problem [3,4], the ordinal string is simpler and more convenient than
non-ordinal string in the operation. Crossover operation of GA using ordinal strings is
far more difficult than one using non-ordinal strings although various crossover operators for GA using ordinal strings such as PMX, OX and CX [5] have been proposed. A Partheno-Genetic Algorithm (PGA) to solve the combinatorial optimization
problem is proposed [6,7]. PGA is a Genetic Algorithm (GA) using ordinal character
strings as chromosomes, and its genetic operation is achieved by the genetic operator
operating in one chromosome only such as gene exchange operator, gene shift operator and gene inversion operator, instead of crossover operators operating between two
chromosomes such as PMX, OX and CX. Compared with the Traditional Genetic

---

* Supported by the University Doctoral Foundation of Chinese State Education Department
(20030533014).

N.R. Pal et al. (Eds.): ICONIP 2004, LNCS 3316, pp. 224–229, 2004.
© Springer-Verlag Berlin Heidelberg 2004

Algorithms (TGA), the genetic operation in PGA is simpler and more efficient, the initial population need not be varied and there is no "immature convergence", while PGA has the basic features of TGA.

2 The Genetic Operators of PGA

Definition 1. Gene exchange operation of PGA is the procedure in which two or several genes (characters) in a chromosome (character string) are exchanged by certain probability p_e. The exchanged genes are randomly selected.

Gene exchange operation can be divided into one-point gene exchange operation and multi-point gene exchange operation. The former exchanges only two genes in a chromosome; the latter takes a definite positive integer u_e at first, and then randomly selects a positive integer $j \in \{1,2,\cdots,u_e\}$, finally exchanges j pair genes in a chromosome.

Definition 2. Gene shift operation of PGA is to shift the genes (characters) in one or several sub-strings in a chromosome (character string) backward in turn and to shift the last gene (character) in this (these) sub-string(s) to the first location, by certain probability p_s. In the gene shift operation, the shifted sub-string(s) and its (their) length are randomly determined.

Gene shift operation can also be divided into one-point gene shift operation and multi-point gene shift operation. The former only shifts the genes in one sub-string in a chromosome; the latter takes a definite positive integer u_s at first, and then randomly selects a positive integer $j \in \{1,2,\cdots,u_s\}$, finally shifts the genes in j sub-strings in a chromosome.

Definition 3. Gene inversion operation of PGA is to invert the genes (characters) in one or several sub-strings in a chromosome (character string) by certain probability p_i. In the gene inversion operation, the inverted sub-string(s) and its (their) length are randomly determined.

Gene inversion operation can also be divided into one-point gene inversion operation and multi-point gene inversion operation. The former only inverts the genes in one sub-string in a chromosome; the latter takes a definite positive integer u_i at first, and then randomly selects a positive integer $j \in \{1,2,\cdots,u_i\}$, finally inverts the genes in j sub-strings in a chromosome.

Definition 4. Gene leap operation of PGA is the procedure in which one or several genes (characters) in a chromosome (character string) to leap to other values in its same order gene set [6], by certain probability p_m. In the gene leap operation the changed genes are randomly selected.

Gene leap operation consists of one-point gene leap operation and multi-point gene leap operation. The former takes only one gene in a chromosome to leap to other values in its same order gene set; the latter takes a definite positive integer u_m at first, and then randomly selects a positive integer $j \in \{1, 2, \cdots, u_m\}$, finally takes j genes in a chromosome to leap to other values in their same order gene set.

Definition 5. Gene recombination operation of PGA is all of the genetic operations that adjust the location of the ordinal genes (characters) in a chromosome such as gene exchange operation, gene shift operation and gene inversion operation.

3 Formula of the Schema Theorem of PGA

Schema theorem is one of fundamental theory of GA. The formula of the schema theorem has relation with the operative process of PGA. This paper analyses the schema theorem of PGA based on the model operative process of PGA that has given out in Reference [6].

Assume gene recombination operator and gene leap operator destroy schemas by the probability p_{rd} and p_{md} respectively, then the number of the schema H in the new individuals generated from population t by gene recombination operator and gene leap operator is

$$m_1(H,t) = m(H,t)(1 - p_d) .$$ (1)

Gene recombination operator and gene leap operator search subspace Λ_e and subspace Λ_m [6] respectively, that is, they process the schemas in the their respective subspace, so that p_d in formula (1) should take p_{rd} or p_{md} respectively, while they do not combine one with the other. The number of the schema H in population $(t+1)$ is

$$m(H, t+1) \geq [m(H,t) + m_1(H,t)] \frac{f(H)}{2\overline{f}} .$$ (2)

where $\overline{f}$ is the average fitness of $2N$ individuals that contain N individuals in population t and N new individuals produced by the genetic operation, where N is the number of all of the individuals in a population; $f(H)$ is the average fitness of the individuals including schema H. Formula (2) is not an equation because we ignore the number of schema H obtained from non-schema H. From formula (1) and formula (2), we have

$$m(H, t+1) \geq m(H,t) \frac{f(H)}{\overline{f}} \left(1 - \frac{p_d}{2}\right) .$$ (3)

That is the formula of the schema theorem of PGA, which gives out the floor level of the number of the schema H in next population.

4 Global Convergence of PGA

It is obvious that all of the population of PGA make up a homogeneous finite Markov chain.

Definition 6. If $A \in R^{n \times n}$, and for any positive integer $i, j (1 \le i, j \le n)$, we have

(1) if $a_{ij} > 0$, then A is called positive matrix, expressed as $A > 0$;

(2) if $a_{ij} \ge 0$, then A is called non-negative matrix, expressed as $A \ge 0$.

Theorem 1. The Markov chain formed by all population which are generated from initial population λ_i by gene leap operators is ergodic.

Proof. Probability transition matrix of gene leap operator $P_m > 0$.

Theorem 2. The Markov Chain formed by all population generated from initial population λ_i by gene exchange operators is ergodic.

Proof. Probability transition Matrix of gene exchange operator $P_e > 0$.

Theorem 3. The Markov Chain formed by all population generated in PGA is ergodic.

Proof. Probability transition Matrix of PGA $P > 0$.

   Theorem 3 indicates that PGA can find global optimal solution, but it does not imply that PGA is global convergent.

Theorem 4. PGA is not global convergent.

Proof. From theorem 3 we know for any states λ_i and λ_j in the population space, there is a positive integer $s \to \infty$, subjecting $P_{ij}^{(s)} > 0$. And $\Pi = \lim P^{(t)}$ is a random Matrix, all the rows of which are identical, where P is the probability transition matrix of PGA. From $\sum_j \Pi_{ij} = 1$ we know that unless there is only one element 1 in each row of Π (this is not possible), we have $0 \le \Pi_{ij} < 1$.

   This implies PGA can find global optimal solution only in a probability less than 1. In other words, PGA is not global convergent.

   The essential case why PGA is not global convergent is that the found optimal solution can be destroyed in the operating procedure.

Definition 7. The operation that reproduces directly the optimal individuals of each population into next population is called optimal maintaining operation. PGA containing optimal maintaining operation is called Optimal Maintaining PGA (OMPGA).

Theorem 5. OMPGA is global convergent.

Proof. Assume that Λ_0 is a set of all the population consisting of one or more optimal individuals, then

$$|\Lambda_0| = \sum_{i=1}^{r} C_r^i g_c^{l(n-i)} (l!)^{n-i}, \tag{4}$$

where r is the number of optimal individuals in string space, $r \geq 1$; g_c is called same order gene number [6]; l is length of a chromosomes; n is the number of chromosomes in a population. The probability transition matrix of OMPGA can be expressed as

$$P = \begin{bmatrix} Q & O \\ T & U \end{bmatrix},$$
(5)

where Q is the probability transition matrix of Λ_0, and Q is a closed-class, while U is a transitive-class.

From theorem 3, we know that any state no in Λ_0 can be transferred into a state in Λ_0 in limited steps. Therefore the probability of there being one or more optimal individuals in population

$$\lim p^{(t)} = 1.$$
(6)

So OMPGA is global convergent.

It is evident that optimal maintaining operation is the key operation to make the PGA global convergent.

5 Conclusion

In order to use genetic algorithm in solving combinatorial optimization problem conveniently, a partheno-genetic algorithm was proposed. PGA maintains the basic features of TGA. Moreover, the genetic operation of PGA is simpler and more efficient than TGA. The initial population need not be varied and there is no "immature convergence" in PGA.

The schema theorem of PGA was thoroughly analyzed. From the point of view of processing schemas, the genetic operators of PGA have their respective function. During the genetic operation, gene recombination operator and gene leap operator continually produce new schemas, while selection operator, on one hand, maintains the excellent schemas with high fitness, but on the other hand, falls into disuse bad schemas with low fitness. Similarly with TGA, through genetic operators processing schemas, the individuals in the population continually move towards optimal individual in PGA, finally the optimal solution can be gained.

The global convergence of PGA was studied. It was also proved that optimal maintaining operation is the key operation to make this algorithm global convergent.

References

1. Lin W., Delgadofiras Y. G., Gause D. C. et al: Hybrid Newton Raphson Genetic Algorithm for the Traveling Salesman Problem. Cybernetics and Systems. 1995, 26(4): 387-412
2. Huang Xiaoyuan, Xiao Sihan, Wu Shulin: Application of Genetic Algorithm in Train Line Holding. Information and Control. 1996, 25(1): 58-63 (in Chinese)
3. Wang Bo, Zhang Qun, Wang Fei, et al: Quantitative Analysis of Infeasible Solution to Job Shop Scheduling Problem. Control and Decision, 2001, 16(1): 33-36 (in Chinese)

4. Imed Kacem, Slim Hammadi, Pierre Borne: Approach by Localization and Multi-objective Evolutionary Optimization for Flexible Job-Shop Scheduling Problems. IEEE trans on Systems, Man and Cybernetics-Part C: Applications and Reviews, 2002, 32(1):1-13
5. Pedro Larranga, Cindy M.H. Kuijipers, Roberto H. Murga, et al: Learning Bayesian Network Structures for the Best ordering with Genetic Algorithms. IEEE trans on Systems Man and Cybernetics-Part A: Systems and Humans. 1996, 26(4): 487-493
6. Li Maojun, Tong Tiaosheng: A Partheno-genetic Algorithm and Its Applications. Journal of Hunan University, 1998, 25(6):56-59
7. Li Maojun, Zhu Taoye, Tong Tiaosheng: Comparison Between Partheno-genetic Algorithm and Traditional Genetic Algorithm. System Engineering, 2001, 19(1):61-65

Evaluation of Comprehensive Learning Particle Swarm Optimizer

Jing J. Liang, A. Kai Qin, Ponnuthurai Nagaratnam Suganthan, and S. Baskar

BLK S-2, School of Electrical and Electronic Engineering
Nanyang Technological University, Singapore 639798
{liangjing,qinkai}@pmail.ntu.edu.sg, SBaskar@ntu.edu.sg
http://www.ntu.edu.sg/home/EPNSugan

Abstract. Particle Swarm Optimizer (PSO) is one of the evolutionary computation techniques based on swarm intelligence. Comprehensive Learning Particle Swarm Optimizer (CLPSO) is a variant of the original Particle Swarm Optimizer which uses a new learning strategy to make the particles have different learning exemplars for different dimensions. This paper investigates the effects of learning proportion P_c in the CLPSO, showing that different P_c realizes different performance on different problems.

1 Introduction

Particle swarm optimizer (PSO) simulates the behaviors of the birds flocking. In PSO, each solution is a point in the search space and may be regarded as a "bird". The bird would find food through its own efforts and social cooperation with the other birds around it. We refer to the *bird* as a *particle* in the algorithm. All particles have fitness values and velocities. The particles fly through the D dimensional problem space by learning from the best experiences of all the particles. Therefore, the particles have a tendency to fly towards better search area over the course of search process. The velocity $V_i(d)$ and position $X_i(d)$ updates of dth dimension of the ith particle are presented below [1,2]:

$$V_i(d) = \omega * V_i(d) + c_1 * rand1() * (Pbest_i(d) - X_i(d)) + c_2 * rand2() * (Gbest(d) - X_i(d)) \quad (1)$$

$$X_i(d) = X_i(d) + V_i(d) \quad (2)$$

where c_1 and c_2 are the acceleration constants representing the weighting of stochastic acceleration terms that pull each particle towards *pbest* and *gbest* positions. $rand1()$ and $rand2()$ are two random functions in the range [0,1]. $X_i = (x_{i1}, x_{i2}, ..., x_{iD})$ is the position of the ith particle; $Pbest_i = (pbest_{i1}, pbest_{i2}, ..., pbest_{iD})$ is the best previous position yielding the best fitness value $pbest_i$ for the ith particle; $Gbest = (gbest_1, gbest_2, ..., gbest_D)$ is the best position discovered by the whole population; $V_i = (v_{i1}, v_{i2}, ..., v_{iD})$ represents the rate of the position change (velocity) for particle i. ω is the inertia weight used to balance between the global and local search abilities. If $|V_i d|$ exceeds a positive constant value Vmax specified by user, then the velocity of that dimension is assigned to be $sign(V_i d)V_{max}$ that is, particles' velocity on each dimension is clamped to a maximum magnitude Vmax.

N.R. Pal et al. (Eds.): ICONIP 2004, LNCS 3316, pp. 230–235, 2004.
© Springer-Verlag Berlin Heidelberg 2004

The PSO algorithm is simple in concept, easy to implement and computationally efficient. Since its introduction in 1995 by Kennedy and Eberhart [1,2], PSO has attracted a lot of attention. Many researchers have worked on improving its performance in various ways and developed many interesting variants [3-10]. A New Learning Strategy is proposed in [11] to improve the original PSO, where each dimension of a particle learns from just one particle's historical best information, while each particle learns from different particles' historical best information for different dimensions for a few generations. This novel strategy ensures that the diversity of the swarm is preserved to discourage premature convergence and at the same time does not introduce any complex computations to the original PSO algorithm. Three versions were discussed demonstrating outstanding performance on solving multimodal problems in comparison to several other variants of PSO. Among those three versions, the Comprehensive Learning Particle Swarm Optimizer (CLPSO) is the best according to the experimental results. Hence, we further investigate the CLPSO in this paper.

This paper is organized as follows. Section 2 describes the Comprehensive Learning Particle Swarm Optimizer. Section 3 defines the benchmark continuous optimization problems used for experimental evaluation of the algorithms, the experimental setting, and discusses the results. Section 4 presents conclusions and directions for future work.

2 Comprehensive Learning Particle Swarm Optimizer

Though there are numerous versions of PSO, premature convergence when solving multimodal problems is still the main deficiency of the PSO. In the original PSO, each particle learns from its *pbest* and *gbest* simultaneously. Restricting the social learning aspect to only the *gbest* in the original PSO appears to be somewhat an arbitrary decision. Furthermore, all particles in the swarm learn from the *gbest* even if the current *gbest* is far from the global optimum. In such situations, particles may easily be attracted and trapped into a local optimum if the search environment is complex with numerous local solutions. As the fitness value of a particle is determined by all dimensions, a particle which has discovered the value corresponding to the global optimum in one dimension may have a low fitness value because of the poor solutions in other dimensions. This good genotype may be lost in this situation. The CLPSO is proposed based on this consideration. In CLPSO, there are three main differences compared to the original PSO:

1) Instead of using *pbest* and *gbest* as the exemplars, other particles' *pbest* are also used as the exemplars to guide a particle's flying direction.
2) Instead of learning from the same exemplars for all dimensions, each dimension of a particle in general learns from a different exemplar for different dimensions for a few generations. Each dimension of a particle could learn from the corresponding dimension of different particle's *pbest*.
3) Instead of learning from two exemplars at the same time in every generation as in the original PSO's eqn (1), each dimension of a particle learns from just one exemplar for a few generations.

In other words, CLPSO learns from the *gbest* of the swarm, the particle's *pbest* and the *pbests* of other particles so that the particles learn from the elite, itself and other particles. In CLPSO, m dimensions are randomly chosen to learn from the *gbest*. Some of the remaining D-m dimensions are randomly chosen to learn from some randomly chosen particles' *pbests* according to a probability P_c, while the remaining dimensions learn from its *pbest*. When m=0, though it seems *gbest* has no use, in fact it is one particle's *pbest* and has equal chance to be learnt by other particles. From the experiments in [11], m=0 gives better performance for complex problems, so CLPSO with m=0 is called the basic CLPSO and is used in the experiments in this paper. In this case, P_c is called learning proportion which means the proportion of the dimensions learnt from other particles. These operations increase the particles' initial diversity and enable the swarm to overcome premature convergence problem. By inspecting the expressions in eqns (1) and (2), we understand that PSO performs variable-wise update and not a vector update of the positions of particles. In other words, each dimension is updated independently. Hence, learning each dimension of a particle from a different *pbest* exemplar is within the spirit of the original PSO.

In order to prevent particles moving out of the search range, we proposed a method to constrain the particles within the range by calculating the fitness value of the particle to update its *pbest* and *gbest* only if the particle is in the range. Because the particles' *pbests* are all within the range, the particle will finally return to the search range. The pseudo code for CLPSO is given in Fig. 1.

Initialize positions and associated velocities of all particles in the population randomly in the D-dimensional search space. Evaluate the fitness values of all particles. Set the current position as *pbest* and the current particle with the best fitness value in the whole population as the *gbest*.
For k=1 to max_iteration

$$\omega(k) = \frac{(\omega_0 - 0.2) \times (\max\_gen - k)}{\max\_gen} + 0.2 \quad and \quad \omega_0 = 0.9 \tag{3}$$

If Mod(k,10)=1   //assign dimensions every 10 generations
For i=1 to ps,   // ps is the population size
rc=randperm(D);   //random permutation of D integers

$$b_i = zeros(1, D) ; b_i = \lceil rand(1, D) - 1 + P_c \rceil \tag{4}$$

$$f_i = \lceil rand(1, D).* ps \rceil \qquad // \lceil \ \rceil \text{ represents ceiling operator} \tag{5}$$

EndFor i
EndIf
For i=1 to ps // updating velocity, position of each particle
For d=1 to D // updating V, X of each dimension
if $b_i(d) == 1$ $V_i(d) = \omega_k * V_i(d) + rand() * (pbest_{fi(d)}(d) - X_i(d))$ (6a)

Else $V_i(d) = \omega_k * V_i(d) + rand() * (pbest_i(d) - X_i(d))$ (6b)

End If
$V_i(d) = \min(V_{max}(d), \max(-V_{max}(d), V_i(d)))$ // Limit the velocity
$X_i(d) = X_i(d) + V_i(d)$ // Update the position   (7)

EndFor d

If $X_i \in [X_{min}, X_{max}]$

Calculate the fitness value of X_i, Update pbest, gbest if needed

EndIf

EndFor i

Stop if a stop criterion is satisfied

EndFor k

Fig. 1. CLPSO's flow chart

3 Experimental Results and Discussions

In order to show the effects of the learning proportion P_c, which determines how many dimensions are chosen to learn from other particles' *pbest*s, experiments were conducted on six benchmark functions and their rotated versions obtained by using Salomon's algorithm [12]. The functions' dimensions are all 10. The number of particles is set 10 and the maximum number of generations is set 3000 and 8000 for unrotated problems and rotated problems respectively. P_c is set 0.05, 0.1, 0.2, 0.3, 0.4 and 0.5 separately and the algorithms are run 30 times for each P_c. Stop criteria are not set and all the experiments are run for full generations. The global optima are shifted in our experiments randomly in the search range, so the global optimum has different corrdinate values in different dimensions. The six benchmark minimization problems used in the experiments are listed below:

$$f_1(x) = \sum_{i=1}^{n} |x_i|^{i+1} \qquad \text{where} \quad -1 \le x_i \le 1 \qquad (8)$$

$$f_2(x) = \sum_{i=1}^{n} x_i^2 \qquad \text{where} \quad -5.12 \le x_i \le 5.12 \qquad (9)$$

$$f_3(x) = \sum_{i=1}^{n-1} [100(x_i^2 - x_{i+1})^2 + (x_i - 1)^2] \qquad \text{where} \quad -2.048 \le x_i \le 2.048 \qquad (10)$$

$$f_4(x) = \sum_{i=1}^{n} \frac{x_i^2}{4000} - \prod_{i=1}^{n} \cos(\frac{x_i}{\sqrt{i}}) + 1 \qquad \text{where} -600 \le x_i \le 600 \qquad (11)$$

$$f_5(x) = \sum_{i=1}^{n-1} (20 + e - 20 e^{-0.2\sqrt{0.5(x_{i+1}^2 + x_i^2)}} - e^{0.5(\cos(2\pi x_{i+1}) + \cos(2\pi x_i))}) \qquad (12)$$
$$\text{where} \ -32.768 \le x_i \le 32.768$$

$$f_6(x) = \sum_{i=1}^{n} (x_i^2 - 10\cos(2\pi x_i) + 10) \qquad \text{where} \quad -5.12 \le x_i \le 5.12 \qquad (13)$$

Among the six benchmark test functions, functions 1 and 2 are unimodal problems while functions 3-6 are multimodal problems. Tables 1 and 2 present the results of CLPSO under different P_c for test functions unrotated and rotated problems respectively. We can observe that in general CLPSO gives better performance on unrotated problems compared to the rotated problems. For unrotated problems, smaller P_c gives

Table 1. The results for the test functions without coordinate rotation

Pc / Func.	0.05	0.1	0.2	0.3	0.4	0.5
f_1	**1.8660e-058**	1.8477e-050	2.0265e-041	4.2851e-034	1.2526e-033	8.8705e-033
f_2	4.6947e-025	**2.2285e-025**	6.7862e-024	6.1037e-023	2.3877e-021	1.8417e-022
f_3	**1.4146e+000**	2.3840e+000	3.9404e+000	5.7626e+000	6.2301e+000	6.3456e+000
f_4	4.0314e-003	**2.3842e-003**	6.9777e-003	9.3566e-003	1.7317e-002	1.7242e-002
f_5	**6.0751e-014**	2.2157e-013	8.5523e-012	1.5495e-010	1.0010e-009	7.8453e-010
f_6	**0**	2.3216e-001	4.3115e-001	1.0613e+000	1.3929e+000	2.2884e+000

Table 2. The results for the test functions with coordinate rotation

Pc / Func.	0.05	0.1	0.2	0.3	0.4	0.5
f_1	1.3869e-008	1.1838e-008	1.0986e-008	**7.7857e-009**	8.3719e-009	2.2811e-008
f_2	**1.2654e-075**	2.2788e-070	6.8253e-064	5.9804e-057	1.1278e-053	3.2754e-054
f_3	**3.6688e+000**	4.4424e+000	5.4370e+000	6.4183e+000	6.7061e+000	6.3345e+000
f_4	**3.0349e-002**	3.4778e-002	3.8025e-002	4.0031e-002	3.7627e-002	4.9053e-002
f_5	9.4059e+000	5.9588e+000	4.6321e-001	**3.4128e-001**	6.0970e-001	1.0495e+000
f_6	6.7488e+000	5.9394e+000	4.0070e+000	**3.5885e+000**	3.7267e+000	3.7130e+000

better performance, but for the rotated problems, the situation is more complex. CLPSO realises the best performance at different P_c values for different functions. Different learning proportion P_c values give similar results for simple unimodal problems but seriously affect CLPSO's performance on multimodal problems.

4 Conclusions

In this paper, we represent further investigation on a novel comprehensive Learning Particle Swarm Optimizer (CLPSO). The effects of its learning proportion P_c are discussed. Experiments are conducted on functions without and with coordinate rotation and conclusions are given. Considering that the CLPSO yields the best performance at different P_c values for different functions, it is clearly that there is a need to adapt P_c, to handle different problems adaptively. Hence, we are in the process of developing adaptive CLPSO, which could learn a proper P_c.

References

1. R. C. Eberhart, J. Kennedy, "A new optimizer using particle swarm theory," *P. 6th Int. Symposium on Micromachine and Human Science,* Nagoya, Japan, pp. 39-43, 1995.
2. J. Kennedy, R. C. Eberhart, "Particle swarm optimization," *P. of IEEE International Conference on Neural Networks,* Piscataway, NJ, pp. 1942-1948, 1995.
3. Y. Shi and R. C. Eberhart, "A modified particle swarm optimizer," *Proc. of the IEEE Congress on Evolutionary Computation (CEC 1998),* Piscataway, NJ, pp. 69-73, 1998.
4. A. Ratnaweera, S. Halgamuge, and H. Watson, Self-organizing hierarchical particle swarm optimizer with time varying accelerating coefficients *IEEE Transactions on Evolutionary Computation,* vol. 8, pp. 240-255, Jun, 2004.
5. J. Kennedy, "Small worlds and mega-minds: effects of neighborhood topology on particle swarm performance," *P. CEC,* Washington DC, pp. 1931-1938, 1999.
6. J. Kennedy and R. Mendes, "Population structure and particle swarm performance," *Proc. of the CEC 2002,* Honolulu, Hawaii USA, 2002.
7. P. N. Suganthan, "Particle swarm optimiser with neighborhood operator," *P. of Congress on Evolutionary Computation,* Washington DC, pp. 1958-1962, 1999.
8. F. van den Bergh and A. P. Engelbrecht, A cooperative approach to particle swarm optimization *IEEE Trans. on Evolutionary Computation,* vol. 8, pp. 225-239, Jun, 2004.
9. M. Lovbjerg, T. K. Rasmussen, T. Krink, "Hybrid particle swarm optimiser with breeding and subpopulations," *P. of Genetic and Evolutionary Computation Conf.,* 2001.
10. T. Peram, K. Veeramachaneni, C. K. Mohan, "Fitness-distance-ratio based particle swarm optimization," *P. IEEE Swarm Intelligence Sym.,* Indiana, USA., pp. 174-181, 2003.
11. J. J Liang, A. K. Qin, P. N. Suganthan, and S. Baskar, 'Particle swarm optimization algorithms with novel learning strategies', *P. IEEE Int. Conf. on Systems Man and Cybernetics,* The Netherlands, October 2004 (http://www.ntu.edu.sg/home/EPNSugan/).
12. R. Salomon, "Reevaluating genetic algorithm performance under coordinate rotation of benchmark functions," *BioSystems,* vol. 39, pp.263-278, 1996.

Evolutionary Learning Program's Behavior in Neural Networks for Anomaly Detection

Sang-Jun Han, Kyung-Joong Kim, and Sung-Bae Cho

Dept. of Computer Science, Yonsei University,
134 Shinchon-dong, Sudaemoon-ku, Seoul 120-749, Korea
{sjhan,kjkim,sbcho}@cs.yonsei.ac.kr

Abstract. Learning program's behavior using machine learning techniques based on system call audit data is effective to detect intrusions. Among several machine learning techniques, the neural networks are known for its good performance in learning system call sequences. However, it suffers from very long training time because there are no formal solutions for determining the suitable structure of networks. In this paper, a novel intrusion detection technique based on evolutionary neural networks is proposed. Evolutionary neural networks have the advantage that it takes shorter time to obtain superior neural network than the conventional approaches because they learn the structure and weights of neural network simultaneously. Experimental results against 1999 DARPA IDEVAL data confirm that evolutionary neural networks are promising for intrusion detection.

1 Introduction

In host-based anomaly detection, the idea of learning program's behavior has been studied and used actively by many researchers. It considers normal behavior from the point of individual program. Profiles for program's behavior are built and the behaviors which deviate from the profile significantly are recognized as attacks. Machine learning methods have been used to profile program's behavior because it can be viewed as a binary classification problem which is one of the traditional problems in pattern classification. Especially, in previous researches, neural network showed the performance superior to other techniques. However, profiling normal behavior requires very long time due to the huge amount of audit data and computationally-intensive learning algorithm. Moreover, to apply neural network to real world problems successfully, it is very important to determine the topology of network, and the number of hidden nodes which are proper for the given problem, because the performance hinges upon the structure of network. Unfortunately, although many works on designing the domain-specific network structure automatically, there is no absolute solution [1] and typically the network structure is designed by repeating trial and error cycles on the basis of the experiences of working on similar problem. A.K. Ghosh who showed the best performance against the pubic benchmark data trained 90 neural networks in total for each program: 10, 15, 20, 25, 30, 35, 40, 50, and 60 hidden nodes

N.R. Pal et al. (Eds.): ICONIP 2004, LNCS 3316, pp. 236–241, 2004.
© Springer-Verlag Berlin Heidelberg 2004

and 10 networks for each number of hidden nodes. Then a neural network which showed best performance against the validation data was selected [2]. Therefore it takes a very long time to build normal behavior model and it is the vital drawback of neural network-based intrusion detection technique.

In this paper, we employ evolutionary neural network (ENN) to overcome the shortcoming of the conventional intrusion detection technique based on neural network. ENN does not require trial and error cycles for designing the network structure and the near optimal structure can be obtained automatically. Due to these advantages of ENN, we can get better classifier in shorter time. We examine the proposed method through experiments with real audit data and compare the result with that of other methods.

2   Intrusion Detection with Evolutionary NNs

Fig. 1 illustrates the overall architecture of ENN-based intrusion detection technique. We use system call-level audit data provided by BSM (Basic Security Module) of Solaris operating system. Preprocessor monitors the execution of specified programs and generates system call sequences by programs. GA modeler builds normal behavior profiles using ENN. One neural network is used per one program. New data are input to the corresponding neural network. If the evaluation value exceeds the pre-defined threshold, the alarm is raised.

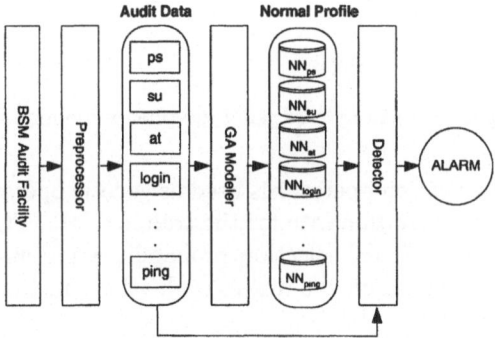

Fig. 1. Overall architecture of the proposed technique.

2.1   Modeling Normal Behavior

Our ENN has L input nodes because the system call sequence S_t which is generated at time t with window length L is used as input. There are two output nodes which represent normal and attack behavior respectively. 10 input nodes are used because we have set the window length as 10. There are 15 hidden nodes among which the connectivity is determined by evolutionary algorithm. Anomaly detector uses only attack-free data in training phase, but to train the supervised learner like neural network, the data labeled as attack are also needed.

For this reason, we have generated the artificial random system call sequences and used them as intrusive data. The training data is generated by mixing real normal sequences and artificial intrusive sequences in the ratio of 1 to 2. In this way, we can obtain the neural network which classifies all system call sequences except given normal sequence as attack behavior.

There are several genotype representations methods for neural network such as binary, tree, linked list, and matrix representation. We have used a matrix-based genotype representation because it is straightforward to implement and easy to apply genetic operators. When N is the total number of nodes in a neural network including input, hidden, and output nodes, the matrix is $N \times N$ whose entries consist of connection links and the corresponding weights. In this model, each neural network uses only forward links. In the matrix, upper right triangle (see Fig. 2) has connection link information and lower left triangle describes the weight values corresponding to the connection link information. The number of hidden nodes can be varied within the maximum number of hidden nodes in the course of genetic operations.

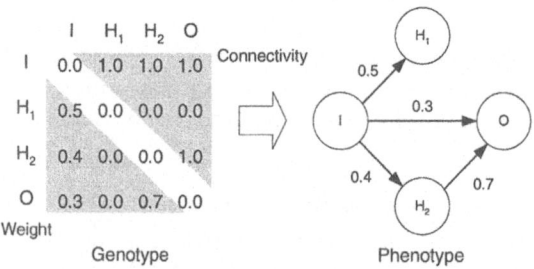

Fig. 2. An example of genotype-phenotype mapping.

Crossover and mutation operator is used as genetic operators and the fitness is calculated as the recognition rate for the training data. The rank-based selection in which the individuals' selection probabilities are assigned according to the individuals' rank based on the fitness evaluation function values is used.

2.2 Anomaly Detection

For accurate intrusion detection, it is important to recognize the temporal locality of abnormalous events, not the fluctuation of the output value [2]. High output values of attack node for very short time should be ignored because it is not sufficient to decide if that process is attack. To do that it is required to consider the previous output values as well as the current output values. For this purpose, we define a new measure of abnormality that has a leaky integrator. When o_t^1 denotes the output value of attack node, o_t^2 denotes the output value of normal node, and w_1, w_2, w_3 denote the weights to these values, the raw evaluation score r_t is calculated as follws:

$$r_t = w_1 \cdot r_{t-1} + w_2 \cdot o_t^1 + w_3 \cdot o_t^2 \tag{1}$$

It retains the evaluation value of past evaluation with decay and we get higher abnormality of current process as the output value of attack node is higher and the output value of normal node is lower. In this way, we can measure the abnormality of program's behavior robustly to short fluctuation and recognize the temporal locality of abnormal behaviors.

We define threshold and check whether its abnormality is exceeds it, to determine whether current process is attack or not. However, the decision boundaries vary from program to program because the different neural network is used to evaluate the different program behavior. Thus, applying a threshold to overall neural network is not feasible. To solve this problem we have normalized the raw evaluation values statistically. First, we test the training data using the trained neural network and we calculate the mean and variance of r_t. Then, under assumption of that r_t is normally distributed, we transform r_t to corresponding value in standard normal distribution R_t. When m is the mean of r_t and d is the standard deviation against the training data, the normalized evaluation value R_t is calculated as follows:

$$R_t = eval(S_t) = \frac{r_t - m}{d} \qquad (2)$$

If R_t exceeds the pre-defined threshold, current process is considered as attack.

3 Experiments

3.1 Experimental Settings

To verify the proposed method, we have used the 1999 DARPA intrusion evaluation data set [4]. In this paper, our experiments are focused on detecting U2R attack attempts to gain root privilege by privileged program misuse. Thus, we monitors only SETUID privileged programs which are the target of most U2R attacks. This data set consists of five weeks of audit data. 1-3 week data are for training and 4-5 week data are for testing. We have used 1 and 3 weeks data which do not contain any attacks for training neural networks and 4 and 5 week data are used for testing. The test data contain 11 instances of 4 types of U2R attacks.

Population size is 20 and the maximum generation number is 100. Crossover rate is 0.3 and mutation rate is 0.08. The neural network which has the highest fitness is selected and used for testing.

3.2 Results

Comparison of Training Time. The time required for training general MLP and ENN is compared. The training program was run on the computer with the dual Intel Pentium Zeon 2.4GHz processor, 1GB RAM, and Sun Solaris 9 operating system and the average time was taken. In the case of MLP, the number of hidden nodes varied from 10 and 60 and for each number of hidden

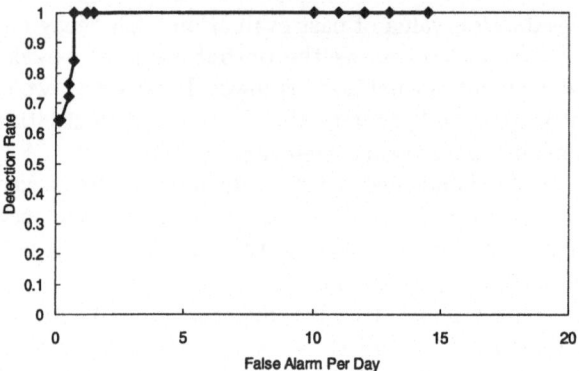

Fig. 3. Intrusion detection performance of ENN.

nodes, 10 networks were trained. Total 90 networks were trained. Error back propagation algorithm was iterated until 5000 epoch. ENN has the maximum 15 hidden nodes and the population of 20 neural networks was evolved to the 100th generation. Both neural network has 10 input nodes and 2 output nodes and are trained with the training data of *login* program which consists of 1905 sequences.

The conventional approach which repeats trial-and-error cycle requires about 17 hours 50 minutes. However, in case of evolutionary neural network, it takes 1 hour 14 minutes. Evolutionary approach can reduce the learning time as well as it has advantage that the near optimal network structure can be obtained.

Comparison of Detection Performance. Fig. 3 depicts the detection/false alarm plot which illustrates the intrusion detection performance of the proposed method. It produces 0.7 false alarms at 100% detection rate. In 1999 DARPA IDEVAL, the method which showed the best performance at detecting U2R attacks is the work of A.K. Ghosh *et at.* that learns system call sequences with Elman recurrent neural network [5]. It showed 3 false alarms at 100% detection rate [3]. The performance of ENN is superior to that of Elman network. This result illustrates that ENN can find more optimal neural network than the conventional neural network which has static and regular structure.

Comparsion of Network Structure. Table 1 compares ENN trained with *ps* program's behavior and general MLP in terms of network structure. Both have the same number of nodes: 10 input nodes, 15 hidden nodes, and 2 output nodes. The total number of connections does not differ much. However, ENN has more various types of connection including connection types which do not exist in MLP such as connections from input node to output node and from hidden node to hidden node. In the work of A.K. Ghosh *et al.*, they improved the performance by retaining context information between samples with recurrent topology. On the other hand, ENN attempts to increase learnable samples by forming non-regular and complex network structure.

Table 1. Comparison of network structure.

(a) ENN

From\To	Input	Hidden	Output
Input	0	86	15
Hidden	0	67	19
Output	0	0	0

(b) MLP

From\To	Input	Hidden	Output
Input	0	150	0
Hidden	0	0	30
Output	0	0	0

4 Conclusion

This paper proposes an evolutionary neural network approach for improving the performance of anomaly detection technique based on learning program's behavior. The proposed method cannot only improve the detection performance, but also reduce the time required for training because it learns the structure and weights of neural network simultaneously. The experimental result against 1999 DARPA IDEVAL which is superior to previous works verifies the proposed method. As future work, it is needed to find the network structure which is good for intrusion detection by analyzing the evolved structures. For more accurate modeling, we can employ multiple expert neural networks which are evolved with speciation and combine them.

Acknowledgement

This paper was supported by Brain Science and Engineering Research Program sponsored by the Korean Ministry of Science and Technology.

References

1. X. Yao, "Evolving Artificial Neural Networks," *Proceedings of the IEEE*, vol. 87, no. 9, pp. 1423-1447, 1999.
2. A. K. Ghosh, A. Schwartzbard, and M. Schatz, "Learning Program Behavior Profiles for Intrusion Detection," *Proceedings of the 1st USENIX Workshop on Intrusion Detection and Network Monitoring*, pp. 51-62, Santa Clara, CA, April, 1999.
3. A. K. Ghosh, C. C. Michael, and M. A. Schatz, "A Real-Time Intrusion Detection System Based on Learning Program Behavior," *Proceedings of the Third International Symposium on Recent Advances in Intrusion Detection*, pp. 93-109, 2000.
4. MIT Lincoln Laboratory, "DARPA Intrusion Detection Evaluation," Available from http://www.ll.mit.edu/IST/ideval/index.html.
5. R. Lippmann, J. Haines, D. Fried, J. Korba, and K. Das, "The 1999 DARPA Off-Line Intrusion Detection Evaluation," *Computer Networks*, vol. 34, no. 4, pp. 579-595, 2000.

Gray and Binary Encoding in the (1+1)-EA

Uday K. Chakraborty

University of Missouri, St. Louis, MO 63121, USA
uday@cs.umsl.edu

Abstract. The expected first passage time to optimality is used to analyze the relative performance of Gray and binary encoding for five variants of the (1+1)-EA.

1 Introduction

In this paper we build on our earlier work on the Gray-binary issue [1]. We
derive the transition probabilities of the Marvov chain models of (1+1)-EA. The
Markov chain is homogeneous.

2 Markov Model of the (1+1)-EA

The following versions of the (1+1)-EA ([2–4]) are analyzed here:

Algorithm 1

1. *Initialization* Choose a point – the *current* point, x_c – at random and evaluate
   it.
2. *Mutation* Mutate the current point by (probabilistically) flipping each bit
   (using a predetermined probability of bit-wise mutation, p_m), obtaining a
   (possibly) new point, x_a, and evaluate it.
3. *Selection* If the new point has a better fitness, accept the new point as the
   current point (that is, $x_c \leftarrow x_a$ with probability 1); otherwise leave the
   current point unaltered.
4. *Iteration* If a predetermined termination condition is not satisfied, go to
   step 2.

In the 1+1-EA the search begins with a single point and proceeds from one
point (state) to another. For an L-bit problem the search space consists of 2^L
points (states). At any single step, the process can move from a given point to
itself or to any one of the better points. A move from a current state i to a better
state j takes place with probability $p_{ij} = p_m^{n_{ij}} \cdot (1 - p_m)^{L-n_{ij}}$ where n_{ij} is the
Hamming distance between the two strings. The process stays in the same state
i with probability $1 - \sum_{k \in A_i} p_{ik}$ where A_i is the set of states that are better
than i. Therefore the entries of the $2^L \times 2^L$ transition probability matrix of the
Markov chain for the 1+1 EA are given by

$$
p_{ij} = \begin{cases} p_m^{n_{ij}} \cdot (1 - p_m)^{L-n_{ij}} & \text{for } j \in A_i \\ 1 - \sum_{k \in A_i} p_m^{n_{ik}} \cdot (1 - p_m)^{L-n_{ik}} & \text{for } i = j \\ 0 & \text{otherwise} \end{cases}
\tag{1}
$$

N.R. Pal et al. (Eds.): ICONIP 2004, LNCS 3316, pp. 242–247, 2004.
© Springer-Verlag Berlin Heidelberg 2004

Algorithm 2

1. *Initialization* Choose a point – the *current* point, x_c – at random and evaluate it.
2. *Mutation* Choose a bit uniformly randomly from among the L bits, and flip that bit, obtaining a new point, x_a, and evaluate it.
3. *Selection* If the new point has a better fitness, accept the new point as the current point (that is, $x_c \leftarrow x_a$) with a predetermined probability p; that is, between the two points x_c and x_a, accept the better with probability p, the other with the complementary probability $1 - p$.
4. *Iteration* If a predetermined termination condition is not satisfied, go to step 2.

In this case, mutation always produces a new string with exactly one bit flipped. The use of the parameter p gives us a range of selection pressures ($0.5 < p <= 1$), $p = 1$ corresponding to elitism. The transition probabilities are given by:

$$p_{ij} = \begin{cases} \frac{1}{L}p & \text{for } j \in H_{i1} \text{ and } j \text{ with a better fitness than } i \\ \frac{1}{L}(1-p) & \text{for } j \in H_{i1} \text{ and } j \text{ with a worse fitness than } i \\ 1 - \sum_{k \in H_{i1}} p_{ik} & \text{for } i = j \\ 0 & \text{otherwise} \end{cases} \quad (2)$$

where H_{i1} stands for the set of Hamming-distance-one neighbors of i.

Algorithm 3

Algorithm 3 is similar to Algorithm 2, except that in step 2 a pre-determined number, n ($n > 1$), of bits are (randomly) chosen to be flipped. Thus equation 2 holds for Algorithm 3 with H_{i1} replaced with H_{in}.

Algorithm 4 [3]

1. *Initialization* Choose a point – the *current* point, x_c – at random and evaluate it.
2. *Mutation* Choose a bit uniformly randomly from among the L bits, and flip that bit, obtaining a new point, x_a, and evaluate it.
3. *Selection* If the new point has a better fitness, accept the new point as the current point (that is, $x_c \leftarrow x_a$) with probability 1; if the new point is worse, accept it with probability $1/L$. If the two points have the same fitness, accept any one uniformly at random.
4. *Iteration* If a predetermined termination condition is not satisfied, go to step 2.

The transition probabilities are given by:

$$p_{ij} = \begin{cases} \frac{1}{L} & \text{for } j \in H_{i1} \text{ and } j \text{ with a better fitness than } i \\ \frac{1}{L}\frac{1}{L} & \text{for } j \in H_{i1} \text{ and } j \text{ with a worse fitness than } i \\ \frac{1}{L}\frac{1}{2} & \text{for } j \in H_{i1} \text{ and } j \text{ and } i \text{ have the same fitness} \\ 1 - \sum_{k \in H_{i1}} p_{ik} & \text{for } i = j \\ 0 & \text{otherwise} \end{cases} \quad (3)$$

244   Uday K. Chakraborty

Algorithm 5 This has been referred to as Algorithm RLS_p in [4].

1. *Initialization* Choose a point – the *current* point, x_c – at random and evaluate it.
2. *Mutation* Choose a bit uniformly randomly from among the L bits, and flip that bit, obtaining a new point, x_a. Now independently flip each of the bits of x_a, except the one flipped earlier, using a bitwise mutation probability of p. Let x_b be the point thus generated. (Note that x_a and x_b may be the same.)
3. *Selection* If the new point x_b has a better fitness, accept the new point as the current point (that is, $x_c \leftarrow x_b$) with probability 1. Otherwise, leave the current point x_c unaltered.
4. *Iteration* If a predetermined termination condition is not satisfied, go to step 2.

Note that this algorithm ensures that the new point will have at least one bit changed. Let H_{ik} represent the Hamming-distance-k neighborhood of point i (that is, the set of points that are at a Hamming distance of k from point i). To calculate the probability with which an L-bit point i will generate as the next move a better (fitter) point $j \in H_{ik}$, we note that i and j differ by exactly k bits (by the definition of H). One of these k bits must be flipped by the first part of the mutation process, and the corresponding probability is $1/L$ (all bits being equally likely to be chosen for flipping). Next, the remaining $k-1$ bits must be flipped (while the $L-1-(k-1)$ bits must not be flipped) – the probability of this event is $p^{k-1}(1-p)^{L-1-(k-1)}$. The calculation can be completed by noting that there are exactly k ways in which this last-mentioned event can occur (corresponding to the k bits that get changed). Therefore The transition probabilities are given by:

$$
p_{ij} = \begin{cases} \frac{1}{L} k p^{k-1}(1-p)^{L-k} & \text{for } j \in H_{ik} \text{ and } j \text{ with a better fitness than } i \\ 1 - \sum_{k=1}^{L}(p_{ij}|j \in H_{ik}) & \text{for } i = j \\ 0 & \text{otherwise} \end{cases}
$$

$$(4)$$

3 Expected First Passage Time to Convergence

For Algorithms 1, 2, 4 and 5 the transition probability matrix, $\mathcal{P}$, has exactly one absorbing state and the other states are transient. Let $\mathcal{Q}$ be the matrix obtained by truncating $\mathcal{P}$ to include only the non-absorbing states. Then $\mathcal{I} - \mathcal{Q}$ gives the "fundamental matrix", and the mean time to absorption, starting from a given transient state, is given by the row-sum of the corresponding row of the matrix $(\mathcal{I} - \mathcal{Q})^{-1}$. The expected value of the expected first passage time to the global optimum is then given by $\mathcal{E} = \frac{1}{2^L} \sum_{i=1}^{2^L} E(T_i)$ where E denotes expectation, and T_i is a random variable for the first passage time, given the start state i. For an absorbing state i, $P(T_i = 0)$ is unity. The expected value $\mathcal{E}$ is computed for both binary and Gray encoding and is used as the basis of comparison. Algorithm 3 does not have a finite mean first passage time. For $n = 2, L = 3$ there are four states which once entered can never be left, that is, starting from any one of these four states, the global optimum can never be reached. Similarly, for $n = 3, L = 3$, the global optimum can be reached from only one state other than the global optimum itself.

4 Results

For L bits, we have 2^L distinct function evaluations, and for any given set of 2^L values, we can permute these 2^L values, creating a total of $(2^L)!$ different functions. Without loss of generality, we consider a minimization problem. By covering all $(2^L)!$ functions, we have included all possible situations.

Table 1. The number of local minima in all possible functions defined over three bits. In the left part, the functions are divided into 4 categories corresponding to 1,2,3 or 4 local minima in the integer neighborhood. The right part shows the *total* number of functions with 1,2,3 or 4 local minima.

Integer		Gray		Binary	
#min	#fn	#min	#fn	#min	#fn
				1	64
1	512	1	512	2	384
				3	64
				1	3056
		1	6144	2	10032
2	14592			3	1360
		2	8448	4	144
				1	4112
		1	1984	2	13296
3	23040	2	16000	3	4336
		3	5056	4	1296
		2	32	1	1408
4	2176	3	704	2	768
		4	1440		
Total	40320		40320		40320

#min	#functions		
	Integer	Gray	Binary
1	512	8640	8640
2	14592	24480	24480
3	23040	5760	5760
4	2176	1440	1440
Total	40320	40320	40320

Performance comparisons are shown in Table 2 where the expected first passage times have been used as the basis of comparison. An encoding is better if it has a smaller expected first passage time to find the global optimum. Note that no EA runs (experiments) were performed; we obtained the first passage times theoretically, via the Markov chain calculations.

Table 2. Performance of Binary and Gray on Algorithms 1, 2, 4, and 5 ($L = 3$).

# Fn.	# Min.			# Winner							
				Algo 1		Algo 2		Algo 4		Algo 5	
	I	G	B	Gray	Bin	Gray	Bin	Gray	Bin	Gray	Bin
64	1	1	1	32	32	36	28	40	24	32	32
384	1	1	2	64	320	384	0	384	0	68	316
64	1	1	3	0	64	64	0	64	0	0	64
768	2	1	1	384	384	312	416	300	428	384	384
5248	2	1	2	432	4816	5248	0	5092	156	448	4800
128	2	1	3	0	128	128	0	128	0	0	128
2288	2	2	1	1988	300	0	2288	16	2272	1964	324
4784	2	2	2	3556	1228	1832	2896	2096	2632	3544	1240
1232	2	2	3	812	420	928	304	920	312	800	432
144	2	2	4	96	48	120	24	120	24	96	48
224	3	1	1	112	112	100	108	100	108	112	112
1568	3	1	2	248	1320	1568	0	1516	52	260	1308
192	3	1	3	0	192	192	0	192	0	0	192
2016	3	2	1	1604	412	0	2016	208	1808	1604	412
9024	3	2	2	4516	4508	4672	4104	4448	4328	4184	4840
3664	3	2	3	2012	1652	2816	848	2664	1000	1896	1768
1296	3	2	4	768	528	1272	24	1272	24	768	528
1872	3	3	1	1872	0	0	1872	0	1872	1872	0
2704	3	3	2	2336	368	176	2528	208	2496	2336	368
480	3	3	3	480	0	88	320	88	320	480	0
16	4	2	1	16	0	0	16	0	16	16	0
16	4	2	2	0	16	8	8	0	16	0	16
432	4	3	1	432	0	0	432	0	432	432	0
272	4	3	2	160	112	0	272	0	272	160	112
960	4	4	1	960	0	0	960	0	960	960	0
480	4	4	2	480	0	0	480	0	480	480	0

A "head-to-head minimax" comparison (Table 2) shows that the two representations produce different numbers of winners, e.g., 23360 for Gray and 16960 for binary in Algorithm 1 ($p_m = 0.8$), G:19944/B:19944 with 432 ties in Algorithm 2 ($p = 0.9$), G:19856/B:20032 with 432 ties in Algorithm 4, and G:22896/B:17424 in Algorithm 5 ($p = 0.7$).

Overall, the results show that it is not necessarily true that fewer local optima make the task easier for the evolutionary algorithm.

Table 3 shows how the relative performance of the two encodings changes with changes in the algorithm parameter. For $p_m = 0.5$ (random search) in Algorithm 1, the performances of Gray and binary should be identical, and this was corroborated by our numerical results. For Algorithm 1, when all 40320 functions are considered, binary is better for $p_m < 0.5$ and Gray is better for $p_m > 0.5$. Above 0.5, the relative advantage of Gray over binary decreases with increasing p_m, and below 0.5, the lower the mutation rate, the less pronounced the edge of binary over Gray.

Table 3. Effect of parameter values on the relative performance of Binary and Gray coding ($L = 3$). The total number of functions is 40320 for Algorithms 1 and 5, and 39888 (432 ties) for Algorithm 2.

Algo 1	# Winner		Algo 2	# Winner		Algo 5	# Winner	
	Gray	Binary		Gray	Binary		Gray	Binary
$p_m = 0.005$	17440	22880	$p = 0.6$	20008	19880	$p = 0.01$	17648	22672
$p_m = 0.05$	17440	22880	$p = 0.7$	20000	19888	$p = 0.05$	17568	22752
$p_m = 0.1$	17288	23032	$p = 0.8$	19744	20144	$p = 0.1$	17304	23016
$p_m = 0.333$	16840	23480	$p = 0.9$	19944	19944	$p = 0.4$	20296	20024
$p_m = 0.4$	16656	23664				$p = 0.5$	22984	17336
$p_m = 0.49$	16384	23936				$p = 0.9$	22584	17736
$p_m = 0.499$	16320	24000						
$p_m = 0.501$	24000	16320						
$p_m = 0.51$	23968	16352						
$p_m = 0.55$	23968	16352						
$p_m = 0.8$	23360	16960						
$p_m = 0.95$	23024	17296						
$p_m = 0.99$	23008	17312						

Because the transition probabilities are independent of the string fitnesses, the results of this paper hold for *any* discrete or discretized fitness function (with distinct fitness values).

5 Conclusions

Over all possible functions there is not much difference between the two representations, but fewer local optima do not necessarily make the task easier for Gray coding. The relative performance of the two encodings depends on the algorithm parameters.

References

1. Chakraborty, U.K., Janikow, C.Z., An analysis of Gray versus binary encoding in genetic search. *Information Sciences* 156, 2003, pp. 253-269.
2. Droste, S., Analysis of the (1+1) EA for a dynamically changing objective function, Tech. Report No. Cl-113/01, Univ. of Dortmund, 2001.
3. He, J., Yao, X., From an individual to a population: An analysis of the first hitting time of population-based evolutionary algorithms, IEEE Trans. Evol. Comput. 6(5), 2002, 495-511.
4. Wegener, I., Witt, C., On the optimization ofmonotone polynomials by the (1+1) EA and randomized local search, Proc. GECCO-2003, pp. 622-633, LNCS 2723, Springer, 2003.

Asymptotic Stability
of Nonautonomous Delayed Neural Networks*

Qiang Zhang[1,2], Xiaopeng Wei[1], Jin Xu[1], and Dongsheng Zhou[2]

[1] Center for Advanced Design Technology, Dalian University,
Dalian, 116622, China
[2] School of Mechanical Engineering, Dalian University of Technology,
Dalian, 116024, China
zhangq26@163.com

Abstract. A delay differential inequality is established in this paper. Based on this inequality, global asymptotic stability of nonautonomous delayed neural networks is analyzed. A new sufficient condition ensuring the global asymptotic stability for this kind of neural networks is presented. This condition is easy to be checked.

1 Introduction

The stability of autonomous delayed neural networks has been deeply studied in the past decades and many important results on the global asymptotic stability and global exponential stability of one unique equilibrium point have been presented, see, for example,[1]-[17] and references cited therein. However, to the best of our knowledge, few studies have considered dynamics for nonautonomous delayed neural networks [18]. In this paper, by using a differential inequality, we discuss the global asymptotic stability of nonautonomous delayed neural networks and obtain a new sufficient condition. We do not require the delay to be differentiable.

2 Preliminaries

The dynamic behavior of a continuous time nonautonomous delayed neural networks can be described by the following state equations:

$$
\begin{aligned}
x_i'(t) = &-c_i(t)x_i(t) + \sum_{j=1}^{n} a_{ij}(t)f_j(x_j(t)) \\
&+ \sum_{j=1}^{n} b_{ij}(t)f_j(x_j(t - \tau_j(t))) + I_i(t).
\end{aligned}
\tag{1}
$$

where n corresponds to the number of units in a neural networks; $x_i(t)$ corresponds to the state vector at time t; $f(x(t)) = [f_1(x_1(t)), \cdots, f_n(x_n(t))]^T \in R^n$

* The project supported by the National Natural Science Foundation of China and China Postdoctoral Science Foundation

N.R. Pal et al. (Eds.): ICONIP 2004, LNCS 3316, pp. 248–253, 2004.
© Springer-Verlag Berlin Heidelberg 2004

denotes the activation function of the neurons; $A(t) = [a_{ij}(t)]_{n \times n}$ is referred to as the feedback matrix, $B(t) = [b_{ij}(t)]_{n \times n}$ represents the delayed feedback matrix, while $I_i(t)$ is an external bias vector at time t, $\tau_j(t)$ is the transmission delay along the axon of the jth unit and satisfies $0 \leq \tau_i(t) \leq \tau$.

Throughout this paper, we will assume that the real valued functions $c_i(t) > 0, a_{ij}(t), b_{ij}(t), I_i(t)$ are continuous functions. The activation functions $f_i, i = 1, 2, \cdots, n$ are assumed to satisfy the following hypothesis

$$|f_i(\xi_1) - f_i(\xi_2)| \leq L_i|\xi_1 - \xi_2| \ , \forall \xi_1, \xi_2. \tag{2}$$

This type of activation functions is clearly more general than both the usual sigmoid activation functions and the piecewise linear function (PWL): $f_i(x) = \frac{1}{2}(|x+1| - |x-1|)$ which is used in [8].

The initial conditions associated with system (1) are of the form

$$x_i(s) = \phi_i(s), \ s \in [-\tau, 0], \ \tau = \max_{1 \leq i \leq n} \{\tau_i^+\} \tag{3}$$

in which $\phi_i(s)$ are continuous for $s \in [-\tau, 0]$.

Lemma 1. *Assume $k_1(t)$ and $k_2(t)$ are nonnegative continuous functions. Let $x(t)$ be a continuous nonnegative function on $t \geq t_0 - \tau$ satisfying inequality (4) for $t \geq t_0$.*

$$x'(t) \leq -k_1(t)x(t) + k_2(t)\bar{x}(t) \tag{4}$$

where $\bar{x}(t) = \sup_{t-\tau \leq s \leq t} \{x(s)\}$. If the following conditions hold

$$\begin{array}{l} 1) \ \int_0^\infty k_1(s)ds = +\infty \\ 2) \ \int_{t_0}^t k_2(s)e^{-\int_s^t k_1(u)du}ds \leq \delta < 1. \end{array} \tag{5}$$

then, we have $\lim_{t \to \infty} x(t) = 0$.

Proof. It follows from (4) that

$$x(t) \leq x(t_0)e^{-\int_{t_0}^t k_1(s)ds} + \int_{t_0}^t k_2(s)e^{-\int_s^t k_1(u)du}\bar{x}(s)ds, \ t \geq t_0 \tag{6}$$

For $t \geq t_0$, let $y(t) = x(t)$, and for $t_0 - \tau \leq t \leq t_0$, $y(t) = \sup_{t_0 - \tau \leq \theta \leq t_0}[x(\theta)]$. From (6), we obtain

$$x(t) \leq x(t_0) + \delta \sup_{t_0 - \tau \leq \theta \leq t}[x(\theta)], \ t \geq t_0 \tag{7}$$

then, we can get

$$y(t) \leq x(t_0) + \delta \sup_{t_0 - \tau \leq \theta \leq t}[y(\theta)], \ t \geq t_0 - \tau \tag{8}$$

Since the right hand of (8) is nondecreasing, we have

$$\sup_{t_0 - \tau \leq \theta \leq t}[y(\theta)] \leq x(t_0) + \delta \sup_{t_0 - \tau \leq \theta \leq t}[y(\theta)], \ t \geq t_0 - \tau \tag{9}$$

and

$$x(t) = y(t) \leq \frac{x(t_0)}{1 - \delta}, \ t \geq t_0 \tag{10}$$

By condition 1), we know that $\lim_{t \to \infty} \sup x(t) = x^*$ exists. Hence, for each $\varepsilon > 0$, there exists a constant $T > t_0$ such that

$$x(t) < x^* + \varepsilon, \ t \geq T \tag{11}$$

From (6) combining with (11), we have

$$\begin{aligned} x(t) &\leq x(T)e^{-\int_T^t k_1(s)ds} + \int_T^t k_2(s)e^{-\int_s^t k_1(u)du}\bar{x}(s)ds \\ &\leq x(T)e^{-\int_T^t k_1(s)ds} + \delta(x^* + \varepsilon), \ t \geq T \end{aligned} \tag{12}$$

On the other hand, there exists another constant $T_1 > T$ such that

$$\begin{aligned} x^* - \varepsilon &< x(T_1) \\ e^{-\int_T^{T_1} k_1(u)du} &\leq \varepsilon \end{aligned} \tag{13}$$

therefore,

$$x^* - \varepsilon < x(T_1) \leq x(T)\varepsilon + \delta(x^* + \varepsilon) \tag{14}$$

Let $\varepsilon \to 0^+$, we obtain

$$0 \leq x^* \leq \delta x^* \tag{15}$$

this implies $x^* = 0$. This completes the proof.

3   Global Asymptotic Stability Analysis

In this section, we will use the above Lemma to establish the asymptotic stability of system (1). Consider two solutions $x(t)$ and $z(t)$ of system (1) for $t > 0$ corresponding to arbitrary initial values $x(s) = \phi(s)$ and $z(s) = \varphi(s)$ for $s \in [-\tau, 0]$. Let $y_i(t) = x_i(t) - z_i(t)$, then we have

$$\begin{aligned} y_i'(t) &= -c_i(t)y_i(t) + \sum_{j=1}^n a_{ij}(t)\left(f_j(x_j(t)) - f_j(z_j(t))\right) \\ &+ \sum_{j=1}^n b_{ij}(t)\left(f_j(x_j(t - \tau_j(t))) - f_j(z_j(t - \tau_j(t)))\right) \end{aligned} \tag{16}$$

Set $g_j(y_j(t)) = f_j(y_j(t) + z_j(t)) - f_j(z_j(t))$, one can rewrite Eq.(16) as

$$y_i'(t) = -c_i(t)y_i(t) + \sum_{j=1}^n a_{ij}(t)g_j(y_j(t)) + \sum_{j=1}^n b_{ij}(t)g_j(y_j(t - \tau_j(t))) \tag{17}$$

Note that the functions f_j satisfy the hypothesis (2), that is,

$$\begin{aligned} |g_i(\xi_1) - g_i(\xi_2)| &\leq L_i|\xi_1 - \xi_2| \ , \forall \xi_1, \xi_2. \\ g_i(0) &= 0 \end{aligned} \tag{18}$$

Theorem 1. *Let*

$$k_1(t) = \min_i \left[2c_i(t) - \sum_{j=1}^{n} \left(L_j \left(|a_{ij}(t)| + |b_{ij}(t)| \right) + \frac{\alpha_j}{\alpha_i} L_i |a_{ji}(t)| \right) \right] > 0 \quad (19)$$

$$k_2(t) = \max_i \sum_{j=1}^{n} \left(\frac{\alpha_j}{\alpha_i} L_i |b_{ji}(t)| \right)$$

where $\alpha_i > 0$ is a positive constant. Eq.(1) is globally asymptotically stable if

$$\begin{aligned} &1) \ \int_0^\infty k_1(s)ds = +\infty \\ &2) \ \int_{t_0}^{t} k_2(s) e^{-\int_s^t k_1(u)du} ds \le \delta < 1. \end{aligned} \quad (20)$$

Proof. Let $z(t) = \frac{1}{2} \sum_{i=1}^{n} \alpha_i y_i^2(t)$, Calculating the time derivative of $z(t)$ along the solutions of (17), we get

$$\begin{aligned}
z'(t) &= \sum_{i=1}^{n} \alpha_i y_i(t) y_i'(t) \\
&= \sum_{i=1}^{n} \alpha_i y_i(t) \left[-c_i(t) y_i(t) + \sum_{j=1}^{n} a_{ij}(t) g_j(y_j(t)) \right. \\
&\qquad\qquad\qquad \left. + \sum_{j=1}^{n} b_{ij}(t) g_j(y_j(t - \tau_j(t))) \right] \\
&= \sum_{i=1}^{n} \alpha_i \left[-c_i(t) y_i^2(t) + \sum_{j=1}^{n} a_{ij}(t) y_i(t) g_j(y_j(t)) \right. \\
&\qquad\qquad\qquad \left. + \sum_{j=1}^{n} b_{ij}(t) y_i(t) g_j(y_j(t - \tau_j(t))) \right] \\
&\le \sum_{i=1}^{n} \alpha_i \left[-c_i(t) y_i^2(t) + \sum_{j=1}^{n} L_j |a_{ij}(t)| |y_i(t)| |y_j(t)| \right. \\
&\qquad\qquad\qquad \left. + \sum_{j=1}^{n} L_j |b_{ij}(t)| |y_i(t)| |\bar{y}_j(t)| \right] \quad (21)
\end{aligned}$$

Recall that the inequality $2ab \le a^2 + b^2$ holds for any $a, b \in R$. Employing this inequality, we can obtain

$$\begin{aligned}
z'(t) &\le \sum_{i=1}^{n} \alpha_i \left[-c_i(t) y_i^2(t) + \sum_{j=1}^{n} \frac{L_j |a_{ij}(t)|}{2} \left(y_i^2(t) + y_j^2(t) \right) \right. \\
&\qquad\qquad\qquad \left. + \sum_{j=1}^{n} \frac{L_j |b_{ij}(t)|}{2} \left(y_i^2(t) + \bar{y}_j^2(t) \right) \right]
\end{aligned}$$

$$= -\frac{1}{2} \sum_{i=1}^{n} \alpha_i \left[2c_i(t) - \sum_{j=1}^{n} \left(L_j \left(|a_{ij}(t)| + |b_{ij}(t)| \right) + \frac{\alpha_j}{\alpha_i} L_i |a_{ji}(t)| \right) \right] y_i^2(t)$$

$$+ \frac{1}{2} \sum_{i=1}^{n} \alpha_i \left(\sum_{j=1}^{n} \frac{\alpha_j}{\alpha_i} L_i |b_{ji}(t)| \right) \bar{y}_i^2(t)$$

$$\leq -k_1(t) z(t) + k_2(t) \bar{z}^2(t) \tag{22}$$

According to Lemma above, if the conditions 1) and 2) are satisfied, then we have $\lim_{t \to \infty} z(t) = \lim_{t \to \infty} \frac{1}{2} \sum_{i=1}^{n} \alpha_i y_i^2(t) = 0$, which implies that $\lim_{t \to \infty} y_i(t) = 0$. This completes the proof.

Remark 1. Note that the criteria obtained here are independent of delay and the coefficients $c_i(t)$, $a_{ij}(t)$ and $b_{ij}(t)$ may be unbounded. Therefore, the results here improve and generalize those obtained in [18].

4   Conclusion

A new sufficient condition ensuring global asymptotic stability for nonautonomous delayed neural networks is given by utilizing a delay differential inequality. Since the condition does not impose the differentiability on delay, it is less conservative than some presented in the earlier references.

References

1. Zhang, Q., Ma, R., Xu, J.: Stability of Cellular Neural Networks with Delay. Electron. Lett. **37** (2001) 575–576
2. Zhang, Q., Ma, R., Wang, C., Xu, J.: On the Global Stability of Delayed Neural Networks. IEEE Trans.Automatic Control **48** (2003) 794–797
3. Zhang, Q., Wei, X.P. Xu, J.: Global Exponential Convergence Analysis of Delayed Neural Networks with Time-Varying Delays. Phys.Lett.A **318** (2003) 537–544
4. Arik, S.: An Improved Global Stability Result for Delayed Cellular Neural Networks. IEEE Trans.Circuits Syst.I. **49** (2002) 1211–1214
5. Arik, S.: An Analysis of Global Asymptotic Stability of Delayed Cellular Neural Networks. IEEE Trans.Neural Networks. **13** (2002) 1239–1242
6. Cao, J., Wang, J.: Global Asymptotic Stability of a General Class of Recurrent Neural Networks with Time-Varying Delays. IEEE Trans.Circuits Syst.I. **50** (2003) 34–44
7. Chen, A., Cao, J., Huang, L.: An Estimation of Upperbound of Delays for Global Asymptotic Stability of Delayed Hopfiled Neural Networks. IEEE Trans.Circuits Syst.I. **49** (2002) 1028–1032
8. Chua, L.O., Yang, L.: Cellular Neural Networks: Theory and Applications. IEEE Trans.Circuits Syst.I. **35** (1988) 1257–1290
9. Feng, C.H., Plamondon, R.: On the Stability Analysis of Delayed Neural Networks Systems. Neural Networks. **14** (2001) 1181–1188
10. Huang, H., Cao, J.: On Global Asymptotic Stability of Recurrent Neural Networks with Time-Varying Delays. Appl.Math.Comput. **142** (2003) 143–154

11. Liao, X., Chen, G., Sanchez, E.N.: LMI-Based Approach for Asymptotically Stability Analysis of Delayed Neural Networks. IEEE Trans.Circuits Syst.I. **49** (2002) 1033–1039
12. Liao, X.X., Wang, J.: Algebraic Criteria for Global Exponential Stability of Cellular Neural Networks with Multiple Time Delays. IEEE Trans.Circuits Syst.I. **50** (2003) 268–274
13. Mohamad, S., Gopalsamy, K.: Exponential Stability of Continuous-Time and Discrete-Time Cellular Neural Networks with Delays. Appl.Math.Comput. **135** (2003) 17–38
14. Roska, T., Wu, C.W., Chua, L.O.: Stability of Cellular Neural Network with Dominant Nonlinear and Delay-Type Templates. IEEE Trans.Circuits Syst.**40** (1993) 270–272
15. Zeng, Z., Wang, J., Liao, X.: Global Exponential Stability of a General Class of Recurrent Neural Networks with Time-Varying Delays. IEEE Trans.Circuits Syst.I. **50** (2003) 1353–1358
16. Zhang, J.: Globally Exponential Stability of Neural Networks with Variable Delays. IEEE Trans.Circuits Syst.I. **50** (2003) 288–290
17. Zhou, D., Cao, J.: Globally Exponential Stability Conditions for Cellular Neural Networks with Time-Varying Delays. Appl.Math.Comput. **131** (2002) 487–496
18. Jiang, H., Li, Z., Teng, Z.: Boundedness and Stability for Nonautonomous Cellular Neural Networks with Delay. Phys.Lett.A **306** (2003) 313–325

A New PID Tuning Technique
Using Differential Evolution for Unstable
and Integrating Processes with Time Delay

Zafer Bingul

Kocaeli University, Mechatronics Engineering, Veziroglu Kampusu,
Kocaeli, Turkey
zaferb@kou.edu.tr

Abstract. In this paper, differential evolution algorithm (DEA), one of the most promising Evolutionary Algorithm's, was employed to tune a PID controller and to design a set-point filter for unstable and integrating processes with time delay. The proposed cost function used in DEA gives the shortest trajectory with minimum time in the phase plane. The results obtained from the proposed tuning method here were also compared with the results of the method used in [1]. A time-domain cost function is deployed in order to obtain good compromise between the input step response and disturbance rejection design. The PID controllers optimized with DE algorithm and the proposed cost function gives a performance that is at least as good as that of the PID tuning method from [1]. With PID tuning method using DEA, a faster settling time, less or no overshoot and higher robustness were obtained. Furthermore, the tuning method used here is successful in the presence of high noise.

1 Introduction

The PID controllers are widely used for industrial control processes, since they have a simple structure and their performances are quite robust for a wide range of operating conditions. The three well known types of time delayed unstable processes used especially in chemical systems are: The first order delayed unstable process (FODUP), the second order delayed unstable process (SODUP) and third order delayed unstable process (TODUP).

A closed-loop control system contains a set-point filter, a PID controller and a plant. The PID controller is composed of three components: a proportional part (k_p) a derivative part (k_d) and an integral part (k_i).

$$C(s) = k_p + \frac{k_i}{s} + k_d s \qquad (1)$$

In the design of a PID controller, these three constants must be selected in such a way that the closed loop system has to give the desired response. The desired response should have minimal settling time with a small or no overshoot in the step response of the closed loop system. For unstable systems, it is common to use a set-point filter $(F(s))$. The first order set-point filter (FOSF) is used in processes having one unstable

N.R. Pal et al. (Eds.): ICONIP 2004, LNCS 3316, pp. 254–260, 2004.
© Springer-Verlag Berlin Heidelberg 2004

pole ($as+1$) and the second order set-point filter (SOSF) is used in processes having two unstable poles (as^2+bs+1).

In this work, differential evolution (DE) was applied to optimize the constants of the PID controller and also the constants of the set-point filter. To show the effectiveness of the proposed method used here, the step responses of closed loop system were compared with that of the paper [1]. Next, the method was tested for the robustness to model errors. In the robustness test, the model is slightly changed by increasing the time delay and adding noise to the model.

2 Application of Differential Evolution to PID Tuning

The differential evolution algorithm (DEA), developed by Storn and Price a few years ago, is one of the most promising new Evolutionary Algorithm's (EA) [2], [3]. DE is a simple, population-based, direct search algorithm for globally optimizing functions defined on totally ordered spaces, including functions with real-valued parameters.

In original DEA, a big population size, NP is needed to maintain the diversity of the population so that the algorithm can converge to global optimum without being stuck in a local minima. In this work, a new random mutation operator is proposed to prevent immature convergence. There are two advantages to use this operator: maintaining the diversity in the population and small population size. This new operator mutates the child vector, after it is created by the original DEA. The probability of a child vector to be mutated is given by the mutation rate (MR). The change of the child vector in the mutation is determined by the mutation power (MP).

Set-point filter and PID controller are applied to the control of unstable systems with time delay. To achieve this, three constants of the PID controller and one or two constants which depend on the number of unstable poles of the system for the set-point filter must be optimized to have desirable input step response. To accomplish this optimization problem, the following cost function was used in unstable processes:

$$J(\rho) = \int_0^{T_S} \sqrt{(\dot{e}^2 + \ddot{e}^2)} \; t^2 dt \tag{2}$$

where $\dot{e}$ is the first derivative of the error, $\ddot{e}$ is the second derivative of the error, T_S is the simulation time and ρ is a vector containing the PID and the set-point filter constants. If the error is initially constant, the cost function is stuck in a trivial equilibrium point ($\dot{e} = 0$, $\ddot{e} = 0$). Using the cost function given above in an unstable process causes no problem since instability of the system diverges the error from this trivial equilibrium point. By adding the square of the error having a small weighting, this cost function can be used not only in unstable systems but also in other systems. Final cost function now takes the form:

$$J(\rho) = \int_0^{T_S} \sqrt{(0.1\,e^2 + \dot{e}^2 + \ddot{e}^2)} \; t^2 dt \tag{3}$$

The DE optimization process using this cost function tries to achieve the minimum path between the initial condition point and the equilibrium point with minimum time. In the optimization process, the parameters of the DE algorithm, were chosen as NP=10, CR =0.8, F=0.8, MR=0.1, and MP=0.1 for 100 generations. These parameters were determined experimentally to obtain the desired result in a reasonable time.

3 Results and Discussion

In this section, the PID controller tuned with differential evolution algorithm (PIDDE) with cost function given in equation (3) is tested with the following systems.

$$G_1(s) = \frac{e^{-0.4s}}{s-1} \tag{4}$$

$$G_2(s) = \frac{e^{-0.5s}}{(5s-1)(2s+1)(0.5s+1)} \tag{5}$$

$$G_3(s) = \frac{2e^{-0.3s}}{(3s-1)(s-1)} \tag{6}$$

$$G_4(s) = \frac{e^{-0.2s}}{s(s-1)} \tag{7}$$

G_1 and G_2 are taken from Huang [4] and G3 and G4 are taken from Lee [1]. Figures 1-3 show the step response of the systems and the corresponding control signals. Unit step payload change was applied to all systems after they reached steady state. Table 1 gives the PID controller constants and the set-point filter constants for the PIDDE tuning method and those of [1].

Table 1. PID and setpoint filter constants for the systems

		K_p	K_i	K_d	a	b
$G_1(s)$	PIDDE	2.624	1.235	0.346	1.985	–
	Lee	2.634	1.045	0.406	2.36	–
$G_2(s)$	PIDDE	4.494	0.552	7.397	5.999	–
	Lee	7.144	1.069	11.82	4.276	–
$G_3(s)$	PIDDE	1.429	0.739	4.091	5.551	1.959
	Lee	2.315	1.298	4.366	3.252	1.715
$G_4(s)$	PIDDE	3.865	2.514	3.484	1.405	1.567
	Lee	0.841	0.254	2.365	8.453	3.361

In order to compare the PIDDE method with [1], in terms of following performance criteria, Integral Absolute Error (IAE), 2% settling time (T_S) and the overshoot percentage (OS) are given in Table 2. As can be seen in Table 2, the PIDDE method produces the best settling time with small or no overshoot.

Table 2. Performance criteria for the controllers of the systems

		IAE	T_s	OS(%)
$G_1(s)$	PIDDE	1.1868	2.18	0
	Lee	1.4131	3.34	0
$G_2(s)$	PIDDE	4.2525	7.9	0
	Lee	3.5579	11.26	3.7324
$G_3(s)$	PIDDE	2.7572	5.48	1.7825
	Lee	2.3054	10.6	4.4247
$G_4(s)$	PIDDE	1.5841	3.42	0
	Lee	3.5278	8.88	2.3532

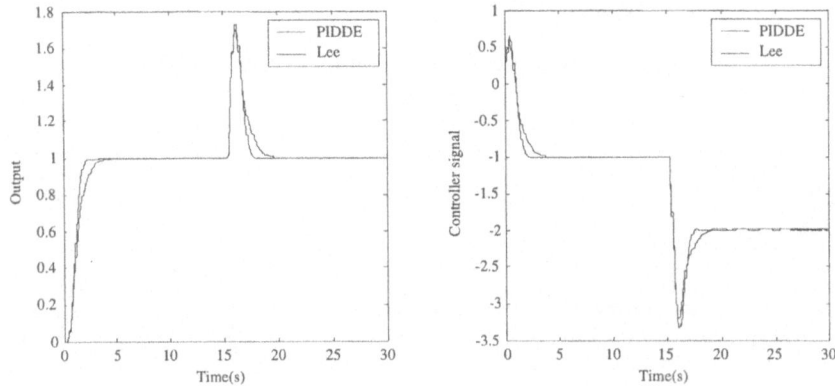

Fig. 1. Step responses for the closed-loop system with $G_1(s)$

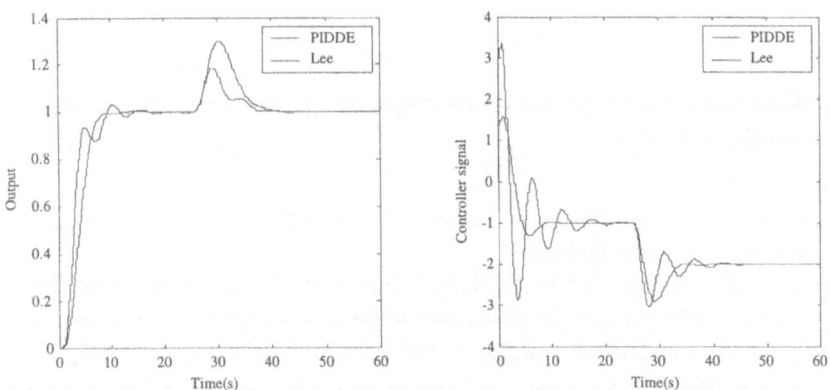

Fig. 2. Step responses for the closed-loop system with $G_2(s)$

Figure 4-a,b illustrates the step response to G_4, the corresponding control signals and the phase plane. As can be seen in Fig 4a, PIDDE method outperforms [1] in terms of rise and settling time. To see the behaviour of G_4 to step response and unit step payload change more clearly, phase plane analysis is made. The controller tuned with PIDDE method converges quickly from initial point to equilibrium point.

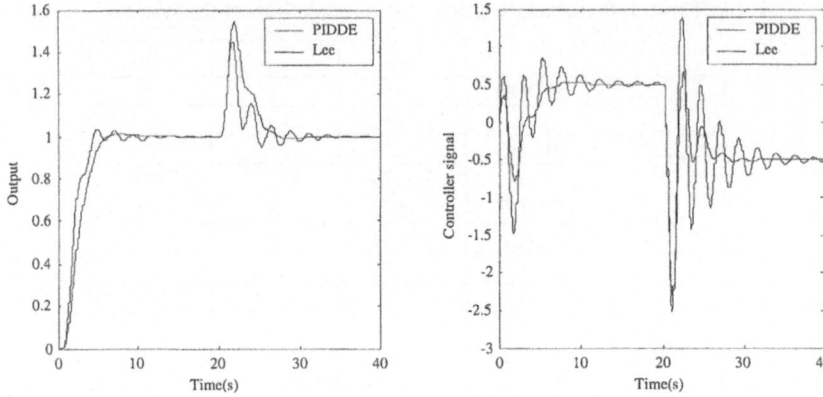

Fig. 3. Step responses for the closed-loop system with $G_3(s)$

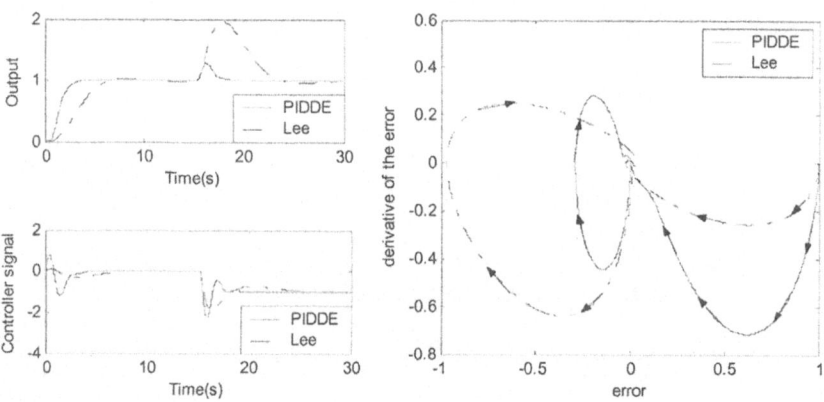

Fig. 4. a) Step response and the control output for the closed-loop system, $G_4(s)$, b) the corresponding phase plot

Moreover, it does not diverge as much as [1] from the equlibrium point as the unit step payload change is applied.

A controller tuning method should be robust to model errors and noise. To test the robustness of the methods, firstly the time delay of the system $G_3(s)$ is increased 15% and secondly, gaussian white noise of 0.05 variance is added to system $G_4(s)$ as input disturbance. Figure 5 shows the input step response of system $G_3(s)$ with added noise. Figure 6 illustrates the step response of system $G_4(s)$ changed time delay, respectively. As can be seen in the figures, controllers tuned with DEA method was robust to the model errors. On the other hand, the controllers tuned by [1] showed oscillatory behaviour under the presence of noise and it was not able to stabilize the system when the time delay of the system is increased by 15%.

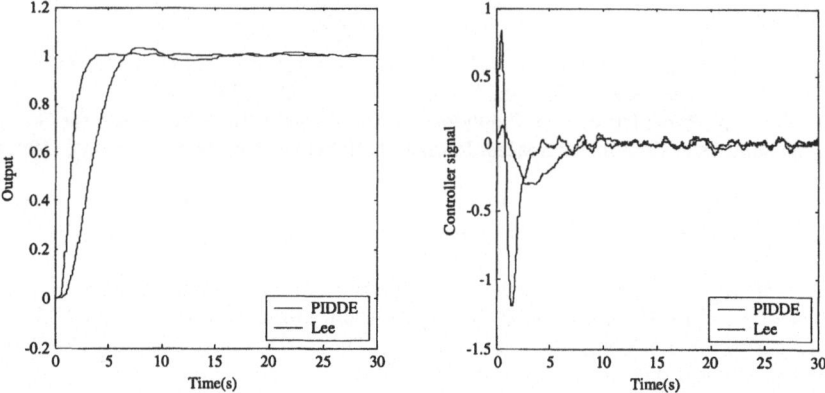

Fig. 5. Step response and the control output for the closed-loop system, $G_4(s)$ with noisy input disturbance signal

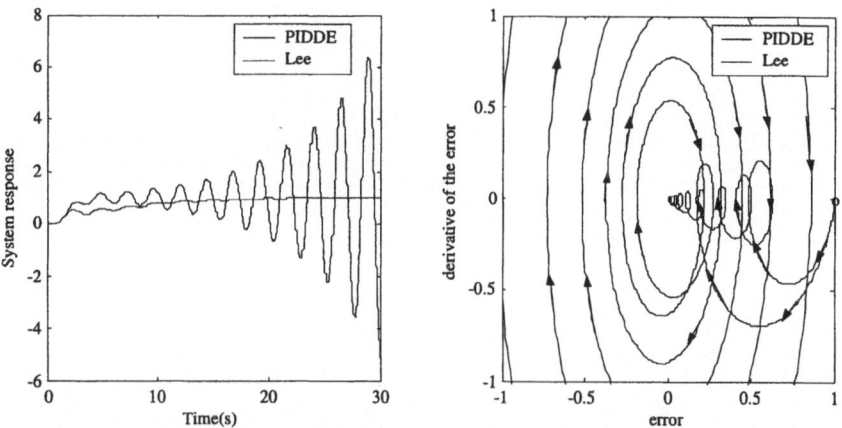

Fig. 6. Step response and the phase plot for the closed-loop system, $G_3(s)$

4 Conclusions

In this study, a new PID tuning process for unstable and integrating processes with time delay based on differential evolution algorithm was developed and compared to other tuning technique [1]. In this paper, two contributions were made: A new mutation operator is proposed in order to maintain the diversity of the population and a time-domain cost function based on the minimum length of the trajectory with minimum time in the phase plane is proposed. DEA using the cost function was employed to tune the constants of the PID controllers and set-point filter for four different unstable plants with time delay. In comparison to [1], the PIDDE method produces the smaller settling time with small or no overshoot. To test the robustness of the DEA tuning technique, noise and time delay changes were applied to the models used here. Even in the presence of high noise and changed time delay, systems tuned with DEA showed good control behavior.

References

1. Y. Lee, J. Lee, S. Park,: PID Controller Tuning for Integrating and Unstable Processes with Time Delay, Chemical Engineering Science, vol. 55, (2000), 3481-3493.
2. R. Storn, K. Price: Differential Evolution - A Simple and Efficient Heuristic for Global Optimization Over Continuous Spaces, Journal of Global Optimization, vol. 11(4), (1997) 341-359.
3. K. Price, R. Storn: Minimizing the Real Functions of the ICEC'96 contest by Differential Evolution, IEEE International Conference on Evolutionary Computation (ICEC'96), (1996) 842–844.
4. H.P. Huang, C.C. Chen,: Control-System Synthesis for Open-Loop Unstable Process with Time Delay, IEE Process-Control Theory and Applications, vol. 144 (4), (1997) 334-336.
5. K.J. Aström and T. Hägglund: PID Controllers: Theory, Design, and Tuning, The Instrmentation, Systems, and Automation Society, (1995).
6. H.A. Varol, Z. Bingul: A New PID Tuning Technique Using Ant Algorithm, American Control Conference, Boston, USA, (2004) 2154-2159.
7. A.B. Rad and W.L. Lo: Predictive PI controller, Int. J. Control, vol. 60(5), (1994) 953-975.

Representation and Identification of Finite State Automata by Recurrent Neural Networks

Yasuaki Kuroe

Center for Information Science, Kyoto Institute of Technology
Matsugasaki, Sakyo-ku, Kyoto 606-8585, Japan
kuroe@dj.kit.ac.jp

Abstract. This paper presents a new architecture of neural networks for representing deterministic finite state automata. The proposed model is capable of strictly representing FSA with the network size being smaller than the existing models proposed so far. We also discuss an identification method of FSA from a given set of input and output data by training the proposed neural networks. We apply the genetic algorithm to determine the parameters of the neural network such that its input and output relation becomes equal to those of a target FSA.

1 Introduction

The problem of representing and learning finite state automata (FSA) with artificial neural networks has attracted a great deal of interest recently. Several models of artificial neural networks for representing and learning FSA have been proposed and their computational capabilities have been investigated [1–5].

In recent years, there have been increasing research interests of hybrid control systems, in which controlled objects are continuous dynamical systems and controllers are implemented as discrete event systems. One of the most familiar model representations of discrete event systems is a model representation by FSA. It has been strongly desired to develop an identification method of unknown FSA with reasonable efficiency and accuracy. It is, therefore, becoming an important problem to investigate what architectures of neural networks are suitable for representing FSA in applications of control problems.

In this paper, we propose a new architecture of neural networks for representing deterministic FSA. The proposed model is a class of recurrent hybrid neural networks, in which two types of neurons, static neurons and dynamic neurons are arbitrarily connected. The proposed model of neural networks is capable of strictly representing FSA with the network size being smaller than the existing models proposed so far. We also discuss an identification method of FSA from a given set of input and output data by training the proposed neural networks. The proposed neural networks make the identification easier than the existing models because of less number of learning parameters.

2 Finite State Automata

In this paper we consider finite state automata (FSA) M defined by

$$M = (Q, q_0, \Sigma, \Delta, \delta, \varphi) \tag{1}$$

N.R. Pal et al. (Eds.): ICONIP 2004, LNCS 3316, pp. 261–268, 2004.
© Springer-Verlag Berlin Heidelberg 2004

where Q is the set of state symbols: $Q = \{q_1, q_2, \cdots, q_r\}$, r is the number of state symbols, $q_0 \in Q$ is the initial state, Σ is the set of input symbols: $\Sigma = \{i_1, i_2, \cdots, i_m\}$, m is the number of input symbols, Δ is the set of output symbols: $\Delta = \{o_1, o_2, \cdots, o_l\}$, l is the number of output symbols, $\delta: Q \times \Sigma \to Q$ is the sate transition function and φ: $Q \times \Sigma \to \Delta$ is the output function.

We suppose that the FSA M operates at unit time intervals. Letting $i(t) \in \Sigma$, $o(t) \in \Delta$ and $q(t) \in Q$ be the input symbol, output symbol and state symbol at time t, respectively, then the FSA M is described by a discrete dynamical system of the form:

$$M : \begin{cases} q(t+1) = \delta(q(t), i(t)), \ q(0) = q_0 \\ o(t) = \varphi(q(t), i(t)) \end{cases} \tag{2}$$

The objective of this paper is to discuss the problem of identification of the FSA described by (1) or (2) by using neural networks.

3   Hybrid Recurrent Neural Networks

We introduce hybrid recurrent neural networks [6] to represent FSA. The neural networks consist of two types of neurons, static neurons and dynamic neurons, which are arbitrarily connected as shown in Fig.1. Let N_d, N_s, M, L and K be the numbers of dynamic neurons, static neurons, external inputs and external outputs in the network, respectively. The mathematical model of the dynamic neurons is given by

$$\begin{cases} u_i^d(t+1) = \sum_{j=1}^{N_d} w_{ij}^{dd} y_j^d(t) + \sum_{j=1}^{N_s} w_{ij}^{ds} y_j^s(t) + \sum_{\ell=1}^{L} w_{i\ell}^{dI} I_\ell(t) + \theta_i^d, \quad u_i^d(0) = u_{i0}^d \\ y_i^d(t) = f_i^d(u_i^d(t)), \quad (i = 1, 2, \cdots, N_d) \end{cases} \tag{3}$$

and that of the static neurons is given by

$$\begin{cases} u_i^s = \sum_{j=1}^{N_s} w_{ij}^{ss} y_j^s + \sum_{j=1}^{N_d} w_{ij}^{sd} y_j^d + \sum_{\ell=1}^{L} w_{i\ell}^{sI} I_\ell + \theta_i^s \\ y_i^s = f_i^s(u_i^s), \quad (i = 1, 2, \cdots, N_s) \end{cases} \tag{4}$$

where u_i^d, y_j^d and θ_i^d are the state, the output and the threshold value of the i-th dynamic neuron, respectively, and u_i^s, y_j^s and θ_i^s are the state, the output and the threshold value of the i-th static neuron, respectively. I_ℓ is the ℓ-th external input, w_{ij}^{dd} is the weight from the j-th dynamic neuron to the i-th dynamic neuron, w_{ij}^{ds} is the weight from the j-th static neuron to the i-th dynamic neuron, $w_{i\ell}^{dI}$ is the weight from the ℓ-th external input to the i-th dynamic neuron, w_{ij}^{ss} is the weight from the j-th static neuron to the i-th static neuron, w_{ij}^{sd} is the weight from the j-th dynamic neuron to the i-th static neuron, $w_{i\ell}^{sI}$ is the weight from the ℓ-th external input to the i-th static neuron. $f_i^d(\cdot)$ and $f_i^s(\cdot)$ are nonlinear output functions of the dynamic and static neurons such as sigmoidal functions. The external outputs O_k are expressed by

$$O_k = \sum_{k=1}^{N_d} \delta_{kj}^{Od} y_j^d + \sum_{k=1}^{N_s} \delta_{kj}^{Os} y_j^s, \quad (k = 1, 2, \cdots, K) \tag{5}$$

where δ_{kj}^{Od} and δ_{kj}^{Os} take values 1 or 0. If the output of the j-th dynamic (static) neuron is connected to the k-th external output, $\delta_{kj}^{Od} = 1$ ($\delta_{kj}^{Os} = 1$), otherwise $\delta_{kj}^{Od} = 0$ ($\delta_{kj}^{Os} = 0$).

4 Neural Network Architectures for Representing FSA

4.1 Recurrent Second-Order Neural Networks

For representing FSA with neural networks, it is important to investigate suitable architectures of neural networks. There have been done several works on the representation of FSA with neural networks or investigation of relationships between neural network architectures and FSA. A typical representative of neural network architectures for representing FSA is a class of second-order neural networks.

Let each state symbol q_i be expressed by r dimensional unit basis vector, that is $q_1 = (1, 0, \cdots, 0), q_2 = (0, 1, \cdots, 0), \cdots, q_r = (0, 0, \cdots, 1)$. Similarly let each input symbol i_i and each output symbol o_i be expressed by m and l dimensional unit basis vectors, respectively. With the use of these expressions the functions δ and φ in (2) can be represented with the products of the state $q(t)$ and input $i(t)$. Then (2) can be rewritten as follows.

$$
\begin{cases}
q_i(t+1) = \sum_{l=1}^{m} i_l(t) \sum_{j=1}^{r} a_{ij}^l q_j(t) \\
o_k(t) = \sum_{l=1}^{L} i_l(t) \sum_{j=1}^{N} c_{kj}^l q_j(t)
\end{cases}
\tag{6}
$$

where a_{ij} and c_{kj} are parameters. From this equation, we can construct a neural network which is capable of representing FSA as shown in Fig. 2. Note that the network has the second-order (product) units, which comes from the products of the inputs and states in (6). The literatures [3] and [5] proposed similar architectures of neural networks and show their capability of representing FSA.

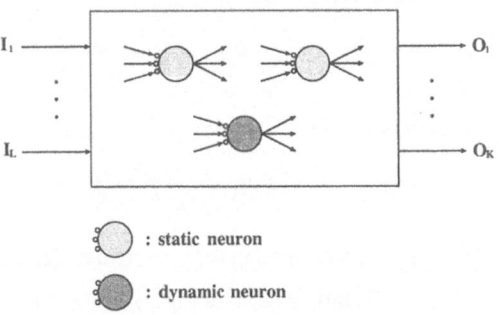

Fig. 1. Hybrid recurrent neural networks.

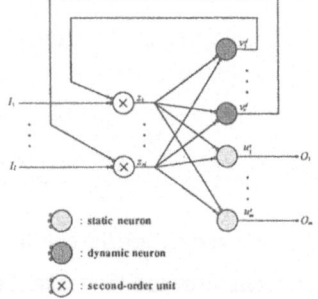

Fig. 2. A recurrent second-order neural network for representing FSA.

4.2 Proposed Neural Network Architectures

In the recurrent second-order neural networks obtained in §4.1 or in [3] and [5], each state of FSA is represented by assigning one neuron individually. Then, as the number of states of a target FSA increases, the number of neurons required for representing the FSA increases, which makes it difficult to identify the FSA because of a large number of network parameters.

We encode all the state symbols q_i ($i = 1, 2, \cdots, r$), input symbols i_i ($i = 1, 2, \cdots, m$) and output symbols o_i ($i = 1, 2, \cdots, \ell$) of FSA as binary variables. Then $q(t)$, $i(t)$ and $o(t)$ in (2) can be expressed as follows.

$$
\begin{aligned}
q(t) &= (s_1(t), s_2(t), \cdots, s_\alpha(t)) &\quad (s_i(t) \in \{0,1\}) \\
i(t) &= (x_1(t), x_2(t), \cdots, x_\beta(t)) &\quad (x_i(t) \in \{0,1\}) \\
o(t) &= (y_1(t), y_2(t), \cdots, y_\gamma(t)) &\quad (y_i(t) \in \{0,1\})
\end{aligned}
$$

where α, β and γ are natural numbers, which depend on r, m and l, respectively, that is, α, β and γ are the minimum natural number satisfying $r \leq 2^\alpha$, $m \leq 2^\beta$ and $l \leq 2^\gamma$, respectively. By using the representation, we can transform (2) into the following equations.

$$
M : \begin{cases}
s_i(t+1) = \delta_i(s_1(t), \cdots, s_\alpha(t), x_1(t), \cdots, x_\beta(t)) & (i = 1, 2, \cdots, \alpha) \\
y_i(t) = \varphi_i(s_1(t), \cdots, s_\alpha(t), x_1(t), \cdots, x_\beta(t)) & (i = 1, 2, \cdots, \gamma)
\end{cases} \quad (7)
$$

where $\delta_i : \{0,1\}^{\alpha+\beta} \to \{0,1\}$ and $\varphi_i : \{0,1\}^{\alpha+\beta} \to \{0,1\}$ are Boolean functions.

It is well known that any Boolean function can be expanded into one of the canonical forms. We represent the Boolean functions δ_i and φ_i in the principal disjunctive canonical form. For simplicity, we introduce new variables z_i, $i = 1, 2, \cdots, n$ ($n = \alpha + \beta$) defined by $z_1 = s_1, z_2 = s_2, \cdots, z_\alpha = s_\alpha, z_{\alpha+1} = x_1, z_{\alpha+2} = x_2, \cdots, z_n = x_\beta$. Define $Z_1(t), Z_2(t), \cdots, Z_{2^n}(t)$ by

$$
\begin{aligned}
Z_1(t) &= z_1(t) \wedge \cdots \wedge z_{n-1}(t) \wedge z_n(t) \\
Z_2(t) &= z_1(t) \wedge \cdots \wedge z_{n-1}(t) \wedge \bar{z}_n(t)
\end{aligned}
$$

$$
\vdots
$$

$$
Z_{2^n}(t) = \bar{z}_1(t) \wedge \cdots \wedge \bar{z}_{n-1}(t) \wedge \bar{z}_n(t),
$$

which are called the fundamental products of $z_1, z_2, \cdots, z_n$. We can rewrite (7) as

$$
M : \begin{cases}
s_i(t+1) = \bigvee_{j=1}^{2^n} a_{ij} Z_j(t) & (i = 1, 2, \cdots, \alpha) \\
y_i(t) = \bigvee_{j=1}^{2^n} b_{ij} Z_j(t) & (i = 1, 2, \cdots, \gamma)
\end{cases} \quad (8)
$$

where a_{ij} and b_{ij} are the coefficients of δ_i and φ_i represented in the principal disjunctive canonical form and they take the values '1' or '0', and $\bigvee_{i=1}^{n} x_i = x_1 \vee x_2 \vee \cdots \vee x_n$.

We now discuss expressions of the logical operations in (8). Let 'true $= 1$' and 'false $= 0$' and define the function $S(\cdot)$ by $S(x) = 1$ for $x \geq 0$ and $S(x) = -1$ for $x < 0$. Then the logical product $x_1 \wedge x_2 \wedge \cdots \wedge x_k$ is given by $y = S(x_1 + x_2 + \cdots + x_k - (k-$

1)), the logical sum $x_1 \vee x_2 \vee \cdots \vee x_k$ is given by $y = S(x_1 + x_2 + \cdots + x_k + (k-1))$ and the not $\bar{x}$ is given by $y = -x$. By using these expressions, (8) can be transformed into the following equation without logical operations.

$$M : \begin{cases} s_i(t+1) = S(\sum_{j=1}^{2^n} a_{ij} Z_j(t) + n_i^s - 1) & (i = 1, 2, \cdots, \alpha) \\ y_i(t) = S(\sum_{j=1}^{2^n} b_{ij} Z_j(t) + n_i^y - 1) & (i = 1, 2, \cdots, \gamma) \end{cases} \tag{9}$$

where

$$Z_1(t) = S(z_1(t) + \cdots + z_{n-1}(t) + z_n(t) - (n-1))$$
$$Z_2(t) = S(z_1(t) + \cdots - z_{n-1}(t) + z_n(t) - (n-1))$$
$$\vdots \tag{10}$$
$$Z_{2^n}(t) = S(-z_1(t) - \cdots - z_{n-1}(t) - z_n(t) - (n-1))$$

and n_i^s and n_i^y are the number of the elements of $\{a_{ij} : a_{ij} = 1, i = 1, 2, \cdots, \alpha, j = 1, 2, \cdots, 2^n\}$ and $\{b_{ij} : b_{ij} = 1, i = 1, 2, \cdots, \gamma, j = 1, 2, \cdots, 2^n\}$, respectively.

We now propose a new architecture of neural networks for representing FSA M. The neural network is constructed based on (9) and (10). The relation of the first equation of (9) can be realized by dynamic neurons, the relation of the second equation of (9) and the relation of (10) can be realized by static neurons as follows.

Consider a hybrid recurrent neural network described by (3),(4) and (5) where we let $N_d = \alpha$, $N_s = 2^{\alpha+\beta} + \gamma$, $L = \beta$, $K = \gamma$ and $f_i^d(\cdot) = S(\cdot)$ and $f_i^s(\cdot) = S(\cdot)$. In the network the state vector of the dynamic neurons $\boldsymbol{u}_i^d = (u_1^d, u_2^d \cdots, u_\alpha^d)$, the external input vector $\boldsymbol{I} = (I_1, I_2, \cdots, I_\beta)$ and the external output vector $\boldsymbol{O} = (O_1, O_2, \cdots, O_\gamma)$ correspond to the state q, the input i and the output o of FSA, respectively, where they are encoded as binary variables. To realize the first equation of (9), α dynamic neurons described by (3) are assigned, in which we let $w_{ij}^{dd} = 0$ for $i, j = 1, 2 \cdots, \alpha$, $w_{ij}^{ds} = 0$ for $i = 1, 2 \cdots, \alpha$, $j = 2^{\alpha+\beta} + 1, 2^{\alpha+\beta} + 2, \cdots, 2^{\alpha+\beta} + \gamma$, and $w_{i\ell}^{dI} = 0$ for $i = 1, 2 \cdots, \alpha$, $\ell = 1, 2, \cdots, \beta$. Note that the values of θ_i^d $(i = 1, 2, \cdots, \alpha)$ can be determined uniquely from the values of w_{ij}^{ds} $(i = 1, 2 \cdots, \alpha, \ j = 1, 2, \cdots, 2^{\alpha+\beta})$ by the definition of n_i^s in (9). To realize (10), $2^{\alpha+\beta}$ static neurons described by (4) are assigned, in which we let $w_{ij}^{ss} = 0$ for $i, j = 1, 2 \cdots, 2^{\alpha+\beta}$. Note that the values of w_{ij}^{sd} $(i = 1, 2 \cdots, 2^{\alpha+\beta}, \ j = 1, 2 \cdots, \alpha)$, $w_{i\ell}^{sI}$ $(i = 1, 2 \cdots, 2^{\alpha+\beta}, \ \ell = 1, 2 \cdots, \beta)$ and θ_i^s $(i = 1, 2 \cdots, 2^{\alpha+\beta})$ can be determined from (10); w_{ij}^{sd} and $w_{i\ell}^{sI}$ take the values of '1' or '0' and $\theta_i^s = -(n-1)$ $(i = 1, 2 \cdots, 2^{\alpha+\beta})$. To realize the second equation of (9), additional γ static neurons described by (4) are assigned, in which we let $w_{ij}^{ss} = 0$ for $i = 2^{\alpha+\beta} + 1, 2^{\alpha+\beta} + 2, \cdots, 2^{\alpha+\beta} + \gamma$, $j = 2^{\alpha+\beta} + 1, 2^{\alpha+\beta} + 2, \cdots, 2^{\alpha+\beta} + \gamma$, $w_{ij}^{sd} = 0$ for $i = 2^{\alpha+\beta} + 1, 2^{\alpha+\beta} + 2, \cdots, 2^{\alpha+\beta} + \gamma$, $j = 1, 2, \cdots, \alpha$ and $w_{i\ell}^{sI} = 0$ for $i = 2^{\alpha+\beta} + 1, 2^{\alpha+\beta} + 2, \cdots, 2^{\alpha+\beta} + \gamma$, $j = 1, 2, \cdots, \beta$. Note that the values of θ_i^s $(i = 2^{\alpha+\beta} + 1, 2^{\alpha+\beta} + 2, \cdots, 2^{\alpha+\beta} + \gamma)$ can be determined uniquely from the values of w_{ij}^{ss} $(i = 2^{\alpha+\beta} + 1, 2^{\alpha+\beta} + 2, \cdots, 2^{\alpha+\beta} + \gamma, \ j = 1, 2, \cdots, 2^{\alpha+\beta})$ by the

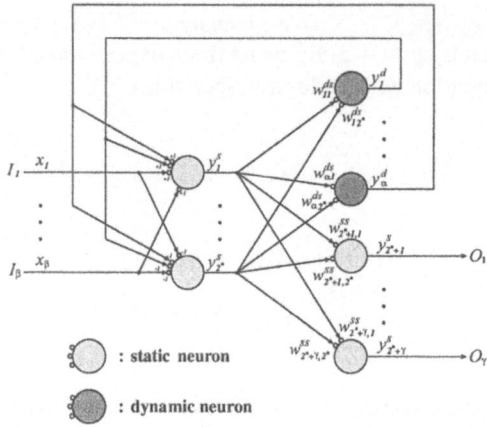

Fig. 3. Hybrid recurrent neural network for representing FSA.

definition of n_i^y in (9). Furthermore the external outputs described by (5) are assigned, in which we let $\delta_{kj}^{Od} = 0$ for $k = 1, 2 \cdots, \gamma$, $j = 1, 2, \cdots, \alpha$ and $\delta_{kj}^{Os} = 0$ for $k = 1, 2 \cdots, \gamma$, $j = 1, 2, \cdots, 2^{\alpha+\beta}$. Figure 3 shows the neural network thus constructed, which consists of α dynamic neurons, and $2^{\alpha+\beta} + \gamma$ static neurons. It can be shown by using (9) and (10) that the network is capable of strictly representing any FSA.

5 Identification of FSA

5.1 Identification Method

In this section we discuss the identification problem of unknown FSA by using the proposed neural network. We formulate the identification problem of FSA as follows. Given a set of data of input and output sequences of a target FSA, determine values of the parameters of the neural network such that its input and output relation becomes equal to that of the FSA. It is proper to assume the following.

A1 *The set of state symbols and the initial state of FSA are unknown.*
A2 *The state transition function δ and output function φ of FSA are unknown.*
A3 *The sets of input symbols Σ and output symbols Δ of FSA are known.*
A4 *A set of data of input sequences $\{i(t)\}$ and the corresponding output sequences $\{o(t)\}$ are available.*

For the identification we construct the hybrid recurrent neural network in the manner discussed in the previous section. Note that, the number of the external inputs and outputs can be determined as $L = \beta$ and $K = \gamma$, on the other hand, the number of the dynamic neurons ($N_d = \alpha$) and the static neurons ($N_s = 2^{\alpha+\beta} + \gamma$) cannot be determined since the number of the states of the FSA is not known (A1). Note also that the initial values of the dynamic neurons $u_i^d(0)$ can not be given a priori because of the assumption A1. Hence the parameters to be determined in the identification problem are the number of the dynamic neurons α, values of the weights w_{ij}^{ds} ($i = 1, 2 \cdots, \alpha$, $j =$

$1, 2, \cdots, 2^{\alpha+\beta})$ and w_{ij}^{ss} $(i = 2^{\alpha+\beta}+1, 2^{\alpha+\beta}+2, \cdots, 2^{\alpha+\beta}+\gamma, j = 1, 2, \cdots, 2^{\alpha+\beta})$ and values of the initial conditions of the dynamic neurons u_{i0}^{d} $(i = 1, 2 \cdots, \alpha)$.

The identification problem can be formulated as a combinatorial optimization problem because all the connection weights of the neural networks take only the values '0' or '1'. We can apply the genetic algorithm to determine the parameters of the hybrid recurrent neural network thus constructed such that its input and output relation becomes equal to that of the target FSA. We have developed a learning method of the neural network based on the genetic algorithm. The details are omitted.

Here we make the comparison between the proposed neural network shown in Fig. 3 and the recurrent second-order neural network, shown in Fig. 2 or proposed in [3,5], from the view of the numbers of parameters to be determined. In the proposed neural network, the total number of parameters is $2^{(\alpha+\beta)} \times (\alpha + \gamma) + \alpha$ for a given α. In the recurrent second-order neural network, the total number is $rm(r + l) + r$ for a given r. Noting that α, β and γ are the minimum natural numbers satisfying $r \leq 2^{\alpha}$, $m \leq 2^{\beta}$ and $\ell \leq 2^{\gamma}$, we can find that the identification by using the proposed neural network requires to determine less number of parameters, compared with the second-order neural networks, which becomes more remarkable as the number of the states of FSA increases.

5.2 Experiments

First we consider a simple FSA whose state transition diagram is shown in Fig. 4. This FSA accepts the sequences consisting of only '1'. The number of state symbols of the FSA is one and $\Sigma = \Delta = \{0, 1\}$. We construct a hybrid recurrent neural network with $\beta = 1$ and $\gamma = 1$ because $\Sigma = \Delta = \{0, 1\}$. However α can not be determined because the number of state symbol is unknown. Supposing that the number of state symbol of the FSA is estimated at most 4 ($r \leq 4$), we can let $\alpha \leq 2$ in the proposed genetic algorithm because $r \leq 4$. It is known that FSA with r state symbols is uniquely identified by using all input sequences of length $2r - 1$. We apply the proposed genetic algorithm by using all sequences of length $2r - 1$ where $r = 4$ ($m^{2r-1} = 128$ patterns) as learning data of the neural network. Figure 5 shows an example of the obtained neural networks when the learning algorithm converged. It can be checked that the operation of the neural network shown in Fig. 5 is the same as that of the target FSA shown in Fig. 4.

Next we consider the FSA whose state transition diagram is shown in Fig. 6, which accepts the sequence $(10)^{*}$. The number of the state symbols of the FSA is three and

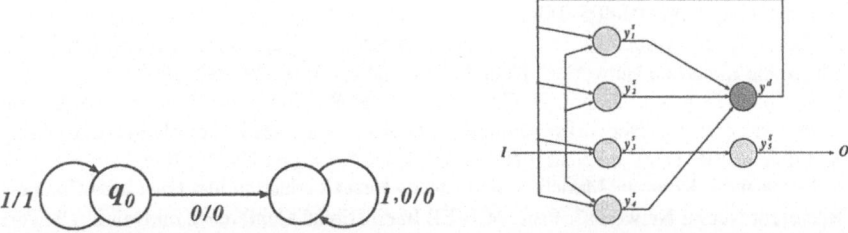

Fig. 4. FSA accepting the sequences consisting of only '1'.

Fig. 5. An example of obtained neural networks by the proposed method for Example 1.

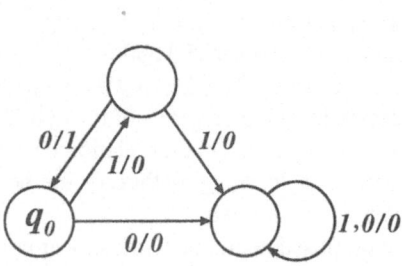

Fig. 6. FSA accepting the sequence $(10)^*$.   **Fig. 7.** An example of obtained neural networks by the proposed method for Example 2.

$\Sigma = \Delta = \{0, 1\}$. Supposing also that the number of state symbol of the FSA is estimated at most 4 ($r \leq 4$), we can let $\alpha \leq 2$, $\beta = 1$ and $\gamma = 1$. Similar to the first example, we apply the proposed genetic algorithm by using all sequences of length $2r - 1$ as learning data. Figure 7 shows an example of the obtained neural networks, from which, we can obtain its state transition diagram that turns out to be equivalent to the target FSA shown in Fig 6.

6   Conclusions

This paper presented a new architecture of neural networks, recurrent hybrid neural networks, for representing deterministic FSA. The proposed neural networks are capable of strictly representing FSA with the network size being smaller than the existing models. We also discussed an identification method of FSA by using the proposed neural networks. It should be noted that the proposed neural networks make the identification easier because of less number of learning parameters, which comes from the smaller network size. The author would like to express his gratitude to Prof. T. Mori and Mr. N. Ono for their valuable discussions.

References

1. M. L. Minsky:"Computation:Finite and Infinite Machines", Prentice-Hall, New York, 1967.
2. N. Alon, A. K. Dewdney and T. J. Ott: "Efficient Simulation of Automata by Neural Nets", Journal of Association for Computing Machinery, vol.38, No.2, April, pp.495-514, 1991.
3. C. L. Giles, C. B. Miller, D. Chen, H. H. Chen, G. Z. Sun and Y. C. Lee: "Learning and Extracting Finite State Automata with Second-Order Recurrent Neural Networks", Neural Computation, 4, pp.393-405, 1992.
4. Z. Zegn, R. M. Goodman and P. Smyth: "Learning Finite State Machines with Self-Clustering Recurrent Networks", Neural Computation, 5, pp.976-990, 1993.
5. C. L. Giles, D. Chen, G. Sun, H. Chen, Y. Lee and W. Goudreau: "Constructive Learning of Recurrent Neural Networks: Limitations of Recurrent Casade Correlation and a Simple Solution", IEEE Trans. on Neural Networks, Vol.6, No.4, pp.829-836, July, 1995.
6. Y. Kuroe and I. Kimura: "Modeling of Unsteady Heat Conduction Field by Using Composite Recurrent Neural Networks", Proc. of IEEE International Conference on Neural Networks, Vol.1, pp.323-328, 1995

Neural Network Closed-Loop Control
Using Sliding Mode Feedback-Error-Learning

Andon V. Topalov* and Okyay Kaynak

Department of Electrical and Electronic Engineering,
Mechatronics Research and Application Center, Bogazici University,
Bebek, 34342 Istanbul, Turkey
{topalov,kaynak}@boun.edu.tr

Abstract. A novel variable-structure-systems-based approach to neuro-adaptive feedback control of systems with uncertain dynamics is proposed. An inner sliding motion is established in terms of the controller parameters. The outer sliding motion is set up on the system under control, the state tracking error vector being driven towards the origin of the phase space. The equivalence between the two sliding motions is shown. The convergence of the on-line learning algorithm is demonstrated and the conditions are given. Results from a simulated trajectory tracking control task for a CRS CataLyst-5 industrial robot manipulator are presented. The proposed scheme can be considered as a further development of the well-known feedback-error-learning method.

1 Introduction

In high performance control applications, robustness is a fundamental requirement. It is well known that Variable structure control (VSC) can effectively be used to robustify the control system against disturbances and uncertainties [1]. A recent tendency in the literature is to exploit the strength of the technique for on-line learning in computationally intelligent systems [2].

Robot manipulators are frequently used as a test bed for evaluation of computationally intelligent control methods since they are hard to control nonlinear systems. A well known approach in robot control is the so called "feedback-error-learning" proposed by Kawato et al. [3]. It, in its original form, is based on the neural network (NN) implementation of the computed torque, plus a secondary proportional-derivative (PD) controller. The output of this conventional feedback controller (CFC) is used as a learning error signal to update the weights of the NN, trained as a feedforward controller. In more recent literature, the method has been extended and applied to learning schemes where NN is used as adaptive nonlinear feedback controller (NNFC) [4].

In the present work the approach in [4] is further extended by using a VSC-based on-line learning algorithm in the NNFC. It establishes an inner sliding motion in terms of the controller parameters, forcing the learning error toward zero. The outer sliding motion is related to the controlled nonlinear system with uncertain dynamics, the state tracking error vector of which is simultaneously forced towards the origin of the phase space.

In the second section of the paper, the proposed sliding mode feedback-error-learning approach is presented. Its performance is evaluated in the third section by the

---

* On leave from Control Systems Dept., TU Sofia, br. Plovdiv, 4000 Plovdiv, Bulgaria.

N.R. Pal et al. (Eds.): ICONIP 2004, LNCS 3316, pp. 269–274, 2004.
© Springer-Verlag Berlin Heidelberg 2004

simulation studies carried out for trajectory control of an experimental manipulator. Finally, section 4 summarizes the findings of this work.

2 The Sliding Mode Feedback-Error-Learning Approach

2.1 Initial Assumptions and Definitions

The proposed control scheme is depicted in fig. 1. A PD controller (the CFC block in fig. 1) is utilized both as an ordinary feedback controller to guarantee global asymptotic stability and as an inverse reference model for the system under control.

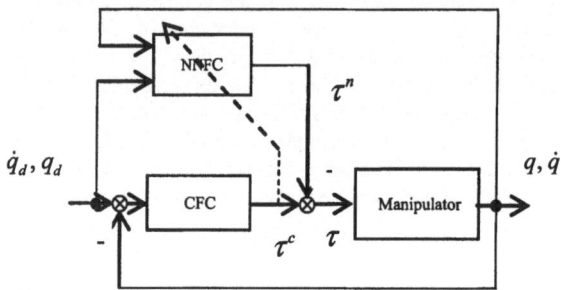

Fig. 1. Block diagram of nonlinear regulator sliding mode feedback-error-learning scheme

Consider a two-layered feedforward NN implemented as NNFC where $X(t) = \left[x_1(t), \ldots, x_p(t) \right]^T$ is the vector of the time-varying input signals augmented by the bias term, $T_H(t) = \left[\tau^n_{H1}(t), \ldots, \tau^n_{Hn}(t) \right]^T$ is the vector of the output signals from the neurons in the hidden layer and $\tau^n(t)$ is the scalar signal representing the network output. The weight matrix of the neurons in the hidden layer is denoted by $W1(t)_{(n \times p)}$, where each element $w1_{i,j}(t)$ means the weight of the connection of the neuron i from its input j. $W2(t)_{(1 \times n)} = \left[w2_1(t), \ldots, w2_n(t) \right]$ is the weight vector for the output node. Both $W1(t)_{(n \times p)}$ and $W2(t)_{(1 \times n)}$ are augmented by including bias weight components. Let $f(\cdot)$ be nonlinear, differentiable, monotonously increasing activation function of the neurons in the hidden layer. The neuron in the output layer is considered to have a linear activation function.

It will be assumed that the input vector of the NNFC and its time derivative are bounded, i.e. $\|X(t)\| \le B_X$ and $\|\dot{X}(t)\| \le B_{\dot{X}}$ $\forall t$ with B_X and $B_{\dot{X}}$ being known positive constants. Due to physical constraints, it is also assumed that the magnitude of the row vectors $W1_i(t)$ constituting the matrix $W1(t)$ and the elements of the vector $W2(t)$ are bounded, i.e. $\|W1_i(t)\| \le B_{W1}$ and $|w2_i(t)| \le B_{W2}$ $\forall t$ for some known constants B_{W1} and B_{W2}, where $i = 1, 2, \ldots, n$. It is assumed that $\tau(t)$ and $\dot{\tau}(t)$ are also bounded signals, i.e. $|\tau(t)| \le B_\tau$, $|\dot{\tau}(t)| \le B_{\dot{\tau}}$ $\forall t$ where B_τ and $B_{\dot{\tau}}$ are positive constants.

2.2 The VSC-Based On-Line Learning Algorithm

In the NNFC, VSC based learning algorithms are used to force the system error and the learning error to zero, the sliding surfaces being defined as $s_p(e,\dot{e})=\dot{e}+\lambda e$ and $s_c(\tau'',\tau)=\tau^c=\tau''+\tau$ respectively where λ is a constant that determines the slope of the sliding surface.

Definition 1. A sliding motion is said to exist on a sliding surface $s_c(\tau'',\tau)=\tau^c(t)=0$, after a hitting time t_h if the condition $s_c(t)\dot{s}_c(t)=\tau^c(t)\dot{\tau}^c(t)<0$ is satisfied for all t in some nontrivial semi open subinterval of time of the form $[t,t_h)\subset(-\infty,t_h)$.

Theorem 1. If the adaptation law for the weights $W1(t)$ and $W2(t)$ of NNFC is chosen respectively as

$$\dot{w1}_{i,j}=-\left(\frac{w2_i x_j}{X^T X}\right)\alpha\, sign(s_c), \quad \dot{w2}_i=-\left(\frac{\tau_{H_i}^n}{T_H^T T_H}\right)\alpha\, sign(s_c) \tag{1}$$

with α being sufficiently large positive constant satisfying $\alpha>nB_A B_{W1}B_{\dot{X}}B_{W2}+B_t$ then, given an arbitrary initial condition $s_c(0)$, the learning error $\tau^c(t)$ converges to zero in a finite time t_h estimated by

$$t_h\le\frac{|s_c(0)|}{\alpha-nB_A B_{W2}B_{W1}B_{\dot{X}}+B_t} \tag{2}$$

and a sliding motion is sustained on $\tau^c=0$ for all $t>t_h$.

Proof. Consider $V_c=\frac{1}{2}s_c^2$ as a Lyapunov function candidate. Then differentiating V_c yields:

$$\dot{V}_c=s_c\dot{s}_c=s_c\left(\dot{\tau}''+\dot{\tau}\right)=s_c\left\{\left[\sum_{i=1}^{n}\dot{w2}_i f\left(\sum_{j=1}^{p}\dot{w1}_{i,j}x_j\right)\right]'+\dot{\tau}\right\}$$

$$=s_c\left[\sum_{i=1}^{n}\dot{w2}_i\tau_{H_i}^n+\sum_{i=1}^{n}w2_i A_i\sum_{j=1}^{p}\left(\dot{w1}_{i,j}x_j+w1_{i,j}\dot{x}_j\right)+\dot{\tau}\right]$$

$$=s_c\left[-\sum_{i=1}^{n}\frac{\tau_{H_i}^n}{T_H^T T_H}\alpha\, sign(s_c)\tau_{H_i}^n+\sum_{i=1}^{n}A_i\sum_{j=1}^{p}\left(-\frac{w2_i x_j}{X^T X}\right)\alpha\, sign(s_c)x_j w2_i+w1_{i,j}\dot{x}_j w2_i\right)+\dot{\tau}\right] \tag{3}$$

$$=s_c\left(-\alpha sign(s_c)-\sum_{i=1}^{n}A_i\alpha w2_i^2 sign(s_c)+\sum_{i=1}^{n}A_i w2_i\sum_{j=1}^{p}w1_{i,j}\dot{x}_j+\dot{\tau}\right)$$

$$=-\left(\alpha+\alpha\sum_{i=1}^{n}A_i w2_i^2\right)|s_c|+\left(\sum_{i=1}^{n}A_i w2_i\sum_{j=1}^{p}w1_{i,j}\dot{x}_j+\dot{\tau}\right)s_c\le-\alpha|s_c|+s_c\left(\sum_{i=1}^{n}A_i w2_i\sum_{j=1}^{p}w1_{i,j}\dot{x}_j+\dot{\tau}\right)$$

$$\le-\alpha|s_c|+\left(nB_A B_{W2}B_{W1}B_{\dot{X}}+B_t\right)|s_c|=|s_c|\left(-\alpha+nB_A B_{W2}B_{W1}B_{\dot{X}}+B_t\right)<0\quad\forall s_c\ne0$$

where $A_i(t)$, $0 < A_i(t) = f'\left(\sum_{j=1}^{p} wl_{i,j}x_j\right) \le B_A$ $\forall i,j$ is the derivative of the activation

function $f(.)$, and B_A corresponds to its maximum value.

The inequality (3) means that the controlled trajectories of the learning error $s_c(t)$ converge to zero in a stable manner. The convergence will takes place in finite time which is estimated by eq. (2) (see the prove in [2]).

2.3 Relation Between the VSC-Based Learning of the Controller and the Sliding Motion in the Behavior of the System

The relation between the sliding line s_p and the zero adaptive learning error level s_c, when λ is taken as $\lambda = \dfrac{k_P}{k_D}$, is determined by the following equation:

$$s_c = \tau^c = k_D\dot{e} + k_Pe = k_D\left(\dot{e} + \frac{k_P}{k_D}e\right) = k_Ds_p \tag{4}$$

where k_D and k_P are the PD controller gains.

The tracking performance of the system under control can be analyzed by introducing $V_p = \dfrac{1}{2}s_p^2$ as a Lyapunov function candidate.

Theorem 2. If the adaptation strategy for the adjustable parameters of the NNFC is chosen as in equation (1) then the negative definiteness of the time derivative of the above Lyapunov function is ensured.

Proof. Evaluating the time derivative of the Lyapunov function V_p yields:

$$\dot{V}_p = \dot{s}_ps_p = \left(\frac{s_c}{k_D}\right)'\frac{s_c}{k_D} = \frac{1}{k_D^2}\dot{s}_cs_c \le \frac{1}{k_D^2}|s_c|\left(-\alpha + nB_AB_{W2}B_{W1}B_{\dot{X}} + B_t\right) < 0, \forall s_c, s_p \ne 0 \tag{5}$$

Remark 1. The obtained results mean that, assuming the VSC task is achievable, utilization of τ^c as the learning error for the NNFC together with the tuning law of (1) enforces the desired reaching mode followed by the sliding regime for the system under control. It is straightforward to prove that the hitting occurs in finite time (see [2]).

3 Trajectory Tracking Control of a Simulated CRS CataLyst-5 Industrial Robot Manipulator

In this section, the effectiveness of the proposed approach is evaluated by simulation studies carried out on an experimental manipulator (CRS CataLyst-5), the control task being the trajectory control of the two consecutive (the second and the third) joints. The manipulator dynamics has been accurately modeled using Matlab SimMechanics toolbox by taking into account the data related to the links frame assignment, distances between the joint axes, default orientations, mass and inertia tensors for each link with respect to the center of gravity, friction dynamics, gear mechanisms and motor transfer functions of each robot joint.

Two identical NNs (one per joint), each with 5 neurons in its hidden layer are used as NNFC. The sampling time and the learning parameter α are taken to be 1 ms and 0.5 respectively. The reference signals to be followed are sinusoidal ones of frequency $\pi/2$ rad/sec and with amplitudes equal to 30 and 45 degrees respectively.

The results are presented on fig. 2. It can be seen that the CFC torque signals are suppressed by the NNFC and the joint outputs closely follow the required trajectories demonstrating a good tracking performance of the control scheme.

4 Conclusion

A novel approach for generating and maintaining sliding motion in the behavior of a system with uncertainties in its dynamics is introduced. The system under control is under a closed-loop simultaneously with a conventional PD controller and an adaptive variable structure neural controller. The presented results from a simulated trajectory tracking control of an industrial manipulator have demonstrated that the predefined sliding regime could be generated and maintained if the NNFC parameters are tuned in such a way that the reaching is enforced. Another prominent feature that should be emphasized is the computational simplicity of the proposed approach.

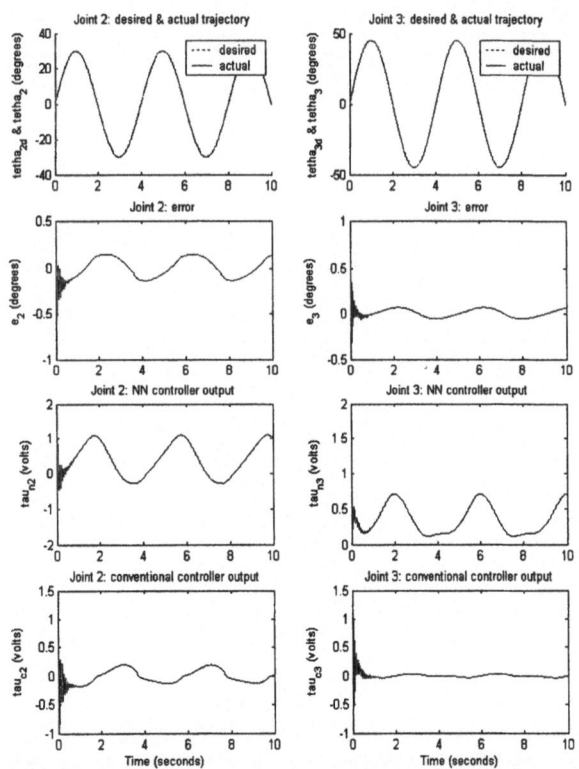

Fig. 2. Simulation results from the trajectory tracking control task

Acknowledgements

The authors would like to acknowledge Bogazici University Research Fund Project No: 03M109.

References

1. Utkin, V. I.: Sliding Modes in Control and Optimization. Springer-Verlag, Berlin Heidelberg New York (1992)
2. Shakev, N. G., Topalov, A. V., and Kaynak, O.: Sliding Mode Algorithm for On-Line Learning in Analog Multilayer Feedforward Neural Networks. In: Kaynak et al. (eds.): Artificial Neural Networks and Neural Information Processing. Lecture Notes in Computer Science, Springer-Verlag, Berlin Heidelberg New York (2003) 1064-1072
3. Kawato, M., Uno, Y., Isobe, M., and Suzuki, R. A.: Hierarchical Model for Voluntary Movement and with Application to Robotics. IEEE Contr. Syst. Mag., Vol. 8, No. 2 (1988) 8-16
4. Gomi, H., and Kawato, M.: Neural Network Control for a Closed-Loop System Using FeedBack-Error-Learning. Neural Networks, Vol. 6 (1993) 933-946

State Estimation and Tracking Problems: A Comparison Between Kalman Filter and Recurrent Neural Networks

S. Kumar Chenna[1,2], Yogesh Kr. Jain[1], Himanshu Kapoor[1], Raju S. Bapi[1], N. Yadaiah[3], Atul Negi[1], V. Seshagiri Rao[4], and B.L. Deekshatulu[1]

[1] Dept. of Computer and Information Sciences, University of Hyderabad, India
{bapics,atulcs,bldcs}@uohyd.ernet.in
[2] Honeywell Technology Solutions, Bangalore
[3] Dept. of Electrical Engineering, JNTU, Hyderabad, India
yadaiahn@hotmail.com
[4] SHAR Computer Facility, ISRO, Sriharikota, AP, India

Abstract. The aim of this paper is to demonstrate the suitability of recurrent neural networks (RNN) for state estimation and tracking problems that are traditionally solved using Kalman Filters (KF). This paper details a simulation study in which the performance of a basic discrete time KF is compared with that of an equivalent neural filter built using an RNN. Real time recurrent learning (RTRL) algorithm is used to train the RNN. The neural network is found to provide comparable performance to that of the KF in both the state estimation and tracking problems. The relative merits and demerits of KF vs RNN are discussed with respect to computational complexity, ease of training and real time issues.

Keywords: Recurrent Neural Network, KF, Real time recurrent learning, Tracking, State estimation

1 Introduction

Traditionally, state estimation and tracking problems are solved using KFs (for example, see [1], [2]). Recurrent neural networks (RNN) have received much research attention because of their powerful capability to represent attractor dynamics and to preserve information through time [3]. KF is a well known recursive, linear technique that works optimally when the system equations are linear and the noises (system and measurement) are uncorrelated and white Gaussian [1]. Extended KF is formulated to deal with simple nonlinearities in the system equations. However, in general when the system and/or noise deviate from Kalman assumptions, the convergence and optimality results of KF are not guaranteed. Since neural networks of appropriate size are known to be capable of approximating a wider class of nonlinear functions [4], it is expected that neural networks, especially RNN, offer a better alternative for KF even when Kalman assumptions are violated. In this paper, we attempt to use RNNs to solve the

N.R. Pal et al. (Eds.): ICONIP 2004, LNCS 3316, pp. 275–281, 2004.
© Springer-Verlag Berlin Heidelberg 2004

estimation and tracking problems. In recent years, a variety of approaches have been proposed for training the RNNs, such as the back propagation through time (BPTT) [5], real time recurrent learning algorithm (RTRL), extended Kalman filter (EKF), etc. In this paper, the RTRL algorithm is used for training the recurrent network. We present an example state estimation problem and a tracking problem to illustrate the application of RTRL trained RNN for these problems. Further, we also simulated KFs for solving the same problems to enable a comparison of the performance of RNN and KF. Relative performance is compared in terms of prediction capability and tracking error. An introduction to KF and RNN are presented first and then the description of the problems is given next. Finally the results are presented and a discussion of relative merits and demerits of these two methods is given.

2   Description of Kalman Filter

The Kalman filter is a technique for estimating the unknown state of a dynamical system with additive noise. The KF has long been regarded as the optimal solution to many tracking and state prediction tasks [1]. The strength of KF algorithm is that it computes on-line. This implies that we don't have to consider all the previous data again to compute the current estimates, we only need to consider the estimates from the previous time step and the current measurement. Popular applications include, state estimation [6], navigation, guidance, radar tracking [2], sonar ranging, satellite orbit computation, etc. These applications can be summarized into various classes such as denoising, tracking and control problems. The basic KF is optimal in the mean square error sense (given certain assumptions), and is the best possible of all filters, if state and measurement inputs are Gaussian vectors and the additive noise is white and has zero mean [1].

We now begin the description of the KF. The block diagram of basic discrete-time kalman filter is shown in Figure 1. We assume that the system can be modelled by the state transition equation,

$$\mathbf{X}_{k+1} = \mathbf{A}\mathbf{X}_k + \mathbf{B}\mathbf{u}_k + \mathbf{w}_k \qquad (1)$$

where $\mathbf{X}_k$ is the state at time k, $\mathbf{u}_k$ is an input control vector, $\mathbf{w}_k$ is additive noise from either the system or the process. $\mathbf{B}$ is the input transition matrix and $\mathbf{A}$ is the state transition matrix. The measurement system can be represented by a linear equation of the form,

$$\mathbf{Z}_k = \mathbf{H}\mathbf{X}_k + \mathbf{v}_k \qquad (2)$$

where $\mathbf{Z}_k$ is the measurement prediction made at time k, $\mathbf{X}_k$ is the state at time k, $\mathbf{H}$ is the observation matrix and $\mathbf{v}_k$ is additive measurement noise. The KF uses a feed-back control for process estimation. The KF algorithm consists of two steps – a prediction step and an update step as described below.

Prediction (time-update): This predicts the state and process covariance at time $k+1$ dependent on information at time k.

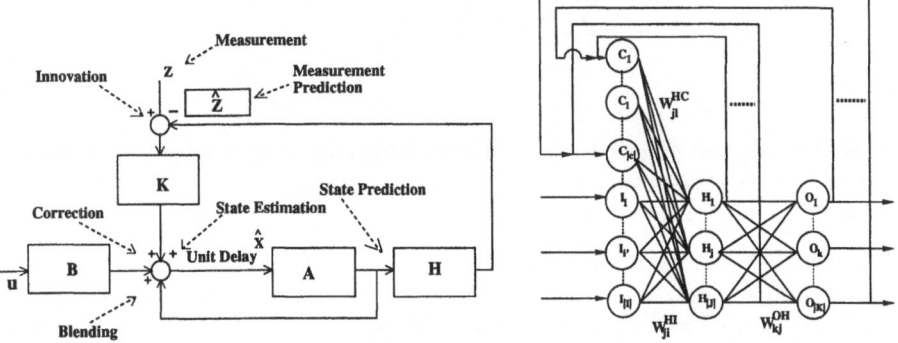

Fig. 1. Block Diagram of Kalman Filter.   **Fig. 2.** RNN Architecture.

Update (measurement update): This updates the state, process covariance and Kalman gain at time $k+1$ using a combination of the predicted state and the observation at time $k+1$. Not all the equations are shown here for want of space but see Brookner [1], for detailed equations.

Together with the initial conditions and the error covariance matrices, the steps listed above define the discrete-time sequential, recursive algorithm for determining the linear minimum variance estimate known as the Kalman filter [1].

3   Recurrent Neural Network Architecture

Recurrent Neural Networks (RNN) form a much wider class of neural networks, as they allow feedback connections between neurons, making them dynamical systems. The behaviour of a recurrent network is dependent on all its previous inputs. Recurrent neural networks have been used in a number of identification and control scenarios [7]. The work reported in this paper differs from a previous attempt at comparison of KF with RNN in several ways [6]. The feedback in our RNN is both from the output and hidden layers to the input layer unlike from only output to input layer in [6] and the learning method adopted here is RTRL as opposed to conjugate gradient method in earlier paper [6]. A simplified and more detailed representation of Recurrent Network is shown in Figure 2.

Units of the input layer **I** and the recurrent layer **H** and the output layer **O** are fully connected through weights $\mathbf{W}^{HI}$ and $\mathbf{W}^{OH}$, respectively. The current output of the recurrent units at time t is fedback to the context units at time $t+1$ through recurrent connections so that $\mathbf{C}^{(t+1)} = \mathbf{H}^{(t)}$. Hence, every recurrent unit can be viewed as an extension of input to the recurrent layer. As they hold contextual information from previous time steps, they represent the memory of the network.

Given the input pattern at time t, $\mathrm{I}^{(t)} = (I_1{}^{(t)}, ..., I_i{}^{(t)}, ...,I_{|I|}{}^{(t)})$, and recurrent activities $\mathrm{H}^{(t)} = (H_1{}^{(t)}, ..., H_j{}^{(t)}, ..., H_{|H|}{}^{(t)})$, the recurrent unit's net input $\tilde{H}$ and output activity $net_i{}^{(t)}$ are calculated as

$$\tilde{H}_i^{(t)} = \sum_j W_{ij}{}^{HI} I_j^{(t)} + \sum_j W_{ij}{}^{HC} H_j^{(t-1)} \tag{3}$$

$$net_i^{(t)} = f(\tilde{H}_i^{(t)}) \tag{4}$$

where $|I|$, $|H|$ and $|O|$ are the number of input, hidden and output units, respectively, and f is the activation function. In this work we are using the symmetrical transfer function $f(x) = \tanh x$.

Learning Algorithm for RNN: There are several algorithms available to train recurrent networks based on streams of input-output data. Perhaps the most widely used are real-time recurrent learning (RTRL) and backpropagation through time (BPTT) [5]. In this paper RTRL algorithm is used for recurrent network training because of its good convergence property and its on-line nature [3]. Real-time recurrent learning (RTRL) has been independently derived by many authors, although the most commonly cited reference for it is Williams and Zipser [3]. This algorithm computes the derivatives of states and outputs with respect to all weights as the network processes the sequence, that is, during the forward step. The supervised learning process uses 'Teacher Forcing' technique [3]. The advantage of using RTRL is the ease with which it may be derived and programmed for a new architecture as it does not involve any unfolding over time as in BPTT.

4   Simulation Experiments

Simulation studies were performed for a) state estimation problem [6] and b) tracking problem. The configuration for the KF and RNN for each of these problems is described below.

System I: *State Estimation Problem* [6]

Kalman: The state update equation and measurement equation are given by:

$$X(t) = 0.9X(t-1) + W(t) \quad \& \quad Z(t) = X(t) + V(t) \tag{5}$$

and the noise sources are white Gaussian noise sequences with zero mean. The process noise covariance Q is 0.1997 and the measurement noise covariance R is 0.1. It can be observed that this is a simple scalar version of the state estimation problem where the state transition and measurement systems are scalar valued, with coefficients taken as 0.9 and 1, respectively.

RTRL: The architecture we took consisted of **1** input node, **3** hidden nodes and **1** output node.

System II: *Tracking Problem*

Kalman: The state can be described as $\mathbf{X}(t) = [\mathbf{x}(t), \dot{\mathbf{x}}(t)]^T$. The state update equation and measurement equation are given by:

$$\mathbf{X}(t) = \mathbf{A}\mathbf{X}(t-1) + \mathbf{W}(t) \quad \& \quad \mathbf{Z}(t) = \mathbf{H}\mathbf{X}(t) + \mathbf{V}(t) \tag{6}$$

$$where \quad \mathbf{A} = \begin{bmatrix} 1 & 1 \\ 0 & 1 \end{bmatrix}, \quad \mathbf{H} = \begin{bmatrix} 1 & 0 \\ 0 & 1 \end{bmatrix}, \quad Q_k = \begin{bmatrix} 1/3 & 1/2 \\ 1/2 & 1 \end{bmatrix} \sigma_q^2, \quad R_k = \sigma_r^2$$

Where Q_k and R_k are the process and observation noise covariance matrices and $\sigma_q^2 = 0.01$; $\sigma_r^2 = 0.1$.

RTRL: The architecture consisted of **2** input nodes, **8** hidden nodes and **2** output nodes.

Simulation for each system is conducted as follows:

A training data set as well as a separate test data set are produced by running the system equations. Each data set contains 100 sequences of 100 I/O pairs (Z(n),X(n)), for a total of 10000 I/O pairs. The data are scaled to the range [-1,1]. The RNN architecture comprises one hidden layer of nodes with non-linear activation function (symmetrical transfer function **tanh(x)**), whereas the output nodes are linear. The state is initialized to some random value and the weights are also initialized to random values, uniformly distributed between -0.05 and +0.05. RTRL is used to train the net and it is trained until error criterion threshold is achieved. Testing is done with a separate data set and the results are reported in Figures 3 and 4. The KF parameters are computed using statistical estimation techniques over the same test data set.

5   Discussion

Figure 3 depicts the simulation of the state estimation problem. Figure 3(a) compares the resulting kalman and neural filter estimates of the true state. Figure 3(b) shows the error plots of KF and RNN. From Figure 3, it is evident that KF and RNN show comparable performance on the estimation problem. Figure 4 depicts the simulation of tracking problem. Figures 4(a) and 4(c) show the performance of KF versus RNN with respect to tracking of position and velocity of a vehicle, respectively. Figures 4(b) and 4(d) depict the corresponding error plots of position and velocity. It is evident from these figures that the difference between the desired and estimated values (tracking error) for RNN are almost zero whereas that with KF is not zero but appears to be a random value with zero-mean. The tracking error behaviour of KF is in expected lines as per the algorithm.Kalman filter is a simple, on-line, optimal algorithm but works only for linear systems with Gaussian noise. RNN is expensive in terms of space and time complexities. However, nonlinear approximation can be achieved and there is no restrictive gaussian assumption with RNNs.

6   Conclusion

A recurrent neural network of the type described in this paper is capable of closely matching a basic KF in performance on state estimation and tracking problems. KF is less expensive computationally both in space and time complexities as compared to RNN trained via RTRL algorithm. However, the attractive

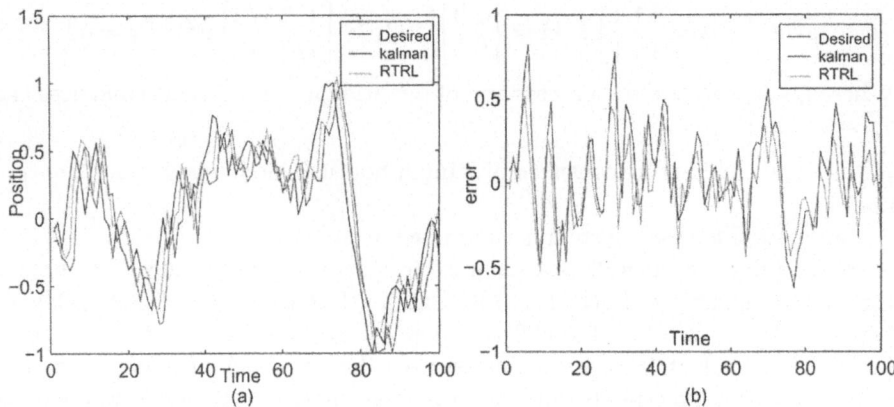

Fig. 3.

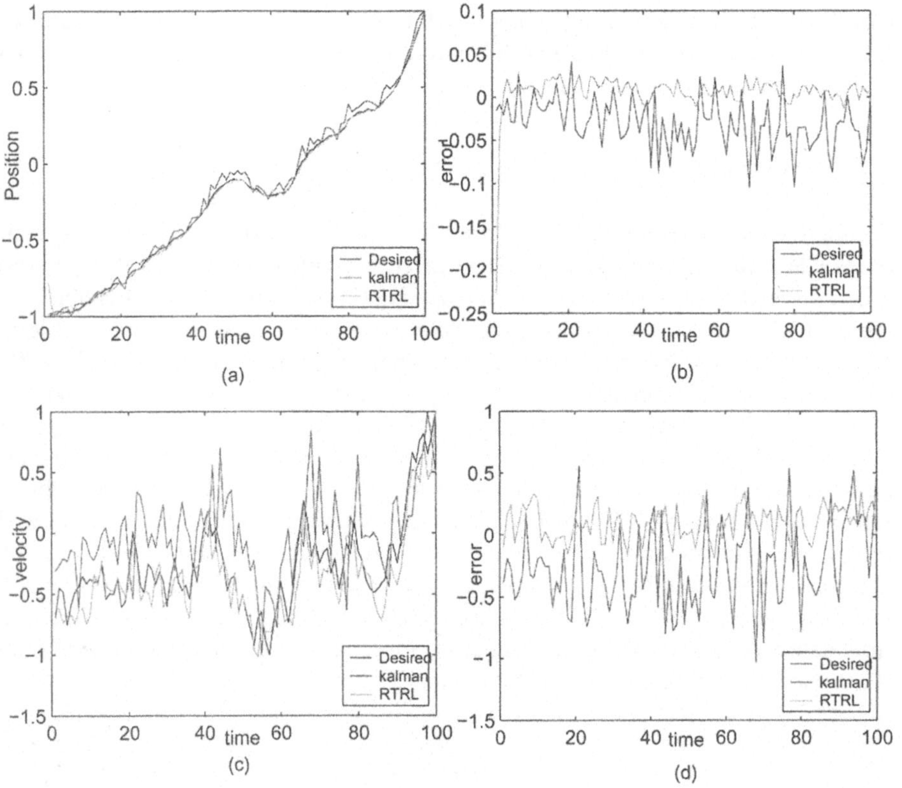

Fig. 4.

feature of RNNs is that the technique works without any significant modifications for nonlinear and non-gaussian cases also. Whereas in KF, whenever these assumptions are violated, the algorithm becomes differently structured and is more complex [4]. Thus, while the RTRL itself may not necessarily be the algorithm of choice for training recurrent networks, it may help provide a basis for both gaining a deeper understanding of existing recurrent network learning techniques and more importantly, creating more computationally attractive algorithms that allow one to optimize the trade-off between computational effort and learning speed. Other methods such as EKF are also available for training RNNs [8] and will be taken up in future, specifically for estimation and tracking problems.

References

1. Brookner, E., *Tracking and kalman filtering made easy*, John Wiley and Sons, Inc, New York, USA, 1998.
2. Pillai, S. K., Seshagiri Rao, V., Balakrishanan, S. S., and Sankar, S., A New Method of Model Compensation in Kalman Filter - Application to Launch Vehicles, *Proceedings of The Fourteenth International Symposium on Space Technology and Science*, Tokyo, 1984.
3. Williams, R. J., and David Zipser, A learning algorithm for continually running fully recurrent neural networks, *Neural Computation*, vol. 1, pp.270–280, 1989.
4. Haykin, S., *Neural Networks*, Second Edition, Pearson Eductation, Inc, New Delhi, India, 1999.
5. Werbos, P. J. Backpropagation Through Time: What it does and how to do it? *Proc. of the ICNN*, SanFrancisco, CA, USA, 1993.
6. DeCruyenaere, J.P. and Hafez, H.M., A comparison between Kalman filters and recurrent neural networks, *Proc. of the IJCNN*, Baltimore, USA, IEEE Press, vol. **IV**, pp.247–251, 1992.
7. Narendra, K. S., Parthasarathy, K., Identification and control of dynamical systems using neural networks, *IEEE Transactions on Neural Networks*, vol. **1**, no. l, pp.4–27, 1990.
8. Puskorius, G. V. and Feldkamp, L. A., Neurocontrol of Nonlinear Dynamical Systems with Kalman Filter Trained Recrrent Networks, *IEEE Transactions on Neural Networks*, vol. **5**, No. 2, pp.279–297, March 1994.

A Connectionist Account
of Ontological Boundary Shifting

Shohei Hidaka and Jun Saiki

Department of Intelligence Science and Technology,
Graduate School of Informatics, Kyoto University
Yoshida-Honmachi, Sakyo-ku, Kyoto, 606-8501, Japan
{hidaka,saiki}@cog.ist.i.kyoto-u.ac.jp
http://www.cog.ist.i.kyoto-u.ac.jp

Abstract. Previous research on children's categorizations has suggested that children use perceptual and conceptual knowledge to generalize object names. In particular,the relation between ontological categories and linguistic categories appears to be a critical cue to learning object categories. However, the mechanism underlying this relation remains unclear. Here we propose a connectionist model for the acquisition of ontological knowledge by learning linguistic categories of entities. The results suggest that linguistic cues help children attend to specific perceptual properties.

1 Introduction

Categorization, an essential cognitive ability, involves the compression of information and is one solution to handling an almost infinite number of entities efficiently. Categorizing entities and learning words is a basic linguistic ability. Quine [7] discussed the difficulty of word learning in situations that have many possible interpretations. This problem is encountered by children when acquiring word meanings in the early stage, as parents' daily words to their children are spoken with many possible interpretations. So how do children learn word meanings in such a situation? Children must logically reject many useless possibilities, and can not acquire word meaning at the first attempt. However, children do not actually consider useless possibilities; instead they acquire temporary word meaning for words presented only once. Landau, Smith and Jones [5] claimed that children learn words so quickly because they use prior knowledge of vocabulary and entities as constraints. They showed that shape is an important property for the categorization of objects and they called this cognitive process 'shape bias'. Colunga & Smith [1] and Samuelson [8] suggested that children attended to perceptual features depending on the solidity of objects. In other words, children recognize the nature of entities and use them to generalize to novel words. In the present study, we focus on how children acquire knowledge about the nature of entities and ontological categories.

Some researchers have suggested a deep relation between ontological categories and linguistic categories. In particular, the relation between count/ mass

N.R. Pal et al. (Eds.): ICONIP 2004, LNCS 3316, pp. 282–287, 2004.
© Springer-Verlag Berlin Heidelberg 2004

noun syntax in English and objects/ substance ontology is typical. Imai & Gentner [3] expanded upon the experiments of Soja, Carey and Spelke [9] to verify the difference between English and Japanese speakers in this regard. As English syntax is compatible with the ontological distinction between objects and substance, but Japanese is not, their comparison reveals the influence of count/ mass syntax on ontological category. Their results suggested a different categorization of simple objects between English and Japanese speakers. Imai & Gentner considered these simple objects to be near the boundary between objects and substance, since they were objects but also resembled substances in that parts of the object were similar to the whole. Their experiments showed the linguistic influence on ontological categories of ambiguous entities.

Japanese expresses animacy in syntax through verb form. For example, in sentences, (1) 'Animates-ga **iru**,' and (2) 'Inanimates-ga **aru**,' 'iru' and 'aru' have essentially the same meaning as the verb 'to be' in English, but an animate subject requires 'iru' and an inanimate one 'aru' (hereinafter reffered to as 'iru'/'aru' syntax). Yoshida & Smith [10] verified the influence of Japanese syntax by using objects simulating animates, and suggested that English and Japanese speakers had different categorical criteria. They proposed the 'boundary shift hypothesis' (BSH). where the linguistic cues influence the ontological boundaries on an 'individuation continuum', which explains ontological categories by individuation [6]. However, the mechanism of boundary shifting remains to be fully elucidate.

1.1 Previous Work

Hidaka & Saiki [2] proposed a computational model explaining BSH. They quantified the common feature space by English and Japanese adults' vocabulary rating (see also Figure 1) by asking adults to rate the applicability of 16 adjective pairs to 48 nouns (e.g., "a monkey is (very dynamic, dynamic, neither, static, very static)."). Furthermore, they estimated English- and Japanese-specific ontological space using a principal component analysis (PCA)-based model including specific syntactical categories (i.e. count/ mass and 'iru'/ 'aru' syntax), and they simulated the experiment of Yoshida & Smith [10] using the results of this estimation. We believe that feature attention learning is sufficiently powerful to change ontological knowledge and explain BSH. Therefore, in the present work, we show that the Attention Learning COVEring map (ALCOVE) [4], which is successfully simulates adult's category learning, can also explain children's attentional shift.

2 Simulation

We simulated Yoshida & Smith [10]'s experiment, known as the "novel word generalization task", that suggests BSH. They conducted three experiments showing the ontological difference between Japanese and English monolingual children. The following is a brief summary of their second experiment, which we simulated.

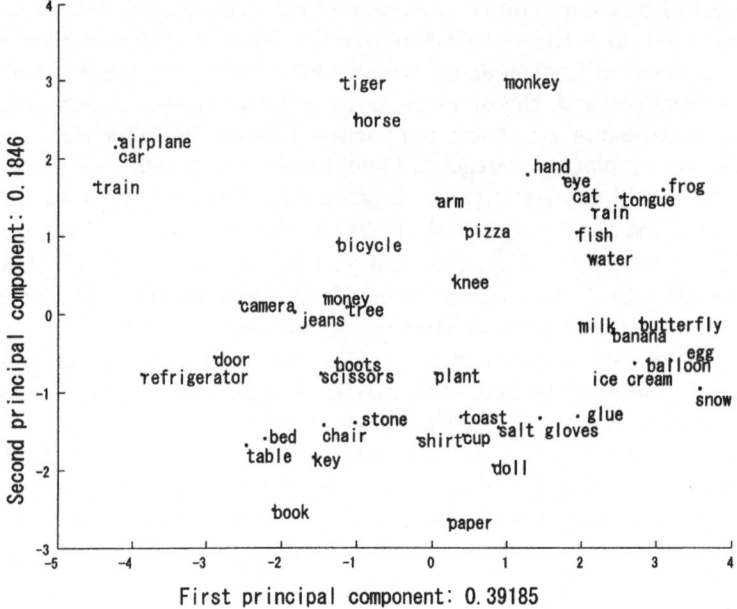

Fig. 1. Result of adults' vocabulary rating (Hidaka & Saiki, 2004): the first two principal components.The first principal component (x axis) was interpreted as 'solidity' or 'size' of objects. The second principal component (y axis) was interpreted as 'animacy' or 'movement' of objects.

Their participants were 3-year-old English and Japanese monolingual children. Experimenters presented them exemplars with pipes resembling animal legs and attached a novel label to it (e.g. in Japanese 'Kore-wa __ dayo', in English 'This is __.'). Experimenters gave no syntactic cues, like use of 'iru/ aru' to indicate the animacy of the label to children. The experimenters then presented them test objects and asked whether it had a novel label (e.g. in Japanese 'Kore-wa __-kana?', in English 'Is this __ ?'). Exemplars and test objects were controlled to be matched or not matched in three perceptual features (Table 1). The results showed different responses between English speakers and Japanese speakers where English speakers tended to generalize novel labels to test objects matched in shape, but Japanese speakers did not.

2.1   Method

In the present experiment, we used ALCOVE [4] to simulate Yoshida & Smith's experiment. ALCOVE is an exemplar-based neural network model that has an input layer which receives attentional modulation, a hidden layer with exemplar units and an output layer with category units. It has an error-driven learning algorithm to optimize its attention and weights between the hidden layer and output layer. In the present simulation, the input layer had 16 units represent-

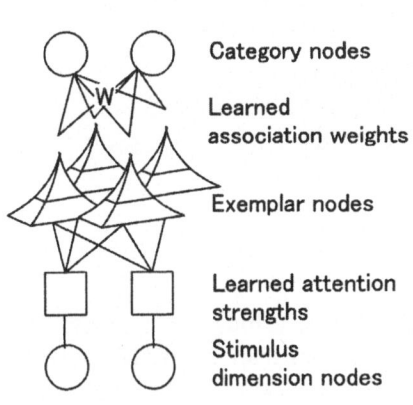

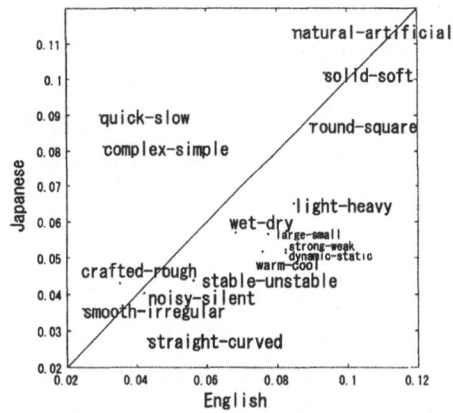

Fig. 2. ALCOVE (revised from Kr-uschke, (1992) [4]).

Fig. 3. The attention weight of English and Japanese condition.

ing the psychological features of Hidaka & Saiki [2] and attentions initialized to one. The hidden layer had 48 exemplar units representing each of the 48 entities by holding the mean value of each category. The output layer had two units representing linguistic category. The output layer represented the count/mass category and 'iru'/ 'aru' category in the English and Japanese condition respectively. The model performed a novel word generalization task simulating Yoshida & Smith's experiment after learning linguistic categories in 40 epochs.

The novel word generalization task in the simulation is to say 'yes' to a test stimulus similar to the exemplar. Three features (shape, color and texture) were manipulated in the behavioral experiment, but we handled only shape and texture in this simulation. We selected the shape and texture dimensions based on the perceptual expressivity [2].The shape dimensions were 'round-square' (.83), 'straight-curved' (.67) and 'large-small' (.63), and the texture dimensions were 'smooth-irregular' (.25), 'complex-simple' (.17) and 'finely crafted-rough hewn' (.13)[1]. At first we presented the model with novel exemplars that have uniform random values as feature dimensions. The model was then presented with a feature-controlled test stimulus and it would classify the stimulus as Being similar to the novel exemplar ('yes') or different ('no'). We defined the probability of a 'yes' response (P_{yes}) based on the Euclidean distance δ between the two output vectors corresponding to the exemplar and the test stimulus (see equation 1). $b > 0$ is the scaling parameter of the conversion from a distance to a similarity.

$$P_{yes} = exp(-b\delta) \tag{1}$$

---

[1] We selected the three most expressive dimensions. Values in parentheses represent expressivity of shape or texture. Range of expressivity is from 1 (most appropriate) to -1 (least appropriate).

Table 1. Experimental conditions of Yoshida & Smith (2003). 'm' represents a feature match between exemplar and test object, and 'N' represents non-match.

condition	S+T+C	S+C	C
shape	m	m	N
texture	m	N	N
color	m	m	m

2.2   Results

We show the learned attention weights of English (learning count /mass category) and Japanese (learning 'iru' /'aru' category) normalized by the total sum of the weights (Figure 3). The result suggested the network in the English condition attended more to shape dimension (e.g. straight-curved, large-small) and that in the Japanese condition, it attended more to material and movement dimension (e.g. smooth-irregular, quick-slow). The results of Yoshida & Smith [10] (Figure 4) were reproduced by our model (Figure 5). Using a Monte Carlo simulation, we estimated that the scaling parameter b is 1.8. In the behavioral experiment, the English speakers categorized the stimuli based on shape and the Japanese speakers categorized them based on multiple features. These results provide evidence for BSH because they suggest the difference of criteria used between English and Japanese. From this perspective, our model fitted the behavioral results well ($R^2 = .96$).

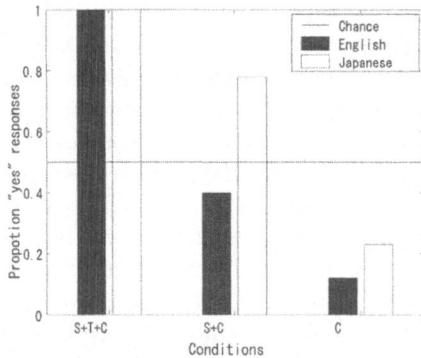

Fig. 4. Results of Yoshida & Smith [10].

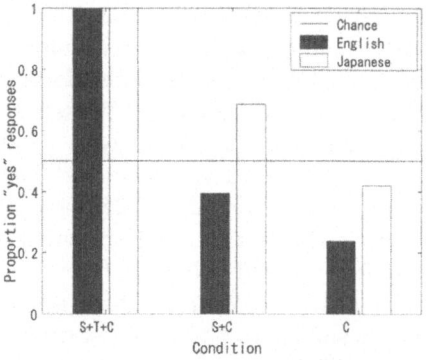

Fig. 5. Results of the simulation.

2.3   Discussion

The present study demonstrated that the connectionist model could simulate behavioral data by learning linguistic categories. Therefore, this work implemented a computational model that expands the BSH proposed by Hidaka & Saiki [2] in the form of connectionist model. One contribution of our work here is to provide associational "learnablity" to the previous computational model, as the model learned language-specific linguistic categories of entities. The results (Figure 3) suggested that linguistic categories influenced learners' attention. In the English condition, the model attended to shape dimension, which is consistent with Colunga & Smith [1] and Samuelson [8] who showed that American children attended more to the shape of objects during object categorization. In contrast, in the Japanese condition, the model attended to material and movement dimensions, which is consistent with Yoshida & Smith [10] who showed that Japanese children attended to multiple features and animacy of objects. In addition to qualitative matches with previous data, our model could make a good quantitative fit to the behavioral data of Yoshida & Smith [10]. We showed here that a general category learning model can account for crosslinguistic differences in object categorization, known as ontological boundary shifting, that is intimately related to children's word learning bias.

Acknowledgments

This work was supported by Grants-in-Aid for Scientific Research from JMEXT (No. 15650046), and the 21st Century COE Program from JMEXT (D-2 to Kyoto University).

References

1. Colunga, E. & Smith, L. (2000) Committing to an Ontology: A Connectionist Account, *The Twenty Second Annual Meeting of the Cognitive Science Society*.
2. Hidaka, S. & Saiki, J. (2004) A mechanism of ontological boundary shifting, *The Twenty Sixth Annual Meeting of the Cognitive Science Society*, 565-570.
3. Imai, M. & Gentner, D. (1997). A cross-linguistic study of early word meaning: universal ontology and linguistic influence., *Cognition*, *62*, 169-200.
4. Kruschke, J. (1992). ALCOVE: An exemplar-based connectionist model of category learning., *Psychological Review*, *99*, 22-44.
5. Landau, B., Smith, L.B. & Jones, S.S. (1988). The importance of shape in early lexical learning, *Cognitive Development*, *3*, 299-321.
6. Lucy, J.A. (1992). Language diversity and thought: A reformulation of the linguistic relativity hypothesis., Cambridge: Cambridge University Press.
7. Quine, W.V.O. (1960). *Word and Object.*, Cambridge, MA:MIT Press,.
8. Samuelson, L.K. (2002) Statistical Regularities in Vocabulary Guide Language Acquisition in Connectionist Models and 15-20 Month Olds., *Developmental Psychology*, *38*, 1016-1037.
9. Soja, N. N. , Carey, S. & Spelke, E. S. (1991). Ontological categories guide young children's inductions of word meanings: object terms and substance terms., *Cognition*, *38*, 179-211.
10. Yoshida, H. & Smith, L. B. (2003) Shifting ontological boundaries: how Japanese- and English- speaking children generalize names for animals and artifacts., *Developmental Science*, *6*, 1-34.

A Neural Network Model for Trace Conditioning

Tadashi Yamazaki and Shigeru Tanaka

Laboratory for Visual Neurocomputing, RIKEN Brain Science Institute,
2-1 Hirosawa, Wako, Saitama 351-0198, Japan
tyam@brain.riken.jp, shigeru@riken.jp

Abstract. We studied the dynamics of a neural network which have both of recurrent excitatory and random inhibitory connections. Neurons started to become active when a relatively weak transient excitatory signal was presented and the activity sustained due to the recurrent excitatory connections. The sustained activity stopped when a strong transient signal was presented or when neurons were disinhibited. The random inhibitory connections modulated the activity patterns of neurons so that the patterns evolved without recurrence with time. Hence, a time passage between the onsets of the two transient signals was represented by the sequence of activity patterns.
We then applied this model to the trace eyeblink conditioning which is mediated by the hippocampus. We assumed this model as CA3 of hippocampus and considered an output neuron corresponding a neuron in CA1. The activity pattern of the output neuron was similar to that of CA1 neurons during the trace conditioning which was experimentally observed.

1 Introduction

It is widely known that the hippocampus plays a critical role in declarative learning and memory. An example of declarative learning is trace eyeblink conditioning [1]. In this paradigm, a conditioned stimulus (CS; e.g., a tone) presented before an unconditioned stimulus (US; e.g., an airpuff) which elicits an automatic conditioned response (CR; e.g., an eyeblink). The offset of the CS and onset of the US do not overlap, creating a off-stimulus interval. The subject is tested for learning an association between the CS and the US, as evidenced by the CR in anticipation of the US. The hippocampus is known to play a necessary role in learning a well-timed CR in anticipation of the US [2]. Thus, neurons in the hippocampus must have a kind of "memory trace" of the CS that bridges the time interval to form a CS-US association, but how?

So far, we have studied a simple random recurrent inhibitory network and found that activity patterns of neurons evolved with time without recurrence due to random recurrent connections among neurons. The sequence of activity patterns was generated by the trigger of an external signal, suggesting that a time passage from the trigger of an external signal could be represented by the sequence of activity patterns [3]. In this paper, we extended this idea and studied the dynamics of a neural network model which have both recurrent

N.R. Pal et al. (Eds.): ICONIP 2004, LNCS 3316, pp. 288–293, 2004.
© Springer-Verlag Berlin Heidelberg 2004

excitatory and random inhibitory connections. When a relatively weak transient signal was presented, neurons in this model started to become active and the activity was sustained even during the off-stimulus period due to the recurrent excitatory connections. The sustained activity stopped when a strong transient signal was presented or when neurons were disinhibited. On the other hand, since the random recurrent connections generated non-recurrent activity patterns of neurons, the time passage was represented.

We then applied this model to the trace eyeblink conditioning which is mediated by the hippocampus. We assumed this model as CA3 of hippocampus and considered an output neuron corresponding a neuron in CA1. The activity pattern of the output neuron was similar to that of CA1 neurons during the trace conditioning which is experimentally observed.

2 Model Description

The model consists of N excitatory neurons and the same number of inhibitory neurons. Let $z_{\mathrm{ex}i}(t)$ and $z_{\mathrm{inh}i}(t)$ be the activities of excitatory neuron i and inhibitory neuron i at time t, respectively. For a neuron type $\mathrm{T} \in \{\mathrm{ex}, \mathrm{inh}\}$, $z_{\mathrm{T}i}(t)$ is defined as

$$z_{\mathrm{T}i}(t) = \begin{cases} u_{\mathrm{T}i}(t) & u_{\mathrm{T}i}(t) > \theta_{\mathrm{T}}, \\ 0 & \text{otherwise.} \end{cases}$$

$u_{\mathrm{ex}i}(t)$ and $u_{\mathrm{inh}i}(t)$ are internal states of excitatory neuron i and inhibitory neuron i at time t, respectively, which are calculated as

$$\tau_{\mathrm{ex}} \dot{u}_{\mathrm{ex}i}(t) = -u_{\mathrm{ex}i}(t) + I_i(t) + \sum_j w_{\mathrm{ex}i \leftarrow \mathrm{ex}j} z_{\mathrm{ex}j}(t) - \sum_j w_{\mathrm{ex}i \leftarrow \mathrm{inh}j} z_{\mathrm{inh}j}(t) \quad (1)$$

$$\tau_{\mathrm{inh}} \dot{u}_{\mathrm{inh}i}(t) = -u_{\mathrm{inh}i}(t) + \sum_j w_{\mathrm{inh}i \leftarrow \mathrm{ex}j} z_{\mathrm{ex}j}(t) - \sum_j w_{\mathrm{inh}i \leftarrow \mathrm{inh}j} z_{\mathrm{inh}j}(t), \quad (2)$$

where for $\mathrm{T}, \mathrm{T}' \in \{\mathrm{ex}, \mathrm{inh}\}$, $w_{\mathrm{T}i \leftarrow \mathrm{T}'j}$ is the weight of the synaptic connection from neuron j of type T' to neuron i of type T, τ_{T} is the time constant, and $I_i(t)$ is the external input to excitatory neuron i at time t.

Synaptic connections are defined as follows. $w_{\mathrm{ex}i \leftarrow \mathrm{ex}j}$ is set at $c_{\mathrm{ex} \leftarrow \mathrm{ex}}$ for any i and j: excitatory connections are all-to-all, $w_{\mathrm{inh}i \leftarrow \mathrm{ex}j}$ is given under the binomial distribution $\Pr(w_{\mathrm{inh}i \leftarrow \mathrm{ex}j} = 0) = \Pr(w_{\mathrm{inh}i \leftarrow \mathrm{ex}j} = c_{\mathrm{inh} \leftarrow \mathrm{ex}}/N) = 0.5$, $w_{\mathrm{ex}i \leftarrow \mathrm{inh}j}$ is set at $c_{\mathrm{ex} \leftarrow \mathrm{inh}}$ if $i = j$ and 0 otherwise, thus each inhibitory neuron inhibits its corresponding excitatory neuron, and $w_{\mathrm{inh}i \leftarrow \mathrm{inh}j}$ is set at $c_{\mathrm{inh} \leftarrow \mathrm{inh}}$ for any i and j for simplicity. When we consider disinhibition of excitatory neurons, $c_{\mathrm{ex} \leftarrow \mathrm{inh}}$ and $c_{\mathrm{inh} \leftarrow \mathrm{inh}}$ are set at the half.

External input signals are given as follows. For any t and i, $I_i(t)$ is set at I_{aff} when $1 \leq i \leq N_{\mathrm{CS}}$ and $t_{\mathrm{CSonset}} \leq t \leq t_{\mathrm{CSoffset}}$, I_{aff} when $1 \leq i \leq N_{\mathrm{US}}$ and $t_{\mathrm{USonset}} \leq t \leq t_{\mathrm{USoffset}}$, and 0 otherwise.

We demonstrate that the activity patterns of neurons generated using Eq. (2) can represent a time passage, that is, the activity pattern at one time step is dissimilar to the pattern at a different time step when the interval between the two steps are large. Therefore, we use the following correlation function as the similarity index.

$$C(t_1, t_2) = \frac{\sum_i z_{exi}(t_1) z_{exi}(t_2)}{\sqrt{\sum_i z_{exi}^2(t_1)} \sqrt{\sum_i z_{exi}^2(t_2)}}. \tag{3}$$

Parameter values are set arbitrarily as follows: $N = 1000$, $T = 1000$, $\tau_{ex} = 20.0$, $\tau_{inh} = 50.0$, $\theta_{ex} = 0.1$, $\theta_{inh} = 0.1$, $c_{ex \leftarrow ex} = 3.0$, $c_{inh \leftarrow ex} = 6.0$, $c_{ex \leftarrow inh} = 20.0$, $c_{inh \leftarrow inh} = 6.0$, $I_{aff} = 1.0$, $N_{CS} = 400$, $N_{US} = 1000$, $t_{CSonset} = 0$, $t_{CSoffset} = 20$, $t_{USonset} = 800$, and $t_{USoffset} = 820$.

3   Results

The left panel in Fig. 1 shows the plot of active states of the first 500 neurons out of N excitatory neurons during T steps. At time t ($t_{CSonset} \leq t \leq t_{CSoffset}$), neurons i ($1 \leq i \leq N_{CS}$) were given external signals I_{aff} and they started to become active. Then, the activity spread out among all neurons through recurrent excitatory connections and was sustained during the off-stimulus period ($t > t_{CSoffset}$). At time t ($t_{USonset} \leq t \leq t_{USoffset}$), neurons i ($1 \leq i \leq N_{US}$) were given external signals but due to the recurrent inhibition not all neurons became active. After the USoffset, neurons gradually became inactive and suddenly stopped activities at $t \approx 920$. We also examined the effect of disinhibition. This time, we did not present the external signals but disinhibited neurons during ($t_{USonset} \leq t \leq t_{USoffset}$). After the disinhibition, the sustained activity suddenly stopped (data not shown). During the off-stimulus period, once a neuron started to become active, its activity continued for several hundreds of steps, and then the neuron became inactive. Some neurons were reactivated after the inactive period. Thus, the active and inactive periods appeared alternately.

The right panel in Fig. 1 shows the similarity index calculated using Eq. (3). Since Eq. (3) takes two arguments of t_1 and t_2, we obtained a $T \times T$ matrix, where the row and the column were specified by t_1 and t_2, respectively. Similarity

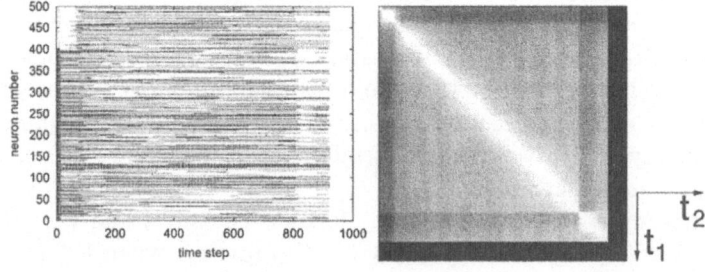

Fig. 1. Raster plot of active states ($z_{exi}(t) > \theta_{ex}$) of the first 500 excitatory neurons (left) and the similarity index (right).

indices were plotted in a gray scale in which black indicated 0 and white 1. A white band appeared diagonally. Since the similarity index at the identical step $(t_2 = t_1)$ takes 1, the diagonal elements of the similarity index appeared white. The similarity index decreased monotonically as the interval between t_1 and t_2 became longer. This result indicates that the activity pattern of neurons changed gradually with time and did not recur.

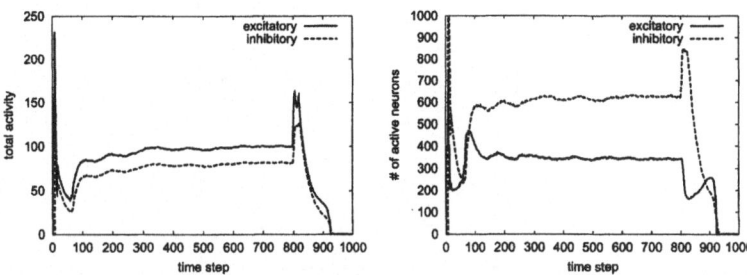

Fig. 2. Total activity of excitatory (solid line) and inhibitory (dashed line) neurons (left) and the number of active excitatory (solid line) and inhibitory (dashed line) neurons (right).

The left panel in Fig. 2 shows the total activity of neurons while the right panel the number of active neurons. As can be seen, both of the total activity and the number of active neurons remained constant during the off-stimulus period. These results suggest that the representation of a time passage was stable.

Then, we examined if this model could make association between the CS and US onsets temporally separated, which were motivated by the trace eyeblink conditioning. We regarded the present model as the hippocampus CA3 because of its recurrent excitatory and inhibitory connections and we considered an output neuron corresponding a neuron in CA1, which was connected with excitatory neurons in the model with synaptic weights representing Schaffer collaterals. We assumed that the output neuron received only the US directly through an another pathway corresponding to the perforant path and the CA3 neurons received only the CS (see Discussion). The output neuron learns to associate the US onset with the CS onset which are disconnected by the off-stimulus interval [4].

We ran the simulation twice. In the first run, we determined active neurons at $t = t_{\text{USonset}}$ and set their synaptic weights to 1 while weights of inactive neurons 0, namely,

$$w_i = \begin{cases} 1 & z_{\text{ex}i}(t_{\text{USonset}}) > \theta_{\text{ex}}, \\ 0 & \text{otherwise}, \end{cases}$$

where w_i represents the synaptic weight of neuron i. This corresponds to the Long-Term Potentiation (LTP) at Shaffer collaterals induced by the conjunctive stimulation of the output neuron by the US and by the signals from our CA3 neurons. In the second run, we calculated the net input to the output neuron as $\sum_i w_i z_{\text{exi}}(t)$. In order to see if the output neuron anticipates the US onset only by the CS stimulation, the US signal was not presented to the output neuron. We regarded the value of the net input as the activity of the output neuron by assuming the linear response of the output neuron. Thus, the output neuron learns to elicit responses at $t = t_{\text{USonset}}$ by the trigger of the CS onset $t = t_{\text{CSonset}}$ followed by the off-stimulus period.

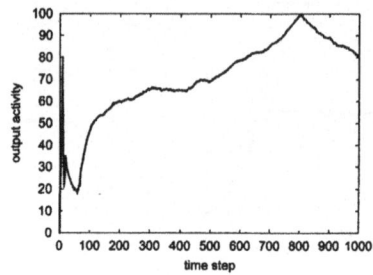

Fig. 3. Plot of the net input $\sum_i w_i z_{\text{exi}}(t)$.

Figure 3 shows the activity of the output neuron. As can be seen, after the CS onset, the activity transiently increased. It reached the maximal value at $t = 10$ and then sharply decreased. After the CS offset the activity increased and converged to constant at $t \approx 300$. Then, at $t \approx 500$ it started to move again and slowly increased towards $t = t_{\text{USonset}}$. At $t = t_{\text{USonset}}$ the activity reached the maximum value and then gradually decreased. As as result, the output neuron could associate the US onset with the CS onset. The initial transient increase and the following slow increase towards $t = t_{\text{USonset}}$ are typical responses of neurons in the hippocampus CA1 during the trace conditioning [2]. Hence, the present model successfully reproduced the hippocampal activity in the trace conditioning.

4   Discussion

We studied the dynamics of a model which have recurrent excitatory and random inhibitory connections. Due to the recurrent excitatory connections individual neurons could generate sustained activity during the off-stimulus period while due to the random recurrent inhibitory connections the neurons exhibited the random repetition of transition between active and inactive states. Hence, the population of active neurons changed gradually with time and did not recur. This property was confirmed by calculating the similarity index.

We then examined if the present model could account for the trace eyeblink conditioning mediated by the hippocampus. We regarded the model as the hippocampus CA3 and incorporated an output neuron corresponding to a neuron in CA1. We calculated the activity of the output neurons and the activity profile was similar to the one observed experimentally [2]. We assumed that the output neuron received only the US directly through an another pathway corresponding to the perforant path. Input signals to the hippocampus first arrive at both the layers 2 and 3 of the entorhinal cortex (EC2, EC3), and neurons in CA1 receive input signals from EC3 through the perforant path and from CA3 through Shaffer collaterals [5]. Activation of perforant path neurons to CA1 do not evoke neuronal activity in CA1 when neurons in CA3 are not activated [6]. Since neurons in CA3 are inactive just before the CS onset, CA1 neurons cannot become active by the CS stimulation only. Hence, we could ignore the CS presented to the output neuron. We also assumed that the CA3 neurons received only the CS. CA1 neurons excite inhibitory neurons in the septum and in turn these inhibitory neurons inhibit inhibitory neurons in CA3 [7]. We hypothesized that this disinhibition stops the sustained activity. Therefore, we ignored the US stimulation to CA3 neurons.

In the present model, the recurrent excitatory connections were assumed to be all-to-all. Rolls has argued that CA3 plays a role of an autoassociation memory [5]. If so, the connections should be symmetric and thus the activity pattern of neurons converges to a steady state. Since all-to-all connections is a variation of symmetric connections, hence, our assumption may not be too simplistic.

Levy and his colleagues have developed a model of the hippocampus CA3 and recently in [8] they have reported that their model successfully reproduced the activity pattern of CA1 neurons in the trace conditioning. Their model connects the CS onset to the US onset by the temporally asymmetric Hebbian learning between pairs of CA3 neurons [9]. In their model, inhibitory neurons are incorporated only to regulate the total activity of CA3 neurons. On the other hand, this study demonstrated that inhibitory neurons could work more: they modulate the activity of excitatory neurons and generate a sequence of activity patterns without recurrence, which can represent a time passage from the CS onset. This study may shed light on roles of inhibitory neurons.

References

1. Christian, K.M., Thompson, R.F. Learn. and Mem. **11** (2003) 427–455
2. McEchron, M.D., Disterhoft, J.F. J. Neurophys. **78** (1997) 1030–1044
3. Yamazaki, T., Tanaka, S. In: Society for Neuroscience Abstract. (2003)
4. McNaughton, B.L. Brain Res. Rev. **16** (1991) 202–204
5. Rolls, E.T. Hippocampus **6** (1996) 601–620
6. Bartesaghi, R., Gessi, T. Hippocampus **13** (2003) 235–249
7. Tóth, K., Borhegyi, Z., Freund, T.F. J. Neurosci. **13** (1993) 3712–3724
8. Rodriguez, P., Levy, W.B. Behav. Neurosci. **115** (2001) 1224–1238
9. Devanne, D., Gähwiler, B.H., Thompson, S.M. J. Physiol. **501** (1998) 237–247

Chunking Phenomenon
in Complex Sequential Skill Learning in Humans

V.S. Chandrasekhar Pammi[1], K.P. Miyapuram[1],
Raju S. Bapi[1], and Kenji Doya[2,3]

[1] Department of Computer and Information Sciences, University of Hyderabad, India
bapics@uohyd.ernet.in
[2] Computational Neuroscience Laboratories, ATR International, Kyoto, Japan
[3] Okinawa Institute of Science and Technology, Gushikawa, Okinawa, Japan
doya@atr.jp

Abstract. Sequential skill learning is central to much of human behaviour. It is known that sequences are hierarchically organized into several chunks of information that enables efficient performance of the acquired skill. We present clustering analysis on response times as subjects learn finger movement sequences of length *24* arranged in two ways — *12* sets of two movements each and *6* sets of four movements each. The experimental results and the analysis point out that greater amount of reorganization of sequences into chunks is more likely when the set-size is kept lower and discuss the cognitive implications of these findings.

Keywords: Chunking, Sequence Learning, Skill Acquisition, Clustering

1 Introduction

Most of the higher-order intelligent behaviour such as problem solving, reasoning and language involve acquiring and performing complex sequence of activities [1]. It is known that acquiring a complex sequential skill involves chaining a number of primitive actions to make the complete sequence. The notion of chunking in the context of limited capacity of short term working memory was introduced by Miller [2]. Hierarchical organization of movement sequences has been suggested by several researchers (for example: [3]). A sequence might consist of several sub-sequences and these sub-sequences in turn can contain sub-sub-sequences. Using a modified mxn visuo-motor sequence task [4,5], we set out to investigate the phenomenon of chunking during learning of a complex sequential skill.

The concept of chunking in sequential behaviour has been previously studied in animals (see [6] for a review) and in humans (for example: see [7]). Using a 2x10 sequence task, Sakai et al. [7] have demonstrated that different subjects chunk the same sequence of movements differently. They have also shown that performance on a shuffled sequence after learning was less accurate and slower when the chunk patterns were disrupted than when they were preserved. This clearly suggests an operational role for the chunks as a single memory unit that facilitates efficient performance of the sequence. The amount of information to be

N.R. Pal et al. (Eds.): ICONIP 2004, LNCS 3316, pp. 294–299, 2004.
© Springer-Verlag Berlin Heidelberg 2004

processed at a time forms a *set*. The current experiment specifically addresses the differences in chunk formation when same amount of information is organized in two different ways. We hypothesized that a smaller set-size would enable spontaneous chunking across several sets, while increasing the set-size will limit the chunk formation to single sets.

2 Materials and Methods

In the mxn task (Figure 1), visual stimuli in the form of m lighted squares (called a set) on a 3x3 grid display are presented. Subjects responded to the visual cues by pressing the corresponding keys on a keypad. The correct order of pressing the m keys (called a set) is to be learnt by trial and error. On successful completion of a set, subjects are presented the next set and so on. Subjects learn to complete n such sets (called a hyperset). If the subjects are not able to complete a set within a specific time period (average of 0.8 sec per keypress) or if they press an incorrect key, a screen-flash appears and the sequence is reset to beginning of the hyperset. We utilized two different mxn tasks in the current study. We arranged a total of *24* finger movements to be learnt as 4x6 and 2x12 tasks. Figure 1 shows the 2x12 task, in which subjects learn 12 sets of 2 movements each. In the 4x6 task, four lighted squares are presented at a time ($m = 4$) and subjects have to learn six such sets ($n = 6$). Hence there is an increased long-range prediction load in the 2x12 task, while there is an increased short-range prediction load in the 4x6 task.

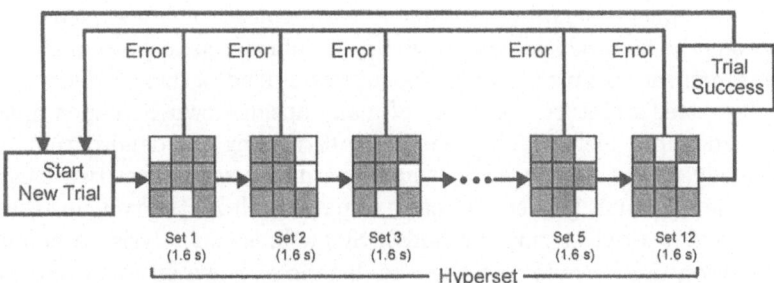

Fig. 1. The 2x12 task procedure. Subjects learn by trial and error to press $m(= 2)$ keys successively for $n(= 12)$ times in response to visual stimuli on a 3x3 grid display. On successful completion of a set, subjects are allowed to progress to the next set and the trial is reset to the beginning of the hyperset upon error.

2.1 Experimental Paradigm

A total of 18 subjects participated in the study. The subjects were explained the task procedure before the experiment. Subjects performed the two experiments (2x12, 4x6) for four sessions each. Further, every session consisted of six blocks of the sequence task. In every sequence block, subjects practiced the sequence for a fixed duration of 36 sec. A random hyperset was generated for each subject that remained fixed during the experiment. To reduce the possibility of any explicit structure or patterns in the sequence, the hyperset was generated such that any repetition or transposition of sets does not occur. To enable smooth performance of the movements, subjects were encouraged to respond as quickly as they could throughout the experiment. Moreover, subjects were instructed to use their index, middle and ring fingers for the three columns of the keypad - left, middle and right columns respectively.

2.2 Data Analysis

Subjects performed several trials in a block and a trial was terminated upon error. We measured the response times for each successful set in a trial and call these the set completion times. The response time (RT) is measured as time taken from presentation of the visual cues to the completion of the set (pressing all corresponding keys in correct order). The focus of the current paper is to examine the hierarchical organization of sequences by controlling the total number of movements to be learnt. To study the chunk formation, we employed two strategies. Firstly, we plotted the stacked graph of set completion times for successfully completed trials. Next, we performed cluster analysis on the cumulative set response times (RT) to identify the pattern of chunking (clusters across the sets). We use a bottom-up hierarchical cluster analysis to identify the hierarchical sequence structure possibly employed by the subjects. The hierarchical clustering methodology is suitable to address the problem of investigating the chunking processes in sequence learning as sequences can consist of sub-sequences, which in turn could be hierarchically organized. A graphical representation of the hierarchical organization thus found is depicted as a dendrogram. A dendrogram is essentially a tree structure consisting of many upside-down U shaped lines connecting nodes in a hierarchical tree. For constructing the dendrogram, single-linkage analysis was performed in which the distance between two clusters is taken as the minimum between all pairs of patterns from the two clusters (refer to [8] for review of clustering methods). Single-linkage analysis on cumulative set-completion times would result in the distances shown on the y-axis of dendrogram reflecting the actual set RTs. The clusters corresponding to the chunks were identified by performing a one-way analysis of variance (ANOVA) on the set completion times between successive sets. A significant pause between sets is identified as the beginning of a new chunk.

3 Results

Subjects showed learning related improvements by successfully acquiring the hyperset. Figure 2 shows the number of sets completed and the average set completion time for all the trials of one subject (WY) for both the 2x12 and 4x6 tasks.

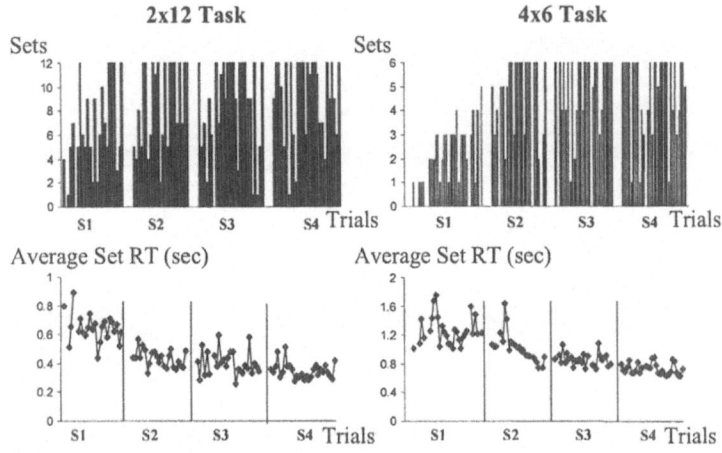

Fig. 2. Learning related improvements observed in the two mxn tasks. Number of sets completed *(top panel)* and average set completion time *(bottom panel)* over all the trials across the four sessions (S1 to S4) are shown. Vertical lines in *bottom panel* (and corresponding gaps in *top panel*) are shown to demarcate sessions.

Out of the 18 subjects that participated in our experiments, in this paper results from three representative subjects (KU, NS and WY) are presented for analysis of the chunking phenomenon. The stacked graphs shown in Figure 3a reveal a clear bunching pattern evolving as training progressed in the 2x12 task. Each bunch represents a chunk. The data of set RTs from last session was used for identifying chunking patterns. ANOVA between successive set RTs revealed significant ($p < 0.05$) pauses for several sets representing the beginning of an ensuing chunk. (KU: set 3, 8; NS: sets 3, 6, 8, 12; WY: set 7). The dendrogram plots show the hierarchical structuring of the sequence and thus reveal subsequences within the cluster identified from the ANOVA. Although ANOVA revealed main cluster patterns, it is interesting that the complete hierarchical sequence structure can be identified by the dendrogram. For example, the nested sequential structure for KU is (1 2) 3 (4 (5 (6 7)) 8 (9 10) (11 12)) and for WY is ((1 2) 3 (4 (5 6)) (7 (8 9)) ((10 11) 12).

Figure 3b shows the chunking phenomenon for the 4x6 task for the same three subjects. The cumulative set RTs (Figure 3b) did not show any bunching pattern across sets. The dendrogram shows that subjects require similar amount of time for completion of each set. The ANOVA results revealed non-significant p values for all pairs of successive set RTs, thus possibly indicating the absence of any significant pauses during sequence acquisition.

4   Conclusions and Future Work

Chunking offers a flexible way of learning. We have demonstrated that subjects employ different strategies for chunking when the same number of finger movements were arranged in two different ways. Our results suggest that when the

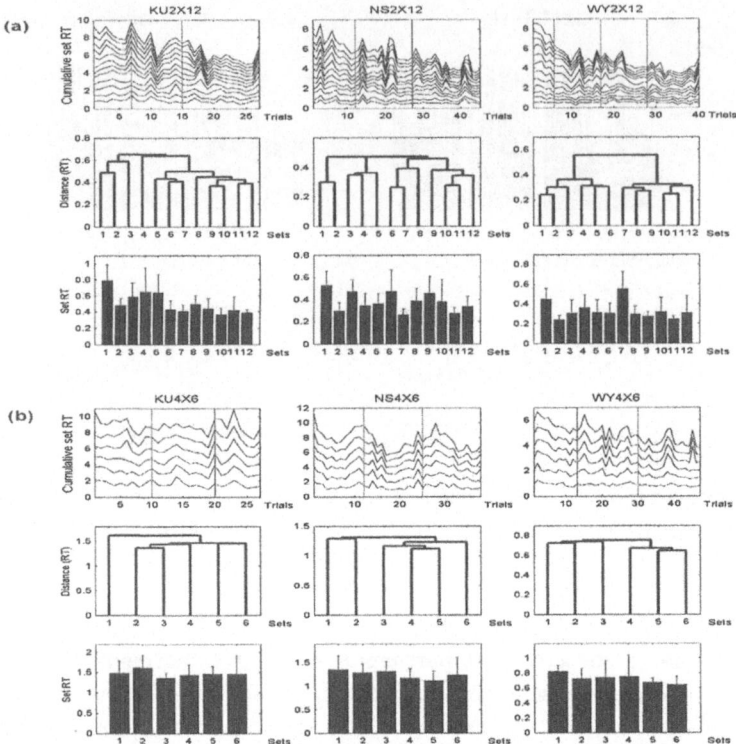

Fig. 3. Chunking phenomenon observed in three subjects (KU, NS, WY) for the 2x12 and 4x6 tasks. *top panel*: The cumulative set RTs for successful trials for the four sessions (delineated by vertical lines), *middle panel*: dendrogram, *bottom panel*: mean set RTs in the last session. (a) 2x12 Task. Cumulative set RTs show a clear bunching pattern for few sets of 2x12 task. The dendrogram shows the hierarchical structure of the sequence acquired by the subject. (b) 4x6 Task. The cumulative set RTs for successful trials does not show any bunching pattern across sets. The dendrogram shows that subjects require similar amount of time for each set possibly indicating optimisation of performance within a set.

set-size is larger as in the 4x6 task, there is less reorganization across sets. Subjects have to process more amount of information in each set and because of the increased short-term cognitive load, it appears that performance optimization was more within the sets but less across the sets. This is consistent with the theory of limited capacity of working memory [2]. On the other hand, when the *set-size* was kept smaller but the number of sets to be processed was increased as in the 2x12 task, we observed remarkable reorganization across the sets. As the number of sets to be internalized (12) is larger than the short-term memory capacity, it appears that subjects compressed the information into a number of chunks. The results from the current behavioural study have implications for the cognitive models of hierarchical sequence learning. The results point out that a model that learns sequences using a limited capacity working memory (WM) would need to optimize in two different ways depending on the amount of information to be processed at any instance of time. If the amount stretches to the limit of WM then optimization process needs to operate within the logical unit (set). If the amount is well within the WM capacity, optimization across the logical units (sets) would facilitate efficient performance. Neural bases for chunking strategy have been suggested in various brain areas including the pre supplementary motor area [9] and the basal ganglia [10]. In the current study, we have also collected fMRI images from all the subjects while they performed the mxn tasks. The analysis of fMRI data comparing the 2x12 and 4x6 tasks, which is in progress, is expected to reveal brain mechanisms underlying different strategies adopted for chunking process.

Acknowledgements

Authors would like to thank Dr. K. Samejima, ATR International for help with conducting the experiments. We also thank Dr. Atul Negi and Mr. Ahmed, University of Hyderabad for helpful discussions on Clustering analysis. The grants from JST, Japan under the ERATO and CREST schemes for conducting the experiments are gratefully acknowledged. Pammi would like to thank CSIR, New Delhi, India for the Senior Research Fellowship.

References

1. R. Sun. Introduction to sequence learning. In R. Sun and C. L. Giles, editors, *Sequence Learning – Paradigms, Applications and Algorithms*, volume 1828, pages 1–10. Springer-Verlag LNAI, 2000.
2. G. A. Miller. The magical number seven plus or minus two: Some limits on our capacity for processing information. *The Psychological Review*, 63:81–97, 1956.
3. D. A. Rosenbaum, S. B. Kenny, and M. A. Derr. Hierarchical control of rapid movement sequences. *Journal of Experimental Psychology: Human Perception and Performance*, 9:86–102, 1983.
4. O. Hikosaka, M. K. Rand, S. Miyachi, and K. Miyashita. Learning of sequential movements in the monkey: Process of learning and retention of memory. *Journal of Neurophysiology*, 74:1652–1661, 1995.
5. R. S. Bapi, K. Doya, and A. M. Harner. Evidence for effector independent and dependent representations and their differential time course of acquisition during motor sequence learning. *Experimental Brain Research*, 132:149–162, 2000.
6. H. Terrace. Chunking and serially organized behavior in pigeons, monkeys and humans. In R. G. Cook, editor, *Avian visual cognition*. Comparative Cognition Press, Medford, MA, 2001.
7. K. Sakai, K. Kitaguchi, and O. Hikosaka. Chunking during human visuomotor sequence learning. *Experimental Brain Research*, 152:229–242, 2003.
8. A. K. Jain, M. N. Murty, and P. J. Flynn. Data clustering: A review. *ACM Computing Surveys*, 31:264–323, 1999.
9. S. W. Kennereley, K. Sakai, and M. F. S. Rushworth. Organization of action sequences and role of the Pre-SMA. *Journal of Neuroohysiology*, 91:978–993, 2004.
10. A. M. Graybiel. The basal ganglia and chunking of action repertoires. *Neurobiology of Learning and Memory*, 70:119–136, 1998.

Cognitive Process of Emotion Under Uncertainty

Ayako Onzo[1,2] and Ken Mogi[1,2]

[1] Sony Computer Science Laboratories, Takanawa Muse Bldg, 3-14-13,
Higashigotanda, Shinagawa-ku, Tokyo, 141-0022, Japan
onzo@olive.ocn.ne.jp, kenmogi@csl.sony.co.jp
[2] Tokyo Institute of Technology, 4259, Nagatsuta-cho, Midori-ku,
Yokohama, 226-8502, Japan

Abstract. One of the missions of the cognitive process of animals, including humans, is to make reasonable judgments and decisions in the presence of uncertainty. The balance between exploration and exploitation investigated in the reinforcement-learning paradigm is one of the key factors in this process. Recently, following the pioneering work in behavioral economics, growing attention has been directed to human behaviors exhibiting deviations from simple maximization of re-ward. Here we study the nature of human decision making under the existence of reward uncertainty, employing a condition where the reward expectancy is constant (flat reward condition). The characteristic behavioral patterns exhibited by subjects reveal the underlying reward-related neural mechanism. The relevance of this result to the functions of dopamine neurons is discussed.

1 Introduction

Animals, including humans, encounter novel stimuli in the course of life, incurring perceptual uncertainty. How animals coordinate their actions in such an uncertain environment is one of the crucial aspects of cognition.

Metacognition is considered to be essential in the robust perception of uncertainty [1][2][3][4]. Hampton reported the metacognitive ability of rhesus monkeys [3]. It was found that the monkey has a "metacognition" of its internal state, i.e., its own assessment of the likelihood of conducting the task successfully.

To keep drawing on one's past experience might prevent us from coming up to the alternative sources of reward, and might work unfavorably for one's survival. The balance between exploration and exploitation has been investigated in the reinforcement learning paradigm [5], where the trade-off between exploration and exploitation is dealt with from the perspective of optimization. The agent has to exploit what it already knows in order to obtain reward, but it also has to explore in order to make better action selections in the future.

In the developmental process, the psychological safe base provided by caretakers is considered to be a necessary basis for the infant's voluntary exploration of novel stimuli [6][7]. Perception of safe base as a basis for exploration is likely to be relevant also in mature humans.

Shultz and his colleagues revealed that dopamine neurons code uncertainty itself [8]. There was a sustained increase in activity that grew from the onset of the conditioned stimulus to the expected time of reward. The peak of the sustained activation occurred at the time of potential reward, which corresponds to the moment of greatest

N.R. Pal et al. (Eds.): ICONIP 2004, LNCS 3316, pp. 300–305, 2004.
© Springer-Verlag Berlin Heidelberg 2004

uncertainty. These results suggest that dopamine neurons might respond to uncertainty itself and uncertainty could be regarded as the secondary reward. The temporal parameters involved in the learning of action-reward association, e.g. the discount rate, and their correlate in the dopamine system [9][10][11], are expected to be important in the metacognition of uncertainty and related cognitive processes.

Here we investigate the human's ability to handle uncertainty in a robust way by studying how subjects behave in the presence of uncertainty, where reward is constant regardless of the action chosen (flat reward condition). This particular paradigm reveals the internal cortical dynamics involved in judgment under uncertainty separate from the conventional tight coupling with the reward structure.

2 Experimental Settings

12 healthy young adults participated in the experiment. In each game, subjects were asked to increase the resources within 20 trials. The initial resource was 5. The probability of winning a reward was constant (p) in every trial. The given reward was x/p in a win (probability =p) and 0 in a loss (probability = 1-p), where x was the amount of betting. The net gain was therefore always 0. The subjects repeated 30 games under one condition. In the experiments reported here, p=0.25.

The amount of betting was restricted to 1 (bet) or 0 (escape). The subjects were given information on the present resources, trial number, and the probability of winning. The subjects made a choice after an interval of 5 seconds. The outcome was presented as "Win", "Lose" or "Escape". The game was over when the resource became 0.

3 Results

We examined how the last outcome influenced the subject's betting behavior. From the objective point of view, there is no reason to assume differential betting behavior depending on the previous outcome (Win, Lose, or Escape) since the winning probability is independent from the previous results.

However, we found differential betting behavior in the subjects (Fig. 1 and 2). In particular, the difference between the betting ratio for the (previously) "Lose" and (previously) "Win" conditions was found to be statistically significant. There is no way to explain this behavior in terms of optimizing objective reward.

In the reinforcement learning paradigm, temporal parameters such as discount rate is important in the temporal coordination of learning, which likely reflect neural mechanisms including the dopamine system. We investigated how the influence of a particular outcome is discounted with the progress of time (Fig. 3). The statistically significant effect of a "Win" result leading to less betting choice (i.e. more escape choice) is observed to decay within a few trials.

4 Discussion

Our results suggest that the dynamics of brain's internal reward system cannot be described by a tight coupling with external reward structure alone, but has a rich internal dynamics of its own.

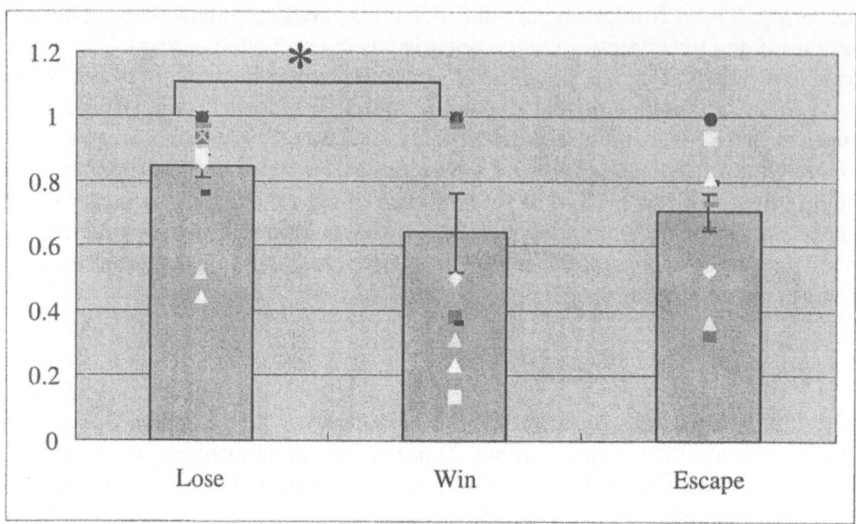

Fig. 1. Influence of previous outcome on present betting behavior. Significant difference is observed between "lose" and "win" (p = 0.022 < 0.05). Each dot represents an individual average and bars are the average of N = 12. Note the considerable difference between individuals

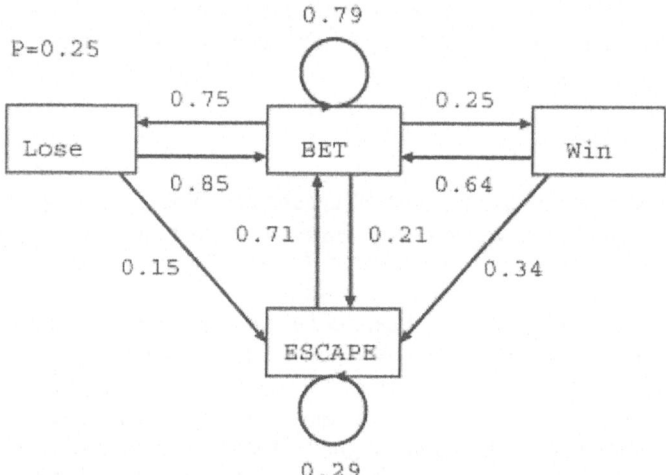

Fig. 2. Finite state transition diagram representation of betting behavior. The numbers represent the probability of transition. N=12

The smaller probability to bet after a "Win" result in Fig.1 and 2 might reflect the dynamics of brain's internal reward system. For instance, a "Win" result might lead to a higher activity of reward related neurons, thus resulting in a smaller tendency to take a risk. Informal interviews with the subjects after the experiment suggested that they were in general unaware of the fact that they were behaving differentially depending on the previous outcome. The reward system involved in the differential behavior seems to be functioning for the most part unconsciously.

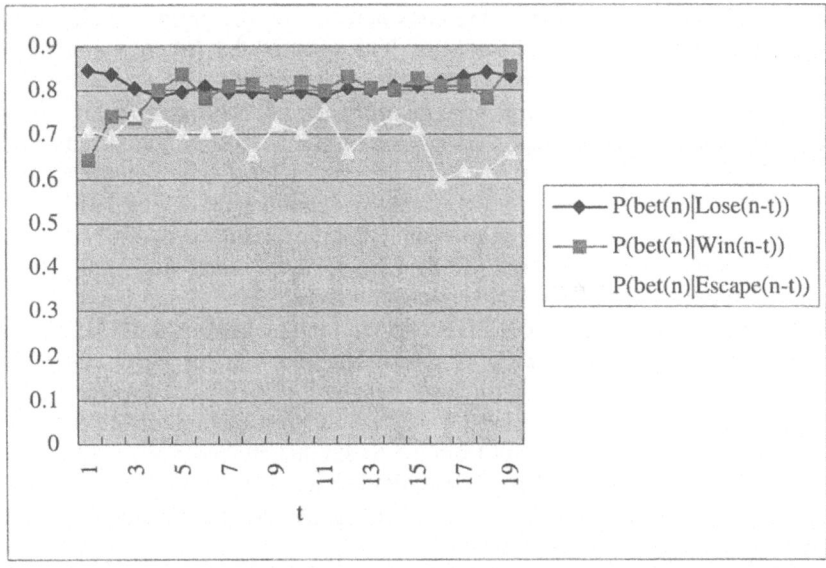

Fig. 3. The discount factor in the effect of previous outcome. Here we plot $P(\text{bet}(n)|\text{Lose}(n\text{-}t))$, $P(\text{bet}(n)|\text{Win}(n\text{-}t))$ and $P(\text{bet}(n)|\text{Escape}(n\text{-}t))$, where n is the present trial and t is the referred number of the trial

Similar results have already been suggested in the behavioral economics. Kahneman and his colleagues studied on the introspective value of lotteries [12][13]. In their experiment, the subjects had to choose between lottery A and B. In A, there is a sure loss of $750. In B, there is a 75% chance to lose $1000 and a 25% chance to lose nothing. It was found that although both lotteries had an identical expected value, a clear majority of respondents preferred B (13% of the subjects chose A and 87% chose B). This result shows that there is a risk seeking preference on this kind of negative choice. They obtained a hypothetical value function by investigated people's decision. The value function is (a) defined on gains and losses rather than on total wealth, (b) concave in the domain of gains and convex in the domain of losses, and (c) considerably steeper for losses than for gains. And Sanfey and his colleagues also showed that people changed their behavior depending on the partner (a computer or a human) in the ultimatum game because of their feeling of unfairness [14].

The differential betting behavior found in our "flat reward condition" suggests the existence of internal reward and evaluation process influenced by, but ultimately independent of, external reward structure. Among the factors possibly involved in these processes are (1) perceived safe base, (2) memory of the result of recent betting, (3) perception of the probability of winning.

Milinski and Wedekind found that the constraints on the working memory of players affect the strategy that the players employ in it iterated prisoner's dilemma [15]. In a prisoner's dilemma game, the subject can take alternative strategies, e.g. Tit-For-Tat, and Pavlovian. The outcome of each round of betting is a function only of the probability and is independent of the result of past betting. However, the memory of the result of past betting, in particular of those in the recent past, might affect the player's decision of how much to bet.

We found that the influence of previous outcome on the present betting behavior decays within a few trials (~10 seconds). It is possible that emotion related neural processing take a certain amount of time to be executed. Rather than realizing a simple stimulus-response relation, in emotional processing the brain might attempt to make a reasonable judgment based on the perceived uncertainty, perceived safe base, perceived nature of reward, etc. Since the decision making based on these elements likely involves complex computation, finally culminating in a winner-take-all type neural mechanism, it is reasonable to assume that processing involving emotion systems take a certain amount of time. This nature of "deep" computation might explain the particular aspect of temporal development observed here.

In the current betting game, constraints on working memory is most likely to affect the player's memory of the result of recent betting. Since the expected reward in this game is constant regardless of the subject's behavior, differential behavior on the part of the subject cannot be explained on the basis of reward optimization. Thus, the current game condition is different from the traditional studied games, e.g. prisoner's dilemma, altruism game, etc. The differential behavior of the subject can only arise from the internal reward modulations, reflecting the brain mechanism for making decisions in an uncertain environment.

Finally, there was a considerable difference of betting behavior among subjects. Such heterogeneity of strategy is typically observed in gaming under the presence of uncertainty, and might reflect a general tendency of the neural system involved in the robust handling of uncertainty.

Acknowledgements

We thank Ruggiero Cavollo, Hisayuki Nagashima, Kei Omata, Takayasu Sekine, Tamami Sudo, Fumiko Tanabe, Fumihiko Taya, Takanori Yokoyama, Toru Yanagawa, and Zhang Qi for their helpful discussions.

References

1. Graham G, Neisser J. Probing for relevance: what metacognition tells us about the power of consciousness. Conscious Cogn. (2000) Jun; 9 (2 Pt 1) 172-7
2. Griffiths D, Dickinson A, Clayton N. Episodic memory: what can animals remember about their past? Trends Cogn Sci. (1999) Feb; 3 (2): 74-80
3. Hampton RR. Resus monkeys know when they remember. PNAS. (2001); 98 (9); 5359-62
4. Koriat A. The feeling of knowing: some metatheoretical implications for consciousness and control. Conscious Cogn. (2000) Jun; 9 (2 Pt 1): 149-71
5. Sutton RS, Barto AG. Reinforcement learning. MIT Press (1998)
6. Ainsworth MD. Object relations, dependency, and attachment: a theoretical review of the infant-mother relationship. Child Dev. (1969) Dec; 40 (4): 969-1025
7. Bowlby J. Attachment. Perseus Books (1982)
8. Fiorillo CD, Tobler PN, Schultz W. Discrete coding of reward probability and uncertainty by dopamine neurons. Science. (2003) Mar 21; 299 (5614): 1898-902
9. Schultz W. Multiple reward signals in the brain.Nat Rev Neurosci. (2000) Dec;1 (3): 199-207. Review.
10. Schultz W, Dayan P, Montague PR. A neural substrate of prediction and reward. Science. (1997) Mar 14; 275 (5306): 1593-9

11. Shizgal P, Arvanitogiannis A. Neuroscience. Gambling on dopamine. Science. (2003) Mar 21;299(5614):1856-8.
12. Kahneman D, Slovic P, Tversky A. Judgment under uncertainty. Cambridge University Press (1982)
13. Kahneman D, Tversky A. Choices, Values, and Flames. Cambridge University Press 2000
14. Sanfey AG, Rilling JK, Aronson JA, Nystrom LE, Cohen JD. The neural basis of economic decision-making in the Ultimatum Game. Science. (2003) Jun 13;300(5626):1755-8.
15. Milinski, M. & Wedekind, C. Working memory constraints human cooperation in the prisoner's dilemma. Proc Natl Acad Sci U S A. (1998) November 10; 95 (23): 13755-8

The Locus of Word Length and Frequency Effect in Comprehending English Words by Korean-English Bilinguals and Americans*

Kichun Nam[1], Yoonhyong Lee[2], and Chang H. Lee[3]

[1] Department of Psychology, Korea University, Korea
kichun@korea.ac.kr
[2] Department of Psychology, University of North Carolina at Chapel Hill, USA
[3] Department of Psychology, Pusan National University, Korea

Abstract. Three experiments on English word recognition have been conducted in order to investigate the locus of word length effect, as well as the locus of the frequency effect. The other aim of this study was to investigate whether the processing of English word recognition for Koreans is similar to that of Americans' or not in the respect of word length and frequency effects. In all experiments, the interacting pattern between length and frequency was examined. If the interaction is additive, length and frequency affect separate stages. Experiment 1 showed that degradation of stimuli by mixing the case had no effect on word length and frequency. The processing patterns between the naming and the lexical decision (Experiment 2), and between the naming and the delayed naming (Experiment 3) were similar for both word length and frequency effects. In addition, Korean bilinguals showed similar length and frequency effect as American. The locus of length effect is located at the lexical access and a post-lexical stage. Because word frequency and length are two most influential lexical variables in word recognition, this result indicates that Korean bilingual access the lexical system of English as much the similar way as American.

1 Introduction

Word recognition is separated into three processing stages: pre-lexical, lexical access, and post-lexical stage. Pre-lexical stage refer to the information transformation from a visually presented word to the representation form which is contained in mental lexicon and lexical access means verifying words in the mental lexicon. Post-lexical processing is all other processes after lexical access. For the word recognition processing, word frequency and word length have been known to be the main lexical variables in word recognition [1]. The word length effect, which is defined as the low accuracy rate and slower recognition speed in recognizing longer words, occurs during word recognition process. The presence of word length effect has been consistently reported in various studies. For example, Chumbley and Balota reported the word length effect in the lexical decision task, when other lexical variables (e.g., frequency) are controlled [2]. In the study related with English word recognition by the bilingual, Korean, word length effect during English word recognition was also reported [3]. For the question about the locus of the word length effect, several researchers disagree on the locus of the effect. Forster and Chamber argued that since

---

* This work was supported by KRF (KRF-2004-074-HM0004).

N.R. Pal et al. (Eds.): ICONIP 2004, LNCS 3316, pp. 306–315, 2004.
© Springer-Verlag Berlin Heidelberg 2004

both naming and lexical decision task shows the word length effect and the size of the effects are similar in both tasks, a pre-lexical or the lexical access stage is the locus of the word length effect [4]. In contrast, Hudson and Bergman suggested since the word length effect varies according to the degrees of orthographic difficulty, it would affect the orthographic judgment, a post-lexical stage, rather than the word recognition itself [5]. Word frequency has been also shown to be the main lexical variable in word recognition [2], [6]. The study examined the effect of word frequency in recognizing and naming English written words by Korean also showed the word frequency effect [7]. The locus of the frequency effect has been proven to be in the lexical access stage or a post-lexical stage. Traditional models for the word frequency effect insist that its locus occurs only in the lexical access stage, which is related with the identification of letter strings in a word [8]. However, several other studies have agreed on the possibility that word frequency also affects post-lexical stages, such as the mapping of letter strings to a naming code or a meaning code, as well as decision components in various tasks [2], [6].

Recently, studies of word recognition turn their eyes to the bilingualism and multilingualism. The main interests are whether there is one unique system to process every language or there are individual systems for each language and if one language system influences other language system automatically when they want to use a language system [9], [10], [11]. As foreign language can be learned based upon mother language, foreign word recognition occurs with the same way with mother language [12]. However, Katz and Frost, Simpson and Kang insisted that word recognition processing will differ according as the word structure [13], [14]. Considering various researches about bilingual word recognition, whether there is one system for one language or there is one system for two languages highly depends on the structure of the languages. Languages with similar grammar and alphabetic structure, such as many European languages, may have a system closely connected with each language. However, totally different languages, such as Chinese and English, may have individual processing systems. Korean language system is quite different from that of English. As they are different in aspects of orthographic, semantic, morphological and phonological rule, it is hard to believe that Korean and English have one common processing system for bilinguals. In this regard, what kinds of processing can work for the Koreans' English word recognition is focused in this study instead of focusing whether Korean and English are same system or not in Bilinguals.

The main purpose of this study is to find out the locus of word length effect among the pre-lexical, the lexical access, and the post-lexical stage. Another purpose is to differentiate information-processing type of Koreans and Americans in terms of word frequency and length effect, when Koreans recognize English words as a foreign language and Americans as their mother language. If there is any difference between Koreans and Americans in their form of responses, it means Koreans use different information processing system from that of Americans. If there would be a gap in their overall reaction time but not in the patterns of responses, it indicates, related with word frequency and length, that Koreans depend on the equal mechanism as the Americans'. Three experiments were conducted, each manipulation reflecting the different stage of word recognition. In experiment 1, degraded the word by mixing case was tested in order to reflect the pre-lexical stage. In experiment 2, the performance between the naming and the lexical decision was tested in order to investigate the influence of the extra decision component in the lexical decision task. Finally, in experiment 3, performance between the normal naming and the delay naming was compared in order to investigate whether the articulatory execution stage affects the performance or not. With all these experiments, we also investigated the effect of word length in English word processing of Koreans and Americans. As Korean students' English fluency is much worse than that of Americans', Koreans would show longer reaction time than American. However, if their reaction patterns are similar to each other, it means that those two groups process in a similar way. That means the time difference between two groups only indicates the difference of processing speed.

2   Experiment 1

The first experiments were designed to investigate whether word length has effect on the pre-lexical processing in the visual word recognition. We conducted the immediate naming experiment by using the visually degraded words by mixing letter cases (e.g., DegRAde) and by using intact words (e.g., intact). Stimulus inputs in the initial stages of word recognition are transformed to an abstract processing unit, either a phonological unit, or an orthographic unit. Because degrading words would interfere with the transformation, the locus of degradation of words is at a prelexical stage.

If the interacting pattern of the word length and degradation is additive, the two variables are believed to affect the different stage. But if there is an interaction between two variable effects, this pattern would indicate that two effects arise in the same stage according to the Stenberg's logic [15], [16]. In other words, we can assume that the word length have some influences on the initial stage of word recognition, if there is difference of word length effects between the degraded condition and the intact condition.

Method

Subjects: Eighty-four Korean subjects were recruited from Introductory Psychology class in Korea University. All of them were right-handed and had normal vision. Forty students were male and forty-four students were female. Their mean age was 21 years old. Forty-six native English speakers at the University of Texas-Austin participated in order to fulfill their experimental credit. Twenty-three students were male and twenty-three students were female. Their mean age was 21 years old.

Materials Each of the two word lists was composed of 42 English words. In each list, half were 4-letter-words and the others were 7-letter-words. At each length, half were high frequency words and the others were low frequency words. The mean of high frequency words was 141.4, and the mean of low frequency word was 4.75 [17]. In the degraded condition, 2 letters from 4-letter-words were capital letters and 3 letters from 7-letter-words were capital letters, mixing the lower and upper case alternatively. Components of both degraded and intact lists were composed of the words which have the same word frequency. Additional 10 words, half of them were normal and the other half were mixed, were also constructed to use as the practice stimuli.

Procedure and Design: Participants sat in front of the 586 pentium computer with a view distance of 60 cm. All stimuli were presented on the center point of the computer screen until the response. The participants were instructed to read aloud the words as fast as possible without sacrificing accuracy. An experimenter sat beside the participant to record the misread, the pronunciation errors, and the other errors (e.g., hesitation). Halves of the Korean and American subjects participated in the intact condition and the other half participated in the degrade condition. Thus, the experimental design was 2(Group) between and 2(Condition) X 2(Length) X 2(Frequency) within Mixed ANOVA.

Results and Discussion

The words with errors were discarded. The error rate of American for each condition was less than 1%. Korean error rate was 2.85% for intact condition and 5.49% for degrade condition. In American's case, The frequency, length, and their interaction were all statistically significant in the intact condition (length: $F(1,22)=26.5$ MSe=2032.5, P<0.01, frequency: $F(1,22)=18.0$ MSe=1895.2, P<0.01, interaction: $F(1,22)=9.9$ MSe=991.2, P<0.01). In the degrade condition, all are also significant (length: $F(1,22)=27.4$ MSe=1341.428, P<0.01, frequency : $F(1,22)=28.5$ MSe=773.5, P<0.01, interaction : $F(1,22)=17.9$ MSe=743.8, P<0.01). In Korean's case, both results of intact and degrade condition are same to American (Intact; length: $F(1,42)=161.8$ MSe=7539.8, P<0.01, frequency : $F(1,42)=272.5$ MSe=3036.2, P<0.01, interaction : $F(1,42)=20.3$ MSe=2420.1, P<0.01, degrade; length: $F(1,42)=305.5$ MSe=4798.1, P<0.01, frequency : $F(1,42)=105.2$ MSe=6022.1, P<0.01, interaction : $F(1,42)=33.8$ MSe=4892.6, P<0.01). In the different tasks, American showed no differences in all conditions (Length & Task; $F(1,44)=0.4$ MSe=1687.0, frequency & Task; $F(1,44)=0.496$ MSe=1334.404, interaction & Task; $F(1,44)=0.15$ MSe=867.5). Korean showed the significant difference in the interaction condition (Length & Task; $F(1,84)=0.9$ MSe=6168.9, frequency & Task; $F(1,84)=1.4$ MSe=4529.1, interaction & Task; $F(1,84)=4.6$ MSe=3656.4, P<0.05). The results show that word length and frequency effects have significant effects in both degraded and intact condition for Koreans and Americans. Importantly, word length and frequency effects were similar across the degraded and the intact condition, which means the length and the frequency effects do not have their loci in the prelexical stage related with degradation. From the similarity between Korean and American data, we can assume that native Korean and native English speakers do a similar type of processing, at least, related with length and frequency.

Table 1. Mean reaction time(RT) and standard deviation (in milliseconds)

| | American | | | | Korean | | | |
| | INTACT | | DEGRADE | | INTACT | | DEGRADE | |
	HIGH	LOW	HIGH	LOW	HIGH	LOW	HIGH	LOW
LONG	480	540	524	579	885	1058	1101	1285
	(61)	(131)	(74)	(104)	(135)	(163)	(193)	(208)
SHORT	453	470	508	515	751	855	979	1038
	(60)	(68)	(62)	(61)	(82)	(117)	(155)	(188)

3 Experiment 2

The second experiments were designed to investigate whether word length has effect on the lexical access stage. We used the immediate naming task and lexical decision task. Lexical access stage and phonological representation stage are involved in the immediate naming task. The lexical access stage and post-access decision stage are involved in the lexical decision task. According to the study comparing the naming task and the lexical decision task by Monsell, Doyle, and Haggard, if there is no difference in word length effect resulted from those two tasks, we can assume that the word length effect have influenced on the common stages of the two tasks, the lexical access stage [6]. Alternatively, the different performance between the naming and the lexical decision would indicate that the decision components of the lexical decision task or the phonological transformation stage of the naming are related with the word length effect.

Method

Subjects: 86 Korean subjects and 46 native English speakers participated. All of them were from the introductory psychology class as experiment 1 but did not attend to experiment 1. Forty-two students were male, and the rest students were female, and their mean age was 21 years old. Forty-six native English speakers at the University of Texas, Austin were participated in order to fulfill their experimental credit. Twenty-three students were male, and twenty-three students were female, and their mean age was 21 years old.

Materials: 4 kinds of words conditions were used, which were long, short and high, low frequency conditions. Short words were consisted of 4 letters and long words were 7 letters. High frequency words' frequency rate were from 52 to 290 based on the Kucera & Francis frequency book and low frequency words' were 6 [17]. Each word condition was assigned to 4 word lists according to a 2x2 design. Overall, 80 words were used. In case of the lexical decision task, 80 control words were also used.

Procedure and Design: For the naming task, the procedure and the design were same as the experiment 1. In LDT, participants were told to decide whether the presented letter strings are real words or not. They were asked to press 'yes' or 'no' button quickly without sacrificing accuracy.

Results and Discussion

The words with errors were discarded. The error rate of American for each condition was less than 1%. Korean error rate was 2.36% for naming condition and 1.79% for LDT condition. In American's case, The frequency, length, and their interaction were all statistically significant in the naming condition (length: $F(1,22)=30.105$ MSe=812.925, P<0.01, frequency : $F(1,22)=68.074$, MSe=568.258, P<0.01, interaction : $F(1,22)=32.680$ MSe=388.166, P<0.01). In the LDT condition, the frequency and length were statistically significant (length: $F(1,22)=14.050$ MSe=2152.497, P<0.01, frequency : $F(1,22)=32.327$ MSe=3048.384, P<0.01, interaction : $F(1,22)=0.001$ MSe=1669.949). In Korean's case, both results of naming and LDT condition are same to American (Naming; length: $F(1,42)=237.352$ MSe=4536.065, P<0.01, frequency : $F(1,42)=158.874$ MSe=2905.206, P<0.01, interaction : $F(1,42)=76.112$ MSe=1805.287, P<0.01, LDT; length: $F(1,42)=111.006$ MSe=7546.140, P<0.01, frequency : $F(1,42)=98.734$ MSe=8203.941, P<0.01, interaction : $F(1,42)=0.064$ MSe=2995.801). In the difference of the tasks, American showed difference in the frequency and interaction conditions (Length & Task; $F(1,44)=0.103$ MSe=1482.711, frequency & Task; $F(1,44)=3.800$ MSe=6872.284, P=0.058, interaction & Task; $F(1,44)=6.409$ MSe=1034.557 P<0.05). Korean showed the same result with American (Length & Task; $F(1,84)=1.239$ MSe=6041.102, frequency & Task; $F(1,84)=4.381$ MSe=5554.574 P<0.05, interaction & Task; $F(1,84)=30.797$ MSe=2400.544, P<0.05).

Table 2. Mean reaction time (RT) and standard deviation for experiment 2

	American				Korean			
	LDT		NAMING		LDT		NAMING	
	HIGH	LOW	HIGH	LOW	HIGH	LOW	HIGH	LOW
LONG	609	674	482	547	914	1049	884	044
	(94)	(93)	(42)	(71)	(132)	(177)	(125)	(151)
SHORT	572	638	474	491	772	912	783	830
	(91)	(84)	(41)	(44)	(98)	(131)	(101)	(120)

Results from table 2 show that word length and frequency have significant effects on the naming and lexical decision times in Korean and American subjects. In addition, in naming condition, the interaction effects of the two variables were both significant for Koreans and Americans, again confirming the similar type of processing across the two population groups as Experiment 1. Most importantly, related with the main purpose of this experiment, we can see there is no task difference in word length effects but there is significant task difference in word frequency effect for both populations (cf. the task difference of the frequency effect for American approached statistical significance (p = .057)). This means that word length affects the common stage of the naming and the lexical decision, the lexical access stage, and word frequency affects the articulation of naming task, or the lexical decision component of the lexical decision task. The results of this experiment are partly consistent with those of Forster and Chamber [4]. Since they argued that the locus of the word length effect would reside in a pre-lexical, or the lexical access stage, the common patterns of the naming and the lexical decision in this experiment supported their argument.

4 Experiment 3

The third experiments were designed to investigate the word length effect on the post-lexical processing. We used the immediate naming task and the delayed naming task. From the delayed naming task, we can examine production stage that comes after lexical access is related with the word length effect. Specifically, if there is significant word length effect in delayed naming task, word length has locus on the phonological production stage. In contrast, if there is no difference in word length effect between the delayed naming and the immediate naming tasks, the locus of the word length effect is not in the phonological production stage, a post-lexical stage.

Method

Subjects: 84 Korean subjects and 38 native English speakers participated in the experiment.

Materials: Four kinds of words conditions were used, which were long, short and high, low frequency conditions. Short words were consisted of 4 letters and long words were 7 letters. High frequency words' frequency rate were from 55 to 394 in the Kucera & Francis frequency book and low frequency words' were 4 or 5 [17]. Each word condition was assigned to 4 word lists according to a 2x2 design and overall, 84 words were used.

Procedure: All the procedure and the design were same as the experiment 1.

Results and Discussion

The words with errors were discarded. The error rate of American for each condition was less than 1%. Korean error rate was 2.62% for non-delay condition and 1.71% for delay condition. In American's case, the frequency, length, and their interaction were all statistically significant in the non-delay condition (length: $F(1,18)=30.105$ MSe=812.925, P<0.01, frequency : $F(1,18)=68.074$, MSe=568.258, P<0.01, interaction : $F(1,18)=32.680$ MSe=388.166, P<0.01). In the delay condition, the frequency and length were statistically significant (length: $F(1,18)=14.050$ MSe=2152.497, P<0.01, frequency : $F(1,18)=32.327$ MSe=3048.384, P<0.01, interaction : $F(1,18)=0.001$ MSe=1669.949). In Korean's case, both results of delay and non-delay condition are same to American (Non-delay; length: $F(1,42)=305.508$ MSe=4798.118, P<0.01, frequency : $F(1,42)=105.242$ MSe=6022.120, P<0.01, interaction : $F(1,42)=33.870$ MSe=4892.699, P<0.01, Delay; length: $F(1,42)=208.575$ MSe=5684.144, P<0.01, frequency : $F(1,42)=268.718$ MSe=3773.882, P<0.01, interaction : $F(1,42)=2.911$ MSe=2902.744). In the difference of the tasks, American showed no difference in all condition (Length & Task; $F(1,36)=1.717$ MSe=597.851, frequency & Task; $F(1,36)=3.492$ MSe=854.417, interaction & Task; $F(1,36)=0.432$ MSe=460.908). Korean showed difference in the frequency and interaction conditions (Length & Task; $F(1,84)=1.417$ MSe=5241.131, frequency & Task; $F(1,84)=4.332$ MSe=4878.001, P<0.05, interaction & Task; $F(1,84)=9.038$ MSe=3897.721 P<0.05).

Table 3. Mean reaction time (RT) and standard deviation for experiment 3

	American				Korean			
	DELAY		NON-DELAY		DELAY		NON-DELAY	
	HIGH	LOW	HIGH	LOW	HIGH	LOW	HIGH	LOW
LONG	328	345	489	550	536	566	882	1063
	(74)	(93)	(41)	(65)	(116)	(124)	(159)	(200)
SHORT	310	333	470	485	528	553	754	865
	(75)	(90)	(40)	(41)	(115)	(117)	(91)	(152)

The word length and frequency variables affected performance in the normal naming and the delay naming in a similar manner, and their performance was not differentiated across Korean bilinguals and Americans. The result of Experiment 3 means that the length affects execution stage because the length is significant in the delayed naming task. These results are consistent with the results of Experiment 1 and 2, in which length does not have its locus in a pre-lexical stage (Exp. 1), and affects the lexical access stage (Exp. 2). The results of this experiment are in the similar vein as Hudson and Bergman, in the sense that the results showed the locus of the length effect is in a post-lexical access stage [5]. As they argued that the word length effects arise in an orthographic spelling stage, the results showed a locus of the word length effect is in the articulation execution stage, a post-lexical stage. The similar pattern of processing for Korean and American in this experiment provided an converging evidence on the argument made in the Experiment 1 and 2 that word frequency and length do not differentiate the two populations. Because word frequency and length are two most influential lexical variables in word recognition, this result indicates that Korean bilingual access the lexical system of English as much the similar way as American.

5 General Discussion

Three experiments in this study have investigated the locus of word length effect by segregating the stages of lexical processing into the three stages, pre-lexical, lexical, and post-lexical stage. In addition, the comparison between American and Korean bilinguals has been done in order to investigate how Korean bilinguals process their second language, English. The interpretation of the experiments in this study has been conducted based on the logic of Sternberg [15]. For native English speakers, Experiment 1 showed that the stimulus degradation has not eliminated the word length effect, indicating that the locus of word length effect is not in a pre-lexical stage, the initial perceptual stage. Experiment 2 showed that the degrees of word length effects were similar across the naming and the lexical decision task. It indicates that the decision components of the lexical decision task would not be related with the locus of word length effects. Instead, the common component of the naming and the lexical decision, the lexical access might be a locus of the word length effect. Finally, experiment 3 showed that the delay naming still produced significant length effects. One of the post-lexical stages, the articulation execution stage is seemed to be a locus of the word length effects.

Another main purpose of this study is to find out how the frequency effect is differentiated from the word length effect and how Korean bilinguals process English. The results clearly indicated that the pattern of the frequency effect in the three experiments was similar to that of the word length effect. The loci of the frequency effect and the length effect seem to be in a same stage of word recognition. In addition, the patterns of the processing between American and Korean bilinguals were also similar each other. The word length effect and frequency effect do not differentiate the two population, indicating English word recognition for Korean bilinguals share some common components with American. The results of Experiment 3 indicate that the locus of word length effect would be in a post-lexical stage of the articulation execution. Retrieving long words from working memory buffer would be more time consuming than retrieving short words from the buffer. Alternatively, the rehearsal components of working memory system would produce such results. Specifically, rehearsing long words in the delay duration would be more demanding rehearsing short words. The duration of delay might not be enough to rehearse the long words in a complete manner.

In the results of word naming task and lexical decision task, Americans responded faster in the naming task than in the lexical decision task, but Koreans were not. It can be interpreted that although Koreans process English words in the similar way to Americans, Koreans feel difficulty in word naming. For the English education for Koreans, these results give a suggestion that making sounds and articulation should be emphasized instead of just memorizing the meaning of words. All the experiments, there are consistency that Korean and English speakers do the same type of processing.

References

1. Whaley, G. P. (1978). Word-nonword classification time. *Journal of Verbal Learning and Verbal Behavior, 17*, 143-154.
2. Chumbley, J. I., & Balota, D. A. (1984). "A word's meaning affects the decision in lexical decision." *Memory & Cognition, 12,* 590-606.
3. Lee, H., Nam, K., & Kim, H. (1996). ERP 에 나타난 한글, 한자, 영어 단어 재인의 차이. *한국 인지과학회,* 7, 111-140.
4. Forster, K. I., & Chambers, S .M. (1973). "Lexical access and naming time." *Journal of Verbal Learning and Verbal Behavior, 12,* 627-635.
5. Hudson, P. T. W., & Bergman, M. W. (1985). "Lexical knowledge in word recognition: Word length and word frequency in naming and lexical decision tasks." *Journal of Memory and Language, 24,* 46-58.
6. Monsell, S., Doyle, M. C., & Haggard, P. N. (1989) Effect of frequency on visual word recognition tasks: Where are they? *Journal of Experimental Psychology: Learning, Memory, and Cognition, 18,* 452-467.
7. Lee, Y., Lee, J., Hwang, Y. Chung. Y., & Nam, K.. (2000). Word regularity effect in English word recognition of Korean. *Korean Journal of foreign language education, 7(1),* 25-44.
8. McClelland, J. L., & Rumelhart, D. E. (1981). An interactive activation model of context effects in letter perception: Part 1. An account of basic findings. *Psychological Review, 88,* 375-407.
9. Beauvillain, C. (1992). Orthographic and lexical constraints in bilingual word recognition. In R.J. Harris (Ed.), *Cognitive Processing in bilinguals, 221-235.* Amsterdam: Elsevier Science Publishers B.V.
10. Heuven, W., Dijkstra, T., & Grainger, J. (1998). Orthographic neighborhood effects in bilingual word recognition. *Journal of Memory and Language, 39,* 458-483.
11. Kroll, J. F. (1993). Accessing conceptual representations words in a second language. In R. Schreuder & B. Weltens (Ed.), *The bilingual lexicon.* 53-81. John Benjamins Publishing Company: Amsterdam/Philadelphia.
12. De Groot, A. M. B. (1993). Word Type Effects in Bilingual Processing Tasks: Support for a Mixed-Representational System. In Robert Schreuder and Bert Weltens(Eds). *The Bilingual Lexicon.* (pp. 27-51). John Benjamins Publishing Company, Amsterdam/Philadelphia.
13. Katz, L., & Frost, R. (1992). The reading process is different for different orthographies: The orthographic depth hypothesis. In R. Frost and L. Katz (Eds.), *Orthography, Phonology, Morphology and Meaning* (pp. 67-84). Northholland: Elsevier Science Publishers B.V.

14. Simpson, G. B., & Kang, H. (1994). The flexible use of phonological information in word recognition. *Journal of Memory and Language, 33,* 319-331.
15. Seidenberg, S. (1969). The discovery of processing stages: Extensions of Donders' method. In W. G. Koster (Ed.), *Attention and Performance II.* Amsterdam: North Holland.
16. Besner, D., & Smith, M. C. (1992). Models of visual word recognition: When obscuring the stimulus yields a clearer view. *Journal of Experimental Psychology: Learning, Memory, and Cognition. 18,* 3, 468-482
17. Kucera, H., & Francis, W. N. (1967). *"Computational analysis of present-day American English."* Brown university press. Providence, Rhode Island.

Cerebral Activation Areas with Respect to Word and Sentence Production by Early and Late Korean-English Bilinguals: Event-Related fMRI Study*

Choong-Myung Kim, Donghoon Lee, and Kichun Nam

Department of Psychology, Korea University, Korea
czykim@korea.ac.kr

Abstract. This study was conducted to investigate the cerebral areas related with word and sentence production shown in event-related fMRI. Specifically the current study has an intention to enlighten the difference between native and foreign language processing by early and late Korean-English bilinguals. Two experiments were performed to confirm the areas related with each level of word and sentence generation. The experimental tasks comprises picture naming and sentence production tasks in which subjects were asked to name the picture and to produce a sentence to describe the picture in Korean or in English. While performing the task, event-related activation areas were confirmed. The results showed that inferior frontal gyrus (IFG; BA 44, 45) activated in both early and late Korean-English bilinguals. However the activation areas were reduced in English-presented condition compared to Korean-presented condition as a native language. Additionally as compared with the activation areas in English-presented condition as a foreign language, the late Korean-English bilinguals have more activation than the early. Such results suggested that the different acquisition time of foreign language could result in the different cerebral activation.

1 Introduction

The processing of language production in human is complex cognitive processes underlying from lexical conceptualizing and syntactic planning to word-form selection and phonological encoding [1]. Many studies on language comprehension and production in bilinguals were rapidly progressed along with the development of the brain imaging technique such as PET, fMRI, MEG and high-resolution EEG.

Of those, fMRI is often used to find a region of interest (ROI) related with a specific cognitive processing. However, recently many studies measuring on-line processing response usually adopt so-called event-related fMRI design. More specifically the event-related experiments paired with a randomized design confirms temporal variation following stimulus-related effect as a result of rapid mental processing such as language comprehension or production.

Kim et al. reported that in the case of post-adolescent bilinguals, late learner of foreign language, centers of activation at Broca's area between native and foreign

---

* This Work was Supported by the Korea Ministry of Science & Technology (M1041300000 8-04N-1300-00811).

N.R. Pal et al. (Eds.): ICONIP 2004, LNCS 3316, pp. 316–320, 2004.
© Springer-Verlag Berlin Heidelberg 2004

language is at a short distance [2]. But it is not the case in pre-adolescent bilinguals exposed at early stage of language acquisition. In contrast, Hernandez et al. failed to investigate the cerebral activation difference between the two language processing conditions in the task of picture naming [3].

On the other hand, Pu et al. measured differential hemodynamic response in middle frontal gyrus and inferior frontal gyrus during verb generation task using event-related fMRI [4]. They also indicated that the hemodynamic difference between the two areas did not occur, and insisted that neural system associated with the two languages are very similar despite late bilinguals. Other than the studies noted above, there is a report that grammatical processing of foreign language by early bilinguals corresponds to a procedural memory, whereas that of foreign language by late bilinguals equivalent to a declarative memory [5].

Most studies to date have focused on the processing in language comprehension excluding language production associated with on-line processing at word and sentence level. Furthermore, the comparative studies dealing with language production were under a bias toward the same cognate. With regard to these points, this study tried to demonstrate the differential areas responsible for grammatical encoding and word retrieval which varied in language family and its time of acquisition.

2 Methods

Participants

A total of 16 participants with a consistent right-handedness from KAIST were recruited by bulletin board on the campus. They were divided into 2 groups with 8 bilinguals in each of early and late learners according to the age of foreign language acquisition. All the early bilinguals had lived at least two years in English country and their proficiency was superior to that of late bilinguals, as assessed by TOEFL. Prior to the image acquisition age of language learning, period of exposure, and degree of fluency were examined as well. All the volunteers had normal or corrected to normal vision and no subject had a known history of neurological or psychiatric disorder. Written informed consent was obtained from all participants.

Image Acquisition and Data Analysis

Subjects lay supine in a ISOL 3.0T forte, their heads being secured to minimize movement artifacts. During fMRI scanning stimuli were presented visually on a computer screen.

A total of 20 parallel axial slices (thickness = 5mm no gap) were acquired across the complete brain volume by means of a EPI-BOLD (Echo Planner Imaging-Blood Oxygen Level Dependent) sequence (TE/TR= 35/3000 ms, flip angle= 80°, FOV=220mm, 64 x 64matrix). Total volume image in each session was obtained 204 phases for 612 seconds. MRI data were processed by the SPM99 software, which allows for realignment of functional images, coregistration with structural images, spatial normalization, smoothing with FWHM 7.5mm Gaussian filter, and statistical inference (z-value cut-off, uncorrected $p < 0.001$). The brain region specified in this paper were found by the Talairach-Tournoux system from the MNI coordinates.

Experimental Materials and Procedure

The materials of the experiment consisted of the picture naming task and sentence production task. The former was the naming task presented with the line drawing object and the latter was the production task using the combination given the pictures. Subjects were asked to respond to the stimuli in a native and foreign language, and the conditions were counter-balanced according to the order of participation.

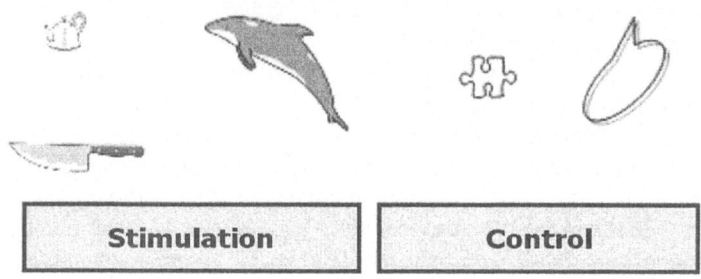

Fig. 1. An example of stimulus used in the experiments

The stimuli are arranged with 2 directions, from top to bottom or from left to right and their size are different in each condition such as the one presented in Fig. 1. Subjects was to produce naming larger picture before smaller one at the first task and to produce the sentence whose subject should be expressed as the larger picture with smaller one included as a component at the second task. Total 240 pictures were used for activation condition of word and sentence production tasks. Meanwhile meaningless pictures of same numbers were presented as control condition to remove the specific brain activation following early visual information and articulation for naming or production. When presented with controlled stimuli, subjects were required to produce and repeat already given babbles. All experimental tasks were conducted to keep utterance covert so as to minimize the movement in acquiring the image.

3   Results and Discussion

Cerebral activation areas responding to the task specified with naming tasks and with producing a sentence in Korean-English bilinguals are provided in Fig. 2 and Fig. 3. Korean and English word generation processes revealed by picture naming tasks showed cortical activation pattern on IFG (BA 44, 45), superior parietal gyrus and occipital cortex (BA 19) in early bilinguals (Fig. 2).

During the Korean and English sentence generation tasks in each group, the activation pattern was similar to word generation tasks, however Korean sentence production tasks produced more activation areas on IFG (BA 44, 45) than Korean word generation. This finding was not consistently observed in English tasks. These experimental results according to the late Korean-English bilinguals are shown well in Fig. 3. Specifically the activation size at left inferior frontal gyrus in picture naming tasks was smaller than in sentence production task using native Korean language. However the difference in activation surrounding IFG between the tasks was not shown in early Korean-English bilinguals.

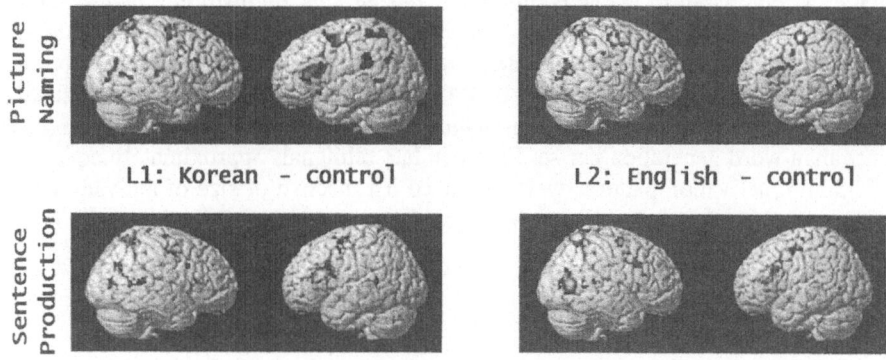

Early bilinguals

Fig. 2. Brain areas activated by early Korean-English bilinguals in each task

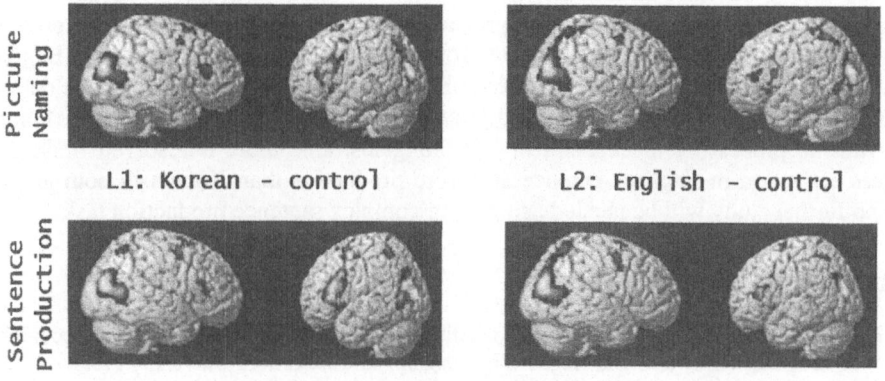

Late bilinguals

Fig. 3. Brain areas activated by late Korean-English bilinguals in each task

Recently a study related to syntactic processing in language comprehension report recruiting dorsal prefrontal cortex (DPFC) [6]. As also shown in this study, the extended DPFC activation together with IFG presumably suggests that additional syntactic processes following sentence production are supported by these areas in conditions responding as Korean language. On the other hand the extensive activation was not investigated in English response condition implying syntactic processing at the stage of foreign language production in late bilinguals was not occurred immediately. This result are also connected with the fact that foreign sentence production on current study needs additional precentral gyrus (BA 6) or anterior cingulate gyrus related with task difficulty and attention instead of DPFC or posterior superior temporal gyrus [7], [8].

In summary, in terms of cross-linguistic view the activation areas in English-presented condition were smaller than in Korean-presented condition as a native lan-

320 Choong-Myung Kim, Donghoon Lee, and Kichun Nam

guage. As compared to the activation areas related with English-presented condition as foreign language stimuli, the late Korean-English bilinguals showed far more activation than the early ones. And in inter-linguistic viewpoint, Korean sentence production tasks produced more extension of IFG activation than Korean word production. This seems that sentence production includes more processes related to building sentence than word generation per se. Lastly in late bilinguals approximately equal contribution of activation patterns were produced irrespective of size of activation areas as compared to the early bilinguals in word and sentence production tasks in two languages.

4 Conclusions

We were able to show that, during language production, word and sentence generation processing of Korean and English in early and late bilinguals activated very similar and overlapping brain regions, inferior frontal, superior parietal and occipital cortex. Particularly extension of activation in inferior frontal gyrus to dorsal prefrontal area during sentence production than words by Korean was observed and this is probably due to syntactic processing related with sentence comprehension or production. However neither of the bilingual groups was consistently represented in English word and sentence production. These results are compatible with the available data in language comprehension domain and provide some further evidence on retardation of syntactic processing mechanism in late bilinguals. One more unresolved finding is less activation in English sentence and word production than Korean in both groups and further study will be needed using more complex sentence production task.

References

1. Levelt, W. J. M. (1999). Language production: a blueprint of the speaker, in Neurocognition of Language (Brown, C and Hagoort, P., eds). pp. 83-122. Oxford University Press
2. Kim, K. H. S., Relkin, N. R., Lee, K. M., & Hirsch, J. (1997). Distinct cortical areas associated with native and second languages. Nature, 388, 171-174
3. Hernandez, A. E., Martinez, A., & Kohnert, K. (2000). In search of the language switch: An fMRI study of picture naming in Spanish-English bilinguals. Brain and Language. 73. 421-431
4. Pu, Y., Liu, H. L., Spinks, J. A., Mahankali, S., Xiong, J., Feng, C. M., Tan, L. H., Fox, P. T., Gao, J. H. (2001). Cerebral hemodynamic response in Chinese (first) and English (second) language processing revealed by event-related functional MRI. Magnetic Resonance Imaging 19, 643-647.
5. Ulman, M., Corkin, S., Coppola, M., Hickok, G., Growdon, J. H., Koroshetz, W. J., & Pinker, S. (1997). A neural dissociation within language: Evidence that mental dictionary is part of declarative memory and that grammatical rules are processed by the procedural system. Journal of Cognitive Neuroscience. 9. 266-276
6. Hashimoto, R. & Sakai, K. L. (2002). Specialization in the left prefrontal cortex for sentence comprehension. Neuron, 35, 589-597
7. Bush, G., Luu, P., and Posner, M. (2000). Cognitive and emotional influences in anterior cingulated cortex. Trends in Cognitive Science, 4, 215-222.
8. Kim, Choong-Myung (2003). Syntactic and semantic integration processes during Korean sentence comprehension: using ERPs as an index of neurophysiology. Doctoral dissertation, Seoul National University, Korea.

Fusion of Dimension Reduction Methods and Application to Face Recognition

Byungjun Son, Sungsoo Yoon, and Yillbyung Lee

Division of Computer and Information Engineering, Yonsei University
134 Shinchon-dong, Seodaemoon-gu, Seoul 120-749, Korea
{sonjun,ssyoon,yblee}@csai.yonsei.ac.kr

Abstract. As dimensionality reduction is an important problem in pattern recognition, it is necessary to reduce the dimensionality of the feature space for efficient face recognition. In this paper, we suggest the fusion of Discrete Wavelet Transform(DWT) and Direct Linear Discriminant Analysis (DLDA) for the efficient dimension reduction. The Support Vector Machines (SVM) and nearest mean classifier (NM) approaches are applied to compare the similarity between the similar and different face data. In the experiments, we show that the proposed method is an efficient way of representing face patterns as well as reducing dimension of multidimensional feature.

1 Introduction

In a recognition system using the biometric features, one may try to use large feature set to enhance the recognition performance. However, the increase in the number of the biometric features has caused other problems. For example, the recognizer using higher dimension feature set requires more parameters to characterize the classifier and requires more storage. Thus, it will increase the complexity of computation and make its real-time implementation more difficult and costly. Furthermore, a larger amount of data is needed for training. To avoid these problems, a number of dimensionality reduction algorithms have already been proposed to obtain compact feature set.

The feature extraction process needs to be effective so that salient features that can differentiate between various classes can be extracted from the face images [1]. Several methods like Principal Component Analysis (PCA) [2] [3], transform the input data so that the features are well separated and classification become easier. Linear transformations are extensively used because they are easy to compute and analytically tractable. Typically, these transformations involve projecting features from a high dimensional space to a lower dimensional space where they are well separated. Linear Discriminant Analysis(LDA) is one such discriminative technique based on Fisher's linear discriminant [4]. Linear Discriminant Analysis (LDA) and Principal Component Analysis (PCA) are two major methods used to extract new features [4]. However, one hopes to obtain new dimensionality reduction method for extracting features with more discriminative power. In this paper, we propose the fusion of wavelet and direct lin-

N.R. Pal et al. (Eds.): ICONIP 2004, LNCS 3316, pp. 321–326, 2004.
© Springer-Verlag Berlin Heidelberg 2004

ear discriminant analysis(DLDA) [5] for reducing dimension of high-dimensional face image.

This paper is organized as follows. In section 2, we overview a multilevel two-dimensional discrete wavelet transform (DWT). Also, we describe the DLDA scheme to linearly transform the subimages of face image obtained by wavelet transform to new feature space with higher separability and lower dimensionality in section 3. Section 4 describes feature matching approach based on SVM. Experimental results and analysis will be stated in section 5, and finally the conclusions are given in section 6.

2   Wavelet Transform

The hierarchical wavelet functions and its associated scaling functions are to decompose the original signal or image into different subbands. The decomposition process is recursively applied to the subbands to generate the next level of the hierarchy. The traditional pyramid-structured wavelet transform decomposes a signal into a set of frequency channels that have narrower bandwidths in the lower frequency region. The DWT was applied for texture classification and image compression because of its powerful capability for multiresolution decomposition analysis. The wavelet decomposition technique can be used to extract the intrinsic features for the recognition of persons by their biometric data. We employ the multilevel 2D Daubechies wavelet transform to extract the face features. Using the wavelet transform, we decompose the image data into four subimages via the high-pass and low-pass filtering with respect to the column vectors and the row vectors of array pixels. Fig. 1 shows the process of pyramid-structured wavelet decomposition.

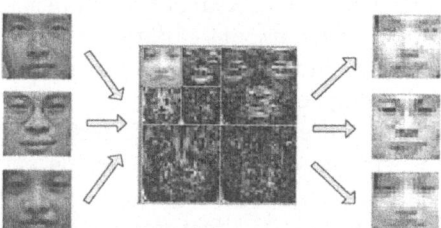

Fig. 1. Example of a 2-level wavelet transform of the face images.

In this paper, we use the statistical features and the two-level lowest frequency subimage to represent robust feature vectors as the initial dimension reduction step, thus statistical features were computed from each subband image. First, we divide the subimages into local windows in order to get robust feature sets against shift and noisy environment (Fig. 2).

Next, we extract first-order statistics features, that is, mean and standard deviation from local windows on the corresponding subimages to represent feature vectors. Generally, the mean extracts low spatial-frequency features and the

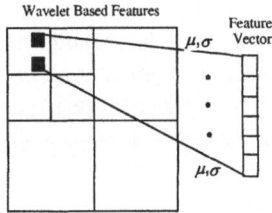

Fig. 2. Arrangement of feature vectors by local windows.

standard deviation can be used to measure local activity in the amplitudes from the local windows. Also, low frequency components represent the basic figure of an image, which is less sensitive to varying images. These feature vectors include both local and global information. The level of low frequency subimage chosen to extract the feature vector depends on size of the image.

3   Direct Linear Discriminant Analysis

After extraction of the face feature vector by wavelet transform, the original face vector x of 3,600 dimensions is transformed to the feature vector y of 241 dimensions. To further reduce the feature dimensionality and enhance the class discrimination, we apply DLDA to convert the feature vector y into a new discriminant vector z with lower dimensions then the feature vector y.

Existing LDA methods first use PCA to project the data into lower dimensions, and then use LDA to project the data into an even lower dimension [6]. The PCA step, however, can remove those components that are useful for discrimination. The key idea of DLDA method is to discard the null space of between-class scatter S_b - which contains no useful information - rather than discarding the null space of S_w , which contains the most discriminative information [5]. Each scatter is given as follows:

$$S_b = \sum_{i=1}^{J} n_i(\mu_i - \mu)(\mu_i - \mu)^T \qquad (n \times n)$$

$$S_w = \sum_{i=1}^{J} \sum_{x \in C_i} (x - \mu_i)(x - \mu_i)^T \qquad (n \times n)$$

where n_i is the number of class i feature vectors, μ_i is the mean of class i, μ is the global mean, and J is the number of classes.

The DLDA method is outlined below. We do not need to worry about the computational difficulty that both scatter matrices are too big to be held in memory because the dimensionality of input data is properly reduced by wavelet transform.

First, we diagonalize the S_b matrix by finding a matrix V such that

$$V^T S_b V = D$$

where the columns of V are the eigenvectors of S_b and D is a diagonal matrix that contains the eigenvalues of S_b in decreasing order. It is necessary to discard eigenvalues with 0 value and their eigenvectors, as projection directions with a total scatter of 0 do not carry any discriminative power at all [5].

Let Y be the first m columns of V (an $n \times m$ matrix, n being the feature space dimensionality),

$$Y^T S_b Y = D_b \qquad (m \times m)$$

where D_b contains the m non-zero eigenvalues of S_b in decreasing order and the columns of Y contain the corresponding eigenvectors.

The next step is to let $Z = Y D^{1/2}$ such that $Z^T S_b Z = I$. Then we diagonalize the matrix $Z^T S_w Z$ such that

$$U^T (Z^T S_w Z) U = D_w \qquad (1)$$

where $U^T U = I$. D_w may contain zeros in its diagonal. We can sort the diagonal elements of D_w and discard some eigenvalues in the high end, together with the corresponding eigenvectors.

We compute the LDA matrix as

$$A = U^T Z^T \qquad (2)$$

Note that A diagonalizes the numerator and denominator in Fisher's criterion.

Finally, we compute the transformation matrix(2) that takes an $n \times 1$ feature vector and transforms it to an $m \times 1$ feature vector.

$$x_{reduced} = D_b^{-1/2} A x \qquad (3)$$

4   SVM-Based Pattern Matching

We only give here a brief presentation of the basic concepts needed. The reader is referred to [7] for a list of applications of SVMs. SVMs are based on structural risk minimization, which is the expectation of the test error for the trained machine. This risk is represented as $R(\alpha)$, α being the parameters of the trained machine. Let β be the number of training patterns and $0 \leq \gamma \leq 1$. Then, with probability $1 - \gamma$ the following bound on the expected risk holds:

$$R(\alpha) \leq R_{emp}(\alpha) + \sqrt{\frac{h(\log(2\beta/h) + 1) - \log(\gamma/4)}{\beta}} \qquad (4)$$

$R_{emp}(\alpha)$ being the empirical risk, which is the mean error on the training set, and γ is the VC dimension. SVMs try to minimize the second term of (4), for a fixed empirical risk.

For the linearly separable case, SVMs provides the optimal hyperplane that separates the training patterns. The optimal hyperplane maximizes the sum of the distances to the closest positive and negative training patterns. This sum is called *margin*. In order to weight the cost of missclassification an additional parameter is introduced. For the non-linear case, the training patterns are mapped onto a high-dimensional space using a kernel function. In this space the decision boundary is linear. The most commonly used kernel functions are polynominals, gaussian, sigmoidal functions.

5   Experimental Results

We used face images from the IIS face database accessible at http://smart.iis. sinica.edu.tw/ [8]. We sampled frontal face images of 100 people from the IIS face database, each person having 10 images with varying expressions. The size of each image is 92×104 pixels, with 256 grey levels per pixel.

We randomly choose five images per person for training, the other five for testing. To reduce variation, each experiment is repeated at least 20 times. We applied LDA, DWT +LDA, DLDA, and DWT +DLDA to a training set. Also, we evaluated the recognition performances using mean classifier(NM) and SVM.

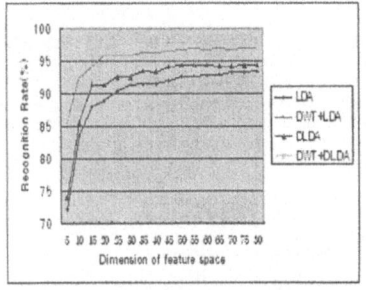
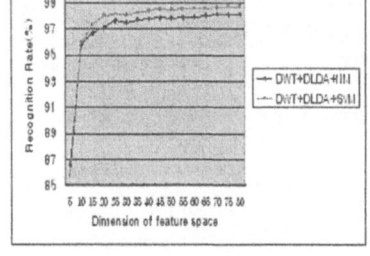

(a) Nearest Mean Classifier   (b) SVM Classifier

Fig. 3. Recognition Rate vs. Dimension of feature space.

Fig. 3 (a) shows the result of recognition rate vs. dimension of feature space. The greatest recognition rate of the LDA approach about the face data is 93.33% with feature vector consisted of 80 components. The DLDA approach about the face achieves 94.35% recognition rate with feature vector consisted of 50 components. The DWT+LDA and DWT+DLDA approach about the face achieve 96.89% and 98.1% with 55 and 75 features, respectively. From Fig. 3 (a), we can also see the DWT+DLDA method achieves the highest recognition rate. In addition, recognition rate of the DWT+DLDA approach is 97.67% when the number of features is 25. It is higher than the best performance of the other methods and has lower dimension than others. As compared in Fig. 3 (b), we also find that DWT+DLDA +SVM for Gaussian kernel($\sigma = 8$) outperforms DWT+DLDA+NM and the best recognition rate of DWT+DLDA+SVM approach is 98.71% when the number of features is 75. This shows that the DWT+DLDA+SVM approach is more efficient method to increase the discriminating power and reduce the feature dimension than the others.

6 Conclusion

In this paper, we have presented effective dimension reduction method for the face recognition. We specifically use the first-order statistics features and two-level lowest frequency subimage obtained by the multiresolution decomposition of 2-D discrete wavelet transform for extracting the robust feature set of low dimensionality. In addition, the DLDA method is used to obtain the feature set with higher discriminative power and lower dimensionality. These methods of feature extraction well suit with face recognition system while allowing the algorithm to be translation and rotation invariant. The experimental results illustrate that the fusion of Discrete Wavelet Transform(DWT) and Direct Linear Discriminant Analysis (DLDA) achieves an effective lower dimensional representation of the high dimensional input data. In addition, when it is used with the SVM, it has the advantage of efficient testing, and good performance compared to other linear classifier.

Acknowledgements

This work was supported in part by the Brain Neuroinformatics Program sponsored by KMST.

References

1. H. Watanabe et al: " Discriminative Metric Design for Robust Pattern Recognition", IEEE Trans. On Signal Processing, vol. 45, PP. 2655-2662, 1997.
2. Ian T. Jollife:"Principal Component Analysis", Springer Verilag, New York, 1986.
3. Richard O. Duda, Peter E. Hart, David G. Stork:"Pattern Classification and Scene Analysis", Wiley Interscience, 2000.
4. W. L. Poston and Marchette, D. J.: "Recursive Dimensionality Reduction Using Fisher's Linear Discriminant," Pattern Recognition, vol. 31, no. 7, PP.881-888, 1988.
5. Jie Yang, Hua Yu: "A Direct LDA Algorithm for High-Dimensional Data - with Application to Face Recognition," Pattern Recognition, vol. 34, no. 10, PP.2067-2070, 2001.
6. D. Swets and J. Weng: "Using discriminant eigenfeatures for image retrieval," IEEE Transactions on Pattern Analysis and Machine Intelligence, vol. 8, no. 8, PP.831-836, 1996.
7. A. Tefas , C. Kotropoulos , and I. Pitas: "Using support vector machines to enhance the performance of elastic graph matching for frontal face authentication," IEEE Transactions on Pattern Analysis and Machine Intelligence, vol. 23, no. 7, 2001.
8. Laboratories of Intelligent Systems, Institute of Information Science. The IIS Face Database. http://smart.iis.sinica.edu.tw/index.html.

A Hardware-Directed Face Recognition System Based on Local Eigen-analysis with PCNN

C. Siva Sai Prasanna, N. Sudha, and V. Kamakoti

VLSI Laboratory
Department of Computer Science and Engineering
Indian Institute of Technology Madras, Chennai - 600 036 India
sivasai@cse.iitm.ernet.in, {sudha,kama}@shiva.iitm.ernet.in

Abstract. A new face recognition system based on eigenface analysis on segments of face images is discussed in this paper. The eigenfaces are extracted using principal component neural networks. The proposed recognition system can tolerate local variations in the face such as expression changes and directional lighting. Further, the system can be easily mapped onto the hardware.

1 Introduction

In a number of situations, the need to determine who the individuals are and the functions they are permitted to perform has become paramount. Automatic identification of a person based on his/her physiological characteristics is gaining importance. One method of identification is using images of human face.

One approach to face recognition is using "Eigenfaces" obtained through Principal component analysis(PCA)[1]. Eigenfaces are robust to slight translation, rotation and scale changes and can tolerate high blurring effects[2]. However they are affected by significant local variations such as expression changes and directional lighting. In this paper, we propose a face recognition method based on local eigenface analysis. The proposed method gives better results for yale database faces with local variations when compared to existing whole eigenface based face recognition method [3–5].

In this paper, principal Component Neural Networks[6,7] are employed to extract eigenfaces. The weights of the network converge to eigenfaces when trained with face images using Generalized Hebbian Learning algorithm[8]. The neural network based proposed method is feasible for hardware mapping due to the properties of PCNN such as linear neurons and simple learning algorithm. Further, the neural networks inherent parallelism, regularity and cascadability are additional features that are suitable for hardware implementation.

A fully digital hardware architecture is described in this paper. The traditional method of implementation of neural network involves the assignment of a processing element for each neuron. A processing element is usually a sequential logic consisting of few registers and combinational logic that implements the learning rule. The proposed architecture for PCNN reuse the same combinational logic for updating the weights and has minimal set of registers for storing M weight vectors, and the learning rate. Hence, it is area-efficient. The architecture has simple combinational logic that realizes simple integer arithmetic operations. Further, concerning the limited number of I/O pins in

N.R. Pal et al. (Eds.): ICONIP 2004, LNCS 3316, pp. 327–332, 2004.
© Springer-Verlag Berlin Heidelberg 2004

VLSI chips, the proposed local eigenface analysis on small parts(with limited number of pixels) of images is favourable to single chip implementation.

The contributions in this paper are two-fold. One is, a person identification system based on local eigenface analysis on parts of face image is proposed. Second one is, the system is robust to significant local variations.An area-efficient architecture for PCNN-based eigenface extraction is presented.

2 Preliminaries

Principal Component Analysis is a method of representing a data set in more compact form. Principal components are the directions along which the data points in the data set have maximum variance. In mathematical terms, these components are the eigenvectors of the data covariance matrix arranged in the descending order of eigenvalues. The neural network that extracts principal components from the input data is called Principal Component Neural Network(PCNN).

PCNN is a feed-forward neural network with single layer of linear neurons[9, 10]. Each neuron receives the elements of the input vector $X = [x_1, x_2, ..., x_N]$. An N-input M-output network has M neurons and performs principal component analysis of N-dimensional input vectors. The connections from inputs to each neuron are associated with weights. w_{ji} is the weight associated with the ith input to the jth neuron. By generalized Hebbian learning algorithm[8], the weight vector of the jth neuron, $W_j = [w_{j1} \ w_{j2} .... w_{jN}]$, converges to j^{th} principal component of input data vectors. The weights are updated incrementally by the learning rule below.

$$\triangle W_{ji} = \mu y_j [x_i - \sum_{k=1}^{j} w_{ki} y_k]$$

when μ is the learning parameter which takes the value in the range (0,1) and y_j is the output of the jth neuron. The above can be rewritten as :

$$\triangle W_j = \mu y_j [X - \sum_{k=1}^{j} W_k y_k] = \mu y_j [X - S_j] = \mu y_j X'$$

The advantages of neural network based computation of principal components over direct computation are as follows:

- The size of the covariance matrix is large for large dimensions of the input vector which can cause problem while computation due to limited computer memory. A neural network extracts the principal components directly from the input data.
- The direct computation gives all the principal components even though only a few components are required in many applications. The number of principal components to be extracted can be restricted in the neural network.
- The extraction of principal components by direct computation is done using a block of data. On the otherhand, neural networks are adaptive and hence the computation is done online for each input data.
- A neural network is feasible for VLSI implementation.

3   Proposed Method

In order to make the eigenface-based recognition method robust to directional lighting and expression variations, we perform PCA on parts of images. Here the whole $N X N$ image is divided into $p = KXL$ sub images. So, the size of each image is given by $\frac{N}{K} X \frac{N}{L}$. In the proposed method we use either p PCNN's each with $\frac{NXN}{KXL}$ inputs and R outputs (or neurons) to get the results in parallel for p parts of the image or a single PCNN of the same size.

The subimages are preprocessed before applying to PCNN. The input to the PCNN must be of zero-mean [11]. Let I_i, i=1 to M, be the training set of images. We do the averaging of sub-images as follows:

$$A_j = \frac{1}{M} \sum_{i=1}^{M} I_{ij} \ for \ j = 1, 2...p$$

where I_{ij} is the jth part of I_i. Now the mean-centering of the imaes is done by subtracting the mean sub-images from the respective sub-images and is given as:

$$X_{ij} = I_{ij} - A_j$$

where i ranges from 1 to M and j ranges from 1 to p. These sub-images are normalized and they are given as the training vectors to the PCNN. The pixel values of the images are normalized to the range $[0, 1]$. The initial weight vectors are initialized to 0.1 and the learning rate is chosen as 0.4. During training, the parts of the training images are given to the corresponding PCNN's iteratively. In the case of single PCNN, all parts are given to that PCNN itself. When the weights of the PCNN converge, the resultant weight vectors constitute local eigenfaces. The weights of every neural network are stored after the training.

After the training is over, the parts of training image are projected to the corresponding weight vectors and the resultant values are the distance of the input training image to the eigenspace. Let the projected outputs be $P_{ij}(r)$, where i varies from 1 to M, j varies from 1 to p and r varies from 1 to R.

Given a test face, it is recognized as one among the training faces as follows. The test image is also divided into p parts similar to training images. The parts are given to the corresponding PCNN's and the outputs are given by $P_j^t(r)$; j=1 to p, r=1 to R. The l_1 norm between the projected output vector of each part of the test image and the corresponding vector of each training image is computed. That is, $\sigma_{ij} = ||P_{ij} - P_j^t||$. Each test part j is recognized as that of face i^* where $\sigma_{i*j} = \min_i\{\sigma_{ij}\}$ for i=1 to M and j=1 to p.

The entire test image is recognized as face i' if maximum number of its parts are recognized as face i'.

3.1   Advantages

- The effect of directional lighting and facial expressions are limited to only certain parts of face image and hence it affects only the principal components pertaining to these parts.

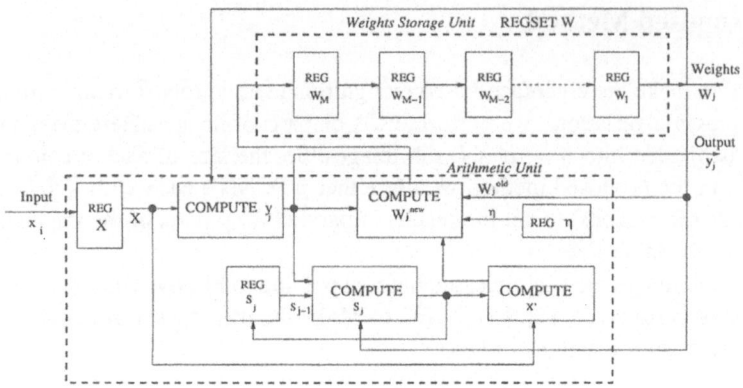

Fig. 1. Hardware architecture of PCNN.

- The size of the input vector for the PCNN is reduced and it satisfies the constraint on the number of I/O pins when it is required to obtain single chip solution for a PCNN.
- In the case of multi-PCNN, high degree of parallelism is achieved.

4   Performance Study

The performance of PCNN-based local eigenface analysis has been evaluated on Yale database consisting of images with varying illumination and expressions. The database has normalized images of 15 persons, each with 11 samples. Five samples of each person have been used for training and the rest has been tested. Ultimately, 75 images constitute training set and 90 images constitute test set.

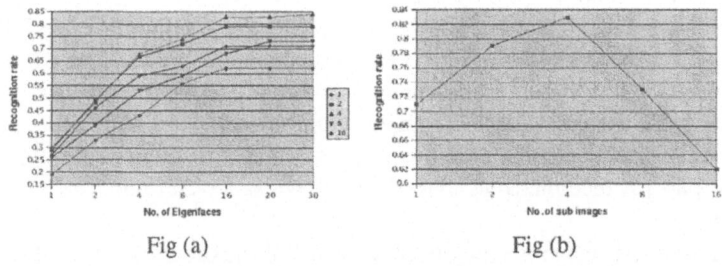

Fig (a)   Fig (b)

Fig. 2. Plots for Single PCNN.

The results are plotted in the figure for both single PCNN and multi-PCNN cases respectively. Figure 2(a) & 3(a) show the plots of recognition rate for varying number of neurons L. The plots are for different values of p. The recognition rate is defined as the ratio of the number of images that were recognized correctly to the number of

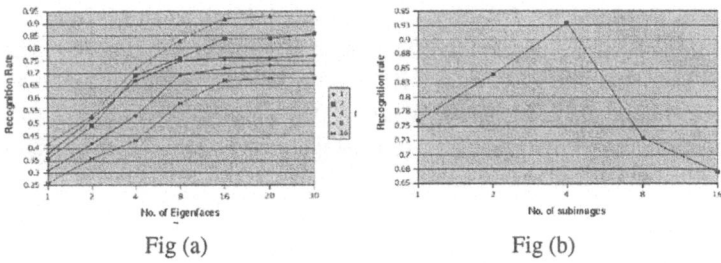

Fig (a)   Fig (b)

Fig. 3. Plots for Multi-PCNN.

the images in the test set. It can be observed that the recognition rate increases as R increases and it reaches saturation for a particular value of R, say R_p for each p. There is no significant improvement in performance for $R > R_p$. The maximum value of R_p is close to 20 in both cases. Figure 2(b) & 3(b) show the plots of recognition rate for varying p and for a fixed value of L. It can be seen that good performance is achieved for p=4. For $p > 4$, the global facial features in the image parts are lost and hence the recognition rate deterioates. Further division of these subimages results in division of important facial features like eyes and hence, it results in degrading performance.

5   Hardware Mapping

The main component of the proposed method is PCNN. It extracts the principal components of subimages for the purpose of recognition. In this section, an area-efficient hardware architecture for PCNN is described. The hardware architecture of PCNN is shown in Figure 1. It has registers to store η and weights W_1 to W_2. There are combinational logic circuits to computer y_j, S_j, X' and weights. The weights are computed iteratively in M clock pulses and they are stored in a pipelined set of registers, REGSET W.

At j^{th} clock pulse, y_j is computed by finding the inner product of input data vector X and the contents of REG W_1 that stores W_j to be updated. S_j is stored in a register for the computation of S_{j+1}. Once S_j is known, X' is computed with S_j and X. W_j is finally updated using X', η and y_j. It is pushed into REG W_M. Simultaneously, the contents of all other weight registers in REGSET W are pushed forward. After M clock pusles, the weight registers store the corresponding updated weights for a given input X.

5.1   Operation of Hardware

Initially REG η is initialized with a value between 0 and 1. REGSET **W** is initialized with random values in the same range. Then the data vectors from the input data set are presented one after another and weights are updated for each data vector. The input data vector X is placed on the input data bus and a start signal S is activated for a period of two clock pulses. When a clock trigger appears during S=1, weights start updating. The weights get updated from W_1 and W_M for the successive M clock pulses. REG S_j is first initialized to 0 and an external counter CNT counts the number of clock pulses.

CNT counts the number of clock pulses. CNT is initialized to M. In any clock pulse j, y_j is computed first followed by the computation of S_j, X' and W_j. The values of S_j and W_j are stored in REG S_j and REG W_j at the clock trigger. Simultaneously, all other register contents in REGSET W are shifted. Once CNT reaches 0, the next input is then placed. The weights converge to principal components of input data space once they are updated with adequate number of input vectors.

6   Conclusions

An eigenface-based new human recognition system has been discussed in this paper. The eigenfaces are extracted for parts of the face image using PCNN. The advantages of the proposed neural network-based local eigenface analysis are,

1. It is robust to local variations such as facial expression changes and directional lighting.
2. It can be easily mapped to hardware.

The performance has been evaluated on Yale database and it is observed that best results can be achieved when face image is divided into 4 parts and the number of eigenfaces equals 20. A hardware architecture for PCNN is also described.

References

1. Mathew A.Turk and Alex P.Pentland, "Eigenfaces for recognition," *Journal of Cognitive Neuroscience*, vol. 3, no. 1, pp. 71–86, 1991.
2. Alexandre Lemieus and Marc Parizeau, "Experiments on eigenfaces robustness," *Proceedings of IEEE International Conference on Pattern Recognition*, vol. 1, pp. 421–424, August 2002.
3. Mathew A.Turk and Alex P.Pentland, "Face recognition using eigenfaces," *Proceedings of IEEE Conference on Computer Vision and Pattern Recognition*, pp. 586–591, June 1991.
4. D.Waldoestl E.Lizama and B.Nickolay, "An eigenfaces-based automatic face recognition system," *Proceedings of IEEE International Conference on Systems, Man and Cybernetics*, pp. 174–177, 1997.
5. Zhujie and Y.L.Yu, "Face recognition with eigenfaces," *Proceedings of IEEE International Conference on Industrial Technology*, pp. 434–438, December 1994.
6. Simon Haykin, *Neural networks : A comprehensive foundation*, Prentice Hall India, 1999.
7. K.I.Diamantaras and S.Y.Kung, *Principal component neural networks:Theory and applications*, Wiley, New York, 1996.
8. T.D.Sanger, "Optimal unsupervised learning in a single layer feedforward neural network," *Neural Networks*, vol. 2, pp. 459–73, 1989.
9. E.Oja, "A simplified neuron model as a principal component analyzer," *Journal of Mathematical Biology*, vol. 15, pp. 267–273, 1982.
10. E.Oja and J.Karhunen, "On stochastic approximation of the eigen vectors and eigen values of the expectation of a random matrix," *Journal of Mathematical Analysis and Applications*, vol. 104, pp. 69–84, 1985.
11. S. Costa and S.Fiori, "Image compression using Principal component neural networks," *Image and Vision Computing Journal*, vol. 19, pp. 649–668, 2001.

The Teager Energy Based Features for Identification of Identical Twins in Multi-lingual Environment

Hemant A. Patil and T.K. Basu

Department of Electrical Engineering, Indian Institute of Technology,
IIT Kharagpur, West Bengal, India-721302
{hemant,tkb}@ee.iitkgp.ernet.in

Abstract. Automatic Speaker Recognition (ASR) is an economic method of biometrics because of the availability of low cost and powerful processors. An important question which must be answered for the ASR system is how well the system resists the effects of determined mimics such as those based on physiological characteristics especially identical twins or triplets. In this paper, a new feature set based on Teager Energy Operator (TEO) and well-known Mel frequency cepstral coefficients (MFCC) is developed. The effectiveness of the newly derived feature set in identifying identical twins has been demonstrated for different Indian languages. Polynomial classifiers of 2nd and 3rd order have been used. The results have been compared with other feature sets such as LPC coefficients, LPC cepstrum and baseline MFCC.

1 Introduction

In this paper, we show the effectiveness of the newly derived feature set for identification of identical twins in multi-lingual environment viz. Marathi, Hindi and Urdu. Training and testing are done for each group for the same language. Identification of identical twins in Marathi language using different features (LPC, LPCC and MFCC) has been reported in [6].

2 The Teager Energy Operator (TEO)

Features derived from a linear speech production models assume that airflow propagates in the vocal tract as a linear plane wave. This pulsatile flow is considered the source of sound production [8]. According to Teager [7], this assumption may not hold since the flow is actually separate and concomitant vortices are distributed throughout the vocal tract. He suggested that the true source of sound production is actually the vortex-flow interactions, which are non-linear and a non-linear model has been suggested based on the energy of airflow.

There are two broad ways to model the human speech production process. One approach is to model the vocal tract structure using a source-filter model. This approach assumes that the underlying source of speaker's identity is coming from the vocal tract configuration of the articulators (i.e., size and shape of the vocal tract) and the manner in which speaker uses his articulators in sound production [5]. An alternative way to characterize speech production is to model the airflow pattern in the vocal tract. The underlying concept here, is that while the vocal tract articulators do move to configure the vocal tract shape (making cues for speaker's identity [5]), it is the

N.R. Pal et al. (Eds.): ICONIP 2004, LNCS 3316, pp. 333–337, 2004.
© Springer-Verlag Berlin Heidelberg 2004

resulting airflow properties which serve to excite those models which a listener will perceive for a particular speaker's voice [7],[8]. Modeling the time-varying vortex flow is a formidable task and Teager devised a simple algorithm which uses a non-linear energy-tracking operator called as Teager Energy Operator (TEO) (in discrete-time) for signal analysis with the supporting observation that hearing is the process of detecting energy. The concept was further extended to continuous-domain by Kaiser [4]. According to Kaiser, energy in a speech frame is a function of amplitude and frequency as well. Let us now discuss this point in brief.

The dynamics and solution (which is a S.H.M.) of mass-spring system are described

$$\frac{d^2x}{dt^2} + \frac{k}{m}x = 0 \Rightarrow x(t) = A\cos(\Omega t + \phi)$$

and the energy is given by

$$E = \frac{1}{2}m\Omega^2 A^2 \Rightarrow E \propto (A\Omega)^2 \qquad (1)$$

From (1), it is clear that the energy of the S.H.M. of displacement signal $x(t)$ is directly proportional not only to the square of the amplitude of the signal but also to the square of the frequency of the signal. Kaiser and Teager proposed the algorithm to calculate the running estimate of the energy content in the signal. (1) can be expressed in discrete-time domain as

$$x(n) = A\cos(\omega n + \phi)$$

By trigonometry,

$$x^2(n) - x(n+1)x(n-1) = A^2\sin^2\omega \approx A^2\omega^2 \approx E_n$$

where E_n gives the running estimate of signal's energy. In continuous and discrete-time, TEO of a signal is $x(t)$ defined by

$$\Psi_c[x(t)] = \left[\frac{dx}{dt}\right]^2 - x(t)\frac{d^2x}{dt^2} \mapsto \Psi_d[x(n)] = x^2(n) - x(n+1)x(n-1) \qquad (2)$$

It is a well known fact that the speech can be modeled as a linear combination of AM-FM signals in some cases [3]. Each resonance or formant is represented by an AM-FM signal of the form

$$x(t) = a(t)\cos(\phi(t)) = a(t)\cos[\int_0^t \omega_i(\tau)d\tau + \phi(0)] \Rightarrow \Psi_c[x(t)] \approx \left(a\frac{d\phi}{dt}\right)^2 \qquad (3)$$

where $a(t)$ is a time varying amplitude signal and $\omega_i(t)$ is the instantaneous frequency given by $\omega_i(t) = d\phi/dt$. This model allows the amplitude and formant frequency (resonance) to vary instantaneously within one pitch period. It is known that TEO can track the modulation energy and identify the instantaneous amplitude and frequency. Motivated by this fact, in this paper a new feature set based on nonlinear model of (3) is developed using the TEO. The idea of using TEO instead of the commonly used instantaneous energy is to take advantage of the modulation energy tracking capability of the TEO. This leads to a better representation of formant information in the feature vector than MFCC [3]. In the next section we will discuss the details of T-MFCC.

3 Teager Energy Based MFCC (T-MFCC)

Traditional MFCC based feature extraction involves pre-processing; Mel-spectrum of pre-processed speech, followed by log-compression of subband energies and finally DCT is taken to get MFCC per frame [2]. In our approach, we employ TEO for calculating the energy of speech signal. Now, one may apply TEO in frequency domain, i.e., TEO of each subband at the output of Mel-filterbank, but there is difficulty from implementation point of view. Let us discuss this point in detail.

In frequency-domain, (2) for pre-processed speech $x_p(n)$ implies,

$$F\{\Psi_c[x_p(t)]\} \mapsto F\{x_p^2(n) - x_p(n+1)x_p(n-1)\} = F\{x_p^2(n)\} - F\{x_p(n+1)x_p(n-1)\} \tag{4}$$

Using shifting and multiplication property of Fourier transform, we have

$$F\{x_p(n+1)x_p(n-1)\} = \frac{1}{2\pi} \int_{2\pi} X_{1p}(\theta)X_{2p}(\omega - \theta)d\theta$$

where $X_{1p}(\omega) = e^{-j\omega}X_p(\omega)$ and $X_{2p}(\omega) = e^{j\omega}X_p(\omega)$. Hence (4) becomes

$$F\{\Psi_c[x_p(t)]\} = \frac{1}{2\pi} \int_{2\pi} \left(1 - e^{j\omega}e^{-2\theta}\right)X_p(\theta)X_p(\omega - \theta)d\theta \tag{5}$$

Thus (5) is difficult to implement in discrete-time and also time-consuming. So we have applied TEO in the time-domain. Let us now see the computational details of T-MFCC.

Speech signal $x(n)$ is first passed through pre-processing stage (which includes frame blocking, hamming windowing and pre-emphasis) to give pre-processed speech signal $x_p(n)$. Next we calculate the Teager energy of $x_p(n)$:

$$\Psi_d[x_p(n)] = x_p^2(n) - x_p(n+1)x_p(n-1) = \psi_1(n)(say)$$

The magnitude spectrum of the TEO output is computed and warped to Mel frequency scale followed by usual log and DCT computation (of MFCC) to obtain T-MFCC.

$$T - MFCC = \sum_{l=1}^{L} \log[\Psi_1(l)]\cos\left(\frac{k(l-0.5)}{L}\pi\right), k = 1, 2, ..., Nc.$$

where $\Psi_1(l)$ is the filterbank output of $F\{\psi_1(n)\}$ and $\log[\Psi(l)]$ is the log of filterbank output and $T - MFCC(k)$ is the kth T-MFCC. T-MFCC differs from the traditional MFCC in the definition of energy measure, i.e., MFCC employs L^2 energy in frequency domain (due to Parseval's equivalence) at each subband whereas T-MFCC employs Teager energy in time domain. Fig. 1 shows the functional block diagram of MFCC and T-MFCC.

4 Experimental Results

Database of 12 twins is prepared from different places in India. Out of 12 twins, database of 8 twins is prepared from different places in Maharashtra in Hindi and Marathi whereas the database of 4 Urdu speaking twins is prepared from Kolkata, Unrao and Umri village (Uttar Pradesh). Text material for recording consisted of isolated words, digits, combination-lock phrases and a contextual speech of approximately 30s duration. Polynomial classifiers of 2nd and 3rd order are used as the basis for all

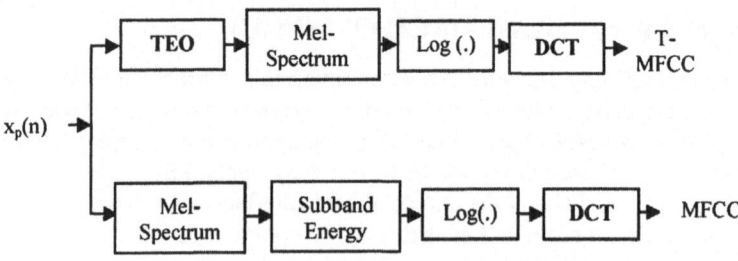

Fig. 1. Block diagram for T-MFCC and MFCC

experiments [1]. Feature analysis was performed using 12th order LPC on a 23.2 ms frame with an overlap of 50%. Each frame was pre-emphasized with the filter 1-0.97z-1, followed by Hamming window and then the mean value is subtracted from each speech frame (similar pre-processing steps were performed for MFCC and T-MFCC except mean removal). We have taken 2 samples more to compute T-MFCC than that for LPC, LPCC and MFCC because of TEO processing. The results are shown in Table 1-6 for different training durations with 2nd and 3rd order polynomial approximation. Some of the observations from the results are as follows:

- For 2nd order polynomial approximation, average success rate for T-MFCC is found to be better than MFCC, LPC and LPCC for 30s training duration whereas it is found to be better than LPC and LPCC for 60s and 90s training durations.
- Average success rates for T-MFC, MFCC, LPC and LPCC go up by more than 16%, 13%, 12% and 15% respectively for 3rd order polynomial approximation as compared to their 2nd order counterparts for various training and testing durations. This is as expected for a classifier of higher order polynomials.
- For 3rd order polynomial approximation, T-MFCC and MFCC performed equally well and these feature sets together outperformed LPC and LPCC for 30s training duration whereas for 60s and 90s training durations, MFCC outperformed T-MFCC, LPC and LPCC.
- It is interesting to note that in majority of the cases of misidentification, the misidentified person is the actual speaker's twin brother/sister (except in those cases where the twin pair is of different sex).

Table 1. Success Rates for 2nd Order Approximation with 30s training

TEST (SEC)	T-MFCC	MFCC	LPC	LPCC
1	83.33	83.33	66.66	62.50
3	75.00	79.16	75.00	75.00
5	79.16	75.00	70.83	70.83
7	75.00	75.00	79.16	70.83
10	83.33	79.16	75.00	70.83
12	83.33	79.16	75.00	70.83
15	83.33	79.16	75.00	66.66
Avg. rate	**80.35**	78.56	73.80	69.64

Table 2. Success Rates for 3rd Order Approximation with 30s training

TEST (SEC)	T-MFCC	MFCC	LPC	LPCC
1	83.33	87.50	79.16	83.33
3	87.50	83.33	83.33	79.16
5	87.50	87.50	79.16	79.16
7	87.50	87.50	79.16	79.16
10	87.50	87.50	75.00	83.33
12	87.50	87.50	75.00	83.33
15	87.50	87.50	79.16	83.33
Avg. rate	**86.90**	**86.90**	78.56	81.54

Table 3. Success Rates for 2ⁿᵈ Order Approximation with 60s training

Test (sec)	T-MFCC	MFCC	LPC	LPCC
1	87.50	87.50	66.66	70.83
3	79.16	79.16	70.83	75.00
5	75.00	83.33	66.66	75.00
7	75.00	87.50	70.83	79.16
10	75.00	87.50	75.00	79.16
12	79.16	87.50	70.83	75.00
15	79.16	87.50	70.83	79.16
Avg. rate	**78.56**	**85.71**	70.23	76.18

Table 4. Success Rates for 3ʳᵈ Order Approximation with 60s training

Test (sec)	T-MFCC	MFCC	LPC	LPCC
1	91.66	95.83	83.33	87.50
3	91.66	95.83	91.66	95.83
5	91.66	**100**	87.50	91.66
7	91.66	**100**	83.33	91.66
10	**100**	**100**	79.16	91.66
12	95.83	**100**	75.00	87.50
15	95.83	**100**	75.00	91.66
Avg. rate	**94.04**	**98.80**	82.14	91.06

Table 5. Success Rates for 2ⁿᵈ Order Approximation with 90s training

Test (sec)	T-MFCC	MFCC	LPC	LPCC
1	83.33	91.66	70.83	66.66
3	79.16	79.16	70.83	79.16
5	83.33	79.16	75.00	79.16
7	87.50	83.33	75.00	75.00
10	91.66	87.50	75.00	75.00
12	87.50	91.66	75.00	75.00
15	87.50	91.66	75.00	79.16
Avg. rate	**85.71**	**86.30**	73.80	75.59

Table 6. Success Rates for 3ʳᵈ Order Approximation with 90s training

Test (sec)	T-MFCC	MFCC	LPC	LPCC
1	95.83	91.66	79.16	83.33
3	91.66	95.83	87.50	91.66
5	91.66	95.83	87.50	87.50
7	91.66	95.83	87.50	87.50
10	**100**	**100**	87.50	87.50
12	**100**	**100**	79.16	83.33
15	**100**	**100**	83.33	91.66
Avg. rate	**95.83**	**97.02**	84.52	87.49

5 Conclusion

In this paper, Teager Energy based MFCC (T-MFCC) features are proposed for speaker identification of identical twins in multilingual environment. Their performance was compared with conventional features and found to be effective.

References

1. Campbell, W. M., Assaleh, K. T., Broun, C. C.: Speaker recognition with polynomial classifiers. IEEE Trans. on Speech and Audio Processing. 10 (2002) 205-212
2. Davis, S. B., Mermelstein, P.: Comparison of parametric representations for monosyllabic word recognition in continuously spoken sentences. IEEE Trans. Acoust., Speech and Signal Processing. 28 (1980) 357-366
3. Jabloun, F., Cetin, A. E., Erzin, E.: Teager energy based feature parameters for speech recognition in car noise. IEEE Signal Processing Lett. 6 (1999) 259-261
4. Kaiser, J.F.: On a simple algorithm to calculate the 'energy' of a signal. Proc. of Int. Conf. on Acoustic, Speech and Signal Processing. 1(1990) 381-384
5. Kersta, L.G.: Voiceprint Identification. Nature 196 (1962) 1253-1257
6. Patil, Hemant A., Basu, T. K.: Text-independent identification of identical twins in Marathi language in noisy environment. In Proc. of 2ⁿᵈ Int. Conf. on Artificial Intelligence in Engg. and Tech, Malaysia (2004) 190-196
7. Teager, H.M.: Some observations on oral air flow during phonation. IEEE Trans. Acoust., Speech, Signal Process. 28(1980) 599-601
8. Zhau, G., Hansen, J.H.L., Kaiser, J. F.: Non-linear feature based classification of speech under stress. IEEE Trans. on Speech and Audio Processing, 9(2001) 201-216

A Fast and Efficient Face Detection Technique Using Support Vector Machine

R. Suguna*, N. Sudha, and C. Chandra Sekhar

Department of Computer Science and Engineering
Indian Institute of Technology Madras, Chennai - 600 036, India
hitec_suguna@hotmail.com, {sudha,chandra}@cs.iitm.ernet.in

Abstract. We present an efficient technique for face detection that filters the face-like regions from a gray image and then detects the faces using Support Vector Machines (SVM) from these potential regions. The technique focusses on extracting the probable eye-regions since eyes are the most prominent features of a face. The probable location of eyes is used to segment a square region constituting face-like region from the image. The extracted regions are verified using a single SVM for making binary decision: face/non-face. Experimental results show that the proposed face detection system is several orders of magnitude faster than the existing SVM-based systems.

1 Introduction

Human beings have no difficulty in locating a face in a photograph even when the face is occluded, rotated or scaled by a certain amount. Images containing faces are essential to intelligent vision-based human computer interaction. To build fully automated systems that analyze the information contained in face images, robust and efficient face detection algorithms are required. The interest in automatic face detection and recognition is mainly driven by the demand of applications such as the video conferences, intelligent human computer interfaces, nonintrusive identification, verification for credit cards, automatic teller machines transactions.

Since face detection can be considered as a two class (face/non-face) pattern recognition problem, various neural network architectures [1–3] have been proposed. One drawback of these methods is that the network architecture has to be extensively tuned (number of layers, number of nodes, learning rates etc.) to get good performance. Further, the methods for training a classifier (eg. Bayesian, multilayer perceptron or radial basis function network) are based on minimizing the training error. However, Support Vector Machines [4] (SVMs) operate on induction principle called structural risk minimization, which aims to minimize the upper bound on the expected generalization error. They implicitly project patterns to a higher dimensional space and then form a linear decision surface between the projected face and non-face patterns. The mapping of input patterns to a higher dimesional space is implemented by a kernel function K. An SVM classifier is a linear classifier where the separating hyperplane is chosen to minimize

---

* R.Suguna is currently with Department of Information Technology, M.N.M. Jain Engineering College, Chennai - 600 096, India.

N.R. Pal et al. (Eds.): ICONIP 2004, LNCS 3316, pp. 338–343, 2004.
© Springer-Verlag Berlin Heidelberg 2004

the expected classification error of the unseen test patterns. This optimal hyperplane is defined by a weighted combination of a small subset of training vectors called support vectors. Estimating the hyperplane is equivalent to solving a linearly constrained quadratic programming problem.

This paper deals with detection of frontal faces in a gray scale image, using SVMs. We extract probable eye regions, which guides further processing in face detection. Face detection can be viewed as a two class recognition problem, in which an image sub-region is classified as being a face or non-face. Support Vector Machines [4] are used for such classification. The objective of this work is to detect frontal faces in an image with variations in scale and position. The detection of a face from an image involves segmentation of possible face regions and verifying for face/non-face. The issues addressed in this work are: (a) Extraction of face-like regions which guide the face detection, (b) determining the scale of these regions, and (c) verifying the extracted region for face/non-face. In the entire study only gray scale images are considered.

2 Related Work

Support vector machines (SVM) were first applied to face detection by Osuna *et al.* [5]. Here, a SVM with second degree polynomial as kernel function is taken. The input images of size 19×19 are preprocessed for masking, lighting correction and histogram equalization before giving to SVM classifier. In some other work [6, 7], SVM is used for detecting face and other objects. The systems used Haar wavelet features as input to a SVM classifier. In all the above mentioned ones, for detecting the faces in a given image, a fixed size window is slided over the image and the trained classifier is used to decide which patterns are the object of interest. This technique is computationally intensive, since it applies exhaustive search on all possible positions of windows in images of different scale. In this paper, we present a speed-up technique which can identify, relatively quickly, regions in an image that are likely to contain human faces.

3 Proposed Face Detection Technique

The SVM-based face detection system can be divided into two main stages : Training phase and Detection phase. The block diagram of the system is shown in Fig. 1.

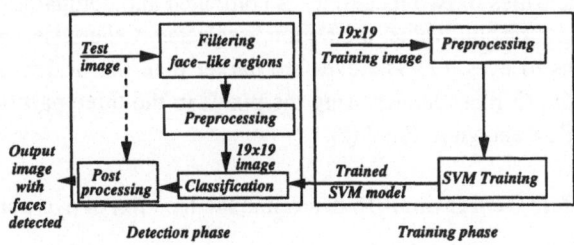

Fig. 1. Block diagram of face detection process.

3.1   Training Phase

The SVM is trained with face and non-face images of fixed size. The images are pre-processed first.

Preprocessing. Three preprocessing steps [1] were applied to the gray images to reduce within-class image variations. The images are first masked, that is the pixels close to the boundary of training images were removed in order to eliminate parts belonging to the background. A best-fit intensity plane was subtracted from the gray values of masked images to compensate for cast shadows, and histogram equalization was applied to remove variation in the image brightness and contrast. The resulting gray values of each image are normalized to [0,1] and given as input to the SVM for training. Appropriate kernel is chosen for SVM.

3.2   Detection Phase

In this phase, the trained SVM model is used to detect faces in a given image of any size. A technique is proposed to quickly select face-like regions in the image. The speed-up technique is based on the fact that, for frontal faces, the eyes are usually a prominent feature of the face. The distance between a pair of eyes gives an indication of the size of the face, and the positions of the eyes can be used to estimate the orientation. Using this premise, the technique identifies the regions that are most likely to contain faces. These regions are then verified, in turn, to check whether a face actually exists. Filtering the regions of interest relies on detecting possible pairs of eyes that may or may not belong to a face. The eye-pairs inherently provide information about the location and scale of potential faces. Square regions around the eye-pairs are used to establish the area that may contain the rest of the face. A suitable face verification technique can then be used to confirm the existence of faces. Our system uses SVM for the final verification. The individual stages of the proposed speed-up technique are described as follows.

Eye-Like Region Extraction. This initial stage receives an image and eye-like regions are filtered based on the following properties: (1) The region of eyes is often darker than the regions of the nose and cheek. This property is evaluated using pixel intensities of two adjacent rectangular windows shown in Fig. 2 (b). The difference between the sum of pixel intensities of two rectangles is computed and compared to a threshold. If it is less than the threshold value, the region is rejected. Otherwise, the upper window probably belongs to eyes. (2) The eyes are darker than the bridge of the nose. This is evaluated in the similar manner using the pixels in the three partitions of the upper rectangular window shown in Fig 2 (c).

Potential Eye-Pair Generation. To determine the location of eye-pairs, the following steps are executed.

1. For the regions in upper rectangle $(left, middle, right)$ the corresponding average intensity values $(avg\_left, avg\_mid, avg\_right)$ are computed.
2. The average intensity value of lower rectangle (avg_low) is computed.

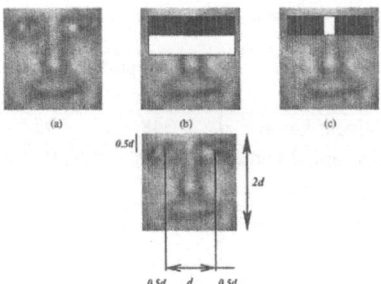

Fig. 2. Top: Eye-like region extraction (a) Sample image (b) Feature 1 (c) Feature 2; Bottom: Face-like region extraction.

3. If $(avg\_left \leq avg\_mid)$ and $(avg\_right \leq avg\_mid)$, find the pixels in left rectangle whose intensities are less than avg_low and compute their centroid which may be the potential left eye center. Similar computation holds for finding the right eye center.

Face-Like Region Extraction. The distance d between the calculated eye centers is computed. Based on d,a square region is extracted as in Fig. 2 which covers the main facial features(eyes, nose, mouth). The extracted face-like regions are resized to the size of the training images, preprocessed and given to SVM model for verification. The output of SVM gives a binary decision, face or non-face. Using these decisions, suitable post processing is done to show the locations of faces in the given image.

The extracted face-like regions are resized to the size of the training images, preprocessed and given to SVM model for verification. The output of SVM gives a binary decision, face or non-face. Using these decisions, suitable post processing is done to show the locations of faces in the given image.

4   Experimental Results

We use MIT-CBCL face database [8] to evaluate the performance of the proposed face detection system. The training dataset consist of 2428 face images and 4548 non-face images. The size of each image is 19 × 19 pixels. After preprocessing, each image is represented by a reduced 301-dimensional vector. This is due to removal of 60 pixels along the boundary of the image while masking. To evaluate the performance of SVM, we used MIT CBCL test set which is the subset of CMU Test set1. The test set consists of 472 face images and 23,573 non-face images of 19 × 19 pixels. Appropriate kernel is chosen by evaluating three common types of kernel functions. They are linear, polynomial and Gaussian kernels.

The SVMs based on these kernels are trained with the training set of images. The trained models are then tested on the test set. The 2nd-degree polynomial kernel seems to provide a good compromise between computational complexity and classification performance. The performance of SVM with Gaussian kernel is slightly better but required more support vectors (773 versus 456) than the polynomial SVM. Hence, the

Table 1. Performance of SVM for different kernels.

Kernel type	# of Support vectors	% of correct classification
linear kernel	348	95.34
Polynomial kernel with degree 2	456	97.99
Gaussian kernel with std dev 5	773	98.08

2nd-degree polynomial kernel is preferred for face detection. Experimental studies for evaluating the performance of SVM are given in Table 1.

Figure 3 shows the results of our face detection system on images with multiple faces and cluttered background. It can be seen that the proposed speed-up technique significantly reduces the number of sub-windows that need preprocessing and verification. This inturn reduces the time taken for processing an image. A comparison of time taken by the existing face detection system in which all possible sub-windows of 19x19 are given to SVM and the proposed system is given in the Table 2.

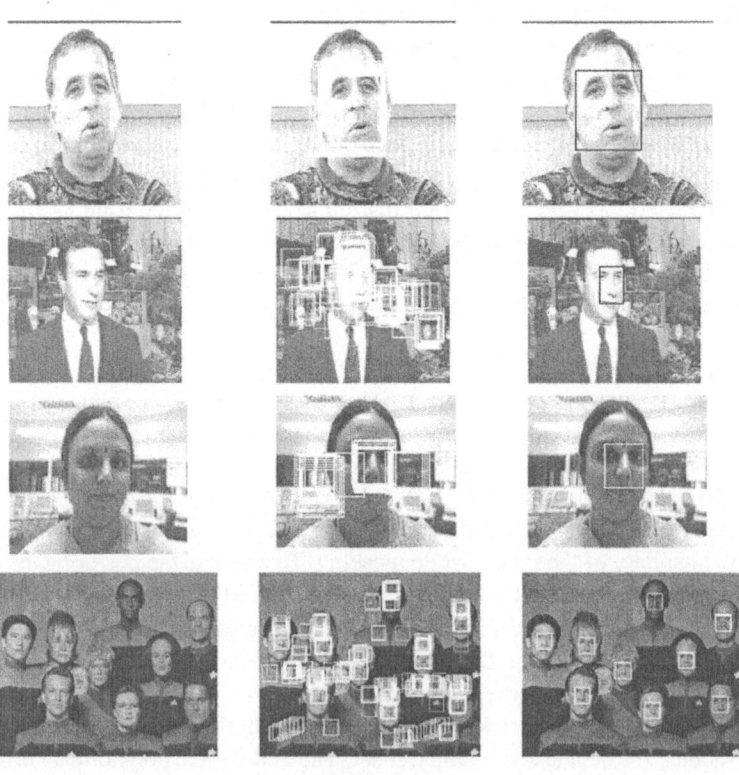

Test images   Extraction of face–like regions   Output images

Fig. 3. Face Detection results.

Table 2. Comparison of time taken for processing by the existing and proposed face detection system.

Test image	Existing system [5]		Proposed system	
	# windows processed	Time taken for detecting faces	# windows processed	Time taken for detecting faces
1	67044	78 min	11	19 sec
2	67044	78 min	261	2 min
3	14484	17 min	17	5 sec
4	243216	4 hrs	1058	6 min

5 Conclusions

In this paper, we presented a speed-up technique to extract face-like regions by searching for probable eye-regions in a given image. Our proposed face detection system incorporating the technique reduces the number of subwindows that need preprocessing and verification. A single support vector machine with 2nd degree polynomial kernel has been applied for face/non-face verification. The proposed system is several times faster than the existing neural network-based detection systems and is suitable for real-time applications. Further, the system performs well for frontal faces in gray scale images with variation in scale and position.

References

1. K.-K. Sung, *Learning and Example Selection for Object and Pattern Recognition*. PhD thesis, MIT, Artificial Intelligence Laboratory and Center for Biological and Computational Learning, Cambridge, MA, 1996.
2. H. Rowley, S. Baluja, and T. Kanade, "Neural network-based face detection," *IEEE Transactions on Pattern Analysis and Machine Intelligence*, vol. 20, pp. 23–38, January 1998.
3. M.-H. Yang, D. J. Kriegman, and N. Ahuja, "Detecting faces in images: A survey," *IEEE Transactions on Pattern Analysis and Machine Intelligence*, vol. 24, pp. 34–58, January 2002.
4. S. Haykin, *Neural Networks*. Prentice-Hall of International Inc., 1999.
5. E. Osuna, R. Freund, and F. Girosi, "Training support vector machine: An application to face detection," in *Proceedings, IEEE Conference on Computer Vision and Pattern Recognition*, (Puerto Rico, USA), pp. 130–136, 1997.
6. C. Papageorgiou and T. Poggio, "A trainable system for object detection," *International Journal of Computer Vision*, vol. 38, pp. 15–33, June 2000.
7. B. Heisele, A. Verri, and T. Poggio, "Learning and vision machines," in *Proceedings of the IEEE*, vol. 90, pp. 1164–1177, July 2002.
8. "CBCL Face Database 1, MIT Center For Biological and Computation Learning." http://www.ai.mit.edu/projects/cbcl.

User Enrollment Using Multiple Snapshots of Fingerprint

Younhee Gil[1], Dosung Ahn[1], Choonwoo Ryu[2], Sungbum Pan[1], and Yongwha Chung[3]

[1] Information Security Research Division
Electronics and Telecommunications Research Institute
{yhgil,dosung,sbpan}@etri.re.kr
http://www.etri.re.kr
[2] Department of Automation Engineering
INHA University
cwryu@vision.inha.ac.kr
[3] Deppartment of Computer and Information Science
Korea University
ychungy@korea.ac.kr

Abstract. As a method of preserving of privacy and the security of sensitive information, biometrics has been studied and used for the past few decades. A number of fingerprint verification approaches have been proposed until now. However, fingerprint images acquired using current fingerprint input devices that have small field of view are from just very limited areas of whole fingertips. Therefore, essential information required to distinguish fingerprints could be missed or extracted falsely. The limited and somewhat distorted information are detected from them, which might reduce the accuracy of fingerprint verification systems. In the systems that verify the identity of two fingerprints using fingerprint features, it is critical to extract the correct feature information. In order to deal with these problems, compensation of imperfect information can be performed using multiple snapshots of enrollee's fingerprints. In this paper, additional fingerprint images are used in enrollment phase. Our experiments using FVC 2002 databases show that the enrollment using multiple snapshots improves the performance of the whole fingerprint verification system.

1 Introduction

Traditionally, verified users have gained access to their property or service via dozens of PIN/password, smart cards and so on. However, these knowledge based, token based security methods have crucial weakness that can be lost, stolen, or forgotten. In recent years, there is an increasing trend of using biometrics. X9.84 [1] standard defines terminology of biometrics as 'A measurable biological or behavioral characteristic, which reliably distinguishes one person from another, used to recognize the identity, or verify the claimed identity, of an enrollee'. The fingerprint is one of widely used biometrics satisfying uniqueness and permanency [2]. Thus a number of fingerprint verification approaches have been proposed until now. Jain et al. [3] presented a minutiae-based verification, which aligns minutiae using Hough transform and performs minutiae matching by bounding box. Ross et al. [4] proposed hybrid matching method of local-based matching and global-based matching to enhance the performance. Pan et al. proposed an alignment algorithm using limited processing

N.R. Pal et al. (Eds.): ICONIP 2004, LNCS 3316, pp. 344–349, 2004.
© Springer-Verlag Berlin Heidelberg 2004

power and memory space to be executed in a smart card, and showed the possibility of match-on-card [5].

Although it is true that technical improvement has been achieved, there still exist challenging problems relating to the quality of fingerprint images and reliability of extracted minutiae. Most of input devices get fingerprint images having fingerprints being pressed on it not rolled, as a result, the area of fingerprint images can not help being very limited. Fingerprint mosaicking [6] uses multiple fingerprint images to generate template, augmenting minutiae sets from plural fingerprint images on enrollment stage. But, it does not check the reliability of each fingerprint images.

We proposed enrollment using multiple fingerprint images to extend enrolled fingerprint image and also guarantee the reliability of each fingerprint image. And we have tested our algorithm on the first FVC 2002 database [7,8].

This paper is organized as follows. Section 2 describes fingerprint verification system and the enrollment using plural fingerprint images briefly. Section 3 explains proposed methods and the experimental results using the methods are shown in Section 4. And we conclude in Section 5.

2   User Verification Using Fingerprint

It is widely known that the fingerprint is unique and invariant with aging, which implies that user authentication can be done comparing two fingerprints [2]. In general, a professional fingerprint examiner relies on details of ridge structures of the fingerprint in order to make fingerprint identifications. And the structural features are composed of the points where ridges end or bifurcate, that are called minutiae. Fig. 1 shows small part of an enlarged fingerprint image and two types of minutiae pointed by square marker, and the branch from minutia represents the direction of the minutiae. Usually, each minutia is described by the coordinate location, the direction ridge flows and the type, whether it is ridge ending or bifurcation.

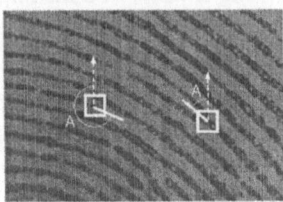

Fig. 1. Fingerprint Minutiae: Ending and Bifurcation, A presents the direction of each minutia

The orientation of the minutiae is represented in degrees, with zero degrees pointing vertical to the up, and increasing degrees proceeding counter-clockwise. The orientation of a ridge ending is determined by measuring the angle between vertical axis and the line starting at the minutia point and running through the middle of the ridge. The orientation of a bifurcation is determined by measuring the angle between the vertical axis and the line starting at the minutia point and running through the middle of the intervening valley between the bifurcating ridges.

Fig. 2 shows a fingerprint verification system, which consists of two phases: enrollment and verification. In the off-line enrollment phase, at first, the fingerprint image of an enrollee is acquired and preprocessed. Then, the minutiae are extracted from the raw image and stored as enrolled template. And in the on-line verification phase, it reads the fingerprint from a claimer, and detects the minutiae information through the same procedure as in the enrollment phase. Then, it estimates the similarity between the enrolled minutiae and the input minutiae. Image preprocessing refers to the refinement of the fingerprint image against the image distortion occurred during the image acquisition and transmission. Minutiae extraction refers to the detection of features in the fingerprint image and finding out of their information, i.e., position, direction and type.

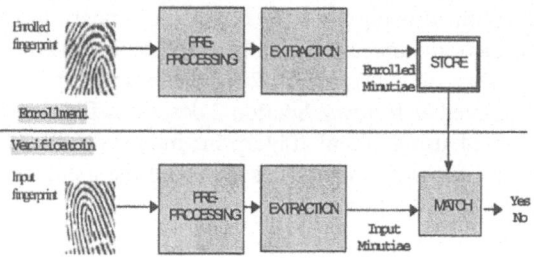

Fig. 2. Fingerprint Verification

Based on the minutiae, the claimed fingerprint is compared with the enrolled fingerprint. Generic minutiae matching is composed of alignment stage and matching stage. In order to match two fingerprints captured with unknown direction and position, the differences of direction and position between two fingerprints should be evaluated, and alignment between them needs to be preceded. In the alignment stage, transformations such as translation and rotation between two fingerprints are estimated, and two minutiae sets are aligned according to the estimated alignment parameters. If the alignment procedure is performed accurately, the remaining matching stage is referred to point matching simply. In matching stage, two minutiae are compared based on their position, direction, and type. Then, a matching score is computed.

Fingerprint Matching Using Minutiae. Minutiae matching is composed of alignment stage and matching stage. In order to make the explanation easier, we define the notation of two minutiae sets extracted from enrolled and claimed fingerprint images two minutiae sets as P and Q respectively.

$$\mathbf{P} = \left\{ \left(p_x^1, p_y^1, \alpha^1 \right), ..., \left(p_x^P, p_y^P, \alpha^P \right) \right\}$$
$$\mathbf{Q} = \left\{ \left(q_x^1, q_y^1, \beta^1 \right), ..., \left(q_x^Q, q_y^Q, \beta^Q \right) \right\} \tag{1}$$

where (p_x^i, p_y^i, α^i) and (q_x^j, q_y^j, β^j) are the three features (spatial position and direction) associated with the ith and jth minutia in the set P and Q respectively, and P and Q are the number of elements in the P and Q set.

The alignment stage gets two minutiae sets, P and Q, as input and estimates how their differences of position and orientation were when the two fingerprints were captured. And then, transforms minutiae set Q for the claimed fingerprint to have same locality as the enrolled fingerprint image according to the estimated difference. For the purpose of proper alignment, the estimation of the rotation and translation parameters must precede finding $(\Delta x, \Delta y)$ and $\Delta \theta$ satisfying formula (2) according to the formula (3).

$$F_{\theta,\Delta x,\Delta y}((q_x,q_y,\beta)^T) = (p_x,p_y,\alpha)^T \tag{2}$$

$$F_{\theta,\Delta x,\Delta y}\begin{pmatrix} x \\ y \\ \theta \end{pmatrix} = \begin{pmatrix} \cos \Delta \theta & \sin \Delta \theta & 0 \\ -\sin \Delta \theta & \cos \Delta \theta & 0 \\ 0 & 0 & 1 \end{pmatrix}\begin{pmatrix} x \\ y \\ \theta \end{pmatrix} + \begin{pmatrix} \Delta x \\ \Delta y \\ \Delta \theta \end{pmatrix} \tag{3}$$

where ($\Delta x, \Delta y$) and $\Delta \theta$ are the translation and rotation parameters; $(p_x,p_y,\alpha)^T$ represents the enrolled minutiae and $(q_x,q_y,\beta)^T$ represents the claimed minutiae.

After alignment step, the comparison of the information of two minutiae sets, P and Q^a, is accomplished by point pattern matching in the polar coordinate system with respect to the center of foreground. More details can be found in [5]

3   Super Template Using Multiple Snapshots

As automatic fingerprint identification and authentication systems rely on the two most prominent minutiae, a minutiae extraction algorithm is critical to the performance of the system. However, even superior extraction process including removal of false minutiae have some false minutiae remained still. And it can miss the true ones. The performance of fingerprint verification is influenced by both of them. Also fingerprint images acquired using current fingerprint input devices that have small field of view are from just very limited areas of whole fingertips, therefore, essential information required to distinguish fingerprints could be missed. In particular, if these happen during enrollment phase and the low quality minutiae are stored as enrolled template, it will be serious problem, because they will affect the matching phase continuously. In the other word, if the system ensures there are neither false minutiae nor missed minutiae, its reliability will be increased.

We suggested using plural fingerprint images on the enrollment phase to discard the false minutiae and compensate the missed minutiae. The system that adopts the enrollment using multiple snapshots is shown in fig. 3.

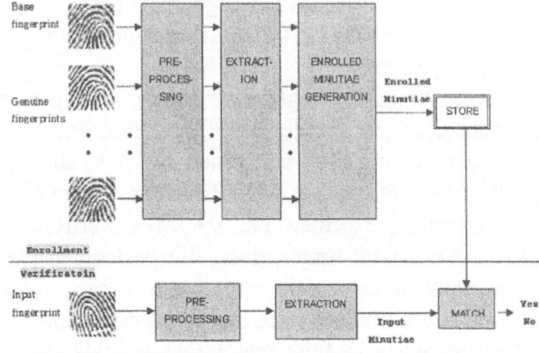

Fig. 3. Enrollment using Multiple Snapshots from One Fingerprint

Fig. 4 shows the generation of super template from three snapshots. In order to make explanation easier, pseudo minutiae images are used. We can see reference minutiae and multiple snapshots in the Fig. 4. In the reference minutiae, the minutiae are represented by squares, and in two multiple snapshots, the minutiae are pointed by black dots and circles respectively.

The super template can be created by selecting true minutiae from the combination of reference minutiae set and two minutiae sets from multiple snapshots. And before that, three minutiae should be aligned. The role of reference minutiae is to produce the baseline on aligning them. The right image in fig.4 is superimposed image of three minutiae sets. According to it, there are three kinds of minutiae. In some position, only one minutia exists, on the other hand, two or three minutiae from different snapshots are plotted in some places simultaneously. We treat the former as false minutiae because they are caused by the temporal fault during preprocess step and they cannot happen continuously and discard them: 6, 7, 8 are to be discarded. Using multiple snapshots has another effect that the region of window can be extended virtually. The gray region in the figure represents the original window size.

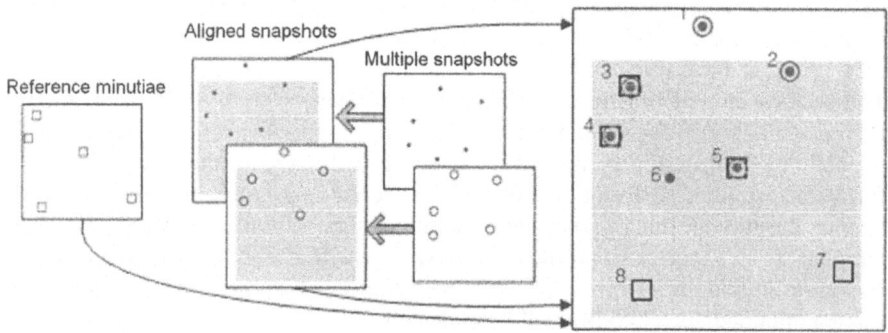

Fig. 4. Generation of Super Template from three snapshots

4   Experimental Results and Conclusion

We have tested our fingerprint verification algorithm using one of the FVC 2002 databases [7,8]. The details about FVC 2002 databases are in [8]. Among them, A set of DB1 was used in our experiment.

For the enrollment, the first four fingerprint images among eight were used, and the remaining four were used for verification. Therefore, when a matching was labeled GENUINE if the matching was performed between fingerprint images from same finger, and IMPOSTER otherwise, 400 GENUINEs were able to be performed. And 9900 IMPOSTERs were performed, i.e., 99 IMPOETERs have been tested for one fingerprint using the other 99 fingerprints. We performed two tests in order to show the effect of the adoption of enrollment using multiple snapshots: 1.enrollment using single impression; 2. enrollment using multiple impression.

Fig. 5 presents the distribution of false match rate and false non-match rate of second experiment. Vertical axis represents the normalized distribution of matching scores, and horizontal axis represents the score ranging from 0 to 100. When using multiple snapshots on enrollment, equal error rate is observed lower by 1.38% comparing to the first experiment. And when false non-match rate is set 1%, in the false matches happen at the rate of about 6.15%.

According to the upper results, we can see that using multiple snapshots of the same fingerprint in case of enrollment corrects the false minutiae, and increase the performance entire system as a result.

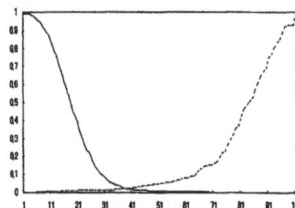

Fig. 5. FNR/FNMR curve when using four fingerprint images on the enrollment

References

1. ANSI web site, http://www.ansi.org/
2. Jain, L.C., Halici, U., Hayashi, I., Lee, S.B., Tsutsui, S.: Intelligent Biometric Techniques in Fingerprint and Face Recognition, CRC Press LLC, (1999)
3. Jain, A., Hong, L., Bolle, R.: On-line Fingerprint Verification. IEEE Trans. on Pattern Analysis and Machine Intelligence, Vol.19, No.4 (1997) 302–313
4. Ross, A., Jain, A. K., Reisman, J.: A Hybrid Fingerprint Matcher, Pattern Recognition, Vol. 36, No. 7, (2003) 1661-1673
5. Pan, S.B., Gil, Y.H., Moon, D., Chung, Y., Park, C.H.: A Memory-Efficient Fingerprint Verification Algorithm using A Multi-Resolution Accumulator Array, ETRI Journal, Vol. 25, No. 3, (2003) 179–186
6. Jain, A. K., Ross, A.: Fingerprint Mosaicking, Proc. ICASSP, (2002)
7. FVC 2002 web site, http://bias.csr.unibo.it/fvc2002
8. Maio, D., Maltoni, D., Cappelli, R., Wayman, J.L., Jain, A.K.: FVC2002:Second fingerprint verification competition, Proc. ICPR, (2002)

Signature Verification
Using Static and Dynamic Features

Mayank Vatsa[1], Richa Singh[1], Pabitra Mitra[1], and Afzel Noore[2]

[1] Department of Computer Science & Engineering
Indian Institute of Technology, Kanpur, India
mayank_richa@yahoo.com, pmitra@iitk.ac.in
[2] Lane Department of Computer Science & Electrical Engineering
College of Engineering and Mineral Resources
West Virginia University, Morgantown, USA
Afzel.Noore@mail.wvu.edu

Abstract. A signature verification algorithm based on static and dynamic features of online signature data is presented. Texture and topological features are the static features of a signature image whereas the digital tablet captures in real-time the pressure values, breakpoints, and the time taken to create a signature. 1D - log Gabor wavelet and Euler numbers are used to analyze the textural and topological features of the signature respectively. A multi-classifier decision algorithm combines the results obtained from three feature sets to attain an accuracy of 98.18%.

1 Introduction

Identification of an individual using behavioral biometrics is becoming prevalent and includes online and offline signature verification. Online verification deals with both static (number of black pixels, length and height of signature, etc) and dynamic features (time taken and the speed of signing, etc) of the signature, while offline verification extracts only the static features.

In this paper, we present an online signature verification algorithm which uses the online and offline features extracted from data tablet. Online features such as pressure values, breakpoints, and time taken to generate a signature are used to compute the matching score. Signature pattern is generated using the data points extracted from the tablet and then the static features, i.e. texture and topological features are analyzed to perform the matching. 1D-log Gabor [1] is used to extract the textural features of the signature pattern and Euler number is used to extract the topological features to compute the matching score for the static features. The weighted sum rule based multi-classifier decision algorithm combines the matching scores of online and offline features. The following sections describe the algorithm in detail and discuss the experimental results.

2 Signature Verification System

The block diagram of the signature verification system is shown in Figure 1. Forgery of signatures can be classified as: a) *Random forgery*, where the forger randomly guesses the signature, b) *Skilled forgery*, where the forger has prior knowledge of the

N.R. Pal et al. (Eds.): ICONIP 2004, LNCS 3316, pp. 350–355, 2004.
© Springer-Verlag Berlin Heidelberg 2004

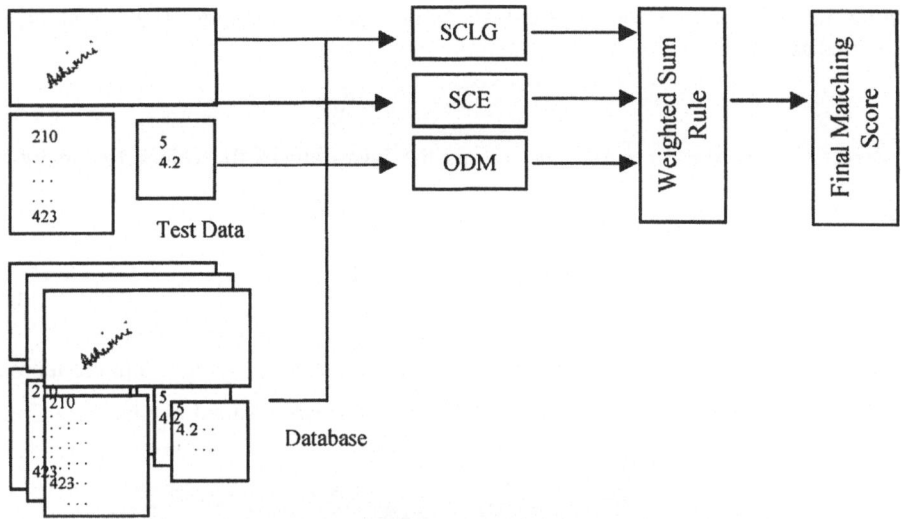

Fig. 1. Signature Verification System

signature and might have practiced in advance, and c) *Tracing*, where a signature instance is used as a reference to attempt forgery. Most systems have high verification rates for random forgery but low rates for skilled forgery and tracing. Our proposed signature verification algorithm combines static and dynamic feature set to obtain a high accuracy for both skilled forgery and tracing.

2.1 Data Preprocessing

The data acquisition process involves reading the reference signature data with the help of a digitizing tablet and obtaining the dynamic parameters (pressure, breakpoints and total time for a signature) and the image of the signature (Figure 1). Next, the input data is preprocessed using a low-pass filter to eliminate spurious noise inherent in the acquisition process [2].

2.2 Extraction of Static Features

The textural and topological features of a signature are extracted using algorithms based on 1D log Gabor and Euler numbers respectively. The resultant image generated by encoding the textural features is called Signature Code - log Gabor (SCLG). Euler numbers give a vector matrix which contains values extracted from the topological behavior of the signature. This vector matrix is called Signature Code – Euler (SCE).

Generating Signature Code – log Gabor: For generating the Signature Code using 1D - log Gabor wavelet, 2D normalized pattern is decomposed into a number of 1D signals. 1D signals are convolved with the 1D log-Gabor wavelets. A Gabor function is a harmonic wave modulated by a Gaussian function. In [6], log-Gabor filters are used for natural textures which often exhibit a linearly decreasing log power spectrum. In the frequency domain, log-Gabor filter bank according to Bigun et al. [1] is defined as:

$$G_{ij}(\omega_r, \omega_\varphi) = G(\omega_r - \omega_{r_i^\alpha}, \omega_{\varphi_i^\alpha})$$ (1)

where (r, φ) are polar coordinates, $\omega_{r_i^\alpha}$ is the logarithm of the center frequency at scale i, $\omega_{\varphi_i^\alpha}$ is the jth orientation and $G(\omega_r, \omega_\varphi)$ is defined as:

$$G_{\omega_r, \omega_\varphi} = \exp(\frac{\omega_r^2}{2\sigma_{ri}^2})\exp(\frac{\omega_\varphi^2}{2\sigma_{\varphi j}^2})$$ (2)

where σ_{ri}^2 and $\sigma_{\varphi j}^2$ are parameters of the Gaussian function. This algorithm produces a bitwise template containing a number of bits of information called as Signature Code – log Gabor. Figure 2 shows an example of SCLG.

Fig. 2. Signature Code of two different Signatures using 1D - log Gabor

Generating Signature Code – Euler: For generating the SCE from Euler numbers [4], four binary images corresponding to the four most significant bits of signature template are extracted from the signature image. Corresponding to each MSB extracted, one binary image is obtained for the whole image. Euler number is then computed for each of these binary images to obtain the topological property of the image i.e. (Number of Connected Components - Number of Holes). The topological property is useful for global description of regions in the image as it is unaffected by deformation or rotation. To generate the SCE, Euler numbers of the four binary images are stored in a vector matrix as its first four elements. The next three values represent the difference between adjacent Euler numbers. Figure 3 shows the SCE of a person at different instances.

	Euler Vector				Difference		
Image 1	49	49	51	52	0	2	1
Image 2	49	49	51	52	0	2	1
Image 3	49	49	51	53	0	2	2

Fig. 3. Image showing the Euler Code for an image

2.3 Online Data Extraction

The parameters obtained from online signatures consist of pressure values, time, x-tilt, y-tilt, x-value, y-value and breakpoints. In our algorithm, we use pressure, time and breakpoints for matching purposes. Although x-tilt and y-tilt are two important online features, these depend on holding style of the pen and orientation of the data tablet. X-values and y-values generate the signature pattern (static) and are therefore not considered again for online data matching.

2.4 Matching

Matching of the signatures includes matching of SCLG, matching of SCE, and matching of online data with stored signature data. For matching SCLG, Hamming distance based matching algorithm [3] is used. Hamming distance (*HD*) for the two SCLG is calculated using the equation below:

$$HD = \frac{1}{N}\sum_{i=1}^{N} A_i \oplus B_i \text{ and } MS_{SCLG} = (1 - HD) \tag{3}$$

where A_i and B_i are the two bit-wise SCLGs to be compared, N is the number of bits represented by each SCLG and $\oplus$ is the XOR operation. *HD* gives the matching score (MS_{SCLG}) for SCLG. For handling rotation, templates are shifted left and right bit-wise and a number of *HD* values are calculated from successive shifts [3]. The bit-wise shifting in the horizontal direction corresponds to rotation of the original signature template at an angle based on the angular resolution.

Fig. 4. SCLG of the same signature at different instances (HD = 0.3219)

For SCE based matching, Directional Difference Matching (DDM) algorithm is used. A comparison matrix is constructed with its elements as binary numbers. The matrix stores the results of comparison of input SCE to the SCE from the database. For comparing Euler numbers and Differences we use the following equation:

$$[X_1, X_2] = Y \pm \varepsilon \tag{4}$$

where Y is the Euler number/Difference from the input Euler code and ε is the tolerant error. If the value of X (Euler number / Difference from the stored Euler code) lies between X_1 and X_2, then we indicate a 1 in the comparison matrix; otherwise we enter a 0. Figure 5 illustrates the completed comparison matrix. In the experiment, FAR-FRR graph shows that maximum accuracy is achieved for $\varepsilon = 3$.

	Euler Vector				Difference		
Input	56	56	59	58	0	3	1
Database	56	56	58	58	0	2	0
Comparison	1	1	1	1	1	1	1

Fig. 5. Illustrating the Generation of Comparison Matrix

For matching, the number of 1's and 0's in the comparison matrix are counted. All 1's, denote a perfect match and all 0's signify a perfect mismatch. Although, ideally two SCE generated from the same signature should have a comparison matrix of all 1's, practically this does not occur because of the errors incurred at various stages. The comparison matrix gives the matching scores (MS_{SCE}) of SCE depending on the number of occurrence of 1's.

$$MS_{SCE} = Non\text{-}Zero\text{-}values\ (Comparison\ Matrix)\ /\ 7 \tag{5}$$

Matching of online data is carried out based on DDM algorithm. Signature tablet gives pressure values A for every signature (depending on the size of the signature, which is generally between 350 – 425 values), number of breakpoints B (number of times user has lifted the pen from the tablet) and total time T (time for generating the signature on the tablet). Thus from every signature which is to be matched using Equation 4, we have $A + 2$ online data, represented by O. After experimenting with different values of ε, for the first A values ε is set to 15 and for the last two values (B and T) ε is 1 in order to achieve optimal accuracy. A comparison matrix is generated by comparing O values of data from the database and input test data. The verification is performed at the threshold of 23 zeros in the comparison matrix. The matching score of online data (MS_{ODM}) is calculated as follows:

$$MS_{ODM} = Non\text{-}Zero\text{-}values\ (Comparison\ Matrix)\ /\ O \tag{6}$$

2.5 Multi-classifier Decision Algorithm

Weighted sum rule is used as the multi-classifier decision algorithm to obtain the final matching result [5]. Based on three matching scores obtained above, the weighted sum rule for fusion is given as,

$$MS = a * MS_{SCLG} + b * MS_{SCE} + c * MS_{ODM} \tag{7}$$

where, a, b and c are the weight factors, and MS is the final matching score of the multi-classifier algorithm. This is the statistical and user specific matching score obtained after experiments. The range of values for MS is given as: $0 \leq MS \leq 1$, where 0 denotes rejection and 1 denotes acceptance. Best matching results are obtained at $a=0.35$, $b=0.20$ and $c=0.45$ and hence threshold of 0.819 is used for selecting any query as genuine based on the characteristics of the FAR-FRR graphs.

3 Experimental Results

The proposed algorithm was tested using the signature database collected by the authors. The database consists of 1,100 images from 110 different individuals. There are 660 genuine signatures (6 per person), 220 images of random forgery (2 per person), 110 images of skilled forgery (1 per person), and 110 images representing forgery by tracing (1 per person). For training purposes, 330 genuine images from each individual were are and the remaining 770 images are used as test data. Thresholds of different values are determined by analyzing the results of the experiments. FAR-FRR graphs are used to determine the optimal thresholds for best performance. The thresholds of the four matching scores (MS_{SCLG}, MS_{SCE}, MS_{ODM} and MS) are found to be 0.650, 0.875, 0.935 and 0.823 respectively. Using these thresholds in the FAR-FRR graph, the best performance and the accuracy of the multi-classifier decision algorithm is determined. It has been found from the FAR-FRR graph that the proposed algorithm gives the maximum accuracy of 98.18%. Table 1 shows the results obtained from these algorithms.

Table 1. Accuracy of the algorithms

Algorithm	Genuine	Random	Skilled & Tracing	Accuracy
Euler	91.49	88.71	71.04	83.75
1D log Gabor	98.21	98.16	91.21	95.86
Online Data	98.62	98.55	96.38	97.85
Weighted Sum Rule	98.75	98.69	97.11	98.18

4 Conclusion

In this paper a signature verification algorithm has been presented which uses the static and dynamic features of the signature data. Signature data includes the signature image, pressure values, number of breakpoints and time. The static behavior of the signature is analyzed using 1D - log Gabor wavelet transform and Euler numbers and matching is performed to obtain the matching scores for the offline features. Online data are matched using statistical method and a matching score is calculated. Finally weighted sum rule is used as multi-classifier decision algorithm for calculating the final matching score which classifies the query data as matched or mismatched. The experimental results show that this algorithm is robust to skilled forgeries and tracing with an overall accuracy of 98.18%.

References

1. Bigun J. and du Buf J. M., N-folded symmetries by complex moments in Gabor space and their applications to unsupervised texture segmentation, IEEE Transactions on PAMI Vol. 16, No. 1, (1994) 80-87.
2. Brault J-J. and Plamondon R., Segmenting Handwritten Signatures at Their Perceptually Important Points, IEEE Transactions on PAMI, Vol. 15, No. 9, (1993) pp. 953-957.
3. Daugman, J. Recognizing Persons by their Iris Patterns, in Biometric: Personal Identification in Networked Society, A. Jain, R. Bolle, S. Pankati, Eds, Kluwer, (1998) 103-121.
4. Gonzalez, Woods, *Digital Image Processing*, Second Edition, Pearson Education.
5. Kittler Josef, Hatef Mohamad, Duin Robert P. W. and Mates Jiri, On combining classifiers, IEEE Transactions on PAMI, Vol. 20, No.3, (1998) 226–239.
6. Rubner Y. and Tomasi C., Coalescing Texture Descriptors. Proceedings of the ARPA Image Understanding Workshop, (1996).
7. Scott D. C., Jain A.K. and Griess F. D., On-line Signature Verification, Pattern Recognition, Vol. 35, No. 12, (2002), 2963–2972.

Face Recognition Using SVM
Combined with CNN for Face Detection

Masakazu Matsugu, Katsuhiko Mori, and Takashi Suzuki

Canon Inc., Intelligent I/F Project., 5-1, Morinosato-Wakamiya, Atsugi, 243-0193 Japan
{matsugu.masakazu,mori.katsuhiko,suzuki.takashi}@canon.co.jp

Abstract. We propose a model for face recognition using a support vector machine being fed with a feature vector generated from outputs in several modules in bottom as well as intermediate layers of convolutional neural network (CNN) trained for face detection. The feature vector is composed of a set of local output distributions from feature detecting modules in the face detecting CNN. The set of local areas are automatically selected around facial components (e.g., eyes, moth, nose, etc.) detected by the CNN. Local areas for intermediate level features are defined so that information on spatial arrangement of facial components is implicitly included as output distribution from facial component detecting modules. Results demonstrate highly efficient and robust performance both in face recognition and in detection as well.

1 Introduction

Face recognition algorithms have been extensively explored [1]-[3], [5]-[7], [9], [12]-[14] and most of which address the problem separately from object detection, which is associated with image segmentation, and many assume the existence of objects to be recognized without background. Some approaches, in the domain of high-level object recognition, address economical use of visual features extracted in the early stage for object detection. However, only a few object recognition algorithms proposed so far explored efficiency in the combined use of object detection and recognition [9].

For example, in the dynamic link matching (DLM) [14], Gabor wavelet coefficient features are used in face recognition and detection as well. However, we cannot extract shape as well as spatial arrangement information on facial components directly from those features since, for a set of nodes of the elastic graph, they do not contain such information. This necessitated to device the graph matching technique, a computationally expensive procedure, which requires quite different processing from feature detection stage. Convolutional neural networks (CNN) [8] have been exploited in face recognition and hand-written character recognition. In [10], we proposed a CNN model for robust face detection. SVM has also been used for face recognition [5]-[7], [9], [13]. In particular, in [6], [7], SVM classification was used for face recognition in the component-based approach.

This study, in the domain of face recognition as a case study for general object recognition with object detection, explores the direct use of intermediate as well as low level features obtained in the process of face detection. Specifically, we explore the combined use of convolutional neural networks (CNN) and support vector machines (SVM), the former used for feature vector generation, the latter for classification.

N.R. Pal et al. (Eds.): ICONIP 2004, LNCS 3316, pp. 356–361, 2004.
© Springer-Verlag Berlin Heidelberg 2004

Proposed algorithm is one of component-based approaches [6], [7] with appearance models represented by a set of local, area-based features. The direct use of intermediate feature distributions obtained in face detection, for face recognition, brings unified and economical process that involves simple weighted summation of signals, implemented both in face detection and recognition.

The outline of the paper is as follows. Section 2 gives a brief overview of convolutional neural networks for face detection. In Section 3, we describe procedures for the extraction of feature vectors for face recognition followed by description about face recognition procedure using one-vs-one multi-class SVM. Section 4 presents experimental results followed by conclusion in Section 5.

2   Convolutional Neural Networks for Face Detection

Convolutional neural networks, with hierarchical feed-forward structure, consist of feature detecting (FD) layers, each of which followed with a feature pooling (FP) layer or sub-sampling layer [2].

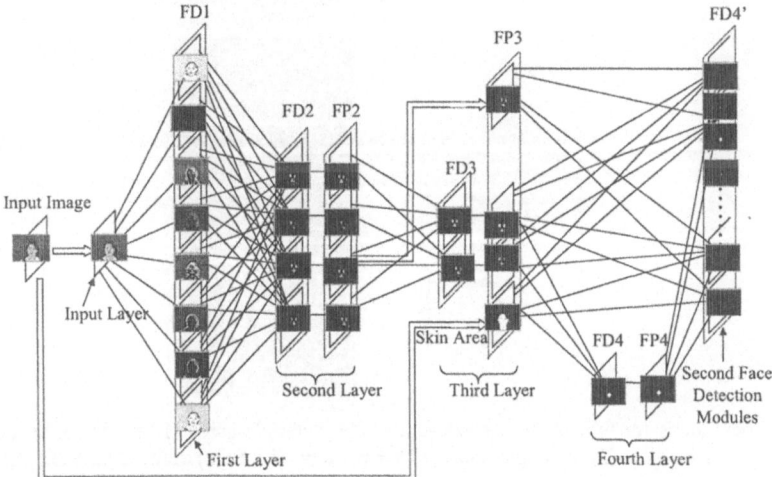

Fig. 1. The CNN model with selective activated modules for face detection [10]

This architecture comes with the property of robustness in object recognition such as translation and deformation invariance as in well-known *neocognitrons* [4], which also have similar architecture. Our CNN model for face detection as shown in Fig.1 is slightly different from traditional ones in that it has only FD modules in the bottom and top layers. Feature pooling (FP) neurons perform either maximum value detection or local averaging in their receptive fields of appropriate size.

The intermediate features detected in FD2 constitute a set of figural alphabets [10] in our CNN. Local features in FD1 are used as basis of figural alphabets such as corner-like structures and elongated blobs, which are used for eye or mouth detection. Face detecting module in the top layer is fed with a set of outputs from facial component (e.g., such as eye, mouth) detectors as spatially ordered set of local features of intermediate complexity.

3   Feature Vectors Extracted from Low and Intermediate CNN Outputs

We describe feature vectors and the procedure for their generation in face recognition. A feature vector, F, used in SVM for face recognition is an N dimensional vector, synthesized from a set of local output distributions, F_1 (as shown in Fig.2(1)), in a module detecting edge-like feature in FD1 layer in addition to output distributions, F_2, (as shown in Fig.2(2)) of two intermediate-level modules detecting eye and mouth in FD2 layer. Thus, $F = (F_1, F_2)$ where $F_1 = (F_{11}, ..., F_{1m})$ and $F_2 = (F_{21}, ..., F_{2n})$ are synthesized vectors formed by component vectors, F_{1k} ($k=1, ..., m$) and F_{2k} ($k=1, ..., n$). Here, m and n are the number of local areas for F_1 and F_2 component vectors, respectively. Each component vector represents possibility or presence of specific class of local feature in an assigned local area. Dimension of a component vector is the area of a rectangular region as in Fig.2. Thus dimension of feature vector, N, is the total summation of respective dimensions of component vectors.

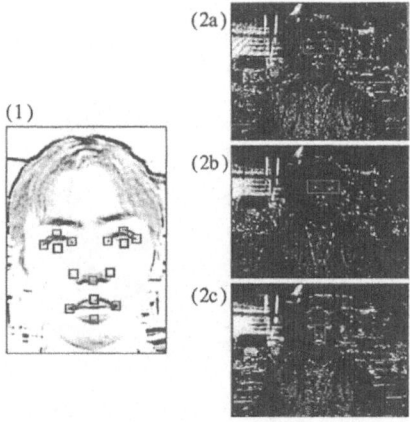

Fig. 2. Local areas for feature vector extraction. (1) rectangle areas (15 x 15) set around eye-corners, mouth corners (2) rectangle areas (125 x 65: a,b; 45 x 65:c) defined for FD2 output

In particular, $F_1 = (F_{11}, F_{12}, ..., F_{1,15})$, and local areas, total number of assigned areas being 15 as in Fig.2 (1), for component vectors are set around eye, nose, and mouth, using the detected eye location from the CNN. F_1 reflects shape information of eye, mouth, and nose. $F_2 = (F_{21}, F_{22}, F_{23})$, and each component vector reflects spatial arrangement of eye or eye and nose, etc., depending on how local areas in FD2 (e.g., positions and size) are set. The procedure for feature vector extraction is as follows. First, we define a set of local areas for outputs in FD1 as well as in FD3 modules based on the position of eyes and mouth detected from FD3 modules. Secondly, feature vector components are extracted. Finally, a feature vector is formed as concatenation of component vectors.

Positions of local areas in FD1 module are set around specific facial components (i.e., eyes, mouth) as illustrated in Fig. 2 (1). The size of respective local areas in the output plane of FD1 module is set relatively small (e.g., 11 x 11) so that local shape information of figural alphabets can be retained in the output distribution, while the local area in the FD2 plane is relatively larger (e.g., 125 x 65) so that information concerning spatial arrangement of facial components (e.g., eye) is reflected in the distribution of FD2 outputs.

4   Results

As in [10], [11], training of the CNN is performed module by module using fragment images as positive data extracted from database (e.g., Softpia Japan) of more than 100 persons. Other irrelevant fragment images extracted from background images are used as negative samples. The size of partial images for the training is set so that only one class of specific local feature is contained. The number of training data set is 14847 including face images and background image for FD4 module, 5290 for FD3, and 2900 for FD2.

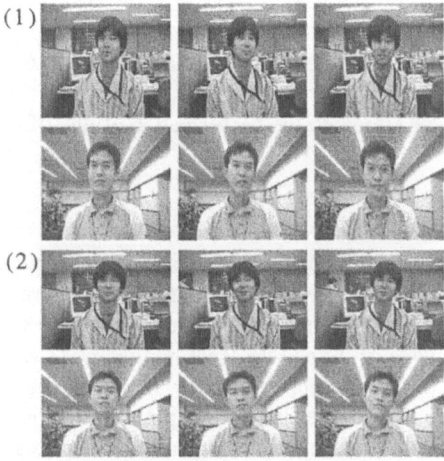

Fig. 3. Training images of faces with varying poses; (1) rotation in depth (2) rotation in plane

For face recognition, we use an array of linear SVMs, each trained for one-against-one multi-class recognition of faces. The SVM library used in the simulation is *libsvm2.5*, available in the public domain. In the SVM training, we used a dataset of feature vectors (FVs) extracted, under varying image capturing conditions, for each person in the way described in Section3.

The size of input image is VGA, and as illustrated in Fig.2, the size of local areas for FVs is 15 x15, 125 x 65, or 45 x 65 depending on the class of local features. As indicated in Fig.2, the number of local areas for FD1 feature and FD2 feature is fourteen and two, respectively. The number of FVs for one person is 30, which are obtained under varying image capturing conditions so that size, pose, facial expression, and lightning conditions of respective faces are slightly different.

The result shown in Fig.4, obtained using test images, different from training data, indicates robustness to size variability from 0.8 to 1.2 (relative size in units of area for reference face), demonstrating 100% recognition with 0% false acceptance rate. Using the same dataset, we compared our model with commercially available software (by Omron Inc.), which is based on DLM [14]. The recognition rate turned out to be

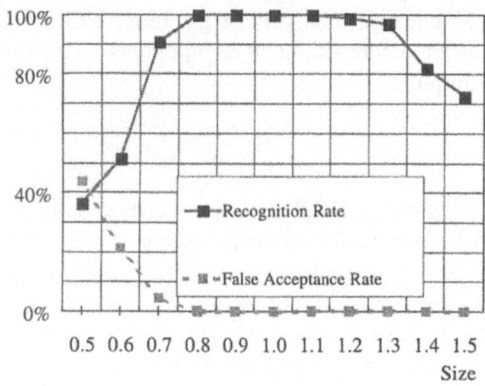

Fig. 4. Robust face recognition performance for 20 people under varying sizes of face

almost the same for the relative size of 0.8 to 1.2, while F.A.R. is slightly inferior to our model (i.e., F.A.R. is not perfectly zero), suggesting that our model involving much simpler operations equals to the performance of one of the best models.

5 Summary

We presented a new model of object recognition with economical use of low or intermediate level features. The preliminary results demonstrated robustness and efficiency in face recognition combined with detection, with 100 % recognition rate and 0% F.A.R. for 600 images of 20 people. Test images were with complex background (e.g., Fig.3), captured under varying conditions, including size and pose variability.

The novelty of proposed model lies in simple and efficient mechanism of object recognition that involves extracting alphabetical local features in CNN for face detection and generating support vectors in SVM from local area-based-feature vectors from intermediate outputs in CNN, both require relatively simple linear operation. This mechanism is in contrast with renowned method [14] in that our proposed model does not require computationally expensive and biologically implausible mechanism, a graph matching in DLM.

The computational procedure proposed mainly involves only weighted summation of inputs, implemented both in CNN and in linear SVM. Thus the approach presented in this study can be described in a common framework of relatively simple neuronal computation, weighted summation of inputs. This can be directed to incorporate as substrate for general model of object recognition and detection.

References

1. Belhumeur, P., Hesoanha, P., Kriegman, D.: Eigenfaces vs fisherfaces: recognition using class specific linear projection. IEEE Trans. on Pattern Analysis and Machine Intelligence **19** (1997) 711-720
2. Brunelli, R., Poggio T.: Face recognition: features versus templates. IEEE Trans. on Pattern Analysis and Machine Intelligence **15** (1993) 1042-1052

3. Turk, M., Pentland, A.: Face recognition using eigenfaces. Proc. IEEE Conf. On Computer Vision and Pattern Recognition (1991) 586-591
4. Fukushima, K.: Neocognitron: a self-organizing neural networks for a mechanism of pattern recognition unaffected by shift in position. Biological Cybernetics 36 (1980) 193-202
5. Guodong, G., Li, S., Kapluk, C.: Face recognition by support vector machines. Proc. IEEE International Conf. On Automatic Face and Gesture Recognition (2000) 196-201
6. Heisele, B.,Ho, P., Poggio, T.: Face recognition with support vector machines: global versus component-based approach. Proc. International Conf. on Computer Vision (2001) 688-694
7. Heisele, B., Koshizen, T.: Components for Face Recognition Proc. IEEE International Conf. on Automatic Face and Gesture Recognition (2004)
8. Le Cun, Y., Bengio, T.: Convolutional networks for images, speech, and time series. In: Arbib, M.A. (ed.): The handbook of brain theory and neural networks, MIT Press, Cambridge (1995) 255-258
9. Li, Y., Gong, S., Liddel, H.: Support vector regression and classification based multi-view face detection and recognition. Proc. IEEE International Conf. on Automatic Face and Gesture Recognition (2000) 300-305
10. Matsugu, M., Mori, K., Ishii, M., Mitarai, Y.: Convolutional spiking neural network model for robust face detection. Proc. International Conf. on Neural Information Processing (2002) 660-664
11. Mitarai, Y.,Mori, K., Matsugu, M.: Robust Face Detection System Based on Convolutional Neural Networks Using Selective Activation of Modules (In Japanese). Proc. Forum in Information Technology (2003) 191-193
12. Moghaddam, B., Wahid, W., Pentland, A.: Beyond eigenfaces: probabilistic matching for face recognition. Proc. IEEE International Conf. on Automatic Face and Gesture Recognition (1998) 30-35
13. Pontil, M., Verri, A.: Support vector machines for 3-d object recognition. IEEE Trans.on Pattern Analysis and Machine Intelligence 20 (1998) 637-646
14. Wiskott, L., Fellous, J.-M., Krüger, N., von der Malsburg, C.: Face recognition by elastic bunch graph matching. IEEE Trans. on Pattern Analysis and Machine Intelligence 19 (1997) 775-779

Face Recognition Using Weighted Modular Principle Component Analysis

A. Pavan Kumar, Sukhendu Das, and V. Kamakoti

Department of Computer Science and Engineering
Indian Institute of Technology Madras, Chennai -6000036
apavan@peacock.iitm.ernet.in, {sdas,kama}@iitm.ernet.in

Abstract. A method of face recognition using a weighted modular principle component analysis (WMPCA) is presented in this paper. The proposed methodology has a better recognition rate, when compared with conventional PCA, for faces with large variations in expression and illumination. The face is divided into horizontal sub-regions such as forehead, eyes, nose and mouth. Then each of them are separately analyzed using PCA. The final decision is taken based on a weighted sum of errors obtained from each sub-region. A method is proposed, to calculate these weights, which is based on the assumption that different regions in a face vary at different rates with expression, pose and illumination.

1 Introduction

Biometry using face recognition is an increasingly important area today. Its applications are becoming more important, as in ATM machines, criminal identification, access restriction, monitoring public areas for known faces. The task of automated face recognition is very difficult due to the similarity of faces in general and large variations in the faces of same person due to expression, pose and illumination.

Various algorithms have been proposed for the automatic face recognition in last few decades, with varying degrees of success. Rama Chellappa et al.[1] gave a detailed survey of face recognition algorithms based on neural network models, statistical models, and feature-based models. Majority of the contributions are based on PCA [2], LDA [3] and SVM [4] techniques. Modular PCA [5, 6] is an improvement proposed over PCA. Most of these AFR algorithms evaluate faces as one unit which leads to problems due to variations in expression, illumination and pose. This neglects the important fact that few facial features are expression invariant and others are more susceptible to the expressions.

In this paper we propose a modified approach, where different parts of face (eyes, nose, lips) are separately analyzed and the final decision is based on the weighted sum of errors obtained from separate modules. We have also proposed a method to calculate these weights using the extent to which each sub-region, of a subject, is spread in the eigenspace. The weights are the measures of intra-person variance of the sub-region.

N.R. Pal et al. (Eds.): ICONIP 2004, LNCS 3316, pp. 362–367, 2004.
© Springer-Verlag Berlin Heidelberg 2004

This paper is organized as follows: Section 2 gives an overview of PCA. Section 3 describes the proposed algorithm. In Section 4, we discuss the experiments and results. Finally Section 5 gives the conclusions and future scope of work.

2   Review of PCA

PCA is a dimensionality reduction technique. Usually a face image of size $N * N$, can be represented as a point in a N^2-Dimension space, termed as *image Space*. Since most faces are similar in nature, face images are not randomly distributed in the image space and fall in a small subspace, called *face Space*. The concept of PCA is to find vectors that best describe the distribution of these faces in the image subspace.

Let the training set be Γ_1, Γ_2, Γ_3, ..., Γ_M where M is the number of faces in the training set. These faces are represented by using column vectors $(N^2 * 1)$ instead of the usual matrix representation $(N * N)$. The average face of the training set, Ψ, is calculated as, $\Psi = \frac{1}{M} \sum_{m=1}^{M} \Gamma_m$. A vector that describes the difference of each face from average face is obtained, as $d_m = \Gamma_m - \Psi$, $m = 1. ..., M$. The covariance matrix is obtained as,

$$C = \frac{1}{M} \sum_{m=1}^{M} d_m d_m^T \tag{1}$$

The eigenvectors of this matrix are computed and the most significant S eigenvectors, $\mu_1, \mu_2, ..., \mu_S$ are chosen as those corresponding to largest corresponding eigenvalues. Given these eigenvectors, each face Γ_m can be expressed as a set of weights, $w_{m,s}$, which are obtained as,

$$w_{m,s} = \mu_s^T(\Gamma_m - \Psi), \quad m = 1, 2, \ldots, M; \; s = 1, 2, \ldots, S; \tag{2}$$

The weights obtained in above equation form the weight vector for the corresponding face m, $\Omega_m = [w_{m,1} \, w_{m,2} \ldots w_{m,S}]^T$, where $m = 1, 2, \ldots, M$. Given a test face Γ_{test}, it is projected on the face space and the weights are obtained as in (2), $w_{test,s} = \mu_s^T(\Gamma_{test} - \Psi)$, where $s = 1, .., S$, which gives the corresponding weight vector, Ω_{test}. The error vector, which is the euclidean distance between the test face Ω_{test} and training faces Ω_m, is obtained as, $e_m = ||\Omega_{test} - \Omega_m||$, where $m = 1, 2, \ldots, M$.

The test face Γ_{test} is said to have best matched with a face, $\Gamma_{m'}$, for which the error vector e_m is minimum. Suitable threshold τ can be used for rejection as $\tau < min(e_m)$.

3   Proposed Methodology

We propose an algorithm based on modular PCA, which modularizes the face into sub-regions and performs recognition on each sub-region individually. Each face is horizontally split into a set of sub-regions such as forehead, eye, nose, mouth, chin. For each sub-region, ρ, of each face, we now compute average sub-region, calculate covariance matrix, eigenvectors and the weight set as mentioned in section 2. All these computations can be implemented in parallel. Finally net error is obtained as a weighted sum of the error vectors of individual sub-regions. The given face is classified as to belong to that class which is at nearest euclidean distance in the face space.

3.1   Training

Let the training set contain L subjects, where each subject is one person. Each person has N different faces. So the training set has $M = LN$ faces. All M faces are divided in to R regions. Hence, each rth partition of nth sample of lth subject is, $\rho_{l,n,r}$, where $l = 1, 2, ..., L$; $n = 1, 2, ..., N$; $r = 1, 2, ..., R$. Thus entire training set can be represented as $T_{set} = \{ \rho_{l,n,r} \mid \forall l, n, r \}$. The following steps are repeated for each sub-region $r = 1, 2, ..., R$. For each rth sub-region, an average sub-region, ψ_r is computed over all faces as, $\psi_r = \frac{1}{LN} \sum_{l=1}^{L} \sum_{n=1}^{N} \rho_{l,n,r}$. This equation can be conveniently rewritten as,

$$\psi_r = \frac{1}{M} \sum_{m=1}^{M} (\Upsilon_m)_r, \quad r = 1, 2, ..., R; \ M = LN;$$

where $(\Upsilon_m)_r$ is the rth sub-region of mth face.

The covariance matrix C_r of rth sub-region is calculated as in (1) and its eigenvectors are computed. The most significant S eigenvectors, $((\mu_1)_r , ..., (\mu_S)_r)$, are considered for each sub-region r as mentioned in section 2. Then each sub-region r of face m can be expressed as a set of weights, $(w_{m,s})_r$, which are calculated as,

$$(w_{m,s})_r = (\mu_s)_r^T ((\Upsilon_m)_r - \psi_r), \quad m = 1, 2, ..., M; \ s = 1, 2, ..., S; \ r = 1, 2, ..., R; \tag{3}$$

Similarly weight vector of each sub-region $(\Omega_m)_r$ is generated from these weights as, $(\Omega_m)_r = [\ (w_{m,1})_r \ \ (w_{m,2})_r \ \ ... \ \ (w_{m,S})_r \ ]$, $m = 1, 2, ..., M$; $r = 1, 2, ..., R$.

3.2   Intra-subject Variance of Each Sub-region

As mentioned, the final decision is based on the weighted sum of error vectors obtained from each sub-region. These weights represent a measure of *the extent of variation in eigenspace for a sub-region of a subject across all samples.* For each sub-region r of each subject, l, average sub-region $(\Phi_l)_r$ is calculated. Then For each sub-region r, the measure of variance for l^{th} subject is,

$$(P_l)_r = \frac{1}{N} \sum_{n=N*(l-1)+1}^{l*N} [(\Omega_n)_r - (\Phi_l)_r]^2, \quad l = 1, 2, ..., L; \ r = 1, 2, ..., R; \tag{4}$$

It may be noted that more compact sub-regions have lesser value of $(P_l)_r$.

3.3   Classification

Given a test face, Γ_{test}, it is split into R horizontal sub-regions as in the training phase. These regions can be represented as, $(\Upsilon_{test})_r$ where $r = 1, 2, ..., R$. These regions are then projected onto face space, weights are calculated as in (3), $(w_{test,s})_r = (\mu_s)_r^T((\Upsilon_m)_r - \psi_r)$. The corresponding weight vector is built as, $(\Omega_{test})_r = [(w_{test,1})_r \ (w_{test,2})_r \ \cdots \ (w_{test,S})_r], \ r = 1, 2, ..., R;$

The error vector for a region r, is the euclidean distance between $(\Omega_{test})_r$ and $(\Omega_m)_r$. It is computed as $(E_m)_r = [(\Omega_{test})_r - (\Omega_m)_r]^2, \ m = 1, 2, ..., M; \ r = 1, 2, ..., R$. For each subject, the sub-region that is more invariant to expressions and illuminations is given more priority in the net error function. This is implemented by multiplying each error of the sub-region with the measure obtained in (4). The net error function for comparing a test image Γ_{test} with Γ_m is,

$$(F_{test})_m = \sum_{r=1}^{R} [(\Omega_{test})_r - (\Omega_m)_r]^2.(P_l)_r, \qquad m = 1, 2, \ldots, M; \qquad (5)$$

where l is the subject of mth sample. The test face is said to have matched with face m', for which $(F_{test})_{m'} = min(F_{test})_m, \forall m$. Suitable threshold is used to reduce false acceptance.

Reconstruction: The sub-regions of the test face can be reconstructed from the eigenvectors, the weight vectors of each sub-region and variance measure of each subject as, $(\rho_r)_{rc} = \psi_r + (P_l)_r[\sum_{i=1}^{S}(w_{test,i})_r(\mu_i)_r]$, where $r = 1, 2, \ldots, R$ and l is the subject into which Γ_{test} is classified as. The test face can be obtained by concatenating these reconstructed sub-regions.

4   Experiments and Results

The algorithm was tested on the Yale Face Database. This database consists of 15 subjects each with 11 different samples with varying expressions and illumination. The training set consists of only 6 images of each subject whereas the other 5 images are used for testing. This choice was done such that both the sets had expression and illumination variations. Figure 1 shows the images used in testing and training phases of the experiment for a subject.

The faces were first cropped horizontally (manually in this experiment), into three sub-regions containing forehead, eyes with nose and mouth as shown in

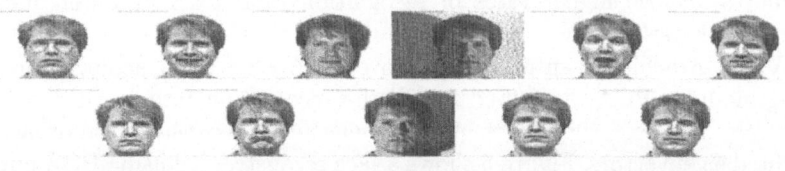

Fig. 1. Some examples of faces used for training (top row) and testing (bottom row).

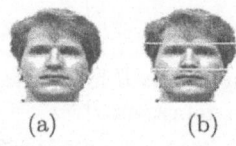

(a)   (b)

Fig. 2. (a) Actual face and (b) the cropped modules of the face from the Yale database.

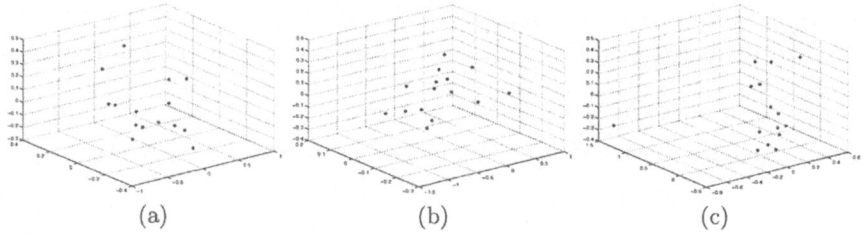

(a)   (b)   (c)

Fig. 3. Distribution of average (a) foreheads (b) eyes and (c) mouths for all subjects in 3-D eigenspace using the first 3 eigenvectors.

Table 1. Intra-person variance of each sub-region, for 10 different subjects.

Subject	1	2	3	4	5	6	7	8	9	10
Forehead	0.5357	0.8212	0.8415	0.8456	1.0000	0.7148	0.7089	0.7924	0.7697	0.6346
Eyes	0.8567	1.0000	0.9986	1.0000	0.9222	1.0000	1.0000	1.0000	1.0000	1.0000
Mouth	1.0000	0.5556	1.0000	0.9508	0.9204	0.7769	0.9810	0.9399	0.9455	0.8648

Fig. 2. The method of training explained in section 3.1 was applied to all these sub-regions and the weight vectors were computed. Then measures of intra-person variance for all the sub-regions were calculated as in section 3.2. Figure 3 shows the distribution of average foreheads, eyes and mouths for all the subjects in 3-D eigenspace using the first three eigenvectors. A set of weights obtained using 20 eigenvectors, for 10 different subjects, are given in Table 1. These weights are normalized for each subject.

We performed PCA on the actual samples and modular PCA, weighted modular PCA (WMPCA) on the partitioned set with varying number of eigenvectors. The recognition rates obtained using PCA, MPCA and WMPCA, for 5, 10, 20, 30, 40 eigenvectors are illustrated in Fig. 4. It can be observed that WMPCA is able to achieve higher rates of recognition than PCA at lower number of eigenvectors itself.

WMPCA achieved an accuracy of over 87% while PCA achieved only 76%. Using modular PCA, described in [6], the recognition rate reached only 80%. There also has been significant improvement in the reconstruction of faces from weighted eigenvectors. Figure 5 shows a face reconstructed using PCA and WMPCA. The recognition rate of WMPCA improved to 89%, if 7 images of each subject were used for training.

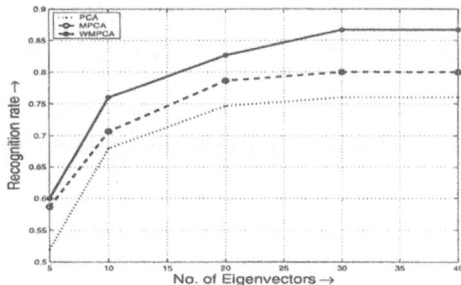

Fig. 4. Results of PCA, MPCA, WMPCA for different number of eigenvectors used in the experiment.

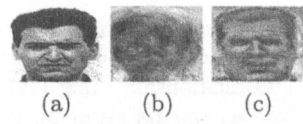

(a)   (b)   (c)

Fig. 5. Reconstruction of faces: (a) Test face, (b) Reconstruction using PCA, (C) using WMPCA.

5   Conclusions

In this paper we assume that different regions of the face vary at different rates due to variations in expression and illumination. We recognize all sub-regions of the face independently and the final decision is a weighted sum of the errors of each sub-region. We also calculate the intra-person variance of each sub-region, which is a measure of how each sub-region of a subject varies over various expressions and illuminations. The results were very promising and the method is suitable for real time applications. The recognition rate shows improvement over PCA and modular PCA in case of faces having variation in expression and illumination.

References

1. Zhao W. and Chellappa R. and Phillips P. J.: Face Recognition: A Literature survey. ACM Computing Surveys **35** (2003) 339–458
2. Turk M. and Pentland A.: Eigen faces for recognition. Journal of cognitive neuroscience **3 No. 1** (1991) 71–86
3. Peter N. Belhumeur and Hesphana P.: Eigenfaces vs. Fisherfaces: Recognition using class specific linear projection. IEEE Trans. on Pattern Analysis and Machine Intelligence **19, No.7** (1997) 711–720
4. Bernde Heisele and Purdy Ho and Tomoso Poggio: Face recognition with support vector machines: Global Vs Component based approach. Proceedings of International Conference on Computer Vision **2** (2001) 688–694
5. Pentland A. and Moghaddam B. and Starner T.: View based and modular eigenspaces for face recognition. In: Proceedings of IEEE Computer Society Conference on Computer Vison and Pattern Recognition. (1994) 84–91
6. Rajkiran G. and Vijayan K. : An improved face recognition technique based on modular PCA approach . Pattern Recognition Letters **25 , No. 4** (2004) 429–436

Self-organizing Relationship (SOR) Network with Fuzzy Inference Based Evaluation and Its Application to Trailer-Truck Back-Up Control

Takanori Koga, Keiichi Horio, and Takeshi Yamakawa

Kyushu Institute of Technology
Graduate School of Life Science and Systems Engineering
Hibikino 2-4, 8080196 Fukuoka, Japan
koga-takanori@edu.brain.kyutech.ac.jp
{horio,yamakawa}@brain.kyutech.ac.jp

Abstract. In this paper, the self-organizing relationship (SOR) network with fuzzy inference based evaluation is proposed. The SOR network can extract a desired I/O relationship using I/O vector pairs and their evaluations. The evaluations can be given by a user or calculated by the evaluation function. However, in many applications, it is difficult to calculate the evaluation using simple functions. It is effective to employ fuzzy inference for evaluating the I/O vector pairs. The proposed system is applied to design the trailer-truck back-up controller, and experimental result is easily realized with some fundamental fuzzy if-then rules.

1 Introduction

The self-organizing maps (SOM)[1][2] is one of the most popular neural networks, and it has been applied to many fields such as pattern recognition, data visualization, data analysis, and so on[3][4]. By modifying the SOM, we proposed the self-organizing relationship (SOR) network which is established to approximate a desired I/O relationship of a target system[5]. The I/O vector pairs of the target system are employed as the learning vectors, and weight vectors which are parameters in SOR network are updated based on the evaluations of I/O vector pairs. There are many systems that the correct I/O relationship is not available but the I/O relationship can be evaluated by evaluation functions or intuition of users and so on. The validity of the SOR network was verified by applying it to design of the control system[6]. However, in some cases, it is difficult to evaluate the I/O relationship of the target system using simple evaluation functions. In this paper, we propose a new evaluation method in which the evaluations are calculated by fuzzy inference.

2 Self-organizing Relationship (SOR) Network

The SOR network consists of the input layer, the output layer and the competitive layer, in which n, m and N units are included, respectively, as shown

N.R. Pal et al. (Eds.): ICONIP 2004, LNCS 3316, pp. 368–374, 2004.
© Springer-Verlag Berlin Heidelberg 2004

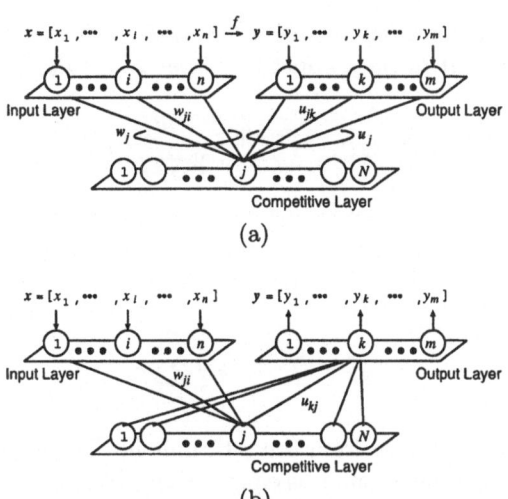

Fig. 1. The structure of SOR network. (a)Learning mode, (b)Execution mode.

in Fig. 1. The j-th unit in the competitive layer is connected to the units in the input and the output layers with weight vectors $w_j = [w_{j1}, \ldots, w_{jn}]$ and $u_j = [u_{j1}, \ldots, u_{jm}]$, respectively. The network can be established by learning in order to approximate the desired function $y = f(x)$. The operation of the SOR network is consist of both learning mode and execution mode.

2.1 Learning Mode of SOR Network

In learning, the random I/O vector pair $I = [x, y] = [x_1, \ldots, x_n, y_1, \ldots, y_m]$ is applied, as the learning vector, to the input and the output layers together with the evaluation E for the I/O vector pair as shown in Fig. 1(a). The evaluation E may be assigned by the network designer, given by the intuition of the user or obtained by examining the system under test. The value of E is positive and negative in accordance with judgment of the designer, preference of the user or score of examination. The positive E causes the self-organization of attraction to the learning vector and the negative one does that of repulsion from the learning vector. It means that the weight vectors are updated by the following equation.

$$v_j^{new} = \begin{cases} v_j^{old} + \alpha(t)E(I - v_j^{old}) & for\ E \geq 0 \\ v_j^{old} + \beta(t)E \exp(-\parallel I - v_j^{old} \parallel)\mathrm{sgn}(I - v_j^{old}) & for\ E < 0 \end{cases} \quad (1)$$

The weight vectors are arranged in area where desired I/O vector pairs exist by learning.

2.2 Execution Mode of SOR Network

After the learning, the SOR network is ready to use as the I/O relationship generator. The operation is referred to as the execution mode and it is illustrated in Fig. 1(b). The actual input vector x^* is applied to the input layer, and the output of the j-th unit in the competitive layer z_j is calculated by:

$$z_j = \exp(-\frac{\|x^* - w_j\|^2}{2\gamma_j^2})$$
(2)

where, γ_j is a parameter representing fuzziness of similarity, and z_j represents the similarity measure between the weight vector w_j and the actual input vector x^*. The output of the k-th unit in the output layer y_k^* is calculated by:

$$y_k^* = \sum_{j=1}^{N} z_j u_{kj} / \sum_{j=1}^{N} z_j$$
(3)

where, u_{kj} is a weight vector from the j-th unit in the competitive layer to the k-th unit in the output layer and it is equal to u_{jk} obtained in the learning mode. The output of the network $y^* = [y_1^*, \ldots, y_m^*]$ represents the weighted average of u_j by the similarity measure z_j.

2.3 Problems of SOR Network

The successful application of the SOR network is a DC motor control[6]. In this application, the evaluation function is defined based on the decrease of error. It means that the controller was designed in consideration of only decrease of the positioning error of the DC motor. In some applications, however, it is very difficult to design evaluation functions because of the complexity of the system. A design of a trailer-truck back-up control system is one of such applications.

3   Fuzzy Inference Based SOR Network

In order to apply the SOR network to design a trailer-truck back-up control system, we propose a new evaluation method employing fuzzy inference.

3.1 Trailer-Truck Back-Up Control

The trailer-truck (semi-trailer type) used in this paper is illustrated in Fig. 2. The trailer-truck goes back at a constant velocity $v(=0.1[m/s])$, and a control objective is to make the trailer-truck follow a target line. There are many papers in which the trailer-truck back-up control system is designed[7]-[9]. Especially, the method using fuzzy logic control is very efficient. However, an expert who knows the characteristics of behavior of the trailer-truck is needed to construct the fuzzy controller.

3.2 Application of the SOR Network
to Trailer-Truck Back-Up Control

A new trailer-truck control system by employing the SOR network is proposed. In the proposed system, the angle between trailer and truck ϕ, the angle of the

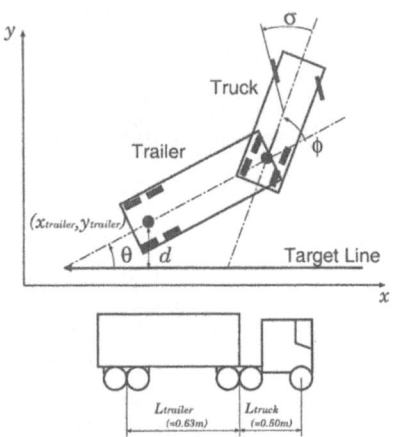

Fig. 2. Trailer-truck used in this paper.

trailer θ and the distance between the trailer-truck and the target line d as shown in Fig.2 are input to the SOR network. Accordingly, the SOR network generates the front wheel angle σ. In order to obtain the data for learning of the SOR network, the trail is done. At first, the state of the trailer-truck $(\phi(l), \theta(l), d(l))$ is randomly generated. Then, front wheel angle $(\sigma(l))$ is randomly generated. These values are elements of the learning vector. As a result of operation, the state at $(l+1)$ is observed. The designer should evaluate the I/O relationship of the system with fuzzy inference by observing the state at (l) and $(l+1)$, where, l represents time sampling, and the sampling interval is 1.0[sec].

3.3   Evaluation Based on Fuzzy Inference

In order to evaluate the trial results, we employed a fuzzy inference (product-sum-gravity method[10]). Fuzzy if-then rules and membership functions are shown in Fig.3. The input variables were defined to have five membership functions each, 'PL', 'PS', 'ZR' 'NS' ,'and 'NL'. At first, 5×5×5=125 fuzzy if-then rules are prepared, then the rules are marged as shown in Fig.3. The antecedent of the rules is the state of the trailer-truck at l, and the consequent of the rules is the decreace of the error between state (l) and $(l+1)$. The decreace of the error is normalized, the range is from -1 to 1. These fuzzy if-then rules are consist of four fundamental evaluation strategies described as follows.

(Strategy 1) If the angle between trailer and truck ϕ is large, it should be decreased. If the angle ϕ becomes large, about 90 degree, the trailer-truck can not be controlled any more. It is called jack-knife phenomenon, and to avoid falling into this phenomenon is above everything else.

(Strategy 2) If the trailer-truck is directed away from the target line, the direction should be corrected to meke the trailer-truck approach the target line.

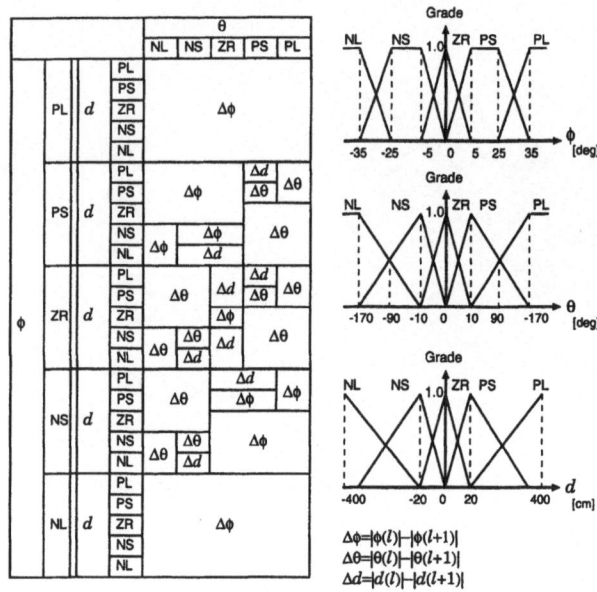

Fig. 3. Fuzzy if-then rules and membership functions used in the fuzzy inference.

(Strategy 3) If the trailer-truck move in a direction opposite to the target direction, the direction should be corrected to meke the trailer-truck move in the target direction.

(Strategy 4) If strategies 1-3 are satisfied, the trailer-truck should be controlled to follow the target line while considering the distance d and the angle θ.

It is not difficult to construct these rules even if a user does not have enough knowledge about dynamics of the trailer-truck.

4   Computer Simulation Results

In order to verify the effectiveness of the proposed method, we show computer simulation results of trailer-truck back-up control. In the simulation, it is assumed that the model of the trailer-truck is given and described by geometrical model[9]. The number of the learning vectors of the SOR network is 50,000, and the learning vectors are evaluated by fuzzy inference with the fuzzy if-then rules shown in Fig. 3. In the learning of the SOR network, the number of learning iterations is 1,000, and the number of units on the competitive layer is 900 (30×30). The initial values of the learning rates $\alpha(0)$ and $\beta(0)$ are 0.4 and 0.025, respectively. In the execution mode of the SOR network, the parameter γ_j is decided in consideration of the distribution of the weight vectors by:

$$\gamma_j = \frac{1}{P} \sum_{p=1}^{P} \|\mathbf{w}_j - \mathbf{w}_j^{(p)}\| \tag{4}$$

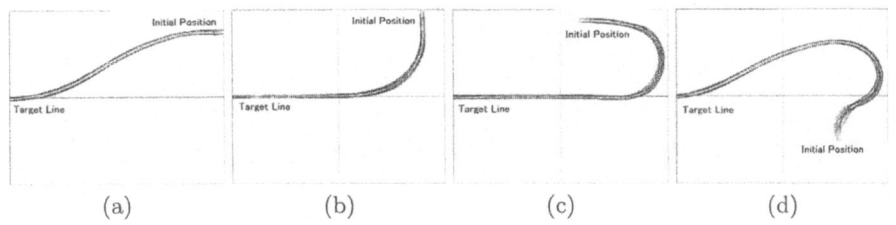

Fig. 4. The simulation results. Initial values: (a) $\phi=0$[deg], $\theta=0$[deg], $d=300$[cm], (b) $\phi=0$[deg], $\theta=90$[deg], $d=300$[cm], (c) $\phi=0$[deg], $\theta=180$[deg], $d=350$[cm], (d) $\phi=30$[deg], $\theta=-120$[deg], $d=-100$[cm].

where, $\mathbf{w}_j^{(p)}$ is the p-th unit according to the distance to the weight vector $\mathbf{w}_j$. In this simulation, $P = 5$. The simulation results are shown is Fig. 4 (a)-(d). In each figure, the lengths of the work space are 1000 cm and 800 cm for the x and y axes, respectively. The trailer-truck is controlled to satisfy the control objective from the various initial states while avoiding jackknife phenomenon . Especially, it is very difficult to control the trailer-truck from large ϕ (Fig. 4(d)). It is known that jack-knife phenomenon is avoided at first and the trailer-truck follows the target line successfully.

5 Experimental Results

The experiments in which the remote control trailer-truck is used are achieved. The flow of processing is as follows. The motion capture system captures the coordinates of three markers attached to the trailer-truck using two CCD cameras, and the coordinates are input to the PC through digital I/O board. In the PC, the angles ϕ, θ and the distance d are calculated and the front wheel angle σ, *i.e.* the output of the SOR network, is calculated. The command is sent to the truck through the A/D converter and the remote control proportional system. The weight vectors of the SOR network used in the experiment are the same to those used in the computer simulation. Fig. 5 shows the experimental result. The black line in the figures is the target line. The initial values of the distance and angles are $\phi = 0$[degree], $\theta = 135$[degree], and $d = 250$[cm] as shown in Fig. 5 (a). The angle of the trailer becomes smaller and the trailer-truck follows the target line finally. In cases of other initial values, it is confirmed that the trailer-truck can follow the target line.

6 Conclusions

In this paper, we proposed the new evaluation algorithm of the learning vectors of the SOR network. The fuzzy inference is employed to realize it. The fuzzy controller which should control trailer-truck directly requires strict knowledge and fine adjustment of parameters. On the other hand, the fuzzy if-then rules used in the proposed algorithm can be constructed without special knowledge

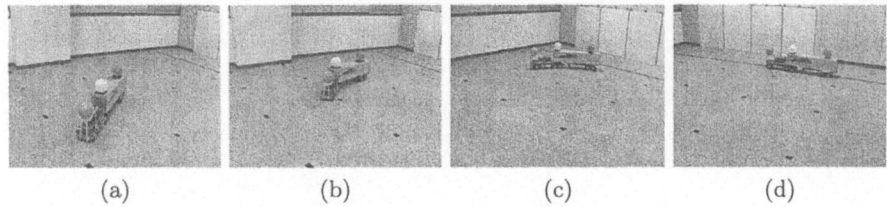

(a)   (b)   (c)   (d)

Fig. 5. The Experimental Result. (a)0sec, (b)4sec, (c)8sec, (d)12sec.

about the characteristics of the trailer-truck. In order to verify the effectiveness of
the proposed algorithm, computer simulation and the experiments using remote
control trailer-truck is achieved. The experimental results show that the control
objective is satisfied without any special knowledge about characteristics of the
trailer-truck by employing the proposed algorithm.

References

1. T. Kohonen, "Self-organizing formation of topologically correct feature map," *Biol. Cybern.*, Vol.43, pp.59-69, 1982.
2. T. Kohonen, *Self-organizing maps*, Springer Verlag, 1995.
3. R. P. Lippmann, "Pattern Classification Using Neural Networks," *IEEE Communication Magazine*, Vol.27, No.11, pp47-50, 1989.
4. H. Tokutaka, K. Fujimura, K. Iwamoto, S. Kishida, K. Yoshihara, "Application of Self-Organizing Maps to a Chemical Analysis," *Proc. of Int. Joint Conf. on Neural Information Processing*, Vol.2, pp.1318-1321,1997.
5. T. Yamakawa and K. Horio, "Self-organizing relationship (SOR) network," *IEICE Trans. on Fundamentals*, Vol.E82-A, pp.1674-1678, 1999.
6. K. Horio and T. Yamakawa, "Adaptive Self-Organizing Relationship Network and Its Application to Adaptive Control," *Proc. of the 6th Int. Conf. on SoftComputing and Information/Intelligent Systems (IIZUKA2000)*, pp.299-304, 2000.
7. D. Nguyen and B. Widrow, "The truck backer-upper: an example of self-learning in neural nerwork," *Proc. of IJCNN'89*, pp.357-363, 1989.
8. S.G. Kong and B. Kosko, "Adaptive fuzzy-systems for backing-up a truck-and-trailer," *IEEE Trans. on Neural Networks*, Vol.3(2), pp.211-223, 1992.
9. K. Tanaka and M. Sano, "A robust stabilization problem of fuzzy control systems and its application to backing up control of a truck-trailer," *IEEE Trans. on Fuzzy Systems*, Vol.2(2), pp.119-133, 1994.
10. M. Mizumoto,"Fuzzy controls by product-sum-gravity method," *Advancement of Fuzzy Theory and Systems in China and Japan,* International Academic Publishers, c1.1-c.1.4., 1990.

In-Vehicle Noise and Enhanced Speech Intelligibility

Akbar Ghobakhlou[1] and Richard Kilgour[2]

[1] KEDRI, Auckland University of Technology, Private Bag 92006, Auckland, New Zealand
akbar@aut.ac.nz
[2] Navman NZ Limited, PO Box 68155, Newton, Auckland, New Zealand
rkilgour@navman.com

Abstract. In-Car speech recognition will be pervasive over the coming years. The goal of speech enhancement is to increase the quality and intelligibility of speech in a noisy environment. The focus of the present research is to evaluate the effect of speech enhancement on the intelligibility of spoken language in a moving vehicle. Here, an ECoS network is used as a model to evaluate the intelligibility. A baseline performance was established using clean speech data. This data was then mixed with various types of in-vehicle noise at several signal-to-noise ratios. Speech enhancement techniques were applied to the noisy speech data. The performance of the ECoS model was evaluated when the noisy and enhanced speech was presented. Several factors were found to affect the recognition rate, including noise type and noise volume.

1 Introduction

As speech recognition becomes pervasive, several challenges need to be overcome. Applications such as telematics and navigation systems can be greatly enhanced by speech recognition, allowing control by drivers while leaving hands and eyes free. Migration of speech recognition technology to mobile devices presents some specific challenges. Typically, such devices have limited resources to dedicate to the speech recognition task, especially for the recognition to be done in real time.

By their very nature, mobile devices are used in a number of varying environments. Each environment presents different noise characteristics. Speech recognition engines need to include some mechanism for compensating for the expected range of noise. The task of the current research is in-vehicle navigation devices. Here, the expected noise can be limited to that which occurs in a vehicle cockpit environment.

The intent of noise reduction algorithms is to increase the intelligibility and/or the quality of the source signal. Quality is a subjective measure, whereas intelligibility is objective [1]. Noise reduction algorithms applied to spoken language are also known as speech enhancement algorithms. The performance of speech enhancement algorithms is measured by quality and intelligibility. Speech recognition algorithms are concerned with the latter. Therefore, even where a speech enhancement algorithm produces a result that sounds worse to a human listener, it can still be seen as successful if the recognition rate increases.

1.1 In-Vehicle Noise

Noise encountered in a moving vehicle is created by a number of factors. The sources of in-vehicle noise are varied. They include the persistent sources such as engine,

N.R. Pal et al. (Eds.): ICONIP 2004, LNCS 3316, pp. 375–380, 2004.
© Springer-Verlag Berlin Heidelberg 2004

road and aerodynamic turbulence, as well as spurious sources such as traffic noise and road surface changes.

Such factors vary between vehicles, but retain several characteristics. Engine noise varies with the revolutions per minute (RPM). The RPM increase during acceleration and the resulting increase in noise is one of the more difficult conditions for speech recognition systems [2]. The majority of the vehicle noise is of frequencies less than 1 kHz.

Engine noise and road noise tends to be less than 1 kHz frequency, while wind noise if greater then 500 Hz. Coarse road conditions do not appear to significantly reduce recognition performance as the frequencies are less than 200 Hz [2].

An assumption is made here that only a single channel is available for speech recognition. With many mobile devices, only a single microphone is available. Some noise reduction techniques, such as blind source separation and beamforming[3], require an array of microphones.

Several standard algorithms are available for single channel noise reduction. The Wiener filter [1] and Spectral Subtraction [1, 4] are well established techniques for reduction of noise.

The primary objective of the current experiments was to evaluate the effect on intelligibility of noisy data before and after speech enhancement. To achieve this, the clean and noisy data were collected independently. This allows the control of several variables that would otherwise be uncontrollable.

The primary advantage of this approach was to allow the two sources to be mixed together to any desired signal-to-noise ratio (SNR). Additionally, different types of noise can be mixed with the same speech signal. This allows the effect of different noise environments to be evaluated independently. This includes conditions such as variations of speed and acceleration, as well as noise data collected from a variety of different vehicles, locations and road conditions.

Studies have shown that in noisy environments, talkers will modify their speaking style in an attempt to compensate for environmental noise (the "Lombard Effect") [3]. If the speech signal was collected in the absence of noise, the talker will not change their manner of speech. The Lombard effect may have a positive or negative effect on the speech recognition process. In any case, it has been eliminated as a variable in the current experiments. A second environmental variable that is removed is the acoustic properties of the car cockpit.

1.2 The ECoS Paradigm

The Evolving Connectionist System (ECoS) paradigm was developed to address several of the perceived disadvantages of traditional connectionist systems [5]. It is a structurally evolving paradigm that modifies the structure of the network as training examples are presented. Although the seminal ECoS architecture was the Evolving Fuzzy Neural Network (EFuNN), several other architectures have been developed that utilise the ECoS paradigm [5, 6]. Human listeners can experience fatigue over testing sessions [1], reducing the objectivity of the intelligibility measure. ECoS is a brain-like method that can be used to measure intelligibility.

The general principles of ECoS are to learn fast from a large amount of data through one-pass training, to adapt in an on-line mode where new data is incrementally accommodated, to "memorise" data exemplars for a further refinement, or for information retrieval, and to learn and improve through active interaction with other systems and with the environment in a multi-modular, hierarchical fashion [5, 6].

ECoS can avoid several problems associated with traditional connectionist structures. They are hard to over-train, they learn quickly, and they are far more resistant to catastrophic forgetting than most of the other models. Conversely, since they deal with new examples by adding nodes to their structure, they rapidly increase in size and can become unwieldy if no aggregation or pruning operations are applied [6].

The Simple Evolving Connectionist Structure. The Simple Evolving Connectionist System (SECoS) is a minimalist implementation of the ECoS paradigm [6]. It is a simplified version of EFuNN for cases where fuzzified inputs are not necessary. There are several advantages to using SECoS. Firstly, the simpler architecture means they are easier to understand and analyse. Secondly, their unfuzzified input space is of a lower dimensionality than a corresponding EFuNN, which allows the SECoS to model the training data with fewer nodes in the evolving layer than an equivalent EFuNN.

2 Experiments

The intent of the first experiment was to see the effect of the speech enhancement algorithms on the intelligibility of the speech. It was expected that the recognition rate would increase after the enhancement. The secondary purpose was to see if the noise types had an effect on the recognition rate. To see the effect of the noise type independent of the SNR, the noise needed to be normalised.

2.1 Data

Clean speech data used in experiments was taken from the Otago Digit Corpus [7]. The speech data was recorded in a quiet room environment to obtain clean speech signals. 23 native New Zealand English speakers participated in the recording sessions using close-mouth microphone. The speech was sampled at 22.05 kHz and quantised to a 16 bit signed number. Each word was uttered five times with distinct pauses in between. Each of these words was then manually segmented and labelled.

Speech data was prepared for the English digits "zero" to "nine". The speech data used in these experiments obtained from two groups of speakers; group A (8 male and 8 female) and group B (3 male and 4 female). The data from group A divided into two sets; training set A, testing set A. The data from group B was used in testing set B.

Noise data was collected from a dashboard mounted microphone inside a moving vehicle. The vehicle used was a 1999 Toyota Cavalier. Several seconds of noise were collected for four noise conditions. These noise conditions are shown in Table 1.

2.2 Method

The noise conditions were mixed into the clean data at various SNR. In all cases, the volume of the noise data was changed, while the signal data remained at a constant

Table 1. Recognition rate over all noise conditions

Noise Type	Description
N1	Vehicle accelerating from 0 to 50 km/h.
N2	Vehicle travelling at a steady 50km/h
N3	Vehicle accelerating from 50 to 100 km/h
N4	Vehicle travelling at a steady 100 km/h

volume. The effect of the noise type and the volume may not be independent. For example, the SNR under acceleration conditions may be the cause of the reduced performance seen in [2], and not the quality of the noise in itself. In this way the noise was normalised, as the final SNR was independent of the original volume of each noise condition.

Training set A with 480 examples was obtained from 3 utterances of each word in group A. The remaining 2 utterances were used in the testing set A for a total of 320 examples. Testing set B with 210 examples was obtained from the 3 utterances of each word in group B. Spectral analysis of the speech signal was performed over 20 ms to make the input vector features [5].

A SECoS was initialised with 100 input nodes and 10 output nodes (one for each word). The network was trained on clean data, and only on the training data set. After the data had been mixed with the noise data, the recall of the network established a baseline condition. The network was then tested after the data had undergone speech enhancement. Two speech enhancement algorithms were used here: Spectral Subtraction and Wiener Filtering.

2.3 Results

The baseline results are shown in Table 2. Here, the recognition rate is averaged over all noise conditions. The recognition performance on the clean data is shown in the last column. The word error rate for clean data was 0.21% for the training set, 8.75% for test A, and 7.51% for test B. Note that even a small amount of noise reduced the recognition rate significantly. This is expected, as the networks had not previously been subjected to noisy data. The word error rate for 9dB SNR data was 50.00% for the training set, 50.94% for test A, and 42.87% for test B.

Tables 3 and 4 show the results of the network after the data had undergone spectral subtraction and Wiener filtering respectively. Even after ignoring the noise type as a variable, both filter types showed a significant increase of recognition performance.

Table 2. Recognition rate as true positive and true negative over all noise conditions

	Train		Test A		Test B	
SNR	Pos%	Neg%	Pos%	Neg%	Pos%	Neg%
0dB	37.81	93.09	34.69	92.74	42.96	93.66
6dB	43.23	93.69	40.42	93.38	47.41	94.16
9dB	50.00	94.44	49.06	94.34	57.13	95.24
12dB	55.31	95.03	56.15	95.13	62.87	95.87
18dB	57.60	95.29	57.29	95.25	64.07	96.01
-	99.79	99.98	91.25	99.03	92.59	99.18

Table 3. Recognition rate as true positive and true negative after Spectral Subtraction across all noise types

	Train		Test A		Test B	
SNR	Pos%	Neg%	Pos%	Neg%	Pos%	Neg%
0dB	44.53	93.84	41.15	93.46	49.44	94.38
6dB	49.58	94.40	47.92	94.21	54.54	94.95
9dB	55.16	95.02	55.00	95.00	62.59	95.84
12dB	59.32	95.48	60.63	95.63	64.54	96.06
18dB	60.89	95.65	61.15	95.68	65.09	96.12

Table 4. Recognition rate as true positive and true negative after Wiener filtering across all noise types

	Train		Test A		Test B	
SNR	Pos%	Neg%	Pos%	Neg%	Pos%	Neg%
0dB	41.98	93.55	39.69	93.30	48.89	94.32
6dB	48.18	94.24	46.25	94.03	54.54	94.95
9dB	55.73	95.08	54.17	94.91	61.67	95.74
12dB	59.74	95.53	60.42	95.60	66.11	96.23
18dB	60.99	95.67	61.56	95.73	68.61	96.51

A MANOVA was performed to see if the noise types had a significant effect on the recognition rates. Even after the SNR was normalised, the noise types had a significant effect on the recognition results ($F<0.025$). Also, the MANOVA confirmed that the SNR had a significant effect on recognition.

The word error rate for 9 dB SNR data after Spectral Subtraction was 44.84% for the training set, 45.00% for Test A, and 37.41% for Test B. After Wiener filtering, the word error rate for 9dB data was 44.27% for the training set, 45.83% for Test A, and 38.33% for Test B. For the 9dB condition, using a speech enhancement resulted in an error rate improvement of between 4.54% and 5.94%. Similar improvements were seen over the other noise levels, and in no condition did the speech enhancement result in a decline in recognition performance.

3 Conclusion

By normalising the noise amplitudes, it was possible to see the effect due to the quality of the noise, independently of SNR. The quality of the noise, reflected in the four noise conditions, was still significant. Thus, any further experiments need to account for this dimension, and SNR alone is not enough to capture the effect of noise.

The quality of the noise is a significant factor for intelligibility. Some noise conditions that appear benign, such as deceleration, may decrease recognition performance if not explicitly accounted for. Further analysis should be performed on more noise conditions, which may also include external factors such as road surface, and density of traffic.

Even a small amount of noise greatly reduces recognition. Noisy data should be included in the training data for the SECoS network. The SECoS will implicitly model the noise. The results should show if the explicit noise reduction performs better than implicit modelling, and whether these techniques can be used together to obtain optimal recognition results in the presence of noise.

The experimental design allows the mixing of clean data with noise. Ideally, further 'clean' data can be collected accounting for both the Lombard effect and the acoustic characteristics of the vehicle cockpit. Once the Lombard effect is included as a variable, any interaction between the quality and/or amplitude of the noise and the style of speaking needs to be analysed.

Acknowledgement

We would like to thank Associate Professor Brett Collins for his assistance in the statistical analysis of the results contained herein.

References

1. Ephraim, Y., Lev-Ari, H. and Roberts, W. J. J.: A Brief Survey of Speech Enhancement. The Electronic Handbook, CRC Press, (2003). to appear
2. Hoshino, H.: Noise-Robust Speech Recognition in a Car Environment Based on the Acoustic Features of Car Interior Noise. The R&D Review of Toyota CRDL 39, 1 (2004)
3. Plucienkowski, J. P., Hansen, J. H. L., and Angkititrakul, P.: Combined Front-End Signal Processing for In-Vehicle Speech Systems. Eurospeech-2001, Aalborg, Denmark (2001) 1573-1576
4. R. Martin.: Spectral subtraction based on minimum statistics. Proc EUSIPCO, Edinburgh (1994) 1182-1185
5. Kasabov, N.: Evolving connectionist systems: Methods and applications in bioinformatics, brain study and intelligent machines, Springer, London (2002)
6. Ghobakhlou, A., Watts, M. and Kasabov, N.: Adaptive Speech Recognition with Evolving Connectionist Systems. Journal of Information Science 156 (2003) 70-83
7. Sinclair, S., Watson, C.: Otago Speech Data Base. Proceedings of NNES'95, Dunedin, IEEE Computer Society Press, Los Alamos (1995)

An Evolving Neural Network Model
for Person Verification Combining Speech and Image

Akbar Ghobakhlou, David Zhang, and Nikola Kasabov

KEDRI, Auckland University of Technology, Private Bag 92006, Auckland, New Zealand
{akbar,dzhang,nkasabov}@aut.ac.nz

Abstract. This paper introduces a method based on Evolving Connectionist Systems (ECOS) for person verification tasks. The method allows for the development of models of persons and their on-going adjustment based on new speech and face images. Some experimental person verification models based on speech and face image features are developed based on this method where speech and face image information are integrated at a feature level to model each person. It is shown that the integration of speech and image features improves significantly the accuracy of the person verification model when compared with the use of only image or speech data.

1 Introduction

Biometric verification can be defined as a process of uniquely identifying a person by evaluating one or more distinguishing biological traits. Unique identifiers include fingerprints, hand geometry, retina, iris patterns, face image and voice. There are many biometric features that distinguish individuals from each other, thus many different sensing modalities have been developed [5]. These identifiers may be used individually, as exemplified by the iris scan system deployed in the banking sector and currently being tested for airport security [6].

Over the past few years, interest has been growing in the use of multiple modalities to solve automatic person identification problems. The motivation for using multiple modalities is multi-fold. In the first instance different modalities measure complementary information and by this virtue multimodal systems can achieve better performance than single modalities. Single feature may fail to be exact enough for identification of individuals.

In this paper we propose person verification module based on Evolving Connectionist Systems (ECoS) [1]. Person verification models developed based on speech face image and integrated features. Each person is modelled by placing nodes during the training process. From previous work [2], it was shown that ECoS could be used to create adaptive speech recognition systems. ECoS use a local learning algorithm where each neuron in the evolving layer of the network represents data in a small region from the problem space.

The following sections describe the method used and the experimental system built to demonstrate the method. First, the pre-processing and feature extraction methods are described and then the classification ECoS principles are presented and illustrated on a simple case problem.

N.R. Pal et al. (Eds.): ICONIP 2004, LNCS 3316, pp. 381–386, 2004.
© Springer-Verlag Berlin Heidelberg 2004

2   Speech and Face Image Signal Processing

2.1   Speech Signal Sampling and Processing

In the speaker verification model, a text-dependent module was built. The speech data was captured using close-mouth microphone. The speech was sampled at 22.05 kHz and quantized to a 16 bit signed number. In order to extract Mel Frequency Cepstrum Coefficients (MFCC) as acoustic features, spectral analysis of the speech signal was performed over 20ms with Hamming window and 50% overlap. Discrete Cosine Transformation (DCT) was applied on the MFCC of the whole word to obtain input feature vectors [2].

2.2   Face Image Processing

In the face verification model, the images were captured using a web-cam with a resolution of 320×240. Once a new image was captured, features were extracted using the composite profile technique. The composite profile features are composed of the average value of the columns in the image followed by the average value of rows in the image. It is a relevant feature to characterize symmetric and circular patterns, or patterns isolated in a uniform background. This feature can be useful to verify the alignment of objects. In order to reduce the number of features, the interpolation technique was applied to the 60 features.

3   ECoS for Dynamic Modelling and Classification

Here we use an implementation of the ECoS models called Evolving Classifier Function (ECF) [1]. The ECF algorithm classifies input data into a number of classes and finds their class centres in the n-dimensional input space by "placing" a rule node in the evolving layer. Each rule node is associated with a class and an influence (receptive) field representing a part of the n-dimensional space around the rule node. Generally such an influence field in the n-dimensional space is a hyper-sphere. Essentially each client is modelled by a number of rule nodes that represent that client.

There are two distinct modes of ECF operation, learning and recognition. The details of the original algorithm of these two operation modes were introduced in [1]. In this paper, the recognition algorithm of ECF was modified for the task of person verification. Accordingly we call it verification algorithm. The verification algorithm consists of the following steps:

- With the trained ECF module, when a new test sample I is presented, first it is checked whether it falls within the influence field of the rule nodes representing the claimed identity of the sample I . This is achieved by calculating the Euclidean distance between this sample and appropriate rule nodes, then comparing this distance D_i with the corresponding influence field Inf_i. The sample I is verified as person i if the relation (1) is satisfied.

$$D_i <= Inf_i \qquad (1)$$

- If the sample I doesn't fall in the influence field of any existing rule node,

○ Find the rule node which has the shortest distance to this sample, note this distance as D_{min}.

○ If this distance D_{min} is less than a pre-set acceptance threshold θ, the sample I is verified as person i. Otherwise, this sample is rejected by this verification module.

This verification algorithm was applied to the speaker, face image and integrated verification modules. Figure 1 illustrates the overall process of adaptive person verification system.

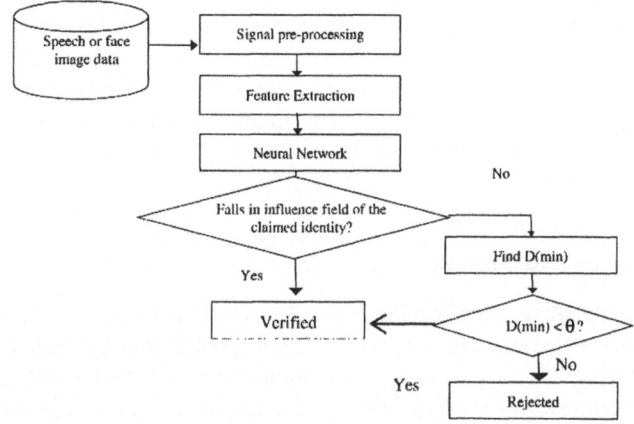

Fig. 1. Overall view of an adaptive connectionist person verification system

4   Integration of Speech and Face Image at the Feature Level

Speech and face image information were used for the person verification task. Individual ECF modules were built for both speech and face image sub-network. In addition, features obtained from speech and face image of clients were merged to form integrated input features. There are various strategies of combining multimodal sources of information. In this approach, speech and face image information were integrated at the feature level. There are 100 input features in a speech sample and 64 input features in a face image sample. These two set of features were concatenated to form the integrated input features.

5   System Implementation and Experimental Results

Person verification system is essentially a two-class decision task where the system can make two types of errors. The first error is a false acceptance, where an impostor is accepted. The second error is false rejection, where a true claimant is rejected. False Acceptance Rate (FAR) and False Rejection Rate (FRR) are calculated according to the following equations:

$$FAR = \frac{I_A}{I_T} \qquad (1) \qquad FRR = \frac{C_R}{C_T} \qquad (2)$$

where I_A the number of impostors classified as true claimants, I_T is the total number of impostor presented, C_R is the number of true claimant classified as impostors and C_T is the total number of true claimant presented. The trades off between these errors are adjusted using the acceptance threshold θ.

5.1 Data Preparation

In this study, speech data were taken from 8 members of the KEDRI institute [3]. As the speech module is text-dependent, all the speakers were requested to say the word "security" for speech-based speaker verification. Five samples from each speaker were collected to form the training dataset. Another 5 samples from each of these speakers were used to form a testing dataset. In a similar fashion the face images of the same peoples were captured to prepare training and testing datasets. Finally, the input features from speech and face image were integrated according to the method described in section 4.

5.2 Experiments and Results

An Adaptive Speaker Verification Module. An ECF neural network engine was built based on the speech training dataset. Each speaker was modelled by allocating rule nodes during training session. The number of rule nodes assigned for each speaker is determined by the maximum influence field. Figure 2 illustrates the performance of ECF on testing dataset.

As shown in Figure 2, the smaller the Maximum Inference Field, the more rule nodes are allocated for each client. This leads to a high correct acceptance rate of 92% and small FRR and FAR errors of 1%.

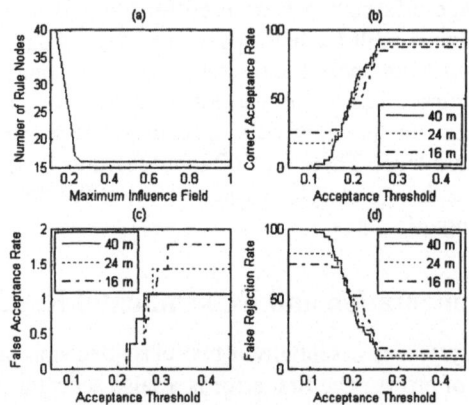

Fig. 2. ECF performance on speaker verification task. (a) Number of rule nodes created vs various influence field values. (b) Correct acceptance rate vs acceptance thresholds. (c) FAR vs acceptance thresholds. (d) FRR vs acceptance thresholds

An Adaptive Face Image Verification Module. Image verification system was built and validated. In a similar fashion to speaker verification model, each person was modelled by allocating rule nodes during training session. The number of rule nodes assigned for each client is determined by the maximum influence field. Figure 3 illustrates the performance of ECF on testing dataset.

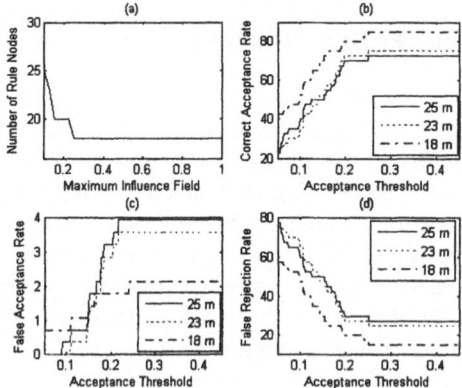

Fig. 3. ECF performance on face image verification task. (a) Number of rule nodes created vs various influence field values. (b) Correct acceptance rate vs acceptance thresholds. (c) FAR vs acceptance thresholds. (d) FRR vs acceptance thresholds

As illustrated in Figure 3, the smaller the Maximum Inference Field, the more rule nodes are allocated for each person. The best ECF performance was achieved with 18 rule nodes with the correct acceptance rate of 85% and FAR error of just over 2%.

A Person Verification Module Based on Integrated Voice and Face Features. The training dataset for this experiments obtained by concatenating the speech training dataset and face image training dataset as described in Section 4. Each integrated sample has 164 input features. An ECF model was built using the integrated training dataset and test on the integrated testing dataset A and B. The test results are shown in Figure 4.

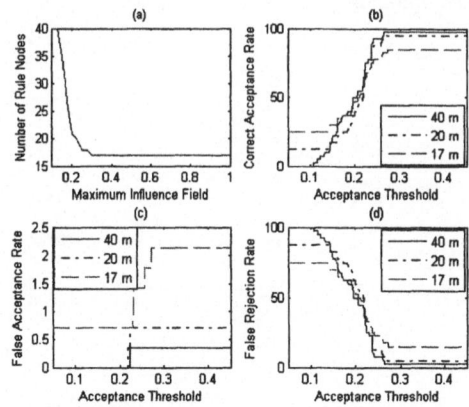

Fig. 4. ECF performance on integrated features. (a) Number of rule nodes created vs various influence field values. (b) Correct acceptance rate vs acceptance thresholds. (c) FAR vs acceptance thresholds. (d) FRR vs acceptance thresholds

The results in figure 4 show that the smaller the Maximum Inference Field, the more rule nodes are allocated for each person. The best ECF performance was achieved with correct acceptance rate of 97% and FAR error of just less than 0.5%.

6   Conclusions and Future Research

This paper presented a method based on evolving connectionist system ECF for person verification tasks. The performance of ECF of individual and integrated modules shows the ECF capability in modelling each person by placing rule nodes to create the person's verification model. The verification module based on integrated voice and face features outperformed both single modules showing improvement in correct acceptance rate and lower FAR and FRR errors. The evolving property of ECF [1] allows for new persons to be added or removed from the system. Further experiments and analysis need to be done to evaluate the performance of this methodology on persons who are not participated in training.

References

1. Kasabov N.: Evolving connectionist systems: Methods and applications in bioinformatics, brain study and intelligent machines, Springer Verlag, 2002
2. Ghobakhlou A., Watts M. and Kasabov N.: Adaptive speech recognition with evolving connectionist systems, Information Sciences 156(2003), 71-83
3. Knowledge Engineering & Discovery Research Institute, Auckland University of Technology, New Zealand, http://www.kedri.info
4. Kasabov N., Postma E.,Herik J. V. D.: AVIS: a connectionist-based framework for integrated auditory and visual information processing, Information Sciences 123 (2000), 127-148
5. Brunelli R., Falavigna D.: Person identification using multiple cues, IEEE Transactions on Pattern Analysis and Machine Intelligence 17(1995), 955-966
6. Luettin J., Thacker N. A., Beet S.W.: Active shape models for visual speech feature extraction, in: D.G. Storck, M.E.Heeneeke(Eds.), Speechreading by Humans and Machines, Springer, Berlin, 1996, 383-390

Adaptive Affine Subspace Self-organizing Map with Kernel Method

Hideaki Kawano, Keiichi Horio, and Takeshi Yamakawa

Kyushu Institute of Technology
2-4 Hibikino Wakamatsu-ku Kitakyushu, 808-0196, Japan
kawano-hideaki@edu.brain.kyutech.ac.jp,
{horio,yamakawa}@brain.kyutech.ac.jp

Abstract. Adaptive Subspace Self-organizing Map (ASSOM) is an evolution of Self-Organizing Map, where each computational unit defines a linear subspace. Recently, its modified version, where each unit defines an affine subspace instead of the subspace, has been proposed. The affine subspace in a unit is represented by a mean vector and a set of basis vectors. After training, these units result in a set of affine subspace detectors. In numerous cases, however, these are not enough to describe a class of patterns because of its linearity. In this paper, the Adaptive Affine Subspace SOM (AASSOM) on the high-dimensional space with kernel method is proposed in order to achieve efficient classification. By using the kernel method, linear affine subspaces in the AASSOM can be extended to nonlinear affine subspaces easily. The effectiveness of the proposed method is verified by applying it to some simple classification problems.

1 Introduction

Adaptive-Subspace Self-Organizing Map (ASSOM)[1] is an evolution of Self-Organizing Map (SOM)[2]. The ASSOM is capable, to a certain degree, of capturing the invariant features of patterns as a linear subspace composed by some bases. Recently, Adaptive Affine Subspace Self-Organizing Map (AAS-SOM) which is a modified version of ASSOM has been proposed[3]. In the AAS-SOM, more general classification problem, e.g., face recognition problem, can be achieved by employing affine subspaces instead of subspaces. In complex situations, however, it does not extract useful features because of its linearity.

On the other hand, kernel method has been focused because it can extend from a linear algorithm to a nonlinear algorithm easily. The idea of the kernel method is that data set is transformed into a nonlinear feature space, which is a very high-dimensional space related to the input space by the implicit nonlinear map. The kernel method was introduced in the context of Support Vector Machine (SVM) for the first time[4]. Then various linear algorithms have been extended to nonlinear algorithms by using the kernel method and successful results have been achieved[5][6].

N.R. Pal et al. (Eds.): ICONIP 2004, LNCS 3316, pp. 387–392, 2004.
© Springer-Verlag Berlin Heidelberg 2004

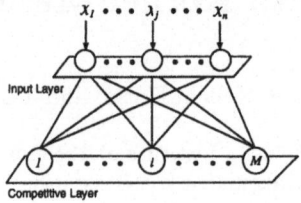

Fig. 1. A structure of the Adaptive Affine Subspace Self-Organizing Map (AASSOM).

In this paper, we propose a nonlinear extended version of the AASSOM by applying kernel method, referred as Kernel Adaptive Affine Subspace Self-Organizing Map (KAASSOM), in order to achieve efficient classification. The KAASSOM could be expected to construct nonlinear affine subspaces so that effective representation of data belonging to the same category is achieved with a small number of bases. The effectiveness of the proposed method is verified by applying it to some simple pattern classification problems.

2   Adaptive Affine Subspace Self-organizing Map (AASSOM)

In this section, we give a brief review of the original AASSOM. Fig.1 shows the structure of the AASSOM. It consists of an input layer and a competitive layer, in which n and M units are included respectively. Suppose $i \in \{1, \cdots, M\}$ is used to index computational units in the competitive layer, the dimensionality of the input vector is n. The i-th computational unit constructs an affine subspace, which is composed of a mean vector $\boldsymbol{\mu}^{(i)}$ and a subspace spanned by H basis vectors $\boldsymbol{b}_h^{(i)}$, $h \in \{1, \cdots, H\}$. First of all, we define the orthogonal projection of a input vector $\boldsymbol{x}$ onto the affine subspace of i-th unit as

$$\hat{\boldsymbol{x}}^{(i)} = \boldsymbol{\mu}^{(i)} + \sum_{h=1}^{H} (\boldsymbol{\phi}^{(i)^T} \boldsymbol{b}_h^{(i)}) \boldsymbol{b}_h^{(i)}, \tag{1}$$

where $\boldsymbol{\phi}^{(i)} = \boldsymbol{x} - \boldsymbol{\mu}^{(i)}$. Therefore the projection error is represented as follows:

$$\tilde{\boldsymbol{x}}^{(i)} = \boldsymbol{\phi}^{(i)} - \sum_{h=1}^{H} (\boldsymbol{\phi}^{(i)^T} \boldsymbol{b}_h^{(i)}) \boldsymbol{b}_h^{(i)}. \tag{2}$$

This model is more general strategy than the ASSOM. To illustrate why this is so, let us consider a very simple case: Suppose we are given two clusters as shown in Fig.2(a). It is not possible to use one dimensional subspaces, that is lines intersecting the origin O, to approximate the clusters. This is true even if the global mean is removed, so that the origin O is translated to the centroid of the two clusters. However, two one-dimensional affine subspaces can easily approximate

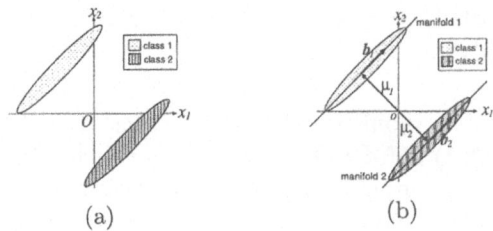

Fig. 2. (a) Clusters in 2-dimensional space: An example of the case which can not be separated without a mean value. (b) Two 1-dimensional affine subspaces to approximate and classify clusters.

the clusters as shown in Fig.2(b), since the basis vectors are aligned in the direction that minimizes the projection error. In the AASSOM, the input vectors are grouped into episodes in order to be presented to the network. For pattern classification, an episode input is defined as a subset of training data belonging to the same category. Assume that the number of elements in the subset is E, then an episode input χ_q in the class q is denoted as $\chi_q = \{x_1, x_2, \cdots, x_E\}, \chi_q \subseteq \Omega_q$, where Ω_q is a set of training patterns belonging to the class q. The set of input vectors of an episode has to be recognized as one class, such that any member of this set and even an arbitrary linear combination of them should have the same winning unit.

The training process has the following steps:

(a) **Winner lookup.** The unit that gives the minimum projection error for an episode is selected. We denote this as the winner, whose index is c. This criterion is represented as follows:

$$c = \arg\min_i \left\{ \sum_{e=1}^{E} \|\tilde{x}_e^{(i)}\|^2 \right\}, \quad i \in \{1, \cdots, M\}. \tag{3}$$

(b) **Learning.** For each unit i, and for each x_e, update $\mu^{(i)}$

$$\mu^{(i)}(t+1) = \mu^{(i)}(t) + \lambda_m(t) h_{ci}(t) \left(x_e - \mu^{(i)}(t) \right), \tag{4}$$

where $\lambda_m(t)$ is the learning rate for $\mu^{(i)}$, $h_{ci}(t)$ is the neighborhood function with respect to the winner c. Both $\lambda_m(t)$ and $h_{ci}(t)$ are monotonic decreasing function w.r.t. t. Then update the basis vectors

$$b_h^{(i)}(t+1) = b_h^{(i)}(t) + \lambda_b(t) h_{ci}(t) \frac{\phi_e^{(i)}(t)^T b_h^{(i)}(t)}{\|\hat{\phi}_e^{(i)}(t)\|\|\phi_e^{(i)}(t)\|}, \phi_e^{(i)}(t), \tag{5}$$

where $\phi_e^{(i)}(t) = x_e - \mu^{(i)}(t+1)$, $\hat{\phi}_e^{(i)}(t) = \sum_{h=1}^{H} (\phi^{(i)}(t)^T b_h^{(i)}(t)) b_h^{(i)}(t)$ and $\lambda_b(t)$ is the learning rate for the basis vectors, which is also monotonic decreasing function w.r.t. t.

3   Kernel Adaptive Affine Subspace Self-organizing Map (KAASSOM)

In the KAASSOM, an input x is transformed into the Hilbert space H. The transformed data $\Phi(x)$ is fed to the network, of which structure is the same as AASSOM. To achieve the implicit mapping to high-dimensional feature space, the mean vector and the basis vectors in the i-th unit are represented by $\Phi(\mu^{(i)})$ $= \sum_{l=1}^{N} \alpha_l^{(i)} \Phi(x_l)$ and $\Phi(b_h^{(i)}) = \sum_{l=1}^{N} \beta_{hl}^{(i)} \Phi(x_l)$, respectively. N is the total number of training samples and $\alpha_l^{(i)}$ and $\beta_{hl}^{(i)}$ are the parameters adjusted by learning. In the KAASSOM, the norm of the orthogonal projection error onto the i-th affine subspace with respect to present input x_p is calculated as follows:

$$
\begin{aligned}
\|\Phi(\tilde{x}_p)^{(i)}\|^2 = {} & K(x_p, x_p) + \sum_{l_1=1}^{N} \sum_{l_2=1}^{N} \alpha_{l_1}^{(i)} \alpha_{l_2}^{(i)} K(x_{l_1}, x_{l_2}) + \sum_{h=1}^{H} P_h^{(i)2} \\
& - 2 \sum_{l=1}^{N} \alpha_l K(x, x_l) + 2 \sum_{h=1}^{H} \sum_{l_1=1}^{N} \sum_{l_2=1}^{N} P_h^{(i)} \alpha_{l_1} \beta_{hl_2} K(x_{l_1}, x_{l_2}) \\
& - 2 \sum_{h=1}^{H} \sum_{l=1}^{N} P_h^{(i)} \beta_{hl} K(x, x_l),
\end{aligned} \tag{6}
$$

where $P_h^{(i)} = \sum_{l=1}^{N} \beta_{hl}^{(i)} K(x_p, x_l) - \sum_{l_1=1}^{N} \sum_{l_2=1}^{N} \alpha_{l_1}^{(i)} \beta_{hl_2}^{(i)} K(x_{l_1}, x_{l_2})$ and $K(x_s, x_t)$ is a kernel function, which means the inner product w.r.t. $\Phi(x_s)$ and $\Phi(x_t)$. Kernel trick in SVM employs a kernel function $K(x_s, x_t)$ instead of the inner product between $\Phi(x_s)$ and $\Phi(x_t)$. In practice, since we do not need an implicit form of Φ, we first determine K that can be decomposed in the form of inner product. From Mercer theorem, the symmetric positive definite kernel K can be decomposed into the inner product form[7]. In this paper, the following function, referred as polynomial kernel,

$$
K(x_s, x_t) = (x_s^T x_t)^d, \tag{7}
$$

is used. The training procedure of KAASSOM has the following steps:

(a) **Winner lookup.** The winner is decided by the same manner as the AAS-SOM as follows:

$$
c = \arg\min_i \left\{ \sum_{e=1}^{E} \|\Phi(\tilde{x}_p)^{(i)}\|^2 \right\}, \quad i \in \{1, \cdots, M\}. \tag{8}
$$

(b) **Learning.** The learning rule for $\alpha_l^{(i)}$ and $\beta_{hl}^{(i)}$ are as follows:

$$
\Delta \alpha_l^{(i)} = \begin{cases} -\alpha_l^{(i)} \lambda_m h_{ci} & for \ l \neq p \\ -\alpha_l^{(i)} \lambda_m h_{ci} + \lambda_m h_{ci} & for \ l = p \end{cases}, \tag{9}
$$

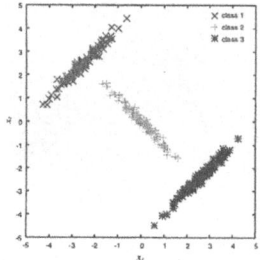

Fig. 3. Distribution of data used in the first experiment.

Table 1. Classification performance results (%).

	AASSOM	KAASSOM
class 1	93	100
class 2	100	100
class 3	92	100

$$\Delta\beta_{hl}^{(i)} = \begin{cases} -\alpha_l^{(i)}\lambda_b h_{ci}T & for \; l \neq p \\ -\alpha_l^{(i)}\lambda_b h_{ci}T + \lambda_b h_{ci}T & for \; l = p \end{cases}, \tag{10}$$

where $T = \dfrac{\Phi(\phi_e^{(i)}(t))^T \Phi(b_h^{(i)}(t))}{||\Phi(\tilde{\phi}_e^{(i)}(t))|| ||\Phi(\phi_e^{(i)}(t))||}$.

4   Experimental Results

To verify the theoretical results and compare the performance of the proposed method with the AASSOM, we apply the proposed method to some simple classification problems.

Experiment 1 Figure 3 shows the distributions of data used in the Experiment 1. In the problem, the goal is to classify the given input coordinates as belonging to one of the three distributions. In the experiment, the parameters are assigned as follows: the numbers of basis vectors are $H = 1$ in AASSOM, $H = 2$ in KAASSOM and $d = 3$ in the polynomial kernel. Table 1 shows the experimental results for the Experiment 1. From the experiment, the KAASSOM outperforms the AASSOM clearly as shown in Table 1.

Experiment 2 Figure 4(a) shows the distributions of data used in the experiment 2. This case is non-separable case in linear method. In the experiment, the parameters are assigned as follows: the numbers of basis vectors are $H = 1$ in AASSOM, $H = 2$ in KAASSOM, $d = 2$ in the polynomial kernel. Figure 4(b) and (c) shows the decision regions learned by AASSOM and KAASSOM, respectively. By this experiment, we can see KAASSOM extracts suitable affine subspaces efficiently.

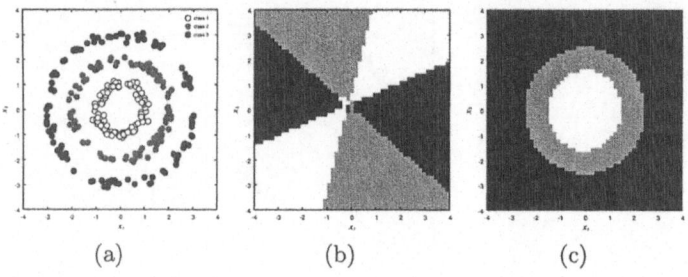

$$\text{(a)} \qquad\qquad \text{(b)} \qquad\qquad \text{(c)}$$

Fig. 4. (a) Training data used in the second experiment. (b) Decision regions learned by AASSOM and (c) KAASSOM.

5   Conclusions

The Kernel Adaptive Affine Subspace Self-Organizing Map was proposed as a new pattern classification method. The proposed method has been extended to a nonlinear method easily from AASSOM by applying the kernel method. The effectiveness of the proposed method were verified by the experiments. The proposed algorithm has highly promising applications of the ASSOM in a wide area of practical problems.

References

1. T. Kohonen, "Emergence of Invariant-Feature Detectors in the Adaptive-Subspace Self-Organizing Map", Biol.Cybern, 75, pp.281-291, 1996.
2. T. Kohonen, "Self-Organizing Maps", Springer-Verlag Berlin Heidelberg New York, 1995.
3. Z. Q. Liu, "Adaptive Subspace Self-Organizing Map and Its Application in Face Recognition", International Journal of Image and Graphics, Vol.2, No.4, pp.519-540, 2002.
4. V. Vapnik, "The Nature of Statistical Learning Theory, Second Edition", Spring-Verlag New York Berlin Heidelberg, 1995.
5. B. Schölkopf, A. J. Smola and K. R. Müler, "Nonlinear Component Analysis as a Kernel Eigenvalue Problem,", Technical Report 44, Max-Planck-Institut fur biologische Kybernetik, 1996.
6. S. Mika, G. Rätsch, J. Weston, B. Schölkopf, K. R. Müler "Fisher discriminant analysis with kernels," Neural Networks for Signal Processing IX, pp. 41-48, IEEE, 1999.
7. B. Schölkopf and A. J. Smola, "Learning with Kernels", The MIT Press, 2002.

Scene Memory on Competitively Growing Neural Network Using Temporal Coding: Self-organized Learning and Glance Recognizability

Masayasu Atsumi

Dept. of Information Systems Sci, Faculty of Eng., Soka University
1-236 Tangi-cho, Hachioji-shi, Tokyo 192-8577, Japan
matsumi@t.soka.ac.jp

Abstract. We have been building the competitively growing neural network using temporal coding for quick one-shot object learning and glance object recognition, which is the core of our saliency-based scene memory model. This neural network represents objects using latency-based temporal coding and grows size and recognizability through learning and self-organization. This paper shows that self-organized learning is quickly performed and glance recognition is successfully performed by our model through simulation experiments of a robot equipped with a camera.

1 Introduction

It is known that a human can learn scenes in almost one shot and also recognize them at a glance. In these processes, spatially circumscribed regions are selected based on saliency-based attention as well as volition-controlled attention before further processing. The former is rapid, bottom-up and task-independent attention and the latter is slow, top-down and task-dependent attention [4]. The attentive higher processing of these regions engraves scene memory that consists of attended objects, complexes of objects and their spatial relation from the egocentric point of view.

We have been building the competitively growing neural network using temporal coding [1], named COGNET (COmpetitively Growing NEural network using Temporal coding), for quick one-shot object learning and glance object recognition, which is the core of our saliency-based scene memory model [2] in which objects in saliency-based attended spots are sequentially encoded to be invariant with respect to position and size by this network and their positions and sizes are encoded simultaneously. In this network, objects are internally represented using latency-based temporal coding. This network enables fast self-organized learning of objects involving recruitment and similarity-based sorting of neurons. It also enables glance recognition of objects based on the latency-based temporal coding.

In this paper, we discuss the fast self-organized learnability and the glance recognizability of objects in scenes through simulation experiments of a robot equipped with a camera.

N.R. Pal et al. (Eds.): ICONIP 2004, LNCS 3316, pp. 393–398, 2004.
© Springer-Verlag Berlin Heidelberg 2004

2 A Model of Scene Memory

A model of saliency-based scene memory is shown in Fig. 1. In the first phase, contrast and opponent color channels of red, green, blue and yellow are computed at each pixel of a scene image [4]. Then the saliency map is produced to represent saliency at every pixel of the image by combining contrast and opponent color channels [3]. Next, connected object regions are segmented using a grow-and-merge method on the saliency map and their bounding boxes are extracted as attended spots. Three or less attended spots are extracted. In the second phase, attended spots are processed in the order of decreasing saliency at a certain interval. In this phase, an object in each attended spot is encoded to be invariant with respect to position and size by the COGNET, and the position and the size of the attended spot are encoded at the same time.

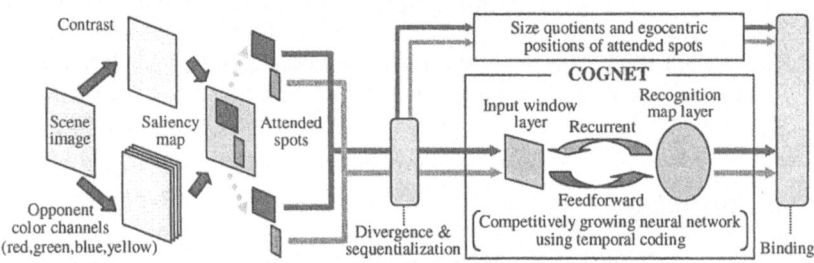

Fig. 1. A model of saliency-based scene memory

The COGNET consists of the input window layer and the recognition map layer. The input window layer consists of $l_w \times l_w \times 5$ neurons that receive normalized values of contrast and four opponent colors at $l_w \times l_w$ sections in height and width on an attended spot. The recognition map layer is a competitively growing layer where neurons can be arranged in a two-dimensional lattice. An object in an attended spot is encoded by the winner neuron in the recognition map layer. The input window layer and the recognition map layer have the entire reciprocal connection by feedforward and recurrent synapses for recognition and recall. Meanwhile, the position of an object is expressed by the center coordinate of its bounding attended spot. The size of an object is expressed by the size quotient $q = \frac{l_s}{l_w}$ where l_s is the larger of height and width of the attended spot.

Finally, consecutive processing results of attended spots in a scene are bound together and a scene memory is given by a set of triplets each of which consists of the winner neuron, the center coordinate and the size quotient for each attended spot. We call this triplet an attended spot code and a set of triplets a scene code.

3 The COGNET

In neural network models, information is internally represented in the form of a rate code or a temporal code. Several temporal coding schemes have been proposed [6]. In the COGNET, latency from trigger such as pulse transmission or stimulation until firing is used as a temporal code to encode information.

Each neuron in the input window layer converts an input value into a latent period so that the larger the input value is, the shorter the latent period is. As a result, a spatial pattern of input strength that codes an object is converted into a spatiotemporal pattern of pulse transmission, and the object is internally represented by this pattern. In neurons in the recognition map layer, membrane potential is computed based on a spatiotemporal pattern of pulse arrival, that is, a pattern of latency from pulse arrival until the competition time. The membrane potential represents how well a neuron encodes an input pulse pattern by a value between 1 and −1. The competition time is controlled by the pre-competition inhibition imposed on neurons in the recognition map layer. At the competition time, if the maximum membrane potential of all neurons is above a threshold of discrimination, a neuron that takes the maximum membrane potential is selected as the winner neuron. Otherwise, a new neuron is recruited as the winner neuron, whose synaptic efficacy is initialized so that it takes the maximum membrane potential to the spatiotemporal pattern of pulse arrival. The winner neuron n_s fires at the competition time. Every neuron n_k whose membrane potential is above a threshold of neighborhood is selected as a neighbor neuron and fires with a lag proportional to difference of membrane potential between n_k and n_s. The winner neuron and its neighbor neurons modulate their synaptic efficacy just after firing. Synaptic modulation is performed for the winner neuron so as to memorize an object that is encoded by the pulse pattern, and for neighbor neurons so as to bring their memory close to the one of the winner neuron. In addition, in order to achieve topology preservation [5], the sorting of neighbor neurons is performed in descending order of their membrane potentials to bring them close to the winner neuron on a two-dimensional lattice of the recognition map layer.

Recurrent synaptic efficacy is modulated to be similar with corresponding feedforward synaptic efficacy based on transmission of pulses to the input window layer. As a result, symmetrical synaptic efficacy is acquired in reciprocal connection, which enables recall of an object's mental image for an external input to a neuron in the recognition map layer.

4 Experimental Results

To evaluate the self-organized learnability and the glance recognizability of objects in scenes, simulation experiments of a robot equipped with a color camera were conducted. Fig. 2 shows an experimental T-maze world, and also shows an example of a saliency map and attended spots computed on a camera image of 64 × 64 pixels in height and width which is processed about once a second. In experiments, the l_w of the input window layer is set to 12. The threshold of discrimination is 0.75 and the threshold of neighborhood is 0.5. Since latency is designed to take a value in the range $[0, 255]$, the length of pre-competition inhibition period is set to 255 in default.

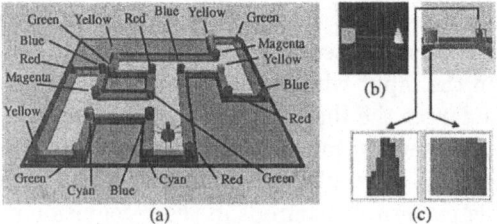

Fig. 2. (a)An experimental world. (b)A saliency map. (c)Attended spots. A total of 20 objects, where three each of red, blue, green and yellow cylinders, two each of magenta and cyan cylinders, and one each of red, blue, green and yellow cones, are arranged

4.1 Self-organized Learnability

As for self-organized learning, performance of one-shot object learning and fast self-organization of object memory were evaluated, and also invariant and discriminative object learning performance and scene learning performance were evaluated. In experiments, a robot slowly moves one lap around the maze while learning scenes, then moves another lap while recognizing scenes and recording scene codes without learning. After that, the robot repeatedly searches for a target scene that is picked out every 25 codes from the recorded sequence.

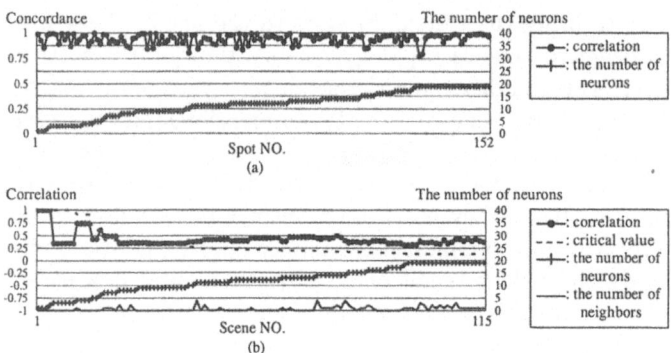

Fig. 3. (a)One-shot learning performance. (b)Fast self-organization performance

Fig. 3(a) shows a series of concordance between attended object images and winner neurons for them and a series of the number of neurons as the number of attended spots increases in the learning lap. The concordance is given by the formula $\sum_j (w_{sj}^F \times e_j)/\sqrt{\sum_j (w_{sj}^F)^2}\sqrt{\sum_j (e_j)^2}$, where w_{sj}^F is a feedforward synaptic efficacy between the winner neuron s and a neuron j in the input window layer and e_j is an input image element to the neuron j. Since the concordance always takes high values that are larger than the threshold of discrimination, we can conclude that attended objects were learned quickly.

Fig. 3(b) shows series of the degree of self-organization in the recognition map layer, the number of neighbors and the number of neurons as the number of scenes increases in the learning lap. The degree of self-organization is obtained as the Kendall's rank correlation coefficient between the similarity distance and the Manhattan distance for all neuron pairs on a two-dimensional lattice. The similarity distance of each neuron pair is calculated as the cosine of the angle between synaptic efficacy vectors. A dotted line in the figure shows the critical value of a positive correlation at the significance level (right-sided probability) of 0.5%. We can observe that the degree of self-organization is quickly recovered though it sometimes falls when a neuron is recruited especially in the first stage. It is also observed that significant and steady self-organization is achieved as the number of neurons increases in response to increase of the object variety.

Fig. 4 shows an example of object encoding in the recognition map layer and discriminative and invariant rates of object recognition in the recording lap. By repeating experiments, it was confirmed that invariant object recognition with respect to position and size was achieved with a high probability. Since target scene search succeeded almost perfectly, it was also confirmed that positions and sizes of objects were encoded suitably enough for scene recognition.

Recognition map layer

ID	Class of objects encoded	Discriminative rate	Invariant rate	ID	Class of objects encoded	Discriminative rate	Invariant rate
R0	Green cyl.	100%	100%	R1	Blue cyl.	93%	100%
R2	Red cyl.	100%	100%	R3	Magenta cyl.	100%	100%
R4	Yellow cyl.	95%	100%	R6	Blue cone	100%	100%
R8	Green cone	100%	100%	R13	Cyan cyl.	100%	100%
R14	Red cone	100%	100%	R16	Yellow cone	100%	100%
R7	Magenta cyl. and yellow cyl. behind	100%	100%	R9	A row of blue, red and green cyl.	100%	100%
R10	Red cyl. and green cyl. behind	100%	100%	R11	Green cyl. and blue cyl. behind	63%	100%
R12	Blue cyl. and green cyl. in front	100%	100%	R17	Blue cyl. and Magenta cyl. in front	100%	100%
R18	Blue cyl., O1* and Magenta cyl. in front	100%	100%	O1*: Yellow cyl. or a row of yellow and green cyl.		

Fig. 4. An example of discriminative and invariant object learning performance. For each neuron, a class of objects is the one that occupies the maximum number of objects encoded to it. The rate of discriminative recognition means the ratio of the number of objects in the class to the number of all objects encoded to the neuron. The rate of invariant recognition means the ratio of the number of objects in the class encoded to the neuron to the number of objects in the class encoded to all neurons

4.2 Glance Recognizability

The glance recognition is considered to be achieved based on quickly capturing partial information that represents distinctive features of objects. In the COGNET, this ability can be realized by shortening the length of the pre-competition inhibition period, which is in default set to 255 at learning, and suppressing pulse transmissions after the competition time.

Fig. 5(a) shows the success rate of the glance recognition when the length of the pre-competition inhibition period is shortened. Fig. 5(b) shows the success rate of the glance recognition against the number of effective pulse transmissions, that is pulse arrivals before the competition time, which is decreased by

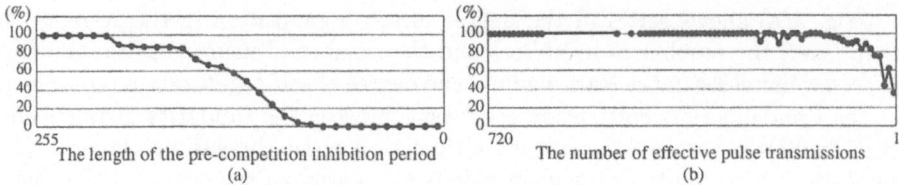

Fig. 5. The success rate of the glance recognition. It was tested whether glance recognition succeeded for 147 object images that the robot had paid attention on the maze. Data are summed up every 8 interval

shortening the length of the pre-competition inhibition period. We can observe that a high recognition success rate is achieved using only a small number of early pulse arrivals by shortening the length of the pre-competition inhibition period. These pulses encode large contrast or opponent color values that capture distinctive feature of objects.

5   Conclusions

We have evaluated self-organized learning and glance recognition performance of our saliency-based scene memory model whose core is the COGNET. As for learnability, it was confirmed that quick one-shot object learning and fast self-organization of object memory were performed, discriminative and invariant object recognition with respect to position and size was achieved and also positions and sizes of objects were encoded suitably enough for scene recognition. As for recognizability, it was confirmed that glance object recognition was achieved using only early pulse arrivals which encoded distinctive feature of objects.

References

1. Atsumi, M.: Scene Learning and Glance Recognizability based on Competitively Growing Spiking Neural Network. Proc. 2004 IEEE IJCNN, pp.2859-2864 (2004)
2. Atsumi, M.: Saliency-based Scene Recognition based on Growing Competitive Neural Network. SMC 2003 Conf. Proc., pp.2863-2870 (2003)
3. Breazeal, C., Scassellati, B.: A Context-dependent Attention System for a Social Robot. Proc. of 16th Int. Joint Conf. on Artificial Intelligence, pp.1146-1151 (1999)
4. Itti, L., Koch, C., Niebur, E.: A Model of Saliency-based Visual Attention for Rapid Scene Analysis. IEEE Trans. on Pattern Analysis and Machine Intelligence, **20**(11), pp.1254-1259 (1998)
5. Kohonen, T.: Self-Organizing Maps, Springer-Verlag (1995)
6. Thorpe, S., Delorme, A., Rullen, R. V.: Spike-based Strategies for Rapid Processing. Neural Networks, **14**, pp.715-725 (2001)

Pulsed Para-neural Networks (PPNN) Based on MEXOR Logic

Andrzej Buller[1], Ismail Ahson[2], and Muzaffar Azim[2]

[1] Advanced Telecommunications Research Institute International (ATR), NIS Labs.
2-2-2 Hikaridai, Keihanna Science City, Kyoto 619-0288 Japan
buller@atr.jp

[2] Jamia Millia University, Okhla, New Delhi 10025 India
mehar@ahson.org, m_azim1@yahoo.com

Abstract. We present Pulsed Para-Neural Networks (PPNN) defined as graphs consisting of processing nodes and directed edges, called axons. We discuss PPNNs in which every node has up to 3 inputs and returns a pulse at clock t if it received one and only one pulse at clock t-1. Axons represent pure delays. We provide theoretical account for MEXOR Logic underlying the behavior of the presented PPNNs. A number of proven theorems and schemes of practical devices shows that MEXOR-based PPNNs may be considered as a step toward a future-generation evolvable hardware for a brain-like computing.

1 Introduction

The research on MEXOR Logic and pulsed para-neural networks (PPNN) aims to develop a future-generation evolvable hardware for brain-like computing. MEXOR, known in switching theory as the elementary symmetric function S^3_1 (Sasao 1999, p. 99), is a three-argument Boolean function returning 1 if one and only one of the arguments equals 1. Currently MEXOR logic does not enjoy a special interest within the mainstream of logic, probably because it is not too attractive from the point of view of available VLSI technology. Nevertheless, PPNNs working based on MEXOR Logic are being developed for future-generation hardware (Buller et al. 2004).

S^n_1-based elements were implemented by Genobyte Inc. in the second half of 90s in the CAM-Brain Machine (CBM) that was to evolve neural-like circuits in a 3-dimensional cellular automata space (Korkin at al. 2000). The circuits included so called "dendritic cells" executing delayed S^5_1, however, according to the, so called, CoDi paradigm (Gers et al. 1997) the only role of the cells was to "collect" signals, while the key processing was to be done in so called "neural cells". Although the CBM failed to evolve anything practical, the research on CoDi-style computation were conducted at the ATR, Kyoto, as well as at the Ghent University. The Kyoto team concentrated on spike-train processing in PPNNs (Buller 2003) and non-evolutionary synthesis of CoDi, which resulted in the development of the *Neuro-Maze*™ *3.0 Pro* (see Liu 2002) – a CAD tool on which, among others, a module for reinforcement learning and various robotic-action drivers has been synthesized (Buller & Tuli 2002; Buller et al. 2002; Buller et al. 2003). The work of the Ghent team resulted in the discovery, that all practical spike-train processing can be done using exclusively the "dendritic cells" (i.e. without the CBM's "neural cells").

N.R. Pal et al. (Eds.): ICONIP 2004, LNCS 3316, pp. 399–408, 2004.
© Springer-Verlag Berlin Heidelberg 2004

Hendrik Eeckhaut & Jan Van Campenhout (2003) provided schemes based exclusively on delayed S^n_1, $n \leq 3$ for all sixteen 2-input Boolean functions, a circuit built on a single plane in which two signal paths can cross without interfering, and other devices. Hence, we replaced 3-dimensional CoDi with 2-dimensional qCA board, eliminated all processing units but the delayed 3-input MEXOR, and separated the problem of CA synthesis into the problem of PPNN synthesis and routing. Some promising results in MEXOR-based PPNN synthesis have been recently reported (Liu & Buller 2004).

This paper summarizes the current state of MEXOR Logic and PPNN theory and provides a draft theoretical framework for the future research in the area. In Section 2 we provide some fundamentals of Boolean logic. Section 3 contains MEXOR-specific notes and lemmas. Sections 4 and 5 present PPNN-related lemmas and practical spike-train processing devices. In Section 6 we speculate about potential application of the results.

2 Basic Concepts and Notations

Let us introduce the following notation for Boolean functions:

Function	Notation
NOT	~[argument]
AND	[argument1][argument2]
OR	[argument1] + [argument2]
XOR	[argument1] $\oplus$ [argument1]
MEXOR	$\square$([argument1],[argument2],[argument3])

NOT works in such a way that $\sim 0 = 1$, $\sim 1 = 0$. As for the next three functions, they work as shown in Table 1.

Table 1. Basic Boolean operations

X	y	xy	x+y	x $\oplus$ y
0	0	0	0	0
0	1	0	1	1
1	0	0	1	1
1	1	1	1	0

Table 2. Selected negation-related Boolean operations

x	Y	~ x	~ x y	((~ x)y) $\oplus$ x
0	0	1	0	0
0	1	1	1	1
1	0	0	0	1
1	1	0	0	1

As for Table 1, it can be noted that:

Note #1: i. $0 \oplus 1 = 1 \oplus 0 = 1$, $1 \oplus 1 = 0$,   therefore $x \oplus 1 = 1 \oplus x = {\sim}x$.
  ii. $0 \oplus 0 = 1$, $1 \oplus 0 = 1 \oplus 0 = 1$,   therefore $x \oplus 0 = 0 \oplus x = x$.

Note #2: $x \oplus y = x({\sim}y) + ({\sim}x)y$.
  Let us also provide four useful lemmas:

Lemma 1 (Sasao 1999, p. 44):
  i. $(x \oplus y) \oplus z = x \oplus (y \oplus z)$.   ii. $xy \oplus xz = x(y \oplus z)$.
  iii. $x \oplus y = y \oplus x$.   iv. $x \oplus x = 0$.

Lemma 2 (Sasao 1999, p. 45): $xy = 0 \Leftrightarrow x + y = x \oplus y$.

Lemma 3 (Hatchel & Somenzi 1999, p. 471): $(x + y) \oplus (x + z) = ({\sim}x)(y \oplus z)$.

Lemma 4: i. $y \oplus (y({\sim}x)) = xy$.   ii. $x \oplus (y({\sim}x)) = x + y$.

Proof: Ad i. According to Lemma 1ii, $yz \oplus yp = y(p \oplus z)$, which for $p = ({\sim}x)$ and $z = 1$ becomes $y1 \oplus y({\sim}x) = y(({\sim}x) \oplus 1)$. According to Note #1, $(({\sim}x) \oplus 1 = {\sim}({\sim}x)$ $= x$. Hence, $y \oplus (y({\sim}x)) = yx = xy$. QED.
Ad ii. Since the fourth, OR-related column of Table 1 and the last column of Table 2, showing the values of $x \oplus (y({\sim}x))$ are identical, this means that for all (x, y), $x \oplus (y({\sim}x))$ is equal to $x + y$. QED

3 MEXOR Logic

MEXOR Logic is a system of data processing that uses only one Boolean function we call MEXOR and denoted "□". In this paper we deal with three-argument MEXOR working as stated in Definition 1.

Definition 1. MEXOR, denoted □, is a 3-argument Boolean function that returns 1 if one and only one of its arguments equals 1.

  MEXOR is not to be confused with a three-argument function $f(x, y, z) = x \oplus y \oplus z$. The difference is shown in Table 3. As it can be seen, if all three arguments are equal to 1, f returns 1, while MEXOR returns 0. This property makes MEXOR very suitable as a basic building block in a synthesis of custom Boolean functions and spike-train processing devices.

Table 3. MEXOR compared with 3-input XOR

x	y	z	$\Box(x, y, z)$	$x \oplus y \oplus z$
0	0	0	0	0
0	0	1	1	1
0	1	0	1	1
0	1	1	0	0
1	0	0	1	1
1	0	1	0	0
1	1	0	0	0
1	1	1	0	1

A number of MEXOR-related notes can be derived directly from Table 3.

Note #3: $\Box(x, 0, 0) = \Box(0, x, 0) = \Box(0, 0, x) = x$.

Note #4: $\Box(x, y, 0) = \Box(x, 0, y) = \Box(0, x, y) = \Box(y, x,0) = \Box(y, 0, x) = \Box(0, y, x) = x \oplus y$.

Note #5: $\Box(x, 1, 0) = \Box(x, 0, 1) = \Box(1, x, 0) = \Box(0, x, 1) = \Box(1, 0, x) = \Box(0, 1, x) = {\sim}x$.

Note #6: $\Box(x, y, y) = \Box(y, x, y) = \Box(y, y, x) = x({\sim}y)$.

Lemma 5. i. $\Box(\Box(x, x, y), y, 0) = xy$. ii. $\Box(\Box(x, x, y), x, 0) = x + y$.

Proof: Let $\Box(\Box(x, x, y), y, 0) = u$, $\Box(\Box(x, x, y), x, 0) = w$.
According to Note #6, $\Box(x\ x\ y) = y({\sim}x)$. Hence, $u = \Box(y({\sim}x), y)$, $w = \Box(y({\sim}x), x)$, which, according to Note #4, gives $u = y({\sim}x) \oplus y$, $w = y({\sim}x) \oplus y$. Hence, according to Lemma 2i, $u = xy$, while, according to Lemma 2ii, $w = x + y$. QED

Lemma 6. $xyz = 0 \Rightarrow x \oplus y \oplus z = \Box(x, y, z)$.

Proof: $xyz = 0 \Leftrightarrow x = 0$ or $y = 0$ or $z = 0$. Let, therefore, check all of the three cases. For $x = 0$, according to Lemma 1i and Note #1, $x \oplus y \oplus z = 0 \oplus (y \oplus z) = y \oplus z$, which, according to Note #4, equals $\Box(0, y, z) = \Box(x, y, z)$. According to the same lemma, for $y = 0$, $x \oplus y \oplus z = x \oplus 0 \oplus z = (x \oplus 0) \oplus z = z \oplus (x \oplus 0) = (z \oplus x) \oplus 0 = z \oplus x = \Box(x, 0, z) = \Box(x, y, z)$, while for $z = 0$, $x \oplus y \oplus 0 = x \oplus y = \Box(x, y, 0) = \Box(x, y, z)$.Hence, the lemma is true for all cases satisfying the statement $xyz = 0$. QED

Lemma 7. $({\sim}y)x + yz = \Box(\Box(\Box(x, x, y), x, 0), \Box(y, z, z), 0)$.

Proof: Let $({\sim}y)x + yz = u$. Note that (i) $xz = 1 \Rightarrow ({\sim}y)x + yz = 1$, (ii) for all y, $({\sim}y)y = 0$, (iii) $(p = 1 \Leftrightarrow q = 1) \Rightarrow p = p + q$, and (iv) for all q, $q + 0 = q$. Thus, $u = ({\sim}y)x + xz + ({\sim}y)y + yz = x(({\sim}y) + z) + y(({\sim}y) + z) = x({\sim}(y({\sim}z))) + y({\sim}(y({\sim}z))) = (x + y)({\sim}(y({\sim}z))) = (x + y)({\sim}(y({\sim}z))) + 0 = (x + y)({\sim}(y({\sim}z))) + 0 = (x + y)({\sim}(y({\sim}z))) + ({\sim}y)y({\sim}x)({\sim}z) = (x + y)({\sim}(y({\sim}z))) + ({\sim}(x + y)y({\sim}z)$. Hence, according to Note #2, $u = (x + y) \oplus y({\sim}z)$. According to Note #4, $u = \Box((x + y), y({\sim}z), 0)$, which, according to Lemma 2ii and Note #6, equals $\Box(\Box(\Box(x, x, y), x, 0), \Box(y, z, z), 0)$. QED

Lemma 8. $({\sim}x)yz + x({\sim}y)z + xy({\sim}z) + xyz = \Box(x, p, q)$, where $p = \Box(x, x, y)$, $q = \Box(\Box(x, y, 0), z, z)$.

Lemma 7 and Lemma 8 show how to have a multiplexer and a majority function built of MEXORs only.

4   Pulsed Para-neural Networks (PPNN)

The next two definitions introduce the notions of pulsed para-neuron and pulsed para-neural network (PPNN).

Definition 2. An *n-input pulsed para-neuron* is a processing element operating on binary time-series $(x_{i,0}, x_{i,1}, ..., x_{i,t}, ...)$, $i = 0, 1, .., n\text{-}1$, such that it returns 1 at clock $t+1$ iff $f(w_{0,t}x_{0,t}, w_{1,t}x_{1,t}, ..., w_{n\text{-}1,t}x_{n\text{-}1,t})$ was equal to 1, where f is a Boolean function, $w_{i,t}$ is a binary value representing a weight of i-th input at clock t.

Definition 3. A *pulsed para-neural network (PPNN)* is a graph consisting of processing nodes and directed edges, where each node is a pulsed para-neuron and each edge represents a specified delay of data transmission between the points it connects.

In this paper we discuss only PPNNs such that for each para-neuron, $f \equiv \square$ and weights are equal to 1 for the period of interest (Fig. 1a). Note that even such a MEXOR-based PPNN meets some definitions of a neural network (eg. Hecht-Nielsen 1990, p. 2-3) and can be interpreted as a modification of the historical formal neuron (Fig. 1b).

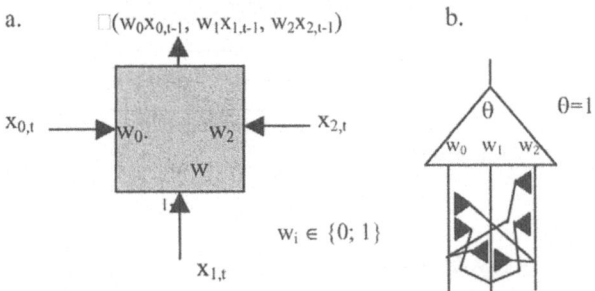

Fig. 1. MEXOR-based para-neuron. (a) a scheme according to Definition 3. (b) an equivalent device based on formal (McCulloch-Pitts) neuron and six pre-synaptic inhibition connections

PPNN graphical notation uses squares representing para-neurons and arrows representing delays. Thick, labeled arrows represent delays that are longer than 1 clock or unknown, thick, non-labeled arrows represent 1-clock delays, and tiny, non-labeled arrows represent connections of zero delay. Two or more points of reference may concern a given arrow: begin, end, and a possible fan-out points. A given label appears as a specified number of clocks or as a symbol of a variable and concerns only an arrow's section between two nearest points of reference (Fig. 2).

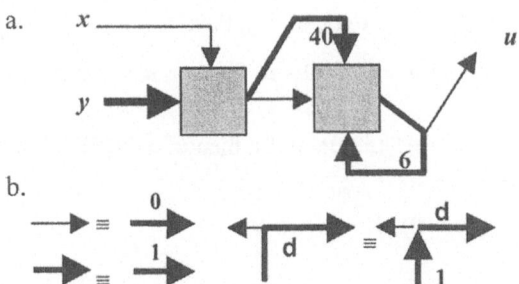

Fig. 2. PPNN graphical notation. a. an example of a network, b. simplifying replacements

As it can be noted, in case of PPNN, it is essential not only what values are provided to a given para-neuron, but also when they are provided. Table 4 presents a notation facilitating an analysis of PPNN behaviors.

If a series provided to a para-neuron's input is **0**, this means that the related weight equals 0 or that the related input is not used. Otherwise, it is assumed that the related weight equals 1. Let us therefore formulate some lemmas:

Lemma 9. i. $\underline{\square}(x, y, z) = \leftarrow\square(x, y, z)$.

ii. $\underline{\square}(a+d\leftarrow x, b+d\leftarrow y, c+d\leftarrow z) = d\leftarrow\underline{\square}(a\leftarrow x, b\leftarrow y, c\leftarrow z)$.

Table 4. PPNN notation

Expression	Interpretation
$x_0x_1...x_t...$	time-series $(x_0, x_1, ..., x_t, ...)$, where t is a given clock
?	undetermined value of x_t
$?^d$	series of d question marks; $?^1 = ?$, $?^2 = ??$, $?^{d+1} = ??^d$
$\leftarrow x \quad (= ?x)$	series x delayed by one clock
$d\leftarrow x \quad (= ?^d x)$	series x delayed by d clocks
0	000000...
1	111111...
$\square(x, y, z)$	A series returned by a para-neuron that are provided with series x, y, and z

Lemma 10. If $u = \square(x, y, 0)$, then $\forall_{t\geq0} u_{t+1} = x_t \oplus y_t$.

Proof: Let $u = u_0u_1u_2...$, $x = x_0x_1x_2...$, $y = y_0y_1y_2...$. According to Lemma 4i, $u = \square(x, y, 0) = \leftarrow\square(x, y, 0)$. Hence, according to Definition 2, $u_0u_1u_2... = ?v_0v_1v_2...$, where for all $t\geq0$, $v_t = \square(x_t, y_t, 0)$. According to Note #5, $\square(x_t, y_t, 0) = x_t \oplus y_t$. Hence, for all $t\geq0$, $u_{t+1} = v_t = x_t \oplus y_t$, QED.

Lemma 11. If $u = \square(x, 1, 0)$, then $\forall_{t\geq0} u_{t+1} = \sim x_t$.

Proof: Lemma 5 states that if $u = \square(x, y, 0)$, then $\forall_{t\geq0} u_{t+1} = x_t \oplus y_t$. Hence, for y = 1, $u_{t+1} = x_t \oplus 1 = \sim x_t$. QED

Lemma 12. If $u = \square(\square(x, x, y), \leftarrow y), 0)$, i≠j, then $\forall_{t\geq0} u_{t+2} = x_t y_t$.

Lemma 13. If $u = \square(\square(x, x, y), \leftarrow x), 0)$, i≠j, then $\forall_{t\geq0} u_{t+2} = x_t + y_t$.

Lemma 14. If $x = 10$ and $u = \square(x, D\leftarrow u, 0)$, then u = 0AAA..., where A is 1 followed by D 0s.

Lemma 15. If $x = 1$ and $u = \square(x, D\leftarrow u, 0)$, then u = 0BBB..., where B is a series of D 1s followed by D 0s.

Lemma 16. $\square(\leftarrow y, \square(y, x, 0), 0) = = \leftarrow\leftarrow x$.

Proof:

$$x' = \square(\leftarrow y, \square(y, x, 0), 0)$$

$= \square(\leftarrow y, \leftarrow\square(y, x, 0), 0)$	/ Lemma 4i
$= \square(\leftarrow y, \leftarrow y \oplus x, 0)$	/ Note #4
$= \leftarrow\square(y, y \oplus x, 0)$	/ Lemma 4ii
$= \leftarrow\leftarrow\square(y, y \oplus x, 0)$	/ Lemma 4i
$= \leftarrow\leftarrow (y \oplus y) \oplus x$	/ Note #4
$= \leftarrow\leftarrow 0 \oplus x$	/ Lemma 1ii
$= \leftarrow\leftarrow x$. QED	/ Note #2

Lemmas 10-13 may serve as tips for PPNN-based synthesis of custom Boolean functions (see Fig. 3). As it may be noted, a para-neuron wired in such a way serves

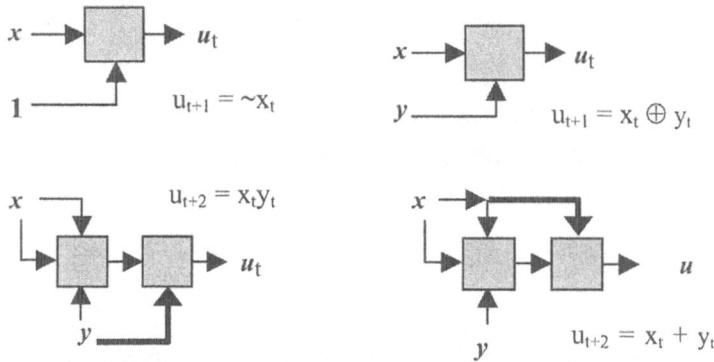

Fig. 3. PPNNs for basic Boolean functions (cf. Lemmas 5, 6, 7& 8)

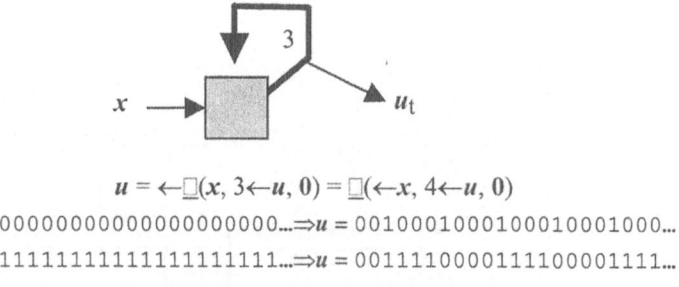

$$u = \leftarrow\Box(x, 3\leftarrow u, 0) = \Box(\leftarrow x, 4\leftarrow u, 0)$$
$$x = 10000000000000000000000...\Rightarrow u = 001000100010001000...$$
$$x = 111111111111111111111...\Rightarrow u = 001111000011110000 1111...$$

Fig. 4. An example of PPNN-based oscillator (cf. Lemmas 14 and 15)

as an oscillator producing a periodic input pattern depending on the input pattern (Fig. 4).

5 PPNN-Based Devices

Existence of PPNN-based NOT and AND Boolean functions ensures that there is a PPNN for any Boolean function. By having axons of defined delays, one can provide generated values to specified PPNN nodes at desired moments in time. This means that any manipulation of spiketrains can be done by using an appropriate PPNN. Moreover, owing to properties of the device shown in Fig. 5, we can build any PPNN-based spike-train processor without crossed edges. In those section we present the "flat" overpass (Fig. 5), erasable pulsed memory (Fig. 6 & 7), para-neuron of the second kind (Fig. 8).

In "Flat overpass" Pl. note that: $x' = \Box(\leftarrow y, \Box(y, x, 0), 0)$, $y' = \Box(\leftarrow x, \Box(x, y, 0), 0)$, hence, $x' = \leftarrow\leftarrow x$, $y' = \leftarrow\leftarrow y$ (Lemma 11).

In Erasable pulsed memory (EPM) a memorized 1 circulates in the loop constituted by the nodes M_1, M_2, M_3, and M_4. It arrives at the input s and comes to the loop through the AND gate built using the nodes A_1, and A_2, if also 1 arrives to the input p at the same clock. Then M1 every fourth clock produces 1 that come to two inputs of D preventing the circulating 1 from being erased by 1s arriving at s later. The memory can be erased via providing 1 to the input r at appropriate clock.

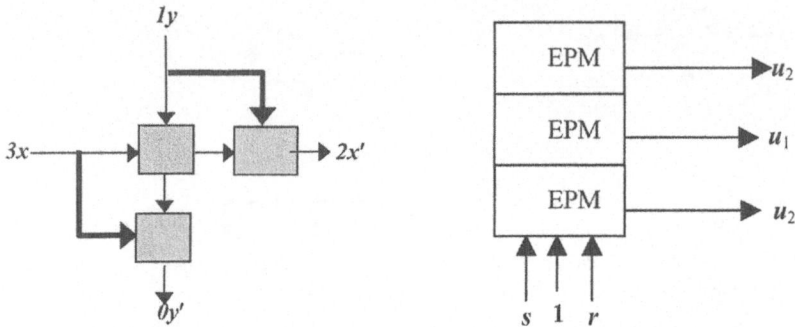

Fig. 5. "Flat overpass"   **Fig. 6.** A Counter made of Erasable Pulsed Memories

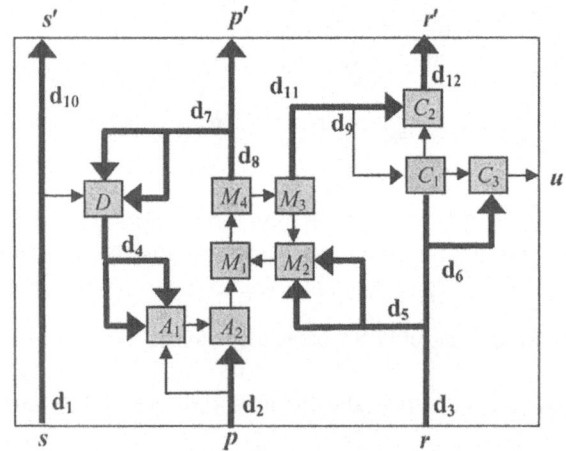

Fig. 7. Erasable pulsed memory (EPM)

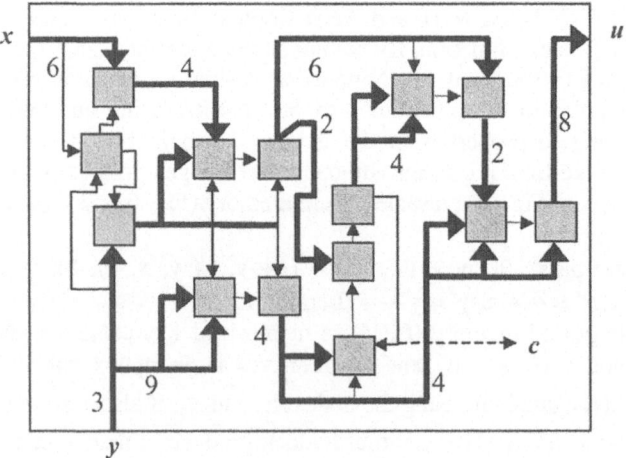

Fig. 8. Pulsed para-neuron of the 2nd kind (PPN2)

In the counter made of erasable pulsed memories when a pulse arrives at s for the first time, the lowest EPM will start producing 1s every fourth clock as signal u_0 and keeping the second EPM ready to work. Then, when another pulse arrives at s at appropriate clock, the second EPM will start producing 1s every fourth clock as signal u_1 and keeping the third EPM ready to work. Consecutive pulses arriving at r at appropriate clocks will switch-off EPMs top to down.

The PPNN-based device works according to formulas:

$$u_{t+32} = x_t y_t + c_{t+16}(x_t + y_t), \quad c_{t+16} = \Box(c_t, x_t, y_t).$$

Owing to this it behaves as an extension of 2-input formal neuron with the threshold equal to 2 and both weights equal to 1, where activation is accumulated in time.

6 Concluding Remarks

The lemmas we proved and schemes of practical devices we presented show that MEXOR logic and MEXOR-based Pulsed Para-Neural Networks (PPNN) may be considered as a candidate for a future-generation evolvable hardware in which artificial brain-like systems will be eventually implemented. In order to make MEXOR/PPNN computing a more attractive sub-area of switching theory, a lot of additional results should be shown. Indeed, further research should, first of all, bring answers to the following questions:

- How to synthesize efficiently arbitrary Boolean functions (and/or spike-train processing devices) using exclusively MEXORs (and/or para-neurons)?
- How many p-n junctions are necessary to build a MEXOR and para-neuron?
- How many para-neurons are necessary to implement associative memories, multipliers, etc.?
- Can a MEXOR's internal scheme be built without crossed wires?
- Can MEXOR (and para-neuron) be built of other substrate than semiconductor?

As it can be noted, the idea of MEXOR logic and PPNN triggers an avalanche of intellectual and technological challenges. Another set of lemmas facilitating a reduction of MEXOR expressions without "mediation" of traditional Boolean functions should be worked out. Also an expansion for direct conversion of arbitrary truth table into possibly short MEXOR expression should be invented.

As for internal structure of MEXOR or para-neuron, it is important to compare the number of required p-n junctions and number of indispensable wire-crossings with traditional NOR-NAND and XOR-AND technologies. Even if the comparison challenges MEXOR, it is a chance that for most practical cases, especially in brain-like systems, excessive junctions in a single MEXOR may be compensated by relatively smaller number of required gates. Anyway, PPNNs may attract VLSI industry owing to their potential to be implemented as 2-dimensional sheets covered by a single-layer of semiconductor, free of crossing paths).

Another chance for MEXOR/PPNN hardware to challenge NOR-NAND- and XOR-AND-based circuits comes a possibility that there is a way of building a MEXOR of a non-semiconductor material as a device simpler than, NAND or XOR built in the same technology.

Acknowledgement

This research was conducted as a part of the *Research on Human Communication* supported by the National Institute of Information and Communications Technology of Japan (NICT).

References

1. A. Buller, CAM-Brain Machines and Pulsed Para-Neural Networks (PPNN): Toward a hardware for future robotic on-board brains, *8th International Symposium. on Artificial Life and Robotics (AROB 8th '03), January 24-26, 2003, Beppu, Japan*, pp. 490-493.
2. A.Buller, H. Hemmi, M. Joachimczak, J. Liu, K. Shimohara & A. Stefanski, "ATR Artificial Brain Project: Key Assumptions and Current State", 9th International Symposium on Artificial Life and Robotics (AROB 9th), Jan. 28-30, 2004, Japan, 2004. pp 166-169.
3. A.Buller & T.S. Tuli, Four-legged robot's behavior controlled by Pulsed para-neural networks (PPNN), Proceedings, 9th International Conference on Neural Information Processing (ICONIP'02), Vol.1. November 18-22, 2002, Singapore, pp. 239-242.
4. A.Buller, H. Eeckhaut & M. Joachimczak, Pulsed para-neural network (PPNN) synthesis in a 3-D cellular automata space, Proceedings, 9th International Conference on Neural Information Processing (ICONIP'02), Vol.1. Nov. 18-22, 2002, Singapore, pp. 581-584.
5. A.Buller, Joachimczak M., Liu J., Stefanski A. Neuromazes: 3-Dimensional Spiketrain Processors, WSEAS Transactions on Computers, Issue 1, Vol 3, 2004, pp. 157-161.
6. H. Eeckhaut and J. Van Campenhout "Handcrafting Pulsed Neural Networks for the CAM-Brain Machine", 8th International Symposium on Artificial Life and Robotics (AROB 8th '03), January 24-26, 2003, Beppu, Japan, pp. 494-498
7. F. Gers, H. de Garis & M. Korkin "CoDi-1Bit: A Simplified Cellular Automata based Neuron Model", Evolution Artificiele 97, Nimes, France, October 22, 1997, pp. 211-229.
8. G.D. Hachtel & F. Somenzi, Logic Synthesis and Verification Algorithms, Kluwer Academic Publ., 1996.
9. R. Hecht-Nielsen, Neurocomputing, Addison-Wesley, 1990.
10. M. Korkin, G. Fehr & G. Jeffrey, Evolving hardware on a large scale, Proceedings, The Second NASA / DoD Workshop on Evolvable Hardware, July 2000, Pasadena, pp. 173-81.
11. W.S. McCulloch & W. Pitts, A logical calculus of the ideas immanent in nervous activity, Bulletin of Math. Bio, 5, 1943, pp. 115-133.
12. J. Liu, NeuroMaze™. User's Guide, Version 3.0. ATR Human Information Science Labs., Kyoto, Japan, 2002.
13. T. Sasao, Switching Theory for Logic Synthesis, Boston: Kluwer Academic Publ., 1999.

Knowledge Reusing Neural Learning System for Immediate Adaptation in Navigation Tasks

Akitoshi Ogawa and Takashi Omori

Graduate School of Information Science and Technology, Hokkaido University
Kita 14 Nishi 9, Sapporo, Japan
akitoshi@complex.eng.hokudai.ac.jp

Abstract. A characteristic feature of conventional intelligent agents is the amount of trials that are required for them to learn. Since the tasks they encounter change depending on the environment, it is difficult for a learning system to compress the learning time using a priori knowledge. In the real world, however, agents confront a whole range of new tasks one by one, and have to solve them one by one without consuming learning time. A serious problem for the real-world agent is the amount of learning time needed. We suppose that one reason for a long learning time is the nonuse of prior knowledge. It is natural to expect that a kind of fast adaptation to tasks would be possible when we reuse knowledge that is acquired from similar past experiences. For this problem, we propose a neural-network based learning system that can immediately or quickly solve new tasks by reusing knowledge already acquired. We adopt a navigation task as an example and show the effectiveness of our method in variations on the task by comparing its performance with other methods.

1 Introduction

One of the major characteristics of intelligent agents in the real world is that a task is unknown in advance and an agent may require a huge number of trials to learn it. The agent faces new unknown tasks in unexpected order one after the other in the real world. The number of required trials, therefore, may become a serious problem for the agent in its effort to survive. One reason for long learning time is the nonuse of prior knowledge that causes a large learning space. If the agent reuses the knowledge acquired from past experiences in a task similar to previous tasks, the agent will finish the learning with much fewer trials. This is fast adaptation to a new task.

In this paper, we propose a new neural learning system that reuses knowledge of previously learned tasks to solve a new task. The system can control the topology of a neural network, construct module-like sub-neural networks in a self-organized manner and form a neural circuit that generates an output action from an input of the new task environment. The system searches a combination of the sub-networks to form a circuit, in which the connection knowledge within the sub-network and the combination of the sub-networks are reused. The agent with

N.R. Pal et al. (Eds.): ICONIP 2004, LNCS 3316, pp. 409–415, 2004.
© Springer-Verlag Berlin Heidelberg 2004

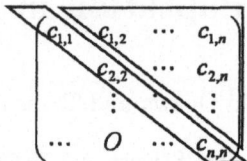

Fig. 1. Structure matrix that represents the network topology between the modules and the learning rules for each of the modules. For example, $c_{6,12}$ designates the connection and learning rule between module No.6 and module No.12.

our system reuses the knowledge of past tasks and exhibits fast adaptation to a new task. If the agent keeps the combinations of sub-networks used in past similar tasks, the agent will be able to adapt to various types of unexperienced tasks in a short time using the combination knowledges. To verify the effectiveness of the proposed system, we apply it to a basic benchmark of robot navigation and compare its performance with other learning methods. The result shows that our system exhibits fast adaptation based on a new learning concept of knowledge reuse that differs much from the learning concept of other neural networks.

2   Knowledge Reusing Neural Learning System

2.1   Network Building Using Structure Matrix

In our system, the combination of modular-type neural networks decides the network processing. This combination is defined by factors of leaning rules, topology between sub networks, transfer functions of neurons and connection weights. These factors decide the rough behavior of the whole neural network. After that, learning according to the input from the environment decides the network's fine processing. We assume that the four factors above are the knowledge for re-solving the tasks.

The modules of inputs, intermediates and outputs are numbered in a sequential order. An output of a module is sent to other modules with a larger number to avoid confusion in the computational order of the modules. We introduce a structure matrix to represent the network structure except for the connection weights. If there are n modules, a $n \times n$ regular matrix (Fig.1) is used for the structure matrix.

We use a genetic algorithm (GA) to search for parameters of the structure matrix appropriate for each task. In our system, a gene directly represents the structure matrix. In the GA method, a set of distributed individuals is brought toward the target point in the search space. If the target structure matrix is within the range of the distribution, the agent can find the target matrix within a few generations. Immediate adaptation is derived from the relative disposition of the distribution of individuals and the target.

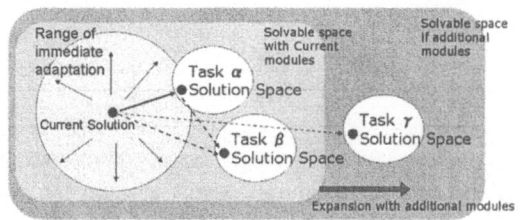

Fig. 2. Range of structural flexibility and immediate adaptation of the network. If the range of immediate adaptation by individuals and their weights covers the new task in the task space, the agent can immediately adapt to the task. However, if the new task is out of the range, the agent needs additional searching and learning.

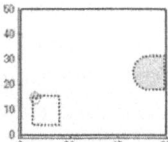 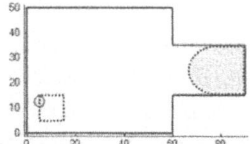

Fig. 3. Task A: starting area is designated as the area on the dotted line. A light is located at (60,25). The goal is the area around the light.

Fig. 4. Task B: starting area is designated as the area on the dotted line. The light is located at (90,25). The goal is the area around the light.

2.2 Immediate Adaptation by Reusing Knowledge

When an agent has finished the searching and learning in task A, the optimum individual has the structure matrix and weights for the task. We then assume that the agent faces task B, which is similar to yet different from task A in some parts. Additional searching and learning based on the knowledge derived from task A can be used to realize fast searching and learning in another task. We can expect if there is less additional searching, the adaptation to task B is faster.

In the initial search of task B, the structure matrixes for task A mutate and became the first generation of new task B. If task B is sufficiently similar to task A, the search time is shortened and immediate adaptation is accomplished (Fig.2).

3 Computer Simulation

3.1 Gradually Making Task Settings More Complicated

We prepared three navigation tasks: task A, task B and task C, which gradually become more complicated. The tasks shown in Fig.3, Fig.4 and Fig.5 are implemented in a Khepera simulator. Adaptation of each individual is judged when the agent reduces the number of steps to a goal and the learning converges.

We evaluated two cases of task changes to verify the efficiency of the knowledge reuse.

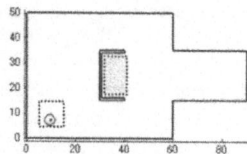

Fig. 5. Task C: starting area is designated as the area on the dotted line. The light is located at (90,25). The goal is the area inside of the walls at the center of the environment.

Case I : The agent acquires knowledge in task A and reuses the knowledge when it encounters task B.

Case II : The agent reuses the knowledge of case I when it encounters task C.

In task B, all the knowledge derived from task A can be reused. Therefore, an agent can adapt to task B if it acquires some additional knowledge. In task C, however, some of the knowledge from task B can be reused, but some knowledge interferes with the adaptation. We show a transition of the fitness value of the optimum individual.

3.2   Detail of the Structure Matrix

We divide the Khepera's front sensors into 6 groups with respect to direction and sort. We prepare six input modules corresponding to the groups and six intermediate modules that have two neurons inside. Three output modules correspond to the actions: move ahead, right turn and left turn. The following two transfer functions for the neurons are prepared ($a = 1, b = 1, c = 2$).

1. $f(x) = \exp(-ax)$
2. $f(x) = \frac{-1}{1+\exp(-bx+c)}$

Weights are initialized with positive small numbers. We prepare the following three learning rules.

1. $\Delta w = \eta(x - w)$ if $x > 0.5$
2. $\Delta w = \eta(x - w)$ if $x \leq 0.5$
3. $\Delta w = \eta(x - w)$
   where η is the learning rate ($\eta = 0.1$).

Weights are updated at every step ($2step/sec$). Input modules output a vector in which a sensory input is normalized to $[0, 1]$. The following equation determines the inputs x of the translation function in the intermediate and the output modules, where the i-th module receives the j-th module's output y_j and the weights of the connection are w_{ij}. The actual action is determined by a competition between the output modules.

$$x_i = \sum_j |y_j - w_{ij}|.$$

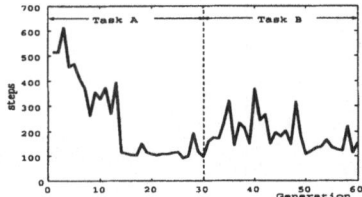

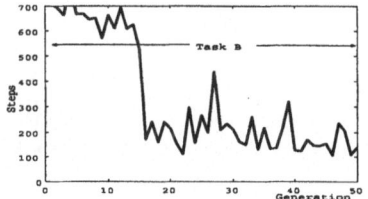

Fig. 6. Result in case I. Task A was performed to the 30th generation, and the task switched to task B at the 31st generation. The navigation agent acquired knowledge for task A at the 14th generation and adapted to task B in the first generation of the task.

Fig. 7. Result of task B not reusing the knowledge for task A. The agent acquired the knowledge for task B at the 16th generation.

3.3   Task Setting for Action Learning and Searching

The fitness of each individual is the average steps of six trials. One trial ends when the agent reaches the goal, collides with an obstacle or reaches 800 timeout steps. The fitness value for the collision is the same as the timeout. Each individual continues learning throughout the trials.

The parameters of GA are as follows: 20 for population, 0.1 for mutation rate and 2 for the number of elites. Individuals in the first generation are generated randomly. One point crossover by the row of the structure matrix is used. Pairs for the crossover are selected randomly. Mutation is applied to an element of the matrix. If the mutation operation removes a connection, the system removes the corresponding weights. If the mutation operation generates a connection, the system generates and initializes a new connection. If there is no mutation, weights are carried over generations.

In case I, task A switches to task B when the 31st generation starts. The system reuses the knowledge derived from task A at the first generation of task B, and the system keeps on searching. The same is true for case II. To show the efficiency of the knowledge reuse, we compare case I with a case in which the same agent starts and continues learning in task B for 50 generations. Fig.6 and Fig.7 show the results of case I, while Fig.8 shows the result of the case II.

3.4   Comparison

We compare our system with the evolutionary robotics(ER) and reinforcement learning (RL) methods. Fig.9 and Fig.10 show results of ER and RL respectively.

Evolutionary Robotics [FU1,NF1] There are various methods of genetic programming for acquiring robot control processing and of real-valued GA for searching weights of a neural network. We adopted the latter for comparison because most ER studies use it.

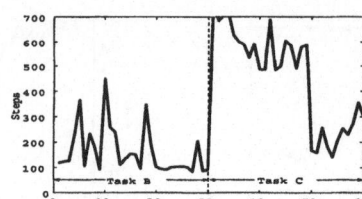

Fig. 8. Result in case II. The agent adapted to task C at the 19th generation after the task switch.

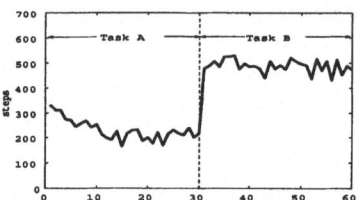

Fig. 9. Result of ER in case I. ER system adapted to task A, but could not adapt to task B.

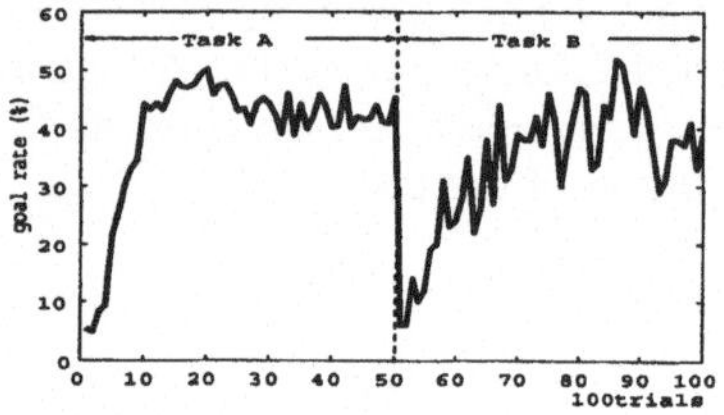

Fig. 10. Result of RL in case I. Task A switched to task B at the 5000th trial. Goal rate converged at about the 2000th trial after the switch.

Reinforcement Learning [SB1] We adopt the actor-critic (λ) method because of its learning speed and usability in a continuous space. Since the raw sensory input raises the perceptual aliasing problem, we used position input in the world coordinates (x, y, θ) because of its ease in learning.

Table 1 shows a comparison of the number of steps until adaptation; the fitness of our system and ER is judged by the number of generations and that of RL is judged by the number of trials. The adaptation speed of our system is distinct in case I. ER and RL failed in case II, but our system could adapt.

Table 1. Comparison of experimental results. Generations or trials until convergence of learning (-: failed to learn task).

	task A	task B	task C	
Our System	14	1	19	[generation]
ER	10	-	-	[generation]
RL	1200	2000	-	[trial]

4   Conclusion

In this paper, we proposed a learning system that reuses knowledge from former tasks by controlling the topology of a neural network and constructing module-like sub-neural networks in a self-organized manner. We compared our learning system with ER and RL and evaluated its efficiency.

Intelligent agents in the real world have insufficient time to adapt to different tasks, and such agents should aggressively reuse knowledge acquired in past tasks. Our system acquires knowledge in the form of functional modules that can be easily reused to immediately or quickly adapt to a new task.

References

[FU1] Floreano, D., Urzelai, J.: Evolutionary robots with on-line self-organization and behavioral fitness. Neural Networks **13** (2000) 431–443

[Ka1] Kasabov, N.: Evolving Connectionist Systems. Springer (2003)

[NF1] Nolfi, S., Floreano, D.: Evolutionary Robotics. MIT PRESS (2000)

[SB1] Sutton, R. S., Barto, A. G.: Reinforcement Learning: An Introduction. MIT PRESS (1998)

Universal Spike-Train Processor for a High-Speed Simulation of Pulsed Para-neural Networks

Michal Joachimczak[1], Beata Grzyb[2], and Daniel Jelinski[1]

[1] Technical University of Gdansk, Dept. of Electronics,
Telecommunications and Informatics ul. Gabriela Narutowicza 11/12, 80-952 Gdansk, Poland
{guhru,danjel}@gabri.pl
[2] Maria Curie-Sklodowska University in Lublin,
Faculty of Mathematics, Physics and Computer Sci. Pl. M. Curie-Skłodowskiej 5,
20-031 Lublin, Poland
bea_jo@gabri.pl

Abstract. We propose a Universal Spiketrain Processor (USP) – a computing device for high-speed execution of Pulsed Para-Neural Networks (PPNN). PPNN is a graph consisting of processing nodes and directed edges, called axons. We provide a description of USP architecture, a USP-specific language of PPNN coding, and methods of conversion of the already existing PPNN code into a USP code. We also compare the USP's performance with the already existing cellular-automata-based tools for PPNN simulation.

1 Introduction

Universal Spiketrain Processor (USP) is a computing device for high-speed execution of *Pulsed Para-Neural Networks* (PPNN) developed at the Advanced Telecommunications Research Institute International (ATR), Kyoto, Japan. PPNN is a graph consisting of processing nodes and directed edges, called axons (Figure 1). Every PPNN node represents a delayed function operating on spike-trains. Every axon transmits spikes with a defined delay. Both nodes and axons receive/return spikes at discrete moments of time (clocks) [1]. Very large PPNNs are intended to be synthesized as artificial brains [2][3]. Smaller PPNNs can be designed to serve as robotic controllers [4] or devices for processing of various signals (including music) [5].

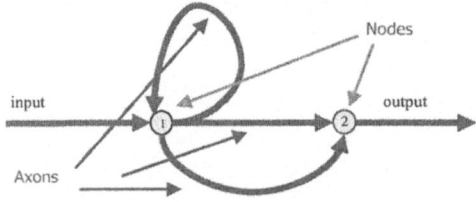

Fig. 1. A simple PPNN graph

In cellular-automatic PPNNs (CA-PPNN) every node takes a form of a cell occupying a defined part of space (e.g. square, cube, or hypercube), while every axon is a chain of cells of the same shape. Every axonic cell serves as a 1-clock delay. Each cell has a certain number of memory bits. Some of the bits code the function realized

N.R. Pal et al. (Eds.): ICONIP 2004, LNCS 3316, pp. 416–421, 2004.
© Springer-Verlag Berlin Heidelberg 2004

by this cell and its current state. Other bits specify from which neighbors signals are received and to which signals are being sent. Input and output points are associated with particular facets of the cell (cf. [6]).

The CAM-Brain Machine (CBM), produced in 90s by the Genobyte, Inc. was the first CA-PPNN-dedicated hardware offering two kinds of nodes: a threshold (neuron-like) element and a multi-input gate returning 1 if one and only one pulse entered it at previous clock. According to the manufacturer, the CBM was to allow its user to define and run a CA-PPNN occupying up to 64,640 3-dimensional modules of 24×24×24 cells [7]. A CBM-compatible software simulator has been written to facilitate a research on PPNNs at the centers that cannot afford to have the expensive CBM. The simulator, together with a related CAD tool constitutes the *NeuroMaze*™ *3.0 Pro* package [8][9]. An example of a CBM-compatible CA-PPNN module is shown in Figure 2.

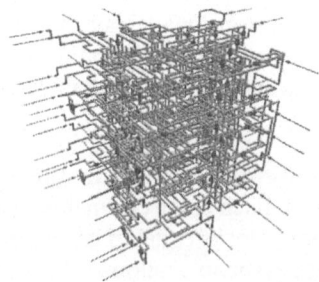

Fig. 2. A typical 3-dimensional CA-PPNN structure

In the CBM the states of every CA cells are updated in parallel by their own separate circuits. The time required to update the state of each cell sequentially on a PC is acceptable but not satisfactory. The need to accelerate the *NeuroMaze*™ *3.0 Pro* motivated our research.

2   Universal Spike-Train Processor (USP)

It has been noticed that in most of practically used CA-PPNN modules only a small fraction of cells is actually processing signals, while the remaining ones broadcast pulses to nodes or remain unused [5]. To take advantage of that, we have developed a different approach where updating all cells in N axons is replaced with moving 2 pointers.

2.1   Basic Assumptions

Let's assume spike-trains to be series of data, sampled at discrete moments of time called clocks. Each sample is represented as n-bits of data coding the amplitude of the signal at the given clock. All the samples inside one USP system use the same, configurable number of bits.

The task of a USP system is to perform a simulation of a given PPNN graph (consisting of processing nodes and axons). After leaving the output of a node, signal travels through an axon and reaches its target node after D clocks, where D is a property of a given axon. Therefore each axon is a buffer of the length D. We call it a delay channel. Each node may perform a single- or multi-valued logic on its inputs and its internal state according to a function that is assigned to it.

PPNN shown in Figure 1 is a very simple graph with 3 axons and 2 nodes. However, to make the graph complete, information about axon delays and functions performed by particular nodes is necessary.

2.2 USP Architecture

USP architecture consists of four basic modules: Spike-train memory, Channel manager, Net manager, Processing units space.

2.2.1 Spike-Train Memory
Spike-train Memory is a one-dimensional array that stores all the samples that are currently traveling through the delay channels and thus, have to be buffered. The total size of this array is equal to the sum of lengths of all delay lines.

2.2.2 Channel Manager
Channel Manager divides Spike-train memory into delay channels and provides access to their inputs and outputs. Each delay channel is a part of Spike-train memory. It delays signals by the number of clocks equal to its length, working as first-input-first-output (FIFO) buffer. As we want to minimize the amount of data that has to be moved, each delay channel has a pointer associated with it. The pointer points to the current state of the channel's output. With the next clock this value is overwritten by the state of the channel's input and the pointer is incremented. When the pointer becomes equal to the length of the delay channel it is reset to 0. This way, to provide desired delays, with every clock we only need to read and write one single sample for each channel, no matter its length.

Channel Manager contains internal memory that stores the information about all the delay channels in an array. One array entry contains following information:

- address in the Spike-train-memory
- length of a channel
- current channel's pointer value
- optionally – some amount of cache memory to reduce the number of accesses to Spike-train memory

2.2.3 Net Manager
Net Manager distributes processed data between outputs and inputs of nodes and channels. Similarly to Channel Manager, it contains internal memory which stores an array containing the list of all connections. Each array entry stores the source type and number as well as destination type and number. The type can be either a channel or a node. When the destination type is a node, additionally an input number is stored.

2.2.4 Processing Units Space
Processing Units Space provides memory space to allocate the network processing nodes. Each node consists of one output, a set of inputs, a small internal memory, and

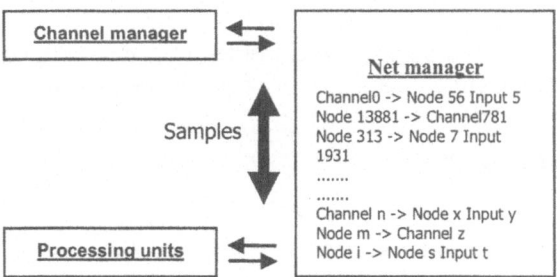

Fig. 3. The USP Net Manager

logic. Memory can be used to store information associated with each input, such as input weight, and internal state of the node. Logic calculates function value from the current state of memory and inputs, and sends the result to the output of the node.

In current C++ implementation, all nodes are descendants of same base class, with virtual function to update node's state, defined by the programmer.

3 PPNN Simulation and Analysis Using USP

3.1 CA-PPNN to PPNN Graph Conversion

Main reason of creating USP was increasing the speed of calculations required to test a PPNN CA. Hence we have prepared several conversion routines of different PPNN automata types to USP.

The conversion algorithm we present is composed of three phases. First all nodes, that is the cells that are processing data in different manner than just copying input and external inputs and outputs, are stored on a list (phase 1). Then, for each node, the lengths and destinations of intra-module delay channels starting from it are examined. (phase 2). Then the connections between modules are converted (phase 3). The implementation of these phases may vary, depending on a particular type of PPNN CA to convert. The conversion from CoDi to USP in current version of qCAEditor is done in the following manner:

Phase 1:
To find nodes, the conversion program sequentially checks all cells in all modules.

Phase 2:
For each node the program examines all its neighbors, to which it sends its signal. If a neighbor is a node, a node-to-node connection is created, and next neighbor is examined. If a neighbor is an axon cell, it becomes the current cell, a structure called channel (corresponding to a sequence of axon cells) and a node-to-channel connection are created. Then all neighbors, to which the current cell sends its signal are examined. If there is one and only one such cell, and it is axon, it becomes the current cell, the channel length is increased by one, and the procedure is repeated. If there is no such cell, the channel is discarded. Otherwise the channel is ended. Considering only the cells to which the current cell sends its signal, if a cell is a node, a channel-to-node connection is created. If a cell is an axon, it starts a new channel, which is recursively examined in the same way as above, and a channel-to-channel connection is created.

Phase 3:
As all module inputs and outputs are converted to nodes, each connection between CBM modules can be simply converted to a node-to-node connection.

3.2 Simulated CA vs. USP

We have compared the performance of software implementations of two CA-PPNN models to the USP simulation of their PPNN graphs. Benchmarks have been made using Pentium 4 2.0 GHz workstation.

3.2.1 CoDi Model

Two structures have been benchmarked. First one, an associative memory module, Memloop (see Figure 2) simulated under NeuroMaze 3.0 was updated with maximum speed of 1724 clocks per second. (cps) When converted into PPNN graph and simulated with USP under qCAEditor a maximum rate of 6818 cps was measured. Second structure, a set of 3 CoDi modules capable of playing TicTacToe game, achieved a maximum update rate of 692 cps in CA mode and 4225 cps in USP mode.

This stands for almost 400% performance gain for Memloop and more than 600% for TicTacToe.

3.2.2 qCA

There are two modes in which qCA is used – default - binary, where all spiketrains have a fixed amplitude and fuzzified, where spikes are represented as Gaussians. Fuzzified mode is used for smoothing the fitness landscape when the qCA structure is being evolved.

A qCA structure consisting of 25152 cells, adding two 16-bit numbers was simulated. In CA mode, a maximum rate of 2174 cps was measured for binary qCA and 152 cps for fuzzified. Switching into USP mode resulted in a 391% and 3600% performance gain, respectively.

4   Concluding Remarks and Future Work

Universal Spiketrain Processor has proven to be an efficient device for Pulsed-Para-Neural-Networks execution, providing speed and memory effective alternative for CA simulation. It can be observed, that the more computationally demanding CA-PPNN model, the higher is the performance gain of switching to USP. For smaller structures, a software implementation of USP running on a PC workstation can even outperform hardware CA implementations. This makes it a vital tool for our PPNN research.

As for a software implementation, a few improvements still need to be done. First, the size of Spiketrain Memory and Channel Manager can be reduced owing to the fact, that all channels starting at a given node share the same data. It would be enough to store only the data of the longest channel starting at each node. This will improve the speed of Channel Manager module equally to the average number of axons leaving each node. Second, an Object Oriented USP Definition Language need to be developed, that would give an ability to design PPNN structures orders of magnitude larger than designed at present time. Finally, a feature allowing to specify a function for each type of a node will become very useful for exploring new PPNN models properties.

A hardware implementation of USP is also possible and is currently being considered.

Acknowledgements

Authors' work was supported by Gdansk Artificial Brain Research Initiative (www.gabri.pl)

References

1. A. Buller, CAM-Brain Machines and Pulsed Para-Neural Networks (PPNN): Toward a hardware for future robotic on-board brains, *Proceedings, 8th International Symposium. on Artificial Life and Robotics (AROB 8th '03), January 24-26, 2003, Beppu, Japan*, pp. 490-493.
2. A. Buller, H. Eeckhaut & M. Joachimczak, Pulsed para-neural network (PPNN) synthesis in a 3-D cellular automata space, *Proceedings, 9th International Conference on Neural Information Processing (ICONIP'02), Vol.1. Nov. 18-22, 2002, Singapore*, pp. 581-584.
3. A. Buller, H. Hemmi, M. Joachimczak, J. Liu, K. Shimohara & A. Stefanski, ATR Artificial Brain Project: Key Assumptions and Current State, *Proceedings, 9th International Symposium on Artificial Life and Robotics (AROB 9th '04), January 28-30, 2004, Beppu, Japan, 2004*.
4. A. Buller & T.S. Tuli, Four-legged robot's behavior controlled by Pulsed para-neural networks (PPNN), *Proceedings, 9th International Conference on Neural Information Processing (ICONIP'02), Vol.1. Nov. 18-22, 2002, Singapore*, pp. 239-242.
5. M. Joachimczak & A. Buller. Universal Spiketrain Processor. ATR HIS, Kyoto, unpublished reseach memo, *2003*.
6. F. Gers, H. de Garis & M. Korkin "CoDi-1Bit: A Simplified Cellular Automata based Neuron Model", *Evolution Artificiele 97, Nimes, France, October 22, 1997*, pp. 211-229.
7. M. Korkin, G. Fehr & G. Jeffrey, Evolving hardware on a large scale, *Proceedings, The Second NASA / DoD Workshop on Evolvable Hardware, July 2000, Pasadena*, pp. 173-181.
8. D. Jelinski & M. Joachimczak, Heuristic-based Computer-Aided Synthesis of Spatial β-type Pulsed Para-Neural networks (3D-βPPNNs), *Proceedings, 8th International Symposium. on Artificial Life and Robotics (AROB 8th '03), January 24-26, 2003, Beppu, Japan*, pp. 499-501.
9. A. Buller A., Joachimczak M., Liu J., Stefanski A. Neuromazes: 3-Dimensional Spiketrain Processors, *WSEAS Transactions on Computers, Issue 1, Vol 3, 2004*, pp. 157-161.

Knowledge Extraction from Artificial Associative Memory for Helping Senile Dementia Patients

JeongYon Shim

Division of General Studies, Computer Science, Kangnam University
San 6-2, Kugal-ri, Kihung-up,YongIn Si, KyeongKi Do, Korea
Tel: +82 31 2803 736
mariashim@kangnam.ac.kr

Abstract. We designed Knowledge Extraction mechanism from Artificial associative memory system for helping memory formation and associative knowledge retrieval of Senile dementia patients. This system has the hierarchical structure. It can learn and perceive the things and has the associative knowledge retrieval function according to the associative keywords. This system has also Episode memory described by Script which is helpful for the simple association of events. We constructed event oriented virtual memory and tested learning and associative knowledge retrieval function with a Senile dementia patient's data.

1 Introduction

The brain scientists classify memory to four parts,namely Episode memory, Learning memory, Proceeding memory and Priming memory. Those parts take charge of the different functions and are linked very complexly. 'Episode memory' stores the events of our living. In the type of this memory, consciousness takes part in and the information about time and location is memorized. 'Learning memory' stores the logical facts and rules[2]. We are conscious of the two types of Episode memory and Learning memory. These memories get the various information from the surrounding environment,arrange the data and allocate them to many parts in the cortex.

We designed Knowledge Extraction mechanism from Artificial associative system with Episode memory for helping memory formation and associative knowledge retrieval of Senile dementia patients.

2 The Design of Associative Memory Frame

2.1 The Structure

As shown in Fig. 1, This system consists of Learning memory, Rule base, and Episode memory. They are related to the others according to their association. The memory is constructed by information acquisition process. The obtained data from the knowledge environment come into Input Interface and are temporarily stored in Temporary memory. They are selected and distributed by the

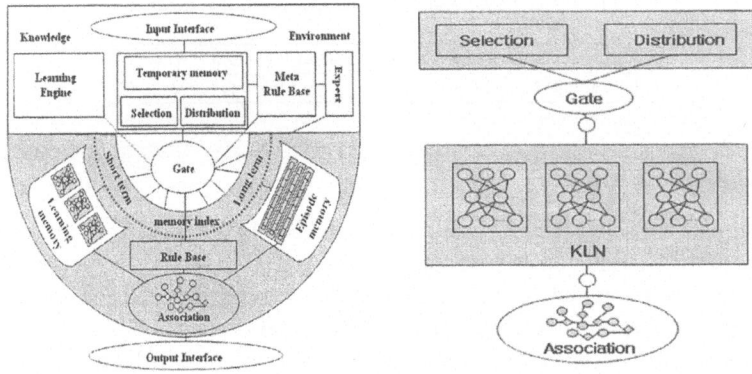

Fig. 1. Associative Memory Frame.   **Fig. 2.** Knowledge Learning Frame.

basic mechanism. For autonomous learning mechanism. Learning engine receives the training data of the special domain and process its learning mechanism. Episode memory stores the event oriented facts with the information of time and location and memorize them according to the flow of events sequentially. Memory Index which composed of Short term memory Index and Long term memory Index is used for efficient knowledge retrieval.

3   Data Acquisition and Storing in the Memory

3.1   Autonomous Learning Mechanism

Knowledge Learning Frame has the Hierarchical structure and also has KLN (Knowledge Learning Net) module which consists of modular Neural networks representing the domain knowledge for the autonomous learning process as shown in Figure 2[1]. These domains are connected to the corresponding areas in the other layers vertically and related with the associative relation in the association module.The observed mixed input data are selected and distributed to the corresponding NN(Neural Network) in the Selection/Distribution module by filtering function F_i, which is the criteria for determining the state of firing.

$$F_i = P(C_i|e_1, e_2, \ldots, e_n) = \prod_{k=1}^{n}(C_i|e_k) \qquad (1)$$

where C_i denotes a hypothesis for disease class and $e_k = e_1, \ldots, e_n$ denotes a sequence of observed data. F_i can be obtained by calculating the belief in C_i. If filtering factor F_i is over the threshold, q_i, $(F_i \geq q_i)$, the corresponding class is fired. The corresponding KLN of the fired class starts the learning or perception mechanism and produces the output. The values of the cells that don't belong to the activated class, are filtered and cleared. The structure of NN is three layered neural network trained by BP learning algorithm[3].

Table 1. DataBase of Episode memory: event.

Title	actor	action	object	reason	location	time
shopping	Kim	bought	apples	party	E-mart	2003-12-01

3.2   Episode Memory

We designed Episode memory as Data Base that is composed of six attributes, namely, actor, action, object, reason, location and time. This system analyzes the events and stores the facts according to the six attributes. Such a distribution can be very important factors for the information retrieval.

We designed a Relational graph for Episode memory which represents not only simple relations but also script of event as shown in Figure 3.

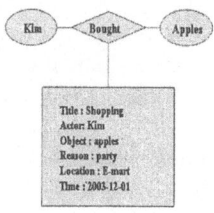

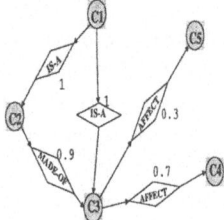

Fig. 3. Relational graph and script for Episode memory.

Fig. 4. Relational graph in Association level.

These data are interpreted to the easier comprehensible form:

Kim bought apples for party in E-mart on 2003-12-01.

For removing the complexity of transferring to the natural language form, we fixed the Tense to Past.

3.3   Association

Association level consists of nodes and their relations. These nodes are connected to their neighbors according to their associative relations horizontally and connected to NN of the previous layer vertically. Their relations are represented by the relational graph as shown in Fig. 4.

The relation between the two nodes is represented by both linguistic term and its associative strength.

    TRANSFORM-TO : transform in time
    AFFECT : partial transform
    IS-A : generalization
    MADE-OF :component

The linguistic terms are represented as TRANSFER-TO, AFFECT, IS-A, MADE-OF, NOT and SELF. Linguistic term denotes the degree of representing the associative relation. It can be transformed to the associative strength that has a real value of [-1,1]. The positive value denotes the excitatory strength and the negative value represents the inhibitory relation between the nodes. It is used for extracting the related facts. The minus sign of the minus value means the opposite directional relation as -TRANSFORM-TO.

The relational graph is transformed to the forms of AM(Associative Matrix) in order to process the knowledge retrieval mechanism. AM has the values of associative strengths in the matrix form.

For example, the relational graph in Fig.3. can be transferred to the associative matrix A.

The associative matrix,A, is:

$$A = \begin{bmatrix} 1.0 & 1.0 & 1.0 & 0.0 & 0.0 \\ -1.0 & 1.0 & 0.9 & 0.0 & 0.0 \\ -1.0 & -0.9 & 1.0 & 0.7 & 0.3 \\ 0.0 & 0.0 & -0.7 & 1.0 & 0.0 \\ 0.0 & 0.0 & -0.3 & 0.0 & 1.0 \end{bmatrix}$$

The matrix, A, has the form of $A = [R_{ij}]$. The associative strength,R_{ij}, between C_i and C_j is calculated by equation (2).

$$R_{ij} = P(a_i|a_j)D \qquad (2)$$

where D is the direction arrow, $D = 1 or -1$, $i = 1, \ldots, n$, $j = 1, \ldots, n$.

Using this Associative Matrix A, this system can extract the related facts by following knowledge retrieval algorithm.

4   Knowledge Retrieval

4.1   Knowledge Retrieval from Learning Memory

After the memory construction is completed by learning mechanism, Knowledge retrieval is available. Knowledge Retrieval mechanism extracts the related data connected to Association level and Episode memory by making the inferential output nodes a starting point. The following algorithm represents the knowledge retrieval process.

Algorithm 1 : knowledge retrieval algorithm

Step 1: Input the data to Input Interface.
Step 2: Calculate the reaction value by filtering function
Step 3: Select the activated class.
Step 4: Propagate the filtered data to KNN of the activated class
Step 5: Calculate the inferential value of the output node by BP algorithm in KNN
Step 6: Data Extraction in Association level:
   Search the associated nodes connected to the inferential output node of KNN in the row of the activated node in AM.
Step 7: IF((not found) AND (found the initial activated node))
   Goto Step 8.
   ELSE
   Output the found fact.
   Add the found fact to the list of inference path.
   Goto Step 6
Step8: STOP

When the class,C_i, in the relational graph is assumed to be activated, from the node, C_i, the inferential paths can be extracted using the knowledge retrieval algorithm. The inferential path, I_i has the following form.

$$I_i = [C_i \ (R_{ij}) \ C_j]$$

where C_i is i-th class node,R_{ij} is the associative strength between C_i and C_j. The following example is the result from the matrix A using the knowledge retrieval mechanism.

$I_1 = [C_1$ IS-A(1.0) C_2 TRANSFER-TO(0.9) C_3 MADE-OF(0.7) $C_4]$
$I_2 = [C_1$ IS-A (1.0) C_2 TRANSFER-TO(0.9)]
$I_3 = [C_3$ MADE-OF(0.7) C_5]
$I_4 = [C_1$ IS-A(1.0) C_3 MADE-OF(0.7) $C_4]$
$I_5 = [C_1$ IS-A(1.0) C_3MADE-OF(0.7)$C_5]$

From the obtained inferential paths, this system can extract the related facts as much as user wants by masking with the threshold, θ. In this step, the connected facts that has the value of the associative strength over the threshold are extracted. In the case of I_1,when the threshold is 0.7, the extracted path is $[C_1$ IS-A(1.0) C_2 TRANSFER-TO(0.9) $C_3]$. The another function of the knowledge retrieval mechanism is to infer the new relations. From the following extracted inferential path, $C_i(R_{ij})C_j(R_{jk})C_k$, we can elicit the new inferred path between C_i and C_k. The new associative strength,R_{ik}, is calculated by equation (3).

$$R_{ik} = R_{ij} * R_{jk} \tag{3}$$

The inferential path, $I_1 : C_1$ IS-A(1.0) C_2 TRANSFER-TO (0.9) C_3 MADE-OF(0.7) C_4, can produce the new relations, $C_1(0.9)$ C_3, C_1 (0.63) C_4 by its mechanism[4].

4.2   Knowledge Retrieval from Episode Memory

Episode memory is logically represented by Script and physically r composed of relational Data Base which has 7 attributes. Knowledge retrieval step is processed by the following form of query.

SELECT * from *table* where *table.id = value*

For example,the query of example in the previous section has the following expression. We can select the necessary attributes by adjusting this query form.

SELECT Title,actor,action,object,reason,location,time from EVENT where EVENT.object = apple

From this query this system produce the output.

Title : shopping
actor: Kim
object: apple
reason: party
location: E-mart
time: 2003-12-01

As we described in the previous section, these output data are interpreted to the easier comprehensible form.

5 Experiments

This system is applied to the helping system for a Senile dementia patients. We tested this system with a Senile dementia patient's data of events occurred during one week. Figure 5 shows the process of classifying three patterns of the patients from the input factors which consists of emotional state,temperature, characteristics, economic state, marriage, disease history and etc.. Figure 6 shows the retrieved script from DB in Episode memory.

```
--------------------------------------
Artificial associative memory testing...
Knowledge Retrieval
1. Associative Frame
2. Episode memory
--------------------------------------

Select? 1
Enter the name of input file? IN.dat
Enter the reaction degree? 0.7

**Input and Reaction stage
The selected group is G2
KMM2 is fired.

   Type1     Type2     Type3
 0.987650 0.345600 0.001234

**Knowledge Retrieval stage

 Type1 Transform_to(0.3) Type2
```

```
Select?2
Knowledge Retrieval from Episode memory.
Input the keyword: time
Input the date: 2004-01-10

Title       actor action  object          reason    location    time
Visiting    Sumi  visited P1              seeing    home        2004-01-10
Shopping    P1    bought  milk,apple,bread eating    supermarket 2004-01-10
watching-TV P1    watched TV              enjoying  home        2004-01-10

Input the Keyword: actor
Input the name of actor: Sumi

Title    actor action  object  reason  location  time
Visiting Sumi  visited P1      seeing  home      2004-01-10
helping  Sumi  helped  P1      walking hospital  2004-01-13
```

Fig. 5. Knowledge Retrieval from Associative frame.

Fig. 6. Knowledge Retrieval from Episode memory.

6 Conclusion

We designed Artificial Associative memory for helping memory formation and associative knowledge retrieval of Senile dementia patients. This system has the hierarchical structure. It can learn and perceive the things and has the associative knowledge retrieval function according to the associative keywords. This system has also Episode memory described by Script which is helpful for the simple association of events. We constructed event oriented virtual memory and tested learning and associative knowledge retrieval function with a Senile dementia patient's data. It is expected that this study will contribute to helping Senile dementia patients.

Acknowledgements

This work was supported by a Kangnam University Research Grant in 2004.

References

1. Jeong-Yon Shim, Knowledge Retrieval Using Bayesian Associative Relation in the Three Dimensional ModularSystem,pp630-635, *Lecture Notes in Computer Science(3007), 2004.*
2. John R. Anderson Learning and Memory *Prentice Hall.*
3. Laurene Fausett Fundamentals of Neural Networks *Prentice Hall.*

Some Experiments on Training Radial Basis Functions by Gradient Descent

Mercedes Fernández-Redondo, Carlos Hernández-Espinosa, Mamen Ortiz-Gómez, and Joaquín Torres-Sospedra

Universidad Jaume I, D. de Ingeniería y Ciencia de los Computadores,
Avda. Vicente Sos Baynat s/n, 12071 Castellón, Spain
espinosa@icc.uji.es

Abstract. In this paper we present experiments comparing different training algorithms for Radial Basis Functions (RBF) neural networks. In particular we compare the classical training which consists of a unsupervised training of centers followed by a supervised training of the weights at the output, with the full supervised training by gradient descent proposed recently in some papers. We conclude that a fully supervised training performs generally better. We also compare *Batch training* with *Online training* of fully supervised training and we conclude that *Online training* leads to a reduction in the number of iterations and therefore increase the speed of convergence.

1 Introduction

A RBF has two layers of neurons. The first layer, in its traditional form, is composed of neurons with a Gaussian transfer function and the second layer has neurons with a linear transfer function. The output of a RBF can be calculated with equation 1 and 2.

$$\hat{y}_{i,k} = \mathbf{w}_i^T \cdot \mathbf{h}_k = \sum_{j=1}^{c} w_{ij} \cdot h_{j,k} \tag{1}$$

$$h_{j,k} = \exp\left(-\frac{\left\| \mathbf{x}_k - \mathbf{v}_j \right\|^2}{\sigma^2} \right) \tag{2}$$

Where v_j are the center of the Gaussian functions (GF), σ control the width of the GF and w_i are the weights among the Gaussian units (GU) and the output units.

As equations 1 and 2 show, there are three elements to design in the neural network: centers and widths of the GU and the weights among the GU and output units.

There are two different procedures to design the network. One is to train the networks in two steps. First we find the centers and widths by using same unsupervised clustering algorithm and after that we train the weights among hidden units and output units by a supervised algorithm. This process is usually fast, and the most important step is the training of centers and widths [1-4].

The second procedure is to train simultaneously the centers and weights in a full supervised fashion, similar to the algorithm Backpropagation (BP) for Multilayer Feedforward. This procedure was not traditionally used because it has the same drawbacks of BP, long training time and high computational cost. However, it has received quite attention recently [5-6].

N.R. Pal et al. (Eds.): ICONIP 2004, LNCS 3316, pp. 428–433, 2004.
© Springer-Verlag Berlin Heidelberg 2004

In [5-6] it is used a sensitivity analysis to show that the traditional GU (called "exponential generator function") of the RBF network has low sensitivity for gradient descent training for a wide range of values of the widths, this parameter should be tuned carefully. As an alternate two different transfer functions are proposed, called in the papers "lineal generator function" and "cosine generator function" with a better sensitivity. Unfortunately, the experiments shown in the papers are performed with only two databases and the RBF networks are compared with equal number of GU. In our opinion the number of GU should be determined by trial and error and cross-validation, and it can be different for a traditional unsupervised training and a gradient descent training.

In contrast, in this paper we present more complete experiments with nine different databases, and include in the experiments four traditional unsupervised training algorithms and a fully gradient descent training with the three transfer functions analysed in papers [5-6]. Furthermore, we also presents experiments with *Batch* and *Online learning*, in the original references the training was performed in *Batch* mode. But as we will comment later *Online training* can greatly reduce the training time.

2 Theory

2.1 Training by Gradient Descent

"Exponential (EXP) Generator" Function. This RBF has the usual Gaussian transfer function described in equations 1 and 2. The equation for adapting the weights is:

$$\Delta \mathbf{w}_p = \eta \cdot \sum_{k=1}^{M} \varepsilon_{p,k}^0 \cdot \mathbf{h}_k \tag{3}$$

Where η is the learning rate, M the number of training patterns and $\varepsilon_{p,k}^0$ is the output error, the difference between the target of the pattern and the output.

The equation for adapting the centers is the following:

$$\Delta v_q = \eta \cdot \sum_{k=1}^{M} \varepsilon_{p,k}^h \cdot (\mathbf{x}_k - \mathbf{v}_q) \tag{4}$$

Where η is the learning rate and $\varepsilon_{p,k}^h$ is the hidden error given by the following equations:

$$\varepsilon_{p,k}^h = \alpha_{q,k} \cdot \sum_{i=1}^{n_o} \varepsilon_{i,k}^0 \cdot w_{iq} \qquad \alpha_{q,k} = \frac{2}{\sigma^2} \cdot \exp\left(-\frac{\|\mathbf{x}_k - \mathbf{v}_q\|^2}{\sigma^2}\right) \tag{5}$$

In the above equations n_o is the number of outputs and these equation are for *Batch training*, the equations for *Online training* are evident.

"Lineal (LIN) Generator" Function. In this case the transfer function of the hidden units is the following:

$$h_{j,k} = \left(\frac{1}{\|\mathbf{x}_k - \mathbf{v}_j\|^2 + \gamma^2}\right)^{\frac{1}{m-1}} \tag{6}$$

Where we have used $m=3$ in our experiments and γ is a parameter that should be determined by trial and error and cross-validation.

The above equations 3, 4 and 5 are the same, but in this case $\alpha_{q,k}$ is different and is given in the following equation:

$$\alpha_{q,k} = \frac{2}{m-1} \cdot \left(\left\| \mathbf{x}_k - \mathbf{v}_q \right\|^2 + \gamma^2 \right)^{\frac{m}{1-m}} \tag{7}$$

"Cosine (COS) Generator" Function. In this case the transfer function is the following:

$$h_{j,k} = \frac{a_j}{\left(\left\| \mathbf{x}_k - \mathbf{v}_j \right\|^2 + a_j^2 \right)^{\frac{1}{2}}} \tag{8}$$

Equations 3 and 4 are the same, but in this case the hidden error is different:

$$\varepsilon_{p,k}^h = \left(\frac{h_{j,k}^3}{a_j^2} \right) \cdot \sum_{i=1}^{no} \varepsilon_{i,k}^0 \cdot w_{iq} \tag{9}$$

The parameter a_j is also adapted during training, the equation is the following:

$$\Delta a_j = \left(\frac{\eta}{a_j} \right) \cdot \sum_{i=1}^{no} h_{j,k} \cdot (1 - h_{j,k}^2) \cdot \varepsilon_{p,k}^h \tag{10}$$

2.2 Training by Unsupervised Clustering

Algorithm 1. This training algorithm is the simplest one. It was proposed in [1]. It uses adaptive k-means clustering to find the centers of the GU.

After finding the centers, we should calculate the widths of the GU. For that, it is used a simple heuristic, we calculate the mean distance between one center and one of the closets neighbors, P, for example, the first $(P=1)$, second $(P=2)$, third $(P=3)$ or fourth $(P=4)$ closest neighbor.

Algorithm 2. It is proposed in reference [2]. The GU are generated incrementally, in stages. A stage is characterized by a parameter δ that specifies the maximum radius for the hypersphere that includes the random cluster of points that is to define the GU, this parameter is successively reduced in every stage k. The GU at any stage are randomly selected, by choosing an input vector x_i from the training set and search for all other training vectors within the δ_k neighborhood of x_i. The training vector are used to define the GU (the mean is the center, and the standard deviation the width). The stages are repeated until the cross-validation error increases. The algorithm is complex and the full description can be found in the reference.

Algorithm 3. It is proposed in [3]. They use a one pass algorithm called *APC-III*, clustering the patterns class by class instead of the entire patterns at the same time. The *APC-III* algorithm uses a constant radius to create the clusters, in the reference this radius is calculated as the mean minimum distance between training patterns multiplied by a constant α.

Algorithm 4. It is proposed in reference [4]. The GU are generated class by class, so the process is repeated for each class. In a similar way to algorithm 2 the GU are generated in stages. A stage is characterized by its majority criterion, a majority criterion of 60% implies that the cluster of the GU must have at least 60% of the patterns belonging to its class. The method will have a maximum of six stages, we begin with a majority criterion of 50% and end with 100%, by increasing 10%. The GU at any stage h are randomly selected, by picking a pattern vector x_i of class k from the training set and expand the radius of the cluster until the percentage of patterns belonging to the class falls below the majority criterion. To define the next GU another pattern x_i of class k is randomly picked from the remaining training set and the process repeated. The successive stage process is repeated until the cross-validation error increases. The algorithm is complex and the full description is in the reference.

3 Experimental Results

We have applied the training algorithms to nine different classification problems from the UCI repository of machine learning databases. They are Balance Scale (BALANCE), Cylinders Bands (BANDS), Liver Disorders (BUPA), Credit Approval (CREDIT), Glass Identification (GLASS), Heart Disease (HEART), the Monk's Problems (MONK'1, MONK'2) and Voting Records (VOTE). The complete data and a full description can be found in the UCI repository (http://www.ics.uci.edu /~mlearn/MLRepository.html).

The first step was to determine the appropriate parameters of the algorithms by trial and error and cross-validation. We have used an extensive trial and error procedure.

After that, with the final parameters we trained ten networks with different partition of the data in training set, cross-validation set and test set, also with different random initialization of centers and weights. With this procedure we can obtain a mean performance in the database (the mean of the ten networks) and an error.

These results are in Table 1, 2 and 3. We have for each database the mean percentage in the test and the mean number of clusters in the network.

Comparing the results of the same algorithm trained by gradient descent in the case of *Batch training* and *Online training*, we can see that the differences in performance are not significant. The unique two cases where there is a difference are in *EXP*, Mok1 (Batch= 94.7 and Online= 98.5) and in *LIN*, Mok2 (Batch= 82.8 and Online= 89.6). The fundamental difference between both training procedures is in the number of iterations and the value of the learning step. For example, 8000 iterations, η=0.001 in *EXP* Batch for Balance and 6000 iterations, η=0.005 in *EXP* Online.

Table 1. Performance of the different algorithms, Radial Basis Functions

DATABASE	TRAINING ALGORITHM							
	Exp Batch		Exp Online		Lineal Batch		Lineal Online	
	Perc.	Cluster	Perc.	Cluster	Perc.	Cluster	Perc.	Cluster
Balance	90.2±0.5	45	90.2±0.5	60	90.1±0.5	45	90.6±0.5	50
Band	74.1±1.1	110	74.0±1.1	40	74.5±1.1	30	73.4±1.0	35
Bupa	69.8±1.1	35	70.1±1.1	40	71.2±0.9	10	69.7±1.3	15
Credit	86.1±0.7	40	86.0±0.8	30	86.2±0.7	10	85.8±0.8	10
Glass	92.9±0.7	125	93.0±0.6	110	91.4±0.8	35	92.4±0.7	30
Heart	82.0±1.0	155	82.0±1.0	20	82.1±1.1	15	81.8±1.1	10
Monk1	94.7±1.0	60	98.5±0.5	30	93.2±0.7	15	94.5±0.7	15
Monk2	92.1±0.7	80	91.3±0.7	45	82.8±1.2	25	89.6±1.2	50
Vote	95.6±0.4	35	95.4±0.5	5	95.6±0.4	25	95.6±0.4	10

Table 2. Performance of the different algorithms, Radial Basis Functions

| DATABASE | TRAINING ALGORITHM | | | | | | | |
| | Cosine Batch | | Cosine Online | | UC Alg. 1 | | UC Alg. 2 | |
	Perc.	Cluster	Perc.	Cluster	Perc.	Cluster	Perc.	Cluster
Balance	89.9±0.5	25	90.0±0.7	40	88.5±0.8	30	87.6±0.9	88.5±1.6
Band	75.0±1.1	120	74.9±1.1	125	74.0±1.5	60	67±2	18.7±1.0
Bupa	69.9±1.1	15	70.2±1.1	40	59.1±1.7	10	57.6±1.9	10.3±1.5
Credit	86.1±0.8	10	86.1±0.8	25	87.3±0.7	20	87.5±0.6	95±14
Glass	93.5±0.8	105	92.6±0.9	15	89.6±1.9	100	79±2	30±2
Heart	82.1±1.0	25	81.9±1.1	15	80.8±1.5	100	80.2±1.5	26±4
Monk1	89.8±0.8	100	90.2±1.0	145	76.9±1.3	90	72±2	93±8
Monk2	87.9±0.8	125	86.6±1.1	45	71.0±1.5	90	66.4±1.7	26±4
Vote	95.6±0.4	20	95.4±0.4	10	95.1±0.6	40	93.6±0.9	53±5

Table 3. Performance of the different algorithms, Radial Basis Functions

| DATABASE | TRAINING ALGORITHM | | | |
| | UC Alg. 3 | | UC Alg. 4 | |
	Perc.	Cluster	Perc.	Cluster
Balance	88.0±0.9	94.7±0.5	87.4±0.9	45±7
Band	67±4	97.2±0.3	65.8±1.4	4.5±1.3
Bupa	60±4	106.2±0.3	47±3	11±5
Credit	87.9±0.6	161.10±0.17	86.4±0.9	32±4
Glass	82.8±1.5	59.9±0.7	81.2±1.8	22±2
Heart	72±4	71.8±0.6	78±3	10±2
Monk1	68±3	97.4±0.6	64±2	23±6
Monk2	66.5±0.8	143±0	71.6±1.5	20±2
Vote	94.1±0.8	120.30±0.15	76±5	5.0±1.1

Table 4. Performance of Multilayer Feedforward with Backpropagation

DATABASE	Number of Hidden	Percentage
BALANCE	20	87.6±0.6
BANDS	23	72.4±1.0
BUPA	11	58.3±0.6
CREDIT	15	85.6±0.5
GLASS	3	78.5±0.9
HEART	2	82.0±0.9
MONK'1	6	74.3±1.1
MONK'2	20	65.9±0.5
VOTING	1	95.0±0.4

Comparing *EXP*, *LIN* and *COS* generator functions, we can see that the general performance is quite similar and the differences are not significant except in the case Monk1 where the performance of *EXP* is clearly better. In other aspects, *EXP* and *LIN* functions need a higher number of trials for the process of trial and error to design the network (*EXP*, Band= 280 trials; *LIN*, Band= 280 trials; *COS*, Band= 112), because cosine generator functions adapt all parameters and the initialization is less important. But in contrast, the number of iterations needed to converge by *COS* functions is usually superior (*EXP*, Band= 10000 iterations; *LIN*, Band= 15.000; *COS*, Band= 75000), so globally speaking the computational cost can be considered similar.

Comparing unsupervised training algorithms among them, it seems clear that the classical algorithm 1, k-means clustering shows the better performance.

Finally, comparing unsupervised training with fully supervised training we can see that the best alternative (under the performance point of view) is supervised training, it achieves a better performance in databases Balance, Bupa, Glass, Heart, Monk1, and Monk2. In 6 of 9 databases. So, gradient descent is the best alternative.

In order to perform a further comparison, we include the results of Multilayer Feedforward with Backpropagaion in Table 4.

We can see that the results of RBF are better. This is the case in all databases except Credit, Heart and Voting.

4   Conclusions

In this paper we have presented a comparison of unsupervised and fully supervised training algorithms for RBF networks. The algorithms are compared using nine databases. Our results show that the fully supervised training by gradient descent may be the best alternative under the point of view of performance. The results of RBF are also compared with the results of Multilayer Feedforward with Backpropagation, the performance of a RBF network is better. So we think it is a better alternative.

References

1. Moody, J., Darken, C.J., "Fast Learning in Networks of Locally-Tuned Procesing Units". Neural Computation. **1**, (1989) 281-294
2. Roy, A., Govil, S., et alt. "A Neural-Network Learning Theory and Polynomial Time RBF Algorithm". IEEE Trans. on Neural Networks. **8** no. 6, (1997 1301-1313
3. Hwang, Y., Bang, S., "An Efficient Method to Construct a Radial Basis Function Neural Network Classifier". Neural Network. **10** no. 8, (1997) 1495-1503
4. Roy, A., Govil, S., et alt., "An Algorithm to Generate Radial Basis Function (RBF)-LikeNets for Classification Problems". Neural Networks. **8** no. 2, (1995) 179-201
5. Krayiannis, N., "Reformulated Radial Basis Neural Networks Trained by Gradient Descent". IEEE Trans. on Neural Networks. **10** no. 3, (1999) 657-671
6. Krayiannis, N., Randolph-Gips, M., "On the Construction and Training of Reformulated Radial Basis Functions". IEEE Trans. Neural Networks. **14** no. 4, (2003) 835-846

Predictive Approaches
for Sparse Model Learning

S.K. Shevade[1], S. Sundararajan[2], and S.S. Keerthi[3]

[1] Dept of Computer Science and Automation, Indian Institute of Science
Bangalore - 560 012, India
[2] Philips Innovation Campus, #1, Murphy Road
Bangalore - 560 008, India
[3] Yahoo! Research Labs, 210 S. DeLacey Avenue, Pasadena, CA-91105, USA

Abstract. In this paper we investigate cross validation and Geisser's sample reuse approaches for designing linear regression models. These approaches generate sparse models by optimizing multiple smoothing parameters. Within certain approximation, we establish equivalence relationships that exist among these approaches. The computational complexity, sparseness and performance on some benchmark data sets are compared with those obtained using relevance vector machine.

1 Introduction

In supervised learning our aim is to a learn a model which describes dependency of the targets $\{t_i\}_{i=1}^{N}$ on the input vectors $\{x_i\}_{i=1}^{N}$. Here, $x_i \in \mathbb{R}^n$ denotes the ith input example in the training set and t_i denotes the ith output. A model is designed with the objective of making accurate predictions for previously unseen values of $\mathbf{x}$. This is also referred to as the generalization ability of the model. The targets might be class labels (for the classification problem) or real valued (for the regression problem). In this paper, we focus on the regression problem. The classification problem will be addressed in a future paper.

Over the past few years, kernel methods like support vector machines (SVMs) have become very popular. These methods have been shown to generalize well on many real-world data sets. However, one of the drawbacks of SVMs is that the number of support vectors grows linearly with the size of the training set. This increases the computational complexity of the inference unless some post-processing is done [1]. Therefore, there is a need to design sparse models which also generalize well.

A more popular approach to model the target function is to use a linear model, $y(\mathbf{x}; \mathbf{w}) = \mathbf{w} \cdot \boldsymbol{\phi}(\mathbf{x})$ where $\mathbf{w} = (w_0, w_1, \ldots, w_M)^T$ is the parameter vector and $\boldsymbol{\phi}(\mathbf{x}) = (\phi_0(\mathbf{x}), \phi_1(\mathbf{x}), \ldots, \phi_M(\mathbf{x}))^T$ is the set of fixed basis functions.

In the real-world data, the presence of noise in the data implies that the model should be designed so as to avoid over-fitting. For simplicity, we assume that $y(x) : \mathbb{R}^n \to \mathbb{R}$. We can express $\mathbf{t}$ as

$$\mathbf{t} = \mathbf{y} + \epsilon = \boldsymbol{\Phi}\mathbf{w} + \epsilon \qquad (1)$$

where ϵ is a error vector which can be modelled as independent zero-mean Gaussian with variance $\sigma^2 (= \frac{1}{\beta})$ and Φ denotes the "design matrix" which is of size N x M+1. In the SVM, $\phi_i(\mathbf{x}) = \mathbf{K}(\mathbf{x}, \mathbf{x_i})$, with $K(.,.)$ a positive definite kernel function and $\mathbf{x}_i$ an example from the training set. The components of $\mathbf{w}$ in (1) can be determined by minimizing a suitable error function. A commonly used error function is

$$\bar{E} = \frac{\beta}{2} \left\| \Phi \mathbf{w} - \mathbf{t} \right\|^2 + \lambda \sum_i w_i^2 \qquad (2)$$

where the hyper-parameter λ controls the balance between fitting the training data and smoothing the output function. In many practical problems, this error function which enforces a uniform smoothness over the entire input space may not be suitable, especially for functions which are smooth only in some subsets of input space. In this paper, we study a more flexible model which associates a regularization parameter α_i with every basis function. The corresponding cost function can be given by

$$E = \frac{\beta}{2} \left\| \Phi \mathbf{w} - \mathbf{t} \right\|^2 + \sum_i \alpha_i w_i^2 \qquad (3)$$

There exist various model selection strategies which can be used to find the optimal hyperparameters $(\boldsymbol{\alpha}, \beta)$. For linear models, Tipping [7] introduced a general Bayesian framework for obtaining sparse solutions for regression as well as classification problems. In this approach, the sparse solution is obtained by maximizing the marginal likelihood. The final model consists of a small number of basis functions, called relevant vectors and hence is known by the name "Relevance Vector Machine" (RVM). Tipping [7] gave an iterative approach to update the hyperparameters associated with the basis functions, starting from all the M basis functions (full model).

In this paper, we propose an approach based on leave-one-out (LOO) error to estimate the optimal values of these hyperparameters and show its equivalence to the generalized cross-validation (GCV) approach [3] under certain approximation.

The paper is organized as follows. Section 2 discusses the way to determine the hyperparameter using the LOO error and its relationship to the GCV error. Equivalence between the GCV and Geisser's Predictive Probability (GPP) [6] approaches is also described. Numerical experiments on some benchmark data sets are described in Section 3. Section 4 concludes the paper.

2   Model Selection Using LOO Error

We now discuss about the computation of the hyperparameters α_i in (3) so as to minimize E. Typically, the parameter λ in (2) is estimated using techniques like cross-validation. This method however becomes intractable for estimating a large number of hyperparameters (as in (3)). In the following, we discuss a way to estimate these parameters systematically.

The stationary condition for the problem (3) is $\nabla_{\mathbf{w}} E = 0 \Rightarrow (\mathbf{A} + \beta \Phi^T \Phi)\mathbf{w} = \beta \Phi^T \mathbf{t}$ where $\mathbf{A} = Diag(\alpha_i)$. Given a choice for the regularization parameters (α, β), the weight vector at which the cost (3) reaches a minimum is given as

$$\mathbf{w}_{\mathrm{M}} = \beta \mathbf{C}^{-1} \Phi^T \mathbf{t}$$

where $\mathbf{C} = (\mathbf{A} + \beta \Phi^T \Phi)$. Our aim is to find the optimal hyperparameters for a given $\mathbf{w}_{\mathrm{M}}$. We do this by minimizing the leave-one-out error. Let us assume that the ith example is left out and the model is trained using the remaining N-1 samples to get the model with the parameter $\mathbf{w}^{(i)}$. Let $y_j^{(i)}$ denote the new output for the jth sample $(j \neq i)$. The output of this new model on the ith example is given by,

$$\hat{y}_i = \mathbf{w}^{(i)} \cdot \phi_i \tag{4}$$

The LOO error is then calculated as

$$E_{\mathrm{LOO}} = \frac{1}{2} \sum_i (\hat{y}_i - t_i)^2 \tag{5}$$

Since it is computationally not feasible to train N different models and estimate $\mathbf{w}^{(i)}$ independently for every model, we use the following result,

$$\mathbf{w}^{(i)} = \mathbf{w} + \beta(y_i - t_i)\mathbf{C}_{-i}^{-1}\phi_i \tag{6}$$

where $\mathbf{C} = \mathbf{C}_{-i} + \beta \phi_i \phi_i^T$ and $\mathbf{C}_{-i}^{-1} = \mathbf{C}^{-1} + \dfrac{\mathbf{C}^{-1}\phi_i \phi_i^T \mathbf{C}^{-1}}{\frac{1}{\beta} - \phi_i^T \mathbf{C}^{-1}\phi_i}$. It is easy to derive this result using the stationary conditions for the models which use $\mathbf{w}$ and $\mathbf{w}^{(i)}$ and using the fact that $y_j^{(i)} = y_j + (\mathbf{w}^{(i)} - \mathbf{w}) \cdot \phi_j$.

Using (4) and (6) we can write,

$$\hat{y}_i = \mathbf{w}^{(i)} \cdot \phi_i = y_i + \beta(y_i - t_i)\frac{\phi_i^T \mathbf{C}^{-1}\phi_i}{1 - \beta\phi_i^T \mathbf{C}^{-1}\phi_i} \tag{7}$$

The LOO error in (5) can therefore be calculated as

$$E_{\mathrm{LOO}} = \frac{1}{2} \sum_{i=1}^{N} \left(\frac{y_i - t_i}{1 - \beta\phi_i^T \mathbf{C}^{-1}\phi_i}\right)^2 \tag{8}$$

It is important to note here that the LOO error can be calculated with the knowledge of $\mathbf{w}_{\mathrm{M}}$ and there is no need to evaluate N different models. For a fixed $\mathbf{w}_{\mathrm{M}}$, it is therefore easy to optimize E_{LOO} w.r.t. the hyperparameters by using a simple gradient-based approach. The new set of hyperparameters are then used to find the new value of $\mathbf{w}_{\mathrm{M}}$. This procedure is repeated till there is no significant improvement in the cost in (3) or there is no significant change in the hyperparameter values. To ensure that the hyperparameters α_j's and β are positive, one can use the transformation, $\kappa_j = \log(\alpha_j)$ and treat κ_j's as the variables for optimization.

The comparative results of the LOO based approach with the relevance vector machine on various datasets are given in Section 3. Note that the sum of squared error at the optimum weight $\mathbf{w}_M$ is $\sum_i (y_i - t_i)^2 = \mathbf{t}^T \mathbf{P}^2 \mathbf{t}$ where $\mathbf{P} = \mathbf{I}_N - \beta \boldsymbol{\Phi} \mathbf{C}^{-1} \boldsymbol{\Phi}^T$. Let us use the following approximation,

$$1 - \beta \boldsymbol{\phi_i}^T \mathbf{C}^{-1} \boldsymbol{\phi_i} = \frac{1}{N} \sum_{j=1}^{N} (1 - \beta \boldsymbol{\phi_j}^T \mathbf{C}^{-1} \boldsymbol{\phi_j}) = \frac{1}{N} \text{trace}(\mathbf{P}) \qquad (9)$$

Using this approximation, the LOO error (8) can be given as

$$\text{GCV}(\boldsymbol{\alpha}, \beta) = \frac{N^2 \mathbf{t}^T \mathbf{P}^2 \mathbf{t}}{2(\text{trace}(\mathbf{P}))^2} \qquad (10)$$

This is also called the generalized cross-validation (GCV) error.

Instead of using the gradient based approach to estimate the hyperparameters $(\boldsymbol{\alpha}, \beta)$, Orr [5] suggested a way to estimate these hyperparameters using a sequential procedure which repeatedly estimates one parameter at a time. For this purpose, the GCV error in (10) can be written as

$$\text{GCV}(\alpha_i, \beta) = \frac{N^2 (a\Delta_i^2 - 2b\Delta_i + c)}{2(\psi\Delta_i - \delta)^2} \qquad (11)$$

where $\mathbf{P}_{-i} = \mathbf{I}_N - \beta \boldsymbol{\Phi}_{-i} \mathbf{C}_{-i}^{-1} \boldsymbol{\Phi}_{-i}^T$, $\Delta_i = \frac{1}{\beta}\alpha_i + \boldsymbol{\phi_i}^T \mathbf{P}_{-i} \boldsymbol{\phi_i}$, $a = \mathbf{t}^T \mathbf{P}_{-i}^2 \mathbf{t}$, $b = \mathbf{t}^T \mathbf{P}_{-i}^2 \boldsymbol{\phi_i} \mathbf{t}^T \mathbf{P}_{-i} \boldsymbol{\phi_i}$, $c = \boldsymbol{\phi_i}^T \mathbf{P}_{-i}^2 \boldsymbol{\phi_i} (\mathbf{t}^T \mathbf{P}_{-i} \boldsymbol{\phi_i})^2$, $\psi = \text{trace}(\mathbf{P}_{-i})$, and $\delta = \boldsymbol{\phi_i}^T \mathbf{P}_{-i}^2 \boldsymbol{\phi_i}$.

For computational purposes, it is convenient to optimize the cost in (11) w.r.t. the parameter $\tilde{\alpha}_i = \frac{\alpha_i}{\beta}$. Taking the derivative of GCV w.r.t. $\tilde{\alpha}_i$ and equating it to zero implies that $g_i \tilde{\alpha}_i + h_i = 0$ where $g_i = b\psi - a\delta$, $h_i = (b\psi - a\delta)d + (b\delta - c\psi)$, and $d = \boldsymbol{\phi_i}^T \mathbf{P}_{-i} \boldsymbol{\phi_i}$.

For notational convenience, we write $\tilde{\alpha}_i$ as α_i. Note that we are always interested in a non-negative optimum. Let α_i^* denote an optimal value. It can be obtained by using the following conditions:

1. $g_i > 0, h_i > 0$: $\alpha_i^* = 0$, 2. $g_i < 0, h_i < 0$: $\alpha_i^* = \infty$ 3. $g_i > 0, h_i < 0$: $\alpha_i^* = -\frac{h_i}{g_i}$, 4. $g_i < 0, h_i > 0$: The solution is at $\alpha_i^* = 0$ or $\alpha_i^* = \infty$ depending upon the end at which the GCV function (11) has a minimum value. 5. $g_i = 0$: $\alpha_i^* = 0$ if $h_i > 0$ or $\alpha_i^* = \infty$ if $h_i < 0$.

In the simulation experiments, for numerical stability, the basis function with optimal hyperparameter value more than a prescribed value is removed. Moreover, the hyperparameter was set to a small value when the optimal value was 0. Sections 3 compares the GCV approach with the LOO approach.

2.1   GCV and GPP

In this section we introduce Geisser's predictive probability (GPP) maximization approach [6] and show its equivalence to the GCV model. The GPP function is given by $\text{GPP}(\alpha, \sigma^2) = -\frac{1}{N} \sum_{i=1}^{N} \log(p(t_i|\mathbf{x}_i, Z_{-i}, \alpha, \sigma^2))$ where Z_{-i} denotes the set of training examples excluding the ith example and $p(t_i|\mathbf{x}_i, Z_{-i}, \alpha, \sigma^2)$ denotes the predictive density function. In this approach, the parameters α and σ^2 are obtained by minimizing the GPP function. Within the Gaussian modeling assumptions [6], the predictive density function is also a Gaussian and therefore we have,

$$\text{GPP}(\alpha, \sigma^2) = \frac{1}{2N} \sum_{i=1}^{N} \left(\frac{((t_i - y(\mathbf{x}_i; \mathbf{w}^{(i)}))^2}{\sigma_i^2} + \log(2\pi\sigma_i^2) \right)$$

where $\sigma_i^2 = \frac{\sigma^2}{1 - \beta\boldsymbol{\phi}_i{}^T C^{-1}\boldsymbol{\phi}_i}$. Let $E_D(\mathbf{w}) = \sum_{i=1}^{N} (t_i - y(x_i; \mathbf{w}))^2$. Using the approximation used in equation (9) the above equation reduces to

$$\text{AGPP}(\alpha, \sigma^2) = \frac{E_D(\mathbf{w})}{2\sigma^2(N - \gamma)} + \frac{1}{2}\log\left(\frac{2\pi\sigma^2 N}{N - \gamma}\right) \tag{12}$$

where $N - \gamma = \text{trace}(\mathbf{P})$. In this case, the optimal noise level is given as $\frac{E_D(\mathbf{w})}{N-\gamma}$. It can be shown that the optimal value of α obtained by solving (12) is same as the one obtained by minimizing the GCV function. Therefore, within the approximation, LOO, GCV and AGPP approaches are equivalent. However, the numerical experiments may show slightly different results because of the approximation and the actual optimization procedure involved.

3   Numerical Experiments

The different approaches for hyperparameter estimation, proposed in this paper, are evaluated on three popular benchmark data sets for regression problem and are compared with RVM. For all the data sets, Gaussian kernel was used and the width parameter was chosen using 5-fold cross-validation. Two of these data sets are generated in a way described in [2]. We refer to these data sets as Friedman Dataset #2 and Friedman Dataset #3. For these data sets, average results for 100 repetitions are quoted, where 200 randomly generated training examples (with input dimension 4 and output dimension 1 for each data set) are utilized, along with 1000 noise-free test examples. The paired t-test was conducted to find the statistical significance (with confidence 0.01) of the errors in all the three cases which are indicated as (RVM, LOO), (RVM, GCV) and (LOO, GCV). The results are reported in Table 1. It is clear that the generalization performance of the GCV approach is better than that of the LOO approach on both the data sets. However, the number of relevant vectors in the latter is less on average. The RVM approach performs better than both the LOO and the GCV approaches on Friedman data set #3. However, on Friedman data set #2, the generalization performance of the GCV approach is better.

We also tested these approaches on Sunspot time series data with input dimension 6 as given in [4]. The test set errors obtained using the RVM, LOO and the GCV approaches were .097, .098 and .091 respectively. The corresponding numbers of relevant vectors were 13, 12 and 12.

Predictive Approaches for Sparse Model Learning   439

Table 1. Normalized mean squared error (NMSE), the number of relevant vectors and
statistical significance results for RVM, LOO and GCV on two Friedman data sets

Friedman Dataset # 2		RVM	LOO	GCV
NMSE	min	.1847	.1764	.1456
	max	.4415	.4449	.4503
	mean	.2840	.2676	.2262
	std	.0568	.0622	.0586
# RV	min	5	9	14
	max	21	30	85
	mean	15.18	20.32	32.51
	std	3.4798	5.0490	10.9853
	p-value	(R, L) .0267	(R, G) 1.15e-9	(L, G) 2.39e-5

Friedman Dataset # 3		RVM	LOO	GCV
NMSE	min	.1433	.1559	.1527
	max	.2400	.3046	.2617
	mean	.1888	.2018	.1930
	std	.0190	.0262	.0239
# RV	min	6	7	8
	max	29	45	39
	mean	15.14	22.09	22.44
	std	3.7820	7.5426	6.8
	p-value	(R, L) 1.0	(R, G) .9118	(L, G) .0068

4 Conclusion

Sparse model learning using the LOO and GCV approaches were studied and
compared with the RVM. Preliminary results suggest that the generalization
performance of the proposed approaches are comparable with that of the RVM;
however, the number of relevant vectors is more. The proposed approaches have
the same computational complexity as that of RVM. More detailed experiments
on large-scale problems need to be conducted to evaluate these approaches further.

References

1. Burges, C. J. C., and Schölkopf, B.: Improving the accuracy and speed of support
   vector machines, Advances in NIPS. 9 (1997) 375–381
2. Friedman, J. H.: Multivariate adaptive regression splines, Annals of Statistics. 19(1)
   (1991) 1–141
3. Golub, G. H., Heath, M. and Wahba, G.: Generalized cross-validation as a method
   for choosing a good ridge parameter, Technometrics. 21(2) (1979) 215–223
4. Naftaly, U., Intrator, N., and Horn, D.: Optimal ensemble averaging of neural networks, Comput. Neural Syst. 8 (1997) 283–296
5. Orr, M. J. L.: Local smoothing of Radial Basis Function Networks, Proceedings of
   International Symposium on Neural Networks, Hsinchu, Taiwan. (1995)
6. Sundararajan, S. and Keerthi, S. S.: Predictive approaches for choosing hyperparameters in Gaussian Processes, Neural Computation. 13(5) (2001) 1103-1118
7. Tipping, M. E.: Sparse Bayesian Learning and the Relevance Vector Machine. Journal of Machine Learning Research. 1 (2000) 211-244

Multiple Instance Learning
with Radial Basis Function Neural Networks

Abdelhamid Bouchachia

University of Klagenfurt, Dept. of Informatics-Systems
Universitätsstrasse 65, A-9020 Klagenfurt, Austria
hamid@isys.uni-klu.ac.at

Abstract. This paper investigates the application of radial basis function neural networks (RBFNs) for solving the problem of multiple instance learning (MIL). As a particular form of the traditional supervised learning paradigm, MIL deals with the classification of patterns grouped into bags. Labels of bags are known but not those of individual patterns. To solve the MIL problem, a neural solution based on RBFNs is proposed. A *classical* application of RBFNs and *bag unit-based* variants are discussed. The evaluation, conducted on two benchmark data sets, showed that the proposed bag unit-based variant performs very well.

1 Introduction

Recently multiple instance learning has raised an increasing interest within the community of machine learning. MIL arises in learning situations where the concept to be learned is recognizable only at the level of sets, called *bags*, of the training samples. The learning algorithm is required to induce a classifier whose classification decision is concerned with bags and not with individual samples and which is able to predict the labels of the unseen bags. Similar but more complicated than traditional supervised learning, in the MIL context the labels of individual training samples are not known. From the practical point of view of MIL, a bag stands for an object which has many possible representations. Each instance or representation in a bag describes the same object. The label of the object depends on that of its alternative representations. MIL is concerned with binary classification problems (i.e. objects are either positive or negative). An object is labeled positive if at least one of its instances is positive, otherwise negative. Various methods have been used to solve the MIL problem. Using the concept of axis-parallel rectangle (APR), Dietterich et al. [4] studied three APR-based algorithms to solve the drug activity prediction and found that algorithms which ignore the MIL aspect during the learning process perform poorly. Based on the same concept of APR, an algorithm, called MULTINST, was proposed in [2]. In [9], the authors applied two K-nearest neighbor (kNN)-based algorithms, namely citation-kNN and Bayesian-kNN. In [8], the authors applied a regression algorithm that assumes each bag hving a representative instance to be searched. In [1, 5], MIL was proposed to deal with scene classification. In [7], a complex neural architecture based on the backpropagation algorithm was proposed to solve the MIL problem.

In this paper, we suggest to apply RBFNs. The idea is to check how well RBFNs perform when applied for such an unconventional supervised learning task. As will be

N.R. Pal et al. (Eds.): ICONIP 2004, LNCS 3316, pp. 440–445, 2004.
© Springer-Verlag Berlin Heidelberg 2004

shown later, two variants of RBFNs architectures are discussed. While the first variant applies RBFNs in a classical way, the second variant involves a modular neural architecture. Each module in this architecture is an RBF network, called *bag unit*. This modular architecture can support a large number of bags. Adding new bags as reference in the system can be done easily without requiring any retraining of the whole neural system since each bag is modeled as a separate network. Thus, the suggested architecture allows incremental learning. The other aspect behind using RBFNs is their instantaneous training which saves much time compared with the multilayer perceptron.

The rest of the paper is organized as follows. Section 2 explains briefly radial basis functions networks. Sections 3 and 4 discuss respectively two variants of applying RBF networks to the MIL problem. The evaluation of these two options on two data sets, Musk1 and Musk2, is presented in section 5. Finally, section 6 concludes the paper.

2 Radial Basis Function Neural Networks

Inspired by research in regions of the cerebral cortex and the visual cortex, RBF networks have been proposed by Moody and Darken [6] as supervised learning neural networks. An RBF network is a two-layer architecture where each unit in the hidden layer represents a radial basis function. These units measures the degree of overlap (or matching) between input vectors and a set of prototypes drawn from the training set. A RBFN is a mapping $M : R^n \rightarrow R^m$ such that each input vector $x_i \in R^n$ is of dimension n and vectors $C_j \in R^n$ $(j = 1..H)$ representing the prototypes of the input vectors. The output space of the mapping is of m-dimensions (i.e., size of the output vectors). The output of each RBF unit (called also receptive field) is given as:

$$\phi_j(x_i) = \phi_j(\|x_i - C_j\|) \tag{1}$$

where $\|.\|$ is the Euclidean norm on the input space to compute the distance between the n-dimensional input i and a hidden unit j. The function ϕ has various forms. Here, the Gaussian function is considered, i.e., ϕ has the following form:

$$\phi_j(x_i) = exp(-\frac{\|x_i - C_j\|^2}{\sigma_j^2}) \tag{2}$$

where σ_j is the width of the jth RBF unit. Note that if $x_i = C_j$, $\phi_j(x_i) = 1$ yielding maximum overlap. The kth output, $y_j(x_i)$, of an RBF network according to the weighted sum option is:

$$y_k(x_i) = \sum_{j=0}^{H} \phi_j(x_i) \cdot w(k,j) \tag{3}$$

where $\phi_0() = 1$, $w(k, j)$ is the weight of the jth receptive field to the kth output and $w(k, 0)$ is the bias of the kth output.

The key problem in RBF networks is the design of the parameters of the receptive fields: the prototypes C_j and the widths σ_j. Generally, prototypes representing the sub-regions (or classes) of the input space are found using clustering algorithms such as k-means and fuzzy k-means. It is important, however, to notice that these algorithms determine clusters independently of any semantical information about the real classes of the input. In this work, the prototypes are determined by taking the mean of the desired region of the input as will be shown later. Using the centers found, the widths, which are the radii of the Gaussian basis functions, are searched. Radii should be set so that the Gaussian from one center overlaps with near centers to a certain extent to ensure smoothness across the input space. The other possibility is to apply isotropic Gaussian functions whose width is fixed according to the spread of the centers. In this work, the width is computed taking into account the maximal spread of samples within the desired region. During the training stage, for each sample x_i, $y_k(x_i)$ is computed. This can be expressed in a matrix form as:

$$Y = \Phi W \tag{4}$$

The goal of the training stage is to find the weight W. This can be done in two ways; either through a repetitive adjustment of the weight using the delta learning rule or by computing W directly in an instantaneous manner, i.e.

$$W = \Phi^{-1} Y \tag{5}$$

provided that Φ is nonsingular. To avoid the singularity problem, a small value λ is added to the diagonal terms.

3   Classical Use of RBF

The first option is that of the conventional utilization of RBF networks for classification. Here, each instance gets the label of the bag to which it belongs. Since in the MIL context, we have a two-class problem (positive, negative), one can generate one or more centers per class as illustrated in Fig. 1a. This means that each class is described as a mixture of Gaussians. Thus, the set of positive bags (resp. negative bags) is considered as one distinct region of the input space. Once the centers of the receptive fields are identified, their widths are determined and the weight matrix W is computed. During the testing process, bags are introduced to the network successively. The actual (inferred) label of each bag is compared against the expected one. The question is how to determine the actual label of a testing bag. For a bag, the output of each of its instances is recorded and the actual computed label of the bag at hand is that of the instance with the highest output.

4   Bag Unit-Based Architecture

To apply RBF netwoks in MIL context efficiently, each bag will be trained on its own RBF network. Thus, the whole setting is a modular architecture of RBF networks, where each network is called *bag unit* (see Fig. 1b). Each bag unit comprises two receptive field units. The first one has the center of the instances of the corresponding bag and a width estimated on the basis of the spread of the instances within that bag. The second receptive field unit is designed for two purposes. The first purpose is to increase the discrimination power of the bag unit and the second purpose is to have sufficient number of receptive units (without causing an overfitting problem). To achieve that, the second receptive field of the bag unit corresponds to the most similar bag from the other class and its parameters, the center and the width, are determined as explained earlier. During the testing process, each instance from a given bag is introduced to the set of bag units and the output of each bag unit is recorded. A testing instance is judged positive (resp. negative) if the winner -the bag unit with the highest output activity- is positive (resp. negative). Then, a testing bag will get a positive label if at least one of its individual instances is assigned a positive label, otherwise negative. Once the label of the testing bag is determined, its match against the expected label is performed to compute the classification performance.

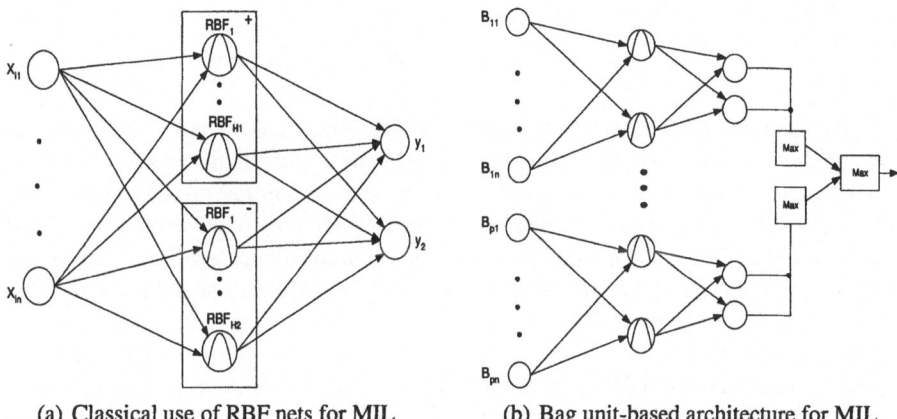

(a) Classical use of RBF nets for MIL   (b) Bag unit-based architecture for MIL

Fig. 1. RBF networks for MIL

5   Evaluation

To evaluate the approach presented here, the two well known data sets *Munsk1* and *Musk2* are used. They constitute the popular benchmark data sets in the context of MIL [4]. In these data sets, each bag describes a molecule for which several instances describing the conformations of that molecule are provided. Each instance consists of 166 features. Table 1 summarizes the characteristics of the two data sets. Because the data sets are relatively small, we apply a 10-fold cross validation to compute the classification performance. The data sets are partitioned into 10 roughly equal-sized subsets with a roughly equal proportion of negative and positive bags in all subsets.

For both experiments, the prototypes of the receptive fields are generated using subtractive clustering which essentially a modified version of the Mountain method [3]. According to this clustering method, each point is seen as a potential cluster center for which the following measure is applied:

$$C_j = \sum_{i=1}^{N} e^{-\alpha ||x_j - x_i||^2} \tag{6}$$

Table 1. Description of the musk data sets

Data set	Musk(positive)	Non-musk(negative)	No.bags	No.instances
1	47	45	92	476
2	39	63	102	6598

where $\alpha = 4/r^2$ and $r > 0$ represents the neighborhood radius for each cluster. The point x_j with maximum score C_j^* will be considered as the first cluster center denoted as x_1^*. To get the next cluster center, a modified mountain function having the form:

$$C_j = C_j - C_j^* \cdot e^{-\beta \|x_i - x_j\|^2} \tag{7}$$

is applied, where β is typically 1.5α. The new cluster center is identified and the process is repeated until the stopping criterion is reached (e.g $C_j^* \leq \epsilon$). The widths σ_j in Eq. 2 are obtained on the basis of the neighborhood radius r. Since Gaussians are applied here, the standard deviation of each cluster is computed by:

$$\sigma_j = r \cdot \frac{max(x_{ij}) - min(x_{ij})}{\sqrt{8}}, \quad i = 1..N \tag{8}$$

For the first option, the classical use of RBFs, one cluster center for the positive bags and two cluster centers for the negative bags were identified. Let r_p be the radius used for positive bags to determine their center, and r_n the radius for negative bags. The best results were obtained when r_p was set to 28 (to obtain one center) and r_n was set to 8 (to obtain two centers). The classification performance achieved with respect to this first option is 0.791 for Musk1 and 0.753 for Musk2. For the second option, the bag unit-based use of RBFs, the number of RBF networks is the total number of positive and negative bags. Each bag is represented as a network, where the hidden layer comprises two hidden RBF units as explained in Sec. 4. The first unit represents the corresponding bag, while the second represents the closest bag in the other class. Here, the similarity between bags is computed as the minimum distance between the bags' centers. Actually this option was tested under two settings. In the first setting, the hidden layer as described is used. In the second setting, the second node in each individual network is ignored. Again, subtractive clustering is used to identify the centers of the receptive field units. Widths were computed with a radius of 28 ($r_p = r_n$). The second setting achieved better classification performance. In fact. The classification accuracy achieved under the first setting is 0.643 for Musk1 and 0.58 for Musk2. The second setting, on the other hand, provides much better results with performance rate of 0.903 for Musk1 and 0.866 for Musk2.

To appreciate the results obtained using radial basis functions networks, a comparison against other approaches is presented in Tab. 2. If the bag unit-based RBFs are considered, one can notice that the approach suggested here achieves better performance compared with the other neural networks-based approaches. Two approaches: iterated APR and Citation-kNN required some further preprocessing of the data and much training time outperform the bag unit-based RBFs. In contrast, RBFNs have an instantaneous training model and, therefore, less training time is required. In addition, the modular architecture based on bag units allows an incremental learning; an aspect which is not available with the other methods. This aspect is important in many applications.

Table 2. Comparison of RBF Networks-based MIL against other methods

Algorithm	Musk1	Musk2
Iterated-Discrima APR [4]	92.4	89.2
Diversity Density [5]	88.9	82.5
Bayesian-kNN [9]	90.2	82.4
Citation-kNN [9]	92.4	86.3
Multinst [2]	76.7	84.0
BP neural networks [4]	75.0	67.7
multiple instance NN [7]	88.0	82.0
RBF Networks (Classical use)	79.1	75.3
RBF Netwoks (bag unit-based)	90.3	86.6

6   Conclusion

This paper discussed the application of RBFNs to solve the problem of multiple instance learning. Two possible variants were introduced and evaluated. The bag unit-based variant outperforms the classical use of RBF networks. The study showed that RBF networks are a good neural-based alternative and achieve better results with less time and no preprocessing of the multiple instance data is involved. As a future work, we intend to apply this approach for character recognition which is a typical multiple instance learning case. The modularity of the bag unit-based architecture is important due to the incremental learning facet involved in character recognition.

References

1. S. Andrews, T. Hofmann, and I. Tsochantaridis. Multiple Instance Learning with Generalized Support Vector Machines. In *Proc. of* 14th *Nat. Conf. on A.I.*, pages 943–944, 2002.
2. P. Auer. On learning from Multi-instance Examples: Empirical Evaluation of a Theoretical Approach. In *Proc. of* 14th *Int. Conf. on Machine Learning*, pages 21–29, 1997.
3. S. Chiu. Fuzzy Model Identification Based on Cluster Estimation. *Journal of Intelligent and Fuzzy Systems*, 2(3), 1994.
4. T. Dietterich, R. Lathrop, and T. Lozano-Perez. Solving the Multiple Instance Problem with Axis-parallel Rectangles. *Artificial Intelligence*, 89:31–71, 1997.
5. O. Maron and A. Ratan. Multiple-Instance Learning for Natural Scene Classification. In *Proc. of* 15th *Int. Conf. on Machine Learning*, pages 341–349, 1998.
6. J. Moody and C. Darken. Fast Learning in Networks of locally-tuned Processing Units. *Neural Computation*, 1:284–294, 1989.
7. J. Ramon and L. De Raedt. Multi-instance Neural Networks. In *Proc. of the Int. Conf. on Machine Learning Workshop on Attribute-Value and Relational Learning*, 2000.
8. S. Ray and O. Page. Multiple Instance Regression. In *Proc. of* 18th *Int. Conf. on Machine Learning*, pages 425–432, 2001.
9. J. Wang and J. Zucker. Solving Multiple-Instance Problem: A Lazy Learning Approach. In *Proc. of* 17th *Int. Conf. on Machine Learning*, pages 1119–1125, 2000.

Leverages Based Neural Networks Fusion

Antanas Verikas[1,2], Marija Bacauskiene[2], and Adas Gelzinis[2]

[1] Intelligent Systems Laboratory, Halmstad University
Box 823, S-301 18 Halmstad, Sweden
antanas.verikas@ide.hh.se
[2] Department of Applied Electronics, Kaunas University of Technology
Studentu 50, LT-3031, Kaunas, Lithuania

Abstract. To improve estimation results, outputs of multiple neural networks can be aggregated into a committee output. In this paper, we study the usefulness of the leverages based information for creating accurate neural network committees. Based on the approximate leave-one-out error and the suggested, generalization error based, diversity test, accurate and diverse networks are selected and fused into a committee using data dependent aggregation weights. Four data dependent aggregation schemes – based on *local variance, covariance, Choquet integral, and the generalized Choquet integral* – are investigated. The effectiveness of the approaches is tested on one artificial and three real world data sets.

1 Introduction

A variety of schemes have been proposed for combining multiple estimators. The approaches used most often include averaging [1], weighted averaging [1, 2], the fuzzy integral [3, 4], probabilistic aggregation [5], and aggregation by a neural network [6]. We can say that a combiner assigns weights of value to neural networks in one way or another. The aggregation weights assigned to neural networks can be the same in the entire data space or can be different – data dependent – in various regions of the space [1, 2]. The use of data-dependent weights, when properly estimated, provides a higher estimation accuracy [2, 7].

Numerous previous works on neural network committees have shown that an efficient committee should consist of networks that are not only very accurate, but also diverse in the sense that the network errors occur in different regions of the input space [8]. Bootstrapping [9], Boosting, AdaBoosting [10], and negative correlation learning [11] are the most often used approaches for data sampling when training neural network committees aiming to create diverse members of the committees.

In this paper, we further study committees designed using data dependent aggregation weights. Leverages of the training data is the main source of information exploited in various stages of the designing process. First, models, highly affected by some influential training data points, are detected and excluded from the set of candidate models. The leverages based information is then utilized in the suggested, generalization error based, diversity test for selecting accurate and

N.R. Pal et al. (Eds.): ICONIP 2004, LNCS 3316, pp. 446–451, 2004.
© Springer-Verlag Berlin Heidelberg 2004

diverse approved networks to be fused into a committee. Finally, the information is exploited to calculate the data dependent aggregation weights for fusing the selected networks. We investigate two ordinary – *averaging* and *variance based* – and four data dependent – *local variance, covariance, the Choquet integral, and the generalized Choquet integral based* – aggregation schemes. The usefulness of the leverages based information for creating accurate neural network committees is demonstrated on one artificial and three real world problems.

2 Training and Selecting Networks to Be Aggregated into a Committee

We use the Levenberg-Marquardt algorithm to train neural networks. At the ith iteration of the algorithm the q-parameter vector $\boldsymbol{\theta}_{(i)}$ is updated according to

$$\boldsymbol{\theta}_{(i)} = \boldsymbol{\theta}_{(i-1)} + (\mathbf{Z}_{(i)}^T \mathbf{Z}_{(i)} + \alpha_{(i)} \mathbf{I}_q)^{-1} \mathbf{Z}_{(i)}^T (\mathbf{y} - \mathbf{f}(\mathbf{X}, \boldsymbol{\theta}_{(i-1)})) \tag{1}$$

where $\mathbf{X} = [\mathbf{x}^1 \mathbf{x}^2 ... \mathbf{x}^N]^T$ is the training data set of N points, the N-vector $\mathbf{f}(\mathbf{X}, \boldsymbol{\theta}) = [f(\mathbf{x}^1, \boldsymbol{\theta}) ... f(\mathbf{x}^N, \boldsymbol{\theta})]^T$ stands for the neural network outputs, $\mathbf{y}$ is the target N-vector, $\alpha_{(i)}$ is the scalar, $\mathbf{I}_q$ is a $(q \times q)$ identity matrix, and $\mathbf{Z}_{(i)} = [\mathbf{z}_{(i)}^1 \mathbf{z}_{(i)}^2 ... \mathbf{z}_{(i)}^N]^T$ is the Jacobian matrix at ith iteration with $\mathbf{z}_{(i)}^k = \frac{\partial f(\mathbf{x}^k, \boldsymbol{\theta})}{\partial \boldsymbol{\theta}} \big|_{\boldsymbol{\theta} = \boldsymbol{\theta}_{(i-1)}}$

The diagonal elements $h^{ii} = \mathbf{z}^{iT} (\mathbf{Z}^T \mathbf{Z})^{-1} \mathbf{z}^i$, $i = 1, ..., N$ of the orthogonal projection matrix $\mathbf{Z} (\mathbf{Z}^T \mathbf{Z})^{-1} \mathbf{Z}^T$, called leverages of the training data, satisfy the following relations: $\sum_{i=1}^N h^{ii} = q$, $0 \leq h^{ii} \leq 1$ [12].

The value of h^{ii} provides an indication of the degree of influence of the ith data point on the model. The closer h^{ii} is to unity, the larger is the influence of the ith data point on the model. We exploit leverages for selecting networks to be aggregated into a committee as well as for estimating the aggregation weights.

Starting from a linear model, we train a series of models with an increasing number of hidden units until the aforementioned relations for h^{ii} are steadily violated. We train each model on B bootstrapping replicates. For each replicate training is repeated from I different initializations. Models do not satisfying the relations are discarded and the approved ones are evaluated by calculating the approximate leave-one-out error [12]

$$E_{LOO} = \sqrt{\frac{1}{N} \sum_{i=1}^N \left(\frac{r^i}{1 - h^{ii}} \right)^2} \tag{2}$$

where $r^i = y^i - f(\mathbf{x}^i, \boldsymbol{\theta}_{LS})$ is the ith residual and $\boldsymbol{\theta}_{LS}$ stands for the least-squares estimate of the parameter vector. Then, for each bootstrap replicate, a model with the smallest value of E_{LOO} is selected. Finally, committee members are selected amongst the candidate models exhibiting satisfactory values of E_{LOO}. The selection is performed so as to minimize the average correlation between the vectors of the approximate leave-one-out error $\mathbf{e}$, $e^i = r^i / (1 - h^{ii})$. By minimizing the correlation we attempt to choose diverse members to be aggregated into a committee.

3   Aggregating Networks into a Committee

We employ four linear – *averaging, variance, data dependent variance, and co-variance based linear combination* – and two non-linear – *the Choquet and generalized Choquet integral based fusion* – aggregation schemes. The Choquet integral based aggregation can be considered as generalization of the linear weighted averaging approach. In all the schemes, leverages based aggregation weights are utilized. The use of leverages allows computing data dependent weights.

When applying a linear aggregation scheme, for a given input vector $\mathbf{x}$ we need to estimate the unknown target variable y by forming a linear combination of the L estimates f_i – outputs of L neural networks

$$\widehat{y} = \sum_{i=1}^{L} w_i f_i = \mathbf{w}^T \mathbf{f} \qquad (3)$$

where $\mathbf{f} = [f_1, ..., f_L]^T$ and $\mathbf{w} = [w_1, ..., w_L]^T$ is the vector of the combination weights, satisfying the constraint $\sum_{i=1}^{L} w_i = 1$, $w_i \geq 0$, $i = 1, ..., L$. The *averaging* scheme assumes all the weights w_i to be equal.

For unbiased and uncorrelated estimators the expected error of a combined estimator is minimized for $w_i^* = 1/var_i$, where var_i is the variance of the ith estimator [1] – *variance based aggregation*. Since we are mainly interested in data dependent combination weights, we use local variance estimates for calculating the weights $w_i(\mathbf{x}) = 1/var(f_i(\mathbf{x}))$ – *data dependent variance based aggregation*.

A number of different methods can be used to estimate the variance of a predictor for a given data point $\mathbf{x}$. The approach adopted in this paper uses the first-order terms of a Taylor's series expansion to approximate f with a linear function

$$var(f_i(\mathbf{x})) = s^2 \mathbf{z}^T (\mathbf{Z}^T \mathbf{Z})^{-1} \mathbf{z} \qquad (4)$$

where $s^2 = \mathbf{r}^T \mathbf{r}/(N - q)$ and $\mathbf{r} = \mathbf{y} - \mathbf{f}(\mathbf{X}, \boldsymbol{\theta}_{LS})$ is the residual N-vector.

One can expect further improving the performance of a combined estimator by exploiting not only variance of predictions but also covariance between predictions of the same estimator – *covariance based aggregation*. This idea, assuming that the data set is split into L subsets, was introduced in the context of kernel-based regression systems [13].

Thus, estimates at other N_Q query data points are used to improve the estimate at a given data point $\mathbf{x}$. Let $\mathbf{f}_i = [f_i(\mathbf{x}^1), ..., f_i(\mathbf{x}^{N_Q})]^T$ be the vector of outputs of the ith network for all the N_Q query points. It is then shown [13] that for unbiased and uncorrelated networks the optimal N_Q-vector of target estimates $\widehat{\mathbf{y}}$ is given by

$$\widehat{\mathbf{y}} = \left(\sum_{i=1}^{L} (cov_i)^{-1} \right)^{-1} \left(\sum_{i=1}^{L} (cov_i)^{-1} \mathbf{f}_i \right) \qquad (5)$$

where cov_i is the covariance of the N_Q predictions of the ith network. In this paper, the estimate of the covariance is based on Eq. (4).

The *Choquet integral based aggregation* is implemented by calculating the integral for all the decision classes. The final decision is made according the maximum value of the integral.

Let g be a λ-fuzzy measure defined on the set Ψ of L neural networks being aggregated. The *Generalized Choquet integral* of a function $f : \Psi \rightarrow \mathbb{R}^+$ with respect to g is defined as [14]

$$C_{g/\beta}(f_1,\ldots,f_L) = \left(\sum_{i=1}^{L} v_i f_{(i)}^{\beta} \right)^{1/\beta} \tag{6}$$

where the indices i have been permuted so that $0 \leq f_{(1)} \leq \ldots f_{(L)} \leq 1$, $f_{(0)} = 0$, $v_i = g(A_i) - g(A_{i+1})$ with $A_i = \{(i), \ldots, (L)\}$, and the parameter $\beta \in [-\infty, \infty]$. By setting $\beta = 1$ the ordinary *Choquet integral* is obtained. The values g_i are called the densities of the λ-fuzzy measure. We use data dependent densities and set the densities to the values of weights w_i estimated as in the data dependent variance based approach without requiring that the weights sum to unity. When g is the λ-fuzzy measure, the values of $g(A_i)$ can be computed recursively: $g(A_1) = g(\{\psi_1\}) = g_1$, $g(A_i) = g_i + g(A_{i-1}) + \lambda g_i g(A_{i-1})$, for $1 < i \leq L$.

4 Experimental Investigations

In all the tests, we used $B = 20$ bootstrap replicates, $I = 7$ different initial values of weights and 10 different partitioning of the data set into <Training> and <Test> sets of equal size. The mean values and standard deviations of the correct classification rate presented in this paper were calculated from these 10 trials. In all the tests, we used the same number, $L = 7$, of committee members. Large committees imply using large values of B. The E_{LOO} value of the least accurate network included into a committee has not exceeded the E_{LOO} value of the most accurate member more than 30%. The value of the parameter β of the generalized Choquet integral was found by cross-validation and was equal to $\beta = 5$. The number of query points N_Q used in the covariance based approach was set to the number of effective degrees of freedom of the given data [13].

4.1 Data Used

To test the schemes studied we used three real world and one artificial problem. The data used are available at: www.dice.ucl.ac.be/pub/neural-net/ and www.ics.uci.edu/~mlearn/.

The United States Congressional voting records data set–CVR consists of the voting records of 435 congressman on 16 major issues in the 98th Congress. The votes are categorized into one of the three types of votes: (1) *Yea*, (2) *Nay*, and (3) *Unknown*. The task is to predict the correct party affiliation of a congressman.

The University of Wisconsin Breast Cancer Data Set–UWBC consists of 699 patterns. Amongst them there are 458 benign samples and 241 malignant ones. Each of these patterns consists of nine measurements taken from fine needle aspirates from a patient's breast.

The Pima Indians Diabetes Data Set contains 768 samples taken from patients who may show signs of diabetes. Each sample is described by eight features. There are 500 samples from patients who do not have diabetes and 268 samples from patients who are known to have diabetes.

The artificial Concentric data are two-dimensional with two classes and uniform concentric circular distributions. The points of the class ω_0 are uniformly distributed into a circle of radius 0.3 centered on (0.5, 0.5). The points of the class ω_1 are uniformly distributed into a ring centered on (0.5, 0.5) with internal and external radii equal to 0.3 and 0.5, respectively. There are 2500 instances, with 1579 in class ω_1 and the remainder in class ω_0. The theoretical error is 0%. The mean error, reported in previous studies is about 2.7%.

4.2   Simulation Results

Table 1 summarizes the test set classification error rate obtained for the databases from the different aggregation schemes. In the Table, we also provide the average error obtained from the most accurate single neural network.

Table 1. Test Set Classification Error Rate for the Different Databases

Scheme\Database	UWBC	CVR	Diabetes	Concentric
Single Net	4.51 (1.61)	5.40 (1.67)	24.82 (1.83)	2.63 (0.98)
Averaging	3.76 (1.12)	3.96 (1.23)	23.33 (1.41)	0.62 (0.29)
Variance	3.86 (1.14)	4.05 (1.25)	23.46 (1.45)	0.61 (0.31)
Variance(x)	2.65 (1.18)	3.54 (1.29)	22.23 (1.53)	0.49 (0.30)
Covariance	2.45 (1.20)	3.56 (1.31)	22.07 (1.61)	0.43 (0.34)
C Integral	2.68 (1.21)	3.68 (1.32)	22.26 (1.78)	0.48 (0.36)
GC Integral	2.70 (1.26)	3.61 (1.30)	22.22 (1.82)	0.47 (0.39)

As can be seen from the Table, the committees built exhibit considerable performance improvements if compared to the most accurate single neural network. On average, the data dependent aggregation schemes created more accurate committees than the simpler averaging and variance based approaches. While the covariance based technique, on average provided the lowest classification error, the difference between the results obtained from the data dependent schemes is not statistically significant. We have not observed any significant difference between the results obtained using the generalized and ordinary Choquet integral. We have also tested committees when, instead of exploiting the diversity test, the first L most accurate networks, as evaluated according to the average E_{LOO}, were used to build a committee. However, such committees were less accurate than those exploiting the suggested diversity test.

5   Discussion and Conclusions

We investigated the usefulness of the leverages based information for creating accurate neural network committees. The results obtained indicate that the information is useful in various contexts. First, by employing the leverages based analysis, models, highly affected by some influential data points, are detected and excluded. The leverages based information is then utilized in the suggested diversity test for selecting accurate and diverse approved networks to be fused into a committee. Finally, the information can be effectively exploited to calculate the data dependent aggregation weights for fusing the selected networks. The covariance based aggregation technique seems to be the most accurate. However, more studies are needed to see if statistically significant improvements can be obtained, if compared to the more simple local variance based approach.

References

1. Taniguchi, M., Tresp, V.: Averaging regularized estimators. Neural Computation **9** (1997) 1163–1178
2. Verikas, A., Lipnickas, A., Malmqvist, K., Bacauskiene, M., Gelzinis, A.: Soft combination of neural classifiers: A comparative study. Pattern Recognition Letters **20** (1999) 429–444
3. Gader, P.D., Mohamed, M.A., Keller, J.M.: Fusion of handwritten word classifiers. Pattern Recognition Letters **17** (1996) 577–584
4. Verikas, A., Lipnickas, A.: Fusing neural networks through space partitioning and fuzzy integration. Neural Processing Letters **16** (2002) 53–65
5. Kittler, J., Hatef, M., Duin, R.P.W., Matas, J.: On combining classifiers. IEEE Trans Pattern Analysis and Machine Intelligence **20** (1998) 226–239
6. Kim, S.P., Sanchez, J.C., Erdogmus, D., Rao, Y.N., Wessberg, J., Principe, J.C., Nicolelis, M.: Divide-and-conquer approach for brain machine interfaces: nonlinear mixture of competitive linear models. Neural Networks **16** (2003) 865–871
7. Woods, K., Kegelmeyer, W.P., Bowyer, K.: Combination of multiple classifiers using local accuracy estimates. IEEE Trans Pattern Analysis Machine Intelligence **19** (1997) 405–410
8. Krogh, A., Vedelsby, J.: Neural network ensembles, cross validation, and active learning. In Tesauro, G., Touretzky, D.S., Leen, T.K., eds.: Advances in Neural Information Processing Systems. Volume 7. MIT Press (1995) 231–238
9. Breiman, L.: Bagging predictors. Technical Report 421, Statistics Departament, University of California, Berkeley (1994)
10. Freund, Y., Schapire, R.E.: A decision-theoretic generalization of on-line learning and an application to boosting. Journal of Computer and System Sciences **55** (1997) 119–139
11. Liu, Y., Yao, X.: Ensemble learning via negatine correlation. Neural Networks **12** (1999) 1399–1404
12. Monari, G., Dreyfus, G.: Local overfitting control via leverages. Neural Computation **14** (2002) 1481–1506
13. Tresp, V.: A Bayesian committee machine. Neural Computation **12** (2000) 2719–2741
14. Yager, R.R.: Generalized OWA aggregation operators. Fuzzy Optimization and Decision Making **3** (2004) 93–107

A Process of Differentiation in the Assembly Neural Network

Alexander Goltsev[1], Ernst Kussul[2], and Tatyana Baidyk[2]

[1]Cybernetics Center of Ukrainian Academy of Sciences, Kiev, Ukraine
agoltsev@adg.kiev.ua
[2]Center of Applied Science and Technological Development
UNAM (National Autonomous University of Mexico)
ekussul@servidor.unam.mx
tbaidyk@aleph.cinstrum.unam.mx

Abstract. An assembly neural network model is described. The network is artificially partitioned into several sub-networks according to the number of classes that the network has to recognize. In the process of primary learning Hebb's neural assemblies are formed in the sub-networks by means of modification of connections' weights. Then, a differentiation process is executed which significantly improves the recognition accuracy of the network. A computer simulation of the assembly network is performed with the aid of which the differentiation process is studied in a set of experiments on a character recognition task using two types of separate handwritten characters: Ukrainian letters and Arabic numerals of MNIST database.

1 Introduction

The present paper considers a neural network model based on Hebb's theory about cell assemblies in the brain [6], [10]. According to this, the neural network is named as an assembly neural network. The assembly neural networks are described in [1–5]. New algorithm of a differentiation procedure is presented in the paper. The differentiation procedure is intended to improve a recognition accuracy of the assembly network with analog connections. The purpose of the paper is experimental study of the differentiation process in the assembly network on a character recognition task. The experiments with a computer model of the assembly network have shown that the differentiation process significantly increases the network's recognition performance.

2 Description of the Assembly Neural Network

In spite of the fact that functioning algorithms of the assembly network have been presented in the literature [3–5], it is necessary to provide here a concise information about these algorithms in order not to refer a reader to other publications for the details that are required for explanation of the algorithms described below.

The assembly network is destined to recognize a pre-defined number of classes of objects. Some number of training samples of each class is given to learn the network. The network task is to classify test samples of the same classes.

The assembly network has the following pre-organization: it is artificially partitioned into several sub-networks according to the number of classes that the network has to recognize so that one sub-network represents one recognized class. Let the

N.R. Pal et al. (Eds.): ICONIP 2004, LNCS 3316, pp. 452–457, 2004.
© Springer-Verlag Berlin Heidelberg 2004

network be intended to recognize M classes. Therefore, the network is partitioned into M sub-networks each of which includes the same number of neurons N and all possible connections between them. The network neurons are numbered separately within each sub-network.

The assembly network falls into such a category of learning machines that requires some pre-processing of an input image by means of a pre-determined set of algorithms in order to extract some features from the image and create some description of the image in terms of these features. Thus, a certain set of features is extracted from every input image (sample) presented to the network as for learning as for recognition. Each set of features extracted from an input sample is encoded into activation of a definite neural pattern within a sub-network for subsequent processing. Let us name a neural encoding of the set of features, which is extracted from the xth input sample, as a pattern of initial neural activity. A binary vector $G(x)$ of N components is used to represent this pattern.

An output characteristic of every network neuron is linear and has no threshold. A state of each neuron is described by its activity (nonnegative integer). An integer-valued vector E^m of N components represents activities of all neurons of the mth sub-network. Let us introduce a separate, integer-valued, two-dimensional connection matrix W of $N \times N$ components for each sub-network. Initially all weights of all connections have zero values in the network.

A process of primary learning in the assembly network is of a supervised learning type. A large number of neural assemblies are formed in the network during the primery learning process. According to Hebb [6], a neural assembly is a group of neurons that are bound into the assembly by means of mutual excitatory connections between them. Let us divide the primary learning process into a number of learning steps. Each presentation of a new training sample, which results in memorization of the corresponding neural encoding of this sample in the network, is the learning step.

The learning process of the mth sub-network is executed as follows. Let us denote the connection matrix of the sub-network which is formed after k previous learning steps by the notation $W^m(k)$. Let the m_x^{th} training sample of the mth class be presented to the network for the next $(k+1)^{th}$ learning step. The pattern of initial neural activity $G(m_x)$ is set in the mth sub-network. Hebb's learning rule is used for modification of connection weights. This means that after the $(k+1)^{th}$ learning step, the connection matrix of the mth sub-network is calculated by the equation

$$W_{i,j}^m(k+1) = W_{i,j}^m(k) + \Delta W\left(G_i(m_x) \wedge G_j(m_x)\right), \qquad (1)$$

where $\wedge$ is conjunction; $i = 1, 2, 3, ..., N$; $j = 1, 2, 3, ..., N$.

Thus, the neural encoding of the set of features extracted from the input training sample is bound into the neural assembly in the corresponding sub-network. The same learning procedures are successively applied to all training samples of the mth class. Then, the analogous series of procedures are carried out to form the neural assemblies in all other sub-networks and the stage of primary learning is completed.

An assemblage of excitatory connections created in each sub-network in the primary learning process, constitute an integral description of the corresponding class.

A recognition process also begins with feature extraction from an input test sample; let it be the xth test sample. Unlike the primary learning stage, the pattern of initial neural activity is set in all sub-networks in parallel to create equal opportunities for all sub-networks to win a competition between them. This initial neural activity spreads (only once) through the connection structure of the network and form some secondary distribution of activities of all network neurons. The output vector $E^m(x)$ for the mth sub-network is calculated by the equation

$$E_i^m(x) = G_i(x) \sum_{j=1}^{N} G_j(x)\, W_{j,i}^m \,. \tag{2}$$

The network uses R-neurons for the recognition process; each R-neuron is a representative of a certain sub-network. Every R^m-neuron summarizes secondary activities of all those neurons that constituted the pattern of initial neural activity in the mth sub-network. An integer-valued vector R of M components represents activity of all R-neurons. The mth component of this vector is calculated by the formula

$$R^m(x) = \sum_{i=1}^{N} E_i^m(x). \tag{3}$$

In order to classify the input test sample, R-neuron with maximum output is found in the network. This R-neuron defines the sub-network winner and, consequently, the class of the test sample presented for recognition.

3   A Procedure of Differentiation

It is possible to tune the integer-valued weights of connections formed in the primary learning stage aiming to increase the network's recognition ability. This tuning process is named as a secondary learning process or a process of differentiation. A prototype of the differentiation procedure presented below is described in [1], [2].

In the procedure of differentiation, the model repeatedly considers all available training samples in turn. Let the xth training sample of the mth class be considered by the model. The model tries to recognize it. Let the recognition result be such that the R^m-neuron has the maximum excitation level among all R-neurons of the network. This means that the network classifies the input training sample to the correct m^{th} class. If activity of the R^m-neuron exceeds activities of all other R-neurons more than by a preassigned value Ω, then the model proceeds to consider the next training sample. If activity of the correct R^m-neuron exceeds activity of some wrong R^v-neuron less than by the value Ω, the following two procedures are fulfilled.

In the first procedure, all weights of connections of the neural assembly that represents the considered training sample are increased by the value ΔW in the correct m^{th} sub-network. In the second procedure, the weights of connections between all neurons of the pattern of initial neural activity of this training sample in the wrong v^{th} sub-network are decreased by the same value ΔW. The former procedure is described by Eq. (1). The latter one is expressed by the formula

$$W_{i,j}^v(k+1) = W_{i,j}^v(k) - \Delta W \left(G_i(m_x) \wedge G_j(m_x) \right), \tag{4}$$

These two procedures increase the competitiveness of the correct m^{th} sub-network and decrease the competitiveness of the wrong v^{th} sub-network.

A successive consideration of all samples of the available training set continues, epoch by epoch, until the network becomes able to recognize successfully each sample of the training set under the above conditions related with the Ω value. Thus, the parameter Ω is an external influence, which artificially counteracts the correct classification of the recognized training sample by the network.

4 Experiments

The differentiation process in the assembly network was experimentally studied on a task of recognition of separate handwritten characters of two types: letters of Ukrainian alphabet and Arabic numerals of MNIST database [11].

The experiments with Ukrainian letters were conducted as follows. A very simple set of features presented in [4] was used for description of the letter shapes. The training set consisted of 80 samples of each letter (2 560 samples, a total). The network performance was measured using the test set which consisted of 20 samples of each letter (640 samples, a total).

At first, a neural assembly structure was formed in the network in the primary learning stage. Then, the differentiation process began. The first step of the differentiation process was done with $\Omega=0$. The network's performance was measured. The second step of the differentiation process was performed with some increment of the Ω value. The recognition accuracy of the network was measured again. The subsequent steps were carried out analogously.

The experiments have shown that the differentiation procedure is effective tool for increasing the recognition ability of the assembly network. It is possible to take as a primary recognition ability of the network such a percentage that is achieved by the network after accomplishment of the differentiation procedure for $\Omega=0$. Then in the above experiments, the differentiation procedure results in growth of the network's relative performance by 37%.

The same growth of the relative recognition ability of the assembly network was recorded in the following experiments that were carried out on recognition of separate handwritten digits of MNIST database [11]. The database contains 60 000 samples in its training set and 10 000 ones in the test set. Some more complicated set of features was used for description of the digits. The experiments were conducted according to the above description. During the process of differentiation (with growth of the parameter Ω), the recognition ability of the network significantly increases as it is expressed by the curve of Fig. 1. In these experiments, the model's recognition rate has achieved 97.93% (207 errors).

The differentiation process in the assembly network takes rather much time. However, the subsequent recognition procedure is quick. In particular, the process of recognition of all 10 000 test samples of MNIST database takes 5 minutes and 25 seconds on PC of 2 000MHz.

5 Discussion and Conclusions

Thus, the experiments have shown that the differentiation procedure considerably increases the recognition ability of the assembly neural network and does it in the same way in case of using different feature sets and/or other recognized objects.

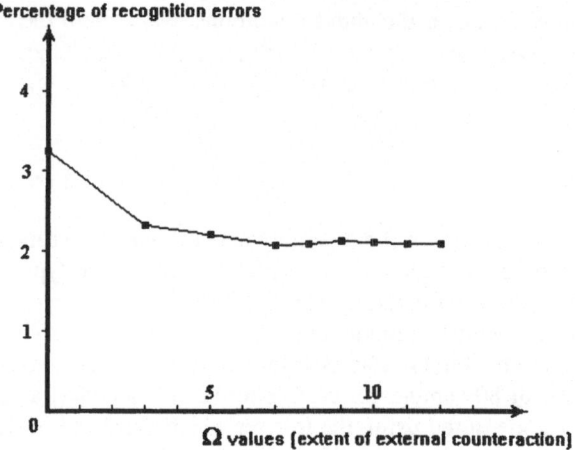

Fig. 1. Performance of the assembly neural network as a function of progress of the differentiation process. Percentage of recognition errors are plotted on the Y-axis. Ω values are plotted on the X-axis. Parameter Ω is measured in percentage to activity of the R-neuron, which represents the correct sub-network during recognizing the training sample under consideration

The differentiation procedure is very reliable in the sense of the network's convergence. Indeed, the convergence was reached in all experiments described above even in the case of the training set of MNIST database, which contains a large amount of ambiguous samples. Such a fact definitely testifies that the assembly network structure is able to be adequately adopted in the process of differentiation so that to form unambiguous descriptions of all recognized classes in the sub-networks.

The differentiation procedure has to be executed in any assembly network with analog connections for its normal functioning because the network without this procedure has rather low recognition ability. At the same time in spite of accomplishment of the differentiation procedure, the recognition rate of the assembly network of 97.93% demonstrated in the above experiments on MNIST database is still low in comparison, for instance, with the rate of 99.58% reported in [9], which is achieved on a simple perceptron-type classifier [7–9]. However, it is worth to note that the best recognition percentages on MNIST database are attained with the aid of multiple expansion of the training set by means of distortion of its digits. Application of this method considerably increases the recognition rate. The result of the present model has been achieved without any artificial expansion of the training set size, which fact should be taken into account for correct appraising of this result. Nevertheless, designing of more adequate feature set continues to be a topical task for a future work.

References

1. Goltsev, A.: An Assembly Neural Network for Texture Segmentation. Neural Networks 9(4) (1996) 643–653
2. Goltsev, A., Wunsch, D.C.: Inhibitory Connections in the Assembly Neural Network for Texture Segmentation. Neural Networks 11(5) (1998) 951–962
3. Goltsev, A: A Process of Generalization in the Assembly Neural Network. Proceedings of ICONIP'2001 (2001) 47–52

4. Goltsev, A., Húsek, D.: Some Properties of the Assembly Neural Networks. Neural Network World 12(1) (2002) 15–32
5. Goltsev, A., Wunsch, D.C.: Generalization of Features in the Assembly Neural Networks. International Journal of Neural Systems (IJNS) 14(1) (2004) 39–56
6. Hebb, D.O.: The Organization of Behavior. John Wiley, New York (1949)
7. Kussul, E.M., Baidyk, T.N., Lukovitch, V.V., Rachkovskij, D.A.: Adaptive Neural Network Classifier with Multifloat Input Coding. Proceedings of 6th Int. Conf. "Neuro-Nimes 93" (1993) 209–216
8. Kussul, E.M., Baidyk, T.N., Lukovitch, V.V., Rachkovskij, D.A.: Neural Classifiers with Distributed Coding of Input Information. Neurocomputers 3–4 (1994) 13–24 (in Russian)
9. Kussul, E., Baidyk, T.: Permutative Coding Technique for Handwritten Digit Recognition System. Proceedings of IJCNN'2003 (2003) 2163–2168
10. Milner, P.M.: The Cell Assembly: Mark 2. Psychological Review 64 (1957) 242–252
11. MNIST database: http://yann.lecun.com/exdb/mnist/

Managing Interference
Between Prior and Later Learning

L. Andrew Coward[1], Tamás D. Gedeon[1], and Uditha Ratnayake[1,2]

[1] Department of Computer Science, Australian National University,
Canberra, ACT 0200, Australia
[2] Dept. of Electrical and Computer Engineering
The Open University of Sri Lanka, PO BOX 21, Nawala, Nugegoda, Sri Lanka

Abstract. A device algorithm which permanently records an expanding portfolio of similar conditions is described, along with an architecture in which this algorithm is used to avoid interference between prior and later learning.

1 Introduction

A barrier to complex learning in artificial neural networks is catastrophic interference between later learning and earlier learning [1]. There have been numerous attempts to address this issue [2] but success has been limited as the complexity of the problem addressed by the network increases. A critical issue for learning complex combinations of capabilities is maintaining adequate meanings for information generated by one part of a network but used for different purposes in many other parts of the network [3]. Avoiding interference in these terms depends on maintaining information meanings, as learning changes information. This paper describes an investigation of the interference between prior and later learning in a connectionist architecture called the recommendation architecture in which explicit management of the maintenance of operational meanings is possible.

2 Implementation of the Recommendation Architecture

Properties of recommendation architecture systems have been investigated in learning problems such as telecommunications network management and document classification [4; 5]. There is a separation in the recommendation architecture between clustering which defines and detects conditions in the information available to the system and competition which associates different combinations of conditions with different behaviours. The device used in clustering is illustrated in figure 1A. The weights of inputs which define conditions to this device are binary. The device detects activity in a set of regular inputs, and produces an output if a subset larger than a threshold is present. The set is defined by converting active provisional inputs to regular if the number of active regular inputs is less than the threshold, the total of active regular and provisional inputs is above the threshold, a signal exciting the recording of conditions is present, and no signal inhibiting the recording of conditions is present. The effect of this algorithm is that once a device has produced an output in response to the activity of a specific set of inputs, it will always in the future produce an output in response to the activity of the same set. This permanence is in contrast with percep-

N.R. Pal et al. (Eds.): ICONIP 2004, LNCS 3316, pp. 458–464, 2004.
© Springer-Verlag Berlin Heidelberg 2004

tron type algorithms standard in ANNs in which the adjustments to individual weights mean that a device may not respond to an exact repetition of a condition which earlier generated a response.

Condition recording devices are arranged in layers. Inputs to devices in the first layer are system inputs, condition defining inputs to devices in the second layer are from first layer devices and so on. The complexity of conditions detected increases through the layers, where complexity is defined as the total number of system inputs which contribute to a condition. There are additional inputs from devices in specific modules to condition recording devices in specific other modules. These additional inputs are from special purpose devices which detect the level of activity of condition recording devices in the source module, and excite or inhibit changes to conditions detected by their target device.

A modular hierarchy is superimposed on the device layers, in which modules have specific functional roles. The first level of module is made up of devices in an area on one layer. The second level is a column made up of corresponding areas on a sequence of four layers. The third level is an array of parallel columns. The functions of these modules can be understood by consideration of two columns in the same array as illustrated in figure 1B. In each column, layer 4 is a single special purpose device that produces a binary output which is 1 if any of the devices in layer 3 of the column are active, otherwise 0. The layer four device is connected to devices in competition.

Initially, all the condition recording devices in a column have randomly selected provisional condition defining inputs. Condition defining connectivity between layers is only within columns. Columns are initiated one at a time, no other column is initiated until the previous column is generating a layer 4 output in response to a significant percentage of input states. Biases are placed on the random selection process for provisional inputs in favour of groups of inputs which have tended to be active in the same system input state in the past. Once some columns have been initiated, groups are defined with a bias in favour of inputs which have tended to be active at the same time in input states for which no previously initiated column has activity above a threshold level in its layer 1.

A high level of input activity into layer 1 of a column excites initiation of a column and inhibits initiation of any other column. The first input state will therefore initiate a column. This initiation means that condition recording devices in layers 1 to 3 will define regular conditions until there is activity in layer 3, at which point further definition will be inhibited. The next input state could have one of several results. One is that regular conditions are detected in all levels 1 through 3 and the column produces an output. Recording of conditions will be inhibited. A second is that there is activity in layer 2 but not in layer 3. In this situation, additional conditions will be recorded in layers 1 through 3 until layer 3 activity inhibits further recording. The third is that there is activity in layer 1. Such activity will inhibit the initiation of another column. When multiple columns have been initiated, there is another possibility. There could be activity in layer 2 of several columns, but inhibition between columns limits recording to the column with the highest proportion of active layer 2 devices.

The effect of a series of input states is that columns are defined each detecting a somewhat different portfolio of conditions. The array definition process thus results in compression of a large number of input characteristics which discriminate weakly between different categories into a much smaller number of portfolios which discriminate much more strongly between the categories. Increase in discrimination means that an individual characteristic can occur in some instances of N different categories, but a column is active in response to instances of n different categories, where in general n << N.

Category identification achieved by competition associating different combinations of portfolio outputs with different categories is therefore easier than with system inputs. Each portfolio output is assigned a weight in favour of each category, and weights are adjusted until high integrity identifications are achieved. This weight adjustment can be guided in two ways. In supervised feedback the competition component corresponding with the correct category is identified. In consequence feedback it is indicated whether the category identification by the system is correct or incorrect.

If the learning process in competition does not converge, the lack of convergence is an indication that the columns provide inadequate discrimination between categories. Such a lack of convergence would be demonstrated, for example, if at different times when the same set of columns was active and generated a category identification, the identification was sometimes correct and sometimes incorrect. This situation triggers generation of additional columns which provide additional discrimination.

The values of the device algorithm and modular hierarchy are that they result in column outputs in response to every input state; they result in consistent column outputs in the sense that once a column has generated an output in response to an input state it will always produce an output in response to an identical input state; they manage condition recording to prevent excessive recording; and they compress a large input space into a smaller output space with better category discrimination.

In the terminology of unsupervised learning in ANNs, columns correspond with heuristically defined clusters [6]. The reference vector for a column is implicit in the conditions recorded in layer 2, and is evolved whenever an input state is added to the column. The critical difference from unsupervised learning is that if an input state contains a condition which has been assigned to a cluster, an exact repetition of that condition in another input state will always be detected by the same cluster.

Competition devices in figure 1B have excitatory inputs from the special purpose devices in layer 4, a maximum of one from each such device. These inputs have continuously variable weights, and devices produce outputs proportional to the sum of the input weights of all active inputs. The category corresponding with the device with the largest output is the selected identification. When the category corresponding with a competition device is indicated by supervision, an input to that device is established from any layer 4 devices in columns with output activity which does not already have such a connection. This new input is given a standard weight. If the total of the weights of all active inputs to the device corresponding with the target behaviour is less than the total for any other device, the weights of the currently active inputs into the target device are all increased by the same proportion until the output from the target device is the largest.

In consequence learning, if the category is correct, the weights of active inputs to the competition device with the largest output are increased by 5% of the standard weight. If the category is incorrect, the weights of those active inputs are decreased by 10% of the standard weight. Under no feedback conditions, the identity of the category is compared with the identity generated by the system, but no feedback is provided.

Interference between later and earlier learning could occur in a number of ways. One is that changes to column portfolio definitions during later learning might dilute the discrimination of the portfolio. This dilution is limited because conditions must be similar to the existing portfolio to be added and if different conditions are needed they will be will be assigned to a new column. Dilution is also limited because as the portfolio expands, the number of devices which must be active in layer 2 for recording to occur increases. Hence the degree of difference from the existing portfolio under which addition of conditions is allowed decreases with learning. Finally, because recognition of a category is recommended by the presence of a number of portfolios, the effect of change to one portfolio will be limited.

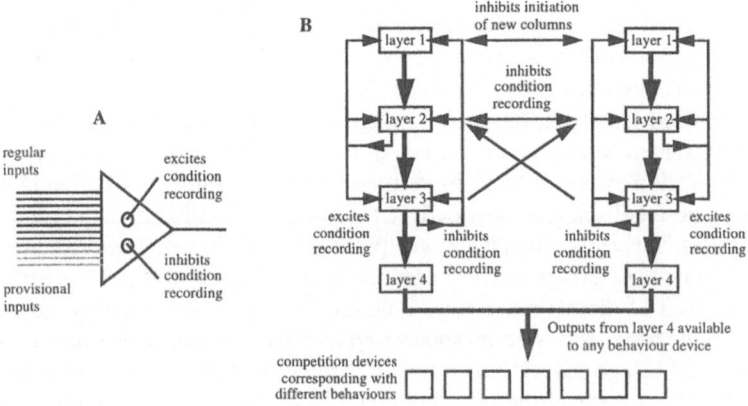

Fig. 1. A. Condition recording device. B. Modules in clustering, and outputs to competition

Any column will typically target a number of different category components in competition. The second way in which interference between later and earlier learning could occur is when the weights assigned as a result of supervision to inputs to a new category device are partially for columns which also have weights into a previously learned category. For some combinations of column outputs in response to an instance of the old category the total weight into the new category could exceed that into the old category. This dilution will be limited by the degree of discrimination between the categories provided by the columns, and also can be controlled by consequence feedback in response to incorrect identifications.

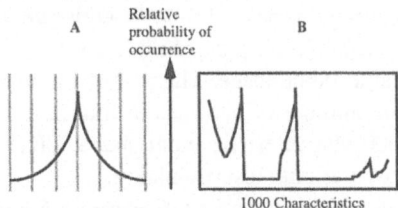

Fig. 2. Statistical generation of category instances. A. Probability distribution shape divided into six segments. One of the six is selected at random to define the relative probability for each 100 characteristic block in a category, plus a 40% chance of selecting zero probability. B. Relative probability of occurrence of characteristics in an example category

The test scenario to investigate the effects of later learning on prior learning was that the system was presented with a series of category instances. The presence or absence of 1000 characteristics could be discriminated in each instance, and the information available to the system from each object was therefore equivalent to a 1000 bit binary vector.

Instances were defined by the set of characteristics which they possessed. Sixty different categories were defined by probability distributions for the occurrence of characteristics in instances of the category. An example distribution is illustrated in figure 2B. 1000 objects were constructed for each category. The construction process was random selection of characteristics from a set with the appropriate distribution for the category. No two objects in the same category were identical. Individual characteristics occurred on average in instances of 26.1 categories or 43.4% of the total. This percentage is a measure of the ability of individual characteristics to provide discrimination between categories of inputs.

In supervised feedback, the identity of the competition component corresponding with the presented category was provided. In consequence feedback, a correct or incorrect indication was provided in response to the category identified. Under test conditions, no feedback was provided. After column definition in response to the first 50 categories, supervised feedback was provided for between 1 and 32 instances of those categories. At this point (P1) identification accuracy was determined for 5 instances of each of the 50 categories. Columns were then extended by experience of the full 60 categories followed by supervised feedback for between 1 and 32 instances of the new 10 categories. At this point (P2) identification accuracy was determined for 5 instances of each of the 60 categories. An alternative learning profile was that consequence feedback on the original 50 categories was provided when supervised feedback was provided on the new categories. At this point (P3) identification accuracy was determined for 5 instances of each of the 60 categories. In the entire process including column definition, supervised and consequence feedback, and testing phases, all instances presented to the network were different.

An array of columns contained on average about 23 columns, including about one column added during learning of the extra 10 categories. One column does not in general correlate unambiguously with one category. On average a column generated an output in response to some instances of about a quarter of the categories. Since on average a characteristic appeared in some instances of about half of the categories, this represents an improvement in operational discrimination by a factor of 2. Category identification accuracy for one array after supervised learning of the original 50 categories was 45%. Accuracy improved to 75% with two sets, 85% with three sets, and over 90% with four sets. Learning capability can thus be improved by additional sets of columns in the same input space. The rest of the results reported here are for four sets of columns. The number of columns in four sets averaged 93, compression by over 10 from the 1000 characteristic input space. Clustering thus achieved both compression and higher operational discrimination.

The first learning process achieved a 91.6% accuracy for instances from 50 categories. In general the variation in recognition accuracy using different object instances but with the same experience generating clustering was ±0.5%. The variation in recognition accuracy for different clustering experiences was ±2.5%.

Table 1. Recognition accuracy of original 50 categories before and after learning of the additional 10 categories, with and without consequence feedback on identifications of the original categories during supervised feedback on the additional categories. Results are averages for five experiments

Number of instances of each category used during supervised training	Final recognition correctness measured on 5 instances of each of 50 original categories at end of process:		
	P1	P2	P3
1	67.6%	26.4%	38.0%
2	72.4%	31.2%	51.6%
4	75.6%	32.8%	70.8%
8	84.4%	40.4%	77.6%
16	88.4%	44.4%	81.6%
32	91.6%	44.8%	79.6%

Table 1 shows the accuracy with which instances of the first 50 categories were identified following the different learning processes. The recognition accuracy for the extra ten categories was greater than 90% at points P2 and P3. For many of the errors in identification, the second choice identification was correct. In other words, the competition device corresponding with the correct identification had the second highest output activity. For example, in table 1 for 16 supervised instances, 88.4% of the instances of the first 50 categories were correctly identified before learning the additional 10 categories (point P1), but only 44.4% of instances after the additional learning (point P2). However, for 32.4% of the instances at P2, the first identification was incorrect but the second choice was correct. This type of information robustness accounts for the way in which identification accuracy was restored to 81.6% if simple consequence feedback on identifications of the original 50 categories was provided during supervised learning of the new 10 categories (point P3).

Table 1 demonstrates that subsequent learning of an extra 10 categories has an effect on prior learning of the 50 categories, but the effect is a gradual degradation not catastrophic destruction. The gradual degradation is the result of slight broadening of the similarity definition for some columns, and of clusters with weights in competition indicating some of the original categories gaining large weights in favour of new categories. Both of these effects can be limited by providing simple consequence feedback for identifications of the original categories during the period in which supervised feedback is provided for the new categories.

3 Conclusions

It has been demonstrated that using a condition recording device algorithm in an architecture called the recommendation architecture, interference between later and prior learning is not catastrophic, and can be managed in a controlled fashion.

References

1. French, R. M. Catastrophic Forgetting in Connectionist Networks. Trends in Cognitive Science Volume 3 (1999)128-135.
2. Robins, A. Catastrophic Forgetting, rehearsal, and pseudorehearsal. Connection Science Volume 7 (1995) 123 - 146.

3. Coward, L. A.. The Recommendation Architecture: lessons from the design of large scale electronic systems for cognitive science. Journal of Cognitive Systems Research Volume 2 (2001) 111-156
4. Coward, L. A., Gedeon, T. D. and Kenworthy, W. Application of the Recommendation Architecture to Telecommunications Network Management. Int. Journal of Neural Systems Volume 11. (2001) 323-327.
5. Ratnayake, U. and Gedeon, T. D. Application of the Recommendation Architecture for Discovering Associative Similarities in Documents. (2002) Proceedings of the 9th International Conference on Neural Information Processing.
6. Gokcay, E. and Principe, J. C. Information Theoretic Clustering. IEEE Transactions on Pattern Analysis and Machine Intelligence Volume 24 (2002) 158-169.

A Neural Learning Rule for CCA Approximation

M. Shahjahan and K. Murase

Department of Human and Artificial Intelligence Systems
University of Fukui, Bunkyo 3-9-1, Fukui 910-8507, Japan
{jahan,murase}@synapse.his.fukui-u.ac.jp

Abstract. A new learning rule is implemented for approximating canonical correlation analysis(CCA) with artificial neural networks. A correlation objective function is maximized in order to find identical or correlated item from several sets of data. A simple weight update rule is derived, that is computationally much more inexpensive than the standard statistical technique. We demonstrate the network capabilities on artificial and real-world data. The experimental results show that this method is a good approximator of CCA as well as correlated item identifier.

1 Introduction

Artificial Neural Networks (ANNs) are well known for their capacity for implementing powerful transformation. Human understandable and extractable information can be separated from these transformation. A very common example is a PCA implementation by ANNs, which extract principle components from the input data (Oja, 1982). The nonlinear extentions of PCA networks (Oja, 1997) have been shown to be capable of more sophisticated statistical technique such as Factor Analysis(FA) (Charles, 1998) and Projection Persuit (Fyfe, 1995).

In this article, we present a new learning rule for canonical correlation analysis(CCA). Canonical correlation analysis (Mardia, 1979) is used when we have two data sets which we believe have some underlying correlation. Say, two sets of input data sets, from which we draw iid (independent and identical distribution) samples to form a pair of input vectors, x_1 and x_2. In classical CCA, we attempt to find the linear combination of the variables that gives us maximum correlation between the combinations, y_1 and y_2. Let $y_1 = \mathbf{w_1}\mathbf{x_1} = \sum_j w_{1j}x_{1j}$, and $y_2 = \mathbf{w_2}\mathbf{x_2} = \sum_j w_{2j}x_{2j}$ as shown in Fig. 1(a).

The goal is to find the those values of $\mathbf{w_1}$ and $\mathbf{w_2}$ that maximize the correlation between y_1 and y_2. Whereas PCA and FA deal with the interrelationship within a set of variables, CCA deals with the relationships between two sets of variables. We implement an iterative learning rule that does not need to view the process as one of finding the best predictor of the set $\mathbf{x_2}$ by the set $\mathbf{x_1}$, and vice versa.

N.R. Pal et al. (Eds.): ICONIP 2004, LNCS 3316, pp. 465–470, 2004.
© Springer-Verlag Berlin Heidelberg 2004

2   The Learning Rule

We propose a correlation objective function to be maximized which in turn
maximizes correlation between y_1 and y_2. The *objective function* is defined as
follows.

$$F(y) = (y_k - \bar{y}) \sum_{j \neq k} (y_j - \bar{y}) \tag{1}$$

where y_k is the output of the output unit k and $\bar{y}$ is the average of the total
output units and can be computed as: $\bar{y} = \frac{1}{M} \sum_{k=1}^{M} y_k$. Where M is the number
of output units. Now differentiating $F(y)$ with respect to weight as a whole.
Considering without any loss of generality that w_i is a weight of CCA network,
it may be from $\mathbf{w_1}$ or $\mathbf{w_2}$. By applying chain rule for differentiation.

$$\eta \Delta w_i = \eta \frac{\partial F(y)}{\partial w_i} = \eta \frac{\partial F(y)}{\partial y_k} \cdot \frac{\partial y_k}{\partial w_i} \tag{2}$$

Now

$$\frac{\partial F(y)}{\partial y_k} = \sum_{j \neq k} (y_j - \bar{y}) = -(y_k - \bar{y}) \tag{3}$$

and

$$\frac{\partial y_k}{\partial w_i} = \frac{\partial \sum_i w_i x_i}{\partial w_i} = x_i \tag{4}$$

We assume $\bar{y}$ is constant with respect to y_i throughout the calculation. Therefore,
weight update rule is:

$$\eta \Delta w_i = -\eta x_i (y_k - \bar{y}) \tag{5}$$

where η is a learning rate parameter. The weight update rule is thus generalized
as follows for any input set k.

$$w_{ki} = w_{ki} + \eta \Delta w_{ki} \tag{6}$$

$$w_{ki} = w_{ki} - \eta x_{ki} (y_k - \bar{y}). \tag{7}$$

3   Experimental Results

3.1   Two Sets of Data

The first experiment comprises an artificial data set: $\mathbf{x_1}$ is a four dimensional
vector and $\mathbf{x_2}$ is three-dimensional vector. The initial weight is selected randomly
within a certain small range. The experiment is done by another set of input
data set which are identical with the experiment done by Lai (Lai, 1999). The
elements in the vectors of $\mathbf{x_1}$ and $\mathbf{x_2}$ are drawn from the zero-mean Gaussian
distribution, $N(0, 1)$. In order to introduce correlation between the two vectors,
$\mathbf{x_1}$, and $\mathbf{x_2}$, an identical sample from $N(0, 1)$ is used in the first elements of each
vector. Thus there is no correlation between the two vectors other than that

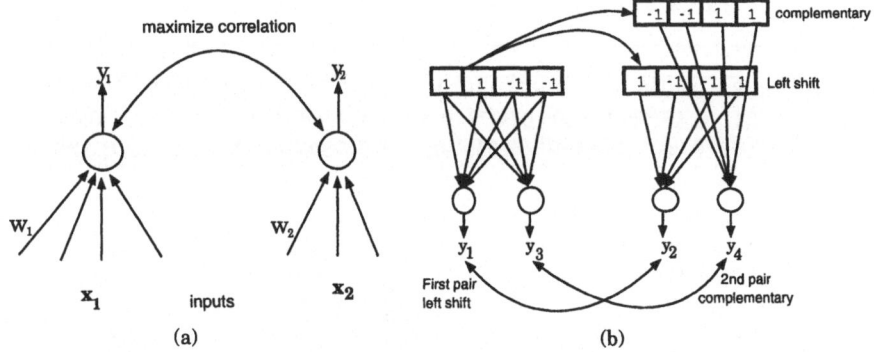

Fig. 1. (a) The CCA network. By adjusting weights, w_1 and w_2, we maximize the correlation between y_1 and y_2. (b) Pair wise correlation maximization. First pair for left bit shift and 2nd pair for bit reversal. The weights should discover this information.

existing between the first element of each vector. The initial learning rate and iteration number were 0.0001 and 1000 respectively. It is clear from Table 1(A) that there is high correlation between the first element of each vector and in fact there is only one correlation between the vectors. The results are coherent with Lai's (Lai, 1999) similar experiment as shown in Table 1(B).

The algorithm is applied to a real-world data, daily currency exchange rate between British pound and Canadian dollar into USD, collected from financial data finder, Ohio state university. They are placed in x_{11} and x_{21}, respectively. The linear correlation between them is 0.84. After training the converged weights listed in Table 1(C) clearly show that there is no correlation except the first two elements. The input exchange rates should be transformed within $N(0,1)$.

Table 1. The converged weight vector $\mathbf{w_1}$ and $\mathbf{w_2}$.

	A	B	C
$\mathbf{w_1}$	**0.434** 0.060 0.014 0.012	**0.679** 0.023 -0.051 -0.006	**0.385** 0.001 0.011 0.001
$\mathbf{w_2}$	**0.435** 0.005 0.011	**0.681** 0.004 0.005	**0.391** 0.010 0.006

3.2   More Than Two Sets of Data

In a similar fashion, we create an artificial data set, which comprises three vectors, $\mathbf{x_1}$, $\mathbf{x_2}$, $\mathbf{x_3}$, each of whose elements is initially independently drawn from $N(0,1)$. In order to introduce correlation between vectors, an identical $N(0,1)$ is used in any one element of the three input vectors. The first element of vector $\mathbf{x_1}$, 2nd element of vector $\mathbf{x_2}$ and 3rd element of vector $\mathbf{x_3}$ are made identical. Table 2(A) clearly shows diagonal elements are highly correlated.

The previous real world data was tested by placing rates of British pound in x_{11}, Canadian dollar in x_{22} and Dutch mark in x_{33}. They are correlated each

Table 2. The converged weight vectors $\mathbf{w}_1$, $\mathbf{w}_2$ and $\mathbf{w}_3$.

	A			B			C		
$\mathbf{w}_1$	**0.4882**	0.0036	0.0001	**0.3039**	0.0139	0.0720	**1.0101**	0.0021	0.0006
$\mathbf{w}_2$	0.0041	**0.4879**	0.0039	0.0078	**0.3861**	0.1750	0.0041	**0.5051**	0.013
$\mathbf{w}_3$	-0.0017	0.0018	**0.4882**	-0.2659	-0.2761	**0.5847**	-0.0003	0.0003	**0.3367**

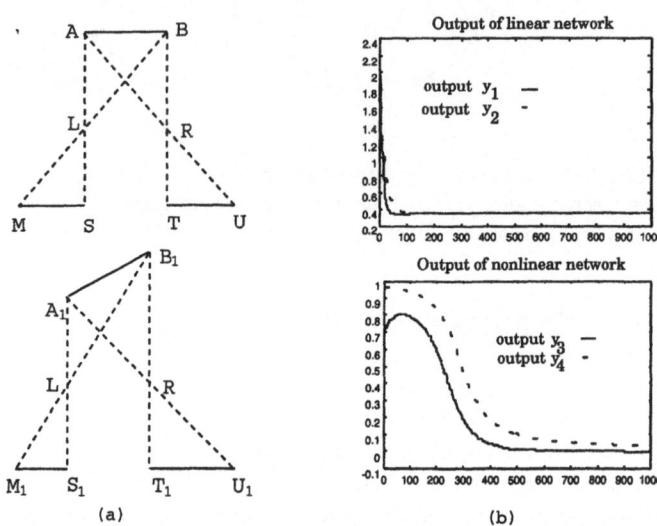

Fig. 2. (a) The top figure represents visual information from a surface AB which is passed through pupils R and L to flat retinas MS and TU. The bottom figure represents the same scene when the external surface $A_1 B_1$ is not parallel to the plane of retinas. (b) The top figure represents the outputs, y_1 and y_2 of linear network and the bottom figure represents the outputs, y_3 and y_4 of nonlinear network. Visual inspection would suggest that nonlinear network has more flexibility than linear network although the convergence is faster in linear network.

other in the positive sense. The network clearly found those correlation by the presence of heavier weights listed in Table 2(B). The method is capable to extract correlated components even from real world data set.

The factor by which two compounds are related such as alkenes ($C_n H_{2n}$) or alkynes ($C_n H_{2n-2}$) related with a factor of 2. We tested the network ability to find out those factor by placing $x_{22} = 2x_{11}$ and $x_{33} = 3x_{11}$. After training, the extracted factors listed in Table 2(C) were found as 2 ($0.5051 \times 2 = 1.0101$) and $3(0.3367 \times 3 = 1.0101)$, respectively.

3.3   Extraction of Bit Shift and Bit Reversal Information

It has been suggested (Lai, 1999) that one of the goals of sensory information processing may be the extraction of common information between different sensors. The idea behind the network of Fig. 1(b) is to extract correlation with bit

Table 3. The converged weight vectors w_1, w_2, w_3, and w_4.

w_1	0.2657	-0.0183	0.0465	0.6389	w_3	-0.2955	-0.2453	0.1679	0.0136
w_2	-0.0183	0.0465	0.6389	0.2657	w_4	0.2955	0.2453	-0.1679	-0.0136

shift and reversal information. In other words, if there is any feature in identical between the two inputs, it should be discovered by the weight representation irrespective of their position in the input vectors.

The four components of each vector are generated randomly and then a left shift version and a complementary version are used in vectors x_2 and x_4, respectively. The corresponding weight vector should discover these information after training. From Table 3, it is clear that first two weight vectors, w_1 and w_2 clearly discover left shift information and the second two weight vectors, w_3 and w_4 clearly discover the bit reversal information.

3.4   Nonlinear Correlations

We know the relationship between the projection of the surface on the retinas is a function of the angle between the plane of the surface and the plane of retinas as shown in Fig. 2(a). We investigate how cortex extracts depth information from surface at any one time. Here, we investigate the general problem of maximizing correlations between two data sets when there may be an underlying nonlinear relationship between the data sets: the data set is generated according to the following equation as done by Lai (Lai, 1999). $x_{11} = \theta - \pi + \mu_1$, $x_{12} = \theta - \pi + \mu_2$, $x_{21} = 1 - sin(\theta) + \mu_3$, and $x_{21} = cos(\theta) + \mu_4$ where θ is drawn from a uniform distribution in $[0, 2\pi]$ and μ_i, $i = 1....4$ are drawn from the zero mean Gaussian distribution $N(0, 0.1)$. x_{11} and x_{12} define a linear manifold within the input space where each manifold is only approximately due to the presence of noise while x_{21} and x_{21} define a circular manifold in the two-dimensional input space.

Therefore, $x_1 = \{x_{11}, x_{12}\}$ lies on a line $x_{11} = x_{12}$ while $x_2 = \{x_{21}, x_{22}\}$ lies on or near the circular manifold $x_{21}^2 + x_{22}^2 = 1$. We wish to investigate whether the network can find nonlinear correlations between the two data sets, x_1 and x_2. To do this experiment, we train two pairs of output neurons: (a) we train one pair of weights w_1 and w_2 using rules for linear units. (b) we train a second pair of outputs, y_3 and y_4 which are calculated after passing through a nonlinear function, say tanh(.). The update law for nonlinear network is thus $w_{kj} = w_{kj} - \eta(1 - (y_3)^2)x_{kj}(y_k - \bar{y})$. We use a learning rate of 0.0001 for all weights and the learning was stopped after 1000 iterative loops. The outputs of two networks are shown in Fig. 2(b).

4   Discussion

The proposed algorithm discovers not only the correlation between the components of multiple input data sets but also important aspects of input data sets. A heavier weight value denotes correlation among the elements of data sets for both artificial and real world data. The network is able to find the linear and nonlinear correlation between data set. The robustness of the algorithm is less dependent on the initial condition. The nonlinear network, shown in Fig. 2(b)(bottom), illustrates that a smooth agreement between the output with some degree of flexibility. It is also interesting that the nonlinear sample correlation(0.98) is higher than the linear correlation(0.89). This is due to the nice 'run together' behavior of outputs.

The proposed method has some advantages over the others. We need one learning rate parameter and short training time, while recent Lai's method (Lai, 1999) requires multiple Lagrange multipliers, multiple learning rates and long training time. The proposed method finds correlated components from several data sets, while standard PCA and ICA (Hyverinen, Karhunen & Oja, 2001) try to find possibly uncorrelated and independent variables.

5   Conclusion

A simple neural learning rule is implemented for approximating canonical correlation analysis. The effectiveness of the algorithm is shown by several artificial data and real world data such as currency exchange rates among several countries. After training, the weight vectors suggest that the learning rule can approximate CCA. The network discovers correlated components, bit shift and bit reversal information from both type of data sets. Its application to a nonlinear network explores its smooth capture of the relationship between two outputs. Extending it to more complex real world problem may be left as a task for the future work.

References

Charles, D., and Fyfe, C., Modeling multiple cause structure using rectification constraints. Networks: Computation in Neural Systems, 9, 167-182, 1998.

Fyfe, C. and Baddeley, R., Non-linear data structure extracting using simple habbian networks. Biological Cybernetics, 72(6), 533-541, 1995.

Hyverinen, A., Karhunen, J., and Oja, E., (2001). Independent Component Analysis. John Wiley & Sons Publishing, USA.

Lai, P. L., and Fyfe, C., A neural implementation of canonical correlation analysis. Neural Networks, 12, 1391-1397, 1999.

Mardia, K. V., Kent, J., and Bibby, Multivariate analysis, New York, Academic Press, 1979.

Oja, E., A simplified neuron model as a principle component analyzer. Journal of Mathematical Biology, 16, 267-273, 1982.

Oja, E., The nonlinear pca learning rule in independent component analysis. Neurocomputing, 17, 25-45, 1997.

Adaptive Learning
in Incremental Learning RBF Networks

T.N. Nagabhushan and S.K. Padma

Department of Information Science & Engineering
Sri Jayachamarajendra College of Engineering, Mysore 570 006, India
{tnn,skp}@sjce.ac.in

Abstract. This paper presents a modification to the incremental learning algorithm originally proposed by Bernd Fritzke.This algorithm is a single stage approach to build a RBF neural network in the training phase. It combines the unsupervised and supervised learning stages to incrementally generate the RBF architecture. The algorithm uses accumulated error information to insert new neurons dynamically in the hidden space. The algorithm maintains constant parameters to control the growth of the network. This may result in non-optimal architectures. Modifications that provide adaptive capabilities to the learning parameters have been proposed in this work. Two benchmark data sets have been used for training. Results show an improved performance over the original incremental learning algorithm.

Keywords: Incremental Learning, Constructive learning, Neural networks, RBF networks, Adaptive learning.

1 Introduction

Constructive learning algorithms for neural networks have been receiving wider attention in recent times [1] [2] [3] [4]. The main goal of such algorithms is to generate an optimal neural network architecture for a given training set. One of the first constructive learning algorithms was the Cascade correlation learning architecture [5]. It uses the gradient descent learning to build a layered network. Platt [6] proposed a Resource Allocating Network (RAN) which builds a RBF network based on the novelty of the input pattern vectors. Fritzke [7] [8] proposed a single stage incremental learning algorithm for RBF networks and reported interesting results. It is observed from the above that the incremental learning algorithms build neural network architectures dynamically during learning maintaining constant adaptation parameters. A compact optimal architecture is very much necessary to exhibit good generalization capabilities. ε_1 and ε_2 serve as constant adaptation parameters that control the growth of the network. For every input pattern presented, ε_1 and ε_2 control the coarse and fine movements of the RBF units in the input space during learning. The idea is to allow the RBF units to follow the input pattern distribution. Due to this, RBF units are subjected to continuous movement while attempting to place

N.R. Pal et al. (Eds.): ICONIP 2004, LNCS 3316, pp. 471–476, 2004.
© Springer-Verlag Berlin Heidelberg 2004

themselves at optimal locations in the input space. Holding ε_1 and ε_2 consant throughout the learning process tends to dislocate the RBF units to non-optimal locations thus resulting in longer training times and bigger size networks at the end of convergence.

We, in our approach, present a scheme that dynamically change the parameter values depending on the accumulated error. The objective is to allow the parameters to acquire new values based on the network output error and aid in developing compact network architecture. We have modified Fritzke's incremental algorithm to have adaptive capabilities for the parameters ε_1, ε_2 and the learning rate η. Bench mark data sets have been used to impart training to the network.

2   Proposed Modifications

The following modifications have been proposed on the incremental learning algorithm to provide adaptive capabilities to the learning parameters:

2.1   Center Adaption

For each presentation of the input, the best matching unit (BMU) is computed and moved towards the input pattern by ε_1 times the distance between the input pattern and BMU. ε_1 is a small positive value which is computed using the error accumulated across the output units. It has been found that ε_1 and the accumulated error have direct correlation. The position of the RBF unit in the input space dictates the error for every input pattern presented since RBF uses Gaussian activation function. The neighbors of the BMU are also moved towards the input pattern by a smaller value ε_2, where ε_2 is made linearly dependent on ε_1. Thus ε_1 and ε_2 control the coarse and fine movements of RBF units in the input space during learning.

2.2   Weight Training

The weights between the RBF units and output units are trained using delta rule. For each presentation of the input, one adaptation of the delta rule is done. The learning rate η is made to vary according to the accumulated error. When the error is high, η can be coarse meaning we can move down the error surface with a bigger step size. As error decreases, η is made smaller to enable finer step sizes during final stages of convergence.

2.3   Choice of Initial Centers

The initial network consists of two RBF units. Instead of choosing any two random points in the input space, two patterns from the training set forming the farthest neighbors have been selected as initial centers.

2.4   Non-linear Output

Since we are experimenting on classification problems, the outputs are binary in nature. Instead of using linear outputs where output units do only a weighted summation over their inputs, we have applied the sigmoidal function on the network output.

3   Algorithm

1. Select two farthest vectors from the input space as initial RBF centers. Connect them by an edge. Set their widths to be equal to the distance between them, that is

$$\sigma_1 = \sigma_2 = \|x_1 - x_2\| \tag{1}$$

where x_1 and x_2 are the coordinates of the two selected RBF units.

2. For a given training pattern (ξ^μ, ζ^μ), calculate the output of the RBF network using

$$O_i^\mu = \sum_{j=1}^{jmax} W_{ij} V_j^\mu \tag{2}$$

where

$$V_j^\mu = exp\left(\frac{-\|\hat{x}_j - \xi_k^\mu\|^2}{\sigma_j^2}\right) \tag{3}$$

Calculate error using

$$e^\mu = \zeta^\mu - O^\mu \tag{4}$$

3. Calculate ε_1 using

$$\varepsilon_1 = sigmoid(e^\mu) \tag{5}$$

Calculate ε_2 using

$$\varepsilon_2 = \varepsilon_1^2 \times \varepsilon_1^2 \tag{6}$$

For every input pattern ξ^μ that is presented, find the best matching unit (BMU) using

$$\|\hat{x}_j - \xi^\mu\| \tag{7}$$

Move the BMU ε_1 times the current distance towards the input pattern using

$$\hat{x}_{jbmu}(new) = \hat{x}_{jbmu}(old) + \varepsilon_1(\hat{x}_j - \xi^\mu) \tag{8}$$

Move all the immediate neighbors of the BMU ε_2 times their current distance towards the input pattern using

$$\hat{x}_{jneighbor}(new) = \hat{x}_{jneighbor}(old) + \varepsilon_2(\hat{x}_j - \xi^\mu) \tag{9}$$

4. Update the weights between the hidden and output units using

$$W_{ij}^{new} = W_{ij}^{old} + \eta[\zeta^\mu - O^\mu]V_j^\mu \tag{10}$$

where η is the learning rate selected using the lookup table (Table 1).

Table 1. Lookup table for choosing η

S.N	If the error e^μ is :	η
1	$((e^\mu > 0.1)\&(e^\mu <= 0.5))$	$\eta=0.30$
2	$((e^\mu > 0.05)\&(e^\mu <= 0.1))$	$\eta=0.26$
3	$((e^\mu > 0.005)\&(e^\mu <= 0.05))$	$\eta=0.22$
4	$((e^\mu > 0.0005)\&(e^\mu <= 0.005))$	$\eta=0.20$
5	$((e^\mu > 0.00005)\&(e^\mu <= 0.0005))$	$\eta=0.16$
6	$((e^\mu > 0.000005)\&(e^\mu <= 0.00005))$	$\eta=0.12$
7	$((e^\mu > 0.0000005)\&(e^\mu <= 0.000005))$	$\eta=0.10$
8	$((e^\mu > 0.00000005)\&(e^\mu <= 0.0000005))$	$\eta=0.05$

5. The width of RBF units are computed using 'Age' information. For each input presented, compute BMU and the next immediate BMU. Connect them by an edge and associate it with an age variable. When an edge is created, its age variable is set to zero. The age of all edges emanating from the BMU is increased at every adaptation step. Edges exceeding an age limit A_{max} are deleted and so also the nodes having no more emanating edges. The insertion of a new RBF unit is based on the squared error accumulated across all the output units.

6. A RBF unit is inserted between a unit which has accumulated maximum error and any of its neighbors. Its weights are set to small random values. Its width is set to the mean distance between the neighboring units.

7. Repeat steps 2 to 6 until classification error for all the patterns falls below a set value.

4 Experiments

Two data sets have been used to train the adaptive incremental learning algorithm. The iris data set has 4 standard features and comprises of 3 classes. The training set consists of 150 patterns. The temp data set comprises of 124 symptoms of protective devices in a power network and 40 different types of faults. Five hundred samples have been considered for training.

Figures 1 and 2 show the growth of the network and fall of error as a function of epochs for the temp training set. The solid lines in the graph show learning characteristics for adaptive parameter learning while dotted lines indicate the learning characteristics for constant parameter learning. The data sets for testing have been obtained by incorporating 1% and 2% noise to the original data sets.

Table 2 shows the performance of both algorithms for the two data sets. With constant learning parameters, the incremental learning algorithm produces a network of size 30 and 76 units while adaptive learning algorithm produces a network of size of 19 and 59 units respectively. The learning time also shows reduction from 1446 and 304 epochs to 880 and 214 epochs respectively. With 1% and 2% noise, adaptive learning network exhibits better generalization compared to the network generated using constant learning parameters.

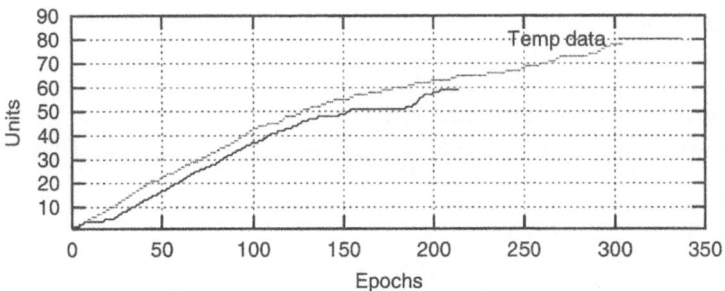

Fig. 1. Units Vs Epochs

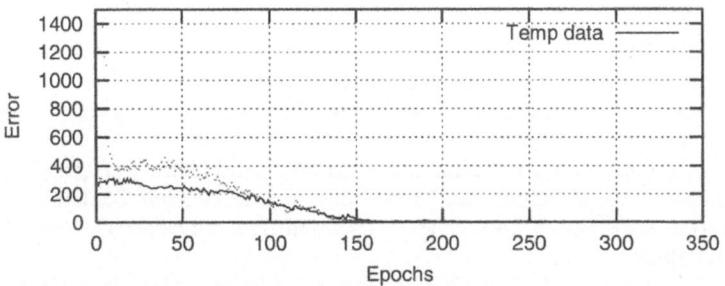

Fig. 2. Error Vs Epochs

Table 2. Results: Adaptive Vs Constant Parameter Learning RBF Network

Data set	Adaptive Parameter Network				Constant Parameter Network			
	Training		Generalization		Training		Generalization	
	Units	Epochs	1% Noise	2% Noise	Units	Epochs	1% Noise	2% Noise
Iris	19	880	1.33	2.0	30	1446	2.0	2.0
Temp	59	214	0.4	0.4	76	304	4.2	4.4

5   Conclusions

We have proposed modifications to the original incremental learning algorithm
to have adaptive capabilities. The learning parameters ε_1, ε_2 and η are allowed
to change as a function of accumulated error. Instead of keeping ε_1, ε_2 and η
constant, they are made to vary according to the local error. Since the RBF units
follow the local error distribution, convergence is much faster thereby generating
a network with less number of RBF units. These modifications have resulted in
significant improvements in the network size, learning time and generalization.

References

1. Broomhead & Lowe, Multivariable Function Interpolation and Adaptive Networks, Complex Systems 2 (1988) 312-355.
2. T J Moody & C J Darken, Fast Learning in Networks of Locally Tuned Processing Units, Neural Computation 1 (1989) 151-160.
3. T Poggio & F Girosi, Networks for Approximation and Learning, Proceedings of the IEEE 78(9) (1990) 1481-1497.
4. J C Platt, Learning by Combining Memorization and Gradient Descent, In. J. Neural Information System-3 (1991).
5. Fahlman S E & C Lebiere, The Cascade-Correlation Learning Architecture, Advances in Neural Information Processing Systems 2 (1990) 524-532.
6. J C Platt, A Resource-Allocating Network for Function Interpolation, Neural Computation 3(2) (1991) 213-225.
7. B Fritzke, Supervised Learning with Growing Cell Structures, Advances in Neural Information Processing Systems 6 (1994) 255-262.
8. B Fritzke, Fast Learning with Incremental RBF Networks, Neural Processing Letters, Vol.1, No.1 (1994) 2-5.
9. T N Nagabhushan & S K Padma, Incremental Neural Learning Algorithms for Classification of Spiral Structures, Proceedings of the National Conference on Recent Trends in Information Technology, Karpagam Arts & Science College, Coimbatore (2002) 19-23.
10. T N Nagabhushan & S K Padma, Extended Tower Algorithm for Multi Category Classification, ICICS-PCM 2003, Nanyang Technological University, Singapore (2003).

Recurrent Neural Networks
for Learning Mixed k^{th}-Order Markov Chains

Wang Xiangrui and Narendra S. Chaudhari

School of Computer Engineering, Block N4-02a-32, Nanyang Avenue
Nanyang Technological University, Singapore 639798
xray@pmail.ntu.edu.sg, asnarendra@ntu.edu.sg

Abstract. Many approaches for Markov chain construction determine only the parameters for the first order Markov chain. In this paper, we present a method for constructing mixed k^{th} order Markov Chains by using recurrent neural networks. In terms of input length n, our method needs $O(n)$ operations. We apply our method for classification on the Splice-junction Gene Sequences Database. Experimental results show that our method has less error rate than the traditional Markov Models.

1 Introduction

In traditional framework of Markov Chains[1], the models are able to capture the classification of highly observed sequence patterns within sequences. In the first-order Markov Chains, only two-length sequence patterns are captured, while in the second-order Markov Chains, three-length sequence patterns are captured. In the same manner, k^{th} –order Markov Chains can and only can capture $(k+1)$-length sequence patterns.

Neural networks are believed to have a representative power to capture structural information. Many approaches have been proposed to learn structural information by neural networks. These approaches need the feedback, so this kind network is called *recurrent neural network* [2-4]. There are some works related Markov models with recurrent neural networks [5], however, their applicability for learning mixed chains is limited.

In practice, it is difficult to determine exact value of parameter k in Markov Chain models. Highly observed patterns may include sequences of different length at the same time. Therefore, a method which uses all important patterns (in terms of classification) and their probability parameters of different length is desired. We give a formulation of this problem.

In this paper, we present our approach to solve the problem of constructing mixed k^{th}–order Markov Chains using recurrent network, which uses explicit construction of neurons to learn Markov Chains from given samples. We use pruning to ensure efficiency.

In the next section, the model is briefly presented. Our method is given in section 3. Section 4 shows some experimental results. Concluding remarks are given in section 5.

N.R. Pal et al. (Eds.): ICONIP 2004, LNCS 3316, pp. 477–482, 2004.
© Springer-Verlag Berlin Heidelberg 2004

2   The Model

Consider the recurrent network in Figure 1. The input I^t is a symbol from the alphabet of a running sequence. At each time t, one symbol is put as the input to the recurrent neural network, and at the next time $t+1$, the following symbol will be put as input. In this way, the sequence is given to the neural network. After taking the input symbol, a corresponding node $h^t_{i,j}$ will be activated. Here, i means the ith order information of the Markov chain, j means the jth node with the same i value.

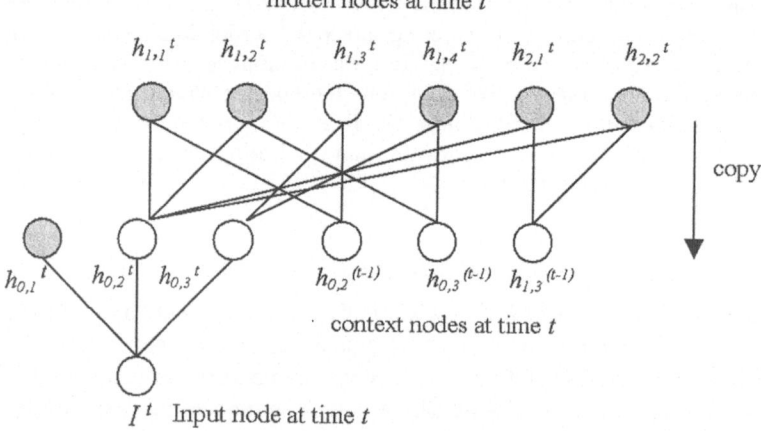

Fig. 1. Recurrent neural network structure for mixed k^{th} order Markov models

We attach a 'purity value' with each node. This purity value, represented by $S_{i,j}$, is given by:

$$S_{i,j} = lg(Pr(h_{i,j}|P)/Pr(h_{i,j}|N)), \qquad (1)$$

where, P is the set of positive samples, and N is the set of negative samples, $Pr(h_{i,j}|P)$ denotes the probability of activating node $h_{i,j}$ in the running of a positive sample sequence, while $Pr(h_{i,j}|N)$ denotes the probability of activating node $h_{i,j}$ in the running of a negative sample sequence. In practice, we use weighted frequencies to represent the probabilities:

$$Pr(h_{i,j}|P) = n_{i,j}/n_P, \qquad (2)$$
$$Pr(h_{i,j}|N) = n_{i,j}/n_N, \qquad (3)$$

where, $n_{i,j}$ is the number of times of activation of the node $h_{i,j}$, n_P and n_N denote the number of positive and negative sequences respectively. This does not promise normalization, but enough for the scoring. The use of logarithm in (1) makes the score addable, and also simplifies the result from some normalization factor.

If a score of a node is larger than a certain threshold tr, we consider the node be a 'pure' one, that is, it is enough for classification, and will need no further check. In Figure 1, the shaded nodes are 'pure' nodes.

The steps involved the classification are as follows:

- put the unclassified sequence from the first symbol, activate the connected nodes in the hidden layer.
- if the activated node is a 'pure' node, calculate the score S for sequence as $S = S + S_{i,j}$, reset all activated nodes to inactivated; else Copy the activating result of the hidden layer to the context layer.
- get the next symbol in the sequence, until reach the end, the sequence is positive, if S is positive, and negative otherwise.

The information of the sequences is captured within the context nodes in our network.

3   Construction of the Model

We now discuss a way to construct the recurrent network. We use a framework of recurrent neural network to first discover highly observed patterns of small length, and later to learn mixed k^{th} order Markov Chains. When a short pattern is considered to be 'pure' enough, the longer sequences containing the short pattern as a sub-pattern are no longer considered, because the short patterns are already enough for classification. The construction of the recurrent neural network contains the following steps:

1. For each symbol in the alphabet Σ, create a node representing it. Suppose $\Sigma = \{a, b, c\}$. Then, the sequences for classification are input to the network. Originally, the hidden nodes in the network are all non-pure nodes. As the sequences are input into the network symbol by symbol, their frequencies are calculated by adding one to the visit number each time the corresponding node is activated(Figure 2). Note that there are two such visit numbers, one for positive and the other for negative. After this, we calculate the scores of the nodes using (1). Then, mark those nodes whose score greater than the threshold tr or less than $-tr$. Suppose this time the node representing symbol a is marked as pure node.

2. From the previous scan, we know which nodes are 'pure' i.e. considered to give enough confidence to make classification. Then, the non-pure nodes will be copied for two length patterns. In the example, b^t and c^t are copied as b^t and c^t, shown in Figure 3.

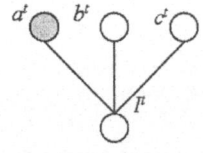

Fig. 2. Recurrent neural network structure after the first scan

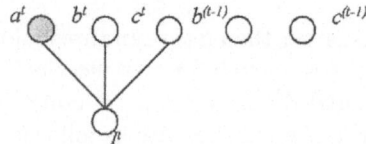

Fig. 3. Recurrent neural network structure constructing 1st order context

3. Further, four nodes are constructed to calculate first order information. Again, the sequences are input to the network. This time, the network uses the inputs in this way: if a 'pure' node is activated, the network resets all activated nodes to inactivated ones. The pruning is done in this way. This pruning results in shorter patterns having higher priority than the longer patterns. If a 'pure' node is not activated, the context nodes are copied from the corresponding hidden nodes. New input is input. If a 'pure' node is activated, the network is reset. If not, the hidden nodes combined the context nodes activate newly constructed nodes. Again, the frequencies are calculated and after the scan, 'pure' nodes are marked. After the 2nd scan, the resulting structure of the recurrent neural network is shown in Figure 4.

4. We repeat steps 2 and 3 at most $(k+1)$ times or until all the newly constructed nodes are "pure". The limit number k is used to avoid too many nodes.

In the example in Figure 4, the network copies bc^t to context nodes. In the next scan, we get the final network (Figure 5).

In the worst case, we may need a node for each pattern. Therefore, the network has $O(|\Sigma|^{(k+1)})$ nodes; in practice, the nodes are reduced by limiting

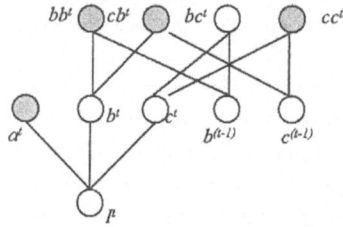

Fig. 4. Recurrent neural network after the second scan

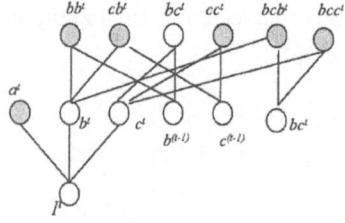

Fig. 5. Final recurrent neural network after scans

k. In the worst case, a sequence may activate all its sub-patterns, requiring $O(n |\Sigma|^{(k+1)})$ operations. We note that k and $|\Sigma|^{(k+1)}$ are constant with respect to input length n, so the time complexity is only linear in n.

4   Experimental Results

We implemented our method on a workstation with Pentium 4 CPU 2.00GHz and 523,280KB RAM running MS Windows 2000. Since classification of gene sequences is one of the main task of Markov Chains like model, in our experiments, we apply our method to do classification on the Splice-junction Gene Sequences Database, which includes gene sequences from classes "ei" (25%), "ie" (25%) and Neither (50%) on http://www.ics.uci.edu/ ∼mlearn/MLSummary.html.

We use 10 fold cross validation to test our method. The following figure shows the accuracy under difference k-levels.

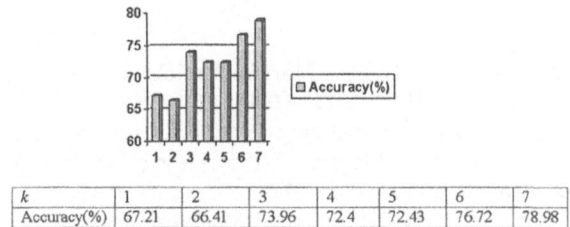

k	1	2	3	4	5	6	7
Accuracy(%)	67.21	66.41	73.96	72.4	72.43	76.72	78.98

Fig. 6. Accuracy under different k - levels

It can be seen in Figure 6 that, while the accuracy varies when k grows, greater k is likely to have more accuracy. Note that, when $k = 1$ and $k = 2$, the model degrades to HMMs and Markov Chains. Therefore, by using our method, the accuracy is shown to be increased from HMMs and Markov Chains.

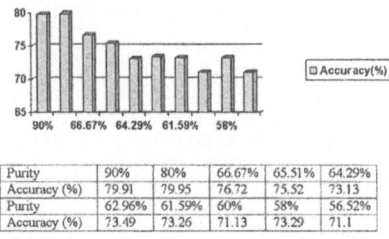

Purity	90%	80%	66.67%	65.51%	64.29%
Accuracy (%)	79.91	79.95	76.72	75.52	73.13
Purity	62.96%	61.59%	60%	58%	56.52%
Accuracy (%)	73.49	73.26	71.13	73.29	71.1

Fig. 7. Accuracy under different k - levels

The Figure 7 shows the accuracy under difference purity thresholds. It shows that when the threshold drops, the accuracy drops. Higher purity limitation means less pruning and therefore means more cost in time and space. Hence, there is a balance between the accuracy and efficiency.

5   Conclusions

We have presented an explicit method to construct recurrent neural networks for mixed k^{th} order Markov Chains, which captures highly observed $(k+1)$-length patterns that are useful for classification.

Our method uses stochastic information. However, unlike existing stochastic inference, the structure of our model is dynamically constructed during the training process instead of an arbitrary one.

Our method also captures mixed k^{th} order relationships in the sample while ignore non-typical relationships. It uses pruning to avoid falling into trivial separations and thus it prevents higher complexity. Running in the form of neural network grants our work could be adapted for parallel environments.

In the experimental results, we show that by using combined longer sequencing information, the accuracy is improved with our method.

References

1. L.R.Rabiner: A Tutorial on Hidden Markov Models and Selected Applications in Speech Recognition. Proc. IEEE 77(2) (1989) 257-286
2. C. Omlin and C.L. Giles: Constructing Deterministic Finite State Automata in Recurrent Neural Networks. Journal of the Association of Computing Machinery(JACM), Vol. 45 No. 6 (1996) 937-972
3. P. Frasconi, M. Gori and G. Soda: Injecting Nondeterministic Finite State Automata into Recurrent Networks. Tech. Rep., Dipartimento di Sistemi e Informatica, University di Firenze, Italy, Florence, Italy (1993)
4. J. Elman: Finding Structure in Time. Cognitive Science, Vol.14 (1990) 179-211
5. Y. Sakakibara and M. Golea: Simple Recurrent Networks as Generalized Hidden Markov Models with Distributed Representations. Proc. IEEE Internat. Conf. On Neural Networks (ICNN'95), IEEE Computer Society Pres, New York (1995) 979-984

An Efficient Generalization
of Battiti-Shanno's Quasi-Newton Algorithm
for Learning in MLP-Networks

Carmine Di Fiore, Stefano Fanelli, and Paolo Zellini

Department of Mathematics, University of Rome "Tor Vergata", Rome, Italy
fanelli@mat.uniroma2.it

Abstract. This paper presents a novel Quasi-Newton method for the minimization of the error function of a feed-forward neural network. The method is a generalization of Battiti's well known OSS algorithm. The aim of the proposed approach is to achieve a significant improvement both in terms of computational effort and in the capability of evaluating the global minimum of the error function. The technique described in this work is founded on the innovative concept of "convex algorithm" in order to avoid possible entrapments into local minima. Convergence results as well numerical experiences are presented.

1 Introduction

It is well known that, in order to determine the global minimum of the error function of a MLP-network, it is necessary to superimpose global optimization techniques in the computational scheme of a backpropagation algorithm. An approach of this type was utilized in [7] to derive a preliminary version of a "pseudo–backpropagation" method and in [8] to refine the global algorithm introduced in [7] thereby dealing with high-dimensional problems more efficiently.

The latter approach is founded on a new definition, involving both the mathematical properties of the error function and the behaviour of the learning algorithm. The corresponding hypotheses, called "non–suspiciousness conditions" (NSC), represent a sort of generalization of the concept of convexity. Roughly speaking, a "non–suspect" minimization problem is characterized by the fact that, under general regularity assumptions on the error function, a "suitable" pseudo-backpropagation algorithm is able to compute the optimal solution, avoiding unfair entrapments into local minima.

In [7] we proved that under the NSC and with a proper choice of the step-sizes the optimal algorithm can be obtained by a classical gradient descent-type scheme.

The present paper has several aims. Firstly, we present a generalization of OSS algorithm [2]. Secondly, we show that second order methods of Quasi-Newton(QN)-type can be implemented in the frame of the NSC theory. Thirdly, we prove that the novel algorithm, named Generalized Battiti (GB), can be successfully applied to large networks.

N.R. Pal et al. (Eds.): ICONIP 2004, LNCS 3316, pp. 483–488, 2004.
© Springer-Verlag Berlin Heidelberg 2004

A fundamental contribution towards the implementation of efficient QN-methods to MLP-networks is based on the utilization of generalized BFGS-type algorithms, named $\mathcal{L}QN$, involving a suitable family of matrix algebras $\mathcal{L}$ (see [5],[6]). The main advantage of these methods is based upon the fact that they have an $O(n \log n)$ complexity per step and that they require $O(n)$ memory allocations, where n is the number of network connections. Moreover, $\mathcal{L}QN$ methods are competitive with the most recent L-BFGS algorithms ([13],[14]), since they perform a sort of optimal second order information compression in the array of the eigenvalues of the best approximation to the Hessian matrix.

In this paper we show that a simplified variant of $\mathcal{L}QN$ methods can be easily applied in the learning process of an MLP-network. The latter variant represents a significant improvement of Battiti's algorithm and is preferable to the L-BFGS methods utilizing a comparable amount of memory.

2 Generalized BFGS-Type Algorithms

Let us consider the optimisation problem:

$$\min_{\mathbf{w} \in \Re^n} E(\mathbf{w}) \tag{1}$$

where $E(\mathbf{w})$ is the error function of an MLP-network.

Let us suppose that (1) has a solution $\mathbf{w}^*$ (global minimum) and let E_{min} denote the corresponding value of the function E.

Denote by $\nabla E(\mathbf{w})$ and by $\nabla^2 E(\mathbf{w})$ the gradient vector and the Hessian matrix of E in $\mathbf{w}$, respectively. The matrix B_{k+1}, replacing $\nabla^2 E(\mathbf{w}_{k+1})$ in the $BFGS$ method, is a rank-2 perturbation of the previous positive definite Hessian approximation B_k, defined in terms of the two current difference vectors $\mathbf{s}_k = \mathbf{w}_{k+1} - \mathbf{w}_k$ and $\mathbf{y}_k = \nabla E(\mathbf{w}_{k+1}) - \nabla E(\mathbf{w}_k)$ by the following formula:

$$B_{k+1} = \varphi(B_k, \mathbf{s}_k, \mathbf{y}_k) = B_k + \frac{1}{\mathbf{y}_k^T \mathbf{s}_k} \mathbf{y}_k \mathbf{y}_k^T - \frac{1}{\mathbf{s}_k^T B_k \mathbf{s}_k} B_k \mathbf{s}_k \mathbf{s}_k^T B_k. \tag{2}$$

The method introduced by Battiti [2], named OSS, has $O(n)$ complexity per step, being a simple memory-less modification of equation (2), in which the identity matrix I is used, instead of B_k, to compute the new approximation B_{k+1}. In [2] it is shown that OSS can be extremely competitive with the original $BFGS$ method to perform optimal learning in a large MLP-network. Unfortunately, by the very nature of the memory-less approach, the amount of second order information contained in OSS is considerably reduced in comparison with the standard $BFGS$ method.

The main problem connected to the calculation of the Hessian approximation B_{k+1} is, in fact, to minimize the computational complexity per iteration, by maintaining a QN rate of convergence.

In [9] we introduced a generalized iterative scheme, named $\mathcal{L}QN$, where the matrix B_{k+1} is defined by utilizing a suitable matrix $\tilde{B}_k$ instead of B_k ($BFGS$) or I (OSS):

$$B_{k+1} = \varphi(\tilde{B}_k, \mathbf{s}_k, \mathbf{y}_k) \tag{3}$$

We point out that in the $\mathcal{L}QN$ methods $\tilde{B}_k$ is in general a *structured dense* matrix. The method considered in [13] is instead obtained from (3) by setting :

$$\tilde{B}_k = \frac{\|\mathbf{y}_k\|^2}{\mathbf{y}_k^T \mathbf{s}_k} I \qquad (4)$$

Notice that (4) is the simplest L-BFGS algorithm (i.e. L-BFGS for $m = 1$ [1]). We study here other possible choices in the family of matrices:

$$\tilde{B}_k = \alpha_k I \qquad (5)$$

It is interesting to observe (see [10] for the rigorous proof) that *the choice of* α_k *given in* (4) *is associated to the minimum value of the condition number of* $B_{k+1} = \varphi(\alpha_k I, \mathbf{s}_k, \mathbf{y}_k)$. In [13],[14] the latter choice was only justified by weaker arguments based on experimental observations.

In this paper, we have implemented in particular the following values of α_k:

$$\alpha_k^* = \left(1 - \frac{1}{n}\right)\alpha_{k-1}^* + \frac{1}{n}\frac{\|\mathbf{y}_{k-1}\|^2}{\mathbf{y}_{k-1}^T \mathbf{s}_{k-1}} \qquad (6)$$

$$\begin{cases} \alpha_k^{**} = \alpha_{k-1}^{**} + \frac{1}{n-1}\left(\frac{\|\mathbf{y}_{k-1}\|^2}{\mathbf{s}_{k-1}^T \mathbf{y}_{k-1}} - \frac{\mathbf{v}_k^T \mathbf{v}_k}{\mathbf{s}_k^T \mathbf{v}_k}\right); \\ \mathbf{v}_k = \alpha_{k-1}^{**}\left(\mathbf{s}_k - \frac{\mathbf{s}_{k-1}^T \mathbf{s}_k}{\|\mathbf{s}_{k-1}\|^2}\mathbf{s}_{k-1}\right) + \frac{\mathbf{y}_{k-1}^T \mathbf{s}_k}{\mathbf{s}_{k-1}^T \mathbf{y}_{k-1}}\mathbf{y}_{k-1} \end{cases} \qquad (7)$$

The formula (6) can be derived by minimizing $\|\alpha_k I - \varphi(\alpha_{k-1}^* I, \mathbf{s}_{k-1}, \mathbf{y}_{k-1})\|_F$, being $\|\cdot\|_F$ the Frobenius norm. *We underline that the resulting matrix* $\alpha_k^* I$ *is the best least squares fit* $\mathcal{L}_{B_k}$ *of* B_k *in the space* $\mathcal{L} = \{zI : z \in C\}$ (see [9]), i.e. (6) gives rise to a $\mathcal{L}QN$ method. The formula (7) is obtained by solving the minimum problem:

$$\min_{\alpha_k} \|\varphi(\alpha_k I, \mathbf{s}_k, \mathbf{y}_k) - \varphi(\varphi(\alpha_{k-1}^{**}I, \mathbf{s}_{k-1}, \mathbf{y}_{k-1}), \mathbf{s}_k, \mathbf{y}_k)\|_F,$$

or, in other words, by minimizing the distance between B_{k+1} and the matrix defining the search direction in the L-BFGS algorithm, for m=2. In Section 4, we shall compare the performances of the algorithms based on the use of (5)-(6) and (5)-(7) with the original OSS and (4).

3   Theoretical Results

The following assumptions, named "non–suspiciousness conditions" were originally introduced in [3], [12] and redefined in a more suitable form in [7].

Definition 1. *The non-suspiciousness conditions hold if* $\exists \lambda_k$:
   1. $\forall \epsilon_a \in \Re^+, \exists \epsilon_s \in \Re^+ : \|\nabla E(\mathbf{w}_k)\| > \epsilon_s$ *during the optimization algorithm, apart from* k: $E(\mathbf{w}_k) - E_{min} < \epsilon_a$;
   2. $\lambda_k \|\nabla E(\mathbf{w}_k)\|^2 \leq \epsilon_a$;
   3. $E \in C^2$ *and* $\exists H > 0 : \|\nabla^2 E(\mathbf{w})\| \leq H$.
   The following convergence result was proved in [12](see also [4]):

Theorem 1. *Let the non-suspiciousness conditions hold for problem* (1). *Then,* $\forall \epsilon_a \in \Re^+, \exists k^{**} : \forall k > k^{**}$:

$$E(\mathbf{w}_k) - E_{min} < \epsilon_a, \tag{8}$$

by using the gradient descent-type scheme: $\mathbf{w}_{k+1} = \mathbf{w}_k - \lambda_k \nabla E(\mathbf{w}_k)$.

Let E.S.W. be the following λ-subset of the classical Strong Wolfe conditions

$$\begin{cases} E(\mathbf{w}_k + \lambda \mathbf{d}_k) \leq E(\mathbf{w}_k) + c_1 \lambda \nabla E(\mathbf{w}_k)' \mathbf{d}_k, \\ c_3 |\nabla E(\mathbf{w}_k)' \mathbf{d}_k| \leq |\nabla E(\mathbf{w}_k + \lambda \mathbf{d}_k)' \mathbf{d}_k| \leq c_2 |\nabla E(\mathbf{w}_k)' \mathbf{d}_k| \end{cases} \tag{9}$$

being $\lambda > 0$ and $0 < c_1 \leq c_3 < c_2 < 1$ proper constants. E.S.W. *is not empty whenever* $\mathbf{d}_k$ *is a descent direction* (see [10],[14]). It can be proved (see again [10]) the fundamental:

Theorem 2. *Let the condition* 1. *of* **Definition 1.** *hold for problem (1). Moreover, assume that* $E \in C^2$ *and* $\exists m, M > 0$:

$$\frac{\|\mathbf{y}_k\|}{\|\mathbf{s}_k\|} \geq m \quad ; \quad \frac{\|\mathbf{y}_k\|^2}{\mathbf{y}_k^T \mathbf{s}_k} \leq M \tag{10}$$

Then, $\forall \epsilon_a \in \Re^+, \exists k^{**} : \forall k > k^{**}$:

$$E(\mathbf{w}_k) - E_{min} < \epsilon_a, \tag{11}$$

by using the QN-type scheme:

$$\begin{cases} \mathbf{w}_{k+1} = \mathbf{w}_k - \lambda_k B_k^{-1} \nabla E(\mathbf{w}_k) \\ B_k = \varphi(\tilde{B}_{k-1}, \mathbf{s}_{k-1}, \mathbf{y}_{k-1}) \end{cases} \tag{12}$$

Theorem 2 shows that QN-type methods can be implemented in the frame of the NSC theory. It is important to emphasize that the scalars λ_k, evaluated by (9) in connection with (10), perform a sort of *experimental translation* of the condition 1. of **Definition 1.** (see once again [10]). Observe that, if E were a convex function, the second inequality in (10) would be satisfied ([9]). Furthermore, the first inequality implies that $\|\nabla^2 E(\xi_k)\| \geq m$, being $\xi_k = \mathbf{w}_k + t(\mathbf{w}_{k+1} - \mathbf{w}_k), 0 < t < 1$. This fact justifies the following:

Definition 2. *A QN-method satisfying the conditions* (10) *is called a convex algorithm*(see [10] for more details).

4   Experimental Results

In this section we study the local convergence properties of the algorithms described in section 2. In particular, it is shown that the novel values of α_k given in (6),(7) are the most competitive. Some well known non-analytical tasks taken from UCI Repository of machine learning databases (IRIS and Ionosphere [11]) are selected as benchmarks. We consider the training of *i-h-o* networks where i, h and o are the number of input, hidden and output nodes, respectively.

Since thresholds are associated with hidden and outer nodes, the total number of connections (weights) is $n = ih + ho + h + o$. In the learning process of IRIS and Ionosphere n= 315, 1408, respectively (for more details see [5]). CPU-time is referred to a Pentium 4-M, 2 GHz with a machine precision of $.1 \times 10^{-18}$.

LN, GB1, GB2 indicate the algorithms utilizing the values of α_k given in (4), (6) and (7), respectively. In all the algorithms we implement the same line-search technique, i.e. the classical efficient Armijo-Goldstein (A.G.) conditions ([7],[8]).

Define $p_k = \frac{\|\mathbf{y}_k\|^2}{\mathbf{y}_k^T \mathbf{s}_k}$

$$
\begin{aligned}
&\mathbf{w}_0 \in \Re^n, \; \mathbf{d}_0 = -\nabla E(\mathbf{w}_0) \\
&For \; k = 0, 1, \ldots : \\
&\left\{
\begin{aligned}
&\mathbf{w}_{k+1} = \mathbf{w}_k + \lambda_k \mathbf{d}_k, \quad \lambda_k \in A.G. \\
&\mathbf{s}_k = \mathbf{w}_{k+1} - \mathbf{w}_k, \; \mathbf{y}_k = \nabla E(\mathbf{w}_{k+1}) - \nabla E(\mathbf{w}_k) \\
&\mathbf{define} \; \alpha_k : \\
&OSS : \; \alpha_k = 1 \\
&LN : \; \alpha_k = p_k \\
&GB1 : \; if(k = 0)\{\alpha_k = 1 \; or \; \alpha_k = p_k\}else\{ \\
&\qquad \alpha_k = \left(1 - \tfrac{1}{n}\right)\alpha_{k-1} + \tfrac{1}{n}\frac{\|\mathbf{y}_{k-1}\|^2}{\mathbf{y}_{k-1}^T \mathbf{s}_{k-1}}; \\
&\qquad \} \\
&GB2 : \; if(k = 0)\{\alpha_k = 1 \; or \; \alpha_k = p_k\}else\{ \\
&\qquad \mathbf{v}_k = \alpha_{k-1}\left(\mathbf{s}_k - \frac{\mathbf{s}_{k-1}^T \mathbf{s}_k}{\|\mathbf{s}_{k-1}\|^2}\mathbf{s}_{k-1}\right) + \frac{\mathbf{y}_{k-1}^T \mathbf{s}_k}{\mathbf{s}_{k-1}^T \mathbf{y}_{k-1}}\mathbf{y}_{k-1}; \\
&\qquad \alpha_k = \alpha_{k-1} + \tfrac{1}{n-1}\left(\frac{\|\mathbf{y}_{k-1}\|^2}{\mathbf{s}_{k-1}^T \mathbf{y}_{k-1}} - \frac{\mathbf{v}_k^T \mathbf{v}_k}{\mathbf{s}_k^T \mathbf{v}_k}\right); \\
&\qquad \} \\
&B_{k+1} = \varphi(\alpha_k I, \mathbf{s}_k, \mathbf{y}_k) \\
&\mathbf{d}_{k+1} = -B_{k+1}^{-1}\nabla E(\mathbf{w}_{k+1})
\end{aligned}
\right.
\end{aligned}
$$

The following table reports the number of iterations (the seconds) required by the algorithms to obtain $E(\mathbf{w}_k) < 10^{-1}$, where E is the error function of the corresponding MLP. Experiments are related to different initial weights of IRIS and Ionosphere networks. Notice that GB1 and GB2 are often more efficient than LN. *OSS in some cases performs less iterations, but it is always dominated in terms of CPU time* . The latter experimental result depends upon the fact that, in order to evaluate $\lambda_k \in$ A.G., OSS requires more computational effort. The general $\mathcal{L}QN$, using *dense and structured matrices* $\mathcal{L}$ ($\mathcal{L}$=Hartley algebra, see [9]), outperforms all the other algorithms.

5   Conclusions

All GB-algorithms examined in Table 1 require $6n$ memory allocations (see [14] for L-BFGS $m = 1$). Moreover, the computational complexity per step is cn, being c a very small constant. Further experimental results have shown that the use of L-BFGS methods, for $m = 2$ or $m = 3$, does not reduce the number of iterations. Since minimizing the error function of a neural network is a nonlinear least squares problem, one could use the well known Levenberg-Marquardt(LM)

Table 1. k (seconds): $f(\mathbf{x}_k) < 10^{-1}$.

Algorithms	iris1	iris2	iono1	iono2	iono3
OSS	49528 (1037)	42248 (863)	3375 (277)	2894 (236)	2977 (245)
LN	41291 (361)	40608 (358)	4054 (165)	5156 (212)	5171 (215)
GB1, $\alpha_0 = 1$	24891 (207)	34433(286)	4415 (174)	3976 (157)	4159 (164)
GB2, $\alpha_0 = 1$	20086 (166)	17057 (142)	3821 (152)	4519 (178)	3794 (149)
General $\mathcal{LQN}$	12390 (112)	15437 (140)	993 (49)	873 (43)	1007 (48)

method. Unfortunately, LM needs at least $O(n^2)$ memory allocations and its implementation requires the utilization of more expensive procedures than GB algorithms (i.e. Givens and Householder for QR factorizations). As a matter of fact, the original algorithm OSS turns out to be much more efficient than LM for large scale problems (see [15]).

References

1. M.Al Baali, Improved Hessian approximations for the limited memory BFGS method, *Numer. Algorithms*, Vol. 22, pp.99–112, 1999.
2. R. Battiti, First- and second-order methods for learning: between steepest descent and Newton's method, *Neural Computation*, Vol. 4, pp. 141–166, 1992.
3. M. Bianchini, S. Fanelli, M.Gori, M.Protasi, Non-suspiciousness: a generalisation of convexity in the frame of foundations of Numerical Analysis and Learning, *IJCNN'98*, Vol.II, Anchorage, pp. 1619–1623, 1998.
4. M.Bianchini, S.Fanelli, M.Gori, Optimal algorithms for well-conditioned nonlinear systems of equations, *IEEE Transactions on Computers*, Vol. 50, pp. 689-698, 2001.
5. A.Bortoletti, C.Di Fiore, S.Fanelli, P.Zellini, A new class of quasi-newtonian methods for optimal learning in MLP-networks, *IEEE Transactions on Neural Networks*, Vol. 14, pp. 263–273, 2003.
6. C.Di Fiore, S.Fanelli, P.Zellini, Matrix algebras in quasi-newtonian algorithms for optimal learning in multi-layer perceptrons, *ICONIP Workshop and Expo*, Dunedin, pp. 27–32, 1999.
7. C.Di Fiore, S.Fanelli, P.Zellini, Optimisation strategies for nonconvex functions and applications to neural networks, *ICONIP 2001*, Vol. 1, Shanghai, pp. 453–458, 2001.
8. C. Di Fiore, S.Fanelli, P.Zellini, Computational experiences of a novel algorithm for optimal learning in MLP-networks, *ICONIP 2002*, Vol. 1, Singapore, pp. 317–321, 2002.
9. C. Di Fiore, S. Fanelli, F. Lepore, P. Zellini, Matrix algebras in Quasi-Newton methods for unconstrained optimization, *Numerische Mathematik*, Vol. 94, pp. 479–500, 2003.
10. C. Di Fiore, S.Fanelli, P.Zellini, Convex algorithms for optimal learning in MLP-networks, *in preparation*
11. R.O. Duda, P.E. Hart, *Pattern Classification and Scene Analysis*, Wiley, 1973.
12. P. Frasconi, S. Fanelli, M. Gori, M. Protasi, Suspiciousness of loading problems, *IEEE Int. Conf. on Neural Networks*, Vol. 2, Houston, pp. 1240–1245, 1997.
13. D.C. Liu, J. Nocedal, On the limited memory BFGS method for large scale optimization, *Math. Programming*, Vol. 45, pp.503–528, 1989.
14. J. Nocedal, S.J. Wright, *Numerical Optimization*. New York: Springer-Verlag, 1999
15. http://www.mathworks.com/access/helpdesk/help/toolbox/nnet/backpr14.html

Incremental Learning and Dimension Selection Through Sleep

Koichiro Yamauchi

Graduate School of Information Science and Technology, Hokkaido University
Kita ku, Kita 13 Jyou Nishi 8 Chou, Sapporo, 060-8628, Japan
yamauchi@complex.eng.hokudai.ac.jp

Abstract. Dimension selection from high-dimensional inputs is very important in improving the performance of machine-learning systems. Dimension selection via learning, however, usually requires a large number of iterations to find an appropriate solution. This means that learning involving dimension-selection is hard to achieve in an incremental manner. An improved version of the previous work to overcome this problem is presented, namely, "Incremental Learning with Sleep"(ILS). This system repeats two learning phases, awake and sleep phases, alternately. The system records new novel instances like instance-based learning during the awake phase, whereas it achieves both an optimization of parameters and dimension selection during the sleep phase. Because the learning of the awake phase is very quick, the system's apparent learning time is very short.

The experimental results show the extended ILS allows learning inputs incrementally but ignores redundant dimensions during the learning.

1 Introduction

Incremental learning is important for systems that have to achieve recognition of known instances and learn unknown instances simultaneously. Normally, however, the systems usually forget past instances due to their learning new instances. To avoid forgetting, the systems have to re-learn past instances. However, the re-learning process sometimes wastes learning time.

However, instance-based learning systems such as k-Nearest Neighbors learn new instances only by appending them to the database so that the systems do not waste learning time. However, the systems waste a huge amount of resources to record all instances.

To achieve a balance between accelerating the learning speed and reducing the amount of resources, several systems have been presented. [1] [2]. These systems memorize unknown instances quickly by using a fast-learning network (F-Net) in the same way as k-NN during the awake phase. Also a slow-learning network (S-Net) learns pseudo patterns generated by the F-Net with a small number of hidden units during the sleep phase.

In this paper, an extension of the ILS to accomplish dimension selection during the sleep phase is presented.

N.R. Pal et al. (Eds.): ICONIP 2004, LNCS 3316, pp. 489–495, 2004.
© Springer-Verlag Berlin Heidelberg 2004

2   Neural Network Used

In this study, the system uses two types of Gaussian Radial Basis Function Networks [3]. The first one consists of isotropic Gaussian-basis functions; the other consists of aerotropy Gaussian-basis functions.

Let $f_i^*(x, \theta)$ be the ith output value of the network to input vector x: $f_i^*(x, \theta) = \sum_\alpha c_{i\alpha} \phi_\alpha(x)$, where θ denotes the parameter vector, which represents whole parameters in the network. $c_{i\alpha}$ denotes the connection strength between the αth hidden unit and the ith output unit. $\phi_\alpha(x)$ is the α-th hidden unit's output. The first and second types of hidden outputs are

$$\phi_\alpha(x) = \exp\left(-\frac{\sum_i (x_i - u_{\alpha i})^2}{2\sigma_\alpha^2}\right) , \ \phi_\alpha(x) = \exp\left(-\frac{\sum_i w_{\alpha i}^2 (x_i - u_{\alpha i})^2}{2\sigma_\alpha^2}\right), \quad (1)$$

respectively. In the second type, $w_{\alpha i}$ denotes the importance of the ith input dimension. $u_{\alpha i}$ and σ_α denote the center and standard deviation. During the learning, $w_{\alpha i}$, $u_{\alpha i}$ and σ_α are modified.

The first type of hidden unit is used only in the F-Net, and the second is used in both the F- and S-Net.

3   System Structure

The new system described here consists of two Gaussian Radial Basis Function networks: an F-Net and an S-Net (Fig. 1). The system has two learning phases, the awake and the sleep phases. During the awake phase, the system recognizes instances presented, and the system output mainly comes from the S-Net. If the S-Net's error in the current instance exceeds a threshold, the F-Net learns the instance quickly, in the same way as k-NN[1], to compensate for the S-Net's error. The F-Net output is added to the S-Net output. As a result, the system output is always close to the desired output.

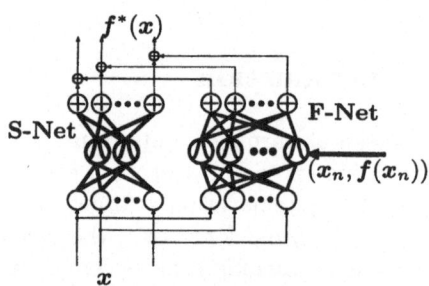

Fig. 1. Structure and Behavior of the system

Note that, the F-Net uses an isotropic Gaussian Basis Function as mentioned in the previous section to achieve the k-NN-like learning.

In contrast, during the sleep phase, the S-Net learns pseudo instances generated by the F-Net to bring its output close to that of the F-Net with a small number of hidden units. The learning method is a modified version of the Minimum Resource Allocating Network (MRAN) [4], which uses a growing strategy

---

[1] If the RBF learns each instance with one hidden unit, the RBF shows k-NN-like behavior.

that prunes redundant cells, as described later. The modified MRAN learns instances normally, even if the distribution of inputs is unstable.

The S-Net continues learning until its outputs become close enough to those of the F-Net. After that, all hidden units in the F-Net are removed. The S-Net uses the aerotropy (the second type of) Gaussian-basis functions as described in the previous section. Through the sleep phase, each hidden unit in the S-Net adjusts the weight parameter $W_{\alpha j}$, namely the importance of each dimension, to ignore redundant dimensions.

Strictly speaking, the S-Net learning process is applied to only some of the hidden units and corresponding weights between the hidden and the output units. In the following text, we call the hidden units "plastic hidden units." The plastic hidden units are units activated during the adjacent awake phase. During the S-Net learning, outputs from non-plastic hidden units are fixed to zero.

Furthermore, prior to S-Net learning, parameters of the plastic hidden units are copied to the F-Net to make the F-Net output agree with the desired S-Net output. Note that the S-Net memory is not interfered with by this learning, since S-Net's non-plastic hidden units do not change its parameters.

F-Net learning is, however, arrested during the S-Net learning. Therefore, the system stops reading new novel instances during the learning period, resulting in alternating learning processes for the S- and F-Net.

In the following text, $f_f^*(x, \theta_f)$ and $f_s^*(x, \theta_s)$ denote the output vectors of the F-Net and S-Net. Here, θ_* and x denote the parameter vector of the two networks and the input vector, respectively.

$$
\begin{aligned}
&\underline{\textbf{foreach }} x_n \\
&\underline{\textbf{if }}\ \min_{S\alpha} \|x_n - u_{S\alpha}\| > \epsilon_D \text{ and} \\
&\quad \min_{F\alpha} \|x_n - u_{F\alpha}\| > \epsilon_D \text{ and} \\
&\quad \|D(x_n) - f^*(x_n)\|^2 > \epsilon_E \quad \underline{\textbf{then}} \\
&\quad \text{allocate new cell } F\alpha \\
&\quad u_{F\alpha} = x_n \\
&\quad c_{F\alpha} = D(x_n) - f^*(x_n) \\
&\quad \sigma_{F\alpha} = \min\left(\lambda_F \min_{S\alpha} \|x_n - u_{S\alpha}\|,\right. \\
&\quad\quad\quad\quad \left. \lambda_F \min_{F\beta \neq F\alpha} \|x_n - u_{F\beta}\|, \sigma_{\max}\right) \\
&\underline{\textbf{else}} \\
&\quad \theta_f := \theta_f - \epsilon \nabla_{\theta_f} \|D(x_n) - f^*(x_n)\|^2
\end{aligned}
$$

Fig. 2. Learning Procedure of F-Net

4   Awake Phase

During the awake phase, the system recognizes known inputs by yielding the sum of the outputs from the F-Net and S-Net while the F-Net learns unknown instances quickly. Let $f^*(x)$ and $D(x)$ be the final output of the system and the desired output to x, where $f^*(x) = f_f^*(x, \theta_f) + f_S^*(x, \theta_S)$. The F-Net learns a new unknown instance $(x_n, D(x_n))$ as shown in Fig. 2. The learning method for the F-Net is based on normal gradient descent learning. The F-Net allocates new basis functions in the same way as a resource allocating network (RAN) [5]. The allocation procedure is similar to that of the RAN; however, the allocation condition is not restricted by the learning epochs. Thus, changing the input distribution becomes possible. Note that the allocation condition depends not only on the distribution of the basis function of the F-Net, but also on that of the S-Net. In Fig. 2, the suffixes 'S' and 'F' indicate the parameters of the S- and F-Net.

5   Sleep Phase

During the sleep phase, the S-Net not only learns the outputs from the F-Net but also prunes redundant hidden units and input connections. However, the F-Net output is not the desired output for the S-Net but is, in fact, S-Net error. Therefore, the system converts the output to the desired output by copying some of the hidden units in the S-Net to the F-Net. The copied hidden units are the "plastic hidden units" that were activated during the adjacent awake phase.

As a result, the output becomes the desired output for the S-Net, which reflects not only the new instances but also the old memories that will be re-learned to avoid forgetting.

Note that the outputs of the non-plastic hidden units are fixed to zero during this learning to ensure consistency. Consequently, the pa-rameters of the non-plastic hidden units of the S-Net are not changed during this learning.

```
for each pseudo instance (x̂_F, f*_F(x̂_F))
  for each S_j-th hidden unit
  if ||x̂ − u_{Sα}|| < σ_{Sα} then
    Calculate the contributing ratio r_{S_j}(x̂_F)
    (see text.)
    if r_{S_j}(x̂_F) < δ for M consecutive x̂_F then
      prune the S_j-th hidden unit;
```

Fig. 3. Pruning Algorithm

The S-Net then begins to learn pseudo patterns $(\hat{x}_F, f_F^*(\hat{x}_F))$ generated by the F-Net, where $\hat{x}_F$ is the pseudo input vector that is generated by adding random noise to the centroid u_{F_j} of each hidden unit.

The system randomly chooses one hidden unit in the F-Net every time one pseudo input is generated. If the system chooses the j-th hidden unit, each element of the pseudo input vector $\hat{x}_F$ is $\hat{x}_{Fi} = u_{Fji} + \sigma_{F_{max_{ji}}}\kappa$, where κ is a random value in the interval [0,1]. Here, $\sigma_{F\,max_{ji}}$ denotes the maximum distance between the ith element of the input vector and that of the centroid of the j-th hidden unit. $\sigma_{F\,max_{ji}}$ is determined during the awake phase as follows. If the j-th hidden unit center is the closest to the current input and $\sigma_{F\,max\,ji} < |u_{Fji}-x_{ni}|$, then $\sigma_{F\,max\,ji}$ is updated to be the current distance: $\sigma_{F\,max\,ji} := |u_{Fji} - x_{ni}|$.

The S-Net's learning method is a modified version of the Minimum Resource Allocating Network (MRAN) learning algorithm [4], which is online learning with a pruning strategy.

> **for** $n = 1$ **to** $N_{optimize}$ **times**
>
>   <u>Get</u> pseudo pattern $\hat{x}_F$ from the F-Net.
>
> $$e_{rmsn} := \sqrt{\frac{\sum_{i=t-M}^{T} \|f_F^*(\hat{x}_{Fi}) - f_S^*(\hat{x}_{Fi})\|^2}{M}}$$
>
> $e_n = \max\{\epsilon_{max}\gamma^n, \epsilon_{min}\}$, where $\gamma < 1$ is a decay
> constant
>
>   **if** $\min_{S\alpha} \|x_F - u_{S\alpha}\| > e_n$ and
> $\|f_F^*(\hat{x}_F) - f_S^*(\hat{x}_F)\| > e_E$ and
> $e_{rmsn} > e'_{rmsn}$ and
>   Number of cells is less than that of F-Net
>   **then**
>   allocate new cell $S\alpha$
> $u_{S\alpha} = \hat{x}_F$, $c_{S\alpha} = f_F^*(\hat{x}_F) - f_S^*(\hat{x}_F)$
> $\sigma_{S\alpha} = \min(\lambda_S \min_{S\beta} \|\hat{x}_F - u_{S\beta}\|, \sigma_{max})$
> $w_{\alpha i} = 1$
>   **else**
> $u_{S\alpha} := u_{S\alpha} - \varepsilon \nabla u_{S\alpha} \|f(\hat{x}_F) - f^*(\hat{x}_F)\|^2$
> $c_{S\alpha} := c_{S\alpha} - \varepsilon \nabla c_{S\alpha} \|f(\hat{x}_F) - f^*(\hat{x}_F)\|^2$
> $\sigma_{S\alpha} := \sigma_{S\alpha} - \varepsilon \nabla \sigma_{S\alpha} \|f(\hat{x}_F) - f^*(\hat{x}_F)\|^2$
> $w_{S\alpha i} := w_{S\alpha i} - \varepsilon \nabla w_{S\alpha i} \|f(\hat{x}_F) - f^*(\hat{x}_F)\|^2$
>   **endif**
> **Execute the pruning algorithm**(see text)
> **endfor**

Fig. 4. S-Net Learning procedure

In the pruning strategy, the S-Net removes any hidden units whose contribution ratio is smaller than a threshold. The contribution ratio of the S_j-th hidden unit is the relative magnitude of outputs, as per $r_j(\hat{x}_{F_k}) = \|c_j \phi_{S_j}(\hat{x}_{F_k})\|/ \max_i \|c_i \phi_{S_i}(\hat{x}_{F_k})\|$, where c_j is the connection strength vector between the S_j-th hidden unit and the output units. The distribution of pseudo inputs is varied due to the change in the number of hidden units. The original M-RAN pruning method, however, usually disposes hidden units that are needed if the distribution of $\hat{x}_F$ inputs is varied. To overcome this shortcoming, the ratio of confidence of each hidden unit is only measured when the unit is activated. Therefore, the contribution ratio $r_j(\hat{x}_F)$ is only estimated when $\|\hat{x} - u_{S\alpha}\| < \sigma_{S\alpha}$, where $\sigma_{S\alpha}$ is the standard deviation of the hidden unit. The summarized pruning algorithm is given in Fig. 3.

Figure 4 shows the whole S-Net learning procedure. In this figure, e_{rmsn} denotes the mean square error of the S-Net, and δ denotes a performance-threshold, which determine the ratio of the pruning of hidden units.

The learning process is repeated $N_{optimize}$ times, where $N_{optimize}$ is set N_o times the number of hidden F-Net units. In the experiment that will be described next, N_o was set to 200. After the system learns with the S-Net, it removes all hidden F-Net units. Then, the awake phase is restarted.

6   Experiment

The performances of the new system (ILS-new) and the conventional ILS (ILS-old), which uses only the isotropic Gaussian-basis functions, were examined experimentally with respect to two datasets.

First of all, the systems were tested using the 1st dataset: $D = 2\exp(-0.8(x_1-3)^2)+2\exp(-0.8(x_2-3)^2)$ in the interval $0 \leq x_* \leq 6$. Note that (x_1, x_2, D) shows the cross-bar-like patterns so that D does not partly depend on x_1 or x_2. During the 1st and 2nd awake phases, ILS-new and -old learned the datasets in $0 \leq x_1 \leq 3$ and $3 \leq x_1 \leq 6$, respectively. Figure 5 shows the outputs of ILS-new after the 1st and 2nd sleep phases. The learning samples during corresponding adjacent awake phase are also plotted in this figure. We can see that ILS-new learned the datasets incrementally. The outputs of ILS-old were almost the same as those of ILS-new.

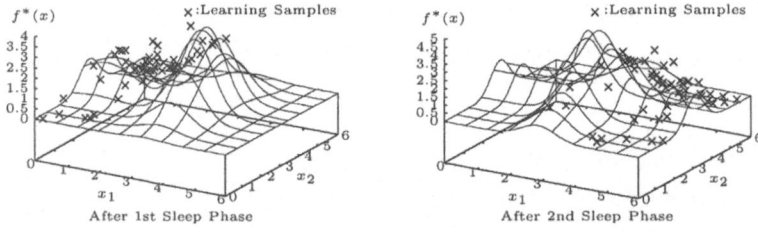

Fig. 5. ILS-new outputs immediately after the 1st and 2nd sleep phases

Figure 6 shows the resulting hidden units of ILS-new and -old after the 2nd sleep phase. We can see that ILS-new learned the datasets with only two hidden units, each of which reduces the weight for x_1 or x_2. However, ILS-old wastes 5 hidden units to learn the dataset.

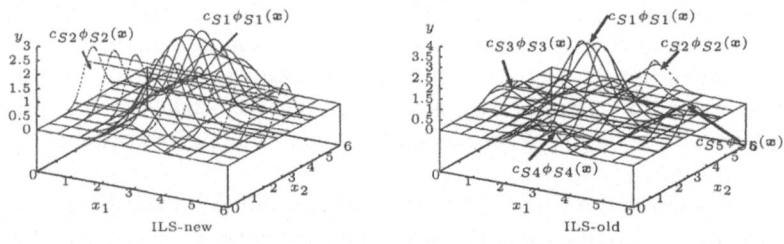

Fig. 6. Resulting S-Net hidden units after 2nd sleep phase of ILS-NEW and -OLD

Next, the performances of the two systems with respect to the servo datasets from the UCI Machine Learning Repository[2] were examined. The performances

---

[2] http://www1.ics.uci.edu/~mlearn/MLRepository.html

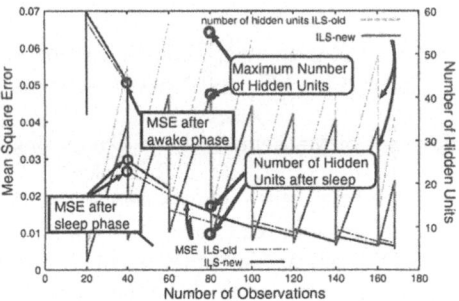

Fig. 7. Mean Square Error(MSE) and Number of Hidden Units of ILS-new and -old to Servo data-set: During each awake phase, the systems learned 20 new samples. MSE was examined after each awake and sleep phases

immediately after every awake and sleep phase were evaluated with $E_{all}(t) = \sum_k^{all} SqureErr(x_k, \theta(t))/$ Data-Size. Note that $E_{all}(t)$ reflects both the ratio of forgetting and the generalization ability.

The number of hidden units was also evaluated. However, the learning tasks require additional resources. So, the maximum and minimum number of hidden units were evaluated. We can see from Fig. 7 that the mean square errors of ILS-new and -old are almost the same, but the number of ILS-new's hidden units after each sleep phase is about half that of ILS-old.

7   Conclusion

An extended version of "Incremental Learning with Sleep (ILS)" was presented. The system achieves dimension-selection in ignoring redundant input dimensions during the sleep phase. Experimental results show that the new system incrementally learns multi-dimensional patterns with fewer hidden units than that of the conventional ILS. This is a very important property in achieving high generalization ability for the incremental learning of high-dimensional patterns.

References

1. K. Yamauchi, S. Itho, and N. Ishii. In *IEEE SMC'99 1999 IEEE System, Man and Cybernetics Conference*, volume III, pages 390–395, October 1999.
2. K. Yamauchi. In *International Joint Conference on Neural Networks, IJCNN2003*, volume 4, pages 2776–2781, July 2003.
3. T. Poggio and F. Girosi. *Proceeding of the IEEE International Conference on Neural Networks*, 78(9):1481–1497, September 1990.
4. Lu Yingwei, N. Sundararajan, and P. Saratchandran. *Neural Computation*, 9:461–478, 1997.
5. J.Platt. *Neural Computation*, 3(2):213–225, 1991.

The Most Robust Loss Function for Boosting

Takafumi Kanamori[1], Takashi Takenouchi[2],
Shinto Eguchi[3], and Noboru Murata[4]

[1] Department of Mathematical and Computing Sciences
Tokyo Institute of Technology
Ookayama 2-12-1, Meguro-ku, Tokyo 152-8552, Japan
kanamori@is.titech.ac.jp
[2] Institute of Statistical Mathematics, 4-6-7, Minami-Azabu
Minato-ku, Tokyo 106-8569, Japan
ttakashi@ism.ac.jp
[3] Institute of Statistical Mathematics, 4-6-7, Minami-Azabu
Minato-ku, Tokyo 106-8569, Japan
eguchi@ism.ac.jp
[4] School of Science and Engineering, Waseda University
3-4-1 Ohkubo, Shinjuku, Tokyo 169-8555, Japan
noboru.murata@eb.waseda.ac.jp

Abstract. Boosting algorithm is understood as the gradient descent
algorithm of a loss function. It is often pointed out that the typical
boosting algorithm, Adaboost, is seriously affected by the outliers. In
this paper, loss functions for robust boosting are studied. Based on a
concept of the robust statistics, we propose a positive-part-truncation
of the loss function which makes the boosting algorithm robust against
extreme outliers. Numerical experiments show that the proposed boosting algorithm is useful for highly noisy data in comparison with other
competitors.

1 Introduction

We study a class of loss functions for boosting algorithms in binary classification problems. Boosting is a learning algorithm to construct a predictor by
combining, what is called, *weak hypotheses*. Adaboost [3] is a typical implementation of boosting and is shown to be a powerful method from theoretical and
practical viewpoints. The boosting algorithm is derived by the gradient method
[4]. Through these studies, the boosting algorithm is viewed as an optimization
process of the loss function. The relation between Adaboost and the maximum
likelihood estimator was also clarified from the viewpoint of the information
geometry [6, 8].

Since a loss function plays an important role in statistical inference, the relation between the loss function and the prediction performance is widely studied
in statistics and machine learning communities. In last decade some useful loss
functions for classification problems have been proposed, for examples, the hinge
loss for support vector machine [11], the exponential loss for Adaboost [4] and so

N.R. Pal et al. (Eds.): ICONIP 2004, LNCS 3316, pp. 496–501, 2004.
© Springer-Verlag Berlin Heidelberg 2004

on. Since these typical loss functions are convex, highly developed optimization techniques is applicable to the global minimization of the loss function.

We study the robustness of boosting. Adaboost puts too much weights on the solitary examples even though these are outliers which should not be learned. For example, outliers may occur in recording data. Some studies revealed that the outlier seriously degrades the generalization performance of Adaboost. Though several improvements to recover the robustness are already proposed, the theoretical understanding is not enough. We measure the influence of the outlier by *gross error sensitivity* [5] and derive the loss function which minimizes the gross error sensitivity. Our main objective is to propose a boosting algorithm that is robust for outliers.

This paper is organized as follows. In section 2, we briefly introduce boosting algorithms from the viewpoint of optimization of loss functions. In section 3, we explain some concepts in the robust statistics and derive robust loss functions. Numerical experiments are illustrated in section 4. The last section is devoted to concluding remarks.

2   Boosting Algorithms

Several studies have clarified that boosting algorithm is derived from the gradient descent method for loss functions [4, 7]. The derivation is illustrated in this section.

Suppose that a set of examples $\{(x_1, y_1), \ldots, (x_n, y_n)\}$ is observed, where x_i is an element in the input space $\mathcal{X}$ and y_i takes 1 or -1 as the class label. We denote the set of weak hypothesis by $\mathcal{H} = \{h_t(x) : \mathcal{X} \to \{1, -1\} \mid t = 1, \ldots, T\}$, where each hypothesis assigns the class label for the given input.

Our aim is to construct a powerful predictor $H(x)$ by combining weak hypotheses, where $H(x)$ is given as the linear combination of weak hypotheses, that is,

$$H(x) = \sum_{t=1}^{T} \alpha_t h_t(x).$$

Using the predictor, we can assign the class label corresponding to x as $\text{sign}(H(x))$, where $\text{sign}(z)$ denotes the sign of z.

Loss functions are often used in the classification problems. The loss of the predictor H given a sample (x, y) is defined as $l(-yH(x))$, where $l : R \to R$ is twice continuously differentiable except finite points. Typically convex and increasing functions are used because of the facility of the optimization. Let us define the *empirical loss* as

$$L_{emp}(H) = \frac{1}{n} \sum_{i=1}^{n} l(-y_i H(x_i)).$$

The minimization of the empirical loss provides the estimator of the predictor.

The gradient method for the empirical loss provides the boosting algorithm as follows.

Boosting Algorithm
Input: Examples $\{(x_1, y_1), \ldots, (x_n, y_n)\}$ and the initial predictor $H_0(x) \equiv 0$.
Do for $m = 1, \ldots, M$

  Step 1. Put the weight as $w(x_i, y_i) \propto l'(-y_i H_{m-1}(x_i)), i = 1, \ldots, n$, where
  l' be the differential of l and the sum of weights is set to be equal to one.
  Step 2. Find $h_m \in \mathcal{H}$ which minimizes the weighted error,

$$\sum_{i=1}^{n} w(x_i, y_i) I(y_i \neq h(x_i)).$$

  Step 3. Find the coefficient $\alpha_m \in R$ which attains the minimum value of

$$\frac{1}{n} \sum_{i=1}^{n} l(-y_i(H_{m-1}(x_i) + \alpha h_m(x_i))).$$

  Step 4. Update the predictor, $H_m(x) \leftarrow H_{m-1}(x) + \alpha_m h_m(x)$.
Output: The final predictor: $\text{sign}(H_M(x))$.

Note that in the above algorithm the exponential loss $l(z) = e^z$ derives Adaboost.

3   Robust Loss Functions

It is often pointed out that Adaboost is much sensitive to outliers. Some alternative boosting algorithms are proposed to overcome this drawback [9]. In this section we derive robust loss functions from the viewpoint of robust statistics.

We define $P(y|x)$ as the conditional probability of the class label for a given input x. Suppose that samples are identically and independently distributed. When the sample size goes to infinity, the empirical loss converges to the expected loss function:

$$L(H) = \int_{\mathcal{X}} \mu(dx) \sum_{y=\pm 1} P(y|x) \, l(-yH(x)),$$

where μ denotes the probability measure on the input space $\mathcal{X}$. Let $H^*(x)$ be the minimizer of $L(H)$. When the loss function l satisfies $l'(z)/l'(-z) = \rho(z)$, where ρ is a monotone increasing function and l' denotes the differential of l, the minimizer of $L(H)$ is given as $H^*(x) = \rho^{-1}\left(\frac{P(1|x)}{P(-1|x)}\right)$. This formula is derived from the variational of $L(H)$.

Even if l_1 and l_2 are different loss functions, $l'_1(z)/l'_1(-z) = l'_2(z)/l'_2(-z)$ could hold. For examples, all of the following loss functions

$$l_{ada}(z) = e^z, \quad l_{logit}(z) = \log(1 + e^{2z}),$$

$$l_{mada}(z) = \begin{cases} z & z \geq 0 \\ \frac{1}{2}e^{2z} - \frac{1}{2} & z < 0 \end{cases}$$

satisfy the formula $l'(z)/l'(-z) = e^{2z}$. The loss functions l_{logit} and l_{mada} are used for Logitboost [4] and Madaboost [2], respectively. Therefore estimators given by these loss functions are identical to $H^*(z) = \frac{1}{2}\log\frac{P(1|x)}{P(-1|x)}$ when the sample size goes to infinity.

We study the influence of outliers to the coefficients α's in boosting algorithm. As studied above, the solution of $\min_{\alpha\in R} L(H^* + \alpha h)$ is $\alpha = 0$. When the outlier $(\tilde{x}, \tilde{y})$ contaminates the data, the probability distribution of the data is perturbed to

$$\tilde{P}(x, y) = (1 - \epsilon)P(y|x)\mu(x) + \epsilon\,\delta_{(\tilde{x},\tilde{y})}(x, y),$$

where $\delta_{(\tilde{x},\tilde{y})}$ denotes the point mass distribution at $(\tilde{x}, \tilde{y})$ and ϵ is the ratio of the outlier included in the data. The influence to the coefficient of $h(x)$ is measured by $\alpha_\epsilon(\tilde{x}, \tilde{y})$ which is the solution of

$$\min_{\alpha\in R} \int_{\mathcal{X}} \sum_{y=\pm 1} \tilde{P}(x, y)l(-y(H^*(x) + \alpha h(x)))$$

$$= \min_{\alpha\in R} (1 - \epsilon)L(H^* + \alpha h) + \epsilon l(-\tilde{y}H^*(\tilde{x}) - \alpha\tilde{y}h(\tilde{x})).$$

The amplitude of $\alpha_\epsilon(\tilde{x}, \tilde{y})$ denotes how much the outlier influences to the coefficient. The *gross error sensitivity* [5] is defined as

$$\gamma(l) = \sup_{\tilde{x}\in\mathcal{X},\tilde{y}=\pm 1} \lim_{\epsilon\to +0} \frac{|\alpha_\epsilon(\tilde{x}, \tilde{y})|}{\epsilon}.$$

The *most B-robust* [5] loss function for ρ is defined as the one which minimizes the gross error sensitivity under the condition of $\frac{l'(z)}{l'(-z)} = \rho(z)$. The form of the most B-robust loss function is specified by the following theorem.

Theorem 1. *Under the regularity conditions, the most B-robust loss function for ρ is given as*

$$l_\rho(z) = \begin{cases} z & z \geq 0 \\ \int_0^z \rho(w)dw & z < 0. \end{cases} \tag{1}$$

The proof is given by a certain extension of Theorem 7 in Murata et al. [8].

Note that the differential of loss functions is proportional to the weight on the sample in boosting algorithms. For the most B-robust loss function, the differential is equal to constant in the region of $z \geq 0$, and thus so much weight is not assigned to the outlier. This is an intuitive reason why the estimator by l_ρ is resistant for the outlier.

Theorem 1 provides the way to make the loss function l robust. Substituting $l'(z)/l'(-z)$ to $\rho(z)$ in (1), we obtain the most B-robust function $l^*(z)$ as

$$l^*(z) = \begin{cases} z & z \geq 0 \\ \int_0^z \frac{l'(w)}{l'(-w)}dw & z < 0. \end{cases} \tag{2}$$

This formula is useful to derive robust estimators from known loss functions. In section 4 we derive the robust loss function from contamination models [1, 10].

4   Numerical Experiments

In the following numerical experiments, we study the two-dimensional binary classification problems. Labeled examples are generated from a fixed probability and a few class labels are flipped as outliers. The detailed setup is as follows. Inputs are uniformly distributed on $[-\pi, \pi] \times [-\pi, \pi]$, and the conditional probability of the class label is defined as $P(y|x_1, x_2; \eta, H) \propto (1 - \eta)e^{yH(x_1,x_2)} + \eta$, where $H(x_1, x_2) = x_2 - 3\sin(x_1)$ and $\eta = 0.5$. Moreover $a\%$ outliers are mixed, where these are randomly chosen from top $10a\%$ examples sorted in the descending order of $|H(x_1, x_2)|$. Hence outliers are scattered far from the decision boundary.

The conditional probability $P(y|x_1, x_2; \eta, H)$ is a sort of contamination models [1]. In the context of boosting, Takenouchi and Eguchi [10] proposed the loss function $l(z; \eta) = (1 - \eta)e^z + \eta z$ for the contamination models. The tuning parameter, η, determines the ratio of the contamination. In their paper the boosting by the loss function, *eta-boost*, is shown to be useful for reducing the influence of *mislabels* which denote the flip of the class label around the decision boundary. We can solve the analytic form of the most B-robust loss function for $l(z; \eta)$ as

$$l^*(z; \eta) = \begin{cases} z & z \geq 0 \\ \frac{1}{\eta^2}\left[(1 - \eta)(e^z - 1)\eta + (2\eta - 1)\log(1 + (e^z - 1)\eta)\right] & z < 0. \end{cases} \quad (3)$$

The boosting by (3) could be robust compatibly for outliers and mislabels. We refer to the boosting algorithm by (3) as *robust eta-boost*.

In this numerical experiments, "stumps" [4] is used as the weak learner in the boosting algorithm. In order to evaluate the performance of estimators, test errors are averaged in 100 different runs for each boosting method. In each run 200 training data are used. The test error is calculated on 5000 test data. Estimators evaluated here are Adaboost, Adaboost$_{Reg}$ [9], Madaboost [2], Logitboost [4], eta-boost [10] and robust eta-boost. Some tuning parameters involved in each boosting method are determined by 10-fold cross validations.

In the table 1 the generalization performance is shown for each level of outliers. When outliers do not exist, Adaboost$_{Reg}$ is better than the other algorithms. As the ratio of outliers increases, boosting algorithms other than robust eta-boost and Madaboost are affected by outliers.

We confirmed that Adaboost$_{Reg}$ with the appropriate tuning parameter gives good generalization performance even under 2% outliers. The estimation of tuning parameter by the cross validations needs a high computational cost because it takes a wide range of values, although the selection of the tuning parameter is important for the generalization ability of Adaboost$_{Reg}$. We think that Adaboost$_{Reg}$ does not show good generalization performance under the existence of outliers because of the limitation of the search region for the tuning parameter. On the other hand we observed that the generalization performance of robust eta-boost is not significantly degraded by the inappropriate value of η. When training data are contaminated by outliers, the averaged results for robust

Table 1. The averaged test errors in % for several boosting algorithms. The bold face denotes the best generalization performance under each ratio of outliers.

outliers	0%	1%	2%
robust eta-boost	14.87	**15.66**	**15.70**
eta-boost	14.70	16.86	16.90
Adaboost$_{Reg}$	**14.57**	15.74	16.09
Adaboost	14.86	16.49	16.50
Madaboost	15.04	15.96	15.98
Logitboost	14.95	16.01	16.01

eta-boost is slightly better than those for Madaboost because robust eta-boost is able to control the ratio of mislabels by η.

5   Concluding Remarks

In this paper we formulated a way of constructing a robust boosting algorithm. We proposed the formula for the most B-robust loss function. Applying the formula to contamination models, we obtain *robust eta-boost* which is resistant to both mislabels and outliers. The usefulness of the proposed algorithm is confirmed by simple numerical experiments.

References

1. Copas, J. (1988). Binary regression models for contaminated data. *J. Royal Statist. Soc. B.*, *50*, 225–265.
2. Domingo, C., & Watanabe, O. (2000). MadaBoost: A modification of AdaBoost. *Proc. of the 13th Conference on Computational Learning Theory, COLT'00*.
3. Freund, Y., & Schapire, R. E. (1997). A decision-theoretic generalization of on-line learning and an application to boosting. *Journal of Computer and System Sciences*, *55*, 119–139.
4. Friedman, J. H., Hastie, T., & Tibshirani, R. (2000). Additive logistic regression: A statistical view of boosting. *Annals of Statistics*, *28*, 337–407.
5. Hampel, F. R., Rousseeuw, P. J., Ronchetti, E. M., & Stahel, W. A. (1986). *Robust statistics. the approach based on influence functions*. John Wiley and Sons, Inc.
6. Lebanon, G., & Lafferty, J. (2002). Boosting and maximum likelihood for exponential models. *Advances in Neural Information Processing Systems*.
7. Mason, L., Baxter, J., Bartlett, P. L., & Frean, M. (1999). Boosting algorithms as gradient descent. *Advances in Neural Information Processing Systems*.
8. Murata, N., Takenouchi, T., Kanamori, T., & Eguchi, S. (2004). Information geometry of u-boost and bregman divergence. *Neural Computation*. To appear.
9. Rätsch, G., Onoda, T., & Müller, K.-R. (2001). Soft margins for adaboost. *Machine Learning*, *42*, 287–320.
10. Takenouchi, T., & Eguchi, S. (2004). Robustifying adaboost by adding the naive error rate. *Neural Computation*, vol. 16, num. 4, pp. 767 – 787.
11. Vapnik, V. (1995). *The Nature of Statistical Learning Theory*. Springer, NY.

An On-Line Learning Algorithm with Dimension Selection Using Minimal Hyper Basis Function Networks

Kyosuke Nishida, Koichiro Yamauchi, and Takashi Omori

Graduate School of Information Science and Technology, Hokkaido University
Kita 14 Nishi 9, Kita, Sapporo, 060-0814, Japan
{knishida,yamauchi,omori}@complex.eng.hokudai.ac.jp

Abstract. In this study, we extend a minimal resource-allocating network (MRAN) which is an on-line learning system for Gaussian radial basis function networks (GRBFs) with growing and pruning strategies so as to realize dimension selection and low computational complexity. We demonstrate that the proposed algorithm outperforms conventional algorithms in terms of both accuracy and computational complexity via some experiments.

1 Introduction

On-line learning with structural adaptation is a very important issue for adaptive systems in unknown environments. Lu et al. proposed a minimal resource-allocating network (MRAN), which allocates hidden units autonomously based on the novelty of input data, and prunes redundant hidden units[1, 2]. MRAN usually realizes high accuracy with low resources; however, it is difficult to say whether MRAN always performs well in terms of both accuracy and computational complexity for high-dimensional input data. In the learning of high-dimensional input space, essential information will usually be contained in a part of the input dimensions. However, MRAN cannot ignore such unnecessary input dimensions since MRAN uses isotropic Gaussian basis functions. As a result, MRAN must allocate a large number of hidden units so as to memorize combination of the essential information and the unnecessary information.

In this paper, we propose a new on-line learning algorithm to resolve such high-dimensional problems, which is named Hyper MRAN (HMRAN). In Section 2, we show the proposed method HMRAN. Experimental results are given in Section 3, followed by the conclusion in Section 4.

2 Proposed Method HMRAN

We propose a new on-line learning algorithm HMRAN to overcome the drawbacks of MRAN and to realize further improvement. This network uses Hyper Basis Function[4] instead of the isotropic Gaussian basis function to ignore

N.R. Pal et al. (Eds.): ICONIP 2004, LNCS 3316, pp. 502–507, 2004.
© Springer-Verlag Berlin Heidelberg 2004

the unnecessary input dimensions, and uses a Localized Extended Kalman Filter (LEKF) to reduce the computational complexity of the high-dimensional problem[5–8]. HMRAN also introduces a merging strategy and improved pruning strategy in addition to the conventional pruning strategy to avoid the generation of redundant hidden units.

2.1 Hyper Basis Function

The output of HMRAN is denoted by $f(x_n) = c_0 + \sum_{k=1}^{K} c_k \phi_k(x_n)$, where c_0 is a bias term, and $\phi_k(x_n)$ denotes a response of the kth hidden unit to inputs x_n; $\phi_k(x_n)$ is a Gaussian function given by $\phi_k(x_n) = \exp\left(-\|x_n - t_k\|^2_{V_k}/\sigma_k{}^2\right)$, where t_k is a center, σ_k is a width of the Gaussian function, and $\|\cdot\|_{V_k}$ denotes the "Weighted norm." The Weighted norm is given by $\|x_n - t_k\|^2_{V_k} = (x_n - t_k)^T V_k^T V_k (x_n - t_k)$, where V_k is square matrix[4].

To extend the algorithm from MRAN to HMRAN, we replace the Euclidean norm with the Weighted norm. In this study, V_k is used with a diagonal matrix to reduce the computational complexity for calculation of the Weighted norm, and hence V_k is described as $V_k = \text{diag}\{v_k\}$, where $v_k = [v_{k_1}, v_{k_2}, ..., v_{k_J}]^T$. Then, $\text{diag}\{v_k\}$ is the matrix that has the diagonal element v_k and all other elements of that are 0. Note that, when the network allocates a new hidden unit, HMRAN initializes V_{K+1} as $V_{K+1} = I$ to set the importance of all dimensions to equal.

In this case, the network parameter vector $w_{n-1} = [c_0^T, c_1^T, t_1^T, v_1^T, \sigma_1, ...,]^T$ and the gradient matrix B_n are described as $B_n = [I, \phi_1 I, (2\phi_1/\sigma_1{}^2) \times c_1(x_n - t_1)^T V_1^T V_1, -(2\phi_1/\sigma_1{}^2)c_1(x_n - t_1)^T \text{diag}\{x_n - t_1\}V_1^T, (2\phi_1/\sigma_1{}^3) \times c_1\|x_n - t_1\|^2_{V_1}, ...]^T$.

2.2 Localized EKF

MRAN and HMRAN basically use original EKF (Global EKF; GEKF) to modify the network parameters; however, GEKF requires high computational complexity. One approach to reduce the computational complexity of MRAN is proposed by Li et al[3]. This Extended MRAN (EMRAN) reduces its computational complexity by omitting parameter modification of hidden units except for the closest hidden unit to current input data. Consequently, its computational complexity is much lower than MRAN, although its accuracy is usually degraded along with the learning process.

To overcome the problem, HMRAN uses Localized EKF (LEKF) instead of GEKF for the learning of high-dimensional data. LEKF for RBFs was proposed by Birgmeier[8]. It reduces the computational complexity by dividing the parameter vector into several sub-vectors of parameters that correspond to each hidden unit.

Birgmeier, however, only shows the algorithm applicable to one-dimensional output network. To make the algorithm support the multi-dimensional outputs network, we extend the Birgmeier's algorithm by applying EKF to each output unit's parameters. Here, suppose K and M are the number of hidden units and the number of parameters per hidden unit, respectively. Then, the number of multiplications and storage required per iteration by GEKF are $O(M^2K^2)$ and $O(M^2K^2)$, respectively. On the other hand, those by LEKF become $O(M^2K) + O(K^2)$ and $O(M^2K) + O(K^2)$. Thus, the computational complexity of LEKF is much lower than that of GEKF for the learning of high-dimensional data. The summarized algorithm of LEKF is shown in Fig. 1. Note that LEKF also does not need to calculate inverse matrix.

Algorithm LEKF

Output $f(x_n) = c_0 + \sum_{k=1}^{K} c_k \phi_k(x_n)$ with error $e_n = y_n - f(x_n)$.
for each kth hidden unit

Set $w_{n-1}^k = [t_k^T, v_k^T, \sigma_k]^T$, $\mu_n^k = (2\phi_k/\sigma_k^2)c_k$,

$\nu_n^k = [(x_n - t_k)^T V_k^T V_k, -(x_n - t_k)^T \text{diag}\{x_n - t_k\} V_k^T, \|x_n - t_k\|_{V_k}^2/\sigma_k]^T$,

$\psi_n^k = P_{n-1}^k \nu_n^k$, $\alpha_n^k = \nu_n^{k^T}\psi_n^k$, $\beta_n^k = \mu_n^{k^T}\mu_n^k$, $K_n^k = \dfrac{1}{\lambda + \alpha_n^k\beta_n^k}\psi_n^k\mu_n^{k^T}$.

Update $w_n^k = w_{n-1}^k + \dfrac{\mu_n^{k^T}e_n}{\lambda + \alpha_n^k\beta_n^k}\psi_n^k$, $P_n^k = P_{n-1}^k - \dfrac{\beta_n^k}{\lambda + \alpha_n^k\beta_n^k}\psi_n^k\psi_n^{k^T} + qI$.

endfor
Set $s_n = [1, \phi_1, \phi_2, ..., \phi_K]^T$.
for each lth output unit

Set $c_{n-1}^l = [c_{0_l}, c_{1_l}, c_{2_l}..., c_{K_l}]^T$, $\xi_n^l = P_{n-1}^l s_n$, $\zeta_n^l = s_n^T\xi_n^l$, $K_n^l = \dfrac{1}{\lambda + \zeta_n^l}\xi_n^l$.

Update $c_n^l = c_{n-1}^l + K_n^l e_n^l$, $P_n^l = P_{n-1}^l - \dfrac{1}{\lambda + \zeta_n^l}\xi_n^l\xi_n^{l^T} + qI$.

endfor

Fig. 1. Algorithm LEKF

2.3 Merging Strategy

Next, we introduce a merging strategy in addition to the conventional pruning strategy. First of all, the network finds two hidden units. The network then merges those two units by generating a new hidden unit, which learns the sum of the two units' outputs. However, if the new hidden unit cannot approximate the sum of the two units' outputs, the merging process is canceled. Note that pseudo patterns are generated from the two hidden units for the learning of the new hidden unit. The summarized algorithm of the Merging Strategy is shown in Fig. 2.

2.4 Improvement of Pruning Strategy

The pruning strategy of MRAN uses normalized output values $r_{k_l} = |o_{k_l}/o_{\max_l}|$, where $o_k(x_n) = c_k \exp\left(-\|x_n - t_k\|^2/\sigma_k^2\right)$ and $o_{\max}(x_n) = [\max_k |o_{k_1}|, ..., \max_k |o_{k_L}|]^T$. Then, if the normalized output values of all output dimensions r_{k_l} ($l = 1, ..., L$) are less than a threshold δ_p for M_p consecutive observations, the kth hidden unit is regarded as a redundant unit and is pruned.

In the above conventional strategy, however, the normalized output values of redundant hidden units are apt to be high in the area where a value of $\|f(x_n)\|$ is close to 0. In such situations, the pruning process wastes large amounts of iteration time to prune such redundant hidden units.

Algorithm Merging Strategy
Find two hidden units:
  Select randomly $m_1 \in \{1, ..., K\}$; Set $m_2 = \arg \min_k ||t_k - t_{m_1}||_{V_{m_1}}$.
  if $||t_{m_2} - t_{m_1}||_{V_{m_1}} > \epsilon_n$ then Finish. endif

Generate pseudo patterns:
  for i from 1 to M_m
    Select randomly $k \in \{m_1, m_2\}$, $\rho_i \in [0, \epsilon_n]$, $u_{ij} \in [-1, 1]$ $(j = 1, ..., J)$.
    Set $x_i^* = t_k + (\rho_i/||u_i||)u_i$, $y_i^* = o_{m_1}(x_i^*) + o_{m_2}(x_i^*)$.
  endfor

Initialize the parameters of unit m:
  Set $c_m = c_{m_1} + c_{m_2}$, $t_m = 0.5(t_{m_1} + t_{m_2})$, $v_m = 0.5(v_{m_1} + v_{m_2})$,
  $\sigma_m = 0.5(\sigma_{m_1} + \sigma_{m_2})$.

do
  Calculate $E_h = 0.5 \sum_{i=1}^{M_m} ||y_i^* - o_m(x_i^*)||^2$.
  Update $w_m = w_m - \eta \nabla_{w_m} E_h$ (learning of the unit m's parameters: w_m).
  if $|E_h - E_{h-1}| < \delta_s$ **then**
    Calculate $e_{rms}^* = \sqrt{(\sum_{i=1}^{M_m} ||y_i^* - o_m(x_i^*)||^2)/M_m}$.
    if $e_{rms}^* < \delta_m$ **then**
      Calculate $o_{rms}^k = \sqrt{(\sum_{i=1}^{M_m} ||o_k(x_i^*)||^2)/M_m}$ $(k = m_1, m_2)$.
      if $o_{rms}^{m_1} > o_{rms}^{m_2}$ **then** Set $k = m_1$, $d = m_2$. **endif**
      else Set $k = m_2$, $d = m_1$. **endif**
      Set $w_k = w_m$ (to inherit the error covariance matrix of EKF).
      Prune the dth hidden unit.
      Adjust the dimensionality of P_n to suit the reduced network.
    endif
    break;
  endif
while (true)

Fig. 2. Algorithm Merging Strategy

To overcome this problem, the network executes the pruning strategy to promote the redundant hidden units only when $||f(x_n)||$ exceeds the threshold f_p. This improvement is effective for the problem in which the network output is apt to become 0, such as the learning of logical functions.

2.5 Whole HMRAN Algorithm

The entire summarized algorithm of HMRAN is shown in Fig. 3.

3 Experimental Result: 16-Input Multiplexer

In this section, we show the performances of HMRAN using LEKF (L-HMRAN) in relation to a benchmark data set: the output set of a 16-input multiplexer. The performance of L-HMRAN is compared with that of the original EMRAN and a real-time learning algorithm for a Multilayered Neural Network (MLNN) based on the Extended Kalman Filter, which is proposed by Iiguni et al.[6], under the same condition. The description of the parameters used in the following experiment is omitted because of space limitations.

A multiplexer is a logical circuit that selects one input from many inputs, and outputs the selected input. In this experiment, for instance, logical function Y of

Algorithm HMRAN

for each n

  Output $f(x_n) = c_0 + \sum_{k=1}^{K} c_k \phi_k(x_n)$ with error $e_n = y_n - f(x_n)$.

  Set $\epsilon_n = \max[\epsilon_{max}\gamma^n, \epsilon_{min}]$, $\quad d_n = \begin{cases} \min_k \|x_n - t_k\|_{V_k} & (K \neq 0) \\ \epsilon_{max} & (K = 0) \end{cases}$,

  $e_{rmsn} = \sqrt{(\sum_{i=n-(M-1)}^{n} \|e_i\|^2)/M}$.

  if $\|e_n\| > \epsilon_{min}$ & $d_n > \epsilon_n$ & $e_{rmsn} > e'_{min}$ then
    Allocate a new hidden unit:
      Set $c_{K+1} = e_n$, $t_{K+1} = x_n$, $V_{K+1} = I$, $\sigma_{K+1} = \kappa d_n$.
    Adjust the dimensionality of P_n to suit the increased network.
  else Update the network parameters using GEKF/LEKF. endif

  Call the pruning strategy:
  if $\|f(x_n)\| > f_p$ then
    Set $o_k(x_n) = c_k \exp\left(-\|x_n - t_k\|_{V_k}^2/\sigma_k^2\right)$ $(k = 1, ..., K)$,

    $o_{max}(x_n) = \left[\max_k |o_{k_1}|, ..., \max_k |o_{k_L}|\right]^T$.
    for each kth hidden unit
      Set $r_{k_l} = |o_{k_l}/o_{max_l}|$ $(l = 1, ..., L)$.
      if $(\forall l)r_{k_l} < \delta_p$ for M_p consecutive observations then
        Prune the kth hidden unit.
        Adjust the dimensionality of P_n to suit the reduced network.
      endif
    endfor
  endif

  Call the merging strategy.
endfor

Fig. 3. Algorithm HMRAN

the multiplexer is expressed by $Y = D_0 \bar{S}_3 \bar{S}_2 \bar{S}_1 \bar{S}_0 + D_1 \bar{S}_3 \bar{S}_2 \bar{S}_1 S_0 + D_2 \bar{S}_3 \bar{S}_2 S_1 \bar{S}_0 + ... + D_{14} S_3 S_2 S_1 \bar{S}_0 + D_{15} S_3 S_2 S_1 S_0$, where $D_0 \sim D_{15}$ denotes the inputs, and $S_0 \sim S_3$ are the data-selection signals.

In this case, the combination of all input patterns is $2^{16+4} = 1,048,576$, and hence it is unrealistic that the network stores all input patterns. Here, the above logical function shows that each logical term does not consist of all input dimensions. If each hidden unit of HMRAN can find necessary input dimensions, HMRAN will realize an on-line learning successfully with few resources.

In the following text, $x = [D_0, ..., D_{15}, S_0, ..., S_3]^T$, and $y = Y$. In this experiment, $\{D_j\}_{j=0}^{15}$, $\{S_j\}_{j=0}^3 \in \{0, 1\}$ are selected randomly to generate a set of $30,000$ training data and a set of $3,000$ test data. Table 1 shows the averaged performances over 50 trials, and Fig. 4 shows one example of the Output Error Rate and the transaction of the number of hidden units, respectively.

Here, Output Error Rate is the error rate in the case that the network output $f(x)$ is converted to the logical output as follows: if $f(x) > 0.5$ then $f(x) = 1$ otherwise 0.

The results suggest that L-HMRAN realizes high accuracy with few resources. L-HMRAN successfully learned the function with the small number of training examples, which was less than 1% of the combination of all input patterns.

Table 1. Performance for 16-input multiplexer. K is the number of hidden units, N_p is the number of network parameters, RMSE is the RMS Error for the test data set, ϵ is the Output Error Rate, and N_0 is the the number of training examples when ϵ first reaches 0. The number of trials in which ϵ becomes 0 are given in parentheses on the column of N_0

method	K	N_p	RMSE	ϵ	N_0
EMRAN	113.08	2488.76	0.407	0.236	– (0)
MLNN $(20-K-1)$	16	353	0.388	0.211	– (0)
	48	1057	0.215	0.0360	– (0)
	80	1761	0.136	0.00791	28756 (1)
	112	2465	0.104	0.00323	29740.5 (2)
L-HMRAN	31.12	1308.04	0.00815	0	6565.92 (50)

4   Conclusion

In this paper, we showed that the proposed method HMRAN overcomes the drawbacks of MRAN. HMRAN is an effective learning method for resolving multidimensional input problems by using the Hyper Basis Function. In addition, HMRAN also realizes both usefulness and high accuracy by employing LEKF which maintains accuracy as well as GEKF but with less computational complexity. Future work will entail applying the proposed method to many real-world problems to verify its performance. A remaining problem is determining clear setting methods for the HMRAM parameters.

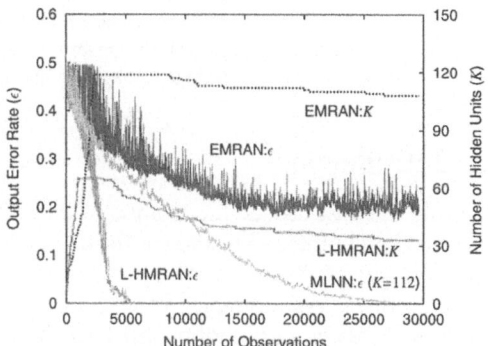

Fig. 4. Output Error Rate and Number of Hidden Units for 16-input multiplexer

References

1. Y.Lu, N.Sundararajan, P.Saratchandran, *Neural Computation* 9, pp.461–478, 1997.
2. N. Sundararajan, P. Saratchandran, Y. Lu, *World Scientific*, 1999.
3. Y. Li, N. Sundararajan, P. Saratchandran, *IEE Proc. Control Theory Appl.*, vol. 147, no. 4, pp. 476–484, July 2000.
4. T. Poggio, F. Girosi, *Proc. IEEE*, vol. 78 no. 9, pp. 1481–1497, September 1990.
5. G. V. Puskorius, L. A. Feldkamp, *Proc. International Joint Conference on Neural Networks*, Seattle, vol. I, pp. 771–777, 1991.
6. Y. Iiguni, H. Sakai, H. Tokumaru, *IEEE Trans. Signal Processing*, vol. 40, no. 4, pp. 959–966, April 1992.
7. S. Shah, F. Palmieri, M. Datum, *Neural Networks*, vol. 5, no. 5, pp. 779–787, 1992.
8. M. Birgmeier, *Proc. 1995 IEEE Int. Conf. Neural Networks*, pp. 259–264, November 1995.

Density Boosting for Gaussian Mixtures

Xubo Song[1], Kun Yang[1], and Misha Pavel[2]

[1] Department of Computer Science and Engineering, OGI School of Science and Engineering
Oregon Health and Science University, Beaverton, Oregon 97006
{xubosong,kyang}@cse.ogi.edu
[2] Department of Biomedical Engineering, OGI School of Science and Engineering
Oregon Health and Science University, Beaverton, Oregon 97006
pavel@bme.ogi.edu

Abstract. Ensemble method is one of the most important recent developments in supervised learning domain. Performance advantage has been demonstrated on problems from a wide variety of applications. By contrast, efforts to apply ensemble method to unsupervised domain have been relatively limited. This paper addresses the problem of applying ensemble method to unsupervised learning, specifically, the task of density estimation. We extend the work by Rosset and Segal [3] and apply the boosting method, which has its root as a gradient descent algorithm, to the estimation of densities modeled by Gaussian mixtures. The algorithm is tested on both artificial and real world datasets, and is found to be superior to non-ensemble approaches. The method is also shown to outperform the alternative bagging algorithm.

1 Introduction

During the past decade the method of ensemble learning was firmly established as a practical and effective technique for difficult supervised learning tasks. An ensemble method predicts the output of a particular example using a weighted vote over a set of base learners. For example, Freund and Shapire's Adaboost algorithm [13,17] and Breiman's Bagging algorithm [9] have been found to give significant performance improvement over algorithms for the corresponding base classifiers, and have lead to the study of many related algorithms [10] [2] [12] [4]. The differences among these ensemble methods exist in several aspects, such as how the inputs are re-weighted to form a base learner, how these base learners are combined, and how the cost functions are defined.

The success of ensemble methods can be explained from various perspectives. Under the bias/variance decomposition formulation, the ensemble method is said to be able to reduce the variance while keeping a minimal bias [19]. Recent theoretical results suggest that these algorithms have the tendency to produce large margin classifiers [14]. The margin of an example, defined as the difference between the total weight assigned to the correct label and the largest weight assigned to an incorrect label, can be interpreted as an indication of the confidence of correct classification. Therefore, loosely speaking, if a combination of classifiers correctly classifies most of the training examples with a large margin, then the error probability will be small. The effectiveness of boosting can also be understood in term of well-known statistical principles, namely additive modeling and maximum likelihood – it is shown that boosting can be viewed as an approximation to additive modeling on the logistic scale using maximum Bernoulli likelihood[12]. Recently, it is shown [4] that "boosting-

N.R. Pal et al. (Eds.): ICONIP 2004, LNCS 3316, pp. 508–515, 2004.
© Springer-Verlag Berlin Heidelberg 2004

type" algorithms for combining classifiers can be viewed as gradient descent on an appropriate cost functional in a suitable inner product space. Regardless how it is interpreted, the benefit of ensemble method is evident. Originally designed for classification problems, ensemble method has also been successfully generalized to regression problems [20].

On an orthogonal front is the classic task of multivariate density estimation. The problem of density estimation is one of great importance in engineering and statistics, playing an especially role in exploratory data analysis, statistical pattern recognition and machine learning. Frequently one must estimate density function for which there is little prior knowledge concerning the shape of the density and for which one wants a flexible and robust estimator. There have been traditionally two principal approaches to density estimation, namely the parametric approach which makes assumptions of the density, and the nonparametric view which is essentially distribution free. In recent years, Gaussian mixture models, a parametric method, have attracted much attention due to their flexibility in modeling multimodal shapes, and the applicability of the EM (Expectation Maximization) learning algorithm that is efficient in finding the optimal parameters.

In this paper we apply ensemble learning methods to the task of density estimation, specifically to density distribution represented as Gaussian Mixture Model. In contrast to the vast popularity of ensemble methods for supervised learning, attempts to apply ensemble methods to unsupervised tasks have been limited. Recently, Rosset and Segal proposed a boosting technique to estimate densities represented as Bayesian Networks [3]. Their boosting technique, in the form of gradient descent algorithm, was originally proposed by Breiman [10], and was formally formulated by Mason *et al* for classification problems [4]. In this paper, we extend the work by Rosset and Segal [3], and apply it to probabilistic densities represented as Gaussian mixtures. We note that a few researchers have considered learning mixture via stacking or bagging [1] [5].

This paper is organized as follows. In Section 2, we provide a review of the principle behind the boosting algorithm as gradient descent for classification problems and its adaptation to density estimation problems. In Section 3, we introduce the Mixture of Gaussian (MOG) as the density model, and address several implementation-specific issues for boosting such density functions. Experimental results are presented in Section 4, and Section 5 offers discussions and future directions.

2 Boosting of Density Estimation: Overview of the Framework

Boosting is a method that incrementally builds linear combinations of weak base learners to form a strong learner. Many researchers have come up with different versions of boosting algorithms. The formulation that casts boosting as a gradient descent algorithm is presented by Mason et al [4] for classification problems. Such a formulation provides a unifying framework for many variants of boosting algorithms. Here we provide an overview of this formulation.

First we introduce some notations. Let (x, y) be the input-output pairs, where x is from the input space (for instance, $\mathbf{R}^N$) and y is from the label space (for instance, $y \in \{-1, 1\}$). Let $F(x)$ be a combined classifier $F(x) = \sum_{t=1}^{T} \alpha_t h_t(x)$, where

$h_t(x)$'s are base classifiers from some fixed class $\mathbf{H}$, and α_t are the weights for these base classifiers. For a given set of data samples $S = \{(x_1, y_1), ..., (x_n, y_n)\}$, a margin cost function $L(F)$ can be defined for $F(x)$ on S where $L(F) = \frac{1}{n}\sum_{i=1}^{n} L(y_i F(x_i))$

Suppose we have a function $F \in lin(\mathbf{H})$, where $lin(\mathbf{H})$ is the set of all linear combinations of functions in $\mathbf{H}$. We would like to find a new function $f \in \mathbf{H}$ such that the cost function $L(F + \varepsilon h)$ decreases for small ε. Expand $L(F + \varepsilon h)$ in an inner product space and ignoring the higher order terms, we have

$$L(F + \varepsilon h) = L(F) + \varepsilon < \nabla L(F), h >.$$

This derivation is doable under proper choice of inner product space and cost function. Therefore, the optimal choice of h is the one that maximizes $-\varepsilon < \nabla L(F), h >$. Since such an h is not guaranteed to be in the function class $\mathbf{H}$, we instead find an h from $\mathbf{H}$ with a large value of $-\varepsilon < \nabla L(F), h >$. Then the new combined classifier will be $F_{t+1} = F_t + \alpha_{t+1} h_{t+1}$ for appropriately chosen α_{t+1}.

The analysis above provides the basic principle that describes boosting algorithms as a gradient descent algorithm. Such a perspective enables a general formulation of ensemble algorithms. Many existing ensemble methods, such as Adaboost [13], LogitBoost [12] and ConfidenceBoost [22], can be shown to be special cases of this abstract formulation with specific choices of margin cost functions and step sizes.

This general formulation, originally presented for classification problems, also provides guidelines for representing density estimation problem. Rosset and Segal presented such an adaptation to density estimation in [3]. For density estimation, the data set is $S = \{(x_1, y_1), ..., (x_n, y_n)\}$. The basis class $\mathbf{H}$ is comprised of probability distributions, and the base learner h is a density distribution. The combined learner $F(x)$ at each boosting iteration t is a probability distribution $F_{t-1}(x) = \sum_{j<t} \alpha_j h_j(x)$. The cost function L of $F(x)$ on the sample set S can be defined as the negative log likelihood

$$L(F) = \sum_i L(x_i, F(x_i)) = \sum_i L(x_i, \sum_j \alpha_j h_j(x_i)) = \sum_i -log(\sum_j \alpha_j h_j(x_i)). \quad (1)$$

We then identify, at iteration t, a base learner $h_t \in \mathbf{H}$ so that it gives the largest improvement in the loss at the current fit. Intuitively, h_t assigns higher probability to samples that receives low probability by the current model F_{t-1}. Formally, if we apply $L(F+\varepsilon h) = L(F) + \varepsilon < \nabla L(F), h >$ to $L(F)$ as the negative log likelihood function, we have $L(F + \varepsilon h) = L(F) - \varepsilon \sum_i \frac{1}{F_{t-1}(x_i)} h(x_i)$.

The optimal base learner $h(x)$ will be the one that maximizes the second term in the equation above. Since the coefficients for $h(x_i)$ is $\frac{1}{F_{t-1}(x_i)}$, it coincides well with our intuition of putting more emphasis on the samples with low probability from the current learner. At this moment, the optimal base learner is added to the current model F_{t-1} to form the new model F_t. We cannot simply augment the model, since $F_{t-1} + \alpha_t h_t$ may no longer be a probability distribution. Rather, we adopt the form $F_t = (1 - \alpha_t)F_{t-1} + \alpha_t h_t$ where $0 \le \alpha \le 1$.

For the implementation of the algorithm, several factors need to be decided: the model of the base weak learners $\mathbf{H}$, the method for searching for the optimal base weak learner h_t at each boosting iteration, and method for searching for the optimal α that combines h_t to the current model.

The resulting generic boosting algorithm for density estimation can be summerized as follows:

1. Set $F_0(x)$ to uniform on the domain of x.

2. For $t = 1$ to T

   a. Set $w_i = \dfrac{1}{F_{t-1}(x_i)}$;

   b. Find $h_t \in \mathbf{H}$ to maximize $\sum_i w_i h_t(x_i)$;

   c. Find $\alpha_t = \arg\min\limits_{\alpha} \sum_i - \log((1 - \alpha_t)F_{t-1}(x_i) + \alpha_t h_t(x_i))$.

3. Output the final model F_T.

3 Boosting of Gaussian Mixture Densities

In this section we apply the boosting algorithm described above to densities represented by mixtures of Gaussians. Gaussian mixture model has wide appeal for its flexibility and ease of implementation. In this case, a base learner h is defined as $h(x) = \sum_k p_k N(\mu_k, \Sigma_k)$, where p_k is the weight for the kth mixture component, and $N(\mu, \Sigma) = \dfrac{1}{(2\pi)^{d/2}|\Sigma|^{1/2}} e^{-\frac{1}{2}(x-\mu)^T \Sigma^{-1}(x-\mu)}$ is a Gaussian with mean μ and covariance matrix Σ. This specifies the basis model $\mathbf{H}$. The unknown parameters are p_k, μ_k, Σ_k for all k. As mentioned above, one of the implementation-specific considerations is the method for searching the optimal base learner h at each boosting iteration. This is accomplished by using the EM algorithm, which estimates the optimal parameters p_k, μ_k, Σ_k for all k. In other words, the EM algorithm searches for p_k, μ_k, Σ_k such that $log(\sum_i w_i h_t(x_i)) = log(\sum_i w_i \sum_k p_k N(\mu_k, \Sigma_k))$ is maximized. Another consideration, namely the method to search for the optimal α, is typically accomplished by using a line search [3]. In this paper, we treat α_t as an unknown parameter, and use the EM algorithm to estimate it. In other words, given the current weak learner h_t, α_t is estimated by the EM algorithm to be the one that maximizes $\sum_i - \log((1 - \alpha_t)F_{t-1}(x_i) + \alpha_t h_t(x_i))$.

In summery, the density boosting algorithm specifically for Gaussian mixture model is as follows:

4. Set $F_0(x)$ to uniform on the domain of x.

5. For $t = 1$ to T

    a. Set $w_i = \dfrac{1}{F_{t-1}(x_i)}$;

    b. Use EM to find p_k, μ_k, Σ_k for all k to maximize $log(\sum_i w_i \sum_k p_k N(\mu_k, \Sigma_k))$;

    The optimal parameters specify $h_t(x) = \sum_k p_k N(\mu_k, \Sigma_k)$.

    c. Use EM to find

$$\alpha_t = \arg \min_\alpha \sum_i - \log((1 - \alpha_t) F_{t-1}(x_i) + \alpha_t h_t(x_i)) \cdot$$

6. Output the final model F_T .

4 Experimental Results

We evaluate the performance of the density boosting algorithm for Gaussian Mixture Models on both artificial and real data. Three distinct data sets are used. The first one is an artificial data set, with data points drawn from a known density $p(z) = .2N(-5., .8) + .5N(0., 1.4) + .3N(8., 2.)$. We use 50 data points for density estimation and 50 for model evaluation. The second data set is the phoneme data collected at the Center for Spoken Language Understanding (CSLU) at OGI. The phonemes come from the OGI stories corpus with multi-speakers and continuous speech. The data is of dimension 10, corresponding to 10 mel-frequency cepstral coefficients. Each entry is the result of the average cepstral coefficient calculated over 7, 30ms frames. The frames overlap by 5ms and the central frame is centered on the phoneme. We use the data for the phoneme /m/. The training set contains 400 examples, and test set contains 100 examples. The third data set is a corporate bond rating data set, provided by a major wall street bank [21]. The data contains bonds issued by 196 industrial firms. Only investment bonds (as opposed to junk bonds) are included in our experiment. The data points have 10 dimensions, corresponding to time-averaged, proprietary financial ratios. The training set contains 100 training examples, and the test set contains 27 examples. A three-component Gaussian mixture is used to model all three data sets. For each data set, the boosting algorithm is repeated 1000 times, and the log likelihood of the test data is averaged over the repetitions.

The results are presented in Fig. 1. The x-axis corresponds to the total number T of weak learners that are combined into the final model (the ensemble size). Therefore, $T = 1$ corresponds to single learner non-boosting case. The y-axis corresponds to the log likelihood of test data. The results of density estimation using boosting algorithm corresponds to the upper blue solid curves in these figures. It can be seen clearly that the log likelihood consistently improves with the ensemble size. Therefore, the boosting algorithm outperforms non-boosting single learner algorithm.

We also compare the boosting algorithm to the bagging algorithm [9]. For bagging, a strong learner is formed by combining multiple weak learners with equal weight. Each weak learner is trained on a bootstrap sample of the training data. The bagging procedure is repeated 1000 times, and the log likelihood of the test data is averaged over the repetitions. The results using bagging algorithm correspond to the lower black dotted curves in each figure. As can be seen, boosting consistently outperforms bagging for equal ensemble size.

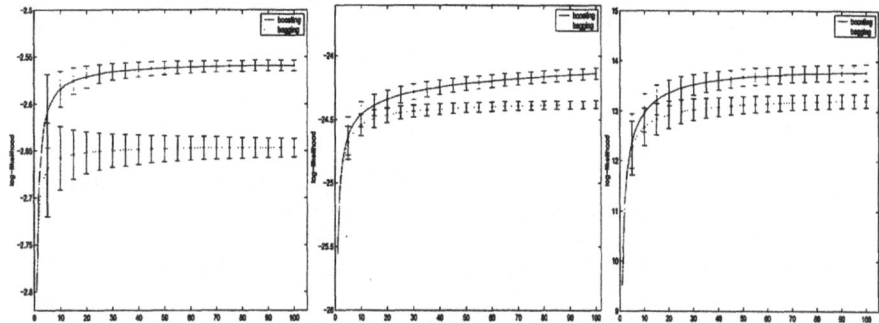

Fig. 1. Results on artificial data (left), on phoneme data (middle) and bond data (right)

5 Discussions

In this paper we extend the boosting methodology to the task of density estimation using Gaussian Mixture Models, and demonstrate its superiority over non-boosting single base learner approaches on both artificial and real world datasets. We also compare the boosting algorithm to an alternative bagging algorithm. We observe that boosting outperformed bagging on all three datasets. We believe that this approach holds promise, and may lead to techniques applicable to other unsupervised learning domains.

One of the assumptions we made in the paper was that the number of components for the Gaussian mixture is known. However, the algorithm presented is not limited to fixed components. In general, the number of components can be treated as one of the parameters, which can be optimized by a data-driven approach such as the EM algorithm. This will also imply that we adopt a non-parametric model for the density distribution.

A key concept in most of the ensemble methods concerns the "weakness" of a base learner. As s matter of fact, "weakness" of the base learners is required in order for a boosting method to be effective. For classification problems, "weakness" is loosely defined as "slightly better than random guessing". It is possible to measure the "weakness" of a model when data are given as labeled pairs. Unfortunately, for unsupervised problems such as density estimation, even thought it is possible to evaluate a model by its likelihood, there is no obvious relationship between the likelihood and the "weakness" of the model. (In our implementation, a weak learner h is identified by applying EM algorithm to the re-weighted training data until the EM converges. This may be interpreted that we don't explicitly impose the weakness constraint.) One possibility is to use the entropy of h as the gauge, and a model is considered "weak" if its entropy is slightly smaller than that of a uniform distribution. We intend to explore this in the future.

Gaussian Mixture Model is chosen in this paper as the representation for a density distribution due to its versatility and ease of applicability. However, we believe that the effectiveness of the boosting technique is not limited to a particular form of density parameterization. Our future work will involve applying the density boosting method to other density representations, including the formulatation for non-parametric density representations.

References

1. P. Smyth and D. Wolpert, An evaluation of Llinearly combining density estimators via stacking, *Machine Learning*, vol 36, 1/2, pp53-89, July 1999.
2. D. Wolpert, Stacked Generalization, *Neural Networks*, 5(2):241-260.
3. S. Rosset and E. Segal, Boosting density estimation, *Advances in Neural Information Processing* 15, 2002.
4. L. Mason, J. Baxter, P. Bartlett and P. Frean, Boosting algorithms as gradient descent in function space, *Advances in Neural Information Processing* 12, pp512-518, 1999.
5. D. Ormoneit and V. Tresp, Improved Gaussian mixture density estimates using bayesian penalty terms and network averaging', *Advances in Neural Information Processing* 8, pp 542-548, 1996.
6. M. Jordan and R. Jacobs, Hierarchical mixtures of experts and the EM algoriths, *Neural Computation*, 6, pp 181-214, 1994.
7. A. Dempster, N. Laird and D. Rubin, Maximum likelihood from imcomplete data via the EM algorithm, *J. Royal Statistical society B*, 1989.
8. B. W. Silverman, Denstity Estimation for Statistics and Data Analysis, Chapman and Hall, NY, 1986.
9. L. Breiman, Bagging predictors, *Machine Learning*, 24, pp 123-140, 1996.
10. L. Breiman, Prediction games and arcing algorithms. Technical Report 504, Department of Statistics, University of California, Berkeley, 1998.
11. B. Efron and R. Tibshirani, *An Introduction to the Boostrap*. Chapman & Hall, 1994.
12. J. Friedman, T. Hastie and R. Tibshirani, Additive logistic regression: a statistical view of boosting, *The Annals of Statistics*, 38(2):337-374, April 2000.
13. Y. Freund and R. Shapire, Experiments with a new boosting algorithm, *Proceedings of the Thirteenth International Conference on Machine Learning*, Bari, Italy, July 3-6, pp 148-156, 1996.
14. R. Shapire, Y. Freund, P. Bartlett and W. Lee, Boosting the margin: a new explanation for the effetiveness of voting methods, *The Annals of Statistics*, 26(5):1651-1686, October, 1998.
15. R. E. Shapire, The boosting approach to machine learning: An overview, *MSRI Workshop on Nonlinear Estimation and Classification*, 2002.
16. Y. Freund and R. E. Shapire, A short introduction to boosting, *Journal of Japanese Society for Artificial Intelligence*, 14(5):771-780, September, 1999.
17. Y. Freund and R. Shapire, A decision-theoretic generalization of on-line learning and an application to boosting, *Journal of Computer and System Sciences*, 55(1):119-139, August 1997.
18. R. Shapire, P. Stone, D. McAllester, M. Littman and J. Csirik, Modeling auction price uncertainty using boosting-based conditional density estimation, *Machine Learning: Proceedings of the Nineteenth International Conference*, 2002.
19. R. Meir, Bias, variance and the combination of estimators; the case of linear least squares, *Advances in Neural Information Processing Systems*, vol. 7, eds. G. Tesauro, D. Touretzky and T. Leen, 1995.

20. R. Zemel and T. Pitassi, A gradient-based algoriths for regression problems, *Advances in Neural Information Processing Systems*, NIPS 2001.
21. J. Moody and J. Utans, Architecture Selection Strategies for Neural Networks: Application to Corporate Bond Rating Prediction, in Refenes A.N. (ed.), *Neural Networks in the Capital Markets*, John Wiley & Sons, 1994.
22. R. Schapire and Y. Singer, Improved boosting algorithms using confidence-rated predictions. *Proceedings of the Eleventh Annual Conference on Computational Learning Theory*, pp80-91, 1998.

Improving kNN Based Text Classification with Well Estimated Parameters*

Heui Seok Lim

Dept. of Software, Hanshin University, Korea
limhs@hs.ac.kr

Abstract. This paper propose a method which improves performance of kNN based text classification by using well estimated parameters. Some variants of the kNN method with different decision functions, k values, and feature sets are proposed and evaluated to find out adequate parameters. Our experimental results show that kNN method with carefully chosen parameters are very significant in improving the performance and reducing size of feature set. We carefully conclude that it is very worthy of tuning parameters of kNN method to increase performance rather than having hard time in developing a new learning method.

1 Introduction

Automatic text categorization is a problem of assigning predefined categories to free text documents based on the likelihood suggested by a training set of labelled texts. Document categorization is one solution to effective retrieval of large amount of textual information by providing good indexing and summarization of document content. Topic spotting for newswire stories and document routing are the most commonly investigated application domains in the text categorization.

An increasing number of approaches to text categorization have been proposed, including nearest neighbor classification[4], [7], [8], [9], Bayes probabilistic approaches[2], [6], decision tree[2] and neural networks[10]. Performances of the methods and cross method comparison were well reported in [8], [9].

k-Nearest Neighbor(kNN) method is a well known statistical approach. The kNN algorithm is quite simple: given a test document, the system finds the k nearest neighbors among the training documents, and uses the categories of the k top-ranking neighbors to predict the categories of the input document. The similarity score of each neighbor document to the new document being classified is used as the weight of each of its categories, and the sum of category weights over the k nearest neighbors are used for category ranking. Due to its simplicity, the kNN method has been intensively applied to many text categorization systems. But a through evaluation of the method has been rarely done so far.

There are some important parameters to be considered for improving performance of the method: decision function, k value for considering nearest neighbors, and size of feature set. We expect that ideally chosen parameters of kNN

---

* This Work was Supported by Hanshin University Research Grant in (2004).

N.R. Pal et al. (Eds.): ICONIP 2004, LNCS 3316, pp. 516–523, 2004.
© Springer-Verlag Berlin Heidelberg 2004

play an very important role in improving performance of kNN method on the same training and test documents. To decide the good parameters, we must know the effect of the individual parameters and the interactions of each parameters.

This paper focuses on comparative evaluation of Korean text categorization system based on kNN method. Experimental results with various parameters will be discussed.

2   kNN Learning

The algorithm of kNN learning consists of simply storing the presented training data. This algorithm assumes all the document corresponds to vectors in n-dimensional vector space. The neighbors of an example are defined in terms of one of distance measurements such as Euclidean distance and cosine measurement. For example, a document x_i is defined as $\overrightarrow{x_i} = (f_{i1}, f_{i2}, f_{i3}, ..., f_{in})$ where f_{ij} means the jth feature value of x_i. Features are elements that represent and model the document. A set of indexing terms[1] are usually used as feature set and an weighting scheme is applied to represent importance of each feature on the document. DF(document frequency), IG(information gain), entropy, and conventional TF/IDF scheme in information retrieval are the frequently used weighting schemes.

Once documents are represented as vectors in which the elements are weights of corresponding features, the distance or similarity between two documents, x_i and x_j is defined as

$$d(\overrightarrow{x_i}, \overrightarrow{x_j}) \equiv \sqrt{\sum_{r=1}^{n} (x_{ir} - x_{jr})^2} \tag{1}$$

when using Euclidean distance measurement or as

$$d(\overrightarrow{x_i}, \overrightarrow{x_j}) \equiv \frac{\overrightarrow{x_i} \cdot \overrightarrow{x_j}}{|\overrightarrow{x_i}| \cdot |\overrightarrow{x_j}|} \tag{2}$$

when using cosine measure. A major advantage of the kNN learning is that it can estimate a target function locally and differently for each new instance to be classified instead of estimating the target function once for the entire instance space[4].

3   Decision Functions of kNN Classification

Decision function in kNN based text classification is a function that assigns a category to a query document using k nearest neighbors. Several decision

---

[1] Only nouns extracted by morphological analyzer are indexing terms in most of Korean systems.

functions may be used: discrete-valued function(DVF), similarity-weighted function(SWF), average-similarity-weighted function(ASWF). Text classification using the DVF function is defined as equation 3.

$$TC(\vec{x_q}) \leftarrow \arg\max_{c_j \in C} \sum_{\vec{d_i} \in kNN}^{k} y(\vec{d_i}, c_j) \tag{3}$$

where $y(\vec{d_i}, c_j) \in \{0, 1\}$ is the classification for document $\vec{d_i}$ with respect to category c_j ($y = 1$ for YES, and $y = 0$ for NO). The DVF assigns c_i which has the most common value of $y(\vec{d_i}, c_i)$ among the k training examples nearest to x_q as a answer categorization. While DVF is very simple, it does not use the contribution of each of the k neighbors according to their distance or similarity to the query document. SWF is one refinement to the DVF which weights the contribution of each of the neighbors according to their similarity to the query point, giving greater weight to closer neighbors. SWF decision function is defined as equation 4.

$$TC(\vec{x_q}) \leftarrow \arg\max_{c_j \in C} \sum_{\vec{d_i} \in kNN}^{k} sim(\vec{x_q}, \vec{d_i}) y(\vec{d_i}, c_j) \tag{4}$$

where $sim(\vec{x_q}, \vec{d_i})$ is the similarity between the query document x_q and the training document d_i. Although many other measures are possible, we use the cosine value to calculate the similarity between the documents. The SWF is robust to noisy training data and quite effective when it is provided a sufficiently large set of training data[4]. The SWF may suffer from some isolated noisy training examples. One possible solution to the problem is smoothing out the impact of such noisy training examples. ASWF can be one of smoothing out method of SWF by taking the weighted average of the k neighbors nearest query point. It is defined as in equation 5.

$$TC(\vec{x_q}) \leftarrow \frac{\arg\max_{c_j \in C} \sum_{\vec{d_i} \in kNN}^{k} sim(\vec{x_q}, \vec{d_i}) y(\vec{d_i}, c_j)}{\sum_{\vec{d_i} \in kNN}^{k} y(\vec{d_i}, c_j)} \tag{5}$$

4   k Value for kNN

The k value in kNN method indicates the number of nearest neighbors. If all training examples are considered when classifying a new query instance, we call the algorithm a *global method*. If only some nearest examples are considered, we call it a *local method*.

If we use the DVF as a decision function, the global method cannot be used because it allows very distant examples have much effect on decision function. In contrast, if we use SWF or ASWF as a decision function, there is really no harm in allowing all training examples to have an influence on the classification, because very distant examples will have very little effect on the decision function. The only disadvantage of the global method is that the classifier will run more slowly.

In this paper, we have experimented both the global and the local methods to show effect of the number of nearest neighbors in classifying query instances. The SWF were used as a decision function in our experiment and the results will be described in experimental results.

5   Size of Feature Set for kNN Learning

In information retrieval or text categorization problem, a document is represented as a n-dimensional vector in a feature space. The feature space usually consists of the unique terms that occur in document set. Usually the unique terms are hundreds of thousands of terms. The enormous number of features is a major difficulty of text categorization problems. So, it is highly desirable to reduce the size of the native feature set. Document frequency(DF) thresholding is the simplest technique for the reduction. DF is the number of documents in which a term occurs in the whole document set . DF thresholding method removes terms in higher rank than predetermined rank or removes terms whose DF is less than some predefined threshold value. The basic assumption is that rare terms are either non-informative for category prediction, or not influential in global performance. However, DF thresholding is not a pricipled criterion for selecting predictive features. It is against a widely received assumption in information retrieval, which is low-DF terms are assumed to be relatively informative and should not be removed. We will examine this assumption with somewhat different DF thresholding strategy, DF thresholding considering Category Frequency(CF) which is called DFCF thresholding in this paper. Category Frequency is the number of categories in which a term occurs. DFCF value of a feature i is defined as follows:

$$CFDF_i = DF_i \times \frac{1}{CF_i} \qquad (6)$$

where DF_i is the number of documents in which the feature i occurs and CF_i is the number of categories in which the feature i occurs. Using CFDF value, terms with high DF value and low CF value may be more likely to be selected than those with high DF value and high CF value. CFDF thresholding removes terms whose CFDF value is in higher rank than some predetermined rank.

6   Experiments

We uses a Korean noun extracting system which extracts every unit nouns and compound nouns in a sentence to get index terms[1]. Index terms are stored in inverted file. Each record of the inverted file has a pointer to the location of posting file which has document id in which a term is occurred and weight of the term. We uses traditional TF/IDF term weighting scheme in information retrieval system[5]. TF/IDF weighting scheme is defined as in equation 7.

$$W_{ij} = \frac{TF_{ij}}{max_l TF_{lj}} \times log \frac{N}{DF_i} \qquad (7)$$

In equation 7, N is the total number of training examples. $max_l\, TF_{lj}$ and N is used to normalize the TF value and DF value, respectively.

We use Korean corpora for this study which consists of 5,037 documents from Korean web sites. The corpora was build carefully so that category distribution was balanced. The total number of unique words in the corpora is about 260,000. The document have been manually assigned a category among 70 categories. We divided the corpora randomly into a training set of 4294 and a test set of 743 documents.

For evaluating performance of text categorization and information retrieval, recall, precision and F_1 measures are usually used. Recall is defined to be the ratio of correct assignments by the system divided by the total number of correct assignments and precision is the ratio of correct assignments by the system divided by the total number of the systems's assignment. F_1 measure is defined as in equation 8.

In equation 8, R and P denotes recall and precision. For evaluating average performance across categories, two conventional methods are used: macro-averaging and micro-averaging. The detail of both methods is described in [8]. We used the standard recall, precision, and macro-averaged F_1 measures to evaluate the various kNN-based text classification systems. Macro-averaged F_1 measure is calculated using macro-averaged recall and precision measures.

$$F_1 = \frac{2RP}{R+P} \qquad (8)$$

Figure 1 shows F_1 scores of systems using the DVF, SWF, and ASWF decision functions. SWF shows the best F_1 scores along the every feature set sizes

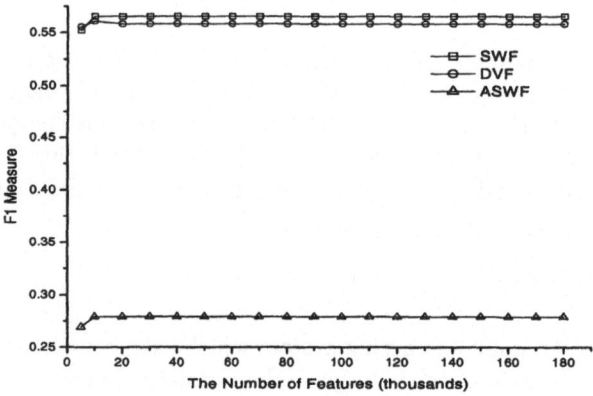

Fig. 1. Performances of different decision functions

while ASWF shows the worst results. It is interesting that ASWF which expected to smooths out noisy training examples shows worst results. This may be due to the fact that our training data are clustered densely to the centroid to each class. Figure 1 also displays the F_1 measures reached with different sizes of feature set. All the methods reached critical point of scores with about 10%(20,000 unique words) of whole features. Increasing the number of features helped improve the performances slightly. This means that up to 90% of the unique terms can be eliminated with no loss in categorization performance. This coincides with the previous results about effect of feature set size[3], [9].

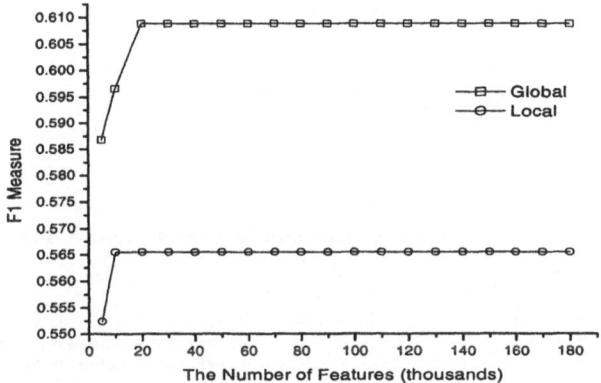

Fig. 2. Performances of global and local methods

Figure 2 is performance results of global method and local method in considering nearest neighbors. In this results, k value of local method was fixed for 30 and the SWF was used as a decision function because this showed the best result in the experiment on decision functions. As figure 2 shows, Global method shows higher F_1 measures along with all the different feature set sizes. We wondered that it was a final conclusion to adopt the global method to acquire higher performance measure even though it required more computational costs than using local method. It may be cased by we used improper k value, 30 in this experiment. To solve the curiosity, we evaluated SWF methods with different k value parameter. Figure 3 shows the F1 scores of SWF methods along different k values with 20,000 features. This results shows the k value greatly affects the performance and the local method with carefully chosen k value, 80 in our result, can show the similar results without much computational costs.

Figure 4 shows the results of performance with DF and CFDF thresholding in feature selection. In this experiment, SWF is was used as a decision function and the k value was 30. As shown in the figure, F_1 scores when using CFDF thresholding are slightly higher than using DF thresholding along every feature set size. This results shows that the proposed CFDF thresholding which considers both document frequency and category frequeycy is more effective than the previous simple DF thresholding to reduce feature set. And it is very interesting

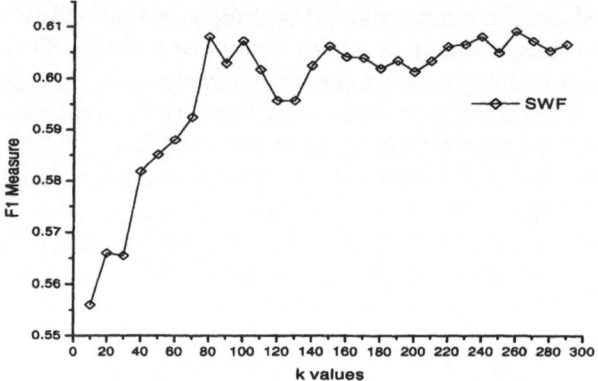

Fig. 3. Performances along different k values

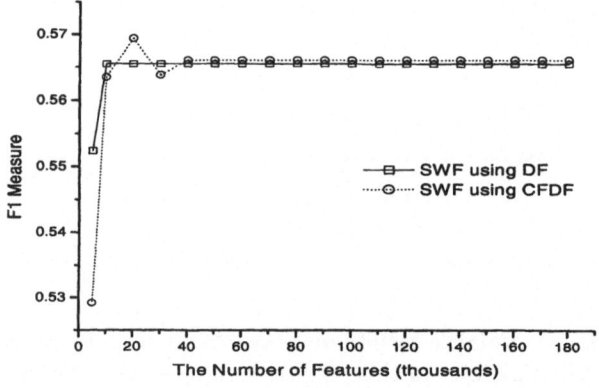

Fig. 4. Performances along different thresholding methods

that the amount of improvement on the set of 20,000 features is much more than on the other feature set sizes. The set of 20,000 features is the set with which the performances are reached at critical point of F_1 score in the figure 2.

7   Conclusion

This paper described a comparative evaluation results of Korean text categorization systems based on kNN learning. The performances with different parameters were evaluated on Korean corpora of 5,037 documents. We used DF thresholding and a new DFCF thresholding proposed in this paper in feature selection. The conventional TF/IDF weighting scheme was used to calculate weights of features in a document and training vectors are indexed in inverted file. SWF which considers weighted similarity between documents showed the best performance and ASWF the worst results. The local method with carefully chosen k value shows similar performance with the global method, while the global method usually shows higher performance than the local method with randomly chosen k value. We also verified that only small amount of features, 10% of the whole features, can be used as a feature set without much loss in Korean text categorization. According to the our experimental results, Korean text categorization system based on kNN learning which uses SWF as a decision function, 80 as k value of kNN, 10% of whole features as feature set is much more effective and efficient than systems which uses randomly chosen decision function, k value, and feature set.

In this study, we focused on comparative evaluations of only kNN based methods while there are many approaches to text categorization. We will continue to find out which approaches are more adequate and which kind of feature is effective for Korean text categorization.

References

1. D. G. Lee. *A High Speed Index Term Extracting System Considering the Morphological Configuration of Noun.* Master thesis of Dept. Computer Science and Engineering, Korea University, (2000)
2. David D. Lewis, Robert E. Schapire, James P. Callan, and Ron Papka. Training algorithms for text categorization. *Proc. of the Third Annual Symposium on Document Analysis and Information Retrieval,* (1994)
3. David D. Lewis. Feature Selection and Feature Extraction of Text Categorization. *Proc. of Speech and Natural Language Workshop,* (1992) 212–217
4. Tom M. Mitchell. *Machine learning.* McGraw Hill, (1996)
5. G. Salton. *Automatic Text Processing: The Transformation, Analysis, and Retrieval of Information by Computer.* Addison-Wesley, (1989)
6. K. Tzeras and S. Hartman. Automatic indexing based on bayesian inference networks. *Proc. of the 16th Annual Int. ACM SIGIR Conference on Research and Development in Information Retrieval,* (1993) 22–34
7. Y. Yang. Expert network: Effective and efficient learning from human decisions in text categorization and retrieval. *17th Annual Int. ACM SIGIR Conf. on Research and Development in Information Retrieval,* (1994) 13–22
8. Y. Yang. An evaluation of statistical approaches to text categorization. *Journal of Information Access,* (1996) 99–95
9. Y. Yang and J.P. Pedersen. A comparative study on feature selection in text categorization. In Jr. D. H. Fisher, editor, *The Fourteenth Int. Conf. on Machine Learning,* Morgan Kaufmann, (1997) 412–420
10. E. Wiener, J.O. Pedersen, and A.S. Weigend. A neural network approach to topic spotting. *Proc. of the Fourth Annual Symposium on Document Analysis and Information Retrieval,* (1995)

One-Epoch Learning for Supervised Information-Theoretic Competitive Learning

Ryotaro Kamimura

Information Science Laboratory, Tokai University
1117 Kitakaname Hiratsuka Kanagawa 259-1292, Japan
ryo@cc.u-tokai.ac.jp

Abstract. In this paper, we propose a new computational method for a supervised competitive learning method. In the supervised competitive learning method, information is controlled in an intermediate layer, and in an output layer, errors between targets and outputs are minimized. In the intermediate layer, competition is realized by maximizing mutual information between input patterns and competitive units with Gaussian functions. One problem is that a process of information maximization is computationally expensive. However, we have found that the method can produce appropriate performance with a small number of epochs. Thus, we restrict here the number of epochs to only one epoch for facilitating learning. This computational method can overcome the shortcoming of our information maximization method. We applied our method to chemical data processing. Experimental results showed that with only one epoch, the new computational method gave better performance than did the conventional methods.

1 Introduction

In this paper, we propose a new computational method for the supervised information-theoretic competitive learning method. In the supervised competitive learning, a process of information maximization is computationally expensive. The new computational method can solve this shortcoming. The new method can contribute to neural computing from three perspectives: (1) this is a new type of information-theoretic competitive learning; (2) the new model is a hybrid model in which information maximization and error minimization are combined; and (3) we can restrict the number of learning epochs to only one to facilitate learning.

First, our method is based upon a new type of information-theoretic competitive learning. We have so far proposed information-theoretic approaches to competitive learning [1], [2], [3]. In the methods, competitive processes have been realized by maximizing mutual information between input patterns and competitive units. When information is maximized, just one unit is turned on, while all the others are off. Thus, we can realize competitive processes by maximizing mutual information. In addition, in maximizing mutual information, the entropy of competitive units must be maximized. This entropy maximization can realize equiprobabilistic competitive units without the special techniques that have so far been proposed in conventional competitive learning [4], [5], [6], [7].

N.R. Pal et al. (Eds.): ICONIP 2004, LNCS 3316, pp. 524–529, 2004.
© Springer-Verlag Berlin Heidelberg 2004

Second, our new model is a hybrid model in which information maximization and error minimization are combined with each other. There have been many attempts to model supervised learning based upon competitive learning. For example, Rumelhart and Zipser [8] tried to include teacher information in competitive learning. They called this method the "correlated teacher learning" method, in which teacher information is included in input patterns. On the other hand, Hechet-Nielsen tried to combine competitive learning directly with error minimization procedures [9], in what are called "counter-propagation networks." In our method, we can use the pseudo-inverse matrix operation to produce outputs, and learning is much faster than with counter-propagation.

Third, we restrict the number of epochs to one to facilitate learning. As above mentioned, information-theoretic supervised competitive learning is a new type of supervised learning method with many good points. One shortcoming is that a process of information maximization is computationally more expensive than conventional competitive learning methods. However, many experiments have suggested that the number of learning epochs can be restricted to a small number of epochs. Because information need not to be maximized for many practical problems, we suppose that only one epoch is enough to show good performance. Experiment results have shown that this one-epoch learning can really give better generalization performance.

2   Information Acquisition

Information is defined as decrease in uncertainty observed in initial stages of learning. Uncertainty decrease or information is defined by Information

$$I = -\sum_{\forall j} p(j) \log p(j) + \sum_{\forall s} \sum_{\forall j} p(s)p(j \mid s) \log p(j \mid s), \tag{1}$$

where $p(j)$, $p(s)$ and $p(j|s)$ denote the probability of firing of the jth unit, the probability of the sth input pattern and the conditional probability of the jth unit, given the sth input pattern, respectively.

Let us present update rules to maximize information content. As shown in Figure 1, a network is composed of input units x_k^s and competitive units v_j^s. The jth competitive unit receives a net input from input units, and an output from the jth competitive unit can be computed by

$$v_j^s = \exp\left(-\frac{\sum_{k=1}^{L}(x_k^s - w_{jk})^2}{2\sigma^2}\right), \tag{2}$$

where L is the number of input units, w_{jk} denote connections from the kth input unit to the jth competitive unit, and σ controls the width of the Gaussian function. The output is increased as connection weights come closer to input patterns. The conditional probability $p(j \mid s)$ is computed by

$$p(j \mid s) = \frac{v_j^s}{\sum_{m=1}^{M} v_m^s}, \tag{3}$$

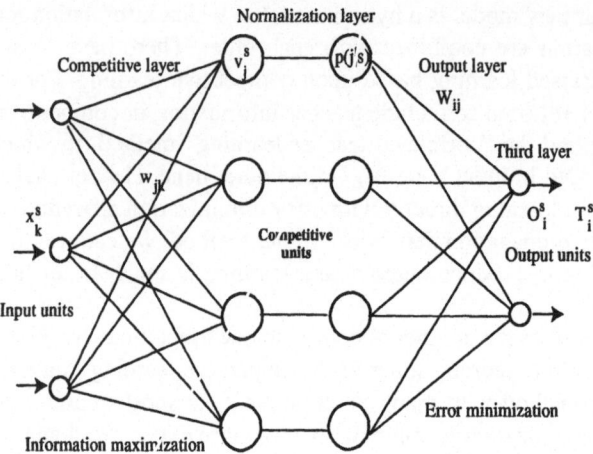

Fig. 1. A network architecture to control information content.

where M denotes the number of competitive units. Since input patterns are supposed to be given uniformly to networks, the probability of the jth competitive unit is computed by

$$p(j) = \frac{1}{S} \sum_{s=1}^{S} p(j \mid s), \tag{4}$$

where S is the number of input patterns. Information I is computed by

$$I = -\sum_{j=1}^{M} p(j) \log p(j) + \frac{1}{S} \sum_{s=1}^{S} \sum_{j=1}^{M} p(j \mid s) \log p(j \mid s). \tag{5}$$

To maximize mutual information, entropy must be maximized, and at the same time conditional entropy must be minimized. When conditional entropy is minimized, each competitive unit responds to a specific input pattern. On the other hand, when entropy is maximized, all competitive units are equally activated on average. As information becomes larger, specific pairs of input patterns and competitive units become strongly correlated. Differentiating information with respect to input-competitive connections w_{jk}, we have

$$\Delta w_{jk} = -\alpha \sum_{s=1}^{S} \left(\log p(j) - \sum_{m=1}^{M} p(m \mid s) \log p(m) \right) Q_{jk}^{s}$$

$$+ \alpha \sum_{s=1}^{S} \left(\log p(j \mid s) - \sum_{m=1}^{M} p(m \mid s) \log p(m \mid s) \right)$$

$$\times Q_{jk}^{s} \tag{6}$$

where α is the learning parameters, and

$$Q_{jk}^s = \frac{(x_k^s - w_{jk})p(j \mid s)}{S\sigma^2}. \tag{7}$$

In the output layer, errors between targets and outputs are minimized. The outputs from the output layer are computed by

$$O_i^s = \sum_{j=1}^{M} W_{ij}p(j|s), \tag{8}$$

where W_{ij} denote connection weights from the jth competitive unit to the ith output unit. Errors between targets and outputs can be computed by

$$E = \frac{1}{2}\sum_{s=1}^{S}\sum_{i=1}^{N}(T_i^s - O_i^s)^2, \tag{9}$$

where T_i^s denote targets for output units O_i^s and N is the number of output units. This linear equation is directly solved by using the pseudo-inverse of the matrices of competitive unit outputs.

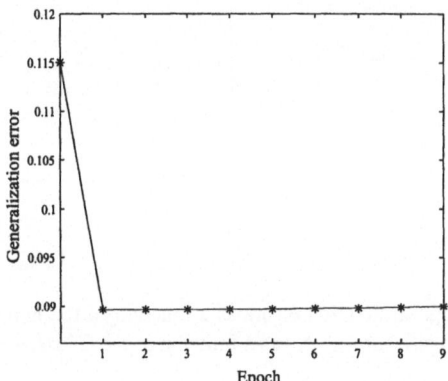

Fig. 2. A typical learning curve for generalization.

3 Chemical Data Processing

We applied the method to chemical data processing. In this experiment, we try to show that the new method shows good performance in generalization. We try to infer subjects' preference on chemical substances. The data was composed of 140 subjects, and their responses to 48 kinds of odors such as aromatic, meaty, sickly as well as their preference. The preference ranges between 1 and 7, and as the number is larger, the rate of preference is smaller. Networks must infer the preference of subjects based upon responses to 48 kinds of odors. In this experiment, the number of input, competitive and output unit were set to 48, 20 and one unit, respectively. We used ten-fold cross validation for evaluating generalization performance.

Figure 2 shows a typical curve for generalization. As can be seen in the figure, with one epoch the generalization error drastically drops, and then remain almost a constant. This figure suggests that information-theoretic learning needs a smaller number of learning epochs. Figure 3(a) shows training and generalization errors as a function of the number of epochs. As shown in the figure, training errors are constantly decreased as the number of competitive units is increased. However, generalization errors are decreased and then increased significantly as the number of competitive units is increased. Figure 3(b) shows training and generalization errors as a function of the Gaussian width σ. Training errors are decreased significantly when the σ is increased from 0.5 to 1. However, the errors remain a constant for larger σ values. On the other hand, generalization errors tend to decrease as the Gaussian width σ is larger. Figure 4 shows the comparison of generalization performance by four methods. Compared with errors by conventional methods (BP, a probabilistic network, LVQ), generalization errors obtained by our method were significantly small.

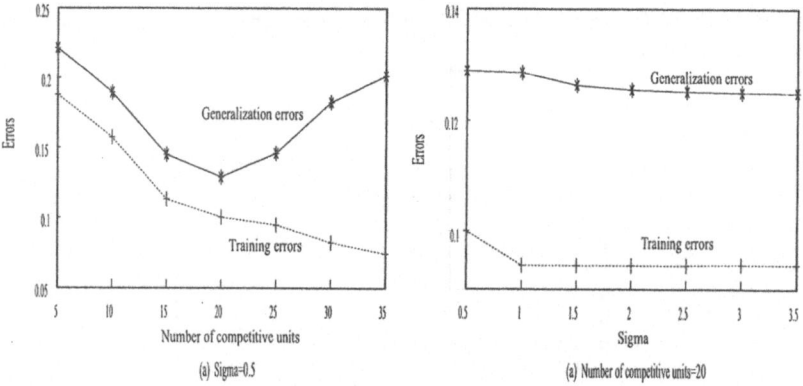

Fig. 3. (a) Training and generalization errors as a function of the number of competitive units; and (b) training and generalization errors as a function of the Gaussian width σ.

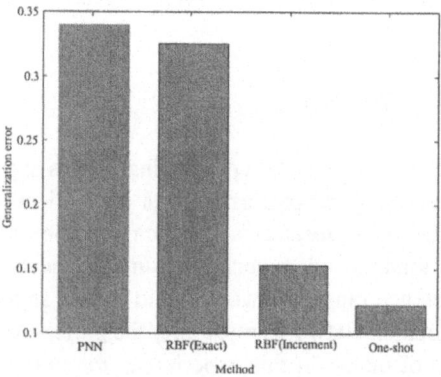

Fig. 4. Comparison of generalization performance by four methods. Errors are the error rates.

4 Conclusion

In this paper, we have proposed a new computational method for supervised infor-mation-theoretic competitive learning. In the paradigm of competitive learning, this is a hybrid model in which unsupervised and supervised learning are combined with each other, which is close to the counter-propagation networks. The difference is that in our method competitive unit outputs are computed by using the Gaussian functions, and in the output layer, the least mean squared error method is used. In the paradigm of the radial-basis function networks, this is a new approach to determine the center of classes. Though our method has many good points, compared with conventional methods, it is computationally expensive, because we need to compute complicated measures such as conditional entropies. However, this paper has shown that the number of epochs needed to show appropriate performance is very small. Thus, we have shown a possibility that our supervised competitive learning method can be practically used. Finally, though some problems remain unsolved, I think that the approach outlined here is a step toward a new information-theoretic approach to neurocomputing.

References

1. R. Kamimura, T. Kamimura, and O. Uchida, "Flexible feature discovery and structural infor-mation," *Connection Science*, vol. 13, no. 4, pp. 323–347, 2001.
2. R. Kamimura, T. Kamimura, and H. Takeuchi, "Greedy information acquisition algorithm: A new information theoretic approach to dynamic information acquisition in neural networks," *Connection Science*, vol. 14, no. 2, pp. 137–162, 2002.
3. R. Kamimura, "Progressive feature extraction by greedy network-growing algorithm," *Complex Systems*, vol. 14, no. 2, pp. 127–153, 2003.
4. D. DeSieno, "Adding a conscience to competitive learning," in *Proceedings of IEEE International Conference on Neural Networks*, (San Diego), pp. 117–124, IEEE, 1988.
5. S. C. Ahalt, A. K. Krishnamurthy, P. Chen, and D. E. Melton, "Competitive learning algorithms for vector quantization," *Neural Networks*, vol. 3, pp. 277–290, 1990.
6. L. Xu, "Rival penalized competitive learning for clustering analysis, RBF net, and curve detection," *IEEE Transaction on Neural Networks*, vol. 4, no. 4, pp. 636–649, 1993.
7. M. M. V. Hulle, "The formation of topographic maps that maximize the average mutual infor-mation of the output responses to noiseless input signals," *Neural Computation*, vol. 9, no. 3, pp. 595–606, 1997.
8. D. E. Rumelhart and D. Zipser, "Feature discovery by competitive learning," *Cognitive Science*, vol. 9, pp. 75–112.
9. R. Hecht-Nielsen, "Counterpropagation networks," *Applied Optics*, vol. 26, pp. 4979–4984, 1987.

Teacher-Directed Learning with Gaussian and Sigmoid Activation Functions

Ryotaro Kamimura

Information Science Laboratory, Tokai University
1117 Kitakaname Hiratsuka Kanagawa 259-1292, Japan
ryo@cc.u-tokai.ac.jp

Abstract. In this paper, we propose a new computational method for information-theoretic competitive learning that maximizes information about input patterns as well as target patterns. The method has been called *teacher-directed learning*, because target information directs networks to produce appropriate outputs. In the previous method, we used sigmoidal functions to activate competitive units. We found that the method with the sigmoidal functions could not increase information for some problems. To remedy this shortcoming, we use Gaussian activation functions to simulate competition in the intermediate layer, because changing the width of the functions accelerates information maximization processes. In the output layer, we used the ordinary sigmoid functions to produce outputs. We applied our method to two problems: an artificial data problem and a chemical data problem. In both cases, we could show that information could be significantly increased with the Gaussian functions, and better generalization performance could be obtained.

1 Introduction

In this paper, we propose a new computational method for teacher-directed learning. In the teacher-directed learning method, teacher information is included in the input layer, and it directs networks to produce appropriate outputs. Because errors between targets and outputs need not be back-propagated, this is a very efficient learning method. However, the method fails to produce appropriate outputs for some problems. To remedy this problem, we propose a new computational method in which the Gaussican activation function is newly used. In the previous methods, we used the sigmoidal activation function [1], [2], [3], [4], [5], [6]. As information is increased, strongly negative connections are generated, and information can easily be increased. However, too strongly negative connections may blur teacher information, and this causes difficulty in decreasing errors between targets and outputs. To remedy this shortcoming, we use Gaussian functions, because we can increase information content by adjusting the width of Gaussian functions. When the width is smaller, competitive units tend to respond to a limited number of input patterns, which should be realized by maximizing mutual information between input patterns and connection weights. Thus, information maximization can be facilitated by adjusting the Gaussian width.

N.R. Pal et al. (Eds.): ICONIP 2004, LNCS 3316, pp. 530–536, 2004.
© Springer-Verlag Berlin Heidelberg 2004

2   Teacher-Directed Information Maximization

Information is defined as decrease in uncertainty from an initial state to a state after receiving input patterns [7], [2]. This uncertainty decrease, or the information, is defined by

$$I(j \mid s) = -\sum_{\forall j} p(j) \log p(j) + \sum_{\forall s} \sum_{\forall j} p(s) p(j \mid s) \log p(j \mid s), \qquad (1)$$

where $p(j)$, $p(s)$ and $p(j|s)$ denote the probability of firing of the jth unit, the sth input pattern and the conditional probability of firing of the jth unit, given the sth input pattern, respectively.

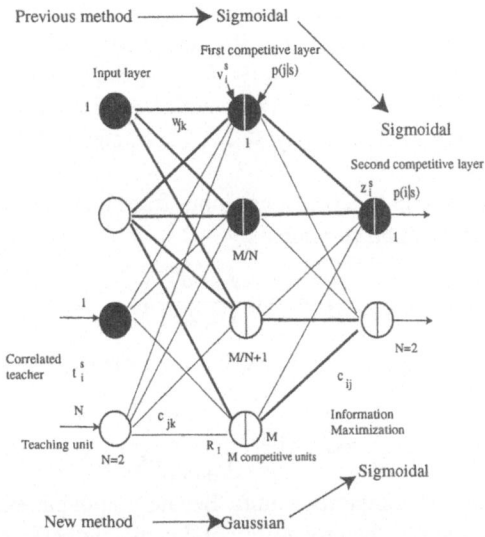

Fig. 1. Multi-layered network architecture for teacher-directed learning.

Then, we attempt to apply the information discussed above to neural networks. As shown in Figure 1, a network is composed of an input, the first competitive and the second competitive layer. In each layer, information is maximized. However, some connections are fixed and do not change throughout learning. Connections from teachers to the first competitive units are fixed, and connections between two competitive layers are also fixed. Thus, connections to be updated are those from training units to the first competitive layer. In a testing phase, no connection weights between correlated teachers and the first competitive layer exist. Thus, networks must infer the final states without teacher information.

Let us present update rules to maximize information content. As shown in Figure 1, a network is composed of L input units and M competitive units. We denote the value of the kth input, given the sth input pattern by x_k^s. For simplicity, the number of competitive units is even, the number of output unit is two and targets are binary. This means that a unit corresponding to a target class is turned on, while all the other units are off. We use weighted distance between input patterns and connection weights to incorporate information on targets. Weighted distance of the jth competitive unit, given the sth input pattern, is defined by

$$d_j^s = \phi_j^s \sum_{k=1}^{L} (x_k^s - w_{jk})^2, \tag{2}$$

where w_{jk} denote connections from the kth input unit to the jth competitive unit, and ϕ_j^s is defined by

$$\phi_j^s = \sum_{k=1}^{N} c_{jk} t_k^s, \tag{3}$$

where t_k^s are targets (supposed to have one or zero), and connection weights c_{jk} are constants. In this paper, the weights c_{j1} for correlated teachers are set to ϵ for $1 \leq j \leq M/2$ and $2 - \epsilon$ for $M/2 + 1 \leq j \leq M$, respectively, where M is the number of competitive units and $0 < \epsilon < 1$. The weights c_{j2} have inverse values: $2 - \epsilon$ for $1 \leq j \leq M/2$ and ϵ for $M/2 + 1 \leq j \leq M$. These connections are not updated and fixed throughout learning.

The jth competitive unit receives a net input from input units, and an output from the jth competitive unit can be computed by

$$v_j^s = \exp\left(-\frac{d_j^s}{2\sigma^2}\right). \tag{4}$$

In modeling competition among units, we use normalized outputs as conditional probabilities:

$$p(j \mid s) = \frac{v_j^s}{\sum_{m=1}^{M} v_m^s}, \tag{5}$$

where M is the number of competitive units. Because input patterns are supposed to be given uniformly to networks, the probability of the jth competitive unit is computed by

$$p(j) = \frac{1}{S} \sum_{s=1}^{S} p(j \mid s), \tag{6}$$

where S is the number of input patterns. Thus, information is computed by

$$I(j \mid s) = -\sum_{j=1}^{M} p(j) \log p(j) + \frac{1}{S} \sum_{s=1}^{S} \sum_{j=1}^{M} p(j \mid s) \log p(j \mid s). \tag{7}$$

Differentiating information I with respect to input-competitive connections w_{jk}, we have the final update rules:

$$\Delta w_{jk} = -\alpha \sum_{s=1}^{S} \left(\log p(j) - \sum_{m=1}^{M} p(m \mid s) \log p(m) \right) Q_{jk}^s$$

$$+ \alpha \sum_{s=1}^{S} \left(\log p(j \mid s) - \sum_{m=1}^{M} p(m \mid s) \log p(m \mid s) \right) Q_{jk}^s, \tag{8}$$

where α is the learning parameter and

$$Q_{jk}^s = \frac{p(j \mid s)(x_k^s - w_{jk})}{S\sigma^2}.$$ (9)

By using this update rule, mutual information is increased as much as possible. In a process of information maximization, units compete with each other, and finally a unit wins the competition. By maximizing information, we can simulate competitive learning.

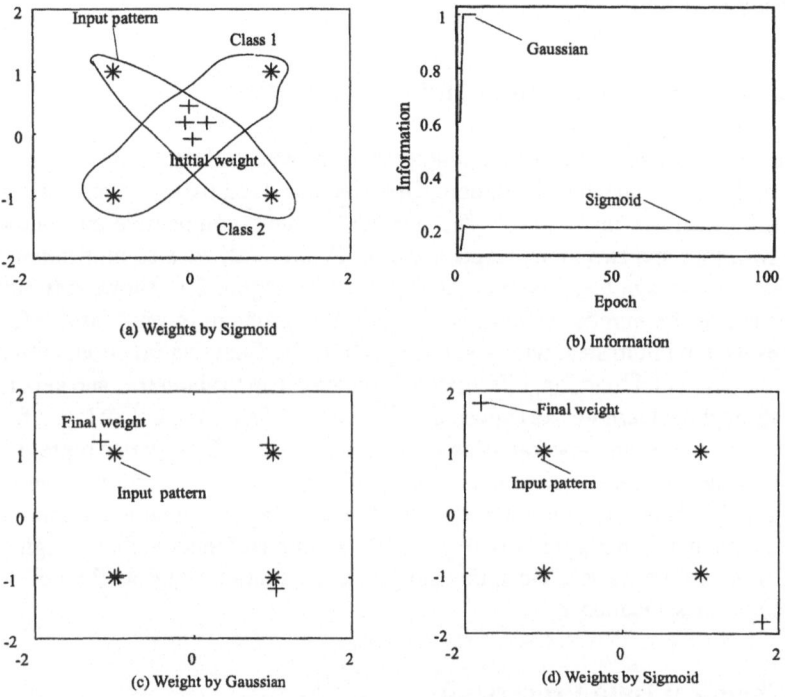

Fig. 2. (a) Input patterns and initial connection weights; (b) information as a function of the number of epochs; (c) final weights by the Gaussian function; and (d) final weights by the sigmoid function.

In the second competitive layer, probabilities are computed in the same way as the previous method. The ith competitive unit in the second competitive layer receives a net input from the first competitive layer, and an output from the ith competitive unit can be computed by

$$z_i^s = \frac{1}{1 + \exp(-\sum_{j=1}^M c_{ij}p(j|s))}$$ (10)

where c_{ij} denotes a connection from the jth competitive unit in the first competitive layer to the ith competitive unit in the second competitive layer. Weights c_{ij} in the second competitive layer are some constants. For example, when the number of competitive units is two, c_{1j} are set to a constant γ for $1 \leq j \leq M/2$ and $-\gamma$ for $M/2 + 1 \leq j \leq M$, respectively. Weights c_{2j} have inverse values[1]. Conditional probabilities are computed by normalized activities:

$$p(i|s) = \frac{z_i^s}{\sum_n^N z_n^s}. \tag{11}$$

Mutual information is computed in the same way. However, this mutual information need not to be updated. By increasing the constant γ, mutual information can be automatically increased.

3   Experiment No.1: Artificial Data Problem

In the first experiment, we use a simple artificial data and try to show how teacher-directed learning can find a solution. The data was composed of four input patterns shown by asterisks in Figure 2. The number of input, competitive and output units were two, four and two units, respectively. In Figure 2(a), we also plot initial random weights (in plus symbols) used in the experiments. Figure 2(b) shows information as a function of the number of epochs. Because this problem is very easy, information reaches its maximum state with just one epoch by the Gaussian function. However, by using the sidmoidal function, information is increased up to about 0.2, and kept to be the same throughout learning. As expected, in the intermediate competitive layer, four input patterns are represented separately as shown in Figure 2(c). By separate representations, networks can solve the problem. This problem can easily be solved by other methods such as LVQ. However, our method seems to reach the final state much more rapidly. On the other hand, in Figure 2(d), by using the sigmoidal function, final weights are far away from input patterns. Thus, this simple demonstration suggests the utility of our method for classification.

4   Chemical Data Processing

In this experiment, we try to show that the new method shows good performance in generalization. In the experiment, we try to infer subjects' preference on chemical substances. The data was composed of 140 subjects' responses (yes or no) to 48 kinds of odors such as aromatic, meaty, sickly as well as their preference. Networks must infer the preference of subjects based upon responses to 48 kinds of odors. In this experiment, the number of input, competitive and output unit were set to 48, 10 and two units, respectively. We used ten-fold cross validation for evaluating generalization performance.

Figure 3(a) shows information as a function of the Gaussian width σ. The number of competitive units was set to six giving the best performance. As the parameter σ is increased, information is naturally decreased. Figure 3(b) shows generalization errors as a function of the width σ. Generalization errors are in the first place decreased, and then increased as the Gaussian width is increased. When the σ is one, the best generalization

---

[1] In the following experiments, this γ was set to 10 as the first approximation.

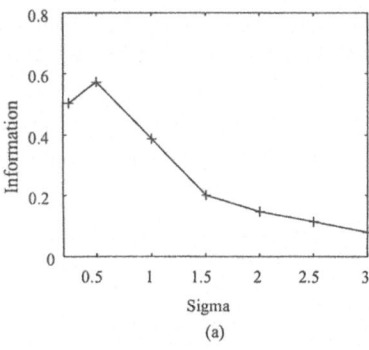 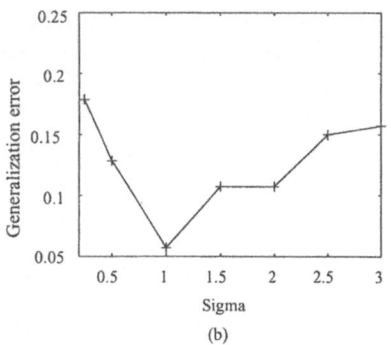

(a)   (b)

Fig. 3. (a) Information as a function of the Gaussian width σ; and (b) generalization errors as a function of the Gaussin width σ.

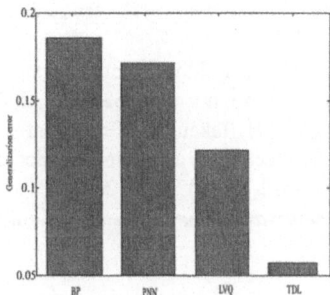

Fig. 4. Comparison of generalization errors by four methods. PNN means a probabilistic neural network.

performance is obtained. Figure 4 shows the comparison of generalization errors by four methods. As can be seen in the figure, generalization error by our method is significantly better than the other conventional methods.

5   Conclusion

In this paper, we have proposed a new computational method for a new type of supervised learning based upon information-theoretic competitive learning. Teacher information is utilized in learning by using weighted distance between input patterns and connection weights. By using teacher information in the input layer, distance between input patterns and connection weights is changed. For example, when distance is large, and teacher information suggests smaller distance, the distance is made smaller by correlated teachers. In competition, we use Gaussian functions of the distance. When weighted distance is smaller, competitive units tend to strongly fire. In this way, we can incorporate teacher information in the input layer. We have applied the method to an artificial data and a chemical data problem. In all problems, we have shown that teacher-direction is very effective in producing outputs. In addition, we have shown the lowest level of generalization errors by our new method. Though some problems must be solved, for example, parameter setting, scalability, it is certain that the new method opens up a new perspective in neural computing.

References

1. R. Kamimura and S. Nakanishi, "Improving generalization performance by information minimization," *IEICE Transactions on Information and Systems*, vol. E78-D, no. 2, pp. 163–173, 1995.
2. R. Kamimura and S. Nakanishi, "Hidden information maximization for feature detection and rule discovery," *Network*, vol. 6, pp. 577–622, 1995.
3. R. Kamimura, "Minimizing α-information for generalization and interpretation," *Algorithmica*, vol. 22, pp. 173–197, 1998.
4. R. Kamimura, T. Kamimura, and T. R. Shultz, "Information theoretic competitive learning and linguistic rule acquistion," *Transactions of the Japanese Society for Artificial Intelligence*, vol. 16, no. 2, pp. 287–298, 2001.
5. R. Kamimura, T. Kamimura, and O. Uchida, "Flexible feature discovery and structural information," *Connection Science*, vol. 13, no. 4, pp. 323–347, 2001.
6. R. Kamimura, T. Kamimura, and H. Takeuchi, "Greedy information acquisition algorithm: A new information theoretic approach to dynamic information acquisition in neural networks," *Connection Science*, vol. 14, no. 2, pp. 137–162, 2002.
7. L. L. Gatlin, *Information Theory and Living Systems*. Columbia University Press, 1972.

Gradient Type Learning Rules for Neural Networks Based on Watcher-Environment Model

M. Tanvir Islam and Yoichi Okabe

Department of Electronic Engineering, University of Tokyo, Hongo 7-3-1,
Bunkyo-Ku, Tokyo 113-8656, Japan
tanvir@is.t.u-tokyo.ac.jp

Abstract. Moderatism, which is a learning rule for ANNs, is based on the principle that individual neurons and neural nets as a whole try to sustain a "moderate" level in their input and output signals. In this way, neural network receives feedback signal from the outside environment, and the principle of connectionism is preserved. In this paper, a potential Moderatism-based local, gradient learning rule based on a watcher-environment model is proposed.

1 Introduction

Artificial neural networks, specially the multi-layer feedforward neural networks have long been used in the field of recognition for their competence in learning. Moderatism [1][2] is a learning model of neurons that models a biological learning characteristic of neither receiving nor transmitting too strong or weak signals. In our previous papers [4][5] we showed that a learning rule called "Error Based Weight Update" (EBWU) shows similar performance as BP in some pattern learning experiments. However, it was seen that when the number of outputs increase, the learning performance deteriorates. To solve this problem, gradient-based learning rules are a good solution. In [6] we showed that the inclusion of the cost of Moderatism in the error function of BP increases the average learning performance. In this paper, we propose one gradient learning rule that are based on Moderatism. Also we show the results of some pattern learning experiments to observe how the new learning rule performs.

2 Learning Rules for Neural Networks

Learning rules for neural networks can be divided into two main categories, namely global learning rules and local learning rules. In global learning rules the learning cost is calculated globally and the weights are adopted according to the global learning rule. In this way, all the synapses of the networks must somehow know their contributions in the global cost and adjust themselves. On the other hand, Local learning rules correspond with connectionism because in local learning rules the synapses do not know the global error, so they only adjust their weights to minimize the local cost. The gradient rule described in this paper is also a local learning rule. In the next sections we shall explain the concept of Moderatism and the gradient rule based on Moderatism.

N.R. Pal et al. (Eds.): ICONIP 2004, LNCS 3316, pp. 537–542, 2004.
© Springer-Verlag Berlin Heidelberg 2004

3   Moderatism

The theoretical background of Moderatism [1][2] comes from the concept of handling feedback signals from the surrounding environment. In general, the feedback signal from the environment to the brain represents either the penalty or the reward for any action ordered by the brain. In either case, the amplitude of the signal can be equally high. Therefore, it is almost impossible to know beforehand what type of feedback signal will come next from the environment. So, in an unknown environment, the best way to survive is to act in way so that the amplitude of the feedback signal from the environment is neither too big nor too small, and holds to a "Moderate" level. In this way it is possible to process signals of penalty and reward in the same manner. From here we shall go through the mathematical model of Moderatism.The dynamics of the neurons is showed in figure 1.

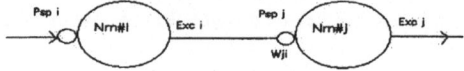

Fig. 1. Neuron-Synapse model

The synaptic weight between neuron #i and #j is w_{ji}, the thresholds of the two neurons are θ_i, θ_j, inputs to the neurons are Psp_i, Psp_j and the outputs are Exc_i, Exc_j respectively. The input signal coming to neuron #j from neuron #i is:

$$Psp_j = w_{ji} \times Exc_i \tag{1}$$

The output of neuron #j is

$$Exc_j = Sigmoid(Psp_j - \theta_j) \tag{2}$$

Where

$$Sigmoid(x) = \frac{1}{(1 + e^{-x})} \tag{3}$$

The variable parameters in this case are the synaptic weights (w_{ji}) of the network. The calculation of signals and cost are given in equation (4).

$$
\begin{aligned}
&Exc = Sigmoid(Input - \theta)\\
&DcPsp = mean(Input)\\
&DcExc = mean(Exc)\\
&AcPsp = (Input - DcPsp)^2\\
&AcExc = (Exc - DcExc)^2\\
&CostPsp = (AcPsp - AcMod)^2\\
&CostExc = (AcExc - AcMod)^2\\
&CostPsp(w_{ji}) = CostPsp_i + CostPsp_j\\
&CostExc(w_{ji}) = CostExc_i + CostExc_j\\
&AcCost(w_{ji}) = CostPsp(w_{ji}) + CostExc(w_{ji})
\end{aligned}
\tag{4}
$$

Here, $mean(x)$ is the average of x over a certain period of time, $AcCost(w_{ji})$ is the cost of weight w_{ji} that needs to be minimized through weight update, and $AcMod$ is the moderate value of the network.

4 Neural Net Structure

The network structure for the new gradient rule is based on a loop structure because a feedback is introduced from the environment. The network structure is based on "watcher-environment" model [7] showed in figure 2. The output of the network is then sent to the "environment", where the learning error is determined. For simplicity, the environment is set to a function that calculates the sum of squares of the learning errors for the training patterns. Then this environment sends back its response (in this case the sum of the squared errors) to a "watcher" neuron, which symbolizes the sensory input from the environment. The output of the watcher neuron is used as the amplitude of a sine wave, which is given as feedback to the input neurons. As the network tries to keep the AC signals to a very small moderate level, the weights are updated in a way so that the sine wave deteriorates, thus minimizing the amplitude of the watcher, and as a result minimizing the training error.

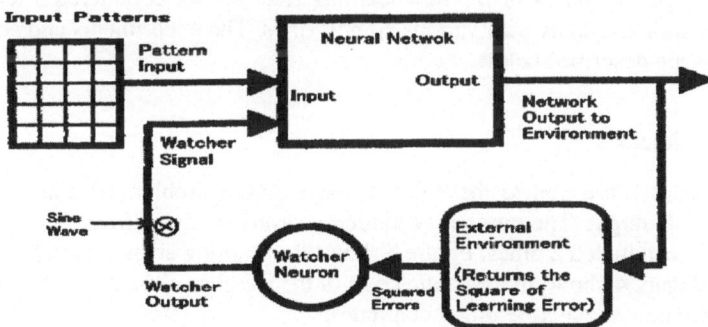

Fig. 2. The "watcher-environment" model

5 Local Learning Rule Based on Moderatism

Based on Moderatism, we propose the following gradient weight-update rule:

$$\Delta w_{ji}^{t} = -\alpha \times \frac{\partial AcCost(w_{ji})^{t}}{\partial w_{ji}} \qquad (5)$$

Where α is the learning rate. We can see from equation (5) that the learning rule is local, unlike BP. Each weight is updated according to the cost that particular weight bears, thus following the principles of connectionism. It is expected that the network will minimize the learning cost as individual weights minimize their own costs, resulting the minimization of the output signal of the watcher neuron.

We thought that Δw can also be linearly related to the gradient of the square of $AcCost(w_{ji})$, so considering the gradient of the square of cost, we get equation (6). The learning rate in this case is different from the learning rate in equation

$$\Delta w'_{ji} = -\alpha' \times AcCost(w_{ji})' \times \frac{\partial AcCost(w_{ji})'}{\partial w_{ji}} \tag{6}$$

Where $\alpha' \neq \alpha$. It should be noticed that the performances of the two learning rules stated in equations (5) and (6) depend on the learning rates. Learning is faster if the learning rate is gradually increased up to a limit, after which the learning rules do not converge. The achieved upper limits of α and α' are 200.0 and 20.0 respectively. To obtain these values we conducted some pattern learning experiments, but in this case we did not use the feedback loop showed in figure 2. Instead we used only the feed-forward part of the structure.

We observed from learning experiments that the learning rule of (6) performs worse than (5). Therefore, from now on, we consider only the learning rule (5).

6 Pattern Learning Experiment

To test the performances of the new learning rule (5), we conducted 3 learning experiments with problems with various complexities. The experiments and corresponding results are described below.

6.1 Experiment 1

In experiment 1, we trained the 2-dimensional EXOR problem to a network with 2 inputs, and 1 output. The number of hidden neurons is varied from 1 to 2 as the experiment is conducted 2 times. Figure 3 shows the learning error curves for the EXOR problem. Figure 4 shows the synaptic costs of the weights. We can see that the synaptic costs are minimized as learning continues.

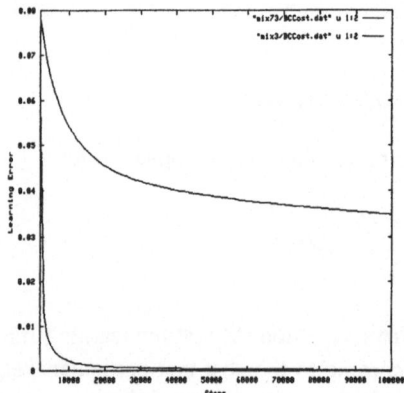

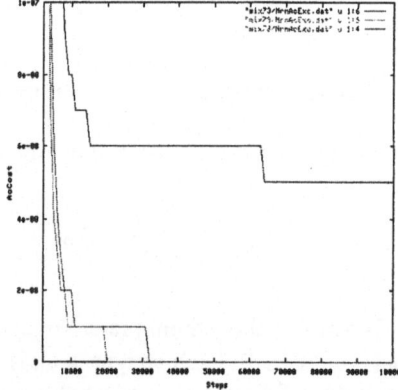

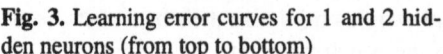

Fig. 3. Learning error curves for 1 and 2 hidden neurons (from top to bottom)

Fig. 4. Synaptic cost curves of the network

6.2 Experiment 2

In experiment 2, we trained a 9 input, 1 output problem to a network with 9 inputs, 4 hidden neurons and 1 output. In each of the input patterns, only one of the 9 values is high and others are low. The teaching outputs are different for each of the outputs. Figure 5 and 6 show the learning error curve and synaptic costs of the network during the training process.

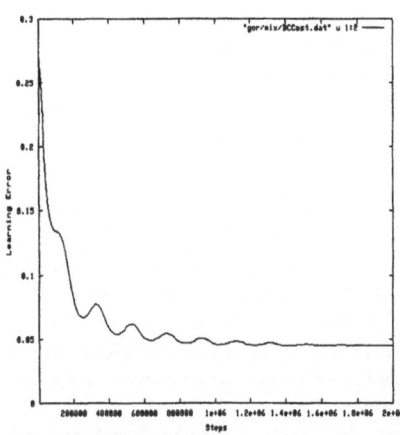

Fig. 5. Learning error curve   **Fig. 6.** Synaptic costs of the network

6.3 Experiment 3

In experiment 3 we trained a more complex 9 input-1 output problem to a network with similar structure as experiment 2. The 9 input patterns are constructed like:

Pattern 1-3: One of the rows has high values, Pattern 4-6: One of the columns has high values, Pattern 7-8: The diagonal pixels have high values, Pattern 9: Pixels 2,4,6,8 have high values. The output values are different for all inputs. Figure 7 and 8 show the learning error curve and synaptic weights of the network during training.

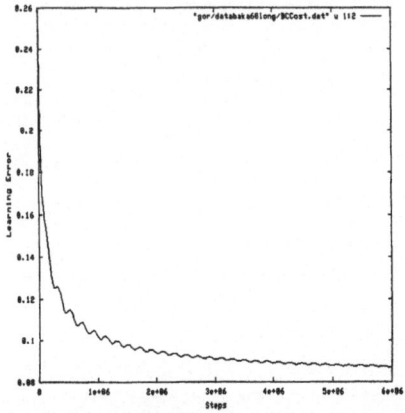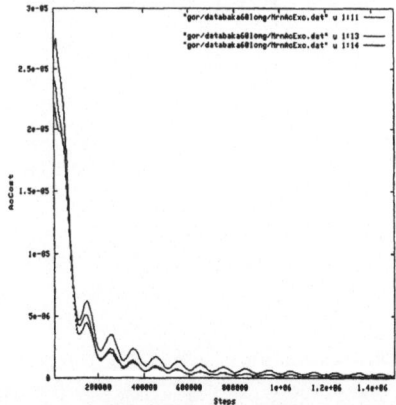

Fig. 7. Learning error curve   **Fig. 8.** Synaptic weights

7 Conclusion

In this paper we have proposed a local gradient learning rule that is theoretically based on Moderatism. The learning rule is implemented on a neural network model that follows the principles of connectionism. We found that the learning rule shows good performance in reducing both learning error and synaptic *AcCost* in the 3 learning experiments. However, we expect that a choice of time variant learning rate and optimum number of hidden neurons will probably cause better performance in experiments 2 and 3. In future we plan to conduct more experiments with bigger networks and datasets to further test the learning capability of the new learning rule.

References

1. Y. Okabe, "Moderationism: Feedback Learning of Neural Networks", *Proceedings of 1988 Intl. Industrial Electronics Conference (IECON'99)*, IEEE, pp. 1028-1033, 1988.
2. T. Kouhara, and Y. Okabe, "Learning algorithm based on Moderationism for multilayer neural networks", *Intl. Joint Conf. On Neural Networks*, vol. 1, pp. 487-490, 1993.
3. C. M. Bishop, "Neural networks for pattern recognition", *Oxford University Press*, 1995.
4. M. Tanvir Islam, and Y. Okabe, "A New Pattern Learning Algorithm For Multilayer Feedforward Neural Networks", *In Proceedings of Intl. Conference on Computer and Information Technology (ICCIT2001)*, pp. 251-255, December 2001.
5. M. Tanvir Islam, and Y. Okabe, "Pattern Learning by Neural Networks Based on Moderatism", *11th Annual Conference of Japanese Neural Network Society, JNNS 2001*, pp. 129-130, September 2001.
6. M. Tanvir Islam, and Y. Okabe, "Pattern Learning by Multilayer Neural Networks Trained by A Moderatism-Based New Algorithm", *In Proceedings of the International Conference On Neural Information Processing (ICONIP02)*, November 2002.
7. M. Tanvir Islam, Yoichi Okabe, "New Gradient Learning Rules for Artificial Neural Nets Based on Moderatism and Feedback Model", *IEEE International Symposium on Signal Processing and Information Technology (ISSPIT2003)*, December 2003 .

Variational Information Maximization for Neural Coding

Felix Agakov[1] and David Barber[2]

[1] University of Edinburgh, 5 Forrest Hill, EH1 2QL Edinburgh, UK
felixa@inf.ed.ac.uk
www.anc.ed.ac.uk
[2] IDIAP, Rue du Simplon 4, CH-1920 Martigny Switzerland
www.idiap.ch

Abstract. Mutual Information (MI) is a long studied measure of coding efficiency, and many attempts to apply it to population coding have been made. However, this is a computationally intractable task, and most previous studies redefine the criterion in forms of approximations. Recently we described properties of a simple lower bound on MI [2]. Here we describe the bound optimization procedure for learning of population codes in a simple point neural model. We compare our approach with other techniques maximizing approximations of MI, focusing on a comparison with the Fisher Information criterion.

1 Introduction

The problem of encoding real-valued stimuli x by a population of neural spikes y may be addressed in many different ways. The goal is to adapt the parameters of any mapping $p(y|x)$ to make a desirable population code for a given set of patterns $\{x\}$. There are many possible desiderata. One could be that *any* reconstruction based on the population should be accurate. This is typically handled by appealing to the Fisher Information which, with care, can be used to bound mean square reconstruction error. Another approach is to bound the probability of a correct reconstruction. Here we consider maximizing the amount of information which the spiking patterns y contain about the stimuli x (e.g. [7], [5]). The fundamental information theoretic measure in this context is the mutual information

$$I(x, y) \equiv H(x) - H(x|y), \tag{1}$$

which indicates the decrease of uncertainty in x due to the knowledge of y. Here $H(x) \equiv -\langle \log p(x) \rangle_{p(x)}$ and $H(x|y) \equiv -\langle \log p(x|y) \rangle_{p(x,y)}$ are marginal and conditional entropies respectively, and the angled brackets represent averages over all variables contained within the brackets.

The principled information theoretic approach to learning neural codes maximizes (1) with respect to parameters of the encoder $p(y|x)$. However, it is easy to see that in large-scale systems exact evaluation of $I(x, y)$ is in general computationally intractable. The key difficulty lies in the computation of the conditional entropy $H(x|y)$, which is tractable only in a few special cases. Standard

N.R. Pal et al. (Eds.): ICONIP 2004, LNCS 3316, pp. 543–548, 2004.
© Springer-Verlag Berlin Heidelberg 2004

techniques often assume that $p(x, y)$ is jointly Gaussian, the output spaces are very low-D, or the channels are invertible [4]. Other methods suggest alternative objective functions (e.g. approximations based on the *Fisher Information* [5]), which, however, do not retain proper bounds on $I(x, y)$. Here we analyze the relation between a simple variational lower bound on the mutual information [2] and standard approaches to approximate information maximization, focusing specifically on a comparison with the Fisher Information criterion.

1.1   Variational Lower Bound on Mutual Information

A simple lower bound on the mutual information $I(x, y)$ follows from non-negativity of the Kullback-Leibler divergence $KL(p(x|y)||q(x|y))$ between the exact posterior $p(x|y)$ and its variational approximation $q(x|y)$, leading to

$$I(x, y) \geq \tilde{I}(x, y) \overset{\text{def}}{=} H(x) + \langle \log q(x|y) \rangle_{p(x,y)}. \tag{2}$$

Here $q(x|y)$ is an arbitrary distribution saturating the bound for $q(x|y) \equiv p(x|y)$. The objective (2) explicitly includes[1] both the encoder $p(y|x)$ (distribution of neural spikes for a given stimulus) and decoder $q(x|y)$ (reconstruction of the stimulus from a population of neural firings). The flexibility of the choice of the decoder $q(x|y)$ makes (2) particularly computationally convenient.

2   Variational Learning of Population Codes

To learn optimal stochastic representations of the continuous training patterns $x_1, \ldots, x_M$ according to (2), we need to choose a continuous density function for the decoder $q(x|y)$. Computationally, it is convenient to assume that the decoder is given by the isotropic Gaussian $q(x|y) \sim \mathcal{N}(Uy, \sigma^2 I)$, where $U \in \mathbb{R}^{|x| \times |y|}$. For simplicity, we limit the discussion to this case only (though other, e.g. correlated or nonlinear cases may also be considered). Then for the empirical distribution $p(x) = \sum_{m=1}^{M} \delta(x - x_m)/M$ we may express the bound (2) as a function of the encoder $p(y|x)$ alone

$$\tilde{I}(x, y) \propto \text{tr} \left\{ \langle xy^T \rangle \langle yy^T \rangle^{-1} \langle yx^T \rangle \right\} + const. \tag{3}$$

Note that the objective (3) is a proper bound for any choice of the stochastic mapping $p(y|x)$. We may therefore[2] use it for optimizing a variety of channels with continuous source vectors.

---

[1] The bound (2) corresponds to the criteria optimized by Blahut-Arimoto algorithms (e.g. [6]); however, we optimize it for both encoder and decoder subject to enforced tractability constraints.

[2] From (3) it is clear that if $\langle yy^T \rangle$ is near-singular, the varying part of the objective $\tilde{I}(x, y)$ may be infinitely large. However, if the mapping $x \mapsto y$ is probabilistic and the number of training stimuli M exceeds the dimensionality of the neural codes $|y|$, the optimized criterion is typically positive and finite.

2.1   Sigmoidal Activations

Here we consider the case of high-dimensional continuous patterns $x \in \mathbb{R}^{|x|}$ represented by stochastic firings of the post-synaptic neurons $y \in \{-1, +1\}^{|y|}$. For conditionally independent activations, we obtain

$$p(y|x) = \prod_{i=1,...,|y|} p(y_i|x) \stackrel{\text{def}}{=} \prod_{i=1,...,|y|} \sigma(y_i(w_i^T x + b_i)) \tag{4}$$

where $w_i \in \mathbb{R}^{|x|}$ is a vector of the synaptic weights for neuron y_i, b_i is its threshold, and $\sigma(a) \stackrel{\text{def}}{=} 1/(1 + e^{-a})$. Optimization of (3) for $W \stackrel{\text{def}}{=} \{w_1, \ldots, w_{|y|}\} \in \mathbb{R}^{|x| \times |y|}$ readily gives

$$\Delta W \propto \sum_{m=1,...,M} \text{cov}(y|x_m) \left(\tilde{D} \lambda_{x_m} + \Sigma_{yy}^{-1} \Sigma_{yx} \left(x_m - \Sigma_{xy} \Sigma_{yy}^{-1} \lambda_{x_m} \right) \right) x_m^T, \tag{5}$$

where $\Sigma_{yy} \stackrel{\text{def}}{=} \langle yy^T \rangle$, $\Sigma_{yx} \equiv \Sigma_{xy}^T \stackrel{\text{def}}{=} \langle yx^T \rangle$ are the second-order moments, $\tilde{D}$ corresponds to the diagonal of $\Sigma_{yy}^{-1} \Sigma_{yx} \left(\Sigma_{yy}^{-1} \Sigma_{yx} \right)^T$, and $\lambda_i(x) \stackrel{\text{def}}{=} \langle y_i \rangle_{p(y_i|x)} = 2\sigma(w_i^T x + b_i) - 1$ is the expected conditional firing of y_i. The update for the threshold Δb has the same form as (5) without the post-multiplication of each term by the training stimulus x_m^T.

From (5) it is clear that the magnitude of each weight update $\Delta w_i \in \mathbb{R}^{|x|}$ decreases with a decrease in the corresponding conditional variance $\text{var}(y_i|x_m)$. Effectively, this corresponds to a variable learning rate – as training continues and magnitudes of the synaptic weights increase, the firings become more deterministic, and learning slows down. One may also obtain a stochastic rule

$$\Delta W \propto \tilde{D} \langle \lambda_x x^T \rangle + \Sigma_{yy}^{-1} \langle \lambda_x x^T \rangle \left(\Sigma_{xx} - \langle x \lambda_x^T \rangle \Sigma_{yy}^{-1} \langle \lambda_x x^T \rangle \right) \tag{6}$$

where $\Sigma_{xx} \stackrel{\text{def}}{=} \langle xx^T \rangle$. Clearly, (6) is decomposable as a combination of the stochastic Hebbian and anti-Hebbian terms, with the weighting coefficients determined by the second-order moments of the firings and input stimuli. Additionally, from (6) one may see that the *"as-if Gaussian"* approximations [7] are suboptimal under the variational lower bound (2) – see [1] for details.

3   Fisher Information and Mutual Information

Let $\hat{x} \in \mathbb{R}^{|x|}$ be a statistical estimator of the input stimulus x obtained from the stochastic neural firings y. It is easy to see that $x \to y \mapsto \hat{x}$ forms a Markov chain with $p(\hat{x}|y) \sim \delta(\hat{x} - \hat{x}(y))$. If $\hat{x}$ is *efficient*, its covariance saturates the Cramer-Rao bound (see e.g. [6]), which results in an upper bound on the entropy of the conditional distribution $H(p(\hat{x}|x))$. From the data processing inequality, one may obtain a lower bound on the mutual information

$$I(x, y) \geq H(\hat{x}) + \langle \log |F_x| \rangle_{p(x)}/2 + const, \tag{7}$$

where $F_x = \{F_{ij}(x)\} \stackrel{\text{def}}{=} -\langle \partial^2 \log p(y|x)/\partial x_i \partial x_j \rangle_{p(y|x)}$ is the Fisher Information matrix. Despite the fact that the mapping $y \mapsto \hat{x}$ is deterministic, exact computation of the entropy of statistical estimates $H(\hat{x})$ in the objective (7) is in general computationally intractable. It was shown that under certain assumptions $H(\hat{x}) \approx H(x)$ [5], leading to the approximation

$$I(x, y) \gtrsim \tilde{I}_F(x, y) \stackrel{\text{def}}{=} H(x) + \langle \log |F_x| \rangle_{p(x)}/2 + const, \tag{8}$$

which is then used as an approximation of $I(x, y)$ independently of the bias of the estimator. Since $H(x)$ is independent of $p(y|x)$, maximization of (8) is equivalent to maximization of (7) where the intractable entropic term is ignored.

For sigmoidal activations (4), the criterion (8) is given by

$$\tilde{I}_F(x, y) \propto \sum_{m=1,\ldots,M} \log |W^T \text{cov}(y|x_m) W| + const. \tag{9}$$

Interestingly, if $|x| = |y|$ then optimization of (9) leads to $\Delta W = 2W^{-T} - \langle \lambda_x x^T \rangle$, which (apart from the coefficient at the inverse weight – *redundancy* term) has the same form as the learning rule of [4] derived for noiseless invertible channels. Notably, the weight update has no Hebbian terms. Moreover, from (9) it is clear that as the variance of the stochastic firings decreases, the objective $\tilde{I}_F(x, y)$ may become infinitely loose. Since directions of low variation swamp the volume of the manifold, neural spikes generated by a fixed stimulus may often be inconsistent. It is also clear that optimization of $\tilde{I}_F(x, y)$ is limited to the cases when $W^T W \in \mathbb{R}^{|x| \times |x|}$ is full-rank, which complicates applicability of the method for a variety of tasks involving relatively low-D encodings of high-D stimuli.

4   Variational Lower Bound *vs.* Fisher Approximation

Since $\tilde{I}_F(x, y)$ is in general not a proper lower bound on the mutual information, it is difficult to analyze its tightness or compare it with the variational bound (2). To illustrate a relation between the approaches, we may consider a Gaussian decoder $q(x|y) \sim \mathcal{N}_x(\mu_y; \Sigma)$, which transforms the variational bound into

$$\tilde{I}(x, y) = -\frac{1}{2} \left\langle \text{tr} \left\{ \Sigma^{-1}(x - \mu_y)(x - \mu_y)^T \right\} \right\rangle_{p(x,y)} + \frac{1}{2} \log |\Sigma^{-1}| + const. \tag{10}$$

Here $\Sigma \in \mathbb{R}^{|x| \times |x|}$ is a function of the conditional $p(y|x)$. Clearly, if the log eigenspectrum of the inverse covariance of the decoder is constrained to satisfy

$$\sum_{i=1,\ldots,|x|} \log l_i(\Sigma^{-1}) = \sum_{i=1,\ldots,|x|} \langle \log l_i(F_x) \rangle_{p(x)}, \tag{11}$$

where $\{l_i(\Sigma^{-1})\}$ and $\{l_i(F_x)\}$ are eigenvalues of Σ^{-1} and F_x respectively, then the lower bound (10) reduces to the objective (8) amended with the average quadratic reconstruction error

$$\tilde{I}(x, y) = -\frac{1}{2} \underbrace{\left\langle \text{tr} \left\{ \Sigma^{-1}(x - \mu_y)(x - \mu_y)^T \right\} \right\rangle_{p(x,y)}}_{\text{reconstruction error}} + \frac{1}{2} \underbrace{\langle \log |F_x| \rangle_{p(x)}}_{\text{Fisher criterion}} + const. \tag{12}$$

Arguably, it is due to the subtraction of the non-negative squared error that (10) remains a general lower bound independently of the parameterization of the model and spectral properties of F_x. Another principal advantage of the variational approach to information maximization is the flexibility in the choice of the decoder [1].

5 Experiments

Variational Information Maximization *vs.* Fisher Criterion

In the first set of experiments we were interested to see how the value of the true MI changed as the parameters were updated by maximising the Fisher criterion $\tilde{I}_F(x, y)$ and the variational bound $\tilde{I}(x, y)$. The dimension $|y|$ was set to be small, so that the true $I(x, y)$ could be computed. Fig. 1 illustrates changes in $I(x, y)$ with iterations of the variational and Fisher-based learning rules, where the variational decoder was chosen to be an isotropic linear Gaussian with the optimal weights (3). We found that for $|x| \leq |y|$ (Fig. 1 (*left*)), both approaches tend to increase $I(x, y)$ (though the variational approach typically resulted in higher values of $I(x, y)$ after just a few iterations). For $|x| > |y|$ (Figure 1 (*right*)), optimization of the Fisher criterion was numerically unstable and lead to no visible improvements of $I(x, y)$ over its starting value at initialization.

Variational IM: Stochastic Representations of the Digit Data

Here we apply the simple linear isotropic Gaussian decoder to stochastic coding and reconstruction of visual patterns. After numerical optimization with an explicit constraint on the channel noise, we performed reconstruction of 196-dimensional continuous visual stimuli from 7 spiking neurons. The training stimuli consisted of 30 instances of digits 1, 2, and 8 (10 of each class). The source variables were reconstructed from 50 stochastic spikes at the mean of the optimal approximate decoder $q(x|y)$. Note that since $|x| > |y|$, the problem could not be efficiently addressed by optimization of the Fisher Information-based criterion (9). Clearly, the approach of [4] is not applicable either, due to its fundamental assumption of invertible mappings between the spikes and the visual stimuli. Fig. 2 illustrates a subset of the original source signals, samples of the corresponding binary responses, and reconstructions of the source data.

6 Discussion

We described a variational approach to information maximization for the case when continuous source stimuli are represented by stochastic binary responses. We showed that for this case maximization of the lower bound on the mutual information gives rise to a form of Hebbian learning, with additional factors depending on the source and channel noise. Our results indicate that other approximate methods for information maximization [7], [5] may be viewed as approximations of our approach, which, however, do not always preserve a proper

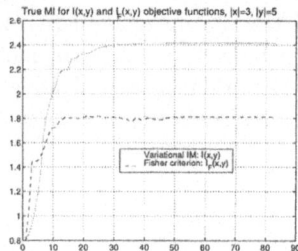 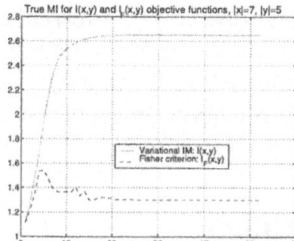

Fig. 1. Changes in the exact mutual information $I(x, y)$ for parameters of the coder $p(y|x)$ obtained by maximizing the variational lower bound and the Fisher information criterion for $M = 20$ training stimuli. *Left:* $|x| = 3$, $|y| = 5$ *Right:* $|x| = 7$, $|y| = 5$.

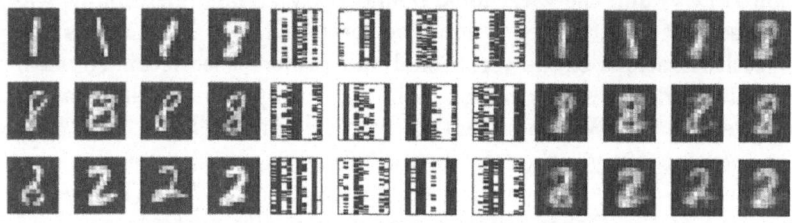

Fig. 2. *Left:* a subset of the original visual stimuli. *Middle:* 20 samples of the corresponding spikes generated by each of the 7 neurons. *Right:* Reconstructions from 50 samples of neural spikes (with soft constraints on the variances of firings).

bound on the mutual information. We do not wish here to discredit generally the use of the Fisher Criterion, since this can be relevant for bounding reconstruction error. However, for the case considered here as a method for maximising information, we believe that our method is more attractive.

References

1. Agakov, F. V. and Barber, D (2004). Variational Information Maximization and Fisher Information. Technical report, UoE.
2. Barber, D. and Agakov, F. V. (2003). The IM Algorithm: A Variational Approach to Information Maximization. In *NIPS*.
3. Barlow, H. (1989). Unsupervised Learning. *Neural Computation*, 1:295–311.
4. Bell, A. J. and Sejnowski, T. J. (1995). An information-maximization approach to blind separation and blind deconvolution. *Neural Computation*, 7(6):1129–1159.
5. Brunel, N. and Nadal, J.-P. (1998). Mutual Information, Fisher Information and Population Coding. *Neural Computation*, 10:1731–1757.
6. Cover, T. M. and Thomas, J. A. (1991). *Elements of Information Theory*. Wiley.
7. Linsker, R. (1989). An Application of the Principle of Maximum Information to Linear Systems. In *NIPS*.

Comparison of TDLeaf(λ) and TD(λ) Learning in Game Playing Domain

Daniel Osman and Jacek Mańdziuk

Faculty of Mathematics and Information Science, Warsaw University of Technology
Plac Politechniki 1, 00-661 Warsaw, Poland
dosman@prioris.mini.pw.edu.pl, mandziuk@mini.pw.edu.pl

Abstract. In this paper we compare the results of applying TD(λ) and TDLeaf(λ) algorithms to the game of give-away checkers. Experiments show comparable performance of both algorithms in general, although TDLeaf(λ) seems to be less vulnerable to weight over-fitting. Additional experiments were also performed in order to test three learning strategies used in self-play. The best performance was achieved when the weights were modified only after non-positive game outcomes, and also in the case when the training procedure was focused on stronger opponents. TD-learning results are also compared with a pseudo-evolutionary training method.

1 Introduction

The Temporal Difference TD(λ) algorithm [1] has been successfully used for learning optimal control in domains with a large state space. Some of the well known applications involving TD(λ) are TDGammon [2] (computer backgammon), KnightCap [3] (computer chess), TDL Chinook [4] (computer checkers) and computer Go [5].

In [6] we have applied the TD(λ) algorithm in the domain of give-away checkers (GAC). The game shares the same rules of playing [7] as regular checkers. The only difference is the goal. In GAC a player wins if no legal move can be made in his turn. The game at first glance may seem trivial or at least not interesting. However a closer look reveals that a strong positional knowledge is required in order to win. A simple piece disadvantage isn't a good estimation of in-game player's performance. Due to the fact that GAC is not a very popular game, we did not concentrate at this point, on creating a master GAC playing program. Our aim was to show that in a domain with a large state space, a control learning program can benefit from the Temporal Difference algorithm even when a relatively simple value function is used (with only 22 weights).

Continuing the work started in [6], we now extend the experiment by testing the TDLeaf(λ) algorithm [3]. TDLeaf(λ) is a modification of TD(λ), enhanced for use in domains, where a d-step look ahead state search is performed in order to choose an action to be executed at a given time step.

We also test a pseudo-evolutionary learning method (EVO) described in [8] and compare it with TD(λ) and TDLeaf(λ).

N.R. Pal et al. (Eds.): ICONIP 2004, LNCS 3316, pp. 549–554, 2004.
© Springer-Verlag Berlin Heidelberg 2004

Several possible training strategies related to Temporal Difference learning are possible in practice. Results presented in [6] show that learning only on non-positive game outcomes i.e. (loss or tie) is much more efficient than learning on all games. In this paper (section 4) we propose and verify a new learning strategy denoted by L3 which consists in playing up to three games in a row against opponents that are stronger than the learning player. The results are very promising.

2   Value Function, TD(λ) and TDLeaf(λ) Algorithms

The value function is used to assign values to states. The value of state s is an approximation of the final game outcome accessible from s. The possible game outcomes are $+100$ for win, -100 for loss and 0 for tie. Each state is defined by a limited number of features (22 in our experiment). The list of features implemented was based on the one used by Samuel in [9]. During the game, the following value function was used:

$$V(s,w) = a \cdot \tanh \left(b \cdot \sum_{k=1}^{K} \omega_k \cdot \phi_k(s) \right), \qquad a = 99, \ b = 0.027, \ K = 22 \qquad (1)$$

where $\phi_1(s), \ldots, \phi_K(s)$ are state features and $w = [\omega_1, \ldots, \omega_K]^T \in \mathbb{R}^K$ is the tunable weight vector. Parameter $a = 99$ to guarantee that $V(s, w) \in (-99; +99)$ and $b = 0.027$ in order to decrease the steepness of the $\tanh(\cdot)$ function.

The TD(λ) and TDLeaf(λ) algorithms are used in order to modify the weights of $V(s, w)$. The goal of this weight correction is to achieve a perfect value function, that is the one that always returns the correct game outcome prediction from any given state $s \in S$. In TDLeaf(λ) [3] the equation for modifying weights is as follows:

$$\Delta w = \alpha \cdot \sum_{t=1}^{N-1} \nabla_w V(s_t^{(l)}, w) \cdot \sum_{i=t}^{N-1} \lambda^{i-t} \cdot d_i \qquad (2)$$

where $s_1, s_2, \ldots, s_N$ are the states observed by the learning player during the entire course of the game and $s_1^{(l)}, s_2^{(l)}, \ldots, s_N^{(l)}$ are the principal variation leaf nodes calculated for these states. Parameter $\alpha \in (0,1)$ is the learning step size and $\lambda \in (0,1)$ is the decay constant. $\nabla_w V(s_t^{(l)}, w)$ is the gradient of $V(s_t^{(l)}, w)$ relative to weights w and $d_i = V(s_{i+1}^{(l)}, w) - V(s_i^{(l)}, w)$ represents the temporal difference in state values obtained after a transition from state s_i to s_{i+1}.

The difference between TD(λ) and TDLeaf(λ) is that in TD(λ) the gradient at time step t in (2) is calculated for $V(s_t, w)$ as opposed to $V(s_t^{(l)}, w)$ in TDLeaf(λ).

3   Experiment Design

Ten learning players were trained in the experiment. Players 1 to 5 played using white pieces (they performed the initial move). Players 6 to 10 played using red pieces. The look ahead search depth was set to $d = 4$. In the first stage, learning players 1 and 6 started with all weights set to zero. The rest of the learning players had their weights initialized with random numbers from interval $(-10, +10)$. Each learning player played in total 10,000 games against a distinct set of 25 random opponents. The opponents in subsequent games were chosen according to some predefined permutation. All 250 opponents were pairwise different and had their weights initialized randomly from interval $(-10, +10)$.

In the second stage, each learning player started out with the weights that were obtained after finishing the first training stage. This time each learning player played 10,000 games against a common set of 25 opponents. 20 of them were randomly picked from the learning players being developed in the first stage at different points in time. The remaining 5 opponents had their weights initialized randomly from interval $(-10, +10)$. The opponents were not modified during training.

Results obtained in the first stage were presented in [6]. Here, we report the results of the second stage of the training phase together with a test (validation) phase which involved matching up every learning player with 100 strong opponents developed in another experiment. These test opponents were not encountered earlier during training. One test phase match was repeated after every 250 training games. There was no weight modification during the test phase. The purpose of this phase was to test the general performance of the learning players. For every win the learning player was rewarded with 1 point, 0.5 for a tie and 0 points for a loss. The percentage results presented in the next section show the fraction of points received by the learning player out of the total number of points available.

In the pseudo-evolutionary learning method (EVO) only one opponent was used during training. This opponent was modified after every two games by adding Gaussian noise to all of its weights. If the learning player lost both games or lost and drawn then its weights were shifted by 5% in the direction of the opponent's weight vector. The two players changed sides after every game. We present the results of the EVO method obtained during training games 1-10,000 since in subsequent games (10,001-20,000) the performance started to gradually decrease.

4   TD(λ), TDLeaf(λ) and EVO Results

Choosing the best learning strategy for Temporal Difference method is still an open question. Many authors, for example [5] and [3], stressed the importance of this aspect. Besides tuning parameters α and λ in (2), it is even more important to properly choose the quality and the number of opponents that the learning player will play against. One must also choose whether the weights of the learning player are to change after each game (this approach will be denoted by LB) or only after games that the learning player lost or drawn (denoted by LL). Another question is whether to play the equal number of times against each opponent or to concentrate on the stronger ones (this strategy will be denoted by L3). Strategy L3, proposed by the authors, is a modification of LL. In L3 the

Table 1. TD, TDLeaf and EVO average results.

(a) Training phase results.

games	TDLeaf+LL	TDLeaf+LB	TD+LL	TD+LB	TD+L3	EVO
1 - 2,500	61.3%	59.0%	61.8%	52.9%	56.7%	64.5%
2,501 - 5,000	66.3%	65.6%	64.3%	54.9%	45.5%	63.5%
5,001 - 7,500	60.1%	63.3%	71.9%	51.8%	40.8%	64.8%
7,501 - 10,000	60.1%	64.6%	62.8%	52.3%	37.5%	63.1%

(b) Test phase results (against strong opponents).

games	TDLeaf+LL	TDLeaf+LB	TD+LL	TD+LB	TD+L3	EVO
1 - 2,500	51.7%	50.4%	53.3%	43.3%	53.3%	30.1%
2,501 - 5,000	52.2%	55.0%	56.7%	44.8%	57.4%	32.9%
5,001 - 7,500	50.6%	56.1%	54.0%	47.9%	56.7%	35.3%
7,501 - 10,000	50.8%	56.0%	54.8%	48.2%	56.1%	36.8%

learning player after losing a game plays the next game with the same opponent (not more than three times in a row, however). This causes the learning process to be concentrated on stronger opponents.

The LB learning strategy for TDLeaf was enhanced (similarly to [3]) by preventing the learning player from learning on the opponents' blunders (more precisely: state transitions that were not predicted by the learning player). This limitation of the number of states that the learning player could learn on was applied only when the learning player won with the opponent. In case of a loss or tie, the learning player learned on all encountered states.

The results of the training phase for all learning strategies presented in Table 1(a) clearly show the superiority of the TD algorithm using the LL learning strategy (TD+LL). The worst performance was obtained by TD+L3, which can be explained in the following way: in L3 the learning player concentrates more on stronger opponents. Since in case of losing the game, the learning player was forced to play against the same opponent again, it turned out that the learning player often lost 3 games in a row when the opponent was really strong. Moreover, due to decrease of α in time the above situation (3 loses in a row) happened more frequently in subsequent training games than at the beginning of the training period. This observation was supported by the following results obtained when the learning coefficient α was kept constant: 56.7%, 45.5%, 44.5%, 44.0%.

In [3] the TDLeaf algorithm showed a faster performance increase than plain TD when learning to play turbo chess. In our experiments indeed TDLeaf achieved the maximum of its performance (66% in games 2,501-5000) earlier than TD (72% in games 5,001-7,500). This is also presented in Figs 1(a) and 1(b). There were a few TDLeaf learning players during games 2,501-5,000 that performed particularly well. However as can be seen in Table 1(a), the overall average training performance of TDLeaf was inferior to TD.

Table 1(b) presents the results of the test phase. This time the results of TDLeaf+LB and TD+LL were comparable. It is important to note that for TDLeaf the LB learning strategy was superior to LL. The most superior results

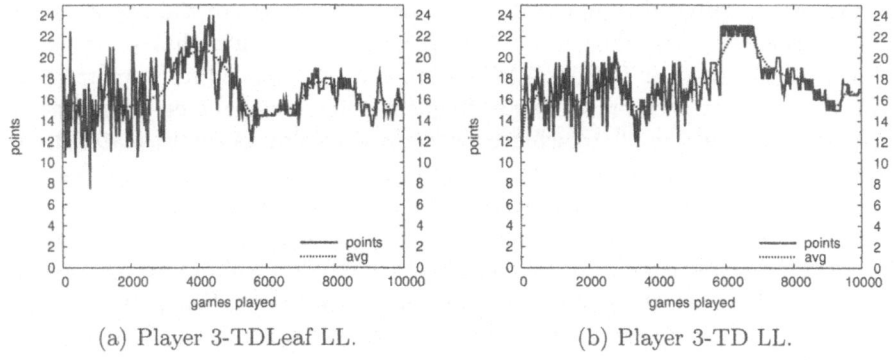

(a) Player 3-TDLeaf LL.   (b) Player 3-TD LL.

Fig. 1. History of performance changes for the best learning players using TDLeaf+LL and TD+LL algorithms during the training phase.

were achieved by TD+L3 method, which confirms the efficacy of L3 learning strategy. The raise in performance from 64.3% to 71.9% by TD+LL in the training phase was not reflected in the test phase where in the same time a fall from 56.7% to 54.0% was observed. This phenomenon of increasing training phase performance along with decreasing test phase results (over-fitting) was only sporadically observed when using the TDLeaf algorithm.

In our experiments the EVO method turned out to be significantly inferior to TD in the test phase (Table 1(b)). Training results for EVO should not be directly compared with the remaining ones in Table 1(a) because of different training strategy used. A similar learning method was used earlier in [8] where its performance was superior to TD. This could have been caused by different characteristics of the games chosen (GAC vs Rummy). Moreover, in our experiments a more sophisticated TD self-play learning strategy had been used enabling better state space exploration due to frequent opponent changing.

In another test phase (not presented in the paper) that consisted in playing against 100 randomly generated opponents the results reached the level of 70-75% for TD and TDLeaf and 60% for EVO. Although the results may not seem very high, they do show the amount of improvement obtained with the Temporal Difference algorithm and promise better performance if a more sophisticated value function were used.

5 Conclusions

The main conclusion from this work is that in the training phase, the TDLeaf algorithm shows overall inferior performance compared to TD. In the test phase however, the results of both algorithms are comparable. The TDLeaf algorithm has an additional benefit of faster performance improvement and seems to be less vulnerable to weight over-fitting which results in good training phase results combined however with an average test phase performance.

The TD learning methods are visibly more efficient than the pseudo-evolutionary technique described in [8]. The results also confirm that self-play can be successfully used in Temporal Difference learning applied in a deterministic game of give-away checkers as long as frequent opponent changing is guaranteed. As it was mentioned in [3], to achieve a good level of play, one must match up the learning player against opponents with strengths similar to that of the learning player. Playing against many different opponents (25 in our case) ensures adequate state space exploration. There is a great chance that once we encounter the same opponent again, the learning player's weights will not be the same as before and therefore the game will take a different course. Secondly in case of many opponents the chance that some of them will share a similar strength of play as the learning player increases.

The L3 training strategy, in which the learning player is more focused on strong opponents was inferior in the training phase but achieved the best score in the test phase. Closer investigation of L3 method is one of our current research goals.

References

1. Sutton, R.: Learning to predict by the method of temporal differences. Machine Learning **3** (1988) 9–44
2. Tesauro, G.: Temporal difference learning and td-gammon. Communications of the ACM **38** (1995) 58–68
3. Baxter, J., Tridgell, A., Weaver, L.: Knightcap: A chess program that learns by combining td(λ) with game-tree search. In: Machine Learning, Proceedings of the Fifteenth International Conference (ICML '98), Madison Wisconsin (1998) 28–36
4. Schaeffer, J., Hlynka, M., Jussila, V.: Temporal difference learning applied to a high-performance game-playing program. In: International Joint Conference on Artificial Intelligence (IJCAI). (2001) 529–534
5. Schraudolph, N.N., Dayan, P., Sejnowski, T.J.: Learning to evaluate go positions via temporal difference methods. In Baba, N., Jain, L., eds.: Computational Intelligence in Games. Volume 62. Springer Verlag, Berlin (2001)
6. Mańdziuk, J., Osman, D.: Temporal difference approach to playing give-away checkers. In Rutkowski, L., Siekmann, J.H., Tadeusiewicz, R., Zadeh, L.A., eds.: 7th International Conference on Artificial Intelligence and Soft Computing (ICAISC 2004), 7th International Conference, Zakopane, Poland. Volume 3070 of Lecture Notes in Computer Science., Springer (2004) 909–914
7. Alemanni, J.B.: Give-away checkers.
   http://perso.wanadoo.fr/alemanni/ give_away.html (1993)
8. Kotnik, C., Kalita, J.K.: The significance of temporal-difference learning in self-play training td-rummy versus evo-rummy. In Fawcett, T., Mishra, N., eds.: Machine Learning, Proceedings of the Twentieth International Conference (ICML 2003), Washington, DC, USA, AAAI Press (2003) 369–375
9. Samuel, A.L.: Some studies in machine learning using the game of checkers. IBM Journal of Research and Development **3** (1959) 210–229

Rule Extraction by Seeing Through the Model

Tuve Löfström[1], Ulf Johansson[1], and Lars Niklasson[2]

[1] University of Borås, Department of Business and Informatics, Borås, Sweden
{tuve.lofstrom,ulf.johansson}@hb.se
[2] University of Skövde, Department of Computer Science, Skövde, Sweden
lars.niklasson@his.se

Abstract. Much effort has been spent during recent years to develop techniques for rule extraction from opaque models, typically trained neural networks. A rule extraction technique could use different strategies for the extraction phase, either a local or a global strategy. The main contribution of this paper is the suggestion of a novel rule extraction method, called Cluster and See Through (CaST), based on the global strategy. CaST uses parts of the well-known RX algorithm, which is based on the local strategy, but in a slightly modified way. The novel method is evaluated against RX and is shown to get as good as or better results on all problems evaluated, with much more compact rules.

1 Introduction

The task to derive transparent formulations from opaque models is often referred to as "rule extraction". Much effort has been spent during the last decades to find effective ways of extracting rules from neural networks and other opaque models, such as committees and boosted decision trees. The reasons for using neural networks in data mining are discussed in [1] and [2] and the benefit of using committees is discussed in [3]. The most important reason for considering opaque learning techniques is the superior ability to generalize; i.e. perform well even on unseen data. Rule extraction allows both a powerful predictive model and an explanation of the relationships found by the model.

Three basic requirements (see e.g. [4]) on extracted rules are accuracy, fidelity, and comprehensibility. Accuracy measures the quality of the predictions on unseen data, with the obvious goal that the accuracy of the rules should be close to that of the opaque model. Fidelity measures how close the predictions of the rules match the predictions of the network. Comprehensibility tries to capture that a rule is easy enough to understand, making it possible to act upon. Comprehensibility is often measured using some complexity measures on the number of conditions in a set of rules. Finally, it is also useful to explore what requirements a rule extraction method imposes on the network and on the format of the data set; see e.g. [1].

Most researchers discuss two major strategies for the rule extraction task; see e.g. [5] - [7]. Craven in [7] uses the terms local and global strategies. Rule extraction techniques using the local strategy extract rules by decomposing a neural network into smaller parts and extracting rules separately for the different parts of the network. Finally, the rule sets from the parts are merged into one rule set that explains the output in terms of the input. When the global strategy is adopted, the opaque model is treated as a function representing the relation between input and output. Thus the task is to directly learn that function and express it as a rule.

N.R. Pal et al. (Eds.): ICONIP 2004, LNCS 3316, pp. 555–560, 2004.
© Springer-Verlag Berlin Heidelberg 2004

The local strategy has some inherent disadvantages in comparison to the global strategy. Techniques using the local strategy can only extract rules from a limited set of opaque models, usually from a limited set of neural network architectures. Techniques using the global strategy, on the other hand, are in general not limited to any specific opaque model.

Several attempts to find effective ways to extract rules have been presented during the years. For a description and comparison of two local methods see [8] - [10]. For two examples using the global strategy see [11] and [12].

The purpose of this paper is to present a new global method for rule extraction, using parts of the RX algorithm presented in [10]. The novel method, is presented and evaluated against the RX algorithm.

2 Problems Used

The problems used are well-known problems previously used within the Machine Learning (ML) community. All the problems are classification problems. The problems have two, five and nineteen dependent classes. The purpose of using problems with so different dimensions is to be able to test the scalability for the used techniques. One of the problems has been used to evaluate the NeuroLinear in previous papers [9]. This problem is used in the same way in this study to enable comparison of the results.

2.1 Problem 1: Cleveland

The Cleveland database has been obtained from the Hungarian Institute of Cardiology, Budapest from Dr. Andras Janosi. It has previously been used in several ML studies; see e.g. [9]. The task is to classify heart disease using 13 attributes, 5 discrete and 8 continuous. The two dependent classes are sick and healthy.

2.2 Problem 2: Annealing

The Annealing database has been donated by David Sterling and Wray Buntine. The task is to classify the results from an annealing process using 38 attributes, 32 discrete and 6 continuous. The dependent classes are 1, 2, 3, 5, U, with about 2/3 of the instances classified as 3.

2.3 Problem 3: Soybean

The Soybean database has been donated by Ming Tan & Jeff Schlimmer. The task is to classify soybeans into one of 19 dependent classes using 35 discrete attributes.

3 Method

NeuroRule presented in [10] has a basic setup containing three steps. The first two steps can be repeated several times.

1. Train a neural network for a specific problem
2. Prune the network
3. Extract rules from the pruned network using the RX algorithm

The RX algorithm clusters the activation values of the hidden nodes and creates the rules with the mean of each cluster instead of the real activation value. A detailed description of the algorithm can be found in [10].

NeuroRule can only extract rules from problems where the variables have been transformed into binary variables. As a consequence, the problems need to be represented with a much larger number of variables than the original problem. A detailed description of how this is done can be found in [10].

Since RX uses the local strategy, it can not operate on a committee of networks but must use only one network. Table 1 shows the benefits of using committees on two of the evaluated problems. There is no single network with a better accuracy on either the training or the test set than the committee, whose output is the average of all the networks. For the annealing problem there is, however, individual networks that are as good as the committee. The soybean problem is rather complex, partly because of the large number of dependent classes, which might be an explanation of the great difference in performance between individual networks and the committee.

Table 1. Improvements with committees

	Annealing		Soybean	
	Train	Test	Train	Test
Individual Net 1-10 (Average result)	93.7 %	94.3 %	75.6 %	73.3 %
Committee	98.8 %	98.3 %	89.6 %	90.1 %

Techniques using the local strategy are limited in their scalability because they can not use for instance committees and boosting. Most of them also have major scalability problems since the complexity of the rule extraction phase, in most cases is exponential; see e.g. [11].

The criterion most difficult to fulfil when using the global strategy, is often assumed to be fidelity; see e.g. [11] - [12]. On the other hand, techniques using the local strategy could be expected to achieve very high fidelity, because of the nature of the strategy. Obviously fidelity is important since it measures how well the extracted rules explain the opaque model.

3.1 The Cluster and See Through (CaST) Method

The number of conditions in the rules or decision trees is often dependant on the complexity of the input patterns. By reducing the number of input values, the number of conditions could probably be reduced as well. To achieve a reduced number of input values the following steps is suggested.

1. Apply a clustering algorithm on the inputs to find clusters of input values which results in (almost) as good accuracy as the original values.
2. Generate rules that describe the network outputs in terms of the clustered inputs.

The RX algorithm has four steps, where in the first step the continuous activation values from the hidden nodes are discretized. In the following steps the final rule set is built stepwise, first from hidden layer to output, then from input to hidden layer, and finally these two rule sets are merged.

By modifying the first step of the RX algorithm slightly, it can be used to cluster input values. The clustering algorithm is presented below:

Clustering Algorithm

```
Initialization: Let F = 1.
For each input variable c = 1,2,...,i:
1. If F ≤ 0 then stop. Let D be the number of clusters of
   input values. Let δ₁ be the value of the first instance of
   the training set.
   Let H(1) = δ₁, count(1) = 1, sum(1) = δ₁, and let D = 1.
   Let ε = F * (max(input(c)) - min(input(c))).
2. For each instance r = 1,2,...,m in the training set:
   a. Let δ be the value of instance r.
   b. If there exists any index j' such that
      |δ-H(j')| = min ⱼ₌₁,...,ₚ |δ-H(j)| and |δ-H(j')| ≤ ε
      Then: Let count(j') = count(j') + 1,
      sum(j') = sum(j') + δ
      Else: Create a new cluster. Let D = D + 1, H(D) = δ,
      count(D) = 1, and sum(D) = δ.
   c. Update the cluster means. H(j) = sum(j)/count(j), j =
      1,2,...,D.
3. If the accuracy is below the preset threshold then reduce
   F and restart at 1, else stop.
```

In the clustering algorithm, i is the number of variables, m is the number of instances and $input(c)$ is the set of all values for variable c occurring in the training instances.

Any transparent technique can be used in step 2 to create the rules. In this experiment, a technique using the heuristics proposed in [13] is used to create the rules.

CaST is evaluated against NeuroRule, using RX, and the standard tool See5 on the problems described above. Only the results with the highest accuracy will be presented, since accuracy usually is the most important criteria. Pruning was used to reduce the problems for both NeuroRule and CaST.

4 Results

In Table 2 the results from the problems are presented. NeuroRule fails to produce a result for the soybean problem. It was not possible to sufficiently prune the network and still preserve an acceptable accuracy when the input variables had to be represented as binary variables. With an accuracy of around 60 %, the network was still too complex to allow rules to be extracted in reasonable time, i.e. within a day or two.

CaST extracts rules from committees while NeuroRule extracts only from single networks. On the Cleveland problem, NeuroRule and CaST obtain equal results, except that the complexity of the NeuroRule results makes the rules all but comprehensible. NeuroRule was, however, able to create a much less complex rule set with a lower accuracy. A probable reason for the large difference in complexity between CaST and NeuroRule is that NeuroRule had to represent the problems with a larger number of variables, while CaST could use clusters of values from the original representation of the problem, i.e. with fewer variables.

Table 2. Results

Problem	Technique	Accuracy	Opaque Model	Fidelity	Complexity
	CaST	82.5 %	82.5 %	100 %	6
Cleveland	NeuroRule	82.5 %	82.5 %	100 %	572
	See5	81.1 %	-	-	13
	CaST	98.3 %	98.3 %	99.8 %	28
Annealing	NeuroRule	97.5 %	97.5 %	100 %	53
	See5	98.5 %	-	-	28
	CaST	89.3 %	90.1 %	98.3 %	122
Soybean	NeuroRule	-	-	-	-
	See5	88.8 %	-	-	92

The results from See5 and CaST are similar regarding both complexity and accuracy for all the problems. For two of the problems, it was possible to obtain better accuracy with the opaque model than with See5, even though the opaque model had been pruned.

As a comparison, the average accuracy on the Cleveland problem for the rules extracted with NeuroLinear presented in [9] is 78.2 %, with a standard deviation of 6.86. The complexity is hard to compare, since only the number of rules is presented, not the number of conditions in the rules, but CaST has got at least as small rules as NeuroLinear on this problem. CaST and NeuroRule have results comparable to NeuroLinear on the Cleveland problem.

The fidelity is very high on all problems, both for NeuroRule and CaST.

5 Conclusions

The limited number of experiments does not allow any far-reaching conclusions. CaST obtains better results than NeuroRule overall, but the only obvious difference is the complexity. CaST can also compete with the standard tool See5 on all the problems, regarding both accuracy and complexity. The fact that it was possible to get better results with pruned opaque models than with See5 implies that it is useful to use rule extraction as a natural technique to create powerful and transparent models. A key result from the experiments is that CaST, on the data sets used, meets all the criterions rather well, with high accuracy and fidelity despite the compact rules.

6 Discussion and Future Work

The problems used show that CaST is able to scale up to complex problems, like the soybean problem. The clustering algorithm used can, however, only create clusters with adjacent values. By using more general clustering algorithms the method could be improved.

Other techniques to extract the rules in the second step should be evaluated. By using other extraction techniques, it would be possible to choose the representation language for the rules. It would also be interesting to study which rule extraction techniques that can be used without losing the good fidelity achieved in the presented results.

A meta learning study, where CaST is evaluated against several other rule extraction and transparent learning techniques, should be performed. Such study would make it possible to draw conclusions regarding the usefulness of the proposed method.

Results in this study indicate that the global method extracts more comprehensible rules than the local method; this should be evaluated in a broader study.

References

1. Craven, M.; Shavlik, J. *Using neural networks for data mining*. Future Genertion Computer Systems, vol 13, 211-229, 1997
2. Lu, H.; Setiono, R.; Liu, H. *Effective Data Mining Using Neural Networks*. IEEE Trans. Knowl. Data Eng. vol 8(6), 957-961, 1996
3. Bishop, C. M. *Neural Networks for Pattern Recognition*. Oxford University Press, Oxford, UK, 1995.
4. Johansson, U., König, R. and Niklasson, L. *Rule Extraction from Trained Neural Networks Using Genetic Programming*. 13th International Conference on Artificial Neural Networks (ICANN), Istanbul, Turkey, pp 13-16, 2003.
5. Tickle, A.; Golea, M.; Hayward, R.; Diederich, J. *The Truth Is in There: Current Issues in Extracting Rules from Feedfoward Neural Networks*. Proceedings of the International Conference on Neural Networks. Volume 4, 2530-2534, 1997
6. Tickle, A.; Andrews, R.; Diederich, J. *A survey and critique of techniques for extracting rules from trained artificial neural networks*. Knowledge-Based Systems. Volume 8, 373-389, 1995
7. Craven, M. *Extracting comprehensible models from trained neural networks*. Ph.D. Thesis, University of Wisconsin, Madison, 1996.
8. Hayashi, Y.; Setiono, R.; Yoshida, K. *A comparison between two nerual network rule extraction techniques for the diagnosis of the hepatobiliary disorders*. Artificial Intelligence in Medicine, vol 20, 205-216, 2000
9. Setiono, R.; Liu, H. *NeuroLinear: From neural networks to oblique decision rules*. Neurocomputing, vol 17, 1-24, 1997
10. Lu, H.; Setiono, R.; Liu, H. *NeuroRule: A Connectionist Approach to Data Mining*. Proceedings of the 21st VLDB Conference, Zürich, Switzerland, 1995.
11. Craven, M.. *Extracting Comprehensible Models form Trained Neural Networks*. PhD Thesis, University of Wisconsin, Madison. 1996.
12. Johansson, U. *Rule Extraction – the Key to Accurate and Comprehensible Data Mining Models*. Licentiate Thesis, University of Linköping. 2004.
13. Saito, K.; Nakano, R. *Medical diagnostic expert system based on PDP model*. Proceedings of the IEEE International Conference on Neural Networks, San Diego, CA, IEEE Press, 255-262, 1988

An Auxiliary Variational Method

Felix V. Agakov[1] and David Barber[2]

[1] University of Edinburgh, 5 Forrest Hill, EH1 2QL Edinburgh, UK
felixa@inf.ed.ac.uk
http://anc.ed.ac.uk
[2] IDIAP, Rue du Simplon 4, CH-1920 Martigny Switzerland
david.barber@idiap.ch

Abstract. An attractive feature of variational methods used in the context of approximate inference in undirected graphical models is a rigorous lower bound on the normalization constants. Here we explore the idea of using augmented variable spaces to improve on the standard mean-field bounds. Our approach forms a more powerful class of approximations than any structured mean field technique. Moreover, the existing variational mixture models may be seen as computationally expensive special cases of our method. A byproduct of our work is an efficient way to calculate a set of mixture coefficients for any set of tractable distributions that principally improves on a flat combination.

1 Introduction

Probabilistic treatment of uncertainty provides a principled way of reasoning in stochastic domains. Unfortunately, mathematical consistency often comes at a price of inherent intractability of many interesting models, such as Boltzmann machines

$$p(\mathsf{x}) = \exp\{-E(\mathsf{x})\}/Z, \quad Z = \sum_{\mathsf{x}} \exp\{-E(\mathsf{x})\}. \tag{1}$$

In general the complexity of evaluating the partition function Z is exponential in the size of the largest clique in the associated junction tree. For dense models the exact evaluations are in general computationally infeasible, and approximations need to be considered. In this paper we focus on computation of lower bounds on Z, which may also be used to approximate formally intractable marginals.

Variational approximations have been widely used in physics and engineering and more recently applied to graphical modeling (e.g. [5]). In this context they are typically used to obtain rigorous (but relatively simple) bounds on the normalizing constant. A popular class of such methods is based on the KL-divergence

$$KL(q(\mathsf{x})\|p(\mathsf{x})) = \langle \log q(\mathsf{x})\rangle_{q(\mathsf{x})} - \langle \log p(\mathsf{x})\rangle_{q(\mathsf{x})} \geq 0, \tag{2}$$

where $\langle\ldots\rangle_{q(\mathsf{x})}$ denotes an average over $q(\mathsf{x})$, and the bound is saturated if and only if $q(\mathsf{x}) \equiv p(\mathsf{x})$. In the case of the Boltzmann distribution (1), non-negativity of (2) yields the well-known class of lower bounds

$$\log Z \geq -\langle \log q(\mathsf{x})\rangle_{q(\mathsf{x})} - \langle E(\mathsf{x})\rangle_{q(\mathsf{x})}, \tag{3}$$

N.R. Pal et al. (Eds.): ICONIP 2004, LNCS 3316, pp. 561–566, 2004.
© Springer-Verlag Berlin Heidelberg 2004

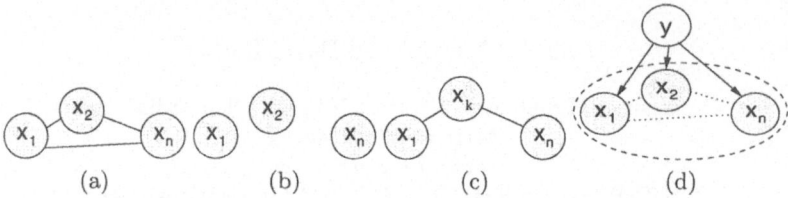

Fig. 1. (a) A fully connected network representing the intractable $p(x)$; (b) standard mean field model $q_{MF}(x)$; (c) structured mean field model $q_{SMF}(x)$; (d) a mixture of mean field models [all the variables x are coupled through the mixture label y.].

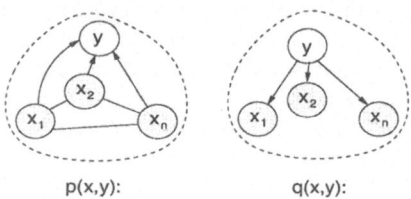

p(x,y):   q(x,y):

Fig. 2. An auxiliary MF model. The target $p(x,y)$ is approximated by $q(x,y)$, which is structured *in the augmented space*. [Note that the marginal $p(x)$ expressed from $p(x,y)$ is identical to the original fully connected pairwise distribution shown on Figure 1 (a).

where $q(x)$ is typically restricted to lie in a tractable family and varied to obtain the tightest bound within the family. Coupled with an upper bound on Z, expression (3) may be used for bounding marginals of $p(x)$. This procedure may also be used to optimize a lower bound on the marginal likelihood in partially observable models, which is a natural generalization of the EM algorithm [7].

1.1 Existing Variational Approximations

The tractability of the bound (3) depends on the choice of the approximating distribution $q(x)$, which is in the simplest case given by the factorized *mean field* (MF) model $q_{MF}(x) = \prod_i q(x_i)$ (see Fig. 1 (a), (b)). Factorized approximations may be simple, but inaccurate when $p(x)$ is strongly coupled; moreover, due to uni-modality they may miss a significant mass contributing to Z. One way to go beyond the factorized assumption for $q(x)$ is to consider a *structured mean field* approximation, which also introduces conditional independencies, but retains some of the structure of $p(x)$. Often it is assumed that $q(x)$ has a sparse graphical representation (e.g. it is a (poly)tree, see Figure 1 (c)), which typically leads to an improvement on the bound at a moderate increase in computational cost. Another approach examined recently [4], [6] uses mixtures of mean field type models (see Figure 1 (d)). This is a powerful extension of factorized approximations, since the resulting $q(x)$ is in general multi-modal and coupled in x (Fig. 1); however, in this case the bound (3) is itself intractable. Known techniques [6], [4], [3] handle the intractability by effectively using the Jensen's bound on top of (3), which is computationally costly and numerically unstable in practice unless *all* the mixture components $q(x|y)$ have the same structure.

2   Auxiliary Variational Method

Intuitively, evaluation of the bound (3) on $\log Z$ for variational mixture approximations requires minimization of the KL-divergence between two fully-connected distributions (see (2)). However, computationally it could be useful to retain a sparser structural form of $q(x, y)$ and use it as an approximation. To do this, we introduce *auxiliary variables* y to the target distribution in such a way that the marginal $\tilde{p}(x)$ of the augmented model $p(x, y)$ has the same graphical structure as the original target $p(x)$ (see Fig. 2). Then we minimize the KL-divergence between $q(x, y)$ and $p(x, y)$ in the joint variable spaces. This case is different from standard structured approximations, as all the variables x of the marginal $\langle q(x|y) \rangle_{q(y)}$ remain fully connected. However, similarly to structured mean field methods, the approximation $q(x, y)$ in the joint space is constrained to be sparse.

Another motivation for this work is the reported success of auxiliary sampling techniques, such as *Hybrid Monte-Carlo* or the *Swendsen-Wang* [8] algorithms. It has been shown that by augmenting the original variable space with auxiliary variables and sampling from joint distributions in the augmented spaces, one can achieve a significant improvement over standard MCMC approaches. The purpose of the auxiliary variables in this context is to capture (structural) information about clusters of correlated variables. It may therefore be hoped that an *auxiliary variational* method performing approximations in the augmented space $\{x, y\}$ may improve on simple approximations in $\{x\}$.

2.1   Optimizing the Auxiliary Variational Bound

Let $p(x, y) = p(x)p(y|x)$ define the joint distribution of the original variables x and auxiliary variables y in the augmented $\{x, y\}$ space. From the divergence $KL(q(x, y) \| p(x, y))$ in the joint space it is easy to obtain an expression for the lower bound on the normalizing constant of (1), which is given by

$$\log Z \geq \sum_{y} q(y) \left[\langle -E(x) - \log q(x|y) \rangle_{q(x|y)} \right] + \tilde{I},$$

$$\tilde{I} = \sum_{x} \sum_{y} q(x, y) \log \frac{p(y|x)}{q(y)} \tag{4}$$

where $p(y|x)$ is an arbitrary *auxiliary conditional* distribution. Clearly, (4) decomposes as a convex sum of the standard lower bounds with approximations $q(x|y)$ and a lower bound $\tilde{I}(x, y)$ on the mutual information. This may be used to improve on a single best (tractable) approximation $q(x|y)$, which is reconstructed by trivially setting $p(y|x) \equiv p(y)$. Note that (4) is tractable as long as $p(y|x)$, $q(y)$, and $q(x|y)$ are constrained to lie in tractable families – there is no need to use further variational relaxations in this case. One tractable choice for the auxiliary mapping $p(y|x)$ is a Gaussian (in this case $q(x|y)$ should also be parameterized as y is real-valued). Another case leading to exact computations is obtained when each node y_i in $p(y|x)$ has a small number of x-parents, and

$q(x, y)$ is a tree. Some other parameterizations may not lead to exact bounds, but may nevertheless result in efficient and practically useful approximations (see [1] for discussions and derivations of the EM algorithms for some of these cases).

2.2   Specific Auxiliary Representations

Here we briefly look at two useful choices of $p(y|x)$ for pairwise Markov networks with $E(x) \stackrel{\text{def}}{=} x^T W x$. As usual, we assume that $q(x, y)$ is tractable, i.e. the only problematic term in (4) is the auxiliary expectation $\langle \log p(y|x) \rangle_{q(x,y)}$.

Parametric Constraints on the Auxiliary Distributions

If the auxiliary space is given by a single multinomial variable $y \in \{1, \ldots, M\}$, a natural choice for $p(y|x)$ is to use a *softmax* type representation

$$p(y_k|x) \propto \exp\left\{ f(x^T u^{(k)} + b^{(k)}) \right\}, \quad U = \{u^{(1)}, \ldots, u^{(M)}\} \in \mathbb{R}^{|x| \times M}, b \in \mathbb{R}^M \quad (5)$$

where $f(x; U, b)$ is some differentiable function and $p(y_k|x)$ is the probability of the auxiliary variable y being in state k. Unfortunately, if the weight vector $u^{(k)}$ is dense, one may need to bound $\langle \log p(y_k|x) \rangle_{q(x,y)}$ (a cheaper alternative is to approximate such terms as $\log p(y_k|\langle x \rangle_{q(x)})$ or use a multivariate factorial representation of the auxiliary variables $p(y|x) = \prod_i p(y_i|x)$ with the *Gaussian field* [2] approximation of $\langle \log p(y_i|x) \rangle_{q(x|y)}$ – see [1] for details). For dense weights such approximations usually do not lead to significant deviations of the objective and are shown to be both accurate and efficient [1]; however, the optimized function is no longer a strict bound on $\log Z$. We now consider a tractable case when $p(y|x)$ has constraints on the parental structure for each auxiliary variable.

Structural Constraints on Auxiliary Distributions

If $\pi_x(y_i)$ and $\pi_y(y_i)$ are x- and y-parents of y_i in the mapping $p(y|x)$, we get

$$\langle \log p(y|x) \rangle_{q(x,y)} = \sum_{i=1,\ldots|y|} \langle \log p(y_i|\pi_x(y_i), \pi_y(y_i)) \rangle_{q(y_i, \pi_y(y_i), \pi_x(y_i))} . \quad (6)$$

The representational complexity of each conditional in (6) in this case is of the order of $s^{|\pi_x(y_i)| + |\pi_y(y_i)|}$, where s is the number of states (for simplicity assumed to be equal for each variable). Since we are free to choose the form of the distribution $p(y|x)$, we can limit its parental structure so that $|\pi_x(y_i)| + |\pi_y(y_i)|$ is small. Clearly, for discrete variables this allows an exact representation of the conditionals. It is also clear that the computational complexity of evaluating (6) is limited by the cost of marginalization of $|x| + |y| - |\pi_x(y_i)| - |\pi_y(y_i)| - 1$ variables from $q(x, y)$, which is tractable as long as $q(x, y)$ is in a tractable family. E.g., in the special case when $q(y, x) = \prod_{l=1}^{|y|} q(y_l) \prod_{j=1}^{|x|} q(x_j|\rho(x_j))$ is a polytree with a small number of y-parents $\rho(x_j)$, the marginalization is exponential in $|\rho(\pi_x(y_i))\setminus\{\pi_y(y_i)\cup y_i\}|$, which is acceptable if both $p(y|x)$ and $q(x|y)$ are sparse.

3   Relation to Variational Mixture Models

The existing variational mixture approaches may be viewed as a special case of the *unconstrained* auxiliary formulation (where $p(y|x) \equiv q(y|x)$). In this case (4) is intractable even for factorized mixtures with a few components (as $|x|$ is large). This *requires* further factorized relaxations, such as the ones used by [4], [6], [3]. Also, unless all the components $q(x|y)$ have identical structures, the optimization of the existing bounds may become numerically unstable (as it requires computing non-factorized summations of exponentially small terms [6]).

In our approach, we optimize the bound on $\log Z$ subject to constraints on the auxiliary conditional. By first constraining $p(y|x)$ to be tractable and then optimizing the bound (4), we essentially incorporate the constrained Blahut-Arimoto algorithm into the variational inference framework. Arguably, our approach generalizes variational mixture approximations similarly to the way that the variational EM [7] generalizes the standard EM algorithm: by allowing a flexibility in the choice of $p(y|x)$, it improves numerical stability and helps to significantly simplify the computations. Moreover, our method suggests a way to extend variational mixture approaches to structured auxiliary spaces, which may be useful for boosting the effective number of mixture states.

4   Experimental Results

Throughout the simulations, it was assumed that $p(x)$ is a pairwise Markov net with the energy $E(x) = x^T W x + x^T b$ and $x \in \{-1, 1\}^{|x|}$ (see [1] for details).

Reweighting Structured Representations: A by-product of our framework is a simple and fast way to re-weight any set of differently structured (tractable) distributions $q(x|y)$, which principally improves on trivial combinations. For a uniformly chosen $W \in \mathbb{I}^{10 \times 10}$, we generated $K = 10$ spanning trees with the weights $W^{(m)}$ such that $W_{ij}^{(m)} = W_{ij}$ for all i, j and $m = 1, \ldots, K$. Then we optimized (4) for $q(y)$ assuming that $q(x|y_k)$ and $p(y_k|x)$ were fixed (see Fig. 3). In this case the bounds were $L_r \approx 8.38$, $L_u \approx 8.87$, $L_b \approx 8.33$, and $\mathbf{L_{av}} \approx \mathbf{9.52}$ for the random, uniform, best single, and auxiliary variational weightings respectively. As expected, the auxiliary method leads to the tightest bound.

Structured Auxiliary Mappings: To investigate an influence of a structure of $p(y|x)$ on (4), we assumed a fixed $q(x, y)$ with $x \in \{-1, 1\}^{|x|}$ and $y \in \{-1, 1\}^{|y|}$. The marginals $q(y_i)$ (flip rate) were fixed to be constant for all the nodes y_i. Fig. 3 (c) shows the improvement $\tilde{I}$ (over the convex combination in (4)) as a function of the flip rate for a model with $|x| = 5$, $|y| = 10$. The total number of x- and y-parents of each y_i was constrained to satisfy $|\pi_x(y_i)| + |\pi_y(y_i)| \leq 4$. We observed that some of the optimal auxiliary mappings were close to the theoretically optimal (but generally intractable) $q(x|y)$, though the choice of the structure proved to be important (see [1] for details).

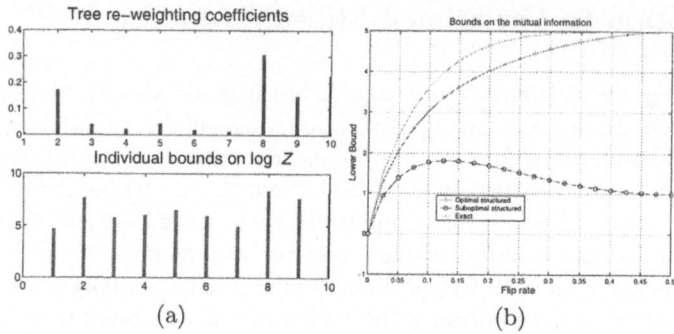

Fig. 3. (a) *Top:* weights of the fixed structured approximations; *Bottom:* bounds on $\log Z$ for each approximation; (b) influence of the structure of $p(y|x)$ on $\tilde{I}$.

5   Summary

We have presented an approach that generalizes the standard KL- variational procedure to the use of auxiliary variables, which provides a systematic improvement over standard structured approximations. We have also showed that the variational mixture approximations could be seen as special and more computationally expensive cases of our approach. Finally, we showed that our method can be easily generalized to factorial and structural state representations. One way to use it in practice is to find weightings for any set of tractable distributions.

References

1. Agakov, F. V. and Barber, D. (2004). An Auxiliary Variational Method. Technical report, EDI-INF-RR-0205, School of Informatics, University of Edinburgh.
2. Barber, D. and Sollich, P. (2000). Gaussian fields for approximate inference. In *Neural Information Processing Systems 12*. The MIT Press.
3. El-Hay, T. and Friedman, N. (2002). Incorporating Expressive Graphical Models in Variational Approximations: Chain-Graphs and Hidden Variables. In *UAI*.
4. Jaakkola, T. S. and Jordan, M. I. (1998). Improving the Mean Field Approximation via the Use of Mixture Distributions. In Jordan, M. I., editor, *Learning in Graphical Models*. Kluwer Academic Publishers.
5. Jordan, M. I., Ghahramani, Z., Jaakkola, T. S., and Saul, L. K. (1998). An Introduction to Variational Methods for Graphical Models. In Jordan, M. I., editor, *Learning in Graphical Models*, chapter 1. Kluwer Academic Publishers.
6. Lawrence, N. D., Bishop, C. M., and Jordan, M. I. (1998). Mixture Representations for Inference and Learning in Boltzmann Machines. In *UAI: Proceedings of the 14th Conference*.
7. Neal, R. M. and Hinton, G. E. (1998). A View of the EM Algorithm that Justifies Incremental, Sparse, and Other Variants. In Jordan, M. I., editor, *Learning in Graphical Models*, chapter 1. Kluwer Academic Publishers.
8. Swendsen, R. and Wang, J.-S. (1987). Nonuniversal critical dynamics in Monte Carlo simulations. *Physical Review Letters*, 58:86–88.

Gaussian Process Regression
with Fluid Hyperpriors

Ramūnas Girdziušas and Jorma Laaksonen

Helsinki University of Technology, Laboratory of Computer and Information Science
P.O. Box 5400, FI-02015 HUT, Espoo, Finland
{Ramunas.Girdziusas,Jorma.Laaksonen}@hut.fi

Abstract. A Gaussian process model can be learned from data by identifying the covariance matrix of its sample values. The matrix usually
depends on some fixed parameters called input length scales. Their estimation is equivalent to finding the corresponding diffeomorphism of the
process inputs. Spatially variable length scales are difficult to estimate in
the absence of good *a priori* values. We suggest a fluid-based nonlinear
map of the process inputs and our experiments validate such a model on
a synthetic problem.

1   Introduction

The Gaussian process (GP) model [6, 8] is an elegant alternative to parametric
approaches such as multilayer perceptron (MLP) neural networks [5]. The GP
model postulates that any collection of predicted variables follows normal distribution with a specific mean and covariance. Usually, the distribution depends
on some key adaptable parameters called spatial length scales of the GP inputs.

According to Bayesian GP theory [3], the length scales can be interpreted
as the model hyperparameters and they can be integrated out or estimated in
order to maximize the Bayesian evidence. However, a fixed-length-scale model
does not take into account that data correlation (smoothness) properties vary
throughout the input space. Models with small length scales over-fit data whereas
large length scales filter out important rapid changes.

The estimation of spatially variable length scales is equivalent to finding a
specific nonlinear diffeomorphism of the input space [3]. The adaptation of the
hyperparameters can be reduced to redistribution of the data points so that the
input regions which contain rapid changes are covered more densely. The problem
is that the nonlinear map estimation requires *a priori* constraints. Parametriza-
tion by using radial basis functions with the weight decay priors [3] prevents
large deformations. As a result, the whole approach does not go hand in hand
with the nonparametric nature of GP.

In this work, we seek for smooth transformations of the input space whose
region compressibility can be controlled by a few parameters. We will show how
such *a priori* information can be incorporated by applying a fluid dynamics approach. In the sequel, we first introduce the Gaussian process regression problem

N.R. Pal et al. (Eds.): ICONIP 2004, LNCS 3316, pp. 567–572, 2004.
© Springer-Verlag Berlin Heidelberg 2004

and state an equivalence between spatially varying length scales and nonlinear input deformations in Section 2. A fluid dynamics-based method is proposed in Section 3. The experiments of Section 4 have been conducted with data that has outliers and regions of different regularity. Section 5 presents our conclusions.

2   Gaussian Process Regression

2.1   The Gaussian Process Model

A zero-mean random process $t(\mathbf{x})$ is Gaussian if a vector of its sample values $\mathbf{t}_M = [t(\mathbf{x}_1), \ldots, t(\mathbf{x}_M)]^T$ given at any $M \in \mathbb{N}$ spatial locations $\mathcal{X}_M = \{\mathbf{x}_n\}|_{n=1}^M$, $\mathbf{x}_n \in \mathbb{R}^L$ is normally distributed [8, 3]:

$$p(\mathbf{t}_M | \mathcal{X}_M, \boldsymbol{\theta}) = (2\pi)^{-\frac{M}{2}} |\mathbf{C}_M|^{-\frac{1}{2}} \exp\left(-\frac{1}{2} \mathbf{t}_M^T \mathbf{C}_M^{-1} \mathbf{t}_M\right). \tag{1}$$

We assume that the elements of the covariance matrix are equal to

$$C_M(m, n; \mathcal{X}_M, \boldsymbol{\theta}) = \theta_1 \exp\left(-\sum_{l=1}^L \frac{(x_m^{(l)} - x_n^{(l)})^2}{r_l^2}\right) + \theta_2 \delta_{mn}. \tag{2}$$

A Gaussian process model given by Eq. (2) states that the nearby located inputs correspond to close output values. A vector of hyperparameters $\boldsymbol{\theta} = \{\theta_1, \theta_2, r_1, \ldots, r_L\}$ includes the normalizing constants $\theta_{1,2}$ and the spatial length scales r_l. The number of hyperparameters in this model barely exceeds the input space dimension L. This quite simple case becomes a difficult one to handle when the length scales r_l vary in space.

An important method to estimate spatially variable length scales introduces a nonlinear diffeomorphism $\mathbf{x} \mapsto \mathbf{f}(\mathbf{x})$ of the input space [3]:

$$f_l(\mathbf{x}_n) = \mathbf{x}_0^{(l)} + \int_0^1 r_l^{-1}(\mathbf{x}_0 + k(\mathbf{x}_n - \mathbf{x}_0)) dk, \tag{3}$$

where the integrals of positive functions $r_l^{-1}(\mathbf{x})$ are evaluated along the lines that join some reference point $\mathbf{x}_0$ and any data point $\mathbf{x}_n$. Therefore, GP with spatially dependent length scales is equivalent to the GP with nonlinear deformation of the input space. The inverse length scales $r_l^{-1}(\mathbf{x})$ can be approximated by using a weighted sum of the radial basis functions [3]:

$$r_l^{-1}(\mathbf{x}) = \sum_{j=1}^J w_j \exp\left(-\sum_{l=1}^L \frac{(x-a)^2}{2\sigma^2}\right), w_j \geq 0. \tag{4}$$

The positiveness of weights is ensured by putting them into the form $w_j = e^{\beta_j}$ and imposing zero-mean flat Gaussian prior onto these parameters.

The above model is computationally attractive, but suffers from several drawbacks. GP is now very much of a parametric model. It is hard to incorporate *a priori* information about the spatial length scales through their positive weights. The approach does not guarantee smooth maps and emphasizes only small displacements.

2.2   Estimation of Model Outputs and Parameters

Suppose we are given data as a set of the inputs $\mathcal{X}_N = \{\mathbf{x}_n\}|_{n=1}^{N}$, $\mathbf{x}_n \in \mathbb{R}^L$ and their outputs $\mathcal{T}_N = \{t_n\}|_{n=1}^{N}$, $t_n \in \mathbb{R}$. Assume the outputs follow Gaussian process. Our main interest lies in the predictive distribution of the GP output t_{N+1} due to its new input $\mathbf{x}_{N+1}$:

$$p(t_{N+1}|\mathcal{X}_{N+1}, \mathcal{T}_N) = \int p(t_{N+1}|\mathcal{X}_{N+1}, \mathcal{T}_N, \theta) p(\theta|\mathcal{X}_N, \mathcal{T}_N) d\theta \qquad (5)$$

$$\approx p(t_{N+1}|\mathcal{X}_{N+1}, \mathcal{T}_N, \theta^{\mathrm{MP}}) . \qquad (6)$$

Eq. (5) represents a Bayesian way to estimate the model output by using full integration over the unknown hyperparameters. Eq. (6) states the approximate method to estimate the predictive density according to the Bayesian Evidence framework [3]. When the number of hyperparameters is small, one could postulate a noninformative (possibly improper) prior over the hyperparameters θ and estimate them by maximizing the log-likelihood (log-evidence) of the data:

$$\mathcal{L}(\theta) = \ln p(\theta|\mathcal{X}_N, \mathcal{T}_N) = -\frac{1}{2}\ln|\mathbf{C}_N| - \frac{1}{2}\mathbf{t}_N^T \mathbf{C}_N^{-1} \mathbf{t}_N - \frac{N}{2}\ln 2\pi. \qquad (7)$$

Here $\mathbf{C}_N \equiv \mathbf{C}_N(\theta)$ and it is formed by using all GP inputs as indicated by the subscript N. The case of spatially dependent length scales $r_l(\mathbf{x})$ significantly increases the dimension of the hyperparameter vector θ and therefore better priors are necessary to estimate the hyperparameters.

3   Fluid Dynamics Approach to Spatial Length Scales

3.1   Gaussian Process with Fluid Hyperpriors

We believe that the notion of spatial length scales is helpful in understanding the possible GP regression outcome, but it is better to incorporate a priori knowledge of the problem directly through the input deformations. Thus, we propose Gaussian process model with the covariance matrix expressed in terms of the displacement field $\mathbf{u}(\mathbf{x})$, i.e. $\mathbf{f}(\mathbf{x}) \equiv \mathbf{x} - \mathbf{u}(\mathbf{x})$:

$$C_N(m, n; \mathcal{X}_N, \theta, \mathbf{f}(\cdot)) = \theta_1 \exp\left(-\sum_{l=1}^{L} \frac{(f_l(\mathbf{x}_m) - f_l(\mathbf{x}_n))^2}{r_l^2}\right) + \theta_2 \delta_{mn} . \qquad (8)$$

Consequently, the log-evidence $\mathcal{L} \equiv \mathcal{L}(\theta, \mathbf{u})$, i.e. it depends on the stationary GP parameters according to Eq. (8) and also on the displacement field $\mathbf{u}(\mathbf{x})$.

Consider the first order approximation of log-evidence given by Eq. (7) w.r.t. the displacement $\mathbf{u}(\mathbf{x})$ and let the later one follow an incremental process

$$\mathbf{u}_{t+1}(\mathbf{x}) = \mathbf{u}_t(\mathbf{x}) + \eta_t(\mathbf{x}), \ \mathbf{u}_0(\mathbf{x})|_\Omega = 0, \ \mathbf{u}_t(\mathbf{x})|_{\partial\Omega} = 0, \ \text{for } \forall\, t \geq 0. \qquad (9)$$

Here Ω and $\partial\Omega$ denote the data domain and its border, respectively. Requirement to have displacements which are zero on the boundary is useful as it can prevent the contraction of the data domain into a single point.

Let us postulate that the displacement increment $\boldsymbol{\eta}_t(\mathbf{x})$ follows the time-dependent Gaussian random field

$$\boldsymbol{\eta}_t \sim p(\boldsymbol{\eta}_t|\nabla_{\mathbf{u}_t}\mathcal{L}, \lambda, \mu) = \frac{1}{Z_{\lambda,\mu}} \exp\left(\boldsymbol{\eta}_t^T \nabla_{\mathbf{u}_t}\mathcal{L} - \mathcal{E}_{\lambda,\mu}(\boldsymbol{\eta}_t)\right). \qquad (10)$$

The normalization factor $Z_{\lambda,\mu}$ makes Eq. (10) a proper density. Its computation is irrelevant considering fixed second level hyperparameters λ and μ. The exponential term $\mathcal{E}_{\lambda,\mu}(\boldsymbol{\eta}_t)$ is the Navier-Stokes energy

$$\mathcal{E}_{\lambda,\mu}(\boldsymbol{\eta}_t) = \int_{\Omega} (\mu\|\nabla\boldsymbol{\eta}_t + \nabla\boldsymbol{\eta}_t^T\|^2 + 2\lambda(\nabla \cdot \boldsymbol{\eta}_t)^2)\mathbf{dx} \,, \qquad (11)$$

$$\mu > 0, \; 2\mu + \lambda > 0. \qquad (12)$$

These equations can be obtained from the probabilistic formulation of the stationary viscous Navier-Stokes equations in the Eulerian reference frame by ignoring the inertial and pressure terms and assuming that the Jacobian matrix $\mathbf{I} - \nabla\mathbf{u}_t \approx \mathbf{I}$ [2, 4].

Such a model is useful for several reasons. Eqs. (9) and (10) express the regularized gradient ascent on the log-evidence w.r.t. the displacement field $\mathbf{u}_t(\mathbf{x})$. The incremental formulation allows large deformations while preserving their smoothness. The prior information of the desired map can be modeled by specifying the second level hyperparameters λ and μ. They define: (1) the displacement smoothness via the symmetrized norm of the $\boldsymbol{\eta}_t$ gradients and (2) the volume compression ratios via the divergence term $\nabla \cdot \boldsymbol{\eta}_t$.

3.2   Estimation of the Nonlinear Map

A skeleton of the algorithm to estimate the deformation map of the input space that maximizes Bayesian evidence is

> *Initialize* ϵ, λ, μ, κ, set $t = 0$, $\mathbf{u}_0(\mathbf{x}) = \mathbf{0}$.
> WHILE $\|\mathcal{L}_t - \mathcal{L}_{t-1}\| > \epsilon$,
> $$\boldsymbol{\theta}^* = \arg\max_{\boldsymbol{\theta}} \mathcal{L}(\boldsymbol{\theta}, \mathbf{u}_t), \qquad (13a)$$
> $$\boldsymbol{\eta}_t = \arg\max_{\boldsymbol{\eta}_t} p(\boldsymbol{\eta}_t|\nabla_{\mathbf{u}_t}\mathcal{L}(\boldsymbol{\theta}^*, \mathbf{u}_t), \lambda, \mu), \qquad (13b)$$
> $$\mathbf{u}_{t+1} = \mathbf{u}_t + \boldsymbol{\eta}_t \frac{\kappa}{\max(\boldsymbol{\eta}_t)} \,,$$
> $$t \leftarrow t + 1 \,.$$
> END.

$\nabla_{\mathbf{u}_t}\mathcal{L}(\boldsymbol{\theta}^*, \mathbf{u}_t)$ is estimated according to Eq. (7) and (8). Eq. (13a) is solved by applying a few iterations of the conjugate gradient method [8, 3]. Methods to solve Eq. (13b) are discussed in [9]. Whenever $\mathbf{x} \in \mathbb{R}^2, \mathbb{R}^3$ one can directly apply the Finite Element method [1]. κ denotes the maximum displacement increment which has to be fixed beforehand. The second-level hyperparameters λ and μ can be chosen so that the flow is either (i) hyperelliptic: $\lambda = 10, \mu = 1$, (ii) Laplacian: $\lambda = -1, \mu = 1$ or (iii) hypoelliptic: $\lambda = -2, \mu = 1$.

4   Experiment: Regression with Outliers

We consider a hill-plateau regression problem which is to estimate the following two-dimensional vector field $\mathbf{z}(\mathbf{x})$:

$$\begin{cases} z_1(x,y) = 4\exp\left(-\frac{(x-2.5)^2+(y-2.5)^2}{0.7}\right) - \frac{2}{1+\exp(-6x)} + 1, \\ z_2(x,y) = z_1(y,-x) \end{cases}$$

from its noisy sample. This problem tests the regressor on its local and global properties because $\mathbf{z}(\mathbf{x})$ is a superposition of a local radial basis field with a log-sigmoid field. A training sample was formed by taking noisy values $\tilde{\mathbf{z}}(\mathbf{x}) = \mathbf{z}(\mathbf{x}) + \boldsymbol{\xi}(\mathbf{x})$, $\boldsymbol{\xi}(\mathbf{x}) \sim \mathcal{N}(0,\mathbf{A})$, $\mathbf{A}^{-1} = \sigma^2\mathbf{I}$, $\sigma = 0.1$. Values were created on a regular 21×21 grid from the interval $(x,y) \in [-5,5]$. We turned 5% of the data points into outliers by adding uniform noise $\upsilon \sim \mathcal{U}(-1,1)$ to $\tilde{\mathbf{z}}(\mathbf{x})$. Fig. 1a displays $z_1(x,y)$ component.

The three different second-level hyperparameter settings described in Section 3.2 were tested. As can be seen in Fig. 1b, the displacements clearly maximize log-evidence in all the three cases. However, when the flow is hypoelliptic,

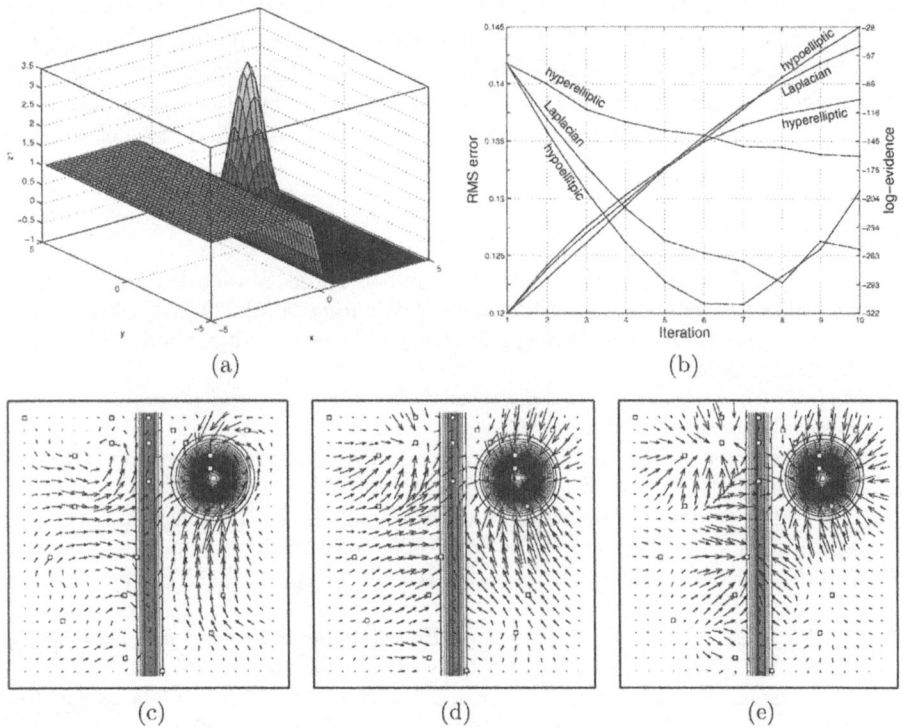

(a)   (b)

(c)   (d)   (e)

Fig. 1. The hill-plateau regression: (a) $z_1(x,y)$, (b) Evolution of RMS error and log-evidence; the displacement field $\mathbf{u}_t$ with different flow regimes: (c) hyperelliptic, (d) Laplacian and (e) hypoelliptic. White squares mark the locations of outliers.

the root mean square (RMS) error attains a minimum around the value 0.121 and starts increasing. Nonetheless, the method improves the standard Gaussian process which gives RMS error varying between $[0.141, 0.16]$ by approximately 15%. Multilayer perceptron with robust cost functions gives the best RMS error equal to 0.14 [5].

The estimated displacement fields $u_t(x)$ are shown in Figs. 1c-e. Notice in Fig. 1c that large $\lambda = 10$ value corresponds to nearly incompressible flow regime and such a hyperelliptic flow produces vortices. That kind of a setting perhaps represents too strong *a priori* knowledge that we do not really possess in this problem, and therefore the log-evidence values are lower and the RMS errors higher than in the other two cases.

5   Conclusions

A Gaussian process with fluid hyperpriors is an alternative to non-stationary covariance matrix models [3, 7]. The use of fluid hyperpriors expresses the belief that the map should be smooth while not imposing the requirement of small displacements. Experiments have shown that the proposed method improves the performance of a GP with a stationary covariance matrix and gives error values comparable to the state of the art approaches such as robust formulations of the MLP network [5]. Second-level hyperparameters λ and μ represent *a priori* knowledge about the ellipticity nature of the map. Here they were fixed constants, but it would be useful to consider their probabilistic model as well.

References

1. J. Alberty, C. Carstensen, S. A.Funken, and R. Klose. Matlab Implementation of the Finite Element Method in Elasticity. *Computing*, 69(3):239–263, 2002.
2. G. Christensen. *Deformable Shape Models for Anatomy*. Ph.d. thesis, Washington University, 1994.
3. M.N. Gibbs. *Bayesian Gaussian Processes for Regression and Classification*. Ph.d. thesis, Cambridge University, 1997.
4. U. Grenander and M. I. Miller. Computational Anatomy: An Emerging Discipline. *Quart. Appl. Math.*, LVI(4):617–694, December 1998.
5. T. Kärkkäinen and E. Heikkola. Robust Formulations for Training Multilayer Perceptrons. *Neural Computation*, 16(4):837–862, April 2004.
6. R.M. Neal. Priors for Infinite Networks. Technical Report CRG-TR-94-1, The University of Toronto, 1994.
7. C.J. Paciorek. *Nonstationary Gaussian Processes for Regression and Spatial Modelling*. Ph.d. thesis, Carnegie Mellon University, 2003.
8. C.E. Rasmussen. *Evaluation of Gaussian Processes and Other Methods for Nonlinear Regression*. Ph.d. thesis, The University of Toronto, 1996.
9. Gert Wollny and Frithjof Kruggel. Computational Cost of Nonrigid Registration Algorithms Based on Fluid Dynamics. *IEEE Trans. on Medical Imaging*, 21(8):946–952, 2002.

Learning Team Cooperation

Ron Sun[1] and Dehu Qi[2]

[1] Cognitive Science Department, Rensselaer Polytechnic Institute, Troy, NY 12180
[2] Department of Computer Science, Lamar University, Beaumont, TX 77710

Abstract. A cooperative team of agents may perform many tasks better than single agents. The question is how cooperation among self-interested agents should be achieved. It is important that, while we encourage cooperation among agents in a team, we maintain autonomy of individual agents as much as possible, so as to maintain flexibility and generality. This paper presents an approach based on bidding utilizing reinforcement values acquired through reinforcement learning. We tested and analyzed this approach and demonstrated that a team indeed performed better than the best single agent as well as the average of single agents.

1   The Model

We developed a multi-agent learning model, named MARLBS. It mixes completely two mechanisms – (1) reinforcement learning and (2) bidding. That is, the learning of individual agents and the learning of cooperation among agents are simultaneous and thus interacting. This model extends existing work, in that it is not limited to bidding alone, for example, not just bidding alone for forming coalitions, or bidding alone as the sole means for learning. Neither is it a model of pure reinforcement learning, without explicit interaction among agents. It addresses the combination and the interaction of the two aspects: reinforcement learning and bidding.

1.1   A Sketch of the Model

A team is composed of a number of member agents. Each member receives environmental information and makes action decisions. In any given state, only one member of the team is in control. The team's action is the active (controlling) member's action. Each team member learns how to take actions in its environment through reinforcement learning when it is active (i.e., when it is in control of the actions of the team). In each state, the controlling agent decides whether to continue or to relinquish control.

If it decides to relinquish control, another member will be chosen to be in control through a bidding process. That is, once the current member in control relinquishes its control, to select the next agent, it conducts a bidding process among members of the team (who are required to bid their best Q values; more later). Based on the bids, it decides which member should take over next from the current point on (as a "subcontractor"). The current member in control then

N.R. Pal et al. (Eds.): ICONIP 2004, LNCS 3316, pp. 573–578, 2004.
© Springer-Verlag Berlin Heidelberg 2004

takes the bid as its own payoff. Thus, the member who is more likely to benefit from being in control, and to benefit the other agents on its team, is likely to be chosen as the new member in control.

1.2   Model Details

The structure of an individual agent is as follows:

- Individual action module Q: Each Q module selects and performs actions, and each learns through Q-learning (Watkins 1989).
- Individual controller module CQ: Each CQ module learns when the agent should continue and when it should relinquish control, through Q-learning (separate from that of the Q module).

Let state s denote the actual observation by a team at a particular moment. We assume reinforcement (payoffs and costs) is associated with current state, denoted as $g(s)$. The overall algorithm is as follows:

1. Observe the current state s.
2. The current active agent on the team (the current agent in control; i.e., the currently active Q/CQ pair) takes control. If there is no active one (when the team first starts), go to step 5.
3. The CQ module of the active agent selects and performs a control action based on $CQ(s, ca)$ for different ca. If the action chosen by CQ is *end*, go to step 5. Otherwise, the Q module of the active agent selects and performs an action based on $Q(s, a)$ for different a.
4. The active agent (both of its Q and CQ modules) performs learning based on the reinforcement received. Go to step 1.
5. The bidding process (which is based on current Q values) determines the next member agent (the next pair of Q/CQ) to be in control. The agent that relinquished control performs learning taking into consideration the expected winning bid (which is its payoff for giving up control).
6. Go to step 1.

In bidding, usually each member of the team submits its bid, and the one with the highest value wins. However, during learning, for the sake of exploration, a stochastic selection of bids is conducted based on the Boltzmann distribution:

$$prob(k) = \frac{e^{bid_k/\tau}}{\sum_l e^{bid_l/\tau}}$$

where τ is the temperature that determines the degree of randomness in bid selection. The higher a bid is, the more likely the bidder will win. The winner will then subcontract from the current member in control. The current member in control takes the chosen bid as its own payoff[1].

---

[1] We do not allow the current member in control who decided to relinquish control to participate in bidding itself, so as to avoid unnecessary calls for bidding.

One crucial aspect of this bidding process is that the bid a team member sub-mits must be its best Q value for the current state. In other words, each member agent is not free to choose its own bids. A Q value resulting from Q-learning (see Watkins 1989) represents the total expected (discounted) reinforcement that an agent may receive based on its own experience. Thus, a bid is fully determined by a agent's experience with regard to the current state: how much reinforcement (payoff and cost) the agent will accrue from this point on if it does its best.

An agent, in submitting a bid (the best Q value for the current state), takes into account both its own reinforcement and the gains from subsequent subcon-tracting to other agents (because it takes the accepted subsequent bid as its own payoff when subcontracting).

This is an "open-book" bidding process in which there is no possibility of intentional over-bidding or under-bidding.

On top of learning, evolution may be used to enhance cooperation.

2   Experiments

We applied MARLBS to learning Backgammon, in order to evaluate the model – the usefulness and performance of MARLBS in complex problem solving (in complex sequential decision tasks).

PubEval is a publicly available computer player (Tesauro 1992), commonly used as a benchmark. We used it as an evaluator of our model.

We played MARLBS against PubEval. The result of 400,000 games was as shown in Figure 1. The maximum, average, and minimum winning percentages were calculated from the 15 teams in each generation. To generate these numbers, each team of each generation played against PubEval for 50 games. The average winning percentage of MARLBS (averaged over all 15 teams) reached 54.8. The maximum winning percentage of MARLBS reached 62.

The performance of individual team members was tested. Recall that each member is an autonomous agent and can play the game by itself. The results are in Table 1 and Table 2. Table 1 shows the team member performance, in terms of the best member, the worst member, and the average of all the members, of the best team (that is, the team that had the highest winning percentage when playing against PubEval). The column "Best vs. Worst" records the performance of the best member of the team playing against the worst. For the sake of com-parison, the performance of the best team is also listed there. All the numbers are winning percentages of playing against PubEval for 50 games, except the column "Best vs. Worst". Table 2 shows the performance of the members of the worst team. The numbers there are similar.

We notice that at the end of training, the best team performed better than any member of the best team (including the best member of the best team). The best team also outperformed its members on average. That is, there is a clear advantage in having a multi-agent team as opposed to having a set of individual agents and choosing a best one. In some way, this fact demonstrates why multi-agent learning is useful, which is due to the synergy within a team, created by

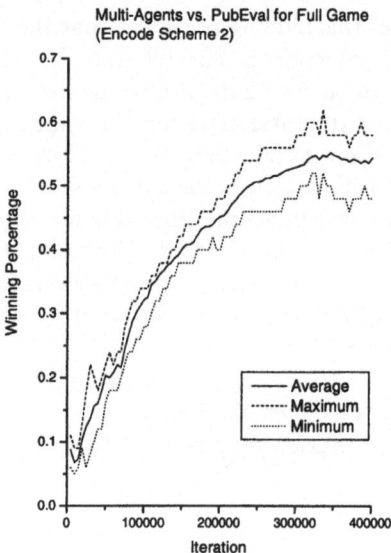

Fig. 1. The winning percentages of MARLBS over time, played against PubEval in full games.

Table 1. Member performance in the best team. All data are winning percentages out of 50 games.

Iteration	Members in the best team				Best Team
	Average	Best	Worst	Best vs. Worst	
100,000	0.32	0.34	0.28	0.66	0.36
200,000	0.42	0.46	0.38	0.54	0.46
300,000	0.49	0.52	0.44	0.52	0.54
400,000	0.50	0.56	0.46	0.52	0.60

Table 2. Member performance in the worst team. All data are winning percentages out of 50 games.

Iteration	Members in the worst team				Worst Team
	Average	Best	Worst	Best vs. Worst	
100,000	0.20	0.24	0.16	0.72	0.20
200,000	0.36	0.38	0.30	0.62	0.28
300,000	0.40	0.42	0.34	0.62	0.40
400,000	0.42	0.46	0.36	0.58	0.42

the emergent cooperation and coordination of team members. However, note also that the worst team did not perform better than all of its members. This fact suggests that the best teams were able to learn to achieve good coordination among its members, while the worst team failed to do so. It also suggests that

Table 3. Members of the best team versus the best team. All data are winning percentages of a member playing against its team in 50 games.

Iteration	Members of the best team playing against the best team		
	Average	Best	Worst
100,000	0.45	0.62	0.38
200,000	0.46	0.52	0.36
300,000	0.45	0.50	0.38
400,000	0.44	0.48	0.40

Table 4. Members of the worst team versus the worst team. All data are winning percentages of a member playing against its team in 50 games.

Iteration	Members of the worst team playing against the worst team		
	Average	Best	Worst
100,000	0.48	0.56	0.32
200,000	0.47	0.58	0.42
300,000	0.46	0.54	0.38
400,000	0.46	0.52	0.36

the best team was the best at least in part because it achieved better cooperation among its members.

We also tested the performance of a team playing against its members. The results are in Table 3 and Table 4. For the best team, the performance of the whole team was not better than that of its best member at the beginning. But after a sufficient number of iterations, the whole team outperformed its best member (and all the other members as well). On the other hand, for the worst team, the performance of the whole team was never as good as its best member. Again, this set of data suggests the following two points: (1) A good team, due to the emergent cooperation of its members, had the advantage of the synergy among agents, and as a result, it performed better than the individual agents on the team; (2) the best team was the best at least in part because of the coordination and cooperation of the team members – these agents learned to cooperate with each other and divided up the task among themselves to achieve a better performance than any individual agent could.

Note that team cooperation improved over time. As indicated by Table 3, early on, the best member of the best team outperformed the team. But after 400,0000 iterations, the team outperformed the agents (including the best agent) on the team. This fact indicated the increasing cooperation of agents on the team over time, which led to the improved team performance against its members (because, without increasing cooperation, a team would never outperform its member agents).

Why does cooperation lead to better performance? Divide-and-conquer is generally a useful approach. In this case, the division of labor among team members makes the learning task faced by each neural network (within each member agent) easier, because each of them can focus on learning a subset of

578 Ron Sun and Dehu Qi

Table 5. Comparisons with other Backgammon players in terms of winning percentage against PubEval.

	MARLBS	TD-Gammon	HC-Gammon
Winning Percentage	62	59.3	45
Iteration	400,000 games	> 1,000,000 games	400,000 games

input/output mappings, rather than treating all of them as equally important. In this way, each of them may learn better and, through their cooperation, the whole team may perform better as well. Cooperation in this model emerges from the interaction of agents, and is not externally determined[2].

The result of experimental comparisons with other Backgammon learning systems, in terms of winning percentage against PubEval, is shown in Table 5. MARLBS compares favorably with other Backgammon learning systems. Compared with these other players, our model achieved a better performance than HC-Gammon, but a roughly comparable performance as TD-Gammon. However, note that MARLBS was trained far less than TD-Gammon.

3 Conclusions

In this work, we developed an approach toward establishing cooperation among a group of self-interested agents, based on bidding utilizing reinforcement values acquired through reinforcement learning. It is a straightforward method that is generic and works in a variety of task domains. Notably, in the method, the autonomy and the self interest of individual agents are maintained as much as possible while cooperation among agents of a team is facilitated.

References

1. J. Pollack and A. Blair, (1998). Co-evolution in the successful learning of Backgammon strategy. *Machine Learning*, 32(3), 225-240.
2. G. Tesauro, (1992). Practical issues in temporal difference learning. *Machine Learning*, 8:257-277, 1992.
3. C. Watkins, (1989). *Learning with Delayed Rewards*. Ph.D Thesis, Cambridge University, Cambridge, UK.

---

[2] An alternative way of describing the advantage that teams have is that different Q functions handle different regions of the state space, which makes the function approximation of each Q function simpler. As a result of appropriate partitioning of the state space, the whole system performs better.

Training Minimal Uncertainty Neural Networks by Bayesian Theorem and Particle Swarm Optimization

Yan Wang, Chun-Guang Zhou, Yan-Xin Huang, and Xiao-Yue Feng

College of Computer Science and Technology, JiLin University ChangChun 130012, China
wy6868@hotmail.com, cgzhou@jlu.edu.cn

Abstract. A new model of minimal uncertainty neural networks (MUNN) is proposed in this article. The model is based on the Minimal Uncertainty Adjudgment to construct the structure, and it combines with Bayesian Theorem and Particle Swarm Optimization (PSO) for training. The model can determine the parameters of neural networks rapidly and efficiently. The effectiveness of the algorithm is demonstrated through the classification of the taste signals of 10 kinds of tea. The simulated results show its feasibility and validity.

1 Introduction

How to determine the structure and the parameters of the neural networks promptly and efficiently has been a difficult point all the time in the field of neural networks research [1][2]. At present the basic idea to solve this problem is to dig the proper information from the data of research, and then guide the construction of neural networks via the previously acquired information, such as successful in constructing neural networks in light of Bayesian Theorem [3][4]. Optimizing neural networks according to Particle Swarm Optimization (PSO) is newly invented in recent years [5][6]. A new model of minimal uncertainty neural networks (MUNN) to construct the neural networks is discussed in this paper. It is derived from Minimal Uncertainty Adjudgment to construct the networks structure, which combines with Bayesian Theorem and Particle Swarm Optimization (PSO) for training. When it is employed in the classification of the taste signals of 10 different kinds of tea [7], the experimental results show its feasibility and validity. And theoretically the minimal uncertainty neural networks can also be applied in other terrains of classification.

2 Minimal Uncertainty Neural Networks (MUNN) Model

Definition 1. Let $X=\{x_1,x_2...x_N\}$ be a N-dimensional input vector, and all attributes $x_1,x_2...x_N$ are separate independent; let $P(y|X)$ denote the probability of event that X comes forth and output y occurs, then $\pi = 1 - P(y|X)$ is called **uncertainty** of X to y.

Theorem 1. As in definition 1, we get the uncertainty of X to y: $\pi = \prod_{i=1}^{N}(1 - P(y|x_i))$.

Proof: Let event $A_i=\{x_i$ comes forth and then y occurs$\}$, $i\in[1,2...N]$, it is obvious that N events $A_1,A_2...A_N$ are separate independent. If event $A=\{X$ comes forth and then y occurs$\}$, we have : $P(A) = P(A_1 \cup A_2 ... \cup A_N)$.

N.R. Pal et al. (Eds.): ICONIP 2004, LNCS 3316, pp. 579–584, 2004.
© Springer-Verlag Berlin Heidelberg 2004

Hence $P(A) = 1 - P(\overline{A}) = 1 - P(\overline{A_1 \cup A_2 \cup ... \cup A_N}) = 1 - P(\overline{A_1} \cdot \overline{A_2} ... \cdot \overline{A_N})$.

Therefore, since $A_1, A_2 ... A_N$ are independent, we have:

$$P(A) = 1 - P(\overline{A_1})P(\overline{A_2}) ... P(\overline{A_N}) = 1 - \prod_{i=1}^{N}(1 - P(A_i)).$$

Transformed $P(A_i)$ and $P(A)$ into conditional probability format $P(A_i) = P(y \mid x_i)$ and $P(A) = P(y \mid X)$. From the above equations, we can easily see that:

$$P(y \mid X) = 1 - \prod_{i=1}^{N}(1 - P(y \mid x_i)). \text{ Then we get: } 1 - P(y \mid X) = \prod_{i=1}^{N}(1 - P(y \mid x_i)).$$

Furthermore, as $\pi = 1 - P(y \mid X)$ is the uncertainty of X to y, we obtain:

$$\pi = \prod_{i=1}^{N}(1 - P(y \mid x_i)). \text{ The proof is completed.}$$

Corollary 1: When $Y = \{y_1, y_2 ... y_M\}$ is a classified collection set, the uncertainty of X to y_j ($j \in [1,2 ... M]$) is determined as: $\pi_j = 1 - P(y_j \mid X) = \prod_{i=1}^{N}(1 - P(y_j \mid x_i))$.

When classified, choose the minimal uncertainty π_j as the finial adjudgment, which is defined as **Minimal Uncertainty Adjudgment**.

From Corollary 1, we have: $\pi_j = \prod_{i=1}^{N}(1 - P(y_j \mid x_i)) = P(y_j)^N \prod_{i=1}^{N}(\frac{1 - P(y_j \mid x_i)}{P(y_j)})$.

Suppose a set of observations A is given, $x_{ii'}$ is the property of attribute x_i, then it becomes: $\pi_j = P(y_j)^N \prod_{x_{ii'} \in A}(\frac{1 - P(y_j \mid x_{ii'})}{P(y_j)})$. Take logarithm of both sides:

$$\log(\pi_j) = N \log P(y_j) + \sum_{x_{ii'} \in A} \log(\frac{1 - P(y_j \mid x_{ii'})}{P(y_j)}) = N \log P(y_j) + \sum_{i=1}^{N} \log(\frac{1 - P(y_j \mid x_{ii'})}{P(y_j)})O_{ii'}$$

Where $O_{ii'} = 1$, if $x_{ii'} \in A$; and $O_{ii'} = 0$, otherwise. This equation is especially suitable for implementation to a neural network. The general formula for signals propagating in neural networks is: $s_j = \beta_j + \sum_i \omega_{ji}o_i$. Then we can make the identifications:

$$\beta_j = N \log P(y_j) \tag{1}$$

$$\omega_{ji} = \log(\frac{1 - P(y_j \mid x_{ii'})}{P(y_j)}) \tag{2}$$

$$\pi_j = \exp s_j \tag{3}$$

Based on the above formulae, the structure of neural networks is determined as shown in Fig.1. The significations of each layers are as the followings:

Layer[A](input samples): A is the observation set, $x_{ii'}$ is the property of x_i.

Layer[B](weights selection): Where $O_{ii'} = 1$, if $x_{ii'} \in A$, and $O_{ii'} = 0$, otherwise.

Layer[C](transmited calculation): From Eq. (1) and Eq. (2) we can get s_j.

Layer [D] (output uncertainty): Using Eq. (3) to obtain the output of π_j.

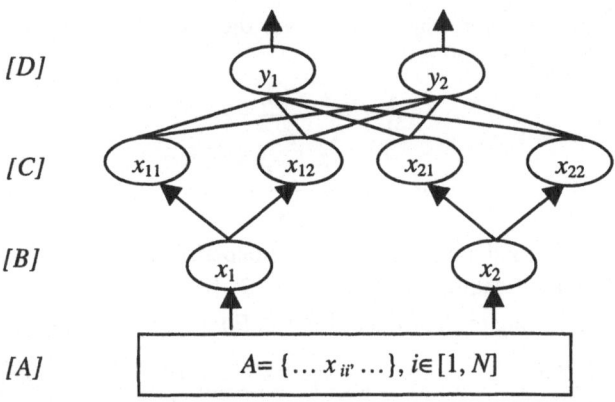

[D]

[C]

[B]

[A]

$A = \{ \dots x_{ii'} \dots \}, i \in [1, N]$

Fig. 1. A simply minimal uncertainty neural network with 2 input attributes and 2 classes($N=2$)

This model is defined as **minimal uncertainty neural networks** (MUNN). Using different approaches in the followed sections, the weights and biases are estimated and then the estimation of the posterior probability of the corresponding class is obtained. Furthermore, we can make decisions according to the Minimal Uncertainty Adjudgment.

3   Determining Weights and Biases of MUNN by Bayesian Theorem

When having N-dimensional input vector $X=\{x_1, x_2 \dots x_N\}$ and M-dimensional output vector $Y=\{y_1, y_2 \dots y_M\}$, we have the conditional probability by Bayesian Theorem:

$$P(y_j|x_i) = P(y_j)\frac{P(x_i|y_j)}{P(x_i)}. \text{ And then: } \frac{P(y_j|x_i)}{P(y_j)} = \frac{P(x_i|y_j)}{P(x_i)} = \frac{P(x_i,y_j)}{P(y_j)P(x_i)}.$$

This implies the terms in Eq. (2) becomes: $\dfrac{1-P(y_j|x_i)}{P(y_j)} = \dfrac{P(x_i)-P(x_i,y_j)}{P(y_j)P(x_i)}.$

Now we deduce Eq. (2) to:

$$\omega_{ji} = \log(\frac{P(x_i)-P(x_i,y_j)}{P(y_j)P(x_i)}) \tag{4}$$

The purpose of training phase is to determine the weights and the biases, which is done by counted all units, to keep track on how many times each attribute value and each pair wise combination of attribute values has occurred. In detail, the counters are calculated as [8]:

$$C = \sum_r k^{(r)} \qquad c_{ii'} = \sum_r k^{(r)} \xi_{ii'}^{(r)} \qquad c_{ii',jj'} = \sum_r k^{(r)} \xi_{ii'}^{(r)} \xi_{jj'}^{(r)}$$

where r is the pattern number in the training set, $\xi_{ii'}^{(r)}$ and $\xi_{jj'}^{(r)}$ indicate the presence of $x_{ii'}$ in $x_i^{(r)}$ and $y_{jj'}$ in $y_j^{(r)}$. $k^{(r)}$ is the "strength" of the pattern, typically is 1. The conditional probability of an event is estimated by the number of observations where that event occurred divided by the total number of observations, that is (Here ii' and jj' are replaced by single indices which range over the possible values of all attributes, to make the notation more simple):

$$\hat{P}_i = \frac{c_i + \alpha / n_i}{C + \alpha} \tag{5}$$

$$\hat{P}_{ij} = \frac{c_{ij} + \alpha /(n_i m_j)}{C + \alpha} \tag{6}$$

α is usually a very small number, and in later experiments we assume that α is 1/C, so the error of the classification is close to $\log 1 / C^2$ [8].

Therefore, according to Eq. (5) and Eq. (6), Eq. (1) and Eq. (4) are deduced to:

$$\beta_j = N \log \frac{c_j + \alpha / m_j}{C + \alpha} \quad ; \quad \omega_{ji} = \log \frac{((c_i + \alpha / n_i) - (c_{ij} + \alpha /(n_i m_j))(C + \alpha))}{(c_i + \alpha / n_i)(c_j + \alpha / m_j)} .$$

4 Determining Weights and Biases of MUNN by PSO

Particle Swarm Optimization (PSO) is a population based optimization strategy introduced by Kennedy and Eberhart in 1995 [9]. PSO is initialized with a group of random particles (solutions) and then updates theirs velocity and positions with the following formulae:

$$v(t+1) = v(t) + c_1 * rand() * (pBest(t) - Present(t)) + c_2 * rand() * (gBest(t) - Present(t))$$
$$Present(t+1) = Present(t) + v(t+1)$$

$v(t)$ is the particle velocity, $Persent(t)$ is the current particle. $pBest(t)$ and $gBest(t)$ are defined as individual best and global best. $rand()$ is a random number between [0 1]. c_1, c_2 are learning factors. Usually $c_1 = c_2 = 2$. To accelerate searching velocity and to avoid oscillation, the improvement of v is utilized in the following experiments [5]:

$$v(t+1) = \chi * (\omega * v(t) + c_1 * rand() * (pBest(t) - Present(t))$$
$$+ c_2 * rand() * (gBest(t) - Present(t)))$$

The parameter settings of $v(t)$ have not had the theoretical basis till now. All the parameters are set according to experience. In our experiments, we set $\chi=0.9$, $\omega=0.8$.

The weights and biases of minimal uncertainty neural networks model can act as the parameters of the particles directly, and the misclassified ones are the fitness values. The final result of training is expected to be explained by the probability of Eq.(1) and Eq.(2).

5 Determining MUNN via Hybrid of Bayesian Theorem and PSO

Training the minimal uncertainty neural networks via the hybrid of Bayesian Theorem and PSO is implemented as follows:

1. A MUNN's weights and biases can be determined by Bayesian Theorem.
2. This group of weights and biases can set one particle of PSO. Then some other particles are generated randomly or around that particle.
3. A better result of classification can be obtained through training this group of particles by means of PSO.

Thus Bayesian accelerate the speed of convergence of PSO, while the global and local search ability of PSO further makes up the possibility of getting local optimum by Bayesian.

6   Experimental Results

To test the validity of our approaches for classification, we classified and tested the 2-dimensional taste signals of 10 kinds of tea (100 samples of each kind, 1,000 training samples in total) [7]. The signals were preprocessed to meet the requirement of the discrete input of the minimal uncertainty neural networks. We isometric compartmentalized the continuous value of attributes into 11 and 13 dispersed areas according to the different dimensional characters of the taste signals. Original taste data and the comparison of the results are illustrated in Fig. 2 and table 1.

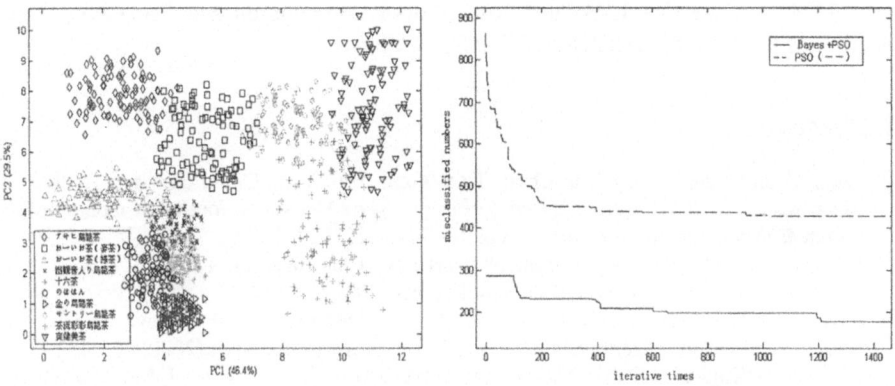

Fig. 2. The left figure shows the taste signals of 10 kinds of tea; the right one gives a comparison of effects of Bayesian, PSO, and Bayesian plus PSO on determining MUNN

Table 1. A comparison of several ways to determine MUNN. PSO takes about 20 seconds every 100 times of iteration (Pentium3 1GHz CPU, 128M RAM, Win2000 OS, VC++6.0)

Total samples 1000	Misclassified numbers	Percentage of correct	Iterative times
Bayesian	287	71.3%	1
PSO	429	57.1%	1500
Bayesian+ PSO	178	82.2%	1501

When to determine the minimal uncertainty neural networks by Bayesian Theorem, we suppose $\alpha=1/C$. The number of wrong identifications is 287 and the percentage of correct identification is 71.3%. The result is preferable.

When through PSO, we assume that there are 20 particles in the group, the iterations of repeated circulations are 1,500, $c1 = c2 = 2$, $\chi=0.9$, $\omega=0.8$. The wrong classified numbers are 429. The result is not good enough.

When via the hybrid of Bayesian Theorem and PSO, all parameters are set as above. Finally, misclassified numbers is 178, and the correct rate is 82.2%. The improvement of identification rate strongly supports the validity of combination of the two methods.

7   Conclusions

A novel networks structure, called minimal uncertainty neural networks, is presented in this article. The new structure reserves the rapidity of minimal uncertainty neural networks trained by Bayesian Theorem and absorbs PSO's advantage of fast correction towards networks. The identification result is encouraged when minimal uncertainty neural networks is introduced into the identification of the taste signals of tea. Future research of this topic is to find a more effective method of dispersing the continuous value of input attributes and to construct the model of multi-layer minimal uncertainty neural networks.

Acknowledgement

This work was supported by the Natural Science Foundation of China (Grant No. 60175024) and the Key Laboratory for Symbol Computation and Knowledge Engineering of the National Education Ministry of China.

References

1. Zhou Chun-Guang, Liang Yan-Chun, Tian Yong, Hu Cheng-Quan, Sun Xi.: A Study of Identification of Taste Signals Based on Fuzzy Neural Networks. Journal of Computer Research & Development (in Chinese), Vol. 36, No.4 (1999) 401-409
2. Johansen, M. M.: Evolving Neural Networks for Classification. Topics of Evolutionary Computation 2002, Collection of Student Reports (2002) 57-63
3. Ninan Sajeeth Philip, K.Babu Joseph.: Boosting the Differences: A Fast Bayesian Classifier Neural Network. Intelligent Data Analysis, 4, IOS press, Netherlands (2000) 463-473
4. Matthew A.Kupinski, Darrin C.Edwards, Maryellen L.Giger, Charles E.Metz.: Ideal Observer Approximation Using Bayesian Classification Neural Networks. IEEE Transactions On Medical Imaging, Vol. 20, NO.9, SEP (2001) 886-899
5. Jakob Vesterstrom, Jacques Riget.: Particle Swarms–Extensions for improved local, multimodal, and dynamic search in numerical optimization. [master thesis]. EVALife Group, Department of Computer Science Ny Munkegade, Bldg,540, University of Aarhus DK-800 Aarhus C.Denmark, May (2002)
6. Gudise, V. G., Venayagamoorthy, G. K.: Comparison of particle swarm optimization and backpropagation as training algorithms for neural networks. Proceedings of the IEEE Swarm Intelligence Symposium 2003 (SIS 2003), Indianapolis, Indiana, USA. (2003) 110-117
7. Shunichiro Watanabe, Akira Yokoyama, Zhang Wenyi, Takeshi Abiko, Hidekazu Uchida, Teruaki Katsube.: Detection Mechanism of Taste Signals with Commercial Ion Sensors. The Journal of the Institute of Electrical Engineers of Japan (in Japanese), Vol.CS-98-30 (1998) 13-18
8. Ander Holst.: The Use of a Bayesian Neural Network Model for Classification Tasks. [PhD dissertation]. Department of Numerical Analysis and Computing Science, Royal Institute of Technology, Stockholm, Sweden. September (1997)
9. J.Kennedy, R.C.Eberhart.: Particle Swarm Optimization. Proc. IEEE int'l conf.on neural networks, Vol. IV, IEEE service center, Piscataway, NJ (1995) 1942-1948

A Forward-Propagation Rule for Acquiring Neural Inverse Models Using a RLS Algorithm

Yoshihiro Ohama, Naohiro Fukumura, and Yoji Uno

Toyohashi University of Technology, Toyohashi Aichi 441-8580, Japan
{ohama,fkm,uno}@system.tutics.tut.ac.jp

Abstract. It has been suggested that inverse models serve feedforward controllers in the human brain. We have proposed a novel learning scheme to acquire a neural inverse model of a controlled object. This scheme propagates error "forward" in a multi-layered neural network to solve a credit assignment problem based on Newton-like method. In this paper, we apply a RLS algorithm to this scheme for the stability of learning. The suitability of the proposed scheme was confirmed by computer simulation; it could acquire an inverse dynamics model of a 2-link arm faster than a conventional scheme based on a back-propagation rule.

1 Introduction

The existence of inverse models as feedforward controllers which have a reverse input-output relation to controlled objects in the human brain has been suggested. Investigating how to acquire such inverse models in the motor control, many learning schemes in which the inverse models are implemented in artificial multi-layered neural networks have been proposed. For updating the parameters of neural networks, most of the learning schemes have used a back-propagation rule[1]. However, the back-propagation channels in biogenic neural networks have not been found yet.

A forward-propagation rule as a novel learning scheme has been proposed[2]. In this scheme, a credit assignment problem is directly solved in multi-layered neural networks. Assuming the closed-loop system to be regarded as an approximately identity map, this problem can be solved by propagating error "forward" in the networks based on Newton-like method. The convergence is very fast but sometimes unstable.

In the current work, a forward-propagation rule is developed for the stability of learning. More specifically, we modify the representation of this scheme and apply a RLS algorithm which means regularization. We also conduct computational experiments to show that the proposed scheme can acquire inverse models. The results show that the convergence is very fast and the stability is provided even if learning parameters are roughly adjusted.

2 Forward-Propagation Learning Rule

Fig.1 shows the proposed scheme which consists of a multi-layered feedforward neural network as a feedforward controller and a controlled object. Given desired

N.R. Pal et al. (Eds.): ICONIP 2004, LNCS 3316, pp. 585–591, 2004.
© Springer-Verlag Berlin Heidelberg 2004

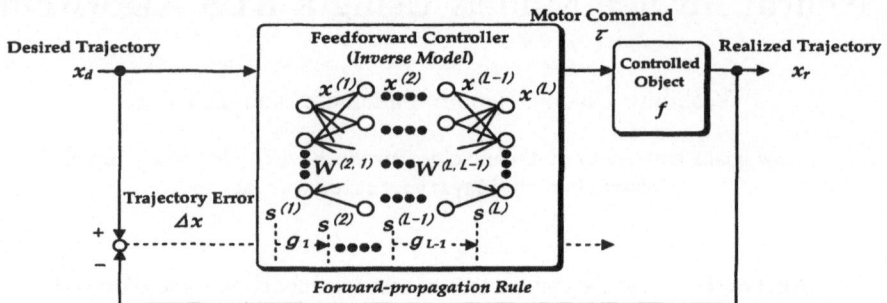

Fig. 1. A forward-propagation learning scheme for a neural inverse model.

trajectory x_d, the neural network produces motor command τ and the controlled object f is driven. The objective of the learning is that trajectory error $\Delta x = x_d - x_r$ comes to zero by acquiring the neural inverse model of f.

The adaptive process of this scheme consists of two kinds of operation performed repeatedly in each iteration; one is to estimate desired signals in the middle layers based on Newton-like method and the other is to update connection weights based on multiple linear regression. In the learning process, Δx is not propagated "backward" but "forward" in the neural network. Moreover, this learning can be performed even if desired torque τ is unknown.

2.1   Estimation of Desired Signals

Let a L-layered neural network be used, the input signal and the neural activation of the i-th layer are described $s^{(i)} \in R^{M^{(i)}}$ and $x^{(i)} : R^{M^{(i)}} \to R^{M^{(i)}}$, where $M^{(i)}$ means a number of the i-th layer neurons. The activation of $x^{(i)}$ is assigned only differentiable function whether or not it is linear. We introduce linear operator g_i to describe mapping from $s^{(i)}$ to $s^{(i+1)}$ as follows:

$$s^{(i+1)} = W^{(i+1,i)T} x^{(i)}(s^{(i)}) \equiv g_i, \qquad (1)$$

$$W^{(i+1,i)} = \begin{bmatrix} w_1^{(i+1,i)} & w_2^{(i+1,i)} & \cdots w_{M^{(i+1)}}^{(i+1,i)} \end{bmatrix}, \qquad (2)$$

$$w_k^{(i+1,i)} = \begin{bmatrix} w_{k1}^{(i+1,i)} & w_{k2}^{(i+1,i)} & \cdots & w_{kM^{(i)}}^{(i+1,i)} \end{bmatrix}^T, \qquad (3)$$

where $W^{(i+1,i)} \in R^{M^{(i)} \times M^{(i+1)}}$ is a matrix to represent connection weights from the i-th layer to the $(i+1)$-th layer and $w_{kj}^{(i+1,i)}$ is a connection weight from the j-th neuron of the i-th layer to the k-th neuron of the $(i+1)$-th layer.

In the learning of multi-layered neural networks, the credit assignment problem, which means how to deliver an error signal to some neural layers, must be solved. We found that this problem can be solved by the forward-propagation error. We consider the estimation of appropriate teaching signal $\hat{s}^{(i)}$. When the controlled object is driven, the network estimates $\hat{s}^{(i)}$ to reduce Δx in the next iteration step. Since Δx is regarded as a function of $s^{(i)}$, the ideal $\hat{s}^{(i)}$ is a solution of equation $\Delta x(s^{(i)}) = 0$. Here, the Damped-Newton method[3] can be applied to find an approximate solution.

$$\Delta x + \frac{\partial \Delta x}{\partial s^{(i+1)}} \Delta s^{(i+1)} = 0, \tag{4}$$

$$\hat{s}^{(i+1)} = s^{(i+1)} + \varepsilon \Delta s^{(i+1)}. \tag{5}$$

$\Delta s^{(i+1)}$ is a solution of Eq.(4) and ε is a learning rate$(0 < \varepsilon \ll 1)$. It should be noticed that Eq.(4) cannot provide an unique solution because of $M^{(1)} \neq M^{(i+1)}$.

To solve such problem, linear operators Φ_i and Ψ_i are defined as

$$\Phi_i = g_i \circ g_{i-1} \circ \cdots \circ g_1, \tag{6}$$

$$\Psi_i = g_{L-1} \circ g_{L-2} \circ \cdots \circ g_{i+1}. \tag{7}$$

Let us assume that the neural network has acquired an approximate inverse model of f before learning. Hence, the closed-loop system is regarded as approximately identity mapping.

$$\Phi_i \circ f \circ \Psi_i \simeq I. \tag{8}$$

Differentiating Eq.(8) about $s^{(i+1)}$, we get an obvious condition as

$$\frac{\partial \Phi_i}{\partial s^{(1)}} \frac{\partial \{f \circ \Psi_i\}}{\partial s^{(i+1)}} \simeq E, \tag{9}$$

where E is an unit matrix. Since $\Delta x = x_d - x_r = x_d - f \circ \Psi_i$, we can apply Eq.(9) to Eq.(4) multiplied by $\partial \Phi_i / \partial s^{(1)}$ from left. Thus, $\Delta s^{(i+1)}$ is given by

$$\begin{aligned}
\Delta s^{(i+1)} &\simeq \frac{\partial \Phi_i}{\partial s^{(1)}} \Delta x \\
&= \frac{\partial g_i}{\partial s^{(i)}} \frac{\partial \Phi_{i-1}}{\partial s^{(1)}} \Delta x \\
&= \frac{\partial g_i}{\partial s^{(i)}} \Delta s^{(i)}. \tag{10}
\end{aligned}$$

Using Eqs.(5) and (10), the calculation can be regarded as Newton-like method.

$$\begin{aligned}
\hat{s}^{(i+1)} &= s^{(i+1)} + \varepsilon \frac{\partial g_i}{\partial s^{(i)}} \Delta s^{(i)} \\
&= s^{(i+1)} + \varepsilon W^{(i+1,i)T} \frac{\partial x^{(i)}}{\partial s^{(i)}} \Delta s^{(i)}. \tag{11}
\end{aligned}$$

Note that $\hat{s}^{(i+1)}$ can be calculated by propagating $\Delta s^{(i)}$ "forward" through the connection weights. In other words, this learning algorithm solves the credit assignment problem by propagating the error signal forward, which can be also described as Newton-like method.

2.2 Update of Connection Weights

The connection weights $W^{(i+1,i)}$ in each layer should be appropriately updated to realize the signal $\hat{s}^{(i+1)}$ after the estimation of desired signal. Let us consider bach-mode learning in which the connection weights are updated after the control of an object. One of desired connection weights $\hat{W}^{(i+1,i)}$ is a solution of the normal equation given by

$$\hat{W}^{(i+1,i)} = \left(X^{(i)T} X^{(i)} \right)^{-1} X^{(i)T} \hat{S}^{(i+1)}, \tag{12}$$

where $X^{(i)} \in R^{N \times M^{(i)}}$ is a set of output signal sequences in the i-th layer, $\hat{S}^{(i+1)} \in R^{N \times M^{(i+1)}}$ is a set of estimated signal sequences in the $(i+1)$-th layer and N is a number of sequences. Since it is not clear that the scalar correlation matrix $X^{(i)T} X^{(i)}$ is a positive definite matrix, Eq.(12) should include a regularization term given by

$$\hat{W}^{(i+1,i)} = \left(X^{(i)T} X^{(i)} + \delta E \right)^{-1} X^{(i)T} \hat{S}^{(i+1)}, \tag{13}$$

where δ is a small positive number. Considering on-line learning, such regularized solution is approximately obtained by a RLS algorithm[4].

$$k^{(i)}(n) = \frac{\lambda^{-1} P^{(i)}(n-1) x^{(i)}(n)}{1 + \lambda^{-1} x^{(i)T}(n) P^{(i)}(n-1) x^{(i)}(n)}, \tag{14}$$

$$\xi_k^{(i)}(n) = \hat{s}_k^{(i+1)}(n) - \hat{w}_k^{(i+1,i)T}(n-1) x^{(i)}(n), \tag{15}$$

$$\hat{w}_k^{(i+1,i)}(n) = \hat{w}_k^{(i+1,i)}(n-1) + k^{(i)}(n) \xi_k^{(i)}(n), \tag{16}$$

$$P^{(i)}(n) = \lambda^{-1} P^{(i)}(n-1) - \lambda^{-1} k^{(i)}(n) x^{(i)T}(n) P^{(i)}(n-1), \tag{17}$$

$$\hat{w}_k^{(i+1,i)}(0) = 0, \quad P^{(i)}(0) = \delta^{-1} E, \tag{18}$$

where λ is a rate of forgetting factor $(0 \ll \lambda \le 1)$, $k^{(i)}(n)$ is a gain vector, $\xi_k^{(i)}(n)$ is a prediction error and $P^{(i)}(n)$ is an estimated inverse correlation matrix. In order to reduce the calculation cost in bach-mode learning, this algorithm, which is regarded as a weighted method of recursive least squares, can be applied.

3   Simulation of Motor Learning

We illustrate the suitability of the proposed scheme by computer simulation. We examined a learning problem for a 3-layered feedforward neural network in an inverse dynamics model of a 2-link arm with two rotation joints in x-y plane as shown in Fig.2(a). The dynamics of the controlled object can be described by the following:

$$\tau = M(\theta)\ddot{\theta} + h(\theta, \dot{\theta}), \tag{19}$$

where τ is the drive torque, θ is the joint angle, M is the inertia matrix of the arm, and h represents the Coriolis and centrifugal force. M and h were calculated according to the values: $m_1 = 1.59\,\text{kg}, m_2 = 1.44\,\text{kg}, l_1 = 0.30\,\text{m}, l_2 = 0.35\,\text{m}, l_{g1} = 0.18\,\text{m}, l_{g2} = 0.21\,\text{m}, I_1 = 0.0477\,\text{kgm}^2, I_2 = 0.0588\,\text{kgm}^2$. Here, m is the mass of the arm, l is the length of the arm, l_g is the center of gravity and I is the inertia of the arm.

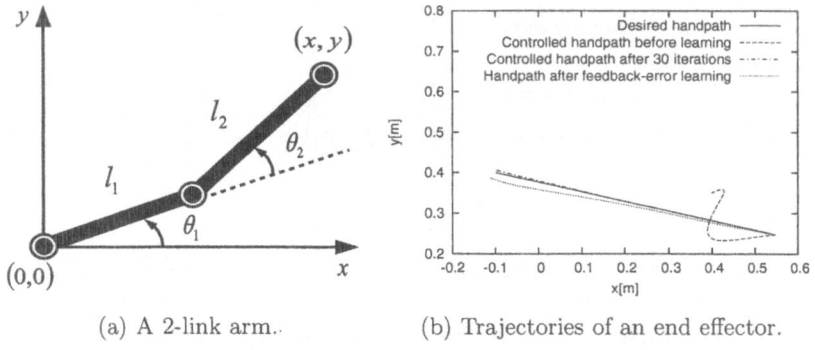

(a) A 2-link arm.   (b) Trajectories of an end effector.

Fig. 2. A controlled object and the result of learning.

The initial values of the connection weights were randomly selected in the uniform distribution on (-0.1,0.1). To satisfy Eq.(8) before learning, we prepared a sequence of the torque $T = [\tau(1), \tau(2), \cdots, \tau(500)]^T \in R^{500 \times 2}$ which is the suitable sin wave and a sequence of the angular trajectory $\Theta \in R^{500 \times 2}$ which is the trajectory of the 2-link arm controlled by T. Note that T and Θ have no relation to the training sequence for learning. Inputting Θ to the neural network, $X^{(2)} = x^{(2)}(g_1(\Theta)) \in R^{500 \times M^{(2)}}$ was derived. An approximate inverse model was obtained by substituting the solution of the following equation for $W^{(3,2)}$ which was solved by the method of recursive least-squares($\lambda = 1.0$).

$$T = X^{(2)} W^{(3,2)}. \tag{20}$$

A 3-layered neural network with 6 hidden units was used for learning the inverse dynamics model of the 2-link arm. The activation functions of input and output layer units were linear but those of hidden units were nonlinear as

$$x_k^{(2)} = \frac{1}{1 + \exp(-s_k^{(2)})}. \tag{21}$$

The desired trajectory of an end effector, which is a minimum jerk trajectory for 1.0 s, is shown in Fig.2(b) by a solid line. This trajectory was sampled at 100 Hz in angular space to obtain a training sequence $X_d \in R^{100 \times 6}$. For updating the connection weights, we applied a method of recursive least squares ($\lambda = 1.0$) in bach-mode learning. The learning results of the forward-propagation learning rule(FP) were simulated with the learning rate $\varepsilon = 0.2$ and the regularization parameter $\delta = 1.0 \times 10^{-4}$. In order to indicate the effectiveness of the learning scheme, we prepared the neural inverse model which was acquired by feedback-error learning(FEL)[5] after 500 iterations with the back-propagation learning rule(BP) based on a steepest decent method. In FEL, the feedback-error signal has been used as an error in the output layer of the neural inverse model. The parameter of the feedback controller was experimentally decided in this simulation.

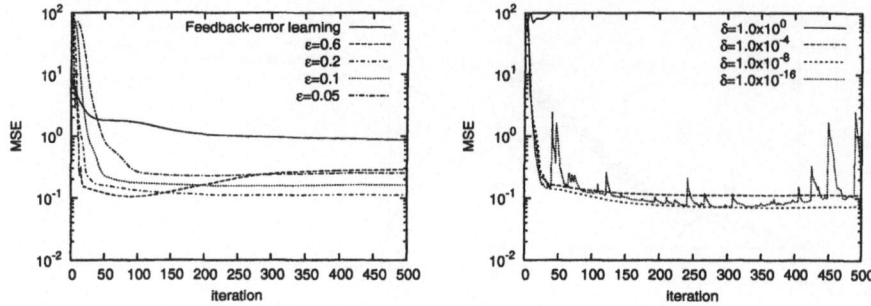

(a) The cases of changing ε and feedback-error learning.   (b) The cases of changing δ at $\varepsilon = 0.2$.

Fig. 3. The learning curves for acquiring the neural inverse models.

Both the realized trajectories of the end effector before and after learning are shown in Fig.2(b) by dashed lines and the realized trajectory by feedback-error learning is shown in Fig.2(b) by a dotted line. Fig.2(b) shows that FP is clearly faster and more correct than the conventional scheme based on BP.

In order to realize stable learning, it was required that the range of distribution for initializing connection weights was carefully adjusted. On the other hand, ε and δ did not need strict adjustment. Fig.3 shows learning curves depending on the values of the parameters. We can see that the speed of convergence with FP was quite fast and the stable learning was provided even if ε and δ were roughly adjusted.

4   Conclusion

We have aimed to the learning scheme which is consistent with physiology. In this paper, we have developed a forward-propagation rule for acquiring inverse models by using a RLS algorithm. The proposed scheme was applied to learning an inverse dynamics problem of a 2-link arm, and its convergence was faster than a feedback-error learning based on BP. Through the learning curves given by the simulation, we confirmed that the proposed scheme provided the stable learning by roughly adjusting the parameters of adaptive algorithm as long as paying attention to the initial states of neural networks.

Acknowledgement

This study was supported by the 21st Century COE Program "Intelligent Human Sensing" from the Japanese Ministry of Education, Culture, Sports, Science and Technology.

References

1. D.E. Rumelhart, G.E. Hinton and R.J. Williams "Learning representations by back-propagation errors," Nature, vol.323, pp.533-536, 1986.
2. K. Nagasawa, N. Fukumura and Y. Uno, "A forward-propagation rule for acquiring inverse models in multi-layered neural networks," (in Japanese), IEICE Transactions, J85-D-II, pp.1066-1074, Jun 2002.
3. J.M. Ortega and W.C. Rheinboldt, Iterative solution of nonlinear equations in several variables, Academic Press, New York, 1970.
4. S. Heykin, Adaptive filter theory, 3rd Edition, Prentice Hall, Englewood Cliff, NJ, 1996.
5. M.Kawato, "Feedback-error-learning neural network for supervised motor learning," in Advanced Neural Computers, ed. R. Eckmiller, pp.365-372, North Holland, Amsterdam, 1990.

Generalization in Learning Multiple Temporal Patterns Using RNNPB

Masato Ito[1] and Jun Tani[2]

[1] Sony Corporation, Tokyo, 141-0001, Japan
masato@pdp.crl.sony.co.jp
[2] Brain Science Institute, RIKEN, Saitama, 351-0198, Japan
tani@brain.riken.go.jp

Abstract. This paper examines the generalization capability in learning multiple temporal patterns by the recurrent neural network with parametric bias (RNNPB). Our simulation experiments indicated that the RNNPB can learn multiple patterns as generalized by extracting relational structures shared among the training patterns. It was, however, shown that such generalizations cannot be achieved when the relational structures are complex. Our analysis clarified that the qualitative differences appear in the self-organized internal structures of the network between generalized cases and not-generalized ones.

1 Introduction

Learning temporal patterns from examples are important problems in various domains including robot learning, adaptive process controls, auditory processing and etc. Recurrent neural network (RNN) [1] has been investigated for this purpose and it has been shown that an RNN is good at learning a single pattern by self-organizing a corresponding attractor [1].

Then a question arises that how multiple temporal patterns can be learned using RNNs. There are two distinct approach for the problem by using local representation and distributed representation. In the local representation scheme, each temporal pattern is learned to be stored in a local module network by utilizing winner-take-all dynamics among the modules. Such examples can be seen in [2, 3]. On the other hand in the distributed representation, multiple temporal patterns are learned in a single network by sharing its neural units and synaptic weights. One possible implementation is the scheme of recurrent neural network with parametric biases (RNNPB) [4] in which the parametric bias (PB) plays a role of modulation parameters of RNN dynamical structures. By modulating the PB, different temporal patterns are generated. The values of PB for each of target patterns are self-organized during learning processes.

An interesting characteristics of the RNNPB is that it can generate not only learned patterns but also varieties of unlearned patterns by modulating the PB values [4]. It is, however, not clear yet that how such unlearned patterns are generated and under what sort of constraints they do. One of the most essential characteristics in the distributed representation scheme is that each

N.R. Pal et al. (Eds.): ICONIP 2004, LNCS 3316, pp. 592–598, 2004.
© Springer-Verlag Berlin Heidelberg 2004

pattern is learned not independently but as embedded in sorts of relational structures among other learned patterns. Therefore, unlearned patterns can be generated as constrained by such relational structures organized in the network. It can be said that the learning of a set of patterns is generalized when such relational structures among patterns are successfully extracted in the network.

The current study examines this generalization characteristics of the RNNPB learning by conducting a set of simulation experiments. In the first experiment, we investigate how generalization is achieved in learning a set of training patterns those share simple constraints. In the second experiment, we investigate a representational case of not achieving the generalization with using training patterns that are posed with more complex constraints. Through these experiments we attempt to clarify the underlying mechanisms of achieving generalizations as well as not achieving of them in learning multiple patterns in the RNNPB.

2 Model

First, here are the basic ideas behind our model. Figure 1 shows the neural net architecture used in the current study.

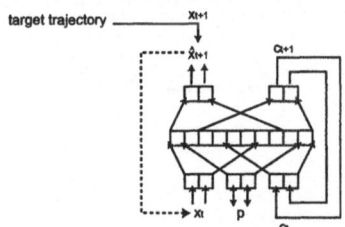

Fig. 1. The RNN associated with the PB inputs.

The architecture employs a Jordan-type [5] recurrent neural network (RNN) associated with the PB nodes. The RNN learns to generate sequence patterns by receiving x_t as inputs and generating predictions of the next step inputs $\hat{x}_{t+1}$ as outputs by using the context state c_t and the PB vector p. Here, x_t, c_t and p are vectors. Note that p is fixed while x_t and c_t change dynamically during generating a temporal pattern. The current context state c_t represented by context nodes in the input layer is mapped to that in the next time step, c_{t+1}, represented by context output nodes. The PB vector plays the role of the pattern modulator which is analogous to bifurcation parameters of nonlinear dynamical systems. Each specific temporal pattern is generated while the PB is clamped to its corresponding value. Specific values of PB for generating each training pattern is self-determined through their prior learning processes (as will be described later). The following subsections will describe the learning processes in detail as well as the pattern generation processes.

2.1   The Learning Process

The idea of learning in the model is to search simultaneously for optimal synaptic weights that are common for N of training sequence patterns and N of optimal PB vectors each of which is specific to one of the training sequence patterns.

p_n as the PB vector values at learning step n is updated iteratively for each training pattern with using back-propagation through time (BPTT) algorithm [5] while the synaptic weights in the network is updated. The BPTT utilizes a window of a working memory that stores computational results in the current learning step for each training sequence. The forward activation sequence by cascade, the error sequence and the delta error sequence are stored. The step length of the sequences stored in the working memory is denoted as L (that is equal to step length of the training sequence pattern). For each learning iteration, L steps of the forward dynamics are computed by cascading the network for each sequence. Once the L steps of the output sequence are generated, the errors between the teaching targets x_{t+1} and its prediction outputs $\hat{x}_{t+1}$ are computed. The error at each step is back-propagated through time in the sequence by which the delta error at PB nodes at each step in the cascaded network is obtained. The summation of this delta error over L steps provides the update direction of the PB vector in order to minimize the total error for the training sequence. The update equations for the ith unit of the PB at learning step n are:

$$\delta\rho_n^i = k_{bp} \cdot \sum_{t=0}^{L} \delta_t^{bp^i} \tag{1}$$

$$\Delta\rho_n^i = \epsilon \cdot \delta\rho_n^i + \eta \cdot \Delta\rho_{n-1}^i \tag{2}$$

$$p_n^i = sigmoid(\rho_n^i) \tag{3}$$

In Eq. (1) the δ force for the update of the internal potential values of the PB ρ is obtained from the summation of the delta error at PB node $\delta_t^{bp^i}$ for L steps. Then, the ρ_n is updated by using the delta force by means of the steepest descent method. The current PB p_n are obtained by means of the sigmoidal outputs of the internal potential values ρ_n.

2.2   The Pattern Generation

Once the synaptic weights in the RNN are determined by the learning process, each of trained sequence pattern can be generated using the so-called closed-loop mode with clamping the PB vector with the values obtained in the learning. In the closed-loop mode, the RNN's forward dynamics proceeds autonomously without receiving the inputs x_t externally. Instead, the prediction outputs $\hat{x}_{t+1}$ is fed-back to the inputs. It is noted that the RNN can generate various dynamic patterns beyond learned ones by arbitrary setting the PB vector, of which characteristics is the main discussion topics in the current paper.

3   Learning Experiments

In the learning experiments, 2 channels sinusoidal patterns are employed for training patterns. In the experiments, two learning sets are used. The set 1 consists of 5 patterns each of which has different amplitude and frequency while other properties, such as phase differences between two channels and offset, are the same. The set 2 consists of 5 patterns each of which has different offset and phase difference between channel 1 and 2 with other properties set as the same. In the first simulations, only training set 1 is learned. In the subsequent simulation, we examine the learning case with these two training sets merged into one training set. The RNNPB used in these experiments has 2 input nodes and 2 prediction output nodes for learning the forward dynamics of patterns. It also had 2 PB nodes, 60 hidden nodes, and 60 context nodes. Parameter settings in Eq. (1), (2) were $k_{bp} = 0.5$, $k_{nb} = 0.4$, $\epsilon = 0.1$, $\eta = 0.9$. In the learning, the RNNPB learns all patterns in the training set simultaneously. The learning is iterated for 100000 steps, starting from randomly set initial synaptic weights. The final root-mean-square error of the output nodes was less than 0.0002 over all learning results.

3.1   Training Set 1: Amplitude and Frequency

In the simulation 1, the RNNPB was trained using the set 1. After it was confirmed that the RNNPB can regenerate all patterns in the set 1, we examined how much the RNNPB can also generate unlearned patterns that are possibly generated by interpolating the training patterns in terms of amplitude and frequency. For this purpose, interpolated patterns are prepared by taking intermediate amplitude and frequency between selected pairs of the training patterns as the targets. Then, the PB values are searched for best mimicking those interpolated patterns.

In Figure 2, it is observed that the RNNPB can generate the interpolated patterns successfully. In Figure 2, the plot (b) shows the outputs generated by the RNNPB, N2-3 that is interpolated between the (a) pattern 2 and (b) 3 in the training set is generated with the PB of $(0.22, 0.20)$. The amplitude of N2-3 is intermediate between the pattern 2 and pattern 3 in the training set. In the same way, other 3 patterns, N2-4, N3-5 and N4-5 have intermediate characteristics among the training patterns. It is noted that those patterns are generated with modulating only for their amplitude and frequency, but not for other properties.

Next, in order to clarify the structure of the mapping between the PB and corresponding temporal pattern characteristics, a two-dimensional phase diagram for the PB space was plotted. More specifically, the amplitude and the periodicity of the generated patterns were plotted with two varying values of the PB nodes. The amplitude and the periodicity for generated patterns was computed while the two values of the PB were gradually changed at 0.01 intervals. The sequences of 1000 step lengths were generated by the forward dynamics in the closed-loop mode, and then the periodicity and the amplitude were calcu-

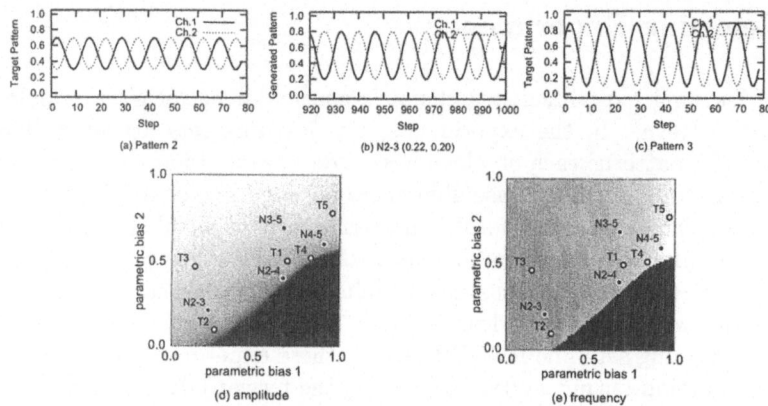

Fig. 2. The generated patterns correspond to intermediate patterns (a) N2-3 between the pattern (a) 2 and (c)3 in the learning sample. The phase plots for (d) the amplitude and (e) the periodicity for the outputs using the values of the parametric biases after learning the training set 1.

lated at each point in the PB space using the sequence from the 600th to the 1000th step in order to exclude the initial transient period.

Figure 2 (d) and (e) show the phase plots generated for (d) amplitude and (e) periodicity by using color grading from black to white. In Figure 2 (d), a white tile denotes that the amplitude is 1.0 while a black one indicates zero amplitude (meaning that the trajectory converges to a fixed point). In Figure 2 (e), a white tile denotes that the periodicity is more than 30 steps, while a black one indicates the convergence to a fixed point. In figure 2 (d) and (e), N labels indicate the PB values that generated the interpolated patterns. And T labels indicate the PB values that regenerated the training patterns.

We observed that the amplitude and periodicity changes smoothly over the PB space except for the region of fixed point dynamics. In the PB space, it is observed that the PB vectors T1, T2, T3, T4 and T5 which can regenerate the training patterns, are widely distributed in the limit cycling dynamics. And it is also observed that each PB vector for the interpolated patterns, N2-3, N2-4 and N3-5, is located in intermediate between T2 and T3, T2 and T4, and T3 and T5, respectively. These observations conclude that the learning of the training set 1 was well generalized by capturing the underlying regularity in the training set 1, that is – in generating sinusoidal patterns their amplitude and periodicity can be changed independently while other profiles are preserved.

Note that we obtained similar results for the set 2 in terms of offset and phase difference.

3.2 Training Set 1 and 2

In the simulation 2, the RNNPB was trained using both of the set 1 and 2. We confirmed that the RNNPB can regenerate all patterns in the set 1 and 2.

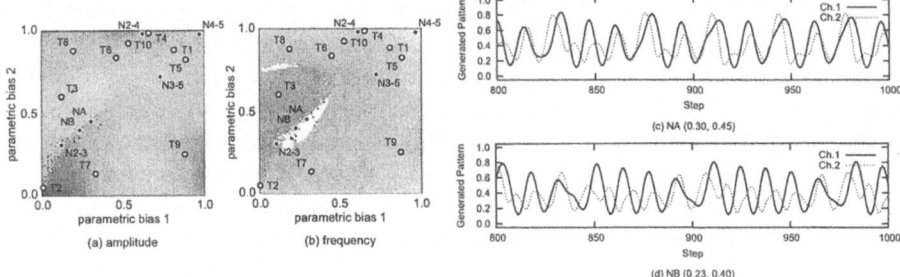

Fig. 3. The phase plots for (a) the amplitude and (b) the periodicity for the outputs using the values of the parametric biases after learning the training set 1 and 2. The diverse patterns (c) and (d) generated with the PB values in the region of complex structures in a relatively small PB space.

Figure 3 shows the phase plots generated for (a) amplitude and (b) periodicity using color grading from black to white.

It is observed that the PB space is self-organized with much more complex structures compared to those in the simulation 1. The PB points those correspond to the regenerations of the training patterns in the set 2 and their interpolated patterns are placed in the orthogonal region from the top-left to the down-right. On the other hand, those points corresponding to the set 1 are placed as divided into two regions of the down-left and the top-right parts. At each of these regions, some of the PB points for the interpolations are found to be way out from intermediate between two PB points that correspond to training patterns to be interpolated.

In examining whole possible patterns generated in the all space of the PB, non-periodic patterns were found frequently especially in the boundary between the regions of the set 1 and the set 2. Figure 3 (c) and (d) show such fluctuated patterns generated diversely, which cannot be explained by simple interpolations among the training patterns. (c) NA was generated with the PB vector $(0.30, 0.45)$ and (d) NB with $(0.23, 0.40)$. It is considered that these fluctuated patterns are generated because the PB mapping in the boundary is highly distorted in a nonlinear way where two distinct functional structures meet each other in a conflicting manner. These observations suggest that generalization in learning becomes much harder when embedding of different structures are attempted in a relatively small PB space.

4   Summary

Our experiments showed that the RNNPB can learn multiple temporal patterns by extracting certain common structures among them. In the successful learning case, it was observed that the self-organized mapping between the PB and the characteristics of generated patterns becomes smooth where patterns interpolated among the training patterns can be generated. This explains the generalization capability of the RNNPB when the training patterns share relatively simple constraints. It was, however, observed that the RNNPB learning cannot be generalized well in more complex situations where two distinct relationships exist among training patterns. It seems that the PB mapping is self-organized with substantial distortions by attempting to embed complex structures in a relatively small PB space. In such situations, diverse fluctuated patterns, which cannot be explained by simple interpolations among the trained patterns, tend to be generated.

References

1. Jordan, M.: Attractor dynamics and parallelism in a connectionist sequential machine. In: Proc. of Eighth Annual Conference of Cognitive Science Society, Hillsdale, NJ: Erlbaum (1986) 531–546
2. Wolpert, D., Kawato, M.: Multiple paired forward and inverse models for motor control. Neural Networks 11 (1998) 1317–1329
3. Tani, J., Nolfi, S.: Learning to perceive the world as articulated: an approach for hierarchical learning in sensory-motor systems. In Pfeifer, R., Blumberg, B., Meyer, J., Wilson, S., eds.: From animals to animats 5. Cambridge, MA: MIT Press. (1998) later published in Neural Networks, vol12, pp1131–1141, 1999.
4. Tani, J., Ito, M.: Self-organization of behavioral primitives as multiple attractor dynamics: a robot experiment. IEEE Trans. on Sys. Man and Cybern. Part A 33 (2003) 481–488
5. Rumelhart, D., Hinton, G., Williams, R.: Learning internal representations by error propagation. In Rumelhart, D., Mclelland, J., eds.: Parallel Distributed Processing. Cambridge, MA: MIT Press (1986) 318–362

Structural Learning of Neural Network for Continuous Valued Output: Effect of Penalty Term to Hidden Units

Basabi Chakraborty and Yusuke Manabe

Faculty of Software and Information Science, Iwate Prefectural University
Iwate 020-0193, Japan
basabi@soft.iwate-pu.ac.jp

Abstract. Multilayer feed forward networks with back propagation learning are widely used for function approximation but the learned networks rarely reveal the input output relationship explicitly. Structural learning methods are proposed to optimize the network topology as well as to add interpretation to its internal behaviour. Effective structural learning approaches for optimization and internal interpretation of the neural networks like structural learning with forgetting (SLF) or fast integration learning (FIL) have been proved useful for problems with binary outputs. In this work a new structural learning method based on modification of SLF and FIL has been proposed for problems with continuous valued outputs. The effectiveness of the proposed learning method has been demonstrated by simulation experiments with continuous valued functions.

1 Introduction

The most popular model of artificial neural network is the multilayer feedforward neural network with back-propagation (BP) learning. Although powerful, backpropagation learning suffers from some serious drawbacks. One of the important problems is that the neural network with BP learning becomes a black box due to use of all of the initial network parameters such as link-weights and hidden units. Therefore, we can not interpret the knowledge embedded in the trained neural network. This problem prevents further application of the neural network to real world problems. In order to solve the above problem, there is a class of learning methods including optimization of neural network structure, called structural learning.

Structural learning with forgetting(SLF) proposed by Ishikawa [1] in which a quasi-linear penalty term is added to the normal BP error function leads to a skeletal network which facilitates discovery of rules or knowledge related to the problem from the training data. However selection of proper forgetting factor, distributed representation of hidden units leading to unnecessary links and large output errors must be taken care of by additional learning phase. Fast integration learning (FIL) [2] has been proposed to accelerate forgetting in SLF and suppress

N.R. Pal et al. (Eds.): ICONIP 2004, LNCS 3316, pp. 599–605, 2004.
© Springer-Verlag Berlin Heidelberg 2004

emergence of distributed representation of hidden units by introducing a different form of penalty term in the backpropagation error function. Both the learning techniques have been successfully applied to the problems with binary outputs such as discovery of boolean function or classification problems but they are not appropriate for approximating continuous valued outputs. In this work a modification of SLF and FIL, suitable for continuous valued outputs has been proposed and its effectiveness compared to SLF and FIL has been studied by simulation with several continuous valued functions.

2   Structural Learning with Penalty Terms

2.1   Structural Learning with Forgetting (SLF)

Structural learning with forgetting contains three components, learning with forgetting, hidden unit clarification and learning with selective forgetting which are applied in order. The three criterion functions J_f, learning with forgetting, J_h, learning for hidden units clarification and J_s, learning with selective forgetting are defined as follows :

$$J_f = J + \varepsilon' \sum_{i,j} |w_{ij}|, \tag{1}$$

$$J_h = J_f + c' \sum_j min\{h_j, 1 - h_j\}, \tag{2}$$

$$J_s = J + \varepsilon' \sum_{|w_{ij}|<\theta_\varepsilon} |w_{ij}| \tag{3}$$

where J is mean squared error, $\sum_{i,j} |w_{ij}|$ is the penalty term for forgetting, $\sum_j min\{h_j, 1 - h_j\}$ is the penalty term for hidden unit clarification, ε' and c' are relative weights of each penalty terms. w_{ij} represents connection weight from ith unit to jth unit, while h_j represents output of jth hidden unit and θ_ε is threshold for acheiving selective forgetting.

The learning by J_f contributes to the removal of redundant connection weights. The learning by J_h contributes to the removal of distributed representation of connection weights on hidden units by forcing them to either active or inactive. The learning by J_s contributes to the minimization of mean square error by decaying the connection weights only whose absolute value are below a threshold while keeping the skeletal structure intact.

2.2   Fast Integration Learning (FIL)

Fast integration learning proposed by Kikuchi et al. [2] is an improvement over SLF in which the three criterion functions to be used in order are defined as follows:

$$J_1 = -\sum_k \{t_k ln o_k + (1 - t_k) ln(1 - o_k)\}, \tag{4}$$

$$J_2 = J_1 + \beta' \sum_j |h_j - 0.5| + \varepsilon' \sum_{ij} |w_{ij}|, \tag{5}$$

$$J_3 = J_1 + \beta' \sum_{|h_j - 0.5| < \theta_\beta} |h_j - 0.5| + \varepsilon' \sum_{|w_{ij}| < \theta_\varepsilon} |w_{ij}| \tag{6}$$

β' and ε' are relative weights of each of the penalty terms, h_j is output of jth hidden unit, w_{ij} is connection weight from ith unit to jth unit and θ_β and θ_ε are threshold parameters. J_1 represents back propagation learning with cross entropy criterion. The learning by J_2 contributes to the suppression of distributed representation on the hidden units. The learning by J_3 contributes to the minimization of training error.

3   Proposed Variance Suppressive Learning (VSL)

We found that for both SLF and FIL learning, the penalty terms for suppression of the distributed representation on the hidden units have an effect of producing near constant output values (i,e. small output variance) for the redundant hidden units. Considering this outcome, a modified learning technique with new penalty terms, specially suitable for problems with continuous valued inputs and outputs, has been proposed in this paper in which the hidden units of near zero output variance are deleted by adding their outputs to the next higher layer neurons as a bias input. VSL comprises of two criterion functions, which are to be applied in order. The two functions are,

$$J_I = J_c + \varepsilon' \sum_{ij} |w_{ij}| + \gamma' \sum_j \frac{net_j^2}{1 + net_j^2} \tag{7}$$

$$J_{II} = J_c + \varepsilon' \sum_{|w_{ij}| < \theta_\varepsilon} |w_{ij}| \tag{8}$$

J_c represents back propagation learning with cross entropy criterion and same as J_1 in Eq. 4 of FIL. ε' and γ' represent relative weights of each of the penalty terms in J_I. net_j represents the total input of jth hidden unit, w_{ij} is connection weight from ith unit to jth unit and θ_ε is threshold parameter.

The learning by J_I achieves structural optimization simulteneously by forgetting (first term) and detecting redundant hidden units by modified integration term (second term) by pushing the outputs of so-called redundant hidden units to a constant value. Finally the hiddent units with near zero variance are deleted by adding its value as the bias to the next higher layer unit as follows [4] [5]:

if $\sigma_{h_j}^2 \simeq 0$ then $b_k^{new} = b_k^{old} + w_{jk} \bar{h}_j$.

$\sigma_{h_j}^2$ represents variance of jth hidden unit, b_k represents bias of the kth unit in the next higher layer and $\bar{h}_j$ represents average output of jth hidden unit. The proper setting of threshold in above equation is also easy as the difference between variance of neccessary units and the unneccessary units is so large that the redundant hidden units can easily be detected. The learning by J_{II} contributes to the minimization or training error for fine tuning of the training similar to SLF and FIL.

4   Simulation Experiments

4.1   Learning of Quadratic Functions

Simulation experiments have been done with the following two simultaneous quadratic functions and the effect of SLF, FIL and VSL for structural optimization of the network have been analyzed.

$$y_1 = x_1 x_2 + x_1 \tag{9a}$$
$$y_2 = x_1 x_2 + x_2 \tag{9b}$$

The values of x_1, x_2 are in the range $(0.1 \approx 0.6)$. Initial neural network structure has 2 input units, 5 hidden units and 2 output units. The networks are trained independently by three learning techniques: Structural Learning with Forgetting(SLF), Fast Integration Learning(FIL) and Variance Suppressive Learning(VSL) and the simulation results are compared. The parameters related to penalty term for hidden unit determination for respective learning rules are varied to examine their effects. The learning epoch for each learning methods are SLF=(J_f, J_h, J_s)=(5000,12500,12500), FIL=(J_1, J_2, J_3)=(2000,14000,14000) and VSL=(J_I, J_{II})=(15000,15000) respectively.

4.2   Simulation Results and Discussions

Fig.1 and Fig.2 represent the structure of the final network after training with SLF or FIL and VSL respectively. The width of the connection links is proportional to the connection weights i,e. the wider links represents higher or stronger connection weights. After learning with SLF and FIL the final networks in each case reflects the functional relationship given by equations 9a and 9b. The network after learning with VSL represents perfectly the functional relationship. Thus it is found that the proposed learning is better than SLF and FIL in obtaining optimized structures. Fig.3, Fig.4 and Fig.5 represents the effect of changing parameters c, β and γ in SLF, FIL and VSL learning for obtaining optimized structure respectively. From the figures it is evident that VSL is least sensitive to parameter tuning among the three structural learning techniques.

4.3   Learning of Henon Map and Logistic Map

The second simulation is learning of Henon map and Logistic map. The objective of this simulation is to analyze the ability of knowledge extraction from learned neural network with proposed learning method VSL. The equations are as follows:

    Henon Map:

$$x(t+1) = 1 - ax(t)^2 + y(t) \tag{10a}$$
$$y(t+1) = bx(t) \tag{10b}$$

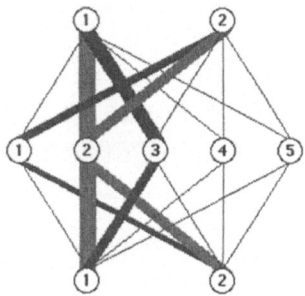

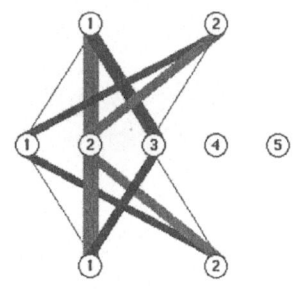

Fig. 1. Network Structure after learning with SLF or FIL

Fig. 2. Network Structure after learning with VSL

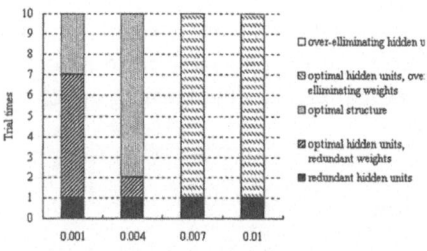

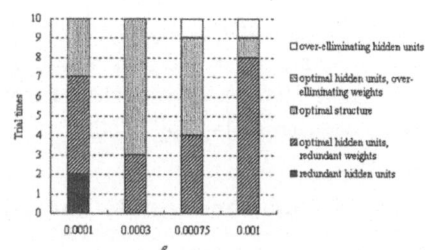

Fig. 3. Effect of c in SLF on structural optimization

Fig. 4. Effect of β in FIL on structural optimization

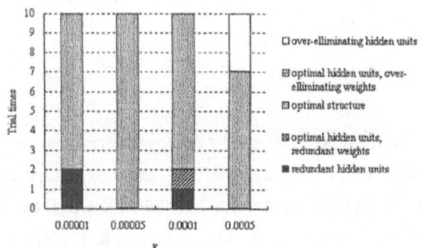

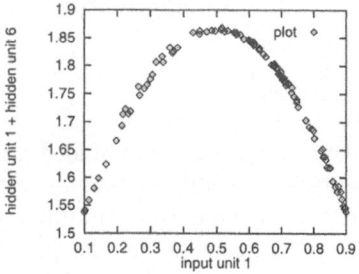

Fig. 5. Effect of γ in VSL on structural optimization

Fig. 6. Extracting Relation between $x(t)$ and $x(t+1)$

Logistic Map:

$$z(t+1) = cz(t)(1 - z(t)) \tag{11}$$

The number of data generated from the above equations for simulation is 100 ($t = 1$ to 101), with the parameters $a = 1.4$, $b = 0.3$, $c = 4.0$, $x(0) = 0.1$, $y(0) = 0.1$, $z(0) = 0.1$. Each neural network structure has initially 3 input units, 7 hiden units and 3 output units. Input units 1, 2 and 3 represent $x(t)$, $y(t)$ and $z(t)$ respectively and output unit 1, 2 and 3 represent $x(t+1)$, $y(t+1)$

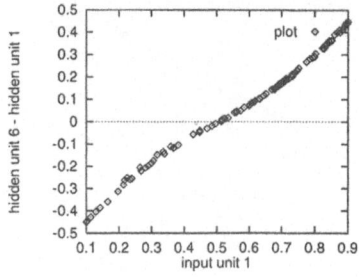
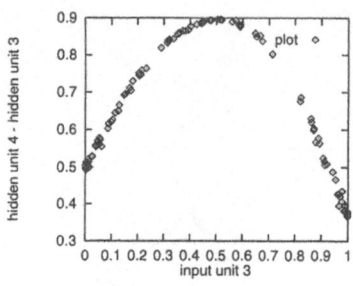

Fig. 7. Extracting Relation between $x(t)$ and $y(t+1)$

Fig. 8. Extracting Relation between $z(t)$ and $z(t+1)$

and z(t+1) respectively. The parameter settings for VSL are $\eta = 0.1$, $\alpha = 0.2$, $\varepsilon = 0.00005$, $\theta_\varepsilon = 1.0$ and $\gamma = 0.0001$. These parameters have been chosen as the respective best parameters through trial and error. The learning epochs for VSL are $(J_I, J_{II})=(7500,7500)$ respectively.

4.4   Simulation Results and Discussion

Fig. 6, Fig. 7 and Fig. 8 represent the relation expressed in Eq.10a,Eq.10b and Eq. 11 perfectly by the connection links and connection weights of the input and hidden unit neurons of the learned network with proposed VSL. Thus it is seen from the simulation results that the proposed structural learning is efficient in representing the input-output relationship explicitly by the trained network.

5   Conclusion

Structural optimization of feed forward networks for proper comprehensible representation of input/output function of the network revealing the underlying structure of the data is an important issue in artificial neural network approach for problem solving. In this paper structural learning of the network by using penalty terms with the error back propagation learning has been investigated and a structural learning algorithm with new penalty terms based on structural learning with forgetting (SLF) and fast integration learning (FIL), an improvement of SLF, has been proposed. SLF and FIL are effective for structural optimization for problems with binary outputs but are not very appropriate for continuous valued problems. The proposed learning algorithm VSL is found to perform better in structural optimization of the network than SLF and FIL in mapping continuous valued problems as shown by simulation experiments. Moreover it has been found from the simulation results that the proposed VSL is very efficient in extracting the underlying functional relationship from the data by representing it explicitly through structurally optimized trained network.

References

1. M. Ishikawa, 'Structural Learning with Forgetting', Neural Networks, Vol 9, no.3, pp. 509–521, 1996.
2. S. Kikuchi and M. Nakanishi, 'Extracting Classification Rules using Modified Structural Learning with Forgetting and Parallel Multi-Layer Network', Trans IEE Japan, Vol.120-C, No. 8/9, pp. 1181-1187,2000. (in Japanese)
3. S. Kikuchi and M. Nakanishi, 'Recurrent Neural Network with Short Term Memory and Fast Structural Learning Method', Trans. IEICE, Vol. J-84-D-II, No. 1, pp. 159–169, 2001. (in Japanese)
4. J. Sietsma and R. J. F. dow, ' Creating Artificial Neural Networks That Generalize', Neural Networks, Vol. 4, pp. 67–79, 1991.
5. T. Masuda et al., 'Compact Structuring of Hierarchical Neural Networks by Combining Extra Hidden Units', Trans. SICE, Vol. 28, No. 4, pp. 519–527, 1992. (in Japanese)

Argumentation Neural Networks

Artur d'Avila Garcez[1,*], Dov Gabbay[2], and Luís C. Lamb[3,**]

[1] Dept. of Computing, City University London, UK
aag@soi.city.ac.uk
[2] Dept. of Computer Science, King's College London, UK
dg@dcs.kcl.ac.uk
[3] Institute of Informatics, UFRGS, Porto Alegre, Brazil
LuisLamb@acm.org

Abstract. While neural networks have been successfully used in a number of machine learning applications, logical languages have been the standard for the representation of legal and argumentative reasoning. In this paper, we present a new hybrid model of computation that allows for the deduction and learning of argumentative reasoning. We propose a *Neural Argumentation Algorithm* to translate argumentation networks into standard neural networks, and prove correspondence between the semantics of the two networks.

1 Introduction

Neural-Symbolic integration concerns the application of symbolic knowledge within the connectionist paradigm [3]. While neural networks have been successfully used in a number of machine learning applications, logical languages have been the standard for the representation of argumentative reasoning [2,5]. In this paper, we present a new hybrid model of computation that allows for the deduction and learning of argumentative reasoning [1]. We do so by using Neural-Symbolic Learning Systems, which use simple, single hidden layer neural networks to represent and learn nonmonotonic, epistemic or temporal symbolic knowledge with the use of off-the-shelf neural learning algorithms [4].

Our goal is to facilitate learning capabilities in value-based argumentation frameworks, as arguments may evolve over time, with certain arguments being strengthened and others weakened. The neural networks are set up by a *Neural Argumentation Algorithm,* introduced in this paper. The proof that the neural network computes the argumentation network is then given, thus showing that the two representations are equivalent. In Section 2, we briefly present the basic concepts of neural-symbolic systems used throughout this paper. In Section 3, we introduce the *Neural Argumentation Algorithm* and prove that the neural network computes the given argumentation network. In Section 4, we conclude and discuss directions for future work.

---

* Partly supported by The Nuffield Foundation UK.
** Partly supported by CNPq and FAPERGS Brazil.

N.R. Pal et al. (Eds.): ICONIP 2004, LNCS 3316, pp. 606–612, 2004.
© Springer-Verlag Berlin Heidelberg 2004

2   Neural-Symbolic Learning Systems

The *Connectionist Inductive Learning and Logic Programming System* (C-ILP) [3] is a massively parallel computational model based on an artificial neural network that integrates inductive learning from examples and background knowledge with deductive learning from logic programming. C-ILP's *Translation Algorithm* maps a logic program $\mathcal{P}$ into a single hidden layer neural network $\mathcal{N}$ such that $\mathcal{N}$ computes the least fixed-point of $\mathcal{P}$. This provides a massively parallel model for logic programming. In addition, $\mathcal{N}$ can be trained with examples using Backpropagation [8], having $\mathcal{P}$ as background knowledge. The knowledge acquired by training can then be extracted, closing the learning cycle [3].

Let us exemplify how C-ILP's *Translation Algorithm* works. Each rule (r_l) of $\mathcal{P}$ is mapped from the input layer to the output layer of $\mathcal{N}$ through one neuron (N_l) in the single hidden layer of $\mathcal{N}$. Intuitively, the *Translation Algorithm* from $\mathcal{P}$ to $\mathcal{N}$ has to implement the following conditions: (c_1) The input potential of a hidden neuron (N_l) can only exceed N_l's threshold (θ_l), activating N_l, when all the positive antecedents of r_l are assigned the truth-value *true* while all the negative antecedents of r_l are assigned *false*; and (c_2) The input potential of an output neuron (A) can only exceed A's threshold (θ_A), activating A, when at least one hidden neuron N_l that is connected to A is activated.

In order to use $\mathcal{N}$ as a massively parallel model for logic programming, we just have to follow two steps: (i) add neurons to the input and output layers of $\mathcal{N}$, allowing it to be recurrently connected; and (ii) add the corresponding recurrent connections with fixed weight $W_r = 1$, so that the activation of output neuron A feeds back into the activation of input neuron A, the activation of output neuron B feeds back into the activation of input neuron B, and so on.

In the case of argumentation networks it will be sufficient to consider definite logic programs (i.e. programs without negation). In this case, the neural network will contain only positive weights (W). We will then expand such a positive network to represent *attacks* using negative weights from the network's hidden layer to its output layer.

3   Argumentation Neural Networks

Let us start by considering a *moral debate* example: *Hal, a diabetic, loses his insulin in an accident through no fault of his own. Before collapsing into a coma, he rushes to the house of Carla, another diabetic. She is not at home, but Hal breaks into her house and uses some of her insulin. Was Hal justified? Does Carla have a right to compensation?* The following are some of the arguments involved in the example. **A:** Hal is justified, he was trying to save his life; **B:** It is wrong to infringe the property rights of another; **C:** Hal compensates Carla; **D:** Hal is endangering Carla's life; **E:** Carla has abundant insulin; and **F:** If Hal is too poor to compensate Carla he should be allowed to take the insulin as no one should die because they are poor.

In [1], arguments and counter-arguments are arranged in an argumentation network, as in Figure 1(a), where an arrow from argument X to argument Y indicates that X *attacks* Y. For example, the fact that it is wrong to infringe Carla's right of property (**B**) attacks Hal's justification (**A**). In the argumentation network of Figure 1(a), some aspects may change as the debate progresses and actions are taken with the strength of an argument in attacking another changing in time. This is a learning process that can be implemented using a neural network (Figure 1(b)) in which the weights encode the strength of arguments. The network is an auto-associative single hidden layer network. Solid arrows represent positive weights and dotted arrows represent negative weights. Arguments are supported by positive weights and attacked by negative ones. Argument **A** (input neuron A), for example, supports itself (output neuron A) with the use of hidden neuron h_1. Similarly, argument **B** supports itself (via h_2), and so on. From the argumentation network, **B** attacks **A**, and **D** attacks **A**. The attacks are implemented in the neural network by the negative weights (see dotted lines in Figure 1(b)) with the use of h_2 and h_4 in the case of the above attacks, and so on.

The network of Figure 1(b) is a standard feedforward neural network that can be trained with the use of a standard neural learning algorithm. Training would change the initial weights of the network (the initial beliefs on the strength of arguments and counter-arguments), according to examples of input/output patterns, i.e. examples of the relationship between arguments. If the absolute value of the weight from neuron h_1 to output neuron A is greater than the sum of the absolute values of the weights from neurons h_2 and h_4 to A, one could say that argument **A** prevails (in which case output neuron A should be *activated* in the neural network).

Let us now implement the above behaviour using C-ILP networks. The *Neural Argumentation Algorithm* introduced below takes an argumentation network as input and produces a C-ILP neural network as output. These networks use a semi-linear activation function $h(x) = (2/(1 + e^{-\beta x})) - 1$ and inputs in $\{-1, 1\}$. We define $A_{min} \in (0, 1)$ as the minimum activation for a neuron to be considered *active* (or *true*), and $A_{max} \in (-1, 0)$ as the maximum activation for a neuron to be considered *non active* (or *false*). We assume, for mathematical convenience and without loss of generality, that $A_{max} = -A_{min}$. We need to define the values of A_{min}, W and θ such that the neural network computes the behaviour of the argumentation network. The values are set by the algorithm below, and come from the proof of Theorem 1, which shows that the neural network indeed computes the argumentation network. The neural network can then be run in parallel to compute the prevailing arguments, as exemplified in the sequel, or trained to represent different argument values.

Definition 1. *An argumentation network has the form $\mathcal{A} = <\alpha, attack, v>$, where α is a set of arguments, $attack \subseteq \alpha^2$ is a relation indicating which arguments attack which other arguments, and v is a function from attack to $\{0, 1\}$, which gives the relative strength of an argument. If $v(\alpha_i, \alpha j) = 1$ then α_i is said to be stronger than α_j. Otherwise, α_i is said to be weaker than α_j.*

Neural Argumentation Algorithm

1. Given an argumentation network $\mathcal{A}$ with arguments $\alpha_1, \alpha_2, ..., \alpha_n$, make $\mathcal{P} = \{r_1 : \alpha_1 \to \alpha_1, r_2 : \alpha_2 \to \alpha_2, ..., r_n : \alpha_n \to \alpha_n\}$;
2. Let $A_{min} > 0$, $W > 0$ and $W' < 0$;
3. Calculate $W \geq (1/\beta A_{\min}) \cdot (ln(1 + A_{\min}) - ln(1 - A_{\min}))$;
4. For each rule r_l of $\mathcal{P}$ $(1 \leq l \leq n)$:
   (a) Add a neuron N_l to the hidden layer of $\mathcal{N}$;
   (b) Connect neuron α_l in the input layer to N_l and set the connection weight to W;
   (c) Connect N_l to neuron α_l in the output layer and set the connection weight to W;
   (d) Set the threshold of all hidden and output neurons to 0;
5. Set $g(x) = x$ as the activation function of the neurons in the input layer of $\mathcal{N}$.
6. Set $h(x)$ as the activation function of the neurons in the hidden and output layers of $\mathcal{N}$.
7. For each $(\alpha_i, \alpha_j) \in attack$, do:
   (a) Connect hidden neuron N_i to ouptut neuron α_j;
   (b) If $v(\alpha_i, \alpha_j) = 0$ then set the connection weight to $W' > h^{-1}(A_{\min}) - WA_{\min}$;
   (c) If $v(\alpha_i, \alpha_j) = 1$ then set the connection weight to $W' < (h^{-1}(-A_{\min})) - W)/A_{\min}$;
8. If $\mathcal{N}$ is to be fully-connected, set all other connections to 0.

Note that the program $\mathcal{P}$ obtained from an argumentation network will always have the form of a set of rules $r_i : \alpha_i \to \alpha_i$. As a result, differently from in the C-ILP *Translation Algorithm* [3], in which rules having more than one antecedent are accounted for, here there is always a single antecedent per rule (α_i). This allows us to take $A_{\min} > 0$ and $\theta = 0$ when using C-ILP's algorithm to calculate W and W'. The fact that $W > 0$ and $W' < 0$ fits very well with the idea of arguments having strengths (W), and attacks also having strengths (W'). Once the constraints imposed on W and W' by the argumentation algorithm are satisfied, their relative values could be defined in a standard way, for example, by an audience using a voting system [1]. In this system, at some point, an accumulation of attacks with different strengths - neither being individually stronger than the argument being attacked - might produce a value $\sum_i W'_i$ that overcomes W. This is precisely the way that neural networks work. All we need to make sure is that the neural network *computes* the prevailing arguments of the argumentation framework, according to the following definition.

Definition 2. *($\mathcal{N}$ computes $\mathcal{A}$) Let $(\alpha_i, \alpha_j) \in attack$. We say that a neural network $\mathcal{N}$ computes the prevailing arguments of an argumentation framework $\mathcal{A}$ if (i) and (ii) below hold. (i) If α_i is stronger than α_j then output neuron α_j will not be activated when input neurons α_i and α_j are both activated, and (ii) If α_i is weaker than α_j then the activation of input neuron α_i will not be individually responsible for output neuron α_j being deactivated when input neuron α_j is activated.*

Theorem 1. *(Correctness of Argumentation Algorithm) For each argumentation network $\mathcal{A}$, there exists a feedforward neural network $\mathcal{N}$ with exactly one hidden layer and semi-linear neurons such that $\mathcal{N}$ computes $\mathcal{A}$.*

Proof. First, we need to show that the neural network containing only positive weights computes $\mathcal{P}$. When $r_i : \alpha_i \rightarrow \alpha_i \in \mathcal{P}$, we need to show that (a) if $\alpha_i \geq A_{\min}$ in the input layer then $\alpha_i \geq A_{\min}$ in the output layer. We also need to show that (b) if $\alpha_i \leq -A_{\min}$ in the input layer then $\alpha_i \leq -A_{\min}$ in the output layer. (a) In the worst case, the input potential of hidden neuron N_i is $WA_{\min}$, and the output of N_i is $h(WA_{\min})$. We want $h(WA_{\min}) \geq A_{\min}$. Then, again in the worst case, the input potential of output neuron α_i will be $WA_{\min}$, and we want $h(WA_{\min}) \geq A_{\min}$. As a result, $W \geq h^{-1}(A_{\min})/A_{\min}$ needs to be verified, which gives $W \geq (1/\beta A_{\min}) \cdot (ln(1 + A_{\min}) - ln(1 - A_{\min}))$, as in the argumentation algorithm. The proof of (b) is analogous to the proof of (a). Now, we need to show that the addition of negative weights to the neural network implements the attacks in the argumentation network. When $v(\alpha_i, \alpha_j) = 1$, we want to ensure that the activation of output neuron α_j is smaller than $-A_{\min}$ whenever both hidden neurons N_i and N_j are activated. In the worst case scenario, N_i presents activation $A_{\min}$ while N_j presents activation 1. We have $h(W + A_{\min}W') < -A_{\min}$. Thus, we need $W' < (h^{-1}(-A_{\min})) - W)/A_{\min}$; this is obtained directly from the argumentation algorithm. Similarly, when $v(\alpha_i, \alpha_j) = 0$, we want to ensure that the activation of output neuron α_j is larger than $A_{\min}$ whenever both hidden neurons N_i and N_j are activated. In the worst case scenario, now N_i presents activation 1 while N_j presents activation $A_{\min}$. We have $h(A_{\min}W + W') > A_{\min}$. Thus, we need $W' > h^{-1}(A_{\min}) - WA_{\min}$; again, this is obtained directly from the argumentation algorithm. $\square$

Our next step is to run the neural network to find out which arguments prevail in a situation. The key to running the network properly is to connect output neurons to their corresponding input neurons using weights fixed at 1, so that the activation of output neuron A, for example, is fed into the activation of input neuron A the next time round. This implements chains such as A attacks B, B attacks C, C attacks D, and so on, by propagating activations around the network, as the following example illustrates.

Example 1. (Moral Debate Neural Network) We apply the *Neural Argumentation Algorithm* to the argumentation network of Figure 1(a), and obtain the neural network of Figure 1(b). From C-ILP's Translation Algorithm, we know that $A_{min} > 0$ and $W > 0$. Let us take $A_{min} = 0.5$ and $W = 5$ (recall that W is the weight of solid arrows in the network). Following [5], we reason about the problem by grouping arguments according to the features of *life, property* and *fact*. Arguments **A**, **D** and **F** are related to the right of life, arguments **B** and **C** are related to property rights, and argument **E** is a fact. We may argue whether *property* is stronger than *life* but facts are always the strongest. If property is stronger than life then $v(B, A) = 1$, $v(D, A) = 1$, $v(C, B) = 1$, $v(C, D) = 1$, $v(E, D) = 1$, and $v(F, C) = 0$. From the *Neural Argumentation Algorithm*, when $v(\alpha_i, \alpha_j) = 0$ we must have $W' > -1.4$, and when $v(\alpha_i, \alpha_j) = 1$ we must have

$W' < -12.2$. From Theorem 1, the network will compute the expected prevailing arguments, as follows: F does not defeat C, C defeats B, E defeats D and, as a result, we obtain {A,C,E} as the acceptable set of arguments. Now, if life is considered stronger than property then $v(F,C) = 1$ instead, and as a result, F defeats C and, since C is defeated, it cannot defeat B, which in turn cannot defeat A. In this case, we obtain the set {A,B,E,F} of acceptable arguments[1]. This shows that two different lines of argumentation will provide the same answer to the question of whether Hall was justified (A), but two different answers to the question of whether Carla has the right to compensation (C).

4   Conclusion and Future Work

In this paper, we have presented a new hybrid model of computation that allows for the deduction and learning of argumentative reasoning. The model combines value-based argumentation frameworks and neural-symbolic learning systems by providing a translation from argumentation networks to C-ILP neural networks, and a theorem showing that such a translation is correct. The model works not only for acyclic argumentation networks but also for circular networks and enables cummulative argumentation through learning. Experiments on learning argumentation neural networks capable of evolving over time are currently being conducted. Complexity issues regarding the parallel computation of arguments in contrast with standard value-based argumentation frameworks are also being investigated. We believe that a neural implementation of this reasoning process may, in fact, be advantageous from a purely computational point of view due to neural networks' parallelism.

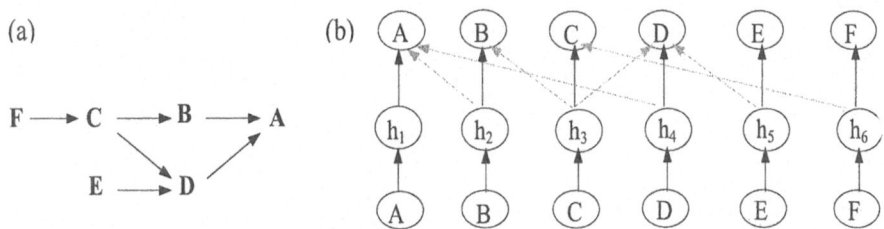

Fig. 1. Moral debate argumentation network (a), and neural network (b)

References

1. T. J. M. Bench-Capon. Persuasion in practical argument using value-based argumentation frameworks. *Journal of Logic and Computation*, 13:429–448, 2003.
2. A. Bondarenko, P. Dung, R. Kowalski, and F. Toni. An abstract, argumentation theoretic approach to default reasoning. *Artificial Intelligence*, 93:63–101, 1997.

---

[1] The complete set of argument values in this case is: $v(B,A) = 0$, $v(D,A) = 1$, $v(C,B) = 1$, $v(C,D) = 0$, $v(E,D) = 1$, and $v(F,C) = 1$. The values of W' are calculated in the same way as before.

3. A. S. d'Avila Garcez, K. Broda, and D. M. Gabbay. *Neural-Symbolic Learning Systems: Foundations and Applications.* Springer-Verlag, 2002.
4. A. S. d'Avila Garcez and L. C. Lamb. Reasoning about time and knowledge in neural-symbolic learning systems. In S. Thrun, L. Saul, and B. Schoelkopf, eds, *Advances in Neural Information Processing Systems 16*, Proc. of NIPS 2003, Vancouver, Canada, pp 921–928, MIT Press, 2004.
5. D. M. Gabbay and J. Woods. The law of evidence and labelled deduction: A position paper. *Phi News*, 4, October 2003.
6. S. Haykin. *Neural Networks: A Comprehensive Foundation.* Prentice Hall, 1999.
7. T. M. Mitchell. *Machine Learning.* McGraw-Hill, 1997.
8. D. E. Rumelhart, G. E. Hinton, and R. J. Williams. Learning internal representations by error propagation. In D. E. Rumelhart and J. L. McClelland, eds, *Parallel Distributed Processing*, volume 1, pp 318–362. MIT Press, 1986.

A Neighbor Generation Mechanism Optimizing Neural Networks

Amanda Lins and Teresa Ludermir

Centro de Informática - Universidade Federal de Pernambuco (UFPE)
Caixa Postal 7851 - 50.732-970, Recife – PE, Brazil
{apsl,tbl}@cin.ufpe.br

Abstract. This paper proposes the utilization of new neighbor generation method in conjunction with search techniques. The proposed mechanism works by adding a random number between -n and +n to the connection weights, where n is the weight value of each respective connection. This value may be multiplied by an adjustable ratio. The present paper shows the results of experiments with three optimization algorithms: simulated annealing, tabu search and hybrid system for the optimization of MLP network architectures and weights. In the context of solving the odor recognition problem in an artificial nose, the proposed mechanism has proven very efficient in finding minimal network architectures with a better generalization performance than the hybrid system mechanism used.

1 Introduction

Multilayer Perceptron (MLP) is the most utilized model in neural network applications using the backpropagation training algorithm. The definition of architecture in MLP networks is a very relevant point is, as a lack of connections may make the network incapable of solving the problem of insufficient adjustable parameters, while an excess of connections may cause an overfitting of the training data.

In experiments with MLP networks, a number of attempts with different architecture are usually made, such as varying the number of hidden units. However, besides taking a considerable amount of time and work, the choice is not always sufficiently adequate.

Most of the existing implementations for optimizing the MLP just codify the architecture. Nonetheless, error evaluation also depends on the weights. Optimizing both architecture and weight makes the cost evaluation more precise.

Thus, it becomes important to develop automatic methods for defining MLP architectures. Global optimization methods such as simulated annealing [1], tabu search and the methodology proposed by Yamazaki [3] (a hybrid system that combines characteristics of simulated annealing, tabu search and a local training algorithm for soft adjusment weights) are addressed in this paper.

The main objective of the paper is to propose a utilization of new neighbor generation method in conjunction with search techniques. The proposed method has been denominated mechanism 2. Through experimental results the paper demonstrates the main advantages of mechanism 2, including a decreased error rate and small complexity.

N.R. Pal et al. (Eds.): ICONIP 2004, LNCS 3316, pp. 613–618, 2004.
© Springer-Verlag Berlin Heidelberg 2004

Section 2 describes the problem and database. Section 3 presents some details of the implementation. Section 4 shows and analyzes the experimental results. The conclusion is presented in Section 5.

2 Problem and Database Descriptions

The addressed problem deals with the odor classification of three different vintages (1995, 1996 and 1997) of commercial red wine (Almadém, Brazil) produced with merlot grapes. Details on the construction of this prototype can be found in [4].

The total size of the data set is 5400 patterns. The present work divided the data in the following way: 50% of the patterns were put in the training set, 25% were delegated to the validation set and 25% to the test set, following the suggestion by *Proben1* [5]. The patterns were normalized in the [-1, +1] interval.

3 Implementation

The simulated annealing and tabu search algorithms can be found in [6] and [7].

In order to minimize the limitations of simulated annealing and tabu search, Yamazaki developed an optimization methodology for MLP networks [3]. In this approach, a set of new solutions is generated at each iteration. The best one is selected according to the cost function as performed by tabu search. However, the best solution is not always accepted since this decision is guided by a probability distribution, which is the same as the one used in the simulated annealing. During the execution of the methodology, the topology and the weights are optimized and the best solution found so far (s_{BSF}) is stored. At the end of this process, the MLP architecture contained in s_{BSF} is kept constant and the weights are taken as the initial ones for training with the backpropagation algorithm in order to perform a fine-tuned local search.

Given a set S of solutions and a real-valued cost function f, the methodology searches for the global minimum s such that $f(s) < f(s')$, for all s' attached S. The search stops after I_{max} epochs and a cooling schedule updates the temperature T_i of epoch i. Algorithm 1 shows the methodology structure.

In order to implement this methodology for a problem, the following aspects must be defined: (1) the representation of solutions; (2) the cost function; (3) the generation mechanism for the new solutions; and (4) the cooling schedule and stopping criteria. In this work, each MLP is specified by an array of connections, and each connection is specified by two parameters: (a) the connectivity bit, which is equal to one if the connection exists, and zero otherwise; and (b) the connection weight, which is a real number. If the connectivity bit is equal to zero, its associated weight is not considered, for the connection does not exist in the network. The cost of each solution is the mean of two important parameters: (1) the classification error for the training set (percentage of incorrectly classified training patterns); and (2) the percentage of connections used by the network. Therefore, the algorithms try to minimize both network performance and complexity. Only valid networks (i.e., networks with at least one unit in the hidden layer) were considered. The generation mechanism for the new solutions are described in the following section.

```
S₀ ← initial solution

T₀ ← initial temperature

Update S_BSF with S₀ (best solution found so far)

For i = 0 to I_max − 1
    If i + 1 is not a multiple of I_T
        T_{i+1} ← T_i
    else
        T_{i+1} ← new temperature
        If stopping criteria are not satisfied
            Stop execution
    Generate a set of k new solutions from S_i
    For i = 0 to k
        a = random number in the interval [-1,1]
        s'_i = s_i + a
    Choose the best solution s' from the set
    If f(s') < f(s_i)
        S_{i+1} ← s'
    else
        S_{i+1} ← s' with probability e^{-[f(s')−f(s_i)]/T_{i+1}}
    Update S_BSF (if f(S_{i+1}) < f(S_BSF))

Keep the topology contained in S_BSF constant and use the weights
as initial ones for training with the backpropagation algorithm.
```

Algorithm 1.

3.1 Generation Mechanism for the Neighbors

The generation mechanism acts as follows: first the connectivity bits for the current solution are changed according to a given probability, which in this work is set at 20%. This operation deletes some network connections and creates new ones. Then, a random number taken from a uniform distribution in [-1.0, +1.0] is added to each connection weight. These two steps can change both topology and connection weights to produce a new neighbor solution. This mechanism is called Mechanism 1 throughout this paper.

The modification proposed and used in the experiments – Mechanism 2 – aims to increase the search space. Thus, the interval was expanded in proportion to its own weight value, that is, a random number was added, which was taken from a uniform distribution in the [-n, +n] interval, where n is the respective weight value connection and it can be multiplied by a particular adjustable rate. The value 5 was used in these experiments. Therefore, the random numbers were generated in the [-5n, +5n] interval. The basic structure is as follows:

Generate a set of k new solutions from s_i

For $i = 0$ to k

   a = random number in the interval [-1,1]

   dw = (2 * weight s_i * a) - weight s_i

   $s'_i = s_i$ + (rate * dw)

4 Experiments and Results

An MLP network with just one hidden layer was used in the experiments. The initial topology has 4 hidden nodes that have all the possible feedforward connections between adjacent layers. For each initial topology, the same 10 random initializations of weights were used, belonging to a uniform distribution between −1.0 e +1.0. For each weight initialization, 30 executions of each algorithm were performed. Based on the classification error for the validation set, the 10 best and 10 worst results were eliminated. The analysis of the results of 10 remaining executions was then considered.

4.1 Simulated Annealing

The average error obtained with simulated annealing for Mechanism 1 with backpropagation (4.6281) was smaller than simulated annealing for mechanism 1 without backpropagation (1.6629). This demonstrates the great influence of refinement occurring through the local algorithm. Without adding the complexity of the backpropagation, the error of Mechanism 2 was even smaller (1.3355).

4.2 Tabu Search

The behavior with the tabu search algorithm was similar to the simulated annealing when analyzing the results obtained using Mechanism 1 without the local algorithm, followed by Mechanism 2 with the local algorithm. The average error obtained with Mechanism 2 was 1.0540, while the error obtained with Mechanism 1 combined with the local algorithm was 1.5022.

4.3 Yamazaki Proposal

The results obtained with Yamazaki Proposal are shown in Table 1. Column 1 display the weights initialization, average and standard deviation. Columns 2 and 3 display respectively the error test set classification and the number of connections for Mechanism 1 without backpropagation. Columns 4 and 5 display respectively the error test set classification and the number of connections for Mechanism 1 with backpropagation. Columns 6 and 7 display respectively the error test set classification and the number of connections for Mechanism 2 without backpropagation.

The error obtained with Mechanism 1 was 0.9774, changing to 0.3688 when refined with backpropagation. Utilizing Mechanism 2, we obtained 0.3413 without using backpropagation. The same behavioral line occurred with Simulated Annealing and Tabu Search algorithms. Using backpropagation with Mechanism 2 does not improve the results. Therefore, its use is unnecessary.

Table 1. Results for Mechanism 1 without backpropagation, Mechanism 1 with backpropagation and Mechanism 2 without backpropagation

Weights Column 1	Error Column 2	Connec. Column 3	Error Column 4	Connec. Column 5	Error Column 6	Connec. Column 7
1	1.7778	9.1	0.3778	9.1	0.5689	9
2	1.2000	9.3	0.7852	9.3	0.6756	8.4
3	0.6370	9.5	0.1852	9.5	0.2370	9.2
4	0.5778	9.8	0.2519	9.8	0.2370	9.6
5	2.0074	8.7	0.1704	8.7	0.2370	8.8
6	**0.3778**	**9.7**	0.8741	9.7	**0.2370**	**9.1**
7	0.4593	8.7	**0.0889**	**8.7**	0.2370	9.2
8	0.5778	8.3	0.3037	8.3	0.3200	9.2
9	0.9259	9.7	0.4148	9.7	0.2370	9.7
10	0.9333	9.5	0.2370	9.5	0.4267	9.1
Average	**0.9474**	**9.23**	**0.3689**	**9.23**	**0.3413**	**9.13**
Stand. Dev.	0.5588	0.5122	0.2620	0.5122	0.1621	0.3683

To compare all three algorithms, we applied t-Student paired tests at a 5% significance level for the test set classification error, with 9 degrees of freedom.

The results obtained by the global method together with Mechanism 2 are better than the global method together with Mechanism 1 at a confidence interval of 95%. Equivalence was also confirmed in the results of the average errors between the global method using Mechanism 1 refined with the local method and the local method alone using Mechanism 2.

5 Conclusion

The objective of this work was to show the results of a new mechanism for generating neighbors, utilizing three optimization algorithms: simulated annealing, tabu search and a hybrid system for the problem of the odor recognition through an artificial nose. T-Student paired tests were used to compare the results.

The proposed mechanism provided a better performance in generalization, as well as being able to produce networks with less time complexity, as it was not necessary to use refining with a local algorithm.

Our future work will test this proposed mechanism on different databases, as well as perform more experiments with the adjustment of parameters, refining the results of the global method used with Mechanism 2 with some local algorithm, verifying the best moment for using the local method, and introducing modifications in the solution and cost function representations.

Acknowledgments

The authors would like to thank CNPq, Capes and Finep (Brazilian agencies) for their financial support.

References

1. S. Kirkpatrick, C. D. Gellat Jr. and M. P. Vecchi: Optimization by simulated annealing, Science, 220:671-680,1983.

2. F. Glover: Future paths for integer programming and links to artificial intelligence, Computers and Operation Research, Vol. 13, pp.533-549, 1986.
3. A. Yamazaki and T.B. Ludermir: Neural Network Training with Global Optimization Techniques, International Journal of Neural Systems, Vol. 13, N. 2, pp. 77-86, 2003.
4. M. S. Santos: Construction of an artificial nose using neural networks (In Portuguese) Ph.D Thesis, Federal University of Pernambuco, 2000.
5. L. Prechelt: Proben 1 - A Set of Neural Network Benchmark Problems and Benchmarking Rules Technical Report 21/94, Fakultat fur Informatik, Universitat Karlsruche, Germany, September, 1994.
6. N. Metropolis, A. W. Rosenbluth, M. N. Rosenbluth, A. H. Teller and E. Teller: Equation of state calculations by fast computing machines, J. of Chem. Phys., Vol. 21, No. 6, pp. 1087-1092, 1953.
7. D.T. Pham and D. Karaboga: Introduction, Intelligent Optimisation Techniques, pp. 1-50, Springer-Verlag, 2000.
8. T.M. Mitchell: Machine Learning, MC-Graww Hill, 1997.

Collaborative Agent Learning Using Neurocomputing

Saulat Farooque[1], Ajith Abraham[2], and Lakhmi Jain[1]

[1] School of Electrical and Information Engineering,
University of South Australia, Adelaide, Australia
saulat.farooque@tenix.com, lakhmi.jain@unisa.edu.au
[2] Department of Computer Science, Oklahoma State University, USA
ajith.abraham@ieee.org

Abstract. This paper investigates the use of Generalised Regression Neural Network (GRNN) to create and train agents capable of detecting face images. This agent would make up the 'Detection Agent' in an architecture comprising of several different agents that collaborate together to detect and then recognise certain images. The overall agent architecture will operate as an Automatic Target Recognition' (ATR) system. The architecture of ATR system is presented in this paper and it is shown how the detection agent fits into the overall system. Experiments and results using the detection agent are also presented.

1 Introduction

The concept of software identities that have the ability (or intelligence) to perform some of the tasks that humans perform has great potential for numerous reasons. Firstly, intelligent agents could be used to provide humans suggestions or make decisions for them in response to a query. Secondly, intelligent agents could be deployed in dangerous situations to make decisions on behalf of humans. Lastly, intelligent agents could be utilised to perform tasks that are too difficult for humans such as complex computation or quickly responding to certain stimuli that humans may be too slow for. Such intelligent agents have great potential in many diverse industries ranging from commerce to defence.

Programming languages often employ controls such as the "if" statement, which acts as a trigger for an event to occur. This represents a form of intelligence, where the software reacts to a certain stimulus. However, this is a static form of intelligence, which never changes with different situations. Dynamic intelligence is more similar to what humans encounter, i.e. having the ability to adapt to different types of situations. This is mainly a result of humans having the ability to learn and recognise different situations [5]. Likewise software can also be written to adapt, recognise and learn from pervious experiences [4].

Neurocomputing is one such method that employs the process of *learning* to mimic the learning process in humans. In this research study we deployed neural networks to train agents to accomplish certain tasks. The role of the agents developed for the case study will be to operate together as an 'Automatic Target Recognition' (ATR) system which should be able to collaborate with each other to detect the presence of faces within an image [1]. Such a technology will exploit the ability of intelligent agents to learn from *previous experiences* (to identify face images) and then combine this knowledge with image processing techniques to extract and identify faces from within

N.R. Pal et al. (Eds.): ICONIP 2004, LNCS 3316, pp. 619–624, 2004.
© Springer-Verlag Berlin Heidelberg 2004

a regular image containing other details. This investigation will explore the ability of intelligent agents to learn knowledge and to accomplish a task by collaborating with other agents within the system. In Section 2 we present the architecture of ATR system followed by experiment results in Section 3. Some conclusions are also provided towards the end.

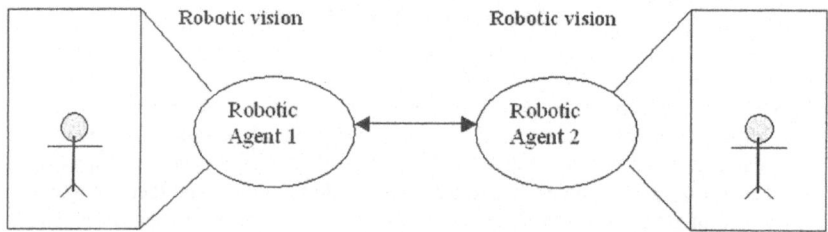

Fig. 1. Robotic Agents

2   Architecture of ATR

It has been suggested that intelligent learning agents could be used in robots for defence purposes. Such robots could be thrown into buildings and then asked to travel though corridors visualising their environment. The aim of these robots would be to seek out objects (i.e. humans, bombs) and inform other robots or humans about knowledge gained. Figure 1 below describes how intelligent learning agents embedded in robots could seek out humans and then share information between another agents. Since the aim of the project is to be able to detect objects (whether it be human, animal or non-living), several different types of sensors could be within the agents. Heat or motion sensors could be embedded into agents to detect the presence of humans and animals. However these sensors may fail if the object is stationary or non-living. Imaging sensors could be a good alternative in such scenarios. A complete and comprehensive intelligent agent could employ all the three sensors (heat, motion and image sensors) to gather information and exchange between other agents to establish certain facts. However for the purpose of this investigation only knowledge gained from imaging sensors is examined.

Before any knowledge can be exchanged between robots an image scanned by a robot must be processed and a decision must be made whether a human is present in the image. This is where intelligence and learning is introduced to these robots. The overall aim of this project is to investigate such a concept. The concept of agents learning from a knowledge base can be tested on an ATR system where agents can automatically detect and recognise images. Such an ATR system could be created to investigate collaborative agent learning. Architecture for the development of the ATR system described above could comprise of two different stages: The Learning Stage and the Detection Stage. The Learning Stage would teach an agent what a person looks like. This process would make up the knowledge base for the ATR system. This could be done using ANN and would utilise a series of face images. The entry point into the ATR System would be via the Detection Stage. This would take images that a robot scans, and tries to detect if there are targets or images of interest within the picture. If possible targets are detected, these images are clipped and a Region Of

Interest (ROI) is determined. This information is then passed onto the recognition agent. The recognition agent would refine the search and would determine whether the selected images are of the required targets (e.g. people's faces). Once it has been established that a person's face has been detected this information could then be shared between any other robot in the system. The Learning Stage could be developed as a learning agent that distributes information to the Detection Stage and the Recognition Stage. Figure 2 describes how the ATR system would work with the two different stages.

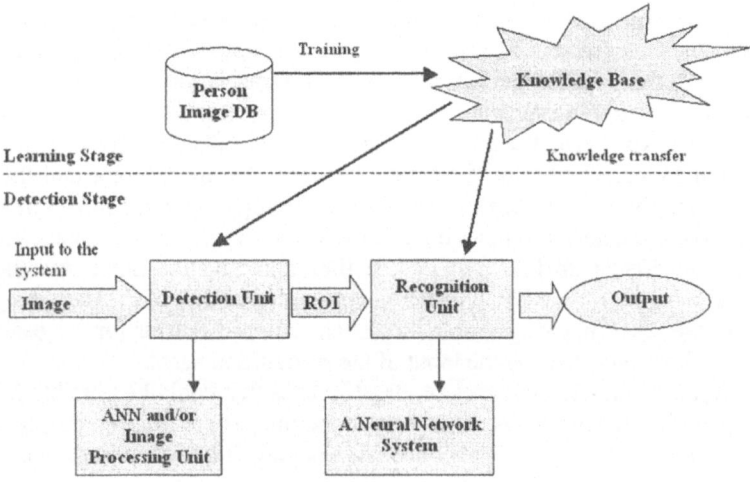

Fig. 2. Architecture of ATR

The Learning Stage comprises of the single learning agent and the Detection Stage comprises of two distinct agents: the detection agent and the recognition agent. This agent architecture would operate by having the learning agent constantly training the detection and recognition agents. This would also comprise of constantly updating the database of images for training purposes. The detection and recognition agents would have a single role of detecting possible regions of interest (detection agent) and then recognising specific images from these regions of interest (recognition agent). This paper is more focused on the development of detection agents.

2.1 Functioning of Detection Agent

The role of the detection agent is to detect its target image, whether it be of a person's face as in Figure 3 (a), or any other image that the agent wishes to detect. The detection agent is there to simply detect the presence of the target image and pass all the probable candidates off to the recognition agent. We modeled the detection agent using a two layer Generalised Regression Neural Network (GRNN). To increase the likelihood of the neural network detecting the target image, the target image is divided into tiles during the training process. The tiles created on the target image form non-overlapping grids, with each region of the grid being made up of binary values of 1 and 0 (since binary images are used). An important factor is determining how many tiles should be used on the image. Too many tiles would result in decreased efficiency due to increased processing time of each tile, and too few tiles would result in the image not being tiled at all. Through experimentation it was discovered that a tiling size of 15 seemed to work best for most instances. The training is then performed on these tiles rather than the target image itself. This ensures a higher success rate if incomplete or distorted images are presented to the network once it has been trained. The number of output neurons for such a network is dependent on the number of target images (in this case just one). Each tile of the target image is assigned the same unique output indicating that it belongs to that face image. The horizontal rows of pixels in each tile are concatenated to create the network training vector. The first layer has the same number of neurons as there are inputs (for a face image tiled as a 15 x 15 grid, there would be 225 inputs). Each input is then subjected to a weight, derived from the transpose of the input vector of training tiles. When new inputs are detected, the GRNN provides an output, based on interpolation of past records. This is worked out by calculating the difference between the input vector and the training vector, which gives the probability value of each output neuron [2][3]. The detection of targets (face images) within an image of a room is performed by once again dividing the room image up into tiles. These tiles make up the input into the trained GRNN. The output of this GRNN determines the probability of the tiles being targets, and those tiles that may be possible targets are cropped out for further examination. The cropped out tiles become the input of the recognition agent.

As a result of the various types of target images that could be possible for the detection agent, no specific physical feature in the images are selected during the training and running of the agent. Rather specific imaging features are target such as number of connected areas present in the image, mean size of the connecting areas, and the standard deviation of the areas.

3   Experiment and Test Results Using GRNN

As discussed in the previous section, the detection agent is trained using a two layer Generalised Regression Neural Network. The agent is trained on a particular face image, such as that in Figure 3(a) [6], which is passed into the system in TIFF format. After training is complete a new image (Figure 3(b)) that contains images similar to that of Figure 3(a) embedded within a larger image is presented to the trained detection agent. It is now the role of the detection agent to detect possible regions of interests within Figure 3(b). These regions of interests would represent regions within Figure 4 where the embedded images may be present. The results obtain from the detection agent (i.e. the images extracted that represent regions of interest from Figure 4) are passed onto the recognition agent to further refine the search and to positively detect and identify the faces. Passing individual images (regions of interest) to the recognition agent may not be very practical as this could possibly greatly hamper the performance of the overall system. A much quicker and effective method that could be used as the input to the recognition agent is to input key features extracted from these regions of interest.

The key feature from the region of interest that would be extracted from the Detection Agent are:

(a)   (b)

Fig. 3. (a) Training image of a persons face (b) Test image

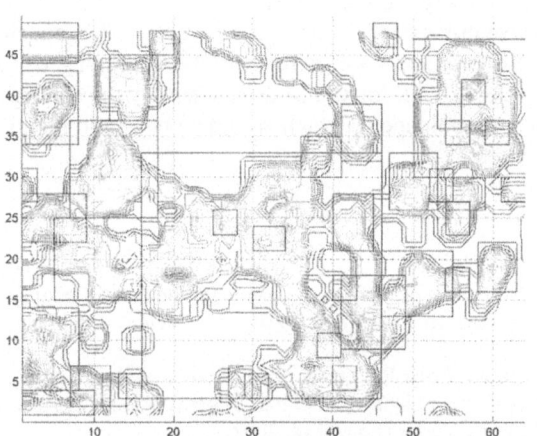

Fig. 4. Contour density plots of ROI

(a)   (b)   (c)

(d)

Fig. 5. (a-d) Extracted image clips

- Number of connected areas present in the image,
- Mean size of the connecting areas, and
- The standard deviation of the areas.

Figure 4 represents the contour density plots of the regions of interest when Figure 3 (b) is passed into the trained detection agent. Each contour density plot refers to an area or a ROI on the test image. Some kind of filtering mechanism is to be used to extract only those contours that may be probable targets. For this purpose a density threshold level is decided and all contour plots above a set density level are extracted. The extracted ROI look like the figures illustrated in Figures 5 (a)-(d). These are the clipped images, which acts as the input to the recognition unit. The extracted features such as the number of connected areas present in the image, the mean size of the connecting areas, and the standard deviation of the areas from images shown in Figure 5 are used as the inputs to the recognition agent.

4 Conclusion

In this paper a detection agent trained to detect face images was created. An image containing face images was presented to the agent and a test was performed to determine whether any positive hits could be achieved. Initial results from the detection stage of the architecture seems quite encouraging. In the result presented in this paper, forty regions of interest were detected. Amongst the forty, there were four hits. These four are shown in Figure 5. Apart from Figure 5 (a), the other three images either shows too much information or too little information. This problem can be addressed by experimenting with the internal parameters of the Generalised Regression Neural Network. Even though some of the results show either too much or too little information, the results obtained have shown that all the three embedded images within Figure 3 (b) have been successfully detected. This is a very positive result. A success rate as close as possible to 100% is required for the success of recognition agent and hence the overall ATR system. If the detection agent fails to detect a possible target, this information would not be passed onto the recognition agent and hence the overall ATR system would fail to positively identify a possible target.

Future work will examine the development of the recognition agent. The recognition agent would perform the same operation as we performed in determining which ROI was a hit and which could be rejected. All forty ROI determined by the detection agent would be passed on to the recognition agent, which would be tasked to identify the positive targets.

References

1. Filippidis A., Jain L.C. and Martin N., Fusion of Intelligent Agents for the Detection of Aircraft in SAR Images, *IEEE Transactions on Pattern Analysis and Machine Intelligence*, Vol. 22, No. 4, 2000.
2. Joshi V., Jain L.C., Seiffert U., Zyga K., Price R.J. Leisch F., Neural Techniques in Logo Recognition, First International Workshop on Hybrid Intelligent Systems, Adelaide, Australia, Advances in Soft Computing, Physica-Verlag, pp. 25-32, 2002.
3. Seiffert U., Logo feature Extraction of Region based properties, Release 1.0, University of Magdeburg, Germany, 2000.
4. Tecuci G., Boicu M., Marcu D., Stanescu B., Boicu C.and Comello J., Training and Using Disciple Agents: A Case Study in the Military Center of Gravity Analysis Domain, in *AI Magazine*, AAAI Press, Menlo Park, California, 2002.
5. Boicu M., Modeling and Learning with Incomplete Knowledge, PhD Thesis in Information Technology, School of Information Technology and Engineering, George Mason University, 2002.
6. The AR Face Database, http://rvl1.ecn.purdue.edu/~aleix/ar.html

Cognitive Routing in Packet Networks

Erol Gelenbe

Dennis Gabor Chair
Dept. of Electrical and Electronic Engineering
Imperial College
London SW7 2BT, UK
e.gelenbe@imperial.ac.uk

Abstract. While the Internet has beome a resounding success for human communication and information gathering and dissemination, Quality of Service (QoS) remains a major subject of concern for users of packet networks. In this paper we show how distributed intelligence and on-line adaptation can be used to enhance QoS for users of packet networks, and present our Cognitive Packet Network approach as a manner to address this important issue. Our presentation is based on ongoing experimental reserach that uses several "Cognitive Packet Network" (CPN) test-beds that we have developed.

1 Introduction

Novel networked systems are emerging to offer user oriented flexible services, using the Internet and LANs to reach different parts of the same systems, and to access other networks, users and services. Examples include Enterprise Networks, Home Networks (Domotics), Sensor Networks, Networks for Military Units or Emergency Services. The example of Home Networks is significant in that a family may be interconnected as a unit with PDAs for the parents and children, the health monitoring devices for the grand-parents, the video cameras connected to the network in the infants' bedroom, with connections to smart home appliances, the home education server, the entertainment center, the security system, and so on. Clearly, such systems must allow for diverse QoS requirements which can be implicit due to the nature of the application (e.g. alarm system connection to the security service, or video-on-demand or voice-over-IP), or which may be explicitly formulated by the end users of the network. Such networks raise interesting issues of intelligence and adaptation to user needs and to the networking environment, including routing requirements, possibly self healing, security and robustness. Thus on-line adaptation, in response to QoS needs, available resources, perceived performance of the different network infrastructure elements, instantaneous traffic loads, and economic costs, would be a very attractive feature of such networks.

2 The Cognitive Packet Network (CPN)

CPN [7, 8, 6, 10] is a packet routing protocol which addresses QoS using adaptive techniques based on on-line measurement. Although most of our work on CPN

N.R. Pal et al. (Eds.): ICONIP 2004, LNCS 3316, pp. 625–632, 2004.
© Springer-Verlag Berlin Heidelberg 2004

concerns wired networks, we have also developed a wireless extension of which can operate seamlessly with wired CPN or IP networks [9].

In CPN, users are allowed to declare QoS Goals such as: "Get me the data object *Ob* via the path(s) of highest bandwidth which you can find", where *Ob* is the handle of some data object [2], or "Find the paths with least power consumption to the mobile user *Mn*", or "Get the video output *Vi* to my PDA as quickly as possible" (i.e. with minimum overall delay). CPN is designed to *Accept Direction*, by inputing Goals prescribed by users. It exploits *Self-Observation* with the help of *smart packets* so as to be aware of the network state including connectivity of fixed or mobile nodes, power levels at mobile nodes, topology, paths and path QoS. It performs *Self-Improvement*, and *Learns from the experience of smart packets* using neural networks and genetic algorithms [11] to determine routing schemes with better QoS. It will *Deduce* hitherto unknown routes by combining or modifying paths which have been previously learned so as to improve QoS and robustness. CPN makes use of three types of packets: smart packets (SP) for discovery, source routed dumb packets (DP) to carry payload, acknowledgements (ACK) to bring back information that has been discovered by SPs. Conventional IP packets tunnel through CPN to seamlessly operate mixed IP and CPN networks. SPs are generated by a user (1) requesting that a path having some QoS value be created to some CPN node, or (2) requesting to discover parts of network state, including location of certain fixed or mobile nodes, power levels at nodes, topology, paths and their QoS.

SPs *exploit the experience of other packets* using random neural network (RNN) based Reinforcement Learning (RL) [3, 7]. RL is carried out using a Goal which is specified by the user who generated a request for the connection. The decisional weights of a RNN are increased or decreased based on the observed success or failure of subsequent SPs to achieve the Goal. Thus RL will tend to prefer better routing schemes, more reliable access paths to data objects, and better QoS. In an extended version of the CPN network which is presented in [11], but that we do not discuss in this paper, the system **deduces** new paths by combining previously discovered paths, and using the estimated or measured QoS values of those new paths to select better paths. This is similar conceptually to a genetic algorithm which generates new entities by combination or mutation of existing entities, and then selects the best among them using a fitness function. These new paths can be tested by probes so that the actual QoS can be evaluated. When a SP arrives to its destination, an ACK is generated and heads back to the source of the request. It updates *mailboxes* (MBs) in the CPN nodes it visits with information which has been discovered, and provides the source node with the successful path to the node. All packets have a life-time constraint based on the number of nodes visited, to avoid overburdening the system with unsuccessful requests or packets which are in effect lost. A node in the CPN acts as a storage area for packets and mailboxes (MBs). It also stores and executes the code used to route smart packets. It has an input buffer for packets arriving from the input links, a set of mailboxes, and a set of output buffers which are associated with output links. CPN software is integrated into the Linux kernel 2.2.x, providing a single application program interface (API) for the programmer to access CPN. CPN routing algorithms also run seamlessly on ad-hoc wireless and wired connections [9], without specific dependence on the nature (wired or wireless) of the links, using QoS awareness to optimize behavior across different connection technologies and wireless protocols.

The SP routing code's parameters are updated at the router using information collected by SPs and brought back to routers by the ACK packets. Since SPs can meander and get lost in the network, we destroy SPs which have visited more than a fixed number of nodes, and this number is set to 30 in the current test-beds. This number is selected based on the fact that it will be two to three times larger than the diameter of any very large network that one may consider in practice. For each incoming SP, the router computes the appropriate outgoing link based on the outcome of this computation. A recurrent random neural network (RNN) [1] with as many "neurons" as there are possible outgoing links, is used in the computation. The weights of the RNN are updated so that decision outcomes are reinforced or weakened depending on how they have contributed to the success of the QoS goal. In the RNN the state q_i of the $i-th$ neuron in the network is the probability that the $i-th$ neuron is excited. Each neuron i is associated with a distinct outgoing link at a node. The q_i satisfy the system of non-linear equations:

$$q_i = \lambda^+(i)/[r(i) + \lambda^-(i)], \tag{1}$$

where

$$\lambda^+(i) = \sum_j q_j w_{ji}^+ + \Lambda_i, \quad \lambda^-(i) = \sum_j q_j w_{ji}^- + \lambda_i, \tag{2}$$

w_{ji}^+ is the rate at which neuron j sends "excitation spikes" to neuron i when j is excited, w_{ji}^- is the rate at which neuron j sends "inhibition spikes" to neuron i when j is excited, and $r(i)$ is the total firing rate from the neuron i. For an n neuron network, the network parameters are these n by n "weight matrices" $\mathbf{W}^+ = \{w^+(i,j)\}$ and $\mathbf{W}^- = \{w^-(i,j)\}$ which need to be "learned" from input data.

RL is used in CPN as follows. Each node stores a specific RNN for each active source-destination pair, and each QoS class. The number of nodes of the RNN are specific to the router, since (as indicated earlier) each RNN node will represent the decision to choose a given output link for a smart packet. Decisions are taken by selecting the output link j for which the corresponding neuron is the most excited, i.e. $q_i \leq q_j$ for all $i = 1, .. , n$. Each QoS class for each source-destination pair has a QoS Goal G, which expresses a function to be minimized, e.g., Transit Delay or Probability of Loss, or Jitter, or a weighted combination, and so on. The reward R which is used in the RL algorithm is simply the inverse of the goal: $R = G^{-1}$. Successive measured values of R are denoted by R_l, $l = 1, 2, ..$; These are first used to compute the current value of the decision threshold:

$$T_l = aT_{l-1} + (1-a)R_l, \tag{3}$$

where $0 < a < 1$, typically close to 1. Suppose that we have now taken the $l - th$ decision which corresponds to neuron j, and that we have measured the $l - th$ reward R_l. We first determine whether the most recent value of the reward is larger than the previous value of the threshold T_{l-1}. If that is the case, then we increase very significantly the excitatory weights going into the neuron that was the previous winner (in order to reward it for its new success), and make a small increase of the inhibitory weights leading to other neurons. If the new reward is not greater than the previous threshold, then we simply increase moderately all excitatory weights leading to all neurons, except for the previous winner, and increase significantly the inhibitory weights leading to the previous winning neuron (in order to punish it for not being very successful this time). Let us denote by r_i the firing rates of the neurons before the update takes place:

$$r_i = \sum_1^n [w^+(i,m) + w^-(i,m)], \tag{4}$$

We first compute T_{l-1} and then update the network weights as follows for all neurons $i \neq j$:

- If $T_{l-1} \leq R_l$
    - $w^+(i,j) \leftarrow w^+(i,j) + R_l$,
    - $w^-(i,k) \leftarrow w^-(i,k) + \frac{R_l}{n-2}$, if $k \neq j$.
- Else
    - $w^+(i,k) \leftarrow w^+(i,k) + \frac{R_l}{n-2}, k \neq j$,
    - $w^-(i,j) \leftarrow w^-(i,j) + R_l$.

Since the relative size of the weights of the RNN, rather than the actual values, determine the state of the neural network, we then re-normalize all the weights by carrying out the following operations. First for each i we compute:

$$r_i^* = \sum_1^n [w^+(i,m) + w^-(i,m)], \tag{5}$$

and then re-normalize the weights with:

$$w^+(i,j) \leftarrow w^+(i,j) * \frac{r_i}{r_i^*},$$
$$w^-(i,j) \leftarrow w^-(i,j) * \frac{r_i}{r_i^*}.$$

Finally, the probabilities q_i are computed using the non-linear iterations (1), (2). The largest of the q_i's is again chosen to select the new output link used to send the smart packet forward. This procedure is repeated for each smart packet for each QoS class and each source-destination pair.

An important question is whether the scheme we have proposed can only function if the number, or percentage, of SPs (and hence ACKs) used is very high. This is a question that we have examined attentively, both by examining the actual length of SPs, and with numerous experiments [6]. The results of one of these experiments for a heavily loaded network, is summarized in Figure 1

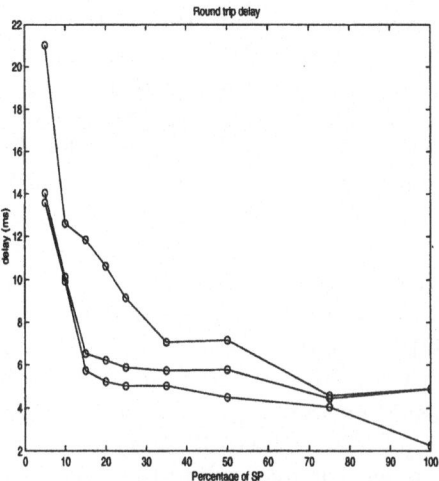

Fig. 1. Average round-trip delay for smart (top) and dumb (bottom) packets, and average delay for all packets (center) as a function of the percentage of smart packets. These measurements were obtained for an end-to-end connection in the presence of obstructing traffic on several links of the network.

where we report the round-trip delay experienced by SPs and DPs, and by all packets, when the percentage of SPs added on top of DP traffic was varied from 5% to 100% in steps of 5%. In these experiments, the user specified QoS Goal is "delay" so that what is being measured is the quantity that the user would like CPN to minimize. The top curve shows the round-trip delay for SPs, while the bottom curve is the round-trip delay for DPs, with the average delay of all packets being shown in the middle. As expected, when there are 100% of SPs, the average delay for SPs is the same as the average delay for all packets. The interesting result we observe is that as far as the DPs are concerned, when we have added some 15% or 20% of SPs we have achieved the major gain in delay reduction. Going beyond those values does not significantly reduce the delay for DPs. In effect, DPs are typically full sized Ethernet packets (as are IP packets in general), while SPs and ACKs are 10% of their size. Thus, if 20% of SP traffic is added, this will result in 14% traffic overhead, if ACKs are generated by both DPs and SPs, and only 4% of traffic overhead if ACKs are only generated in response to SPs. Note also that ACKs and DPs do not require a next hop to be computed at each node, contrary to IP packets. Both in CPN and IP we can of course reduce next-hop computations using appropriate caching and hardware acceleration.

One of the major requirements of CPN is that it should be able to start itself with <u>no</u> initial information, by first randomly searching, and then progressively improving its behaviour through experience. Since the major function of a network is to transfer packets from some source S to some destination D, CPN must be able to establish a path from S to D even when there is no prior information available in the network. The network topology we have used in these experiments is shown in Figure 2 (left), with the source and destinations at the top left and bottom right ends of the diagram. The network contains 24 nodes, and each node is connected to 4 neighbours. Because of the possibility of repeatedly visiting the same node on a path, the network contains an unlimited number of paths from S to D. However, the fact that SPs are destroyed after they visit 30 nodes, does limit this number though it still leaves a huge number of possible paths. In this set of experiments, the network is always started with empty mailboxes, i.e. with no prior information about which output link is to be used from a node, and with neural network weights set at identical values, so that the neural network decision algorithm at nodes initially will produce a random choice. Each point shown on the curves of Figures 2 and 3 are a result of 100 repetitions of the experiment under identical starting conditions.

Let us first discuss the curve in Figure 2 (right). An abscissa value of 10 indicates that the number of SPs used was 10, and – assuming that the experiment resulted in an ACK packet coming back to the source – the ordinate gives the average time (over the 100 experiments) that it elapse between the instant that the first SP was sent out, and the first ACK comes back. Note that the first ACK will be coming back from the correct destination node, and that it will be bringing back a valid forward path that can be used by the subsequent useful traffic. We notice that the average set-up time decreases significantly when we go from a few SPs to about 10, and after that, the average set-up time does not improve appreciably. Its value somewhere between 10 and 20 milliseconds actually corresponds to the round-trip transit time through the hops. This does not mean that it suffices to have a small number of SPs at the beginning, simply because the average set-up time is only being measured for the SPs which are *successful*; unsuccessful SPs are destroyed after 30 hops. Figure 3 gives a more complete understanding of what is happening. Again for an x-axis value of over 10 packets, we see that the probability (right figure) of successfully setting up a path is 1, while with a very small number of packets this figure drops down to

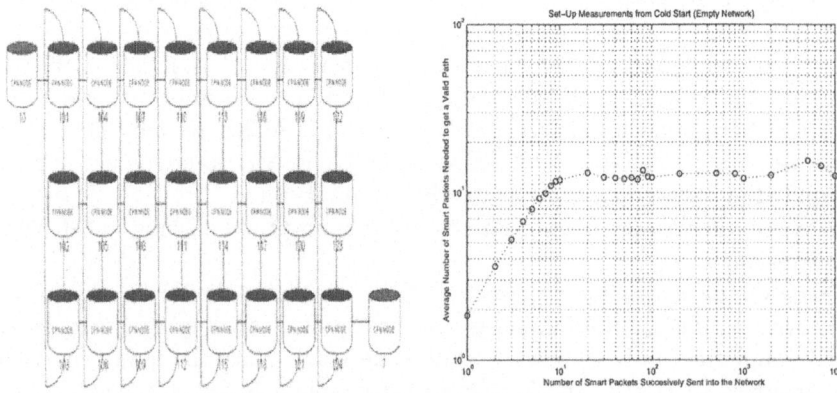

Fig. 2. CPN Network Topology for Cold Start Experiments (Left) and Average Number of SPs Needed to Obtain a Path to the Destination (Right).

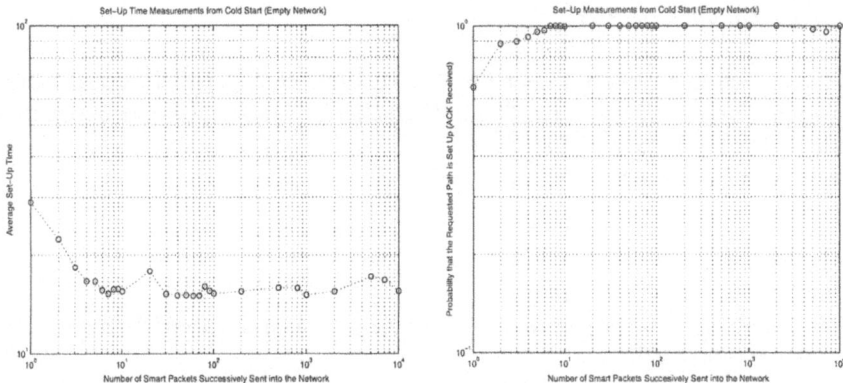

Fig. 3. Average Network Set-Up Time (Left) and Probability of Successful Connection from Cold Start, both from Cold Start, as a Function of the Initial Number of Smart Packets.

about 0.65. These probabilities must of course be understood as the empirically observed fraction of the 100 tests which result in a successful connection. The conclusion from these two data sets is that to be safe, starting with an empty system, a fairly small number of SPs, in the range of 20 to 100, will provide almost guaranteed set-up of the connection, and the minimum average set-up time. Since the previous curves show that connections are almost always established with as few as 10 SPs, and because the average round-trip connection establishment time is quite short, we would expect to see that a connection will generally be established before all the SPs are sent out by the source. This is exactly what we observe on Figure 3 (left). The x axis shows the number of SPs sent into the network, while the y axis shows the average number sent in (over the 100 experiments) before the first ACK is received. For small numbers of SPs sent out by the source, until the value 10 or so, the relationship is linear. However as the number of SPs being inserted into the network increases, we see that after (on the average) 13 packets or so have been sent out, the connection is already established (i.e. the first ACK has returned to the source). This again indicates that a fairly small number of SPs suffice to establish a connection. In addition to the experiments on the test-bed, simulations have been conducted for a 1000 node network with results which are significantly similar.

3 Conclusions

We have discussed an experimental packet network architecture that is controlled by neural networks using reinforcement learning to offer QoS to its users. Our ongoing research examines how a CPN sub-system can be inserted into the Internet to carry out traffic engineering functions and protection from denial of service attacks, and the use of CPN in ad hoc wireless networks.

Acknowledgements

This research is supported by EPSRC, the Engineering and Physical Sciences Research Council of the UK. The author thanks his former and current PhD students Dr. Ricardo Lent, Dr Zhiguang Xu, Mike Gellman, Peixiang Liu, Arturo Nunez, and Pu Su, for their important contributions to this project.

References

1. E. Gelenbe. "Learning in the recurrent random neural network", *Neural Comp.* 5(1), 154–164, 1993.
2. R. E. Kahn, R. Wilensky "A framework for digital object services", *c.nri.dlib/tn95-01.*
3. U. Halici, "Reinforcement learning with internal expectation for the random neural network" *Eur. J. Opns. Res.*, 126 (2) 2, 288-307, 2000.
4. E. Gelenbe, E. Şeref, Z. Xu. "Simulation with learning agents", *Proc. IEEE*, Vol. 89 (2), 148–157, 2001.
5. E. Gelenbe, R. Lent, Z. Xu, "Towards networks with cognitive packets", Opening Invited Paper, *International Conference on Performance and QoS of Next Generation Networking*, Nagoya, Japan, November 2000, in K. Goto, T. Hasegawa, H. Takagi and Y. Takahashi (eds), "Performance and QoS of next Generation Networking", Springer Verlag, London, 2001.
6. E. Gelenbe, R. Lent, Z. Xu, "Design and performance of cognitive packet networks", *Performance Evaluation*, 46, pp. 155-176, 2001.
7. E. Gelenbe, R. Lent, Z. Xu "Measurement and performance of Cognitive Packet Networks", *J. Comp. Nets.*, 37, 691–701, 2001.
8. E. Gelenbe, R. Lent, Z. Xu "Networking with Cognitive Packets", *Proc. ICANN.*, Madrid, August 2002.
9. E. Gelenbe, R. Lent "Mobile Ad-Hoc Cognitive Packet Networks", *Proc. IEEE ASWN*, Paris, July 2-4, 2002.
10. E. Gelenbe et al. "Cognitive packet networks: QoS and performance", Keynote Paper, *IEEE MASCOTS Conference*, San Antonio, TX, Oct. 14-16, 2002.
11. E. Gelenbe, P. Liu, J. Lainé "Genetic algorithms for route discovery", SPECTS'03, Summer Simulation Multiconference, Society for Computer Simulation, Montreal, 20-24 July, 2003.
12. E. Gelenbe, K. Hussain "Learning in the multiple class random neural network", *IEEE Trans. on Neural Networks* 13 (6), 1257–1267, 2002.

TWRBF – Transductive RBF Neural Network with Weighted Data Normalization

Qun Song and Nikola Kasabov

Knowledge Engineering & Discovery Research Institute
Auckland University of Technology
Private Bag 92006, Auckland 1020, New Zealand
{qsong,nkasabov}@aut.ac.nz

Abstract. This paper introduces a novel RBF model – *Transductive Radial Basis Function Neural Network* with *Weighted Data Normalization* (TWRBF). In transductive systems a local model is developed for every new input vector, based on some closest to this vector data from the training data set. The *Weighted Data Normalization* method (WDN) optimizes the data normalization range individually for each input variable of the system. A gradient descent algorithm is used for training the TWRBF model. The TWRBF is illustrated on two case study prediction/identification problems. The first one is a prediction problem of the Mackey-Glass time series and the second one is a real medical decision support problem of estimating the level of renal functions in patients. The proposed TWRBF method not only gives a good accuracy for an individual, "personalized" model, but depicts the most significant variables for this model.

1 Introduction: Transductive Modeling and Weighted Data Normalization

Most of learning models and systems in artificial intelligence, developed and implemented so far, are based on *inductive* methods, where a model (a function) is derived from data representing the problem space and this model is further applied on new data. The model is usually created without taking into account any information about a particular new data vector (test data). An error is measured to estimate how well the new data fits into the model. The inductive learning and inference approach is useful when a global model of the problem is needed even in its very approximate form. In contrast to the inductive learning and inference methods, *transductive* methods estimate the value of a potential model (function) only in a single point of the space (the new data vector) utilizing additional information related to this point. This approach seems to be more appropriate for clinical and medical applications of learning systems, where the focus is not on the model, but on the individual patient. Each individual data vector may need an individual, local model that best fits the new data, rather then a global model, in which the new data is matched without taking into account any specific information about this data.

A transductive method is concerned with the estimation of a function in single point of the space only [13]. For every new input vector x_i that needs to be processed for a prognostic/classification task, the N_i nearest neighbours, which form a data subset D_i, are derived from an existing data set D or/and generated from an existing

N.R. Pal et al. (Eds.): ICONIP 2004, LNCS 3316, pp. 633–640, 2004.
© Springer-Verlag Berlin Heidelberg 2004

model M. A new model M_i is dynamically created from these samples to approximate the function in the point x_i – Fig. 1 and Fig. 2. The system is then used to calculate the output value y_i for this input vector x_i.

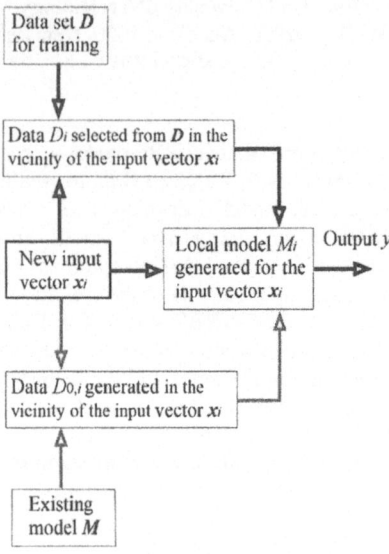

Fig. 1. A block diagram of a *transductive* model. An individual model M_i is trained for every new input vector x_i with data use of samples D_i selected from a data set D, and data samples $D_{0,i}$ generated from an existing model (formula) M (if such a model is existing). Data samples in both D_i and $D_{0,i}$ are similar to the new vector x_i according to defined similarity criteria

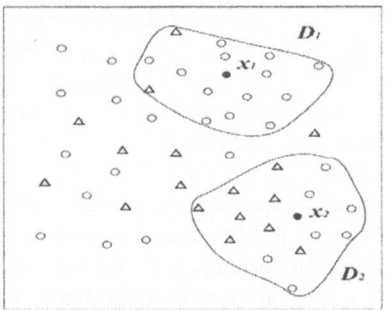

Fig. 2. The new data vector is in the centre of a transductive model (here illustrated with two of them – x_1 and x_2), and is surrounded by a fixed number of nearest data samples selected from the training data D and possibly generated from an existing model M

In many neural network models and applications, raw (not normalized) data is used. This is appropriate when all the input variables are measured in the same units. Normalization, or standardization, is reasonable when the variables are in different units, or when the variance between them is substantial. However, a general normalization means that every variable is normalized in the same range, e.g. [0, 1] with the assumption that they all have the same importance for the output of the system.

For many practical problems, variables have different importance and make different contribution to the output. Therefore, it is necessary to find an optimal normalization and assign proper importance factors to the variables. Such a method can also be used for feature selection, or for reducing the size of the input vectors through keeping the most important ones [12]. This is especially applicable to a special class of models – the clustering based neural networks or fuzzy systems (or also: distance-based; prototype-based) such as RBF [6, 10] and ECOS [3, 4]. In such systems, distance between neurons or fuzzy rule nodes and input vectors are usually measured in *Euclidean distance*, so that variables with a higher upper bound of the range will have more influence on the learning process and on the output value, and vice versa.

The paper is organized as follows: Sect. 2 presents the structure and algorithm of the TWRBF model. Sect. 3 and 4 illustrate the approach on two case study problems. Conclusions are drawn in Sect. 5.

2 Transductive RBF Neural Networks with Weighted Data Normalization: Structure and Learning Algorithm

The origin of the radial basis function (RBF) models is in the approximation theory and in the regularization theory [10]. They deal with the general approximation problem common to supervised learning neural networks. Standard RBF networks have three layers of neurons: input layer - that represents input variables; hidden layer – representing centers of clusters of data in the input space; and output layer – that represents the output variables. Although RBF networks usually need more neurons in their hidden layer than standard feed-forward back-propagation networks, such as MLP, they train much faster than MLP networks and can be used to extract meaningful rules based on clusters of data. RBF models work very well when there is enough training data.

A TWRBF network has an additional layer – the WDN layer for weighted data normalization. *Gaussian* kernel functions are used as activation functions for each neuron in the hidden layer. The TWRBF structure is shown in Fig. 3.

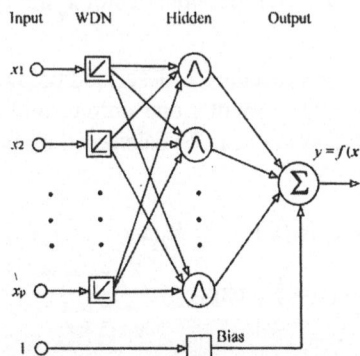

Fig. 3. The Structure of the proposed TWRBF Network

For each new data vector x_q, the TWRBF learning algorithm performs the follow-ing steps:

1. Normalize the training data set (the values are between 0 and 1) with the initial input variable weights.
2. Search in the training data set in the input space with weighted normalized *Euclid-ean* distance, defined as Eq.1, to find N_q training examples that are closest to x_q. The value for N_q can be pre-defined based on experience, or – optimized through the application of an optimization procedure. Here we assume the former approach.

$$\|x - y\| = \frac{1}{P} \left[\sum_{j=1}^{P} w_j^2 |x_j - y_j|^2 \right]^{\frac{1}{2}} \tag{1}$$

where: $x, y \in R^P$ and w_j are weights.

3. Calculate the distances d_i, $i = 1, 2, ..., N_q$, between each of these data samples and x_q. Calculate the vector weights $v_i = 1 - (d_i - \min(d))$, $i = 1, 2, ..., N_q$, $\min(d)$ is the minimum value in the distance vector $d = [d_1, d_2, ... , d_{Nq}]$.
4. Use the *ECM* clustering algorithm [5, 11] (or some other clustering methods) to cluster and partition the input sub-space that consists of N_q selected training sam-ples.
5. Create *Gaussian* kernel functions and set their initial parameter values according to the *ECM* clustering procedure results; for each cluster, the cluster centre is taken as the centre of a *Gaussian* function and the cluster radius is taken as the width.
6. Apply the gradient descent method to optimize the parameters of the system in the local model M_q that include the individual variable normalization weights (the up-per bound of the normalization intervals, the lower bound being 0) (see Eq. 2-9 below).
7. Search in the training data set to find N_q samples (the same to Step 2), if the same samples are found as the last search the algorithm turns to Step 8, otherwise, Step 3.
8. Calculate the output value y_q for the input vector x_q applying the trained model.
9. End of the procedure.

The parameter optimisation procedure is described below:
Consider the system having P inputs, one output, and M neurons in the hidden layer, the output value of the system can be calculated on input vector $x_i = [x_1, x_2, ..., x_P]$ as follows:

$$y = f(x_i) = b_0 + b_1 R_1(x_i) + b_2 R_2(x_i), ... , + b_M R_M(x_i) \tag{2}$$

$$\text{here,} \quad R_l(x_i) = \prod_{j=1}^{P} \exp\left[-\frac{w_j^2 (x_{ij} - m_{lj})^2}{2\sigma_{lj}^2} \right] \tag{3}$$

m_{lj} and σ_{lj} are parameters of *Gaussian* functions, w_j are weights of input variables.

In the TWRBF learning algorithm, the following indexes are used:

- Training Data pairs: $i = 1, 2, ..., N$;
- Input Variables: $j = 1, 2, ..., P$;
- Neurons in the hidden layer: $l = 1, 2, ..., M$;
- Learning Iterations: $k = 1, 2, ...$

The TWRBF learning algorithm minimizes the following objective function (an error function):

$$E = \frac{1}{2} \sum_{i=1}^{N} v_i [f(x_i) - t_i]^2 \quad (v_i \text{ are defined in Step 3}) \tag{4}$$

The gradient descent algorithm (back-propagation algorithm) is used on the training data pairs, $[x_i, t_i]$, to obtain the recursions for updating the parameters b, m, σ and w such that E of Eq. 4 is minimized:

$$b_0(k+1) = b_0(k) - \eta_b \sum_{i=1}^{N} v_i [f(x_i) - t_i] \tag{5}$$

$$b_l(k+1) = b_l(k) - \eta_b \sum_{i=1}^{N} \{ v_i R_l(x_i)[f(x_i) - t_i] \} \tag{6}$$

$$m_{lj}(k+1) = m_{lj}(k) - \eta_m \sum_{i=1}^{N} \left\{ v_i R_l(x_i) b_l[f(x_i) - t_i] \frac{w_j^2 (x_{ij} - m_{lj})}{\sigma_{lj}^2} \right\} \tag{7}$$

$$\sigma_{lj}(k+1) = \sigma_{lj}(k) - \eta_\sigma \sum_{i=1}^{N} \left\{ v_i R_l(x_i) b_l[f(x_i) - t_i] \frac{w_j^2 (x_{ij} - m_{lj})^2}{\sigma_{lj}^3} \right\} \tag{8}$$

$$w_j(k+1) = w_j(k) - \eta_w \sum_{l=1}^{M} \sum_{i=1}^{N} \left\{ v_i R_l(x_i) b_l[f(x_i) - t_i] \frac{w_j(m_{lj} - x_{ij})^2}{\sigma_{lj}^2} \right\} \tag{9}$$

here, η_b, η_m, η_σ and η_w are learning rates for updating the parameters b, m, σ and w respectively.

3   Case Study Example of Applying the TWRBF for a Time-Series Prediction

In this paper, the TWRBF network is applied to a time series prediction. The TWRBF learning is demonstrated on the Mackey-Glass (MG) time series prediction task [8]. The MG time series is generated with a time-delay differential equation as follows:

$$\frac{dx(t)}{dt} = \frac{0.2 \, x(t - \tau)}{1 + x^{10}(t - \tau)} \tag{10}$$

To obtain this time series values at integer points, the *fourth-order Runge-Kutta method* was used to find the numerical solution to the above MG equation. Here we used the following parameter values: a time step of 0.1, $x(0) = 1.2$, $\tau = 17$ and $x(t) = 0$ for $t < 0$. From the input vector $[x(t - 18) \; x(t - 12) \; x(t - 6) \; x(t)]$, the task is to predict the value $x(t + 85)$. In the experiments, 3000 data points, from t = 201 to 3200, were extracted as training data; 500 data points, from t = 5001 to 5500 were taken as testing data. For each of these testing data sample a local transductive model is created and tested on this data.

Table 1. Experimental results on MG data

Model	Neurons or rules	Training iterations	Testing NDEI	WDN - Input variable			
				w1	w2	w3	w4
MLP	60	500	0.022	1	1	1	1
ANFIS	81	500	0.025	1	1	1	1
DENFIS	58	100	0.017	1	1	1	1
RBF	128	200	0.031	1	1	1	1
TWRBF	6.2 (average)	200	0.012	0.95	1	0.74	0.89

To compare the performance of the TWRBF method we conducted the following experiments. We applied some well-known inductive models, such as MLP [9], RBF [9], ANFIS [1, 2] and DENFIS [5], on the same data. The experimental results are listed in Table 1 that includes the number of neurons in the hidden layer (for MLP and RBF) or fuzzy rules (for ANFIS and DENFIS), training iterations and testing NDEI – *non-dimensional error index* which is defined as the *root mean square error* (RMSE) divided by the standard deviation of the target series, the WDN upper bound for each input variable. It is seen that variable x2 is at average the most significant when this method is applied.

The proposed method, TWRBF, evolves individual, "personalized" models and performs better at average than the other inductive, global models. This is a result of the fine tuning of each local, individual model in TWRBF for each simulated example. The finely tuned local models achieve a better local generalization.

4 Case Study Example of Applying the TWRBF for a Medical Decision Support Problem

A real data set from a medical institution is used here for experimental analysis. The data set has 447 samples, collected at hospitals in New Zealand and Australia. Each of the records includes six variables (inputs): age, gender, serum creatinine, serum albumin, race and blood urea nitrogen concentrations, and one output - the Glomerular Filtration Rate (GFR) value. All experimental results reported here are based on 10-cross validation experiments with the same model and parameters and the results are averaged. In each experiment 70% of the whole data set is randomly selected as training data and another 30% as testing data.

For comparison, several well-known methods are applied on the same problem, such as the MDRD function [7], MLP and RBF neural network [9], adaptive neural fuzzy inference system (ANFIS) [1] and dynamic evolving neural fuzzy inference system (DENFIS) [5] along with the proposed TWRBF and results are given in Table 2. The results include the number of fuzzy rules (for ANFIS and DENFIS), or neurons in the hidden layer (for RBF and MLP), the testing RMSE (root mean square error), the testing MAE (mean absolute error) and the average weights of the input variables representing their average significance for the GFR prediction. It is seen that Sreatenine and Age are the most important variables at average.

Table 2. Experimental results on GFR data

Model	Neurons or rules	Testing RMSE	Testing MAE	Weights of input variables					
				Age w1	Sex w2	Scr w3	Surea w4	Race w5	Salb w6
MDRD	—	7.74	5.88	1	1	1	1	1	1
MLP	12	8.44	5.75	1	1	1	1	1	1
ANFIS	36	7.49	5.48	1	1	1	1	1	1
DENFIS	27	7.29	5.29	1	1	1	1	1	1
RBF	32	7.22	5.41	1	1	1	1	1	1
TWRBF	5.1 (average)	7.10	5.15	0.92	0.63	1	0.78	0.37	0.45

The proposed TWRBF method not only gives a good accuracy at average, but provides a good accuracy for an individual, "personalized" model, and also depicts the most significant variables for this model as it is illustrated on fig.3 on a single, randomly selected sample from the GFR data. The importance of variables for a particular patient may indicate potential specific problems and a personalized treatment.

Table 3. A TWRBF personalised model of a single patient (one sample from the GFR data)

Values for the Input variables of a selected patient	Age 58.9	Sex Female	Scr 0.28	Surea 28.4	Race White	Salb 38
Weights of input variables (TWRBF) for this patient	0.84	0.62	1	0.83	0.21	0.41
Results	GFR (desired) 18.0		MDRD 14.9		TWRBF 16.2	

5 Conclusions

This paper presents a transductive RBF neural network with weighted data normalization method – TWRBF. The TWRBF results in a better local generalization over new data as it develops an individual model for each data vector that takes into account the new input vector location in the space, and it is an adaptive model, in the sense that input-output pairs of data can be added to the data set continuously and immediately made available for creating transductive models. This type of modelling can be called "personalised", and it is promising for medical decision support systems. As the TWRBF creates a unique sub-model for each data sample, it usually needs more performing time than inductive models, especially in the case of training and simulating on large data sets.

Further directions for research include: (1) TWRBF system parameter optimization such as optimal number of nearest neighbors; and (2) applications of the TWRBF method for other decision support systems, such as: cardio-vascular risk prognosis; biological processes modeling and prediction based on gene expression micro-array data.

Acknowledgement

The research presented in the paper is funded by the New Zealand Foundation for Research, Science and Technology under grant NERF/AUTX02-01. We thank Dr

Mark Marshall and Ms Maggie Ma from the Middlemore hospital in Auckland for their help with the clinical data.

References

1. Fuzzy System Toolbox, MATLAB Inc., Ver.2 (2002)
2. Jang, R., "ANFIS: adaptive network-based fuzzy inference system", *IEEE Trans. on Syst.,Man, and Cybernetics*, vol. 23, No.3, (1993) 665 – 685 1993.
3. Kasabov, N., "Evolving fuzzy neural networks for on-line supervised/unsupervised, knowledge–based learning," *IEEE Trans. SMC* – part B, Cybernetics, vol.31, No.6 (2001) 902-918
4. Kasabov, N., Evolving connectionist systems: Methods and Applications in Bioinformatics, Brain study and intelligent machines, Springer Verlag, London, New York, Heidelberg, (2002)
5. Kasabov, N. and Song, Q., "DENFIS: Dynamic, evolving neural-fuzzy inference systems and its application for time-series prediction," *IEEE Trans. on Fuzzy Systems*, vol. 10, (2002) 144 – 154
6. Lee, Y., "Handwritten digit recognition using K nearest-neighbor, radial basis function, and back-propagation neural networks," *Neural Computation*, vol.3 No.3 (1991) 440 – 449
7. Levey, A. S., Bosch, J. P., Lewis, J. B., Greene, T., Rogers, N., Roth, D., for the Modification of Diet in Renal Disease Study Group, " A More Accurate Method To Estimate Glomerular Filtration Rate from Serum Creatinine: A New Prediction Equation", *Annals of Internal Medicine*, vol. 130 (1999) 461 – 470
8. Mackey, M. C., and Glass, L., "Oscillation and Chaos in Physiological Control systems", *Science*, vol. 197 (1977) 287 – 289
9. Neural Network Toolbox user's guide, The MATH WORKS Inc., Ver.4 (2002)
10. Poggio, F., "Regularization theory, radial basis functions and networks" In: *From Statistics to Neural Networks: Theory and Pattern Recognition Applications*. NATO ASI Series, No.136 (1994) 83 – 104
11. Song, Q. and Kasabov, N., "ECM - A Novel On-line, Evolving Clustering Method and Its Applications", *Proceedings of the Fifth Biannual Conference on Artificial Neural Networks and Expert Systems (ANNES2001)*, (2001) 87 – 92
12. Song, Q. and Kasabov, N., "Weighted Data Normalizations and feature Selection for Evolving Connectionist Systems Proceedings", *Proc. of The Eighth Australian and New Zealand Intelligence Information Systems Conference* (ANZIIS2003), (2003) 285 – 290
13. V. Vapnik, Statistical Learning Theory, John Wiley & Sons, Inc. (1998)

An Incremental Neural Network
for Non-stationary Unsupervised Learning

Shen Furao[1] and Osamu Hasegawa[1,2]

[1] Tokyo Institute of Technology, Yokohama, Japan
furaoshen@isl.titech.ac.jp
[2] PRESTO, Japan Science and Technology Agency (JST)

Abstract. A new online learning mechanism is proposed for unsupervised classification and topology representation. It can represent the topological structure of unsupervised online non-stationary data, report the reasonable number of clusters, eliminate noise, and give typical prototype patterns of every cluster without priori conditions such as a suitable number of nodes or a good initial codebook.

1 Introduction

Traditional clustering methods such as k-means [1] and LBG [2] suffer from the dependence on initial starting conditions and predetermining of number of clusters. Learning of topology structure for nonstationary data distribution is difficult for some self-organizing neural networks such as self-organizing map [3] and neural gas [4] for the use of decaying adaptation parameters will lead to "frozen network". Growing neural gas (GNG) [5] suffers from the permanent increase in number of nodes and GNG-U [6] will destroy the learned old prototype patterns. Hamker's life-long learning model [7] only suits for some supervised learning tasks.

In this work, our goals are: (1) to process online and nonstationary data. (2) without priori conditions such as a suitable number of nodes or a good initial codebook or knowing how many classes there are, to conduct unsupervised learning, report a suitable number of classes and represent the topological structure of input probability density. (3) to separate classes with low-density overlap and detect the main structure of clusters that are polluted by noise.

2 Proposed Algorithm

Suppose we say that two samples belong to the same cluster if the Euclidean distance between them is less than threshold distance T. To obtain "natural" clusters, T must be greater than typical within-cluster distance and less than typical between-cluster distance [8].

For within-cluster insertion, we insert a node between node q with maximum accumulated error and node f, which is among the neighbors of q with maximum accumulated error. To ensure insertion of a new node leads to a decrease in error,

N.R. Pal et al. (Eds.): ICONIP 2004, LNCS 3316, pp. 641–646, 2004.
© Springer-Verlag Berlin Heidelberg 2004

and to avoid catastrophic allocation of new nodes, when a new node is inserted, we evaluate insertion by a utility parameter to judge if insertion is successful.

We separate clusters by removing nodes located in a region with low probability density. If the number of input signals generated so far is an integer multiple of a parameter, remove those nodes with no or only one topological neighbor. The idea is based on the consideration that if a node has no or only one neighbor, it means during a period, the accumulated error of that node has very low chance to become maximum and the insertion of new nodes near that node is difficult, i.e., the probability density of the region the node lies in is very low. Unlike some other techniques in [9] and [6], this proposed strategy works well for removing nodes in low density regions without additive computation load and avoids the use of special parameters. In addition, this technique periodically removes nodes caused by noise.

Algorithm 2.1: Proposed algorithm

1. Initialize node set A to contain two nodes, c_1 and c_2 with weight vectors chosen randomly from the input pattern. Initialize connection set C, $C \subset A \times A$, to the empty set.
2. Input new pattern $\xi \in R^n$.
3. Search the nearest node (winner) s_1, and the second-nearest node (second winner) s_2 by $s_1 = \arg\min_{c \in A} \|\xi - W_c\|$, $s_2 = \arg\min_{c \in A \setminus \{s_1\}} \|\xi - W_c\|$. If the distance between ξ and s_1 or s_2 is greater than similarity threshold T_{s_1} or T_{s_2}, the input signal is a new node, add it to A and go to Step2 to process the next signal. The threshold T is calculated by *Algorithm 2.2*.
4. If a connection between s_1 and s_2 does not exist, create it. Set the age of the connection between s_1 and s_2 to zero.
5. Increment the age of all edges emanating from s_1 by 1.
6. Add Euclidian distance between input signal ξ and the winner to a local accumulated error E_{s_1}, $E_{s_1} = E_{s_1} + \|\xi - W_{s_1}\|$.
7. Add 1 to a local accumulated number of signals M_{s_1}, $M_{s_1} = M_{s_1} + 1$.
8. Adapt the weight vectors of the winner and its direct topological neighbors by fraction $\epsilon_1(t)$ and $\epsilon_2(t)$ of the total distance to the input signal,
   $\Delta W_{s_1} = \epsilon_1(t)(\xi - W_{s_1})$
   $\Delta W_i = \epsilon_2(t)(\xi - W_i)$ for all direct neighbors i of s_1
   We adopt a scheme like k-means to adapt the learning rate over time by $\epsilon_1(t) = 1/t$ and $\epsilon_2(t) = 1/100t$.
9. Remove edges with an age greater than a predefined threshold age_{dead}. If this results in nodes having no more emanating edges, remove them as well.
10. If the number of input signals generated so far is an integer multiple of parameter λ, insert a new node as follows:
    (a) Determine node q with maximum accumulated error E.
    (b) Determine among neighbors of q node f with maximum accumulated error.
    (c) Add new node r to the network and interpolate its weight vector from q and f, i.e., $W_r = (W_q + W_f)/2.0$.

(d) Interpolate accumulated error E_r, accumulated number of signal M_r and inherited error-radius R_r from E_q, E_f, M_q, M_f and R_q, R_f by $E_r = (E_q + E_f)/6.0$, $M_r = (M_q + M_f)/4.0$, and $R_r = (R_q + R_f)/4.0$. Here, error-radius of node i is defined by mean of accumulated error, E_i/M_i. R_i serves as memory for the error-radius of node i at the moment of insertion. It is updated at each insertion, but only for affected nodes.

(e) Decrease accumulated error of q and f by $E_q = 2E_q/3.0$, $E_f = 2E_f/3.0$.

(f) Decrease accumulated number of signal of q and f by $M_q = 3M_q/4.0$, $M_f = 3M_f/4.0$.

(g) Judge whether insertion is successful or not. If the error-radius is larger than inherited error-radius R_i ($\forall i \in \{q, r, f\}$), i.e., insertion is not able to decrease the mean error of this local area, insertion is not successful; else, update the inherited error-radius, i.e., if $E_i/M_i > R_i$ ($\forall i \in \{q, r, f\}$), insertion is not successful, new node r is removed; else, $R_q = E_q/M_q$, $R_f = E_f/M_f$, and $R_r = E_r/M_r$.

(h) If insertion is successful, insert edges connecting new node r with nodes q and f, and remove the original edge between q and f.

(i) For all nodes in A, search for nodes having no neighbor or only one neighbor, then remove them.

11. After long constant time period LT, use *Algorithm 2.3* to classify nodes into different clusters. Report the number of clusters, output all nodes belonging to different clusters.

12. Go to Step2 to continue unsupervised online learning.

The similarity threshold T (in *Algorithm 2.1*, step3) must be greater than within-cluster distance and less than between-cluster distance. Based on this idea, we propose *Algorithm 2.2* to calculate similarity threshold T_i.

Algorithm 2.2: Calculation of similarity threshold T

1. Initial the similarity threshold of node i to $+\infty$ when node i is generated as a new node.

2. When node i is a winner or second winner, update similarity threshold T_i:
   - If the node has direct topological neighbors, T_i is updated as the maximum distance between node i and all of its neighbors,
     $T_i = \max_{c \in N_i} \|W_i - W_c\|$, N_i is the neighbor set of node i.
   - If node i has no neighbors, T_i is updated as the minimum distance of node i and all other nodes in A, $T_i = \min_{c \in A \setminus \{i\}} \|W_i - W_c\|$.

If there is a series of nodes $x_i \in A$, $i = 1, 2, \ldots, n$, makes (i, x_1), (x_1, x_2), $\ldots$, (x_{n-1}, x_n), $(x_n, j) \in C$. We say there is a "path" between node i and node j.

Algorithm 2.3: Classify nodes to different clusters

1. Initialize all nodes as unclassified.

2. Randomly choose one unclassified node i from node set A. Mark node i as classified and label it as class C_i.

3. Search A to find all unclassified nodes connected to node i with a "path." Mark these nodes as classified and label them as the same class as node i.

4. If there are unclassified nodes, go to Step2 to continue classification until all nodes are classified.

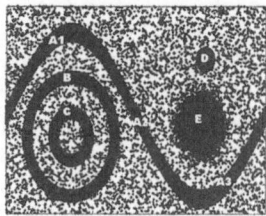 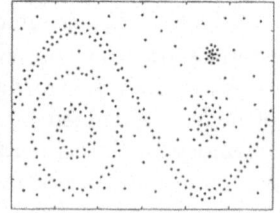 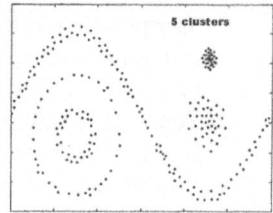

Fig. 1. Original data set   **Fig. 2.** GNG results   **Fig. 3.** Proposed results

3   Experiment

We conducted our experiment on the data set shown in Fig. 1. An artificial 2-D data set is used to take advantage of its intuitive manipulation, visualization, and resulting insight into system behavior. The data set is separated into five parts, i.e., A, B, C, D, and E. A is shaped like a sinusoid and separated into A1, A2, and A3 to show incremention property. The B and C data set is a famous single-link example. D is a circular area. E is a data set satisfied 2-D Gaussian distribution. We add random noise (5% of useful data) to the data set to simulate real-world data. In Fig. 1, there are overlaps between clusters, and noise is distributed over the whole data set.

In this experiment, we compare our proposed method with GNG to show the advantage of our method. In all experiments, we set parameter $\lambda = 100$, and $age_{dead} = 100$. For GNG, the maximum number of nodes is predefined as 300.

3.1   Experiment in Stationary Environment

First, we use Fig. 1 as a stationary data set, 100,000 patterns are randomly chosen from areas A, B, C, D, and E. Topological results of GNG and proposed method are shown in Fig. 2 and Fig. 3. For stationary data, GNG can represent the topological structure, but it is affected by noise and all nodes are linked to form one cluster. The proposed method efficiently represents the topology structure and gives the number of clusters and typical prototype nodes of every cluster.

3.2   Experiment in Nonstationary Environment

We simulate online learning by using a paradigm as follows: from step 1 until 20,000, patterns are randomly chosen from area A1. At step 20,001, the environment changes and patterns from area A2 are chosen. At step 40,001, the environment changes again, etc. Table 1 details specifications of the test environment. The environment changes from I to VII. In each environment, areas used to generate patterns are marked with "1," and other areas are marked with "0." For every environment, we add 5% noise to test data and noise is distributed over the whole data space.

The results for GNG and GNG-U in Fig. 4 and Fig. 5 show that GNG cannot represent the topological structure of online nonstationary data well, GNG-U deletes all old learned patterns and only represents the structure of new input patterns. Neither method eliminates noise. In GNG-U results, for example, nodes beyond area E are all caused by noise distributed over the whole space (Fig. 5).

Table 1. Nonstationary

Area	Environment						
	I	II	III	IV	V	VI	VII
A1	1	0	1	0	0	0	0
A2	0	1	0	1	0	0	0
A3	0	0	1	0	0	1	0
B	0	0	0	1	1	0	0
C	0	0	0	0	1	0	0
D	0	0	0	0	0	1	0
E	0	0	0	0	0	0	1

Fig. 4. GNG results   **Fig. 5.** GNG-U results

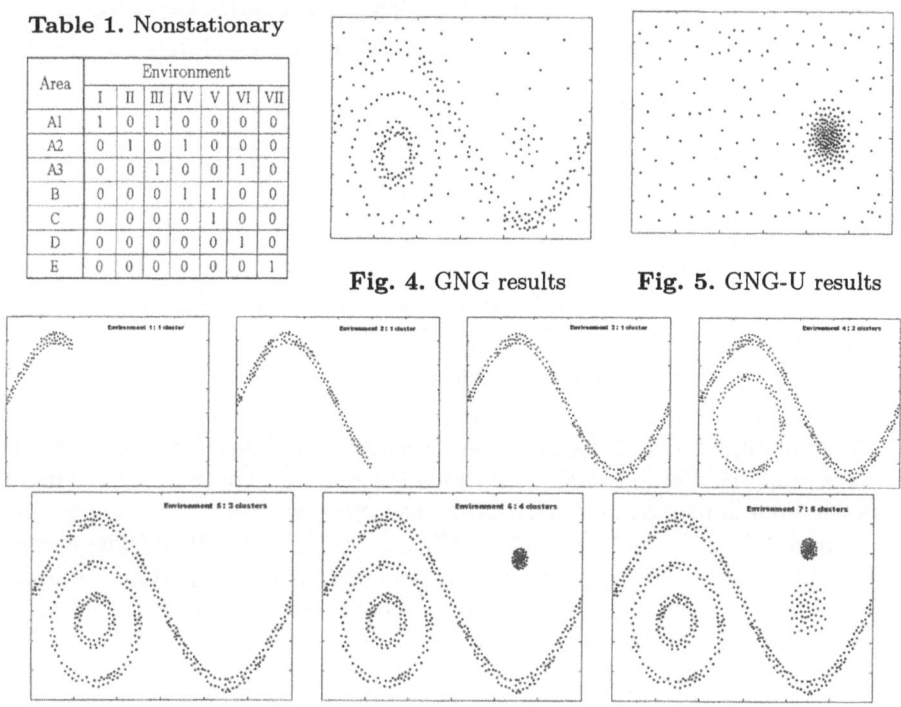

Fig. 6. Nonstationary environments, results of proposed method

Fig. 6 shows the results of proposed method. After learning in one environment, we report intermediate topological structures. Environments I, II, and III test a complicated artificial shape (area A). A is separated into three parts and data comes to the system sequentially. We find nodes of area A increasing following the change of environment from I to II and III, but all nodes linked to form the same class. In environment I, the data set of area A1 is tested. If probability changes to zero in some regions, such as in environment II, remaining nodes of area A1 preserve the knowledge of previous situations for future decisions. In the future, the reappearance of area A1 (environment III) does not raise error and knowledge is completely preserved, so nearly no insertion happens and most nodes remain at their positions. Environments IV and V test for a difficult situation (data sets such as areas B and C). The system also works well, removing noise between areas and separating the B and C. Environments VI and VII test an isolated data set: area D (circular) and area E (Gaussian distribution).

Fig. 3 and Fig. 6 show that the proposed method processes stationary and nonstationary environments well. It detects the main structure from original data, which is polluted by noise. It controls the number of nodes needed for representation of topology and reports the number of clusters.

Fig. 7 shows how the number of nodes changes during online learning. When an input signal comes from a new area (see Table 1, environment changes from I to VII), the number of nodes increases. In the same environment, after some learning steps, the increase in nodes stops and converges to a constant because further insertion cannot lead to decreasing of error. Noise leads to the system frequently inserting and deleting nodes; thus, there is a small fluctuation in the number of nodes in Fig. 7.

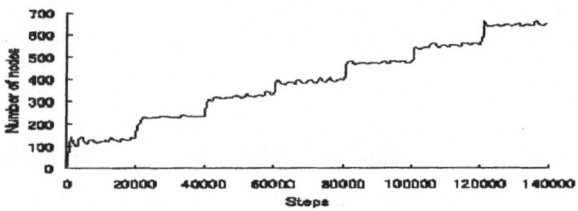

Fig. 7. Number of nodes during online learning

In addition, we tested up to 5000-dimensional non-linear artificial data set and the proposed method worked very well. For real world data, we tested digits data set from Japan ETL6 handwriting character database, it is composed of digits from '0' to '9', and there are 1,383 samples for every digit. The system reported that there were 10 classes, and then the prototype patterns of every class were also reported. Due to lack of space, we can not report the detail of the experiment for real data in this paper.

References

1. Likas, A., Vlassis, N., & Verbeek, J.J. (2003). "The global k-means clustering algorithm," Pattern Recognition, Vol. 36, 451-461.
2. Patane, G., & Russo, M. (2001). "The enhanced LBG algorithm," Neural Networks, Vol.14, 1219-1237.
3. Kohonen, T. (1982). "Self-organized formation of topologically correct feature maps," Biological Cybernetics, Vol 43, 59-69.
4. Martinetz, T.M., Berkovich, S.G., & Schulten, K.J. (1993). ""Neural-gas" network for vector quantization and its application to time-series prediction," IEEE Transactions on Neural Networks, Vol. 4, No. 4, 558-569.
5. Fritzke, B. (1995). "A growing neural gas network learns topologies," In Advances in neural information processing systems, 625-632.
6. Fritzke, B. (1997). "A self-organizing network that can follow non-stationary distributions," In Proceedings of ICANN-97, 613-618.
7. Hamker, F.H. (2001). "Life-long learning cell structures – continuously learning without catastrophic interference," Neural Networks, Vol. 14, 551-573.
8. Duda, R.O., Hart, P.E., & Stork, D.G. (2001). "Pattern classification, 2nd ed.," A Wiley-Interscience Publication.
9. Fritzke, B. (1994). "Growing cell structures – a self-organizing network for unsupervised and supervised learning," Neural Networks, Vol. 7, 1441-1460.

Computing Convex-Layers by a Multi-layer Self-organizing Neural Network

Amitava Datta and Srimanta Pal

Indian Statistical Institute,
203 B. T. Road, Calcutta - 700 108
amitava@isical.ac.in
http://www.isical.ac.in/~amitava

Abstract. A multi-layer self-organizing neural network model has been proposed for computation of the convex-layers of a given set of planar points. Computation of convex-layers has been found to be useful in pattern recognition and in statistics. The proposed network architecture evolves in such a manner that it adapts itself to the hull-vertices of the convex-layers in the required order. Time complexity of the proposed model is also discussed.

1 Introduction

The convex-hull of a given planar set of points is defined as the smallest convex polygon containing all points in the set. The convex-layers of a set of points is a natural extension of its convex-hull. The convex-hull $C(S)$ for a given (finite) set S of planar points is the minimal convex set containing all points of S. Let $V(S)$ be the set of hull vertices of S. The convex-layers $CL(S)$ of the set S is the set of convex polygons $\{C(S), C(S'), C(S''), \ldots\}$, where S' is obtained by removing the hull points of S, i.e., $S' = S - V(S)$; similarly, $S'' = S' - V(S')$ and so on. The concept of convex-layers has applications in many areas including pattern recognition and statistics [3, 7, 9]. For example, in robust estimation, a common problem is to evaluate an unbiased estimator that is not very sensitive to outliers. In two-dimensional case, this can be accomplished by removing the outliers of a point set by peeling the convex-layers iteratively until a predetermined fraction of the original point set remains [7].

There are many sequential algorithms for the convex-layers problem with $O(n^2)$ [4, 9] and $O(n \log^2 n)$ [8] time complexity, where n is the size of the input. The most efficient method, to our knowledge, is Chazelle's algorithm [2] which takes $O(n \log n)$ time. We propose here a multi-layered neural network with $O(n)$ processors for computation of the convex-layers of a given planar set. The complexity of the model is $O(n)$.

2 Computational Model

To find the convex-layers, we need to find repeatedly the convex-hulls of a set S of n points representing the input vectors (the signals), i.e.

$$S = \{P_i \mid i = 1, 2, \cdots, n\}. \tag{1}$$

N.R. Pal et al. (Eds.): ICONIP 2004, LNCS 3316, pp. 647–652, 2004.
© Springer-Verlag Berlin Heidelberg 2004

The point P_i is defined by $\mathbf{X_i} = (x_i, y_i)$. For S, the following results hold good:

(1) The convex-hull of S is a convex polygon.
(2) Each edge of the polygon is a hull-edge.
(3) Each vertex of the polygon is a hull-vertex.
(4) Every hull-edge partitions the plane into two half-planes such that one contains all points of S and other contains none.

Suppose $\Pi = \{\pi_i \mid i = 1, 2, \cdots, n\}$ is a set of n processors. Each processor π_i is associated with the points P_i and stores a vector $(x_i, y_i, w_i^{(x)}, w_i^{(y)}) = (X_i, W_i)$, where $X_i = (x_i, y_i)$ is the coordinate of P_i and $W_i = (w_i^{(x)}, w_i^{(y)})$ is its weight vector (equivalent to *direction ratio*) for $i = 1, 2, \cdots, n$. Each processor π_i is connected with $\pi_j \; \forall \; i, j = 1, 2, \cdots, n$ to form a complete network. These processors are termed as *point-processors*. During learning, some of the point-processors are labeled as *hull-processors* depending on certain criterion. When the learning is complete, as we shall see later, the hull-vertices of the respective convex-layers are mapped as the hull-processors of the network in an orderly fashion.

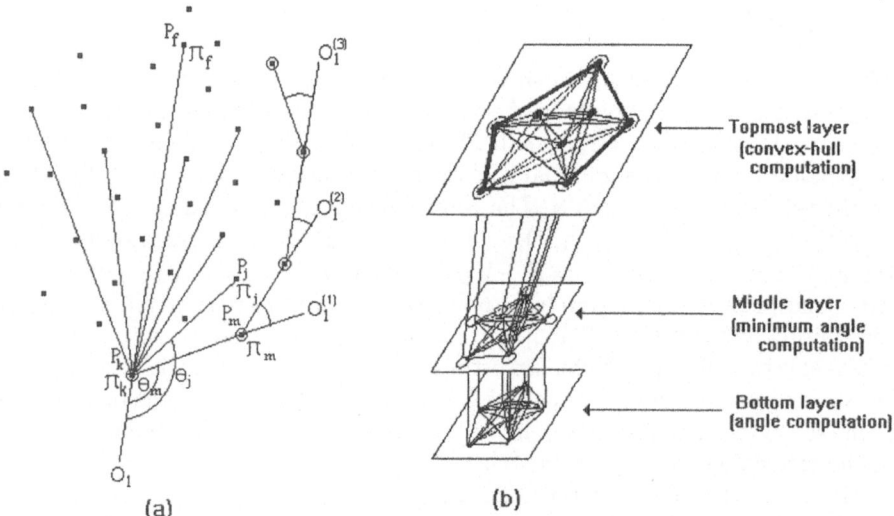

Fig. 1. (a) Point-processors and hull-processors (circled). (b) The proposed three layered architecture.

Definition. We will call two processors π_i and π_j equivalent if their coordinates are the same i.e., if $X_i = X_j$.

Initialization

First we select the farthest processor $\pi_k \in \Pi$ from a randomly chosen processor $\pi_f \in \Pi$ (known as *seed-processor*) using a 'ranknet'-like network [10]. Thus π_k is chosen as

$$\|X_k - X_f\| = \max_i \|X_i - X_f\|. \tag{2}$$

Clearly π_k is a hull-processor [5]. We term π_k as a *mother* (first mother) hull-processor. The initially assigned weights for the neurons are as follows:

$$\mathbf{W}_j(0) = \begin{cases} X_j - X_k & \text{if } j \neq k \\ X_k - X_f & \text{if } j = k \end{cases} \tag{3}$$

for $j = 1, 2, \cdots, n$.

The convex-layer computation algorithm has three major steps: angle calculation, finding the minimum angle and weight update [10]. These steps are repeated iteratively until all the processors are labeled as hull-processors.

Angle Computation

Compute the smallest angle θ_j, between the vectors $\overline{P_k P_j}$ and $\overline{P_k O_1}$ with respect to the mother hull-processor π_k (see Fig. 1(a)) as

$$\theta_j = \cos^{-1} \frac{<W_k, W_j>}{||W_k||\cdot||W_j||}$$

$$= \cos^{-1} \frac{w_k^{(x)} w_j^{(x)} + w_k^{(y)} w_j^{(y)}}{\sqrt{(w_k^{(x)})^2 + (w_k^{(y)})^2}\sqrt{(w_j^{(x)})^2 + (w_j^{(y)})^2}}. \tag{4}$$

$j = 1, 2, \cdots, n \ (j \neq k)$.

A 'ranknet'-like network [10] is used to find the minimum angle θ_m using (5).

$$\theta_m = \min_j \theta_j. \tag{5}$$

The processor π_m is selected as the *winner* in the above competition by the 'ranknet'. π_m is declared as a hull-processor and as a *daughter* of the mother hull-processor π_k. The mother hull-processor π_k now becomes *nonproductive* and the newly created daughter π_m becomes a *productive* mother hull-processor. The idea is to allow only one processor in the network to reproduce so that all angles can be computed with respect to it. A mother can reproduce a daughter only once after which it becomes nonproductive (and hence no more remains a mother) and the daughter becomes the mother. A processor may reproduce a second daughter if it wins the competition again. Moreover, a mother processor, except the first mother, is made *idle* or *inactive* after reproducing a daughter so that it (the mother) does not participate in the angle computation and in the competition process (that is, in the process of finding the minimum angle) in the successive iterations. An idle processor can be looked as if it is not present in the network.

Updating the Weights

In this network, weights are updated using (6) after finding the minimum angle.

$$W_j(t+1) = \begin{cases} W_j(t) - W_m(t) & \text{for } j = 1, 2, \cdots, n \\ & \text{and } j \neq m, k \\ W_j(t) & \text{for } j = m \\ -W_m(t) & \text{for } j = k. \end{cases} \tag{6}$$

The above process of angle calculation and finding the processor with the minimum angle (that is, the process of reproduction) is repeated with respect to the productive mother hull-processor. We declare a *phase* is complete when the most recent daughter processor becomes equivalent to the first mother hull-processor. Thus the first phase generates the outermost convex-layer. The computation continues to the next step with the most recent daughter π_r (which is equivalent to the first mother). The most recent daughter π_r being productive, it in turn produces a daughter π_h, which is labelled as a hull-processor. Note that this newly generated hull-processor, π_h, does not belong to the set of hull-processors obtained in the first phase since all hull-processors, other than the first mother, are set idle. Now the first mother, π_r becomes idle and its most recently created daughter π_h is declared as the first mother of the second phase. This *chain of reproduction* is continued until all the processors in the network become hull-processors.

The algorithm for computation of the convex-layers can be implemented on a three (bottom, middle and topmost) layered network (see Fig. 1(b)) as follows:

Algorithm:

Step 1.   [**Initialization**] Select the seed-processor π_f at random. Select the first mother hull-processor π_k satisfying Eqn. (2) and declare it as the first mother hull-processor. Set the initial weights as using (3).

Step 2.   [**Reproduction**] The mother hull-processor creates a daughter processor as follows:

Step 2(a) [**Angle computation**] All processors, except the mother, in the bottom layer compute the angle (with respect to the mother hull-processor) using the activation function in Eqn. (7). If no processor responds in angle computation (that is, all the processors are exhausted and are labeled as hull-processors) then make the first mother of the current phase idle and go to Step 8.

$$o_j = f(W_j, W_k) = \cos^{-1} \frac{<W_k, W_j>}{||W_k|| \; ||W_j||}$$

$$= \cos^{-1} \frac{w_k^{(x)} w_j^{(x)} + w_k^{(y)} w_j^{(y)}}{\sqrt{(w_k^{(x)})^2 + (w_k^{(y)})^2} \sqrt{(w_j^{(x)})^2 + (w_j^{(y)})^2}}. \qquad (7)$$

where, π_k is a mother hull-processor and o_j is angle θ_j in (4).

Step 2(b) [**Minimum angle computation**] The minimum angle is computed in the middle layer, where the mother hull-processor does not participate. The processor corresponding to the minimum angle is declared as the daughter processor.

Step 2(c) [**Report to topmost layer**] The winner information is sent to the topmost layer to set the link 'ON' from the mother to the daughter.

Step 3.   [**Weight updating**] Update the weights using Eqn. (6).

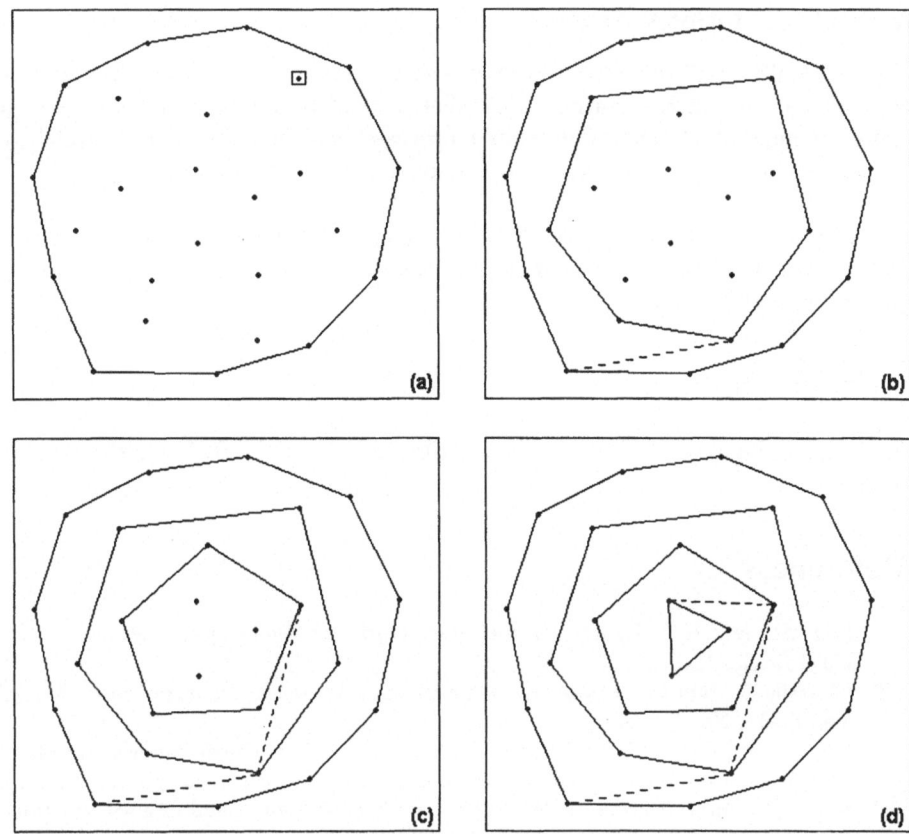

Fig. 2. (a)-(d) The results after phases 1, 2, 3 and 4 respectively (dot enclosed by a box in (a) indicates the seed-processor). The links from the first mother of one phase to that of the next phase (denoted by dashed lines) are reset 'OFF' in the Step 7.

Step 4. Make the mother nonproductive and make the daughter a productive mother. Also make the (old) mother idle if it is not the first mother.

Step 5. Repeat Steps 2 through 4 until the most recent daughter processor is equivalent to the first mother.

Step 6. **[End of Phase]** Repeat Steps 2 through 4 to get the first mother of the next phase.

Step 7. Make the first mother of the current phase idle and reset the corresponding link (between the first mother of the current phase and the first mother of the next phase) 'OFF' in the topmost layer. Go to Step 2.

Step 8. Activate all the processors and stop.

On completion of the above algorithm, the convex-layers are obtained in the topmost layer of the network. Figure 2 shows one such example.

2.1   Computational Aspects

After the initial hull-processor is created every processor in the network computes its own output independently (in parallel) [1]. Thus this computation takes a constant amount of time. The winner processor can be selected in parallel by a 'ranknet'-like layer. Hence, computational complexity of this step will not depend on size of the input data set [10]. It is easy to see that the network needs to compute the output (in the bottom layer) and then select the winner (in the middle layer) n times to exhaust all the point-processors. Thus, the whole process takes $O(n)$ time.

Acknowledgement

The authors highly acknowledge the contribution of Prof. N. R. Pal in the present work.

References

1. S.G. Akl and K.A. Lyons, *Parallel Computational Geometry*, Englewood Cliffs, NJ: Prentice-Hall, 1993.
2. Chazelle B., "On the convex layers of a Planar set", IEEE Trans. on Info. Theory, Vol. IT-31, No. 4, 1985.
3. Chazelle B., Guibas L. and Lee D.T.,"The power of geometric duality," in proc. 24th IEEE Ann. Symp. Foundations of Computer Science, 1983, pp.217-225.
4. Green P.J. and Silverman B.W., "Constructing the convex hull of a set of points in the plane," Comput. J., vol 22, pp. 262-266, 1979.
5. G. Hadley, *Linear Programming*, Addison-Wesley Publishing Company, 1962.
6. Haykin Simon, *Neural networks*, New Jersey: Prentice Hall, 1994.
7. Huber P. J., "Robust statistics: a review," Ann. Math. Statist., vol.43, no. 3, pp.1041-1067, 1972.
8. Overmars M.H. and Leeuwen J. van, "Maintenance of configurations in the plane", J. Comput. Syst. Sci., vol. 23, pp.166-204, 1981.
9. F.P. Preparata and M.I. Shamos, *Computational Geometry: An Introduction.* New York: Springer-Verlag, 1985.
10. S. Pal, A. Datta and N. R. Pal, "A multi-layer self-organizing model for convex-hull computation", *IEEE Trans. Neural Network*, vol. 12, pp. 1341-1347, 2001.

Cost-Sensitive Greedy Network-Growing Algorithm with Gaussian Activation Functions

Ryotaro Kamimura[1] and Osamu Uchida[2]

[1] Information Science Laboratory, Tokai University
1117 Kitakaname, Hiratsuka, Kanagawa 259-1292, Japan
ryo@cc.u-tokai.ac.jp
[2] Department of Human and Information Science
School of Information Technology and Electronics, Tokai University
1117 Kitakaname, Hiratsuka, Kanagawa, 259-1292, Japan
o-uchida@tokai.ac.jp

Abstract. In this paper, we propose a new network-growing algorithm which is called the *cost-sensitive greedy network-growing algorithm*. This new method can maximize information while controlling the associated cost. Experimantal results show that the cost minimization approximates input patterns as much as possible, while information maximization aims to extract distinctive features.

1 Introduction

A number of network growing approaches have been proposed in neural networks, because constructive approaches are economical in computation and they have a good chance of finding smaller networks for given problems. For example, in supervised learning, one of the most popular constructive methods is the cascade correlation network that grows a network by maximizing the covariance between the outputs of newly recruited hidden units and the errors of the network outputs [1], [2]. In the popular backpropagation method, a method called *constructive backpropagation* was developed in which connections into a newly recruited hidden unit are only updated [3].

We have developed a new type of information-theoretic network growing algorithm called *greedy network-growing* [4], [5]. These methods are completely different from previous approaches, because information maximization is used to increase complexity of networks. In this algorithm, we suppose that a network attempts to absorb as much information as possible from outer environment. To absorb information on the outer environment, the networks gradually increase their complexity until no more complexity is needed. When no more additional information can be obtained, the networks recruit another unit, and then it again tries to absorb information maximally.

One of the problems of the method is that we do not know to what extent the algorithm can extract representations faithful to input patterns. It happens sometimes that information maximization does not necessarily produce representations faithful to input patterns. To produce faithful representations, we introduce a cost function. We call this information maximization with the cost function *cost-sensitive information maximization*.

N.R. Pal et al. (Eds.): ICONIP 2004, LNCS 3316, pp. 653–658, 2004.
© Springer-Verlag Berlin Heidelberg 2004

2   Cost-Sensitive Greedy Network-Growing Algorithm

The general idea of networks with greedy network-growing algorithm has been realized in neural networks in the previous paper[6]. Figure 1 shows an actual network architecture for greedy algorithm. Figure 1(M=1) represents an initial state of information maximization, in which only two competitive units are used. We need at least two competitive units, because our method aims to make neurons compete with each other. First, information is increased as much as possible with these two competitive units. When it becomes difficult to increase information, the first cycle of information maximization is finished. In this case, just one unit wins the competition, while the other loses. Then, a new competitive unit is added, as shown in Figure 1(M=2). These processes continue until no more increase in information content is possible.

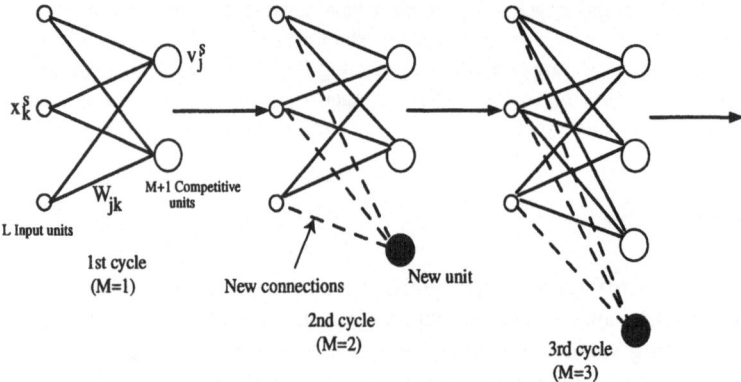

Fig. 1. A process of network growing.

Now, we can compute information content in a neural system. We consider information content stored in competitive unit activation patterns. For this purpose, let us define information to be stored in a neural system. Information stored in the system is represented by decrease in uncertainty [9]. Uncertainty decrease, that is, information $I(t)$ at the tth epoch, is defined by

$$I(t) = -\sum_{\forall j} p(j;t) \log p(j;t) + \sum_{\forall s} \sum_{\forall j} p(s)p(j \mid s;t) \log p(j \mid s;t), \qquad (1)$$

where $p(j;t)$, $p(s)$ and $p(j|s;t)$ denote the probability of firing of the jth unit in a system at the tth learning epochs, the probability of the sth input pattern and the conditional probability of the jth unit, given the sth input pattern at the tth epoch, respectively. Let us define a cost function

$$C = \sum_{\forall s} p(s) \sum_{\forall j} p(j \mid s)C_j^s, \qquad (2)$$

where C_j^s is a cost of the jth unit for the sth input pattern.

Let us now present update rules to maximize information content in every stage of learning. For simplicity, we consider the Mth growing cycle, and t denotes the cumulative learning epochs throughout the growing cycles. As shown in Figure 1, a network at the tth epoch is composed of L input units and $M+1$ competitive units. The jth competitive unit receives a net input from input units, and an output from the jth competitive unit $v_j^s(t)$ can be computed by the Gaussian type function, that is,

$$v_j^s(t) = \exp\left(-\frac{\sum_{k=1}^{L}(x_k^s - w_{jk}(t))^2}{2\sigma^2} \right), \tag{3}$$

where L is the number of input units, and w_{jk} denote connections from the kth input unit to the jth competitive unit [7]. The output is increased as connection weights are closer to input patterns. The conditional probability $p(j \mid s; t)$ at the tth epoch is computed by

$$p(j \mid s; t) = \frac{v_j^s(t)}{\sum_{m=1}^{M+1} v_m^s(t)}, \tag{4}$$

where M denotes the Mth growing cycle. Since input patterns are supposed to be given uniformly to networks, the probability of the jth competitive unit is computed by

$$p(j; t) = \frac{1}{S}\sum_{s=1}^{S} p(j \mid s; t). \tag{5}$$

Information $I(t)$ is defined by

$$I(t) = -\sum_{j=1}^{M+1} p(j; t) \log p(j; t) + \frac{1}{S}\sum_{s=1}^{S}\sum_{j=1}^{M+1} p(j \mid s; t) \log p(j \mid s; t). \tag{6}$$

In this paper, we define a cost function at the tth epoch by the average distortion between input patterns and connection weights, that is,

$$C(t) = \frac{1}{S}\sum_{s=1}^{S}\sum_{j=1}^{M+1} p(j \mid s; t) \sum_{k=1}^{L}(x_k^s - w_{jk}(t))^2. \tag{7}$$

As the function $J(t) = I(t) - C(t)$ becomes larger, specific pairs of input patterns and competitive units become strongly correlated. Differentiating the function $J(t)$ with respect to input-competitive connections $w_{jk}(t)$, we have

$$\frac{\partial J(t)}{\partial w_{jk}(t)} = -\sum_{s=1}^{S} Q_{jk}^s(t) \log p(j; t) + \sum_{s=1}^{S}\sum_{m=1}^{M+1} Q_{jk}^s(t) p(m \mid s; t) \log p(m; t)$$

$$+ \sum_{s=1}^{S} Q_{jk}^s(t) \log p(j \mid s; t) - \sum_{s=1}^{S}\sum_{m=1}^{M+1} Q_{jk}^s(t) p(m \mid s; t) \log p(m \mid s; t)$$

$$+ \sum_{s=1}^{S} p(j \mid s; t)(x_k^s - w_{jk}(t)), \tag{8}$$

where

$$Q_{jk}^s = \frac{x_k^s - w_{jk}}{S\sigma^2 \sum_{m=1}^{M+1} v_m^s} \exp\left(-\frac{\sum_{k=1}^{L}(x_k^s - w_{jk}(t))^2}{2\sigma^2}\right). \tag{9}$$

Thus, update rules of the cost-sensitive greedy network-growing algorithm is given by

$$\begin{aligned}
\Delta w_{jk}(t) = &-\alpha \sum_{s=1}^{S} Q_{jk}^s(t) \log p(j;t) + \alpha \sum_{s=1}^{S} \sum_{m=1}^{M+1} Q_{jk}^s(t) p(m \mid s;t) \log p(m;t) \\
&+\beta \sum_{s=1}^{S} Q_{jk}^s(t) \log p(j \mid s;t) - \beta \sum_{s=1}^{S} \sum_{m=1}^{M+1} Q_{jk}^s(t) p(m \mid s;t) \log p(m \mid s;t) \\
&+\gamma \sum_{s=1}^{S} p(j \mid s;t)(x_k^s - w_{jk}(t)),
\end{aligned} \tag{10}$$

where α, β and γ are parameters.

By changing three parameters, we can have different types of learning methods. For example, when β was set to zero, this method becomes an entropy maximization method with cost minimization.

Finally, we should state how to add a new competitive unit. We define a relative increase in $J(t)$ by

$$R(t) = \frac{|J(t) - J(t-1)|}{J(t-1)}, \tag{11}$$

where $t = 1, 2, 3, \cdots$. If $R(t)$ is less than a certain point ϵ for three consecutive epochs, the networks recruit a new competitive unit.

3   Experimental Results

In this experiment, we attempt to show that the greedy network-growing algorithm with the cost function can extract input patterns faithfully with much less learning epochs. The data was an artificial one shown in Figure 2. The number of competitive units was increased gradually up to five, with four growing cycles. The number of input units was 30.

Figure 3 shows information as a function of the number of epochs t and final connection weights by three methods. As shown in Figure 3(a1), information is gradually increased up to 1.5 by the information maximization method (without cost minimization) using the inverse of the Euclidean distance as the activation function [8], that is,

$$v_j^s(t) = \frac{1}{\sum_{k=1}^{L}(x_k^s - w_j^k(t))^2}. \tag{12}$$

We can see that final connection weights capture perfectly input patterns (Figure 3(a2)). Figure 3(b1) shows information as a function of the number of epochs by the information maximization method with Gaussian activation functions. Information is rapidly

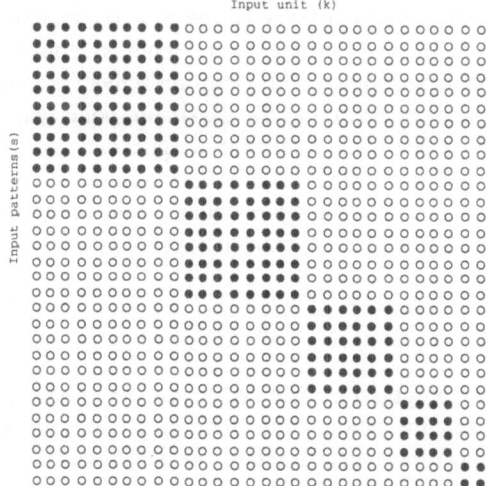

Fig. 2. Artificial data for the first experiment.

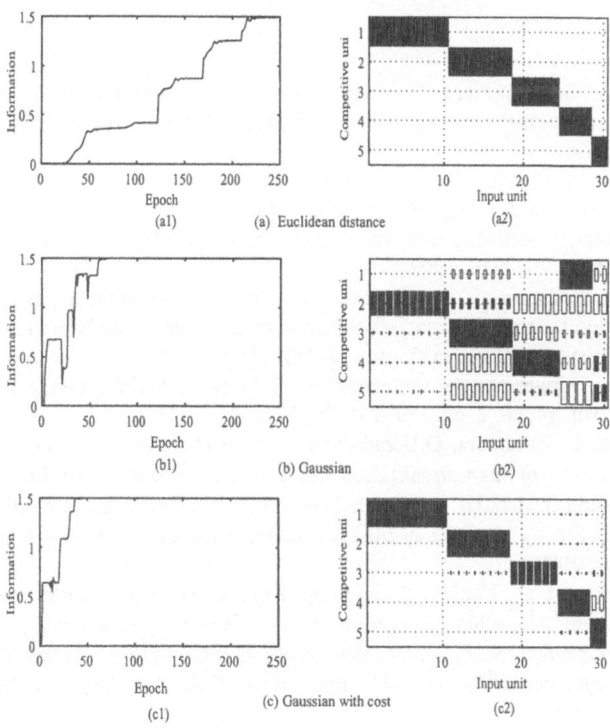

Fig. 3. Information and final connection weights by three methods.

increased up to 1.5 with about 60 epochs. However, as shown in Figure 3(b2), final connection weights are significantly different from input patterns. Finally, when the cost-sensitive information maximization with Gaussian activation functions is used (Figure 3(c1)), information is increased up to the maximum point with less than 50 epochs. In addition, we can see that final connection weights can capture almost perfectly input patterns as the Euclidean distance activation functions do (Figure 3(c2)). Thus, we can say that the information maximization with the cost minimization can significantly accelerate learning with good feature extraction performance.

4   Conclusion

In this paper, we have proposed a novel greedy network-growing algorithm. This new method can maximize information while controlling the associated cost. We have applied the method to an artificial data problem. We have found that the cost minimization approximates input patterns as much as possible, while information maximization aims to extract distinctive features.

References

1. S. E. Fahlman and C. Lebiere, "The cascade-correlation learning architecture," in *Advances in Neural Information Processing*, vol. 2, (San Mateo: CA), pp. 524–532, Morgan Kaufmann Publishers, 1990.
2. T. A. Shultz and F. Rivest, "Knowledge-based cascade-correlation: Using knowledge to speed learning," *Connection Science*, vol. 13, pp. 43–72, 2001.
3. M. Lehtokangas, "Modelling with constructive backpropagation architecture," *Neural Networks*, vol. 12, pp. 707–716, 1999.
4. R. Kamimura, T. Kamimura, and H. Takeuchi, "Greedy information acquisition algorithm: A new information theoretic approach to dynamic information acquisition in neural networks," *Connection Science*, vol. 14, no. 2, pp. 137–162, 2002.
5. R. Kamimura, "Progressive feature extraction by greedy network-growing algorithm," *Complex Systems*, vol. 14, no. 2, pp. 127–153, 2003.
6. R. Kamimura, T. Kamimura, O.Uchida and H. Takeuchi, "Greedy information acquisition algorithm," in *Proc. of International Joint Conference on Neural Networks*, 2002.
7. R. Kamimura and O. Uchida, "Accelerated Greedy Network-Growing Algorithm with application to Student Survey," in *Proc. of IASTED International Conference on Applied Simulation and Modelling*, 2003.
8. R. Kamimura and O. Uchida, "Improving Feature Extraction Performance of Greedy Network-Growing Algorithm," *Intelligent Data Engineering and Automated Learning - Lecture Notes in Computer Sciences*, Vol.2690, pp.1056-1061, Springer-Verlag, 2003.
9. L. L. Gatlin, *Information Theory and Living Systems*. Columbia University Press, 1972.

An Efficient Skew Estimation Technique
for Binary Document Images Based
on Boundary Growing and Linear Regression Analysis

P. Shivakumara, G. Hemantha Kumar, D.S. Guru, and P. Nagabhushan

Department of Studies in Computer Science, University of Mysore
Mysore – 570006, Karnataka
hudempsk@yahoo.com

Abstract. Skew angle estimation is an important component of an Optical Character Recognition (OCR) and Document Analysis Systems (DAS). In this paper, a novel and efficient (in terms of accuracy and computations) method to estimate skew angle of a scanned document image is proposed. The proposed technique works based on fixing a boundary for connected components and growing boundaries. The technique uses Linear Regression Analysis to estimate skew angles. However, the technique works based on the assumption that the space between the two adjacent text lines is greater than the space between the successive characters present in a text line. The proposed method is compared with other existing methods. The experimental results are also presented.

1 Introduction

Today most information is saved, used and distributed by electronic means. Scanners convert paper documents into a format suitable for a computer. The increasing applications of digital document analysis system have resulted in the development of digital text processing systems. The main purpose of these systems is to transform text images into recognized ASCII characters, which is mainly performed by OCR systems. An OCR system often consists of a preprocessing stage, a document layout understanding and segmentation stage, a feature extraction stage and classification stage. Many of these stages function well if the document is not skewed during the scanning process and its text lines are strictly horizontal. Although there are some techniques for character segmentation, which work on, skewed documents too, but are ineffective and involve great computational cost [7-11]. It is therefore preferable, in the preprocessing stage, to determine the skew angles of the digitized documents.

The literature survey specific to skew estimation has revealed that there exist methods for skew detection, based on Projection Profile (PP) [3], Hough Transform (HT) [4], Fourier Transform (FT) [13], Nearest Neighbor Clustering (NNC) [5], and Interline Cross Correlation (ICC) techniques [12].

The authors [1– 2] have proposed skew detection techniques for estimation of skew angle of skewed text binary documents based on linear regression analysis. The techniques are very simple compared to the above methods since the methods do not involve Hough transform and others expensive methods to determine skew angle for a skewed document. However, the skew angle is limited to only $\pm 10^0$ for which accuracy is preserved.

N.R. Pal et al. (Eds.): ICONIP 2004, LNCS 3316, pp. 659–665, 2004.
© Springer-Verlag Berlin Heidelberg 2004

The implementation of few existing methods [3,4,5] has revealed that some are accurate but expensive and others are inexpensive but less accurate. Hence there is a need for new inexpensive and accurate methods.

In section 2, we describe the proposed methodology. Experimentation and Comparative Study are given in section 3. Conclusion is given in section 4.

2   Proposed Methodology

We present the boundary growing and liner regression analysis based method for skew detection for binary document images. This approach is proposed to extract relevant coordinates of pixels in the text line. These coordinates are subjected to linear regression formula given in Equation (1) to estimate skew angle for each text line. This procedure is repeated for two or three text lines to confirm the skew angle.

2.1   Boundary Growing Approach (BGA)

Method starts with labeling the connected components in the document. The connected components are labeled by scanning document column-wise as shown in Fig. 1 and 2. Identify the black pixel of a character of a line in the document. The boundary for a character of text line is fixed and the centroid of bounding box of a character is computed. The method allows the boundary to grow until it reaches next pixel of the nearest neighboring component. This procedure is continued till the text line ends. Threshold for deciding end of the text line is based on the space between the characters and words. The method calculates height of each rectangle of the characters in the text line. The average height is determined as sum of heights of all rectangles of connected components divided by number of connected components. The characters with larger heights are reduced to average height. Some characters whose height is less than average height are ignored. However, the average height depends on the number of connected components present in the text line. This procedure is repeated for all the text lines present in the document. This is depicted in Fig. 3. This method brings all the connected components to one particular size resulting in improvement in accuracy. The smaller sized characters such as dots commas, hyphens do not participate in the process. This method uses only two to three text lines to estimate skew angle of whole document causing reduction in processing time. The method works for all types of documents containing different languages, different fonts and etc. Hence, this method is accurate and inexpensive to estimate skew angle.

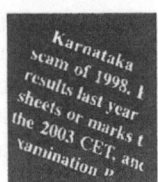   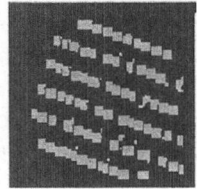   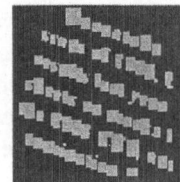

Fig. 1. Skewed image   **Fig. 2.** Labeled components   **Fig. 3.** Uniform sized components

```
Algorithm: Boundary Growing Approach Begins (BGA)
Input: Skewed document
Output: Skew angle of skewed document
Method:
```

Step1: Identify the black pixel of character of any text line in the document by scanning column wise.

Step2: Fix the boundary for the character using (X_{min} and Y_{min},) and (X_{max}, Y_{max}) coordinates of characters.

Step3: Compute centroid for rectangle using X and Y coordinates of rectangle

$$c[x] = (X_{min} + X_{max}) / 2 \text{ and } c[y] = (Y_{min} + Y_{max}) / 2,$$

where C[x] and C[y] are the coordinates of centroids.

Step4: Draw one more rectangle by incrementing one unit from centroid (growing a boundary)

Step5: Allow this boundary to grow until it reaches next pixel of neighbor character.

Step6: Calculate height of the character using the X and Y coordinates of text line.

Step7: Define average height of the character in the text line.

Step 8: Reduce the size of the characters whose height is greater than the average height.

Step9: Ignore characters whose height is smaller than the average height.

Step10: Extract lowermost, uppermost coordinates and coordinates of centroids of characters of text line.

Step11: Use these coordinates to determine skew angle for the text line by applying LRA.

Step12: Repeat this procedure for two or three text lines to get average of skew angle.

Method ends

2.2 Linear Regression Analysis for Skew Estimation (BGA-LRA)

In order to estimate skew angle we have used the BGA to extract the required coordinates. The BGA gives only uppermost coordinates of characters of individual text lines to estimate skew angle. This method is called BGA-LRA-UP. BGA-LRA-CC is another method to estimate skew angle by extracting centroids of characters of text lines. BGA-LRA-LP is one more method to estimate by extracting lowermost coordinates of pixels of characters of text lines. Skew angle is estimated using Equation (1) and Equation (2).

$$B = \frac{n\sum_{i=1}^{n} x_i y_i - \left(\sum_{i=1}^{n} x_i\right)\left(\sum_{i=1}^{n} y_i\right)}{n\sum x_i^2 - \left(\sum_{i=1}^{n} x_i\right)^2} \tag{1}$$

$$T\theta = \tan^{-1}(B) \tag{2}$$

Where B is the slope of the skew used in estimating actual degree of the skew ($T\theta$). The three methods BGA-LRA-UP, BGA-LRA-CC and BGA-LRA-LP are compared with respect to Means of skew angles (M), Standard Deviation (SD) and Mean Time (MT) for estimating skew angles for 20 documents to select an efficient method among three.

From Table 1, it is observed that, BGA-LRA-CC gives better accuracy with respect to M, SD and MT than BGA-LRA-LP and BGA-LRA-UP since these methods are sensitive to ascenders and descenders present in the text line. However, BGA-LRA-CC considers coordinates of centroids and does not much depend on the ascenders and descenders. Hence the BGA-LRA-CC is less sensitive to ascenders and descenders compared to BGA-LRA-LP and BGA-LRA-UP. The small variations are taken care by LRA. It is concluded that the BGA-LRA-CC exhibits better accuracy than BGA-LRA-UP and BGA-LRA-LP with respect to M, SD and MT and hence it is compared with the existing methods.

Table 1. Comparative study of BGA-LRA based methods

True Angle	BGA-LRA-UP			BGA-LRA-CC			BGA-LRA-LP		
	M	SD	MT	M	SD	MT	M	SD	MT
3	2.98	0.32	0.52	3.124	0.28	0.52	2.956	0.312	0.52
5	5.31	0.21	0.48	5.03	0.32	0.48	5.2	0.189	0.48
10	10.12	0.31	0.52	10.19	0.28	0.52	10.2	0.29	0.52
15	15.14	0.29	0.56	15.16	0.25	0.56	15.2	0.31	0.56
20	20.16	0.4	0.52	20.07	0.41	0.52	20	0.371	0.52
30	30	0.57	0.55	30.05	0.39	0.55	30	0.439	0.55
40	39.8	0.72	0.51	40	0.61	0.51	39.88	0.609	0.51
45	45	0.69	0.49	45.2	0.49	0.49	45.01	0.513	0.49

3   Experimentation and Comparative Study

We present experimental results to evaluate the performance of the proposed methods by comparing the results of existing methods. We have considered more than hundred document images from different books, magazines, journals, textbooks, newspaper cuttings, different fonts, different language documents, different resolution documents and noisy document images. The documents are tested by pre specified angles varying between 0 and 45^0 and are considered as true skew angles. The documents are subjected to all proposed and existing methods. Further, to establish the superiority of the proposed methods, we have considered the special cases comprising of different types of documents and different resolutions. In order to show that the proposed method is accurate and inexpensive, we have derived two more parameters such as average standard deviations (ref. Fig. 4) and average mean time (ref. Fig. 5). We implemented the methods on the system with configuration 128 MB RAM, 6 GB and 650 MHZ.

From Table 2, it is observed that the proposed method BGA-LRA-CC exhibits better accuracy compared to all other methods with respect to M. This is the advantage offered by BGA.

From Table 3, it is observed that, the BGA-LRA-CC gives consistent results compared to other methods with respect to SD.

From Table 4, it is observed that the proposed method takes less MT compared to other methods.

Table 2. Comparative study of proposed method with existing methods with respect to M

True Angle	HPP	VPP	BGA-LRA-CC	HTWI	HTULP	NNCM	HTLMP
3	7.3	3.15	3.124	3.2	3.128	3.54	3.15
5	8.04	5.28	5.03	5.59	5.68	4.8	5.35
10	12.19	11.2	10.19	12.88	10.17	8.05	10.12
15	15.9	14.76	15.16	15.72	15.72	15.82	15.18
20	22.6	19.14	20.07	18.9	20.11	20.78	20.17
30	32.7	27.5	30.06	30.02	30.32	27.8	30.5
40	40.2	36.2	40	40.9	40.42	43.19	40.21
45	45.9	46.16	45.2	45.3	45.2	45.95	45.15

HPP: Horizontal Projection Profile based Method
VPP: Vertical Projection Profile based Method
HTULP: Hough Transform for Upper and Lower Pixels of Characters
HTWI: Hough Transform for Whole Image
NNCM: Nearest Neighboring Clustering based Method
HTLMP: Hough Transform for Lowermost Pixels of Characters
BGA-LRA-CC: Proposed Method.

Table 3. Comparative study of proposed method with existing methods with respect to SD

True Angle	HPP	VPP	BGA-LRA-CC	HTWI	HTULP	NNCM	HTLMP
3	2.86	0.76	0.28	0.3	0.34	0.54	0.365
5	2.12	1.54	0.32	0.548	0.45	0.64	0.426
10	1.761	1.94	0.286	1.456	0.42	1.24	0.416
15	1.44	2.13	0.258	0.826	0.51	1.14	0.496
20	1.562	3.13	0.411	0.749	0.72	0.90	0.398
30	1.769	2.25	0.393	0.994	1.42	1.46	0.856
40	2.02	2.86	0.617	0.618	0.71	1.55	0.358
45	0.99	3.25	0.49	0.462	0.86	0.64	0.926

Table 4. Comparative study of proposed method with existing methods with respect to MT

True Angle	HPP	VPP	BGA-LRACC	HTWI	HTULP	NNCM	HTLMP
3	1.16	1.28	0.52	13.66	2.12	2.2	2.32
5	1.16	1.28	0.48	14.06	1.98	2.15	2.31
10	1.16	1.28	0.52	13.13	2.38	1.83	2.21
15	1.16	1.28	0.56	11.19	2.12	1.95	2.42
20	1.16	1.28	0.52	12.04	2.38	1.96	2.10
30	1.16	1.28	0.55	11.36	2.18	2.03	2.61
40	1.16	1.28	0.51	11.17	2.17	2.01	2.68
45	1.16	1.28	0.49	10.5	2.32	2.03	2.59

From the Fig. 4 and Fig. 5 it is evident that the proposed method BGA-LRA-CC takes less variation and time in estimating skew angles compared to the existing methods. With this we conclude that BGM-LRA-CC (C) is the best method.

Special Cases

We have presented experimental results on different document images with 10^0 skew. Different language documents, document images from journals, document images of different resolutions, synthetic images, images of newspaper cuttings, images from textbooks and sample of noisy images shown in Fig. 6 are considered for the experimentation.

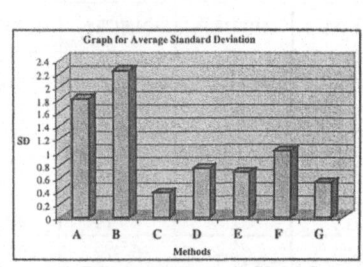
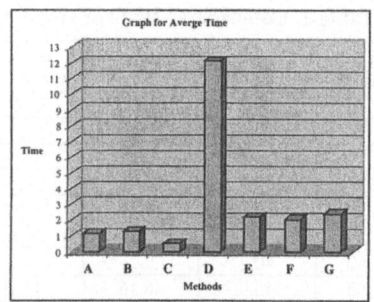

Note: A: HPP, B: VPP, C: BGM-LRA-CC, D: HTWI, E: HTULP, F: NNCM, G: HTLMP.

Fig. 4. Methods v/s average SD   **Fig. 5.** Methods v/s average MT

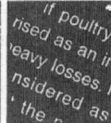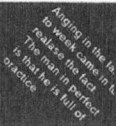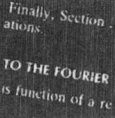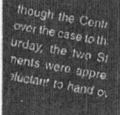

Fig. 6. Different samples considered for experimentation

From Table 5, it is noticed that, all the methods fail to handle the text with picture documents including BGA-LRA-CC. For all other cases, the methods give accuracy including BGA-LRA-CC. For different languages: NNCM, HPP, VPP methods fail to give accuracy. Projection profile based methods work based on the space between the text lines. NNCM also fail since it requires connected components. For synthetic images: all methods give accuracy. For noise image: NNCM fails since it labels noise also as connected components. For text books images: NNCM, HPP, VPP, methods fail to give results. For journal images: NNCM, HPP, VPP methods fail to give results.

Table 5. Comparative study for different samples of 10^0 skewed documents

Cases	HPP	VPP	BGA-LRA-CC	HTWI	HTULP	NNCM	HTLMP
Different Language	39.1	43	9.935	11	11.5	24	11.5
Synthetic	45	12	9.526	10	10	11	10.5
Noise	27	31	1.06	11	9.5	0	9.5
Text Book	29	25	10.69	10.5	10.5	12.1	11
Journal	3.58	2.7	9.936	12	11	35	10.5
Text with Picture	71	81	26	45	38	0	1.7

From Table 6, it is observed that the NNCM, HPP and VPP fail to give results. However, all other methods work well with all resolutions.

Table 6. Comparative study for different dpi of 23^0 skewed documents

dpi	HPP	VPP	BGA-LRA-CC	HTWI	HTULP	NNCM	GVLC	HTLMP
75	22.1	23	22.9	18.5	23.5	90	25	23
100	46	45	22.3	17.5	23.5	21	23.5	22.5
150	29	38	22.8	4.5	23	21	24.9	23
300	41	44	22.63	22.5	23	19	22	23
400	75	90	22.77	48	23.5	0	21.9	22.5

4 Conclusions

We have implemented some existing well-known methods in order to bring out the efficacy of our proposed methods. We have considered projection profile, nearest neighbor clustering, Hough transform based methods for considering different samples of data sets. We have shown that our proposed method based on BGA gives better results than other methods with respect to accuracy and time and computations. However, all the methods including existing methods fail to give results for text with picture document images. Extension of BGA with the help of fuzzy technique for estimating efficiently skew angle for the complex document images such as document containing pictures, graphs, tables and etc would be our future work.

Acknowledgment

The authors acknowledge the support extended by Prof. Basavaraj S Anami, Mr. Manjunath Aradya V and Mr. Noushath S.

References

1. Shivakumara P et al, Statistical Methodology for Skew Detection in Binary Text Document Images for Document Image Mosaicing. Journal of the Society of Statistics, Computer and Applications, Vol.1, Nos 1 and 2, 2003 (New series), PP 81-90.
2. Shivakumara. P et al, Text-Skew Detection Through Contour Following in Document Image, Proceedings of National Workshop on Computer Vision, Graphics and Image Processing –WVGIP- 2002, 15th and 16th of February, 2002, pp 39-44.
3. Baird H.S, The Skew Angle of Printed Documents, Proceedings of Conference Society of Photographic Scientists and Engineers, 1987, pp 14-21.
4. Srihari S. N and Govindaraju V, Analysis of textual images using the Hough Transform, Machine Vision and Applications, Vol. 2, 1989, 141-153.
5. Hashizume A et al, A Method of Detecting the Orientation of Aligned Components, Pattern Recognition Letters, Vol. 4, April 1986, pp 125-132.
6. Yue Lu and Chew Lim Tan, A nearest-neighbor chain based approach to skew estimation in document images, Pattern Recognition Letters 24, 2003, pp 2315-2323.
7. Pal U and Chaudhuri B. B, An Improved document skew angle estimation technique, Pattern Recognition Letters 17, 1996, pp 899-904.
8. Amin A and Fischer S, A Document Skew Detection Method using the Hough Transform, Pattern Analysis and Applications, 2000, pp 242-253.
9. Kwag H.K et al, Efficient skew estimation and correction algorithm for document images, Image and Vision Computing 20, 2002, pp 25-35.
10. Gatos B et al, Skew detection and text line position determination in digitized documents, Pattern Recognition, Vol. 30, No. 9, 1997, pp 1505-1519.
11. Liolios et al, On the generalization of the form identification and skew detection problem, Pattern Recognition 35, 2002, pp 253-264.
12. Yan, H. Skew correction of document images using interline cross-correlation, Computer Vision, Graphics, and Image Processing 55, 1993, pp 538-543
13. Postl. W Detection of liner oblique structure and skew scan in digitized documents. In Proc. of International Conference on Pattern Recognition, 1986, pp 687-689.

Segmenting Moving Objects
with a Recurrent Stochastic Neural Network

Jieyu Zhao

The CST Research, Ningbo University, Ningbo, 315211, China
zhao_jieyu@hotmail.com

Abstract. Moving object segmentation with a neural network is a challenging task. In this paper, we have proposed an effective optimization scheme for dynamic image segmentation and computer vision using a large scale recurrent stochastic binary neural network. The model is formally defined and can be described as a Markov Random Field. Related theoretical results and Bayesian interpretation are presented. Experimental results on real world video images illustrate the effectiveness of the approach.

1   Introduction

Segmenting interesting objects from a complex background is a challenging task in image processing and computer vision. Numerous attempts have been made in recent years to solve this hard problem [2]. Probabilistic approaches, which allow us to handle uncertainties in a systematic way, are widely adopted [4,5]. Among those successful ones are Bayesian methods [7], Markov random fields [8], and stochastic graphical models [10,14].

Most research on image segmentation using probabilistic approaches only deals with still images. Work on video image segmentation is less common than still-based image due to the poor video quality, constantly changing background and significant illumination variation. The time-consuming computational process is also a major obstacle for the application of stochastic models.

For video object segmentation, a significant clue is provided by the motion of the object. Humans can easily discern an independently moving object without knowing its identity. The computational mechanism for this remarkable brain function still remains unclear. Recent research on visual perception suggests a "biased competition hypothesis" [10].

In this paper we propose a feasible framework for video object segmentation using a large scale recurrent stochastic binary neural network. The partitioning of the image is implemented through an unsupervised optimization process. The recurrent network has a potential to outperform finite memory model and the clusters in the network can reflect meaningful information processing states [12]. The detailed introduction of the stochastic mode and related Bayesian interpretation are presented. Experimental results show the successful object segmentation on real world video images with a complex background.

2   Stochastic Models

Stochastic computing model is expected to provide a powerful technology for image processing and pattern recognition. It is very promising for global optimization since the model is able to avoid the system being trapped in local minima.

N.R. Pal et al. (Eds.): ICONIP 2004, LNCS 3316, pp. 666–672, 2004.
© Springer-Verlag Berlin Heidelberg 2004

The stochastic model we adopted here is a recurrent stochastic binary network. It is a parallel distributed information processing structure in the form of a directed graph where the nodes of the graph are stochastic binary unit. We formally define the model as follows:

Definition 1. A *recurrent stochastic binary network B(V,W,U)* is a pseudo-graph with vertex set V having state $S \in \{-1,+1\}^n$ and edge set W of real value defined on a neighborhood structure N. U is a dynamic updating mechanism which randomly selects the vertices (or units) and changes their states with updating rule $S_i = F(\sum_{j \in N} w_{ij}S_j)$, where F is a random activation function.

In this paper we only use a sequential and symmetric model in which the state of the unit is updated sequentially, and all the connecting edge of are symmetric, that is, $w_{ij}=w_{ji}$ for all unit i and j.

The random activation function F of the unit can be chosen to have different forms depending on the application. Two commonly used random functions are the stochastic cumulative Gaussian function and the stochastic logistic function.

Definition 2. A k-input *stochastic Gaussian unit* is a binary state unit which takes value 1 with probability $p(W,V,\theta)$ or value -1 with probability $1-p(W,V,\theta)$, $p(W,V,\theta)$ is a cumulative Gaussian distribution function:

$$p(W,V,\theta) = \int_{\theta}^{\infty} \frac{1}{\sqrt{2\pi}\sigma} e^{-\frac{(x-\mu)^2}{2\sigma^2}} dx \tag{1}$$

where $W = (w_1,w_2,...,w_k)$ is the weight vector of the unit, $V = (v_1,v_2,...,v_k)$ is the corresponding input vector, $\mu = \sum_{i=1}^{k} w_i v_i$ is the sum of the weighted inputs into the unit, θ is the threshold and σ a control parameter for the randomness.

According to the law of large numbers the distribution of the sum of the noisy inputs into a stochastic unit will obey a Gaussian distribution if the number of the inputs is large. In this case the activation function can be described as a cumulative Gaussian distribution function. Thus the network of stochastic Gaussian unit is a very generic model of stochastic computation.

Another very useful random activation function is the stochastic logistic function.

Definition 3. A k-input *stochastic logistic unit* is a binary state unit which takes value 1 with probability $p(h)$ or value -1 with probability $1- p(h)$, where

$$p(h) = \frac{1}{1+e^{-h/T}}, \qquad h = (\sum_{i=1}^{k} w_i v_i) - \theta. \tag{2}$$

T is the "temperature" which controls the randomness within the network, and θ is the threshold.

A recurrent network of stochastic logistic units is actually a partially connected Boltzmann machine [1]. The only difference between these two models is that the recurrent stochastic binary network has a predefined neighborhood structure N while the Boltzmann machine is fully connected.

If the number of the input k is large and there are no extremely large weight values, the cumulative Gaussian function and the logistic function become very similar. We can practically use these two functions alternatively if we set $\sigma = 1.699T$ [13].

The structure of a stochastic binary network can be feed-forward or recurrent. The computation in a feed-forward network is straightforward. The input is propagated through the network and produces the output right away. In contrast, a recurrent network always needs some kind of relaxation to reach stable 'equilibrium states' that is then read out as the final output of the network. In this paper we only discuss the application of a recurrent structure stochastic binary network.

There is a very strong relationship between the recurrent stochastic binary networks and the Markov Random Fields (MRF). A family of random variables F on set S is defined to be an MRF if and only if the following two conditions are satisfied: (a) $P(f) > 0$, for all f ; (b) $P(f_i | f_{S-\{i\}}) = P(f_i | f_{N_i})$ where N_i is the set of neighbors of vertex i. The condition (b) states that the assignment of a state to a vertex is conditionally dependent on the assignment to other vertices only through its neighbors. This is a totally local property. The global behavior of an MRF is described as a Gibbs distribution, which takes the form:

$$P(f) = \frac{e^{-E(f)/T}}{\sum_{f \in F} e^{-E(f)/T}} \tag{3}$$

where $E(f)$ is the energy function and T is the temperature. The energy is a sum of clique potentials over all possible cliques C: $E(f) = \sum_{c \in C} Vc(f)$. For a recurrent stochastic binary network, the energy function is simplified as $E = -\frac{1}{2}\sum_{i,j} w_{ij} S_i S_j$.

The Hammersley-Clifford theorem [3] establishes the equivalence of the Markov random field and Gibbs random field (GRF). The theorem states that F is an MRF on S with respect to N if and only if F is a GRF on S with respect to N.

It is well known that a sequential Boltzmann machine is an MRF [1]. The recurrent network of stochastic Gaussian unit can also be treated as an MRF.

Theorem 1. *The stationary distribution of a sequential recurrent network of stochastic Gaussian unit uniquely exists and the stationary distribution is approximately a Gibbs distribution if the connections are symmetric and the size of the network is not too small.*

The proof can be achieved by using the theory of Markov chains with some approximation [13].

With the above theorem we are able to describe quantitatively both the local property (the Markovianity) and the global property (the Gibbs distribution) of the stochastic model.

3   Bayesian Interpretation

Using the Bayesian framework, the task of stochastic computation is to find the optimal estimator which minimizes the following risk

$$R(f^*) = \int_{f \in F} C(f^*, f) P(f | d) df \tag{4}$$

where $C(f^*, f)$ is the cost function of estimator f when the truth is f^*, d is the observation and $P(f | d)$ is the posterior distribution. According to the Bayesian rule,

$P(f \mid d) = p(d \mid f)P(f)/p(d)$ where $P(f)$ is the prior probability of f, $p(d \mid f)$ is the conditional pdf of the observation d, and $p(d)$ is the density of d. If we use the following cost function

$$C(f^*,f) = \begin{cases} 0 & \text{if } |f^*-f| \leq \delta \\ 1 & \text{otherwise} \end{cases} \tag{5}$$

where $\delta > 0$ is a small constant, the Bayes risk is

$$R(f^*) = \int_{f:|f^*-f|>\delta} P(f \mid d)df = 1 - \int_{f:|f^*-f|\leq\delta} P(f \mid d)df. \tag{6}$$

The risk can be approximated by $R(f^*) = 1 - kP(f \mid d)$ where k is the volume of space for which $|f^*-f| \leq \delta$. Thus minimizing the Bayes risk is equivalent to maximizing the posterior probability $f^* = \arg \max_{f \in F} P(f \mid d)$, For a fixed d, $p(d)$ is a constant.

Thus $P(f \mid d) \propto P(d \mid f)P(f)$.

Therefore, the optimal estimator is

$$f^* = \arg \max_{f \in F} \{p(d \mid f)P(f)\}. \tag{7}$$

If the prior distribution $P(f)$ is flat, maximizing the posterior probability is simplified to the maximizing of the likelihood.

The maximization of the posterior probability in the Bayesian framework is equivalent to minimization of the posterior energy function of a MRF, that is, the minimization of the energy in a stochastic binary network [4].

To successfully apply the above *maximum a posterior* (MAP) approach in image processing, one needs to define a neighborhood structure and derive the posterior energy function to define the MAP solution to an optimization problem, or map the problem onto a recurrent stochastic network by an appropriate setting of weights such that solution to the problem is corresponding to the minimization of the energy function.

4   Experimental Setup

We evaluate the effectiveness of the optimization approach using a recurrent stochastic network on real outdoor video images with a complex and noisy background. The first video clip was taken at a busy traffic site where many cars were moving in both directions (Fig. 1). The second clip was the ocean front scene (Fig. 4). It can be seen the background of the scenes contains many non static components such as swaying trees and bushes, and dynamic ocean waves. The original video clip was in DV format (720x576). We reduced its resolution to 360x288 for a less noise level. Our target is to segmenting the moving cars and pedestrians from the complex background.

A widely used technique for moving object segmentation is the background subtraction [9,11]. Many background subtraction methods work by comparing color or intensities of pixels in the video frames to a reference background image. Significant differences in intensity from the reference image are attributed to motion of objects. We use a simple method to constantly estimate the mean of the dynamic background. For the current input X_t, the mean is calculated by $B_t = (1-\alpha)B_{t-1} + \alpha X_t$. The difference between the current frame and the mean of the background is shown in Fig. 2 (left) and Fig. 4 (right).

Many motion analysis approaches make use of optical flow of video images. Optical flow is the velocity field in the image plane caused by the motion of the observer and objects in the scene. It contains important information about the directions of the moving objects and how fast they are moving. However, the computation of optical flow is based on the assumption of constant intensity and smoothness, which is valid only in some limited circumstance. In the case of real world video images taken from a noisy environment, the optical flow calculated with the Horn-Schunck's method [6] is far from ideal (Fig. 2 and Fig. 5).

Our approach is to setup an optimization scheme which maps the moving object segmentation problem onto a recurrent stochastic binary network. The key issue is the proper design of the energy function such that the global minima of the energy function corresponds to our desired solution to the moving object segmentation. The energy function should contain all necessary information of the moving cars and the background. Indeed, we do not know exactly which form of the energy function corresponds to the *optimal* solution of the segmentation problem. The validation of the stochastic model needs to be assessed by the outcome of the segmentation results.

The recurrent stochastic binary network we used has a simple structure of 360x288 units, arranged on a two dimensional plane and each unit is connected to its neighboring 120 (11x11-1) units (except to itself). The energy function of the network is $E = -\frac{1}{2}\sum_{i,j} w_{ij}S_iS_j$, and the weight values are set as follows:

$$w_{ij} = \mathrm{sgn}\big((tmpr(i)-t)\times(tmpr(j)-t)\big)\times\left\{\alpha[opt(i)+opt(j)]+\beta[|\,tmpr(i)\,|+|\,tmpr(j)\,|]+\gamma\frac{itn(i)+itn(j)}{1+|\,itn(i)-itn(j)\,|}\right\} \quad (8)$$

where $itn()$ is the intensity, $opt()$ is the optical flow and $tmpr()$ is the temporal difference between the current input and the mean, α,β,γ are parameters to control the balance among corresponding terms, and t is a threshold value.

The simulated annealing process is controlled by the temperature in the stochastic network. By gradually lowering the temperature from a high initial value to a low point with a certain cooling schedule, the energy function in the network will hopefully reach a global minimum. We use a simplified cooling schedule $T_n = T_0/(n+1)$ in the experiment. Mean field annealing method is faster but it may easily converge to a local minimum. The whole process takes about 30 seconds on a Pentium 4 Mobile 2G laptop. The segmentation result is shown in Fig. 3 (right) and Fig. 5 (right).

Fig. 1. Two consecutive image frames from the first clip, note the leftmost car is moving slowly and waiting to turn left.

Fig. 2. The difference between the current frame and the mean of background (left) and the optical flow calculated with the two consecutive frames (right)

Fig. 3. The output of the stochastic model at an early stage of the simulated annealing process (left) and the final segmentation result (right)

 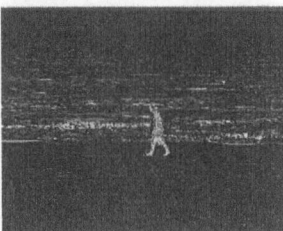

Fig. 4. The ocean front clip (left) and the difference between the input and the mean (right)

Fig. 5. The optical flow of two consecutive image frames (left) and the final segmentation results (right) (we use the white color for the foreground here)

5   Conclusions

We have proposed a recurrent stochastic network model for object segmentation. Experimental results on real world video images have illustrated the effectiveness of the model. The computation of this model is completely local and biologically plausible. Future work will be focused on the acceleration of the algorithm and the establishment of a unified mechanism for both supervised and unsupervised learning.

Acknowledgements

This research was supported by the National Natural Science Foundation of China (NSFC60273094), the Natural Science Foundation of Zhejiang Province and the Ministry of Education.

References

1. Ackley, D.H., Hinton, G.E., Sejnowski, T.J.: A learning algorithm for Boltzmann machines. *Cognitive Science*, Vol. 9 (1985) 147-169
2. Aggarwal, J.K., Cai, Q.: *Human Motion Analysis: A Review,* Computer Vision and Image Understanding, Vol. 73 (1999) 428-440
3. Dougherty, E. R.: *Random Process for Image and Signal Processing*, IEEE Press, New York (1999)
4. Geman, S., Geman, D.: Stochastic relaxation, Gibbs distributions, and the Bayesian restoration of images. *IEEE Transactions on Pattern Analysis and Machine Intelligence*, Vol. **6** (1984) 721–741
5. Hammersley, J.M., Clifford, P.: Markov field on finite graphs and lattices, unpublished (1971)
6. Horn, B.K.P., Schunck, B.G.: *Determining optical flow*, Artificial Intelligence Vol. 17 (1981) 185-203
7. Lee, T.S., Mumford, D.: Hierarchical Bayesian inference in the visual cortex, *Journal of Optical Society of America, A.* Vol. 20 (2003) 1434-1448
8. Li, S.Z.: *Markov Random Field Modeling in Image Analysis*, Springer-Verlag, Tokyo (2001)
9. Monnet, A., Mittal, A., Paragios, N., Ramesh, V.: Background Modeling and Subtraction of Dynamic Scenes, IEEE International Conference on Computer Vision (2003) 1305- 1312
10. Rolls, E.T., Deco, G.: *Computational Neuroscience of Vision*, Oxford University Press, New York (2002)
11. Seki, M., Wada, T., Fujiwara, H., Sumi, K.:Background Subtraction based on Cooccurrence of Image Variations, Proceedings of the IEEE Computer Society Conference on Computer Vision and Pattern Recognition, Vol. II (2003) 65-72
12. Tino, P., Cernansky, M., Benuskova, L.: Markovian Architectural Bias of Recurrent Neural Networks, IEEE Transactions on Neural Networks, Vol. 15, No. 1 (2004) 6-15
13. Zhao, J.: A Recurrent Stochastic Binary Network, *Science in China*, Ser. F, Vol. 44, No. 5, (2001) 376-388
14. Zhu, S. C.: Statistical Modeling and Conceptualization of Visual Patterns, *IEEE Trans. on Pattern Analysis and Machine Intelligence*, Vol. 25, No. 6 (2003) 691-712

Real-Time Gaze Detection via Neural Network

Kang Ryoung Park

Division of Media Technology, SangMyung University, Seoul, Republic of Korea
parkgr@smu.ac.kr

Abstract. Human gaze can provide important information in man machine interaction. To overcome the disadvantages of previous gaze detecting researches, we propose a new method with a wide and an auto
panning/tilting/focusing narrow view camera. In order to enhance the
performance of detecting facial features, we use a SVM (Support Vector Machine) and the eye gaze position on a monitor is computed by a
multi-layered perceptron.

1 Introduction

Gaze detection system is important in many applications such as computer interface for the handicapped. The gaze detection researches can be classified into
4 categories. First one is that focused on 2D/3D head motion estimation [2][11].
Second one is that for the facial gaze detection [3-9][12][13][15] and the third
one is the eye gaze detection [10][14]. And last one is that considering both head
and eye movement has been researched. Ohmura and Ballard et al.[5][6]'s methods and Rikert et al.[9]'s method has the constraints that the user's Z distance
should be measured manually and take much time to compute the gaze position.
The researches of [3][4][16] show the facial gaze detection methods and have the
limits that the gaze errors are increased in case that the eye movements happen. To overcome such problems, the research of [17] shows the facial and eye
gaze detection, but uses only one wide view camera. In such case, the eye image
resolution is too low and the fine movements of user's eye cannot be exactly detected. Wang et al.[1]'s method provides the advanced approaches that combines
head pose and eye gaze estimation by a wide view camera and a panning/tilting
narrow camera (showing small angular gaze error of below 1 degree). However,
in order to compute the gaze position, their method supposes that they know
the 3D distance between two eyes, that between both lip corners and the 3D
diameter of eye ball. Also, they suppose there is no individual variation for the
3D distances and diameter. However, our preliminary experiments show that
there are much individual variations for the 3D distances/3D diameter and such
cases can increase much gaze errors (the angular error of more than 5 degree).
To overcome above problems, we propose the new method for detecting gaze position. In order to enhance the performance of detecting facial features, we use
a SVM and the eye gaze position on a monitor is computed by a multi-layered
perceptron.

N.R. Pal et al. (Eds.): ICONIP 2004, LNCS 3316, pp. 673–678, 2004.
© Springer-Verlag Berlin Heidelberg 2004

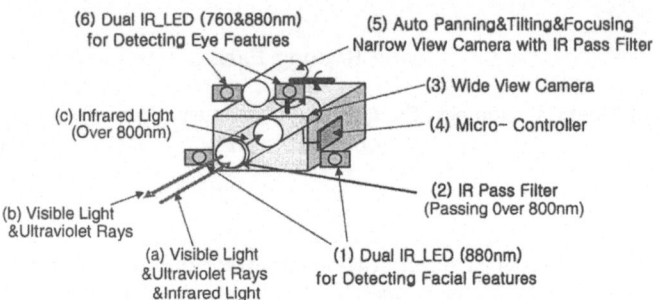

Fig. 1. The gaze detecting system

2   Localization of Facial Features in Wide View Image

In order to detect gaze position on a monitor, we first locate facial features in wide view images. To detect facial features robustly, we implement a gaze detection system as shown in Fig. 1. The IR-LED(1) is used to make the specular reflections on eyes. When a user starts our gaze detection system, the micro-controller(4) turns on the illuminator(1) synchronized with the even field of CCD signal and turns off it synchronized with the next odd field of CCD signal, successively [17]. From that, we can get a difference image between the even and the odd image and the specular reflection points on both eyes can be easily detected because their image gray level are higher than other regions [17]. In addition, we use the Red-Eye effect and the method of changing Frame Grabber decoder value in order to detect more accurate eye position [17]. In general, the NTSC signal from camera has high resolution $(0 \sim 2^{10} - 1)$, but the range of A/D conversion by conventional decoder of the Frame Grabber is low resolution $(0 \sim 2^8 - 1)$. So, the NTSC signal in high saturated range such as the corneal specular reflection on eye and the some reflection region on facial skin can be represented as same image grey level $(2^8 - 1)$, which makes it difficult to discriminate the corneal specular reflection. However, the NTSC signal level of corneal specular reflection is higher than that of other reflection due to the reflectance rate. So, if we make the decoder brightness value lower, there is no high saturated range and the corneal specular reflection and the other reflection can be discriminated, easily. Around the detected corneal specular reflection points, we determine the eye candidate region of 30*30 pixels and locate the accurate eye (iris) center by the circular edge detection method. After that, we detect the eye corner by using eye corner shape template and SVM (Support Vector Machine) [17]. SVMs have been recently proposed as a new technique for solving pattern recognition problems. SVMs perform pattern recognition between two point classes by finding a decision surface determined by certain points of the training set, termed as Support Vectors (SV) and SVs are regarded as data which are difficult to be classified among training. At the same time, the decision surface found tends to have the maximum distance between two classes. In general, it is reported that its classification performance is superior to that of MLP (Multi-Layered Perceptron). Especially, when plenty of positive and negative data are not obtained and input data is much noisy, the MLP cannot show the reliable classification results. In addition, MLP requires many initial parameter settings and it usually is performed by user heuristic experience. In this paper, we use a polynomial kernels of degree 5 for SVM in order to solve non-linearly separable problem. That is why the dimension of input data is big, so we use the polynomials of high degree. In this case, the problem is defined as 2 class problem. The first class shows the genuine eye corner and the second one does the imposter eye corner. It is reported that the other inner products such as RBF, MLP, Splines and B-Splines do not affect the generation of support vector. Our experimental results comparing the polynomial kernel to MLP for SVM kernel show the same results. The C factor affects the generalization of SVM and we use 10,000 as C factor, which is selected by experimental results. We get 2000 successive image frames for SVM training and additional 1000 images are used for testing. Experimental results show the classification error for training data is 0.11% and that for testing data is 0.2%. The classification time of SVM is 8 ms in Pentium-III 866MHz. After locating eye centers and eye corners, the positions of nostrils can be detected by anthropometric constraints in a face and SVM. In order to reduce the effect by the facial expression change, we do not use the lip corners. Experimental results show that RMS error between the detected feature positions and the actual positions are 1 pixel (of both eye centers), 2 pixels (of both eye corners) and 4 pixels (of both nostrils) in 640×480 pixels image. From those, we use 5 feature points (left/right eye corners of left eye, left/right eye corners of right eye, nostril center) in order to detect facial gaze position.

3 4 Steps for Computing Facial Gaze Position

After feature detection, we take 4 steps in order to compute a gaze position [3][4][17]. At the 1st step, when a user gazes at 5 known positions on a monitor, the 3D positions (X, Y, Z) of initial 5 feature points (detected in the section 2) are computed automatically including Z distance between the user and the monitor [3][4]. At the 2nd step and 3rd step, when the user rotates/translates his head in order to gaze at one position on a monitor, the new (changed) 3D positions of those 5 features can be computed by 3D motion estimation. At the 4th step, one facial plane is determined from the new (changed) 3D positions of the 5 features and the normal vector (whose origin exists in the middle of the forehead) of the plane shows a gaze vector by head (facial) movements. The gaze position on a monitor is the intersection position between a monitor and the gaze vector [3][4][17].

4 Auto Panning/Tilting/Focusing of Narrow View Camera

Based on the new (changed) 3D positions of the 5 feature points (which are computed at the 2nd and 3rd step as mentioned in section 3), we can pan and tilt the narrow view camera in order to capture the eye image. For that, we also perform the coordinate conversion between monitor and narrow view camera [3]. Conventional narrow view camera has small DOF and there is the limitation of increasing the DOF with the fixed focal camera. So, we use the auto focusing narrow view camera. For auto focusing, the Z distance between the eye and the camera is required and we can obtain the Z distance at the 2nd and 3rd step (as mentioned in section 3). In the case that the surface of glasses can make the specular reflection which covers the whole eye image, the eye region is not detected and we cannot compute the eye gaze position. So, we turn on the both sided illuminator alternately like Fig. 1(6).

5   Localization of Eye Features in Narrow View Image

After we get the focused eye image, we perform the localization of eye features. J. Wang et al. [1] uses the method that detects the iris outer boundary by elliptical fitting. However, the upper and lower region of iris outer boundary tend to be covered by eyelid and inaccurate iris elliptical fitting happens due to the lack of iris boundary pixels. In addition, their method computes eye gaze position by checking the shape change of iris when a user gazes at monitor positions. However, our experimental results show that the shape change amount of iris is very small and it is difficult to detect the accurate eye gaze position with that. So, we use the positional information of both pupil and iris. Also, we use the information of shape change of pupil, which does not tend to be covered by eyelid. In general, the IR-LED of short wavelength (700nm $\sim$ 800nm) makes the high contrast between iris and sclera and that of long wavelength (800nm $\sim$ 900nm) makes the high contrast between pupil and iris. Based on that, we use the IR-LED illuminator of multi-wavelength (760nm and 880nm) as shown in Fig. 1(6). The shapes of iris and pupil are almost ellipse, when the user gazes at a side position of monitor and we use the canny edge operator to extract edge components and a 2D edge-based elliptical Hough transform. In order to detect the eye corner position, we detect the eyelid and we use the region-based eyelid template deformation and masking method. Here, we use 2 deformable templates (parabolic shape) for upper and lower eyelid detection, respectively. Experimental results show that RMS errors between the detected eye feature positions and the actual ones are 2 pixels (of iris center), 1 pixel (of pupil center), 4 pixels (of left eye corner) and 4 pixels (of right eye corner). Based on the detected eye features, we select the 22 feature values and we can compute eye gaze position on a monitor.

6   Detecting the Gaze Position on a Monitor

In section 3, we explain the gaze detection method only considering head movement. However, when a user gazes at a monitor position, both the head and eyes tend to be moved simultaneously. So, we compute the additional eye gaze position with the detected 22 feature values (as mentioned in section 5) and a neural network (multi-layered perceptron). The numbers of input nodes, hidden nodes and output nodes are 11, 8 and 2, respectively. For output function of neural network, we use a limited logarithm function, which shows better performance than that in case of using other functions like a linear, sigmoid etc. That is because the narrow view camera is above the wide view camera as shown in Fig. 1 and the 2D eye movement resolution is decreased in case of gazing at the lower positions of the monitor. The continuous output values of neural network represent eye gaze position on a monitor. Here, the input values for neural network are normalized by the distance between the iris/pupil center and the eye corner, which are obtained in case of gazing at monitor center at the 1st step of section 3. That is because we do not use a zoom camera and the distance is changed according to the user's Z position. After detecting eye gaze position, we can determine a final gaze position based on the vector summation of facial and eye gaze position [17].

7 Performance Evaluations

The gaze detection error of our method is compared to that of our previous methods [3][4][15][17]. The test data are acquired when 95 users gaze at 23 gaze positions on a 19" monitor. Here, the gaze error is the RMS error between the actual gaze positions and the computed ones. In the 1st experiment, the gaze errors are calculated in two cases. The case I shows the gaze error about test data including only head movements. The RMS error of each method is following(unit is cm); that of Linear Interpolation [15] is 5.1, that of Single Neural Net [15] is 4.23, that of Combined Neural Nets [15] is 4.48, that of [3] method is 5.35, that of [4] method is 5.21, that of [17] method is 3.40 and that of proposed method is 2.24. The case II shows the gaze error about test data including head and eye movements. The RMS error of each method is following(unit is cm); that of Linear Interpolation [15] is 11.8, that of Single Neural Net [15] is 11.32, that of Combined Neural Nets [15] is 8.87, that of [3] method is 7.45, that of [4] method is 6.29, that of [17] method is 4.8 and that of proposed method is 2.89. At the 2nd experiment, the points of radius 5 pixels are spaced vertically and horizontally at 1.5" intervals on a 19" monitor with monitor resolution of 1280×1024 pixels as such Rikert's research [9]. The RMS error between the real and calculated gaze position is 2.85 cm and it is superior to Rikert's method (almost 5.08 cm). Our gaze error is correspondent to the angular error of 2.29 degrees on X axis and 2.31 degrees on Y axis. In addition, we tested the gaze errors according to user's Z distance. The RMS errors are 2.81cm at 55cm, 2.85cm at 60cm, 2.92cm at 65cm and the performance of our method is not affected by the user's Z position change. Last experiment for processing time shows that our gaze detection process takes about 100ms in Pentium-III 866MHz and it is much smaller than Rikert's method (1 minute in alphastation 333MHz).

8 Conclusions

This paper describes a new gaze detecting method. In future works, we will use the method of capturing higher resolution eye image with zoom lens and it will increase the accuracy of gaze detection. In addition, the method to increase the auto panning/tilting/focusing speed should be researched to decrease total processing time.

References

1. J. Wang and E. Sung, 2002. Study on Eye Gaze Estimation, IEEE Trans. on SMC, Vol.32, No.3, pp.332-350
2. A. Azarbayejani., 1993, Visually Controlled Graphics. IEEE Trans. PAMI, Vol.15, No.6, pp.602-605
3. K. R. Park et al., Apr 2000, Gaze Point Detection by Computing the 3D Positions and 3D Motions of Face, IEICE Trans. Inf.&Syst.,Vol.E.83-D, No.4, pp.884-894
4. K. R. Park, Oct 1999, Gaze Detection by Estimating the Depth and 3D Motions of Facial Features in Monocular Images, IEICE Trans. Fund., Vol.E.82-A, No.10, pp.2274-2284
5. K. OHMURA et al., 1989. Pointing Operation Using Detection of Face Direction from a Single View. IEICE Trans. Inf.&Syst., Vol.J72-D-II, No.9, pp.1441-1447
6. P. Ballard et al., 1995. Controlling a Computer via Facial Aspect. IEEE Trans. on SMC, Vol.25, No.4, pp.669-677
7. A. Gee et al., 1996. Fast visual tracking by temporal consensus, Image and Vision Computing. Vol.14, pp.105-114
8. J. Heinzmann et al., 1998. 3D Facial Pose and Gaze Point Estimation using a Robust Real-Time Tracking Paradigm. Proceedings of ICAFGR, pp.142-147
9. T. Rikert, 1998. Gaze Estimation using Morphable Models. ICAFGR, pp.436-441
10. A.Ali-A-L et al., 1997, Man-machine Interface through Eyeball Direction of Gaze. Proc. of the Southeastern Symposium on System Theory, pp.478-82
11. J. Heinzmann et al., 1997. Robust Real-time Face Tracking and Gesture Recognition. Proc. of the IJCAI, Vol.2, pp.1525-1530
12. Matsumoto-Y, et al., 2000, An Algorithm for Real-time Stereo Vision Implementation of Head Pose and Gaze Direction Measurement. the ICAFGR. pp.499-504
13. Newman-R et al., 2000, Real-time Stereo Tracking for Head Pose and Gaze Estimation. Proceedings the 4th ICAFGR 2000. pp.122-8
14. Betke-M et al., 1999, Gaze Detection via Self-organizing Gray-scale Units. the Proc. of IWRATFG. pp.70-76
15. K. R. Park et al., 2000. Intelligent Process Control via Gaze Detection Technology. EAAI, Vol.13, No.5, pp.577-587
16. K. R. Park et al., 2002. Gaze Position Detection by Computing the 3 Dimensional Facial Positions and Motions. Pattern Recognition, Vol.35, No.11, pp.2559-2569
17. K. R. Park, 2002, Facial and Eye Gaze detection. LNCS, Vol.2525, pp.368-376

CA Based Document Compression Technology

Chandrama Shaw, Biplab K. Sikdar, and N.C. Maiti

Department of Computer Science and Technology, B. E. College
Howrah 711 103, India
chandrama_shaw@yahoo.co.in, {biplab,nm}@cs.becs.ac.in

Abstract. This paper proposes a new technology for document compression developed around the sparse network of Cellular automata (CA). It identifies the different class of segments in a scanned document - such as text, image, and background. The compression of the document is then performed on each segment employing the CA based compression technique specially designed for that class of segment. It results in a significant improvement in online document compression with no visible artifacts and ideally suits for low cost high speed hardware realization.

1 Introduction

The voluminous data of different types (text, image, figure etc.) are transmitted as a document over communication channels. A high speed data transmission, over the network, demands an efficient document-compression[1] technique to reduce the data rate. For text and line in a document graphics we need lossless compression [5], whereas for the images and background we can tolerate lossy reconstruction.

This fact motivates the design of a Cellular Automata (CA) based compression technology for the documents, considered for online network transmission. It segments blocks of pixels into different classes such as text, image and background as shown in *Fig.1*. The blocks of each class are then compressed around a codebook [2] specifically designed for that class with different allowances for information loss.

The basic codebook design algorithm(s) [3], performs pattern matching and exploits the concept of Vector Quantization (VQ). CA based implementation of VQ for image compression (TSVQ) [2] makes it more speedy and robust against misclassification [4]. The experimentation, reported in *Section 5*, establishes that the proposed technology achieves higher compression ratio and faster compression/decompression compared to state-of-the-art techniques. Overview of the design follows.

2 Overview of Document Compression

The proposed compression technique, consists of two major components - segmentation and compression. The first step of compression operation is to design

---

[1] This work is supported by AICTE: File No: 8020/RID/TAPTEC-46/2001-2002.

N.R. Pal et al. (Eds.): ICONIP 2004, LNCS 3316, pp. 679–685, 2004.
© Springer-Verlag Berlin Heidelberg 2004

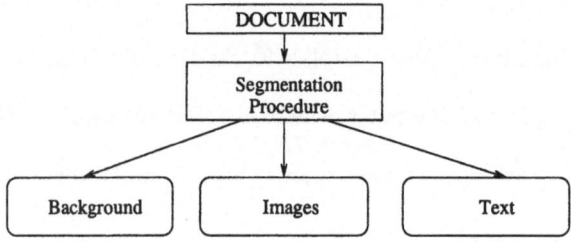

Fig. 1: Block Diagram of Segmentation method

the codebook for each class (text, image, background) of blocks. The well known *Vector Quantization* (VQ) [2] method has been applied to generate the *codebook* and addresses the following inherent drawbacks of VQ:

– Difficulty in generation of *codebook* that would ensure good quality of image with high compression ratio; and
– The high processing overhead at encoding stage.

Rather than developing general compression scheme, we concentrate on a specific type of document to improve the quality of compression as well as the *compression ratio*. While designing the codebook for a class, the proposed scheme extracts the domain knowledge. The encoding time is reduced substantially by employing CA technology. The *CA* has been used as an *implicit memory* to store the *codebook*. The next section introduces the basics of CA.

3   Cellular Automata (CA)

The next state of a two state (0/1) 3-neighborhood CA, cell depends on its present state and the present states of its right and left neighbors. An n-cell linear CA is characterized by an $n \times n$ characteristic matrix T [1], Where

$$T[i,j] = \begin{cases} 1, \text{ the next state of the } i^{th} \text{ cell depends on the present state of the } j^{th} \text{ cell} \\ 0, \text{ otherwise,} \end{cases}$$

The state transition behavior of such a CA is characterized by

$$S_{t+1} = T \times S_t + F \tag{1}$$

where S_t & S_{t+1} represent the CA states at t^{th} and $(t+1)^{th}$ instant of time.
Non Group CA: For a non-group CA, det[T]=0. The set of non-cyclic states in a non-froup CA forms an tree rooted at the cyclic state (attractor), where leaf nodes are the non-reachable states. The states rooted at attractor α form α-basin. The depth d is the number of Steps required to reach the nearest cyclic state from a non-reachable state. In example of CA (T) of Fig. 2, there are four attractors (00000, 01111, 10000, and 11111). Its depth d = 3. The states (00100, 00101,.., 00000) form the 00000-basin.

A non-group CA with multiple (single cycle) attractors is called MACA. The CA of Fig.2 is an MACA. In an n-cell MACA with k = 2^m attractors, there

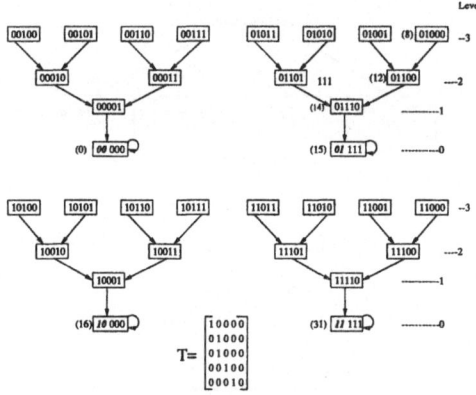

Fig. 2. State transition diagram of a 5-cell non-group CA

exists 'm' bit positions referred to as *PEF* at which the attractors generate pseudo-exhaustive 2^m patterns. The MACA of Fig.2 has 4 attractors with two MSB as the PEF. The MACA is employed to design the compression technology.

4   CA Based Document Compression

The proposed technology implements *Vector quantization* (VQ) that maps the n dimensional vectors into a finite set of vectors, called the codebook. Each vector of the codebook is a codevector. The compression technique encodes each block (16 × 16, 8 × 8, 4 × 4 and so on) of input file with the index of a *codevector* in the *codebook* that gives minimum deviation. The index of *codevector* is sent to the destination end. The *decoder* at destination, on receiving the index, replaces the entry associated with the index *(codevector)* found from the *codebook*. The codebook is kept in destination site. However, at the encoding end it is replaced by an MACA.

4.1   Codebook Design

For the current application, separate codebooks are designed for text, image and background blocks of a document. For image blocks, we design 3-layer codebooks based on TSVQ [2] technique. It considers the blocks of size 16X16, 8X8, and 4X4. Similarly, for text, the five layer codebooks are designed considering 16 × 16, 8 × 8, 4 × 4, 2 × 2 and single pixel blocks.

During compression, whenever the match between a higher dimension block (say 16 × 16) and the entries of the codebook fails, it tries to find out the best match in a lower dimension (8 × 8, 4 × 4 or so on) within a threshold limit.

In a design, the training set (Fig.3) contains the 16 × 16 blocks of a particular type (say text), segmented from the document. The 16 × 16 training set is then partitioned into different sets (8 × 8, 4 × 4, 2 × 2) on the basis of some threshold values. While compressing the text segment, if an incoming 2 × 2 block does not

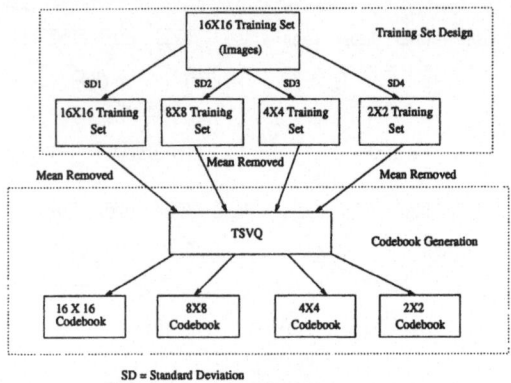

SD = Standard Deviation
TSVO = Tree Structure Vector Quantization

Fig. 3. Block diagram of codebook generation scheme

match with the entries of 2×2 codebook within the threshold limit, it is then treated as the single pixel. The details are reported in [4].

4.2 MACA as a Codebook

In the current work, the function of a codebook is implemented with MACA based 2-class classifier. The example MACA of Fig.2 can act as a 2-class classifier, where Class I is represented by the set of attractor basins [I]= {01111, 10000 & 11111} and for Class II it is [II] = {00000}. If the CA is loaded with a pattern (say 10100) and is allowed to run for 'd' (3) cycles, then it reaches to an attractor (10000). The PEF (10) of the attractor identifies the class (Class I) of the pattern. The PEF yields the memory address of the class (Fig.4). distinguish the two classes.

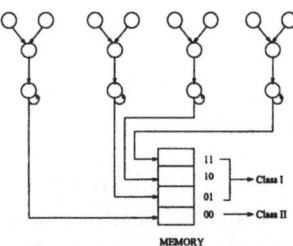

Fig. 4. $MACA$ based classification strategy

Fig.5 illustrates the design of a set of MACA for the codebook. Let assume the codevectors $C = \{\{S_0\}, \{S_1\}, \{S_2\}, \{S_3\}\}$ is to be classified into four classes. At the first level, C is divided into two classes $C_0 = \{\{S_0\}, \{S_1\}\}$ and $C_1 = \{\{S_2\}, \{S_3\}\}$. The MACA (T_0) is designed to classify C_0 and C_1 (Fig.5). The same process is then applied for C_0 and C_1 to isolate $\{S_0\}$ & $\{S_1\}$ and $\{S_2\}$ & $\{S_3\}$ respectively. It results in a multi-class classifier with 2-class classifiers - T_0, T_1 and T_2 that is equivalent to the $TSVQ$.

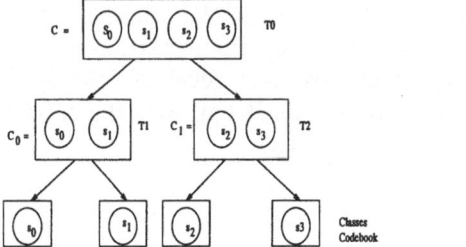

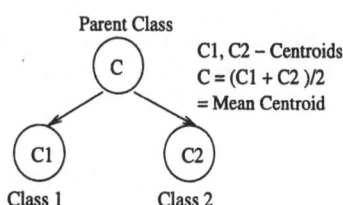

Fig. 5. Logical structure of multi-class classifier equivalent to TSVQ

Fig. 6. Discrete value of centroid

4.3 Searching of Codebook Designed with MACA

In order to identify the best match in binary TSVQ scheme, an input vector is compared with the two centroids of two vector clusters at each level of the tree. A sequence of comparisons is to be done in the subsequent levels till the leaf node is reached. The MACA based 2-class classifier is designed to model the comparison operation at each node of TSVQ binary tree. For a given codevector P_j ($P_j \in S_1$), if the $MACA$ (T_0) is loaded with P_j and allowed to run, it returns class C_0. At the next level, T_1 is loaded with P_j that outputs the class of P_j as S_1.

4.4 Requirements for the Design

In the testing phase, when a vector $say\ V = \begin{pmatrix} 0001 \\ 0011 \\ 0111 \\ 1111 \end{pmatrix}$ traverses through a tree

based on MACA classifiers, it always gets attracted to the class that has centroid with lesser hamming distance [6]. If following are the centroids of a design,

$$\begin{pmatrix} ClassI & ClassII \\ centroid & centroid \\ 0000 & 1001 \\ 0001 & 1010 \\ 0111 & 1100 \\ 1111 & 0100 \end{pmatrix}$$

then it is expected that the vector V will fall in Class I.

The inherent tendency of a classifier, in testing phase, is to follow the hamming distance. For example, 7 and 8 are very close if we consider Euclidean distance. However according to their hamming distance, these two are far away. We realize discretization on real data that maintains the minimum hamming distance if the vectors are of same class.

Discretization: Each n-dimension data of Class I or Class II is represented by an n-bit number (0/1) depending on its value in a dimension. For Example, let $C1 = \{12, 7, -2, 6, 9\}$ and $C2 = \{3, 18, 10, 26, -3\}$ are the centroids of Class

I and Class II. The mean centroid C = (C1+C2)/2 = {8, 13, 4, 16, 3}. If V = {10, 10, 6, 19, 1} is a vector in real field, then after discretization V = {1, 0, 1, 1, 0} (1 represents value > mean-centroid). We have developed the T's for the MACA based classifiers on these discretized data. It improves the classification quality at the cost of marginal processing overhead.

5   Experimental Results

The training set for designing the codebooks of different types of blocks is formed from 10 scanned documents.

Table 1. Execution Time

Block Size	Full Search (ms)	TSVQ (ms)	MACA (ms)
4 × 4	0.0121	0.00824	0.00562
8 × 8	0.0473	0.03312	0.01367
16 × 16	0.1941	0.13192	0.04102

Table 2. CR & PSNR

Document title	CR (%) CA	CR (%) JPEG 2K	PSNR (db) CA	PSNR (db) JPEG 2K
doc1	89.50	89.53	45.55	44.54
doc2	74.42	74.51	45.24	43.35
doc3	85.60	85.60	44.99	43.62
doc4	62.50	62.22	45.47	43.67
doc5	74.98	74.84	45.50	43.34
doc6	71.95	71.95	44.35	43.06
doc7	75.65	75.66	44.62	43.64

Table 3. Compression ratio for different block types

Doc Name	Size in (KB)	% of text Original	% of text Comp	% of Image Original	% of Image Comp	% of Background Original	% of Background Comp	Overall %
doc1	441	22.00	41.00	4.00	91.25	74.00	99.51	89.50
doc2	467	46.00	39.10	8.00	92.18	46.00	99.40	74.42
doc3	453	20.00	38.00	7.00	88.00	73.00	99.45	85.60
doc4	460	64.00	35.80	9.00	87.74	27.00	99.11	62.50
doc5	461	46.00	37.57	4.00	89.00	49.00	99.42	74.98
doc6	463	49.00	37.57	4.00	86.65	47.00	99.49	71.95
doc7	464	43.00	37.00	5.00	88.69	52.00	99.43	75.65

The efficiency of the MACA based search engine is compared with that of the full search and TSVQ [2] in Table 1. Column I depicts the size of a block. The columns II, III, and IV report the average search times to find closest codevector for a block. The experimention is done on Intel P-IV, 1.2 MHz platform. The results of Table 1 confirm that the search time in the proposed design (software version) is significantly lesser than that of full search and TSVQ. However, the major gain could be derived through hardwired implementation of MACA tree [1]. Table 2 gives the results of proposed scheme & JPEG 2000 [7] in respect of compression ratio (CR) and PSNR. It is established that the CA based document compression technology achieves higher PSNR. The compression gain for each segment type (text, image and background) has been shown in Table 3.

6   Conclusion

The proposed document compression scheme is a unique in application of CA. The CA is employed for low cost hardware implementation of compression technique to support high speed data transmission.

References

1. P. P. Chaudhuri, D. R. Chowdhury, S. Nandi, and S. Chatterjee, "Additive cellular automata, theory and applications, vol. 1," *IEEE Computer Society Press, Los Alamitos, California*, no. ISBN-0-8186-7717-1, 1997.
2. Allen Gresho and Robert M. Gray, *"Vector Quantization and Signal Compression"*, Kluwer Academy Publishers.
3. Mark Nelson, *"The Data Compression Book"*, M&T Books.
4. Chandrama Shaw, Debashis Chatterji, Pradipta Maji, Subhayan Sen, B. N. Roy, P Pal Chaudhuri, *"A Pipeline Architecture For Encompression (Encryption + Compression) Technology"*, in *Proc.* of 17th *Int. Conf.* on *VLSI* Design, India, pp. 303-314, January 2003.
5. J. Ziv and A. Lempel, *"A universal algorithm for sequential data compression,"* IEEE Trans. on Infor. Theory IT-23 (1977) 337-343.
6. N. Ganguly, P. Maji, S. Dhar, B. K. Sikdar and P. Pal Chaudhuri, *"Evolving Cellular Automata as Pattern Classifier"*, Proceedings of Fifth International Conference on Cellular Automata for Research and Industry, ACRI 2002, Switzerland.
7. Diego Santa-Cruz, Touradj Ebrahimi, Joel Askelf, Mathias Larsson and Charilaos Christopoulos. *"JPEG 2000 still image coding versus other standards."*, In SPIE's 46th annual meeting, Applications of Digital Image Processing XXIV , Proc. of SPIE, volume 4472, pages 267-275, San Diego, California, Jul. 29-Aug. 3, 2001.

Size-Independent Image Segmentation by Hierarchical Clustering and Its Application for Face Detection

Motofumi Fukui, Noriji Kato, Hitoshi Ikeda, and Hirotsugu Kashimura

Fuji Xerox Corporate Research Laboratory, 430 Sakai, Nakai-machi,
Ashigarakami-gun, Kanagawa 259-0157, Japan
{motofumi.fukui,noriji.kato,hitoshihi.ikeda,kashimura.hirotsugu}
@fujixerox.co.jp

Abstract. In this paper, we introduce a technique to detect a target object quickly. Our idea is based on onservation on the clusters into which an image is divided by hierarchical k-means clustering with space feature and color feature. This clustering method has the advantage of extracting the region of an object with some varied size. We insist that our idea should lead to detect a target object quickly, because it is not necessary to search the locations containing no targets. First, we evaluate our clustering method and second, we demonstrate that our method is effective on an object detection by applying to our face detection system. We show that the detection time can be reduced by 24%.

1 Introduction

Many researchers and engineers are interested in constructing the basic technology of recognizing what kind of objects there are and counting how many they are in a natural image, as human beings can do effortlessly. Especially face recognition technology is one of the most popular topics of research in the fields of image processing. It also has some applications such as human-computer interfaces, image retrieval, and security systems. To detect some faces quickly and precisely in an original image is the entrance of a face recognition system. In computer vision systems, this task is important. To address this task and realize a face recognition system, some ideas for face detection are proposed [1,2].

Face detection systems that can detect faces precisely have been developed by adopting the template matching method [3], but this method has the disadvantage of taking much time until all faces in target images are detected. So the template matching method has a risk of searching the locations including no faces obviously.

In the case of human beings, we can detect all faces in target images effortlessly with moving our eyes hardly at all. Many psychologists support the claim that human covert visual attentional mechanism plays an important role in detecting an object we want to see and an object we get to be used to see, like a face, instantaneously [4]. In face detection, some pseudo-attentional systems have been used to limit search locations and detect all faces quickly [5,6]. However face feature spaces have such a wide variability that it is difficult to gaze the locations of faces by using only low-level features derived from a priori information of the face. Still more to cope with the other objects besides faces, it is desirable to develop the system that does not depend on specific features for a target object.

N.R. Pal et al. (Eds.): ICONIP 2004, LNCS 3316, pp. 686–693, 2004.
© Springer-Verlag Berlin Heidelberg 2004

The work reported on this paper addresses a technique to segment an image by using k-means clustering and its application for face detection. By introducing the hierarchical clustering algorithm, an original image can be divided into some sub-clusters with some varied size. If an object detection system has only to search around the centroid of their sub-cluster, object candidates can be reduced and consequently it can lead to detect target objects quickly. In Section 2, we describe the hierarchical k-means clustering algorithm. In Section 3 and 4, we describe how to evaluate our clustering method and additionally we evaluate our idea by really applying to our face detection system [7,8]. Finally, Section 5 is devoted to conclusion.

2 Hierarchical K-means Clustering

Clustering analysis is frequently used for exploring the basic structure of a given image. This tool can be made use of grouping image pixels with similar feature into a cluster. K-means clustering is one of the most popular method. Many styles of this clustering have been proposed and are shown to be effective for many computer vision systems due to the easy computer-friendly algorithms [9,10,11].

2.1 K-means Clustering

Now we consider a k-cluster problem for image segmentation. K-clustering is a partition of the n-pixels set in a given image I into k disjoint sets $S_1,...,S_k$, named by clusters such that a cost function is minimized. This function is defined to be the sum of squared distance between the centroid d-dimensional feature vector s_i of each cluster S_i and each feature vector p_i of an pixel belonging to the cluster. On the tth trial, by Eq. (1) D-value is calculated to each feature vector p_i.

$$D(p_i,t) = \min_j \left\{ \left\| p_i - s_j^{(t-1)} \right\|^2 \right\}, \quad j \in \{1,..., k\} \tag{1}$$

Each pixel P_i is assigned to the cluster $S_j^{(t)}$ such that this cost function is minimized. Subsequently each cluster $S_j^{(t)}$ is updated and the centroid of all pixels currently in $S_j^{(t)}$ is represented as Eq. (2).

$$s_j = \frac{\sum_i p_i}{|S_j|}, \quad j \in \{1,..., k\}, \quad P_i \in S_j \tag{2}$$

Here $|S|$ denotes the number of pixels in S. This cost function has been calculated until it does not change significantly. There are many cases that the centroids of initialized clusters $(t=0)$ are selected randomly from feature vectors of all pixels. We locate these initialized centroids evenly all over the image.

In this paper we regard p as $(x,y,a,b)^T$ in R^4. Here $(x,y)^T$ denotes coordinates of a pixel P and $(a,b)^T$ denotes $(a$-value,b-value$)^T$ of a pixel P which is represented in lab-space. Usually only color feature is utilized by k-means clustering for image segmentation, but we use coordinates of P, too. It is because most pixel belonging to an object is located closely in the 2D-space and two approximate pixels each other should be often forced to be grouped into the same cluster. Furthermore we use lab-space but not rgb-space, because it is said to reflect on human quality of their visual system.

$$D(p,t) = \min_{j}\{(x - x_j)^2 + (y - y_j)^2 +$$
$$w * (a - a_j)^2 + w * (b - b_j)^2\}, \quad j \in \{1,...,k\} \tag{3}$$

D-value is calculated by Eq. (3). Here $(x_j,y_j,a_j,b_j)^T$ denotes the feature vector s_j and w denotes a weight parameter between location space and color space. Figure 1(b)-(d) show an example of how the clusters obtained by Eq. (3) depend on w. In the case that w has a small value, there are many clusters as blobs, while in the case that w has a large value, some pixels with similar color feature form the same cluster one another and many broken clusters are generated on the original image. We select w adequately by trial.

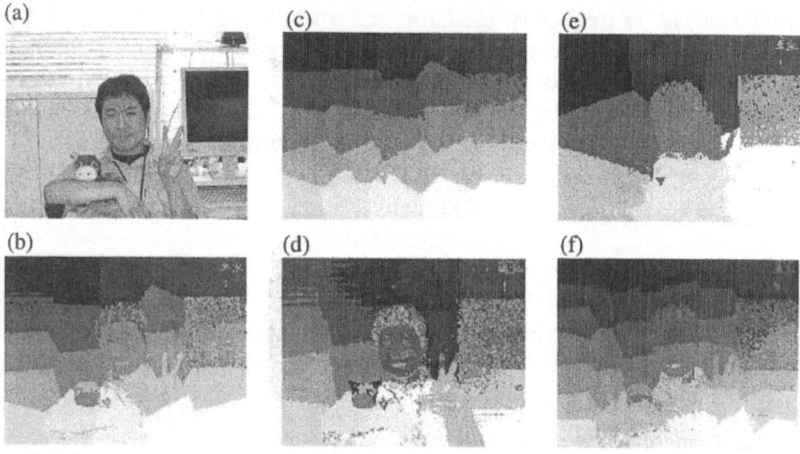

Fig. 1. Image segmentation by k-means clustering. (a) shows an original color image. (b)-(f) show the example of image segmentation. In (b)(e)(f), a weight value w is set up as 1.0, in (c) 10.0, and in (d) 100.0. In (b)-(d) these three images are divided into 25 clusters, in (e) 9 clusters, and in (f) 49 clusters

K-means clustering is an effective tool for image segmentation but the number of clusters k must be fixed in advance. Ideally k is equal to the number of objects in the image. If a priori information to the image is unknown, it is very difficult to decide the value k in advance. Figure 1(b)(e)(f) show an example of how the clusters depend on k. The selection of inadequate k occasionally leads to generate some mistaken clusters. To cope with these annoying problems, we use a hierarchical clustering method.

2.2 Hierarchical K-means Clustering

A general hierarchical clustering is used in carrying out an agglomeration at any stage of the clustering [12]. We use this kind of clustering aimed at dividing a cluster on a level into sub-clusters on the next one. Figure 2 shows how a $g^{(z)}$-tuple cluster can arise in the algorithm. Here let z be the level of hierarchy and g be the number of clusters on the level. When an original image is divided into q sub-clusters on the top level and a cluster on the $(z-1)^{th}$ level is divided into m sub-clusters on the zth level ($z \geq 2$), $g^{(z)}$ is equal to $qm^{(z-1)}$. In the case of $q=m=2$, 256 clusters are generated by the 8th level. Some clusters with varied size can be used on the stage of object detection by means of compiling sub-clusters generated on each level within the identical time. In addition to the advantage, the time required for generating these clusters is about $O(mzt)$, faster than the one by normal k-means clustering ($\doteqdot O(nkdt)$).

Applying all sub-clusters generated on each level to object detection, generation of some discarded clusters leads to excessive detection time. To get rid of these clusters and reduce the total number of clusters, we set up a threshold value $Th^{(z)}$. If the minimum weighted Euclidian distance among the centroids of sub-clusters generated on each level z is below $Th^{(z)}$, we group these two sub-clusters into one cluster.

While hierarchical k-means clustering has the advantage of the computational time and wide-ranged size variation relative to normal k-means clustering, it has the disadvantage of the degree of accuracy. Of course We apply our clustering method to change a weight value $w^{(z)}$ every level z of hierarchy. We evaluate this effect by introducing into our face detection system and detecting faces on the images belonging to our image database in Section 3 and 4.

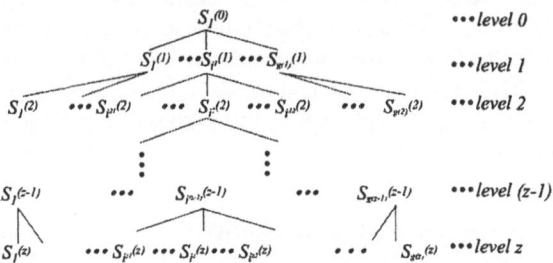

Fig. 2. Hierarchical k-means clustering. Level 0 is the top level and level z is the bottom level. $Sj^{(l)}$ is the jth sub-cluster generated on the level l ($1 \leq j \leq g^{(l)}$). $Si^{(k-1)(k-1)}$ on the level $k-1$ is divided into ($i^{k2}-i^{k1}+1$) ($=m$) sub-clusters on the next level k

3 Evaluation Method

To evaluate accuracy of generated clusters, we introduce a coincidence value. Additionally we apply our clustering method to our developed face detection system by the kernel-based non-linear subspace model [7,8].

3.1 Coincidence Value

This value indicates a reference mark on how similar to the region of objects a generated cluster is. Now T_O is defined to the minimizing square-shaped template covering an object O, which is usually determined by hand for evaluation. The template vector $U = (u_1,...,u_n)^T$ and the cluster vector $V_i = (v_{i1},...,v_{in})^T$ corresponding to a cluster S_i denote the following Eq. (4). And the coincidence value $Co(O)$ of a cluster S_i to an object O is defined as (U,V_i). Here (A,B) denotes inner product of vectors A and B.

Maximum coincidence value is equal to 1.0 and in this case T_O synchronizes exactly with a cluster. We estimate accuracy of clustering by means of finding a cluster of which the coincidence value is the largest to the object included in a image. For example, in Figure 1, the maximum coincidence value is estimated as (b)0.59, (c)0.54, and (d)0.52 respectively. This result shows that Figure 1(b) is the best clustering of the three.

$$u_j = 1/\sqrt{|T_o|} \cdots P_j \in T_o, \; otherwise \; 0$$
$$v_{ij} = 1/\sqrt{|S_i|} \cdots P_j \in S_i, \; otherwise \; 0$$

(4)

3.2 Evaluation by Our Face Detection System

Our system realizes invariant pattern matching by re-normalizing the image with three feature units, consisted of a scaling unit, a rotation unit, and a translation unit. On our face detection system, a square-shaped template constructed by unsupervised learning is scanned serially on an original image, so the time required for detecting target faces fluctuate in line with the number of locations where this template is scanned and the kind of sizes of it. The following procedure shows how to fix this window format after a generated cluster S. The determination procedure is as follows; (1)estimate the centroid $(x,y)^T$ of S, (2)estimate the square measure T^2 of S, (3)set up a square-shaped window, (4)input this window to our engine. As a result, we implement maximum k face candidates to our engine, when k is the number of generated clusters by our method. If the number of k can be reduced, we can detect a face quickly.

4   Experiment, Result, and Discussion

To evaluate the performance of clustering, we prepare a color-image database which contains 254 frontal faces with some varied size. The distribution of the relative size to the image which includes them is from 0.03 to 0.57. Moreover we divide these faces into 5 groups according to the relative size. Descending order of the relative size is from group1 to group5. Each group has 50 or 51 faces.

4.1   Evaluation of Coincidence Value

We set up the bottom level as 8, q as 2, m as 2, and $Th^{(z)}$ as 0. These parameters are set up in view of the minimum relative size. Furthermore a weight value $w^{(z)}$ on the level z should be inverse proportion to z., because a 2D-distance between two pixels belonging to the same cluster comes near as descending the level. On our clustering, we set up it as $10.0/2.0^{(z-1)}$. By using this parameter, we can generate 511 clusters in total per one image. In Figure 3, we show the frequency distribution of this coincidence value of each group. Each maximum average Co is estimated as 0.62, 0.61, 0.58, 0.51, and 0.43. These results indicate that maximum Co is reduced as the relative face size becomes small, especially a face in group5. We suspect that this results might be caused by the influence of a failure to clustering on a level, on clustering on the rest level. This influence may represent that an error of image segmentation is accumulated with lowering the level, in the case of hierarchical clustering.

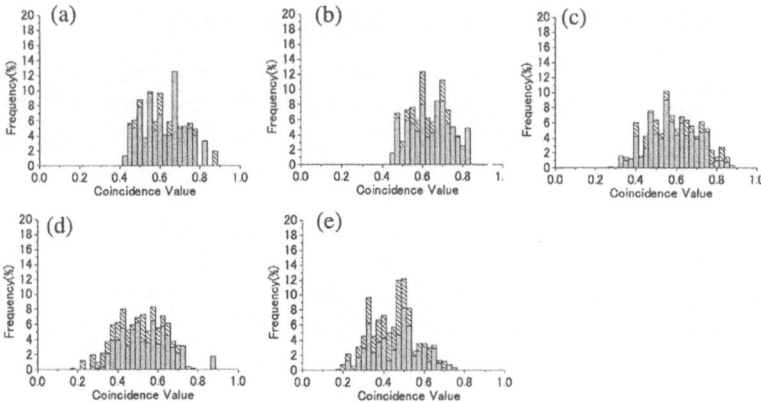

Fig. 3. Frequency distribution of the maximum coincidence value. (a),...,(e) shows the results of group1,..., group5 respectively. Filled color shows the frequence of detected faces, while shaded portion shows undetected faces

4.2 Relationship Between Coincidence Value and Detection Rate

Subsequently, we analyze the relationship between the coincidence value and the detection rate of faces. We prepare normal k-means clustering by which an image with faces belonging to group5 is divided into 256 clusters, in order to analysis this relationship by using clusters with almost same size. In Figure 4, we show the results by two cases: using ab-space and lab-space in clustering. In either case, we get the results of high correlation between these two parameters. These results demonstrate that this coincidence value is the good mark on accuracy of clustering.

In the case of using lab-space, the detection rate is reduced as w is large, because maybe there is large change of intensity in a natural scene. While, in the case of using ab-space, the detection rate is maintained even if w changes. This advantage is valid for the hierarchical clustering such that w would be changed by degree of the level, because an adequate value of w may be estimated without difficulty. Additionally The maximum detection rate between these cases is almost same.

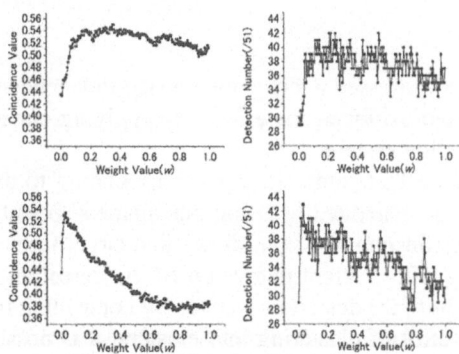

Fig. 4. A comparison between the coincidence value and the number of detected faces on group5. The number of clusters is set up as 256. Upper graphs show the results by clustering with ab-space, lower graphs with lab-space

4.3   Analysis for the Number of Face Candidates and Speed

On hierarchical clustering, the same parameter is used. In the case with Th, $Th^{(z)}$ should be inverse proportion to z. We set up it as $(4000.0)^{1/2}/z$. Pentium IV 2.8 GHz is used.

Finally, we analysis the speed of face detection. The total time is the sum of the one for clustering and the one for detecting in our engine. The detecting time is almost proportional to the number of face candidates by our system. By introducing our idea of clustering, we try to reduce the number of face candidates which are input in our engine. Table 1 shows the comparisons among 3 methods of face detection. In practice, we use 8 sub-windows with the same area as the main window, which are located adjacent to it.

Table 1. Detection Rate, Speed and the Number of Candidates

method	full scanning	hierarchical k-means clustering	hierarchical k-means clustering with Th
detection rate(%)	83.1	77.6	75.6
detection time(ms)	5000	3830	3530
total number of candidates	5800	3972	3670

By using the hierarchical clustering, the time required for clustering can be reduced ($\doteqdot$400ms). Furthermore we can reduce face candidates (=441). Consequently the total detection time can be reduced by about 24%. By introducing $Th^{(z)}$ which is set up as a smaller value according to the level z and reducing the number of face candidates, the total detection time can be moreover reduced. While in the case of using normal k-means clustering, we have much time for detection (=7350ms). It is because it takes much time for clustering with many clustering size k. However compared with the full scanning, the detection rate calculated by our method deteriorates in small measure. In future work, we want to correct this problem.

5   Conclusion

In this paper, we have addressed a technique on size-independent image segmentation for object detection and exhibited the effects by applying our idea to our modulared view-based face detection system.

First, we have shown that some clusters corresponding to the objects with varied size in an image can be generated by using our image segmentation technique based on k-means clustering. Second, we have shown that target faces can be detected faster than by full scanning with a slight detraction of the detection rate, by applying our clustering method to our face detection system. By controlling the segmentation without accumulating an error of clustering into clusters, it is possible to realize a faster face detection system. In addition to this suggestion, we do not use a priori information of the face in clustering, so our image segmentation technique would be effective on a general object detection system.

References

1. Hjelmas E, Low B (2001) Face detection:a survey. Comp. Vis. Img. Und. 83:236-274
2. Yang M, Kriegman D, Ahuja N (2002) Detecting faces in images:a survey. IEEE Trans. Patt. Anal. Mach. Intell. 24:34-58
3. Rowley H, Baluja S, Kanade T (1998) Neural network-based face detection. IEEE Trans. Patt. Anal. Mach. Intell. 20:23-38
4. Pashler H(ed) (2001) Attention. Psychology Press
5. Wang C, Brandstein M (1999) Multi-source face tracking with audio and visual data. Proc. IEEE 3rd Workshop on Multimedia Signal Processing pp.169-174
6. Jacquin A, Eleftheriadis, A (1995) Automatic location tracking of faces and facial features in video sequences. Proc. 1st Intl. Workshop on Automatic Face and Gesture Recognition pp.142-147
7. Ikeda H, Kato N, Kashimura H, Shimizu M (2003) Scale, rotation, and translation invariant fast face detection system. In Proc. of the 5th IASTED Intl. Conf. on Signal and Image Processing. pp.146-151
8. Kato N, Ikeda H, Kashimura H, Shimizu M (2003) Scaling, rotation, and translation invariant image recognition using competing multiple subspaces. Proc. INNS-IEEE Intl. Joint Conf. on Neural Networks pp.1268-1273
9. Su M, Chou C (2001) A modified verison of the k-means algorithm with a distance based on cluster symmetry. IEEE Trans. Patt. Anal. Mach. Intell. 6:674-680
10. Matousek J (2000) On approximate geometric k-clustering. Discrete & Comp. Geometry 24:61-84
11. Tailor J, Gisler G (1997) A contiguity-enhanced k-means clustering algorithm for unsupervised multispectral image segmentation. Proc. SPIE 3159 pp.108-118
12. Murtagh F (1983) A survey of recent advances in hierarchical clustering algorithms. The Comp. J. 26:354-359

Lecture Notes in Computer Science 3316

Commenced Publication in 1973
Founding and Former Series Editors:
Gerhard Goos, Juris Hartmanis, and Jan van Leeuwen

Editorial Board

David Hutchison
  Lancaster University, UK
Takeo Kanade
  Carnegie Mellon University, Pittsburgh, PA, USA
Josef Kittler
  University of Surrey, Guildford, UK
Jon M. Kleinberg
  Cornell University, Ithaca, NY, USA
Friedemann Mattern
  ETH Zurich, Switzerland
John C. Mitchell
  Stanford University, CA, USA
Moni Naor
  Weizmann Institute of Science, Rehovot, Israel
Oscar Nierstrasz
  University of Bern, Switzerland
C. Pandu Rangan
  Indian Institute of Technology, Madras, India
Bernhard Steffen
  University of Dortmund, Germany
Madhu Sudan
  Massachusetts Institute of Technology, MA, USA
Demetri Terzopoulos
  New York University, NY, USA
Doug Tygar
  University of California, Berkeley, CA, USA
Moshe Y. Vardi
  Rice University, Houston, TX, USA
Gerhard Weikum
  Max-Planck Institute of Computer Science, Saarbruecken, Germany

Nikhil R. Pal   Nikola Kasabov
Rajani K. Mudi   Srimanta Pal
Swapan K. Parui (Eds.)

Neural
Information Processing

11th International Conference, ICONIP 2004
Calcutta, India, November 22-25, 2004
Proceedings

 Springer

Volume Editors

Nikhil R. Pal
Srimanta Pal
Indian Statistical Institute
Electronics and Communications Sciences Unit
203 B. T. Road, Calcutta 700 108, India
E-mail: {nikhil,srimanta}@isical.ac.in

Nikola Kasabov
Auckland University of Technology
Knowledge Engineering and Discovery Research Institute (KEDRI)
Private Bag 92006, Auckland, New Zealand
E-mail: nik.kasabov@aut.ac.nz

Rajani K. Mudi
Jadavpur University
Department of Instrumentation and Electronics Engineering
Salt-lake Campus, Calcutta 700098, India
E-mail: rkmudi@iee.jusl.ac.in

Swapan K. Parui
Indian Statistical Institute
Computer Vision and Pattern Recognition Unit
203 B. T. Road, Calcutta 700 108, India
E-mail: swapan@isical.ac.in

Library of Congress Control Number: 2004115128

CR Subject Classification (1998): F.1, I.2, I.5, I.4, G.3, J.3, C.2.1, C.1.3, C.3

ISBN 978-3-540-23931-4   ISBN 978-3-540-30499-9 (eBook)
DOI 10.1007/978-3-540-30499-9

This work is subject to copyright. All rights are reserved, whether the whole or part of the material is concerned, specifically the rights of translation, reprinting, re-use of illustrations, recitation, broadcasting, reproduction on microfilms or in any other way, and storage in data banks. Duplication of this publication or parts thereof is permitted only under the provisions of the German Copyright Law of September 9, 1965, in its current version, and permission for use must always be obtained from Springer. Violations are liable to prosecution under the German Copyright Law.

springeronline.com

© Springer-Verlag Berlin Heidelberg 2004

Originally published by Springer-Verlag Berlin Heidelberg New York in 2004

Typesetting: Camera-ready by author, data conversion by Olgun Computergrafik
Printed on acid-free paper   SPIN: 11359166   06/3142   5 4 3 2 1 0

Preface

It is our great pleasure to welcome you to the *11th International Conference on Neural Information Processing* (ICONIP 2004) to be held in Calcutta. ICONIP 2004 is organized jointly by the Indian Statistical Institute (ISI) and Jadavpur University (JU). We are confident that ICONIP 2004, like the previous conferences in this series, will provide a forum for fruitful interaction and the exchange of ideas between the participants coming from all parts of the globe. ICONIP 2004 covers all major facets of computational intelligence, but, of course, with a primary emphasis on neural networks. We are sure that this meeting will be enjoyable academically and otherwise.

We are thankful to the track chairs and the reviewers for extending their support in various forms to make a sound technical program. Except for a few cases, where we could get only two review reports, each submitted paper was reviewed by at least three referees, and in some cases the revised versions were again checked by the referees. We had 470 submissions and it was not an easy task for us to select papers for a four-day conference. Because of the limited duration of the conference, based on the review reports we selected only about 40% of the contributed papers. Consequently, it is possible that some good papers are left out. We again express our sincere thanks to all referees for accomplishing a great job. In addition to 186 contributed papers, the proceedings includes two plenary presentations, four invited talks and 18 papers in four special sessions. The proceedings is organized into 26 coherent topical groups. We are proud to have a list of distinguished speakers including Profs. S. Amari, W.J. Freeman, N. Saitou, L. Chua, R. Eckmiller, E. Oja, and T. Yamakawa. We are happy to note that 27 different countries from all over the globe are represented by the authors, thereby making it a truly international event.

We are grateful to Prof. A.N. Basu, Vice-Chancellor, JU and Prof. K.B. Sinha, Director, ISI, who have taken special interest on many occasions to help the organizers in many ways and have supported us in making this conference a reality. Thanks are due to the Finance Chair, Prof. R. Bandyopadhyay, and the Tutorial Chair, Prof. B.B. Chaudhuri.

We want to express our sincere thanks to the members of the Advisory Committee for their timely suggestions and guidance. We sincerely acknowledge the wholehearted support provided by the members of the Organizing Committee. Special mention must be made of the organizing Co-chairs, Prof. D. Patranabis and Prof. J. Das for their initiative, cooperation, and leading roles in organizing the conference. The staff members of the Electronics and Communication Sciences Unit of ISI have done a great job and we express our thanks to them. We are also grateful to Mr. Subhasis Pal of the Computer and Statistical Services Center, ISI, for his continuous support. Things will remain incomplete unless we mention Mr. P.P. Mohanta, Mr. D. Chakraborty, Mr. D.K. Gayen and Mr. S.K. Shaw without whose help it would have been impossible for us to make this conference a success. We must have missed many other colleagues and friends who have helped us in many ways; we express our sincere thanks to them also.

We gratefully acknowledge the financial support provided by different organizations, as listed below. Their support helped us greatly to hold this conference on this scale.

Last, but surely not the least, we express our sincere thanks to Mr. Alfred Hofmann and Ms. Ursula Barth of Springer for their excellent support in bringing out the proceedings on time.

November 2004

Nikhil R. Pal
Nikola Kasabov
Rajani K. Mudi
Srimanta Pal
Swapan K. Parui

Funding Agencies

- Infosys Technologies Limited, India
- IBM India Research Lab, India
- Department of Science and Technology, Govt. of India
- Council of Scientific and Industrial Research, Govt. of India
- Reserve Bank of India
- Department of Biotechnology, Govt. of India
- Defence Research and Development Organization, Govt. of India
- Department of Higher Education, Govt. of West Bengal, India
- Jadavpur University, Calcutta, India
- Indian Statistical Institute, Calcutta, India

Organization

Organizers
Indian Statistical Institute, Calcutta, India
Jadavpur University, Calcutta, India
Computational Intelligence Society of India (CISI), India

Chief Patrons
K.B. Sinha, Indian Statistical Institute, India
A.N. Basu, Jadavpur University, India

Honorary Co-chairs
S. Amari, Riken Brain Science Institute, Japan
T. Kohonen, Neural Networks Research Centre, Finland

General Chair
N.R. Pal, Indian Statistical Institute, India

Vice Co-chairs
E. Oja, Helsinki University of Technology, Finland
R. Krishnapuram, IBM India Research Lab, India

Program Chair
N. Kasabov, University of Otago, New Zealand

Organizing Co-chairs
D. Patranabis, Jadavpur University, India
J. Das, Indian Statistical Institute, India

Joint Secretaries
S.K. Parui, Indian Statistical Institute, India
R.K. Mudi, Jadavpur University, India
S. Pal, Indian Statistical Institute, India

Tutorials Chair
B.B. Chaudhuri, Indian Statistical Institute, India

Finance Chair
R. Bandyopadhyay, Jadavpur University, India

Technical Sponsors
Asia Pasific Neural Network Assembly (APNNA)
World Federation on Soft Computing (WFSC)

Advisory Committee

A. Atiya, Cairo University, Egypt
Md.S. Bouhlel, National Engineering School of Sfax, Tunisia
S. Chakraborty, Institute of Engineering and Management, India
G. Coghill, University of Auckland, New Zealand
A. Engelbrecht, University of Pretoria, South Africa
D. Fogel, Natural Selection Inc., USA
K. Fukushima, Tokyo University of Technology, Japan
T. Gedeon, Australian National University, Australia
L. Giles, NEC Research Institute, USA
M. Gori, Università di Siena, Italy
R. Hecht-Nielsen, University of California, USA
W. Kanarkard, Ubonratchathani University, Thailand
O. Kaynak, Bogazici University, Turkey
S.V. Korobkova, Scientific Centre of Neurocomputers, Russia
S.Y. Lee, Korea Advanced Institute of Science and Technology, Korea
C.T. Lin, National Chiao-Tung University, Taiwan
D.D. Majumder, Indian Statistical Institute, India
M. Mitra, Jadavpur University, India
D. Moitra, Infosys Technologies Limited, India
L.M. Patnaik, Indian Institute of Science, India
W. Pedrycz, University of Alberta, Canada
S.B. Rao, Indian Statistical Institute, India
V. Ravindranath, National Brain Research Centre, India
J. Suykens, Katholieke Universiteit Leuven, Belgium
A.R. Thakur, Jadavpur University, India
S. Usui, Neuroinformatics Lab., RIKEN BSI, Japan
L. Wang, Nanyang Technological University, Singapore
L. Xu, Chinese University of Hong Kong, Hong Kong
T. Yamakawa, Kyushu Institute of Technology, Japan
Y.X. Zhong, University of Posts and Telecommunications, China
J. Zurada, University of Louisville, USA

Track Chairs

Quantum Computing	E. Behrman (USA)
Bayesian Computing	Z. Chan (New Zealand)
Bio-informatics	J.Y. Chang (Taiwan)
Support Vector Machines and Kernel Methods	V.S. Cherkassky (USA)
Biometrics	S.B. Cho (Korea)
Fuzzy, Neuro-fuzzy and Other Hybrid Systems	F.K. Chung (Hong Kong)
Time Series Prediction and Data Analysis	W. Duch (Poland)
Evolutionary Computation	T. Furuhashi (Japan)
Neuroinformatics	I. Hayashi (Japan)
Pattern Recognition	R. Kothari (India)
Control Systems	T.T. Lee (Taiwan)
Image Processing and Vision	M.T. Manry (USA)
Robotics	J.K. Mukherjee (India)
Novel Neural Network Architectures	M. Palaniswami (Australia)
Brain Study Models	V. Ravindranath (India)
Brain-like Computing	A. Roy (USA)
Learning Algorithms	P.N. Suganthan (Singapore)
Cognitive Science	R. Sun (USA)
Speech and Signal Processing	H. Szu (USA)
Computational Neuro-science	S. Usui (Japan)
Neural Network Hardware	T. Yamakawa (Japan)

Organizing Committee

K. Banerjee	U. Garai	P. Pal
J. Basak	Karmeshu	S. Raha
U. Bhattacharya	R. Kothari	Baldev Raj
B.B. Bhattacharya	S. Kumar	K. Ray
B. Chanda	K. Madhanmohan	K.S. Ray
N. Chatterjee	K. Majumdar	B.K. Roy
B.N. Chatterji	A.K. Mandal	K.K. Shukla
B. Dam	M. Mitra	B.P. Sinha
A.K. De	D.P. Mukherjee	B. Yegnanarayana
K. Deb	K. Mukhopadhyay	
U.B. Desai	P.K. Nandi	
B.K. Dutta	U. Pal	

Reviewers

S. Abe
A. Abraham
M. Alcy
A.S. Al-Hegami
N.M. Allinson
E. Alpaydin
L. Andrej
A. Arsenio
A. Atiya
M. Atsumi
M. Azim
T. Balachander
J. Basak
Y. Becerikli
R. Begg
L. Behera
Y. Bengio
U. Bhattacharya
C. Bhattacharyya
A.K. Bhaumik
A. Biem
Z. Bingul
S. Biswas
S.N. Biswas
M. Blumenstein
S. Buchala
A. Canuto
A.C. Carcamo
F.A.T. Carvaho
K.-M. Cha
D. Chakraborty
Debrup Chakraborty
S. Chakraborty
U. Chakraborty
C.-H. Chan
B. Chanda
H. Chandrasekaran
K. Chang
B.N. Chatterjee
B.B. Chaudhuri
H.-H. Chen
J. Chen
K. Chen
T.K. Chen

V. Cherkassky
W. Cheung
E. Cho
S. Cho
S.B. Cho
K. Chokshi
N. Chowdhury
A.A. Cohen
R.F. Correa
S. Daming
J. Das
N. Das
C.A. David
A.K. Dey
R. De
K. Deb
W.H. Delashmit
G. Deng
B. Dhara
G. Dimitris
M. Dong
Y. Dong
T. Downs
K. Doya
W. Duch
A. Dutta
D.P.W. Ellis
M. Er
P. E'rdi
M. Ertunc
A. Esposito
E.C. Eugene
X. Fan
Z.-G. Fan
O. Farooq
S. Franklin
M. Fukui
T. Furuhashi
M. Gabrea
M. Gallagher
U. Garain
A. Garcez
S.S. Ge
T. Gedeon

T.V. Geetha
A. Ghosh
S. Ghosh
B.G. Gibson
K. Gopalsamy
K.D. Gopichand
R. Gore
M. Grana
L. Guan
C. Guang
H. Guangbin
A. Hafez
M. Hagiwara
M. Hattori
I. Hayashi
G. Heidemann
G.Z. Chi
S. Himavathi
P.S. Hiremath
A. Hirose
L. Hongtao
C.-H. Hsieh
W.-Hsu
B. Huang
H.-D. Huang
Y.K. Hui
M.F. Hussin
S. Ikeda
T. Inoue
H. Ishibuchi
P. Jaganathan
M. Jalilian
M. Jalili-Kharaajoo
G. Ji
X. Jiang
L. Jinyan
T. Kalyani
T. Kambara
C.-Y. Kao
K. Karim
N. Kasabov
U. Kaymak
O. Kaynak
S.S. Khan

J.Y. Ki
J. Kim
J.-H. Kim
K. Kim
K.B. Kim
H. Kita
A. Koenig
M. Koppen
K. Kotani
R. Kothari
R. Kozma
K. Krishna
R. Krishnapuram
S.N. Kudoh
C. Kuiyu
A. Kumar
A.P. Kumar
S. Kumar
M.K. Kundu
Y. Kuroe
S. Kurogi
J. Kwok
H.Y. Kwon
J. Laaksonen
S. LaConte
A. Laha
D. Lai
S. Laine
R. Langari
J.-H. Lee
K.J. Lee
V.C.S. Lee
J. Li
P. Li
S. Lian
C. Lihui
C.-J. Lin
C.T. Lin
C. Liu
J. Liu
P. Lokuge
R. Lotlikar
M. Louwerse
C.L. Lu
T. Ludermir

C.K. Luk
P.-C. Lyu
Y. Maeda
S. Maitra
S.P. Maity
K. Majumdar
M. Mak
F.J. Maldonado
A.K. Mandal
J. Mandziuk
N. Mani
D.H. Manjaiah
M. Matsugu
Y. Matsuyama
B. McKay
O. Min
M. Mitra
S. Mitra
B.M. Mohan
P.P. Mohanta
R.K. Mudi
S. Mukherjea
A. Mukherjee
D.P. Mukherjee
J.K. Mukherjee
K. Mukhopadhyaya
A.K. Musla
W. Naeem
P. Nagabhushan
H. Najafi
T. Nakashima
P.K. Nanda
P. Narasimha
M. Nasipura
V.S. Nathan
G.S. Ng
A. Nijholt
G.S. Nim
D. Noelle
A. Ogawa
E. Oja
H. Okamoto
P.R. Oliveira
T. Omori
Y. Oysal

G.A.V. Pai
N.R. Pal
S. Pal
U. Pal
S. Palit
R. Panchal
J.-A. Park
K.R. Park
S.K. Parui
M.S. Patel
S.K. Patra
M. Perus
T.D. Pham
A.T.L. Phuan
H. Pilevar
M. Premaratne
S. Puthusserypady
S. Qing
M. Rajeswari
K.S. Rao
F. Rashidi
M. Rashidi
V. Ravindranath
B.K. Rout
A. Roy
P.K. Roy
R. Roy
J. Ruiz-del-Solar
S. Saha
S. Saharia
A.D. Sahasrabudhe
S. Sahin
J.S. Sahmbi
M. Sakalli
A.R. Saravanan
S.N. Sarbadhikari
P. Sarkar
P.S. Sastry
A. Savran
H. Sawai
A. Saxena
C.C. Sekhar
A. Sharma
C. Shaw
B.H. Shekar

P.K. Shetty
Z. Shi
C.N. Shivaji
P. Shivakumara
K.K. Shukla
A.P. Silva Lins
M.J. Silva Valenca
J.K. Sing
R. Singh
S. Singh
S. Sinha
M. Sirola
G. Sita
K.R. Sivaramakrishnan
J. Sjoberg
K. Smith
X. Song
M.C.P. de Souto
A. Sowmya
R. Srikanth
B. Srinivasan
P.N. Suganthan
C. Sun
R. Sun
E. Sung
V. Suresh
J. Suykens

R. Tadeusiewicz
P.K.S. Tam
H. Tamaki
C.Y. Tang
E.K. Tang
P. Thompson
K.-A. Toh
A. Torda
V. Torra
D. Tran
J. Uchida
S. Usui
P. Vadakkepat
B. Valsa
D. Ventura
A. Verma
B. Verma
J. Vesanto
E. Vijayakumar
D. Wang
J. Wang
L. Wang
S. Wang
J. Watada
O. Watanabe
J. Wei
A. Wichert

R.H.S. Wong
K.-W. Wong (Kevin)
B. Xia
C. Yamaguchi
Y. Yamaguchi
T. Yamakawa
L. Yao
S. Yasui
Z. Yeh
D.C.S. Yeung
H. Yigit
C.-G. Yoo
N.M. Young
C. Yu
M. Yunqian
C. Zanchettin
Z. Zenn Bien
B.-T. Zhang
D. Zhang
L. Zhang
Q. Zhang
Y. Zhang
L. Zhiying
S. Zhong
D. Zhou
H. Zujun

Table of Contents

Computational Neuroscience

Evolutionary Computation

Control Systems

Cognitive Science

Biometrics

Adaptive Intelligent Systems

Brain-Like Computing

Learning Algorithms

Novel Neural Networks

Image Processing

Pattern Recognition

Neuroinformatics

Fuzzy Systems

Neuro-fuzzy Systems

Hybrid Systems

Ant Colony

Neural Network Hardware

Robotics

Signal Processing

Support Vector Machine

Time Series Prediction

Bioinformatics

Human-Like Selective Attention Model
with Reinforcement and Inhibition Mechanism

Sang-Bok Choi[1], Sang-Woo Ban[2], and Minho Lee[2]

[1] Dept. of Sensor Engineering, Kyungpook National University,
1370 Sankyuk-Dong, Puk-Gu, Taegu 702-701, Korea
csv@palgong.knu.ac.kr
[2] School of Electronic and Electrical Engineering, Kyungpook National University,
1370 Sankyuk-Dong, Puk-Gu, Taegu 702-701, Korea
swban@palgong.knu.ac.kr, mholee@knu.ac.kr

Abstract. In this paper, we propose a trainable selective attention model that can not only inhibit an unwanted salient area but also reinforce an interesting area. The proposed model was implemented by the bottom-up saliency map model in conjunction with the top-down attention mechanism. The bottom-up saliency map model generates a salient area, and human supervisor decides whether the selected salient area is inhibited or reinforced. The fuzzy adaptive resonance theory (Fuzzy-ART) network can generate an inhibit signal or a reinforcement signal so that the sequence of attention areas is modified to be a desired scan path. Computer simulation results show that the proposed model successfully generates the plausible scan path of salient region.

1 Introduction

The selective attention mechanism makes the human-vision system very effective in processing high dimensional image data with great complexity. If we apply the human-like selective attention function to the active vision system, an efficient and intelligent active vision system can be developed. Considering the human-like selective attention function, top-down or task dependent processing can affect how to determine the saliency map as well as bottom-up or task independent processing [1].

As a previous work, several models have been proposed by many researchers [1-3]. However, the weight values of these feature maps for constructing the saliency map are still determined artificially. Moreover, all of these models are non-interactive with environment and human being, and resultantly it is insufficient to give confidence of the selected salient area whether the selected area is interesting.

In this paper, we propose a new selective attention model to mimic human-like selective attention mechanism not only with truly bottom-up process but also with interactive process to inhibit an unwanted area and to reinforce attention to a desired area in subsequent visual search process. In order to implement such a selective attention model, we use the bottom-up saliency map (SM) model in conjunction with the fuzzy adaptive resonant theory (Fuzzy ART) network. The proposed model has two process modes. In training process, the Fuzzy ART network learns about uninteresting areas and desired areas that are decided by human supervisor interactively, which is different from the conventional Fuzzy ART network. In test mode, the vigilance parameter in the Fuzzy ART network determines whether the new input area is inter-

N.R. Pal et al. (Eds.): ICONIP 2004, LNCS 3316, pp. 694–699, 2004.
© Springer-Verlag Berlin Heidelberg 2004

esting or not, because the Fuzzy ART network memorizes the characteristics of the unwanted salient areas or desired salient areas. If the vigilance value is larger than a threshold, the Fuzzy ART network for inhibition inhibits the selected area in the bottom-up saliency map model so that the area should be ignored in the subsequent visual search process, and also the Fuzzy ART network for reinforcement modifies the scan path like human vision system so that the interesting area is to be the most salient area.

In Section 2, we explain our developed bottom-up saliency map model. Section 3 shows the proposed trainable selective attention model using ART network. Simulation results and conclusion will be followed.

2   The Summary of Bottom-Up Saliency Map Model

The simple biological visual pathway for bottom-up processing is from retina to visual cortex through the LGN [4]. In order to implement a human-like visual attention function, we consider the SM model. In our approach, we use SM model that reflects the functions of retina cells, LGN and visual cortex. Since the retina cells can extract the edge and intensity information as well as color opponency, we use the edge, intensity and color opponent coding as the basis features of SM model [5]. In order to consider the function of LGN and the primary visual cortex to detect a shape of an object, we consider the symmetry feature as additional basis. The symmetry information is obtained by noise tolerant general symmetry transform (NTGST) method [6]. The ICA can be used for modeling the roles of the primary visual cortex for redundancy reduction according to Barlow's hypothesis and Sejnowski's result [7, 8]. Barlow's hypothesis is that human visual cortical feature detectors might be the end result of a redundancy reduction process [8], and Sejnowski's result is that the ICA is the best way to reduce redundancy [7]. After the convolution between channel of the feature maps and the filters obtained by ICA learning, the saliency map is computed by summation of all feature maps for every location [5].

3   Trainable Selective Attention Model

Although the proposed bottom-up saliency map model generates plausible salient areas and scan path, the selected areas may not be an interesting area for human because the saliency map only uses the primitive features such as intensity, edge, color and symmetry information. In order to implement more plausible selective attention model, we need an interactive procedure together with the bottom-up information processing.

Human ignores uninteresting area even if it has salient primitive features, and can memorize the characteristics of the unwanted area. We do not give an attention to a new area with similar characteristics of the previously learned unwanted area. Also, human can pay an attention to an interesting area even if it does not have salient primitive features, or it is less salient than any other area. We propose a new selective attention model to mimic such a human-like selective attention mechanism considering not only the primitive input features but also interactive property with environment. Moreover, human brain can learn and memorize many new things without catastrophic forgetting of existing ones. It is well known that the Fuzzy ART network can be easily trained for additional input pattern and also can solve the stability-plasticity dilemma in conventional multi-layer neural network [9]. Therefore, we use the Fuzzy ART network together with the bottom-up saliency map model to implement a trainable selective attention model that can interact with human supervisor. During the training process, the Fuzzy ART network learns and memorizes the characteristics of the uninteresting areas and/or interesting areas selected by the bottom-up saliency map model. The uninteresting or interesting areas to be trained by Fuzzy ART are decided by human supervisor, which is different from the general Fuzzy ART training mechanism. After successful training of the Fuzzy ART network, an unwanted salient area is inhibited and a desired area is reinforced by the vigilance value ρof the Fuzzy ART network. The feature maps, which are used as input of the Fuzzy ART network, consist of continuous real values. Thus, we considered the Fuzzy ART network that can process the real values like ART 2 network. Moreover, Fuzzy ART network has more or less simple structure to implement than ART 2 network and also shows suitable performance for analog pattern clustering. Any ART-type net can be characterized by its preprocessing, choice, match and adaptation rule, where choice and match define the search circuit for a fitting prototype. The detail description for these roles of the Fuzzy ART is referred in [9].

Fig. 1 shows the architecture of the trainable selective attention model during training process.

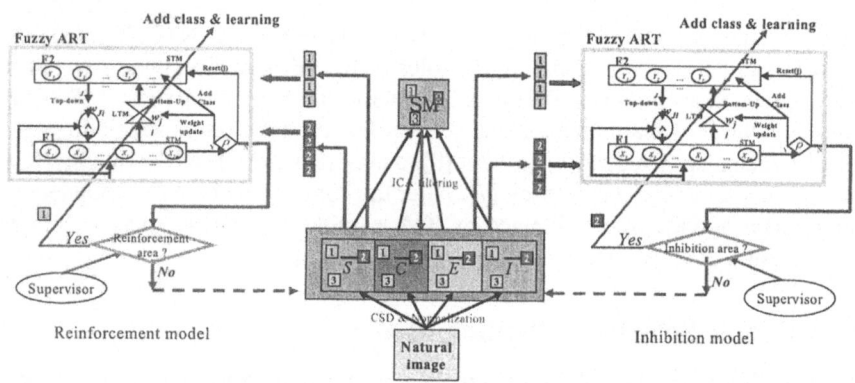

Fig. 1. The architecture of the proposed trainable selective attention model using Fuzzy ART network during training mode ($\overline{I}$: intensity feature map, $\overline{E}$: edge feature map, $\overline{S}$: symmetry feature map, $\overline{C}$: color feature map, ICA : independent component analysis, SM : saliency map). Square blocks 1 and 3 in the SM are interesting areas, but block 2 is uninteresting area

As shown in Fig. 1, the attention area is decided by the bottom-up saliency map model and the four feature maps corresponding to the attention area are used as input of the Fuzzy ART model, and then a supervisor decides whether it is a reinforcement area or an inhibition area. If the selected area is unwanted area even though it has salient features, the inhibition part of the Fuzzy ART model trains and memorize that area to be ignored in later processing. If the selected area is an interesting area, then it is not involved in training process of the Fuzzy ART model for inhibition. If some area is decided to be reinforced by the supervisor, that area is trained by the reinforcement part of the Fuzzy ART model.

Fig. 2 shows the architecture of the proposed model during test mode. After training process of the Fuzzy ART model is successfully finished, it memorizes the characteristics of unwanted areas and desired area. If a salient area selected by the bottom-up saliency map model of a test image has similar characteristics of Fuzzy ART memory, it is ignored by inhibiting that area in the saliency map or magnified by reinforcing that area in the saliency map. In the proposed model, the vigilance value in the inhibition part of the Fuzzy ART model is used as a decision parameter whether the selected area is interesting or not. When the 4 feature maps of an unwanted salient area inputs to the inhibition part of the Fuzzy ART model, the vigilance value is higher than a threshold, which means that it has similar characteristics with the trained unwanted areas. Therefore, the inhibition part of the trainable selective attention model inhibits those unwanted salient areas not so as to give an attention to them. In contrast, when a desired salient area inputs to the reinforcement part of the Fuzzy ART model, the vigilance value becomes higher than a threshold, which means that such an area is interesting area. As a result, the proposed model can focus on a desired area, but it does not focus on a salient area with unwanted feature.

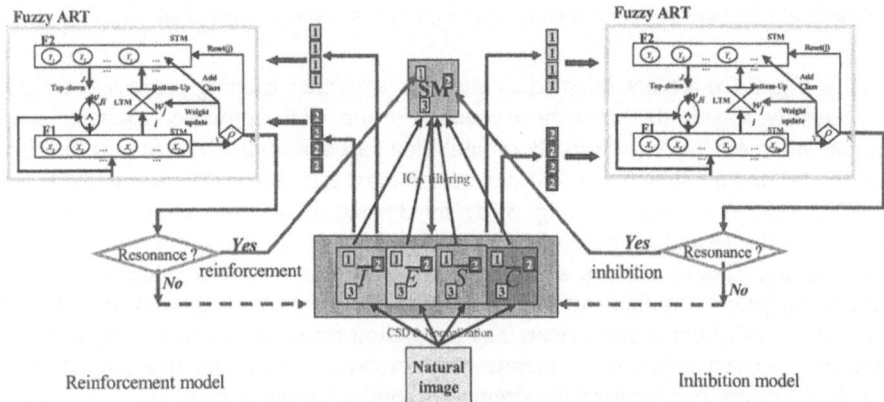

Fig. 2. The architecture of the proposed trainable selective attention model using Fuzzy ART network during test mode ($\overline{I}$: intensity feature map, $\overline{E}$: edge feature map, $\overline{S}$: symmetry feature map, $\overline{C}$: color feature map, ICA : independent component analysis, SM : saliency map) Square blocks 1 and 3 in the SM are trained interesting areas, but block 2 is trained uninteresting area

4 Computer Simulation and Experimental Results

Fig. 3 (a) and (b) show the simulation results of the proposed bottom-up attention model. The numbers in the Fig. 3 (b) represent the scan path according to the degree of saliency.

Fig. 3 (c) and (d) shows the simulation result of the proposed trainable selective attention model. In Fig. 3 (b), the 3rd and 4th salient area are decided as unwanted Salient region by human supervisor. As shown in Figs. 3 (c), inhibited 3rd and 4th salient

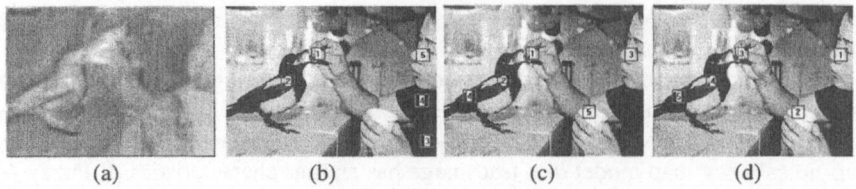

Fig. 3. Simulation results: (a) bottom-up saliency map of input image, (b) scan path generated by attention model, (c) scan path generated by attention model with inhibition, (d) scan path generated by attention model with both inhibition and reinforcement

Fig. 4. Simulation results in natural scenes (First column: scan path generated by bottom-up attention model, Second column: scan path generated by top-down trainable attention model in conjunction with bottom-up attention model): (a) outdoor scene, (b) indoor scene

area in Fig. 3 (b) are not selected as attention area after training process is finished successfully. Fig. 3 (d) shows the simulation results of the proposed trainable selective attention model considering both inhibition part and reinforcement part. In Fig. 3 (c), the 3rd and 5th salient area are decided as most interesting salient regions by human supervisor. As shown in Fig. 3 (d), the 3rd and 5th salient area in Fig. 3 (c) are selected as the 1st and 2nd salient areas after reinforcement training process is finished successfully. This simulation shows that the proposed trainable selective attention model can properly inhibit an unwanted area and reinforce a desired area through interaction with human supervisor. The simulation result shown in Fig. 3 (d) shows that the proposed model can generate very reasonable scan path like human visual attention system. Fig. 4 shows the simulation results for two scenes.

As shown in Fig. 4 (a), the proposed model can properly pay an attention to the cars in the outdoor parking lot like human vision system. Moreover, our attention model can decide well meaningful salient area in the indoor environment as shown in Fig. 4 (b). In Fig. 4, the first column images shows the selected attention areas using only the bottom-up saliency map model, and the second column shows the selected attention areas using the proposed trainable selective attention model using Fuzzy ART. As shown in Fig. 4, the simulation results show that our proposed model can generate human-like scan path by considering the top-down trainable inhibition and reinforcement mechanism as well as the bottom-up visual processing mechanism.

5 Conclusion

We proposed a trainable selective attention model that can not only inhibit an un-wanted salient area but also reinforce an interesting area in a static natural scene. The proposed model was implemented using a Fuzzy ART network in conjunction with a biologically motivated bottom-up saliency map model. Computer simulation results show that the proposed method gives a reasonable salient region and generates scan path that can be modified to desired ones through interaction with human supervisor.

Acknowledgement

This work was supported by Korea Research Foundation Grant. (KRF-2002-D120303-E00047).

References

1. Itti, L., Koch, C., Niebur, E.: A model of saliency-based visual attention for rapid scene analysis. IEEE Trans. Patt. Anal. Mach. Intell. Vol. 20. 11 (1998) 1254-1259
2. Koike, T., Saiki, J.: Stochastic Guided Search Model for Search Asymmetries in Visual Search Tasks. BMCV 2002, Lecture Notes in Computer Science, Vol. 2525, Springer-Verlag, Heidelberg (2002) 408-417
3. Sun, Y., Fisher, R.: Hierarchical Selectivity for Object-Based Visual Attention. BMCV 2002, Lecture Notes in Computer Science, Vol. 2525, Springer-Verlag, Heidelberg (2002), 427-438
4. Bruce Goldstein E.: Sensation and Perception. 4th edn. An international Thomson publishing company, USA (1995)
5. Park, S. J., An, K. H., and Lee, M.: Saliency map model with adaptive masking based on independent component analysis. Neurocomputing, Vol. 49, (2002) 417-422
6. Seo, K.S., Park, C.J., Cho, S.H., Choi, H.M.: Context-Free Marker-Controlled Watershed Transform for Efficient Multi-Object Detection and Segmentation. IEICE Trans. Vol. E84-A. 6 (2001) 1066-1074
7. Bell, A.J., Sejnowski, T.J.: The independent components of natural scenes are edge filters. Vision Research. 37 (1997) 3327-3338
8. Barlow, H.B., Tolhust, D.J.: Why do you have edge detectors? Optical society of America Technical Digest. 23 (1992) 172
9. Frank, T., Kraiss, K. F., and Kuklen, T.: Comparative analysis of Fuzzy ART and ART-2A network clustering performance. IEEE Trans. Neural Networks Vol. 9 No. 3 May (1998) 544-559

Genetic Algorithm for Optimal Imperceptibility in Image Communication Through Noisy Channel

Santi P. Maity[1], Malay K. Kundu[2], and Prasanta K. Nandi[3]

[1] Bengal Engineering College (DU), P.O.-Botanic Garden, Howrah, India, 711 103
[2] Machine Intelligence Unit, 203 B. T. Road, Kolkata, India, 700 108
[3] Bengal Engineering College (DU), P.O.-Botanic Garden, Howrah, India, 711 103

Abstract. Data embedding in digital images involves a trade off relationship among imperceptibility, robustness, data security and embedding rate etc. Genetic Algorithms (GA) can be used to achieve optimal solution in this multidimensional nonlinear problem of conflicting nature. The use of the tool has been explored very little in this topic of research. The current paper attempts to use GA for finding out values of parameters, namely reference amplitude (A) and modulation index (μ) both with linear and non linear transformation functions, for achieving the optimal data imperceptibility. Results on security for the embedded data and robustness against linear, non linear filtering, noise addition, and lossy compression are reported here for some benchmark images.

1 Introduction

The properties of data hiding techniques in digital images such as perceptual transparency, higher capacity, statistical invisibility or security of the hidden data, and robustness to some types of attacks are related in a conflicting manner and the design tradeoff depends on the applications (see [1]and [2]). Genetic Algorithms (GAs) can be used to optimize the conflicting requirements of data hiding problem but the use of the tool has been explored very little in this topic of research although there are many problems in the area of pattern recognition and image processing [3] where GAs perform an efficient search in complex spaces in order to achieve an optimal solution.

In the present paper GA is used to find two parameter values, namely reference amplitude (A) and modulation index (μ) in linear and non linear transformation functions used to modulate the auxiliary message. The cover object (image) is chosen in which the message is well hidden unlike in watermarking applications where the message must be hidden in the cover work to which it refers and no other. Image regions with relatively low information content and having pixel values in the lower and upper portion of the dynamic range are used for data hiding as the characteristics of human visual system (HVS) are less sensitive to the change at these two ends.

The paper is organized as follows. Sections 2 describes transformation functions for message modulation and how to calculate the range of parameters.

N.R. Pal et al. (Eds.): ICONIP 2004, LNCS 3316, pp. 700–705, 2004.
© Springer-Verlag Berlin Heidelberg 2004

Genetic algorithm for data embedding is presented in section 3 with embedding and recovery process in section 4. Sections 5 and 6 present some results for highlighting the use of GA and conclusions respectively.

2   Modulation Function and Calculation of Parameters

The cover and the auxiliary message are chosen as gray level images and the auxiliary message is modulated using suitable transformation functions to match the characteristics of the cover image regions. Distortions are then created over stego image to simulate the behavior of a noisy channel.

The power-law function $X' = A(X + \varepsilon)^\mu$ which is widely used for image enhancement operation is considered as modulation function. Here X denotes the pixel value in auxiliary message and transformation function modulates X to X', the pixel value of the cover image selected for embedding. Two other transformation functions, One is linear transformation function of the form $X' = A(1 + \mu X)$ and other one is parabolic function of the form $X' = A(1 + \mu\sqrt{X})$, are compared for their suitability on imperceptibility, security and robustness issues of data embedding. The following sub sections describe how to calculate the range of parameter values.

2.1   Calculation of A

The modulation function is as follows:

$$X' = A(X + \varepsilon)^\mu \tag{1}$$

Differentiating X' with respect to X, we get

$$dX'/dX = A\mu(X + \varepsilon)^{\mu-1} \tag{2}$$

Here dX'/dX is positive provided $A > 0$, $\mu > 0$ and $(X + \varepsilon) > 0$, which implies X' increases monotonically with X. The upper (U') and lower (L') bound of the modulated pixel values are

$$U' = X_{max}' = A(X_{max} + \varepsilon)^\mu \tag{3}$$
$$L' = X_{min}' = A(X_{min} + \varepsilon)^\mu \tag{4}$$

The range (Ψ) of the modulated pixel values is given as follows:

$$\Psi = U' - L' = A[(X_{max} + \varepsilon)^\mu - (X_{min} + \varepsilon)^\mu] \tag{5}$$

The relation shows that for large A value, the span of the modulated pixel values (Ψ) will be large leading to smaller probability of matching between the modulated message and embedding regions. This in turn suggests to select lower value of A for better imperceptibility. The small span (Ψ) is also possible for large A value provided very small value is selected for μ. But it is shown in the detection process that small value of μ will make the auxiliary message vulnerable to elimination in noisy transmission media. Similar argument also holds good for the value of A. The value of A depends on selection of the auxiliary message as well as regions selected for embedding. As rule of thumb A is selected as

$$A = X'_{mode}/(X_{mode} + \varepsilon)^{\mu} \qquad (6)$$

where X'_{mode} and X_{mode} respectively denote the mode of the gray values for the embedding regions and the auxiliary messgaes.

2.2   Calculation of μ

Power-law transformation suggests if μ value is taken small ($\mu < 1.0$) keeping A constant, auxiliary message is mapped into a narrow range of gray values. This fact is also supported by equation (5). Confinement of gray values inside a narrow range increases the probability of matching between the modulated message and the data embedding region. But very small value of μ makes the detection of message impossible even after a very small image distortion. The upper and lower value of μ are calculated as follows:

It is found that X' is a monotonically increasing function of X. The value of ε, acts as offset value in image display, is set to (~ 0.01). From equation (1), we write $X'_{max} = A(255 + 0.01)^{\mu} = A(255.01)^{\mu}$.

The maximum X value is taken 255 for monochrome gray level image. The corresponding μ value is designated as μ_{max} and is related with X_{max} and A as follows: $\mu_{max} \simeq \frac{\log X'_{max} - \log A}{\log 255}$

Similarly, μ_{min} value can be written as follows:

$$\mu_{min} \simeq \frac{\log X'_{min} - \log A}{\log 0.01} = \frac{\log A - \log X'_{min}}{2.0}$$

The value of μ will be positive if A lies between X_{max} and X_{min} where the later values represent the maximum and minimum gray values of the embedding regions respectively.

3   GA for Data Hiding

Let us analyze the use of GA in data hiding problem (any problem) and the main steps are as follows [4]:

(1) Chromosomal representation of the parameter values, namely the reference amplitude (A) and modulation index (μ) associated with the problem.

(2) Creation of an initial population of chromosomes for possible solutions by using random binary strings of length pq where p represents the number of parameters and q are the bits assigned for each parameter.

(3) To quantify the closeness measure among pixel values over sub image or image, average Euclidean distance is considered here as fitness function. The value of pay-off function can be expressed mathematically as follows:

$$F(A, \mu) = \frac{\sum_{i=0}^{N-1} \sum_{j=0}^{N-1} [S_{ij} - e_{ij}(A, \mu)]}{N^2} \qquad (7)$$

where S_{ij} is the gray value of (i, j)-th pixel of the embedding region, $e_{ij}(A, \mu) = A(X_{ij} + c)^{\mu}$, is the gray value of (i, j)-th pixel of the message after modulation and N^2 is the total number of pixels in the embedding regions as well as in the auxiliary message.

(4) According to De Jong's elitist model [5], the best fit string obtained is included in the new population to preserve the best structure in the population.

(5) Although the high mutation probability leads to exhaustive search which may results in better imperceptibility at higher computation cost. Here moderate value of mutation probability is chosen in order to achieve imperceptibility at comparatively lower cost.

Size of the population is 20, Number of generations are 800, Probability of crossover per generation is 0.95, probability of mutation is 0.005.

4   Message Hiding and Recovery

The particular cover image can be selected from its histogram on the basis of higher frequency of occurrence in pixel values in either end and the auxiliary message is modulated using the transformation functions. The message modulated by transformation function using GA replaces the selected regions of the cover image.

Data recovery process uses inverse transformation function that maps X' into X and thus message is recovered. If power-law, linear and parabolic functions are used for message modulation, extracted message can be represented respectively as follows:

$$X = (X'/A)^{1/\mu} - \varepsilon \tag{8}$$

$$X = (1/\mu A)(X' - A) \tag{9}$$

$$X = 1/(\mu A)^2 (X' - A)^2 \tag{10}$$

Differentiating equations (8),(9) and (10) with respect to X' the following equations are obtained respectively,

$$dX/dX' = (1/\mu)(1/A)^{1/\mu}(X')^{1/\mu - 1} \tag{11}$$

$$dX/dX' = 1/\mu A \tag{12}$$

$$dX/dX' = 2(X' - A)/(\mu A)^2 \tag{13}$$

dX/dX' denotes the change of X with respect to the change of X' i.e. a measure of noise immunity in the detection process. The large values of A and μ are preferable for reliable decoding whereas small values of the same are desirable for better imperceptibility. Lower value of dX/dX' indicates better reliability in detection process.

5   Results and Discussions

The efficiency of the proposed algorithm is tested by embedding different messages in several cover images like Cameraman (results reported), Black bear,

Fig. 1. (a) Cover image; (b), (c), (d) Stego images using power-law (PSNR=41.77 dB), parabolic function (PSNR=51.58 dB), linear function (PSNR= 47. 45 dB) respectively; (e) Auxiliary message; (f), (g), (h) Extracted messages from (b), (c), (d) respectively; (i), (j), (k) Extracted messages from mean filtered stego images when power law, parabolic, linear functions are used respectively; (l), (m), (n) Extracted messages from median filtered stego images when power law, parabolic, linear functions are used respectively; (o), (p) Extracted messages from compressed (JPEG) stego images (PSNR=31.06 dB) and (PSNR=31.88 dB) using power law and linear function respectively.

Bandron [6] etc. and the messages are extracted from various noisy version of the stego images. Peak signal to Noise Ratio (PSNR), relative entropy distance (Kulback Leibler distance)[7], and mutual information [8] are used as representative objective measures of data imperceptibility, security and robustness respectively. Table 1 shows the robustness results where 1, 2, and 3 used in the objective measures indicate the results for power-law, linear and parabolic functions respectively.

Robustness efficiency against mean, median filtering and lossy compression are shown in respective figures. Poor robustness in the case of power law function is supported by eqn. (11) where small values of A and μ causes change in X' manifold even for the small change in X. Linear transformation function offers better imperceptibility as large range of message gray values can be mapped to smaller range by choosing the small slope i.e. the product of A and μ. At the same time better resiliency is achieved since dX/dX' (equation 12) is no way dependent on X' although very small values of A and μ will affect the detection process. Best data imperceptibility and security result is possible in case of parabolic function, as small values of A and μ map wide range of message

Table 1. Performance results of different modulation functions

Generation number	PSNR (dB)$_1$	ε value (1)	$I(X;Y)$ value(1)	PSNR (dB)$_2$	ε value (2)	$I(X;Y)$ value(2)	PSNR (dB)$_3$	ε value (3)	$I(X;Y)$ value(3)
50	40.56	0.046	0.28	43.49	0.037	0.48	45.42	0.037	0.38
150	40.79	0.045	0.21	44.36	0.040	0.44	46.74	0.037	0.39
400	40.90	0.041	0.20	47.90	0.036	0.42	48.73	0.036	0.34
600	41.50	0.039	0.19	48.36	0.030	0.39	50.37	0.025	0.35
800	41.77	0.038	0.15	47.45	0.034	0.40	51.53	0.023	0.32

gray values to the narrow region in the lower range of pixel values of the cover image. Detection reliability in such case is also satisfactory since dX/dX' does not contain X' with power term of A and μ like power law transformation function, although small values of the parameters affect detection process little more compared to linear transformation function.

6   Conclusions

The paper proposes an invisible image-in-image communication through noisy channel where linear, power-law and parabolic functions are used to modulate the auxiliary messages. GA is used to find the optimal parameter values, viz. reference amplitude (A) and modulation index (μ) for data imperceptibility. Experimental results show that parabolic function offers higher visual and statistical invisibility and reasonably good robustness, whereas, linear function offers higher robustness with reasonably good invisibility. Power-law function neither provides good resiliency nor imperceptibility. Future work can be directed to develop better form of modulation functions in order to improve the robustness performance against various types of distortions in stego images along with higher embedding rate.

References

1. Macq, B.:Special issue on identification and protection of multimedia information. Proc. IEEE. **87** (1997) 1673–1687
2. Vleeschouwer, C., Delaigle, J., Macq,B.: Invisibility and application functionalities in perceptual watermarking-an overview. Proc. IEEE. **90** (2002) 64–77
3. Ankerbrandt, A., Unckles, B., Petry, F.: Scene recognition using genetic algorithms with semantic nets. Pattern Recognition Lett. **11** (1990) 285-293
4. Pal, S., Bhandari, D., Kundu, M.: Genetic algorithms for optimal image enhancement. Pattern Recognition Lett. **15** (1994) 261–271
5. Goldberg, D.: Genetic Algorithms: Search, Optimization and Machine Learning. Addison-Wesley, Reading, M.A., (1989)
6. http://www.cl.cam.ac.uk/~fapp2/watermarking
7. Cachin, C.: An information theoretic model for steganography. Proceedings of 2nd Workshop on Information Hiding. D. Aucsmith (Eds.). Lecture Notes in Computer Sciences, Springer-verlag, USA, **1525** (1998)
8. Lathi, B.: Modern Digital and Analog Communication Systems (Third Edition). Oxford University Press, New Delhi, India (1999)

High Speed Extraction Model
of ROI for Automatic Logistics System

Moon-sung Park[1], Il-sook Kim[2], Eun-kyung Cho[3], and Young-hee Kwon[3]

[1] ETRI, Postal Technology Research Center, 161 Ga-jung-dong, Daejeon, 305-350, Korea
mspark@etri.re.kr
[2] Paichai University, 439-6 Do-ma-dong Seo-gu, Daejeon, 302-735, Korea
kimis@pcu.ac.kr
[3] DaeDuk College, Department of Computer, Internet & Information,
48 Jang-dong You-sung-gu, Daejeon, 305-715, Korea
{ekcho,yhkwon}@ddc.ac.kr

Abstract. This paper deals with a model for the high-speed extraction of Region of Interest (ROI) during the process of logistics transported on conveyor belt. The objective of this paper is to extract various ROIs of the large size image of logistics more than 4,096 by 4,096. For this purpose, we propose a wavelet extension and a texture extension method, and verified it by the experiments. Our experiments show that the proposed methods extract ROIs nearly 100 percent.

1 Introduction

In proportion of the rapid growth of off-line logistics with the development of e-commerce, it is absolutely needed to develop technology for more effective classification and conveyance [1]. These day logistics is mostly distributed in a rectangular parallelepiped shape for effective load and protection of the product. Presently logistics is transported on conveyor belt, with its addressee on the top side. The operator checks the area code with his own eyes and then inputs the classification information for machine distribution. And the system is to convey the logistics by interpreting the barcode and to deliver them to their destination [1,2]. The system efficiency is mainly relied on the performance of the core information extraction from the logistics. So, it is absolutely needed to develop an effective extraction technology to acquire it.

This paper deals with the extraction of Region-of-Interest (ROI) at high-speed from the large size image and verified it by applying the regular pattern. To extract ROI, we propose new methods by enlarging both of the texture-ROI and the wavelet-based ROI. Especially it examined the feature value of barcode divided by 3 steps with minute block unit, and it applied the labeling system so that it may filter the minute blocks to be erroneously detected or excluded at examination. It also minimized the outer information of logistics so that the examination range can be reduced to the minimum. In the process of examination at each step, the mini buffer is utilized to minimize the access of the original image.

N.R. Pal et al. (Eds.): ICONIP 2004, LNCS 3316, pp. 706–713, 2004.
© Springer-Verlag Berlin Heidelberg 2004

2 Texture-Based ROI Extraction

To extract only the barcode ROI among the various information of the logistics image, it divides whole region into the minute blocks and extract information such as barcode, character strings and logistics appearance, not examining the information in regular sequence. Existing information of each block is applied with the consideration of the features such as edge and ripple in order to satisfy the conditions above, and the acquired image accordingly is divided by 32 by 32 pixels. The dynamic critical value is calculated from the remainder of maximum and minimum values of each block. And also, in order to eliminate the image of the conveyor belt, the logistic image which the remainder of maximum and minimum value is under the dynamic critical value is excluded from the inspection.

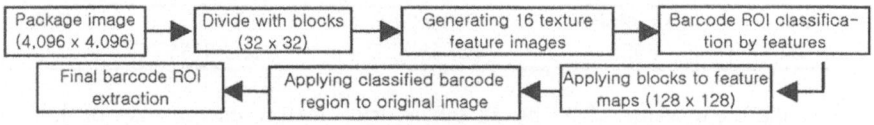

Fig. 1. General flow of texture method

Fig. 1 is the flow diagram for the barcode ROI extraction method. The original image is divided by the block of 32 by 32 pixels (N, M), and each pixel value is subtracted the mean of blocks in order to eliminate the influence of illumination created by the slight window movement occurred near the image. Then the average pixel intensity becomes nearly 0, creating a preprocessing image. Also the mean value of blocks can be computed and the remainder image is acquired by using the Formula 1. In Fig. 2, (a) are the original image, (b) the remainder and (c) the histogram equalized image of the remainder. These transformed images are used to extract the texture feature. In this section, Law's texture energy feature is applied and the following 4 vectors turn to 16 feature vectors.

(a) Original image   (b) Remainder image   (c) Histogram equalized

Fig. 2. The image subtracted the mean value

Here L5 vector indicates the partial average centroid. E5, S5 and R5 extracts the edge, the spot, and the ripple accordingly. Four feature vectors presented are from the study results of Law's texture energy.

$$S(x, y) = \left. \sum_{x, y = 0}^{N, M} I(x, y) \middle/ N \times M \right. \tag{1}$$

L5 (Level) = [1 4 6 4 1], E5 (Edge) = [-1 -2 0 2 1]
S5 (Spot) = [-1 0 2 0 -1], R5 (Ripple) = [1 -4 6 -4 1] .

Also 2-D convolution mask applied to block images is computed by the outer product of vector pairs. For example, mask E5L5 can be computed externally as follows: Each mask of sixteen 5 by 5 pixel creates 16 filtered images when applied to the proposed images. Let $F_k[i,j]$ denote the filtered result of the kth mask in pixel [i, j]. Then in filter k, texture energy map E_k is defined as Formula 2 [3,4].

$$\begin{bmatrix} -1 \\ -2 \\ 0 \\ 2 \\ 1 \end{bmatrix} \times \begin{bmatrix} 1 & 4 & 6 & 4 & 1 \end{bmatrix} = \begin{bmatrix} -1 & -4 & -6 & -4 & -1 \\ -2 & -8 & -12 & -8 & -2 \\ 0 & 0 & 0 & 0 & 0 \\ 2 & 8 & 12 & 8 & 2 \\ 1 & 4 & 6 & 4 & 1 \end{bmatrix} , \quad E_k[r,c] = \sum_{j=c-16}^{c+16} \sum_{i=r-16}^{r+16} |F_k[i,j]| \cdot \qquad (2)$$

Table 1. 9 Energy maps

L5E5/E5L5	L5R5/R5L5	E5S5/S5E5	S5S5	R5R5
L5S5/S5L5	E5E5	E5R5/R5E5	S5R5/R5S5	

With the mean value of 32 by 32 pixels, each texture energy map of 4,096 by 4,096 pixel image is reduced into the 128 by 128 mask image. Once sixteen energy maps are generated, the definite symmetrical pair is combined to the creation of nine last maps replacing each pair by its mean. For example, E5L5 measures the horizontal edge and L5E5 the vertical edge. The mean of these two maps measures the general edge. The resulted nine energy maps are as follows in Table 1. Fig. 3 shows the energy map on block images of barcode, letters and object. Here (a) indicates the original image, (b) E5E5, (c) S5S5, (d) R5R5, (e) E5L5, (f) S5L5, (g) R5L5, (h) S5E5, (i) R5E5, (j) R5S5, (k) L5E5, (l) L5S5, (m) L5R5, (n) E5S5, (o) S5R5.

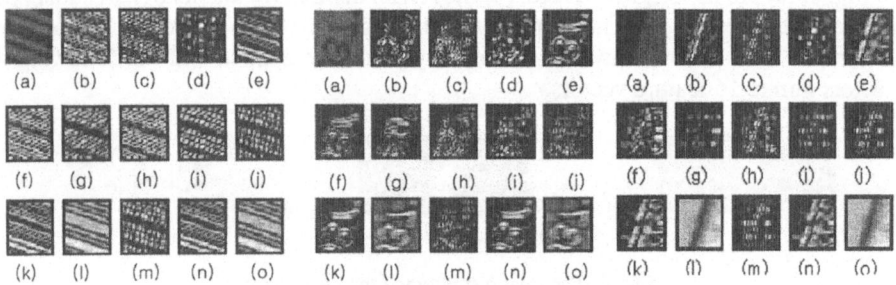

(a) (b) (c) (d) (e)   (a) (b) (c) (d) (e)   (a) (b) (c) (d) (e)
(f) (g) (h) (i) (j)   (f) (g) (h) (i) (j)   (f) (g) (h) (i) (j)
(k) (l) (m) (n) (o)   (k) (l) (m) (n) (o)   (k) (l) (m) (n) (o)

Fig. 3. Block images of barcode, letters and object

Table 2. 9 Feature vectors

	E5E5	S5S5	R5R5	E5L5	S5L5	R5L5	S5E5	R5E5	R5E5
Barcode	61.5	19.79	50.9	298.7	239.7	408	64.39	131.4	53.27
Letters	51.61	18.95	57.72	220.2	130	160.5	33.54	42.21	26.79
Object	18.6	3.6	15.25	75.18	20.38	28.19	6.09	8.22	5.41

Table 2 refers to mean value of barcode, letters and object block of nine feature vectors accordingly. In section 2, only 3 vectors out of 16 are chosen to be used with the creation of the image map in order to extract barcode region only. Since the repetition elements have a larger value in R5L5 according to the feature of energy map, the barcode can be easily identified. Moreover R5E5 represents the repetition of edge and S5L5 emphasizes the feature of spot. Fig. 5 shows a whole image and a image that show three elements. In this section we created 128 by 128 images by using the critical value of the 3 feature vectors above. In order to eliminate some noise nearby and to extract the exact barcode region from the created image, a labeling method and a simple expansion algorithm are applied. It was experimented about 4,096 by 4,096 pixel image in CPU 1GHz. The image is acquired by Metrox line CCD scan camera. The Table 3 presents the block processing time and the general processing time accordingly. We observed that it takes quite long time for the process generally. Also with the processing of the images with letters and pictures in the background, it shows far better results in terms of the confidence as shown in Fig. 4. Afterwards, it is possible to classify the letter region and the picture region if the texture information is more utilized.

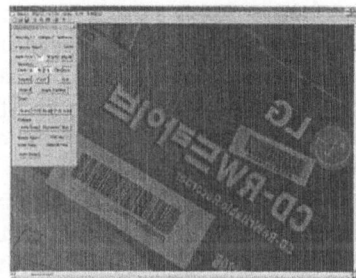 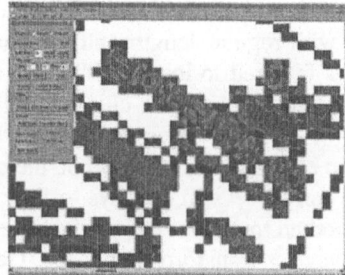

Fig. 4. The original image and the result of R5E5, R5L5, S5L5

Table 3. Block texture procedure time and performance

Class	Block procedure	General procedure
General feature vector	6msec	33,000msec
3 feature vector	1.2msec	2,300msec

It reduces a great amount of time to calculate the vector only needed for the barcode ROI extraction (766.7msec). Also if the background region in both conveyor belt and logistics image is eliminated, the computation time can be reduced significantly. In case of the maximum size logistics, it is added the time for barcode region labeling and interpretation to the feature vector extraction time. However the logistics ROI extraction time should be done within 600 msec including image acquisition time, ROI extraction time and barcode interpretation time.

3 Wavelet-Based ROI Extraction Method

In this paper we propose a barcode ROI extraction method out of the wavelet transformed images. Mostly the barcode used in general logistics such as package delivery is composed of straight lined bars. According to the direction of barcode label its direction varies to horizontal, vertical and diagonal. In the proposed method, the wavelet transformation which has a characteristic of using each featured directions independently well is applied to its best use. It inspects whether they are within the

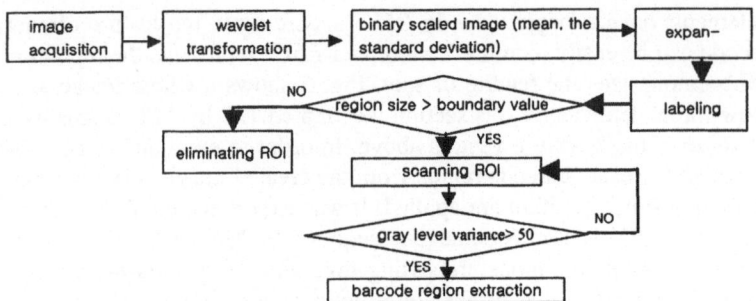

Fig. 5. General flow of wavelet transformation method

barcode region for each frequency range and detects some candidates for barcode region. The barcode candidate regions take the final step to be distinguished about the length and direction of barcode through scan and inspection procedure. In barcode, the bars with regular length and direction are arranged in parallel. And one bar in barcode is featured to locate within the very close distance with another bar nearby. This feature gives the basic clue not only to interpret it but also to extract the candidate region. Therefore there needs a method to detect the candidate region composed of a group of straight bars, based on the basic feature above [7,8].

There can be several of ways to find the candidate region, but in section 3 it presents a method to extract the candidate region by extracting regions including straight bars in wavelet transformed images. Generally in wavelet transformed images, the original image is down sampling to be 1/2. The bar code region of it reacts strongly and widely among regions in each frequency of horizontal, vertical and diagonal images.

Therefore, after the wavelet transformation, the largest region is selected to be the candidate among its horizontal, vertical and diagonal image after the binary scale and region labeling process. These candidate regions are scanned in the direction of their length and they are finally examined to be the barcode region according to their transformation extent of gray level. The flow chart of the presented idea is shown in Fig. 5 Wavelet transformation is to transmit the given function f(x) to wavelet surface and analyze it into a number of resolution. Wavelet space is composed of basis functions with different resolution. In wavelet transformation, the basis functions are generated by expanding and transforming the original wavelet.

$$\phi=\left(\frac{t}{2}\right)=-h_3\phi(t+2)+h_2\phi(t+1)-h_1\phi(t)+h_0\phi(t-1) \quad h_0=\frac{1+\sqrt{3}}{4}, \ h_1=\frac{3+\sqrt{3}}{4}, \ h_2=\frac{3-\sqrt{3}}{4}, \ h_3=\frac{1-\sqrt{3}}{4} \cdot \quad (3)$$

In this part Daubechies is used as a generating function of wavelet. Formula 3 presents the generating function of Daubechies and includes four variables. As barcode ROI has its straight line arrangement existed in horizontal, vertical and diagonal region, after wavelet transformation it applies the binary scale and expansion. It also performs labeling in only three of high frequency regions. Fig. 6 (a) shows the classification to indicate the analysis results of the image signals. Like in Fig. 6 (c), wavelet transformed image is applied by the threshold method using the mean and the standard deviation dispersion in the process of binary scale. The feature of binary scaled barcode region is formed with regular internals between bars.

LL	LH (Horizontal)
HL (Vertical)	HH (Diagonal)

(a)   (b)   (c)

Fig. 6. (a) Resolution result of image signals, (b) Original image, (c) Wavelet transformed image

$$\tau = k_1\mu + k_2\sigma \ . \tag{4}$$

Formula 4 is to calculate the critical value. In the Formula above, □ indicates the mean value, σ the standard deviation, k_1 and k_2 use 1. Since the linear barcode is arranged with regular intervals unlike other characters and noises, so when the interval gets bigger in process of binary scale, the barcode region may be divided. In experiment image, as the interval between bars is from 1 to 5 pixels, if it is covered up 3 by 3 masks, the division of barcode region can be prevented.

Fig. 7 (a) shows the result image of the experiment above by covering up with the 3 by 3 mask. But unfortunately, this has a shortcoming that is not applicable when there is a change of the gray level value and the pixel interval value. The barcode feature presents at least one outstanding feature out of three high frequency region of wavelet. And also in the process of binary scale and expansion, the letters region and the noise region are increased as well. To eliminate the noise region created here, the size of labeling region is measured by labeling method. And the region that has less size of region than the critical value is considered as noise and is eliminated. The rest is processed by the procedure to verify as candidate region. In this experiment, the region under the size of 100 was considered as noise region and was eliminated.

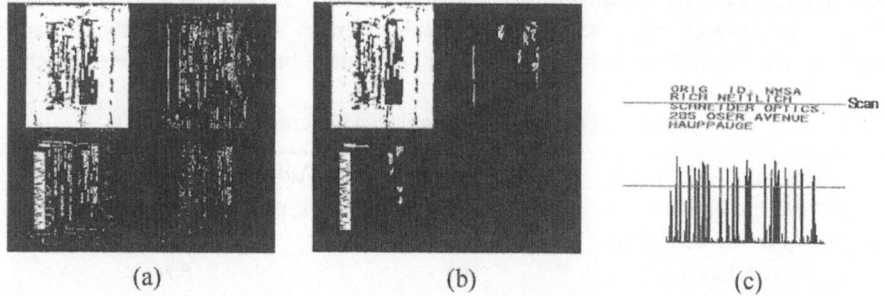

(a)   (b)   (c)

Fig. 7. (a) Expansion of barcode region by 3 by 3 mask, (b) Inspection of candidate region, (c) Labeling

Fig. 7 (b) presents the result image after labeling. Barcode region has tens of black bars and white intervals arranged with a regular rule. This type of barcode region has more change than other letters or noises. Accordingly, to verify the extracted candidate region, it is scanned orthogonally to the length of the region. If the summation of gray level change is more than the critical value, the candidate region is extracted as barcode region. Fig. 7 (c) is the result to scan the median line from the extracted candidate barcode region. In this part the critical value for the change is fixed to 50, considering the mean of image and the average critical value of barcode region value.

Table 4. Average Wavelet algorithm procedure time

Algorithm	Wavelet transformation	Critical value & dilation	Labeling	Barcode verification	Total time consumed
Procedure time (msec)	300	100	450	50	900

On the experiment of barcode extraction from 40 post packages, more than 90% was analyzed to be accurate by down sampling 4,096 by 4,096 pixel logistics image to 1,024 by 1,024 pixel. The processing time on general image takes about 900 msec averages in Pentium 1 GHz. Table 4 indicates the average time taken in partial algorithm.

4   Performance of ROI Extraction Method

The time to acquire logistics is not calculated to compare the efficiency of ROI extraction because major delay to access logistics image is 4,096 by 64 pixel image access time. This means the access is available in 6.24 msec after the image is stored. The length of logistics above case is 7.8 mm. The test logistics are designed in various sizes and also with from simple information to complex images and letters mark on the surface. Logistics is designed in 150 types with a number of different types of barcode, such as linear barcode or four kinds of 2-D barcodes with and without logistics label. The average processing time applied by wavelet and texture method is presented in Table 5. As shown in Table 5, it is proper to use the wavelet transformation method, but when adjusting the test level used in texture method or acquiring barcode and object only, the whole processing can be shortened greatly. It can be applied as a practical system only when labeling time is improved in its performance.

Table 5. Performance comparison with formal method

	Wavelet procedure time	Texture method procedure time
Wavelet transformation	300	2,300 (object, barcode, letters)
Binary scale and expansion	100	
Labeling	500	500
Candidate region analysis	10	10
Total time consumption	910 (msec)	2,810 (msec)

The ROI extraction step during this image processing is presented that the letters, shape, barcode, linear element, or the boundary of image region is included and the masks with little change quantity are eliminated. During this there are some non-extracted regions which should have been classified as ROI candidates.

The outer logistics region with similar level value comparing to the surface of conveyor belt are not partially extracted. And the performance has been analyzed under the real-time testing environments with its specially designed logistics devices and the image acquisition mock-up. It is estimated that a variety of information can be drawn simultaneously with proper steps, after selecting the prospective ROI and adjusting on those non-extracted regions.

5 Conclusions

As mentioned in the paper, it is concluded that the wavelet method shows the better performance. But it requires some more improvements on labeling process time in order to be practically applicable, because it can save significant time to acquire the barcode and the object only after adjusting the examination level used in texture method. And it will be applied as the automatic logistics identification system, but this needs the consideration of the time for the selection of the needed barcode among those extracted. Only 95% of primarily extracted is the accurate barcode and with the process of inspection it becomes 100%. Therefore it should successfully extract more than 99% from the first extraction and also under the situation that the barcode region is damaged or misinterpreted. The solution to these problems should be answered quickly and additionally by the study on the extraction of ROI in 3-D logistics images extracting the barcode region, detecting the error of input direction and extracting ROI of barcode positioning.

References

1. Korea Post, "2001 Yearly Report on Postal Service"consulted http://www.koreapost.go.kr.
2. UPU, "Technical Standards Manual", International Bureau of the Universal Postal Union, 1998. 7.
3. J. R Parker "Algorithms for image processing and computer vision", pp. 250 - 274, 1999.
4. R. C. Gonzalez and R. E. Woods, "Digital Image Processing", Prentice Hall, 2002.
5. ISO/IEC/JTC1/SC31, "Bar Coding Symbology Specification Code 128," 13 March 1999.
6. J. R Parker "Algorithms for image processing and computer vision", pp.250 - 274, 1999.
7. Y.P. Zhou, C.L. Tan "Hough - based model for recognizing bar charts in document images", Proceedings of SPIE Document Recognition and Retrieval VIII, Vol. 4307, 20ders only).
8. Joohyung Cho. Tae-wan Kim, and Kunwoo Lee "Surface Fairing with Boundary Continuity Based on the Wavelet Transform", ETRI Journal, Volume 23, Number2, pp. 85-95, June 2001.

Using Biased Support Vector Machine to Improve Retrieval Result in Image Retrieval with Self-organizing Map

Chi-Hang Chan and Irwin King

Department of Computer Science and Engineering
The Chinese University of Hong Kong
Shatin, N.T., Hong Kong
{chchan,king}@cse.cuhk.edu.hk

Abstract. The relevance feedback approach is a powerful technique in content-based image retrieval (CBIR) tasks. In past years, many intra-query learning techniques have been proposed to solve the relevance feedback problem. Among these techniques, Support Vector Machines (SVM) have shown promising results in the area. More specifically, in relevance feedback applications the SVMs are typically been used as binary classifiers with the balanced input data assumption. In other words, they do not consider the imbalanced dataset problem in relevance feedback, i.e., the non-relevant examples outnumbered the relevant examples. In this paper, we propose to apply our Biased Support Vector Machine (BSVM) to address this problem. Moreover, we apply our Self-Organizing Map-based inter-query technique to reorganize the feature vector space, in order to incorporate the information provided by past queries and improve the retrieval performance for future queries. The proposed combined scheme is evaluated against real world data with promising results demonstrating the effectiveness of our proposed approach.

1 Introduction

The goal of relevance feedback is to learn user's preference from their interaction, and it is a powerful technique to improve the retrieval result in CBIR. Under this framework, a set of images is presented to the user according to the query. The user marks those images as either relevant or non-relevant and then feeds back this into the system. Based on these feedback information, the system presents another set of images to the user. The system learns user's preference through this iterative process, and improves the retrieval performance.

Most of the current relevance feedback systems are based on the intra-query learning approach [1–3]. In this approach, the system refines the query and improves the retrieval result by using feedback information that the user provided. Recently, researchers introduced the regular SVM [1] and one-class SVM [2] into the relevance feedback process, and it shows that SVM-based techniques are more promising and effective techniques than other intra-query approaches. The regular SVM [4] technique treats the relevance feedback problem as a strict binary classification problem. However, this technique does not consider the imbalanced dataset problem, in which the number of non-relevant images are significantly larger than the relevant images. This imbalanced dataset problem will lead the positive data (relevant images) be overwhelmed by the negative data (non-relevant images). The one-class SVM [5] uses only the relevant images in the learning process, and treats the problem as a density estimation problem. The one-class SVM technique seems to avoid the imbalanced dataset problem. However, it cannot work well without the help of negative information.

Recently, researchers propose the use of inter-query information to further improve retrieval result [6, 7]. In the inter-query approach, feedback information from past queries are accumulated to train the system to determine what images are of the same semantic meaning. These approaches show that the retrieval performance can be benefited from the inter-query learning.

In this paper, we propose a relevance feedback technique to incorporate both inter-query and intra-query information for modifying the feature vector space and estimating the users' target. Self-Organizing Map (SOM) [8] is used to cluster and index the images in the database. We apply our SOM-based inter-query technique [6] to modify the feature vector space, in which the SOM of images is stored. This allows for transforming the images distributions and improving their organization in the modified vector space. Moreover, we propose to apply our BSVM [9] technique to capture the user's individual preferences in the relevance feedback process, and address the imbalanced dataset problem in relevance feedback process.

2   Biased Support Vector Machine

Our Biased Support Vector Machine (BSVM) is derived from 1-SVMs, and the objective of BSVM is to describe the data by employing a pair of sphere hyperplanes in which the inner one captures most of the relevant data while the outer one pushes out the non-relevant data. Moreover, the distance between these two sphere hyperplanes have to be maximized. The task can be formulated as an optimization problem and the mathematical formulation of our technique is given as follows.

The objective function for finding the optimal sphere hyperplane can be formulated below:

$$\min_{R \in \mathbb{R}, \xi \in \mathbb{R}, \rho \in \mathbb{R}} \quad bR^2 - \rho + \frac{1}{n\nu} \sum_{i=1}^{n} \xi_i, \tag{1}$$

$$s.t. \quad y_i(||\Phi(\mathbf{x_i}) - \mathbf{c}||^2 - R^2) \leq -\rho + \xi_i, \tag{2}$$

$$b \geq 0, \xi_i \geq 0, \rho \geq 0, 0 \leq \nu \leq 1, \tag{3}$$

where ξ_i are the slack variables for margin error, $\Phi(\mathbf{x_i})$ is the mapping function, $\mathbf{c}$ and R are the center and radius of the optimal hypersphere, ρ is the width of the margin, b is a parameter to control the bias, and $\nu \in [0, 1]$ is a parameter to control the tradeoff between the number of support vectors and margin errors.

The optimization task can be solved by introducing the Lagrange multipliers. The dual of the primal optimization can be shown to take the form

$$\max_{\alpha} \sum_i \alpha_i y_i k(\mathbf{x_i}, \mathbf{x_i}) - \frac{1}{b} \sum_{i,j} \alpha_i \alpha_j y_i y_j k(\mathbf{x_i}, \mathbf{x_j}) \tag{4}$$

$$s.t. \sum_i \alpha_i y_i = b, 0 \leq \alpha_i \leq \frac{1}{n\nu}, \sum_i \alpha_i \geq 1 \tag{5}$$

This dual problem can be solved with quadratic programming techniques.

The way we construct the BSVM is efficient for solving the imbalanced dataset problem in relevance feedback. Since the weight allocated to the positive support vectors in BSVM will be larger than the negative ones when setting a positive bias factor b. This can be useful for solving the imbalanced dataset problem. However, regular SVMs treat the two classes without any bias which is not effective enough to model the relevance feedback problem. Moreover, we fully utilize both positive and negative data in the training process.

3   Proposed Algorithm

3.1   Preprocessing

In the preprocessing procedure, the system performs feature extraction on the images I in the database, and uses a SOM to represent the distribution of the data. We perform a low-level feature extraction on the set of images in the database, and each image is then represented by a feature vector $\mathbf{x}_i \in \mathbb{R}^d$ in a high dimensional vector space. We construct and train a SOM M with feature vectors extracted from the images. After the SOM training, the model vectors in the neurons of M are arranged to match the distribution of the feature space. The model vectors $\mathbf{m}_i \in M$ of neurons in the SOM are used to partition the feature vector space based on the minimum distance classifier, each image I_i is classified into different groups represented by $\mathbf{m}_i$.

3.2   Intra-query Learning

In the intra-query learning process, the system presents a set of images D_t to the user in each iteration t, and the user gives response A_t by marking them as either relevant or non-relevant. The information provided in the k-th query at iteration t is represented by $q_t^k = \{D_1, A_1, \ldots, D_t, A_t\}$, and the system uses it to refine the query. We define D_R and D_N as the set of relevant images and the set of non-relevant images marked by the user from first iteration to the current iteration respectively. The sets D_R and D_N are then represented by the corresponding model vector set M_R and M_N. The BSVM in Section 2 is used to train a decision boundary to classify this two sets of data.

In order to retrieve images from the database, we need to construct a evaluation function to output the relevance value of the neurons, and it is defined by,

$$g(\mathbf{m}_i) = R^2 - ||\Phi(\mathbf{m}_i) - \mathbf{c}||^2 \tag{6}$$

where the radius R and center $\mathbf{c}$ can be solved by a set of support vectors, and it can be done by solving the Eq. (4). The neurons with higher scores will be more likely to be chosen as the targets. The relevance score between an image and its corresponding neutron is measured by their Euclidean distance. Thus, we can rank the images in the database by combining it with the function $g(\mathbf{m}_i)$.

3.3   Inter-query Learning

In most of the relevance feedback systems, only the intra-query feedback information is used to learn the user's preference. However, a small training data set is difficult to provide enough statistical information for achieving this goal and providing good retrieval result. In order to address this problem, we use the inter-query information to modify the feature vector space and cluster the neurons with similar images together, so that the neutrons are organized in a way that ease the process of intra-query learning.

In order to update the similarity measure based on the inter-query feedback information, we modify the model vectors $\mathbf{m}_i$ in the new SOM, such that neurons contain similar images as indicated in the feedback are moved closer to each others. Consider that there are K past queries stored in the system, and inter-query information provided to the system is represented by $\{q^1, \ldots, q^K\}$. Each past query q^k is used to reorganize the vector space of the SOM, and improve the structure of data. Assume in the k-th query, the user marked a set of relevant images D_R^k and a set of non-relevant images D_N^k during the whole retrieval process, M_R^k and M_N^k are the corresponding sets of model vectors respectively. Let $\mathbf{c}'^k$ be the model vector with highest relevance score in Eq. (6), and it is most likely to be the user's target for that query. In our SOM-based inter-query learning, neurons represent relevant images are moved closer to the estimated user's target and those represent non-relevant images are moved away from the estimated user's target. For a long run, the vector space will be modified, in which neurons represent the same image concept are clustered together.

In a SOM, the nearby neurons in the topology are representing similar units, so that the learning process can be improved by moving also the neurons that near to the neurons in the sets M_R^k and M_N^k. The idea of this process is similar to the SOM training process. The equations for modifying the model vectors are defined by,

$$\forall \mathbf{m} \in N(M_R^k), \mathbf{m} = \mathbf{m} + h_{Ri}^k(\mathbf{c}'^k - \mathbf{m}), \tag{7}$$

$$\forall \mathbf{m} \in N(M_N^k), \mathbf{m} = \mathbf{m} + h_{Ni}^k(\mathbf{m} - \mathbf{c}'^k), \tag{8}$$

where $N(M)$ is the set of nearby neurons for M in the SOM topology. h_{Rci}^k and h_{Nci}^k are the neighborhood functions; they are monotonic decreasing function with k and the distance with the corresponding model vector in M_R^k or M_R^k.

4   Experiment

Here, we present the setup of our experiments on various relevance feedback systems for CBIR. The relevance feedback systems involved in the experiment are

MARS [3], 1-SVM [2], ν-SVM [10] and our proposed method. For the purpose of objective measure of performance, we assume that the query judgement is defined on the image categories. The metric of evaluation is the average precision. The dataset used in the experiment is the real-world images chosen from the COREL image collection. The datasets is with 50 categories. Each category includes 100 images belonging to a same semantic class. We extract three different features to represent the images: color, shape and texture. We combine this three different features as a 36-dimensional feature space.

In the first experiment, we evaluate the retrieval performance of four intra-query methods. A category is first picked from the database randomly, and this category is assumed to be the user's query target. The system then improves retrieval results by relevance feedbacks. In each iteration of the relevance feedback process, five images are picked from the database and labelled as either relevant or non-relevant based on the ground truth of the database. The precision of each method is then recorded, and the whole process is repeated for 200 times to produce the average precision in each iteration for each method. Fig. 1 shows the evaluation result on the top-30 average precision.

In the second experiment, we evaluate the retrieval performance of our SOM-based inter-query learning technique by applying it to those intra-query learning techniques in the first experiment. In this experiment, the feature vectors of images in the database are used to train a SOM of dimension of 30×30. The 200 queries generated in the first experiment are used to further train the SOM with our proposed inter-query learning technique. Finally, we generated another 200 queries and recorded the precision of each method. We evaluate the performance of our inter-query learning by comparing the retrieval result with the first experiment. We list the top-30 average precision at various iteration in Table 1.

From the experiment, the BSVM outperforms the other three intra-query techniques. Typical approaches by SVMs without considering the bias in the retrieval tasks is not reasonable and good enough to solve the relevance feedback problem. We also see that regular one-class SVMs do not consider the negative information which cannot learn the feedback well. Moreover, all four intra-query techniques perform better when our SOM-based inter-query learning technique is

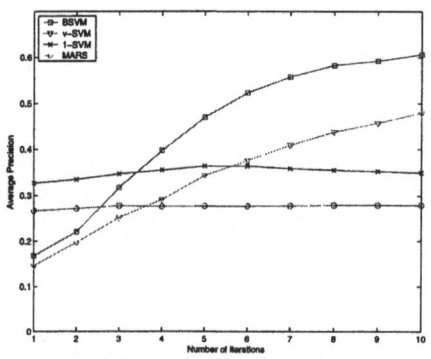

Fig. 1. Top 30 Average Precision

Table 1. Top 30 Average Precision with SOM

Iteration	1	3	5	7	9
MARS	0.267	0.277	0.277	0.277	0.279
MARS	0.285	0.296	0.298	0.298	0.301
1-SVM	0.327	0.347	0.363	0.358	0.352
1-SVM	**0.365**	**0.376**	0.379	0.366	0.371
ν-SVM	0.146	0.252	0.343	0.409	0.458
ν-SVM	0166	0.291	0.376	0.451	0.465
BSVM	0.168	0.317	0.470	0.558	0.592
BSVM	0.191	0.359	**0.515**	**0.602**	**0.611**

applied, and it performs the best when incorporate with our BSVM. It shows that the SOM-based inter-query learning can help the intra-query learning process and improve the retrieval result.

5 Conclusion

In this paper, we investigate SVMs techniques for solving the relevance feedback problems in CBIR. We address the imbalanced dataset problem in relevance feedback and propose a novel relevance feedback technique with Biased Support Vector Machine. The advantages of our proposed techniques are explained and demonstrated compared with traditional approaches. Moreover, we propose a SOM-based inter-query learning technique to reorganize the feature vector space, such that the information provided in past queries is utilized and the retrieval result is improved. We perform experiments on real-world image datasets. The experimental results demonstrate that our BSVM-based relevance feedback algorithm and SOM-based inter-query learning technique are effective and promising for improving the retrieval performance in CBIR.

Acknowledgement

This work is supported in part by the RGC Research Grant Direct Allocation #2050259.

References

1. Tong, S., Chang, E.: Support vector machine active learning for image retrieval. In: Proceedings of the ninth ACM international conference on Multimedia, ACM Press (2001) 107–118
2. Chen, Y., Zhou, X., Huang, T.S.: One-class svm for learning in image retrieval. In: Proceedings of the IEEE international conference on Image Processing. (2001) 34–37
3. Rui, Y., Huang, T.S., Ortega, M., Mehrota, S.: Relevance feedback: A power tool for interactive content-based image retrieval. IEEE Transactions on Circuits and Video Technology (1998) 644–655
4. Vapnik, V.N.: The Nature of Statistical Learning Theory. Springer (1995)
5. Tax, D.M.J., Duin, R.P.W.: Support vector domain description. Pattern Recignition Letters 20 (1999) 1191–1199
6. Chan, C.H., Sia, K.C., King, I.: Utilizing inter and intra-query relevance feedback for content-based image retrieval. In: Special Session of Joint 13th International Conference on Artificial Neural Network and 10th International Conference on Neural Information Processing. (2003)
7. He, X., King, O., Ma, W.Y., Li, M., Zhang, H.J.: Learning a semantic space from user's relevance feedback for image retrieval. IEEE Transactions on Circuits and Systems for Video Technology 13 (2003) 39–48
8. Oja, E., Kaski, S.: Kohonen Maps. Hardbound (1999)
9. Hoi, C.H., Chan, C.H., Huang, K., Lyu, M.R., King, I.: Biased support vector machine for relevance feedback in image retrieval. In: Proceedings of the International Joint Conference on Neural Networks (IJCNN'04) accepted. (2004)
10. Scholkopf, B., Smola, A., Williamson, R., Bartlett, P.L.: New support vector algorithms. Neural Computation 12 (2000) 1207–1245

A Fast MPEG4 Video Encryption Scheme
Based on Chaotic Neural Network

Shiguo Lian, Jinsheng Sun, Zhongxin Li, and Zhiquan Wang

Department of Automation, Nanjing University of Science & Technology,
Nanjing, 210094, P.R. China
sg_lian@163.com

Abstract. In this paper, a fast video encryption scheme is proposed, which combines encryption process with MPEG4 encoding. This scheme encrypts video object layers (VOLs) selectively. That is, the motion vectors, subbands, code blocks or bit-planes are encrypted partially. A stream cipher based on chaotic neural network is used in all these encryption processes, which is of high security and low cost. This encryption scheme keeps compression ratio and file format unchanged, supports direct bit-rate control, and keeps the error-robustness unchanged. These properties make it suitable for real-time applications, such as video-on-demand system, video-conference system, mobile or wireless multimedia, and so on.

1 Introduction

With the development of network technology and multimedia technology, multimedia data are used more and more widely in human's life. In order to protect multimedia data on politics, economics or military, multimedia encryption algorithms have been studied [1-6]. Since the past decade, video encryption has been a topic of great interest. According to the relationship between compression and encryption, these algorithms may be classified into three types: direct-encryption algorithms, partial-encryption algorithms and compression-encryption algorithms. The first type encrypts compressed data directly with traditional or modified cryptographies [1-2]. In spite of high security, it is often of high complexity and changes the file format. Thus, it is more suitable for secure video storing than for real-time transmission. The second type encrypts video data partially or selectively [3-4]. It satisfies real-time requirement by encrypting less data, and realizes encryption during compression process. Thus, it often keeps file format unchanged, supports bit-rate control or recompression in some extent. But some algorithms change compression ratio for they change the statistical characteristics of DCT coefficients. Thus, this type is more suitable for real-time applications with direct bit-rate control required, such as wireless multimedia network or multimedia transmission over narrow bands. The third type combines encryption process with compression process, and realizes compression and encryption at the same time [5-6]. It avoids the shortcoming that the partial-encryption algorithm changes compression ratio greatly. Additionally, it is of low-cost and is easy to be realized. For these advantages, it is suitable for real-time required applications such as video transmission or video access.

Recently, a novel video compression standard MPEG4 [7-8] is announced, and now widely used. Due to the features of high compression ratio, progressive recovery

N.R. Pal et al. (Eds.): ICONIP 2004, LNCS 3316, pp. 720–725, 2004.
© Springer-Verlag Berlin Heidelberg 2004

and random access, it is suitable for applications in Internet, web browsing, document videoing and so on. In order to enhance its applications, we propose a video encryption scheme combining with MPEG4 codec. This scheme encrypts bitsteam selectively and progressively. It is of high security, low-cost and supports direct bit-rate control.

The rest of the paper is arranged as follows. In Section 2, the proposed encryption scheme is proposed. And such performances as security, computational complexity, bit-rate control or error-robustness are analyzed in Section 3. Finally, some conclusions are drawn in Section 4.

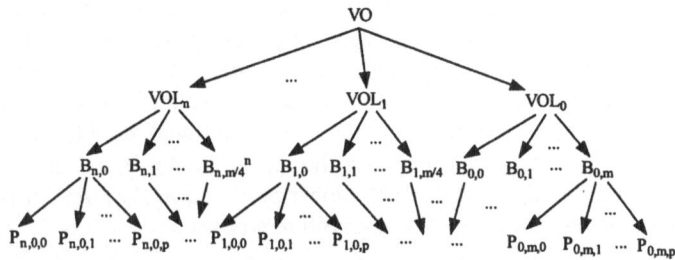

Fig. 1. Progressive components. Each code block is composed of some bit-planes, each VOL is composed of many code blocks, and each VO is composed of several VOLs

2   The Proposed Encryption Scheme

In MPEG4 codec, video data are divided into many VSs (video sessions) and then encoded one by one. Where, each VS includes three layers: VO (video object), VOL (video object layer) and VOP (video object plane). Here, VOP is encoded by the embedded block coding with optimized truncation (EBCOT) [9]. And, VOL encoding and EBCOT encoding are in relation with progressive property. The relationship between them is shown in Fig. 1. Where, each VOL represent a level of resolution. The base VOL is composed of $M \times N$ sized VOPs. And each I-VOP is split into $X \times Y$ sized code blocks, and each code block is encoded from the most significant bit-plane to the least significant one. VOL_i is the i-th VOL in each VO (i=n, n-1, ..., 0), $B_{i,j}$ is the j-th code block in the VOL_i (j=0, 1, ..., $m/4^i$), and $P_{i,j,k}$ is the k-th bit-plane of code block $B_{i,j}$ (k=0, 1, ..., p). The VOLs are numbered from basic layer to the highest layer, and bit-planes are numbered from the least significant one to the most significant one. Thus, the bit-planes can consist of a set in progressive order, $P_{set}=\{P_{n,0,0}, P_{n,0,1}, ..., P_{n,1,0}, P_{n,1,1}, ..., P_{n-1,0,0}, P_{n-1,0,1}, ..., P_{1,0,0}, P_{1,0,1}, ..., P_{0,0,0}, P_{0,0,1}, ..., P_{0,m,p}\}$. Based on this set, selective encryption can be constructed by encrypting only some significant bit-planes. That is, the video produced by encrypting the later bit-planes (for example, from $P_{0,0,0}$ to $P_{0,0,p}$) are more chaotic than the one produced by encrypting the former bit-planes (for example, form $P_{n,0,0}$ to $P_{n,0,p}$). In order to obtain high security and high speed, we propose the following encryption principles:

1) File format information. The file format information, such as file header, packet header, and so on, should be left unencrypted in order to support such operation as image browsing or bit-rate control.

2) VO encryption. Only VOL$_0$ is encrypted greatly, other VOLs are only encrypted by motion vector encryption. In VOL$_0$, I-VOP is encrypted greatly, while P-VOP, B-VOP or S-VOP is only encrypted by motion vector encryption. For each I-VOP, only the code blocks in the lowest subband are completely encrypted, and other ones are only encrypted by selective encryption that encrypts only q of the most significant bit-planes. Through various experiments, we recommend to select q=5, which can get good tradeoff between security and time-efficiency.

3) Cipher selection. Considering that the encoding-passes are often of variable size, block ciphers with fixed plaintext-size are not suitable here. This is because that the bit-rate control and error resilience are both based on encoding-passes that can be cut directly without affecting on the decoding process. Therefore, stream ciphers are more suitable here. However, no suitable random generators can obtain really random sequence now. The ones based on chaotic system and neural network [10,11] are regarded to have higher security compared with traditional pseudo-random number generators. Therefore, we propose to use the stream cipher based on chaotic neural network that is modified from the block cipher proposed in [12]. And it is often used to encrypt motion vector's signs. The encryption process is

$$c_i = f(\omega_i p_i + \omega_{i-1} c_{i-1} + \theta_i). \tag{1}$$

Here, p$_i$ ('0' or '1') and c$_i$ ('0' or '1') are the i-th plaintext and ciphertext respectively, f(x) is the function whose value is 1 if $x \geq 0$ and 0 otherwise, and the parameters ω_i and θ_i are determined by a chaotic binary sequence $B = \{b(0), b(1), b(2), \cdots\}$. That is

$$\omega_i = \begin{cases} 1 & b(2i)=0, \\ -1 & b(2i)=1. \end{cases} \text{ and } \theta_i = \begin{cases} \dfrac{1}{2} & b(2i+1) = 0, \\ -\dfrac{1}{2} & b(2i+1) = 1. \end{cases} \tag{2}$$

Here, the chaotic binary sequence is generated based-on Logistic map

$$x(k+1) = \mu x(k)[1 - x(k)]. \tag{3}$$

Here, we take $\mu = 4$ and $k \in \{0, 1, \cdots, n-1\}$. Initial state x(0) is the key, and the chaotic sequence $x(0), x(1), \cdots, x(n)$ is produced through iterated chaotic map. If $x(i) = 0.b(0)b(1)b(2)b(3)\cdots$, then binary chaotic sequence $b(0), b(1), \cdots, b(m)$ may be constructed by extracting the first m bits of x(i). The bigger m is, the higher the key-sensitivity is, while the higher the computer resolution is required. Through various experiments, we recommend 7<m<17.

The decryption process is symmetric to the encryption one, that is, the plaintext p$_i$ is replaced by the ciphertext c$_i$, and the ciphertext c$_i$ by the plaintext p$_i$. Thus, the cipher is a symmetric one, in which, the initial value x(0) is key that is assigned 128 bits.

4) Pass encryption. Each encoding-pass is encrypted with different key. That is, for each encoding-pass, the chaotic binary sequence is generated from different initial-condition. Thus, if one encoding-pass cannot be synchronized because of transmission errors, the other ones can still be decrypted correctly.

3 Performance Analysis

3.1 Security Analysis

In this encryption scheme, some sensitive code blocks and bit-planes are encrypted, which make the decoded images too chaotic to be understood. Thus, the encryption scheme can get high security in perception. Taking various videos for example, the encryption results are shown in Fig. 2. Where, the number of bit-planes to be encrypted is q=5. It's obvious that, the encrypted videos cannot be understood.

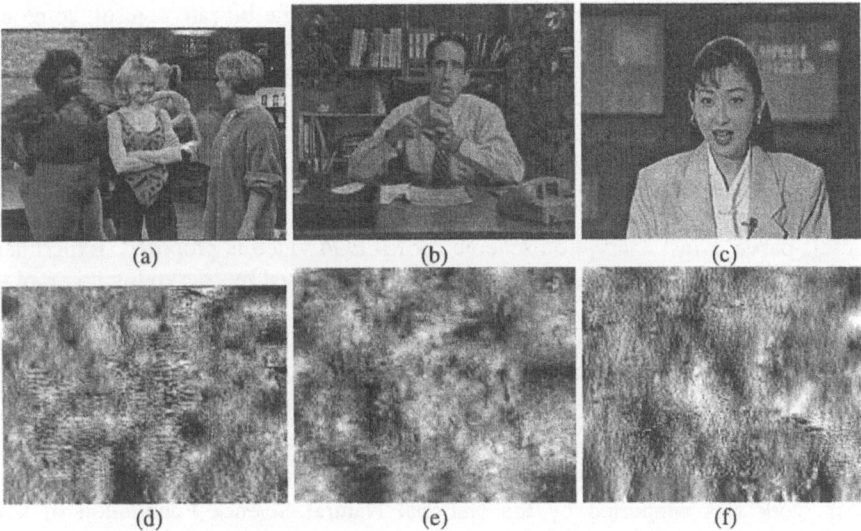

Fig. 2. Encrypted video frames. The ones shown in (a), (b) and (c) are the original videos. The ones shown in (d), (e) and (f) are the encrypted videos

In this cryptosystem, the difficulty of brute-force attack, statistic attack or differential attack is determined by the adopted stream cipher. As we know that, the chaotic system's pseudo-randomness and the neural network's complexity together determine the security of the cryptosystem. The random-similarity of the encrypted bit sequence makes statistic attack difficult to be realized, the high key-sensitivity and plaintext-sensitivity keep it secure against differential attack, and the key space is large enough to make brute-force attack unpractical. Additionally, different encoding-passes are encrypted with different keys, which keep the cryptosystem of high computing security [13]. For example, the brute-force space for each encoding-pass is 2^{128}, and the one for S different encoding-passes is 2^{128S}, which is large enough to keep high security.

3.2 Computational Complexity

In this encryption scheme, I-VOP frames are encrypted by selective bit-plane encryption, while other frames are encrypted by motion vector encryption. For I-VOP frames, the computational complexity of encryption process is in relation with the number of bit-planes to be encrypted: q. Taking various videos with different sizes for example, we test the time ratio between encryption process and the summation of encryption and encoding process. Where, the number of encrypted bit-planes is q=5. The encryption time ratio is also not bigger than 10%, which shows that the encryption process is of low cost.

3.3 Direct Bit-Rate Control

EBCOT codec encodes code blocks into progressive bit-streams. The encryption scheme encrypts compressed bit-streams directly. While bit-rate control, some code passes will be cut directly. The proposed encryption algorithm assures that the changed bit-stream can still be decrypted correctly, which benefits from its advantage that the encoding passes are encrypted independently.

4 Conclusions

In this paper, a fast encryption scheme for MPEG4 video is proposed. Experimental results show that, high security and speed can be obtained by encrypting no more than 10% of the whole bit stream. These properties make it suitable for real-time applications with different security or bit-rate requirement, such as VOD system, pay-TV, wireless or mobile multimedia, and so on.

Acknowledgements

This work was supported by the National Natural Science Foundation of China through the grant number 60374066.

References

1. Qiao L., Nahrstedt K.: A New Algorithm for MPEG Video Encryption. In: Proceeding of the First International Conference on Imaging Science, Systems and Technology (CISST'97). Las Vegas, Nevada (1997) 21-29
2. Qiao L., Nahrstedt K.: Comparison of MPEG Encryption Algorithm. International Journal on Computers and Graphics, Special Issue: Data Security in Image Communication and Network. Vol. 22, No. 3, Permagon Publisher (1998)
3. Tang L.: Methods for Encrypting and Decrypting MPEG Video Data Efficiently. In: Proceedings of the Fourth ACM International Multimedia Conference (ACM Multimedia'96). Boston, MA (1996) 219-230
4. Zeng W.J., Lei S.M.: Efficient Frequency Domain Selective Scrambling of Digital Video. IEEE Trans. on Multimedia (2002)
5. Tosun A.S., Feng W.C.: On Error Preserving Encryption Algorithms for Wireless Video Transmission. In: Proceedings of the ACM International Multimedia Conference and Exhibition n IV. Ottawa, Ont (2001) 302-308
6. Wu C.P., Jay Kuo C.C.: Efficient Multimedia Encryption Via Entropy Codec Design. In: Proceedings of SPIE International Symposium on Electronic Imaging 2001, Vol. 4314. San Jose, CA, USA (2001)
7. ISO/IEC JTC1/SC29/WG11 Document N2552: MPEG-4 Video Verification Model Version 12. 1. (1998)

8. ISO/IEC JTC1/SC29/WG11 N3342: Overview of the MPEG4 Standard. (2000)
9. Taubman D.: High Performance Scalable Image Compression with EBCOT. IEEE Trans. on Image Processing. Vol. 9, No. 7 (2000) 1158-1170
10. Karras D.A., Zorkadis V.: On Neural Network Techniques in the Secure Management of Communication Systems through Improving and Quality Assessing Pseudorandom Stream Generators. Neural Networks, Vol. 16, No. 5-6 (2003) 899-905
11. Chan C.K., Cheng L.M.: Pseudorandom Generator Based on Clipped Hopfield Neural Network. In: Proceedings of the 1998 IEEE International Symposium on Circuits and Systems (ISCAS'98), Vol. 3 (1998) 183–186
12. Yen J.C., Guo J.I.: A Chaotic Neural Network for Signal Encryption/Decryption and Its VLSI Architecture. In: Proceeding of the 10th Taiwan VLSI Design/CAD Symposium. (1999) 319-322
13. Shannon C.: Communication Theory of Secrecy Systems. Bell System Technical Journal. Vol. 28 (1949) 656–715

Content-Based Video Classification
Using Support Vector Machines

Vakkalanka Suresh, C. Krishna Mohan,
R. Kumara Swamy, and B. Yegnanarayana

Speech and Vision Laboratory
Department of Computer Science and Engineering
Indian Institute of Technology Madras, Chennai-600 036, India
{suresh,ckm,kswamy,yegna}@cs.iitm.ernet.in

Abstract. In this paper, we investigate the problem of video classification into predefined genre. The approach adopted is based on spatial and temporal descriptors derived from short video sequences (20 seconds). By using support vector machines (SVMs), we propose an optimized multi-class classification method. Five popular TV broadcast genre namely cartoon, commercials, cricket, football and tennis are studied. We tested our scheme on more than 2 hours of video data and achieved an accuracy of 92.5%.

1 Introduction

Due to significant improvement in processing technologies, network subsystems, and availability of large storage systems, the amount of video data has grown enormously in recent years. In order to make efficient use of this data it should be labeled or indexed in some manner. Also, with the advent of digital TV broadcasts of several hundred channels and the availability of large digital video libraries, it is desirable to classify and categorize video content automatically so that end users can search, choose or verify a desired program based on the semantic content thereof.

There are many approaches to content-based classification of video data. At the highest hierarchy level, video collections can be categorized into different program genres such as cartoon, sports, commercials, news and music. In a recent approach [1], Li et al used PCA to reduce the dimensionality of the features (low-level audio and visual) of video and used Gaussian Mixture Models (GMMs) to model the video classes. In an another approach [2], Truong et al used semantic aspects of a video genre such as editing, motion and color features and C4.5 decision tree algorithm to build the classifier.

At the next level of hierarchy, domain videos such as sports can be classified into different sub-categories. In [3], Xavier et al classify sports video into four sub categories (ice hockey, basketball, football and soccer) by using motion and color features and HMM based models for the video classes. In an another approach [4], by using the statistical analysis of camera motion patterns such as fix, pan, zoom and shake, sports videos are categorized into sumo, tennis, baseball, soccer and football.

N.R. Pal et al. (Eds.): ICONIP 2004, LNCS 3316, pp. 726–731, 2004.
© Springer-Verlag Berlin Heidelberg 2004

At a finer level, a video sequence itself can be segmented and each segment can then be classified according to its semantic content. In [5], sports video segments are first segmented into shots and each shot is then classified into playing field, player, graphic, audience and studio shot categories. Parsing and indexing of news video [6] and semantic classification of basketball segments into goal, foul and coroud categories [7] by using edge-based features are some of the other works carried out at this level.

In this paper, we address the problem of video genre classification for five classes: cartoon, commercials, cricket, football and tennis. In particular, we examine a set of features that would be useful in distinguishing between the classes. We concentrate on features that can be extracted only from the visual content of a video. Although features from multiple modalities (audio, visual and text) have been reported recently in the literature, individual modalities still need to be explored. Support vector machines (SVMs) have been used to perform the classification task.

The rest of this paper is organized as follows: In Section 2, the extraction of spatial and temporal visual features inherent in a video class is described. In Section 3, the system modules for video content modeling and classification are discussed. Section 4 describes experiments on video classification on five TV genre and discusses the performance of the system. Section 5 summarizes/concludes the study.

2   Feature Extraction

A feature is defined as a descriptive parameter that is extracted from an image or a video stream. The effectiveness of any classification scheme depends on the effectiveness of attributes in content representation. We extract set of visual features such as color (features, 1 to 3), shape (features, 4 to 11), motion (feature, 12) and two other visual features (features, 13-14), that provide the discriminatory information useful for high-level classification. The definitions of these features and their intuitive meanings are discussed in the following subsections.

2.1   Color Features

Color is an important attribute for image representation. Color histogram, which represents the color distribution in an image, is one of the most widely used color feature. Since it is not feasible to consider the complete histogram, we have considered the first three moments of the color histogram. From the probability theory we know that a probability distribution is uniquely characterized by its moments. Thus, if we interpret the color distribution of an image as probability distribution, then the color distribution can be characterized by its moments as well [8]. Furthermore, because most of the information is concentrated in the lower-order moments, only the first three moments, mean, variance and skewness are used. Let N be the number of quantized colors and P_i be the number of pixels of the ith color, then the first, second and third moments of the color histogram are given by

$$E = \frac{1}{N} \sum_{i=1}^{N} P_i \tag{1}$$

$$\sigma^2 = \frac{1}{N} \left[\sum_{i=1}^{N} (P_i - E)^2 \right] \tag{2}$$

$$S = \frac{1}{N} \left[\sum_{i=1}^{N} \left(\frac{P_i - E}{\sigma} \right)^3 \right] \tag{3}$$

The RGB(888) color space is quantized into 64 colors by considering only the 2 most significant bits from each plane. For each frame of the video sequence, color histogram is obtained with 64 bins. Then the mean, variance and skewness of the color histogram are obtained as described above.

2.2   Shape Features

Low-level shape based features can be formed from the edges in the image. A histogram of edge directions is translation invariant and it captures the general shape information in the image. Edge histogram descriptor was one of the recommended descriptors for the MPEG-7 standard content description for an image or video. We approximate the horizontal and vertical derivatives by separately filtering each from with the horizontal and vertical sobel filters. Let $\frac{\partial I(x,y)}{\partial x}$ and $\frac{\partial I(x,y)}{\partial y}$ denote the horizontal and vertical derivatives of frame I at point (x, y). Then the angle of the gradient at (x, y) is given by

$$\theta(x, y) = tan^{-1} \left(\frac{\frac{\partial I(x,y)}{\partial y}}{\frac{\partial I(x,y)}{\partial x}} \right) \tag{4}$$

The domain of the edge directions $(0 - 180)$ can be divided into 8 bins. Finally, the 8-dimensional edge direction histograms are calculated by counting the edge pixels in each direction and normalizing with the total number of pixels.

2.3   Motion Feature

Motion is an important attribute of video. Different video genre present different motion patterns. The motions in a video sequence are caused by two different sources which are camera and object(s). In this paper, we use a simple and effective technique where motion is extracted by pixel-wise differencing of consecutive frames using the equation:

$$M(t) = \frac{1}{w * h} \sum_{x=1}^{w} \sum_{y=1}^{h} P_t(x, y) \tag{5}$$

$$\text{where } P_t(x, y) = \begin{cases} 1 & \text{if } |I_t(x, y) - I_{t-1}(x, y)| > \beta \\ 0 & \text{otherwise} \end{cases} \tag{6}$$

where $I_t(x, y)$ and $I_{t-1}(x, y)$ are the pixel values at pixel location (x, y) in t^{th} and $(t-1)^{th}$ frames, respectively. β is the threshold and w and h are width and height of the image respectively.

2.4   Other Visual Features

Along with the color, shape and motion features, we have taken two other features: 1) Ratio of number of pixels with brightness greater than 0.5 to the total number of pixels and 2) ratio of edge pixels to the total number of pixels in an image.

3   Video Content Modeling and Classification

After feature extraction, the next step in video classification task is video content modeling. Many effective modeling techniques have been proposed in the literature. The effectiveness of the classification task depends on the classifier chosen. In this work, we have chosen support vector machines (SVMs) to model the video content wherein, the learning of model involves discrimination of each class against all other classes.

Support vector machines [9] for pattern classification are built by mapping the input patterns into a higher dimensional feature space using a nonlinear transformation (kernel function), and then optimal hyperplanes are built in the feature space as decision surfaces between classes. Nonlinear transformation of input patterns should be such that the pattern classes are linearly separable in the feature space. According to Cover's theorem, nonlinearly separable patterns in a multidimensional space, when transformed into a new feature space are likely to be linearly separable with high probability, provided the transformation is nonlinear, and the dimension of the feature space is high enough [10]. The separation between the hyperplane and the closest data point is called the margin of separation, and the goal of a support vector machine is to find a particular hyperplane for which the margin of separation is maximized. The support vectors constitute a small subset of the training data that lie closest to the decision surface, and are therefore the most difficult to classify.

The performance of the pattern classification problem depends on the type of kernel function chosen. Possible choices of kernel function include; polynomial, Gaussian and sigmoidal. In this work, we have used Gaussian kernel, since it was empirically observed to perform better than the other two. SVMs are originally designed for two class classification problems. In our work, multi-class ($M = 5$) classification task is achieved using one-against-rest approach, where an SVM is constructed for each class by discriminating that class against the remaining $(M - 1)$ classes.

4   Experimental Results

The experiments were carried out on more than 2 hours of video data (≈ 400 video clips each of 20 seconds captured at 25 frames per second) comprising of cartoon, commercial, cricket, football and tennis video categories. The data is collected from different TV channels on different dates and at different times to ensure the variety of data. For each video genre, half of the total number of clips were used for training and remaining half were used for testing. A Gaussian kernel based SVM was constructed for each class by using the one against the rest approach.

During testing phase, given a pattern vector to an SVM model, the result will be a measure of the distance of the pattern vector from the hyper plane constructed as a decision boundary between this class and rest of the classes. A positive value represents that the pattern belongs to the target class and vice versa. Based on the outputs from all the SVMs for a given pattern vector, the class label corresponding to the model giving the highest positive value can be assigned to the pattern vector. In order to make the decision at video clip level, two different approaches can be followed: 1) For each model, averaging the values of all pattern vectors belonging to a particular clip and assign the class label to the clip, based on which model gives the highest average positive value, and 2) count the number of positive outputs per model and assign the class label to the clip corresponding to the model with highest count. In this paper, we have used the first method, since it was empirically observed to perform better than the other. The performance of SVM based classifiers is given in Table 1.

Table 1. Confusion matrix of video classification results using SVM.

	Cartoon	Commercial	Cricket	Football	Tennis
Cartoon	90.325%	6.45%	0%	0%	3.225%
Commercial	4.1%	95.9%	0%	0%	0%
Cricket	0%	2.56%	92.31%	5.13%	0%
Football	0%	2.63%	5.26%	86.85%	5.26%
Tennis	0%	5%	5%	0%	90%

5   Conclusion

We have presented a novel approach to video classification based on color, shape and motion features using support vector machine models. A video database of TV broadcast program containing five popular genre namely cartoon, commercial, cricket, football and tennis is used for training and testing the models. A correct classification rate of 92.5% percent has been achieved. Experimental results indicate that the considered set of features can provide useful information for semantic content understanding and provide discriminating information among the classes considered. Also, it has been shown that SVM based content modelling is an effective method to bridge the gap between the low level features and the high level semantic conceptions. However, inorder to achive better classification performance, evidence from visual features alone may not be sufficient. Evidence from other modalities in a video like audio and text are to be combined with the visual evidence, which will be our future work.

References

1. Xu, L.Q., Li, Y.: Video classification using spatial-temporal features and PCA. In: Int. Conf. Multimedia and Expo, Baltimore, MD, USA (2003)
2. Truong, B.T., Venkatesh, S., Dorai, C.: Automatic genre identification for content-based video categorization. In: Proc. of Int. conf. Pattern Recognition, Barcelona, Spain (2000)
3. Gibert, X., Li, H., Doermann, D.: Sports video categorizing using HMM. In: Int. Conf. Multimedia and Expo, Baltimore, MD, USA (2003)
4. Takagi, S., Hattori, S., Yokoyama, K., Kodate, A., Tominaga, H.: Sports video categorizing method using camera motion parameters. In: Int. Conf. Multimedia and Expo, Baltimore, MD, USA (2003)
5. Assflag, J., Bertini, M., Colombo, C., Bimbo, A.D.: Semantic annotation of sports videos. IEEE Multimedia **9** (2002) 52–60
6. Wactlar, H.D., Kanade, T., Smith, M.A.: Intelligent access to digital video: Informedia project. IEEE Comput. Mag. **29** (1996) 45–52
7. Lee, M.H., Nepal, S., Srinivasan, U.: Edge-based semantic classification of sports video sequences. In: Int. Conf. Multimedia and Expo, Baltimore, MD, USA (2003)
8. Swain, M.J., Ballard, D.H.: Color indexing. Int. Journal of Computer Vision **7** (1991) 11–32
9. Vapnik, V.: Statistical Learning Theory. John Wiley and Sons, New York (1998)
10. Haykin, S.: Neural Networks - A Comprehensive Foundation. Prentice Hall (1999)

Fast Half Pixel Motion Estimation
Based on Spatio-temporal Correlations

HyoSun Yoon, GueeSang Lee*, SooHyung Kim, and Deokjai Choi

Department of Computer Science, Chonnam National University
300 Youngbong-dong, Buk-gu, Kwangju 500-757, Korea
estheryoon@hotmail.com, {shkim,dchoi,gslee}@chonnam.ac.kr

Abstract. A fast half pixel motion estimation algorithm based on spatio-temporal correlations is proposed to reduce the computational complexity. According to spatially and temporally correlated information, the proposed method decides whether half pixel motion estimation is skipped or not for the current block. Experimental results show that the proposed method outperforms most of current methods in computational complexity by reducing the number of search points with little degradation in image quality. When compared to full half pixel search method, the proposed algorithm achieves the search point reduction up to 95% with only $0.01 \sim 0.06$ (dB) degradation of image quality.

1 Introduction

Motion estimation (ME) and motion compensation techniques are an important part of video encoding systems, since it could significantly affect the compression ratio and the output quality. But, ME is very computational intensive part.

Generally, ME is made of two parts, integer pixel motion estimation and half pixel motion estimation. For the first part, integer pixel motion estimation, many search algorithms such as Diamond Search (DS) [1, 2], Three Step Search (TSS) [3], New Three Step Search (NTSS) [4], Four Step Search (FSS) [5], Two Step Search (2SS) [6], Two-dimensional logarithmic search algorithm [7], HEXagon-Based Search (HEXBS) [8], Motion Vector Field Adaptive Search Technique (MVFAST) [9] and Predictive MVFAST (PMVFAST) [10] have been proposed to reduce the computational complexity. Some algorithms among these algorithms can find an integer pixel Motion Vector (MV) by examining less than 10 search points. For the second part, half pixel motion estimation, Full Half pixel Search Method (FHSM) examines eight half pixel points around the integer motion vector. This method takes nearly half of the total computations in the ME that uses fast algorithms for integer pixel motion estimation. Therefore, it becomes more important to reduce the computational complexity of half pixel motion estimation. For these reasons, Horizontal and Vertical Direction as Reference (HVDR) [11], the Parabolic Prediction-based, Fast Half Pixel Search algorithm (PPHPS) [12]and Chen's Fast Half Pixel Search algorithm (CHPS)[13] have been

---

* Corresponding author.

N.R. Pal et al. (Eds.): ICONIP 2004, LNCS 3316, pp. 732–737, 2004.
© Springer-Verlag Berlin Heidelberg 2004

proposed. Since these algorithms do not have any information on the motion of the current block, they always perform half pixel motion estimation to find a half pixel motion vector.

In this paper, we propose a fast algorithm based on spatio-temporal correlations for half pixel motion estimation. According to the information of spatially and temporally correlated motion vector, the proposed method decides whether half pixel motion estimation is skipped or not for the current block.

This paper is organized as follows. Section 2 describes the previous works. The proposed method is described in Section 3. Section 4 reports the simulation results and conclusions are given in Section 5.

2   The Previous Works

In Motion Estimation, half pixel motion estimation is used to reduce the prediction error between the original image and the predicted image. FHSM that is a typical method, is made of two steps. At the first step, the bilinear interpolation for the half pixel vlaues is performed. The formulas for the interpolation are illustrated in Fig.1(a). At the second step, eight half pixel points around the integer motion vector 'C' illustrated in Fig. 1(b) are examined. The point with the minimum cost function value among these points is the half pixel motion vector. To reduce the computational complexity of FHSM, some fast algorithms have been proposed.

In HVDR, 4 neighboring half pixel points in vertical direction and horizontal direction around 'C' illustrated in Fig. 1(b) are examined to decide the best matching point in each direction. Then, a diagonal point between these two best matching points is also examined. The point having the minimum cost function value among these 5 points and 'C' is decided as a half pixel motion vector.

CHPS examines 4 horizontal and vertical half pixel points '2','4','5','7' shown Fig. 1(b).. The point having the minimum cost function value among these 4 points and the point 'C' is decided as a half pixel motion vector.

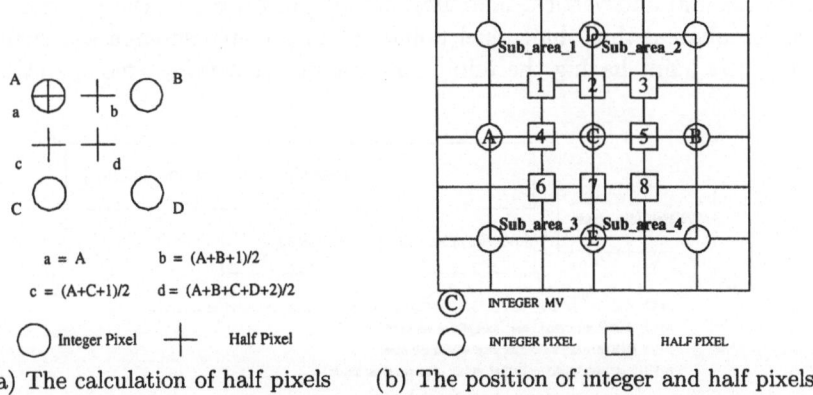

$$a = A \qquad b = (A+B+1)/2$$
$$c = (A+C+1)/2 \qquad d = (A+B+C+D+2)/2$$

(a) The calculation of half pixels   (b) The position of integer and half pixels

Fig. 1. The position and calculation of half pixels

PPHPS predicts the possible optimal half pixel point by using the cost function values of 5 integer pixel points 'A','B','C','D','E' shown Fig. 1(b). The cost function values of the predicted possible optimal half pixel point and its nearest points are calculated to find the best matching point. The point of the minimum cost function value is decided as a final half pixel MV between this best matching point and the point 'C'.

CHPS can not consider the diagonal direction motion in half pixel motion estimation, it reusts in the degradation of image quality. HVDR checks more half pixel points than other fast half pixel motion estimation methods. PPHPS uses complex equations to predict the half pixel motion, it cause a little computational complexity.

3   The Proposed Method

In order to reduce more computational complexity and predict the half pixel motion in half pixel motion estimation, the proposed method exploits spatio-temporal correlations among half pixel motion vectors. In other words, the proposed method exploits spatially and temporally correlated half pixel motion vectors depicted in Fig.2 uses the following obesvations in Table 1, where it can be seen that the highest spatial and temporal correlations are 88%, the lowest correlations are 28%. Spatial and Temporal correaltion in Table 1 is definded as the percentage that MV1_Half, MV2_Half and MV3_Half are equal to MV0_Half. If MV1_Half, MV2_Half and MV3_Half are equal to MV0_Half, the proposed method stops the half pixel motion estimation. Otherwise, Yoon's Fast Half Pixel Search algorithm (YFHPS) is performed to find a half pixel motion vector. YFHPS proposed in this paper predicts the possible subarea by using the cost function values of integer pixel points. According to the position of the possible subarea, three half pixel points in its possible subarea are examined to find a half pixel motion vector. At first, YFHPS decides the best horizontal matching point between 2 horizontal integer pixel points 'A','B' depicted in Fig.1(b) and the best vertical matching point between 2 vertical integer pixel points 'D','E' depicted in Fig.1(b). And then, the possible subarea is selected by using the best horizontal and vertical matching points. According to the position of the possible subarea, three half pixel points in its possible subarea are examined. Finally, the point having the minimum cost function value among these three

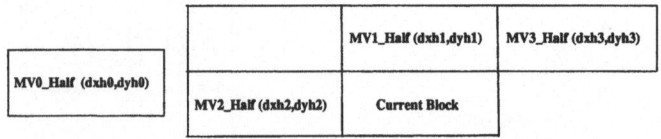

Fig. 2. Blocks for Spatio-Temporal Correlation Information

Table 1. Spatial and Temporal Correlations

	MV1_Half==MV2_Half==MV3_Half==MV0_Half
Akiyo	88%
Claire	62%
Foreman	30%
News	73%
Salesman	81%
Silent	65%
Suzie	28%

half pixel points and the point 'C' Fig. 1(b) is decided as a half pixel motion vector. The propsoed method is summarized as follows.

Step 1 If MV1_Half (dxh1, dyh1), MV2_Half (dxh2, dyh2) and MV3_Half (dxh3, dyh3) are equal to MV0_Half (dxh0, dyh0), go to Step 2. Otherwise, go to Step 3.

Step 2 MV0_Half (dxh0, dyh0) is decided as a half pixel MV of the current block.

Step 3 YFHPS is performed to find a half pixel motion vector.

4   Simulation Result

In this section, we show experimental results for the proposed method. The proposed method has been evaluated in the H.263 encoder. Nine QCIF test sequences are used for the experiment. The mean square error (MSE) distortion function is used as the block distortion measure (BDM). The quality of the predicted image is measured by the peak signal to noise ratio (PSNR), which is defined by

$$MSE = \left(\frac{1}{MN}\right) \sum_{m=1}^{M} \sum_{n=1}^{N} [x(m,n) - \hat{x}(m,n)]^2 \tag{1}$$

$$PSNR = 10 \ log_{10} \frac{255^2}{MSE} \tag{2}$$

In Eq. (1), $x(m,n)$ denotes the original image and $\hat{x}(m,n)$ denotes the motion compensated prediction image. For integer pixel motion estimation, Full Search algorithm is adopted. For half pixel motion estimation, we compared FHSM, HVDR, CHPS, PPHPS and YFHPS to the proposed method in both of image quality and search speed. The simulation results in Table 2 and 3 show that the search speed of the proposed method is faster than the other methods (FHSM, HVDR, CHPS, PPHPS and YFHPS) while its PSNR is similar to them except for FHSM. In other words, the proposed method can achieves the search point reduction up to 95% with only 0.01 ~ 0.06 (dB) degradation of image quality When compared to FHSM.

736 HyoSun Yoon et al.

Table 2. Average PSNR for half pixel motion estimation algorithms

Integer-pel ME method	Full search					
Half-pel ME method	FHSM	HVDR	CHPS	PPHPS	YFHPS	Proposed
Akiyo	34.5	34.41	34.46	34.43	34.5	34.41
Claire	35.05	35.02	35.03	35.05	35.05	35.04
Foreman	29.54	29.52	29.50	29.51	29.51	29.48
M&D	31.54	31.50	31.54	31.52	31.54	31.48
News	30.59	30.49	30.54	30.57	30.57	30.53
Salesman	32.7	32.64	32.67	32.70	32.70	32.66
Silent	31.81	31.80	31.76	31.79	31.80	31.76
Stefan	23.89	23.85	23.86	23.87	23.87	23.83
Suzie	32.19	32.17	32.15	32.19	32.19	32.18

Table 3. The Number of Search points per half pixle MV

	FHSM	HVDR	CHPS	PPHPS	YFHPS	Proposed
Akiyo	8	5	4	3	3	0.39
Claire	8	5	4	3	3	1.55
Foreman	8	5	4	3	3	2.68
M &D	8	5	4	3	3	1.39
News	8	5	4	3	3	0.94
Salesman	8	5	4	3	3	0.62
Silent	8	5	4	3	3	1.12
Stefan	8	5	4	3	3	2.32
Suzie	8	5	4	3	3	2.26

5   Conclusion

Based on spatio-temporal correlations among half pixel MVs, a fast method for half pixel motion estimation is proposed in this paper. According to spatially and temporally correlated information, the proposed method decides whether half pixel motion estimation is skipped or not for the current block. As a result, the proposed method could reduce the computational complexity significantly. Experimental results show that the speedup improvement of the proposed method over FHSM can be up to 4 ~ 25 times faster with a little degradation of the image quality.

Acknowledgement

This work was supported by grant of the Fundamental Technology Research Project, Institute of Information Technology Assessment, Republic of korea (No. 04-fu-072).

References

1. Tham, J.Y., Ranganath, S., Kassim, A.A.: A Novel Unrestricted Center-Biased Diamond Search Algorithm for Block Motion Estimation. IEEE Transactions on Circuits and Systems for Video Technology. **8(4)** (1998) 369–375
2. Shan, Z., Kai-kuang, M.: A New Diamond Search Algorithm for Fast block Matching Motion Estimation. IEEE Transactions on Image Processing. **9(2)** (2000) 287–290
3. Koga, T., Iinuma, K., Hirano, Y., Iijim, Y., Ishiguro, T.: Motion compensated interframe coding for video conference. In Proc. NTC81. (1981) C9.6.1–9.6.5
4. Renxiang, L., Bing, Z., Liou, M.L.: A New Three Step Search Algorithm for Block Motion Estimation. IEEE Transactions on Circuits and Systems for Video Technology. **4(4)** (1994) 438–442
5. Lai-Man, P., Wing-Chung, M.: A Novel Four-Step Search Algorithm for Fast Block Motion Estimation. IEEE Transactions on Circuits and Systems for Video Technology. **6(3)** (1996) 313–317
6. Yuk-Ying, C., Neil, W.B.: Fast search block-matching motion estimation algorithm using FPGA. Visual Communication and Image Processing 2000. Proc. SPIE. **4067** (2000) 913–922
7. Jain, J., Jain, A.: Dispalcement measurement and its application in interframe image coding. IEEE Transactions on Communications. **COM-29** (1981) 1799–1808
8. Zhu, C., Lin, X., Chau, L.P.: Hexagon based Search Pattern for Fast Block Motion Estimation. IEEE Transactions on Circuits and Systems for Video Technology. **12(5)** (2002) 349–355
9. Ma, K.K., Hosur, P.I.: Report on Performance of Fast Motion using Motion Vector Field Adaptive Search Technique. ISO/IEC/JTC1/SC29/WG11.**M5453** (1999)
10. Tourapis, A.M., Liou, M.L.: Fast Block Matching Motion Estimation using Predictive Motion Vector Field Adaptive Search Technique. ISO/IEC/JTC1/SC29/WG11.**M5866** (2000)
11. Lee, K.H.,Choi, J.H.,Lee, B.K., Kim. D.G.: Fast two step half pixel accuracy motion vector prediction. Electronics Letters **36(7)**(2000) 625–627
12. Cheng, D., Yun, H., Junli, Z.: A Prabolic Prediction-Based, Fast Half Pixel Serch Algorithm for Very Low Bit-Rate Moving Picture Coding. IEEE Transactions on Circuits and Systems for Video Technology. **13(6)** (2003) 514–518
13. Cheng, D., Yun, H.: A Comparative Study of Motion Estimation for Low Bit Rate Video Coding. SPIE **4067(3)**(2000) 1239–1249

Local and Recognizable Iso Picture Languages

T. Kalyani[1], V.R. Dare[2], and D.G. Thomas[2]

[1] Department of Mathematics
St. Joseph's College of Engineering
Chennai - 119, India
[2] Department of Mathematics
Madras Christian College
Tambaram, Chennai - 59, India
rchristian@eth.net

Abstract. In the context of a syntactic approach to pattern recognition, there have been several studies in the last few decades on theoretical models for generating or recognizing two-dimensional objects, pictures and picture languages. Motivated by these studies we introduce a new notion of recognizability for a class of picture languages called iso picture languages. We first introduce a notion of local iso picture language and then define a recognizable iso picture language as a projection of a local iso picture language. We prove certain closure properties of this family of languages. We present a learning algorithm LOC for local iso picture languages.

1 Introduction

The first attempt on formalizing the concept of finite state recognizability for two-dimensional languages can be attributed to Blum and Hewitt who have introduced the notion of a four-way automaton moving on a two-dimensional tape as the natural extension of a one-dimensional two-way finite automaton [1]. Since this work, several papers have been devoted to the study of the families of picture languages recognized by four-way automata and several other models of machines that read two dimensional tapes have been defined [2–4, 7].

Motivated by these studies in this paper, we introduce the notion of local and recognizable iso picture languages. We prove certain closure properties for recognizable iso picture languages and present a learning algorithm LOC for local iso picture languages.

2 Preliminaries

In this section we recall the notions of iso triangular tiles and iso pictures [5].

Let $\Sigma = \left\{ \begin{array}{c} a_1\triangle a_3 \\ a_2 \end{array}, \begin{array}{c} b_2 \\ b_3\nabla b_1 \end{array}, \begin{array}{c} c_3 \\ c_1 \end{array}c_2, d_2\begin{array}{c} d_1 \\ d_3 \end{array} \right\}$ be a finite set of labeled isoce-

les right angled triangular tiles of dimensions $\frac{1}{\sqrt{2}}, \frac{1}{\sqrt{2}}$ and 1 unit, obtained

N.R. Pal et al. (Eds.): ICONIP 2004, LNCS 3316, pp. 738–743, 2004.
© Springer-Verlag Berlin Heidelberg 2004

by intersecting a unit square by its diagonals [5]. Gluable rules of tile A are as follows: Tiles which can be glued with A are B, C and D by the rules $\{(a_1, b_1), (a_2, b_2), (a_3, b_3)\}, \{(a_3, c_1)\}, \{(a_1, d_3)\}$. In a similar way the gluable rules can be defined for the remaining tiles.

Definition 1. *[5] An iso array of size $m(m \geq 1)$ is an isosceles right-angled triangular arrangement of elements of Σ, whose equal sides are denoted as S_1 and S_3 and the unequal side as S_2, and it consists of m tiles along side S_2 and it contains m^2 gluable elements of Σ.*

Iso arrays can be classified as U-iso array, D-iso array, R-iso array and L-iso array, if tiles A, B, D and C are used in side S_2 respectively.

Iso arrays of same size can be concatenated using the following concatenation operations. *Horizontal concatenetion* $\ominus$ is defined between U and D iso arrays of same size. *Right diagonal concatenation* $\oslash$ is defined between any two gluable iso arrays of same size. This concatenation includes the following:

   a) $D \oslash U$ (b) $U \oslash R$ (c) $D \oslash L$ (d) $R \oslash L$

In a similar way vertical $\odot$ and left diagonal $\oslash$ concatenations can be defined.

Definition 2. *[5] Let Σ be a finite alphabet of iso triangular tiles. An iso picture of size $(n, m), n, m \geq 1$ over Σ is a picture formed by concatenating n-iso arrays of maximum size m. The number of tiles in any iso picture of size (n, m) is nm^2.*

Any two iso pictures of sizes (n_1, m) and $(n_2, m), n_1, n_2, m \geq 1$ can be concatenated using the rules of concatenation of iso arrays, provided the sides of iso pictures are gluable.

An element of an iso picture p of size (n, m) is represented as $p(i, j, k)$, where i is the ith iso array of the picture and j is the jth row of the ith iso array and k is the kth element of jth row of the ith iso array.

3   Local Iso Picture Languages

The set of all iso pictures over the alphabet Σ is denoted by Σ_I^{**}. An iso picture language L over Σ is a subset of Σ_I^{**}.

Definition 3. *Let p be an iso picture of size (n, m). We denote by $B_{n', m'}(p)$, the set of all sub iso pictures of p of size (n', m'), where $n' \leq n, m' \leq m$.*

Definition 4. *Let p be an iso picture over Σ, then $\hat{p}$ is an iso picture obtained by surrounding p with a special boundary symbols $\{$ $\} \notin \Sigma$.*

3.1   The Family ILOC

Definition 5. *An iso picture language $L \subseteq \Sigma_I^{**}$ is called local if there exists a finite set θ of iso arrays of size 2 over $\Sigma \cup \{$ $\}$ such that $L = \{p \in \Sigma_I^{**} / B_{1,2}(\hat{p}) \subseteq \theta\}$ and is denoted by $L(\theta)$.*

The family of local iso picture languages will be denoted by ILOC. We now give an example of a local iso picture language.

Example 1. Let

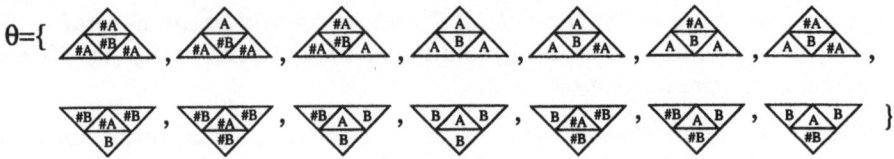

Then the language $L = L(\theta)$ is the local iso picture language of parallelograms of size $(2, m), m \geq 1$ over $\Sigma = \{$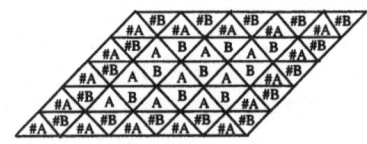$\}$, an element of which is shown in the following figure.

The iso picture language of rhombuses of size $(2, m), m \geq 1$ over the tiles A, B is non local.

3.2   The Family IREC

We now introduce the family of recognizable iso picture languages.

Definition 6. *Let* $p \in \Sigma_I^{**}$ *be an iso picture. Let* Σ *and* Γ *be two finite alphabets and* $\pi : \Gamma \to \Sigma$ *be a mapping which we call, a projection. The projection by mapping* π *of picture* p *is the picture* $p' \in \Sigma_I^{**}$ *such that* $p'(i, j, k) = \pi(p(i, j, k))$ *for all* $1 \leq i \leq n, 1 \leq j \leq m, 1 \leq k \leq 2j - 1$, *where* (n, m) *is the size of the iso picture. In this case* $p' = \pi(p)$.

Definition 7. *Let* $L \subset \Gamma_I^{**}$ *be an iso picture language. The projection by mapping* π *of* L *is the language* $L' = \{p'/p' = \pi(p), \forall p \in L\} \subseteq \Sigma_I^{**}$.

We will denote by $\pi(L)$ the projection by mapping π of an iso picture language L.

Definition 8. *Let* Σ *be a finite alphabet. An iso picture language* $L \subseteq \Sigma_I^{**}$ *is recognizable if there exists a local iso picture language* L' *over an alphabet* Γ *and a mapping* $\pi : \Gamma \to \Sigma$ *such that* $L = \pi(L')$, *where* $L' = L(\theta)$ *and* L *is represented by* (Γ, θ, π).

Example 2.

Let θ={  }

Then $L' = L'(\theta)$ is the local iso picture language of rhombuses, where the diagonals are represented by the tiles ◺A2◿ and ◹B2◹ and the tiles in the remaining positions are represented by tiles ◺A1◿ and ◹B1◹, a member of which is shown in the following figure.

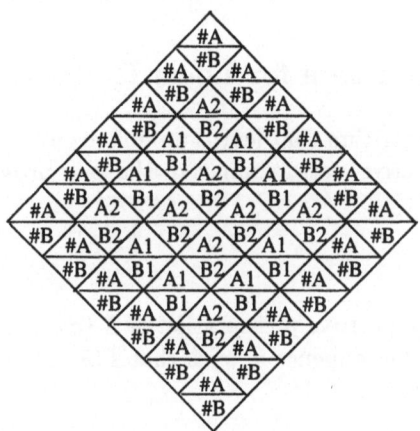

Applying the projection $\pi(A1) = \pi(A2) = A$ and $\pi(B1) = \pi(B2) = B$ in Example 2, we get $L = \pi(L')$. Therefore L, the language of rhombuses is a recognizable iso picture language.

The family of all recognizable iso picture languages will be denoted by IREC. We prove some closure properties for recognizable iso picture languages.

Theorem 1. *The family IREC is closed under projection.*

Proof. Let $L_1 \subseteq \Sigma_{1I}^{**}, L_2 \subseteq \Sigma_{2I}^{**}$ be iso picture languages such that $L_2 = \phi(L_1)$ where $\phi : \Sigma_1 \rightarrow \Sigma_2$. We have to prove that if L_1 is recognizable then L_2 is recognizable.

If L_1 is recognizable then there exists a local language $L' \subseteq \Gamma_I^{**}$ and a projection $\pi : \Gamma \rightarrow \Sigma_1$ such that $L_1 = \pi(L')$. Hence $L_2 = \phi(L_1) = \phi \circ \pi(L')$. i.e., $L_2 = \psi(L')$ where $\psi = \phi \circ \pi : \Gamma \rightarrow \Sigma_2$. Hence L_2 is recognizable.

Theorem 2. *The family IREC is closed under right diagonal, left diagonal, horizontal and vertical concatenation operations.*

Sketch of the Proof. Let L_1 and L_2 be recogniable languages over Σ and let $L = L_1 \oslash L_2$. By definition of right diagonal concatenation, a picture $p \in L$ is constituted by a pair of pictures $p_1 \in L_1$ and $p_2 \in L_2$ with same size of the iso arrays such that the right most segment of P_1 is glued to the left most segment of P_2.

Let $(\Gamma_1, \theta_1, \pi_1)$ and $(\Gamma_2, \theta_2, \pi_2)$ be two representations for L_1 and L_2 respectively, where Γ_1 and Γ_2 are disjoint. We can define a representation (Γ, θ, π) for L as follows: Let $\Gamma = \Gamma_1 \cup \Gamma_2$. The set θ contains all the elements from θ_1 but those corresponding to the right borders and all the elements from θ_2 but those corresponding to the left borders. Moreover some internal tiles corresponding to the two segments where the gluing is made have to be added. Such tiles contain pieces of the right border of pictures in L_1 in the left side and pieces of the left border of pictures in L_2 in the right side.

Similar procedure can be given for other catenation operations.

4   Learning Algorithm for Local Iso Picture Languages

We now present an algorithm that learns an unknown local iso picture language in the limit from positive data, similar to the one presented in [6], for local picture languages.

Algorithm LOC

Input: A sequence of positive presentation of L (members of L).
Output: An increasing sequence θ_i such that $L(\theta_i)$ are local iso
        picture languages.
Procedure
    Initialize E_0 to ϕ
    Construct the initial $\theta_0 = \phi$
    Repeat (for ever)
        Let θ_i be the current conjecture
        Read the next positive example p
        Scan p to obtain $B_{1,2}(\hat{p})$
        $\theta_{i+1} = \theta_i \cup B_{1,2}(\hat{p})$
        $E_{i+1} = E_i \cup \{p\}$
    Output θ_{i+1} as new conjecture.

Lemma 1. *Let $\theta_0, \theta_1, \ldots, \theta_i, \ldots$ be the sequence of conjectures by the algorithm LOC. Then*

1. *for all $i \geq 0, L(\theta_i) \subseteq L(\theta_{i+1}) \subseteq L$ and*
2. *there exists $r \geq 0$ such that for all $i \geq 0, L(\theta_r) = L(\theta_{r+i}) = L$.*

Theorem 3. *Given an unknown local iso picture language L, the algorithm LOC learns, in the limit a set θ_i such that $L(\theta_i) = L$.*

Conclusion

D. Gimmarresi and A. Restivo [2] have introduced the concept of recognizable picture languages using rectangular tiles. In this paper we introduce recogniable iso picture languages, which include other geometrical patterns, and investigate their properties.

Acknowledgements

The first author would like to thank the management of St. Joseph's College of Engineering and Prof.V. Vallinayagam, Head of the Department of Mathematics for the encouragement and constant support to pursue the research work.

References

1. M. Blum and C. Hewitt, Automata on a 2-dimensional tape, IEEE Symposium on Switching and Automata theory, 1967.
2. D. Giammarresi and A. Restivo, Two-dimensional finite state recognizability, Fundamenta Informatica 25 (3, 4): 399-422, 1966.
3. D. Giammarresi and A. Restivo, Hand book of Formal languages, Volume 3, Springer-Verlag, Berlin, 1997.
4. K. Inoue and I. Takanami, A survey of two-dimensional automata theory, Lecture Notes in Computer Science, Vol. 381, Springer-Verlag, Berlin, 1990, 72-91.
5. T. Kalyani, K. Sasikala and V.R. Dare, Generation of pictures by iso array grammar, Proceedings of 2nd National Conference on Mathematical and Computational Models, NCMCM 2003, 260-266, 2003.
6. Rani Siromoney, Lisa Mathew, K.G. Subramanian and V.R. Dare, Learning of Recognizable Picture Languages, International Journal of Pattern Recognition and Artificial Intelligence, Vol. 8, No. 2, 627-639, 1994.
7. A. Rosenfeld and R. Siromoney, Picture languages - A survey. Languages of design, 1: 229-244, 1993.

Multilayer Feedforward Ensembles
for Classification Problems

Mercedes Fernández-Redondo, Carlos Hernández-Espinosa,
and Joaquín Torres-Sospedra

Universidad Jaume I. Dept. de Ingeniería y Ciencia de los Computadores
Avda Vicente Sos Baynat s/n, 12071 Castellon, Spain
{redondo,espinosa}@icc.uji.es

Abstract. As shown in the bibliography, training an ensemble of networks is an interesting way to improve the performance with respect to a single network. However there are several methods to construct the ensemble and there are no complete results showing which one could be the most appropriate. In this paper we present a comparison of eleven different methods. We have trained ensembles of 3, 9, 20 and 40 networks to show results in a wide spectrum of values. The results show that the improvement in performance above 9 networks in the ensemble depends on the method but it is usually marginal. Also, the best method is called "Decorrelated" and uses a penalty term in the usual Backpropagation function to decorrelate the network outputs in the ensemble.

1 Introduction

The most important property of a neural network (NN) is the generalization capability. The ability to correctly respond to inputs which were not used in the training set.

One technique to increase the generalization capability with respect to a single NN consist on training an ensemble of NN, i.e., to train a set of NNs with different weight initialization or properties and combine the outputs of the different networks in a suitable manner to give a single output.

It is clear from the bibliography that this procedure in general increases the generalization capability [1,2].

The two key factors to design an ensemble are how to train the individual networks and how to combine the different outputs to give a single output.

Among the methods of combining the outputs, the two most popular are *voting* and *output averaging* [3]. In this paper we will normally use *output averaging* because it has no problems of ties and gives a reasonable performance.

In the other aspect, nowadays, there are several different methods in the bibliography to train the individual networks and construct the ensemble [1-3], [5-10].

However, there is a lack of comparison among the different methods and it is not clear which one can provide better results.

In this paper, we present a comparison among eleven different methods.

2 Theory

In this section we briefly review the different ensemble methods.

Simple Ensemble: A simple ensemble can be constructed by training different networks with the same training set, but different random weight initialization.

N.R. Pal et al. (Eds.): ICONIP 2004, LNCS 3316, pp. 744–749, 2004.
© Springer-Verlag Berlin Heidelberg 2004

Bagging: This ensemble method is described in reference [5]. It consists on generating different datasets drawn at random with replacement from the original training set. After that, we train the different networks in the ensemble with these different datasets, we use one dataset for each network.

Bagging with Noise (BagNoise): It was proposed in [2], we use in this case datasets of size 10·N (number of training points) generated in the same way of Bagging, where N is the number of training points of the initial training set. Also we introduce a random noise in every selected training point drawn from a normal distribution.

Boosting: This ensemble method is reviewed in [3]. It is conceived for a ensemble of only three networks. The first network is trained with the whole training set. After this training, we pass all patterns through the first network and we use a subset of them, which has 50% of patterns incorrectly classified and 50% classified correctly. With this new training set we train the second network. After that, the original patterns are presented to both networks. If the two networks disagree in the classification, we add the training pattern to the third training set.

CVC: It is reviewed in [1]. In k-fold cross-validation, the training set is divided into k subsets. Then, k-1 subsets are used to train the network and results are tested on the subset that was left out. Similarly, by changing the subset that is left out of the training process, one can construct k classifiers. This is the technique used in this method.

Adaboost: We have implemented the algorithm "Adaboost.M1" in the reference [6]. In the algorithm the successive networks are trained with a training set selected at random, but the probability of selecting a pattern changes depending on the correct classification of the pattern and on the performance of the last trained network. The algorithm is complex and the full description should be looked for in the reference.

Decorrelated (Deco): This ensemble method was proposed in [7]. It consists on introducing a penalty added to the usual error function. The term for network j is:

$$Penalty = \lambda \cdot d(i, j)(y - f_i) \cdot (y - f_j) \qquad (1)$$

Where λ determines the strength of the penalty term and should be found by trial and error, y is the target of the training pattern and f_i and f_j are the outputs of networks number i and j in the ensemble. The term $d(i,j)$ is in equation 2.

$$d(i, j) = \begin{cases} 1, & if \ i = j - 1 \\ 0, & otherwise \end{cases} \qquad (2)$$

Decorrelated2 (Deco2): It was proposed also in reference [7]. It is basically the same method of "Decorrelated" but with a different term $d(i,j)$ in the penalty:

$$d(i, j) = \begin{cases} 1, & if \ i = j - 1 \ \ and \ \ i \ is \ even \\ 0, & otherwise \end{cases} \qquad (3)$$

Evol: This ensemble method was proposed in [8]. In each iteration (presentation of a training pattern), it is calculated the output of the ensemble for the input pattern by voting. If the output is correctly classified we continue with the next iteration. Otherwise, the network with an erroneous output and lower MSE is trained in this pattern until the output of the network is correct. This procedure is repeated for several networks until the vote of the ensemble correctly classifies the pattern.

Cels: It was proposed in [9]. This method also uses a penalty term added to the usual error function. In this case the penalty term for network number i is in equation 4.

$$Penalty = \lambda \cdot (f_i - y) \cdot \sum_{j \neq i} (f_j - y) \qquad (4)$$

Where y is the target and f_i and f_j the outputs of networks i and j.

Ola: This ensemble method was proposed in [10]. In this method, first, several data-sets are generated by using bagging. Every network is trained in one of this datasets and in *virtual data*. The *virtual data* for network i is generated by selecting randomly samples for the original training set and perturbing the sample with a random noise drawn from a normal distribution with small variance. The target for this new virtual sample is calculated by the output of the ensemble without network number i for this sample. For a full description of the procedure see the reference.

3   Experimental Results

We have applied the eleven ensemble methods to ten different classification prob-lems. They are from the UCI repository of machine learning databases. Their names are Cardiac Arrhythmia Database (Aritm), Dermatology Database (Derma), Protein Location Sites (Ecoli), Solar Flares Database (Flare), Image Segmentation Database (Image), Johns Hopkins University Ionosphere Database (Ionos), Pima Indians Dia-betes (Pima), Haberman's survival data (Survi), Vowel Recognition (Vowel) and Wiscosin Breast Cancer Database (Wdbc).

We have constructed ensembles of a wide number of networks, in particular 3, 9, 20 and 40 networks in the ensemble. In this case we can test the results in a wide set of situations.

We trained the ensembles of 3, 9, 10 and 40 networks. We repeated this process of training an ensemble ten times for different partitions of data in training, cross-validation and test sets. With this procedure we can obtain a mean performance of the ensemble for each database (the mean of the ten ensembles) and an error in the per-formance calculated by standard error theory. The results of the performance are in Table 1 and 2 for the case of ensembles of three networks and in Table 3 and 4 for the case of nine, we omit the results of 20 and 40 networks by the lack of space by the improvement of increasing the number of networks is in general not important.

Table 1. Results for the ensemble of three networks

	ARITM	DERMA	ECOLI	FLARE	IMAGEN
Single Net.	75.6 ± 0.7	96.7 ± 0.4	84.4 ± 0.7	82.1 ± 0.3	96.3 ± 0.2
Adaboost	71.8 ± 1.8	98.0 ± 0.5	85.9 ± 1.2	81.7 ± 0.6	96.8 ± 0.2
Bagging	74.7 ± 1.6	97.5 ± 0.6	86.3 ± 1.1	81.9 ± 0.6	96.6 ± 0.3
Bag_Noise	75.5 ± 1.1	97.6 ± 0.7	87.5 ± 1.0	82.2 ± 0.4	93.4 ± 0.4
Boosting	74.4 ± 1.2	97.3 ± 0.6	86.8 ± 0.6	81.7 ± 0.4	95.0 ± 0.4
Cels_m	73.4 ± 1.3	97.7 ± 0.6	86.2 ± 0.8	81.2 ± 0.5	96.82 ± 0.15
CVC	74.0 ± 1.0	97.3 ± 0.7	86.8 ± 0.8	82.7 ± 0.5	96.4 ± 0.2
Decorrelated	74.9 ± 1.3	97.2 ± 0.7	86.6 ± 0.6	81.7 ± 0.4	96.7 ± 0.3
Decorrelated2	73.9 ± 1.0	97.6 ± 0.7	87.2 ± 0.9	81.6 ± 0.4	96.7 ± 0.3
Evol	65.4 ± 1.4	57 ± 5	57 ± 5	80.7 ± 0.7	77 ± 5
Ola	74.7 ± 1.4	91.4 ± 1.5	82.4 ± 1.4	81.1 ± 0.4	95.6 ± 0.3
Simple Ens.	73.4 ± 1.0	97.2 ± 0.7	86.6 ± 0.8	81.8 ± 0.5	96.5 ± 0.2

Table 2. Results for the ensemble of three networks

	IONOS	PIMA	SURVI	VOWEL	WDBC
Single Net.	87.9 ± 0.7	76.7 ± 0.6	74.2 ± 0.8	83.4 ± 0.6	97.4 ± 0.3
Adaboost	88.3 ± 1.3	75.7 ± 1.0	75.4 ± 1.6	88.43 ± 0.9	95.7 ± 0.6
Bagging	90.7 ± 0.9	76.9 ± 0.8	74.2 ± 1.1	87.4 ± 0.7	96.9 ± 0.4
Bag_Noise	92.4 ± 0.9	76.2 ± 1.0	74.6 ± 0.7	84.4 ± 1.0	96.3 ± 0.6
Boosting	88.9 ± 1.4	75.7 ± 0.7	74.1 ± 1.0	85.7 ± 0.7	97.0 ± 0.4
Cels_m	91.9 ± 1.0	76.0 ± 1.4	73.4 ± 1.3	91.1 ± 0.7	97.0 ±0.4
CVC	87.7 ± 1.3	76.0 ± 1.1	74.1 ± 1.4	89.0 ± 1.0	97.4 ± 0.3
Decorrelated	90.9 ± 0.9	76.4 ± 1.2	74.6 ± 1.5	91.5 ± 0.6	97.0 ± 0.5
Decorrelated2	90.6 ± 1.0	75.7 ± 1.1	74.3 ± 1.4	90.3 ± 0.4	97.0 ± 0.5
Evol	83.4 ± 1.9	66.3 ± 1.2	74.3 ± 0.6	77.5 ± 1.7	94.4 ± 0.9
Ola	90.7 ± 1.4	69.2 ± 1.6	75.2 ± 0.9	83.2 ± 1.1	94.2 ± 0.7
Simple Ens.	91.1 ± 1.1	75.9 ± 1.2	74.3 ± 1.3	88.0 ± 0.9	96.9 ± 0.5

By comparing the results of Table 1, 2, 3 and 4 with the results of a single network we can see that there the improvement by the use of the ensemble methods depends clearly on the problem. For example in databases Aritm, Flare, Pima and Wdbc there is not a clear improvement.

In the rest of databases there is an improvement, perhaps the most important one is in database Vowel.

Table 3. Results for the Ensemble of nine networks

	ARITM	DERMA	ECOLI	FLARE	IMAGEN
Adaboost	73.2 ± 1.6	97.3 ± 0.5	84.7 ± 1.4	81.1 ± 0.7	97.3 ± 0.3
Bagging	75.9 ± 1.7	97.7 ± 0.6	87.2 ± 1.0	82.4 ± 0.6	96.7 ± 0.3
Bag_Noise	75.4 ± 1.2	97.0 ± 0.7	87.2 ± 0.8	82.4 ± 0.5	93.4 ± 0.3
Cels_m	74.8 ± 1.3	97.3 ± 0.6	86.2 ± 0.8	81.7 ± 0.4	96.6 ± 0.2
CVC	74.8 ± 1.3	97.6 ± 0.6	87.1 ± 1.0	81.9 ± 0.6	96.6 ± 0.2
Decorrelated	76.1 ± 1.0	97.6 ± 0.7	87.2 ± 0.7	81.6 ± 0.6	96.9 ± 0.2
Decorrelated2	73.9 ± 1.1	97.6 ± 0.7	87.8 ± 0.7	81.7 ± 0.4	96.84 ± 0.18
Evol	65.9 ± 1.9	54 ± 6	57 ± 5	80.6 ± 0.8	67 ± 4
Ola	72.5 ± 1.0	86.7 ±1.7	83.5 ± 1.3	80.8 ± 0.4	96.1 ± 0.2
Simple Ens	73.8 ± 1.1	97.5 ± 0.7	86.9 ± 0.8	81.6 ± 0.4	96.7 ± 0.3

Table 4. Results for the ensemble of nine networks

	IONOS	PIMA	SURVI	VOWEL	WDBC
Adaboost	89.4 ± 0.8	75.5 ± 0.9	74.3 ± 1.4	94.8 ± 0.7	95.7 ± 0.7
Bagging	90.1 ± 1.1	76.6 ± 0.9	74.4 ± 1.5	90.8 ± 0.7	97.3 ± 0.4
Bag_Noise	93.3 ± 0.6	75.9 ± 0.9	74.8 ± 0.7	85.7 ± 0.9	95.9 ± 0.5
Cels_m	91.9 ± 1.0	75.9 ± 1.4	73.4 ± 1.2	92.7 ± 0.7	96.8 ± 0.5
CVC	89.6 ± 1.2	76.9 ± 1.1	75.2 ± 1.5	90.9 ± 0.7	96.5 ± 0.5
Decorrelated	90.7 ± 1.0	76.0 ± 1.1	73.9 ± 1.3	92.8 ± 0.7	97.0 ± 0.5
Decorrelated2	90.4 ± 1.0	76.0 ± 1.0	73.8 ± 1.3	92.6 ± 0.5	97.0 ± 0.5
Evol	77 ± 3	66.1 ± 0.7	74.8 ± 0.7	61 ± 4	87.2 ± 1.6
Ola	90.9 ± 1.7	73.8 ± 0.8	74.8 ± 0.8	88.1 ± 0.8	95.5 ± 0.6
Simple Ens	90.3 ± 1.1	75.9 ± 1.2	74.2 ± 1.3	91.0 ± 0.5	96.9 ± 0.5

There is, however, one exception in performance in the method Evol. This method did not work well in our experiments. In the original reference the method was tested in the database Heart. The results for a single network were 60%, for a simple ensemble 61.42% and for Evol 67.14%. We have performed some experiments with this database and our results for a simple network are 82.0 ± 0.9, clearly different.

Now, we can compare the results of Tables 1, 2, 3 and 4 for an ensemble of different number of networks. We can see that the results are in general similar and the improvement of training an increasing number of networks, for example 20 and 40, is in general marginal. Taking into account the computational cost, we can say that the best alternative for an application is an ensemble of three or nine networks.

We have also calculated the percentage of error reduction of the ensemble with respect to a single network. We have used equation 5 for this calculation.

$$PorError_{reduction} = 100 \cdot \frac{PorError_{single\ network} - PorError_{ensemble}}{PorError_{single\ network}} \qquad (5)$$

The value of the percentage of error reduction ranges from 0%, where there is no improvement by the use of a particular ensemble method with respect to a single network, to 100%. There can also be negative values, which means that the performance of the ensemble is worse than the performance of the single network.

This new measurement is relative and can be used to compare more clearly the different methods. Furthermore we can calculate the mean performance of error reduction across all databases this value is in Table 5 for ensembles of 3, 9, 20 and 40 networks.

Table 5. Mean percentage of error reduction for the different ensembles

	Ensemble 3 Nets	Ensemble 9 Nets	Ensemble 20 Nets	Ensemble 40 Nets
Adaboost	1.33	4.26	9.38	12.21
Bagging	6.86	12.12	13.36	12.63
Bag_Noise	-3.08	-5.08	-3.26	-3.05
Boosting	-0.67			
Cels_m	9.98	9.18	10.86	14.43
CVC	6.18	7.76	10.12	6.48
Decorrelated	9.34	12.09	12.61	12.35
Decorrelated2	9.09	11.06	12.16	12.10
Evol	-218.23	-297.01	-375.36	-404.81
Ola	-33.11	-36.43	-52.53	-47.39
Simple Ens	5.58	8.39	8.09	9.72

According to this global measurement *Ola*, *Evol* and *BagNoise* performs worse than the *Simple Ensemble*. The best methods are *Bagging, Decorrelated* and *Decorrelated2*. In total there are only four methods which perform better than the *Simple Ensemble*.

Also in this table, we can see the effect of increasing the number of networks in the ensemble. There are two methods (*Adaboost* and *Cels*) where the performance seems to increase slightly with the number of networks in the ensemble. But other methods like *Bagging, CVC, Decorrelated, Decorrelated2* and *Simple Ensemble* does not increase the performance beyond 9 or 20 networks in the ensemble. The reason can be that the new networks are correlated to the first ones or that the combination method (the average) does not exploit well the increase in the number of networks.

4   Conclusions

In this paper we have presented experimental results of eleven different methods to construct an ensemble of networks, using ten different databases. We trained ensembles of 3, 9, 20 and 40 networks in the ensemble. The results showed that in general the improvement by the use of the ensemble methods depends clearly on the database, in some databases there is an improvement but in other there is not improvement at all. Also the improvement in performance from three or nine networks in the ensemble to a higher number of networks depends on the method. Taking into account the computational cost, an ensemble of nine networks may be the best alternative for most of the methods. Finally, we have obtained the mean percentage of error reduction over all databases. According to the results of this measurement the best methods are "Decorrelated", "Bagging" and "Cels" and there are only four or five methods which perform better than the "Simple Ensemble".

References

1. Tumer, K., Ghosh, J., "Error correlation and error reduction in ensemble classifiers". Connection Science. **8** nos. 3 & 4, (1996) 385-404
2. Raviv, Y., Intrator, N., "Bootstrapping with Noise: An Effective Regularization Technique". Connection Science. **8** no. 3 & 4, (1996) 355-372
3. Drucker, H., Cortes, C., Jackel, D., et alt., "Boosting and Other Ensemble Methods". Neural Computation. **6**, (1994) 1289-1301
4. Verikas, A., Lipnickas, A., et alt., "Soft combination of neural classifiers: A comparative study". Pattern Recognition Letters. **20**, (1999) 429-444
5. Breiman, L., "Bagging Predictors". Machine Learning. **24**, (1996) 123-140
6. Freund, Y., Schapire, R., "Experiments with a New Boosting Algorithm". Proceedings of the Thirteenth International Conference on Machine Learning. (1996) 148-156
7. Rosen, B., "Ensemble Learning Using Decorrelated Neural Networks". Connection Science. **8** no. 3 & 4, (1996) 373-383
8. Auda, G., Kamel, M., "EVOL: Ensembles Voting On-Line". Proc. of the World Congress on Computational Intelligence. (1998) 1356-1360
9. Liu, Y., Yao, X., "A Cooperative Ensemble Learning System". Proc. of the World Congress on Computational Intelligence. (1998) 2202-2207
10. Jang, M., Cho, S., "Ensemble Learning Using Observational Learning Theory". Proceedings of the International Joint Conference on Neural Networks. **2**, (1999) 1281-1286

Performance Advantage of Combined Classifiers in Multi-category Cases: An Analysis

Xubo Song[1] and Misha Pavel[2]

[1] Department of Computer Science and Engineering, OGI School of Science and Engineering
Oregon Health and Science University, Beaverton, Oregon 97006
xubosong@cse.ogi.edu
[2] Department of Biomedical Engineering, OGI School of Science and Engineering
Oregon Health and Science University, Beaverton, Oregon 97006
pavel@bme.ogi.edu

Abstract. One problem in the field of machine learning is that the performance on the training and validation sets lack robustness when applied in real-life situations. Recent advances in ensemble methods have demonstrated that robust behavior can be improved by combining a large number of weak classifiers. The key insight of this paper is that the performance enhancement due to combining multiple classifiers is considerably greater in multi-category situations than in binary classifications, as long as their errors are conditionally independent. This paper provides some experimental and theoretical analysis of the performance using majority vote, paying special attention to the effect of several parameters that include the number of combined classifiers, weakness of the combined classifiers and the number of classes. These insights can provide guidance for the analysis and design of multi-classifier systems.

1 Introduction

Although many pattern recognition and intelligent signal processing systems have achieved considerable success over the past fifty years, most of them lack robustness in that they exhibit undue sensitivity to irrelevant environmental conditions in novel, unpredictable, and dynamic environments. In contrast, biological systems seem to exhibit considerably higher resilience to the irrelevant variability. In most successful biological systems, the input signals to be classified are decomposed into a large number of interdependent streams and the classification results are obtained by combining the partial results [9]. The potential of this approach has attracted much attention in the field of machine learning. Its capability to improve classification rate over a single classifier has been firmly established as an effective and practical technique for difficult learning tasks. The insight presented in this paper is that the performance enhancement due to combining multiple classifiers is considerably greater in multi-category situations than in binary classifications.

Techniques for multiple classifier combination generally have two stages, namely the base classifier formation stage, and the classifier combination stage. The goal of classifier formation stage is to come up with base classifiers that are maximally diversified and complementary to one another, and the goal of the classifier combination stage is to optimally combine these base classifiers.

N.R. Pal et al. (Eds.): ICONIP 2004, LNCS 3316, pp. 750–757, 2004.
© Springer-Verlag Berlin Heidelberg 2004

The classifiers can be combined using a variety of strategies. Typical strategies include fixed rules (e.g., simple averaging, weighted averaging, majority vote, maximum, median) [1-3] and trained rules (e.g., modular networks, Bayesian combination, Dempster-Shafer) [4-5]. Among them, majority vote is by far the simplest to implement, and yet it has been found to be just as effective as more complicated schemes in improving classification performance [5]. Most of the reported work on classifier combination has been experimental in nature. Recently, some theoretical analysis has been carried out to study the behavior and performance of majority vote. In [3] and [7], the authors studied the two-class voting scheme. In [10], the authors initiated using majority vote for multi-class cases, and applied it to ensemble of neural networks. Contrary to the assumptions held by many investigators, extension of binary classifications to multiple classes is not trivial and we are not aware of any general analysis on the performance of majority voting for multiple classes.

In this paper, we analyze the performance of combined multiple classifiers using majority vote with arbitrary number of classes. Specifically, we study this performance as a function of several factors – the number of classes, the strength of the individual base classifiers and the number of base classifiers being combined. We present simulation results as well as theoretical analysis of the asymptotic behavior of combined classifiers. Conditions and assumptions are also described that make the analysis tractable.

2 Problem Specification

We first introduce some notations. Let x be the input feature from the input space X, and $c(x)$ be its true class label from the label space $\Omega = \{1, 2, ..., K\}$. We assume a set of M classifiers $\{h_1(x), h_2(x), ..., h_M(x)\}$, where $h_i(x): X \rightarrow \Omega$. Each classifier $h_i(x)$ is characterized by a confusion matrix $\{p_{vw}{}^i(x) = \Pr(h_i(x) = v \mid c(x) = w)\}$ for a given input x. For simplicity, we assume $p_{vw}{}^i(x) = p$ if $v = w$ (correct classification) and $p_{vw}{}^i(x) = q$ if $v \neq w$ (incorrect classification) for all classifier $h_i(x)$ and for all input x. Apparently $p + (K-1)q = 1$. We assume that the classifiers are better than random guesses and therefore have $p > q$.

Let P_c be the probability that the combined classifier is correct. To evaluate the performance of combining M classifiers using majority vote, we compare P_c with the probability of a single classifier being correct, namely p. We are especially interested in how P_c behaves as a function of the parameters M, K, p and q.

Assume that $\Pr(h_i(x) \mid h_j(x), c(x)) = \Pr(h_i(x) \mid c(x))$ for all i and j. Based on this assumption, all classifiers will label a given x independently, resulting in M class labels from K classes. Denote z_i as the count from class i. Clearly $\sum_1^K z_k = M$. Then the joint distribution of all class counts $\Pr(z_1, z_2, ..., z_K)$ is multinomial:

$$\Pr(z_{1,}z_{2,}...,z_K) = \frac{M!}{z_1!z_2!...z_K!}p^{z_1}q^{z_2}q^{z_3}...q^{z_k} = \frac{M!}{z_1!z_2!...z_K!}p^{z_1}q^{M-z_1} \tag{1}$$

Without loss of generality, we assume the true label for the given input x is class 1, i.e., $c(x)=1$. According to the majority rule, the true class for x is recovered if and only if $z_1 > z_k$ for all $k>1$. Then the probability of success using majority vote is

$$P_c = \Pr(z_1 > z_{k,}\forall k > 1) = \sum_{z_1}\sum_{z_2}...\sum_{z_K} p(z_1,z_2,...,z_k) \tag{2}$$

where the summation is taken over all $\{z_{1,}z_{2,}...,z_K\}$ combinations subject to that $z_1 > z_K, \forall k > 1$ and that $\sum_1^K z_k = M$.

Equation 2 provides the formula for calculating the success rate P_c of combining M classifiers using majority vote for K-category problems. However, it is difficult to obtain an exact closed form for Equation 2 for arbitrary K and M. We will study the general behavior of P_c without obtaining a closed form. We will introduce approximations that simplify the computation of P_c. Such approximations will be validated by Monte Carlo simulation. We will also study the asymptotic behavior of P_c for large M.

3 Analysis of Performance P_c for Multi-category Case

Majority vote has been a much studied subject since its origin in the Condorcet Jury Theorem (CJT) [8], which states the following for the two-class case:

Condorcet Jury Theorem: For odd $M > 3$, the following are true:

(1) If $p > 0.5$, then P_c is monotonically increasing in M and $P_c \to 1$ as $M \to \infty$.

(2) If $p < 0.5$, then P_c is monotonically decreasing in M and $P_c \to 0$ as $M \to \infty$.

(3) If $p = 0.5$, then $P_c = 0.5$ for all M.

We conjecture the analogy to CJT for the multi-category case as follows: Assume $M \bmod K \neq 0$. Then,

(1) If $p > \frac{1}{K}$, then P_c is monotonically increasing in M, and $P_c \to 1$ as $M \to \infty$.

(2) If $p < \frac{1}{K}$, then P_c is monotonically decreasing in M, and $P_c \to 0$ as $M \to \infty$.

(3) If $p = \frac{1}{K}$, then $P_c = \frac{1}{K}$ for all M.

Figures 1 and 2 show the Monte Carlo simulation results which illustrate our conjecture, demonstrating the effects of various variables. For all figures, unless otherwise noted, the y-axis corresponds to the success rate P_c. In all case, we choose better-than-random-guess classifiers so that $p > q$. Figure 1(a) shows the effect of changing M. For fixed $p = 0.101$ and $K = 10$, we can easily see that

$q = 0.099889$. Even for classifiers so weak, the probability of success increases as M becomes large, and eventually approaches 1. Figure 1(b) illustrates the effect of changing K that results in stronger base classifiers. In this case, p is fixed at 0.101, therefore q decreases as K increases. As a result, the base classifiers become stronger for larger K. For a given M, P_c approaches 1 as the classifiers become stronger. The rate of P_c approaching 1 is faster for larger M. This is indicated by the difference between the upper curve (M=5000) and lower curve (M=500). Figure 2 demonstrates the effect of changing K for weak classifiers. The upper curve corresponds to P_c, and the lower curve corresponds to p. In this case, we choose $p = \frac{1}{K}*1.02$, which indicates the base classifiers are always only slightly better (by 2% in this case) than random guess for any K. For a fixed M, the advantage of P_c compared to single classifier success rate p diminishes as K becomes large, and both approach zero for large K.

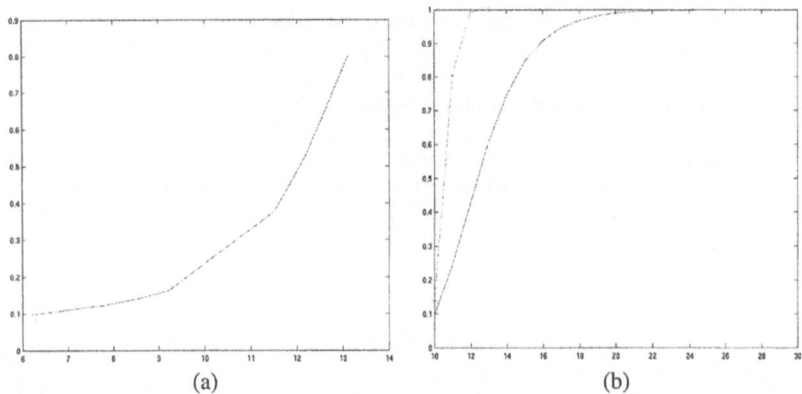

(a)   (b)

Fig. 1. (a) Performance P_c curve for K =10, p=0.101 and q=0.099889. The x-axis corresponds to log M, (b) Performance P_c curve for p=0.101 and q = (1-p)/(K-1). The x-axis corresponds to K. The upper curve is for M=5000 and the lower curve is for M=500

4 Approximation of P_c

In an effort to obtain some closed form for P_c, we investigate the feasibility of certain approximations that would make its computation tractable. We can rewrite Equation 2 as follows:

$$P_c = \Pr(z_1 > z_k, \forall k > 1) = \Pr(z_1 > \max(z_2,...,z_k)) \tag{3}$$

Let $y = \max(z_2,...,z_K)$. Then it follows

$$P_c = \sum_{t=M/2}^{M} \Pr(z_1 = t) + \sum_{t=M/K}^{M/2} \Pr(y < t \mid z_1 = t)\,\Pr(z_1 = t) \tag{4}$$

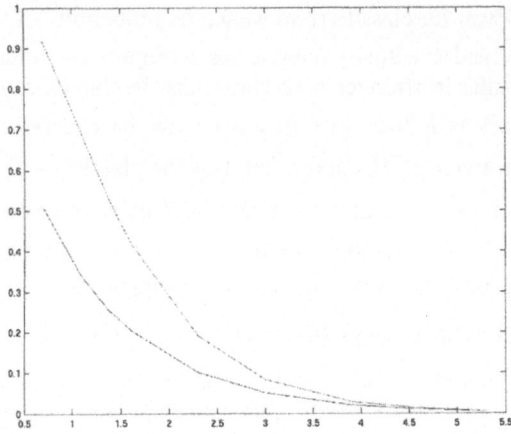

Fig. 2. Performance curves for M = 5000 and $p = \frac{1}{K}(1+.02)$. The x-axis corresponds to $\log K$. The upper curve is for P_c and the lower curve is for p

Continue with Equation 4, we can upper bound P_c by

$$P_c \leq \sum_{t=M/2}^{M} \mathrm{Pr}(z_1 = t) + \sum_{t=M/K}^{M/2} \mathrm{Pr}(y < t)\,\mathrm{Pr}(z_1 = t) \qquad (5)$$

where

$$\mathrm{Pr}(z_1 = t) = \frac{M!}{t!(M-t)!} p^t (1-p)^{M-t}$$

and

$$\mathrm{Pr}(y < t) = \mathrm{Pr}(\max(z_2,...,z_K) < t) = \{ \sum_{y=0}^{t} \frac{M!}{y!(M-y)!} q^y (1-q)^{M-y} \}^{K-1} .$$

Equation 5 provides a computable approximation to P_c by assuming that z_2, ..., z_K are independent. By applying Stirling's formula, Equation 5 can be computed, even for large M. The approximation of P_c using Equation 5 is very close to the Monte Carlo simulation results (see Figure 3), which doesn't make any assumption and are numerically accurate. Such closeness implies that the independence assumption of z_2, ..., z_K is reasonable for large M and large z_2, ..., z_K. Experiments also show that for large M, the first term in Equation 5 is negligible compared to the second term.

Notice that in our analysis a vote is considered successful only if z_1 is strictly larger than z_k. No tie is counted in computing P_c. This is solely for convenience considerations, and doesn't affect the overall conclusions. Modifications can be made to accommodate ties with appropriate weighting of the tie events. The probability of tie events decreases as M becomes large.

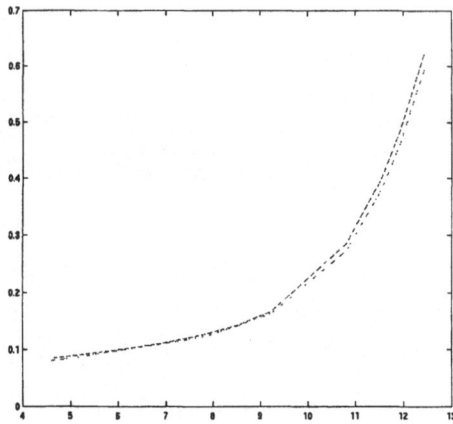

Fig. 3. Comparison of the Monte Carlo simulation of P_c (lower line) and the computation of P_c using Equation (6) (upper line). The x-axis corresponds to $\log M$ .Here the parameters are $K = 10$, p = 0.101, and q=0.099889

5   Asymptotic Behavior of P_C

Equation 1 provides the multinomial joint distribution for $\Pr(z_1, z_2, ..., z_K)$, from which we can deduce the expectations $E(z_1)=pM$, and $E(z_2)=...=E(z_k)=qM$. This implies that in the limit, it is true that $z_1 > z_k$ for any K and therefore P_c =1. In fact, according to Law of Large numbers, we have $\Pr(|\frac{z_1}{M}-p|<\varepsilon) \to 1$ and $\Pr(|\frac{z_k}{M}-q|<\delta) \to 1$ for $K>1$ as M becomes large. Here we provide an analytical approximation of P_c for large M . According to the Central Limit Theorem, the limiting distributions of $z_1, z_2, ..., z_K$ are Gaussians. Specifically,

$$\Pr(|\frac{z_1}{M}-p|<\varepsilon) \to \Re(s\varepsilon)-\Re(-s\varepsilon), \text{ where } s=\sqrt{\frac{M}{pq}}, \text{ and } \Re(x)=\frac{1}{2\pi}\int_{-\infty}^{x}\exp(-\frac{1}{2}y^2)dy$$

is the normal distribution. Then we can approximate P_c by

$$P_c \approx \Pr(|\frac{z_1}{M}- p|<\varepsilon \text{ and } |\frac{z_k}{M}- q|<\delta \text{ for } \forall k > 1)$$

(Choose $\delta=p-q-\varepsilon$)

$$\approx \Pr(|\frac{z_1}{M}- p|<\varepsilon)\prod_{k=2}^{K}\Pr(|\frac{z_k}{M}- q|<\delta)$$

$$\approx (2\Re(s\varepsilon)-1)(2\Re(s\delta)-1)^{K-1} \approx (1-\frac{2}{s\varepsilon}\exp(-\frac{1}{2}s^2\varepsilon^2))(1-\frac{2}{s\delta}\exp(-\frac{1}{2}s^2\delta^2))^{K-1}$$

Omitting the intermediate steps, it turns out that

$$P_c \approx 1 - \frac{8}{\sqrt{\dfrac{M(p-q)^2}{pq}}} \exp\{-\frac{1}{8}\frac{M(p-q)^2}{pq}\} = 1 - \frac{8}{\sqrt{\dfrac{M(pK-1)^2}{p(1-p)(K-1)}}} \exp\{-\frac{1}{8}\frac{M(pK-1)^2}{p(1-p)(K-1)}\} \tag{6}$$

Equation 6 provides asymptotic approximations for P_c. We can see that as $M \to \infty$, P_c approaches 1. For fixed p, the larger K, the larger P_c, since q becomes smaller and thus the base classifiers become stronger. For very weak classifiers where $pK-1$ equals a small positive constant (i.e., p is only slightly larger than $\frac{1}{K}$), the large the K, the weaker the classifiers, and thus the smaller P_c. All these facts coincide with our intuitions. Equation 6 also provides an upper bound for the probability of the majority being correct for multi-category classification problems when M is large.

6 Conclusions

Our empirical as well as theoretical analysis have established that for a large number of weak classifiers, where the probability of correct classification for each is only slightly larger than the probability of incorrect classification, the overall probability of success will approach 1 as the number of classifier increases. This insight can provide guidance for the design of real classification systems. The proposed approach is to increase the diversity of the classifier rather than focus on efficient information representation. Thus each classifier doesn't need to have a lot of discriminative power, as long as it is better than random guess where $p = \frac{1}{K}$. Having a large number of classifiers can compensate for the weakness of the individual classifiers.

The critical assumption underlying this work is that the classifiers are conditionally independent, and in particular, the erroneous responses need to be uncorrelated. To what extent this is achievable remains an empirical question. Our preliminary investigations in several realistic areas, e.g., speech recognition [11,12], suggest that this assumption can in practice be sufficiently well satisfied to yield the desired results. Our current efforts are directed towards developing methodologies to train classifiers with enhanced degree of conditional independence.

References

1. Breiman, L. (1996) Bagging Predictors. *Machine Learning*, 24(2), 123-140.
2. Kittler, J. (1998) On Combining Classifiers. *IEEE Trans. Pattern Analysis and Machine Intelligence*, 20(3):226-239.
3. Lam, L. & Suen, C. (1997) Application of majority voting to pattern recognition: An analysis of its behavior and performance. *IEEE Trans. on Systems, Man and Cybernetics, Part A: Systems and Humans*, 27(5): 553-568.

4. Dietterich, T.G. (2000) Ensemble Methods in Machine Learning. In Kittler, J. and Roli, F. (eds.), *Multiple Classifier Systems, First International Workshop.* Lecture Notes in Computer Science, Vol. 1857, pp 16-29. Springer Verlag. New York.
5. Freund, Y. & Schapire, R.E. (1996) Experiments with a new boosting algorithm. In *Proc. 13th International Conference on Machine Learning*, pp148-156. Morgan Kaufmann.
6. Drucker, H., Cortes, C., Jackel, L., LeCun, Y. & Vapnik, V. (1994) Boosting and other ensemble methods. *Neural Computation,* 6:1289-1301.
7. Kuncheva, L.I., Whitaker, C.J., Shipp, C.A. & Duin, R.P.W. (2003) Limits on the majority vote accuracy in classifier fusion. *Pattern Analysis and Applications*, 6(1): 22-31.
8. de Condorcet, N.C. (1785) Essai sur l'application de l'analysis a la probabilite des decisions rendues a la pluralite des voix. Imprimerie Royale, Paris.
9. Fletcher, H. (1953) *Speech and Hearing in Communication.* New York: Krieger.
10. Ji, C. & Ma, S. (1997) Combination of weak classifiers. *IEEE Trans. Neural Networks*, 8 (1): 32-42.
11. Pavel, M. and Hermansky, H. (1997) Information Fusion by Human and Machine, *First European Conference on Signal Analysis and Prediction*, Strahov Monastery, Prague, Czech Republic.
12. Hermansky, H., Tibrewala, S. & Pavel, M. (1996) Towards ASR on Partially Corrupted Speech, *the Fourth International Conference on Spoken Language Processing*, Philadelphia, PA, USA.

Web Documents Categorization Using Neural Networks

Renato Fernandes Corrêa[1,2] and Teresa Bernarda Ludermir[1]

[1] Polytechnic School, Pernambuco University
Rua Benfica, 445, Madelena, Recife – PE, Brazil, 50.750-410
[2] Center of Informatics – Federal University of Pernambuco
P. O. Box 7851, Cidade Universitária, Recife – PE, Brazil, 50.732-970
{rfc,tbl}@cin.ufpe.br

Abstract. This paper shows, through experimental results, that artificial neural networks are good classifiers for the text categorization task. The paper compares the results of experiments on text categorization using Multilayer Perceptron, Self-organizing Maps, C4.5 decision tree and PART decision rules. The experiments were carried out with K1 collection of web documents.

1 Introduction

Information overloading has become a serious problem due to the ever increasing of volume of electronically stored documents available mainly by means of the Internet. Innumerable Information Retrieval Systems (IRS), called search engines, have been implemented to assist users in finding documents that contain information relevant to their particular needs. Despite the aid of these systems, finding necessary information is often difficult and time-consuming due to vast amount of information available.

Improved IRS performance is achieved through organizing text documents into categories according to subject. This is known as text categorization.

Some search engines, such as Yahoo, have provided a hierarchy of categories in which documents are classified manually. However, manual text categorization is a difficult and laborious task even for experienced professionals.

An alternative for transforming text categorization into a viable task is the construction of Text Categorization Systems. This system utilizes machine-learning techniques in order to automate the classification of documents, thus allowing wide scale classification to be carried out quickly and concisely.

The objective of this work is to show that Artificial Neural Networks is good for text categorization. The paper presents and compares the results of experiments on text categorization using the Multilayer Perceptron (MLP) [1] and Self-organizing Maps (SOM) [2] artificial neural networks, as well as symbolic machine-learning algorithms [3]: the C4.5 decision tree and PART decision rules. The experiments were carried out with the K1 collection of web documents [4].

2 Text Categorization

Text Categorization systems typically have to process an influx of new documents – some times in real time –, assigning them individually to none, one or a number of different categories. The set of categories into which the documents are classified is typically predefined by the designer or maintainer of the system and generally re-

N.R. Pal et al. (Eds.): ICONIP 2004, LNCS 3316, pp. 758–762, 2004.
© Springer-Verlag Berlin Heidelberg 2004

mains unchanged over long period of time. Three stages in text categorization are considered: representation, preprocessing, and classification.

The representation stage consists of defining documents and their respective categories in such a way that they may be used as the input and output of the machine learning algorithms. For documents, the vectorial representation is generally adopted. In this representation, terms or words become indices and the respective values represent the importance of the term in the document. These terms consist of isolated words or groups of related words. The assumption adopted here is that the semantics of the document can be expressed by a set of indexation terms. Categories may be associated to the document vectors as labels or a boolean vector can be used, where each position indicates the relevancy of the document to a category.

The pre-processing stage consists of reducing the number of terms used to represent documents. This stage involves decisions on how the lexical analysis of documents will be carried out, which algorithms will be used to remove word affixes, which words are irrelevant in the categorization process, and which algorithm will be used for the selection or extraction of the final set of indexation terms.

In the classification stage, the partition of document vectors in training and test sets are determined. The training set is used to train the classifiers. The test set is for evaluating their capacity for generalization. The machine learning algorithms and respective training parameters are also chosen in this stage, as well as the metrics used to evaluate the test set system.

3 Methodology

The experiments consisted of the binary text categorization of the K1 collection [4]. This set consists of a collection of 2,340 English-language web documents manually classified in one of 20 news categories at Yahoo: Health, Business, Sports, Politics, Technology and 15 subclasses of Entertainment (film, people, television, review, etc.). The document of the collection had been categorized manually. Thus, the learning and performance analysis of the generated classifiers was done on the basis of their proximity to the human-categorization.

Each of the collection documents was represented individually as a boolean vector, based on term frequency, with 410 positions, where each position indicates the presence or the absence of a particular term in the document. It was used mean document frequency for pruning insignificant or too generic words (words that occur less than 0.01 or more than 0.10 times on average respectively) [7], followed of document Relevance Score (20 terms more positive and 10 more negative for category) [8].

Two separate stages were carried out: the training and testing of the classifiers. Distinct sets of documents, the training set and the test set, from the collection were used in these two stages. The document vectors were randomly partitioned into the training and testing sets, each one with 1,170 documents.

The performance of the machine-learning algorithms in the text categorization task was measured by comparing the classification error on test set of the best classifier generated by each algorithm.

For the SOM networks, the boolean vectors were transformed into unitary vectors. The categories were codified and associated with the respective document vectors as a boolean vector for MLP and as labels for SOM and symbolic algorithms.

The neural classifiers were constructed and tested using the MATLAB programming environment from The MathWorks Incorporation.

MLP Networks were trained with the Backpropagation learning algorithm with momentum [1]. The topologies had one hidden layer. The number of neurons in the input layer was 410 (dimension of the document vector). The number of neurons in the output layer was 20 (number of categories). Both the number of neurons in the hidden layer and the learning rate were determined experimentally. All the neurons used the logistic sigmoid as the activation function. The momentum used in all the experiments was 0.8. Part of the training set was used as the validation set, and the criteria for stopping the training included the loss of generalization [5] surpassing 5% and reaching the point of the maximum number of 1000 epochs. The evaluation of the output of the networks [5] was 1-of-N (simple categorization).

Topologies with different numbers of neurons in the hidden layer were tested: 2, 4, 8, 16 and 32. For each topology, 10 initial configurations were created with random weight initializations. Each initial configuration was trained 5 times with the respective learning rates: 0.1, 0.05, 0.01, 0.005 and 0.001.

In the experiments with the SOM networks, the maps used had a rectangular structure with a hexagonal neighborhood to facilitate visualization. The dimension of the map and the number of training epochs were obtained experimentally. The Gaussian neighborhood function was used. The algorithm used for training SOM networks was the unsupervised batch-map SOM [2] for being the quickest and having few adjustable parameters. Topologies with different map dimensions were tested: 8x5, 10x8 e 14x12. For each topology, 10 initial configurations were created with random weight initializations. For each topology, the initial neighborhood size was equal to half the number of neurons within the largest dimension. The final neighborhood size was always 1. Each initial configuration was trained by 20 and 50 epochs; the number of epochs determines how mild the decrease of neighborhood size will be, since it is linearly decreasing with the number of epochs.

The text categorization was carried out assigning to each neuron the category of the majority of documents mapped in the training set. Each document is only represented by one neuron – the one where the vector model is most similar to the respective document vector.

The document vectors of the test set received the category assigned to the neuron where they were mapped. For the SOM and MLP networks, having found the best likely neural architecture, the instance of this architecture generating the least classification errors in the test set was chosen as the best respective classifier.

For C4.5 and PART techniques, the classifiers were constructed using the Weka tool [3]. Differently from the neural networks, the C4.5 and PART do not need an initial search point and, therefore, they were executed just once. The value of the training parameters of the C4.5 and PART algorithms were: pruning confidence threshold equal to 0.25 and minimum number of instances per leaf equal to 2.

In the Table 1, the performance of the algorithms is in ascending order according to the classification error in the test for the best and worse performance. A Ranking column was generated from the classification error and indicates how many times the error found for a classifier was greater than the least amount of error found for the collection.

4 Results

The classifiers had difficulties representing the Entertainment subclasses within the K1 collection due to the similarity of the documents and the small number documents in 6 of these subclasses. Nevertheless, the results obtained for this collection proved to be satisfactory in view of the range of classification errors, on the part of the classifiers, equal or inferior to 38.38% (cf. Table 1).

Table 1 shows that MLP classifier obtained the least classification error on test set. This performance was the consequence of a better representation of the Entertainment subclasses. SOM networks were in second place with a classification error 1.30 times bigger than that one got by MLP networks. The C4.5 and PART classifiers had practically the same performance, classification errors were about 1.47 times bigger than that one got by MLP networks.

The best MLP network had 32 neurons in the hidden layer and was trained through 5000 epochs with a learning rate of 0.001 and momentum of 0.8. The average classification error of 10 MLP networks with this architecture was of 28.02% with a standard deviation of 1.25%.

The best SOM network for this collection consisted of a 14x12 unit map trained through 50 epochs with initial ray of 7 and a final ray equal to 1. The average classification error of the 10 SOM networks with this architecture was of 35.79% with a standard deviation of 1.31%.

The two main characteristics of the SOM maps were verified: the preservation of the proximity relations between similar documents and consequently related categories, and the allocation of units for each category in accordance with the frequency of occurrence [6]. Thus, these networks can be used as support tools for a manual document categorizer, as well as a way to explore a document collection, having as its initial interface the map generated and labeled with the most numerous categories in each neuron.

Table 1. Performance of the best classifiers for K1 collection

Algorithm	Classification Error	Ranking
MLP	25.90	1.00
SOM	33.68	1.30
C4.5	37.95	1.47
PART	38.38	1.48

5 Conclusions

In this work, the evaluation of the performance of connectionist and symbolic machine-learning algorithms in a text categorization task was presented by means of experiments with K1 web documents collection. MLP and SOM were the most promising algorithms in the solving of text categorization problem. The SOM networks, despite learning in an unsupervised manner, had a good performance in the solving of text categorization problem and got a better performance than symbolic algorithms.

References

1. Rumelhart, D.E., Hinton, G.E. and Williams, R.J.: 'Learning internal representations by error propagation'. Parallel Distributed Processing (Edited by D.E. Rumelhart and J.L. McClelland), Vol. 1, pp. 318-362, Cambridge, MIT Press, 1986
2. Kohonen, T., Kaski, S., Lagus, K., Salojärvi, J., Honkela, J., Paatero, V. and Saarela, A.: 'Self Organization of a Massive Document Collection'. IEEE Transaction on Neural Networks, v. 11, n. 3, May 2000, pp. 574-585
3. Witten, I. H. and Frank, E.: 'Data Mining: Practical Machine Learning Tools and Techniques with Java Implementations'. Morgan Kaufmann, 2000
4. Boley, D., Gini, M., Gross, R., Han, E., Hastings, K., Karypis, G., Kumar, V., Mobasher, B. and Moore, J.: 'Partitioning-based clustering for web document categorization'. Decision Support Systems, v.27, 1999, pp. 329-341
5. Prechelt, L.: 'Proben1 – A Set of Neural Network Benchmark Problems and Benchmarking Rules'. Technical Report 21/94, Fakultät für Informatik, Universität Karlsruhe, Germany, 1994
6. Lin, X., Soergel, D. and Marchionini, G.: 'A self-organizing semantic map for information retrieval'. Proceedings of the Fourteenth Annual International ACM/SIGIR Conference on Research and Development in Information Retrieval, Chicago, IL, 1991, pp. 262-269
7. Strehl, A., Ghosh, J., Mooney, R: 'Impact of Similarity Measures on Web-page Clustering'. Proc. of the 17th National Conference on Artificial Intelligence: Workshop of Artificial Intelligence for Web Search (AAAI 2000), July 2000, Austin, Texas. pp. 58–64
8. Wiener, E., Pedersen, J., Weigend, A. 'A Neural Network Approach to Topic Spotting'. In Proceedings of the Fourth Annual Symposium on Document Analysis and Information Retrieval (SIDAIR'95), pp. 317–332, Nevada, Las Vegas, 1995. University of Nevada, Las Vegas.

Gender Classification of Face Images:
The Role of Global and Feature-Based Information

Samarasena Buchala[1], Neil Davey[1], Ray J. Frank[1], Tim M. Gale[1,2],
Martin J. Loomes[1], and Wanida Kanargard[1]

[1] Department of Computer Science, University of Hertfordshire,
College Lane, Hatfield, AL10 9AB, UK
{S.Buchala,N.Davey,R.J.Frank,T.Gale,M.J.Loomes,W.Pensuwon}
@herts.ac.uk
[2] Department of Psychiatry, QEII Hospital, Welwyn Garden City, AL7 4HQ, UK

Abstract. Most computational models of gender classification use global information (the full face image) giving equal weight to the whole face area irrespective of the importance of the internal features. Here we use a two-way representation of face images that includes both global and featural information. We use dimensionality reduction techniques and a support vector machine classifier and show that this method performs better than either global or feature based representations alone.

1 Introduction

Most computational models of gender classification use whole face images, giving equal weight to all areas of the face, irrespective of the importance of internal features for this task. In this paper we evaluate the importance of global and local features based on experimentation. Global processing largely deals with coarse information like shape and configuration of internal features, while featural processing involves more detailed representation (e.g. eyes, mouth etc). We use these two representations and use a Support Vector Machine (SVM) classifier for gender classification. As the face images data have a very high dimensionality, we also use dimensionality reduction techniques before classification.

The remainder of the paper is organised as follows. Related work is discussed in the next Section. Section 3 discusses the methodology used for this study. Section 4 discusses the experimental results. We conclude with some discussion in Section 5.

2 Related Work

The gender classification problem has attracted researchers from the fields of Psychology and Computer Science. While the research in Psychology [1], [2], [3] is in the context of human vision and identifying differing features in males and females, the Computer Science research [4], [5], [6], [7], [8] is mostly from the perspective of face recognition. The computational models range from using pixel-based information to representations obtained from geometric measurements. Studies also vary as in the size of training sets used and in the type of features present or absent (for example, some studies use hair information while others do not). Nevertheless, most mod-

N.R. Pal et al. (Eds.): ICONIP 2004, LNCS 3316, pp. 763–768, 2004.
© Springer-Verlag Berlin Heidelberg 2004

els, specifically the pixel-based, use whole face images, where all features carry equal weight. These can be termed as global models.

3 Methodology

We use a two way representation of face images which embodies both global and featural information. From a 128 × 128 face image three sub-images are obtained as illustrated in Fig.1. A 32 × 64 pixel strip pertaining to the eyes region, taking the midpoint between the two eyes as a reference point, and a 32 × 64 pixel strip pertaining to the mouth region, taking midpoint of the mouth as a reference point are extracted from each face image. These sub-images account for the featural information. The third sub-image is a 64 × 64 reduced resolution version of the original image and this represents global information. A similar type of face representation was also used by Luckman et al [9] for their computational model of familiar face recognition.

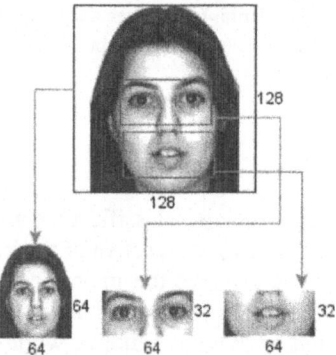

Fig. 1. Three sub-images are obtained from the original 128 × 128 image. A 32 × 64 image pertaining to the eye region and a 32 × 64 image pertaining to the mouth region are extracted from the original image. The third sub-image is a 64 × 64 reduced resolution version of the original image.

As the face images data have a very high dimensionality and due to "curse of dimensionality" [10], we apply dimensionality reduction techniques before applying an SVM for classification.

Principal Component Analysis (PCA) [11] is a popular dimensionality reduction technique and linearly transforms a D dimensional dataset X to a d dimensional dataset Y, without significant loss of information, where $d \leq D$.

Self Organising Map (SOM) [12] is a nonlinear method and learns a mapping from a D dimensional input space X to a d dimensional output space Y by using principles of Vector Quantization and Topological Mapping.

Curvilinear Component Analysis (CCA) [13], a recent technique, has the ability to reduce the dimensionality of strongly-nonlinear data. The output is a free space which takes the shape of the submanifold of the data. CCA minimizes the following error function:

$$E = \frac{1}{2}\sum_{i=1}^{N}\sum_{j=1}^{N}\left(d_{i,j}^{X} - d_{i,j}^{X}\right)^{2} F_{\lambda}\left(d_{i,j}^{Y}\right) \quad \forall \; j \neq i. \tag{1}$$

Where $d_{i,j}^{X}$ and $d_{i,j}^{Y}$ are the Euclidean distances between points i and j in the input space X and output space Y respectively. $F_{\lambda}\left(d_{i,j}^{Y}\right)$ is the neighbourhood function. The idea of CCA is to match distances in the input and output spaces. However, preservation of larger distances may not be possible in the case of nonlinear data, as a global unfolding of the manifold is required to reduce the dimension. In this case, it is important that at least local (smaller) distances should be preserved. For this reason CCA uses the neighbourhood function which ensures the condition of distance matching is satisfied for smaller distances while it is relaxed for larger distances. For details of the update rule, the reader is referred to [13].

The classification is performed using an SVM. The SVM [14] is a recently developed learning method, for pattern classification and regression. The basic idea of the SVM is to find the optimal hyperplane that has the maximal margin of separation between the classes, while having minimum classification errors.

Given a set of examples and their labels $\{(x_1, y_1),(x_2, y_2),\ldots, (x_N, y_N)\}$ where $y_i \in \{-1,1\}$, the optimal hyperplane is given as:

$$f(X) = \sum_{i=1}^{N}\alpha_i y_i k(X, X_i) + b. \tag{2}$$

Constructing the optimal hyperplane is equivalent to finding α_i with nonzero values. The examples corresponding to the nonzero α_i are called support vectors. $K(x, x_i)$ is a kernel function, which implicitly maps the example data points into a high dimensional feature space, and takes inner product in that feature space. The potential benefit of a kernel function is that the data is more likely to be linearly separable in the high dimensional feature space, and also the actual mapping to the higher-dimensional space is never needed. We used an RBF kernel in our experiments.

4 Experiments

Experiments are carried out using 400 frontal face (200 females and 200 males) grey scale images. The faces are from the following databases: FERET [15], AR [16], and BioId [17]. Three sub-images, as explained in the previous Section, are extracted for each of the 400 faces. Histogram equalization is then applied on all three sub-images to normalize for different lighting conditions. We use five-fold cross validation, with 320 faces (160 females and 160 males) for each training set and 80 faces (40 females and 40 males) for each test set, and report average classification rates using a SVM classifier, with RBF kernel. Before applying classification, dimensionality reduction techniques discussed in Section 3 are applied on the sub-images data. For PCA reduction we use the first few principal components, which account for 95% of the total variance of the data, and project the data onto these principal components. As CCA has the ability to reduce the dimensionality of strongly-nonlinear data, we use an

Intrinsic Dimension[1] estimation technique, the Correlation Dimension [18] and reduce the data dimension to this Intrinsic Dimension. For SOM reduction, the subspace dimensionality is chosen as 64 (8 × 8 output grid) for the whole face and 36 (6 × 6 output grid) for eyes and mouth sub-images.

First we present classification results on the sub-images data. As shown in Table 1, all three sub-images produced high classification rates, indicating a surprisingly high amount of gender information in each of them. The figures in the parentheses indicate the subspace dimensionality. Classification is performed on the composite data, obtained by combining the data from the three sub-images. It can be seen from Table 2 that PCA performed marginally better than CCA and SOM. However, CCA uses far fewer variables (70) than PCA (759). For a comparison, we also report the classification rates of the data of the original 128 × 128 faces. It can be seen from Table 2 that the composite data, which includes both global and featural information, performed significantly better than the purely global model. It can be seen from Fig.2. that the composite data outperformed all other data representations.

Table 1. Average classification rates of the sub-images by SVM classifier. Figures in the parentheses are the number of variables obtained after dimensionality reduction.

Feature	PCA	CCA	SOM
Eyes	85.5% (250)	82.75% (22)	80.25% (36)
Mouth	81.25% (253)	81.55% (22)	80.25% (36)
Full Face	87.5% (256)	87% (26)	83.25% (64)

Table 2. Classification rates of the composite image and original image data by SVM classifier. Figures in the parentheses are the number of variables obtained after dimensionality reduction.

Feature	PCA	CCA	SOM
Composite	92.25% (759)	91.5% (70)	89.75% (136)
Original Full Face	86.5% (253)	85.5% (26)	83.25% (64)

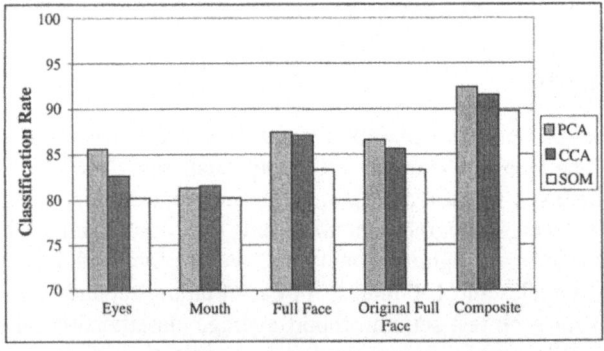

Fig. 2. Average classification rates on different features.

---

[1] Due to correlations, linear and nonlinear, a D dimensional data may actually lie in a d dimensional space. This true dimension d is called Intrinsic Dimension, where $d \leq D$. As PCA accounts only linear correlations, it is unable to reduce the data dimension to its intrinsic dimension, when the correlations are nonlinear.

5 Discussion and Conclusion

Hair, especially for females, forms a major part of the image and has a dominating affect on the classification. Many males with long hair and females with short hair were misclassified when the original full face images are used. The global and feature based model largely solved this problem, by reducing the affect of misleading hairstyles, while not removing important hair information. Fig.3. shows examples of individual faces that are misclassified when the original full face images are used and classified correctly by the global and feature based model.

Fig. 3. Examples of the faces that are misclassified due to hair style of the individuals.

The global and feature based model for gender classification presented here performs significantly better than the global and featural models individually. This model allows inspection of facial data at various component levels and the results presented suggest that all components carry high levels of gender information. We believe that this type of representation also acts as a weighting factor of information, where highly variable discriminatory information (like hair) alone does not affect classification.

We also investigated three dimensionality reduction techniques. The Performance of CCA, a nonlinear technique, is comparable to PCA, while it uses far fewer variables than PCA.

References

1. Bruce, V., et al., "Sex discrimination: how do we tell the difference between male and female faces?," Perception, 1993. **22**: p. 131-152.
2. Burton, A.M., V. Bruce, and N. Dench, "What's the difference between men and women? Evidence from facial measurement.," Perception, 1993. **22**: p. 153-176.
3. Abdi, H., D. Valentin, B. Edelman, and J.A. O'Toole, "More about the difference between men and women: evidence from linear neural networks and the principal component approach.," Perception, 1995. **24**: p. 539-562.
4. Golomb, B., A.,, D. Lawrence, T., and T. Sejnowski, J., "Sexnet: A neural network identifies sex from human faces.," Advances in Neural Information Processing Systems, 1991. **3**: p. 572-577.
5. Brunelli, R. and T. Poggio. "HyperBF networks for gender classification,". in DARPA Image Understanding Workshop. 1992.
6. Tamura, S., H. Kawai, and H. Mitsumoto, "Male/female identification form 8 × 6 very low resolution images by neural network.," Pattern Recognition, 1996. **29**no.(2): p. 331-335.
7. Moghaddam B. and M.-H. Yang, "Gender classification with support vector machines.," Technical Report : TR-2000-01, Mitsubishi Electric Research Laboratory. January, 2000.

8. Sun, Z., X. Yuan, G. Bebis, and S. Louis, J. "Neural-Network-based gender classification using genetic search for eigen-feature selection,". in IEEE international joint conference on neural networks. 2002.

9. Luckman, A., N.M. Allinson, A.M. Ellis, and B.M. Flude, "Familiar face recognition: A comparative study of a connectionist model and human performance.," Neurocomputing, 1995. **7**: p. 3-27.

10. Bellman, R.E., Adaptive control processes: A guided tour. 1961: Princeton University Press.

11. Jolliffe, I., T., Principal component analysis. 1986, New York: Springer-Verlag.

12. Kohonen, T., Self organizing maps. 3rd ed. 2001: Springer-Verlag.

13. Demartines, P. and J. Herault, "Curvilinear component analysis: A self-organizing neural network for nonlinear mapping of data sets.," IEEE Transactions on Neural Networks, 1997. **8**no.(1): p. 148-154.

14. Cortes, C. and V. Vapnik, "Support-vector networks.," Machine Learning, 1995. **20**: p. 273-297.

15. Phillips, P.J., H. Wechsler, J. Huang, and P. Rauss, "The FERET database and evaluation procedure for face recognition algorithms.," Image and Vision Computing, 1998. **16**no.(5): p. 295-306.

16. Martiniz, A.M. and R. Benavente, "The AR face database.," Technical Report : 24, CVC. June, 1998.

17. Jesorsky, O., K. Kirchberg, and R. Frischholz. "Robust face detection using the hausdorff distance,". in International Conference on Audio- and Video-based Biometric Person Authentification. 2001. Halmstad, Sweden.

18. Grassberger, P. and I. Proccacia, "Measuring the strangeness of strange attractors.," Physica D, 1983. **9**: p. 189-208.

Classification of SAR Images Through a Convex Hull Region Oriented Approach

Simith T. D'Oliveira Junior,
Francisco de A.T. de Carvalho, and Renata M.C.R. de Souza

Centro de Informatica - CIn / UFPE, Av. Prof. Luiz Freire,
s/n - Cidade Universitaria, CEP: 50740-540 - Recife - PE, Brasil
{stdj,fatc,rmcrs}@cin.ufpe.br

Abstract. This paper presents a new symbolic classifier based on a region oriented approach. Concerning the learning step, each class is described by a region (or a set of regions) in $\Re^p$ defined by the convex hull of the objects belonging to this class. In the allocation step, the assignment of a new object to a class is based on a dissimilarity matching function which compares the class description (a region or a set of regions) with a point in $\Re^p$. To show the usefulness of this approach, experiments with simulated SAR images were considered. The evaluation of the proposed classifier is based on the prediction accuracy and it is achieved in the framework of a Monte Carlo experience.

1 Introduction

Symbolic Data Analysis (SDA) [1] has been introduced as a new domain related to multivariate analysis, pattern recognition and artificial intelligence in order to extend classical exploratory data analysis and statistical methods to symbolic data. Symbolic data allows multiple (sometimes weighted) values for each variable, and it is why new variable types (interval, categorical multi-valued and modal variables) have been introduced.

Ichino et al. [4] introduced a symbolic classifier as a region oriented approach for quantitative, categorical, interval and categorical multi-valued data. At end of the learning step, the symbolic description of each group is obtained through the use of an approximation of a *Mutual Neighborhood Graph* (*MNG*) and a symbolic join operator. In the allocation step, an observation is assigned to a particular group based on a dissimilarity matching function.

Souza et al. [8] and De Carvalho et al. [2] have proposed another *MNG* approximation to reduce the complexity of the learning step without losing the classifier performance in terms of prediction accuracy. Concerning the allocation step, a classification rule, based on new similarity and dissimilarity measures, have been introduced to carry out the assignment of an individual to a class.

In this paper, we present a new symbolic classifier based on a region oriented approach. Here, the description of each class is a region (or a set of regions) in $\Re^p$ defined by the convex hull of the objects belonging to this class which is obtained through a suitable approximation of a *Mutual Neighborhood Graph*

N.R. Pal et al. (Eds.): ICONIP 2004, LNCS 3316, pp. 769–774, 2004.
© Springer-Verlag Berlin Heidelberg 2004

(*MNG*). This approach aims to reduce the over-generalization that is produced when each class is described by a region (or a set of regions) defined by the hyper-cube formed by the objects belonging to this class and then to improve the accuracy performance of the classifier. The assignment of a new object to a class is based on a dissimilarity matching function which compares the class description with a point in $\Re^p$.

2   Basic Concepts

Let $\Omega = \{\omega_1, \cdots, \omega_n\}$, be a set of n individuals described by p quantitative features $X_j (j = 1, \ldots, p)$. Each individual ω_i $(i = 1, \ldots, n)$ is represented by a quantitative feature vector $\mathbf{x}_i = (x_{i1}, \ldots, x_{ip})$, where x_{ij} is a *quantitative feature value*. A quantitative feature value may be either a continuous value (e.g., $x_{ij} = 1.80$ meters of height) or an interval value (e.g., $x_{ij} = [0, 2]$ hours, the duration of a student evaluation).

Let $C_k = \{\omega_{k1}, \ldots, \omega_{kN_k}\}, k = 1, \ldots, m$, be a class of individuals with $C_k \cap C_{k'} = \emptyset$ if $k \neq k'$ and $\cup_{k=1}^m C_k = \Omega$. The individual $\omega_{kl}, l = 1, \ldots, N_k$, is represented by the continuous feature vector $\mathbf{x}_{kl} = (x_{kl1}, \ldots, x_{klp})$.

A symbolic description of the class C_k can be obtained by using the join operator (Ichino et al (1996)).

Definition 1. The join between the continuous feature vectors $\mathbf{x}_{kl}$ ($l = 1, \ldots, N_k$) is an interval feature vector which is defined as $\mathbf{y}_k = \mathbf{x}_{k1} \oplus \ldots \oplus \mathbf{x}_{kN_k} = (x_{k11} \oplus \ldots \oplus x_{kN_k1}, \ldots, x_{k1j} \oplus \ldots \oplus x_{kN_kj}, \ldots, x_{k1p} \oplus \ldots \oplus x_{kN_kp})$, where $x_{k1j} \oplus \ldots \oplus x_{kN_kj} = [min\{x_{k1j}, \ldots, x_{kN_kj}\}, max\{x_{k1j}, \ldots, x_{kN_kj}\}](j = 1, \ldots, p)$.

Moreover, we can associate to each class C_k two regions in $\Re^p$: one spanned by the join of its elements and another spanned by the convex hull of its elements.

Definition 2. The *J-region* associated to class C_k is a region in $\Re^p$ which is spanned by the join of the objets belonging to class C_k and it is defined as $R_J(C_k) = \{\mathbf{x} \in \Re^p : min\{x_{k1j}, \ldots, x_{kN_kj}\} \leq x_j \leq max\{x_{k1j}, \ldots, x_{kN_kj}\}, j = 1, \ldots, p\}$. The volume associated to the hyper-cube defined by the region $R_J(C_k)$ is $\pi(R_J(C_k))$.

Definition 3. The *H-region* associated to class C_k is a region in $\Re^p$ which is spanned by the convex hull formed by the objects belonging to class C_k and it is defined as $R_H(C_k) = \{\mathbf{x} = (x_1, \ldots, x_j, \ldots, x_p) \in \Re^p : \mathbf{x}$ is inside the envelop of the convex hull defined by the continuous feature vectors $\mathbf{x}_{kl} = (x_{kl1}, \ldots, x_{klp}), l = 1, \ldots N_k\}$. The volume associated to the internal points inside the convex hull envelop defined by $R_H(C_k)$ is $\pi(R_H(C_k))$.

The *mutual neighborhood graph (MNG)* (Ichino et al (1996)) yields information about interclass structure.

Definition 4. The objects belonging to class C_k are each one *mutual neighbors* (Ichino et al (1996)) if $\forall \omega_{k'l} \in C_{k'}$ $(k' \in \{1, \ldots, m\}, k' \neq k), \mathbf{x}_{k'l} \notin R_J(C_k)$ $(l = 1, \ldots, N_{k'})$. In this case, the *MNG* of C_k against $\overline{C_k} = \cup_{\substack{k'=1 \\ k' \neq k}}^m C_{k'}$, which is constructed by joining all pairs of objects which are mutual neighbors, is a complete graph.

If the objects belonging to class C_k are not each one mutual neighbors, we look for all the subsets of C_k where its elements are each one mutual neighbors and which are a *maximal clique* in the *MNG*, which, in that case, is not a complete graph. To each of these subsets of C_k we can associate a *J-region* and calculate the volume of the corresponding hyper-cube defined by it.

In this paper we introduce another definition to the *MNG*.

Definition 5. The objects belonging to class C_k are each one *mutual neighbors* if $\forall \omega_{k'l} \in C_{k'}, k' \in \{1, \ldots, m\}, k' \neq k, \mathbf{x}_{k'l} \notin R_H(C_k)(l = 1, \ldots, N_{k'})$. The *MNG* of C_k against $\overline{C_k} = \cup_{\substack{k' \neq k \\ k'=1}}^{m} C_{k'}$, defined in this way is also a complete graph.

If the objects belonging to class C_k are not each one mutual neighbors, again, to each *maximal clique* (subset of C_k) we can associate a *H-region* and calculate the volume of the corresponding convex-hull defined by it.

3 Symbolic Classifier

This section introduces the learning and allocation steps of the symbolic classifier based on a convex hull region oriented approach.

3.1 Learning Step

The idea of this step is to provide a description of each class through a region (or a set of regions) in $\Re^p$ defined by the convex hull formed by the objects belonging to this class, which is obtained through a suitable approximation of a Mutual Neighborhood Graph (*MNG*).

Concerning this step, we have two basic remarks. When the *MNG* of a class C_k is not complete, it is necessary to construct of an approximation of the *MNG* because the computational complexity in time to find all maximal cliques on a graph is exponential. The second concerns the kind of region that is suitable to to describe a class C_k. The class description based on a *J-region* over-generalizes the class description given by a *H-region*. It is why we selected this last option.

The construction of the *MNG* for the classes C_k $(k = 1, \ldots, m)$ and the representation of each class by a *H-region* (or by a set of *H-region*) is accomplished in the following way:

For $k = 1, \ldots, m$ do
1. Find the the region $R_H(C_k)$ (according to *definition 3*) associated to class C_k and verify if the objects belonging to this class are each one mutual neighbors according to the *definition 5*
2. If it is the case, construct the *MNG* (which is a complete graph) and stop
3. If it is not the case (*MNG* approximation) do:
   3.1. choose an object of C_k as a seed according to the lexicographic order of these objects in C_k; do $t = 1$ and put the seed in the set C_k^t; remove the seed from C_k

     3.2. add the next object of C_k (according to the lexicographic order) to C_k^t if all the objects belonging now to C_k^t remains each one mutual neighbors according to the *definition 5*; if this is true, remove this object from C_k

     3.3. repeat step 3.2) for all remaining objects in C_k

     3.4. Find the region $R_H(C_k^t)$ (according to *definition 3*) associated to C_k^t)

     3.5. if $C_k \neq \emptyset$, do $t = t + 1$ and repeat steps 3.1) to 3.4) until $C_k = \emptyset$

4. construct the *MNG* (which now is not a complete graph) and stop

At the end of this algorithm it is computed the subsets $C_k^1, \ldots, C_k^{n_k}$ of class C_k and it is obtained the description of this class by the *H-regions* $R_H(C_k^1), \ldots, R_H(C_k^{n_k})$.

3.2 Allocation Step

The aim of the allocation step is to associate a new object to a class based on a dissimilarity matching function that compares the class description (a region or a set of regions) with a point in $\Re^p$.

Let ω be a new object, which is candidate to be assigned to a class $C_k (k = 1, \ldots, m)$, and its corresponding description given by the continuous feature vector $\mathbf{x} = (x_1, \ldots, x_p)$. Remember that from the learning step it is computed the subsets $C_k^1, \ldots, C_k^{n_k}$ of C_k.

The *classification rule* is defined as follow: ω is affected to the class C_k if

$$\delta(\omega, C_k) \leq \delta(\omega, C_h), \forall h \in \{1, \ldots, m\} \tag{1}$$

where $\delta(\omega, C_h) = min\{\delta(\omega, C_h^1), \ldots, \delta(\omega, C_h^{n_h})\}$.

In this paper, the dissimilarity matching function δ is defined as

$$\delta(\omega, C_h^s) = \frac{\pi(R_H(C_h^s \cup \{\omega\})) - \pi(R_H(C_h^s))}{\pi(R_H(C_h^s \cup \{\omega\}))}, \quad s = 1, \ldots, n_h \tag{2}$$

4   The Monte Carlo Experiences

In order to show the usefulness of the approach proposed in this paper, a special kind of SAR simulated image is classified in this section.

4.1   SAR Simulated Images

Synthetic Aperture Radar (SAR) is a system that possesses its own illumination and produces images with a high capacity for discriminating objects. It uses coherent radiation, generating images with speckle noise. Data SAR possesses a random behaviour that is usually explained by a multiplicative model [3]. Different kinds of detection (intensity or amplitude format) and types of regions can be modelled by different distributions associated to the return signal. The homogeneous (e.g. agricultural fields), heterogeneous (e.g. primary forest) and extremely heterogeneous (e.g. urban areas) region types are considered in this work.

In this paper, two situations of images are considered ranging in classification from moderate (here named situation 1) to greatly difficult (here named situation 2). We generate the distribution associated to each class in each situation by using an algorithm for generating gamma variables. Moreover, the *Lee filter* [6] was applied to the data before segmentation in order to decrease the effect of speckle noise. The segmentation was obtained using the region growing technique [5], based on the t-student test (at the 5% significance level) for the merging of regions.

4.2 Experimental Evaluation

The evaluation of our method, called here *H-region approach*, where class representation, MNG approximation and dissimilarity matching function are based on *H-regions*, was performed based on prediction accuracy in comparison with the approach where class representation, MNG approximation and dissimilarity matching function are based on *J-regions* (called here *J-region approach*). The Monte Carlo experience with 100 replications was performed for images of sizes $64 \times 64, 128 \times 128$ and 256×256, taking into consideration situations 1 and 2. The prediction accuracy of the classifier was measured through the error rate of classification obtained from a test set. The estimated error rate of classification corresponds to the average of the error rates found for these replications.

The comparison according to the average of the error rate was achieved by a paired Student's t-test at the significance level of 5%. Table 3 shows the average error rate, suitable (null and alternative) hypothesis and the observed values of the test statistic for various sizes and the two image situations. In this table, the test statistics follow a Student's t distribution with 99 degrees of freedom and μ_1 and μ_2 are, respectively, the average error rate for the *H-region* approach and the *J-region* approach.

Table 1. Comparison between the classifiers according to the average error rate.

Simulated SAR images	H-region approach	J-region approach	$H_0 : \mu_2 \geq \mu_1$ $H_1 : \mu_2 < \mu_1$
64×64 situation 1	5.78	8.29	-5.19
64×64 situation 2	24.83	24.92	-0.15
128×128 situation 1	2.68	3.42	-5.03
128×128 situation 2	16.52	16.89	-1.45
256×256 situation 1	1.39	1.87	-8.38
256×256 situation 2	13.67	14.34	-4.57

From Table 1, we can conclude that in all cases (image size and situations corresponding to difficulty degree of classification) the average error rate for *H-*

region approach is lower than that for the *J-region* approach. Also, from the test statistics we can conclude that the *H-region* approach outperforms the *J-region* approach.

5   Concluding Remarks

A new classifier based on a region oriented approach is presented in this paper. Concerning the learning step, each class is described by a region (or a set of regions) in $\Re^p$ defined by the convex hull formed by the objects belonging to this class, which is obtained through a suitable approximation of a Mutual Neighborhood Graph (*MNG*). To show the usefulness of this approach, SAR simulated images are considered presenting situations ranging from moderate to great difficult of classification. In the allocation step, to assign a segment to a region, a dissimilarity matching function was introduced.

The evaluation of the proposed classifier, which uses in the learning and allocation steps the *H-region*, was based on error rate of classification calculated in the framework of a Monte Carlo experience with 100 replications in comparison with the classifier which uses in the learning and allocation steps the *J-region*. The experiments showed that the *H-region approach* furnishes better results.

Acknowledgments

The authors would like to thank CNPq (Brazilian Agency) for its financial support.

References

1. Bock, H.H. and Diday, E.: Analysis of Symbolic Data: Exploratory Methods for Extracting Statistical Information from Complex Data. Springer, Berlin Heidelberg (2000)
2. De Carvalho, F.A.T., Anselmo, C.A.F. and Souza, R.M.C.R.: Symbolic approach to classify large data sets, In: Data Analysis, Classification, and Related Methods, Kiers, H.A.L. et al (Eds.), Springer, (2000) 375–380
3. Frery, A.C. Mueler, H.J. Yanasse, C.C.F. and Sant'ana, S.J.S.: A model for extremely heterogeneous clutter. *IEEE Transactions on Geoscience and Remote Sensing*, 1, (1997) 648–659
4. Ichino, M., Yaguchi, H. and Diday, E.: A fuzzy symbolic pattern classifier In: Diday, E. et al (Eds.): Ordinal and Symbolic Data Analysis. Springer, Berlin, (1996) 92–102
5. Jain, A. K.: Fundamentals of Digital Image Processing. Englewood Cliffs. Prentice Hall International Editions (1988)
6. LEE, J. S.: Speckle analysis and smoothing of synthetic aperture radar images. *Computer Graphics and Image Processing*, 17, (1981) 24–32
7. O'Rourke, J.: Computational Geometry in C (Second Edition), Cambridge University Press (1998)
8. Souza, R. M. C. R., De Carvalho, F. A. T. and Frery, A. C.: Symbolic approach to SAR image classification. IEEE 1999 International Geoscience and Remote Sensing Symposium, Hamburgo, (1999) 1318–1320

Clustering of Interval-Valued Data
Using Adaptive Squared Euclidean Distances

Renata M.C.R. de Souza, Francisco de A.T. de Carvalho, and Fabio C.D. Silva

Centro de Informatica - CIn / UFPE, Av. Prof. Luiz Freire,
s/n - Cidade Universitaria, CEP: 50740-540 - Recife - PE, Brasil
{rmcrs,fatc,fcds}@cin.ufpe.br

Abstract. This paper presents a clustering method for interval-valued
data using a dynamic cluster algorithm with adaptive squared Euclidean
distances. This method furnishes a partition and a prototype to each
cluster by optimizing an adequacy criterion that measures the fitting
between the clusters and their representatives. To compare a class with
its representative, the method uses an adaptive version of a squared Euclidean distance to interval-valued data. Experiments with real and artificial interval-valued data sets shows the usefulness of the this method.

1 Introduction

Clustering is concerned with the summarizing and extracting information on
data sets. *Symbolic Data Analysis* (SDA) [1] is a new domain related to multivariate analysis, pattern recognition and artificial intelligence, that aims to provide
suitable methods (clustering, factorial techniques, decision tree, etc.) to manage
symbolic data. A cell of a symbolic data table may contain a set of categories,
an interval, or a weight (probability) distributions. Concerning partitioning clustering methods, SDA has provided suitable tools to cluster interval-valued data
(see, for example, [5]).

The standard adaptive dynamic cluster algorithm [3] is a two-step relocation
algorithm involving the construction of the clusters and the identification of a
representative or prototype of each cluster by locally minimizing a criterion. This
algorithm uses a different distance to compare each cluster with its representative
and it is able to find clusters of different shapes and sizes in a given set of objects.

In this paper, we present a dynamic cluster method based on adaptive squared
Euclidian distances for partitioning a set of individuals each one described by an
interval-valued data vector. The proposed method is the adaptive version of a
dynamic cluster algorithm presented in [2]. Experiments with real and artificial
interval valued data sets are considered and an evaluation of the clustering results
based on the computation of an external cluster validity index in the framework
of the Monte Carlo experience is accomplished. To compare the clustering results
furnished by the proposed algorithm with the clustering results furnished by its
non-adaptive version, we use the paired t-Student test.

N.R. Pal et al. (Eds.): ICONIP 2004, LNCS 3316, pp. 775–780, 2004.
© Springer-Verlag Berlin Heidelberg 2004

2   Adaptive Dynamic Cluster Algorithm

The standard adaptive dynamic cluster algorithm [3] starts from a set of representatives or an initial partition and interactively applies an allocation step in order to assign the individuals to the clusters, according to their proximity from the class prototypes, followed by a representation step where the class prototypes are updated according to the assignment of the individuals until convergence is achieved. Moreover, at each iteration there is a different distance associated with each cluster, i.e., the distance is not determined once and for all, furthermore is different from one class to another.

Let $E = \{s_1, \ldots, s_n\}$ be a set of n objects described by p interval-valued variables. Each object s_i $(i = 1, \ldots, n)$ is represented as a vector of intervals $\mathbf{x}_i = (x_i^1, \ldots, x_i^p)$, where $x_i^j = [a_i^j, b_i^j] \in I = \{[a, b] : a, b \in \Re, a \le b\}$ $(j = 1, \ldots, p)$. Let P be a partition of E into K clusters $\{C_1, \ldots, C_K\}$, where each cluster C_k $(k = 1, \ldots, K)$ has a prototype L_k that is also represented as a vector of intervals $\mathbf{y}_k = ([\alpha_k^1, \beta_k^1], \ldots, [\alpha_k^p, \beta_k^p])$.

Our method searches for a partition $P = (C_1, \ldots, C_K)$ of E in K classes, its corresponding set of K class prototypes $L = (L_1, \ldots, L_K)$ and a set of K different distances $d = (d_1, \ldots, d_K)$ associated with the clusters which locally minimizes an adequacy criterion that is usually stated as

$$W(P, L, d) = \sum_{k=1}^{K} \sum_{i \in C_k} d_k(\mathbf{x}_i, \mathbf{y}_k) \tag{1}$$

where $d_k(\mathbf{x}_i, \mathbf{y}_k)$ is an adaptive dissimilarity measure between an object $s_i \in C_k$ and the class prototype L_k of C_k.

2.1   Adaptive Distances Between Two Vectors of Intervals

In [3] an adaptive distance d_k is defined according to the structure of a cluster C_k and is described by a vector of coefficients $\lambda_k = (\lambda_k^1, \ldots, \lambda_k^p)$ with $\lambda_k^j > 0$ $(j = 1, \ldots, p)$ e $\prod_{j=1}^{p} \lambda_k^j = 1$. In this paper, we define an adaptive distance between the two vectors of intervals $\mathbf{x}_i$ and $\mathbf{y}_k$ as

$$d_k(\mathbf{x}_i, \mathbf{y}_k) = \sum_{j=1}^{p} \lambda_k^j \phi_E(x_i^j, y_k^j) \tag{2}$$

where

$$\phi_E(x_i^j, y_k^j) = (a_i^j - \alpha_k^j)^2 + (b_i^j - \beta_k^j)^2 \tag{3}$$

is the sum of the squares of the differences between the lower bounds and the upper bounds of the intervals $x_i^j = [a_i^j, b_i^j]$ and $y_k^j = [\alpha_k^j, \beta_k^j]$.

The distance in equation (3) corresponds to represent an interval $[a, b]$ as a point $(a, b) \in \Re^2$, where the lower bounds of the intervals are represented in the x-axis, and the upper bounds in the y-axis, and then compute the squared L_2 distance between the points (a_i^j, b_i^j) and (α_k^j, β_k^j). Therefore, the distance function in equation (2) is a weighted version of the squared L_2 (Euclidean) distance to interval-valued data.

2.2 The Optimizing Problem

The optimizing problem is stated as follows: find the class prototype L_k of the class C_k and the adaptive squared Euclidian distance d_k associated to C_k which minimizes an adequacy criterion measuring the dissimilarity between this class prototype L_k and the class C_k according to d_k. Therefore, the optimization problem has two stages:

a) The class C_k and the distance d_k $(k = 1, \ldots, K)$ are fixed. We look for the prototype L_k of the class C_k which locally minimizes

$$\Delta(\mathbf{y}_k, \boldsymbol{\lambda}_k) = \sum_{i \in C_k} d_k(\mathbf{x}_i, \mathbf{y}_k) = \sum_{j=1}^{p} \lambda_k^j \sum_{i \in C_k} \left[(a_i^j - \alpha_k^j)^2 + (b_i^j - \beta_k^j)^2 \right] \quad (4)$$

The criterion $\Delta(\mathbf{y}_k, \boldsymbol{\lambda}_k)$ being additive, the problem becomes to find for $j = 1, \ldots, p$, the interval $y_k^j = [\alpha_k^j, \beta_k^j]$ which minimizes $\sum_{i \in C_k} [(a_i^j - \alpha_k^j)^2 + (b_i^j - \beta_k^j)^2]$. Using the *method of least square*, we found that $\hat{\alpha}_k^j$ is the average of the set $\{a_i^j, i \in C_k\}$, the set of lower bounds of the intervals $x_i^j = [a_i^j, b_i^j], i \in C_k$, and $\hat{\beta}_k^j$ is the average of the set $\{b_i^j, i \in C_k\}$, the set of upper bounds of the intervals $x_i^j = [a_i^j, b_i^j], i \in C_k$.

b) The class C_k and the prototype L_k $(k = 1, \ldots, K)$ are fixed. We look for the vector of weights $\boldsymbol{\lambda}_k = (\lambda_k^1, \ldots, \lambda_k^p)$ with $\lambda_k^j > 0$ $(j = 1, \ldots, p)$ e $\prod_{j=1}^{p} \lambda_k^j = 1$ that minimizes the criterion $\Delta(\mathbf{y}_k, \boldsymbol{\lambda}_k)$. According to Diday and Govaert [3], the coefficient λ_k^j which minimizes the criterion $\Delta(\mathbf{y}_k, \boldsymbol{\lambda}_k)$ given in equation 4 is obtained by the Lagrange multipliers method and it is:

$$\hat{\lambda}_k^j = \frac{\left[\prod_{h=1}^{p} (\sum_{i \in C_k} (a_i^j - \alpha_k^j)^2 + (b_i^j - \beta_k^j)^2) \right]^{\frac{1}{p}}}{\sum_{i \in C_k} (a_i^j - \alpha_k^j)^2 + (b_i^j - \beta_k^j)^2} \quad (5)$$

3 Experiments

To show the usefulness of these methods, experiments with two artificial interval-valued data sets of different degrees of difficulty to be clustered (clusters of different shapes and sizes, etc) and an application with a real data set are considered in this section. The evaluation of the clustering results is based on the corrected Rand (CR) index [4]. The CR index assesses the degree of agreement (similarity) between an a priori partition (in our case, the partition defined by the seed points) and a partition furnished by the clustering algorithm.

Initially, we considered two standard quantitative data sets in $\Re^2$. Each data set has 450 points scattered among four clusters of unequal sizes and shapes: two clusters with ellipsis shapes and sizes 150 and two clusters with spherical shapes of sizes 50 and 100. The data points of each cluster in each data set were drawn according to a bi-variate normal distribution with non-correlated components.

Data set 1, showing well-separated clusters, is generated according to the following parameters:

a) Class 1: $\mu_1 = 28$, $\mu_2 = 22$, $\sigma_1^2 = 100$, $\sigma_{12} = 0$, $\sigma_2^2 = 9$;
b) Class 2: $\mu_1 = 65$, $\mu_2 = 30$, $\sigma_1^2 = 9$, $\sigma_{12} = 0$, $\sigma_2^2 = 144$;
c) Class 3: $\mu_1 = 45$, $\mu_2 = 42$, $\sigma_1^2 = 9$, $\sigma_{12} = 0$, $\sigma_2^2 = 9$;
d) Class 4: $\mu_1 = 38$, $\mu_2 = -1$, $\sigma_1^2 = 25$, $\sigma_{12} = 0$, $\sigma_2^2 = 25$;

Data set 2, showing overlapping clusters, is generated according to the following parameters:

a) Class 1: $\mu_1 = 45$, $\mu_2 = 22$, $\sigma_1^2 = 100$, $\sigma_{12} = 0$, $\sigma_2^2 = 9$;
b) Class 2: $\mu_1 = 65$, $\mu_2 = 30$, $\sigma_1^2 = 9$, $\sigma_{12} = 0$, $\sigma_2^2 = 144$;
c) Class 3: $\mu_1 = 57$, $\mu_2 = 38$, $\sigma_1^2 = 9$, $\sigma_{12} = 0$, $\sigma_2^2 = 9$;
d) Class 4: $\mu_1 = 42$, $\mu_2 = 12$, $\sigma_1^2 = 25$, $\sigma_{12} = 0$, $\sigma_2^2 = 25$;

Each data point (z_1, z_2) of the data set 1 and 2 is a seed of a vector of intervals (rectangle): $([z_1 - \gamma_1/2, z_1 + \gamma_1/2], [z_2 - \gamma_2/2, z_2 + \gamma_2/2])$. These parameters γ_1, γ_2 are randomly selected from the same predefined interval. The intervals considered in this paper are: $[1, 8], [1, 16], [1, 24], [1, 32]$, and $[1, 40]$. Figure 1 shows these artificial interval-valued data sets.

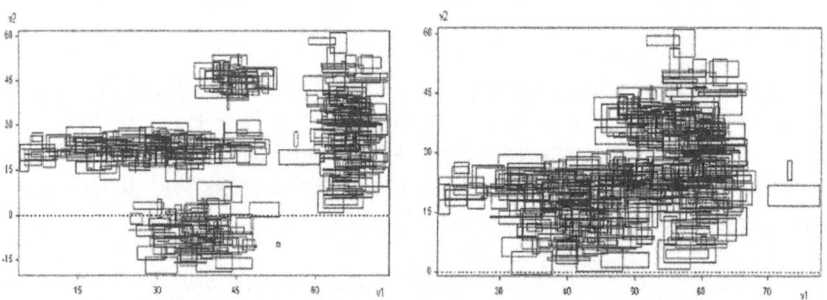

Fig. 1. Symbolic data showing well-separated classes and overlapping classes

In order to compare the adaptive dynamic cluster algorithm proposed in the present paper with the non-adaptive version of this algorithm, this section presents the clustering results furnished by these methods according to artificial interval-valued data sets 1 and 2. The non-adaptive dynamic cluster algorithm uses a suitable extension of the squared L_2 (Euclidian) metric to compare the vectors of intervals $\mathbf{x}_i$ and $\mathbf{y}_k$:

$$d(\mathbf{x}_i, \mathbf{y}_k) = \sum_{j=1}^{p} \left[(a_i^j - \alpha_k^j)^2 + (b_i^j - \beta_k^j)^2 \right] \qquad (6)$$

For artificial data sets, the CR index is estimated in the framework of a Monte Carlo experience with 100 replications for each interval data set as well as for each predefined interval. In each replication a clustering method is run 50 times and the best result, according to the corresponding adequacy criterion is selected. The average of the corrected Rand (CR) index among these 100 replications is calculated.

Table 1 shows the values of the average CR index according to adaptive and non-adaptive methods, as well as artificial interval-valued data sets 1 and 2. From these results it can be observed that the average CR indices for the adaptive method are greater than those for the non-adaptive method.

Table 1. Comparison between the clustering methods according to the average of the correct Rand index

Range of values of $\gamma_i\ i=1,2$	Symbolic Data Set 1		Symbolic Data Set 2	
	Adaptive Method	Non-Adaptive Method	Adaptive Method	Non-Adaptive Method
[1, 8]	0.944	0.710	0.523	0.404
[1, 16]	0.934	0.711	0.496	0.408
[1, 24]	0.887	0.705	0.473	0.404
[1, 32]	0.823	0.620	0.385	0.405
[1, 40]	0.781	0.716	0.397	0.394

The comparison between the proposed clustering methods is achieved by a paired Student's t-test, at the level of 5% of significance. Table 2 shows the suitable (null and alternative) hypothesis and the observed values of the test statistic following a Student's t distribution with 99 degrees of freedom. In this table, μ and μ_1 are, respectively, the average of the CR index for the non-adaptive and adaptive method. From these results, we can reject the hypothesis that the average performance of the adaptive method is inferior to the non-adaptive method.

Table 2. Statistics of paired Student's t-tests comparing the methods

Range of values of $\gamma_i\ i=1,2$	$H_0 : \mu_1 \leq \mu$ $H_1 : \mu_1 > \mu$			
	Symbolic data set 1	Decision	Symbolic data set 2	Decision
$\gamma \in [1, 8]$	70.17	Reject H_0	23.20	Reject H_0
$\gamma \in [1, 16]$	55.15	Reject H_0	19.61	Reject H_0
$\gamma \in [1, 24]$	25.04	Reject H_0	13.43	Reject H_0
$\gamma \in [1, 32]$	15.08	Reject H_0	9.27	Reject H_0
$\gamma \in [1, 40]$	11.42	Reject H_0	0.37	Accept H_0

A data set with 12 fish species, each specie being described by 13 interval variables and 1 categorical variable is used as an application. These species are grouped into four a priori clusters of unequal sizes according to the categorical variable: two clusters (Carnivorous and Detritivorous) of sizes 4 and two clusters of sizes 2 (Omnivorous and Herbivorous).

The CR indices obtained from the comparison between the a priori partition and the partitions given by the adaptive and non-adaptive methods are, respectively, 0.340 and -0.02. Therefore, the performance of the adaptive method is superior to the non-adaptive method concerning also this interval-valued data set.

4   Concluding Remarks

A partitioning cluster method for interval-valued data using a dynamic cluster algorithm with adaptive squared Euclidean distances was presented in this paper. The algorithm locally optimizes an adequacy criterion which measures the fitting between the classes and its representatives (prototypes). To compare the individuals and the class and prototypes, adaptive distances based on a weighted version of the squared L_2 (Euclidean) distance well adapted to interval-valued data have been introduced. These adaptive distances are parameterized according the intra-class structure of the partition and they are able to recognize clusters of different shapes and sizes.

Experiments carried out with real and artificial interval-valued data sets showed the usefulness of this adaptive clustering method. The accuracy of the results furnished by this clustering method is assessed by the CR index and compared with the results furnished by the non-adaptive version of clustering method. This CR index is calculated in the framework of a Monte Carlo experience with 100 replications. Concerning artificial data sets, statistic tests support the evidence that the average behaviour of this index for the adaptive method is superior to the non adaptive method. The adaptive method also outperforms the non adaptive one concerning the fish interval valued data set.

References

1. Bock, H.H. and Diday, E.: Analysis of Symbolic Data: Exploratory Methods for Extracting Statistical Information from Complex Data. Springer, Berlin Heidelberg (2000)
2. De Carvalho, F. A. T. Brito, P. and Bock H. H., Dynamical Clustering for symbolic quantitative data. In: Workshop in Symbolic Data Analysis, Cracow, (2002)
3. Diday, E. and Govaert, G.: Classification Automatique avec Distances Adaptatives. R.A.I.R.O. Informatique Computer Science, **11** (4) (1977) 329–349
4. Hubert, L. and Arabie, P.: Comparing Partitions. Journal of Classification, **2** (1985) 193–218
5. Souza, R.M.C.R. and De Carvalho, F. A. T.: Clustering of interval data based on city-block distances. Pattern Recognition Letters, **25** (3) (2004) 353–365

A Two-Pass Approach to Pattern Classification

Subhadip Basu[1], C. Chaudhuri[2], Mahantapas Kundu[2],
Mita Nasipuri[2], and Dipak Kumar Basu[2]

[1] Computer Sc. & Engg. Dept., MCKV Institute of Engineering, Howrah-711204, India
[2] Computer Sc. & Engg. Dept., Jadavpur University, Kolkata-700032, India

Abstract. A two-pass approach to pattern recognition has been described here. In this approach, an input pattern is classified by refining possible classification decisions obtained through coarse classification of the same. Coarse classification here is performed to produce a group of possible candidate classes by considering the entire input pattern, whereas the finer classification is performed to select the most appropriate one from the group by considering features only from certain group specific regions of the same. This makes search for the true pattern class in the decision space more focused or guided towards the goal by restricting the finer classification decision within a smaller group of possible candidate classes in the second pass. The technique has been successfully applied for optical character recognition (OCR) of handwritten Bengali digits. It has improved the classification rate to 93.5% in the second pass from 90.5% obtained in the first pass.

1 Introduction

The recent trend for improving the performance of a pattern recognition system is to combine the complementary information provided by multiple classifiers [1-3]. Classifier combination becomes useful particularly when the concerned classification schemes are all different. This can be achieved either by using different representations of the input pattern with multiple classifiers or by using the same representation of the input with multiple versions of the same classifier. A typical example in the latter case, as given in [1], may include multiple k-nearest neighbor classifiers, each using the same representation of the input pattern, but different classifier parameters such as the value of k and the distance metric used for determining the nearest neighbors in the feature space. Another example in this case may be a set of neural network classifiers, all of the same type but with different weight sets obtained through different training strategies. An important issue in classifier combination approaches is how the individual classifier's decisions can be combined. If the classifiers' decisions are available in forms of class labels then the final decision can be made on the basis of majority votes. In some cases, the classifiers' decisions may be available in forms of some measurement values, each representing certain posteriori probability, membership value or some degree of belief, indicating how closely the input pattern is related to some pattern class. For combining decisions of such classifiers, fuzzy rules, Bayesian and Dempster Shafer approaches have been applied.

Success of the *classifier combination* approach can be explained as follows. In this approach, weakness of one classifier is complemented with the strength of another classifier. And also insufficiency of one feature set in discriminating certain pattern characteristics is complemented with another feature set. In classifier combination

N.R. Pal et al. (Eds.): ICONIP 2004, LNCS 3316, pp. 781–786, 2004.
© Springer-Verlag Berlin Heidelberg 2004

approach, all constituent classifiers work in parallel and no classifier can help the other to search for the true pattern class of the input pattern in the decision space. Considering this, a *two-pass approach* is introduced in the present work. In this approach, it is possible that one classifier's decision can help another to make its search for the true pattern class more focused or guided towards the goal, improving the recognition rate of the pattern recognition system at the same time.

The two-pass classifier described here first performs a *coarse classification* on the input pattern by restricting the possibility of classification decision within a group of classes, smaller than the original group of classes considered initially. In the second pass, the classifier refines its earlier decision by selecting the true class of the input pattern from the group of candidate classes selected in the first pass. In doing so, unlike the previous pass, the classifier concentrates only on certain regions of the input pattern, specific to the group of classes selected in the earlier pass. The group of candidate classes formed in the first pass of classification is determined by the *top choice* of the classifier in the same pass. There is a high chance that an input pattern classified into a member class of such group originally belongs to some other member class of the same group. By observing the *confusion matrix* on the *training data*, all such necessary groups of candidate classes can be formed for a particular application. The groups are formed on the basis of the statistical information obtained through the application of the classifier on the training data with the same features selected for the first pass. Secondary choices of the classifier are not considered in selection or for formation of a group.

The work presented here also embodies results of an investigation carried out to establish authenticity of the proposed technique by experimenting with the *handwritten digit recognition* problem. Handwritten digit recognition is a realistic benchmark problem of pattern recognition. It represents the core problem of many potential applications related to reading amounts from bank cheques, extracting numeric data from filled in forms, interpreting handwritten pin codes from mail pieces and so on. The digit patterns considered here consist of samples of handwritten digits of *Bengali* script. Popularity wise, Bengali stands 2nd after Hindi, both as a script and a language in Indian subcontinent. Compared to Chinese, Japanese, Korean, it has received little attention as a subject of OCR research [4] until recently.

For conducting experimentation for the present work, Multi Layer Perceptrons (MLPs) have been selected as pattern classifiers [5] for their superb *learning* and *generalization* abilities.

2 Feature Selection

Success of pattern recognition systems mostly depends on how best the discriminatory properties or features of the sample patterns have been identified. Features are so selected that their values remain close to each other for the same class of patterns and differ appreciably for different classes of patterns. Typical digit patterns of first ten natural numbers (0 to 9) taken from Bengali script are shown in Fig. 1.

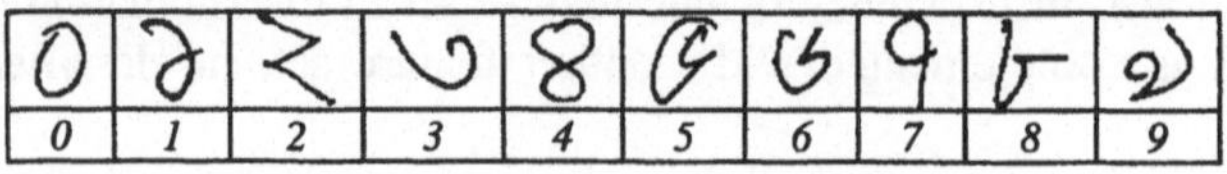

Fig. 1. The decimal digit set of Bengali script

2.1 Features for Coarse Classification

In this work, 36 features have been selected for coarse classification of handwritten Bengali digits. Images of handwritten digit samples are each scaled to 32x32 pixels size to ease the feature extraction process. The features for coarse classification are illustrated below.

2.1.1 Shadow Features

Shadow features [6] are computed by considering the lengths of projections of the digit images, as shown in Fig. 2, on the four sides and eight octant dividing sides of the minimum size boxes enclosing the same. 16 shadow features are considered in all for the present work. For the bars, on which more than one projections are taken, the sum of the lengths of all such projections on each bar is to be considered. Each value of the shadow feature so computed is to be normalized by dividing it with the maximum possible length of the projections on the respective bar.

2.1.2 Centroid Features

Coordinates of centroids of black pixels in all the 8 octants of a digit image are considered to add 16 features in all to the feature set. Fig. 3(a-b) show approximate locations of all such centroids on two different digit images. It is noteworthy how these features can be of help to distinguish the two images.

  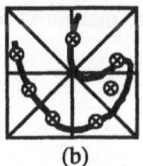 

(a)   (b)

Fig. 2. Shadow features   **Fig. 3(a-b).** Centroid features   **Fig. 4.** Diagonal distance features

2.1.3 Diagonal Distance Features

Distances of the first black pixels from the four corners of each digit image are considered as diagonal distance features. It is illustrated in Fig. 4.

2.2 Features for Finer Classification

At the stage of finer classification, different feature sets are used for three pre fixed groups of candidate classes. The candidate classes of the Bengali digit patterns are grouped for this as {1,2,9}, {3,6} and {0,4,5}. The groups will be referred as group #1, group #2 and group #3 respectively. The features selected for these groups are illustrated below.

2.2.1 Features for Group #1

73 features have been selected in all for this group. Out of these, one is a distance feature represented by the average of the distances of the first black pixels from the bottom of the digital image. It is illustrated in Fig. 5. The other 72 features are all bar features [7] computed in the lower rectangular half of the digital image. The pixel positions of the two opposite corners of this half are chosen as (16,0) and (31,31).

Bar features, which supply directional information, are computed with *binary images*. The images need not be size normalized. From each pixel position on an image, bar features are computed separately along the four directions *viz.*, east, northeast, north and northwest, by measuring lengths of the longest bars that fit equicolour pixels in the respective directions. For computation of bar features, an entire image is divided into a number of rectangular overlapping regions. For each of these regions, eight bar features are computed by averaging the background and foreground feature values for the four specified directions.

In the present work, the lower half of the image is divided into 9 overlapping rectangular regions, for computing 72 bar features. Each of these regions is of size $h/2$ x $w/2$, where h and w are the height and the width of the lower half of the image frame respectively. The top left corners of the 9 overlapping regions are at positions $\{(r,c) \mid r = 0, h/4, 2h/4$ and $c = 0, w/4, 2w/4\}$.

2.2.2 Features for Group #2
74 features have been selected for this group. Out of these, 72 bar features have been selected for a rectangular region with its two opposite corners at the pixel positions (0,16) and (24,31) on the digital image. 73rd feature is selected as the average of distances from topside of the first quadrant, on the image frame, to the first black pixels in the same quadrant. It is illustrated in Fig. 6(a). 74th feature is selected as the average of distances from right side of the same quadrant to the first black pixels as found within it. How these distances are measured for an image is shown in Fig. 6(b).

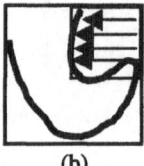

(a)   (b)

Fig. 5. Distance features for group #1   **Fig. 6(a-b).** Distance features for group #2

2.2.3 Features for Group #3
79 features have been selected for this group. Out of these, 72 bar features have been selected for a rectangular region with its two opposite corners at the pixel positions (8,0) and (24,31). The average of the distances from left side of the digital image to the first black pixels of the image and the average of the distances from right side of the same to the first black pixels of the image constitute 73rd and 74th features respectively. How these distances are computed is illustrated in Fig. 7(a). 75th feature is represented by the average of distances from a line, joining the mid points of the two opposite vertical sides of the image frame, to the first black pixels in upward direction. 76th feature is represented by computing the distances in the same way from the same line as before but in the reverse direction. Fig. 7(b) illustrates these two features. Features 77 to 79 represent the distances of the first black pixels from the center of the image frame in three directions *viz.*, northeast, east and southeast respectively as shown in Fig. 7(c).

(a)

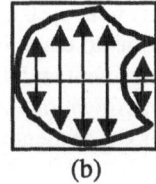
(b)

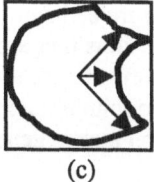
(c)

Fig. 7(a-c). Distance features for group #3

3   Experimental Results and Discussion

The training and the test sets for the work consist of 300 and 200 randomly selected samples of handwritten digit patterns respectively. In each set, samples from 10 digit classes are included in equal numbers. For coarse classification, an MLP (36-12-10) is designed after training it for 10,000 iterations with the learning rate (η) and the momentum term (α) tuned to values of 0.6 and 0.7 respectively. The percentage recognition rate, as observed with this MLP on the test data, is **90.5%**. A confusion matrix formed from this experimentation is shown in Table 1. Each $c_{i,j}{}^{th}$ element of this matrix shows the number of digit patterns from the ith class misclassified into j^{th} class, for $i \neq j$. It can be observed from the confusion matrix that the 9th digit pattern has been once misclassified into class 1 and thrice misclassified into class 2. The reverse, though not observed in all cases with the dataset under consideration, can also be assumed to be a possibility with extension of the same. So, an input pattern classified as 1, 2 or 9 requires finer classification to rule out other two possibilities before confirmation of this classification decision. To ensure this, pattern classes 1,2 and 9 are grouped together for finer classification.

It can also be observed from the confusion matrix that the 3rd digit pattern has been classified into class 6 for four times. Occurrences of the other instances of misclassification of the same have been observed in much lesser numbers. Again, the 6th digit pattern has been misclassified into class 3 and class 1 for once each. Considering all these, classes 3 and 6 are grouped together for finer classification. With the same line of consideration, digit classes 0, 4 and 5 have also been grouped together for the same.

There is no instance of misclassification found for the test samples of classes 7 and 8. But a few samples from the two other classes 3 and 9 have been misclassified into these classes. The number of misclassified samples for any such class pair does not exceed 1. All these initially lead to inclusion of classes 7 and 8 to the group {3,6}, but due to the lack of an appropriate feature set to distinguish the samples of pattern classes 3,6,7 and 8, the original group is restored finally. For samples classified into class 7 or 8, some *belief composition* method may be tried to combine classification decisions from more than one sources before arriving at the final decision. This is how the class groups are formed for finer classification of input patterns.

The groups mainly help to concentrate on some selected regions of the image frame for refinement of coarse classification decisions on the basis of certain group specific *locally salient features*. The MLPs designed for recognition of patterns belonging to groups #1, #2 and #3 are (73-30-3) trained for 20,000 iterations with $\eta=0.65$ and $\alpha=0.7$, (74-3-2) trained for 10,000 iterations with $\eta=0.7$ and $\alpha=0.7$ and

(74-3-3) trained for 10,000 iterations with $\eta=0.6$ and $\alpha=0.7$ respectively. Classification decisions that assign input patterns to class 7 or 8 after coarse classification cannot be refined with the present arrangement and are assumed to be final. The percentage recognition rate after finer classification is improved to **93.5%** finally. The confusion matrix generated from this experiment is shown in Table 2.

Table 1. The confusion matrix after coarse classification

	0	1	2	3	4	5	6	7	8	9
0	20	0	0	0	0	0	2	0	0	0
1	0	19	0	0	0	0	0	1	0	1
2	0	0	19	0	0	0	0	0	0	3
3	0	0	0	13	0	0	1	0	0	0
4	0	0	0	0	20	1	0	0	0	0
5	0	0	1	1	0	17	0	0	0	0
6	0	0	0	4	0	0	18	0	0	0
7	0	0	0	1	0	0	0	20	0	0
8	0	0	0	1	0	0	0	0	20	1
9	0	1	0	0	0	0	0	0	0	15

Table 2. The confusion matrix after finer classification

	0	1	2	3	4	5	6	7	8	9
0	20	0	0	1	0	0	0	0	0	0
1	0	20	0	0	0	0	0	1	0	1
2	0	0	19	0	0	0	0	0	0	1
3	0	0	0	14	0	0	1	0	0	0
4	0	0	0	0	20	1	0	0	0	0
5	0	0	1	0	0	19	0	0	0	0
6	0	0	0	3	0	0	18	0	0	0
7	0	0	0	1	0	0	0	20	0	0
8	0	0	0	1	0	0	0	0	20	1
9	0	0	0	0	0	0	0	0	0	17

Acknowledgements

Authors are thankful to the CMATER and the SRUVM project, C.S.E. Department, Jadavpur University, for providing necessary infrastructural facilities during the progress of the work. One of the authors, Mr. S. Basu, is thankful to the authorities of MCKV Institute of Engineering for kindly permitting him to carry on the research work.

References

1. Joseph Kittler *et al.*, "On combining classifiers", IEEE Trans. PAMI, vol. 20, no. 3, Mar. 1998, pp. 226-239.
2. Y.S. Huang, C.Y. Suen, "A method of combining multiple experts for the recognition of unconstrained handwritten numerals", IEEE Trans. PAMI, vol. 17,no.1,Jan. 1995, pp. 90-94.
3. Tin Kam Ho, Jonathan J. Hull, Sargur N. Srihari, "Decision combination in multiple classifier syatems", IEEE Trans. PAMI, vol. 16, no. 1, Jan. 1994, pp. 66-75.
4. B.B. Chaudhuri and U. Pal, "A complete printed *Bangla* OCR system", Pattern Recognition, vol. 31, no. 5, pp. 531-549.
5. K. Roy *et al.*, "An application of the multi layer perceptron for handwritten digit recognition", CODEC 04, Jan.1-3, 2004, Kolkata.
6. D.J. Burr, "Experiments on neural net recognition of spoken and written text," IEEE Trans. Acoust., Signal Process., vol. 36, no. 7, pp 1162-1168, 1988.
7. Paul Gader, Magdi Mohamed and Jung Hsien Chiang, "Comparison of crisp and fuzzy character neural networks in handwritten word recognition", IEEE Trans. Fuzzy Systems, vol. 3, no. 3, Aug. 1995, pp. 357-363.

A Long Memory Process Based Parametric Modeling and Recognition of PD Signal

Pradeep Kumar Shetty

Dept. of HVE, Indian Institute of Science, Bangalore, India

Abstract. We address the problem of recognition and retrieval of relatively weak industrial signal such as Partial Discharges (PD) buried in excessive noise. The major bottleneck being the recognition and suppression of stochastic pulsive interference (PI) which has similar time-frequency characteristics as PD pulse. Therefore conventional frequency based DSP techniques are not useful in retrieving PD pulses. We employ statistical signal modeling based on combination of long-memory process and probabilistic principal component analysis (PPCA). An parametric analysis of the signal is exercised for extracting the features of desired pules. We incorporate a wavelet based bootstrap method for obtaining the noise training vectors from observed data. The procedure adopted in this work is completely different from the research work reported in the literature, which is generally based on deserved signal frequency and noise frequency.

1   Introduction

PD analysis has been the indispensable, non-destructive, sensitive and most powerful diagnostic tool for on-line, on-site condition monitoring of high power, high voltage equipments. A major constrain encountered with on-line digital PD measurements is the coupling of external interferences that directly affect the sensitivity and reliability of the acquired PD data. The more important of them being, discrete spectral interferences (DSI), periodic pulse shaped interferences, external random pulsive interferences and random noise generic to measuring system itself. In most of the cases, external interferences yield false indications, there-by reducing the credibility of the PD as a diagnostic tool. Many researchers, have proposed signal processing techniques to suppress the different noise component such as, FFT thresholding, adaptive digital filter, IIR notch filter, wavelet based methods with varying degree of success [1]. Due to the inherent difficulties involved in "on-line" recognition of PD data, general methods have not been reported in the literature which forms the subject matter of this paper.

1.1   Problem Enunciation

DSI can be identified and eliminated in frequency domain as they have a distinct narrow-band frequency spectrum concentrated around the dominant frequency, whereas, PD pulses have relatively a broad band spectrum. Periodic pulse shaped interferences can be gated-off in time domain (any PD occurring in that time interval is lost). But, it is very difficult to identify and suppress PI, as they have many characteristics in common

N.R. Pal et al. (Eds.): ICONIP 2004, LNCS 3316, pp. 787–793, 2004.
© Springer-Verlag Berlin Heidelberg 2004

(both in time and frequency domain) with PD pulses. Also, PI is a random occurrence like PD, which aggravates the process of separation. Thus, PI continues to pose serious problems for reliable on-line, on-site PD measurement.

A statistical signal modeling for estimation of the desired signal has been undertaken in this paper. Locating the PD/PI pulses are the first step in further analysis of the signal. In this regard, we enhance the observed noisy signal using wavelet based soft thresholding method and the pulses are detected using simple peak-detector. Further analysis of the signal is undertaken around the detected location. An innovative model based on long memory process and PPCA is employed for obtaining the pdf of the signal. A Gaussian parametric model has been implemented for feature extraction of the desired pulse and the PD data is classified using a simple nearest neighbour method. Since the PD signal is combination of different sinusoidal and random noises, this method is quite realistic, also, the long-range dependence of this natural signal is effectively modeled by fBm process. We derive the noise process from the observed signal using wavelet based bootstrap process.

1.2   PD/PI Pulse Detection

It has been observed that, PD and PI pulses randomly occur in time. Therefore detection of the pulses is a primary requirement in further analysis of the signal. The signal-to-noise ratio of the PD signal is generally less (around -25dB) and it is difficult to visualize the location and the form of pulses in the observed noisy signal. In this regard, we denoise the noisy observed signal using wavelet based *sureshrink* soft thresholding method and make use of a simple peak detector to detect the location of pulsive activity. A minimum of S scale discrete wavelet transform is taken, where, $S = \lfloor \frac{log(F_s) - log(F_d)}{log(2)} - 1 \rfloor$. Here, F_s is the sampling frequency and F_d is the upper cutoff frequency of the PD detector. A windowed signal of appropriate size is taken around the detected location for further analysis.

2   Probabilistic PCA

Principal Component Analysis (PCA) is a widely used tool for data analysis. Given a set of $d-$dimensional data vector y, the q principal axes U_j, $j = 1, 2, ..., q$, are those onto which the retained variance under projection is maximal. These principal axes are the q eigenvectors corresponding to the q dominant eigenvalues of the sample covariance matrix of the data y. The analysis using PCA does not involve any probability model for the data. Tipping and Bishop [2] showed that by assuming a latent variable model for the data vectors, the data vectors can be represented in terms of its principal components. This approach is very useful because, we not only represent the data in terms of its principal components, but also a probability model for the data can be derived. This model in turn can be used for the tasks like estimation and detection of signals.

2.1   Probability Model for PCA

Any $d-$dimensional data vector y can be related to $q-$ dimensional ($q < d$) latent variables z as:

$$y = h + Lz + \gamma \tag{1}$$

where, γ and z are independent random processes. h is the mean of the data vectors. By defining a prior pdf to z, the above equation induces a corresponding pdf to y. If we assume $z \sim N(0, I_q)$ and $\gamma \sim N(0, C_\gamma)$, then, y is also a Gaussian with, $y \sim N(h, LL^T + C_\gamma)$, where, I_q and I are $q \times q$ and $d \times d$ identity matrices. With the above pdf's for z and γ, we can show that the columns of L are the rotated and scaled principal eigenvectors of the covariance matrix of the data vector y. In the above model, the observed vector y is represented as the sum of systematic component (Lz) and random noise component (γ). It is shown in [2] that the ML estimate of L and σ^2 are given by, $L = U_q (\Lambda_q - \sigma^2 I)^{1/2} R$. Where, the q column vectors in U_q are the eigenvectors of the covariance matrix of the data with the corresponding eigenvalues in the diagonal matrix Λ_q. R is an arbitrary rotation matrix. The energy in the remainining $(d - q)$ eigen vectors is given by σ^2. The model order (q) is estimated using Akaike information criterion (AIC) as explained in [3], which is found to be two.

3   A Probability Model for PD/PI Time Series

We propose a model for the analysis of PD/PI pulses buried in noise, as:

$$y(t) = \sum_{t=0}^{d} x(t - k)h(k) + w(t), \quad t = 0....d - 1 \qquad (2)$$

where, y is the observed time series, x is the system impulse response and w is the noise component. We model h by non parametric model based on smooth FIR filter. The Eqn. 2 can be written in matrix form as: $y = Xh + w$. Here, X is the convolution matrix, which is identity matrix I_d. The noise w can be represented using the latent variable model defined in the section 2, as, $w = Lz + \gamma$, where z is a q-dimensional ($q < N$) latent variable and γ is a random noise component. This method is quite realistic in modeling the PD signal, since the observed PD signal is combination of pulses, DSI and other random components. The matrix L is called as systematic noise matrix, which characterizes the systematic noise component by considering q principal components of w, corresponding to the first q dominant eigenvalues. Being a natural signal, PD exhibits long range dependence. Therefore, the random noise component γ is modeled using fBm process, which is explained in section 4. Assuming Gaussian pdf models as described in the section 2, the pdf of noise can be given as, $w \sim N(0, C_y)$, where, $C_y = LL^T + C_\gamma$. Finally, the observed time series can be represented as,

$$y = Xh + Lz + \gamma \qquad (3)$$

Therefore, the probability model for the observed PD/PI time-series y for a given h is distributed as, $y|h \sim N(Xh, C_y)$.

4   Modeling Noise by fBm Process

The physical phenomena like PD exhibits long-term dependencies and $1/f$ type of behaviour over wide range of frequencies [4]. Also, the natural signals are non-stationary in nature. Therefore, the standard assumption of indepedence and normality of noise random variables are not valid in modeling the natural signals. One well-known model of long-memory processes proposed by Mandelbrot and VanNess [5] is fractional Brownian motion. Among others, self-similarity property makes wavelet transform, a preferred tool for analysis of the fBm processes. The noise model is $\gamma(t) = \gamma_d(t) + \gamma_i(t)$, where, $\gamma_d(t)$ describe the long-memory process and $\gamma_i(t)$ represent the independent random noise. By taking DWT of scale m, we have $W\gamma = W\gamma_d + W\gamma_i$. The $1/f$ type of signals exhibit Karuhunen-Loeve like properties in wavelet domain and therefore $W\gamma_d$ are independent of $W\gamma_i$ and uncorrelated. The variance of the wavelet coefficients in each scale is given by: $var(W\gamma) = \sigma_\gamma^2 = \sigma_d^2\beta^{-m} + \sigma_i^2$. The parametr β is related to the Hurst component H, which completely describes the long-memory process. The covariance function of the self-similar process with Hurst component H is given as $R_d(t,s) = \frac{\sigma_H^2}{2}(|s|^{2H} + |t|^{2H} - |t-s|^{2H})$, where, $\sigma_H^2 = \Gamma(1-2H)cos(\pi H)/(\pi H)$. The parameter set $\Theta = [H, \sigma_d^2, \sigma_i^2]$ has to be estimated for modeling the signal. This is achieved by employing the ML technique in wavelet domain. The likelihood function L is given by,

$$L(\Theta) = p(W\gamma\,;\Theta) = \prod_{m,n\in R} \frac{1}{\sqrt{2\pi\sigma_\gamma^2}} exp\left(-\frac{(W\gamma_n^m)^2}{2\sigma_\gamma^2}\right) \qquad (4)$$

where, n represents the number of wavelet coefficients in scale m. The covariance matrix (C_γ) of γ, obtained using: $C_\gamma = C_d + C_i$. Where, C_d is estimated using H and C_i is estimated as: $C_i = \sigma_i^2 I_d$.

5   A Bootstrap Based Covariance Matrix Estimation

A set of training data for noise (i.e. non-pulsive region of the data) is needed to estimate C_y. But this requires the prior knowledge of absence of PD/PI pulse in the observed data. Also, the noise characteristics of physical signals change with respect to time. Hence, estimate of noise covariance matrix at one location cannot be used for another location. To over come this problem, we use a bootstrap method in wavelet domain to extract the noise process from the observed signal.

Wornell proved that [6], wavelet transform decorrelates a large class of physical signals. Hence, the detailed coefficients in the wavelet domain can be assumed to be independent and identically distributed (iid) processes. Therefore, these coefficients can be exchanged (resampled without replacement) in wavelet domain. In [4], it was shown that, resampling of the detailed coefficients do not alter the covariance structure of the signal. Thus, we can generate multiple processes from a single process by resampling the detailed wavelet coefficients.

As expalined in section 1.2, the detailed coefficients in S level decomposition of signal do not consist of PD signal components. Therefore, we can generate the noise process by resampling the detailed coefficients. We propose following algorithm for generating the noise process and estimation of covariance matrix.

1. Take a S-level Wavelet Transform of windowed signal y, where, S is as given in Sec. 1.2.
2. Set scaling coefficients to zero.

3. Resample each detailed coefficient without replacement and take inverse wavelet transform to get a realization of the noise process (y_p).
4. Repeat the above steps for a required number of realizations (M).
5. Estimate the covariance matrix, $C = \sum_{p=1}^{M} \mathbf{y_p^t y_p}$, where, $\mathbf{y_p}$ is a realization of the time-series. We use M equal to two times the length of the time-series to get a reliable estimate of C.
6. Estimate the systematic noise covariance matrix (LL^T), as explained in section 2.1
7. To extract the random noise component (γ) from y_p, form a projection matrix $B = U_q U_q^T$ (as explained in Sec.2.1). Obtain γ by projecting y_p onto B and secure the random noise component (γ).
8. Estimate C_f using γ, as explained in Sec. 4.
9. Obtain C_y, using: $C_y = LL^T + C_\gamma$

6   Parametric Modeling of PD/PI Pulse Buried in Noise

Feature extraction of the pattern is the basic step in any pattern recognition exercise. Therefore the characterization of the PD pulses are essential for discriminating the PD pulses from PI. In classical signal processing approach, frequency information is used to separate signal from noise. In our case, both PD and noise are pulsive, both have broad band frequency spectrum which overlaps in frequency domain. Therefore a need arises of investigating different characteristics of the signal, for effective separation. Recent literature [8] suggest that parameters like, rise time, pulse width, proximity, pulse magnitude and repetition rate can serve as indices of PD pulse. The qualitative assessment of the parameters are a sufficient index and actual numerical value is not of great interest. Also, PD pulses assume different shapes depending on the source of PD. Therefore a feature extraction methodology which is sensitive to shape of the pulses are employed for further analysis.

The task of completely representing PD and PI pulses in a functional form is extremely difficult, as they assume wide variety of shapes. The complexity is further enhanced by oscillatory nature of some of the pulses. Therefore, to extract the features, we intend to perform one lobe analysis (dominant lobe) and model h using a Gaussian function, which is given by,

$$h(t) = \eta \, e^{\left(\frac{t-\nu}{\beta}\right)^2} \tag{5}$$

The parameter set $\theta = [\eta \ \nu \ \beta]$, in which, η, is the pulse height, ν represents the pulse center and β is the function of pulse width.

6.1   ML Parameter Estimation

Let $p(y; \theta)$ represent the probability density function (pdf) of y. Then,

$$p(y; \theta) = \frac{1}{(2\pi)^{\frac{k}{2}}|C_y|^{\frac{1}{2}}} exp[-\frac{1}{2}(y - Xh)^T C_y^{-1}(y - Xh)] \tag{6}$$

In the above equation $|C_y|$ represents determinant of C_y which is obtained as explained in section 5. The log-likelihood function, (l), is given by,

$$l = K - \frac{1}{2}(y - Xh)^T C^{-1}(y - Xh) \tag{7}$$

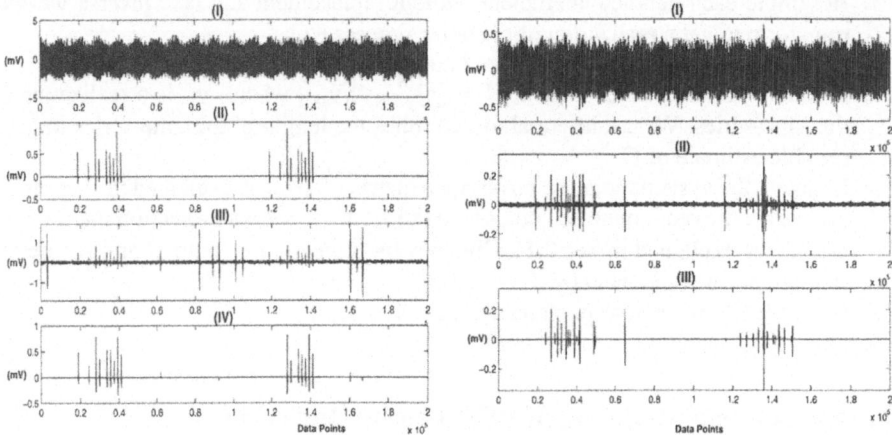

(a) (I) Observed (simulated) signal (II) Location of added PD pulses (III) Enhanced Signal (both PD and PI pulses can be seen) (IV) Retrieved PD pulses

(b) (I) Observed (real) signal (II) Enhanced Signal (both PD and PI pulses can be seen) (III) Retrieved PD pulses

Fig. 1. Output of non-parametric method considering real and simulated data

where, K is a constant. The likelihood function l is maximized w.r.t parameter set θ, to obtain ML estimates[9]. Alternatively this is equivalent to minimizing the equation,

$$l \propto \frac{1}{2}(y - Xh)^T C^{-1}(y - Xh) \tag{8}$$

Sample values of the estimated parameters are:

$\theta_{PD1} = [0.45, 150, 15.8]$ $\qquad$ $\theta_{PI1} = [1.21, 151, 25.0]$

$\theta_{PD2} = [0.94, 148, 14.2]$ $\qquad$ $\theta_{PI2} = [0.75, 150, 29.1]$

Apparently, the parameter β seems to discriminate the PD and PI pulses. This is due to the fact that, the PD pulse has relatively broader band of frequencies compared to pulsive interference, thus it is a sharply rising pulse in time domain.

A binary classifier known as weighted nearest neighbor (WNN) classifier, based on the Eucledian distance from the unlabeled instance (i.e. estimated parameters of test pulse) to the training set (parameters of standard pulse) has been implemented, wherein, one class represents the PD pulses and the other one represents the PI. In WNN classifier, parameter β rendered more weight for having strong discriminatory feature compared to other parameters. The detected pulse was retained, if found to be PD.

The overall output of the procedure is shown in figure 1. In figure 11(a), the observed (simulated) data, the location of the added PD pulses, the enhanced data and the final output is shown. All PI pulses have been suppressed and most of the PD pulses have been retrieved as shown in figure 11(a)(IV). It can be seen that, sixteen PD pulses out of eighteen have been retrieved without much shift in pulse position. The effectiveness of the method in dealing with the real data is shown in figure 11(b). In this case, all

PD pulses have been retrieved with one noise pulse being misclassified as PD pulse. A small amount of reduction of pulse height was observed in the retrieved pulses, which is attributed to filtering operation.

7   Conclusion

The problem of on-line recognition of PD signal is approached in a different perspective than conventional DSP techniques and theory to model the noisy signal has been developed. The time series modeling based on long-memory process is realistic and models the signal reasonably accurately. The performance of the parametric model is found to be reasonably good in recognition and retrieving PD pulse. The methods proposed is completely automatic and there is no user interference in PD measurement.

References

1. Satish, L., Nazneen, B.:Wavelet denoising of PD signals buried in excessive noise and interference. IEEE Transaction on DEI. Vol. 10. No. 2. April 2003. pp 354–367.
2. Tipping, M.E., Bishop, C. M.: A hierarchical latent variable model for data visualization. IEEE trans. PAMI. Vol. 20. no-3. 1998. pp.25-35. 281-293.
3. M. H Hayes "Statistical Digital Signal Processing and Modelling", John Wlley and Sons inc. 1996, Chap. 8, pp. 445-447.
4. Flandrin, P.: Wavelet analysis and synthesis of fractional Brownian motion. IEEE transaction on Information Theory. Vol. 38. no-2. 1992. pp.910-917.
5. Wornell, G: Signal Processing with Fractals: A Wavelet Based Approach. Prentice Hall PTR. Newjersy. 1996. Chap. 3. pp. 30-46.
6. G. W. Wornell   "A Karhunen-Loeve-like expansion for 1/f processes via wavelets", IEEE, Trans. Inform. Theory. vol. 36, July 1990, pp. 859-861.
7. Wornell, G. W.: A Karhunen-Loeve-like expansion for 1/f processes via wavelets. IEEE. Trans. Inform. Theory. vol. 36. July 1990. pp. 859-861.
8. Stone, G. C.: Practical techniques to measure PD in operating equipment, Proc. 3rd Int. Conf. on Properties and Application of Dielectric Materials, Tokyo, Japan, 1991, pp 1–17.
9. Kay, S. M.: Fundamentals of Statistical Signal Processing-Estimation Theory, Prentice Hall PTR,Newjersy, 1998, Chapter 7, 10,11,12, pp. 157-214, 309-415.

A Fusion of Neural Network Based Auto-associator and Classifier for the Classification of Microcalcification Patterns

Rinku Panchal and Brijesh Verma

Faculty of Informatics and Communication
Central Queensland University, Rockhampton, QLD 4702, Australia
{r.panchal,b.verma}@cqu.edu.au

Abstract. This paper presents a novel approach to combine a neural network based auto-associator and a classifier for the classification of microcalcification patterns. It explores the auto-associative and classification abilities of a neural network approach for the classification of microcalcification patterns into benign and malignant using 14 image structure features. The proposed technique used combination of two neural networks; auto-associator and classifier for classification of microcalcification. It obtained 88.1% classification rate for testing dataset and 100% classification rate for training dataset.

1 Introduction

Each year around the world million of women develop breast cancer during their lifetime. The stage of development depends upon the detection time, early detection prevents patient to pass through high stage traumatic treatment and increases chance of survival. There has been a considerable decline in mortality from breast cancer in women because of the Breast Screening programs [1]. Thus prevention and an early detection of breast cancer tumors are immediate demand of society.

Mammography continues to be regarded as one of the most useful techniques for early detection of breast cancer tumor. The presence of microcalcifications in breast tissue is one of the main features for its diagnosis [2]. Microcalcifications are mammographic hallmarks of early breast cancer [3]. Due to their subtlety, detection and classification (from benign to malignant) are two key issues. In many cases of screening database, microcalcifications exhibit both class (benign and malignant) characteristics. Interpretation of such cases often produce screening errors; either to miss malignant cases or more unnecessary biopsies. A higher prognostic rate is anticipated by combining the mammographer's interpretation and computer analysis [4]. A Computer-Aided Diagnosis (CAD) system can serve as a 'vital second reader' to the radiologists to improve overall classification.

Current image processing techniques makes even smaller microcalcification detection easier, though classification of malignant and benign microcalcifications remains a challenging issue for researchers. It's clear from previously proposed methods [2-14] that the selection of significant features and the type of classifier and topography for particular classifiers are most important factors in pattern classification process. Still this area needs enhancement to use digital mammography as clinical tool in every day practice.

N.R. Pal et al. (Eds.): ICONIP 2004, LNCS 3316, pp. 794–799, 2004.
© Springer-Verlag Berlin Heidelberg 2004

Selection and extraction of significant type(s) of features, which characterize each pattern uniquely is very important to achieve optimum classification. Features found in literature are region-based features [5], shape based features [6], image structure features [5, 7, 8], texture based features [9, 10], and position related features. A feature selection method often used to determine an "optimal" subset of features to use with a particular classifier and performance of classifier considered as an evaluating criteria [11]. Features containing irrelevant or redundant information may have detrimental effects on classifier performance. Sequential forward/backward selection [12] and genetic algorithms [13] have been used for optimal feature(s) subset selection for mammographic abnormalities classification. Many different methods have been used to classify microcalcifications in digital mammograms. Most common are statistical methods [3, 7] and artificial neural networks (ANN) [2, 4 -14]. The learning ability of neural network from given pattern attributes and to classify each pattern into appropriate class using acquired knowledge make neural network more popular in the field of pattern recognition.

Shen et al [6] achieved 94% classification for benign calcifications and 87% for malignant calcifications using one-hidden layer Back-propagation ANN. A comparative study of statistical methods and ANN for classification carried out by woods et al [7]. They reported that back-propagation learning algorithm takes long time to train network. It was difficult to determine the learning rate by which the weights were changed and updated. Qian et al [14] used a multilayer feed-forward neural network with back-propagation learning algorithm. They identified a problem [7] with the error generated during network training and modified the algorithm adding Kalman gain to weights during weight adjustment to minimize the error.

The main objective of this research work is to investigate the auto-associative and classification abilities of a neural network approach for the classification of microcalcification patterns into benign and malignant using 14 image structure features. Two neural networks such as an auto-associator and a classifier are combined to classify benign and malignant microcalcifications. The remainder of paper is broken down into 4 sections. In section 2 we discussed our research methodology. Experimental results are presented in section 3. In section 4, we discussed and analyzed the experimental results. The conclusions and future research directions are stated in Section 5.

2   Research Methodology

2.1 Digital Mammograms

Digital mammograms for this research work are taken from Digital Database for Screening Mammography (DDSM) established by University of South Florida. The main purpose of this resource is to aid researches to evaluate and compare their research work with others. DDSM provides mammograms with ground truth and other information. This is free available database and can be downloaded from the University of South Florida's online digital mammography database website http://marathon.csee.usf.edu/Mammogrphy/Database.html.

Dataset contains total of 126 calcification areas: 84 (43 benign and 41 malignant) areas in training set and 42 (21 benign and 21 malignant) areas in testing set.

2.2 Feature Extraction

Feature extraction process is divided into two parts: 1) extract suspicious areas from already marked mammograms; 2) feature extraction from the extracted suspicious areas. Suspicious areas are already marked in all digital mammograms of DDSM by at least two expert radiologists. For area extraction, first boundary of each abnormalities of mammogram is defined by solving chain code values available in ".OVERLAY" file. For easiness each abnormality area is resized. Grey level values of each suspicious area and respective surrounding boundary area are extracted to calculate feature values. 14 image structure features are number of pixels, histogram, average grey, average boundary grey, contrast, difference, energy, modified energy, entropy, modified entropy, standard deviation, modified standard deviation, skew and modified skew [8, 13].

2.3 Classifier

The proposed research method use fusion of two neural networks: auto-associator for benign and malignant class patterns association and classifier for classification of associative patterns into benign and malignant classes. Both networks have single hidden layer architecture (figure 1) and use back-propagation learning algorithm.

Auto-association discovers redundancies in a set of input patterns, which lead to more efficient featural representations. It encourages the network to discover any structure present in the input so the input can be represented more abstractly and compactly. A trained auto-associative network returns a composite of its learned patterns most consistent with the new information. Auto-associative neural network (AANN) takes input and output the same feature vector i.e. 14 image structure features. Hidden neuron values of trained AANN are set as input to classifier network. Classifier network has two output nodes in output layer, which represents each class. The desired output set as (0.9, 0.1) for malignant class and (0.1, 0.9) for benign class.

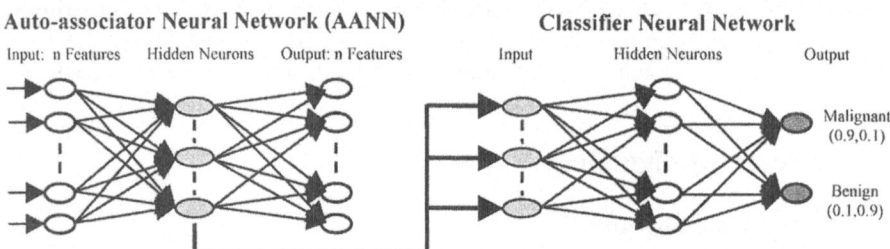

Hidden neurons values of hidden layer of trained AANN as input feature vector to Classifier NN

Fig. 1. Architecture of two neural networks

3   Experimental Results

The proposed technique is implemented using C language on UNIX platform. Five modules have been defined separately and integrated together to perform entire classification process. The five modules are: 1) area extraction, 2) feature extraction, 3) feature normalization, 4) auto-association of microcalcification patterns and 5) classification. Both networks trained extensively with various network parameters: AANN to regenerate the input patterns more consistent with new information and classifier network to produce the optimum classification with regenerated associative patterns. RMS error is considered as evaluating criteria for AANN. While final classification results are used to evaluate classifier network performance on associative patterns.

Table 1 shows number of experimental results using proposed technique with 14 image structure features. Learning rate 0.1 and momentum 0.1 were set for both networks. It attained 100% training classification rate. The highest testing classification rate 88.1% with the corresponding 96.4% training classification rate was obtained using 10 hidden neurons for both networks and 50000 iterations for AANN and 20000 iterations for classifier NN. While 85.7% testing classification rate was attained in many experiments. To attain the optimum classification on associative patterns, NN classifier was trained for more iteration. Table 2 shows the results of experiments carried out on associative patterns with 10 hidden neurons and different number of iterations for NN classifier.

Table 1. The highest classification rate obtained using different network topologies

AANN		Classifier NN			
Hidden Neurons	Iterations	Hidden Neurons	Iterations	Training (%)	Testing (%)
8	10000	10	20000	98.8	81
10	50000	6	10000	94	85.7
10	50000	6	70000	100	73.8
10	50000	10	20000	96.4	88.1
10	50000	10	50000	100	69
14	10000	6	50000	98.8	83.3
14	20000	8	50000	100	81

4 Discussions and Analysis

Many experiments were run with proposed technique using 14 image structure features to classify microcalcification patterns into benign and malignant classes. It is observed during classifier training that with increase in number of iterations improved the training classification rate gradually. While for testing dataset, it improved initially but after attaining the optimum classification it started dropping (table 2). This is because the network is over-trained on the training data and rather learning it has remembered the solution, which decreases its ability to recognize the data it had not been trained with.

It is of prime importance for any microcalcification classification system to produce the optimum classification with low false positive and false negative errors. Table 2 results show that classification of benign microcalcifications is higher than malignant microcalcifications on both training and testing dataset. This may be because of both class patterns are very identical and during classification one class takes over the other.

It is clear from the results (tables 1 and 2) that the associative patterns obtained from AANN produces good classification with minimal classifier training. Associative patterns obtained from AANN trained with 10 hidden neurons produced optimum classification 88.1% on testing dataset. These results showed that neural-association represents the input patterns more abstractly and compactly, i.e. 10 hidden neurons represent 14 features.

Table 2. Performance of classifier network with increase in iterations in training

AANN : 10 hidden neurons; 50000 iterations & Classifier NN : 10 hidden neurons						
Iterations	Training			Testing		
	Malignant	Benign	Total (%)	Malignant	Benign	Total (%)
5000	32	42	88.1	16	20	85.7
10000	37	43	95.2	17	17	81
20000	38	43	96.4	18	19	88.1
30000	39	43	97.6	16	18	81
40000	40	43	98.8	14	18	76.2
50000	41	43	100	14	15	69

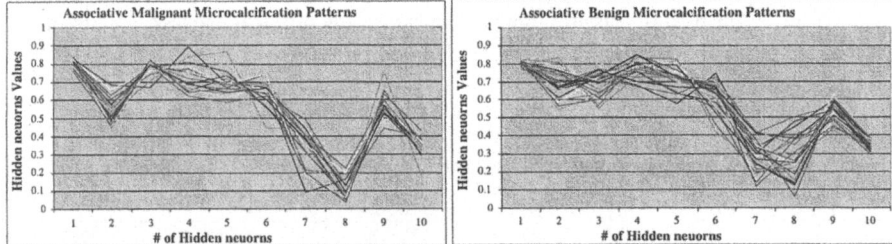

Fig. 2. Associative training patterns

Figure 2 shows the graphical representations of a number of associative patterns of benign and malignant classes, which attained the highest testing classification rate 88.1% i.e. associative patterns obtained from AANN trained with 10 hidden neurons for 50000 iterations. Graphical representation of associative patterns and experimental results show that auto-association improves the overall featural representation of both class patterns. Auto-association effectively draws each class characteristics reducing the total number of features (compactly) exploring each feature of all extracted features (abstractly).

5 Conclusions and Further Research

In this paper we proposed a technique, which uses a fusion of two neural networks: auto-associator and classifier for the classification of microcalcification patterns. We obtained highest classification rate 88.1% on testing dataset and 100% on training dataset. 85.7% classification rate for testing dataset was consistent with many network configurations for both networks. The results obtained with proposed methodology motivate further to explore the auto-associative abilities of neural network to regenerate the input patterns most consistent with new information. For such patterns some non-iterative classifiers can be used for classification process, to save both time and cost on network training.

References

1. National Breast Cancer Institute. www.nbcc.org.au
2. S. Halkiotis, J. Mantas and T. Botsis, Computer-Aided Detection of Clustered Micro-calcification in Digital Mammograms, Proc. of the 5th European Systems Science Congress, 2002.
3. M. F. Salfity, G. H. Kaufmann, P. Granitto and H. A. Ceccatto, Automated Detection and Classification of Clustered Microcalcifications using Morphological Filtering and Statistical Techniques, IWDM-2000, Medical Physics Pub., pp. 253-258, 2001.
4. L. Kinnard, S-C. B. Lo, P. Wang, Separation of Malignant and Benign Masses using Image and Segmentation Features. http://www.imac.georgetown.edu
5. Y. Chitre, A. P. Dhawan and M. Moskowitz, Artificial Neural Network Based Classification of Mammographic Microcalcifications using Image Structure Features, State of The Art in Digital Mammographic Image Analysis, World Sci. Pub., Vol. 9, pp. 167-197, 1994.
6. L. Shen, R. M. Rangayyan and J. E. Leo Desautels, Detection and Classification of Mammographic Calcifications, State of The Art in Digital Mammographic Image Analysis, Int. Journal of Pattern Recognition and Artifical Intelligence, Vol. 7, pp. 1403-16, 1993.
7. K. S. Woods, J. L. Solks, C. E. Priebe, Comparative Evaluation of Pattern Recognition Techniques for Detection of Microcalcifications in Mammography, State of The Art in Digital Mammographic Image Analysis, World Sci. Pub., Vol. 9, pp. 213-231, 1994.
8. B. Verma and J. Zakos, A Computer-Aided Diagnosis System for Digital Mammograms Based on Fuzzy-Neural and Feature Extraction Techniques, IEEE Trans. on IT in Biomedicine, Vol. 5, pp.46-54, 2001.
9. K. Bovis and S. Singh, Detection of Masses in Mammograms using Texture Measures, Proc.15th IEEE-ICOPR, Vol. 2, pp. 267-270, 2000.
10. L. Christoyianni, E. Dermatas and G. Kokkinakis, Fast Detection of Masses in Computer-Aided Mammography, IEEE Signal Processing, Vol. 17, pp. 54 –64, 2000.
11. M. A. Kupinski and M. L. Giger, Feature Selection with limited datasets, Medical Physics, Vol. 26, pp. 2176-2182, 1999.
12. S. Yu and L. Guan, A CAD System for the Automatic Detection of Clustered Microcalcifications in Digitized Mammogram Films, IEEE Trans. on Medical Imaging, Vol. 19, pp. 115-126, 2000.
13. P. Zhang, B. Verma and K. Kumar, Neural vs. Statistical Classifier in Conjunction with Genetic Algorithm Feature Selection in Digital Mammography, IEEE-CEC, Vol. 2, pp. 1206-13, 2003.
14. W. Qian and X. Sun, Digital Mammography: Wavelet Transform and Kalman-Filtering neural network in Mass Segmentation and Detection, Acad. Radiol., Vol. 8, pp. 1074-82, 2001.

Time Series Classification for Online Tamil Handwritten Character Recognition – A Kernel Based Approach

K.R. Sivaramakrishnan[1] and Chiranjib Bhattacharyya[2]

[1] Dept. of Electrical Engineering
Indian Institute of Science, Bangalore, India
sivaram@ee.iisc.ernet.in
[2] Dept. of Computer Science & Automation
Indian Institute of Science, Bangalore, India
chiru@csa.iisc.ernet.in

Abstract. In this paper, we consider the problem of time series classification. Using piecewise linear interpolation various novel kernels are obtained which can be used with Support vector machines for designing classifiers capable of deciding the class of a given time series. The approach is general and is applicable in many scenarios. We apply the method to the task of Online Tamil handwritten character recognition with promising results.

1 Introduction

In a large variety of applications it is important to discriminate between various time series. In this paper we address the problem of time series classification and apply our results to recognize Online Tamil Handwritten characters. The proposed approach should find ready applications in various devices like tablet PCs, PDAs etc.

Time series classification can be approached in various ways. In the recent past [9,10] propose to pre-process the time series in a suitable way so that the data can be handled by conventional classifiers e.g Support Vector Machines(SVMs). There is also substantial litreature on designing specialized classifiers for time series data [5–8,12,11], which doesn't involve any preprocessing step.

The approach proposed in the paper consists of interpolating a given time series by sum of piecewise linear basis functions. The obtained function is re-sampled at equal intervals and the sampled values are used by a SVM classifier to give a decision on the class of the time series. The entire process can be described in a reproducing kernel Hilbert space(RKHS) setting. The RKHS setting gives an elegant kernel interpretation. The efficacy of the proposed approach is evaluated on Online Tamil Handwritten Character Recognition problem. The remainder of the paper is organized as follows. Section 2 briefly describes the time series classification problem. Section 3 describes the proposed approach while Section 4 details the experiments and results. Section 5 summarizes the results and discusses future directions.

N.R. Pal et al. (Eds.): ICONIP 2004, LNCS 3316, pp. 800–805, 2004.
© Springer-Verlag Berlin Heidelberg 2004

2   The Problem of Time Series Classification

Consider a dataset $\mathcal{D} = \{(\mathcal{F}_k, y_k) | \mathcal{F}_k \in \mathbb{R}^{n_k}, y_k \in \{1, -1\}, 1 \leq k \leq N\}$ specified by N tuples where each tuple consists of a time series $\mathcal{F}_k$ and its label y_k. The kth time series

$$\mathcal{F}_k = \{\mathcal{F}_k(t_{1_k}), \mathcal{F}_k(t_{2_k}), \ldots, \mathcal{F}_k(t_{n_k})\} \tag{1}$$

consists of observations $\mathcal{F}(t_{i_k}) \in \mathbb{R}$ at time instants $t_{1_k} < t_{2_k} < \ldots < t_{n_k}$. In a general setting the difference between adjacent time instants $(t_{i_k} - t_{(i-1)_k})$ and the length of time series (n_k) may be arbitrary. We wish to learn a classifier from the dataset $\mathcal{D}$ which would predict the label of a given time series.

In the recent past Support Vector Machines(SVMs) have emerged as a powerful tool for binary classification problems. However they cannot be readily applied to time series data due to variable length of time series. To get over this problem one can compute a suitable statistic which can be used for discrimination. Another approach could be re-sampling the time series at specified instants of time and feed these re-sampled values to a SVM classifier.

Online handwritten character data is one of the examples of time series data as specified in (1). Each character is represented by x and y coordinates of the pen at various time instants. Each character is of variable length and is sampled at variable time instants, see figure 1b. We propose to interpolate the given time series by piecewise linear functions. The interpolated function can be used as surrogate to re-sample the values.

3   The Proposed Approach

Let $f : [0, 1] \rightarrow \mathbb{R}$ and $g : [0, 1] \rightarrow \mathbb{R}$ be two differentiable functions with the dot product

$$(f, g) = f(0)g(0) + \int_0^1 f'(s)g'(s) \, ds \tag{2}$$

Consider piecewise linear functions, defined as follows (see figure 1a)

$$R_t(s) = 1 + \min(s, t) \; s, t \in [0, 1] \tag{3}$$

the dot product (2) is computed by $(R_{t_i}, R_{t_j}) = 1 + \min(t_i, t_j)$ [1]. The piecewise linear functions (3) can be interpreted as a reproducing kernel in a suitable Hilbert space $\mathcal{H}$ with the dot product defined as in (2) [1].

Let $f = \{f(t_1), f(t_2), \ldots, f(t_n)\}$ be a function specified at n points, where $f(t_i)$ is the function evaluation at t_i and $0 \leq t_1 < t_2 \ldots < t_n \leq 1$. We seek to approximate f by a sum of piecewise linear functions $R_{t_j}(t)$

$$\bar{f}(t) = \sum_{j=1}^{n} c_j R_{t_j}(t) \tag{4}$$

---

[1] $\min(u, v) = u$ if $u < v$ else $\min(u, v) = v$.

The fixed basis functions $R_{t_j}(t)$ are defined with respect to t_j, $j = 1, 2, \ldots, n$. while the scalars $c_j \in \mathbb{R}$ are unknown. The goal is to choose the coefficients c_j such that $\bar{f}$ interpolates f at the given arguments $t_j, j = 1, 2, \ldots, n$.

It is easy to see that for any differentiable function $g : [0,1] \rightarrow \mathbb{R}$ the dot product with R_t is given by $(g, R_t) = g(t)$. Using this property one can choose the coefficients such that it satisfies[2]

$$(\bar{f} - f, R_{t_j}) = 0, \forall j = 1, 2, \ldots, n$$

Substituting the expression for $\bar{f}$ from (4) and after some algebra we obtain

$$\sum_{j=1}^{n}(R_{t_j}, R_{t_i})\, c_j = (f, R_{t_j}) = f(t_i), \forall i = 1, 2, \ldots, n \tag{5}$$

The function resulting from a such a choice of c can be shown to be optimal in the sense that it satisfies $\min_{\bar{f}}\|f - \bar{f}\|_{\mathcal{H}}$ [1].

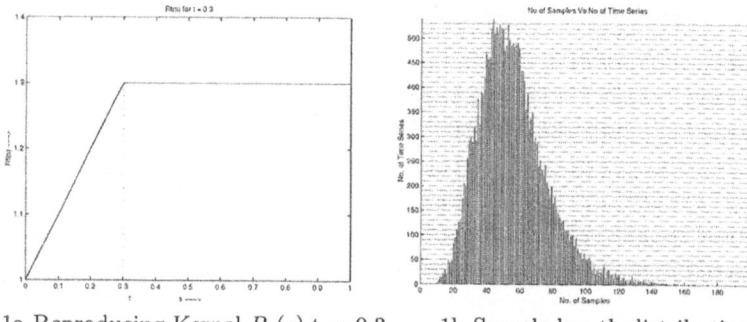

1a Reproducing Kernel $R_t(s)$,t $= 0.3$ 1b Sample length distribution

Fig. 1.

Choosing the coefficients $c_1, c_2, \ldots, c_n$ as in (5) the $\bar{f}$ in (4) is evaluated at time instants $s_1, s_2, \ldots, s_m$ where $s_{i+2} - s_{i+1} = s_{i+1} - s_i$, $\forall i = 1, 2, \ldots, m - 2$. This function is now sampled at equally spaced time instants and of length m and can be classified by SVMs. The re-sampling process can also be described by dot products as follows

$$\bar{f}(t) = (\bar{f}, R_t) = \sum_{j=1}^{n} c_j(R_{t_j}, R_t) \tag{6}$$

The process of interpolation (5) and re-sampling (6) can be described more compactly as

$$F = R_{m \times n} R_{n \times n}^{-1} f_n \tag{7}$$

where $f_n = [f(t_1), f(t_2), \ldots, f(t_n)]^T$ is the specified vector of function evaluations at time instants $t_1, t_2, \ldots, t_n$. The kernel matrix $R_{n \times n}(i, j) = (R_{t_i}, R_{t_j})$ contain the dot products of piecewise linear basis functions at various time in-

---

[2] Use $g(t) = \bar{f}(t) - f(t)$.

Algorithm 1 TRAIN(Training Data $\mathcal{D}$, No of re-sampled points m)

---

1. Rescale the time axis to ([0,1]) for each time series $\mathcal{F}_i$ so that
2. Find the coefficients C_i for each $\mathcal{F}_i$ by solving (5).
3. Find re-sampled function values F_i for the ith time series at time instants $s_k = \frac{k}{m}$ by solving (6).
4. Use the m dimensional vector F_i's to train a SVM classifier.

---

stants $\{t_1, t_2, ..., t_n\}$. The matrix $R_{m \times n}(i, j) = (R_{s_i}, R_{t_j})$ contain the dot products with the basis functions for the re-sampling time instants $\{s_1, s_2, ..., s_m\}$. We are now ready to state our algorithm The computational time complexity in steps 1-3 is dominated mainly by step 2 which has a complexity of $O(n^3)$. To predict the class of a time series one computes steps $1 - 3$ to re-sample from the time series and then the re-sampled values are fed to a SVM classifier.

4   Experiments and Results

4.1   Tamil Online Handwritten Characters Dataset

The data set consists of 162 tamil characters. Each character is written by a writer at 10 different times, total number of writers being 15. The data consists of 162 classes and each class has 150 samples. Each sample is of variable length and sample is sampled at varying instants of time. The distribution of the sample size is shown in the figure 1b. A sample set of characters are shown in figure 2.

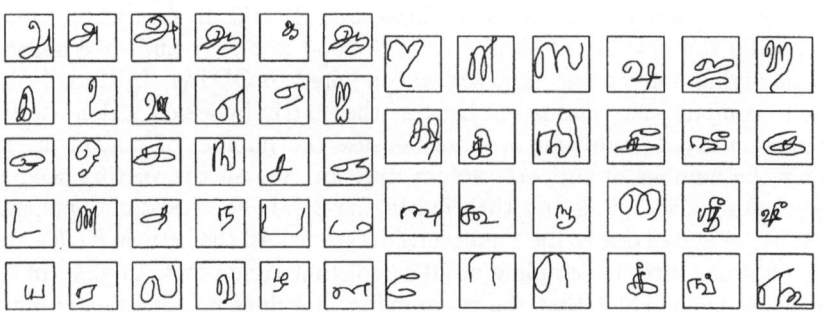

Fig. 2. Sample Tamil Characters

4.2   Results

The above mentioned dataset was used to evaluate the proposed method. As a benchmark we compare our method with Gaussian Dynamic Time Wrapping(DTW) proposed in [2]. They use DTW instead of euclidean distance to define a kernel $\left[exp\left(-\frac{DTW(\mathcal{F}_x^1, \mathcal{F}_y^2)}{\sigma^2} \right) \right]$ and use it with SVMs to discriminate between different time series.

Table 1. Accuracy and Timings

Kernel/Method	Accuracy	Timing (per test sample in seconds)
GDTW	70.4	0.78
K_1	74.8	0.45
K_2	78.5	0.46
K_3	78.2	0.37

The approaches described in the previous section can be seen as a kernel defined on two time series $\mathcal{F}_1$ and $\mathcal{F}_2$ of length n_1 and n_2 respectively. More precisely using (7) we can write

$$K_1(\mathcal{F}_1, \mathcal{F}_2) = F_1^T F_2 = \mathcal{F}_1^T R_{n_1 \times n_1}^{-1} R_{n_1 \times m} R_{m \times n_2} R_{n_2 \times n_2}^{-1} \mathcal{F}_2$$

We have also experimented with the following kernels

$$K_2(\mathcal{F}_1, \mathcal{F}_2) = \exp(-\frac{1}{2\sigma^2} K_1(\mathcal{F}_1, \mathcal{F}_2)) \quad K_3(F_1, F_2) = \exp(-\frac{1}{2\sigma^2} \|F_1 - F_2\|^2)$$

The data was randomly partitioned into 40% training and 60% test data set. The training data was used to train the SVM using different kernels and the test data was used to measure the generalization error for each kernel. This was done for 20 times and the average accuracy is reported. Apart from accuracy it is important that the classifier should quickly predict the class of the time series. Keeping this in mind the average timing statistics for each test sample is also reported. In our implementation we have used SVM[3, 4]. As can be seen from the results the new set of kernels provide good accuracy and take lesser time for classifying a given test time series, compared to GDTW. In Kernel K_3, the time to compute the class is less because the matrix inversion is done only once whereas in K_1, K_2 the time is more because the number of matrix inversions equals the number of support vectors. But for K_3, all the vectors need to be stored after re-sampling and then fed to the SVM. For real-time applications, K_3 is better suited due to its timing considerations. All the kernels K_1, K_2 and K_3 were obtained after re-sampling at 20 equidistant time points in [0,1],($m = 20$). Our experiments show that the re-sampling rate does not drastically influence the classifier.

5 Conclusions and Future Work

We have proposed a novel time series classification algorithm applicable to a large class of problems. Currently we are investigating approximations to step 2 in algorithm 1 to reduce the computational complexity involved in solving (6). The work presented here easily generalizes to piecewise polynomials. In future we would like to examine the efficacy of the following $R_t^q(s) = \sum_{j=0}^{q-1} \frac{s^j t^j}{(j!)^2} + \int_0^{min(s,t)} \frac{(s-u)^{q-1}(t-u)^{q-1} du}{((q-1)!)^2}$ basis functions. For $q = 1$ it reduces to (3). These basis functions are different from splines, and it would be useful to evaluate their performance on real life applications. Empirical results on Online Tamil Handwritten Character recognition task show that the method has promise. The method is general and is applicable to any language. Future work will consist in applying the method to other Indian languages.

References

1. R. E. Moore, Computational Functional Analysis, 1985.
2. Claus Bahlmann, Bernard Haasdonk, and Hans Burkhardt. On-line handwriting recognition with support vector machines–a kernel approach. In Proc. of the 8th IWFHR, pages 49-54, 2002.
3. Junshui Ma, Yi Zhao and Stanley Ahalt,OSU SVM Classifier Matlab Toolbox (ver 3.00).
4. T. Joachims, Making large-Scale SVM Learning Practical. Advances in Kernel Methods - Support Vector Learning, B. Schalkopf and C. Burges and A. Smola (ed.), MIT-Press, 1999.
5. Juan J. Rodríguez and Carlos J. Alonso González. Time series classification by Boosting Interval Based Literals. In *Inteligencia Artificial, Revista Iberoamericana de Inteligencia Artificial. No.11 (2000), pages 2–11.*
6. Mohammed Waleed Kadous. Temporal Classification: Extending the classification paradigm to multivariate time series, PhD Thesis, School of Computer Science and Engineering, University of New South Wales.
7. D.Eads *et al*, Genetic Algorithms and Support Vector Machines for Time Series Classification, *Proc SPIE* **4787**(2002) pages 74–85.
8. Vladimir Pavlovic, Brendan Frey and Thomas S. Huang, Time Series Classification using Mixed-State Dynamic Bayesian Networks, *IEEE Conf. Computer Vision and Pattern Recognition*, Ft. Collins, CO, 1999, pages 609–617.
9. Eamonn J. Keogh and Michael J. Pazzani, An enhanced representation of time series which allows fast and accurate classification, clustering and relevance feedback, *Proc. of the 4th Int'l Conference on Knowledge Discovery and Data Mining*, pages 239 – 241, 1998.
10. Geurts P., Pattern extraction for time series classification, *Proc of Principles of Data Mining and Knowledge Discovery*, 5th European Conference, Freiburg, Germany, pages 115–127, 2001.
11. William H. Hsu and Sylvian R. Ray, Construction of Recurrent Mixture Models for Time Series Classification,*Proc. of Int'l Joint Conference on Neural Networks (IJCNN-99)*, Washington, DC, 1999, Vol 3, pages 1574–1579.
12. C. Dietrich, F. Schwenker and G. Palm, Classification of Time Series Utilizing Temporal and Decision Fusion, MCS 2001, LNCS 2096, pp. 378 - 387, 2001.
13. Vladimir Vapnik, The Nature of Statistical Learning Theory, 1995.

Tamil Handwriting Recognition
Using Subspace and DTW Based Classifiers

Niranjan Joshi[1], G. Sita[1],
A.G. Ramakrishnan[1], and Sriganesh Madhvanath[2]

[1] Dept. of Electrical Engg., Indian Institute of Science, Bangalore, India
{niranjan,sita,ramkiag}@ee.iisc.ernet.in
[2] Hewlett-Packard Laboratories, Bangalore, India
srig@hplabs.com

Abstract. In this paper, we report the results of recognition of on-line handwritten Tamil characters. We experimented with two different approaches. One is subspace based method wherein the interactions between the features in the feature spate are assumed to be linear. In the second approach, we investigated an elastic matching technique using dynamic programming principles. We compare the methods to find their suitability for an on-line form-filling application in writer dependent, independent and adaptive scenarios. The comparison is in terms of average recognition accuracy and the number of training samples required to obtain an acceptable performance. While the first criterion evaluates effective recognition capability of a scheme, the second one is important for studying the effectiveness of a scheme in real time applications. We also perform error analysis to determine the advisability of combining the classifiers.

1 Introduction

Handwriting recognition is a desirable attribute for real time operation of hand held systems where the resources are limited and the devices are too small to have full sized keyboards. Cursive English script recognition is already an inbuilt feature in pocket sized Personal Digital Assistants (PDA) with very high recognition accuracies. For a good review of online handwriting recognition, see [1]. Online handwriting recognition is especially very relevant in Indian scenario, as symbols requiring long key stroke sequences are very common in Indian languages. It also eliminates the need to adapt to any complex key stroke sequences and handwriting input is faster compared to any other text input mechanism for Indian languages. Given the complexity of entering the Indian scripts, using a keyboard, handwriting recognition has the potential to simplify and thereby revolutionize data entry for Indian languages.

The challenges posed by Indian languages are different from English. In addition, there has been very little research on machine recognition of Indian scripts. Consequently, exhaustive experimentation is necessary in order to get a good insight into the script from machine recognition point of view. In this paper, we address the problem of online handwriting recognition of Tamil which is a

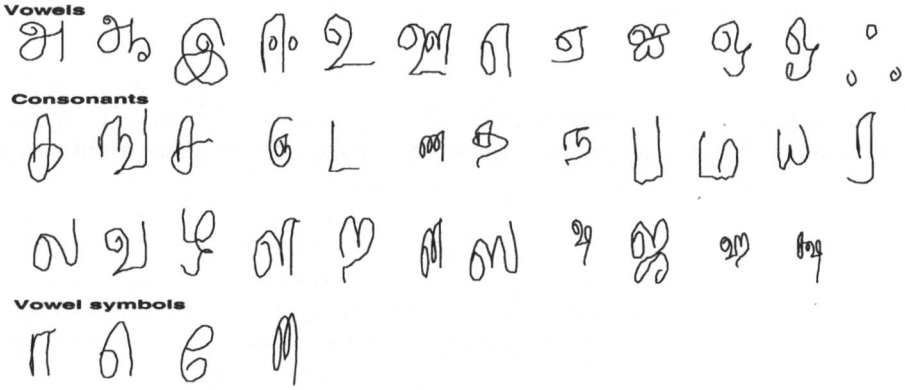

Fig. 1. Tamil character set

popular South Indian language and also one of the official languages in countries such as Singapore, Malaysia, Sri Lanka.

There are 156 distinct symbols/characters in Tamil of which 12 are pure vowels and 23 are pure consonants. This set of 35 characters are the basic character units of the script and the remaining character classes are vowel-consonant combinations. They are composed of two parts, namely the basic character and a modifier symbol corresponding to each of the basic character. Although in most of the cases, the basic character and the modifier are written in separate strokes, in the present work, we consider them to be written in single stroke. Fig. 1 presents the basic Tamil character set. In south Indian scripts such as Tamil, Malayalam, Kannada, and Telugu, characters are written in isolation. Hence, in the current work, each character is considered as a separate class for recognition. The input is a temporally ordered sequence of (x, y) pen coordinates corresponding to an isolated character obtained from the digitizer.

We experimented with two different approaches for character recognition. One is principal component analysis based method wherein each character class is modeled as a subspace. A consequence of this is that whenever a core pattern and its variations occur, all the linear combinations of these patterns are treated as members of the class. This is equivalent to synthesizing patterns by taking linear combinations. The second approach uses dynamic programming principles for recognition and uses elastic matching . A comparison of both the methods is carried out to find their suitability for an online form filling application in writer dependent, independent and adaptive scenarios. In writer dependent case, the character model is built using the individual writer's data only. In writer independent case, the data of the writer under consideration is not a part of the training set. In writer adaptive case, the training set consists of the data of all other writers in addition to a part of the current writer's data. In all the three cases, the training data is different from the test data.

2   Preprocessing

We use a Pocket PC for dynamic capture of the handwritten characters. The input from the digitizer corresponding to a handwritten character is a sequence of points of the form (x_i, y_i) with embedded pen-up and pen-down events when multiple strokes are involved. Pre-processing is required in order to compensate for variations in time and scale, and can be classified into two steps - smoothing and normalization. Smoothing is performed to reduce the amount of high frequency noise in the input resulting from the digitizer or tremors in writing.In our scheme, each stroke is smoothed independently using a 5-tap Gaussian low-pass filter. Normalization is carried out to account for variability in character size and pen velocity. The details of these operations are given in [5].

3   Methods

3.1   Subspace Based Classification

It is essentially a linear transformation of the feature space. By selecting the principal directions in which variance is significant, the feature space can be approximated by a lower order space. We use this method to model each character class as a subspace. As a consequence of this, whenever we have a core pattern and its variations, all the linear combinations of these patterns are treated as members of the class. The method is briefly desscribed below. For more details, refer [3].

Let the N training vectors of a particular class be $(\mathbf{x}_1, \cdots, \mathbf{x}_N)$. The correlation matrix is defined as,

$$\mathbf{R_x} = \frac{1}{N} \sum_{i=1}^{N} \mathbf{x}_i \mathbf{x}_i'$$

For finding the principal components, we solve the eigen value equation,

$$\lambda \mathbf{v} = \mathbf{R}_x \mathbf{v}$$

In this fashion, the basis vectors for each class k are computed as a set of N eigen vectors $\mathbf{v}_j^k, j = 1, \cdots, N$. Each eigen vector is normalized so that the basis is orthonormal. For a given test vector $\mathbf{x}_{test}$, its projection distance to the subspaces spanned by individual character classes is used as a measure to recognize its correct class lable. In this work, subspace spanned by the first 11 eigen vectors is used after experimentation.

An advantage of subspace method is its ability to approximate the feature vector in a low dimensional space which leads to a reduction in the running time in real time applications. This is possible as normally the smallest eigen values correspond to spurious variations in the character. Therefore, selecting a subset of the original subspace increases the accuracy of classification.

3.2 DTW Based Classification

Dynamic time warping (DTW) is a elastic matching technique. It allows nonlinear alignment of sequences and computes a more sophisticated similarity measure. This is especially useful to compare patterns in which rate of progression varies non-linearly which makes similarity measures such as euclidean distance and cross-correlation unusable. Classifiers using DTW-based distances have been shown to be well suited for handwriting recognition task by several researchers [1, 4] .

Suppose we have two time series $Q = (q_1, \cdots, q_n)$ and $C = (c_1, \cdots, c_m)$, of length n and m respectively. To align two sequences using DTW, we construct an n-by-m matrix where the $(i, j)^{th}$ element is the Euclidean distance $d(q_i, c_j) = (q_i - c_j)^2$ between the two points q_i and c_j. Each matrix element (i, j) corresponds to the alignment between the points q_i and c_j. A warping path, W is a contiguous set of matrix elements that defines a mapping between Q and C and is written as $W = w_1, \cdots, w_K$ where $max(m, n) \leq K < m + n - 1$. The warping path is typically subject to several constraints such as, boundary conditions, continuity, monotonicity, and windowing [4]. The DTW algorithm finds the point-to-point correspondence between the curves which satisfies the above constraints and yields the minimum sum of the costs associated with the matchings of the data points. There are exponentially many warping paths that satisfy the above conditions. The path that minimizes the warping cost is,

$$DTW(Q, C) = min\{\sqrt{\sum_{k=1}^{K} w_k / K}$$

The warping path can be found very efficiently using dynamic programming to evaluate the following recurrence which defines the cumulative distance $\gamma(i, j)$ as the distance $d(i, j)$ found in the current cell and the minimum of the cumulative distances of the adjacent elements.

In order to resolve the confusion among character classes and reduce computational time, in this method classification stage is divided into two steps. In the first pre-classification step, we use Euclidean distance as a measure for obtaining the possible character candidates. This is followed by fine classification using (x, y) coordinates as features on the output classes given by pre-classification step. Both the steps use DTW as the distance measure. More details of this method can be found in [6].

4 Database

The database is collected from 15 native Tamil writers using a custom application running on a Pocket PC (Compaq iPAQ 3850). It contains vowels, consonants, vowel-consonant combinations and special characters totalling 156 symbols. This set covers all the discrete symbols that make up all characters in Tamil. Ten samples of each of the symbols under consideration were collected from each

Table 1. Samples used in WD,WI and WA modes

Mode	Training samples	Test samples
WI	19440	4860
WD	1134	486
WA	23814	486

writer totalling 23400 samples. The ten datasets from each writer are collected at different times to avoid fatigue in the writers which would reflect in the hand writing. In the user interface of the training mode, the character to be written is displayed and the user has to write in a given writing area. In the testing mode, the user writes in boxes obviating the need for character segmentation.

5   Experimental Results

The objective of the current investigation is to evaluate the performance of both subspace based and DTW based methods so as to combine the advantages of the two schemes to formulate subsequently a hybrid scheme for handwriting recognition. The comparison is carried out for all the three modes, namely, writer dependent (WD), writer independent (WI) and writer adaptive (WA). The difference in the three modes is only in the training dataset. For writer independent recognition, we use "leave one out" strategy. In this strategy, out of the 15 writers data, data from 12 writers is used for training the recognizer. The remaining 3 writers data is used as the test data. The recognizer is trained to recognize the variations of a specific writer only in the case of WD recognition. Hence, out of the ten data sets of a particular class of a given writer, seven are used as training set to model a class and the remaining three datasets are used for testing. In WA case, the recognizer is trained to incorporate a much larger variability in the writing style by including other writers' data along with the specific writer's data. Table 1 presents the number of samples used in all the three modes for training as well testing the recognizer's performance. All the experiments are run using a Pentium IV processor with 512 MB RAM.

In subspace method, we experimented using different number of principal components to approximate the chracter classes under conisderation. The mean accuracy of a given classification scheme is computed by averaging the accuracies computed across all writers. This is compared across WD, WI and WA modes. It is found that the gain in recognition accuracy is not very significant if we further increase the number of principal components beyond 11. Hence, in the rest of the investigation, we considered only first 11 principal components for subspace based studies.

The comparison is carried out with respect to the average class recognition accuracy for different number of training samples. Average recognition accuracy is found out by dividing number of correctly recognized test patterns with total number of test patterns. This leads to the minimum number of training samples required for an acceptable recognition performance for each of the schemes. Fig. 2

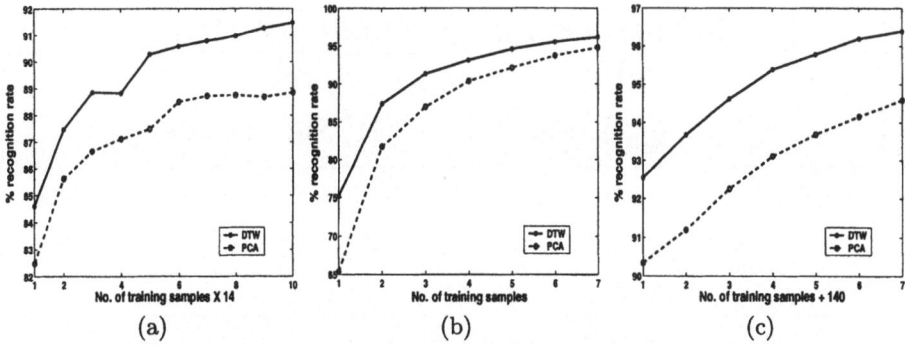

Fig. 2. Recognition for (a) WI case (b) WD case (c) WA case

shows that in each of the modes, the DTW method outperforms subspace method in terms of recognition accuracy. In writer adaptive case, the performance of both methods is good as a larger writing style variability is incorporated in the model by including other writers data in the training set.

6 Error Analysis

In this section, we present the analysis of confused characters in both the methods. For this, we manually check the structure of the misrecognized test samples. Only those errors are considered, which occur frequently. Fig. 3 shows set of confusion pairs/triplets/quadruples. *Group 1* confusions are observed with both classifiers. Confusions of type 1 occur due to a "loop" getting confused with either a "cusp" or a "straight line". Confusions of type 2 occur because of a sharp "corner" getting confused with a "curved" part. It is apparent that structures of most of the characters involved in these types of confusions are very much similar. Therefore both the methods which provide global (dis)similarity measure get easily confused within these characters. However this observation provides some important clues for further modifications. Confused characters shown in the figure along with rest of the characters similar to them can be grouped together for first level of classification. Further classification can be performed by giving importance to local features. Rest two groups of errors shown in Fig. 3 are specific only to a certain method. In Table 2, a group-wise error comparison of the methods in each of the modes of operation, namely, WI,WD,WA is presented. The second and third columns of the table present the percentage errors obtained with each of the methods and for different modes. In the other

Table 2. Comparison of error rates for subspace and DTW based methods

Mode	PCA	DTW	Common
WI	11.15	8.52	3.56
WD	5.23	3.30	1.47
WA	5.41	3.60	1.20

Group 1: Confusions common to both methods	
Type 1	(எரி னி),(எ ன ன ணை),(எரீ னீ),(ன் ள்), (நூ ஜா),(ளு நு),(ா எ),(ஏ ர),(ரு ரூ)
Type 2	(க்ஷீ க்ஷு),(ஹீ ஹு),(ஸீ ஸு),(ஷு வ்ஷீ), (ஜீ ஜு),(வ ல),(வீ லீ),(வி லி)
Group 3: Confusions specific to subspace method	
(ங நு),(ட பீ),(ஆ ஆ),(ந் த்),(ந ற),(சூ ஆ), (ஏ எ அ),(ல ஸ), ,(ரு ஓ),(யி பி)	
Group 3: Confusions specific to DTW method	
(ஃ),(ஷ வி),(மூ மூ மு மூ),(ஷ க்ஷ),(ந ர), (ஜ ஜ),(றா னா),(உ ட),(னு று)	

Fig. 3. Confusion set

columns, group wise errors are presented. Failure of subspace method is possibly due to estimated subspaces for confused characters are very "near" or overlapping. Errors in DTW method are mainly because its elastic matching capability overfits the template. However since these set of errors are of non-overlapping in nature a classifier combination scheme combining these two classifiers could prove helpful for improving overall accuracy.

7 Conclusions

The suitability of two different schemes, namely subspace based method and dynamioc time warping(DTW) based methods, for online handwriting recognition of Tamil script is investigated in three different writer modes. Although, the performance of DTW based method is marginally better, in terms of speed subspace based method wins over. To overcome the data dependency of subspace method, and use the advantage of elastic and nonlinear matching capability of DTW, hierarchical classification schemes are currently being investigated. Although both the methods are studied for a specific real time application, at present recognition speed is not being compared as essentially the objective is to reap the advantages of both the methods. Although DTW based method is computationally expensive, it can be overcome by using prototype selection/reduction methods.

References

1. Charles C. Tappert, Ching Y. Suen, and Tory Wakahara, "The state of the art in on-line handwriting recognition," *IEEE Trans. Pattern Analysis and Machine Intelligence*, vol. 12(8), 1990, pp. 787-808.
2. C. S. Sundaresan and S. S. Keerthi, "A study of representations for pen based handwriting recognition of Tamil characters," *Fifth International Conference on Document Analysis and Recognition*, Sep. 1999, pp.422-425.
3. Deepu V., "On-line writer dependent handwriting character recognition," Master of Engineering project report, *Indian Institute of Science*, India, Jan. 2003.
4. E. Keogh and M. Pazzani, " Derivative dynamic time warping," *First SIAM International Conference on Data Mining (SDM'2001)*, Chicago, USA, 2001.
5. X. Li, D.Y. Yeung, "On-line handwritten alphanumeric character recognition using dominant points in strokes," *Pattern Recognition*, 30(1), 1997, pp. 31-44.
6. Niranjan Joshi, G. Sita, A.G. Ramakrishnan, and Sriganesh Madhvanath, "Comparison of elastic matching algorithms for on-line Tamil handwriting recognition," ICONIP'04, Kolkatta, 2004.

Recognition of Bangla Handwritten Characters Using an MLP Classifier Based on Stroke Features

T.K. Bhowmik, U. Bhattacharya, and Swapan K. Parui

CVPR Unit, Indian Statistical Institute, Kolkata, 108, India
swapan@isical.ac.in

Abstract. A recognition scheme for handwritten basic Bangla (an Indian script) characters is proposed. No such work has been reported before on a reasonably large representative database. Here a moderately large database of Bangla handwritten character images is used for the recognition purpose. A handwritten character is composed of several strokes whose characteristics depend on the handwriting style. The strokes present in a character image are identified in a simple fashion and 10 certain features are extracted from each of them. These stroke features are concatenated in an appropriate order to form the feature vector of a character image on the basis of which an MLP classifier is trained using a variant of the backpropagation algorithm that uses self-adaptive learning rates. The training and test sets consist respectively of 350 and 90 sample images for each of 50 Bangla basic characters. A separate validation set is used for termination of training of the MLP.

1 Introduction

OCR systems are now available commercially at affordable prices and can recognize many fonts. Even so it is important to note that in certain situations these commercial packages are not always satisfactory and problems still exist with unusual character sets, fonts and with documents of poor quality. However, research now focuses more on handprinted and handwritten character recognition. Unfortunately, the success of OCR could not extend to handwriting recognition due to large variability in people's handwriting styles [1]. Diverse schemes for handwritten character recognition are discussed in [2]. Among the Indian scripts some works [3], [4] are found on on-line/off-line recognition of Devnagari characters. However, not much work has been done on off-line recognition of handwritten characters of Bangla, the second-most popular language and script in the Indian subcontinent and the fifth-most popular language in the world. There are 50 basic characters (11 vowels and 39 consonants) in Bangla apart from the numerals. The difficulty in automatic recognition of these handwritten Bangla characters arises because this is a moderately large symbol set and they are usually extremely cursive even when written separately. Only a few works on handwritten Bangla character recognition are available [5], [6]. However, the results reported in there are based on small databases collected in laboratory environments. It is now well established that a scheme for recognition of handwriting must be trained and tested on a reasonably large number of samples. A few works report high recognition accuracies on moderately large databases of handprinted Bangla numerals [7], [8], [9]. However, it requires extensive research work for an efficient scheme for off-line recognition of handwritten basic Bangla characters.

N.R. Pal et al. (Eds.): ICONIP 2004, LNCS 3316, pp. 814–819, 2004.
© Springer-Verlag Berlin Heidelberg 2004

In this paper, the dominant vertical and horizontal strokes present in a handwritten character image are identified in the form of digital curves with 1-pixel width and 10 certain features are extracted from each such curve. These features contain shape, size and position information of a curve or stroke and are normalized. The feature vectors of the strokes obtained from a character image are concatenated in a particular order to form the feature vector of the character image. Such a feature vector has length 100 and retains the essential information of the character image. An MLP classifier is designed using these feature vectors. The proposed method is quite general in the sense that it can be applied to other scripts also where the nature of the strokes may be different. The present approach is robust in the sense that it is independent of several aspects of input shape such as thickness, size etc.

2 Stroke Features

The gray level image of a Bangla character is first median filtered and then thresholded into a binary image. Let A be a binary Bangla character image. The aim now is to identify the vertical and horizontal strokes that are present in A. Such a stroke will be represented as a digital curve which is one-pixel thick and in which all the pixels except two have exactly two 8-neighbours, the other two pixels being the end pixels. In order to get the digital curves representing the vertical and horizontal strokes, two directional view based binary images from A are created. Let E be a binary image consisting of object pixels in A whose right or east neighbour is in the background. In other words, E is formed by the object pixels of A that are visible from the east (Fig.1a). Similarly, S is a binary image consisting of object pixels in A whose bottom or south neighbour is in the background (Fig.1b). The connected components in E represent strokes that are vertical, that is, where the pen movement is upward or downward. Similarly, the connected components in S represent strokes that are horizontal. These components are digital curves whose shape, size and position information will be used for classification. Only the sufficiently long curves in E and S are considered.

(a)   (b)

Fig. 1. Image of a Bangla character. (a) Dark and gray pixels indicate E and A images respectively. (b) Dark and gray pixels indicate S and A images respectively.

2.1 Extraction of Stroke Features

From each digital curve in E and S, ten scalar feature values are extracted. These features indicate the shape, size and position information of a digital curve with respect to the character image. A curve C in E is traced from bottom upward. Suppose the bottom most and the top most pixel positions in C are P_0 and P_6 respectively. The five points $P_1, P_2, ..., P_5$ on C are found such that the curve distances between P_{i-1} and P_i (i=1, ..., 6) are equal [10], [11]. Let θ_i , i=1,2,...,6 be the angles that the lines $\overrightarrow{P_{i-1}P_i}$ make with the x-axis. Since the digital curve here is a vertical stroke, $45^0 \leq \theta_i \leq 135^0$. Note that the angles θ_i are features that are invariant under scaling or size of the curve and represent only its shape. Another shape feature of the curve is defined as L_1 = (Euclidian distance between two vertices P_0 and P_6)/(curve length of C). L_1 represents the degree of linearity of the curve where $0 < L_1 \leq 1$.

Let H be the height of the binary character image. That is, $H = row_2 - row_1 + 1$ where row_1 is the first row in the image having an object pixel and row_2 is the last such row. Let L_2 = (curve length of C) / H. L_2 is the normalized size of the curve. The position features of C are given by $(\overline{X}, \overline{Y})$ which is the centre of gravity of the positions of pixels in C. Note that the entries in $(\overline{X}, \overline{Y})$ are also divided by H for normalization purpose. The centre of gravity is useful in finding the relative positions of two digital curves present in a character image. The *feature vector* of a vertical stroke C is defined as $(\theta_1, \theta_2, \theta_3, \theta_4, \theta_5, \theta_6, L_1, L_2, \overline{X}, \overline{Y})$. For example, the feature vectors of the vertical strokes present on the left sides in the image in Fig.1 is (135.02, 124.22, 90.04, 99.42, 80.58, 55.47, 0.8389, 0.9435, 0.1429, 0.4286).

The features extracted from a horizontal stroke representing curve C in S are similar. Here C is traced from west to east. Suppose the west most and the east most pixel positions in C are Q_0 and Q_6 respectively. The five points $Q_1, ..., Q_5$ on C are such that the curve distances between Q_{i-1} and Q_i ($i=1, ..., 6$) are equal. These points on C are found in the same way as P_i. Let α_i, $i=1,2,...,6$ be the angles that the lines $\overrightarrow{Q_{i-1}Q_i}$ make with the x-axis. Since the curve here represents a horizontal stroke, $-45^0 \leq \alpha_i \leq 45^0$. Note that the angles α_i, like θ_i, are invariant under scaling. The size and position features are the same as in the case of curves in E. The *feature vector* of a horizontal stroke C is defined as $(\alpha_1, \alpha_2, \alpha_3, \alpha_4, \alpha_5, \alpha_6, L_1, L_2, \overline{X}, \overline{Y})$. For example, the feature vector of the bottom horizontal stroke in the image in Fig.2 is (-34.19, -29.57, -15.41, 15.41, 17.76, 44.98, 0.8284, 0.5748, 0.4048, 0.9048). To make the above features suitable as input to an MLP classifier, we normalize them as follows. Each of the 10 features of a vertical/horizontal stroke is normalized between 0 and 0.9 for reasons that will be clear later. Each angle β (θ_i or α_i) is normalized as ($\beta + 45$)/200 as $-45^0 \leq \beta \leq 135^0$. L_1 is normalized as $0.9 * L_1$ and so on.

It has been observed that the number of significant strokes present in a character image varies between 1 and 5 in both vertical and horizontal directions. Thus, a character image has at most 10 feature vectors. Now we define below a feature vector $U = (u_1, \cdots, u_{100})$ of length 100 of a character image so that the first 50 entries come from the vertical strokes and the rest from the horizontal strokes present in the image. The vertical strokes are arranged from left to right on the basis of $\overline{X}$ and the horizontal strokes are arranged from top to bottom on the basis of $\overline{Y}$. The feature vectors of the vertical strokes are concatenated in the above order to form $u_1, \cdots, u_{50}$. In case the number of vertical strokes is less than 5, the last empty entries are filled with unity. Similarly, the feature vectors of the horizontal strokes are concatenated to form $u_{51}, \cdots, u_{100}$. Thus, the empty entries, if any, contain 1 while the valid entries contain values between 0 and 0.9.

2.2 MLP Classifiers

Use of ANN in handwritten character recognition tasks has become very popular because ANN tools (say, MLP classifiers) perform efficiently when input data are often affected by noise and distortions. Also, the parallel architecture of a connectionist network model and its adaptive learning capability are added advantages. In our approach, we feed the feature vector of length 100 to suitable MLP classifiers for the final classification purpose. Below, we provide a brief account of a few necessary aspects of the use of MLP classifiers. Since the selection of a suitable learning rate has an important role towards the success of backpropagation training algorithm, we consider a modified backpropagation algorithm [12] in which the learning rates are dynamically adapted based on the gradients of the system error.

Network Architecture: During simulations we consider MLPs with one hidden layer. However, the selection of the optimal number of nodes in the hidden layer is not straightforward and involves conflicting interests [13]. We have made an exhaustive search and found that when the number of hidden nodes is approximately 50% of the input layer size, the best recognition performance is achieved on the test set.

Criterion for Termination of Training: Overtraining is a common problem of the backpropagation algorithm and deciding how much training should be given to an MLP classifier is very crucial. There are various approaches for termination of training and in the present paper, we consider a validation set [14] of samples other than the training and test samples for this purpose. This validation set has been used as an indicator for termination of training. Usually, during training, the system error on both the training and validation sets decreases monotonically till a certain amount of training is imparted to the concerned MLP classifier. However, afterwards the error on the validation set starts increasing although the same on the training set keeps reducing if further training is continued. This is the time when overtraining starts and the classifier starts losing its generalization capability. We stop further training when it is found that system error on the validation set fails to improve for three consecutive sweeps for the first instance. The connection weight values before this error starts increasing are considered final.

3   Experimental Results

We have simulated the present recognition scheme on a moderately large representative database of handwritten Bangla characters. To the best of our knowledge, there does not exist any standard database for handwritten Bangla character images and so we generated one such database. This database has been developed neither in a laboratory environment nor it is non-uniformly distributed over different classes. This database includes 25000 samples collected from different sections of the population of West Bengal, the Bangla speaking state of India. The whole set of samples of this database has been randomly grouped into training, validation and test sets consisting of 17500, 3000 and 4500 samples respectively. A few samples are shown in Fig. 2 below.

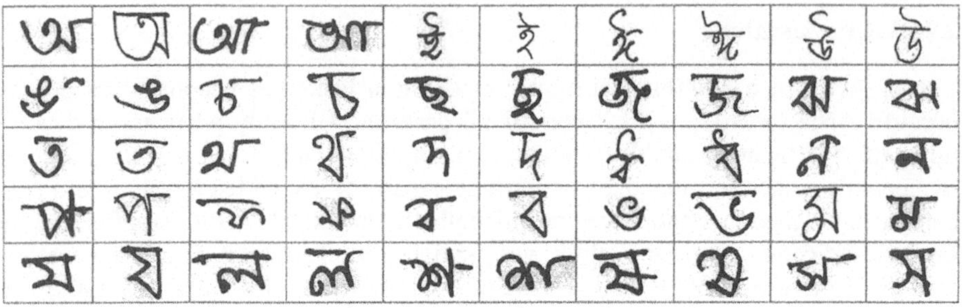

Fig. 2. Samples of Bangla handwritten characters.

Above we have considered vertical strokes computed from the East view. However, sometimes the vertical stroke from the East gets disconnected and disappears after thresholding on stroke length. In such a case the vertical stroke from the West view may be useful. Similarly, the horizontal strokes computed from the North view may also be useful. When strokes are computed from all the four directions (East, South, West, North), the length of the feature vector of the character image becomes 200. We consider both these cases and make two different simulations - one using 100 input features (extracted from east and south projections) and another using 200 input features (extracted from east, west, north and south projections). For each of these two cases, we have made several simulation runs varying the number of nodes in the hidden layer of the respective MLP classifiers and have observed that recognition accuracy on the test set of samples can be improved by considering the hidden layer size approximately 50% of that of the input layer (feature vector size). In Table 1, we present these recognition results on both the training and the test set. The validation set of samples has been used only to determine how much training should be given to the concerned MLP network so that it may have good generalization capability.

Table 1. Recognition results.

Input feature vector size: 100			Input feature vector size: 200		
Hidden Nodes	Recognition Accuracy(%)		Hidden Nodes	Recognition Accuracy(%)	
	Training set	Test set		Training set	Test set
25	79.87	75.24	50	85.82	80.16
35	83.42	79.67	70	87.42	82.91
45	85.13	80.58	90	88.78	84.11
50	85.16	80.58	100	89.11	84.27
55	85.16	80.56	110	89.43	84.33
60	86.18	79.69	130	90.38	82.93
70	86.96	77.71	150	90.71	81.40

References

1. Senior, A W.: Off-Line Cursive Handwriting Recognition using NNs. *PhD Dissertation*, University of Cambridge, England, (1994)
2. Plamondon, R., Srihari, S. N.: On-Line and Off-Line Handwriting Recognition: A Comprehensive Survey. IEEE Trans. Patt. Anal. and Mach. Intell. 22 (2000) 63-84
3. Connell, S. D., Sinha, R. M. K., Jain, A. K.: Recognition of Unconstrained On-Line Devnagari Characters. Proc. of the 15th Int. Conf. on Patt. Recog., Bercelona, Spain (2000), 368-371
4. Sethi, K.: Machine Recognition of Constrained Handprinted Devnagari. Pattern Recognition. 9 (1977) 69-75
5. Datta, A., Chaudhury, S.: Bengali Alpha-Numeric Character Recognition using Curvature Features. Pattern Recognition. 26 (1993) 1757–1770
6. Rahman, F. R., Rahman, R., Fairhurst, M. C.: Recognition of Handwritten Bengali Characters: A Novel Multistage Approach. Pattern Recognition. 35 (2002) 997-1006
7. Bhattacharya, U., Das, T. K., Datta, A., Parui, S. K., Chaudhuri, B. B.: A Hybrid Scheme for Handprinted Numeral Recognition Based on a Self-organizing Network and MLP Classifiers. International Journal for Pattern Recognition and Artificial Intelligence. 16 (2002) 845-864
8. Bhattacharya, U., Chaudhuri, B. B.: A Majority Voting Scheme for Multiresolution Recognition of Handprinted Numerals. Proc. of the 7th ICDAR, Edinburgh, Scotland (2003)
9. Pal, U., Chaudhuri, B. B.: Automatic Recognition of Unconstrained Off-Line Bangla Handwritten Numerals. Advances in Multimodal Interfaces. Springer Verlag Lecture Notes on Computer Science (LNCS-1948), Eds. T. Tan, Y. Shi and W. Gao., 2000, 371-378
10. Parui, S. K, Majumder, D. Dutta: Shape Similarity Measures for Open Curves. Pattern Recognition Letters, 3 (1983), 129-134
11. Parui, S. K., Bhowmik, T. K., Bhattacharya, U.: A Novel Scheme for Extraction of Shape Descriptions from Handwritten Bangla Characters. Proc. of 2nd WCVGIP, Gwalior (2004) 1-4
12. Bhattacharya, U., Parui, S. K.: Self-Adaptive Learning Rates in Backpropagation Algorithm Improved its Function Approximation Performance. Proceedings of the IEEE International Conference on Neural Networks, Australia, 1995 (San Diego:IEEE), 2784-2788
13. Kruschke, J. K.: Creating Local and Distributed Bottlenecks in Hidden Layers of Backpropagation Networks. Proc. of the 1988 Connectionist Models Summer School, Edited by D. Touretzky, G. Hinton, and T. Sejnowski, (San Mateo, CA: Morgan Kaufmann), 120-126
14. Bhattacharya, U., Parui, S. K.: An Improved Backpropagation Neural Network for Detection of Road-Like Features in Satellite Imagery. Int. J. Remote Sensing, 18 (1997) 3379 – 3394

Elastic Matching Algorithms
for Online Tamil Character Recognition

Niranjan Joshi[1], G. Sita[1], A.G. Ramakrishnan[1], and Sriganesh Madhvanath[2]

[1] Dept. of Electrical Engg., Indian Institute of Science, Bangalore, India
{niranjan,sita,ramkiag}@ee.iisc.ernet.in
[2] Hewlett-Packard Laboratories, Bangalore, India
srig@hplabs.com

Abstract. We present a comparison of elastic matching schemes for writer dependent on-line handwriting recognition of isolated Tamil characters. Three different features are considered namely, preprocessed x-y co-ordinates, quantized slope values, and dominant point co-ordinates. Seven schemes based on these three features are compared using an elastic distance measure. The comparison is carried out in terms of recognition accuracy, recognition speed, and number of training templates. In this paper, we present the results obtained using these schemes along with some possible grouping strategies. Error analysis is also presented in brief.

1 Introduction

On-line handwriting recognition means machine recognition of the writing as user writes on a sensitive screen. Here the writing is stored as a sequence of points in time as against an image in the case of off-line handwriting recognition. On-line handwriting recognition is a desirable attribute for hand held systems, where resources are limited and the devices are too small to have full sized keyboards. Given the complexity of entering the Indian scripts using a keyboard, hand-writing recognition has the potential to simplify and thereby revolutionize data entry for Indian languages. For a good review of on-line handwriting recognition, see [1, 2].

There are totally 156 commonly used, distinct symbols in Tamil. Of these, 12 are pure vowels, 23 are pure consonants, and the remaining are vowel-consonant combinations. The pure vowels and consonants are shown in Fig. 1. The present work is based on template based elastic matching algorithms [3–5] for writer de-pendent recognition. We present the results of our experiments on three different features and seven different recognition schemes based on dynamic time warping (DTW). The choice of these schemes and their comparison is motivated by the factors like number of training samples used, recognition accuracy, and recogni-tion speed. Rest of the paper details of the methods used and the experimental results obtained.

N.R. Pal et al. (Eds.): ICONIP 2004, LNCS 3316, pp. 820–826, 2004.
© Springer-Verlag Berlin Heidelberg 2004

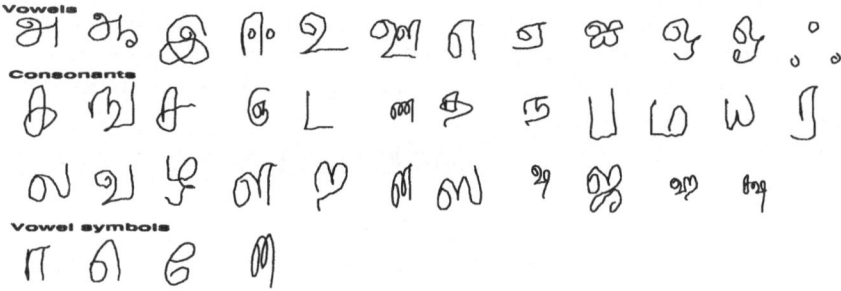

Fig. 1. Basic Tamil character set

2 Database Collection

The iPAQ pocket PC (ARM processor running on Win CE operating system) is used for data collection. Sampling rate for this device is 90 points per second. A database of 20 writers, each writing all 156 characters ten times, is collected. Each character is written in a separate box, which obviates the need for character segmentation. In this work, we are attempting writer dependent recognition. Therefore, out of the ten data samples of a particular class for a given writer, up to seven samples are used as training set to model a class and the rest are used for testing.

3 Preprocessing and Normalization

Pen-up and pen-down information is recorded as an integral part of the data-acquisition. This facilitates us to track the number of strokes and their order in each character. The initial number of pen-down points varies roughly between 50 and 200 points for different characters.

The input data, which is uniformly spaced in time, is resampled to obtain a constant number of points uniformly sampled in space. The total length of trajectory is divided by the number of intervals required after resampling. The resampled characters are represented as a sequence of points $[x_n, y_n]$, regularly spaced in arc length. Uniform re-sampling results in 60 points for all characters. Special care has been taken for re-sampling characters with multiple strokes in order to preserve the proportion of the strokes. To reduce the effect of noise, the X and Y coordinate sequences are separately smoothed, using a 5-tap low pass Gaussian filter [4]. Each stroke is smoothed independently. The characters are normalized by centering and rescaling. Features obtained from these pre-processed characters are used for building the recognizer.

4 Feature Extraction

We experimented with three different features. Preprocessed $\{(x, y)\}$ co-ordinates are one of the features used. As the preprocessed character is normalized to 60 points, the feature dimension is 120 for this case. Another feature used is quantized slopes. This is also referred to as direction primitives. In order to get these features, we first find the slope angle of the segment between two consecutive points of P as,

$$\theta_i = \tan^{-1}\left(\frac{y_{i+1} - y_i}{x_{i+1} - x_i}\right)$$

The slope angle obtained is quantized uniformly into 8 levels to make this feature invariant to noise. The dimension of this feature is 60. The third feature experimented with is the set of dominant points. Dominant points of a character P are those points where the slope values q_i change noticeably. A point p_i of P is said to be a dominant point, if the following two conditions are satisfied:

$$(q_{i+1} - q_i + 8) \% 8 \geq CT \quad \text{and} \tag{1}$$

$$(q_i - q_{i+1} + 8) \% 8 \geq CT \tag{2}$$

where, CT is *Curvature Threshold*, $2 \leq i \leq 59$, and % is *modulo* operator. Fig. 2 illustrates this concept. CT is the threshold for retaining any point as a dominant point; it can take any value from the set $\{0, \ldots, 4\}$. By default, the first and last points of P are considered dominant points. As CT increases, the structure of the character becomes coarser and vice versa. The dimension of this feature varies from one template to another, according to the inherent complexity of the character and how it has been written. This derived feature implicitly uses slope information along with the explicit use of (x, y) co-ordinate information.

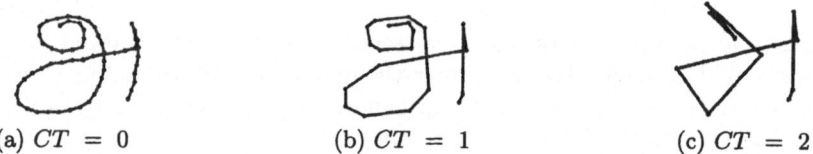

(a) $CT = 0$ (b) $CT = 1$ (c) $CT = 2$

Fig. 2. Dominant points of a character for different FI values

5   Character Recognition Schemes

This section describes seven distinct recognition schemes based on DTW. Being an elastic matching technique, DTW allows to compare two sequences of different lengths. This is especially useful to compare patterns in which rate of progression varies non-linearly, which makes similarity measures such as Euclidean distance and cross-correlation unusable. Here, time alignment is carried out using dynamic programming concepts [6]. The DTW dissimilarity measure (or *distance*) between two sequences A and B is given by,

$$D(A, B) = \min_{\mathcal{W}}\left[\sum_{k=1}^{K} w_k \Big/ K\right] \tag{3}$$

where, $\mathcal{W}$ is a set of all possible warping paths under given constraints. The minimization in Eqn. 3 is carried out using dynamic programming techniques.

5.1 Basic Schemes

The following four schemes employ single stage procedures for recognition and hence we refer to them as *basic schemes*. The choice of these schemes is motivated by their possible use in real time applications.
• **Scheme 1:**
This scheme uses preprocessed x-y co-ordinates as features. Euclidean distance is used as *cost measure* of dissimilarity between two points of feature vector. DTW distance between test pattern and templates is used in a nearest neighbor classifier. Computational complexity of this method is $O(M^2 N)$ where M is the length of the sequence and N is the total number of training templates across all the classes.
• **Scheme 2:**
This scheme uses quantized slope values as features. A fixed cost matrix, given by Table 1, is used to find out *cost measure* of dissimilarity between two quantized slopes. DTW distance measure and nearest neighbor classifier are used for classification. Even though the computational complexity is the same as that for *scheme 1*, Euclidean distance calculation is replaced by the simple table look-up operation and hence overall speed is expected to improve.

Table 1. Cost measure matrix for quantized slope values

	0	1	2	3	4	5	6	7
0	0	0.4	0.7	1	1	1	0.7	0.4
1	0.4	0	0.4	0.7	1	1	1	0.7
2	0.7	0.4	0	0.4	0.7	1	1	1
3	1	0.7	0.4	0	0.4	0.7	1	1
4	1	1	0.7	0.4	0	0.4	0.7	1
5	1	1	1	0.7	0.4	0	0.4	0.7
6	0.7	1	1	1	0.7	0.4	0	0.4
7	0.4	0.7	1	1	1	0.7	0.4	0

• **Scheme 3:**
This scheme uses dominant point x-y co-ordinates as features. CT is set to 1. The rest of the procedure is same as that for *scheme 1*. Computational complexity for this scheme is $O(\hat{M}^2 N)$ where $\hat{M} < N$. As a result, considerable improvement in the recognition speed is expected. This scheme is completely equivalent to *scheme 1* if we set CT to 0.
• **Scheme 4:**
This scheme uses preprocessed x-y co-ordinates as features and Euclidean distance as *cost measure*. The difference between this scheme and *scheme 1* is that here warping path is forced to follow diagonal path of the warping matrix. Hence this is a one to one or *rigid* matching scheme. Nearest neighbor classifier is used for classification purposes. Computational complexity of this scheme is $O(MN)$. Hence this is the fastest scheme among four basic schemes.

5.2 Hybrid Schemes

The other three schemes are combinations of the above four basic schemes. Each of the following three *hybrid schemes* accomplishes the recognition task in two stages. The first stage is a *pre-classification* stage with low computational complexity and selects top 5 choices as its output. Second stage selects the output from these 5 choices and provides *post-classification*.

• **Scheme 5:**

This scheme uses quantized slope based classifier described in *scheme 2* at the pre-classification stage and preprocessed x-y co-ordinate based classifier described in *scheme 1* at the post-classification stage.

• **Scheme 6:**

This scheme uses dominant point co-ordinates based method described in *scheme 3* at both of its stages. In the pre-classification stage, CT is set to 2 and in the post-classification stage, it is reduced to 1. We refer to this scheme as *hierarchical* dominant point based scheme, where dominant points are selected by gradually reducing CT value. At high values of CT, computational complexity is low but the structure of the character is also coarse, which suits the first stage of classification.

• **Scheme 7:**

This scheme uses rigid matching scheme based on preprocessed x-y co-ordinates (*scheme 4*) at its pre-classification stage. Post-classification uses elastic matching scheme based on preprocessed x-y co-ordinates (*scheme 1*). Higher computational complexity of DTW is due to its elastic matching capability. The idea here is therefore to gradually switch from rigid matching to elastic matching.

6 Experimental Results and Discussion

We study the performance of these schemes with respect to three criteria: average recognition accuracy, average recognition speed, and number of training templates used per class. While the first criterion evaluates effective recognition capability of a scheme under consideration, the remaining two are important for studying effectiveness of that scheme in real time applications. All the results are obtained on a machine with Intel P4 processor (1.7 GHz) and 256 MB RAM.

First we present the results for the basic schemes (*schemes 1-4*). Our main intention to study these methods is to find their suitability for hybrid schemes. In this experiment, we use 7 training templates per class. Table 2 gives the results for the basic schemes. We notice that the recognition accuracy for *scheme 1* is the highest. However, recognition speed is the lowest. This fact is attributed to the inherent computational complexity of DTW distance based methods. Low accuracy of *scheme 2* implies that quantized slope values by themselves do not work as a feature for the problem at hand, though they provide small gain in recognition speed. However, high accuracy and improved recognition speed of *scheme 3* suggest that implicit use of quantized slope values is useful for recognition. *Scheme 4* exhibits fairly low accuracy, which may be due to the fact that Tamil characters are more curved in nature and only rigid matching is not

Table 2. Recognition results for basic schemes (20 writers; 7 training and 3 test patterns /writer)

Scheme	top 1 (%)	top 2 (%)	top 3 (%)	top 4 (%)	top 5 (%)	Speed (chs/s)
1	96.30	99.22	99.50	99.64	99.69	1.69
2	88.22	96.21	97.85	98.51	98.87	3.31
3	94.89	98.51	98.93	99.09	99.15	5.83
4	90.65	96.37	97.69	98.20	98.49	67.39

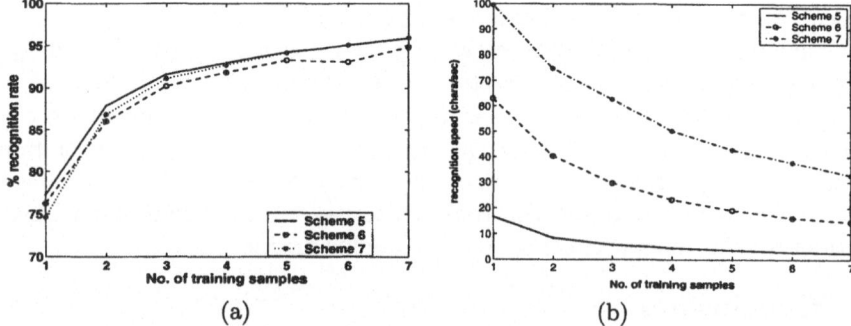

(a)   (b)

Fig. 3. (a) Recognition speed vs. number of training samples for the hybrid schemes (b) Average number of dominant points vs. character class

sufficient for their recognition. Hence *schemes 2-4* can be used as the first stage of a two stage classification strategy.

Next, we present the results for the hybrid schemes (*schemes 5-7*). Figures 3 (a) & (b) show the performance of hybrid schemes in terms of accuracy and recognition speed, respectively, against number of training templates used. We notice that *schemes 5* and *7* perform equally well in the region where training templates are more, with the maximum accuracy of 95.89%. However recognition speed for *scheme 5* is much lower than that of *scheme 7*. On the other hand, *scheme 6* shows lower recognition accuracies than the other two methods. This is possibly because the process of extraction of dominant points may result in removal of some useful information, which in turn leads to increased error rate. But it has better recognition speed than *scheme 5* and hence it is more useful for real time application purposes. Graphs in Fig. 3 show remarkable decrease in recognition accuracy below 3 training templates per class. A possible reason for this may be that most of the writers adopt two or more writing styles for some characters. This variability is completely disregarded when less number of training templates are used. From the experiments, we note that even though the preprocessed characters contain 60 points, the number of extracted dominant points from each character is much smaller. Fig. 4 illustrates few example characters with corresponding dominant points shown.

In general, it is observed that average number of dominant points a character contains is around 25. Since dominant points are the points with "high information" content, the number of dominant points present in a character can serve as a feature for a grouping strategy which will help in improving the recognition

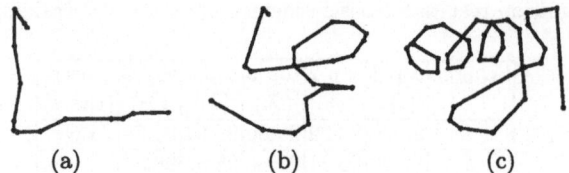

(a)   (b)   (c)

Fig. 4. Characters and corresponding number of dominant points. 14 in (a), 21 in (b) & 35 in (c)

speed. We conclude this section by listing some of the prominent confusion pairs (or triplets). They include (ன, ஈ, ன), (ஜ்ⁿ, ஜ்), (ௌ, வ), (டூ, டூ, டூ) and (ஓ, ர). These errors are observed irrespective of the scheme used. While some of them look visually similar such as, (ன, ஈ, ன) and (ௌ, வ), others occur due to elastic matching capability of DTW viz. (டூ, டூ, டூ) and (ஓ, ர). A possible solution to overcome this problem could be to extract structural features such as loops and cusps. Another possible solution could be to use these schemes in combination with other classifiers producing non-overlapping errors.

7   Conclusions

We have described implementation details of an on-line Tamil handwriting recognition system. A comparison of seven different schemes based on three different features and dynamic time warping based distance measure has been presented. Our results show that dominant points based two-stage scheme (*scheme 6*), and combination of rigid and elastic matching schemes (*scheme 7*) perform better than rest of the schemes, especially from the point of view of implementing them in a real time application. Efforts are underway to devise character grouping schemes for hierarchical classification, and classifier combination schemes so as to obtain a computationally more efficient recognition scheme with improved accuracy.

References

1. Charles C. Tappert, Ching Y. Suen, and Tory Wakahara, "The state of the art in on-line handwriting recognition," *IEEE Trans. Pattern Analysis and Machine Intelligence*, vol. 12, no. 8, Aug. 1990, pp. 787-808.
2. R. Plamondon and S. N. Srihari, "On-line and off-line handwriting recognition: A comprehensive survey," *IEEE Trans. Pattern Analysis and Machine Intelligence*, vol. 22, no. 1, Jan. 2000, pp. 63-74.
3. S. D. Connell and A. K. Jain, "Template-based on-line character recognition," *Pattern Recognition*, 34(2001), pp. 1-14.
4. X. Li and D. Y. Yeung, "On-line handwritten alphanumeric character recognition using dominant points in strokes," *Pattern Recog.*, vol. 30, no. 1, pp. 31-44, 1997.
5. S. Masaki, M. Kobayashi, O. Miyamoto, Y. Nakagawa, and T. Matsumoto, "An on-line handwriting character recognition algorithm RAV (Reparamentrized Angle Variations)," *IPSJ journal*, vol. 41, no. 9, 1997, pp. 919-925.
6. E. Keogh and M. Pazzani, "Derivative dynamic time warping," *First SIAM International Conference on Data Mining (SDM'2001)*, Chicago, USA, 2001.

Automated Classification of Industry and Occupation Codes Using Document Classification Method*

Heui Seok Lim[1] and Hyeoncheol Kim[2]

[1] Dept. of Software, Hanshin University, Korea
limhs@hs.ac.kr
http://nlp.hs.ac.kr
[2] Computer Science Education, Korea University, Korea
hkim@comedu.korea.ac.kr

Abstract. This paper describes development of the automated industry and occupation coding system for the Korean Census records. The purpose of the system is to convert natural language responses on survey questionnaires into corresponding numeric codes according to standard code book from the Census Bureau. We employ kNN(k Nearest Neighbors)-based document classification method and information retrieval techniques to index and to weight index terms. In order to solve the description inconsistency of many respondents, we use nouns and phrases acquired from past census data. Using the data, we could estimate the nouns or phrases frequently used to describe a certain code. The Experimental results show that the past census data plays an important role in increasing code classification accuracy.

1 Introduction

The Korean National Statistical Office(KNSO) collects industry and occupation information of individuals in annual census. This task has been usually done by human coders who read the description written in informal natural language by individuals and decide corresponding codes with the help of coding guidelines and standard code book. This manual coding is laborious, time-consuming, and error prone.

There has been several researches on automated coding systems since early 1980s in U.S., France, Canada, Japan[1], [3], [5], [10]. AIOCS(Automated Industry and Occupation Coding System) was developed and used for the 1990 U.S. Census [1], [5]. Since then, the AIOCS has been improved with new approaches: Eli Hellerman algorithm, a self-organizing neural network, the holograph model, nearest neighbor and fuzzy search techniques, etc. [1], [5]. Memory Based Reasoning system was one of the successful system among the efforts to improve the AIOCS [4]. The ACTR(Automated Coding by Text Retrieval) system is the

---

* This Work was Supported by the Korea Ministry of Science & Technology (M10413000008-04N1300-00811).

N.R. Pal et al. (Eds.): ICONIP 2004, LNCS 3316, pp. 827–833, 2004.
© Springer-Verlag Berlin Heidelberg 2004

generalized automated coding system developed by Statistics Canada. It is based on the Eli Hellerman Algorithm[10], similar to AIOCS and is also designed for a wide range of coding application. ACTR is however unsuitable for Census Bureau industry and occupation coding since it is designed to assign a code to a single text string and weighting scheme cannot be altered. At this time, many of the systems are at preliminary stages and the Census Bureau is conducting research on a wide range of different automated coding and computer-assisted coding systems.

In this paper, we propose an Automated Korean Industry and Occupation Coding System(AKIOCS) using information retrieval technologies: indexing the contents of code book and retrieving a set of highly ranked codes. We abstract the coding problem as a code retrieval problem that the most highly ranked codes are classified into Korean Standard Industry and Occupation codes.

2   Overview

Figure 1 shows an overview of the system. The proposed system consists of two main modules: Indexing module and Code Generation module. The indexing module extracts index terms by using morphological analyzer and noun extractor[8], calculates weight of each term using TF/IDF weighting scheme weights and makes an index database of which structure is inverted file. Code generation module forms a query from an input written in natural language and retrieves all the candidate codes. The candidate codes are ranked by similarity in descending order and k most highly ranked codes are assigned as a classification results. The details of each module are explained in the following sections.

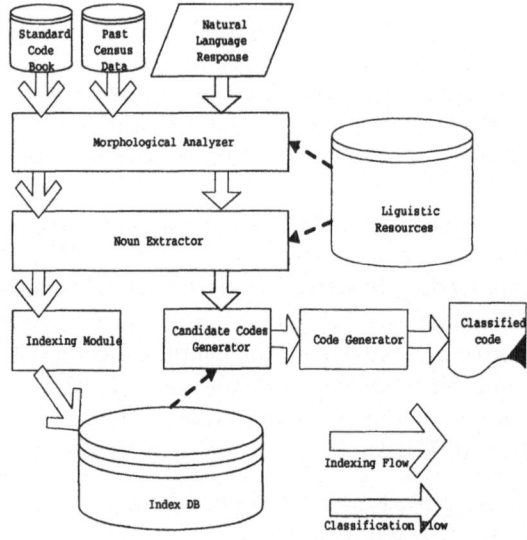

Fig. 1. The overall architecture of the proposed system

3   Augmenting Code Description

The standard code book provided by the KNSO describes each classification code with four fields: code name, short description of the code, several examples and exceptional cases. The classification codes are organized hierarchically in five levels as shown in table 1. There are 1,837 codes in industry code book and 2,070 codes in occupation code book.

Table 1. Industry(Occupation) classification code structure

Code	Sections (level1)	Divisions (level2)	Groups (level3)	Classes (level4)	Sub-Classes (level5)
Digits	1	2	3	4	5
	alphabetic	numeric	numeric	numeric	numeric
Example	A	01	012	0121	01211
# of codes	17(11)	63(46)	194(162)	442(447)	1121(1404)

The code description for each code provides very limited information of only 40 to 50 word size. Another problem is that many individual responses usually contain ambiguous, incorrect and ill-formed data. Variability of the terms and expressions between many respondents is also very serious: same occupation can be described in so many ways with so many different terms while the standard code description book contains very small number of fixed terms. Human experts obviously use his/her own knowledge and experience to classify the descriptions into codes, in addition to the help of coding guidelines provided. To alleviate the problem of limited information available in the standard code book, it is desirable to use Korean thesaurus which is not available for the present. Instead, we employ a method to augment code description by adding nouns and phrases occurred in past census data. Using the data, we could augment each code description with nouns or phrases which are frequently used to describe the code by many respondents.

Even though we augmented the code description by adding the past census data, the system still suffers from limited information problem. That is, the terms are not consistent with the ones from respondents. To solve the problem, we add predefined index terms to each code . The predefined index terms are nouns expected to be more probable to be used in describing the code by the respondents. KNSO provided us with 23,431 records for industry and 17,184 records for occupation classification.

4   Indexing

In our system, we take each code as a document in document classification. The system indexes the standard code book and predefined index term list extracted from the data provided by KNSO[6], [7].

The indexing module extracts index terms by using morphological analyzer and noun extractor[8] and calculating weight for each indexing term. Then, it makes an inverted file including code ID, starting position of the posting file, and the number of the posting files. Code ID is a classification code in a code book. The posting file includes the code ID in which a term is occurred, weight for the term. The starting position of the posting file is a starting address in the posting file for a term.

Weight for a term indicates how the term is discriminative and it is calculated by the following modified TF/IDF weighting scheme[9], [2].

$$w_{ij} = \frac{f'_{ij}}{max_l f_{lj}} \times log \frac{N}{n_i} \tag{1}$$

where w_{ij} is the weight of the ith term in the jth code, N is the total number of codes in a code book, and n_i is the number of code descriptions in which the term appeared. Modified frequency, f'_{ij} is equal to $5^{1} \times f_{ij}$ if term i occurred in past census data, otherwise, equal to the conventional term frequency, f_{ij}. Through indexing and weighting terms, each code is represented as vector in the n-dimensional term space. The dimension of the code vector is the total number of the valid terms. Value of each component of the vector is the weight of the corresponding term. Later in code generation step, input record is also represented as a vector and distance between the vectors is calculated as a measure of similarity.

5   Code Generation

The individual response is converted into query which is set of nouns of the company name, business type and job description. The code generator retrieves corresponding codes by *vector space model*. The vector space model uses vector representation for each code description and user query, and retrieves codes by calculating cosine similarity between the code vectors and query vector. Cosine similarity between j^{th} code, $\overrightarrow{C_j} = (w_{1j}, w_{2j}, \ldots, w_{tj})$ and query, $\overrightarrow{Q} = (w_{1q}, w_{2q}, \ldots, w_{tq})$ is defined as follows:

$$sim(C_j, Q) = \frac{\overrightarrow{D_j} \cdot \overrightarrow{Q}}{|\overrightarrow{C_j}| \times |\overrightarrow{Q}|} = \frac{\sqrt{\sum_{i=1}^{t} w_{ij} \times w_{iq}}}{\sqrt{\sum_{i=1}^{t} w_{ij}^2} \times \sqrt{\sum_{i=1}^{t} w_{iq}^2}} \tag{2}$$

In the equation (2) , $\sqrt{\sum_{i=1}^{t} w_{ij}^2}$ is calculated in indexing time for each code and $\sqrt{\sum_{i=1}^{t} w_{iq}^2}$ is not calculated because it is same for all the codes.

We use modified kNN algorithm as target function of code generator. The target function is defined as in 3.

$$\hat{f}(x_q, l)_p \leftarrow argmax_{c \in C_l} \sum_{i=1}^{k} \zeta(c, candidate_i) \tag{3}$$

$$where\ \zeta(a, b) = 1\ if\ a_l = b_l,\ 0\ otherwise$$

---

[1] The constant 5 is acquired heuristically through many experiments.

Table 2. Production rate of Industry Code in p mode(%)

	p	level2	level3	level4	level5
before augmenting	1	76.03	70.16	60.23	57.08
	2	81.04	80.91	75.02	67.23
	3	83.52	81.10	77.02	72.98
	10	84.55	83.02	80.70	78.38
after augmenting	1	87.08	82.46	73.12	66.08
	2	95.16	90.91	85.14	77.01
	3	97.22	94.16	90.03	82.65
	10	99.10	98.22	95.80	92.88

Table 3. Production rate of Job Code in p mode(%)

	p	level2	level3	level4	level5
before augmenting	1	66.03	60.25	55.83	47.30
	2	71.04	65.72	58.45	57.23
	3	73.52	70.84	62.71	61.72
	10	74.55	88.56	78.80	73.27
after augmenting	1	71.08	65.36	73.12	66.08
	2	84.67	78.56	85.14	77.01
	3	89.14	83.65	90.03	82.65
	10	95.20	92.86	95.80	92.88

The target function takes two parameters, x_q and l. In equation 3, a_l represents a code which consists of l substrings from the first digits to lth digits of the code a. The target function, $\hat{f}$ generates p best codes of which length is l. $p(\geq 1)$ is parameter which indicates the mode of code generator. If p is 1, the system works fully automatic mode which makes one best code. If $p \geq 1$, the system makes p best probable candidates.

6   Experimental Results

We used 10-fold cross-validation for performance evaluation. The evaluation is done with a production rate as in equation 4. We assume a code is correctly assigned in p mode if the system makes p best candidate codes which include correct code for the input case.

$$PR_p = \frac{\sharp\ of\ correctly\ assigned\ cases}{\sharp\ of\ input\ cases} \times 100 \tag{4}$$

Table 2 and 3 shows production rates of generating industry code and job code respectively. The first column represents indexed code sets whether they are augmented by adding predefined index terms and past census data. The production rate of industry code is rather higher than that of job code for most cases. This is because the average number of words included in input for industry code is more larger than that for job code. This means the input for industry code has many distinguishable index terms and information between different codes in both industry code and job code classification. It is very promising that the production rate with augmented code set are much improved. This means that the augmenting method of code sets with predefined index terms and past census data are very effective to increase production rate. The improvement came from using the past census data of 46,762 instances for industry and 36,286 for occupation code.

The production rate of fully automatic mode is rather low in every level for both industry and job code classification. But the production rate is rapidly increased as the p value is increased. The production rates are above 92% in both code classification tasks. This illustrates that the proposed system is premature as a fully automated coding system but it may be used as a semi-automatic code classification which minimizes search space of code set for human coder.

7 Conclusions

This paper describes development of the automated industry and occupation coding system for the Korean Census using information retrieval technique and kNN learning. We proposed a method to use nouns and phrases acquired from the past census data to solve discrepancy between the terms of the standard code book and the terms from the individual responses.

We performed the experiment to see how code level and candidate set size change system performance, and how much the past census data improves system performance. The experimental results show that using the past census data is very successful in increasing production rate. For example, the production rate for industry level4 code increases from 60.23% to 73.12% and from 80.70% to 95.80% in both $p = 1$ and $p = 10$ modes respectively.

The average production rate as a fully automated coding system is not satisfactory yet. This unsatisfactory result resulted from variability of the terms and expressions used to describe same code and spacing errors in input data. We proposed a solution to alleviate the variability of the terms by using past census data and predefined index terms and showed the effectiveness of the method. One of possible solutions of spacing errors in input is to use n-gram index terms by using n-gram stemming algorithm[2]. We will try to improve the proposed system by using n-gram index technique in near future.

Even though the system performance is not good as a fully automatic system, it can be successfully used as a computer-assisted clerical system that generates several candidate codes in a fixed size of candidates set. The computer-assisted clerical system is used for residual coding of cases which a fully automated system could not decide.

References

1. Apeel, M. V. and Hellerman, E.: Census Bureau Experiments with Automated Industry and Occupation Coding. Proceedings of the American Statistical Association (1983) 32–40

2. Baeza-Yates and Ribeiro-Neto.: Modern Information Retrieval. Addison-Wesley, 1999.
3. Chen, B., Creecy, R. H., and Appel, M.: On Error Control of Automated Industry and Occupation Coding. Journal of Official Statistics, Vol. 9, No. 4. Korean Statistical Office (1993) 729–745
4. Creecy, R. H., Masand, B. M., Smith, S.J., and Walts, D. L.: Trading MIPS and Memory for Knowledge Engineering. Communications of the ACM, Vol. 35, No. 8, (1992) 48–64
5. Gilman, D. W. and Appel, M. V.: Automated Coding Research At the Census Bureau. U.S. Census Bureau, http://www.census.gov/srd/papers/pdf/rr94-4.pdf
6. Korean Standard Industrial Classification. Korean Statistical Office, (2000)
7. Korean Standard Classification of Occupations. Korean Statistical Office, (2000)
8. Lee, D.G.: A High Speed Index Term Extracting System Considering the Morphological Configuration of Noun. M.S. Thesis, Dept. of Computer Science and Engineering, Korea Univ., Korea, (2000.)
9. Salton, G. and McGill, M.J.: Introduction to Modern Information Retrieval. McGraw-Hill, New York, (1983)
10. Rowe, E. and Wong, C.: An Introduction to the ACRT Coding System. Bureau of the Census Statistical Research Report Series No. RR94/02 (1994)

Abnormality Detection in Endoscopic Images Using Color Segmentation and Curvature Computation

P.S. Hiremath, B.V. Dhandra, Ravindra Hegadi, and G.G. Rajput

Department of P.G. Studies and Research in Computer Science,
Gulbarga University, Gulbarga-585106 Karnataka, India
hiremathps@hotmail.com, ravindrahegadi@rediffmail.com

Abstract. In this paper, a method for detecting possible presence of abnormality in the endoscopic images of lower esophagus is presented. The pre-processed endoscopic color images are segmented using color segmentation based on 3σ-intervals around mean RGB values. The zero-crossing method of edge detection is applied on the gray scale image corresponding to the segmented image. For the large contours, the Gaussian smoothing is performed for eliminating the noise in the curve. The curvature for each point of the curve is computed considering the support region of each point. The possible presence of abnormality is identified, when curvature of the contour segment between two zero crossing has the opposite curvature signs to those of such neighboring contour segments on the same edge contours. The experimental results show successful abnormality detection in the test images using this proposed method.

1 Introduction

The technique of endoscopy has expanded the understanding of numerous gastrointestinal diseases since from its wide spread use in the late 1960s. The careful inspection of mucosal surface has greatly improved the ability to care the affected patients. As the video endoscope containing the intensity light source, suction equipment, guided camera, etc, passes under direct vision, through the esophagus and the stomach into a portion of duodenum, it transmits the video clipping of tissues for the display, the storage and the analysis [1]. Endoscopy of lower gastrointestinal system provides real time image information about colorectal mucosa and is being used increasingly to identify abnormalities and disorders of the colon [2]. Colonic polypoid lesions are the most common pathology found during endoscopy. Detecting and promoting the deal with the polyps is important, because the presence of an adenoma is associated with synchronous and meta synchronous polyps and cancer [3]. The abnormality of polyps and tumors are mainly detected when the surface of the lipoma is eroded or irregular in contrast to a smooth surface. Normally, the creases of colon haustra, which are seen as contours in the endoscopic image, are smooth and are of arc shapes. However, the presence of polyps or tumors will lead to the shape of these contours being seen as distorted. Such distorted shape is reflected by the change of curvature sign along a normal smooth contour of the same curvature sign. Thus the possible presence of abnormality can be detected, if the contour's curvature is analyzed. This approach is used by Krishnan et.al.[9] for the intestinal abnormality detection from Endoscopic images based on the Canny's method for edge detection followed by curvature analysis. In the present paper, a method to detect the possible presence of abnormality us-

N.R. Pal et al. (Eds.): ICONIP 2004, LNCS 3316, pp. 834–841, 2004.
© Springer-Verlag Berlin Heidelberg 2004

ing color segmentation of the image based on 3σ-interval method [5] for edge detection followed by curvature analysis for the image contours is presented. The experimental results show that the proposed method is successful in abnormality detection at a lesser computational cost as compared to the method in [9].

2 Methods

The edge contours corresponding to the image contours of haustra creases are extracted using RGB color segmentation and edge detection by zero crossings method. The contours are then smoothed along its length to make them suitable for curvature computation, since curvature is represented by second order differential making it very noise sensitive. Abnormality in the image is detected through curvature analysis along the contours.

2.1 RGB Color Segmentation

The endoscopic images of lower esophagus are smoothed using average filter to reduce the effect of small bright spots created by light reflection and small lumen regions created in the image. The RGB color image contains 3 times more data than a gray scale image. However, the three maps should not be processed independently because it appears that a strong spatial and chromatic correlation exists. The average filter is considered for smoothing multiple image [R(x,y), G(x,y), B(x,y)] componentwise. For the regions of interest in the test images, the mean and standard deviation of RGB values are determined using the statistical sampling technique [5]. This information is stored as the knowledge base for automatic segmentation of any given input image. Table 1 shows the estimated mean and standard deviation of RGB values for the region of interest.

Table 1. Mean and standard deviation of RGB values for the region of interest.

	Abnormal	Region	Values
Variables	R	G	B
Mean	120.93	56.41	53.65
Standard Deviation	17.39	14.45	16.03

The segmentation is carried out using 3σ-intervals around mean RGB values stored in the knowledge base. If there is any region in the image containing the bright spots due to the reflection of light sources, the pixel lying in that region will have RGB values nearer to 256. The pixels corresponding to the lumen regions will have RGB values nearer to zero. Such regions are not considered as the region of interest during color classification based on interval of RGB values. Hence, the region containing bright spots and lumen regions are avoided during further processing steps. If the bright spot due to reflection is in the centre of ROI, it will give rise to contours after the edge detection of the segmented region, which needs to be processed. This will increase the processing time unnecessarily, which can be avoided by eliminating the bright spots by replacing the corresponding pixels with the mean RGB values. This process eliminates the bright spots and thus minimizing the misdetection of the abnormal region as the normal region.

2.2 Edge Detection

After the color segmentation the extracted color image is converted into gray scale image. The edge detection scheme is applied on this image based on zero crossings method. An edge detection technique based on the zero crossings of second derivatives [10] explores the fact that step edge corresponds to an abrupt change in the image function. So, we can notice that for an image function $f(x)$

$$\frac{d^2 f(x)}{dx^2} = 0 \qquad (1)$$

is the condition for the edge detection by zero crossing. The output of this process is a series of edge contour.

2.3 Contour Curvature Computation

After edge detection using zero crossing technique, large contours are used for curvature analysis. Due to the discrete boundary representation and quantization errors, false local concavities and convexities along a contour are formed. This noisy nature of binary contours must be taken into account to obtain reliable estimates of contour curvature. Hence, a Gaussian filter is used to smooth the contour points to reduce the noise effect [6]. However, the width of Gaussian filter, w, that controls the degree of smoothing has to be chosen carefully. A large value of w will remove all small details of the contour curvature, while a small value will permit false concavities and convexities to remain making it to difficult to choose an appropriate w. To solve this problem a support region is employed which will dynamically determine the Gaussian parameter.

2.4 Determination of Support Region

The support region concept can be explained using the figure 1. The support region for each point on the curve is the number of points obtained from the implementation of the algorithm 1.

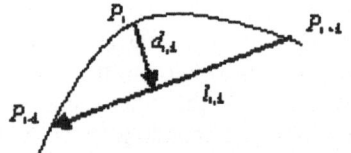

Fig. 1. Representation of Support region.

Algorithm 1
(i) Determine the length of the chord joining the points $P_{i,k-1}, P_{i,k+1}$ as

$$l_{i,k} = \left| P_{i,k-1} P_{i,k+1} \right| \qquad (2)$$

(ii) Let $d_{i,k}$ be the perpendicular distance from P_i to the line joining $\overline{P_{i,k-1}P_{i,k+1}}$

Start with k=1, compute $l_{i,k}$ and $d_{i,k}$ until one of the following conditions hold

(a) $l_{i,k} \geq l_{i,k+1}$ (3)

(b) $\dfrac{d_{i,k}}{l_{i,k}} \geq \dfrac{d_{i,k+1}}{l_{i,k+1}}$ for $d_{i,k} \geq 0$

Now, the region of support of P_i is the set of points, which satisfy either condition (a) or condition (b), that is,

D(P_i) = { $P_{i-k,}$ ...., P_{i-1}, $P_{i,}$ P_{i+1}, ..., P_{I+k} |Condition (a) or condition (b)}   (4)

2.5 Gaussian Smoothing

A planar curve can be defined in parametric form as $(x(t),\ y(t)) \in R^2$, where t is the path length along the curve. Smoothing is done by convolving $x(t)$ and $y(t)$ with the Gaussian filter. A one dimensional Gaussian filter is defined as

$$\eta(t,w) = \frac{1}{\sqrt{2\pi w^2}} e^{\left(-t^2/2w^2\right)}$$ (5)

where, w is the width of the filter, which needs to be determined. The smoothed curve is denoted by set of points $(X(t,w),\ Y(t,w))$, where,

$$x(t,w) = x(t) \otimes \eta(t,w)\ ,\quad x(t,w) = y(t) \otimes \eta(t,w)$$ (6)

where $\otimes$ denotes the convolution.

The measurement of the curvature of the point should be based on the local properties within its region of support, and the length of Gaussian smooth filter can be proportional to the region of support [7]. This implies that the neighboring points closer to the point of interest should have higher weights than those points further away. This method is less sensitive to noise. The Gaussian filter applied here will have the following window length and width [9]:

Window Len =2xSupp. RegionD(P_i)+1, Width w =Support Region D(P_i) / 3   (7)

After finding the support region, the curvature for each point on the curve is calculated in the following way.

2.6 Curvature Computation

For the continuous curve C, expressed by $\{x(s),\ y(s)\}$, where s is the arc length of the edge point, the curvature can be expressed as:

$$k(s) = \frac{\dot{x}\ddot{y} - \ddot{x}\dot{y}}{\left(\dot{x}^2 + \dot{y}^2\right)^{3/2}}$$ (8)

where, $\dot{x} = dx/ds$, $\ddot{x} = d^2x/ds^2$, $\dot{y} = dy/ds$, $\ddot{y} = d^2y/ds^2$.

For digital implementation, the coordinate functions $x(s)$ and $y(s)$ of the curvature are represented by a set of equally spaced Cartesian grid samples. The derivatives in the equation (8) are calculated by finite differences as:

$$\dot{x}_i = x_i - x_{i-1}, \quad \dot{y}_i = y_i - y_{i-1}, \quad \ddot{x}_i = x_{i-1} - 2x_i + x_{i+1}, \quad \ddot{y}_i = y_{i-1} - 2y_i + y_{i+1} \qquad (9)$$

The modified algorithm for curvature computation is presented in the following algorithm 2.

Algorithm 2

(i) Find region of interest using color segmentation based on 3σ-intervals around mean RGB values.

(ii) Convert the region of interest into gray scale image and find edge contours by edge detection using zero crossings method.

(iii) Determination of support region for each contour point.

(iv) Smooth the contour by a Gaussian filter with the width proportional to the support region.

(v) Compute curvature for each point on the Gaussian smoothed curve using equation (8) and (9).

While performing curvature computation we come across two special conditions for which the alternate solutions need to be given. They are:

(i) When the edge point is on a straight line, the curvature for that point is assigned to zero.

(ii) When the support region for an edge point is 1, this point will not be smoothed. So, the smoothing on this point is performed using the following equation:

$$(\hat{x}_i, \hat{y}_i) = \frac{1}{2}(x_i, y_i) + \frac{1}{4}[(x_{i-1}, y_{i-1}) + (x_{i+1}, y_{i+1})] \qquad (10)$$

where, $(\hat{x}_i, \hat{y}_i)$ is the smoothed point of (x_i, y_i).

2.7 Abnormality Detection in Endoscopic Image

The detection of possible presence of abnormality is performed by analyzing the curvature change along each edge contour. The curvature of each edge point on the edge contour is computed using the algorithm 2. Two thresholds, c_{th} and n_{th} are used in the analysis. c_{th} is the curvature value threshold, and n_{th} is number of edge points in a segment. Along the edge contour, if the absolute curvature of the point is bigger than c_{th}, the point counting starts until the absolute curvature value of the point is less than c_{th}. If the point count is bigger than n_{th}, an edge segment is formed. The possible presence of abnormality is detected when the curvature of a segment has opposite sign to those of such neighboring segments on the same edge contour. Also such a segment is bounded by two significant zero crossings.

3 Results and Discussion

The experiment is carried out with the six test images obtained from the medical expert. The images are collected from Olympus Endoscopic equipment. The Matlab 5.2 software is used for implementation of the algorithm. The figure 2(a) shows an abnormal endoscopic image. This image is smoothed using the average filter and segmented using 3σ-interval around mean RGB values stored in knowledge base. The

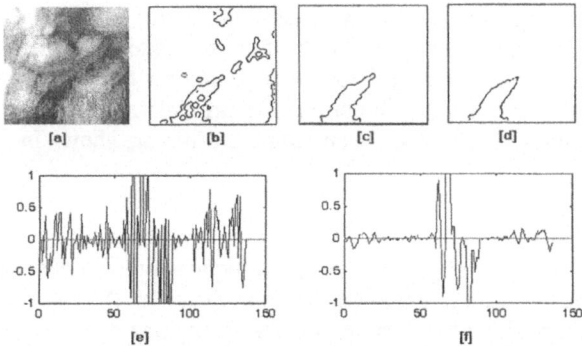

Fig. 2. Results for an abnormal image [a] Color image showing abnormality, [b] Edge detected by zero crossing method, [c] A large edge contour corresponding to the image, [d] Edge after smoothing, [e] Curvature profile of the edge, [f] Smoothed curvature profile of the edge.

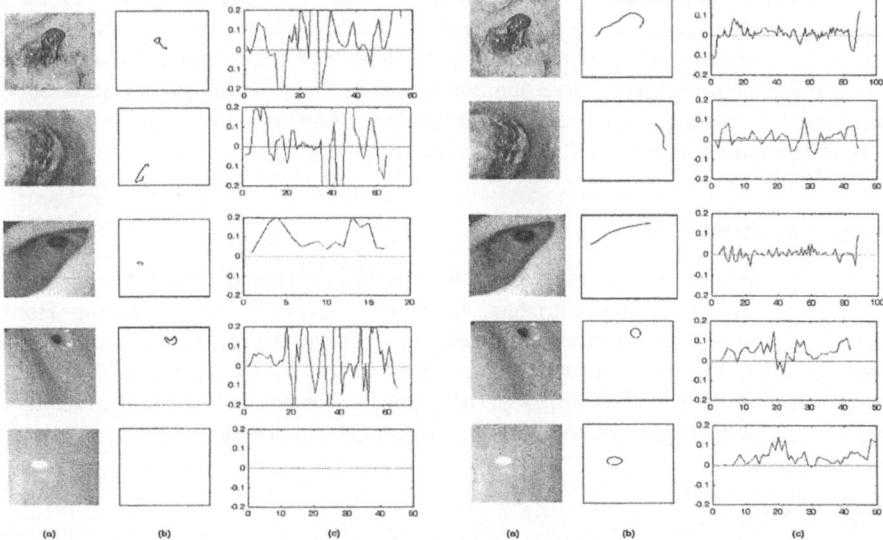

Fig. 3. Curvature profile of the images based on proposed method (a) Original Color image, (b) Large smoothed edge and (c) Smoothed curvature for the corresponding edge.

Fig. 4. Curvature profile of the images based on the method in [2]. (a) Original Color image, (b) Large smoothed edge and (c) Smoothed curvature for the corresponding edge.

edges corresponding to the segmented image are found using the zero crossings method as shown in figure 2(b). The figure 2(c) shows a large edge contour for the image. This edge contour is smoothed using Gaussian filter as shown in figure 2(d). The width of the filter depends on the support region for each point on the edge contour. The curvature profile along an edge contour is given in figure 2(e). It can be observed that the curvature has many rugged zero crossings. In order to analyze the signals effectively and automatically by the computer, the curvature is smoothed by a one-dimensional Gaussian filter, with standard deviation $\sigma=4$. The smoothed curve profile is shown in figure 2(f). Figure 3 shows the curvature profile obtained for 5 sample images. Fig. 3(a) shows the endoscopic color image. Fig. 3(b) shows the smoothed large edge contour obtained after the edge detection by the proposed method. Fig. 3(c) is the smoothed curvature profile for the image using the proposed method. For the same images, the curvatures are determined using the method proposed by Krishnan et.al.[9], and the curvature profiles are shown in the figure 4.

4 Conclusion

Endoscopic images are rich in color. The technique adopted in this paper for segmenting the abnormal portion is based on the color features. The miss detection of bright spots in the image due to reflection and lumen regions as the regions of interest is avoided thus reducing the time required for further image processing. However, this is not avoided in the method adopted by Krishnan et. al. [2], since, it is based on edge detection of intensity image. Even though the present method requires more time for pre-processing the color image [RGB components], but generates large continuous edge contours, as compared to the method proposed by Krishnan et. al. [2]. Large continuous edge contours are more advantageous than shorter ones in so far as the curvature computation and determination of number of zero-crossings is concerned. The present method located more number of zero-crossings in the edge contour curvatures and hence is a good indicator of the presence of abnormalities in the color image than the method proposed by Krishnan et. al. [2].

Acknowledgement

Authors are grateful to the referees for their useful comments. The authors are also grateful to Dr. M. K. Ramakrishna, MS., Sri Lakshminarayana Nursing Home, Raichur, for providing endoscopic images and rendering manual segmentation for the present study. The authors are also indebted to Dr. P. Nagabushan, Dr. G. Hemanthkumar and Dr. D.S. Guru, Dept. of Studies in Computer Science, University of Mysore, for their helpful discussions and encouragement during this work.

References

1. Guillou C.Le. et.al.: Knowledge Representation and case Indexing in Upper Digestive Endoscopy, Proceedings of 22nd Annual EMBS International Conference, Chicago IL (2000)
2. Krishnan S.M., et.al.: A neural Network based approach for Classification of Colon Abnormality, Intl. Symposium on Intelligent Robotic Systems (1998)
3. Silverstein F., Tytget G.: Atlas of Gastrointestinal Endoscopy, Gower Med. Pub.
4. Refael Gonzalez, Richard E. Woods: Digital Image Processing, Pearson Edition Asia, 2nd Edition (2002).
5. Hiremath P.S., Dhandra B.V., Iranna Humnabad, Ravindra Hegadi, Rajput G.G.: Detection of esophageal Cancer (Necrosis) in the Endoscopic images using color image segmentation, Proceedings of second National Conference on Document Analysis and Recognition (NCDAR-2003), Mandya, India (2003)
6. Torre V., Poggio T.A.: On Edge Detection, IEEE Transactions on Pattern Analysis and Machine Intelligence, Vol. 8 (1986) 147-163
7. Ansari N., Huang K.W.: Non-parametric Dominant Point Detection, Pattern Recognition, Vol. 24 (1991) 849-862

8. Tech C., Chin R.T.: On the Detection of Dominant points on Digital Curves, IEEE Trans. on Pattern Analysis and Machine Int., Vol. 11, 859-872, (1989).
9. Krishnan S.M., et.al.: Intestinal Abnormality Detection from Endoscopic Images, The 20th Annual International Conference of IEEE EMBS 98, Hongkong (1998).
10. Marr D., Hildreth E.: Theory of Edge detection, in kasturi R. and Jain R.C. Editors, Computer Vision, IEEE Los Alamitos, CA (1991) 77-107
11. Milan Sonka, Vaclav, Hlavac, Roger Boyle: Image Processing and Machine Vision, PWS Publishing Company (2001)

Fault Diagnosis for Industrial Images Using a Min-Max Modular Neural Network

Bin Huang and Bao-Liang Lu

Department of Computer Science and Engineering, Shanghai Jiao Tong University
1954 Hua Shan Rd., Shanghai, P.R. China 200030
uu.hb0@public1.sz.js.cn, blu@cs.sjtu.edu.cn

Abstract. This paper presents a new fault diagnosis method for industrial images based on a Min-Max Modular (M^3) neural network and a Gaussian Zero-Crossing (GZC) function. The most important advantage of the proposed method over existing approaches such as radial-basis function network and support vector machines is that our classifier has locally tuned response characteristics and the misclassification rate of faulty product images can be controlled as small as needed by turning two parameters of the GZC function while the correct rate can be influenced to some extend. The experimental results on a real-world fault diagnosis problem of industrial images indicate that the effectiveness of the proposed method.

1 Introduction

The aim of developing a fault diagnosis system for industrial images described in this paper is to exactly classify all the CCD sensing images acquired from a product line into two categories: qualified product and faulty product. A strict requirement for the classifier used in fault diagnosis systems is that the classifier is not allowed to mistakenly classify images of faulty products into the category of qualified products. Although various image classification methods have been developed in the last few years [3], many of these methods, however, are not suitable to fault diagnosis for industrial images because they lack locally tuned response characteristics and the misclassification rate on faulty product images can not be controlled by a user.

In our previous work we have proposed a min-max modular (M^3) neural network [1] for dealing with large-scale pattern classification problem. As a kind of divide-and-conquer technique, the main idea of M^3 network is to divide a complex multi-class problem into a series of smaller and simpler two-class subproblems and combine the individual solutions of all the subproblems into a solution to the original problem. An attractive feature of the M^3 networks is that any multi-class problem can be decomposed into a number of linearly separable subproblems.

For linearly separable problems, we have proposed a Gaussian Zero-Crossing (GZC) function. The combination of the M^3 network and the GZC function has two important advantages over existing classifiers. One is that fast learning

N.R. Pal et al. (Eds.): ICONIP 2004, LNCS 3316, pp. 842–847, 2004.
© Springer-Verlag Berlin Heidelberg 2004

without error can be achieved to any complex pattern classification problems; the other is that the locally tuned response features of the GZC function are maintained in the entire M^3 network. In this paper, we apply the M^3 network with the GZC function to a fault diagnosis problem of industrial images.

2   Feature Extraction with Wavelet Transform

2.1   Image Processing

A real-world image set used in this paper consists of CCD sensing images acquired from a product line at a leading manufacturing company. In this paper, all the images are in the form of 256 gray-scales bitmap and are classified into two classes: *correct image* and *faulty image*. The size of the images is 128*128 in pixels. Two kinds of enlarged images are illustrated in Fig. 1, in which (a), (b) belong to *correct image* and (c) is classified into *faulty image*.

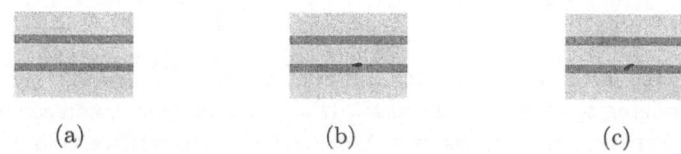

(a)   (b)   (c)

Fig. 1. Industrial CCD sensing images for qualified products (a) and (b), and defective product (c)

2.2   Feature Extraction Using Wavelet Transform

We used wavelet transform techniques to extract the features of images. The image data were convolved by using the Daubechies wavelets [6], which is orthogonal and with compact support. In the experiments below, the order N was set to 2 with the encoding level 1. After wavelet transforming, the dimension of the feature vector representing an image is 4225.

3 M^3 Network and GZC Function

3.1 M^3 Network

The proposed method is based on a min-max modular neural network model [1]. With the task decomposition method, a K-class problem can be uniquely partitioned into a number of linearly separable problems, each of which includes only two different training data that belong to two classes. The number of linearly separable problems is given by

$$\sum_{i=1}^{K-1} \sum_{j=i+1}^{K} L_i \cdot L_j \tag{1}$$

where L_i denotes the number of training data belonging to class C_i.

After training individual component modules assigned to learn associated two-class sub-problems, all the trained component modules are integrated into a min-max modular neural network with the MAX, MIN, or/and INV units according to two module combination laws [1].

3.2   Gaussian Zero-Crossing Discriminant Function

A fatal weakness of linear discriminant functions is that they lack locally tuned response characteristics [2]. This deficiency may lead classifiers to mistakenly produce proper output even when an unknown input is presented.

To overcome the deficiency of linear discriminant functions, we have proposed a Gaussian Zero-Crossing (GZC) function for solving linearly separable problems in our previous work [2]. The definition of the Gaussian zero-crossing discriminant function is given by:

$$f_{ij}(x) = \exp\left[-\left(\frac{\|x - c_i\|}{\sigma}\right)^2\right] - \exp\left[-\left(\frac{\|x - c_j\|}{\sigma}\right)^2\right] \tag{2}$$

where $x \in \mathbf{R}^n$ is the input vector, $c_i \in \mathbf{R}^n$ and $c_j \in \mathbf{R}^n$ are the given training inputs belonging to class C_i and class C_j $(i \neq j)$, respectively, and are used as two different receptive field centers, $\sigma = \lambda \|c_j - c_i\|$ is the receptive field width, λ is a user-defined constant $(0 < \lambda)$, and the norm $\|z\|$ is the Euclidean norm of vector z [2]. An important advantage of the GZC discriminant function over existing linear discriminant functions is its locally tuned response characteristics.

The output for the M^3 network with GZC discriminant function is defined by

$$g(x) = \begin{cases} 1 \text{ if } y_i(x) > \theta^+ \\ \text{Unknown if } \theta^- \leq y_i(x) \leq \theta^+ \\ -1 \text{ if } y_i(x) < \theta^- \end{cases} \tag{3}$$

where θ^+ and θ^- are the upper and lower threshold limits of the M^3 network, respectively; and $y_i(x)$ denotes the transfer function of the M^3 network for class C_i.

One attractive feature of the GZC discriminant function is that the proper generalization performance of the M^3 network can be easily controlled by tuning θ^+ and θ^-. From (2), we see that the receptive field width of the GZC function is determined by two factors: the constant λ defined by a user and the distance between two different receptive field centers.

4   Experiment Results

To evaluate the effectiveness of the proposed method and compare it with support vector machines, we carry out simulations on a real-world fault diagnosis problem of industrial images. The training data set consists of 500 images and the test data set contains 83 images. Table 1 shows the distributions of the training and test data.

Table 1. Distributions of The Training and Test Data

	No. Data	
Class	Training	Test
Images for qualified products	400	100
Images for faulty products	6	23

Table 2. Performance Comparison of M³–GZC with SVMs

Methods	SVs	Success Rate(%)		Error No. in Testing	
		Training	Test	Correct→Fault[a]	Fault→Correct[b]
M³–GZC	N/A	100.00	96.75	0	4
SVMs(σ=2.0)	137	100.00	96.75	0	4
SVMs(σ=1.4)	206	100.00	96.75	0	4
SVMs(σ=1.0)	328	100.00	92.69	0	9
SVMs(σ=0.7)	438	100.00	83.74	0	20
SVMs(σ=0.5)	460	100.00	81.30	0	23

[a] The number that the Correct are misclassified to the Fault.
[b] The number that the Fault are misclassified to the Correct.

4.1 Experiment 1

In this experiment, we compare our method with standard SVMs. Both θ^+ and θ^- were set to 0.01 for the M³ network with the GZC function and C for SVMs [4] was selected as 4, 8, 16, 32. The experimental results are shown in Table 2.

From Table 2, we see that our method has the same recognition rate on the whole test data as the standard SVMs. The results demonstrate that the M³ neural network with the GZC function has a well-performed structure for classification problems. When the radius (σ) of the kernel function[1] in SVMs is narrowed, however, the corresponding recognition rate become worse, especially, more number of images for defective products are mistakenly classified as the category of qualified products. From this result, we can point out that it is difficult for SVMs to reduce the misclassification rate on faulty images by shrinking the radius of the kernel function.

4.2 Experiment 2

In this experiment, we demonstrate that the rate of misclassifying the images of faulty products to the category of qualified products can be reduced to 0 with the use of the GZC discriminant function, while the correct recognition rate can be maintained in certain level. The results of the experiment is illustrated in Fig. 2. Here, $\theta^- = -\theta^+$. The unknown recognition rate means the percentage of the M³ network producing 'I don't know' outputs. From Fig.2 (a) and (b), we can see that when $|\theta|$ is greater than or equal to 0.6147, the error recognition

[1] $\exp(\frac{\|X-X_i\|^2}{-2\sigma^2})$.

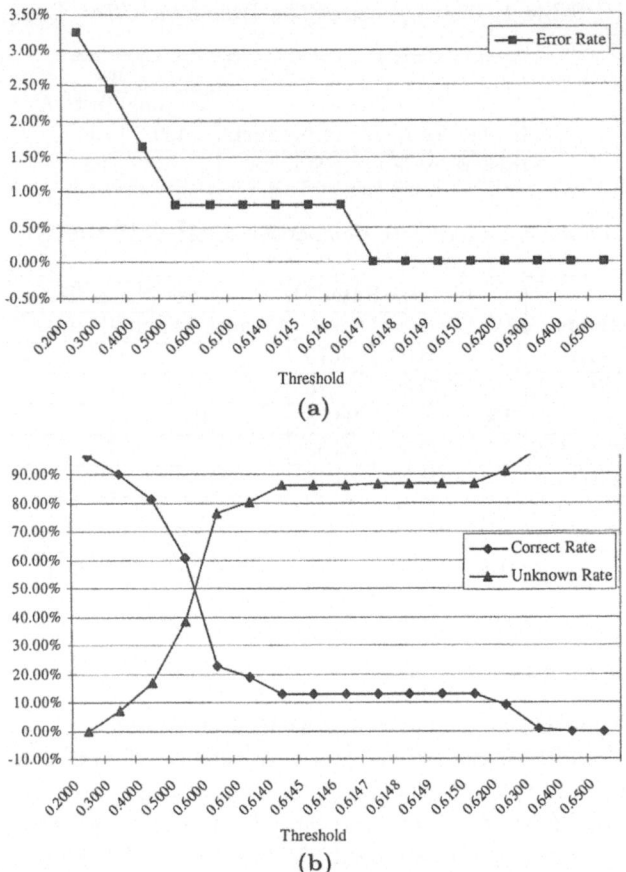

Fig. 2. Correct recognition rate and unknown recognition rate (a), and error recognition rate (b), as a function of the value of threshold of the M^3 network

rate is reduced to 0%. Meanwhile, the correct recognition rate is kept at 13.01%. Although this correct recognitin rate is highly unsatisfactory for practical applications, the proposed method provides us with a promising approach to dealing with fault diagnosis problems of industrial images.

5 Conclusions

We have presented a fault diagnosis method based on the combination of the M^3 neural network and the Gaussian Zero-Crossing discriminant function. The experiment results indicate that the proposed method have the advantage to reduce the misclassification rate on faulty products that SVMs are lack of in our experiments. As to future work, we will refine the internal classification structure for improving correct recognition rate while reducing the rate of misclassifying faulty product images.

Acknowledgements

The authors would like to thank Mr. Kai-An Wang for the help on simulations. This work was partially supported by the National Natural Science Foundation of China via the grant NSFC 60375022.

References

1. Lu, B.L., Ito, M.: Task Decomposition and Module Combination Based on Class Relations: A Modular Neural Network for Pattern Classification. IEEE Trans. on Neural Networks, Vol. 10 (1999) 1244-1256
2. Lu, B.L., Ichikawa, M.: A Gaussian Zero-Crossing Discriminant Function for Min-Max Modular Neural Networks. Proc. of 5th International Conference on Knowledge-Based Intelligent Information Engineering Systems & Allied Technologies, (2001) 298-302
3. Barla, A., Odone, F., Verri, A.: Old Fashioned State-of-the-art Image Classification. Proceedings of 12th International Conference on Image Analysis and Processing (2003) 566-571
4. Cristianini, N., Taylor, J.S.: An Introduction to Support Vector Machines and Other Kernel-based Learning Methods. Cambridge University Press (2000)
5. Richard O. Duda, R.O., Hart, P.E., Stock, D.G.: Pattern Classification (2nd Ed.). John Wiley & Sons, Inc. (2001)
6. Daubechies, I.: Ten Lectures on Wavelets. SIAM (1992)
7. Hayashi, S., Asakura, T., Zhang, S.: Study of Machine Fault Diagnosis System Using Neural Networks. Proc. of IJCNN '02, Vol.1 (2002) 956 - 961
8. Asakura, T., Kobayashi, T., Xu, B.J., Hayashi, S.: Fault Diagnosis System for Machines Using Neural Networks. Trans. JSME, Vol.43, No.2 (2000) 364-372

Cellular Automata Based Pattern Classifying Machine for Distributed Data Mining

Pradipta Maji and P Pal Chaudhuri

Department of Computer Science and Engineering & Information Technology
Netaji Subhash Engineering College, Kolkata, India 700 152
pradiptamaji@hotmail.com, palchau@vsnl.net

Abstract. In this paper, we present the design and application of a pattern classifying machine (PCM) for distributed data mining (DDM) environment. The PCM is based on a special class of sparse network referred to as Cellular Automata (CA) The desired CA are evolved with an efficient formulation of Genetic Algorithm (GA). Extensive experimental results with respect to classification accuracy and memory overhead confirm the scalability of the PCM to handle distributed datasets.

1 Introduction

The meaningful interpretation of different distributed sources of voluminous data is increasingly becoming difficult. Due to high response time, lack of proper use of distributed resources, and its inherent characteristics, conventional centralized data mining algorithms are not suitable for distributed environment. Consequently, researchers, practitioners, entrepreneurs from diverse fields are focusing on development of sophisticated techniques for knowledge extraction, which leads to the promising field of distributed data mining (DDM). Most DDM algorithms are designed for parallel processing on distributed data. Similar algorithms are applied on each distributed data source concurrently, producing one local model per source. All local models are next aggregated to produce the final model.

In this paper, we present design and application of a PCM to address the problem of DDM. At each local site, we design a CA based PCM as a base classifier that classifies the dataset available in a site. The classifier design is based on a special class of CA, termed as Multiple Attractor CA (MACA) [1, 2]. The desired structure of MACA for a particular dataset is obtained by employing GA. The GA significantly reduces the search space. On completion of this training phase, the set of PCMs are copied at central location. In the testing phase, in order to identify the class of a data element, the PCMs in central location are operated. The majority voting scheme is next implemented to aggregate the results of base classifiers and predict the class of data element.

2 Multiple Attractor CA (MACA)

The pattern classifier proposed in this paper employs a special class of linear CA referred to as MACA [1, 2]. The state transition graph of an MACA consists

N.R. Pal et al. (Eds.): ICONIP 2004, LNCS 3316, pp. 848–853, 2004.
© Springer-Verlag Berlin Heidelberg 2004

of a number of cyclic and non-cyclic states. The set of non-cyclic states of an MACA forms inverted trees rooted at the cyclic states. The cyclic states with self loop are referred to as attractors. Fig. 1 depicts the state transition diagram of a 5-cell MACA with four attractors $\{00000(0), 00011(3), 00100(4), 00111(7)\}$. The states of a tree rooted at the cyclic state α forms α-basin. The detailed characterization of MACA is available in [1, 2]. A few fundamental results for an n-cell MACA having k number of attractor basins are next outlined.

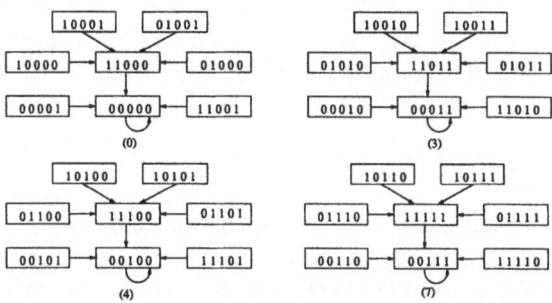

Fig. 1. State transition diagram of a 5-cell MACA

Definition 1 *An m-bit field of an n-bit pattern set is said to be pseudo-exhaustive if all possible 2^m patterns appear in the set.*

Theorem 1 *In an n-cell MACA with $k = 2^m$ attractors, there exists m-bit positions at which the attractors generate pseudo-exhaustive 2^m patterns.*

Result I: An n-bit MACA with 2-attractor basins can be represented by an n-bit binary string, termed as Dependency Vector (DV). If DV is an n-bit Dependency Vector and $\mathcal{P}$ is an n-bit pattern, then the modulo-2 sum (XOR) of the dependent variables of $\mathcal{P}$ (where DV contains 1's) is equal to zero if $\mathcal{P}$ belongs to zero basin; otherwise 1. That is,

$$DV \cdot \mathcal{P} = \begin{cases} 0, \text{ if } \mathcal{P} \in \text{ zero basin} \\ 1, \text{ if } \mathcal{P} \in \text{ non-zero basin} \end{cases} \tag{1}$$

Result II: An n-bit MACA with 2^m-attractor basins can be represented by an n-bit Dependency String (DS). An n-bit Dependency String DS is produced through concatenation of m number of Dependency Vectors of length $n_1, n_2, \cdots, n_m$ respectively, where $n_1 + n_2 + \cdots + n_m = n$ and $\mathcal{P}$ is an n-bit pattern whose attractor basin is to be identified. For each DV_i (of length n_i), the dependent variables of the corresponding n_i bits of $\mathcal{P}$ (say $\mathcal{P}_i$) gives either 0 or 1, - that is,

$$DV_i \cdot \mathcal{P}_i = \begin{cases} 0, \text{ if } \mathcal{P}_i \in \text{ zero basin of } DV_i \\ 1, \text{ if } \mathcal{P}_i \in \text{ non-zero basin of } DV_i \end{cases}$$

which indicates the value of ith pseudo-exhaustive bit. Finally, a string of m binary symbols can be obtained from m number of DVs. This m-bit binary string is the pseudo-exhaustive field (PEF) of the attractor basins where the pattern $\mathcal{P}$ belongs. That is, the PEF of the attractor basin of $\mathcal{P}$ is given by

$$PEF = DS \cdot \mathcal{P} = [DV_1 \cdot \mathcal{P}_1][DV_2 \cdot \mathcal{P}_2] \cdots [DV_m \cdot \mathcal{P}_m] \qquad (2)$$

where DS and $\mathcal{P}$ - both are n-bit vectors. So, the complexity of identifying the PEF of an attractor basin is $O(n)$. This specific result is of significant importance for our design. It enables the scheme to identify the class of an input element with linear complexity. GA formulation to arrive at the desired MACA realizing this specific objective has been proposed in [1] with $O(n^3)$ complexity. The design reported in the next section achieves classification with linear complexity.

3   Two Stage Pattern Classifier

To enhance the classification accuracy of the machine, we have refined the approach reported in [1] and report the design of a CA based classifier with $O(n)$ complexity. Multi-class classifier is built by recursively employing the concept of two class classifier.

3.1   MACA Based Two Stage Classifier (TSC)

The design of MACA based classifier for two n-bit pattern sets S_1 and S_2 should ensure that elements of one class (say S_1) are covered by a set of attractor basins that do not include any member from class S_2. Any two n-bit patterns $\mathcal{P}_1 \in S_1$ and $\mathcal{P}_2 \in S_2$ should fall in different basins. Let, an n-bit MACA with 2^m-attractor basins can classify two n-bit pattern sets S_1 and S_2. That is,

$$DS \cdot \mathcal{P}_1 \neq DS \cdot \mathcal{P}_2 \qquad (3)$$

where DS is an n-bit Dependency String consisting of m number of Dependency Vectors. Here, the total number of attractor basins is 2^m and the pseudo-exhaustive field (PEF) (Theorem 1) of each attractor basin is an m-bit binary pattern/string. Let, k_1 and k_2 be two m-bit pattern sets consisting of pseudo-exhaustive bits of attractors of two n-bit pattern sets S_1 and S_2 respectively. Then, k_1 and k_2 can also be regarded as two m-bit pattern sets for two class classification. So, we synthesize an MACA based two class classifier such that one class (say k_1) belongs to one attractor basin and another attractor basin houses the class k_2. Any two m-bit patterns $p_1 \in k_1$ and $p_2 \in k_2$ should fall in different attractor basins, – that is,

$$DV \cdot p_1 \neq DV \cdot p_2 \qquad (4)$$

where DV is an m-bit Dependency Vector.

Fig 2 represents the architecture of Two Stage Classifier. It consists of three layers - input, hidden and output layers denoted as x_i $(i = 1, 2, \cdots, n)$, y_j $(j = 1, 2, \cdots, m)$, and o_k $(k = 1)$ respectively. While the first classifier (Classifier#1

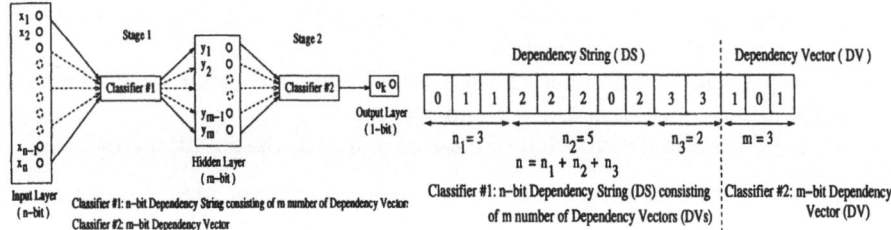

Fig. 2. Two Stage Classifier (TSC)

of Fig 2) maps an n-bit pattern of the input layer into an m-bit pattern (PEF) of the hidden layer, the second classifier (Classifier#2) maps that m-bit pattern into a single bit (either 0 or 1) of the output layer. That is, Classifier#1 provides an appropriate mappings of patterns of input layer into PEF (pseudo-exhaustive field) of the hidden layer and the Classifier#2 implements the classification of the PEFs rather than original patterns. In order to evolve TSC (two MACAs for Stage 1 and 2) realizing this design objective, we have employed GA.

3.2   Genetic Algorithm (GA) for Two Stage Classifier

The structure of GA revolves around the concept of encoding a solution in string format referred to as chromosome and evolving solutions according to its fitness.

Chromosome. Rather than conventional bit string, the proposed scheme employs a chromosome which consists of two parts: (i) a DS for Classifier#1 - a symbol string of numerical digits; and (ii) a DV for Classifier#2 - a binary string as defined in Results I and II of Section 2. The length of a chromosome is equal to $(n + m)$ where n is the number of bits in a pattern and m is number of PEF bits of Classifier #1 attractors. The m-bit patterns are classified by Classifier #2. Fig 2 represents a chromosome corresponding to 10-bit MACA with 2^3-attractor basins. It consists of a 10-bit $(= n)$ DS [011|22202|33] (Classifier #1) and a 3-bit $(= m)$ DV < 101 > (Classifier #2). The DS has partition points at 3rd and 8th positions and the DVs are of length 3 $(= n_1)$, 5 $(= n_2)$ and 2 $(= n_3)$ respectively.

Fitness Function. In classification, the fitness F of a particular chromosome in a population is determined by two factors: (i) the capability of the evolved DS (Classifier #1) for classifying the given input pattern set S_1 and S_2 into separate set of attractor basins - this is referred to as the factor F_1; and (ii) capability of the evolved DV (Classifier #2) for classifying the pseudo-exhaustive field (PEF) set k_1 and k_2 into different attractor basins - this is referred to as F_2.

Determination of F_1 and F_2. For ease of subsequent discussions the following terminologies are introduced.

- k denotes number of attractor basins of an n-cell MACA in which the dataset S $(S = S_1 \cup S_2)$ is to be distributed.
- N_{ij} represents the number of elements of class j covered by i^{th} attractor basin, where $i = 1, 2, 3, 4, \cdots, k$ and $j = 1, 2$.
- M_i indicates the distribution of class elements in the ith attractor basin

The diversity of ith attractor basin of an n-cell k-attractor MACA is given by

$$M_i = \frac{\max\{N_{i1}, N_{i2}\}}{N_{i1} + N_{i2}} \tag{5}$$

The ith $(i = 1, 2, 3, 4, \cdots, k)$ attractor basin indicates class j $(j = 1, 2)$ for which N_{ij} is maximum. The fitness function F_1 of DS is determined by the percentage of patterns which are correctly classified into different attractor basins. That is,

$$F_1 = \frac{1}{k} \sum_{i=1}^{k} M_i \tag{6}$$

Similarly, F_2 has been calculated for two pattern set k_1 and k_2. That is,

$$F_2 = \frac{1}{2}(M_1 + M_2) \tag{7}$$

The fitness F of a particular chromosome is given by

$$F = F_1 \cdot F_2 \tag{8}$$

The experimental results reported next, confirm that this relation, although evolved empirically, provides desired direction to arrive at the best solution.

4   Performance Analysis

To evaluate the efficiency of proposed classifier, we perform extensive experiments for different values of n (number of attributes) and t (size of datasets). Table 1 represents the efficiency of the proposed Two Stage Classifier. The GA has been evolved for maximum 100 generations. The following conclusions can be derived from this experimental results: (i) the classifier has high classification accuracy (Column III) irrespective of number of attributes (n) and size of the datasets (t); (ii) the memory overhead of the proposed classifier, as per Column IV, is independent of the size of datasets (t); and (iii) generation and retrieval time, as the results of Columns V and VI indicate, are linear in nature.

Next we provide a brief analysis of performance of MACA based PCM. Details are available in [3].

Table 2 reports detail result of cluster detection. The experimental results confirm the following facts: (i) the accuracy in DDM is better than that of centralized PCMs; and (ii) as the number of sites (N) increases, classification accuracy for DDM environment significantly increases over centralized algorithm. All results establish that the classification accuracy of MACA based classifiers in distributed environments are superior than that of centralized environment.

Table 1. Efficiency of Two Stage Classifier

Topology $(n : m : 1)$	Size of Dataset	Classification Accuracy	Memory Overhead (Byte)	Generation Time (ms)	Retrieval Time (ms)
200:7:1	5000	98.03	207	1102	301
	10000	97.83	207	1139	578
400:10:1	5000	96.93	410	1789	606
	10000	97.03	410	1809	1125
500:10:1	5000	96.43	510	2344	710
	10000	96.71	510	2339	1403

Table 2. Clusters Detection by MACA, $n = 500$ and $t = 5000$

No of Sites (N)	Value of m	Classification Accuracy	
		Centralized DM	DDM
20	5	98.13	99.23
		98.07	99.26
		98.23	99.37
30	5	98.13	99.62
		98.07	99.68
		98.23	99.82

5   Conclusion

The paper presents the detailed design and application of an efficient PCM for DDM environments. The proposed model is built around a special class of sparse network referred to as CA. The excellent classification accuracy and low memory overhead figures establish the CA as an efficient classifier for DDM environments.

Acknowledgement

This research work is supported by Dept. of CST, B. E. College (DU), India. We acknowledge BEC, India.

References

1. N. Ganguly, P. Maji, S. Dhar, B. K. Sikdar, and P. P. Chaudhuri, "Evolving Cellular Automata as Pattern Classifier," *Proceedings of Fifth International Conference on Cellular Automata for Research and Industry, ACRI 2002, Switzerland*, pp. 56–68, October 2002.
2. P. Maji, C. Shaw, N. Ganguly, B. K. Sikdar, and P. P. Chaudhuri, "Theory and Application of Cellular Automata For Pattern Classification," *Accepted for publication in the special issue of Fundamenta Informaticae on Cellular Automata*, 2004.
3. P. Maji, B. K. Sikdar, and P. P. Chaudhuri, "Cellular Automata Evolution For Distributed Data Mining," *Proceedings of Sixth International Conference on Cellular Automata for Research and Industry, ACRI 2004*, October 2004.

Investigating the Use of an Agent-Based Multi-classifier System for Classification Tasks

Anne M. Canuto, Araken M. Santos, Marjory C. Abreu,
Valéria M. Bezerra, Fernanda M. Souza, and Manuel F. Gomes Junior

Informatics and Applied Mathematics Department
Federal University of Rio Grande do Norte, Brazil
anne@dimap.ufrn.br

Abstract. This paper proposes NeurAge, an agent-based multi-classifier system for classification tasks. This system is composed of several neural classifiers (called neural agents) and its main aim is to overcome some drawbacks of multi-classifier systems and, as a consequence, to improve performance of such systems.

1 Introduction

The main idea of using multi-classifier systems is that the combination of classifiers can lead to an improvement in the performance of a pattern recognition system in terms of better generalisation and/or increased efficiency and clearer design [4]. However, the choice of a combination method which is more suitable for an application is a difficult process, having to execute exhaustive testing to choose the best combination method. In some situations, small changes in the structure of the multi-classifier system, for instance, can drastically change the performance of the combination method and, as a consequence, of the multi-classifiers system. One alternative way to smooth out this problem is to transform the classifiers into agents, which are able to make its own decision in a more autonomous and flexible way. The main aim of this paper is to propose a multi-neural agent system capable of being used for classification tasks, such as Data Mining [2] and character classification, among others.

This paper is organised as follows. Section 2 describes multi-neural classifiers systems (MCNS). The proposed system is presented in Section 3, describing the general architecture of a neural agent and an action plan for the negotiation process. Section 4 shows an experimental work, using a numeric character database. Section 5 presents the final remarks of this work.

2 Multi-classifiers Systems

Multi-classifiers systems (MCS) use the idea that different classifier system can provide complementary information about a certain input pattern, improving, in this way, the overall classification performance of the system. When a MCS system is composed of neural classifiers, it is called Multi-neural classifiers system (MCNS)[9]. Although combined classifiers have some potential advantages over single ones, implementation is not an easy task. The main problem lies in the determination of the

N.R. Pal et al. (Eds.): ICONIP 2004, LNCS 3316, pp. 854–859, 2004.
© Springer-Verlag Berlin Heidelberg 2004

combination strategy. Usually, the choice of the best combination method needs the execution of exhaustive testing. In some situations, small changes in the structure of the multi-classifier system, in input information or in the confidence of one classifier can drastically change the performance of the combination method and, as a consequence, of the multi-classifiers system. This problem has been addressed by several authors, such as in [31,10].

The ideal is the use of a multi-classifier system more dynamic which is capable of adjusting more easily to changes in its structure, confidence or environment. In this case, an input is clamped into the system and all classifiers produce their output. However, instead of providing the outputs to a combination method, all classifiers would communicate to each other in other to reach a common output for the system. In this sense, the process of providing the overall solution for the system is not a centralized one, in which an agreement must be reached for all classifiers. Also, even if one classifier entity does not provide an output, the other entities will communicate and will produce an overall output. Finally, the decrease of confidence in one classifier can be noted by the other classifiers and this can be used during the negotiation process, allowing the system to treat in a more flexible way changes in the classifier confidences.

3   Multi Agent Systems for Classification

As it is desired to have entities which work in an efficient and flexible way, entities of the proposed system can be seen as a agents. In this sense, neural classifiers will be transformed into neural agents, which are able to make its decision in a more autonomous and flexible way. In this sense, a multi-classifiers system becomes a multi-agent system. In this case, two main aspects have to be seen in more details, which are: the internal architecture of the neural agent and some action plans during the negotiation process of this systems.

3.1   Neural Classifier Agent

The main idea behind the functioning of a neural agent is that once an input pattern is provided, the controller passes the needed information to the decision making module, which accesses the neural network module to produce its output. Then, controller can decide to communicate with other agents in order to reach a common result. During the negotiation process, it might be necessary for the agent to change its opinion about its current output or to perform a new decision making process. Also, an agent may decide to perform the decision making process one more time, analyzing other criteria or pattern features.

Figure 1 shows the architecture of a neural agent. As the main goal of all agents is the same, the general structure for all agents is the same. This agent has four main modules, which are:

- Controller: It receives the user queries and defines the activation order of its internal processes. For instance, this module decides, based on the negotiation result, if it is important for the agent to change its existing result in order to reach a common result.

- Decision making: It is responsible for reasoning about its knowledge in order to define the best output for a neural network classifier. The main idea of this module is to search for a result, eliminating those who do not fit the existing conditions. Then, it ranks the results in a decreasing order, according to a set of evaluation criteria. Finally, it picks the first (best) one and defines it as the correct result.
- Negotiation: It is responsible for the communication with other agents in order to reach a common result. It builds an action plan for negotiation or uses one action plans previously taken by it. During the negotiation process, an agent can be suggested to change its result. However, it has autonomy to decide whether to change or to confirm its current result.
- Neural Network: It is responsible for executing the neural network method of the agent. It is aimed that each agent can have one neural network, but it could be performed using different parameters (topology, input features and so on), providing, in this sense, different results for an input pattern.

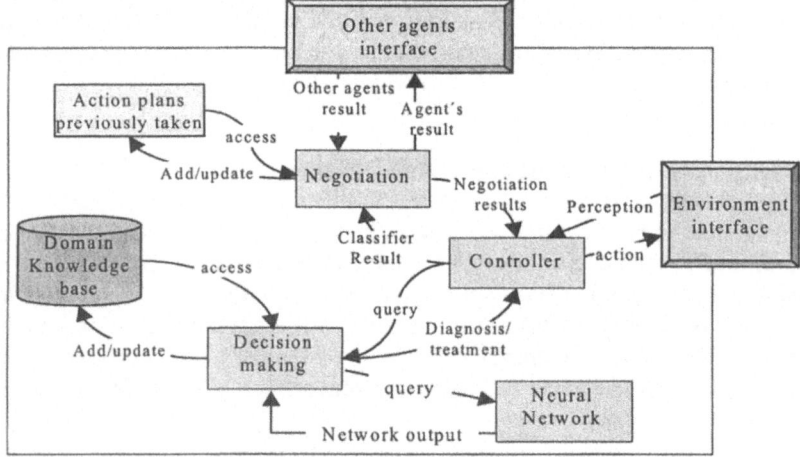

Fig. 1. Internal Architecture of a Neural Classifier Agent.

3.2 An Action Plan to Negotiate

As already mentioned, the controller module of an agent builds an action plan for the negotiation protocol, which, for instance, can be based on techniques of game theory [8]. Alternatively, an agent can choose to use one action plans previously taken by it. In this section, an action plan is presented, which can be used for the neural agents of the proposed system.

A way for the agents to negotiate a common result is to try to decrease the confidence degree of the other agents during the negotiation process. An agent can check some problems of the other agents in order to decrease their confidence. This can be done through the use of sensitive analysis. Therefore, in order to start negotiating, an agent should calculate a sensitivity analysis to all input attributes. This can be done excluding and/or varying the values of an input attribute and analysing the variation in the performance of its neural network method. Then, the functioning of a neural agent has the following steps:

1. Allow its neural network method to train the set of patterns;
2. Calculate the sensibility analysis for all attributes. This process is done for all classes of the problem. Also, it is important to calculate the training mean for all attributes of each class;
3. Start the negotiation process trying to show the other agents that their results are not good ones. The best way to convince the other agents is to decrease their confidence, which can be done in the following way:
   3.1. Calculate the distance (for example: Euclidean distance) from the input attribute and the training mean of that attribute for all classes;
   3.2. Rank the attributes in a decreasing order of the distance (from the least similar attribute to the most similar one);
   3.3. For the other agent, check the class assigned by it. For the first N attributes (distance higher than a threshold).
       3.3.1. Check the sensibility of the neural network to this attribute. Send a message to the other agent suggesting a punishment to the confidence degree of that agent.

   It is important to emphasize that once one agent sends a suggestion to punish the other agent, the other agent will also send a suggestion to punish the first one. Every time that both agents suggest punishments, it is called a round. This process proceeds until all attributes are analyzed or when one of the agents keeps a negative confidence for a number of rounds.
4. After the negotiation process, the neural agent with the highest confidence degree is said to be the most suitable one and its output is considered as the overall output.

The main idea behind this process is that the more distant one attribute is from the training mean, the highest is the probability that a sensitive classification method is wrong. In this sense, this is used to suggest a decrease the confidence degree of an agent. The punishment value is calculated taking into account the ranking of the attribute, the magnitude of the difference with the training mean along with the magnitude of the sensibility of the neural network. Finally, the sensitivity analysis and the training mean, along with some environmental information are transformed into rules and compose the knowledge base of the neural agent.

4 Experimental Work

In order to investigate the performance of the proposed multi-neural agent system, an empirical study was conducted. The chosen task was character recognition and the database was developed by the Centre of Excellence for Document Analysis and Recognition (CEDAR) of the State University of New York at Buffalo [7]. The set of alphanumeric characters was extracted from ZIP codes and is composed of 16x24 binary patterns.

Two neural agents were used in this experimental work, one using a multi-layer perceptron [6], with a 384-150-10 topology and a learning rate of 0.3. The other neural agent uses a RePART neural network, which is a more developed version of the simpler Fuzzy ARTMAP [4]. Both neural networks were trained using the same training set, 2000 patterns, and tested with 1000 different patterns. This process was performed ten times.

In order to calculate the sensitivity analysis, input patterns were divided into nine regions and each region was considered as an attribute. Then, for each region, ten different variations were presented to the neural network. After that, the variation in performance was calculated and transformed into percentage. These values compose the sensitivity analysis of the neural networks. Also, the training mean is calculated to all nine regions of all ten classes. For the negotiation process, an Euclidean distance measure was used to calculate the difference between the input attribute and the training mean. Only distances higher than 5% were considered. When the confidence degree reaches 0.15, the agent is suggested to change its result. If the agent accepts to change its results and still provides a different result from the other, the negotiation process starts again. If an agent keeps a negative confidence for two rounds, its result is not considered anymore and the negotiation process is finished.

4.1 Analysis of the Implementation Results

Table I shows error mean and standard deviation of some classification methods. The first two columns present the performance of individual classifiers, the ones which composed the neural agents of NeurAge. The last column shows the performance of the proposed system. Finally, the third and fourth columns illustrates the performance of two well-known centralized combination methods, which are: Sum and Borda Count [9]. As it can be seen from Table I, when using centralized combination method, there was a decrease in the error mean when compared to individual classifier. However, this improvement can be considered small. Another important fact from Table I is that the proposed system had the lowest error mean and standard deviation of all methods. The improvement in the error mean reached 6,7% when compared to the RePART network.

Table 1. Recognition rate of two individual classification methods, two combination methods and the proposed multi-agent system.

Classification Method	MIP Network	RePART network	Sum Combination	Borda Count	NeurAge
Error Mean	10,91	11,57	8,94	8,33	4,87
St Deviation	3,02	2,84	2,54	2,6	2,03

It is important to emphasize that both neural networks produced wrong outputs to 5,2%, on average, of the total of testing patterns. In these cases, it would be difficult for any combination system to change the overall result. However, NeurAge changed their results in almost 20% of these cases (1,01% out of 5,2%), which was done through suggestions to the agents to change its result. It is a very important result, not only for the error mean, but also to show the capability of negotiation for the agents.

In analyzing the action plan for negotiation, it could be observed that an average of five rounds were necessary for the agents to finish the negotiation process and in all cases an agreement was reached. It is a good result since an agreement was always reached and in a relative small number of rounds. However, one problem of the negotiation problem is that it is more comfortably used between two agents. In this simpler version of the system, it is not a problem because it will always be a negotiation between two agents. However, for future versions of the proposed system, further improvements in the negotiation protocol has to be done to allow more agents during the negotiation process.

5 Final Remarks

In this paper, an agent-based multi-classifier system was proposed. This system is composed of several neural agents which negotiate in order to reach an agreement of a common result for an input pattern. Also, an action plan was presented which allow the agents to negotiate. An Experimental work has been done using a character database and using a system composed of two neural agents. Initial experiments have been executed and satisfactory results were reached in the performance of the proposed systems, decreasing the error mean in almost 7%, when comparing to individual classification methods. It also produced a lower error mean than two combination methods (Sum and Borda count). However, some improvements have to be performed in order to provide all functionalities of the proposed system. Mainly in the negotiation protocols, further adjustment have to be done and are currently under way.

References

1. Ben-Yacoub, S, Abdeljaoued, Y, and Mayoraz, E. "Fusion of Face and Speech data for Person Identity Verification", IEEE Trans on Neural Networks, 10(5), pp. 1065-1075, 1999.
2. Bezerra, Valéria; Canuto, Anne and Campos, André. "A Multi-agent System for Extracting Knowledge in Image Databases". To appear in 8th Brazilian Symposium on Artificial Neural Networks (SBRN), October, 2004.
3. Breiman, L. Combining Predictors. In Combining Artificial Neural Nets: Ensemble and Modular Multi-net Systems, (Ed) A. J. C. Sharkey. Spring-Verlag, pp. 31-49, 1999.
4. A Canuto.. Combining neural networks and fuzzy logic for applications in character recognition. PhD thesis, University of Kent, 2001.
5. S-B Cho. Pattern. Recognition with Neural Networks combined by Genetic Algorithm. Fuzzy Sets and Systems, 103: 339-347, 1999.
6. S. Haykin. Neural Networks, A Comprehensive Foundation. Prentice Hall.1998
7. J J Hull. A database for handwritten text recognition. IEEE Transactions on Pattern Analysis and Machine Intelligence,5(16):550–554, 1994.
8. Osborne, M. "An Introduction to Game Theory". Oxford University Press, 2003.
9. Sharkey, A. Multi-net System. Em Combining Artificial Neural Nets: Ensemble and Modular Multi-net Systems, (Ed) A. J. C. Sharkey. Spring-Verlag, pag. 1-30, 1999.
10. Yamashita, Y and Komori, H and Suzuki, M. Running multiple neural networks for process trend interpretation. Journal of Chemical Engineering of Japan, 32(4): 552-556, 1999.

A New MDS Algorithm
for Textual Data Analysis

Manuel Martín-Merino[1] and Alberto Muñoz[2]

[1] University Pontificia of Salamanca, C/Compañía 5, 37002 Salamanca, Spain
mmerino@ieee.org
[2] University Carlos III, C/Madrid 126, 28903 Getafe, Spain
alberto.munoz@uc3m.es

Abstract. MDS algorithms are data analysis techniques that have been
successfully applied to generate a visual representation of multivariate
object relationships considering only a similarity matrix. However in high
dimensional spaces the concept of proximity become meaningless due to
the data sparsity and the maps generated by common MDS algorithms
fail often to reflect the object proximities.
In this paper, we present a new MDS algorithm that overcomes this
problem transforming the dissimilarity matrix in an appropriate manner. Besides a new dissimilarity is proposed that reflects better the local
structure of the data. The connection between our model and a kernelized
version of the Kruskal MDS algorithm is also studied.
The new algorithm has been applied to the challenging problem of word
relation visualization. Our model outperforms several alternatives proposed in the literature.

1 Introduction

Visualization algorithms are useful multivariate analysis techniques that help to
discover the underlying structure of high dimensional data [7]. A large variety
of neural based techniques have been proposed to this aim, such as Kohonen
Maps [11], PCA based algorithms [13] or multidimensional scaling algorithms
(MDS) [8]. In particular, MDS algorithms have been applied to a broad range
of practical problems [7] even as a complementary tool to Kohonen maps [11].

Let δ_{ij} be the dissimilarity matrix made up of object proximities. MDS algorithms look for object coordinates in a low dimensional space (usually $\mathbb{R}^2$
for visualization purposes) such that the inter-pattern dissimilarities (δ_{ij}) are
approximately preserved. This visual representation is a valuable help to understand the inherent structure of the data and to analyze the clustering tendency
of the objects.

However, the object relations induced by common MDS algorithms become
frequently meaningless when a large percentage of similarities (s_{ij}) are close to
zero [6]. This usually happens when the dimension of the vector space representation is high and the object relationships are local [2]. In this case, s_{ij} is non
zero only for the first nearest neighbors [1] and even those similarities show a
bias toward small values due to the 'curse of dimensionality' [5, 9].

N.R. Pal et al. (Eds.): ICONIP 2004, LNCS 3316, pp. 860–867, 2004.
© Springer-Verlag Berlin Heidelberg 2004

A wide range of practical applications such as the textual data analysis comply with the features mentioned above [16, 15, 4]. Consider for instance, the problem of word relation visualization. A large number of terms (specific terms [3]) have non-null relation only with a small group of terms related with the same semantic concept. Besides, due to the high dimension of the vector space representation, non zero similarities get close to zero [5].

MDS algorithms fail often to reflect the local relationships (smaller distances) when a large percentage of similarities are nearly zero [6]. In this paper we propose a new MDS algorithm that focuses on the preservation of the smaller distances and avoids that the larger dissimilarities distort the maps. A new metric is also defined that reflects better the proximities inside local neighborhoods. Finally the new algorithm is compared with a widely known alternatives including neural based techniques such as Sammon [17] and SOM [11].

This paper is organized as follows: In section 2 we discuss the major challenges that poses the visualization of textual data. In section 3 a new MDS algorithm is presented. In section 4 the algorithm is applied to the problem of word relation visualization. Finally in section 5 we get conclusions and outline future research trends.

2   Visualizing Textual Data with MDS Algorithms

MDS algorithms are helpful techniques to visualize multivariate object relationships considering only a similarity matrix (s_{ij}). In this section we first discuss some peculiar properties of the similarities usually considered to model textual data. Next we comment the impact that this kind of similarities have on the performance of most MDS algorithms.

Let x_i, $x_j \in \mathbb{R}^p$ denote the vectorial representation of two terms. The textual data exhibit two relevant properties. First, the vector space dimension is very high because it depends on the number of documents in the database [3]. Second, the vectorial representation is very sparse (small percentage of non-zero components) [4]. That is, most of the terms group into semantic subspaces of much smaller dimension than p. The above properties justify that the similarities considered to model textual data follow a Zipf law [16]. That is, most of the similarities are close to zero.

To understand this problem, consider for instance a widely used similarity such as the Jaccard [8]. This similarity is proportional to the number of variables simultaneously non-zero for both objects. Due to the sparsity of the vector space representation, J_{ij} will be larger than 0 only for a small fraction of terms that share non-zero components [2]. Besides that, non-zero similarities are often affected by the 'curse of dimensionality' becoming close to zero [5, 1]. This explains why the Jaccard similarity histogram follows a Zipf law. Notice that a large variety of alternative similarities are somewhat affected by the same problem [5, 1] including certain popular dissimilarities such as the χ^2 [12].

It has been shown in [6] that the performance of any algorithm based on distances is strongly degraded when a large percentage of similarities are nearly zero. In particular the maps show a geometrical distribution for the objects that has nothing to do with the underlying structure of the dataset. Besides the objects are put together in the map because they are far away from the the same subset of patterns. Hence small distances in the map could suggest false object relationships.

Therefore any MDS algorithm proposed to deal with textual data, should avoid the negative effect of the similarities that are nearly zero and improve the preservation of the smaller distances. Besides, new dissimilarities should be defined that reflect more accurately the local structure of the data [1] and that help to alleviate the 'curse of dimensionality'.

3   A New MDS Algorithm

In this section a new MDS algorithm is presented that improves the preservation of the smaller dissimilarities keeping under control the influence of the larger ones. A new metric is also defined that models appropriately the local relationships and that helps to alleviate the 'curse of dimensionality'. Finally it is shown that the proposed MDS algorithm can be interpreted as a kernel method that transforms the original dissimilarities in an appropriate manner.

Let (δ_{ij}) be a dissimilarity matrix. The kruskal MDS algorithm [8] looks for an object configuration in $\mathbb{R}^d$ $(d = 2)$ that minimizes the following Stress function (originally by a gradient descent technique):

$$E = \frac{\sum_i \sum_{i<j} (\delta_{ij} - d_{ij})^2}{\sum_i \sum_{i<j} \delta_{ij}^2}, \tag{1}$$

where d_{ij} denotes the inter-pattern distances in $\mathbb{R}^d$. Notice that a large variety of MDS algorithms optimize a similar quadratic error up to normalization factors [8]. Therefore the ideas proposed in this paper are applicable to a wide range of MDS algorithms.

First, we introduce a new Stress function that allow us to control, for each object, the percentage of nearest neighbors that contributes to the error function. Obviously, this feature will allow to drop the larger dissimilarities that distort the map and focus on the preservation of the smaller distances. This will help to reflect accurately the non zero similarities. The new Stress function is defined as:

$$E = \frac{\sum_i \sum_{j \in V_i} (\delta_{ij} - d_{ij})^2}{\sum_i \sum_{j \in V_i} \delta_{ij}^2}, \tag{2}$$

where V_i is the local neighborhood corresponding to the first $N(t)$ nearest neighbors of object i in iteration t. $N(t)$ is updated in each iteration by means of the following rule $N(t) = N_i + \frac{N_i - N_f}{N_{iter}} t$ (see [11] for alternatives), where N_i and N_f determine the number of nearest neighbors considered in the first and last iterations respectively. Notice that the computation of the nearest neighbors is done in output space ($\mathbb{R}^d$). In this way we overcome the problematic determination of neighbors in high dimensional spaces [5].

The error function (2) can be minimized through an iterative algorithm as in the classic version [8]. Notice that in the first iterations ($N(t) \approx N_i$) and the algorithm behaves similarly to the classic version. Therefore, only the position of dense vectors (broad terms) is interpretable in the map, because they contribute with a large number of non-zero similarities (see [14] for details). Hence their position in the map is hardly distorted by similarities that are almost zero.

In the last iterations ($N(t) \approx N_f$) and only a small number of nearest neighbors are considered in the error function. This feature will allow to improve the position of sparse vectors (specific terms) that have non-zero similarities with a small number of nearest neighbors. Therefore, two objects will appear together in the map just because they are related according to the dissimilarity considered.

We have mentioned in section 2 that another relevant feature that distorts the MDS maps is the small variance of non zero similarities. To avoid this problem a modification of the Euclidean distance is introduced that models more accurately the local relations and helps to alleviate the 'curse of dimensionality'. The new dissimilarity gives more weigh to the more relevant variables and reduces the effect of noisy features following the recommendations of [1]. However, the discriminant power of the variables is determined locally considering only the objects that belong to the same neighborhood V_i. Assuming that the variables with large L_1 norm have low discriminant power [12], the new dissimilarity can be defined as:

$$\delta_{ij} = \sum_s \frac{1}{f_{si}} (x_{is} - x_{js})^2, \qquad (3)$$

where f_{si} denotes the L_1 norm for the s variable considering only the nearest neighbors of i in the current iteration. Obviously, those variables with local L_1 norm equal to 0 are drop from the sum.

Finally we discuss the connection between our algorithm and a kernel version of the Kruskal MDS.

First, notice that considering only the smaller distances in the error function (2) is analogous to transform the map distances in equation (1) through the following neighborhood function,

$$H(d_{ij}, \eta) = \begin{cases} d_{ij} & d_{ij} < \eta \\ 1 & \text{otherwise} \end{cases} \qquad (4)$$

where η determines the number of neighbors considered in the function error and it is assumed that δ_{ij} is normalized to 1. The neighborhood (4) may be approximated by a continuous nonlinear function: $h(d_{ij}; \eta) = [1 - \exp(-d_{ij}/\eta)]$. Now transforming the map distances through this nonlinear function, the error for the Kruskal MDS algorithm takes the following form:

$$E = \frac{\sum_i \sum_{j<i} \left[\delta_{ij}^2 - [1 - \exp(-d_{ij}^2/\eta^2)] \right]^2}{\sum_i \sum_{j<i} \delta_{ij}^4} \qquad (5)$$

We have substituted in (5) the distances by their square because the square is a monotonic function and does not change the position of the local minima [8].

Notice that given a Gaussian kernel, there exists a non-linear map ϕ such that $k(\boldsymbol{x}_i, \boldsymbol{x}_j) = \phi(\boldsymbol{x}_i)^T \phi(\boldsymbol{x}_j)$ [18]. Therefore the equation (5) may be written in terms of a nonlinear transformation ϕ as

$$E = \frac{\sum_i \sum_{j<i} \left[\delta_{ij}^2 - (\phi(\boldsymbol{x}_i) - \phi(\boldsymbol{x}_j))^T (\phi(\boldsymbol{x}_i) - \phi(\boldsymbol{x}_j))\right]^2}{\sum_i \sum_{j<i} \delta_{ij}^4}, \qquad (6)$$

where we have considered that for a Gaussian kernel, the Euclidean distance in feature space can be written as $\|\phi(\boldsymbol{x}_i) - \phi(\boldsymbol{x}_j)\|^2 = 2[1 - \exp(-d_{ij}^2/\eta^2)]$ [18].

The error function (6) suggests that our model can be considered a kernel version of the Kruskal MDS algorithm that transforms nonlinearly the objects in the map to a feature space [18]. The non-linear mapping ϕ will help to approximate better the smaller dissimilarities avoiding the negative effect of the larger ones.

4   Experimental Results

In this section the MDS algorithm proposed is applied to the problem of word relation visualization and compared with some alternatives. To this aim we have built a document collection made up of 2000 documents recovered from three commercial databases "LISA", "INSPEC" and "Sociological Abstracts". The thesaurus available has induced a classification of words into seven main topics and guarantee the result's objectivity.

To evaluate the mapping algorithms, we check if words belonging to the same cluster according to the map, are assigned to the same group by the thesaurus. To this aim, the word map is first clustered into seven groups using a clustering algorithm such as PAM [10]. Then, different features of the partition induced by the map are evaluated through the following objective measures: The F measure [3] determines if words clustered together in the map are related in the thesaurus. It is a measure of the overall word map quality. The entropy measure E [19] gives an idea of the map overlapping between the different topics induced by the thesaurus. Smaller values are preferred. Finally, the mutual information I [19] is a nonlinear correlation measure between the classification induced by the thesaurus and by the word map respectively. This measure is more sensible to the position of less frequent terms (sparse vectors).

The terms are codified using the vector space model [3] and normalized by the L_2 norm [11]. All the mapping algorithms have been initialized by an SVD [4, 13] to avoid that no one gets trapped in a local minima.

Table 1 compares empirically the MDS proposed in this paper (row 4) with three widely used alternatives presented in the literature; the Sammon's mapping [17] (row 3), the Kruskal multidimensional scaling [8] (row 2) and the self organizing maps (SOM) [11] (row 1). Row (4a) corresponds to the kernel MDS algorithm presented in section 3 while row (4b) shows the results for a variant in which V_i is a spherical neighborhood of decreasing radius d_i. From the analysis of table 1 we report the following conclusions:

Table 1. Experimental results for the kernel MDS (rows [4]) presented in this paper versus the classic alternatives (rows [2,3]) and SOM.

	F	E	I
[1] SOM	0.70	0.38	0.23
[2] *Kruskal MDS*	*0.57*	*0.51*	*0.19*
[3] Sammon	0.53	0.51	0.18
[4a] Kernel MDS $N_i =$ 1200 (70%), $N_f = 150$ (8%)	**0.68**	**0.45**	**0.23**
[4b] Kernel MDS $d_i =$ 0.97 (60%), $d_f = 0.4$ (15%)	0.66	0.46	0.22

Parameters: [2,3,4] *$Niter = 90$, $\alpha = 0.38$.* [1] *Nneur = 88, niter = 30, $\sigma_i = 30$, $\sigma_f = 2$*

The kernel MDS algorithms (4) outperform significantly both, the Sammon mapping (3) and the Kruskal MDS algorithm (2). In particular, the new algorithm (4a) improves the Mutual Information (I) up to 21% which suggests that the position of the less frequent terms (sparse vectors) is significantly improved in the map. Therefore our model helps to avoid the tendency of classic MDS algorithms to put together specific terms in the map regardless of their semantic meaning (see section 2). Hence the relations between specific terms and their nearest neighbors are reflected better by the MDS proposed and the visualization of the local data structure is substantially improved. Moreover, the Entropy reduction of 12% suggests that the previous feature helps to minimize the overlapping between terms (particularly the more specific) assigned to different groups by the thesaurus. Finally, it is worth noting that our method achieves a remarkable improvement of the overall word map quality ($\Delta F = 19\%$).

The second version (4b) that assumes spherical neighborhoods for V_i achieves similar results.

The SOM performs similarly to the kernel MDS algorithm (4a). However, the MDS proposed exhibits two interesting properties for many practical problems. First, the computational complexity can be reduced to be linear with the number of patterns. To this aim, we can use for instance any of the methods proposed in [15] to represent the dataset with a reduced number of prototypes. Notice that the computational complexity of SOM is quadratic with the number of patterns which hampers the application to large datasets. Second, the Kernel MDS algorithm is able to work considering only a dissimilarity matrix.

Finally figure 1 illustrates the performance of our MDS algorithm. For the sake of clarity only a small subset of terms have been shown. Terms assigned to different groups by the thesaurus are plotted in different gray scale and are split by dashed line. Notice that associations involving specific terms are identified correctly by the map. Moreover, the map allow us to discover easily the different clusters (topics) of the database.

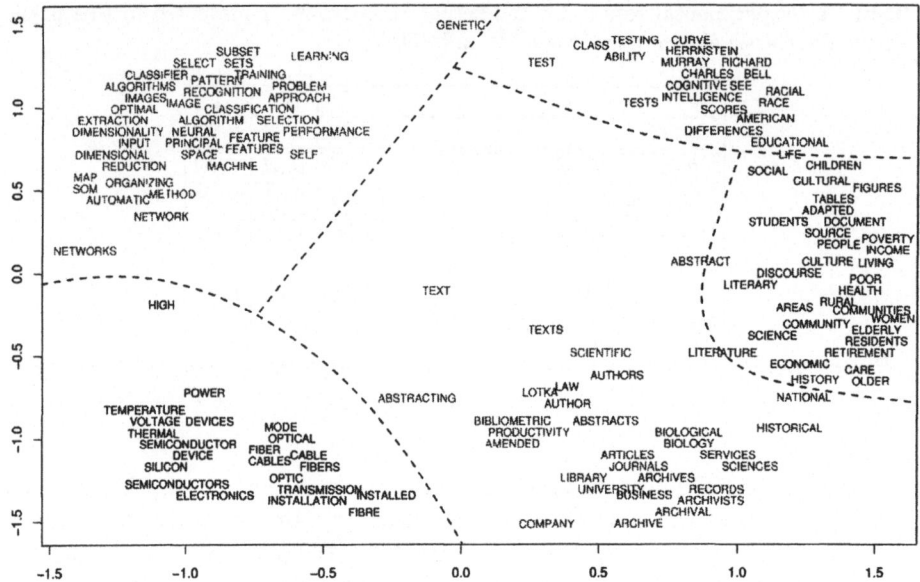

Fig. 1. Word map generated by the kernel MDS algorithm for a subset of terms.

5   Conclusions and Future Research Trends

In this paper we have presented a new MDS algorithm that is able to handle sparse high dimensional data. The new model transforms the dissimilarities in an appropriate manner and incorporates a new metric that reflects better the local proximities. The resulting algorithm is shown to be related to a kernelized version of the Kruskal MDS algorithm. Finally, the new algorithm has been applied to the challenging problem of word relation visualization.

The experimental results show that our algorithm outperforms two widely used nonlinear projection techniques such as Sammon's mapping and Kruskal MDS algorithm. Particularly, the position of non frequent terms (specific terms) is significantly improved by the new model as well as the clustering structure induced by the map. Besides the MDS proposed has revealed as an interesting alternative to SOM that is able to work directly from a dissimilarity matrix.

Future research will focus on the development of new dimension reduction techniques based on the ideas proposed in this paper.

References

1. C. C. Aggarwal. Re-designing distance functions and distance-based applications for high dimensional applications. In Proc. of SIGMOD-PODS, 1, pages 13-18, 2001.
2. C. C. Aggarwal and P. S. Yu. Redefining clustering for high-dimensional applications. IEEE Transactions on Knowledge and Data Engineering, 14(2):210–225, March/April 2002.

3. R. Baeza-Yates and B. Ribeiro-Neto. Modern Information Retrieval. Addison-Wesley, New York, 1999.
4. M. W. Berry and Z. Drmac and E. R. Jessup. Matrices, Vector Spaces and Information Retrieval. SIAM review. 41(2), 335-362, 1999.
5. K. Beyer, J. Goldstein, R. Ramakrishnan, and U. Shaft. When is "nearest neighbor" meaningful? In Proc. of the (ICDT), Lecture Notes in Computer Science (1540), pages 217-235, Jerusalem, Israel, 1999. Springer Verlag.
6. A. Buja, B. F. Logan, J. A. Reeds, and L. A. Shepp. Inequalities and positive-definite functions arising from a problem in multidimensional scaling. The Annals of Statistics, 22(1):406–438, 1994.
7. A. Buja, D. Swayne, M. Littman and N. Dean. XGVIS: Interactive data visualization with multidimensional scaling.Submitted to the Journal of Computational and Graphical Statistics. 2003. http://www.research.att.com/~andreas
8. T. F. Cox and M. A. A. Cox. Multidimensional Scaling. Chapman & Hall/CRC Press, New York, second edition, 2001.
9. K. Fukunaga. Statistical Pattern Recognition. Academic Press, UK., $2^{\underline{a}}$ ed., 1990.
10. L. Kaufman and P. J. Rousseeuw. Finding groups in Data; An Introduction to Cluster Analysis. John Wiley and Sons, USA, 1990.
11. T. Kohonen. Self-Organizing Maps. Springer Verlag, Berlin, second edition, 1995.
12. L. Lebart, A. Morineau, and J. F. Warwick. Multivariate Descriptive Statistical Analysis. John Wiley, New York, 1984.
13. J. Mao and A. K. Jain. Artificial neural networks for feature extraction and multivariate data projection. IEEE Transactions on Neural Networks, 6(2), March 1995.
14. M. Martín-Merino and A. Muñoz. Self organizing map and Sammon mapping for asymmetric proximities. In Lecture Notes on Computer Science (2130), 429-435, Vienna, 2001. Springer Verlag.
15. A. Muñoz and M. Martín-Merino. Visualizing asymmetric proximities with MDS models. Proc. of the European Symposium on Artificial Neural Networks ESANN'03, Bruges, Belgium, 2003, 51-58.
16. A. Muñoz. Compound key word generation from document databases using a hierarchical clustering art model. Journal of Intelligent Data Analysis, 1(1):25-48, 1997.
17. J. W. Sammon. A nonlinear mapping for data structure analysis. IEEE Transactions on Computers, C-18:401–409, May 1969.
18. Scholkopf, B. and A.J. Smola. Learning with Kernels, MIT Press, Cambridge, 2002.
19. A. Strehl, J. Ghosh, and R. Mooney. Impact of similarity measures on web-page clustering. Workshop of Artificial Intelligence for Web Search , Austin, Texas, USA, 58-64, July 2000.

Chaotic Behavior in Neural Networks and FitzHugh-Nagumo Neuronal Model

Deepak Mishra, Abhishek Yadav, and Prem K. Kalra

Department of Electrical Engineering, IIT Kanpur, 208 016 India
{dkmishra,ayadav,kalra}@iitk.ac.in

Abstract. There has been a lot of discussion about whether chaos theory can be applied to the dynamics of the brain. It is obvious that nonlinear mechanisms are crucial in neural systems and also it has been established from various experiments that neuronal activities and EEG recordings show many characteristics of chaotic behavior. Chaotic behavior of a three-dimensional neural network representing nonlinear dynamics of local-averaged spike-rate in neurons is investigated. The same study has been carried out to investigate chaos in FitzHugh-Nagumo neuronal model. It was found that these models exhibit chaotic dynamics at some parametric values.

1 Introduction

A neuron communicates to other neurons via electrical impulses, also called potentials, and chemical secretions called neurotransmitters. These changes appear as wriggling lines along the time axis in a typical EEG record[8]. For example, Freeman and colleagues have developed mathematical models for EEG signals generated by the olfactory system in rabbits. These investigators have suggested that the learning and recognition of novel odors, as well as the recall of familiar odors can be explained through the chaotic dynamics of the olfactory cortex's electrical activity[1,7,8,10]. Activity of such a model is determined by three nonlinear differential equations. Each neuron has a nonlinearity represented in terms of transfer function[7,8]. This paper presents numerical simulations to explain how the stability of the system passes from stable to chaotic regime and a sudden qualitative change in chaotic dynamics takes place. Das *et.al.* has investigated presence of chaos in three dimensional hopfield like neural network[4]. In this work, the biological significance of the same neural network is seeked. It was found that FHN model exhibits chaotic dynamics at some parametric values[3]. Gao *et.al.*[5] and Jing *et.al.*[14] have investigated chaotic behavior of FHN neuronal model considering integration time step as bifurcation parameter. This parameter is related to the numerical method and not to the model and therefore it reflects the characterstics of numerical method and not of the model[2,12]. In section 2, we discussed the change in behavior of the firing rate activities of a three-dimensional dynamic neural network with the variation in parameters. Chaotic behavior in FitzHugh-Nagumo model is discussed in Section 3. In Section 4, some conclusions drawn from this investigation are discussed.

N.R. Pal et al. (Eds.): ICONIP 2004, LNCS 3316, pp. 868–873, 2004.
© Springer-Verlag Berlin Heidelberg 2004

2   Chaos in Three-Dimensional Model of Neural Network

A small neural network consisting of three neurons is considered whose activity is determined by three nonlinear differential equations[6]. It has been found that with the variation in parameters, model exhibits different types of responses, i.e. stable, periodic and chaotic. The nonlinear differential equations for the above model can be encapsulated as:

$$\frac{\partial x(t)}{\partial t} = f_1(w_{12}y(t) + w_{13}z(t)) - \alpha_1 x(t) \tag{1}$$

$$\frac{\partial y(t)}{\partial t} = f_2(w_{21}x(t) + w_{23}z(t)) - \alpha_2 y(t) \tag{2}$$

$$\frac{\partial z(t)}{\partial t} = f_3(w_{31}x(t) + w_{32}y(t)) - \alpha_3 z(t) \tag{3}$$

Where,

$$f_i(s) = (1 + exp(-\beta_i(s - \theta_i)))^{-1} \tag{4}$$

where $x(t)$, $y(t)$ and $z(t)$ are the short-time averaged firing-rate activities of the neurons 1, 2 and 3 respectively. θ_i and β_i are the threshold and slope of transfer function of neuron i respectively. α_i is the decay rate of the neuron i. The extent to which the input of neuron i is driven by the output of neuron j is characterized by its synaptic weight w_{ij}. Different types of analyses have been carried out in order to find the behavioral change of the dynamics of the system with variations in parameters.

2.1   Eigenvalue Analysis for Three-Dimensional Nonlinear Dynamic Neural Network Model

The analytically calculated eigenvalues for the linearized three dimensional model are shown in Table 1. These eigenvalues are corresponding to different values of parameter w_{13}. It is found that at a particular value of parameter ($w_{13} = -5.64$), real part of at least one complex eigenvalue of linearised model is almost zero. This indicates the presence of behavioral change in the system at $w_{13} = -5.64$. Fig. 1(a) shows the maximum lyapunov exponent for the given system, plotted against the bifurcation parameter w_{13}. From Fig. 1(a), it is found that maximum lyapunov exponent is positive between $w_{13} = -300$ and $w_{13} = -5.64$. The bifurcation diagram is drawn in Fig. 1(b). It is noticeable that at $w_{13} = -4$, period doubling is observed and chaotic bifurcation starts at $w_{13} = -5.64$.

2.2   Time Response and Phase Portrait of the Model

The time response and phase portrait for the three dimensional neural network model for different values of w_{13} are shown in Fig. 2. It is observed that for $w_{13} = -5.64$ neural network model exhibits the chaotic behavior. The other parameters are $\beta_1 = 7$, $\beta_2 = 7$, $\beta_3 = 13$, $\theta_1 = 0.5$, $\theta_2 = 0.3$, $\theta_3 = 0.7$, $\alpha_1 = 0.65$, $\alpha_2 = 0.42$, $\alpha_3 = 0.1$, $w_{12} = 1$, $w_{21} = 1$, $w_{23} = 0.1$, $w_{31} = 1$ and $w_{32} = 0.02$.

Table 1. Eigenvalues of linearized model of neuaral network at different values of parameter, w_{13}.

$At(w_{13} = -11)$	$At(w_{13} = -5.64)$	$At(w_{13} = -1)$
Equillibrium Points		
$x^* = 0.32$	$x^* = 0.39$	$x^* = 0.56$
$y^* = 1.34$	$y^* = 1.65$	$y^* = 2.28$
$z^* = 0.09$	$z^* = 0.23$	$z^* = 1.86$
Eigenvalues		
$\lambda_1 = -1.4532811$	$\lambda_1 = -1.17$	$\lambda_1 = -0.44$
$\lambda_2 = 0.14 - j0.55$	$\lambda_2 = 0.56 \times 10^3 - j0.79$	$\lambda_2 = -0.36 - j1.63$
$\lambda_3 = 0.14 + j0.55$	$\lambda_3 = 0.56 \times 10^3 + j0.79$	$\lambda_3 = -0.36 + j1.63$

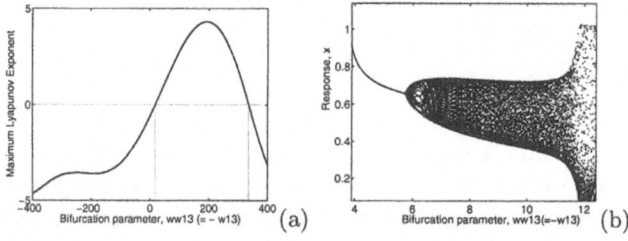

Fig. 1. Plots of Maximum Lyapunov exponent and bifurcation diagram for three-dimensional nonlinear dynamic neural network model, against bifurcation parameter w_{13}. (a)Maximum Lyapunov exponent; (b)Bifurcation diagram.

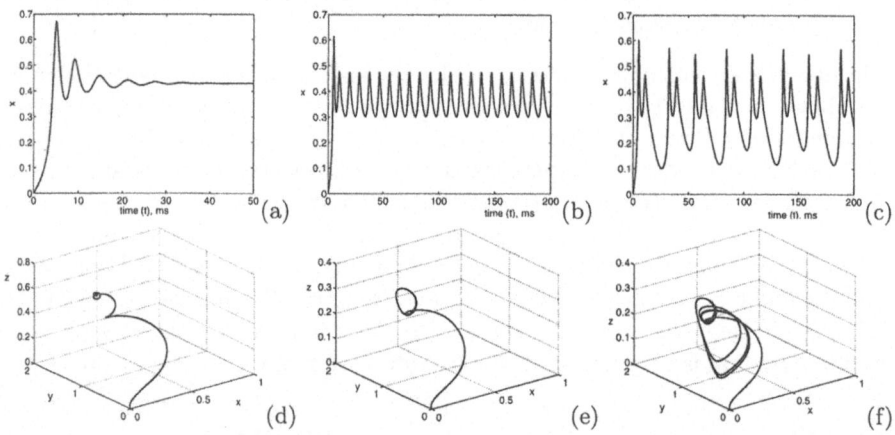

Fig. 2. Time responses and phase portraits for neural network dynamics. (a) Time response at $w_{13} = -3.2$ (converging); (b) Time response at $w_{13} = -4$ (oscillatory); (c) Time response at $w_{13} = -5.64$ (chaotic); (d) Phase portrait at $w_{13} = -3.2$ (converging); (e) Phase portrait at $w_{13} = -4$ (oscillatory); (f) Phase portrait at $w_{13} = -5.64$ (chaotic).

3   Chaos in FitzHugh-Nagumo (FHN) Model

In the mid-1950's, FitzHugh sought to reduce the Hodgkin-Huxley model [9] to a two variable model for which phase plane analysis can be applied. His general observation was that the gating variables n and h have slow kinetics relative to m [11,13]. Moreover, for the parameter values specified by Hodgkin and Huxley, $n+h$ is approximately 0.8. This led to a two variable model, called the fast-slow phase plane model:

$$C_m \frac{dV}{dt} = -\overline{G}_K n^4 (V - V_K) - \overline{G}_{Na} m^3 h (V - V_{Na}) - G_l (V - V_l) + I_{appl} \quad (5)$$

$$n_w(V) \frac{dn}{dt} = (n_\infty(V) - n) \quad (6)$$

A polynomial model reduction of the FHN model is

$$\frac{dV}{dt} = V(V - \alpha)(1 - V) - W + I \quad (7)$$

$$\frac{dW}{dt} = \varepsilon(V - \gamma W) \quad (8)$$

3.1   Analysis of the FHN Model

The FHN neuron model is analysed with I and γ as bifurcation parameters. Here the effect of parameter variation to its dynamical behavior has been explored.

3.2   Analysis of the FHN Model with Current as a Parameter

We have investigated the behavioral change in the dynamics with respect to I by plotting real part of complex eigenvalues, Lyapunov exponent, bifurcation diagram, phase portrait and the time response (Fig. 3 and 4).

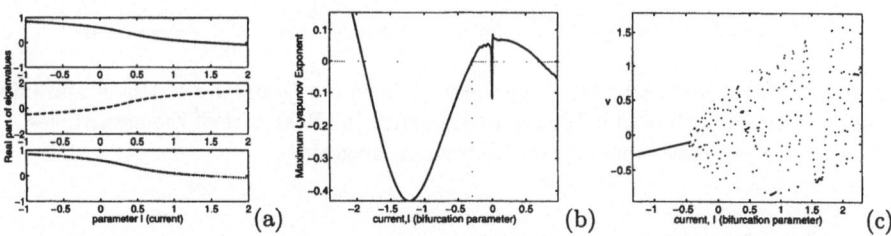

Fig. 3. Plots of real part of the eigenvalues, Maximum Lyapunov exponent and bifurcation diagram with current, I as bifurcation parameter. (a) Real part of the eigenvalues; (b) Maximum Lyapunov exponent; (c) Bifurcation diagram.

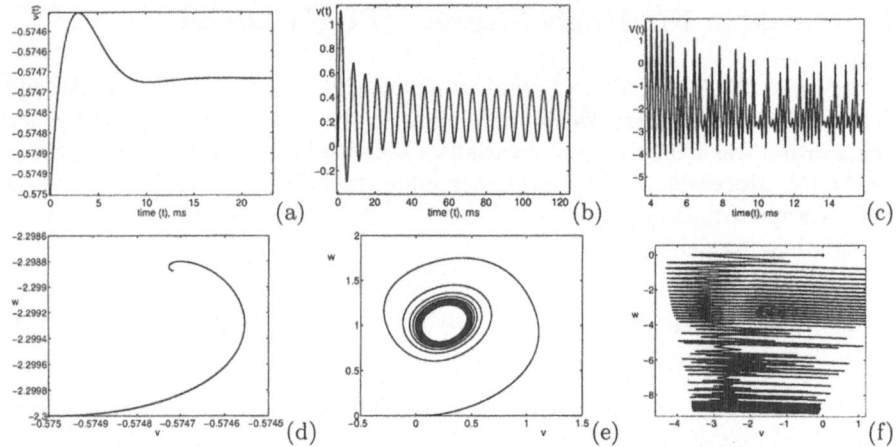

Fig. 4. Time responses and phase portraits for FHN model. (a) Time response at $I = -3$ (converging); (b) Time response at $I = -1$ (oscillatory); (c) Time response at $I = -0.35$ (chaotic); (d) Phase portrait at $I = -3$ (converging); (e) Phase portrait at $I = -1$ (oscillatory); (f) Phase portrait at $I = -0.35$ (chaotic).

3.3   Analysis of the FHN Model with γ as a Parameter

We have investigated the behavioral change in the dynamics of FHN model with respect to γ by plotting real part of complex eigenvalues, Lyapunov exponent, bifurcation diagram and time responses (Fig. 5 and 6). It is observed that FHN model exhibits stable, periodic and chaotic behavior for different values of γ.

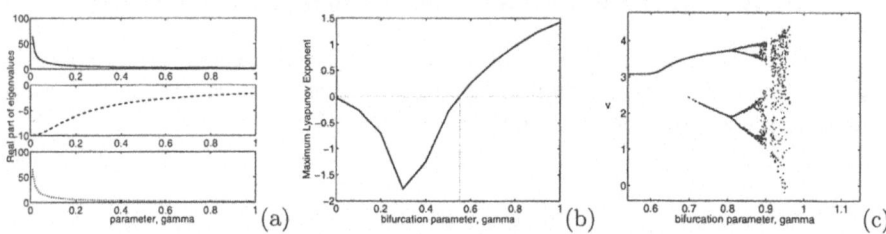

Fig. 5. Plots of real part of the eigenvalues, Maximum Lyapunov exponent and bifurcation diagram with γ as bifurcation parameter. (a) Real part of the eigenvalues; (b) Maximum Lyapunov exponent; (c) Bifurcation diagram.

4   Conclusion

As an effort towards seeking the possibility of presence of chaos in FitzHugh-Nagumo neuron model and biologically motivated neural networks we investigated the effect of parameter variations on the behavior of these models. It is observed that the firing-rate activities in neural network can exhibit chaotic

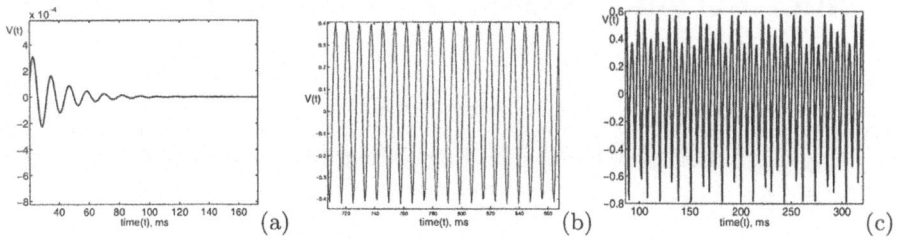

Fig. 6. Time responses for FHN model: (a) at $\gamma = 0.25$ (converging); (b) at $\gamma = 0.5$ (oscillatory); (c) at $\gamma = 0.9$ (chaotic).

behavior at some parametric values. The results of FitzHugh-Nagumo neuron model depict the presence of chaotic behavior in this model also. Thus chaos theory can be applied to explain the dynamics of the brain in order to explain many characteristics of the EEG data.

References

1. Koch, C.: Biophysics of Computation,Oxford University Press, (1999)
2. Zak, S.: Systems and Control,Oxford University Press, (2003)
3. Strogatz, S.H.: Nonlinear Dynamics and Chaos,Westview Press, (2000)
4. Haykin, S.: Neural Networks: A Comprehensive Foundation, Second Edition, Pearson Education Asia, (2002)
5. Gao, Y.: Chaos and bifurcation in the space-clamped FitzHugh-Nagumo system, Academy of Mathematics and System Sciences, Chinese Academy of Sciences, Chaos Solutions and Fractals, (2004) 943–956
6. Das, A., Roy, A.B., Das, P.: Chaos in three dimensional neural networks, Applied Math. Modeling, (2000) 511–522
7. Das, A., Roy, A.B., Das, P.: Chaos in three dimensional general model of neural network, Int. J. Bifurcation and Chaos, (2001)
8. Freeman, W.J., Cybern, B.: Simulation of chaotic EEG patterns with a dynamic model of the olfactory system, Biological Cybernetics, (1987) 139–150
9. Hodgkin, A.L., Huxley, A.F.: Aquantitative description of membrane current and its applications to conduction and excitation in nerve, J. Physiol. (Lond.), 116 (1952) 500–544
10. Guckenheimer, J., Oliva, R.A.: Chaos in the Hodgkin-Huxley Model,Siam Journal on Applied Dynamical Systems, Vol. 1, No.1, (2002) 105–114
11. FitzHugh, R.: Impulses and physiological states in theoretical models of nerve membranes. Biophys. J.1 (1961) 445–466
12. Wiggins, S.: Introduction to applied nonlinear dynamical systems and chaos, Springer-Verlag (1990)
13. Nagumo, J.S., Arimoto, Yoshizawa, S.: An active pulse transmission line simulating nerve axon. Proc. IRE 50 (1962) 2061–2071
14. Jing, Z., Chang, Y., Guo, B.: Bifurcation and chaos in discrete FitzHugh-Nagumo system, Chaos, Solitons and Fractals, Volume 21, Issue 3, (2004) 701–720

Snap-Shots on Neuroinformatics and Neural Information Processing Research in Singapore

Lipo Wang[1,2]

[1] College of Information Engineering, Xiangtan University,
Xiangtan, Hunan, China
[2] School of Electrical and Electronic Engineering,
Nanyang Technology University,
Block S1, Nanyang Avenue, Singapore 639798
elpwang@ntu.edu.sg

Abstract. This paper summarizes some of the key research areas in neuroinformatics and neural networks that have recently been or are being carried out in Singapore. Researchers in Singapore have been proposing novel algorithms for various neural networks, i.e., radial basis function networks, fuzzy neural networks, multilayer perceptrons, and support vector machines. Neural networks have been applied to solving a wide variety of difficult problems in bioinformatics, multimedia, data mining, and communications. Researchers in Singapore are also working with neurophysiologists on functional brain imaging, brain atlases, and brain disease analysis.

Keywords: RBF, fuzzy, neural network, brain, neuroinformatics

1 Introduction

Despite its small geographical size, Singapore has been making significant contributions to the research and development in neuroinformatics and neural networks over the past decades. Active research are being carried out in many fronts, including novel neural network algorithms, applications of neural networks, as well as investigations in functional brain imaging, brain atlases, and brain disease analysis.

Most of the academic research activities concentrate in the two major universities in Singapore, that is, the National University of Singapore (NUS) and the Nanyang Technological Unviersity (NTU), whereas a number of national research institutes (RIs) mainly carry out applied research. The Agency for Science, Technology, and Reasearch (A*STAR) is the main source of scientific and technological research funding in Singapore, and manages all national RIs in Singapore. A number of RIs are active in neural network and related research, especially Singapore Institute of Manufacturing Technology (SIMTech), the Institute of Infocomm Research (I^2R), and the BioInformatics Institute (BII). The Ministry of Education also funds a part of the research in the two universities, for example, with funding to cover scholarships for graduate students.

N.R. Pal et al. (Eds.): ICONIP 2004, LNCS 3316, pp. 874–879, 2004.
© Springer-Verlag Berlin Heidelberg 2004

As suggested by its title, the present paper by no means reflects an exhaustive collection of *all* research activities in neuroinformatics and neural networks in Singapore.

2 Proposing Novel Neural Network Algorithms

Researchers at the NTU Neural Network Research Group (NNRG) led by Lipo Wang have recently developed an importance ranking technique called "separability-correlation measure" [1] for feature selection. After eliminating irrelevant input features, the complexity of a classifier can be reduced and the classification performance improved. They also proposed a modified method for efficient construction of an RBF classifier [1], by allowing for large overlaps between clusters corresponding to the same class label. This approach significantly reduces the structural complexity of the RBF network and improves the classification performance.

The NNRG also proposed a noisy chaotic neural network, which combines the best properties of conventional simulated annealing and efficient chaotic search [2]. They have successfully applied this method to a variety of optimization problems, such as broadcast scheduling in packet radio networks, channel assignment in mobile communications [3], noisy and blurred image restoration, and image segmentation [4].

Lu and Rajapakse at the NTU School of Computer Engineering incorporated constraints or prior knowledge into learning algorithm while improving independent component analysis (ICA). They have developed a method to eliminate indeterminacy in ICA [6]. Further, they have developed the ICA with Reference (ICA-R) to incorporate input stimuli in fMRI experiments into the analysis directly.

Phua and Ming at the NUS Department of Computer Science proposed the use of parallel quasi-Newton (QN) optimization techniques to improve the rate of convergence of the training process for multilayer perceptron (MLP) neural networks [7]. Simulations with nine benchmark problems show that the proposed algorithms outperform other existing methods. Guan and Li [8] at the NUS Department of Electrical and Computer Engineering proposed a method to divide the original problem into a set of subproblems, each of which is composed of the whole input vector and a fraction of the output vector. By means of a monotonic transformation, Toh [5] at the I^2R derived a sufficient condition for global optimality of a network error function, based on which a penalty-based algorithm was derived directing the search towards possible regions containing the global minima.

Lee, Keerthi, Ong, and DeCoste [10] from the NUS Mechanical Engineering Department proposed an efficient method for computing the leave-one-out (LOO) error for support vector machines (SVMs) with Gaussian kernels quite accurately. The new method often leads to speedups of 10C50 times compared to standard LOO error computation. Keerthi [9] proposed an efficient iterative techniques for computing radius and margin in SVMs.

3   Applications of Neural Networks

LimSoon Wong, Valdimir Brusic, Vladimir Brajic, and Li Jinyang of the Knowledge Discovery Group of the I^2R used neural networks for a number of bioinformatics applications, such as prediction of immunogenic peptides [25] and recognition of promoters [26]. In addition, they have used neural networks to find translation initiation sites and transcription start sites from genomic sequences.

Lo and Rajapakse combined hidden Markov models (HMMs) and neural networks in order to capture compositional properties of complex genes with minimal error rate and high correlation coefficient. Splice site detection was also examined by a higher-order Markov model implemented with a neural network [17].

Wang et al are applying novel feature selection techniques and a dynamically-generated fuzzy neural network [18][19] that they proposed earlier to cancer classification with microarray gene expression data. They [20] were able to obtain highly accurate results using a much smaller number of genes compared to other results reported in the literature. It is useful to reduce the number of genes required for cancer classification because: (a) it means lighter computational burden and lower "noise" arising from irrelevant gene data; (b) in some cases it even becomes possible to extract simple rules for doctors to make accurate diagnosis without the need for any classifiers; (c) it can simplify gene expression tests to include only a small number of genes rather than thousands of genes, which may bring down the cost for cancer testing; and (d) it calls for further investigations into possible biological relationship between these small numbers of genes and cancer developmenttreatment.

Corne, Fogel, Rajapakse, and Wang are editing a special issue for the Soft Computing journal on "Soft Computing for Bioinformatics and Medical Informatics", featuring several novel neural network applications.

Cao and Tay [11] from the NUS Mechanical Engineering Department applied SVMs to financial time series prediction. Five real futures contracts collated from the Chicago Mercantile Market were used as the data sets. Song et al from the NTU School of Electrical and Electronic Engineering [22] applied a robust SVM for bullet hole image classification.

Wang edited a book titled "Support Vector Machines: Theory and Applications" to be published by Springer-Verlag in 2004 [12]. This book consists of about 20 chapters contributed by active researchers in the SVM area covering the most recent advances in both theoretical studies and SVM applicaions to a variety of challenging problems. A recent book edited by Rajapakse and Wang [13] entitled "Neural Information Processing: Research and Development" includes numerous examples on latest models and applications of neural networks. Two edited volumes by Halgamuge and Wang concentrate on recent work on clustering, prediction, and data mining [14][15]. Another book [16] covers a range of neural network applications to multimedia processing, including image and video processing.

Zhang et al [31] of the Media Division of I^2R recently developed a high performance face recognition system by combining Gabor wavelet networks for visual pattern representation and kernel associative memories for nonlinear concept learning. Werner et al [32] of the Communication and Devices Division proposed a neural network approach for adaptive routing in dynamic topology networks, e.g., LEO satellite networks and mobile ad hoc networks. Suganthan at the NTU School of Electrical and Electronic Engineering used [21] self-organizing maps for shape indexing.

K.K. Tan and colleagues at the Department of Electrical and Computer Engineering at the NUS developed neural network-based nonlinear controllers and tested these controllers in simulation and laboratory work. Excellent performance and transient response under different operating conditions were established. These controllers will be applied in the power electronic industry [23].

Srinivasan leads a three-year project applying neural networks to real-world electrical load forecasting. Neural network-based prediction models have also been developed for congestion prediction in dynamic street traffic network [24]. As the traffic patterns vary from day to day, a classifier is used to group the input data into a number of clusters. Feedforward neural networks are then individually tuned to specify the input-output relationship for each cluster. The neural network architecture is optimized using evolution algorithms. Another funded project involves development, simulation and testing of neural network-based systems for incident detection and network re-routing for traffic networks. A novel constructive probabilistic neural network has been developed and excellent performance on real data obtained from Singapore expressways has been obtained.

4   Research in Neuroinformatics

The Biomedical Imaging Lab of BII, led by Wieslaw Nowinski, carries out research and development in a variety of areas, including construction of anatomical, functional and vascular brain atlases; development of atlas-based applications for neurosurgery, neuroradiology, human brain mapping [27]; rapid and automated brain segmentation based on domain knowledge and image processing [28] [29]; identification and localization of brain pathology; multi-modal and model-to-data registration; geometric and physical modeling; brain databases, and virtual reality in brain intervention. BII has constructed the Cerefy Brain Atlas Database with gross anatomy, brain connections, subcortical structures, and sulcal patterns containing 1,000 structures and 400 sulcal patterns.

Neuroinformatics lab at I^2R, led by Guan Cuntai, focuses on the investigation and development of effective mathematical framework and learning algorithms for the analysis of brain signals, with emphasis on EEG (electroencephalographic) signal.

The NTU neuroinformatics group concentrates on the analysis of both structural and functional MRI in order to segment different tissue classes of the brain or identify subcortical structures of the brain. They developed techniques for the detection of activation from functional MR images and modeling human brain surface by active NURBS surfaces [30].

5 Conclusions

Despite the small geographic size and the absence of national consorted programs in neuroinformatics or neural networks like those in Japan, Korea, and China, there has been very active research in both neuroinformatics and neural networks in Singapore. We expect that these efforts will continue in the future and more exciting research results will be reported by researchers from Singapore.

References

1. Fu, X., Wang, L.: Data Dimensionality Reduction with Application to Simplifying RBF Network Structure and Improving Classification Performance. IEEE Transactions on System, Man, Cybern, Part B - Cybernetics, Vol. 33. (2003) 399–409
2. Wang, L., Li, S., Tian, F., Fu, X.: A Noisy Chaotic Neural Network for Solving Combinatorial Optimization Problems: Stochastic Chaotic Simulated Annealing. IEEE Trans. System, Man, Cybern, Part B - Cybernetics (2004)
3. Wang, L.: Soft Computing in Communications. Springer-Verlag, Berlin (2003) 131–145
4. Yan, L., Wang, L., Yap, K.H.: A noisy chaotic neural network approach to image denoising. Proc. IEEE Intern. Conf. Image Processing (2004)
5. Toh, K.A.: Deterministic Global Optimization for FNN Training. IEEE Trans. Systems, Man and Cybernetics, Part B, Vol. 33 (2003) 977 - 983
6. Lu., W., Rajapakse, J.C.: Eleminating Indeterminacy in ICA. Neurocomputing, Vol. 50 (2003) 271–290
7. Phua, P.K.H., Ming, D.: Parallel Nonlinear Optimization Techniques for Training Neural Networks. IEEE Trans. Neural Networks, Vol. 14 (2003) 1460–1468
8. Guan, S., Li, S.: Parallel Growing and Training of Neural Networks Using Output Parallelism. IEEE Trans. Neural Networks, Vol. 13 (2002) 542–550
9. Keerthi, S.S.: Efficient Tuning of SVM Hyperparameters Using Radius/margin Bound and Iterative Algorithms. IEEE Trans. Neural Networks, Vol. 13. (2002) 1225–1229
10. Lee., M.M.S., Keerthi, S.S., Chong., J., DeCoste, D.: An Efficient Method for Computing Leave-one-out Error in Support Vector Machines with Gaussian Kernels. IEEE Transactions on Neural Networks, Vol. 15 (2004) 750–757
11. Cao, L.J., Tay, F.E.H.: Support Vector Machine with Adaptive Parameters in Financial Time Series Forecasting, IEEE Trans. Neural Networks, Vol. 14 (2003) 1506–1518
12. Wang, L.: Support Vector Machines: Theory and Applications. Springer-Verlag (2004)
13. Rajapakse, J.C., Wang, L.: Neural Information Processing: Research and Development. Springer-Verlag (2004)
14. Halgamuge, S., Wang, L.: Computational Intelligence for Modelling and Predictions. Springer-Verlag (2004)
15. Halgamuge, S., Wang, L.: Classification and Clustering for Knowledge Discovery, Springer-Verlag (2004)
16. Tan, Y.P., Yap, K.H., Wang, L.: Intelligent Multimedia Processing with Soft Computing. Springer-Verlag (2004).
17. Ho, S.L., Rajapakse, J.C.: Splice Site Detection with a Higher-order Markov Model Implemented on a Neural Network. Genomic Informatics (2003)

18. Frayman, Y., Wang, L.: A Dynamically-constructed Fuzzy Neural Controller for Direct Model Reference Adaptive Control of Multi-input-multi-output Nonlinear Processes. Soft Computing, Vol. 6 (2002) 244–253
19. Frayman, Y., Wang, L.: A Fuzzy Neural Approach to Speed Control of an Elastic Two-mass System. Proc. 1997 International Conference on Computational Intelligence and Multimedia Applications (1997) 341–345
20. Liu, B., Wan, C., Wang, L.: Unsupervised gene selection via spectral biclustering. Proc. IJCNN (2004)
21. Suganthan, P.N.: Shape indexing using self-organizing maps. IEEE Trans. Neural Networks. vol.13 (2002) 835 - 840
22. Song, Q., Hu, W., Xie, W.: Robust Support Vector Machine with Bullet Hole Image Classification. IEEE Trans. Systems, Man and Cybernetics, Part C, Vol. 32 (2002) 440–448
23. Huang, S.N., Tan, K.K., Lee, T.H.: Further Results on Adaptive Control for a Class of Nonlinear Systems Using Neural Networks. IEEE Trans. Neural Networks, Vol. 14 (2003) 719–722
24. Srinivasan, D., Xin J., Cheu, R.L.: Evaluation of Adaptive Neural Network Models for Freeway Incident Detection. IEEE Trans. Intelligent Transportation Systems, Vol. 5 (2004) 1–11
25. Tatsumi T.: Mage-6 Encodes Hla-drbeta1*0401-presented Epitopes Recognized by Cd4+t Cells from Patients with Melanoma or Renal Cell Carcinoma. Clinical Cancer Research, Vol. 9 (2003) 947–954
26. Bajic, V.B., Seah, S.H.: Dragon Gene Start Finder: An Advanced System for Finding Approximate Location of the Start of Gene Transcriptional Units. Genome Research, Vol. 13 (2003) 1923–1929
27. Nowinski, W.L., Thirunavuukarasuu, A.: The Cerefy Clinical Brain Atlas. Theime (2004)
28. Xia, Y., Hu, Q., Nowinski, W.L.: A Knowledge-driven Algorithm for a Rapid and Automatic Extraction of the Human Cerebral Ventricular System from mr Neuroimages. NeuroImage, Vol. 21 (2004) 269–283
29. Hu, Q., Nowinski, W.L.: A Rapid Algorithm for Robust and Automatic Extraction of the Midsagittal Plane of the Human Cerebrum from Neuroimages Based on Local Symmetry and Outlier Removal. NeuroImage, Vol. 20 (2003) 2154–2166
30. Meegama, R.G.N., Rajapakse, J.C.: Nurbs-based Segmentation of the Brain in Medical Images. International Journal of Pattern Recognition and Artificial Intelligence, Vol. 17 (2003) 995–1009
31. Zhang, H.,Zhang, B., Huang, W., Tian, Q.: Gabor Wavelet Associative Memory for Face Recognition. IEEE Transactions on Neural Networks (2004)
32. Werner, M.: A Neural Network Approach to Distributed Adaptive Routing of Leo Intersatellite Link Traffic. VTC98. (1998) 1498–1502

Deciphering the Genetic Blueprint
of Cerebellar Development
by the Gene Expression Profiling Informatics

Akira Sato, Noriyuki Morita, Tetsushi Sadakata, Fumio Yoshikawa,
Yoko Shiraishi-Yamaguchi, JinHong Huang, Satoshi Shoji,
Mineko Tomomura, Yumi Sato, Emiko Suga, Yukiko Sekine,
Aiko Kitamura, Yasuyuki Shibata, and Teiichi Furuichi[*]

Laboratory for Molecular Neurogenesis, RIKEN Brain Science Institute
2-1 Hirosawa, Wako, Saitama 351-0198, Japan
tfuruichi@brain.riken.jp
http://www.brain.riken.go.jp/labs/lmn/index.html

Abstract. The brain is the ultimate genetic system to which a large number of genes are devoted. To extract and visualize biological information from such large data sets accumulated in the post-sequencing era, the use of bioinformatics would be a very powerful means. To understand the genetic basis of mouse cerebellar postnatal development, we have analyzed the whole transcription or gene expression (transcriptome) during the developmental stages on a genome-wide basis, and have systematized the spatio-temporal gene expression profile information in a comprehensive database (Cerebellar Development Transcriptome [CDT] database) from a bioinformatics point of view. This CDT database would open up a new field for deciphering the genetic blueprint for cerebellar development.

1 Introduction

Recently, the genome sequencing projects suggest that the mouse and the human genomes each seem to contain in the neighborhood of 30,000 protein coding genes (1, 2). The brain with the structural and functional complexity is thought to express a large number of genes (3, 4, 5). In the post-sequencing era, it is time to address how many and what kinds of genes are involved in its development, wiring up, and function.

The postnatal development of mouse cerebellum is accomplished by a series of cellular events (cell proliferation and migration, dendrogenesis and axogenesis, synaptogenesis, myelination, etc.) that are genetically coded (Fig. 1). These cytogenetic and morphogenetic events occur on schedule within the first three weeks of life. Although many researchers have paid attention to this hereditary plan, it is not understood completely. The decoding states of such genetic codes for cerebellar development can be captured by analyzing all the gene expressions (transcriptions), *"transcriptome"* during the developmental stages; in other words, the spatio-temporal gene expression profiling on a genome-wide basis precisely through the developmental stages would underlie the deciphering of the genetic blueprint for cerebellar development.

---

[*] To whom correspondence should be addressed.

N.R. Pal et al. (Eds.): ICONIP 2004, LNCS 3316, pp. 880–884, 2004.
© Springer-Verlag Berlin Heidelberg 2004

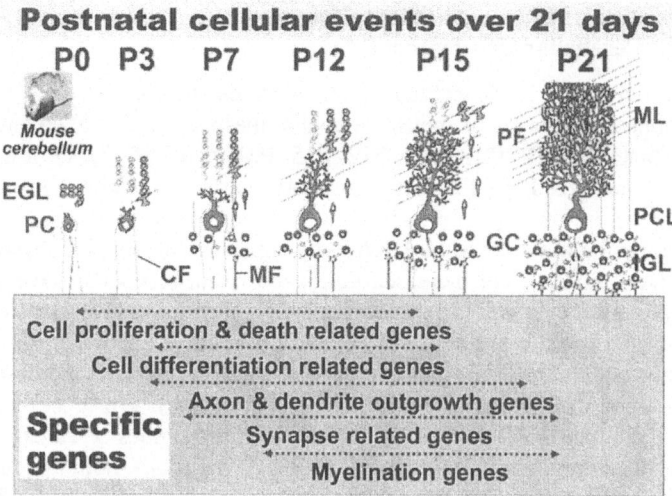

Fig. 1. Differential expression of specific gene groups responsible for a series of magnificent developmental events of mouse cerebellum within the first three weeks of life. EGL, external germinal layer; ML, molecular layer; PCL, Purkinje cell layer; IGL, internal granular layer; PC, Purkinje cell; GC, granule cell; CF, climbing fiber; MF, mossy fiber; PF, parallel fiber. GCs are generated by vigorous cell proliferation in the EGL, extend their PF axons, and migrate downward to the IGL where the MF-GC synapses are formed. PCs undergo robust outgrowth of dendrites and form elaborate arborization with numerous synapses with the PFs and CFs.

We are exploring the differential gene expression by utilizing genome-wide analysis approaches, fluorescence differential display (FDD), cDNA microarray, and GeneChip analysis. Then, spatial (cellular) and temporal (developmental) specificities in their expression patterns are being analyzed with *in situ* hybridization (ISH) brain histochemistry and reverse transcription-polymerase chain reaction (RT-PCR) analysis, respectively. To extract and visualize biological meanings from such large data sets accumulated, the use of bioinformatics would be a very powerful means. We systematize all these lines of information (spatio-temporal expression profiles [RT-PCR gel patterns, ISH brain images, GeneChip and microarray curves], gene annotation and clustering, links to other relevant databases, etc.) in the CDT database[1] from a bioinformatics point of view.

2   Results and Discussion

To understand the genetic blueprint for mouse cerebellar postnatal development, we are attempting to clarify the cerebellar development transcriptome (CDT) by exploring the gene expression responsible for the developmental stage by the genome-wide analysis approaches, and to systematize these lines of gene expression information in an integrative CDT database.

---

[1] The CDT database will be online accessible (announce at our web site: http://www.brain.riken.go.jp/labs/lmn/index.html).

882   Akira Sato et al.

2.1 Exploration of Differential Gene Expression during Cerebellar Development

First, by utilizing the FDD technique, we analyzed differential gene expression at eight developmental stages of mouse cerebella; these stages included embryonic day (E) 18, postnatal day (P) 0, P3, P7, P12, P15, P21, and P56 (3). Of a total of about 12,000 bands analyzed, 83.1% were constantly expressed throughout the postnatal stages, whereas the other (16.9%, about 2,000 bands) showed differential patterns. It was showed by the previous RNA-DNA hybridization analysis (4, 5, 6) that complexity of poly(A)+ RNAs were increased during postnatal cerebellar development. The FDD analysis also indicates that the complexity of gene expression (in number of genes) reaches a peak in the first and second postnatal weeks when the cytogenetic and morphogenetic events are concentrated (about 20% increase in comparison with earlier stage, and then about 12 % decrease to later stage). We succeeded in cloning 2,194 non-redundant FDD clones (as of 2002). These clones were categorized into four groups: known genes (1,843 clones, 85.3% of the total), homologous sequences (57 clones, 2.7%), expressed sequence tags (ESTs) (162, clones, 7.5%), and genome/unknown/etc. (141 clones, 6.5%) (as of August, 2004).

Second, the GeneChip (Affymetrix GeneChip MuU74, 12,654 genes) showed that most of them (10,321 genes, 81.6%) were differentially expressed at least in one of the five postnatal stages tested (E18, P7, P14, P21, and P56) (7). We made the underlying data publicly available at our web site (http://www.brain.riken.go.jp/labs/lmn/GeneChipCblDev.html) and also through the Gene Expression Omnibus (http://www.ncbi.nih.gov/geo). Of these 10,321 genes, 897 genes (8.7%) showed a differential expression with more than two-fold changes. Functional clustering (34 gene clusters) of these genes indicates a developmental transcriptomic feature that many genes identified are categorized into the clusters related to cell cycle and proliferation, cell growth and differentiation, transcriptional regulation, and signal transduction (7).

Third, we generated the custom-made cDNA microarray containing about 2,400 cDNA sequences (FDD clones plus about 250 cerebellar genes that are known to be indispensable for cerebellar development and function). We have applied this cDNA microarray to the gene expression profiling during cerebellar development.

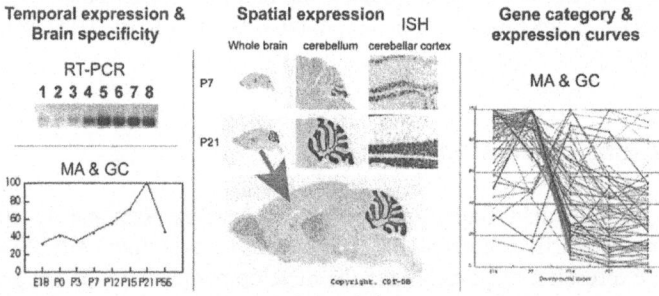

Fig. 2. Spatio-temporal gene expression profiling data. Temporal patterns and brain specificities are defined by RT-PCR. GeneChip (GC) and microarray (MA) curves are also compiled for temporal expression profiling. Spatial patterns are defined by ISH. ISH images of whole brains, cerebella, and cerebellar cortices (including magnified views). 34 gene categories and expression curves (GC and MA data) within each category are compiled.

2.2 Spatio-temporal Gene Expression Profiling

We are intensively analyzing the spatio-temporal expression profiles of the genes identified as described above. The *temporal*, developmental expression patterns are analyzed by RT-PCR using RNAs isolated from E18, P0, P3, P7, P12, P15, P21, and P56 cerebella. Then, the *spatial*, cellular expression patterns are analyzed with ISH histochemistry of sagittal sections of P7 and P21 brains (3, 8). The *brain specificity* (tissue distribution patterns) is determined by RT-PCR using RNAs prepared from eight different tissues (brain, thymus, heart, lung, liver, kidney, spleen, and testis). All the data of RT-PCR gel banding patterns and ISH brain staining images are acquired and complied as digital data with image scanners and CCD camera-equipped microscopes, respectively.

2.3 Cerebellar Development Transcriptome (CDT) Database

We integrate the gene expression profile information of about 2,600 genes (gene annotations, functional clusters, RT-PCR patterns of about 1,500 genes, ISH brain images of about 1,000 genes, expression plots and relative values of about 1,100 genes) (as of August, 2004) in the CDT database. The CDT database is an online database (OS, Linux; database software, Oracle; operation software, Internet Explorer or Netscape) with various search functions (expression profiles [temporal patterns, spatial patterns, brain distribution patterns, and brain specificity], gene cluster, keyword [gene name, protein domain and function, cellular function, etc.], and ID number) and multiple links to public bioinformatics databases (Jackson Lab.-*MGI*, EBI/Sanger - *Ensembl*, NCBI-*Entrez Gene*, -*GEO*, -*Nucleotide*, -*OMIM*, -*UniGene*, -*PubMed*) for easy access to additional information.

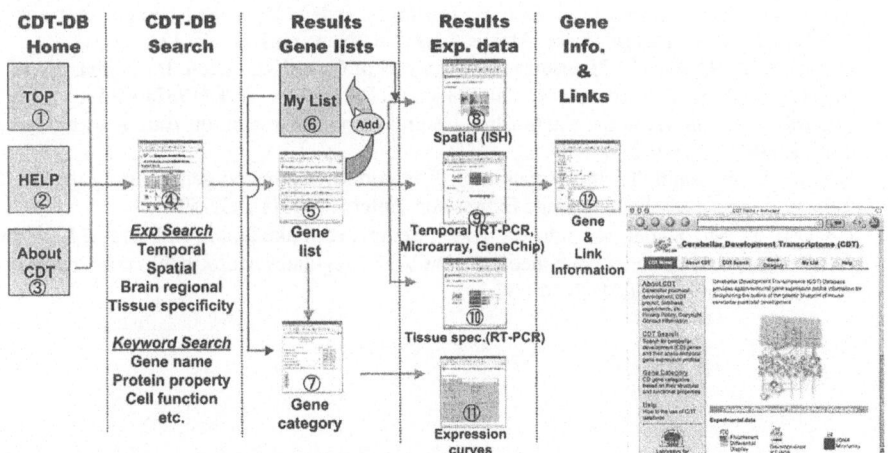

Fig. 3. The outline of the Cerebellar Development Transcriptome (CDT) database (left) and a view of CDT database home page (right).

3 Conclusions

For deciphering the genetic blueprint of mouse cerebellar development, we generate the database in which the CDT information is integrated. The CDT database provides us a clue to extract and visualize an outline of the genetic basis of cerebellar development, and would be a product due to the coming of neuroinformatics era. Further studies on relational networks of the genes profiled will shed light on the genetic blueprint for cerebellar development.

Acknowledgements

This study was supported by Grants-in-Aid for Scientific Research from the Japanese Ministry of Education, Culture, Sports, Science and Technology (MEXT) and the Japan Society for the Promotion of Science (JSPS) and by the Institute of Physical and Chemical Research (RIKEN).

References

1. International Human Genome Sequencing Consortium: Initial sequencing and analysis of the human genome. Nature (2001) 409: 860-921.
2. Mouse Genome Sequencing Consortium: Initial sequencing and comparative analysis of the mouse genome. Nature (2002) 420: 520-562.
3. Shiraishi, Y., Mizutani, A., Bito, H., Fujisawa, K., Narumiya, S., Mikoshiba, K., Furuichi, T.: Cupidin, an isoform of Homer/Vesl, interacts with the actin cytoskeleton and activated Rho family samall GTPases and is expressed in developing mouse cerebellar granule cells. J. Neuosci. (1999) 19: 8389-8400.
4. Kaplan, B.B.: Current Approaches to the Study of Gene Expression in the Adult and Developing Brain. In: Rassin, D., Harber, B, Brujan B. (eds.): Basic and Clinical Aspects of Nutrition and brain Development. Alan R. Liss, Inc., New York (1987) 131-155.
5. Kaplan, B.B.: RNA-DNA Hybridization: Analysis of Gene Expression. In: Lajtha, A. (ed.): The Handbook of Neurochemistry. Vol. 2. Plenum Press, New York (1982) 1-26.
6. Sutcliffe, J.G.: mRNA in the Mammalian Central Nervous System. In: Ann. Rev. Neurosci. Vol. 11. (1988) 157-198.
7. Kagami, Y., Furuichi, T.: Investigation of differentially expressed genes during the development of mouse cerebellum. Gene Expression Patterns (2001) 1: 39-59.
8. Sadakata, T., Mizoguchi, A., Sato, Y., Katoh-Semba, R., Fukuda, M., Mikoshiba, K., Furuichi, T.: Secretory granule-associated protein CAPS2 regulates neurotrophin release and cell survival. J. Neurosci. (2004) 24: 43-52.

Korean Neuroinformatics Research Program: From the Second Phase to the Third Phase

Soo-Young Lee

Brain Science Research Center, Department of BioSystems,
and Department of Electrical Engineering and Computer Science
Korea Advanced Institute of Science and Technology
373-1 Guseong-dong, Yuseong-gu, Daejeon 305-701, South Korea
sylee@kaist.ac.kr

Abstract. The second phase of the Korean Brain Neuroinformatics Research Program had completed by June 2004, and the 3rd phase had started from July 2004 for 4 years. Since the 3rd phase is regarded as the final phase of Korean brain national research program started from November 1998 for 10, the 3rd phase program aims for the system integration of the developed technologies as well as continuation of previous research efforts on basic technologies for artificial vision, auditory, inference, and behaviour systems. The system integration will come as a "digital brain" to combine all the 4 functions, and an integrated demonstration system, i.e., "artificial secretary" alias "OfficeMate" with exceptional human-like information processing capabilities. Researches on measurement and signal analysis for neuroscience are also included as the infrastructure. It is a joint effort of researchers from many different disciplines including neuroscience, cognitive science, mathematics, physics, electrical engineering, computer science, etc.

1 Introduction

The Korean Brain Research Program has been sponsored by Korean Ministry of Science and Technology for 10 years from November 1998, and includes research programs on neurobiology, neuroinformatics, and biomedical applications on brain diseases. [1][2] The Neuroinformatics program has two goals, i.e., to understand information processing mechanisms in biological brains and to develop intelligent machines with human-like functions based on the mechanism. [3][4] It is a joint effort of researchers from many different academic disciplines, including cognitive neuroscience, mathematics, electrical engineering, and computer science. The second phase had been completed by June 2004, and the Program got into its third and final phase for 4 years from July 2004. [5] In this paper the results of the second phase and the goals of the third phase will be presented.

2 Results of the Second Phase

At the second phase the Brain Neuroinformatics Research Program has 4 functional groups for the 4 brain functional module, i.e., artificial vision system, artificial auditory system, cognition & inference system, and human behaviour system. Also, the

N.R. Pal et al. (Eds.): ICONIP 2004, LNCS 3316, pp. 885–890, 2004.
© Springer-Verlag Berlin Heidelberg 2004

fifth group is devoted to measurements of brain signals such as fMRI and EEG. Each group has a fully-multidisciplinary research team. [3][4] As shown in Figure 1, for each scientific issue to be studied, one need start from measured data, propose mathematical models, and develop functional systems. If measured data are not available, one need develop measurement technology and conduct the measurement by oneself. To propose mathematical models from the measured data one may utilize information pro-cessing algorithms, and the mathematical models may propose other measurements to confirm the models. To develop functional systems one may need specific hardware implementations, and the developed functional systems may verify the mathematical models. Therefore, a "system approach" is required, and an interdisciplinary research team had been formed for each functional module to integrate efforts of researchers from many different academic disciplines. Also, one team is devoted to measurement technology for brain signals, mainly fMRI and EEG signals.

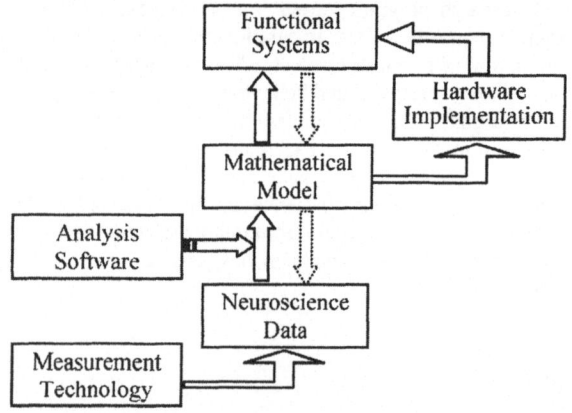

Fig. 1. Systems Approach of Brain Neuroinformatics Researches.

For example, research activities focusing on the auditory module are based on the simplified diagram of the human auditory central nerve system. The mechanical vibration of the eardrum is converted into neural signals at inner hair cells (IHCs) through the basilar membrane in the cochlea. Each IHC signal represents an acoustic input signal with specific frequency filtering and nonlinear characteristics. Hence it is believed that the "front end" of the human ear is a nonlinear spectral analyzer. The IHC signals from left and light cochleae are combined at superior olivery complexes (SOCs), and further go to auditory cortexes through inferior colliculus (IC) and MGB. This binaural signal processing at SOCs conducts sound localization and noise reduction. Although earlier auditory signal processing mechanisms at the cochlea and possibly up to the SOC-level are relatively well understood, the signal processing mechanism between the SOC and auditory cortex is less understood, and therefore represented as dotted lines. It is also known that some neurons at the auditory cortex layer respond to specific sound components with complex time-frequency characteristics. Speech recognition and language understanding take place at the higher-level brain. It is worth noticing that, in addition to forward signal paths, there also exist backward signal paths. This backward path is responsible to top-down attention, which filters out irrelevant components from noisy input speeches.

Detail functions currently under modelling are summarized in Fig. 2. The object path, or "what" path, includes nonlinear feature extraction, time-frequency masking, and complex feature formation from cochlea to auditory cortex. These are the basic components of speech feature extraction for speech recognition. The spatial path, alias the "where" path, consists of sound localization and noise reduction with binaural processing. The attention path includes both bottom-up (BU) and top-down (TD) attention. However, all of these components are coupled together. Especially, the combined efforts of both BU and TD attentions control the object and spatial signal paths.

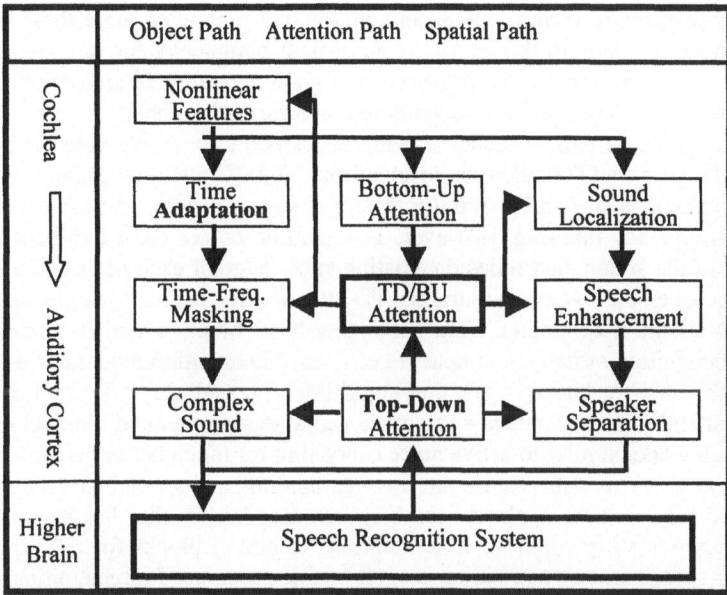

Fig. 2. Three functional paths and their relationship in an artificial auditory system. The red and blue lines denote forward and backward signal pathways, respectively, and the latter is controlled by top-down selective attention based on previous knowledge acquired into the speech recognition system in the brain.

The nonlinear feature extraction model is based on cochlear filter bank and logarithmic nonlinearity. The cochlear filter bank consists of many bandpass filters, of which center frequencies are distributed linearly in logarithmic scale. The quality factor Q, i.e., ratio of center frequency to bandwidth, of bandpass filters is quite low, and there are overlaps in frequency characteristics. The logarithmic nonlinearity provides wide dynamic range and robustness to additive noises. Time-frequency masking is a psychoacoustic phenomenon, where a stronger signal suppresses weaker signals in nearby time and frequency domains. Frequency masking is modelled by lateral inhibition in the frequency domain, which also helps to increase frequency selectivity with overlapping filters. Time masking is also implemented as lateral inhibition, but only forward (progressive) time masking is incorporated. Another important research subject is the representation (or coding) of the speech signals. Simple frequency features at the inner-hair-cells and complex features such as onset/offset and frequency modulation at the auditory cortex are modelled based on sparse-coding or independent component analysis.

For the binaural processing at the spatial path conventional models estimate interaural time delay, i.e., time-delay between signals from left and right ears, based on cross-correlation, and utilize the time-delay for sound localization and noise reduction. Interaural intensity difference is also utilized for advanced models. However, these models assume only direct sound paths from a sound source to two ears, which is not valid for many real-world environments with multipath reverberation and multiple sound sources (e.g., speech inside an automobile with external road and wind noise, and reverberation of speech mixed with music from the audio system). Therefore, it is required to incorporate deconvolution and separation algorithms in the binaural processing. Due to the increased number of parameters for time-delayed components the simple correlation measure is not good enough, and an extended binaural processing model has been developed based on information theory.

For the attention path, a model is being developed to combine both the bottom-up (BU) and top-down (TD) attention mechanisms. The BU attention usually incurs from strong sound intensity and/or rapid intensity changes in time, and closely related to the time-frequency masking. However, TD attention comes from familiarity and importance of the sound, and relies on existing knowledge of each person. For example, a word or a person's voice may trigger TD attention for relevant people only. Therefore, TD attention originates from the higher-level brain, which is modelled by a speech recognition system. A simple yet efficient TD attention model has been developed based on error back-propagation algorithm with multi-layer Perceptron classifiers. TD attention also provides a reference signal to the extended binaural processor, which then works similar to active noise cancelling for much better performance.

Many auditory models require intensive computing, and special hardware has been developed for real-time applications. A speech recognition chip has been developed as a System-On-Chip solution, which consists of circuit blocks for A/D conversion, nonlinear speech feature extraction, programmable processor for recognition, and D/A conversion. Also, the extended binaural processing model has been implemented at the FPGA level, and an ASIC version will be introduced in the near future.

All three functional paths are coupled together, and models of each path only are expected to contribute slightly to the overall performance improvement. All these processing mechanisms are being understood, analyzed, and implemented by integrated chips. It is interesting to note that the human auditory processing mechanism utilizes information theory in many respects.

3   Goals of the Third Phase

The third phase program will consist of 9 projects in 4 areas, i.e., brain-like systems, artificial vision and auditory technology, artificial cognition and human behaviour, and brain signal measurement and analysis. [5]

In the brain-like systems we will have 2 projects, i.e., "Artificial Brain" project and "Active Vision System" project. In the artificial brain project we will develop an integrated hardware and software platform for the brain-like intelligent systems, which combine all the technologies developed for the brain functions in the second phase. With 2 microphones, 2 cameras (or retina chips), and one speaker, the artificial brain will have the functions of vision, auditory, cognition, and behaviour. The sound localization, speech enhancement, visual attention, face recognition, lip reading, and emotion recognition and representation are included. It will have self-learning and user modelling capabilities for practical user interfaces. Also, with this platform, we plan to develop a testbed application, i.e., "artificial secretary" alias "office mate." The "Active Vision System" is a part of the "Artificial Brain," but a small team will be assigned to this important building block.

Table 1. Project Names of the Third Phase Korean Brain Neuroinformatics Research Programm.

Area	Project Name
Brain-like System	Artificial Brain and its Application (Office Mate)
	Active Vision System
Brain-based Audio-Visual Technology	Stereo-Vision and Space Perception
	Sound Source Separation and Recovery
	Continuous Speech Understanding
Brain-based Cognition and Behavior Technology	Memory and Cognition Model
	Sensory Information Fusion and Human Behavior Model
	Brain-based Active Learning
Brain Signal Measurement	Brain Signal Measurement and Analysis

We will also have 6 projects for the 2 future technologies, which require more understanding on the brain mechanism for the mathematical modeling and engineering applications for the brain-like systems. For these projects it is strongly recommended to form interdisciplinary research teams among cognitive scientists and information engineers.

In the human-like vision and auditory technology we will focus on 3 technologies, which are required for the artificial brain but need more development during the third phase. The stereo vision is important for the perception of 3-dimensional objects and spaces. Although the binaural processing had been studied in the context of independent component analysis and will be incorporated into the artificial brain, it needs further development for the distributed or more sources. Also, it is important to find the basic differences between the recognition of isolated and continuous speeches, and utilizes it to the robust continuous speech recognition tasks.

In the cognitive and human behaviour technology we will work on mathematical models of memory and cognitive inference. We will also work on mathematical models of sensory fusion and human behaviour, which include the evaluation of the emotional states and their relationship to the human behaviour. The last technology project is devoted to the learning and evaluation program, which is based on the learning mechanism of the human being.

The team for the brain signal measurement and analysis will continue to work on new technologies for the simultaneous measurement of fMRI and EEG signals, DT-MRI, and ms-MRI technologies.

4 Conclusion

The goal of the Korean Brain Neuroinformatics Research Program is to make the future human society as productive and enjoyable as possible. Intelligent machines will serve humans with human-like 5 sensors and self-learning inference capability. Those machines will work like your friends and colleagues. With the help of these intelligent machines human productivity will be greatly increased, which will result in the Third Revolution after the Industrial Revolution and Computer Revolution.

With the reduced working time and less efforts for the living itself, human will be able to put more efforts on "more human-like" jobs. Works requiring creativity will still remain as the sole jobs of human beings. Science, engineering, and art may be examples of those works. Much faster progress will be obtained in science and engineering, which will again serve for the prosperity of human beings.

Intelligence to machine! Freedom to mankind!

Acknowledgment

The author would like to thank Korean Ministry of Science and Technology to support the Korean Brain Neuroinformatics Research Program.

References

1. S.Y. Lee, *et al.*, *A Study on Brain Research Programs*, Research Report to Korean Ministry of Science and Technology, Sept. 1997.
2. Korean Ministry of Science and Technology, *Basic Plan for Brain Research and Development Program – Braintech21*, Sept. 1997.
3. Y. Kim and S.Y. Lee, *Initiatives on Auditory Pathway Modelling for Robust Speech Recognition at KAIST*, SpeechTek, Nov. 2002.
4. S.Y. Lee, *Korean Initiatives on Brain-like Information Processing Systems: From Biology to Functional Systems*, Neural Network Society Newsletter, July 2003.
5. S.Y. Lee, *Korean Brain Neuroinformatics Research Program: The 3rd Phase*, IJCNN2004, Budapest, Hungary, July 2004.

A Guided Tour of Neuroinformatics Research in India

Prasun Kumar Roy and Nandini Chatterjee Singh

National Brain Research Centre, Manesar, Gurgaon, Haryana 122 050, India
{pkroy,nandini}@nbrc.ac.in

Abstract. The study of neuroinformatics and computational neuroscience in India encompasses research on neural information processing, neural networks, information theory, neuroimaging, processing of EEG signals, speech production and neuronal recordings. The computational neuroscience division of NBRC is developing computational techniques to investigate speech production and their disorders in children, with special reference to autism. Stochastic Resonance approach is being used to enhance neuroimaging findings with application to brain tumour diagnosis and contrast accentuation during scanning. The National Facility of Functional Brain Imaging is being set up at NBRC to serve as India's first research venture on high-field MRI/fMRI/MRS/MRSI. Research work pursued in other institutions in the country is also briefly outlined, implying the presence of a vibrant research community. Efforts are also being undertaken to develop a National Neuroinformatics Initiative.

1 Introduction

The National Brain Research Centre (NBRC) is the first institute of its kind in India devoted to the development and promotion of fundamental research on Neuroscience across its entire gamut: Computational Neuroscience and Neuroimaging, Molecular and Cellular Neuroscience, and Systems and Cognitive Neuroscience. The NBRC has been established as an autonomous institute of the Department of Biotechnology (DBT), Ministry of Science and Technology, Government of India. Since its inception in 1999, NBRC has been steadily working towards basic research and technological advances that translated into better diagnostic tools and rational therapies for health problems related to neurosciences and related fields. The mandate of NBRC is to carry out research on brain function in health and disease. Furthermore, NBRC has been awarded 'Deemed University' status by the Ministry of Human Resources Development and is the only institute of DBT to achieve this status. The Centre has a vibrant activity of MSc, PhD and post-doctoral programmes in the various branches of Neuroscience, and also collaborates with institutes, laboratories and industry on specific areas.

2 Neuroinformatics Research

As an emergent discipline issuing out of the cross-fertilization of neuroscience and information technology, the field of Neuroinformatics endeavours to develop and apply advanced computing tools, artificial systems and database approaches essential for understanding and integrating brain structure and function. NBRC investigators address a variety of problems spanning mathematical and theoretical neuroscience,

N.R. Pal et al. (Eds.): ICONIP 2004, LNCS 3316, pp. 891–897, 2004.
© Springer-Verlag Berlin Heidelberg 2004

neuroengineering and neurocomputing. In contrast to stand-alone research laboratories, research at NBRC has been shaped along two mutually symbiotic components: extramural and intramural. In the former, the Centre's investigators interacts or collaborates with neuroscientists across the nation, whereas in the latter the investigators independently lead intensive research programmes at the laboratories of the Centre. Some of the salient aspects of research in both the components are highlighted below.

2.1 Feature Extraction in Human Speech

For extracting the features in natural sound and human speech, a novel computational approach has been initiated [1]. An objective of this research is to study the phonological structure of different Indian languages and characterize their similarities and differences. A new technique called the modulation spectrum separates the amplitude and frequency modulations in a speech signal. This will be used along with Independent Component Analysis (ICA) to identify new features for speech recognition in humans.

2.2 Computational Approaches to Speech Disorders in Children

To communicate ideas and express our wants and needs, language is the code that we learn to use. Reading, writing, speaking, and some gesture systems are all forms of language. Speech is the spoken form of language. The most basic unit of any language is the phoneme, the smallest unit of sound that differentiates meaning. As phonemes are physically complex acoustic stimuli, the role that complex auditory processing plays in the development of phonological systems have been a topic of increased research concentration. It is known that human speech signals have a time-varying frequency structure and thus require joint representations of time and frequency.

Towards building such a picture, scientists have begun to "visualize" sound in terms of temporal and modulations of the amplitude envelope. In this framework sounds are characterized by broadband spectra with energy at many different frequencies and elaborate spectral and temporal structure. We have developed a modulation spectrum which presents a representation where any sound can be broken into a series of temporal and spectral modulations [2]. Thus we have a joint representation of both temporal and spectral components. Among the language disorders prevalent in Indian population, an important one is known as specific language impairment, or SLI. Often there is no obvious explanation for a child's language difficulties: hearing is normal, nonverbal ability is sufficient, the child does not have any apparent physical or psychiatric disability and comes from a normal home background.

There is a theoretical approach that envisions that although children with SLI have normal hearing, they may have difficulty in distinguishing sounds that are brief or rapid. Since the modulation spectrum is a representation of sound in terms of temporal and spectral modulations, one would expect that a modulation spectrum of the speech sounds of children with SLI would reflect the modulations missing in their speech versus the speech of normal children. Such an approach could also test the hypothesis that children with LLI (Language Learning Impairment) often require longer time periods between acoustic events to discriminate them as compared to normal children. The Speech and Language Laboratory (SALLY) at NBRC headed by Dr. Nandini C Singh addresses these different issues and in the process is also hoping to obtain some clues on early detection of this impairment. Based on the results they would like to use the results from this analysis to design some subsequent therapy. Collateral research is also being done on the therapeutic implications of melodic sounds and music for the human brain [3].

2.3 Non-equilibrium Information Theory and Spatiotemporal Processing in Neurons

Probably the most fundamental challenge in neuroscience is understanding the neural code, how the neural system encodes and communicates information. On the other hand, the basic code in bioscience had long been known for over half a century, namely the genetic code comprising a template of four different pyrimidine base molecules. In contrast, the template of neural information transmission is the same, the spikes are all alike. Herein lies the fundamental difference between neuroscience and bioscience. We know that the same spike can transmit different types of information depending on inter-spike arrangement and its statistical distribution. We study this mechanism using communication theory and multiplexing operation where multiple entities of information can flow through the same medium (neuron) using the same symbol (spike), under real-life biological condition of environmental and biophysical noise. Using nonlinear dynamic approach, we focus on the constructive role of fluctuations or noise (stochastic transition and resonance) in information transmission at various neurobiological levels, from single neurons, to ensembles, to the full neurocognitive system. Our approach extends realistically the classical Shannon-Nyquist communication model, 'the stochastic binary channel with noise'.

The Computational Neuroscience and Neuroimaging laboratory led by Dr Prasun Roy is elucidating a new generalized theory of information transmission and processing beyond the Shannon paradigm and use the theory to understand and modulate information processing in high throughput conditions of neuronal systems, as in intensive adaptation, plasticity, and epileptic seizures [4]. The group has found out that information transmission occurs via two multiplexing modes, namely the equilibrial mode (first-order mode) vis-a-vis the non-equilibrium mode (second-order mode). At an optimal level of fluctuations, there is a transition point, the "iso-dissipation point", where first and second order modes equalizes. For such balanced or equimodal information transmission situation, there is an optimal level of system fluctuations, thus indicating a process of stochastic activation and resonance. Conventional models cannot satisfactorily analyze information transmission during such conditions as high intensity adaptation, plasticity, habituation, or in clinical conditions as epilepsy, migraine or tinnitus. The formalism presented here endeavours to open a path in this direction.

2.4 Application of Stochastic Resonance and Stability Analysis for Brain Imaging and Therapy: Using Noise to Defeat Noise

A major recent advancement in biophysics is the incisive phenomenon of the Stochastic Resonance Effect (SRE), whereby optimized perturbation or random-wave fluctuation is used to impressively enhance the behaviour or sensitivity of a system to an input signal or parameter. SRE occurs because the positive peak of the weak signal, under specific conditions, adds up with positive peaks of the noise, resulting in a considerable stronger signal or signal:noise ration (SNR). SR has been used to enhance optical, electron-paramagnetic-resonance, microscopic and raman images or spectra. Here we aim to apply the SRE paradigm for increasing the efficiency of neuroimaging, in MRI, fMRI and diagnostic or therapeutic radiology (tomography and tomotherapy). NBRC researchers have designed the pathways by which perturbation can be administered during the process of MRI or computed tomography and tomotherapy, and have devised a preliminary algorithm to induce stochastic resonance in the system [5]. To induce SRE, the power of the perturbation is varied until one obtains a peak enhancement, as observed on monitoring the output at each stage where SRE is used. SRE is being harnessed to distinguishing between brain tumour recurrence and post-radiotherapy necrosis, whose problematic differential diagnosis is an ordeal for conventional imaging procedure. The stochastic augmentation is also being explored for enhancing the diagnosis of Alzheimer's disease using Dempster-Shaefer evidence theory.

Using computational models of the brain and tumours therein (digital radiological phantoms), the research group is investigating the amplified effect of perturbation of radiotherapetic inflow (flux rate) on brain tumours. Indeed, system perturbations can produce striking changes on population dynamics of target cells. Rather than the usual practice of administering a steady radiotherapetic beam exposure to tumour, the approach being developed computationally fluctuates the flux rate of the photon radiation so that one can induce a resonant enhancement of the antitumour efficiency in radiotherapy [6]. This project work is being done at NBRC with cooperation from Institute of Nuclear Medicine and Allied Sciences, Ministry of Defence, New Delhi.

2.5 Brain Image Section Detection Using ICA and Support Vector Machine Analysis Joint Project Between NBRC and Indian Institute of Science, Bangalore

Neurocomputing has been an important area of research at I.I.Sc.-Bangalore [7, 8]. A crucial area in Neuroinformatics is to generating imaging databases for use in brain mapping, this furnishes a reference source to research laboratories and newer findings could be incorporated as progress in investigations are made. The project involves the construction of a 3-D atlas with associated ontologies and relationships for the mouse. It is visualized as a programme to develop neuroanatomical core ontology of a secure Web-accessible database for the rapid analysis of scientific information about the brain. Initial development has focused on a coordinated adaptation of topographic brain atlases of the brain of the mouse to provide an interactive on-line environment for the analysis of all forms of neuroscientific information, which can be mapped, by an atlas.

Some glimpses of research in computational neuroscience in different parts of the country are offered in the following sections.

2.6 Modelling of Receptive Field Formation and Cortex-Retina Interaction

The Department of Electrical Engineering at Indian Institute of Technology – Delhi, has an intensive research team led by Dr B Bhaumik, working in the problem of simulating and analyzing the orientation and direction selectivity of simple cells as well as studying a developmental model for receptive field formation in these cells [9]. To obtain the response of cortical cells to sinusoidal grating input, the retinal layer has been modeled as a 2-D sheet of retinal ganglion cells (RGC) lying one over the other, the first sheet corresponding to 'on' centre RGCs and the other to 'off' centre RGCs. The stimuli used in the visual space consists of a moving sinusoidal grating of a given spatial and temporal frequency with varying orientation, this stimulus traverses in the up-and-down direction for each orientation of the grating. The response of cortical cell at a given time is calculated using spike-response model (SRM). This computational experiment shows that about 18% cells of the simulated cortex are direction-selective, and this model has been a pioneering one for receptive field formation in directionally selective cells [10].

2.7 Investigation of Aspects of Serial Order in Cognition

An active area initiated by Dr Bapi Raju, of the Dept. of Computer and Information Sciences at the University of Hyderabad, pursues investigations on theory and applications of Neural Networks, Brain Modelling and problems in Cognitive Science [11]. The researchers have recently been investigating sequence processing underlying various human activities such as speech, language, skill learning, planning etc.

2.8 Analysis of fMRI Activation Data

Over the years, the Indian Statistical Institute-Kolkata has been a major focus of research in neurocomputing and brain image-processing. Here Dr S Purkayastha and his team has initiated investigations on functional neuroimaging [12]. Studies of estimation of hemodynamic response function, with special focus on Bayesian methods, are being pursued. They also propose to actively use the software fMRIstat (developed by Keith J. Worsley of McGill University, Montreal, Canada). Across a multidisciplinary setting comprising of neuropsychologists, neurologists, radiologists and statisticians, an appreciable endeavour on experimental work on fMRI has been undertaken at Bangur Institute of Neurology, Calcutta University.

2.9 Computational Exploration of Neurotransmission in Smooth Muscle

Computational analysis and *in silico* experimentation of neurochemical processes are fast developing into a new research focus in the neuroscience arena. Dr. R. Manchanda at the Bio-engineering department at the Indian Institute of Technology in Mumbai, pioneers a research unit that has implemented a discrete electrical model of 3-D syncy-tium representing smooth muscle [13]. Theoretical analyses can now be carried out of the effects on smooth muscle electrical activity, measured as syncytial synaptic potential of imposed conditions. The conditions imposed are chosen such as to mimic experi-mental conditions or protocols. Experimental application of presumptive uncoupling agents is modelled as alteration in relevant electrical parameters of the system. Alteration of function at the bundle, rather than cellular, level, can be explored [14].

3   National-Level Programs

Some schemes, funded by Dept of Biotechnology, Govt. of India, have been entrusted to NBRC and these programmes have been envisioned to help develop the research expertise and scientific manpower. Two of these undertakings deserve special mention.

3.1   National Facility on Functional Brain Imaging

The country's first high-field 3 Tesla MRI/fMRI/MRS/MRSI system dedicated to brain research is being set up as the National Facility of Functional Brain Imaging in NBRC. Available to neuroscientists across universities and institutes, the facility is being developed as a model centre for basic and applied studies on the research field that would play a pivotal role among neuroscientist across the region of South Asia. Neurocognitive investigations would encompass primate studies, analysis of normal human subjects as well as subjects having neurological or behavioral deficits, with special reference to the Indian milieu.

3.2   National-Level Neuroscience Network

As a promising illustration of the power of information technology to foster neuroscience research, one of the major goals of NBRC is to network the existing neuroscience groups/institutions in the country and promote multidisciplinary investigations in neuroscience. This facilitates sharing of expertise and available infrastructure for mutual benefit. Networking also helps to bring together researchers from varying backgrounds to pursue common objectives that may be beyond the capacity of an individual investigator, group or institution. This is important since the major achievements in neuroscience are being made through a multidisciplinary approach bringing together scientists working in different disciplines into the main stream of neuroscience and brain research activity. The networking is facilitated by information sharing through electronic network and identifying "Collaborating" centres for mutual interaction. Currently 43 centres throughout India are networked to NBRC.

3.3   National Neuroinformatics Initiative

Reference may also been made to the endeavour being undertaken to develop a National Initiative in Neuroinformatics by involving various Institutes to develop newer models for language and speech processing, analysis of images in various conditions of normality and disease, as well as various other areas of research. Students at postgraduate and the Ph.D. level would undertake projects and rotate through various Institutes depending on the available expertise and thus also facilitate more interaction and develop better networks.

4   Conclusion

Given India's pre-eminence in I.T., the novel field of Neuroinformatics (NI), the interface of information engineering and neuroscience, has already received a fillip in the country. Not only does NI help to unravel the mystery of working of the normal and abnormal brain, the discipline is also a forceful locomotive for the soft computing, computer vision, pattern recognition and signal processing industry as also for the defence establishment; hence the critical promise of NI for the I.T. sector which is being increasingly appreciated across the country. A collective collaborative effort among neuroscientists, computing researchers and electronic engineers is anticipated for ushering in this promising cognitive, clinical and technological modality of NI. There is ample reason to share this optimism as a number of Indian institutions together with NBRC are developing the critical mass, in both manpower and resources, to harness this collaboration, whose direct beneficiary would be the billion-plus Indian populace.

References

1. Singh, N *et al*: Modulation spectra of natural sound, J. Acoust. Soc. Amer., 114 (2003) 3394-34
2. Singh, L *et al*: Statistics of sound, In: Singh N: Proc Int Conf Theor Neuro, NBRC Gurgaon (2003)
3. Singh, N: Healing sound, In: Bagchi, K: Music, mind and health, Soc Geront Res., Delhi (2003)
4. Roy, P *et al*: Control analysis of neural processing, Lecture Notes Artif Intel. 2275 (2002) 191-202
5. Roy, P: Stochastic resonance imaging and therapy: Brain as stochastic processor. This volume
6. Roy, P *et al*: Neurocomputation: fluctuation instability. IEEE Trans Evol Comp 6 (2002) 292-305
7. Kumar, J *et al*: Mapping ANN in message system, IEEE Trans Systems Cyb-B, 26 (1996) 822-35
8. Rajan, K *et al*: High speed computation for PET image, IEEE Trans Nuclear Sci, 41 (1994) 1721-28
9. Bhaumik, B: Cooperation based on cell receptive field, J. Comput. Neurosci., 14 (2003) 211-2
10. Bhaumik, B: Receptive field of simple cell, J. Inst. Electronics & Telecom Engr 49 (2003) 87-96
11. Raju, B: Effector independent representation: motor learning, Exp Brain Res, 132 (2000) 149-62
12. Palit, S: Wavelet identifies activation, In: Proc Indo-US Symp Brain Res, NBRC Gurgaon (2003)
13. Turale, N *et al*: Simulation for physiological network, Med Biol Engg Comp, 41 (2003) 589-9
14. Vaidya, P *et al*: Synaptic analysis using wavelet, IEEE Trans. Biomed. Engg., 47 (2000) 701-8

CMAC with Fuzzy Logic Reasoning

Daming Shi, Atul Harkisanka, and Chai Quek

School of Computer Engineering, Nanyang Technological University, Singapore 639798
asdmshi@ntu.edu.sg, aharki@pmail.ntu.edu.sg,
ashcquek@ntuix.ntu.ac.sg

Abstract. This paper proposes a fuzzy CMAC model with truth value restriction inference scheme, which provide the original CMAC with a firm and intuitive fuzzy logic reasoning framework. Our proposed model smoothens the network output and increases the approximation ability, as well as reduces the memory requirement. Moreover, the membership functions and the fuzzy rules used in the fuzzy CMAC have clear semantic meaning. Our experiments are conducted on some benchmark datasets, and the results show that our method outperforms the existing representative techniques. The high learning capability of our model results from the ability·to handle uncertainty in the inference process.

1 Introduction

The cerebellar model articulation controller (CMAC), presented by Albus [1], is an autoassociative memory feed-forward neural network. CMAC is a table-lookup technique and has local generalization abilities that are dependent on the overlap of the association vectors. CMAC stores information and learns locally which means that only a small number of weights are updated or altered for a training vector, and therefore, only a few computations per step are needed. CMAC has attractive characteristics of local generalization, output superposition and incremental learning. It also has a faster learning time because of the limited number of computations per cycle and is thus easily realizable in hardware. However, CMAC suffers from inherent disadvantages like its inefficiency in storing data storage, since the memory size required grows exponentially with respect to the number of input variables and its weak performance in classifying inputs which are similar and highly overlapping.

Some researchers have introduced fuzzy logic to CMAC. As we know, neural fuzzy systems greatly improve the transparency of neural networks and enable a better understanding of their inner working. Existing fuzzy CMAC models, such as [2] and [3], whose weights are fuzzy by using membership functions, have considerable advantages over CMAC. Fuzzy membership in respective fields is introduced in FCMAC to increase the learning capability for the model. The FCMAC structure reduces the memory requirement by a great deal as compared to the original CMAC. The outputs of the model are given by the corresponding fuzzy weights.

The neural fuzzy systems proposed so far in literature can be broadly classified into two categories. The first category, such as ANFIS network [4] and the ARIC model [5], is parameter tuning with an initial rule base. The second category, such as Falcon-ART network [6] and POPFNN [7-9], can automatically formulate a set of fuzzy rules from the numerical training data. As the latter provide more powerful learning ability, it attracts more and more interest.

N.R. Pal et al. (Eds.): ICONIP 2004, LNCS 3316, pp. 898–903, 2004.
© Springer-Verlag Berlin Heidelberg 2004

Falcon-ART [6] is a hybrid connectionist system that can operate in a highly autonomous way. However, Falcon-ART suffers from inconsistency of the rule-base due to the inherent clustering techniques of the adaptive resonance theory (ART) [10]. This makes the fuzzy rules extracted from a neural fuzzy network meaningless and obscure. The POPFNN network [7-9] has a strong correspondence to the inference steps in the truth value restriction (TVR) [11] method in fuzzy logic that gives it a strong theoretical basis. There are three main disadvantages with POPFNN, namely, (1) it lacks the flexibility to incorporate new knowledge after training has completed, (2) it is susceptible to noisy and spurious data since it employs prototype-clustering schemes, (3) it requires prior knowledge such as the number of clusters to be computed and are poor at solving non-partitionable problems. The GenSoFNN [12] networks overcome the above weaknesses by automatically formulating the fuzzy rules from the numerical training data. In spite of all the strengths, GenSoFNN is a highly computationally intensive model and its performance is good only for a small set of features.

To address the shortcomings of the above existing neural fuzzy systems, a new model is proposed in this paper. The structure of this paper is outlined as follows. In section 2, the truth value restriction scheme is described. In Section 3, the original CMAC is fuzzified with TVR inference scheme. Section 4 gives the experimental analysis using our proposed model. Conclusion is presented in Section 5.

2 Truth Value Restriction Inference Scheme

Truth value restriction uses implication rules to derive the truth values of the consequents from the truth value of the antecedents [11]. In the *TVR* methodology, the degree to which the actual given value of A' of a variable x agrees with the antecedent value A in the proposition *"IF x is A THEN y is B"* is represented as a fuzzy subset of truth space. The TVR inference scheme uses the *inverse truth function modification* (ITFM) process to compute a function $\tau_{A\tilde{A}}$ that would transform the fuzzy proposition p to p'. The function $\tau_{A\tilde{A}}$ is in fact a fuzzy set in the truth space and is defined as:

$$\tau_{A\tilde{A}} = \begin{cases} \sup\{\mu_{\tilde{A}}(x) \mid x \in \mu_A^{-1}(a)\}, & \text{if } \mu_A^{-1}(a) \neq \phi \\ 0, & \text{otherwise} \end{cases} \quad (1)$$

Where μ_A and $\mu_{\tilde{A}}$ are the respective membership functions of the fuzzy set A and $\tilde{A}$ defined on the universe of discourse $\mathbf{X}$; x denotes a value in $\mathbf{X}$; a is the membership value of x in fuzzy set $\mathbf{A}$; $\mu_A^{-1}(a)$ is the set of values of x in $\mathbf{X}$ that take membership value a in the fuzzy set $\mathbf{A}$; and ϕ denotes the empty set or null set.

When the sup-T operation is used to resolve the composition of the observed input and the fuzzy rule, the truth-value $\tau_{B\tilde{B}}$ of the consequent "y is $\mathbf{B}$" can be computed using the following equation:

$$\tau_{B\tilde{B}}(b) = \sup\{m_I[\tau_{A\tilde{A}}(I(a,b))]\} \quad (2)$$

Where m_I is the forward reasoning function (usually a T-norm operation); and I is the implication rule.

3 Fuzzy CMAC with TVR

In this research work, the computed truth-values of the antecedents can be effectively propagated in the hybrid structure of a neural fuzzy system. The truth-value of the proposition in the antecedent is computed and allowed to propagate through the network. It is this value that is used to calculate the proposition in the consequent. This treatment makes the TVR a viable inference scheme for implementation in a neural fuzzy system.

The fuzzy CMAC can be viewed as a 6-layer hierarchical structure: (1) Input space. (2) Fuzzification layer. (3) Sensor Layer S, (4) Association Layer, (5) Post association Layer P and (6) Output Layer R.

Implementing the TVR scheme provides the network with a consistent rule base and a strong theoretical foundation. As compared to the CRI scheme for a multiple-input-single-output (MISO) system TVR scheme is more logical and intuitive to the human reasoning process. In CRI, the conjunction of fuzzy sets could result in an un-normalized and irregularly shaped fuzzy set defined on the Cartesian space. The proposed model also uses discrete incremental clustering (DIC) [13] for its self-organization phase and the back-propagation learning algorithm for the parameter-learning phase.

The simplified fuzzy reasoning and CMAC are explained briefly for formulating CMAC-fuzzy system. Given the input data x^o, the output of the simplified fuzzy reasoning y^τ is derived by the following equation:

$$\omega^p = \prod_{j=1}^{n} \mu_{A_j\,p}(x_j^o), \quad \text{for } j=1,2,\ldots,n \tag{3}$$

$$y^\tau = \frac{\displaystyle\sum_{p=1}^{N} \omega^p \times w^p}{\displaystyle\sum_{p=1}^{N} \omega^p} \quad \text{for } p=1,2,\ldots,N \tag{4}$$

where $\mu_{A_j\,p}(x_j^o)$ is a membership value, and ω^p is the total membership value of the antecedent part. To find the $\mu_{A_j\,p}(x_j^o)$ in equation (3), a clustering approach should be adopted. Clustering is a process to partition a data space or a given data set into different classes/groups so that the intra-class data points are more similar than the inter-class points. This similarity is measured using a metric of which the Euclidean distance is commonly used. Clustering is an exploratory approach to analyze a given numerical data set by creating a structural knowledge representation of the data set. This structural knowledge representation is the grouping of the data points into classes.

In this research, the discrete incremental clustering technique is used to specify the fuzzy sets. The DIC technique uses raw numerical values of a training data set with no pre-processing. It computes trapezoidal-shaped fuzzy sets and each fuzzy label (fuzzy set) belonging to the same input/output dimension has little or no overlapping of kernel with its immediate neighbors. Also, the number of fuzzy sets (fuzzy labels) for each input/output dimension need not be the same. In DIC technique, if the fuzzy label (fuzzy set) for a particular input/output dimension already exists, then it is not 'recreated'. The DIC technique has five parameters. They are fuzzy set support parameter SLOPE, plasticity parameter β, tendency parameter TD, input threshold and output threshold. The detailed description of DIC algorithm can be seen in [13].

Some techniques, such as hashing, can be used to map a large table from the logical memory to a smaller table in the physical memory [14]. Once we can obtain the correct set of weights for a particular set of inputs we can calculate the output. It is noticed that the weighted average mean approach is used to calculate the output rather than just the simple mean.

The weights between the post-association layer and the response unit are initialized to zero and then updated through training. The output for the input data x_i, is calculated by

$$y_i^\tau = \sum_{k=1}^{M} y_i^{\tau(k)} \tag{5}$$

where $y_i^{\tau(k)}$ is the output of the kth layer for the input data x_i.

4   Experimental Results

Our experiments were conducted on three benchmark data sets: Phoneme data set (Phoneme), Wisconsin breast cancer data set (WDBC), traffic data set (Traffic). (1) **Wisconsin Diagnostic Breast Cancer (WDBC).** University of Wisconsin Hospital compiles the WDBC database for accurately diagnosing breast masses based solely on Fine Needle Aspiration (FNA) test [15]. This dataset is obtained from [16]. (2) **Phoneme Data (Phoneme).** Phoneme database [17] developed by Dominique Van Cappel distinguishes nasal and oral vowels. The classification is done based on 5 different attributes chosen to characterize vowels. The attributes are the amplitudes of the first five harmonics, normalized by the total energy. (3) **Traffic Data (Traffic).** Function approximation means finding hypothesis (function) that can predict the values of function to be approximated that are yet to be seen, given examples of that function. Traffic flow prediction is one such example of universal approximation. The raw traffic flow data is obtained from [12]. The data were collected at a site located at exit 15 along the east bound Pan Island Expressway (PIE) in Singapore using loop detectors embedded beneath the road surface.

Table 1 shows the comparison among 5 neural fuzzy models on the above 3 benchmark datasets. The five neural fuzzy models are our proposed FCMAC-TVR, FCMAC-CRI, FALCON-AARt, POP-TVR and GENSO-TVR respectively. From table 1, we can see that FCMAC-TVR outperforms all the other models except some particular cases. The advantage of FCMAC-TVR is its leaning scheme. As TVR is more logical and intuitive to the human reasoning process, it is more suitable for CMAC.

Table 1. Comparison of various neural fuzzy models on benchmark datasets

Datasets	Fuzzy Neural Network Models				
	FCMAC-TVR	FCMAC-CRI	FALCON-AART	POP-TVR	GENSO-TVR
WDBC	91.1%	90.4%	N/A	89.2%	87.9%
Phoneme	87.3%	89.7%	83.9%	87.2%	87.8%
Traffic	90.0%	87.3%	84.3%	81.2%	79.9%

5 Conclusion

A FCMAC-TVR network is proposed in this paper. This neural network model has the characteristic of high learning speed and localization. TVR inference scheme makes the system more transparent and less rigid. It also gives the network a consistent rule base and a strong theoretical foundation. The proposed model uses DIC for self-organizing phase and the back-propagation learning algorithm for the parameter-learning phase. Our experiments are conducted on 4 benchmark datasets, and the results show that the performance of FCMAC-TVR is comparable to those of representative neural fuzzy systems such as FALCON-AART, POP-TVR and GENSO-TVR, etc.

References

1. J. S. Albus, *"A new approach to manipulator control: The cerebellar model articulation controller (CMAC)"*, Transaction of the ASME, Dynamic Systems Measurement and Control, vol. 63, pg 220-227, 1995.
2. J. Ozawa, I. Hayashi and N. Wakami, *"Formulation of CMAC-Fuzzy system"*, IEEE international Conference on Fuzzy Systems - Fuzzy-IEEE, San Diego, CA, pp. 1179-1186, 1992.
3. K. Zhang and F. Qian, *"Fuzzy CMAC and Its Application"*, Proceedings of the 3rd World Congress, Page(s): 944 -947 vol.2, 2000.
4. J. S. R. Jang, *"ANFIS: Adaptive-Network-Based Fuzzy Inference Systems"*, IEEE Transaction Systems, Man & Cybernetics, vol. 23: pp.665-685, 1993.
5. H. R. Berenji and P. Khedkar, *"Learning and Tuning Fuzzy logic Controllers through Reinforcements"*, IEEE Transaction, Neural Networks, vol. 3: pp 724-740, 1992.
6. C. J. Lin and C. T. Lin, *"An ART-Based Fuzzy Adaptive Learning Control Network"*, IEEE Transaction Systems, 5(4): pp. 477-496, 1997.
7. C. Quek and R. W. Zhou, *"POPFNN: A Pseudo Outer-Product Based Fuzzy Neural Network"*, IEEE Transaction, Neural Networks, 9(9): pp. 1569-1581, 1996.
8. C. Quek and R. W. Zhou, *"POPFNN-AARS(S): A Pseudo Outer-Product Based Fuzzy Neural Network"*, IEEE Transaction Systems, Man & Cybernetics, 29(6):859-870, 1999.
9. K. Ang, C. Quek and M. Pasquier, *"POPFNN-CRI(S): Pseudo Outer Product based Fuzzy Neural Network using the Compositional Rule of Inference and Singleton Fuzzifier"*, IEEE Transactions on Systems, Man and Cybernetics, 2004.
10. S. Grossberg, *"Adaptive pattern classification and universal recoding: II. Feedback, expectation, olfaction, illusions"*, Biological Cybernetics, 23: pp187-202, 1976.
11. R. L. Mantaras, *"Approximate reasoning models"*, Ellis Horwood Limited, 1990.
12. W. L. Tung and C. Quek, *"GenSoFNN: A Generic Self-organizing Fuzzy Neural Network"*, IEEE Transaction, Neural Networks, vol. 13, pg 1075-1086, 2002.

13. W. L. Tung and C.Quek, *"DIC: A Novel Discrete Incremental Clustering Technique for the Derivation of Fuzzy Membership Functions"*, Proceedings of the 7th Pacific Rim International Conference on Artificial Intelligence: Trends in Artificial Intelligence, 2002.
14. Z. Q. Wang, J. L. Schiano and M. Ginsberg, *"Hash-Coding in CMAC Neural Networks"*, IEEE International Conference on Neural Networks, Page(s): 1698 -1703 vol.3, 1996.
15. O. L. Mangasarian, W. N. Street and W. H. Wolberg, *"Pattern Recognition via linear Programming: Theory and Application"*, Medical Diagnosis, 22-31, 1990.
16. Wisconsin Diagnostic Breast Cancer Database. URL: http://www.cs.wisc.edu/~olvi/uwmp/cancer.html
17. ELENA. URL: http://www.dice.ucl.ac.be/neural-nets/Research/Projects/ELENA/elena.htm.

A Fuzzy Multilevel Programming Method for Hierarchical Decision Making

Bijay Baran Pal* and Animesh Biswas

Department of Mathematics, University of Kalyani, Kalyani 741235, India
Tel. +913325825439
bbpal18@hotmail.com

Abstract. This paper describes a fuzzy programming method for multilevel programming problems in a large hierarchical decision making organization. In the proposed approach, first the objectives at different decision making units are described fuzzily by setting an imprecise aspiration level to each of them. Then the defined fuzzy goals for the objective functions and the control vectors of the upper-level units are characterized by membership functions for measuring the degree of satisfaction of the decision makers located at different hierarchical levels.

In the solution process, a linear programming formulation of the problem for minimizing the group regret of degree of dissatisfaction of the decision makers using the distance function is considered first. Then the linear programming problem is transformed into a priority based fuzzy goal programming problem to achieve the most satisfactory decision for overall benefit of the organization.

To illustrate the approach, a numerical example studied [1] previously is solved and compared the solution with the conventional [1, 11] approaches.

1 Introduction

Multilevel programming (MLP) can be viewed as an extension of bilevel programming (BLP) for solving large and complex organizational planning problems with multiplicity of objectives, in which one decision maker (DM) is located at each of the hierarchical decision making levels and controls separately a decision vector for optimizing his / her own objective. In the decision making situation, although the execution of decision is sequential from an upper-level to a lower-level, the decision for optimizing the objective of an upper-level DM is often affected by the reaction of a lower-level DM due to his / her dissatisfaction with the decision, because the objectives at different levels often conflict each other. In such case, the problem for proper distribution of decision powers to the DMs is often encountered in most of the hierarchical decision situations.

During 1960's, several solution approaches for MLP problems as well as BLP problems as a special case have been deeply investigated [1–4] by the pioneer

---

* Corresponding Author: (B.B. Pal).

N.R. Pal et al. (Eds.): ICONIP 2004, LNCS 3316, pp. 904–911, 2004.
© Springer-Verlag Berlin Heidelberg 2004

researchers in the field from the view point of their potential use to different real-life hierarchical decision systems. But the use of a classical approach often leads to the paradox that the decision power of a lower-level DM dominates that of a higher-level DM. To overcome this situation, an ideal point dependent solution approach based on the preferences of the DMs for achievement of their objectives has been introduced by Wen and Hsu [12] in 1991. But their method does not always provide a satisfactory decision in highly conflicting hierarchical decision situation.

In order to overcome the shortcomings of the classical approaches, a fuzzy programming (FP) approach to hierarchical decision problems has been introduced by Lai [6] in 1996. Lai's solution concept has been further extended by Shih et al. [9] and Shih and Lee [10] to make a reasonable balance of decision powers to the DMs. The main difficulty with a conventional FP approach is that re-evaluation of the problem again and again by re-defining the elicited membership values of the objectives is involved to reach a satisfactory decision. To avoid such a computational difficulty, goal satisficing method in GP [5] for minimizing the regrets of the DMs in case of a BLP problem has been studied by Moitra and Pal [7] in the recent past. However, extensive study on fuzzy MLP problems is at an early stage.

In the present paper, the fuzzy goal programming (FGP) solution approach [7] to BLP problems is extended to solve MLP problem. Here, in the decision process, minimization of the under-deviational variables of the defined membership goals with highest membership value (unity) as the aspiration levels of them on the basis of pre-emptive priority as well as the relative weights of importance of those which are at the same priority level for achieving the aspired levels of the fuzzy goals is taken into consideration in the decision making context.

2   Formulation of MLP Problem

Let the vector of variables $\mathbf{X} = (x_1, x_2, \cdots, x_n)$ be involved in the hierarchical decision system, and let F_l and $\mathbf{X}_l$ be the objective function and control vector of the decision variables of the l-th level DM, where $l = 1, 2, \cdots L; L \leq n$ and where $\bigcup_l \{\mathbf{X}_l | l = 1, 2, \cdots, L\} = \mathbf{X}$.

Then the generic form of a MLP in a hierarchical nested structure can be presented as:

Find $\mathbf{X}(\mathbf{X}_1, \mathbf{X}_2, \cdots, \mathbf{X}_L)$ so as to

$$\max_{\mathbf{X}_1} F_1(\mathbf{X}) = \sum_{l=1}^{L} \mathbf{C}_{1l} \mathbf{X}_l \qquad \text{(top-level)}$$

where, for given $\mathbf{X}_1, \mathbf{X}_2, \mathbf{X}_3, \cdots, \mathbf{X}_L$ solve

$$\max_{\mathbf{X}_2} F_2(\mathbf{X}) = \sum_{l=1}^{L} \mathbf{C}_{2l} \mathbf{X}_l \qquad \text{(second-level)}$$

....

where, for given $\mathbf{X}_1, \mathbf{X}_2, \cdots, \mathbf{X}_{L-1}, \mathbf{X}_L$ solves

$$\max_{\mathbf{X}_L} F_L(\mathbf{X}) = \sum_{l=1}^{L} \mathbf{C}_{Ll} \mathbf{X}_l \qquad \text{(L-th level)}$$

subject to

$$(\mathbf{X_1}, \mathbf{X_2}, \cdots, \mathbf{X_L}) \in S = \{\sum_{l=1}^{L} \mathbf{A}_l \mathbf{X}_l \leq \mathbf{b}\} \mathbf{X}_l \geq 0, l = 1, 2, \cdots, L \qquad (1)$$

where $\mathbf{C}_{l'l}(l' = 1, 2, \cdots, L)$ and $\mathbf{b}$ are constant vectors, $\mathbf{A}_l(l = 1, 2, ..., L)$ are constant matrices. It is also assume that $S(\neq \Phi)$ is bounded.

Now, in the classical approaches to MLPs, it is to be noted that the decision $\mathbf{X}_{l+1}$ is made by a lower-level DM subject to the decision $\mathbf{X}_l$ made by the respective higher-level DM in the order of their hierarchy. But in such a hierarchical execution process for making decision, the lower-level DMs are always found to be dominated by the higher-level DMs. As a consequence, decision deadlock arises frequently in most of the decision organizations.

To overcome the above situation, relaxation of individual decision as well as the objective value upto certain tolerance limits of each of the higher-level DMs are needed essentially to execute the decision power of the DMs properly and thereby making overall benefit of the organization.

2.1   FP Problem Formulation

To formulate the FP model of the problem (1), the objective functions $F_l(l = 1, 2, \cdots, L)$ and decision vectors $\mathbf{X}_l(l = 1, 2, \cdots, L - 1)$ are to be transformed into fuzzy goals by means of assigning imprecise aspiration levels to them, and then they are to be characterized by the associated membership functions.

Construction of Membership Functions. Since each DM is interested in optimizing his / her own objective function, the individual best objective value can be reasonably considered as the aspiration level of the corresponding objective. Let $(\mathbf{X}_1^l, \mathbf{X}_2^l, \cdots, \mathbf{X}_L^l; F_l^M)$ be the optimal decision of the l-th level DM. Then the fuzzy objective goals appear as $F_l \gtrsim F_l^M (l = 1, 2, \cdots, L)$.

Again in a hierarchical decision situation, since the benefit of a lower-level DM depends on the relaxation of the decision of the higher-level DM, the fuzzy goals for the control vectors can be defined as $\mathbf{X}_l \gtrsim \mathbf{X}_l^l (l = 1, 2, \cdots, L - 1)$.

Now, since execution of decision is sequential from an upper- to lower-level, the lower tolerance limits of the stated fuzzy goals can be determined as follows: Let the value $F_l^{l+1}[= F_l(\mathbf{X}_1^{l+1}, \mathbf{X}_2^{l+1}, \cdots, \mathbf{X}_L^{l+1})] < F_l^M, (l = 1, 2, \cdots, L - 1)$ be the lower tolerance limits of the successive upper-level DMs, which actually represent the least possible objective values of them by giving full relaxation for achievement of the objectives to the successive lower-level DMs. Again, since the L-th level DM is at the bottom position, the lower tolerance limit for the L-th objective can be considered as

$$F_L^m[= min(F_l(\mathbf{X}_1^l, \mathbf{X}_2^l, \cdots, \mathbf{X}_L^l); l = 1, 2, \cdots, L - 1)] < F_L^M.$$

Now relaxation of decision of a higher-level DM is needed for benefit of a lower one and that depends on the decision making context. Let $\mathbf{X}_l^m(\mathbf{X}_l^{l+1} < \mathbf{X}_l^m < \mathbf{X}_l^l); l = 1, 2, \cdots, L-1$ be the lower tolerance limits of the decision vectors of the upper-level DMs.

Note 1. If alternative solutions are found in optimizing the objective function at a higher-level individually, then to give relaxation in executing decision power to the extent possible to the next lower-level DM, the lower tolerance value of the objective at the l-th level would have to be considered as

$$F_l^{l+1} = min[F_l(\mathbf{X}_{1t}^{l+1}, \mathbf{X}_{2t}^{l+1}, \cdots, \mathbf{X}_{Lt}^{l+1})|t = 1, 2, \cdots, T], l = 1, 2, \cdots, L-1$$

where $\mathbf{X}_{lt}^{l+1}$ is renamed for the actual $\mathbf{X}_l^{l+1}$ to designate it as the t-th alternative optimal solution, and where T is the total of it.

Again, since the l-th level DM has a motivation to give relaxation to make satisfactory decision by the $(l+1)$-th level DM, the existence of alternative optimal solutions $\mathbf{X}_{lt}^l (t = 1, 2, \cdots, T)$, leads to assign the fuzzy aspiration level of the decision vector $\mathbf{X}_l$ as $\mathbf{X}_{lv}^l$ which will satisfy the following:

$$F_{l+1}^{l+1}(\mathbf{X}_{lv}^l) \geq F_{l+1}^{l+1}(\mathbf{X}_{lt}^l) \ \forall\{\mathbf{X}_{lt}^l|t = 1, 2, \cdots, T \ and \ v \neq t\}; \ l = 1, 2, \cdots, L-1.$$

However, in the solution identification, if tie occurs then any one of the tied solutions can be considered there.

The Membership Functions. The membership functions for the defined fuzzy goals can be constructed as [8]:

$$\mu_{F_l}[F_l(\mathbf{X})] = \begin{cases} 1 & \text{if } F_l(\cdot) \geq F_l^M \\ \frac{F_l(\cdot)-F_l^{l+1}}{F_l^M-F_l^{l+1}} & \text{if } F_l^{l+1} \leq F_l(\cdot) < F_l^M \\ 0 & \text{if } F_l(\cdot) < F_l^{l+1}, \quad l = 1, 2, \cdots, L-1, \end{cases} \quad (2)$$

$$\mu_{F_L}[F_L(\mathbf{X})] = \begin{cases} 1 & \text{if } F_L(\cdot) \geq F_L^M \\ \frac{F_L(\cdot)-F_L^m}{F_L^M-F_L^m} & \text{if } F_L^m \leq F_L(\cdot) < F_L^M \\ 0 & \text{if } F_L(\cdot) < F_L^m, \end{cases} \quad (3)$$

$$\mu_{\mathbf{X}_l}[\mathbf{X}_l] = \begin{cases} 1 & \text{if } \mathbf{X}_l \geq \mathbf{X}_l^l \\ \frac{\mathbf{X}_l-\mathbf{X}_l^m}{\mathbf{X}_l^l-\mathbf{X}_l^m} & \text{if } \mathbf{X}_l^m \leq \mathbf{X}_l < \mathbf{X}_l^l \\ 0 & \text{if } \mathbf{X}_l < \mathbf{X}_l^m, \quad l = 1, 2, \cdots, L-1. \end{cases} \quad (4)$$

Now, the primary aim of each of the DMs is to achieve the highest membership value of his / her own objective. But, it is not possible in a practical decision situation due to limitations of the resources. In such a case, the concept of distance function can be effectively used for minimizing the regrets of the DMs in the decision making context.

2.2 Use of Distance Function

The distance function can be presented as [13]:

$$D_p(u(x)) = [\textstyle\sum_{i=1}^n (u_i^* - u_i(x))^p]^{\frac{1}{p}}, p \geq 1, \text{ where } D_p(u(x)) \text{ represents the}$$
distance between the utopia point u_i^* and the actual utilities resulting from the decision x. The parameter p takes the value depending on the decision situation.

In the present decision situation, u_i^* designates the highest membership value 1, and $u_i(x)$ represents the achieved membership value. Now when $p=1$, $D_1(u(x))$ represents the sum of individual regret, which may be interpreted as the group regret.

The FP model for minimizing the group regret can be presented as:

$$min\ D_1 = \sum_{l=1}^{L}[1 - \mu_{F_l}(F_l(\mathbf{X}))] + \sum_{l=1}^{L-1}[\mathbf{I_1} - \mu_{\mathbf{X}_l}(\mathbf{X}_l)]$$
$$\text{subject to } \mu_{F_l}(F_l(\mathbf{X})) \leq 1, \quad l = 1, 2, \cdots, L$$
$$\mu_{\mathbf{X}_l}(\mathbf{X}_l) \leq \mathbf{I_2}, \quad l = 1, 2, \cdots, L-1$$
$$\text{and the given system constraints in (1)} \qquad (5)$$

where $\mathbf{I_1}$ and $\mathbf{I_2}$ are respectively the row and column vectors with all elements equal to 1 and the dimension of each depends on $\mathbf{X}_l, l = 1, 2, \cdots, L$.

The problem (5) can be recast as:

$$max\ D_1' = \sum_{l=1}^{L} \mu_{F_l}(F_l(\mathbf{X})) + \sum_{l=1}^{L-1} \mu_{\mathbf{X}_l}(\mathbf{X}_l)]$$
$$\text{subject to the given constraints in (5)} \qquad (6)$$

It is to be noted here that the problem (6) is similar to the additive FGP model proposed by Tiwari *et al* [11]. However, to arrive at a most satisfactory decision, the FGP formulation in the conventional form of GP is considered here.

3 FGP Formulation

It is to be noted that maximization of D_1' in problem (6) means achievement of membership values of the fuzzy goals to the extent possible. So, achievement of the aspired level 1 of each of the membership function can be taken into account and thereby an equivalent FGP model can be constructed.

Now, since a considerable number of fuzzy goals are involved with the problem and they usually conflict among themselves, it seems that the FGP formulation of the problem on the basis of priority of DMs' needs and desires is effective in the present decision situation.

The priority based FGP can be presented as [8]:

Find $\mathbf{X}(x_1, x_2, \cdots, x_n)$ so as to
$$min\ Z = [P_1(\mathbf{d}^-), P_2(\mathbf{d}^-), \cdots, P_k(\mathbf{d}^-), \cdots, P_K(\mathbf{d}^-)]$$
$$\text{and satisfy } \frac{F_l(\cdot) - F_l^{l+1}}{F_l^M - F_l^{l+1}} + d_l^- - d_l^+ = 1, \quad l = 1, 2, ..., L-1$$
$$\frac{F_L(\cdot) - F_L^m}{F_L^M - F_L^m} + d_L^- - d_L^+ = 1$$
$$\frac{\mathbf{X}_l - \mathbf{X}_l^m}{\mathbf{X}_l^l - \mathbf{X}_l^m} + \mathbf{d_r^-} - \mathbf{d_r^+} = \mathbf{I}_l, r = L + l, \text{ where } l = 1, 2, \cdots, L-1$$
$$d_l^+, d_l^- \geq 0; \mathbf{d_r^+}, \mathbf{d_r^-} \geq \mathbf{0},$$
$$\text{and subject to the given system constraints in (1)} \qquad (7)$$

where Z represents the vector of K priority achievement function and d_l^+, d_l^- represent the over- and under- deviational variables, respectively, and $\mathbf{d_r^+}, \mathbf{d_r^-}$ are the vectors of over- and under- deviational variables associated with the respective goals. $P_k(\mathbf{d}^-)$ is the function of weighted under deviational variables, and which is of the form: $P_k(\mathbf{d}^-) = \sum_{l=1}^{L} w_{kl}^- d_{kl}^- + \sum_{r=L+1}^{2L-1} \mathbf{w}_{kr}^- \mathbf{d}_{kr}^-$, where d_{kl}^- and $\mathbf{d}_{kr}^-$ are renamed for d_l^- and $\mathbf{d}_r^-$ to represent them at the k-th priority level. $w_{kl}^-(\geq 0)$ and $\mathbf{w}_{kr}^-(\geq \mathbf{0})$ are the numerical weights and the vector of numerical weights, respectively, and they are determined as:

$$w_{kl}^- = \frac{1}{(F_l^M - F_l^{l+1})_k}, l = 1, 2, \cdots, L-1; \quad w_{kL}^- = \frac{1}{(F_L^M - F_L^m)_k},$$
$$\mathbf{w}_{kr}^- = \frac{1}{(\mathbf{X}_l^l - \mathbf{X}_l^m)_k}, \quad r = l + L, \text{ where } l = 1, 2, \cdots, L-1,$$

where the suffix $'k'$ to the RHS values is used to represent them at the k-th priority level.

Also, the priorities have the relationship $P_1 >>> P_2 >>> \cdots >>> P_k >>> \cdots >>> P_K$, which implies that the goals at the highest priority level (P_1) are achieved to the extent possible before the set of goals at the second priority level (P_2) is considered, and so forth.

4   Illustrative Example

The trilevel programming problem studied by Anandalingam [1] is solved.

Find (x_1, x_2, x_3) so as to
$$\max_{x_1} F_1 = 7x_1 + 3x_2 - 4x_3 \qquad \text{(top-level)}$$
where, for given x_1, x_2 *and* x_3 solve
$$\max_{x_2} F_2 = x_2 \qquad \text{(second-level)}$$
where, for given x_1 *and* x_2, x_3 solves
$$\max_{x_3} F_3 = x_3 \qquad \text{(third-level)}$$
subject to $x_1 + x_2 + x_3 \leq 3, x_1 + x_2 - x_3 \leq 1, x_1 + x_2 + x_3 \geq 1,$
$-x_1 + x_2 + x_3 \leq 1, x_3 \leq 0.5; x_1, x_2, x_3 \geq 0 \qquad (8)$

The individual optimal solutions of the three successive levels are $(x_1^1, x_2^1, x_3^1; F_1^M)$ $= (1.5, 0, 0.5; 8.5)$, $(x_1^2, x_2^2, x_3^2; F_2^M) = (0, 1, 0; 1)$ and $(0.4, 1, 0.4; 1)$, and $(x_1^3, x_2^3, x_3^3; F_3^M) = (0, 0.5, 0.5; 0.5)$, respectively. Then, following the procedure, the fuzzy goals are obtained as: $F_1 \gtrsim 8.5, F_2 \gtrsim 1, F_3 \gtrsim 0.5$ and $x_1 \gtrsim 1.5, x_2 \gtrsim 1$.

The lower tolerance limits of the fuzzy objective goals are obtained as $F_1^2 = 3, F_2^3 = 0.5, F_3^m = 0$.

Again following the procedure, $x_1^m = 1.4(x_1^2 < 1.4 < x_1^1)$ and $x_2^m = 0.75(x_2^3 < 0.75 < x_2^2)$ are taken into consideration as the lower tolerance limits of the decisions x_1 and x_2, respectively.

Using the above numerical values, the membership functions for the defined fuzzy goals can be obtained by (2), (3) and (4).

Then, the executable FGP model can be designed under a given priority structure as:

Find (x_1, x_2, x_3) so as to
$$\min Z = [P_1(\tfrac{1}{5.5}d_1^-), P_2(\tfrac{1}{0.5}d_2^- + \tfrac{1}{0.5}d_3^-), P_3(\tfrac{1}{0.1}d_4^- + \tfrac{1}{0.25}d_5^-)]$$
and satisfy: $\mu_{F_1} : \frac{7x_1 + 3x_2 - 4x_3 - 3}{5.5} + d_1^- - d_1^+ = 1,$
$\mu_{F_2} : \frac{x_2 - 0.5}{0.5} + d_2^- - d_2^+ = 1,$

$\mu_{F_3} : \frac{x_3}{0.5} + d_3^- - d_3^+ = 1, \quad \mu_{x_1} : \frac{x_1 - 1.4}{0.1} + d_4^- - d_4^+ = 1,$
$\mu_{x_2} : \frac{x_2 - 0.75}{0.25} + d_5^- - d_5^+ = 1, \quad d_j^+, d_j^- \geq 0 , j = 1, 2, \ldots, 5.$
subject to the given system constraints in (8). $\hfill (9)$

The *software* LINGO (version 6.0) is used to solve the problem.

The achieved solution is $(x_1, x_2, x_3) = (1.5, 0, 0.5)$ with $(F_1, F_2, F_3) = (8.5, 0, 0.5)$.

The resulting membership values of the fuzzy objective goals are $\mu_{F_1} = 1, \mu_{F_2} = 0, \mu_{F_3} = 1$.

Note 2. The solution of the problem obtained by Anandalingam [1] using conventional crisp approach is $(x_1, x_2, x_3) = (0.5, 1, 0.5)$ with $(F_1, F_2, F_3) = (4.5, 1, 0.5)$.

Again if the additive FGP approach of Tiwari *et al.* [11] is used to the problem (6) directly, the solution is then obtained as $(x_1, x_2, x_3) = (1, 0.5, 0.5)$ with $(F_1, F_2, F_3) = (6.5, 0.5, 0.5)$ and $(\mu_{F_1}, \mu_{F_2}, \mu_{F_3}) = (0.636, 0, 1)$.

It is to be observed that the conventional FP solution is better than the crisp solution from the view point of achieving the aspired goal levels of the objectives in the order of hierarchy.

However, a comparison of the model solution with the above results shows that a better decision is achieved here. Also, it is worthy to note here that the top-level DM has a leading position for the proposed solution of the problem in the decision making context.

From the above view points it may be claimed that the model solution is much acceptable with regard to optimizing overall benefit of the organization.

5 Conclusion

The proposed solution approach can be extended to multiobjective MLP problems without involving any computational difficulty. An extension of the approach for fuzzy nonlinear MLP problem may be a problem in future studies. Finally it is hoped that the approach presented here may open up a new look into the way of solving real-life hierarchical decision problems.

Acknowledgement

The authors are grateful to the anonymous referees whose constructive comments have improved the presentation of the paper. The second author is also grateful to Council for Scientific and Industrial Research (CSIR), New Delhi, India for providing financial support in pursuing the research work.

References

1. Anandalingam, G.: A Mathematical Programming Model of Decentralized Multi-level Systems. J. Opl. Res. Soc. **11** (1988) 1021 – 1033

2. Bard, J. F., Falk, J. E.: An Explicit Solution to the Multi-Level Programming Problem. Comp. and Ops. Res. **9** (1982) 77 – 100
3. Bialas, W. F., Karwan, M. H.: On Two Level Optimization. IEEE Trans. on Auto. Control. **27** (1982) 211 – 214
4. Burton, R. M.: The Multilevel Approach to Organizational Issues of the Firm-Critical Review. Omega **5** (1977) 457 – 468
5. Ignizio, J.P.: Goal Programming and Extensions. Lexington, D. C. Health (1976)
6. Lai,Y.J.: Hierarchical Optimization. A Satisfactory Solution. Fuzzy Sets and Syst. **77** (1996) 321–335
7. Moitra, B. N., Pal, B. B.: A Fuzzy Goal Programming Approach For Solving Bilevel Programming Problems.: In Pal, N.R., Sugeno, M. (eds.): Advances in Soft Computing, AFSS-2002. LNAI. vol. **2275**. Springer, Berlin (2002) 91 – 98
8. Pal, B.B.,Moitra, B.N.: A Goal Programming Procedure for Solving Problems with Multiple Fuzzy Goals Using Dynamic Programming. Euro. J. Opl. Res. **144** (2002) 480–491
9. Shih, H. S., Lai, Y. J., Lee, E S.: Fuzzy approach for multilevel programming problems. Comp. and Ops. Res. **23** (1996) 73 – 91
10. Shih, H. S., Lee, S.: Compensatory fuzzy multiple level decision making. Fuzzy Sets and Sys. **14** (2000) 71 – 87
11. Tiwari, R.N., Dharmar, S., Rao, J.R.: Fuzzy goal programming - An additive model. Fuzzy Sets and Sys. **24** (1987) 27 – 34
12. Wen, U. P., Hsu, S. T.: Efficient solution for the linear bilevel programming problem. Euro. J. Opl. Res. **62** (1991) 354 – 362
13. Yu, P.L.: A Class of Solutions for Group Decision Problems. Mgmt. Sci. **19** (1973) 936–946

Fuzzy Rule-Based Systems
Derived from Similarity to Prototypes

Włodzisław Duch[1,2] and Marcin Blachnik[3]

[1] Department of Informatics, Nicholaus Copernicus University, Grudziądzka 5, Toruń, Poland
[2] School of Computer Engineering, Nanyang Technological University, Singapore
[3] Division of Computer Methods, Department of Electrotechnology, The Silesian University of Technology, ul. Krasińskiego 8, 40-019 Katowice, Poland

Abstract. Relations between similarity-based systems, evaluating similarity to some prototypes, and fuzzy rule-based systems, aggregating values of membership functions, are investigated. Fuzzy membership functions lead to new types of similarity measures and similarity measures, including probabilistic distance functions that are applicable to symbolic data, lead to new types of membership functions. Optimization of prototype-based rules is an interesting alternative to neurofuzzy systems. As an illustration simple prototype-based rules are found for leukemia gene expression data.

1 Introduction

Investigation of relationships between fuzzy systems and similarity based systems is quite fruitful, leading to new methods in both fields. Fuzzy models usually start from membership functions (MFs) defining linguistic variables. In most applications MFs should be derived from data together with logical rules, optimizing the performance of the system. Some neurofuzzy systems can do this [1–3]. Networks based on separable basis functions (i.e. calculating products of one-dimensional MFs) perform essentially optimization of fuzzy MFs with a specific aggregation of rule conditions to form a conclusion. In both network and fuzzy approaches the form of the MFs is fixed, only their parameters, and sometime their number (if constructive networks are used) are optimized. The system designer selects from a few types of elementary functions that determine the type of decision borders that the system will provide, and thus determine the complexity of the final system. In similarity-based methods [4] the training set provides the reference examples and the similarity of (or distance to) new cases is used for evaluation.

Similarity-based methods may be used in more general situation than neural or fuzzy methods because they do not require numerical representation of inputs with fixed number of components. Similarity between complex objects (such as proteins, texts, software or financial institutions) may be determined using various quantitative and qualitative procedures. Such methods may also be presented in a network form [5]. Prototype cases that allow for reliable classification using a few features in the distance function may be used to formulate similarity-based rules, providing an interesting alternative to neurofuzzy approach. In the simplest case using Euclidean distance metric

N.R. Pal et al. (Eds.): ICONIP 2004, LNCS 3316, pp. 912–917, 2004.
© Springer-Verlag Berlin Heidelberg 2004

and a single prototype per class hyperplane decision borders are created, leading to a linear classification machine. How to set fuzzy systems that has identical decision borders? What type of similarity measures correspond to the typical fuzzy functions and vice versa? Are prototype-based rule systems always equivalent to fuzzy rule systems? Although some theoretical work has been devoted to understanding fuzzy sets in terms of similarity such practical questions remain unanswered.

In the next section relation between similarity and fuzzy systems are presented, and probabilistic, data-dependent similarity measures leading to prototype rules that have no simple equivalents in fuzzy rules introduced. Some examples illustrating the relations between fuzzy rules and prototype-based systems are shown in the third section. Very simple prototype-based rules are found for leukemia gene expression data.

2 Probability, Similarity and Fuzzy Sets

Fuzzy set $\mathcal{F}$ is defined by the universe $\mathcal{X}$ and the MFs $\chi_{\mathcal{F}}(X)$, specifying the degree to which elements of this universe belong to the set $\mathcal{F}$. This degree may be understood as estimation of $X \in \mathcal{X}$ similarity to typical (or $\chi_{\mathcal{F}}(X) \approx 1$) elements of $\mathcal{F}$. In fuzzy modeling each feature $\mathbf{X}_i$ of an object $\mathbf{X}$ is filtered through a large receptive field $\mathcal{F}_{ij}$, defined by a MF $\mu_{F_j}(X_i)$. Simple functions, such as triangular, trapezoidal or Gaussian are used to model the degree to which some value X_i belongs to the receptive field $\mathcal{F}_{ij}$. Several features are combined together to evaluate similarity to known objects. A general form of prepositional classification rule is:

$$\text{IF } \mathbf{X} \sim \mathbf{O}^{(i)} \text{ THEN Class membership is } \chi_{C_j}(\mathbf{X}) \tag{1}$$

The operator $\sim$ represents similarity determining the membership values $\chi_{C_j}(\mathbf{X})$ in classes to which $\mathbf{O}^{(i)}$ objects typical $\chi_{C_i}(\mathbf{O}^{(i)}) \approx 1$ for this class belong. Rules partition the feature space into areas where $\chi_{C_i}(\mathbf{X}) > \chi_{C_j}(\mathbf{X})$, that is similarity to objects from different classes dominates. In fuzzy logic the overall similarity is calculated as a T-norm (frequently a product) of MFs $\mu_{ji}(X_j)$ for relevant features. The crisp form of logical rules (L-rules) is obtained when rectangular MFs are used, partitioning feature space into hyper-rectangles. Fuzzy rules (F-rules) with popular triangular or Gaussian MFs provide more complex decision borders. An alternative way to partition the feature space and classify the data is to use a set of prototype-based rules (P-rules) and find minimal distance:

$$\text{IF } P = \arg\min_{\mathbf{P}'} D(\mathbf{X}, \mathbf{P}') \text{ THEN Class}(\mathbf{X}) = \text{Class}(\mathbf{P}) \tag{2}$$

where $D(\mathbf{X}, \mathbf{P})$ is a dissimilarity function (usually a distance function). If the minimal distance rule is used to find the nearest prototype, the decisions borders will have polyhedral shapes. The goal here is to find a small number of prototypes $\mathbf{P}$ and a simple similarity functions that can give accurate classification and understanding of the problem. Similarity functions based on Minkovsky's distance are very useful:

$$D(\mathbf{X}, \mathbf{P})^{\alpha} = \sum_{i=1}^{N} W_i |X_i - P_i|^{\alpha} \tag{3}$$

where W_i are feature scaling factors, taken from standardization or treated as adaptive parameters. For large exponents α contours of constant distance become rectangular. In the limit Chebyshev (or L_∞ norm) distance function $D_\infty(\mathbf{X},\mathbf{P}) = \max_i |X_i - P_i|$ has rectangular contours of constant values. Introducing thresholds d_P, rules of the form: IF $D_\infty(\mathbf{X},\mathbf{P}) \leq d_P$ THEN C, are equivalent to conjunctive crisp rules: IF $X_1 \in [P_1 - d_{P1}/W_1, P_1 - d_{P1}/W_1] \wedge ... \wedge [P_k - d_{Pk}/W_k, P_k - d_{Pk}/W_k]$ THEN C. These rules may not cover the whole feature space, while minimal distance rules always partition the whole space.

Any T-norm, for example $S(\mathbf{X},\mathbf{P}) = \prod_{i=1} \mu(X_i - P_i)$, may be used as a similarity function and be related to distance functions by exponential (or other) transformations, $S(\mathbf{X},\mathbf{P}) = \exp(-D(\mathbf{X},\mathbf{P}))$. Additive distance functions are then converted to the multiplicative similarity factors (MFs). In particular Euclidean distance function $D_2(\mathbf{X},\mathbf{P})^2 = \sum_i W_i(X_i - P_i)^2$ is equivalent to a Gaussian similarity function $S_2(\mathbf{X},\mathbf{P}) = \exp(-||\mathbf{X}-\mathbf{P}||^2)$ centered at $\mathbf{P}$ with ellipsoidal contours of constant values $||\mathbf{X}-\mathbf{P}|| =$const, equal to a product of Gaussian MFs $S_2(\mathbf{X},\mathbf{P}) = \prod_i G(X_i, P_i)$.

F-rules may be replaced by P-rules with appropriate similarity functions, but the reverse does not hold in general. Manhattan distance function $D_1(\mathbf{X},\mathbf{P}) = \sum_{i=1} |X_i - P_i|$, Canberra distance: $D_{Ca}(\mathbf{X},\mathbf{Y}) \sum_{i=1} |X_i - Y_i|/|X_i + Y_i|$, used below. and many other distance measures (see [6]) are not equivalent to commonly used MFs and T-norms. More general form of rules are obtained if more than one prototype is used in the rule condition: IF among k most similar prototypes P_i class C is dominating than $C(\mathbf{X}) = C$. Such rules should be useful in approximation problems, but for classification they are rather difficult to understand and require more prototypes (at least k) per class.

An interesting group of distance measures is based on the Value Distance Metrics (VDM) [6]. A value difference for feature X_j in a K-class problem is defined as:

$$d_V(X_j, Y_j)^q = \sum_{i=1}^{K} |p(C_i|X_j) - p(C_i|Y_j)|^q \tag{4}$$

where $p(C_i|X_j) = N_i(X_j)/N(X_j)$ is the number of times $N_i(X_j)$ the value X_j of feature j occurred in vectors belonging to class C_i, divided by $N(\mathbf{X}_j)$, the number of times the value $\mathbf{X}_j$ occurred for any class. The distance between two vectors $\mathbf{X}, \mathbf{Y}$ with discrete (nominal, symbolic) elements is computed as a sum of value differences $D_V(\mathbf{X},\mathbf{Y})^q = \sum_{j=1}^{N} d_V(\mathbf{X}_j, \mathbf{Y}_j)^q$. Distance is defined here via a data-dependent matrix with the number of rows equal to the number of classes and the number of columns equal to the number of all attribute values. The probabilities may be replaced by mutual information between the value of a feature and the class label.

For continuous inputs probabilities are computed either by discretization (Discrete Value Difference Metric, DVDM), or via interpolation (Interpolated Value Difference Metric, IVDM) (see [6, 7] where other VDM functions are also presented). VDM distance functions are useful especially for symbolic attributes, where typical distance functions may not work so well. Fuzzy MFs are also difficult to define for such data. P-rules based on VDM distances may still be replaced by F-rules by creating MFs that approximate their decision borders. Distance functions will be converted to similarity functions and replaced by products of MFs (only product T-norm is used in this paper).

3   Illustrative Examples

Two-dimensional models are used here to visualize decision borders in order to understand P-rule systems. Similarity functions corresponding to the Minkowski's distance measures $D(\mathbf{X}, \mathbf{P})^\alpha$ for different α after exponential transformation become products of MFs centered on selected prototype.

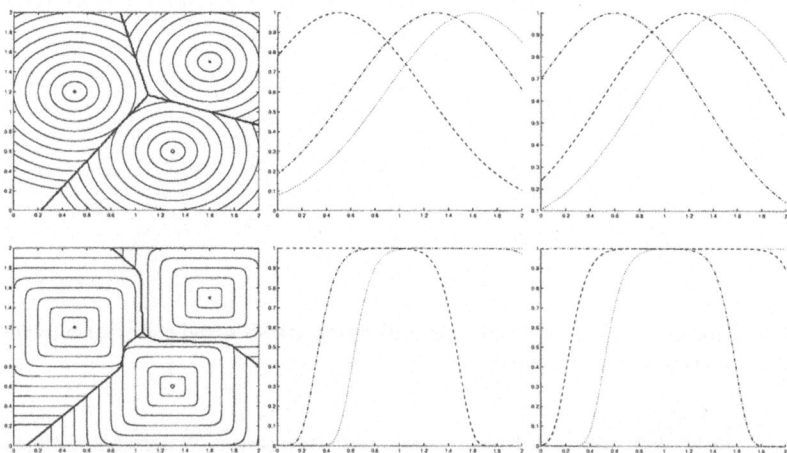

Fig. 1. Contours and decision borders for a 3-class problem using Minkowski distance function with α=2, 10, and the MFs for the two dimensions derived from these distances.

Square of the Euclidean distance leads to Gaussian MFs (1, top row), increasing α leads to more steep MFs (1, bottom row), with $\alpha \to \infty$ giving rectangular contours for crisp logic. For small α MFs have exponential character with sharp peak. The number of unique MFs for each dimension is equal to the number of prototypes. Positions of the prototypes are reflected in the position of the MFs. The width of these functions result from scaling factors W_i that are used for each feature. These factors may be used as adaptive parameters; small values of W_i correspond to a very flat MF that covers the whole data, and thus provides a rule condition that is always true and may be deleted.

Fuzzy logic systems use typical MFs such as triangular, trapezoidal or Gaussian. Transformation of distance functions into similarity functions leads to new types of MFs of different shapes. For example, using Canberra distance function (popular in the nearest neighbor pattern recognition methods) asymmetric MFs are created (Fig. 2).

The inverse transformation, going from MFs to distance functions, is also interesting. Gaussian MFs lead to the square of Euclidean distance function, and all bell-shaped MFs will show similar behavior. Membership functions of triangular or trapezoidal shapes do not have a simple counterpar in the common distance function. Products of triangular functions $T_3(x - x_m; \Delta x)$ equal to zero outside $x_m \pm \Delta x$ interval and to 1 for $x = x_m$, correspond to sums of distance functions $D_3(x - x_m; \Delta x)$ that are infinite outside this interval and behave like $- \ln(1 + (x - x_m)/\Delta)$ for $x \in [x_m - \Delta x, x_m]$ and $- \ln(1 - (x - x_m)/\Delta)$ for $x \in [x_m, x_m - \Delta x]$. Thus fuzzy logic rules with tri-

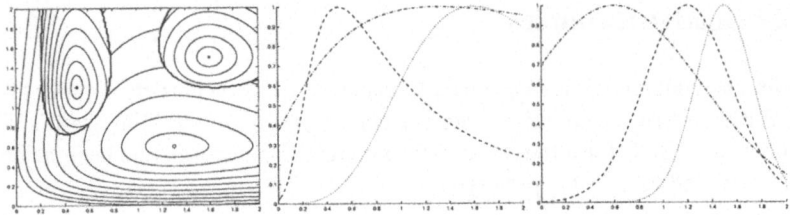

Fig. 2. Contours of a 3 class problem for Canberra distance function for different α=2 and the MFs derived from these distances.

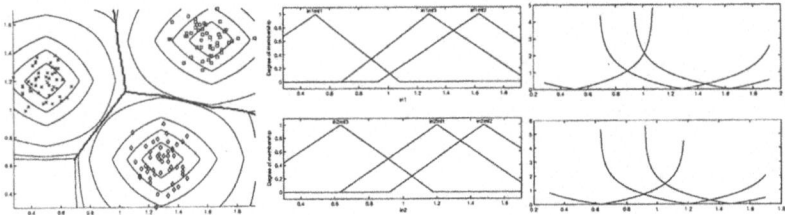

Fig. 3. Decision borders of fuzzy logic rules with triangular MFs, and corresponding prototypes and distance functions for each prototype.

angular MFs are equivalent to P-rule system with prototypes centered at the maxima of the products of MFs and $D_3(x - x_m)$ distance functions, as illustrated in Figure 3.

Creating MFs from data-dependent VDM distance functions allows for a discovery of a "natural" MFs for problems with symbolic or nominal attributes. VDM measure can be used with continuous attributes after discretization of the training data features.

The Leukemia gene expression data [8] has been analyzed looking for best prototype for each of the two classes, acute lymphoblastic leukemia (ALL) and acute myeloid leukemia (AML). Distinguishing between these two leukemias is considered to be challenging because their appearance is highly similar [8]. 7129 gene expression features from microarray experiments are given for each case. The "neighborhood analysis" method developed in the original paper finds 1100 genes that are correlated with ALL-AML class distinction. Prediction is based on a rather complex method that assigns weights to the most useful 50 genes and than calculates "prediction strenghts" (PS) as a sum of votes with threshold 0.3. Training was done on 38 samples (27 ALL and 11 AML), using the leave-one-out method to set parameters, and testing was done on 34 samples (20 ALL and 14 AML). As a result 36 samples were correctly predicted and for two samples PS was below the critical 0.3 threshold. 29 of 34 test samples had large correct PS (median 0.77).

Using logical rules with a single condition based on threshold value for each of the 7129 features identified one that makes no errors on the training data, two features that give a single error, and 14 features that make 2 errors. Since the best feature has quite narrow gap between the two classes 3 best features were taken to generate prototypes, optimizing them using the LVQ approach. Resulting P-rules use VDM metric and one prototype per class; that make no errors on the training data, and only 3 errors on the test data.

4   Conclusions

Rule-based classifiers are useful only if rules are reliable, accurate, stable and sufficiently simple to be understood [9]. Prototype-based rules seem to be a useful addition to the traditional ways of data explanation based on crisp or fuzzy logical rules. They may be helpful in cases when logical rules are too complex or difficult to obtain. Using data-dependent, probabilistic distance functions may lead to natural membership functions that may be difficult to derive in other way. A small number of prototype-based rules with specific similarity functions associated with each prototype may provide complex decision borders that are hard to approximate using logical systems. Such simple rules have been generated for medical datasets using heterogeneous decision tree [10]. Results obtained here for the gene expression data confirm the usefulness of this approach.

Systematic investigation of various membership functions, T-norms and co-norms, their relation to distance function, and the algorithms to discover good prototypes, is under way.

References

1. Nauck, D., Klawonn, F., Kruse, R. (1997): *Foundations on Neuro-Fuzzy Systems*. Wiley, Chichester
2. Pal, S.K., Mitra S. (1999): *Neuro-Fuzzy Pattern Recognition*. J. Wiley, New York
3. Duch, W., Diercksen, G.H.F. (1995): Feature Space Mapping as a universal adaptive system. Computer Physics Communic. **87**, 341–371
4. Duch, W. (2000): Similarity based methods: a general framework for classification, approximation and association, Control and Cybernetics **29**, 937–968
5. Duch, W., Adamczak, R., Diercksen, G.H.F. (2000): Classification, Association and Pattern Completion using Neural Similarity Based Methods. Applied Mathematics and Computer Science **10**, 101–120
6. Duch W, Jankowski N. New neural transfer functions. Neural Computing Surveys 2 (1999) 639-658
7. Wilson, D.R., Martinez, T. R. (1997): Improved Heterogeneous Distance Functions. Journal of Artificial Intelligence Research **6**, 1–34
8. Golub, T.R. et al. (1999): Molecular Classification of Cancer: Class Discovery and Class Prediction by Gene Expression Monitoring. Science **286**, 531-537
9. Duch W, Adamczak R, Grabczewski K, A new methodology of extraction, optimization and application of crisp and fuzzy logical rules. Trans. on Neural Networks 12 (2001) 277-306
10. Grąbczewski, K., Duch, W. (2002) Heterogenous forests of decision trees. Springer Lecture Notes in Computer Science Vol. 2415, pp. 504-509.

Generalized Rule-Based Fuzzy Cognitive Maps: Structure and Dynamics Model

Vadim V. Borisov and Alexander S. Fedulov

Smolensk Branch of the Moscow Power Engineering Institute (Technical University)
Department of Computer Engineering and Electronic,
Energeticheskiy proezd 1, 214013 Smolensk, Russia
borvv@sci.smolensk.ru, fedulov_a@mail.ru

Abstract. Generalized Rule-Based Fuzzy Cognitive Maps (GRFCM) are Fuzzy Cognitive Maps that use completely the fuzzy approach to the analysis and modeling of complex qualitative systems. All components (concepts, interconnections) and mechanisms (causality influence, causality accumulation, system dynamics) of the GRFCM are fuzzy. The offered dynamics model for GRFCM allows to describe and analyze essential features of complex qualitative system's behavior.

1 Introduction

The desire to use the advantages of fuzzy sets theory has resulted in creation of Fuzzy Cognitive Maps in which fuzzy rule-based systems are used for description of influence between the concepts. Rules-Based Fuzzy Cognitive Maps (RBFCM) is a kind of Fuzzy Cognitive Maps [1], [2].

In the RBFCM the value of each concept can be represented as a fuzzy (linguistic) variable. The influence between two concepts is represented as a linguistic fuzzy rule-based system that deals with the changes of fuzzy values of concepts.

The special fuzzy operation – Fuzzy Carry Accumulation (FCA) – is used for accumulation of influences from several input-concepts into the output-concept. This operation of accumulation allows to process fuzzy sets on two systems axes (Y-Axis: value of membership functions – X-Axis: Universe of discourse) and to take into account additive character of separate concepts causalities. However FCA-mechanism seems rather arbitrary.

RBFCM operates only with the changes of fuzzy concepts values. To increase flexibility of models of nonlinear dynamics of complex systems it is possible to use not only changes but also absolute levels of fuzzy concept values.

2 Generalized Rule-Based Fuzzy Cognitive Maps

Generalized Rule-Based Fuzzy Cognitive Maps (GRFCM) are the development of Fuzzy Cognitive Maps which use completely the fuzzy approach for the analysis and modeling of complex qualitative systems [3], [4]. GRFCM is a fuzzy causal map of the following kind:

$$G = (C, W),$$

where $C = \{C_1, C_2, ..., C_p\}$ – set of the concepts, $W = \{w_{ij}\}$ – set of the interconnections between the concepts.

N.R. Pal et al. (Eds.): ICONIP 2004, LNCS 3316, pp. 918–922, 2004.
© Springer-Verlag Berlin Heidelberg 2004

Each concept C_i ($i \in I = \{1, 2, ..., p\}$) contains several membership functions (mbf) $T_i = \{T_1^i, T_2^i, ..., T_{m_i}^i\}$. These mbfs represent the concept's values (typical states).

The weights of influence w_{ij} ($i, j \in I = \{1, 2, ..., p\}$)) between typical states of two concepts are represented also as membership functions $T_{w_{ij}} = \{T_{11}^{w_{ij}}, ..., T_{zl}^{w_{ij}}\}$, $z \in Z = \{1, 2, ..., m_i\}$, $l \in L = \{1, 2, ..., m_j\}$).

All components (concepts, interconnections) and mechanisms (causal influence, causal accumulation, system dynamics) of the GRFCM are fuzzy.

Fig. 1 shows the structure of Generalized Rule-Based Fuzzy Cognitive Map.

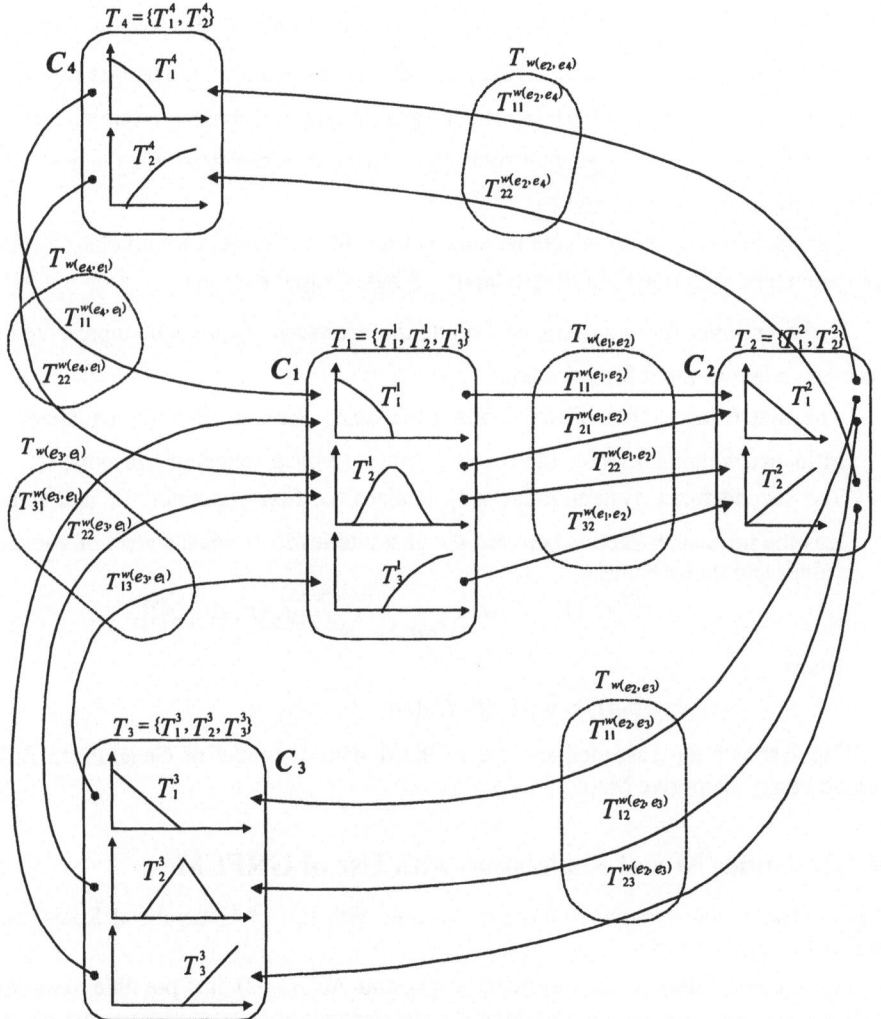

Fig. 1. GRFCM structure.

3 Dynamics Model of Fuzzy Cognitive Maps

This paper presents *a dynamics model of fuzzy cognitive maps*. This model allows to take into account nonlinear behavior of complex qualitative systems due to a joint account of influence of levels and changes of values of the GRFCM concepts.

$$\begin{cases} \tilde{C}_j(t+1) = \tilde{C}_j(t) \,\tilde{+}\, [\underset{i=1,\,2,\,...,\,N}{\tilde{+}} \Delta \tilde{C}_{ij}(t+1)], \\ \Delta \tilde{C}_{ij}(t+1) = \tilde{f}_{ij}[\tilde{C}_i(t), \tilde{C}_j(t), \Delta \tilde{C}_i(t)], \end{cases}$$

or

$$\tilde{C}_j(t+1) = \tilde{C}_j(t) \,\tilde{+}\, \{\underset{i=1,\,2,\,...,\,N}{\tilde{+}} \tilde{f}_{ij}[\tilde{C}_i(t), \tilde{C}_j(t), \Delta \tilde{C}_i(t)]\},$$

where t, $t+1$ – discrete values of time; $j = 1, 2,..., P$ – index of output-concept; P – number of GRFCM concepts; $\tilde{C}_i$, $\Delta \tilde{C}_{ij}$ – fuzzy sets describing values and changes of input-concept C_i; $\tilde{C}_j$, $\Delta \tilde{C}_j$ – fuzzy sets describing values and changes of output-concept C_j; N – number of the input-concepts directly influencing a output-concept; $\tilde{+}$ – operation of fuzzy algebraic sum; $\tilde{f}_{ij}$ – fuzzy operator describing causal influence between concepts C_i and C_j.

The operation of fuzzy algebraic sum is used for the correct accumulation of the output values $\Delta \tilde{C}_{ij}(t+1)$ of Single Input – Single Output systems.

The three-input fuzzy system, describing the operation $\tilde{f}_{ij}$, may be represented by two-cascade two-input fuzzy system:

- The first fuzzy system $\tilde{\varphi}_{ij}(t)$ realizes the fuzzy operator $\tilde{w}_{ij}$ and actualizes the influence weights between the absolute values (typical states) of concepts;
- The second fuzzy system $\Delta \tilde{C}_{ij}(t+1)$ realizes the fuzzy operator $\tilde{f}_{ij}$ and actualizes the influence weights between the absolute levels (typical states) of concepts values and their changes.

$$\tilde{C}_j(t+1) = \tilde{C}_j(t) \,\tilde{+}\, \{\underset{i=1,\,2,\,...,\,N}{\tilde{+}} \tilde{f}_{ij}[\tilde{\varphi}_{ij}(t), \Delta \tilde{C}_i(t)]\},$$

where

$$\tilde{\varphi}_{ij}(t) = \tilde{w}_{ij}[\tilde{C}_i(t), \tilde{C}_j(t)]$$

Fig. 2 shows the cascaded scheme of the dynamics model of Generalized Rule-Based Fuzzy Cognitive Map.

4 Dynamics Model Realization with Use of GRFCM

Let's consider the dynamics model realization with use of Generalized Rule-Based Fuzzy Cognitive Map.

For representation of fuzzy systems $\tilde{\varphi}_{ij}(t)$ and $\Delta \tilde{C}_{ij}(t+1)$ it is possible to use linguistic fuzzy rule-based systems. These systems allow to present the antecedents and consequents of fuzzy rules as membership functions.

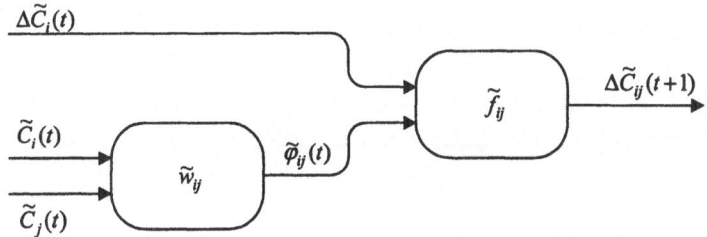

Fig. 2. Cascade scheme of dynamics model of GRFCM.

Let's consider an example of interaction of input-concept $\tilde{C}_i$ (which contains membership functions $T_i = \{ T_1^i, T_2^i, T_3^i \}$) with output-concept $\tilde{C}_j$ (which contains membership functions $T_j = \{ T_1^j, T_2^j \}$).

The first fuzzy system $\tilde{\varphi}_{ij}(t)$ realizes the fuzzy operator $\tilde{w}_{ij}$. This operator $\tilde{w}_{ij}$ is described by set of membership functions $T_{w_{ij}} = \{ T_{11}^{w_{ij}}, T_{21}^{w_{ij}}, T_{22}^{w_{ij}}, T_{32}^{w_{ij}} \}$. Fuzzy system $\tilde{\varphi}_{ij}(t)$ itself is defined by the fuzzy rules of the following kind:

$$\textbf{IF } \tilde{C}_i = T_1^i \textbf{ AND } \tilde{C}_j = T_1^j, \textbf{ THEN } \tilde{\varphi}_{ij}(t) = T_{11}^{w_{ij}},$$

$$\textbf{IF } \tilde{C}_i = T_2^i \textbf{ AND } \tilde{C}_j = T_1^j, \textbf{ THEN } \tilde{\varphi}_{ij}(t) = T_{21}^{w_{ij}},$$

$$\textbf{IF } \tilde{C}_i = T_2^i \textbf{ AND } \tilde{C}_j = T_2^j, \textbf{ THEN } \tilde{\varphi}_{ij}(t) = T_{22}^{w_{ij}},$$

$$\textbf{IF } \tilde{C}_i = T_3^i \textbf{ AND } \tilde{C}_j = T_2^j, \textbf{ THEN } \tilde{\varphi}_{ij}(t) = T_{32}^{w_{ij}}.$$

The second fuzzy system $\Delta\tilde{C}_{ij}(t+1)$ realizes the fuzzy operator $\tilde{f}_{ij}$ and can be represented, for example, in the following kind:

$$\textbf{IF } \tilde{\varphi}_{ij}(t) = T_{11}^{w_{ij}} \textbf{ AND } \Delta\tilde{C}_i(t) = \text{``Insignificant negative''},$$

$$\textbf{THEN } \Delta\tilde{C}_{ij}(t+1) = T_{11}^{w_{ij}},$$

$$\textbf{IF } \tilde{\varphi}_{ij}(t) = T_{21}^{w_{ij}} \textbf{ AND } \Delta\tilde{C}_i(t) = \text{``Insignificant negative''},$$

$$\textbf{THEN } \Delta\tilde{C}_{ij}(t+1) = T_{11}^{w_{ij}},$$

$$\textbf{IF } \tilde{\varphi}_{ij}(t) = T_{11}^{w_{ij}} \textbf{ AND } \Delta\tilde{C}_i(t) = \text{``Middle positive''},$$

$$\textbf{THEN } \Delta\tilde{C}_{ij}(t+1) = T_{21}^{w_{ij}},$$

$$\cdots$$

$$\textbf{IF } \tilde{\varphi}_{ij}(t) = T_{22}^{w_{ij}} \textbf{ AND } \Delta\tilde{C}_i(t) = \text{``Significant positive''},$$

$$\textbf{THEN } \Delta\tilde{C}_{ij}(t+1) = T_{32}^{w_{ij}}.$$

The small number of membership functions $T_{w_{ij}} = \{ T_{11}^{w_{ij}}, T_{21}^{w_{ij}}, T_{22}^{w_{ij}}, T_{32}^{w_{ij}} \}$ allows to reduce the number of the rules of two-cascade fuzzy system considerably.

Resulting value of output-concept change $\Delta \widetilde{C}_{ij\,\text{res}}(t+1)$ is computed by means of the consequents composition of all fuzzy rules.

The above mentioned procedure should be executed for all input-concepts C_i ($i = 1, 2,\ldots, N$) in relation to a considered output-concept C_j ($j = 1, 2, \ldots, P$).

Then all received fuzzy subsets $\Delta \widetilde{C}_{ij\,\text{res}}(t+1)$ ($i = 1, 2, \ldots, N$) are accumulated using fuzzy algebraic sum.

$$\Delta \widetilde{C}_{j\,\text{res}}(t+1) = \underset{i=1,\,2,\,\ldots,\,N}{\widetilde{+}}\ \Delta \widetilde{C}_{ij\,\text{res}}(t+1).$$

Then the next value $\widetilde{C}_j(t+1)$ of the output-concept is defined. For this purpose the operation of fuzzy sum over the fuzzy sets $\widetilde{C}_j(t)$ and $\Delta \widetilde{C}_{j\,\text{res}}(t+1)$ is carried out.

$$\widetilde{C}_j(t+1) = \widetilde{C}_j(t) \,\widetilde{+}\, \Delta \widetilde{C}_{j\,\text{res}}(t+1).$$

After each step of modeling the identification of the absolute values (typical states) of concept C_j will be carried out at time ($t+1$). This identification is done, for example, on the basis of the maximal degree of fuzzy equality with one of typical fuzzy states (mbf) $T_j = \{\, T_1^j, T_2^j, \ldots, T_{m_i}^j \,\}$ of this concept.

All of the above mentioned actions are carried out for all concepts $C = \{C_1, C_2, \ldots, C_p\}$ of GRFCM during modeling.

5   Conclusion

This paper presents Generalized Rule-Based Fuzzy Cognitive Maps. GRFCM are Fuzzy Cognitive Maps which use completely the fuzzy approach for the analysis and modeling of complex qualitative systems. All components (concepts, interconnections) and mechanisms (causality influence, causality accumulation, system dynamics) of the GRFCM are fuzzy.

The dynamics model of fuzzy cognitive maps is offered. This model allows to take into account nonlinear behavior of complex qualitative systems due to a joint account of influence of values and changes of the GRFCM concepts.

References

1. Carvalho, J. P., Tomé, J. A.: Rule-based fuzzy cognitive maps and fuzzy cognitive maps – a comparative study. In Proc. of the 18th International Conference of the North American Fuzzy Information Processing Society, NAFIPS'99, New York (1999) 115–119.
2. Carvalho, J. P., Tomé, J. A.: Rule-based fuzzy cognitive maps – expressing time in qualitative system dynamics. In Proc. of the FUZZ-IEEE'2001, Melbourne, Australia, (2001) 280–283.
3. Borisov, V. V., Fedulov, A. S.: Computer support of complex qualitative systems, Publishing House: Goryachaya Liniya – Telecom, Moscow (2002) 176 p. (in Russian).
4. Borisov, V. V., Fedulov, A. S.: Generalized fuzzy cognitive map. Neurocomputers: Design and Application, Begell House, Moscow – NY, no. 4 (2004) 3–20.

Development of Adaptive Fuzzy Based Multi-user Detection Receiver for DS-CDMA

Sharmistha Panda and Sarat Kumar Patra

Department of Electronics and Instrumentation Engineering,
National Institute of Technology, Rourkela, India-769 008
Tel: +91 661 2475922; Fax: +91 661 2472926
{20207308,skpatra}@nitrkl.ac.in

Abstract. This paper investigates the problem of multiuser detector (MUD) for direct sequence code division multiple access (DS-CDMA) system. A radial basis function (RBF) receiver provides the optimum receiver performance. We propose a fuzzy implementation of the RBF receiver. This fuzzy receiver provides considerable computational complexity reduction with respect to RBF receivers. The fuzzy receiver provides exactly the same bit error rate performance (BER) as the RBF receiver. Extensive simulation studies validate our finding.

1 Introduction

The demand for increased capacity on mobile communication system such as GSM has led to newer technologies like code division multiple access (CDMA), wide band CDMA systems. It is believed that the capacity of CDMA technique is much higher then that of the established TDMA system [1]. CDMA allows frequency re-use in the neighboring cells and even distribution of the workload among the cells, and user transparent soft hand-off as the call is re-routed from one cell to another. With this CDMA technology has been used in voice, data and network communications.

CDMA systems suffer from interference from other users in the cell. It is also affected by channel multipath interference with fading in presence of additive white Gaussian noise (AWGN). Mitigation of these effects in receivers require high computational complexity. Instead of attempting to cancel the interference from other users in the system, the principle of multiuser detector [2] uses multiple access interference (MAI) as additional information to obtain a better estimate of the intended data. The multiuser detector (MUD) processes the signal at bit rate derived from the bank of matched filters. These processed signal are then processed by different types of receivers. Since the optimal decision boundary in DS-CDMA is non-linear [3], it can be optimally implemented by radial basis function (RBF) network [2,4], at an expense of increased computational complexity. The complexity in terms of center calculation grows exponentially with the number of users. Considering DS-CDMA a non-linear classification problem, it has been shown that the non-linear receivers always outperform the conventional linear receivers. Existing non-linear receivers based on artificial neural network (ANN), multiple layer perceptron (MLP), polynomial series, recurrent networks can approximate the decision boundary well and possess superior performance, but at an expense of higher computational complexity and larger and complex training technique

N.R. Pal et al. (Eds.): ICONIP 2004, LNCS 3316, pp. 923–928, 2004.
© Springer-Verlag Berlin Heidelberg 2004

and therefore difficult for practical implementation.Thus considerable investigation is underway in this regard. Fuzzy systems have been extensively used for many non-linear applications including pattern classification. The close relationship between the fuzzy and the RBF [5] prompted us to use adaptive fuzzy systems as a candidate for DS-CDMA MUD receiver.

This paper is organized as follows, Following this section, DS-CDMA system model is outlined first. The next section provides a discussion on adaptive fuzzy filter and its implementation for MUD receiver for DS-CDMA. The performance of the proposed receiver with other standard receivers is discussed next. The last section provides the concluding remarks.

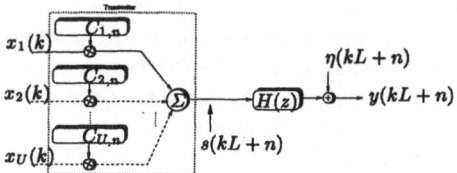

Fig. 1. DS-CDMA down link transmitter for U transmitting users.

2 DS-CDMA System Model

The system model considered in this paper is presented in Fig.1. It shows the down link scenario where the mobile unit receives signal $y(kL + n)$ from the base station. The information bits corresponding to one of U users are denoted as $x_i(k)$. $x_i(k)$ takes the value $+1/-1$ with equal probability and k denotes the time index of user transmitted symbols. The information bits transmitted by each user are convolved with each of their mutually orthogonal spreading sequences $C_{i,n}$. Gold code, convolution codes, Pseudo-noise (PN) codes [6] are some of the coding techniques used. With this the BW of $x_i(k)$ is enhanced. The processing gain (PG) of the system is defined as $PG = \frac{W}{B}$ where, W denotes the spreaded signal bandwidth (BW) and B is the unspreaded signal BW. The spreaded signal from each of the user are combined to form

$$s(kL + n) = \sum_{i=1}^{U} x_i(k)C_{i,n} \tag{1}$$

which is transmitted through the channel $H(z)$. The channel corrupts the signal with inter symbol interference (ISI) and effects of fading. AWGN also gets added to the signal. With this the received signal $y(kL + n)$ can be denoted as

$$y(kL + n) = H(z) \otimes s(kL + n) + \eta(kL + n) \tag{2}$$

where $\otimes$ denotes the convolution and $\eta(kL + n)$ is the AWGN component at chip rate. The job of the receiver is to estimate $x_i(k)$ of the desired user using the information content in $y(kL + n)$. The input is sampled at chip rate n and process the signal at sample rate k. This is called chip level based receiver (CLB). Due to high computational complexity of nonlinear CLB receivers multiuser detection is used [4]. The structure of a MUD receiver using RBF is shown in Fig.2. The output vector of the preprocessor $\tilde{\mathbf{x}}(k) = [\tilde{x}_1(k), \ldots, \tilde{x}_U(k)]^T$ is fed to a RBF network. The output of the RBF can be denoted as

$$t(k) = \sum_{j=1}^{2^U} w_j \exp\left(\frac{-\|\tilde{\mathbf{x}}(k) - \mathbf{c}_j\|^2}{2\sigma^2}\right) \tag{3}$$

where, the RBF has 2^U centres of dimension U, σ is the centre spread parameter and w_j denotes the weight associated with each centre. The RBF output $t(k)$ is passed through a hard limiter to provide $\hat{x}_i(k)$, the estimate of the transmitted symbol of the desired user $x_i(k)$. As the number of transmitting users increases, the computational complexity of the RBF receiver also increases in terms of number of centres.

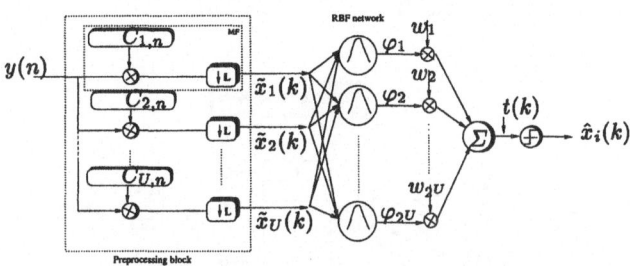

Fig. 2. RBF receiver with preprocessing stage.

3 Fuzzy Adaptive Filter for DS-CDMA

3.1 Adaptive Fuzzy Filters

Fuzzy logic system uses linguistic informations to process it's input. The fuzzifier converts the real world crisp input to a fuzzy output described by the membership function. The inference engine provides the relationship between the fuzzy input in terms of membership functions and the fuzzy output of the controller using a set of IF ... THEN ... rules derived from the rule base. The defuzzifier converts the inferences to provide the crisp output. Generally in a fuzzy system the rule base is generated in advance with expert knowledge of the system under consideration. In [7], online learning properties was introduced which provided scope for training the fuzzy system.

Wang et. al. presented fuzzy basis functions (FBF) and used them as a fuzzy filter [8] for channel equalization. Later on the fuzzy implementation of MAP equalizer was investigated [5]. It was shown that the fuzzy equalizer can provide the MAP decision function like RBF. These equalizers address some of the problems associated with the previously reported fuzzy equalizers. In this paper we implement a modification of these fuzzy filters for MUD in DS-CDMA scenario.

3.2 Fuzzy Filter for DS-CDMA Multi-user Detection Receiver

The RBF receiver decision function in (3) discussed in the previous section can also be represented as

$$t(k) = \sum_{j=1}^{2^U} w_j \left\{ \prod_{i=1}^{U} \exp\left(\frac{-\|\tilde{x}_{j,i}(k) - c_{j,i}\|^2}{2\sigma^2}\right) \right\} \tag{4}$$

where $1 \leq i \leq U$ constitute the ith components of the RBF centre and the RBF input. The inner product of $\exp(.)$ of vector has been replaced by product of $\exp(.)$ of scalar terms of the vector. The function presented in (4) can be represented by a fuzzy system shown in Fig.3. The output of the preprocessing block, feeds the fuzzy filer. The fuzzy filter consists of fuzzifier with Gaussian membership function. The centres of the membership function are located at -1 and $+1$. There are 2^U rules in the rule base. The product inference block provides 2^U outputs generated with product rule. The defuzzifier provides a weighted sum of it's input from inference block with it's set of weights. The receiver so designed is presented in Fig.3. This receiver can be considered as an alternative implementation of RBF receiver [5]. This fuzzy receiver proposed here can be trained with gradiant search algorithm like LMS.

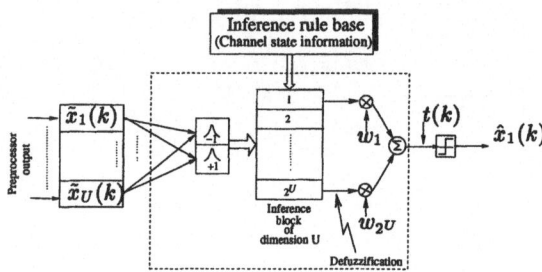

Fig. 3. Fuzzy implementation of RBF receiver.

An example is considered to describe the details of the fuzzy receiver discussed here. If the number of users in the scenario discussed here is $U = 2$, there will be $2U = 4$ fuzzified inputs to the inference engine from a total of 2 input scalars constituting the input vector. The number of rule base is $2^U = 4$ and the output defuzzifier combines these 4 inference outputs with suitable weights. If the number of active user increases to 6 the number of fuzzy inputs will be $2U = 12$ and number of inference rule will be $2^U = 64$.

Table 1. Computational complexity for MUD receivers using RBF and Fuzzy.

U	Tech- que	Centres/ Rule	Multiplication.	Addition/ Subtraction/ Comparison	$exp(.)$
2	RBF	4	12	8	4
	Fuzzy1	4	12	8	4
7	RBF	128	1024	896	128
	Fuzzy1	128	910	142	14

This receiver proposed provides considerable computational complexity reduction compared to RBF receiver. The computational complexity comparison between RBF and fuzzy receiver when 2 and 7 users are active is presented in Table.1. From the table it can be seen that, the fuzzy based MUD receiver provides the RBF implementation of MUD receiver with considerable computational complexity reduction in terms of multiplication, addition and exp(.) calculations.

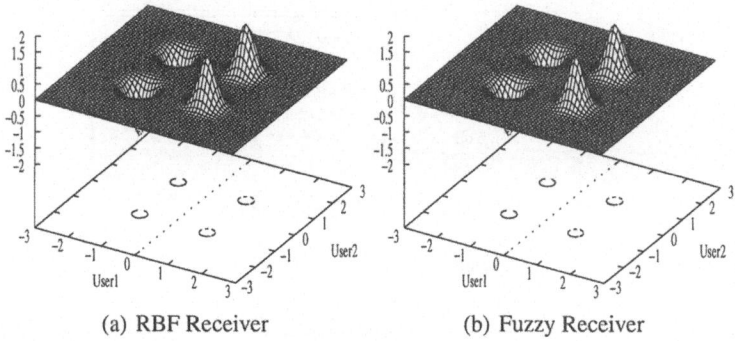

(a) RBF Receiver   (b) Fuzzy Receiver

Fig. 4. Surface plot and decision boundary of RBF and Fuzzy MUD receivers at $E_b/N_o = 10dB$.

4 Simulation Results

Extensive simulation studies were conducted to validate the proposed fuzzy MUD receiver for DS-CDMA application. The results obtained were compared with MUD receivers using RBF network and simple linear receiver using LMS training. During the training period the receiver parameters were optimized/trained with 1000 random samples and parameters so obtained were averaged over 50 experiments. The parameters of the receiver were fixed after the training phase. The RBF and fuzzy receiver decision surface along with their decision boundaries for a two user case is plotted in Fig.4. From here it can be seen that the fuzzy MUD receiver provides a decision boundary exactly same as the RBF receiver.

In the next phase of simulation studies, bit error rate (BER) was considered as the performance index. Monte Carlo simulations were conducted to estimate the BER performance of fuzzy MUD receiver and was compared with RBF and linear MUD receivers. A total of 10^7 bits were transmitted by each user and a minimum 100 errors were recored. The tests were conducted for different levels of E_b/N_o and varying number of users active in the cell.

The BER performance of the three types of receivers with 2 users and 7 users active in the system is shown in Fig.5. Fig.5(a) shows the performance for channel $H(z) = 0.5 + z^{-1}$ and Fig.5(b) shows the performance for the channel $0.3482 + 0.8704z^{-1} +$

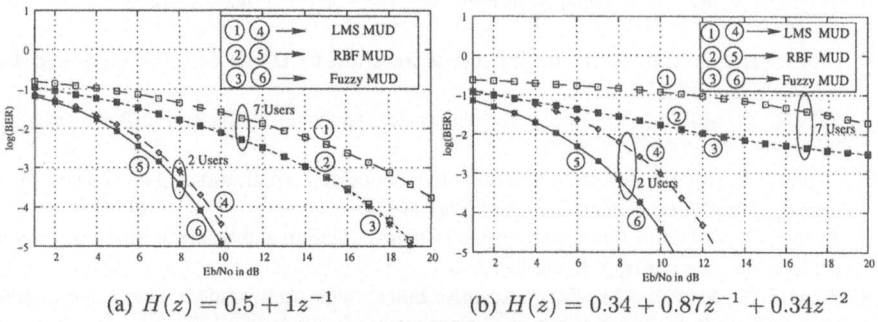

(a) $H(z) = 0.5 + 1z^{-1}$ (b) $H(z) = 0.34 + 0.87z^{-1} + 0.34z^{-2}$

Fig. 5. BER performance for varying E_b/N_o.

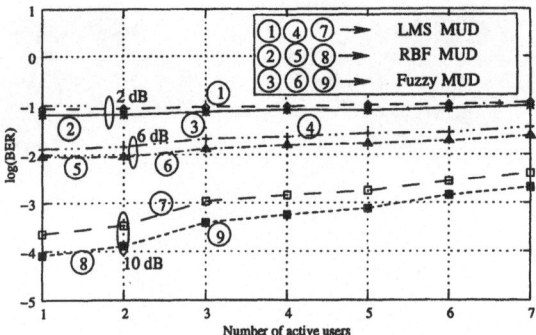

Fig. 6. BER performance for varying no of users at different values of E_b/N_o.

$0.3482z^{-2}$. From the BER performance it can be seen that the fuzzy receiver provides a performance which is exactly same as RBF receiver. Following this, performance of the fuzzy receiver was tested for varying levels of users active in the system for a fixed value of E_b/N_o in the channel. The channel used for the test is characterized by $0.407 - 0.815z^{-1} - 0.407z^{-2}$. The BER performance for E_b/N_o of 2dB, 6dB and 10dB is shown in Fig.6. The simulation studies show that the proposed fuzzy receiver performs exactly same as optimal RBF MUD receiver.

5   Conclusion

In this paper the RBF based MUD receiver has been implemented with fuzzy system. This fuzzy receiver proposed uses Gaussian membership function, product inference and center of gravity defuzzifier. This receiver provides computational complexity reduction over the optimal RBF receiver. Simulation studies show that the performance of the receiver proposed is exactly similar to RBF receiver.

References

1. Shiung, D., Jin-Fu Chang: Enhancing the Capacity of DS-CDMA System Using Hybrid Spreading Sequences. IEEE Transactions on Communications **52** (2004) 372–375
2. Mitra, U., Poor, H.V.: Neural Network Techniques for Adaptive Multiuser Demodulation. IEEE Transactions on Selected Areas in Communications **12** (1994) 1460–1470
3. Mulgrew, B.: Applying Radial Basis Functions. IEEE Signal Processing Magazine **13** (1996) 50–65
4. Cruickshank, D.G.M.: Radial Basis Function Receivers for DS-CDMA. IEE Electronics Letter **32** (1996) 188–190
5. Patra, S.K., Mulgrew, B.: Fuzzy Techniques for Adaptive Nonlinear Equalization. Signal Processing **80** (2000) 985–1000
6. Dixon, R.C.: Spread Spectrum System with Commercial Applications. 3 edn. Wiley- Interscience, John Wiley & Sons, Inc., New York (1994)
7. Wang, L.X.: Adaptive Fuzzy Systems and Control : Design and Stability Analysis. Prentice Hall, Englewood Cliffs, N.J., USA (1994)
8. Wang, L.X., Mendel, J.M.: Fuzzy Adaptive Filters, with Application to Non-linear Channel Equalization. IEEE Transactions on Fuzzy Systems **1** (1993) 161–170

A Partitioning Method for Fuzzy Probabilistic Predictors*

Marcelo Andrade Teixeira[1,2] and Gerson Zaverucha[1]

[1] Systems Engineering and Computer Science - COPPE
Federal University of Rio de Janeiro (UFRJ)
P.O. Box 68511, ZIP 21945-970, Rio de Janeiro, Brazil
{mat,gerson}@cos.ufrj.br
[2] Electric Power Research Center (CEPEL)
P.O. Box 68007, ZIP 21944-970, Rio de Janeiro, Brazil
mat@cepel.br

Abstract. We present a new partitioning method to determinate fuzzy regions in fuzzy probabilistic predictors. Fuzzy probabilistic predictors are modifications of discrete probabilistic classifiers, as Naive Bayes Classifier and Hidden Markov Model, in order to enable them to predict continuous values. Two fuzzy probabilistic predictors, Fuzzy Markov Predictor and the Fuzzy Hidden Markov Predictor, are applied to the task of monthly electric load single-step forecasting using this new partitioning and successfully compared with two Kalman Filter Models, and two traditional forecasting methods, Box-Jenkins and Winters exponential smoothing. The employed time series present a sudden significant changing behavior at their last years, as it occurs in an energy rationing.

1 Introduction

In statistics and pattern recognition, the typical approach for the handling of continuous variables is to use a parametric family of distributions, as in Kalman Filter Models [5], [17], which makes strong assumptions about the nature of data: the induced model can be a good approximation of the data, if these assumptions are warranted. Machine learning, on the other hand, deal with continuous variables by discretizing them, which can suffer from loss of information [3]. When it is used to predict continuous variables this approach is known as regression-by-discretization [2], [16] and it motivated the creation of three discrete probabilistic predictors: the Naives Bayes for Regression (NBR) [2], the Markov Model for Regression (MMR) [12], [14] and the Hidden Markov Model for Regression (HMMR) [14]. Through a generalization of the regression-by-discretization approach named fuzzification [7], we developed three fuzzy probabilistic predictors: the Fuzzy Bayes Predictor (FBP) [13], the Fuzzy Markov Predictor (FMP) [13], and the Fuzzy Hidden Markov Predictor (FHMP) [14].

Indeed, the discretization and the fuzzification in probabilistic predictors are estimating the distribution of a continuous variable [13], [10]. In both, the simplest approach is to make a uniform partitioning of the continuous space. This partitioning is

---

* The authors are partially financially supported by the Brazilian Research Agencies CAPES and CNPq, respectively.

N.R. Pal et al. (Eds.): ICONIP 2004, LNCS 3316, pp. 929–934, 2004.
© Springer-Verlag Berlin Heidelberg 2004

not necessarily the best one to approximate a probability density function. Density trees [15] are an alternative way to make the discrete approximation without the uniform restriction. In this work, it is a fuzzy version of the density tree growing algorithm that divides the continuous space in fuzzy regions.

The paper is organized as follows. In section 2, FHMP is reviewed. In section 3, it is discussed how discretization can approximate distributions of continuous variables. In section 4, it is presented the new partitioning method to determinate fuzzy regions in fuzzy probabilistic predictors. In section 5, FHMP and FMP, using this new partitioning method, are applied to task of monthly electric load single-step forecasting where the employed time series present a sudden significant changing behavior at their last years. They are compared to Kalman Filter Models and traditional forecasting methods. Finally, in section 6, some conclusions and future work are discussed.

2 Fuzzy Hidden Markov Predictor

Hidden Markov Model (HMM) [9] is a particular Dynamic Bayesian Network (DBN) [4]. A DBN is a Bayesian Network (BN) that represents a temporal probability model: S_t is a set of state variables (hidden) and E_t is a set of evidence variables (observed). In the HMM, S_t is a singleton discrete random variable.

Hidden Markov Model for Regression (HMMR) [14] is the HMM applied to regression by discretization of the space of observations: for each continuous observation there is a corresponding discrete value representing the interval that contains the continuous value. In the Fuzzy Hidden Markov Predictor (FHMP), the space of observations is divided into fuzzy regions. The membership function $m_r(v)$ returns a real number in [0, 1] for a region r and a continuous value v (Fig. 1).

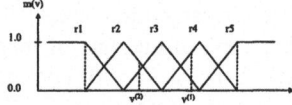

Fig. 1. Fuzzy regions and the membership function

Fuzzy probabilities [18] arise when random variables are fuzzy. Following [18], the probability of a fuzzy variable B equal to rk is the expectation E(.) of the membership function of rk. $E(m_{rk})$ can be estimated from a sample (training examples):

$$p(B = rk) = E(m_{rk}) \approx (\textstyle\sum_{x \in \text{sample}} m_{rk}(x)) / \text{size\_of\_sample} \qquad (1)$$

We assume that S_t and $E_{t,j}$ ($1{\leq}j{\leq}m$) are fuzzy random variables. For each continuous attribute ($a_{t,j}$) there is a corresponding set of fuzzy values ($e_{t,j}$) representing the fuzzy interval that contains this continuous value. Considering the conjunction of attributes (a_t), d_t represents the set of conjunctions of fuzzy intervals (e_t) with $m_{e_t}(a_t) > 0$:

$$d_t = \{e_t | m_{e_t}(a_t) > 0\}, \text{ where } e_t = (e_{t,1},..., e_{t,m}) \text{ and } m_{e_t}(a_t) = m_{e_{t,1}}(a_{t,1}) \times ... \times m_{e_{t,m}}(a_{t,m}) \quad (2)$$

Since S_t is hidden in the training data, the estimation of fuzzy probabilities (parameters) $p(S_t|S_{t-1})$ and $p(E_t|S_t)$ must be made by the EM algorithm [4]:

$$p(S_0) = N(S_0) / N, \quad p(S_{t+1}|S_t) = N(S_{t+1},S_t) / N(S_t), \quad p(E_{t,j}|S_{t,j}) = N(E_{t,j},S_t) / N(S_t) \quad (3)$$

where N is the total number of training examples, N(.) and N(., .) are computed by:

$$N(S_t) = \Sigma^T_{t=1} \; p(S_t|d_{1:t}), \; N(S_{t+1},S_t) = \Sigma^T_{t=1} \; p(S_{t+1},S_t|d_{1:t}), \quad (4)$$
$$N(E_{t,j},S_t) = \Sigma^T_{t=1} \; p(E_{t,j},S_t|d_{1:t})$$

where T is the last time slice, $p(S_{t+1},S_t|d_{1:t})$ and $p(E_{t,j},S_t|d_{1:t})$ are infered by:

$$p(S_{t+1},S_t|d_{1:t}) = p(S_{t+1}|S_t).p(S_t|d_{1:t}), \quad p(E_{t,j},S_t|d_{1:t}) = m_{E_{t,j}}(a_{t,j}).p(S_t|d_{1:t}) \quad (5)$$

Filtering (calculating $p(S_t|d_{1:t})$) is given by the two formulas:

$$p(S_t|d_{1:t}) = (\Sigma_{each \; e_t \in d_t} \; m_{e_t}(a_t).p(S_t|d_{1:t-1},e_t)) / (\Sigma_{each \; e_t \in d_t} \; m_{e_t}(a_t)) \quad (6a)$$

$$p(S_t|d_{1:t-1},e_t) = \alpha.\{\Pi_j p(e_{t,j}|S_t)\}.(\Sigma_{s_{t-1}} \; p(S_t|s_{t-1}).p(s_{t-1}|d_{1:t-1})) \quad (6b)$$

EM is an iterative procedure: in order to compute the parameters we have to calculate N(.) and N(., .) by use of inference; and inference is done through the use of the current parameters. This process repeats itself until it reaches a stopping condition.

A future fuzzy observation $E_{t+1,j}$ is predicted by the computation of $p(E_{t+1,j}|d_{1:t})$:

$$p(E_{t+1,j}|d_{1:t}) = (\Sigma_{s_{t+1,j}} \; p(E_{t+1,j}|s_{t+1}).(\Sigma_{s_t} \; p(s_{t+1}|s_t).p(s_t|d_{1:t}))) \quad (7)$$

This future fuzzy observation is converted to a continuous value $v_{FHMP\,t,j}$:

$$v_{FHMP\,t,j} = \Sigma_{e_{t+1,j} \in \; dom(E_{t+1,j})}\{ \; \bar{v}_{e_{t+1,j}}.p(e_{t+1,j}|d_{1:t}) \; \}, \quad (8)$$

where $\bar{v}_{e_{t+1,j}}$ is the center value of $e_{t+1,j}$

3 Density Estimation by Discretization

It is possible to approximate the distributions of continuous variables by discretizing them. Since fuzzification could be considered a generalization of discretization, continuous variables distributions can be approximated by fuzzification. For example, consider the case where we want to estimate the probability density function $f(v \mid a_1, a_2, \dots, a_m)$ and calculate its mean to use as a prediction of v when we know the continuous values $a_1, a_2, \dots, a_m$. The NBR executes this task by discretizing $a_1, a_2, \dots, a_m$ into intervals $e_1, e_2, \dots, e_m$ and then it makes a prediction of v by

$$v_{NBR} = \Sigma_{s \in \; dom(S)}\{m(s).p(s|e_1, \dots, e_m)\} = \Sigma_{s \in \; dom(S)}\{m(s).\alpha.p(s).\Pi_j p(e_j|s)\} \quad (9)$$

where $m(s)$ is the mean of the interval s, α is a normalizing constant, $p(s)$ and $p(e_j|s)$ are calculated by simple counting of discretes values (intervals) and their conjunctions from the discretization of the training examples:

$$p(S) = N(S) / N, \quad p(E_j|S) = N(E_j , S) / N(S) , 1 \le j \le m \quad (10)$$

Indeed, the NBR is estimating the probability density function

$$f(v|a) = f(v \mid a_1, a_2, \dots, a_m) = f(v).\Pi_j f(a_j|v) / \int(f(v).\Pi_j f(a_j|v))dv \quad (11)$$

employing the following approximations [10] through discretization:

$$f(v) = p(s) / h_s \, , v \in s; \qquad f(a_j, v) = p(e_j \, , s) / (h_{e_j} . h_s) \, , a_j \in e_j \text{ and } v \in s \qquad (12a)$$

$$f(a_j|v) = f(a_j, v) / f(v) = p(e_j|s) / h_{e_j} \, , a_j \in e_j \text{ and } v \in s \qquad (12b)$$

where h_{e_j} and h_s are the sizes of the intervals e_j and s, respectively. So, we have:

$$f(v|a) = \alpha.p(s).\Pi_j p(e_j|s) / h_s \, , v \in s \text{ and } e_1 \in a_1 \text{ and } ... e_m \in a_m \qquad (13)$$

and the prediction is given by the expected value of v (that is, its mean):

$$v_{NBR} = \int v.f(v|a)dv = \Sigma_s m(s).\alpha.p(s).\Pi_j p(e_j|s) \qquad (14)$$

4 Fuzzy Partitioning in Fuzzy Probabilistic Predictors

The previous section showed that we can use discretization or fuzzification of a continuous random variable as a means to estimate its probability density function. Until now all our fuzzy probabilistic predictors employed triangular membership functions that are uniformly distributed in the continuous space. This uniform fuzzification is not necessarily the best one to approximate a probability density function f. Density trees [15] are an alternative way to make this approximation without the uniform restriction. They discretize the continuous space in intervals in the following way:

- start with N continuous samples in an interval that covers the entire domain of f;
- split the current interval into two equally sized intervals if it has at least $\sqrt{N}$ samples and its distance from the root node does not exceed $\lfloor(\log_2 N)/4\rfloor$;
- repeat this process for each new interval while the conditions are satisfied.

The partitioning made by the density tree is similar to the one made by the Binary Space Partitioning (BSP) in neuro-fuzzy systems [11]. Our fuzzy version of the density tree growing algorithm works in the same manner of the original one for the division of intervals. The difference is in the placing of fuzzy regions on these intervals:

- the starting interval contains two uniform triangular fuzzy regions whose maximums and minimums (memberships equal to 1 and 0) are the limits of the interval, and the intersection of the fuzzy regions is the middle point of the interval;
- each current interval contains part of two triangular fuzzy regions whose maximums and minimums (memberships equal to 1 and 0) are the limits of the interval, and the intersection of the fuzzy regions is the middle point of the interval;
- when the current interval is splitted into two equally sized intervals (if the conditions are satisfied), the minimums (memberships equal to 0) of the two previous fuzzy regions are changed to the middle point of the original interval, and a new triangular fuzzy region is inserted with maximum (membership equal to 1) equal to the middle point of the original interval and minimums (memberships equal to 0) equal to the limits of the original interval;
- this process is repeated for each new interval while the conditions are satisfied.

An example of this partitioning is shown in Fig. 2. We start with 2 fuzzy regions, r1 and r2, and the algorithm places new fuzzy regions r3, r4 and r5 in this order.

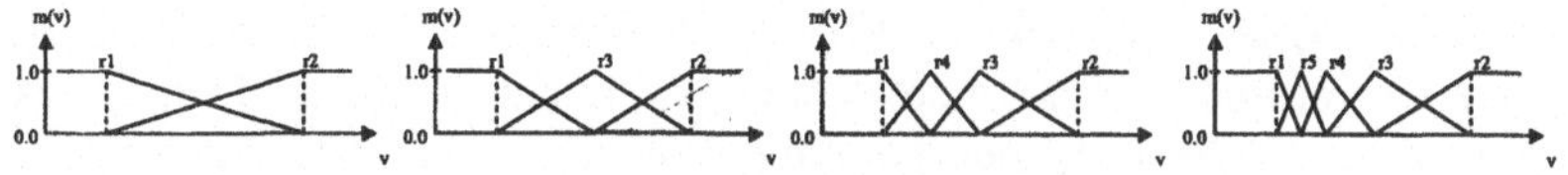

Fig. 2. Fuzzy partitioning

5 Experimental Results

FMPtree and FHMPtree (the new partitioning) are applied to the task of monthly electric load single-step forecasting and compared with FMP and FHMP (uniform partitioning), two Kalman Filter Models, STAMP [5] and BATS [17], and two traditional forecasting methods, Box-Jenkins [1] and Winters exponential smoothing [8].

The forecast errors (Table 1) are from three series of monthly electric load (3×12 months of test data). These series were obtained from Brazilian utilities and present a sudden significant changing behavior at their last years, as it occurs in an energy rationing. The error metric used was MAPE (Mean Absolute Percentage Deviation):

$$MAPE = (\sum_{i=1}^{n} |e_i|) / n \qquad (15)$$

where $e_i = ((actual_i - forecast_i) / actual_i) * 100\%$, and n = number of examples.

All the systems employed the last 3 years of the series as the test set, and the preceding 5 years as a training data to make the single-step prediction of the next month. So, for each month of the test set the systems are retrained with the preceding 5 years. Forward validation [6] was utilized for the selection of the number of fuzzy regions (uniform partitioning) and the number of attributes.

Table 1. MAPE errors

	Series 1	Series 2	Series 3
FMP	4.03%	5.10%	3.46%
FHMP	2.75%	5.04%	3.28%
FMPtree	2.71%	4.97%	3.42%
FHMPtree	2.75%	5.36%	3.12%
STAMP	3.12%	4.65%	3.54%
BATS	7.28%	8.93%	7.48%
Box-Jenkins	3.29%	4.58%	3.60%
Winters	2.79%	6.06%	3.33%

6 Conclusion and Future Work

We presented a new partitioning method to determinate fuzzy regions in fuzzy probabilistic predictors. FMP and FHMP using this new partitioning method were applied to the task of monthly electric load forecasting where the employed time series present a sudden significant changing behavior at their last years.

Analyzing Table 1, we can see that FMP and FHMP using the new partitioning method obtained competitive results when compared with FMP and FHMP using uniform partitioning, two Kalman Filter Models and two traditional forecasting methods.

As future work, we would like to utilize the new partitioning method with FBP, test other time series, and explore more complex versions of the FHMP, for example, using smoothing in the EM, incorporating structure in the state and evidence variables.

References

1. G.E.P. Box, G.M. Jenkins and G.C. Reinsel, Time Series Analysis: Forecasting & Control, Prentice Hall, 1994.
2. E. Frank, L. Trigg, G. Holmes and I.H. Witten, "Naive Bayes for regression," Machine Learning, Vol.41, No.1, pp.5-25, 1999.
3. N. Friedman, M. Goldszmidt and T.J. Lee, "Bayesian network classification with continuous attributes: getting the best of both discretization and parametric fitting," 15th Inter. Conf. on Machine Learning (ICML), pp. 179-187, 1998.
4. Z. Ghahramani, "Learning dynamic Bayesian networks," in Adaptive Processing of Sequences and Data Structures, Lecture Notes in Artificial Intelligence, C.L. Giles and M. Gori (eds.), Berlin, Springer-Verlag, 1998, pp. 168-197.
5. A.C. Harvey, Forecasting, structural time series models and the Kalman filter, Cambridge University Press, 1994.
6. J.S.U. Hjorth. Computer Intensive Statistical Methods. Validation Model Selection and Bootstrap. Chapman & Hall. 1994.
7. J.M. Mendel, "Fuzzy logic systems for engineering: a tutorial," Proceedings of the IEEE, vol.83, pp.345-377, 1995.
8. D.C. Montgomery, L.A. Johnson and J.S. Gardiner, Forecasting and Time Series Analysis, McGraw-Hill Companies, 1990.
9. L.R. Rabiner, "A tutorial on hidden Markov models and selected applications in speech recognition," Proc. of the IEEE, vol. 77, no. 2, pp. 257-286, 1989.
10. D.W. Scott, "Density Estimation",. in P. Armitage & T. Colton, editors, Encyclopedia of Biostatistics, pp. 1134-1139. J. Wiley & Sons, Chichester, 1998.
11. F.J. Souza, M.M.R. Vellasco and M.A.C. Pacheco. "Hierarchical Neuro-Fuzzy Quadtree Models", Fuzzy Set and Systems, IFSA, Vol. 130(2), pp. 189-205, 2002.
12. M.A. Teixeira, K. Revoredo and G. Zaverucha, "Hidden Markov model for regression in electric load forecasting," ICANN/ICONIP, Istanbul, Turkey, pp. 374-377. 2003.
13. M.A. Teixeira and G. Zaverucha, "Fuzzy Bayes and Fuzzy Markov Predictors" Journal of Intelligent and Fuzzy Systems. IOS Press, vol. 13, numbers 2-4, pp. 155-165, 2003.
14. M.A. Teixeira and G. Zaverucha, "Fuzzy hidden Markov predictor in electric load forecasting," International Joint Conference on Neural Networks, Vol. 1, pp. 315-320, 2004.
15. S. Thrun, J. Langford, and D. Fox. "Monte Carlo hidden Markov models: Learning nonparametric models of partially observable stochastic processes". Proc. of the International Conference on Machine Learning (ICML), pp. 415-424, 1999.
16. L. Torgo and J. Gama, "Regression using classification algorithms," Intelligent Data Analysis, Vol.1, pp. 275-292, 1997.
17. M. West and J. Harrison, Bayesian Forecasting and Dynamic Models, Springer, 1997.
18. L.A. Zadeh, "Probability measures of fuzzy events," Jour. Math. Analysis and Appl. 23, pp. 421-427, 1968.

Fuzzy Compactness Based Adaptive Window Approach for Image Matching in Stereo Vision

Gunjan and B.N. Chatterji

Department of Electronics and Electrical Communication Engineering
Indian Institute of Technology, Kharagpur, 721302 India

Abstract. The central problem in any window based approach for image matching is the determination of correct window size. While the window should be small enough to have same disparity level throughout, it must be large enough to contain variations that can be matched. Thus it is obvious that a fixed window size is not enough.

There have been many approaches to vary the window size that is used for matching. Most of them work by modeling the disparity variation within the window and then developing a cost function which determines whether or not to reduce the window size at that location.

This paper presents a simple fuzzy based approach to vary the window size locally as per the intensity variation. This uses a measure called the fuzzy compactness to determine whether or not the intensity variation in the window is enough to split the window.

1 Introduction

The central problem in stereo vision is image matching where corresponding points in the image pair are to be obtained. This is illustrated in figure 1. There have been many approaches proposed which can be broadly divided into two parts - the feature based [8] and the area based approaches [7].

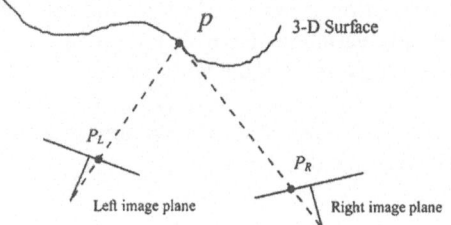

Fig. 1. Basic concept of stereo vision.

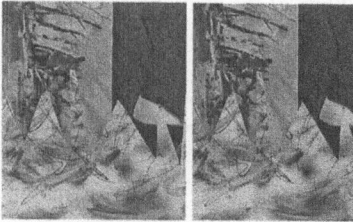

Fig. 2. The Sawtooth image pair showing large areas of low intensity variation.

The feature based approaches are two step process – extraction of features in the images followed by matching between the set of features obtained. Some of the common approaches have been based on edge extraction and matching of features like lines, corners and contours. Though these approaches are accurate because of the many parameters that can be associated with the features and fast because of the highly reduced search space, they are not able to give a dense disparity map. This is so because they match only selected points that are covered in features.

N.R. Pal et al. (Eds.): ICONIP 2004, LNCS 3316, pp. 935–940, 2004.
© Springer-Verlag Berlin Heidelberg 2004

The area based approaches also known as the window based approaches rely on statistical measures as the sum of squared difference (SSD), correlation and sum of absolute difference (SAD). A window is selected in one of the images and based on any of the above methods, an appropriate match is searched for the same in the other window. For example, the use of correlation will search corresponding neighborhood in the other image for maximum value.

The method works well for areas of constant disparity but fails at the areas of disparity variations. The reason is the basic assumption in this approach – the disparity is constant in the window.

So the window needs to be small enough that the disparity within the window is constant. However taking a very small window leads to the case where there is very low intensity variation within the window. In such a case the problem faced is ambiguity in matching because there is not enough intensity variation to achieve a unique match.

This problem occurs most commonly in the areas of image pair that constitute the background. This is illustrated in figure 2 which shows many areas to be very much similar with small intensity variations.

To overcome the problems of a fixed window, an adaptive window approach was first suggested by Kanade and Okutomi [1]. The landmark paper suggested modeling the disparity variation within a window as a Gaussian function and evaluating a window cost function that decides whether to change the window size or not.

Other works as [1],[2],[3] and [4] on adaptive window approaches have concentrated on modeling the disparity within the window and evaluating two sets of cost functions – one based on intensity variation and the other on disparity variation.

The scheme suggested in this paper is based on a fuzzy measure of the intensity variation – Fuzzy Compactness which is explained in the next section. The motivation for using only an intensity variation measure is the fact the areas of disparity changes are marked by intensity variations in the form of occurrences of edges and intensity steps. This implies that areas of higher intensity variations have very high chance of having disparity variations as well and hence justify varying the window size.

The idea presented here is computationally less expensive as only one cost has to be evaluated for each window. From the results it is clear that even without modeling the disparity variation it is possible to obtain accurate dense maps.

2 Fuzzy Compactness

Natural images have many features that have ambiguity or fuzziness associated with them. Many authors have used fuzzy set concepts for image processing. Pal and Rosenfeld [6] and Rosenfeld and Haber [9] have introduced several fuzzy geometric properties of images. They have used a fuzzy measure of image compactness for enhancement and thresholding.

A fuzzy subset of a set S is a mapping μ from S into [0,1]. For any element $p \, \varepsilon \, S$, $\mu(p)$ is called the degree of membership of p in μ or more generally the membership function.

A conventional crisp set is the special case where $\mu(p)$ can take on the values 0 or 1 only. Extending this to an image, the image can be interpreted as an array of fuzzy singletons. Each pixel has a membership value depending on its intensity value relative to some level l, where $l=1, 2, 3, ..., L-1$.

Such a relation can be expressed as

$$\mu_x(X_{ij}) = \mu_{ij} / X_{ij} \tag{1}$$

Pal and Rosenfeld [6] extended the concepts of image geometry to fuzzy subsets and generalized some of the standard geometric properties among regions to fuzzy sets. A few of them are mentioned to understand the concepts. $\mu_x(X_{ij})$ is referred to as μ for sake of simplicity.

The area of μ is defined as

$$a(\mu) = \int \mu \, d\mu \tag{2}$$

The integral is taken over any region outside which μ is defined to be zero. For the case of digital images where μ is constant over a small unit namely the pixel, the area is defined as

$$a(\mu) = \sum \sum \mu(i,j) \tag{3}$$

where the double summation signifies the summation over a region.

For a piecewise constant case, the perimeter of μ is defined as

$$P(\mu) = \sum \sum \sum |\mu_i - \mu_j| \, |A_{ijk}| \tag{4}$$

This is the weighted sum of the lengths of the arcs A_{ijk} along which the i-th and j-th regions having constant values of μ, μ_i and μ_j respectively meet, weighted by the absolute difference of the values. In the case of an image, the pixels can be considered as the piecewise constant regions. This leads to a further simplification of the equation to the one below

$$P(\mu) = \sum \sum |\mu_{ij} - \mu_{ij+1}| + |\mu_{ij} - \mu_{i+1j}| \tag{5}$$

This takes into the pixels as shown in figure 3, namely the east and south neighbors.

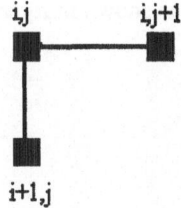 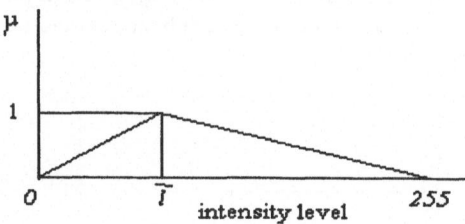

Fig. 3. Representation of adjacent pixels in the image.   **Fig. 4.** Pictorial representation of the membership function.

Following the area and perimeter, Compactness of μ is defined as the ratio of area and square of the perimeter as below

$$Comp(\mu) = a(\mu) / P^2(\mu) \tag{6}$$

Having defined the fuzzy terms to be used we now define the membership values to the pixels. For the purpose of matching, windows which are rectangular in shape are considered. In such a case a membership function which looks like the one shown in figure 4 is used. For this purpose first the average intensity level in the window is calculated. The pixels with intensity level equal to this average value are assigned a membership value of 1 while others are given a linearly decreasing value thereafter.

A region that has very low intensity variations will have many values of μ which are same or very similar. This means that regions of similar intensity levels will have very low perimeter values and hence a high value of compactness.

On the other hand a region with more intensity variations will have varying values of μ and hence the value of compactness will be low. This is exactly as per what is expected from the definition of compactness.

3 Algorithm for Matching

The algorithm proposed in this paper consists of two steps:

1. Partitioning one of the image into regions of varying sized windows.
2. Searching the second image for matching window of corresponding size.

To start with the right image is taken. This is then portioned by an iterative procedure. The use of rectangles implies that the partitioning procedure includes splitting of the rectangles when needed.

The idea used in this paper is that locations of disparity discontinuity or change is marked by change in intensity levels as well and those regions should have a smaller window size than what is used elsewhere. So the partitioning algorithm works as given below for some assumed threshold value, th.

1. Take window with upper-left co-ordinates (a,b) and lower-right co-ordinates (c,d) and compute the membership values μ as per local intensity distribution in the window.
2. Compute the area, perimeter and compactness of the region.
3. If compactness > th, the region has low levels of intensity variation and need not be split further.
4. If compactness < th, the region has high intensity variations that may represent a variation in disparity and hence should be further split as shown in figure 5.

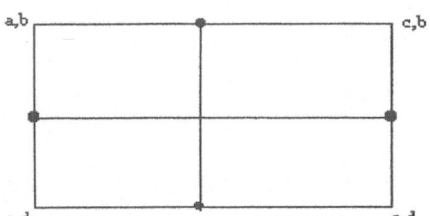

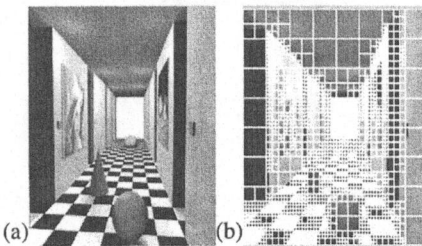

Fig. 5. Splitting of window, dark points are mid points of respective sides.

Fig. 6. (a) Right image form the corridor image pair (b) result of partitioning.

Figure 6(b) shows the result of partitioning applied to the corridor image which is shown in figure 6(a). It can be seen that areas like the ceiling and side walls in the front that have almost same intensity levels have a larger window. On the other hand, areas like the images on side walls or top left corner have smaller windows.

Figure 7 shows how the number and size of windows vary with threshold. Figure 7(a) has the lowest threshold and hence lesser number of windows. However, figure 7(d), having highest threshold among the four cases, has more number of windows.

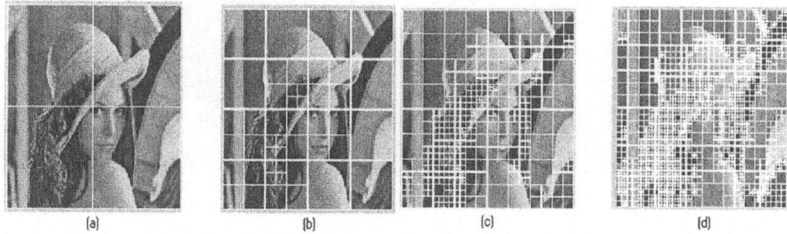

Fig. 7. Variation in the number of splitting with threshold ; threshold increases from (a) to (d).

The next step in the matching algorithm is the search for matching window in the second image. For making the process easier, a simple measure is used for matching. As mentioned earlier, a range of statistical quantities can be used to define the degree of match. However initially, SAD was used for the process of matching.

4 Experimental Results

The algorithm was tested on stereo image pairs obtained from the website of Middlebury which gives stereo image pairs along with disparity calculated by using ground truth.

Shown in figure 8(a) is the disparity as calculated from the ground truth and given for comparing results. Figure 8(b) shows the result of applying the partitioning process to the right image in the image pair. This process was followed by matching which resulted in a disparity map as shown in figure 8(c). It is seen that there is very small error which is summarized in table 1.

Figure 9(a) shows the partitioning result on the right image for a threshold value of 1.0. The corresponding disparity map is shown in figure 9 (b).

This shows a improvement as the error decreases drastically. The disparity map is shown as the top view of the 3-D plot.

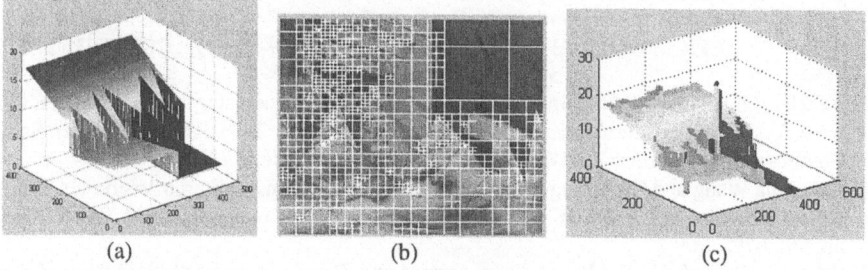

(a)   (b)   (c)

Fig. 8. (a) Disparity from ground truth (b) partitioning for th=0.1 (c) disparity obtained by proposed algorithm.

The average error shows improvement by increasing the threshold value:

Table 1. Variation in average error with threshold value.

Threshold value	Average error (%)
0.1	2.26
1.0	0.49

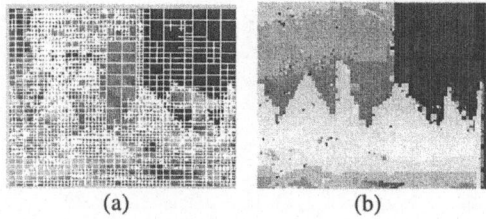

Fig. 9. (a) partitioning for th=1.0 (b) view of the disparity obtained by algorithm.

This is expected because with higher threshold more smaller rectangles appear at the edges leading to better match. However, this comes at the cost of more computation and noises in the disparity map in form of spikes.

5   Conclusions

The proposed algorithm is an efficient and accurate method that uses fuzzy set concepts to compute disparity. Given the simple cost function used to vary the window size, the method is computationally fast and accurate results are obtained by varying the threshold value.

Improvements are possible by changing the shape of the window.

References

1. Kanade T. and Okutomi M. : A stereo matching algorithm with an adaptive window: theory and experiment. IEEE Trans.Pattern Anal. Machine Intell., 16, 920-932. 1994
2. Scherer, S.; Andexer, W.: Pinz, A. :Robust adaptive window matching by homogeneity constraint and integration of descriptions ,Pattern Recognition, 1998. Proceedings. Fourteenth International Conference on, Volume: 1, 16-20 Pages:777 - 779 vol.1 Aug. 1998
3. Lotti, J.-L.; Giraudon, G. : Adaptive window algorithm for aerial image stereo Pattern Recognition, 1994. Vol. 1 - Conference A: Computer Vision & Image Processing., Proceedings of the 12th IAPR International Conference on, Volume: 1, 9-13 Pages:701 - 703 vol.1 Oct. 1994
4. Veksler, O. : Fast variable window for stereo correspondence using integral images, Computer Vision and Pattern Recognition, 2003. Proceedings. 2003 IEEE Computer Society Conference on, Volume: 1, 18-20 Pages:I-556 - I-561 vol.1 June 2003
5. Jung-Hua Wang and Chih-Ping Hsiao; On Disparity Matching in Stereo Vision via a Neural Network Framework ,Proceedings of the National Science Council ROC(A) Vol. 23, No. 5, pp. 665-678, 1999
6. Pal, S.K., Rosenfeld, A. : Image Enhancement and Thresholding by Optimization of Fuzzy Compactness, PRL(7), pp. 77-86., 1988
7. Marapane, S. B. and M. M. Trivedi : Region-based stereo analysis for robotic applications. IEEE Trans. Syst., Man, Cybern.,19, pp1447-1464, 1989.
8. Nasrabadi, N. M. and C. Y. Choo : Hopfield network for stereo vision correspondence. IEEE Trans. Neural Networks, 3, pp5-13, 1992
9. Rosenfeld, A. and Haber, S :The Perimeter of a fuzzy set, Pattern Recognition, Vol. 18, pp. 125-130, 1985

BDI Agents Using Neural Network and Adaptive Neuro Fuzzy Inference for Intelligent Planning in Container Terminals

Prasanna Lokuge and Damminda Alahakoon

School of Business Systems, Monash University, Australia
{Prasanna.Lokuge,Damminda.Alahakoon}@infotech.monash.edu.au

Abstract. Vessel berthing operations in a container terminal is a very complex application since environmental changes should be considered in assigning a right berth for a vessel. Dynamic planning capabilities would essentially enhance the quality of the decision making process in the terminals. Limitations in the social ability and learning capabilities of the generic BDI execution cycle have been minimized in the proposed architecture. Paper describes the use of Belief-desires-intention (BDI) agents with neural network and adaptive neuro fuzzy inference system in building the intelligence especially in planning process of the agent. Previous knowledge and the uncertainty issues in the environment are modeled with the use of intelligent tools in the hybrid BDI agent model proposed in the paper. This would essentially improve the adaptability and autonomy features of the BDI agents, which assures better planning, scheduling and improved productivity of the terminal.

1 Introduction

Shipping applications are heterogenous, distributed, complex, dynamic, and large, which essentially requires cutting edge technology to yield extensibility and efficiency. One of the important applications in container terminal operations is the berthing system of a container terminal. System requires to determine expected berthing time (ETB), expected completion time (ECT) of the vessels, a birth, allocation of cranes, labour, trucks for the stevedoring (loading and discharging) of containers assuring maximum utilization of resources and finally guaranteeing the high productivity of the terminal.

Agent oriented systems are based on practical reasoning system, which perhaps use philosophical model of human reasoning and have been used in achieving optimal solutions for many business application in the recent past. A number of different approaches have emerged as candidates for the study of agent-oriented systems [1], [2]. Belief –Desires-Intention (BDI) model has come to be possibly the best-known and best-studied model of practical reasoning agents [3]. But it has shown some limitations in certain business applications. In particular, the basic BDI model appears to be inappropriate for building complex systems that must learn and adapt their behaviors and such systems are becoming increasingly important in today's context in the business applications. The berthing system in a container terminal can be considered as one such a complex system, which requires learning, and adaptability in making rational decision promptly. We propose a hybrid BDI agent model for the operations in container terminals, which essentially extend the learning and adaptability features of

N.R. Pal et al. (Eds.): ICONIP 2004, LNCS 3316, pp. 941–946, 2004.
© Springer-Verlag Berlin Heidelberg 2004

the current BDI agents. In this paper, we describe how hybrid BDI agent architecture coupled with Neural Network and Adaptive Neuro Fuzzy Inference system (ANFIS) in the berthing application of container terminal operations could improve the decision making process in a complex, dynamic environment.

The research is carried out at the School of Business Systems, Monash University, Australia, in collaboration with the Jaya Container Terminal at the port of Colombo, Sri Lanka and Patrick Terminals in Melbourne, Australia. The rest of the paper is organized as follows: Section 2 describes multi agent system in a berthing application of a container terminal. Section 3 describes Proposed hybrid BDI architecture in berthing system. Section 4 describes a vessel berthing test case scenario in a container terminal and concluding remarks are provided in Section 5.

2 Multi Agents in a Berthing System

Operations in a berthing application have been grouped in to three main areas in the proposed system, namely Vessel related operations, operations in a berth and scheduling tasks of the terminals. Multi agent architecture has been proposed in handling the above tasks in the terminal using hybrid agents model. Hybrid Beliefs-desires-intention (BDI) agent model [2] has been proposed in handling the various berthing related operations mentioned above in container terminals.

Tasks related to berths, vessels and scheduling are being proposed to be handled with three different types of agents namely, VESSEL-AGENT (VA), SCHEDULE AGENT (SA) and BERTH-AGENT (BA). VA is primarily responsible for informing the vessel details to other agents. SA in the proposed hybrid BDI model is the main agent who handles many important tasks such as berth-assignments, rescheduling vessels, vessel shifting, etc. BA is responsible for all the operations at berths and makes every effort to improve the productivity and achieve set targets by SA. Main agents in the system are shown in figure 1.

Some of important data used in the proposed hybrid agent model in the container terminals are: CTY^v - Cargo type of the new vessel v, could be, normal, dangerous, and perishable. NOB^v - Number of boxes in the new vessel v. VBD^v -Vessel berthing requirement of the new vessel v. VSD^v - Vessel sailing draft requirement of the new vessel v. VCR^v - Crane requirement of the new vessel v. BDR_b - Berth Draft of the berth b. COR_b - Crane outreach of the berth b. GCP_b - Gross crane productivity of the berth b. SKL_b – Skill level of labor in berth b. ETA^v - Expected arrival time of the new vessel v. ETB^v - Expected time of berth of the new vessel v. ETC^v_b - Expected time of completion of the vessel v in berth b. STV^v_b – Sailing time of the vessel v in berth b. $GBP_b{}^v$- Berth productivity for vessel v in berth b, in moves per hour(mph). NOC^v_b - Number of cranes allocation for vessel v in berth b. Vx^i_b - Left side distance of the vessel V in berth b. Vx^j_b - Right side distance of the vessel V in berth b. W^t - Rating of the weather conditions are denoted with a range of 1 to 5. Vx^i_b, Vx^j_b - left and right side distance kept in the vessel V in berth b. C^v_b – Capacity of the vessel in berth b. l_b- Length of the berth b. $v_1{}^b$, $v_2{}^b$ - Left and right positions of vessel v in berth b. $w^b{}_1$, $w_2{}^b$ - Left and right positions of vessel w in berth b.

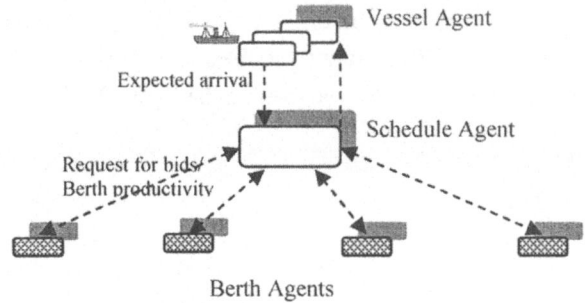

Fig. 1. Main Agents in the proposed system.

Vessel-Agents, Schedule-agent and Berth-Agents shown in the figure 1 need to be implemented as intelligent agents with the ability to make decisions considering the dynamic changes in the environment. Generic BDI execution cycle given below does not consider previous knowledge and dynamic changes in the environment in making rational decisions. Therefore in the next section we describe our proposed hybrid BDI model that essentially overcome the above drawback in the generic BDI model and use intelligence in vessel scheduling and planning.

> *Initialise-state ();*
> *Repeat*
>> *Options:=option-generated(event-queue);*
>> *Selected-options:=deliberate(options);*
>> *Update-intentions (selected-options);*
>> *Execute ();*
>> *Get-new-external-events ();*
>> *Drop-successful-attitudes ();*
>> *Drop-impossible-attitudes ();*
> *End repeat*

3 Proposed Hybrid BDI Architecture in a Berthing System

Two modules have been introduced in the proposed architecture. *"Generic BDI module" (GDM)* handles the execution cycle with simple well-defined plans and " *Knowledge Acquisition module" (KAM)* provides the required knowledge and leaning in handling uncertainty and vague situations in the terminal with the use of supervised neural network and adaptive neuro fuzzy systems. Simple plans such as Check-outreach-of cranes (VCR^v, COR_i), Sailing-draft-requirement (VSD^v, BDR_i) are being executed in *GDM* module of the hybrid agent model. *KAM* module of the agent is described in the next section.

3.1 Knowledge Acquisition Module *(KAM)* of the BDI Agent

Supervised neural network architecture is being trained to produce the expected GBP^n_b of individual berth in the terminal. Set of beliefs and desires given in the figure 2 are being taken as inputs to the neural network model. Expected Sailing time of the new vessel (STV^n_b) in individual berths is produced using 5 layered adaptive neuro fuzzy inference systems (ANFIS). Input variables to ANFIS are: W^t, Vx^i_b, Vx^j_b, and C^v_b. Architecture of the *KAM* module is given in the figure 2.

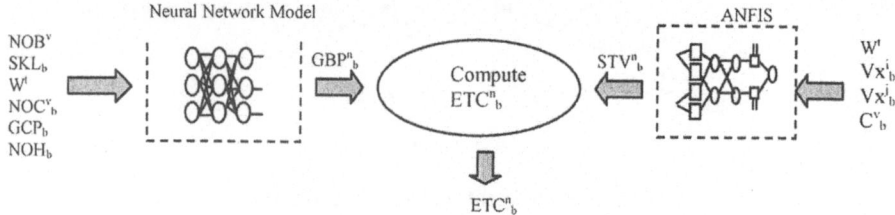

Fig. 2. *KAM* intelligence Architecture.

Multilayered feedforward neural network with back propagation is used to obtain the GBP^n_b in the first part of the *KAM* module. Five layered Adaptive neuro fuzzy inference (ANFIS) [21] system is the second part of the KAM module, which produces STV^n_b, The input parameters used in the ANFIS are given in the figure 2. Summary of the Layers used in the ANFIS is described as follows. Membership function of the layer one is given as $O^1_i = \mu_{ai}(x)$, where

$$\mu_{ai}(x) = \frac{1}{1 + \left[\left(\frac{x - c_i}{a_i}\right)^2\right]^{b_i}},$$

a_i, b_i, c_i are parameters. Layer 2, node function uses T-norm operator and sends the product as $w_i = \mu_{ai}(x) \otimes \mu'_{bi}(y)$. Layer 3 calculates the firing strength of the ith rule as $\overline{w} = \frac{w_i}{w_1 + w_2}$. Output of the 4th layer is given as $O^4_i = \overline{w}_i f_i$

$= \overline{w}\left(p_i x + q_i + r_i\right)$. Where p_i, q_i, r_i is the parameter set used in the 4th layer. Finally the single node in the 5th layer computes the overall output as the summation of all income signals as $O^5_i = \sum_i \overline{w} f_i = \frac{\sum_i w_i f_i}{\sum_i w_i}$.

Proposed architecture in the BDI agents facilitates learning from past data and approximate reasoning, as well as rule extraction and insertion. Therefore the required knowledge in computing the expected gross berth productivity and sailing time for the new vessel in individual berth is being modeled using the above intelligent tools.

4 A Vessel Berthing Test Case Scenario

GDM module in proposed architecture executes the simple plans in the intention structure of the agent. Required knowledge in predicting GBP^n_b and STV^n_b of the new vessel are produced with the introduction of intelligent tools. Timely changes in the beliefs are regularly considered before the execution of the next plans in the intention structure ensuring accurate results. Intelligent tools in the proposed *KAM* module have shown a remarkable improvement in making rational decision in the agent model. Few membership functions and decision surfaces produced by ANFIS are shown in the figure 3.

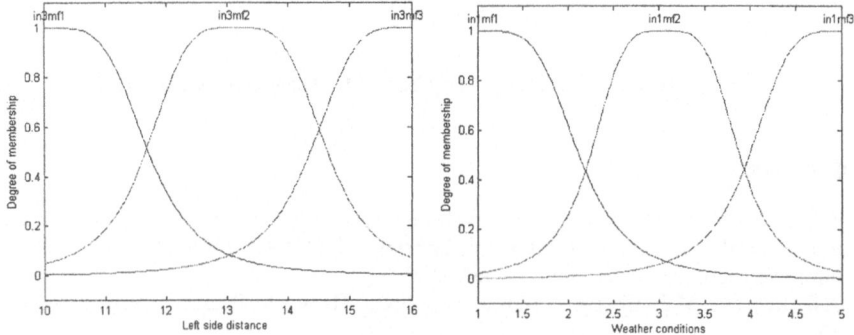

Fig. 3a. Membership functions.

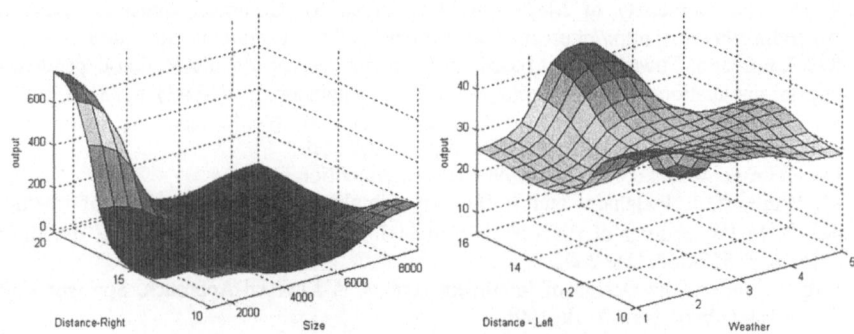

Fig. 3b. Decision surfaces.

5 Conclusion

Knowledge acquisition module with intelligent tools in the proposed agent architecture has assured more accurate results in the berthing system.

Beliefs and desires in the environment have been used as inputs to the ANFIS and trained neural network in the *KAM* module to produce berthing related decisions. Results produced from the *KAM* module is being analyzed before the execution of plans in the intention structure of the agent model. Actual time of completion of vessels and the results obtained using a hybrid BDI agent model in a container terminal is shown in the figure 4.

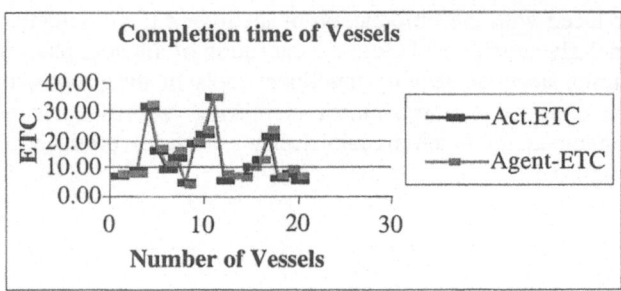

Fig. 4. ETC of vessels produced by Schedule-Agent and Actual values.

References

1. A.S. Rao and M. Georgeff, BDI agents: from theory to practice. In proceedings of the first international conference on Multi agents systems (ICMAS-95), 1995 url : http://www.citeseer.nj.nec.com/rao95bdi.html
2. A.S. Rao and M.P. Georgeff. BDI agents: From Theory to Practice. Technical note 56, 1995.
3. Paolo busetta and Kotagiri Ramamohanarao, Technical Report 97/16, The University of Melbourne, Department of Computer Science, Melbourne, Australia, 1997.
4. P. Busetta and Kotagiri R. An Architecture for Mobile BDI Agents. Technical Report 97/16, The University of Melbourne, Department of Computer Science, Melbourne, Australia,1997. url: http://citeseer.nj.nec.com/article/busetta97architecture.html
5. S.C. Laufmann, Towards agent-based software engineering for information-dependent enterprise applications. IEE Proceedings – Software Engineering, V144(1), p. 38-50, 1997.
6. M. Georgeff, B. Pell, M. Pollack, M. Tambe, and M. Wooldridge, The Belief-Desire-Intention Model of Agency,SpringerPublishers,1998. url: http://citeseer.nj.nec.com/georgeff99beliefdesireintention.html
7. M. Winikoff, L Padgham, and J. Harland. Simplifying the development of intelligent agents. In Proceedings of the 14th Australian Joint Conference on Artificial Intelligence (AI'01), p. 557-568, Dec.2001.
8. Jorg P. Muller, The Design of Intelligent Agents, A Layered Approach, Springer-Verlag Berlin Heidelberg, New York, 1996
9. Nicholas R. Jennings, Intelligent Agents VI, Agent Theories Architecture, and Languages, 6th International workshop; /proceedings ATAL'99,Florida, 1999.
10. George Bojadziew and Maria Bojadziev, Fuzzy sets, fuzzy logic, applications, World Scientific publishing Co Pte Ltd, Singapore, 1995.

A Neuro-fuzzy Approach for Predicting the Effects of Noise Pollution on Human Work Efficiency

Zaheeruddin[1] and Garima[2]

[1] Department of Electrical Engineering, Faculty of Engineering and Technology,
Jamia Millia Islamia (A Central University),
New Delhi-110025, India
zaheer_2k@hotmail.com
[2] Department of Computer Science, Galgotia College of Engineering & Technology,
UP Technical University, Greater Noida-201308, U.P., India
gari_ag1@rediffmail.com

Abstract. In this paper, an attempt has been made to develop a neuro-fuzzy model for predicting the effects of noise pollution on human work efficiency as a function of noise level, type of task, and exposure time. Originally, the model was developed using fuzzy logic based on literature survey. So, the data used in the present study has been synthetically generated from the previous fuzzy model. The model is implemented on Fuzzy Logic Toolbox of MATLAB© using adaptive neuro-fuzzy inference system (ANFIS). ANFIS discussed in this paper is functionally equivalent to Sugeno fuzzy model. Out of the total input/output data sets, 80% was used for training the model and 20% for checking purpose to validate the model.

1 Introduction

The traditional equation based techniques for the solution of real world problems are not suitable for modeling non-linearity in the complex and ill-defined systems. During the last three decades, model free techniques such as fuzzy logic and neural networks have provided an attractive alternative to accommodate the non-linearity and imprecise information found in real world for modeling the complex systems. The term fuzzy logic is used in two different senses. In narrow sense, fuzzy logic (FLn) is a logical system – an extension of multi-valued logic that intends to serve as logic of approximate reasoning. In a wider sense, fuzzy logic (FLw) is more or less synonymous with fuzzy set theory; the theory of classes with un-sharp boundaries. In this perspective, FL=FLw, and FLn is merely a branch of FL Today, the term fuzzy logic is used predominantly in its wider sense [1]. Zadeh suggested a linguistic approach for modelling complex and ill-defined systems [1-4]. The fuzzy systems employing fuzzy if-then rules can model human knowledge in form of easily understandable linguistic labels but these systems lack learning capabilities and depend entirely on knowledge of human experts. Neural networks, on the other hand, possess good learning capabilities but still not widely used for addressing real world problems as they are unable to explain the user about a particular decision in a human-comprehensible form. To have both the capabilities of learning and interpretability in a single system, hybridization of neural networks and fuzzy logic is the basic need of future intelligent systems.

N.R. Pal et al. (Eds.): ICONIP 2004, LNCS 3316, pp. 947–952, 2004.
© Springer-Verlag Berlin Heidelberg 2004

The effects of noise pollution on human beings are studied through social surveys based on questionnaires. These questionnaires are generally words and propositions drawn from natural language. For example, noise level, type of task, and exposure time may be represented by the words (*low, medium, high*), (*simple, moderate, complex*), and (*short, medium, long*) respectively. However, measurements of these parameters are done with the help of some scientific instruments, which provides the numerical values to the researchers. Hence, the study of noise pollution is the unique combination of linguistic and numerical values. It is in this context that an attempt has been made in this paper to develop a neuro-fuzzy system for predicting the effects of noise pollution on human work efficiency.

The paper is organized as follows. Section 2 is devoted to the description of effects of noise pollution on human beings and Section 3 introduces the neuro-fuzzy modeling aspects. In Section 4, implementation details and architecture is presented. Results are discussed in Section 5 followed by the conclusion in Section 6.

2   Noise Pollution and Human Work Efficiency

In the past, many studies have been conducted to determine the effects of noise pollution on human performance involving variety of tasks. One of the most important parameter for evaluating the human performance is noise level. The level of noise, which produces adverse effects, is greatly dependent upon the type of task. Simple routine tasks usually remain unaffected at noise levels as high as 115 dB or above, while more complex tasks are disrupted at much lower levels [5]. Noise hinders the performance of subjects involving audio-visual task [6], sentence-verification, vowel-consonant recognition [7], proof reading and solving challenging puzzles [8]. There are ample evidences showing negative associations between exposure to high air and road traffic noise in reading acquisition among children [9]. Office noise is found to disrupt the performance on memory for prose and mental arithmetic tasks [10]. The effects of noise on human performance have also been investigated by researchers based on other factors such as sex [11] and age [12] but these factors have very little effects. Depending on the nature of the task, human performance gets affected differently under the impact of different noise levels and duration of exposure only.

3   Neuro-fuzzy Computing

Neuro-fuzzy computing is a judicious integration of the merits of neural and fuzzy approaches. This incorporates the generic advantages of artificial neural networks like massive parallelism, robustness, and learning in data-rich environments into the system. The modeling of imprecise and qualitative knowledge as well as the transmission of uncertainty is possible through the use of fuzzy logic. In the last decade, various neuro-fuzzy systems have been developed [13-17]. Some of the well known neuro-fuzzy systems are ANFIS [18], DENFIS [19], SANFIS [20], FLEXNFIS [21] and others. Our present study is based on Adaptive Neuro-Fuzzy Inference System (ANFIS). An ANFIS is a multilayer feedforward network consisting of nodes and directional links through which nodes are connected. Moreover, part or all the nodes are adaptive, which means that their outputs depend on the incoming signals and on the parameter(s) pertaining to these nodes [18]. ANFIS provides a method for fuzzy model to learn information about a data set. It fine tunes the membership function parameters associated with fuzzy inference system using either backpropagation algorithm [22] alone or in combination with a least squares type of method [18].

4   Implementation and Architecture

In order to predict the effects of noise pollution on human work efficiency, the output (reduction in work efficiency) is taken as a function of noise level, type of task, and exposure time. For system identification, Takagi-Sugeno-Kang (TSK) fuzzy inference system is used, as it is simple and computationally efficient [23,24]. TSK model uses fuzzy if-then rules where antecedents are defined by a set of nonlinear parameters and consequents are linear combination of input variables and constant terms. For tuning these parameters, ANFIS architecture available in Fuzzy Logic Toolbox of MATLAB© [25] is used. The corresponding ANFIS architecture functionally equivalent to TSK fuzzy inference system is shown in Fig. 1.

Layer 1 (Input Layer): Inputs are noise level, type of task, and exposure time.

Layer 2 (Fuzzification Layer): This layer represents the membership functions of each of the input variables and is given by $\mu_{Aj}(x_j)$. Here,

$$\mu_{Aj}(x_j) = \cfrac{1}{1 + \left[\left(\cfrac{x_j - c_j}{a_j}\right)^2\right]^{b_j}} \tag{1}$$

where x_j is the number of inputs and A_j is their corresponding linguistic labels and (a_j, b_j, c_j) is the parameter set. The gbell shaped membership function is chosen because of its smooth and concise notation. Fig. 2 shows the membership functions for input noise level.

Layer 3 (Inference Layer): The output of each neuron in this layer represents the firing strength (w_i) of each rule using multiplicative inference method. For instance,

R1: IF noise level is *high* AND type of task is *complex* AND exposure time is *long*
    THEN reduction in work efficiency is approximately 96%

$$w_1 = \mu_{\text{high}}(noiselevel) \times \mu_{\text{complex}}(type\ of\ task) \times \mu_{\text{long}}(exposure\ time) \tag{2}$$

Layer 4 (Normalization Layer): The output of ith node of inference layer is normalized as

$$\overline{w_i} = \frac{w_i}{w_1 + w_2 + ... + w_R} \tag{3}$$

where $i = 1,2,.....,R$ and R is total number of rules.

Layer 5 (Output Layer): This layer generates the consequent of each rule depending on the normalized firing strength and is given by

$$O_i = \overline{w_i} f_i \tag{4}$$

where f_i is constant (for example 96% in the above rule). If f_i is linear function of input variables then it is called first order Sugeno fuzzy model and if f_i is a constant (as in our present model) then it is called zero order Sugeno fuzzy model.

Layer 6 (Defuzzification Layer): This layer computes the weighted average of output signals of the output layer and is given by

$$O = \sum_i \overline{w}_i f_i$$

(5)

For training the above ANFIS model to fine tune the parameters, 80% data sets are provided. In forward pass, the functional signals keep going forward till the error measure is calculated at the output layer. The consequent parameters are identified using least square estimator (LSE) method and updated. In the backward pass, the error rates are then propagated from output layer to input layer using gradient descent method and nonlinear premise parameters are updated. Fig. 3 shows the training root mean square error (RMSE) curve for the model.

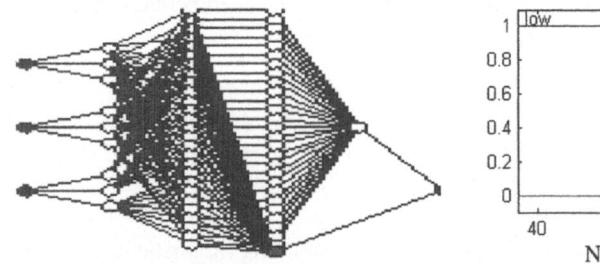

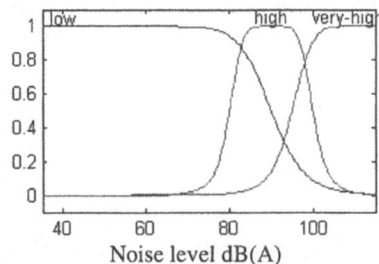

Noise level dB(A)

Fig. 1. ANFIS architecture.   **Fig. 2.** Membership function (input 1).

5 Results and Discussion

The ANFIS used here contains 27 rules with three gbell shaped membership functions being assigned to each input, total number of fitting parameters is 54 which are composed of 27 premise parameters and 27 consequent parameters while in our original fuzzy model [26,27], the membership functions used were triangular and total number of rules were 81. The flexibility of this model is that the parameters can be tuned to give a more realistic representation of the system. The results can be obtained either in 3-D or 2-D forms. The 3-D representation is not comprehensible while 2-D representation is more interpretable and understandable. As an example, one graph depicting the reduction in work efficiency versus noise-level with long exposure time for simple, moderate, and complex tasks is shown in Fig. 4. It is evident from this figure that there is no reduction in work efficiency up to the noise level of 75 dB(A) irrespective of the type of task. Further, reduction in work efficiency is negligible up to the noise level of 90 dB(A) for simple and moderate tasks while it is about 40% for complex task. The work efficiency starts reducing after 90 dB(A) even for simple and moderate tasks. At 100 dB(A), work efficiency reduces to 36%, 58%, and 76% for simple, moderate, and complex tasks, respectively. There is significant reduction in work efficiency after 100 dB(A) for all type of tasks. When noise level is in the interval of 110-115 dB(A), it is 56% for simple, 90% for moderate, and 96% for complex tasks.

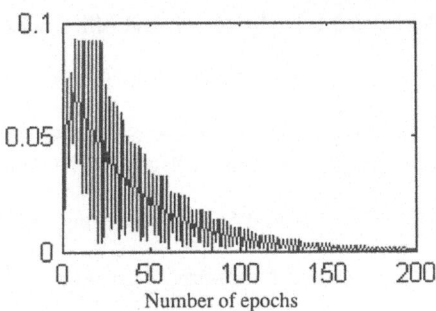
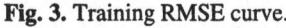

Number of epochs

Fig. 3. Training RMSE curve.

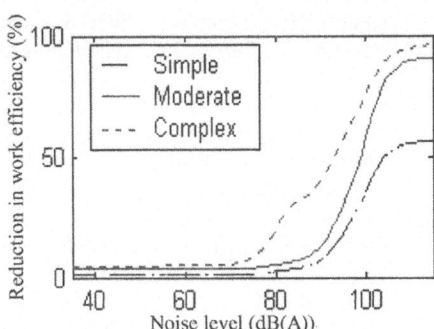

Noise level (dB(A))

Fig. 4. Reduction in work efficiency for long exposure time.

6 Conclusion

The main thrust of the present work has been to develop a neuro-fuzzy model for the prediction of work efficiency as a function of noise level, type of tasks and exposure times. It is evident from the graph that the work efficiency, for the same exposure time, depends to a large extent upon the noise level and type of task. Similarly, graphs for reduction in work efficiency versus noise levels for simple, moderate, and complex tasks at different exposure times can be obtained. It has also been verified that simple tasks are not affected even at very high noise level while complex tasks get significantly affected at much lower noise level. Moreover, minor changes are observed in the shape of the membership functions after training the fuzzy model using ANFIS. This is because of the close agreement between the knowledge provided by the expert and input-output data pairs. However, if the available expert knowledge about the system is not very accurate even then the appropriate results can be obtained by training the model using neuro-fuzzy approach.

References

1. Zadeh, L.A.: Soft Computing and Fuzzy Logic. IEEE Software. November (1994) 48–56
2. Zadeh, L.A.: Fuzzy Sets. Information and Control. 8 (1965) 338-353
3. Zadeh, L.A.: Fuzzy Algorithm. Information and Control. 12 (1968) 94-102
4. Zadeh, L.A.: Outline of a new approach to the analysis of complex systems and decision processes. IEEE Transactions on Syst., Man, and Cyber. SMC-3 (1973) 28-44
5. Suter, A.H.: Noise and its effects. This is available online at http://www.nonoise.org/library/suter/suter.html. (1991)
6. Arnoult, W.D., Voorhees, J.W.: Effects of aircraft noise on an intelligibility task. Human Factors. 22 (1980) 183-188
7. Fu, Q.J., Shanno, R.V., Wang, X.: Effects of noise and spectral resolution on vowel and consonant recognition: acoustic and electric hearing. J. Acoust. Soc. Am. 6 (1998) 3586-3596

8. Percival, L., Loeb, M.: Influence of noise characteristics on behavioural after-effects. Human Factors. 22 (1980) 341-352
9. Hygge, S., Evans, G.W., Bullinger, M.: The Munich Airport Noise Study: Psychological, cognitive, motivational, and quality of life effects on children. In: Vallet, M. (ed.): Noise as a Public Health Problem. INRETS, France (1993) 301-308
10. Banbury, S., Berry, D.C.: Disruption of office related tasks by speech and office noise. Brit. J. Psychol. 89 (1998) 499-517
11. Herrmann, D.J., Crawford, M., Holdsworth, M.: Gender-linked differences in everyday memory performance. Brit. J. Psychol. 83 (1992) 221-231
12. Westerman, S.J., Davies, D.R., Glendon, A.I., Stammers, R.B., Matthews, G.: Ageing and word processing competence: compensation or compilation? Brit. J. Psychol. 89 (1998) 579-597
13. Lin, Y., Cunningham, G.A.: A New Approach to Fuzzy-Neural System Modelling. IEEE Trans. Fuzzy Syst. 3 (1995) 190-198
14. Figueiredo, M., Gomide, F.: Design of Fuzzy Systems Using Neuro-fuzzy Networks. IEEE Trans. Neural Networks. 10 (1999) 815-827
15. Chakraborty, D., Pal, N.R.: Integrated Feature Analysis and Fuzzy Rule-Based System Identification in a Neuro-Fuzzy Paradigm. IEEE Trans. Syst., Man, and Cyber. 31 (2001) 391-400
16. Chakraborty, D., Pal, N.R.: A Neuro-Fuzzy Scheme for Simultaneous Feature Selection and Fuzzy Rule-Based Classification. IEEE Trans. Neural Networks. 15 (2004) 110-123
17. Jang, J.-S.R., Sun, C.-T., Mizutani, E.: Neuro-Fuzzy and Soft Computing. First Indian Reprint. Pearson Education, New Delhi (2004)
18. Jang, J.-S.R.: ANFIS: Adaptive-network-based fuzzy inference system. IEEE Trans. Syst., Man, and Cyber. 23 (1993) 665-685
19. Kasabov, N.: DENFIS: Dynamic evolving neural-fuzzy inference system and its application for time-series prediction. IEEE Trans. Fuzzy Syst. 10 (2002) 144-154
20. Wang, J.S., Lee, C.S.G.: Self-adaptive neuro-fuzzy inference systems for classification applications. IEEE Trans. Fuzzy Syst. 10 (2002) 790-802
21. Rutkowski, L., Cpalka, K.: Flexible Neuro-Fuzzy Systems. IEEE Trans. Neural Networks. 14 (2003) 554-574
22. Rumelhart, D.E., Hinton, G.E., Williams, R.J.: Learning internal representations by error propagation. In: Rumelhart, D.E., McClelland, J.L. (eds.): Parallel distributed processing: explorations in the microstructure of cognition. MIT Press, Cambridge (1986) 318-362
23. Sugeno, M., Kang, G.T.: Structure Identification of Fuzzy Models. Fuzzy Sets and Systems. 28 (1988) 15-33
24. Takagi, T., Sugeno, M: Fuzzy Identification of Systems and its Applications to Modelling and Control. IEEE Trans. Syst., Man, and Cyber. 15 (1985) 116-132
25. Fuzzy Logic Toolbox for use with MATLAB®. The MathWorks Inc., USA (2000)
26. Zaheeruddin, Singh, G.V., Jain, V.K.: Fuzzy Modelling of Human Work Efficiency in Noisy Environment. Proc. The IEEE Internat. Conf. Fuzzy Systems, Vol. 1 (2003) 120-124
27. Zaheeruddin, Jain, V.K.: A Fuzzy Approach for Modelling the Effects of Noise Pollution on Human Performance. Internat. Journal of Advance Computational Intelligence and Intelligent Informatics. 8(2004) to appear

Evolving Fuzzy Neural Networks
Applied to Odor Recognition

Cleber Zanchettin and Teresa B. Ludermir

Information Technology Center – Federal University of Pernambuco
P.O. Box 7851, Cidade Universitária, Recife – PE, Brazil, 50.732-970
{cz,tbl}@cin.ufpe.br

Abstract. This paper presents the use of Evolving Fuzzy Neural Networks as pattern recognition system for odor recognition in an artificial nose. In the classification of gases derived from the petroliferous industry, the method presented achieves better results (mean classification error of 0.88%) than those obtained by Multi-Layer Perceptron (13.88%) and Time Delay Neural Networks (10.54%).

1 Introduction

An artificial nose is a sensing device capable of detecting and classifying odors, vapors, and gases automatically. The artificial nose consists of a sensor system and an automated pattern recognition system. The sensor system is an array of several elements, where each sensor measures a different property of the odor. This device can identify a wide of odorants with high sensitivity and recognize substances by the combination and relative proportions of compounds. This is achieved by combining a set of sensing elements of broad and overlapping selectivity profiles. Each odorant substance presented to the sensor system, usually an odor-reactive polymer sensor, generates a pattern of resistance values that characterizes the odor. This pattern is often preprocessed first and then given to the pattern recognition system, which in turn classifies the odorant stimulus.

A wide range of odorants needs to be identified in many sectors, like food industry, environmental monitoring and medicine [1]. For this reason, artificial noses are currently being developed as systems for automatic recognition of substances. Such devices are extremely important in applications like inspection of food quality, control of cooking processes, detection of gas leaks in environmental protection, diagnosis of medical conditions and many other situations.

In this work we propose the use of an adaptive and on-line learning mechanism in the pattern recognition system of the artificial nose. The method, Evolving Fuzzy Neural Networks (EFuNNs) [2] is a connectionist feedforward architecture that facilitates learning from data, reasoning over fuzzy rules, aggregation, insertion and rule extraction.

To evaluate the performance of the Evolving Fuzzy Neural Network in the classification of odor patterns, the results are compared with two other classifiers: Multi-Layer Perceptron Neural Networks (MLP) [3], the type of neural network most commonly used for odor classification in artificial noses; and Time-Delay Neural Networks (TDNN) [4] that has displayed excellent results in the classification of odor patterns. The experiments presented in this paper aim to classify gases derived from petroleum, such as ethane, methane, butane, propane and carbon monoxide.

N.R. Pal et al. (Eds.): ICONIP 2004, LNCS 3316, pp. 953–958, 2004.
© Springer-Verlag Berlin Heidelberg 2004

The remainder of this paper is divided into five sections. Next section presents connectionist classifiers applied in odor recognition. Section 3 presents the adaptive and on-line learning method. Section 4 and Section 5 concentrates the experiments, tests and results of the classifiers. In Section 6 some final remarks are presented.

2 Connectionist Odor Pattern Recognition Systems

Artificial neural networks have been widely applied as pattern recognition systems in artificial noses. Some advantages of this approach are: (1) the ability to handle non-linear signals from the sensor array; (2) adaptability; (3) fault and noise tolerance; and (4) inherent parallelism, resulting in fast operation.

The most commonly used Artificial Neural Network for odor classification in artificial noses has been the Multi-Layer Perceptron, together with the backpropagation learning algorithm [3]. Amongst the neural architectures, the type of network that has presented most promising results to odor classification is the Time Delay Neural Networks [4]. The TDNN consists of a pattern recognition system able to analyze the temporal features of the signals generated by the sensors of the artificial nose. This system considers the variation of the signals generated by the sensors along the time interval in which the data acquisitions were done.

Interesting results had been also found with the use of hybrid architectures in the pattern recognition system of the artificial nose. Hybrid applications, as the use of Wavelet Analysis in the preprocessing the coming data of the sensor system [5], use of neuro-fuzzy networks for selection of the sensor ones more important in the odor classification, and extraction of classification rules of the sensors of the artificial nose [6]. Despite the good performance of these systems, new applications in artificial noses need dynamic structures to consider new odors or changes that can happen in the characteristics of the odors patterns or environment in your operation.

The artificial neural networks and hybrid systems usually utilized in artificial noses present deficiencies in the treatment of dynamics data, the structures of the networks are fixed and do not change with the addition of new data. For example, the network uses a fixed set of input features, neurons and classes. In contrast with the patter recognition systems for artificial noses in use, this work propose the use of a model that is able to accommodate new data as they become available and adapt its structure and parameters continuously in an on-line mode.

3 Evolving Fuzzy Neural Network

The adaptive networks EFuNNs are neural networks that realize a set of fuzzy rules and a fuzzy inference machine in a connectionist way, and evolve according to the ECOS principles [7]. An EFuNN is a connectionist system that facilitates learning from data, reasoning over fuzzy rules, aggregation, rule insertion, rule extraction. EFuNNs operate in an on-line mode and learn incrementally through locally tuned elements. They grow as data arrive, and regularly shrink through pruning of nodes, or through node aggregation. The EFuNN is an architecture that can classify multiple classes. In addition, if through training a new class is added the EFuNN can automatically evolve a new output to reflect the change in the data set.

The interest in the application of EFuNN in the treatment of the odor patterns appeared from three of your functionalities: (1) possibility of extraction of the knowledge of the neural network in classification rules; (2) incremental learning; and (3) on-line training. In a device that looks for efficiency, speed, plasticity and size these characteristics are very appreciated. In EFuNN, all nodes are created during (possibly one-pass) learning. The nodes representing membership functions (fuzzy label neurons) can be modified during learning. The model learns relationships between input and output data in an iterative and on-line way, and fuzzy rules may then be extracted to explain what the network has learned.

4 Experiments

In this research the aim is to classify five gases: ethane, methane, butane, propane and carbon monoxide from the petroliferous industry (Petrobrás, Brazil). A prototype of an artificial nose was used to acquire the data. The sensor systems have been often built with polypyrrol-based gas sensors. Some advantages of using such kind of sensors are [8]: (1) rapid absorption kinetics at environment temperature; (2) low power consumption, as no heating element is required; (3) resistance to poisoning; and (4) the possibility of building sensors tailored to particular classes of chemical compounds.

The prototype is composed of eight distinct gas sensors, built by electrochemical deposition of polypyrrol using different types of dopants. The data were obtained with nine data acquisitions for each one of the gases, by recording the resistance value of each sensor at every 20 seconds during 40 minutes. In the acquisition phase, each sensor has obtained 120 resistance values for each gas. A pattern is a vector of eight elements representing the values of the resistances recorded by the sensor array.

In MLP experiments, the data set for training and test of the network were divided into: training set, containing 50% of the total amount of patterns; validation set, containing another 25%; and test set, which contains the remaining 25%. The patterns were normalized to the range [-1, +1], the processing units were implemented by *hyperbolic tangent activation function* [9]. The network was trained with only a single hidden layer, with five different topologies (4, 8, 12, 16 and 20 hidden units). The training algorithm used is a version of the Levenberg-Maquardt method, described in [10]. For each topology, 30 runs were performed with different random weight initializations. Training was stopped if: (i) the GL_5 criterion defined in Proben1 [9] was satisfied twice; (ii) the training progress criterion was met, with $P_5(t)<0.1$ [9]; or (iii) a maximum number of 1000 epochs was achieved.

For TDNN training, the architecture needs to receive complete curves generated by the sensors during the data acquisition. Thus, only complete data acquisitions can be used as training, validation and test sets. For experiments, one of the data acquisitions (960 patterns of each gas) was used as training set; in the same way, other two acquisitions (with the same amount of data) were used as validation and test sets. The choice of the acquisitions was done in an arbitrary way, by using the three first data acquisitions. The same data normalization, network topologies, training algorithm and training criteria adopted for the MLP experiments was used in TDNN experiments.

In EFuNN experiments, eight input processing units were used, corresponding to the values of the eight sensors. Each input/output processing unit contains two pertinence functions. The data set for training and test of the network were divided into: training set, containing 75% of patterns; and test set, which contains the remaining 25%. The training mode used was one pass training. For each experiment, 30 runs were performed with different partitions of the data, to compare the results with MLP and TDNN experiments. The patterns are presented to the network without normalization. The aspects observed at the end of training were: Root Mean Squared Error percentage of the training set, network scale and the classification error of the test set.

5 Results

In the MLP and TDNN experiments, the error measures analyzed were the RMS Error, for training, validation and test sets, and classification error of test set. The best result of MLP neural network was obtained in topology with 8 nodes in the hidden layer, a classification error of 13.88%. In TDNN experiments the best result was obtained in topology with 8 nodes in the hidden layer too, a error of 10.54% in test set.

In EFuNNs experiments, some key parameters of the EFuNN configuration and the average results for the 30 partitions are shown in Table 1. Four experiments are performed: (1) without pruning and aggregation; (2) with rule insertion; (3) with both pruning and aggregation; and (4) with insertion of a new class in on-line training.

Table 1. Parameters and results of EFuNN experiments. **MF.No.:** the fuzzy membership function number; **Errthr:** error threshold; **Pruning:** prune or not prune; **Pru.A.:** age of a node that increases one after the network processes an example; **nAgg.:** the number of example, after that an aggregation will be done; **R.Field:** maximum receptive radius; **Node:** number of node in rule layer; **RMSE:** Root Mean Square Error of test set; **Class.:** classification error of test set.

		Parameters					Results		
No.	MF.No.	Errthr	Pruning	Pru.A.	nAgg.	R.Field	Node	RMSE	Class.
1	2	0.1	No		10	0.5	134	0.3651	0.0088
2	2	0.9	Yes	3000	100	0.5	102	0.4275	0.0282
3	2	0.6	Yes	1000		0.5	79	0.5601	0.0467
4	2	0.9	No			0.5	161	0.3785	0.0126

In the first experiment, pruning or aggregation operations in node rules were not performed. As can be observed in Table 1 (row 1), the EFuNN got a good generalization in the odor classification. The obtained classification error was of 0.88%.

In the second experiment a new EFuNN architecture was created. In this new architecture 10 rules were inserted, remainders of previous trainings and new odor patterns were presented to this network. The classification error it was of 2.82% (Table 1, row 2) and the amount of rules was of 102 rules.

In third experiment, presented in Table 1 (row 3), all nodes rules were created dynamically during the training. However, during this training it is accomplished the pruning and node aggregation. With the application of these techniques, the amount of nodes rules in the network decreased and for consequence the amount of extracted rules of the neuro-fuzzy network also decreased. However, at the same time an increase of classification error is verified.

The fourth experiment seeks to verify the on-line training and the dynamic insertion of classes in the network. As it is shown in the Table 1 (row 4), the fact of the fifth gas be presented later on the training of the others gases, it did not harm the generalization of the model for the five gases. The classification error of this simulation was of 1.26%.

The Students' t-test [11] was used to perform the statistical analysis in the results. The analytical goal relates to verifying the null hypothesis (NH). The NH hypothesis assumes that the average performance of the both classifiers is the same. The alternative hypothesis (AH) is assumed to be true if NH is refuted and in this case indicates that the average performance of the first classifier is superior to the last classifier.

(1) NH: $\mu_x = \mu_y$, i.e. expected mean performance of the classifiers are the same. AH: $\mu_x \geq \mu_y$: Reject NH if $|t| > t_{\alpha/2,f}$ where f is the degree of freedom and α is the significance level. The value of $t_{\alpha/2,f}$ is obtained from a statistical table.

The x and y values represent respectively the mean error of classifiers in the odor recognition experiments. The results obtained are summarized in Table 2.

Table 2. Data Analysis.

General Results			
	Mean	Variance	Observations
MLP	0.1388	0.1874	30
TDNN	0.1054	0.0891	30
EFuNN	0.0088	0.0029	30
Statistics of paired students' t-test			
Classifiers	t stat	t Critical two-tail ($t_{\alpha/2,f}$)	Conclusion
MLP – TDNN	0.8812	2.756	$|t| < t_{\alpha/2,f}$ therefore NH is not reject
MLP – EFuNN	3.7957	2.756	$|t| > t_{\alpha/2,f}$ therefore NH is reject
TDNN – EFuNN	5.9326	2.756	$|t| > t_{\alpha/2,f}$ therefore NH is reject

The value $t_{\alpha/2,f}$ can be seen in statistical tables and in this case represents the critical value for the two-tailed t-test ($t_{\alpha/2,f} = 2.756$ for $\alpha = 5\%$ and $f = 29$) [11]. Therefore $|t| < t_{\alpha/2,f}$ and the null hypothesis (NH) is not reject for the first comparation, but is reject for the last two. These results support the hypothesis that the average performance of the EFuNN is superior to the MLP and TDNN in odor classification.

6   Final Remarks

In this work, results for a pattern recognition system in an artificial nose have been presented. A comparison among the three pattern classification algorithms, MLP, TDNN and EFuNN, was made. The results show that the EFuNN achieved a better generalization performance than those obtained by MLP and TDNN neural networks to odor recognition. While the classification error of EFuNN was 0.88%, this error for MLP and TDNN, in the best performances, was respectively 13.88% and 10.54% in test sets.

Besides the good performance, the EFuNN presents some characteristics unexplored in the construction of artificial noses, as the incremental learning, and the possibility of on-line change in the number of sensors. Another advantage is no necessity of preprocessing in the data, which can reduce the training time of the device and to increase the speed in the identification of the gases detected by the sensor system.

For simple tasks currently we have technologies for odor recognition, future research needs to consider other functionalities in the construction of artificial noses, as the automatic acquisition of knowledge, automatic configuration of the devices, and the versatility of the artificial noses.

References

1. Gardner J. W., Hines E. L.: Pattern Analysis Techniques. Handbook of Biosensors and Electronic Noses: Medicine, Food and the Environment, CRC Press (1997) 633–652.
2. Kasabov N.: Evolving Fuzzy Neural Networks for Supervised/Unsupervised On-Line, Knowledge-Based Learning. IEEE Trans. on Systems, Man and Cybernetics, Part B: Cybernetics, Vol. 31, No. 6, December (2001) 902–918.
3. Rumelhart D. E., Hinton G. E., Williams R. J.: Learning Representations by Backpropagation Errors, Nature, No. 323 (1986) 533–536.
4. Yamazaki A., Ludermir T.B.: Classification of Vintages of Wine by an Artificial Nose with Neural Networks. In: International Conference on Neural Information Processing, Vol. 1, China (2001) 184-187.
5. Zanchettin C., Ludermir T. B.: Wavelet Filter for Noise Reduction and Signal Compression in an Artificial Nose. In: Hybrid intelligent System, Melbourne, Austrália (2003) 907–916.
6. Zanchettin C., Ludermir T. B.: A Neuro-Fuzzy Model Applied to Odor Recognition in an Artificial Nose. In: Hybrid intelligent System, Melbourne, Australia (2003) 917–926.
7. Kasabov N.: The ECOS framework and the "eco" training method for evolving connectionist systems. In: Journal of Advanced Computational Intelligence, Vol. 2, No. 6, (1983) 1–8.
8. Persaud K. C., Travers P. J.: Arrays of Broad Specificity Films for Sensing Volatile Chemicals. In: Handbook of Biosensors and Electronic Noses: Medicine. Food and the Environment (Edited by E. Kress-Rogers), CRC Press (1997) 563–592.
9. Prechelt L.: Proben1 – A Set of Neural Network Benchmark Problems and Benchmarking Rules. Technical Report 21/94, Universität Karlsruhe, Germany (1994).
10. Fletcher R.: Pratical Methods of Optimization. Wiley (1987).
11. Johson R. A., Wichern D. W.: Applied Multivariate Statistical Analysis. Prentice Hall, (1999) 767p.

Differential Evolution Based On-Line Feature Analysis in an Asymmetric Subsethood Product Fuzzy Neural Network

C. Shunmuga Velayutham and Satish Kumar

Dept. of Physics and Computer Science
Faculty of Science
Dayalbagh Educational Institute
Dayalbagh, Agra 282005 India
skumar_db@ieee.org

Abstract. This paper proposes a novel differential evolution learning based online feature selection method in an asymmetric subsethood product fuzzy neural network (ASuPFuNIS). The fuzzy neural network has fuzzy weights modeled by asymmetric Gaussian fuzzy sets, mutual subsethood based activation spread, product aggregation operator that works in conjunction with volume defuzzification in a differential evolution learning framework. By virtue of a mixed floating point-binary genetic coding and a customized dissimilarity based bit flipping operator, the differential evolution based asymmetric subsethood product network is shown to have online feature selection capabilities on a synthetic data set.

1 Introduction

Synergistic fuzzy neural models, which combine the merits of connectionist and fuzzy approaches, possess the ability to refine initial domain knowledge and are able to operate and adapt in both numeric as well as linguistic environments [3–5]. They also possess the ability to perform system identification (SI) implicitly. It is well known that feature analysis plays an important role in SI [6], for not all the features may be required to understand the function underlying input-output relationship, some features may be redundant and some may be indifferent to the system output thereby causing a derogatory effect on the performance of the system. While most of the fuzzy rule-based system identification methods either ignore feature analysis or do it offline, very few fuzzy neural models have been proposed in the literature for the task of simultaneous feature extraction and system identification [7, 8].

In this paper, we propose a differential evolution learning based online feature selection method in an asymmetric subsethood product fuzzy neural network (ASuPFuNIS) introduced in [1],[2]. ASuPFuNIS has fuzzy weights modeled by asymmetric Gaussian fuzzy sets, mutual subsethood based activation spread, product aggregation operator that works in conjunction with volume defuzzifi-

N.R. Pal et al. (Eds.): ICONIP 2004, LNCS 3316, pp. 959–964, 2004.
© Springer-Verlag Berlin Heidelberg 2004

cation in a differential evolution (DE) learning framework. In [1],[2] the ASuP-FuNIS network learnt network parameters using gradient descent. However, in the present work, the network is trained by differential evolution [9]. Differential evolution is a population based, stochastic function minimizer that uses vector differences to perturb a population of vectors. A mixed floating point-binary genetic coding along with a customized bit flipping operator has been introduced for online feature selection. The feature selection capability of this DE learning based ASuPFuNIS has been evaluated on a synthetic data set [6].

The organization of this paper follows: Section 2 describes the asymmetric subsethood product fuzzy neural network, Section 3 deals with the differential evolution algorithm, Section 4 describes the simulation experiments and presents the results and Section 5 concludes the paper.

2   Asymmetric Subsethood Product Network

The ASuPFuNIS architecture [2] shown in Fig. 1. Nodes in the input and output layer represent features and target variables. Each hidden node represents a fuzzy rule; input-hidden node connections represent fuzzy rule antecedents and hidden-output node connections represent fuzzy rule consequents. A connection from node i to node j is represented by the triplet $w_{ij} = (c_{ij}, \sigma_{ij}^l, \sigma_{ij}^r)$: the center, left spread and right spread of an asymmetric Gaussian fuzzy set. The asymmetric subsethood network can simultaneously admit the numeric as well as linguistic inputs. A numeric input is fuzzified by treating it as the center of an asymmetric Gaussian membership function with tunable left and right spreads and is represented by the triplet $x_i = (c_i, \sigma_i^l, \sigma_i^r)$. Linguistic inputs are also modeled using a similar triplet.

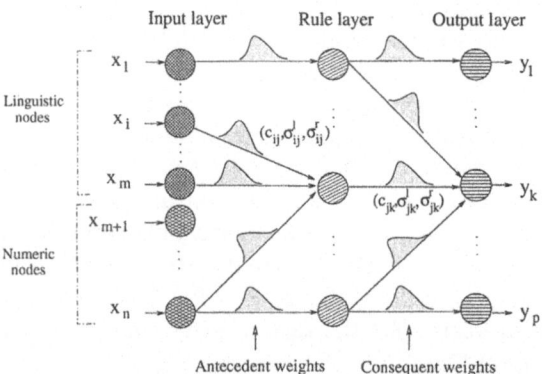

Fig. 1. Architecture of the asymmetric subsethood product fuzzy neural network.

Signal transmission along input-hidden node connections is computed using a fuzzy mutual subsethood measure [10]. This quantifies the net signal value as the extent of overlap between the input fuzzy signals $s_i = (c_i, \sigma_i^l, \sigma_i^r)$ and fuzzy antecedent weights $w_{ij} = (c_{ij}, \sigma_{ij}^l, \sigma_{ij}^r)$:

$$\mathcal{E}(s_i, w_{ij}) = \frac{\mathcal{C}(s_i \cap w_{ij})}{\mathcal{C}(s_i) + \mathcal{C}(w_{ij}) - \mathcal{C}(s_i \cap w_{ij})} \tag{1}$$

where $\mathcal{C}(\cdot)$ denotes the cardinality of a fuzzy set. A subsethood based product aggregation operator at rule nodes aggregates all the transmitted signals to compute the rule firing strength z_j as

$$z_j = \prod_{i=1}^{n} \mathcal{E}_{ij} . \tag{2}$$

The signal of each output node y_k is determined using standard volume based centroid defuzzification [10]

$$y_k = \frac{\sum_{j=1}^{q} z_j \left(c_{jk} + \frac{\sigma_{jk}^r - \sigma_{jk}^l}{\sqrt{\pi}} \right) (\sigma_{jk}^l + \sigma_{jk}^r)}{\sum_{j=1}^{q} z_j (\sigma_{jk}^l + \sigma_{jk}^r)} \tag{3}$$

where q is the number of rule nodes and $(c_{jk}, \sigma_{jk}^l, \sigma_{jk}^r)$ represents the center, left and right spreads of consequent fuzzy weights.

3   Differential Evolution Coding and Operation

Differential Evolution (DE) is a novel population based parallel search method [9]. The crucial ideal behind DE, that differentiates it from other population based search methods, is a new scheme for generating trial vectors: adding the weighted difference vector between two random population members to a third member [9]. If the resulting trial vector yields a lower objective function value than a predetermined population member, it replaces the vector with which it was compared. Crossover operation involves choosing a substring of the trial vector and shuffling it with a substring of a random vector or the best member of the current generation.

The genetic coding employed in this paper uses both floating point and binary numbers. The floating point part codes the fuzzy input and weight triplets of antecedent and consequent connections subsethood product network. In the binary part, each bit encodes the presence or absence of an antecedent connection in the network. This is different from the conventional procedure of coding the feature or rule node count for enable bits. Together with a properly designed fitness function, this coding scheme lends the flexibility to select features, learn the antecedent connectivity pattern and the rule node count along with the network parameters using DE.

We employ a variant of the DE where the weighted difference between two random population members always perturbs the best vector of the current generation. The weighted difference and the trial vector generation operations on the enable bits of the randomly chosen members are modeled by a *dissimilarity based* bit flipping operation. This operator preserves structural information of the network. A dissimilarity in the random vector enable bits triggers a bit flipping with a certain probability. Crossover in the floating point part involves shuffling a substring between the trial vector and a predetermined population member while the binary part employs a uniform crossover operation.

4   Simulation Results

The efficacy of the differential evolution learning for ASuPFuNIS is tested on a synthetic data set [6] generated by the following nonlinear static system with two inputs x_1, x_2 and a single output y:

$$y = (1 + x_1^{-2} + x_2^{-1.5})^2, \ 1 \le x_1, x_2 \le 5. \tag{4}$$

Using this nonlinear system, 50 input-output data were generated by randomly picking 50 pairs of points from $1 \le x_1, x_2 \le 5$. To demonstrate the feature selection capability of the proposed methodology, two random variables x_3 and x_4, in the range $[1, 5]$ were added as dummy inputs so that these features would be indifferent to the output of the system. To measure the performance of ASuP-FuNIS, we employed the following performance index (PI) measure as defined in [11]:

$$\text{PI} = \frac{\sqrt{\sum_{k=1}^{50} (y_k' - y_k)^2}}{\sum_{k=1}^{50} |y_k|} \tag{5}$$

where y_k and y_k' respectively denotes the desired and the actual outputs of the network.

Parameter and Antecedent Learning. To demonstrate the learning capability of the DE based asymmetric subsethood product network, we initially trained and tested the network on the data set without any dummy inputs i.e., on the two input, single output system as described by (4). The fitness function used a simple sum of squared error (SSE) measure to evolve the network parameters. A 3-rule ASuPFuNIS gave a root mean squared error (RMSE) of 0.0432 and a performance index (PI) of 0.0024. This is to be compared with FLEXNFIS in [12] that obtained an RMSE of 0.0739 with 4 rules while [11] obtained a PI of 0.0035 with three rules.

To learn the optimal antecedent connectivity pattern along with the network parameters, the fitness function was modified to include the ratio of the antecedent connections used in the network as calculated from the enable bits and the total connections in the fully connected network. The differential evolution learning pruned a rule by deleting both the antecedent connections to the third rule, and obtained a performance index measure of 0.0076. Note that although the fitness function aims to learn the optimal antecedent connectivity pattern, rule structure gets evolved implicitly.

Feature Selection. Here, the fitness function is a weighted sum of two terms: the SSE and the ratio of features used in the network total number of features i.e.,

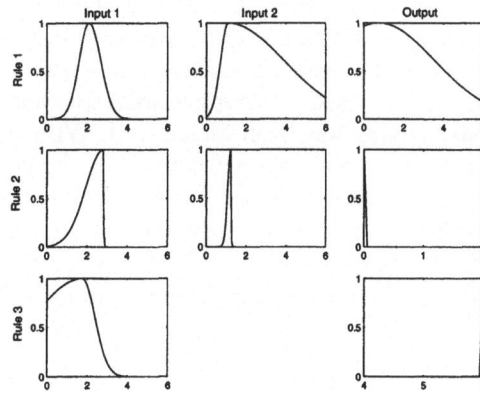

Fig. 2. Rule plots of the asymmetric subsethood product fuzzy neural network.

four. A 3-rule ASuPFuNIS successfully removed the indifferent features x_3 and x_4 and obtained a high performance index measure of 0.0039. This performance is comparable with that of the 3-rule network used in the parameter learning thereby demonstrating the robustness of the DE based online feature analysis. Fig. 2 shows the rule plots of the 3-rule network. Interestingly, as the figure shows, one of the antecedents got pruned during learning. This implicit learning stems from the coding scheme which uses antecedent connectivity information in enable bits rather than the input features or rule nodes count. Table 1 compares the performance of ASuPFuNIS with other models on the synthetic data set for feature analysis. As can be seen in the table, the differential evolution based asymmetric subsethood product network outperformed the other methods by achieving a high performance. It is worth mentioning that [8] did not tune the membership functions during feature analysis. The comparison shows the high performance as well as the robustness of the differential evolution based online feature analysis of ASuPFuNIS.

Table 1. Performance of 3-rule ASuPFuNIS against other methods on synthetic data.

Method	Rule counts	PI
Sugeno and Yasukawa [6]	6	0.01
Chakraborty and Pal [8]	16	0.01
ASuPFuNIS [2]	3	0.0039

5 Conclusions

A novel scheme for online feature analysis in an asymmetric subsethood product fuzzy neural network (ASuPFuNIS) using differential evolution learning has been proposed in this paper. The differential evolution method uses a mixed floating point-binary coding to represent respectively the fuzzy weights and antecedent connectivity pattern of the network. Unlike the conventional coding of input features or rule node counts, the present work uses antecedent connection information which provides the network with implicit structural learning abilities. A customized dissimilarity based bit flipping operator that preserves structural information in binary coding has been introduced. With a simple fitness function, the asymmetric subsethood product network is shown to reject indifferent features successfully on a synthetic data set. The differential evolution learning based asymmetric subsethood product fuzzy neural network is currently being extended to perform feature analysis and structural evolution in an integrated manner. This will be reported as part of future work.

References

1. Velayutham, C.S., Kumar, S.,: Some applications of an asymmetric subsethood product fuzzy neural inference system. Proceedings of 12th IEEE International Conference on Fuzzy Systems (FUZZ-IEEE 2003) St. Louis, Missouri, USA (May 2003) 202-207
2. Velayutham, C.S., Kumar, S.,: Asymmetric Subsethood Product Fuzzy Neural Inference System (ASuPFuNIS). IEEE Transactions on Neural Networks, forthcoming.
3. Mitra, S., Hayashi, Y.: Neuro-fuzzy rule generation: survey in soft computing framework. IEEE Transactions on Neural Networks 11 (May 2000) 748–768
4. Kasabov, N.: Neuro-fuzzy techniques for intelligent processing. Physica Verlag (1999)
5. Lin, C., Lee, C.S.G.: Neural fuzzy systems: A neuro-fuzzy synergism to intelligent systems. Upper Saddle River NJ: Prentice Hall P T R (1996)
6. Sugeno, M., Yasukawa, T.,: A fuzzy-logic-based approach to qualitative modeling. IEEE Transactions on Fuzzy Systems 1(1) (February 1993) 7-31
7. Chakraborty, D., Pal, N.R.,: A neuro-fuzzy scheme for simultaneous feature selection and fuzzy rule-based classification. IEEE Transactions on Neural Networks 15(1) (January 2004) 110-123
8. Chakraborty, D., Pal, N.R.,: Integrated feature analysis and fuzzy rule-based system identification in a neuro-fuzzy paradigm. IEEE Transactions on Systems, Man and Cybernetics Part B: cybernetics 15(1) (June 2001) 391-400
9. Storn, R., Price, K.,: Differential Evolution - a simple and efficient adaptive scheme for global optimization over continuous spaces. Technical Report TR-95-012, ICSI (March 1995).
10. Kosko, B.: Fuzzy engineering Englewood Cliffs: Prentice Hall (1997)
11. Lin, Y., Cunningham III, G.A.,: A new approach to fuzzy-neural system modeling. IEEE Transactions on Fuzzy Systems 3(2) (May 1995) 190-198
12. Rutkowski, L., Cpalka, K.,: Flexible neuro-fuzzy systems. IEEE Transactions on Neural Networks 14(3) (May 2003) 554-574

Neuro-fuzzy System for Clustering of Video Database

Manish Manori A.[1], Manish Maheshwari[2], Kuldeep Belawat[3],
Sanjeev Jain[1], and P.K. Chande[4]

[1] Samrat Ashok Technological Institute, Vidisha (M.P.)
manishmanoria@rediffmail.com, dr_sanjeevjain@hotmail.com
[2] Makhanlal Chaturvedi National University, Bhopal (M.P.)
manishbhom@yahoo.com
[3] Lakshmi Narain College of Technology, Bhopal (M.P.)
kuldeepbelawat@yahoo.com
[4] Director, Moulana Azad National Institute of Technology, Bhopal (M.P.)
pkchande@yahoo.com

Abstract. Due to poor and non-uniform lighting conditions of the object, imprecise boundaries and color values, the use of fuzzy systems makes a viable addition in image analysis. Given a continuous video sequence V, the first step in this framework for mining video data is to parse it into discrete frames. This is an important task since it preserves the temporal information associated with every frame. A database of images is created, from which features are extracted for each image and stored in feature database. This framework focuses on color as feature and considers HLS color space with color quantization into eight colors. Using fuzzy rules, fuzzy histogram of all these eight colors is calculated and stored in feature database. A Radial Basis Function (RBF) Neural Network is trained by the fuzzy histogram of random images and similarity measure is calculated with all other frames. Frames, which have distance between ranges specified for clustering, are clustered into one cluster.

1 Introduction

The drastic advances in both hardware and software technologies are making digital multimedia application technically and economically feasible. This forms a culture that becomes increasingly visual, relying more and more on non-textual formats for learning, entertainment and communication. The inherent features of the visual data are "imprecision", "partial", and "user preferences"[4]. Fuzzy logic is based on the theory of fuzzy sets and, unlike classical logic, it aims at modeling the imprecise (or inexact) modes of reasoning and thought processes (with linguistic variables) that play an essential role in the remarkable human ability to make rational decisions in an environment of uncertainty and imprecision [13]. The imprecision in an image contained within color value can be handled using fuzzy sets. The notations like "good contrast" or "sharp boundaries", "light red", "dark green", etc. used in image enhancement by fuzzy logic are termed as linguistic variables. These variables can be perceived qualitatively by the human reasoning. As they lack in crisp and exhaustive quantification, they may not be understood by machine. To overcome this limitation to a large extent, fuzzy logic tools empower a machine to mimic human reasoning.

N.R. Pal et al. (Eds.): ICONIP 2004, LNCS 3316, pp. 965–970, 2004.
© Springer-Verlag Berlin Heidelberg 2004

Neural Network implementation of fuzzy systems called fuzzy neural networks or neural network based fuzzy (neuro-fuzzy) system will possess the advantages of both types of systems and overcome the difficulties of each type of system. In fact, the resulting systems not only support numerical mathematical analysis, hardware implementation, distributed parallel processing and self learning but capable of dealing with difficulties arising from uncertainty, imprecision and noise.

We proposed a neuro-fuzzy based system used to cluster the feature in large image and video data sets, while preserving their temporal information.

2 Related Works

Clustering is a method of grouping data into different groups, so the data in each group share similar trends and patterns. Clusters of objects are formed so that objects within a cluster have high similarity in comparison to one another, but are very dissimilar to objects in other clusters [1], [10]. There are two main groups of cluster analysis methods - the first is hierarchical [8], [9] and the second non-hierarchical or partitioning [3].

In Russo and Ramponi [11]. IF.. THEN.. ELSE. fuzzy rules directives (much similar to humane-like reasoning) for image enhancement. Here, a set of neighborhood pixels forms the antecedent part of the rule and the pixel to be enhanced is changed by the consequent part of the rule. In Pal and King, [12] an image can be considered as an array of fuzzy singletons having a membership value that denotes the degree of some image property in the range {0,1}. S. Chen et al. [5], R. Cucchiara et al.[6] , and D. Dailey et al. [7] proposed Multimedia data mining frameworks for traffic monitoring system.

3 Proposed Technique

Clustering of Video database involve the following scheme:

3.1 Video Parsing

Parsing is the process of temporal structure segmentation that involves the detection of temporal boundaries and identification of meaningful components of videos thus extracts structural information of video. A video sequence is viewed as a set of "group of frames" or GOFs. In the proposed work AVI video file is used for experiments. In this sample AVI video file display rate is 12 frames/second. As near by frames are similar in color contents, it is time-consuming process to compare each & every frame. It is more beneficial if we capture one frame among all the similar frames. Therefore we take one frame in one second. In this module AVI video file as input and get still images or frames in JPEG format as output.

3.2 Feature Extraction

Feature extraction is the process of interacting with images & video data and performs extraction and recognition of meaningful information of images with descriptions based on properties that are inherent in the images themselves. An image is a matrix of pixel in which each pixel is represented by color intensity. This RGB values for each are converted to HLS values color model. Since RGB components are highly correlated and chromatic information is the not directly fit for use, This RGB values for each are converted to HLS (Hue-Lightness-Saturation) values color model.

3.2.1 Fuzzification of the HLS Values. The hue values range from 0 to 360 degrees and hue represents the dominant color of a pixel. The fuzzification of hue is done in such a way that the non-crisp boundaries between the colors can be represented much better. Six symbols are used in order to characterize the hue values at the distance of 60 degree

Hue = {RED, YELLOW, GREEN, CYAN, BLUE, MAGENTA]]}

The saturation & lightness values ranges from 0 to 1. Three symbols are used to characterize these quantities

Saturation = { Small, Medium, Large }

Lightness = { Small, Medium, Large }

The lightness values from 0 to 1. The values run as , 0 appears black (no light) while 1 is full illumination, which washes out the color (it appears white).

3.2.2 Fuzzy Histogram. Color histogram as a set of bins where each bin donates the probability of pixels in the image being of a particular color.

The simplest approach is to normalize the histogram by the value of its largest bin, in such way that the most probable color will have a membership degree of 1 within the fuzzy set "image" [2]. The most predominant color can be thus considered as the most typical for the given image and the constructed fuzzy histogram measures the typicality of a color within the image.

In the proposed work hue and Lightness values of each pixel are considered as the input for the calculation of histogram.

If lightness is small
      Then   Color is black
If lightness is large
      Then   Color is white
If lightness is medium and Hue is

$$\begin{bmatrix} \text{Red} \\ \text{Magenta} \\ \text{Blue} \\ \text{Yellow} \\ \text{Cyan} \\ \text{Green} \end{bmatrix}$$

      Then   Color is of Hue

The output is the pixel belongs to which color histogram bin i.e. the color histogram is incremented by one. Thus we get histogram of eight colors for each image or frame. Finally fuzzy histogram of each color is obtained by dividing each histogram bin by the largest number of histogram bin in that image. Fig.1 shows sample image and its fuzzy histogram.

	Black	Red	Green	Yellow	Blue	Magenta	Cyan	White
Image.jpg	0.3818	0.471	0	0.1744	1	0.2424	0	0.5043

Fig. 1. Fuzzy histogram of sample image

3.3 Cluster Formation

In the proposed work the fuzzy histogram of each color in an image is input to the Radial Basis Function neural network (RBF) i.e. 8 neurons in input layer. Inputs to these neurons are feature vector of randomly chooses image. With these feature vector neural network is trained for the output value 1 and error 0. The no. of training epocs is set to 50.

These video and frame descriptions serve as a natural form of "meta data" upon which conceptual clustering techniques can be applied so that semantically meaningful groups (so-called cluster) can be constructed. Feature i.e. fuzzy histogram from feature database of each frame are given to trained RBF neural network and obtain output. This output is the distance of image from the training image. A range is specified for clustering, all the frames having distance between a particular range are grouped into one cluster.

4 Experimental Results

In order to evaluate performance, a prototype is implemented in this work. The training frames of neural network is prepared by considering eight inputs as fuzzy histogram of images. Then the feature database of each frame are given to traning neural network and based on their distance different clusters are formed. In random order two frames are selected as training frames. Video frames are retrieved from database and their features from feature database of every frame are given to trained RBF neural network and calculate its output. This output shows the distance of frame from training frame. If two frames have same distance from training frame then both the frame are similar. All the frames, having distance between a range specified for clustering, are clustered into one cluster as shown in figure 2.

5 Conclusion

This work presented a color feature based clustering approach for the fuzzy categorization of database images. This clustering mechanism can categorize the images into different clusters based on their distances from training image, while preserving their temporal information.

Training image	Cluster images
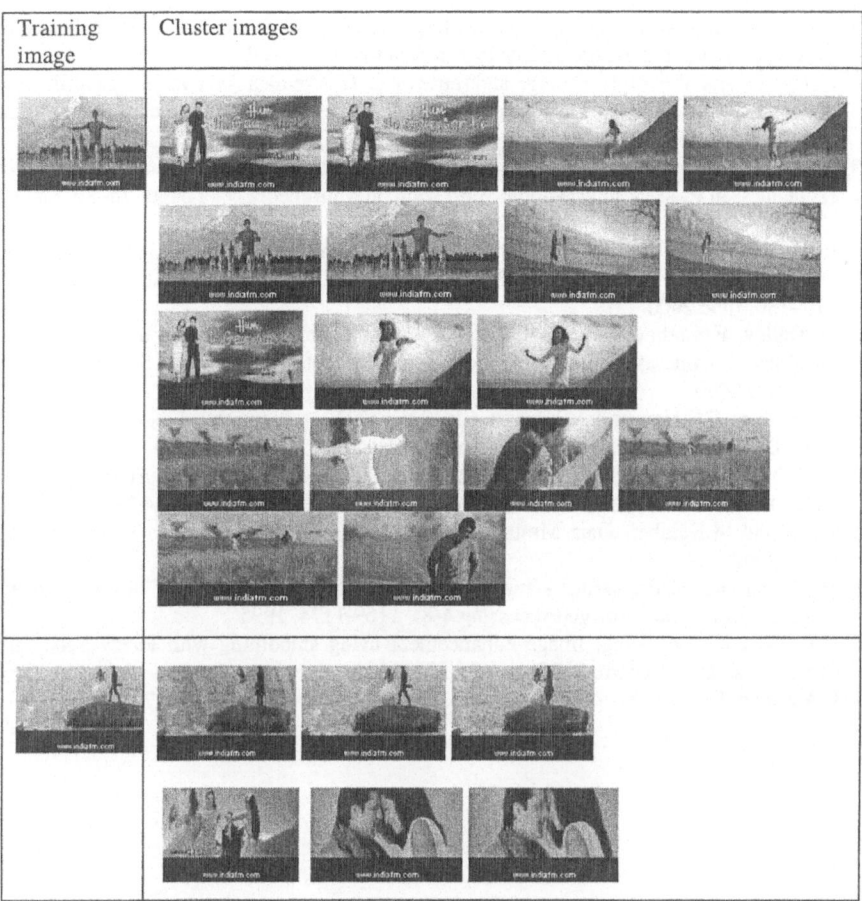	

Fig. 2. Cluster Images

This paper is a general framework for the video data mining to perform the fundamental tasks which are temporal division of video sequences into frames, feature extaction and clustering of frames. For feature extraction fuzzy histogram of eight color using HLS model is prepared.

Although our experimental data set are limited, the results are showing that the proposed framework is performing the fundamental tasks effectively and efficiently. The main advantage of this technique is that it is based on hue value of image pixel, this scheme works nicely in day and night (dark) images/video.

References

1. R. Agrawal, J. Gehrke, D. Gunopuios and P. Raghavan.: Automatic subspace clustering of high dimensional data for data mining application. Proc. ACM-SIGMOD, pp. 94-105, 1998
2. J.C.Bezdek: Fuzzy models - what are they and why? , IEEE trans. On Fuzzy systems. 1(1): 1-5, Feb 1993

3. P. Bradley, U. Fayyad and C.Reina: Scaling clustering algorithms to large databases. proc. Fourth int. conf. knowledge discovery and data mining, pp 9-15, 1998
4. C. Carson and V.E. Ogle: Storage and retrieval of feature data for a very large online image collection. Bulletin of the IEEE Computer society Technical committee on Data Engineering, vol. 19, pp 19-25, 1996
5. S. Chen, M. Shyu, C. Zhang, and J. Strickrott: Multimedia data mining for traffic video sequences. In Proc. of International Workshop on Multimedia Data pages 78–86, San Francisco, CA, August 2001
6. R. Cucchiara, M. Piccardi, and P. Mello: Image analysis and rule-based reasoning for a traffic monitoring system. IEEE Transactions on Intelligent Transportation Systems, 1(2): 119–130, June 2000
7. D. Dailey, F. Cathey, and S. Pumrin: An algorithm to estimate mean traffic speed using uncalibrated cameras. IEEE Transactions on Intelligent Transportation Systems, 1(2): 98–107, June 2000
8. G.Karypis, E.H.Han and V.Kumar: CHAMELEON: a hierarchical clustering algorithm using Dynamic modeling. Computer, vol.32, no. 8, pp. 68-75, Aug.1999
9. S. Guha, R. Rastogi, and K.Shim: CURE: An efficient algorithm for clustering large databases. Proc. ACM-SIGMOID Int. conf. on management of data, pp 73-84 1998
10. J.Han and M.Kamber: Data Mining concepts and Techniques. Morgan Kaufmann Publishers, 2002
11. M. Russo and G. Ramponi: A fuzzy operator for the enhancement of blurred and noisy images. IEEE Trans. Image Processing 4(8), 1169-1174, 1995.
12. S.K. Pal and R.A. King: Image enhancement using smoothing with Fuzzy Sets. IEEE Trans. Sys. Man Cybern. SMC-11, 494-501, 1981.
13. L.A.Zadeh: Fuzzy logic, neural networks, and soft computing. ACM 37:77-84, 1994

Dynamic Neuro-fuzzy Inference and Statistical Models for Risk Analysis of Pest Insect Establishment

Snjezana Soltic[1, 2], Shaoning Pang[2], Nikola Kasabov[2],
Sue Worner[3], and Lora Peackok[3]

[1] Department of Electrical & Electronic Engineering, Manukau Institute of Technology,
Manukau City, New Zealand
ssoltic@manukau.ac.nz
[2] Knowledge Engineering & Discovery Research Institute
Auckland University of Technology, Auckland, New Zealand
spang@aut.ac.nz
[3] Center for Advanced Bio-protection Technologies,
Ecology and Entomology Group
Soil, Plant and Ecological Science Division
Lincoln University, Canterbury, New Zealand
Worner@lincoln.ac.nz

Abstract. The paper introduces a statistical model and a DENFIS-based model for estimating the potential establishment of a pest insect. They have a common probability evaluation module, but very different clustering and regression modules. The statistical model uses a typical K-means algorithm for data clustering, and a multivariate linear regression to build the estimation function, while the DENFIS-based model uses an evolving clustering method (ECM) and a dynamic evolving neural-fuzzy inference system (DENFIS) respectively. The predictions from these two models were evaluated on the meteorological data compiled from 454 worldwide locations, and the comparative analysis shows advantages of the DENFIS-based model as used for estimating the potential establishment of a pest insect.

1 Introduction

A variety of methods have been designed to predict the likelihood of pest establishment upon a species introduction into an area [1], [2], [3], [4], [5], [6], [7]. It is observed that, (1) a number of methods have been developed specifically for problems at hand, and therefore have relatively narrow applicability, and (2) usually only one method was applied to a data set, and therefore there is a lack of comparative analysis that show advantages and disadvantages of using different methods on the same data set.

The analysis of the response of a pest to influential environmental variables is often so complex that traditional methods are not very successful. Artificial neural networks have been studied as a promising tool for decision support in ecological research [8], [9]. The studied neural networks are mainly of a multilayer perceptron type that have some drawbacks such as absence of incremental learning, no facility for extracting knowledge (rules) and often, not good generalization [8]. This research

N.R. Pal et al. (Eds.): ICONIP 2004, LNCS 3316, pp. 971–976, 2004.
© Springer-Verlag Berlin Heidelberg 2004

describes and compares two models for predicting the potential establishment of a pest in new locations using *Planocuccus citri* (Risso), the citrus mealybug, as a case study. The software environment NeuCom (www.kedri.info) was used in the paper for the analysis and the prediction.

2 Experiments

2.1 Data Set

In the experiment, meteorological data compiled from 454 worldwide locations where *Planocuccus citri* (Risso) has been recorded as either present (223 locations) or considered absent (232 locations), were used. Each location is described using a 16-dimensional vector and a class label (present/absent). Note that, the class label for a number of locations from the absent class might be false absent. The pest species may be absent at a location simply because it may never have reached it, and not because the climate is unsuitable for its establishment.

2.2 Problem Definition

The assessment of the establishment potential of any species (response variable) can be formulated by the following: Given a problem space: $D = \{X_1, X_2, \cdots, X_k, Y\}$, where $X_i (i = 1, \cdots, k)$ are data examples from D, and $Y = y_1, y_2, \cdots, y_k$ is the vector under estimation. Suppose $X = x_1, x_2, \cdots, x_l$. The target is to predict Y in terms of X by modeling an estimation function $Y = f(X)$. The estimation function f is then used to make spatial predictions of the response, e.g., to predict the establishment of a pest in a new area following entry.

2.3 Models

Two models are introduced and discussed in this paper: (1) a statistical model, and, (2) a dynamic evolving neural-fuzzy inference system (DENFIS)-based model, which are denoted as Model I and Model II respectively.

These two models have a common probability evaluation module, but very different clustering and regression modules. Model I uses a typical K-means algorithm for data clustering [10], and a multivariate linear regression to build the estimation function. Model II clusters data using an evolving clustering method (ECM) [10] and estimates f by a dynamic evolving neural-fuzzy inference system (DENFIS). The details of the DENFIS can be referenced in [10], [11]. Both models fit response surfaces as a function of predictors in environmental space $E = \{X_1, X_2, \cdots X_k\}$, where $X_i (i = 1, \cdots, k)$ are data examples from D and then use the spatial pattern of predictor surfaces to predict the response in geographical space $G = \{g_1, g_2, \cdots, g_k\}$, where the examples are of type $g_i = (latitude_i, longitude_i)$. Model II is incrementally trainable on new data in contrast to Model I.

We implemented the statistical model to predict the establishment potential as follows.

1. Apply a clustering algorithm to data from the problem space D.
2. Suppose $\{C_1, C_2, \cdots, C_\xi\}$, are clusters from the clustering module. For each cluster $C_i \in \{C_1, C_2, \cdots, C_\xi\}$ calculate the mean vector and establishment potential using:

$$X_i^c = \frac{\sum_{j=1}^{|C_i|} X}{|C_i|}, \quad p_i^c(Y \mid x_1, x_2, \cdots, x_k) = \frac{\sum_{j=1}^{|C_i|} p(y \mid x_1, x_2, \cdots, x_k)}{|C_i|}, i = 1, \cdots \xi. \quad (1)$$

3. Use $\mathbf{P}^c$ and $\mathbf{X}^c$ to build the estimation function f.

4. Use f to make spatial predictions of the response (e.g., estimate the establishment potential for each location given in the original data set D).

Note that the regression is performed among clusters C, instead of among samples in D. This enables the model to estimate probability without losing the key information among clusters.

The above procedure was repeated using both models. In Model I the K-means module was used for clustering of the original data set D where the number of clusters, iterations and replicates was set to 20, 100 and 5 respectively. In Model II ECM was used for partitioning data D into 20 clusters (the number of clusters can be and was controlled by selecting the maximum distance, *MaxDist*). Thereafter, the multiple linear regression model was used to build the estimation function (Model I):

$$y = 0.78017 - 0.52528x_1 - 0.1023x_2 + 4.262e - 005x_3 + 0.030326x_4 + 0.0020693x_5$$
$$+ 1.0084x_6 - 1.748x_7 + 1.9414x_8 - 0.13537x_9 - 1.1652x_{10} + 0.87642x_{11}$$
$$- 0.08011x_{12} - 0.96676x_{13} - 0.078018x_{14} + 1.9266x_{15} - 1.2633x_{16}$$

In Model II DENFIS was applied to $\mathbf{P}_{ecm}^c$ and $\mathbf{X}_{ecm}^c$. Consequently, we obtained 15 rules, each of them representing the 15 rule nodes created during learning. Those rules cooperatively function as an estimate that can be used to predict the establishment potential of the citrus mealybug at each location.

The first rule extracted is as follows:

Rule 1: if x_1 is f(0.20 0.75) & x_2 is f(0.20 0.70) & x_3 is f(0.20 0.10) &
x_4 is f(0.20 0.53) & x_5 is f(0.20 0.33) & x_6 is f(0.20 0.73) &
x_7 is f(0.20 0.75) & x_8 is f(0.20 0.76) & x_9 is f(0.20 0.76) &
x_{10} is f(0.20 0.72) & x_{11} is f(0.20 0.71) & x_{12} is f(0.20 0.69) &
x_{13} is f (0.20 0.69) & x_{14} is f(0.20 0.71) & x_{15} is f(0.20 0.72) &
x_{16} is f(0.20 0.71) then
$$y = -2.45 - 27.88x_1 - 150.94x_2 - 1.27x_3 - 4.04x_4 + 4.65x_5 - 59.0x_6 + 85.32x_7$$
$$- 19.85x_8 - 29.54x_9 + 72.0x_{10} + 45.41x_{11} - 129.34x_{12} + 203.15x_{13}$$
$$+ 11.39x_{14} + 12.75x_{15} - 6.59x_{16}$$

3 Results

In Table 1, we compared the DENFIS-based model with the statistical model on the establishment potential prediction of the citrus mealybug at 24 locations. The first 12 locations were chosen because they were given establishment potential estimates greater than 0.7 by Model I. The second 12 locations were given estimates greater than 0.7 by Model II. Each location is described by a pair of geographic coordinates (latitude, longitude), which is given in column 2. The predictions by the statistical and the DENFIS-based are presented in column 3 and column 4, respectively. For the purpose of the comparison, column 5 records the known establishment status of the pest (presence: 1/absence: 0).

Table 1. Results for 24 selected locations. The correct matches are shown in bold

Location	(Latitude, Longitude)	Model I	Model II	Label
Shaam, Selenge	(50.1, 106.2)	1	**0.45**	0
Saran-Paul', Russia	(64.28, 60.88)	0.87	**0.55**	0
Nape, Laos	(18.3, 105.1)	0.80	**0.42**	0
Bangladesh	(24, 90)	**0.80**	0.65	1
Hacienda Santa Elena	(22.52, -99)	0.75	**0.47**	0
Seoul	(37.6, 127)	**0.74**	0.48	1
Tamanrasset, Algeria	(22.78, 5.52)	0.74	**0.39**	0
Najaf, Iraq	(31.98, 44.32)	0.74	**0.55**	0
Dhubri, India	(26.02, 89.98)	0.73	**0.63**	0
Thailand	(16, 102)	**0.73**	0.55	1
Asuncion, Paraguay	(-25.3, -57.7)	**0.73**	0.60	1
Monclova, Coah.	(26.88, -101.42)	0.72	**0.41**	0
Valencia	(39.5, -0.4)	0.49	**1**	1
Lima	(-12.1, -77)	0.16	**0.87**	1
Torit, Sudan	(4.4, 32.5)	0.42	**0.84**	1
Juba, Sudan	(4.87, 31.6)	0.42	**0.83**	1
Ghana	(8, -1)	0.49	**0.75**	1
Ibadan, Nigeria	(7.4, 3.9)	0.41	**0.75**	1
Rwanda	(-2, 30)	0.41	**0.74**	1
Uganda	(2, 32)	0.47	**0.73**	1
Zhejiang (Chekiang)	(29, 120)	0.27	**0.71**	1
Trinidad	(21.48, -80)	0.29	**0.71**	1
Fujian / Fukien	(26, 118)	0.36	**0.71**	1
Dakar, Senegal	(14.7, -17.5)	0.36	**0.71**	1

Given a threshold value, P_{thr}, for scores greater than P_{thr}, set $P = 1$ representing the pest presence, otherwise set $P = 0$ and the pest is absent. Given a location $g_i = (latitude_i, longitude_i)$ in column 2, if the prediction P_i equals to the true value from the 5th column, then the prediction is matched. As can be seen, Model I gives 4 matches in 24 locations, while Model II gives 20 matches.

In Fig. 1 we carried out another comparison, where establishment potentials of citrus mealybugs from 454 worldwide locations are estimated by the above two prediction models, and their performances were measured by match-degree/threshold-value plots. The match-degree, defined as a ratio between the number of locations with a match and the total number of locations, was assessed over the range $P_{thr} \in [0.4, 0.8]$. As can be seen, although both models have similar accuracy predicting the absence of the pest, Model II slightly outperforms Model I. In the case of the presence of the pest, Model II is better than Model I in that Model II achieves more matches than Model I for each $P_{thr} \in [0.4, 0.8]$. Particularly, when $P_{thr} \geq 0.6$, the two models give a significant difference in accuracy, where Model II accuracy increases to 100% while the accuracy of the Model I drop down to 0%.

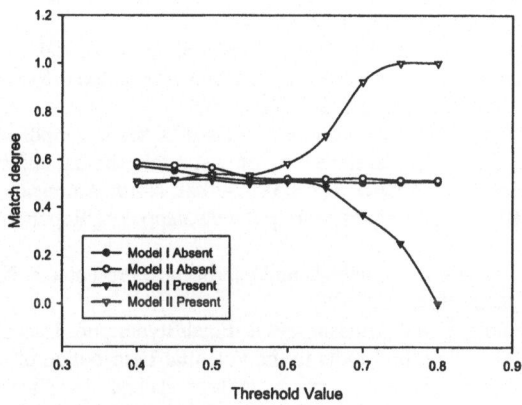

Fig. 1. The accuracy of the models predicting the pest presence or absence at 454 locations expressed in terms of match-degree/threshold-value plots

4 Conclusions

In this paper, we introduced and compared a statistical model and a DENFIS-based model for estimating the potential establishment of pest insects. We used both models in a case study to predict the establishment of the citrus mealybug.

The DENFIS-based model is recommended for on-line prediction applications. If new, yet unseen data becomes available DENFIS will adapt its structure and produce output to accommodate the new input data. During learning, this model creates rules that are useful to researchers who study pest-environmental relationships. The model is preferred because it employs local rather than global clustering, thus the information about pest locations is better conserved in the estimation. This comparative analysis clearly illustrates the advantages of the DENFIS-based model when used for estimating the establishment potential of this particular species of pest insect, and therefore it is a possible new solution for general pest risk assessment.

Acknowledgments

Snjezana Soltic wish to acknowledges the support of this work by the Research Committee of the Department of Electrical and Electronic Engineering at the Manukau Institute of Technology, through the Departmental Research Fund.

References

1. Sutherst R.W., Maywald, G.F. and Bottomley, W.: From CLIMEX to PESKY, a generic expert system for pest risk assessment. *EPPO Bulletin* (1991) **21**:595-608
2. Dentener, P.R., Whiting D.C., Connoly, P.G.: Thrips palmi karny (Thysanoptera: Thripidae): Could it survive in New Zealand? In: Proc. of 55th Conference of New Zealand Plant Protection Society Incorporated (2002) 18-24
3. Dobesberger, E.J.: Multivariate techniques for estimating the risk of plant pest establishment in new environments. Presented at NAPPO International Symposium on Pest Risk Analysis, Puerto Vallarta, Mexico, (2002) Available:
   http://www.nappo.org/PRA-Symposium/PDF-Final/Dobesberger.pdf , December 2003
4. Dobesberber, E.: Climate based modelling of pest establishment and survival in support of rest risk assessment., In: Annual report 1999-2000, North American Plant Protection Organization (2000) 35-36, Available: http://www.nappo.org/Reports/AnnRep-99-00-e.pdf, December 2003
5. Stynes, B.: Pest risk analysis: methods and approaches. Presented at NAPPO PRA Symposium, Puerto Vallarta, Mexico, (2002)
   http://www.nappo.org/PRA-Symposium/PDF-Final/Stynes.pdf , December 2003.
6. Baker, R.H.A.: Predicting the Limits to the Potential Distribution of Alien Crop Pests. In: Halman G., Schwalbe, C.P. (eds.): Invasive arthropods and agriculture: problems and solutions. Science Publisher Inc., Enfield, New Hampshire (2002) 208-241
7. Cohen, S.D.: Evaluating The Risk of Importation of Exotic Pests Using Geospatial Analysis and Pest Risk Assessment Model. First International Conference on Geospatial Information in Agriculture and Forestry, Lake Buena Vista, Florida, USA, (1998)
   http://www.aphis.usda.gov/ppd/evaluating.pdf December 2003
8. Worner, S.P. et. al.: Neurocomputing for decision support in ecological research. Conference on Neurocomputing and Evolving Intelligence, Auckland, New Zealand, 20-21 November 2003 (2003)
9. Gevrey, M., Dimopoulus, I., Lek, S.: Review and comparison of methods to study the contribution of variables in artificial neural network models. In: Ecological Modelling 160 (2003) 249-264
10. Kasabov, N.: Evolving connectionist systems: Methods and applications in bioinformatics, brain study and intelligent machines. Springer-Verlag (2002)
11. Kasabov, N., Song, Q.: Dynamic Evolving Neural-Fuzzy Inference System and Its Application for Time-Series Prediction. In: IEEE Trans. on Fuzzy Systems, vol. 10. (2002) 144-154

An Enhanced Fuzzy Multilayer Perceptron

Kwang Baek Kim[1] and Choong Shik Park[2]

[1] Dept. of Computer Engineering, Silla University, Busan, Korea
gbkim@silla.ac.kr
[2] Dept. of Computer Engineering, Youngdong University, Youngdong, Korea

Abstract. Error back-propagation algorithm of the multilayer perceptron may result in local-minima because of the insufficient nodes in the hidden layer, inadequate momentum set-up, and initial weights. In this paper, we proposed the fuzzy multilayer perceptron which is composed of the ART1 and the fuzzy neural network. The proposed fuzzy multilayer perceptron using the self-generation method applies not only the ART1 to create the nodes from the input layer to the hidden layer, but also the winner-take-all method, modifying stored patterns according to specific patterns, to adjustment of weights. The proposed learning method was applied to recognize individual numbers of student identification cards. Our experimental result showed that the possibility of local-minima was decreased and the learning speed and the paralysis were improved more than the conventional error back-propagation algorithm.

1 Introduction

The Error back-propagation (EBP) algorithm uses gradient descent as the supervised learning rule to minimize the cost function defined in terms of the error value between the output vector and the target one for an given input [1]. The idea is to minimize the network total error by adjusting the weights. Each weight may be thought of as a dimension in an N-dimensional error space. In error space the weights act as independent variable and the shape of the corresponding error surface is determined by the error function in combination with the training set. However, the algorithm has the drawback that the convergence speed of learning is slower and the possibility of falling into the local minima is induced by the insufficient number of nodes in the hidden layer and the unsuitable initial connection weights [2]. During the learning process, the algorithm uses credit assignment for propagating error value of the output layer's nodes backward to the nodes in the hidden layer. As a result, paralysis can be induced in the hidden layer. Generally, the recognition algorithms using the EBP are plagued by the falling-off of recognition rate caused by the empirical determination of the number of hidden layer nodes and the credit assignment procedure [3][4]. If the hidden layer has too many nodes, the redundant nodes, which have no effect on discriminative performance, result in a longer learning time. If the hidden layer does not have sufficient nodes, the possibility of placing the connection weights in local minima may be increased [4][5]. In this paper, we proposed the fuzzy multilayer perceptron, which is composed of the ART1 and fuzzy neural network, for the solving problem of setting the number of nodes of the hidden layer in the EBP algorithm.

N.R. Pal et al. (Eds.): ICONIP 2004, LNCS 3316, pp. 977–982, 2004.
© Springer-Verlag Berlin Heidelberg 2004

2   Related Researches

There are two approaches for combining the fuzzy theory and the neural network theory. The first approach is to combine the strong points of both theories: the fuzzy logic is represented as a rule-like form and the neural network can classify the patterns by the learning algorithm [6]. The second approach is to combine their similar characteristics of both theories [7]. Min operation between fuzzy variables of antecedents in inference rules and input values corresponds to multiplying operation between connection weights and inputs given to neurons. Furthermore, Max operation of the conclusion parts of inference rules corresponds to adding operations of the multiplication between connection weights and inputs given to neurons. Max-Min neural network as used the second approach uses fuzzy logic to update the weights in a multilayer perceptron rather than the delta rule, which uses multiplication and addition operations [8].

3   Enhanced Fuzzy Multilayer Perceptron

Fig.1 shows the proposed learning architecture to self-generate nodes of the hidden layer. The proposed network is presented with a large number of patterns and each hidden layer neuron represents the cluster center. The prototype pattern for each cluster is represented by the weights from the hidden neuron to the input neuron. Vigilance criterion is used to achieve unsupervised learning which determines the actual number of clusters. In the proposed architecture, the connection structure between input layer and hidden layer is similar to structure of the modified ART1. A node of hidden layer represents each class. The nodes in hidden layer are fully connected to nodes in input and output layers. In the case of backward propagation by comparing target vector with actual output vector, we adapt a winner-take-all method to modify weighting factor of only the synapse that is connected to the neuron representing the winner class. The adaptation of weight of synapses between output layer and hidden layer is accomplished by Max-Min neural network.

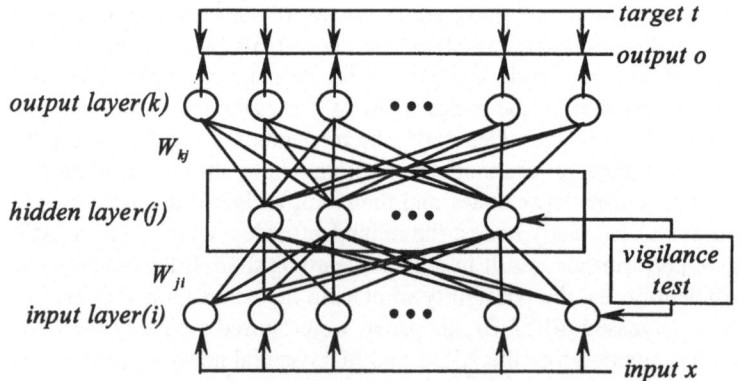

Fig. 1. Enhanced fuzzy multilayer perceptron architecture

The creation for organizing the hidden layer nodes is based on the number of determining classes based on input patterns. Based on ART1, we assume the number of maximum initial nodes of the hidden layer as the number of classes. Starting with one node, we allocate related classes to the initially suggested pattern from the input layer of the node. Next input patterns choose a winner from the nodes in the present state. If all the existing nodes fail to choose a winner, one node is added and allocated to the class for the presented pattern. In this way, patterns are sequentially presented and the nodes for the class are created dynamically. If the stored pattern of the winner node is similar to the input pattern, it becomes the winner. Otherwise, classification is repeated until we get a winner. If an existing node is found to be a winner node, all the weights linking that node to the input layer are updated to reflect the accommodation of that input pattern into the representative class.

The proposed algorithm uses a winner-take-all method instead of the EBP learning to change weights. We should adjust the weights connected to the winner node from the hidden layer to the input layer. To reflect target vector for the input pattern to the actual output vector by the representative class, we change only the connection weights related to the output layer node and its representative class. The proposed fuzzy multilayer perceptron is trained using self-generation method as follows:

Step 1. Initialize the bottom-up weight w_{ji} and the top-down weight t_{ji} between the input layer and the hidden layer. Assign a random value to w_{kj} and θ_k, where $i(i=1,...,m)$ is the input layer, $j(j=1,...,n)$ is the hidden layer, and $k(k=1,...,p)$ is the output layer. ρ is the vigilance parameter, which determines how close an input has to be to correctly match a stored pattern.

$$t_{ji}(0) = 1, \ w_{ji}(0) = \frac{1}{m+1}$$

Set ρ, where $0 < \rho \le 1$

Step 2. Set the input vector x_i and the target vector t_k.

Step 3. Calculate the output vector o_j of the hidden layer.

$$o_j = \sum_{j=1}^{n} w_{ji} \times x_j \qquad (1)$$

Step 4. Select a winner node $O_{j^\bullet}$.

$$o_{j^\bullet} = Max[o_j] \qquad (2)$$

Step 5. Compare the similarity.

If $\dfrac{\|T \bullet X\|}{\|X\|} \ge \rho$, go to *step 7*. Else, go to *step 6*.

Step 6. Reassign zero to o_{j*} in the winner node and go to *step 4*.

Step 7. Adjust the top-down and bottom-up weights of the winner node.

$$t_{j^*_i}(n+1) = t_{j^*_i}(n) \times x_i \tag{3}$$

$$w_{j^*_i}(n+1) = \frac{t_{j^*_i}(n+1) \times x_i}{0.5 + \sum_{i=1}^{m} w_{j^*_i} \times x_i} \tag{4}$$

Step 8. Calculate NET using the representative class o_{j^*} of the hidden layer and the connection weight w_{kj^*} of output layer. Calculate output vector o_k of the output layer using NET and bias term θ_k.

$$NET = \left\{ o_{j^*} \circ w_{kj^*} \right\} \tag{5}$$

$$o_k = NET \vee \theta_k \tag{6}$$

where " $\circ$ " denotes max-min composition.
Step 9. Adjust the connection weights and bias term.

$$w_{kj^*}(n+1) = w_{kj^*}(n) + \alpha \Delta w_{kj^*}(n+1) + \beta \Delta w_{kj^*}(n) \tag{7}$$

$$\theta_k(n+1) = \theta_k(n) + \alpha \Delta \theta_k(n+1) + \beta \Delta \theta_k(n) \tag{8}$$

where α is the learning rate and β is the momentum.

$$\Delta w_{kj^*} = \sum_{k=1}^{p}(t_k - o_k)\frac{\partial o_k}{\partial w_{kj^*}}, \Delta \theta_k = \sum_{k=1}^{p}(t_k - o_k)\frac{\partial o_k}{\partial \theta_k} \tag{9}$$

$$if \ o_k = w_{kj}, \ \frac{\partial o_k}{\partial w_{kj^*}} = 1, \ otherwise, \frac{\partial o_k}{\partial w_{kj^*}} = 0 \tag{10}$$

$$if \ o_k = \theta_k, \ \frac{\partial o_k}{\partial \theta_k} = 1, \ otherwise \ , \frac{\partial o_k}{\partial \theta_k} = 0 \tag{11}$$

Step 10. For all training pattern pair, if TSS is larger than error criteria, go to *Step 3.* Otherwise, learning ends.

4 Experiments and Performance Evaluation

For performance evaluation, we have compared the EBP algorithms and the proposed algorithm using 10 number patterns extracted from student identification cards as learning data. Table 1 (error criteria = 0.01) shows the number of epoch and TSS according to various momentums by applying 10 number patterns to the EBP and the proposed method. In the EBP algorithm, experiments were executed from 5 nodes of the hidden layer to 10 nodes. The experiment of the 10 nodes of the hidden layer showed the most short learning time and good convergence of learning. Therefore,

Table 1 is the result of the experiment with the 10 nodes of hidden layer. The proposed algorithm using ART1 has generated 10 nodes of the hidden layer after learning. As shown Table 1, the proposed method is lesser insensitive, takes smaller learning time, and has lesser TSS than the EBP algorithm.

Table 1. Comparison of the learning speed between EBP and enhanced fuzzy multilayer perceptron

Methods	Momentum	# of Epoch	TSS
EBP	0.1	6320	0.009964
	0.5	6034	0.009725
	0.9	5410	0.009252
Enhanced Fuzzy Multilayer Perceptron	0.1	204	0.009861
	0.5	57	0.002735
	0.9	57	0.002735

Table 2 shows the number of the success convergence by applying 50 number patterns to the EBP and the proposed method. In the EBP algorithm, the initial learning rate is 0.3 and the momentum is 0.5. In the proposed algorithm, the vigilance parameter is 0.9. In the criterion of success convergence, the number of epoch is limited to 20,000 and TSS is 0.04. As shown Table 2, the number of success convergence of the proposed method is larger and the average number of epoch of that is smaller than the EBP algorithm.

Table 2. Comparison of the learning convergence between EBP and enhanced fuzzy multilayer perceptron

Methods	# of trials	# of success	# of the nodes of hidden layer	average # of Epoch
EBP	10	4	12	10952
Enhanced Fuzzy Multilayer Perceptron	10	10	15	571

In conclusion, the experiment for performance evaluation shows that the proposed fuzzy multiplayer perceptron has less short learning time and more stable convergence than the previous learning algorithm. The reason is that the adjustment of weights by the winner-take-all method decreases the amount of computation, and adjusting only the related weights decreases the competitive stages as a premature saturation. Therefore, there is less possibility of the paralysis and local minima in the proposed method.

5 Conclusion

To improve the problem of setting the size of node in the hidden layer, we proposed the fuzzy multilayer perceptron which is composed of the ART1 algorithm and the fuzzy neural network. The proposed learning structure applied ART1 to the connection structure between the input layer and the hidden layer and applied the output layer of ART1 to the hidden layer of the proposed structure. Therefore the learning structure is generally fully connected. However when its learning algorithm applied to the winner-take-all method which is to backpropagate only the connection weight which is connected to a representative class. The adjustment of the connection weights from the output layer to the hidden layer is applied with the fuzzy neural network. The proposed algorithm applied the winner-take-all method adjusts the weights. Because the information of the patterns which is effective to the class in the hidden layer can be stored, the paralysis is decreased due to the credit assignment of the hidden layer, and the learning time and convergence of learning are improved. In the experiment for performance evaluation using number patterns, the proposed method is quite robust with respect to minor change in the momentum parameter, takes smaller learning time, and is more convergent than the EBP algorithm. But the number of nodes in the hidden layer is increased or decreased according to the setting-up vigilance parameter. Our future work will be to make improvement on this problem.

References

1. James, A., Freeman, A.: Neural Networks: Algorithm, Application and Programming Techniques. Addison-Wesley (1991)
2. Hirose, Y., K. Yamashita, K., Hijiya, S.: Back-Propagation Algorithm Which Varies the Number of Hidden Units. Neural Networks, Vol.4, (1991) 61-66
3. Kavuri, S. N., Ventatasubramanian, V.: Solving the Hidden Node Problem in Neural Networks with Ellipsoidal Units and Related Issues. Proceedings of IJCNN, Vol. 1, (1992) 775-780
4. Kim, K. B., Kang, M. H., and Cha, E. Y.: A Fuzzy Self-Organized Backpropagation using Nervous System. Proceeding of SMC, Vol.2, (1997) 1457-1462
5. Kim K. B., Kim Y. J.: Recognition of English Calling Cards by Using Enhanced Fuzzy Radial Basis Function Neural Networks. IEICE Trans. Fundamentals, Vol.E87-A, No.6, (2004) 1355-1362
6. Gupta, M. M. and Qi, J.: On Fuzzy Neuron Models. Proceedings of IJCNN, Vol.2, (1991) 431-435
7. Saito, T., and Mukaidono, M.: A Learning algorithm for Max-Min Network and its Application to Solve Relation Equations. Proceedings of IFSA, (1991) 184-187
8. Czogala, E., and Buckley, J. J.: Fuuzy Neural Controller. Proceedings of IEEE Fuzzy Systems, Vol. 1, (1992) 197-202

Intelligent Multi-agent Based Genetic Fuzzy Ensemble Network Intrusion Detection

Siva S. Sivatha Sindhu, P. Ramasubramanian, and A. Kannan

Department of Computer Science and Engineering
Anna University, Chennai 600025, India
sivatha127@yahoo.com, suryarams@cs.annauniv.edu, kannan@annauniv.edu

Abstract. This paper proposes a distributed superior approach for preventing malicious access to corporate information system. It is to identify a foolproof system to obviate the manual analysis and breaches in the networking system by using a distributed approach and a technique of genetic algorithm that automates the generation of fuzzy rules. The experimental study is performed using audit data provided by MIT Lincoln labs. In order to reduce single point of failures in centralized security system, a dynamic distributed system has been designed in which the security management task is distributed across the network using Intelligent Multi-Agents.

1 Introduction

Security has become a major issue in many organizations but most systems still rely on user ID and password systems to provide user authentication and validation. Many other mechanisms and technologies like firewalls, encryption, authentication, vulnerability checking, access control policies can offer security but it is still susceptible for attacks from hackers who takes advantage of system flaws and social engineering tricks. In addition, computer systems with no connection to public networks remain vulnerable to disgruntled employees or other insiders who misuse their privileges. This observation results in the fact that much more emphasis has to be placed on Intrusion Detection Systems(IDSs).

This paper proposes a distributed approach to network security using agents and a genetic algorithm to generate fuzzy rules instead of manual design that is used to check the user's on-line profile to detect abnormal behavior of the user.

2 Related Works

Majority of early IDSs was designed to detect attacks upon a single host [2]. These methods have a central focal point for security which could itself become the focus of an attack [3]. Deploying host agent to protect the security of distributed environment reduces the traffic disposed by the core agent, alleviates the core agent's load, complexity of maintenance work, facilitates the management and enhances the whole system's efficiency. Unfortunately, most security

N.R. Pal et al. (Eds.): ICONIP 2004, LNCS 3316, pp. 983–988, 2004.
© Springer-Verlag Berlin Heidelberg 2004

systems [2][3] in distributed environments are based on passive and static software components. Most software is now written with multiple goals, and the larger the software becomes the more bugs it tends to have [4]. To optimize the distribution of software components in distributed systems, our system is based around Agents. This is because agents can be written so that they only have a single goal, and the code can be designed around that code. Agents are robust and highly adaptable software entities and have minimum impact on the use of computational power and memory usage of the system [6]. Agents are independently running entities; they can be added and removed from a system without altering other components. Thus, there is no need to restart the network intrusion detection system when there is a system change. Furthermore, agents may provide mechanisms for reconfiguring them at run time without even having to restart them. In this paper, we introduce a newly implemented multi-agent architecture, which can be reused, easily extended and can be flexibly adapted to future changes in agent technology or other environments.

3   Architecture

The general architectural framework for an Intelligent Multi-Agent based Distributed Network IDS system is illustrated in Fig. 1. It has been implemented by using Aglets Software Development Kit(ASDK) [5], and API Java Aglet(J-AAPI) developed by IBM Tokyo Research Laboratory. It consists of two main components. They are 1. Core Agent 2. User Agent.

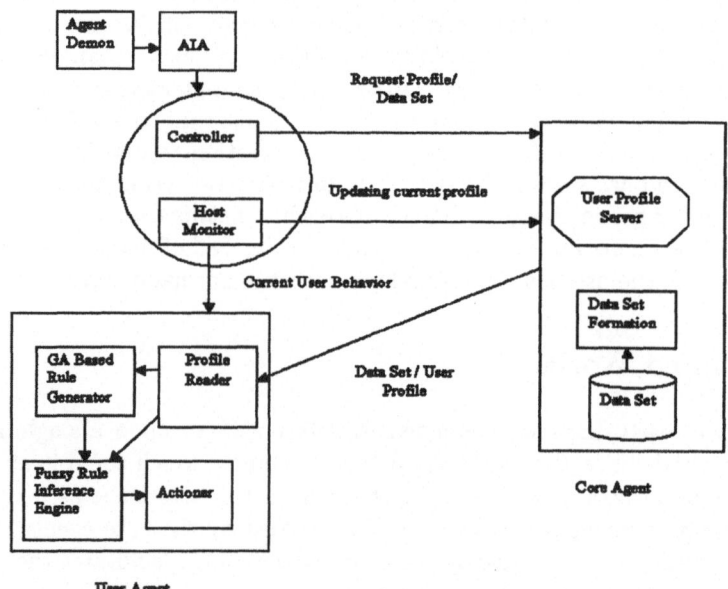

Fig. 1. Intelligent Multi-Agent based Network Intrusion Detection System.

3.1 Core Agent

Core Agent acts as a data processing unit and as a data repository for the User Agents. The Core Agent is responsible for storing user logs for all users that have access to the protected network. It is also responsible for giving the user log to the User agent, each time that a user log is requested by the controller. It consists of 1. User Profile Server 2. Data Set Formation.

- **Host Monitor(Mobile Agent):** In distributed environment, the performance of each host has to be monitored constantly so that performance drop or failure of any node can be detected. Based on that corrective measures can be taken to maintain the overall performance level of the network. When an Information Agent is created, it sends a monitor agent to every host in the network. The monitor agent then starts monitoring the performance as soon as it reaches the host at regular intervals and this interval can be programmed.
- **User Profile Server:** In the User Profile Server the information about each user is maintained and is transmitted to the User Agent upon request by the Controller in the Core Agent.
- **Data Set Formation:** This step involves formatting the data in the proper format that can be used by the ID engine to detect patterns of behavior.

3.2 User Agent

An User Agent resides on every host on the protected distributed environment. This agent is responsible for monitoring each user that logs on to the host and it is started during the login process. User Agents are individual software entities that act autonomously, by monitoring the user behavior. It consists of two components namely 1. Agent Demon 2. AIA (Agent Identification and Authentication) 3. Controller 4. Profile Reader 5. GA based Fuzzy Rule Generation 6. Fuzzy Rule Inference Engine 7. Actioner.

- **Agent Demon:** This static agent is responsible for the input traffic preprocessing. It monitors traffic and extracts User ID.
- **AIA:** This static agent is used for identification and authentication of user and performs their conventional functionalities.
- **Controller:** This mobile agent requests the Core Agent for User Profile and Data Set. Based on the request ID the Core Agent responds with the required information.
- **Profile Reader:** This mobile agent is responsible to read and transmit the user profile to the Rule Inference Engine.
- **GA Based Fuzzy Rule Generation:** It is used to automate the generation of fuzzy rules with the user profile obtained from the core agent.
- **Fuzzy Rule Inference Engine:** This is the decision making component for the IDS. This component decides whether the user behavior is a normal or abnormal one.

- **Actioner:** Actioner's(Static Agent) role is to take necessary actions when an intrusion is detected. When an attack is detected exactly, the Actioner does one of the following operations to terminate the attack: 1. Warn the system administrator 2. Warn the end user 3. Kill the specific application that has caused invalid behavior 4. Prevent the end user from running any further application. Case 2, 3 & 4 can be achieved locally at the client workstation.

4   Experimental Results

4.1   Data Source

The data that has been used in the proposal is the data set prepared and maintained by the MIT Lincoln labs [1]. In this dataset, forty-two attributes that usually characterize network traffic behavior compose each record. In this data set we use four main intrusion classes with samples sizes like 0.01%, 0.23%, 79.5%, 0.83% and 0.59%. The data set also contain 22 different types of attacks within the 4 main intrusion classes. The normal class staying at 19.3%. The test thus conducted using the 10% of Knowledge Discovery and Data Mining(KDD) cup '99 data set proved the proposed approach.

4.2   Feature Selection

Feature selection is used to find features that are most indicative of intrusions and can be used to classify intrusions. Table. 2 lists the features selected for each class.

4.3   Training and Testing Data

In order to conduct an experimental setting, a 20% of each class was chosen and used for testing and the remaining 80% was used for training. The combined proportion of samples from the normal class and the denial of service(DoS) class is almost 99% of the data set, so more number of samples was kept for these two classes in the training data set. Each genetic search was initialized with a random population of individuals and a random number of evolutions. Each individual in the population uses a number of samples from the data set that is proportional to the number of samples of that class present in the dataset. The proportion of samples from each class is given by the proportion of samples of each class in the training data set. Thus a uniform number of samples were considered for each of these classes.

4.4   Experimental Topology

The reported results were obtained using a population size of 200 and a maximum of 20 iterations. The proposed approach was able to generate simple rules. The best solution contained the following rule:

Class	Feature Description
Normal	Hot indicators: Number of "hot" indicators Destination bytes: Number of bytes sent from the destination system to the host system Source bytes: Number of bytes sent from the host system to the destination system Compromised conditions: Number of compromised conditions Dst_host_rerror_rate: % of connections that have REJ errors from a destination host
Probe	Dst_host_diff_srv_rate: % of connections to different services from a destination host Rerror_rate: % of connections that have REJ errors Srv_diff_host_rate: % of connections that have same service to different hosts Logged in: binary decision Service: type of service
Denial of Service (DoS)	Count: Number of connections made to the same host system in a given interval of time Compromised conditions: Number of compromised conditions Wrong_fragments: no of wrong fragments Land: 1 if connection is from/to the same host/port; 0 otherwise Logged in: 1 if successfully logged in; 0 otherwise
User to Root(U2Su)	Root shell: 1 if root shell is obtained; 0 otherwise Dst_host_srv_serror_rate: % of connections to the same service that have SYN errors from a destination host No of file creations: no of file creation operations Serror_rate: % of connections that have SYN errors Dst_host_same_src_port_rate: % of connections to same service ports from a destination host
Root to User (R2L)	Guest login: 1 if the login is a "guest" login; 0 otherwise No of file access: no of operations on access control files Destination bytes: Number of bytes sent from the destination system to the host system Failed logins: no of failed login attempts Logged in: binary decision

Fig. 2. Feature Selection.

- If Src_bytes > 122 And If Dst_bytes < 9926 And If Hot indicators are 0 And If Compromised conditions are 0 And If dst_host_rerror_rate is 0.0 Then it is normal.
- If Count < 10 And If Compromised condition < 10 And If Wrong_Fragment is 0 And If Land is false And If Logged in is true Then it is DoS.
- If diff_srv_rate < 4 And If srv_diff < 5 And If logged is true And If service is < 3 And If rerror_rate is 0.0 Then it is Probe.
- If Rootshell is false And If dst_host_srv_serror_rate is < 7 And If No: of file creations is < 3 And If Serror_rate is 0 And dst_host_same_src_port_rate is 0 Then it is U2S.
- If Guest login is false And If No: of file access is < 10 And If Destination bytes is < 10 And If failed login is 0 And logged in is true Then it is R2L.

Table. 1 summarizes false alarm rate and detection rate.

Table 1. Receiver Operating Characteristic(ROC).

Dataset	Detection Rate	False Positive Alarm	False Negative Alarm
Normal	0.9625	0.0375	-
U2Su	0.8703	-	0.1296
R2L	0.95	-	0.05
DoS	0.916	-	0.08
Probe	0.4736	-	0.526

5   Conclusions and Future Works

In this paper, an intelligent multi-agent based anomaly intrusion prediction system has been implemented in order to detect future intrusions in networks. The significance of the approach is that, the user agent at the client workstation takes all decision and actions over the invalid user and thus reduces the burden of the server. The evolved rules allow characterization of the normal and abnormal behaviors in a simple way and they are not complex as no more than five attributes are used in each rule. Future research in this direction could be the evaluation of network intrusion detection model by training the GA with various other dataset such as DARPA, WINE etc.

References

1. KDD-cup data set, Available at URL
   http://kdd.ics.uci.edu/databases/ kddcup99/kddcup99.html (2004)
2. Michael, C.C., Anup Ghosh.: Simple, State-Based Approaches to Program-Based Anomaly Detection. In ACM Transactions on Information and System Security. **5** (2002) 203–237.
3. Nong Ye, Sean Vilbert and Qiang Chen.: Computer Intrusion Detection Through EWMA for Autocorrelated and Uncorrelated Data. In Proceedings of IEEE Transactions on Reliability. **52** (2003) 75–82.
4. Pikoulas, J., Buchanan, W.J., Manion, M., Triantafyllopoulos, K.: An intelligent agent intrusion system. In Proceedings of the 9th IEEE International Conference and Workshop on the Engineering of Computer Based Systems - ECBS, IEEE Comput. Soc., Luden, Sweden. (2002) 94–102.
5. Java Aglet, IBM Tokyo Research Laboratory, Available at URL
   http://www.trl.ibm.co.jp/aglets (2004)
6. Triantafyllopoulos, K., Pikoulas, J.: Multivariate Bayesian regression applied to the problem of network security. Journal of Forecasting. **21** (2002) 579–594.

Genetic Algorithm Based Fuzzy ID3 Algorithm

Jyh-Yeong Chang, Chien-Wen Cho, Su-Hwang Hsieh, and Shi-Tsung Chen

Department of Electrical and Control Engineering
National Chiao Tung University
1001 Ta Hsueh Road, Hsinchu, Taiwan 300, R.O.C.
jychang@mail.nctu.edu.tw

Abstract. In this paper, we propose a genetic algorithm (GA) based fuzzy ID3 algorithm to construct a fuzzy classification system with both high classification accuracy and compact rule base size. This goal is achieved by two key steps. First, we optimize by GA the parameters controlling the means and variances of fuzzy membership functions and leaf node conditions for tree construction. Second, we prune the rules of the tree constructed by evaluating the effectiveness of the rule, and the remaining rules are retrained by the same GA proposed. Our proposed scheme is tested on various famous data sets, and its results is compared with C4.5 and IRID3. Simulation result shows that our proposed scheme leads to not only better classification accuracy but also smaller size of rule base.

1 Introduction

Decision tree classifiers (DTC's), playing important roles in machine learning field, are successfully and widely used to extract knowledge from existing data in many areas such as radar signal classification, character recognition, remote sensing, medical diagnosis, expert systems, speech recognition, image processing, *etc.* [1]. The most important feature of DTC's is the capability to break down a complex decision-making process into a collection of simpler decisions; thus they provide a solution which is often easier to interpret. Among various DTC's [1], ID3 [2] was proposed by Quinlan and has become one of the most popularly adopted DTC's. It is an efficient method for decision making for classification of symbolic data by adding the crucial idea of selecting featuring using the information gain. However, in the case of dealing with numerical data, ID3 cannot work without further modifications. To overcome this limitation, CART [3] and C4.5 [4] discretize numerical attributes by partitioning and dynamically compute associated thresholds according to the condition along each path; therefore the accuracy is raised up at the cost of loss of comprehensibility. There are two typical methods used to partition continuous attributes: one is to partition the attribute range into intervals using a threshold [5], and the other is to partition the attribute domain into intervals using a set of cut points [6].

Fuzzy decision tree classifier (FDTC), based on fuzzy sets, is another approach to deal with continuous valued attributes. Except for a modified information evaluation, it adopts almost the same steps as what is done in traditional DTC's [7], [8]. To increase comprehensibility and avoid the misclassification due to sudden class change

N.R. Pal et al. (Eds.): ICONIP 2004, LNCS 3316, pp. 989–995, 2004.
© Springer-Verlag Berlin Heidelberg 2004

near the cut points of attributes, FDTC represents attributes with linguistic variables and partitions continuous attributes into several fuzzy sets. Details about techniques for the design of fuzzy decision trees can be found in [7].

Among various fuzzy DTC's, fuzzy ID3, based on selecting best features according to the concept of gain in entropy, is one of the most important fuzzy decision trees [9]. The construction of a fuzzy ID3 consists of three main steps: generating the root node having the set of all data, generating and testing new nodes to see if they are leaf nodes by some criteria, and breaking the non-leaf nodes into branches by best selection of features according to entropy calculation. Pal *et al.* proposed a DTC called RID3 [11], which calculates membership function using the fuzzy c-means with best tuned fitness function of thresholds by GA. Furthermore, Pal and Chakraborty also proposed IRID3 [12], which is an improved version of RID3 with the control of number of nodes (features) by refining the fitness function and pruning process to remove useless nodes.

In this paper, we improve the fuzzy ID3 algorithm proposed by Umano *et al.* [9] in both accuracy and the size of the tree through two key steps. First, we optimize the thresholds of leaf nodes and the mean and variance of fuzzy numbers involved. Second, we prune the rules of the tree by evaluating the effectiveness of the rules, and then the reduced tree is retrained by the same GA proposed. On testing the data sets used, our proposed method are found to be consistently better than C4.5 and IRID3.

2 Fuzzy ID3 Algorithm

A brief introduction of fuzzy ID3 method will be given below. In FID3 algorithm, we assign each data point a unit membership value. To reduce data points, we combine data points which have the same value for all feature, and normalize membership value of all data points. For feature ranking, fuzzy ID3 follows ID3 in selecting the feature based on the maximum information gain which is computed by the probability of ordinary data, but FID3 is evaluated by the membership value of the data point [9].

Assume that we have a training set D, where each data point has l features A_1, A_2, ..., A_l, and n decision classes C_1, C_2, ..., C_n, and m fuzzy sets F_{i1}, F_{i2}, ..., F_{im} for feature A_i. Let D^{C_k} be a fuzzy subset in D whose decision class is C_k, $|D|$ be the cardinality of D, and $|D^{C_k}|$ be the cardinality of D^{C_k}. Here the algorithm to generate a fuzzy decision tree is shown in the following:

1) Generate the root node that has a set of all data points, i.e., a fuzzy set of all data point with the unit membership value.
2) If a node t with a fuzzy set of data D satisfies the following conditions then it is a leaf node and we record the certainty values

$$\frac{\left|D^{C_k}\right|}{|D|} \tag{1}$$

of the node.
   i)   The proportion of a data set of a class C_k, is greater than or equal to a threshold θ_r, that is,

$$\frac{\left|D^{C_k}\right|}{|D|} \geq \theta_r. \tag{2}$$

ii) The number of a data set is smaller than a threshold θ_n, that is,

$$|D| < \theta_n .$$ (3)

iii) There are no attributes for more classifications.

3) If it does not satisfy the above conditions, it is not a terminal node, and the branch node is generated as follows:

i) For feature A_i ($i = 1, 2, ..., l$), calculate the information gains $G(A_i, D)$ [10], and select the branch feature by decreasing $G(A_i, D)$ gradually.

ii) Divide D into fuzzy subsets $D_1, ..., D_m$ according to the result of feature ranking, and calculate membership value of each data point in D_j, which is the product of the membership value of each data point in D and the membership value of corresponding every subset of selected feature A_i.

iii) Generate new node $t_1, t_2 ..., t_m$ for fuzzy subsets $F_{\max j}$ to edges that connect between the nodes t_j and t.

iv) Replace D by D_j ($j=1, 2,...,m$), and repeat from 3 recursively until the destination of all paths are terminal nodes.

3 Fuzzy ID3 Using Genetic Algorithm

From the description above, the structure of FID3 scheme is determined by the thresholds θ_r, θ_n, and the membership functions of the various feature values. A good selection of fuzzy rule base, θ_r, θ_n, and the membership functions are best matched to database to be processed, would greatly improve the accuracy of the decision tree. To this end, any optimization algorithms seem appropriate for this purpose. In particular, GA-based scheme is highly recommended since a gradient computation for conventional optimization approach is usually not feasible for a decision tree. This is because condition-based decision path is nonlinear in nature, and hence its gradient is not defined. With this concept in mind, we will introduce, in this section, genetic algorithm to search for the best θ_r, θ_n, and membership functions of all feature values for the design of FID3. We use GA [13], [14] to tune the threshold θ_r, θ_n, and the parameters of the membership functions of feature values. The membership function for each feature is assumed to be Gaussian-type and is given by

$$m(x) = \exp(-\frac{(x-\mu)^2}{2\sigma^2}) ,$$ (4)

where x is the corresponding feature value of the data point with mean μ and standard deviation σ. Thus for each membership function, we have two parameters μ and σ to tune. Adopted from [6], to minimize the rule number and maximize the accuracy, let the fitness function

$$f = A + \frac{\eta}{R} ,$$ (5)

where A is the accuracy of the classification, R is the total rule number, and η is the influence of the rule number. In the beginning of the tuning process, we set η to a value such that η / R is greater than A. This means that reduction of classification rule number receives a higher priority over improvement of the accuracy. As GA evolves, we gradually continue to decrease the value of η so that the improvement of the accuracy starts dominating. That is, we reduce η to zero in k equal steps. After k steps, η is always zero. In other words, we focus on the improvement of the accuracy after η becomes zero. Thus we can decrease the rule number with the same performance.

3.1 Fuzzy Decision Tree Inference

According to the rule base, inference in the decision tree starts from the root node and iteratively tests each node indicated by the rule until reach at a leaf node. Note that we have recorded the membership value $|D^{C_k}|$ of the leaf node as mentioned above and it represents the certainty of each class of the corresponding rule.

Since we get the $|D^{C_k}|$ value of each leaf node, the node is assigned by all class names with $|D^{C_k}|$. On the other hand, every leaf node has all class with certainty. With the leaf node, the rule produced can classify the data with certainty of all class but not directly classify the data to a specific class. For example, the rule firing strength, membership degree, from this node of a data point looks like as follows:

If x_1 is F_{12} And x_2 is F_{23} ...
Then Class 1 with certainty 0.7 and Class 2 with certainty 0.2 ...

Here x_1 is the first feature value of the data point and F_{12} is the fuzzy set whose membership function is defined by the μ and σ associated with node 2 of the first feature of the tree. 0.7 and 0.2 are the certainties of the class 1 and class 2, respectively. The steps of using the rule base to classify a data are shown in the following:

1) According to the rule, we multiply the membership value of the test data of the corresponding subset from the root to the leaf node sequentially. That is the firing strength

$$\prod_{i=1}^{l} m_i(x).$$ (6)

2) Multiply the final production of the membership value in step 1) and the certainty of the class of the leaf node, then we get $J(n)$ of this rule ($n=1, 2, ..., $ class number). Here we use the normalized certainty of the class.
3) Repeat step 1) and 2) until that all rules have classified the test data.
4) Sum up class membership, firing strength, of all the rules.
5) The test data is assigned to the class that has the maximum value in step 4).

3.2 Pruning the Rule Base

We have used the GA to improve the performance of classification task and decrease the rule number as well. Here we propose a rule pruning method to further minimize the number of rules as follows:

1) For each rule, when any data point is classified, we maintain the production value of the membership value and the certainty of each class, $J(n)$.
2) $J(n)$ corresponding to the correct class of the data point gets positive sign and the others get negative sign.
3) Sum $J(1), J(2),...$ for all classes of $J(n)$ then we get the credit of the rule to classify this data point.

4) Repeat from 1) until all data points are classified by this rule and we get the final credit of this rule.

5) Remove the rules whose final credits are less than certain threshold and/or have big drops.

The final credit of each rule computed above represents the effectiveness of the rule in performing the classification task. If the rule is essential in classification then it would get high credit value. On the contrary, if the credit is small, for example, less than zero, this rule could be an insignificant or redundant rule. We will describe the reason why as follows.

The rule that classifies the data to its true class or classify the data to the wrong class will be cumulatively counted. In this way, we can prune the insignificant or inconsistent rules to obtain a smaller and efficient rule base set. After finishing rule pruning, we retune the parameters again by GA in the pruned rule-base constructs.

As mentioned in 5), we prune the rule that the rule credit is less than some threshold. Here we propose a simple method to select the threshold. For a typical example for instance, first we get the cumulative rule credits of all the rules in the rule base. We sort and plot the rule credit values of the rules we have generated as shown in Fig. 1. We find that around the rule number 32, the slope of the credit curve takes a visible drop and then up to 36 it pulls down rapidly. It means that there is a credit gap after rule 32 and this rule may be insignificant or redundant. Hence we can select the credit values of rule 32 as the threshold to prune the rule base.

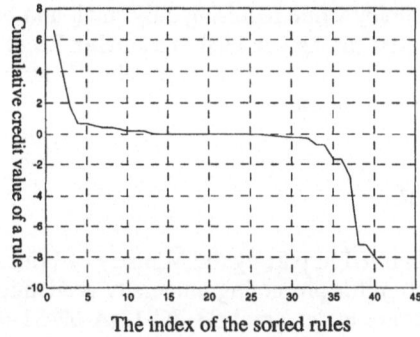

The index of the sorted rules

Fig. 1. The cumulative credit values of sorted rules

4 Experiment

Here five well known data sets, including Iris [15], Crude oil [16], Norm4 [10], Myo electric [17], and Glass [18], are used for testing. The performance comparison between the proposed GA based FID3, IRID3 [12], and C4.5 [4] is shown is Table 1. For Iris data set, the classification accuracy is better than those of the other two algorithms. The number of rules of GA based FID3 is fewer than C4.5 by one. For Crude oil data set, the accuracy and rule numbers are equal to C4.5; however, they are better than those of IRID3. For Norm4 data set, IRID3 and C4.5 have similar performance, whereas GA based FID3 achieves a better 95.1% accuracy and requires 12 rules which is the same as IRID3 and C4.5. In addition, for Myo data set, the GA based

FID3 also outperforms in the size of rules. Only two rules is needed in our proposed method to achieve the same accuracy of IRID3 and C4.5; however, they need five and six rules, respectively. Finally, for Glass data set, it is evident that GA based FID3 still demonstrates the best performance not only in the accuracy but also in the size of rules required.

Table 1. Accuracy and Size of Rules Comparison

Data set	GA+FID3		IRID3		C4.5	
	Accuracy	Size	Accuracy	Size	Accuracy	Size
Iris	98.6	4	98.0	4	98.0	5
Crude oil	92.9	5	91.1	7	92.9	5
Norm4	95.1	12	94.8	12	94.6	12
Myo	98.6	2	98.6	5	98.6	6
Glass	76.2	12	65.4	30	72.9	13

5 Conclusion

In this paper, we proposed a genetic algorithm based fuzzy ID3 algorithm to construct a fuzzy classification system with high classification accuracy. Our rule extraction scheme is quite powerful. We tried to optimize the rule parameters in the tuning process by genetic algorithm. In addition, we formulated a pruning method to obtain a more efficient rule base. On testing to some famous data sets, we have obtained very high classification accuracy while requiring only small number of rules. It is remarked that the decision tree after tuning can lead to a smaller fuzzy rule base and the pruned rule base can usually retain or even improve the classification performance despite the reduction of the number of the rules.

Acknowledgement

This research was supported in part by the Ministry of Education under grant EX-91-E-FA06-4-4, the program for promoting university academic excellence, by the Ministry of Economic Affairs under grant 93-EC-17-A-02-S1-032, and by the National Science Council under Grant NSC 92-2213-E-009-112, Taiwan, R.O.C.

References

1. Safavian, S. R., Landgrebe, D.: A survey of decision tree classifier methodology. IEEE Trans. Syst., Man, Cybern., Vol. 21. (1991) 660–674
2. Quinlan, J. R.: Induction on decision trees. Machine Learning, Vol. 1. (1986) 81–106
3. Breiman, L. et al.: Classification and Regression Trees. Monterey, CA: Wadsworth and Brooks/Cole (1984)
4. Quinlan, J. R.: C4.5, Programs for Machine Learning. San Mateo, CA: Morgan Kauffman (1993)
5. Fayyad, U. M., Keki, B. I.: On the handing of continuous-valued attributes in decision tree generation. Machine Learning. (1992) 87–102

6. Fayyad, U. M., Keki, B. I.: Multi-interval discretization of continuous valued attributes for classification learning. IJCAI-93. (1993) 1022–1027
7. Janikow, C. Z.: Fuzzy decision trees: issues and methods. IEEE Trans. Syst., Man, Cybern. B, Vol. 28. (1998) 1-14
8. Peng, Y. H., P. A. Flach, P. A.: Soft discretization to enhance the continuous decision tree induction. IDDM-2001. Germany (2001)
9. Ichihashi, H. et al., Neural fuzzy ID3: A method of inducing fuzzy decision trees with linear programming for maximizing entropy and algebraic method. Fuzzy Sets Syst., Vol. 81. No.1. (1996) 157–167
10. Pal, N. R., Bezdek, J. C.: On cluster validity for the fuzzy c-means model. IEEE Trans. Fuzzy Syst., Vol. 3. (1995) 370-379
11. Pal, N. R., Chakraborty, S., Bagchi, A.: RID3, an ID3-like algorithm for real data. Inf. Sci., Vol. 96. (1997) 271–290
12. Pal, N. R., Chakraborty, S.: Fuzzy rule extraction from ID3-type decision trees for real data. IEEE Trans. Syst., Man, Cybern. B, Vol. 31. (2001) 745-754
13. Lin, C. T., Lee, C. S. G.: Neural Fuzzy Systems: A Neural-Fuzzy Synergism to Intelligent Systems. Prentice-Hall,Upper Saddle River, New Jersey (1996)
14. Goldberg D.: Genetic Algorithms in search optimization and machine learning. Addison-Wesley, Reading, MA (1989)
15. Anderson, E.: The irises of the gaspe peninsula. Bull. Amer. IRIS Soc., Vol. 59. (1935) 2-5
16. Gerrid, P. M., Lantz, R. J.: Chemical analysis of 75 crude oil samples from Pliocene sand units. Elk oil fields, California. U.S. Geologic. Surv. Open-File Report (1969)
17. Mehrota, K. et al.: Elements of Artificial Neural Networks. MIT Press, Cambridge, MA (1996)
18. Holte, R. C.: Very simple classification rules perform well on most commonly used data set. Mach. Learn., Vol. 11. (1993) 63-91

Neural-Evolutionary Learning in a Bounded Rationality Scenario

Ricardo Matsumura de Araújo and Luís C. Lamb

Institute of Informatics
Federal University of Rio Grande do Sul
Porto Alegre, 91501-970, Brazil
{rmaraujo,lamb}@inf.ufrgs.br

Abstract. This paper presents a neural-evolutionary framework for the simulation of market models in a bounded rationality scenario. Each agent involved in the scenario make use of a population of neural networks in order to make a decision, while inductive learning is performed by means of an evolutionary algorithm. We show that good convergence to the game-theoretic equilibrium is reached within certain parameters.

1 Introduction

Classical economics makes the assumption that economic agents involved in any system are perfectly rational. This usually means that each agent has knowledge of all relevant aspects of the environment, a logical and coherent preference, and enough computational power to process all this information in order to choose the best course of action to attain the highest, optimal point in his or her preference scale [13]. This assumption facilitates the use of analysis tools within game theory to predict the outcome of interactions of multiple agents. However, it is not the case that real agents behave in a perfect rational sense: they are usually endowed with *bounded rationality* [12].

This term, coined by Herbert Simon in the 1950s, refers to an approach now widely used for modeling reasoning in economic scenarios. Systems endowed with bounded rational agents might offer quite different behavior from ones with rational agents. Some systems, on the other hand, seem to allow analysis *as if* agents were rational [11], even though this may not be the case. Such systems are of particular interest, since they allow the use of traditional, tractable techniques for behavioral analysis [6].

Arthur [1] has proposed *the El Farol problem* in order to provide insights in systems of interacting bounded rational agents in a simplified market model. Since then, this model has been widely discussed together with other evolutionary game scenarios, such as the Minority Game [16]. The El Farol problem is as follows. *There are N agents; each agent has to decide whether or not to go to the El Farol Bar at some week; an agent will go if he or she expects that at most aN agents are going, where $a \in [0, 1]$; otherwise the bar would be overcrowded and the agent will not go. The only source of information available to the agents is a global history of past weeks attendance, and no explicit communication is allowed between the agents[1].* The interest in modeling this

---

[1] Some papers report experiments with explicit communication, see [14] for an example.

N.R. Pal et al. (Eds.): ICONIP 2004, LNCS 3316, pp. 996–1001, 2004.
© Springer-Verlag Berlin Heidelberg 2004

kind of problem is the assumption that the utility gain of each agent depends directly on the decision of all other agents, leading to an interesting paradox: if many agents believe that El Farol will be overcrowded, few will go; if they all believe nobody will go, all will go.

In [1], an experiment was set by allowing each agent to choose over an internal set of fixed strategies in order to predict the next week attendance. Each strategy used only a history window to make the prediction (e.g. the same as the previous week, the average over 5 last weeks, a fixed number etc.) and the most accurate predictor was chosen at each time step. With a total of 100 agents and setting the bar to be overcrowded if 60 or more agents attended ($a = 0.6$), the simulations showed a convergence to a mean attendance of 60. Game theory tells us that a Nash equilibrium exists using mixed-strategy and 60 is the expected mean attendance [1], the same results observed in Arthur's simulations.

In response to this experiment, [6] suggested that if agents could be *creative*, in the sense that they could come up with *new* strategies other than those pre-specified (as was the case in Arthur's simulations), the system would not show a convergence, but rather it would behave in a more chaotic manner, thus showing that game-theoretic expectation would be of little use to predict the behavior of such systems. Learning was conducted using an evolutionary paradigm, where each strategy is represented by an autoregressive equation, as a function of past weeks attendances, which parameters were mutated to give birth to new, possibly better, generations of predictors. In the present paper, we refer to "dynamic learning" as the model of learning that allows the creation of new models, in contrast to "static learning" where learning is only made on choices over a fixed set of pre-specified models. Even though in [6] dynamic learning is explicitly included in the model, the predictions made are only able to capture linear patterns, which is a rather strong assumption on the system.

We believe that it is plausible that the study of other classes of machine learning algorithms applied to the same problem may show a different behavior, possibly closer to the real world market behavior that the problem tries to model. In this paper we present empirical results from the simulation of the El Farol problem using agents capable of dynamic learning through a population of neural networks, evolved by an evolutionary algorithm. By doing this, we aim to further understand the role of learning in the dynamics of economic scenarios and multi-agent systems in general. We then show that the system shows a better convergence to the game-theoretic equilibrium when compared to the proposed setup in [6].

Section 2 provides the architecture used to model the agents; Section 3 presents detailed results of the experiments; Section 4 the concludes the paper and discusses directions of future work.

2 The Hybrid Agent Architecture

An agent is defined as a system that receives a vector A of length M, representing a history window of past weeks attendance, and outputs a single bit of information, namely "0" if it will not attend to the bar in the current week and "1" if it will. The vector A is here considered external to the agent since it is meant to be a perfect information available to every agent. We do not consider the case where agents have different perceptions on this information. Next, we describe the composition of an agent.

In multi-agent decision problems it is common the use of the concept of "mental models" [1, 4, 3], where each agent has a population of predictors and one is chosen to make the decision at each simulation step. This concept is used in this experiment as well. Each agent is internally equipped with a population of K models, represented as neural networks, chosen due to its ability to capture non-linear patterns [7]. Neural networks have been widely used in economic modeling (see e.g. [8]). In [9] neural networks were applied to the Minority Game, a variant of the El Farol Problem, but agents were composed of a single neural network, differing from the approach taken here, and [2] has presented an application of genetic learning over neural networks within an alternative economic context.

We use multi-layer perceptrons (MLP) [7] composed of an input layer, one hidden layer and one single output unit. The input layer is composed of M input units, each receiving one unique value from $\mathcal{A}$. This way the number of input units effectively represents the agent's memory size. The hidden layer is composed of H units. The value of H roughly controls the capacity of the neural network to process the information received [15] and is the same to every neural network in the system. The ouput of the output unit is taken as a prediction of attendance based on the input. All nodes in a layer are fully connected to the next layer and use a sigmoidal activation function. Fig. 1 shows the topology adopted.

All networks have their prediction accuracy evaluated, at each simulation step, through a *fitness function* and the best performing network is chosen by each agent to make its current week's prediction. The agent's decision is then: Output "0" if predicted attendance is greater than aN; Output "1" otherwise.

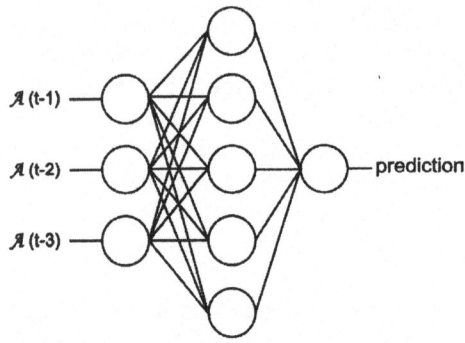

Fig. 1. Example topology for $M = 3, H = 5$.

On top of the population of predictors runs an evolutionary algorithm. The model of learning used here closely follows the one described by Fogel in [5]. For every agent, the algorithm generates one offspring for each neural network by adding a zero-mean random gaussian value $\sigma \in [-1, 1]$ to each weight i:

$$w^i_{offspring} = w^i_{parent} + \sigma$$

The operation results in a total of $2K$ neural networks. All of them are evaluated with the fitness function and the best K replace the current population. Through this proceeding, it is ensured that new strategies are created all the time and put in competition with previous ones, generating possibly better ones.

3   Simulation Setup and Results

The simulations described here were made using the following parameters: $N = 100$, $K = 10$, $M = 12$, $H = 50$, $a = 0.6$. Except for H, which was chosen for having presented good results in trials, all other parameters follow from those used in [1] and [6]. The fitness function of the system was taken as being the sum of squared errors of tests through 10 different history windows.

Figure 2 (a) shows a typical attendance over 500 weeks *without* the evolutionary learning applied. Thus, each agent can only choose among the fixed set of models created at the start of the simulation. It is interesting to note that randomly initialized neural networks do not contain any explicit strategies as was the case in Arthur's model. Despite this, the system's behavior is very close to that of [1], showing a mean convergence near Nash equilibrium of 60 with minor fluctuations, with a mean of 59.02 and standard deviation of 4.13. These values were calculated in the "steady-state" region, where the transient fluctuations due the random initialization are surpassed, which we take as starting around week 50.

By allowing the agents to learn using the evolutionary algorithm we reach the typical results depicted in Fig. 2 (b). Despite a small increase to 4.25 in standard deviation, convergence is still observed. The mean attendance, after the transient stage, is 58.98. We show that agents are effectively learning in Fig. 3, where the average fitness of the agents is shown over 500 weeks of a typical simulation.

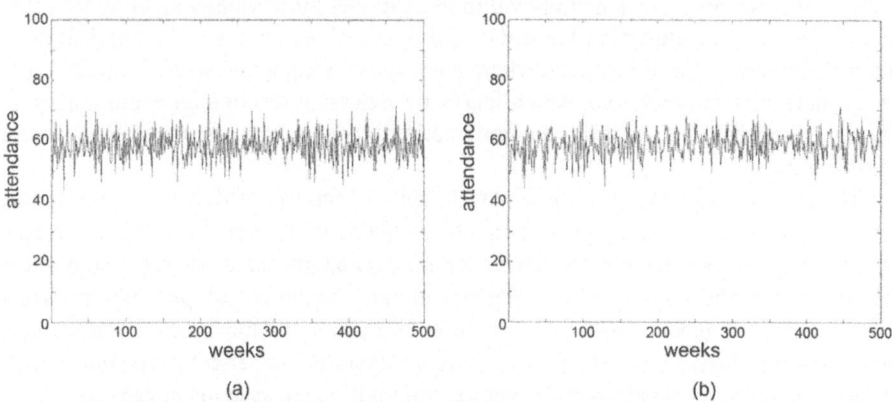

Fig. 2. Typical weekly attendance: (a) without evolutionary learning; (b) with evolutionary learning.

Even though the results with static learning are very similar to Arthur's in [1], the results with the evolutionary (dynamic) learning applied differ qualitatively from those presented in [6]. The mean attendance gets closer to the game-theoretic expectation and

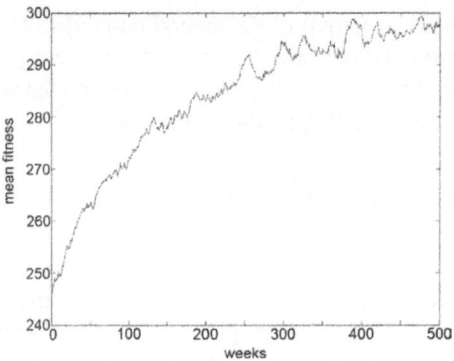

Fig. 3. Mean fitness averaged over all 100 agents at each week.

standard deviation in a typical trial is much smaller. In fact, although dynamic learning is clearly taking place, the system presents almost identical behaviour compared to the case where only static learning is acting. Thus, the overall behavior is predicted in both cases, to some extent, by classical game-theory. This better converging behavior might be explained by the use of neural networks, which are known to be able to capture non-linear patterns, in contrast to the linear predictors used in [6]. It is known that the behaviour of this kind of evolutionary game is very dependent on the memory size of the agents [10], being important to notice that this comparisons were made using the same memory size.

4   Conclusions and Future Work

In this paper we proposed a neural-evolutionary model for learning in a bounded rationality scenario. We illustrated the use of our approach by means of an application to the well-known El Farol Problem, where each agent is equipped with a population of neural networks as predictors, which learns by induction through an evolutionary algorithm. Empirical results showed a good convergence to the game-theoretic expected equilibrium.

The results presented here show that, in spite of the use of dynamic learning, the problem does not necessarily present chaotic behavior as suggested in [6]. This corroborates the hypothesis that the underlying learning paradigm seems to play a substantial role in the rate and stability of convergence in our case study and, possibly, in evolutionary games in general. Although we make no claims that the behavior shown here represents real market behavior, these results are important to better understand the role of learning in bounded rationality scenarios and multi-agent systems in general.

Our experiments have also suggested that the stability (mean and standard deviation) is highly influenced by the *complexity* of the neural networks i.e. simulations using networks of different sizes (number of nodes and layers) presented qualitatively different behaviors. As future work we plan to study such variations by analysing the role of complexity and computational power in the emergent properties of evolutionary games.

Acknowledgements

This work has been partly supported by CNPq and FAPERGS.

References

1. W. Brian Arthur. Inductive reasoning and bounded rationality. *American Economic Review (Papers and Proceedings)*, 84:406–411, 1994.
2. Gianluca Baldassarre. Neural networks and genetic algorithms for the simulation models of bounded rationality theory - an application to oligopolistic markets. *Rivista di Politica Economica*, 12:107–146, 1997.
3. Bruce Edmonds. Modelling bounded rationality in agent-based simulations using the evolution of mental models. In T. Brenner, editor, *Computational Techniques for Modelling Learning in Economics*, pages 305–332. Kluwer, 1999.
4. Bruce Edmonds and Scott Moss. Modelling bounded rationality using evolutionary techniques. *Proceedings AISB'97 workshop on Evolutionary Computation*, pages 31–42, 1997.
5. David Fogel. *Evolutionary Computation: Toward a New Philosophy of Machine Intelligence*. IEEE Press, 2a. edition, 2000.
6. David Fogel, Kumar Chellapilla, and Peter Angeline. Inductive reasoning and bounded rationality reconsidered. *IEEE Transactions on Evolutionary Computation*, 3(2):142–146, July 1999.
7. Simon Haykin. *Neural Networks: A Comprehensive Foundation*. Prentice Hall, 2nd edition, 1998.
8. Ralf Herbrich, Max Keilbach, Thore Graepel, Peter Bollmann-Sdorra, and Klaus Obermayer. Neural networks in economics. *Advances in Computational Economics*, (11):169–196, 1999.
9. W. Kinzel, R. Metzler, and I. Kanter. Dynamics of interacting neural networks. *Physica A*, 33(14):141–147, April 2000.
10. Esteban Moro. The minority game: an introductory guide. In Elka Korutcheva and Rodolfo Cuerno, editors, *Advances in Condensed Matter and Statistical Physics*. Nova Science Publishers, Inc., 2004.
11. Scott Moss. Boundedly versus procedurally rational expectations. In H. Hallet and P. McAdam, editors, *New Directions in Macro Economic Modelling*. Kluwer, 1997.
12. Ariel Rubinstein. *Modeling Bounded Rationality*. Zeuthen Lecture Book Series. The MIT Press, Cambridge, Massachussets, 1998.
13. Herbert Simon. A behavioral model of rational choice. *The Quarterly Journal of Economics*, LXIX, February 1955.
14. Frantisek Slanina. Social organization in the minority game model. *Physica A*, (286):367–376, 2000.
15. Joseph Wakeling and Per Bak. Intelligent systems in the context of surrounding environment. *Physica Review E*, 64(051920), October 2001.
16. Yi-Cheng Zhang. Modeling market mechanism with evolutionary games. *Europhysics News*, March/April 1998.

Rule Extraction Framework
Using Rough Sets and Neural Networks

Yi Xu and Narendra S. Chaudhari

School of Computer Engineering, Block N4-02a-32, Nanyang Avenue,
Nanyang Technological University, Singapore, 639798
xuyi@pmail.ntu.edu.sg, asnarendra@ntu.edu.sg

Abstract. This paper deals with the simplification of classification rules for data mining using rough sets theory combined with neural networks. In the attribute reduction process, the proposed approach generates minimal reduct and minimum number of rules with high accuracy. Experimental results with sample data sets in UCI repository show that this method gives a good performance in getting concise and accurate rules.

1 Introduction

With the increasing of the amount of data stored in various forms, the difficulties in searching useful information from such a large volume of data also grow. Many researchers have proposed different approaches to extract the meaningful knowledge from a lot of data.

One issue in data mining is to classify the data into groups for better understanding. However, the result of data mining should be explicit and understandable and the classification rules be short and clear.

Rough sets theory has been used to acquire some sets of attributes for classification [7]. It offers opportunities to discover useful information in training examples. Several works and extensions of rough sets have been proposed.

Yasdi [1] uses rough sets for the design of knowledge-based networks in the rough-neuro framework. This methodology consists of generating rules from training examples by using rough sets concepts and mapping them into a single layer of connection weights of a four-layered neural network. In this rough-neuro framework, rough sets were used to speed up or simplify the process of using neural network.

A hybrid intelligent system, which is proposed by B. S. Ahn et al. [3] combines rough set approach with neural network to predict the failure of firms based on the past financial performance data. When a new object is predicted by rule set, it is fed into the neural network if it does not match any of the rules. Thus they use rough sets as a preprocessing tool for neural networks. Though this approach can get high classification accuracy, some knowledge in neural networks is still hidden and not comprehensible for user.

Pabitra Mitra et al. [2] designed a hybrid decision support system for detecting the different stages of cervical cancer. This system includes the evolution

N.R. Pal et al. (Eds.): ICONIP 2004, LNCS 3316, pp. 1002–1007, 2004.
© Springer-Verlag Berlin Heidelberg 2004

of knowledge-based subnetwork modules with genetic algorithms using rough sets theory and the ID3 algorithm. The performance of this methodology with modular network is superior in terms of classification score, training time, and network sparseness. S. K. Pal et al. [4] proposed other methodology for evolving a rough-fuzzy multi-layer perceptron with modular concept using a genetic algorithm to obtain a structured network suitable for both classification and rule extraction.

Most of these previous works use rough sets as a preprocessing tool of neural classifier. As rough sets provide useful technology to reduce redundant attributes from information systems, it is mostly applied as a tool as feature selector. However, the accuracy is low because the classification region defined by rough sets is relatively simple. Thus it is natural to combine these two methods for their complementary features. By removing the noise attribute by neural network, rough sets can accelerate the training time and improve its classification accuracy.

Another problem for rough sets is that it can generate more than one reduct during the process of attribute reduction. Though some algorithms [10] are proposed for attribute reduction, it is usually difficult to achieve optimal attribute reduct and generate an efficient rule set for large-scaled data.

Our approach is to obtain the minimal attribute reduct based on information gain. In this paper, we present a simple model of rough sets and backpropagation neural network to extract the minimal rule sets from data. The paper is organized as follows: Section 2 gives basic concepts of rough sets. Section 3 formally describes our model for rule extraction. Some experiment results are shown and analyzed in section 4. Finally concluding remarks are included in section 5.

2   Preliminaries

Rough sets theory was introduced by Pawlak in 1982 as a mathematical approach to deal with vagueness and uncertainty of information [5]. Reduction of knowledge by rough set is used in inducing decision rules according to a specific class [6]. The reduction of knowledge eliminates condition attributes without affecting a decision attribute. After analyzing the relation of attribute and removing dispensable attribute, it induces a set of minimal rules.

2.1   Indiscernibility

An information system (IS) consists of a 4-tuple as follows: $S =< U, Q, V, f >$, where U is the universe consisting of a finite set of n objects $\{x_1, x_2, ..., x_n\}$, Q is a finite set of attributes, $V = \bigcup_{q \in Q} V_q$, where V_q is a domain of the attribute q, and $f : U \times Q \rightarrow V$ is the total decision function called the information function.

Some objects (let them be x and y, where $x, y \in U$) in S cannot be distinguished in terms of a set of attributes (say, a set A, where $A \subseteq Q$). For a given subset of attributes $A \subseteq Q$, a binary relation $IND(A)$ is defined as follows: $IND(A) = \{(x, y) \in U : \text{for all } a \in A, f(x, a) = f(y, a)\}$. This binary relation $IND(A)$ is an equivalence relation on the set U. We say that the objects x, y are "indiscernable" (by a set of attributes A) iff $(x, y) \in IND(A)$.

2.2   Approximations

Two basic approximations used in rough sets theory are defined as follows:

$$A_*(X) = \bigcup_{x \in U} \{A(x) : A(x) \subseteq X\}, \ A^*(X) = \bigcup_{x \in U} \{A(x) : A(x) \cap X \neq \phi\}$$

They are called A-lower and A-upper approximation of X, respectively. The set $AN_A(X) = A^*(X) - A_*(X)$ is referred to as the A-boundary region of X. If the boundary region of X is empty set, X is crisp with respect to A; otherwise, if $AN_A(X) = \phi$, the set X is referred to as rough with respect to A.

2.3   Reduct and Core

Some attributes in IS may be redundant and can be eliminated without the loss of information. Core and reduct are fundamental rough sets concepts, which are used for knowledge reduction. A reduct is the essential part of an IS, which captures the information represented by "discernable" set of attributes. A core is a common part of all reducts. After attribute reduction, redundant attributes, called superfluous attributes are removed. An attribute c_i is identified as superfluous attribute iff $IND(A - c_i) = IND(A)$, where A is a set of attributes.

3   Rule Extraction Framework with Rough Sets

As described in Fig. 1, our rule extraction framework consists of four major steps:

1. Data is preprocessed by entropy-based discretization.
   Before proceeding on data sets, the raw data must be preprocessed in order to deal with continuous variables. We apply minimal entropy partitioning to discretize continuous attributes [8]. We use the class information entropy of candidate partitions to select bin boundaries for discretization. As a result, the training data are stored into the decision tables.

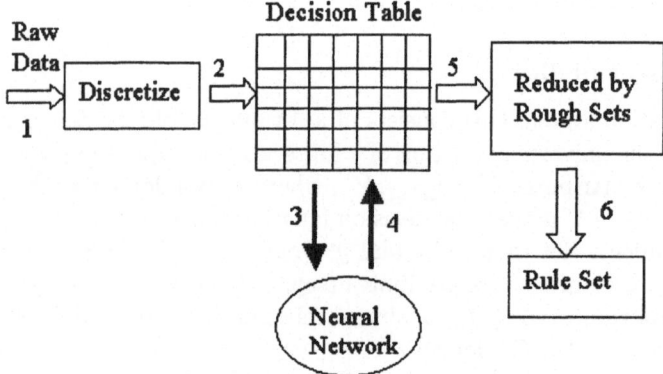

Fig. 1. Framework of rule extraction.

2. Reduced data is fed into neural network.

   In this phase, we employ backpropogation (BP) network to do further attribute reduction. The noisy input attributes are removed after the learning. This approach enhances the classification ability of rough sets. Making use of the robustness to noise and generalization ability of BP network, the noisy attributes can be removed from the decision table.

3. Apply rough sets to get optimal reduct.

   Rough sets can find the relationship for the attributes for different classes. There may exist more than one reduct of condition attributes for a decision table. However, these reducts may require more memory, which make the performance not practical. Commonly, the optimal reduct with the fewest attributes can help to generate more efficient rule sets. So we need to find an optimal reduct with minimal number of attributes. We assume the decision table is the only information source.

   First we find a set of attribute reducts using judgment of attribute (RJ) algorithm in [9]. Next we find the minimal reduct using information gain. we can express value of information of a set of objects S as entropy as follows:

$$E(S) = -\sum_{i=1}^{n} P_i log_n P_i \tag{1}$$

   where n is the number of objects S. S_i is a set of objects, which belong to the ith class value. $P_i = |S_i|/|S|$, is the ratio of the number of elements in set S_i divided by the total number of elements $S(S_i \subseteq S)$. When attribute A_i has m distinct attribute values, the information of attribute A_i can be defined as follows:

$$E(A_i) = \sum_{i=1}^{m} W_i * E(S_i) \tag{2}$$

   Note that S_i is a set of examples of low level, and W_i is defined as:

$$W_i = \frac{number\ of\ examples\ in\ S_i}{number\ of\ examples\ in\ S} \tag{3}$$

   Thus the information gain of A_i is obtained as follows:

$$gain(A_i) = 1 - E(A_i)/E(S) \tag{4}$$

   We assume that the information gain of each reduct can be obtained by adding the information gain of each attribute in the reduct. This assumption, although simplistic, gives good results for rule extraction for practical data sets.

   The attribute with the highest information gain is selected as a "useful" candidate in our reduct. Based on this heuristic approach, our algorithm of attribute reduction module (called ARM) is described as follows:

   Input: a decision table $T = <U, Q, V, f>$.

   Output: Our reduct (ORed) of T.

    1: Generate a set of attribute reducts using RJ algorithm on T and save them to reduct pool

    2: Compute the information gain for all the attributes

    3: Initialize the information gain of ORed, denoted as R_m is 0

    4: **while** reduct pool is **NOT** empty **do**

    5:    Select one reduct from the reduct pool

    6:    Compute the information gain for this reduct, denoted as R_i

    7:    **if** $R_i > R_m$ **then**

    8:      Set this reduct as the member of ORed

    9:      Set $R_m = R_i$

    10:    **end if**

    11: **end while**

    12: Return the ORed

4. Rule generation.

A reduced table can be seen as a rule set, where each rule corresponds to one object of the table. After the removal of the redundant condition attributes, the rules are derived from the objects.

4   Experiments

We use algorithm ARM explained in Section 3 to extract rules from some real-life data sets available in the UCI repository. These rules are compared with rough set (RS) based approach in [11]. We randomly separate each data set to two parts: two thirds as training set and the rest one third as testing set. The continuous data are initially discretized using the entropy-based partitioning method. We also tested 20 times for each case and the average results are presented in Table 1.

Table 1. Comparison of Test Results.

Datasets	Datasets Sizes	Algorithm	Avg. Selected Attributes	Avg. No. of Rules	Avg. Accuracy (%)	Uncovered Region (%)
breast	683	ARM	1.5	6.2	96.4	2.89
		RS	1.6	7.8	92.3	3.10
diabetes	768	ARM	1.2	4	86.1	2.61
		RS	1.5	6	73.2	2.93
glass	214	ARM	1.9	20.4	76.2	4.20
		RS	2.2	24.5	60.4	5.24
iris	150	ARM	1.2	2.44	96.0	1.91
		RS	1.29	3.55	95.1	2.05

The results indicate that the accuracy of rules of our approach is better than that of the RS algorithm. As we generate rules from the optimal reduct, rules are less in number using our approach than the rough sets method.

5 Conclusions

We have discussed a combination of machine learning techniques, namely rough sets and neural networks for rule extraction. An approach to generate the reduct from decision table, that we have presented in this paper, can effectively reduce the number of rules and promise the cover region of rules. This is in contrast with previous methods to generate all reducts. Experiments show our approach can generate more concise and more accurate rules when compared with rough set based method.

References

1. R. Yasdi: Combining Rough Sets Learning and Neural Learning Method to Deal with Uncertain and Imprecise Information. *Neurocomputation* **7** (1995) 61–84
2. Pabitra Mitra, Sankar K. Pal: Staging of Cervical Cancer with Soft Computing. *IEEE Trans. on Biomedical Engineering* **47** (2000) No. 7, 934-940
3. B. S. Ahn, S. S. Cho,C. Y. Kim: The Integrated Methodology of Rough Set Theory and Artificial Neural Network for Business Failure Prediction. *Expert Systems with Applications* **18** (2000) 65–74
4. S. K. Pal, S. Mitra, and P. Mitra: Rough-Fuzzy MLP: Modular Evolution, Rule Generation, and Evaluation. *IEEE Trans. on Knowledge and Data Engineering* **15** (2003) No. 1 14-25
5. Z. Pawlak, J. Grzymala-Busse, R. Slowinski,W. Ziarko: Rough Sets. *Communications of the ACM* **38** (1995) No.11 88–95
6. X. Hu, N. Cercone, W. Ziarko: Generation of Multiple Knowledge from Databases Based on Rough Sets Theory. Rough Sets and Data Mining, Kluwer (1997) 109-121
7. Andrew Kusiak: Rough Set Theory: a Data Mining Tool for Semiconductor Manufacturing. *IEEE Trans. on Electronics Packaging Manufacturing* **24** (2001) No.1 44-50
8. U. M. Fayyad and K. B. Irani: Multi-interval Discretization of Continuous-valued Attributes for Classification Learning. *Proc. of the 13th International Joint Conference on Artificial Intelligence*, Morgan Kaufmann (1993) 1022-1027
9. Dan Pan, Qi-Lun Zheng, An Zeng, and Jin-Song Hu: A Novel Self-Optimizing Approach for Knowledge Acquisition. *IEEE Trans. on Systems, Man. and Cybernetics-Part A: Systems and Humans* **32** (2002) No. 505-514
10. J. Starzyk, D. Nelson and K. Sturtz: A Mathematical Foundation for Improved Reduct Generation in Information Systems. *Knowledge and Information Systems* **2** (2000), 131-146
11. X. Chen, S. Zhu, and Y. Ji: Entropy Based Uncertainty Measures for Classfication Rules with Inconsistency Tolerance. *Proc. of the IEEE International Conference on Systems, Man and Cybernetics* (2000), 2816-2821

A Fusion Neural Network
for Estimation of Blasting Vibration

A.K. Chakraborty[1], P. Guha[2], B. Chattopadhyay[2], S. Pal[3], and J. Das[3]

[1] Central Mining Research Institute, 3rd Floor,MECL Complex
Dr. Ambedkar Bhavan, Seminary Hills, Nagpur - 440 006, India
cmrirc@satyam.net.in
[2] Institute of Engineering & Management,Y 12, Sector-V
Saltlake Electronics Complex, Calcutta - 700 091
[3] Electronics and Communication Sciences Unit
Indian Statistical Institute, 203 B T Road, Calcutta 700 108
jdas@isical.ac.in

Abstract. This paper presents the effectiveness of multilayer perceptron (MLP) networks for estimation of blasting vibration using past observations of various blasting parameters. Here we propose a fusion network that combines several MLPs and *on-line* feature selection technique to obtain more reliable and accurate estimation over the empirical models.

1 Introduction

In the nature, future events need to be predicted on the basis of past history when there is no concrete relationship among others. In such cases, knowledge of underlying laws governing the process can be a very powerful and accurate means of prediction. The discovery of empirical relationship within the parameters for a given system can be useful for prediction. However, the laws underlying the behavior of a system are not easily discovered and the empirical regularities are not always evident and can often be masked by environmental hazards.

The prediction and control of ground vibration from blasting has been a subject of in-depth research for the mining and environmental scientists during the last three decades [1]-[5]. Blasts, if not properly designed, may result in ground motions of sufficient intensity to damage the nearby structures. Over the years a number of guidelines have been emerged relating to ground motion and structural damage. Comprehensive regulations in this regard have been developed in various countries like USA, UK, Australia, Sweden, India, France and others based on various observations and research. The regulations are based on the threshold peak particle velocity (PPV) in most of the cases.

Here we focus on estimation of blasting vibration peak particle velocity (v) based on past measurements of various blasting parameters. Normally, blasting parameters are measured for each blast. We have collected the following blasting parameters from a blast in *manganese, limestone, muscovite schist, coal bearing sand stone* and *granite* mines from 10 open cast excavations: (1) *hole diameter* (d), (2) *burden* (B_d), (3) *spacing* (S_d), (4) *effective burden* (B_e), (5) *effective spacing* (S_e), (6) *subgrade drilling* (d_{sg}), (7) *number of decking* (N_d), (8) *charge per hole* (W_h), (9) *specific charge* (Q), (10) *charge per round* (W_r), (11) *maximum charge per delay* (W_{max}), and (12) *distance of seismograph from blast site* (D). We have different observations on these variables for different quality of rocks.

Our objective is to design a suitable model for the estimation of ground vibration from blasting, so as to reduce the estimation error as well as it should have good generalization ability.

This paper is organized as follows: first we study the effectiveness of multilayer perceptron networks for prediction of the blasting vibration along with different empirical model. Next we select few features with the help of domain expert to study the performance of multi-layer perceptron networks. Then we shall study the on-line feature selection method for these data and then propose a fusion network which uses several MLPs to realize a much better prediction system with higher reliability.

2   Different Models for Vibration Estimation

Prediction models for blasting vibration (v) are of two types:(1) empirical model and (2) neural network model.

Empirical Models: There are a number of empirical models mentioned in the literature [6]. Some of them are given below:

Model 1:   USBM :
$$v = K\left(D / \sqrt{W_{max}}\right)^{-n}.$$

Model 2:   Langefors-Kihlstrom :
$$v = K\left(\sqrt{W_{max}} / D^{2/3}\right)^{n}.$$

Model 3:   Ambraseys-Hendron :
$$v = K\left(D / W_{max}^{1/3}\right)^{-n}.$$

Model 4:   Indian Standard Predictor :
$$v = K\left(W_{max} / D^{2/3}\right)^{n}.$$

Model 5:   CMRI Predictor :
$$v = n + K\left(D / \sqrt{W_{max}}\right)^{-1}.$$

where, K and n are constants.

Result: The performance of these models are shown in table 1 for a given set of data, and it is found that correlation is less than 0.5 in each case.

Neural Network Models: Neural network like *multi-layer perceptron* (MLP), *radial-basis function* (RBF), etc. are used for prediction [6,7]. It consists of several layers of neurons of which the first one is the input layer and the last one is the output layer, remaining layers are called hidden layers. There are complete connections between the nodes in successive layers but there is no connection within a layer. Every node, except the input layer nodes, computes the weighted sum of its inputs and apply a sigmoidal function to compute its output, which is then transmitted to the nodes of the next layer [7]. The objective of the MLP learning is to set the connection weights such that the error between the network output and the target output is the minimum. The network weights may be updated by several methods of which the backpropagation technique is most popular one. In this work we have used the backpropagation learning. It is known that a single hidden layer is sufficient for a multilayer perceptron to compute an uniform approximation of a given training set (represented by the set of inputs and a desired (target) output) [7]. Hence in this study we restrict ourselves to a three-layer network i.e., one hidden layer.

Data Preparation for MLP: For a blast, we have observed p variables, let us denote them by $x \in R^p$. Now assume that the blasting vibration (v) of an observation is determined by other parameters of that observation. Thus we attempt to predict blasting vibration (v) using $x \in R^p$. Let N be the total number of available observations. So we can construct the input x) and output v) data for training. In order to train the network we will use input-output pairs (x, v). After obtaining the set of observations (X,V), we partition X (and also V) randomly as $X_{tr}(V_{tr})$ and $X_{te}(V_{te})$ such that $X_{te} \cup X_{tr} = X$, $X_{te} \cap X_{tr} = \phi$. X_{tr} is then used for training the system and X_{te} is used to test the system. In our data set N=470, and p=12. A typical partition is $|X_{tr}| = 423$ and $|X_{te}| = 47$.

Results: We have made several runs of the MLP net with different number of hidden nodes (n_h) like n_h =10, 12 and 15. Table1 reports some typical performance in terms of the coefficient of correlation among all predictors. It exhibits that MLP models are better than empirical models (Table 1). Although the results of some of the MLPs are quite satisfactory but some are not good. Few possible reasons are: (1) process of MLP training, (2) weight initialization, (3) network architecture, (4) selection of a trained MLPs for the estimation model, and (5) the input features.

3 Feature Analysis

Feature Computation: We have observed that a number of empirical relations for the prediction/estimation of blasting vibration are proposed by different researchers in Model 1 through Model 5 and it depends on two parameters: *maximum charge per delay* and *distance of seismograph*. Therefore these empirical relations are the expert knowledge. In this context, we assume these are the important features to improve the accuracy. Here we are considering the following derived features for our analysis:

$$v_1 = D/\sqrt{W_{max}}, \quad v_2 = \sqrt{W_{max}/D^{2/3}}, \quad v_3 = D/W_{max}^{1/3}, \quad v_4 = W_{max}/D^{2/3} \text{ and } v_5 = \sqrt{W_{max}}/D.$$

Online Feature Selection: We have observed that all features are not responsible for the estimation of vibration. Therefore, we do *on-line* feature selection task to choose good features for improved estimation of blast vibration. The basic concept of *on-line* feature selection technique is as follows: The *on-line* feature selection net (FSMLP) [8] has two parts: First part is the *gateing* part where each input is attached with a gate i.e., weight value. Initially it is almost zero. The second part is the standard MLP part. The gate is realized by a function known as *gate function*. Suppose a p input feature selection net, weights γ_i are initialized with values that make $F(\gamma_i) = 1/(1 + e^{-\gamma_i})$ close to 0 for all input node i. Consequently, for the *i*th input $x_i, x_i F(\gamma_i)$ is small at the beginning of training, so the FSMLP *allows only a very small "fraction"' of each input feature value to pass into the standard part of the MLP.* As the network trains, it selectively allows only important features to be active by increasing their attenuator weights (and hence, increasing the multipliers of $\{x_i\}$ associated with these weights) as dictated by the gradient descent. The training can be stopped when the network has learnt satisfactorily, i.e., the mean squared error is low or the number of iteration reaches a maximum limit. Features with *low* attenuator weights are eliminated from the feature set.

Table 1. Performance analysis among different models.

Models	Correlation(r^2)	Max Deviation	Avg Deviation	Std.Deviation
Model1	0.4165	62.57	4.09	7.60
Model2	0.3291	94.16	4.77	9.35
Model3	0.2266	103.24	4.86	9.19
Model4	0.3291	94.17	4.58	8.39
Model5	0.4572	61.68	4.27	6.59
MLP				
h=10	0.797	43.65	2.81	4.00
h=12	0.802	37.95	2.87	3.94
h=15	0.821	31.73	2.74	3.74
FUSED	0.822	31.37	2.74	3.59
MLP with computed features				
h=10	0.843	32.10	2.43	3.69
h=12	0.867	27.43	2.32	3.33
h=15	0.872	27.44	2.21	3.25
FUSED	0.884	25.92	2.22	3.13
FSMLP				
h=10	0.881	25.96	2.07	3.06
h=12	0.872	26.51	2.20	3.18
h=15	0.882	27.56	2.09	3.05
FUSED	0.884	25.15	2.05	3.00

Results: In order to select the good features, we train the FSMLP using the entire data set. And after the features are selected, we train the fusion net with those features.

Here only 12 out of 17 features are important for estimation of blasting vibration. The network accepts the following features: *hole diameter, burden, spacing, effective burden, subgrade drilling, number of decking, charge/round, maximum charge/delay, distance of seismograph from blast site*, v_1, v_2, and v_4. And it rejects the following features: *effective spacing, charge/hole, specific charge*, v_3 and v_5.

It is very reasonable to expect that the *max charge/delay* can have influence on the blasting vibration variation but not the *charge/hole*. The network can capture those information. Among the accepted features, FSMLP has given the maximum importance to the distance of seismograph from blast site (D), the next important feature is the maximum charge per delay (W_{max}). These two are very logical as we are predicting the ground vibration The third important feature selected by the network, as a Geologist will expect, is the effective burden (B_e) and subgrade drilling (d_{sg}) is the next important feature.

4 Fusion Network

Architecture: The architecture of this fusion network with p input nodes and K output nodes is shown in Fig. 1. It has two major layers: MLP layer and Fusion layer. m MLPs constitute the first major layer where each MLP has p input nodes and K output nodes. The fusion layer is a network with Km input nodes from MLP layer and K output nodes.

Training: The fusion network is trained as follows:

MLP layer: First we train m MLP's $M_1, M_2, \cdots, M_m$ with normalized training data. Note that, each MLP is trained in an off-line mode.

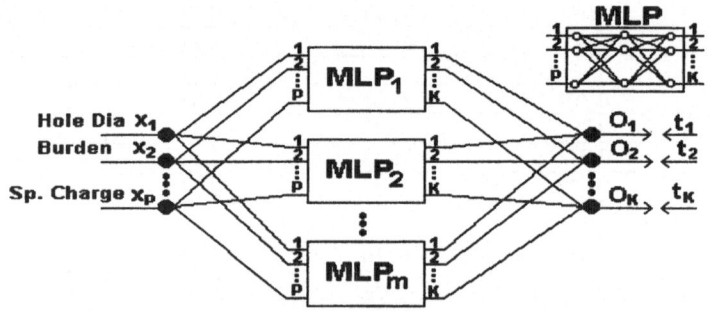

Fig. 1. A Fusion neural network for ground vibration estimation.

Fusion layer: An input vector $\mathbf{x} \in R^p$ is applied to the MLP layer which contains m trained MLPs. Therefore Km outputs will be produced by this layer. These outputs will be the input to the fusion layer. Suppose MLP M_i in the MLP layer accepts $\mathbf{x}$ as normalized input and it produce an output O_i, $\forall i = 1,2,\cdots,K$. The output O_i of the output node i of the fusion network is calculated as $O_i = (\sum_{1 \le j \le m} w_{ji} O_{ji}^{q_i})^{1/q_i}$, $\forall i = 1,2,\ldots,K$ where O_{ji} is the value of the output node i of the MLP j and w_{ji} is the connection weight between the output node i of the MLP j and output node i of the Fusion layer. Now the computed output O_i is compared with the corresponding target output V_i. This produced error helps to adjust the weights in the fusion layer but the weights of the MLP layer are not updated. We obtain the weight updation formula after minimizing the sum of square errors with respect to the weight vector. This training process continue as an MLP. At the end of the training, this weight vectors are freezed for future use. In this case we have $p=12, K=1$ and $m=3$.

Prediction: At the end of the training of different layers like each MLP in MLP layer and fusion layer, we obtain different weight matrices which are dependent on the training data set. Now we shall use this composite fusion network along with these weight matrices for prediction. We have tested this network with our test data set. The predicted vibration is computed based on forward pass and is shown in table 1.

5 Discussion

The proposed fusion network consistently performs better than the conventional MLP network. The reliability of an estimation model based on a fusion network is quite high rather than an individual MLP based model. In fusion network, an individual MLP fails to estimate a reasonably good result even worst, but in this situation the fusion network produce a reliable and good estimate. Feature selection is an important factor for better estimation of blasting parameters. In this regards FSMLP turns out to be an excellent tool that can select good features while learning the estimation. Therefore the combined use of FSMLP and fusion net results in an excellent paradigm for estimation of blasting parameters. There are couple of other areas where we need to do experiments. For example, we plan to use fusion for prediction of other blasting parameters.

References

1. Duvall, W.I and Fogleson, D.E., 1962, Review Criteria for Estimating damage to Residences from Blasting Vibration, *USBM*-I,5968.
2. Langefors, U. and Kihlstrom, B., 1978, *The Modern Techniques of Rock blasting*. Wiley And Sons, inc., New York, 438.
3. Ambraseys, N.R. and hendron, A.J., 1968, Dynamic Behaviour of Rock Mass. Rock Mechanics in Engineering Practices (Ed. Stagg K.G. and Zienkiewics, O.C.), John Wiley And Sons, London, 203-207.
4. Indian Standard, 1973, Criteria for safety and design of structures subjected to under ground blast. *ISI.*, IS-6922.
5. Pal Roy, P. (1993), Putting Ground Vibration Prediction into Practices. *Colliery Guardian.* U.K., Vol. 241, No. 2, pp.63-67
6. Rai R, Maheshwari M, Sohane N, Ranjan A, "Prediction of Maximum Safe Charge per Delay by Application of Artificial Neural Network" *National Seminar On Rock Fragmentation*, 23-24 January 2004, B.H.U, India.
7. Haykin S., *Neural Networks A comprehensive Foundation*. Macmillan College Publishing Co., New York, 1994.
8. N.R. Pal and K. Chintalapudi, "A connecionist system for feature selection", *Neural, parallel and scientific computation*, Vol. 5, No. 3, pp. 359-381, 1997.

Nonlinear Feature Extraction
Using Evolutionary Algorithm

E.K. Tang[1], Ponnuthurai Nagaratnan Suganthan[1], and Xin Yao[2]

[1] School of Electrical and Electronic Engineering
Nanyang Technological University, Singapore 639798
tangke@pmail.ntu.edu.sg
http://www.ntu.edu.sg/home/EPNSugan
[2] School of Computer Science, University of Birmingham, Birmingham, B15 2TT, UK
X.Yao@cs.bham.ac.uk

Abstract. We propose a method of nonlinear feature extraction for 2-class problems. A simple sigmoid function is used to extract features that are negatively correlated to each other. To evaluate the effectiveness of the proposed method, we employ linear and non-linear support vector machines to classify using the extracted feature sets and the original feature sets. Comparison on 4 datasets shows that our method is effective for nonlinear feature extraction.

1 Introduction

When solving a pattern classification problem, it is common to apply a feature extraction method as a pre-processing technique, not only to reduce the computation complexity, but possibly also to obtain better classification performance by removing irrelevant and redundant information in the data. A class of feature extraction procedures can be defined by a transformation $\mathbf{Y}=T(\mathbf{X})$, where $\mathbf{X} \in R^D$, $\mathbf{Y} \in R^d$ and the transformation T is obtained by optimizing suitable objectives. According to this transformation function T, feature extraction can be roughly categorized into linear feature extraction and non-linear feature extraction. In many classical linear feature extraction methods, special objective functions are selected so that optimal solutions have a closed form, such as Fisher's linear discriminant analysis [1] and its variants [2][3][4]. Obviously, the closed-form solutions have the computational advantage. However, in many real-world problems, closed form methods are likely to yield suboptimal solution. Another problem is, a closed form solution can be easily obtained in linear feature extraction, but if the expected transformation function is nonlinear, it will be very hard or even impossible to obtain a closed-form solution. Therefore, many iterative methods have also been developed in the past. In these methods, iterative optimization algorithms, such as the expectation-maximization (EM) algorithm, simulated annealing and the evolutionary-based algorithms (EA), are employed to find the optimal solution for the objective function.

As a kind of powerful optimization algorithms, different evolutionary algorithms have been applied to solve both linear and nonlinear feature extraction problems [5][6]. In the evolution procedure, the population is composed of different feature sets. One most straightforward way to apply the EA to feature extraction is to employ the classification accuracy on training set that can be achieved from the extracted

N.R. Pal et al. (Eds.): ICONIP 2004, LNCS 3316, pp. 1014–1019, 2004.
© Springer-Verlag Berlin Heidelberg 2004

feature set, as in [5]. But as the dimensionality of extracted feature set increases, it will be more difficult to evolve a suitable solution in acceptable time. Sometimes, even the most powerful EA may not yield a good result. Therefore, we present a method named negatively correlated feature extraction (NCFE) to nonlinearly extract features for two-class classification problems in this paper. In this algorithm, a different fitness function is employed so that it is not necessary to extract a number of features at the same time, but one by one, while high classification accuracy can still be achieved from the extracted features.

The organization of this paper is as follows: In section 2, we discuss how negative correlation can be employed to extract useful features for 2-class problems. In section 3, we describe the details of the design of the fitness function and NCFE. Section 4 compares the performance of NCFE+Linear-SVM with a nonlinear SVM on 4 UCI 2-class datasets [7]. In section 5 we present conclusions.

2 Negatively Correlated Features for Classification

When a set of features is extracted, both the class separability of each feature and the relationship between the features may influence the classification accuracy. Based on the results reported in [8] [9], we expect that if the extracted features are negatively correlated while possessing relevant information for classification, the overall classification accuracy may improve. Fig.1.a and Fig.1.b.show the distribution of the two classes in a dataset with two feature dimensions f_1 and f_2. In each feature, the two classes overlap on each other. We use the same feature set to construct the datasets, the means and variances of the two features in Fig.1.a are same with the features in Fig.1.b, but the separability between the two classes are obviously different.

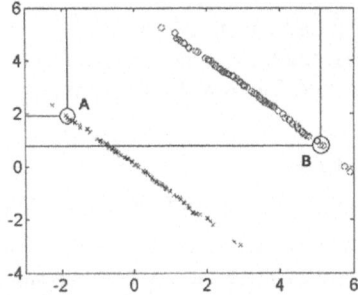
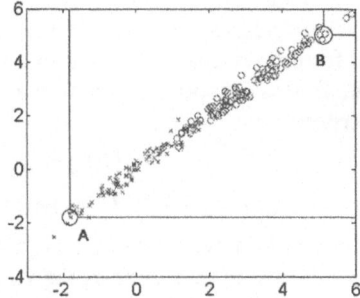

Fig. 1a. Data distribution in 2-d.   **Fig. 1b.** Data distribution in 2-d.

What is the difference between the two datasets? We can consider the region A and region B in the figures to illustrate the difference. In Fig.1.b, distances between the two regions are approximately same on the two features, while in Fig.1.a, the two regions are close to each other on one feature, and distant on the other. If we define a variable d_i as the distance between two samples on feature i (Eg. the distance between points in region A and points in region B on one dimension), a small d_1 between A and B will correspond to a large d_2 in Fig.1.a, while a large d_1 will correspond to a large d_2 in Fig.1.b. Similar example can be observed for any other regions in the figures. Therefore, we can conclude that, the combination of features f_1 and f_2 is likely to increase separability between the two classes if d_1 and d_2 are negatively correlated. This strategy can be understood more easily in Liu et al's framework of negatively correlated classifiers ensemble [8], where classifier ensemble is applied to the dataset, and different classifiers are required to classify different samples of the dataset. In our method, each feature can be viewed as a classifier by setting a threshold value to it, so it is the features instead of the classifiers that should be negatively correlated. In this paper, to demonstrate the advantages of our method, the number of features is selected as 2.

3 Nonlinear Negatively Correlated Feature Extraction Using EAs

Unlike linear feature extraction, we do not know which nonlinear functions are suitable to extract features for a specific problem. Hence, the most straightforward strategy is to make use of a pool of nonlinear functions. By using evolution programming (EP) or genetic programming (GP) algorithms optimal combination of different nonlinear functions and their parameters can be evolved for each problem. However, the main purpose of our investigation is to understand the influence of negatively correlated features on the classification accuracy. Therefore, a simple sigmoid function $f = \dfrac{1}{1 + exp(-Xa)}$ is employed, where f is the extracted feature, X is the d-dimensional dataset and a is the coefficient vector to be evolved. To make the experimental results illustrate the influence of our method more efficiently, we choose to extract only two features.

In NCFE, the individuals in evolution are the coefficient vectors a in sigmoid function, and the fitness value of each individual is computed from the corresponding extracted features. Once a suitable solution is found for a, it is used to extract the final feature set that will be fed to a classifier. To implement the negative correlation as the fitness function, we employ a commonly used framework of optimizing multiple criteria.

$$J(f) = J_1(f) - rJ_2(f) \tag{1}$$

In the expression (1), J_1 is the class separability measure of feature f_i, J_2 is the measure of how well the two features are negatively correlated and r is a coefficient. As we discussed in section 1, the two features are evolved one by one instead of simultaneously. Hence, J_2 is only necessary to be considered when evolving the second feature. In other words, we first evolve a single feature that optimize J_1, and then use the whole expression (1) to evolve the second feature. We employ the Fisher's criterion as the J_1:

$$J_1(f) = S_{bf} / S_{wf} \tag{2}$$

where S_{bf} is the between-class variance and S_{bf} is the within-class variance on the feature f. For a single feature, the value of this criterion is maximized if the samples of the same class are as close as possible while at the same time the samples of different classes are as distant as possible.

For a dataset with n samples, a feature f can be represented by an n-dimensional vector, and J_2 is defined as following:

$$J_2 = \sum_{i=1}^{n} abs(sign((f_{i1} - m_1)(f_{i1} - m_{k1})) + sign((f_{i2} - m_2)(f_{i2} - m_{k2}))) \qquad (3)$$

where f_{ij} is the value of feature f_j in ith sample, m_j is the mean of f_j, and m_{kj} is the mean of of feature f_j in class k (k=1,2). The Eq. (3) can be illustrated as follows:

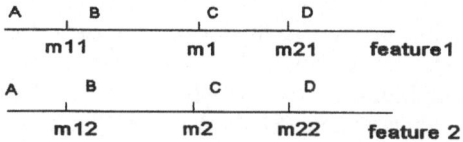

Fig. 2. Illustration of two negatively correlated features.

In the Fig.2, m_1 is mean of feature 1 in the whole dataset, m_{11} and m_{21} are means of the two classes respectively. m_2, m_{12} and m_{22} are the values of feature 2. For a single feature, expression $sign((f_{ij} - m_j)(f_{ij} - m_{kj}))$ only takes either –1 or 1. When the ith sample is close to the mean of the whole dataset (i.e. lies in regions B or C), it is more difficult to classify, and the product of $(f_{ij}-m_j)$ and $(f_{ij}-m_{kj})$ is negative, so the value of the expression is -1, and it is 1 if the product of $(f_{ij}-m_j)$ and $(f_{ij}-m_{kj})$ is positive. By combining expressions (2) and (3) together, we obtain the fitness function for the evolution procedure. Based on the first feature that has the largest value of expression (2), the fitness function of the second feature (expression (1)) is maximized when expression (2) is maximized while expression (3) is minimized.

As our objective is to investigate the suitability of the fitness function, we employ a simple evolution strategy algorithm in NCFE. The major steps are given as follows.

1) Randomly generate an initial population M of coefficient vectors a, calculate their fitness values.
2) Use all M vectors as parents to create n_b offspring vectors by Gaussian mutation.
3) Add the offspring vectors to the population and calculate the fitness values of the n_b vectors, compare fitness value of all $M+ n_b$ individuals in the population and prune it to the M fittest vectors.
4) Go to step 2 until the maximum number of generations has been reached. Then the fittest vector in the last generation is chosen as the optimal coefficient vector.

4 Experiments

In order to test the discriminant ability of the negatively correlated features, we performed experiments on 4 UCI 2-class datasets [7]. They are Australian Credit Card, Pima Diabetes, Heart Disease and Wisconsin-Breast Cancer datasets. The feature sets extracted by NCFE is applied to a linear support vector machine, and the classification accuracy is compared with the performance of a nonlinear SVM (N-SVM). Here, we choose the linear SVM (L-SVM) as the linear classifier, and value of r is set at 20. M and n_b of the Evolution Strategy are 10 and 20 respectively, and number of generations is 5000.

All datasets are randomly split into a testing and a training set. The testing set contains approximately 25 percent of the data, while the train set contains the remaining 75 percent. For each pair of training and testing sets, we run the program for 30 times independently. The whole procedure is repeated for 5 times and the average result of 150 runs on each dataset is used to analyze the final results of NCFE.

Firstly, if the negative correlation has positive effects on the feature sets, the higher the fitness value, the higher the CR should be. For each dataset, since we run the program for 150 times, the corresponding 150 results are evenly divided into 2 groups according to the fitness values, namely results with large fitness value (LFV) and results with small fitness value (SFV). Average CRs of the two groups are compared in the first two rows of Table 1. It obviously demonstrates our conclusion.

General performances of the algorithms are compared. The last three rows of Table 1 compare the average CR of N-SVM with NCFE+L-SVM and NCFE+N-SVM. NCFE+L-SVM performs better on Australian Credit Card dataset and Pima datasets than N-SVM. We observe that NCFE+N-SVM achieves better results on Diabetes, Credit Card and Heart Disease datasets. This is reasonable because a nonlinear feature extraction algorithm may not always find all the relevant nonlinear properties of a dataset. Since no algorithms can be optimal for all real-world problems, the result demonstrates that our method is generally comparable to the N-SVM.

Table 1. Experimental results.

	Pima Diabetes	Credit card	Heart disease	Breast Cancer
SFV	75.85	87.44	80.64	95.86
LFV	76.49	88.18	81.13	96.67
N-SVM	74.38	86.22	83.2	**96.84**
L-ACR	76.49	88.18	81.13	96.67
N-ACR	**76.55**	**89.1**	**83.42**	96.16

In Table 1, SFV is the average CR on the feature sets with small fitness value. LFV is the average CR on the feature sets with large fitness value. N-SVM is the average CR achieved by N-SVM. L-ACR is the average CR achieved by NCFE+L-SVM and N-ACR is the average CR achieved by NCFE+N-SVM.

5 Conclusions and Discussions

We proposed a novel nonlinear feature extraction algorithm for 2-class classification problems. A specific fitness function motivated by the concept of negative correlation is designed for the evolutionary algorithm. Experimental results on 4 UCI datasets show that applying a single linear SVM on our extracted feature set can yield classification accuracy that comparable with achieved by a well tuned nonlinear SVM on the original feature set. In our preliminary work, only 2-class problems are considered. This is because that when applying the negative correlation framework to multi-class problem, the position relationship between points in the feature space will become more complex, and some further modifications should be made to generalize our method. Furthermore, our method can also be generalized to multi-class problem by decomposing the problem into a set of 2-class problems. The combination strategies that can be employed here have been well developed in the past [10].

References

1. R. A. Fisher, "The Statistical Utilization of Multiple Measurements," *Ann. Eugenics*, vol. 8, pp. 376-386, 1938.
2. R. Lotlikar and R. Kothari, "Fractional-step Dimensionality Reduction," *IEEE Trans. Pattern Analysis and Machine Intelligence,* vol. 22, pp. 623-627, 2000.
3. L. Chen, H. Liao, M. Ko, J. Lin, and G. Yu, "A New LDA-based Face Recognition System Which Can Solve the Small Sample Size Problem," *Pattern Recognition*, vol. 33, pp. 1713-1726, 2000.
4. T. Hastie and R. Tibshirani, "Discriminant Analysis by Gaussian Mixtures," *J. Royal Statistical Soc., B,* vol. 58, pp. 155-176, 1996.
5. M. L. Raymer, W. F. Punch, E. D. Goodman, L. A. Kuhn and A. K. Jain, "Dimensionality Reduction Using Genetic Algorithms," *IEEE T. Evolutionary Computation*, 4:164-171, 2000.
6. M. Kotani, M. Nakai and K. Akazawa, "Feature extraction Using Evolutionary Computation," *Proc. 1999 Congress on Evolutionary Computation,* 1999.
7. C. L. Blake and C. J. Merz, *UCI Repository of Machine Learning Databases*, Univ. of California, Irvine, 1996. http://www.ics.uci.edu/~mlearn/MLRepository.html
8. Y. Liu, X. Yao and T. Higuchi, "Evolutionary Ensembles with Negative Correlation Learning," *IEEE T. Evolutionary Computation*, 4:380-387, 2000.
9. G. Brown, X. Yao, J. Wyatt, H. Wersing and B. Sendhoff, "Exploiting Ensemble Diversity for Automatic Feature Extraction," *P. 9th Int Conf. on Neural Information Processing,* 2002.
10. T. F. Wu, C. J. Lin and R. C. Weng, "Probability Estimates for Multi-class Classification by Pairwise Coupling," *J of Machine Learning Research, vol. 5, pp. 975-1005,* 2004.

Hybrid Feature Selection
for Modeling Intrusion Detection Systems

Srilatha Chebrolu, Ajith Abraham, and Johnson P. Thomas

Department of Computer Science, Oklahoma State University, USA
ajith.abraham@ieee.org, jpt@okstate.edu

Abstract. Most of the current Intrusion Detection Systems (IDS) examine all data features to detect intrusion or misuse patterns. Some of the features may be redundant or contribute little (if anything) to the detection process. We investigated the performance of two feature selection algorithms involving Bayesian Networks (BN) and Classification and Regression Trees (CART) and an ensemble of BN and CART. An hybrid architecture is further proposed by combining different feature selection algorithms. Empirical results indicate that significant input feature selection is important to design an IDS that is lightweight, efficient and effective for real world detection systems.

1 Introduction and Related Research

IDS have become important and widely used for ensuring network security. Since the amount of audit data that an IDS needs to examine is very large even for a small network, analysis is difficult even with computer assistance because extraneous features can make it harder to detect suspicious behavior patterns [4][7]. Complex relationships exist between the features and IDS must therefore reduce the amount of data to be processed. This is very important if real-time detection is desired. Reduction can occur by data filtering, data clustering and feature selection. In complex classification domains, features may contain false correlations, which hinder the process of detecting intrusions. Extra features can increase computation time, and can have an impact on the accuracy of IDS. Feature selection improves classification by searching for the subset of features, which best classifies the training data [8]. In the literature a number of work could be cited wherein several machine learning paradigms, fuzzy inference systems and expert systems, were used to develop IDS [4][5]. Authors of [8] have demonstrated that large number of features is unimportant and may be eliminated, without significantly lowering the performance of the IDS. IDS task is often modeled as a classification problem in a machine-learning context.

2 Feature Selection and Classification Using AI Paradigms

2.1 Bayesian Learning and Markov Blanket Modeling of Input Features

The Bayesian Network (BN) is a powerful knowledge representation and reasoning algorithm under conditions of uncertainty. A Bayesian network $B = (N, A, \Theta)$ is a Directed Acyclic Graph (DAG) (N, A) where each node n $\in$ N represents a domain variable (e.g. a dataset attribute or variable), and each arc $a \in$ A between nodes represents a probabilistic dependency among the variables, quantified using a conditional probability distribution (CP table) $\theta_i \in \Theta$ for each node n_i. A BN can be used

N.R. Pal et al. (Eds.): ICONIP 2004, LNCS 3316, pp. 1020–1025, 2004.
© Springer-Verlag Berlin Heidelberg 2004

to compute the conditional probability of one node, given values assigned to the other nodes. Markov Blanket (MB) of the output variable T, is a novel idea for significant feature selection in large data sets [9]. *MB (T)* is defined as the set of input variables such that all other variables are probabilistically independent of T. A general BN classifier learning is that we can get a set of features that are on the Markov blanket of the class node. The Markov blanket of a node n is the union of n's parents, n's children and the parents of n's children [2]. This subset of nodes shields n from being affected by any node outside the blanket. When using a BN classifier on complete data, the Markov blanket of the class node forms feature selection and all features outside the Markov blanket are deleted from the BN.

2.2 Classification and Regression Trees Learning and Modeling Input Features

The Classification and Regression Trees (CART) methodology is technically called as binary recursive partitioning [1]. The key elements of CART analysis are a set of rules for splitting each node in a tree; deciding when tree is complete and assigning a class outcome to each terminal node. As an example, for the DARPA intrusion data set [3] with 5092 cases and 41 variables, CART considers up to 5092 times 41 splits for a total of 208772 possible splits. For splitting, Gini rule is used which essentially is a measure of how well the splitting rule separates the classes contained in the parent node. Splitting is impossible if only one case remains in a particular node or if all the cases in that node are exact copies of each other or if a node has too few cases. Feature selection is done based on the contribution the input variables made to the construction of the decision tree. Feature importance is determined by the role of each input variable either as a main splitter or as a surrogate. Surrogate splitters are defined as back-up rules that closely mimic the action of primary splitting rules. Suppose that, in a given model, the algorithm splits data according to variable '*protocol_type*' and if a value for '*protocol_type*' is not available, the algorithm might substitute '*service*' as a good surrogate. Variable importance, for a particular variable is the sum across all nodes in the tree of the improvement scores that the predictor has when it acts as a primary or surrogate (but not competitor) splitter. Example, for node i, if the predictor appears as the primary splitter then its contribution towards importance could be given as $i_{importance}$. But if the variable appears as the nth surrogate instead of the primary variable, then the importance becomes $i_{importance} = (p^n) * i_{improvement}$ in which p is the 'surrogate improvement weight' which is a user controlled parameter set between (0-1).

3 Experiment Setup and Results

The data for our experiments was prepared by the 1998 DARPA Intrusion Detection Evaluation program by MIT Lincoln Labs [6]. The data set contains 24 attack types that could be classified into four main categories namely *Denial of Service (DOS)*, *Remote to User (R2L)*, *User to Root (U2R)* and *Probing*. The data set has 41 attributes for each connection record plus one class label. Some features are derived features, which are useful in distinguishing normal connection from attacks. These features are either nominal or numeric. Some features examine only the connections in the past two seconds that have the same destination host as the current connection, and calculate statistics related to protocol behavior, service, etc. These are called same host features. Some features examine only the connections in the past two seconds that have the same service as the current connection and are called same service features. Some other connection records were also sorted by destination host, and features were constructed using a window of 100 connections to the same host instead of a time window. These are called host-based traffic features. Some features that look for suspicious behavior in the data packets like number of failed logins etc. are called content features. Our experiments have three phases namely data reduction, training phase and testing phase. In the data reduction phase, important variables for real-time intrusion detection are selected by feature selection. In the training phase, the Bayesian neural network and classification and regression trees constructs a model using the training data to give maximum generalization accuracy on the unseen data. The test data is then passed through the saved trained model to detect intrusions in the testing phase. The data set for our experiments contains randomly generated 11982 records having 41 features [3]. The 41 features are labeled in order as *A, B, C, D, E, F, G, H, I, J, K, L, M, N, O, P, Q, R, S, T, U, V, W, X, Y, Z, AA, AB, AC, AD, AF, AG, AH, AI, AJ, AK, AL, AM, AN, AO* and the class label is named as *AP*. This data set has five different classes namely *Normal, DOS, R2L, U2R* and *Probes*. The training and test comprises of 5092 and 6890 records respectively. All the IDS models are trained and tested with the same set of data. As the data set has five different classes we perform a 5-class binary classification. The *Normal* data belongs to class 1, *Probe* belongs to class 2, *DOS* belongs to class 3, *U2R* belongs to class 4 and *R2L* belongs to class 5. All experiments were performed using an AMD Athlon 1.67 GHz processor with 992 MB of RAM.

Table 1. Performance of Bayesian Belief Network.

Attack Class	41 variables			17 variables		
	Train (sec)	Test (sec)	Accuracy (%)	Train (sec)	Test (sec)	Accuracy (%)
Normal	42.14	19.02	99.57	23.29	11.16	**99.64**
Probe	49.15	21.04	**99.43**	25.07	13.04	98.57
DOS	54.52	23.02	**99.69**	28.49	14.14	98.16
U2R	30.02	15.23	**64.00**	14.13	7.49	60.00
R2L	47.28	12.11	**99.11**	21.13	13.57	98.93

3.1 Modeling IDS Using Bayesian Network

We selected the important features using the Markov blanket model and found out that 17 variables of the data set forms the Markov blanket of the class node as explained in Section 2.1. These 17 variables are *A, B, C, E, G, H, K, L, N, Q, V, W, X, Y, Z, AD* and *AF*. Further Bayesian network classifier is constructed using the training data and then the classifier is used on the test data set to classify the data as an attack or normal. Table 1 depicts the performance of Bayesian belief network by using the original 41 variable data set and the 17 variables reduced data set. The training and testing times for each classifier are decreased when 17 variable data set is used. Using the 17 variable data set there is a slight increase in the performance accuracy for *Normal* class compared to the 41 variable data set.

Table 2. Performance of classification and regression trees.

Attack Class	41 variable data set			12 variable data set		
	Train (sec)	Test (sec)	Accuracy (%)	Train (sec)	Test (sec)	Accuracy (%)
Normal	1.15	0.18	99.64	0.80	0.02	**100.00**
Probe	1.25	0.03	**97.85**	0.85	0.05	97.71
DOS	2.32	0.05	**99.47**	0.97	0.07	85.34
U2R	1.10	0.02	48.00	0.45	0.03	**64.00**
R2L	1.56	0.03	90.58	0.79	0.02	**95.56**

Table 3. Performance of Bayesian and CART using reduced datasets.

Attack Class	Bayesian with 12 variables			CART with 17 variables		
	Train (sec)	Test (sec)	Accuracy (%)	Train (sec)	Test (sec)	Accuracy (%)
Normal	20.10	10.13	98.78	1.03	0.04	**99.64**
Probe	23.15	11.17	99.57	1.15	0.13	**100.00**
DOS	25.19	12.10	98.95	0.96	0.11	**99.97**
U2R	11.03	5.01	48.00	0.59	0.02	**72.00**
R2L	19.05	12.13	**98.93**	0.93	0.10	96.62

Table 4. Performance of CART and Bayesian network using 19 variables.

Class	Bayesian	CART
Normal	**99.57**	95.50
Probe	96.71	**96.85**
DOS	**99.02**	94.31
U2R	56.00	**84.00**
R2L	**97.87**	97.69

Table 5. Performance of ensemble approach using different data sets.

Class	Number of Variables		
	12	17	41
Normal	**100.00**	99.64	99.71
Probe	99.86	**100.00**	99.85
DOS	99.98	**100.00**	99.93
U2R	**80.00**	72.00	72.00
R2L	**99.47**	99.29	99.47

3.2 Modeling IDS Using Classification and Regression Trees

We decided the important variables depending on the contribution of the variables for the construction of the decision tree. Variable rankings were generated in terms of percentages. We eliminated the variables that have 0.00% rankings and considered only the primary splitters or surrogates as explained in Section 2.2. This resulted in a reduced 12 variable data set with $C, E, F, L, W, X, Y, AB, AE, AF, AG$ and AI as variables. Further the classifier is constructed using the training data and then the test data is passed through the saved trained model. Table 2 compares the performance of CART using the 41 variable original data set and the 12 variable reduced data set. Furthermore, the accuracies of classes U2R and R2L have increased by using the 12 variable reduced data set. Further, we used the Bayesian reduced 17 variable data set (Section 3.1) to train CART and the CART reduced 12 variable dataset (Section 3.2) to train Bayesian network. As illustrated in Table 3 except R2L all other classes were classified well by the CART algorithm. Moreover, training and testing time for each class are greater for Bayesian network classifier compared to CART algorithm.

3.3 Feature Ranking Using Support Vector Machines

We also attempted to evaluate the performance of CART and Bayesian network using the reduced dataset (same input variables) given in [8]. Table 4 shows the performance comparisons of CART and Bayesian network using 19 variables. Except *U2R*, the 17 and 12 variable dataset performs well for all the other classes.

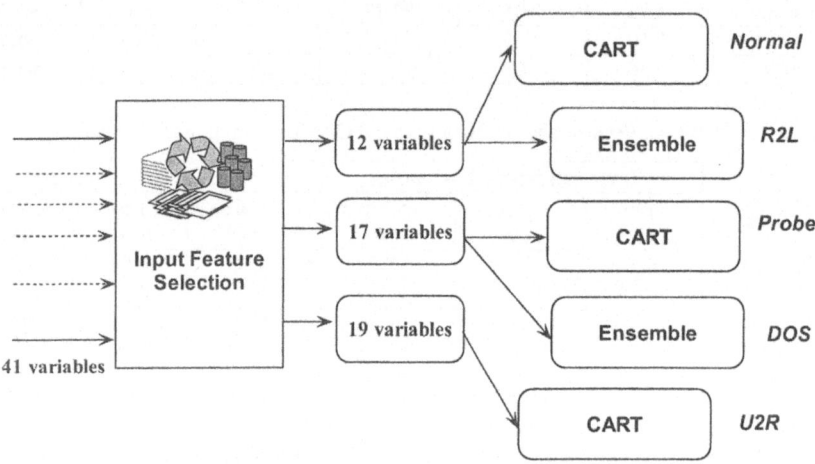

Fig. 1. Developed IDS model for different attack classes.

3.4 Ensemble Approach Using Reduced Data Sets

In this approach we first construct the Bayesian network classifier and CART models individually to obtain a very good generalization performance. The ensemble approach is used for 12, 17 and 41 variable dataset. In the ensemble approach, the final outputs were decided as follows: Each classifier's output is given a weight (0-1 scale) depending on the generalization accuracy as given in Section 3.1-3.2. If both classifiers agree then the output is decided accordingly. If there is a conflict then the decision given by the classifier with the highest weight is taken into account. Table 5 illustrates the ensemble results using the different data sets. From the results, we can conclude that ensemble approach gives better performance than the two individual separately used models. The ensemble approach basically exploits the differences in misclassification (by individual models) and improves the overall performance. Since the *U2R* class is best detected by CART using 19 variables a final ensemble model is constructed as illustrated in Figure 1. By using the ensemble model in Figure 1, *Normal*, *Probe* and *DOS* could be detected with 100% accuracy and *U2R* and *R2L* with 84% and 99.47% accuracies respectively.

4   Conclusions

In this research, we have investigated new techniques for intrusion detection and performed data reduction and evaluated their performance on the benchmark intrusion data. Our initial experiments using PCA/ICA to compress data was not successful (due to space limitations the results are not reported in this paper). We used the feature selection method using Markov blanket model and decision tree analysis. Following this, we explored general Bayesian Network (BN) classifier and Classification and Regression Trees (CART) as intrusion detection models. We have also demonstrated performance comparisons using different reduced data sets. The proposed ensemble of BN and CART combines the complementary features of the base classifiers. Finally, we propose a hybrid architecture involving ensemble and base classifiers for intrusion detection. From the empirical results, it is evident by using the hybrid model *Normal, Probe* and *DOS* could be detected with 100% accuracy and *U2R* and *R2L* with 84% and 99.47% accuracies respectively. Our future research will be directed towards developing more accurate base classifiers particularly for the detection of *U2R* type of attacks.

References

1. Brieman L., Friedman J., Olshen R. and Stone C., Classification of Regression Trees. Wadsworth Inc., 1984.
2. Cheng J., Greiner R., Kelly J., Bell D.A. and Liu W., Learning Bayesian Networks from Data: an Information-Theory Based Approach, The Artificial Intelligence Journal, Volume 137, Pages 43-90, 2002.
3. KDD cup 99 Intrusion detection data set
   <http://kdd.ics.uci.edu/databases/kddcup99/kddcup.data_10_percent.gz>
4. Lee W., Stolfo S. and Mok K., A Data Mining Framework for Building Intrusion Detection Models, In Proceedings of the IEEE Symposium on Security and Privacy, 1999.
5. Luo J. and Bridges S. M., Mining Fuzzy Association Rules and Fuzzy Frequency Episodes for Intrusion Detection, International Journal of Intelligent Systems, John Wiley & Sons, Vol. 15, No. 8, pp. 687-704, 2000.
6. MIT Lincoln Laboratory. <http://www.ll.mit.edu/IST/ideval/>
7. Mukkamala S., Sung A.H. and Abraham A., Intrusion Detection Using Ensemble of Soft Computing Paradigms, Third International Conference on Intelligent Systems Design and Applications, Springer Verlag Germany, pp. 239-248, 2003.
8. Sung A.H. and Mukkamala S., Identifying Important Features for Intrusion Detection Using Support Vector Machines and Neural Networks, Proceedings of International Symposium on Applications and the Internet (SAINT 2003), pp. 209-217, 2003.
9. Tsamardinos I., Aliferis C.F.and Statnikov A., Time and Sample Efficient Discovery of Markov Blankets and Direct Causal Relations, 9th ACM SIGKDD International Conference on Knowledge Discovery and Data Mining, USA, ACM Press, pages 673-678, 2003.

Feature Selection for Fast Image Classification with Support Vector Machines

Zhi-Gang Fan, Kai-An Wang, and Bao-Liang Lu

Departmart of Computer Science and Engineering, Shanghai Jiao Tong University,
1954 Hua Shan Road, Shanghai 200030, China
{zgfan,kaianwang}@sjtu.edu.cn, blu@cs.sjtu.edu.cn

Abstract. According to statistical learning theory, we propose a feature
selection method using support vector machines (SVMs). By exploiting
the power of SVMs, we integrate the two tasks, feature selection and
classifier training, into a single consistent framework and make the feature selection process more effective. Our experiments show that our
SVM feature selection method can speed up the classification process
and improve the generalization performance of the classifier.

1 Introduction

Pattern classification is a very active research field in recent years. As a result
of statistical learning theory, support vector machines (SVMs) is an effective
classifier for the problems of high dimension and small sample sets. This is a
very meaningful breakthrough for machine learning and pattern classification
because both high dimension and small sample set problems are too difficult to
be solved by classical paradigms. According to the principle of structural risk
minimization, SVMs can guarantee a high level of generalization ability. SVMs
can obtain an optimal separating hyperplane as a trade-off between the quality of
empirical risk and the complexity of the classifier. Furthermore, SVMs can solve
linearly non-separable problems using kernel functions, which map the input
space into a high-dimensional feature space where a maximal margin hyperplane
is constructed [1].

In fact, SVMs are not only a good classification technique but also a good
feature selection method. The problem of feature selection is well known in machine learning. Data overfitting arises when the number of features is large and
the number of training samples is comparatively small. This case is very common especially in image classification. Therefore, we must find a way to select
the most informative subset of features that yield best classification performance
for overcoming the risk of overfitting and speeding up the classification process.
By investigating the characteristics of SVMs, it can be found that the optimal
hyperplane and support vectors of SVMs can be used as indicators of the important subset of features. Therefore, through these indicators, the most informative
features can be selected effectively.

N.R. Pal et al. (Eds.): ICONIP 2004, LNCS 3316, pp. 1026–1031, 2004.
© Springer-Verlag Berlin Heidelberg 2004

2   SVMs and Feature Ranking

2.1   Support Vector Machines

Support vector machine is a machine learning technique that is well-founded in statistical learning theory. Statistical learning theory is not only a tool for the theoretical analysis but also a tool for creating practical algorithms for pattern recognition. This abstract theoretical analysis allows us to discover a general model of generalization. On the basis of the VC dimension concept, constructive distribution-independent bounds on the rate of convergence of learning processes can be obtained and the structural risk minimization principle has been found. The new understanding of the mechanisms behind generalization not only changes the theoretical foundation of generalization, but also changes the algorithmic approaches to pattern recognition.

As an application of the theoretical breakthrough, SVMs have high generalization ability and are capable of learning in high-dimensional spaces with a small number of training examples. It accomplishes this by minimizing a bound on the empirical error and the complexity of the classifier, at the same time. With probability at least $1 - \eta$, the inequality

$$R(\alpha) \leq R_{emp}(\alpha) + \Phi\left(\frac{h}{l}, \frac{-\log(\eta)}{l}\right) \tag{1}$$

holds true for the set of totally bounded functions. Here, $R(\alpha)$ is the expected risk, $R_{emp}(\alpha)$ is the empirical risk, l is the number of training examples, h is the VC dimension of the classifier that is being used, and $\Phi(\cdot)$ is the VC confidence of the classifier.

According to equation (1), we can find that the uniform deviation between the expected risk and empirical risk decreases with larger amounts of training data l and increases with the VC dimension h. This leads us directly to the principle of structural risk minimization, whereby we can attempt to minimize at the same time both the actual error over the training set and the complexity of the classifier. This will bound the generalization error as in (1). This controlling of both the training set error and the classifier's complexity has allowed SVMs to be successfully applied to very high dimensional learning tasks.

We are interesting in linear SVMs because of the nature of the data sets under investigation. Linear SVMs uses the optimal hyperplane

$$(w \cdot x) + b = 0 \tag{2}$$

which can separate the training vectors without error and has maximum distance to the closest vectors. To find the optimal hyperplane one has to solve the following quadratic programming problem: minimize the functional

$$\Phi(w) = \frac{1}{2}(w \cdot w) \tag{3}$$

under the inequality constraints

$$y_i[(x_i \cdot w) + b] \geq 1, \quad i = 1, 2, \ldots, l. \tag{4}$$

where $y_i \in \{-1, 1\}$ is class label. We can obtain the functional

$$W(\alpha) = \sum_{i=1}^{l} \alpha_i - \frac{1}{2} \sum_{i,j}^{l} \alpha_i \alpha_j y_i y_j x_i^T x_j \tag{5}$$

It remains to maximize this functional under the constraint

$$\sum_{i=1}^{l} \alpha_i y_i = 0, \quad \alpha_i \geq 0, \quad i = 1, \ldots, l \tag{6}$$

Once the optimization problem has been solved, we can obtain w as follows:

$$w = \sum_{i=1}^{l} \alpha_i y_i x_i \tag{7}$$

It is usually the case that most of the parameters α_i are zero. The decision hyperplane therefore only depends on a smaller number of data points with non-zero α_i; these data points are called support vectors. So we can change the equation (7) as

$$w = \sum_{i \in SV} \alpha_i y_i x_i \tag{8}$$

As a result, equation (2) can be obtained and the SVM classifier has been built.

2.2 Feature Selection and Classification

According to the hyperplane as shown in equation (2), the linear discriminant function can be constructed for SVMs classifier as follows:

$$D(x) = (w \cdot x) + b \tag{9}$$

The inner product of weight vector $w = (w_1, w_2, \ldots, w_n)$ and input vector $x = (x_1, x_2, \ldots, x_n)$ determines the value of $D(x)$. Fig.1 shows that the $|w_k|$ of a SVMs example with R^{4096} input space has obvious variance. Intuitively, the input features in a subset of $(x_1, x_2, \ldots, x_n)$ that are weighted by the largest absolute value subset of $(w_1, w_2, \ldots, w_n)$ influence most the classification decision. If the classifier performs well, the input features subset with the largest weights should correspond to the most informative features [4]. Therefore, the weights $|w_k|$ of the linear discriminant function can be used as feature ranking coefficients. However, this way for feature ranking is a greedy method and we should look for more evidences for feature selection. In [7], support vectors have been used as evidence.

Assume the distance between the optimal hyperplane and the support vectors is Δ, the optimal hyperplane can be viewed as a kind of Δ-margin separating hyperplane which is located in the center of margin $(-\Delta, \Delta)$. According to [3],

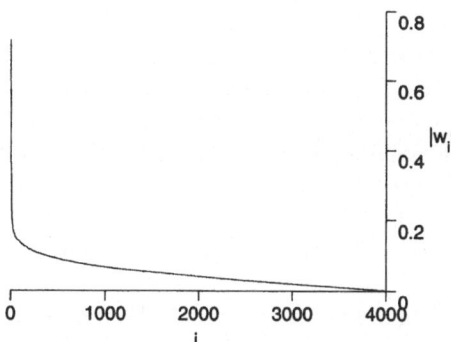

Fig. 1. $|w_i|$ ordered decreasingly in a linear SVMs example with R^{4096} input space.

the set of Δ-margin separating hyperplanes has the VC dimension h bounded by the inequality

$$h \leq \min \left(\left[\frac{R^2}{\Delta^2} \right], n \right) + 1 \tag{10}$$

where R is the radius of a sphere which can bound the training vectors $x \in X$.

Inequality (10) points out the relationship between margin Δ and VC dimension: a larger Δ means a smaller VC dimension. Therefore, in order to obtain high generalization ability, we should still maintain margin large after feature selection. However, because the dimensionality of original input space has been reduced after feature selection, the margin is always to shrink and what we can do is trying our best to make the shrink small to some extent. Therefore, in feature selection process, we should preferentially select the features which make more contribution to maintaining the margin large. This is another evidence for feature ranking. To realize this idea, we introduce a coefficient given by

$$c_k = \left| \frac{1}{l_+} \sum_{i \in SV_+} x_{i,k} - \frac{1}{l_-} \sum_{j \in SV_-} x_{j,k} \right| \tag{11}$$

where SV_+ denotes the support vectors belong to positive samples, SV_- denotes the support vectors belong to negative samples, l_+ denotes the number of SV_+, l_- denotes the number of SV_-, and $x_{i,k}$ denotes the kth feature of support vector i in input space R^n.

The larger c_k indicates that the kth feature of input space can make more contribution to maintaining the margin large. Therefore, c_k can assist $|w_k|$ for feature ranking. The solution is that, combining the two evidences, we can order the features by ranking $c_k|w_k|$. We present below an outline of the feature selection and classifier training algorithm.

- Input:
  Training examples

$$X_0 = [x_1, x_2, \ldots x_l]^T$$

- Initialize:

  Indices for selected features: $s = [1, 2, \ldots n]$

  Train the SVM classifier using samples X_0
- For $t = 1, \ldots, T$:
    1. Compute the ranking criteria $c_k |w_k|$ according to the trained SVMs
    2. Order the features by decreasing $c_k |w_k|$, select the top M_t features, and eliminate the other features
    3. Update s by eliminating the indices which not belong to the selected features
    4. Restrict training examples to selected feature indices

$$X = X_0(:, s)$$

    5. Train the SVM classifier using samples X
- Outputs:

  The final SVM classifier and features selected by SVMs

Usually, the iterative loop in the algorithm should be terminated before the training samples can not be separated by a hyperplane. Clearly, this algorithm can integrate the two tasks, feature selection and classifier training, into a single consistent framework and make the feature selection process more effective.

3   Experiments

In order to verify the effect of our SVM feature selection method, we use the SVMs without feature selection and the SVMs with feature selection respectively in our experiments for comparison study. Two other feature selection methods (proposed in [4] , [7]) have been compared with our method. The data set used in the first experiment has totally 3433 samples which are all industrial images from a manufacturing company and 2739 samples were selected as training set, the other 694 samples were selected as test set.

In the second experiment, we use the ORL face database of Cambridge University. The non-face images (negative samples) are obtained from the Ground Truth database of Washington University and the total sample size is 2551. Table 1 and Table 2 show the test results after training. Through these results, we see that the success rate can be improved and classification speed increases rapidly at the same time in the test phase using our method.

4   Conclusion and Future Work

On the basis of statistical learning theory, we have presented a feature selection method using SVMs. Our experiments show that this method can remarkably speed up the classification process and improve the generalization performance of the classifier at the same time. In the future work, we will enhance this method and apply it to face classification.

Table 1. Test result on industrial images.

Methods	No. features	Success rate (%)	Test time (s)	Speedup
No selection	4096	96.83	69.2	-
SVM RFE in [4]	500	97.98	2.5	27.68
Selection method in [7]	500	97.55	1.8	38.44
Our method	500	98.27	2.1	32.95

Table 2. Test result on ORL face database.

Methods	No. features	Success rate (%)	Test time (s)	Speedup
No selection	10304	97.43	320.9	-
SVM RFE in [4]	4000	97.62	52.6	6.10
Selection method in [7]	4000	97.33	52.8	6.08
Our method	4000	97.71	51.5	6.23

Acknowledgements

This work was partially supported by the National Natural Science Foundation of China via the grant NSFC 60375022. The authors thank Mr. Bin Huang for the help on preprocessing the training and test data sets.

References

1. Vapnik, V. N.: Statistical Learning Theory. Wiley, New York (1998)
2. Vapnik, V. N.: The Nature of Statistical Learning Theory. Springer-Verlag, New York (2000)
3. Vapnik, V. N.: An Overview of Statistical Learning Theory, IEEE Trans. Neural Networks. vol. 10, no.5, (1999) 988-999
4. Guyon, I., Weston, J., Barnhill, S., Vapnik, V. N.: Gene Selection for Cancer Classification using Support Vector Machines. Machine Learning, vol. 46, (2002) 389-422
5. Mao, K. Z.: Feature Subset Selection for Support Vector Machines Through Discriminative Function Pruning Analysis. IEEE Trans. Systems, Man, and Cybernetics, vol. 34, no. 1, (2004) 60-67
6. Evgeniou, T., Pontil, M., Papageorgiou, C., Poggio, T.: Image Representations and Feature Selection for Multimedia Database Search. IEEE Trans. Knowledge and Data Engineering, vol. 15, no. 4, (2003) 911-920
7. Heisele, B., Serre, T., Prentice, S., Poggio, T.: Hierarchical classification and feature reduction for fast face detection with support vector machine. Pattern Recognition, vol. 36, (2003) 2007-2017

Dimensionality Reduction by Semantic Mapping in Text Categorization

Renato Fernandes Corrêa[1,2] and Teresa Bernarda Ludermir[2]

[1] Polytechnic School, Pernambuco University
Rua Benfica, 455, Madalena, Recife - PE, 50.750-410, Brazil
[2] Center of Informatics – Federal University of Pernambuco
P.O. Box 7851, Cidade Universitária, Recife – PE, 50.732-970, Brazil
{rfc,tbl}@cin.ufpe.br

Abstract. In text categorization tasks, the dimensionality reduction become necessary to computation and interpretability of the results generated by machine learning algorithms due to the high-dimensional vector representation of the documents. This paper describes a new feature extraction method called semantic mapping and its application in categorization of web documents. The semantic mapping uses SOM maps to construct variables in reduced space, where each variable describes the behavior of a group of features semantically related. The performance of the semantic mapping is measured and compared empirically with the performance of sparse random mapping and PCA methods and shows to be better than random mapping and a good alternative to PCA.

1 Introduction

When the data vectors are high-dimensional it is computationally infeasible to use data analysis or pattern recognition algorithms which repeatedly compute similarities or distances in the original data space [1], as well as interpret the results and mining knowledge from models generated by machine learning algorithms.

In text categorization tasks, the documents are normally represented by high-dimensional data vectors with length equals to the number of distinct terms in the vocabulary of the corpus. Thus, methods of dimensionality reduction are essential to implementation of effective text categorization systems.

The objective of this paper is to show a application of a new feature extraction method called semantic mapping in text categorization. The semantic mapping was derived of sparse random mapping method [2] and uses self-organizing maps [3] to cluster terms semantically related (i.e. terms that refer to the same topic).

This paper is organized as follows. In Section 2 and 3 are described the sparse random mapping method and the semantic mapping method respectively. Section 4 describes the metodology and results of the experiments on text categorization [4]. Section 5 contains the conclusions and future works.

2 Sparse Random Mapping

The sparse random mapping method (SRM) [2] is a variation of random mapping (RM) [1], both generated and used in the context of WEBSOM project. WEBSOM is a method for organizing textual documents onto a two-dimensional map display using

N.R. Pal et al. (Eds.): ICONIP 2004, LNCS 3316, pp. 1032–1037, 2004.
© Springer-Verlag Berlin Heidelberg 2004

SOM, the maps are a meaningful visual background providing: an overview of the topics present in the collection and means to make browsing and content-addressable searches.

In the SRM, the original n-dimensional data vector, denoted by x, is multiplied by a random matrix R. The mapping

$$y = R\,x\,, \tag{1}$$

results in a d-dimensional vector y. R is constructed as a sparse matrix, where a fixed number k of ones (tipically 5 or 3 or 2) was randomicaly generated in each column (determining in which extracted features each original feature will participate), and the others elements remained equal to zero.

Experimentally, SRM shows to be a method generally applicable that approximately preserves the mutual similarities between the data vectors [2]. The SRM was used successfully also in [5].

3 Semantic Mapping

The basis of SRM is group orginal features in clusters of co-ocorrence or semantically related features and force the correspondence between clusters and extracted features in the construction of the matrix of projection R.

Initially, the vectors in training set are used as meta-features to describe the original features. In text categorization this description has direct interpretation because the semantics or means of a term can be deduced analyzing the context where this is applied, i.e., the set of documents (or document vectors) where it occurs [6].

The original features are grouped in semantic clusters training a self-organizing map (SOM). In SOM maps, similar training vectors are mapped in the same node or neighboring nodes [7], as co-occurrent terms are represented by similar vectors, clusters of co-occurrent terms are formed. In text categorization these clusters typically correspond to topics or subjects treated in documents and probably contain semantic related terms. The formed maps are called semantic maps. The number of nodes in semantic map must be equals to the number of extracted features wanted.

After the training of semantic map, the matrix R is constructed. Each semantic cluster corresponds to one extracted feature, and a original feature will participate of it if the corresponding cluster is among the k best representative clusters of that original feature. Thus, while in SRM the position of the ones in each column of R is determined randomicaly, in the method of semantic mapping the position of the ones in each column of R is determined in accordance with the semantic clusters where each original feature was mapped. The set of matrices of projection generated by SM is a subset of that generated by SRM, thus SM too approximately preserves the mutual similarities between the data vectors after projection to reduced dimension.

The mapping of the vectors of data for the reduced dimension is done using the Equation (1).

The computational complexity of the SM method is $O(nd(iN + k))$ that is the complexity of the construction of the semantic map with d units by SOM algorithm from n vectors (original characteristics) with N dimensions (number of vectors in training set) for i epochs plus the superior complexity of the construction of the matrix of mapping with k ones in each column. This complexity is smaller than the complexity of PCA ($O(Nn^2) + O(n^3)$ [1]), and still linear to the number of characteristics in the original space as the SRM ($O(nd)$).

4 Experiments

The experiments consist of the application of semantic mapping (SM), sparse random mapping (SRM) and principal component analysis (PCA) to a problem of text categorization [4]. The performances achieved by SRM and PCA were used as references to evaluate the performance of SM.

4.1 Methodology

The performance of the projection methods was measured by the mean classification error generated by five SOM maps in the categorization of the projected document vectors of a test set, trained with the respective projected document vectors of a training set. Four dimensions of projection were used: 100, 200, 300 and 400 extracted features.

Thirty matrices of projection were generated for SRM and SM methods. The number of ones in each column in the projection matrix was: 1, 2, 3 and 5.

The PCA method [3] involves the use of SVD algorithm in the extraction of the principal components of the matrix of correlation of the characteristics in the training set. Four matrices of projection were mounted, one for each dimension, taking the 100, 200, 300 and 400 first components respectively.

The matrices of projection generated by the three methods were applied on boolean vector-documents, where each position indicates the presence or the absence of determined term in the document, thus forming the projected vectors in the reduced dimensions. The motivation in projecting boolean document vectors is to test the methods when the minimum of information is supplied. The projected vectors of the training set and test had been normalized and used to construct the document maps and to evaluate the performance of these respectively.

The classification error for a SOM map is the percentual of documents incorrectly classified when each map unit is labeled according to the category of the document vectors in training set that dominated the node. Each document is mapped to the map node with the closest model vector in terms of Euclidean distance.The document vectors of the test set received the category assigned to the node where they were mapped.

The SOM maps are used here as classifiers because the low sensitivity of the training SOM algorithm to the distortions of similarity caused by random mapping [1]. These SOM maps are denominated document maps.

The documents categorized belong to K1 collection [8]. This collection consists of 2340 Web pages classified in one of 20 news categories at Yahoo: Health, Business, Sports, Politics, Technology and 15 subclasses of Entertainment (without subcategory, art, cable, culture, film, industry, media, multimedia, music, online, people, review, stage, television, variety). The document vectors of the collection were mounted using the vector space model [9]. These vectors were preprocessed eliminating generic and non-informative terms [8]; the final dimension of the vectors was equal to 2903 terms. After preprocessing, the document vectors was divided randomicaly for each category in half for training set and half for test set; the length of each set was 1170 document vectors. The categories were codified and associated to document vectors as labels.

The algorithm used for training SOM maps was batch-map SOM [2] because it is quick and have few adjustable parameters. The SOM maps used to construct the semantic maps and document maps had a rectangular structure with a hexagonal neighborhood to facilitate visualization. The Gaussian neighborhood function was used. For each topology, the initial neighborhood size was equal to half the number of nodes with the largest dimension plus one. The final neighborhood size was always 1. The number of epochs of training was 10 in rough phase and 20 in the fine-tuning phase. The number of epochs determines how mild the decrease of neighborhood size will be, since it is linearly decreasing with the number of epochs. The dimensions of document maps were 12x10 units (as sugested in WEBSOM project [1]) with the model vectors with 100, 200, 300 and 400 features. Because there is no prior knowledge in word clustering, the semantic maps had the most squared possible topologies: 10x10, 20x10, 20x15 and 20x20, with the model vectors with 1170 features. For all SOM maps topology, randomly initialized configurations were obtained using the som_randinit function of somtoolbox.

4.2 Results

The first step was the evaluation of the number of ones needed in each column of the matrices of projection generated by SRM and SM in order to minimize the errors of classification in the test set. The t-test of combined variance [10] was used to compare the peformances of the methods with different numbers of ones, it was applied on the average and the standard deviation of the errors of classification achieved by each method in the test set.

The semantic mapping and SRM generate one better representation of documents in all the dimensions when 2 ones was used in each column of the projection matrix, minimizing the errors of classification. SRM was lesser sensible to the number of ones in the columns of the projection matrix than SM, fact already expected due to the purely random nature of SRM in extract features.

Figure 1 shows the averaged classification error in the test set generated by SRM, SM and PCA in function of the mapping dimensions. It was used 2 ones in each column of the matrices of projection for methods SRM and SM. The bars denote one standard deviation over 150 experiments for SRM and SM (combination of the 30 matrices of projection with the 5 document maps) and 5 for PCA (combination of the matrix of projection with the 5 document maps).

In Figure 1, PCA had the best performance, followed by SM, that generated classification errors smaller than SRM for all the dimensions. The difference between the performances of PCA and SRM is strong significant but lesser than 10%, this fact becomes SRM a good alternative due to high computational cost of the PCA.The SRM's and SM classification errors decreases significantly with increasing of the dimension of projection. In contrast to SRM, the SM and PCA preserve practically the same mutual similarity between document vectors in different dimensions.

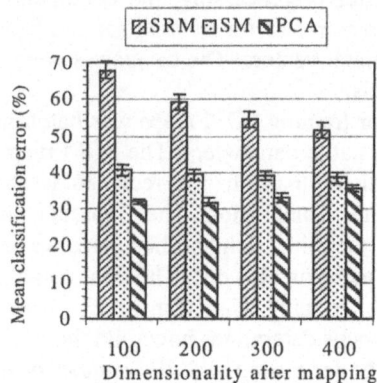

Fig. 1. Classification error as function of reduced dimension of document vectors.

Table 1 shows the best results achieved by each method. All the differences between the performance of the methods are strong significant.

Table 1. Better results generated by method.

Method	Dimension	Trn mean err	Trn std. dev.	Tst mean err	Tst std. dev.
PCA	200	25,93	0,82	31,91	1,08
SM	400	31,46	0,96	38,46	1,43
SRM	400	43,69	1,72	51,72	2,13

5   Conclusions

Theorically and experimentally, the characteristics extracted for the method of semantic mapping (SM) shown to be more representative of the content of the documents and better interpretable that those generated by sparse random mapping (SRM).

SM showed to be a viable alternative to PCA in the dimensionality reduction of high-dimensional data due to the performance relatively close to PCA and the computational cost linear to the number of characteristics in the original space as SRM.

As future works it is intended to test SM method on term-frequency based representation of document-vectors and to modify SM to use weights attributed to each word instead of simple ones in each column with the goal of improve performance.

References

1. Kaski, S.: Dimensionality Reduction by Random Mapping: Fast Similarity Computation for Clustering. Proc. IJCNN'98 Int. Joint Conf. Neural Networks, Vol. 1. (1998) 413-418
2. Kohonen, T., Kaski, S., Lagus, K., Salojärvi, J., Honkela, J., Paatero, V., Saarela, A.: :Self Organization of a Massive Document Collection. IEEE Transaction on Neural Networks, Vol. 11, No. 3, May 2000. IEEE Press (2000)

3. Haykin, S.: Neural Networks: a Comprehensive Foundation – 2nd ed. Prentice-Hall, (1999)
4. Sebastiani, F.: Machine Learning in Automated Text Categorization. Proc. ACM Computing Surveys, Vol. 34, No. 1, March 2002. (2002) 1-47
5. Bingham, E., Kuusisto, J., Lagus, K.: ICA and SOM in Text Document Analysis. Proc. SIGIR'02. Tampere, Finland (2002)
6. Siolas, G., d'Alché-Buc, F.: Mixtures of Probabilistic PCAs and Fisher Kernels for Word and Document Modeling. Artificial Neural Networks - ICANN 2002, International Conference, Madrid, Spain, August 28-30, 2002, Proceedings. Lecture Notes in Computer Science 2415, ISBN 3-540-44074-7. Springer (2002) 769-776.
7. Lin, X., Soergel, D., Marchionini, G.: A Self-organizing Semantic Map for Information Retrieval. Proceedings of the Fourteenth Annual International ACM/SIGIR Conference on Research and Development in Information Retrieval. Chicago, IL (1991) 262-269.
8. Boley, D., Gini, M., Gross, R., Han, E., Hastings, K., Karypis, G., Kumar, V., Mobasher, B., Moore, J.: Partitioning-based Clustering for Web Document Categorization, Decision Support Systems, Vol. 27. (1999) 329-341.
9. Salton, G., McGill, M. J.: Introduction to Modern Information Retrieval. McGraw-Hill, New York (1983)
10. Spiegel, M. R.: Schaum's Outline of Theory and Problems of Estatistics. McGraw-Hill, (1961)

Non-linear Dimensionality Reduction
by Locally Linear Isomaps

Ashutosh Saxena[1], Abhinav Gupta[2], and Amitabha Mukerjee[2]

[1] Department of Electrical Engineering,
Indian Institute of Technology Kanpur, Kanpur 208016, India
Ashutosh.Saxena@ieee.org
[2] Department of Computer Science and Engineering,
Indian Institute of Technology Kanpur, Kanpur 208016, India
{Abhigupt,Amit}@cse.iitk.ac.in

Abstract. Algorithms for nonlinear dimensionality reduction (NLDR) find meaningful hidden low-dimensional structures in a high-dimensional space. Current algorithms for NLDR are Isomaps, Local Linear Embedding and Laplacian Eigenmaps. Isomaps are able to reliably recover low-dimensional nonlinear structures in high-dimensional data sets, but suffer from the problem of short-circuiting, which occurs when the neighborhood distance is larger than the distance between the folds in the manifolds. We propose a new variant of Isomap algorithm based on local linear properties of manifolds to increase its robustness to short-circuiting. We demonstrate that the proposed algorithm works better than Isomap algorithm for normal, noisy and sparse data sets.

1 Introduction

Nonlinear dimensionality reduction involves finding low-dimensional structures in high-dimensional space. This problem arises when analyzing high-dimensional data like human faces, speech waveforms, handwritten characters and natural language. Previous algorithms like Principal Component Analysis, Multidimensional scaling and Independent Component Analysis fail to capture the hidden non-linear representation of the data [1, 2]. These algorithms are designed to operate when the manifold is embedded almost linearly in the high-dimensional space. There are two approaches to solve this problem: Global (Isomaps [3, 4]) and Local (Local Linear Embedding [5] and Laplacian Eigenmaps [6]).

Tenenbaum [3] describes an approach that uses easily measured local metric information to learn the underlying global geometry of a data set based on isomaps. It attempts to preserve geometry at all scales, by mapping nearby points on the manifold to nearby points in low-dimensional space, and faraway points to faraway points. Since, the algorithm aims to find correct geodesic distances by approximating them with a series of euclidean distances between neighborhood points, it gives correct representation of the data's global structure.

Local approaches (LLE [5] and Laplacian Eigenmaps [6]) try to preserve the local geometry of the data. By approximating each point on the manifold with a

N.R. Pal et al. (Eds.): ICONIP 2004, LNCS 3316, pp. 1038–1043, 2004.
© Springer-Verlag Berlin Heidelberg 2004

linear combination of its neighbors, and then using the same weights to compute a low-dimensional embedding. LLE tries to map nearby points on the manifold to nearby points in the low-dimensional representation. In general, local approaches are computationally efficient, but Landmark Isomaps [4] achieve computational efficiency equal to or in excess of existing local approaches. Local approaches have good representational capacity, for a broader range of manifolds, whose local geometry is close to Euclidean, but whose global geometry may not be. Conformal Isomaps [4], an extension of Isomaps, are capable of learning the structure of certain curved manifolds. However, Isomap's performance exceeds the performance of LLE, specially when the data is sparse.

In presence of noise or when the data is sparsely sampled, short-circuit edges pose a threat to Isomaps and LLE algorithms [7]. Short-circuit edges occur when the folds in the manifolds come close, such that the distance between the folds of the manifolds is less than the distance from the neighbors. In this paper, we propose an algorithm which increases the robustness of the Isomaps. Locally Linear Isomaps (LL-Isomaps), a variant of Isomaps are proposed which use the local linearity properties of the manifold to choose neighborhood for each data point. We demonstrate that the proposed algorithm works better than Isomap algorithm for normal, noisy and sparse data sets.

In Section 2 we discuss Tenenbaum's approach using Isomaps and the Roweis approach using Local Linearity to solve the problem. The proposed algorithm has been described in Section 3. In Section 4, results are discussed, followed by conclusion in Section 5.

2   Current Approaches

2.1   Isometric Feature Mapping (Isomaps)

NLDR algorithm reduces the dimensionality of high-dimensional data and hence only the local structure is preserved. This implies that the euclidean distance is meaningful between the nearby points only. Tenenbaum et. al [3] proposed an algorithm that measures the distance between two far-away points on the manifold (called the geodesic distance) and tries to obtain a low-dimensional embedding using these distances.

Isomap algorithm can be described in three steps:

1. Neighbors of each point are determined. The neighbors are chosen as points which are within the ϵ distance or using K-nearest neighbor approach. These neighborhood relations are represented as a weighted graph G over the data points, with edges of weight $d_X(i, j)$ between neighboring points.
2. Isomap estimates the geodesic distances $d_M(i, j)$ between all pairs of points on the manifold M by computing their shortest path distances $d_G(i, j)$ in the graph G. The shortest path can be found by using Floyd-Warshall's algorithm or Dijasktra algorithm.
3. Reduce the dimensionality of the data by using MDS algorithm on the computed shortest path distance matrix.

The residual error of the MDS algorithm determines the performance of the Isomap algorithm. A zero error implies that the computation of the geodesic distance was correct. The dimensionality of a manifold is determined by decrease in the error as dimensionality of low-dimensional embedding vectors Y is increased. The correct low-dimensional embedding is obtained when the error goes below a certain threshold.

2.2 Local Linear Embedding

The LLE algorithm proposed by Roweis et. al [5] uses the fact that the data point and its neighbors lie on a linear patch whose local geometry is characterized by linear coefficients that construct the point. This characterization is also valid in lower dimensions. Suppose the data consist of N real-valued vectors X_i, each of dimensionality n, sampled from some underlying manifold and let Y_i represent global internal coordinates on the manifold (coordinates in low-dimensional space). The algorithm can described in three steps below

1. Assign neighbors to each data point X_i using K-nn approach.
2. Compute the weights W_{ij} that best linearly reconstruct X_i from its neighbors.
3. Compute the low-dimensional embedding vectors Y_i best reconstructed by W_{ij}.

3   Proposed Algorithm (K_{LL} Isomaps)

There is a serious problem in the Isomap algorithm which is referred to as Short Circuiting [7]. When the distance between the folds is very less or there is noise such that a point from a different fold is chosen to be a neighbor of the point, the distance computed does not represent geodesic distance and hence the algorithm fails (Fig. 2(a)).

We propose an algorithm which uses local linearity property of the manifolds to determine the neighborhood of a data point as opposed to using a K-nearest neighbor or ϵ-map. This results in a better neighborhood of the point, which in turn gives lower residual variance and robustness. The problem with the previous algorithms is that they consider only the distance for determining the neighborhood and they fail when the folds of manifold come close to each other. This approach not only overcomes the problem of short-circuiting but also produces better estimates of geodesic distances and hence the residual error is less than the Tenenbaum's algorithm.

The proposed algorithm first finds a candidate neighborhood using K-nearest neighbor (K-nn) approach. The linear combination of the candidate neighbors are used to reconstruct the data-point. The weight for each neighbor can be estimated by reducing the reconstruction error:

$$\epsilon(W) = \sum \left| X_i - \sum_{j \neq i} W_{ij} X_j \right|^2 \tag{1}$$

Now $K_{LL} \leq K$ neighbors are chosen based on the values of reconstruction weights. The neighbors whose Euclidean distance is less and those lying on the locally linear patch of the manifold get higher weights, and hence are selected preferably. These K_{LL} (same for every point) neighbors are used in the rest of the Isomap algorithm (Section 2.1) to calculate geodesic distances and the low-dimensinal embedding. The proposed algorithm has two parameters $\{K, K_{LL}\}$.

4   Results and Discussion

We compare the results of our algorithm for following classes of data:

1. Sparsely Sampled data
2. Noisy Data
3. Dense data without noise

Swiss-roll data set (n=3, d=2) and Synthetic face data set (n=4096, d=3) [3] were used. The quality metric for comparing the proposed algorithm with the Tenenbaum's algorithm is the residual variance at the expected Manifold dimension d. We show that the proposed algorithm not only overcomes the problem of short-circuiting but also gives less residual variance.

In sparsely sampled data sets, the Euclidean distance between points in neighborhood becomes larger as compared to the distance between different folds of the manifold. Tenenbaum's algorithm either faces the problem of short-circuiting (Fig. 1(a)), or has to choose a very low value of K, which gives a large residual variance. The proposed algorithm uses the same number of neighbors (=5) as Tenenbaum's algorithm, but is able to find the correct dimensionality (=3). Fig. 1(b) shows plot of residual variance of proposed K_{LL}-Isomaps and Tenenbaum's K-Isomaps, using best performances of both. The proposed algorithm works with higher value of K_{LL}, even with sparse data, hence gives a much lower value of residual variance. In worst case, when $K_{LL} = K$, the proposed algorithm performs as good as Tenenbaum's algorithm.

Additive-White-Gaussian-Noise (AWGN) with SNR of 10 dB is added to the original Swiss-roll data with 1000 sample points. Tenenbaum's algorithm with $K = 6$ (which works with noiseless data) fails due to short-circuit edges (Fig. 2(a)). Using the proposed method, the problem of short-circuiting is easily removed as shown in Fig. 2(b), and the correct low-dimensional embedding is found more robustly with $K_{LL} = 6$. For noisy data, Tenenbaum's algorithm has to choose a lower value of K to avoid short-circuiting. In Fig. 3(b), the best possible results with Tenenbaum ($K = 5$), and our algorithm ($K = \{7, 4\}$) are shown. It can be seen that our algorithm out-performs Tenenbaum's algorithm.

Even for dense Synthetic Face data (without noise), our algorithm gives better residual variance as compared to Tenenbaum's algorithm (Fig. 3(a)).

5   Conclusion

The Isomap algorithm, with its broad appeal, opened up new frontiers by its various applications; but was not robust to short-circuiting, resulting in drastically

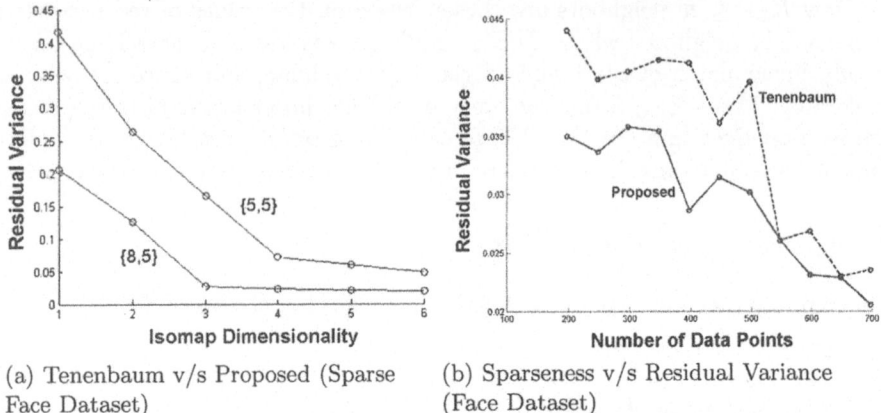

(a) Tenenbaum v/s Proposed (Sparse Face Dataset)

(b) Sparseness v/s Residual Variance (Face Dataset)

Fig. 1. Comparison of Tenenbaum's algorithm with the proposed algorithm. (a) For $N = 349$, Tenenbaum's algorithm is represented by {5,5}, and it predicts the manifold dimensionality to be 4 because of a short-circuit edge. This problem can be overcome by reducing K but this leads to a high residual variance. Proposed K_{LL}-Isomap gives smaller error for the *same* number of neighbors and the dimensionality is correctly predicted to be 3. (b) Comparison of Tenenbaum's Isomap with K_{LL}-Isomap for varying level of sparseness. The number of sample data-points was varied and the error in both the algorithms (with their best case) was computed. The K_{LL}-Isomap outperforms Isomap in all the cases, except two where the errors are same in both the algorithms.

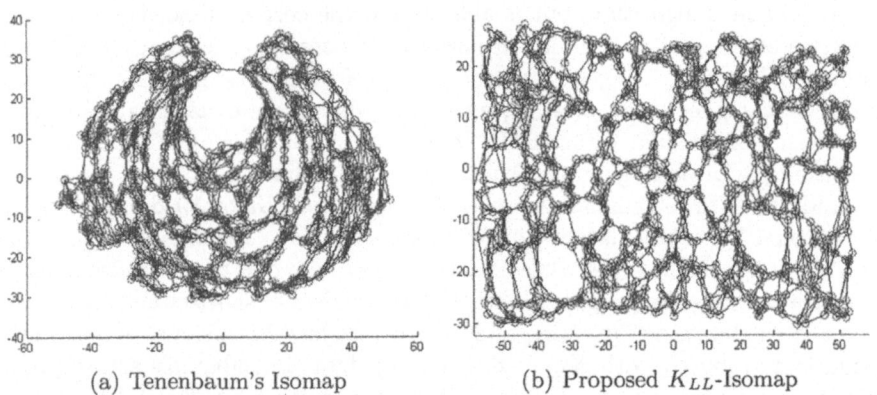

(a) Tenenbaum's Isomap

(b) Proposed K_{LL}-Isomap

Fig. 2. Noisy swiss roll embeddings in two dimensions as obtained by Tenenbaum's Isomap and K_{LL}-Isomaps. The swiss roll dataset consisted of 1000 points. The Isomap algorithm had a short-circuit edge and hence gave incorrect embedding.

different (and incorrect) low-dimensional embedding. We proposed a new variant of Isomaps based on local linearity properties of the manifolds to increase the its robustness to short-circuiting. We demonstrated that the proposed algorithm works better than Isomap algorithm for normal, noisy and sparse data sets.

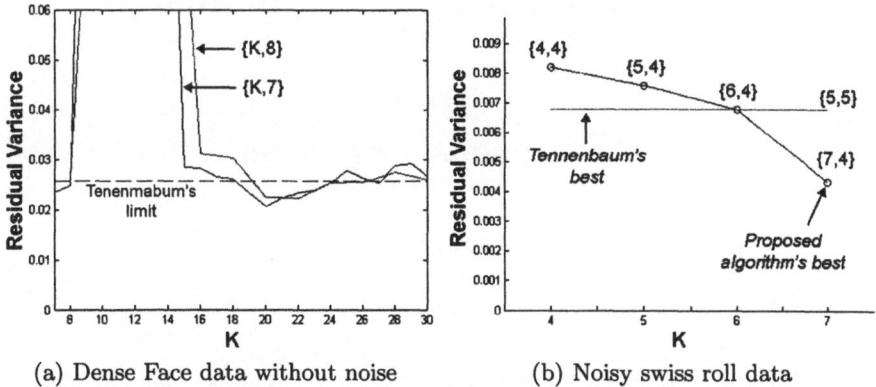

(a) Dense Face data without noise   (b) Noisy swiss roll data

Fig. 3. Accuracy as a function of parameter K. Increasing K gives more choice to the proposed algorithm to choose K_{LL} neighbors on the basis of weights and hence the performance improves (Residual Error decreases).

References

1. Murase, H., Nayar, S.: Visual learning and recognition of 3d objects from appearance. International Journal Computer Vision **14** (1995)
2. J.W. McClurkin, L.M. Optican, B.R., Gawne, T.: Concurrent processing and complexity of temporally encoded neuronal messages in visual perception. Science **253** (1991) 675–657
3. Tenenbaum, J.B., Silva, V.d., Langford, J.C.: A Global Geometric Framework for Nonlinear Dimensionality Reduction. Science **290** (2000) 2319–2323
4. Silva, V.d., Tenenbaum, J.B.: Global versus local methods in nonlinear dimensionality reduction. In S. Becker, S.T., Obermayer, K., eds.: Advances in Neural Information Processing Systems 15. MIT Press, Cambridge, MA (2003) 705–712
5. Roweis, S.T., Saul, L.K.: Nonlinear Dimensionality Reduction by Locally Linear Embedding. Science **290** (2000) 2323–2326
6. Belkin, M., Niyogi, P.: Laplacian eigenmaps and spectral techniques for embedding and clustering. In Dietterich, T.G., Becker, S., Ghahramani, Z., eds.: Advances in Neural Information Processing Systems 14, Cambridge, MA, MIT Press (2002)
7. Balasubramanian, M., Schwartz, E.L., Tenenbaum, J.B., Silva, V.d., Langford, J.C.: The Isomap Algorithm and Topological Stability. Science **295** (2002) 7a

Applications
of Independent Component Analysis

Erkki Oja

Helsinki University of Technology,
Neural Networks Research Centre,
P.O.B. 5400, 02015 HUT, Finland
erkki.oja@hut.fi

Abstract. Blind source separation (BSS) is a computational technique
for revealing hidden factors that underlie sets of measurements or signals.
The most basic statistical approach to BSS is Independent Component
Analysis (ICA). It assumes a statistical model whereby the observed
multivariate data are assumed to be linear or nonlinear mixtures of some
unknown latent variables with nongaussian probability densities. The
mixing coefficients are also unknown. By ICA, these latent variables can
be found. This article gives the basics of linear ICA and reviews the
efficient FastICA algorithm. Then, the paper lists recent applications of
BSS and ICA on a variety of problem domains.

1 Introduction

Blind source separation (BSS) is a computational technique for revealing hidden
factors that underlie sets of measurements or signals. This problem is very general, and many kinds of solutions may be suggested. The most basic statistical
approach to BSS is *Independent Component Analysis* (ICA)[14]. It assumes a
statistical model whereby the observed multivariate data, typically given as a
large database of samples, are assumed to be linear or nonlinear mixtures of
some unknown latent variables. The mixing coefficients are also unknown. The
latent variables are nongaussian and mutually independent, and they are called
the independent components of the observed data. By ICA, these independent
components, also called sources or factors, can be found. Thus ICA can be seen
as an extension to Principal Component Analysis and Factor Analysis. ICA is
a much richer technique, however, capable of finding the sources when these
classical methods fail completely.

In many cases, the measurements are given as a set of parallel signals or time
series. Typical examples are mixtures of simultaneous sounds or human voices
that have been picked up by several microphones, brain signal measurements
from multiple EEG sensors, several radio signals arriving at a portable phone,
or multiple parallel time series obtained from some industrial process. The term
blind source separation is used to characterize this problem.

N.R. Pal et al. (Eds.): ICONIP 2004, LNCS 3316, pp. 1044–1051, 2004.
© Springer-Verlag Berlin Heidelberg 2004

2   Independent Component Analysis: Theory and Algorithms

Assume that we have a set of observations, given as a sample $(\mathbf{x}(1), ..., \mathbf{x}(t),$ $...\mathbf{x}(T))$ of m-dimensional vectors. Typically, in signal and image separation, the index t would stand for the time or spatial location index (for images or 3D voxel arrays, a row-by-row scanning will produce a one-dimensional array). The elements of vector $\mathbf{x}(t)$ would be the signal amplitudes or pixel/voxel gray levels of our m measurement signals or images at the temporal or spatial location t.

ICA is a statistical technique in which we assume that this is an i.i.d. sample from a random vector, denoted by $\mathbf{x}$. In the simplest form of ICA, we assume the following *latent variable model* for our random observation vector $\mathbf{x}$ (for references and details see the text-books [18, 12, 14, 8])

$$\mathbf{x} = \mathbf{As}. \tag{1}$$

There the n -dimensional vector $\mathbf{s}$ consists of *statistically independent* elements s_j called the *sources*. It can be assumed that both $\mathbf{x}$ and $\mathbf{s}$ are zero mean; if not, the sample can always be normalized to zero. The observations x_i are now linear combinations or mixtures of the sources s_j. The matrix $\mathbf{A}$ is called in ICA the *mixing matrix*. A further assumption on the sources s_j is that they are *nongaussian* (except at most one of them). This is in sharp deviation from classical nonparametric techniques like Factor Analysis, in which everything is assumed gaussian, and this is in fact the key property giving the much enhanced power to ICA as compared to the classical statistical expansions.

We may further assume that the dimensions of $\mathbf{x}$ and $\mathbf{s}$ are the same. If originally $\dim \mathbf{x} < \dim \mathbf{s}$, or there are more sources than observed variables, then the problem becomes quite difficult [14]. If, on the other hand, $m = \dim \mathbf{x} > \dim \mathbf{s} = n$, then model (1) implies that there is redundancy in $\mathbf{x}$ which is revealed and can be removed by performing Principal Component Analysis on $\mathbf{x}$. In fact, another related step called *whitening* is very useful as a preprocessing stage in ICA, and it can be combined into the dimensionality reduction. In whitening, $\mathbf{x}$ is linearly transformed into another n-dimensional vector that has unit covariance matrix. Whitening can be always performed because we only need the covariance matrix of observation vectors $\mathbf{x}$, which can be readily estimated from a sample. Let us assume in the following that whitening has always been performed in the model, and denote simply by $\mathbf{x}$ the whitened observation vector whose dimension is the same as that of the source vector $\mathbf{s}$.

Whitening has another desirable side-effect: it can be shown that then $\mathbf{A}$ is an *orthogonal* matrix, for which $\mathbf{A}^{-1} = \mathbf{A}^T$. So, if we knew matrix $\mathbf{A}$, we could directly solve the unknown source vector $\mathbf{s}$ from the model by

$$\mathbf{s} = \mathbf{A}^T\mathbf{x}.$$

It is an interesting finding that very few assumptions suffice for solving the mixing matrix and, hence, the sources. All we need is the assumption that the sources s_i are statistically independent and nongaussian, except at most one [10].

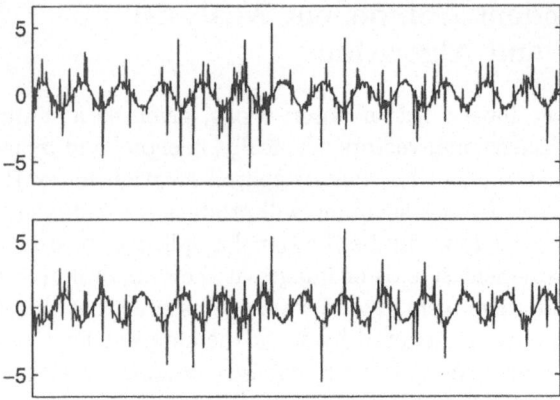

Fig. 1. Mixed signals.

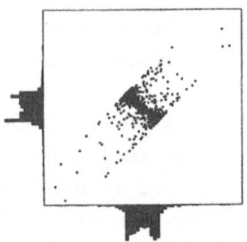

Fig. 2. Histogram of the two amplitudes of the mixed signals x_1, x_2.

Consider the following simple example: we have two signals, shown in Fig. 1, that are linear combinations or mixtures of two underlying independent nongaussian source signals. This example is related to model (1) in such a way that the elements x_1, x_2 of the random vector $\mathbf{x}$ in (1) are the amplitudes of the two signals in Fig. 1. The signals provide a sample $\mathbf{x}(1), \ldots \mathbf{x}(T)$ from this random vector. The joint histogram of the sample vectors is plotted in Fig. 2; each point in the scatter plot corresponds to one time point in Fig. 1. The vector $\mathbf{x}$ is now white in the sense that x_1 and x_2 are zero mean, uncorrelated, and have unit variance. This may not be apparent from the histogram but can be verified by estimating the covariance matrix of all the points.

The example suggests a method that in fact is highly useful and forms the basis of some practical ICA algorithms. Consider a line passing through the origin at the center of the data cloud in Fig. 2. Denote a unit vector defining the direction of the line by $\mathbf{w}$. Then the projection of a data point $\mathbf{x}$ on the line is given by $y = \mathbf{w}^T \mathbf{x}$. This can be considered as a random variable whose density is approximated by the histogram of the projections of all the data points in the cloud on this line. No matter what is the orientation of the line, it always holds that y has zero mean and unit variance. The unit variance is due to $\mathrm{E}\{y^2\} = \mathrm{E}\{(\mathbf{w}^T \mathbf{x})^2\} = \mathbf{w}^T \mathrm{E}\{\mathbf{x}\mathbf{x}^T\}\mathbf{w} = \mathbf{w}^T \mathbf{w} = 1$ where we have used the facts that $\mathbf{x}$ is white ($\mathrm{E}\{\mathbf{x}\mathbf{x}^T\} = \mathbf{I}$) and $\mathbf{w}$ has unit norm ($\mathbf{w}^T \mathbf{w} = 1$).

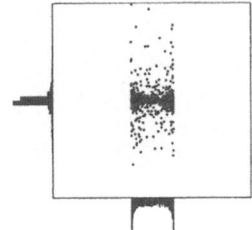

Fig. 3. Histogram of the two amplitudes of the separated signals y_1, y_2.

However, it is easy to see from Fig. 2 that the density of y will certainly vary as the orientation of the line varies, meaning that all the moments of y cannot stay constant. In fact, no other moment than the first and second ones is constant. What is most important is that any such moment, say, $E\{y^3\}$ or $E\{y^4\}$ or in fact $E\{G(y)\}$, with $G(y)$ a nonlinear and non-quadratic function, will attain a number of maxima and minima when the orientation of the line goes full circle, and some of these extrema coincide with orientations in which the 2-dimensional density factorizes into the product of its marginal densities - meaning independence.

In Fig. 3, the coordinate system has been rotated so that the fourth moment $E\{y^4\}$ is maximal in the vertical direction and minimal in the horizontal direction. We have found two new variables $y_1 = \mathbf{w}_1^T \mathbf{x}$ and $y_2 = \mathbf{w}_2^T \mathbf{x}$, with $\mathbf{w}_1, \mathbf{w}_2$ orthonormal, that satisfy

$$p(y_1, y_2) = p(y_1)p(y_2)$$

with $p(.)$ the appropriate probability densities. The variables are thus independent and it holds

$$\mathbf{y} = \mathbf{W}\mathbf{x}$$

where $\mathbf{W} = (\mathbf{w}_1 \mathbf{w}_2)^T$. We have solved the inverse of the model (1) and obviously found the mixing matrix: $\mathbf{A} = \mathbf{W}^T$.

Fig. 4 shows $y_1(t), y_2(t)$ again arranged in their correct time order. It is seen that they form two signals, one a random nongaussian noise and the other one a deterministic sinusoid. These were in fact the original signals that were used to make the artificial mixtures in Fig. 1. In the context of separating time series or signals, the ICA technique is an example of blind signal separation.

The above illustrative example can be formalized to an efficient mathematical algorithm. What we need is a numerical method to maximize, say, the fourth moment $E\{y^4\}$ in terms of a unit norm weight vector $\mathbf{w}$. Mathematically, the criterion is then

$$\max J^{kurt}(\mathbf{w}) = E\{y^4\} = E\{(\mathbf{w}^T \mathbf{x})^4\}, \ \|\mathbf{w}\| = 1. \tag{2}$$

A possibility for maximizing this is gradient ascent. The gradient of $E\{y^4\}$ with respect to $\mathbf{w}$ is $4E\{y^3 \mathbf{x}\} = 4E\{(\mathbf{w}^T \mathbf{x})^3 \mathbf{x}\}$. We could build a simple gradient ascent algorithm on this. However, gradient methods are notoriously slow. A better idea is a fast algorithm with higher-order convergence speed. Such a method

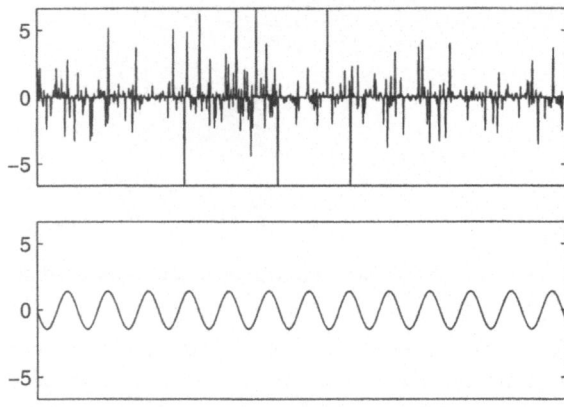

Fig. 4. Separated signals.

is provided by the FAstICA algorithm. For finding one independent component (one weight vector **w**), the algorithm is as follows:

1. Choose the initial value randomly for the weight vector **w**.
2. Repeat Steps 3,4 until the algorithm has converged:
3. Normalize **w** to unit norm.
4. Update **w** by

$$\mathbf{w} \leftarrow \mathrm{E}\{(\mathbf{w}^T\mathbf{x})^3\mathbf{x}\} - 3\mathbf{w}. \tag{3}$$

This algorithm was introduced in Ref. [15] and further extended and analyzed in Ref. [13]; for a detailed review, see Ref. [14]. The FastICA algorithm is available in public-domain software [11] from the author's web pages. The algorithm can be run either in a deflation mode, in which the orthogonal weight vectors (columns of the mixing matrix **A**) can be found one at a time, or in a parallel mode, in which all the independent components and the whole matrix **A** are solved in one iteration.

The above method of fourth order moment maximization can be shown[13, 14] to be an example of a powerful criterion of finding *maximally nongaussian orthogonal directions* through the multidimensional density $p(\mathbf{x})$. Cost functions like maximum likelihood, minimization of marginal entropies, or minimal mutual information are shown to be intimately related to this basic criterion. Other classical algorithms to solving the basic linear ICA model have been reported e.g. in Refs. [2, 4, 5, 7, 9, 8, 10, 17], as reviewed in Ref. [14].

As shown in Ref. [16], maximizing the nongaussianity of a projection $y = \mathbf{w}^T\mathbf{x}$ can be achieved by looking at the extrema of a generic cost function $\mathrm{E}\{G(y)\} = \mathrm{E}\{G(\mathbf{w}^T\mathbf{x})\}$ over the unit sphere $\|\mathbf{w}\| = 1$. For the generic cost function, the FastICA algorithm is otherwise the same as for kurtosis, but the central updating rule

$$\mathbf{w} \leftarrow \mathrm{E}\{(\mathbf{w}^T\mathbf{x})^3\mathbf{x}\} - 3\mathbf{w} \tag{4}$$

must now be replaced by

$$\mathbf{w} \leftarrow \mathrm{E}\{g(\mathbf{w}^T\mathbf{x})\mathbf{x}\} - \mathrm{E}\{g'(\mathbf{w}^T\mathbf{x})\}\mathbf{w}. \tag{5}$$

It is easy to see that for $g(y) = y^3$, (5) becomes (4) because $E\{y^2\} = 1$. For more details on the general FastICA algorithm, see Ref. [14] and the software documentation[11].

When the additive noise cannot be assumed to be zero in the ICA model, we have the noisy ICA model, also termed independent factor analysis [3]. This is due to the fact that it is otherwise similar to the factor analysis model, with the difference that the factors y_i are not uncorrelated (thus independent) gaussians, but rather independent nongaussians. Some solution methods are reviewed in the textbook Ref. [14].

3 Applications of ICA

For applications of ICA and Blind source separation, the most comprehensive literary source are the Proceedings of the 4 Workshops on BSS and ICA, held since 1999 (Refs. [6, 20, 19, 1]). A complete review of application papers even from these sources is outside the scope of this article. The following categories and topics can be listed. By far the most voluminous applications are speech and audio separation as well as the analysis of biomedical signals and images. In the talk, some applications will be covered in detail.

- Speech and audio separation
  - Separation of speech signals
  - Speaker detection
  - Speech enhancement in car environment
  - Separation of musical instruments
  - Extraction of drum tracks
- Biomedical signal and image processing
  - Artefact removal and source extraction from electroencephalography (EEG)
  - BSS in Magnetoencephalography (MEG)
  - BSS in Electrocardiography (ECG)
  - BSS in Magnetocardiography (MCG)
  - BSS in Electrogastrograms (EGG)
  - BSS in Electromyography (EMG)
  - Atrial fibrillation analysis
  - Spatio-temporal analysis of fMRI images
  - Finding spatial signal patterns in brain optical imaging
  - Finding spatial IC's from myocardial PET images
- Telecommunications and antenna arrays
  - ICA in adaptive beamforming
  - Signal separation in CDMA mobile communications
  - Multiuser detection
  - Interference cancellation
  - BSS in contactless identification systems
  - Radar detection

- Image processing
  - Finding non-redundant spatial image filters
  - ICA in lossless image coding
  - Segmentation of textured images
  - Separating sources in astrophysical images
  - Analysis of the cosmic microwave background
  - Terrain classification in multispectral satellite images
  - Detecting faces from videos
  - Digital image watermarking
  - Separation of reflections in images
- Text and document processing
  - Clustering of text documents
  - ICA in multimedia modelling
  - Finding topics in text collections
  - Web image retrieval
- Industrial applications
  - Rotating machine vibration analysis
  - Acoustical machine monitoring
  - Identification of components in NMR spectroscopy and other spectra
  - ICA in infrared imaging
  - Testing of metal slabs
  - ICA in chemical reactions
- Environmental issues
  - Seismic analysis
  - Analysis of telluric current data
  - Analysis of volcanic explosion quakes
  - Analysis of weather and climate patterns
  - IC's of odour signals in electronic nose
- Financial time series analysis
  - Preprocessing for exchange rate time series prediction
  - Finding independent stock portfolios
  - Independent Factor model in finance
- Bioinformatics
  - Gene classification from microarray data
  - IC's of DNA substrings.

References

1. S. Amari, A. Cichocki, S. Makino, and N. Murata, editors. *Proc. of the 4th Int. Workshop on Independent Component Analysis and Signal Separation, Nara, Japan, April 1-4, 2003*. Brain Science Institute, Riken, Tokyo, 2003.
2. S.I. Amari, A. Cichocki, and H.H. Yang. A new learning algorithm for blind source separation. In *Advances in Neural Information Processing Systems 8*, pages 757–763. MIT Press, 1996.
3. H. Attias. Independent factor analysis. *Neural Computation*, 11(4):803–851, 1999.

4. A.J. Bell and T.J. Sejnowski. An information-maximization approach to blind separation and blind deconvolution. *Neural Computation*, 7:1129–1159, 1995.
5. J.-F. Cardoso. Source separation using higher order moments. In *Proc. IEEE Int. Conf. on Acoustics, Speech and Signal Processing (ICASSP'89)*, pages 2109–2112, Glasgow, UK, 1989.
6. J. F. Cardoso, C. Jutten, and P. Loubaton, editors. *Proc. of the 1st Int. Workshop on Independent Component Analysis and Signal Separation, Aussois, France, January 11–15, 1999*. INPG, Grenoble, 1999.
7. J.-F. Cardoso and B. Hvam Laheld. Equivariant adaptive source separation. *IEEE Trans. on Signal Processing*, 44(12):3017–3030, 1996.
8. A. Cichocki and S. Amari. *Adaptive Blind Signal and Image Processing*. Wiley, New York, 2002.
9. A. Cichocki and R. Unbehauen. Robust neural networks with on-line learning for blind identification and blind separation of sources. *IEEE Trans. on Circuits and Systems*, 43(11):894–906, 1996.
10. P. Comon. Independent component analysis – a new concept? *Signal Processing*, 36:287–314, 1994.
11. The FastICA MATLAB package. Available at
    http://www.cis.hut.fi/projects/ica/fastica/.
12. M. Girolami, editor. *Advances in Independent Component Analysis*. Springer, London, 2000.
13. A. Hyvärinen. Fast and robust fixed-point algorithms for independent component analysis. *IEEE Trans. on Neural Networks*, 10(3):626–634, 1999.
14. A. Hyvärinen, J. Karhunen, and E. Oja. *Independent Component Analysis*. Wiley, New York, 2001.
15. A. Hyvärinen and E. Oja. A fast fixed-point algorithm for independent component analysis. *Neural Computation*, 9(7):1483–1492, 1997.
16. A. Hyvärinen and E. Oja. Independent component analysis by general nonlinear Hebbian-like learning rules. *Signal Processing*, 64(3):301–313, 1998.
17. C. Jutten and J. Herault. Blind separation of sources, part I: An adaptive algorithm based on neuromimetic architecture. *Signal Processing*, 24:1–10, 1991.
18. T.-W. Lee. *Independent Component Analysis – Theory and Applications*. Kluwer, 1998.
19. T.-W. Lee, T.-P. Jung, S. Makeig, and T. Sejnowski, editors. *Proc. of the 3rd Int. Workshop on Independent Component Analysis and Signal Separation, San Diego, CA, December 9–13, 2001*. Salk Institute, CA, 2001.
20. P. Pajunen and J. Karhunen, editors. *Proc. of the 2nd Int. Workshop on Independent Component Analysis and Blind Signal Separation, Helsinki, Finland, June 19–22, 2000*. Otamedia, Espoo, 2000.

Supervised Independent Component Analysis with Class Information

Manabu Kotani, Hiroki Takabatake, and Seiichi Ozawa

Kobe University, Kobe 657-8501, Japan
ozawasei@kobe-u.ac.jp

Abstract. Independent Component Analysis (ICA) is a method to transform from mixed signals into independent components. ICA has been so far applied to blind signal separation problems such as sound, speech, images, and biological signals. Recently, ICA is applied to feature extraction for face, speech, and image recognitions. Since ICA is an unsupervised learning, extracted independent components are not always useful for recognition purposes. In this paper, we propose a new supervised learning approach to ICA using class information to enhance the separability of features. The proposed method is implemented by a three-layered feedforward network in which target signals are given to the output units. The defined objective function is composed of the following two terms: one is for evaluating independency of hidden outputs and the other is for evaluating errors between output signals and their targets. Simulations are performed for some datasets in the UCI repository to evaluate the effectiveness of the proposed method. In the proposed method, we obtain higher recognition accuracies as compared with a conventional unsupervised ICA algorithm.

1 Introduction

Independent Component Analysis (ICA) has been mainly applied to blind signal separation and has used to recover independent signals from mixture signal such as speech, images, and biological signals. Applications of ICA to feature extractions have been also a recent topic of research interest. There are some studies about feature extractions using ICA for images and sounds [1]-[5]. Olshausen and Field [1] have shown the characteristics of basis functions extracted from natural scenes by a sparse coding algorithm. The characteristics of basis functions are similar to the response properties of neurons in a primary visual cortex. Similar results were also obtained in the other ICA algorithms [2]-[3].

On the other hand, there are several researches using extracted features for the pattern recognition. Bartlett and Sejnowski [6] have applied to recognition of human faces and shown that features obtained by the infomax algorithm [7] were better than features obtained by Principal Component Analysis (PCA) with regard to recognition accuracy. Ozawa et al. [8] have applied FastICA [9] algorithm to Japanese hand-written Hiragana character and performed recognition experiments with the extracted features. However, the recognition accuracy

N.R. Pal et al. (Eds.): ICONIP 2004, LNCS 3316, pp. 1052–1057, 2004.
© Springer-Verlag Berlin Heidelberg 2004

using ICA features was almost the same as that of PCA features. Thus, this result suggests that ICA is not always superior to PCA and it is not easy to obtain high-performance features using ICA.

To overcome this difficulty, there are some studies about introducing class information into ICA. Ozawa et al. have proposed a supervised ICA in which linear separability of extracted features projected on each axis is maximized based on an objective function used in the linear discriminant analysis [10]. Umeyama et al.[11] have proposed another Supervised ICA (SICA) by maximizing correlation with control signals. Although SICA is a powerful approach to extracting effective features, the problem is how proper control signals are given in advance.

In this paper, we propose a new type of supervised ICA without such control signals. The proposed method uses a model of three-layered neural networks whose target signals are class information. The objective function consists of two components: one is for increasing independency of hidden outputs and the other is for reducing the squared error between network outputs and their targets. That is to say, the learning algorithm maximizes the independence of hidden outputs and minimizes output errors. We apply the proposed method to various standard databases in the UCI machine learning repository [12].

2   SICA Using Class Information

The proposed supervised ICA is implemented by a three-layer feedforward networks whose outputs correspond to class information. Activation functions of the hidden and the output units are the sigmoid functions, $f(u) = 1/(1+\exp[-u])$. The numbers of input and hidden units are Ns and the number of output units is M that is the number of classes.

Hidden outputs, y_p^h, for the pth input are given as follows:

$$y_p^h = f\left(W^h x_p + \theta^h\right),\tag{1}$$

where $x_p = [x_{p1}, ..., x_{pN}]^T$ is the pth input vector, $[\cdot]^T$ is the transpose of $[\cdot]$, $\theta^h = [\theta_1^h, ..., \theta_N^h]^T$ is a N dimensional threshold vector of hidden layer, $W^h = [w_1^h, ..., w_N^h]^T$ is a $N \times N$ weight matrix between input and hidden layers, and w_i^h is a N dimensional weight vector between the input layer and the ith unit in the hidden layer.

On the other hand, M dimensional output vector, $y_p^o = [y_{p1}^o, ..., y_{pM}^o]^T$, in the output layer is defined as follows:

$$y_p^o = f\left(W^o y_p^h + \theta^o\right),\tag{2}$$

where θ^o is a M dimensional threshold vector in the output layer, W^o is a $M \times N$ weight matrix between the hidden and the output layers, and w_i^o is a M dimensional weight vector between the hidden layer and the ith unit in the output layer.

The objective function, J, is defined as follows:

$$J = J_B + \beta J_I,\tag{3}$$

where J_B is the objective functions for the error minimization and J_I is the objective functions for the independence between hidden outputs, and β is a constant value.

J_B is defined as

$$J_B = \frac{1}{2M}E\{(y_p^o - t_p)^T(y_p^o - t_p)\}, \tag{4}$$

where $E\{\cdot\}$ means expectation and t_p is a target vector for the pth input. Only one element of this target vector is set to one to express the class information; that is, the dimensions of a target vector is the same as the number of classes.

On the other hand, J_I is defined as follows:

$$J_I = (\kappa^L - \kappa^I)^2, \tag{5}$$

where κ^L is an average of absolute values of kurtosis for the hidden units and κ^I is an average of absolute values of kurtosis for independent components when the original ICA is performed for the input pattern. The average of absolute values of kurtosis is defined as

$$\kappa = \frac{\sum_{j=1}^{N}|kurt_j|}{N} \tag{6}$$

$$kurt_j = \frac{E\{(y_{pj}^I)^4\}}{\left(E\{(y_{pj}^I)^2\}\right)^2} - 3, \tag{7}$$

where $y_{pj}^I = (w_j^h)^T x_p$.

The gradient of Eq. (3) with respect to weight vectors, W^o and W^h, are derived as following equations:

$$\Delta W^o = -\frac{\partial J}{\partial W^o} = E\{\delta_p^o(y_p^h)^T\} \tag{8}$$

$$\Delta W^h = -\frac{\partial J}{\partial W^h} = E\{\delta_p^h(x_p)^T\} + \beta D_p^I, \tag{9}$$

and

$$\delta_{pk}^o = (t_{pk} - y_{pk}^o)y_{pk}^o(1 - y_{pk}^o) \tag{10}$$

$$\delta_{pj}^h = (\sum_k \delta_{pk}^o w_{kj}^o)y_{pj}^h(1 - y_{pj}^h) \tag{11}$$

$$D_{pji}^I = \frac{4\,\text{sign}(\kappa_j^L)}{N(E\{(y_{pj}^I)^2\})^3}\left(E\{x_{pi}(y_{pj}^I)^3\}E\{(y_{pj}^I)^2\}\right.$$
$$\left. - E\{(y_{pj}^I)^4\}E\{x_{pi}y_{pj}^I\}\right), \tag{12}$$

where $\text{sign}(u)$ is a sign function whose output is 1 for $u > 0$ and -1 for $u \leq 0$.

The learning algorithm of the weight matrix is shown as follows:

$$W(t+1) = W(t) + \eta(\Delta W(t) + \gamma \Delta W(t-1)) \tag{13}$$

where η and γ are constants, t is a learning iteration, and W means W^h and W^o.

Table 1. Specifications of five databases from UCI repository.

Database	Samples	Dimensions	Classes
Bupa	345	6	2
Pima	768	8	2
Iris	150	4	3
Tae	151	5	3
Credit	690	14	2

The learning procedures are described as follows:

1. Dimension reduction: dimensions of input patterns are reduced to N dimensions according to magnifications of eigenvalues using principal component analysis.
2. Whitening: average and variance of the data are normalized to 0 and 1, respectively.
3. Initial condition: FastICA[9] is applied to the whitening data and κ^I in Eq. (5) is calculated from the independent components obtained by FastICA. W^h is set to an unit matrix and W^o is set to random values of $-0.2 \sim +0.2$.
4. Learning: W is trained using Eq. (13).

3   Experiments

We applied to 5 databases from UCI machine learning database repository[12] to evaluate the effectiveness of the proposed method. These databases are composed of only numerical data and have no missing data. Table 1 shows the number of samples, dimensions, and classes in each database.

Recognition accuracies were calculated using 10-fold cross validation method. In the initial condition of the learning procedure, we used the symmetric FastICA and the nonlinear function was $G(u) = -\exp(-u^2/2)$. The target patterns for the output units were 1 for the kth output unit and 0 for the other output units when the input pattern were labelled in the kth class. Experiments were performed in various combinations of $\beta = 10^{-4} \sim 10^{-1}$, $\eta = 10^{-2} \sim 1$, and $\gamma = 0.1$. The number of input units, which was the reduced dimensions, was also varied from $N = 1$ to the original dimension of each database listed in Table 1.

After the learning was over, we evaluated the separability of extracted features, which correspond to hidden outputs, through several recognition accuracy tests. In this experiment, we adopted k-Nearest Neighbor (k-NN)as a classifier. This is because we want to see the potential effectiveness of our proposed scheme introducing supervisor into ICA independent of classifiers. We applied two kinds of distance measures to the k-NN: Euclid distance measure, D_{euc}, and cosine distance measure, D_{cos}, are defined as follows:

$$D_{euc}(\boldsymbol{y}_p^I, \tilde{\boldsymbol{y}}_p^I) = \sqrt{\sum_i (y_{pi}^I - \tilde{y}_{pi}^I)^2} \tag{14}$$

Table 2. Recognition accuracy using PCA, ICA, and the proposed method for all databases.

Database	PCA	ICA	ICAC
Bupa	69.0	72.2	75.5
Pima	75.4	75.1	78.7
Iris	96.7	93.3	98.7
Tae	65.3	67.3	70.0
Credit	86.8	87.8	88.0

$$D_{cos}(\boldsymbol{y}_p^I, \tilde{\boldsymbol{y}}_p^I) = \frac{\sum_i y_{pi}^I \tilde{y}_{pi}^I}{\sqrt{\sum_i (y_{pi}^I)^2}\sqrt{\sum_i (\tilde{y}_{pi}^I)^2}} \tag{15}$$

where $\boldsymbol{y}_p^I$ and $\tilde{\boldsymbol{y}}_p^I$ are features vectors of the template data and the test data, respectively. Recognition accuracies were calculated using these distance measures and $k = 1 \sim 15$ of the k-NN.

Table 2 shows the highest recognition accuracy among various results for each database. Table 2 also shows results using PCA, FastICA, and the proposed method. Results using PCA and FastICA are calculated as well as the case in which the recognition accuracies using the proposed method are calculated. Features in PCA and FastICA are principal components and independent components, respectively. "PCA" means the results using PCA, "ICA" means those using FastICA, and "ICAC" means those using the proposed method. These results show that the performance using ICA is better than that using PCA for three databases: Bupa, Tae, and Credit. On the other hand, the performance using the proposed method is better than those using PCA and ICA for all databases. Furthermore, we performed 5% one-sided T-test against these results and obtained that there were significant differences for Pima and Iris between the ICAC and the ICA.

4   Conclusions

We presented a new approach to feature extraction in pattern recognition tasks. Independent Component Analysis (ICA) was extended in a supervised learning fashion in order to extract useful features with good class separability. The proposed method was implemented by a three-layered feedforward network, in which pattern features are obtain from hidden outputs and the class information is given to output units as their targets. An objective function to be minimized includes two components: one is for increasing independency of hidden outputs and the other is for minimizing errors between network outputs and their target signals. The evaluation was carried out for five datasets of UCI repository: Bupa, Pima, Iris, Tae, and Credit. In the experiment, some promising results were obtained as a feature extraction method. The recognition performance of the proposed method was higher than the performance of an unsupervised ICA algorithm. This suggests that not only independent property of features but also the enhancement in class separability is needed to extract good features.

However, we often need long training time to converge on a fixed point, which result from two different objective functions to be defined. In some cases, these two objective functions would be competitive, and then the convergence speed might be slow. The current version of our proposed method is based on a standard optimization technique, steepest descent method. Hence, some other optimization techniques could be applied to our supervised ICA scheme in order to improve the convergence speed. This is left as our future works.

References

1. B. A. Olshausen and D. J. Field, "Emergence of Simple-Cell Receptive Field Properties by Learning a Sparse Code for Natural Images," *Nature*, vol. 381, pp. 607–609, 1996.
2. A. J. Bell and T. J. Sejnowski, "The 'Independent Components' of Natural Scenes are Edge Filters," *Vision Research*, vol. 37, pp. 3327–3338, 1997.
3. A. Hyvärinen and P. Hoyer, "Emergence of Phase and Shift Invariant Features by Decomposition of Natural Images into Independent Feature Subspaces," *Neural Computation*, vol. 12, pp. 1705–1720, 2000.
4. A. J. Bell and T. J. Sejnowski, "Learning the Higher-order Structure of a Natural Sound," *Network: Computation in Neural Systems*, vol. 7, pp. 261–266, 1996.
5. M. S. Lewicki and T. J. Sejnowski, "Learning Nonlinear Overcomplete Representations for Efficient Coding," *Advances in Neural Information Processing Systems*, vol. 10, pp. 556–562, 1998.
6. M. S. Bartlett, J. R. Movellan, and T. J. Sejnowski, "Face Recognition by Independent Component Analysis," *IEEE Trans. on Neural Networks*, vol. 13, no. 6, pp. 1450–1464, 2002.
7. A. J. Bell and T. J. Sejnowski, "An Information Maximization Approach to Blind Separation and Blind Deconvolution," *Neural Computation*, vol. 7, pp. 1129–1159, 1995.
8. S. Ozawa, and M. Kotani, "A Study of Feature Extraction and Selection Using Independent Component Analysis," *Proc. of International Conference on Neural Information Processing*, CD-ROM, 2000.
9. A. Hyvarinen, "Fast and Robust Fixed-Point Algorithms for Independent Component Analysis," *IEEE Trans. on Neural Networks*, vol. 10, no. 3, pp. 626–634, 1999.
10. Y. Sakaguchi, S. Ozawa, and M. Kotani, "Feature Extraction Using Supervised Independent Component Analysis by Maximizing Class Distance," *Proc. of Int. Conf. on Neural Information Processing 2002*, vol. 5, pp. 2502–2506, 2002.
11. S. Umeyama, S. Akaho, and Y. Sugase, "Supervised Independent Component Analysis and Its Applications to Face Image Analysis," *Technical Report of IEICE*, vol. NC99, no. 2, pp. 9–16, 1999.
12. http://www1.ics.uci.edu/~mlearn/

Automated Diagnosis of Brain Tumours Using a Novel Density Estimation Method for Image Segmentation and Independent Component Analysis Combined with Support Vector Machines for Image Classification

Dimitris Glotsos[1], Panagiota Spyridonos[1], Panagiota Ravazoula[2],
Dionisis Cavouras[3], and George Nikiforidis[1]

[1] Medical Image Processing and Analysis Laboratory, Department of Medical Physics,
School of Medicine, University of Patras, 26500, Patras, Greece
dimglo@med.upatras.gr
[2] Department of Pathology, University Hospital of Patras, 26500, Rio, Greece
[3] Department of Medical Instrumentation Technology, Technological Institute of Athens,
12210, Aigaleo, Athens, Greece

Abstract. A computer-aided system was developed for the automatic diagnosis of brain tumours using a novel density estimation method for image segmentation and independent component analysis (ICA) combined with Support Vector Machines (SVM) for image classification. Images from 87 tumor biopsies were digitized and classified into low and high-grade. Segmentation was performed utilizing a density estimation clustering method that isolated nuclei from background. Nuclear features were quantified to encode tumour malignancy. 46 cases were used to construct the SVM classifier. ICA determined the most important feature combination. Classifier performance was evaluated using the leave-one-out method. 41 cases collected from a different hospital were used to validate the systems' generalization. For the training set the SVM classifier gave 84.9%. For the validation set classification performance was 82.9%. The proposed methodology is a dynamic new alternative to computer-aided diagnosis of brain tumours malignancy since it combines robust segmentation and high effective classification algorithm.

1 Introduction

In diagnosing brain tumours astrocytomas, the determination of tumor malignancy grade is a critical step that determines patient management [1-2]. However, the doctor subjective interpretation in grade assessment has been shown to influence diagnostic accuracy [3]: 36% to 62% agreement among different pathologists. The necessity was generated to objectify the diagnostic process. Towards this direction, computer-aided diagnosis (CAD) systems were introduced.

Previous studies proposing CAD systems in astrocytomas grading investigated supervised classification techniques with examples linear discriminant analysis [4], neural networks [5-6], self-editing nearest neighbor nets [7], decision tree and nearest neighbor models [8-9]. However, these studies were focused in processing specialized processed microscopic images. CAD systems utilizing daily clinical protocols, are the only useful to support the regular diagnostic procedure followed by the pa-

N.R. Pal et al. (Eds.): ICONIP 2004, LNCS 3316, pp. 1058–1063, 2004.
© Springer-Verlag Berlin Heidelberg 2004

thologists [10]. Little effort has been made to construct CAD systems able to process images obtained from routine stained Hematoxylin-Eosin (HE) biopsies because these images are difficult to be segmented.

In this work, we propose a novel application of Support Vector Machines (SVM) combined with independent component analysis (ICA) for classification of astrocytomas as low or high-grade according to routine clinical protocols [11-12]. The possibility that SVM may generalize ensuring good performance even with limited training samples, made the selection of SVM most attractive. HE routine stained images were segmented using a novel clustering algorithm recently proposed by the authors [13] that is based on the concept of density estimation in an autocorrelation feature space. In contradiction to other approaches [4-9], the proposed system's generalization to unseen clinical data was evaluated.

2   Methods and Materials

Clinical material comprised 87 biopsies of astrocytomas collected from the University Hospital of Patras, (46/87) Greece and the METAXA Hospital, (41/87) Greece. Tumour grade was defined as low (30/87) or high-grade (57/87) according the WHO system by 2 independent pathologists. Microscopic images from biopsies were digitized (768x576x8 bit) at a magnification of x400 using a light microscopy imaging system consisting of a Zeiss KF2 microscope and an Ikegami color video camera.

2.1   Image Segmentation and Feature Extraction

A segmentation algorithm was performed (figure 1) to separate nuclei from surrounding tissue in order to quantify diagnostic features from cell nuclei. The algorithm takes as input the grey level of each pixel. Subsequently, maps the input space in an autocorrelation feature space. Then, calculates non-parametrically the PDF of data in that space using a modification of the Probabilistic Neural Network (PNN) [20]. Subsequently, seeks the two most prominent peaks of the PDF. These peaks are characterized as cluster centroids and based on a k-nearest heuristic, data are assigned either belonging to the nuclei cluster or surrounding tissue cluster. A binary image is produced, with nuclei pixels lighted white and the remaining background lighted black. Small noisy regions are corrected with size filtering and fill holes operations. The final segmentation derives by superimposing the original image to the binary filtered one.

The original data space is transformed into an autocorrelation feature space because the differences between data belonging to different clusters are more prominent in that space [13]. The transformation is defined as follows: Assuming we have N d-component feature vectors $x_1,...,x_N$, we derive two-component autocorrelation feature vectors $a_1...,a_N$. For simplicity, we assume N is even and define for all $i=1,...,N$

$$B_i = \sum_{j=1, j \neq i} x_i^T x_j \text{ and } a_i = \left[(1 - B_i)B_i, (1 - B_i)^2 B_i \right]$$ (1)

for i=1,...N, with a_i the autocorrelation features.

The modified PNN based algorithm presents 5 layers [13], instead of four as originally proposed for supervised classification [20]: 1) The input layer with one node for each data sample. 2) The pattern where the Probability density function (PDF) based on each data sample is calculated by using a Gaussian kernel as activation function. 3) The summation layer where the PDF of all data is computed by adding up the PDF estimates based on individual data samples. 4) The clustering layer where clusters centroids are defined as the peaks of the PDF. 5) In the output layer data belonging to clusters are detected based on a k-nearest neighbor heuristic.

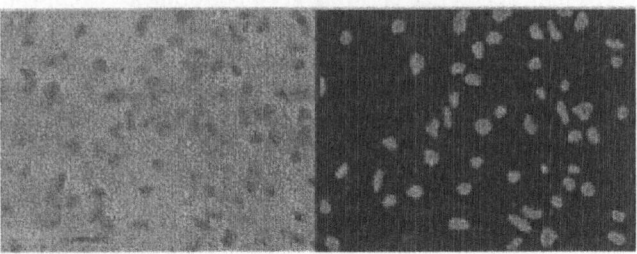

Fig. 1. Example of a typical HE stained brain tumour biopsy and the resulted segmented image

Two kinds of features were generated, selecting at least 50 non-overlapping nuclei per case (patient): 18 related to the size and shape of the cell nucleus and 22 textural features (first-order, co-occurrence, run-length based) that encoded chromatin distribution and nuclear DNA content. Details for the calculation of these features can be found elsewhere [14]. After feature generation, each distinct case was represented by a 40-dimension feature vector.

2.2 Feature Selection and Classification

46 cases collected from the PATRAS hospital were used to construct the SVM classifier. There are many feature selection methods for preserving class separability [24-25]. In the current study, the most important feature combination was determined using ICA. With ICA, those features that are statistically independent can be extracted from a population of variables. In this work we utilized the FastICA MATLAB package originally proposed in [15]. Exhaustive search could not be used due to the high complexity of SVM setting. Classifier performance was evaluated using the leave-one-out method [16]. 41 cases collected from the METAXA hospital were used to validate the systems' generalization retaining the vector combination derived from ICA and the same parameters for the SVM classifier. The results for both clinical datasets are presented in table 1.

By mapping input vectors into a higher dimension feature space and defining the hyperplane that has the maximum distance from the closest training data, the SVM [17] can be utilized for binary classification problems with discriminant as follows:

$$g(\mathbf{x}) = \text{sign}\left(\sum_{i=1}^{N} \alpha_i y_i K(\mathbf{x}, \mathbf{x}_i) + b \right) \tag{2}$$

where $\mathbf{x}_i$ training data belonging to either class $y_i \in \{+1,-1\}$, N the number of training samples, α_i, b weight coefficients and K the transformation or kernel function [19]. Kernel functions utilized were the radial basis function (RBF) kernel with value of

$\gamma = 1 / (2\sigma^2)$ set equal to 0.5 after testing values from 0.005 to 6 and polynomial kernel of degree 1 and 2. The adjustable parameter C that specifies the importance of misclassifications, was experimentally determined equal to 10.

$$K_{RBF}(x,x_i)=\exp\left(\tfrac{-|z-z_i|^2}{2\sigma^2}\right), \sigma=spread, \quad K_{POLYNMIAL}(x,x_i)=\left(\left(x^T x_i\right)+1\right)^d, d=degree \qquad (3)$$

The optimization problem of calculating parameters a_i was solved by using the routine quadprog as suggested in [18] provided with the MATLAB optimization toolbox.

3 Results

SVM with polynomial kernel of degree 2 optimized classification performance in the training set resulting in 84,9% accuracy (table 1). The best vector consisted of 5 textural features (inertia, inverse different moment, energy and correlation derived from the co-occurrence matrix and long run length from the run length matrix) and 3 morphological (area, roundness and concavity). Retaining the best vector combination and the same parameters for the SVM classifier, the system ability to generalize was evaluated. The overall classification accuracy was 82,9%.

Table 1. Truth Table demonstrating classification results for both clinical datasets

	PATRAS training set			METAXA validation set		
Diagnosis	Low-grade	High-grade	accuracy	Low-grade	High-grade	accuracy
Low-grade	14	5	73,7%	8	3	72,7%
High-grade	2	25	92,6%	4	26	86,7%
accuracy			84,9%			82,9%

4 Discussion and Conclusions

To evaluate the performance of the segmentation algorithm, it was not practical to make synthetic images, since modeling of nuclei images is not easy. The most reliable method proposed so far [21], is to compare the results with the reference of a manual segmentation performed by the experts. Under this perspective, the physicians concluded on the success on boundary detection in terms of correct and wrongly delineated nuclei. On average 93% of all nuclei were correctly delineated. The misclassification result of 7% may be considered of limited significance since the number of nuclei ranged from 50 to 120 for each image.

The results obtained with the SVM classifier, are comparable to those obtained by cross-validation discriminant analysis [22], and better than those obtained by the nearest-neighbor approach [8] and fuzzy logic models [23] proposed for automatic grading of astrocytomas. However it has to be stressed that the staining method employed in the present work is the one adopted in every day clinical practice and it is not as accurate in staining nuclei as the specialized methods used in other studies. Additionally, in contrast to previous work [4-9], the introduced methodology was validated for clinical material collected from two hospitals.

The proposed methodology is a dynamic new alternative to computer aided diagnosis of brain tumours malignancy since it combines robust segmentation and high effective classification algorithm. Thus, it can provide a powerful tool to support the regular diagnostic procedure followed by the pathologists.

Acknowledgements

We thank European Social Fund (ESF), Operational Program for Educational and Vocational Training II (EPEAEK II) and particularly the Program IRAKLEITOS for funding the above work.

References

1. DeAngelis, L.M.: Brain tumors. New England Journal of Medicine, 344 (2001) 114-123
2. Shapiro, W.: Biology and treatment of malignant gliomas. Oncology 12 (1998) 233-240
3. Mittler, M., Walters, B., Stopa, E.: Observer reliability in histological grading of astrocytoma. Journal of Neurosurgery 85 (1996) 1091-1094
4. Scarpelli, M., Bartels, P., Montironi, R., Thompson, D.: Morphometrically assisted grading of astrocytomas. Analytical and quantitative Cytology and Histology 16 (1994) 351-356
5. Kolles, H., v.Wangenheim, A.: Automated grading of astrocytomas based on histo- morphometric analysis of ki-67 and Fleugen stained paraffin sections. Analytical cellular pathology 8 (1995) 101-116
6. McKeown, M., Ramsay, A.: Classification of Astrocytomas and Malignant Astrocytomas by Principal Component analysis and a Neural Net. Neuropathology & Experimental Neurology 55 (1996) 1238-1245
7. Kolles, H., v. Wangenheim, A., Rahmel, J., Niedermayer, I., Feiden, W.: Data-Driven Approaches for decision making in Automated Tumour Grading. An example in Astrocytoma Grading. Analytical and Quantitative Cytology and Histopathology, 18 (1996) 298-304
8. Decaestecker, C., Salmon, I., Dewitte, O., Camby, I., Van Ham, P.: Nearest-neighbor classification for identification of aggressive versus nonaggressive astrocytic tumours by means of image cytometry-generated variables. Journal of Neurosurgery 86 (1997) 532-537
9. Sallinen, P., Sallinen S., Helen, T., Rantala, I., Helin, H., Kalimo, H.: Grading of diffusely infiltrating astrocytomas by quantitative histopathology, cell proliferation and image cytometric DNA analysis. Neuropathology and Applied Neurobiology 26 (2000) 319-331
10. Glotsos, D., P., Petalas, P., Cavouras, D., Ravazoula, I., Dadioti, P., Lekka, I., Nikiforidis, G.: Computer-Based Malignancy Grading of Astrocytomas Employing a Support Vector Machine Classifier, the WHO Grading System and the Regular Hematoxylin-Eosin Diagnostic Staining Procedure. Analytical and Quantitative Cytology and Histopathology 26 (2003) 77-83
11. Kleihues, P., Burger P., Scheithauer, B.: Histological typing of tumours of the central nervous system. Geneva: World Health Organization (1993)
12. Spyridonos, P., Ravazoula, P., Cavouras, D., Nikiforidis, G.: Neural Network based segmentation and classification system for the automatic grading of histopathological sections of urinary bladder carcinoma. Analytical and Quantitative Cytology Histopathology 24 (2002) 317-324
13. Glotsos, D., Tohka, J., Soukka, J., Ruotsalainen, U.: A new Approach to Robust Clustering by Density Estimation in an Autocorrelation Derived Feature Space. Proceedings of the 6th NORDIC signal processing symposium (2004) 296-299

14. Spyridonos, P., Ravazoula, P., Cavouras, D., Berberidis, K., Nikiforidis, G.: Computer-based grading of haematoxylin-eosin stained tissue sections of urinary bladder carcinomas. Medical Informatics & The Internet in Medicine 26 (2001) 179-190
15. Deniz, O., Castrillon, M., Hernadez M.: Face recognition using independent component analysis and support vector machines. Pattern recognition letters 24 (2003) 2153-2157
16. Theodoridis, S., Koutroubas K.: Pattern recognition. Academic Press (1999) 342
17. Kechman, V.: Learning and Soft Computing, MIT, USA (2001) 121-189
18. Christianini, N., Taylor, J.S.: An introduction to support vector machines and other kernel-based learning methods. Cambridge University Press, UK (2000) 135-136
19. Schad, L., Schmit, H., Lrenz, W., Scarpelli, M., Bartels, R.: Numerical grading of astrocytomas. Medical informatics 12 (1987) 11-22
20. Specht, D.: Probabilistic neural networks. Neural Networks 3 (1990) 109-118
21. Lee, K-M., Street, N.: A fast and robust approach for automated segmentation of breast cancer nuclei. Proceedings of the IASTED International Conference on Computer Graphics and Imaging (1999) 42-47.
22. Reinhold, N., Schlote, W.: Topometric Analysis of Diffuse Astrocytomas. Analytical and Quantitative Cytology and Histopathology 25 (2003) 12-18
23. Belacel, N., Boulassel, M.: Multicriteria fuzzy assignment method: a useful tool to assist medical diagnosis. Artificial intelligence in medicine 21 (2001) 201-207
24. Yi Lu Murphey, Hong Guo: Automatic feature selection - a hybrid statistical approach. Proceedings of the 15th International Conference on Pattern Recognition, (2000) 382 – 385
25. Bressan, M., Vitria, J.: On the selection and classification of independent features. IEEE Transactions on Pattern Analysis and Machine Intelligence 25 (2003) 1312 - 1317

Temporal Independent Component Analysis for Separating Noisy Signals*

Liqing Zhang

Department of Computer Science and Engineering
Shanghai Jiaotong University, Shanghai 200030, China
zhang-lq@cs.sjtu.edu.cn

Abstract. In this paper, we formulate the problem of blind source separation into temporal independent component analysis. The main purpose of such a formulation is to use both high-order statistics and temporal structures of source signals. In order to accelerate the training process, we employ the conjugate gradient algorithm for updating the demixing model. Computer simulations are presented to show the separation performance of the proposed algorithm. It is observed that the proposed approach has advantages in separating correlated signals or noisy signals in short time windows. Thus it is promising to overcome the over-separating problem by using both the temporal structures and high-order statistics.

1   Introduction

Independent component analysis (ICA) has been accepted as a standard data analysis tool in the neural network and signal processing societies [1, 2]. However, there still exist a number of problems in dealing with real world data using ICA. In many applications, the problem usually does not satisfy the basic assumptions of ICA model. One typical application of ICA is electroencephalographic (EEG) data analysis. EEG usually is very noisy and its mixing model is time-variable. One challenging problem is to extract and localize evoked potentials from EEG measurements in a very short time window. Still another problem is the over-separating problem [3, 4]. In order to tackle the problems, we suggest to explore both the high order statistics and temporal structures of source signals. The main idea is to formulate the blind separation into a framework of independent residual analysis. By analyzing the mutual independence of the residual signals, we can derive learning algorithms for the demixing model and the temporal structures.

When we consider the separation problem in a short time window, the sample data is not sufficient to test the statistics. For example, the empirical correlation of two independent variables may not close to zero due to insufficient data. This makes it difficult for many high-order statistics algorithm to separate sources from measurements in a short time window. The independent residual analysis

---

* This work was supported by the National Natural Science Foundation of China under grant 60375015.

N.R. Pal et al. (Eds.): ICONIP 2004, LNCS 3316, pp. 1064–1069, 2004.
© Springer-Verlag Berlin Heidelberg 2004

explores both the temporal structures and high order statistics, providing a possible solution to blind separation in a short time window.

2   Problem Formulation

Assume that s_i, $i = 1, \cdots, n$ are mutually spatially independent source signals, of which each temporally correlated with zero mean. Suppose that source $s_i(k)$ is modelled by a stationary AR model,

$$s_i(k) = \sum_{p=1}^{N} a_p^i s_i(k - p) + \varepsilon_i(k), \tag{1}$$

where N is the degree of the AR model and $\varepsilon_i(k)$ is zero-mean, independently and identically distributed (that is, white) time series called the residual. For the sake of simplicity, we use the notation $A_i(z) = 1 - \sum_{p=1}^{N} a_p^i z^{-p}$, z is the z-transform variable. Since in the blind separation setting, the source signals are unknown, we need to impose some constraints on the linear filters. We assume that the linear filters $A_i(z)$ are minimum phase throughout this paper. Suppose that sensor signals are instantaneous mixtures of the source signals. Let $\mathbf{x}(k) = [x_1(k), \cdots, x_n(k)]^T$ be the set of linearly mixed signals,

$$\mathbf{x}(k) = \mathbf{H}\mathbf{s}(k). \tag{2}$$

Here, $\mathbf{H} = (H_{ij})$ is an $n \times n$ unknown nonsingular mixing matrix. Blind source separation problem is to find a linear transform which transforms the sensor signals into maximally mutually independent components, which are considered as the estimates of source signals. Let $\mathbf{W}$ be an $n \times n$ nonsingular matrix which transforms the observed signals $\mathbf{x}(k)$ to

$$\mathbf{y}(k) = \mathbf{W}\mathbf{x}(k). \tag{3}$$

The general solution to the blind separation problem is to find a matrix $\mathbf{W}$ such that $\mathbf{W}\mathbf{A} = \mathbf{\Lambda}\mathbf{P}$, where $\mathbf{\Lambda} \in \mathbf{R}^{n \times n}$ is a nonsingular diagonal matrix and $\mathbf{P} \in \mathbf{R}^{n \times n}$ is a permutation matrix.

3   Cost Function

In this section, we introduce the mutual information of residual signals as a criterion for training the demixing matrix and temporal structure parameters. The residual independent analysis provides us a new way to explore both the temporal structures and high-order statistics of source signals. From the source model, we have $\varepsilon(k) = \mathbf{A}(z)\mathbf{s}(k)$, where $\mathbf{A}(z)$ can be estimated via the linear prediction method if the source signals $\mathbf{s}(k)$ are known. When the temporal structure $\mathbf{A}(z)$ and the demixing matrix $\mathbf{W}$ is not well-estimated, the residual signals

$$\mathbf{r}(k) = (r_1(k), \cdots, r_n(k))^T = \mathbf{A}(z)\mathbf{W}\mathbf{x}(k) \tag{4}$$

are not mutually independent. Therefore, it provides us a new criterion for training the demixing model and temporal structures to make the residuals $\mathbf{r}(k)$ spatially mutually independent and temporally identically independently distributed.

Assume $q(\mathbf{r})$ is the probability density function of $\mathbf{r}$ and $q_i(r_i)$ is the marginal probability density function of r_i, $i = 1, \cdots, n$. Now we introduce the mutual information rate $I(\mathbf{r})$ between a set of stochastic processes $r_1, \cdots, r_n$ as

$$I(\mathbf{r}) = -H(\mathbf{r}) + \sum_{i=1}^{n} H(r_i), \tag{5}$$

where $H(r_i)$ and $H(\mathbf{r})$ are the entropies of random variables r_i and $\mathbf{r}$ respectively. For blind deconvolution problem, Amari et al [5] and Pham [6] simplify the first term of cost function (5) and derive a cost function as follows

$$l(\mathbf{W}, \mathbf{A}(z)) = -\frac{1}{2\pi j} \oint_\gamma \log |det(\mathbf{A}(z)\mathbf{W})| z^{-1} dz - \frac{1}{L} \sum_{k=1}^{L} \sum_{i=1}^{n} \log q_i(r_i(k)). \tag{6}$$

where j is the imaginary unit of complex numbers, and the path integral is over the unit circle γ of the complex plane. The first term of right side of equation (6) is introduced to prevent the filter $\mathbf{W}$ from being singular. To simplify the cost function, we calculate the first term of the right side of equation (6) as follows

$$\log |det(\mathbf{A}(z)\mathbf{W})| = \log |det(\mathbf{W})| + \log |det(\mathbf{A}(z))| \tag{7}$$

Because the temporal filters $\mathbf{A}(z)$ is causal and minimum phase, we can easily verify

$$\frac{1}{2\pi j} \oint_\gamma \log |det(\mathbf{A}(z))| z^{-1} dz = 0. \tag{8}$$

Now combining equations (7), (8) with (6), we obtain a simplified cost function for independent residual analysis

$$l(\mathbf{W}, \mathbf{A}(z)) = -\log(|det(\mathbf{W})|) - \frac{1}{L} \sum_{k=1}^{L} \sum_{i=1}^{n} \log q_i(r_i(k)). \tag{9}$$

Independent residual analysis can be formulated into the semiparametric model [7]. The probability density function q and the temporal filter $\mathbf{A}(z)$ are seen as the nuisance parameters in the semiparametric model. The demixing matrix $\mathbf{W}$ is called as the parameters of interest. The semiparametric approach suggests using an estimating function to estimate the parameter of interest, regardless of the nuisance parameters. In this paper, we suggest to estimate the nuisance parameters in order to have better separating performance of the algorithm.

4   Conjugate Gradient Algorithm

In this section, we derive a learning algorithm based on the conjugate gradient descent approach for the demixing matrix. We assume that the probability density functions and the temporal filters are known for a moment during the derivation of a learning algorithm for the demixing matrix. To describe the conjugate gradient method for minimizing cost function, we need first to calculate the natural gradient

$$\nabla l(\mathbf{W}, \mathbf{A}(z)) = (-\mathbf{I} + \frac{1}{L} \sum_{k=1}^{L} \sum_{p=0}^{N} \mathbf{A}_p \left[\varphi\left(\mathbf{r}(k)\right) \mathbf{y}^T(k-p) \right]) \mathbf{W}. \qquad (10)$$

where $\varphi(\mathbf{r}) = (\varphi_1(r_1), \cdots, \varphi_n(r_n))^T$ is the vector of activation functions, defined by $\varphi_i(r_i) = -\frac{q_i(r_i)}{q_i(r_i)}$.

Given an initial value $\mathbf{W}_0$ and $k = 1$, the conjugate gradient algorithm starts out by searching in the steepest descent direction (negative of the gradient) on the first iteration.

$$\mathbf{H}_0 = -\nabla l(\mathbf{W}_0, \mathbf{A}(z)). \qquad (11)$$

Now we perform one-dimensional search algorithm to find the minimum point of the cost function $l(\mathbf{W}, \mathbf{A}(z))$

$$\mathbf{W}_k = \exp(t_* \mathbf{H}_{k-1} \mathbf{W}_{k-1}^{-1}) \mathbf{W}_{k-1}, \ t_* = \arg\min_t l(\mathbf{W}_{k-1}(t)), \qquad (12)$$

along the geodesic: $\mathbf{W}_{k-1}(t) = \exp(t \mathbf{H}_{k-1} \mathbf{W}_{k-1}^{-1}) \mathbf{W}_{k-1}$. The new search direction $\mathbf{H}_k$ is defined by the following equation

$$\mathbf{H}_k = -\nabla l(\mathbf{W}_k) + \gamma_k \tau \mathbf{H}_{k-1} \qquad (13)$$

where $\tau \mathbf{H}_{k-1}$ is the parallel translation from $\mathbf{W}_{k-1}$ to $\mathbf{W}_k$, i.e.

$$\tau \mathbf{H}_{k-1} = \mathbf{H}_{k-1} \mathbf{W}_{k-1}^{-1} \mathbf{W}_k. \qquad (14)$$

The value γ_k in equation (13) is evaluated by

$$\tau_k = <\mathbf{G}_k - \tau \mathbf{G}_{k-1}, \tau \mathbf{G}_k> / <\tau \mathbf{G}_{k-1}, \tau \mathbf{G}_{k-1}> . \qquad (15)$$

For the geometrical structures, such as the geodesic and Riemannian metric of nonsingular matrices, refer to [8]. The conjugate gradient algorithm search the minimum point along the geodesic, which produces generally faster convergence than steepest descent directions. Both theoretical analysis and computer stimulations show that the conjugate gradient algorithm has much better learning performance than the natural gradient does.

Here we briefly introduce learning algorithms for adapting the nuisance parameters in the semiparametric ICA model. From stability analysis, we see that that temporal structures might also affect the leaning performance of the natural gradient algorithm. By using the gradient descent approach, we obtain the learning algorithm for the filter coefficients a_k^i

$$\Delta a_p^i(k) = -\eta_k' \frac{1}{L} \sum_{k=1}^{L} \varphi_i \left\{ r_i(k) \right\} y_i(k-p), \qquad (16)$$

where η_k' is the learning rate. For activation function adaptation, refer to [9].

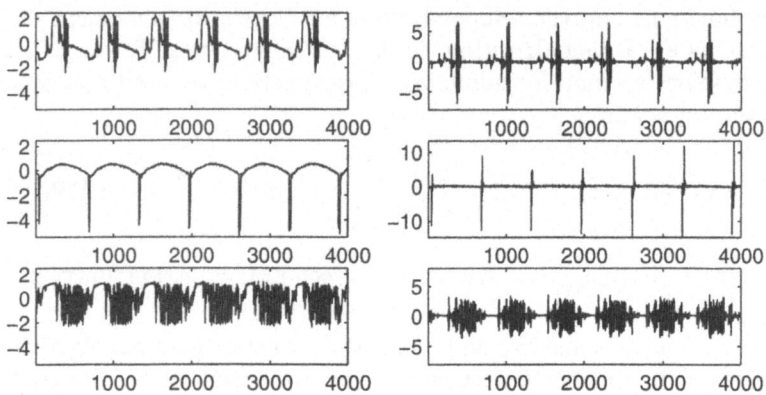

Fig. 1. First row illustrates three source signals sampled from three images, and second row shows their residuals.

5   Computer Simulations and Conclusions

In this section, we present computer simulations to demonstrate the performance of our proposed independent residual analysis (IRA) algorithm. To evaluate its performance, we employ the multichannel inter-symbol interference [10],

$$M_{ISI} = \sum_{i=1}^{n} \left[\frac{|\sum_{j=1}^{n} |g_{ij}| - \max_j |g_{ij}|}{\max_j |g_{ij}|} + \frac{|\sum_{j=1}^{n} |g_{ji}| - \max_j |g_{ji}|}{\max_j |g,_{ji}|} \right].$$

where $\mathbf{G} = (g_{ij}) = \mathbf{W} * \mathbf{H}$. In our simulation, the source signals are sampled from three building images. Figure 1 plots the source signals and their residuals. Due to temporal structure in the images, the empirical correlations between the three source signals are not close to zero. However, the empirical correlations between the three residual signals become much smaller then those of the source signals. This motivates us to use the residual signals to identify the demixing model and to recover source signals as well. We apply different ICA algorithms, such as IRA, JADE, and SOBI, to a random mixture of the three signals. Additive noises are added to the mixing model with signal-to-noise ratios, varying from 5db to 30db. The averaged separation performances of 20 trials are showed in figure 2. Because the source signals are not mutually independent, Algorithms JADE and SOBI cannot separate well from their mixture. Computer simulations indicated that the proposed independent residual analysis (IRA) achieves much better separation performance than the other algorithms do.

In conclusion, exploring the temporal structures and high order statistics of the source signals provides us a new dimension for blind source separation. Both theoretical analysis and computer simulations show the independent residual analysis works well ever when the empirical correlations of source signals are not close to zero. This suggests that the independent residual analysis approach provide a possible solution to blind separation in a short time window and the over-separating problem in separation with insufficient data.

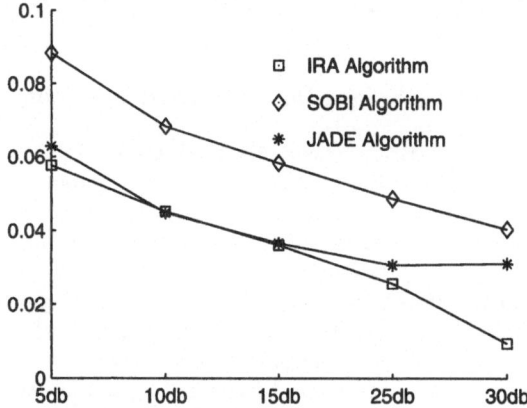

Fig. 2. The inter-symbol interference index comparisons between different algorithms, where the horizontal axe indicates the signal-to-noise ratios, and the vertical axe indicates the inter-symbol interference index.

References

1. Hyvärinen, A., Karhunen, J., Oja, E.: Independent Component Analysis. John Wiley & Sons, Chichester, UK (2001)
2. Cichocki, A., Amari, S.: Adaptive Blind Signal and Image Processing. John Wiley, Chichester, UK (2003)
3. Hyvarinen, A., Sarela, J., Vigario, R.: Spikes and bumps: Artefacts generated by independent component analysis with insufficient sample size. In: Proc. ICA'99, Aussois, France (1999) 425–429
4. Särelä, J., Vigário, R.: Overlearning in marginal distribution-based ica: Analysis and solutions. J. Mach. Learn. Res. **4** (2003) 1447–1469
5. Amari, S., Cichocki, A.: Adaptive blind signal processing– neural network approaches. Proceedings of the IEEE **86** (1998) 2026–2048
6. Pham, D.: Mutual information approach to blind separation of stationary sources. In: Proc. ICA'99, Aussois, France (1999) 215–220
7. Amari, S., Cardoso, J.F.: Blind source separation– semiparametric statistical approach. IEEE Trans. Signal Processing **45** (1997) 2692–2700
8. Zhang, L.: Geometric structures and unsupervised learning on manifold of nonsingular matrices. Neurocomputing (2004) Accepted
9. Zhang, L., Amari, S., Cichocki, A.: Self-adaptive blind source separation based on activation function adaptation. IEEE Transactions on Neural Networks **15** (2004) 1–12
10. Amari, S., Cichocki, A., Yang, H.: A new learning algorithm for blind signal separation. In Tesauro, G., Touretzky, D., Leen, T., eds.: Advances in Neural Information Processing Systems 8 (NIPS*95). (1996) 757–763

Blind Dereverberation of Single-Channel Speech Signals Using an ICA-Based Generative Model

Jong-Hwan Lee[1], Sang-Hoon Oh[2], and Soo-Young Lee[3]

[1] Brain Science Research Center
and Department of Electrial Engineering & Computer Science
Korea Advanced Institute of Science and Technology, Daejeon 305-701, Korea
Phone: +82-42-869-5431, Fax: +82-42-869-8490
jhlee@neuron.kaist.ac.kr
[2] Department of Information Communication Engineering
Mokwon University, Daejeon, 302-729, Republic of Korea
[3] Brain Science Research Center
Department of BioSystems and Dept. of Electrical Engineering & Computer Science
Korea Advanced Institute of Science and Technology, Daejeon 305-701, Korea

Abstract. In this paper, an adaptive blind dereverberation method based on speech generative model is presented. Our ICA-based speech generative model can decompose speeches into independent sources. Experimental results show that the proposed blind dereverberation model successfully performs even in non-minimum phase channels.

1   Introduction

In real room environments, sounds are corrupted with delayed versions of themselves reflected from walls. This room reverberation severely degrades intelligibility of speeches and performance of automatic speech recognition system [1]. In some applications, it is necessary to recover an unknown source signal using only observed signal through an unknown convolutive channel. This problem is called the blind deconvolution and also known as the blind dereverberation when convolving channels are room impulse responses. Almost every methods for the blind deconvolution are developed under the assumption that a source signal is independent identically distributed (IID) and non-Gaussian [2–8]. When an IID non-Gaussian source signal is convolved with a multi-path channel, the probability density function (p.d.f.) of the received signal approaches to Gaussian due to the central limit theorem. Deconvolution can then be accomplished by adapting a deconvolution filter which makes the p.d.f. of deconvolved signal away from Gaussian [2–4]. When sources are not IID such as speeches, the existing algorithms cannot be directly applied.

In this paper, we make a generative model of speeches, which linearly decompose them into independent components. In the first stage of our model, we extract independence transform matrix using independent component analysis (ICA) of natural human speech signals. Using the independence transformation of speeches, we derive blind dereverberation learning rules based on the Least Square (LS) method [6, 7].

N.R. Pal et al. (Eds.): ICONIP 2004, LNCS 3316, pp. 1070–1075, 2004.
© Springer-Verlag Berlin Heidelberg 2004

2   ICA-Based Speech Generative Model

We adopt ICA algorithms to find efficient representations of speech signals such that their sample by sample redundancy is reduced significantly. This redundancy reduction leads nonstationary correlated speech signals to IID-like signals.

ICA assumes a source vector $\mathbf{s}$ whose components $s_i (i = 1, \cdots, N)$ are mutually independent. We can only observe linear combinations

$$\mathbf{x} = \mathbf{A}_I \mathbf{s} \tag{1}$$

where $\mathbf{A}_I$ is an $N \times N$ mixing matrix and their columns are called as basis vectors. After ICA adaptation which minimizes the mutual information among unknown sources [3, 5], estimated sources will be as independent as possible. If the observation vector is a frame of speech, we can find an independent signal vector and related basis vectors. Here we will call $\mathbf{W}_I = \mathbf{A}_I^{-1}$ as the "independence transform matrix".

To learn $\mathbf{W}_I$ from natural human speech signals, we used 10 sentences from one speaker (mcpm0), which corresponds to DR1 New England dialect of the train set in the TIMIT continuous speech corpus. 8kHz sampling was used to reduce computation time. We assumed 16 basis vectors for the ICA-based speech generative model and each speech frame were composed of 16 samples, i.e. 2ms time interval. Figure 1 shows a diagram of the speech generative model. A part of mcpm0's sentence, 'she had your dark suit', can be generated with independent sources through trained 16 basis vectors.

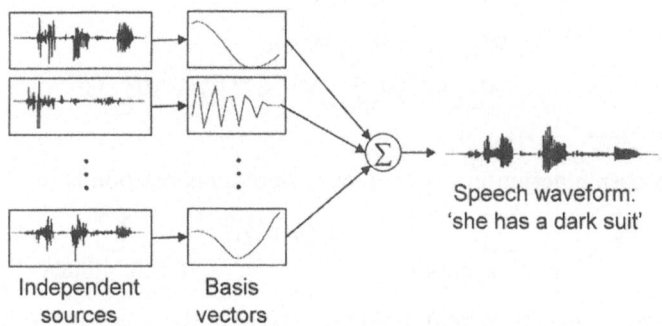

Independent   Basis
sources   vectors

Fig. 1. Diagram of speech generative model with 16 trained basis vectors.

To check the independence transform property of $\mathbf{W}_I$, joint p.d.f. of two adjacent samples was estimated. Figure 2 (a) shows the contour-plots of joint p.d.f. of two adjacent samples for the mcpm0's sentences. Although adjacent samples in natural human speech signals are highly correlated, their dependencies are very much reduced when the independence transform matrix is applied as shown in Fig.2 (b).

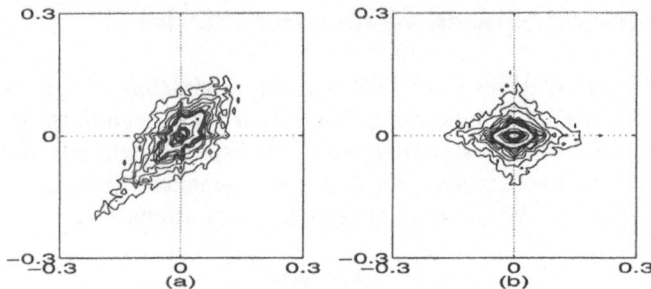

Fig. 2. Contour-plots of joint p.d.f. for mcpm0's sentences. (a) two adjacent samples in original unprocessed speech signal, (b) 1st and 2nd components transformed with $\mathbf{W}_I$.

3 Learning Rule for Nonminimum-Phase Channels

Now, we derive the algorithm for non-minimum phase channel based on the LS measure [6, 7]. In the dereverberation block of Fig.3, let's define $\hat{\mathbf{U}}^F \equiv \mathbf{W}_{\text{fft}} \, \hat{\mathbf{u}}$, $\mathbf{X}^F \equiv \mathbf{W}_{\text{fft}} \, \mathbf{x}$, and $\mathbf{W}^F \equiv \mathbf{W}_{\text{fft}} \, \mathbf{w}$, where $\mathbf{W}_{\text{fft}}$ denotes discrete Fourier transform matrix and the superscript ^{F} means frequency domain representation. Now the dereverberated speech signal $\hat{\mathbf{u}}$ can be expressed in the frequency domain as,

$$\hat{\mathbf{U}}^F = \mathbf{W}^F \otimes \mathbf{X}^F, \tag{2}$$

where $\otimes$ means component by component multiplication. IID-like signal $\mathbf{u}$ can be expressed in the frequency domain as,

$$\begin{aligned}
\mathbf{U}^F = \mathbf{W}_{\text{fft}} \, \mathbf{u} &= \mathbf{W}_{\text{fft}} \mathbf{W}_I \, \hat{\mathbf{u}} \\
&= \mathbf{W}_{\text{fft}} \mathbf{W}_I \mathbf{W}_{\text{fft}}^{-1} \, (\mathbf{W}^F \otimes \mathbf{X}^F) = \mathbf{W}_\alpha^F \, \hat{\mathbf{U}}^F,
\end{aligned} \tag{3}$$

where $\mathbf{W}_\alpha^F \equiv \mathbf{W}_{\text{fft}} \mathbf{W}_I \mathbf{W}_{\text{fft}}^{-1}$.

The LS cost function in the frequency domain corresponds to

$$\mathbf{J}_{LS} \equiv \sum_{\text{all fft points}} |\mathbf{U}^F - \text{fft}\{g(\mathbf{u})\}|^2 = \sum_{\text{all fft point } i} |e_i|^2 \tag{4}$$

where $g(\cdot)$ is the Fisher score function [5] and e_i is the i-th component of $(\mathbf{U}^F - \text{fft}\{g(\mathbf{u})\})$. We can obtain the update rule by minimizing $\mathbf{J}_{LS}$ with respect to $\mathbf{W}^F$. That is, in matrix formulation,

$$\frac{\partial \mathbf{J}_{LS}}{\partial \mathbf{W}^{F\,*}} = \{\mathbf{W}_\alpha^{F\,H}(\mathbf{U}^F - \text{fft}\{g(\mathbf{u})\}\} \otimes \mathbf{X}^{F\,*} \tag{5}$$

where superscript ^{H} denotes the Hermitian operator and * is the complex conjugate. Finally, using the relative gradient ([8]),

$$\begin{aligned}
\Delta \mathbf{W}^F &\propto -\frac{\partial \mathbf{J}_{LS}}{\partial \mathbf{W}^{F\,*}} \otimes \mathbf{W}^{F\,*} \otimes \mathbf{W}^F \\
&= -\{\mathbf{W}_\alpha^{F\,H}(\mathbf{U}^F - \text{fft}\{g(\mathbf{u})\})\} \otimes \hat{\mathbf{U}}^{F\,*} \otimes \mathbf{W}^F.
\end{aligned} \tag{6}$$

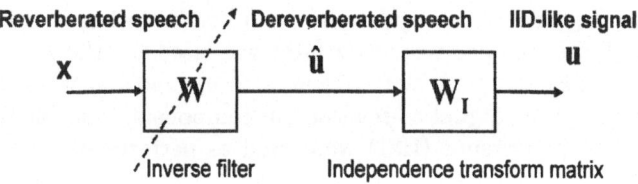

Fig. 3. Proposed blind dereverberation method with speech generative model.

4 Experimental Results

We conducted blind dereverberation experiments using simulated room impulse response. During this deconvolution phase the independence transform matrix is fixed to the previously-trained values. To get the simulated room impulse response we used the commercial software 'Room Impulse Response v2.5' which assumes a rectangular enclosure with a source-to-receiver impulse response calculated using a time-domain image expansion method [9].

We assumed that the room dimensions are $4\ m \times 5\ m \times 3\ m$, a sound speed of $345\ m/s$, and reflection coefficients for 4 walls are 0.9, ceiling and floor are 0.7. Volume of the room is $60\ m^3$, and the reverberation time is $0.56\ s$.

Three different reverberant channels regarding the position of the source and receiver were used for experiments. The position of the source was fixed at $(2\ m, 2\ m, 1\ m)$, and the positions of three receivers were at $(2\ m, 1.7\ m, 1\ m)$, $(2\ m, 1.5\ m, 1\ m)$ and $(2\ m, 1\ m, 1\ m)$. The length of room impulse response was truncated by 512 samples. Fig.4 shows obtained three different room impulse responses. Channel distortions are much heavier as the distances are increased.

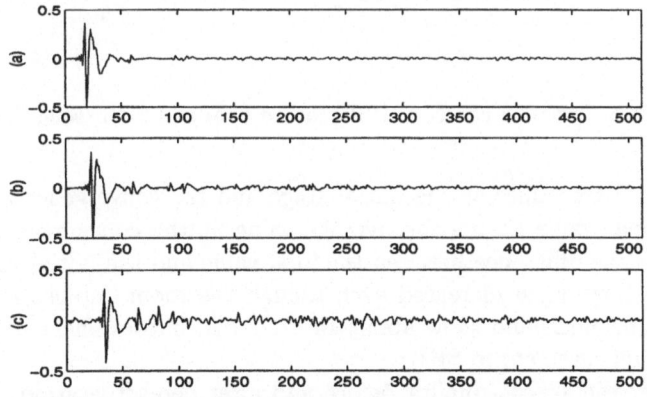

Fig. 4. Three different simulated room impulse responses. The distances between the source and receiver are (a) $0.3\ m$ (channel 1), (b) $0.5\ m$ (channel 2) and (c) $1\ m$ (channel 3).

Equation (6) was used to update inverse filter **W** in Fig.3. 1024-tap delayed causal FIR (finite impulse response) filter was used for the inverse filter, and the delay was 512 samples. Ten sentences of mcpm0's speaker were used for blind dereverberation. Signal-to-reverberant component ratio (SRR) and inverse of inter-symbol interference (IISI) were used as performance measure. SRR is defined as:

$$\text{SRR (dB)} = 10 \ \log \left(\frac{\sum_n \hat{s}_n^2}{\sum_n (\hat{s}_n - \hat{u}_n)^2} \right) \tag{7}$$

where $\hat{s}$ is unknown clean speech signal and $\hat{u}$ is dereverberated signal. IISI is a measure of how close the dereverberated impulse response to the delta function. IISI is defined as:

$$\text{IISI (dB)} = 10 \ \log \left(\frac{\sum_k |t_k|^2 - \max_k |t_k|^2}{\max_k |t_k|^2} \right) \tag{8}$$

where **t** is the convolution of the reverberant channel and the estimated inverse filter of the channel. Higher SRR and IISI show better result.

Figure 5 show the learning curves for the channels. Dashed and solid lines show the resulting IISI and SRR values respectively. Totally 7000 sweeps were performed, and training converged at about 1000 sweeps.

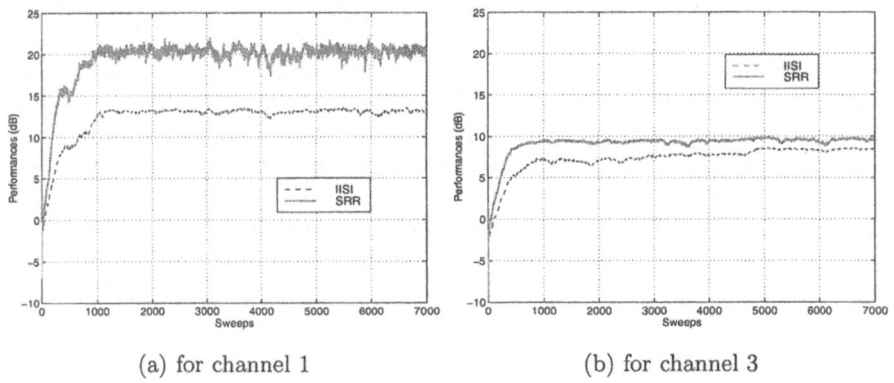

(a) for channel 1   (b) for channel 3

Fig. 5. Learning curves of IISI (dashed line) and SRR (solid line).

IISI and SRR values at the initial stage and the convergence are shown in Table 1. Final value means the average value at the convergence, and increment means the difference between the final value and the initial value. Performances are very much increased even though the room impulse responses are non-minimum phase and show about 15 ~17 (dB) improvement in IISI and 20 ~ 27 (dB) improvement in SRR.

To verify the speech quality before and after dereverberation, and predict the performance improvement in the automatic speech recognition system we compared the spectrograms. Spectrogram of reverberated speeches is blurred by the room impulse response especially in the mid and high frequency ranges.

Table 1. IISI and SRR values at the initial stage and the convergence.

	IISI (dB)			SRR (dB)		
	Initial	Final	Increment	Initial	Final	Increment
Channel 1	-4.1	13.0	17.1	-7.0	20.0	27.0
Channel 2	-4.6	12.8	17.4	-7.3	20.0	27.3
Channel 3	-6.8	8.5	15.3	-9.9	9.6	19.5

Those corrupted frequency structure could be recovered after dereverberation process and we can expect that speech recognition rate would be improved.

5 Conclusion

In this paper, a method for blind dereverberation based on speech generative model was proposed and LS-based learning rule was derived. Proposed blind dereverberation method was successfully applied to the simulated room impulse responses even though it is non-minimum phase and shows around 20 (dB) improvement in SRR and IISI.

Acknowledgment

This research was supported as a Brain Neuroinformatics Research Program by Korean Ministry of Science and Technology.

References

1. Haas, H.: The influence of a single echo on the audibility of speech. Journal of the Audio Engineering Society. **20**(2) (1972) 146–159
2. Shalvi, O., Weinstein, E.: New criteria for blind deconvolution of nonminimum phase systems (channels). IEEE Trans. on Information Theory. **36**(2) (1990) 312–321
3. Bell, A. J., Sejnowski, T. J.: An information-maximization approach to blind separation and blind deconvolution. Neural Computation. **7**(6) (1995) 1004–1034
4. Cichocki, A., Amari, S.: Adaptive blind signal and image processing - Learning algorithms and applications. John Wiley & Sons, Ltd. (2002)
5. Lee, T. W.: Independent component analysis - Theory and applications. Boston: Kluwer Academic Publisher. (1998)
6. Bellini, S.: Bussgang techniques for blind deconvolution and equalization. In Blind Deconvolution (S. Haykin, ed.). Englewood Cliffs, New Jersey: Prentice Hall. (1994) 8–52
7. Godfrey, R., Rooca, F: Zero memory non-linear decvonolution. Geophysical Prospecting. **29** (1981) 189–228
8. Cardoso, J. F., Laheld, B. H.: Equivariant adaptive source separation. IEEE Trans. on Signal Processing. **44**(12) (1996) 3017–3030
9. http://www.dspalgorithms.com/room/room25.html

Permutation Correction of Filter Bank ICA Using Static Channel Characteristics

Chandra Shekhar Dhir[1,3], Hyung Min Park[1,3], and Soo Young Lee[1,2,3]

[1] Department of Biosystems
Korea Advanced Institute of Science and Technology, Daejeon 305-701, Korea
[2] Department of Electrical Engineering and Computer Science
Korea Advanced Institute of Science and Technology, Daejeon 305-701, Korea
[3] Brain Science Research Center
Korea Advanced Institute of Science and Technology, Daejeon 305-701, Korea
Phone: +82-42-869-5431, Fax: +82-42-869-8490
shekhardhir@neuron.kaist.ac.kr

Abstract. This paper exploits static channel characteristics to provide a precise solution to the permutation problem in filter bank approach to Independent Component Analysis (ICA). The filter bank approach combines the high accuracy of time domain ICA and the computational efficiency of frequency domain ICA. Decimation in each sub-band helps in better formulation of the directivity patterns. The nulls of the directivity patterns are dependent on the location of the source signals and this property is used for resolving the permutation problem. The experimental results, show a good behavior with reduced computational complexity and do not require non-stationarity of the signals.

1 Introduction

In real world situation, we often have undesired signals in addition to the signal of primary interest. Separation of these unwanted signals is of vital importance to many applications such as noise robust telecommunication, speech enhancement, bio-medical signal processing,etc. Traditionally, signal separation was performed using spatial filtering techniques governed by principles of beamforming [1]. On the other hand, ICA is a signal processing approach which estimates the individual source signals from just the mixtures that are linear summation of convolved independent source signals [2]. ICA has, therefore, found its wide application in many fields such as Blind Source Separation (BSS).

Time domain ICA shows a good performance, but this method is computationally complex for signal separation. On the other hand, frequency domain approach can considerably reduce the computational complexity but its separation performance is inferior to time domain ICA [3,4]. It also faces the permutation and scaling problem at every frequency of interest [8]. Over-sampled filter bank approach to ICA proposed by Park et. al. utilizes time domain FIR filters to divide the complex separation problem into many simpler problems without the block effects of frequency domain approaches. It shows better separation performance than the other methods but still suffers from the permutation problem at

N.R. Pal et al. (Eds.): ICONIP 2004, LNCS 3316, pp. 1076–1081, 2004.
© Springer-Verlag Berlin Heidelberg 2004

every sub-band [4]. Permutation problem in filter bank ICA was resolved using the similarity between the envelopes of the separated signals [5]. This similarity can give a wrong measure if the envelopes of the source signals are similar to each other. In this paper, we propose a computationally less complex and more precise method to solve the permutation problem of filter bank ICA.

In section 2 and section 3, we briefly review filter bank approach to ICA and beamforming, respectively. The proposed algorithm to resolve permutation problem of filter bank ICA is presented in Section 4. Experimental results using artificially generated acoustic room response and real office room situation are summarized in Section 5, which is followed by conclusions in Section 6.

2   Filter Bank Approach to ICA

If observations are a linear sum of convolved independent signals, each observation is defined as [2]

$$x_i(n) = \sum_{j=1}^{N} \sum_{l=0}^{L-1} a_{ij}(l)s_j(n-l), \tag{1}$$

where a_{ij} denotes the mixing filter of length L between the source s_j and the observation x_i and N is the total number of source signals.

The ICA network in each sub-band uses feedback architecture and entropy maximization algorithm to learn the adaptive filters $w_{ij}(l)$, which forces the output $u_i(n)$ to reproduce original source signals. The learning rules for adapting the unmixing filters are [2]

$$\Delta w_{ii}(0) \propto 1/w_{ii}^* - \varphi(u_i(n))x_i^*(n),$$
$$\Delta w_{ii}(l) \propto -\varphi(u_i(n))x_i^*(n-l),\ l \neq 0,$$
$$\Delta w_{ij}(l) \propto -\varphi(u_i(n))u_j^*(n-l),\ i \neq j. \tag{2}$$

To deal with complex data, the non-linear score function was re-defined as [6]

$$\varphi(u_i) = -\frac{\frac{\partial p(|u_i|)}{\partial(|u_i|)}}{p(|u_i|)} exp(j.\angle u_i). \tag{3}$$

The estimated signals in every sub-band still suffer from permutation and scaling indeterminacy. The unmixing matrix obtained by ICA satisfies the condition

$$(\mathbf{W} \cdot \mathbf{A})_k = (\mathbf{P} \cdot \mathbf{D})_k, \tag{4}$$

where $\mathbf{W}, \mathbf{A}$, $\mathbf{P}$ and $\mathbf{D}$ are the unmixing, mixing, permutation and diagonal matrix, respectively in the kth sub-band. The scaling problem can be resolved by choice of direct filters as scaling factor 1. After the permutation ambiguity is solved, the desired source signals are obtained by synthesis [4].

3   Beamforming and Directivity Pattern

Beamforming is a spatial filtering technique for separation of the wanted signal from observations. It is multiple input-single output network, which considers instantaneous mixing condition under far field assumption. The phase response of source signal s_j sampled by a linear sensor array on the observation x_i is given as [1]

$$a_{ij} = exp\left(-\frac{j2\pi f d_i \sin\theta_{ij}}{c}\right), \tag{5}$$

where f is the frequency, d_i is the sensor location, θ_{ij} is the incident angle of the source s_j on sensor x_i and c is the velocity of sound.

For a 2×2 ICA network, the case can be analyzed as a cascade of two beamformers which jams one signal and allows the other to pass [7]. The angular location of the signal to be jammed or its direction-of-arrival (DOA) can be estimated by finding the angle corresponding to the minima of the directivity patterns. This knowledge about the location of the source signals can be used in fixing permutation. The directivity patterns are defined as [8]

$$\varphi_1 = w_{11} + w_{12}exp\left(-\frac{j2\pi f d \sin\theta_2}{c}\right), \tag{6}$$

$$\varphi_2 = w_{21} + w_{22}exp\left(-\frac{j2\pi f d \sin\theta_1}{c}\right). \tag{7}$$

4   Proposed Method

Among different methods to solve permutation problem, Park et al. used the similarity between the envelopes of the separated signal [4, 5]. However, Murata's method is dependent on non-stationarity of the signals for fixing permutation. To overcome this constraint, we propose a new method that uses the stationarity of the mixing environment, i.e., both the speakers and microphones are located at fixed points. The proposed method is independent of any assumption on the signal characteristics.

The input mixtures in every sub-band are analyzed in the frequency domain. Due to high stop band attenuation by the analysis filter in over-sampled filterbank case, aliasing can be neglected [4]. Therefore, the Fourier transform of the input mixture x_i to the kth sub-band ICA network is given as

$$X_i\left(\frac{f}{M}\right) = H_k\left(\frac{f}{M}\right)\sum_{j=1}^{N} A_{ij}\left(\frac{f}{M}\right) S_j\left(\frac{f}{M}\right), \tag{8}$$

where A_{ij} is the channel response between signal s_j and the observation x_i and H_k is the frequency response of the analysis filter in the kth sub-band.

Decimation makes a weaker reverberant condition and results in the reduction of the length of mixing filters by factor M. In many real situations, the filter coefficient corresponding to the direct path has the largest magnitude and the following filter coefficients at every M^{th} tap can be comparatively neglected. Thus, the response of the channel between source s_j and observation x_i after decimation can be approximately expressed as

$$A_{ij}\left(\frac{f}{M}\right) \approx a_{ij}(0)exp\left(-\frac{j2\pi fd\sin\theta_{ij}}{Mc}\right) \tag{9}$$

which is equivalent to phase response of instantaneous mixing condition.

For a 2×2 mixing system, we can consider the two observations as a linear array separated by distance d. Using equation (4),(6) and (7) we obtain the directivity patterns for the kth sub-band of the filter bank as

$$\varphi_{1k} = H_k\left(\frac{f}{M}\right)\left(1 + W_{12k}\left(\frac{f}{M}\right)exp\left(-\frac{j2\pi fd\sin\theta_2}{Mc}\right)\right), \tag{10}$$

$$\varphi_{2k} = H_k\left(\frac{f}{M}\right)\left(W_{21k}\left(\frac{f}{M}\right) + exp\left(-\frac{j2\pi fd\sin\theta_1}{Mc}\right)\right). \tag{11}$$

The permutation indeterminacy can be resolved by the knowledge of DOA of the source signals at every sub-band. The DOAs are calculated in the frequency range of the anti-aliasing filter and are defined as

$$\theta_{1k} = arg_{\theta_1}min(abs(\varphi_{2k})), \tag{12}$$

$$\theta_{2k} = arg_{\theta_2}min(abs(\varphi_{1k})), \tag{13}$$

in the kth sub-band. Considering the order $[\theta_1 \ \theta_2]$ in the first sub-band as no permutation, a decision about permutation in k^{th} sub-band can be made taking the order of first sub-band as reference. If the values of DOAs are interchanged, the outputs of the sub-band are permuted and this ambiguity can be resolved by simply interchanging the rows of the unmixing matrix before synthesis.

5   Experiments

We performed BSS experiments for mixtures of speech signals in reverberant conditions. Two Korean sentences of 5 seconds length at 16kHz sampling rate were recorded as speech signals. For a 2×2 system, signal-to-interference (SIR) ratio at the output is given as

$$SIR = \frac{1}{2}\left|10\log\left(\frac{< (u_{1,s_1}(n))^2 > \cdot < (u_{2,s_2}(n))^2 >}{< (u_{1,s_2}(n))^2 > \cdot < (u_{2,s_1}(n))^2 >}\right)\right|, \tag{14}$$

where $u_{j,s_i}(n)$ denotes the jth output of the cascaded mixing/unmixing system when only $s_i(n)$ is active.

The recorded observations were obtained by artificially generating acoustic impulse response using Image method [9]. The virtual rectangular room used for simulation is shown in figure 1. The wall, ceiling and floor reflection coefficients of

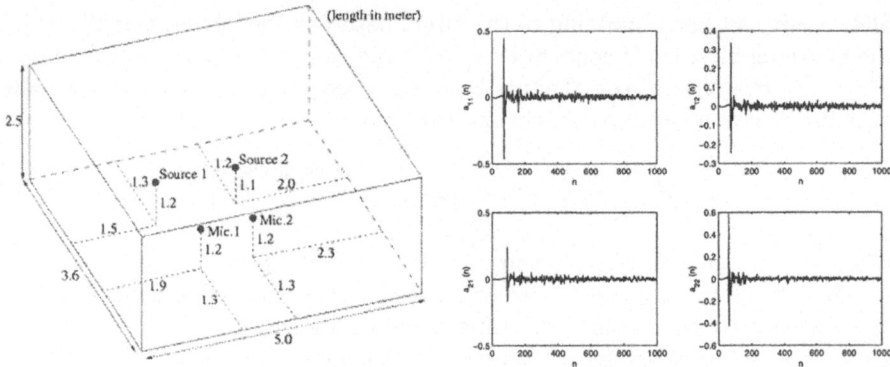

Fig. 1. Virtual room to simulate response of 2 speakers to 2 microhone points.

Fig. 2. Impulse response of real room recording in normal office condition.

the virtual room were set as 0.4, 0.5 and 0.25, respectively. The mixture powers were appropriately normalized to avoid mismatch between the non-linear score function and the recovered signal levels.

For a more realistic evaluation of the proposed method, we mixed the two speech signals using the mixing conditions of a normal office room. The impulse responses of the normal office room for a 2×2 mixing system are shown in figure 2. Figure 3 shows the learning curves of filter bank approach when the speech signals are mixed using the virtual room and the real normal office like situation.

Figure 4 shows the comparison between Murata's method and the proposed algorithm when two stationary signals are separated from its mixture in real office room situation using filter bank architecture. The stationary signals, voice babble and f-16 fighter noise, were obtained from NOISEX-92 CD-ROMs. The figure shows very preliminary results for a few sweeps and it can be seen that the proposed algorithm is able to correct permutation at every iteration step more efficiently than the Murata's method.

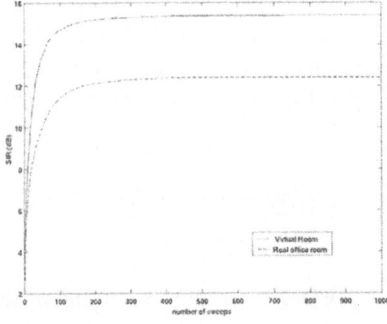

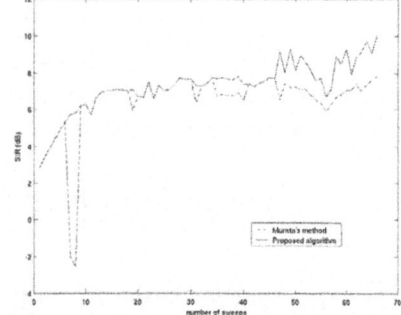

Fig. 3. Learning curve of proposed algorithm for virtual room and real office condition.

Fig. 4. Comparison results for mixture of stationary signals.

6   Conclusion

A new method to solve the permutation problem in filter bank ICA is proposed. This approach does not require non-stationarity of signals and exploits the static nature of the channels. Instantaneous mixing assumption used for formulating directivity patterns and estimating DOA is more appropriate due to decimation. The proposed algorithm could be used to separate mixtures in real-world situations and is computationally more efficient. Preliminary experiments on mixtures of stationary signals give better permutation correction in comparison to the Murata's method. Further work is in progress to improve the convergence of the learning curves.

Acknowledgment

This research was supported as a Brain Neuroinformatics Research Program by Korean Ministry of Science and Technology.

References

1. Godara, L. C.: Application of Antenna Arrays to Mobile Communications, Part II: Beam-Forming and Direction-of-Arrival Considerations. Proc. of IEEE, vol. 85, No. 8 (1997) 1195–1245
2. Lee, T. W.: Independent component analysis - Theory and applications. Boston: Kluwer Academic Publisher. (1998)
3. Araki, S., Makino, S., Nishikawa, T., Saruwatari, H.: Fundamental limitations of frequency domain blind source separation for convolved mixture of speech. Proc. Int. Conf. ICA and BSS (2001) 132–137.
4. Park, H. M., Oh, S. H., Lee, S. Y.: A uniform over-sampled filter bank approach to independent component analysis. Proc. Int. Conf. Neural Networks and Signal processing, Vol. 2 (2003) 1354–1357
5. Murata, N., Ikeda, S., Ziehe, A.: An approach to blind source separation based on temporal structure of speech signals. Neurocomputing, vol. 41 (2001) 1–24
6. Sawada, H., Mukai, R., Araki, S., Makino, S.: A polar coordinate based activation function for frequency domain blind source separation. Proc. Int. Conf. ICA and BSS (2001) 663–668
7. Araki, S., Makino, S., Mukai, R., Hinamoto, Y., Nishikawa, T., Saruwatari, H.: Equivalence between frequency domain blind source separation and frequency domain adaptive beam-forming. Proc. of ICASSP, vol. 2 (2002) 1785–1788
8. Sawada, H., Mukai, R., Araki, S., Makino, S.: A robust and precise method for solving permutation problem of frequency-domain blind source separation. Proc. 4th Int. symposium on ICA and BSS (2003) 505–510.
9. Allen, J.B., Berkly, D.A.: Image method for efficiently simulating small-room acoustics. Journal of the Acoustic Society of America, vol. 65, no. 4 (1979) 943–950

Minimal Addition-Subtraction Chains with Ant Colony

Nadia Nedjah and Luiza de Macedo Mourelle

Department of Systems Engineering and Computation,
Faculty of Engineering, State University of Rio de Janeiro, Brazil
{nadia,ldmm}@eng.uerj.br

Abstract. Addition-subtraction chains (AS-chains) consist of a sequence of integers that allow one to efficiently compute power T^E, where T varies but E is constant. The shorter the AS-chain is, the more efficient the computation. Solving the minimisation problem that yields the shortest addition-subtraction is NP-hard. There exists some heuristics that attempt to obtain reduced AS-chains. In this paper, we compute minimal addition- subtraction chains using the ant-colony methodology.

1 Introduction

The performance of cryptosystems [1] is then primarily determined by the implementation efficiency of the modular multiplication, division and exponentiation. As the operands, i.e the plain text of a message or the cipher (possibly a partially ciphered) text are usually large (1024 bits or more), and in order to improve time requirements of the encryption/decryption operations, it is essential to attempt to minimise the number of modular operations performed.

The straightforward method to compute $C = T^E \mod M$ requires more multiplications than necessary. For instance, to compute T^{31}, it needs 30 multiplications. However, T^{31} can be computed using only 7 multiplications: $T \to T^2 \to T^3 \to T^5 \to T^{10} \to T^{11} \to T^{21} \to T^{31}$. But if division is allowed, T^{31} can be computed using only 5 multiplications and one division: $T \to T^2 \to T^4 \to T^8 \to T^{16} \to T^{32} \to^- T^{31}$, where $\to^-$ denotes a division.

The basic question is: what is the fewest number of multiplications and divisions to compute T^E, given that the only operations allowed is multiplying or dividing two already computed powers of T? Answering this question is NP-hard, but there are several efficient heuristics that can find a near optimal ones.

Ant systems [2] are distributed multi-agent systems [3] that simulate real ant colony. Each agent behaves as an ant within its colony. Despite the fact that ants have very bad vision, they always are capable to find the shortest path from their nest to wherever the food is. To do so, ants deposit a trail of a chemical substance called *pheromone* on the path they use to reach the food. On intersection points, ants tend to choose a path with high amount of pheromone. Clearly, the ants that travel through the shorter path are capable to return quicker and so the pheremone deposited on that path increases relatively faster than that deposited

N.R. Pal et al. (Eds.): ICONIP 2004, LNCS 3316, pp. 1082–1087, 2004.
© Springer-Verlag Berlin Heidelberg 2004

on much longer alternative paths. Consequently, all the ants of the colony end using the shorter way.

In this paper, we exploit the ant colony methodology to obtain an optimal solution to addition-subtraction chain (AS-chain) minimisation NP-complete problem. In order to clearly report the research work performed, we subdivide the rest of this paper into five sections. In Section 2, we clearly state the minimisation problem we are focusing on and describe the AS-chain-based methods. Thereafter, in Section 3, we provide and comment the ant colony-based model including the data structures required for the bookkeeping of the reached (partial) AS-chain as well as rules applied to represent and update the pheromone level. Subsequently, in Section 4, we expose the results obtained by the ant system and compare them to those evolved using genetic algorithms as well as to those exploited by traditional methods such as m-ary and sliding window methods.

2   AS-Chain Minimisation

The AS-chain-based methods use a sequence of positive integers such that the first number of the chain is 1 and the last is the exponent E, and in which each member is the sum or the difference of two previous members of the chain. For instance, the AS-chains used in the introduction are $(1, 2, 3, ..., E-2, E-1, E)$, $(1, 2, 3, 5, 10, 11, 21, 31)$ and $(1, 2, 4, 8, 16, 32, 31)$. A formal definition of an AS-chain of length l for an positive integer n is a sequence of positive integers $(a_1, a_2, \ldots, a_l)$ such that $a_1 = 1, a_l = n$ and $a_k = a_i + a_j$ or $a_k = a_i + a_j$, wherein $1 \leq i \leq j < k \leq l$.

The algorithm used to compute the modular exponentiation $C = T^E \bmod M$ based on a given non-redundant AS-chain, is specified in Algorithm 1, wherein PoT stands for the array of *Powers of T*.

Algorithm 1. AddSubChainBasedMethod(T, M, E)
1:    Let $(a_1 = 1, a_2, \ldots, a_l = E)$ be the AS-chain;
2:    $PoT[0] := T \bmod M$;
3:    for $k := 1$ to l do
4:        Let $a_k := a_i \pm a_j | i < k$ and $j < k$;
5:        if $a_k := a_i + a_j$ then $PoT[k] := PoT[i] \times PoT[j] \bmod M$;
6:        else $PoT[k] := PoT[i] \div PoT[j] \bmod M$;
7:    return $PoT[l]$;
end.

Finding a minimal AS-chain for a given number is NP-hard [4]. Therefore, heuristics were developed to attempt to approach such a chain. The most used heuristics is based an addition-only chain and consist of scanning the digits of E binary representation from the less significant to the most significant digit and grouping them in partitions P_i [4], [5]. Modular exponentiation methods based on constant-size partitioning of the exponent are usually called m-ary, where m is a power of two and $\log_2 m$ is the size of a partition while methods based on variable-size windows are usually called *sliding window* [4], [5].

It is perfectly clear that the shorter the addition-sibtraction chain is, the faster Algorithm 1. Consequently, the AS-chain minimisation problem consists of finding a sequence of numbers that constitutes an addition- subtraction chain for a given exponent. The sequence of numbers should be of a minimal length. The heuristics used the m-ary and sliding window methods [4], [5] generate a relatively short AS-chains. However, the performance of modular exponentiation can be further improved if the AS-chain is much shorter. In previous research work, we applied genetic algorithms to evolve a minimal AS-chains. Indeed, the application of genetic algorithms produced much shorter AS-chains, compared with those of the m- ary and sliding window methods. Interested readers can find further details in [5]. In this paper, we describe an ant system that applies ant colony principles to the AS-chain minimisation problem. We show that this ant system finds shorter addition chains, compared with the ones based on the heuristics used in the m-ary and sliding window methods as well to those evolved by the genetic algorithms.

3   AS-Chain Minimisation Using Ant System

Ant systems can be viewed as multi-agent systems [3] that use a shared memory (SM) through which the agents communicate and a local memory LM_i for each agent A_i to bookkeep the locally reached problem solution. Mainly, the shared memory holds the pheromone information while the local memory LM_i keeps track of the solution (possibly partial) that agent A_i reached so far.

The behaviour of an artificial ant colony is summarised: The first step consists of activating N distinct artificial ants that should work in simultaneously. Every time an ant conclude its search, the shared memory is updated with an amount of pheromone, which should be proportional to the quality of the reached solution. This called *global* pheromone update. When the solution yield by an ant's work is suitable then all the active ants are stopped. Otherwise, the process is iterated until an adequate solution is encountered.

In this section, we concentrate on the specialisation of the ant system to the AS-chain minimisation problem.

Ant System Shared Memory. It is a two-dimension array. The array has E rows. The number of columns depends on the row. It can be computed as in Eq. 1, wherein NC_i denotes the number of columns in row i.

$$NC_i = \begin{cases} 2^{i-1} - i + 1 & \text{if } 2^{i-1} < E \\ 1 & \text{if } i = E \\ E - i + 3 & \text{otherwise} \end{cases} \qquad (1)$$

An entry $SM_{i,j}$ of the shared memory holds the pheromone deposited by ants that used exponent $i + j$ as the i th. member in the built addition chain. Note that $1 \leq i \leq E$ and for row i, $0 \leq j \leq NC_i$. Fig. 1 gives an example of the shared memory for exponent 17. In this example, a table entry is set to show the exponent corresponding to it. The exponent $E_{i,j}$ corresponding to entry $SM_{i,j}$

1											
2											
3	4										
4	5	6	7	8							
5	6	7	8	9	10	11	12	13	14	15	16
6	7	8	9	10	11	12	13	14	15	16	17
7	8	9	10	11	12	13	14	15	16	17	
8	9	10	11	12	13	14	15	16	17		
9	10	11	12	13	14	15	16	17			
10	11	12	13	14	15	16	17				
11	12	13	14	15	16	17					
12	13	14	15	16	17						
13	14	15	16	17							
14	15	16	17								
15	16	17									
16	17										
17											

Fig. 1. Example of the shared memory content for $E = 17$.

should be obtainable from exponents from previous rows. Eq. 2 formalise such a requirement.

$$\begin{array}{ll} E_{i,j} = E_{k_1,l_1} + E_{k_2,k_2} | & 1 \leq k_1, k_2 < i, 0 \leq l_1, l_2 \leq j, k_1 = k_2 \Longleftrightarrow l_1 = l_2 \\ E_{i,j} = E_{k_1,l_1} - E_{k_2,k_2} | & 1 \leq k_1, k_2 < i, j < l_1 \leq NC_i, 0 \leq l_2 < j, \\ & k_1 \neq k_2, k_2 = i - 1 \Longleftrightarrow l_2 \neq j + 1 \end{array} \quad (2)$$

Note that, in Fig. 1, the exponents in the shaded entries are not valid exponents as for instance exponent 7 of row 4 can is not obtainable from the sum of two previous different stages, as described in Eq. 2. The computational process that allows us to avoid these exponents is of very high cost. In order to avoid using these few exponents, we will penalise those ants that use them and hopefully, the solutions built by the ants will be almost all valid addition chains.

Ant Local Memory. Each ant is endowed a local memory that allows it to store the solution or the part of it that was built so far. This local memory is divided into two parts: the first part represents the (partial) AS-chain found by the ant so far and consists of a one-dimension array of E entries; the second part holds the *characteristic* of the solution. It represents the solution fitness to the objective of the optimisation.

AS-Chain Characteristics. The fitness evaluation of AS-chain is performed with respect to two aspects: *(a)* how much a given chain adheres to the Definition 1, i.e. how many of its members cannot be obtained summing up or subtracting two previous members of the chain; *(b)* how far the AS-chain is reduced, i.e. what is the length of the chain. Eq. 3 shows how to compute the fitness of an AS-chain, wherein $\mathbb{I} = \{k \mid 3 \leq k \leq n, \forall i, j, 1 \leq i, j < k, a_k \neq a_i \pm a_j\}$

$$Fitness(E, (a_1, a_2, \ldots, a_n)) = \frac{E \times (n-1)}{a_n} + \sum_{k \in \mathbb{I}} penalty \quad (3)$$

Pheromone Trail and State Transition Function. There are three situations wherein the pheromone trail is updated: *(a)* when an ant chooses to use exponent $F = i \pm j$ as the ith. member in its solution, the shared memory cell $SM_{i,j}$ is incremented with a constant value of pheromone $\Delta\phi$, as in the first line of Eq. 4; *(b)* when an ant halts because it reached a complete solution, say $A = (a_1, a_2, \ldots, a_n)$, all the shared memory cells $SM_{i,j}$ such that $i + j = a_i$ are incremented with pheromone value of $1/Fitness(A)$, as in the second line of Eq. 4. Note that the better is the reached solution, the higher is the amount of pheromone deposited in the shared memory cells that correspond to the addition chain members. *(iii)* The pheromone deposited should evaporate. Priodically, the pheromone amount stored in $SM_{i,j}$ is decremented in an exponential manner [6] as in the last line of Eq. 4.

$$SM_{i,j} := SM_{i,j} + \Delta\phi, \quad \text{every time } a_i = i \pm j \text{ is chosen}$$
$$SM_{i,j} := SM_{i,j} + 1/Fitness((a_1, a_2, \ldots, a_n)), \quad \forall i,j \mid i \pm j = a_i \quad (4)$$
$$SM_{i,j} := (1 - \rho)SM_{i,j} \mid \rho \in (0, 1], \quad \text{periodically}$$

An ant, say A that has constructed partial AS-chain $(a_1, a_2, \ldots, a_i, 0, \ldots, 0)$ for exponent E, is said to be in *step i*. In step $i + 1$, it may choose exponent a_{i+1} as follows: *(i)* $a_i + 1$, $a_i + 2$, ..., $2a_i$, if $2a_i \leq E$. That is, ant A may choose one of the exponents that are associated with the shared memory cells SM_{i+1,a_i-i}, SM_{i+1,a_i-i+1}, ..., $SM_{i+1,2a_i-i-1}$. Otherwise (i.e. if $2a_i > E$), it may only select from exponents $a_i + 1$, $a_i + 2$, ..., $E + 2$. In this case, ant A may choose one of the exponent associated with $SM_{i+1,a_i-i}, SM_{i+1,a_i-i+1}, \ldots, SM_{i+1,E-i+1}$; *(b)* i, $i + 1$, ..., $a_i - 1$, if $a_i > i + 1$. That is ant A may choose one of the exponents that correspond to the shared memory cells $SM_{i+1,0}$, $SM_{i+1,1}$, ..., SM_{i+1,a_i-i-2}. Otherwise (i.e. if $a_i \leq i + 1$), ant A cannot use a subtraction in step $i + 1$.

4   Performance Comparison

The ant system was implemented using Java as a multi-threaded ant system. Each ant was simulated by a thread that implements the artificial ant computation. A Pentium IV-HT^{TM} of a operation frequency of 1GH and RAM size of 2GB was used to run the ant system and obtain the performance results. We compared the performance of the recoding m-ary methods to the genetic algorithm and ant system-based methods. The average lengths of the AS-chains for different exponents obtained by using these methods are given in Table 1. The exponent size is that of its binary representation (i.e. number of bits). The ant system-based method always outperforms all the others, including the genetic algorithm-based method [7].

5   Conclusion

In this paper we applied the methodology of ant colony to the addition chain minimisation problem. We implemented the ant system described using muti-threading (each ant of the system was implemented by a thread). We compared

Table 1. Average length of addition chain for binary, quaternary and octal method vs. genetic algorithm and ant system-based methods.

Size	R-Binary	R-Quaternary	R-Octal	Genetic Algorithms	Ant System
128	169	170	168	158	142
256	340	341	331	318	273
512	681	682	658	629	561
1024	1364	1365	1313	1279	1018

the results obtained by the ant system to those of m-ary methods (binary, quaternary and octal methods). Taking advantage of the a previous work on evolving minimal addition chains with genetic algorithm, we also compared the obtained results to those obtained by the genetic algorithm. The ant system always finds a shorter addition chain and gain increases with the size of the exponents.

References

1. Rivest, R., Shamir, A. and Adleman, L., A method for Obtaining Digital Signature and Public-Key Cryptosystems, Communications of the ACM, (21):120–126, 1978.
2. Dorigo, M. and Gambardella, L.M., Ant Colony: a Cooperative Learning Approach to the Travelling Salesman Problem, IEEE Transaction on Evolutionary Computation, (1)1:53–66, 1997.
3. Feber, J., Multi-Agent Systems: an Introduction to Distributed Artificial Intelligence, Addison-Wesley, 1995.
4. Downing, P. Leong B. and Sthi, R., Computing Sequences with Addition Chains, SIAM Journal on Computing, (10)3:638–646, 1981.
5. Nedjah, N., Mourelle, L.M., Efficient Parallel Modular Exponentiation Algorithm, Lecture Notes in Computer Science, Springer-Verlag, (2457):405–414, 2002.
6. Stutzle, T. and Dorigo, M., ACO Algorithms for the Travelling Salesman Problems, Evolutionary Algorithms in Engineering and Computer Science, John-Wiley, 1999.
7. Nedjah, N. and Mourelle, L.M., Minimal addition-subtraction chains using genetic algorithms, Lecture Notes in Computer Science, Springer-Verlag, (2457):303–313, 2002.

TermitAnt: An Ant Clustering Algorithm Improved by Ideas from Termite Colonies

Vahid Sherafat[1], Leandro Nunes de Castro[2], and Eduardo R. Hruschka[2]

[1] State University of Campinas (Unicamp)
sherafat@dca.fee.unicamp.br
[2] Catholic University of Santos (UniSantos)
{lnunes,erh}@unisantos.br

Abstract. This paper proposes a heuristic to improve the convergence speed of the standard ant clustering algorithm. The heuristic is based on the behavior of termites that, when building their nests, add some pheromone to the objects they carry. In this context, pheromone allows artificial ants to get more information, at the local level, about the work in progress at the global level. A sensitivity analysis of the algorithm is performed in relation to the proposed modification on a benchmark problem, leading to interesting results.

1 Introduction

Several species of animals and insects benefit from sociality in various ways, usually resulting in greater survival advantages. Social behaviors have also inspired the development of several computational tools for problem-solving, which compose the field known as *swarm intelligence* [1,5], such as ant colony optimization algorithms [4], collective robotics [6], ant clustering algorithms [7], and others.

This paper explores an approach from the swarm intelligence field, inspired by the clustering of dead bodies and nest cleaning in ant colonies. In particular, we improve the standard ant clustering algorithm introduced by Lumer and Faieta [7]. Motivated by the observation that termites add pheromone to soil pellets when building a nest [2] (this serves as a sort of reinforcement signal to other termites placing more pellets on the same portion of the space), we added a pheromone heuristic function to the standard ant clustering algorithm [7], here for brevity called SACA, in order to improve the convergence speed of the algorithm. A sensitivity analysis is performed to study the influence of the pheromone heuristics in the standard algorithm.

2 Termite Nest Building: A Useful Behavior

During the construction of a nest, each termite places somewhere a soil pellet with a little of oral secretion containing attractive pheromone. This pheromone helps to coordinate the building process during its initial stages. Random fluctuations and heterogeneities may arise and become amplified by positive feedback, giving rise to the final structure (mound). Each time one soil pellet is placed in a certain part of the space, more likely another soil pellet will be placed there, because all the previous pellets contribute with some pheromone and, thus, attract other termites. There are, however, some negative feedback processes to control this snowballing effect, for instance, the depletion of soil pellets or a limited number of termites available on the

N.R. Pal et al. (Eds.): ICONIP 2004, LNCS 3316, pp. 1088–1093, 2004.
© Springer-Verlag Berlin Heidelberg 2004

vicinity. It is also important to note that the pheromone seems to loose its biological activity or evaporate within a few minutes of deposition [2].

This behavior is interesting from a clustering perspective, because it allows the environment to provide some reinforcement signals to the clustering agents based on the density of objects on that region of the space. This means that regions of the space with greater density of data promote a greater attraction to the deposition of more data. Combined with an appropriate tuning of the standard ant clustering algorithm, this proposal can result in faster convergence than SACA.

3   TermitAnt: A Modified Ant Clustering Algorithm

3.1   The Standard Ant Clustering Algorithm (SACA)

Lumer and Faieta [7] introduced a method for organizing datasets into clusters based on the model of Deneubourg et al. [3], in which ant-like agents move at random on a 2D grid where objects are scattered at random. Each ant-like agent can either pick up an object from the grid or drop it on the grid. The probability of picking up an object decreases with both the density of other objects and the similarity with other objects within a given neighborhood. By contrast, the probability of dropping an object increases with the similarity and the density of objects within a local region. This led to the algorithm here referred to as SACA (Standard Ant Clustering Algorithm).

Independently of the dimension of the input data, each datum is randomly projected onto one of the cells on the grid. Thus, a cell or patch is responsible for hosting the index of one input pattern, and this indicates the relative position of the datum in the grid. The general idea is to have items, which are similar in their original N-dimensional space, in neighboring regions of the grid. Therefore, the indexes of neighbor data on the grid must indicate similar patterns in their original space.

In SACA, each site or cell on the grid can be occupied by at most one object. At each step of the algorithm, an ant is selected at random and can either pick up or drop an object at its current location, according to probabilistic rules. Assume that $d(i,j)$ is the Euclidean distance between objects i and j in their N-dimensional space. The density dependent function for object i, is defined by the following expression:

$$f(i) = \begin{cases} \dfrac{1}{s^2}\Sigma_j(1-d(i,j)/\alpha) & \text{if } f(i) > 0 \\ 0 & \text{otherwise.} \end{cases}$$

(1)

where s^2 is the number of sites in the surrounding area of i, and α is a constant that scales the dissimilarities between objects. The maximum value for $f(i)$ is obtained if, and only if, all sites in the neighborhood are occupied by equal objects. For the density function given by Eq.(1), the probability of picking up and dropping an object i is given by Eqs. (2) and (3) respectively. The parameters k_p and k_d are threshold constants equal to 0.1 and 0.15 respectively, and $f(i) \in [0,1]$.

$$P_{pick}(i) = \left(\frac{k_p}{k_p + f(i)}\right)^2$$

(2)

$$P_{drop}(i) = \begin{cases} 2f(i) & \text{if } f(i) < k_d; \\ 1 & \text{otherwise.} \end{cases}$$

(3)

3.2 TermitAnt: Adding Pheromone to SACA

One important drawback of SACA relates to the lack of mechanisms by which artificial ants could get more information, at the local level, about the *work in progress* at the global level. A simple way to achieve such information is to create a local variable $\phi(i)$ associated with each bi-dimensional position i on the grid such that the quantity of pheromone in that exact position can be determined. Inspired by the way termites use pheromone to build their nests, the artificial agents in the modified ant clustering algorithm will add some pheromone to the objects they carry. In a more abstract sense, one can consider that the pheromone is indirectly deposited in the position in which each object is laid. During each cycle, the artificial pheromone $\phi(i)$ evaporates at a fixed rate and it is thus diffused on the environment. More specifically, in order to accommodate the addition of pheromone to the objects, we propose some variations on the picking and dropping probability functions of SACA, which are now given by:

$$P_{pick}(i) = (1 - Phe(\min, \max, P, \phi(i))) \times \left(\frac{k_p}{k_p + f(i)}\right)^2 . \tag{4}$$

$$P_{drop}(i) = (1 + Phe(\min, \max, P, \phi(i))) \times \left(\frac{f(i)}{k_d + f(i)}\right)^2 . \tag{5}$$

$$Phe(\min, \max, P, \phi(i)) = \frac{2.P}{\max - \min}\phi(i) - \frac{2.P.\max}{\max - \min} + P , \tag{6}$$

where,

- max: current largest amount of pheromone perceived by agent i;
- min: current smallest amount of pheromone perceived by agent i;
- P: the maximum influence of the pheromone in changing the probability of picking and dropping data elements;
- $\phi(i)$: the quantity of pheromone in the current position i.

In Eq.(6), the function $Phe(\cdot)$, which is depicted in Fig. 1, gives a value in the range $[-P, P]$, and represents the local relative amount of pheromone perceived by each artificial ant. The more pheromone in a certain position (a value closer to P), the greater the value of function $Phe(\cdot)$, which results in a decrement of the probability of picking up elements or in an increment of the probability of dropping an element. The probabilities P_{pick} and P_{drop} are always bounded to the interval $[0,1]$.

The rate by which the pheromone evaporates is a preset parameter as in Eq. (7). Therefore, a region with a high quantity of pheromone is probably both a large cluster and a cluster under construction. Each artificial ant has a memory of the maximum and minimum values of pheromone, $\phi(i)$, perceived on the grid, but it also forgets this value with a fixed rate as described by Equations (8) and (9):

$$\phi(i) \leftarrow \phi(i) \times 0.99 . \tag{7}$$

$$\max \leftarrow \max \times 0.99 \tag{8}$$

$$\min \leftarrow \min \times 1.01 \tag{9}$$

Ramos et al. [9] also proposed a pheromone-based approach aimed to reduce random explorations of the grid. In short, their approach is based on the fact that ants also communicate by means of pheromone, i.e., when they leave their nests to search for food, they lay a trail of pheromone on their path. Thus, the number of ants that has traveled on the path determines the strength of the pheromone trail, and the ants that travel the shortest path reinforce this path with more amount of pheromone, helping others to follow them. In essence, our *TermitAnt* algorithm allows ants to focus more on the clusters being formed (ants *perceive* the grid positions in which objects were manipulated by other ants) instead of on the paths by which they travel on the grid.

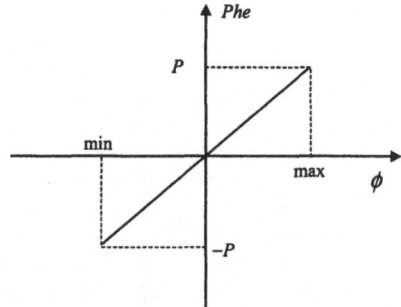

Fig. 1. Function $Phe(\text{max},\text{min},P,\phi_i)$.

4   The Influence of Pheromone on SACA

To assess the sensitivity of the algorithm to the proposed variation, we performed an experiment with a modified version of the well-known four classes data set proposed by Lumer and Faieta [7] to study SACA, which corresponds to four distributions of 25 data points each, defined by Gaussian probability density functions with various means μ and fixed standard deviation $\sigma = 1.5$, $G(\mu,\sigma)$, as follows:

$$A = [x \propto G(0,1.5), y \propto G(0,1.5)]; B = [x \propto G(0,1.5), y \propto G(8,1.5)];$$
$$C = [x \propto G(8,1.5), y \propto G(0,1.5)]; D = [x \propto G(8,1.5), y \propto G(8,1.5)].$$

In the experiments to be described here, the adopted performance measure was the convergence rate of the algorithm (how many cycles it takes to converge) and the classification error. In the present context, the algorithm is said to have converged after it was capable of correctly identifying the four clusters available in the data set. This is possible because the input data was intentionally generated so as to present four well separated clusters. A cycle here corresponds to 10,000 steps of an ant. The algorithm was implemented using StarLogo® [8]. The grid has a dimension of 25×25 and 10 ants are allowed to explore it.

To evaluate the sensitivity of the algorithm in relation to the proposed pheromone function, the parameters k_p and k_d were kept fixed at 0.20 and 0.05, respectively, based on previous experimentation. Fig. 2 presents the average percentage classification error, $E(\%)$, and the number of cycles for convergence, C, as a function of P. The parameter P was varied from 0 to 1.3 in 0.1 steps. Thus, the algorithm was analyzed from a case where there is no pheromone at all to a case in which a very high amount of pheromone is added. Note that for $P > 1.0$ the influence of pheromone is greater than the influence of the similarity among data, and thus becomes detrimental, because the algorithm tends to result in incorrect groupings.

Fig. 3 presents the average pheromone of the patches that contain no item (dotted line), the average pheromone of the patches that contain an item (solid line), and the average pheromone on the grid (dashed line). The plot uses a log scale on the y-axis. Note that the average pheromone on the empty cells stabilizes with a value around 1, a value around 2.5 on the whole grid, and a value around 13 on the occupied cells. The scale on the x-axis corresponds to the simulation time in seconds. The algorithm is stopped after it finds all clusters.

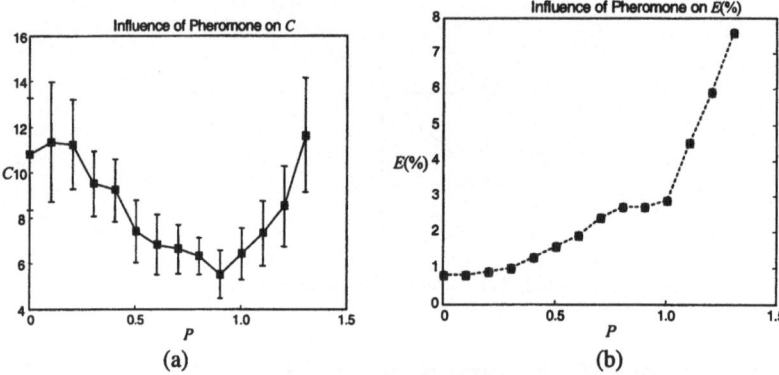

(a)   (b)

Fig. 2. The influence of the parameter P on the convergence and error rates of SACA. The parameter P was varied from 0 to 1.3 in 0.1 steps ($P = 0:0.1:1.3$). The values presented are the average over 12 runs $\pm$ the standard deviation. (a) Sensitivity of the algorithm in terms of number of cycles for convergence. (b) Sensitivity of the algorithm in terms of percentage error rate.

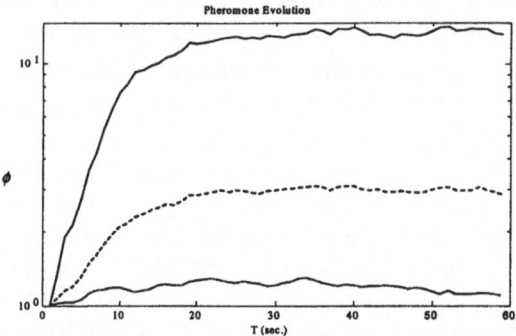

Fig. 3. Evolution of pheromone (ϕ) on the grid along the adaptation. Bottom curve (dotted line): average pheromone on the patches without any item; Middle curve (dashed line): average pheromone on the grid; Top curve (solid line): average pheromone on the patches with an item.

5 Conclusions

In relation to the sensitivity analysis of the algorithm, it could be observed that the addition of pheromone provides the ants with some form of global information about the current configuration of the grid, and this may lead to faster convergence rates and a more appropriate separation of the clusters contained in the input dataset. It was also interesting to observe that, to some extent, the use of pheromone works like a progressive vision, because as it diffuses to the environment, ants reasonably far from a cluster can be attracted to ('see') it in a stigmergic fashion. As in natural ants, smelling is a very important sense that provides general information about the environment.

Although the sensitivity analysis presented here and the proposed modification of the standard algorithm are relevant for aiding in the practical application of the algorithm, much research still has to be performed to validate the potential of this technique, including the proposed modification. Thus, we are starting to investigate the application of the modified algorithm to benchmark and real-world problems.

Acknowledgments

The authors acknowledge both CNPq and FAPESP for their financial support.

References

1. Bonabeau, E., Dorigo, M. and Théraulaz, G., *Swarm Intelligence from Natural to Artificial Systems*, Oxford University Press (1999).
2. Camazine, S., Deneubourg, J.-L., Franks, N. R., Sneyd, J., Theraulaz, G. and Bonabeau, E., *Self-Organization in Biological Systems*, Princeton University Press (2001).
3. Deneubourg, J. -L., Goss, S., Franks, N., Sendova-Franks, A., Detrain, C. and Chrétien, L., "The Dynamics of Collective Sorting: Robot-Like Ant and Ant-Like Robot", In J. A. Meyer and S. W. Wilson (eds.) *Simulation of Adaptive Behavior: From Animals to Animats*, pp. 356-365, Cambridge, MA, MIT Press/Bradford Books (1991).
4. Dorigo, M., *Optimization, Learning and Natural Algorithms*, (in Italian), Ph.D. Thesis, Dipartimento di Elettronica, Politecnico di Milano, IT (1992).
5. Kennedy, J., Eberhart, R. and Shi. Y., *Swarm Intelligence*, Morgan Kaufmann Publishers (2001).
6. Kube, C. R., Parker, C. A. C., Wang, T., Zhang, H., "Biologically Inspired Collective Robotics", In L. N. de Castro & F. J. Von Zuben, *Recent Developments in Biologically Inspired Computing*, Idea Group Inc., Chapter 15 (2004).
7. Lumer, E. D. and Faieta, B., "Diversity and Adaptation in Populations of Clustering Ants", In D. Cliff, P. Husbands, J. A. Meyer, S.W. Wilson (eds.), *Proc. of the 3rd Int. Conf. on the Simulation of Adaptive Behavior: From Animals to Animats*, 3, MIT Press, pp. 499-508 (1994).
8. Resnick, M., *Turtles, Termites, and Traffic Jams: Explorations in Massively Parallel Microworlds*, Cambridge, MA: MIT Press (1994).
9. Ramos, V., Muge, F., Pina, P. Self-Organized Data and Image Retrieval as a Consequence of Inter-Dynamic Synergistic Relationships in Artificial Ant Colonies. In J. Ruiz-del-Solar, A. Abrahan and M. Köppen Eds., Soft-Computing Systems - Design, Management and Applications, Frontiers in Artificial Intelligence and Applications: IOS Press, v. 87, 500-509, Amsterdam (2002).

Definition of Capacited p-Medians
by a Modified Max Min Ant System with Local Search

Fabrício Olivetti de França[1], Fernando J. Von Zuben[1], and Leandro Nunes de Castro[2]

[1] DCA/FEEC/Unicamp
State University of Campinas (Unicamp)
Caixa Postal 6101, 13083-852 – Campinas/SP, Brazil
[2] Research and Graduate Program on Informatics
Catholic University of Santos (UniSantos)
R. Dr. Carvalho de Mendonça, 144
11070-906, Santos/SP, Brazil

Abstract. This work introduces a modified MAX MIN Ant System (MMAS) designed to solve the Capacitated p-Medians Problem (CPMP). It presents the most relevant steps towards the implementation of an MMAS to solve the CPMP, including some improvements on the original MMAS algorithm, such as the use of a density model in the information heuristics and a local search adapted from the uncapacitated p-medians problem. Extensions to a recently proposed updating rule for the pheromone level, aiming at improving the MMAS ability to deal with large-scale instances, are also presented and discussed. Some simulations are performed using instances available from the literature, and well-known heuristics are employed for benchmarking.

1 Introduction

The capacitated p-medians problem (CPMP), also known as capacitated clustering problem, is a combinatorial programming task that can be described as follows: given a graph with n vertices (clients), find p centers (medians) and assign the other vertices to them minimizing the total distance covered, limited to a capacity restriction. This problem is a special case of the "capacitated plant location problem with single source constraints" and many other combinatorial problems [1]. As such, the CPMP can be proved to be NP-complete [2]. Its practical use varies from industrial and commercial planning to every cluster related problem, like data mining, pattern recognition, vehicle routing and many others.

Ant Systems (AS) were first proposed in [3] as an attempt to use the ant foraging behavior as a source of inspiration for the development of new search and optimization techniques. By using the pheromone trail as a reinforcement signal for the choice of which path to follow, ants tend to find "minimal" routes from the nest to the food source. The system is based on the fact that ants, while foraging, deposit a chemical substance, known as pheromone, on the path they use to go from the food source to the nest. The standard system was later extended [4], giving rise to the so-called Max Min Ant System (MMAS). The main purpose of the max-min version was to improve the AS's capabilities by combining exploitation with exploration of the search space, and by imposing bounds to the pheromone level, thus avoiding stagnation.

N.R. Pal et al. (Eds.): ICONIP 2004, LNCS 3316, pp. 1094–1100, 2004.
© Springer-Verlag Berlin Heidelberg 2004

This paper presents the application of an ant colony optimization (ACO) algorithm to the capacitated p-medians problem (CPMP). In particular, it describes one form of applying a modified Max Min Ant System (MMAS) to the CPMP problem that includes a local search heuristics and combines the MMAS with a framework from the literature in order to improve the performance of the algorithm. Therefore, the contributions of this work are twofold: 1) the application of an ACO algorithm to a new problem; and 2) the presentation of a modified algorithm which demonstrated robustness in solving large instances of the CPMP problem.

This paper is organized as follows. Section 2 provides a mathematical description of the problem. Section 3 describes the MMAS framework and presents the adaptations to be incorporated so that it can be used to solve the CPMP. Section 4 outlines the presentation of the new framework. Experimental results are then reported in Section 5, and Section 6 concludes the work and provides avenues for future research.

2 Problem Formulation

On a complete graph, given n nodes with a predefined capacity and demand, the purpose is to choose $p < n$ among them as capacitated medians and to attribute each one of the remaining $n-p$ nodes, denoted clients, to one of the chosen medians, so that the capacity of each median is not violated by the cumulated demand, and the sum of the distances from each client to the corresponding median is minimal. Every node is a candidate to become a median, and the solution will consider demand and capacity of medians, and only demand of clients.

Defining an $n{\times}n$ matrix $\mathbf{X}$, with components $x_{ij} \in \{0,1\}$, $i,j=1,...,n$, and an n-dimensional vector $\mathbf{y}$, with components $y_j \in \{0,1\}$, $j=1,...,n$, the following associations will be imposed:

$$x_{ij} = \begin{cases} 1, \text{if node } i \text{ is allocated to median } j \\ 0, \text{otherwise} \end{cases} \qquad y_j = \begin{cases} 1, \text{if node } j \text{ is a median} \\ 0, \text{otherwise} \end{cases}$$

Under these conditions, the CPMP formulation as an integer programming problem can be made as follows:

$$\min_{X,y} f(X) = \sum_{i=1}^{n}\sum_{j=1}^{n} d_{ij}x_{ij} \cdot \tag{1}$$

subject to,

$$\sum_{i=1}^{n} x_{ij} = 1, j = 1,...,n, \ x_{ij} \le y_j, i,j = 1,...,n, \ \sum_{j=1}^{n} y_j = p,$$

$$\sum_{i=1}^{n} x_{ij} \cdot a_i \le c_j, \text{ for } j \text{ such that } y_j = 1. \tag{2}$$

where:

n = number of nodes in the graph a_i = demand of node i

c_j = capacity of median j d_{ij} = distance between nodes i and j

p = number of medians to be allocated

3   Ant System

The basic AS [3] is conceptually simple, and can be described as follows:

```
While it < max_it do
        for each ant do:
            build_solution();
            update_pheromone();
        endfor
    end
```

In the algorithm described above, function build_solution() builds a solution to a problem based on a pheromone trail and some information heuristics (optional). Each ant k traverses one node per iteration step t and, at each node, the local information about its pheromone level, τ_i, is used by the ant such that it can probabilistically decide the next node to move to, according to the following rule:

$$p_i^k(t) = \begin{cases} \dfrac{[\tau_i(t)]^\alpha . [\eta_i]^\beta}{\sum_{l \in J^k} [\tau_l(t)]^\alpha . [\eta_l]^\beta} & \text{if } i \in J^k \\ 0 & \text{otherwise} \end{cases}, \tag{3}$$

where $\tau_i(t)$ is the pheromone level of node i, η_i is the information heuristic of node i, and J^k is the tabu list of nodes still to be visited by ant k. The parameters α and β are user-defined and control the relative weight of trail intensity $\tau_i(t)$ and visibility η_i.

While visiting a node i, ant k deposits some pheromone on it, and the pheromone level of node i is updated according to the following rule:

$$\tau_i \leftarrow \rho.\tau_i + \Delta\tau_i, \tag{4}$$

where $\rho \in (0,1]$ is the pheromone decay rate. In minimization problems, the pheromone increment is given by

$$\Delta\tau_i = \begin{cases} 1/f(S), & \text{if } i \in S \\ 0, & \text{otherwise} \end{cases}. \tag{5}$$

where S is the solution used to update the trail and $f(S)$ is the objective function.

3.1   Max Min Ant System (MMAS)

An improvement to the Ant System, called Max Min Ant System (MMAS), was introduced in [4]. On this implementation, the pheromone trail is updated only on the global best and/or local best solutions, instead of on solutions created by every ant, thus promoting a better exploitation of the search space. Another peculiarity is the inclusion of upper and lower bounds to the pheromone level (τ_{max} and τ_{min}), thus helping to avoid stagnation. Initially all trail is set to the upper bound in order to favor exploration. As in [5], the upper bound is usually chosen to be

$$\tau_{max} = \frac{1}{1-\rho} . \frac{1}{Fbest} . \tag{6}$$

where $Fbest$ is the objective function of the best solution found so far, and the lower bound is set to $\tau_{min} = \tau_{max}/2n$.

3.2 MMAS Applied to the Capacitated *p*-Medians Problem

To avoid having to update τ_{max} every time a new best solution is found, its value is fixed on the upper bound of the objective function. In this case, it is set to *n×gdist*, where *gdist* is the greatest distance between two nodes of a graph containing *n* nodes.

The ant structure is composed of an array of *p* elements chosen according to Eq. (3). The information heuristic (η) proposed here is a density model for the CPMP based on [6]. The idea is to calculate an optimistic density of a cluster if a given node was to be chosen as the median. The computation is made as follows:

```
For i = 1 to n do,
     sorted_nodes = sort_nodes(i);
     [all_nodes, distance] = allocate(i,sorted_nodes);
                all_nodes
     dens(i)= ──────────
                distance
End
```

Function `sort_nodes()` sorts all nodes based on their distance to node *i*; and function `allocate()` assigns each node in `sorted_nodes` to *i* until the capacity is reached, it returns `all_nodes`, associated with the number of nodes allocated, and `distance` which accounts for the distance between these points and node *i*. When all *p* medians are chosen, the CPMP becomes a Generalized Assignment Problem (GAP), so that the constructive solution method proposed in [1] can be promptly adopted.

After a solution is built, one iteration of local search is performed based on [7] and [8], and described in [9]. It consists of making the best possible swap of a median with a client. The pheromone trail is updated just for the global and local best solutions, by an increment of 1/*Fbest* and 1/*Flocal_best* on medians contained in the global and local best solutions, respectively. After a total of 30% of the number of iterations without any improvement (defined empirically), all pheromone trails are reinitialized.

4 Improving the MMAS Algorithm

A well-known problem with the AS is that of scaling the objective function to update the pheromone trail. If not appropriately done, the performance of the algorithm tends to be unsatisfactory for large instances. In [10] and [11] the authors proposed a framework for the AS that can also be applied to its variations like MMAS. The main idea is to normalize the bounds of the pheromone trails in the range [0,1], and consequently normalize the fitness function $f(\cdot)$.

To apply the MMAS to the capacitated *p*-medians problem using the new updating rule, some modifications had to be introduced to take full advantage of all the problem information available. First, τ_{min} and τ_{max} were set to 0.001 and 0.999, respectively, and the pheromone trail was initialized to 0.5. The update rule for pheromone was calculated using the global and the current best solutions, giving a better quality function as follows:

$$\Delta\tau_i = \begin{cases} \frac{gbest-lbest}{gbest}, & \text{if gbest} \geq \text{lbest and } i \in \{gbest, lbest\} \\ 1 - \frac{lbest-gbest}{lbest}, & \text{if gbest} < \text{lbest and } i \in \{gbest, lbest\} \\ 0, & \text{otherwise} \end{cases} \qquad (7)$$

where *gbest* and *lbest* are the global and local best solutions, respectively. In this equation, every time the local best is better than the global one, the pheromone is updated proportionally to the difference between them. Otherwise, the complementary value is taken; thus, the closer the local best from the global best, the closer $\Delta\tau_t$ becomes to one.

It was also adopted a convergence control for this algorithm in order to restart every time the algorithm stagnates. For this problem, it is intuitive that when the pheromone trail converges, p points (number of medians) will be at the upper bound and the remaining will be at the lower bound, so every time the sum of all pheromone follows Eq. (8), the algorithm is said to have stagnated and is thus restarted.

$$\sum_i \tau_i = p \cdot \tau_{max} + (n-p) \cdot \tau_{min}. \qquad (8)$$

5 Experimental Results

The experiments were performed to compare the MMAS with the improved MMAS (MMAS.IMP) presented here. For each algorithm, 2,000 iterations were run on an Athlon XP 2000+, 512 MB RAM running Linux Slackware 9.1, compiled with gcc 3.2, not optimized at compilation. Additionally, on the first set of instances (Osman) the performance of the modified algorithm was compared with the results presented in [1], referred as HSS.OC, which is an implementation of a hybrid composed of Simulated Annealing and Tabu Search. The results are summarized in Table 1.

Table 1. MMAS, MMAS.IMP and HSS.OC results for Osman's set of instances, "Avg." is the average relative percentage deviation from best for all 20 instances.

Osman	MMAS	MMAS.IMP	HSS.OC
Avg. (%)	0.133	0.042	0.049

Table 2 shows the set of larger instances created by Lorena in [12]. In this case, the MMAS.IMP presents a superior performance when compared with the simple MMAS algorithm. Furthermore, it was capable of finding better solutions than the best known solution to date.

6 Conclusions

This paper presents the application of an ant colony optimization algorithm to the capacitated p-medians problems (CPMP). In particular, it describes one form of applying the Max Min Ant System (MMAS) to the CPMP problem that includes a local search heuristics, and combines the MMAS with a new updating rule from the literature in order to improve the performance of the algorithm, mainly when large instances are considered.

Table 2. MMAS and MMAS.IMP for Lorena set of instances, the "Best Known" column is the best known solution found so far, results in bold are the best found among the algorithms, and "Time" is the execution time in seconds.

		MMAS			MMAS.IMP		
Lorena	Best Known	Sol.	%	Time	Sol.	%	Time
n=100 p=10	17288	17288	0.00	295.95	17252	**−0.21**	90.26
n=200 p=15	33395	33254	−0.42	540.96	33207	**−0.56**	615.15
n=300 p=25	45364	45251	−0.25	8109.64	45245	**−0.26**	3726.32
n=300 p=30	40635	40638	0.01	7818.13	40521	**−0.28**	3766.40
n=402 p=30	62000	62423	0.68	12701.24	62020	**0.03**	9299.55
n=402 p=40	52641	52649	0.02	10500.15	52492	**−0.28**	9295.11
Avg.			0.006			**−0.261**	

With the extensions proposed here, based on the framework presented in [11], the final results demonstrated to be competitive and even better than other heuristics found in the literature. For future research, it must be investigated an adaptive distribution of importance factors given for the local and global bests. Also, a better GAP local search or even a constructive heuristics can be implemented to assign clients to medians.

Acknowledgements

The authors would like to thank Capes, Fapesp and CNPq for the financial support.

References

1. Osman, I. H.; Christofides, N., Capacitated Clustering Problems by Hybrid Simulated Annealing and Tabu Search. Pergamons Press, England, 1994, Int. Trans. Opl. Res. Vol. 1, No. 3, pp. 317-336, 1994.
2. Garey, M.R. and Johnson, D. S., Computers and Intractability: a Guide to the Theory of NP-Completeness. San Francisco: Freeman. 1979.
3. Dorigo M. Optimization, Learning and Natural Algorithms. Ph.D.Thesis, Politecnico di Milano, Italy, in Italian, 1992.
4. Stützle, T. and Hoos, H.H. The MAX-MIN ant system and local search for the traveling salesman problem. In T. Bäck, Z. Michalewicz, and X. Yao, editors, Proceedings of the IEEE International Conference on Evolutionary Computation (ICEC'97), IEEE Press, Piscataway, NJ, USA, pp. 309-314, 1997.
5. Stützle T. and Dorigo M. ACO algorithms for the Quadratic Assignment Problem. In D. Corne, M. Dorigo, and F. Glover, editors, New Ideas in Optimization, pages 33-50. McGraw-Hill, 1999.
6. Ahmadi, S. and Osman, I.H., Density based problem space search for the capacitated clustering problem. Annals for Operational Research, 2004 (in press).
7. Resende, G.C.M. and Werneck, F. R. On the implementation of a swap-based local search procedure for the p-median problem. Proceedings of the Fifth Workshop on Algorithm Engineering and Experiments (ALENEX'03), Richard E. Ladner (Ed.), SIAM, Philadelphia, pp. 119-127, 2003.

8. Teitz, M. B.; Bart, P., Heuristic Methods for Estimating the Generalized Vertex Median of a Weighted Graph. Operation Research, 16(5):955-961, 1968.
9. De França, F.O., Von Zuben, F.J. and De Castro, L.N., A Max Min Ant System Applied to the Capacitated Clustering Problem, IEEE International Workshop on Machine Learning for Signal Processing, São Luis, Brazil, 2004.
10. Blum, C., Roli, A. and Dorigo, M. Hc-Aco: The hyper-cube framework for Ant Colony Optimization. In Proceedings of MIC'2001 - Meta-heuristics International Conference, volume 2, Porto, Portugal, pp. 399-403, 2001.
11. Blum, C. and Dorigo, M., The Hyper-Cube Framework for Ant Colony Optimization. IEEE Transactions on Systems, Man and Cybernetics, Part B, 34(2): 1161-1172, 2004.
12. Lorena, L.A.N. and Senne, E.L.F. Local Search Heuristics for Capacitated P-Median Problems, Networks and Spatial Economics 3, pp. 407-419, 2003.

Investigations into the Use of Supervised Multi-agents for Web Documents Categorization

Siok Lan Ong[1], Weng Kin Lai[1], Tracy S.Y. Tai[1],
Choo Hau Ooi[2], and Kok Meng Hoe[1]

[1] MIMOS, Technology Park Malaysia, 57000 Kuala Lumpur, Malaysia
[2] University of Malaya, 50603 Kuala Lumpur, Malaysia

Abstract. The self-organization behavior exhibited by ants may be modeled to solve real world clustering problems. The general idea of artificial ants walking around in search space to pick up, or drop an item based upon some probability measure has been examined to cluster a large number of World Wide Web (WWW) documents [1]. In this paper, we present a preliminary investigation on the direct application of a Gaussian Probability Surface (GPS) to constrain the formation of the clusters in pre-defined areas of workspace with these multi-agents. We include a comparison between the clustering performances of supervised ants using GPS against the typical ants clustering algorithm. The performance of both supervised and unsupervised systems will be evaluated on the same dataset consisting of a collection of multi-class web documents. Finally, the paper concludes with some recommendations for further investigation.

1 Introduction

Social insects make up 2 % of all species of living organisms that live in this world [2], with ants forming by far the largest group - 50% of these social insects are ants. Within the ant colonies, there is specialization in the tasks that need to be performed. Many of these simple but yet important tasks are very similar to some of the real world problems for humans. For example, the foraging behavior of ants has shown to be a useful computing paradigm for solving discrete optimization problems [3]. Similarly, the self-organizing behavior of ants may be used to model intelligent application such as clustering. In this paper, we focus on the task performed by the specialized worker ants that include nest and cemetery maintenance through clustering, particularly in clustering the fast growing source of online text documents.

Similar to any typical document clustering, web documents clustering may generally be seen as dividing the documents into homogeneous groups with the main purpose that documents within each cluster should be similar to one another while those within different clusters should be dissimilar [4]. Unfortunately, the sheer size of the World Wide Web makes it difficult to manually categorize the documents. In order to automate the process, different well-established clustering approaches have been widely applied to effectively organize the documents based on the above principle in terms of processing time, quality of clustering and spatial distribution. The straightforward model which ants move randomly in space to pick up and deposit items on the basis of local information has also been explored to cluster web documents [1].

This paper examines the direct implementation of a *Gaussian Probability Surface* (GPS) to supervise these homogeneous multi-agents to form clusters within a speci-

N.R. Pal et al. (Eds.): ICONIP 2004, LNCS 3316, pp. 1101–1109, 2004.
© Springer-Verlag Berlin Heidelberg 2004

fied dropping zone. In addition, the results will also be compared with those obtained through unsupervised multi-agents clustering. Basically, the main idea is building on the concept of *"self-organisation along a template"*, whereby a template mechanism is combined with the self-organisation mechanism. More specifically, it involves mapping each pixel in a workspace layer to a pixel in another surface within the same relative spatial location. Combining the underlying self-organizing mechanisms of the algorithm with templates allows all items be deposited in some particular regions of space[5].

2 Ant Colony Models

The ability of insects such as ants living in a colony has fascinated many in the scientific community and this has led to more detailed studies on the collective behavior of these creatures. Even though these insects may be small in size and live by simple rules, but yet they are able to survive well within their colony. Recently, scientists have found that this behavior could be borrowed to solve complex tasks such as text mining, networking etc. Deneubourge et al. had developed this concept further by modeling the ant's action in organizing their nests for data classification. Assuming each of these multi-agents carries one item at a time and there is only one item type, the probabilistic functions, P are as shown below:

$$\text{Picking-up probability, } P_p = \left(\frac{k_1}{k_1 + f}\right)^2 \tag{1}$$

$$\text{Dropping probability, } P_d = \left(\frac{f}{k_2 + f}\right)^2 \tag{2}$$

f denotes the fraction of similar items in the neighbourhood of the agent, while k_1 and k_2 are threshold constants. When $f \to 1$, there are many similar items in the neighbourhood. This indicates that there is a high possibility that the multi-agent will put down the item it is carrying, as P_d is high. Similarly, the agent is not likely to pick up the item when P_p is low. This will happen when most of the items in the neighbourhood are dissimilar, indicated by $f \to 0$. Essentially, there is a high possibility of picking up items which are isolated and transporting them to another region where they are now more of its kind in the neighbourhood. The possibility of dropping the item will be low when $P_d \to 0$. Lumer & Faieta (LF) [6] had reformulated Deneubourg et al.'s [7] model to include a distance function, d between data objects for the purpose of exploratory data analysis. The binary distance between objects i and j, $d(o_i, o_j)$, is assgined 1 for dissimilar objects and 0 for similar objects. The fraction of items in the neighbourhood, f in equation (1) and (2) is replaced with the local density function, $f(o_i)$ which measures the average similarity of object i with other objects j in its neighbourhood, N. Given a constant, α, and the cells in a neighbourhood, $N(c)$, $f(o_i)$ may be defined as:

$$f(o_i) = \begin{cases} \dfrac{1}{\left|N(c)^2\right|} \sum_{o_j \in N(c)} \left[1 - \dfrac{d(o_i,o_j)}{\alpha}\right] & \text{if } f > 0 \\[3em] 0 & \text{otherwise} \end{cases} \tag{3}$$

3 Supervised Ant Colony Models with Gaussian Probability Surface (GPS)

In several species of ants, the worker ants are known to perform corpse aggregation and brood sorting where the clusters formed is at arbitrary locations [8]. However, there are other species, like the *Acantholepsis custodiens* ants that are known to perform self-organization which are constrained by templates [9]. A template is a pattern that is used to construct another pattern. In the case of the ants in Nature, they utilize the information related to the temperature and humidity gradients in their surroundings to build their nests ants to spatially distribute their brood [9]. This concept of self-organizing with templates has been used by *Dorigo & Theraulaz* for data analysis and graph partitioning[5].

With such mechanisms, the result is that the final structures would closely follow the configuration defined by the templates. However, this is only useful in applications where the numbers of clusters are known beforehand. The template we have used here, in the form of a *Gaussian Probability Surface* (GPS) guides the multi-agents to form clusters within a *toroidal* working space. The GPS is shown in equation 4 below.

$$P(x,y) = P_{max} \sum_{i=1}^{n} \left[e^{\left(-\left(\frac{(x-x_{0i})^2 + (y-y_{0i})^2}{\sigma^2}\right)\right)} \right] + \delta \tag{4}$$

where $0 \le P_{max} \le 1$ and $0 \le P(x,y) \le 1$

δ offset value.

σ^2 constant defined by user that also determines the steepness of the Gaussian probability surfaces.

x_{oi}, y_{oi} Coordinates of the centre of each dropping zone (i.e. the peak of the humps).

x, y Coordinates on any single point in the workspace.

i number of humps, $1 \le n \le 5$.

The probability surface is two dimensional and isotropic (circular symmetry). Figure 1 shows an example of the GPS model with four humps. As the height of the surface increase, the probability of dropping the document by the multi-agents is higher. Hence, more similar documents are expected to be clustered in the area under an area defined by each hump. This will enhance the quality of clustering by having clusters with similar document types in the specified dropping zone instead of forming in non-deterministic region of the workspace.

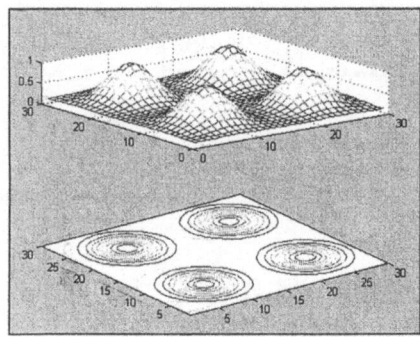

Fig. 1. The Gaussian Probability Surface (GPS) superimposed onto the toroidal working space.

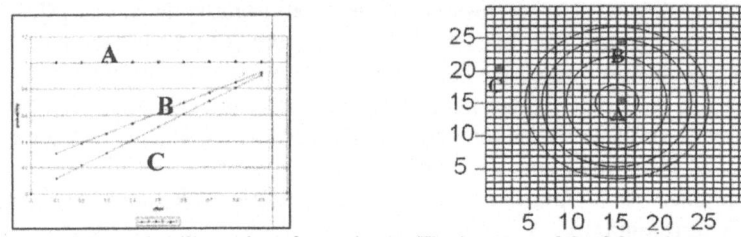

(a) The plot of the probability values for each   (b) The location of the 3 locations monitored.
of the 3 points monitored.

Fig. 2. The probability values of 3 locations for various offsets.

Figure 2(a) shows how the probability values with different values of the offset δ. It may be seen that the probability for depositing an item increases linearly with an increase in the offset (δ) values. The probability is close to 1 at the peak of the humps and there was only a slight increase for any increments of the offset δ. However, there is a significant change in the probability for points at the lower portion of the surface for different values of the offset. In addition, the dropping probability distribution in the regions between the contour lines does not vary much for higher offsets. This implies that there are actually more space for the multi-agents to unload the documents for higher offsets.

The multi-agents can only move one step in any direction at each time unit from its existing location to an *unoccupied* adjacent cell. Only a single agent and/or a single item is allowed to occupy any one cell at a time. An agent occupying any cell, c on the clustering space immediately perceives a neighbourhood of 8 adjacent cells i.e. $N(c) = 8$. The decision of an *unladen* agent to either pick up or ignore an item o_i at cell c is dictated by a probability P_p that is based on a local density function, $g(o_i)$. This local density function determines the similarity between o_i and other items o_j, where $j \in N(c)$. If an agent *laden* with item o_i lands on an empty cell c, it will calculate a probability P_d based on the same function $g(o_i)$ and decides whether to drop o_i or keep on carrying it. Unlike f (see Eq. 3) which uses a distance measure and an additional parameter α, the function $g(o_i)$ uses a similarity measure which may be defined as follows:

$$g(o_i) = \frac{1}{N(c)} \sum_{o_j} S(o_i, o_j) \qquad (5)$$

where $S(o_i, o_j)$ is a measure of the similarity between objects o_i and o_j. To model the inherent similarity within documents, we used the *cosine* measure,

$$S_{cos}(doc_i, doc_j) = \frac{\sum_{k=1}^{r} (f_{i,k} \times f_{j,k})}{\sqrt{\sum_{k=1}^{n} (f_{i,k})^2} \times \sqrt{\sum_{k=1}^{m} (f_{j,k})^2}} \qquad (6)$$

where r is the number of common terms in doc_i and doc_j, n and m represent the total number of terms in doc_i and doc_j respectively. $f_{a,b}$ is the frequency of term b in doc_a. A useful property of S_{cos} is that it is invariant to large skews in the weights of document vectors, but sensitive to common concepts within documents.

In this paper, we employ the Gaussian Probability Distribution to guide the multi-agents to drop the documents onto a specified dropping zone in a two dimensional workspace. This model requires large samples and repeated measurements with random errors distributed according to the Gaussian probability [10].

4 Experimental Set-Up

The 84 web pages used in the experiment came from four different categories – Business, Computer, Health and Science that were randomly retrieved from the *Google* web directory. These were then pre-processed to extract a representative set of features. The main purpose of this feature extraction process is to identify a set of most descriptive words of a document. For these text documents, we have augmented these words with weights, while disregarding the linguistic context variation at the morphological, syntactical, and semantic levels of natural language [11]. The extracted *word-weight* vectors are usually of high dimensions. In our case, the collection of web documents yielded 6,976 distinct words.

We investigated the classification of the dataset above using supervised and unsupervised multi-agents within a 30x30 toroidal grid and 15 homogeneous agents with threshold constants, $k_1 = 0.01$ and $k_2 = 0.15$. Both supervised and unsupervised approaches were set to run at a maximum of iterations, t_{max} of 140,000. As there are four document categories in this experiment, we have specified a similar number of humps to be used.

5 Results and Discussion

This section depicts the experimental results of supervised and unsupervised multi-agents clustering with the parameters setting as described in the previous section.

Figure 3(a) shows how the web documents were initially scattered on the two dimensional workspace at time $t=0$. After 140,000 iterations, four clusters of mixed classes of documents were formed. We evaluated the quality of the results through the measures of purity and entropy. Table 1 below shows the purity (measures the similarity) and entropy (which measures the distribution of various (actual) categories of documents within a cluster) values of the clusters [12]. A high value of purity suggests that the cluster is a *pure* subset of the dominant class. Similarly, an entropy value of 0 means the cluster is comprised entirely of one class. The overall entropy value is the weighted sum of the individual values for each cluster which takes into account the size of each cluster formed. The same applies for the overall purity value.

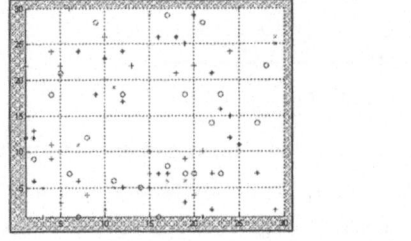
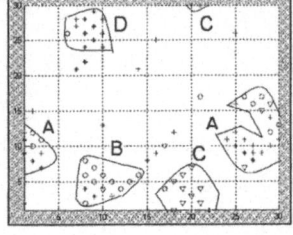

Key: o – Business,   ∇ - Computer,   + - Health,   * - Science

Fig. 3a. The random distribution of the web documents on the workspace at time, *t=0*.   **Fig. 3b.** Clusters of documents formed at *t = 140,000.*

Table 1. The purity and entropy values for different clusters of documents and the overall result.

Cluster	Entropy	Purity	Majority Class
A	0.9297	0.4000	Health
B	0.4732	0.7857	Business
C	0.7419	0.6429	Computer
D	0.5204	0.7500	Science
Overall	**0.6088**	**0.4881**	-

Fig. 4. Graphical representation of the differences in the size of the clusters formed.

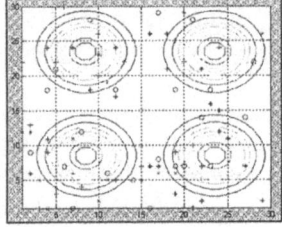
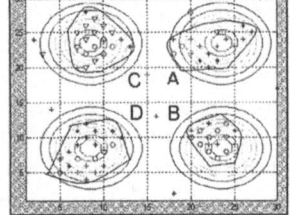

Key: o – Business,   ∇ - Computer,   + - Health,   * - Science

Fig. 5a. The random distribution of the web documents on the workspace at *t =0*.   **Fig. 5b.** Four clusters with each containing a majority of different classes were formed at *t = 80,000.*

Figure 5(a) shows the initial placement of the documents which were scattered on the workspace with the contour plots of the GPS superimposed upon it. After 80,000 iteration it was obvious that the multi-agents had sorted the documents into four different clusters. Most of the documents in the contour regions were closely placed near to the centre of each cluster. In addition, there were only nine documents found scattered at the base of the probability surfaces (indicated by the areas outside the contours). The purity and entropy values obtained from this aprpoach are depicted in Table 2 below:

Table 2. The purity and entropy values for different
clusters of documents and the overall result.

Cluster	Entropy	Purity	Majority Class
A	0.8061	0.5000	Health
B	0.7960	0.5625	Business
C	0.8390	0.5263	Computer
D	0.7243	0.5455	Science
Overall	0.7038	0.4762	–

Fig. 6. Graphical representation
of the differences in the size of
the clusters formed.

In comparison, although both approaches produce clusters which have nearly similar purity values, however the entropy for the supervised approach using the GPS was approximately 10% lower when compared to the unsupervised approach. Moreover, we also found that it was difficult to identify the clusters if GPS was not employed. In other words, the spatial distribution between clusters was uneven without GPS. Conversely, if GPS was adopted, we could easily identify the clusters in the contour areas because the spaces between the clusters were more distinct. In addition, the clusters formed by the GPS were neat and more tightly coupled whereas those without GPS were loose as shown in Figure 2(b) and 3(b). A graphical representation of the difference in the size of the cluster formed is shown in figures 4 & 6. These depict the differences in the size of the clusters formed. Both graphical representations were drawn on the same scale. Clearly, there is also a greater uniformity in the size of the clusters generated when the clusters were formed with GPS.

In terms of the stability of the clusters, with GPS, the multi-agents were able to move most documents into the cluster itself and seldom went beyond the specified regions. Any document could be easily moved around the workspace when the agents were fully unsupervised. Hence, we suggest that the GPS was actually guiding the multi-agents to cluster the documents and constraint the size of clusters in certain regions. More importantly, there was also an improvement in processing time required. With the supervised approach, the clusters were formed at 80,000 iterations, as compared with 140,000 for the unsupervised approach. This would be very useful for the retrieval and access of high dimension web documents.

6   Concluding Remarks and Future Directions

In this paper, we presented the findings of an extended study on using a multi-agent system based on the collective behavior of social insects i.e. ants, to cluster web documents retrieved from a popular search engine. Unlike earlier work, we introduced the direct application of a Gaussian Probability Surface (GPS) to constraint the formation of the clusters in pre-defined areas in the workspace. The experimental results showed that the proposed multi-agent system is able to induce clusters with better clusters than those obtained without this probability surface. We are aware that these results may only be marginally better. Obviously, more extensive experimental results need to be obtained but we are encouraged with the results obtained so far. In addition, we believe that with the right refinements, the system may be able to produce significantly better results. Visually it is clear that the clusters are better formed than those obtained when there is no GPS. If we examine figure 3(b) again, we can clearly identify two smaller clusters. Essentially, unlike the GPS-driven clustering approach, without specifying the exact number of clusters to be formed, the unsupervised approach has formed a total of 4 large clusters and 2 smaller ones. The results obtained, although not on par with the classification ability of human experts, do demonstrate the potential of ant-like multi-agent systems in handling complex and high-dimension data clusters.

In conclusion, there are still several significantly important areas in the current system that has to be improved before it can measure up against other well-established document clustering methods. We have also noticed that the offsets do have a profound effect on the quality of the clusters formed. They are also known to affect the speed of convergence of the multi-agent system. Obviously, if the offset of the GPS is high, the multi-agents have a higher freedom to drop the web documents that they may be carrying over a wider area. A low offset has the opposite effect. We would be exploring whether a non-stationary offset can produce better clusters in the future. In addition, we would also like to explore a larger perceivable time-dependent neighbourhood for agents and a better formulation of a stopping criterion based on homogeneity and spatial distribution of clusters. Lastly, our observation on the initial random distribution of data points on the workspace, which can significantly affect the clustering results, would also be a good area for future research.

Acknowledgements

The authors would like to thank *C.C. Loy* for his help in generating the contour graphs.

References

1. Hoe K. M., Lai W.K., & Tracy Tai, *"Homogeneous Ants for Web Document Similarity Modeling and Categorization"*, Proceedings of the Third International Workshop on Ant Algorithms, pp 256 – 261, September 12th – 14th, 2002, Brussels, Belgium.
2. M.Dorigo, *Artificial Life: "The Swarm Intelligence Approach"*, Tutorial TDI, Congress on Evolutionary Computing, Washington, DC. (1999).
3. Engelbrecht, A.P., *"Computational Intelligence: An Introduction"*, John Wiley & Sons Ltd (2002), ISBN: 0-470-84870-7.
4. J.Handl, J.Knowles and M.Dorigo, *"Ant Based Clustering: a comparative study of it's relative performance with respect to k-means, average link and ld-som"*, http://wwwcip.informatik.uni-erlangen.de/~sijuhand/TR-IRIDIA-2003-24.pdf, March 24th 2004.
5. Bonabeau, E., Dorigo, M., and Theraulaz, G., *"Swarm Intelligence: From Natural to Artificial Systems"*, University Press, Oxford (1999), pp 199.
6. Lumer, E.D. and Faieta, B., *"Diversity and Adaptation in Populations of Clustering Ants"*, Int. Conf. Simulation of Adaptive Behavior: Fr. Animals to Animats. MIT, MA (1994).
7. Deneubourg, J. L., Goss, S., Franks, N.R., Sendova-Franks, A., Detrain, C., and Chretien, L., *"The Dynamics of Collective Sorting: Robot-like Ants and Ant-like Robots"*, Int. Conf. Simulation of Adaptive Behaviour: Fr. Animals to Animats. MIT, MA (1990).
8. Bonabeau, E., Dorigo, M., and Theraulaz, G., *"Swarm Intelligence: From Natural to Artificial Systems"*, University Press, Oxford (1999), pp 149.
9. Bonabeau, E., Dorigo, M., and Theraulaz, G., *"Swarm Intelligence: From Natural to Artificial Systems"*, University Press, Oxford (1999), pp 184.

10. Department of Physic and Astronomy, *"Physics and Astronomy: The Gaussian Distribution"*, http://physics.valpo.edu/courses/p310/ch2.3_gaussian/, March 24th 2004.
11. Baeza-Yates, R.. and Ribeiro-Yates, B., *"Modern Information Retrieval"*, ACM, NY (1999).
12. Steinbach, M., Karypis, G., and Kumar, V., *"A Comparison of Document Clustering Techniques"*, KDD Workshop on Text Mining (2000).

OrgSwarm – A Particle Swarm Model of Organizational Adaptation

Anthony Brabazon[1], Arlindo Silva[2,3], Tiago Ferra de Sousa[3], Michael O'Neill[4], Robin Matthews[5], and Ernesto Costa[2]

[1] Faculty of Commerce, University College Dublin, Ireland
anthony.Brabazon@ucd.ie
[2] Centro de Informatica e Sistemas da Universidade de Coimbra, Portugal
ernesto@dei.uc.pt
[3] Escola Superior de Tecnologia, Instituto Politecnico de Castelo Branco, Portugal
arlindo@est.ipcb.pt
[4] Dept. Of Computer Science &Information Systems, University of Limerick, Ireland
Michael.ONeill@ul.ie
[5] Centre for International Business Policy, Kingston University, London, UK

Abstract. This study extends the particle swarm metaphor to include the domain of organizational adaptation. A simulation model, *OrgSwarm*, is constructed in order to examine the impact of strategic inertia on the adaptive potential of a population of organizations. The key finding is that a degree of strategic inertia, in the presence of an election mechanism, can assist rather than hamper adaption of the population.

1 Introduction

Following a long-established metaphor of adaptation as search, strategic adaptation is considered in this study as an attempt to uncover fitness peaks on a high-dimensional strategic landscape. Some strategic configurations produce high profits, others produce poor results. The search for good strategic configurations is difficult due to the vast number of configurations possible, uncertainty as to the nature of topology of the strategic landscape faced by an organization, and changes in the topology of this landscape over time. Despite these uncertainties, the search process for good strategies is not blind. Decision-makers receive feedback on the success of their current and historic strategies, and can assess the payoffs received by the strategies of their competitors. Hence, certain areas of the strategic landscape are illuminated. A key characteristic of the framework which integrates the search heuristics examined in this study, is that organizations do not adapt in isolation, but interact with each other. Their efforts at strategic adaption are guided by 'social' as well as individual learning. The work of [4] & [8], drawing on a swarm metaphor, has emphasized similar learning mechanisms. We extend this work into the organizational domain.

N.R. Pal et al. (Eds.): ICONIP 2004, LNCS 3316, pp. 1110–1116, 2004.
© Springer-Verlag Berlin Heidelberg 2004

2   OrgSwarm Model

This study constructs a novel simulation model (*OrgSwarm*) to examine the impact of strategic inertia on the rate of strategic adaptation of a population of organizations. This paper presents initial results from this model, and future work is planned to extend these results further. The model can be classed as a *multi-agent system* (MAS). MASs focus attention on collective intelligence, and the emergence of behaviors through the interactions between the agents. MASs usually contain a world (environment), agents, definition of relations between the agents, a set of activities that the agents can perform, and changes to the environment as a result of these activities. The key components of the simulation model, the landscape generator (environment), and the adaption of the basic Particle Swarm algorithm to incorporate the activities and interactions of the agents (organizations), are described next.

2.1   Strategic Landscape

In an organizational setting, a strategy can be conceptualized as being the choice of what activities an organization will perform, and the subsequent choices as to how these activities will be performed [10]. These choices define the strategic configuration of the organization. Recent work by [9] and [11] has recognized that strategic configurations consist of interlinked individual elements (decisions), and have applied general models of interconnected systems such as Kauffman's NK model [2,3] to examine the implications of this for processes of organizational adaptation. It is noted *ab initio* that application of the NK model to define a strategic landscape is not atypical and has support from existing literature in organizational science [9,11,1]. The NK model considers the behavior of systems which are comprised of a configuration (string) of N individual elements. Each of these elements are in turn interconnected to K other of the N elements (K<N). In a general description of such systems, each of the N elements can assume a finite number of states. If the number of states for each element is constant (S), the space of all possible configurations has N dimensions, and contains a total of $\prod_{i=1}^{N} S_i$ possible configurations. In Kauffman's operationalization of this general framework [3], the number of states for each element is restricted to two (0 or 1). Therefore the configuration of N elements can be represented as a binary string. The parameter K, determines the degree of fitness interconnectedness of each of the N elements and can vary in value from 0 to N-1. In one limiting case where K=0, the contribution of each of the N elements to the overall fitness value (or worth) of the configuration are independent of each other. As K increases, this mapping becomes more complex, until at the upper limit when K=N-1, the fitness contribution of any of the N elements depends both on its own state, and the simultaneous states of all the other N-1 elements, describing a fully-connected graph. Altering the value of K affects the ruggedness of the described landscape (graph), and consequently impacts on the difficulty of search on this landscape [2,3]. In this study, the strategy of an organization is characterized as consisting of N attributes [9]. Each of these attributes represents a strategic decision or

policy choice, that an organization faces. Hence, a specific strategic configuration s, is represented as a vector $s_1, \ldots, s_N$ where each attribute can assume a value of 0 or 1 [11]. The vector of attributes represents an entire organizational form, hence it embeds a choice of markets, products, method of competing in a chosen market, and method of internally structuring the organization [11]. Good consistent sets of strategic decisions - configurations, correspond to peaks on the strategic landscape.

2.2 Particle Swarm Algorithm

This section provides an introduction to the canonical Particle Swarm Algorithm (PSA)[1], and outlines how the algorithm is adapted for this study. A fuller description of the particle swarm algorithm and the cultural model which inspired it is provided in [4, 8, 5]. Under the particle swarm metaphor, a swarm of particles (entities) are assumed to move (fly) through an n-dimensional space, typically looking for a function optimum. Each particle is assumed to have two associated properties, a current position and a velocity. Each particle also has a memory of the best location in the search space that it has found so far (*pbest*), and knows the location of the best location found to date by all the particles in the population (*gbest*). At each step of the algorithm, particles are displaced from their current position by applying a velocity vector to them. The size and direction of this velocity is influenced by the velocity in the previous iteration of the algorithm (simulates 'momentum'), and the current location of a particle relative to its *pbest* and *gbest*. Therefore, at each step, the size and direction of each particle's move is a function of its own history (experience), and the social influence of its peer group. The mechanisms of the Particle Swarm model bear *prima facie* similarities to those of the domain of interest, organizational adaptation. It embeds concepts of a population of entities which are capable of individual and social learning. However, the model requires modification before it can employed as a plausible model of organizational adaptation. In this study, we adapt the basic PSA to embed inertia or 'anchoring', and an election operator.

Organizations do not have complete freedom to alter their current strategy. Their adaptive processes are subject to inertia. Inertia springs from the organization's culture, history, and the mental models of its management. These forces serve to anchor an organization to its strategic history. Inertia could be incorporated into the PSA in a variety of ways. We have chosen to incorporate it into the velocity update equation, so that the velocity and direction of the particle at each iteration is also a function of the location of its 'strategic anchor'. Therefore for the simulations, the usual PSA velocity update equation is altered by adding an additional *anchor* or 'inertia' term:

$$v_i(t+1) = v_i(t) + R_1(y_i - x_i(t)) + R_2(\hat{y} - x_i(t) + R_3(a_i - x_i(t)) \qquad (1)$$

---

[1] The term PSA is used in place of PSO (Particle Swarm Optimization) in this paper, as the object is not to develop a tool for 'optimizing', but to adapt and apply the swarm metaphor as a component of a model of organizational adaptation.

where a_i represents the position of the anchor on the strategic landscape for organization i, R_3 represents a weight coefficient, $x_i(t)$ represents the current position of organization i on the strategic landscape, and y_i & $\hat{y}$ represent its pbest and the gbest. The anchor can be fixed at the initial position of the particle (organization) at the start of the algorithm, or it can be allowed to 'drag', thereby being responsive to the recent adaptive history of the particle.

Real-world organizations do not usually intentionally move to 'poorer' strategies. Hence, an 'election' or *ratchet* operator operator is implemented, whereby position updates which would worsen an organization's strategic fitness are discarded. In these cases, an organization remains at its current location on the strategic landscape. An economic interpretation of the election operator, is that strategists carry out a mental simulation or 'thought experiment'. If the expected fitness of the new strategy appears unattractive, the 'bad idea' is discarded. We also adapt the basic PSA for the binary nature of the strategic landscape used in this study, and use the BinPSO algorithm of [6].

3   Results

This section provides the results from our simulation study. All reported fitnesses are the average population fitnesses across 30 separate simulation runs at the conclusion of 5,000 iterations of the algorithm. On each simulation run, the NK landscape is specified anew, and the positions and velocities of particles are randomly initialized at the start of each run. The simulations employ a population of 20 particles, with a circular neighborhood of size 18. Values of N=96 and K=0,4 & 10 are selected in defining the landscapes in this simulation, representing landscapes with differing degrees of fitness inter-connectivity. Table 1 provides the results for each of ten distinct PSA 'variants', at the end of 5,000 iterations, across three static landscape 'scenarios' (K=0,4 & 10). In each scenario, the same series of simulations are undertaken. Initially, a basic (binary) PSA is employed, without an anchor term or an election operator. This simulates a population of organizations searching a strategic landscape, where members of the population have no strategic anchor (no inertia), and where organizations do not utilize a ratchet (conditional move) operator in deciding whether to alter their position on the strategic landscape. The basic PSA is then supplemented by a series of strategic anchor formulations, ranging from a fixed position (fixed at the organization's initial position) anchor which does not change position during the simulation, to one which adapts after a time-lag (moving anchor). In both the initial and moving anchor experiments, a weight value of 1 (the three weight coefficients, $R_1, \ldots, R_3$, in equation 1 are constrained to sum to four) is attached to the inertia term in the velocity equation, and a time-lag of 20 periods (therefore, the location of the strategic anchor for an organization at t_n is the position of the organization at t_{n-20}) is used for the moving anchor. In the experiments concerning the affect of error when assessing the future payoffs to potential strategies, three values of *error* are examined.

Examining the results in table 1 suggests that the basic PSA, without inertia or ratchet operators, performs poorly on a static landscape, even when there is no error in assessing the 'payoffs' to potential strategies. The average populational fitness (averaged across all 30 simulation runs) obtained after 5,000 iterations is no better than random search, suggesting that unfettered adaptive efforts, based on 'social communication' between organizations (gbest), and a memory of good past strategies (pbest) is not sufficient to achieve high levels of populational fitness, even when organizations can make error-free assessments of the 'payoff' of potential strategies. When a ratchet operator is added to the basic PSA (Ratchet PSA-No Anchor), a significant improvement (statistically significant at the 5% level) in both average populational, and average environment best fitness is obtained across landscapes of all K values, suggesting that the simple decision heuristic of *only abandon a current strategy for a better one* leads to notable increases in populational fitness.

Table 1. Average populational fitness after 5,000 iterations, K=0, 4 & 10.

Algorithm	Fitness (N=96, K=0)	(N=96, K=4)	(N=96, K=10)
Basic PSA	0.4641	0.5002	0.4991
Ratchet PSA-No Anchor	0.5756	0.6896	0.6789
Ratchet-No Anchor, e=0.05	0.4860	0.6454	0.6701
Ratchet-No Anchor, e=0.20	0.4919	0.5744	0.5789
Ratchet-Initial Anchor, w=1	0.6067	0.6991	0.6884
Ratchet-Initial Anchor, w=1, e=0.05	0.5297	0.6630	0.6764
Ratchet-Initial Anchor, w=1, e=0.20	0.4914	0.5847	0.5911
Ratchet-Mov. Anchor (20,1)	0.6692	0.7211	0.6976
Ratchet-Mov. Anchor (20,1, e=0.05)	0.5567	0.6675	0.6770
Ratchet-Mov. Anchor (20,1, e=0.20)	0.4879	0.5757	0.5837

In real-world organizations, assessments of the payoffs to potential strategies are not error-free. *A priori* we do not know whether this could impact positive or negatively on the evolution of populational fitness, as permitting errorful assessments of payoff could allow an organization to escape from a local optimum on the strategic landscape, and possibly therefore to uncover a new 'gbest'. In essence, an errorful assessment of payoff may allow a short-term 'wrong-way' move (one which temporarily reduces an organization's payoff), but which in the longer-term leads to higher payoffs. Conversely, it could lead to the loss of a promising but under-developed strategy, if an organization is led away from a promising part of the strategic landscape by an incorrect payoff assessment. To examine the impact of errorful payoff assessment, results are reported for the Ratchet PSA-No Anchor, for values of error of 0.05 and 0.20. In applying these errors, the fitness estimate used by the inventor in applying the election operator is *actual fitness of the new strategy * (1+ 'error')*, where *error* is drawn from a normal distribution with a mean of zero and a standard deviation of 0.05 and 0.20. Hence, despite the election operator, a strategist may sometimes choose a 'bad' strategy because of an incorrect *ex-ante* assessment of its fitness. Table 1 shows that these produce lower results (statistically significant at 5%) than the error-free case. As the size of the error increases, the average populational fitness declines, suggesting that the utility of the ratchet operator decreases as the level of error in assessing the payoff to potential strategies rises.

The experiments implementing strategic anchoring/inertia (initial anchor with weight=1, and moving anchor on a 20-lag period with weight=1) for each of the three values of the error ratio generally indicate that the addition of strategic inertia enhances average populational fitness. Comparing the results for the two forms of strategic inertia indicates that a moving anchor performs better when organizations can make error-free assessments of the payoff to potential strategies, but when these payoffs are subject to error neither form of strategic anchoring clearly dominates the other in terms of producing the higher average populational fitness.

In summary, the results for the three static landscape scenarios do not support a hypothesis that errorful assessments of payoffs to potential strategies are beneficial for populations of organizations. In addition, the results suggest that strategic inertia, when combined with an election operator, produces higher average populational fitness, but the benefits of this combination dissipates when the level of error in assessing *ex-ante* payoffs becomes large.

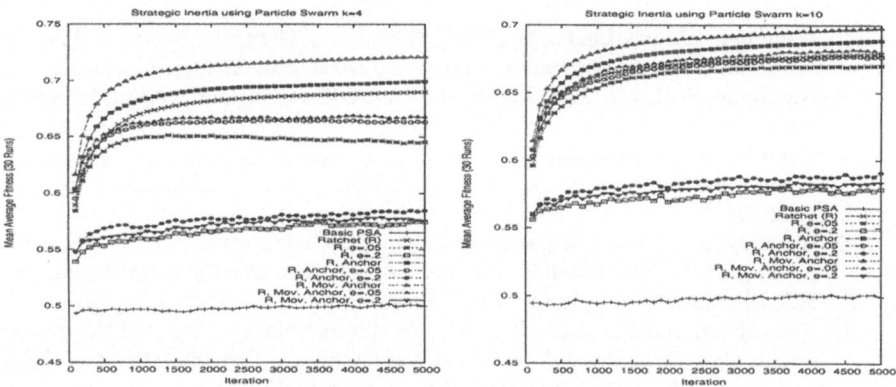

Fig. 1. Plot of the mean average fitness on the static landscape where k=4 (left) and K=10 (right).

4   Conclusions

In this paper a novel synthesis of a strategic landscape, defined using a combination of the NK model and a Particle Swarm metaphor, is used to model the strategic adaption of organizations. The results suggest that a degree of strategic inertia, in the presence of an election operator, can generally assist rather than hamper the adaptive efforts of populations of organizations in static strategic environments, when organizations can accurately assess payoffs to future strategies. The results also suggest that errorful assessments of the payoffs to potential strategies leads to a deterioration in populational fitnesses as the degree of error increases. It is also noted that despite the claim for the importance of social learning in populations of agents, the results suggest that social learning is not always enough, unless learnt lessons can be maintained by means of an election mechanism. The benefits of anchoring come at a price. The affect of gbest, pbest and anchoring terms, is to 'pin' each organization to a region of the strategic landscape. To the extent that the entire population of organizations have converged to a relatively small region of the strategic landscape, they may find it impossible to migrate to a new high-fitness region if that region is far away from their current location. In real-world environments, this is compensated for by the birth of new organizations.

It is noted that the concept of inertia / anchoring developed in this paper is not limited to organizations, but is plausibly a general feature of social systems. Hence, the extension of the social swarm model to incorporate inertia may prove useful beyond this study.

References

1. Gavetti, G. and Levinthal, D. (2000). Looking Forward and Looking Backward: Cognitive and Experiential Search, *Administrative Science Quarterly*, 45:113-137.
2. Kauffman, S. and Levin, S. (1987). Towards a General Theory of Adaptive Walks on Rugged Landscapes, *Journal of Theoretical Biology*, 128:11-45.
3. Kauffman, S. (1993). *The Origins of Order*, Oxford,England: Oxford University Press.
4. Kennedy, J. and Eberhart, R. (1995). Particle swarm optimization, *Proceedings of the IEEE International Conference on Neural Networks*, December 1995, pp. 1942-1948.
5. Kennedy, J. (1997). The particle swarm: Social adaptation of knowledge, *Proceedings of the International Conference on Evolutionary Computation*, pp. 303-308, Piscataway, New Jersey: IEEE Press.
6. Kennedy, J. and Eberhart, R. (1997). A discrete binary version of the particle swarm algorithm, *Proceedings of the Conference on Systems, Man and Cybernetics*, pp. 4104-4109, Piscataway, New Jersey: IEEE Press.
7. Kennedy, J. (1999). Minds and Cultures: Particle Swam Implications for Beings in Sociocognitive Space, *Adaptive Behavior*, 7(3/4):269-288.
8. Kennedy, J., Eberhart, R. and Shi, Y. (2001). *Swarm Intelligence*, San Mateo, California: Morgan Kauffman.
9. Levinthal, D. (1997). Adaptation on Rugged Landscapes, *Management Science*, 43(7):934-950.
10. Porter, M. (1996). What is Strategy?, *Harvard Business Review*, Nov-Dec, 61-78.
11. Rivkin, J. (2000). Imitation of Complex Strategies, *Management Science*, 46(6):824-844.

Analysis of Synchronous Time in Chaotic Pulse-Coupled Networks

Hidehiro Nakano[1] and Toshimichi Saito[2]

[1] Dept. of Computer Science and Media Eng., Musashi Institute of Technology
1-28-1, Tamazutsumi, Setagaya, Tokyo, 158–8557, Japan
[2] Dept. of Electronics, Electrical and Computer Eng., Hosei University
3-7-2, Kajino-cho, Koganei, Tokyo, 184–8584, Japan
nakano@ic.cs.musashi-tech.ac.jp, tsaito@k.hosei.ac.jp

Abstract. This paper studies pulse-coupled networks of chaotic oscillators. The networks have a simple local connection structure and can exhibit global synchronization of chaos. Adjusting parameters appropriately, the synchronization can be achieved rapidly. We investigate synchronous time of the networks for various network parameters and topologies. The dynamics of the networks can be simplified into a return map which is given analytically. Using the return map, fast calculation for the synchronous time is possible.

1 Introduction

Pulse-coupled networks (PCNs) of integrate-and-fire neurons [1] can exhibit various synchronous phenomena [2]-[5]. Adjusting parameters appropriately, synchronization in the PCNs can be achieved rapidly [5]. The rapid synchronous phenomena can be applied into effective information processing systems including image processing [5]. In the published literatures, periodic PCNs have been the main focus. On the other hand, we have proposed a chaotic pulse-coupled network (CPCN) [6]-[8]. The CPCN can exhibit interesting grouping synchronization of chaos such that some of chaos synchronous groups appear partially in the network. Using the grouping synchronization, we have applied the CPCN to image segmentation [7][8].

In this paper, we study chaos synchronous phenomena of CPCNs. The CPCN consists of chaotic spiking oscillators (CSOs, [6]) and has a simple local connection structure. The CPCN can exhibits global synchronization of chaos such that all the CSOs exhibit chaos synchronization. Adjusting parameters appropriately, synchronization in the CPCNs can be achieved rapidly. We investigate synchronous time of the networks for various network parameters and topologies. The dynamics of the networks can be simplified into a return map which is given analytically. Using the return map, trajectories of each oscillator can be calculated fast and accurately; fast calculation for the synchronous time is possible. The results provide useful information for the development of flexible and efficient engineering systems.

N.R. Pal et al. (Eds.): ICONIP 2004, LNCS 3316, pp. 1117–1122, 2004.
© Springer-Verlag Berlin Heidelberg 2004

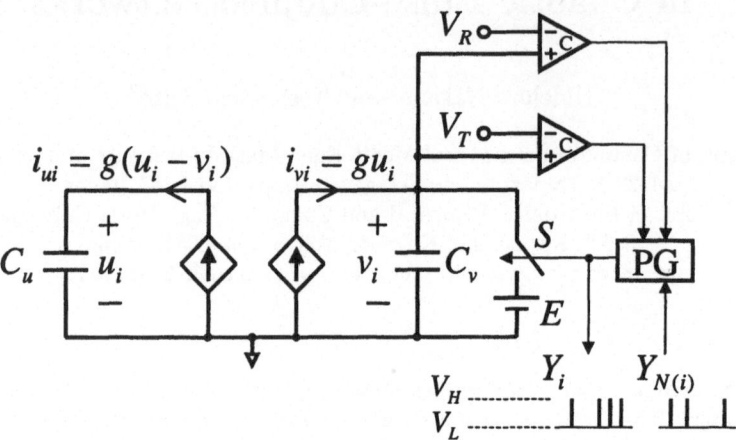

Fig. 1. A chaotic spiking oscillator (CSO): A unit circuit in a chaotic pulse-coupled network (CPCN).

2 Chaotic Pulse-Coupled Network

Fig. 1 shows a chaotic spiking oscillator (CSO) which is a unit circuit in a chaotic pulse-coupled network (CPCN). The circuit dynamics is described by

$$\frac{d}{dt}\begin{bmatrix} C_v\, v_i \\ C_u\, u_i \end{bmatrix} = \begin{bmatrix} 0 & g \\ -g & g \end{bmatrix}\begin{bmatrix} v_i \\ u_i \end{bmatrix}, \quad \text{if } Y_i(t) = V_L, \tag{1}$$

$$\begin{bmatrix} v_i(t^+) \\ u_i(t^+) \end{bmatrix} = \begin{bmatrix} E \\ u_i(t) \end{bmatrix}, \quad \text{if } Y_i(t) = V_H, \tag{2}$$

$$Y_i(t) = \begin{cases} V_H & \text{if } v_i(t) = V_T \text{ or } (Y_{N(i)}(t) = V_H \text{ and } v_i(t) > V_R), \\ V_L & \text{otherwise}, \end{cases} \tag{3}$$

where $Y_i(t)$ denotes an output spike-train of each CSO, and $i \in \{1, 2, \cdots, N\}$ and $N(i) \in \{1, 2, \cdots, N\}$ denote indexes of the ith CSO and its neighbor CSOs, respectively. We assume that Equation (1) has unstable complex characteristic roots $\delta\omega \pm j\omega$, where

$$\delta\omega = \frac{g}{2C_u} > 0, \quad \omega^2 = \frac{g^2}{C_v C_u} - \left(\frac{g}{2C_u}\right)^2 > 0.$$

In this case, the state vector (v_i, u_i) can vibrate divergently below the threshold voltage V_T. Using dimensionless variables and parameters:

$$\tau = \omega t, \quad x_i = \frac{v_i}{V_T}, \quad y_i = \frac{\delta}{V_T}\left(-v_i + \frac{2C_u}{C_v}u_i\right), \quad z_i = \frac{1}{V_H - V_L}\left(Y_i(\tfrac{\tau}{\omega}) - V_L\right),$$

$$\delta = \frac{1}{\sqrt{4\frac{C_u}{C_v} - 1}}, \quad q = \frac{E}{V_T}, \quad a = \frac{V_R}{V_T},$$

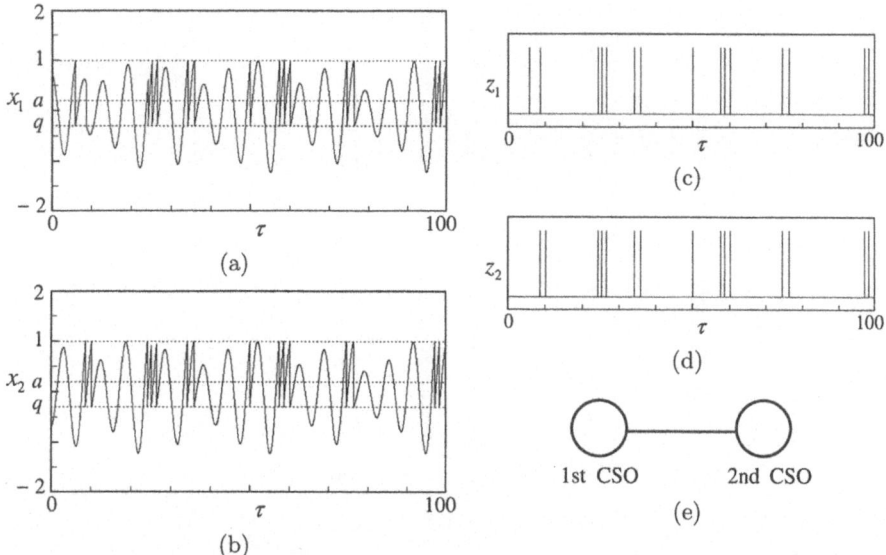

Fig. 2. A typical phenomenon for a CPCN of 2 CSOs: $N(1) = 2$, $N(2) = 1$ ($\delta = 0.07$, $q = -0.3$, $a = 0.5$). Initial states are $(x_1(0), y_1(0)) = (0.7, 0)$ and $(x_2(0), y_2(0)) = (-0.7, 0)$. (a) and (b): Time-domain waveforms of x_1 and x_2. (c) and (d): Output spike-trains z_1 and z_2. (e) Network Topology.

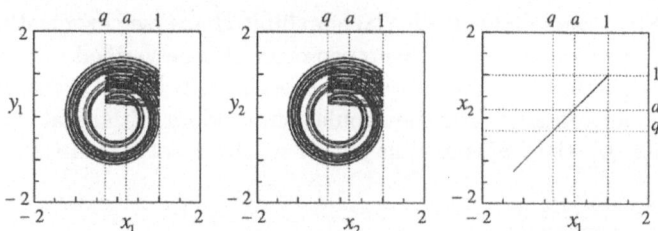

Fig. 3. Phase space trajectories in steady state for the simulation in Fig. 2.

Equations (1) to (3) are transformed into

$$\frac{d}{d\tau}\begin{bmatrix} x_i \\ y_i \end{bmatrix} = \begin{bmatrix} \delta & 1 \\ -1 & \delta \end{bmatrix}\begin{bmatrix} x_i \\ y_i \end{bmatrix}, \quad \text{if } z_i(\tau) = 0,$$

$$\begin{bmatrix} x_i(\tau^+) \\ y_i(\tau^+) \end{bmatrix} = \begin{bmatrix} q \\ y_i(\tau) + \delta(x_i(\tau) - q) \end{bmatrix}, \quad \text{if } z_i(\tau) = 1, \tag{4}$$

$$z_i(\tau) = \begin{cases} 1 & \text{if } x_i(\tau) = 1 \text{ or } (z_{N(i)}(\tau) = 1 \text{ and } x_i(\tau) > a), \\ 0 & \text{otherwise.} \end{cases}$$

For simplicity, we fix parameters as the following: $\delta = 0.07$, $q = -0.3$. In this case, each CSO without coupling exhibits chaos [6]. Figs. 2 and 3 show a typical phenomenon for a CPCN of 2 CSOs. For the simulation, different initial states

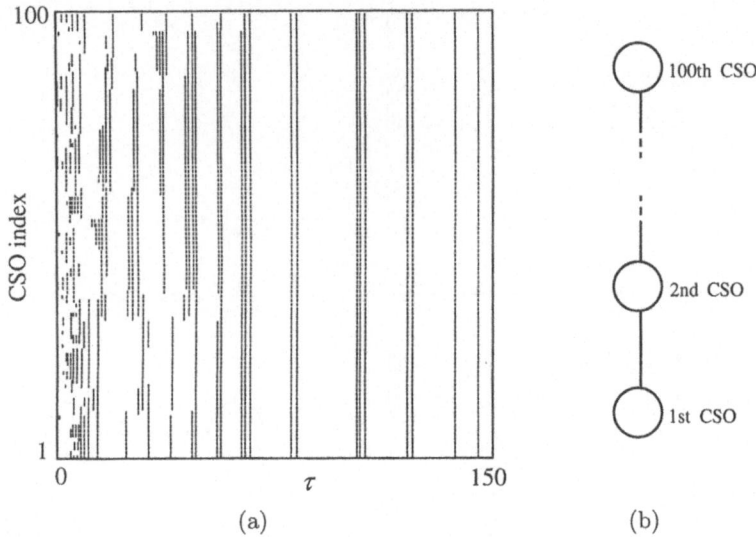

(a)   (b)

Fig. 4. A typical phenomenon for a CPCN of 100 CSOs having a ladder topology: $N(1) = 2$, $N(i) = \{i-1, i+1\}$ for $2 \leq i \leq 99$, $N(100) = 99$ ($\delta = 0.07$, $q = -0.3$, $a = 0.2$). (a) Output spike-trains. The solid lines represent output spikes of each CSO. (b) Network topology.

are assigned to each CSO. Both CSOs exhibit chaos synchronization via transient state. Such synchronous phenomena can also be verified in the laboratory experiments [8]. Fig. 4 shows a typical phenomenon for a CPCN of 100 CSOs having a ladder topology. For the simulation, random initial states are assigned to each CSO. All the CSOs exhibit global synchronization via transient state.

3   Analysis

In this section, we investigate synchronous time of the CPCNs. In order to more efficiently calculate the synchronous time, we introduce a mapping procedure. First, we define the following objects.

$$L = L_1 \cup L_2 \cup \cdots \cup L_N, \quad L_i = \{ \mathbf{x} \mid x_i = q \}, \quad i \in \{1, 2, \cdots, N\}, \tag{5}$$
$$\mathbf{x} \equiv (x_1, y_1, x_2, y_2, \cdots, x_N, y_N).$$

Let us consider trajectories starting from $\mathbf{x}(\tau_0^+) \in L$. As the trajectories start from $\mathbf{x}_0 \in L$, either trajectory of the CSOs must reach the threshold at some finite time $\tau = \tau_1$, and the trajectories return to $\mathbf{x}(\tau_1^+) \in L$. We can then define a return map

$$F: L \to L, \quad \mathbf{x}(\tau_0^+) \mapsto \mathbf{x}(\tau_1^+). \tag{6}$$

This map is given analytically by using exact piecewise solutions [8]. Therefore, by using the return map trajectories of each CSO can be calculated fast and

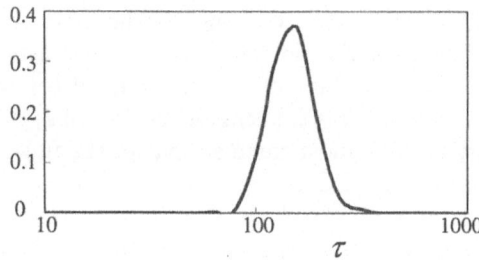

Fig. 5. Distribution of synchronous time for 500 random initial states. The CPCN consists of 100 CSOs having a ladder topology ($\delta = 0.07$, $q = -0.3$, $a = 0.2$).

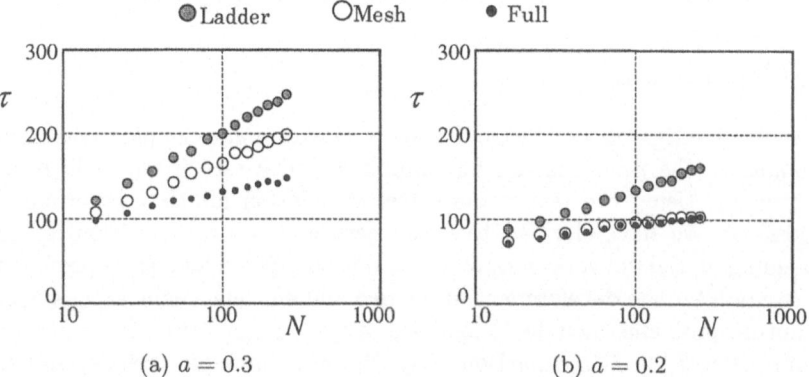

    (a) $a = 0.3$ (b) $a = 0.2$

Fig. 6. Mean synchronous time ($\delta = 0.07$, $q = -0.3$). Each plot denotes mean time for 500 random initial states of CPCNs.

accurately. Let τ_n be synchronous time if Equation (7) is satisfied just after the time τ_n:

$$\max_{i \neq j} \Delta_{ij}(\tau_n^+) < \varepsilon,$$

$$\Delta_{ij}(\tau_n^+) \equiv \sqrt{(x_i(\tau_n^+) - x_j(\tau_n^+))^2 + (y_i(\tau_n^+) - y_j(\tau_n^+))^2}, \tag{7}$$

where ε denotes a sufficiently small value. Since trajectories of each CSO can be calculated fast and accurately, the time τ_n can be calculated fast.

Fig. 5 shows normalized distribution of synchronous time for 500 trials for a CPCN of 100 CSOs having a ladder topology. For each trial, uniform random numbers are assigned as initial states. Note that for all the trials synchronization is achieved at some finite time. Each statistics value for the trials is the following:

	Min	Max	Ave	SD
Synchronous time	58.7	290.0	135.4	33.0

Fig. 6 shows mean synchronous time for the following three kinds of topologies: (1) Nearest-neighbor-coupled ladder topology of N CSOs, (2) Nearest-neighbor-coupled mesh topology of $M \times M$ CSOs ($N = M^2$), and (3) Full-coupled topology of N CSOs. As the value N (the number of CSOs) or a (refractory threshold) increases, the synchronous time increases for every topology. For $a = 0.3$, the synchronous time of the full-coupled topology is the shortest and that of the ladder topology is the longest. For $a = 0.2$, the synchronous time of the mesh topology is the almost same as that of the full-coupled topology.

Coupling method of our CPCN is different from general coupled systems in terms of the following. In our CPCN, the first state variables x_i are equalized to the same value q instantaneously if $x_i > a$. If the parameter a is sufficiently small, a spike of either CSO must propagate to all the CSOs and resets all the states to the same value; the coupling corresponds to full-coupling. This means that the CPCN with a simple local connection structure has efficient functionality and may be developed into flexible information processing systems.

4   Conclusions

We have investigated synchronous time of chaotic pulse-coupled networks. The dynamics of the networks can be simplified into a return map which is given analytically. Using the return map, fast calculation for the synchronous time is possible. We have clarified that the networks can be synchronized rapidly depending on the network parameters and topologies. The results provide useful information for the development of flexible and efficient engineering systems.

Future problems include 1) analysis of synchronous time for grouping synchronization [7][8], 2) comparison of synchronous time for periodic and chaotic networks, and 3) analysis of bifurcation phenomena.

References

1. Keener, J.P., Hoppensteadt, F.C., Rinzel, J.: Integrate-and-fire models of nerve membrane response to oscillatory input. SIAM J. Appl. Math., vol. 41, (1981) 503-517
2. Mirollo, R.E., Strogatz, S.H.: Synchronization of pulse-coupled biological oscillators. SIAM J. Appl. Math., vol. 50, (1990) 1645-1662
3. Catsigeras, E., Budelli, R.: Limit cycles of a bineuronal network model. Physica D, vol. 56, (1992) 235-252
4. Izhikevich, E.M.: Weakly pulse-coupled oscillators, FM interactions, synchronization, and oscillatory associative memory. IEEE Trans. Neural Networks, vol. 10, no. 3, (1999) 508-526
5. Campbell, S.R., Wang, D., Jayaprakash, C.: Synchrony and desynchrony in integrate-and-fire oscillators. Neural Comput., vol. 11, (1999) 1595-1619
6. Nakano, H., Saito, T.: Basic dynamics from a pulse-coupled network of autonomous integrate-and-fire chaotic circuits. IEEE Trans. Neural Networks, vol. 13, no. 1, (2002) 92-100
7. Nakano, H., Saito, T.: Synchronization in a pulse-coupled network of chaotic spiking oscillators. Proc. of MWSCAS, vol. I, (2002) 192-195
8. Nakano, H., Saito, T.: Grouping synchronization in a pulse-coupled network of chaotic spiking oscillators. IEEE Trans. Neural Networks (to appear)

A Spiking Oscillator with Quantized State and Its Pulse Coding Characteristics

Hiroshi Hamanaka, Hiroyuki Torikai, and Toshimichi Saito

EECE Dept., HOSEI University, Tokyo, 184-8584 Japan
hamanaka@nonlinear.k.hosei.ac.jp

Abstract. This paper studies a *quantized spiking oscillator* that can be implemented by a simple electronic circuit. The oscillator can have huge variety of stable periodic spike-trains and generates one of them depending on an initial state. Using a spike position modulation, the output spike-train can be coded by a digital sequence. We then clarify basic characteristics of the co-existing spike-trains and the pulse coding.

1 Introduction

This paper studies *quantized spiking oscillator* (QSO) that can be implemented by a simple electronic circuit [1]. The QSO has a quantized state and can have rich dynamics, e.g., the QSO can have huge variety of co-existing spike-trains and generates one of them depending on an initial state. The dynamics of the QSO can be described by a quantized spike position map (Qmap). Adjusting a system parameter, the QSO can realize various Qmaps different pulse codings, e.g., binary and Gray codings [2]. Among the Qmaps, in this paper, we focus on a tent-shaped one (*Tent Qmap*). We then clarify basic characteristics of the Tent Qmap: the number of the co-existing spike-trains, period of each spike-train and attraction basins of each spike-train. Also, we introduce a spike interval modulation [2] and clarify the encoding characteristics. This paper gives the first in-depth analysis for the Tent Qmap. A simple QSO circuit and basic laboratory experiments can be found in [2]. We note that the QSO with the adjustable parameter might be suitable for implementation by reconfigurable circuits like FPGA.

The QSO has been studied as a simplified artificial spiking neuron model and has been used to construct pulse-coupled artificial neural networks [4]-[7]. Pulse-coupled networks can exhibit various spatio-temporal phenomena including periodic and chaotic synchronizations. Based on such phenomena, several engineering applications have been proposed, e.g., image segmentation, dynamics associative memory and spike-based communications [5]-[8]. On the other hand, digital (or quantized state) dynamical systems can have interesting dynamics and their applications have been investigated [9]-[11]. Hence the results of this paper may be fundamental to develop a digital pulse-coupled network and to consider its interesting dynamics.

N.R. Pal et al. (Eds.): ICONIP 2004, LNCS 3316, pp. 1123–1128, 2004.
© Springer-Verlag Berlin Heidelberg 2004

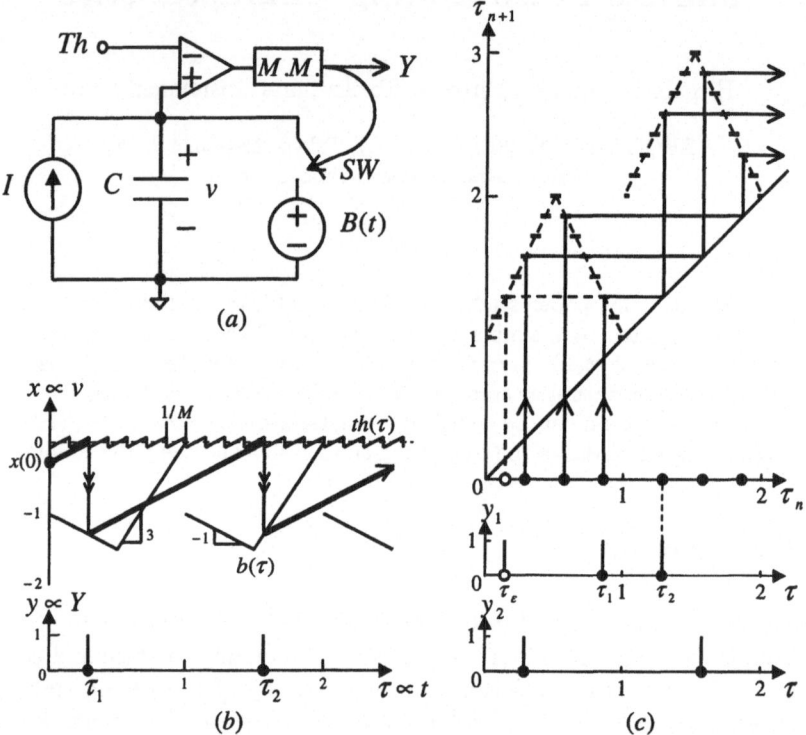

Fig. 1. (a) Quantized spiking oscillator. M.M. represents monostable multi-vibrator. (b) Basic dynamics ($M = 7$). $x, y, \tau, b(\tau)$ and $th(\tau)$ are proportional to $v, Y, t, B(t)$ and $Th(t)$, respectively. "•" represents SPSP and "○" represents EPSP. (c) The broken lines show spike position map f. The stepwise function show Qmap g corresponding to (b).

2   Quantized Spiking Oscillator

Fig.1 shows a quantized spiking oscillator (QSO) with its behavior. In the figure, $B(t)$ is a periodic piecewise-linear base, $Th(t)$ is a periodic sawtooth threshold:

$$B(t) = \begin{cases} A_1 t - B_1, \text{ for } 0 \leq t < \frac{T}{2}, \\ A_2 t - B_2, \text{ for } \frac{T}{2} \leq t < T, \end{cases} B(t + T) = B(t),$$

$$Th(t) = \frac{I}{C}(t - \frac{T}{2M}), \quad Th(t + \frac{T}{M}) = Th(t),$$

(1)

where M is a positive integer. The base and the threshold are synchronized; and the parameters A_1, A_2, B_1 and B_2 are chosen to satisfy $B(t) < Th(t)$. Below the threshold $Th(t)$, the state v increases by integrating the current $I > 0$ and an output is $Y = -E$. If the state v hits the threshold $Th(t)$, the comparator triggers the monostable multi-vibrator to produce a spike $Y = E$. The spike closes the switch SW, and the state v is reset to the base $B(t)$[1]. Repeating in this manner, the QSO generates a spike-train $Y(t)$. Using the following dimensionless variables and parameters

$$\tau = \tfrac{t}{T}, \ x = \tfrac{Cv}{IT}, \ y = \tfrac{Y+E}{2E}, \ b(\tau) = \tfrac{C}{IT}B(T\tau), \ th(\tau) = \tfrac{C}{IT}Th(T\tau),$$
$$\alpha_1 = \tfrac{C}{I}A_1, \ \alpha_2 = \tfrac{C}{I}A_2, \ \beta_1 = \tfrac{C}{IT}B_1, \ \beta_2 = \tfrac{C}{IT}B_2, \tag{2}$$

the dynamics is described by

$$\begin{cases} \dot{x} = 1 \ \text{and} \ y(\tau) = 0, & \text{for } x(\tau) < th(\tau), \\ x(\tau^+) = b(\tau^+) \ \text{and} \ y(\tau) = 1, & \text{if } x(\tau) = th(\tau), \end{cases}$$

$$th(\tau) = \tau - \tfrac{1}{2M}, \ \text{for} \ 0 \le \tau < \tfrac{1}{M}, \ th(\tau + \tfrac{1}{M}) = th(\tau), \tag{3}$$

$$b(\tau) = \begin{cases} \alpha_1\tau - \beta_1, \text{for } 0 \le \tau < 0.5, \\ \alpha_2\tau - \beta_2, \text{for } 0.5 \le \tau < 1.0, \end{cases} \ b(\tau+1) = b(\tau), \ b(\tau) < th(\tau)$$

where $\dot{x} \equiv dx/d\tau$. In this paper, we select M as the control parameter and fix the other four parameters $(\alpha_1, \alpha_2, \beta_1, \beta_2) = (-1, 3, 1, 3)$ as shown in Fig.1.

First, as a preparation, we consider the case of $th(\tau) = 0$. Let τ_n denote the n-th spike position. The spike position τ_n is governed by a *spike position map*:

$$\tau_{n+1} = f(\tau_n) \equiv \tau_n - b(\tau_n), \ \ f : \mathbf{R}^+ \to \mathbf{R}^+ \tag{4}$$

where $\mathbf{R}^+$ denotes the positive reals. The broken lines in Fig.1(c) show a spike position map f that generates a chaotic spike-train [12]. We note that the QSO can realize various spike position maps by adjusting the base $b(\tau)$.

Next, we consider the case where the threshold $th(\tau)$ is a sawtooth signal in Equation (3). In this case, the state x can hit the threshold $th(\tau)$ only at discontinuity points $\tau \in \{0, \tfrac{1}{M}, \tfrac{2}{M}, \ldots\}$. Then the spike position is quantized (i.e., $\mathbf{R}^+ \ni \tau_1 \mapsto \tau_2 \in \mathbf{L} \equiv \{\tfrac{0}{M}, \tfrac{1}{M}, \tfrac{2}{M}, \ldots\}$), and the spike-train $y(\tau)$ is super-stable[2] for the initial continuous state τ_1. Then the spike position is governed by a *quantized spike position map* (Qmap, see Fig.1(b)):

$$\tau_{n+1} = g(\tau_n) \equiv \tfrac{1}{M}\text{Int}(Mf(\tau_n) + \tfrac{1}{2}), \ \ g : \mathbf{L} \to \mathbf{L} \tag{5}$$

where $\text{Int}(\tau)$ gives the integer part of τ. Hereafter we refer to M as *quantization frequency*. Introducing a spike phase $\theta_n \equiv \tau_n \pmod 1$, we obtain the following return map F and quantized return map G:

$$\theta_{n+1} = F(\theta_n) \equiv f(\theta_n) \pmod 1, \ \ F : [0,1) \to [0,1),$$
$$\theta_{n+1} = G(\theta_n) \equiv g(\theta_n) \pmod 1, \ \ \mathbf{L}_0 \to \mathbf{L}_0 \equiv \{\tfrac{0}{M}, \ldots, \tfrac{M-1}{M}\}. \tag{6}$$

---

[1] For simplicity, we consider an ideal switching: the state v is reset to the base $B(t)$ instantaneously without delay.

[2] Even if an spike position τ_n is perturbed as $\tau_n + \delta$, the next spike position τ_{n+1} is restricted on $\mathbf{L}$.That is, the perturbation δ is reset to zero by the quantization and the spike-train $y(\tau)$ is said to be super-stable[1].

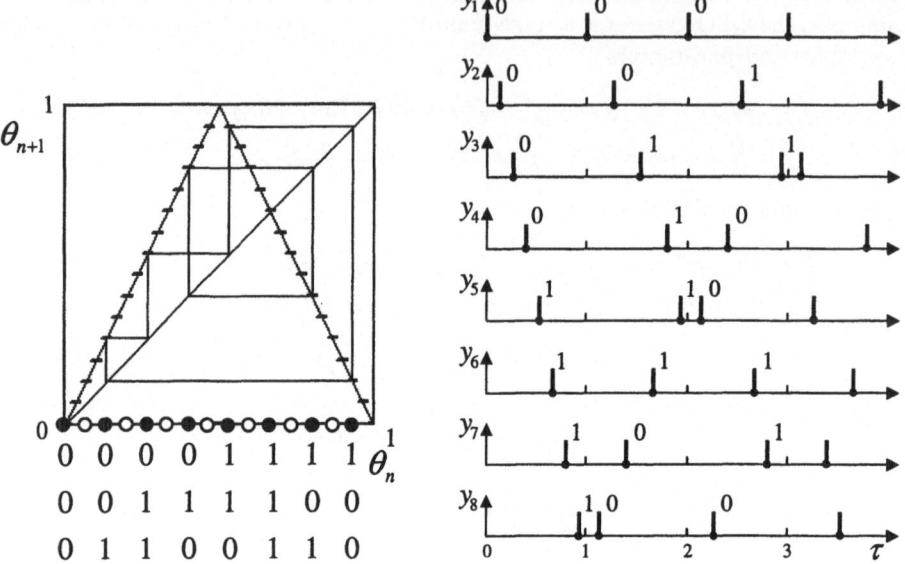

Fig. 2. Quantized return map G ($M = 15$). The SSPTs are coded by 3-bit Gray codes.

The quantized return map G and corresponding spike-trains are shown in Fig.2. Because the return map F in Fig.2 is the Tent map [12], we refer to g in Fig.1(c) as *Tent Qmap*. Now, let us consider dynamics of the Qmap g. τ_* is said to be a *super-stable periodic spike position* (SPSP) with period P and cycle Q if P and Q are the minimum integers such that $g^Q(\tau_*) - \tau_* = P$, where g^Q denotes the Q-fold composition of g. y_* is said to be a *super-stable periodic spike-train* (SSPT) with period P if y_* is represented by the SPSPs $(\tau_*, g(\tau_*), \ldots, g^{Q-1}(\tau_*))$. As shown in Fig.1(c), the Qmap g has co-existing SSPTs and generates one of them depending on the initial state τ_1. τ_ϵ is said to be an *eventually periodic spike position* (EPSP) if τ_ϵ is not an SPSP but $g^l(\tau_\epsilon)$ is an SPSP for some integer l. An EPSP is shown in Fig.1(c). The Tent Qmap satisfies $i + 1 \leq g(\tau_n) < i + 2$ for $i \leq \tau_n < i + 1$, where i is a nonnegative integer. Hence a spike-train $y(\tau)$ has one spike $y = 1$ in each unit interval $[i, i+1)$, and the initial spike position τ_1 can be restricted on $\mathbf{L_0}$. Let N denote the number of co-existing SSPTs. The characteristic of N is shown in Fig.3(a). In order to characterize EPSP, let N_e denote the number of EPSPs in $\mathbf{L_0}$ and let N_p denote the number of SPSPs in $\mathbf{L_0}$. We define an EPSP ratio

$$\rho = \frac{N_e}{N_p + N_e}. \tag{7}$$

The characteristic of ρ is shown in Fig.3(b). We have the following theorem.

Theorem 1: Let M be an odd integer. Let us divide the set $\mathbf{L_0}$ into two disjoint subsets: $\mathbf{L_0} = \mathbf{L}_{even} \cup \mathbf{L}_{odd}$, $\mathbf{L}_{even} \equiv \{\frac{0}{M}, \frac{2}{M}, \ldots, \frac{M-1}{M}\}$ and $\mathbf{L}_{odd} \equiv \{\frac{1}{M}, \frac{3}{M}, \ldots, \frac{M-2}{M}\}$. Then we can give the following: a spike-train $y(\tau)$ starting

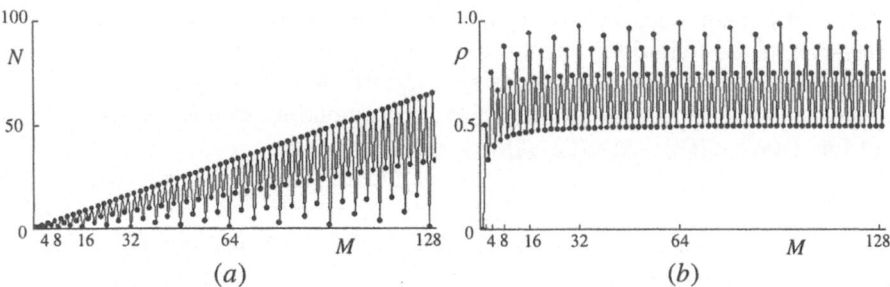

Fig. 3. (a) Number N of the co-existing SSPTs. (b) EPSP ratio ρ.

from $\tau_1 \in \mathbf{L}_{even}$ is an SSPT, and a spike position $\tau_1 \in \mathbf{L}_{odd}$ is an EPSP. Also the number of co-existing SSPTs N and the EPSP ratio ρ are give by

$$N = \frac{M+1}{2}, \quad \rho = \frac{M-1}{2M}. \tag{8}$$

In Fig.2, we can confirm $N = 8$ and $\rho = \frac{7}{15}$. Concerning the period, we have presented the following theorem in [2].

Theorem 2[2]: Let $M = 2^k - 1$ or $2^k + 1$, where k is a positive integer. Then the period of each SSPT is k or a measure of k.

In Fig.2, we can confirm the period of each SSPT is 4, 2 or 1. From further numerical experiments, we may be able to extend Theorem 2 as the following.

Conjecture: Let M be an odd integer. Let an initial states $\frac{m}{M} \in \mathbf{L}_{even}$ of an SSPT be reduced into an irreducible fraction $\frac{m'}{M'}$. Then the period of the SSPT is given by the minimum integer in $\{l \in \{1, 2, \ldots, M' - 1\}|2^l - 1 = 0 \ (\mathrm{mod}\ M')$ or $2^l + 1 = 0 \ (\mathrm{mod}\ M')\}$.

In order to consider encoding characteristics, we define a *spike position modulation*:

$$\omega(\theta_n) = \begin{cases} 0, & \text{for } 0 \le \theta_n < 0.5, \\ 1, & \text{for } 0.5 \le \theta_n < 1.0. \end{cases} \tag{9}$$

That is, a spike which lies on the first (second) half interval is coded by "0" ("1") as shown in Fig.2. Using this modulation, an SSPT is coded by a $(k-1)$-bit digital sequence $(\omega(\theta_1), \omega(\theta_2), \ldots, \omega(\theta_{k-1}))$. Then we have the following theorem.

Theorem 3: Let $M = 2^k - 1$. Then an SSPT starting from $\tau_1 \in \mathbf{L}_{even}$ is coded by a digital sequence $(\omega_1, \omega_2, \ldots, \omega_{k-1})$ given by

$$\begin{cases} \omega_1 = H_1, \\ \omega_{n+1} = H_n \oplus H_{n+1}, \end{cases} \quad \sum_{n=1}^{k-1} 2^{k-(n+1)} H_n = \frac{(2^k - 1)}{2} \tau_1, \tag{10}$$

where $\oplus$ represents the exclusive OR. As a result, the co-existing spike-trains correspond to all the $(k - 1)$-bit Gray codes. Proofs of theorem 1 and 3 will be given in a fully developed version.

This theorem suggests that the QSO can be applied to a spike-based A/D converter as the following. The initial state $x(0)$ corresponds to an analog input, and gives the first spike position as $\tau_1 = \frac{1}{M}\text{Int}(-Mx(0) + \frac{1}{2})$. According to the input, the QSO generates an SSPT $y(\tau)$ corresponding to a Gray coded digital output $(\omega(\theta_1), \omega(\theta_2), \ldots, \omega(\theta_{k-1}))$.

Adjusting shape of the base $b(\tau)$, the QSO can realize not only Gray coding but also various codings [2]. The QSO with adjustable base may be well suited for implementation by reconfigurable circuits like FPGA. A preliminary discrete-component-based QSO circuit and basic laboratory experiments can be found in [2].

3   Conclusions

We have introduced the quantized spiking oscillator (QSO) and analyzed its basic characteristics. We have also shown that the QSO can realize the Gray coding for the output spike-train and can be applied to the spike-based A/D converter. Future problems include: (a) analysis of Qmap from information coding/processing perspectives, (b) detailed analysis of the A/D conversion characteristic, and (c) synthesis and analysis of a network of Qmaps having interesting functions.

References

1. H.Torikai and T.Saito, Analysis of a quantized chaotic system, Int. J. Bif. and Chaos, 12, 5, 1207-1218, 2002.
2. H.Hamanaka, H.Torikai and T.Saito, Spike position map with quantized state and its application to algorithmic A/D converter, Proc. of IEEE/ISCAS, 2004 (accepted).
3. S.Signell, B.Jonsson, H.Stenstrom and N.Tan, New A/D converter architectures based on Gray coding, Proc. of IEEE/ISCAS, pp.413-416, 1997.
4. H.Torikai and T.Saito, Synchronization phenomena in pulse-coupled networks driven by spike-train inputs, IEEE Trans. Neural Networks, 15, 2, 2004.
5. G. Lee & N.H. Farhat, The bifurcating neuron network 2, Neural networks, 15, pp.69-84, 2002.
6. S.R.Campbell, D.Wang & C.Jayaprakash, Synchrony and desynchrony in integrate-and-fire oscillators, Neural computation, 11, pp.1595-1619, 1999.
7. Special issue on Pulse Coupled Neural Network, IEEE Trans. Neural Networks, 10, 3, 1999.
8. G.M.Maggio, N.Rulkov and L.Reggiani, Pseudo-chaotic time hopping for UWB impulse radio, IEEE Trans. CAS-I, 48, 12, pp.1424-1435, 2001.
9. S.Wolfram, Universality and complexity in cellular automata, Pysica D, 10, pp.1-35, 1984.
10. A.C.Davies, Digital counters and pseudorandom number generators from a perspective of dynamics, in Nonlinear Dynamics of Electronic Systems, NDES, pp.373-382, 1994.
11. D.R.Frey, Chaotic digital encoding: an approach to secure communication, IEEE Trans. CAS-II, 40, 10, pp.660-666, 1993.
12. E.Ott, Chaos in Dynamical Systems, Cambridge Univ. Press, 1993.

Concurrent Support Vector Machine Processor
for Disease Diagnosis

Jae Woo Wee and Chong Ho Lee

Department of Electrical Engineering, Inha University, Yonghyun-dong,
Nam-gu, Incheon, 402-751, Korea
wizet2000@empal.com

Abstract. The Concurrent Support Vector Machine processor (CSVM) that performs all phases of recognition process including kernel computing, learning, and recall on a chip is proposed. The classification problems of bio data having high dimension are solved fast and easily using the CSVM. Hardware-friendly support vector machine learning algorithms, kernel adatron and kernel perceptron, are embedded on a chip. Concurrent operation by parallel architecture of elements generates high speed and throughput. Experiments on fixed-point algorithm having quantization error are performed and their results are compared with floating-point algorithm. CSVM implemented on FPGA chip generates fast and accurate results on high dimensional cancer data.

1 Introduction

Recent advent of DNA microarray chip has brought our attention on the efficient analytical methods for high dimensional diagnostic data, such as gene expression profiles. The Support Vector Machine (SVM) has shown promises on a variety of biological classification tasks, including gene expression microarrays [1].

Parallelism is often employed for fast computation of such high dimensional data. The hardware architectures of SVM are published [2, 3, 4], where analog or digital hardware are developed. Kerneltron [2], performing internally analog and massively parallel kernel computation, deals with real-time applications of object detection and recognition successfully. This analog hardware carries out recall processes on-chip, leaving the learning process off-chip. A digital hardware of SVM [3, 4] proposes hardware-friendly SVM learning algorithm and performs the learning process on-chip. However this digital hardware performs kernel computation off-chip.

Our CSVM provides fast on-chip kernel computing and learning process using shared data bus in parallel. Quadratic programming in original SVM training algorithm is not suitable for hardware implementation, due to its complexity and large memory consumption. Thus we use kernel adatron (KA) [5] and kernel perceptron (KP) [4] learning algorithms can be implemented on a silicon chip since these algorithms make use of recursive-updating equations instead of quadratic programming. Kernel scaling method [6] that restricts kernel value within certain range is also utilized in order to reduce the data width and area of the hardware. The detailed architecture of CSVM is described and its performance on high dimensional data is demonstrated.

N.R. Pal et al. (Eds.): ICONIP 2004, LNCS 3316, pp. 1129–1134, 2004.
© Springer-Verlag Berlin Heidelberg 2004

2 Kernel Adatron Algorithm

KA algorithm uses the gradient ascent routine to maximize the margin in feature space. This algorithm is easy to implement and theoretically guaranteed to converge in a finite number of steps to the maximal margin.

Consider a two-class classification problem with training set $(\underline{x}_1, y_1),...,(\underline{x}_n, y_n)$, where $\underline{x}_i \in \mathfrak{R}^d$ and $y_i = \{+1, -1\}$. The kernel function $K(\underline{x}_i, \underline{x}_j)$ realizes a dot product in the feature space. KA algorithm is described as follows.

{Initialization} Set $\alpha_i = 0$ for all i and select learning rate parameter η
{Learning loop}
repeat
    for i =1,...n
$$z_i = \sum_{j=1}^{l} \alpha_j y_j K(\underline{x}_i, \underline{x}_j), \quad \Delta\alpha_i = \eta(1 - y_i z_i)$$
      if $(\alpha_i + \Delta\alpha_i) > 0$ **then** $\alpha_i \leftarrow \alpha_i + \Delta\alpha_i$ **else** $\alpha_i \leftarrow 0$
    end for
{Termination condition}
until (maximum number of iterations reached or
$\gamma = 0.5 \times [\min_{\{i|y_i=+1\}} (z_i) - \max_{\{i|y_i=-1\}} (z_i)] \approx 1$)

3 Kernel Perceptron Algorithm

KP algorithm changes the domain of work from the vector $\underline{w}$ of perceptron to $\underline{\alpha}$ of support vector machine. KP algorithm is described as follows.

{Initialization} Set $\alpha_i = 0$ for all i, $b = 0$
{Learning loop}
repeat
    for $i =1,...,n$
$$o_i = \sum_{j=1}^{n} \alpha_j q_{ij} + b, \quad q_{ij} = y_i y_j K(\underline{x}_i, \underline{x}_j)$$
      if $o_i \leq 0$ **then** update $\alpha_i = \alpha_i + 1$, $b = b + y_i$
    end for
{Termination condition}
until (maximum number of iterations reached or $o_i > 0$ for all i)

4 CSVM Architecture

4.1 System Overview

The hardware architecture composed of Support Vector Elements (SVEs) holding support vector streams is shown in Fig. 1. All SVEs are connected via shared data bus and parallel operation of SVEs generates high throughput. Extension of SVEs using

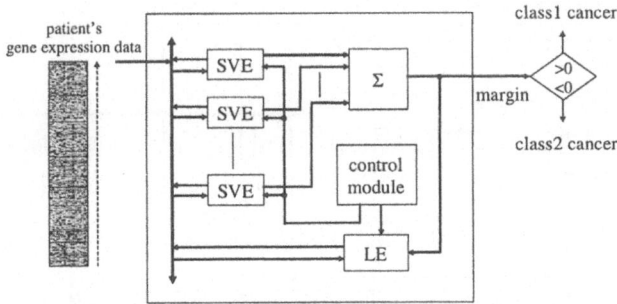

Fig. 1. CSVM architecture for cancer classification.

multi-chips is possible when training patterns are numerous. The hardware architecture is effectively packed in a small area. The similarity of SVM learning and recall algorithms - KA and KP - enables us to embed both algorithms on a chip.

Operating phases of the hardware consists of loading, kernel computation, learning, and recall. The computation of kernel function requires data from one SVE sent to other SVEs via shared data bus and the kernel computations are simultaneously done at all SVEs. The kernel computations are performed faster than those done sequentially by n times. In the learning phase, alphas and bias value are generated in the Learning Element (LE) and the updated parameters are sent to corresponding SVEs. The sign of margin that determines classes of the test data is computed in parallel during recall phase. Also recall phase by parallel operation is faster than by sequential operation by n times.

4.2 Building Blocks

The SVE shown in Fig. 2 consists of memory, computing module and interface with shared data bus. The multiplication by y_i is not performed during KA learning and recall phase, and is omitted in computing z_{ij}. Two's complementer is a substitution of a multiplier because y is 1 or -1. When a SVE sends data to other elements, the enable signal of SVE_i is activated.

Kernel computing module includes linear, polynomial, and RBF kernel functions: $K(\underline{x}_i, \underline{x}_j) = \underline{x}_i^T \underline{x}_j$ (linear), $K(\underline{x}_i, \underline{x}_j) = (\underline{x}_i^T \underline{x}_j + p)^2$ (polynomial) and $K(\underline{x}_i, \underline{x}_j) = exp(-c \| \underline{x}_i - \underline{x}_j \|^2)$, where $c = 1/2\sigma^2$ (RBF). To reduce the chip area, kernel computing module uses two multipliers for three kinds of kernel functions. As shown Fig. 3, one multiplier located next to the accumulator computes $\underline{x}_i^T \underline{x}_j$ for linear and polynomial kernel and computes $\| \underline{x}_i - \underline{x}_j \|^2$ for RBF kernel. The other multiplier performs $(\underline{x}_i^T \underline{x}_j + p)(\underline{x}_i^T \underline{x}_j + p)$ for polynomial kernel and $c \| \underline{x}_i - \underline{x}_j \|^2$ for RBF kernel. A lookup table (LUT) generates the exponential function of RBF kernel. The proper number of address bits of LUT is determined by experiments. It is useful to scale the argument of the kernel, so that the numerical values do not overflow the module as the dimension of the input data increases. kernel scaling method is realized by shifting a bit string.

Kernel values computed during kernel computing phase are used for learning. Parameters required by LE sent and received to SVE_i via shared data bus. When termi-

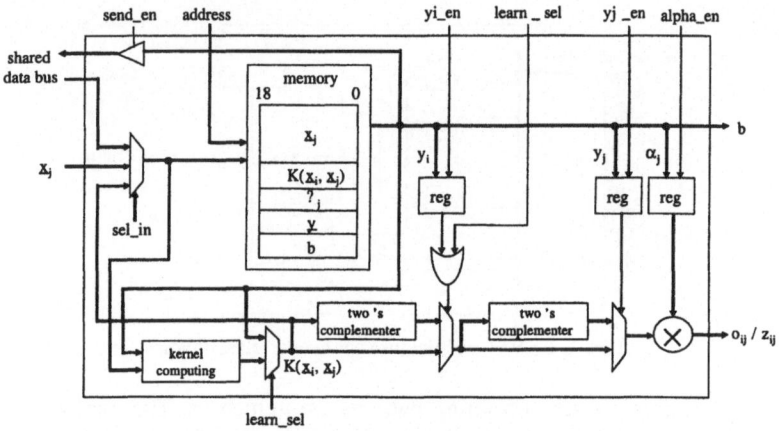

Fig. 2. Support Vector Element.

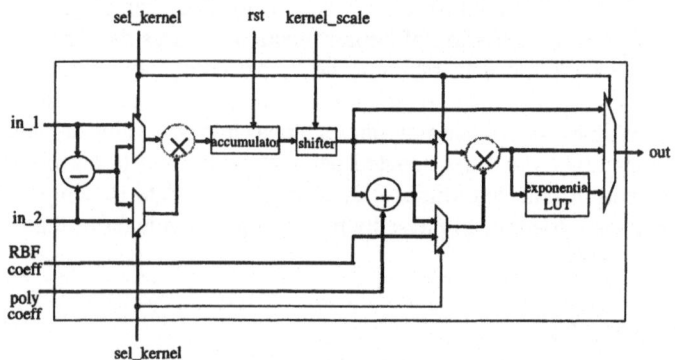

Fig. 3. Kernel Computing Module.

nation condition of learning is satisfied, finally obtained parameters are saved in memory of SVE_i.

5   Results

5.1   Floating- and Fixed-Point Experiments

Experiments are made to verify the efficiency of the CSVM on sonar data and cancer data. Sonar dataset consists of 104 learning patterns and 104 test patterns of 60 features each. The classification of such data is known to be a difficult problem due to its high dimensions and low degree of overlap between training and test sets. Floating-point algorithm of KP learning and RBF kernel function misclassifies 7 test patterns among 104 patterns, which outperforms the results of Anguita et al. [4]. Results of KP are superior to those of KA that misclassifies 9 test patterns at best. Table 1 summarizes the results of the fixed-point algorithm for different data widths and address bit sizes of LUT. When data widths are over 22bits and address bits of LUT is over 14bits, the same error rates as those from the floating-point algorithm are obtained.

Table 1. Percentages of misclassified test patterns on sonar dataset for different data widths and address bits sizes of LUT.

Data widths	Number of address bits of LUT			
	12	14	16	18
20	14.4%	19.2%	15.4%	15.4%
22	15.4%	6.7%	6.7%	6.7%
24	9.6%	6.7%	6.7%	6.7%

Table 2. Test errors on leukemia dataset for different data widths and kernel scales.

Data widths	Kernel scales				
	1	4	16	64	256
14	14	15	17	13	1
16	20	18	17	1	1
18	20	17	1	1	1
20	16	1	1	1	1

Leukemia dataset [7] that is gene expression profiles obtained from DNA microarray chip consists of 7129 genes for each of 72 patients. Among the 72 samples, 38 samples are used for training data and 34 samples for evaluation in our experiments to classify the leukemia type. The preprocessed data [8] having 3571 genes are actually used. The results of floating-point algorithm of KP and RBF kernel misclassify only 1 test pattern. Table 2 shows the number of test errors using fixed-point algorithm for different data widths and kernel scales. Though the data size reduces to as small as 14bits, the test error decreases to 1 when scaling down the kernel value by 256. Note that the kernel scale must be set larger as dimensional size of data becomes bigger. The kernel scaling method on sonar data having 60 features is not as effective as on cancer data.

5.2 Implementation on FPGA

The CSVM is designed by Verilog HDL and implemented on Xilinx Virtex2 FPGA XC2V8000. The chip contains 40 SVEs and 4000-dimensional data that are large enough to handle the leukemia dataset. The CSVM chip runs at 25MHz with throughput of 1.67×10^6 data/s for recall process. The numbers of cycles needed for each phase are as follows: 135 779 cycles for data loading, 136 154 cycles for kernel computing, 18 cycles for each iteration of learning, and 3586 cycles for recall. It takes merely 11 ms for all phases of KP learning on leukemia data.

6 Conclusion

We propose efficient hardware architecture for SVM employing parallelism on shared data bus and on-chip kernel computing. The computationally complex kernel algorithms are quickly performed on a hardware module together with learning and recall operations of SVM. The CSVM performing all phases of SVM is suitable for on-line and stand-alone applications. The CSVM with kernel scaling method has a great capability to classify bio data having thousands in dimension.

The SVE, the basic element of CSVM, consists of several multipliers and memories occupying larger area of the chip. In order to deal with the real-world problems with large data samples, an area-efficient design of SVE will be necessary.

References

1. Furey T.S., Cristianini N., Duffy N., Bednarski D.W., Schummer M., Haussler D.: Support vector machine classification and validation of cancer tissue samples using microarray expression data. Bioinformatics, Vol. 16, No. 10 (2000) 906-914
2. Genov, R., Cauwenberghs, G.: Kerneltron: support vector "machine" in silicon. IEEE Transactions on Neural Networks. Vol. 14, No. 5 (2003) 1426-1434
3. D. Anguita, A. Boni, S. Ridella: A digital architecture for support vector machines: theory, algorithm and FPGA implementation. IEEE Transactions on Neural Networks. Vol. 14, No. 5 (2003) 993-1009
4. D. Anguita, A. Boni, S. Ridella: Digital kernel perceptron. Electronics Letters. Vol. 38, No. 10 (2002) 445-446
5. T.T. Friess, N. Cristianini, C. Campbell: The kernel-adatron algorithm: a fast and simple learning procedure for support vector machines. 15th International Conference on Machine Learning, Wisconsin, USA (1998) 188-196
6. B. Scholkopf, A. J. Smola: Learning with kernels: support vector machines, regularization, optimization, and beyond. MIT Press (2002)
7. T. R. Golub, D. K. Slonim, P. Tamayo, C. Huard, M. Gaasenbeek, J. P. Mesirov, H. Coller, M. L. Loh, J. R. Downing, M. A. Caligiuri, C. D. Bloomfield, E. S. Lander: Molecular classification of cancer: class discovery and class prediction by gene expression monitoring. Science. Vol. 286, No. 5439 (1999) 531-537
8. S. Dudoit, J. Fridlyand, T. P. Speed: Comparison of discrimination methods for the classification of tumors using gene expression data. JASA. Vol. 97, No. 457 (2002) 77-87

Towards the Unification of Human Movement, Animation and Humanoid in the Network

Yasuo Matsuyama[1], Satoshi Yoshinaga[1,2], Hirofumi Okuda[1],
Keisuke Fukumoto[1], Satoshi Nagatsuma[1], Kazuya Tanikawa[1],
Hiroto Hakui[1], Ryusuke Okuhara[1], and Naoto Katsumata[1]

[1] Department of Computer Science, Waseda University, Tokyo 169-8555, Japan
{yasuo,hijiri,hiro,keisuke,nagatyo,tanikawa-k,h891,oku-ryu,katsu}
@wizard.elec.waseda.ac.jp
[2] IT & Mobile Solutions Network Company, Sony Co. Tokyo 108-6201, Japan
Satoshi.Yoshinaga@jp.sony.com

Abstract. A network environment that unifies the human movement, animation and humanoid is generated. Since the degrees of freedom are different among these entities, raw human movements are recognized and labeled using the hidden Markov model. This is a class of gesture recognition which extracts necessary information transmitted to the animation software and to the humanoid. The total environment enables the surrogate of the human movement by the animation character and the humanoid. Thus, the humanoid can work as a *moving computer* acting as a remotely located human in the ubiquitous computing environment.

1 Introduction

Recent advancement of computing power accompanied by the microminiaturization has promoted sophisticated human interfaces. Another social progress caused by this cost effective enhancement is the networking for the ubiquity. To be compatible with such trends, this paper presents the unification of human movements, animation characters and humanoids in the network computing environment. It is important for this purpose to incorporate various levels of learning algorithms on the machine intelligence.

The degree of the freedom of the human movement is around a few hundred. Humanoids available as contemporary consumer electronics have the freedom of its one tenth. Animation characters as software agents have the order of somewhere in the middle according to the software's sophistication. Because of such differences in the freedom, human movements are modeled first by a Hidden Markov Model (HMM). This problem is a class of gesture recognition which extracts the information transmitted to the animation software and to the humanoid.

The rest of the paper is organized as follows. Chapter 2 is devoted to the generation of the data structure compatible with our purpose. In Chapter 3, an HMM recognizer is designed using the training movements. The learned model is utilized for controlling an animation character and a humanoid called HOAP-2 [1]. Chapter 4 describes the realization of the humanoid movement mimicking the human. Chapter 5 gives concluding remarks including the next step.

N.R. Pal et al. (Eds.): ICONIP 2004, LNCS 3316, pp. 1135–1141, 2004.
© Springer-Verlag Berlin Heidelberg 2004

2   Data Acquisition and Transformation for Human Body Movement

2.1   Measured Raw Data

Human body's movements are measured in real time by the MotionStar™ [2] which uses the direct-current magnetic field. Eleven sensors are used for our measurement. Each sensor measures a 3×1 position vector and a 3×3 rotation matrix. Therefore, human movements give 11 time series of $3 + 3 \times 3 = 12$-dimensional vector-data as numerals. Such raw data *per se* do not have any spatial structure for the body movement. Therefore, we have to specify relationships among these time series.

2.2   Bone Frame Expression

Bones are connected. This connection can be expressed precisely by the tree structure in Figure 1. The root element is selected to represent the Hips. Sub-elements are LeftHip, RightHip, Chest, each of which has further sub-elements. These data are expressed by the BVH format (Bio Vision Hierarchical data) [3].

It is important to maintain the independence of the personal physique. For the sake of this demand, we give the following comments in advance.

(a) Data from 11 sensors are expanded to 17 time series by the interpolation according to the tree structure of Figure 1. This is because sensors fixed at some joints may irregularly move to create inaccurate data. Therefore, for instance, movements of two elbows and two knees are computed by using nearby sensors' data and normalized bones. Such a process gives $17 \times 3 = 54$-row data.

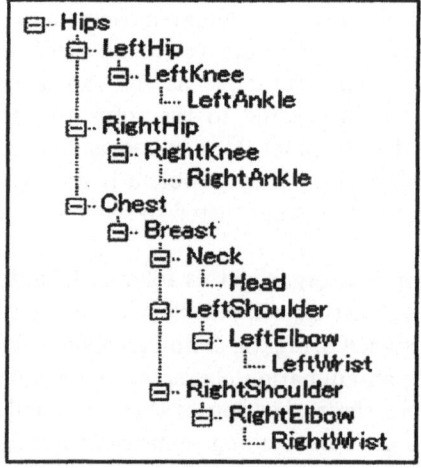

Fig. 1. Tree structure of bones.

(b) Relative rotation angles is found better than absolute ones for the portability to a wide range of humans.

(c) As will be explained in the experiment in Section 3.4, the original data set is further expanded to 69-row data.

(d) The time-frame is selected to be 50 ms. Because of the network communication and the humanoid movement computation, the time-frame needs to be long enough. But, this can not be too much. Thus, the time-frame of 50 ms was selected so that the movement can be tracked and reproduced as an animation smoothly enough based upon the experience of the speech recognition whose typical case is 20 ms.

(e) In the case of the speech recognition, each time-frame is expressed by only 25 rows or so. Therefore, the body movement recognition, or the gesture recognition, can not be a direct adaptation of the well-established speech recognition.

3   Recognizer Design by HMM Learning

3.1   Recognition System

Given the input of the 69-row data stream, it is necessary for the recognizer to categorize human body movement. The Hidden Markov Model (HMM) is a viable learning algorithm for this purpose. HMM's transitions correspond to the labels. The HMM software can be anything if it has a flexible input/output interface. We chose the HTK (Hidden Markov Model Toolkit) [4] since the modification of the I/O style matching with it does not require heavy tasks. Figure 2 illustrates our configuration of the total gesture recognition system. As is usual in learning systems, the model is fixed after the training.

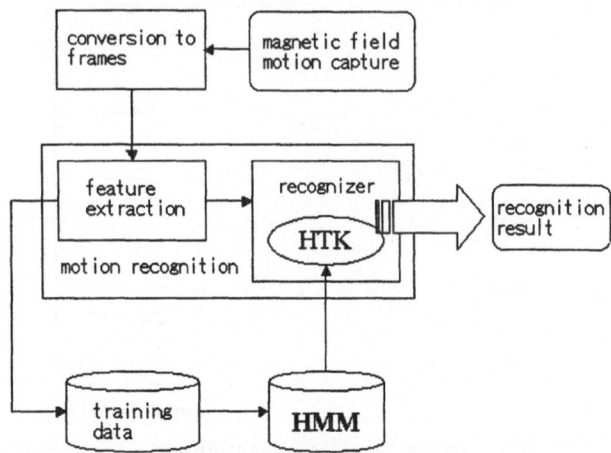

Fig. 2. Recognition system.

3.2 Tasks and Associated Data Preparation

As is expressed by the tree structure of Figure 1, movements of four limbs besides the hips are the most important for the gesture recognition. Therefore, we prepared eight labels for the movements as in Table 1. For this experiment, we prepared training and test data sets as follows.

(a) 10 sets of 8 patterns generated by a single person for training the recognizer $(10 \times 8 = 80$ patterns).
(b) 80 patterns by the same person in a different environment for testing.
(c) Different 8 persons' patterns for testing $(8 \times 8 = 64$ patterns).

Table 1. Recognition labels for movements.

	ArmUp	ArmDown	LegUp	LegDown
Left	LAU	LAD	LLU	LLD
Right	RAU	RAD	RLU	RLD

3.3 Number of States

The first step is to identify an appropriate number of states. There are theoretical criteria for this purpose such as the MDL (Minimum Description Length), however, repeated experiments on real data are essential to decide the actual best number. Therefore, we have to test various number of states by measuring the recognition performance. Figure 3 compares the average log-likelihood which reflects the performance of the recognition by the HMM models[1]. By this test, the number of states was judged to be 5 or 6.

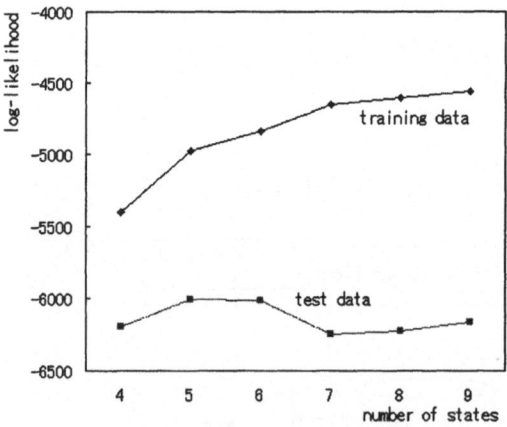

Fig. 3. Average log-likelihood.

---

[1] These data correspond to the case (D) of Section 3.4.

Table 2. Difference to the Runner-up.

	number of states					
	4	5	6	7	8	9
First	-6192.33	-6004.40	-6011.06	-6250.39	-6226.51	-6162.67
Second	-8838.18	-9888.39	-10993.00	-13262.22	-14196.33	-15780.30
difference	2645.86	3884.00	4981.94	7011.83	7969.82	9617.63

The next test is to see the difference between the best model and its runner-up. Table 2 shows the difference in the log-likelihood. This result indicates that the more the number of states is, the larger the difference is. Therefore, we chose the number of states to be 6.

3.4 Selected Features

In parallel to the state number selection, we checked to see which form of the input data is best for the recognition task. We prepared four types of input data:

(A) Use $17 \times 3 = 51$-row rotation data,
(B) Use 51-row *rotation difference* data and 3-row root position (Hips),
(C) Use 105-row data by adding (A) and (B),
(D) Use 69-row data by adding (B) and 5 leaves.

Table 3 summarizes the results of the recognition on the data outside the training data. By this result, the difference of the rotation angle is better than the rotation angle *per se.*

Table 3. Recognition performance.

	method A	method B	method C	method D
subject 1 (48 patterns)	25%(12)	100%(48)	27%(13)	**98%(47)**
subject 2 (46 patterns)	37%(17)	87%(40)	43%(20)	**98%(45)**
total	31%(29)	94%(88)	35%(33)	**98%(92)**

3.5 Animation for Monitoring

There are a few commercially available animation tools for BVH data. But, we had to develop our own display tool. This is because, as in Figure 2, our system needs to be designed including the recognizer and the controller for the succeeding system, the humanoid.

Figure 4 illustrates the course of the LeftLegUp. Thus, the label of LeftLegUp stands for such series of motions, not a still pose of the left-leg-up. This will be related to the humanoid motion of Section 4.

4 Humanoid Motion

4.1 Transmission of Recognized Label

Characters in the animation can behave more sophisticatedly according to the level of the software. But, humanoids can behave only less flexible. Contemporary

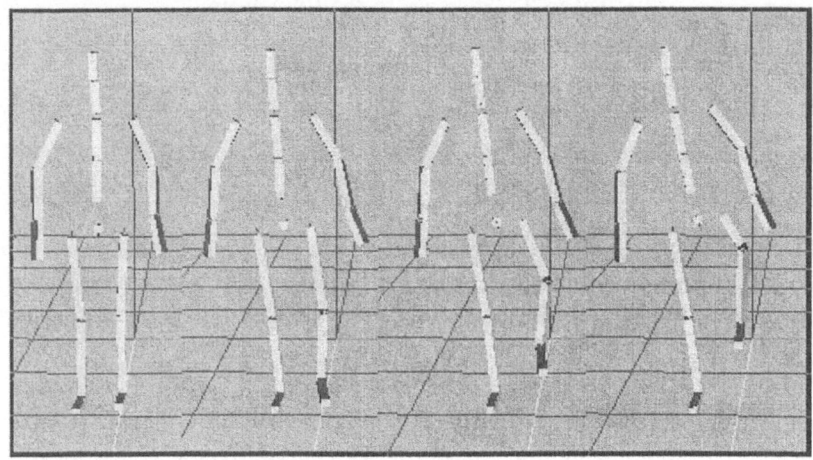

Fig. 4. A series of movements for LeftLegUp.

humanoids, even though they have made a great advance, are mostly composed of metallic materials and powered by motors. Body balances of humans and humanoids are very different. HOAP-2 appearing in this paper has 12 joints with 21 degrees of the movement freedom. Considering this ability, we transmit the recognition results as commands. Transmitting the BVH data directly leads to malfunctioning of humanoid motions. Imagine standing on one leg as is illustrated in Figure 4. This is possible by HOAP-2, however, its duration needs to be shorter than actual human movement.

4.2 Execution of Transmitted Labels

The recognition and the labeling of human motions given in Section 3 have the role of ameliorating the discrepancy between the differences of the freedom and *muscle* powers. Thus, the obtained labels for the motion can be used as commands to the humanoid. The humanoid is controlled by the built-in real-time Linux. Figure 5-left shows LeftLegUp by the humanoid. Figure 5-right illustrates LeftLegUp by the animation character, which is closer to actual human motion.

5 Concluding Remarks

The technical purpose of this paper included

(a) the recognition of human motions,
(b) the utilization of the recognition results for controlling the humanoid,
(c) imbedding the humanoid to a network environment as a movable computing node.

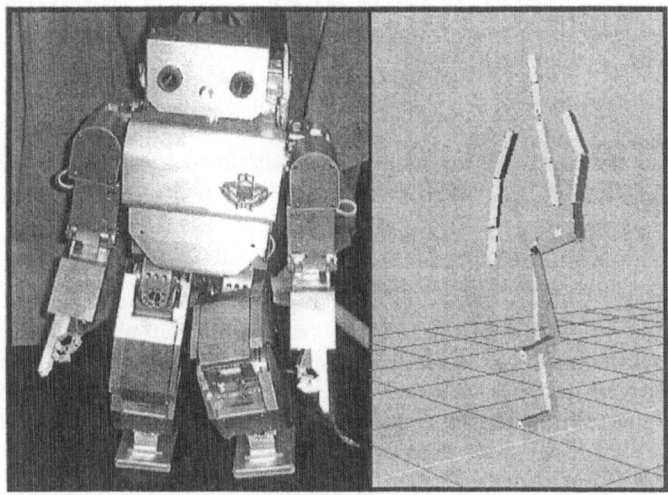

Fig. 5. LeftLegUp by HOAP-2 and the animation character.

As the initiative attempt, these items were satisfied. The use of the recognized label together with lower level data, including biological ones, can enhance the sophistication of the role of the humanoid in the network. This includes the surrogate of a remote human. This is the step connected to this paper's study.

Acknowledgment

This work was supported in part by the Grant-in-Aid for Scientific Research by MEXT, #15300077. The authors are grateful to Mr. Goro Kazama for his early contributions to the motion capturing and animation coding.

References

1. Fujitsu Automation Co.: HOAP-2 Reference Manual (2003)
2. Ascention Technology Co.: http://www.ascention-tech.com/
3. Meredith, M. and Maddock S.: Motion Capture File Formats Explained, Department of Computer Science, University of Sheffield (2001)
4. Young, S., et al.: The HTK Book, Cambridge University Engineering Department, Speech Group and Entropic Research Laboratory Inc. (1989)

A Dual Neural Network for Bi-criteria Torque Optimization of Redundant Robot Manipulators

Shubao Liu and Jun Wang

Department of Automation and Computer-Aided Engineering
Chinese University of Hong Kong, Shatin, N.T., Hong Kong
{sbliu,jwang}@acae.cuhk.edu.hk

Abstract. A dual neural network is presented for the bi-criteria joint torque optimization of kinematically redundant manipulators, which balances between the total energy consumption and the torque distribution among the joints. Joint torque limits are also incorporated simultaneously into the proposed optimization scheme. The dual neural network has a simple structure with only one layer of neurons and is proven to be globally exponentially convergent to the optimal solution. The effectiveness of dual neural network for this problem is demonstrated by simulation with the PUMA560 manipulator.

1 Introduction

Kinematically redundant manipulators are those having more degrees of freedom (DOFs) than required to perform a given task. These redundant DOFs can be utilized to optimize various performance criteria, while performing the given motion task. Among performance criteria, the optimization of joint torques is an appealing one since it is equivalent to effective utilization of actuator powers. Neungadi and Kazerounian [1] presented an approach that locally minimizes joint torques weighted by the inverse of inertia matrix. This optimization criterion corresponds to global kinetic energy minimization, and the local solutions are thus optimal and internally stable. The minimum-effort solution was also proposed to explicitly minimize the largest torque. This solution is consistent with physical limits and enables a better direct monitoring and control of the magnitude of individual joint torques than other norms of joint torques [2]. It is thus more desirable in applications where low individual joint torques is of primary concern. Recently, neural network approaches have been developed for the optimization of redundant manipulators. Tang and Wang [3] and Zhang and Wang [4] proposed recurrent neural networks (Lagrangian, primal-dual and dual neural network) for the torque optimization. In [5], a dual neural network was presented by Zhang et al. for kinematic control for redundant manipulators by minimizing the bi-criteria of Euclidean and infinity norm of joint velocities. In this paper we will extend the results in [5] from the kinematic control to the dynamic control. We will apply the dual neural network to minimize both the weighted norm and infinity norm of the joint torques to get a balance between the total energy consumption and the joint torque distribution, and at the same time take into account of the joint torque limits.

N.R. Pal et al. (Eds.): ICONIP 2004, LNCS 3316, pp. 1142–1147, 2004.
© Springer-Verlag Berlin Heidelberg 2004

2   Problem Formulation

Consider the forward kinematics relations between the joint variables and the pose of the end-effector in Cartesian space

$$r = f(\theta), \quad \dot{r} = J(\theta)\dot{\theta}, \quad J(\theta)\ddot{\theta} = \ddot{r} - \dot{J}(\theta)\dot{\theta} \tag{1}$$

where $\theta \in R^n$ is the joint variable vector, $r \in R^m$ is the pose vector of the end-effector in the Cartesian space ($m < n$ in redundant manipulators), $f(\cdot)$ is a smooth nonlinear function, $J(\theta) \in R^{m \times n}$ is the Jacobian matrix defined as $J(\theta) = \partial f(\theta)/\partial \theta$.

It is well known that the revolute joint robot dynamics is

$$\tau = H(\theta)\ddot{\theta} + c(\theta, \dot{\theta}) + g(\theta) \tag{2}$$

where $H(\theta) \in R^{n \times n}$ is the symmetric positive definite inertia matrix; $c(\theta, \dot{\theta}) \in R^n$ is the component of the torque depending on Coriolis centrifugal forces; $g(\theta) \in R^n$ is the component depending on gravity forces.

As shown in [1] by using calculus of variations, the local optimization of joint torque weighted by inertia results in resolutions with global characteristics; that is, the solution to the local minimization problem

$$\text{minimize } \tau^T H^{-1} \tau, \quad \text{subject to } J(\theta)\ddot{\theta} + \dot{J}(\theta)\dot{\theta} - \ddot{r} = 0 \tag{3}$$

also minimizes the kinetic energy $\int_{t_0}^{t_f} \dot{\theta}^T H \dot{\theta}/2$.

Inverting (2), we have the joint acceleration for given joint torques $\ddot{\theta} = H^{-1}(\tau - c - g)$. Substituting it into (1), the joint torque can be expressed in terms of $\ddot{r}$ as

$$JH^{-1}\tau = JH^{-1}(c+g) + \ddot{r} - \dot{J}\dot{\theta}. \tag{4}$$

Equation (4) can be simplified by introducing two terms $\ddot{r}_\tau = JH^{-1}(c+g)+\ddot{r}-\dot{J}\dot{\theta}$ and $J_\tau = JH^{-1}$. Hence, we have a linear torque-based constraint

$$J_\tau \tau = \ddot{r}_\tau. \tag{5}$$

The inertia inverse weighted joint torque optimization problem (3) can thus be reformulated to a time-varying quadratic program subject to the linear torque-based constraints (5) as

$$\text{minimize } \tau^T H^{-1} \tau, \quad \text{subject to } J_\tau \tau = \ddot{r}_\tau. \tag{6}$$

Further incorporating the infinity-norm optimization and joint torque limits, the bi-criteria torque optimization can be formulated as

$$\text{minimize } \alpha \tau H^{-1}\tau + (1-\alpha)\|\tau\|_\infty^2, \quad \text{subject to } J_\tau \tau = \ddot{r}_\tau, \quad \tau^- \le \tau \le \tau^+. \tag{7}$$

where $\|\tau\|_\infty$ denotes the infinity norm, $\alpha \in (0,1)$ the weight coefficient. τ^- and τ^+ denote respectively upper and lower limits of torque.

Next, let us convert the minimum infinity-norm part of (7) into a quadratic program. By defining $s = ||\tau||_\infty$, the minimization of $(1-\alpha)||\tau||_\infty^2$ can be rewritten equivalently [5] as

$$\text{minimize} \quad (1 - \alpha)s^2, \quad \text{subject to} \quad \begin{bmatrix} I & -e \\ -I & -e \end{bmatrix} \begin{bmatrix} \tau \\ s \end{bmatrix} \leq \begin{bmatrix} 0 \\ 0 \end{bmatrix} \tag{8}$$

where $e := [1, 1, \ldots, 1]^T$ and $0 := [0, 0, \ldots, 0]^T$ are vectors, repsectively, of ones and zeros with appropriate dimensions, I is the identity matrix.

Thus, by defining the variable vector $x = [\tau, s]^T \in R^{n+1}$, the bi-criteria torque optimization problem (7) can be expressed as the following quadratic program:

$$\text{minimize} \quad x^T Q x, \quad \text{subject to} \quad Ax \leq b, \; Cx = d, \; x^- \leq x \leq x^+ \tag{9}$$

where the coefficient matrices and vectors are

$$Q := \begin{bmatrix} \alpha H^{-1} & 0 \\ 0 & (1-\alpha) \end{bmatrix} \in R^{(n+1)\times(n+1)}, \quad A := \begin{bmatrix} I & -e \\ -I & -e \end{bmatrix} \in R^{2n\times(n+1)},$$

$$C := \begin{bmatrix} J_\tau & 0 \end{bmatrix} \in R^{m\times(n+1)}, \quad b := 0 \in R^{2n}, \quad d := \ddot{r}_\tau \in R^m,$$

$$x^- := \begin{bmatrix} \tau^- \\ 0 \end{bmatrix}, \quad x^+ := \begin{bmatrix} \tau^+ \\ \max_{1 \leq j \leq n} |\tau_j^\pm| \end{bmatrix} \in R^{n+1}.$$

Since the objective function in the formulation (9) is strictly convex (due to $0 < \alpha < 1$ and Q is positive definite) and the feasible region of linear constraints is a closed convex set, the solution to the bi-criteria quadratic program (7) is unique and satisfies the Karush-Kuhn-Tucker optimality conditions. Hence the continuity of the bi-criteria solution is guaranteed. As $\alpha \to 0$, the bi-criteria solution reaches the infinity-norm solution and $\alpha \to 1$, the bi-criteria solution becomes the inertia matrix weighted norm solution, which illustrates that the proposed bi-criteria optimization scheme is much more flexible than a single-criterion optimization scheme.

3   Dual Neural Network Model

In this section, a dual neural network [5] is discussed for bi-criteria torque optimization of redundant manipulators. Let us reformulate the constrained quadratic program into a unified form. That is, to treat equality and inequality constraints as special cases of bound constraints, we define

$$\xi^- := \begin{bmatrix} b^- \\ d \\ x^- \end{bmatrix}, \quad \xi^+ := \begin{bmatrix} b \\ d \\ x^+ \end{bmatrix}, \quad E := \begin{bmatrix} A \\ C \\ I \end{bmatrix} \in R^{(3n+m+1)\times(n+1)}.$$

where $b^- \in R^{2n}$, and $\forall j \in \{1, \ldots, 2n\}, b_j^- \ll 0$ sufficiently negative to represent $-\infty$. Then, (9) is rewritten in the following form:

$$\text{minimize} \quad x^T Q x \quad \text{subject to} \quad \xi^- \leq Ex \leq \xi^+. \tag{10}$$

In the above formulation, the generalized feasibility region $[\xi^-, \xi^+]$ is constructed as a closed convex set to facilitate the design and analysis of the dual neural network via the Karush-Kuhn-Tucker condition and the projection operator.

At any time instant, the constrained quadratic programming problem (9) may be viewed as a parametric optimization problem. It follows from the Karush-Kuhn-Tucker condition that x is a solution to (10) if and only if there exists $u \in R^{3n+m+1}$, such that $Qx - E^T u = 0$ and

$$
\begin{cases}
(Ex)_i = \xi_i^-, & \text{if } u_i > 0 \\
(Ex)_i = \xi_i^+, & \text{if } u_i < 0 \\
\xi_i^- \le (Ex)_i \le \xi_i^+, & \text{if } u_i = 0.
\end{cases}
\tag{11}
$$

The complementary condition (11) is equivalent to the system of piecewise linear equation $Ex = g(Ex - u)$. The vector-valued function $g(v) = [\tilde{g}(v_1), \ldots, \tilde{g}(v_{3n+m+1})]^T$ is defined as

$$
\tilde{g}(v_i) =
\begin{cases}
\xi_i^-, & \text{if } v_i < \xi_i^- \\
v_i, & \text{if } \xi_i^- \le v_i \le \xi_i^+, \quad i = 1, \ldots, 3n+m+1 \\
\xi_i^+, & \text{if } v_i > \xi_i^+.
\end{cases}
\tag{12}
$$

Therefore, x is a solution to (10) if and only if there exists a dual decision vector u such that $Qx - E^T u = 0$ and $Ex = g(Ex - u)$; i.e.,

$$
\begin{cases}
x = A^{-1} E^T u \\
g(EQ^{-1} E^T u - u) = EQ^{-1} E^T u.
\end{cases}
\tag{13}
$$

The above optimality condition yields a dual neural network model for solving (10) with the following dynamical equation and output equation:

$$
\begin{aligned}
\dot{u} &= \eta\{g(EQ^{-1}E^T u - u) - EQ^{-1}E^T u\} \\
x &= Q^{-1} E^T u
\end{aligned}
\tag{14}
$$

where $\eta \in R$ is a positive design parameter to scale the convergence of the dual network. For superior online performance, the parameter η, is set as large as hardware permits.

The block diagram of the dual neural network is depicted in Fig. 1, from which we can see that the neural network is composed of only one layer of no more than $3n + m + 1$ neurons and without using any analog multiplier or penalty parameter. Consequently, the dual neural network can be implemented much easier on VLSI. According to [5], the dual neural network is globally convergent to optimal solution for the convex program. Fig. 2 shows the data flow of the neural network based torque optimization. The desired motion of end-effector in cartesian space $\ddot{r}$ is input into the dual neural network, and the network outputs the actuator torque τ, which is the input of the manipulator dynamics.

4   Simulation Results

The Unimation PUMA560 manipulator (as shown in Fig. 3) has six joints. When the pose of the end-effector is considered, PUMA560 is not a redundant manipulator. However, if we consider only the position of the end-effector,

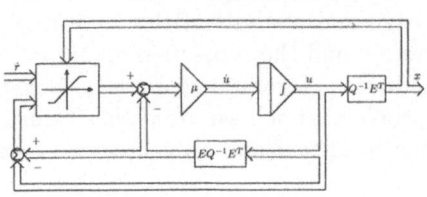

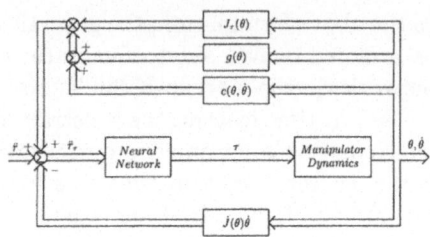

Fig. 1. Block diagram of dual neural network architecture.

Fig. 2. Block diagram of neural network based torque optimization of manipulators.

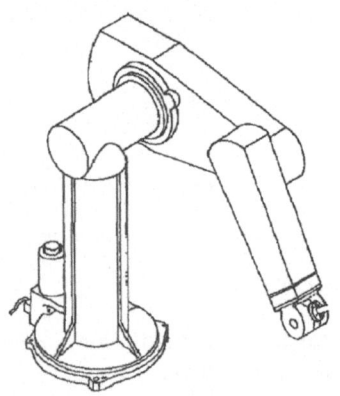

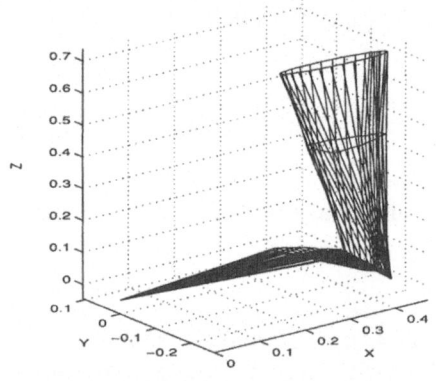

Fig. 3. PUMA560 robot manipulator.

Fig. 4. Motion trajectory of PUMA560 manipulator while tracking a circle.

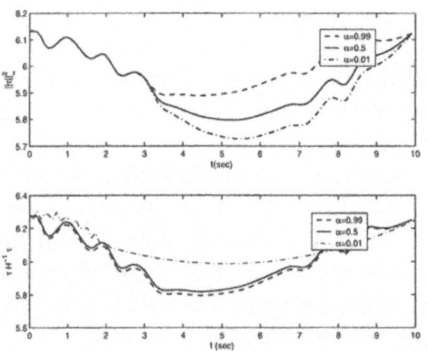

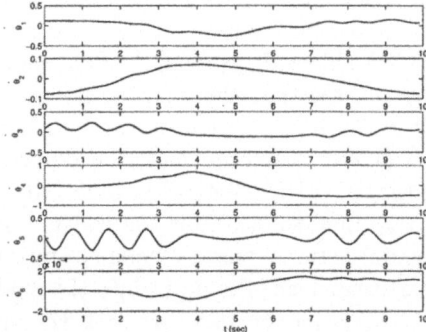

Fig. 5. Comparison of norms of torque for $\alpha = 0.01, 0.5, 0.99$.

Fig. 6. Joint rotation velocities of the PUMA560 when $\alpha = 0.5$.

PUMA560 becomes a redundant manipulator with the associated Jacobian matrix $J(\theta) \in R^{3 \times 6}$.

In this section, we discuss the bi-criteria torque optimization of the PUMA 560 when its end-effector tracks circular paths, by means of the proposed dual neural network.

The desired motion of the end-effector is a circle of radius $r = 10cm$ with the revolute angle about the x axis $\pi/6$. The task time of the motion is $10s$ and the initial joint variables $\theta(0) = [0\ 0\ 0\ 0\ 0\ 0]^T$. Fig. 4 illustrates the simulated motion of the PUMA 560 manipulator in the 3D workspace, which is sufficiently close to the desired one. Fig. 5 shows the infinity norm and the weighted norm when $\alpha = 0.01, 0.5, 0.99$ respectively. As described in Section 3, when $\alpha = 0.01$, the bi-criteria solution is approximate to the infinity-norm solution; while $\alpha = 0.99$, the bi-criteria solution becomes nearly the inertia matrix weighted norm solution. From Fig. 5, we can see that the bi-criteria solution always make a balance between the infinity-norm solution and the weighted-norm solution. Fig. 6 shows the angular velocity of the six joints while $\alpha = 0.5$. In view of discontinuity of pure infinity-norm solution, the bi-criteria solution is smooth, which implies no sudden change of torque. Hence, compared with single-criterion torque optimization, the bi-criteria scheme and the dual neural network are much more flexible in the sense that it can yield any combination of the minimum-effort and minimum-power solutions as needed, at the same time avoid the discontinuity of minimum-effort solution.

5   Concluding Remarks

In this paper, a dual recurrent neural network is applied for bi-criteria torque optimization of the redundant robot manipulators. The dual neural network is globally convergent to the optimal solution. Simulation results show that the neural network for the bi-criteria torque optimization is effective and efficient in balancing the energy consumption and the torque distribution among the joints.

References

1. Nedungadi, A., Kazerouinian, K.: A local solution with global characteristics for joint torque optimization of a redundant manipulator. J. Robot. Syst. **6** (1989) 631–654
2. Shim, I., Yoon, Y.: Stabilized minimum infinity-norm torque solution for redundant manipulators. Robotica **16** (1998) 1993–205
3. Tang, W., Wang, J.: Two recurrent neural networks for local joint torque optimization of kinematically redundant manipulators. IEEE Trans. Syst. Man Cyber. **30** (2000) 120–128
4. Zhang, Y., Wang, J.: A dual neural network for constrained torque optimization of kinematically redundant manipulators. IEEE Trans. Syst. Man Cyber. **32** (2002) 654–662
5. Zhang, Y., Wang, J., Xu, Y.: A dual neural network for bi-criteria kinematic control of redundant manipulators. IEEE Trans. Robot. Automat. **18** (2002) 923–931

A Genetic Approach to Optimizing the Values of Parameters in Reinforcement Learning for Navigation of a Mobile Robot

Keiji Kamei[1] and Masumi Ishikawa[1]

Department of Brain Science and Engineering
Graduate School of Life Science and Systems Engineering
Kyushu Institute of Technology
Kitakyushu, Fukuoka 808-0196, Japan
kamei-keiji@edu.brain.kyutech.ac.jp
ishikawa@brain.kyutech.ac.jp
http://www.brain.kyutech.ac.jp/~ishikawa/

Abstract. Reinforcement learning is a learning framework that is especially suited for obstacle avoidance and navigation of autonomous mobile robots, because supervised signals, hardly available in the real world, can be dispensed with. We have to determine, however, the values of parameters in reinforcement learning without prior information. In the present paper, we propose to use a genetic algorithm with inheritance for their optimization. We succeed in decreasing the average number of actions needed to reach a given goal by about 10-40% compared with reinforcement learning with non-optimal parameters, and in obtaining a nearly shortest path.

1 Introduction

Reinforcement learning [1] has frequently been used in autonomous mobile robots, because both supervised signals and information on the environment can be dispensed with [2]-[4]. Mobile robots acquire proficiency in obstacle avoidance and navigation to a given goal by trial and error based on reward signals from the environment. This is a big advantage for mobile robots in the real world.

In the reinforcement learning, we have to specify the values of parameters such as a discount rate and a learning rate. We do not, however, have prior information on these parameters in general, hence search in a high dimensional space becomes necessary.

A genetic algorithm is good at global search in a high dimensional space [5]. Genetic operations such as selection, crossover and mutation make probabilistic and systematic search possible. For this reason, we propose in this paper to introduce a genetic algorithm for the determination of the values of parameters in reinforcement learning.

Pettinger et al. proposed to improve the performance of a genetic algorithm by reinforcement learning [6]. Their proposal is to iteratively modify the values of parameters in a genetic algorithm such as probabilities of selection, crossover and mutation by reinforcement learning.

N.R. Pal et al. (Eds.): ICONIP 2004, LNCS 3316, pp. 1148–1153, 2004.
© Springer-Verlag Berlin Heidelberg 2004

Caldernoi et al. and Lee et al. proposed to optimize actions generated by a genetic algorithm by reinforcement learning [7] [8].

To the best of our knowledge, proposals which combine reinforcement learning and a genetic algorithm are only for the improvement of a genetic algorithm with the help of reinforcement learning [9]-[11]. We propose, on the contrary, to introduce a genetic algorithm into reinforcement learning to improve its performance by optimizing the values of parameters in reinforcement learning. Additional idea is to take advantage of inheritance in a genetic algorithm. In other words, Q-values in reinforcement learning in the previous generation are used as the initial Q-values in the following generation to speed up the learning.

Section 2 describes the autonomous mobile robot used here. Section 3 presents reinforcement learning. Section 4 explains the details of our proposal on the optimization of the values of parameters in reinforcement learning with the help of a genetic algorithm. Section 5 presents experimental results. Section 6 concludes the paper.

2   Autonomous Mobile Robot

Fig. 1(a) illustrates the mobile robot, *TRIPTERS mini,* and Fig. 1(b) depicts the positions of sensors. The mobile robot has 1 free wheel and 2 independent driving wheels. It cannot rotate on the spot, because the axle between the 2 driving wheels does not pass through the center of the robot. An action of turning actually moves slightly backward due to this property, which is adequately implemented in the simulator.

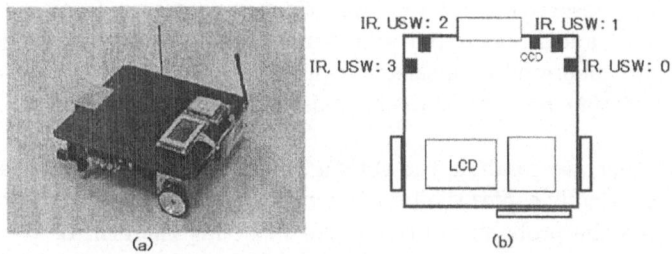

Fig. 1. Mobile robot, (a) Overview of *TRIPTERS mini,* (b) positions of sensors.

In computer experiments in section 5, we assume 3 primitive actions, i.e., moving forward by $100mm$, turning right by $10°$, and turning left by $10°$. A multiple of $100mm$ or $10°$ can easily be realized by a sequence of the corresponding primitive. As these primitive values become smaller, the resulting path becomes more precise and the computational cost increases. Taking this tradeoff into account, we adopt the above 3 primitive actions.

The *TRIPTERS mini* has ultrasonic sensors and infrared(IR) sensors. The ultrasonic sensors on *TRIPTERS mini* can accurately measure the distance to an obstacle not exceeding $800mm$. In contrast, outputs of IR sensors on *TRIPTERS mini* are binary; the output is 1 if the distance is less than 700mm, and 0 otherwise. We use only ultrasonic sensors here, because of the ability of measuring the distance.

3   Reinforcement Learning

Q-learning is one of reinforcement learning methods. Q-learning estimates a value, $Q(s,a)$, as a function of a pair of a state and an action, which we think is suited for a mobile robot. Because of this we adopt the Q-learning here. The Q-learning iteratively updates a value, $Q(s,a)$, as,

$$Q(s,a) \leftarrow Q(s,a) + \alpha \left[r' + \gamma \max_{a'} Q(s',a') - Q(s,a) \right] \tag{1}$$

where s' is the next state, a' stands for the corresponding action, α is a learning rate, γ is a discount rate and r' is the reward from the environment. A penalty, which is a negative reward, is also referred to as a reward for simplicity.

We propose to directly reduce the values, $Q(s,a)'s$, near an obstacle detected by ultrasonic sensors. We restrict the range of reduction of a value function; only the region where the distance to an obstacle is less than a given threshold is subject to reduction. This threshold is expected to further improve the performance of reinforcement learning.

4   Genetic Algorithm

A genetic algorithm is inspired by evolution of living things; search involves genetic operations such as selection, crossover and mutation in a probabilistic way frequently observed in the real evolutionary processes. Each individual has its chromosome and is evaluated by a fitness function. Individuals in the next generation are generated by a selection procedure. A crossover operation and a mutation operation are applied to chromosomes randomly with a predetermined probability.

In this paper, we combine the elitism and the roulette wheel selection. The former automatically copies the best individuals to the next generation, and the latter modifies the probability of selection reflecting the fitness. We also adopt multi-point crossover and conventional mutation.

A chromosome is coded in binary. It is composed of a discount rate, a learning rate, an ε in the ε-greedy policy, a threshold of modification of Q-values based on sensory signals, rewards for actions, i.e., moving forward and turning, and rewards from the environment, i.e., a goal reached, collision and detection of an obstacle. The length of a chromosome is 54 bits, with 6 bits for each parameter. A discount rate is coded in a logarithmic scale as,

$$\gamma = 1 - 10^{-kx} \tag{2}$$

where γ is a discount rate, x is an integer from 0 to 63, and k is a scaling parameter. All other parameters are coded in a linear scale.

Let the probability of the selection of each gene locus be 10%, that of crossover be 10%, and the value of k be 0.1. In this paper, 50 individuals are generated initially, for each of which the fitness is evaluated. We then generate 25 new individuals in addition to the original 50 individuals. Out of 75 individuals, 50 individuals with higher fitness are selected. The resulting 50 individuals constitute the next generation.

The value of fitness of each individual in the initial generation is evaluated by additional learning of 500-episode starting from the values of $Q(s, a)$ at 2000-episode learning. In later generations a child individual is evaluated by additional learning of 500-episode starting from the final values of $Q(s, a)$ of the individual with the best matching chromosome in the previous generation. This is an idea of inheritance we adopted in a genetic algorithm. The fitness function of an individual is expressed as,

$$f = \frac{N_g}{N_E} \times 2.0 + \left(1.0 - \frac{N_{acts}}{N_{\max}}\right) \tag{3}$$

where N_{acts} is the number of actions in successful episodes, $N_{\max}$ is the upper bound of the number of actions in a episode multiplied by the number of successful episodes, N_E is the number of total episodes and N_g is the number of successful episodes.

5   Computer Experiments

We use 2 kinds of environment in Fig.2: the simple environment(Env-S) and the complex one(Env-C). The area of the environment is $4m \times 4m$ and is composed of 20×20 grids, each of which is $20cm \times 20cm$. The state of the mobile robot is defined by its location (one of 20×20 grids) and orientation (one of 8 sectors). The corridor in the Env-S is wider than that in the Env-C, hence the former is considered to be easier to go through than the latter.

An assumption adopted here is that the mobile robot knows its randomly selected state, i.e., location and orientation, at the start of each episode. It is also assumed that the mobile robot knows its state thereafter based on odometry information. Although the state of the mobile robot is discretized, its exact state is also preserved for their calculation at later steps. An episode terminates, provided a mobile robot reaches a goal, collides with obstacles, or the number of actions reaches the upper limit of 250.

Table 1 indicates that in case of Env-S $\gamma = 0.9995$, which is very close to 1, is selected, because long-term viewpoint is required to efficiently reach the goal. The threshold, θ, becomes small to prevent detour. Table 2 indicates that the optimization of parameters decreases the average number of forward actions, that of turning, and that of the total actions by about 40%, 20% and 30%, respectively. Fig.5 illustrates that the resulting path is close to the shortest one.

In case of Env-C, Table 2 indicates that the optimization of parameters decreases the average number of forward actions, that of turning, and that of the total actions by about 40%, 10% and 25%, respectively. Fig.2 illustrates that the resulting path is also close to the shortest one.

Table 1. Parameters in reinforcement learning, their values and their intervals. "best fit" stands for the best fit individual generated by a genetic algorithm and "non-optimal" stands for parameters a priori determined.

	reward for collision	reward for obstacle	reward for goal	discount rate γ
search intervals	[-200.0 0.0]	[-100.0 0.0]	[-15.0 15.0]	[0.8 1.0]
best fit(Env-S)	-162.9	-16.7	1.7	0.9995
best fitl(Env-C)	-128.8	-29.3	-5.0	0.9987
non-optimal	-250.0	-1.0	20.0	0.9900

	reward for forward	reward for rotation	threshold of sensors	learning rate α	ε-greedy policy ε
intervals	[-15.0 -5.0]	[-15.0 -5.0]	[200.0 600.0]	[0.30 0.63]	[0.0050 0.050]
best fit(Env-S)	-5.5	-5.5	200.0	0.32	0.0057
best fit(Env-C)	-6.8	-6.8	466.7	0.35	0.0086
non-opt(Env-S)	-7.5	-7.5	600.0	0.40	0.010
non-opt(Env-C)	-7.5	-7.5	300.0	0.40	0.010

Table 2. "#forward" stands for the number of forward actions, "#rotation" stands for the number of rotation actions, "#actions" stands for the total number of actions. "best fit" stands for the best fit individual generated by a genetic algorithm, and "non-optimal" is the number of total actions averaged over the last 500 episodes.

		# forward	# rotation	# actions	# goals
Env-S	best fit	23.81	19.35	43.16	499
	non-optimal	39.16	23.09	62.25	500
Env-C	best fit	35.93	30.74	66.67	500
	non-optimal	57.44	32.86	90.30	500

6   Conclusion and Discussions

In this paper, we have proposed to combine a genetic algorithm and reinforcement learning. Our key idea is to optimize the values of parameters in reinforcement learning with the help of a genetic algorithm with inheritance.

Computer simulation demonstrates that the average number of actions needed to reach a given goal decreases by about 10-40% compared with reinforcement learning with non-optimal parameters, and a nearly shortest path is successfully obtained.

To clarify the relation between a fitness function and the performance of reinforcement learning such as the length of a resulting path and the number of goals reached, we will try fitness functions with different parameters. This is left for further study. Evaluation using the real *TRIPTERS mini* is also left for future study.

Acknowledgment

This research was supported by the 21st Century COE(Center of Excellence) Program and by Grant-in-Aid for Scientific Research(C)(15500140) both from the Ministry of Education, Culture, Sports, Science and Technology(MEXT), Japan.

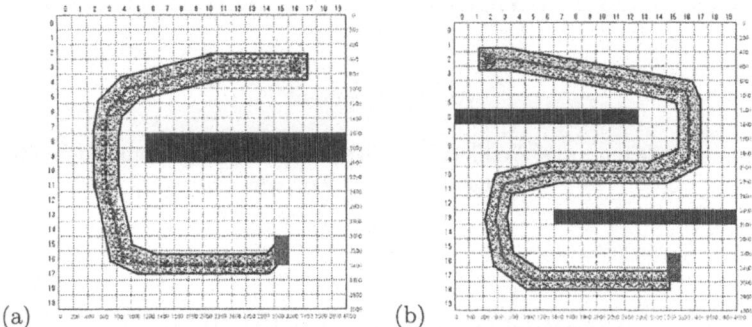

(a)   (b)

Fig. 2. The environment and examples of the resulting paths. (a) Simple environment(Env-S). The number of forward actions and that of left turns are 71 and 20, respectively. (b) Complex environment(Env-C). The number of forward actions, that of left turns and that of right turns are 101, 18 and 18, respectively. A black rectangles are obstacles and a grey one is the goal. The shaded area signifies the passage of the robot body.

References

1. R. S. Sutton and A. G. Barto, "Reinforcement Learning," MIT Press, 1998
2. C. Unsal, P. Kachroo, and J. S. Bay, "Multiple Stochastic Learning Automata for Vehicle Path Control in an Automated Highway System," *Proceedings of the 1999 IEEE Trans. Systems, Man*, and *Cybernetics*, Part A: Systems and Humans, 1999, vol. 29, pp. 120–128.
3. M. M. Svinin, K. Yamada and K. Ueda, "Emergent synthesis of motion patterns for locomotion robots," Artificial Intelligence in Engineering, Elsevier Science, 2001, vol. 15, No. 4, pp. 353–363.
4. C. Balkenius and J. Moren, "Dynamics of a classical conditioning model," Autonomous Robots, 1999, 7, pp. 41–56.
5. R. Pfeifer and C. Scheier, "Understanding Intelligence," MIT Press, 1999
6. James E. Pettinger and Richard M. Everson, "Controlling Genetic Algorithms with Reinforcement Learning," Department of Computer Science, School of Engneering and Computer Science, University of Exeter. EX4 4QF. UK, 2003
7. S. Calderoni and P. Marcenac, "MUTANT: a MultiAgent Toolkit for Artificial Life Simulation," IEEE. Published in the Proceedings of TOOLS-26'98, August 3-7, 1998 in Santa Barbara, California.
8. M. R. Lee and H. Rhee, "The effect of evolution in artificial life learning behavior," Journal of intelligent and robotic systems, 2001, vol. 30, pp. 399–414.
9. R. Abu-Zitar and A. M. A. Nuseirat, "A theoretical approach of an intelligent robot gripper to grasp polygon shaped objects," Journal of intelligent and robotic systems, 2001, vol. 31, pp. 397–422.
10. C. T. Lin and C. P. Jou, "GA-based fuzzy reinforcement learning for control of a magnetic bearing systems," *Proceedings of the 2002 IEEE Trans. Systems, Man*, and *Cybernetics*, Part B: Cybernetics, 2000, vol. 30, pp. 276–289.
11. A. Stafylopatis and K. Blekas, "Autonomous vehicle navigation using evolutionary reinforcement learning," European journal of operational research, 1998, vol. 108, pp. 306–318.

On the Use of Cognitive Artifacts for Developmental Learning in a Humanoid Robot

Artur M. Arsenio

MIT Computer Science and Artificial Intelligence Laboratory, Cambridge 02139, USA
arsenio@csail.mit.edu
http://www.ai.mit.edu/people/arsenio

Abstract. The goal of this work is to boost the robot's object recognition capabilities through the use of learning aids. We describe methods to enable learning on a humanoid robot using learning aids such as books, drawing material, boards or other children toys. Visual properties of objects are learned and inserted into a recognition scheme, which is then applied to acquire new object representations – we propose learning through developmental stages. We present experimental evaluation to corroborate the theoretical framework.

1 Introduction

Teaching a humanoid robot information concerning its surrounding world is a difficult task, which takes several years for a child, equipped with evolutionary mechanisms stored in its genes, to accomplish. Learning aids are often used by human caregivers to introduce the child to a diverse set of (in)animate objects, exposing the latter to an outside world of colors, forms, shapes and contrasts, that otherwise could not be available to a child (such as the image of a Panda). Since these learning aids help to expand the child's knowledge of the world, they are a potentially useful tool for introducing new informative percepts to a robot.

This paper proposes strategies which enable the robot to learn from books and other learning aids.Such strategies rely heavily in human-robot interactions. It is essential to have a human in the loop to introduce objects from a book to the robot (as a human caregiver does to a child). A more effective and complete human-robot communication interface results from adding other aiding tools to the robot's portfolio (which facilitate as well the children' learning process).

Embodied vision methods will be demonstrated with the goal of simplifying visual processing. This is achieved by selectively attending to the human actuator (*Hand* or *Finger*). Indeed, primates have specific brain areas to process the hand visual appearance [6]. A human-robot interactive approach was therefore implemented to introduce the humanoid robot to new percepts stored in books, as described in Section 2. Such percepts are then converted into an useful format through an object recognition scheme (presented in Section 3), which enables the robot to recognize an object in several contexts or to acquire different object representations. Section 4 describes learning from educational activities, such as

N.R. Pal et al. (Eds.): ICONIP 2004, LNCS 3316, pp. 1154–1159, 2004.
© Springer-Verlag Berlin Heidelberg 2004

painting or drawing in paper or jelly boards. Relevant experimental results for each learning activity are presented in each section. Finally, Section 5 draws the conclusions and describes current research directions.

2 Learning from Books

Although a human can interpret visual scenes perfectly well without acting on them, such competency is acquired developmentally by linking action and perception. Actions are not necessary for standard supervised learning, since off-line data is segmented manually. But whenever a robot has to autonomously acquire object categories using its own (or another's) body to generate percepts, actions become indeed rather useful (such as tapping on books).

During developmental phases, children's learning is often aided by the use of audiovisuals and especially, books. Humans often paint, draw or just read books to children during the early months of childhood. A book can be equally a useful learning tool for humans to teach robots different object representations or to communicate properties of unknown objects to them. Aiming at improving the robot's perception and learning capabilities through the use of books, we propose a human aided object segmentation algorithm to tackle the figure (object) – ground (book page) segregation problem. Indeed, a significant amount of contextual information may be extracted from a periodically moving actuator. This can be framed as the problem of estimating $p(o_n|v_{B_{p,\epsilon}}, act_{p,S}^{per})$, the probability of finding object o_n given a set of local, stationary features v on a neighborhood ball B of radius ϵ centered on location p, and a periodic human actuator on such neighborhood with trajectory points in the set $S \subseteq B$. The following algorithm implements the estimation process (see Figure 1):

1. A standard color segmentation [3] algorithm is applied to a stationary image (stationary over a sequence of consecutive frames).
2. A human actor taps with a finger on the object to be segmented.
3. The motion of skin-tone pixels is tracked over a time interval (using the Lucas-Kanade algorithm). The energy per frequency content – using Short-Time Fourier Transform (STFT) – is determined for each point's trajectory.
4. Periodic, skin-tone points are grouped together into the arm mask [1].
5. The trajectory of the arm's endpoint describes an algebraic variety [5] over N^2 (N represents the set of natural numbers). The target object's template is given by the union of all bounded subsets (the color regions of the stationary image) which intersect this variety.

Periodic detection is applied at multiple scales. since the movement might not appear periodic at a coarser scale, but appear as such at a finer scale. If a strong periodicity is not found at a larger scale, the window size is halved and the procedure is repeated again. Periodicity is estimated from a periodogram determined for all signals from the energy of the STFTs over the spectrum of frequencies. These periodograms are processed by a collection of narrow bandwidth band-pass filters. Periodicity is found if, compared to the maximum filter output, all remaining outputs are negligible.

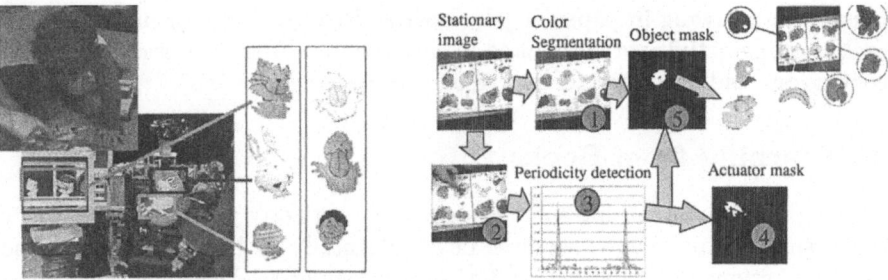

Fig. 1. (left) A human actor teaching the humanoid robot Cog from a fabric book (right) A standard color segmentation algorithm computes a compact cover of color clusters for the image. The actuator's periodic trajectory is used to extract the object's compact cover – the collection of color cluster sets which composes the object.

The algorithm consists of grouping together the colors that form an object. This grouping works by having periodic trajectory points being used as seed pixels. The algorithm fills the regions of the color segmented image whose pixel values are closer to the seed pixel values, using a 8-connectivity strategy. Therefore, points taken from tapping are used to both select and group a set of segmented regions into the full object. Clusters grouped by a single trajectory might either form or not form the smallest compact cover which contains the object (depending on intersecting or not all the clusters that form the object). After the detection of two or more temporally and spatially closed trajectories this problem vanishes. This algorithm was successfully applied to extract templates for fruits, clothes, geometric shapes and other elements from books, under varying light conditions (see Figure 2).

Fig. 2. (left-top) Statistical analysis for object segmentation from books. Errors are given by (template area - object's real area)/(real area). Positive/negative errors stand for templates with larger/smaller area than the real area. Total errors stand for both errors (left-bottom) Segmentation errors. The watermelon, banana and bed have a region with similar color – white – to its background, for which no differentiation is possible, since the intersection of the object's compact cover of color regions with the background is not empty. High variability on the elephant gray levels create grouping difficulties (the compact cover contains too many sets – hard to group). The cherries reflect another problem - small images of objects are hard to segment. (right) Templates for several categories of objects were extracted from dozens of books.

3   Matching Multiple Representations

Object representations acquired from a book are inserted into a database, so that they become available for future recognition tasks. However, object descriptions may came in different formats - drawings, paintings, photos, etc. Hence, methods were developed to establish the link between an object representation in a book and *real* objects recognized from the surrounding world using an object recognition technique.

3.1   Object Recognition

The object recognition algorithm consists of three independent algorithms. Each recognizer operates along orthogonal directions to the others over the input space [2]. This approach offers the possibility of priming specific information, such as searching for a specific object feature (color, shape or luminance) independently of the others [2]. The set of input features are:

Color: Groups of connected regions with similar color
Luminance: Groups of connected regions with similar luminance
Shape: A Hough transform algorithm is applied to a contour image (which is the output of a Canny edge detector). Line orientation is determined using Sobel masks. Pairs of oriented lines are then used as input features

Geometric hashing [7] is a rather useful technique for high-speed performance. In this method, invariants (or quasi-invariants) are computed from training data in model images, and then stored in hash tables. Recognition consists of accessing and counting the contents of hash buckets. An Adaptive Hash table [2] (a hash table with variable-size buckets) was implemented to store affine color, luminance and shape invariants (view-independent for small perspective deformations).

3.2   Linking Representational Cues

A sketch of an object contains salient features concerning the its shape, and therefore there are advantages on learning, and linking, these different representations. Figure 3 shows the recognition of real objects from their representation in a book. Except for a description contained in the book, the robot had no other knowledge concerning the visual appearance or shape of such objects.

Additional possibilities include linking different object descriptions in a book, such as a drawing, which demonstrates the advantages of object recognition over independent input features: the topological color regions of a square drawn in black ink are easily distinguished from a yellow square. But they share the same geometric contours. This framework is also a useful tool for linking other object descriptions in a book, such as a painting, a photo or a printing. Computer generated objects are yet another feasible description (also shown in Figure 3).

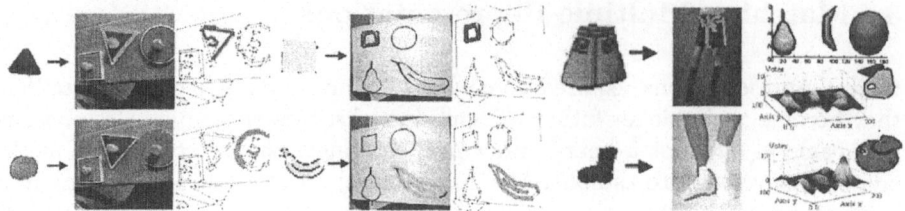

Fig. 3. Object descriptions extracted from books are used to recognize the geometric shapes of (from left to right) real objects; manual drawings; pictures of objects in catalogues; and computer generated objects.

4   Learning from Educational Activities

A common pattern of early human-child interactive communication is through activities that stimulate the child's brain, such as drawing or painting. Children are able to extract information from such activities while they are being performed on-line, which motivated the development of an algorithm that selectively attends to the human actuator (*Hand* or *Finger*), for the extraction of periodic signals from its trajectory. This algorithm operates at temporal, pyramidal levels with a maximum time scale of 16 seconds, according to the following steps:

1. A skin detector extracts skin-tone pixels over a sequence of images
2. A blob detector then groups and labels the skin-tone pixels into five regions
3. Non-periodic blobs are tracked over the time sequence are filtered out
4. A trajectory if formed from the oscillating blob's center of mass over the temporal sequence

Trajectories are also computed by two other parallel processes which receive input data from different sources: an attentional tracker [4], which tracks the attentional focus and is attracted to a new salient stimulus; and from a multi-target tracking algorithm, implemented to track simultaneously multiple targets.

Whenever a repetitive trajectory is detected from any of these parallel processes, it is partitioned into a collection of trajectories, being each element of such collection described by the trajectory points between two zero velocity points with equal sign on a neighborhood. As shown in Figure 4, the object recognition algorithm is then applied to extract correlations between these sensorial signals perceived from the world and geometric shapes present in such world, or on the robot object database.

This framework is being also applied to extract object boundaries from human cues. Indeed, human manipulation provides the robot with extra perceptual information concerning objects, by actively describing (using human arm/hand/finger trajectories) object contours or the hollow parts of objects, such as a cup. A similar strategy has been actively pursued for tactile perception of objects from the robot grasping activities.

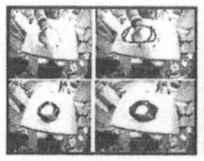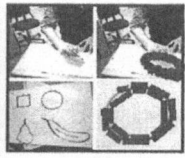

Fig. 4. (left) Learning activities, such as drawing on paper or boards (center) A human is painting a black circle on a sheet of paper with a ink can. The circle is painted multiple times. The hand trajectory is shown, together with edge lines on the background image matched to such trajectory (right) A human draws a circle on a sheet of paper with a pen, which is matched into a circle drawn previously and stored in the robot's database.

5   Conclusions

This paper presented a developmental learning approach to boost the robot's perception and learning capabilities through the use of books. A frequency domain technique was presented to extract appearance templates for a variety of objects, such as animals, clothes, plants, utilities, fruits, furniture, among others. An object recognition scheme incorporates such templates to identify common features along several objects' representations, such as paintings, drawings, photos, computer generated models or real objects. Finally, we introduced an algorithm to learn from other educational activities, such as drawing and painting, to detect geometric shapes in the world or else stored in a database.

Current work is underway to teach the robot simple language skills using learning aids, by exploiting correlations between spoken words and visual motions produced by a human. This way, a human actor will be able to introduce the robot both to an object's appearance and the set of phonemes used on a specific language to describe it. We are also investigating the feasibility of the use of other learning aids, such as television, to introduce the robot to a larger collection of objects, including dynamical systems.

References

1. Arsenio, A. M.: Embodied Vision - Perceiving Objects from Actions. IEEE International Workshop on Human-Robot Interactive Communication (2003)
2. Arsenio, A. M.: Teaching a Humanoid Robot from Books. International Symposium on Robotics (2004)
3. Comaniciu, D. and Meer, P.: Robust Analysis of Feature Spaces: Color Image Segmentation. IEEE Conference on Computer Vision and Pattern Recognition (1997)
4. Paul Fitzpatrick, P.: From First Contact to Close Encounters: A Developmentally Deep Perceptual System for a Humanoid Robot, MIT PhD Thesis (2003)
5. Harris, J.: Algebraic Geometry: A First Course (Graduate Texts in Mathematics, 133). Springer-Verlag, January (1994)
6. Perrett, D., Mistlin, A., Harries, M. and Chitty, A.: Understanding the visual appearance and consequence of hand action. Vision and action: the control of grasping. Ablex (1990) 163–180
7. Wolfson,, H. and Rigoutsos, I.: Geometric hashing: an overview. IEEE Computational Science and Engineering (1997) 10–21

Visual Servo Control for Intelligent Guided Vehicle

J.K. Mukherjee

Advanced Robotics Section DRHR BARC, Mumbai 400085. India
jkmukh@magnum.barc.ernet.in

Abstract. Intelligent Guided Vehicles can attain superior navigation capability through visual servo techniques. Vision guided wheeled mobile robots that autonomously track visible paths are non-holonomic systems with *high non linearity*. Their control is *not amenable* to time invariant state feed back. Vision based control includes visual sensing of track nature, spatial synchronization of track state to control stream for junction resolving and handling of vehicle mechanism's constraints through kinematics behavior model. Development of visual-servo processor and the encompassing test bed environment that generates *model view* of optimized tracking process is discussed.

1 Introduction

Mobile robots have a wide application and have great practical value in defense and industry. They receive great attention from the international academic and industrial society [1], [2], [3] as they are typical non-holonomic systems with high non-linearity. They can not be asymptotically stabilized or tracked via time-invariant state feedback and so vision based corrective control is *more attractive*. In addition, it is *difficult* to exactly model the complete system. Therefore, it has great theoretical and practical value to solve the control problem of mobile robots by using approximate models for impetus computing and correction of associated residual error by outer vision loop. 'Visual Servo control' driven vehicles that follow track lay-outs having junctions, branches and crossovers has been addressed as typical application. Visual control has to follow tracks and execute turns by spatially linking these actions with 'end to end path execution stream'. An indigenously developed non-holonomic model based visual servo control environment is described.

2 System Requirements for Visual Servo Frame-Work

2.1 Track Lay-Out and Its Visual Sensing

Track layout designs are guided by vehicle dynamics and space constraints. Branches and junctions are arcs and combinations of these patterns occur (fig.1). On board camera sees tracks but non parallelism of camera image plane and factory floor need appropriate image correction based on projection parameters and quality of images. View preprocessing techniques [5] are useful. Multiple pass processing [6] can give better result but should be avoided in real time loop.

N.R. Pal et al. (Eds.): ICONIP 2004, LNCS 3316, pp. 1160–1165, 2004.
© Springer-Verlag Berlin Heidelberg 2004

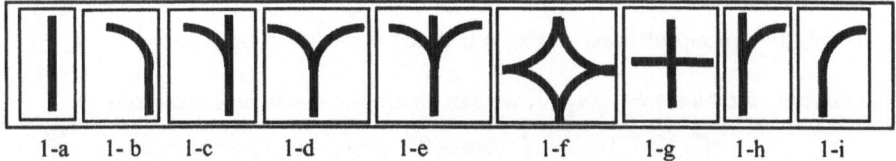

| 1-a | 1- b | 1-c | 1-d | 1-e | 1-f | 1-g | 1-h | 1-i |

Fig. 1. Track descriptions: a, b, h, and i are non branching turns. c, d, e, f are branching tracks and g is cross over (no turn). In real situation some combinations can occur.

2.2 Mechanism Control

'At place frame rotation' is necessary for look around at initial start-up and track recovery after view loss. Differentially steered type δ (2,0) [1] mobile robot (fig -3) is suitable for meeting these needs. It consists of one front caster wheel and two independent rear wheels driven by two dc motors. The motors generate toques T1 and T2 on wheels of radius R that are at distance D from central axis. The body has mass 'm', inertia I_0 and general orientation θ to axis X. Force balancing yields :

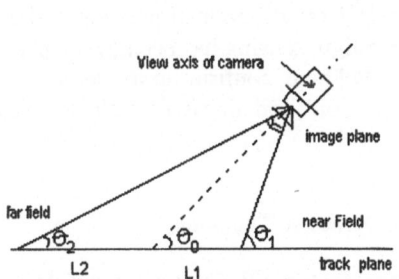

Fig. 2. Inclined view of track floor.θ_1 is 50-45^0 and θ_2 is 20-15^0. L1 < L2.

Fig. 3. Mobile robot type δ (2, 0).

$$\begin{cases} mx\ddot{} = 1/R\ (T_1 + T_2)\ \cos\theta - \lambda\sin\theta \\ my\ddot{} = 1/R\ (T_1 + T_2)\ \sin\theta - \lambda\cos\theta \\ I_0\ \theta\ddot{} = D/R\ (T_1 - T_2) \end{cases} \tag{1}$$

$$\dot{x}\cdot\sin\theta - \dot{y}\cdot\cos\theta = 0 \tag{2}$$

(1) is dynamic equation system, (2) represents controlled 'non-holonomic' motion The model can also be expressed in the form of generalized mechanical system with nonholonomy constraint. Using q for $(x, y, \theta)^T$; A for $(-\sin\theta\ \cos\theta\ 0)$;

$$M = \begin{bmatrix} m & 0 & 0 \\ 0 & m & 0 \\ 0 & 0 & I_0 \end{bmatrix} \quad ; \qquad B = 1/R \begin{bmatrix} \cos\theta & \cos\theta \\ \sin\theta & \sin\theta \\ D & -D \end{bmatrix} \quad ;$$

$$M q\ (q\ddot{}) = B(q)T + A^T(q)\ \lambda \dots (3): \quad \& \quad a(q)q\ddot{} = 0 \tag{4}$$

Choosing one group of basis of the null space of A (q) to be S (q) = $\begin{bmatrix} \text{Cos}\theta & 0 \\ \text{Sin}\theta & 0 \\ 0 & 1 \end{bmatrix}$;

and control variables $v = (v_1 \ v_2)^T$, we can obtain the kinematics equations as

$$\begin{bmatrix} \dot{x} \\ \dot{y} \\ \dot{\theta} \end{bmatrix} = \begin{bmatrix} \text{Cos}\theta & 0 \\ \text{Sin}\theta & 0 \\ 0 & 1 \end{bmatrix} \begin{bmatrix} v_1 \\ v_2 \end{bmatrix} \tag{5}$$

The new control inputs have definite physical meaning, that is, the linear velocity and angular velocity of the wheels that are controlled by the servo loops are related

$$\begin{cases} v_1 = R\,(\omega_1 + \omega_2)/2 & \ldots\ldots \text{ along vehicle axis.} \\ v_2 = R\,(\omega_1 - \omega_2)/2D & \ldots\ldots \text{around body center} \end{cases} \tag{6}$$

This *signifies* that since for reaching a point not on current straight path, ω_1, ω_2 are unequal. The vehicle develops finite orientation error on reaching such targets.

3 Visual Servo Control

An experimental cart with on board DSP control for differential steering and elec-tronic architecture suitable to support multi-paradigm sensing has been developed [4]. The simulated vehicle control model, implemented here, confirms to the servo control model of this vehicle. Controlled 'in-place-rotation' and linear as well as circular paths are generated by the vehicle.

3.1 Effect of Vehicle Control Approach on Dynamic Tracking

Basic motion control mechanism decides several factors [2]. For track types depicted in fig 1, several view scenarios are possible. View axis, vehicle locus as per current control inputs and track direction of related segment of track are the three significant parameters. View axis is tangent to locus at vehicle's location for fixed camera con-figuration. Various situations occur as in fig. (4). Double arrowed dashed line shows view axis. Solid line or curve with single arrow shows track as seen and dashed line or arc without arrow show current path being executed.

Block 2A shows orientation error (linear segment). Vehicle should stop and realign but stopping is avoided to minimize shock. An approach like that in block (2B) with nominal 'K' is attempted to achieve second order smoothness in velocity. Next cor-rection attempt may see *widely varied* condition. If seen while cart is at P' i.e. prior to reaching P, then situation is as shown in 2D. If seen at P" i.e. past the point P then the case is as shown in 2C. The view from camera (shaded part) misses the track alto-gether. Case-3 shows offset- error with correct direction. Attempt to correct it leads to 3B that is similar to 2D. Here algebraic curve fit is needed for smooth movement [7]. In current implementation arc fit is applied in keeping with control mechanism's *native* nature (3C). It is evident that source of complexity lies in equation (6). The vehicle may be controlled to reach a point in motion but to simultaneously ensure precisely desired orientation also is very difficult [3].

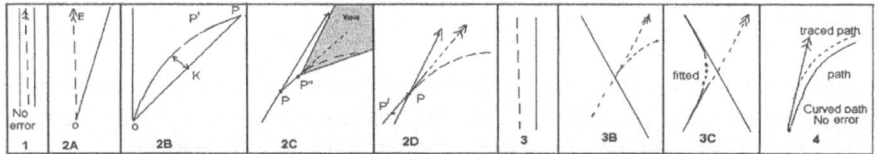

Fig. 4. Possible cases (1 to 4) depending on various dynamic situations of vehicle.

3.2 Prediction-Correction Based Control Approach

Corrections for track following errors are predicted based on the approximate dynamic model of robot, the driving motor dynamics and non-slip kinetic constraints. The visual sensing system evaluates effectiveness of control prior to start of next control loop execution. Vehicle state, offset errors and orientation errors are retrieved from various zones of same frame. All junctions and crossovers appear first in far field and then in intermediate field on later (temporally) frames. Nature of track is assessed from far field. The shape-based patterns form a finite set of combinations. Their occurrence over a sequence of frames forms a finite state set. The vision-processor computes steering parameter using the path seen and the vehicle's model (section 2.2). Delay in servo loop (caused by motor inertia and integrated reflected inertia of vehicle and pay load) and several other factors cause error in track following. Vision system samples track-view and absorbs the orientation error in subsequent track traverses of small size before reaching target by fine modification.

4   Optimization Issues

Speed of travel must be maximized for good throughput. Since accumulated error increases with distance traversed, the frequency of visual processing should be high for higher speeds. On curves the control corrections to be applied per unit distance traveled increases considerably. Consequently negotiation of curved segments of track should be at lower speeds to gain time margins. Incidentally this is *not contradictory* to the speed constraints arising from stable dynamics considerations. Hence predictive control is applied on vehicle's speed too.

5   Integrated Behavior Assessment Environment

The test bench environment is graphics supported (refer fig 5). Test path selection, control file binding and dynamic graphic presentation of car depicting its position and orientation are offered in the composite environment. Camera view window presents the track as scene by system from each new position. Simulation processes have associated seamless loop status review and recovery capabilities. The constant time interval logs create effective temporal behavior display in spatial form. This results in denser occurrence of vehicle positions at lower speed of travel by vehicle. On loss of track view the control stops the vehicle and maintains state for easy analysis.

6 Results

Elaborate test with variety of path patterns and parameters have been carried out on the systems. Range of steering angle, acceleration rate, deceleration rate and speed are accepted as tunable system parameters. The sequence of turn selection is treated as command stream. For brevity, a few cases are detailed here. Fig.(5) shows successful tracking of a complex path. Camera views in (fig.6) are instantaneous views seen by vehicle crossing positions 1 to 7 respectively. *Note* the appearances of middle trifurcation as it appears first in R3, then in inset on top part of fig. 5 and finally in R4 when right turn decision has been taken. In all camera views, chosen path appears in alignment with view axis (vertical centerline of camera view frame) as the cases pertain to a successful tracking behavior. The vehicle deviates more from centre of track on curved path segments *conforming* to the theoretical analysis (case 2B section 3.1). The system exhibits *speed control*. The cart positions are shown at constant time interval. On curves the speed reduces. On top straight section the speed gained is highest as expected. Fig 7a shows successful following on central section and establishes the path programmability aspect. Fig. (7b) shows test with high acceleration and lower deceleration for behavior test. As a consequence the cart fails to make steep turn and track view is lost as expected by theoretical analysis in case 2C of section 3.1. Correction is inadequate at high speed.

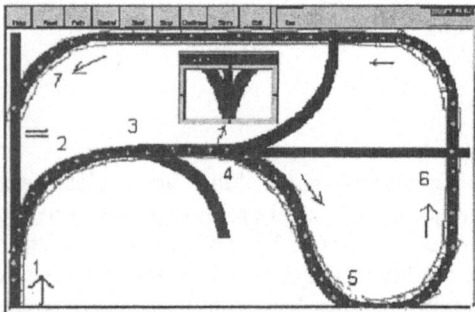

Fig. 5. The console of 'Integrated Visual Servo Environment' with specimen test track.

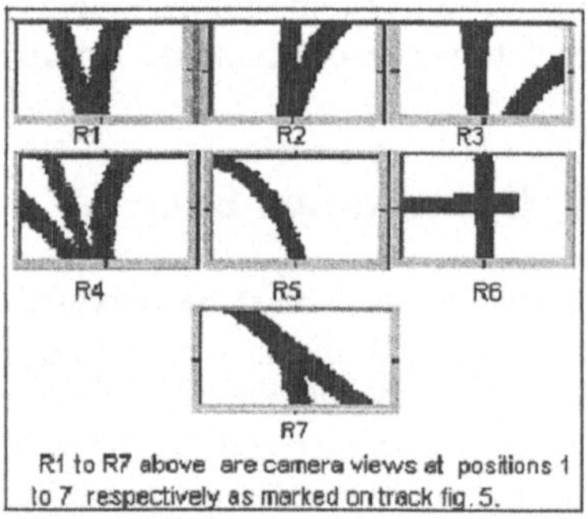

Fig. 6. shows camera views seen at various positions (1 to 7).

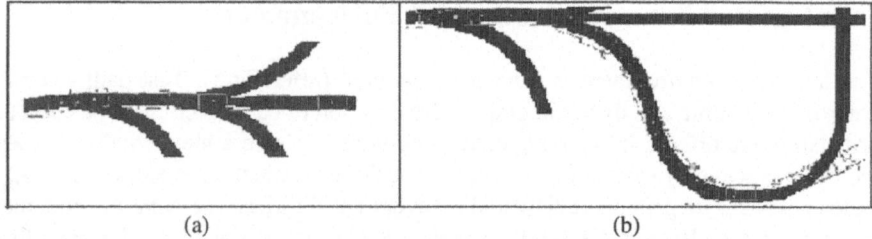

(a)   (b)

Fig. 7. (a) shows track following on straight path at central section of path in fig. 5 while, (b) shows failure with same settings as for the left case on other segment of track.

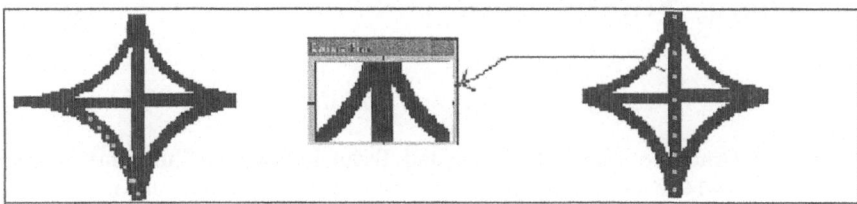

Fig. 8. Combination of 4 way junction. The crossover serves for crossing the zone straight without turning. Note the slow left turn and straight crossover with acceleration.

Figure 8 shows tracking of complex track segment (combination of those in fig 1) as per commanded turn options. AI system uses the track view at junction exit (inset – fig.8) to update internal state and maintain spatial synchronization with track.

7 Conclusion

In the approach developed by author, the control first predicts a steering angle based on forward view of track. After a finite time interval it detects the actual accumulated offset errors from near field of view frame. It also determines desired steering angle change based on the next predicted value derived from far and intermediate fields of instantaneous track view and estimates total correction. Based on nature of track, speed is also modified. The system *serves* for design time assessment of tracks and vehicle control parameters economically. An *indirect fallout* is that track sequence programs also can be tested on this system and 'end to end following' can be verified.

Acknowledgement

The author is thankful to Padmashri G. Govindarajan, Director A&M and EI groups, BARC for constant encouragement and support.

References

1. C.Canudas de Wit et al. 'Theory of Robot Control' Springer-Verlag: London, 1998
2. J.M.Yang, J.K.Kim, 'Sliding mode control for trajectory tracking of nonholonomic wheeled mobile robots', IEEE Transactions on Robotics and Automation, 1999, pp.578-587
3. N.Sarkar et al 'Control of mechanical systems with rolling constraints: application of dynamic control of mobile robots' International Journal of Robotics Res., 1994, p.55-69
4. J.K.Mukherjee et al. 'Remote Guided Vehicle For Sensing and Visual Assessment' National symposium On Nuclear Instrumentation 2004, BRNS-DAE Feb. 2004, pp 641-647
5. J.K.Mukherjee 'Vision Studio 3.0-An Engineering Workstation For Developing Robot Vision' International Symposium on Intelligent Robotic System ISIRS-Nov 1995.
6. J.K.Mukherjee. 'Computer Aided visual Metrics System' National Symposium On Nuclear Instrumentation 2004 - BRNS, DAE Feb. 2004 pp 599 – 606.
7. G. Tanbin et al. 'Parameterized families of polynomials for bounded algebraic curve fitting' IEEE Transaction Pattern Analysis and Machine Intelligence, 1994, 6 : 287 – 303.

A Basilar Membrane Model
Using Simulink for Hearing-Aid Systems

Tetsuya Tsukada[1] and Yoshifumi Sekine[2]

[1] Graduate School of Science and Technology, Nihon University
1-8-14, Kandasurugadai, Chiyoda-ku, Tokyo 101-8308, Japan
ttsukada@hippo.ecs.cst.nihon-u.ac.jp
[2] College of Science and Technology, Nihon University
7-24-1, Narashinodai, Funabashi-Shi, Chiba 274-8501, Japan
ysekine@ecs.cst.nihon-u.ac.jp

Abstract. Our purpose is to apply a basilar membrane model (BMM) to hearing-aid systems to solve problems in existing hearing-aid systems. In this study, we construct the BMM using Simulink in hearing-aid systems. Also, we examine response characteristics of the BMM using Simulink. As a result, we show that the BMM using Simulink can emphasize characteristic frequencies, and the model probably has the effect of hearing compensation. These results suggest that the BMM using Simulink is useful for hearing-aid systems.

1 Introduction

Acoustic information is very important for voice communication. If there is a hearing impairment, then it is hard to understand information in voice communication. Generally, a hearing-aid is used in order to alleviate a hearing impairment. However, more than half of hearing-aid users have felt that existing hearing-aids are uncomfortable, because it is necessary to change the switch according to the situation and/or to adjust the characteristic of a hearing-aid based on an individual user's hearing characteristics.

G. von Békésy [3] analyzed the function of the basilar membrane and clarified that the basilar membrane responded selectively to frequencies of speech sounds based on the basilar membrane's position. B. M. Johnstone *et al.* [4] measured the displacement of the basilar membrane to the stimuli of various sound pressure levels and made clear that the quality factor (Q) of resonance in the basilar membrane varied depending on the sound pressure of an input. J. L. Flanagan [5] derived mathematical models based on Békésy's data for approximating basilar membrane displacement.

Until now, we have constructed the basilar membrane model (BMM) based on Flanagan's mathematical models in consideration of Johnstone's experimental data. Also, we have examined the feature extraction function of our model in order to apply the hearing function to engineering model [6].

Now, we are studying how to apply the BMM to hearing-aid systems, in order to solve problems in existing hearing-aid systems, because we think that hearing

N.R. Pal et al. (Eds.): ICONIP 2004, LNCS 3316, pp. 1166–1171, 2004.
© Springer-Verlag Berlin Heidelberg 2004

impairments should be alleviated by a system with characteristics, which are close to living body characteristics.

When the BMM is applied to hearing-aid systems, deploying the BMM using DSP is an effective method for the following reasons:

1. to cope with individual users requirements
2. to improve convenience
3. to cope with problems such as low power consumption and limited mounting area.

Generally, a system deployed using DSP is described as an algorithm using programming languages. This method can efficiently process, while the method is self-correcting, self-testing, and self-programming [7].

System-level design environments, which can design and verify functions, are proposed in order to reduce the period of design and verification. The MathWorks's MATLAB/Simulink [8] used for system-level design environments, which can visually design and verify based on block diagrams using the graphical user interface (GUI), can directly convert the Simulink model into C code using this exclusive tool. Thus, we can further simplify the system to use on DSP [7] [9].

In this paper, we construct the BMM using Simulink in order to apply the hearing-aid systems. Also, we examine response characteristics of our model and hearing compensation. Firstly, we demonstrate that the composition of the BMM and the correspondence of the characteristics of our model to data of a living body. Secondly, we show the composition of the BMM using Simulink and the response characteristics of our model. Lastly, we show the improved effect of hearing compensation of our model using the modeled hearing loss sound made from the output voice of our model.

2   The Basilar Membrane Model (BMM)

In this section we describe the BMM that can simulate the Q characteristic and the adaptive gain according to the input sound pressure, and the abrupt cutoff property in the high region.

The transfer function $F(s)$ of the BMM is represented as follows:

$$F(s) = \frac{C\omega_0^2}{s^2 + \dfrac{\omega_0}{Q_L}s + \omega_0^2} \cdot \frac{C\dfrac{\omega_0}{Q_B}s}{s^2 + \dfrac{\omega_0}{Q_B}s + \omega_0^2} \cdot \frac{s^2 + \omega_0'^2}{s^2 + \dfrac{\omega_0'}{Q_N}s + \omega_0'^2} \quad (1)$$

where ω_0 is the resonance frequency of the LPF and BPF, Q_L and Q_B are quality factors of the LPF and BPF respectively, ω_0' is the resonance frequency of the BEF, Q_N is a quality factor of the BEF, and C is a constant.

Figure 1 shows frequency response characteristics of $F(s)$ for various input levels compared with Johnstone's data. The horizontal axis shows frequency in Hz, and the vertical axis shows amplitude of $F(s)$ in dB. Where, Q_L was set to 36.3, 7.7, 1.2, and 0.2, when the input level was 20 dB, 40 dB, 60 dB, and 80 dB, respectively (dB re. 20 μPa). $\omega_0/2\pi$ for the BPF and LPF were 18 kHz, $\omega_0'/2\pi$ for the BEF was 20.6 kHz. C for the LPF was 110, C for the BPF was 1.0, and Q_B and Q_N were equal to 3.6 and 0.9, respectively. The results show that $F(s)$ can approximate Johnstone's data well, i.e.

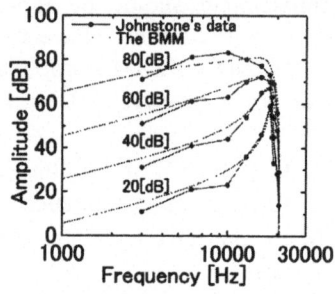

Fig. 1. Frequency responses of $F(s)$ compared with Johnstone's data.

our model can simulate Q and the variation of the gain according to a difference in the input sound pressure, and the abrupt cutoff property in the high region.

3   The BMM Using Simulink

Figure 2 shows the block diagram of the BMM using Simulink. The whole BMM consists of single-channel BMMs connected in parallel, because the basilar membrane has the property that selectively responds to the frequency of that specific position. The single-channel of our model consists of the filter part that can simulate the vibration characteristic of the basilar membrane, and the Q control part (QC) that controls the Q and gain depending on the input sound pressure,

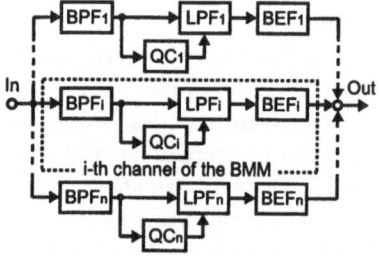

Fig. 2. The blockdiagram of the BMM using Simulink.

corresponding to the function of the outer hair cell [10] in the human auditory system. The input to the QC was taken from the output of the BPF. The Q of the LPF was controlled by the Q characteristic approximated to Johnstone's data. Furthermore, the previous BMM [6] was controlled by the Q of the LPF

using feed-back from the BPF output. In contrast, our model was controlled by the Q of the LPF using feed-forward from the BPF output in order to stabilize the operation, when the model is constituted using Simulink. Our model has the same characteristic as the previous model.

Figure 3 shows Q characteristics of the QC to input levels compared with Johnstone's data. The horizontal axis shows sound pressure in Pa, and the vertical axis shows Q for LPF. The results show that the characteristics of the QC can well approximate Johnstone's data.

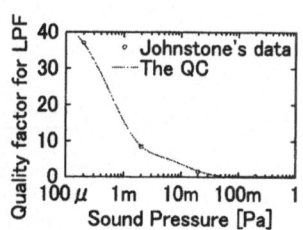

Fig. 3. Characteristics of the QC.

4   Response Characteristics

In this section we describe some properties of the BMM using Simulink to the voice in order to clarify the control function of the Q and gain of our model.

In this study, the number of channels of the BMM using Simulink was 19. Characteristic frequencies of i ($i = 1 \sim 19$) channel was set up at $250 \times 2^{(i-2)/4}$ Hz so that the feature is extracted every 1/4 octave in the frequency band in the range of 200 Hz to 4000 Hz in the existing main constituents of formant frequencies, which is important in discriminating Japanese vowels.

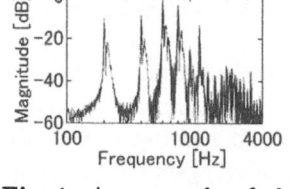

Fig. 4. An example of the spectrum of the Japanese vowel /a/.

Figure 4 shows an example of the spectrum of the Japanese vowel /a/ used for the input. The horizontal axis shows frequency in Hz, and the vertical axis shows normalized magnitude in dB.

Figure 5 shows responses of our model for the input of the Japanese vowel /a/. The horizontal axis shows frequency in Hz, and the vertical axis shows normalized magnitude in dB. Figure 5(a) shows an example of the spectrum of the output voice of our model. This figure shows that our model can emphasize other frequency constituents as against the frequency constituent

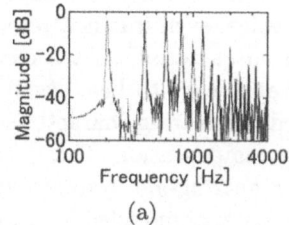

(a)

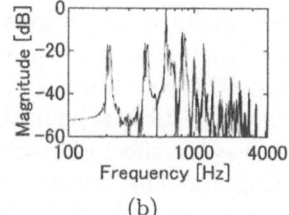

(b)

Fig. 5. Responses of the BMM using Simulink and the fixed Q filter bank for the input of the Japanese vowel /a/. (a) An example of the spectrum of the output voice of the BMM using Simulink for the input of the Japanese vowel /a/. (b) An example of the spectrum of the output voice of the fixed Q filter bank for the input of the Japanese vowel /a/.

with highest level of the vowel. Figure 5(b) shows an example of the spectrum of the output voice of the fixed Q filter bank with the QC in our model of Fig. 2 removed. This figure shows that the fixed Q filter bank can not emphasize other frequency constituents as against the frequency constituent with highest level of the vowel. As a result, our model can emphasize characteristic frequencies of the voice.

5   The Effect of Hearing Compensation

In this section we describe the possibility of the application to hearing-aid systems of the BMM using Simulink. We examine the effect of hearing compensation of our model using modeled hearing loss sounds. Modeled hearing loss sounds were made to decrease the voice emphasized characteristic frequencies with our model, the original voice, and the output voice of the fixed Q filter bank with the hearing characteristic of modeled hearing loss that assumed the hearing characteristic of hearing-impaired people.

Figure 6 shows an example of the audiogram of the modeled hearing loss. The horizontal axis shows frequency in Hz, and the vertical axis shows hearing level in dBHL. This figure shows the characteristic that simulated the hearing characteristic of a sloping hearing loss that assumed most mixed hearing loss as the hearing characteristic of hearing-impaired people.

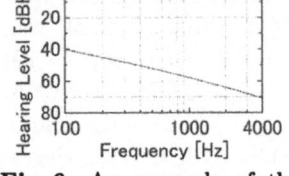

Fig. 6. An example of the audiogram of the modeled hearing loss.

Figure 7 shows modeled hearing loss sounds. The horizontal axis shows frequency in Hz, and the vertical axis shows normalized magnitude in dB. Figure 7(a) shows an example of the spectrum of the modeled hearing loss sound made from the output voice of the BMM using Simulink. This figure shows the output voice of our model can hold the level of characteristic frequencies similar to the original voice, even if it is attenuated with the characteristic that simulated the hearing characteristic. As a result, the output voice of our model become the voice that is easy to hear, even if it is attenuated with the characteristic that simulated the hearing characteristic, that is, our model probably has the effect of hearing compensation. These results suggest that the BMM using Simulink is useful for hearing-aid systems. In contrast, Figure 7(b), (c) show an example of the spectrum of modeled hearing loss sounds made from the original voice, and the output voice of the fixed Q filter bank. These figures show levels of characteristic frequencies over 1000 Hz that the modeled hearing loss sound made from the voice of origin and the modeled hearing loss sound made from the output voice of the fixed Q filter bank decrease more than 10 dB with the characteristic that simulated the hearing characteristic of a sloping hearing loss as compared with the voice.

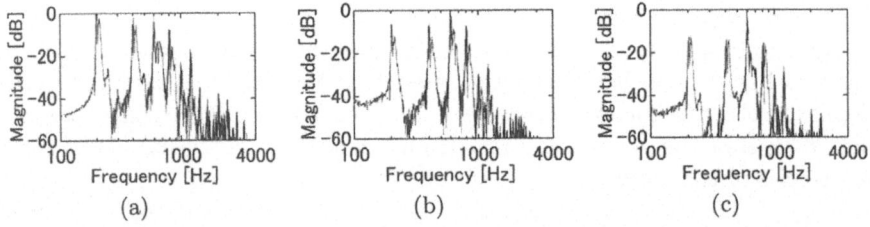

Fig. 7. Modeled hearing loss sounds. (a)An example of the spectrum of the modeled hearing loss sound made from the output voice of the BMM using Simulink. (b)An example of the spectrum of the modeled hearing loss sound made from the original voice. (c)An example of the spectrum of the modeled hearing loss sound made from the output voice of the fixed Q filter bank.

6   Conclusions

In this paper, we constructed the BMM using Simulink in order to apply the BMM to hearing-aid systems. Also, we examined response characteristics of the BMM using Simulink and the effect of hearing compensation.

As a result, we showed that the BMM using Simulink can emphasize characteristic frequencies of the voice. Furthermore, we showed that our model probably has the effect of hearing compensation, because the output voice of our model can hold the level of characteristic frequencies similar to the original voice, even if it is attenuated with the characteristic that simulated the hearing characteristic. These results suggest that the BMM using Simulink is useful for hearing-aid systems.

In future work, we will deploy our model using DSP and carry out listening experiments with hearing-impaired people.

Acknowledgment

This work was supported in part by a grant from the Futaba Electronics Memorial Fundation and Grant-in-Aid #14550334 of the Ministry of Education, Science, Sports and Culture of Japan.

References

1. Y. Ymada, "Sensory aids for the hearing impaired," *IEICE Tech. Rep. (in Japanese)*, SP93-48, pp.31-38, July 1993.
2. Information and Culture Center for the Deaf, *The Questionnaire Survey Reort (in Japanese)*, pp.82-89, Sept 1995.
3. G. von Békésy, *Experiments in Hearring*, McGrawHill, 1960.
4. M. B. Johnstone, R. Patuzzi and G. K. Yates, "Basilar membrane measurements and the travelling wave," *Hearing Research*, vol.22, pp.147-153, 1986.
5. J. L. Flanagan, "Models for Approximating Basilar Membrane Displacement," *Bell Syst. Tech. J.*, vol.39, pp.1163-1191, Sept 1960.
6. S. Takahashi, H. Nakamura and Y. Sekine, "A Hardware Model Based on the Physiological Characteristics of Basilar Membrane and Its Application to Feature Extraction," *IEICE Trans. on Electronics (in Japanese)*, vol.J85-C, no.7, pp.549-556, July 2002.
7. K. Nonami, H. Nishimura and M. Hirata, *Control Systems Design Using MATLAB (in Japanese)*, Tokyo Denki University Press, pp.219-222, May 1998.
8. The MathWorks, Inc., MATLAB, Simulink,
   http://www.mathworks.com
9. K. Hashimoto, M. Sakuragi, K. Tanaka, T. Sato and I. Arita, "Design of Hardware/Software Co-Design Enviroments Based on Simulink," *DA Symposium 2003 (in Japanese)*, pp.163-168, July 2003.
10. B. C. J. Moore, *An Introduction to the Psychology of Hearing (in Japanese)*, K. Ogushi, Seishinshobo, pp.30, April 1994.

Cluster and Intrinsic Dimensionality Analysis of the Modified Group Delay Feature for Speaker Classification

Rajesh M. Hegde and Hema A. Murthy

Department of Computer Science and Engineering
Indian Institute of Technology, Madras, Chennai
{rajesh,hema}@lantana.tenet.res.in

Abstract. Speakers are generally identified by using features derived from the Fourier transform magnitude. The Modified group delay feature(MODGDF) derived from the Fourier transform phase has been used effectively for speaker recognition in our previous efforts.Although the efficacy of the MODGDF as an alternative to the MFCC is yet to be established, it has been shown in our earlier work that composite features derived from the MFCC and MODGDF perform extremely well. In this paper we investigate the cluster structures of speakers derived using the MODGDF in the lower dimensional feature space. Three non linear dimensionality reduction techniques The Sammon mapping, ISOMAP and LLE are used to visualize speaker clusters in the lower dimensional feature space. We identify the intrinsic dimensionality of both the MODGDF and MFCC using the Elbow technique. We also present the results of speaker identification experiments performed using MODGDF, MFCC and composite features derived from the MODGDF and MFCC.

1 Introduction

The most relevant engineering approach to the problem of speaker identification is to represent a speaker by the space which he or she occupies. Indeed there exists a multi-dimensional parameter space in which different speakers occupy different regions. Speakers tend to cluster in this space as points or trajectories at different locations and can occupy more than one region in the entire parameter space. Parameters or features must be chosen such that the clusters are small and well separated. The multidimensional feature space in which speakers position themselves makes pattern recognition difficult, as each observation is made up of a large number of features. Further distances cannot be measured reliably as the covariances of features is difficult to establish. This leads us to investigate effective dimensionality reduction techniques that preserve linear and non linear cluster structures. The issues of cluster analysis and identification of intrinsic dimensionality of the feature set used are crucial. Features like the MFCC which are derived from the Fourier transform magnitude only, by ignoring the phase spectrum, may not be capturing the entire information contained in the signal

N.R. Pal et al. (Eds.): ICONIP 2004, LNCS 3316, pp. 1172–1178, 2004.
© Springer-Verlag Berlin Heidelberg 2004

acquired from each speaker. In this context features derived from phase like the MODGDF [1–3] and composite features derived by combining the MODGDF and MFCC are very relevant. We briefly discuss the MODGDF and use both MODGDF and the traditional MFCC to parametrically represent speakers in this paper. Further cluster structures of speakers in the lower dimensional space derived using non linear dimensionality reduction techniques like Sammon mapping [4] and unsupervised learning algorithms based on manifold learning like Isometric mapping(ISOMAP) [5] and the Locally linear embedding(LLE) [6], have been investigated in this work. Intrinsic dimensionality analysis is carried out using ISOMAP. The Intrinsic dimensionality is identified using the *Elbow* technique from the residual variance curve and its implications in the context of speaker identification are discussed. Finally the classification results using the MODGDF, MFCC and composite features using a GMM based baseline system are listed.

2   The Modified Group Delay Feature

The group delay function [3], defined as the negative derivative of phase, can be effectively used to extract various system parameters when the signal under consideration is a minimum phase signal. The group delay function is defined as

$$\tau(\omega) = -\frac{d(\theta(\omega))}{d\omega} \tag{1}$$

where $\theta(\omega)$ is the unwrapped phase function. The group delay function can also be computed from the speech signal as in [3] using

$$\tau_x(\omega) = \frac{X_R(\omega)Y_R(\omega) + Y_I(\omega)X_I(\omega)}{|X(\omega)|^2} \tag{2}$$

where the subscripts R and I denote the real and imaginary parts of the Fourier transform. $X(\omega)$ and $Y(\omega)$ are the Fourier transforms of $x(n)$ and $nx(n)$, respectively. The group delay function requires that the signal be minimum phase or that the poles of the transfer function be well within the unit circle for it to be well behaved. This has been clearly illustrated in [2] and [3]. It is also important to note that the denominator term $|X(\omega)|^2$ in equation 2 becomes zero, at zeros that are located close to the unit circle. The spiky nature of the group delay spectrum can be overcome by replacing the term $|X(\omega)|^2$ in the denominator of the group delay function with its cepstrally smoothed version, $S(\omega)^2$. Further it has been established in [1] that peaks at the formant locations are very spiky in nature. To reduce these spikes two new parameters γ and α are introduced. The new modified group delay function as in [3] is defined as

$$\tau_m(\omega) = \left(\frac{\tau(\omega)}{|\tau(\omega)|}\right) (|\tau(\omega)|)^\alpha \tag{3}$$

where

$$\tau(\omega) = \left(\frac{X_R(\omega)Y_R(\omega) + Y_I(\omega)X_I(\omega)}{S(\omega)^{2\gamma}}\right) \tag{4}$$

where $S(\omega)$ is the smoothed version of $|X(\omega)|$. The new parameters α and γ introduced vary from 0 to 1 where $(0< \alpha \leq 1.0)$ and $(0< \gamma \leq 1.0)$. The algorithm for computation of the modified group delay function is explicitly dealt with in [3]. To convert the modified group delay function to some meaningful parameters, the group delay function is converted to cepstra using the Discrete Cosine Transform (DCT).

$$c(n) = \sum_{k=0}^{k=N_f} \tau_x(k) \cos(n(2k+1)\pi/N_f) \tag{5}$$

where N_f is the DFT order and $\tau_x(k)$ is the group delay function. The second form of the DCT, DCT-II is used, which has asymptotic properties to that of the *Karhunen Loeve Transformation* (KLT) as in [3]. The DCT acts as a linear decorrelator, which allows the use of diagonal co-variances in modeling the speaker vector distribution.

3 Speaker Cluster Analysis with Sammon Mapping

Speaker classification researchers are usually confronted with the problem of working with huge databases and a large set of multidimensional feature vectors, which exerts a considerable load on the computational requirements. Typically Principal Component Analysis (PCA) and Linear Discriminant Analysis (LDA) are used for dimensionality reduction in the speech context, despite the fact that they may not be optimum for class discrimination problems. We therefore use the Sammon mapping technique[4] for dimensionality reduction of the MODGDF and MFCC as it preserves the inherent structure of the underlying distribution. In order to visualize the cluster structure of individual speakers, we first compute 16 dimensional vector quantization (VQ) codebooks of size 64 by concatenating six sentences of that particular speaker picked from the training set of the NTIMIT [7] database. Each codebook is transformed into a two dimensional codebook of size 64 using Sammon mapping [4]. Sammon mapping, which belongs the class of Multidimensional scaling techniques (MDS) minimizes the following error function to extract lower dimensional information from high dimensional data using gradient descent technique:

$$\varepsilon_{sam} = \frac{1}{\sum_{i=1}^{i=N-1} \sum_{j=i+1}^{i=N} D_{ij}} \sum_{i=1}^{i=N-1} \sum_{j=i+1}^{i=N} \frac{(d_{ij} - D_{ij})^2}{D_{ij}} \tag{6}$$

where d_{ij} is the distance between two points i,j in the d-dimensional output space, and D_{ij} is the distance between two points i, j in the D-dimensional input space, N is the number of points in the input or output space. The results of cluster analysis for two speakers in the two dimensional space is shown in *figure 1(a) and 1(b)*. It is evident that MODGDF clearly separates the two speakers in the low dimensional feature space compared to MFCC.

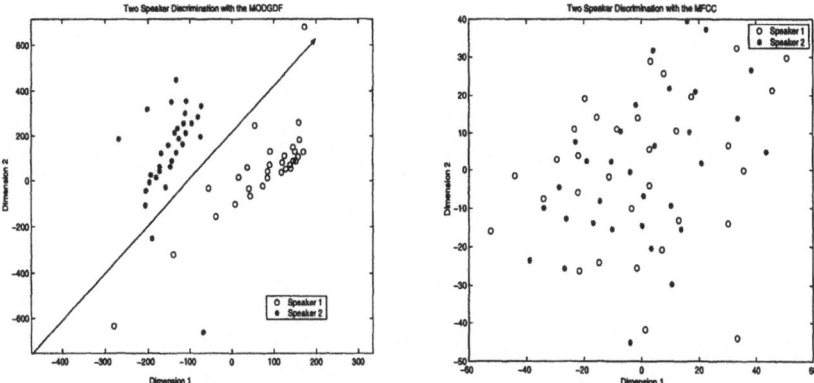

Fig. 1. Two speaker discrimination with the MODGDF (left) and MFCC (right).

4   Intrinsic Dimensionality Analysis Using Unsupervised Learning Algorithms

From the results of sammon mapping, one can be tempted to hypothesize that the projections of MFCC resulted in much greater error than those of the MODGDF. We therefore identify the intrinsic dimensionality of the MODGDF and MFCC using unsupervised learning algorithms like the ISOMAP and LLE and then visualize speakers in the lower dimensional space. Although both ISOMAP and LLE can also be used for identifying the intrinsic dimensionality of any feature set by detecting the dimension at which the error bottoms down, LLE may fail for feature sets twisted and folded in the high dimensional input space. But ISOMAP is guaranteed to asymptotically converge and recover the true dimensionality of even such feature sets. Hence we use ISOMAP and the Elbow technique to identify the true dimensionality of the feature set in this work.

4.1   Isometric Mapping (ISOMAP) and The *Elbow* Technique [5]

The ISOMAP has three steps, The first step determines which points are neighbors on the manifold M, based on the distances $d_x(i,j)$ between pairs of points i, j in the input space X. Two simple methods are to connect each point to all points within some fixed radius e, or to all of its K nearest neighbors. These neighborhood relations are represented as a weighted graph G over the data points, with edges of weight $d_x(i,j)$ between neighboring points. In its second step, Isomap estimates the geodesic distances $d_m(i,j)$ between all pairs of points on the manifold M by computing their shortest path distances $d_g(i,j)$ in the graph G using an appropriate shortest path finding technique like the Dijkstra algorithm. The final step applies classical MDS to the matrix of graph distances $D_G = d_G(i,j)$, constructing an embedding of the data in a d-dimensional euclidean space Y that best preserves the manifolds estimated intrinsic geometry. Further the intrinsic dimensionality of the feature set can be estimated by looking for the *Elbow* at which the curve showing the relationship between residual variance and the

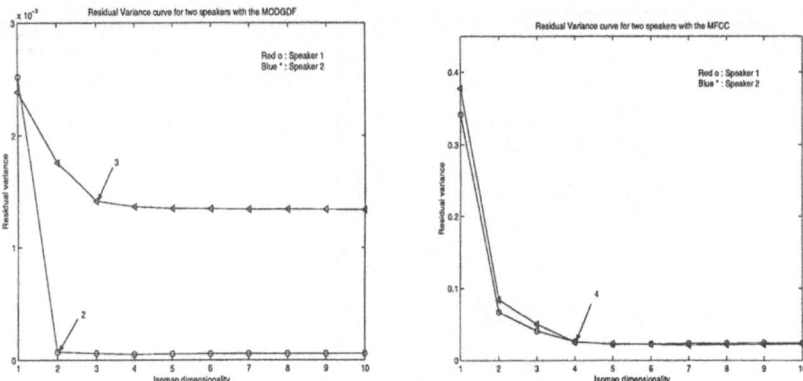

Fig. 2. Residual variance for two Speakers with the MODGDF (left) and MFCC (right) using ISOMAP.

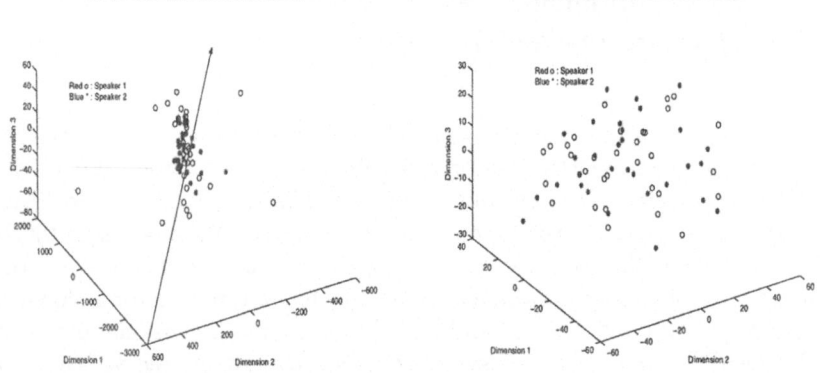

Fig. 3. Two Speaker cluster structure in 3 dimensions with the MODGDF (left) and MFCC (right) using ISOMAP.

number of dimensions of the feature set ceases to decrease significantly which is called the *Elbow* technique. It is important to note that residual variance is the amount of variance in the feature set remaining after the first n principal components have been accounted for. The residual variance curves for two speakers using MODGDF and MFCC are illustrated in *figures 2(a) and 2(b)* respectively. It is interesting to note that MODGDF has a intrinsic dimensionality (2 and 3) while MFCC exhibits an intrinsic dimensionality of 4 with respect to this pair of speakers. The 3 dimensional visualization of codebooks of two speakers with the MODGDF and MFCC using ISOMAP are illustrated in *figures 3(a) and 3(b)*.

4.2   Locally Linear Embedding (LLE) [6]

The LLE is an unsupervised learning algorithm that computes low-dimensional, neighborhood-preserving embeddings of high-dimensional inputs. LLE maps its inputs into a single global coordinate system of lower dimensionality, and its optimizations do not involve local minima. The LLE algorithm, for mapping high dimensional data points, X_i, to low dimensional embedding vectors, Y_i can be summarized in three steps. The first step computes the neighbors of each data point, X_i. In the next step the weights W_{ij} are computed that best reconstruct each data point X_i from its neighbors, minimizing the cost in

$$\varepsilon(W) = \sum_i \left| X_i - \sum_j W_{ij} X_j \right|^2 \qquad (7)$$

by constrained linear fits. The final step computes the vectors best reconstructed by the weights, minimizing the quadratic form

$$\Phi(W) = \sum_i \left| Y_i - \sum_j W_{ij} Y_j \right|^2 \qquad (8)$$

by its bottom nonzero eigenvectors. The 3 dimensional visualization of two speakers with the MODGDF and MFCC using LLE are illustrated in *figures 4(a) and 4(b)*.

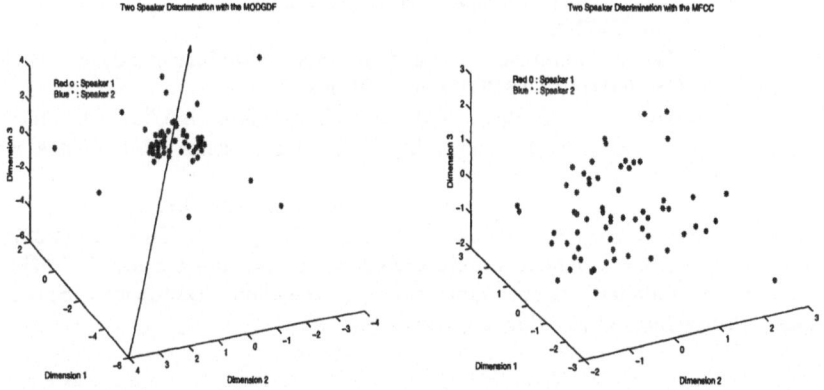

Fig. 4. Two Speaker cluster structure in 3 dimensions with the MODGDF (left) and MFCC (right) using LLE.

5   Classification Results and Conclusions

The MODGDF gave a recognition percentage of 98.5% and 96.5% while MFCC gave 98% and 97% for 100 and 200 speakers on the TIMIT database, using a GMM based baseline system. Composite features derived from MFCC and MODGDF performed at 50% for NTIMIT data. We also noticed that the intrinsic dimensionality was around 2 and 3 for MODGDF for 90% of speakers from the NTIMIT [7] database, while MFCC intrinsic dimensionality was equal to or higher than 4. But it can be concluded from the clustering and intrinsic dimensionality analysis that MODGDF is capable of discriminating speakers in a lower dimensional space while MFCC requires a higher dimensional representation. We investigated the intrinsic dimensionality of a large number of speakers from the NTIMIT database and noticed from the cluster plots that speaker clusters are well separated only at the intrinsic dimensionality of their parametric representations. We therefore intend to identify the intrinsic dimensionality of speakers first and then use this crucial information for automatic speaker identification tasks in our future efforts. This can reduce the computational overhead and also lead us to various other possibilities is speech recognition tasks.

References

1. Rajesh M.Hegde, Hema A.Murthy and Venkata Ramana Rao Gadde: Application of the Modified Group Delay Function to Speaker Identification and Discrimination. Proceedings of the ICASSP 2004, May 2004, Vol 1, pp. 517-520
2. Rajesh M.Hegde and Hema A.Murthy: Speaker Identification using the modified group delay feature. Proceedings of The International Conference on Natural Language Processing-ICON 2003,December 2003, pp. 159-167
3. Hema A. Murthy and Venkata Ramana Rao Gadde: The Modified group delay function and its application to phoneme recognition. Proceedings of the ICASSP, April 2003, Vol.I, pp. 68-71
4. Sammon, Jr., J. W.: A Nonlinear Mapping for Data Structure Analysis. IEEE Transactions on Computers **C-18(5)** (1969) 401-409
5. Joshua B. Tenenbaum, Vin de Silva, and John C. Langford: A Global Geometric Framework for Nonlinear Dimensionality Reduction. Science *www.science.org* **290(5500)** (2000) 2319-2323
6. Sam T. Roweis and Lawrence K. Saul: Nonlinear Dimensionality Reduction by Locally Linear Embedding. Science *www.science.org* **290(5500)** (2000) 2323-2326
7. Charles Jankowski, Ashok Kalyanswamy, Sara Basson, and Judith Spitz: NTIMIT: A Phonetically Balanced, Continuous Speech, Telephone Bandwidth Speech Database. Proceedings of ICASSP-90, April 1990.

Two-Stage Duration Model for Indian Languages Using Neural Networks

K. Sreenivasa Rao, S.R. Mahadeva Prasanna, and B. Yegnanarayana

Speech and Vision Laboratory,
Department of Computer Science and Engineering,
Indian Institute of Technology Madras, Chennai-600 036, India
{ksr,prasanna,yegna}@cs.iitm.ernet.in

Abstract. In this paper we propose a two-stage duration model using neural networks for predicting the duration of syllables in Indian languages. The proposed model consists of three feedforward neural networks for predicting the duration of syllable in specific intervals and a syllable classifier, which has to predict the probability that a given syllable falls into an interval. Autoassociative neural network models and support vector machines are explored for syllable classification. Syllable duration prediction and analysis is performed on broadcast news data in Hindi, Telugu and Tamil. The input to the neural network consists of a set of phonological, positional and contextual features extracted from the text. From the studies it is found that about 80% of the syllable durations are predicted within a deviation of 25%. The performance of the duration model is evaluated using objective measures such as mean absolute error (μ), standard deviation (σ) and correlation coefficient (γ).

1 Introduction

Modeling syllable durations by analyzing large databases manually is a tedious process. An efficient way to model syllable durations is by using features of neural networks. Duration models help to improve the quality of Text-to-Speech (TTS) systems. In most of the TTS systems durations of the syllables are estimated using a set of rules derived manually from a limited database.

Mapping a string of phonemes or syllables and the linguistic structures (positional, contextual and phonological information) to the continuous prosodic parameters is a complex nonlinear task [1998v]. This mapping has traditionally been done by a set of sequentially ordered rules derived based on introspective capabilities and expertise of the individual research workers. Moreover, a set of rules cannot describe the nonlinear relations beyond certain point. Neural networks are known for their ability to generalize and capture the functional relationship between the input-output pattern pairs [1999]. Neural networks have the ability to predict, after an appropriate learning phase, even patterns they have never seen before. For predicting the syllable duration, Feedforward Neural Network (FFNN) models are proposed [1990]. The existing neural network based duration models consists of single neural network for predicting the durations of

N.R. Pal et al. (Eds.): ICONIP 2004, LNCS 3316, pp. 1179–1185, 2004.
© Springer-Verlag Berlin Heidelberg 2004

all sound units. With this the sound units around the mean of the distribution will be predicted better, and for other (long and short) sound units prediction will be poor [1998v][1990].

This paper proposes a two-stage model for predicting the syllable duration. The first stage consists of syllable classifier, which classify the syllable into one of the groups based on duration range. The second stage constitutes three neural network models, which are meant for predicting the duration of syllable in the specific intervals. The paper presents the duration analysis of broadcast news data for three Indian languages (Hindi, Telugu and Tamil) using syllables as basic units.

The paper is organized as follows: Section 2 describes the proposed two-stage model and the performance of duration models intended for specific intervals. The first stage in the proposed duration model is a syllable classifier, which is discussed in Section 3. Evaluation of the proposed duration model is presented in section 4. Final section discusses about the issues to be addressed further.

2   Two-Stage Duration Model

The block diagram of the proposed two-stage duration model is shown in Fig. 1(a). The first stage consists of syllable classifier which groups the syllables based on their duration. The second stage is for modeling the syllable duration which consists of specific models for the given duration interval. In the database, most of the syllable durations are varying from 40-300 ms. We have chosen three successive duration intervals (40-100, 100-150 and 150-300 ms) such that they will cover the entire syllable duration range.

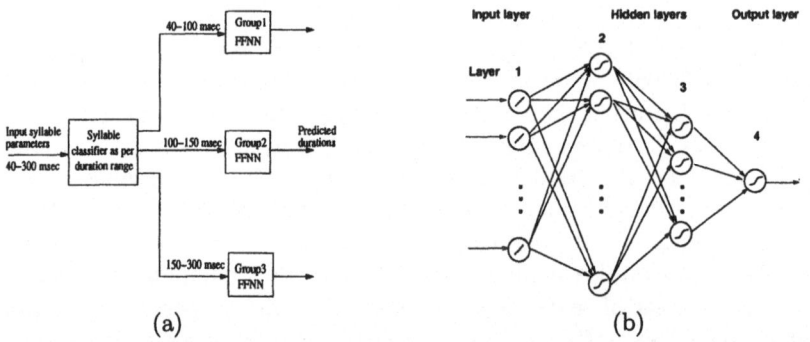

(a)   (b)

Fig. 1. (a) Two-stage duration model (b) Four layer Feedforward neural network.

2.1   Neural Network Structure

For modeling syllable durations, we employed a four layer feedforward neural network whose general structure is shown in Fig. 1(b). The first layer is the input layer which consists of linear elements. The second and third layers are hidden layers, and they can be interpreted as capturing some local and global features in the input space [1999]. The fourth layer is the output layer having one unit representing the syllable duration. For better generalization, several network structures are experimentally verified. The optimum structure arrived is $22L\,44N\,11N\,1N$, where L denotes a linear unit, N denotes a nonlinear unit and the integer value indicates the number of units used in that layer. The nonlinear units use $tanh(s)$ as the activation function, where s is the activation value of that unit. All the input and output parameters were normalized to the range [-1 to +1] before applying to the neural network. The standard backpropagation learning algorithm is used for adjusting the weights of the network to minimize the mean squared error for each syllable duration.

2.2   Speech Database

The database consists of 15 Hindi, 20 Telugu and 25 Tamil news bulletins. In each language these news bulletins are read by male and female speakers. Total durations of speech in Hindi, Telugu and Tamil are around 3.25, 4.5 and 4 hours, respectively. The speech utterances were segmented and labeled manually into syllable-like units. Each bulletin is organized in the form of syllables, words, orthographic text representations of the utterances and timing information in the form of sample numbers. The total database consists of 46222 syllables in Hindi, 81630 syllables in Telugu and 69811 syllables in Tamil.

2.3   Features for Developing Neural Network Model

The features considered for modeling syllable duration are based on positional, contextual and phonological information. The list of features and the number of nodes in a neural network needed to represent the features are given in Table 1.

Table 1. List of the factors affecting the syllable duration, features representing the factors and the number of nodes needed for neural network to represent the features.

Factors	Features	# Nodes
Syllable position in the phrase	1. Position of syllable from beginning of the phrase 2. Position of syllable from end of the phrase 3. Number of syllables in a phrase	3
Syllable position in the word	1. Position of syllable from beginning of the word 2. Position of syllable from end of the word 3. Number of syllables	3
Syllable identity	Segments of syllable	4
Context of syllable	1. Previous syllable 2. Following syllable	4 4
Syllable nucleus	1. Position of the nucleus 2. Number of segments before nucleus 3. Number of segments after nucleus	3
Gender identity	Gender	1

2.4   Performance of the Models in Specific Duration Range

Initially syllables of each of the three languages (Hindi, Telugu and Tamil) are manually classified into 3 groups (40-100, 100-150 and 150-300 ms) based on the duration. For each language three FFNN models are used for predicting the syllable durations in the specific duration intervals. For each syllable the phonological, positional and contextual features are extracted and a 22 dimension input vector is formed. The extracted input vectors are given as input and the corresponding syllable durations are given as output to the FFNN models, and the networks are trained for 500 epochs. The duration models are evaluated with the corresponding syllables in the test set. The deviation of predicted duration from the actual duration is estimated. The number of syllables with various deviations from actual syllable durations are presented in Table 2. In order to objectively evaluate the prediction accuracy, between predicted values and actual duration values, standard deviation of the difference (σ) and linear correlation coefficient (γ) were computed. The standard deviation of the difference between predicted and actual durations is found to be about 13.2 ms and the correlation between predicted and actual durations is found to be 0.91 across the languages in specific duration intervals.

Table 2. Number of syllables having predicted duration within the specified deviation from actual syllable duration for different duration intervals from each of the three languages Hindi, Telugu and Tamil.

Language	Duration range	Training # syls	Testing # syls	# Syllables within deviation			
				< 10%	10-25%	25-50%	> 50%
Hindi	40-100	10000	3057	1611	1155	244	47
	100-150	13000	4112	2462	1641	9	-
	150-300	12000	4053	1989	1875	189	-
Telugu	40-100	19670	5000	1802	2304	692	202
	100-150	24011	6000	3324	2556	120	-
	150-300	20949	6000	2718	2656	622	4
Tamil	40-100	15000	4260	1570	2292	313	85
	100-150	20000	7156	4177	2834	145	-
	150-300	18000	5395	2834	2242	319	-

3   Syllable Classification

In the proposed two-stage duration model, first stage consists of a syllable classifier, which divides the syllables into three groups based on their duration. In this paper Autoassociative Neural Network (AANN) models and Support Vector Machine (SVM) models are explored for syllable classification. The block diagram of syllable classification model is shown in Fig. 2(a).

3.1   AANN Models

Autoassociative neural network models are feedforward neural networks performing an identity mapping of the input space, and are used to capture the

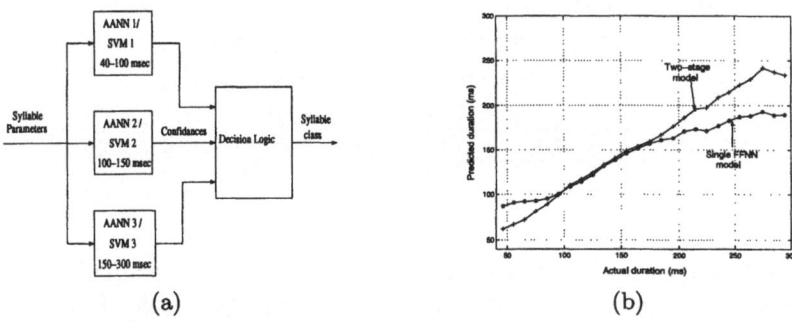

(a)   (b)

Fig. 2. (a) Syllable classification model (b) Prediction performance of two-stage and single FFNN models.

distribution of the input data [1999]. The optimum structures arrived for the study in Hindi, Telugu and Tamil are $22L\ 30N\ 14N\ 30N\ 22L$, $22L\ 30N\ 16N\ 30N$ $22L$ and $22L\ 30N\ 10N\ 30N\ 22L$, respectively. For each language three AANN models are prepared for the duration intervals 40-100, 100-150 and 150-300 ms. For classification task, the syllable parameters are given to each of the model. The output of each model is compared with the input to compute the square error. The error (e) is transformed into a confidence (c) value by using the equation $c = exp(-e)$. The confidence values are given to a decision logic, where the highest confidence value among the models is used for classification. The classification performance of the AANN models are shown in Table 3.

3.2 SVM Models

Support vector machines provide an alternate approach to the pattern classification problems. SVMs are initially designed for two-class pattern classification. Multiclass (n-class) pattern classification problems can be solved using a combination of binary support vector machines. Here we need to classify the syllables into three groups based on duration. An SVM is constructed for each class by discriminating that class against the remaining two classes. The classification system consists of three SVMs. The set of training examples $\{\{(\mathbf{x}_i, k)\}_{i=1}^{N_k}\}_{k=1}^{n}$ consists of N_k number of examples belonging to k^{th} class, where the class label $k \in \{1, 2, \ldots, n\}$. The SVM for the class k is constructed using a set of training examples and their desired outputs, $\{\{(\mathbf{x}_i, y_i)\}_{i=1}^{N_k}\}_{k=1}^{n}$. The desired output y_i for a training example $\mathbf{x}_i$ is defined as follows:

$$y_i = \begin{cases} +1 & : \quad If \quad \mathbf{x}_i \in k \\ -1 & : \quad otherwise \end{cases}$$

The examples with $y_i = +1$ are called positive examples, and those with $y_i = -1$ are called negative examples. An optimal hyperplane is constructed to separate positive examples from negative examples. The separating hyperplane (margin) is chosen in such a way as to maximize its distance from the closest training examples of different classes [1999][1998b]. For a given test pattern $\mathbf{x}$,

Table 3. Classification performance of AANN and SVM models.

Language	% of syllables correctly classified	
	AANN models	SVM models
Hindi	74.68	81.92
Telugu	79.22	80.17
Tamil	76.17	83.26

the evidence is obtained from each of the SVMs, and the maximum evidence is hypothesized as the class of the test pattern. The performance of the classification model using SVMs is shown in Table 3.

4   Evaluation of the Two-Stage Duration Model

For modeling the syllable duration using the proposed two-stage model, syllable parameters are given to all the classification models. Here SVM models are used for syllable classification. The decision logic followed by classification models route the syllable parameters to one of the three FFNN models for predicting the syllable duration. The prediction performance of the two-stage model is presented in Table 4. For comparison purpose, syllable durations are estimated using single FFNN model and its performance is presented in Table 4. Prediction performance of single FFNN and two-stage models for Tamil data is shown in Fig. 2(b). Performance curves in the figure show that short and long duration syllables are better predicted in the case of proposed two-stage duration model. Table 4 and Fig. 2(b) shows that the proposed two-stage model predicts the durations of syllables better compared to single FFNN model.

Table 4. Number of syllables having predicted duration within the specified deviation from actual syllable duration and objective measures for the languages Hindi, Telugu and Tamil using two-stage and single FFNN duration models.

Duration models	Language # Syllables	# Syllables within deviation				Objective measures		
		< 10%	10-25%	25-50%	> 50%	Avg. Err	Std. dev.	Corr.
Two-stage model	Hindi(11222)	4002	4676	2242	302	26.04	20.42	0.81
	Telugu(17000)	6277	6955	2923	842	23.44	23.28	0.82
	Tamil(16811)	7283	6687	2251	590	20.70	21.34	0.85
Single FFNN Model	Hindi(11222)	3312	4012	2875	1023	32.39	25.55	0.74
	Telugu(17000)	4810	5911	4230	2049	28.64	23.92	0.77
	Tamil(16811)	5580	6695	3709	827	25.69	22.56	0.82

5   Conclusions

A two-stage neural network model for predicting the duration of the syllable was proposed in this paper. The performance of the proposed model is shown to be superior compared to the single FFNN model. The performance of the two-stage model may be improved by appropriate syllable classification model and the selection criterion of duration intervals. The performance can be further improved by including the accent and prominence of the syllable in the feature vector. Weighting the constituents of the input feature vectors based on the linguistic and phonetic importance may further improve the performance. The accuracy of labeling, diversity of data in the database, and fine tuning of neural network parameters, all of these may also play a role in improving the performance.

References

[1998v] Vainio M., and Altosaar T.: Modeling the microprosody of pitch and loudness for speech synthesis with neural networks, Proc. Int. Conf. Spoken Language Processing, (Sidney, Australia), Sept. 1998.

[1999]   Haykin S.: Neural Networks:, A Comprehensive Foundation, New Delhi, India: Pearson Education Aisa, Inc., 1999.

[1990]   Campbell W. N.: Analog i/o nets for syllable timing, Speech Communication, vol. 9, pp. 57-61, Feb. 1990.

[1998b] Burges C. J. C.: A tutorial on support vector machines for pattern recognition, Data Mining and Knowledge Discovery, vol. 2, no. 2, pp. 121-167, 1998.

Multichannel Blind Deconvolution of Non-minimum Phase System Using Cascade Structure

Bin Xia and Liqing Zhang*

Department of Computer Science and Engineering,
Shanghai Jiaotong University Shanghai, China
xbin@sjtu.edu.cn

Abstract. Filter decomposition approach has been presented for multichannel blind deconvolution of non-minimum phase systems [12]. In this paper, we present a flexible cascade structure by decomposing the demixing filter into a casual finite impulse response (FIR) filter and an anti-causal scalar FIR filter. Subsequently, we develop the natural gradient algorithms for both filters. Computer simulations show good learning performance of this method.

1 Introduction

Blind deconvolution is to retrieve the independent source signals from sensor outputs by only using the sensor signals and certain knowledge on statistics of the source signals. A number of methods have been developed to deal with the blind deconvolution problem. These methods include the Bussgang algorithms [6, 7], higher order statistical approach (HOS) [3, 5] and the second-order statistics approach (SOS) [9, 10]. When the mixing model is a minimum phase system, we can build a causal demixing system to recover the source signals. Many algorithms will work well for the minimum phase systems. In the real world, the mixing model would be a non-minimum phase system generally. It is a difficult to recover the source signals from measurements which are mixed by a non-minimum phase system.

It is known that a non-minimum phase system can be decomposed into a cascade form of a minimum phase sub-system and a corresponding maximum phase sub-system. Labat et al.[8] presented a cascade structure for single channel blind equalization by decomposed the demixing model. Zhang et al [13] provided a cascade structure to multichannel blind deconvoluiton. Waheed et al [11] discussed several cascade structures for blind deconvolution problem. Zhang et al [12] decomposed a doubly FIR filter into a causal FIR filter and an anti-causal FIR filter. Such a decomposition enables us to simplify the problem of blind deconvolution of non-minimum phase systems. In this paper, we modify the model structure in [12] by decomposing the demixing filter into a causal matrix filter and a scalar anti-causal filter. The two filters in new structure are permutable because the anti-causal filter is scalar and it will be helpful to develop simple learning algorithm. One purpose of decomposition is that we can apply the natural gradient algorithm for training one-sided FIR filters efficiently. The natural gradient, developed by Amari et al [2], is improved learning efficiency in blind separation and

---

* The Project 60375015 supported by National Natural Science Foundation of China.

© Springer-Verlag Berlin Heidelberg 2004

blind deconvolution [1]. Another purpose is to keep the demixing filter stable during training. After introducing the decomposition structure, we can develop the natural gradient algorithms for causal and anti-causal filters independently.

2   Problem Formulation

Consider a convolutive multichannel mixing model, linear time-invariant (LTI) and non-causal systems of form

$$\mathbf{x}(k) = \mathbf{H}(z)\mathbf{s}(z), \qquad (1)$$

where $\mathbf{H}(z) = \sum_{p=-\infty}^{\infty} \mathbf{H}_p z^{-p}$, z is the delay operator, $\mathbf{H}_p$ is a $n \times n$-dimensional matrix of mixing coefficients at time-lag p, which is called the impulse response at time p, $s(k) = [s_1(k), \cdots, s_n(k)]^T$ is an n-dimensional vector of source signals with mutually independent components and $x(k) = [x_1(k), \cdots, x_n(k)]^T$ is the vector of the sensor signals. The objective of multichannel blind deconvolution is to retrieve the source signals using only the sensor signals $x(k)$ and certain knowledge of the source signal distributions and statistics. We introduce a multichannel LTI systems as a demixing model

$$\mathbf{y}(k) = \mathbf{W}(z)\mathbf{x}(k), \qquad (2)$$

where $\mathbf{W}(z) = \sum_{p=-\infty}^{\infty} \mathbf{W}_p z^{-p}$, $\mathbf{y}(k) = [y_1(k), \cdots, y_n(k)]^T$ is an n-dimensional vector of the outputs and $\mathbf{W}_p$ is an $n \times n$-dimensional coefficient matrix at time-lag p.

In blind deconvolution problem, there exist scaling ambiguity and permutation ambiguity because some prior knowledge of source signals are unknown. We can rewrite (2) as

$$\mathbf{y}(k) = \mathbf{W}(z)\mathbf{x}(k) = \mathbf{W}(z)\mathbf{H}(z)\mathbf{s}(k) = \mathbf{P}\mathbf{\Lambda}\mathbf{D}(z)\mathbf{s}(k), \qquad (3)$$

where $\mathbf{P} \in \mathbf{R}^{n \times n}$ is a permutation matrix, $\mathbf{\Lambda} \in \mathbf{R}^{n \times n}$ is a nonsingular diagonal scaling matrix. Then the global transfer function is defined by $\mathbf{G}(z) = \mathbf{W}(z)\mathbf{H}(z)$. The blind deconvolution task is to find a demixing filter $\mathbf{W}(z)$ such that

$$\mathbf{G}(z) = \mathbf{W}(z)\mathbf{H}(z) = \mathbf{P}\mathbf{\Lambda}\mathbf{D}(z), \qquad (4)$$

where $\mathbf{D}(z) = diag\{z^{-d_1}, \ldots, z^{-d_n}\}$.

In order to seek a stable demixing filter, Zhang et al [12] decompose the doubly FIR filter into one causal FIR filter and another anti-causal filter. Based on such a decomposition, we present a new simple cascade form.

3   Filter Decomposition and Learning Algorithm

The main purpose of the filter decomposition is to split one difficult task into several, but easier, subtasks. In order to avoid the errors back propagation which is time consuming [12], we decompose the demixing filter $\mathbf{W}(z)$ into a causal FIR filter and an anti-causal scalar FIR filter. The difference is that the anti-causal matrix filter is replaced by an anti-causal scalar filter. Then the model becomes much simpler where two filters are permutable. Here we stress on the advantage of such decomposition that two sub-filters in the deconvolution model are permutable due to the scalar anti-causal filter. This

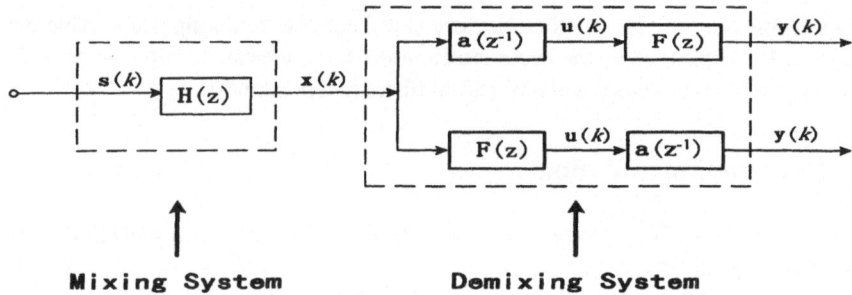

Mixing System   Demixing System

Fig. 1. Illustration of filter decomposition for blind. deconvolution.

property enables us to develop more efficient and simpler algorithms for the demixing model. The decomposition is described as

$$W(z) = a(z^{-1})F(z) \quad \text{or} \quad W(z) = F(z)a(z^{-1}), \tag{5}$$

where $a(z^{-1}) = \sum_{p=0}^{N} a_p z^p$ is a non-causal scalar FIR filter and $F(z) = \sum_{p=0}^{N} F_p z^{-p}$ is a causal FIR filter. The coefficients of three filters satisfy the following relations

$$W_k = \sum_{p-q=k, 0 \le, q \le N} a_p F_q, \text{for} \quad k = -N, \cdots, N. \tag{6}$$

Figure 1 illustrates the permutable cascade form which has two parallel equivalent paths. The permutable property of the model is helpful to derive the efficient algorithm for both sub-filters.

It is obvious that the demixing model $W(z)$ is the pseudoinverse of $H(z)$. For decomposing the demixing model, We directly analyze the structure of the inverse filter $H^{-1}(z)$. The determinant of $H(z)$ can be expressed as

$$det(H(z)) = det(H_0) \prod_{p=1}^{L_1} (1 - b_p z^{-1}) \prod_{p=1}^{L_2} (1 - d_p z^{-1}), \tag{7}$$

where L_1 and L_2 are certain natural numbers, $0 < \| b_p \| < 1$, for $p = 1, \cdots, L_1$ and $\| d_p \| > 1$ for $p = 1, \cdots, L_2$. The b_p, d_p are referred to the zeros of the FIR filter $H(z)$. Using the matrix theory, the inverse of $H(z)$ can be calculated by

$$H^{-1}(z) = c^{-1} z^{L_0 + L_2} \prod_{p=1}^{L_2} (-d_p)^{-1} a(z^{-1}) F(z), \tag{8}$$

where $F(z) = \sum_{r=0}^{\infty} F_r z^{-r} = H^{\sharp}(z) \prod_{p=1}^{L_1} (1 - b_p z^{-1})^{-1}$ is a causal FIR filter and $a(z^{-1}) = \sum_{r=0}^{\infty} a_r z^r = \prod_{p=1}^{L_2} \sum_{q=0}^{\infty} d_p^{-q} z^q$ is an anti-causal FIR filter. It is obviously $\| a_r \|$ and $\| F_r \|$ decay exponentially to zero as r tends to infinity. Hence, the decomposition of demixing filter is reasonable. After decomposing into this cascade form, we can use two one-sided FIR filters to approximate filters $F(z)$ and $a(z^{-1})$, respectively. But the lengthes of $F(z)$ and $a(z^{-1})$ are infinity. In practice, we have to use finite-length filter to approximate them.

$$\mathbf{F}(z) = \sum_{p=0}^{N} \mathbf{F}_p z^{-p} \quad \text{and} \quad \mathbf{a}(z^{-1}) = \sum_{p=0}^{N} \mathbf{a}_p z^p \tag{9}$$

where N is a given positive integer. This approximation will cause a model error in blind decovolution. If we choose a appropriate filter length N, the model error will become negligible.

Natural gradient algorithm, which was developed by Amari [1], is an efficient method for blind signal processing. To introduce the natural gradient for doubly FIR filters, the geometrical structures of FIR filters should be discussed. For further information, the reader is directed to paper [12].

The Kullback-Leibler Divergence has been used as a cost function for blind deconvolution [3, 12] to measure the mutual independence of the output signals. They introduced the following simple cost function for blind deconvolution

$$l(\mathbf{y}, \mathbf{W}(z)) = -\log|det(\mathbf{F}_0)| - \sum_{i=1}^{n} \log p_i(y_i). \tag{10}$$

where the output signals $y_i = \{y_i(k), k = 1, 2 \cdots\}, i = 1, \cdots, n$ as stochastic processes and $p_i(y_i(k))$ is the marginal probability density function of $y_i(k)$ for $i = 1, \cdots, n$ and $k = 1, \cdots, T$. The first term in the cost function is introduced to prevent the matrix $\mathbf{F}_0$ from being singular.

By minimizing (10) and using natural gradient method, we obtain algorithms for both filters.

$$\triangle \mathbf{F}_p = -\eta \sum_{q=0}^{p} (\delta_{0,q} \mathbf{I} - \varphi(\mathbf{y}(k)) \mathbf{y}^T (k - q)) \mathbf{F}_{p-q} \tag{11}$$

$$\triangle \mathbf{a}_p = -\eta \sum_{q=0}^{p} (\varphi^T(\mathbf{y}(k)) \mathbf{y}(k + q)) \mathbf{a}_{p-q} \tag{12}$$

where $\varphi(\mathbf{y}) = (\varphi_1(y_1), \ldots, \varphi_n(y_n))^T$ is the vector of non-linear activation functions, which is defined by $\varphi_i(y_i) = -\frac{p_i'(y_i)}{p_i(y_i)}$. In blind deconvolution problem, activation function φ is unknown. Under the semi-parameter theory, the φ can be regarded as nuisance parameter. It is not necessary to estimate precisely. But, if we choose a better φ, it is helpful for performance of the algorithm. One important factor in determining the activation functions is that the stability conditions of the learning algorithm must be satisfied [4, 12]. The cubic function is good activation function for QAM signals. For further information about activation function selection, see Amari et al. [4].

4 Simulation

In this section, we present computer simulations to illustrate the performance of this algorithm. To show the effectiveness of this algorithm for multichannel blind deconvolution with non-minimum phase systems, we build a non-minimum phase multichannel, which is generated by using state-space method.

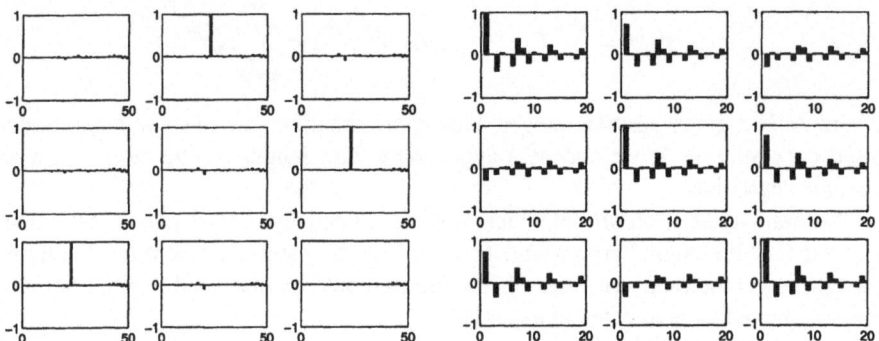

Fig. 2. Coefficients of $\mathbf{G}(z)$ after convergence.   **Fig. 3.** Coefficients of $\mathbf{F}(z)$.

There are two steps during the training in this example. Step 1: we initiate the co-efficients of the $\mathbf{a}(z^{-1})$ and $\mathbf{F}(z)$ by the top line in Figure 1. And then, we can use the output signals $\mathbf{y}$ to train coefficients of the $\mathbf{F}(z)$. Step 2: the coefficients of the $\mathbf{F}(z)$ and $\mathbf{a}(z^{-1})$ are initiated again by the bottom line in Figure 1. Here, the coefficients of $\mathbf{F}(z)$ are trained in step 1. In step 2, we only train coefficients of $\mathbf{a}(z^{-1})$ using $\mathbf{y}$. Figure 2 illustrates the coefficients of the global transfer function $\mathbf{G}(z) = \mathbf{W}(z)\mathbf{H}(z)$ after convergence. Figure 3 and 4 show the coefficients of the causal filter $\mathbf{F}(z)$ and the anti-causal filter $\mathbf{a}(z^{-1})$. It is easy to see that the coefficients of both filters decay as the delay number p increases. Compared with the algorithm in [12], this algorithm avoids error back-propagation and reduces the computational complexity by using a scalar anti-causal FIR filter.

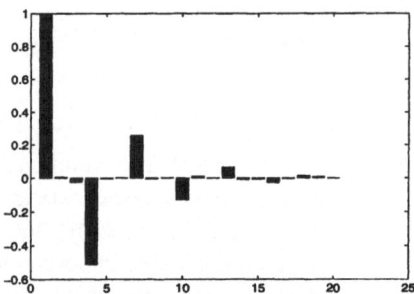

Fig. 4. Coefficients of $\mathbf{a}(z^{-1})$.

5   Conclusion

In this paper we have presented a simple cascade form for multichannel blind deconvo-lution with non-minimum phase system. Under decomposing the demixing anti-causal FIR filter into two sub-demixing FIR filters, we decompose the blind deconvolution problem into some easy sub-tasks. The structure of demixing model is permutable be-cause an anti-causal scalar FIR filter was used. Natural gradient-based algorithms can be easily developed for two one-sided filters. The simulation results show the performance of the proposed decomposition approach is better than conventional algorithms.

References

1. S. Amari. Natural gradient works efficiently in learning. *Neural Computation*, 10(2):251–276, 1998.
2. S. Amari, A. Cichocki, and H. H. Yang. A new learning algorithm for blind signal separation. In G. Tesauro, D. S. Touretzky, and T. K. Leen, editors, *Advances in Neural Information Processing Systems 8(NIPS 95)*, pages 757–763, Cambridge, MA, 1996. The MIT Press.
3. S. Amari, S. Douglas, A. Cichocki, and H. Yang. Novel on-line algorithms for blind deconvolution using natural gradient approach. In *Proc. 11th IFAC Symposium on System Identification, SYSID'97*, pages 1057–1062, Kitakyushu, Japan, July 1997.
4. S. Amari, Tian ping Chen, and A. Cichocki. Stability analysis of learning algorithms for blind source separation. *Neural Networks*, 10(8):1345–1351, 1997.
5. Anthony J. Bell and Terrence J. Sejnowski. An information-maximization approach to blind separation and blind deconvolution. *Neural Computation*, 7(6):1129–1159, 1995.
6. A. Benveniste, M. Goursat, and G. Ruget. Robust identification of a nonminimum phase system: blind adjustment of a linear equalizer in data communication. *IEEE Trans. Automatic Control*, (25):385–399, 1980.
7. D. N. Godard. Self-recovering equalization and carrier tracking in two-dimensional data communication systems. *IEEE Trans. Comm*, (28):1867–1875, 1980.
8. J. Labat, O. Macchi, and C. Laot. Adaptive decision feedback equalization: Can you skip the training period. *IEEE Trans. on communication*, (46):921–930, 1998.
9. L. Tong, G. Xu, and T. Kailath. Blind identification and equalization base on second-order statistics: A time domain approach. *IEEE Trans. Information Theory*, (40):340–349, 1994.
10. J. K. Tugnait and B. Huang. Multistep linear predictors-based blind identification and equalization of multiple-input multiple-output channels. *IEEE Trans. on Signal Processing*, (48):26–28, 2000.
11. K. Waheed and F. M. Salam. Cascaded structures for blind source recovery. In *45th IEEE int'l Midwest Symposium on Circuits and Systems*, volume 3, pages 656–659, Tulsa, Oklahoma, 2002.
12. L. Q. Zhang, S. Amari, and A. Cichocki. Multichannel blind deconvolution of non-minimum phase systems using filter decomposition. *IEEE Trans. Signal Processing*, 2004. In press.
13. L. Q. Zhang, A. Cichocki, and S. Amari. Multichannel blind deconvolution of nonminimum phase systems using information backpropagation. In *Proceedings of ICONIP'99*, pages 210–216, Perth, Australia, Nov.16-20 1999.

A Comparative Study of Feature Extraction Algorithms on ANN Based Speaker Model for Speaker Recognition Applications

Goutam Saha, Pankaj Kumar, and Sandipan Chakroborty

Indian Institute of Technology, Kharagpur, Kharagpur-721302, West Bengal, India
{gsaha,sandipan}@ece.iitkgp.ernet.in
u0ec1005@iitian.iitkgp.ernet.in

Abstract. In this paper we present a comparative study of usefulness of four of the most popular feature extraction algorithm in Artificial Neural Network based Text dependent speaker recognition system. The network uses multi-layered perceptron with backpropagation learning. We show the performance of the network for two phrases with a population of 25 speakers. The result shows normalized Mel Frequency Cepstral Coefficients performing better in false acceptance rate as well as in size of the network for an admissible error rate.

1 Introduction

Automatic Speaker Recognition (ASR) involves recognizing a person from his spoken words[1]-[3]. The goal is to find a unique voice signature to discriminate one person from another. It has found several applications in recent past; primarily to provide biometric feature based security. The recognition process may be text dependent or text independent. In text dependent ASR, speaker is asked to utter a specific string of words both in Enrolment and Recognition phase whereas in Text Independent case systems, ASR recognizes the speaker irrespective of any specific phrase utterance. ASR systems can be open set or closed set. In closed set Recognition systems, the speaker is known a priori to be member of a set of finite speakers. In open system ASR there is also an additional possibility of speaker being an outsider i.e. not from the set of already defined speakers.

The enrolment phase of an ASR system consists of feature extraction module followed by speaker model developed from extracted features. In the verification phase features are extracted similarly and sent to the speaker model to obtain a match score that helps in verifying a speaker from an open set or identifying a speaker from a closed set. The features extracted in feature extraction must exhibit large inter speaker variability and small intra speaker variability for proper recognition. Of several feature extraction algorithms four widely used methods are Linear Predictive Coefficients (LPC)[4], Linear Predictive Cepstral Coefficients(LPCC)[2], Mel Frequency Cepstral Coefficients (MFCC)[2] and Human Factor Cepstral Coefficient (HFCC)[5]. The features extracted by these methods are fed to speaker model for feature based classification of speakers.

N.R. Pal et al. (Eds.): ICONIP 2004, LNCS 3316, pp. 1192–1197, 2004.
© Springer-Verlag Berlin Heidelberg 2004

Several techniques like Hidden Markov Model, Vector Quantization, Multilayer Perceptrons, Radial Basis Functions and Genetic Algorithm have been used for speaker model in ASR before. ASR with 95% accuracy has been achieved using Adaptive Component weight cepstrum in a 10 speaker identification problem[6]. The neural network used here is $10 - 6 - 10$ for vowel phoneme identification using back propagation algorithm. For text dependent case 94% accuracy with MFCC for 14 speaker has been achieved[7] using Recurrent Neural Network architecture $13 - 25 - 14$. Using a gamma neural network[8] an identification system with a 93.6% has been realized for 25 speakers in text dependent case.

In this paper, we investigate the ASR performance using multilayer perceptron which is supposed to be enabled to extract higher order statistics[9] with modified delta rule of learning, unlike conventional delta rule where synaptic weights are updated only on error gradient, modified delta rule uses previous synaptic value as well[5]. This method not only helps in giving extra stability to the system, but also helps in achieving convergence at a faster rate. The work also presents a comparative study of different features like LPC, LPCC, MFCC and HFCC in Neural network based Speaker Model for ASR application in the same framework.

2 Method

2.1 Neural Network Structure

We have developed a Speaker Model based on Neural network. The network uses Multilayer perceptron mechanism with Back Propagation algorithm. MLP has been successfully used to solve complex and diverse classification problems[10],[11]. In our case, the problem is to classify the speech sample feature vectors into several speaker classes. The number of nodes in the input layer equals the feature dimension whereas number of nodes in output layer is same as the number of speakers in the database. The number of nodes in the hidden layer is adjusted empirically for superior performance of the system. The network uses nonlinear hyperbolic tangent activation function.

2.2 Training and Authentication Phases

The whole system can be divided into two phases, Training phase and Authentication phase. In training phase, the feature vectors are fed to the input layer of the network and synaptic weights are adjusted according to Back propagation algorithm. The correction in synaptic is directly proportional to the gradient of total error energy and is given by (1).

$$\Delta\omega_{ji}(n) = -\eta\frac{\delta\varepsilon(n)}{\delta\omega_{ji}(n)} \tag{1}$$

where ω_{ji} is the synaptic weight from i^{th} neuron to j^{th} neuron, η is the learning rate and ε is the total error energy. The index n in the bracket indicates the

n^{th} iteration. In our training phase we have adjusted the synaptic weights by modified delta rule given by (2)

$$\Delta\omega_{ji}(n) = \alpha\Delta\omega_{ji}(n-1) - \eta\frac{\delta\varepsilon(n)}{\delta\omega_{ji}(n)} \tag{2}$$

where α is momentum term. The advantage of this kind of weight update compared to equation 2 is that it allows us to train the network with high learning rate without going into oscillations. Moreover, the learning parameter η is not constant and is adjusted according to the error gradient. If gradient is showing same algebraic sign for consecutive iteration, it implies we are moving towards convergence and hence learning rate should be increased. On the other hand, if we find that gradient sign is changing from one iteration to another, it implies the system is oscillating and learning rate should be decreased. Since we are using supervised learning in training phase, corresponding to each input vector, the node that is the representative of actual speaker is made 1 and rest all are made 0.

Once the training phase is over, authentication of speaker identity is done in Authentication phase. A test feature vector is applied to the input layer and network is simulated to generate outputs at the output node. The one having the maximum value at the output layer is declared as the speaker recognized.

2.3 Normalization of Feature Vector

For faster convergence of the neural network feature vectors should be normalized[12]. We have first the removed the mean values of particular features from each feature vector for all speakers. After that, the obtained features are divided by the maximum magnitude value of the respective features.

Let ρ be the ith feature for speaker j, then the normalized feature corresponding to it is given by (3)

$$\text{Normalized } \rho_{ij} = \frac{\rho_{ij} - \frac{\sum_{i=1}^{N}\rho_{ij}}{N}}{\max|\rho_{ij}|} \tag{3}$$

Thus we have a feature vector set where each value lies between +1 and −1. If normalization were not done, the higher magnitude features resulting in improper learning would bias synaptic weight update.

3 Results

3.1 Database Used

We have used a speech database of 25 speakers, 17 male and 8 female. Since ours is a text dependent ASR, speakers were asked to utter each phrase 10 times. Recording was done in a relatively noise free environment with a sampling frequency of 8KHz. The used phrases are:
(a) A combination lock number "24-32-75"
(b) A codeword "Indian Institute of Technology"

Five utterances of the phrase are used for training phase whereas other five are used for the authentication phase. Note that we used a frame size of 256 samples with 50 percent overlap and window applied is Hamming window function for smoothing spectral distortion at edges. Features are often extracted from each frame. After extraction, the corresponding features are added together and finally these added values are divided by the number of frames to obtain the final feature vector for a utterance.

3.2   Features Used

We use four popular feature extraction techniques mentioned in the introductory part of this paper. In LPC[4] we take 14 coefficients per phrase while in the derived feature LPCC[2] use 20 coefficients are taken into consideration. For both MFCC[2] and HFCC[5] we use 19 of the first 20 coefficents($2-20$) ignoring the first one that represents D.C bias.

3.3   Network Parameters

Table 1 represents the learning parameters for the network with a mean square error(MSE) goal of 0.01. Note that 'learning inc' denotes the scale by which the learning rate should be increased when error gradient is having same algebraic sign for consecutive iterations. 'Learning dec' denotes the factor by which learning rate should be decreased when system is going in oscillations. The number of input nodes as explained in sect. 2.1 are equal to the number of coefficients each feature extraction algorithm produces. Therefore it is 14, 20, 19 and 19 respectively for LPC, LPCC, MFCC and HFCC. We considered two different cases while deciding number of nodes in the hidden layer. In the first case we have twice the number of input nodes in the hidden layer. Table 2 presents the comparison result for this for combination lock phrase as well as the phrase "Indian Institute of Technology". Note that in both the cases number of output nodes is equal to the number of speakers i.e. 25 and total number of samples used for testing $25 \times 5 = 125$. The column training epochs refer to number of iterations to achieve training MSE of 0.01. It is seen that normalized MFCC performing better than others both in terms identification error as well as getting the network trained faster. LPCC and HFCC come next with 3 incorrect identification out of 125. It also shows improvement in both the account or at least in one.

Table 1. Learning Parameters.

Learning rate, η	0.05
Learning inc	1.01
Learning dec	0.9
Momentum term, α	0.9

Table 2. Identification Results for "24-32-75" and "Indian Institute of Technology".

Feature Used	Network Structure	"24 − 32 − 75" Training Epochs	Incorrect Identification	"Indian Institute of Technology" Training Epochs	Incorrect Identification
LPC	14 − 28 − 25	2738	5	2995	15
Normalized LPC	14 − 28 − 25	1263	2	1173	13
LPCC	20 − 40 − 25	1190	2	1454	7
Normalized LPCC	20 − 40 − 25	621	2	656	4
MFCC	19 − 38 − 25	1732	3	2208	6
Normalized MFCC	19 − 38 − 25	869	0	662	5
HFCC	19 − 38 − 25	1013	3	1352	4
Normalized HFCC	19 − 38 − 25	929	3	958	4

In the second case we compare these algorithms in reduction of network structure with a particular admissible accuracy of the system as target to be achieved. Table 3 presents the reduced network structure in each of these cases where the target was more than 95 percent (119 out of 125) of the identification result to be correct. Here, normalized MFCC and HFCC both require 21 nodes in hidden layer while the former require less epochs to get trained. The result for phrase "Indian Institute of Technology" shows HFCC performing better than others and MFCC and LPC coming next. Comparing results of the two phrases we find the combination lock number utterance has more speaker specific information as it gives less error as well as takes less number of epochs to get trained to a specified MSE.

Table 3. Reduced Network Structure for "24-32-75" and "Indian Institute of Technology" with False Acceptance Rate below 5%.

Feature Used	"24 − 32 − 75" Reduced Network Structure	Training Epochs	"Indian Institute of Technology" Reduced Network Structure	Training Epochs
LPC	14 − 28 − 25	2295	14 − 32 − 25	2162
Normalized LPC	14 − 24 − 25	1411	14 − 32 − 25	1651
LPCC	20 − 30 − 25	1950	20 − 42 − 25	1054
Normalized LPCC	20 − 22 − 25	1494	20 − 38 − 25	2004
MFCC	19 − 31 − 25	2344	19 − 36 − 25	2344
Normalized MFCC	19 − 21 − 25	1757	19 − 32 − 25	2850
HFCC	19 − 25 − 25	1829	19 − 32 − 25	1625
Normalized HFCC	19 − 21 − 25	2640	19 − 30 − 25	1153

4   Conclusion

The work gives a comparative assessment of feature extraction algorithms used widely in speaker recognition system on neural network based speaker model. It shows normalized MFCC performing relatively better than others with HFCC coming next. The work also shows a method of comparing suitability of different phrases in text dependent speaker recognition applications. The neural network framework and the comparative scores presented both in terms of error rate and complexity can be used as basis to assess any new development on feature extraction algorithms.

References

1. Furui, S.: An overview of speaker recognition technology. Proceedings of Workshop on Automatic Speaker Recognition, Identification and Verification, Martigny. (1994) 1–9
2. Campbell, J. P.: Speaker Recognition: A Tutorial. Proceedings of the IEEE. **85(9)** (1997) 1437–1462
3. Saha G., Das, M.: On use of Singular Value Ratio Spectrum as Feature Extraction Tool in Speaker Recognition Application, CIT-2003, Bhubaneshwar. (2003) 345–350
4. Atal, B. S.: Effectiveness of linear prediction characteristics of the speech wave for automatic speaker identification and verification. Journal of the Acoustical Society of America. **55** (1974) 1304–1312
5. Skowronski, M. D., Harris, J. G: Human Factor Cepstral Coefficients. Journal of Acoustical Society of America. **112(5)** (2002) 2305
6. Badran ,Ehab: Speaker Recognition based on Artificial Neural Networks Based on vowel phonems. Proceedings of ICSP. (2000)
7. Mueeen,F: Speaker recognition using Artificial Neural Network. Students Conference, ISCON '02. Proceedings. IEEE. **1** (2002) 99-102
8. Wang, C. Xu, D. Principe, J.C.: Speaker verification and identification using gamma neural networks. IEEE International Conference on Neural Networks. (1997)
9. Simon, H.: Neural Networks, A Comprehensive Foundation, Second Edition, chap. 1, (2003), 1–45
10. Archer,N. P., Wang, S.: Fuzzy set representation of neural network classification boundaries. IEEE Transactions on Systems, Man and Cybernetics. **21(4)** (1991) 735–742
11. Sauvola, J., Kauniskangas, H. and Vainamo, K.: Automated document image preprocessing management utilizing grey-scale image analysis and neural network classification. Image Processing and Its Applications, 1997., Sixth International Conference on. **2** (1997)
12. Leuen, Y.: Efficient learning and second order methods, Tutorial at NIPS. Denvar. (1993)

Development of FLANN Based
Multireference Active Noise Controllers
for Nonlinear Acoustic Noise Processes

Debi Prasad Das[1], Ganapati Panda[2], and Sanghamitra Sabat[3]

[1] Silicon Institute of Technology, Bhubaneswar, Orissa, India
[2] National Institute of Technology, Rourkela, Orissa, India
[3] C.V. Raman College of Engineering, Bhubaneswar, Orissa, India

Abstract. In this paper attempts have been made to design and develop multireference active noise control structures using functional link artificial neural network. Two different structures for multiple reference controllers are presented. Computer simulation justifies the effectiveness of the algorithm.

1 Introduction

Active noise control (ANC) [1]-[5] has gained a lot of research interest due to increasing awareness of industrial noise pollution and emerging stricter standards. Although ANC is based upon the simple principle of destructive interference, it has now become enriched with a lot of advanced digital signal processing (DSP) and soft computing techniques. The schematic diagram of a simple feedforward control system for a long, narrow duct is illustrated in Fig. 1.

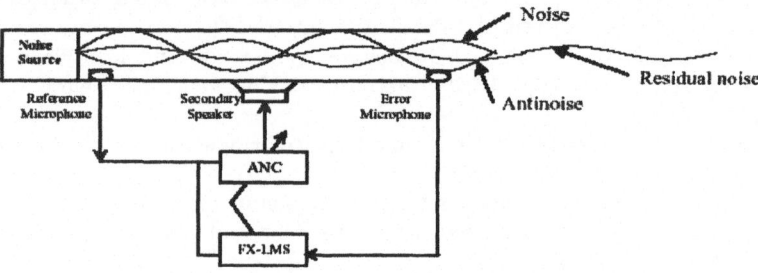

Fig. 1. Schematic diagram of a simple ANC system.

A reference signal is sensed by an input microphone close to the noise source before it passes the canceling loudspeaker. The noise canceller uses the reference input signal to generate a signal of equal amplitude but 180° out of phase. This antinoise signal is used to drive the loudspeaker to produce a canceling sound that attenuates the primary acoustic noise in the duct. The error microphone measures the residual error signal, which is used to adapt the filter coefficients to minimize this error.

In applications concerning the control of various practical acoustic noises, multiple reference controllers have gained an increased interest, especially in applications such as industry, boats and aircraft. Multiple reference signals may be obtained from multiple sources or for different part of the same source but with separated frequency

N.R. Pal et al. (Eds.): ICONIP 2004, LNCS 3316, pp. 1198–1203, 2004.
© Springer-Verlag Berlin Heidelberg 2004

contents. Studies have been carried out [4] on the linear multireference ANC which show satisfactory performance for linear noise processes. But there are several situations [2], [3], [5] where the ANC needs nonlinear control rather than conventional linear control using FXLMS algorithm. In [2], [3], [5] single channel nonlinear ANC systems using multi-layer neural networks are proposed. In this paper two novel Functional Link Artificial Neural Network (FLANN) based multireference ANC structures are developed.

2 Functional Link Artificial Neural Network (FLANN)

The functional link artificial neural network (FLANN) [5] is a useful alternative to the multilayer artificial neural network (MLANN). It has the advantage of involving less computational complexity and a simple structure for hardware implementation.

The conventional MLANN involves linear links. An alternative approach is also possible, whereby the functional link concept is used, which acts on an element of a pattern or the entire pattern itself and generates a set of linearly independent functions. By this process, no new ad hoc information is inserted into the process but the representation gets enhanced. In the functional link model an element x_k, $1 \le k \le N$ is expanded to $f_l(x_k), 1 \le l \le M$. The representative examples of functional expansion are power series expansion, trigonometry expansion, and tensor or outer product [5].

Let us consider the problem of learning with a flat net, which is a net with no hidden layers. Let $\mathbf{X}$ be the Q input patterns each with N elements. Let the net configuration have one output. For the qth pattern, the input components are $x_i^{(q)}$, $1 \le i \le N$ and the corresponding output is $y^{(q)}$. The connecting weights are w_i, $1 \le i \le N$ and the threshold is denoted by α. Thus $y^{(q)}$ is given by

$$y^q = \sum_{i=1}^{N} x_i^q w_i + \alpha, \qquad\qquad q = 1, 2, ..., Q. \qquad\qquad (1)$$

or in matrix form,

$$\mathbf{y} = \mathbf{X}\mathbf{w}. \qquad\qquad (2)$$

The dimension of $\mathbf{X}$ is $Q\times(N+1)$.

If $Q = (N+1)$ and $\mathrm{Det}(\mathbf{X}) \ne 0$, then

$$\mathbf{w} = \mathbf{X}^{-1}\mathbf{y} \qquad\qquad (3)$$

Thus, finding the weights for a flat net consists of solving a system of simultaneous linear equations for the weights $\mathbf{w}$.

If $Q < (N+1)$, $\mathbf{X}$ can be partitioned into a functional matrix $\mathbf{X}_F$ of dimension $Q\times Q$. By setting $w_{Q+1} = w_{Q+2} = ... = w_N = \alpha = 0$, and since $\mathrm{Det}(\mathbf{X}_F) \ne 0$, $\mathbf{w}$ may be expressed as

$$\mathbf{w} = \mathbf{X}_F^{-1}\mathbf{y} \qquad\qquad (4)$$

Equation (4) yields only one solution. But if the matrix $\mathbf{X}$ is not partitioned explicitly, then (3) yields a large number of solutions, all of which satisfy the given constraints.

But if $Q > (N+1)$, then we have

$$\mathbf{X}\,\mathbf{w} = \mathbf{y} \tag{5}$$

where $\mathbf{X}$, $\mathbf{w}$, $\mathbf{y}$ are of dimensions $Q \times (N+1)$, $(N+1) \times 1$ and $Q \times 1$, respectively.

By the functional expansion scheme, a column of $\mathbf{X}$ is enhanced from $(N+1)$ elements to M, producing a matrix $\mathbf{S}$ so that $M \geq Q$. Under this circumstance, we have

$$\mathbf{S}\,\mathbf{w}_F = \mathbf{y} \tag{6}$$

where $\mathbf{S}$, $\mathbf{w}_F$, $\mathbf{y}$ are of dimensions $Q \times M$, $M \times 1$, $Q \times 1$, respectively. If $M = Q$ and Det $(\mathbf{S}) \neq 0$, then

$$\mathbf{w}_F = \mathbf{S}^{-1}\mathbf{y} \tag{7}$$

Equation (7) is an exact flat net solution. However, if $M > Q$, the solution is similar to that of (4). This analysis indicates that the functional expansion model always yields a flat net solution if a sufficient number of additional functions is used in the expansion.

Figure 2 represents a simple FLANN structure, which is essentially a flat net with no hidden layer. In FLANN, N inputs are fed to the functional expansion block to generate M functionally expanded signals which are linearly combined with the M-element weight vector to generate a single output. In this paper, the trigonometric functional expansion is chosen for the FLANN structure because of the below-mentioned reason.

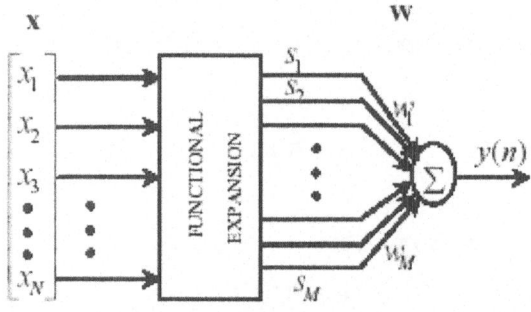

Fig. 2. The structure of a FLANN.

Basis of Using Trigonometric Functional Expansion

Of all the polynomials of Pth order with respect to an orthogonal system, the best approximation in the metric space is given by the Pth partial sum of its Fourier series with respect to the system. Therefore, the trigonometric polynomial basis functions given by $\mathbf{S} = \{x, \sin(\pi x), \cos(\pi x), \sin(2\pi x), \cos(2\pi x), ..., \sin(P\pi x), \cos(P\pi x)\}$ provide a compact representation of the function in the mean square sense.

3 Multireference Nonlinear ANC Structures

Two different control structures are developed for linear control of multireference ANC system.

3.1 Structure 1: A Single Controller (SC)

The two reference signals are added together to form a single reference signal, which is fed to the ANC. The block diagram of the same is shown in Fig. 3. The mathematical analysis and the weight update equations for both are presented as follows. If $d(n)$ is the disturbance signal in Fig. 2, the error signal $e(n)$ is

$$e(n) = d(n) + y(n) = d(n) + S^T(n)W(n) \tag{8}$$

where S is functionally expanded vector containing finite histories of the sum of the two reference signals $x_1(n)$ and $x_2(n)$. The filter weight vector W has a time index n, indicating that their vector is also updated as a per sample basis. The updating algorithm for this structure would be

$$W(n+1) = W(n) - 2\mu e(n)S(n) \tag{9}$$

3.2 Structure 2: Individual Controller (IC)

In the structure shown in Fig. 4, each reference signal is filtered with a separate controller. The control error is calculated as

$$e(n) = d(n) + y(n) = d(n) + S_1^T(n)W_1(n) + S_2^T(n)W_2(n) \tag{10}$$

where W_1 and W_2 are the adaptive filter weight vectors and S_1 and S_2 are the functionally expanded vector of $x_1(n)$ and $x_2(n)$ respectively. This is similar to equation (8), except that now each input reference has an individual weight vector. The filter weights can be updated using a standard (single reference) LMS algorithm. Although the filters are updated using individual algorithms, the weight update schemes are not completely isolated, since they use the same control error. When controlling signals with only a small frequency separation, the individual controller tends to have better convergence performance with fewer filter weights than the other two structures discussed above. One clear disadvantage with this configuration is that the implementation "cost" is effectively doubled. This is particularly important in multi-channel filtered-x implementations, where the filtered-x calculations are very time-consuming.

4 Computer Simulations

To investigate the effectiveness of the two multireference nonlinear control structure computer simulations were carried out. Four different types of experiments are carried out.

Experiment-I: In this experiment the primary the example of nonlinear primary path of [5] is used in the simulation. $d(n) = t(n-2) + 0.08t^2(n-2) - 0.04t^3(n-1)$ where

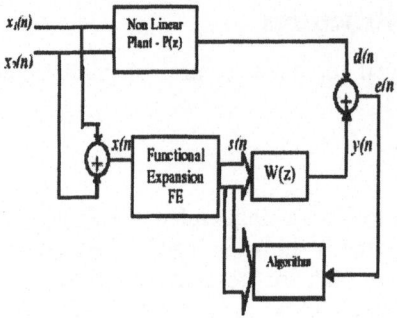
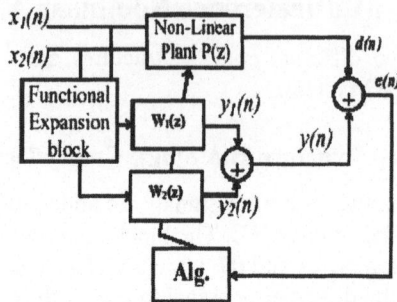

Fig. 3. Structure -1. A Single Control with FLANN used to control the nonlinear noise processes.

Fig. 4. Structure -2. An Individual Control with FLANN used to control the nonlinear noise processes.

$t(n) = x(n) * f(n)$, * denotes convolution and $F(z) = z^{-3} - 0.3z^{-4} + 0.2z^{-5}$. The reference signal $x_1(n)$ and $x_2(n)$ are sinusoidal wave of 100 Hz and 200 Hz, sampled at the rate of 1000 samples/s. The structure-1 is used for the noise cancellation. Fig. 5 shows the convergence characteristics of both LMS and FLANN based controller. From the Fig. 5 it is clearly seen that the FLANN based proposed controller outperforms the conventional LMS algorithm.

Experiment-II: In this experiment the frequency of the reference signal are made 100 Hz and 104 Hz respectively. Fig. 6 shows the convergence characteristics of both LMS and FLANN based single controller. From the Fig. 6 it is clearly seen that neither FLANN based single controller nor the LMS based single controller works well under this situation.

Experiment-III: In this experiment the frequency of the reference signals are taken as 100 Hz and 200 Hz respectively. But the controller is made as individual control. Fig. 7 shows the convergence characteristics of both LMS and FLANN based individual controller. It is clearly seen that the FLANN based proposed controller outperforms the conventional LMS algorithm.

Experiment-IV: In this experiment the frequency of the reference signal are made 100 Hz and 104 Hz respectively. But the controller is selected as individual controller as shown in Fig. 4. Fig. 8 shows the convergence characteristics of both LMS and FLANN based single controller. It is seen that the FLANN based proposed individual controller shows better performance than the conventional LMS algorithm.

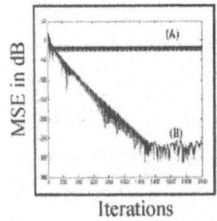
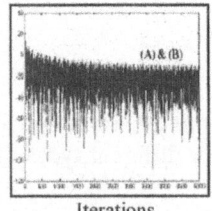
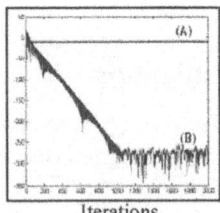
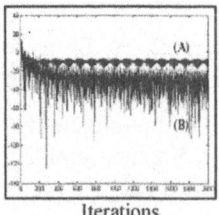

Fig. 5. Experiment I (A) LMS (B) FLANN.

Fig. 6. Experiment II (A) LMS (B) FLANN.

Fig. 7. Experiment III (A) LMS (B) FLANN.

Fig. 8. Experiment IV (A) LMS (B) FLANN.

5 Conclusions

This paper suggests the multireference active noise controller using functional link artificial neural network for nonlinear noise processes. Two different structures for multiple reference controllers are presented. The advantages and disadvantages of different structures are presented in terms of convergence characteristics for two different cases: 1) where the reference signals are pure tones with frequencies 100 Hz & 104 Hz., 2) where the frequencies are 100 Hz & 200 Hz. Exhaustive computer simulation study justifies the effectiveness of the algorithms.

References

1. Kuo S. M. and Morgan D. R.: Active Noise Control Systems – Algorithms and DSP Implementations. New York: Wiley (1996)
2. Strauch P., Mulgrew B.: Active Control of Nonlinear Noise Processes in A Linear Duct in IEEE Transactions on Signal Processing, vol. 46. No. 9, September (1998) 2404-2412
3. Bouchard M., Paillard B. and Dinh C. T. L.: Improved Training of Neural Networks for the Nonlinear Active Noise Control of Sound and Vibration in IEEE Transactions on Neural Networks, vol. 10. No. 2. March (1999) 391-401
4. Sjosten P., Johnsson S., Claesson I. and Lago T. L.: Multireference controllers for active control of noise and vibration in Proceedings of ISMA 21, vol. 1, Leuven, September (1996) 295-303
5. Das D. P. and Panda G.: Active Mitigation of Nonlinear Noise Processes Using Filtered-s LMS Algorithm in IEEE Transactions on speech and audio processing, vol 12 No. 3, May (2004) 313-322

Phase Space Parameters
for Neural Network Based Vowel Recognition

P. Prajith, N.S. Sreekanth, and N.K. Narayanan

School of Information Science & Technology
Kannur University, Thalassery campus, Palayad 670 661
csirc@rediffmail.com

Abstract. This paper presents the implementation of a neural network with error back propagation algorithm for the speech recognition application with Phase Space Point Distribution as the input parameter. By utilizing nonlinear or chaotic signal processing techniques to extract time domain based phase space features, a method is suggested for speech recognition. Two sets of experiments are presented in this paper. In the first, exploiting the theoretical results derived in nonlinear dynamics, a processing space called phase space is generated and a recognition parameter called Phase Space Point Distribution (PSPD) is extracted. In the second experiment Phase Space Map at a phase angle $\pi/2$ is reconstructed and PSPD is calculated. The output of a neural network with error back propagation algorithm demonstrate that phase space features contain substantial discriminatory power.

1 Introduction

Conventional speech signal processing techniques are predicted on linear systems theory, where the fundamental processing space is the frequency domain. Traditional acoustic approaches assume a source-filter model where the vocal tract is modeled as a linear filter. Although the features based on these approaches have demonstrated excellent performance over the years, they are, nevertheless, rooted in the strong linearity assumptions of the underlying physics. Current systems are far inferior to humans, and there are many factors that severely degrade recognition performance. As an alternative to the traditional techniques, interest has emerged in studying speech as a nonlinear system. State of the art speech recognition systems typically use Cepstral coefficient features, obtained via a frame-based spectral analysis of the speech signal [1]. However, recent work in Phase Space Reconstruction Techniques [2] for nonlinear modeling of time-series signals has motivated investigations into the efficacy of using dynamical systems models in the time-domain for speech recognition. In theory, reconstructed Phase Spaces capture the full dynamics of the underlying system, including nonlinear information not preserved by traditional spectral techniques, leading to possibilities for improved recognition accuracy.

The classical technique for phoneme classification is Hidden Markov Models (HMM), often based on Gaussian Mixture Model (GMM) observation probabilities. The most common features are Mel Frequency Cepstral Coefficients (MFCCs).

In contrast, the reconstructed Phase Space is a plot of the time-lagged vectors of a signal. Such Phase Spaces have been shown to be topologically equivalent to the original system, if the embedding dimension is large enough [3], [5]. Structural patterns occur in this processing space, commonly referred to as trajectories or attractors,

N.R. Pal et al. (Eds.): ICONIP 2004, LNCS 3316, pp. 1204–1209, 2004.
© Springer-Verlag Berlin Heidelberg 2004

which can be quantified through invariant metrics such as correlation dimension or Lyapunov exponents or through direct models of the Phase Space distribution [4-8].

Phase Space reconstructions are not specific to any particular production model of the underlying system, assuming only that the dimension of the system is finite. We would like to be able to take advantage of our knowledge about speech production mechanisms to improve usefulness of Phase Space models for speech recognition in particular. The present work is an attempt to recognize the five fundamental vowel units of Malayalam ഇ /i/, ഉ /I/, ഉ /u/, എ /ae/, and ഒ /o/.

2 Parameter Extraction

From a dynamical system analysis point of view a large stream of data is wholly un-suitable, since the dynamics themselves are undergoing continual change and must therefore be non-stationary. A better analysis can be achieved by focusing the analysis onto individual phonemes, single unambiguous sounds that form the building blocks of any language. Vowels play key role in articulated sound language, which is com-monly known as speech. They represent the steady continuant part of the quasi-periodic speech wave and are the most stable elementary speech units.

In the present study isolated Malayalam vowels uttered by a single male speaker is used. The data was collected from the speaker at different occasions. Speech signals are low pass filtered at 4 kHz to remove high frequency components, which are more depended on speaker than on vowel quality. A typical reconstructed phase space map for a Malayalam vowel is shown in Fig.(1). From this map Phase Space Point Distri-bution parameter is extracted as explained below.

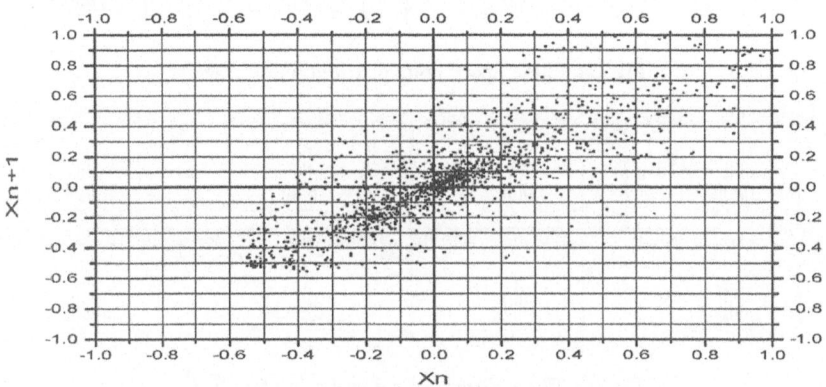

Fig. 1. Phase space map (Vowel ഇ /i/)

The phase space map is generated by plotting X(n) versus X(n+1) of a normalized speech data sequence of a vowel speech segment. The phase space map is divided into grids with 20 x 20 boxes. The box defined by co-ordinates (-1, .9), (-.9, 1) is taken as location 1. Box just right side to it is taken as location 2 and it is extended towards X direction, with the last box in the row as (.9, .9),(1, 1). This is repeated for all rows. Number of points in each location is estimated. A typical plot is given in Fig.2. This operation is repeated for the same vowel uttered at different occasions.

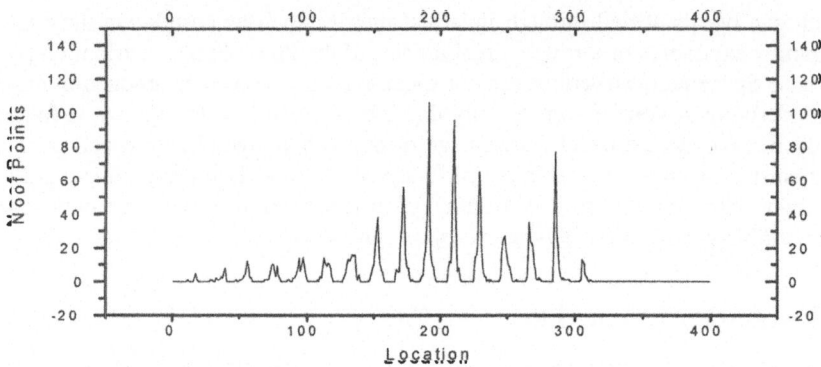

Fig. 2. Phase space point distribution of (▬/Λ/)

Fig.3a - 3e shows the phase space point distribution graph for each vowel uttered at different occasions. The graph thus plotted for different vowels shows the identity for a vowel as regard to pattern. Therefore this technique can be effectively utilized for speech recognition applications.

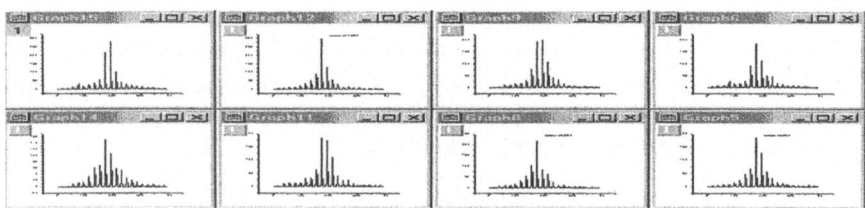

Fig. 3a. Phase Space point distribution for /▬/Λ/

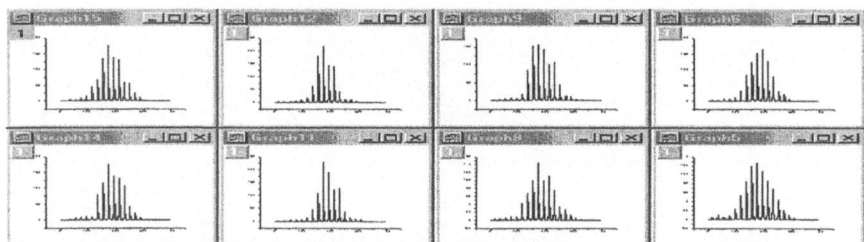

Fig. 3b. Phase Space point distribution for ☺ /I/

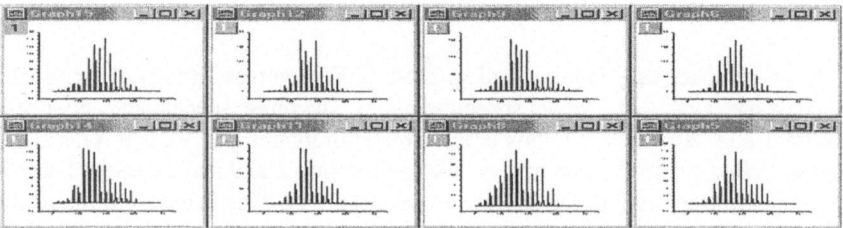

Fig. 3c. Phase Space point distribution for ☺ /u/

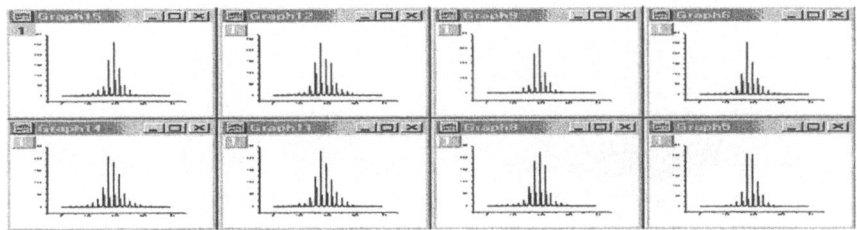

Fig. 3d. Phase Space point distribution for ⌐)/ æ /

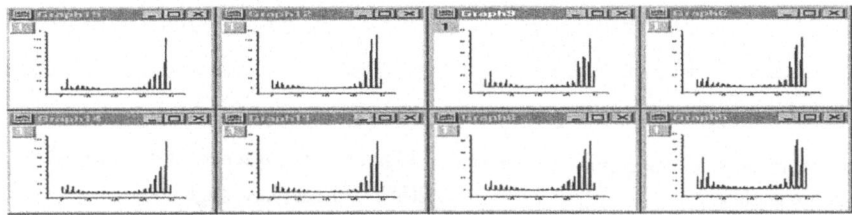

Fig. 3e. Phase Space point distribution for ଓ/O/

These Phase Space Point Distribution parameters are used as input pattern for training the single layer neural network as explained later.

3   PSPD from the Phase Space Map at a Phase Angle π/2

In Xn vs Xn+1 plot the time delay introduced is 1, for a periodic complex signal the displacements at two points with a phase difference of 2π would have same values. This implies that in the Phase Space diagram the points representing such pairs would be lying on a straight line with a slope of $\pi/4$ to the axes. It may be seen that as the phase lag decreases the points are scattered over a broad region. The scattered region reaches a maximum width for a phase difference of $\pi/2$. It collapses into a straight line when the phase difference becomes 2π [9]. Fig.4a shows the Phase Space Map of vowel ⌐/Λ/ at a phase angle $\pi/2$(corresponds to a time lag of T/4, where T is the pitch period) and Fig.4 b shows the corresponding Phase Space Point Distribution.

4   Experimental Results

Neural networks are simulated in C++. Each layer represents a class and the network is simulated as an array of layer objects. Input vector, output vector, weight matrix and error signal vector are the class members. In this present work, the input PSPD parameter is an array of dimension 400 (20 X 20). Since we are focusing a speech recognition system we have to reduce the size of the input vector. For this purpose we take the average distribution of each 20 row. Resultant vector is a single dimensional array of size 20.

Malayalam language is rich with 56 phonemes including short vowels, long vowels, diphthongs and consonants. Vowel recognition is the major concern in designing any practical, efficient speech recognition system Vowels are produced by exciting a

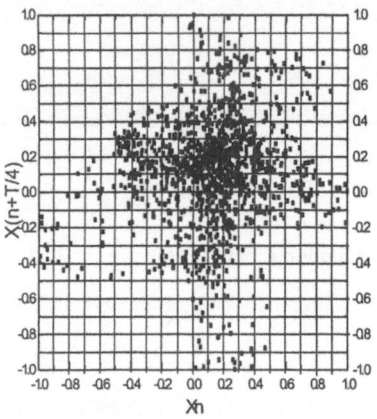 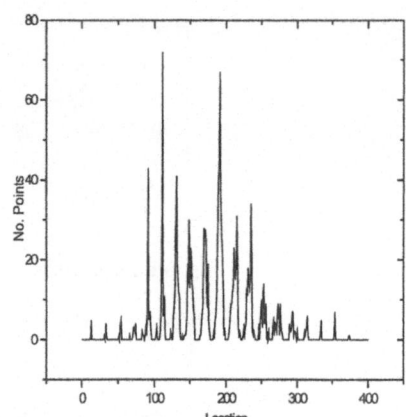

Fig. 4a. Phase space map of (Vowel ⬤/Λ/)   **Fig. 4b**. Phase space point distribution
at phase angle π/2   (⬤/Λ/)   at phase angle π/2

fixed vocal tract with quasi-periodic pulses of air caused by vibration of the vocal cords. The 5 vowels ⬤/Λ/, ⬤/I/, ⬤ /u/, ⬤/ae/, and ⬤/o/. are taken for the recognition experiment.

Multilayer neural network using Error Back Propagation Training algorithm is trained and tested with database file. Recognition accuracies for five Malayalam vowel units are shown bellow in Table 1. When the Phase Space Point Distribution from the Phase Space Map at a phase angle π/2 is taken as the input parameter, the recognition accuracy is shown in the Table 2. It is clear that in the second case recognition accuracy increases considerably.

From the above recognition analysis of Malayalam vowel units, we conclude that phase space distribution is a reliable parameter for speech recognition application using Artificial Neural Network.

Table 1.

Vowel unit	Relative Recognition Accuracy
⬤ /Λ/	95%
⬤/I/	73.33 %
⬤ / ae /	65 %
⬤ / o /	55 %
⬤ / u /	70 %
Average	71.6 %

Table 2.

Vowel unit	Relative Recognition Accuracy
⬤ / Λ /	100 %
⬤/I/	92%
⬤ / ae /	68%
⬤ / o /	76%
⬤ / u /	72 %
Average	81.6 %

5 Conclusion

The Phase Space Point Distribution analysis is an attractive research avenue for increasing speech recognition accuracy. The method has a strong theoretical justification provided by the nonlinear dynamics literature, and represents a fundamental philosophical shift from the frequency domain to the time domain. We are presenting an entirely different way of viewing the speech recognition problem, and offering an opportunity to capture the nonlinear characteristics of the acoustic structure. The experimental results using neural network presented here affirm the discriminatory strength of this approach and future work will determine its overall feasibility and long-term success for both isolated and continuous speech recognition applications.

References

1. Dellar, J.R., Hansen, J.H.L. and Proakis, J.G., "Discrete- Time processing of Speech Signals.", IEEE Press, New York, 908 pp, 2000.
2. Abarbanel, H.D.I., "Analysis of observed chaotic data." Springer, New York, xiv, 272 pp, 1996.
3. Sauer, T., Yorke, J.A and Casdagli, M., "Embedology." Journal of Statistical Physics, 65(3/4): 576-616, 1991.
4. N.K.Narayanan and C.S Sridhar," Parametric Representation of Dynamical Instabilities and Deterministic Chaos in Speech" *Proceedings Symposium on Signals, Systems and Sonars*; NPOL, Cochin,pp.B4.3/1, 1988
5. M.T.Johnson, A.C.Lindgren, R.J.Povinelli, and X. Yuan, "Performance of nonlinear speech enhancements using Phase Space Reconstruction", presented at IEEE International conference on Acoustics, Speech and Signal Processing, Hong Kong, China. 2003
6. N.K.Narayanan, PhD thesis, CUSAT, 1990.
7. Michael Banbrook, Stephan McLaughlin and Iain Mann, "Speech Characterization and Synthesis by Nonlinear Methods", IEEE Transactions on Speech and Audio Processing, vol.7, No. 1, January 1999
8. Takens, F., "Detecting strange attractors in turbulence", Dynamical Systems and Turbulence, 898: 366-381, 1980.
9. Asoke Kumar Datta "A time domain approach to on-line Re-Synthesis of Continuous Speech", JASI, Vol.XXX, pp 129-134, 2002

Speaker Segmentation Based on Subsegmental Features and Neural Network Models

N. Dhananjaya, S. Guruprasad, and B. Yegnanarayana

Speech and Vision Laboratory
Department of Computer Science and Engineering
Indian Institute of Technology Madras, Chennai-600 036, India
{dhanu,guru,yegna}@cs.iitm.ernet.in

Abstract. In this paper, we propose an alternate approach for detecting speaker changes in a multispeaker speech signal. Current approaches for speaker segmentation employ features based on characteristics of the vocal tract system and they rely on the dissimilarity between the distributions of two sets of feature vectors. This statistical approach to a point phenomenon (speaker change) fails when the given conversation involves short speaker turns (< 5 s duration). The excitation source signal plays an important role in characterizing a speaker's voice. We use autoassociative neural network (AANN) models to capture the characteristics of the excitation source that are present in the linear prediction (LP) residual of speech signal. The AANN models are then used to detect the speaker changes. Results show that excitation source features provide better evidence for speaker segmentation as compared to vocal tract features.

1 Introduction

Given a multispeaker speech signal, the objective of speaker segmentation is to locate the instants at which a speaker change occurs. Speaker segmentation is an important preprocessing task for applications like speech recognition, audio indexing and 2-speaker detection. Human beings perceive speaker characteristics at different (signal) levels, which, based on the duration of analysis, can be grouped into segmental (10-50 ms), subsegmental (1-5 ms) and suprasegmental (> 100 ms) features. Most of the current methods for speaker segmentation use the distribution of short-time (segmental) spectral features relating to the vocal tract system, estimated over five or more seconds of speech data, to detect speaker changes. However, these methods cannot resolve speaker changes over shorter durations of data (< 5 s), owing to their dependence on the statistical distribution of the spectral features.

The objective of this study is to explore features present in the source of excitation, to the vocal tract system, for speaker segmentation. In section 2, we give a review of the current approaches to speaker segmentation and bring out their limitations in detecting speaker changes due to short (< 5 s) speaker turns. Section 3 describes the use of autoassociative neural network (AANN) models in

N.R. Pal et al. (Eds.): ICONIP 2004, LNCS 3316, pp. 1210–1215, 2004.
© Springer-Verlag Berlin Heidelberg 2004

characterizing a speaker from the subsegmental features present in the excitation source signal. In section 4 we propose a speaker segmentation algorithm using excitation source features. The performance of the proposed method in speaker segmentation is discussed in section 5. Section 6 summarizes the work and lists a few issues still to be addressed.

2   Need for Alternate Approaches to Speaker Segmentation

Current methods for speaker segmentation use features representing the vocal tract system of a speaker. Two adjacent regions of speech are compared for dissimilarity in the statistical distributions of the feature vectors. Mel-frequency cepstral coefficients (MFCC) or linear prediction cepstral coefficients (LPCC) are used as feature vectors. Some widely used dissimilarity measures include the delta-Bayesian information criterion (dBIC) [1] [2] and Kullback-Leibler distance [2]. In [3], generalized likelihood ratio is used as distance measure to separate out a dominant speaker from other speakers in an air traffic control application. In [2], a multipass algorithm for detecting speaker changes is presented, which uses various window sizes and different dissimilarity measures over different passes. In all these studies, large (> 5 s) speaker turns are hypothesized, while the short turns do not receive attention owing to the application under consideration.

To illustrate the inadequacy of spectral features for speaker change detection, the performance of BIC approach is studied on two types of 2-speaker data, one with long speaker turns and the other with short speaker turns, and is shown in Fig. 1 and Fig. 2, respectively. 19-dimensional weighted LPCCs, obtained from a 12th order LP analysis, are used as feature vectors, and dBIC is used as the dissimilarity measure. It is seen from Fig. 1 that the evidence for speaker change reduces drastically as the window size is reduced, while Fig. 2 illustrates the inability of BIC method in picking the speaker changes with short speaker turns.

3   Speaker Characterization Using Subsegmental Features

Linear prediction (LP) analysis of speech signal gives a reasonable separation of the vocal tract information (LP coefficients) and the excitation source information (LP residual) [4, 5]. If the LP residual of a voiced segment of speech is replaced by a train of impulses separated by one pitch period, and speech is synthesized using the same LP coefficients, it is observed that many of the speaker characteristics are lost. Thus, it is hypothesized that the voiced excitation has significant speaker-specific characteristics. An autoassociative neural network (AANN) model can be used to capture the higher order relations among the samples of the LP residual signal [6]. Blocks of samples of the LP residual (derived over voiced regions) are presented as input to the AANN model. These blocks are presented in a sequence, with a shift of one sample. The blocks are

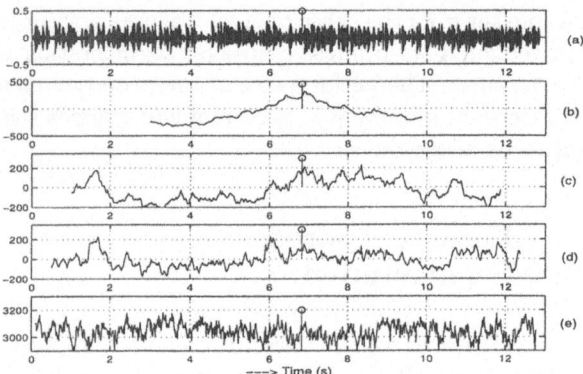

Fig. 1. Case 1: Long (> 5 s) speaker turns. Deteriorating evidence with reducing window sizes. (a) 2-speaker speech signal. (b) to (e) dBIC plots for window sizes of 3 s, 1 s, 0.5 s and 0.1 s respectively. True speaker change is marked by a vertical pole.

typically less than a pitch period in size (subsegmental) and are normalized to unit magnitude before presenting to the AANN model. Once an AANN model is trained with the samples of the LP residual, blocks of samples from a test signal can be presented in a manner similar to the training data. The error between the actual and desired output is obtained, and is converted to a confidence score using the relation, $c = exp(-error)$. The AANN model gives a high confidence scores if the test signal is from the same speaker.

4   Proposed Method for Speaker Segmentation

The algorithm for speaker change detection has two phases, a model building phase and a change detection phase.

Model Building Phase:
An AANN model is trained from approximately 2 sec of contiguous voiced speech which is hypothesized to contain only one speaker. In a casual conversational speech it is not guaranteed that a single random pick of 2 sec data contains only one speaker. In order to circumvent this problem, M (about 10) models are built from M adjacent speech segments of 2 sec, with an overlap of 1 sec. The possibility of at least two pure segments (of a single speaker) is thereby increased. The entire conversation is tested through each of the models to obtain M confidence plots. The cross-correlation coefficients between all possible pairs of confidence plots are computed. N (2 or 4) out of M models are picked which give high correlation coefficient value with each other. The entire process of model building and selection is depicted in Figure 3.

Change Detection Phase:
This phase involves combining evidence from the chosen N confidence plots after model selection. An absolute difference $\triangle\mu$, of average confidence scores from two adjacent window segments (500 ms) is computed to obtain the $\triangle\mu$ plot by

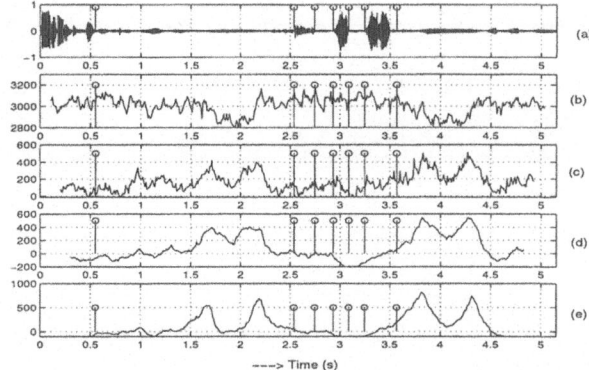

Fig. 2. Case 2: Short (< 5 s) speaker turns. Illustrating lack of evidence for speaker change detection. (a) 2-speaker speech signal. (b) to (e) dBIC plots for window sizes of 0.1 s, 0.2 s, 0.5 s and 1 s respectively. True speaker changes are marked by vertical poles.

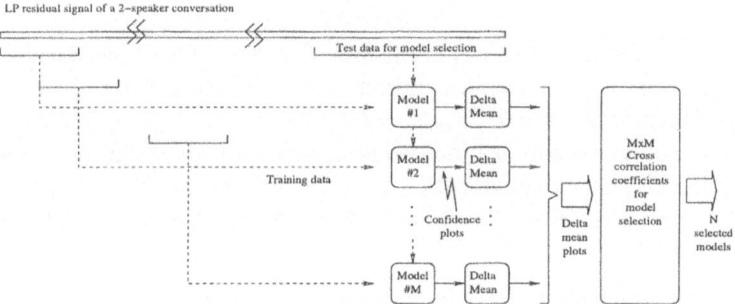

Fig. 3. Model building and selection process.

shifting the pair of windows by 5 ms. Figure 4(b), (c), (d) and (e) show the evidence for the chosen four AANN models. The four evidences are combined using *AND* logic and the result is shown in Figure 4(f). The dBIC plot for the same 2-speaker data, given in Figure 4(g), shows relatively poorer evidence when vocal tract features are used.

5   Performance of the Proposed Approach

Performance Metrics:
The performance of speaker segmentation is evaluated using the false acceptance or alarm rate (FAR) and the missed detection rate (MDR). FAR is the number of false speaker changes, while MDR is the number of missed speaker changes, both expressed as a percentage of the actual number of speaker changes. An ideal system should give an FAR of 0% and an MDR of 0%. The performance of the segmentation is also measured in terms of the segmentation cost function given by, $C_{seg} = 1 - T_c/T_t$,where T_c is the total duration of voiced speech (in

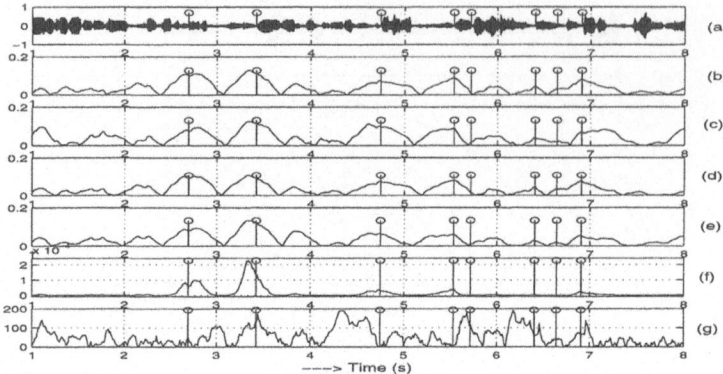

Fig. 4. Combining evidence for speaker change detection. (a) 2-speaker signal with short speaker turns. $\triangle\mu$ plots for (b) model 1, (c) model 2, (d) model 3, (e) model 4, (f) combined evidence and (g) dBIC for vocal tract features. Solid lines indicate the actual speaker change points.

time) correctly segmented and T_t is the total duration of the voiced speech in the conversation. The cost function is normalized by a factor $C_{default}$, to obtain a normalized segmentation cost $C_{norm} = C_{seg}/C_{default}$. $C_{default}$ is the minimum segmentation cost that can be obtained even without processing the conversation (by assigning the entire conversation to either of the speaker). A good system should give a C_{norm} value close to zero, and a value close to one is as good as not processing the conversation.

Data Set for Performance Evaluation:
A total of 10 different 2-speaker conversations each of duration 5 minutes are used to evaluate the performance of speaker segmentation system. The 2-speaker speech signals are casual telephonic conversations and are part of the NIST-2003 database for speaker recognition evaluation [7]. Out of the 10 conversations, 5 are male-male conversations and 5 are female-female conversations. The data set has a total of 1047 actual speaker changes (manually marked). A five layered AANN model with a structure $40L60N12N60N40L$ is used in the experiments and the residual samples are fed to the neural network in blocks of 5 ms. The FAR, MDR and C_{norm} values for the vocal tract based system and the proposed system based on excitation source are compared in Table 1.

Table 1. Speaker segmentation performance of the vocal tract and excitation source based systems ($C_{default} = 0.39$).

System based on	FAR	MDR	C_{seg}	C_{norm}
Vocal tract features	52%	64%	0.35	0.90
Excitation source features	37%	48%	0.27	0.69

6 Summary

In this paper, we have shown the effectiveness of subsegmental features for speaker change detection. Experiments with current approaches indicate that speaker segmentation methods based on statistical distribution of feature vectors do not perform satisfactorily when speaker turns are short (< 5 s). Excitation source features present in the LP residual of speech signal are useful for segmentation. The features can be extracted using AANN models. The results indicate that the subsegmental features from the excitation source signal perform better than the features representing the vocal tract. Combining evidences from multiple AANN models is still an issue and more exploration on this part may lead to improved performance.

References

1. Chen, S., Gopalakrishnan, P.: Speaker, environment and channel change detection and clustering via the Bayesian information criterion. In: Proceedings of DARPA Broadcast News Transcription and Understanding Workshop. (1998) 127–132
2. Delacourt, P., Wellekens, C.J.: DISTBIC: A speaker-based segmentation for audio data indexing. Speech Communication **32** (2000) 111–126
3. H. Gish, M. Siu and R. Rohlicek: Segregation of speakers for speech recognition and speaker identification. In: Proceedings of the International Conference on Acoustics Speech and Signal Processing. Volume 2. (1991) 873–876
4. Makhoul, J.: Linear prediction: A tutorial review. Proceedings of the IEEE **63** (1975) 561–580
5. Rabiner, L., Juang, B.H. In: Fundamentals of Speech Recognition. Prentice-Hall Inc., (Englewood Cliffs, New Jersey, USA)
6. B. Yegnanarayana and K. Sharat Reddy and S. P. Kishore: Source and system features for speaker recognition using aann models. In: Proceedings of the International Conference on Acoustics Speech and Signal Processing. Volume 1. (2001) 409–412
7. Yegnanarayana, B., et. al.: IIT Madras Speaker Recognition system. In: Proc. NIST Speaker Recognition Workshop, Baltimore, Maryland, USA (2003)

Morozov, Ivanov and Tikhonov Regularization Based LS-SVMs

Kristiaan Pelckmans, Johan A.K. Suykens, and Bart De Moor

KULeuven - ESAT - SCD,
Kasteelpark Arenberg 10, B - 3001 Leuven (Heverlee), Belgium
Tel. +32 - 16 - 32 11 45, Fax +32 - 16 - 32 19 70
{kristiaan.pelckmans,johan.suykens}@esat.kuleuven.ac.be
http://www.esat.kuleuven.ac.be/sista/lssvmlab

Abstract. This paper contrasts three related regularization schemes for kernel machines using a least squares criterion, namely Tikhonov and Ivanov regularization and Morozov's discrepancy principle. We derive the conditions for optimality in a least squares support vector machine context (LS-SVMs) where they differ in the role of the regularization parameter. In particular, the Ivanov and Morozov scheme express the trade-off between data-fitting and smoothness in the trust region of the parameters and the noise level respectively which both can be transformed uniquely to an appropriate regularization constant for a standard LS-SVM. This insight is employed to tune automatically the regularization constant in an LS-SVM framework based on the estimated noise level, which can be obtained by using a nonparametric technique as e.g. the differogram estimator.

1 Introduction

Regularization has a rich history which dates back to the theory of inverse ill-posed and ill-conditioned problems [9, 13, 15] inspiring many advances in machine learning [6, 16], support vector machines and kernel based modeling techniques [10, 7, 8]. Determination of the regularization parameter in the Tikhonov scheme is considered to be an important problem [16, 7, 2, 11]. Recently [3], this problem was approached from an additive regularization point of view: a more general parameterization of the trade-off was proposed generalizing different regularization schemes. Combination of this convex scheme and validation or cross-validation measures can be solved efficiently for the regularization trade-off as well as the training solutions.

This paper considers three classical regularization schemes [13, 15, 17] in a kernel machine framework based on LS-SVMs [8] which express the trade-off between smoothness and fitting in respectively the noise level and a trust region. It turns out that both result into linear sets of equations as in standard LS-SVMs where an additional step performs the bijection of those constants into an appropriate regularization constant of standard LS-SVMs. The practical relevance of this result is mainly seen in the exact derivation of the translation of prior knowledge as the noise level or the trust region. The importance of the noise level for (nonlinear) modeling and hyper-parameter tuning was already stressed in [11, 4]. The Bayesian framework (for neural networks,

N.R. Pal et al. (Eds.): ICONIP 2004, LNCS 3316, pp. 1216–1222, 2004.
© Springer-Verlag Berlin Heidelberg 2004

Gaussian processes as well as SVMs and LS-SVMs, see e.g. [19, 8]) allows for a natural integration of prior knowledge in the derivations of a modeling technique, though it often leads to non-convex problems and computationaly heavy sampling procedures. Nonparameteric techniques for the estimation of the noise level were discussed in e.g. [4, 5] and can be employed in the discussed Morozov scheme.

This paper is organized as follows: Section 2 compares the primal-dual derivations of LS-SVM regressors based on Tikhonov regularization, Morozov's discrepancy principle and an Ivanov regularization scheme. Section 3 describes an experimental setup comparing the accuracy of the schemes in relation to classical model-selection schemes.

2 Tikhonov, Morozov and Ivanov Based LS-SVMs

Let $\{x_i, y_i\}_{i=1}^N \subset \mathbb{R}^d \times \mathbb{R}$ be the training data where $x_1, \ldots, x_N$ are deterministic points (fixed design) and $y_i = f(x_i) + e_i$ with $f : \mathbb{R}^d \to \mathbb{R}$ an unknown real-valued smooth function and $e_1, \ldots, e_N$ uncorrelated random errors with $E[e_i] = 0$, $E[e_i^2] = \sigma_e^2 < \infty$. The model for regression is given as $f(x) = w^T \varphi(x) + b$ where $\varphi(\cdot) : \mathbb{R}^d \to \mathbb{R}^{n_h}$ denotes a potentially infinite ($n_h = \infty$) dimensional feature map. In the following, the Tikhonov scheme [9], Morozov's discrepancy principle [15] and Ivanov Regularization scheme are elaborated simultaneously to stress the correspondences and the differences. The cost functions are given respectively as

- Tikhonov [8]

$$\min_{w,b,e_i} \mathcal{J}_T(w, e) = \frac{1}{2} w^T w + \frac{\gamma}{2} \sum_{i=1}^N e_i^2 \text{ s.t. } w^T \varphi(x_i) + b + e_i = y_i, \quad \forall i = 1, ..., N.$$

(1)

- Morozov's discrepancy principle [15], where the minimal 2-norm of w realizing a fixed noise level σ^2 is to be found:

$$\min_{w,b,e_i} \mathcal{J}_M(w) = \frac{1}{2} w^T w \text{ s.t. } \begin{cases} w^T \varphi(x_i) + b + e_i = y_i, \forall i = 1, \ldots N \\ N\sigma^2 = \sum_{i=1}^N e_i^2. \end{cases}$$

(2)

- Ivanov [13] regularization amounts at solving for the best fit with a 2-norm on w smaller than π^2. The following modification is considered in this paper:

$$\min_{w,b,e_i} \mathcal{J}_I(e) = \frac{1}{2} e^T e \text{ s.t. } \begin{cases} w^T \varphi(x_i) + b + e_i = y_i, \forall i = 1, \ldots N \\ \pi^2 = w^T w. \end{cases}$$

(3)

The use of the equality (instead of the inequality) can be motivated in a kernel machine context as these problems are often ill-conditioned and result in solutions on the boundary of the trust region $w^T w \leq \pi^2$.

The Lagrangians can be written respectively as

$$\mathcal{L}_T(w, b, e_i; \alpha_i) = \frac{1}{2}w^T w - \frac{\gamma}{2}\sum_{i=1}^{N} e_i^2 - \sum_{i=1}^{N} \alpha_i(w^T x_i + b + e_i - y_i).$$

$$\mathcal{L}_M(w, b, e_i; \alpha_i, \xi) = \frac{1}{2}w^T w - \xi(\sum_{i=1}^{N} e_i^2 - N\sigma^2) - \sum_{i=1}^{N} \alpha_i(w^T x_i + b + e_i - y_i).$$

$$\mathcal{L}_I(w, b, e_i; \alpha_i, \xi) = \frac{1}{2}e^T e - \xi(w^T w - \pi^2) - \sum_{i=1}^{N} \alpha_i(w^T x_i + b + e_i - y_i).$$

The conditions for optimality are

Condition	Tikhonov	Morozov	Ivanov
$\frac{\partial \mathcal{L}}{\partial w} = 0$	$w = \sum_{i=1}^{N} \alpha_i \varphi(x_i)$	$w = \sum_{i=1}^{N} \alpha_i \varphi(x_i)$	$2\xi w = \sum_{i=1}^{N} \alpha_i \varphi(x_i)$
$\frac{\partial \mathcal{L}}{\partial b} = 0$	$\sum_{i=1}^{N} \alpha_i = 0$	$\sum_{i=1}^{N} \alpha_i = 0$	$\sum_{i=1}^{N} \alpha_i = 0$
$\frac{\partial \mathcal{L}}{\partial e_i} = 0$	$\gamma e_i = \alpha_i$	$2\xi e_i = \alpha_i$	$e_i = \alpha_i$
$\frac{\partial \mathcal{L}}{\partial \alpha_i} = 0$	$w^T \varphi(x_i) + b + e_i = y_i,$	$w^T \varphi(x_i) + b + e_i = y_i,$	$w^T \varphi(x_i) + b + e_i = y_i$
$\frac{\partial \mathcal{L}}{\partial \xi} = 0$	$-$	$\sum_{i=1}^{N} e_i^2 = N\sigma^2$	$w^T w = \pi^2$

$$(4)$$

for all $i = 1, \ldots, N$. The kernel-trick is applied as follows: $\varphi(x_k)^T \varphi(x_l) = K(x_k, x_l)$ for an appropriate kernel $K : \mathbb{R}^D \times \mathbb{R}^D \rightarrow \mathbb{R}$ in order to avoid explicit computations in the high dimensional feature space. Let $\Omega \in \mathbb{R}^{N \times N}$ be such that $\Omega_{kl} = K(x_k, x_l)$ for all $k, l = 1, \ldots, N$. The Tikhonov conditions result in the following set of linear equations as classical [8]:

$$\textbf{Tikhonov}: \quad \left[\begin{array}{c|c} 0 & 1_N^T \\ \hline 1_N & \Omega + \frac{1}{\gamma}I_N \end{array}\right] \left[\begin{array}{c} b \\ \hline \alpha \end{array}\right] = \left[\begin{array}{c} 0 \\ \hline y \end{array}\right]. \quad (5)$$

Re-organizing the sets of constraints of the Morozov and Ivanov scheme results in the following sets of linear equations where an extra nonlinear constraint relates the Lagrange multiplier ξ with the hyper-parameter σ^2 or π^2

$$\textbf{Morozov}: \quad \left[\begin{array}{c|c} 0 & 1_N^T \\ \hline 1_N & \Omega + \frac{1}{2\xi}I_N \end{array}\right] \left[\begin{array}{c} b \\ \hline \alpha \end{array}\right] = \left[\begin{array}{c} 0 \\ \hline y \end{array}\right] \quad \text{s.t. } N\sigma^2 = \alpha^T \alpha, \quad (6)$$

and π^2

$$\textbf{Ivanov}: \quad \left[\begin{array}{c|c} 0 & 1_N^T \\ \hline 1_N & \frac{1}{2\xi}\Omega + I_N \end{array}\right] \left[\begin{array}{c} b \\ \hline \alpha \end{array}\right] = \left[\begin{array}{c} 0 \\ \hline y \end{array}\right] \quad \text{s.t. } \pi^2 = \alpha^T \Omega \alpha. \quad (7)$$

2.1   Formulation in Terms of the Singular Value Decomposition

This subsection rephrases the optimization problem (2) in terms of the Singular Value Decomposition (SVD) of Ω [12]. For notational covenience, the bias term b is omitted from the following derivations. The SVD of Ω is given as

$$\Omega = U \Gamma U^T \quad \text{s.t.} \quad U^T U = I_N, \tag{8}$$

where $U \in \mathbb{R}^{N \times N}$ is orthonormal and $\Gamma = \text{diag}(\gamma_1, \ldots, \gamma_N)$ with $\gamma_1 \geq \cdots \geq \gamma_N$. Using the orthonormality [12], the conditions (6) can be rewritten as

$$
\left\{
\begin{array}{ll}
\alpha & = U \left(\Gamma + \frac{1}{2\xi} I_N \right)^{-1} p \qquad\qquad\qquad\qquad\quad (a) \\[2mm]
N\sigma^2 & = \frac{1}{4\xi^2} p^T \left(\Gamma + \frac{1}{2\xi} I_N \right)^{-2} p = \sum_{i=1}^{N} \left(\frac{p_i}{2\xi\gamma_i+1} \right)^2, \quad (b)
\end{array}
\right. \tag{9}
$$

where $p = U^T y \in \mathbb{R}^N$. Rewriting the Ivanov scheme (7) yields

$$
\left\{
\begin{array}{ll}
\alpha & = U \left(\frac{1}{2\xi} \Gamma + I_N \right)^{-1} p \qquad (a) \\[2mm]
\pi^2 & = \sum_{i=1}^{N} \frac{\gamma_i p_i^2}{(\frac{1}{2\xi}\gamma_i+1)^2}. \qquad\quad (b)
\end{array}
\right. \tag{10}
$$

One refers to the equations (9.b) and (10.b) as secular equations [12, 17].

The previous derivation can be exploited in practical algorithms as follows. As the secular equation (9.b) is strictly monotone in the Lagrange multiplier ξ, the roles can be reversed (the inverse function exist for a nontrivial positive interval): once a regularization constant ξ is chosen, a unique corresponding noise level σ^2 is fixed. Instead of translating the prior knowledge σ^2 or π^2 using the secular equation (which needs an SVD), one can equivalently look for a ξ value resulting respectively in exactly the specified σ^2 or π^2. This can be done efficiently in a few steps by using e.g. the bisection algorithm [12]. The previous derivation of the monotone secular equations states that one obtains not only a model resulting in respectively the specified noise level or trust region, but one gets in fact the optimal result in the sense of (2) and (3).

Figure 1.a illustrates the training and validation performance for Morozov based LS-SVMs for a sequence of strictly positive noise levels. The figure indicates that overfitting on the training comes into play as soon as the noise level is underestimated. The error bars were obtained by a Monte-Carlo simulation as described in the next Section. Figure 1.b shows the technique for model-free noise variance estimation using the differogram [4, 5]. This method is based on a scatterplot of the differences Δ_x of any two input points and the corresponding output observations Δ_y. It can be shown that the curve $E[\Delta_y | \Delta_x = \delta]$ gives an estimate of the noise level at the value where it intersects the Y-axis ($\Delta_x = 0$).

3   Experiments

The experiments focus on the choice of the regularization scheme in kernel based models. For the design of a Monte-Carlo experiment, the choice of the kernel and kernel-parameter should not be of critical importance. To randomize the design of the underlying functions in the experiment with known kernel-parameter, the following class of functions is considered

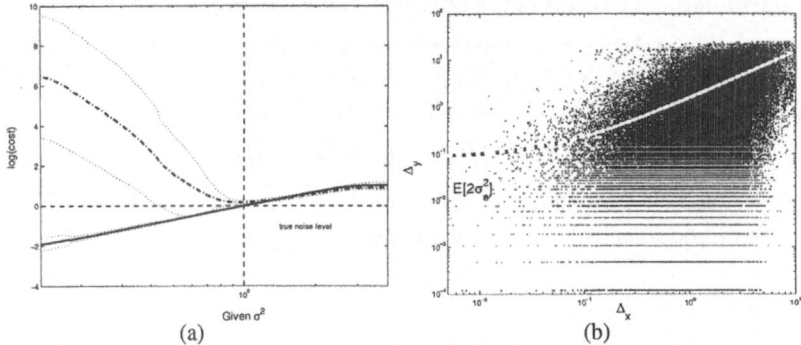

(a)   (b)

Fig. 1. (a) Training error (**solid line**) and validation error (**dashed-dotted line**) for the LS-SVM regressor with the Morozov scheme as a function of the noise level σ^2 (the dotted lines indicate error-bars by randomizing the experiment). The **dashed lines** denote the true noise level. One can see that for small noise levels, the estimated functions suffer from overfitting. (**b**) Differogram cloud of the Boston Housing Dataset displaying all differences between two inputs (Δ_x) and two corresponding outputs (Δ_y). The location of the curve passing the Y-axis given as $E[\Delta_y|\Delta_x = 0]$ results in an estimate of the noise variance.

$$f(\cdot) = \sum_{i=1}^{N} \bar{\alpha}_i K(x_i, \cdot) \qquad (11)$$

where x_i is equidistantly taken between 0 and 5 for all $i = 1, \ldots, N$ with $N = 100$ and $\bar{\alpha}_i$ is an i.i.d. uniformly randomly generated term. The kernel is fixed as $K(x_i, x_j) = \exp(-\|x_i - x_j\|_2^2)$ for all $i, j = 1, \ldots, N$. Datapoints were generated as $y_i = f(x_i) + e_i$ for $i = 1, \ldots, N$ where e_i are N i.i.d. samples of a Gaussian distribution. Although no true underlying regularization parameter is likely to exist for (1), the *true* regularization parameter $\bar{\gamma}$ is estimated by optimizing w.r.t. a noiseless test set of size 10000.

The experiment tests the accuracy of the regularization constant tuning for Morozov's discrepancy principle (see Table 1). It compares results obtained when using exact prior knowledge of the noise level, a model-free estimate of the noise level using the differogram and using data-driven model selection methods as L-fold Cross-Validation (CV), leave-one-out CV, Mallows C_p statistic [14] and Bayesian inference [8]. An important remark is that the method based on the differogram is orders of magnitudes faster than any data-driven method. This makes the method suited as a method for picking a good starting-value for a local search typically associated with a more powerful and computationally intensive generalization criterion. Experiments on the higher dimensional Boston housing data (with standardized inputs and outputs) even suggest that the proposed measure can be sufficiently good as a model selection criterion on its own. For this experiment, one third of the data was reserved for test purposes, while the remaining data were used for the training and selection of the regularization parameter. This procedure was repeated 500 times in a Monte-Carlo experiment.

Table 1. Numerical results on testdata of the experiments as described in Section 3.

	Morozov	Differogram	10-fold CV	leaveoneout	Bayesian	C_p	"true"
	Toy example: 25 datapoints						
mean(MSE)	0.4238	0.4385	0.3111	0.3173	0.3404	1.0072	0.2468
std(MSE)	1.4217	1.9234	0.3646	1.5926	0.3614	1.0727	0.1413
	Toy example: 200 datapoints						
mean(MSE)	0.1602	0.2600	0.0789	0.0785	0.0817	0.0827	0.0759
std(MSE)	0.0942	0.5240	0.0355	0.0431	0.0289	0.0369	0.0289
	Boston Housing Dataset						
mean(MSE)	-	0.1503	0.1538	0.1518	0.1522	0.3563	0.1491
std(MSE)	-	0.0199	0.0166	0.0217	0.0152	0.1848	0.0184

4   Conclusions

This paper compared derivations based on regularization schemes as Morozov discrepancy principle, Ivanov and Tikhonov regularization schemes. It employs these to incorporate prior or model-free estimates of the noise variance for tuning the regularization constant in LS-SVMs.

Acknowledgements

This research work was carried out at the ESAT laboratory of the Katholieke Universiteit Leuven. It is supported by grants from several funding agencies and sources: Research Council KU Leuven: Concerted Research Action GOA-Mefisto 666 (Mathematical Engineering), IDO (IOTA Oncology, Genetic networks), several PhD/postdoc & fellow grants; Flemish Government: Fund for Scientific Research Flanders (several PhD/postdoc grants, projects G.0407.02 (support vector machines), G.0256.97 (subspace), G.0115.01 (bio-i and microarrays), G.0240.99 (multilinear algebra), G.0197.02 (power islands), research communities ICCoS, ANMMM), AWI (Bil. Int. Collaboration Hungary/ Poland), IWT (Soft4s (softsensors), STWW-Genprom (gene promotor prediction), GBOU-McKnow (Knowledge management algorithms), Eureka-Impact (MPC-control), Eureka-FLiTE (flutter modeling), several PhD grants); Belgian Federal Government: DWTC (IUAP IV-02 (1996-2001) and IUAP V-10-29 (2002-2006) (2002-2006): Dynamical Systems and Control: Computation, Identification & Modelling), Program Sustainable Development PODO-II (CP/40: Sustainibility effects of Traffic Management Systems); Direct contract research: Verhaert, Electrabel, Elia, Data4s, IPCOS. JS and BDM are an associate and full professor with K.U.Leuven Belgium, respectively.

References

1. S. Boyd and L. Vandenberghe. *Convex Optimization*. Cambridge University Press, 2004.
2. O. Chapelle, V. Vapnik, O. Bousquet, and S. Mukherjee. Choosing multiple parameters for support vector machines. *Machine Learning*, 46(1-3):131–159, 2002.
3. K. Pelckmans, J.A.K. Suykens, and B. De Moor. Additive regularization: fusion of training and validation levels in kernel methods. *Internal Report 03-184, ESAT-SCD-SISTA, K.U.Leuven (Leuven, Belgium)*, 2003, submitted for publication.
4. K. Pelckmans, J. De Brabanter, J.A.K. Suykens, B. De Moor. Variogram based noise variance estimation and its use in Kernel Based Regression. *in Proc. of the IEEE Workshop on Neural Networks for Signal Processing*, 2003.
5. K. Pelckmans, J. De Brabanter, J.A.K. Suykens, B. De Moor. The differogram: nonparametric noise variance estimation and its use in model selection. *Internal Report 04-41, ESAT-SCD-SISTA, K.U.Leuven (Leuven, Belgium)*, 2004, submitted for publication.
6. T. Poggio and F. Girosi. Networks for approximation and learning. In *Proceedings of the IEEE*, volume 78, pages 1481–1497. Proceedings of the IEEE, septenber 1990.
7. B. Schölkopf and A. Smola. *Learning with Kernels*. MIT Press, Cambridge, MA, 2002.
8. J.A.K. Suykens, T. Van Gestel, J. De Brabanter, B. De Moor, and J. Vandewalle. *Least Squares Support Vector Machines*. World Scientific, 2002.
9. A.N. Tikhonov and V.Y. Arsenin. *Solution of Ill-Posed Problems*. Winston, Washington DC, 1977.
10. V.N. Vapnik. *Statistical Learning Theory*. Wiley and Sons, 1998.
11. V. Cherkassky and F. Mulier. *Learning from Data*. Wiley, New York, 1998.
12. G.H. Golub and C.F. Van Loan. *Matrix Computations*. The John Hopkins University Press, 1989.
13. V.V. Ivanov. *The Theory of Approximate Methods and Their Application to the Numerical Solution of Singular Integral Equations*. Nordhoff International, 1976.
14. C.L. Mallows. Some comments on Cp. *Technometrics*, 40, 661-675, 1973.
15. V.A. Morozov. *Methods for Solving Incorrectly Posed Problems*. Springer-Verlag, 1984.
16. G. Wahba. *Splines Models for Observational Data*. Series in Applied Mathematics, 59, SIAM, Philadelphia, 1990.
17. A. Neumaier. Solving ill-conditioned and singular linear systems: A tutorial on regularization. *SIAM Review*, 40, 636-666, 1988.
18. V. Cherkassky. Practical selection of svm parameters and noise estimation for svm regression. *Neurocomputing, Special Issue on SVM*, 17(1), 113-126, 2004.
19. D.J.C. MacKay. Bayesian interpolation. *Neural Computation*, 4(3), 415-447, 1992.

A Study for Excluding Incorrect Detections of Holter ECG Data Using SVM

Yasushi Kikawa and Koji Oguri

Aichi Prefectural University Graduate School of Information Science and Technology
Nagakute-cho, Aichi 480-1198, Japan
im021011@ist.aichi-pu.ac.jp

Abstract. The inspection of arrhythmia using the Holter ECG is done by automatic analysis. However, the accuracy of this analysis is not sufficient, and the results need to be correct by clinical technologists. During the process of picking up one heartbeat in an automatic analysis system, an incorrect detection, whereby a non-heartbeat is picked up as a heartbeat, may occur. In this research, we proposed the method to recognize this incorrect detection by use of a Support Vector Machine (SVM). When the learning results were evaluated on the ECG wave data from the one hundred subject's heartbeats, this method correctly recognized a maximum of 93% as incorrect detections. These results should dramatically increase the work efficiency of clinical technologists.

1 Introduction

The Holter ECG appeared in the mid-1900s. Since then, it was gone through much research and development. The Holter ECG was recorded over a long period of time; in Japan, it is generally recoded for 24 hours. It is suited to find arrhythmias because of the long duration of its recording. Arrhythmia is caused by stress and increasing age, and it is difficult to predict when and where the arrhythmias take place. Medical specialists are required to spend much effort and time to study a long-term ECG. Therefore, they are generally aided by the automatically analysis of the Holter ECG data using computers beforehand. However, the accuracy of the computer analysis is not sufficient. So, after the analysis, clinical technologists must correct the analyzed results. In the automatic analysis of the Holter ECG, a process which pick up the heartbeat individually, is performed. The R wave is used as the marker to determine the position of one heartbeat on it. However, the process of picking up one heartbeat has mistakes. They are not to be able to find an R wave and to find an incorrect part which is not an R wave. The latter case is addressed in this research, and this case defined incorrect detection. In this research, the above-mentioned incorrect detection is excluded by SVM, which is one of the Pattern Recognition methods. SVM is applied to the fields of object recognition, handwritten digit recognition and others, all of which obtain high evaluations[1].

N.R. Pal et al. (Eds.): ICONIP 2004, LNCS 3316, pp. 1223–1228, 2004.
© Springer-Verlag Berlin Heidelberg 2004

2 Support Vector Machine (SVM)

SVM is one of the Pattern Recognition methods, and is proposed by V.Vapnik and his co-workers[2][3].

SVM separates an input example $\mathbf{X} = (x_1, ..., x_d)$ of dimension d into two classes. A decision function of SVM separates two classes by $f(\mathbf{X}) > 0$ or $f(\mathbf{X}) < 0$. The size of training set N is $(y_i, \mathbf{X}_i)$, $i = 1, ..., N$. Where $\mathbf{X}_i \in R^n$ is the input pattern for the ith example, and $y_i \in \{-1, 1\}$ is the class label. Support Vector classifiers implicitly map $\mathbf{X}_i$ from input space to a higher dimensional feature space which depended on a nonlinear function $\Phi(\mathbf{X})$. A separating hyperplane is optimized by maximization of the margin. Then SVM is solved as the following quadratic programming problem,

$$Maximize : \sum_{i=1}^{n} \alpha_i - \frac{1}{2} \sum_{i,j=1}^{n} \alpha_i \alpha_j y_i y_j K(\mathbf{X}_i, \mathbf{X}_j), \tag{1}$$

$$Subject\ to : 0 \leq \alpha_i \leq C(i = 1, \ldots, n), \sum_{i=1}^{n} \alpha_i y_i = 0. \tag{2}$$

Where $\alpha \geq 0$ are Lagrange multipliers. When the optimization problem has solved, many α_i will be equal to 0, and the others will be Support Vectors. C is positive constant which chosen empirically by the user. This parameter expresses degree of loosing constraint. A larger C can classify training examples more correctly. $K(\mathbf{X}, \mathbf{X}')$ is the kernel function which is inner-product defined by $K(\mathbf{X}, \mathbf{X}') = \Phi(\mathbf{X}) \cdot \Phi(\mathbf{X}')$. Then the SVM decision function is

$$f(\mathbf{X}) = \sum_{\mathbf{X}_i \in SV} \alpha_i y_i K(\mathbf{X}_i, \mathbf{X}) + b. \tag{3}$$

Typical kernel functions are the Gaussian kernel

$$K(\mathbf{X}, \mathbf{X}') = exp\left(-\frac{\|\mathbf{X} - \mathbf{X}'\|^2}{\sigma^2}\right), \tag{4}$$

and others. In this research, The Gaussian kernel is used as the kernel function.

3 Proposed Method

This recognizing method by SVM is proposed in this research. This method recognizes whether the detected place is correct as R wave by R wave detection. By this recognition, if non-R wave position is detected as R wave, it is excluded as incorrect detection. In short, our method can reduce incorrect detections. This method can be applied to ECG analyzing systems which perform R wave detection, and with significant results.

There are often a plural number of leads in the Holter ECG, generally two. However, only one channel is in use, and the other not. In this research, the number of incorrect detections is reduced by the use of two channels. Here we use logical addition, and by doing so, if a detected place was recognized as an R wave in either of the two channels, the place is left.

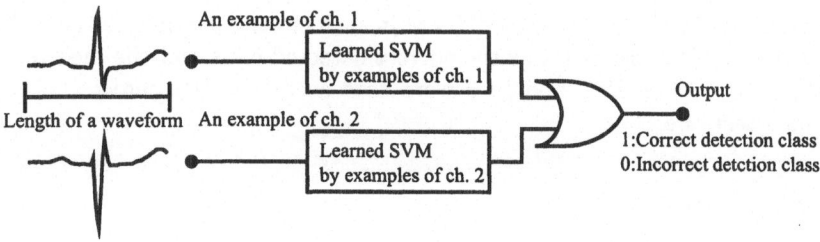

Fig. 1. Proposed method: First, detected places are recognized by SVM at waves of each channel. Next, those output go through OR gate.

3.1   Learning Data

ECG waveform data were used as learning data and estimating data for SVM in this research. This method are supposed to apply to a existing Holter ECG analyzing system. So, The data made use of clinical Holter ECG waveform data. The number of the data is one hundred, and the length of the data is about 24hr. The data's sampling frequency is 125Hz. The part of leads are CM5 as CH1 and NASA as CH2.

SVM has the ability that recognize two classes. So, the data need to be categorized into two classes. First, the Holter ECG analyzing system analyzes the data. In this system, a process which specifies one heartbeat is performed. The results of the process indicate positions that detect correctly as an R wave and those as non-R wave. Next, the positions of incorrect detection are removed by clinical technologists from the results. After comparison of these results, the data are categorized into two groups: the one is correct detection as R waves and the other is incorrect detection. It is difficult to know whether or not the peek of the R wave is indicate after human alteration rightly. For comparison, an allowable range was set up between the peek of the R wave in the analyzed results and altered results. In the altered results, the peek of the R wave is only one between two channels. However, the data have two channels. Furthermore, the peek of the R wave is slightly different between ch.1 and ch.2. So, an allowable range have also to set up between ch.1 and ch.2. The data were categorized to 7 patterns, the numbers of the places on R wave being obtained by comparing the results in Table 1.

Table 1. Results of comparison between the automatic analysis and the altered analysis. 1 express existing an R wave position at Certain times with an allowable range.

Pattern ID	Pt.1	Pt.2	Pt.3	Pt.4	Pt.5	Pt.6	Pt.7
Result of ch.1	1	1	0	1	1	0	0
Result of ch.2	1	0	1	1	0	1	0
Altered Result	1	1	1	0	0	0	1
Number of Examples	9,702,975	776,185	7,702	0	58,501	420,646	3,919

In this research, for learning of SVM in each channel, waveforms of ch.1 were learned using Pt.1 and Pt.5 as the correct detected class and the incorrect one. In the case of ch.2, Pt.1 and Pt.6 were used as the correct detected class and the incorrect one.

4 Experiment

This experiment was performed to conduct learning by changing the number of examples in one class N, the dimension of examples d, the positive constant C and parameter of the Gaussian kernel σ^2, for the SVM of two channels. The conditions for each of the parameters are as follows:

$N : 100, 200$
$d : 15, 19, ..., 71, 75$
$C : 1, 10, 100, 1000, 10000$
$\sigma^2 : 0.01, 0, 1, 1, 10, 100$

N pieces of examples were extracted from each class at random. The example dimension is the converted value from 0.12 second to 0.6 second. 0.12 second is the normal width of the QRS complex in the ECG. 0.6 second is the normal width of the R-R interval in ECG. Learning and estimating were performed 5 times with all combinations of all conditions. Estimated examples were all examples belonging to the correct detected class and the incorrect one. For the estimation of this method, a combination of parameters which was the highest result of each channel's evaluation was used.

5 Results and Discussion

The correct and incorrect detected classes are connected with parameter d, C and σ^2 in a trade-off relationship. So, both recognition rates are simultaneously estimated as the following,

$$R_T(CH, d, C, \sigma^2) = \sqrt{\frac{\bar{R}_A(CH, d, C, \sigma^2)^2 + \bar{R}_B(CH, d, C, \sigma^2)^2}{2}}. \qquad (5)$$

$R_T(CH, d, C, \sigma^2)$ is the evaluation index of the recognition rates, $\bar{R}_A(CH, d, C, \sigma^2)$ is the average recognition rate of correct detected class, and $\bar{R}_B(CH, d, C, \sigma^2)$ is the average recognition rate of incorrect detected class.

Parts of the results which were estimated are shown on Fig. 2. The recognition rates for the correct detected class and incorrect one, are shown in Table 2. When the parameters of Table 2 were applied to the proposed method, the recognition rates were as shown in Table 3.

Application of the proposed method using two channels at the same time with logical addition shown higher recognition rates than that using only single channels in the correct detected class. Before applying the proposed method, correct detected examples were recognized as incorrect detected examples between about 3% and about 4%. These examples correspond to between about

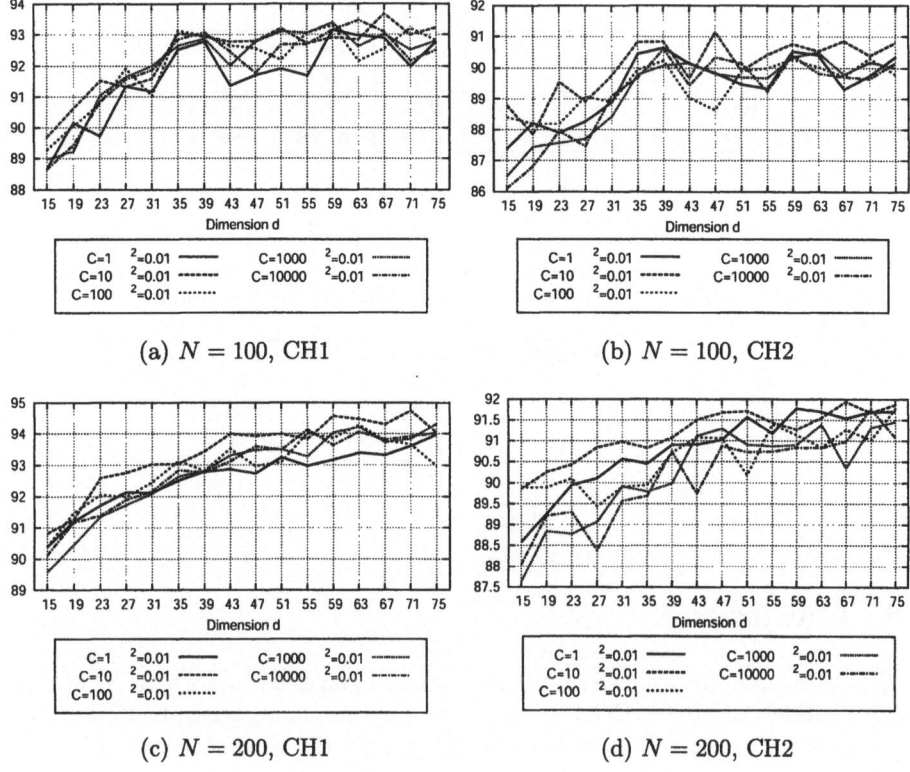

(a) $N = 100$, CH1   (b) $N = 100$, CH2

(c) $N = 200$, CH1   (d) $N = 200$, CH2

Fig. 2. Results of estimation on the experiment.

300,000 and about 400,000 examples. Examples which originally belonged to the correct detected class should be recognized as belonging to the correct detected class as much as possible. So, this method using logical addition is more effective than that using only single channels.

Examples which belong to the incorrect detected class will be subjected to the excluding process when clinical technologists will correct the analyzed results by computer. This proposed method can recognized more than 93%, the number being about 450,000 examples. This is improvement. Between about 300,000 and 400,000 examples which are originally detected as the R wave, are incorrectly recognized as examples which belong to the incorrect detected class. So, the accuracy become lower in the sense of how many of the detected examples are those which should be detected, and which will need to be picked up by clinical technologists. However, clinical technologists will only need to pick out about 300,000 examples of correct detection from about 750,000 examples which are doubted as incorrect detection. This will lighted their load than picking out about 450,000 examples from about 10,000,000 examples.

1228 Yasushi Kikawa and Koji Oguri

Table 2. The recognition rates in the highest evaluation index.

N	CH	d	C	σ^2	Recognition Rates[%] $\bar{R}_A$	$\bar{R}_B$
100	1	67	10	0.01	89.40	97.75
100	2	47	10	0.01	87.74	94.42
200	1	71	10	0.01	91.64	97.75
200	2	67	10	0.01	87.72	95.97

Table 3. The recognition rates of the proposed method.

N	Recognition Rates[%] A	B	Number of Recognized Examples A	B
100	96.68	93.52	9,380,836	448,075
200	96.01	95.08	9,315,826	455,585

A:Correct Detected Class
B:Incorrect Detected Class

6 Conclusion

In this research, we proposed a method that recognizes a waveform by SVM using two channels of Holter ECG at the same time for more accurate Holter ECG automatic analysis and more higher efficiency for the clinical technologists who will be altering the analyzed result. As a result of this, incorrect detection was reduced by more than 93%. The number is about 450,000 examples. However, a loss that non-incorrect detected examples of about 300,000 are recognized as incorrect detected examples, exists. In the future, it will be important to reduce this loss of about 300,000 to as low as possible. We aim at a more accurate system of Holter ECG automatic analysis.

References

1. Hyeran Byun and Seong-Whan Lee: "Applications of Support Vector Machines for Pattern Recognition: A Survey" SVM 2002, LNCS 2388, pp.213-236, 2002.
2. Vladimir N. Vapnik: "The Nature of Statistical Learning Theory 2nd edn" Springer-Verlag, (1999)
3. Nello Cristianini and John Shawe-Taylor: "An Introduction to Support Vector Machines" Cambridge University Press, (2000)

Semi-supervised Kernel-Based Fuzzy C-Means

Daoqiang Zhang, Keren Tan, and Songcan Chen

Department of Computer Science and Engineering
Nanjing University of Aeronautics and Astronautics
Nanjing 210016, P.R. China
{dqzhang,s.chen}@nuaa.edu.cn

Abstract. This paper presents a semi-supervised kernel-based fuzzy c-means algorithm called S^2KFCM by introducing semi-supervised learning technique and the kernel method simultaneously into conventional fuzzy clustering algorithm. Through using labeled and unlabeled data together, S^2KFCM can be applied to both clustering and classification tasks. However, only the latter is concerned in this paper. Experimental results show that S^2KFCM can improve classification accuracy significantly, compared with conventional classifiers trained with a small number of labeled data only. Also, it outperforms a similar approach S^2FCM.

1 Introduction

Recently, semi-supervised learning has attracted much attention in machine learning community. One reason is that in many learning tasks, there is a large supply of unlabeled data but insufficient labeled data because the latter is much more expensive to obtain than the former. In other words, labeled data is accurate but the number is few, and unlabeled data is not accurate whereas their amount is huge. To break away from that dilemma, semi-supervised learning combines labeled and unlabeled data together during training to improve performance. Typically, semi-supervised learning is applicable to both clustering and classification. In semi-supervised clustering, some labeled data is used along with the unlabeled data to obtain a better clustering. However, in semi-supervised classification, additional unlabelled data are exploited with labeled data to obtain a good classification function. A lot of semi-supervised learning algorithms have been proposed to date [1]-[4]. Among them, most semi-supervised clustering algorithms originate from classical clustering algorithms and are for clustering tasks, whereas most semi-supervised classification algorithms originate from classical classification algorithms and are for classification tasks.

In this paper, we present a semi-supervised kernel-based fuzzy c-means algorithm called S^2KFCM, which is based on our previously proposed kernel-based fuzzy c-means clustering algorithm (KFCM) [5][6]. S^2KFCM is the semi-supervised KFCM, and here our goal is to use S^2KFCM not for clustering but for classification tasks. We made comparisons between S^2KFCM and classical classifiers trained with a small number of labeled data, e.g. k-nearest neighbor classifier (k-NN) [7] and support vector machines (SVM) [7]. Comparisons of classification performances are also made between S^2KFCM and another similar algorithm S^2FCM which originated from

N.R. Pal et al. (Eds.): ICONIP 2004, LNCS 3316, pp. 1229–1234, 2004.
© Springer-Verlag Berlin Heidelberg 2004

fuzzy c-means algorithm (FCM) and was for clustering. Experimental results demonstrate the advantages of the proposed approach over other algorithms. In Section 2, we first review the KFCM algorithm. In Section 3 the detailed S²KFCM algorithm is proposed. Section 4 presents the experimental results. Conclusions are made in Section 5.

2 KFCM

Given $X=\{x_1,...,x_n\}$ where x_k in R^s, the original FCM algorithm partitions X into c fuzzy subsets by minimizing the following objective function [8]

$$J_m(U,V) = \sum_{i=1}^{c} \sum_{k=1}^{n} u_{ik}^m \parallel x_k - v_i \parallel^2 .$$ (1)

Here c is the number of clusters, n is the number of data points, and u_{ik} is the membership of x_k in class i takes value in the interval [0,1] such that $\sum_i u_{ik}=1$ for all k. By optimizing the objective function of FCM, one can obtain the alternate iterative equations.

By using the popular 'kernel method', we constructed a kernel version of FCM in our early works, where the original Euclidian distance in FCM is replaced with the following kernel-induced distance measures [5]

$$d(x,y) = \parallel \Phi(x) - \Phi(y) \parallel = \sqrt{K(x,x) - 2K(x,y) + K(y,y)} .$$ (2)

Here Φ is a nonlinear function mapping x_k from the input space X to a new space F with higher or even infinite dimensions. $K(x,y)$ is the kernel function which is defined as the inner product in the new space F with: $K(x,y) = \Phi(x)^T\Phi(y)$, for x, y in input space X.

An important fact about kernel function is that it can be directly constructed in original input space without knowing the concrete form of Φ. That is, a kernel function implicitly defines a nonlinear mapping function. There are several typical kernel functions, e.g. the Gaussian kernel: $K(x,y)=exp(-\parallel x-y\parallel^2/\sigma^2)$, and the polynomial kernel: $K(x,y)=(x^Ty + 1)^d$. Especially for Gaussian kernel, we have $K(x,x)=1$ for all x. For simplicity, we only consider the Gaussian kernel in this paper. By replacing the Euclidean distance in Eq. (1) with the Gaussian kernel-induced distance, we the objective function of KFCM as follows

$$J_m(U,V) = 2\sum_{i=1}^{c} \sum_{k=1}^{n} u_{ik}^m \left(1 - K(x_k,v_i)\right) .$$ (3)

By optimizing Eq. (3), we have the following alternative iterative equations for the Gaussian kernel

$$u_{ik} = \frac{\left(1/\left(1-K(x_k,v_i)\right)\right)^{1/(m-1)}}{\sum_{j=1}^{c} \left(1/\left(1-K(x_k,v_j)\right)\right)^{1/(m-1)}} ,$$ (4)

and

$$v_i = \frac{\sum_{k=1}^{n} u_{ik}^m K(x_k, v_i) x_k}{\sum_{k=1}^{n} u_{ik}^m K(x_k, v_i)} \, . \tag{5}$$

3 S²KFCM

Now we are position to present the semi-supervised KFCM. Assume there are a mall number of labeled data and a large amount of unlabeled data. Each data can be represented as a vector in R^s. Thus all the labeled and unlabeled data can be denoted in a whole matrix form as follows

$$X = \left\{ \underbrace{x_1^l, \ldots, x_{n_l}^l}_{labeled} \,\middle|\, \underbrace{x_1^u, \ldots, x_{n_u}^u}_{unlabeled} \right\} = X^l \bigcup X^u \, . \tag{6}$$

Here the superscript l and u indicate the labeled or unlabeled data respectively, and n_l and n_u denote the number of labeled and unlabeled data respectively. The total number of data is represent with $n = n_l + n_u$. In conventional approach to classifier design, e.g. k-nearest neighbor classifier, only X^l is used to train the classification function, and then use that function to label X^u.

Similarly, a matrix representation of the fuzzy c-partition of X in Eq. (6) has the following form

$$U = \left\{ \underbrace{U^l = \left\{ u_{ik}^l \right\}}_{labeled} \,\middle|\, \underbrace{U^u = \left\{ u_{ik}^u \right\}}_{unlabeled} \right\} . \tag{7}$$

Here the value of the component u_{ik}^l in U^l is known beforehand and typically is set to 1 if the data x_k is labeled with class i, and 0 otherwise. From U^l, we can obtain an initial set of cluster centers or prototypes as follows

$$v_i^0 = \frac{\sum_{k=1}^{n_l} (u_{ik}^l)^m x_k^l}{\sum_{k=1}^{n_l} (u_{ik}^l)^m}, 1 \leq i \leq c \, . \tag{8}$$

Consequently, the membership u_{ik}^u in U^u is updated as follows

$$u_{ik}^u = \frac{\left(1/\left(1 - K(x_k^u, v_i) \right) \right)^{1/(m-1)}}{\sum_{j=1}^{c} \left(1/\left(1 - K(x_k^u, v_j) \right) \right)^{1/(m-1)}}, 1 \leq i \leq c, 1 \leq k \leq n_u \, . \tag{9}$$

Finally, the cluster centers are updated by calculating

$$v_i = \frac{\sum_{k=1}^{n_l} (u_{ik}^l)^m K(x_k^l, v_i) x_k^l + \sum_{k=1}^{n_u} (u_{ik}^u)^m K(x_k^u, v_i) x_k^u}{\sum_{k=1}^{n_l} (u_{ik}^l)^m K(x_k^l, v_i) + \sum_{k=1}^{n_u} (u_{ik}^u)^m K(x_k^u, v_i)} \, . \tag{10}$$

We summarize the above discussion by formalizing the developed algorithm.

The proposed S^2KFCM algorithm

Step 1 Fix c, and select parameters, t_{max}, $m > 1$ and $\varepsilon > 0$ for some positive constant.

Step 2: Initilize $U_0 = [U^d | U^u_0]$.

Step 3: Compute initial prototypes using Eq. (8).

Step 4: For $t = 1, 2, \ldots, t_{max}$.

    (a) Compute the membership u^u_{ik} in U^u using Eq. (9).

    (b) Compute $E_t = \| U^u_t - U^u_{t-1} \|$.

    (c) If $E_t \leqslant \varepsilon$, then stop; else compute the prototypes using Eq. (10), next t.

4 Experiments

In this section, we make numerical comparison between the proposed S^2KFCM and other algorithms including S^2FCM, nearest neighbor (1-NN) and unsupervised FCM and KFCM on some benchmark data sets. We use the Gaussian kernel for S^2KFCM and KFCM and the parameter σ is computed as follows

$$\sigma = \frac{1}{c} \left(\sqrt{\frac{\sum_{j=1}^{n} \| x_j - m \|^2}{n}} \right). \tag{11}$$

Here c is number of clusters, x_j is the labeled or unlabeled data, n is the total number of labeled and unlabeled data and m is the centroid of the n data. In all the experiments, we set the parameters $m=2$, $\varepsilon = 0.001$ and $t_{max}=50$.

The first benchmark data is the well-known Iris data set [9]. It contains 3 clusters with 50 samples each. We choose from the total 150 data one portion as the labeled data set and the other as the unlabeled data set. 1-NN uses only labeled data set for training; FCM and KFCM use only unlabeled data set for clustering, while S^2FCM and S^2KFCM both use labeled and unlabeled data set for better performance. Table 1 shows the numbers of misclassified data of the five algorithms when different sizes of labeled data set are used.

Table 1. Number of misclassified data on Iris data set (#: number of labeled data)

#	1-NN	FCM	KFCM	S^2FCM	S^2KFCM
45	5	11	11	8	6
60	6	9	8	7	5
75	4	6	5	5	4
90	2	3	3	1	1

Table 2. Number of misclassified data on Wine data set (#: number of labeled data)

#	1-NN	FCM	KFCM	S^2FCM	S^2KFCM
45	43	38	36	38	37
60	41	34	34	33	32
75	29	25	26	24	24
90	22	23	22	21	18

From Table 1, we see that in nearly all cases S^2KFCM achieves the best performance. Also we know that the semi-supervised algorithms are superior to the corresponding unsupervised one for classification. Finally, as the labeled data increases, the numbers of misclassified data of all algorithms decrease. It was reported in [10] that when 60 labeled data is used for training, the classification accuracy of SVM is 94.0%, which is very close to that of S^2KFCM.

The second example is the Wine data set [9]. It contains 3 clusters with 59, 71 and 48 samples respectively. We choose from the total 178 data one portion as the labeled data set and the other as the unlabeled data set, and Table 2 gives the results. Clearly, S^2KFCM also shows good advantage over other algorithms.

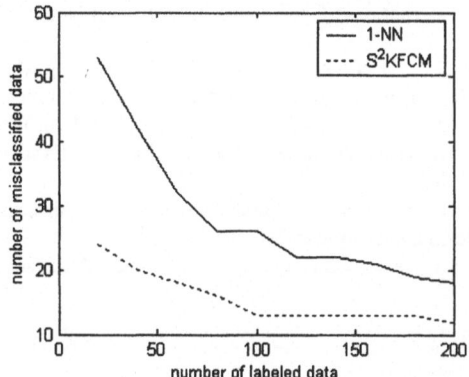

Fig. 1. Comparisons of number of misclassified data on Wisconsin Breast-Cancer data set of 1-NN and S^2KFCM

Finally, we made comparisons between 1-NN and S^2KFCM on Wisconsin Breast-Cancer data set [9], which contains 2 clusters with 444 and 239 data respectively. Fig. 1 shows the results. From Fig. 1, we see that there is a significant increase in the performance of S^2KFCM by using the unlabeled data for classification. When 200 labeled data is used, S^2KFCM has 12 misclassified data, which is comparative to SVM [10]. As number of the labeled data increases, the difference between 1-NN and S^2KFCM decrease, which support our claim that when a small number of labeled data is available, S^2KFCM can obtain better performance compared with classical classifiers.

5   Conclusions

A semi-supervised fuzzy kernel clustering algorithm called S^2KFCM is proposed, which can utilize unlabeled data together with labeled data for classification. Experiments are carried out on benchmark data sets and the results show that by combing both labeled and unlabeled data in the learning process, S^2KFCM achieves good classification results.

References

1. Saint-Jean, C., Frelicot, C.: A Robust Semi-Supervised EM Clustering Algorithm with a Reject Option. In: Proceedings of the International Conference on Pattern Recogntion, 2002
2. Bensaid, A.M., Hall, L.O., Bezdek, J.C., Clarke, L.P.: Partial Supervised Clustering for Image Segmentation. Pattern Recognition 29 (1996) 859-871
3. Pedrycz, W., Waletzky, J.: Fuzzy Clustering with partial supervision. IEEE Trans. on Systems, Man, and Cybernetics, Part B-Cybernetics 27 (1997) 787-795
4. Bennett, K., Demiriz, A.: Semi-Supervised Support Vector Machines. Advances in Neural Information Processing Systems 11 (1999) 368-374
5. Zhang, D.Q., Chen, S.C.: Kernel-based Fuzzy and Possibilistic C-Means Clustering. In: Proceedings of the International Conference on Artificial Neural Networks, Istanbul, Turkey, 2003, 122-125
6. Zhang, D.Q., Chen, S.C.: A Novel Kernelised Fuzzy C-Means Algorithm with Application in Medical Image Segmentation. Artificial Intelligence in Medicine, in press (2004)
7. Duda, R.O., Hart, P.E., Stork, D.G.: Pattern Classification. John Wiley (2001)
8. Bezdek, J.C.: Pattern Recognition with Fuzzy Objective Function Algorithms. Plenum Press, New York (1981)
9. UCI Repository of Machine Learning Databases, University of California, Irvine, available from: http://www.ics.uci.edu/~mlearn/MLRository.html
10. Chen, S.C., Yang, X.B.: Alternative Linear Discriminant Analysis. Pattern Recognition, in press (2004)

Use of Autocorrelation Kernels in Kernel Canonical Correlation Analysis for Texture Classification

Yo Horikawa

Faculty of Engineering, Kagawa University
761-0396 Takamatsu, Japan

Abstract. Kernel canonical correlation analysis (KCCA) with autocorrelation kernels is applied to invariant texture classification. The autocorrelation kernels are the inner products of the autocorrelation functions of original data and effectively calculated with the cross-correlation functions. Classification experiment shows the autocorrelation kernels perform better than the linear and Gaussian kernels in KCCA. Further, it is shown that the generalization ability is degraded as the order of the autocorrelation kernels increases, since relative values of the kernels of different data tend to zero.

1 Introduction

Nonlinear kernel-based statistical analysis for pattern recognition has attracted much attention [1], [2], e. g., support vector machines (SVMs), kernel Fisher discriminant analysis, kernel principal component analysis (KPCA), spectral clustering and kernel canonical correlation analysis (KCCA). In the kernel methods inner products of feature vectors used in classical multivariate analyses is replaced to nonlinear kernel functions, through which nonlinear mappings of original feature vectors to high-dimensional spaces are performed in an implicit manner. SVMs search an optimal linear discriminant function and KPCA finds optimal directions in transformed feature spaces. Recently, SVMs and KPCA using raw gray-scaled values of image data as input were applied to texture classification and segmentation problems and showed good performance comparable with conventional feature extraction methods [3], [4].

Kernel Canonical correlation analysis (KCCA) is a kernel version of canonical correlation analysis [5], [6], [7], [8]. Canonical correlation analysis, proposed by H. Hotelling in 1935, finds linear transformations that yield maximum correlation between two variables. Then KCCA were applied to image data, e.g., pose estimation of objects from their appearance images [9] and content-based retrieval of image database [10].

Autocorrelation kernels, in which the inner product of the autocorrelations of original image data is used for a kernel function, were introduced in SVMs for the detection of face images [11]. The inner product of the autocorrelations is obtained directly through the convolution of original images and large computational costs necessary for the calculation of higher-order correlation functions can be avoided [12]. SVMs with the autocorrelation kernels was applied to texture classification and showed better performance than SVMs with conventional kernels [13].

In this study KCCA with the autocorrelation kernels is applied to texture classification and its performance is evaluated. Formulation of KCCA with the autocorrelation

N.R. Pal et al. (Eds.): ICONIP 2004, LNCS 3316, pp. 1235–1240, 2004.
© Springer-Verlag Berlin Heidelberg 2004

kernels is explained in Sect. 2. The classification experiment on texture images are shown in Sect. 3. The effects of the order of the autocorrelation kernels are discussed in Sect. 4.

2 KCCA and Autocorrelation Kernels

Let (x_i, y_i), $(1 \leq i \leq n)$ are pairs of feature vectors of n sample objects, which describe different aspects of the objects. The feature vectors are transformed to $(\varphi(x_i), \theta(y_i))$ in another feature spaces with nonlinear mappings φ and θ. We assume that the transformed features are centered, i.e., $\sum_{i=1}^{n} \varphi(x_i) = \sum_{i=1}^{n} \theta(y_i) = 0$, for simplicity. (The mean centering can be done in the calculation of kernel functions [14].) Kernel matrices Φ and Θ are defined by $\Phi_{ij} = <\varphi(x_i), \varphi(x_j)>$ and $\Theta_{ij} = <\theta(y_i), \theta(y_j)>$, $(1 \leq i \leq n)$, where $<\cdot, \cdot>$ denotes the inner product. Then we obtain the eigenvector $(f^T, g^T)^T$ of the following generalized eigenproblem.

$$\begin{pmatrix} 0 & \Phi\Theta \\ \Theta\Phi & 0 \end{pmatrix} \begin{pmatrix} f \\ g \end{pmatrix} = \lambda \begin{pmatrix} \Phi^2 + \gamma_x I & 0 \\ 0 & \Theta^2 + \gamma_y I \end{pmatrix} \begin{pmatrix} f \\ g \end{pmatrix} \tag{1}$$

It has been recommended that small multiples of the identity matrix $\gamma_x I$ and $\gamma_y I$ (γ_x, $\gamma_y \geq 0$) are added for the regularization, i.e., avoiding the overfitting to sample data and the singularity of the kernel matrices.

The n-dimensional vectors $f^T = (f_1, \cdots, f_n)$ and $g^T = (g_1, \cdots, g_n)$ are the coefficients of the linear expansions of the projections w_φ and w_θ that maximize the correlation between $u = <w_\varphi, \varphi(x)>$ and $v = <w_\theta, \theta(y)>$ in terms of the transformed sample features.

$$w_\varphi = \sum_{i=1}^{n} f_i \varphi(x_i), \quad w_\theta = \sum_{i=1}^{n} g_i \theta(y_i) \tag{2}$$

The canonical components u and v of (x, y) of a new object are then obtained by

$$u = \sum_{i=1}^{n} f_i <\varphi(x_i), \varphi(x)>, \quad v = \sum_{i=1}^{n} g_i <\theta(y_i), \theta(y)> \tag{3}$$

Under Mercer's condition, the inner products are calculated directly from original feature vectors through kernel functions without evaluating φ and θ (the kernel trick).

$$<\varphi(x_i), \varphi(x)> = k_\varphi(x_i, x), \quad <\theta(y), \theta(y_j)> = k_\theta(y_j, y) \tag{4}$$

The Gaussian kernel function is one of kernels of wide use.

$$k_G(x_i, x_j) = \exp(-\mu \|x_i - x_j\|^2) \tag{5}$$

Autocorrelation kernels are derived by exploiting the kth-order autocorrelation of original feature vector x as the mapping φ [11]. In the following, we consider 2-dimensional image data $x(l, m)$, $(1 \leq l \leq L, 1 \leq m \leq M)$ as the first feature vector x. The kth-order autocorrelation $r_x(l_1, l_2, \cdots, l_{k-1}, m_1, m_2, \cdots, m_{k-1})$ of $x(l, m)$ is defined by

$$r_x(l_1, l_2, \cdots, l_{k-1}, m_1, m_2, \cdots, m_{k-1}) = \sum_l \sum_m x(l, m) x(l+l_1, m+m_1) \cdots x(l+l_{k-1}, m+m_{k-1}) \tag{6}$$

The inner product of the autocorrelations r_{xi} and r_{xj} of image data $x_i(l, m)$ and $x_j(l, m)$ is calculated by

$$<r_{xi}, r_{xj}> = \sum_{l1=0}^{L1-1} \sum_{m1=0}^{M1-1} \{ \sum_{l=1}^{L-l1} \sum_{m=1}^{M-m1} x_i(l,m) x_j(l+l_1, m+m_1)/(LM) \}^k /(L_1 M_1) \tag{7}$$

which corresponds to the sum of the kth power of the cross-correlation of image data. (Note that the definition of the order k is based on a popular use and is different from that in [12].) Computational costs are reduced in the practical order even for large values of k and large data size L and M since the calculation of explicit values of the autocorrelations are avoided. Equation (7) is employed as the kth-order autocorrelation kernel function $k_\varphi(x_i, x_j)$ of image data in KCCA.

To apply KCCA for classification problems we can use an indicator vector as the second feature vector y [8]. When sample objects are categorized into C classes, $y = (y_1, \cdots, y_C)$ corresponding to x is defined by

$$y_c = 1 \quad \text{if } x \text{ belongs to class } c, \qquad y_c = 0 \quad \text{otherwise} \tag{8}$$

Then a mapping θ of y is not adopted, i.e., the linear inner product $y_i{}^T y_j$ is used as the kernel function $k_\theta(y_i, y_j)$. The number of non-zero eigenvalues of the generalized eigenproblem (Eq. (1)) is C-1 under the mean centering. The canonical components u_i ($1 \le i \le C$-1) for an unknown object are calculated with the eigenvectors f_i ($1 \le i \le C$-1) corresponding to them and a feature vector x. Classification methods, e.g., the nearest-neighbor method, the discriminant analysis and SVMs can be applied in the canonical component space.

3 Experiment

3.1 Method

Kernel canonical correlation analysis with autocorrelation kernels is applied to invariant texture classification. Gray-scaled values in [0, 1) are calculated from 8bit black-and-white images scanned from the Brodatz album [15]. Four images of 512×512 pixels shown in Fig. 1 are employed: D5 (Expanded mica), D92 (Pigskin) D4 (Pressed cork) and D84 (Raffia looped to a high pile).

A hundred sample and test images of 50×50 pixels with random transformations and/or additive Gaussian noise are made for each image. The sample images and the test images (Test 1) are taken from the original images of 512×512 pixels at random positions. As well as the random shift, the Gaussian noise with the mean 0 and SD 0.1 is added to each pixel in Test 2. (The SN rates are between 5.5 and 7.0 dB). The applicability to similarity transformations is also tested in Test 3 (random shift [0, 400], scaling [×0.5, ×1.0] and rotation [0, 2π]). All the images for input to KCCA are linearly normalized with the mean 0 and SD 1.0, so that they are indistinguishable by differences only in the mean and SD.

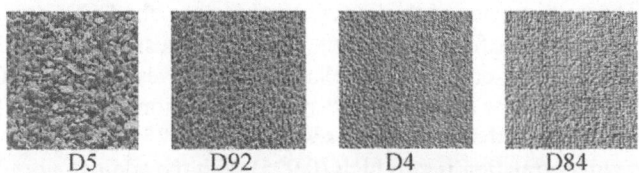

D5   D92   D4   D84

Fig. 1. Texture images from the Brodatz album

The eigenvectors f_i $(1 \leq i \leq C\text{-}1$ $(C = 4))$ corresponding to the non-zero eigenvectors of Eq. (1) are calculated with the sample images. The canonical components u_i $(1 \leq i \leq 3)$ of the test images are calculated with Eq. (3). The following three kinds of the kernel functions are used as $k_\varphi(x_i, x_j)$ $(= \langle\varphi(x_i), \varphi(x)\rangle)$ for the image data: (a) linear kernel $(x_i^T x_j)$; (b) Gaussian kernel (Eq. (5)); (c) autocorrelation kernel (Eq. (7)). The value of μ in the Gaussian kernel is heuristically taken to be 2.0. In the calculation of the autocorrelation kernels, L_1 and M_1 are set to be 10, i.e., the correlations within 10×10 pixels are used in the images of 50×50 pixels. To avoid numerical instability, the elements Φ_{ij} of the kernel matrix Φ are divided by $E\{\Phi_{ij}\}$ in Eq. (1). Values of the regularization parameters are set to be $\gamma_x = \gamma_y = 0.1n$ $(n = 100\times4 = 400)$.

3.2 Results

First, the canonical components u_1 and u_2 of the test images (Test 1) are plotted in Fig. 2. In the linear (a) and Gaussian (b) kernels, the canonical components of four classes are widely overlapped and less effective for classification. In the 2nd-order autocorrelation kernel (c), however, the canonical components of the different classes are well separated. As the order of the autocorrelation kernels increases (d), (e), they are distributed on lines through the center. (The center is slightly shifted from the origin since the mean centering of the test images is done with the kernels of the sample images.)

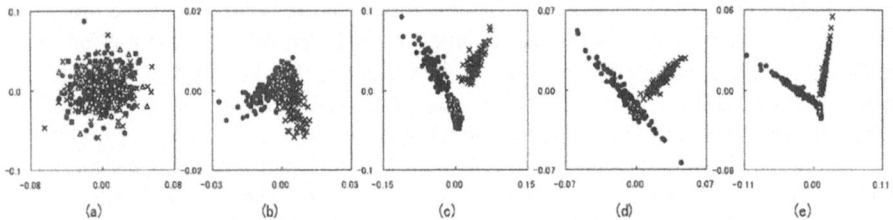

Fig. 2. Canonical components (u_1, u_2) with the linear kernel (a), the Gaussian kernel (b), the 2nd, 3rd, 4th-order correlation kernels (c)-(e). Plotted are circle (●) for D5, triangle (Δ) for D92, square (■) for D4 and cross (×) for D84

The test images are classified with the simple nearest-neighbor method in the canonical component space (u_1, u_2, u_3). The correct classification rates (CCRs) for Tests 1-3 are shown in Fig 3. (Classification with linear SVMs gives similar CCRs though not shown.) The performance of the linear kernel is poor: the CCRs are just over 0.25 (the rate of random classification). When using the Gaussian kernel, the CCRs increase up to 0.6. The 2nd-order autocorrelation kernel gives the highest CCR 0.97 for Test 1. While the CCRs with the 3rd-order autocorrelation kernels are low, the performance is degraded as the order increases (CCR $\approx$ 0.25, for $k = 10$). The CCR of the 2nd-order autocorrelation is still high (0.925) with the additive noise (Test 2), but drops to 0.7 under the scaling and rotation (Test 3) as expected.

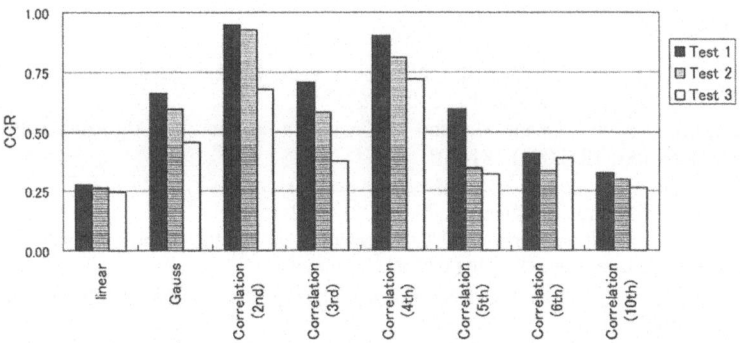

Fig. 3. Correct classification rate (CCR) for the 4-class classification: D4, D84, D5, D92

4 Discussion

In this study, it was shown that KCCA with the autocorrelation kernels, in which the spatial correlations of images are taken into account, gives higher CCRs than the linear and Gaussian kernels in texture classification. The classification performance of KCCA with the 2nd-order autocorrelation kernel competes with the results of the comparative study using LVQ classifiers with various kinds of features [16].

The performance of KCCA with the autocorrelation kernels depends on the order k of the autocorrelations. In KCCA with the autocorrelation kernels, the canonical components of the test images tend to lie on the lines through the center when the order k increases, as shown in Fig. 2(c)-(e). This can be explained as follows. The sum of the kth power of the cross-correlations ($\sum \{cc(\cdot)\}^k$) in the kth-order autocorrelation kernel (Eq. (7)) is approximated by the max norm (max $cc(\cdot)$) for large k [12]. Then the relative values of the autocorrelation kernels $<r_{xi}, r_{xj}>$ of different data ($x_i, x_j, (i \neq j)$) tend to zero as the order k increases ($<r_{xi}, r_{xj}>/<r_{xi}, r_{xi}> \to 0$ ($k \to \infty$, $i \neq j$)). Since each image data is normalized, the kernel matrix Φ tends to be proportional to I. Consequently, the elements of the eigenvectors of Eq. (1) corresponding to sample data of the same classes take the same values. i.e., $(f_1, \cdots, f_n) = (p_1 \cdots p_1, p_2 \cdots p_2, \cdots,$ $p_C \cdots p_C)$ since Θ is a block matrix. The canonical components of sample data x_i belonging to the cth class take the value $<r_{xi}, r_{xi}>p_c$, which depends only on the class. The canonical components of sample data of the same class then lie on a line through the origin of the same direction, i.e. $u_{xi} = <r_{xi}, r_{xi}>p_c$ ($p_c = (p_{c,1}, \cdots, p_{c,C-1})$). For test data x, the canonical components lie on the line if one of sample data is similar to it, but tend to be located at the origin in general since the relative values of the autocorrelation kernels of different data tend to be zero ($<r_{xi}, r_x>/<r_{xi}, r_{xi}> \to 0$).

As a result, the generalization ability and robustness tend to be lost as the order k increases. This effect is the same as the Gaussian kernel with large μ. The choice of the order of the autocorrelation kernels adequate for the objects is then necessary. It has been shown that CCRs generally increase as the order of the kernels increases in the face detection experiments with SVMs [11]. However, the results of this study and the experiment with SVMs [13] suggest that the kernels of low-orders ($k = 2, 3, 4$) are preferable for texture classification.

References

1. Ruiz, A., López-de-Teruel, P. E.: Nonlinear Kernel-Based Statistical Pattern Analysis. IEEE Trans. Neural Networks. 12 (2001) 16-32
2. Müller, K.-R., et al.: An Introduction to Kernel-Based Learning Algorithms. IEEE Trans. Neural Networks. 12 (2001) 181-201
3. Kim, K. I., et al.: Support Vector Machines for Texture Classification. IEEE Trans. Pattern Analysis and Machine Intelligence. 24 (2002) 1542-1550
4. Kim, K. I., et al.: Kernel Principal Component Analysis for Texture Classification. IEEE Signal Processing Letters. 8 (2001) 39-41
5. Lai, P. L., Fyfe, C.: Kernel and Nonlinear Canonical Correlation Analysis. Int. J. Neural Systems. 10 (2000) 365-377
6. Akaho, S.: A Kernel Method for Canonical Correlation Analysis. Proc. Int. Meeting on Psychometric Society (IMPS2001). (2001)
7. Bach, F. R., Jordan, M. I.: Kernel Independent Component Analysis. J. Machine Learning Research. 3 (2002) 1-48
8. Kuss, M., Graepel, T.: The Geometry of Kernel Canonical Correlation Analysis. Technical Report 108. Max Plank Institute for Biological Cybernetics (2002)
9. Melzer, T., Reiter, M., Bischof, H.: Appearance Models Based on Kernel Canonical Correlation Analysis. Pattern Recognition. 36 (2003) 1961-1971
10. Hardoon, D. R., Szedmak, S., Shawe-Taylor, J.: Canonical Correlation Analysis; An Overview with Application to Learning Methods. Technical Report CSD-TR-03-02. Dept. of Computer Science, University of London (2003)
11. Popovici, V., Thiran, J. P.: Higher Order Autocorrelations for Pattern Classification. Proc. IEEE 2001 International Conference on Image Processing (ICIP 2001). (2001) 724-727
12. McLaughlin, J. A., Raviv, J.: Nth-Order Autocorrelations in Pattern Recognition. Information and Control. 12 (1968) 121-142
13. Horikawa, Y.: Comparison of Support Vector Machines with Autocorrelation Kernels for Invariant Texture Classification. Submitted to 17th Int. Conf. Pattern Recognition (ICPR 2004), to appear.
14. Schölkopf, B., Smola, A., Müller K.-R.: Nonlinear Component Analysis as a Kernel Eigenvalue Problem. Neural Computation. 10 (1998) 1299-1319
15. Brodatz, P.: Textures: A Photographic Album for Artists and Designers. Dover, New York (1966)
16. Randen, T., Husøy, J. H.: Filtering for Texture Classification: A Comparative Study. IEEE Trans. Pattern Analysis and Machine Intelligence. 21 (1999) 291-310

Phoneme Transcription by a Support Vector Machine

Anurag Sahajpal[1], Terje Kristensen[1], and Gaurav Kumar[2]

[1] Department of Computer Engineering
Bergen University College
N-5020 Bergen, Norway
{asah,tkr}@hib.no

[2] Indian Institute of Technology, Delhi
Hauz Khas, Delhi-110016
India
gkiitd@rediffmail.com

Abstract. In this paper a support vector machine program is developed that is trained to transcribe Norwegian text to phonemes. The database consists of about 50,000 Norwegian words and is developed by the Norwegian Telecom Research Centre. The transcription regime used is based on SAMPA for Norwegian. The performance of the system has been tested on about 10,000 unknown words.

1 Introduction

The basic abstract symbol representing speech sound is the phoneme. In both text-to-speech systems or in automatic speech recognition systems the phonetic analysis is an important component. Its primary function is to convert the text into the corresponding phonetic sequence. This process is called transcription.

In many systems in use today the phonetic rules are compiled by phonetic experts. However, great effort and expertise are required when setting up such a system for a new language or a specific task. By using an automatic approach to the transcription process much effort may be spared.

Artificial neural network (ANN) [3] may be used to automate such a task. Phoneme transcription by ANN was first introduced by Sejnowski and Rosenberg in their classical paper from 1987 [9]. The main advantage of applying an ANN regime to such a problem is its flexible use for any language or dialect - only the patterns vary. Such an approach will also usually perform well on transcription of words not encountered before. This depends of course on how well the neural network has been trained.

In earlier papers [4],[5],[6] we have shown that a neural network is capable of transcribing Norwegian text fairly well. The ANN approach to the problem of phoneme transcription of a given language can be described as a pattern matching technique where an ANN is trained on a set of transcription patterns to determine how words are transcribed into phonemes. The focus of this paper, however, is to demonstrate how a support vector machine (SVM) [1],[2],[3] can be used to solve the same task.

N.R. Pal et al. (Eds.): ICONIP 2004, LNCS 3316, pp. 1241–1246, 2004.
© Springer-Verlag Berlin Heidelberg 2004

2 SAMPA

The transcription scheme in this paper is an approximation of SAMPA (Speech Assessment Methods Phonetic Alphabet) for Norwegian [8], which is a machine-readable phonetic alphabet and constitutes today's best basis of encoding of machine-readable phonetic notation.

In the transcription scheme for Norwegian based on SAMPA that we have used, the consonants and the vowels are classified into different subgroups defined as follows:

Plosives (6)
Fricatives (7)
Sonorant consonants (5)

where the number in brackets indicate the number of each type.

The vowels are likewise defined by the subgroups:

Long vowels (9)
Short vowels (10)
Diphthongs (7)
Allophones (7)
Allophonic variants (5)

In our phoneme database, ordinary and double stress marks are indicated by ! and !! instead of " and ", that is used by the Norwegian Telecom. Figure 1 shows some words with location of stress, transcribed by the SAMPA notation.

3 Theory

Support Vector Machines is a computationally efficient learning technique that is now being widely used in pattern recognition and regression estimation problems. This approach has been derived from the ideas of statistical learning theory [10],[11] regarding controlling the generalization abilities of a learning machine. In this approach the machine learns an optimum hyperplane that classifies the given pattern. The input feature space, by use of non-linear kernel functions, can be transformed into a higher dimensional space, where the optimum hyperplane can be learnt. This gives a flexibility of using one of many learning models by changing the kernel functions.

The idea of SVM is to explicitly map the input data to some higher-dimensional space, where the data is linearly separable. We can use a mapping:

$$F : \Re^N \to \Im \tag{1}$$

where N is the dimension of the input space, and $\Im$ the higher-dimensional space, termed feature space. If a non-linear mapping F is used, the resulting hyperplane in input-space will be non-linear. Thus, a mapping from input-space to feature-space can be achieved via a substitution of the inner product with:

$$x_i \cdot x_j \to F(x_i) \cdot F(x_j) \tag{2}$$

Now, the training of the network is to maximize the below mentioned Lagrangian function, L_D:

$$L_D = \sum_{i=1}^{N} \alpha_i - \frac{1}{2} \sum_{i,j}^{N} \alpha_i \alpha_j y_i y_j K(\mathbf{x_i}, \mathbf{x_j}) \tag{3}$$

subject to the constraint:

$$\sum_{i=1}^{N} \alpha_i y_i = 0 \tag{4}$$

where $\alpha_i \geq 0$ for $i = 1 \ldots N$ and y_i is the target value.

The choice of the kernel function depends on the application. If the kernel is chosen as Gaussian functions,

$$K(\mathbf{x_i}, \mathbf{x_j}) = e^{-\|\mathbf{x_i} - \mathbf{x_j}\|^2 / 2\sigma^2} \tag{5}$$

then the SVM classifier turns into its equivalent RBF neural network [3]. In this paper, we use the RBF kernel, as defined in 5.

4   Grapheme to Phoneme Transcription

The training data consists of words that are constructed from all the letters (29) of Norwegian. In addition, space is included. Only words with stress indication are included in our training database. Some character combinations have been treated in a special way in the transcription algorithm that we have been using, for example double consonants. A double tt, for instance, is transcribed to a single t. Text units consisting of one or more letters in combination, corresponding to a phoneme, is often called a grapheme.

According to our transcription scheme, the Norwegian language consists totally of 56 different graphemes. 30 of the graphemes are one character long and the rest consists of two or three characters. The second group includes double consonants such as bb, ff, pp, rr, ss etc. diphthongs such as ei, au etc. and allophones such as rd, rl etc.

The graphemes are mapped into corresponding 56 phoneme units. This mapping between graphemes to phonemes is not one-to-one. In other words, a particular grapheme can map, for instance, to two different phonemes, depending upon its context. Figure 1 shows some words of the training file with location of stress, transcribed by the SAMPA notation.

Words	Transcription
apene	!!A:p@n@
lønnsoppgaver	!l2nsOpgA:v@r
politiinspektørene	!puliti:inspk!t2:r@n@
regjeringspartiet	re!jeriNspArti:@
spesialundervisningen	spesi!A:l}n@rvi:sniN@n

Fig. 1. Some words in the training database with different stress placements.

5   Transcription Method

An actual word is considered and a letter pattern is generated by using a window segment on it. The phoneme to be predicted is in the middle of the window. The Norwegian language following the transcription regime that we have used, totally consists of 56 phonemes. Therefore our SVM needs to classify the text into 56 different classes, each for a separate phoneme.

A certain view (or window) is given to the different words, as shown in figure 2. The window size selected in the experiments is seven letters. At any given time the letter in the middle of the window is active. In figure 2 this is r. When r is active, all the other letters will be inactive.

To handle the beginning and end of each word, and to fill up the space of seven letters, a star (*) is used. The desired output of the network is the correct phoneme associated with the centre or the fourth letter in the window. The other six letters - three on either side of the middle letter - provide the context.

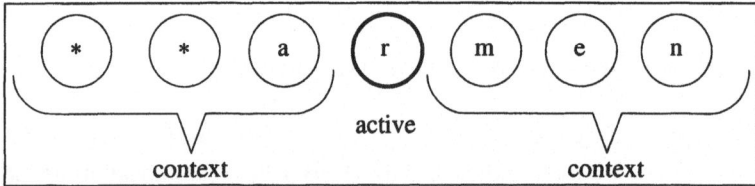

Fig. 2. The Norwegian word "armen" in a seven-letter window. The associated phoneme class of the letter r will be the output of SVM.

The words are stepped through the window letter by letter. At each step the SVM classfies the phoneme associated with the letter in the middle of the window. The parameters of the separating hyperplane are learnt correspondingly.

6   Preprocessing and Coding

The original training data consists of letter patters of words and their transcription. Each such letter pattern has to be preprocessed before it is fed into the network. A separate program has been developed to do this preprocessing. An internal coding table has been defined in the program to represent each letter and its corresponding phoneme.

A pattern file is generated, consisting of a huge amount of patterns conforming to an input format, such as the one given in figure 3. Figure 3 illustrates the pattern corresponding to the word '***ape*'. The target pattern is given by '!!A:', which represents double stress and the phoneme 'long A vowel'. '*' is represented by a 0 in the input.

Input vector : [0 0 0 1 16 5 0]
Target :   [2 34]

Fig. 3. The input format.

The letter 'a' is expressed by a '1' as the 4th attribute. In the target, the corresponding phoneme 'A:' is mapped to the correct position ie. 34 and the double stress is indicated by 2.

Each input pattern (or feature vector) in the file, such as the one given in figure 3, has seven components, one for each letter in the window segment. The components represent the letters of the Norwegian language and, at any given time, the component in the middle is active, for which the phoneme is produced. The three components on either side of the active component provide the context. The size of the preprocessed file is about 17 Mbytes by which the SVM is trained.

The problem of mapping a letter to its phoneme is divided into two classification problems, one for the *stress* classification (ordinary, double or no stress - 3 classes) and other for the *phoneme* classification (56 classes). The stress classification is done separately from phoneme classification. In the final phase of classification, the two output files are merged into a single file.

7 Experiments and Results

All the experiments have been done on a 2.4 GHz Pentium PC with a 256 Mbytes memory. The results produced by the SVM and the target were then compared. The classification model used for multi class patterns was one versus rest. The kernel function used for experimentation is RBF kernel. The cost function parameter C for optimal training was set to 200. The gamma ($\frac{1}{2}\sigma^2$) for the RBF kernel was set as 0.142.

The traning data file consists of about 35,000 Norwegian words. The test file consists of about 10,000 unseen Norwegian words. The individual SVMs (56 different SVM problems - one for each class) are trained. The training time for the experiment was about 15 hours to 7 days, depending upon the value of the different parameters. The program developed used LIBSVM [1] as the toolbox for training and testing. A program was written to combine the results from all the SVM files and make a single file predicting the occurrence of particular phonemes.

The performance of the SVM has been tested on about 10,000 unseen words and the accuracy obtained was 84% after string alignment. An edit distance method was used to perform the string alignment. A sample of Norwegian test words and produced transcriptions are given in table 1.

Table 1. A comparison between target and produced transcriptions for a sample of Norwegian test words produced by the SVM.

Text	Target	Produced
assisterte	Asi!ste:rt@	Asi!ste:rt@
assortert	Asurt!e:rt	As}rte:rt
attesterte	At@!ste:rt@	At@!ste:rt@
proletariat	prul@tAri!A:t	prul@tAri!A:t
representantene	repr@s@n!tAnt@n@	repr}s@n!tAnt@n@
sivilisasjonene	sivilisA!su:n@n@	sivilisA!su:n@n@

8 Conclusion

A support vector machine has been developed to transcribe Norwegian text to phonemes. The SVM has been trained on a phoneme database of about 35,000 Norwegian words. The transcription scheme is based on SAMPA notation for Norwegian. The SVM has been trained on words constructed from all the letters of Norwegian with arbitrary length and different types of stress included.

We have tested two multi class strategies for the training namely, *One vs One* [7] and the *One vs Rest* methods. Because of highly unbalanced data we come to a conclusion that a higher accuracy is achieved by the One vs Rest method. Compared to the accuracy obtained by the MLP network [4],[6], this is a better result. An added advantage is that the SVM approach has a lesser training time. The experiments, so far, indicates that a SVM method is the preferred one to use to transcribe Norwegian text.

References

1. Chih-Chung Chang and Chih-Jen Lin. LIBSVM : a library for Support Vector Machines. available online at http://www.csie.ntu.edu.tw/ cjlin/libsvm, 2001.
2. C.J.C Burges. A Tutorial on Support Vector Machines for Pattern Recognition. *Knowledge Discovery and Data Mining*, 2(2), 1998.
3. Haykin H. *Neural Netowrks - a Comprehensive Foundation*. Prentice Hall, 1999.
4. Kristensen T. Two Neural Network Paradigms of Phoneme Transcription - a Comparison. In *I.E.E.E International Joint Conference on Neural Computing (IJCNN)*, Budapest, Hungary, 2004.
5. Kristensen T. and Treeck B. and Falck-Olsen R. Phoneme Transcription based on Sampa for Norwegian. In *I.E.E.E International Joint Conference on Neural Computing (IJCNN)*, Portland , Oregon, USA, 2003.
6. Kristensen T. and Treeck B. and Falck-Olsen R. Phoneme Transcription of Norwegian Text. In *13th International Conference on Artificial Neural Network and International Conference on Neural Information Processing, ICANN/ICONIP*, Istanbul, Turkey, 2003.
7. Salomon Jesper and King Simon and Osbourne Miles. Framewise phone classification using support vector machines. In *International Conference on Spoken Language Processing*, Denver, USA, 2002.
8. SAMPA. available online at http://www.phon.ucl.ac.uk/home/sampa/norweg.htm.
9. Sejnowski T. J. and Rosenberg C. R. Parallel Networks that Learn to Pronounce English Text. Complex Systems Publications Inc., 1987.
10. Vapnik V. N. *Statistical Learning Theory*. Wiley, New York, 1998.
11. Vapnik V. N. An overview of statistical learning theory. *IEEE Transactions on Neural Networks*, 10, sep 1999.

A Comparison of Pruning Algorithms for Sparse Least Squares Support Vector Machines

L. Hoegaerts, J.A.K. Suykens, J. Vandewalle, and B. De Moor

Katholieke Universiteit Leuven, ESAT-SCD-SISTA
Kasteelpark Arenberg 10, B-3001 Leuven (Heverlee), Belgium
{luc.hoegaerts,johan.suykens}@esat.kuleuven.ac.be

Abstract. Least Squares Support Vector Machines (LS-SVM) is a proven method for classification and function approximation. In comparison to the standard Support Vector Machines (SVM) it only requires solving a linear system, but it lacks sparseness in the number of solution terms. Pruning can therefore be applied. Standard ways of pruning the LS-SVM consist of recursively solving the approximation problem and subsequently omitting data that have a small error in the previous pass and are based on support values. We suggest a slightly adapted variant that improves the performance significantly. We assess the relative regression performance of these pruning schemes in a comparison with two (for pruning adapted) subset selection schemes, -one based on the QR decomposition (supervised), one that searches the most representative feature vector span (unsupervised)-, random omission and backward selection on independent test sets in some benchmark experiments[1].

1   Introduction

In kernel based classification and function approximation the sparseness (i.e. limited number of kernel terms) of the approximator is an important issue, since it allows faster evaluation of new data points. The remaining points are often called support vectors. In Vapnik's SVM [1] the sparseness is built-in due to the ϵ-insensitive loss function that outrules errors of points inside a 'tube' around the approximated function. This results in a quadratic programming problem.

---

[1] This research work was carried out at the ESAT laboratory of the KUL, supported by grants from several funding agencies and sources: Research Council KUL: GOA-Mefisto 666, GOA-Ambiorics, several PhD/postdoc & fellow grants; FWO: PhD/postdoc grants, projects, G.0240.99, G.0407.02, G.0197.02, G.0141.03, G.0491.03, G.0120.03, G.0452.04, G.0499.04, research communities (ICCoS, AN-MMM, MLDM); AWI: Bil. Int. Collaboration Hungary/ Poland; IWT: PhD Grants, GBOU (McKnow); Belgian Federal Government: Belgian Federal Science Policy Office: IUAP V-22 (2002-2006), PODO-II (CP/01/40: TMS and Sustainability); EU: FP5-Quprodis; ERNSI; Eureka 2063-IMPACT; Eureka 2419-FliTE; Contract Research/agreements: ISMC/IPCOS, Data4s, TML, Elia, LMS, IPCOS, Mastercard; BOF OT/03/12, Tournesol 2004 - Project T2004.13. LH is a PhD student with IWT. JS is an associate professor at KUL. BDM and JVDW are full professors at KUL.

N.R. Pal et al. (Eds.): ICONIP 2004, LNCS 3316, pp. 1247–1253, 2004.
© Springer-Verlag Berlin Heidelberg 2004

In LS-SVM [2] a quadratic loss function is used instead, and the optimization problem reduces to solving a linear set of equations. But at the same time the sparseness is lost and must be imposed.

A simple approach to introduce sparseness is based on the sorted support value spectrum (the solution of the set of equations) [3]. From the LS-SVM solution equations follows a reasonable choice for pruning away points with a low error contribution in the dual optimization objective. Another recent paper [4] sophisticates the pruning mechanism by weighting the support values. The datapoint with the smallest error introduced after its omission is selected then. This pruning method is claimed to outperform the standard scheme of [3], but the extent of the comparison was limited to one example where noise was filtered out. We additionally suggest here an improved selection of the pruning point based on their derived criterion. Other methods for achieving sparse LS-SVMs is via Fixed-Size approaches [5] which employ entropy based subset selection in relation to kernel PCA density estimation. This has been successfully applied to a wide class of problems for subspace regression in feature space [6]. For a general overview of pruning we refer to [7].

Pruning is closely related to subset selection, choosing relevant datapoints or variables in order to build a sparse model. Pruning assumes that the model on the full set is iteratively downsized, in a backward manner, while subset selection usually proceeds in a forward manner. Many subset selection schemes can be distinguished that organise their search in a greedy fashion [8]. In particular we focus on two such schemes. A supervised method is based on the QR decomposition [9, 10], that we can employ as omitting points of which the orthogonalized components have least correlation with the output. Furthermore, an unsupervised approach is based on a best fitting span [11], that we can employ as omitting points that have least similarity to that span.

In this paper we aim at a comparison of regression performance between the two pruning procedures and the two subset selection procedures by performing a set of experiments with (i) evaluation on the independent test set and (ii) including random and backward pruning to have an objective measure.

2   LS-SVM for Function Estimation

Assume training data $\{\{(\mathbf{x}_i, y_i)\}_{i=1}^n \in \mathbb{R}^p \times \mathbb{R}\}$ have been given where $\mathbf{x}_i$ are the input data and y_i the target or output values for sample i. The goal of function approximation is to find the underlying relation between input and target values.

LS-SVM [5] assumes an underlying linear model in the weight parameters $\mathbf{w}$ with a bias term b: $y = \mathbf{w}^T \varphi(\mathbf{x}) + b$, where the feature map $\varphi : \mathbb{R}^p \to \mathbb{H}_k$ is a function into the r-dimensional Reproducing Kernel Hilbert space (RKHS) [12] $\mathbb{H}_k$ with an associated kernel $k : \mathbb{R}^p \times \mathbb{R}^p \to \mathbb{R} : (\mathbf{x}_i, \mathbf{x}_j) \mapsto k(\mathbf{x}_i, \mathbf{x}_j)$. A common choice is $k(\mathbf{x}_i, \mathbf{x}_j) = \exp(-\|\mathbf{x}_i - \mathbf{x}_j\|_2^2 / h^2)$, where h is a kernel width parameter. The mapping k provides a similarity measure between pairs of data points, should fulfill the Mercer condition of positive definiteness and is supposed to capture the nonlinearity, while the model remains linear in the parameters [13]. The weights in $\mathbf{w}$ and bias b are to be estimated by minimizing a primal space error cost function $\min_{\mathbf{w},b,\mathbf{e}} J(\mathbf{w}, b, \mathbf{e}) = \mathbf{w}^T\mathbf{w} + \gamma \sum_{i=1}^{n} e_i^2$ s.t. $y_i = \mathbf{w}^T\varphi(\mathbf{x}_i) + b + e_i$, where $i = 1,\ldots,n$. The objective consists of a smallest sum of squares error term (to fit to the training data) and a regularization term to smoothen the approximation (to compensate overfitting). Working with the explicit expression for φ is avoided by considering the dual formulation of this cost function in feature space H_k. The optimization objective becomes the Lagrangian $L(\mathbf{w}, b, \mathbf{e}; \alpha) = J(\mathbf{w}, b, \mathbf{e}) - \sum_{i=1}^{n} \alpha_i(\mathbf{w}^T\varphi(\mathbf{x}_i) + b + e_i - y_i)$, where the α_i's are Lagrange multipliers. One solves by deriving the optimality conditions $\partial L/\partial \mathbf{w} = \partial L/\partial b = \partial L/\partial e_i = \partial L/\partial \alpha_i = 0$. Elimination of the variables $\mathbf{e}, \mathbf{w}$ through substitution naturally leads to a solution expressed solely in terms of inner products $\varphi(\mathbf{x}_i)^T\varphi(\mathbf{x}_j) = k(\mathbf{x}_i, \mathbf{x}_j)$, which results in a linear system [2]:

$$\begin{bmatrix} 0 & \mathbf{1}^T \\ \hline \mathbf{1} & K + \gamma^{-1}I \end{bmatrix} \begin{bmatrix} b \\ \alpha \end{bmatrix} = \begin{bmatrix} 0 \\ \mathbf{y} \end{bmatrix}, \tag{1}$$

where $\mathbf{1}$ is a vector column vector of ones, and $\mathbf{y}$ a vector with target values. The entries of the symmetric positive definite kernel matrix K equal $K_{ij} = k(\mathbf{x}_i, \mathbf{x}_j)$. The role of the (potentially infinite-dimensional) $r \times 1$ weight vector $\mathbf{w}$ in primal space is conveniently taken over by a directly related $n \times 1$ weight vector α in the dual space. Typically there is a model selection (e.g. cross-validation) procedure required for the determination of two hyperparameters (γ, h^2). Once these are fixed, the LS-SVM approximator can be evaluated at any point $\mathbf{x}$ by $\hat{y}(\mathbf{x}) = \mathbf{w}^T\varphi(\mathbf{x}_i) + b = \sum_{i=1}^{n} \alpha_i\varphi(\mathbf{x})^T\varphi(\mathbf{x}) + b = \sum_{i=1}^{n} \alpha_i k(\mathbf{x}_i, \mathbf{x}) + b$. Related models of regularization networks and Gaussian processes have been considered without the bias term b. LS-SVMs have been proposed as a class of kernel machines with a primal-dual formulation for KFDA, KRR, KPCA, KPLS, KCCA, recurrent networks and optimal control [5].

3   Pruning Methods

Pruning methods that are compared in this paper are:

1. Support Values. A simple way of imposing sparseness on the LS-SVM is by pruning those terms of the kernel expansion that have the smallest absolute value [3]. The motivation comes from the fact that the LS-SVM support values are proportional to the errors at the datapoints, namely $\alpha_i = \gamma e_i$. To omit the points that contribute least to the training error is a direct and cheap way to impose sparseness (denoted with 'sv' in our experiments).

2. Weighed Support Values ($\gamma = \infty$). Recently a sophistication of the above pruning scheme has been reported [4] that omits the sample that itself bears least error *after* it is omitted. The derivation makes a distinct criterion corresponding to the value of γ. When no regularization is applied, thus $\gamma = \infty$, one proposes to omit the sample that has the smallest absolute value of α_i, divided by the diagonal element (i, i) of the kernel matrix K. Compared to [3], the extension of [4] comes with a more expensive computation since the kernel matrix needs to

be inverted. It also claims to outperform the standard method and an example
is given where the *training error* is indeed systematically lower.

3. Weighed Support Values ($\gamma \neq \infty$). In case $\gamma \neq \infty$, [4] proposes to
omit the sample that has the smallest absolute value of the ith component
of $\left(AA_\gamma^{-1}e_ie_i^TA_\gamma^{-1}c\right)/\left(e_i^TA_\gamma^{-1}e_i\right)$, were $A = [0,1^T;1,K]$, $c = [0;y]$, $A_\gamma = A + \gamma^{-1}I_{(n+1)}$, and e_i a column vector with value 1 on element $i+1$. Both
cases are included in the experiments. So the alpha's need to be weighed, which
resembles the formula of optimal brain surgeon in which the inverse of the Hessian of the error surface of the model parameters appears in the denominator
[14]. In [4], no examples, nor any comparison was however made for the case
$\gamma \neq \infty$. In this paper we complement this result with experiments.

4. Weighed Support Values Sum ($\gamma \neq \infty$). As an extension to the work of
[4] we propose in the case $\gamma \neq \infty$ to omit the sample such that the *sum* of all
errors introduced in *every* point is smallest. Because omitting a point introduces
error at all points, it makes sense to look at the global increase in error over all
points and exclude the point that minimizes this measure, at no extra cost.

5. Orthogonal Error Minimizing Regressors. A subset selection procedure
[9,10] is motivated from the expression of the sum of squares errors on the training data that is obtained at the *optimal* LS solution. If the regressors are made
orthogonal through e.g. a Gram-Schmidt or QR like decomposition, their impact on reducing the error can be termwise expressed. It then turns out that
choosing orthogonalized regressors that are most coincident with the independent variable, contribute most to the error reduction. This ranks the points and
for pruning the least error-influencing point will be omitted here.

6. Approximating representers. A second subset selection procedure [11] is
unsupervised in nature and aims at finding a span of feature vectors that are
most similar to the other remaining ones. The similarity is measured by the
distance between the remaining vector and an arbitrary linear combination of
previously found vectors (gathered in the set S): $\min_\beta \|\varphi(x_i)-\Phi_S^T\beta\|^2/\|\varphi(x_i)\|^2$.
This criterion selects points by which all the remaining features can be best
approximated. The datapoints are again ranked in importance and for pruning
the last one will be omitted here (denoted 'span' in the experiments).

4   Experiments

In this section we describe the experimental comparison of the above 6 pruning
schemes. The prediction performance on an *independent test set* will serve as a
measure. For reference we include pruning of a randomly selected point (as an
upper bound) and pruning of a backward selected point, i.e. the one that yields a
model with least error after omitting and retraining the model on the remaining
points (as a lower bound). Backward selection can be expected to perform well,
but is an overly expensive method. We applied the algorithms to an artificial
sinc function estimation problem and several benchmarks from the UCI machine
learning repository [15]. In all experiments we standardized the data to zero

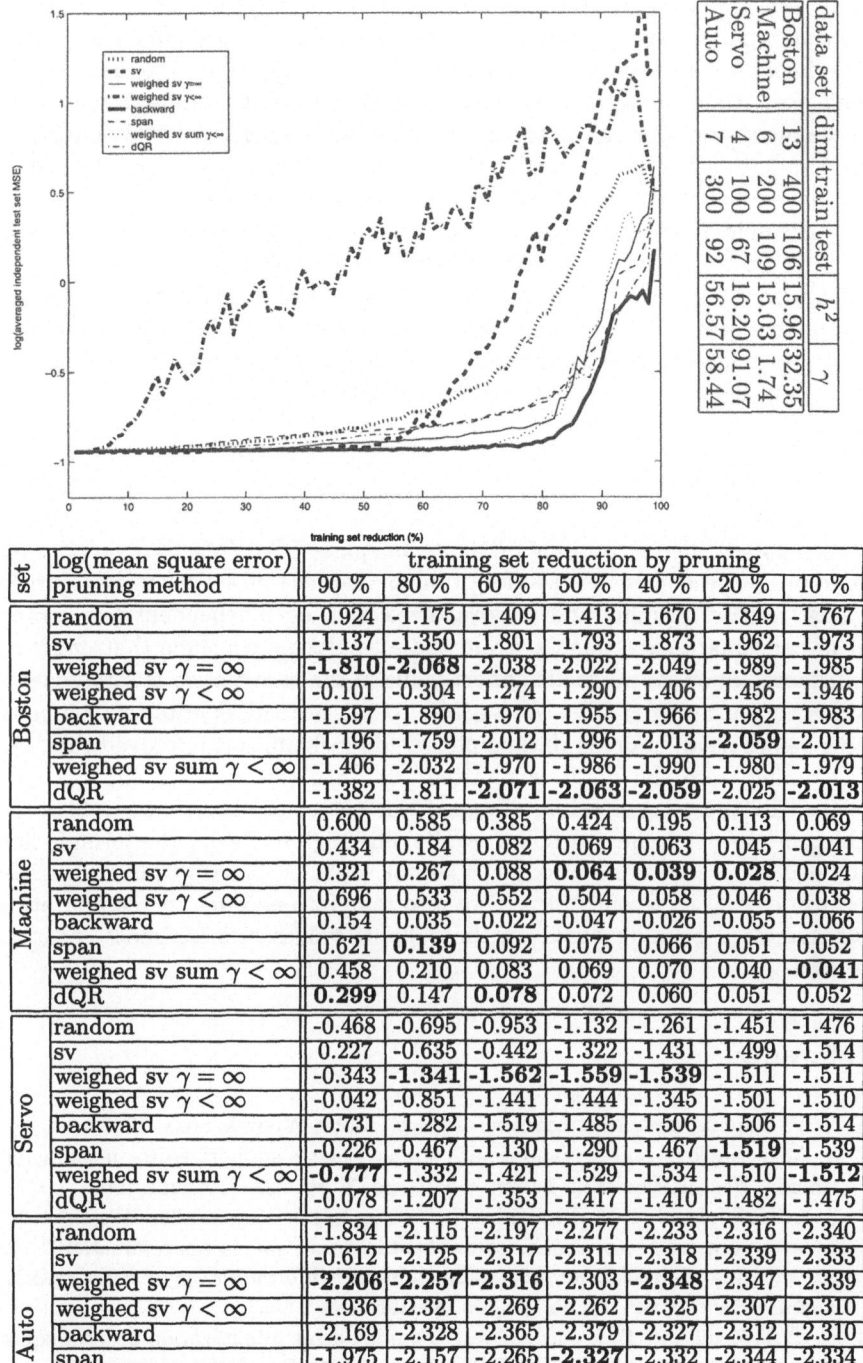

data set	dim	train	test	h^2	γ
Boston	13	400	106	15.96	32.35
Machine	6	200	109	15.03	1.74
Servo	4	100	67	16.20	91.07
Auto	7	300	92	56.57	58.44

set	log(mean square error) pruning method	training set reduction by pruning						
		90 %	80 %	60 %	50 %	40 %	20 %	10 %
Boston	random	-0.924	-1.175	-1.409	-1.413	-1.670	-1.849	-1.767
	sv	-1.137	-1.350	-1.801	-1.793	-1.873	-1.962	-1.973
	weighed sv $\gamma = \infty$	**-1.810**	**-2.068**	-2.038	-2.022	-2.049	-1.989	-1.985
	weighed sv $\gamma < \infty$	-0.101	-0.304	-1.274	-1.290	-1.406	-1.456	-1.946
	backward	-1.597	-1.890	-1.970	-1.955	-1.966	-1.982	-1.983
	span	-1.196	-1.759	-2.012	-1.996	-2.013	**-2.059**	-2.011
	weighed sv sum $\gamma < \infty$	-1.406	-2.032	-1.970	-1.986	-1.990	-1.980	-1.979
	dQR	-1.382	-1.811	**-2.071**	**-2.063**	**-2.059**	-2.025	**-2.013**
Machine	random	0.600	0.585	0.385	0.424	0.195	0.113	0.069
	sv	0.434	0.184	0.082	0.069	0.063	0.045	-0.041
	weighed sv $\gamma = \infty$	0.321	0.267	0.088	**0.064**	**0.039**	**0.028**	0.024
	weighed sv $\gamma < \infty$	0.696	0.533	0.552	0.504	0.058	0.046	0.038
	backward	0.154	0.035	-0.022	-0.047	-0.026	-0.055	-0.066
	span	0.621	**0.139**	0.092	0.075	0.066	0.051	0.052
	weighed sv sum $\gamma < \infty$	0.458	0.210	0.083	0.069	0.070	0.040	**-0.041**
	dQR	**0.299**	0.147	**0.078**	0.072	0.060	0.051	0.052
Servo	random	-0.468	-0.695	-0.953	-1.132	-1.261	-1.451	-1.476
	sv	0.227	-0.635	-0.442	-1.322	-1.431	-1.499	-1.514
	weighed sv $\gamma = \infty$	-0.343	**-1.341**	**-1.562**	**-1.559**	**-1.539**	-1.511	-1.511
	weighed sv $\gamma < \infty$	-0.042	-0.851	-1.441	-1.444	-1.345	-1.501	-1.510
	backward	-0.731	-1.282	-1.519	-1.485	-1.506	-1.506	-1.514
	span	-0.226	-0.467	-1.130	-1.290	-1.467	**-1.519**	-1.539
	weighed sv sum $\gamma < \infty$	**-0.777**	-1.332	-1.421	-1.529	-1.534	-1.510	**-1.512**
	dQR	-0.078	-1.207	-1.353	-1.417	-1.410	-1.482	-1.475
Auto	random	-1.834	-2.115	-2.197	-2.277	-2.233	-2.316	-2.340
	sv	-0.612	-2.125	-2.317	-2.311	-2.318	-2.339	-2.333
	weighed sv $\gamma = \infty$	**-2.206**	**-2.257**	**-2.316**	-2.303	**-2.348**	-2.347	-2.339
	weighed sv $\gamma < \infty$	-1.936	-2.321	-2.269	-2.262	-2.325	-2.307	-2.310
	backward	-2.169	-2.328	-2.365	-2.379	-2.327	-2.312	-2.310
	span	-1.975	-2.157	-2.265	**-2.327**	-2.332	-2.344	-2.334
	weighed sv sum $\gamma < \infty$	-2.149	-2.227	-2.304	-2.312	-2.339	-2.337	-2.336
	dQR	-2.086	-2.177	-2.289	-2.311	-2.316	-2.363	**-2.345**

mean and unit variance, we used the common Gaussian kernel and determined the tuning parameters h^2 and γ with standard 10-fold cross-validation. In every experiment we measured the mean square error on an independent test set, while every time one training sample has been left out according to the criterion of the algorithm. For the mexican hat function we performed 25 times a 100-sized uniformly spaced sample of a sinc with added Gaussian noise with standard deviation $\sigma = 0.2$. We show in the (prototypical) figure the averaged mean square error on an independent test set versus training set reduction (in %). From UCI we used Boston (housing prices), Machine (rel. cpu performance), Servo (rise times) and Auto (fuel consumption). All data set characteristics, selected tuning parameters and pruning results have been reported in the overview table for several pruning reduced training sets. The best results are typeset in bold face (excluding the performance of backward selection).

5   Conclusions

We compared different pruning methods for sparse LS-SVMs: four support value based pruning schemes and two (for pruning adapted) subset selection algorithms against random pruning and backward selection on independent test sets in a set of benchmark experiments. From these results we conclude that omitting a point based upon its weighed support value in the case $\gamma \neq \infty$ is rarely yielding satisfactory pruning results. We suggest to use instead as criterion the *sum* of these values at *all* training datapoints, yielding significant improvements.

Pruning were the point is omitted that has the smallest support value, weighed by the corresponding diagonal element of the inverted kernel matrix, achieved overall excellent pruning results, although in theory the formula is not intended for cases were $\gamma \neq \infty$. The subset selection based pruning algorithms perform overall as second best. If we also take into account the computational cost, standard pruning based on the support values remains most cheap and cheerful.

References

1. V. N. Vapnik, *Statistical Learning Theory*.   John Wiley & Sons, 1998.
2. J. A. K. Suykens and J. Vandewalle, "Least squares support vector machine classifiers," *Neural Processing Letters*, vol. 9, no. 3, pp. 293–300, june 1999.
3. J. A. K. Suykens, J. De Brabanter, L. Lukas, and J. Vandewalle, "Weighted least squares support vector machines : robustness and sparse approximation," *Neurocomputing, Special issue on fundamental and information processing aspects of neurocomputing*, vol. 48, no. 1-4, pp. 85–105, Oct 2002.
4. B. J. de Kruif and T. J. A. de Vries, "Pruning error minimization in least squares support vector machines," *IEEE Transactions on Neural Networks*, vol. 14, no. 3, pp. 696–702, May 2003.
5. J. A. K. Suykens, T. Van Gestel, J. De Brabanter, B. De Moor, and J. Vandewalle, *Least Squares Support Vector Machines*.   World Scientific, Singapore, 2002.

6. L. Hoegaerts, J. A. K. Suykens, J. Vandewalle, and B. De Moor, "Subset based least squares subspace regression in RKHS," *Accepted for publication in Neurocomputing, Special issue*, 2004.

7. R. Reed, "Pruning algorithms – a survey," *IEEE Transactions on Neural Networks*, vol. 4, no. 5, pp. 740–747, 1993.

8. A. J. Smola and B. Schölkopf, "Sparse greedy matrix approximation for machine learning," in *Proc. 17th International Conf. on Machine Learning*. Morgan Kaufmann, San Francisco, CA, 2000, pp. 911–918.

9. A. J. Miller, *Subset Selection in Regression*. Chapman & Hall, 1990.

10. S. Chen, C. Cowan, and P. Grant, "Orthogonal least squares learning algorithm for radial basis function networks," *IEEE Trans. Neural Networks*, vol. 2, pp. 302–309, March 1991.

11. G. Baudat and F. Anouar, "Kernel based methods and function approximation," in *IJCN*, Washington DC, July 2001, pp. 1244–1249.

12. N. Aronszajn, "Theory of reproducing kernels," *Transactions of the American mathematical society*, vol. 686, pp. 337–404, 1950.

13. B. Schölkopf and A. Smola, *Learning with kernels*. MIT Press, 2002.

14. B. Hassibi, D. G. Stork, and G. Wolf, "Optimal brain surgeon and general network pruning," in *Proceedings of the 1993 IEEE International Conference on Neural Networks*, San Francisco, CA, Apr 1993, pp. 293–300.

15. C. Blake and C. Merz, "UCI repository of machine learning databases," 1998. [Online]. Available: http://www.ics.uci.edu/~mlearn/MLRepository.html

Support Vector Machines Approach to Pattern Detection in Bankruptcy Prediction and Its Contingency

Kyung-shik Shin[1], Kyoung Jun Lee[2], and Hyun-jung Kim[1]

[1] Ewha omans University, College of Business Administration
11-1 Daehyun-Dong, Seodaemun-Gu, Seoul 120-750, Korea
ksshin@ewha.ac.kr, charitas@empal.com
[2] School of Business, Kyung Hee University
Hoegi-Dong, Dongdaemun-Ku, Seoul, Korea
klee@khu.ac.kr

Abstract. This study investigates the effectiveness of support vector machines (SVM) approach in detecting the underlying data pattern for the corporate failure prediction tasks. Back-propagation neural network (BPN) has some limitations in that it needs a modeling art to find an appropriate structure and optimal solution and also large training set enough to search the weights of the network. SVM extracts the optimal solution with the small training set by capturing geometric characteristics of feature space without deriving weights of networks from the training data. In this study, we show the advantage of SVM approach over BPN to the problem of corporate bankruptcy prediction. SVM shows the highest level of accuracies and better generalization performance than BPN especially when the training set size is smaller.

1 Introduction

Early techniques of bankruptcy prediction include statistical techniques such as multiple discriminant analysis, logit and probit. Recently artificial intelligence techniques such as neural networks have been an alternative method for the classification problems and numerous theoretical and experimental studies reported the usefulness of the back-propagation neural network (BPN) in classification studies. However, back-propagation neural network models have several limitations in building the model. First, finding an appropriate neural network model among numerous candidates is an artistic work because there are large numbers of controlling parameters and processing elements in the model. Second, its gradient descent search process for computing the synaptic weights can converge to a local minimum that is only a good fit for the training examples. Third, its empirical risk minimization principle seeking the minimization of the training error does not guarantee good generalization performance. Finally, the size of the training set is also the main issue to be resolved because the sufficiency and efficiency of the training set is one of most influencing factors.

In this paper, we investigate the effectiveness of support vector machines (SVM) approach, as an alternative BPN, to corporate failure prediction tasks. SVM classification exercise finds hyperplanes in the possible space for maximizing the distance from the hyperplane to the data points, which is equivalent to solving a quadratic

N.R. Pal et al. (Eds.): ICONIP 2004, LNCS 3316, pp. 1254–1259, 2004.
© Springer-Verlag Berlin Heidelberg 2004

optimization problem [9,10]. The solution of strictly convex problems for SVM is unique and global. In SVM, the structural risk minimization (SRM) principle is implemented and known to have high generalization performance. As the numbers of support vectors increase, SVM model is constructed through the trade-off between decreasing the number of training errors and increasing the risk of overfitting the data. SVM captures geometric characteristics of feature space without a need to derive weights of networks from the training data, therefore it can extract the optimal solution only with the small training set size. There are several arguments and observations supporting the high accuracy of SVM in the small training set size as well as the results showing that the accuracy and generalization performance of SVM is better than that of the standard BPN.

Byun and Lee [2] present a comprehensive survey on applications of Support Vector Machines (SVMs) for pattern recognition. Since SVMs show good generalization performance on many real-life data and the approach is properly motivated theoretically, it has been applied to wide range of applications. The support vector machines (SVMs) has been adopted for predicting bankruptcies [7] and compared with NN, MDA and learning vector quantization (LVQ) [5]. SVM obtained the best results, followed by NN, LVQ and MDA. Van Gestel et al. [8] also reports on the experiment with least squares support vector machines, a modified version of SVMs, and shows significantly better results in bankruptcy prediction when contrasted with the classical techniques.

The structure of this paper is organized as follows: The next section provides a brief description of several superior points of the SVM algorithm compared with BPN. The third section describes the research data and experiments. The fourth section summarizes and analyzes empirical results. The final section discusses the conclusions and future research issues.

2   SVM Versus BPN

Compared with the limitations of the BPN, the major advantages of SVM are as follows:

Few Controlling Parameters: Because a large number of controlling parameters in BPN such as the number of hidden layers, the number of hidden nodes, the learning rate, the momentum term, epochs, transfer functions and weights initialization methods are selected empirically, it is a difficult task to obtain an optimal combination of parameters that produces the best prediction performance. On the other hand, SVM has only two free parameters, namely the upper bound and kernel parameter.

Existence of Unique and Global Solution: Because the gradient descent algorithm optimizes the weights of BPN so that the sum of square error is minimized along a steepest slope of the error surface, the result from training can be massively multi modal, leading to non-unique solutions, and be in the danger of getting stuck in a local minima. On the other hand, training SVM is equivalent to solving a linearly constrained quadratic programming therefore SVM guarantees the existence of unique, optimal and global solution.

Good Generalization Performance: SVM implement the structural risk minimization (SRM) principle that is known to have a high generalization characteristic. SRM is the approach to trading off empirical error with the capacity of the set called VC (Vapnik-Chervonenkis) dimension, which seeks to minimize an upper bound of the generalization error rather than minimize the training error. In order to apply SRM, the hierarchy of hypothesis spaces must be defined before the data is observed. In SVM, the data is first used to decide which hierarchy to use and then subsequently to find a best hypothesis from each. Therefore the good generalization performance of SVM is not always attributable to SRM since the result of SVM is obtained from a data dependent SRM [1]. There exists no explicitly established theory that good generalization performance is guaranteed for SVM. However, it seems plausible that performance of SVM is more robust than that of BPM in that the two measures in terms of the margin and number of support vectors give information on the relation between the input and target function according to different criteria, either of which is sufficient to indicate good generalization. On the other hand, BPN trained based on minimizing a squared error criterion at the network output tends to produce a classifier with the only large margin measure. In addition, flexibility from choosing training data is likely to occur with weights of BPN model, but the maximum hyperplane of SVM is relatively stable and gives little flexibility in the decision boundary.

Training by Geometric Characteristics with the Small Data Set Size: SVM learns through capturing geometric picture corresponding to the kernel function, therefore it is constructed from just a subset of the training data, the support vectors. Moreover, no matter how large the training set size is, SVM has infinite VC dimension. That is why SVM is capable of extracting the optimal solution with the small training set size. On the other hand, as for the case of BPN containing a single hidden layer and used as a binary classifier, the number of training examples should be approximately 10 times the number of weights in the network. With the 10 input and hidden nodes, the learning algorithm will need more than 1000 training set size that is sufficient for a good generalization [6]. In most practical applications there can be a huge gap between the actual size of the training set needed and that is available. Utilizing the feature space images by the kernel function SVM is applicable in circumstances that have proved difficult or impossible for BPN where data in the plane is randomly scattered and where the density of the data's distribution is not even well defined [4].

3 Research Data and Experiments

The research data we employ is provided by Korea Credit Guarantee Fund in Korea, and consists of externally non-audited 2,320 medium-size manufacturing firms, which filed for bankruptcy (1,160 cases) and non-bankruptcy (1,160 cases) from 1996 to 1999. We select 1,160 non-bankrupt firms randomly from among all solvent firms, so the choice covers the whole spectrum from healthy to borderline firms in order to avoid any selection bias. The data set is arbitrarily split into two subsets; about 80% of the data is used for a training set and 20% for a validation set. The training data for SVM is totally used to construct the model and for BPN is divided into 60% training set and 20% test set. The validation data is used to test the results with the data that is not utilized to develop the model.

Using this first data set, 7 different datasets are constructed that differ in the number of cases included in the training and test subsamples. The validation set of all datasets consisting of 464 cases is identical, which ensures that the obtained results from validation data are not influenced by the fluctuation of data arrangement.

We apply two stages of the input variable selection process. At the first stage, we select 52 variables among more than 250 financial ratios by independent-samples t-test between each financial ratio as an input variable and bankrupt or non-bankrupt as an output variable. In the second stage, we select 10 variables using a MDA (multivariate discriminant analysis) stepwise method to reduce dimensionality. We select input variables satisfying the univariate test first and then select significant variables by the stepwise method for refinement.

In this study, the radial basis function is used as the kernel function of SVM. Since SVM does not have a general guidance for determining the upper bound C and the kernel parameter $\delta 2$, this study varies the parameters to select optimal values for the best prediction performance. The MATLAB support vector machine toolbox version 0.55 beta executes these processes [3]. In order to verify the applicability of SVM, we also design BPN as the benchmark with the following controlling parameters. The structure of BPN is standard three-layer with the same number of input nodes in the hidden layer and the hidden and output nodes use the sigmoid transfer function. For stopping the BPN training, test set that is not a subset of the training set is used, but the optimum network for the data in the test set is still difficult to guarantee generalization performance.

4 Results and Analysis

To investigate the effectiveness of the SVM approach trained by small data set size in the context of the corporate bankruptcy classification problem, the results obtained from various data set sizes are compared with those of BPN. The test results for this study are summarized in Table 1. Each cell of Table 1 contains the accuracy of the classification techniques. The results in Table 1 show that the overall prediction performance of SVM on the validation set is consistently good as the number of training set size decreases. Moreover, the accuracy and the generalization using a small size of data set (5th, 6th, and 7th set) are even better than those using a large size of data set (3rd and 4th set). Especially the performance of 5th and 6th set is similarly excellent compared to that of the 1st set.

The experimental result also shows that the prediction performance of SVM is sensitive to the various kernel parameter $\delta 2$. The accuracy on the training set of the most data set decreases as $\delta 2$ increases; on the other hand, the accuracy on the validation set shows a tendency to increases with increasing $\delta 2$. In addition, however, the prediction performance of the validation set is more stable and insensitive than that of training set. This indicates that a small value for $\delta 2$ has an inclination to overfit the training data and an appropriate value for $\delta 2$ plays an important role on the generalization performance of SVM. The results of SVM in the range of $\delta 2$ from 25 to 75 show the best prediction performances. While the results of BPN are comparable with SVM in large size of data set (1st and 2nd set), in case that a training set size is less

Table 1. Classification accuracies (%) of various data set sizes in which C=100

data set	number of set	SVM accuracy (δ^2)					BPN number of set	accuracy
		1	25	50	75	100		
(a) 1st set								
	train (1,856)	95,7	75,4	75,2	74,5	74,2	train (1,392)	71,7
	val (464)	60,3	73,9	76,5	76,3	76,3	val (464)	72,2
	total (2,320)						test (464)	74,1
(b) 2nd set (SVM training set of (a)*25%)								
	train (464)	99,1	80,6	79,7	79,3	78,7	train (370)	75,7
	val (464)	59,9	71,8	72,6	73,7	73,7	val (464)	71,6
	total (928)						test (94)	72,3
(c) 3rd set (SVM training set of (b)*50%)								
	train (232)	100,0	73,2	71,8	70,9	71,8	train (184)	57,1
	val(464)	57,1	65,5	66,2	67,5	67,0	val (464)	56,0
	total(696)						test (48)	56,3
(d) 4th set (SVM training set of (c)*50%)								
	train (116)	100,0	84,0	79,2	75,7	70,8	train (92)	55,4
	val (464)	54,5	67,0	60,8	59,5	57,5	val (464)	54,5
	total (580)						test (24)	54,2
(e) 5th set (SVM training set of (d)*50%)								
	train (58)	100,0	91,4	87,9	84,5	81,0	train (46)	47,8
	val (464)	59,9	72,6	75,2	73,3	73,5	val (464)	50,9
	total (522)						test (12)	50,0
(f) 6th set (SVM training set of (e)*50%)								
	train (28)	100,0	100,0	92,9	89,3	89,3	N/A	
	val (464)	58,8	70,0	73,3	70,9	70,3		
	total (492)							
(g) 7th set (SVM training set of (f)*50%)								
	train (14)	100,0	100,0	92,9	92,9	92,9	N/A	
	val (464)	54,3	62,1	64,0	63,8	62,9		
	total (478)							

than 200, it is hard (3rd, 4th, and 5th set) or even impossible (6th and 7th set) to learn the model.

5   Conclusions

From the results of experiments, we can conclude that SVM is the better approach to learn a small size of data patterns as opposed to ordinary BPN, which confirm existing researches [2,5,7,8]. In addition, our research confirms that SVM has the highest level of accuracies and better generalization performance than BPN as the training set size is getting smaller sets. In this study, we show that the proposed classifier of SVM approach outperforms BPN to the problem of corporate bankruptcy prediction. In addition, we investigate and summarize the several superior points of the SVM algorithm compared with BPN.

Our study has the following limitations that need further research. First, in SVM, the choice of the kernel function and the determination of optimal values of the parameters have a critical impact on the performance of the resulting system. We also need to examine the effect of other factors that is fixed in the experiment such as various values of the upper bound C and the kernel function. The second issue for future research relates to the generalization of SVM on the basis of the appropriate level of the training set size and gives a guideline to measure the generalization performance.

References

1. Burges, C., A Tutorial on Support Vector Machines for Pattern Recognition, Data Mining and Knowledge Discovery, 2(2):955-974, 1998.
2. Byun, H. and Lee, S., Applications of Support Vector Machines for Pattern Recognition: A Survey, Lecture Notes in Computer Science 2388:213-236, 2002.
3. Cawley, G., Support Vector Machine Toolbox, University of East Anglia, School of Information Systems, http://theoval.sys.uea.ac.uk/~gcc/svm/toolbox/, 2000.
4. Friedman, C., Credit Model Technical White Paper, Standard & Poor's, New York, McGraw-Hill, 2002.
5. Fan, A. and Palaniswami, M., A new approach to corporate loan default prediction from financial statements, in Proc. Computational Finance/Forecasting Financial Markets Conf. CF/FFM-2000, London (CD), UK, May 2000.
6. Haykin, S., Neural Networks: A comprehensive foundation. Macmillan College Publication, New York, 1994.
7. Häardle, W., Moro, R., Schäafer, D., Predicting Corporate Bankruptcy with Support Vector Machines, Working Slide, Humboldt University and the German Institute for Economic Research, available on http://ise.wiwi.hu-berlin.de/~isedoc03/slides.pdf, 2003.
8. Van Gestel T., Baesens B., Suykens J., Espinoza M., Baestaens D.E., Vanthienen J., DeMoor B., Bankruptcy Prediction with Least Squares Support Vector Machine Classifiers, Proceedings of the IEEE International Conference on Computational Intelligence for Financial Engineering (CIFEr2003), Hong Kong, pp. 1-8, March 2003.
9. Vapnik, V., "The Nature of Statistical Learning Theory", Springer-Verlag, New York, 1995.
10. Vapnik, V., Statistical Learning Theory, John Wiley & Sons, New York, 1998

Outliers Treatment in Support Vector Regression for Financial Time Series Prediction

Haïqin Yang, Kaizhu Huang, Laiwan Chan, Irwin King, and Michael R. Lyu

Department of Computer Science and Engineering
The Chinese University of Hong Kong
Shatin, N.T. Hong Kong
{hqyang,kzhuang,lwchan,king,lyu}@cse.cuhk.edu.hk

Abstract. Recently, the Support Vector Regression (SVR) has been applied in the financial time series prediction. The financial data are usually highly noisy and contain outliers. Detecting outliers and deflating their influence are important but hard problems. In this paper, we propose a novel "two-phase" SVR training algorithm to detect outliers and reduce their negative impact. Our experimental results on three indices: Hang Seng Index, NASDAQ, and FSTE 100 index show that the proposed "two-phase" algorithm has improvement on the prediction.

1 Introduction

Recently, due to the advantage of the generalization power with a unique and global optimal solution, the Support Vector Machine (SVM) has attracted the interest of researchers and has been applied in many applications, e.g., pattern recognition [1], and function approximation [8]. Its regression model, the Support Vector Regression (SVR), has also been successfully applied in the time series prediction [5], especially in the financial time series forecasting [2]. This model, using the ε-insensitive loss function, can control the sparsity of the solution and reduce the effect of some unimportant data points. Extending this loss function to a general ε-insensitive loss function with adaptive margins has shown to be effective in the prediction of the stock market [9, 10].

In modelling the financial time series, one key problem is its high noise, or the effect of some data points, called outliers, which differ greatly from others. Learning observations with outliers without awareness may lead to fitting those unwanted data and may corrupt the approximation function. This will result in the loss of generalization performance in the test phase. Hence, detecting and removing the outliers are very important. Specific techniques, e.g., a robust SVR network [4] and a weighted Least Squares SVM [6] have been proposed to enhance the robust capability of SVR. These methods would either involve extensive computation or would not guarantee the global optimal solution.

In this paper, we propose an effective "two-phase" SVR training algorithm to detect outliers and reduce their effect for the financial time series prediction. The basic idea is to take advantage of the general ε-insensitive loss function with a non-fixed margin, which can reduce the effect of some data points by enlarging the ε-margin width.

N.R. Pal et al. (Eds.): ICONIP 2004, LNCS 3316, pp. 1260–1265, 2004.
© Springer-Verlag Berlin Heidelberg 2004

The paper is organized as follows. We introduce the SVR with a general ε-insensitive loss function and state the method of detecting and reducing outliers in Section 2. We report experimental results in Section 3. Lastly, we conclude the paper in Section 4.

2   Outliers Detection and Reduction in Support Vector Regression

In this section, we first introduce the Support Vector Regression (SVR) in the time series prediction. We then propose a general ε-insensitive loss function for applying the adaptive margins. Next, we describe our method to detect the outliers and reduce their influence.

2.1   Support Vector Regression for Time Series Prediction

Time series data can be abstracted as $(\mathcal{X}, \mathcal{Y})$ pairs, where $\mathcal{X} \in \mathbb{R}^d$ denotes the space of input patterns, $\mathcal{Y} \in \mathbb{R}$ corresponds to the target value. Usually, the sample is finite and observed in a successive time interval. An N-instance sample series is described as $(\mathcal{X}, \mathcal{Y}) = \{(\mathbf{x}_t, y_t) \mid \mathbf{x}_t \in \mathbb{R}^d, \ y_t \in \mathbb{R}, \ t = 1, \ldots, N\}$. In the financial time series, it may be assumed that all the information can be condensed in the price. Hence, y_t usually represents the price at time t and $\mathbf{x}_t$ represents the p-previous days' prices as $\mathbf{x}_t = (y_{t-p}, \ldots, y_{t-1})$. To analyze this series, one may evaluate a function, f,

$$y_t = f(\mathbf{x}_t) + \sigma_t,$$

from the given N-instance sample series, where σ_t is the noise at time t. The SVR is a currently popular technique to learn the data with good generalization [7].

Typically, the SVR estimates a linear function

$$f(\mathbf{x}) = \mathbf{w}^T \phi(\mathbf{x}) + b, \tag{1}$$

in a feature space, $\mathbb{R}^f$, by minimizing the following regression risk:

$$R_{reg}(f) = \frac{1}{2}\mathbf{w}^T\mathbf{w} + C\sum_{i=1}^{N} l(f(\mathbf{x}_i) - y_i), \tag{2}$$

where the superscript T denotes the transpose, ϕ is a mapping function in the feature space, and b is an offset in $\mathbb{R}$. The term $\frac{1}{2}\mathbf{w}^T\mathbf{w}$ is a complexity term determining the flatness of the function in $\mathbb{R}^f$, C is a regularized constant, and l is a cost function.

Generally, the ε-insensitive loss function is used as the cost function [7]. This function does not consider data points in the range of ε-margin, i.e, $\pm\varepsilon$. It can therefore reduces the effect of those data points lying in the ε-margin to the approximation function and controls the sparsity of solution. The ε-insensitive loss function is defined as $l_\varepsilon(f(\mathbf{x}) - y) = \max(|y - f(\mathbf{x})| - \varepsilon, 0)$.

2.2   SVR with a General ε-Insensitive Loss Function

In the above, the ε-margin is fixed and symmetrical. This setting may lack the flexibility to efficiently model the volatility of the stock market and can not prefer one-side prediction. In order to overcome these problems, we propose a general ε-insensitive loss function. This function divides the margin into two separate parts, up margin, ε^u, and down margin, ε^d, with each part changing adaptively as formulated below:

$$l_{\varepsilon_2}(f(\mathbf{x}_i) - y_i) = \begin{cases} 0, & \text{if } -\varepsilon_i^d < y_i - f(\mathbf{x}_i) < \varepsilon_i^u \\ y_i - f(\mathbf{x}_i) - \varepsilon_i^u, & \text{if } y_i - f(\mathbf{x}_i) \geq \varepsilon_i^u \\ f(\mathbf{x}_i) - y_i - \varepsilon_i^d, & \text{if } f(\mathbf{x}_i) - y_i \geq \varepsilon_i^d \end{cases} \tag{3}$$

The main contribution of proposing this loss function is that we can adopt adaptive margin with non-fixed and asymmetrical characteristics. This would benefit the stock market prediction, e.g., reflecting the volatility of the stock market or avoiding the down side risk.

Minimizing the regression risk of (2) with the cost function of (3) by the Lagrange method, we obtain the following Quadratic Programming (QP) problem:

$$\min_{\alpha, \alpha^*} \frac{1}{2} \sum_{i=1}^{N} \sum_{j=1}^{N} (\alpha_i - \alpha_i^*)(\alpha_j - \alpha_j^*) K(\mathbf{x}_i, \mathbf{x}_j) + \sum_{i=1}^{N} (\varepsilon_i^u - y_i)\alpha_i + \sum_{i=1}^{N} (\varepsilon_i^d + y_i)\alpha_i^*,$$

$$\text{s.t. } \sum_{i=1}^{N} (\alpha_i - \alpha_i^*) = 0, \quad \alpha_i, \; \alpha_i^* \in [0, C], \quad i = 1, \ldots, N,$$

where α_i and α_i^* are the corresponding Lagrange multipliers used to push and pull $f(\mathbf{x}_i)$ towards the outcome of y_i, respectively. $K(\mathbf{x}_i, \mathbf{x}_j) = \phi(\mathbf{x}_i)^T \phi(\mathbf{x}_j)$, the inner product of the mapping function, is the kernel function which satisfies the Mercer's condition.

The above QP problem has a similar form to the original QP problem in the SVR and can be easily implemented or solved by e.g., a commonly used SVM library, LIB-SVM [3]. After solving the above QP problem, we obtain the corresponding Lagrange multipliers α_i and α_i^*, and the weight $\mathbf{w} = \sum_{i=1}^{N} (\alpha_i - \alpha_i^*)\phi(\mathbf{x}_i)$; therefore, we get the approximation function as $f(\mathbf{x}) = \sum_{i=1}^{N} (\alpha_i - \alpha_i^*)K(\mathbf{x}, \mathbf{x}_i) + b$, where the offset b is calculated by exploiting the Karush-Kuhn-Tucker (KKT) conditions (details in [3]).

2.3   Outliers Detection and Reduction

From the KKT conditions, we have

$$\alpha_i(\varepsilon_i^u + \xi_i - y_i + f(\mathbf{x}_i)) = 0, \quad i = 1, \ldots, N, \tag{4}$$
$$\alpha_i^*(\varepsilon_i^d + \xi_i^* + y_i - f(\mathbf{x}_i)) = 0, \quad i = 1, \ldots, N,$$

and

$$(C - \alpha_i)\xi_i = 0, \quad i = 1, \ldots, N, \tag{5}$$
$$(C - \alpha_i^*)\xi_i^* = 0, \quad i = 1, \ldots, N,$$

where ξ_i and ξ_i^* are slack variables used to measure the error of up side and down side, respectively (see Fig. 1(a)).

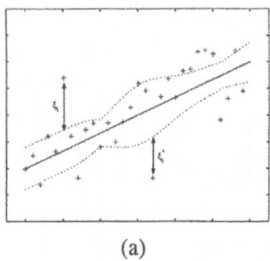 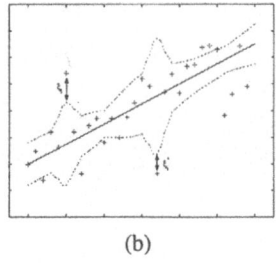

(a)   (b)

Fig. 1. An illustration of detecting and reducing the effect of outliers in the feature space

The KKT conditions in (5) indicate that if $\alpha_i \in [0, C)$, then $\xi_i = 0$; likewise for α_i^* and ξ_i^*. This means that the corresponding data points lie in, or on the ε-margin, i.e., either the ε^u-margin or the ε^d-margin, but not both. Moreover, for $\alpha_i = C$ or $\alpha_i^* = C$, we have

$$\xi_i = y_i - f(\mathbf{x}_i) - \varepsilon_i^u, \quad \forall \alpha_i = C, \quad i = 1, \dots, N,$$
$$\xi_i^* = f(\mathbf{x}_i) - y_i - \varepsilon_i^d, \quad \forall \alpha_i^* = C, \quad i = 1, \dots, N.$$

The above formulae show that increasing ε_i^u and ε_i^d will decrease the corresponding ξ_i and ξ_i^* in the same constructed function f. This will therefore reduce the error caused by the corresponding data points. In addition, if we fulfill the objective in (2), a normal point or a non-outlier will not contain a very large error, i.e., ξ_i or ξ_i^*. Based on these observations, we propose the criterion of detecting outliers, i.e., if ξ_i or ξ_i^* is larger than a threshold, the corresponding data point would be an outlier and we could enlarge the corresponding margin width to reduce its effect (see Fig. 1(b)). This motivates us to propose the following "two-phase" procedure:

Phase 1: Train the SVR model with the ε_i^u and ε_i^d margin setting.
Phase 2: Detect and reduce the effect of the outliers. If $\xi_i > \tau \varepsilon_i^u$, $\varepsilon_i^{u'} = \tau \varepsilon_i^u$; similarly we have $\varepsilon_i^{d'} = \tau \varepsilon_i^d$ for $\xi_i^* > \tau \varepsilon_i^d$. Re-train the SVR model by using the updated margin setting, $\varepsilon_i^{u'}$ and $\varepsilon_i^{d'}$.

Here, τ is a pre-specific constant to denote the suitable threshold.

3   Experiments

In this section, we implement the above "two-phase" procedure and perform the experiments on three indices: Hang Seng Index (HSI), NASDAQ and FTSE 100 index (FTSE). The data are selected from the daily closing prices of the indices from September 1st to December 31th, 2003 (three months' data). The beginning four-fifth data are used for training and the rest one-fifth data are used in the one-step ahead prediction. The experimental performance is evaluated by the Root Mean Square Error (RMSE) and the Mean Absolute Error (MAE), which are frequently used as the statistical metrics.

In the experiments, the input pattern is constructed as a four-day's pattern: $\mathbf{x}_t = (y_{t-4}, y_{t-3}, y_{t-2}, y_{t-1})$. This is based on the assumption that (non)linear relationship occurs in sequential five days' prices. A commonly used function, the Radial Basis

Table 1. Experiment Results

Dataset	Phase	RMSE	MAE	Phase	RMSE	MAE
HSI	I	140.36	116.38	II	140.28	116.26
NASDAQ	I	24.49	20.36	II	22.78	19.49
FTSE	I	59.97	44.74	II	57.90	42.32

Function $K(\mathbf{x}_i, \mathbf{x}_j) = \exp(-\gamma \|\mathbf{x}_i - \mathbf{x}_j\|^2)$, is selected as the kernel function. The margin for time t is set as $\varepsilon_t^u = \varepsilon_t^d = 0.5\rho(\mathbf{x}_t)$, where $\rho(\mathbf{x}_t)$ is the standard deviation of the input pattern at day t as justified in [9]. The parameter pair (C, γ) is set to $(4, 1)$ for HSI, $(2^5, 2^{-6})$ for NASDAQ, and $(2, 1)$ for FTSE, which are tuned by the cross-validation method. In the first phase, we construct the approximation function $f(\mathbf{x}_t)$ by performing the SVR algorithm on the normalized training data using the above settings.

After obtaining the approximation function, we observe that some training data points actually differ largely from the predictive values. We therefore in the second phase, update the corresponding ε_i^u and ε_i^d based on the proposed algorithm. The parameter τ is set to 2 for all three indices. Hence, we can deflate the influence of those differing points. A reason of τ being not so large is that the outliers still contain some useful information for constructing the approximation function and thus we cannot completely ignore them. We report the results in Table 1. The results indicate in the second phase, the prediction performance has improved on all the three indices, especially we obtain 3.45% and 5.41% improvement on the FTSE index for the RMSE and MAE criterion, respectively.

We also plot the results of NASDAQ in Fig. 2. The result of Phase I is illustrated in Fig. 2(a), while that of Phase II is in Fig. 2(b). If comparing these two figures, one can find that the approximation function (the solid line) in Fig. 2(b) is smoother than that in Fig. 2(a). Especially, the highlighted point A is a peak in Fig 2(a), but it is lowered and smoothed in Fig. 2(b). The other highlighted point B is a valley in Fig. 2(a), but it is now lifted in Fig. 2(b). This demonstrates that enlarging the margin width to the outliers can reduce their negative impact.

In addition, in some situations, one may prefer to predict the stock market conservatively, i.e, he would intend to under-predict the boost of stock prices for avoiding the down side risk. To meet this objective, we adopt an asymmetrical margin setting. Concretely, we pick out the corresponding up side support vectors and update their up margin and down margin by $\varepsilon_t^u = 3.8\rho(\mathbf{x}_t)$ and $\varepsilon_t^d = 0.2\rho(\mathbf{x}_t)$, respectively. Here, we use this relatively extreme setting to demonstrate the change and the difference. The graphic result is in Fig. 2(c). It can be observed that the peak A is still lowered, but the valley B is not lifted. Overall, the approximation function maintains the lower predicative values but decreases the higher predictive values, which would be highly valuable in the stock market prediction.

4   Conclusion

In the paper, a novel "two-phase" SVR training procedure is proposed to detect and deflate the influence of outliers. This idea motivates from the phenomenon that enlarging the adaptive margin width in the general ε-insensitive loss function will reduce the ef-

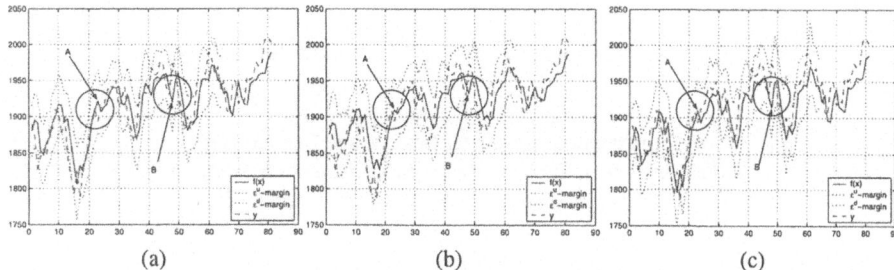

| (a) | (b) | (c) |

Fig. 2. A demonstration of the experimental results in NASDAQ. (a) is the result of Phase I. (b) is the result of Phase II with an enlarged symmetrical margin setting for the outliers detection and reduction. (c) is the result of Phase II with an enlarged asymmetrical margin setting to avoid the down side risk. The solid line is the result of the approximation function. The dashed line is the original time series. The dotted lines correspond to up margin and down margin and they are shifted away from their original places by 30, respectively, in order to make the result clear

fect of the corresponding data points. The experimental results on three indices indicate that this "two-phase" method has improvement on the prediction.

Acknowledgement

The work described in this paper was fully supported by two grants from the Research Grants Council of the Hong Kong Special Administrative Region, China (Project No. CUHK4182/03E and Project No. CUHK4351/02).

References

1. C. Burges. A Tutorial on Support Vector Machines for Pattern Recognition. *Data Mining and Knowledge Discovery*, 2(2):121–167, 1998.
2. L. Cao. Support Vector Machines Experts for Time Series Forecasting. *Neurocompt.*, 51:321–339, 2003.
3. C.-C. Chang and C.-J. Lin. LIBSVM: a Library for Support Vector Machines, 2004.
4. C.-C. Chuang, S.-F. Su, J.-T. Jeng, and C.-C. Hsiao. Robust support vector regression networks for function approximation with outliers. *IEEE Transactions on Neural Networks*, 13:1322 – 1330, 2002.
5. S. Mukherjee, E. Osuna, and F. Girosi. Nonlinear Prediction of Chaotic Time Series Using Support Vector Machines. In J. Principe, L. Giles, N. Morgan, and E. Wilson, editors, *IEEE Workshop on Neural Networks for Signal Processing VII*, pages 511–519. IEEE Press, 1997.
6. J. A. K. Suykens, J. De Brabanter, L. Lukas, and J. Vandewalle. Weighted Least Squares Support Vector Machines: Robustness and Sparse Approximation. *Neurocompt.*, 2001.
7. V. N. Vapnik. *The Nature of Statistical Learning Theory*. Springer, New York, 1995.
8. V. N. Vapnik, S. Golowich, and A. Smola. Support Vector Method for Function Approximation, Regression Estimation and Signal Processing. In M. Mozer, M. Jordan, and T. Petshe, editors, *NIPS*, volume 9, pages 281–287, Cambridge, MA, 1997. MIT Press.
9. H. Yang, L. Chan, and I. King. Support Vector Machine Regression for Volatile Stock Market Prediction. *IDEAL 2002*, volume 2412 of *LNCS*, pages 391–396. Springer, 2002.
10. H. Yang, I. King, L. Chan, and K. Huang. Financial Time Series Prediction Using Non-fixed and Asymmetrical Margin Setting with Momentum in Support Vector Regression. *Neural Information Processing: Research and Development*, pages 334-350. Springer-Verlag, 2004.

Kernel Based Clustering for Multiclass Data

D. Srikrishna Satish and C. Chandra Sekhar

Speech and Vision Laboratory
Department of Computer Science and Engineering
Indian Institute of Technology Madras, Chennai - 600 036 India
{satish,chandra}@cs.iitm.ernet.in

Abstract. In this paper we address the issues in construction of discrete hidden Markov models (HMMs) in the feature space of Mercer kernels. The kernel space HMMs are suitable for complex pattern recognition tasks that involve varying length patterns as in speech recognition. The main issues addressed are related to clustering in the kernel feature space for large data sets consisting of the data of multiple classes. Convergence of kernel based clustering method [1] is slow when the size of the data set is large. We consider an approach in which the multiclass data set is partitioned into subsets, and clustering for the data in each subset is done separately. We consider two methods for partitioning the multiclass data. In the all-class-data method, the partitioning is done in such a way that each subset contains a part of the total data set of each class. In the class-wise-data method, a subset contains the total data of only one class. We study the performance of the two methods on kernel based clustering used to build discrete HMMs in the kernel feature space for recognition of spoken utterances of letters in E-set of English alphabet.

1 Introduction

Development of discriminative training based classification models for varying length patterns is important for speech recognition. Hidden Markov models are suitable for classification of varying length patterns. The performance of hidden Markov models in complex pattern classification tasks is limited because the models are not trained using discriminative methods. Classification models such as multilayer perceptrons and support vector machines are trained using discriminative methods. However, these models are suitable for classification of fixed length patterns only. Complex pattern classification tasks typically involve construction of decision surfaces for nonlinearly separable patterns of varying length. According to Cover's theorem on the separability of patterns, an input space made up of nonlinearly separable patterns may be transformed into a feature space where the patterns are linearly separable with high probability, provided the transformation is nonlinear and the dimensionality of the feature space is high enough [6]. Mercer kernels can be used for nonlinear transformation from the input space to a high-dimensional feature space. If the nonlinear transformation is smooth and continuous, then the topographical ordering of data in the input space will be preserved in the feature space. Because of the easier separability of classes in the feature space, the non-discriminative training based models such as hidden Markov models can be used to solve complex pattern classification tasks involving varying length patterns.

N.R. Pal et al. (Eds.): ICONIP 2004, LNCS 3316, pp. 1266–1272, 2004.
© Springer-Verlag Berlin Heidelberg 2004

The mapping from the input space to a higher dimensional feature space induced by a Mercer kernel can be explicit as in the case of a polynomial kernel or implicit as in the case of a Gaussian kernel [3]. It is possible to construct the discrete hidden Markov models (DHMMs) in the polynomial kernel feature space, by clustering and vector quantization of the explicit feature vectors corresponding to the input space vectors. However, for construction of DHMMs in the feature spaces of kernels that perform implicit mapping, it is necessary to perform clustering and vector quantization in the feature space using kernel functions only. Recently, a kernel based clustering method for the Gaussian kernel feature space has been proposed in [1]. We extend the method for kernel based clustering in polynomial kernel feature space, so that the performance of explicit and implicit clustering methods can be compared. Implementation of this method for large data sets is computationally intensive. We propose two methods suitable for clustering the large data sets consisting of the data belonging to multiple classes. We use the proposed methods for clustering in the feature space to construct the DHMMs for speech recognition.

The organization of the paper is as follows: In the next section, we present a method for clustering in the feature space of a Mercer kernel. In Section 3, we explain two methods of data partitioning for reducing the computational complexity of kernel based clustering of multiclass data. In Section 4, we present the studies on recognition of E-set of English alphabet using discrete HMMs constructed in the kernel feature space.

2 Clustering in Feature Space of a Mercer Kernel

In this section, we first present the criterion for partitioning the data into clusters in the input space using the K-means clustering algorithm. Clustering in the explicitly represented feature space of Mercer kernels such as polynomial kernel can also be realized using the K-means clustering algorithm. Then we present the method for clustering in the feature space of implicit mapping kernels.

Consider a set of N data points in the input space, $\mathbf{x}_i$, $i = 1, 2 \ldots N$. Let the number of clusters to be formed is K. The commonly used criterion for partitioning of the data into K clusters is to minimize the trace of the within-cluster scatter matrix, $\mathbf{S}_w$, defined as follows [1]:

$$\mathbf{S}_w = \frac{1}{N} \sum_{k=1}^{K} \sum_{i=1}^{N} z_{ki}(\mathbf{x}_i - \mathbf{m}_k)(\mathbf{x}_i - \mathbf{m}_k)^T \tag{1}$$

where $\mathbf{m}_k$ is the center of the kth cluster, C_k, and z_{ki} is the membership of data point $\mathbf{x}_i$ to the cluster C_k. The membership value $z_{ki} = 1$ if $\mathbf{x}_i \in C_k$ and 0 otherwise. The number of points in the kth cluster is given as $N_k = \sum_{i=1}^{N} z_{ki}$.

The optimal partitioning of the data points involves determining the indicator matrix, $\mathbf{Z}$, with the elements as z_{ki}, that minimizes the trace of the matrix $\mathbf{S}_w$. This method is used in the K-means clustering algorithm for linear separation of the clusters. For nonlinear separation of clusters of data points, the input space is transformed into a high-dimensional feature space using a smooth and continuous nonlinear mapping, Φ, and the clusters are formed in the feature space. The optimal partitioning in the feature space is based on the criterion of minimizing the trace of the within-cluster scatter matrix in the feature space, $\mathbf{S}_w^\Phi$. The feature space scatter matrix is given by:

$$\mathbf{S}_w^{\Phi} = \frac{1}{N} \sum_{k=1}^{K} \sum_{i=1}^{N} z_{ki} (\Phi(\mathbf{x}_i) - \mathbf{m}_k^{\Phi})(\Phi(\mathbf{x}_i) - \mathbf{m}_k^{\Phi})^T \tag{2}$$

where $\mathbf{m}_k^{\Phi}$, the center of the kth cluster in the feature space, is given by:

$$\mathbf{m}_k^{\Phi} = \frac{1}{N_k} \sum_{i=1}^{N} z_{ki} \Phi(\mathbf{x}_i) \tag{3}$$

The trace of $\mathbf{S}_w^{\Phi}$ can be computed using the innerproduct operations as given below:

$$Tr(\mathbf{S}_w^{\Phi}) = \frac{1}{N} \sum_{k=1}^{K} \sum_{i=1}^{N} z_{ki} (\Phi(\mathbf{x}_i) - \mathbf{m}_k^{\Phi})^T (\Phi(\mathbf{x}_i) - \mathbf{m}_k^{\Phi}) \tag{4}$$

When the feature space is explicitly represented, as in the case of mapping using polynomial kernels, the K-means clustering algorithm can be used to minimize the trace given in Eq.(4). However, for Mercer kernels with implicit mapping used for transformation, it is necessary to express the trace in terms of kernel function. The Mercer kernel function in the input space corresponds to the innerproduct operation in the feature space, i.e.,

$$\mathcal{K}_{ij} = \mathcal{K}(\mathbf{x}_i, \mathbf{x}_j) = \Phi^T(\mathbf{x}_i)\Phi(\mathbf{x}_j) \tag{5}$$

It is shown in [1] that Eq.(4) can be rewritten as

$$Tr(\mathbf{S}_w^{\Phi}) = \frac{1}{N} \sum_{k=1}^{K} \sum_{i=1}^{N} z_{ki} D_{ki} \tag{6}$$

where

$$D_{ki} = \mathcal{K}_{ii} - \frac{1}{N_k} \sum_{j=1}^{N} z_{kj} \mathcal{K}_{ij} \tag{7}$$

The term D_{ki} is the penalty associated with assigning $\mathbf{x}_i$ to the kth cluster in the feature space. However, for a polynomial kernel, D_{ki} may take a negative value because the magnitude of $\mathcal{K}_{ij}$ can be greater than that of $\mathcal{K}_{ii}$. To avoid D_{ki} taking negative values, we replace $\mathcal{K}_{ij}$ in Eq.(7) with the normalized value $\mathcal{K}'_{ij}$ defined as

$$\mathcal{K}'_{ij} = \frac{|\mathcal{K}_{ij}|}{\sqrt{\mathcal{K}_{ii}}\sqrt{\mathcal{K}_{jj}}} \tag{8}$$

A stochastic method for finding the optimal values of elements of $\mathbf{Z}$ that minimizes the expression of the trace in Eq.(10) leads to following iterative procedure [1].

$$\langle z_{ki} \rangle = \frac{\alpha_k \exp(-2\beta D_{ki}^{new})}{\sum_{k'=1}^{K} \alpha_{k'} \exp(-2\beta D_{k'i}^{new})} \tag{9}$$

where

$$D_{ki}^{new} = \mathcal{K}_{ii} - \frac{1}{\langle N_k \rangle} \sum_{j=1}^{N} \langle z_{ki} \rangle \mathcal{K}_{ij} \tag{10}$$

$$\alpha_k = \exp\left(-\beta \frac{1}{\langle N_k{}^2\rangle} \sum_{i=1}^{N}\sum_{j=1}^{N} \langle z_{ki}\rangle \langle z_{kj}\rangle \mathcal{K}_{ij}\right) \tag{11}$$

The parameter β controls the softness of the assignments during the optimization. The terms $\langle z_{ki}\rangle$ and $\langle N_k\rangle$ denote the estimate of the expected values of z_{ki} and N_k respectively. The iterative procedure in Eq.(9) is continued until there is a convergence, i.e., there is no significant change in the values of elements of the indicator matrix Z.

It is important to note that the evaluation of α_k in Eq.(11) requires the computation of the kernel function for every pair of vectors in the data set. When the value of N is large, the computational complexity of the stochastic method is high and the convergence of the method is slow. In the next section we propose an approach for kernel based clustering of large data sets by partitioning the data set into subsets and clustering the data in each subset separately.

3   Partitioning of Multiclass Data

Consider a multiclass data set consisting of the data of M classes. Let N be the size of the multiclass data set. We consider two methods for partitioning the multiclass data set. In the all-class-data method, the data set is partitioned into P subsets in such a way that any subset contains a part of the total data of each class. Therefore each subset contains the data of all the classes. Let K be the number of clusters to be formed for the data in a subset. Let $n_p, p = 1, 2, ..., P$, be the number of vectors in the pth subset. Kernel based clustering for the data of the pth subset involves estimating the values of $K * n_p$ elements of the indicator matrix. Computation of α_k involves $\mathrm{O}(n_p{}^2)$ rather than $\mathcal{O}(N^2)$ kernel operations leading to reduced complexity and fast convergence of the stochastic optimization method. However, since each subset includes some data of all the classes, there will be similarity among the clusters for the data of different subsets. The $K * P$ clusters of all the P subsets are merged into K clusters using the following method. Let C_{p_i} be the ith cluster formed from the data of pth subset, and C_{q_j} be the jth cluster formed from the data of qth subset. Then the similarity of mean vectors of the two clusters is defined as:

$$S_{p_i q_j} = \frac{1}{n_{p_i}}\frac{1}{n_{q_j}} \sum_{l=1}^{n_{p_i}}\sum_{m=1}^{n_{q_j}} \mathcal{K}(\mathbf{x}_{p_{il}}, \mathbf{x}_{q_{jm}}) \tag{12}$$

where $\mathbf{x}_{p_{il}}$ is the lth vector in the ith cluster of pth subset. The cluster C_{p_i} is merged with the maximum similarity cluster C_{q_r} where $r = \arg\max_j(S_{p_i q_j})$. The merging of the clusters in the different subsets is carried out to obtain K clusters for the total data of all the classes. Thus the kernel based clustering using all-class-data method involves computation of a total of $K * N$ elements in P indicator matrices, and merging of $K * P$ clusters into K clusters.

In the class-wise-data method, the partitioning is done in such a way that a subset contains the data of one class only. Let $N_i, i = 1, 2, ..., M$, be the number of vectors in the data of ith class. Since the variability among the data of a class is significantly smaller compared to the variability in data of all the classes, the number of clusters in the data of one class, will be smaller than the number of clusters in the data of all the classes. Let K_c be the number of clusters to be formed in the data of one class. Kernel based clustering for the data of the ith class involves estimation of $K_c * N_i$ elements of the indicator matrix. Since the data in the subsets of different classes are expected to be different, there is no need for merging the clusters formed from the data in different subsets. Therefore, kernel based clustering for the class-wise-data method involves estimation of $\sum_{i=1}^{M} K_c * N_i = K_c * N$ elements in the M indicator matrices. Since $K_c << K$, the computational complexity of the class-wise-data method is significantly less than that of the all-class-data method. We study the performance of the two methods for partitioning the large multiclass data sets on a speech recognition task.

4   Speech Recognition Using DHMMs in Kernel Feature Space

We study the performance of discrete HMMs in the input space and in the kernel feature space for a task in speech recognition. The task involves recognition of spoken utterances of a highly confusable subset of letters in English alphabet, namely, E-set. The E-set includes the following 9 letters: {B,C,D,E,G,P,T,V,Z}. The OGI spoken letter database [2] is used in the study on recognition of E-set. The training data set consists of 240 utterances from 120 speakers for each letter, and the test data set consists of 60 utterances from 30 speakers per letter. For the speech signal of each utterance, short-time analysis of speech is performed using a frame size of 25msec with a shift of 10msec resulting in a sequence of data vectors. Each data vector consists of 12 mel-frequency cepstral coefficients (MFCC), energy, their first order derivatives and their second order derivatives resulting in a dimension of 39. The number of data vectors in the training data set of all the classes is 123806.

In our studies on recognition of E-set, a 5-state, left-to-right, discrete HMM is constructed for each class(letter). For input space DHMMs, a codebook of size 64 is constructed in the 39-dimensional space by clustering the data vectors of all the classes into 64 clusters using K-means algorithm. For construction of DHMMs in the feature space of polynomial kernel, we consider the explicit clustering method and the implicit clustering method. In the explicit clustering method, a codebook of size 64 is constructed in the explicitly represented feature space of polynomial kernel of degree 2. The feature vector includes the monomials of order 0,1 and 2 derived from the 39-dimensional input space vector. Therefore the dimension of the feature space is 820. The K-means clustering algorithm is used for explicit clustering.

In the implicit clustering method, the stochastic method is used for determining the elements of the indicator matrix. For the all-class-data method of partitioning, the total data set of 9 classes is divided into 8 subsets. Each subset contains the data of one utterance per letter from each of 30 speakers. Kernel based clustering method is used for forming 64 clusters for the data in each subset. The 512 clusters for the data of 8 subsets are merged to form 64 clusters for the total data of all the 9 letters. For the class-wise data method of partitioning, each of subset consists of the total training data of one letter. For the data in each of 9 subsets, 8 clusters are formed using kernel based clustering method. Thus there are a total of 72 clusters formed for the data of all the classes. Implicit clustering using the all-class-data method and the class-wise-data method is carried out for the polynomial kernel of degree 2 and for the Gaussian kernel.

Table 1. Classification accuracy (in %) of DHMMs in the input space and the kernel space for recognition of spoken letters in E-set.

Models	Clustering method	Accuracy	No. of clusters
Input space DHMMs	K-means clustering	65.19	64
Polynomial kernel space DHMMs	Explicit, all-class-data clustering	67.96	64
	Implicit, all-class-data clustering	66.11	64
	Implicit, class-wise-data clustering	64.26	72
Gaussian kernel space DHMMs	Implicit, all-class-data clustering	65.37	64
	Implicit, class-wise-data clustering	69.63	72

Vector quantization in the kernel feature space is carried out using the method given in [5].

The classification performance on the test data set for DHMMs built using different methods of kernel based clustering and for different kernels is given in Table 1. The performance is compared with that of the input space DHMMs. It is seen that the explicit clustering in the feature space of polynomial kernel gives a marginally better performance. The performance of class-wise-data clustering method for Gaussian kernel gives a significantly better performance compared to the input space HMMs. It is interesting to note that the performance of the implicit clustering method for Gaussian kernel is better than that of the explicit clustering method for the polynomial kernel. For Gaussian kernel, the class-wise-data method for partitioning gives a better performance than the all-class-data method. However, for polynomial kernel the all-class-data method gives a better performance. The recognition performance of the Support Vector Machines (SVMs) on the same set is 68.52. All-class-data method using Gaussian kernel performs better than SVMs. The time taken for clustering using the class-wise-data method is about 1/4th of the time taken by the all-class-data method. However, the reason behind the variations in the performance is yet to be analysed.

5 Summary and Conclusions

In this paper, we have addressed the issues in clustering in the kernel feature space for multiclass data. Discrete HMMs in kernel feature space are built using kernel based clustering and vector quantization. We have proposed two methods for partitioning of large multiclass data sets. The methods are helpful in the fast convergence of the stochastic optimization method for clustering. Effectiveness of the proposed methods is demonstrated in recognition of spoken letters in E-set of English alphabet. Construction of continuous density HMMs in the kernel feature space will need developing techniques for density estimation in the feature space of Mercer kernels.

References

1. M. Girolami, "Mercer Kernel-Based Clustering in Feature Space," IEEE Transactions on Neural Networks, vol. 13, no. 3, pp. 780–784, May 2002.
2. ISOLET Corpus, Release 1.1, Center for Spoken Language Understanding, Oregon Graduate Institute, July 2000.

3. K. Muller, S. Mika, G. Ratsch, K. Tsuda and B. Schölkopf, "An Introduction to Kernel-Based Learning Algorithms," IEEE Transactions on Neural Networks, vol. 12, no. 2, pp. 181–201, March 2001.
4. L. R. Rabiner and B. H. Juang, Fundamentals of Speech Recognition, Prentice-Hall, 1993.
5. D. S. Satish and C. C. Sekhar, "Discrete hidden Markov models in kernel feature space for speech recognition," in Proceedings of the International Conference on Systemics, Cybernautics and Informatics, Hyderabad, India, Feb 2004, pp. 653–658.
6. Simon Haykin, Neural networks: A Comprehensive Foundation, New Jersey: Prentice-Hall International, 1999.

Combined Kernel Function for Support Vector Machine and Learning Method Based on Evolutionary Algorithm

Ha-Nam Nguyen, Syng-Yup Ohn, and Woo-Jin Choi

Department of Computer Engineering
Hankuk Aviation University, Seoul, Korea
{nghanam,syohn,ujinchoi}@hau.ac.kr

Abstract. This paper proposes a new combined kernel function and its learning method for support vector machine which results in higher learning rate and better classification performance. A set of simple kernel functions are combined to create a new kernel function, which is trained by a learning method employing evolutionary algorithm. The learning method results in the optimal decision model consisting of a set of features as well as a set of the parameters for combined kernel function. The new kernel function and the learning method were applied to obtain the optimal decision model for classification of proteome patterns, and in the comparison with other kernel functions, the combined kernel function showed a higher convergence rate and a greater flexibility in learning a problem space than single kernel functions.

1 Introduction

Support vector machine (SVM) is a learning method that uses a hypothesis space of linear functions in a high dimensional feature space [1-3, 5, 6]. This learning strategy, introduced by Vapnik [2], is a principled and powerful method and outperformed most of the classification algorithms in many applications. However, the computational power of linear learning machines is limited in the cases of the feature space with nonlinear characteristics. It can be easily recognized that real-world applications require more extensive and flexible hypothesis space than linear functions. By using a proper kernel function, we can overcome the nonlinearity of feature space [1, 2, 5, 6]. Also recent improvements in the area of SVM make it easy to implement and overcome the computation time problem due to a large training set [5].

Evolutionary algorithm (EA) is an optimization algorithms based on the mechanism of natural evolutionary procedure [7-9]. Most of evolutionary algorithms share a common conceptual base of simulating the evolution of individual structures through the processes of selection, mutation, and reproduction. In each generation, a new population is selected based on fitness value of previous generation by somehow. Then some members of the population are given the chance to undergo alterations by means of crossover and mutation to form new individuals. In this way, EA performs a multi-directional search by maintaining a population of potential solutions and encourages the formation and the exchange of information among different directions. EAs are generally applied to the problems with a large search space. They are different from random algorithms since they combine the elements of directed and stochastic search. Furthermore, EA is also known to be more robust than directed search methods.

In this paper, we propose a new learning method using GA and SVM. In the new learning method, GA is employed to derive the optimal *decision model* for the classi-

N.R. Pal et al. (Eds.): ICONIP 2004, LNCS 3316, pp. 1273–1278, 2004.
© Springer-Verlag Berlin Heidelberg 2004

fication of patterns, which consists of the optimal set of features and parameters of combined kernel function. SVM is used to evaluate the fitness of the newly generated decision models by measuring the hit ratio of the classification based on the models. We applied GA to obtain the optimal set of features and parameters of combined kernel at the same time. In the comparison with single kernel-based learning methods by extensive experiments on the clinical dataset, proposed learning method achieved faster convergence while searching the optimal decision model and better classification performance.

This paper is organized as follows. In section 2, our new combined kernel and its learning method are presented in detail. In section 3, we compare the performances of combined kernel functions and other individual kernel functions based on case study on proteome pattern samples. Finally, section 4 is our conclusion.

2   Combined Kernel Function and Learning Method

A kernel function provides a flexible and effective learning mechanism in SVM, and the choice of a kernel function should reflect prior knowledge about the problem at hand. However, it is often difficult for us to employ the prior knowledge on patterns to choose a kernel function, and it is an open question how to choose the best kernel function for a given data set. According to *no free lunch theorem* [4] on machine learning, there is no superior kernel function in general, and the performance of a kernel function rather depends on applications.

In our case, a new kernel function is created by combining the set of kernel functions. The combined kernel function has the form of

$$K_{Combined} = (K_1)^{e_1} \circ \cdots \circ (K_m)^{e_m} \tag{1}$$

where $\{K_i \mid i = 1, \ldots, m\}$ is the set of kernel functions to be combined, e_i is the exponent of i-th kernel function, and $\circ$ denotes an operator between two kernel functions. In our case, three types of the kernel functions listed in Table 1 are combined, and multiplication or addition operators are used to combine kernel functions. The parameters in a kernel function play the important role of representing the structure of a sample space. The set of the parameters of a combined kernel function consists of three part - i) the exponents of individual kernel functions, ii) the operators between kernel functions, iii) the coefficients in each kernel function. The optimal set of the parameters maximizing the classification performance can be selected by a machine learning method. In a learning phase, the structure of a sample space is learned by a kernel function, and the knowledge of a sample space is contained in the set of parameters. Furthermore, the optimal set of features also should be chosen in the learning phase. In our case, EA technique is employed to obtain the optimal set of features as well as the optimal combined kernel function.

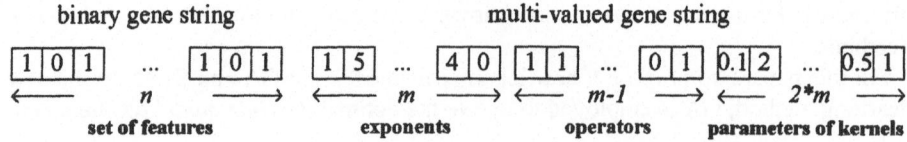

Fig. 1. Structure of a chromosome used in EA procedure

The challenging issue of EA is how to map a real problem into a *chromosome*. In our learning method, we need to map a feature space, the set of the parameters for kernels, and the set of operators combining kernels. Firstly, the set of features is encoded into a *n*-bit binary string to represent an active or non-active state of *n* features. Then the exponents of *m* individual kernel functions, the operators between individual kernel functions, and the coefficients in each individual kernel function are encoded into a multi-valued gene string. The combination of the two gene string forms a chromosome in EA procedure which in turn serves as a *decision model* (see Fig. 1). In learning phase, simulating a genetic procedure, EA creates improved decision models containing a combined kernel function and a set of features by the iterative process of reproduction, evaluation, and selection process. At the end of learning stage, the optimal decision model consisting of a combined kernel function and the set of features is obtained, and the optimal decision model is contained in a classifier to be used classify new pattern samples.

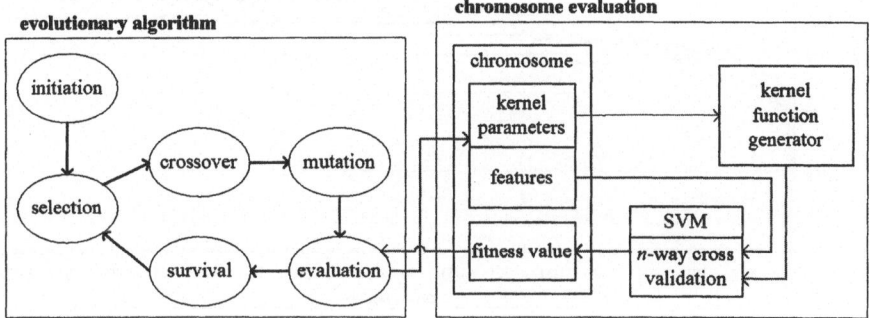

Fig. 2. EA-SVM learning method for combined kernel function

The framework of our learning method is illustrated in Fig. 2. EA creates new chromosomes and searches the *optimal decision model* based on the fitness values obtained from SVM classifier. A chromosome is decoded into a set of features and a kernel function, which are used by SVM classifier. The SVM component is used to evaluate the performance of a *decision model* represented by a chromosome. In SVM classifier, *n*-way cross-validation is used to prevent overfitting, and the classification ratios from *n* tests are averaged to obtain a fitness value.

3   Case Study: Classification of Proteome Patterns

In this section, we apply the new kernel function and learning method to classify the proteome patterns for the identification of breast cancer, compare the performance of the combined kernel function with those of other kernel functions. The proteome pattern samples were provided by Cancer Research Center at Seoul National University in Seoul, Korea.

3.1   Comparison of the Learning Rates of Kernel Function

Our combined kernel function and three other kernel functions (see Table 1) are trained by EA, and their learning rates are compared. Furthermore, the classifiers based on the optimal decision models are evaluated by hit rates. Fig. 3 depicts the maximum hit rates achieved at each generation during the learning phase. While EA was executed for 800 generations, the combined kernel function showed the fastest learning and achieved the highest hit rate among the kernel functions compared. Thus, combined kernel function has the highest convergence rate and the greatest flexibility in learning the structure of sample space. The combined kernel function reached hit rate of 84.5% after 800 generations and the optimal kernel function obtained is

$$K_{Combined} = (K_{Radial})^4 + (K_{Neural})^0 + (K_{Inverse\ multi-quadric})^1$$

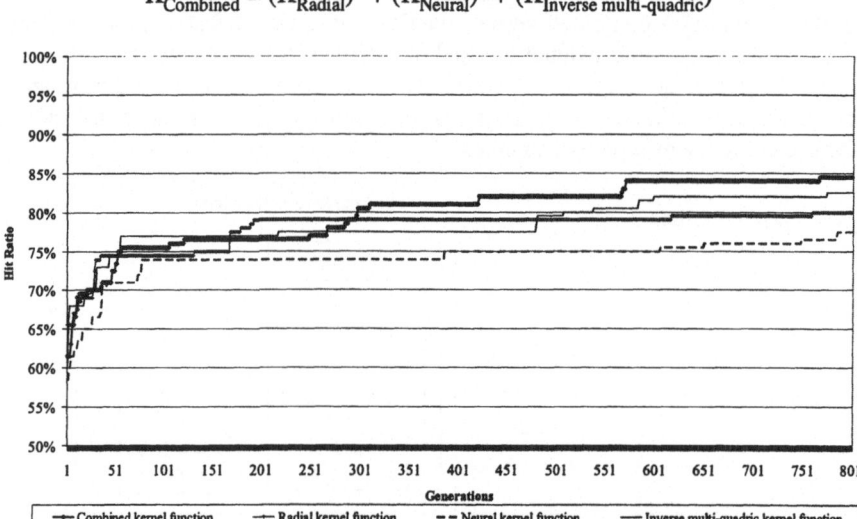

Fig. 3. The comparison of hit ratio rates of kernel functions in learning phase

3.2 Validation Error and Classification

In the next experiment, we validate the optimal *decision model* that was extracted from the learning phase. The "testing on the training set" methodology showed that the training error decreases monotonically during training for most problems [4]. However, the error on the validation set decreases first and then increases. It means the classification model may be overfitting the training data. To prevent the overfitting problem, it was suggested that the training phase should stop at the first minimum of the validation error to obtain the optimal decision model [4] (see Fig. 4). The classification rates by the optimal decision models are evaluated on an independent test set and the results are shown in Table 4. While the validation error rates are higher than those of training error rates in the learning phase due to overfitting, the combined kernel function also outperforms the other kernel functions.

Table 1. Result of classification stage using the *decision model* obtained from valid training

	Combination	Radial	Neural	Inverse multiquadric
Generation	250	100	800	800
Hit rate	84%	74%	77%	69%

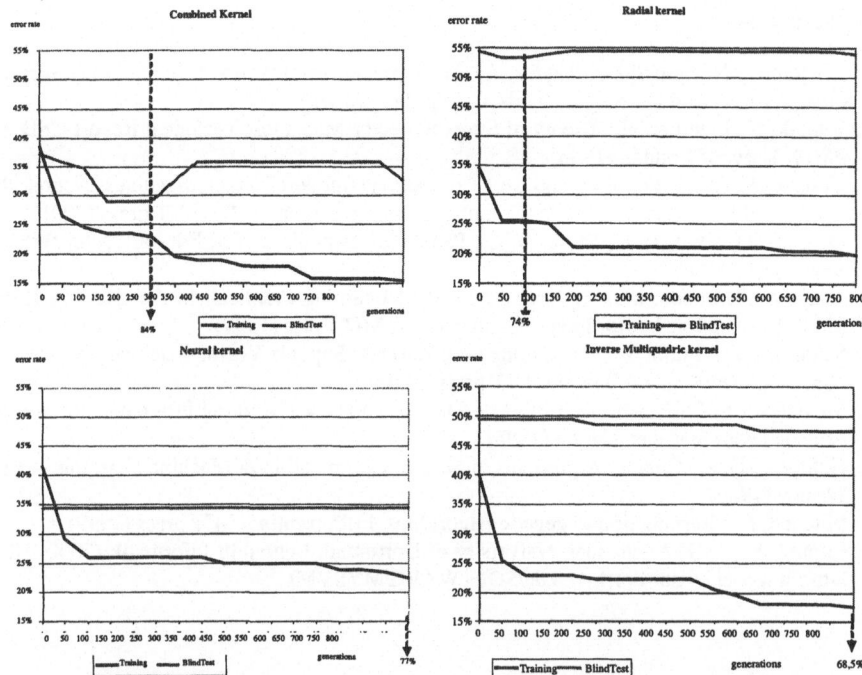

Fig. 4. Estimation error rate to find *decision model* for classification stage

4 Conclusion

In this paper, we proposed a new combined kernel function for SVM and its learning method employing EA technique to obtain the optimal decision model for classification. A kernel function plays an important role of mapping the problem feature space into a new feature space so that the performance of a linear classifier is improved. The combined kernel function and the learning method are applied to classify the clinical proteome patterns. In the comparison of the classifications by combined kernel and other three kernels, the combined kernel shows faster convergence in learning phase and results in the optimal decision model with higher hit rate. Thus, our combined kernel function has greater flexibility in representing a problem space.

Acknowledgement

This research was supported by IRC (Internet Information Retrieval Research Center) in Hankuk Aviation University. IRC is a Kyounggi-Province Regional Research Center designated by Korea Science and Engineering Foundation and Ministry of Science & Technology.
   The authors acknowledge that mySVM [10] was used for the implementation of the proposed method in this paper.

References

1. Cristianini, N. and Shawe-Taylor, J.: An introduction to Support Vector Machines and other kernel-based learning methods. Cambridge (2000)
2. Vapnik, V.N. and et. al.: Theory of Support Vector Machines, Technical Report CSD TR-96-17. Univ. of London (1996)
3. Vojislav Kecman: Learning and Soft Computing: Support Vector Machines, Neural Networks, and Fuzzy Logic Models (Complex Adaptive Systems). The MIT press (2001)
4. Duda, R. O., Hart, P. E., Stork, D. G.: Pattern Classification (2nd Edition). John Wiley & Sons Inc. (2001)
5. Joachims, Thorsten: Making large-Scale SVM Learning Practical. In Advances in Kernel Methods - Support Vector Learning, chapter 11. MIT Press (1999)
6. Schökopf, B., Smola, A. J.: Learning with Kernels: Support Vector Machines, Regularization, Optimization, and Beyond. MIT press (2002)
7. Michalewicz, Z.: Genetic Algorithms + Data structures = Evolution Programs. 3rd rev. and extended edn, Springer-Verlag (1996)
8. Goldberg, D. E.: Genetic Algorithms in Search, Optimization & Machine Learning. Adison Wesley (1989)
9. Mitchell, M.: Introduction to genetic Algorithms. Fifth printing. MIT press (1999)
10. Rüping, S.: mySVM-Manual. University of Dortmund, Lehrstuhl Informatik (2000) URL: http://www-ai.cs.uni-dortmund.de/SOFTWARE/MYSVM/

Neural Network Classification Algorithm for the Small Size Training Set Situation in the Task of Thin-Walled Constructions Fatigue Destruction Control

A.I. Galushkin, A.S. Katsin, S.V. Korobkova, and L.S. Kuravsky

Scientific Center of Neurocomputers, RusAvia

Abstract. High level loads in the acoustic frequency range are the reason for the aviation constructions' elements fatigue destruction and death of the airborne equipment. Acoustic loads have most influence on the thin-walled elements of the aircraft construction. In this work we define the task of fault detection and present the method for solving this problem. Also we present the experiment results of the vibroacoustic system state detection using neural network method for the small size sample set situation.

1 Physical Task Definition

The interest in the investigations in the field of thin-walled constructions fatigue destruction control appeared in the middle 1950s. It was sparkled by the mass fatigue destruction of the aircraft constructions from the acoustic loads, which appeared with the increase in flight speed and shift from reciprocators to turbo-jet and turboprop engines.

Lately the interest in the acoustic vibrations has increased due to the design of the new generation of the passenger aircrafts and hypersonic aircrafts. Experimental analysis of the acoustic vibrations carried out at the special simulators and during the ground and flight tests are usually costly, laborious and require much time. The difficulties in producing acoustic vibrations encourage using computer modeling and computer data processing. One of the approaches to the vibroacoustic destruction control is combining the model and actual results. Usually the data obtained during the actual tests is enough to make a decision using the combined method.

In this paper we present an algorithm for the construction faults classification by the actual test data. The initial data for this task is the set of 10-dimentional vectors; each of them should be classified as belonging to one of the following 7 classes: OK, LEFT, BOUND, CENTER, L_C, R_C, RIGHT.

The first of these classes corresponds to the normal state of the construction, and the other 6 classes are considered to be different types of the construction faults.

2 Mathematical Task Definition

There is a set of numeric characteristics X (the parameters of the analysis subject), which influence the values of functions $F_i(X)$ (functions $F_i(X)$ show if analysis subject belongs to class i). M is the number of vectors for the different analysis subject, the values of $F_i(X)$ for which we know. X is the set of vectors that can be described by the following formula $X = \{X^n, n = 1..M\}$. The number of dimensions

N.R. Pal et al. (Eds.): ICONIP 2004, LNCS 3316, pp. 1279–1284, 2004.
© Springer-Verlag Berlin Heidelberg 2004

for each vector X'' is N. Function $F_i(X)$ can take on the values 1 or 0. It takes on the value of 1 if the vector belongs to the fixed class i $(i=1..k)$, and takes on the value of 0 otherwise. For any given vector X^* we should count the values of $F_i(X^*)$ for each one of the seven classes.

In this paper we present the description and efficiency analysis of the neural network algorithm for $k=7$, $N=10$, $M=7 \cdot 24$.

3 The Algorithms of Dealing with Small Size Training Sets

In our work we present the list of methods for dealing the small sized training sets for adjusting the neural networks with fixed structure:

1. One of the efficient methods of increasing the reliability of the decision made by the neural network is the procedure of multiple iteration of the original small sized sample set. The reliability of the decision is the probability of the correct recognition of the pattern or mean square error in the task of function approximation or any other estimation, which should be designed according to the task which should be solved with the multilayer neural network.
2. One of the possible ways for the neural network generalization is artificial generation of the additional training sets with the average of distribution equal to the components of the original small size sample set and with different dispersion. This dispersion can vary from one experiment to another according to the experiment plan.
3. The problem of choosing the initial conditions before training multilayer neural networks has influence both on the computation speed in the neural network logical basis (the choice of the initial conditions can influence the rate of convergence of the adaptation algorithms to one of the local or to the global extremum of optimization functional) and on the quality of the solution. If we use different methods of choosing initial conditions for the neural network adaptation the results will differ. The mean of the results is a kind of estimation of the quality of the solution produced by the neural network. For this task the initial conditions can be produced by generation of random initial values of the weights of the neural network and also using the adaptation of the variable structured neural network on the original small size training set.
4. The division of the original small training set into even smaller ones and training neural network using the methods described in 2 and 3 and averaging the results over the small parts of the original train set is the additional method of the neural network generalization ability analysis.
5. The last step of this methodology is averaging adaptation results over the variants produced on steps 2, 3 and 4. The resultant function of quality distribution will have its average distribution and dispersion. The average distribution of quality is the basic characteristic of the neural network performance in the small size training set situation. The dispersion of the quality is the estimation of the inauthenticity of the previous characteristic. If the value of the dispersion is large, measures should be taken to improve the quality of the neural network response. One of these measures is complication of the neural network structure.

4   The Algorithm of Forming the Training Set Using the Original Small Size Sample Set

The input of our algorithm is the set of 10-dimensional ($N = 10$) vectors, which consists of 7 classes with 24 vectors in each of them. It is obvious that the training set in this task is insufficient for obtaining trustful results. The idea of our algorithm is to "erode" each pattern so that the resulting training set reaches the necessary size. So we take one of these vectors and make a new set of vectors with normal distribution, the dispersion and average distribution of which are σ and m. They should equal the original pattern. We used Box-Muller formula to solve this problem.

Suppose we have one pattern with $N = 2$. We should choose voluntary U_1, U_2, U_3, U_4 from (0,1]. Then:

$$X1 = \sqrt{-2\log U_1}\,\cos(2\pi U_2);$$
$$X2 = \sqrt{-2\log U_3}\,\cos(2\pi U_4); \tag{1}$$

The values X1 and X2 are added to the original coordinates of the pattern. This operation is applied to 10-dimensional original data vectors.

5   Training the First Layer of the Neural Network with Variable Structure

After obtaining the training set of the appropriate size the neural network should be trained. We used the single-layer neural network with variable structure to classify the vectors of the eroded training set. Figure 2 shows the result of classification of the patterns into 2 classes.

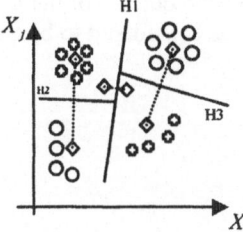

Fig. 1. Classification using hyper planes

Figure 1 shows the projection of the N-dimensional characteristic space onto the plane (X_i, X_j). So, there are 2 classes of samples: daggers and circles. First the geometrical centers of these two classes should be found. They are the closest diamonds on figure 3. The line, connecting these diamond is drawn and the hyperplane H1 separating the samples goes through the center of the segment between the diamonds and transversely to it. After that we suppose that all samples above this hyperplane are of the first class and below the hyperplane are of the second class. At this moment we have the samples of the first class classified to belong to the second class and vice versa. The error for both parts of the space is calculated. We choose the part, where this error is greater and proceed in that part. In our case it is the part above the hyperplane H1. Then the same operation is done to the samples below and above H1. After this we have the space divided into classes correctly. These operations can be represented by the "Logical tree". (fig. 2)

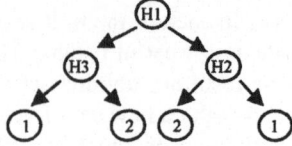

Fig. 2. Logical tree

The division of the space with hyperplanes stops when the total error is 0.

In our task we have 7 classes. To classify the samples in our task we will have 7 different classifiers each of them should classify all the samples as belonging or not to one of the 7 classes. So the fixed class will be considered as the first class, and all other classed will be regarded as the second class.

For the experiments we should specify the values of dispersion σ for producing the training set of the sufficient size. It will be measured in the percentage of the half of the minimal distance between the samples belonging to a fixed class and the samples of the other classes ($L_{min}/2$). The σ values 30%, 50%, 70% were taken for the experiments.

6   Experiment Results

A single-layer neural network with variable number of neurons in the first layer and a block of logical operations over the outputs of the neural network was used for classification. The neural network was considered to be trained when the sample classification error was zero.

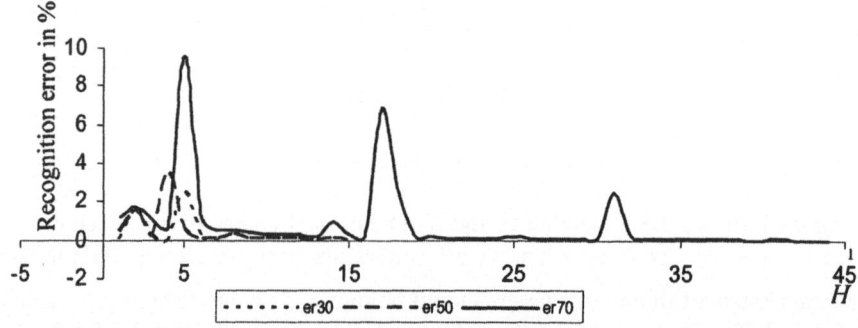

Fig. 3. The classification error dependence on the number of the neurons in the first layer H_1 for OK class; Solid line $\sigma = 70\%$; long-dotted line $\sigma = 50\%$; short-dotted line $\sigma = 30\%$

Figure 4 shows the dependence of the number of neurons in the first layer of the neural network on the value of dispersion for all classes – OK, LEFT, BOUND, CENTER, L_C, R_C, RIGHT.

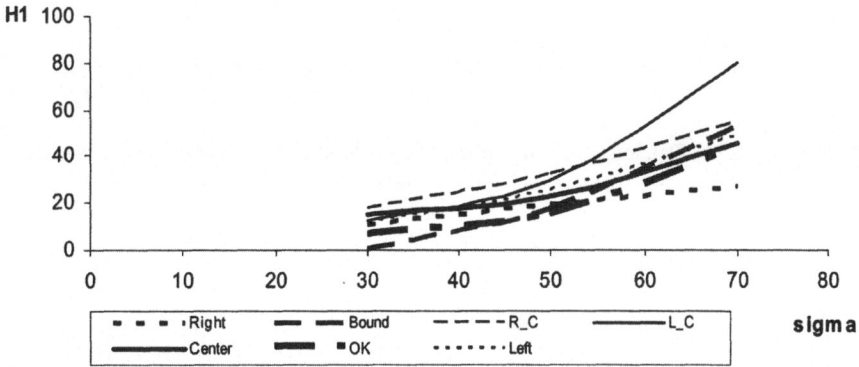

Fig. 4. The dependence of the number of neurons in the first layer of the neural network on the value of dispersion σ for all classes – OK, LEFT, BOUND, CENTER, L_C, R_C, RIGHT

7 Conclusions

Using this method we can get the factors of the trained neural network more adequate of the recognition task than the factors of the neural network trained on the insufficient set of vectors. As the value of σ increases the number of neurons in the first layer of the neural network with variable structure needed to classify the samples increases. This is shown on the figure above. For different classes this dependence differs. As the number of neurons in the first layer of the neural network increases the time needed for training the neural network also increases. The optimal erosion level σ should be chosen for each task specifically concerning the cost of the errors of the first and the second kind.

References

1. H. Yoshiko, U. Shunji, K. Taiko evaluation of artificial neural network classifiers in small sample size situations, Pros. Int. It. Conf. Neural Networks, Nagoya. Oct. 25-29, 1993, (1731-1735).
2. Archer Norman P., Wang Shonhong Learning bias in neural networks and an approach to controlling its effects in monotonic classifications, IEEE Trans. Pattern Anal. And Mach. Intell., 1993, 15, v9 (962-966).
3. Rosenblatt F. The principles of neurodynamics, "Mir", 1965.
4. Galushkin A.I. The synthesis of multilayer systems for pattern recognition. M, "Energy", 1974.
5. Galushkin A.I. Neural Network Theory. M., IPRGR, 2000.
6. Baranov S.N., Kuravsky L.S. Acoustic vibrations in thin-walled constructions: modeling, optimization and analysis. RusAvia, 2001.
7. Galushkin A.I. Neural networks and the problem of small training set. V All-Russia conference Neurocomputers and their applications. 1999.
8. Zador Anthony M., Pearlmutter Barak A. VC dimension of an integrate-and-fireneuron model. Neural Comput. 1996, 8, N°3 -624.
9. Bartlett Peter L., Williamson Robert C. The VC dimension and pseudo dimension of two-layer neural networks with discrete inputs. Neural Comput. 1996, 8, ▯3 – 625-628.
10. Hole A. Some variants on a theorem of Vapnik. Prepr. Ser. Pure Math., Inst. Math. Univ. Oslo 1995, N°30 – 1-11.

11. Hole Arne. Vapnik-Chervonenkis generalization bounds for real valued neural networks. Neural Comput. 1996, 8, N°6 – 1277-1299.
12. Chakraborty Goutam, Shiratori Norio, Noguchi Shoichi. Optimization of overtraining and overgeneralization. Proc. Int. Jt. Conf. Neural Networks, Nagoya, Oct. 25-29, 1993. Vol. 3, 2257-2262.
13. Tzafestas S.G., Dalianis P.J., Anthopoulos G.C. On the overtraining phenomenon of back-propagation neural networks. Math. And Comput. Simul. 1996, 40, N°5 – 6, 507 – 520.

Wavelet-Based Estimation
of Hemodynamic Response Function

R. Srikanth, R. Muralishankar, and A.G. Ramakrishnan

Dept of Electrical Engineering, Indian Institute of Science, Bangalore-560012, India

Abstract. We present a new algorithm to estimate hemodynamic response function (HRF) and drift component in wavelet domain. The
HRF is modeled as a gaussian function with unknown parameters. The
functional Magnetic resonance Image (fMRI) noise is modeled as a fractional brownian motion (fBm). The HRF parameters are estimated in
wavelet domain since wavelet transform with sufficient number of vanishing moments decorrelates a fBm process. Due to this decorrelating
property of wavelet transform, the noise covariance matrix in wavelet
domain can be assumed to be diagonal whose entries are estimated using sample variance estimators at each scale. We study the influence
of sampling time and shape assumption on the estimation performance.
Results are presented by adding synthetic HRFs on null fMRI data.

1   Introduction

Functional Magnetic Resonance Imaging (fMRI) is a noninvasive imaging technique that can be used to study the function of brain. fMRI measures changes
in blood oxygenation and blood volume that result from neural activity and this
response is called as Blood Oxygen Level Dependent (BOLD) response. One
common way of modeling the fMRI time-series is by a convolution model [4]:
The observed time-series at each voxel is the output of a linear filter with the
boxcar function, representing the design paradigm, as the input. The impulse
response of the filter is called as Hemodynamic Response Function (HRF).

Significant work has been done so far [1], [2] in identifying the activated regions for a given task. With the increase in time resolution of fMRI data, there
is a significant interest in estimating the temporal characteristics of fMRI data
[3]. Estimation of HRF is important in not only identifying the regions of activation but also in finding the relative time of activation of different brain regions.
This can be achieved by determining the delay time of HRF at various brain
regions. Amplitudes of the response at different locations give the strength of
response at these locations for a given task. Estimation of HRF also helps in
event-related design paradigms [3] where HRFs will be different for events and
the task. BOLD signal in fMRI time-series is corrupted by physiological noise
due to cardiac and respiratory cycles and their aliased versions, thermal and
scanner noise. Physiological noise components appear as trends or fluctuations
in the voxel time-series. It is observed that fMRI time-series under null conditions exhibit long term dependence (or $1/f$ property) which can be modeled

N.R. Pal et al. (Eds.): ICONIP 2004, LNCS 3316, pp. 1285–1291, 2004.
© Springer-Verlag Berlin Heidelberg 2004

by fractional brownian motion (fBm) [1]. In this work, We model HRF by a gaussian function whose parameters are estimated using Maximum a Posterior Estimation (MAP) and the noise is modeled by fBm. We study the influence of sampling time (TR) and shape assumption on the estimation performance.

2 Probability Model for fMRI Time-Series

The BOLD response of the brain for a given task can be modeled as convolution of the HR function and the input task [4]. The input task $x(t)$ is considered as a binary function of time which has a value 1 during the period of task and a 0 during the rest period.

The observed fMRI time-series at any voxel can be represented as:

$$\mathbf{y} = \mathbf{Xh} + \beta + \mathbf{w} \tag{1}$$

where, $\mathbf{y}$ is an observed time series of size $(N - K \times 1)$, X is a $(N - K \times K)$ convolution matrix, $\mathbf{h}$ is the HRF of length K and β is the known drift component. We model the noise component $\mathbf{w}$ as a fBm process. A fBm process is a zero-mean, non stationary, and non differentiable function of time whose covariance between the process at two times t and u is given by

$$r_w(t, u) = 0.5\sigma_f^2(|t|^{2H} + |u|^{2H} - |t - u|^{2H})$$
$$where, \sigma_f^2 = \Gamma(1 - 2H)cos\pi H/\pi H \tag{2}$$

where, H is called Hurst component with values between 0 and 1 for fBm process. Therefore, the probability of $\mathbf{y}$ given $\mathbf{h}$ is

$$\mathbf{y}|\mathbf{h} \sim \mathbf{N}(\mathbf{Xh} + \beta, \mathbf{C}) \tag{3}$$

where C is the covariance matrix whose elements are given by (2). Wornell et al., [5] observed that wavelet transform with sufficient number of vanishing moments (at least four) approximately decorrelates fBm process. This implies that the correlation of wavelet coefficients within and across the scales is very small. In this work, we use dabauchies wavelet with four vanishing moments since they are compact for a given number of vanishing moments. Multiplying WT matrix W (or applying discrete wavelet transform) on both sides of (1)

$$W\mathbf{y} = \mathbf{WXh} + \mathbf{W}\beta + \mathbf{Ww} \tag{4}$$

where, $W\mathbf{y}$, $W\beta$ and $W\mathbf{w}$ are, respectively wavelet coefficients of observed signal, drift and noise. Let $\mathbf{y_w} = \mathbf{Wy}$, $X_w = WX$, $\beta_\mathbf{w} = \mathbf{W}\beta$ and $\epsilon = \mathbf{Ww}$. Then the above equation can be written compactly as

$$\mathbf{y_w} = \mathbf{X_w h} + \beta_\mathbf{w} + \epsilon \tag{5}$$

For example, for J scales, $\mathbf{y_w}$ is

$$\mathbf{y_w} = [a\mathbf{y}_1^J, d\mathbf{y}_1^J, d\mathbf{y}_1^{J-1}, .., d\mathbf{y}_{2-JN}^{J-1}, ..., d\mathbf{y}_1^1, .., d\mathbf{y}_{2-1N}^1]' \tag{6}$$

where, ay_k^j and dy_k^j (subscript k denotes time index and superscript j indicates scale) are approximate and detail coefficients respectively. X_w can be obtained by applying wavelet transform to each column of X. Now, the probability model for $\mathbf{y_w}$ given $\mathbf{h}$ is

$$\mathbf{y_w}|\mathbf{h} \sim \mathbf{N}(\mathbf{X_w h} + \beta_{\mathbf{w}}, \Lambda) \qquad (7)$$

where, Λ is a diagonal matrix assuming that WT decorrelates the noise process $\mathbf{w}$. The diagonal elements of Λ are variances of wavelet coefficients at each scale.

$$\Lambda = diag[\sigma_J^2, \sigma_J^2, ..., \sigma_1^2] \qquad (8)$$

3   Estimation of HRF and Drift

Hemodynamic response (HR) refers to the local change in blood oxygenation as an effect of increased neuronal activity. This response can be modeled as an output of a linear filter for an input of unit impulse. In this section, we model the system response (HRF) by a Gaussian function, where the parameters give a physiological interpretation [3]. The HRF is represented as:

$$h(t) = \eta \exp(-(t.TR - \mu)^2/\sigma^2) \qquad (9)$$

where, μ is the time lag from the onset of the stimuli to the peak of HR; σ reflects the rise and decay time and η is the amplitude of the response and TR is the sampling period. Let $\theta = [\mu, \sigma, \eta]$ denote the unknown parameters of the HRF.

Now, the estimation of HRF boils down to estimating the unknown parameter θ, for which we use MAP estimation. For this, the parameter θ is modeled as a random variable with a known priori pdf. This method allows to incorporate the prior knowledge of the parameter through the prior pdf. If the prior is properly chosen, one can expect a better estimate of θ. We model μ, σ and η as independent Gaussian random variables.

$$f(\theta) = p(\mu, \sigma, \eta) = f(\mu)f(\sigma)f(\eta) = N(m_\theta, V_\theta) \qquad (10)$$

where m_θ is the vector of means of μ, σ and η and V_θ is a diagonal matrix whose entries are variances of the above random variables.

The prior probability parameters m_θ and V_θ are chosen using the prior knowledge of the HRF. It is observed that when a stimulus of small duration (unit impulse) is applied, the HRF follows after some delay and attains peak after 2-6 seconds of application of the stimulus. Also, HRF lasts for a duration between 7 to 12 seconds. It is also observed that the amplitude of the response η is around 3-5% of the signal intensity. Accordingly, the priors for μ, σ and η can be chosen as [3]:

$$p(\theta) \sim N(m_\theta, V_\theta) \qquad (11)$$

where, $m_\theta = [6, 2, 4]^t$ and $V_\theta = diag(3, 5, 5)$.

The unknown parameters to be estimated are θ and β which is a low frequency signal. Therefore, wavelet coefficients of β at lower scales (high frequencies) should be negligible and can be assumed to be zero. Hence the wavelet coefficients for the drift component can be written as:

$$\beta_w = [a\beta_1^J, d\beta_1^J, dy_1^{J-1}, .., dy_{2^{-J}N}^{J-1}, .., dy_{Jo}^{Jo}, .., dy_{2^{-Jo}N}^{Jo}, 0, 0......0]' \qquad (12)$$

where, Jo is the lowest scale up to which the drift component is significant. We observed from the null data of fMRI that the drift component is significant only at first two higher scales. Therefore we assume $Jo = 2$. We use the following iterative algorithm to estimate the unknown parameters:

1. Find a least square estimate of h, $\hat{h}_{ls} = (X_w' X_w)^{-1} X_w' y_w$.
2. Remove the signal component from the time-series, $\tilde{y}_w = y_w - X_w \hat{h}_{ls}$.
3. Estimate the drift component $\hat{\beta}_w$ by equating the wavelet coefficients to zero from scale $Jo + 1$ to 1.
4. Now use the estimated drift in the wavelet domain and find the MAP estimate of θ as: $\theta_{MAP} = arg\ max_\theta\ p(y_w|h)p(\theta)$

$$\theta_{MAP} = arg\ min_\theta\ (y_w - X_w h(\theta) - \hat{\beta}_w)' (\Lambda)^{-1} (y_w - X_w h(\theta) - \hat{\beta}_w) + \theta' V_\theta^{-1} \theta \qquad (13)$$

   Let $\hat{h}_{MAP} = h(\theta_{MAP})$.
5. Repeat steps (2)-(4) by replacing $\hat{h}_{ls}$ by $\hat{h}_{MAP}$ until convergence.

This algorithm converges in 2-3 iterations and is at each voxel time-series.

The covariance matrix Λ can be estimated by estimating the diagonal elements of Λ which are nothing but variance of wavelet coefficients at each scale. These are related to the Hurst component as $\sigma_j^2 = \sigma_b^2 2^{-(2H+1)j}$, where j is the scale and σ_b is an unknown parameter. Therefore variance of wavelet coefficients can be estimated by estimating H and σ_b different algorithms in the literature [5]. But these algorithms are found to be not reliable for $H > 0$ [6]. Another way of estimating the variance at each scale is to use simple sample variance estimator [7]. fMRI time-series are typically of length 128 or less. The length of wavelet coefficients at scale j will be $N/2^j$. Hence, sample variance estimator will be unreliable due to small data lengths. In [7], it is shown that variance estimators will be more reliable in undecimated wavelet transforms (UWT) compared to the conventional non-redundant discrete wavelet transform (DWT). In UWT, the length of wavelet coefficients will be N irrespective of the scale. In this work, we use DWT for estimating the unknown parameters θ and β and UWT for estimating the variances at each scale.

4   Results and Discussion

We test the algorithms with real fMRI noise obtained from a null experiment, in which data is acquired when the subject is in rest condition. The above algorithms should be robust to the shape of HRF. We add synthetic HRFs (poisson, gamma and gaussian) to this data and estimate them using the above algorithm. We test the influence of shape and sampling times (TR) on the estimation of important parameters like time to peak and amplitude of the response.

The initial TR is fixed at 1 sec. Data at higher TRs (TR=2,3) can be obtained by down sampling the above data. Let the subject be at rest between time $t = 0$ to $t = 20$ seconds, followed by a task between $t = 21$ to $t = 40$ seconds and again rest between $t = 41$ to $t = 80$ seconds. The second rest condition is longer to allow sufficient time for the HRF to decay completely. The entire time duration is called as a cycle. Hence the input, x, to the linear system can be considered as a boxcar function with zeros during the rest condition and ones during the task condition. We add the convolution of x with each one of parametric models for HRF to the null data.

4.1   Performance Measures

We use the following performance indices to characterize the performance of the algorithm.

1. Time to peak t_p is an important parameter of HRF. It characterises the delay in response for a given task. This parameter can be used to find the relative time of activation of various regions. The percentage error in the estimation of time to peak is defined as: $\delta t_p = |\hat{\mu} - \mu|/\mu \times 100$. Where, $\hat{\mu}$ is the estimated time to peak, μ is the actual time to peak. Mean of the gaussian function and parameter of the Poisson model respectively characterize the time to peak. In the Gamma model, the ratio of gamma function parameters characterizes the time to peak.

2. The amplitude of the response (η) to a given stimulus is also an important parameter which characterizes the strength of the response at each voxel. The percentage error in the estimation of this parameter is defined as: $\delta\eta = |\hat{\eta} - \eta|/\eta \times 100$. Where, $\hat{\eta}$ is the estimated amplitude and η is the actual amplitude of HRF.

3. The overall error in the estimation of the HRF is characterized by the sample mean square error (MSE) which is defined as: $MSE = (\hat{h} - h)^t(\hat{h} - h)/K$. Where, $\hat{h}$ and h are, respectively, the estimated and actual HRF and K is the length of the response.

4. The sample correlation ρ between the estimated $(\hat{h})$ and actual HRF (h) measures the match between the estimated and actual HRFs. The range of ρ is from 0 to 1, where 1 corresponds to a perfect match.

4.2   Simulation Results

Our simulations show that the assumed gaussian model is able to recover the parametric functions like gamma, poisson and gaussian. Figure 1 shows the performance of gaussian HRF model at TR = 1, 2 and 3 seconds. At TR = 1 sec, the error in the estimation accuracy of time to peak (δt_p) is less than 13 percent for all the three simulated HRFs. Accuracy reduces with the increase in TR. The sample correlation ρ between the estimated HRF and actual HRF is about 0.85 for all the three HRF models at TR=1second and it decreases with increase in TR. MSE also increases with increase in TR. Hence, one can conclude that for

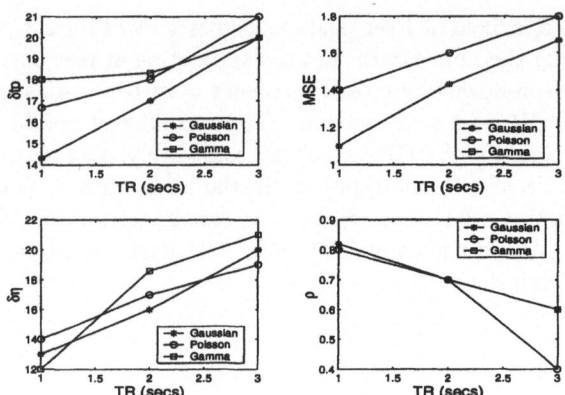

Fig. 1. Performance measures for parametric modeling of HRF at TR = 1, 2 and 3 secs. (a) Percentage error in the estimation of time to peak ($\delta t_p \times 100$) (b) MSE vs TR (c) Percentage error in the estimation of amplitude of HRF ($\delta \eta \times 100$) vs TR (d) Correlation between estimated and actual HRF (ρ).

a reliable estimate of HRF function, low sampling intervals (1 sec and less) are required. The gaussian function model for HRF is adequate to estimate important parameters like time to peak and is also robust to variations in actual HRF shapes.

5   Conclusions

A new wavelet based algorithm to estimate Hemodynamic Response Function (HRF) is presented. fMRI noise is modeled as a fractional Brownian Motion (fBm) and the decorrelating property of fBm process by wavelet transform is used. The influence of sampling time (TR) on the HRF estimation performance is studied. We also studied the influence of HRF shape on the estimation of important parameters like time to peak and amplitude of the function.

Acknowledgements

We thank Prof.P.N.Jayakumar of National Institute of Mental Health and Neuro Science, Bangalore, India for providing the required data and introducing us to the field of fMRI.

References

1. Ed Bullmore, Chris Long, John Suckling, Jalal Fadili, Gemma Calvert, Fernando Zelaya T.Adrian Carpenter and Mick Brammer: Colored Noise and Computational Inference in Neurophysiological (fMRI) Time Series Analysis: Resampling Methods in Time and Wavelet Domain. *Human Brain Mapping* 61–78 2001.
2. Francois Meyer: Wavelet-based Estimation of a Semiparametric Generalized Linear Model of fMRI Time-Series. *IEEE trans. Medical Imaging* vol.22 no.3 pp.315–322 March 2003.

3. Markus Svensen, Frithjof Kruggel and D.Yves von Cramon: Probabilistic Modeling of Single-trial fMRI Data. *IEEE trans. Medical Imaging* vol.19 no.1 pp.25–35 2000.
4. Geoffrey M.Boyton, Stephan A. Engel, Gary H.Glover and David J.Heeger: Linear Systems Analysis of Functional Magnetic Resonance Imaging in Human V1. *Journl of Neuroscience* 4207–4221 July 1996.
5. Gregory W. Wornell: Signal Processing with Fractals: A wavelet-based Approach. New Jersey:Printice Hall INC., 1996.
6. Brett Ninness: Estimation of 1/f Noise. *IEEE trans. on IT* vol.44 Jan 1998 32–46.
7. Donald P.Percival: On Estimation of wavelet variance. *Biometrika* vol.82 3 619–631.

Neural Networks for fMRI Spatio-temporal Analysis

Luo Huaien and Sadasivan Puthusserypady

Department of Electrical and Computer Engineering
National University of Singapore
4 Engineering Drive 3
Singapore 117576
{g0305766,elespk}@nus.edu.sg

Abstract. Most of the analysis techniques applied to functional magnetic resonance imaging (fMRI) consider only the temporal information of the data. In this paper, a new method combining temporal and spatial information is proposed for the fMRI data analysis. The nonlinear autoregressive with exogenous inputs (NARX) model realized by radial basis function (RBF) neural network is used to model the fMRI data. This new approach models the fMRI waveform in each voxel as a regression model that combines the time series of neighboring voxels together with its own. Both simulated as well as real fMRI data were tested using the proposed algorithm. Results show that this new approach can model the fMRI data very well and as a result, can detect the activated areas of human brain successfully and accurately.

1 Introduction

Human brain is the organ (a highly complex nonlinear system) for information processing through the interconnection of millions of neurons. How the brain works exactly is still an enigma to us. Recently, functional magnetic resonance imaging (fMRI) has arisen as an important tool for studying/understanding the human brain activities. fMRI measures the small changes of blood oxygenation level dependent (BOLD) signals due to neural activities. The process underlying the neural activities is believed to be a nonlinear complex process [1]. Modeling the measured fMRI signal is important to understand the properties of fMRI time series. The simplest model is the boxcar model, which uses the square-wave to describe the BOLD response. A more realistic model is the convolutive model [2], which assumes the brain and MR-scanner as a linear system. In [3], Buxton proposed balloon model to examine the dynamics of the BOLD responses.

Most fMRI analysis methods explore the temporal nature of fMRI data. For example, the general linear model (GLM) [4] estimates the voxelwise parameters, such as t-test value. These methods analyze each voxel's time series independently, which can be called univariate time series analysis methods. However, fMRI data has some spatial relationships. The activation is more likely to occur in several millimeters (*i.e.* a few voxels) of the brain region. Hence, modeling spatio-temporal properties of fMRI data will give us a more accurate estimation of the brain activated areas.

Neural network is a good methodology for input-output modeling given a set of sparse data points. In this paper, we proposed the recurrent networks for input-output modeling of fMRI data. In addition to use the time series in a single voxel, we incorporated the time series of the neighboring voxels as well. This method was tested on

N.R. Pal et al. (Eds.): ICONIP 2004, LNCS 3316, pp. 1292–1297, 2004.
© Springer-Verlag Berlin Heidelberg 2004

both simulated as well as real fMRI data. The results show that spatio-temporal analysis of fMRI data by neural network could model the fMRI signal well, which greatly improves the ability of detecting brain activity compared with the conventional t-test.

2 Methodology

The main objective of fMRI data analysis is to extract relevant information from the spatio-temporal data. From the perspective of neural networks, this problem is equivalent to learning an input-output mapping from a set of available time series, $i.e.$ a system identification problem. Normally, system identification can be solved by AR, MA or ARMA models. These models, which rely on linear methods, are well studied in time series analysis. For fMRI analysis, considering the nonlinearity behind the time series, we extended the linear methods to nonlinear ones and use nonlinear autoregressive with exogenous inputs (NARX) model [5] to fMRI data analysis.

2.1 Network Architecture

NARX network, with one or more feedbacks, could function as an input-output mapping network. Considering the spatial clustering features of fMRI data, it is desirable to regress the time series of a single voxel on its own and the neighboring voxles as well. Therefore, the estimated input-output mapping of a specified voxel is best fitted by the time series of its own and the neighboring voxels. Thus, the estimated output $\hat{y}(n+1)$ is not only regressed on previous actual outputs $y(n), \cdots, y(n-p+1)$ of the same specified voxel of interest, but also regressed on the output of the neighboring voxels. This input-output mapping takes the form (for k^{th} voxel):

$$\hat{y}_k(n+1) = \mathbf{F}^{NN}(\mathbf{y}_k(n), \mathbf{y}_1(n), \cdots, \mathbf{y}_{k-1}(n), \mathbf{y}_{k+1}(n), \cdots, \mathbf{y}_m(n), \mathbf{x}(n)) \qquad (1)$$

where $\mathbf{y}_k(n) = [y_k(n), \cdots, y_k(n-p+1)]^T$ is the delayed vector formed by the time series of the voxel of interest (k^{th} voxel), $\mathbf{y}_1(n), \cdots, \mathbf{y}_{k-1}(n), \mathbf{y}_{k+1}(n), \cdots, \mathbf{y}_m(n)$ are the $(m\text{-}1)$ delayed vectors formed by the time series of the neighboring voxels, $\mathbf{x}(n)$ is the input vector, m is the number of total voxels under consideration, including the voxel of interest and the neighboring voxels. This model includes multiple time series, one is from the voxel of interest, and others are from the neighboring voxels. We call it a spatio-temporal regression model realized by neural network. Figure 1 shows the modified architecture of the NARX model for the fMRI data modeling.

2.2 Radial Basis Function (RBF) Network

Radial Basis Function (RBF) network can uniformly approximate any continuous function provided that enough number of hidden units are available. Suppose there are N input data, then the estimated output $\hat{y}_k(n+1)$ of the mapping is taken to be a linear combination of the basis functions, $i.e.$:

$$\hat{y}_k(n+1) = F^{RBF}(\mathbf{p}) = \sum_{i=1}^{M} w_i \exp\left(-\frac{\|\mathbf{p}-\mathbf{c}_i\|}{2\sigma^2}\right) \qquad (2)$$

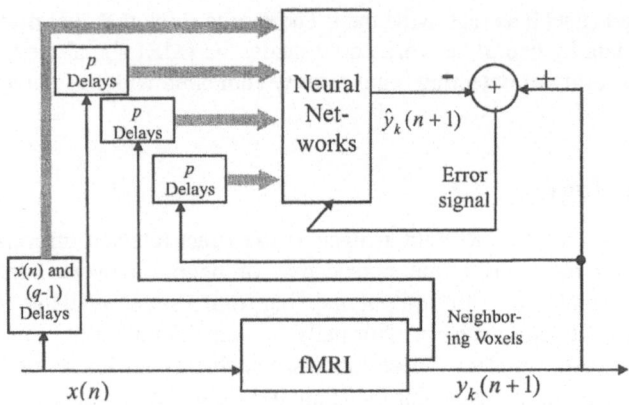

Fig. 1. Extended NARX model for fMRI data analysis

where M is number of hidden units, $\mathbf{p}$ is the vector formed by the input vector $\mathbf{x}(n)$ and the vectors $\mathbf{y}_k(n), \mathbf{y}_1(n), \cdots, \mathbf{y}_{k-1}(n), \mathbf{y}_{k+1}(n), \cdots, \mathbf{y}_m(n)$, $\mathbf{c}_i$ is the center of the radial basis function, w_i's are the output linear weights and σ is the width of radial basis function.

For ill-posed inverse problems, regularization is necessary. It is common to assume that the input-output mapping function is smooth, which means that similar inputs produce similar outputs [5]. The solution to a regularized RBF network is:

$$\mathbf{w} = (\mathbf{\Phi}^T\mathbf{\Phi} + \lambda\mathbf{I})^{-1}\mathbf{\Phi}^T\mathbf{y} \qquad (3)$$

where $\lambda > 0$ is the regularization parameter, which controls the balance between fitting the data and regularization, $\mathbf{\Phi}$ is an $N \times M$ interpolation matrix and $\mathbf{y}$ is the vector formed by output signals. In our method, we use a Bayesian trained RBF network to find the regularization parameter automatically [6-8].

In our method, it is important to determine how many and which neighboring voxels should be included in the regression model. The simplest way is to use a single voxel model, which assumes that the active brain areas consist of isolated voxels. Another method is to use the four nearest neighbors to the center voxel. In our work, we use 3×3 model (including all the 8 voxels surrounding the center voxel) and the recovered signals from each voxels are averaged over this 3×3 window.

As in the usual situation, we assume the input signal to be on-off boxcar function. This boxcar function represents the task block structure of the trial, alternating experimental and control blocks, with value 0 when stimulant is OFF and value 1 when stimulant is ON. This input signal with several delays is added to our regression model, serving as the exogenous inputs.

3 Results

3.1 Simulated Data

One slice (spatial size 79×95) from the fMRI data set was used to form the background and Gaussian noise was added to form the 2D time series, each of which contains 90 time points. A boxcar function was added to the specified areas to simulate the active brain areas with the signal to noise ratio (SNR) = -6dB. The single voxel model, four nearest voxels model and 3×3 model were used to train the network. The *t*-test values were calculated for the original time course (conventional *t*-test) and recovered time courses (our method). Through the simulation, we found that single voxel model and four nearest voxels model did not perform as well as the conventional *t*-test method. This is because the spatial information in these two models is not enough to achieve a better result.

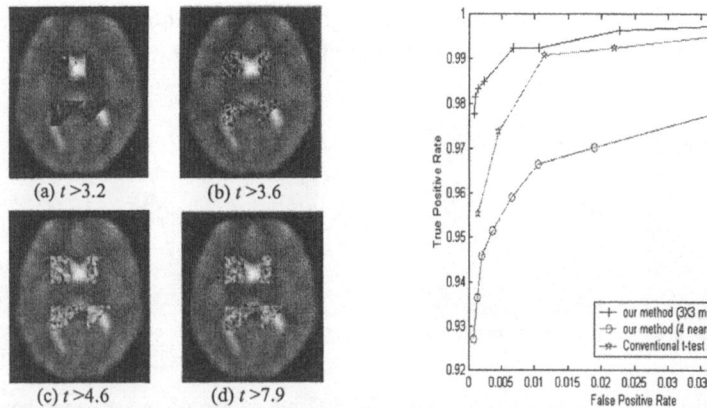

Fig. 2. The detection results of different models with $p<0.001$: (a) Conventional *t*-test; (b) our method with single voxel model; (c) our method with 4 nearest voxels model; (d) our method with 3×3 spatial model

Fig. 3. ROC curves for simulated noisy data with different processing methods

Figure 2 shows the results of the detected area under conventional *t*-test method and our methods with different spatial model. The *t*-test values were chosen when most of the activation areas were detectable, which are displayed under each graph respectively. From Figure 2, we find that our method with 3×3 spatial model not only enhances the *t*-test values of the activation areas, but also decreases the number of voxels incorrectly detected. Thus, this model is able to detect the simulated activation areas reliably compared with the conventional *t*-test method.

We also compared our method with the conventional *t*-test method by using receiver operator characteristic (ROC) analysis [9]. The ROC curve, which is a plot of true activation rate versus false activation rate for different threshold values, describes the ability of the different processing methods. Observing the ROC curves in Figure 3, we could find that our method with 3×3 spatial model produces larger areas under the ROC curve. That means this method performs better than the conventional *t*-test method and the method with four nearest voxels spatial model.

3.2 Real fMRI Data

We also validated our spatio-temporal regression network method by testing it on a block-design real fMRI data. This real fMRI experiment was designed for visuospatial processing task, judgement of line orientation. The details of the experiment can be found in [10]. Before using spatio-temporal regression neural network for modeling, we preprocessed the raw data by SPM (Statistical Parametric Mapping) software for registration, normalization and smoothing [11].

The number of data points of this data set is 100, with the experimental block length and control block length 10. The first ten volumes of data are discarded due to T1 effects [12] in the initial scan of an fMRI time series acquisition. Figure 4 shows the results with different processing methods. The t-test thresholds of respective methods are shown under each graph. From Figure 4, we could see that our method can detect the activation areas reliably with higher t-test threshold. The original and estimated time course of one of these activated voxels is shown in Figure 5.

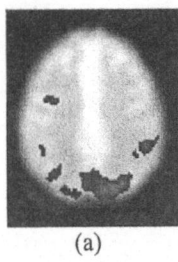
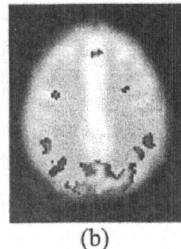
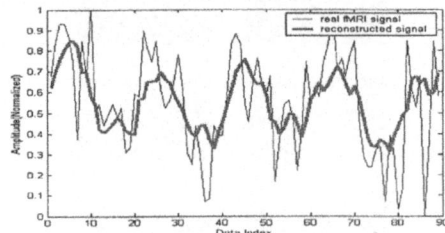

(a)   (b)

Fig. 4. Results of different methods for real fMRI data processing: (a) Conventional t-test ($p<0.001$, $t > 2.5$); (b) our method with 3×3 spatial model ($p<0.001$, $t >7.1$)

Fig. 5. Time course of an activation voxel for the real fMRI activation detection shown in Figure 4 with our method

4   Conclusion and Discussions

In this paper, a spatio-temporal regression model realized by neural network was proposed. This model is an extension to the nonlinear autoregressive with exogenous inputs (NARX) model, which utilizes the information of specified single voxel and neighboring voxels together. The proposed spatio-temporal regression method not only examines the temporal relationship of fMRI time series, but also takes the spatial information of brain activation areas into account. Hence, this model is believed more robust by considering that true fMRI activation is more likely to occur in clusters of several contiguous voxels than in a single voxel. The neural network function was realized by regularized RBF network trained by Bayesian learning. We tested this method over both simulated as well as real fMRI signals. The results show that this spatio-temporal regression method is able to model the fMRI signals well and as a result can realize the sensitive and reliable detection of clustered brain activation.

In our method, we assume that the neighboring voxels affect the specified voxel with identical weight. It is hoped that usage of different weights to the different neighboring voxels should give a more reliable and accurate result. Besides, in our method, we only considered the neighboring voxels in 2D. Inclusion of neighboring voxels of three dimensions will improve the regression and detection results.

References

1. Laird, A.R., Rogers, B.P., Meyerand, M.E.: Investigating the nonlinearity of fMRI activation data. Proc. of the 2nd Joint, EMBS/BMES Conference, Vol.1 (2002) 11–12.
2. Friston, K., Jezzard, P., Turner, R.: Analysis of functional MRI time-series. Human Brain Mapping, Vol.1 (1994) 153-171.
3. Buxton,R., Wong, E., Frank,L.: Dynamics of blood flow and oxygenation changes during brain activation: the balloon model. Mag. Res. in Med., Vol.39(6) (1998) 855-864.
4. Friston, K.J., et al.: Statistical parametric maps in functional imaging: A general linear approach. Hum. Brain Mapping, Vol.2 (1995) 189-210.
5. Simon H.: Neural Networks: A comprehensive foundation. (2nd Ed.). Prentice Hall (1999).
6. Tipping, M.E.: Sparse Bayesian learning and relevance vector machine. Journal of machine learning Research, Vol.1 (2001) 211-244.
7. MacKay, D.J.C.: Bayesian interpolation. Neural Computation, Vol.4(3) (1992) 415-417.
8. Erhard R.: Application of Bayesian trained RBF networks to nonlinear time-series modeling. Signal Processing, Vol.83 (2003) 1393-1410.
9. Constable, R.T., Skudlarski, P., Gore, J.C.: An ROC approach for evaluating functional brain MR imaging and postprocessing protocols. Magnetic Resonance in Medicine, Vol.34 (1995) 57-64.
10. Ng, V. W. K., et al: Identifying rate-limiting nodes in large-scale cortical networks for visuospatial processing: an illustration using fMRI. Journal of Cognitive Neuroscience, Vol.13 (2001) 537-546.
11. Friston, K.J., et al: SPM 97 Course Notes, Wellcome Department of Cognitive Neurology, University of College London (1997).
12. Jezzard P., Matthews P.M., Smith S. M.: Functional MRI: an introduction to methods. Oxford University Press, (2001).

Modeling Corrupted Time Series Data via Nonsingleton Fuzzy Logic System

Dongwon Kim, Sung-Hoe Huh, and Gwi-Tae Park*

Department of Electrical Engineering, Korea University, 1, 5-Ka Anam-Dong,
Seongbuk-Gu, Seoul 136-701, Korea
{upground,sungh,gtpark}@korea.ac.kr

Abstract. This paper is concerned with the modeling and identification of time series data corrupted by noise using nonsingleton fuzzy logic system (NFLS). Main characteristic of the NFLS is a fuzzy system whose inputs are modeled as fuzzy number. So the NFLS is especially useful in cases where the available training data, or the input data to the fuzzy logic system, are corrupted by noise. Simulation results of the Box-Jenkin's gas furnace data will be demonstrated to show the performance. We also compare the results of the NFLS approach with the results of using only a traditional fuzzy logic system. Thus it can be considered NFLS does a much better job of modeling noisy time series data than does a traditional fuzzy logic system.

1 Introduction

A time series is a sequence of observations taken sequentially in time [1]. Many time series data sets can be found in our life. The basic characteristic of a time series is that adjacent values are dependent. Generally speaking, past values in the sequence influence future values [2]. Modeling time series data is to analyze this dependency which is generally believed to be represented by nonlinear relationships. During the past few years, fuzzy modeling techniques [3-7] have become an active research area due to their successful applications to complex, ill-defined and uncertain systems in which conventional mathematical model fails to give satisfactory results. The most widely used fuzzifier in these fuzzy methods is the singleton fuzzifier [9], mainly because of its simplicity and lower computational requirement. However, this kind of fuzzifier may not always be adequate, especially in cases where noise is present in the training data. A different approach is necessary to account for uncertainty in the data, which is why we pay attention to the nonsingleton fuzzifier.

Nonsingleton fuzzifiers have been used successfully in a variety of applications. However, there still exist other applications not yet evaluated. Investigating applicability of the nonsingleton fuzzy logic system (NFLS) to noisy time series data modeling is highly demanded. In this study, nonsingleton fuzzy logic system [8,11] is first applied to model noisy Box-Jenkin's gas furnace data. The Box-Jenkin's gas furnace data will be demonstrated to show the performance of the NFLS. We also compare the results of the NFLS approach with the results of using only a singleton fuzzy logic system (SFLS). Thus it can be considered NFLS is much more successful in producing accurate ability of that series than is a comparable SFLS.

---

* Corresponding author.

N.R. Pal et al. (Eds.): ICONIP 2004, LNCS 3316, pp. 1298–1303, 2004.
© Springer-Verlag Berlin Heidelberg 2004

2 Nonsingleton Fuzzy Logic System

Since the NFLS is first applied to Box-Jenkin's gas furnace time series data [1], its fundamentals are briefly explained. The detailed descriptions and formulations of the NFLS can be found in [8,11]. The overall structure of the NFLS is shown in Fig. 1.

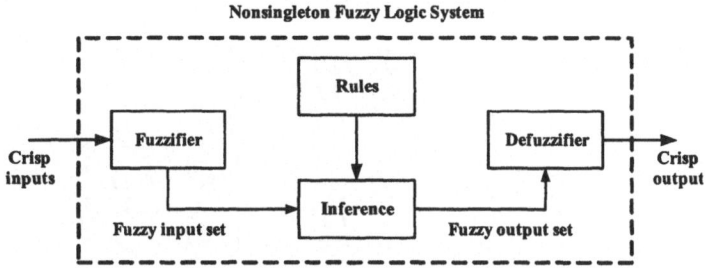

Fig. 1. Structure of the nonsingleton fuzzy logic system

Fuzzy sets can be viewed as membership functions μ_X that map each elements x of the universe of discourse, U, to a number $\mu_X(x)$ in the interval $[0,1]$:

$$\mu_X : U \to [0,1] \tag{1}$$

A fuzzifier maps a crisp point $x \in U$ into a fuzzy set X, whose domain of support is a subset of U. The nonsingleton fuzzifier maps the point $x \in U$ into a fuzzy set X with support x_i, where μ_X achieves maximum value at $x_i = x$ and decreases while moving away from $x_i = x$. Conceptually, the nonsingleton fuzzifier implies that the given input value x is the most likely value to be the correct one from all the values in its immediate neighborhood, however, because the input is corrupted by noise, neighboring points are also likely to be the correct values, but to a lesser degree. So nonsingleton fuzzification is especially useful in cases where the available data contain any kind of statistical, or non-statistical uncertainty. NFLS is not only a generalization of single-ton fuzzy logic system but also provide a reconciliation between fuzzy logic techniques and statistical methods for handling uncertainty [11].

Consider a fuzzy logic system with a rule base of M rules, and let the lth rule be denoted by R^l. Let each rule have n antecedents and one consequent, i.e., it is of the general form

$$R^l : \text{IF } u_1 \text{ is } F_1^l \text{ and } u_2 \text{ is } F_2^l \text{ and} \cdots u_n \text{ is } F_n^l \text{ Then } v \text{ is } G^l \ l = 1, \ldots, M \tag{2}$$

Where u_k, $k=1, \ldots, n$, and v are the input and output linguistic variables, respectively. Each F_k^l and G^l are subsets of possibly different universes of discourse. Let $F_k^l \subset U_k$ and $G^l \subset V$. Each rule can be viewed as a fuzzy relation R^l from a set U to a set V where U is the Cartesian product $U = U_1 \times \cdots \times U_n$. R^l itself is a subset of the Cartesian product $U \times V = \{(x, y) : x \in U, y \in V\}$, where $x \equiv (x_1, x_2, \cdots, x_n)$, and x_k and y are the points in the universes of discourse U_k and V of u_k and v.

Using the rule R^l, NFLS is constructed and its parameters are tuned as follows.

$$y(\mathbf{x}^{(i)}) = f_{ns}(\mathbf{x}^{(i)}) = \sum_{l=1}^{M} \bar{y}^l \phi_l(\mathbf{x}^{(i)}) = \frac{\sum_{l=1}^{M} \bar{y}^l \prod_{k=1}^{n} \mu_{Q_k^l}(x_{k,\max}^{l,(i)})}{\sum_{l=1}^{M} \prod_{k=1}^{n} \mu_{Q_k^l}(x_{k,\max}^{l,(i)})}$$

$$= \frac{\sum_{l=1}^{M} \bar{y}^l \prod_{k=1}^{n} \exp\left(-\frac{1}{2}\left[\dfrac{(x_k^{(i)} - m_{F_k^l})^2}{\sigma_X^2 + \sigma_{F_k^l}^2}\right]\right)}{\sum_{l=1}^{M} \prod_{k=1}^{n} \exp\left(-\frac{1}{2}\left[\dfrac{(x_k^{(i)} - m_{F_k^l})^2}{\sigma_X^2 + \sigma_{F_k^l}^2}\right]\right)} \tag{3}$$

$$m_{F_k^l}(i+1) = m_{F_k^l}(i) - \alpha_m[f_{ns}(\mathbf{x}^{(i)}) - y^{(i)}][\bar{y}^l(i) - f_{ns}(\mathbf{x}^{(i)})]$$
$$\times [\frac{x_k^{(i)} - m_{F_k^l}(i)}{\sigma_X^2(i) + \sigma_{F_k^l}^2(i)}]\phi_l(\mathbf{x}^{(i)}) \tag{4}$$

$$\bar{y}^l(i+1) = \bar{y}^l(i) - \alpha_{\bar{y}}[f_{ns}(\mathbf{x}^{(i)}) - y^{(i)}]\phi_l(\mathbf{x}^{(i)}) \tag{5}$$

$$\sigma_{F_k^l}(i+1) = \sigma_{F_k^l}(i) - \alpha_\sigma[f_{ns}(\mathbf{x}^{(i)}) - y^{(i)}][\bar{y}^l(i) - f_{ns}(\mathbf{x}^{(i)})]$$
$$\times \sigma_{F_k^l}(i)[\frac{x_k^{(i)} - m_{F_k^l}(i)}{\sigma_X^2(i) + \sigma_{F_k^l}^2(i)}]^2 \phi_l(\mathbf{x}^{(i)}) \tag{6}$$

$$\sigma_X(i+1) = \sigma_X(i) - \alpha_X[f_{ns}(\mathbf{x}^{(i)}) - y^{(i)}][\bar{y}^l(i) - f_{ns}(\mathbf{x}^{(i)})]$$
$$\times \sigma_X(i)[\frac{x_k^{(i)} - m_{F_k^l}(i)}{\sigma_X^2(i) + \sigma_{F_k^l}^2(i)}]^2 \phi_l(\mathbf{x}^{(i)}) \tag{7}$$

where $y(\mathbf{x})$, $\bar{y}^l$, ϕ_l and $\mu_{Q_k^l}$ are output for nonsingleton fuzzy system, center of the consequent MF, fuzzy basis function, and function whose supremum can be evaluated, respectively. $m_{F_k^l}$, $\sigma_{F_k^l}$, and σ_X are uncertain mean and standard deviation of kth antecedent fuzzy set and input fuzzy set, respectively. α_m, α_σ, and α_X are the learning parameters of the $m_{F_k^l}$, $\sigma_{F_k^l}$, σ_X, respectively.

3 Simulation Results

We evaluate the performance of NFLS applying it to the modeling of noisy Box-Jenkin's gas furnace time series. In addition, we compare the performance of NFLS with that of singleton FLS. Box and Jenkin's gas furnace is a famous example of system identification. The well-known Box-Jenkins data set consists of 296 input-output observations, where the input u(t) is the rate of gas flow into a furnace and the output y(t) is the CO_2 concentration in the outlet gases. The delayed terms of u(t) and y(t) such as u(t-2), u(t-1), y(t-2), and y(t-1)are used as input variables to the NFLS. The actual system output y(t) is used as target output variable for this model. The performance index (PI) is defined as the root mean squared error

$$PI = \sqrt{\frac{1}{m}\sum_{i=1}^{m}(y_i - \hat{y}_i)^2}$$ (8)

where y_i is the actual system output, $\hat{y}_i$ is the estimated output of each node, and m is the number of data.

Simulation of NFLS is conducted in choosing max-product composition, product implication, height defuzzification, and Gaussian membership function. In Fig. 2, we present the noise-free data which is four input variables of the Box-Jenkin's gas furnace time series. In the figure, (a) and (b) are the delayed terms of methane gas flow rate u(t) and carbon dioxide density y(t), respectively. We also depict the one realization of 5dB uniformly distributed noise data which will corrupt the noise free signals in Fig. 2 (c). Owing to the noise data shown in Fig. 2 (c), input data will be corrupted as shown in Fig. 2 (d-e). In Fig. 2 (d) and (e), we plot the noise corrupted data of Box-Jenkin's gas furnace time series. The corrupted data is employed as input variables to the NFLS. We assigned two or three membership functions to each input variable. So the number of fuzzy rules varies from 16 to 36. The learning parameter α is 0.1485 for the SFLS and NFLS. Fig. 2 (f) shows the modeling results of the SFLS which contains 36 fuzzy rules with two MFs for gas flow rate and three MFs for carbon dioxide density are employed for SFLS. However, the model outputs do not follow the actual output very well so the SFLS is unable to handle the noise.

Modeling results of the nonsingleton fuzzy logic system are shown in Fig. 3. In the Fig. 3 (a), 36 fuzzy rules are used. Here, two MFs for gas flow rate u(t) and three MFs for carbon dioxide density y(t) are employed. In the same way, when three MFs for gas flow rate and two MFs for carbon dioxide density are considered the estimated result is shown in Fig. 3 (b). As can be seen from the Fig. 3 (b), the model output follows actual output well. Therefore, the NFLS does a much better job of modeling a noisy time series data than does a SFLS. The value of the performance index of the NFLS is equal to 0.8493.

4 Conclusions

We have presented modeling of corrupted time series data via a nonsingleton fuzzy logic system. Because the data is corrupted by measurement noise, the most commonly used fuzzifier, singleton fuzzifier, is not adequate. To handle uncertainty in the

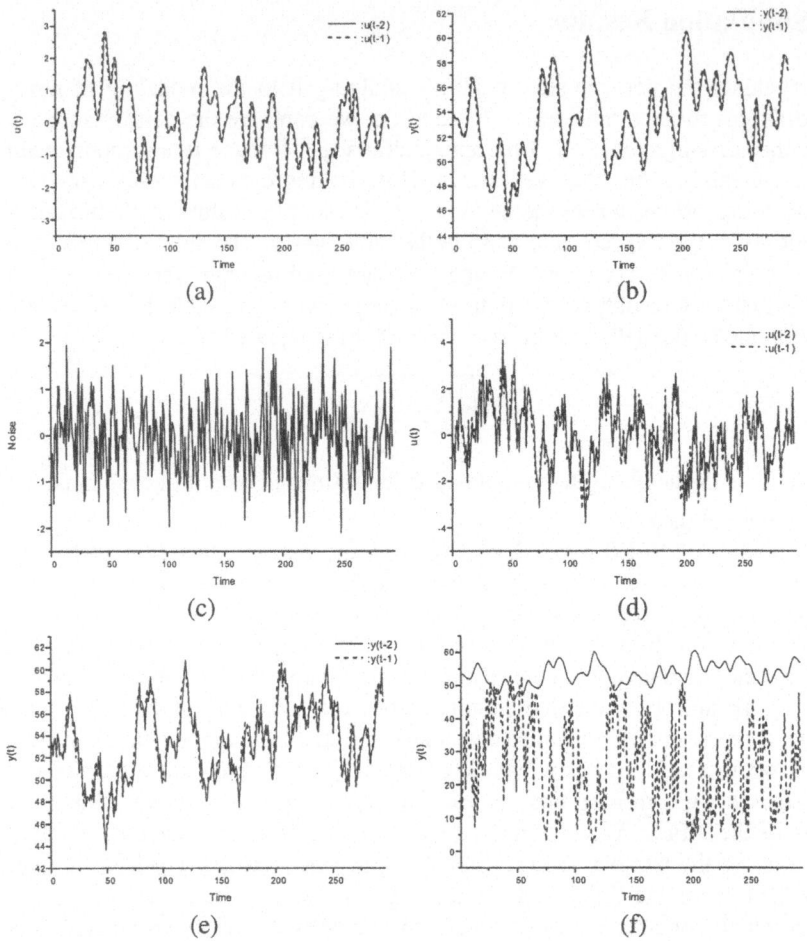

Fig. 2. (a-b): Noise-free Box-Jenkin's gas furnace time series, (c): one realization of 5dB uniformly distributed noise data, (d-e): Noise corrupted Box-Jenkin's gas furnace time series to be employed as input variables for FLS, (f): Modeling results of the singleton fuzzy logic system with 36 fuzzy rules assigned two MFs for gas flow rate $u(t)$ and three MFs for carbon dioxide density $y(t)$ (solid line: actual time series, dotted line: estimated time series)

data, nonsingleton fuzzy logic system (NFLS) is first applied to the corrupted Box-Jenkin's gas furnace time series data.

Comparisons between singleton fuzzy logic system and nonsingleton fuzzy logic system are given. Simulation results of the Box-Jenkin's gas furnace data was demonstrated to show the performance of the NFLS. As we seen from the simulation results, the NFLS provide a way to handle knowledge uncertainty. Meanwhile, SFLS is unable to directly handle uncertainty. Thus it can be considered NFLS does a much better job of modeling a noisy time series data than does a traditional fuzzy logic system.

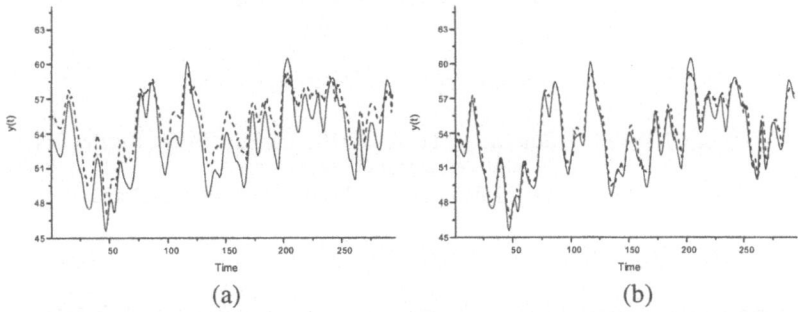

(a)   (b)

Fig. 3. Modeling results of the nonsingleton fuzzy logic system with (a) 36 fuzzy rules assigned two MFs for gas flow rate u(t) and three MFs for carbon dioxide density y(t) and (b) 36 fuzzy rules assigned three MFs for gas flow rate u(t) and two MFs for carbon dioxide density y(t) (solid line: actual time series, dotted line: estimated time series)

References

1. Box, G.E.P., Jenkins, G.M., Reinsel, G.C.: Time Series Analysis: Forecasting and Control 3rd ed. Prentice-Hall (1994)
2. Cho, D.Y, Zhang, B.T.: Bayesian Evolutionary Algorithms for Evolving Neural Tree Models of Time Series Data. Proceedings of the 2000 Congress on Evolutionary Computation. 2 (2000) 1451-1458
3. Chen, J.Q., Chen, L.J.: An on-line identification algorithm for fuzzy systems. Fuzzy Sets and Sytems. 64 (1994) 63-72
4. Jang, J.R.: Fuzzy controllers based on temporal back propagation. IEEE Trans. Neural Networks. 3 (1992) 714-723
5. Takagi, T., Sugeno, M.: Fuzzy Identification of Systems and Its Applications to Modeling and Control. IEEE Trans. Syst. Man Cybern. 15 (1985) 116-132
6. Wang, L.X., Mendel, J.M.: Fuzzy basis functions, universal approximation, and orthogonal least-squares learning. IEEE Trans. Neural Networks. 3 (1992) 807-814
7. Gupta, M.M, Rao, D.H.: On the principles of fuzzy neural networks. Fuzzy Sets Syst. 61 (1994) 1-18
8. Mendel, J.M.: Uncertain Rule-Based Fuzzy Logic Systems: Introduction and New Directions. Prentice Hall, Upper Saddle River NJ (2001)
9. Kosko, B.: Fuzzy systems as universal approximators. IEEE Trans. Comput. 43 (1994) 1329-1333
10. Hayashi, Y., Buckley, J.J., Czogala, E.: Fuzzy neural network with fuzzy signals and weights. Int. J. Intell. Syst. 8 (1993) 527-537
11. Mouzouris, G.C., Mendel, J.M.: Nonsingleton Fuzzy Logic Systems: Theory and Application. IEEE Trans. Fuzzy Syst. 5 (1997) 56-71

Hydrological Forecasting and Updating Procedures for Neural Network

Mêuser Valença[1] and Teresa Ludermir[2]

[1] Chesf/UNIVERSO, Rua Gaspar Peres 427/104, CEP 5670-350, Recife, Brazil
meuser@chesf.gov.br
[2] UFPE – Universidade Federal de Pernambuco, Recife, Brazil
tbl@cin.ufpe.br

Abstract. A non-linear Auto-Regressive Exogenous-input model (NARXM) for river flow forecasting by an output-updating procedure is presented. This updating procedure is based on the structure of a constructive neural network (NSRBN – A Non-linear Sigmoidal Regression Blocks Networks). The NARXM-neural network updating procedure is tested using the daily discharge forecasts of the routing (SSARR – Streamflow Synthesis And Reservoir Regulation) conceptual model operating on the São Francisco River having different discharge conditions. The performance of the NARXM-neural network updating procedure is compared with that of the linear Auto-Regressive Exogenous-input (ARXM) model updating procedure, the latter being a generalisation of the widely used Auto-Regressive (AR) model forecast error updating procedure. The results of the comparison indicate that the NARXM procedure performs better than the ARXM procedure.

1 Introduction

The present paper deals with the application of the neural network technique as an updating procedure for river flow forecasts and with the assessment of its performance by comparison with some of the well established updating procedures. In the context of river flow forecasting, the estimated discharge hydrographs of the conceptual models normally differ, often substantially, from the corresponding observed discharge hydrographs. These differences (i.e. the discharge forecast errors) are due to many factors which include: (1) inadequacy of the structure of the conceptual models to represent the constituent hydrological processes (i.e. model inadequacy or error), (2) poor estimation of the model parameters (due to inefficiency or failure of the optimisation procedures) and (3) both systematic and random errors in the model input and output data [1]. The errors between the simulated and the observed discharge hydrographs can be categorized into three types [2], namely, volume (amplitude) errors, phase errors and shape errors.

In the present paper, the neural network technique is investigated as an alternative model-output updating procedure to that of the ARXM[3][4].

The present paper is organized in the following manner; firstly, a description of the ARXM and the NARXM updating procedures is given. Secondly, the structure of the SSARR conceptual model is outlined, this model having been selected for convenience as the substantive routing conceptual simulation model. Finally, a description is

N.R. Pal et al. (Eds.): ICONIP 2004, LNCS 3316, pp. 1304–1309, 2004.
© Springer-Verlag Berlin Heidelberg 2004

provided of the application of the ARXM and the NARXM updating procedures to the estimated daily discharge series of the SSARR model, for the São Francisco River, and the performances of these two updating procedures are compared.

2 The Auto-regressive Exogenous Input Model (ARXM) Updating Procedure

The Auto-Regressive Exogenous-input Model (ARXM) [5] is a linear input-output model which enables the forecasting of the future values of a time series on the basis of its recent past values and on the basis of the values of one or more exogenous input times series. The one-step-ahead ARXM output updating procedure, can be expressed mathematically as;

$$Q_{i+1} = \sum_{k=1}^{p} a_k Q_{i-k+1} + \sum_{k=0}^{q} b_k Q_{i-k+1} + e_{i+1}^{ARXM} \tag{1}$$

In this equation, Q_{i+1} denotes the as-yet-unmeasured discharge for the $(i+1)^{st}$ time step, the updated one-step ahead discharge forecast $\hat{Q}_{i+1/i}^{ARXM}$ made at the current i^{th} time step having the form

$$\hat{Q}_{i+1/i}^{ARXM} = Q_{i+1} - e_{i+1}^{ARXM} = \sum a_k Q_{i-k+1} + b_0 \hat{Q}_{i+1/i} + \sum b_k \hat{Q}_{i-k+1} \tag{2}$$

where the Q_{i-k+1} are the current and most recently observed discharges, $Q_{i+1/i}$ is the one-step-ahead simulation mode discharge forecast of the substantive routing model, $\hat{Q}_{i-k+1}$ values are the current and recent outputs of that model, p and q are the orders of the auto-regressive and the exogenous input parts of the ARXM, respectively, a_k and b_k being the corresponding coefficient parameters of these two parts.

3 The Non-linear Auto-regressive Exogenous Input Model (NARXM) Updating Procedure

The neural network technique is applied in the present study as an alternative procedure to that of the ARXM for updating the discharge forecasts of the conceptual model. The one-step-ahead NARXM-neural network updating procedure may be expressed as [7]

$$Q_{i+1} = h(Q_i, Q_{i-1}, ..., Q_{i-p+1}, \hat{Q}_{i+1/i}, \hat{Q}_i, ..., \hat{Q}_{i-q+1} + e_{i+1}^{NARXM}) \tag{3}$$

in which h denotes a non-linear functional relation and e_{i+1}^{NARXM} is the residual error of the corresponding updated discharge-forecast $\hat{Q}_{i+1/i}^{NARXM}$, where $\hat{Q}_{i+1/i}^{NARXM} = Q_{i+1} - e_{i+1}^{NARXM}$.

3.1 Network Architecture

The goals here are to present practical methods to realize compact networks using the model with hidden units with sigmoidal blocks activation functions [6]. The activation function is

$$Actv_h(x) = (\sigma_{net(h)} + \theta_h) \tag{4}$$

where h *is* the order of the block (= number of hidden units), θ_h is a bias and $\sigma_{net(h)}$ is the hyperbolic tangent function. The first design step is to divide f (x) up into blocks of equal-degree terms, as in Figure 1. That is

$$f(x) = f_1(x) + f_2(x) + ... + f_d(x) \tag{5}$$

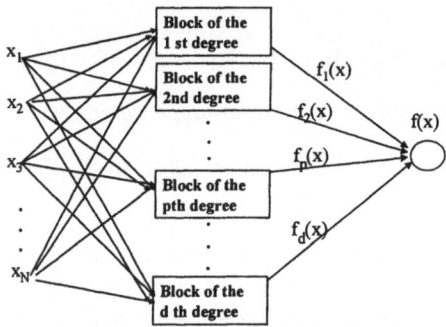

Fig. 1. Network architecture (NSBRN)

The block approach is to realize all terms in $f_p(x)$ functions at the same time, as in Figure 2.

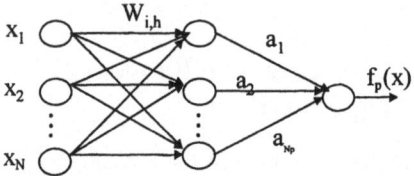

Fig. 2. Block of degree p architecture

The input **x** is an N dimensional vector and x_i is the i-th component of **x**. The inputs are weighted and fed to a layer of h hidden units, where h is the order of the block. Let $f_p(x)$ be the output of the block of degree p. Then,

$$f_p(x) = a_1.(Actv_1(x))^1 + a_2.(Actv_2(x))^2 + + a_3.(Actv_3(x))^3 + ... + a_p.(Actv_p(x))^p \tag{6}$$

and a_h is the weight between h-th hidden unit to output unit and h=1,2,...,p.

4 SSARR Model (Streamflow Synthesis and Reservoir Regulation)

The SSARR Model is a watershed simulation model developed by US Army Corps of Engineers, North Pacific Division, around 1960.

For channel routing, the following storage function is used in SSARR. The change of storage in the channel with respect to that of discharge $dS/dQ = T_S$ is a parameter showing the hydrograph regression due to the channel storage and the time delay of the peak of the hydrograph.

$$O_2 = O_1 + \left[\frac{I_m - O_1}{T_S - t/2}\right], \quad T_S = KTS/Q^n \tag{7}$$

Where: I_m = average inflow; $O_{1,2}$ = outflow at the beginning and end of the unit time; t= unit time of calculation; T_S = channel storage parameter; KTS = a constant for a time lag; Q = discharge; n = a constant for channel routing.

5 Application of the NSRBN – Neural Network and ARXM Updating Procedures

The NARXM-neural network and the ARXM updating procedures are applied to the daily simulation-mode estimated discharges of the SSARR model on the São Francisco River. These calibration and verification periods are non-overlapping, with the first years of the available records being used for calibration, the following years being used for verification purposes. The data obtained correspond to the period between January 1978 and December 2002.

The forecasts were made with one-step-ahead prediction horizon. The data sets were divided into two subsets: the first, corresponding to the period from January 1978 to December 1990, was selected for training; the second subset, corresponding to the period from January 1991 to December 2002, was selected for testing.

The performance of the two updating procedures is evaluated quantitatively using the following numerical indices: (1) The R^2 criterion [3], which is related to the sum of the squares of the differences, F, between the estimated and observed discharges. where F_0 is the sum of the squares of differences between the observed discharge and the mean discharge. (2) The RMSE ("*Root Mean Square Error*") metric penalizes higher errors. Table 1 shows that updating procedures by the he NSRBN has substantially improved the respective R^2 and RMSE values of the different periods.

Table 1. The improvement in the performance of the NSRBN updating procedure over that of the ARXM

Numerical Indices	Calibration period 03/1978 to 12/1991		Verification period 01/1992 to 12/2002	
	ARXM	NSRBN	ARXM	NSRBN
R^2	0,981	**0,988**	0,956	**0,969**
RMSE	307	**241**	314	**276**

The Figures 3 and 4 shows, for example, the tracking behavior of the predicted values by the NSRBN, predicted values by the ARXM model for the calibration period.

6 Conclusions

In the present study, a non-linear model updating procedure for the simulation mode discharge forecasts of the substantive routing conceptual model is developed. The NSRBN model has an in-built mechanism that allows for automatic on-line updating of forecasts.

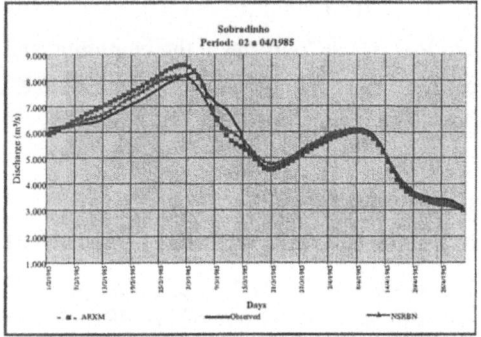

Fig. 3. Comparisons of the observed discharges and the updated discharges for lead time of 1 day

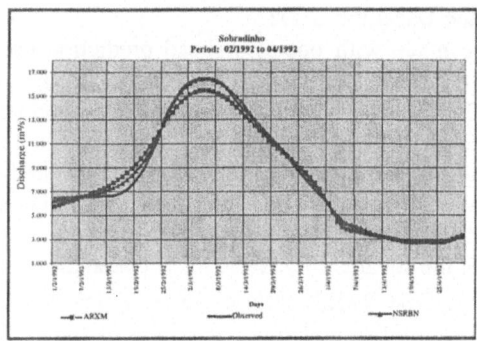

Fig. 4. Comparisons of the observed discharges and the updated discharges for lead time of 1 day

The NSRBN updating procedure is tested using the simulation discharge forecasts of the SSARR model on the São Francisco River. The results of the NSRBN procedure are compared with those of the linear Auto-Regressive Exogenous-input (ARXM) updating procedure for which the well-known autoregressive (AR) model output updating procedure is a special limiting case [4]. In terms of the values obtained for the model efficiency, these comparisons indicate that the NSRBN generally performs better than the ARXM for updating the discharge forecasts of the SSARR routing model.

References

1. Becker, A. and Serban, P., 1990. *Hydrological models for waterresources system design and operation.* Oper. Hydrol. Rep. 34, WMO No. 740, Geneva.
2. Serban, P., and Askew, A.J., 1991. Hydrological forecasting and updating procedures. *IAHS Pub. no.* **201**, 57–369.
3. Ahsan, M. and O'Connor, K.M., 1994. A simple non-linear rainfall-runoff model with a variable gain factor. *J. Hydrol.*, **155**, 151–183.
4. Shamseldin, A.Y. and O'Connor, K.M., 1999. A real-time combination Method for the outputs of different rainfall-runoff models. *Hydrol. Sci. J.*, 44, 895–912.
5. Peetanonchai, B., 1995. *A modification of the AR model for updating of model forecasts.* Final Report for the 6th International advanced Course/Workshop on River Flow Forecasting, 24-30th June, 1995, Department of Engineering Hydrology, National University of Ireland, Galway. (Unpublished).
6. M. J. S. Valença. Analysis and Design of the constructive neural networks for complex systems modeling (in portuguese). Ph.D Theses, UFPE, Brazil, 1999.
7. Bomberger, J.D. and Seborg, D.E., 1998. Determination of model order for NARX models directly from input – output data. *J. Process Control*, **8**, 459–468.

Modeling Gene Regulatory Network in Fission Yeast Cell Cycle Using Hybrid Petri Nets

Ranjith Vasireddy and Somenath Biswas

Dept. of Computer Science and Engineering, IIT Kanpur
Uttar Pradesh, India - 208016
{vasiredy,sb}@cse.iitk.ac.in

Abstract. The complexity of models of gene regulatory network stems from the fact that such models are required to represent continuous, discrete, as well as stochastic aspects of gene regulation. Hybrid stochastic Petri nets as models can fairly incorporate all these aspects while keeping the model as simple as possible. This paper constructs an hybrid Petri net model of the fission yeast cell cycle regulation mechanism. Based on our simulation of the Petri net on an existing tool, we draw some conclusions about the regulation mechanism under study. We discuss the kinds of biologically significant questions that can be answered using hybrid Petri nets as models for genetic regulatory mechanisms.

1 Introduction

A great challenge of post-genomic era is to combine the several understood parts of a cell into a working model of a living, responding, and a reproducing cell. Various kinds of models have been studied to model gene regulatory mechanisms such as Boolean networks [3], graph based models, differential equation models [2,6] and Petri nets [5]. It is well known that among these models, differential equation models most accurately represent gene regulatory networks. The differential equation model by Tyson et al. [2] of the fission yeast cell cycle regulation is an example of such an accurate modelling. Some authors have also incorporated stochasticity into differential equation models, see, e.g., [6]. Differential equation models of gene regulation are, however, difficult to construct and to reason with, mainly because these models fail to capture the way in which we intuitively understand gene regulation networks. Modelling using stochastic Petri nets (as, e.g., [4]), on the other hand, are both relatively easy to construct and to verify. Stochastic Petri nets, however, cannot deal with those aspects which deal with continuous values, such as concentration of mRNA, or of a protein. To overcome this problem, Matsuno et. al. [5] have used hybrid Petri nets in modelling gene regulatory networks. In this work we have also used an hybrid Petri net as a model of the gene regulatory network in fission yeast cell cycle, we shall see that in our model it is fairly easy to reason about biologically significant aspects of fission yeast cell cycle.

N.R. Pal et al. (Eds.): ICONIP 2004, LNCS 3316, pp. 1310–1315, 2004.
© Springer-Verlag Berlin Heidelberg 2004

2   Hybrid Petri Nets

Hybrid Petri net [1] is an extension of Petri net that allows us to handle continuous values. We use the definition of Matsuno et al.[4] to define hybrid Petri nets. Formally, an hybrid Petri net is a six-tuple $Q = (P,T,h,Pre,Post, M_0)$, where P $= p_1, p_2, ..., p_m$, is a finite set of *places* and $T = t_1, t_2, ..., t_n$ is a finite set of *transitions*. $h : P \cup T \rightarrow \{D, C\}$ indicates for every place or transition whether it is a *discrete* or *continuous* one. A non-negative integer called the *number of tokens* is always associated with a discrete place, i.e., when $h(P_i)=D$, and a non-negative real number called the *mark* is always associated with a continuous place, i.e., when $h(P_i) = C$. $Pre(P_i,T_j)$ (respectively, $Post(P_i,T_j)$) is a function that defines the arc from a place P_i (a transition T_j) to a transition T_j(a place P_i), where the arc has a weight of non-negative integer (non-negative real number) if $h(P_i)$ $= D$ ($h(P_i) = C$). We assign a variable d_{T_j}, called the delay time of T_j, to each discrete transition T_j and assign a variable v_{T_j}, called the speed of T_j, to each continuous transition T_j. M_0 is a mapping from the set of places to the set of non-negative integers called the *initial marking*.

A discrete place is represented in graphical form using single circle, and a continuous place using two concentric circles. A discrete transition is represented as a filled in rectangular box, and a continuous transition is represented as a hollow rectangular box. A continuous transition fires continuously and its firing speed is given as a function of values in the places of the model. Without loss of generality, we are using dotted arcs to indicate that firing of the corresponding transition will not lead to a decrease in the concentration of input places.

3   Fission Yeast Cell Cycle Regulation

The cell cycle is an ordered series of events leading to replicating of cells. In eukaryotic cells, the cell cycle consists of two basic processes: DNA synthesis (S phase) and the mitosis phase (M phase). During the S phase, double stranded DNA molecules are replicated to produce pairs of 'sister chromatids', held together by proteins called cohesions. M phase consists of four subphases. Prophase is the first phase when chromosomes condense into compact structures. Metaphase is the next phase when chromosomes are aligned on the mid plane of the mitotic spindle. In anaphase, cohesions are degraded and finally in telophase, daughter nuclei form and the cell begins to divide. S and M phases are separated in time by two gap phases (G1 and G2), constituting the generic cell cycle: G1-S-G2-M.

3.1   Fission Yeast Cell Cycle Engine

The most important components of the eukaryotic cell cycle engine are cyclin-dependent protein kinases, heterodimers consisting of a catalytic subunit (a Cdk) and a regulatory subunit (a cyclin). Cdks are active only in complex with a cyclin partner. The basic cell cycle engine in fission Yeast as explained by Tyson

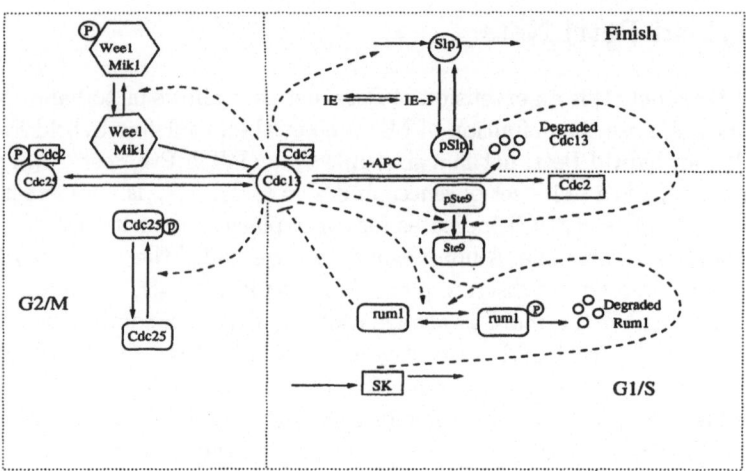

Fig. 1. Cell Cycle Engine of fission Yeast (from Tyson et al.[7])

et al. [7] is given in figure 1. Cdks exert their action by phosphorylating other proteins. Their protein-kinase activity is required to start both DNA replication and mitosis. Lower eukaryotes use only one essential Cdk subunit (Cdc2).

In fission yeast, complexes between Cdk1 and B-type cyclins play the major roles in cycle regulation. Two B-type cyclins Cig1 and Cig2, known as starter kinases, are involved in DNA synthesis. For the cell to enter into mitosis phase, cdc2-cdc13 concentration must be high and to enter into synthesis phase the cdc2-cdc13 concentration must be low. Destruction of Cdk activity as cells exit mitosis is the job of the anaphase promoting complex (APC). Slp1 targets the cohesion complex for disassembly, and both Slp1 and Ste9 present Cdc13 to the APC for ubiquitinaton. Proper timing of these events is controlled by phosphorylation and dephosphorylation of Slp1 and Ste9, by Cdk/Cyclin complexes and the phosphatases that oppose them. There are several complex feedback mechanisms inside a cell, these are easily discernible through the use of our model.

4   Modeling Using Hybrid Petri Nets

In this section, we construct hybrid Petri net pathways for regulation mechanism of both S phase promoting factor and for M-phase promoting factor. Then, we combine the models of S phase and M phase to arrive at a working model for yeast cell cycle. In all transitions (represented using Ti), formulas that are used to describe reaction rates are assigned. However, we do not give these formulae here, in order to keep the model description simple.

The left dotted box in figure 2 shows the constructed model for the S phase promoting factor regulation. In Fission yeast, S phase promoting factor is a cdc2-cig2 complex formed from Cig2 and Cdc2. This is illustrated through the reaction that happens at transition T1. This complex helps in fighting S phase

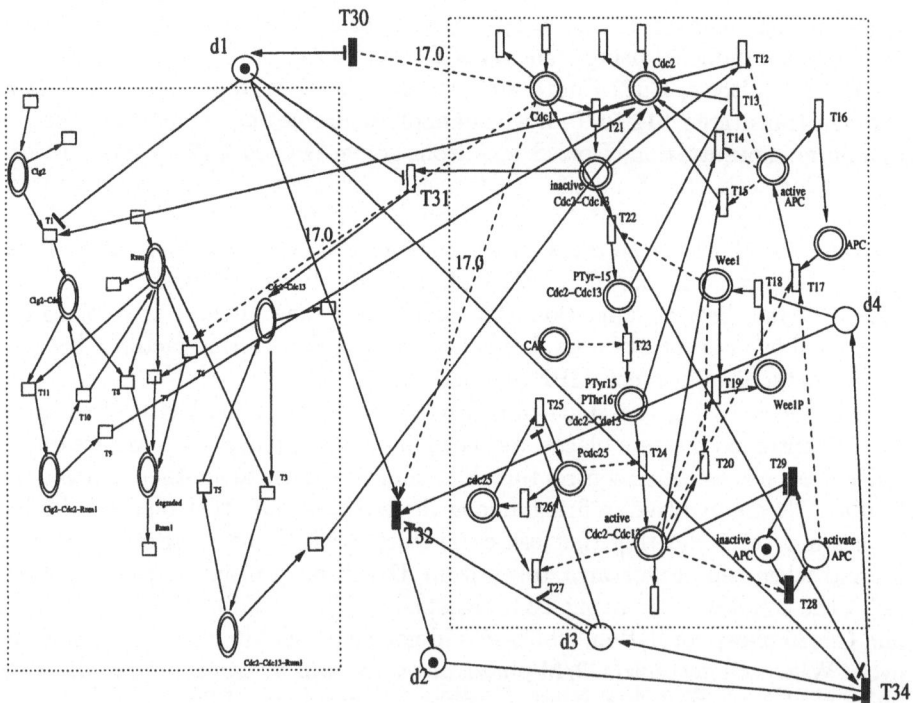

Fig. 2. Hybrid Petri net model of cell cycle regulation in Fission Yeast. (Here left and right dotted boxes indicate S-phase and M-phase respectively)

enemies in the cell cycle. During G1 phase, cells contain a protein, Rum1, which binds to and inhibits any Cdc2:Cdc13 dimers that may be present. This is indicated through transition T3 in the Figure. The trimer can be disrupted by the degradation of Rum1 as indicated through the transition T5. Cig2-Cdc2 dimer helps to prevent cdc2-cdc13 complex from becoming inactive. Rum1 also inactivates S phase promoting factor by forming a trimer Cig2-Cdc2-Rum1 complex that has no kinase activity. This is indicated by the transition T11. However, active Cdc2-Cdc13 helps Cig2-Cdc2-Rum1 complex to break in S phase promoting factor dimer by degrading Rum1. Cig2-Cdc2 and Cdc2-Cdc13 dimers help each other forming a positive feedback loop.

M-phase promoting factor (MPF) is a dimer of cyclin dependent protein kinase (Cdc2) and a cyclin-B (Cdc13). The right dotted box in figure 2 shows the hybrid Petri net model for M phase regulation. The activity of Cdc2-cdc13 depends on the phosphorylation at the sites of specific amino acids Thr-167 and Tyr-15. Cdc2 and Cdc13 form an inactive dimer. MPF dimer is active when its Tyr-15 site is in unphosphorylated state and its Thr-167 site is in phosphorylated state. Wee1 is an enemy of Cdc2-Cdc13 at this stage, which phosphorylates Tyr-15, making it more inactive. However, protein cyclin activating kinase that is present at constant level in cell helps in phosphorylating Thr-167 site. Finally, PCdc25 helps in the dephosphorylation of Tyr-15 amino acid site. These activities are modeled using transitions T22 and T24.

To exit mitosis, Cdc2-Cdc13 activity must be destroyed. This is achieved through transitions T12 to T15, when anaphase promoting complex has certain minimum concentration. Dotted arcs from activeAPC to T12 through T15 in figure 2 indicates that firing of these transitions will not cause the concentration of source place (here active APC) to decrease. Anaphase promoting complex is activated by Cdc2-Cdc13 dimer, when it reaches a certain concentration level. This is achievable in the model through discrete places inactiveAPC and activateAPC and discrete transitions T28 and T29. Firing of these transitions are controlled by concentration level of active M-phase promoting factor. This produces antagonism between MPF and activeAPC.

We are modeling the shift from S phase to M phase and vice-versa using Cdc13 controlled discrete places. As shown in figure 2, places d1 to d4 are essentially those places which control this shift from one phase to another. Transitions T30 to T34 are involved in firing among these places and other places. Initially, d1 has a token indicating that cell cycle starts with S phase (after G1 phase). As and when cell cycle starts, token from d2 will be transferred to d3 and d4, as all the conditions for firing T34 are satisfied. These inhibit Wee1 production and Cdc25 phosphorylation, which should not be present in initial phases of cell cycle. When cell gets into G2/M phase, one token will be transferred to d2. This phase shift is controlled by monitoring the concentration of Cdc13.

5   Simulation Results

We simulated our hybrid Petri net model using the simulator developed by Matsuno et. al. [5]. Dynamics of protein concentrations that are obtained by simulating the hybrid Petri net model described here are shown in figure 3. These results are consistent with known biological facts, as well as results obtained through differential equation models [2,6]. We also simulated the G2 check point control by stopping the production of phosphorylated Cdc25. PCdc25 production stops when DNA replication is not proper. When Pcdc25 production is limited in our cell cycle model, then cell is unable to move from S phase to M phase because of the lack of required MPF activity.

From the graphs shown in Figure 3, it can be easily seen that when MPF concentration reaches more than a certain value, active APC concentration increases. It can also be seen from the graphs that, when active APC concentration reaches a certain peak value, the concentration of MPF starts decreasing. It starts increasing when APC falls below a certain value. From this we can see that MPF is trying to help its enemy APC while APC ubiquitinates active Cdc2-Cdc13. It can be seen from the graph that mutual antagonism exists between MPF and Wee1. These observations are consistent with the known biological facts.

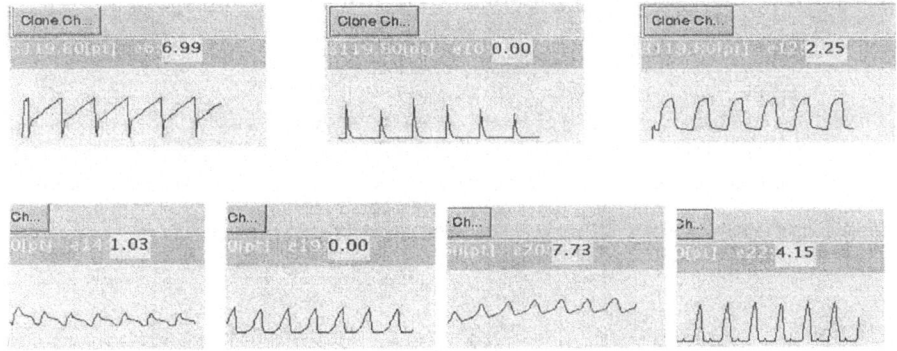

Fig. 3. Concentration graphs of some proteins and their dimers. (MPF, active APC, Wee1, Rum1, Cig2, Cig2-Cdc2 dimer, inactive MPF in that order from top-left to bottom-right)

6   Discussion

In this paper, we have developed a model for fission yeast cell cycle regulation. In our model, we did not include stochasticity. However, stochasticity can be easily incorporated into the model. The type of questions that can be answered using hybrid Petri nets as models are the following. (a) If an activity is inhibited, what other activities can take place? Will it lead to deadlock or live-lock? (b) Does an inhibitor inhibit an entire high-level process? (c) What other activities are in parallel with some process X? (d) How long does process X last when compared to another process Y? Our future work will be to focus on such questions for fission Yeast cell cycle.

References

1. Alla H., Dravid R. Continuous and hybrid Petri nets. Journal of Circuits, Systems and Computers, 8:159-188. (1998).
2. John J Tyson. Modeling the cell division cycle: cdc2 and cyclin interactions, Proceedings of North Atlantic Academic Science, Vol 88, 7328-7332. (1991)
3. Kauffman S. A. Gene Regulatory Networks: A theory for their global structure and behaviours. Currents Topics in Developmental Biology, Vol 6, 145-182, Academic Press, New York.
4. Matsuno, Doi, Nagasaki, Miyano, Adams. Hybrid Petri Net Representation of Gene Regulatory Network. Proceeding Pacific Symposium on Biocomputing, 338-349. (2000).
5. Matsuno, Tanaka Y., Aoshima H., Doi A., Matsui M.,and Miyano S. Bio-pathways representation and simulation on hybrid functional Petri net. In Silico Biology, 3(3), 389-404, 2003.
6. Sveiczer, Novak. A stochastic model of the fission yeast cell cycle. ACH-Models Chem. 133, 299-311. (1996).
7. Tyson, Attila Csikasz-Nagy, and Bela Novak. The dynamics of Cell Cycle Regulation. BioEssays 24: 1095-1109, Wiley Periodicals Inc. (2002).

Protein Metal Binding Residue Prediction Based on Neural Networks

Chin-Teng Lin[1], Ken-Li Lin[1,2], Chih-Hsien Yang[1], I-Fang Chung[3],
Chuen-Der Huang[4], and Yuh-Shyong Yang[5]

[1] Dept. of Electrical and Control Engineering, Chiao-Tung University, HsinChu, Taiwan
ctlin@mail.nctu.edu.tw,
{kennylin.ece90g,blent91.ece91g}@nctu.edu.tw
[2] Computer Center, Chung-Hua University, HsinChu, Taiwan
kennylin@chu.edu.tw
[3] Inst. of Medical Science, University of Tokyo, Tokyo, Japan
cifdmy@hotmail.com
[4] Dept. of Electrical Engineering, HsiuPing Institute of Technology, Dali, Taichung, Taiwan
cdhuang@mail.hit.edu.tw
[5] Inst. of Bioinformatics, Chiao-Tung University, HsinChu, Taiwan
ysyang@cc.nctu.edu.tw

Abstract. It is known that over one-third of protein structures contain metal ions, and they are the necessary elements in life system. Traditionally, structural biologists used to investigate properties of metalloproteins (proteins which bind with metal ions) by physical means and interpret the function formation and reaction mechanism of enzyme by their structures and observation from experiments in vitro. Most of proteins have primary structures (amino acid sequence information) only; however, the 3-dimension structures are not always available. In this paper, a direct analysis method is proposed to predict protein metal-binding amino acid residues only from its sequence information by neural network with sliding window-based feature extraction and biological feature encoding techniques and it can successfully detect 15 binding elements in protein, and 6 binding elements in enzyme.

1 Introduction

It is very interesting that more than one-quarter of the elements in periodic table are required for life, and most of them are metal ions. Many enzymes incorporate metal divalent cations and transition metal ions within their structures to stabilize the folded conformation of protein or to directly participate in the chemical reactions catalyzed by the enzyme.

Metal also provides a template for protein folding, as in the zinc finger domain of nucleic acid binding proteins, the calcium ions of calmodulin (a protein molecule that is necessary for many biochemical process, including muscle contraction and the release of a chemical that carries nerve signals), and the zinc structural center of insulin. Metal ions can also serve as redox centers for catalysis, such as heme-iron centers, copper ions and non-heme irons. Other metal ions can serve as electrophilic reactants in catalysis, as in the case of active site zinc ions of the metalloprotease.

N.R. Pal et al. (Eds.): ICONIP 2004, LNCS 3316, pp. 1316–1321, 2004.
© Springer-Verlag Berlin Heidelberg 2004

The metal-binding occurs in the site, one layer of the protein structure hierarchy. In the descending order by their size, the 4 layers are protein, chain, site and ligand. The top level is protein which may contain one or several chains, and each chain is represented as one polypeptide chain belonged to one protein in nature. Every site contains the coordinate information about entire metal center binding site, just like shown in Fig. 1. Structure biologists are interested in how many atoms (ligands) participate in the site, which residue the ligand belongs, and what these binding ligands are. In this paper, the metal binding residue prediction is focused on.

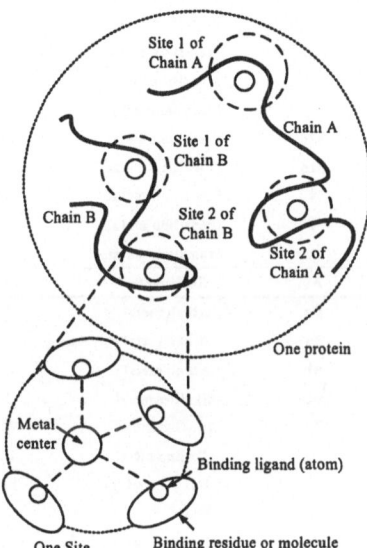

Fig. 1. Binding structure and hierarchy

2 Dataset and Biological Resources

The main data resources come from two web sites, one is the metalloprotein database and browser (MDB) of metalloprotein structure and design program of the Scripps Research Institute (http://metallo.scripps.edu) where all proteins with binding metal can be entirely extracted and the binding site is also defined by nearby amino acid residues and compounds [1]. Another one is Protein Data Bank (PDB, http://www. rcsb.org/pdb/), which provides general information about every protein structure. Hence, by combining these data sets, the detail description of metalloprotein can be driven. For simplicity, the PDB information can be replaced by another compacted data – PDBFinder (http://www.cmbi.kun.nl/gv/pdbfinder/).

There are 43 kinds of metal concerned in MDB. After cross querying by PHP script language from local integrated MySQL database, 41 and 36 metal types can be found in protein and enzyme respectively (Table 1). Each instance in integrated database presents one chain of protein in real world. From binding information of MDB, every position in protein chain sequence can be marked as binding or non-binding and used as the training target.

Table 1. List of elements in metal binding site prediction. Biological level is the classification of life element in [2]. The element type is the element classification from periodic table.

Biological Level	Name	Element Type	chains in Protein	chains in Enzyme
bulk element	Na	Alkali metal	864	484
bulk element	K	Alkali metal	442	234
bulk element	Ca	Alkaline metal	2589	1106
bulk element	Mg	Alkaline metal	1999	863
trace element	I	Halogen	78	33
trace element	Se	Non-metal	225	110
trace element	Co	Transition metal	192	110
trace element	Cr	Transition metal	7	6
trace element	Cu	Transition metal	581	216
trace element	Fe	Transition metal	2893	861
trace element	Mn	Transition metal	1003	434
trace element	Mo	Transition metal	128	70
trace element	Ni	Transition metal	208	101
trace element	V	Transition metal	26	12
trace element	Zn	Transition metal	2433	1087
possibly essential trace element	As	Semi-metal	111	64
N/A	Cs	Alkali metal	7	4
N/A	Li	Alkali metal	3	2
N/A	Rb	Alkali metal	1	0
N/A	Ba	Alkaline metal	3	2
N/A	Be	Alkaline metal	24	3
N/A	Sr	Alkaline metal	13	3
N/A	Al	Basic metal	82	41
N/A	In	Basic metal	1	0
N/A	Pb	Basic metal	31	16
N/A	Tl	Basic metal	18	18
N/A	Eu	Rare Earth	2	1
N/A	Gd	Rare Earth	16	0
N/A	Ho	Rare Earth	7	1
N/A	La	Rare Earth	5	0
N/A	Sm	Rare Earth	20	3
N/A	Yb	Rare Earth	17	7
N/A	Te	Semi-metal	4	2
N/A	Ag	Transition metal	3	1
N/A	Au	Transition metal	14	2
N/A	Cd	Transition metal	379	82
N/A	Hg	Transition metal	236	117
N/A	Pt	Transition metal	8	3
N/A	Tb	Transition metal	1	0
N/A	U	Transition metal	80	16
N/A	W	Transition metal	46	4

3 Machine Learning Scheme and Features

The learning scheme used, in our experiments, is Multi-layer Perceptron (MLP) neural network with back-propagation (BP) algorithm and one hidden layer is used. There are two input coding methods used here: one is direct one-hot coding which presents every amino acid as one 20-bits array. Only one bit in array is '1' and other bits in array are '0'. In this way, every type of natural amino acid can be indicated by the position of the only "1" bit. Owing to the unknown type (usually use the symbol 'X' in sequence) of amino acid in protein sequence, add one bit to record this condition. This is the so-called non-biological coding for amino acid. Another coding method is done by referencing five biological features of amino acid – the probability of occurrence from statistics of NCBI (National Center for Biotechnology Information) database, secondary structures (helix, strand, and turn) propensities in [3] and metal-binding frequencies from integrated database. These five biological features are illustrated in Table 2.

Table 2. Biological features for amino acid coding. First column is one-letter code of amino acid. Second column is the occurrence probability. Column three to five are the propensity of secondary structures. Last column is normalized metal-binding frequency.

Amino Acid Symbol	Occurrence (%)	Helix	Strands	Turn	Metal-Binding Frequency (Normalized)
A	7.49	1.41	0.72	0.82	0.032
C	1.82	0.66	1.40	0.54	0.658
D	5.22	0.99	0.39	1.24	1.000
E	6.26	1.59	0.52	1.01	0.659
F	3.91	1.16	1.33	0.59	0.035
G	7.10	0.43	0.58	1.77	0.120
H	2.23	1.05	0.80	0.81	0.967
I	5.45	1.09	1.67	0.47	0.040
K	5.82	1.23	0.69	1.07	0.033
L	9.06	1.34	1.22	0.57	0.043
M	2.27	1.30	1.14	0.52	0.063
N	4.53	0.76	0.48	1.34	0.198
P	5.12	0.34	0.31	1.32	0.017
Q	4.11	1.27	0.98	0.84	0.075
R	5.22	1.21	0.84	0.90	0.021
S	7.34	0.57	0.96	1.22	0.092
T	5.96	0.76	1.17	0.90	0.117
V	6.48	0.90	1.87	0.41	0.072
W	1.32	1.02	1.35	0.65	0.009
Y	3.25	0.74	1.45	0.76	0.081

Because the binding behavior of central metal is influenced by the surrounding environment in nature, it is necessary to observe in wider scope than single one amino acid so as to decide whether the metal-binding phenomena happens or not. Accordingly, each input vector applied to learning machine is extracted from one segment of entire chain by the concept – sliding window. In our experiments, we take the window size as 17. Each sliding window is centered by the "target" amino acid. And the rest of the amino acids in window are the "neighbors" of the target. Fig. 2 shows the model of encoding and learning scheme when window size is 5.

4 Experimental Results

In the experiments, there are two major sets – protein and enzyme. Each type of element has its own neural network for prediction and five-fold cross validation is used

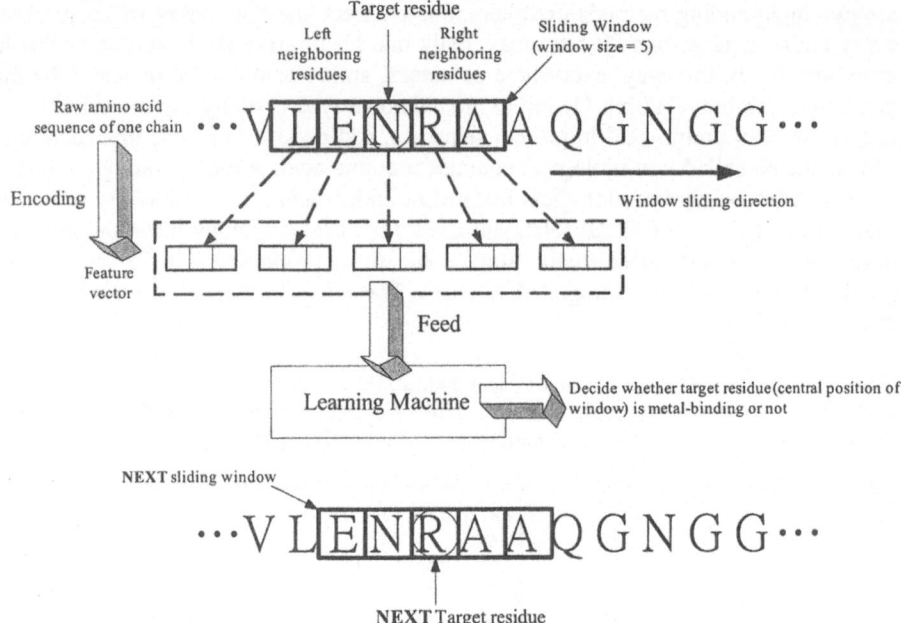

Fig. 2. Feature extraction, learning scheme and sliding window

to evaluate the performance. In Table 3, there are three performance indexes listed – Q_{total} (1), $Q_{predicted}$ (2) and $Q_{observed}$ (3), which are calculated by the value of true positive (TP), true negative (TN), false positive (FP), and false negative (FN). Also, in the first column, coding methods are summarized and in 2nd column "set", P and E represent Protein and Enzyme respectively.

$$Q_{total} = \frac{TP+TN}{TP+TN+FP+FN}(accuracy) .\tag{1}$$

$$Q_{predicted} = \frac{TP}{TP+FP}(postive\ predicitve\ value) .\tag{2}$$

$$Q_{observed} = \frac{TP}{TP+FN}(true\ positive\ rate) .\tag{3}$$

5 Conclusions

In Table 3, we get several observations as follows. First, neural network can detect more types of binding element in protein than in enzyme no matter which coding method is applied. Second, biological coding can perform better than non-biological coding. Especially, the biological coding can detect more meaningful biological elements in metal binding prediction, such as Calcium (Ca), Chromium (Cr), Copper (Cu) and Zinc (Zn). Totally, the experiment shows that neural network successfully predict 15 kinds (4 of them are important life elements) of binding elements in protein, and 6 binding elements in enzyme.

Table 3. The comparison of performance between non-biological and biological coding.

Coding Method	Set	Element	Biological Level	Q-total	Q-predicted	Q-observed
One-Hot Coding (21 bits per amino acid)	P	Cu	trace	98.26%	83.98%	34.09%
		Ho	N/A	98.44%	73.08%	73.08%
		La	N/A	99.68%	91.67%	78.57%
		Ag	N/A	100.00%	100.00%	100.00%
		Cd	N/A	98.11%	73.49%	10.33%
		Tb	N/A	100.00%	100.00%	100.00%
	E	Ho	N/A	95.45%	100.00%	50.00%
		Ag	N/A	100.00%	100.00%	100.00%
Biological Coding (5 attributes per amino acid) 1. Occurrence Probabilities of amino acids 2. Secondary structure Propensities 3. Metal-binding Frequencies	P	Ca	bulk	98.03%	100.00%	0.01%
		Cr	trace	99.06%	100.00%	13.64%
		Cu	trace	97.62%	54.24%	7.28%
		Zn	trace	98.24%	67.47%	0.52%
		Rb	N/A	99.30%	100.00%	25.00%
		Be	N/A	99.61%	85.71%	12.00%
		Tl	N/A	98.85%	100.00%	16.33%
		Ho	N/A	96.12%	8.33%	4.00%
		La	N/A	99.03%	57.14%	30.77%
		Yb	N/A	99.20%	75.00%	6.52%
		Te	N/A	99.54%	100.00%	16.67%
		Ag	N/A	96.03%	78.57%	100.00%
		Cd	N/A	97.96%	38.36%	1.83%
		Tb	N/A	99.38%	100.00%	88.89%
		U	N/A	98.81%	100.00%	1.69%
	E	Cu	trace	98.08%	58.18%	4.32%
		Li	N/A	99.54%	100.00%	33.33%
		Tl	N/A	98.80%	92.86%	13.27%
		Ho	N/A	99.40%	100.00%	85.71%
		Ag	N/A	100.00%	100.00%	100.00%
		Hg	N/A	98.29%	53.33%	1.46%

References

1. Castagnetto, J. M., Hennessy, S. W., Roberts, V. A., Getzoff, E. D., Tainer, J. A., Pique, M.E.: MDB: the Metalloprotein Database and Browser at The Scripps Research Institute. Nucleic Acids Res. ,Vol. 30, No.1 (2002) 379-382
2. Kendrick, M. J., May, M. T., Plishka, M. J., Robinson, K. D.: Metals in Biological System. Ellis Horwood Limited, England (1992) 11-48
3. Wu, C.H., McLarty, J.W.: Neural Networks and Genome Informatics. Elsevier Science Ltd, UK (2000) 67-86

Assessment of Reliability of Microarray Data Using Fuzzy C-Means Classification

Musa Alci[1] and Musa H. Asyali[2]

[1] Department of Electrical and Electronics Engineering,
Ege University, Bornova, Izmir 35100, Turkey
alci@bornova.ege.edu.tr
[2] Department of Biostatistics, Epidemiology, and Scientific Computing,
King Faisal Specialist Hospital and Research Center, Riyadh 11211, Saudi Arabia
asyali@kfshrc.edu.sa

Abstract. A serious limitation in microarray analysis is the unreliability of the data generated from low signal intensities. Such data may produce erroneous gene expression ratios and cause unnecessary validation or post-analysis follow-up tasks. Therefore, elimination of unreliable signal intensities will enhance reproducibility and reliability of gene expression ratios produced from the microarray data. In this study, we applied Fuzzy c-Means classification method to separate microarray data into low (or unreliable) and high (or reliable) signal intensity populations. We compared results of fuzzy classification with that of classification based on normal mixture modeling. Both approaches were validated against reference sets of biological data consisting of only true positives and negatives. We observed that both methods performed equally well in terms of sensitivity and specificity. However, a comparison of the computation times indicated that the fuzzy approach is computationally more efficient.

1 Introduction

DNA microarray technology provides a powerful and efficient means for measuring relative abundance of expressed genes in a variety of applications such as profiling of gene expression between normal and diseased subjects. A comprehensive review of the biological and technical aspects of the microarray technology can be found in [1], [2]. In any microarray experiment, only a small fraction of the genes become expressed as a result of the investigated conditions. Thus, a large portion of the microarray data is comprised of low signal intensities that cause variability or impair reproducibility of the measured ratios between control and experimental samples. There are also other situations that give rise to low signal values such as the deposition of suboptimal amounts of the probes, quality of probes, or incorrect segmentation of the spots. In a recent study Asyali et al. described [3] a classification method based on univariate and bivariate Normal Mixture Modeling (NMM) [4] for the reliability analysis of microarray data. They utilized Expectation Maximization (EM) algorithm [5], [6] to estimate the parameters of the NMM and the class posterior probabilities. They subsequently used Bayesian decision theory [7] to find the optimal decision boundaries that discriminate between the reliable and the unreliable (low) signal intensity populations, based on the estimated class posterior probabilities. In this study, as an alternative to the classification based on NMM, we propose the use of Fuzzy c-Means (FCM) classification [8], [9], compare the results of both approaches against some reference sets, and also crudely evaluate their computational complexity.

N.R. Pal et al. (Eds.): ICONIP 2004, LNCS 3316, pp. 1322–1327, 2004.
© Springer-Verlag Berlin Heidelberg 2004

2 Methods

2.1 Experimental Data

We used data from three independent experiments of microarray gene expression from the same cell system (monocytic leukemia cell line, THP-1, induced by the endotoxin, LPS) [10], in order to test and compare different classification approaches. We used complementary DNA (cDNA) microarray, which contained ~2000 cDNA distinct probes and a total of ~4000 elements. For details of microarray preparation, image acquisition and intensity extraction please see [3]. Our data consisted of fluorescence signal intensities generated from two different channels, Cy3 (green) and Cy5 (red). After background-subtraction and normalization, data were natural log-transformed, as commonly done in microarray data analysis. The log-transform also helps bring the distribution of the data closer to normal density, which helps fitting NMMs. Table 1 shows summary statistics for the two channel data in the three datasets, including mean, standard deviation (SD), and the correlation between the channels ($\rho_{Cy3,Cy5}$), and the number of data points (n).

Table 1. Summary statistics for the three microarray expression datasets. The numbers in the parenthesis are the number of elements in the reference sets

	Dataset 1		Dataset 2		Dataset 3	
	Cy3	Cy5	Cy3	Cy5	Cy3	Cy5
Mean±SD	6.28±1.08	6.16±1.16	6.21±1.05	6.08±1.15	6.24±1.15	5.90±1.56
$\rho_{Cy3,Cy5}$	0.9255		0.9239		0.8672	
n	3027 (53)		3040 (54)		6455 (43)	

During the classification microarray data, it is possible to analyze each channel separately and combine the classification results using AND or OR rules [3]. However, as indicated by the high correlation values in Table 1, bivariate analysis where both channels considered together is more suitable [3].

For each dataset, a reference set of about 50 expressed genes including both endotoxin-induced and housekeeping genes was compiled, based on their expression status in human monocytes. The expression status was obtained from literature and from our previous large-scale microarray expression data [10]. The signal intensities in the reference sets were identified as true positives if the microarray gene expression data agree with the prior knowledge about the expression status, while true negatives were those for which their microarray gene expression data were not consistent with the expected expression or inducibility of these genes. (For further details about the construction of the reference sets and a list of the genes used, please see [3].)

2.2 Fuzzy C-Means Classification (FCM)

Cluster analysis [7], [8] is based on partitioning a collection of data points into a number of subgroups, where the objects in a particular cluster show a certain degree of closeness or similarity. (If the number of clusters are known apriori, as in or case, clustering problem turns into a classification problem.) The similarity measure is generally taken as the Euclidean distance between the data points [7]. Hard clustering, also known as k-Means, assigns each data point to one and only one of the clusters, therefore the degree of membership for each data point to a particular class is either 0 or 1. There are several applications in which the clusters have no clear or well-defined boundaries [11]. In fuzzy clustering, each data point may belong to any class with a certain possibility or degree of membership, a value between 0 and 1. As it will be noted shortly, this concept is similar to the posterior probabilities in the case of mixture models. The rationale behind the fuzzy clustering lies in the reality that an object or data point could be assigned to different classes. That is, if an object does not clearly fit into either of two clusters, this knowledge, expressed by the degree of membership, can be captured. The FCM algorithm is first proposed by Bezdek [8], [9] and briefed here for convenience. Below, c (the number of clusters) is 2, $i = 1, 2$ is the class, $k = 1, 2, ..., n$ is data point, and $l = 1, 2, ..., L$ is iteration index. The $\| \ \|$ refers to the Euclidian norm for vectors and Frobenius norm for matrices. Following the common practice, we selected the exponent parameter m as 2; the maximum number of iterations (L) and termination criterion (ε) were taken as 100 and 10^{-5} respectively.

Step 1. For given dataset $X = \{x_1, x_2, ..., x_n\}$, $x_k \in R^2$, set $l = 1$ and initialize $n \times 2$ partition or membership matrix $U^{(l)}$ with elements $u_{ki}^{(l)}$, such that $0 \le u_{ki}^{(l)} \le 1$, $\sum_{i=1}^{c} u_{ki}^{(l)} = 1, \forall k$.

Step 2. Compute the c-mean vectors (fuzzy centroids) $v_i^{(l)}$s as:

$$v_i^{(l)} = \sum_{k=1}^{n} (u_{ki}^{(l)})^m x_k \Big/ \sum_{k=1}^{n} (u_{ki}^{(l)})^m .$$

Step 3. Compute the degree of membership of all data points for all clusters and update the partition matrix, i.e. obtain $U^{(l+1)}$, as:

$$u_{ki}^{(l+1)} = \left[\sum_{j=1}^{c} \left(\left\| x_k - v_i^{(l)} \right\| \Big/ \left\| x_k - v_j^{(l)} \right\| \right)^{2/(m-1)} \right]^{-1} .$$

Step 4. Check for convergence: Stop, if $\left\| U^{(l+1)} - U^{(l)} \right\| < \varepsilon$ or $l = L$, otherwise, set $l \leftarrow l + 1$ and go to Step 2.

Usually convergence occurs rather rapidly. There is guaranteed convergence in a finite number iterations [12], however, the algorithm may converge to a local minimum. The membership matrix U was initialized with random numbers (normalized to make row sums equal to 1) from the uniform distribution. We performed all the computations using the FCM routine [8] of the Matlab Fuzzy Logic Toolbox (The MathWorks Inc., Natick, MA).

2.3 Classification Using Normal Mixture Modeling

Mixture modeling is a widely used technique for probability density function estimation [6], [7] and found significant applications in various biological problems [13].

We modeled the probability density function (*pdf*) of the microarray data with two bivariate normal *pdf*s as follows:

$$f(x) = \pi N(x; \mu_1, \Sigma_1) + (1 - \pi) N(x; \mu_2, \Sigma_2).$$

Here, $N(x; \mu_i, \Sigma_i) = (2\pi)^{-1} \det(\Sigma_i)^{-1/2} \exp\{-(x - \mu_i)^T \Sigma_i^{-1}(x - \mu_i)/2\}$, $i = 1, 2$ is a bivariate normal *pdf* with mean $\mu_i \in R^2$ and 2×2 covariance matrix Σ_i. The $\pi_i (\geq 0)$ denotes the weight of $N(x; \mu_i, \Sigma_i)$. For each component, we have 2 parameters for the mean vector and 3 parameters for the covariance matrix (because of its symmetry), to estimate. (In addition, we have only one weight to estimate, as $\pi_1 + \pi_2$ must be 1, for $f(x)$ to be proper *pdf*.) The weighted bivariate normal *pdf*s or components, i.e. $\pi_1 N(x; \mu_1, \Sigma_1)$ and $(1 - \pi_1) N(x; \mu_2, \Sigma_2)$, correspond to the class posterior probabilities. By equating the class posterior probabilities and solving for x, we obtain gives the decision boundary, which is a quadratic curve in our case [3], [7].

We used EM algorithm [6], [7] to estimate the mixture parameters. We started the algorithm with an initial estimate of the parameters obtained by the k-Means algorithm [6], [7] and iterated the Expectation and Maximization steps until the changes in the parameters were less than a small preset tolerance (0.0001) or a certain number of iterations (300) were reached. We performed all the computations using an in-house software developed under Matlab (The MathWorks Inc., Natick, MA). Similar to FCM, depending on the initial conditions, EM algorithm may also converge to a local solution, i.e. to a local maximum of the likelihood function.

3   Results

Fig. 1 shows a sample classification result obtained using FCM and NMM, for dataset 2. We observe that the classification boundaries of both approaches are very close and they correctly classify all the points the reference set. The classification performance, in terms of sensitivity and specificity, of the both approaches against the reference sets for the three cases are reported in Table 2. "Reliable" and "unreliable" refer to the results of classification done by the algorithms, whereas the "true positive" and "true negatives" refer to the true or actual class information available in the reference sets.

The time (in seconds) spent by the central processing unit (CPU) to run the algorithms and the number of iterations is also shown in Table 2. (All the computations were done on a personal computer with 1.5 MHz Pentium IV processor and 396 MB of memory, running under Windows-2000™ operating system.)

4   Discussion and Conclusion

The microarray technology lets the biologist study the expression or activity of thousands of the genes at the same time but only a fraction of genes are expressed and low signal intensities constitute a big portion of the data. Such low signal intensities may give rise to erroneous gene expression ratios or false positives. Therefore, filtering of such signals before the subsequent steps of the analysis is required. Some ad-hoc techniques [14] have been suggested in literature to address this issue. Asyali et al. [3]

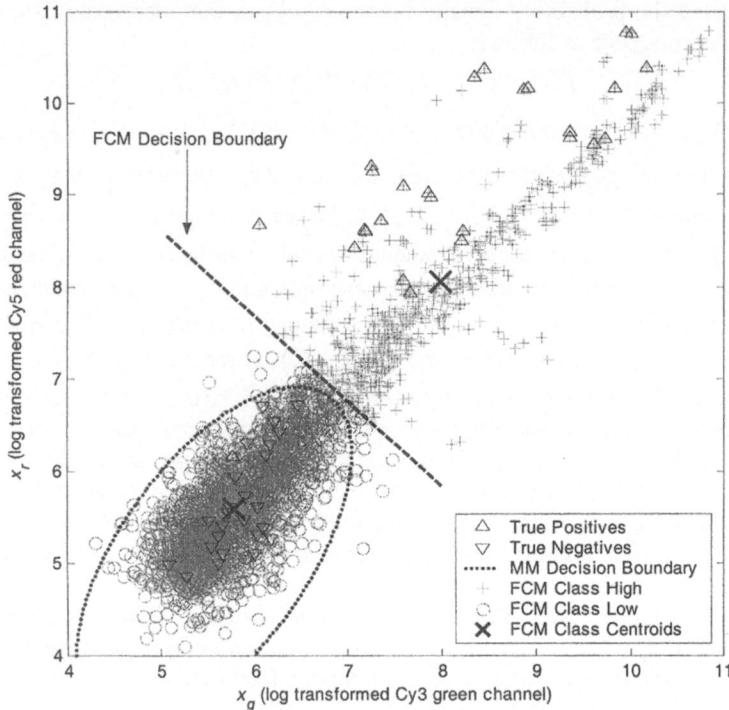

Fig. 1. Fuzzy c-means (FCM) and normal Mixture Modeling (MM) based classification results for dataset 2. (The "high" and "low" refer to the classes of reliable and unreliable to the data points, respectively)

Table 2. Classification results using FCM and NMM

		Fuzzy C-Means		Normal Mixture Modeling	
		Reliable	Unreliable	Reliable	Unreliable
	True positive	26	0	26	0
	True negative	0	27	0	27
Dataset 1	Sensitivity (%)	100.0		100.0	
	Specificity (%)	100.0		100.0	
	CPU time (seconds)	0.250		0.718	
	# of iterations	27		40	
	True positive	27	0	27	0
	True negative	0	27	0	27
Dataset 2	Sensitivity (%)	100.0		100.0	
	Specificity (%)	100.0		100.0	
	CPU time (seconds)	0.250		0.704	
	# of iterations	27		39	
	True positive	27	2	29	0
	True negative	0	14	1	13
Dataset 3	Sensitivity (%)	93.1		100.0	
	Specificity (%)	100.0		92.9	
	CPU time (seconds)	0.406		2.078	
	# of iterations	21		61	

suggested an NMM based approach and successfully demonstrated its advantages over the existing techniques. In this study, we have explored the possibility of accomplishing same signal classification goal using FCM. We have applied both algorithms on our three datasets and assessed their performance by checking the classification decisions against the reference sets.

Our results show that, for all the three cases, FCM performs equally well and as indicated by the CPU times, is computationally more efficient than the NMM. We should note that, this computational efficiency comparison is rather crude, as the tolerance settings of the algorithms were different. However, we recognize that, in general, the update of posterior probabilities in the EM algorithm is a much more costlier operation than the update of the U matrix in the FCM. Therefore, the faster performance of the FCM that we have observed is not surprising. This speed advantage may be important factor in the case of batch processing of larger datasets.

References

1. Nguyen, D. V., Arpat, A. B., Wang, N. and Carroll, R. J. (2002) DNA microarray experiments: biological and technological aspects. Biometrics, 58, 701–717.
2. Golub, T. R., Slonim, D. K., Tamayo, P., Huard, C., Gaasenbeek, M., Mesirov, J. P., Coller, H., Loh, M. L., Downing, J. R., Caligiuri, M. A., Bloomfield, C. D. and Lander, E. S. (1999) Molecular classification of cancer: class discovery and class prediction by gene expression monitoring. Science, 286, 531–537.
3. Asyali, M. H., Shoukri, M. M., Demirkaya O, and Khabar, K. S. A. (2004) Estimation of Signal Thresholds for Microarray Data Using Mixture Modeling. Nucleic Acids Research, 32(7), 1–13.
4. McLachlan, G. J. and Basford, K. E. (1989) Mixture Models, Inference and Applications to Clustering, Marcel Dekker, New York.
5. Redner, R. and Walker, H. (1984) Mixture densities, maximum likelihood and the EM algorithm. SIAM Review, 26, 195–202.
6. Martinez, W. L. and Martinez, A. R. (2001) Computational Statistics Handbook with MATLAB, CRC Press, Boca Raton.
7. Duda, R., Hart, P. and Stork, D. (2000) Pattern Classification, 2nd Edition Ed., Wiley, New York.
8. Jang, J.-S.R., Sun, C.-T., and Mizutani, E. (1997) Neuro-Fuzzy and Soft Computing, Prentice Hall, New Jersey.
9. Bezdek, J.C. (1981) Pattern Recognition with Fuzzy Objective Function Algorithm, Plenum Press, New York.
10. Murayama, T., Ohara, Y., Obuchi, M., Khabar, K. S., Higashi, H., Mukaida, N. and Matsushima, K. (1997) Human cytomegalovirus induces interleukin-8 production by a human monocytic cell line, THP-1, through acting concurrently on AP-1- and NF-kappaB-binding sites of the interleukin-8 gene. J Virol, 71, 5692–5695.
11. Karlik, B., Osman Tokhi, and Alci, M. (2003). Fuzzy clustering neural network architecture for multifunction upper-limb prosthesis. IEEE Trans Biomed Eng, 50(11), 1255–1261.
12. Bezdek, J.C., Hataway, R.J., Sabin, M.J. and Tucker, W.T. (1987) Convergence theory for fuzzy c-means: Counterexamples and repairs. SMC, 17(5), 873–877.
13. McLachlan, G. J., Bean, R. W. and Peel, D. (2002) A mixture model-based approach to the clustering of microarray expression data. Bioinformatics, 18, 413–422.
14. Fielden, M. R., Halgren, R. G., Dere, E. and Zacharewski, T. R. (2002) GP3: GenePix post-processing program for automated analysis of raw microarray data. Bioinformatics, 18, 771–773.

DNA Sequence Pattern Identification Using a Combination of Neuro-Fuzzy Predictors

Horia-Nicolai Teodorescu[1, 2] and Lucian Iulian Fira[1]

[1] Faculty of Electronics and Communications, Technical University of Iasi,
B-dul Carol I, no. 11, Iasi, Romania
{hteodor,lfira}@etc.tuiasi.ro
[2] Romanian Academy, B-dul Carol I, no. 7, Iasi, Romania

Abstract. We address the prediction of the gene structure using a new method and tools, involving the sequence of distances between bases and neuro-fuzzy predictors. The method is tested on the HIV virus genome and the results look promising compared to other methods. We suggest that new, global prediction methods based on implicit, not explicit knowledge, may be as strong as the current, largely explicit knowledge based prediction methods.

1 Introduction

DNA analysis has seen last years a tremendous development. DNA includes huge amounts of data that must be interpreted. The operations to be performed include the recognition of the type of organism to which the gene belongs – when the genetic material comes from different sources, the identification of specific segments in the DNA sequence – sections that represent the genes – and the identification of the proteins the genes code – the prediction of the genes expression. Because of the huge amount of data in the genetic material, the operations must be automated. The first two tasks above are somewhat similar and may use similar tools. They are both essentially related to the identification of patterns in the genetic code. Several tools have been developed for these purposes, including hidden Markov models (HMM) [1], statistical methods, and fuzzy [2-4] and neural models that inherently reflect the statistics. Here, continuing [5-9], we present results using a novel approach in dealing with the genetic sequence prediction, based on its decomposition into four "distance series" and on the use of neuro-fuzzy predictors in a hierarchical pattern identification system.

The syntagma "gene prediction" has several meanings, relating to the object of prediction and depending on the research context. One meaning is to predict the splice sites [10]; another one is to determine what the gene expression result would be (the synthesized proteins).

2 The Pattern Detector

We approach the prediction task from a fresh point of view and propose a new type of gene prediction with a view to classify the genetic sequences, to determine the right splice sites and to classify the information they carry. For this purpose, we preprocess the original, raw base sequence and first produce four base sequences, for the four bases, A, C, G, and T respectively (Fig. 1).

N.R. Pal et al. (Eds.): ICONIP 2004, LNCS 3316, pp. 1328–1333, 2004.
© Springer-Verlag Berlin Heidelberg 2004

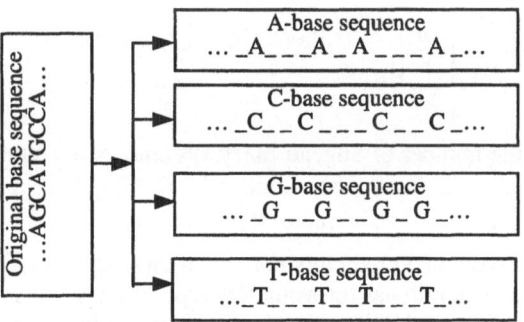

Fig. 1. Processing the base sequence

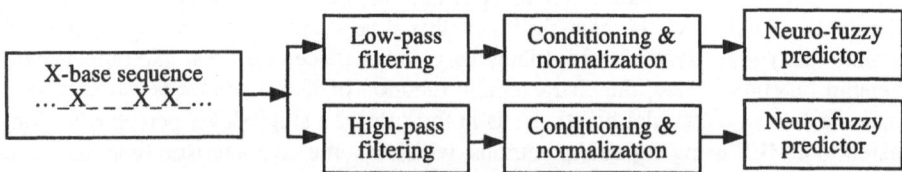

Fig. 2. Details of the processing of the individual sequences

Each of the four series is then preprocessed to derive the "trend" and the "fast varying" components. The low-pass filtering is performed by a 3-step MA filter, according to the formula:

$$y[n] = (x[n-1] + x[n] + x[n+1])/3 .\tag{1}$$

Each of the two components is then predicted and the results are summed to generate the final prediction. Because the neuro-fuzzy system used in the prediction needs normalized values (values in the [-1, 1] interval), the distance series are normalized according to

$$y[n] = 2 \cdot [x[n] - (x_{max} + x_{min})/2]/(x_{max} - x_{min}) .\tag{2}$$

where x_{max}, x_{min} are the maximal and minimal elements in the series.

To improve the prediction outcome, exceptional events (large distances, far out of the spreading range), if any, may first be eliminated. The sketch of the processing of the individual base sequence prediction is shown in Fig. 2.

We have used a neuro-fuzzy predictor, which has been developed and tested in our group in previous years. The architecture for one step predictive system is explained below (see [2]). The predictor is a multi-fuzzy system network with inputs represented by the delayed samples, and the fuzzy cells are Sugeno type 0, with Gauss input membership functions. The formula $\mu(x) = exp(-(x-a)^2 / \sigma)$ represents the membership degree μ of the input x; a is the center and σ is the spreading of the input membership function. Equation (3) stands for the input-output function of the neuro fuzzy predictor.

$$Y = \sum_{k=0}^{M} w_k \cdot \left[\sum_{l=1}^{N} \beta_{kl} \cdot e^{-\frac{(x_{n-k} - a_{kl})^2}{\sigma}} \middle/ \sum_{l=1}^{N} e^{-\frac{(x_{n-k} - a_{kl})^2}{\sigma}} \right] \qquad (3)$$

We denote by M the number of Sugeno fuzzy systems, N the number of membership functions for each Sugeno fuzzy system, index $k = 0 \div M$, index $l = 1 \div N$, a_{kl} – the centers of the Gauss type membership functions, β_{kl} – the singletones and w_k – the weights.

Notice that the individual predictors may be of any type, but FNN predictors have several advantages. The rational for using this type of predictor is based on the complexity of the input-output function of this predictor, on its large number of parameters, and on the possibility to use various algorithms to train it, as explained below.

The class of predictor is given by the input-output function of the predicting system. This function is a ratio with sums of exponentials at the nominator and the denominator. Therefore, the capabilities of this type of predictor are higher than for simple fuzzy logic systems with triangular or other piecewise input and output membership functions. Also, the characteristic function of this predictor is more intricate than the sum of sigmoidal functions, as in the case of a single layer perceptron. Compared to a MLP using sigmoidal neurons, which has the characteristic function represented essentially by composed sigmoidal functions, still the characteristic function of this predictor is still more intricate, because it is a ratio of nonlinear functions. Compared to a MLP using Gaussian RBF neurons, which has the characteristic function represented essentially by composed Gaussian RBF functions, the characteristic function of this predictor is still more intricate, for similar reasons as above.

Regarding the number of parameters involved, showing the adaptation flexibility of the predictor, this predictor is similar to the ones discussed above. Moreover, by adding belief degrees to the rules of the TSK-fuzzy systems representing the neurons, the number of parameters is easily increased.

Regarding the possibility to use various learning algorithms, including gradient-type algorithms, gradient algorithms are easily adaptable to this predictor, in contrast to classic fuzzy systems with piecewise input and output membership functions, which do not accept classic gradient algorithms (because of the non-derivability of the function).

The training of the predictors refer to the adaptation of several sets of parameters, namely of the weights w_i, output singletons, and parameters of the input membership functions, a_k, σ_k. Two basic versions of the training algorithm can be used. According to the first, separate loops are used to adapt each of the above mentioned sets of parameters. We will name this type of algorithm "internal loops algorithm." The second algorithm uses a single loop to adapt all the parameters.

The overall multi-predictor system includes four individual predictors and a decision block (see [9]). This block fuses the results provided by the individual predictors to generate the recognition decision. Notice that, while current prediction methods use explicit knowledge on the gene sequences, system described here uses implicit information, in the sense that the information is included in the predictors, after appropriate training.

3   Results

In this section, we briefly present several results obtained by training the individual fuzzy predictors on the ENV gene (data source [11]), for the distance series corresponding to the A, C, G and T series. The summary of the results are shown in Figures 3 and 4, including the train results on the ENV gene, as well as the test results, both on ENV and on other HIV genes, moreover on genes from other viruses. In these figures, MSE denotes the mean square error and NMSE the normalized MSE.

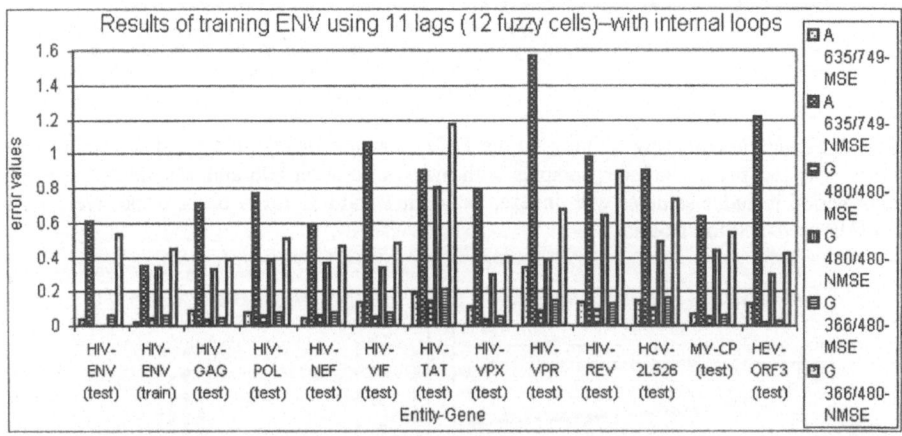

Fig. 3. Results obtained by training on the ENV sequence, using an algorithm with inner (partial) loops. The test period includes samples with indexes 636 to 749 for A basis, respectively 367-480 for G basis. The train period: samples #1 to 635 (A basis), respectively 1-480 for G, 1-366 for G basis

The results obtained by the training with either algorithm indicate that most genes are identified by the predictor for the A-base series. In the Fig. 3, this predictor confuses the ENV with the NEF gene when only the NMSE is used, but behaves correctly for the MSE (MSE for NEF > MSE for ENV–test.) Also, there is an almost miss for the CP gene of the Mosaic virus (NMSE 0.634, vs. 0.609 for ENV–HIV, but MSE-CP=0.071 vs. MSE-ENV=0.034.)

All the other genes are correctly rejected by the NMSE values alone for the A-series predictor, and all are correctly rejected by a combination of MSE and NMSE of this predictor. Excellent results are also obtained with the C-series predictor, with the exception, again, for the CP gene of the Mosaic virus. Poorer results are obtained with the predictors for the T-series, which, however, correctly rejects the CP gene of the Mosaic virus. These results show that a combination of predictors is able to differentiate and even identify the genes.

The error evolution during training shows a correct learning process with steadily decreasing error (see example in Fig. 5). The error histogram show almost-Gauss distributions, meaning that the information in the series has been well extracted (see an example in Fig. 6).

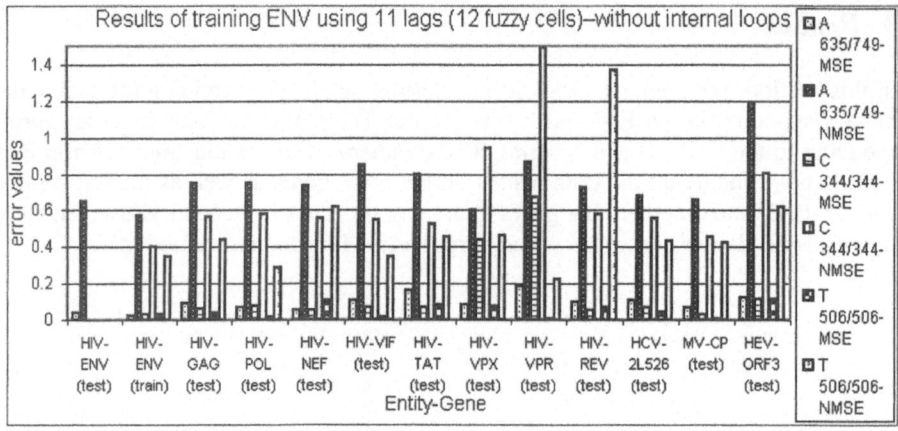

Fig. 4. Results obtained by training on the ENV sequence, using an algorithm without inner loops. The test period includes samples with indexes between 636 and 749 for A basis. The train period includes samples with indexes between 1 and 635 for A basis, respectively 1-344 for C basis, 1-506 for T basis

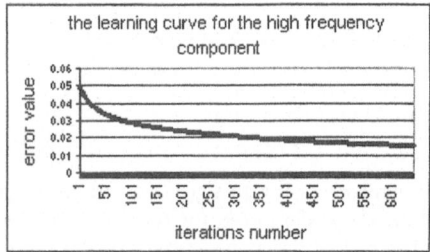

Fig. 5. Error evolution during training on the A series (ENV gene)

Fig. 6. Error histogram for the training results on the ENV gene, A basis, for the high-frequency component

4 Comments and Conclusions

The fact that several base sequences yield similar results in prediction, when a predictor trained on a specific sequence is used, may have several interpretations. The first and easiest interpretation is that the predictor has not been trained well enough. We have to reject this hypothesis, because the training leads to small enough normalized errors, while the error values histogram shows a Gauss-like distribution.

The Gauss-like histogram is not, of course, an indication that the signal (remaining error) is pure white noise, but it is an indication that it might be and a necessary condition for being white noise. Moreover, it is an indication that there is no strong global correlation in the remaining signal. A Gauss histogram, however, is no guarantee that the signal is a noise. Therefore, we need other tools to determine the degree of information extraction from the signal. Such a tool is the self-correlation function.

The second possible interpretation might be that the series carry similar information and represent the almost same generation process. If true, then some of the base sequences from different genes might belong to the same class and then, we could determine classes of genes that, according to the prediction criterion are similar in the same class and dissimilar in different classes. The classes would be information-specific, while members of the same class still may look quite different. This finding may shed new light on the genetic processes and may have important consequences both in biology and bio-informatics.

References

1. Genie: Gene Finder Based on Generalized Hidden Markov Models.
   www.fruitfly.org/seq_tools/genie.html
2. Gasch, A.P., Eisen, M.B.: Exploring the Conditional Coregulation of Yeast Gene Expression through Fuzzy K-Means Clustering. Genome Biology 2002, 3(11): research0059.1 – 0059.22. http://rana.lbl.gov/papers/Gasch_GB_2002.pdf
3. Guthke, R., Schmidt-Heck, W., Hahn, D., Pfaff, M.: Gene Expression Data Mining for Functional Genomics using Fuzzy Technology.
   www.biochem.oulu.fi/BioStat/Guthke_Kluwer2002.pdf
4. Pasanen, T.A., Vihinen, M.: Formulating Gene Regulatory Patterns with Fuzzy Logic, http://www.ki.se/icsb2002/pdf/ICSB_179.pdf
5. Fira, L.I., Teodorescu, H.N.: Genome Bases Sequences Characterization by a Neuro-Fuzzy Predictor, Proc. IEEE-EMBS 2003 Conference, Cancun, Mexico, 3555-3558
6. Teodorescu, H.N.: The Dynamics of the Words. Invited Plenary Lecture, 11th Conf. Applied and Industrial Mathematics, 29-31 May, 2003. University of Oradea, Romania, http://caim2003.rdsor.ro/
7. Teodorescu, H.N., Fira, L.I.: Predicting the Genome Bases Sequences by Means of Distance Sequences and a Neuro-Fuzzy Predictor. F.S.A.I., Vol. 9, Nos. 1–3, (2003), 23-33
8. Teodorescu, H.N.: Genetics, Gene Prediction, and Neuro-Fuzzy Systems–The Context and a Program Proposal. F.S.A.I., Vol. 9, Nos. 1–3, (2003), 15–22
9. Teodorescu, H.N., Fira, L.I.: A Hybrid Data-Mining Approach in Genomics and Text Structures, Proc. The Third IEEE International Conference on Data Mining ICDM '03, Melbourne, Florida, USA, November 19 - 22, (2003), pp. 649-652
10. Thanaraj, T.A.: A Clean Data Set of EST-Confirmed Splice Sites from Homo Sapiens and Standards for Clean-up Procedures. Nucleic Acids Research 1999, vol. 27, no. 13, 2627-2637
11. Los Alamos National Laboratory:
    http://hiv-web.lanl.gov/content/hiv-db/align_current/align-index.html

Genetic Mining of DNA Sequence Structures for Effective Classification of the Risk Types of Human Papillomavirus (HPV)

Jae-Hong Eom[1], Seong-Bae Park[2], and Byoung-Tak Zhang[1]

[1] Biointelligence Lab., School of Computer Science and Engineering,
Seoul National University, Seoul 151-744, South Korea
{jheom,btzhang}@bi.snu.ac.kr
[2] Language & Information Processing Lab., Dept. of Computer Engineering,
Kyungpook National University, Daegu 702-701, South Korea
seongbae@knu.ac.kr

Abstract. Human papillomavirus (HPV) is considered to be the most common sexually transmitted disease and the infection of HPV is known as the major factor for cervical cancer. There are more than 100 types in HPV and each HPV has two risk types, low and high. In particular, high risk type HPV is known to the most important factors in medical judgment. Thus, the classifying the risk type of HPV is very important to the treat of cervical cancer. In this paper, we present a machine learning approach to mine the structure of HPV DNA sequence for effective classification of the HPV risk types. We learn the most informative subsequence segment sets and its weights with genetic algorithm to classify the risk types of each HPV. To resolve the problem of computational complexity of genetic algorithm we use distributed intelligent data engineering platform based on active grid concept called "IDEA@Home." The proposed genetic mining method, with the described platform, shows about 85.6% classification accuracy with relatively fast mining speed.

1 Introduction

Cervical cancer is a leading cause of cancer deaths in women worldwide. It is well established that persistent infection with the Human Papillomavirus (HPV) is associated cervical cancer. Large studies have shown that HPV is present in up to 90% of cervical cancers [1][2]. Since the main etiologic factor for cervical cancer is known as high-risk Human Papillomavirus infection [3], it is now largely a preventable disease [4]. This Human Papillomavirus is one of the most common sexually transmitted diseases and the infection of HPV is still known as the major factor for cervical cancer. This HPV is a double-strand DNA tumor virus that belongs to the papovavirus family (papilloma, polyoma, and simian vacuolating viruses). More than 100 human types are specific for epithelial cells including skin, respiratory mucosa, or the genital tract. And the genital tract HPV types are classified into two or three types by their relative malignant potential as low-, intermediate-, and high-risk types [5]. The common, unifying oncogenic feature of the vast majority of cervical cancers is the presence of HPV, especially high-risk type HPV.

N.R. Pal et al. (Eds.): ICONIP 2004, LNCS 3316, pp. 1334–1343, 2004.
© Springer-Verlag Berlin Heidelberg 2004

Since the HPV classification is important in medical judgments and it is becoming more important, there have been many approaches to classify the risk types of HPVs. Bosch *et al.* [1] investigated whether the association between HPV infection and cervical cancer is consistent worldwide in the context of geographic variation in the distribution of HPV types. Burk *et al.* [6] inspected the risk factors for HPV infection in 604 young college women and they detected various factors of HPV infection (e.g., age, ethnicity, number of lifetime male vaginal sex partners, etc.) through L1 consensus primer polymerase chain reaction and Southern blot hybridization. Park *et al.* [4] used text mining technique to discriminate the risk types of HPVs and they predicted the risk types of several HPVs whose risk types were have been unknown. Muñoz *et al.* [7] classified the risk types with practical experiments based on risk factor analysis. They collected real data from 1,900 cervical cancer patients and analyzed it by PCR (polymerase chain reaction) based assays.

Practical analysis with experiment is the most accurate analysis process. However, it is not easy to conduct every experimental trial when we have many cases to analyze. One alternative for this problem is exploiting computational power to the analysis. The "systems biology" is the field which makes use of this approach to analyze and understand biological phenomena through systems approaches.

In this paper, we present a machine learning approach to mine the structure of HPV DNA sequence for effective classification of the HPV risk types. We learn the most informative subsequence segments and its weights with genetic algorithm to classify the risk types of each HPV. Also we use new data engineering platform based on active grid computing, called "IDEA@Home," to alleviate the problem of computational complexity of genetic algorithm [8].

The remainder of the paper is organized as follows. In Section 2, we simply describe the concept of the distributed intelligent data engineering platform which was used for proposed mining algorithm. Section 3 represents the genetic mining method for learning substructure and its weights of HPV DNA subsequences. Section 4 presents the experimental results. Finally, Section 5 draws conclusions.

2 IDEA@Home: Intelligent Data Engineering Platform

As the data size is becoming increasingly large, more powerful computational ability is needed. One possible alternative for this problem is utilizing distributed computing and there have been many attempts to employ this approach for analyzing high dimensional data which called "Grid Computing." Grid computing is distributed computing, in which a network of computers taps into a main computer server that stores software and data.

One of the most famous grid computing projects is the SETI@Home which aims to analyze radio telescope data to search extraterrestrial intelligence. Grid computing for biological analysis also started. The FightAIDS@Home by the Olson laboratory at the Scripps Research Institute is the first biomedical distributed computing project based on grid computing to discover new drugs, using the growing knowledge of the structural biology of AIDS. The Folding@Home from Stanford University is designed to understand protein folding and related diseases. The "Screensaver Lifesaver" projects from research group of W. Graham Richards has developed and applied computational methods for drug design, molecular similarity analysis and protein structure prediction, and performed simulations of enzyme reaction mechanisms, DNA recognition, and lipid bilayers [9][10].

But, many conventional grid systems are designed to solve only domain-specific problems. Usually they have one master node controlling the whole network and it distributes all information needed for computation. However, the platform IDEA@ Home proposed by Eom *et al.* [8] allows *n* master nodes and each master divides the entire network into *n* sub-networks. The platform also supports computational flexibility through the concept of operation object and data object. The operation object represents the task-specific operation which will be applied to the data object for the analysis and the data object presents the task-specific data. This data object also includes some constraint information which will be referenced by operation object [8].

The GA based method proposed in this paper, actually, is suitable for both standalone computing platform and distributed computing platforms. But, in the stand alone computing platform, general approach of GA require lots of computing time for the analysis of relatively big and complex data. And the data from biological domain usually have these complex characteristics. Thus the grid platform, such as "IDEA@Home," is somewhat useful for these kinds of fields.

In this paper, since the proposed genetic mining method requires relatively heavy computation, we use the IDEA@Home grid computing platform for computational efficiency.

3   Genetic Mining of HPV Sequence Structures

In this paper, we use genetic algorithms (GAs) for mining discriminative sequence sets of HPVs and classifying the risk types of HPVs. Although there are many fast methods (e.g., sequence alignment, SVM and NN, etc.) and GAs are still somewhat time-consuming, GAs are easily implementable in distributed fashion and they also have great potentials for improved search performance of solution space search. Thus, here we use GAs for sequence structure mining (other method will be considered in the future works).

Encoding

Genetic algorithm is a probabilistic search method based on the mechanism of natural selection and genetics [11]. Genetic search is characterized by the fact that N potential solutions for an optimization problem simultaneously sample the search space. These solutions are called *individuals* or *chromosomes* and denoted as $J_i \in \mathbf{J}$, where $\mathbf{J}$ represents the space of all possible individuals. The *population* $\bar{J} = \{ J_1, J_2, ..., J_N \} \in \mathbf{J}^N$ is modified according to the natural evolutionary process with predefined modification rates (i.e., crossover and mutation rates). After the initial population is generated arbitrarily, selection $\omega: \mathbf{J}^N \Rightarrow \mathbf{J}^N$ and variation $\Xi: \mathbf{J}^N \Rightarrow \mathbf{J}^N$ are applied in a loop until some termination criterion is satisfied. Each run of the loop is called generation [12].

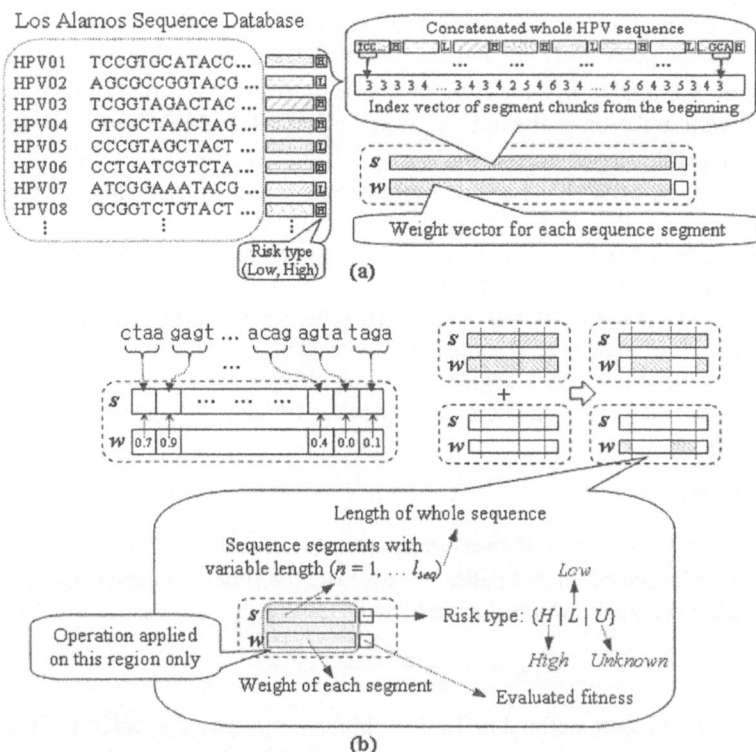

Fig. 1. The encoding scheme of individual chromosome on HPV sequences. Figure (a) shows the encoding scheme of each individual. Each individual in the population has different index vector and its corresponding weight vector. S is a vector of subsequences and W is a vector of weights of subsequences in the segment vector S. An encoding example is described in (b). In this example, all segment length is equally set to 3. Figure (b) shows the schematic working mechanism of 3-point crossover.

The chromosome for HPV classification is defined as a pair of two vectors: a vector for subsequence segments and the other for their weights. The weight of chromosome is a real number ranging from 0.0 to 1.0 and the weight represents the importance of corresponding segment when classifying HPVs. This chromosome is represented as:

$$\mathbf{J} = \begin{pmatrix} j_{s_1}, j_{s_2}, \ldots, j_{s_i}, \ldots, j_{s_L} \\ j_{w_1}, j_{w_2}, \ldots, j_{w_i}, \ldots, j_{w_L} \end{pmatrix} \tag{1}$$

where j_{si} denotes the i-th subsequence segment of DNA sequence of HPVnn and j_{wi} denotes the weight of corresponding subsequence segment. 'nn' corresponds to the index of HPV. L is the number of segments to be considered for classification and is determined by dividing the length of sequence n by the length of segment ρ. Figure 1 shows the representation of chromosome and the mechanism of simple n-point crossover ($n=3$). Generally, GAs use crossover and mutation as variation functions $\Xi: \mathbf{J}^N \Rightarrow \mathbf{J}^N$. For crossover, only the fractions of weights are switched in alternative way based on n-cutting points for selected chromosome pairs. Cutting points are chosen with crossover probability p_c and the uniform random variable sampled again for each point. For mutation, the weight of randomly selected position of chromosome is changed with mutation probability p_m ranging from 0.0 to 1.0. These two operations maintain the diversity of the population.

In this paper, each chromosome represents a subsequence segment with various length and its weights. This is for classifying the corresponding HPV to high or low risk types. The index vector of segment chunks in an individual represents the granularity of sub sequences. For example, an index vector "3 3 4" represents that the chromosome is constructed with successive subsequences with length 3, 3, and 4. In addition, each offspring represents the ordering information of sequence segments through the weights of each segment. A weight represents consideration priority or the importance of a segment for the classification of HPV risk types.

Fitness Function

The fitness function $f: \mathbf{J} \Rightarrow \mathbf{R}$ measures the fitness of a chromosome in terms of classification performance. In this paper, we evaluate the fitness of each offspring with its classification accuracy, which is defined as following:

$$Accuracy = \frac{a+d}{a+b+c+d} \cdot 100\% \tag{2}$$

where a, b, c and d are defined in Table 1. In the proposed approach, the fitness function represents the classification accuracy of each offspring. Answer set was obtained from Los Alamos sequence database [13]. The risk type described in this database is used as an answer set to calculate the classification accuracy. The calculation of the classification accuracy in Equation 2 is based on this information. That is; "answer should be..." are measured according to this answer and the "test results" are the classification result according to the encoding of each individual.

The tag of risk types of each individual is assigned at training step to find the most informative subsequence segments in the whole sequence and to evaluate the fitness value. The estimated fitness value is stored in another tag and this value is used when we select appropriate offspring in the selection procedure of GAs.

The overall procedure of proposed genetic mining method is described in Figure 2. The genetic mining procedure described in Figure 2 is a stand-alone system version.

4   Experimental Results

Datasets

In this paper, we use the HPV sequence database in Los Alamos National Laboratory as a dataset [13]. This database includes HPV compendiums published in 1994–1997 and provides the complete list of 'papillomavirus types and hosts' and the records for each unique papillomavirus types.

Table 1. The contingency table to evaluate the classification performance.

		Test results	
		Low	High
Risk type	Answer should be *Low*	a	b
	Answer should be *High*	c	d

Do following procedures with given parameters:
- Chromosome set $\mathbf{J} = \{ J_1, J_2, ..., J_N \}$
- Crossover probability p_c, Mutation probability p_m
- Number of maximum generation *gmax*
- Number of population to select in each generation M

1. **Set** all chromosomes with randomly generated initial values.
   - Initialize sequence index vectors.
   - Initialize sequence weight vectors.

2. **Repeat** Step 2.1 to 2.3 for all i, i = 1 to *gmax*.
   2.1 **Evaluate** all chromosomes by fitness function f.
   2.2 **Repeat** Step 2.2.1 to 2.2.3 for all j, j = 1 to M.
       2.2.1 **Select** two chromosomes J_a and J_b.
       2.2.2 **Set** offspring[j] = Crossover (J_a, J_b).
       2.2.3 **Set** offspring[j] = Mutation (offspring[j]).
   2.3 **Replace** M chromosomes by offspring generated from Step 2.2.

3. **Return** the optimal chromosome: $\mathbf{J}_{opt}$

Fig. 2. The general procedure of genetic mining. Step 2 represents one generation of generational GAs and it repeats until the population satisfies some predefined convergence criterion.

Settings

To measure the fitness of each individual in genetic mining we use the table of manually classified HPVs as a correct answer, which was previously constructed by Park *et al.* [4] (Table 2) and we used the parameters for genetic mining described in Table 3. The HPVs with "don't know" types are classified according to the discovered subsequence segments and its weights learned from the remaining 72 HPVs. The classification accuracy is measured by Equation 2. This experimental result was calculated through 10-fold cross validation test. For cross validation, the whole dataset is divided into disjoint 10 bins and used as train and test datasets (leave-one-out validation was used; 9 as train, 1 as test dataset).

Results

Muñoz *et al.* classified the types of HPV based on epidemiologic classification method [7]. They pooled data from more than 1,900 patients who have cervical cancer. They detected HPV DNA and assigned type by polymerase-chain-reaction-based

Table 2. The manually classified risk types of 76 HPVs. The "D/K" mean "don't know." There are 18 HPVs with high risk types and 4 HPVs with "don't know" types. This classification of total 76 HPVs is based on the 1997 version of HPV compendium. The classifications of HPV [46, 71, 78, 79] are missing from the table due to the lack of its research data (Park *et al.* [4]).

HPVs	Type	HPVs	Type	HPVs	Type	HPVs	Type
HPV01	Low	HPV20	Low	HPV39	High	HPV59	High
HPV02	Low	HPV21	Low	HPV40	Low	HPV60	Low
HPV03	Low	HPV22	Low	HPV41	Low	HPV61	High
HPV04	Low	HPV23	Low	HPV42	Low	HPV62	High
HPV05	Low	HPV24	Low	HPV43	Low	HPV63	Low
HPV06	Low	HPV25	Low	HPV44	Low	HPV64	Low
HPV07	Low	*HPV26*	D/K	HPV45	High	HPV65	Low
HPV08	Low	HPV27	Low	HPV47	Low	HPV66	High
HPV09	Low	HPV28	Low	HPV48	Low	HPV67	High
HPV10	Low	HPV29	Low	HPV49	Low	HPV68	High
HPV11	Low	HPV30	Low	HPV50	Low	HPV69	Low
HPV12	Low	HPV31	High	HPV51	High	*HPV70*	D/K
HPV13	Low	HPV32	Low	HPV52	High	HPV72	High
HPV14	Low	HPV33	High	HPV53	Low	HPV73	Low
HPV15	Low	HPV34	Low	*HPV54*	D/K	HPV74	Low
HPV16	High	HPV35	High	HPV55	Low	HPV75	Low
HPV17	Low	HPV36	Low	HPV56	High	HPV76	Low
HPV18	High	*HPV57*	D/K	HPV77	Low
HPV19	Low	HPV38	Low	HPV58	High	HPV80	Low

Table 3. Parameters for genetic mining.

	Parameters				
	Pop. size N	Mutation rate P_m	Crossover rate P_c	Replace rate M	Segment length (ρ)
Value	2,000	0.3	0.5	50%	4

assays which are considered as a relatively accurate detection method in current literatures. In their experiment, both HPV54 and HPV70 were classified as low risk types. From our genetic mining method, we also classified these two HPVs as low risk types. This result is identical to the result of Muñoz *et al.* for HPV [54, 70]. Currently, there are insufficient research results on the HPVs which have undetermined risk types, HPV [26, 54, 57, 70], to decide whether the classification result is correct or not. But, from the experimental result of Muñoz *et al.* in comparison with the result of Park *et al.*, we can conclude that our genetic mining method has more ability than previous text mining method of Park *et al.* [4] for capturing the genetic and the structural sequence characteristics of HPVs when we assume that the results of Muñoz *et al.* are correct.

Table 4. Predicted risk types of the HPVs whose risk types are known as "don't know." Note that the HPV70 is classifyed as low risk type which was classified as high-risk type in the previous research result of Park *et al.* [4].

Type	HPV26	HPV54	HPV57	HPV70
Risk	Low	Low	Low	Low

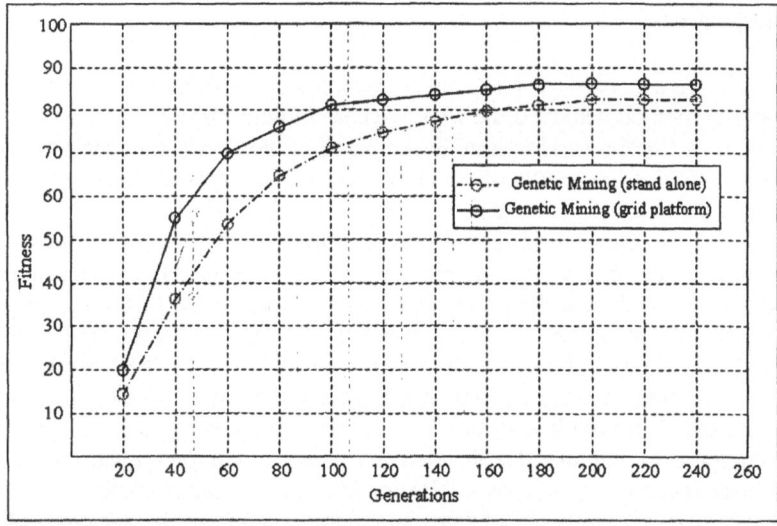

Fig. 3. The average fitness of genetic mining of HPV sequence structures. This figure shows the tendency that genetic mining on the grid platform converges relatively faster than the genetic mining on a single platform (represented as "stand alone"). The maximum fitness of genetic mining on the grid platform was 85.6%. The fitness means the classification accuracy of HPV risk types on sequence data. Whole experiment was conducted five times and averaged for each plot of the graph.

The proposed genetic mining method showed about 85.6% classification accuracy for 72 HPVs in Table 2. This result outperforms that of decision tree based text mining (it was about 81.14%) of Park *et al.* [4] by 4.5%. The population was converged to this accuracy at least within 250 generations which takes about 8.1 hours on described grid platform. Figure 3 shows the fitness convergence graph of proposed genetic mining method on both stand-alone and grid based platform. The graph of genetic mining on the grid platform shows relatively fast converge tendency.

To evaluate the relative success of the proposed genetic mining approach, we compared the classification performance with the result of our previous works [4] which use decision tree as a HPV classification method. And we also implemented Naïve Bayes classifier as a baseline model to compare the classification results. These classification results are described in Table 5. Both classifications with GA in stand-alone system and with GA in grid platform show improved classification accuracy than the previous decision tree based classification result. These GA based method outperform the baseline model, Naïve Bayes, by about 6 to 8 percents.

Table 5. Accuracy of the baseline model (Naïve Bayes), decision tree, and proposed approaches. Decision tree was used in the text ming approach of Part *et al.* [4] as a base classifier. Genetic mining based approches outperformed other two methods in term of classifiaction accuracy.

Method	Naïve Bayes	Decision Tree	GA (stand alone)	GA (GRID)
Accuracy	77.18	81.14	83.93	85.64

5 Conclusions

In this paper, we proposed genetic mining method to classify the risk types of HPVs. The proposed method achieved the improvement in both classification accuracy and processing time (the processing time is compared to the stand-alone system). In particular, the classification results of the HPVs whose risk types were known as "don't know" show coincidence with the recent HPV classification results based on practical experiments. This result based on the genetic mining is different from the previous classification results based on decision tree on the text description of each HPV. Thus, we can conclude that the proposed method is appropriate to analyze biological sequences for classification. Also, the proposed platform cut down the computing time from 22.9 (in stand-alone system) hours to just 8.1 hours. Also, the convergence time of GA is relatively faster in the grid platform than in stand-alone system. We assume that this is maybe due to the 'randomness' in applying genetic operators in distributed platform and also due to relatively powerful computational power. More detailed investigation on this issue will be conducted in the future works.

Moreover, HPVs have many mutants and classification on the risk types of these mutated HPVs is important for medical remedy. Thus, study on the new method for classifying these mutants and study on the more efficient encoding scheme to exploit the proposed genetic mining method with theoretical analysis remain as future works.

Acknowledgements

This research was supported by the Korean Ministry of Science and Technology under the NRL Program and the Systems Biology Program.

References

1. Bosch, F.X., Manos, M.M., Muñoz, N., Sherman, M., Jansen, A.M., Peto, J., Schiffman, M.H., Moreno, V., Kurman, R., Shah, K.V.: Prevalence of human papillomavirus in cervical cancer: a worldwide perspective. *J Natl Cancer Inst* **87** (11) (1995) 796-802.
2. Furumoto, H. and Irahara, M.: Human Papillomavirus (HPV) and Cervical Cancer. *Journal of Medical Investigation* **49** (2002) 124-33.
3. Schiffman, M., Bauer, H., Hoover, R., Glass, A., Cadell, D., Rush, B., Scott, D., Sherman, M., Kurman, R., and Wacholder, S.: Epidemiologic evidence showing that Human Papillomavirus infection causes most cervical intraepithelial neoplasis. *Journal of the National Cancer Institute* **85** (1993) 958-64.
4. Park, S.-B., Hwang, S.-H., and Zhang, B.-T.: Classification of the risk types of human papillomavirus by decision trees. In *Proceedings of the 4th International Conference on Intelligent Data Engineering and Automated Learning* (2003) 540–44.
5. Janicek, M.F. and Averette, H.E.: Cervical cancer: prevention, diagnosis, and therapeutics. *Cancer Journals for Clinicians* **51** (2001) 92-114.
6. Burk, R.D., Ho, G.Y., Beardsley, L., Lempa, M., Peters, M., and Bierman, R.: Sexual behavior and partner characteristics are the predominant risk factors for genital human papillomavirus infection in young women. *J Infect Dis.* **174**(4) (1996) 679-89.

7. Muñoz, N., Bosch, F.X., Sanjosé, S., Herrero, R., Castellsagué, X., Shah, K.V., Snijders, P.J.F., and Meijer, C.J.L.M.: Epidemiologic classification of human papillomavirus types associated with cervical cancer. *The New England Journal of Medicine* **348**(6) (2003) 518-27.

8. Eom, J.-H. and Zhang, B.-T.: IDEA@home: The flexible active grid computing platform based on P2P and network segmentation. *Technical Report BI-04-01*, School of Computer Sci.&Eng., Seoul National Univ., Seoul, Korea, February (2004).

9. Richards, W.G.: Virtual screening using grid computing: the screensaver project. *Nature Reviews Drug Discovery* **1** (2002) 551-55.

10. Davies, E.K., Glick, M., Harrison, K.N., and Richards, W.G.: Pattern recognition and massively distributed computing. *Journal of Computational Chemistry* **23**(16) (2002) 1544-50.

11. Bäck, T., Evolutionary algorithms in theory and practice, *Oxford University Press*. (1996)

12. Kim, S. and Zhang B.-T.: Genetic mining of HTML structures for effective web-document retrieval. *Applied Intelligence* **18** (2003) 243–56.

13. The HPV sequence database in Los Alamos laboratory. http://hpv-web.lanl.gov/stdgen/virus/ hpv/index.html.

Gene Regulatory Network Discovery from Time-Series Gene Expression Data – A Computational Intelligence Approach

Nikola K. Kasabov[1], Zeke S.H. Chan[1], Vishal Jain[1],
Igor Sidorov[2], and Dimiter S. Dimitrov[2]

[1] Knowledge Engineering and Discovery Research Institute (KEDRI),
Auckland University of Technology, Private Bag 92006, Auckland, New Zealand
{nkasabov,shchan,vishal.jain}@aut.ac.nz

[2] National Cancer Institute, Frederick, Washington DC, National Institute of Health, USA
{sidorovi,dimitrov}@ncifcrf.gov

Abstract. The interplay of interactions between DNA, RNA and proteins leads to genetic regulatory networks (GRN) and in turn controls the gene regulation. Directly or indirectly in a cell such molecules either interact in a positive or in repressive manner therefore it is hard to obtain the accurate computational models through which the final state of a cell can be predicted with certain accuracy. This paper describes biological behaviour of actual regulatory systems and we propose a novel method for GRN discovery of a large number of genes from multiple time series gene expression observations over small and irregular time intervals. The method integrates a genetic algorithm (GA) to select a small number of genes and a Kalman filter to derive the GRN of these genes. After GRNs of smaller number of genes are obtained, these GRNs may be integrated in order to create the GRN of a larger group of genes of interest.

1 Introduction

Gene regulatory network is one of the two main targets in biological systems because they are systems controlling the fundamental mechanisms that govern biological systems. A single gene interacts with many other genes in the cell, inhibiting or promoting directly or indirectly, the expression of some of them at the same time. Gene interaction may control whether and how vigorously that gene will produce RNA with the help of a group of important proteins known as transcription factors. When these active transcription factors associate with the target gene sequence (DNA bases), they can function to specifically suppress or activate synthesis of the corresponding RNA. Each RNA transcript then functions as the template for synthesis of a specific protein. Thus the gene, transcription factor and other proteins may interact in a manner that is very important for determination of cell function. Much less is known about the functioning of the regulatory systems of which the individual genes and interaction form a part [6], [8], [15], [20]. Transcription factors provide a feedback pathway by which genes can regulate one another's expression as mRNA and then as protein [3], [5].

The discovery of gene regulatory networks (GRN) from time series of gene expression observations can be used to: (1) Identify important genes in relation to a disease

N.R. Pal et al. (Eds.): ICONIP 2004, LNCS 3316, pp. 1344–1353, 2004.
© Springer-Verlag Berlin Heidelberg 2004

or a biological function, (2) Gain an understanding on the dynamic interaction between genes, (3) Predict gene expression values at future time points. The major approaches that deals with the modelling of gene regulatory networks involve differential equations [14], stochastic models [16], evolving connectionist systems [13], boolean networks [18], generalized logical equations [21], threshold models [19], petri nets [11], bayesian networks [9], directed and undirected graphs.

We propose here a novel method that integrates Kalman Filter [4] and Genetic Algorithm (GA) [10], [12]. The GA is used to select a small number of genes, and the Kalman filter method is used to derive the GRN of these genes. After GRNs of smaller number of genes are obtained, these GRNs may be integrated in order to create the GRN of a larger group of genes of interest. The goal of this work is develop a method for GRN discovery from multiple and short time series data of a large number of genes. The secondary goal is to apply the method as to identify the genes that co-regulate telomerase from the extracts of the U937 plus and minus series obtained in NCI, NIH. Each series contains the time-series expression of 32 pre-selected candidate genes that have been found potentially relevant, as well as the expression of the telomerase. Both the plus series and minus series contains four samples recorded at the (0, 6, 24, 48) [th] hour. Discovering GRN from these two series is challenging in two aspects: first, both series are sampled at irregular time intervals; second, the number of samples is scarce (only 4 samples). A third potential problem is that the search space grows exponentially in size as more candidate genes are identified in the future. Several GRNs of 3 most related to the telomerase genes are discovered, analysed and integrated. The results and their interpretation confirm the validity and the applicability of the proposed method. The integrated method can be easily generalized to extract GRN from other time series gene expression data. This paper reports the methodology and the experimental findings.

2 Modelling GRN with First-Order Differential Equations, State-Space Representation and Kalman Filter

2.1 Discrete-Time Approximation of First-Order Differential Equations

Our GRN is modelled with the discrete time approximation of first-order differential equations, given by:

$$\mathbf{x}_{t+1} = \mathbf{F}\mathbf{x}_t + \varepsilon_t \qquad (1)$$

where $\mathbf{x}_t = (x_1, x_2, \ldots x_n)'$ is the gene expression at the t-th time interval and n is the number of genes modelled, $\Box_t$ is a noise component with covariance E=cov (ε_t), and F=(f_{ij}) i=1,n, j=1,n is the transition matrix relating x_t to x_{t+1}. It is related to the continuous first-order differential equations $dx/dt = \Psi x + e$ by $\mathbf{F} = \tau\Psi + \mathbf{I}$ and $\varepsilon_t = \tau e$ where τ is the time interval {note the subscript notation (t+k) is actually the common abbreviation for (t+k τ)}[7]. We work here with a discrete approximation instead of a continuous model for the ease of modelling and processing the irregular time–course data (with Kalman filter). Besides being a tool widely used for modelling biological processes, there are two advantages in using first-order differential equations.

First, gene relations can be elucidated from the transition matrix $\mathbf{F}$ through choosing a threshold value (ζ; $1 > \zeta > 0$). If $|f_{ij}|$ is larger then the threshold value ζ, $x_{t,j}$ is assumed to have significant influence on $x_{t+1,i}$. A positive value of f_{ij} indicates a positive influence and vice-versa. Second, they can be easily manipulated with KF to handle irregularly sampled data, which allow parameter estimation, likelihood evaluation and model simulation and prediction.

The main drawback of using differential equations is that it requires the estimation of n^2 parameters for the transition matrix $\mathbf{F}$ and n $(n-1)/2$ parameters for the noise covariance $\mathbf{E}$. To minimize the number of model parameters, we estimate only $\mathbf{F}$ and fix $\mathbf{E}$ to a small value. Since both series contain only 4 samples, we avoid over-parameterization by setting n to 4, which is the maximum number of n before the number of parameters exceeds the number of training data {It matches the number of model parameters (the size of $\mathbf{F}$ is $n^2=16$) to the number of training data ($n \times 4$ samples $=16$)}. Since in our case study one of the n genes must be telomerase, we can search for a subset of size $K=3$ other genes to form a GRN.

To handle irregularly sampled data, we employ the state-space methodology and the KF. We treat the true trajectories as a set of unobserved or hidden variables called the *state variables*, and then apply the KF to compute their optimal estimates based on the observed data. The state variables that are regular/complete can now be applied to perform model functions like prediction, parameter estimations instead of the observed data that are irregular/incomplete. This approach is more superior to interpolation methods as it prevents false modelling by trusting a fixed set of interpolated points that may be erroneous.

2.2 State-Space Representation

To apply the state-space methodology, a model must be expressed in the following format called the *discrete-time state space representation*

$$\mathbf{x}_{t+1} = \mathbf{\Phi}\mathbf{x}_t + \mathbf{w}_t \tag{2}$$

$$\mathbf{y}_t = \mathbf{A}\mathbf{x}_t + \mathbf{v}_t \tag{3}$$

$$\text{cov}(\mathbf{v}_t) = \mathbf{R} \qquad \text{cov}(\mathbf{w}_t) = \mathbf{Q} \tag{4}$$

where, $\mathbf{x}_t$ is the system state; $\mathbf{y}_t$ is the observed data; $\mathbf{\Phi}$ is the state transition matrix that relates $\mathbf{x}_t$ to $\mathbf{x}_{t+1}$; $\mathbf{A}$ is the linear connection matrix that relates $\mathbf{x}_t$ to $\mathbf{y}_t$; $\mathbf{w}_t$ and $\mathbf{v}_t$ are uncorrelated white noise sequences whose covariance matrices are $\mathbf{Q}$ and $\mathbf{R}$ respectively. The first equation is called the *state equation* that describes the dynamics of the state variables. The second equation is called the *observation equation* that relates the states to the observation.

To represent the discrete-time model in the state-space format, we simply substitute the discrete-time equation (1) into the state equation (2) by setting $\mathbf{\Phi}=\mathbf{F}$, $\mathbf{w}_t=\varepsilon_t$ and $\mathbf{Q}=\mathbf{E}$ and form a direct mapping between states and observations by setting $\mathbf{A}=\mathbf{I}$. The state transition matrix $\mathbf{\Phi}$ (functional equivalent to $\mathbf{F}$) is the parameter of interest as it relates the future response of the system to the present state and governs the dynamics of the entire system. The covariance matrices $\mathbf{Q}$ and $\mathbf{R}$ are of secondary interest and are fixed to small values to reduce the number of model parameters.

2.3   Kalman Filter (KF)

KF is a set of recursive equations capable of computing optimal estimates (in the least-square sense) of the past, present and future states of the state-space model based on the observed data. Here we use it to estimate gene expression trajectories given irregularly sampled data. To specify the operation of Kalman filter, we define the conditional mean value of the state $\mathbf{x}_t^s$ and its covariance $\mathbf{P}_{tu}^s$ as:

$$\mathbf{x}_t^s = E(\mathbf{x}_t \mid \mathbf{y}_1, \mathbf{y}_2, ..., \mathbf{y}_s) \tag{5}$$

$$\mathbf{P}_{tu}^s = E\left[(\mathbf{x}_t - \mathbf{x}_t^s)(\mathbf{x}_u - \mathbf{x}_u^s)' \mid \mathbf{y}_1, ..., \mathbf{y}_s\right] \tag{6}$$

For prediction, we use the KF forward recursions to compute the state estimates for ($s<t$). For likelihood evaluation and parameter estimation, we use the KF backward recursions to compute the estimates called the smoothed estimates based on the entire data, i.e. ($s=T$; $T>t$ is the index of the last observation), which in turn are used to compute the required statistics.

2.4   Using GA for the Selection of a Gene Subset for a GRN

The task is to search for the genes that form the most probable GRN models, using the model likelihood computed by the KF as an objective function. Given N the number of candidates and K the size of the subset, the number of different gene combinations is $N!/K!(N-K)!$. In our case study, $N=32$ is small enough for an exhaustive search. However, as more candidates are identified in the future, the search space grows exponentially in size and exhaustive search will soon become infeasible. For this reason a method based on GA is proposed. The strength of GA is twofold:

1. Unlike most classical gradient methods or greedy algorithms that search along a single hill-climbing path, a GA searches with multiple points and generates new points through applying genetic operators that are stochastic in nature. These properties allow for the search to escape local optima in a multi-modal environment. GA is therefore useful for optimizing high dimensional functions and noisy functions whose search space contains many local optima points.

2. A GA is more effective than a random search method as it focuses its search in the promising regions of the problem space.

2.5   GA Design for Gene Subset Selection

In the GA-based method for gene subset selection proposed here, each solution is coded as a binary string of N bits. A "1" in the ith bit position denotes that the ith gene is selected and a "0" otherwise. Each solution must have exactly K "1"s and a repair operator is included to add or delete "1"s when this is violated. The genetic operators used for crossover, mutation and selection are respectively the standard crossover, the binary mutation and the (μ, λ) selection operators. Since there are two series – the plus and the minus series of time-course gene expression observations in our case study, a new fitness function is designed to incorporate the model likelihood in both series. For each solution, the ranking of its model likelihood in the plus series and in the minus series are obtained and then summed to obtain a joint fitness ranking. This favors convergence towards solutions that are consistently good in both the plus and the minus series. The approach is applicable to multiple time series data.

2.6 Procedures of the GA-Based Method for Gene Subset Selection

Population Initialization. Create a population of μ random individuals (genes from the initial gene set, e.g. of 32) as the first generation parents.

Reproduction. The goal of reproduction is to create λ offspring from μ parents. This process involves three steps: crossover, mutation and repair.

- *Crossover.* The crossover operator transfers parental traits to the offspring. We use the uniform crossover that samples the value of each bit position from the first parent at the crossover probability p_c and from the second parent otherwise. In general, performance of GA is not sensitive to the crossover probability and it is set to a large value in the range of [0.5, 0.9] [1]. Here we set it to 0.7.

- *Mutation.* The mutation operator induces diversity to the population by injecting new genetic material into the offspring. For each bit position of the offspring, mutation inverts the value at a small mutation rate p_m. Performance of GA is very sensitive to the mutation probability and it usually adapts a very small value to avoid disrupting convergence. Here we use $p_m=1/N$, which has been shown to be both the lower bound value and the optimal value for many test functions [17], [1], providing an average of one mutation in every offspring.

- *Repair.* The function of the repair operator is to ensure that each offspring solution has exactly K "1" to present the indices of the K selected genes in the subset. If the number of "1"s is greater than K, invert a "1" at random; and vice-versa. Repeat the process until the number of "1"s matches the subset size K.

Fitness Evaluation. Here λ offspring individuals (solutions) are evaluated for their fitness. For each offspring solution, we obtain the model likelihood in the both the plus and the minus series and compute their ranking (lower the rank, higher the likelihood) within the population. Next, we sum the rankings and use the negated sum as fitness estimation so that the lower the joint ranking, the higher the fitness.

Selection. The selection operator determines which offspring or parents will become the next generation parents based on their fitness function. We use the (μ, λ) scheme that selects the fittest μ of λ offspring to be the next generation parents. It is worth comparing this scheme to another popular selection scheme $(\mu+\lambda)$ that selects the fittest μ of the joint pool of μ parents and λ offspring to be the next generation parents, in which the best-fitness individuals found are always maintained in the population, convergence is therefore faster. We use the (μ, λ) scheme because it offers a slower but more diversified search that is less likely to be trapped in local optima.

Test for Termination. Stop the procedure if the maximum number of generations is reached. Otherwise go back to the reproduction phase.

Upon completion, GA returns the highest likelihood GRNs found in both the plus and the minus series of gene expression observations. The proposed method includes running the GA-based procedure over many iterations (e.g. 50) thus obtaining different GRN that include possibly different genes. Then we summarize the significance of the genes based on their frequency of occurrence in these GRNs and if necessary we put together all these GRNs thus creating a global GRN on the whole gene set.

3   Experiments and Results

The integrated GA-KF method introduced above is applied to identify genes that regulate telomerase in a GRN from a set of 32 pre-selected genes. Since the search space is small (only C_3^{32}=4960 combinations), we apply exhaustive search as well as GA for validation and comparative analysis.

The experimental settings are as follows. The expression values of each gene in the plus and minus series are jointly normalized in the interval [-1, 1]. The purpose of the joint normalization is to preserve the information on the difference between the two series in the mean. For each subset of n genes defined by the GA, we apply KF for parameter estimation and likelihood evaluation of the GRN model. Each GRN is trained for at least 50 epochs (which is usually sufficient) until the likelihood value increases by less than 0.1. During training, the model is tested for stability by computing the eigenvalues of $(\Phi-I)$ [2], [7]. If any of the real part of eigenvalues is positive, the model is unstable and is abandoned.

For the experiments reported in this paper a relatively low resource settings are used. Parent and offspring population sizes (μ, λ) are set to (20, 40) and maximum number of generations is set to 50. These values are empirically found to yield consistent results over different runs. We run it for 20 times from different initial population to obtain the cumulated results. The results are interpreted from the list of 50 most probable GRNs found in each series (we can lower this number to narrow down the shortlist of significant genes). The frequencies of each gene being part of the highest likelihood GRNs in the plus and in the minus series are recorded. Next, a joint frequency is calculated by summing the two frequencies. The genes that have a high joint frequency are considered to be significant in both minus and plus series.

For exhaustive search, we simply run through all gene combinations of 3 genes plus the telomerase; then evolve through KF a GRN for each combination and record the likelihood of each model in both the plus and minus series. A similar scoring system as GA's fitness function is employed. We obtain a joint ranking by summing the model likelihood rankings in the plus series and the minus series, and then count the frequency of the genes that belong to the best 50 GRNs in the joint ranking. The top ten highest scoring genes obtained by GA and exhaustive search are tabulated in Table 1.

Table 1. Significant genes extracted by GA and through an exhaustive search from 32 selected genes.

Rank	Indices of significant genes found by GA (Freq. of occurrence in Minus GRNs, Freq. of occurrence in Plus GRNs) and their accession numbers in Genbank	Indices of significant genes found by exhaustive search (gene Index)
1	27 (179,185) X59871	20 M98833
2	21 (261,0) U15655	27 X59871
3	12 (146, 48) J04101	32 X79067
4	32 (64, 118) X79067	12 J04101
5	20 (0, 159) M98833	6 AL021154
6	22 (118, 24) U25435	29 X66867
7	11 (0, 126) HG3523-HT4899	5 D50692
8	5 (111, 0) D50692	22 U25435
9	18 (0, 105) D89667	10 HG3521-HT3715
10	6 (75, 0) AL021154	13 J04102

The results obtained by GA and exhaustive search are strikingly similar. In both lists, seven out of top ten genes are common (genes 27, 12, 32, 20, 22, 5, 6) and four out of top five genes are the same (genes 27, 12, 32 and 20). The similarity in the results supports the applicability of a GA-based method in this search problem and in particular, when the search space is too large for an exhaustive search. An outstanding gene identified is gene 27, TCF-1. The biological implications of TCF-1 and other high scoring genes are currently under investigation.

The identified GRNs can be used for model simulation and prediction. The GRN dynamics can also be visualized with a network diagram using the influential information extracted from the state transition matrix. As an example, we examine one of the discovered GRN of genes (33, 8, 27, 21) for both the plus and minus series, shown in Fig. 1 and Fig. 2 respectively. The network diagram shows only the components of Φ whose absolute values are above the threshold value $\zeta=0.3$.

For the plus series, the network diagram in Fig. 1 (a) shows that gene 27 has the most significant role regulating all other genes (note that gene 27 has all its arrows

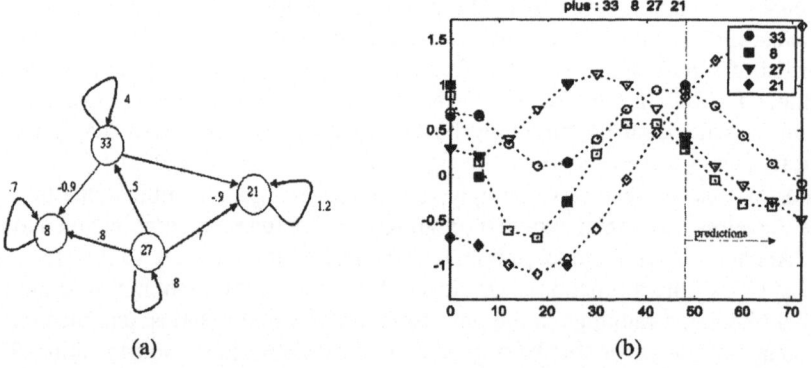

(a)   (b)

Fig. 1. The identified best GRN of gene 33 (telomerase) and genes 8, 27 and 21 for the plus series: (a) The network diagram (b) The network simulation and gene expression prediction over future time. Solid markers represent observations.

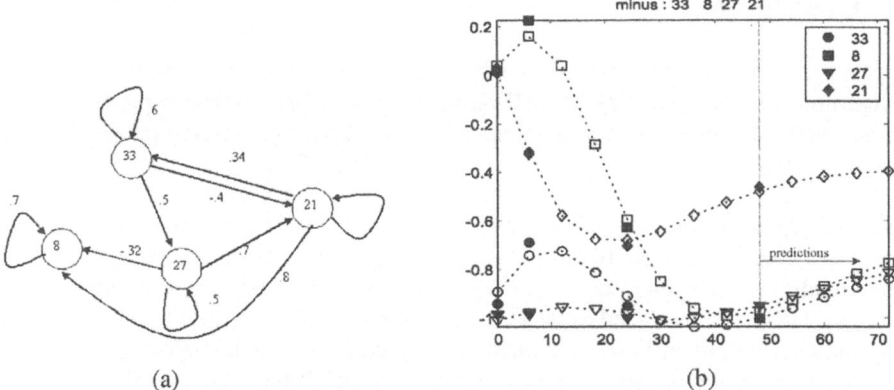

(a)   (b)

Fig. 2. The identified best GRN of gene 33 (telomerase) and genes 8, 27 and 21 for the minus series: (a) The network diagram (b) The network simulation and gene expression prediction over future time. Solid markers represent observations.

out-going). The network simulation, shown in Fig. 1 (b) fits the true observations well and the predicted values appear stable, suggesting that the model is accurate and robust. For the minus series, the network diagram in Fig. 2 (a) shows a different network from that of the plus series. The role of gene 27 is not as prominent. The relationship between genes is no more causal but interdependent, with genes 27, 33 and 21 simultaneously affecting each other. The difference between the plus and minus models is expected. Again, the network simulation result shown in Fig. 2 (b) shows that the model fits the data well and the prediction appears reasonable.

3.1 Building a Global GRN of the Whole Gene Set Out of the GRNs of Smaller Number of Genes (Putting the Pieces of the Puzzle Together)

After many GRNs of smaller number of genes are discovered, each involving different genes (with a different frequency of occurring), these GRNs can be put together to create a GRN of the whole gene set. Representation and illustration for the top five (fittest) GRNs from our experiment are shown in Fig3 and Table 2 respectively.

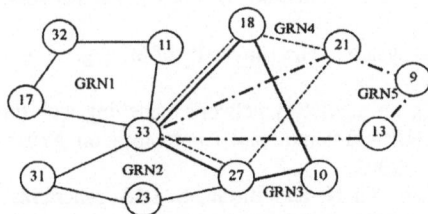

Fig. 3. The five highest likelihood GRN models found by GA in the plus series are put together.

Table 2. Illustration of top five fittest GRNs (plus series).

GRN Number	GRN identified			
1	(33	32	17	11)
2	(33	31	27	23)
3	(33	27	18	10)
4	(33	27	21	18)
5	(33	21	13	9)

4 Conclusions

In this work, we propose a novel method that integrates Kalman Filter and Genetic Algorithm for the discovery of GRN from gene expression observations of several time series (in this case they are two) of small number of observations. As a case study we have applied the method for the discovery of GRN of genes that regulate telomerase in two sub-clones of the human leukemic cell line U937. The time-series contain 12,625 genes, each of which sampled 4 times at irregular time intervals, but only 32 genes of interest are dealt with in the paper. The method is designed to deal effectively with irregular and scarce data collected from a large number of variables (genes). GRNs are modelled as discrete-time approximations of first-order differential equations and Kalman Filter is applied to estimate the true gene trajectories from the irregular observations and to evaluate the likelihood of the GRN models. GA is applied to search for smaller subset of genes that are probable in forming GRN using the model likelihood as an optimization objective. The biological implications of the identified networks are complex and currently under investigation.

References

1. Baeck, T., D. B. Fogel, et al   Evolutionary Computation I and II. Advanced algorithm and operators. Bristol, Institute of Physics Pub (2000)
2. Bay, J. S. (ed.): Fundamentals of Linear State Space Systems, WCB/McGraw-Hill (1999)
3. Bolouri, J. M. B. a. H. (eds.): Computational modelling of Genetic and Biochemical Networks. London, The MIT Press (2001)
4. Brown, R. G. (ed.): Introduction to Random Signal Analysis and Kalman Filtering, John Wiley & Son (1983)
5. Brownstein, M. J., Trent, J.M., and Boguski, M.S., Functional genomics In M. Patterson and M. Handel (eds.): Trends Guide to Bioinformatics (1998) 27-29
6. Collado-Vides, J.: A transformational-grammar approach to study the regulation of gene expression, J. Theor. Biol. **136** (1989) 403-425
7. Dorf, R. and R. H. Bishop: Modern Control Systems, Prentice Hall (1998)
8. Fields, S., Kohara, Y. and Lockhart, D. J.: Functional genomics. Proc Natl. Acad. Sci USA 96 (1999) 8825-8826
9. Friedman, L., Nachman, Pe'er: Using Bayesian networks to analyze expression data. Journal of Computational Biology **7** (2000) 601-620
10. Goldberg, D. E. (1989). Genetic Algorithms in Search, Optimization and machine Learning Reading, MA, Addison-Wesley
11. Hofestadt, R. a. M., F.: Interactive modelling and simulation of biochemical networks Comput. Biol Med. **25** (1995) 321-334
12. Holland. H.: Adaptation in natural and artificial systems, The University of Michigan Press, Ann Arbor, MI (1975)
13. Kasabov, N. and D. Dimitrov: A method for gene regulatory network modelling with the use of evolving connectionist systems. ICONIP - International Conference on Neuro-Information Processing, Singapore, IEEE Press (2002)
14. Likhoshvai, V. A., Matushkin, Yu G., Vatolin, Yu N. and Bazan, S. I: A generalized chemical kinetic method for simulating complex biological systems. A computer model of lambda phage ontogenesis." computational technol. **5**, issue 2 (2000) 87-89
15. Loomis, W. F., and Sternberg, P.W.: Genetic networks. Science **269** (1995) 649

16. Mc Adams, H. H. a. A. A.: Stochastic mechanism in gene expression. Proc. Natl. Acad. Sci. USA **94** (1997) 814-819
17. Muhlenbein, H.: How genetic algorithms really work: I. mutation and hillclimbing. Parallel Problem Solving from Nature 2. B. Manderick. Amsterdam, Elsevier (1992)
18. Sanchez, L., van Helden, J. and thieffry, D.: Establishment of the dorso-ventral pattern during embryonic development of Drosophila melanogaster. A logical analysis. J. Theor. Biol. **189** (1997) 377-389
19. Tchuraev, R. N.: A new method for the analysis of the dynamics of the molecular genetic control systems. I. Description of the method of generalized threshold models. J. Theor. Biol. **151** (1991) 71-87
20. Thieffry, D.: From global expression data to gene networks. BioEssays **21** issue 11 (1999) 895-899
21. Thieffry, D. a. T. R.: Dynamical behaviour of biological regulatory networks-II. Immunity control in bacteriophage lambda. Bull. Math. Biol **57** (1995) 277-297

Sequence Variability
and Long-Range Dependence in DNA:
An Information Theoretic Perspective

Karmeshu[1] and A. Krishnamachari[2]

[1] School of Computer and Systems Sciences, JNU, New Delhi, India
karmeshu@mail.jnu.ac.in
[2] Bioinformatics Center, JNU, New Delhi, India
chari@mail.jnu.ac.in

Abstract. Investigation of the symbolic DNA sequence in terms of its structure and organization is a challenging problem. Inherent uncertainty and sequence variability makes information theoretic framework eminently suitable for applying to variety of computational biology problems. This paper highlights the properties of parametric and nonparametric entropy measures and focusses on few applications where entropic measures have been used. The link between Tsallis entropy and power-law is drawn using maximum entropy principle in capturing the well-known long-range dependence in DNA sequences.

1 Introduction

The statistical analysis of DNA sequence, which can be regarded as a symbolic strings of four nucleotides, is required for understanding the structure and function of genomes. The characteristic of DNA sequences is that they are not statistically homogenous and depict inherent variability. Biological features such as $G+C$ content, C_pG islands, the three base periodicity, $A+T$ strand asymmetry, protein binding motifs, origin of replication are known to exhibit variability in the form of statistical fluctuations along the sequence. For gaining insight in to the issues of intrinsic variability, researchers have widely adopted the information theoretic framework[15, 24]. Information theory though developed in the context of communication by Shannon [25], finds wide applications in a variety of disciplines. This is due to the fact that information theoretic entropy provides a quantitative measure for uncertainty. Thus entropy measure and related concepts are a natural choice for dealing with symbolic sequences like DNA[21].

Information theoretic framework has been widely adopted to study statistical dependence between nucleotides in genomic sequences. One salient features which has drawn the attention of several researchers is in connection with the existence of short and long range correlation in DNA[7, 8, 11, 16, 17, 20]. This feature is not surprising as genomic sequence comprises both information(coding) and regulatory(non-coding) segments. Numerous studies have been carried out to understand and model the underlying mechanisms for generation of long-range correlation structure.It needs to be underlined that the term "long-range

N.R. Pal et al. (Eds.): ICONIP 2004, LNCS 3316, pp. 1354–1361, 2004.
© Springer-Verlag Berlin Heidelberg 2004

correlation" has been used differently in the literature dealing with DNA sequences e.g (i) longer than 3-6 bases (ii) more than 800 bases (iii) 1-10kb [17]. This long-range correlation is generally believed to persist on account of mixture of several length scales in DNA sequences which leads to 1/f noise phenomenon [20]. Long-range dependence leads to power-law behaviour affecting the correlations to decay like a power-law instead of short-range correlations that decay exponentially.

Attempts have been made to generate long-range correlations in terms of stationary and non-stationary fractional Brownian motion process [3]. Another promising approach which can be useful for capturing power law behaviour is based on maximum entropy framework due to Jaynes [13]. Parametric entropy measures such as Rènyi, Havrada-Charvat-Tsallis, [13, 14, 27] are found to be useful in mimicking power law behaviour.

Tsallis entropy though similar to Havrada-Charvat measure, brought to focus the applicability of parametric entropy measures to a wide range of problems across disciplines. This paper highlights the importance of parametric entropy measures to sequence variability and long-range dependence in DNA. To gain better insight, we also give examples of few problems in computational biology where Shannon's entropic measure has been employed.

2 Entropy, Mutual Information and Kullback-Leibler Divergence

Information theory is based on an important quantitative measure called 'entropy' which provides the average information associated with the random variable X. Assuming that the probability mass function $P(X = x_i) = p_X(i), i = 1, 2 \ldots, n$ is known, Shannon [25] defined the entropy as

$$H_S(X) = -\sum_{i=1}^{n} p_X(i) \log p_X(i) \tag{2.1}$$

$$= -E[\log p_X] \tag{2.2}$$

For impossible events $0 \log(0)$ is taken as 0. The Shannon's entropy function possesses several properties such as non-negativity, expansibility, recursivity and additivity. For details see [13, 14]. It is easy to generalise the definition of entropy function in Eq.(2.1) to multivariate case. Writing $p_{i,j} = p(x_i, y_j) = P(X = x_i, Y = y_j); i = 1, 2, \ldots, n$ and $j = 1, 2, \ldots, m$, the joint entropy of the probabilistic system described by a bivariate distribution is

$$H_S(X, Y) = -\sum_{i=1}^{n} \sum_{j=1}^{m} p_{ij} \log p_{ij} \tag{2.3}$$

When X and Y are independent random variables, then

$$H_S(X, Y) = H_S(X) + H_S(Y), \tag{2.4}$$

exhibiting the additivity property of Shannon's entropy function.

In the context of communication theory, Shannon introduced the notion of mutual information for defining the channel capacity. The mutual information between random variables X and Y is defined as

$$I(X, Y) = \sum_i \sum_j p_{XY}(x_i, y_j) \log \left[\frac{p_{XY}(x_i, y_j)}{p_X(x_i) q_Y(y_j)} \right] \qquad (2.5)$$

which is symmetric and non-negative.

Another quantity which is widely used refers to relative entropy or Kullback-Leibler(KL) divergence (distance) between the two probability distributions with mass functions $p_X(x)$ and $q_X(x)$ respectively. KL measure is defined as

$$D_{KL}(P\|Q) = \sum_x p_X(x) \log \frac{p_X(x)}{q_X(x)} \qquad (2.6)$$

It may be noted that KL measure is not symmetric and $D_{KL}(P\|Q) \neq D_{KL}(Q\|P)$. Based on KL measure, it is possible to define a measure of symmetric cross entropy or symmetric divergence as

$$J(P : Q) = D_{KL}(P\|Q) + D_{KL}(Q\|P) \qquad (2.7)$$

which is referred to as J-divergence [13, 12]. A more generalised Jensen-Shannon measure that is bounded and symmetric in (π_1, P) and (π_2, Q) is given as

$$JS_\pi(P, Q) = \pi_1 D_{KL}(P\|\pi_1 P + \pi_2 Q) + \pi_2 D_{KL}(Q\pi_1 P + \pi_2 Q), \qquad (2.8)$$

where $\pi_1 + \pi_2 = 1$ and $\pi_i > 0$ i=1,2.

3   Rènyi and Tsallis Entropy Measures

Using different set of postulates, further generalizations of Shannon's measure have been obtained by changing the requirements of information measure leading to some-well known Rènyi and Havrada-Charvat parametric entropies. Tsallis rediscoverd Havrada-Charvaat entropy in a different context seeking probablistic description of multi fractals geometries[27]. This led Tsallis to develop non-extensive thermodynamics applicable to a wide ranging problems[27].

Rènyi's entropy [19] for a probability distribution $p_1, p_2, \ldots p_n, p_i \geq 0, \sum p_i = 1$ is defined as

$$H_{R,\alpha}(P) = \frac{1}{1-\alpha} \log \sum_i p_i^\alpha \qquad \alpha > 0, \qquad \alpha \neq 1 \qquad (3.1)$$

It may be noted that one recovers Shannon entropy as the parameter α tends to 1. Rènyi's measure of directed divergence is given by

$$D_{R,\alpha}(P\|Q) = \frac{1}{\alpha - 1} \log(\sum p_i^\alpha q_i^{1-\alpha}), \qquad \alpha > 0, \alpha \neq 1 \qquad (3.2)$$

Rènyi's entropy has found wide applications in coding theory, image reconstruction, feature selection [26]. Recently Krishnamchari *et al*[2] have employed this entropy measure to discriminate DNA binding site from background noise.

Tsallis introduced non-extensive entropy which is defined as

$$H_T^q(P) = k \frac{1 - \sum p_i^q}{q - 1}; \qquad \sum p_{i=1} \tag{3.3}$$

where k is a positive constant and q is referred to as non-extensivity parameter. Shannon's entropy is recovered when $q \to 1$. It may be pointed out that Tsallis and Rènyi entropies are closely related. One salient property of Tsallis entropy is that it is not additive. Thus for two independent systems one finds

$$H_T^q(A, B) = H_T^q(A) + H_T^q(B) + (1 - q)H_T^q(A)H_T^q(B) \tag{3.4}$$

The attractive feature of Tsallis entropy [1,6] is that it can be uniquely identified by the principles of thermodynamics[5]. As observed by Abe [1] the thermodynamically exotic complex systems occupy non-equilibrium states for significantly long periods with preserving scale invariant and hierarchical structures [1]. He further observes that "the phase spaces are generically homogenous and accordingly the naive additivity requirement may not be satisfied any more"[1]. Based on this, it can be argued that non-extensive statistical mechanisms become a relevant framework for building statistical mechanics aspects in DNA sequence analysis. [6, 27]

4   Maximum Entropy Principle, Tsallis Entropy and Powerlaw

There are situations where only partial or incomplete information, say in terms of a first few moments is available. In such cases, there could be infinitely large number of distributions which are consistent with the given information. A pertinent question in this regard is to choose the 'most objective' probability distribution consistent with the given moments. The principle of maximum entropy as enunciated by Jaynes' states that the most objective probability distribution is one which has maximum entropy subject to the given constraints[13].

For the purpose of illustration, we consider a case where the first moment A is known. The most objective probability distribution consistent with the first moment is to be obtained. Employing Tsallis entropic framework, the problem can be stated as

$$Max H_T^q(P) = k \frac{1 - \sum p_i^q}{q - 1} \tag{4.1}$$

subject to

$$\sum_i i p_i = A \qquad and \sum p_i = 1 \tag{4.2}$$

Maximization of $H_T^q(p)$ yields

$$p_i = \frac{[1 + \beta(1 - q)i]^{\frac{1}{(q-1)}}}{\sum_i [1 + \beta(1 - q)i]^{\frac{1}{(q-1)}}}, \qquad q > 0 \tag{4.3}$$

where β is Lagrange parameter. It is straight forward to see that as $q \to 1$, one finds

$$p_i = \frac{e^{-\beta i}}{(\sum_i e^{-\beta i})},\tag{4.4}$$

which corresponds to Boltzmann-Gibbs statistics.

The limiting behaviour of p_i as given in (4.3) for large i yields

$$p_i \sim i^{\frac{1}{q-1}},\tag{4.5}$$

leading to power law distribution. Several problems in computational biology depict power-law like behaviour.

5   Applications of Entropic Measures to DNA Sequences

Information theoretic measures have been extensively applied to both DNA and protein sequences, but here only few applications dealing with DNA sequences are highlighted. The general framework prescribed for DNA can also be easily be extended to Protein sequences as well.

5.1   Sequence Variability and Pattern OR Motif Extraction

It is a general practice to perform multiple alignment of closely related sequences and then deduce a consensus pattern for that family of sequences. This is done by considering only the dominant character or base in each column of the multiple alignment and the consensus motif is prescribed. This method has several problems [23]. It is possible to consider the individual contributions of nucleotides and can be quantitatively [24]shown as sequence logos or represent the alignment by a position weight matrix(PWM) using Shannon's entropy measure [22]. The weights are proportional to the conservancy of each base at each column. By employing this type of profile or weight matrix, it is also possible to detect weak but related sequence patterns.

This approach is particularly useful for predicting transcriptor and ribosome binding sites, promoters, and other signals. Since the range of the binding sites are relatively short (from 12 to 20 bases approx.)weight matrix methods are simple, an ideal choice, than employing computationally intensive methods.

5.2   Rènyi Entropy and DNA Binding Sites

The parameter α in Rènyi's entropy provides the flexibility to choose an optimal value in delineating the binding sites from the background. Suitability of Renyi's entropy [19] to few *E.coli* binding sites has been studied [2].

Binding site regions comprises of both signal(s)(binding site) and noise (background). Studies have shown that the information content is above zero at the exact binding site and in the vicinity the it averages to zero [24]. The important question is how to delineate the signal or binding site from the background. One possible approach is to treat the binding site (signal) as an outlier from the surrounding (background) sequences. For a set of characterized known binding site sequences, the critical value α is chosen using an outlier criterion. For the ribosome binding site(RBS)data both Shannon and Renyi entropy measure has been used and the redundancy has been computed [2]. It is evident from the figures 1 and 2 that Renyi measure has flexibility and delineation power compared to Shannon [2].

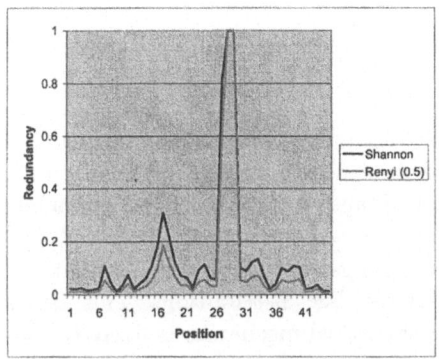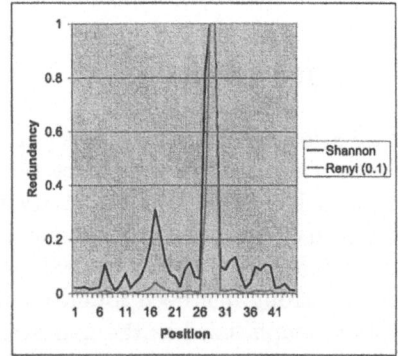

Fig. 1. R-Plot for the RBS site with $\alpha = .5$ **Fig. 2.** R-Plot for the RBS site with $\alpha = .1$

5.3   Modelling Dependency in Protein Binding Sites

Computation of "uncertainty" at each nucleotide position is carried out under the assumption that nucleotides are uniformly and independently distributed as i.i.d variates. Based on these assumptions, position weight matrix and profiles are constructed and are used to search for potential binding sites. The important question is to test this assumption. A theoretical framework based on Jaynes' entropy concentration theorem [13] has been proposed to investigate the distribution of nucleotides in the vicinity of few binding sites. [18]. and analysis with the few experimental data shows that the independent model does not hold good [2].

5.4   Predicting Gene Coding Sequences

Many statistical patterns have been found that distinguishes clearly the coding and non-coding DNA. However they differ from species to species and hence learning models have to be built for classification problems.

Ivo Grosse et al [10]proposed a measure called "Average mutual information" (AMI) to identify coding segments. Individual codon positions are considered while computing AMI. The performance of this approach is comparable with other methods.

5.5 Segmenting Genomic DNA

Segmentation is a procedure that partitions a given sequence into domains of homogenous composition. It is a known fact that the statistical properties of DNA are not homogenously distributed along the sequence [9]. The segmentation procedure employing Jensen-shannon divergence has been used to find domains, isochores, C_pG islands, replication origin and terminus, coding and non-coding sequence borders etc[4, 4, 9]. This simple entropic based distance method is powerful in detecting many biological features.

6 Conclusions

Information theoretic measures partcularly Shannon's entropy has been widely used as it captures the sequence variability in a variety of problems. Very little attention has been given to study the advantage and suitability of parametric entropic measures for symbolic DNA and protein sequences. These entropic measures are ideal choice to deal with short range and long-range correlations. Tsallis entropy has been found to mimic power law distributions. This aspect has been emphasized as the non-extensive statistical mechanics is likely to play a role in computational molecular biology.

References

1. S. Abe, *Tsallis entropy: How unique?*, preprint (2004).
2. A.Krishnamachari, Vijnan moy Mandal, and Karmeshu, *Study of dna binding sites using the renyi parametric entropy measure*, Journal of theoretical biology **227** (2004), 429–436.
3. P. Allergrini, M. Buiatti, and P. Grigolini aand BJ. West, *Fractional brownian motion as a nonstationary process: An alternative paradigm for dna sequences*, Physicaal Review E **57** (1998), no. 4, 4558–4567.
4. RK. Azad, JS. Rao, W. Li, and R. Ramaswamy, *Simplifying the mosaic description of dna sequences*, Phys Rev E **66** (2002), 31913–31918.
5. C. Beck and F. Schlögl, *Thermodynamics of chaotic systems - an introduction*, Cambridge University Press, 1993.
6. M. Buiatti, P. Grigolini, and L. Palatella, *A non extensive approach to the entropy of symbolic sequences*, Physica A **268** (1999), 214–224.
7. C.K.Peng and S.V.Buldyrev, *Long-range correlation in nucleotide sequences*, Nature **356** (1992), 168–170.
8. P. Doukhaan, G. Oppenheim, and MS. Taqqu (eds.), *Theory and applications of long-range dependence*, Birkhöuser, 2002.
9. PB. Galvan, I. Grosse, P. Carpena, JL. Oliver, RR. Roldán, and HE. Stanley1, *Finding borders between coding and noncoding dna regions by entropic segmentation method*, Phys.Rev.Lett **85** (2000), no. 1342-1345.
10. I. Grosse, H. Herzel, SV. Buldyrev, and HE Stanley, *Species independence of mutual information in coding and noncoding dna*, Phys Rev E **61** (2000), 5624–5629.
11. H. Herzel and Große, *Measuring correlations in symbol sequences*, Physica A (1995), 518–542.

12. I.S.Dhillon, S.Mallela, and R.Kumar, *A divisive information theoretic feature clustering algorithm for text classificaation*, Journal of machine learning research **3** (2003), 1265–1287.
13. J.N.Kapur and H.K.Kesavaan, *Entropy optimisation principles with applications.*, Academic press, 1992.
14. Karmeshu and N.R.Pal, *Uncertainty, entropy and maximum entropy entropy principle - an overview*, Entropy Measures, Maximum Entropy Principle and Emerging Applications, Springer, 2003.
15. L.Gatlin, *Information theory and the living system*, Columbia University Press, 1972.
16. W. Li, TG. Marr, and K. Kaneko, *Understanding long-range correlations in dna sequences*, Physica D **75** (1994), 392–416.
17. Wentian Li, *The study of correlation structures of dna sequences:a criticl review*, Computer & Chemistry **21** (1997), no. 4, 252–272.
18. Karmeshu DN. Rao and A. Krishnamachari, *Statistical distribution of nucleotides in the vicinity of the binding sites*, pre-print (2004).
19. Rènyi, *On measures of entropy and information*, vol. 1, Fourth Berkley Symposium on Mathematics, Statistics and Probability, 1961, pp. 547–561.
20. R.F.Voss, *Evolution of long-range fractal correlations and 1/f noise in dna baase sequences*, Phys.Rev.Lett. **68** (1992), no. 25, 3805–3808.
21. Ramon Roman-Rolda'N, Pedro Bernaola-Galvan, and Jose L. Oliver, *Applications of information theory to dna sequence analysis: A review*, Pattern recognition **29** (1996), no. 7, 1187–1194.
22. T.D. Schneider, *Information content of individual genetic sequences*, J.Theor.Biol **189** (1997), no. 4, 427–441.
23. TD. Schneider, *Consensus sequence zen*, Applied Bioinformatics **1** (2002), no. 3, 111–119.
24. T.D. Schneider, G.D. Stormo, L. Gold, and A. Ehrenfeucht, *Information content of binding sites on nucleotide sequences*, J.Mol.Biol **188** (1986), 415–431.
25. C.E. Shannon, *A mathematical theory of communication*, Bell System Tech. J. **27** (1948), 379–423, 623–659.
26. Renata Smolokova, Mark P. Wachowiak, and Jacek Zurada, *An information-theoretic approach to estimaating ultrasound backscatter characteristics*, Computers in Biology and Medicine **34** (2004), 355–370.
27. C. Tsallis, *Theoretical, experimental and computational evidences and connections*, Brazilian Journal of Physics **29** (1999), no. 1.

Author Index

Lecture Notes in Computer Science

For information about Vols. 1–3213

please contact your bookseller or Springer

Vol. 3262: M.M. Freire, P. Chemouil, P. Lorenz, A. Gravey (Eds.), Universal Multiservice Networks. XIII, 556 pages. 2004.

Vol. 3261: T. Yakhno (Ed.), Advances in Information Systems. XIV, 617 pages. 2004.

Vol. 3260: I.G.M.M. Niemegeers, S.H. de Groot (Eds.), Personal Wireless Communications. XIV, 478 pages. 2004.

Vol. 3258: M. Wallace (Ed.), Principles and Practice of Constraint Programming – CP 2004. XVII, 822 pages. 2004.

Vol. 3257: E. Motta, N.R. Shadbolt, A. Stutt, N. Gibbins (Eds.), Engineering Knowledge in the Age of the Semantic Web. XVII, 517 pages. 2004. (Subseries LNAI).

Vol. 3256: H. Ehrig, G. Engels, F. Parisi-Presicce, G. Rozenberg (Eds.), Graph Transformations. XII, 451 pages. 2004.

Vol. 3255: A. Benczúr, J. Demetrovics, G. Gottlob (Eds.), Advances in Databases and Information Systems. XI, 423 pages. 2004.

Vol. 3254: E. Macii, V. Paliouras, O. Koufopavlou (Eds.), Integrated Circuit and System Design. XVI, 910 pages. 2004.

Vol. 3253: Y. Lakhnech, S. Yovine (Eds.), Formal Techniques, Modelling and Analysis of Timed and Fault-Tolerant Systems. X, 397 pages. 2004.

Vol. 3252: H. Jin, Y. Pan, N. Xiao, J. Sun (Eds.), Grid and Cooperative Computing - GCC 2004 Workshops. XVIII, 785 pages. 2004.

Vol. 3251: H. Jin, Y. Pan, N. Xiao, J. Sun (Eds.), Grid and Cooperative Computing - GCC 2004. XXII, 1025 pages 2004.

Vol. 3250: L.-J. (LJ) Zhang, M. Jeckle (Eds.), Web Services. X, 301 pages. 2004.

Vol. 3249: B. Buchberger, J.A. Campbell (Eds.), Artificial Intelligence and Symbolic Computation. X, 285 pages. 2004. (Subseries LNAI).

Vol. 3246: A. Apostolico, M. Melucci (Eds.), String Processing and Information Retrieval. XIV, 332 pages. 2004.

Vol. 3245: E. Suzuki, S. Arikawa (Eds.), Discovery Science. XIV, 430 pages. 2004. (Subseries LNAI).

Vol. 3244: S. Ben-David, J. Case, A. Maruoka (Eds.), Algorithmic Learning Theory. XIV, 505 pages. 2004. (Subseries LNAI).

Vol. 3243: S. Leonardi (Ed.), Algorithms and Models for the Web-Graph. VIII, 189 pages. 2004.

Vol. 3242: X. Yao, E. Burke, J.A. Lozano, J. Smith, J.J. Merelo-Guervós, J.A. Bullinaria, J. Rowe, P. Tiño, A. Kabán, H.-P. Schwefel (Eds.), Parallel Problem Solving from Nature - PPSN VIII. XX, 1185 pages. 2004.

Vol. 3241: D. Kranzlmüller, P. Kacsuk, J.J. Dongarra (Eds.), Recent Advances in Parallel Virtual Machine and Message Passing Interface. XIII, 452 pages. 2004.

Vol. 3240: I. Jonassen, J. Kim (Eds.), Algorithms in Bioinformatics. IX, 476 pages. 2004. (Subseries LNBI).

Vol. 3239: G. Nicosia, V. Cutello, P.J. Bentley, J. Timmis (Eds.), Artificial Immune Systems. XII, 444 pages. 2004.

Vol. 3238: S. Biundo, T. Frühwirth, G. Palm (Eds.), KI 2004: Advances in Artificial Intelligence. XI, 467 pages. 2004. (Subseries LNAI).

Vol. 3236: M. Núñez, Z. Maamar, F.L. Pelayo, K. Pousttchi, F. Rubio (Eds.), Applying Formal Methods: Testing, Performance, and M/E-Commerce. XI, 381 pages. 2004.

Vol. 3235: D. de Frutos-Escrig, M. Nunez (Eds.), Formal Techniques for Networked and Distributed Systems – FORTE 2004. X, 377 pages. 2004.

Vol. 3234: M.J. Egenhofer, C. Freksa, H.J. Miller (Eds.), Geographic Information Science. VIII, 345 pages. 2004.

Vol. 3233: K. Futatsugi, F. Mizoguchi, N. Yonezaki (Eds.), Software Security - Theories and Systems. X, 345 pages. 2004.

Vol. 3232: R. Heery, L. Lyon (Eds.), Research and Advanced Technology for Digital Libraries. XV, 528 pages. 2004.

Vol. 3231: H.-A. Jacobsen (Ed.), Middleware 2004. XV, 514 pages. 2004.

Vol. 3230: J.L. Vicedo, P. Martínez-Barco, R. Muñoz, M. Saiz Noeda (Eds.), Advances in Natural Language Processing. XII, 488 pages. 2004. (Subseries LNAI).

Vol. 3229: J.J. Alferes, J. Leite (Eds.), Logics in Artificial Intelligence. XIV, 744 pages. 2004. (Subseries LNAI).

Vol. 3226: M. Bouzeghoub, C. Goble, V. Kashyap, S. Spaccapietra (Eds.), Semantics of a Networked World. XIII, 326 pages. 2004.

Vol. 3225: K. Zhang, Y. Zheng (Eds.), Information Security. XII, 442 pages. 2004.

Vol. 3224: E. Jonsson, A. Valdes, M. Almgren (Eds.), Recent Advances in Intrusion Detection. XII, 315 pages. 2004.

Vol. 3223: K. Slind, A. Bunker, G. Gopalakrishnan (Eds.), Theorem Proving in Higher Order Logics. VIII, 337 pages. 2004.

Vol. 3222: H. Jin, G.R. Gao, Z. Xu, H. Chen (Eds.), Network and Parallel Computing. XX, 694 pages. 2004.

Vol. 3221: S. Albers, T. Radzik (Eds.), Algorithms – ESA 2004. XVIII, 836 pages. 2004.

Vol. 3220: J.C. Lester, R.M. Vicari, F. Paraguaçu (Eds.), Intelligent Tutoring Systems. XXI, 920 pages. 2004.

Vol. 3219: M. Heisel, P. Liggesmeyer, S. Wittmann (Eds.), Computer Safety, Reliability, and Security. XI, 339 pages. 2004.

Vol. 3217: C. Barillot, D.R. Haynor, P. Hellier (Eds.), Medical Image Computing and Computer-Assisted Intervention – MICCAI 2004, Part II. XXXVIII, 1114 pages. 2004.

Vol. 3216: C. Barillot, D.R. Haynor, P. Hellier (Eds.), Medical Image Computing and Computer-Assisted Intervention – MICCAI 2004, Part I. XXXVIII, 930 pages. 2004.

Vol. 3215: M.G.. Negoita, R.J. Howlett, L.C. Jain (Eds.), Knowledge-Based Intelligent Information and Engineering Systems, Part III. LVII, 906 pages. 2004. (Subseries LNAI).

Vol. 3214: M.G.. Negoita, R.J. Howlett, L.C. Jain (Eds.), Knowledge-Based Intelligent Information and Engineering Systems, Part II. LVIII, 1302 pages. 2004. (Subseries LNAI).